NOBEL PRIZES AWARDED FOR RESEARCH IN GENETICS OR GENETICS-RELATED AREAS

YEAR	RECIPIENTS	NOBEL PRIZE	RESEARCH TOPIC
2002	S. Brenner H. R. Horvitz J.E. Sulston	Medicine or Physiology	Discovery of the genetic regulation of organ development and programmed cell death
2001	L. Hartwell P. Nurse T. Hunt	Medicine or Physiology	Discovery of genes and regulatory molecules controlling the cell cycle
1999	G. Blobel	Medicine or Physiology	Discovery that proteins contain genetically encoded sequences that guide their transport within cells
1997	S. Prusiner	Medicine or Physiology	Discovery of prions—a new biological principle of infection
1995	E. B. Lewis C. Nusslein-Volhard E. Wieschaus	Medicine or Physiology	Genetic control of early development in *Drosophila*
1993	R. Roberts P. Sharp	Medicine or Physiology	RNA processing of split genes
	K. Mullis M. Smith	Chemistry	Development of polymerase chain reaction (PCR) and site-directed mutagenesis (SDM)
1989	J. M. Bishop H. E. Varmus	Medicine or Physiology	Role of retroviruses and oncogenes in cancer
	T. R. Cech S. Altman	Chemistry	Ribozyme function during RNA splicing
1987	S. Tonegawa	Medicine or Physiology	Genetic basis of antibody diversity
1985	M. S. Brown J. L. Goldstein	Medicine or Physiology	Genetic regulation of cholesterol metabolism
1983	B. McClintock	Medicine or Physiology	Mobile genetic elements in maize
1982	A. Klug	Chemistry	Crystalline structure analysis of significant complexes, including tRNA and nucleosomes
1980	P. Berg W. Gilbert F. Sanger	Chemistry	Development of recombinant DNA and DNA sequencing technology
1978	W. Arber D. Nathans H. O. Smith	Medicine or Physiology	Recombinant DNA technology using restriction endonuclease technology
1976	D. C. Gajdusek	Medicine or Physiology	Elucidation of prion-based human diseases, kuru and Creutzfeldt-Jakob dementia
1975	D. Baltimore R. Dulbecco H. Temin	Medicine or Physiology	Molecular genetics of tumor viruses

Year	Recipients	Nobel Prize	Research Topic
1972	G. M. Edelman R. R. Porter	Medicine or Physiology	Chemical structure of immunoglobulins
	C. Anfinsen	Chemistry	Relationship between primary and tertiary structure of proteins
1970	N. Borlaug	Peace Prize	Genetic improvement of Mexican wheat
1969	M. Delbruck A. D. Hershey S. E. Luria	Medicine or Physiology	Replication mechanisms and genetic structure of bacteriophages
1968	H. G. Khorana M. W. Nirenberg	Medicine or Physiology	Deciphering the genetic code
	R. W. Holley	Medicine or Physiology	Structure and nucleotide sequence of transfer RNA
1966	P. F. Rous	Medicine or Physiology	Viral induction of cancer in chickens
1965	F. Jacob A. M. L'woff J. L. Monod	Medicine or Physiology	Genetic regulation of enzyme synthesis in bacteria
1962	F. H. C. Crick J. D. Watson M. H. F. Wilkins	Medicine or Physiology	Double helical model of DNA
	J. C. Kendrew M. F. Perutz	Chemistry	Three-dimensional structure of globular proteins
1959	A. Kornberg S. Ochoa	Medicine or Physiology	Biological synthesis of DNA and RNA
1958	G. W. Beadle E. L. Tatum	Medicine or Physiology	Genetic control of biochemical processes
	J. Lederberg	Medicine or Physiology	Genetic recombination in bacteria
	F. Sanger	Chemistry	Primary structure of proteins
1954	L. Pauling	Chemistry	Alpha helical structure of proteins
1946	H. J. Muller	Medicine or Physiology	X-ray induction of mutations in *Drosophila*
1933	T. H. Morgan	Medicine or Physiology	Chromosomal theory of genetics
1930	K. Landsteiner	Medicine or Physiology	Discovery of human blood groups

Concepts of Genetics

BRONZE SCULPTURE OF THE WATSON-CRICK MODEL OF DOUBLE HELICAL DNA.

Concepts of Genetics

Eighth Edition

WILLIAM S. KLUG
The College of New Jersey

MICHAEL R. CUMMINGS
Illinois Institute of Technology

CHARLOTTE A. SPENCER
University of Alberta

With contributions by

Sarah M. Ward, Colorado State University

PEARSON
Prentice Hall

Pearson Education International

Executive Editor: Gary Carlson
Editorial Assistant: Jennifer Hart
Production Editor and Composition: Prepare, Inc.
Executive Managing Editor: Kathleen Schiaparelli
Managing Editor, Science Media: Nicole Jackson
Senior Media Editor: Patrick Shriner
Marketing Manager: Andrew Gilfillan
Project Manager: Crissy Dudonis
Assistant Manufacturing Manager: Michael Bell
Manufacturing Buyer: Alan Fischer
Developmental Editor: Anne Scanlan-Roher
Art Director: Maureen Eide
Director of Creative Services: Paul Belfanti
Illustrations: Argosy, Imagineering, and Artworks
Senior Managing Editor AV: Patty Burns
Production Manager: Ronda Whitson
Manager, Production Technologies: Matt Haas
Art Editors: Jay McElroy, Sean Hogan, Denise Keller
Production Assistant: Nancy Bauer
Director, Image Resource Center: Melinda Reo
Manager, Rights and Permissions: Zina Arabia
Interior Image Specialist: Beth Boyd-Brenzel
Cover Image Specialist: Karen Sanatar
Cover Designer: JMG Graphics
Interior Designer: Joseph Sengotta
Photo Researcher: Truitt and Marshall
Photo Coordinator: Debbie Hewitson
Cover Image Credits: *Fruit fly: R. Calentine/Visuals Unlimited; Corn maize: Keith Weller/ARS-USDA; Arabidopsis thaliana: © Holt Studios Int./Photo Researchers, Inc.; Caenorhabditis Elegans: © Sinclair Stammers Science Photo Library/ Photo Researchers, Inc.; Saccharomyces Cerevisiae: Dr. Stanley Flegler/Visuals Unlimited; White mouse: Martin Barraud/ Getty Images, Inc.–Stone Allstock*

PEARSON Prentice Hall

Published by Pearson Education, Inc.
Pearson Prentice Hall
Pearson Education, Inc.
Upper Saddle River, NJ 07458

Printed in the United States of America

10 9 8 7 6 5 4 3 2 1

ISBN 0-13-196894-7

Pearson Education LTD., London
Pearson Education Australia PTY, Limited, Sydney
Pearson Education Singapore, Pte. Ltd.
Pearson Education North Asia Ltd, Hong Kong
Pearson Education Canada, Ltd, Toronto
Pearson Educación de Mexico, S.A. de C.V.
Pearson Education—Japan, Tokyo
Pearson Education Malaysia, Pte. Ltd
Pearson Education, Upper Saddle River, New Jersey

Dedication

TO THOSE WHO MEAN THE VERY MOST,

Captains and Kings may rule the world, but it is the presence of those we love and the memory of those we have lost, but who continue to live in our hearts, which bring beauty to living and make all that we do worthwhile.

About the Authors

William S. Klug is currently Professor of Biology at The College of New Jersey (formerly Trenton State College) in Ewing, New Jersey. He served as Chair of the Biology Department for 17 years, a position to which he was first elected in 1974. He received his B.A. degree in Biology from Wabash College in Crawfordsville, Indiana, and his Ph.D. from Northwestern University in Evanston, Illinois. Prior to coming to The College of New Jersey, he was on the faculty of Wabash College as an Assistant Professor. His research interests have involved ultrastructural and molecular genetic studies of oogenesis in *Drosophila*. He has taught the genetics course as well as the senior capstone seminar course in human and molecular genetics to undergraduate biology majors for each of the last 35 years. In 2002, he was the recipient of the initial teaching award given at The College of New Jersey granted to the faculty member who most challenges students to achieve high standards. He also received the 2004 Outstanding Professor Award from the Sigma Pi International, and in the same year, he was nominated as the Educator of the Year, an award given by the Research and Development Council of New Jersey.

Michael R. Cummings is currently Research Professor in the Department of Biological, Chemical and Physical Sciences at Illinois Institute of Technology, Chicago, Illinois. For more than 25 years, he was a faculty member in the Department of Biological Sciences and in the Department of Molecular Genetics at the University of Illinois at Chicago. He has also served on the faculties of Northwestern University and Florida State University. He received his B.A. from St. Mary's College in Winona, Minnesota, and his M.S. and Ph.D. from Northwestern University in Evanston, Illinois. In addition to this text and its companion volumes, he has also written textbooks in human genetics and general biology for nonmajors. His research interests center on the molecular organization and physical mapping of the heterochromatic regions of human acrocentric chromosomes. At the undergraduate level, he teaches courses in Mendelian and molecular genetics, human genetics, and general biology, and has received numerous awards for teaching excellence given by university faculty, student organizations and graduating seniors.

Charlotte A. Spencer is currently an Associate Professor in the Department of Oncology at the University of Alberta in Edmonton, Alberta, Canada. She has also served as a faculty member in the Department of Biochemistry at the University of Alberta. She received her B.Sc. in Microbiology from the University of British Columbia and her Ph.D. in Genetics from the University of Alberta, followed by postdoctoral training at the Fred Hutchinson Cancer Research Center in Seattle, Washington. Her research interests involve the regulation of RNA polymerase II transcription in cancer cells, cells infected with DNA viruses and cells traversing the mitotic phase of the cell cycle. She has taught courses in Biochemistry, Genetics, Molecular Biology and Oncology, at both undergraduate and graduate levels. She has contributed Genetics, Technology and Society essays for several editions of *Concepts of Genetics* and *Essentials of Genetics*. In addition, she has written booklets in the Prentice-Hall Exploring Biology series, which are aimed at the undergraduate nonmajors level.

Brief Contents

Contents

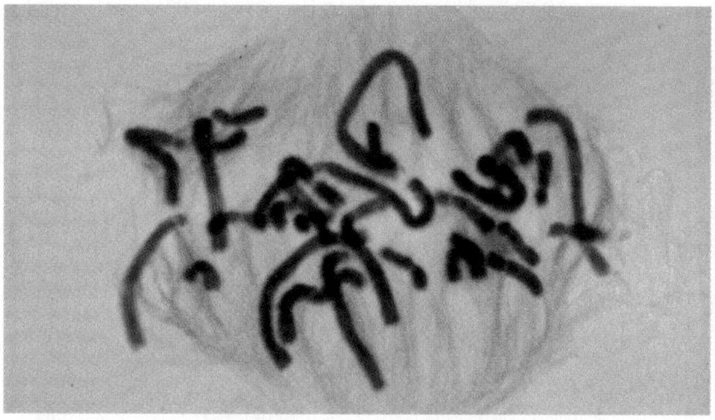

5 Chromosome Mapping in Eukaryotes 100

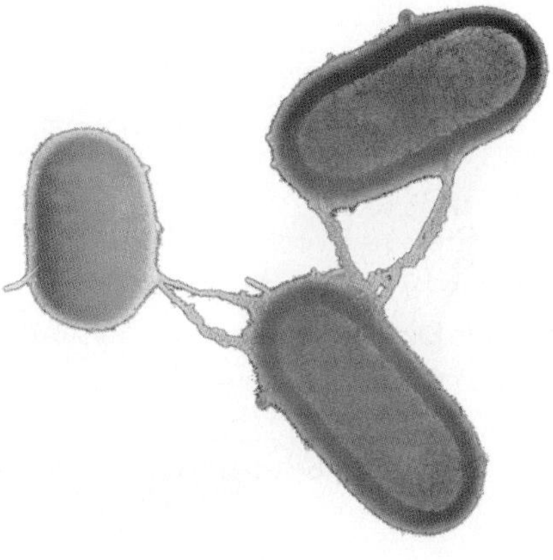

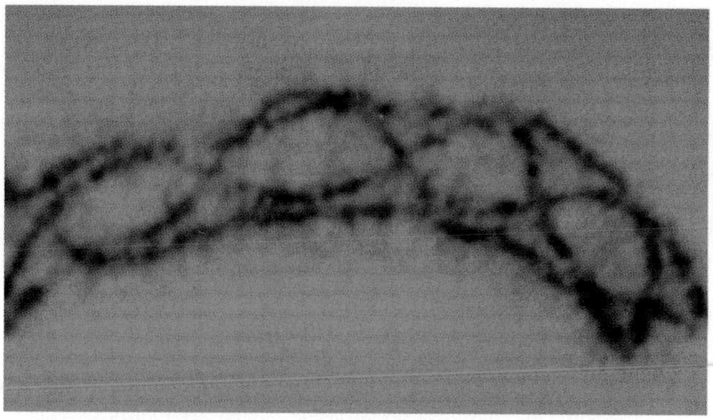

6 Genetic Analysis and Mapping in Bacteria and Bacteriophages 137

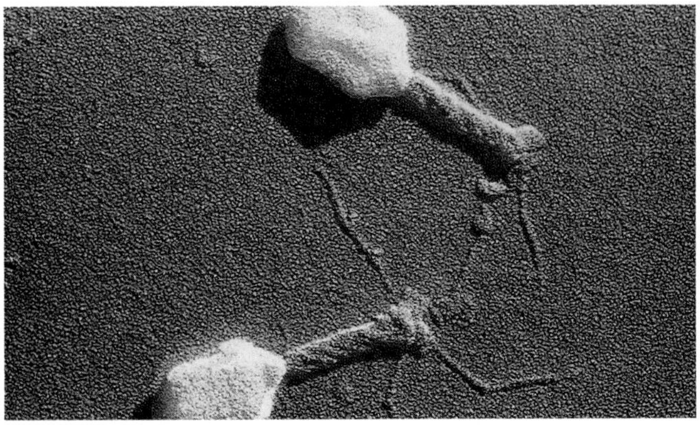

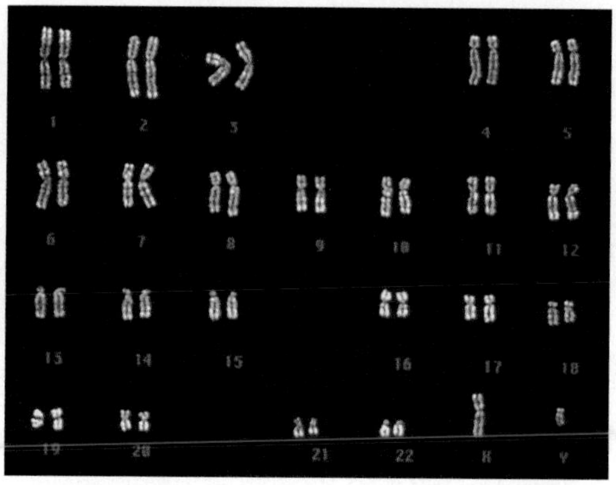

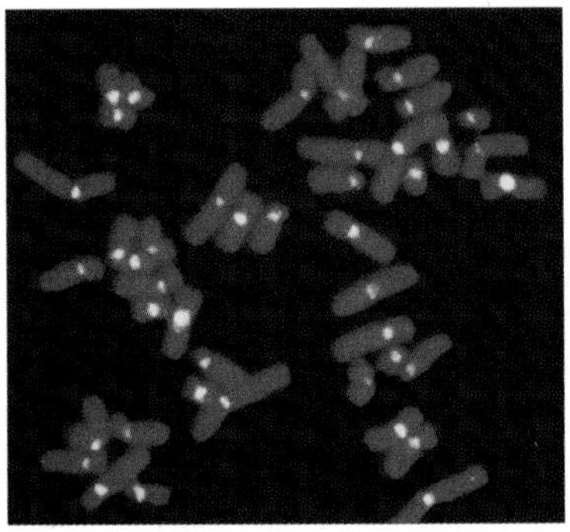

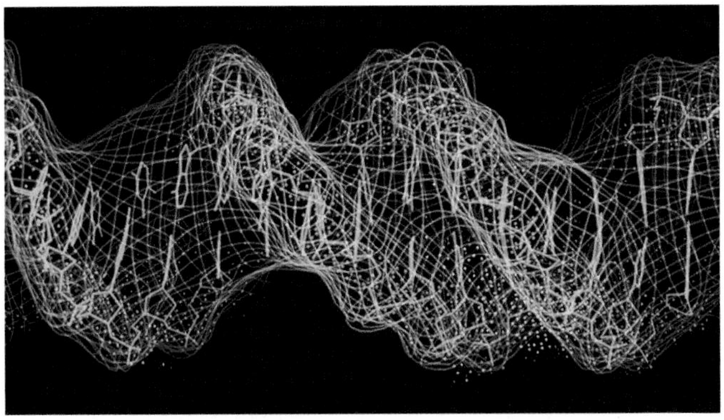

11 DNA Replication and Recombination 263

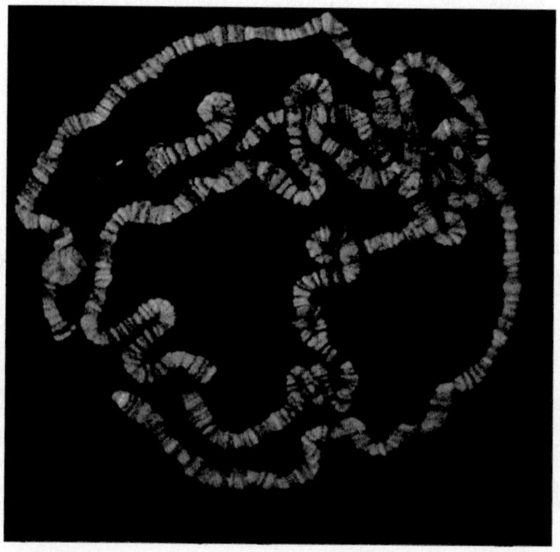

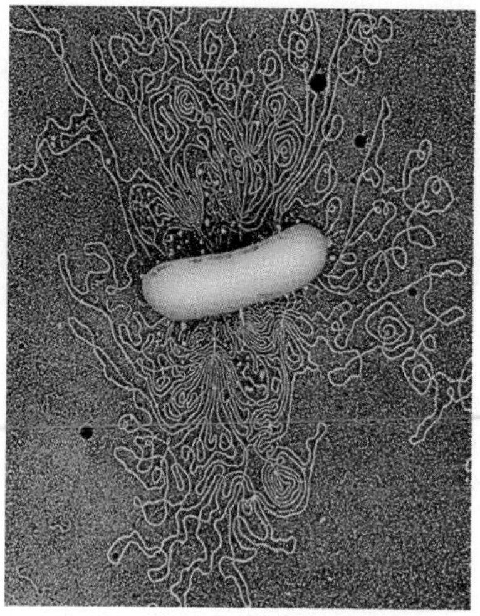

Part Three
Expression and Regulation of Genetic Information

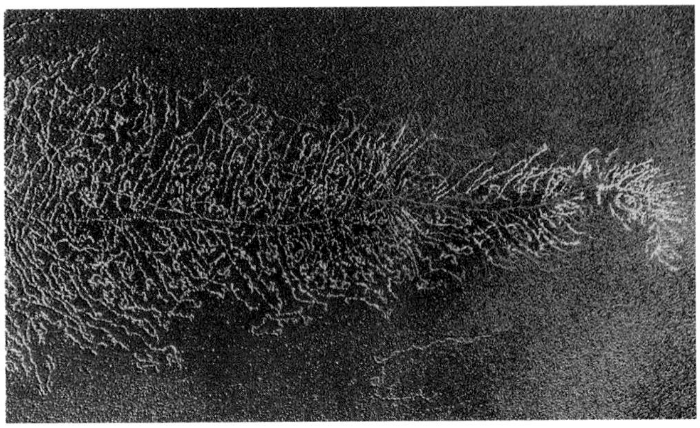

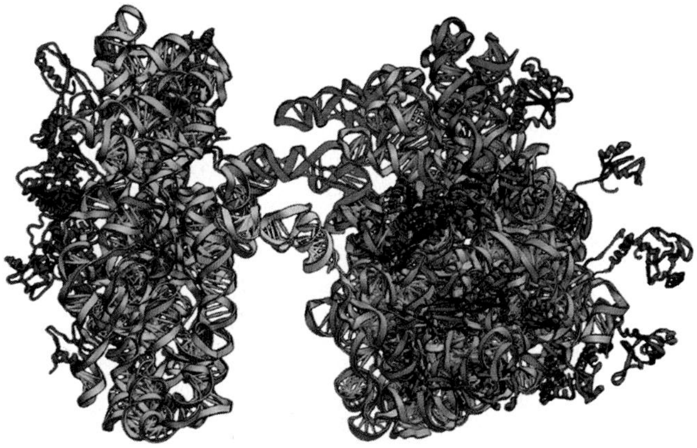

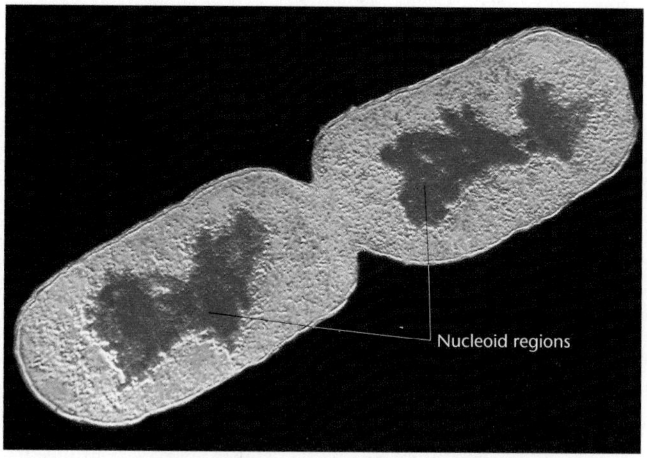
Nucleoid regions

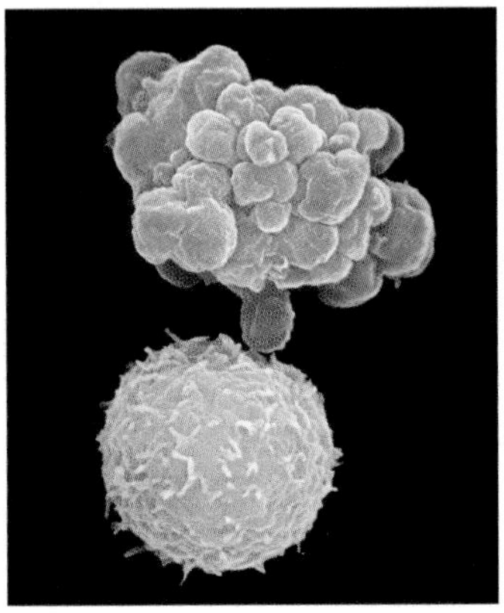

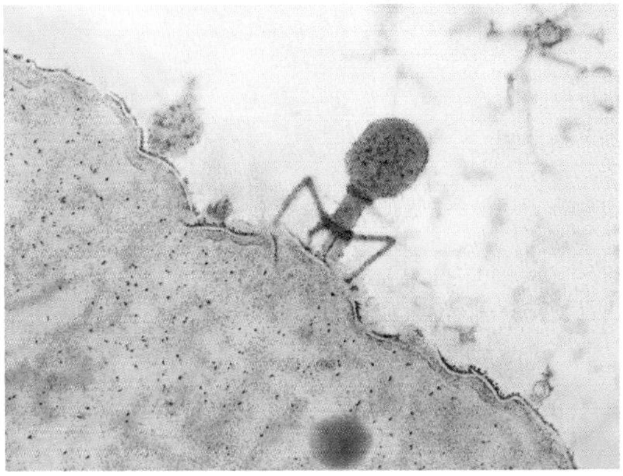

Part Four
Genomic Analysis

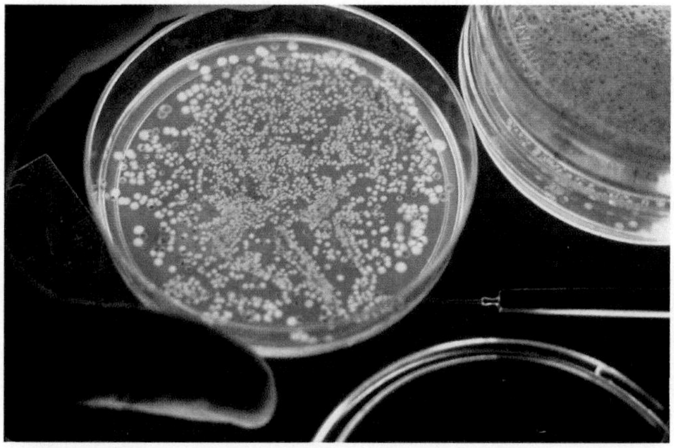

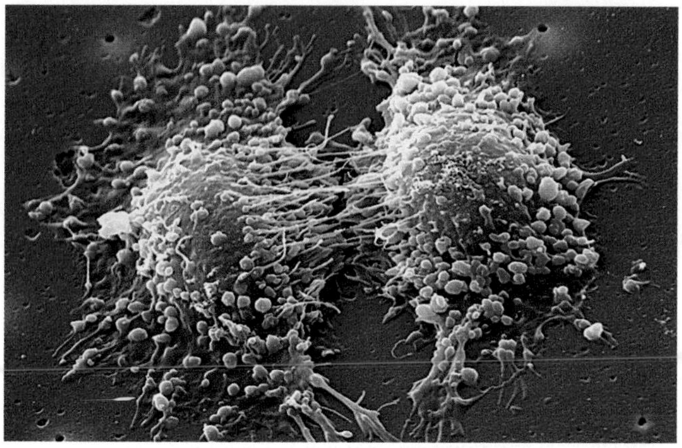

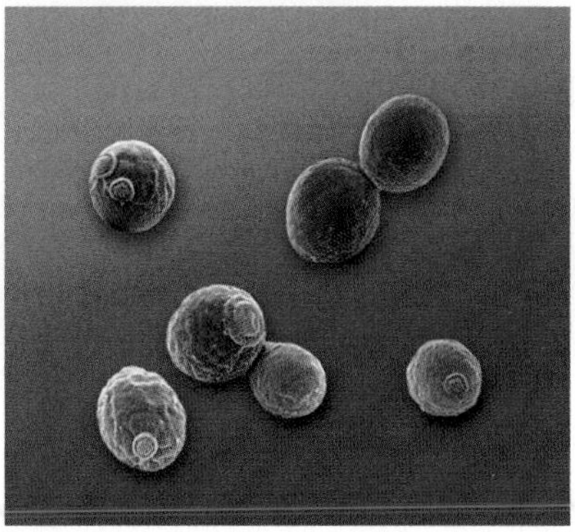

Preface

At times it is useful for textbook authors to step back and look with fresh eyes at their work, how it represents the body of information in its field, and how the material is structured, both scientifically and pedagogically. The publication of the 8th edition of *Concepts of Genetics*, now in its third decade of providing support to students, presents an occasion for a fresh look. The field of genetics has grown tremendously since the book was first published, both in what we know and what we want beginning students to comprehend. In creating this edition, we sought not only to familiarize students with the most important discoveries of the past 150 years, but also to help them relate this information to the underlying genetic mechanisms that explain cellular processes, biological diversity, and evolution. Further, we have also emphasized connections that link transmission genetics, molecular genetics, genomics, and proteomics.

In the early years of this new millennium, discoveries in genetics continue to be numerous and profound. As students of genetics, the thrill of being part of this era must be balanced by a strong sense of responsibility and careful attention to the many scientific, social, and ethical issues that have already arisen, and others that will undoubtedly arise in the future. Policy makers, legislators, and an informed public will increasingly depend upon knowledge of the details of genetics in order to address these issues. As a result, there has never been a greater need for a genetics textbook that clearly explains the principles of genetics.

Goals

In the 8th edition of *Concepts of Genetics*, as with all past efforts, we have six major goals. Specifically, we seek to:

- Emphasize the basic concepts of genetics;
- Write clearly and directly to students in order to provide understandable explanations of complex, analytical topics;
- Establish a careful organization within and between chapters;
- Maintain constant emphasis on science as a way of illustrating how we know what we know;
- Propagate the rich history of genetics that so beautifully elucidates how information is acquired within the discipline as it develops and grows;
- Create inviting, engaging, and pedagogically useful full-color figures enhanced by equally helpful photographs to support concept development.

These goals collectively serve as the cornerstone of *Concepts of Genetics*. This pedagogic foundation allows the book to be used in courses with many different approaches and lecture formats. While the book presents a coherent table of contents that represents one approach to offering a course in genetics, chapters are nevertheless written to be independent of one another, allowing instructors to utilize them in various sequences. We believe that the varied approaches embodied in these goals together provide students with optimal support for their study of genetics.

Writing a textbook that achieves these goals and having the opportunity to continually improve on each new edition has been a labor of love for us. The creation of each of the eight editions is a reflection of not only our passion for teaching genetics, but also the constructive feedback and encouragement provided by adopters, reviewers, and our students over the past three decades.

Features of This Edition

- Organization—A new organization better reflects how genetics is taught in the era of genomics. A new introductory chapter (Chapter 1) gives an overview of molecular biology as a way to connect the early transmission genetics chapters to molecular topics. Enhanced coverage of model organisms is woven throughout the chapters, but is especially prominent in the new chapter on dissection of gene function using mutational analysis (Chapter 21).
- Pedagogy—Two new features appear several times in each chapter. The first, "How Do We Know?," asks the student to reflect on the experimental basis of an important finding to more fully understand rather than memorize the conclusion drawn from the body of research. The second feature, "Now Solve This," directs the student to a problem found at the end of the chapter that is related to the previous text discussion. In each case, a pedagogic hint has been provided. Each feature displays an appropriate icon for ease of identification. It is our hope that the use of these features throughout each chapter will challenge the student to think more deeply about the information he or she has just finished studying.

- New/Revised Chapters—In keeping with our intent to increase the coverage of model organisms utilized in genetic study and to elaborate on the use of mutations to study gene function, we have created a new chapter entitled "Dissection of Gene Function: Mutational Analysis in Model Organisms" (Chapter 21). This chapter gathers together information previously spread throughout many chapters of the text and extends it considerably. The assembled information reflects the sophisticated, modern approaches used to study gene function and provides insights into the role of "model organisms" in genetic study. In addition, we have given particular attention to our coverage of quantitative genetics, population genetics, and evolutionary genetics (Chapters 24, 25, and 26). Coverage in these chapters has been extensively reviewed, and their revision reflects the best thinking of many colleagues whose specialized training is in these fields.
- Modernization of Topics—While we have updated each chapter to reflect the most current and significant findings in genetics, we have paid particular attention to the cutting edge topics of comparative genomics, bioinformatics, and proteomics (Chapter 20) because of their impact on future studies and on society. Also clearly evident are the discussions of the current applications of DNA biotechnology and the ethical issues arising from them (Chapter 22). Additionally, we have further updated and extended our presentation of the rapidly advancing field of cancer genetics (Chapter 18). An-in depth consideration of conservation genetics (Chapter 27) remains as another hallmark of our modern genetic coverage. This field, which attempts to assess and maintain genetic diversity in endangered species, remains at the forefront of genetic studies.
- New Illustrations Throughout—The 8th edition features a whole new color palette as well as the redrafting of almost all figures to enhance their pedagogic value and artistic quality. Many figures feature "flow diagrams" that visually guide a student through experimental protocols and techniques.
- New Photographs—An even greater number of photographs illustrate and enhance this edition.
- Section Numbers—All major sections are numbered, making it easier to assign and locate topics within chapters.
- New/Revised "Genetics, Technology, and Society" Essays—As with every new edition, we have revised many of these essays and added new ones that reflect recent findings in genetics and their impact on society. Among the new essays are those that consider Tay–Sachs disease (Chapter 3), Fragile Chromosome Sites and Cancer (Chapter 8), Quorum Sensing (Chapter 16), and Gene Deregulation and Human Disease (Chapter 17). These new additions supplement essays from past editions, including those that discuss edible vaccines, human sex selection, genetically modified foods, gene therapy, and endangered species such as the Florida panther, among many other topics.
- Emphasis on Problem Solving—Over 200 new problems have been added to the section "Problems and Discussion Questions" at the ends of chapters. Most are "high end" problems found in the subsection that we call "Extra-Spicy Problems." Many of these are based on data derived from the primary literature of genetics. The "Insights and Solutions" sections found at the ends of chapters continue to guide students in learning to think analytically about problems.
- Instructor and Student Media Address Real Needs—Support for lecture presentations and other teaching responsibilities has been increased, including electronic access to more text photos and tables and a greater variety of PowerPoint offerings on the book's Instructor Resource Center on CD/DVD. Media found on the revamped Companion Website reflect the growing awareness that today's students must use their limited study time as wisely as possible.

Emphasis on Concepts

Concepts of Genetics, as its title implies, emphasizes the conceptual framework of genetics. Our experience with this book, reinforced by the many adopters with whom we have been in contact over the years, demonstrates quite conclusively that students whose primary focus is on concepts more easily comprehend and take with them to succeeding courses the most important ideas in genetics as well as an analytic view of biological problem solving.

To aid students in identifying conceptual aspects of a major topic, each chapter begins with a new section called "Chapter Concepts," which outlines the most important ideas about to be presented. Within each chapter, as noted above, the "How Do We Know?" feature asks the student to connect concepts to experiments. Then, the "Now Solve This" feature asks students to link conceptual understanding in a more immediate way to problem solving. Additionally, each chapter ends with a "Chapter Summary," which enumerates the five to ten key points that have been covered. Collectively, these features help to ensure that students focus on and understand concepts as they confront the extensive vocabulary and the many important details of genetics. Carefully designed figures also support this approach throughout the book.

Problem Solving and Insights and Solutions

To optimize the opportunities for student growth in the important areas of problem solving and analytical thinking, each chapter integrates "Now Solve This" questions with accompanying pedagogic hints and concludes with an extensive collection of "Problems and Discussion Questions." These are designed with several levels of difficulty, with the most challenging (those called Extra-Spicy Problems) located at the end of each section. Brief answers to half the problems are in Appendix B. The *Student Handbook* answers every problem and is available to students when faculty decide that it is appropriate. As the reader familiar with previous editions will see, over 200 new problems appear at chapter ends.

As an aid to the student in learning to solve problems, the "Problems and Discussion Questions" section of each chapter is preceded by what has become an extremely popular and successful section called "Insights and Solutions." In this expanded section we stress:

Problem solving
Quantitative analysis
Analytical thinking
Experimental rationale

Problems or questions are posed and detailed solutions or answers are provided. This feature primes students for moving on to the "Problems and Discussion Questions" section that concludes each chapter.

The Genetics MediaLab section is available on the Companion Website. Each MediaLab contains several Web-linked problems designed to enhance and extend the topics presented in the chapter. To complete these problems students must actively participate in the exercises and virtual experiments. For reference, the estimated time required to solve the problem is noted at the beginning of the exercise.

Alternative Edition

To accommodate instructors who wish to organize their course to emphasize molecular genetics before transmission genetics there exists an alternative version of this book entitled *Genetics: A Molecular Perspective*. It begins with DNA and establishes the basis of genetic function at the molecular level before returning to the consideration of the historically important discoveries that fall under the descriptor of "transmission genetics."

Acknowledgments

Contributors

We begin with special acknowledgments to those who have made direct contributions to this text. We particularly thank Sarah Ward at Colorado State University for creating Chapter 27 on Conservation Genetics, and also for providing revised drafts of the chapters involving quantitative and population genetics. Additionally, Amanda Norvell revised several sections emphasizing eukaryotic molecular genetics and Janet Morrison revised numerous aspects of our coverage of evolutionary genetics. Katherine Uyhazi wrote the essay on quorum sensing in bacteria (Chapter 16) and helped revise several other essays. The latter three individuals are colleagues from The College of New Jersey. Additionally, we thank Mark Shotwell at Slippery Rock University for contributing several "Genetics, Technology, and Society" essays. As with previous editions, Elliott Goldstein from Arizona State University was always readily available to consult with us concerning the most modern findings in molecular genetics. We also express special thanks to Harry Nickla at Creighton University. In his role as author of the *Student Handbook* and the *Instructor's Manual*, he has reviewed and edited the problems at the end of each chapter. He also provided the brief answers to selected problems that appear in Appendix B.

We are grateful to all of the above contributors not only for sharing their genetic expertise, but for their dedication to this project as well as the pleasant interactions they provided.

Proofreaders

Proofreading a manuscript of a 784 page text deserves more thanks than words can offer. Our utmost appreciation is extended to the four individuals who confronted this task with patience, diligence, and good humor:

Tamara Horton Mans, Princeton University
Arlene Larson, University of Colorado, Denver
Virginia McDonough, Hope College
Janice Rumph, Montana State University

Reviewers

All comprehensive texts are dependent on the valuable input provided by many reviewers. While we take full responsibility for any errors in this book, we gratefully acknowledge the help provided by those individuals who reviewed the content and pedagogy of this and the previous edition:

Laurel F. Appel, Wesleyan University
Ruth Ballard, California State University, Sacramento
George Bates, Florida State University
Sidney L. Beck, DePaul University
Peta Bonham-Smith, University of Saskatchewan
Paul J. Bottino, University of Maryland
Philip Busey, University of Florida
Alan H. Christensen, George Mason University
Jim Clark, University of Kentucky
Diane Dalo, The College of New Jersey
Garry Davis, University of Alaska, Anchorage
Johnny El-Rady, University of South Florida
Bert Ely, University of South Carolina
Lloyd M. Epstein, Florida State University
Dale Fast, Saint Xavier University
Donald Gailey, California State University, Hayward
George W. Gilchrist, Clarkson University
Sandra Gilchrist, University of South Florida
Thomas J. Glover, Hobart & William Smith Colleges
Elliott S. Goldstein, Arizona State University
Douglas Harrison, University of Kentucky
Don Hauber, Loyola University
Vincent Henrich, University of North Carolina, Greensboro
Philip L. Hertzler, Central Michigan University
Margaret Hollingsworth, SUNY–Buffalo
Rebecca Jann, Queens College
Mitrick A. Johns, Northern Illinois University
Paul F. Lurquin, Washington State University
Clint Magill, Texas A&M University
John McDonald, University of Delaware
Janet Morrison, The College of New Jersey

Harry Nickla, Creighton University
Amanda Norvell, The College of New Jersey
Berl R. Oakley, Ohio State University
Marcia O'Connell, The College of New Jersey
John C. Osterman, University of Nebraska–Lincoln
Dennis T. Ray, University of Arizona
Joseph Reese, Pennsylvania State University
Janice Rumph, Montana State University
Thomas F. Savage, Oregon State University
Cathy Schaeff, American University
Rodney Scott, Wheaton College
Thomas P. Snyder, Michigan Technological University
Paul Spruell, University of Montana
Christine Tachibana, University of Washington
Marty Tracey, Florida International University
Albrecht von Arnim, University of Tennessee
Laurence von Kalm, University of Central Florida
Tracy Whitford, East Stroudsburg University
Janice Rumph, Montana State University

Media reviewers include:

Peggy Brickman, University of Georgia
Carol Chihara, University of San Francisco
Karen Hughes, University of Tennessee
Cheryl Ingram-Smith, Clemson University
David Kass, Eastern Michigan University
John Kemner, University of Washington
Arlene Larson, University of Colorado
John C. Osterman, University of Nebraska, Lincoln
Eric Stavney, DeVry University

Special thanks go to Mike Guidry of LightCone Interactive and Karen Hughes of the University of Tennessee for their original contributions to the media program.

As the above acknowledgments make clear, a text such as this is a collective enterprise. All of the above individuals deserve to share in any success this text enjoys. We want them to know that our gratitude is equaled only by the extreme dedication evident in their efforts. Many, many thanks to them all.

Editorial and Production Input

At Prentice Hall, we express appreciation and high praise for the editorial guidance of Sheri Snavely and Gary Carlson, whose ideas and efforts have helped to shape and refine the features of this and the previous editions of the text. They have worked tirelessly to provide us with reviews from leading specialists who are also dedicated teachers, and to ensure that the pedagogy and design of the book are at the cutting edge of a rapidly changing discipline. We were most fortunate to benefit from valuable developmental editing provided by Anne Scanlan-Roher. We also appreciate the production efforts of those at Preparé Inc., whose quest for perfection is reflected throughout the text. In particular, Fran Daniele and Lorenza Compagnone provided an essential measure of sanity to the otherwise chaotic process of production. Without their work ethic and dedication, the text would never have come to fruition. Shari Meffert and Andrew Gilfillan have professionally and enthusiastically managed the marketing of the text. Crissy Dudonis, biology project manager, has worked tirelessly to ensure that the book, supplements, and cutting-edge interactive media all work together seamlessly. Patrick Shriner managed the media to ensure a stronger emphasis on addressing the needs of today's instructors and students. Finally, the beauty and consistent presentation of the art work is the product of Argosy, Artworks of York, PA, and Imagineering of Toronto. We particularly thank Patricia Burns, Jay McElroy, and Heidi Bertignoll for their efforts.

For the Student

Companion Website—www.prenhall.com/klug

Web Tutorial

Respect for the students' increasingly valuable study time is reflected by the features of the Companion Website, which have been designed to enable users of the Eighth Edition to focus on those chapter sections and topics where they need review or further explanation. The Online Study Guide provides students with a focused, section-by-section review of topic coverage that features concise summary points accompanied by key illustrations and probing review questions that offer hints and feedback. The Web Tutorials offer today's learners the opportunity to quickly and conveniently visualize complex topics and dynamic processes—or to simply refamiliarize themselves with concepts they may have learned earlier but are encountering for the first time in the context of a genetics course. Several new Web Tutorials have been added to the Eighth Edition to keep the media current with the book. The media's strict adherence to both the principles and specific lessons of the textbook means that students and instructors can be assured that study time isn't being squandered on media that confuses students and emphasizes extraneous topics. The media tab on the outside margin of this page appears throughout the book to let you know when there is a Web Tutorial on a topic related to the coverage in the book.

In addition, a Media Lab for each chapter is offered on the Companion Website for those who want to explore genetics beyond the boundaries of a book through the vast array of genetics-related resources available through the Web.

Student Handbook and Solutions Manual

Harry Nickla, Creighton University (0-13-149008-7)
This valuable handbook provides a detailed step-by-step solution or lengthy discussion for every problem in the text. The handbook also features additional study aids including extra study problems, chapter outlines, vocabulary exercises, and an overview of how to study genetics.

New York Times Themes of the Times: Genetics and Molecular Biology

Compiled by Harry Nickla, Creighton University
(0-13-186604-4)
This exciting supplement brings together recent genetics and molecular biology articles from the pages of the *New York*

Times. This free supplement, available through your local representative, encourages students to make the connections between genetic concepts and the latest research and breakthroughs that are making headlines in the field. This resource is updated regularly.

"Research Navigator" www.researchnavigator.com

Prentice Hall's Research Navigator ™ gives your students access to the most current information available for a wide array of subjects via EBSCO's Content Select ™ Academic Journal Database, The New York Times Search by Subject Archive, "Best of the Web" Link Library, and information on the latest news and current events. This valuable tool helps students find the most useful articles and journals, cite sources, and write effective papers for research assignments.

For the Instructor

Instructor Resource Center on CD/DVD

(0-13-149011-7)

The Instructor Resource Center on CD/DVD for the Eighth Edition offers adopters convenient access to the most comprehensive and innovative set of lecture presentation and teaching tools offered by any genetics textbook. Developed to meet the needs of veteran and newer instructors alike, these resources include:

- The JPEG files of all text line drawings with labels individually enhanced for optimal projection results (as well as unlabeled versions) and all text tables
- Most of the text photos, including all photos with pedagogical significance, as JPEG files
- The JPEG files of line drawings, photos, and tables preloaded into comprehensive PowerPoint® presentations for each chapter
- A second set of PowerPoint® presentations consisting of a thorough lecture outline for each chapter augmented by key text illustrations
- An impressive series of concise instructor animations adding depth and visual clarity to the most important topics and dynamic processes described in the text
- The instructor animations preloaded into PowerPoint® presentation files for each chapter
- PowerPoint® presentations containing a comprehensive set of in-class Classroom Response System (CRS) questions for each chapter
- In Word files, a complete set of the assessment materials and study questions and answers from the Testbank, the text's in-chapter text questions, and the student media practice questions, as well as files containing the entire *Instructor's Manual and Solutions Manual*
- Finally, to help instructors keep track of all that is available in this media package, a printable Media Integration Guide in PDF format that lists each chapter's media offerings

Instructor's Resource Manual with Testbank

(0-13-149007-9)

This manual and test bank contains over 1000 questions and problems for use in preparing exams. The manual also provides optional course sequences, a guide to audiovisual supplements, and a section on searching the Web. The testbank portion of the manual is also available in electronic format.

TestGen EQ Computerized Testing Software

(0-13-149010-9)

In addition to the printed volume, the test questions are also available as part of the TestGen EQ Testing Software, a text-specific testing program that is networkable for administering tests. It also allows instructors to view and edit questions, export the questions as tests, and print them out in a variety of formats.

Transparencies

(0-13-149019-2)

Two hundred seventy-five figures from the text are included in the transparency package: 200 four-color transparencies from the text plus 75 transparency masters. The font size of the labels has been increased and boldfaced for easy viewing from the back of the classroom.

Introduction to Genetics

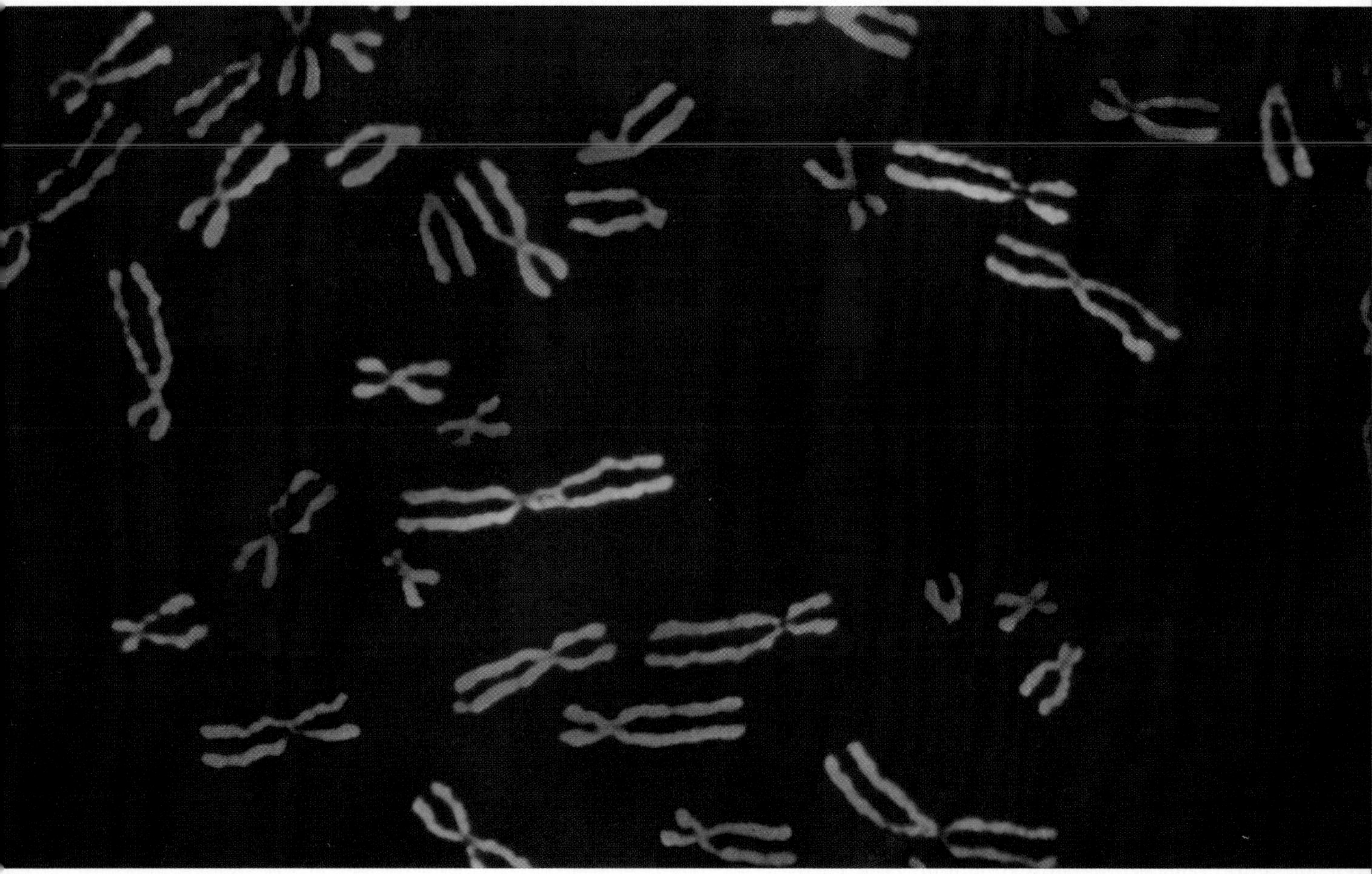

Human metaphase chromosomes, each composed of two sister chromatids joined at a common centromere.

Chapter Concepts

- Using pea plants, Mendel revealed the fundamental principles of transmission genetics. Work by others showed that genes are on chromosomes, and that mutant strains can be used to map genes on chromosomes.
- The discovery that DNA encodes genetic information and solving the structure of DNA and the mechanism of gene expression form the foundation of molecular genetics.
- The development of recombinant DNA technology revolutionized genetics and is the foundation for genome sequencing, including the Human Genome Project.
- Biotechnology uses recombinant DNA technology to produce goods and services in a wide range of areas, including agriculture, medicine, and industry. The use of biotechnology has raised many legal and ethical issues involving the patenting of genetically modified organisms, and the use of gene therapy.
- Model organisms have been used in genetics since the early part of the twentieth century. The extensive genetic knowledge gained from these organisms coupled with recombinant DNA technology and genomics makes these organisms useful as models to study human diseases.
- Genetic technology is developing faster than policy, law, and convention in the use of this technology. Education and participation are key elements in the wise use of this technology.

From our perspective early in the twenty-first century, we can look back and ask when the interaction between genetic technology and society began to affect our lives. Did it begin when recombinant DNA technology was used to produce insulin, when the first food produced by genetic engineering reached the marketplace, or when gene therapy was first used to treat a genetic disorder? While each of these is an incremental step in using genetic knowledge in ways affecting society, we will focus on a case where the use of genetic technology directly affects the citizens of an entire country. This case captures how genetics has a significant impact on society and provides a glimpse of future implications as more advances and applications are developed.

In December 1998, a controversy affecting the 270,000 residents of the island nation of Iceland was coming to a head. Following months of heated debate, the Icelandic Parliament passed a law granting deCODE, a biotechnology company with headquarters in Iceland, a license to create and operate a database called IHD (Icelandic Health Sector Database) containing detailed information drawn from coded (to ensure anonymity) medical records of all Iceland residents. The law also allows deCODE to cross-reference medical information from the IHD with a comprehensive genealogical database from the National Archives. In addition, deCODE can correlate information in these two databases with results of DNA profiles collected from Icelandic donors. The combination of disease, genealogical, and genetic information is a powerful resource available exclusively to deCODE, which can market this information to researchers and companies for a period of 12 years.

This is not a scenario from a movie such as *Gattaca*, but a real example of the interaction underway between genetics and society as we begin a new century. The development and use of these databases in Iceland has spurred the creation of similar projects in other countries as well. The largest is the "UK Biobank" effort launched in Great Britain in 2003. There, a huge database containing the genetic information of 500,000 Britons will be compiled from an initial group of 1.2 million residents. The database will be used to search for susceptibility genes that control complex traits.

Similar projects to develop nationwide databases have been announced in Estonia, Latvia, Sweden, Singapore, and the Kingdom of Tonga, illustrating the global impact of genetic technology.

In the United States, smaller-scale programs involving tens of thousands of individuals are underway at the Marshfield Clinic in Marshfield, Wisconsin; Northwestern University in Chicago, Illinois; and Howard University in Washington, D.C.

Why did deCODE select Iceland for such a project? Because for several reasons, the people of Iceland represent a unique case of genetic uniformity seldom seen or accessible to scientific investigation. This high degree of genetic relatedness results from the peopling of Iceland about 1,000 years ago by a small founding population drawn mainly from Scandinavian and Celtic sources, periodic population reductions by disease and natural disasters that further reduced genetic diversity and, until the last few decades, a lack of immigrants bringing new genes into the population. Thus, for geneticists trying to identify genes that control complex disorders, the Icelandic population is a tremendous asset. Because of the state-supported health care system, medical records exist for all residents as far back as the early 1900s. Genealogical information is available in the National Archives and church records for almost every resident and for more than 500,000 of the estimated 750,000 individuals who have ever lived in Iceland. In spite of the associated controversies, the project already has a number of successes to its credit. Scientists at deCODE have identified genes associated with more than 25 of the most common diseases, including asthma, heart disease, stroke, and osteoporosis.

On the flip side of these successes are issues of privacy, consent, and commercialization—issues at the heart of many controversies arising from the applications of genetic technology. Scientists and nonscientists alike are considering the fate and control of genetic information as it is acquired, and the role of law and society in decisions about how and when genetic technology is used. For example, how will knowledge of the complete nucleotide sequence of the human genome be used? More than at any other time in the history of science, addressing the ethical questions surrounding an emerging technology is now as important as the information gained from that technology.

As you launch your study of genetics, remain sensitive to questions and issues like those just described. There has never been a more exciting time to be part of this science, but never has the need for caution and awareness of social issues been more apparent. This text will enable you to achieve a thorough understanding of modern-day genetics and its underlying principles. Along the way, enjoy your studies, but take your responsibilities as a novice geneticist very seriously.

1.1 From Mendel to DNA in Less Than a Century

Because genetic processes are fundamental to the comprehension of life itself, the discipline of genetics is thought by many to sit at the center of biology. Genetic information directs cellular function, largely determines an organism's external appearance, and serves as the link between generations in every species. As such, knowledge of genetics is essential to the thorough understanding of other disciplines, including molecular biology, cell biology, physiology, evolution, ecology, systematics, and behavior. Genetics therefore unifies biology and serves as its core. Thus, it is not surprising that genetics has a long, rich history. Our starting point for this history is a monastery garden in central Europe in the 1860s.

Mendel's Work on Transmission of Traits

In this garden (Figure 1–1) Gregor Mendel, an Augustinian monk, conducted a decade-long series of experiments using pea plants. In his work, Mendel showed that traits are passed from parents to offspring in predictable ways. From his work, he concluded that traits in pea plants, such as height and flower color, are controlled by discrete units of inheritance we now call **genes**. He further concluded that genes controlling a trait exist in pairs, and that members of a gene pair separate from each other during gamete formation. His work was published in 1866 but was largely ignored until it was partially duplicated and cited in papers by Carl Correns and others around 1900. Having been confirmed by others, Mendel's findings became

FIGURE 1–1 The monastery garden where Gregor Mendel conducted his experiments with garden peas. In 1866, Mendel put forward the major postulates of transmission genetics.

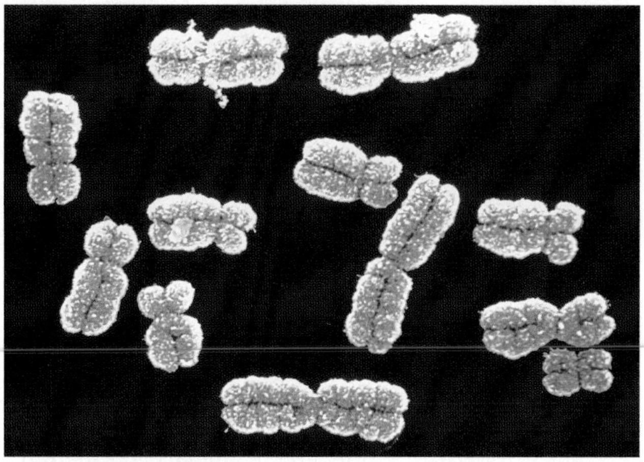

FIGURE 1–2 Colorized image of human mitotic chromosomes as visualized under the scanning electron microscope.

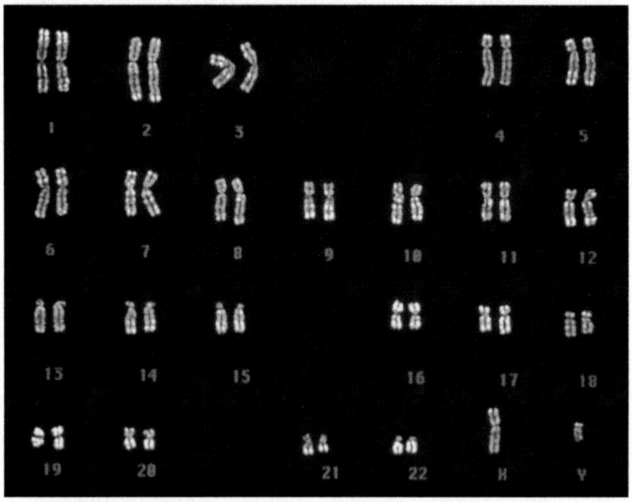

FIGURE 1–3 A colorized image of the human male chromosome set. Arranged in this way, the set is called a karyotype.

recognized as the basis for the transmission of traits in pea plants and all other higher organisms. His work forms the foundation for **genetics**, which is defined as the branch of biology concerned with the study of heredity and variation.

The Chromosome Theory of Inheritance: Uniting Mendel and Meiosis

Mendel did his work before the structure and role of chromosomes was known. About twenty years after his work, advances in microscopy allowed researchers to identify chromosomes (Figure 1–2), and to establish that in eukaryotic organisms (those with a nucleus and cellular membrane systems), each species has a characteristic number of chromosomes called the **diploid number ($2n$)**. For example, humans have a diploid number of 46 (Figure 1–3). Chromosomes in diploid cells exist in pairs, called **homologous chromosomes**. Members of a pair are identical in size and location of the centromere, a structure to which spindle fibers attach during division.

In addition, researchers in the last decades of the nineteenth century described the behavior of chromosomes during two forms of cell division, **mitosis** and **meiosis**. In mitosis (Figure 1–4), chromosomes are copied and distributed so that the two resulting daughter cells each receive a diploid set of

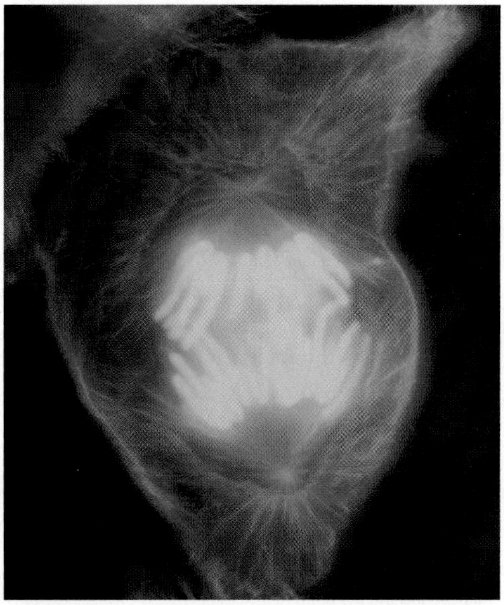

FIGURE 1–4 A stage in mitosis (anaphase) when the chromosomes (stained blue) move apart.

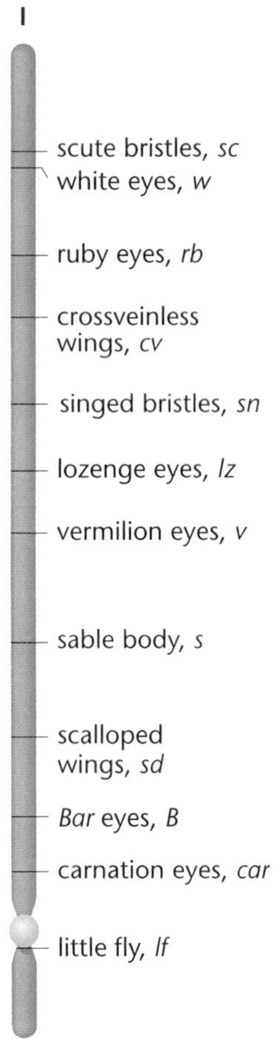

FIGURE 1–5 Chromosome I (the X chromosome) of *D. melanogaster*, showing the location of many genes. Chromosomes can contain hundreds of genes.

chromosomes. Meiosis is a form of cell division associated with gamete formation in animals and spore formation in most plants. Cells produced by meiosis receive only one copy of each chromosome, called the **haploid (*n*) number** of chromosomes. This reduction in chromosome number is essential if the offspring arising from two gametes are to maintain a constant number of chromosomes characteristic of their parents and other members of their species.

Early in the twentieth century, Walter Sutton and Theodore Boveri independently noted that genes and chromosomes have properties in common and that the behavior of chromosomes during meiosis is identical to the behavior of genes during gamete formation. For example, genes and chromosomes exist in pairs and members of a gene pair and members of a chromosome pair separate from each other during gamete formation. Based on these parallels, they each proposed that genes are carried on chromosomes (Figure 1–5). This proposal is the basis of the **chromosome theory of inheritance**, which states that inherited traits are controlled by genes residing on chromosomes that are faithfully transmitted through gametes, maintaining genetic continuity from generation to generation.

Genetic Variation

About the same time that the chromosome theory of inheritance was proposed, scientists began studying the inheritance of traits in the fruit fly, *Drosophila melanogaster*. Soon after, a white-eyed fly (Figure 1–6) was discovered in a bottle containing normal (wild-type) red-eyed flies. This variation was produced by **mutation**, an inherited change in the gene controlling eye color. Chromosomal mutations affect the number and structure of chromosomes. Mutations, whether genetic or chromosomal are defined as any heritable change, and are the source of all genetic variation.

The variant gene discovered in *Drosophila* represents an **allele** of the eye color gene. Alleles are defined as alternative forms of a gene. Different alleles may produce differences in the observable features, or **phenotype** of an organism. The set of alleles for a given trait carried by an organism is called the **genotype**. Using mutant genes as markers, geneticists were able to map the location of genes on chromosomes.

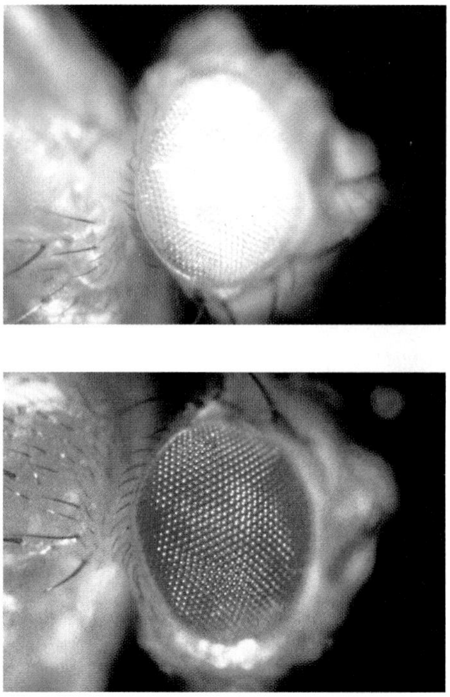

FIGURE 1–6 The normal red eye color in *Drosophila* (bottom) and the white-eyed mutant (top).

The Search for the Chemical Nature of Genes: DNA or Protein?

Work on white-eyed *Drosophila* showed that the mutant trait had a pattern of inheritance that could be traced to a single chromosome, confirming the idea that genes are carried on chromosomes. Once this was established, investigators turned their attention to identifying which chemical component of chromosomes carried genetic information. By the 1920s, DNA and proteins were identified as the major chemical components of chromosomes. Proteins are the most abundant component in cells. There are a large number of different proteins and because of their universal distribution in the nucleus and cytoplasm, many researchers thought proteins would be shown to be the carrier of genetic information.

In 1944, Avery, MacLeod, and McCarty, three researchers at the Rockefeller Institute in New York, provided experimental evidence that DNA was the carrier of genetic information in bacteria. This evidence, although clear-cut, failed to convince many influential scientists. Additional evidence for the role of DNA as a carrier of genetic information came from other researchers who worked with viruses that infect and kill cells of the bacterium *Escherichia coli* (Figure 1–7). One of these viruses, called a **bacteriophage**, or **phage** for short, consists of a protein coat surrounding a DNA core. These experiments showed that the protein coat of the virus remains outside the cell, while the DNA enters the cell and directs the synthesis and assembly of more phage. This work was more proof that

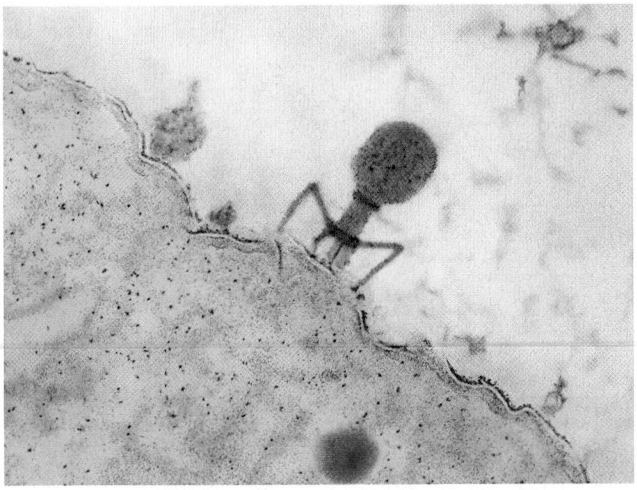

FIGURE 1–7 An electron micrograph showing T phage infecting a cell of the bacterium *E. coli.*

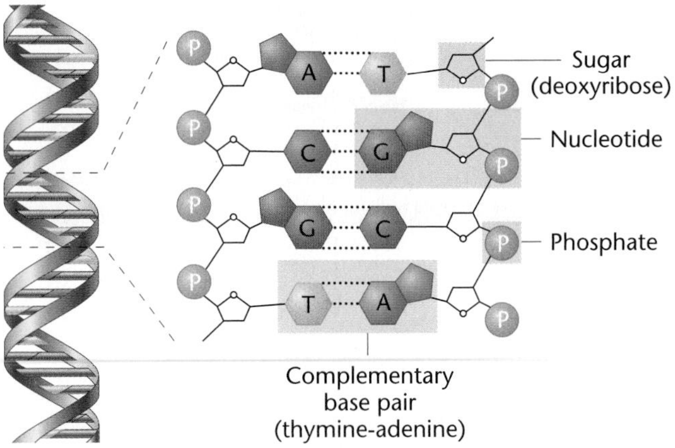

FIGURE 1–8 Summary of the structure of DNA, illustrating the nature of the double helix (on the left) and the chemical components making up each strand (on the right).

DNA carries genetic information. Additional experiments over the next few years provided solid proof that DNA, not protein, is the genetic material, setting the stage for work to establish the structure of DNA.

1.2 Discovery of the Double Helix Launched the Recombinant DNA Era

Once it was accepted that nucleic acid in the form of DNA carries genetic information, efforts were focused on deciphering the structure of DNA, and the mechanism by which information stored in this molecule is expressed to produce an observable phenotype. In the years after this was accomplished, researchers learned how to isolate and make copies of specific regions of DNA molecules, opening the way for the era of recombinant DNA technology.

The Structure of DNA and RNA

DNA is a long, ladder-like molecule that forms a double helix. Each strand of the helix is a linear molecule made up of subunits called **nucleotides**. In DNA, there are four different nucleotides. Each DNA nucleotide contains one of four nitrogenous bases—A (adenine), G (guanine), T (thymine), or C (cytosine). These four bases comprise the genetic alphabet, or **genetic code**, which in various combinations ultimately specify the amino acid sequence of proteins. One of the great discoveries of the twentieth century was made in 1953 by James Watson and Francis Crick, who established that the two strands of DNA are exact complements of one another, such that the rungs of the ladder in the double helix always consist of either A═T or G≡C base pairs. As we shall see in a later chapter, this **complementary relationship** between adenine and thymine and between guanine and cytosine is critical to genetic function. This relationship serves as the basis for both the replication of DNA and for the basis of gene expression. During both processes, DNA strands serve as templates for the synthesis of complementary molecules. Two depictions of the structure and components of DNA are shown in Figure 1–8.

RNA, another nucleic acid, is chemically similar to DNA. RNA contains a different sugar (ribose vs. deoxyribose) in its nucleotides, and contains the nitrogenous base uracil in place of thymine. Additionally, in contrast to the double helix of DNA, RNA is generally single stranded. Importantly, it can form complementary structures with a strand of DNA.

Gene Expression: From DNA to Phenotype

As noted earlier, complementarity is the basis for steps in gene expression. This process begins with the **transcription** of the chemical information in DNA into RNA (Figure 1–9). Once an RNA molecule complementary to one strand of DNA is transcribed, the RNA directs the synthesis of proteins. This is accomplished when the RNA—called **messenger RNA**, or **mRNA**, for short—binds to a **ribosome**. The synthesis of proteins under the direction of mRNA is called **translation** (bottom part of Figure 1–9). Proteins, as the end product of genes, are polymers made up of amino acid monomers. There are 20 different amino acids in living organisms.

How can information contained in mRNA direct the insertion of specific amino acids into protein chains as they are synthesized? The answer is now quite clear. The **genetic code** consists of a linear series of triplet nucleotides present in mRNA molecules. Each triplet reflects the information stored in DNA and specifies the insertion of a specific amino acid into the growing protein chain. This is accomplished by the action of adapter molecules called **transfer RNA (tRNA)**. Within the ribosome, tRNAs recognize the information encoded in the mRNA triplets and specify the proper amino acid for insertion into the protein during translation.

As the preceding discussion shows, DNA makes RNA, which most often makes protein. These processes, known as the **central dogma** of genetics, occur with great specificity. Using

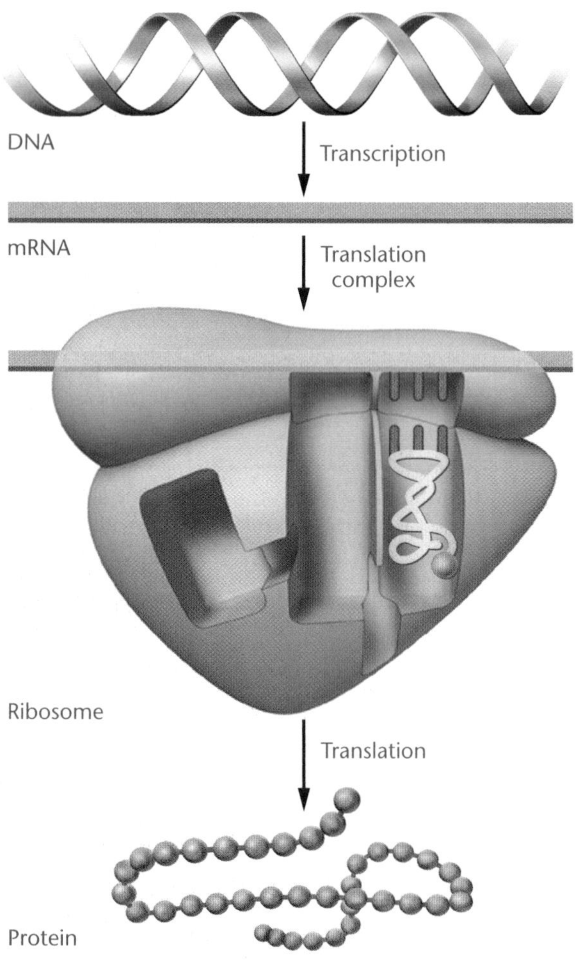

FIGURE 1–9 Gene expression involves transcription of DNA into mRNA (top) and the translation (center) of mRNA on a ribosome into a protein (bottom).

an alphabet of only four letters (A, T, C, and G), genes direct the synthesis of highly specific proteins that collectively serve as the basis for all biological function.

Proteins and Biological Function

As we have mentioned, proteins are the end products of gene expression. These molecules are responsible for imparting the properties that we attribute to living systems. The diverse nature of biological function rests with the fact that proteins are made from 20 different amino acids. In a protein chain that is just 100 amino acids in length, at each position there can be any one of 20 amino acids; the number of different 100 amino acid proteins, each with a unique sequence, is equal to

$$20^{100}$$

Because 20^{10} exceeds 5×10^{12}, or more than 5 trillion, imagine how large 20^{100} is! Obviously, evolution has seized on a class of molecules with the potential for enormous structural diversity to serve as the mainstay of biological systems.

The largest category of proteins includes **enzymes** (Figure 1–10). These molecules serve as biological catalysts, essentially allowing biochemical reactions to proceed at rates that

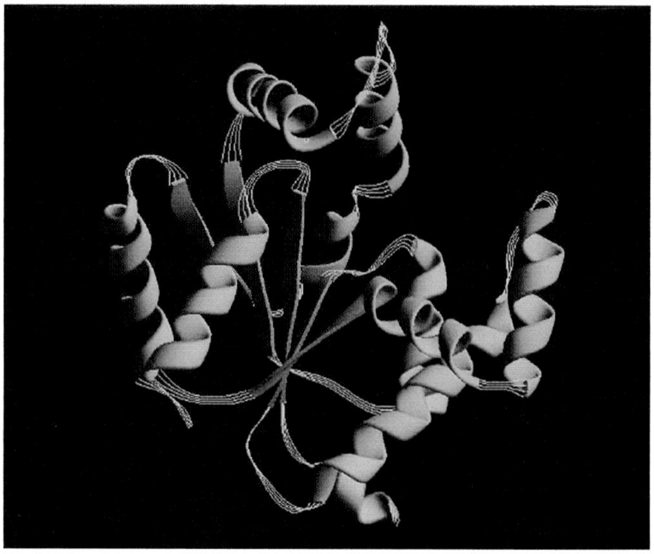

FIGURE 1–10 The three-dimensional conformation of a protein. The amino acid sequence of the protein is depicted as a ribbon.

sustain life under the conditions that exist on Earth. By lowering the energy of activation in reactions, metabolism is able to proceed under the direction of enzymes at body temperature.

There are countless proteins other than enzymes that are critical components of cells and organisms. These include **hemoglobin**, the oxygen-binding pigment in red blood cells; **insulin**, the pancreatic hormone; **collagen**, the connective tissue molecule; **keratin**, the structural molecule in hair; **histones**, the proteins integral to chromosome structure in eukaryotes; **actin** and **myosin**, the contractile muscle proteins; and **immunoglobulins**, the antibody molecules of the immune system. The potential for such diverse functions rests with the enormous variations in three-dimensional conformation of proteins. This conformation is determined by the linear sequence of amino acids constituting the molecule. To come full circle, this sequence is dictated by the stored information in the DNA of a gene that is transferred to RNA, which then directs the synthesis of a protein. DNA makes RNA that then makes protein.

Linking Genotype to Phenotype: Sickle-Cell Anemia

Once a protein is made, its action or location in a cell plays a role in producing a phenotype. When mutation alters a gene, it may abolish or alter that protein's function, and cause an altered phenotype. To trace the chain of events leading from the synthesis of a protein to a phenotype, we will examine sickle-cell anemia, a human genetic disorder. Sickle-cell anemia is caused by a mutant form of hemoglobin, the protein that transports oxygen from the lungs to cells in the body (Figure 1–11). Hemoglobin is a composite molecule made up of two different proteins, α-globin and β-globin, each encoded by a different gene. Each functional hemoglobin molecule contains two α-globin and two β-globin proteins. In sickle-cell anemia, a mutation in the gene encoding β-globin causes an amino acid substitution in 1 of the 146 amino acids in the protein. Figure 1–12 shows part of the DNA sequence,

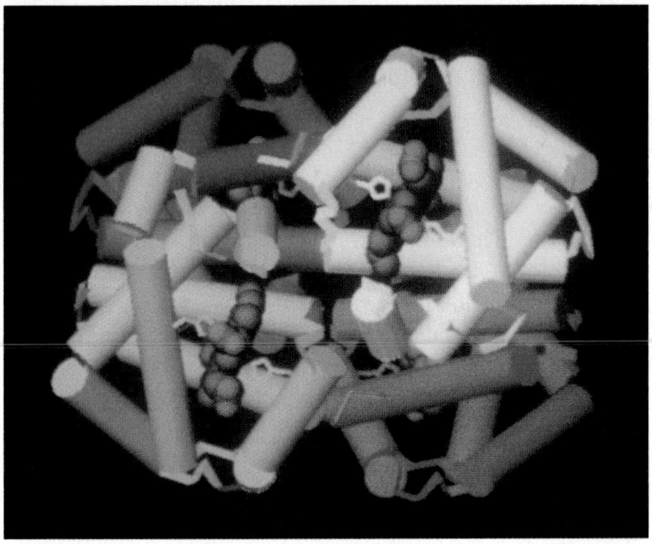

FIGURE 1–11 The hemoglobin molecule, showing the two alpha chains and the two beta chains. A mutation in the gene for the beta chain produces abnormal hemoglobin molecules and sickle cell anemia.

mRNA codons, and amino acid sequence for the normal and mutant forms of β-globin. Notice that the mutation in sickle-cell anemia involves a change in one DNA nucleotide, leading to a change in codon 6 in the mRNA from GAG to GUG, which in turn changes amino acid number 6 in β-globin from glutamic acid to valine. The other 145 amino acids in the protein are not changed by this mutation.

Individuals with two mutant copies of the β-globin gene have sickle-cell anemia. The mutation causes hemoglobin molecules in red blood cells to polymerize when oxygen concentration is low, forming long chains that distort the shape of red blood cells (Figure 1–13). When the blood cells are sickle shaped, they block the flow of blood in capillaries and small blood vessels, causing severe pain and damage to tissues including the heart, brain, muscles, and kidneys. Sickle-cell anemia can result in heart attacks and stroke, and can be fatal if left untreated. In addition, the deformed blood cells break easily, causing anemia by reducing the number of red cells in circulation. Thus,

NORMAL β-GLOBIN

DNA..........................	TGA	GGA	CTC	CTC...........
mRNA.......................	ACU	CCU	GAG	GAG...........
Amino acid..............	thr	pro	glu	glu ....

MUTANT β-GLOBIN

DNA..........................	TGA	GGA	CAC	CTC...........
mRNA.......................	ACU	CCU	GUG	CTC...........
Amino acid..............	thr	pro	val	glu ....

FIGURE 1–12 A single nucleotide change in the DNA encoding the β-globin gene (CTC ⟶ CAC) leads to an altered mRNA codon (GAG ⟶ GUG) and the insertion of a different amino acid (glu ⟶ val), producing an altered version of the β-globin protein, causing sickle cell anemia.

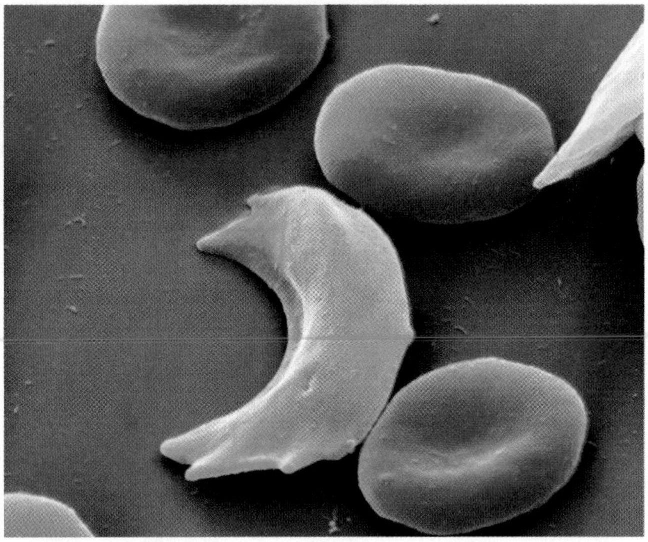

FIGURE 1–13 Normal red blood cells (round) and sickled red blood cells. The sickled cells block capillaries and small blood vessels.

all the symptoms of this disorder are caused by a change in a single nucleotide in a gene that changes one amino acid out of 146 in the β-globin molecule, emphasizing the close relationship between genotype and phenotype.

The β-globin gene is not expressed until a few days after birth, so the mutant protein cannot be detected prenatally. However, by using recombinant DNA technology, the mutant *gene* can be detected prenatally. In addition, the genotypes of family members and others can also be determined, making it possible for people to know if they carry a mutant copy of the gene and are at risk for having an affected child.

1.3 Genomics Grew Out of Recombinant DNA Technology

The era of recombinant DNA began in the early 1970s when researchers discovered that bacteria protect themselves from viral infection by making enzymes that restrict or prevent infection by cutting viral DNA at specific sites. When cut, the viral DNA cannot direct the synthesis of more phage particles, which when released, kill the infected bacterial cell. Scientists quickly realized that such enzymes, called **restriction enzymes**, could be used to cut DNA from any organism at specific nucleotide sequences, producing a reproducible set of fragments. This set the stage for the development of cloning, or making large numbers of copies of these DNA fragments.

Making Recombinant DNA Molecules and Cloning DNA

Soon after it was discovered that restriction enzymes could be used to produce specific DNA fragments, methods were developed to insert these fragments into carrier DNA molecules called vectors and transfer the combined vector and DNA fragment (a **recombinant DNA** molecule) into bacterial cells where hundreds or thousands of copies, or **clones**, of the vector and

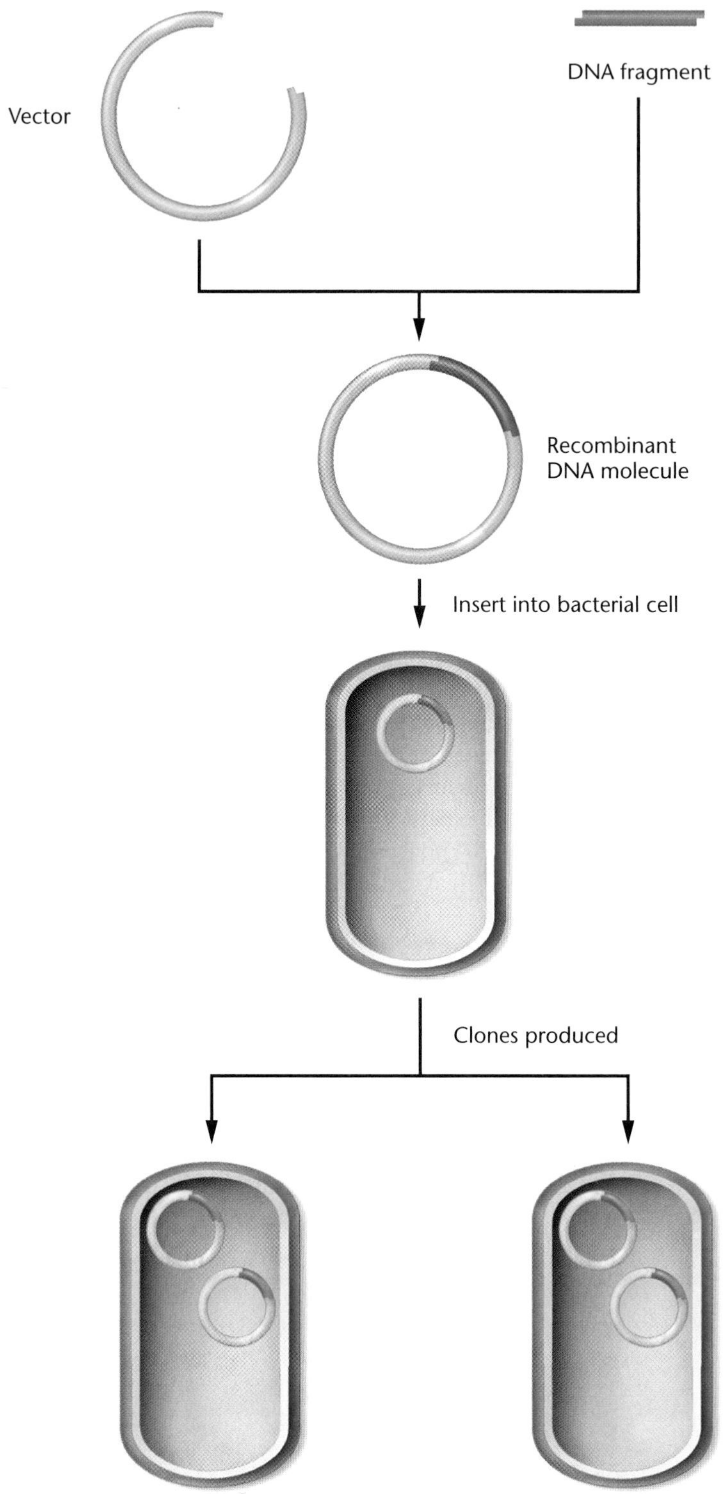

FIGURE 1–14 In cloning, a vector and a DNA fragment produced by cutting with a restriction enzyme are joined to produce a recombinant DNA molecule that is transferred into a bacterial cell, where it is cloned into many copies by replication of the recombinant molecule and by division of the bacterial cell.

DNA fragments are made (Figure 1–14). These cloned copies can be recovered from the bacterial cells and large amounts of the cloned DNA fragment can be isolated. Once large amounts of specific DNA fragments became available by cloning, they were used in many different ways: to isolate genes, to study their organization and expression, and to study their nucleotide sequence and evolution. In addition to preparing large amounts of specific DNA for research, recombinant DNA techniques were the foundation for the biotechnology industry (described in the next section of this chapter).

As techniques became more refined, it became possible to clone larger and larger DNA fragments, paving the way to clone an organism's **genome**, which includes all the DNA carried by that organism. Collections of clones that contain an entire genome are called genomic libraries. Genomic libraries are now available for hundreds of organisms.

Sequencing Genomes: The Human Genome Project

Once genomic libraries became available, scientists began to consider ways to sequence all the clones in a genomic library in an organized way to obtain the nucleotide sequence of an organism's genome. The Human Genome Project began in 1990 as a federally sponsored international effort to sequence the human genome and the genomes of several model organisms used in genetics research. At about the same time, other genome projects, sponsored by industry, got underway. The first genome from a free-living organism, a bacterium (Figure 1–15), was sequenced and reported in 1995 by scientists at a biotechnology company.

In 2001, the publicly funded Human Genome Project and a private genome project undertaken by Celera Corporation reported the first draft of the human genome sequence, covering about 96 percent of the gene-containing portion of the genome. In 2003, the remaining portion of the gene-coding sequence was completed and published. Work is now focused on sequencing the noncoding regions of the genome. Along the way, the genomes of five organisms used in genetic research, *Escherichia coli* (bacterium), *Saccharomyces cerevisiae* (yeast), *Caenorhabditis elegans* (a roundworm), the fruit fly (*D. melanogaster*), and the mouse (*Mus musculus*) were also sequenced.

As genome projects multiplied and more genome sequences were deposited in databases, a new discipline called **genomics**, the study of genomes, came into existence. Genomics uses nucleotide sequence information in databases to study the structure, function, and evolution of genes and genomes. Genomics is drastically changing biology from a laboratory-based science to one combining lab experiments with information technology. Geneticists and other biologists can use information in databases containing nucleic acid sequences, protein sequences, and gene interaction networks to answer experimental questions in a matter of minutes instead of months and years.

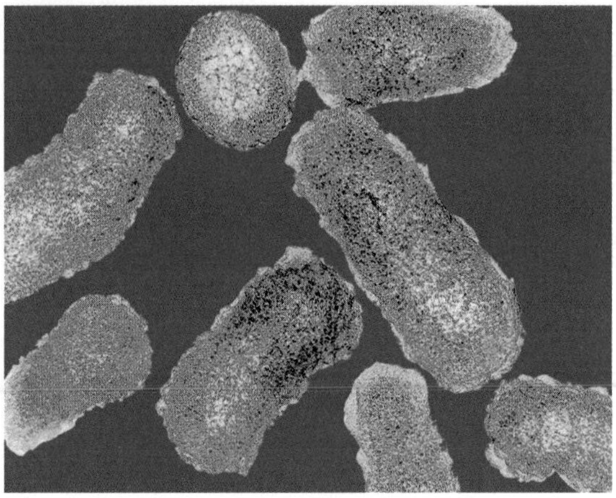

FIGURE 1–15 A colorized electron micrograph of *Haemophilus influenzae*, a bacterium that was the first free-living organism to have its genome sequenced. This bacterium causes respiratory infections and bacterial meningitis in humans.

Recombinant DNA technology has not only greatly accelerated the pace of research, generating new fields of study including genome projects and genomics, but has also given rise to the biotechnology industry, which has grown over the last 25 years to become a major component of the U.S. economy.

1.4 The Impact of Biotechnology Is Growing

Quietly and without much notice in the United States, biotechnology products and services have moved into and revolutionized many aspects of everyday life. Humans have used microorganisms, plants, and animals for thousands of years, but the development of recombinant DNA technology and associated techniques allows us to genetically modify organisms in new ways and use them or their products to enhance our lives. Biotechnology is the commercial use of these modified organisms or their products. It is found at the supermarket, in doctors' offices, drug stores, department stores, hospitals and clinics, on farms, in orchards, law enforcement, court-ordered child support, and even industrial chemicals. We will examine the impact of biotechnology on a small cross-section of everyday life.

Plants, Animals, and the Food Supply

The genetic modification of crop plants is one of the most rapidly expanding areas of biotechnology. Attention has been focused on traits such as resistance to herbicides, insects, and viruses, enhancement of oil content, and others (Table 1.1). Currently, over a dozen genetically modified crop plants have been approved for commercial use in the United States, with dozens more in field trials. Herbicide-resistant corn and soybeans were first planted in the mid-1990s, and now about 40 percent of the corn crop and 80 percent of the soybean crop is genetically modified. In addition, more than 60 percent of the canola crop and 70 percent of the cotton crop are grown from genetically modified strains. It is estimated that more than 60 percent of the processed food in the United States contains ingredients from genetically modified crop plants.

TABLE 1.1 SOME GENETICALLY ALTERED TRAITS IN CROP PLANTS

Herbicide Resistance
Corn, Soybeans, Rice, Cotton, Sugarbeets, Canola
Insect Resistance
Corn, Cotton, Potato
Virus Resistance
Potato, Yellow Squash, Papaya
Altered Oil Content
Soybeans, Canola
Delayed Ripening
Tomato

This agricultural transformation is not without controversy. Critics are concerned that the use of herbicide-resistant crop plants will lead to dependence on chemical weed management and may eventually lead to herbicide-resistant weeds. Others are concerned that traits in genetically engineered crops could be transferred to wild plants in a way that leads to irreversible changes in the ecosystem. We will examine these concerns in Chapter 22.

Biotechnology is also being used to nutritionally enhance crop plants. More than one third of the world's population uses rice as a dietary staple, but most varieties of rice contain little or no vitamin A. Vitamin A deficiency causes more than 500,000 cases of blindness in children each year. A genetically engineered strain, called golden rice, has high levels of two compounds that the body converts to vitamin A. Golden rice is now in testing, and should be available for planting in the near future, its aim to reduce or eliminate this burden of disease. Other crops including wheat, corn, beans, and cassava are also being modified to enhance nutritional value by increasing their vitamin and mineral content.

Livestock such as sheep and cattle have been commercially cloned for more than 25 years, mainly by a method called embryo splitting. This method is used to produce two prize animals instead of one. In 1996, Dolly the sheep (Figure 1–16) was cloned by a new method in which the nucleus of a differentiated adult cell was transferred into an egg that had its nucleus removed. This nuclear transfer method makes it possible to produce hundreds or even thousands of offspring with desirable traits. Cloning by nuclear transfer has many applications in agriculture, sports, and medicine. Some desirable traits, such as high milk production, or speed in race horses, do not appear until adulthood; animals with these traits can now be cloned using differentiated cells. In medical applications, researchers have transferred human genes into animals so that as adults, they produce human proteins in their milk. By selecting and cloning animals with high levels of human protein production, biopharmaceutical companies can produce a herd with uniformly high rates of protein production. Human proteins are used as drugs, and proteins from transgenic animals are now being tested as treatments for diseases such as emphysema. If successful, these proteins will soon be commercially available.

FIGURE 1–16 Dolly, a Finn Dorset sheep cloned from the genetic material of an adult mammary cell, shown next to her first-born lamb, Bonnie.

Who Owns Transgenic Organisms?

Once produced, can a transgenic plant or animal be patented? The answer is yes. The United States Supreme Court ruled in 1980 that living organisms can be patented, and the first organism modified by recombinant DNA technology was patented in 1988 (Figure 1–17). Since then, dozens of plants and animals have been patented. The ethics of patenting living organisms is a contentious issue. Supporters of patenting argue that without the ability to patent the products of research to recover their costs, biotechnology companies will not invest in large-scale research and development. They further argue that patents represent an incentive to develop new products because companies will reap the benefits from taking risks to bring new products to market. Critics argue that patents for organisms such as crop plants will concentrate ownership of food production in the hands of a small number of biotechnology companies, making farmers economically dependent on seeds and pesticides produced by these companies, and reducing the genetic diversity of crop plants as farmers discard local crops that might harbor important genes for resistance to pests and disease. To resolve these and other issues about biotechnology and its uses, a combination of public awareness, education, enlightened social policy, and legislation are needed.

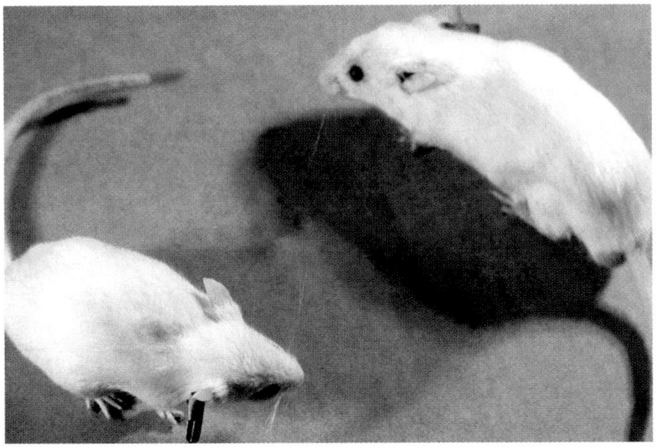

FIGURE 1–17 The first genetically altered organism to be patented, mice from the *onc* strain, genetically engineered to be susceptible to many forms of cancer. These mice were designed for studying cancer development and the design of new anticancer drugs.

Biotechnology in Genetics and Medicine

Biotechnology in the form of genetic testing and gene therapy has already become an important part of medicine. This technology will shape medical practice in the twenty-first century. The importance of developing tests and treatments for genetic diseases is underscored by the estimate that more than 10 million children or adults in the United States suffer from some form of genetic disorder and that every childbearing couple stands an approximately 3 percent risk of having a child with some form of genetic anomaly. The molecular basis for hundreds of genetic disorders is now known (Figure 1–18). For example, the genes for disorders such as sickle-cell anemia, cystic fibrosis, hemophilia, muscular dystrophy, phenylketonuria, and many other metabolic disorders have been cloned. These cloned genes are used for the prenatal detection of affected fetuses. In addition, parents can also learn of their status as "carriers" of a large number of inherited disorders. The combination of genetic testing and genetic counseling gives couples objective information on which they can base informed decisions about childbearing. At present, genetic testing is available for several hundred inherited disorders, and this number will grow as more genes are identified, isolated, and cloned. The use of genetic testing and other technologies, including gene therapy, have raised ethical concerns that have yet to be resolved.

Instead of testing one gene at a time to discover whether someone carries a mutant gene that can produce a disorder in his or her offspring, technology is being developed that will allow screening of an individual's genome to determine the person's risk of developing a genetic disorder or of having a child with a genetic disorder. This technology uses devices called **DNA microarrays** or **DNA chips** (Figure 1–19). Each chip contains thousands of fields, each carrying a different gene. In fact, chips carrying the human genome are now commercially available, making it possible to scan someone's entire genome to see which genetic diseases the individual carries or may develop. DNA chip technology has many other applications as well, such as testing for gene expression in cancer cells to develop therapies tailored to specific forms of cancer.

In addition to testing for genetic disorders, clinicians can transfer normal genes into individuals affected with genetic disorders in a procedure known as **gene therapy**. Although initally successful, therapeutic failures and patient deaths have slowed its development. Recent advances in methods of gene transfer may reduce the risks involved, and it seems certain that gene therapy will become an important tool in treating inherited disorders. In fact, as more is learned about the molecular basis of human diseases, more therapies can be developed. Much of the present-day research on human genetic disorders involves the use of model organisms.

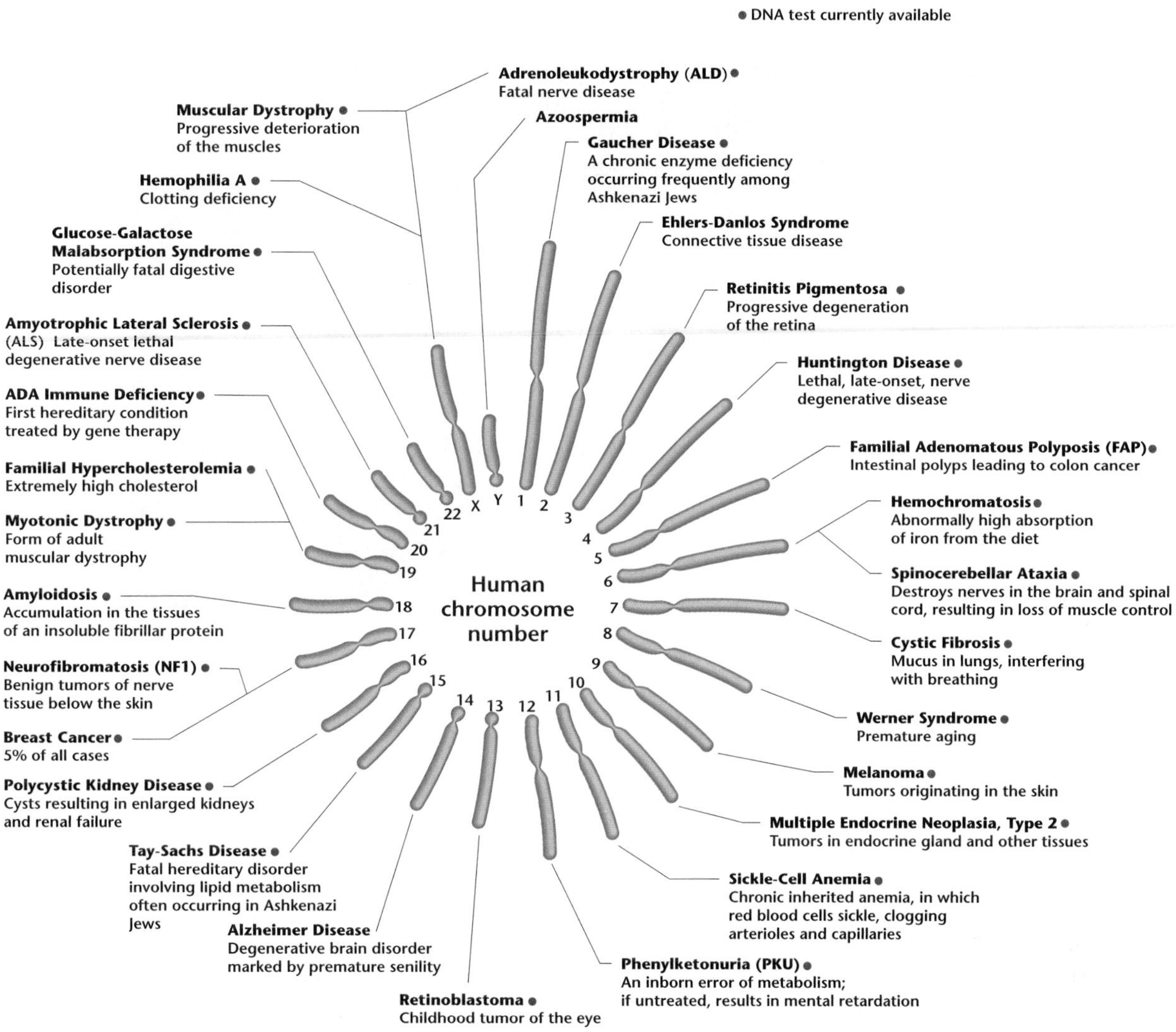

FIGURE 1–18 Diagram of the human chromosome set, showing the location of some genes whose mutant forms cause hereditary diseases. Conditions that can be diagnosed using DNA analysis are indicated by a red dot.

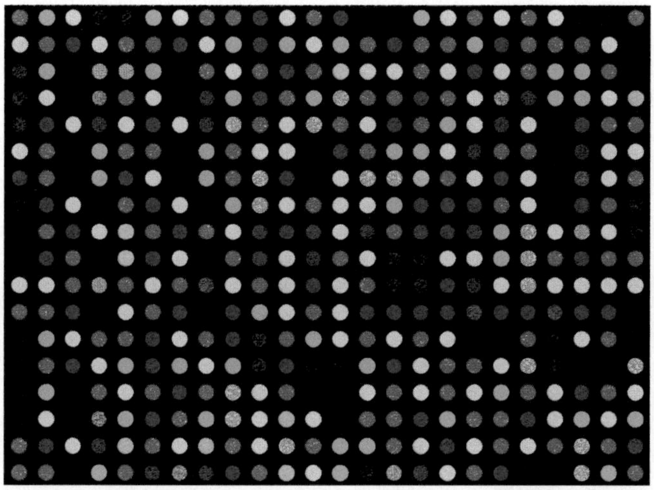

FIGURE 1–19 A DNA microarray. The glass plate in the array contains thousands of fields to which DNA molecules are attached. Using this microarray, DNA from an individual can be tested to detect mutant copies of genes.

1.5 Genetic Studies Rely On the Use of Model Organisms

After the rediscovery of Mendel's work in 1900, genetic research on a wide range of organisms confirmed that the principles of inheritance he described were of universal significance among plants and animals. Although work on the genetics of many different organisms continued, geneticists gradually centered their attention on a small number of organisms, including *Drosophila*, the mouse (*Mus musculus*), and corn (*Zea mays*) (Figure 1–20). These organisms became popular for two main reasons: First, it was clear that genetic mechanisms were the same in most organisms, and second, these species had several advantages for genetic research. They were easy to grow, had relatively short life cycles, produced many offspring, and genetic analysis was fairly straightforward. Over time, researchers created a large catalog of mutant strains for each species. These mutations were carefully studied, characterized, and mapped. Because of their well-developed genetics, these species became **model organisms**, which we define as organisms used for the study of basic

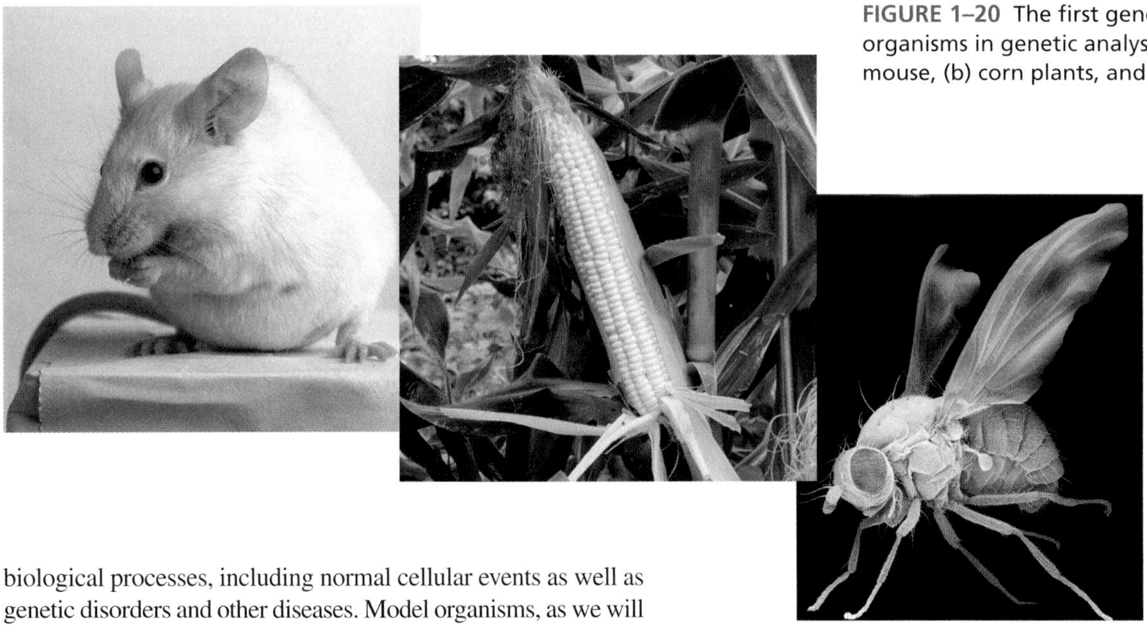

FIGURE 1–20 The first generation of model organisms in genetic analysis included (a) the mouse, (b) corn plants, and (c) the fruit fly.

biological processes, including normal cellular events as well as genetic disorders and other diseases. Model organisms, as we will see in later chapters, are used to study many aspects of biology, including aging, cancer, the immune system, and behavior.

The Modern Set of Genetic Model Organisms

Gradually, other species also became model organisms for research in genetics and modern biology. In the middle years of the twentieth century, viruses (such as the T phages and lambda phage) and microorganisms (including the bacterium *Escherichia coli*, the yeast *Saccharomyces cerevisiae*, and the fungus *Neurospora crassa*) became models (Figure 1–21). Some of these were chosen for the reasons outlined above, while others were selected because they allowed certain aspects of genetics to be studied more easily.

In the last part of the century, three additional organisms were selected and developed as model organisms. Each began as a system used to study some aspect of embryonic development. To study the nervous system and its role in behavior, the nematode *Caenorhabditis elegans* [Figure 1–22(a)] was chosen as a model system. It is small, easy to grow, has a nervous system with only a few hundred cells, and has an unvarying program of cell specification during development. *Arabidopsis thaliana* [Figure 1–22(b)] is a small plant with a short life cycle that can be grown in the laboratory. It was first used to study flower development but has become a model organism for the study of many other aspects of plant biology. The zebrafish, *Danio rerio*, [Figure 1–22(c)] has several advantages for the study of vertebrate development; it is small, reproduces rapidly, and the egg, embryo, and larvae are all transparent. In each of these species, geneticists collected large numbers of mutants, making these organisms useful as model to study not only development but a wide range of other biological processes in plant and animal biology.

Some of the early model organisms are now used to study a narrow range of problems, or have been replaced by other model organisms. *Neurospora*, once a central organism in

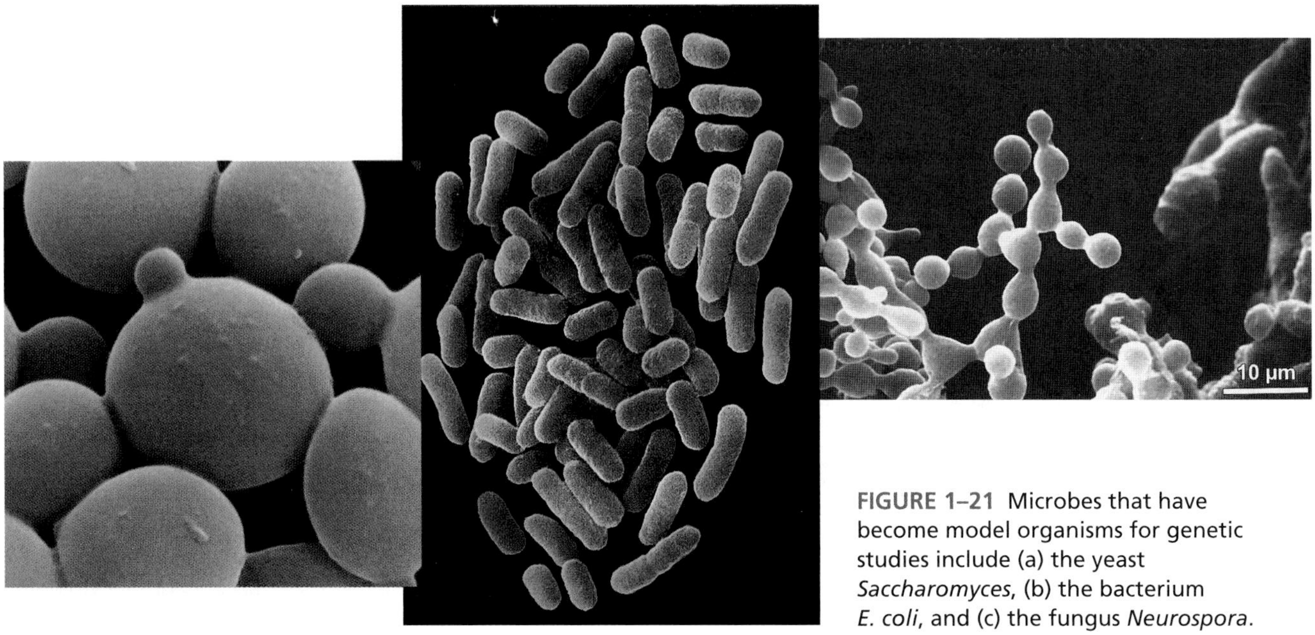

FIGURE 1–21 Microbes that have become model organisms for genetic studies include (a) the yeast *Saccharomyces*, (b) the bacterium *E. coli*, and (c) the fungus *Neurospora*.

genetics, has been displaced by yeast and is now used mainly for research on specialized topics such as circadian rhythms. Corn has been largely replaced by *Arabidopsis* as a model organism for the study of flowering plants.

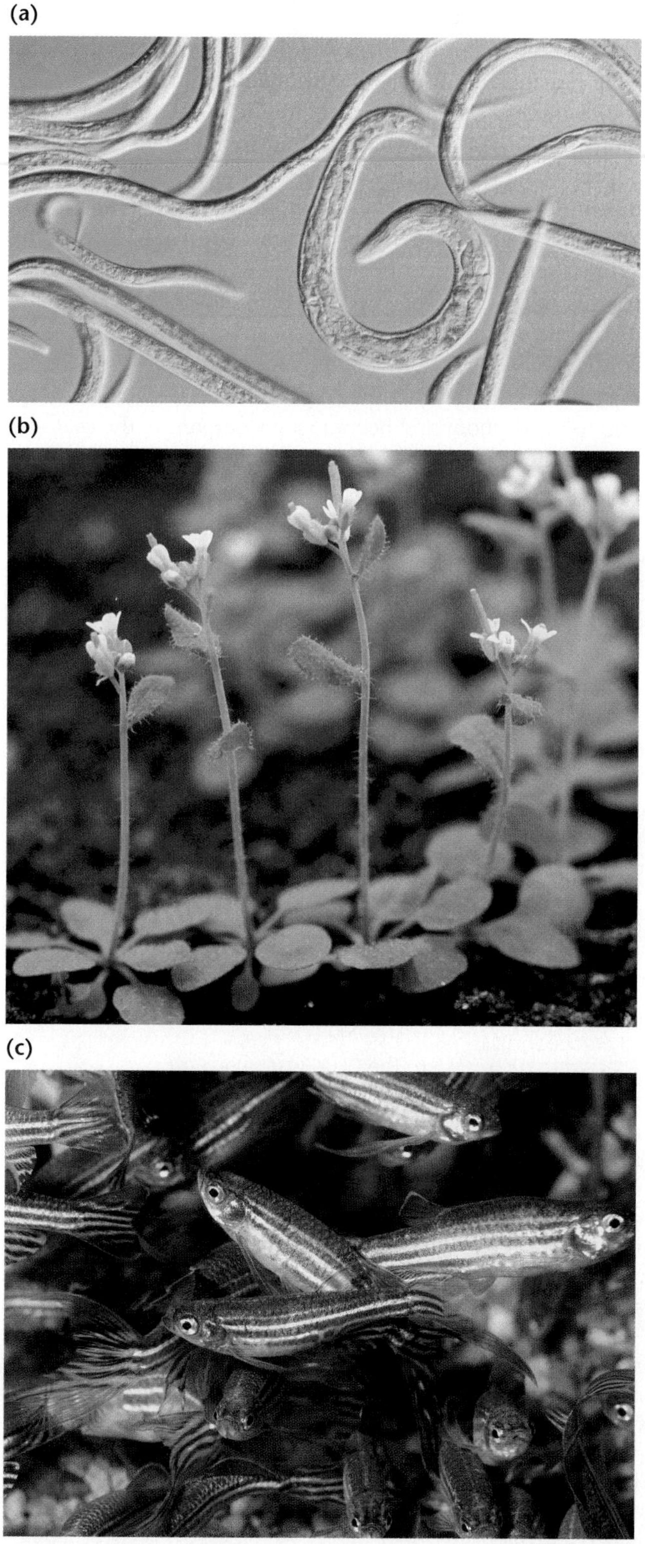

FIGURE 1–22 The third generation of model organisms in genetics includes (a) the roundworm *C. elegans*, (b) the plant *Arabidopsis*, and (c) the zebrafish.

Model Organisms and Human Diseases

The development of recombinant DNA technology and the results from genome sequencing projects have confirmed that all life has a common origin, and as a result, genes with similar functions in different organisms are similar or identical in structure and DNA sequence. In addition, the ability to transfer genes across species has made it possible to develop models of human diseases in organisms ranging from bacteria, fungi, plants, and animals (Table 1.2). For these reasons, the Human Genome Project incorporated projects to sequence the genomes of five model organisms in addition to the human genome. Other genome projects have sequenced the genomes of the remaining model organisms, as well as the genomes of hundreds of other organisms.

It may seem strange to study a human disease such as colon cancer by using *E. coli*, but the basic process of DNA repair (the DNA is defective in some forms of colon cancer) is the same in both organisms, and the gene involved (*mutL* in *E. coli* and *MLH1* in humans) is the same in both organisms. More important, *E. coli* has the advantage of being easier to grow (the cells divide every 20 minutes) and it is easy to create and study new mutations in the *mutL* gene to help understand how it works. This knowledge may eventually lead to the development of drugs and other therapies to treat colon cancer in humans.

Other model organisms, including the fruit fly, *Drosophila melanogaster*, are being used to study specific human diseases. Over several decades, many mutant genes have been identified in *Drosophila* that produce phenotypes with abnormalities of the nervous system, including abnormalities of brain structure, adult-onset degeneration of the nervous system, and visual defects such as retinal degeneration. The information from genome sequencing projects indicates that almost all these genes have human counterparts. As an example, genes involved in a complex human disease of the retina called retinitis pigmentosa are identical to *Drosophila* retinal degeneration genes such as *rdgB* and *rdgC*. Study of these mutations in *Drosophila* is helping to dissect this complex disease and identify the function of the genes involved.

Using recombinant DNA technology, *Drosophila* is being used as a model to study diseases of the human nervous system by transferring a human disease gene into flies. In this way, it is possible to create models for specific human diseases. Flies carrying human genes are used to study the effects of these mutant genes

TABLE 1.2 MODEL ORGANISMS USED TO STUDY HUMAN DISEASES

Organism	Human Diseases
E. coli	DNA repair; colon cancer and other cancers
Yeast	Cell cycle; cancer, Werner syndrome
Drosophila	Cell signaling; cancer
C. elegans	Cell signaling; diabetes
Zebrafish	Developmental pathways; cardiovascular disease
Mouse	Gene expression; Lesch-Nyhan disease, cystic fibrosis, fragile-X syndrome, and many other diseases

on the development and function of the nervous system and its components. In addition to studying the mutant gene itself, the model system can be used to study genes affecting the expression of the human disease genes, and to test the effects of therapeutic drugs on the action of these genes, studies that are difficult or impossible to do in humans. This gene transfer approach in *Drosophila* is being used to study almost a dozen human neurodegenerative disorders, including Huntington disease, Machado-Joseph disease, myotonic dystrophy, and Alzheimer's disease.

As you read through the text, you will encounter these model organisms again and again in the genetic analysis of basic biological processes. Remember, they not only have a rich history in genetics, but are at the forefront in the study of human genetic disorders and infectious diseases. Keep in mind that understanding how a gene controls a process in yeast is relevant to understanding the same gene and the same process in normal human cells.

The development and use of model organisms is only one of the ways genetics and biotechnology are rapidly changing many aspects of everyday life. As discussed in the next section, we have yet to reach a consensus on how and when this technology is acceptable and useful.

1.6 We Live in the "Age of Genetics"

Genetics is no longer just a laboratory science in which researchers study fruit flies or yeast to learn about basic processes in cell function or development. As stated in the beginning of this chapter, it is the core of biology, and the method of choice in dissecting and understanding the functions and malfunctions of biological systems. As knowledge has increased, genetics has become involved in many social issues. Genetics and its applications in the form of biotechnology are now developing much faster than social convention, public policy, and law. Although other scientific disciplines are also expanding in knowledge, none has paralleled the growth of information occurring in genetics. While there has never been a more exciting time to be immersed in the study of genetics, the potential impact of this discipline on society has never been more profound. We are confident that by the end of this course you will agree that the present truly represents the "Age of Genetics," and we urge you to think about and become a participant in the dialogue about genetics and its applications in society.

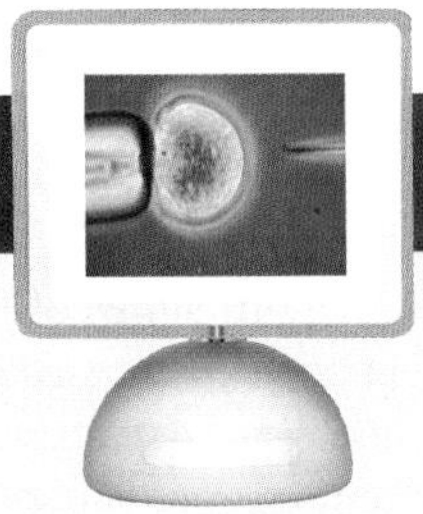

GENETICS, TECHNOLOGY, AND SOCIETY

As you embark on your study of Genetics, we want you to be aware of a special feature of this text found at the conclusion of most chapters: **Genetics, Technology and Society** essays. In these essays we provide coverage of a variety of topics derived from genetics that impact on the lives of each of us, and thus on society in general. Genetics touches all aspects of modern life. Genetic technologies are rapidly changing how we approach medicine, agriculture, law, the pharmaceutical industry, and biotechnology. We now use hundreds of genetic tests to diagnose and predict the course of disease and to detect genetic defects *in utero*. DNA based methods allow scientists to trace the path of evolution taken by many species, including our own. We now design disease resistant and drought resistant crops, as well as more productive farm animals, using gene transfer techniques. We apply DNA profiling methods to paternity testing and murder investigations. Biotechnologies base on information from genomics research have had dramatic effects on industry. The biotechnology industry doubles in size every decade, generating over 700,000 jobs and $50 billion in revenue each year.

Alond with these rapidly changing gene-based technoligies comes a challenging array of ethical dilemmas. Who owns and controls genetic information? Are gene-enhanced agricultural plants and animals safe for humans and the environment? Do we have the right to patent organism and profit from their commercialization? How can we ensure that genomic technologies will be available to all and not just to the wealthy? What are the social issues that accompany the new reproductive technoligies? It is a time when everyone needs to understand genetics in order to make complex personal and societal choices.

The goal of the **Genetics, Technology and Society** essays is to introduce topics that interface with society, including some of the new technologies based on genetics and genomics. As well, we will explore the relevant social and ethical issues. It is our hope that these essays will act as entry points for your exploration of the myriad applications and societal implications of modern genetics. Below, we list the topics that serve as the basis of many of these essays (including the chapter in which each is found). Even should your Genetics course not cover all chapters, we hope that you will find the essays in those chapters of interest. Good reading!

Chapter Summary

1. Mendel's work on pea plants established the principles of the transmission of genes from parents to offspring and established the foundation for the science of genetics.
2. Genes and chromosomes are the fundamental units in the chromosomal theory of inheritance, which explains the transmission of genetic information controlling phenotypic traits.
3. Molecular genetics, based on the central dogma that DNA makes RNA—which makes protein—serves as the underpinnings of Mendelian genetics, referred to as transmission genetics.
4. Recombinant DNA technology allows genes from one organism to be spliced into vectors and cloned, serving as the basis for a far-reaching technology used in molecular genetics.
5. Genomics is one application of recombinant DNA technology in which the entire genetic makeup of an organism is sequenced and the structure and function of its genes are explored. The Human Genome Project is one example of genomics.
6. Biotechnology has revolutionized agriculture, the pharmaceutical industry, and medicine. It has made possible the mass production of medically important gene products. Genetic testing and gene therapy allow detection of individuals with genetic disorders and those at risk of having affected children.
7. The use of model organisms in genetics has advanced our basic understanding of genetic mechanisms and, coupled with recombinant DNA technology, has been used to develop models of human genetic diseases.
8. Genetic technology is affecting many aspects of society. The development of policy and legislation is lagging behind the innovations and uses of biotechnology.

Problems and Discussion Questions

1. Describe Mendel's conclusions about how traits are passed from generation to generation.
2. What is the chromosome theory of inheritance and how is it related to Mendel's findings?
3. Define genotype and phenotype and describe how they are related.
4. What are alleles? If individuals carry genes in pairs, is it possible for more than two alleles of a gene to exist?
5. Given the state of knowledge at the time, why was it difficult for some scientists to accept that DNA is the carrier of genetic information?
6. Contrast chromosomes and genes.
7. How is genetic information encoded in a DNA molecule?
8. Describe the central dogma of molecular genetics and how it serves as the basis of modern genetics.
9. How many different proteins, each with a unique amino acid sequence, are possible in a protein that is five amino acids long?
10. Outline the roles played by restriction enzymes and vectors in cloning DNA.
11. What impact has biotechnology had on crop plants in the United States?
12. Summarize the arguments for and against patenting genetically modified organisms.
13. We all carry 25,000–30,000 genes in our genome. So far, patents have been issued for more than 6,000 of these genes. Do you think that companies or individuals should be able to patent human genes? Why or why not?
14. How has the use of model organisms advanced our knowledge of the genes that control human diseases?
15. If you knew that a devastating late-onset inherited disease runs in your family and you could be tested for it at the age of 20, would you want to know if you are a carrier? Would your answers be likely to change when you reach age 40?

SELECTED READINGS

Barnum, S.R. 2005. *Biotechnology*, 2d ed. Belmont, CA: Brooks-Cole.

Dale, P.J., Clarke, B., and Fontes, E.M.G. 2002. Potential for the environmental impact of transgenic crops. Nature Biotech. 20:567–574.

Fortini, M. and Bonini, N.M. 2000. Modeling human neurodegenerative diseases in *Drosophila. Trends Genet.* 16:161–167.

Lurquin, P. 2002. *High Tech Harvest*. Boulder, CO: Westview Press.

Potter, C.J., Turenchalk, G.S., and Xu, T. 2000. *Drosophila* in cancer research: An expanding role. *Trends Genet.* 16:33–39.

Pray, C.E., Huang, J., Hu, R., and Rozelle, S. 2002. Five years of Bt cotton in China—The benefits continue. *The Plant Journal* 31:423–430.

Primrose, S.B., and Twyman, R.M. 2004. *Genomics: Applications in Human Biology.* Oxford, Blackwell Publishing.

Weinberg, R.A. 1985. The molecules of life. *Sci. Am.* (Oct.) 253:48–57.

Wisniewski, J-P., Frange, N., Massonneau, A., and Dumas, C. 2002. Between myth and reality: Genetically modified maize, an example of a sizeable scientific controversy. *Biochimie* 84:1095–1103.

Yang, X., Tian, X.C., Dai, Y., and Wang, B. 2000. Transgenic farm animals: Applications in agriculture and biomedicine. *Biotechnol. Annu. Rev.* 5:269–292.

Mitosis and Meiosis

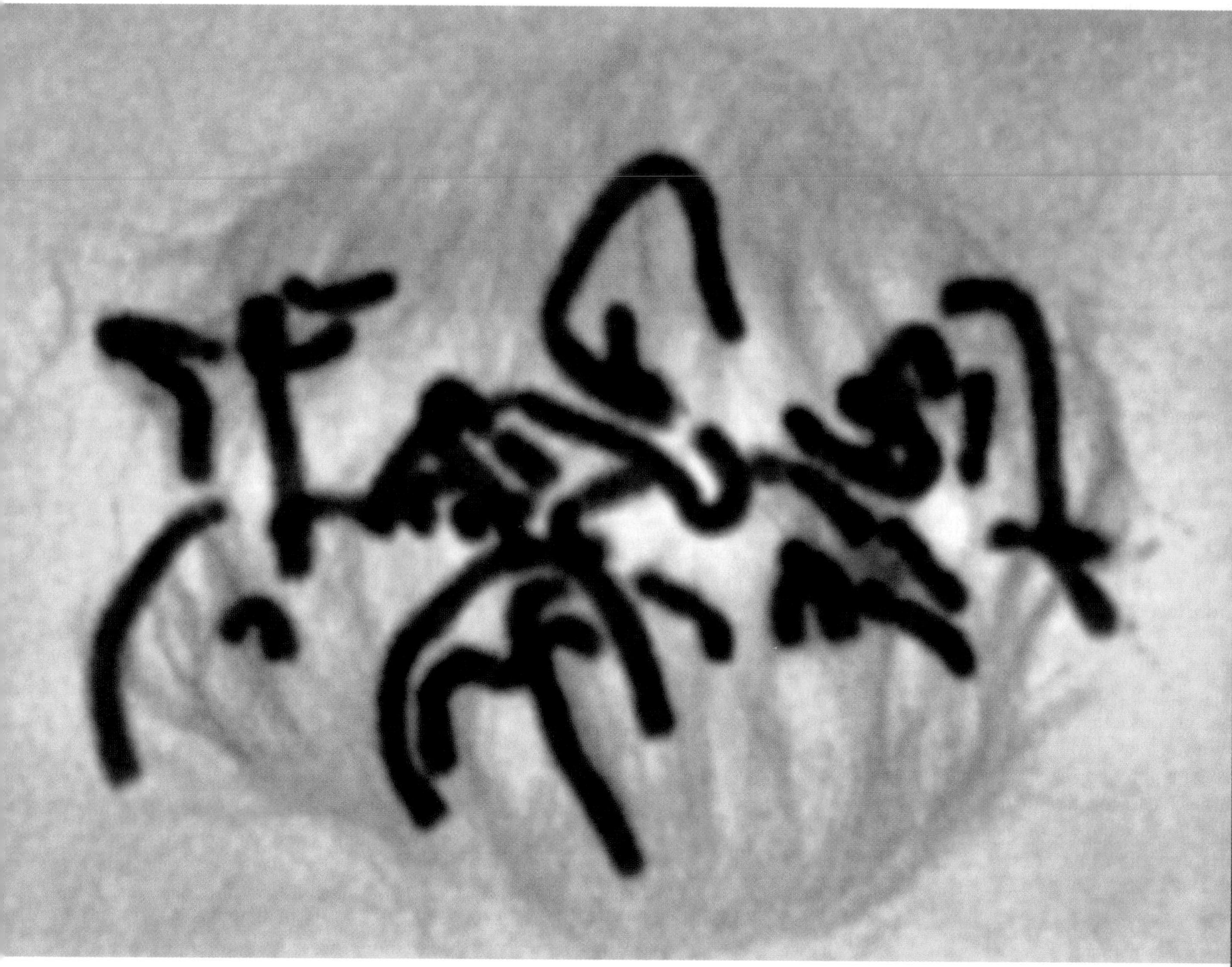

Chromosomes in the prometaphase stage of mitosis, derived from a cell in the flower of Haemanthus.

Chapter Concepts

- Genetic continuity between cells and between organisms of sexually reproducing species is maintained through the processes of mitosis and meiosis, respectively.
- Diploid eukaryotic cells contain their genetic information in pairs of homologous chromosomes, with one member of each pair being derived from the maternal parent and one from the paternal parent.
- Mitosis provides a mechanism to distribute chromosomes that have duplicated into progeny cells during cell reproduction.
- Mitosis converts a diploid somatic cell into two diploid daughter cells.
- The process of meiosis distributes one member of each homologous pair of chromosomes into each gamete or spore, thus reducing the diploid chromosome number to the haploid chromosome number.
- Meiosis generates genetic variability by distributing various combinations of maternal and paternal members of each homologous pair of chromosomes into gametes or spores.
- It is during the stages of mitosis and meiosis that the genetic material has been condensed into discrete structures called chromosomes.
- Plants alternate between a diploid sporophyte stage and a haploid gametophyte stage.

In every living thing there exists a substance referred to as the genetic material. Except in certain viruses, this material is composed of the nucleic acid DNA. A molecule of DNA is organized into units called genes, the products of which direct the metabolic activities of cells. DNA, with its array of genes, is organized into structures called chromosomes, which serve as vehicles for transmitting genetic information. The manner in which chromosomes are transmitted from one generation of cells to the next, and from organisms to their descendants, must be exceedingly precise. In this chapter we consider exactly how genetic continuity is maintained between cells and organisms.

Two major processes are involved in eukaryotes: **mitosis** and **meiosis**. Although the mechanisms of the two processes are similar in many ways, the outcomes are quite different. Mitosis leads to the production of two cells, each with the same number of chromosomes as the parent cell. Meiosis, on the other hand, reduces the genetic content and the number of chromosomes by precisely half. This reduction is essential if sexual reproduction is to occur without doubling the amount of genetic material at each generation. Strictly speaking, mitosis is that portion of the cell cycle during which the duplicated chromosomes are precisely and equally divided into daughter cells. Meiosis is part of a special type of cell division that leads to the production of sex cells: **gametes** or **spores**. This process is an essential step in the transmission of genetic information from an organism to its offspring.

Normally, chromosomes are visible during a cell's life only during mitosis and meiosis. When cells are not undergoing division, the genetic material making up chromosomes unfolds and uncoils into a diffuse network within the nucleus, generally referred to as chromatin. We will briefly review the structure of cells, emphasizing the components that are of particular significance to genetic function. Then we will devote the remainder of the chapter to the behavior of chromosomes during cell division.

2.1 Cell Structure Is Closely Tied to Genetic Function

Before describing mitosis and meiosis, a brief review of the structure of cells will be helpful. As we shall see, many cell components, such as the nucleolus, ribosome, and centriole, are involved directly or indirectly with genetic processes. Other components, the mitochondria and chloroplasts, contain their own unique genetic information. It is also useful to compare the structural differences between the prokaryotic bacterial cell and the eukaryotic cell. Variation in the structure and function of cells is dependent on specific genetic expression by each cell type.

Before 1940, our knowledge of cell structure was limited to what we could see with the light microscope. Around 1940, the transmission electron microscope was in its early stages of development, and by 1950, many details of cell ultrastructure had emerged. Under the electron microscope, cells were seen as highly organized, precise structures. A new world of whorling membranes, organelles, microtubules, granules, and filaments was revealed. These discoveries revolutionized thinking in the entire field of biology. We will be concerned with the aspects of cell structure that relate to genetic study. The typical animal cell shown in Figure 2–1 illustrates most of the structures we will discuss.

Cell Boundaries

All cells are surrounded by a **plasma membrane**, an outer covering that defines the cell boundary and delimits the cell from its immediate external environment. This membrane is not passive; instead, it actively controls the movement of materials into and out of the cell. In addition to this membrane, plant cells have an outer covering called the **cell wall**. One major component of this rigid structure is a polysaccharide called **cellulose**.

Bacterial cells also have a cell wall, but its chemical composition is quite different from that of the plant cell wall, the major component being a complex macromolecule called a **peptidoglycan**. As its name suggests, the molecule consists of peptide and sugar units. Long polysaccharide chains are cross-linked with short peptides, which impart great strength and rigidity to the bacterial cell. Some bacterial cells have still another covering, a **capsule**. This mucuslike material protects these bacteria from phagocytic activity by the host during their pathogenic invasion of eukaryotic organisms. The presence of the capsule is under genetic control. In fact, as we will see in Chapter 9, its loss due to mutation in the pneumonia-causing bacterium *Diplococcus pneumoniae* provided the underlying basis for a critical experiment, proving that DNA is the genetic material.

Many, if not most, animal cells have a covering over the plasma membrane, referred to as the **cell coat**. Consisting of glycoproteins and polysaccharides, its chemical composition differs from comparable structures in either plants or bacteria. The cell coat, among other functions, provides biochemical identity at the surface of cells. These forms of cellular identity at the cell surface are under genetic control. For example, various antigenic determinants, such as the **AB** and **MN antigens**, are found on the surface of red blood cells. In other cells, **histocompatibility antigens**, which elicit an immune response during tissue and organ transplants, are present. A variety of **receptor molecules** are also important components at the surface of cells. These constitute recognition sites that transfer specific chemical signals across the cell membrane into the cell.

The Nucleus

The presence of a nucleus and other membranous organelles characterizes eukaryotic cells. The **nucleus** houses the genetic material, DNA, which is complexed with an array of acidic and basic proteins into thin fibers. During nondivisional phases of the cell cycle, these fibers are uncoiled and dispersed into **chromatin**. As we will soon discuss, during mitosis and meiosis, chromatin fibers coil and condense into structures called **chromosomes**. Also present in the nucleus is the **nucleolus**, an amorphous component where ribosomal RNA is synthesized and where the initial stages of ribosomal assembly occur. The areas of DNA encoding rRNA are collectively referred to as the **nucleolus organizer region**, or the **NOR**.

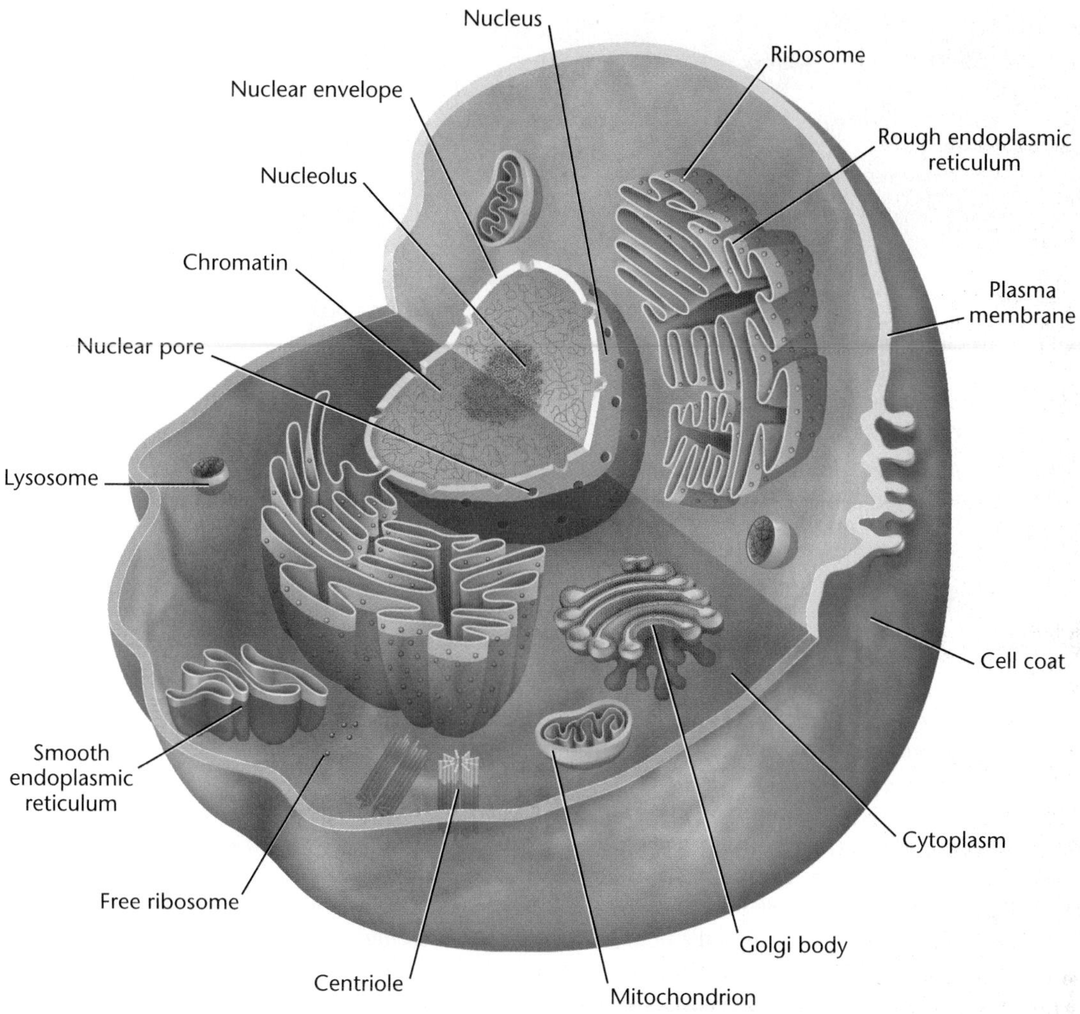

FIGURE 2–1 A generalized animal cell. The cellular components discussed in the text are emphasized here.

The lack of a nuclear envelope and membraneous organelles is characteristic of prokaryotes. In bacteria such as *Escherichia coli*, the genetic material is present as a long, circular DNA molecule that is compacted into an area referred to as the **nucleoid** area. Part of the DNA may be attached to the cell membrane, but in general the nucleoid constitutes a large area throughout the cell. Although the DNA is compacted, it does not undergo the extensive coiling characteristic of the stages of mitosis where, in eukaryotes, chromosomes become visible. Nor is the DNA in these organisms associated as extensively with proteins as is eukaryotic DNA. Figure 2–2, which shows two bacteria forming during cell division, illustrates the nucleoid regions that house the bacterial chromosome. Prokaryotic cells do not have a distinct nucleolus, but do contain genes that specify rRNA molecules.

The Cytoplasm and Cellular Organelles

The remainder of the eukaryotic cell enclosed by the plasma membrane, excluding the nucleus, is composed of **cytoplasm** and all associated cellular organelles. Cytoplasm consists of a nonparticulate, colloidal material referred to as the **cytosol**, which surrounds and encompasses the cellular organelles. Beyond these components, an extensive system of tubules and filaments comprising the cytoskeleton provides a lattice of support structures within the cytoplasm. Consisting primarily of **tubulin-derived microtubules** and **actin-derived microfilaments**, this structural framework maintains cell shape, facilitates cell mobility, and anchors the various organelles.

One organelle, the membranous **endoplasmic reticulum (ER)**, compartmentalizes the cytoplasm, greatly increasing the surface area available for biochemical synthesis. The ER may appear smooth, in which case it serves as the site for synthesizing fatty acids and phospholipids, or it may appear rough because it is studded with ribosomes. Ribosomes serve as sites where genetic information contained in messenger RNA (mRNA) is translated into proteins.

Three other cytoplasmic structures are very important in the eukaryotic cell's activities: mitochondria, chloroplasts, and centrioles. **Mitochondria** are found in both animal and plant cells and are the sites of the oxidative phases of cell respiration. These chemical reactions generate large amounts of adenosine triphosphate (ATP), an energy-rich molecule. **Chloroplasts** are found in plants, algae, and some protozoans. These organelles are associated with photosynthesis, the major energy-trapping process on Earth. Both mitochondria and chloroplasts contain a type of DNA that is distinct from that found in the nucleus. Furthermore, these organelles can

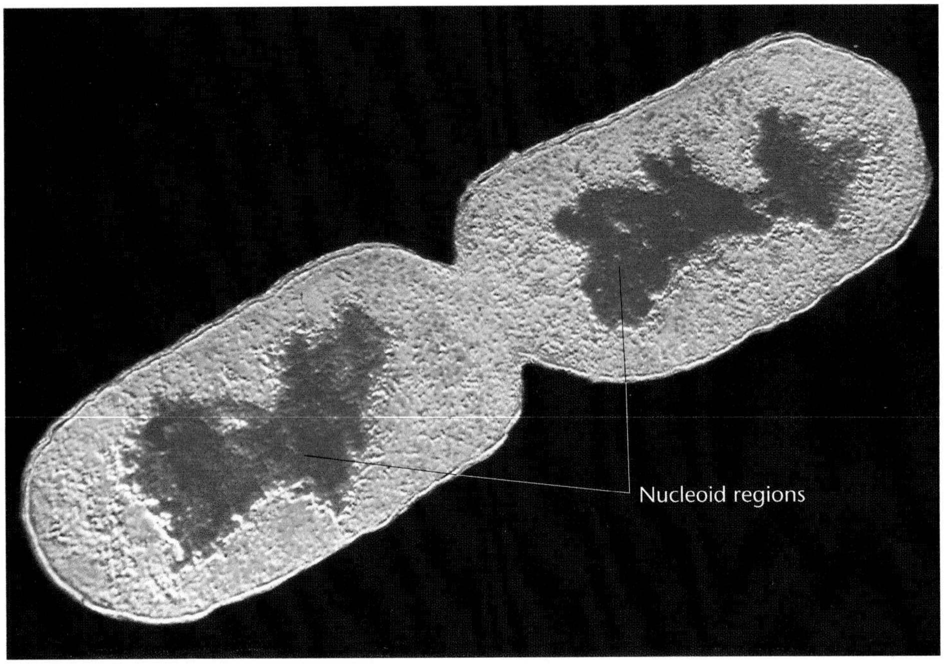

FIGURE 2–2 Color-enhanced electron micrograph of *E. coli* undergoing cell division. Particularly prominent are the two chromosomal areas (shown in red), called nucleoids, that have been partitioned into the daughter cells.

duplicate themselves and transcribe and translate their genetic information. It is interesting to note that the genetic machinery of mitochondria and chloroplasts closely resembles that of prokaryotic cells. This and other observations have led to the proposal that these organelles were once primitive free-living organisms that established a symbiotic relationship with a primitive eukaryotic cell. This theory, which describes the evolutionary origin of these organelles, is called the **endosymbiont hypothesis**.

Animal cells and some plant cells also contain a pair of complex structures called the **centrioles**. These cytoplasmic bodies, located in a specialized region called the centrosome, are associated with the organization of spindle fibers that function in mitosis and meiosis. In some organisms, the centriole is derived from another structure, the basal body, which is associated with the formation of cilia and flagella. Over the years, research has suggested that centrioles and basal bodies contain DNA, which could be involved in the replication of these structures. Currently, this is thought not to be the case.

The organization of **spindle fibers** by the centrioles occurs during the early phases of mitosis and meiosis. Composed of arrays of microtubules, these fibers play an important role in the movement of chromosomes as they separate during cell division. The microtubules consist of polymers of polypeptide subunits of the protein tubulin.

2.2 In Diploid Organisms, Chromosomes Exist in Homologous Pairs

As we discuss the processes of mitosis and meiosis, it is important that you understand the concept of homologous chromosomes. Such an understanding will also be of critical importance in our future discussions of Mendelian genetics. Chromosomes are most easily visualized during mitosis. When they are examined carefully, they are seen to take on distinctive lengths and shapes. Each contains a condensed or constricted region called the **centromere**, which establishes the general appearance of each chromosome. Figure 2–3 shows chromosomes with centromere placements at different points along their lengths. Extending from either side of the centromere are the arms of the chromosome. Depending on the position of the centromere, different arm ratios are produced. As Figure 2–3 illustrates, chromosomes are classified as **metacentric**, **submetacentric**, **acrocentric**, or **telocentric** on the basis of the centromere location. The shorter arm, by convention, is shown above the centromere and is called the **p arm** (p stands for "petite"). The longer arm is shown below the centromere and is called the **q arm** (q being the next letter in the alphabet).

When studying mitosis, several other observations are of particular relevance. First, all somatic cells derived from members of the same species contain an identical number of chromosomes. In most cases, this represents the **diploid number (2*n*)**. When the lengths and centromere placements of all such chromosomes are examined, a second general feature is apparent. Nearly all of the chromosomes exist in pairs with regard to these two criteria. The members of each pair are called **homologous chromosomes**. For each chromosome exhibiting a specific length and centromere placement, another exists with identical features. There are exceptions to the rule of chromosomes in pairs. Bacteria and viruses have but one chromosome, and organisms such as yeasts and molds, and certain plants such as bryophytes (mosses) spend the predominant phase of the life cycle in the haploid stage. That is, they contain only one member of each homologous pair of chromosomes during most of their lives.

Figure 2–4 illustrates the physical appearance of different pairs of homologous chromosomes. There, the human mitotic chromosomes have been photographed, cut out of the print, and matched up, creating a **karyotype**. As you can see, humans have a 2*n* number of 46 and, on close examination, the chromosomes exhibit a diversity of sizes and centromere placements. Note also that each of the 46 chromosomes is clearly a double structure consisting of two parallel **sister chromatids** connected by a common centromere. Had these chromosomes been allowed to continue dividing, the sister chromatids, which are replicas of one another, would have separated into the two new cells as division continued.

Centromere location	Designation	Metaphase shape	Anaphase shape
Middle	Metacentric		
Between middle and end	Submetacentric		
Close to end	Acrocentric		
At end	Telocentric		

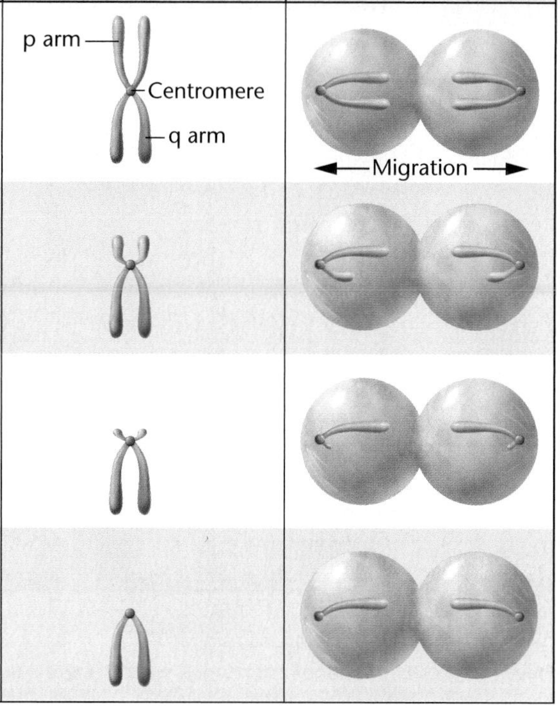

FIGURE 2–3 Centromere locations and designations of chromosomes based on centromere location. Note that the shape of the chromosome during anaphase is determined by the position of the centromere.

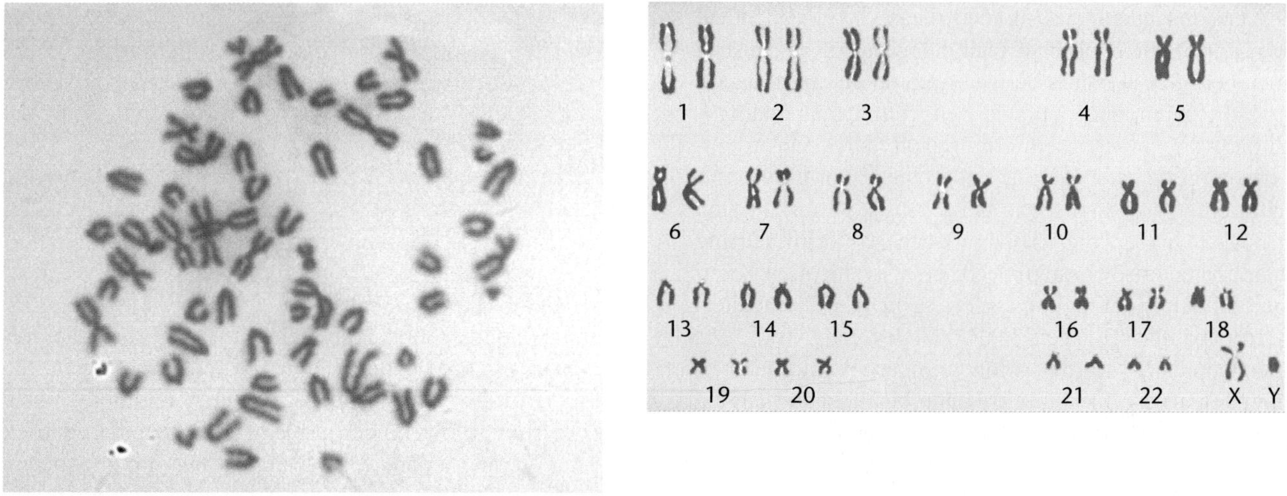

FIGURE 2–4 A metaphase preparation of chromosomes derived from a dividing cell of a human male (left), and the karyotype derived from the metaphase preparation (right). All but the X and Y chromosomes are present in homologous pairs. Each chromosome is clearly a double structure, constituting a pair of sister chromatids joined by a common centromere.

Table 2.1 THE HAPLOID NUMBER OF CHROMOSOMES FOR A VARIETY OF ORGANISMS

Common Name	Scientific Name	Haploid Number	Common Name	Scientific Name	Haploid Number
Black bread mold	*Aspergillus nidulans*	8	House mouse	*Mus musculus*	20
Broad bean	*Vicia faba*	6	Human	*Homo sapiens*	23
Cat	*Felis domesticus*	19	Jimson weed	*Datura stramonium*	12
Cattle	*Bos taurus*	30	Mosquito	*Culex pipiens*	3
Chicken	*Gallus domesticus*	39	Mustard plant	*Arabidopsis thaliana*	5
Chimpanzee	*Pan troglodytes*	24	Pink bread mold	*Neurospora crassa*	7
Corn	*Zea mays*	10	Potato	*Solanum tuberosum*	24
Cotton	*Gossypium hirsutum*	26	Rhesus monkey	*Macaca mulatta*	21
Dog	*Canis familiaris*	39	Roundworm	*Caenorhabditis elegans*	6
Evening primrose	*Oenothera biennis*	7	Silkworm	*Bombyx mori*	28
Frog	*Rana pipiens*	13	Slime mold	*Dictyostelium discoidium*	7
Fruit fly	*Drosophila melanogaster*	4	Snapdragon	*Antirrhinum majus*	8
Garden onion	*Allium cepa*	8	Tobacco	*Nicotiana tabacum*	24
Garden pea	*Pisum sativum*	7	Tomato	*Lycopersicon esculentum*	12
Grasshopper	*Melanoplus differentialis*	12	Water fly	*Nymphaea alba*	80
Green alga	*Chlamydomonas reinhardi*	18	Wheat	*Triticum aestivum*	21
Horse	*Equus caballus*	32	Yeast	*Saccharomyces cerevisiae*	16
House fly	*Musca domestica*	6	Zebrafish	*Danio rerio*	25

The haploid number (n) of chromosomes is equal to one-half the diploid number. Collectively, the total set of genes contained in a haploid set of chromosomes constitutes the **genome** of the species. The examples listed in Table 2.1 demonstrate the wide range of n values found in plants and animals.

Homologous pairs of chromosomes have important genetic similarities. They contain identical gene sites along their lengths, each called a **locus** (pl. **loci**). Thus, they are identical in their genetic potential. In sexually reproducing organisms, one member of each pair is derived from the maternal parent (through the ovum) and one is derived from the paternal parent (through the sperm). Therefore, each diploid organism contains two copies of each gene as a consequence of **biparental inheritance**. As we shall see in the chapters on transmission genetics, the members of each pair of genes, while influencing the same characteristic or trait, need not be identical. In a population of members of the same species, many different alternative forms of the same gene, called **alleles**, can exist.

The conceptual issues of haploid number, diploid number, and homologous chromosomes are important in understanding the process of meiosis. During the formation of gametes or spores, meiosis converts the diploid number of chromosomes to the haploid number. As a result, haploid gametes or spores contain precisely one member of each homologous pair of chromosomes—that is, one complete haploid set. Following fusion of two gametes in fertilization, the diploid number is reestablished; that is, the zygote contains two complete haploid sets of chromosomes. The constancy of genetic material is thus maintained from generation to generation.

How Do We Know?

With the initial appearance of this feature, a brief introduction is in order. Throughout the text, several times in each chapter, we will identify important research questions that have been answered as a result of genetic experimentation. You are asked to relate each question to the previous discussion and review how we acquired the information elucidating the scientific finding. It is our hope that these simple exercises will stimulate and fine-tune your analytical thinking skills.

How do we know that chromosomes exist in homologous pairs?

There is one important exception to the concept of homologous pairs of chromosomes. In many species, one pair, the **sex-determining chromosomes**, is often not homologous in size, centromere placement, arm ratio, or genetic content. For example, in humans, females carry two homologous X chromosomes, while males carry one Y chromosome in addition to one X chromosome (Figure 2–4). The X and Y chromosomes are not strictly homologous. The Y is considerably smaller and lacks most of the gene sites contained on the X. Nevertheless, in meiosis they behave as homologs so that gametes produced by males receive either one X or one Y chromosome.

2.3 Mitosis Partitions Chromosomes into Dividing Cells

The process of mitosis is critical to all eukaryotic organisms. In some single-celled organisms, such as protozoans and some fungi and algae, mitosis (as a part of cell division) provides the basis for asexual reproduction. Multicellular diploid organisms begin life as single-celled fertilized eggs called **zygotes**. The mitotic activity of the zygote and the subsequent daughter cells is the foundation for the development and growth of the organism. In adult organisms, mitotic activity is prominent in wound healing and other forms of cell replacement in certain tissues. For example, the epidermal skin cells of humans are continuously sloughed off and replaced. Cell division also results in the continuous production of reticulocytes that eventually shed their nuclei and replenish the supply of red blood cells in vertebrates. In abnormal situations, somatic cells may lose control of cell division, forming a tumor.

The genetic material is partitioned into daughter cells during nuclear division or **karyokinesis**. This process is quite complex and requires great precision. The chromosomes must first be exactly replicated and then accurately partitioned. The end result is the production of two daughter nuclei, each with a chromosome composition identical to that of the parent cell.

Karyokinesis is followed by cytoplasmic division, or **cytokinesis**. The less complex division of the cytoplasm requires a mechanism that partitions the volume into two parts, then encloses both new cells in a distinct plasma membrane. Cytoplasmic organelles either replicate themselves, arise from existing membrane structures, or are synthesized *de novo* (anew) in each cell. The subsequent proliferation of these structures is a reasonable and adequate mechanism for reconstituting the cytoplasm in daughter cells.

Following cell division, the initial size of each new daughter cell is approximately one-half the size of the parent cell. However, the nucleus of each new cell is not appreciably smaller than the nucleus of the original cell. Quantitative measurements of DNA confirm that there is an amount of genetic material in the daughter nuclei equivalent to that in the parent cell.

Interphase and the Cell Cycle

Many cells undergo a continuous alternation between division and nondivision. The events that occur from the completion of one division until the beginning of the next division constitute the **cell cycle** (Figure 2–5). We will consider the initial stage of the cycle, called **interphase**, as the interval between divisions. It was once thought that the biochemical activity during interphase was devoted solely to the cell's growth and its normal function. However, we now know that another biochemical step critical to the ensuing mitosis occurs during interphase: *the replication of the DNA of each chromosome*. Occurring before the cell enters mitosis, this period during which DNA is synthesized is called the **S phase**. The initiation and completion of synthesis can be detected by monitoring the incorporation of radioactive precursors into DNA.

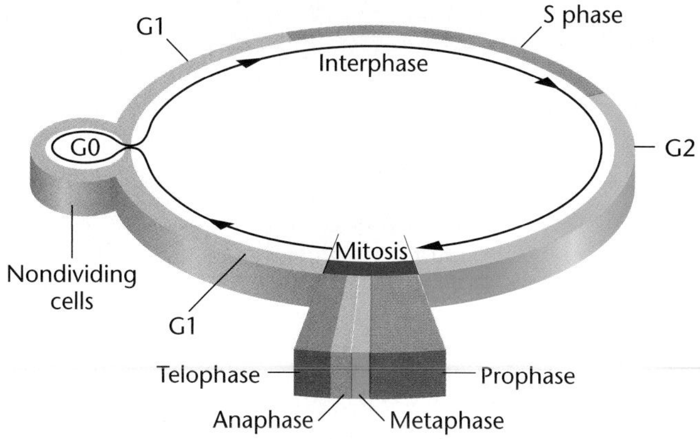

FIGURE 2–5 The intervals comprising an arbitrary cell cycle. Following mitosis, cells enter the G1 stage of interphase, initiating a new cycle. Cells may become nondividing (G0) or continue through G1, where they become committed to begin DNA synthesis (S) and complete the cycle (G2 and mitosis). Following mitosis, two daughter cells are produced and the cycle begins anew for both cells.

Investigations of this nature show two periods during interphase when no DNA synthesis occurs, one before and one after S phase. These are designated **G1 (gap I)** and **G2 (gap II)**, respectively. During both of these intervals, as well as during S, intensive metabolic activity, cell growth, and cell differentiation occur. By the end of G2, the volume of the cell has roughly doubled, DNA has been replicated, and mitosis (M) is initiated. Following mitosis, continuously dividing cells then repeat this cycle (G1, S, G2, M) over and over, as shown in Figure 2–5.

HOW DO WE KNOW?

What experimental approach was used to demonstrate when DNA is duplicated during interphase?

Much is known about the cell cycle based on *in vitro* (Latin for "in glass," meaning in a test tube) studies. While the total length of the cell cycle varies among cells *in vivo* (in living organisms), when grown *in vitro* (in culture), many cells traverse the complete cycle in about 16 hours. The actual process of mitosis occupies only a small part of the overall cycle, often less than an hour. The lengths of the S and G2 phase of interphase are fairly consistent among different cell types. Most variation is seen in the length of time spent in the G1 stage. Figure 2–6 shows the relative length of these intervals in a human cell in culture.

G1 is of great interest in the study of cell proliferation and its control. At a point late in G1, all cells follow one of two paths. They either withdraw from the cycle, become quiescent and enter the **G0 stage** (see Figure 2–5), or they become committed to initiating DNA synthesis and completing the cycle. Cells that enter G0 remain viable and metabolically active but

Interphase			Mitosis
G1	S	G2	M
5	7	3	1

Hours

Pro	Met	Ana	Tel
36	3	3	18

Minutes

FIGURE 2–6 The time spent in each phase of one complete cell cycle of a human cell in culture. Times vary according to cell types and conditions.

do not proliferate. Cancer cells apparently avoid entering G0 or pass through it very quickly. Other cells enter G0 and never reenter the cell cycle. Still others remain in G0, but they can be stimulated to return to G1, and thereby reenter the cell cycle.

Cytologically, interphase is characterized by the absence of visible chromosomes. Instead, the nucleus is filled with chromatin fibers that have formed as the chromosomes have uncoiled and dispersed following the previous mitosis. This is diagrammed in Figure 2–7(a). Once G1, S, and G2 are completed, mitosis is initiated. Mitosis is a dynamic period of vigorous and continual activity. For discussion purposes, the entire process is subdivided into discrete stages, and specific events are assigned to each one. These stages, in order of occurrence, are **prophase**, **prometaphase**, **metaphase**, **anaphase**, and **telophase**. Like interphase, these stages are also diagrammed in Figure 2–7. A photograph of each stage is shown along with each diagram.

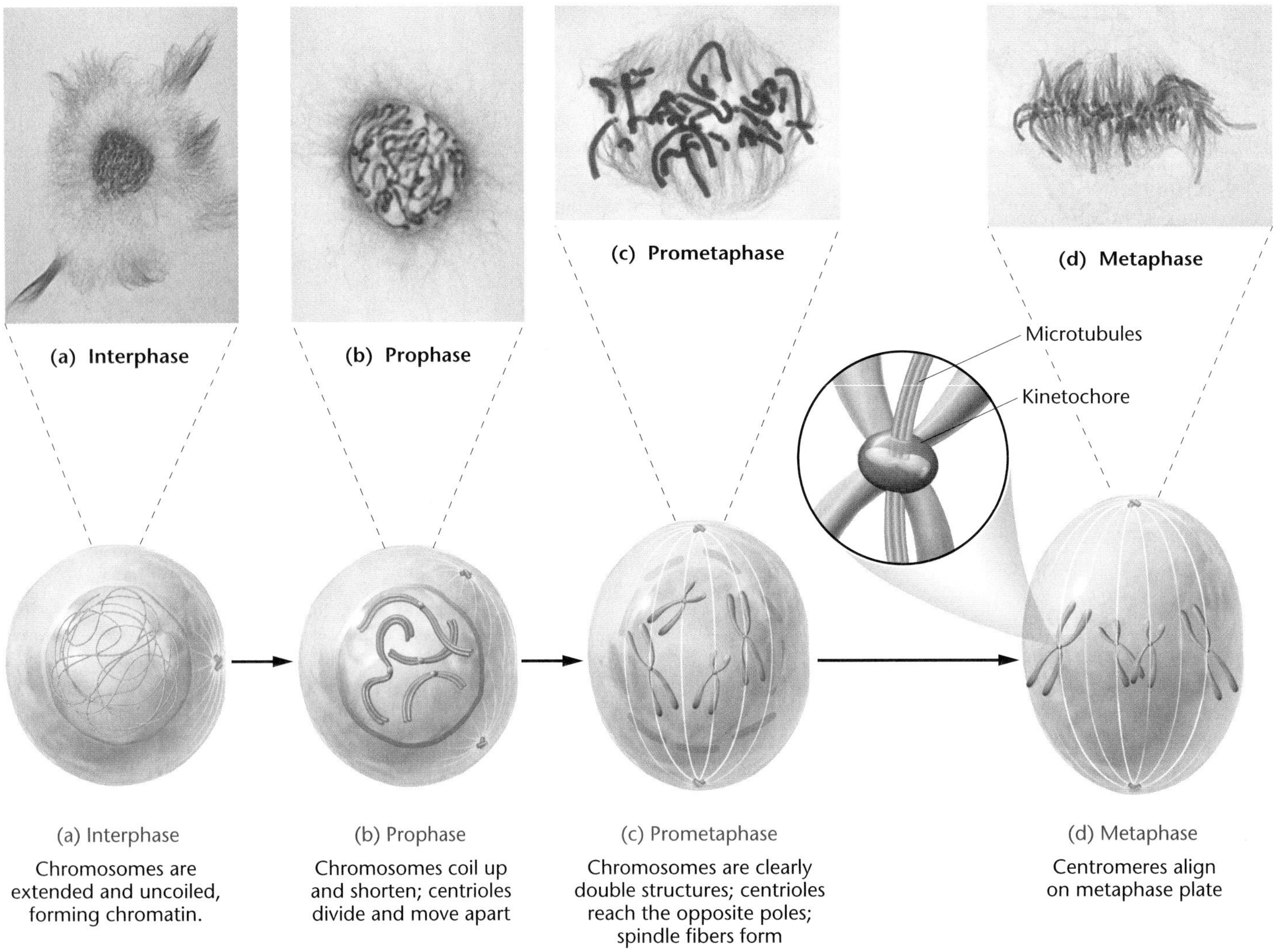
(a) Interphase
(b) Prophase
(c) Prometaphase
(d) Metaphase
Microtubules
Kinetochore
(a) Interphase
Chromosomes are extended and uncoiled, forming chromatin.
(b) Prophase
Chromosomes coil up and shorten; centrioles divide and move apart
(c) Prometaphase
Chromosomes are clearly double structures; centrioles reach the opposite poles; spindle fibers form
(d) Metaphase
Centromeres align on metaphase plate

FIGURE 2–7 Mitosis in an animal cell with a diploid number of 4. The events occurring in each stage are described in the text. Of the two homologous pairs of chromosomes, one contains longer, metacentric members and the other shorter, submetacentric members. The maternal chromosome and the paternal chromosome of each pair are shown in different colors. In (f), the late telophase stage in a plant cell illustrates the formation of the cell plate and lack of centrioles. The light micrographs illustrating the stages of mitosis are derived from the flower of *Haemanthus*.

Web Tutorial 2.1 Mitosis and the Cell Cycle

Prophase

Often, over half of mitosis is spent in prophase [Figure 2–7(b)], a stage characterized by several significant activities. One of the early events in prophase of all animal cells involves the migration of two pairs of centrioles to opposite ends of the cell. These structures are found just outside the nuclear envelope in an area of differentiated cytoplasm called the **centrosome**. It is thought that each pair of centrioles consists of one mature unit and a smaller, newly formed centriole.

The direction of migration of the centrioles is such that two poles are established at opposite ends of the cell. Following their migration, the centrioles are responsible for organizing cytoplasmic microtubules into a series of **spindle fibers** that are formed and run between these poles. This creates an axis along which chromosomal separation occurs. Interestingly, cells of most plants (with a few exceptions), fungi, and certain algae seem to lack centrioles. Spindle fibers are nevertheless apparent during mitosis. Therefore, centrioles are not universally responsible for the organization of spindle fibers.

As the centrioles migrate, the nuclear envelope begins to break down and gradually disappears. In a similar fashion, the nucleolus disintegrates within the nucleus. While these events are taking place, the diffuse chromatin fibers begin to condense, continuing until distinct threadlike structures, or chromosomes, become visible. It becomes apparent near the end of prophase that each chromosome is actually a double structure split longitudinally except at a single point of constriction, the centromere. The two parts of each chromosome are called **chromatids**. Because the DNA contained in each pair of chromatids represents the duplication of a single chromosome, these chromatids are genetically identical. Therefore, they are called **sister chromatids**. In humans, with a diploid number of 46, a cytological preparation of late prophase will reveal 46 chromosomes randomly distributed in the area formerly occupied by the nucleus.

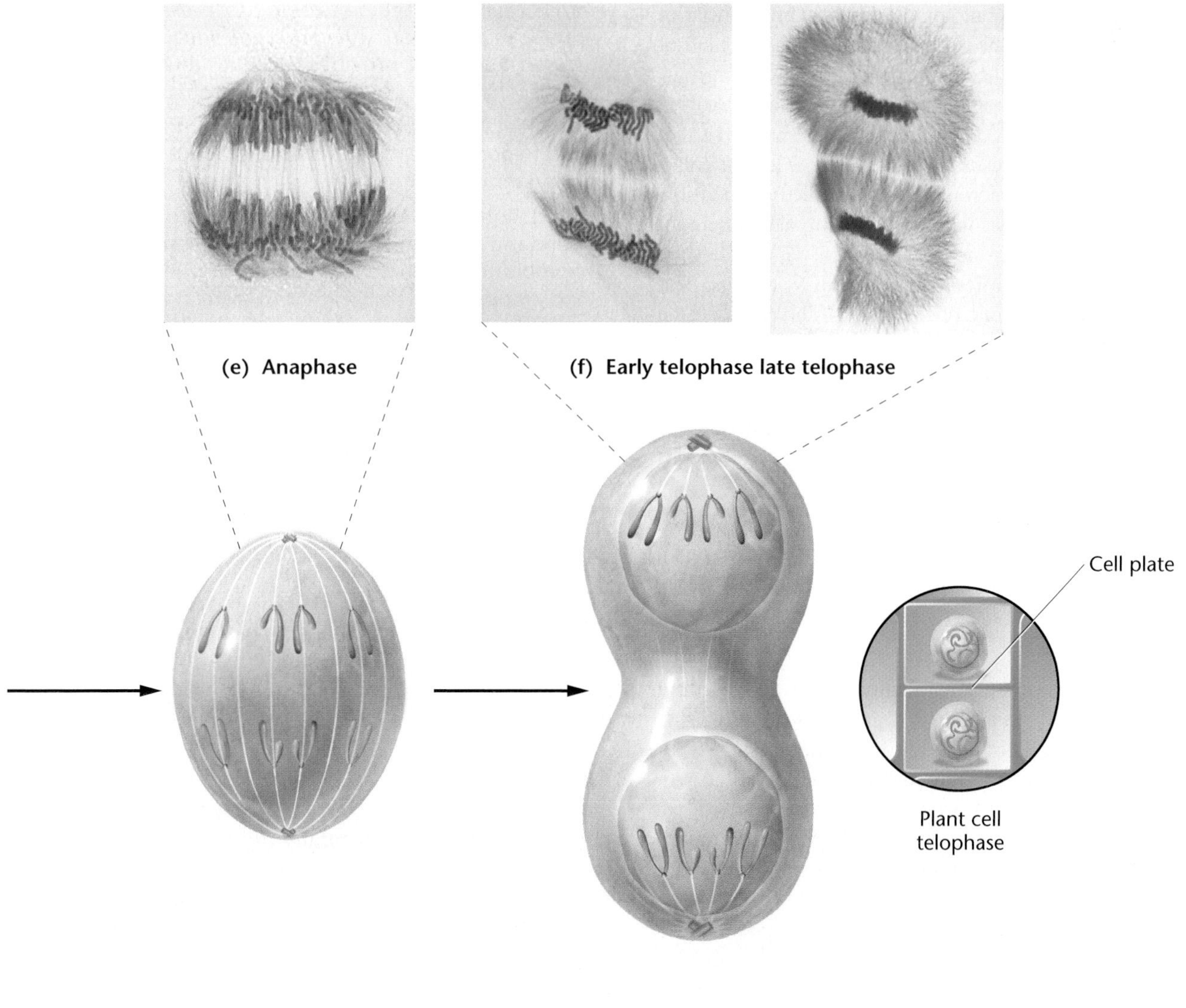

(e) Anaphase

Centromeres split and daughter chromosomes migrate to opposite poles

(f) Telophase

Daughter chromosomes arrive at the poles; Cytokenesis commences

Prometaphase and Metaphase

The distinguishing event of the ensuing stages is the migration of each chromosome, led by the centromeric region, to the equatorial plane. In some descriptions, the term **prometaphase** refers to the period of chromosome movement, and **metaphase** is applied strictly to the chromosome configuration following migration, as depicted in Figures 2–7(c). The equatorial plane, also referred to as the metaphase plate, is the midline region of the cell, a plane that lies perpendicular to the axis established by the spindle fibers.

Migration is made possible by the binding of spindle fibers to a structure associated with the centromere of each chromosome called the **kinetochore**. This structure, consisting of multilayered plates of proteins, forms on opposite sides of each centromere, intimately associating with the two sister chromatids of each chromosome. Once attached to microtubules making up the spindle fibers, the sister chromatids are now ready to be pulled to opposite poles during the ensuing anaphase stage.

We know a great deal about spindle fibers. They consist of **microtubules**, which themselves consist of molecular subunits of the protein **tubulin**. Microtubules seem to originate and "grow" out of the two centrosome regions (containing the centrioles) at opposite poles of the cell. They are dynamic structures that lengthen and shorten as a result of the addition or loss of polarized tubulin subunits. The microtubules most directly responsible for chromosome migration make contact with, and adhere to, kinetochores as they grow from the centrosome region. They are referred to as **kinetochore microtubules** and have one end near the centrosome region (at one of the poles of the cell) and the other anchored to the kinetochore. Note that the number of microtubules that bind to the kinetochore varies greatly between organisms. Yeast (*Saccharomyces*) have only a single microtubule bound to each platelike structure of the kinetochore. Mitotic cells of mammals, at the other extreme, reveal 30 to 40 microtubules bound to each portion of the kinetochore.

At the completion of metaphase, each centromere is aligned at the plate with the chromosome arms extending outward in a random array. This configuration is shown in Figures 2–7(d).

Anaphase

Events critical to chromosome distribution during mitosis occur during the shortest stage of mitosis, **anaphase**. During this phase, sister chromatids of each chromosome *disjoin* (separate) from each other and migrate to opposite ends of the cell. For complete disjunction to occur, each centromeric region must be split in two. This event signals the initiation of anaphase. Once it occurs, each chromatid is referred to as a **daughter chromosome**.

Movement of daughter chromosomes to the opposite poles of the cell is dependent upon the centromere–spindle fiber attachment. Recent investigations reveal that chromosome migration results from the activity of a series of specific proteins, generally called motor proteins. These proteins use the energy generated by the hydrolysis of ATP, and their activity is said to constitute **molecular motors** in the cell. These motors act at several positions within the dividing cell, but all are involved in the activity of microtubules and ultimately serve to propel the chromosomes to opposite ends of the cell. The centromeres of each chromosome *appear* to lead the way during migration, with the chromosome arms trailing behind. The location of the centromere determines the shape of the chromosome during separation. (See Figure 2–3.)

The steps occurring during anaphase are critical in providing each subsequent daughter cell with an identical set of chromosomes. In human cells, there would now be 46 chromosomes at each pole, one from each original sister pair. Figure 2–7(e) shows anaphase prior to its completion.

Now solve this

With the initial appearance of the feature we call "Now Solve This" a short introduction is in order. Occurring several times in this and all ensuing chapters, each entry identifies a problem from the Problems and Discussion Questions section at the end of the chapter. Each selection is related to the discussion just presented. A comment is made about the problem and then a Hint is offered. Each hint provides you with analytical insight that will be useful as you solve the problem.

Problem 2.5 on page 37 involves an understanding of what happens to each pair of homologous chromosomes during mitosis.

Hint: The major issue in solving this problem is to understand that throughout mitosis, members of each homologous pair do not pair up, but instead behave individually.

Telophase

Telophase is the final stage of mitosis and is depicted in Figure 2–7(f). At its beginning, there are two complete sets of chromosomes, one at each pole. The most significant event is **cytokinesis**, the division or partitioning of the cytoplasm. Cytokinesis is essential if two new cells are to be produced from one. The mechanism differs greatly in plant and animal cells. In plant cells, a **cell plate** is synthesized and laid down across the region of the metaphase plate. Animal cells, however, undergo a constriction of the cytoplasm in much the same way a loop of string might be tightened around the middle of a balloon. The end result is the same: Two distinct cells are formed.

It is not surprising that the process of cytokinesis varies among cells of different organisms. Plant cells, which are more regularly shaped and structurally rigid, require a mechanism for depositing new cell wall material around the plasma membrane. The cell plate, laid down during telophase, becomes the **middle lamella**. Subsequently, the primary and secondary layers of the cell wall are deposited between the cell membrane and middle lamella on both sides of the boundary between the two daughter cells. In animals, complete constriction of the cell membrane produces the **cell furrow** characteristic of newly divided cells.

Other events necessary for the transition from mitosis to interphase are initiated during late telophase. They represent a general reversal of events that occurred during prophase. In each new cell, the chromosomes begin to uncoil and become diffuse chromatin once again, while the nuclear envelope reforms around them. The nucleolus gradually reforms and becomes visible in the nucleus during early interphase. The spindle fibers also disappear. At the completion of telophase, the cell enters interphase.

How Do We Know?

How did we learn about the various stages making up mitosis?

2.4 Meiosis Reduces the Chromosome Number from Diploid to Haploid in Germ Cells and Spores

The process of meiosis, unlike mitosis, reduces the amount of genetic material by one half. Whereas in diploids mitosis produces daughter cells with a full diploid complement, meiosis produces gametes or spores with only one haploid set of chromosomes. During sexual reproduction, gametes then combine in fertilization to reconstitute the diploid complement found in parental cells. Figure 2–8 compares the two processes by following two pairs of homologous chromosomes.

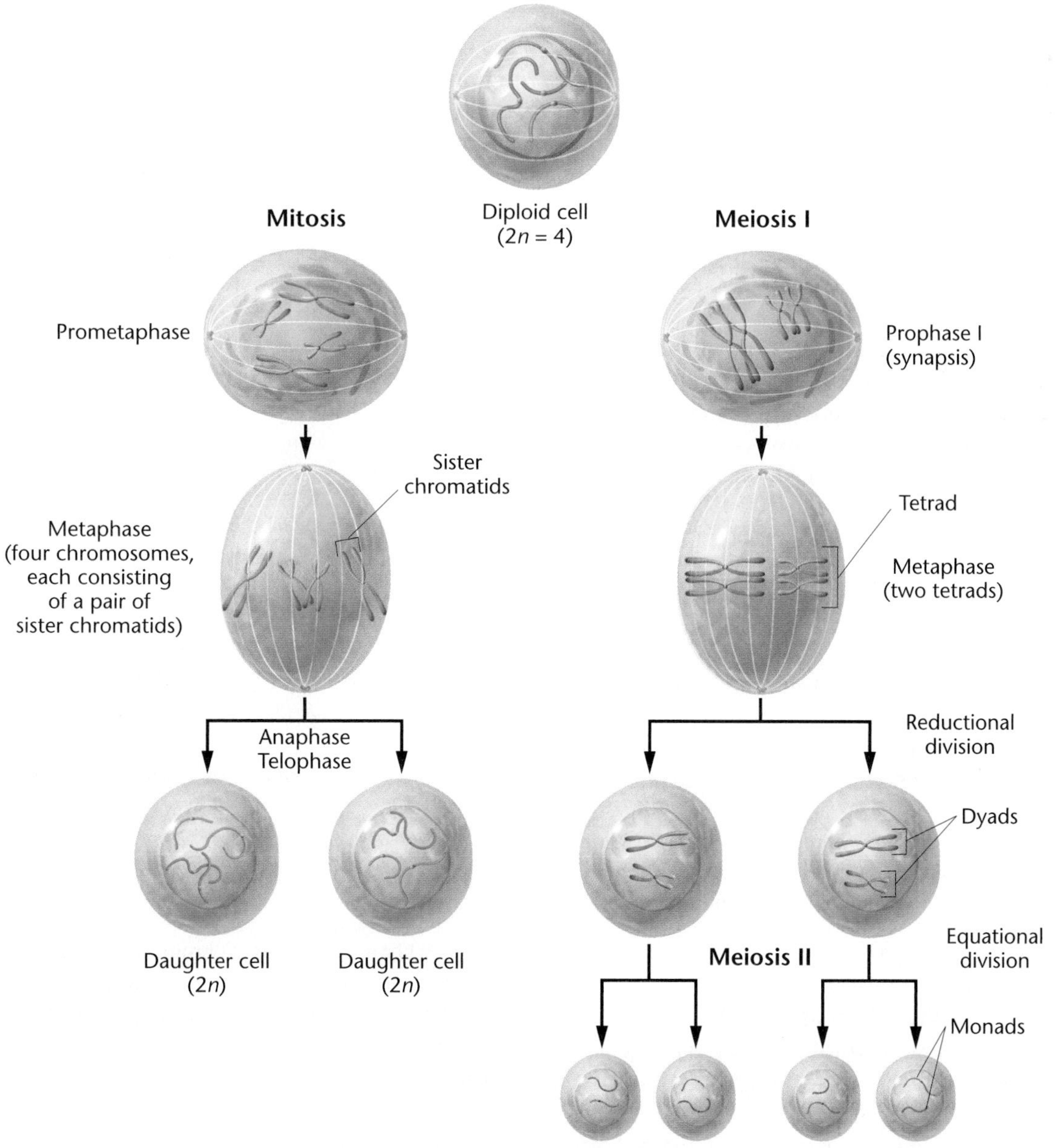

FIGURE 2–8 Overview of the major events and outcomes of mitosis and meiosis. As in Figure 2–7, two pairs of homologous chromosomes are followed.

Meiosis must be highly specific because, by definition, haploid gametes or spores contain precisely one member of each homologous pair of chromosomes. Successfully completed, meiosis ensures genetic continuity from generation to generation. The process of sexual reproduction also ensures genetic variety among members of a species. As you study meiosis, you will see that this process results in gametes with many unique combinations of maternally and paternally derived chromosomes among the haploid complement. With such a tremendous genetic variation among the gametes, a large number of chromosome combinations are possible at fertilization. Furthermore, we shall see that the meiotic event referred to as **crossing over** results in genetic exchange between members of each homologous pair of chromosomes. This creates intact chromosomes that are mosaics of the maternal and paternal homologs from which they are derived, further enhancing the potential genetic variation in gametes and the offspring derived from them. Sexual reproduction therefore reshuffles the genetic material, producing offspring that often differ greatly from either parent. This process constitutes the major form of genetic recombination within species.

An Overview of Meiosis

In the preceding discussion, we established what might be considered the goals of meiosis. Before we consider the phases of this process systematically, we will briefly examine how diploid cells give rise to haploid gametes or spores. You should refer to the meiotic portion of Figure 2–8 during the following discussion.

You have seen that in mitosis each paternally and maternally derived member of any given homologous pair of chromosomes behaves autonomously during division. By contrast, early in meiosis, homologous chromosomes form pairs; that is, they **synapse**. Each synapsed structure is initially called a **bivalent**, which eventually gives rise to a unit, the **tetrad**, consisting of four chromatids. The presence of four chromatids demonstrates that both homologs (making up the bivalent) have, in fact, duplicated. Therefore, in order to achieve haploidy, two divisions are necessary. In meiosis I, described as a **reductional division** (because the number of centromeres, each representing one chromosome, is *reduced* by one half following this division), components of each tetrad—representing the two homologs—separate, yielding two **dyads**. Each dyad is composed of two sister chromatids joined at a common centromere. During meiosis II, described as an **equational division** (because the number of centromeres remains *equal* following this division), each dyad splits into two **monads** of one chromosome each. Thus, the two divisions potentially produce four haploid cells.

The First Meiotic Division: Prophase I

We turn now to a detailed account of meiosis. As in mitosis, meiosis is a continuous process. We name the parts of each stage of division only to facilitate discussion. From a genetic standpoint, three events characterize the initial stage, prophase I (Figure 2–9). First, as in mitosis, chromatin present in interphase thickens and coils into visible chromosomes. Second, unlike mitosis, members of each homologous pair of chromosomes undergo synapsis. Third, crossing over, an exchange process, occurs between synapsed homologs. Because of the complexity of these genetic events, this stage of meiosis has been further divided into five substages: leptonema,* zygonema,* pachynema,* diplonema,*

Web Tutorial 2.2
Overview of Meiosis

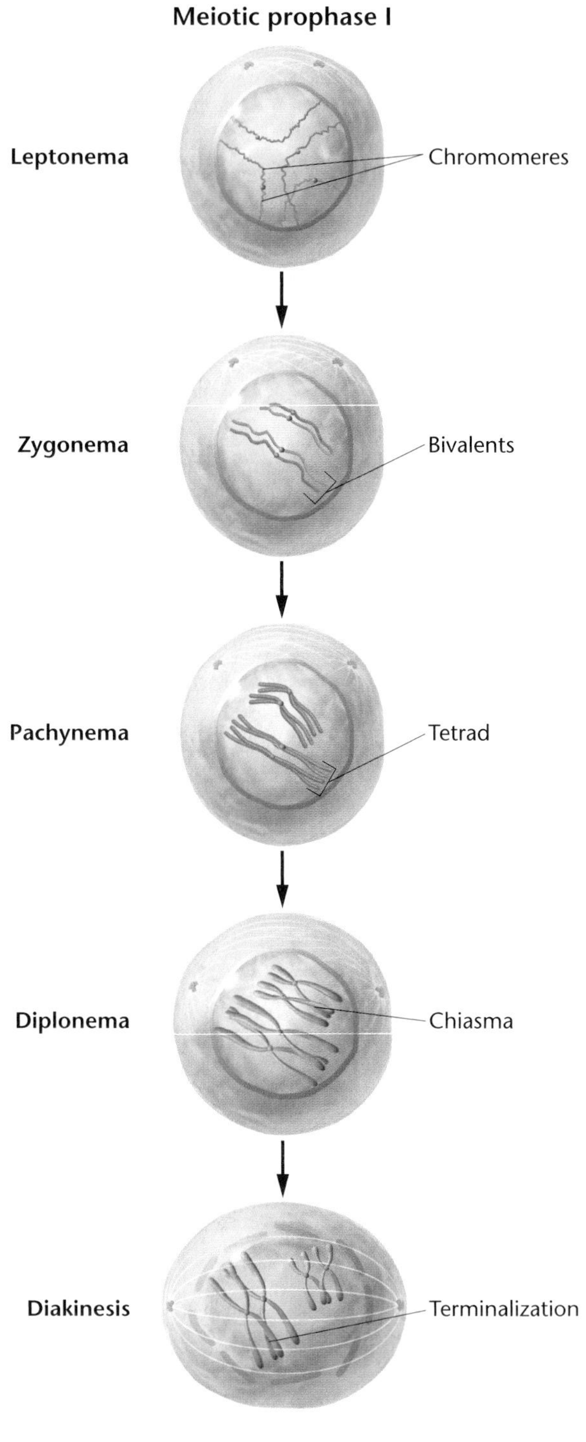

FIGURE 2–9 The substages of meiotic prophase I for the chromosomes depicted in Figure 2–8.

*These are the noun forms of these substages. The adjective forms (leptotene, zygotene, pachytene, and diplotene) are also used in the text.

and diakinesis. As we discuss these substages, be aware that, even though it is not immediately apparent in the earliest phases of meiosis, the DNA of chromosomes has been replicated during the prior interphase.

Leptonema During the **leptotene stage**, the interphase chromatin material begins to condense, and the chromosomes, although still extended, become visible. Along each chromosome are **chromomeres**, localized condensations that resemble beads on a string. Recent evidence suggests that a process called **homology search**, which precedes and is essential to the initial pairing of homologs, begins during leptonema.

Zygonema The chromosomes continue to shorten and thicken during the **zygotene stage**. During the process of homology search, homologous chromosomes undergo a loose alignment with one another, which is complete by the end of zygonema. As meiosis proceeds, an ultrastructural component (one visible only under the electron microsope) called the **synaptonemal complex** is formed between the homologs. We discuss this meiotic component later in the chapter.

At the completion of zygonema, the paired homologs are referred to as **bivalents**. Although both members of each bivalent have already replicated their DNA, it is not yet visually apparent that each member is a double structure. The number of bivalents in each species is equal to the haploid (n) number.

Pachynema In the transition from the zygotene to the **pachytene stage**, coiling and shortening of chromosomes continues, and further development of the synaptonemal complex occurs between the two members of each bivalent. This leads to a more intimate pairing referred to as synapsis. Compared to the rough-pairing characteristic of zygonema, homologs are now separated by only 100 nm.

During pachynema, each homolog is first evident as a double structure, providing visual evidence of the earlier replication of the DNA of each chromosome. Thus, each bivalent contains four member chromatids. As in mitosis, replicates are called sister chromatids, while chromatids from maternal and paternal members of a homologous pair are called nonsister chromatids. The four-membered structure is also referred to as a **tetrad**, and each tetrad contains two pairs of sister chromatids.

Diplonema During the ensuing **diplotene stage**, it is even more apparent that each tetrad consists of two pairs of sister chromatids. Within each tetrad, each pair of sister chromatids begins to separate. However, one or more areas remain in contact where chromatids are intertwined. Each such area, called a **chiasma** (pl. **chiasmata**), is thought to represent a point where nonsister chromatids have undergone genetic exchange through the process referred to above as crossing over. Although the physical exchange between chromosome areas occurred during the previous pachytene stage, the result of crossing over is visible only when the duplicated chromosomes begin to separate. Crossing over is an important source of genetic variability. As indicated earlier, new combinations of genetic material are formed during this process.

Diakinesis The final stage of prophase I is **diakinesis**. The chromosomes pull farther apart, but nonsister chromatids remain loosely associated via the chiasmata. As separation proceeds, the chiasmata move toward the ends of the tetrad. This process, called **terminalization**, begins in late diplonema, and is completed during diakinesis. During this substage period of prophase I, the nucleolus and nuclear envelope break down, and the two centromeres of each tetrad attach to the recently formed spindle fibers. By the completion of prophase I, the centromeres of each tetrad structure are present on the equatorial plate of the cell.

Metaphase, Anaphase, and Telophase I

The remainder of the meiotic process is depicted in Figure 2–10. Following the first meiotic prophase, steps similar to those of mitosis occur. In the metaphase of the first division (**metaphase I**), the chromosomes have maximally shortened and thickened. The terminal chiasmata of each tetrad are visible and appear to be the only factor holding the nonsister chromatids together. Each tetrad interacts with spindle fibers, facilitating movement to the metaphase plate. The alignment of each tetrad prior to this first anaphase is random. One half of each tetrad is pulled to one or the other pole at random, and the other half then moves to the opposite pole.

During the stages of meiosis I, a single centromere holds each pair of sister chromatids together. It does *not* divide. At **anaphase I**, one half of each tetrad (one pair of sister chromatids—called a **dyad**) is pulled toward each pole of the dividing cell. This separation process is the physical basis of what we refer to as **disjunction**, the separation of chromosomes from one another. Occasionally, errors in meiosis occur and separation is not achieved, as we will see later in this chapter. The term **nondisjunction** describes such an error. At the completion of the normal anaphase I, a series of dyads equal to the haploid number is present at each pole.

If crossing over had not occurred in the first meiotic prophase, each dyad at each pole would consist solely of either paternal or maternal chromatids. However, the exchanges produced by crossing over create mosaic chromatids of paternal and maternal origin.

In many organisms, **telophase I** reveals a nuclear membrane forming around the dyads. Next, the nucleus enters into a short interphase period. In other cases, the cells go directly from the first anaphase into the second meiotic division. If interphase occurs, the chromosomes do not replicate because they already consist of two chromatids. In general, meiotic telophase is much shorter than the corresponding stage in mitosis.

The Second Meiotic Division

A second division, referred to as **meiosis II**, is essential if each gamete or spore is to receive only one chromatid from each original tetrad. The stages characterizing meiosis II are shown in the bottom half of Figure 2–10. During **prophase II**, each dyad is composed of one pair of sister chromatids

Web Tutorial 2.3
Meiosis

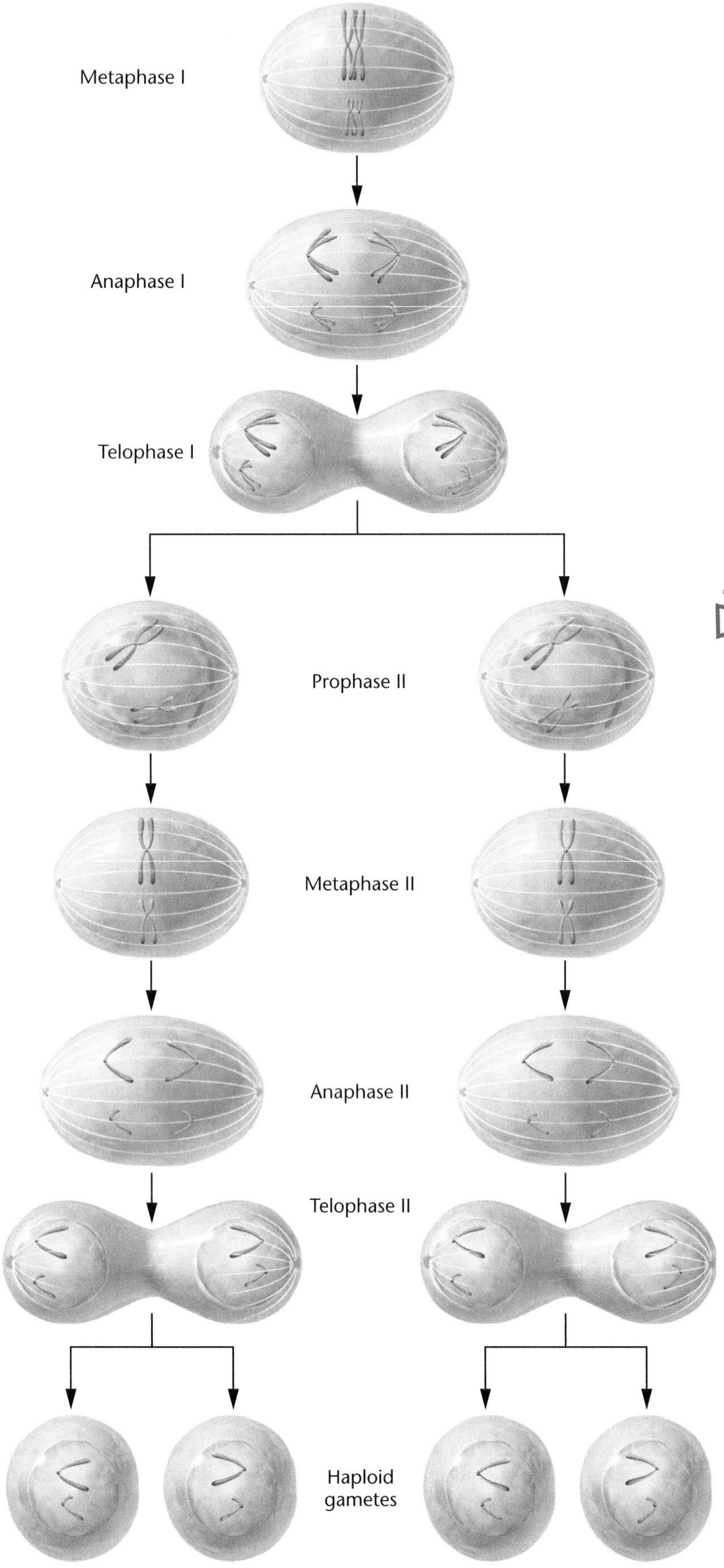

FIGURE 2–10 The major events in meiosis in an animal with a diploid number of 4, beginning with metaphase I. Note that the combination of chromosomes in the cells produced following telophase II is dependent on the random alignment of each tetrad and dyad on the equatorial plate during metaphase I and metaphase II. Several other combinations, which are not shown, can also be formed. The events depicted here are described in the text.

attached by a common centromere. During **metaphase II**, the centromeres are positioned on the equatorial plate. When they divide, **anaphase II** is initiated, and the sister chromatids of each dyad are pulled to opposite poles. Because the number of dyads is equal to the haploid number, **telophase II** reveals one member of each pair of homologous chromosomes present at each pole. Each chromosome is referred to as a **monad**. Following cytokinesis in telophase II, four haploid gametes may result from a single meiotic event. At the conclusion of meiosis, not only has the haploid state been achieved, but if crossing over has occurred, each monad is a combination of maternal and paternal genetic information. As a result, the offspring produced by any gamete will receive a mixture of genetic information originally present in his or her grandparents. Meiosis thus significantly increases the level of genetic variation in each ensuing generation.

Now solve this

Problem 2.9 on page 37 involves an understanding of what happens to the maternal and paternal members of each pair of homologous chromosomes during meiosis.

Hint: The major issue in solving this problem is to understand that maternal and paternal homologs synapse during meiosis. Once it is evident that each chromatid has duplicated, creating a tetrad in the early phases of meiosis, each original pair behaves as a unit and leads to two dyads during anaphase I.

How Do We Know?

How do we know that meiosis has a different final outcome than mitosis?

2.5 The Development of Gametes Varies during Spermatogenesis and Oogenesis

Although events that occur during the meiotic divisions are similar in all cells participating in gametogenesis in most animal species, there are certain differences between the production of a male gamete (spermatogenesis) and a female gamete (oogenesis). Figure 2–11 summarizes these processes.

Spermatogenesis takes place in the testes, the male reproductive organs. The process begins with the expanded growth of an undifferentiated diploid germ cell called a **spermatogonium**. This cell enlarges to become a **primary spermatocyte**, which undergoes the first meiotic division. The products of this division, called **secondary spermatocytes**, contain a haploid number of dyads. The secondary spermatocytes then undergo the second meiotic division, and each of these cells produces two haploid **spermatids**. Spermatids go through a series of developmental changes, **spermiogenesis**, and become highly specialized, motile **spermatozoa**, or **sperm**. All sperm cells produced during spermatogenesis receive equal amounts of genetic material and cytoplasm.

Spermatogenesis may be continuous or may occur periodically in mature male animals, with its onset determined by the nature of the species' reproductive cycle. Animals that reproduce year-round produce sperm continuously, whereas those

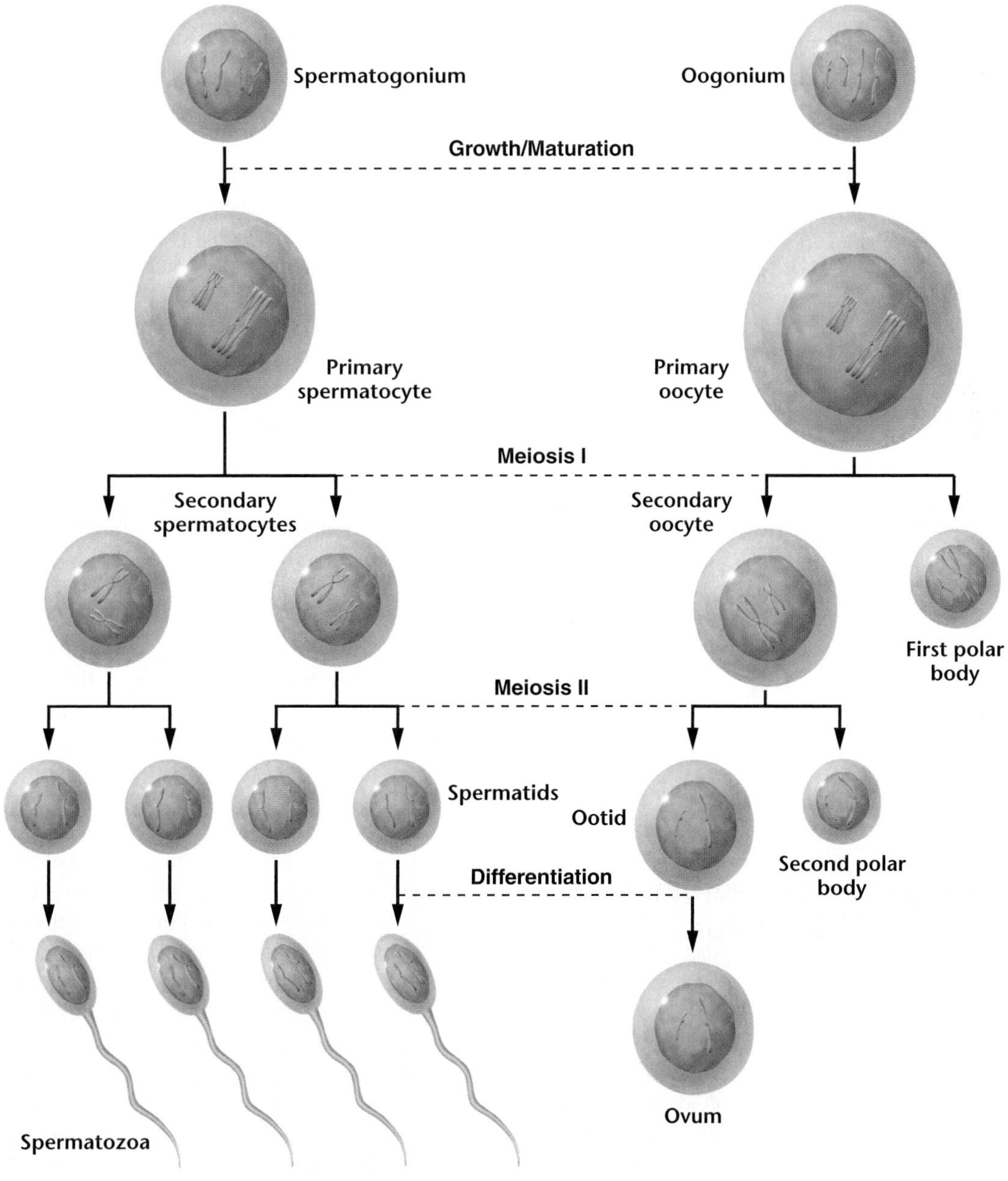

FIGURE 2–11 Spermatogenesis and oogenesis in animal cells.

whose breeding period is confined to a particular season produce sperm only during that time.

In animal **oogenesis**, the formation of **ova** (sing. **ovum**), or eggs, occurs in the ovaries, the female reproductive organs. The daughter cells resulting from the two meiotic divisions receive equal amounts of genetic material, but they do *not* receive equal amounts of cytoplasm. Instead, during each division, almost all the cytoplasm of the **primary oocyte**, itself derived from the **oogonium**, is concentrated in one of the two daughter cells. The concentration of cytoplasm is necessary because a major function of the mature ovum is to nourish the developing embryo following fertilization.

During the first meiotic anaphase in oogenesis, the tetrads of the primary oocyte separate, and the dyads move toward opposite poles. During the first telophase, the dyads present at one pole are pinched off with very little surrounding cytoplasm to form the **first polar body**. The other daughter cell produced by this first meiotic division contains most of the cytoplasm and is called the **secondary oocyte**. The first polar body may or may not divide again to produce two small haploid cells. The mature ovum will be produced from the secondary oocyte during the second meiotic division. During this division, the cytoplasm of the secondary oocyte again divides unequally, producing an **ootid** and a **second polar body**. The ootid then differentiates into the mature ovum.

Unlike the divisions of spermatogenesis, the two meiotic divisions of oogenesis may not be continuous. In some animal species, the two divisions may directly follow each other. In others, including humans, the first division of all oocytes begins in the embryonic ovary, but arrests in prophase I. Many years later, meiosis resumes in each oocyte just prior to its ovulation. The second division is completed only after fertilization.

Now solve this

Problem 2.14 on page 38 involves an understanding of meiosis during oogenesis.

Hint: To answer this question, you must take into account that crossing over occurred during meiosis I between each pair of homologs.

2.6 Meiosis Is Critical to the Successful Sexual Reproduction of All Diploid Organisms

The process of meiosis is critical to the successful sexual reproduction of all diploid organisms. It is the mechanism by which the diploid amount of genetic information is reduced to the haploid amount. In animals, meiosis leads to the formation of gametes, whereas in plants haploid spores are produced, which in turn lead to the formation of haploid gametes.

Furthermore, the mechanism of meiosis is the basis for the production of extensive genetic variation among members of a population. As we have learned, each diploid organism contains its genetic information in the form of homologous pairs of chromosomes, one member of each pair derived from the maternal parent and one member from the paternal parent. Following the reduction to haploidy, gametes or spores contain either the paternal or the maternal representative of every homologous pair of chromosomes. During sexual reproduction, this process has the potential of producing huge quantities of genetically dissimilar gametes. As the number of homologous chromosomes (the haploid number) increases, the possibilities of different combinations of maternal and paternal chromosomes in any given gamete increase. An organism can produce $2n$ number of combinations, where n represents the haploid number. For example, an organism with a haploid number of 10 will produce 2^{10}, or 1024, combinations. Now calculate the number of different combinations of sperm or eggs in our own species: 2^{23}. When you arrive at the answer, you cannot help but be impressed with the potential for genetic variation resulting from meiosis.

The process of crossing over during meiotic prophase I further reshuffles the genetic information between the maternal and paternal members of each homologous pair. As a result, endless varieties of each homolog may occur in gametes, ranging from either intact maternal or paternal chromosomes, where no exchange occurred, to any mixture of maternal and paternal components, depending on where one or more exchanges occurred during crossing over.

In sum, the two most significant points about meiosis are that the process is responsible for

1. the maintenance of equivalent genetic information between generations, and
2. extensive genetic variation within populations.

It is important to touch briefly on the significant role that meiosis plays in the life cycles of fungi and plants. In many fungi, the predominant stage of the life cycle consists of haploid vegetative cells. They arise through meiosis and proliferate by mitotic cell division. In multicellular plants, the life cycle alternates between the diploid **sporophyte stage** and the haploid **gametophyte stage**. While one or the other predominates in different plant groups during this "alternation of generations," the processes of meiosis and fertilization constitute the "bridge" between the sporophyte and gametophyte generations (Figure 2–12). Therefore, meiosis is an essential component of the life cycle of plants.

Finally, we can ask what happens when meiosis fails to achieve the normal outcome. In rare cases during meiosis I or meiosis II, separation, or disjunction, of the chromatids of a tetrad or dyad fails to occur. Instead, both members move to the same pole during anaphase. Such an event is called **nondisjunction**, because the two members fail to disjoin. Nondisjunction during meiosis I or meiosis II leads to gametes with abnormal numbers of chromosomes compared to the haploid number. If such a gamete participates in fertilization, abnormal offspring often result. This shall be a major topic discussed in Chapter 8.

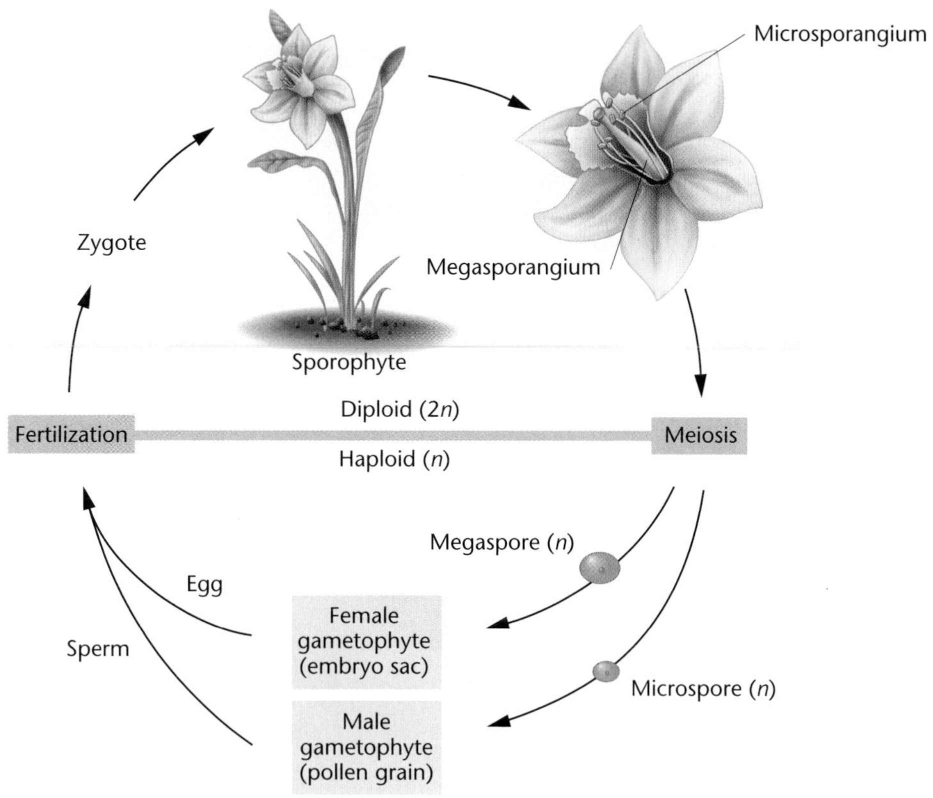

FIGURE 2–12 Alternation of generations between the diploid sporophyte (2n) and the haploid gametophyte (n) in a multicellular plant. The processes of meiosis and fertilization bridge the two phases of the life cycle. This is an angiosperm, where the sporophyte stage is the predominant phase.

Now solve this

Problem 2.20 on page 38 concerns the probability that any particular mixture of maternal or paternal homologs end up together in a gamete.

Hint: You must apply the rules of probability here for multiple events occurring independently of one another. This involves the product law where the probability of many events occurring simultaneously is equal to the product of the individual probabilites.

2.7 Electron Microscopy Has Revealed the Cytological Nature of Mitotic and Meiotic Chromosomes

Thus far in this chapter, we have focused on mitotic and meiotic chromosomes, emphasizing their behavior during cell division and gamete formation. You might wonder why chromosomes are invisible during interphase but visible during the various stages of mitosis and meiosis. Studies using electron microscopy clearly show why chromosomes are visible only during division stages.

Chromatin and Chromosomes

During interphase, only dispersed chromatin fibers are present in the nucleus [Figure 2–13(a)]. Once mitosis begins, however, the fibers coil and fold, condensing into typical mitotic chromosomes [Figure 2–13(b)]. If the fibers comprising the mitotic chromosome are loosened, areas of greatest spreading reveal individual fibers similar to those seen in interphase chromatin [Figure 2–13(c)]. Very few fiber ends seem to be present, and in some cases, none can be seen. Instead, individual fibers always seem to loop back into the interior. Such fibers are obviously twisted and coiled around one another, forming the regular pattern of the mitotic chromosome. Starting in late telophase of mitosis and continuing during G1 of interphase, chromosomes then unwind to form the long fibers characteristic of chromatin, which consist of DNA and associated proteins, particularly proteins called histones. It is in this physical arrangement that DNA can most efficiently function during transcription and replication.

Electron microscopic observations of mitotic chromosomes in varying states of coiling led Ernest DuPraw to postulate the **folded-fiber model**, shown in Figure 2–13(d). During metaphase, each chromosome consists of two sister chromatids joined at the centromeric region. Each arm of the chromatid appears to consist of a single fiber wound much like a skein of yarn. The fiber is composed of tightly coiled double-stranded DNA and protein. An orderly coiling–twisting–condensing process appears to be involved in the transition of the interphase chromatin to the more condensed, mitotic chromosomes. It is estimated that during the transition from interphase to prophase, a 5000-fold contraction occurs in the length of DNA within the chromatin fiber! This process must be extremely precise, given the highly ordered nature and consistent appearance of mitotic chromosomes in all eukaryotes. Note particularly in the micrographs the clear distinction between the sister chromatids constituting each chromosome. They are joined only by the common centromere that they share prior to anaphase.

How Do We Know?

How do we know that mitotic chromosomes are derived from interphase chromatin?

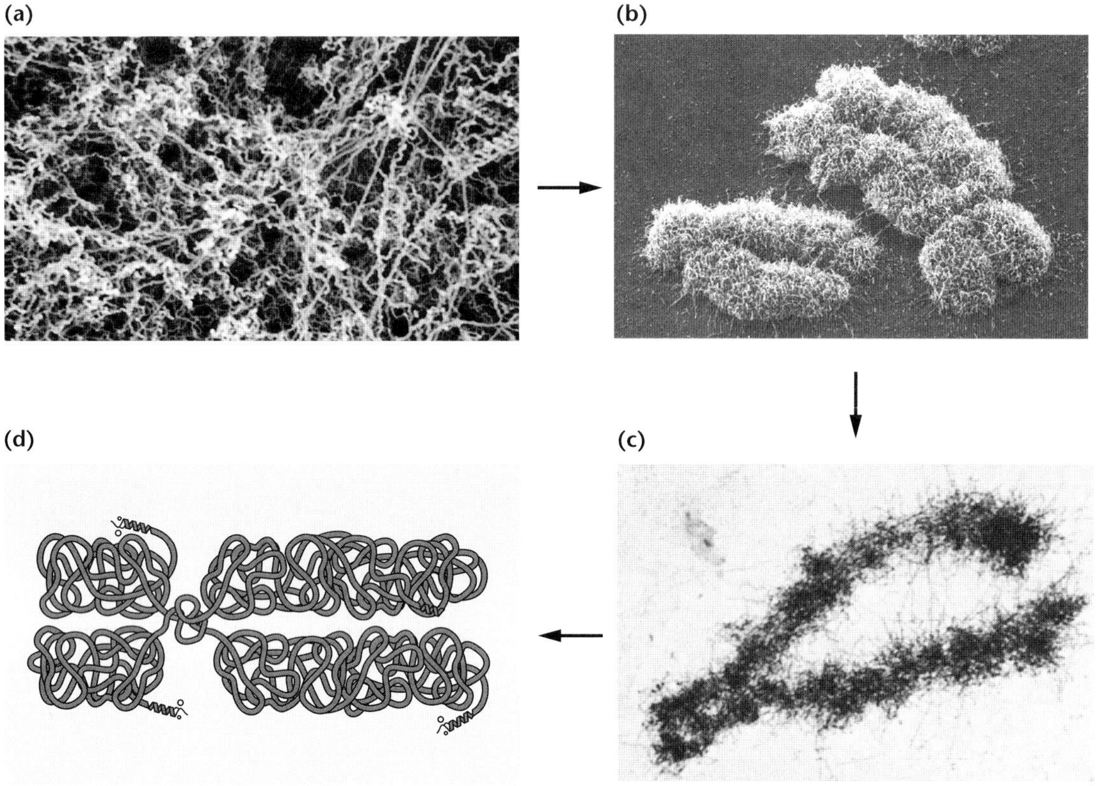

FIGURE 2–13 Comparison of (a) the chromatin fibers characteristic of the interphase nucleus with (b) and (c) metaphase chromosomes that are derived from chromatin during mitosis. Part (d) diagrams the mitotic chromosome and its various components, showing how chromatin is condensed into it. Parts (a) and (c) are transmission electron micrographs, while part (b) is a scanning electron micrograph.

The Synaptonemal Complex

The electron microscope has also been used to visualize another ultrastructural component of the chromosome found only in cells undergoing meiosis. This structure, first introduced during our earlier discussion of the first meiotic prophase stage, is found between synapsed homologs and is called the **synaptonemal complex**.* In 1956, Montrose Moses observed this complex in spermatocytes of crayfish, and Don Fawcett saw it in pigeon and human spermatocytes. Because there was not yet any satisfactory explanation of the mechanism of synapsis or of crossing over and chiasmata formation, many researchers became interested in this structure. With few exceptions, the ensuing studies revealed the synaptonemal complex to be present in most plant and animal cells visualized during meiosis.

As you can see in the electron micrograph in Figure 2–14(a), the synaptonemal complex is a tripartite structure. The central element is usually less dense and thinner (100–150 Å) than the two identical outer elements (500 Å). The outer structures, called lateral elements, are intimately associated with the synapsed homologs on either side. Selective staining has revealed that these lateral elements consist primarily of DNA and protein, suggesting that chromatin is an essential part of them. Some DNA fibrils traverse these lateral elements, making connections with the central element, which is composed primarily of protein. Figure 2–14(b) provides a diagrammatic interpretation of the electron micrograph consistent with the foregoing description.

The formation of the synaptonemal complex begins prior to the pachytene stage. As early as leptonema of the first meiotic prophase, lateral elements are seen in association with sister chromatids. Homologs have yet to associate with one another and are randomly dispersed in the nucleus. As we saw earlier, by the next stage, zygonema, homologous chromosomes begin to align with one another in what is called rough pairing, but they remain distinctly apart by some 300 nm. Then, during pachynema, the intimate association between homologs, characteristic of synapsis, occurs as formation of the complex is completed. In some diploid organisms, this occurs in a zipperlike fashion, beginning at the ends of the chromosomes, which may be attached to the nuclear envelope.

The synaptonemal complex is the vehicle for the pairing of homologs and their subsequent segregation during meiosis.

* An alternative spelling of this term is synaptinemal complex.

where no synaptonemal complexes are formed. Thus, it is possible that the function of this structure may go beyond its involvement in the formation of bivalents.

In certain instances where no synaptonemal complexes are formed during meiosis, synapsis is not complete and crossing over is reduced or eliminated. For example, in male *Drosophila melanogaster*, where synaptonemal complexes are not usually seen, meiotic crossing over rarely, if ever, occurs. This observation suggests that the synaptonemal complex may be important in order for chiasmata to form and crossing over to occur.

The study of *zip1*, a mutation in the yeast *Saccharomyces cerevisiae*, has provided further insights into chromosome pairing. Cells bearing this mutation can undergo the initial alignment stage (rough pairing) and full-length central and lateral element formation, but fail to achieve the intimate pairing characteristic of synapsis. It has been suggested that the gene product of the *zip1* locus is a protein component of the central element of the synaptonemal complex, since it is absent in mutant cells. This observation further suggests that a complete and intact synaptonemal complex is essential during the transition from the initial rough alignment stage to the intimate pairing characteristic of synapsis.

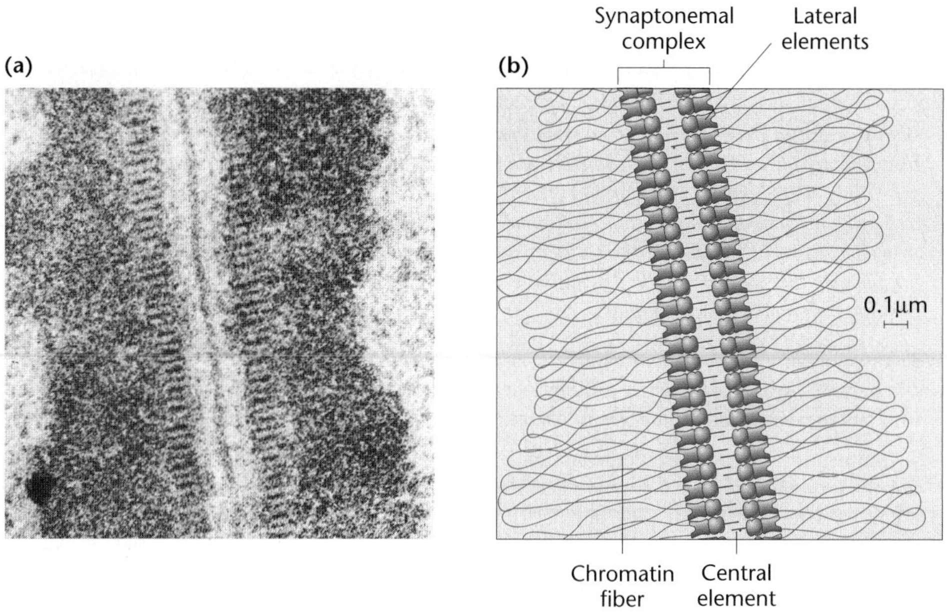

FIGURE 2–14 (a) Electron micrograph of a portion of a synaptonemal complex found between synapsed bivalents of *Neotiella rutilans*. (b) Schematic interpretation of the components making up the synaptonemal complex. The lateral elements, central element, and chromatin fiber are labeled. *D. von Wettstein/Annual Reviews, Inc./With permission, from "Annual Review of Genetics," Volume 6. ©1972 by Annual Reviews, Inc. www.AnnualReviews.org. Photo from D. Von Wettstein.*

Chapter Summary

1. The structure of cells is elaborate and complex. Many components of cells are involved directly or indirectly with genetic processes.
2. In diploid organisms, chromosomes exist in homologous pairs. Each pair shares the same size, centromere placement, and gene sites. One member of each pair is derived from the maternal parent and one is derived from the paternal parent.
3. Mitosis and meiosis are mechanisms by which cells distribute genetic information contained in their chromosomes to progeny cells in a precise, orderly fashion.
4. Mitosis, or nuclear division, is part of the cell cycle and is the basis of cellular reproduction. Daughter cells are produced that are genetically identical to their progenitor cell.
5. Mitosis may be subdivided into discrete stages: prophase, prometaphase, metaphase, anaphase, and telophase. Condensation of chromatin into chromosome structures occurs during prophase. During prometaphase, chromosomes appear as double structures, each composed of a pair of sister chromatids. In metaphase, chromosomes line up on the equatorial plane of the cell. During anaphase, sister chromatids of each chromosome are pulled apart and directed toward opposite poles. Telophase completes daughter cell formation and is characterized by cytokinesis, the division of the cytoplasm.
6. Meiosis, the underlying basis of sexual reproduction, results in the conversion of a diploid cell to a haploid gamete or spore. As a result of chromosome duplication and two subsequent divisions, each haploid cell receives one member of each homologous pair of chromosomes.
7. A major difference exists between meiosis in males and females. Spermatogenesis partitions cytoplasmic volume equally and produces four haploid sperm cells. Oogenesis, on the other hand, accumulates the cytoplasm in one egg cell and reduces the other haploid sets of genetic material to polar bodies. The extra cytoplasm contributes to zygote development following fertilization.
8. Meiosis results in extensive genetic variation by virtue of the exchange during crossing over between maternal and paternal chromatids and their random segregation into gametes. In addition, meiosis plays an important role in the life cycles of fungi and plants, serving as the bridge between alternating generations.
9. Mitotic chromosomes are produced as a result of the coiling and condensation of chromatin fibers characteristic of interphase.

Insights and Solutions

With this initial appearance of "Insights and Solutions," it is appropriate to describe its value to you as a student. This section precedes the "Problems and Discussion Questions" in each chapter and provides sample problems and solutions that demonstrate approaches useful in genetic analysis. The insights you gain by working through this section will help you arrive at correct solutions to ensuing problems in each chapter.

1. In an organism with diploid number of 6, how many individual chromosomal structures will align on the metaphase plate during (a) mitosis, (b) meiosis I, and (c) meiosis II? Describe each configuration.

Solution: (a) In mitosis, where homologous chromosomes do not synapse, there will be 6 double structures, each consisting of a pair of sister chromatids. The number of structures is equivalent to the diploid number. (b) In meiosis I, the homologs have synapsed, reducing the number of structures to 3. Each is called a tetrad and consists of two pairs of sister chromatids. (c) In meiosis II, the same number of structures exist (3), but in this case they are called dyads. Each dyad is a pair of sister chromatids. When crossing over has occurred, each chromatid may contain parts of one of its nonsister chromatids, obtained during exchange in prophase I.

2. Consider two pairs of chromosomes, one larger and metacentric and the other smaller and metacentric. Draw all possible alignment configurations that can occur during metaphase of meiosis I.

Solution: As shown in the diagram below, four configurations are possible when $n = 2$.

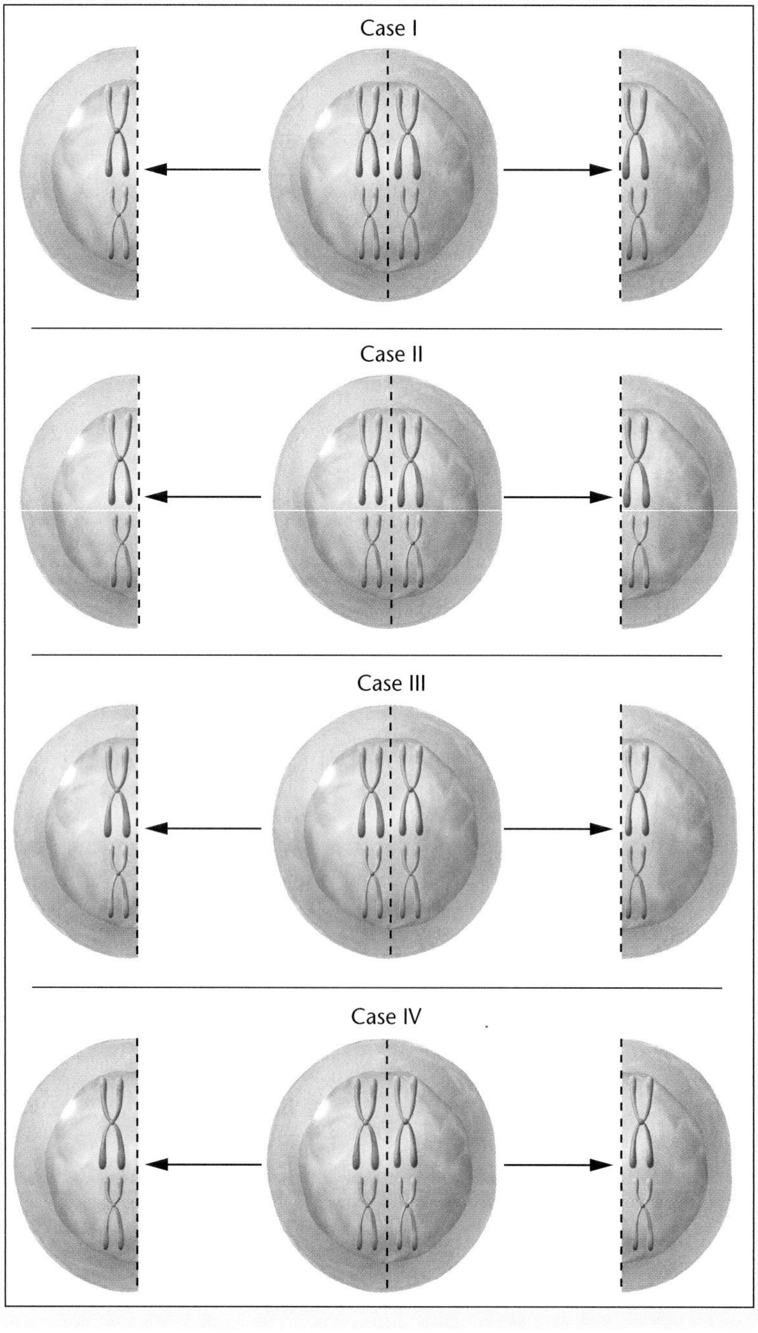

3. For the genes and chromosomes in the previous problem, assume one gene is present on both of the larger chromosomes with two alleles, *A* and *a*, as shown. Also assume a second gene with two alleles (*B*, *b*) is present on the smaller chromosomes. Calculate the probability of generating each gene combination (*AB*, *Ab*, *aB*, *ab*) following meiosis I.

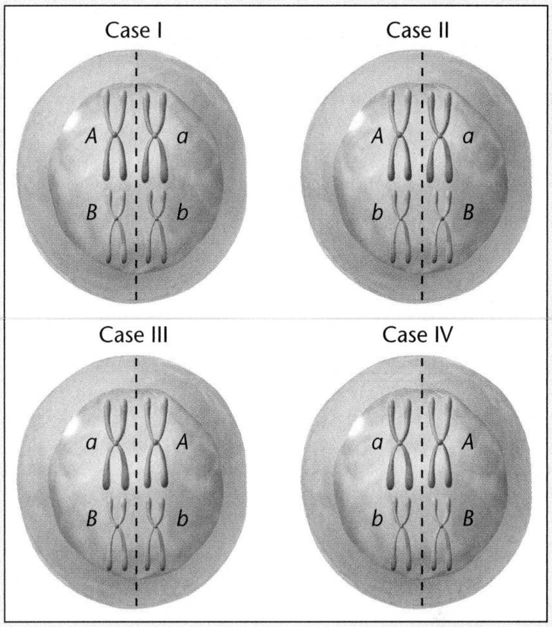

Solution:

Case I	*AB* and *ab*	
Case II	*Ab* and *aB*	
Case III	*aB* and *Ab*	
Case IV	*ab* and *AB*	
Total:	$AB = 2$	$(p = 1/4)$
	$Ab = 2$	$(p = 1/4)$
	$aB = 2$	$(p = 1/4)$
	$ab = 2$	$(p = 1/4)$

4. How many different chromosome configurations can occur following meiosis I if three different pairs of chromosomes are present ($n = 3$)?

Solution: If $n = 3$, then eight different configurations would be possible. The formula 2^n, where *n* equals the haploid number, will allow you to calculate the number of potential alignment patterns. As we will see in the next chapter, these patterns are produced as a result of the Mendelian postulate called *segregation*, and they serve as the physical basis of the Mendelian postulate of *independent assortment.*

5. Describe the composition of a meiotic tetrad as it exists during prophase I, assuming no crossover event has occurred. What impact would a single crossover event have on this structure?

Solution: Such a tetrad contains four chromatids, existing as two pairs. Members of each pair are replicas of one another and are called sister chromatids. They are held together by a common centromere. Members of one pair are maternally derived, whereas members of the other are paternally derived. Maternal and paternal members are called nonsister chromatids. A single crossover event has the effect of exchanging a portion of a maternal and a paternal chromatid, leading to a chiasma, where the two involved chromatids overlap physically in the tetrad. The process of exchange is referred to as crossing over.

Problems and Discussion Questions

1. Explain the role the following cellular components play in the storage, expression, or transmission of genetic information: (a) chromatin, (b) nucleolus, (c) ribosome, (d) mitochondrion, (e) centriole, (f) centromere.
2. Discuss the concepts of homologous chromosomes, diploidy, and haploidy. What characteristics are shared between two chromosomes considered to be homologous?
3. If two chromosomes of a species are the same length and have similar centromere placements yet are not homologous, what is different about them?
4. Describe the events that characterize each stage of mitosis.
5. If an organism has a diploid number of 16, how many chromatids are visible at the end of mitotic prophase? How many chromosomes are moving to each pole during anaphase of mitosis?
6. Describe how chromosomes are named on the basis of their centromere placement.
7. Contrast telophase in plant and animal mitosis.
8. Describe the phases of the cell cycle and the events that characterize each phase.
9. An organism has a diploid number of 16 in a primary oocyte. (a) How many tetrads are present in the first meiotic prophase? (b) How many dyads are present in the second meiotic prophase? (c) How many monads migrate to each pole during the second meiotic anaphase?
10. Contrast the end results of meiosis with those of mitosis.
11. Define and discuss these terms: (a) synapsis, (b) bivalents, (c) chiasmata, (d) crossing over, (e) chromomeres, (f) sister chromatids, (g) tetrads, (h) dyads, (i) monads.
12. Contrast the genetic content and the origin of sister versus nonsister chromatids during their earliest appearance in prophase I of

meiosis. How might the genetic content of these change by the time tetrads have aligned at the equatorial plate during metaphase I?

13. Given the end results of the two types of division, why is it necessary for homologs to pair during meiosis and not desirable for them to pair during mitosis?

14. Examine Figure 2–11, which shows oogenesis in animal cells. Will the genotype of the second polar body (derived from meiosis II) always be identical to that of the ootid? Why or why not?

15. Contrast spermatogenesis and oogenesis. What is the significance of the formation of polar bodies?

16. Explain why meiosis leads to significant genetic variation while mitosis does not.

17. A diploid cell contains three pairs of homologous chromosomes designated C1 and C2, M1 and M2, and S1 and S2. No crossing over occurs. What possible combinations of chromosomes will be present in (a) daughter cells following mitosis? (b) the first meiotic metaphase? (c) haploid cells following both divisions of meiosis?

18. Considering the preceding problem, predict the number of different haploid cells that will occur if a fourth chromosome pair (W1 and W2) is considered in addition to the C, M, and S chromosomes.

19. During oogenesis in an animal species with a haploid number of 6, one dyad undergoes nondisjunction during meiosis II. Following the second meiotic division, the involved dyad ends up intact in the ovum. How many chromosomes are present in (a) the mature ovum and (b) the second polar body? (c) Following fertilization by a normal sperm, what chromosome condition is created?

20. What is the probability that, in an organism with a haploid number of 10, a sperm will be formed that contains all 10 chromosomes whose centromeres were derived from maternal homologs?

21. During the first meiotic prophase, (a) when does crossing over occur? (b) when does synapsis occur? (c) during which stage are the chromosomes least condensed? (d) when are chiasmata first visible?

22. Describe the role of meiosis in the life cycle of a vascular plant.

23. Contrast the chromatin fiber with the mitotic chromosome. How are the two structures related?

24. Describe the "folded fiber" model of the mitotic chromosome.

25. You are given a metaphase chromosome preparation (a slide) from an unknown organism that contains 12 chromosomes. Two are clearly smaller than the rest, appearing identical in length and centromere placement. Describe all that you can about these chromosomes.

Extra-Spicy Problems

As part of the "Problems and Discussion Questions" section in each chapter, we shall present a number of "Extra-Spicy" genetics problems. We have chosen to set these apart in order to identify problems that are particularly challenging. You may be asked to examine and assess actual data, to design genetics experiments, or to engage in cooperative learning. Like genetic varieties of peppers, some of these experiences are just spicy and some are very hot. Hopefully, all of them will leave an aftertaste that is pleasing to those who indulge themselves.

For Questions 26–31 at the right, consider a diploid cell that contains three pairs of chromosomes designated AA, BB, and CC. Each pair contains a maternal and a paternal member (e.g., A^m and A^p). Using these designations, demonstrate your understanding of mitosis and meiosis by drawing chromatid combinations as requested. Be sure to indicate when chromatids are paired as a result of replication and/or synapsis. You may wish to use a large piece of brown manila wrapping paper or a cut-up paper grocery bag and work with another student as you deal with these problems. Such cooperative learning may be a useful approach as you solve problems throughout the text.

26. In mitosis, what chromatid combination(s) will be present during metaphase? What combination(s) will be present at each pole at the completion of anaphase?

27. During meiosis I, assuming no crossing over, what chromatid combination(s) will be present at the completion of prophase? Draw all possible alignments of chromatids as migration begins during early anaphase.

28. Are there any possible combinations present during prophase of meiosis II other than those that you drew in Problem 27? If so, draw them. If not, then proceed to Problem 29.

29. Draw all possible combinations of chromatids during the early phases of anaphase in meiosis II.

30. Assume that during meiosis I none of the *C* chromosomes disjoin at metaphase, but they separate into dyads (instead of monads) during meiosis II. How would this change the alignments that you constructed during the anaphase stages in meiosis I and II? Draw them.

31. Assume that each gamete resulting from Problem 30 participated in fertilization with a normal haploid gamete. What combinations will result? What percentage of zygotes will be diploid, containing one paternal and one maternal member of each chromosome pair?

SELECTED READINGS

Alberts, B., et al. 2002. *Molecular biology of the cell*, 4th ed. New York: Garland.

Brachet, J., and Mirsky, A.E. 1961. *The cell: Meiosis and mitosis*, Vol. 3. Orlando, FL: Academic Press.

DuPraw, E.J. 1970. *DNA and chromosomes*. New York: Holt, Rinehart & Winston.

Glover, D.M., Gonzalez, C., and Raff, J.W. 1993. The centrosome. *Sci. Am.* (June) 268:62–68.

Golomb, H.M., and Bahr, G.F. 1971. Scanning electron microscopic observations of surface structures of isolated human chromosomes. *Science* 171:1024–26.

Hartwell, L.H., and Karstan, M.B. 1994. Cell cycle control and cancer. *Science* 266:1821–28.

Hartwell, L.H., and Weinert, T.A. 1989. Checkpoint controls that ensure the order of cell cycle events. *Science* 246:629–34.

Mazia, D. 1961. How cells divide. *Sci. Am.* (Jan.) 205:101–20.

———. 1974. The cell cycle. *Sci. Am.* (Jan.) 235:54–64.

McIntosh, J.R., and McDonald, K.L. 1989. The mitotic spindle. *Sci. Am.* (Oct.) 261:48–56.

Westergaard, M., and von Wettstein, D. 1972. The synaptinemal complex. *Annu. Rev. Genet.* 6:71–110.

Mendelian Genetics

Gregor Johann Mendel, who in 1866 put forward the major postulates of transmission genetics as a result of experiments with the garden pea.

Chapter Concepts

- Inheritance is governed by information stored in discrete factors called genes.
- Genes are transmitted from generation to generation on vehicles called chromosomes.
- Chromosomes, which exist in pairs, provide the basis of biparental inheritance.
- During gamete formation, chromosomes are distributed according to postulates first described by Gregor Mendel, based on his nineteenth-century research with the garden pea.
- Mendelian postulates prescribe that homologous chromosomes segregate from one another and assort independently with other segregating homologs during gamete formation.
- Genetic ratios, expressed as probabilities, are subject to chance deviation and may be evaluated statistically.

Although inheritance of biological traits has been recognized for thousands of years, the first significant insights into the mechanisms involved occurred almost 140 years ago. In 1866, Gregor Johann Mendel published the results of a series of experiments that would lay the foundation for the formal discipline of genetics. Although Mendel's work went largely unnoticed until about 1900, following the rediscovery of his work, the concept of the gene as a distinct hereditary unit was established. Ways in which genes, as members of chromosomes, are transmitted to offspring and control traits were clarified. Research has continued unabated throughout the twentieth century. Indeed, studies in genetics, most recently at the molecular level, have remained at the forefront of biological research since the early 1900s.

When Mendel began his studies of inheritance using *Pisum sativum*, the garden pea, there was no knowledge of chromosomes or of the role and mechanism of meiosis. Nevertheless, he determined that discrete **units of inheritance** exist and predicted their behavior during the formation of gametes. Subsequent investigators, with access to cytological data, related their own observations of chromosome behavior during meiosis to Mendel's principles of inheritance. Once this correlation was made, Mendel's postulates were accepted as the basis for the study of what is known as **Mendelian** or **transmission genetics**. These principles describe how genes are transmitted from parents to offspring and were derived directly from Mendel's experimentation. Even today, they serve as the cornerstone of the study of inheritance. In this chapter, we focus on the development of Mendel's principles.

3.1 Mendel Used a Model Experimental Approach to Study Patterns of Inheritance

Johann Mendel was born in 1822 to a peasant family in the Central European village of Heinzendorf. An excellent student in high school, he studied philosophy for several years afterward, and in 1843 was admitted to the Augustinian Monastery of St. Thomas in Brno, now part of the Czech Republic. As a monk he took the name of Gregor. In 1849, he was relieved of pastoral duties and received a teaching appointment that lasted several years. From 1851 to 1853, he attended the University of Vienna, where he studied physics and botany. In 1854, he returned to Brno, where for the next 16 years he taught physics and natural science. Mendel received support from the monastery for his studies and research throughout his life.

In 1856, Mendel performed his first set of hybridization experiments with the garden pea. The research phase of his career lasted until 1868, when he was elected abbot of the monastery. Although he retained his interest in genetics, his new responsibilities demanded most of his time. In 1884, Mendel died of a kidney disorder. The local newspaper paid him the following tribute: "His death deprives the poor of a benefactor, and mankind at large of a man of the noblest character, one who was a warm friend, a promoter of the natural sciences, and an exemplary priest."

Mendel first reported the results of some simple genetic crosses between certain strains of the garden pea in 1865. Although his was not the first attempt to provide experimental evidence pertaining to inheritance, Mendel's success where others had failed can be attributed, at least in part, to his elegant model of experimental design and analysis.

Mendel showed remarkable insight into the methodology necessary for good experimental biology. First, he chose an organism that was easy to grow and to hybridize artificially. The pea plant is self-fertilizing in nature, but it is easy to crossbreed experimentally. The plant reproduces well and grows to maturity in a single season. Mendel then chose to follow seven visible features (unit characters), each represented by two contrasting forms or traits (Figure 3–1). For the character stem height, for example, he experimented with the traits *tall* and *dwarf*. He selected six other contrasting pairs of traits involving seed shape and color, pod shape and color, and pod and flower arrangement. From local seed merchants, Mendel obtained true-breeding strains, those in which each trait appeared unchanged generation after generation in self-fertilizing plants.

In addition to his choice of a suitable organism, several factors led to Mendel's success. He restricted his examination to one or very few pairs of contrasting traits in each experiment. He also kept accurate quantitative records, a necessity in genetic experiments. From the analysis of his data, Mendel derived certain postulates that have become the principles of transmission genetics.

The results of Mendel's experiments went unappreciated until the turn of the century, well after his death. Once Mendel's publications were rediscovered by geneticists investigating the function and behavior of chromosomes, however, the implications of his postulates were immediately apparent. He had discovered the basis for the transmission of hereditary traits!

3.2 The Monohybrid Cross Reveals How One Trait Is Transmitted from Generation to Generation

Mendel's simplest crosses involved only one pair of contrasting traits. Each such breeding experiment is called a **monohybrid cross**. A monohybrid cross is made by mating individuals from two parent strains, each of which exhibits one of the two contrasting forms of the character under study. Initially, we examine the first generation of offspring of such a cross, and then we consider the offspring of **selfing** or self-fertilizing individuals from the first generation. The original parents are called the $\mathbf{P_1}$ or **parental generation**, their offspring are the $\mathbf{F_1}$ or **first filial generation**, and the individuals resulting from the selfing of the F_1 generation are the $\mathbf{F_2}$ or **second filial generation**.

The cross between true-breeding pea plants with tall stems and dwarf stems is representative of Mendel's monohybrid crosses. *Tall* and *dwarf* represent contrasting forms or traits of the character of stem height. Unless tall or dwarf plants are crossed together or with another strain, they will undergo self-fertilization and breed true, producing their respective trait generation after generation. However, when Mendel crossed tall plants with dwarf plants, the resulting F_1 generation consisted of only tall plants. When members of the F_1 generation were selfed, Mendel observed that 787 of 1064 F_2 plants were tall,

Web Tutorial 3.1
Monohybrid Cross and Punnett Squares

Character	Contrasting traits		F_1 results	F_2 results	F_2 ratio
Seeds	round/wrinkled		all round	5474 round 1850 wrinkled	2.96:1
	yellow/green		all yellow	6022 yellow 2001 green	3.01:1
Pods	full/constricted		all full	882 full 299 constricted	2.95:1
	green/yellow		all green	428 green 152 yellow	2.82:1
Flower color	violet/white	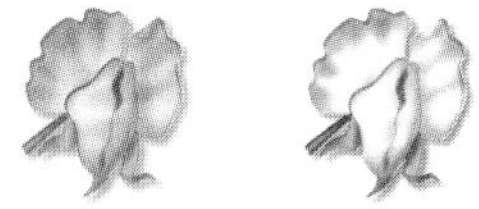	all violet	705 violet 224 white	3.15:1
Flower position	axial/terminal	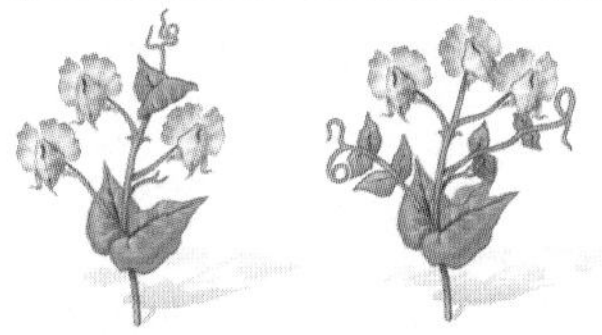	all axial	651 axial 207 terminal	3.14:1
Stem length	tall/dwarf		all tall	787 tall 277 dwarf	2.84:1

FIGURE 3–1 A summary of the seven pairs of contrasting traits and the results of Mendel's seven monohybrid crosses of the garden pea (*Pisum sativum*). In each case, pollen derived from plants exhibiting one trait was used to fertilize the ova of plants exhibiting the other trait. In the F_1 generation, one of the two traits (dominant) was exhibited by all plants. The contrasting trait (recessive) then reappeared in approximately one-fourth of the F_2 plants.

while 277 of 1064 were dwarf. Note that in this cross (Figure 3–1), the dwarf trait disappeared in the F_1 generation, only to reappear in the F_2 generation. Mendel made similar crosses between pea plants exhibiting each of the other pairs of contrasting traits. Results of these crosses are also shown in Figure 3–1. In every case, the outcome was similar to the tall/dwarf cross.

Genetic data are usually expressed and analyzed as ratios. In this particular example, many identical P_1 crosses were made and many F_1 plants—all tall—were produced. Of the 1064 F_2 offspring, 787 were tall and 277 were dwarf—a ratio of approximately 2.8:1.0, or about 3:1. Three-fourths appeared like the F_1 plants, while one-fourth exhibited the contrasting trait, which had disappeared in the F_1 generation.

It is important to point out one further aspect of the monohybrid crosses. In each cross, the F_1 and F_2 patterns of inheritance were similar regardless of which P_1 plant served as the source of pollen (sperm) and which served as the source of the ovum (egg). The crosses could be made either way—that is, pollen from the tall plant pollinating dwarf plants, or vice versa. These are called **reciprocal crosses**. Therefore, the results of Mendel's monohybrid crosses were not sex dependent.

To explain these results, Mendel proposed the existence of what he called **particulate unit factors** for each trait. He suggested that these factors serve as the basic units of heredity and are passed unchanged from generation to generation, determining various traits expressed by each individual plant. Using these general ideas, Mendel proceeded to hypothesize precisely how such factors could account for the results of the monohybrid crosses.

Mendel's First Three Postulates

Using the consistent pattern of results in the monohybrid crosses, Mendel derived the following three *postulates*, or principles, of inheritance:

1. UNIT FACTORS IN PAIRS

   Genetic characters are controlled by unit factors existing in pairs in individual organisms.

   In the monohybrid cross involving tall and dwarf stems, a specific unit factor exists for each trait. Each diploid individual receives one factor from each parent. Because the factors occur in pairs, three combinations are possible: two factors for tallness, two factors for dwarfness, or one of each factor. Every individual possesses one of these three combinations, which determines stem height.

2. DOMINANCE/RECESSIVENESS

   When two unlike unit factors responsible for a single character are present in a single individual, one unit factor is dominant to the other, which is said to be recessive.

In each monohybrid cross, the trait expressed in the F_1 generation results from the presence of the dominant unit factor. The trait that is not expressed in the F_1, but which reappears in the F_2, is under the genetic influence of the recessive unit factor. Note that this dominance–recessiveness relationship pertains only when unlike unit factors are present together in an individual. The terms **dominant** and **recessive** are also used to designate the traits. In the aforementioned case, the trait tall stem is said to be dominant to the recessive trait, dwarf stem.

3. SEGREGATION

During the formation of gametes, the paired unit factors separate, or segregate, randomly so that each gamete receives one or the other with equal likelihood.

These postulates provide a suitable explanation for the results of the monohybrid crosses. Let's use the tall/dwarf cross to illustrate. Mendel reasoned that P_1 tall plants contained identical paired unit factors, as did the P_1 dwarf plants. The gametes of tall plants all received one tall unit factor as a result of segregation. Likewise, the gametes of dwarf plants all received one dwarf unit factor. Following fertilization, all F_1 plants received one unit factor from each parent, a tall factor from one and a dwarf factor from the other, reestablishing the paired relationship. Because tall is dominant to dwarf, all F_1 plants were tall.

When F_1 plants form gametes, the postulate of segregation demands that each gamete randomly receives *either* the tall *or* dwarf unit factor. Following random fertilization events during F_1 selfing, four F_2 combinations will result in equal frequency:

(1) tall/tall

(2) tall/dwarf

(3) dwarf/tall

(4) dwarf/dwarf

Combinations 1 and 4 will clearly result in tall and dwarf plants, respectively. According to the postulate of dominance–recessiveness, combinations 2 and 3 will both yield tall plants. Therefore, the F_2 is predicted to consist of three-fourths tall and one-fourth dwarf plants, or a ratio of 3:1. This is approximately what Mendel observed in his cross between tall and dwarf plants. A similar pattern was observed in each of the other monohybrid crosses (Figure 3–1).

How Do We Know?

- What experimental findings led Mendel to postulate that unit factors that control the inheritance of traits exist in pairs?
- What observations led Mendel to propose that unit factors behave either dominantly or recessively?
- What were the critical observations that led Mendel to propose that unit factors segregate from one another during gamete formation?

Modern Genetic Terminology

To illustrate the monohybrid cross and Mendel's first three postulates in a modern context, we must first introduce several new terms as well as a set of symbols for the unit factors. Traits such as tall or dwarf are physical expressions of the information contained in unit factors. We now call the physical expression of a trait the **phenotype** of the individual.

Mendel's unit factors represent units of inheritance called **genes** by modern geneticists. For any given character, such as plant height, the phenotype is determined by different combinations of alternative forms of a single gene called **alleles**. For example, tall and dwarf are alleles determining the height of the pea plant.

Geneticists use several different conventions involving gene symbols to represent genes. In Chapter 4, we will review some of these, but for now, we will adopt one that we can use consistently throughout this chapter. According to this convention, the first letter of the recessive trait is chosen to symbolize the character in question. The lowercase form of the letter designates the allele for the recessive trait, and the uppercase letter designates the allele for the dominant trait. These gene symbols are also italicized. Therefore, *d* stands for the dwarf allele and *D* represents the tall allele. When alleles are written in pairs to represent the two unit factors present in any individual (*DD*, *Dd*, or *dd*), these symbols are referred to as the **genotype.** This term reflects the genetic makeup of an individual, whether it is haploid or diploid. By following the principle of dominance and recessiveness, we can tell the phenotype of the individual from its genotype: *DD* and *Dd* are tall, and *dd* is dwarf. When identical alleles constitute the genotype (*DD* or *dd*), the individual is said to be **homozygous** or a **homozygote**; when alleles are different (*Dd*), we use the term **heterozygous** or **heterozygote**. Figure 3–2 illustrates the complete monohybrid cross, using modern terminology.

Now solve this

Problem 3.6 on page 62 involves a Mendelian cross where you must determine the mode of inheritance and the genotypes of the parents in a number of instances.

Hint: The first step is to determine how many genes are involved. To do so, convert the data to ratios that are characteristic of Mendelian crosses. In the case of this problem, ask first whether any of the F_2 ratios match Mendel's 3:1 monohybrid ratio.

Mendel's Analytical Approach

What led Mendel to deduce that unit factors exist in pairs? Because there were two contrasting traits for each character, it seemed logical that two distinct factors must exist. However, why does one of the two traits or phenotypes disappear in the F_1 generation? Observation of the F_2 generation helps to answer this question. The recessive trait and its unit factor do not actually disappear in the F_1; they are merely hidden or masked, only to reappear in one fourth of the F_2

offspring. Therefore, Mendel concluded that one unit factor for tall and one for dwarf were transmitted to each F_1 individual; but because the tall factor or allele is dominant to the dwarf factor or allele, all F_1 plants are tall. Given this information, we can ask how Mendel explained the 3:1 F_2 ratio. As shown in Figure 3–2, Mendel deduced that the tall and dwarf alleles of the F_1 heterozygote segregate randomly into gametes. If fertilization is random, this ratio is predicted. If a large population of offspring is generated, the outcome of such a cross should reflect the 3:1 ratio.

Since Mendel operated without the hindsight that modern geneticists enjoy, his analytical reasoning must be considered a truly outstanding scientific achievement. On the basis of rather simple, but precisely executed, breeding experiments, he not only proposed that discrete **particulate units of heredity** exist, but he also explained how they are transmitted from one generation to the next!

Punnett Squares

The genotypes and phenotypes resulting from the recombination of gametes during fertilization can be easily visualized by constructing a **Punnett square**, named after Reginald C. Punnett, who first devised this approach. Figure 3–3 illustrates this method of analysis for the $F_1 \times F_1$ monohybrid cross. Each of the possible gametes is assigned to an individual column or a row, with the vertical column representing those of the female parent and the horizontal row those of the male parent. After we enter the gametes in rows and columns, we can predict the new generation by combining the male and female gametic information for each combination and entering the resulting genotypes in the boxes. This process represents all possible random fertilization events. The genotypes and phenotypes of all potential offspring are ascertained by reading the entries in the boxes.

The Punnett square method is particularly useful when first learning about genetics and how to solve problems. In Figure 3–3, note the ease with which the 3:1 phenotypic ratio and the 1:2:1 genotypic ratio may be derived in the F_2 generation.

The Testcross: One Character

Tall plants produced in the F_2 generation are predicted to have either the *DD* or the *Dd* genotypes. Is there a way to distinguish the genotype of a plant expressing the dominant phenotype? Mendel devised a rather simple method that is still used today in breeding procedures of plants and animals: the **testcross**. The organism of the dominant phenotype, but unknown genotype, is crossed to a **homozygous recessive individual**. For example, as shown in Figure 3–4(a), if a tall plant of genotype *DD* is testcrossed to a dwarf plant, which must have the *dd* genotype, all offspring will be tall phenotypically and *Dd* genotypically. However, as shown in Figure 3–4(b), if a tall plant is *Dd* and is crossed to a dwarf plant (*dd*), then one-half of the offspring will be tall (*Dd*) and the other half will be dwarf (*dd*). Therefore, a 1:1 ratio of tall/dwarf phenotypes demonstrates the heterozygous nature of the tall plant of unknown genotype. The results of testcrosses reinforced Mendel's conclusion that separate unit factors control the tall and dwarf traits.

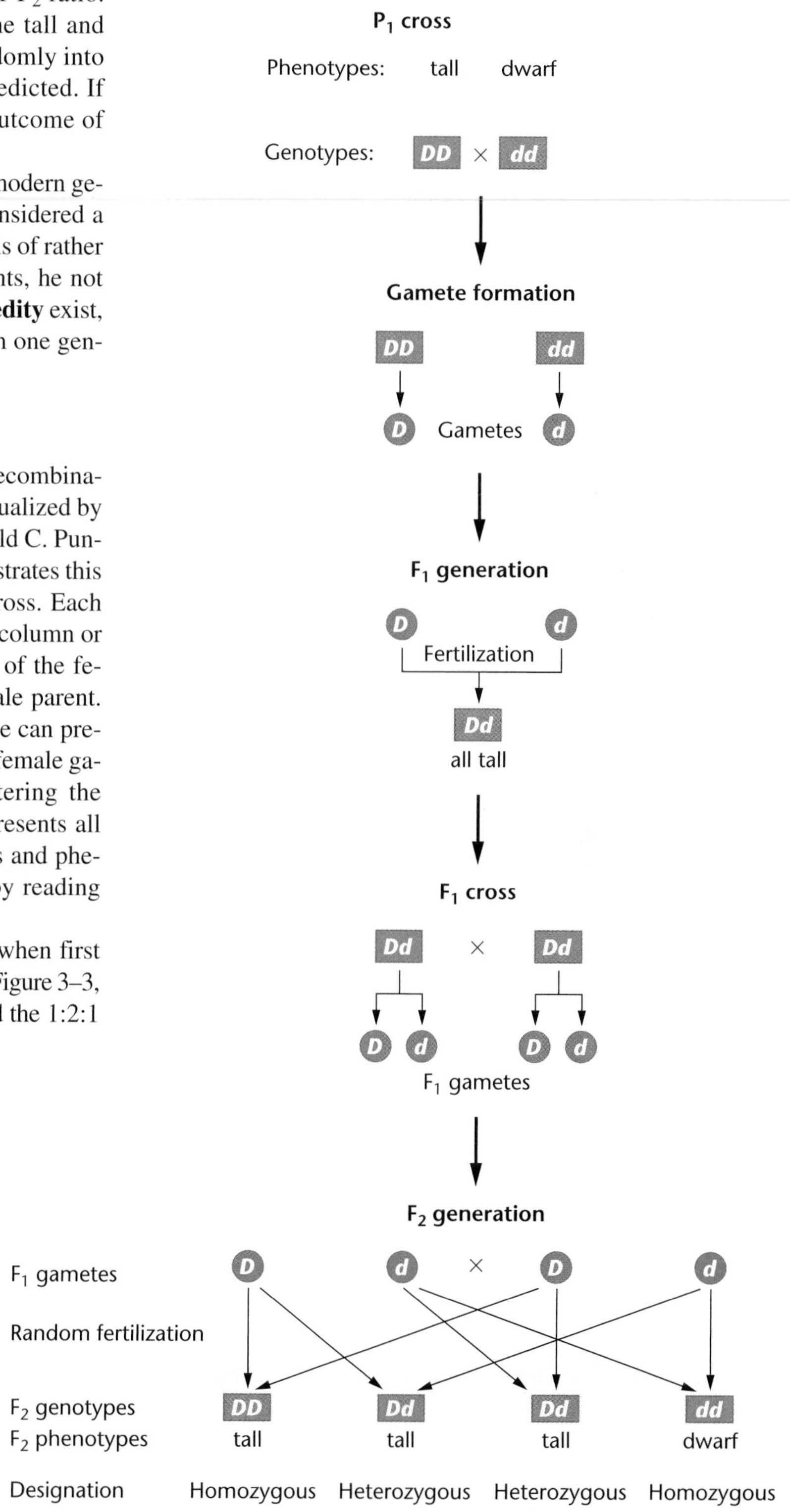

FIGURE 3–2 The monohybrid cross between tall and dwarf pea plants. The symbols *D* and *d* designate the tall and dwarf unit factors, respectively, in the genotypes of mature plants and gametes. All individuals are shown in rectangles, and gametes are shown in circles.

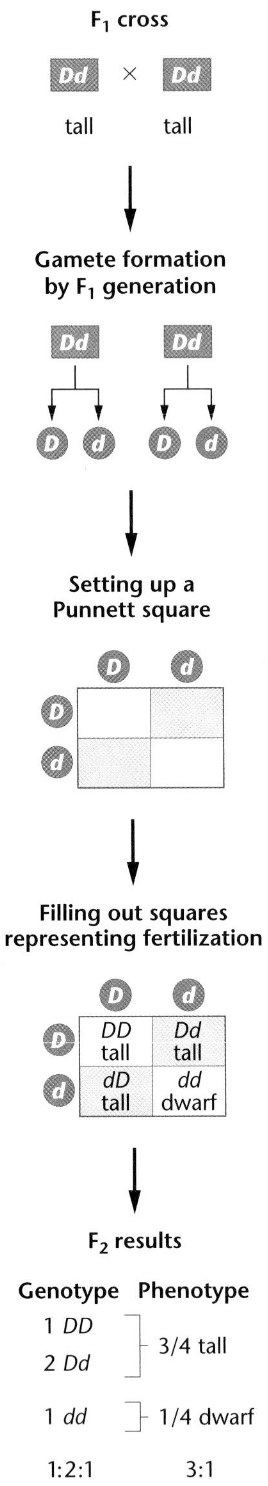

FIGURE 3–3 The use of a Punnett square in generating the F_2 ratio from the $F_1 \times F_1$ cross shown in Figure 3–2.

How Do We Know?

How are geneticists able to experimentally determine whether an organism expressing a dominant trait is homozygous or heterozygous?

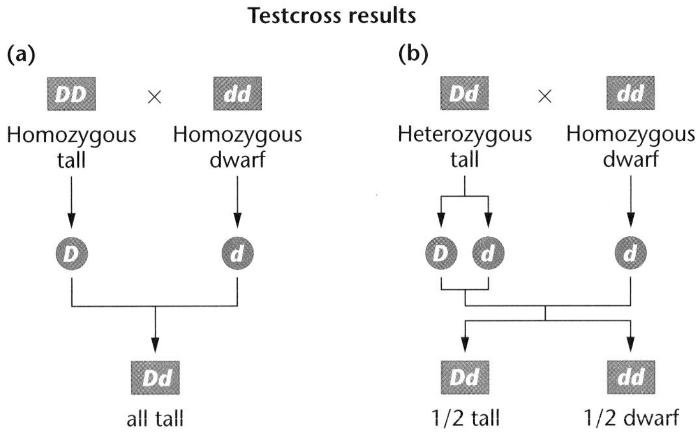

FIGURE 3–4 Testcross of a single character. In (a), the tall parent is homozygous. In (b), the tall parent is heterozygous. The genotype of each tall parent can be determined by examining the offspring when each is crossed to the homozygous recessive dwarf plant.

3.3 Mendel's Dihybrid Cross Revealed His Fourth Postulate: Independent Assortment

As a natural extension of the monohybrid cross, Mendel also designed experiments in which he examined two characters simultaneously. Such a cross, involving two pairs of contrasting traits, is called a **dihybrid cross**, or a **two-factor cross**. For example, if pea plants having yellow seeds that are also round were bred with those having green seeds that are also wrinkled, the results shown in Figure 3–5 will occur. The F_1 offspring are all yellow and round. It is therefore apparent that yellow is dominant to green, and that round is dominant to wrinkled. In this dihybrid cross, the F_1 individuals are selfed, and approximately 9/16 of the F_2 plants express yellow and round, 3/16 express yellow and wrinkled, 3/16 express green and round, and 1/16 express green and wrinkled.

A variation of this cross is also shown in Figure 3–5. Instead of crossing one P_1 parent with both dominant traits (yellow, round) to one with both recessive traits (green, wrinkled), plants with yellow wrinkled seeds are crossed with those with green round seeds. Despite the change in the P_1 phenotypes, both the F_1 and F_2 results remain unchanged.

Independent Assortment

We can most easily understand the results of a dihybrid cross if we consider it theoretically as consisting of two monohybrid crosses conducted separately. Think of the two sets of traits as being inherited independently of each other; that is, the chance of any plant having yellow or green seeds is not at all influenced by the chance that this plant will also have round or wrinkled seeds. Because yellow is dominant to green, all F_1 plants in the first theoretical cross would have yellow seeds. In the second theoretical cross, all F_1 plants would have round seeds, because round is dominant to wrinkled. When Mendel examined the F_1 plants of his dihybrid cross, all were yellow and round, as predicted.

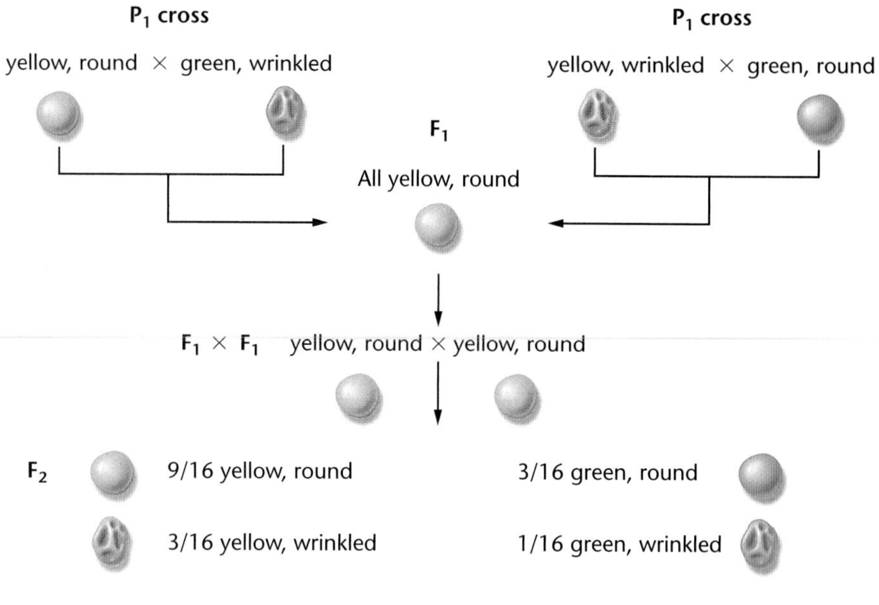

FIGURE 3–5 F_1 and F_2 results of Mendel's dihybrid crosses between yellow, round and green, wrinkled pea seeds and between yellow, wrinkled and green, round pea seeds.

The predicted F_2 results of the first cross are 3/4 yellow and 1/4 green. Similarly, the second cross should yield 3/4 round and 1/4 wrinkled. Figure 3–5 shows that in the dihybrid cross, 12/16 of all F_2 plants are yellow, while 4/16 are green, exhibiting the 3:1 (3/4:1/4) ratio. Similarly, 12/16 of all F_2 plants have round seeds, while 4/16 have wrinkled seeds, again revealing the 3:1 (3/4:1/4) ratio.

Because it is evident that the two pairs of contrasting traits are inherited independently, we can predict the frequencies of all possible F_2 phenotypes by applying the **product law** of probabilities: *When two independent events occur simultaneously, the combined probability of the two outcomes is equal to the product of their individual probabilities of occurrence.* For example, the probability of an F_2 plant having yellow and round seeds is (3/4)(3/4) or 9/16, because 3/4 of all F_2 plants should be yellow and 3/4 of all F_2 plants should be round.

In a like manner, the probabilities of the other three F_2 phenotypes can be calculated: yellow (3/4) *and* wrinkled (1/4) are predicted to be present together 3/16 of the time; green (1/4) *and* round (3/4) are predicted 3/16 of the time; and green (1/4) *and* wrinkled (1/4) are predicted 1/16 of the time. These calculations are illustrated in Figure 3–6. It is now apparent why the F_1 and F_2 results are identical if the parents of the initial cross are yellow and round bred with green and wrinkled or if they are yellow and wrinkled bred with green and round. In both crosses, the F_1 genotype of all plants is identical. Each plant is heterozygous for both gene pairs. As a result, the F_2 generation is also identical in both crosses.

On the basis of similar results in numerous dihybrid crosses, Mendel proposed a fourth postulate:

4. INDEPENDENT ASSORTMENT

During gamete formation, segregating pairs of unit factors assort independently of each other.

This postulate stipulates that any pair of unit factors segregate independently of all other unit factors. Remember that, as a result of segregation, each gamete receives one member of every pair of unit factors. For one pair, whichever unit factor is received does not influence the outcome of segregation of any other pair. Thus, according to the postulate of **independent assortment**, all possible combinations of gametes will be formed in equal frequency.

The Punnett square in Figure 3–7 shows how independent assortment works in the formation of the F_2 generation. Examine the formation of gametes by the F_1 plants. Segregation prescribes that every gamete receives either a *G* or *g* allele and a *W* or *w* allele. Independent assortment stipulates that all four combinations (*GW*, *Gw*, *gW*, and *gw*) will be formed with equal probabilities.

In every $F_1 \times F_1$ fertilization event, each zygote has an equal probability of receiving any of the four combinations from each parent. If a large number of offspring are produced, 9/16 are yellow and round, 3/16 are yellow and wrinkled, 3/16 are

FIGURE 3–6 Computation of the combined probabilities of each F_2 phenotype for two independently inherited characters. The probability of each plant bearing yellow or green seeds is independent of the probability of it bearing round or wrinkled seeds.

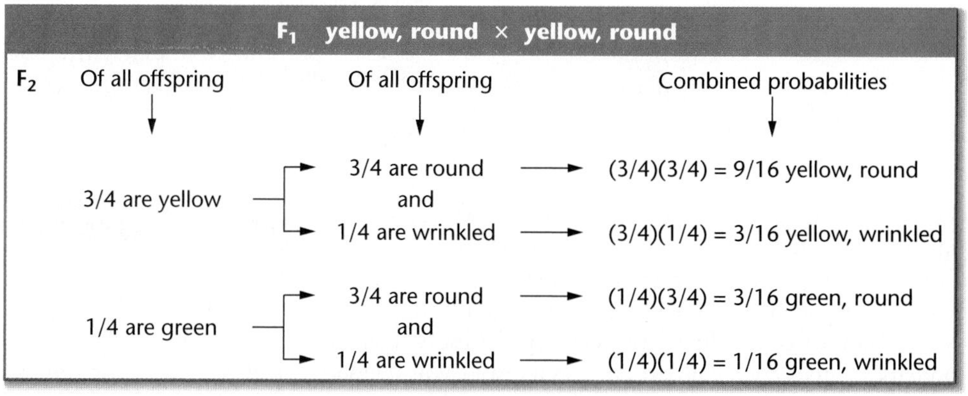

Web Tutorial 3.2
Independent Assortment

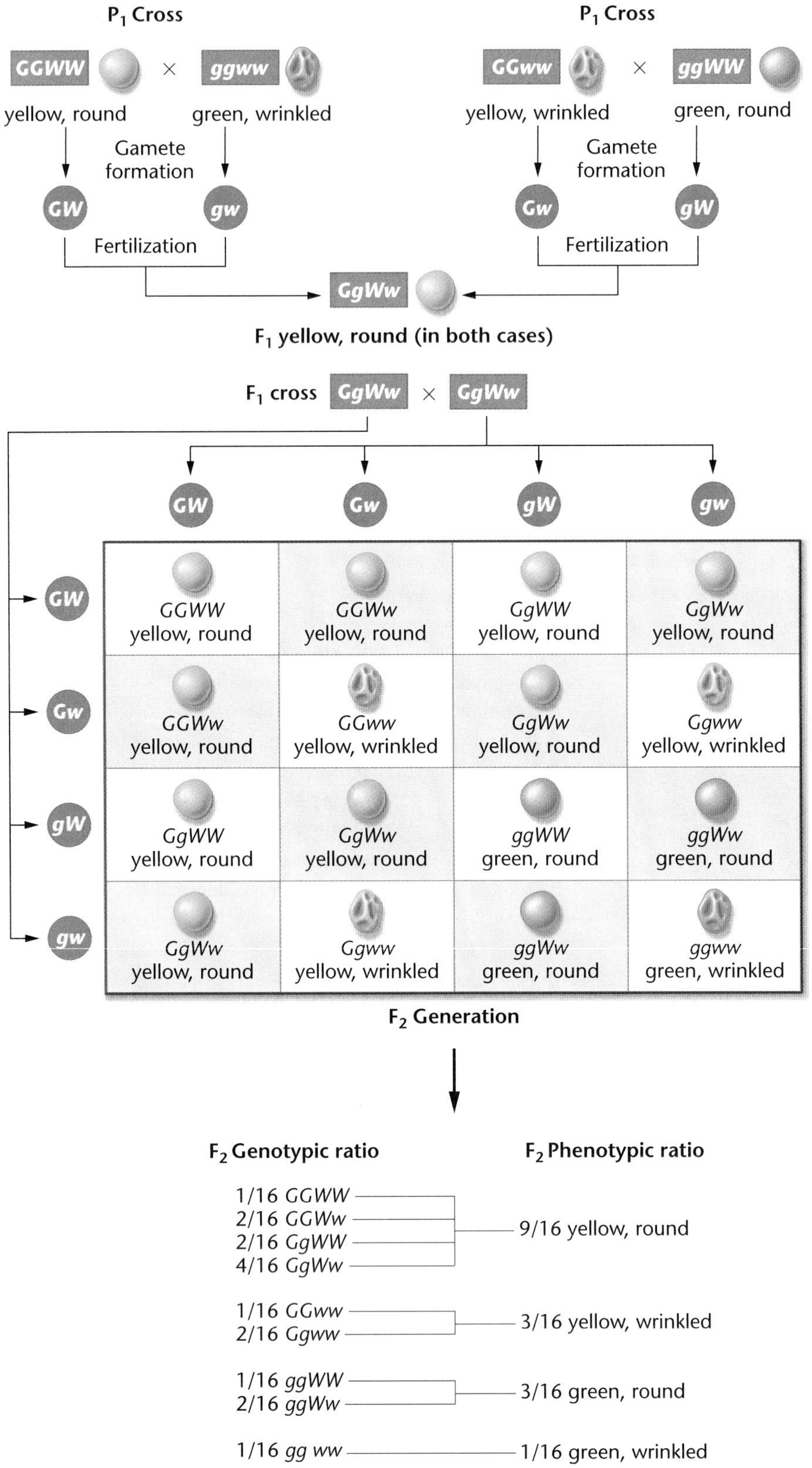

FIGURE 3–7 Analysis of the dihybrid crosses shown in Figure 3–5. The F_1 heterozygous plants are self-fertilized to produce an F_2 generation, which is computed using a Punnett square. Both the phenotypic and genotypic F_2 ratios are shown.

green and round, and 1/16 are green and wrinkled, yielding what is designated as **Mendel's 9:3:3:1 dihybrid ratio**. This is an ideal ratio based on probability events involving segregation, independent assortment, and random fertilization. Because of deviation due strictly to chance, particularly if small numbers of offspring are produced, actual results will seldom match the ideal ratio exactly.

HOW DO WE KNOW?

What experimental observations led Mendel to postulate that each pair of segregating unit factors assorts independently with other segregating pairs of unit factors during gamete formation?

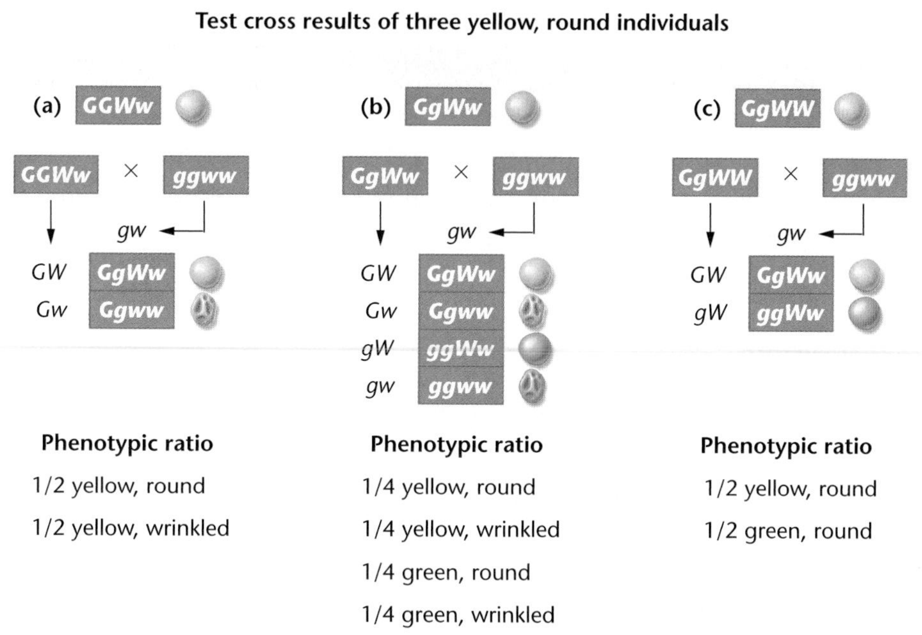

FIGURE 3–8 The testcross illustrated with two independent characters.

The Testcross: Two Characters

The testcross may also be applied to individuals that express two dominant traits, but whose genotypes are unknown. For example, the expression of the yellow round phenotype in the F_2 generation just described may result from the *GGWW*, *GGWw*, *GgWW*, and *GgWw* genotypes. If an F_2 yellow round plant is crossed with the homozygous recessive green, wrinkled plant (*ggww*), analysis of the offspring will indicate the exact genotype of that yellow round plant. Each of these genotypes will result in a different set of gametes and, in a testcross, a different set of phenotypes in the resulting offspring. Three cases are illustrated in Figure 3–8.

Now solve this

Problem 3.9 on page 63 involves a series of Mendelian dihybrid crosses where you must determine the genotypes of the parents in a number of instances.

Hint: In each case, write down everything that you know for certain. This reduces the problem to its bare essentials, clarifying what you need to determine. For example, the wrinkled, yellow plant in case (b) must be homozygous for the recessive wrinkled alleles and bear at least one dominant allele for the yellow trait. Having established this, you need only determine the remaining allele for cotyledon color.

3.4 The Trihybrid Cross Demonstrates That Mendel's Principles Apply to Inheritance of Multiple Traits

Thus far, we have considered inheritance of up to two pairs of contrasting traits. Mendel demonstrated that the processes of segregation and independent assortment also apply to three pairs of contrasting traits in what is called a **trihybrid cross**, also referred to as a **three-factor cross**.

Although a trihybrid cross is somewhat more complex than a dihybrid cross, its results are easily calculated if the principles of segregation and independent assortment are followed. For example, consider the cross shown in Figure 3–9 where the gene pairs representing theoretical contrasting traits are symbolized *A/a*, *B/b*, and *C/c*. In the cross between *AABBCC* and *aabbcc* individuals, all offspring are heterozygous for all three gene pairs. Their genotype, *AaBbCc*, results in the phenotypic expression of the dominant *A*, *B*, and *C* traits. When individuals are parents, each produces eight different gametes in equal frequencies. At this point, we could construct a Punnett square with 64 separate boxes and read out the phenotypes. Because such a method is cumbersome in a cross involving so many factors, another approach, the forked-line method, has been devised to calculate the predicted ratio.

The Forked-Line Method, or Branch Diagram

It is much less difficult to consider each contrasting pair of traits separately and then to combine these results by using the **forked-line method**, which was first illustrated in Figure 3–6. This method, also called a **branch diagram**, relies on the simple application of the laws of probability established for the dihybrid cross. Each gene pair is assumed to behave independently during gamete formation.

When the monohybrid cross $AA \times aa$ is made, we know that

1. All F_1 individuals have the genotype *Aa* and express the phenotype represented by the *A* allele, which is called the *A* phenotype in the discussion that follows.

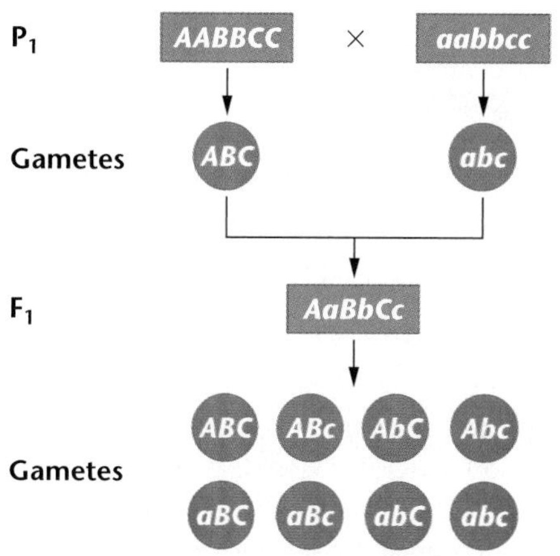

FIGURE 3–9 Formation of P_1 and F_1 gametes in a trihybrid cross.

2. The F_2 generation consists of individuals with either the *A* phenotype or the *a* phenotype in the ratio of 3:1.

The same generalizations apply to the $BB \times bb$ and $CC \times cc$ crosses. Thus, in the F_2 generation, 3/4 of all organisms will express phenotype *A*, 3/4 will express *B*, and 3/4 will express *C*. Similarly, 1/4 of all organisms will express phenotype *a*, 1/4 will express *b*, and 1/4 will express *c*. The proportions of organisms expressing each phenotypic combination can be predicted by assuming that fertilization, following the independent assortment of these three gene pairs during gamete formation, is a random process. We once again simply apply the product law of probabilities.

The phenotypic proportions of the F_2 generation calculated by using the forked-line method are illustrated in Figure 3–10. They fall into the trihybrid ratio of 27:9:9:9:3:3:3:1. The same method can be applied when solving crosses involving any number of gene pairs, *provided* that all gene pairs assort independently from each other. We shall see later that this is not always the case. However, it appeared to be true for all of Mendel's characters.

Note that in Figure 3–10, only phenotypic ratios of the F_2 generation have been derived. It is possible to generate genotypic ratios as well. To do so, we again consider the *A/a*, *B/b*, and *C/c* gene pairs separately. For example, for the *A/a* pair, the F_1 cross is $Aa \times Aa$. Phenotypically, an F_2 ratio of 3/4 *A*:1/4 *a* is produced. Genotypically, however, the F_2 ratio is different; 1/4 *AA*:1/2 *Aa*:1/4 *aa* will result. Using Figure 3–10 as a model, we would enter these genotypic frequencies on the left side of the calculation. Each would be connected by three lines to 1/4 *BB*, 1/2 *Bb*, and 1/4 *bb*, respectively. From each of these nine designations, three more lines would extend to the 1/4 *CC*, 1/2 *Cc*, and 1/4 *cc* genotypes. On the right side of the completed diagram, 27 genotypes and their frequencies of occurrence would appear.

Generation of F_2 trihybrid phenotypes

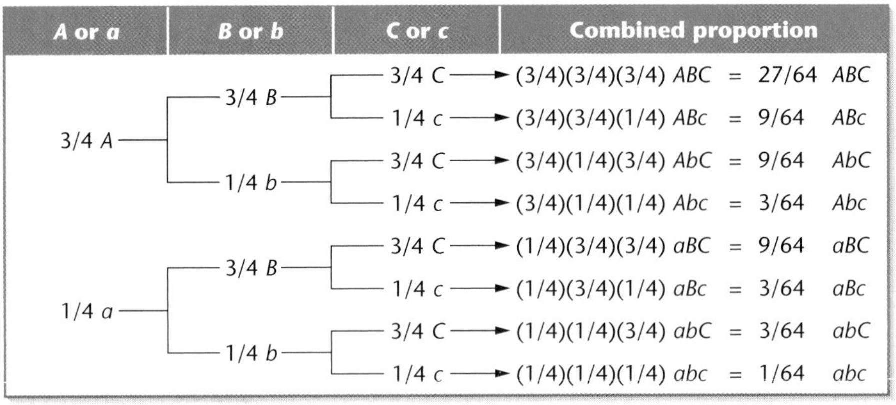

FIGURE 3–10 The generation of the F_2 trihybrid ratio using the forked-line, or branch diagram method, which is based on the expected probabability of occurrence of each phenotype.

Now solve this

Problem 3.17 on page 63 asks you to use the forked-line method to determine the outcome of a number of trihybrid crosses.

Hint: In using the forked-line method, consider each gene pair separately. For example, in this problem, first predict the outcome of each cross for *A/a* genes, then for the *B/b* genes, and finally, for the *C/c* genes. Then you are prepared to pursue the outcome of each cross using the forked-line method.

In crosses involving two or more gene pairs, the calculation of gametes and genotypic and phenotypic results is quite complex. Several simple mathematical rules will enable you to check the accuracy of various steps required in working genetic problems. First, you must determine the number of *heterozygous* gene pairs (n) involved in the cross. For example, where $AaBb \times AaBb$ represents the cross, $n = 2$; for $AaBbCc \times AaBbCc$, $n = 3$; for $AaBBCcDd \times AaBBCcDd$, $n = 3$ (because the *B* genes are not heterozygous). Once n is determined, 2^n is the number of different gametes that can be formed by each parent; 3^n is the number of different genotypes that result following fertilization; and 2^n is the number of different phenotypes that are produced from these genotypes.

TABLE 3.1 SIMPLE MATHEMATICAL RULES USEFUL IN WORKING GENETICS PROBLEMS

Crosses between Organisms Heterozygous for Genes Exhibiting Independent Assortment

Number of Heterozygous Gene pairs	Number of Different Types of Gametes Formed	Number of Different Genotypes Produced	Number of Different Phenotypes Produced*
n	2^n	3^n	2^n
1	2	3	2
2	4	9	4
3	8	27	8
4	16	81	16

*The fourth column assumes that dominance and recessiveness are operational for all gene pairs.

Table 3.1 summarizes these rules, which may be applied to crosses involving any number of genes, *provided that they assort independently from one another*.

3.5 Mendel's Work Was Rediscovered in the Early Twentieth Century

Mendel's work, initiated in 1856, was presented to the Brünn Society of Natural Science in 1865 and published the following year. However, his findings went largely unnoticed for about 35 years. Many reasons have been suggested to explain why the significance of his research was not immediately recognized.

First, Mendel's adherence to mathematical analysis of probability events was quite an unusual approach in that era and may have seemed foreign to his contemporaries. More important, his conclusions drawn from such analyses did not fit well with the existing hypotheses involving the source of variation among organisms. Students of evolutionary theory, stimulated by the proposals developed by Charles Darwin and Alfred Russel Wallace, believed in **continuous variation**, where offspring were a blend of their parents' phenotypes. By contrast, Mendel hypothesized that heredity was due to discrete or particulate units, resulting in **discontinuous variation**. For example, Mendel proposed that the F_2 offspring of a dihybrid cross are merely expressing traits produced by new combinations of previously existing unit factors. Thus, Mendel's hypotheses did not fit well with the evolutionists' preconceptions about causes of variation.

It is also likely that Mendel's contemporaries failed to realize that Mendel's postulates explained *how* variation was transmitted to offspring. Instead, they may have attempted to interpret his work in a way that addressed the issue of *why* certain phenotypes survive preferentially. It was this latter question that had been addressed in the theory of natural selection, but it was not addressed by Mendel. The collective vision of Mendel's scientific colleagues may have been obscured by the impact of this extraordinary theory of organic evolution.

How Do We Know?

What findings confirmed that Mendel's unit factors in pairs were actually homologous pairs of chromosomes?

3.6 The Correlation of Mendel's Postulates with the Behavior of Chromosomes Formed the Foundation of Modern Transmission Genetics

In the latter part of the nineteenth century, a remarkable observation set the scene for the rebirth of Mendel's work: Walter Flemming's discovery of chromosomes in 1879. Flemming was able to describe the behavior of these threadlike structures in the nuclei of salamander cells during cell division. As a result of the findings of Flemming and many other cytologists, the presence of a nuclear component soon became an integral part of ideas surrounding inheritance. In this setting, scientists began to reexamine Mendel's findings.

In the early twentieth century, research led to further interest in Mendel's work. Hybridization experiments similar to Mendel's were independently performed by three botanists, Hugo DeVries, Karl Correns, and Erich von Tschermak. De Vries's work, for example, had focused on unit characters, and he demonstrated the principle of segregation in his experiments with several plant species. He had apparently searched the existing literature and found that Mendel's work had anticipated his own conclusions. Correns and Tschermak had also reached conclusions similar to Mendel's.

In 1902, two cytologists, Walter Sutton and Theodor Boveri, independently published papers linking their discoveries of the behavior of chromosomes during meiosis to the Mendelian principles of segregation and independent assortment. They pointed out that the separation of chromosomes during meiosis could serve as the cytological basis of these two postulates. Although they thought that Mendel's unit factors were probably chromosomes, rather than genes on chromosomes, their findings reestablished the importance of Mendel's work, which served as the foundation of ensuing genetic investigations.

Based on their studies, Sutton and Boveri are credited with initiating the **chromosomal theory of heredity**. As we will see in subsequent chapters, work by Thomas H. Morgan, Alfred H. Sturtevant, Calvin Bridges, and others established beyond a reasonable doubt that Sutton's and Boveri's hypothesis was correct.

Unit Factors, Genes, and Homologous Chromosomes

Because the correlation between Sutton's and Boveri's observations and Mendelian principles is the foundation for the modern interpretation of transmission genetics, we examine it in some detail here.

As we pointed out in Chapter 2, every species possesses a specific number of chromosomes in each somatic (body) cell nucleus characteristic of that species. In diploid organisms, during the formation of gametes, the number is precisely halved (n), and when two gametes combine during fertilization, the diploid number is reestablished. During meiosis, however, the chromosome number is not reduced in a random manner. It was apparent to early cytologists that the diploid number of chromosomes is composed of homologous pairs identifiable by their morphological appearance and behavior. The gametes contain one member of each pair. The chromosome complement of a gamete is quite specific in that the number of chromosomes in each gamete is equal to the haploid number.

With this basic information, we can see the correlation between the behavior of unit factors and chromosomes and genes. Figure 3–11 shows three of Mendel's postulates and the chromosomal explanation of each. Unit factors are really genes located on homologous pairs of chromosomes [Figure 3–11(a)]. Members of each pair of homologs separate, or segregate, during gamete formation [Figure 3–11(b)]. Two different alignments are possible, both of which are shown.

To illustrate the principle of independent assortment, it is important to distinguish between members of any given homologous pair of chromosomes. One member of each pair is derived from the **maternal parent**, while the other comes

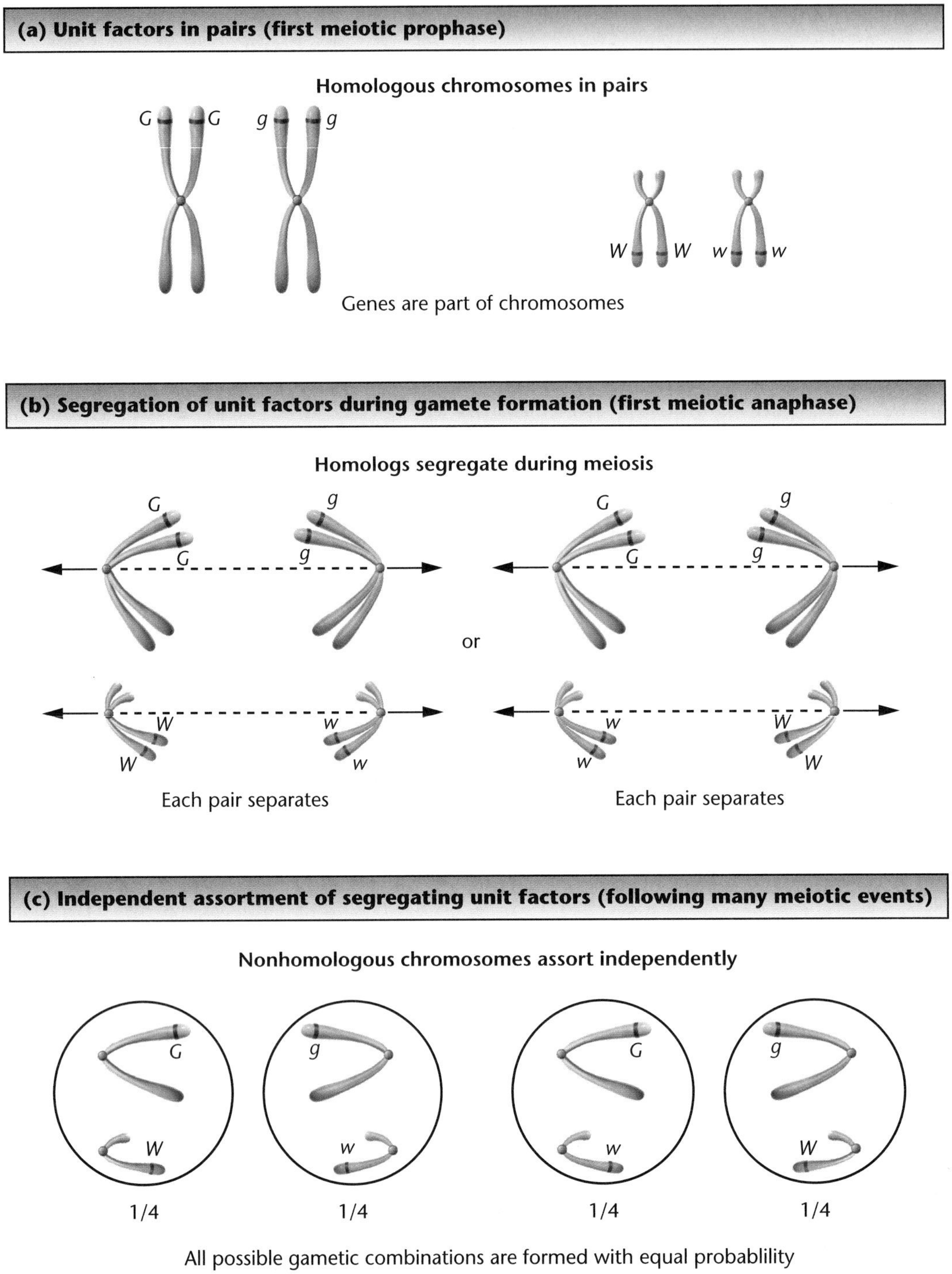

FIGURE 3–11 The correlation between the Mendelian postulates of (a) unit factors in pairs, (b) segregation, and (c) independent assortment, and the presence of genes located on homologous chromosomes and their behavior during meiosis.

from the **paternal parent**. We represent different parental origins by different colors. As shown in Figure 3–11(c), the two pairs of homologs undergo segregation independently of one another during gamete formation. Each gamete receives one chromosome from each pair. All possible combinations are formed. If we add the symbols used in Mendel's dihybrid cross (*G*, *g* and *W*, *w*) to the diagram, we see why equal numbers of the four types of gametes are formed. The independent behavior of Mendel's pairs of unit factors (*G* and *W* in this example) was due to the fact that they were on separate pairs of homologous chromosomes.

From observations of the phenotypic diversity of living organisms, we can logically conclude that there are many more genes than chromosomes. In fact, each chromosome is composed of a large number of linearly ordered genes. Mendel's unit factors (which determine tall or dwarf stems, for example) actually constitute a pair of genes located on one pair of homologous chromosomes. The location on a given chromosome where any particular gene occurs is called its **locus** (pl. **loci**). The different forms taken by a given gene, the alleles (*G* or *g*), contain slightly different genetic information (green or yellow) that determines the same character (seed color). Alleles, therefore, are alternative forms of the same gene. Although we have only discussed genes with two alleles, most genes have *more* than two alternative allelic forms. We discuss the concept of multiple alleles in Chapter 4.

We conclude this section by reviewing the criteria necessary to classify two chromosomes as a homologous pair:

1. During mitosis and meiosis, when chromosomes are visible as distinct structures, both members of a homologous pair are the same size and exhibit identical centromere locations.
2. During early stages of meiosis, homologous chromosomes pair together, or synapse.
3. Homologs contain the same linearly ordered gene loci.

3.7 Independent Assortment Leads to Extensive Genetic Variation

One of the major consequences of independent assortment is the production by an individual of genetically dissimilar gametes. Genetic variation results because the two members of any homologous pair of chromosomes are rarely, if ever, genetically identical. Therefore, because independent assortment leads to the production of all possible chromosome combinations, extensive genetic diversity results.

We have seen that for any individual, the number of possible gametes, each with different chromosome compositions, is 2^n, where n equals the haploid number. Thus, if a species has a haploid number of 4, then 2^4 or 16 different gamete combinations can be formed as a result of independent assortment. Although this number is not high, consider the human species, where $n = 23$. If we calculate 2^{23}, we find more than 8×10^6, or more than 8 million, different types of gametes are possible. Because fertilization represents an event involving only one of approximately 8×10^6 possible gametes from each of two parents, each offspring represents only one of $(8 \times 10^6)^2$ or 64×10^{12} potential genetic combinations. This number of combinations of chromosomes is far greater than the number of humans who have ever lived on Earth! It is no wonder that, except for identical twins, each member of the human species demonstrates such a distinctive appearance and individuality. Genetic variation resulting from independent assortment has been extremely important to the process of evolution in all organisms.

3.8 Laws of Probability Help to Explain Genetic Events

As mentioned earlier, genetic ratios are most properly expressed as probabilities—for example, 3/4 tall:1/4 dwarf. These values predict the outcome of each fertilization event, such that the probability of each zygote having the genetic potential for becoming tall is 3/4, whereas the potential for becoming dwarf is 1/4. Probabilities range from 0, where an event *is certain not to occur*, to 1.0, where an event *is certain to occur*. In this section, we consider the relation of probability to genetics.

The Product Law and the Sum Law

When two or more events occur independently of one another, but at the same time, we can calculate the probability of possible outcomes when they occur together. This is accomplished by applying the **product law**. As mentioned in our earlier discussion of independent assortment (see p. 44), the law states that the probability of two or more events occurring simultaneously is equal to the *product* of their individual probabilities. Two or more events are independent of one another if the outcome of each one does not affect the outcome of any of the others under consideration.

To illustrate the use of the product law, consider the possible results of an event where you toss a penny (*P*) and a nickel (*N*) at the same time and examine all combinations of heads (*H*) and tails (*T*) that can occur. There are four possible outcomes:

$$(P_H{:}N_H) = (1/2)(1/2) = 1/4$$
$$(P_T{:}N_H) = (1/2)(1/2) = 1/4$$
$$(P_H{:}N_T) = (1/2)(1/2) = 1/4$$
$$(P_T{:}N_T) = (1/2)(1/2) = 1/4$$

The probability of obtaining a head or a tail in the toss of either coin is 1/2 and is unrelated to the outcome of the other coin. All four possible combinations are predicted to occur with equal probability.

If we were interested in calculating the probability of a generalized outcome that can be accomplished in more than

Web Tutorial 3.3
Probability

one way, we would apply the **sum law** to the individual mutually exclusive outcomes, as illustrated below. For example, what is the probability of tossing our penny and nickel and obtaining one head and one tail? In such a case, we do not care whether it is the penny or the nickel that comes up heads, provided that the other coin has the alternative outcome. There are two ways in which the desired outcome can be accomplished $[P_H{:}N_T$ and $P_T{:}N_H]$, each with a probability of 1/4. The sum law states that the probability of obtaining any single outcome, where that outcome can be achieved in two or more events, is equal to the sum of the individual probabilities of all such events. Thus, according to the sum law, the overall probability in our example is equal to

$$(1/4) + (1/4) = 1/2$$

One-half of all such tosses are predicted to yield the desired outcome.

These simple probability laws will be useful throughout our discussions of transmission genetics and as you solve genetics problems. In fact, we already applied the product law earlier when we used the forked-line method to calculate the phenotypic results of Mendel's dihybrid and trihybrid crosses. When we wish to know the results of a cross, we need only to calculate the probability of each possible outcome. The results of this calculation then allow us to predict the proportion of offspring expressing each phenotype or each genotype.

There is a very important point to remember when dealing with probability. Predictions of possible outcomes are usually realized only with large sample sizes. If we predict that 9/16 of the offspring of a dihybrid cross will express both dominant traits, it is very unlikely that, in a small sample, exactly 9 of every 16 will do so. Instead, our prediction is that of a large number of offspring, approximately 9/16 will express this phenotype. The deviation from the predicted ratio in small sample sizes is attributed to chance, a subject we examine in our discussion of statistics in the next section. As we will see, the impact of deviation due strictly to chance is diminished as the sample size increases.

Conditional Probability

Sometimes we may wish to calculate the probability of an outcome that is dependent on a specific condition related to that outcome. For example, in the F_2 of Mendel's monohybrid cross involving tall and dwarf plants, what is the probability that a tall plant is heterozygous (and not homozygous)? The condition we have set is to consider only tall F_2 offspring since we know that all dwarf plants are homozygous.

Because the outcome and specific condition are not independent, we cannot apply the product law of probability. The likelihood of such an outcome is referred to as a **conditional probability**. In its simplest terms, we are asking what is the probability that one outcome will occur, given the specific condition upon which this outcome is dependent. Let us call this probability p_c.

To solve for p_c, we must consider both the probability of the outcome of interest and that of the specific condition that includes the outcome. These are (a) the probability of an F_2 plant being heterozygous as a result of receiving both a dominant and a recessive allele (p_a) and (b) the probability of the condition under which the event is being assessed, that is, being tall (p_b).

p_a = plant inheriting one dominant and one recessive allele (i.e., being a heterozygote)

$= 1/2$

p_b = probability of an F_2 plant of a monohybrid cross being tall

$= 3/4$

To calculate the conditional probability (p_c), we divide by p_a by p_b:

$$\begin{aligned} p_c &= p_a/p_b \\ &= (1/2)/(3/4) \\ &= (1/2)(4/3) \\ &= 4/6 \\ p_c &= 2/3 \end{aligned}$$

The conditional probability of any tall plant being heterozygous is two-thirds (2/3). On the average, two thirds of the F_2 tall plants will be heterozygous. We can confirm this calculation by reexamining Figure 3–3.

Conditional probability has many applications in genetics. During genetic counseling, for example, it is possible to calculate the probability (p_c) that an unaffected sibling of a brother or sister expressing a recessive disorder is a carrier of the disease-causing allele (i.e., a heterozygote). Assuming that both parents are unaffected (and are therefore carriers), the calculation of p_c is identical to the preceding example. The value of $p_c = 2/3$.

The Binomial Theorem

Finally, probability can be used for cases where one of two alternative outcomes is possible during each of a number of trials. By applying the **binomial theorem**, we can rather quickly calculate the probability of any specific set of outcomes among a large number of potential events. For example, in families of any size, we can calculate the probability of any combination of male and female children. In a family of four, for example, we can calculate the probability of having two children of one sex and two children of the other sex.

The expression of the binomial theorem is

$$(a + b)^n = 1$$

n	Binomial	Expanded Binomial
1	$(a + b)^1$	$a + b$
2	$(a + b)^2$	$a^2 + 2ab + b^2$
3	$(a + b)^3$	$a^3 + 3a^2b + 3ab^2 + b^3$
4	$(a + b)^4$	$a^4 + 4a^3b + 6a^2b^2 + 4ab^3 + b^4$
5	$(a + b)^5$	$a^5 + 5a^4b + 10a^3b^2 + 10a^2b^3 + 5ab^4 + b^5$
	etc.	etc.

where a and b are the respective probabilities of the two alternative outcomes and n equals the number of trials.

As the binomial is expanded for each value of n, Pascal's triangle, shown in Table 3.2, is useful in determining the numerical coefficient of each term in the binomial equation. In this triangle, each number is the sum of the two numbers immediately above it.

To expand any binomial, the various exponents (e.g., a^3b^2) are determined by using the pattern

$$(a + b)^n = a^n, a^{n-1}b, a^{n-2}b^2, a^{n-3}b^3, \ldots, b^n$$

The numerical coefficients preceding each expression can be most easily determined by using Pascal's triangle. Notice that all numbers other than the 1's are equal to the sum of the two numbers directly above them.

Using these methods, we find that the initial expansion of $(a + b)^7$ is

$$a^7 + 7a^6b + 21a^5b^2 + 35a^4b^3 + \ldots + b^7$$

Applying the binomial theorem, we can return to our original question: *What is the probability that in a family of four children, two are male and two are female?*
First, assign initial probabilities to each outcome:

$$a = \text{male} = 1/2$$
$$b = \text{female} = 1/2$$

Then locate the appropriate term in the expanded binomial, where $n = 4$,

TABLE 3.2 **PASCAL'S TRIANGLE**

n	Numerical Coefficients
n	1
1	1 1
2	1 2 1
3	1 3 3 1
4	1 4 6 4 1
5	1 5 10 10 5 1
6	1 6 15 20 15 6 1
7	1 7 21 35 35 21 7 1
etc.	etc.

* Notice that all numbers other than the 1's are equal to the sum of the two numbers directly above them.

$$(a + b)^4 = a^4 + 4a^3b + 6a^2b^2 + 4ab^3 + b^4$$

In each term, the exponent of a represents the number of males, and the exponent of b represents the number of females. Therefore, the correct expression of p is

$$\begin{aligned} p &= 6a^2b^2 \\ &= 6(1/2)^2(1/2)^2 \\ &= 6(1/2)^4 \\ &= 6(1/16) \\ &= 6/16 \\ p &= 3/8 \end{aligned}$$

Thus, the probability of families of four children having two boys and two girls is 3/8. Of all families with four children, 3 out of 8 are predicted to have two boys and two girls.

Before examining one other example, we should note that a single formula can be applied in determining the numerical coefficient for any set of exponents,

$$n!/(s!t!)$$

where

n = the total number of events

s = the number of times outcome a occurs

t = the number of times outcome b occurs

Therefore, $n = s + t$.

The symbol ! denotes a **factorial**, which is the product of all the positive integers from 1 through some positive integer. For example,

$$5! = (5)(4)(3)(2)(1) = 120.$$

Note that when using factorials, $0! = 1$.

Using the formula, let's determine the probability, in a family of seven, that five males and two females will occur. Thus, $n = 7$, $s = 5$, and $t = 2$. We first extend our equation to include five events of outcome a and two events of outcome b. The appropriate term is

$$\begin{aligned} p &= \frac{n!}{s!t!}a^sb^t \\ &= \frac{7!}{5!2!}(1/2)^5(1/2)^2 \\ &= \frac{(7)\cdot(6)\cdot(5)\cdot(4)\cdot(3)\cdot(2)\cdot(1)}{(5)\cdot(4)\cdot(3)\cdot(2)\cdot(1)\cdot(2)\cdot(1)}(1/2)^7 \\ &= \frac{(7)\cdot(6)}{(2)\cdot(1)}(1/2)^7 \\ &= \frac{42}{2}(1/2)^7 \\ &= 21(1/2)^7 \\ &= 21(1/128) \\ p &= 21/128 \end{aligned}$$

Of families with seven children, on the average, 21/128 are predicted to have five males and two females.

Calculations using the binomial theorem have various applications in genetics, including the analysis of polygenic traits (Chapter 24) and in population equilibrium studies (Chapter 25).

3.9 Chi-Square Analysis Evaluates the Influence of Chance on Genetic Data

Mendel's 3:1 monohybrid and 9:3:3:1 dihybrid ratios are hypothetical predictions based on the following assumptions: (1) each allele is dominant or recessive, (2) segregation occurs normally, (3) independent assortment occurs, and (4) fertilization is random. The last three assumptions are influenced by chance events and therefore are subject to random fluctuation. This concept, called **chance deviation**, is most easily illustrated by tossing a single coin numerous times and recording the number of heads and tails observed. In each toss, there is a probability of 1/2 that a head will occur and a probability of 1/2 that a tail will occur. Therefore, the expected ratio of many tosses is 1:1. If a coin were tossed 1000 times, we would usually expect *about* 500 heads and 500 tails to be observed. Any reasonable fluctuation from this hypothetical ratio (e.g., 486 heads and 514 tails) would be attributed to chance.

As the total number of tosses is reduced, the impact of chance deviation increases. For example, if a coin were tossed only four times, you wouldn't be too surprised if all four tosses resulted in only heads or only tails. But, for 1000 tosses, 1000 heads or 1000 tails would be most unexpected or thought impossible. Actually, the probability of all heads or all tails in 1000 tosses can be predicted to occur with a probability of only $(1/2)^{1000}$. Because $(1/2)^{20}$. is equivalent to less than 1 in 2 million times, an event occurring with a probability as small as $(1/2)^{1000}$ would be virtually impossible.

Two major points are significant here:

1. The outcomes of segregation, independent assortment, and fertilization, like coin tossing, are subject to random fluctuations from their predicted occurrences as a result of chance deviation.

2. As the sample size increases, the average deviation from the expected results decreases. Therefore, a larger sample size diminishes the impact of chance deviation on the final outcome.

In genetics, being able to evaluate observed deviation is a crucial methodology. When we assume that data will fit a given ratio such as 1:1, 3:1, or 9:3:3:1, we establish what is called the **null hypothesis (H_0)**. It is so named because the hypothesis assumes that there is no *real difference* between the **measured values** (or ratio) and the **predicted values** (or ratio). The *apparent* difference can be attributed purely to chance. The null hypothesis is evaluated using statistical analysis. On this basis, the null hypothesis may either (1) be rejected or (2) fail to be rejected. If it is rejected, the observed deviation from the expected is *not* attributed to chance alone; the null hypothesis and the underlying assumptions leading to it must be reexamined. If the null hypothesis fails to be rejected, any observed deviations *are* attributed to chance.

One of the simplest statistical tests devised to assess the goodness of fit of the null hypothesis is **chi-square (χ^2) analysis**. This test takes into account the observed deviation in each component of an expected ratio as well as the sample size and reduces them to a single numerical value. The χ^2 value is then used to estimate how frequently the observed amount of deviation, or more, can be expected to occur strictly as a result of chance. The formula used in chi-square analysis is

$$\chi^2 = \Sigma \frac{(o - e)^2}{e}$$

In this equation,

o = the observed value for a given category

e = the expected value for that category

and

Σ = sum of the calculated values for each category in the ratio

Because $(o - e)$ is the deviation (d) in each case, the equation can be reduced to

$$\chi^2 = \frac{d^2}{e}$$

Table 3.3(a) shows the steps in the χ^2 calculation for the F_2 results of a hypothetical monohybrid cross. If you were analyzing these data, you would work from left to right, calculating and entering the appropriate numbers in each column. Regardless of whether the calculated deviation $(o - e)$ is initially positive or negative, it becomes positive after the number is squared. Table 3.3(b) illustrates the analysis of the F_2 results of a hypothetical dihybrid cross. Based on your study of the calculations involved in the monohybrid cross, check to make certain that you understand how each number was calculated in the dihybrid example.

The final step in chi-square analysis is to interpret the χ^2 value. To do so, you must initially determine the value of the **degrees of freedom (df)**, which is equal to $n - 1$, where n is the number of different categories into which each datum point may fall. For the 3:1 ratio, $n = 2$, so df $= 1$. For the 9:3:3:1 ratio, df $= 3$. Degrees of freedom must be taken into account because the greater the number of categories, the more deviation is expected as a result of chance.

Once the number of degrees of freedom is determined, we can interpret the χ^2 value in terms of a corresponding **probability value (p)**. Because this calculation is complex, we usually take the p value from a standard table or graph. Figure 3–12 shows the wide range of χ^2 and p values for numerous degrees of freedom in both a graph and a table. We will use the graph to explain how to determine the p value. The caption for Figure 3–12(b) explains how to use the table.

TABLE 3.3 CHI-SQUARE ANALYSIS

(a) Monohybrid Cross

Expected Ratio	Observed (*o*)	Expected (*e*)	Deviation (*o* − *e*)	Deviation^2	d^2/e
3/4	740	3/4(1000) = 750	740 − 750 = −10	$(-10)^2 = 100$	100/750 = 0.13
1/4	260	1/4(1000) = 250	260 − 250 = +10	$(+10)^2 = 100$	100/250 = 0.40
	Total = 1000				$\chi^2 = 0.53$
					$p = 0.48$

(b) Dihybrid Cross

Expected Ratio	*o*	*e*	*o* − *e*	d^2	d^2/e
9/16	587	567	+20	400	0.71
3/16	197	189	+8	64	0.34
3/16	168	189	−21	441	2.33
1/16	56	63	−7	49	0.78
	Total = 1008				$\chi^2 = 4.16$
					$p = 0.26$

(a)

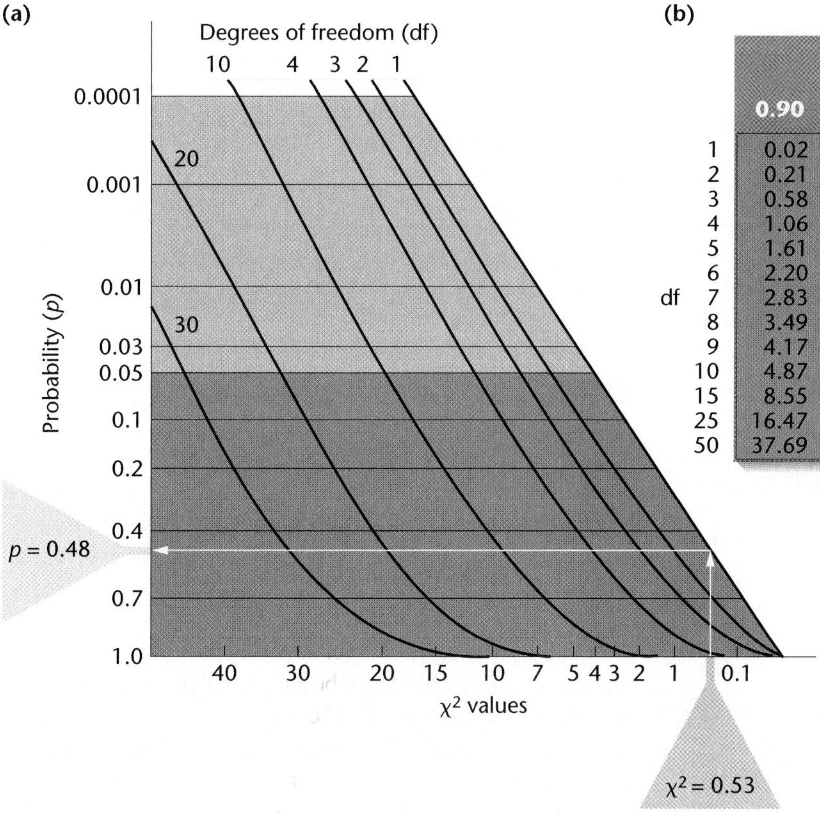

(b)

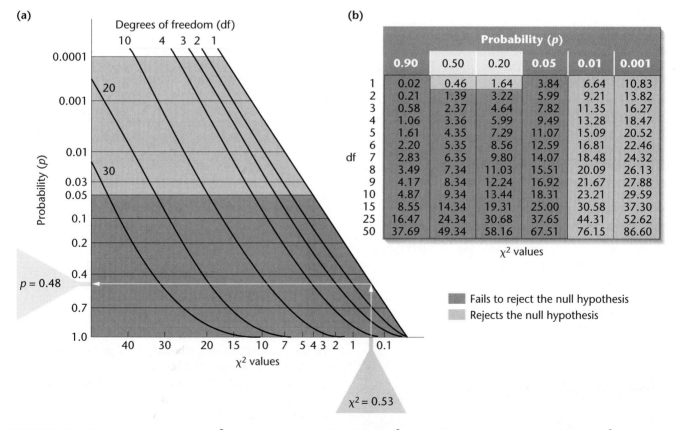

df	Probability (p) 0.90	0.50	0.20	0.05	0.01	0.001
1	0.02	0.46	1.64	3.84	6.64	10.83
2	0.21	1.39	3.22	5.99	9.21	13.82
3	0.58	2.37	4.64	7.82	11.35	16.27
4	1.06	3.36	5.99	9.49	13.28	18.47
5	1.61	4.35	7.29	11.07	15.09	20.52
6	2.20	5.35	8.56	12.59	16.81	22.46
7	2.83	6.35	9.80	14.07	18.48	24.32
8	3.49	7.34	11.03	15.51	20.09	26.13
9	4.17	8.34	12.24	16.92	21.67	27.88
10	4.87	9.34	13.44	18.31	23.21	29.59
15	8.55	14.34	19.31	25.00	30.58	37.30
25	16.47	24.34	30.68	37.65	44.31	52.62
50	37.69	49.34	58.16	67.51	76.15	86.60

χ^2 values

Fails to reject the null hypothesis

Rejects the null hypothesis

FIGURE 3–12 (a) Graph for converting χ^2 values to p values. (b) Table of χ^2 values for selected values of df and p. χ^2 values *greater than those shown at* $p = 0.05$ justify failing to reject the null hypothesis, while χ^2 values *less than those at* $p = 0.05$ justify rejecting the null hypothesis. In our example, $\chi^2 = 0.53$ for 1 degree of freedom is converted to a p value between 0.20 and 0.50. The graph in (a) provides an estimated p value of 0.48 by interpolation. In this case, we fail to reject the null hypothesis.

To determine p, execute the following steps:

1. Locate the χ^2 value on the horizontal or X axis.
2. Draw a vertical line from this point up to the line on the graph representing the appropriate df.
3. Extend a horizontal line from this point to the left until it intersects the vertical or Y axis.
4. Estimate, by interpolation, the corresponding p value.

Using our first example (the monohybrid cross) in Table 3.3, we estimate the p value of 0.48 in this manner [Figure 3–12(a)]. For the dihybrid cross, use this method to see if you can determine the p value. The χ^2 value is 4.16 and df equals 3. The approximate p value is 0.26. Using the table, rather than the graph, confirms that both p values are between 0.20 and 0.50. Examine the table in Figure 3–12(b) to confirm this.

Interpreting χ^2 Calculations

Thus far, we have been concerned only with determining p. The most important aspect of χ^2 analysis is understanding what the p value actually means. We will use the example of the dihybrid cross $p = 0.26$ to illustrate. In these discussions, it is simplest to think of the p value as a percentage (e.g., $0.26 = 26$ percent). In our example, the p value indicates that, if we repeat the same experiment many times, 26 percent of the trials would be expected to exhibit chance deviation as great as or greater than that seen in the initial trial. Conversely, 74 percent of the repeats would show less deviation as a result of chance than initially observed.

The preceding discussion of the p values reveals that a hypothesis (e.g., a 9:3:3:1 ratio) is never proved or disproved absolutely. Instead, a relative standard must be set that serves as the basis for either rejecting or failing to reject the null hypothesis. This standard is most often a p value of 0.05. When applied to chi-square analysis, a p value less than 0.05 means the probability is less than 5 percent that the observed deviation in the set of results could be obtained by chance alone. Such a p value indicates that the difference between the observed and predicted results is substantial and enables us *to reject the null hypothesis*.

On the other hand, p values of 0.05 or greater (0.05 to 1.0) indicate that the observed deviation will be obtained by chance alone 5 percent or more of the time. In such cases, this enables us *to fail to reject the null hypothesis*. Thus, the p value of 0.26, assessing the hypothesis that independent assortment accounts for the results, fails to be rejected. Therefore, the observed deviation can be reasonably attributed to chance.

Now solve this

Problem 3.23 on page 63 asks you to apply χ^2 analysis to a set of data and determine whether the data fit several ratios.

Hint: In calculating χ^2, first determine the expected outcomes using the predicted ratios. Then follow a stepwise approach, determining the deviation in each case, and calculating d^2/e for each category.

A final note is relevant here concerning the case where the null hypothesis is rejected, that is, $p < 0.05$. Suppose, for example, the null hypothesis being tested was that the data represented a 9:3:3:1 ratio, indicative of independent assortment. If the null hypothesis is rejected, what are alternative interpretations of the data? Researchers first reassess the many assumptions that underlie the null hypothesis. In our case, we assumed that segregation operates faithfully for both gene pairs. We also assumed that fertilization is random and that the viability of all gametes is equal irrespective of genotype—that is, all gametes are equally likely to participate in fertilization. Finally, following fertilization, we assumed that all preadult stages and adult offspring are equally viable, regardless of their genotype.

An example will clarify this: Suppose our null hypothesis is that a dihybrid cross between fruit flies will result in 3/16 mutant wingless flies (the proportion of mutant zygotes that may, in fact, occur at fertilization). However, these mutant embryos may not survive as well during their preadult development, or as young adults, compared with flies whose genotype gives rise to wings. As a result, when the data are gathered, there will be fewer than 3/16 wingless flies. Rejection of the null hypothesis is not cause for us to disregard the validity of the postulates of segregation and independent assortment, because other factors affect the result.

The foregoing discussion serves to point out that statistical information must be assessed carefully on a case-by-case basis. When we reject a null hypothesis, we must examine all underlying assumptions. If there is no concern about their validity, then we must consider other alternative hypotheses to explain the results.

How Do We Know?

In examining genetic ratios, how do we know whether the observed deviation is the result of chance or of some other variable that we failed to account for in predicting the ratio?

3.10 Pedigrees Reveal Patterns of Inheritance in Humans

We now explore how to determine the mode of inheritance of phenotypes in humans, where designed crosses are not possible and where relatively few offspring are available for study. The traditional way to study inheritance has been to construct a family tree, indicating the presence or absence of the trait in question for each member of each generation. Such a family tree is called a **pedigree**. Figure 3–13 shows common conventions in human pedigrees. By analyzing a pedigree, we may be able to predict how the trait under study is inherited—for example, is it due to a dominant or recessive allele? When many pedigrees for the same trait are studied, we can often ascertain the mode of inheritance.

Pedigree Conventions

Figure 3–13 shows standard pedigree conventions: circles represent females and squares designate males. Parents are connected by a single horizontal line and vertical lines lead to their offspring. If the parents are related **(consanguineous)**, such

Web Tutorial 3.4
Analysis of Human Pedigrees

as first cousins might be, they are connected by a double line. Offspring are called **sibs** (short for **siblings**) and are connected by a horizontal **sibship line**. Sibs are placed from left to right according to birth order, and are labeled with Arabic numerals. Each generation is indicated by a Roman numeral. If the sex of an individual is unknown, a diamond is used. When a pedigree traces only a single trait, the circles, squares, and diamonds are shaded if the phenotype being considered is expressed and unshaded if not. In some pedigrees, those individuals that fail to express a recessive trait, but are known with certainty to be a heterozygous carrier, have a shaded dot within their unshaded circle or square. If an individual is deceased and the phenotype is unknown, a diagonal line is placed over the circle or square.

Twins are indicated by diagonal lines stemming from a vertical line connected to the sibship line. For **identical** (or **monozygotic**) **twins**, the diagonal lines are linked by a horizontal line. **Fraternal** (or **dizygotic**) **twins** lack this connecting line. A number within one of the symbols represents numerous sibs of the same or unknown phenotypes. The individual whose phenotype first brought attention to the investigation and construction of the pedigree is called the **proband** and is indicated by an arrow connected to the designation **p**. This term applies to either a male or a female.

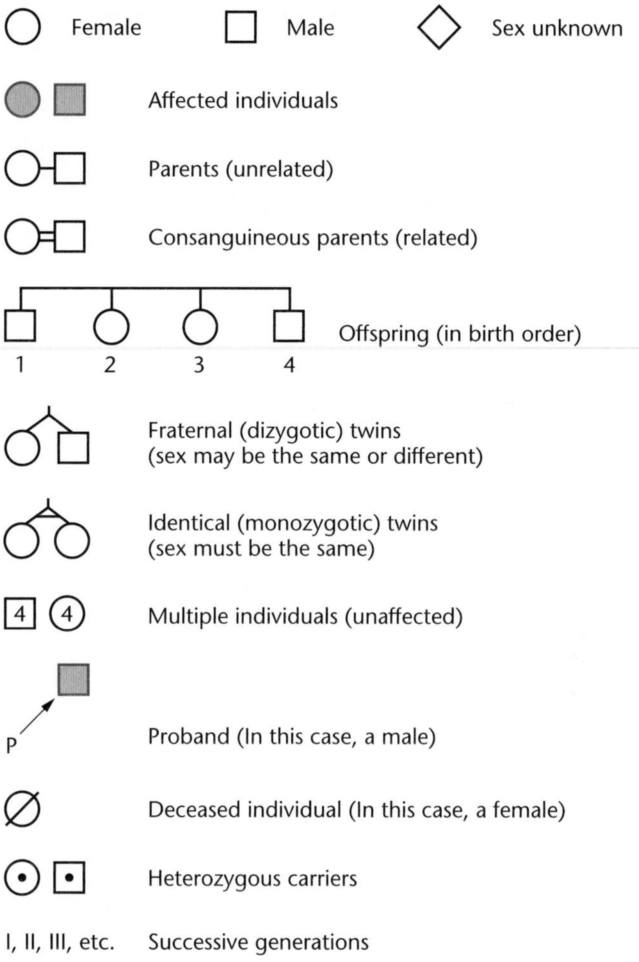

FIGURE 3–13 Conventions commonly encountered in human pedigrees.

Pedigree Analysis

In Figure 3–14, two pedigrees are shown. The first illustrates a representative pedigree for a trait that demonstrates autosomal recessive inheritance, such as albinism. The male parent of the first generation (I-1) is affected. Characteris-

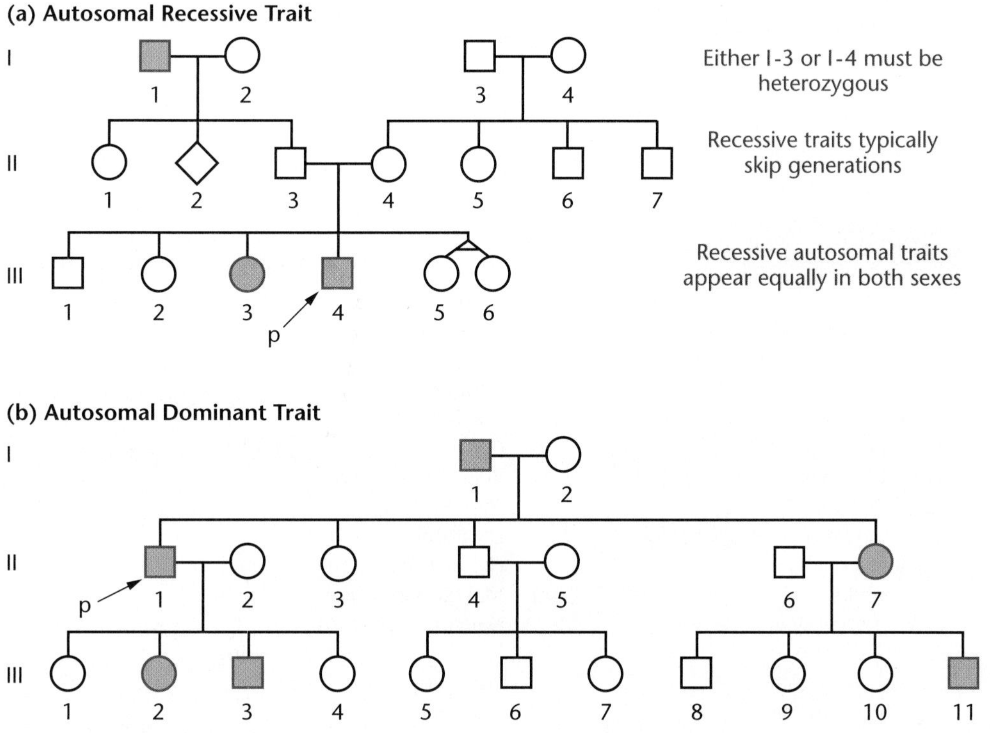

FIGURE 3–14 (a) A representative pedigree for an autosomal recessive trait followed through three generations. (b) A representative pedigree for an autosomal dominant trait followed through three generations.

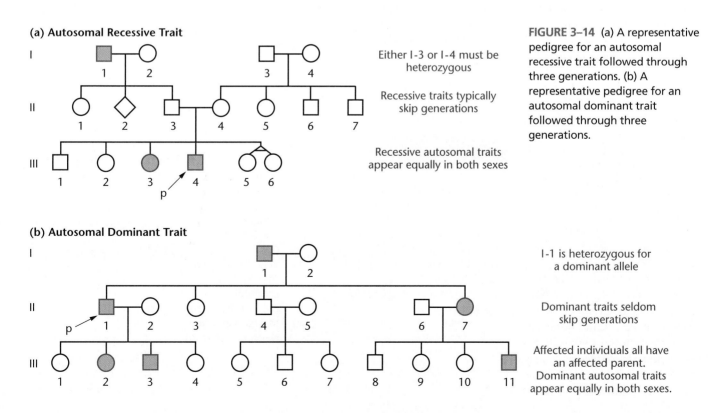

tic of a rare recessive trait with an affected parent, the trait "disappears" in the offspring of the next generation. Assuming recessiveness, we might predict that the unaffected female parent (I-2) is a homozygous normal individual because none of the offspring show the disorder. Had she been heterozygous, one half of the offspring would be expected to exhibit albinism, but none do. However, such a small sample (three offspring) prevents us knowing for certain.

Further evidence supports the prediction of a recessive trait. If albinism were inherited as a dominant trait, individual II-3 would have to express the disorder in order to pass it to his offspring (III-3 and III-4), but he does not. Inspection of the offspring constituting the third generation (row III) provides still further support for the hypothesis that albinism is a recessive trait. If it is, parents II-3 and II-4 are both heterozygous, and approximately one fourth of their offspring should be affected. Two of the six offspring do show albinism. This deviation from the expected ratio is not unexpected in crosses with few offspring. Once we are confident that albinism is inherited as an autosomal recessive trait, we could portray the II-3 and II-4 individuals with a shaded dot within their larger square and circle. Finally, we can note that, characteristic of pedigrees for autosomal traits, both males and females are affected with equal probability. In Chapter 4, we will examine a pedigree representing a gene located on the sex-determining X chromosome. We will see certain limitations imposed on the transmission of X-linked traits, such as that these traits are more prevalent in male offspring and are never passed from affected fathers to their sons.

The second pedigree illustrates the pattern of inheritance for a trait such as Huntington disease, which is caused by an autosomal dominant allele. The key to identifying such a pedigree that reflects a dominant trait is that all affected offspring will have a parent that also expresses the trait. Further, if the sample of offspring within each generation is not particularly small, then the trait is unlikely to skip a generation, as does a rare recessive trait. Like recessive traits, provided that the gene is autosomal, both males and females are equally affected.

When autosomal dominant diseases are rare within the population, and most are, then it is highly unlikely that affected individuals will inherit a copy of the mutant gene from both parents. Therefore, in most cases, affected individuals are heterozygous for the dominant allele. As a result, approximately one half of the offspring inherit it. This is borne out in the second pedigree in Figure 3–14. Further, if a mutation is dominant, and a single copy is sufficient to produce a mutant phenotype, homozygotes are likely to be even more severely affected, perhaps even failing to survive. An illustration of this is the dominant gene for **familial hypercholesterolemia**. Heterozygotes display a defect in their receptors for low density lipoproteins, the so-called LDLs. As a result, too little cholesterol is taken up by cells from the blood, and elevated plasma levels of LDLs result. Such heterozygous individuals have heart attacks during the fourth decade of their life, or before. While heterozygotes have LDL levels about double that of a normal individual, rare homozygotes have been detected. They lack LDL receptors altogether, and have LDL levels nearly ten times above the normal range. They are likely to have a heart attack very early in life, even before age five, and almost inevitably before they reach the age of 20.

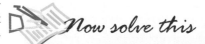

Now solve this

Problem 3.27 on page 64 asks you to examine a pedigree for myopia and predict whether the trait is dominant or recessive.

Hint: One of the first things to look for are individuals who express the trait, but neither of whose parents also express the trait. Such an observation makes it highly unlikely that the trait is dominant.

Pedigree analysis of many traits has historically been an extremely valuable research technique in human genetic studies. However, the approach does not usually provide the certainty in drawing conclusions afforded by designed crosses yielding large numbers of offspring. Nevertheless, when many independent pedigrees of the same trait or disorder are analyzed, consistent conclusions can often be drawn. Table 3.4 lists numerous human traits and classifies them according to their recessive or dominant expression. The genes controlling some of these traits are located on the sex-determining chromosomes. We will discuss pedigrees for X-linked traits in Chapter 4.

How Do We Know?

How do geneticists determine whether a human trait is inherited as the result of a recessive allele or a dominant allele?

TABLE 3.4 **REPRESENTATIVE RECESSIVE AND DOMINANT HUMAN TRAITS**

Recessive Traits	Dominant Traits
Albinism	Achondroplasia
Alkaptonuria	Brachydactyly
Ataxia telangiectasia	Congenital stationary night blindness
Color blindness	Ehler–Danlos syndrome
Cystic fibrosis	Hypotrichosis
Duchenne muscular dystrophy	Huntington disease
Galactosemia	Hypercholesterolemia
Hemophilia	Marfan syndrome
Lesch–Nyhan syndrome	Neurofibromatosis
Phenylketonuria	Phenylthiocarbamide tasting
Sickle-cell anemia	Porphyria (some forms)
Tay–Sachs disease	Widow's peak

Genetics, Technology, and Society

Tay–Sachs Disease: A Recessive Molecular Disorder in Humans

Tay–Sachs disease (TSD) is an autosomal recessive disorder that causes progressive destruction of the central nervous system. Infants with TSD appear normal at birth and seem to develop normally until they are six months old, then gradually lose mental and physical abilities. Afflicted infants become blind, deaf, mentally retarded, and paralyzed within only a year or two, and most do not live beyond age five. Named for Warren Tay and Bernard Sachs, who first described the symptoms and associated them with the disorder in the late 1800s, Tay–Sachs disease results from the loss of activity of the enzyme hexosaminidase A (Hex-A). This enzyme is normally found in lysosomes, organelles that break down large molecules for recycling by the cell. Hex-A is needed to break down the ganglioside GM2, a lipid component of nerve cell membranes. Without functional Hex-A, gangliosides accumulate in neurons in the brain and cause deterioration of the nervous system. Heterozygous carriers of TSD with one normal copy of the gene produce only half the normal amount of Hex-A but show no symptoms of the disorder.

The gene responsible for Tay–Sachs disease is located on chromosome 15 and codes for the alpha subunit of the Hex-A enzyme. Since the gene was isolated in 1985, more than 50 different mutations have been identified that lead to TSD. Although the most common form of the disease is the infantile form, where no functional Hex-A is produced, there is also a rare late-onset form that occurs in patients with greatly reduced Hex-A activity. Late-onset TSD is not detectable until patients are in their twenties or thirties, and it is generally much less severe than the infantile form. Symptoms include hand tremors, speech impediments, muscle weakness, and loss of balance.

Tay–Sachs disease is almost a hundred times more common in Ashkenazi Jews—Jews of central or eastern European descent—than in the general population, and also has a higher incidence in French Canadians and in members of the Cajun population in Louisiana. In the United States, approximately one in every 27 Ashkenazi Jews is a carrier of TSD. By contrast, the carrier rate in the general population and in Jews of Sephardic (Spanish or Portuguese) origin is approximately one in every 250. Although there are currently no effective treatments for TSD, recent advances in carrier screening have helped to reduce the prevalence of the disorder in high-risk populations. Carriers can be identified by tests that measure Hex-A activity or by DNA-based tests that detect specific gene mutations. In addition, ganglioside synthesis inhibitors and Hex-A enzyme replacement therapy are currently being investigated as potential treatments for Tay–Sachs disease.

References

Fernandes, F. and Shapiro, B. 2004. Tay-Sachs Disease. Arch. Neurol. 61:1466–68.

Chapter Summary

1. More than a century ago, Gregor Mendel studied inheritance patterns in the garden pea, establishing the principles of transmission genetics.
2. Mendel's postulates help describe the basis for the inheritance of phenotypic expression. He showed that unit factors, later called alleles, exist in pairs and exhibit a dominant–recessive relationship in determining the expression of traits.
3. Mendel postulated that unit factors must segregate during gamete formation such that each gamete receives only one of the two factors with equal probability.
4. Mendel's postulate of independent assortment states that each pair of unit factors segregates independently of other such pairs. As a result, all possible combinations of gametes will be formed with equal probability.
5. Mendel designed the testcross to determine the exact genotype of peas expressing dominant traits.
6. The discovery of chromosomes in the late 1800s and subsequent studies of their behavior during meiosis led to the rediscovery of Mendel's work, linking the behavior of his unit factors to that of chromosomes during meiosis.
7. The Punnett square and the forked-line methods are used to predict the probabilities of phenotypes and genotypes from crosses involving two or more gene pairs.
8. Genetic ratios are expressed as probabilities. Deriving outcomes of genetic crosses relies on an understanding of the laws of probability, particularly the sum law, the product law, conditional probability, and the binomial theorem.
9. In genetics, variations from the expected ratios due to chance deviations can be anticipated. Statistical analysis is used to test how well a hypothesis is supported by experimental data.
10. Chi-square analysis allows us to assess the null hypothesis; namely, that there is no real difference between the expected and observed values. As such, it tests the probability of whether observed variations can be attributed to chance deviation.
11. Pedigree analysis provides a method for analyzing the inheritance patterns in humans over several generations. Such analysis often provides the basis for determining the mode of inheritance of human traits and disorders.

Insights and Solutions

As a student, you will be asked to demonstrate your knowledge of transmission genetics by solving genetics problems. Success at this task requires not only comprehension of theory, but also its application to more practical genetic situations. Most students find problem-solving in genetics to be challenging, but rewarding. This section is designed to provide basic insights into the reasoning essential to this process.

Genetics problems are in many ways similar to algebraic word problems. The approach taken should be identical: (1) analyze the problem carefully; (2) translate words into symbols, defining each one first; and (3) choose and apply a specific technique to solve the problem. The first two steps are the most critical. The third step is largely mechanical.

The simplest problems are those that state all necessary information about the P_1 generation and ask you to find the expected ratios of the F_1 and F_2 genotypes and/or phenotypes. Always follow these steps when you encounter this type of problem:

a. Determine insofar as possible the genotypes of the individuals in the P_1 generation.

b. Determine what gametes may be formed by the P_1 parents.

c. Recombine gametes by the Punnett square or the forked-line methods, or, if the situation is very simple, by inspection. From the genotypes of the F_1 generation, determine the phenotypes. Read the F_1 phenotypes directly.

d. Repeat the process to obtain information about the F_2 generation.

Performing the aforementioned steps requires an understanding of the basic theory of transmission genetics. For example, consider the following problem:

A recessive mutant allele, *black*, causes a very dark body in *Drosophila melanogaster* when homozygous. The normal wild-type color is described as gray. What F_1 phenotypic ratio is predicted when a black female is crossed to a gray male whose father was black?

To work this problem, you must understand dominance and recessiveness as well as the principle of segregation. Furthermore, you must use the information about the male parent's father. Here is one way to work this problem:

a. Because the female parent is black, she must be homozygous for the mutant allele (*bb*).

b. The male parent is gray; therefore, he must have at least one dominant allele (*B*). Since his father was black (*bb*) and he received one of the chromosomes bearing these alleles, the male parent must be heterozygous (*Bb*).

From here, solving the problem is straightforward:

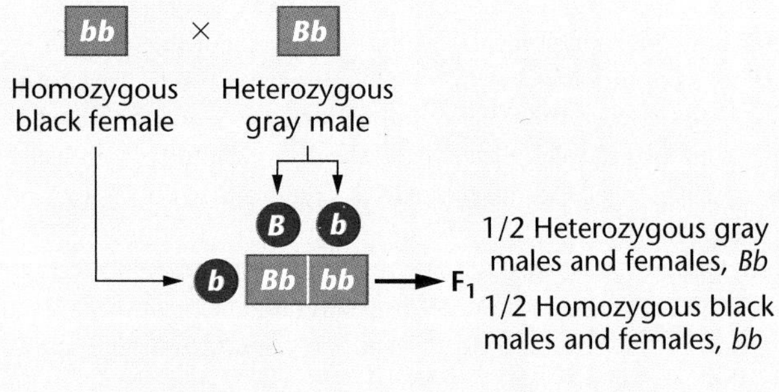

Apply the approach we just studied to the following problems.

1. In his work, Mendel found that full pea pods are dominant to constricted pods, whereas round seeds are dominant to wrinkled seeds. One of his crosses was between full, round plants and constricted, wrinkled plants. From this cross, he obtained an F_1 that was all full and round. In the F_2, Mendel obtained his classic 9:3:3:1 ratio. Using this information, determine the expected F_1 and F_2 results of a cross between homozygous constricted, round and full, wrinkled plants.

Solution: First, define gene symbols for each pair of contrasting traits. Select the lowercase forms of the first letter of the recessive traits to designate those traits, and use the uppercase forms to designate the dominant traits. For example, use *C* and *c* to indicate full and constricted, and use *W* and *w* to indicate the round and wrinkled phenotypes, respectively.

Now, determine the genotypes of the P_1 generation, form gametes, reconstitute the F_1 generation, and determine the F_1 phenotype(s):

P_1	*ccWW* constricted, round	×	*CCww* full, wrinkled
	↓		↓
Gametes	*cW*		*Cw*
F_1		*CcWw* full, round	

You can see immediately that the F_1 generation expresses both dominant phenotypes and is heterozygous for both gene pairs. We can expect that the F_2 generation will yield the classic Mendelian ratio of 9:3:3:1. Let's work it out anyway, just to confirm it, using the forked-line method. Because both gene pairs are heterozygous and can be expected to assort independently, we can predict the F_2 outcomes from each gene pair separately and then proceed with the forked-line method.

Every F_2 offspring is subject to the following probabilities:

$Cc \times Cc$		$Ww \times Ww$	
↓		↓	
CC, *Cc*, *cC*	full	*WW*, *Ww*, *wW*	round
cc	constricted	*ww*	wrinkled

The forked-line method then allows us to confirm the 9:3:3:1 phenotypic ratio. Remember that this represents proportions of 9/16:3/16:3/16:1/16. Note that we are applying the product law as we compute the final probabilities:

3/4 full
- 3/4 round $\xrightarrow{(3/4)(3/4)}$ 9/16 full, round
- 1/4 wrinkled $\xrightarrow{(3/4)(1/4)}$ 3/16 full, wrinkled

1/4 constricted
- 3/4 round $\xrightarrow{(1/4)(3/4)}$ 3/16 constricted, round
- 1/4 wrinkled $\xrightarrow{(1/4)(1/4)}$ 1/16 constricted, wrinkled

2. Determine the probability that a plant of genotype *CcWw* will be produced from parental plants of the genotypes *CcWw* and *Ccww*.

Solution: Because the two gene pairs independently assort during gamete formation, we need only calculate the individual probabilities of the two separate events (*Cc* and *Ww*) and apply the product law to calculate the final probability.

$$Cc \times Cc \rightarrow 1/4\,CC{:}1/2\,Cc{:}1/4\,cc$$

$$Ww \times ww \rightarrow 1/2\,Ww{:}1/2\,ww$$

$$p = (1/2\,Cc)(1/2\,Ww) = 1/4\,CcWw$$

3. In another cross, involving parent plants of unknown genotype and phenotype, the following offspring were obtained.

3/8 full, round

3/8 full, wrinkled

1/8 constricted, round

1/8 constricted, wrinkled

Determine the genotypes and phenotypes of the parents.

Solution: This problem is more difficult and requires keener insights because you must work backward. The best approach is to consider the outcomes of pod shape separately from those of seed texture.

Of all plants, 6/8 (3/4) are full and 2/8 (1/4) are constricted. Of the various genotypic combinations that can serve as parents, which will give rise to a ratio of 3/4:1/4? Because this ratio is identical to Mendel's monohybrid F_2 results, we can propose that both unknown parents share the same genetic characteristic as the monohybrid F_1 parents: They must both be heterozygous for the genes controlling pod shape and thus are

Cc

Before accepting this hypothesis, let's consider the possible genotypic combinations that control seed texture. If we consider this characteristic alone, we can see that the traits are expressed in a ratio of 4/8 (1/2) round: 4/8 (1/2) wrinkled. In order to generate such a ratio, the parents *cannot* both be heterozygous or their offspring would yield a 3/4:1/4 phenotypic ratio. They *cannot* both be homozygous or all offspring would express a single phenotype. Thus, we are left with testing the hypothesis that one parent is homozygous and one is heterozygous for the alleles controlling texture. The potential case of *WW* × *Ww* will not work, because it would also yield only a single phenotype. Hence, we are left with the potential case of *ww* × *Ww*. Offspring in such a mating will yield 1/2 *Ww* (round): 1/2 *ww* (wrinkled), exactly the outcome we are seeking.

Now, let's combine our hypotheses and predict the outcome of crossing. In our solution, because we are only predicting phenotypes, we will use the dash symbol (–) to indicate that the second allele may be either dominant or recessive:

3/4 *C*–
- 1/2 *Ww* → 3/8 *C–Ww* full, round
- 1/2 *ww* → 3/8 *C–ww* full, wrinkled

1/4 cc
- 1/2 *Ww* → 1/8 *ccWw* constricted, round
- 1/2 *ww* → 1/8 *ccww* constricted, wrinkled

As you can see, this cross produces offspring in accordance with the information provided, thus solving the problem. Note that in the solution, we have used *genotypes* in the forked-line method, in contrast to the use of *phenotypes* in the solution of the first problem.

4. In the laboratory, a genetics student crossed flies with normal long wings to flies with mutant dumpy wings, which she believed was a recessive trait. In the F_1, all flies had long wings. In the F_2, the following results were obtained:

792 long-winged flies

and

208 dumpy-winged flies

The student tested the hypothesis that the *dumpy* wing is inherited as a recessive trait by performing χ^2 analysis of the F_2 data.

(a) What ratio did the student hypothesize?

(b) Did the χ^2 analysis support the hypothesis?

(c) What do the data suggest about the *dumpy* mutation?

Solution:

(a) The student hypothesized that the F_2 data (792:208) fit Mendel's 3:1 monohybrid ratio for recessive genes.

(b) The initial step in χ^2 analysis is to calculate the expected results (*e*) by assuming a ratio of 3:1. Then calculate the deviations (*d*) between the expected values and observed values (the actual data):

Ratio	o	e	d	d^2	d^2/e
3/4	792	750	42	1764	2.35
1/4	208	250	– 42	1764	7.06
Total = 1000					

$$\chi^2 = \sum \frac{d^2}{e}$$
$$= 2.35 + 7.06$$
$$= 9.41$$

Consulting Figure 3–12 allows us to determine the probability (*p*). This value will let us determine whether the deviations from the null hypothesis can be attributed to chance. There are two possible outcomes (*n*), so the degrees of freedom (df) $= n - 1$, or 1. The table in Figure 3–12 shows that $p = 0.01$ to 0.001. The graph gives an estimate of about 0.001. That *p* is less than 0.05 leads us to reject the null hypothesis. The data do not statistically fit a 3:1 ratio.

When we accept Mendel's 3:1 ratio as a valid expression of the monohybrid cross, we make numerous assumptions. One of these may explain why the null hypothesis was rejected. We must assume that *all genotypes are equally viable*. That is, at the time the data are collected, genotypes yielding long wings are equally likely to survive from fertilization through adulthood as the genotype yielding *dumpy* wings. Further study would reveal that *dumpy* flies are somewhat less viable than normal flies. As a result, we would expect *less* than 1/4 of the total offspring to express dumpy. This observation is borne out in the data, although we have not proved it.

5. If two parents, both heterozygous carriers of the autosomal recessive gene causing cystic fibrosis, have five children, what is the probability that exactly three will be normal?

Solution: First, the probability of having a normal child during each pregnancy is

$$p_a = \text{normal} = 3/4$$

while the probability of having an afflicted offspring is

$$p_b = \text{afflicted} = 1/4$$

Then apply the formula

$$\frac{n!}{s!t!}a^s b^t$$

where $n = 5$, $s = 3$, and $t = 2$,

$$p = \frac{(5)\cdot(4)\cdot(3)\cdot(2)\cdot(1)}{(3)\cdot(2)\cdot(1)\cdot(2)\cdot(1)}(3/4)^3(1/4)^2$$

$$= \frac{(5)\cdot(4)}{(2)\cdot(1)}(3/4)^3(1/4)^2$$

$$= 10(27/64)\cdot(1/16)$$

$$= 10(27/1024)$$

$$= 270/1024$$

$$p = \sim 0.26$$

PROBLEMS AND DISCUSSION QUESTIONS

When working genetics problems in this and succeeding chapters, always assume that members of the P_1 *generation are homozygous, unless the information given, or the data presented, indicates or requires otherwise.*

1. In a cross between a black and a white guinea pig, all members of the F_1 generation are black. The F_2 generation is made up of approximately 3/4 black and 1/4 white guinea pigs.
   (a) Diagram this cross, showing the genotypes and phenotypes.
   (b) What will the offspring be like if two F_2 white guinea pigs are mated?
   (c) Two different matings were made between black members of the F_2 generation, with the following results.

Cross	Offspring
Cross 1	All black
Cross 2	3/4 black, 1/4 white

   Diagram each of the crosses.
2. Albinism in humans is inherited as a simple recessive trait. For the following families, determine the genotypes of the parents and offspring. (When two alternative genotypes are possible, list both.)
   (a) Two normal parents have five children, four normal and one albino.
   (b) A normal male and an albino female have six children, all normal.
   (c) A normal male and an albino female have six children, three normal and three albino.
   (d) Construct a pedigree of the families in (b) and (c). Assume that one of the normal children in (b) and one of the albino children in (c) become the parents of eight children. Extend the pedigree to include them, predicting their phenotypes (normal or albino).
3. Which of Mendel's postulates are illustrated by the pedigree in Problem 2? List and define these postulates.
4. Discuss how Mendel's monohybrid results served as the basis for all but one of his postulates. Which postulate was not based on these results? Why?
5. What advantages were provided by Mendel's choice of the garden pea in his experiments?
6. Pigeons may exhibit a checkered or plain pattern. In a series of controlled matings, the following data were obtained.

	F_1 Progeny	
P_1 Cross	Checkered	Plain
(a) checkered × checkered	36	0
(b) checkered × plain	38	0
(c) plain × plain	0	35

   Then F_1 offspring were selectively mated with the following results. (The P_1 cross giving rise to each F_1 pigeon is indicated in parentheses.)

	F_2 Progeny	
$F_1 \times F_1$ Crosses	Checkered	Plain
(d) checkered (a) × plain (c)	34	0
(e) checkered (b) × plain (c)	17	14
(f) checkered (b) × checkered (b)	28	9
(g) checkered (a) × checkered (b)	39	0

   How are the checkered and plain patterns inherited? Select and define symbols for the genes involved, and determine the genotypes of the parents and offspring in each cross.

7. Mendel crossed peas having round seeds and yellow cotyledons (seed leaves) with peas having wrinkled seeds and green cotyledons. All the F_1 plants had round seeds with yellow cotyledons. Diagram this cross through the F_2 generation, using both the Punnett square and forked-line, or branch diagram, methods.

8. Based on the preceding cross, in the F_2 generation, what is the probability that an organism will have round seeds and green cotyledons *and* be true breeding?

9. Based on the same characters and traits as in Problem 7, determine the genotypes of the parental plants involved in the crosses shown here by analyzing the phenotypes of their offspring.

Parental Plants	Offspring
(a) round, yellow × round, yellow	3/4 round, yellow 1/4 wrinkled, yellow
(b) wrinkled, yellow × round, yellow	6/16 wrinkled, yellow 2/16 wrinkled, green 6/16 round, yellow 2/16 round, green
(c) round, yellow × round, yellow	9/16 round, yellow 3/16 round, green 3/16 wrinkled, yellow 1/16 wrinkled, green
(d) round, yellow × wrinkled, green	1/4 round, yellow 1/4 round, green 1/4 wrinkled, yellow 1/4 wrinkled, green

10. Are any of the crosses in Problem 9 testcrosses? If so, which one(s)?

11. Which of Mendel's postulates can only be demonstrated in crosses involving at least two pairs of traits? State the postulate.

12. Correlate Mendel's four postulates with what is now known about homologous chromosomes, genes, alleles, and the process of meiosis.

13. What is the basis for homology among chromosomes?

14. Distinguish between homozygosity and heterozygosity.

15. In *Drosophila*, *gray* body color is dominant to *ebony* body color, while *long* wings are dominant to *vestigial* wings. Assuming that the P_1 individuals are homozygous, work the following crosses through the F_2 generation, and determine the genotypic and phenotypic ratios for each generation.

(a) gray, long × ebony, vestigial
(b) gray, vestigial × ebony, long
(c) gray, long × gray, vestigial

16. How many different types of gametes can be formed by individuals of the following genotypes: (a) *AaBb*, (b) *AaBB*, (c) *AaBbCc*, (d) *AaBBcc*, (e) *AaBbcc*, and (f) *AaBbCcDdEe?* What are the gametes in each case?

17. Using the forked-line, or branch diagram, method, determine the genotypic and phenotypic ratios of these trihybrid crosses: (a) *AaBbCc* × *AaBBCC*, (b) *AaBBCc* × *aaBBCc*, and (c) *AaBbCc* × *AaBbCc*.

18. Mendel crossed peas with green seeds to those with yellow seeds. The F_1 generation produced only yellow seeds. In the F_2, the progeny consisted of 6022 plants with yellow seeds and 2001 plants with green seeds. Of the F_2 yellow-seeded plants, 519 were self-fertilized with the following results: 166 bred true for yellow and 353 produced a 3:1 ratio of yellow:green. Explain these results by diagramming the crosses.

19. In a study of black and white guinea pigs, 100 black animals were crossed individually to white animals and each cross was carried to an F_2 generation. In 94 of the cases, the F_1 individuals were all black and an F_2 ratio of 3 black:1 white was obtained. In the other 6 cases, half of the F_1 animals were black and the other half were white. Why? Predict the results of crossing the black and white F_1 guinea pigs from the 6 exceptional cases.

20. Mendel crossed peas with round green seeds to ones with wrinkled yellow seeds. All F_1 plants had seeds that were round and yellow. Predict the results of testcrossing these F_1 plants.

21. Thalassemia is an inherited anemic disorder in humans. Affected individuals exhibit either a minor anemia or a major anemia. Assuming that only a single gene pair and two alleles are involved in the inheritance of these conditions, is thalassemia a dominant or recessive disorder?

22. The following are F_2 results of two of Mendel's monohybrid crosses.

(a)	Full pods	882
	Constricted pods	299
(b)	Violet flowers	705
	White flowers	224

State a null hypothesis to be tested using χ^2 analysis. Calculate the χ^2 value and determine the p value for both. Interpret the p values. Can the deviation in each case be attributed to chance or not? Which of the two crosses shows a greater amount of deviation?

23. In one of Mendel's dihybrid crosses, he observed 315 round yellow, 108 round green, 101 wrinkled yellow, and 32 wrinkled green F_2 plants. Analyze these data using χ^2 test to see if
(a) they fit a 9:3:3:1 ratio.
(b) the round:wrinkled data fit a 3:1 ratio.
(c) the yellow:green data fit a 3:1 ratio.

24. In assessing data that fell into two phenotypic classes, a geneticist observed values of 250:150. She decided to perform a χ^2 analysis by using the following two different null hypotheses: (a) the data fit a 3:1 ratio, and (b) the data fit a 1:1 ratio. Calculate the χ^2 values for each hypothesis. What can be concluded about each hypothesis?

25. The basis for rejecting any null hypothesis is arbitrary. The researcher can set more or less stringent standards by deciding to raise or lower the p value used to reject or fail to reject the hypothesis. In the case of the chi-square analysis of genetic crosses, would the use of a standard of $p = 0.10$ be more or less stringent in failing to reject the null hypothesis? Explain.

26. Consider the following pedigree.

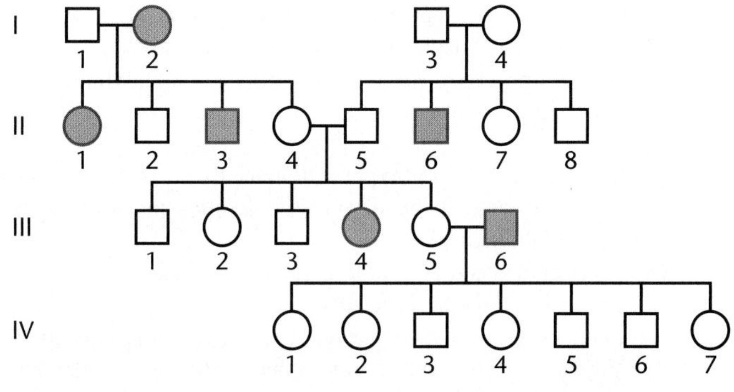

Predict the mode of inheritance and the most probable genotypes of each individual. Assume that the alleles *A* and *a* control the expression of the trait.

27. The following pedigree is for myopia (nearsightedness) in humans.

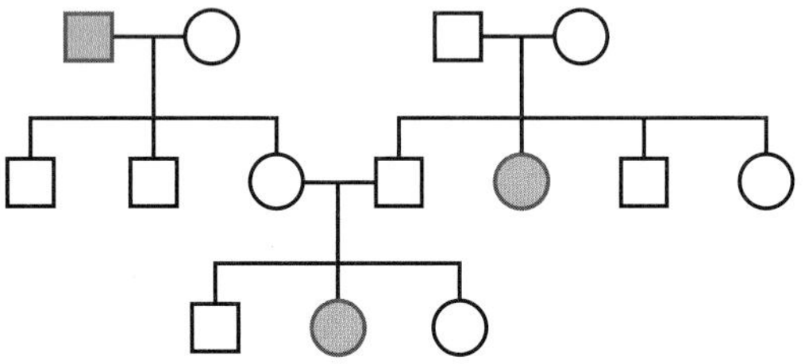

Predict whether the disorder is inherited as the result of a dominant or recessive trait. Determine the most probable genotype for each individual based on your prediction.

28. Consider three independently assorting gene pairs, A/a, B/b, and C/c. What is the probability of obtaining an offspring that is $AABbCc$ from parents that are $AaBbCC$ and $AABbCc$?
29. What is the probability of obtaining a triply recessive individual from the parents shown in Problem 28?
30. Of all offspring of the parents in Problem 28, what proportion will express all three dominant traits?
31. When a die (one of a pair of dice) is rolled, it has an equal probability of landing on any of its six sides.
    (a) What is the probability of rolling a 3 with a single throw?
    (b) When a die is rolled twice, what is the probability that the first throw will be a 3 and the second will be a 6?
    (c) When a die is rolled twice, what is the probability that one throw will result in a 3 and the other throw will result in a 6?
    (d) If two dice are rolled together, what is the combined probability that one will be a 3 and the other will be a 6?
    (e) If one die is rolled and it comes up as an odd number, what is the probability that it is a 5?
32. Consider the F_2 offspring of Mendel's dihybrid cross. Determine the conditional probability that F_2 plants expressing both dominant traits are heterozygous at both loci.
33. Draw all possible conclusions concerning the mode of inheritance of the trait denoted in each of the following limited pedigrees. (Each case is based on a different trait.)

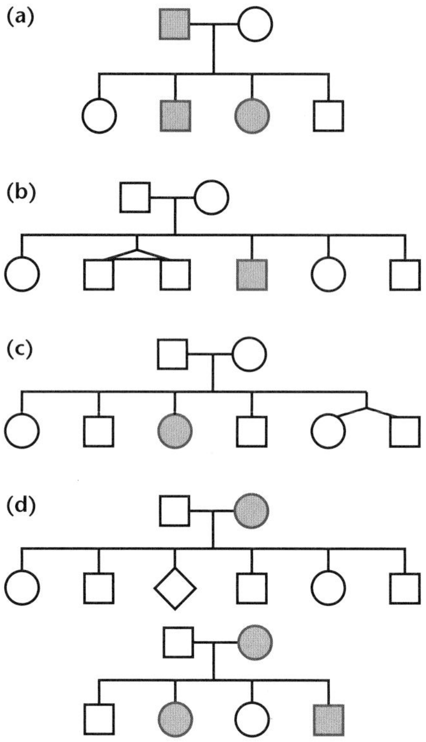

34. Cystic fibrosis is an autosomal recessive disorder. A male whose brother has the disease has children with a female whose sister has the disease. It is not known if either the male or the female is a carrier. If the male and female have one child, what is the probability that the child will have cystic fibrosis?
35. In a family of five children, what is the probability that
    (a) all are males?
    (b) three are males and two are females?
    (c) two are males and three are females?
    (d) all are the same sex?
    Assume that the probability of a male child is equal to the probability of a female child ($p = 1/2$).
36. In a family of eight children, where both parents are heterozygous for albinism, what mathematical expression predicts the probability that six are normal and two are albinos?

Extra-Spicy Problems

37. Two true-breeding pea plants were crossed. One parent is round, terminal, violet, constricted, while the other expresses the respective contrasting phenotypes of wrinkled, axial, white, full. The four pairs of contrasting traits are controlled by four genes, each located on a separate chromosome. In the F_1, only round, axial, violet, and full were expressed. In the F_2, all possible combinations of these traits were expressed in ratios consistent with Mendelian inheritance.
    (a) What conclusion about the inheritance of the traits can be drawn based on the F_1 results?
    (b) In the F_2 results, which phenotype appeared most frequently? Write a mathematical expression that predicts the probability of occurrence of this phenotype.
    (c) Which F_2 phenotype is expected to occur least frequently? Write a mathematical expression that predicts this probability.
    (d) In the F_2 generation, how often is either of the P_1 phenotypes likely to occur?
    (e) If the F_1 plants were testcrossed, how many different phenotypes would be produced? How does this number compare with the number of different phenotypes in the F_2 generation test just discussed?
38. Tay–Sachs disease (TSD) is an inborn error of metabolism that results in death, often by the age of two. You are a genetic counselor, and you interview a phenotypically normal couple, where the male had a female first cousin (on his father's side) who died from TSD, and where the female had a maternal uncle with TSD. There are no other known cases in either of the families, and none of the matings were or are between related individuals. Assume that this trait is very rare.
    (a) Draw a pedigree of the families of this couple, showing the relevant individuals.

(b) Calculate the probability that both the male and female are carriers for TSD.
(c) What is the probability that neither of them is a carrier?
(d) What is the probability that one of them is a carrier and the other is not? (*Hint:* The p values in (b), (c), and (d) should equal 1.)

39. The wild-type (normal) fruit fly, *Drosophila melanogaster*, has straight wings and long bristles. Mutant strains have been isolated that have either curled wings or short bristles. The genes representing these two mutant traits are located on separate autosomes. Carefully examine the data from the following five crosses shown below (running across both columns).
(a) For each mutation, determine whether it is dominant or recessive. In each case, identify which crosses support your answer.
(b) Define gene symbols and, for each cross, determine the genotypes of the parents.

	Number of Progeny			
Cross	straight wings, long bristles	straight wings, short bristles	curled wings, long bristles	curled wings, short bristles
1 straight, short × straight, short	30	90	10	30
2 straight, long × straight, long	120	0	40	0
3 curled, long × straight, short	40	40	40	40
4 straight, short × straight, short	40	120	0	0
5 curled, short × straight, short	20	60	20	60

40. An alternative to using the binomial expansion and Pascal's triangle in determining probabilities of phenotypes in a subsequent generation when the parents' genotypes are known is to use the following equation:

$$\frac{n!}{s!t!}a^s b^t$$

where n is the total number of offspring, s is the number of offspring in one phenotypic category, t is the number of offspring in the other phenotypic category, a is the probability of occurrence of the first phenotype and b is the probability of the second phenotype. Using this equation, determine the probability of a family of 5 offspring having 2 children afflicted with sickle-cell anemia (an autosomal recessive disease) when both parents are heterozygous for the sickle-cell allele.

41. Considering the information in Problem 40, to what do you suppose the following mathematical expression applies?

$$\frac{n!}{s!t!u!}a^s b^t c^u$$

Can you think of a genetic example where it might have application?

42. To assess Mendel's law of segregation using tomatoes, a true-breeding tall variety (*SS*) is crossed with a true-breeding short variety (*ss*). The heterozygous tall plants (*Ss*) were crossed to produce two sets of F_2 data, as follows.

Set I	Set II
30 tall	300 tall
5 short	50 short

(a) Using the χ^2 test, analyze the results for both data sets. Calculate χ^2 values and estimate the p values in both cases.
(b) From the above analysis, what can you conclude about the importance of generating large data sets in experimental settings?

Selected Readings

Bennett, R.L. et. al. 1995. Recommendations for standardized human pedigree nomenclature. *Am. J. Hum. Genet.* 56:745–52.

Carlson, E.A. 1987. *The gene: A critical history*, 2nd ed. Philadelphia, PA: Saunders.

Dunn, L.C. 1965. *A short history of genetics*. New York: McGraw-Hill.

Henig, R.M. 2001. *The monk in the garden: The lost and found genius of Gregor Mendel, the father of genetics*. New York: Houghton-Mifflin.

Miller, J.A. 1984. Mendel's peas: A matter of genius or of guile? *Sci. News* 125:108–109.

Olby, R.C. 1985. *Origins of Mendelism*, 2d ed. London: Constable.

Orel, V. 1996. *Gregor Mendel: The first geneticist*. Oxford: Oxford University Press.

Peters, J., ed. 1959. *Classic papers in genetics*. Englewood Cliffs, NJ: Prentice-Hall.

Sokal, R.R., and Rohlf, F.J. 1987. *Introduction to biostatistics*, 2nd ed. New York: W.H. Freeman.

Stern, C., and Sherwood, E. 1966. *The origins of genetics: A Mendel source book*. San Francisco: W.H. Freeman.

Stubbe, H. 1972. *History of genetics: From prehistoric times to the rediscovery of Mendel's laws*. Cambridge, MA: MIT Press.

Sturtevant, A.H. 1965. *A history of genetics*. New York: Harper & Row.

Tschermak-Seysenegg, E. 1951. The rediscovery of Mendel's work. *J. Hered.* 42:163–72.

Welling, F. 1991. Historical study: Johann Gregor Mendel 1822–1884. *Am. J. Med. Genet.* 40:1–25.

Chapter Four

Extensions of Mendelian Genetics

Walnut

Pea

Rose

Single

Inherited variation in comb shape of chickens, controlled by two pairs of genes.

Chapter Concepts

- While alleles are transmitted from parent to offspring according to Mendelian principles, they often do not display the clear-cut dominant/recessive relationship observed by Mendel.
- In many cases, in contrast to Mendelian genetics, two or more genes are known to influence the phenotype of a single characteristic.
- Still another exception to Mendelian inheritance is the presence of genes on sex chromosomes, whereby one of the sexes contains only a single member of that chromosome.
- Phenotypes are often the result of both genetics and the environment within which genes are expressed.
- The result of the various exceptions to Mendelian principles is the occurrence of phenotypic ratios that differ from those resulting from standard monohybrid, dihybrid, and trihybrid crosses. Collectively, these exceptions are referred to as "extensions of Mendelian genetics."

In Chapter 3, we discussed the fundamental principles of transmission genetics. We saw that genes are present on homologous chromosomes and that these chromosomes segregate from each other and assort independently with other segregating chromosomes during gamete formation. These two postulates are the basic principles of gene transmission from parent to offspring. Once an offspring has received the total set of genes, however, it is the expression of genes that determines the organism's phenotype. When gene expression does not adhere to a simple dominant/recessive mode, or when more than one pair of genes influences the expression of a single character, the classic 3:1 and 9:3:3:1 F_2 ratios are usually modified. In this and the next several chapters, we consider more complex modes of inheritance. Although more complex modes of inheritance result, the fundamental principles set down by Mendel still hold true in these situations.

In this chapter, we restrict our initial discussion to the inheritance of traits controlled by only one set of genes. In diploid organisms, which have homologous pairs of chromosomes, two copies of each gene influence such traits. The copies need not be identical since alternative forms of genes, **alleles**, occur within populations. How alleles act to influence a given phenotype will be our primary focus. We will then turn to **gene interaction**, a situation in which a single phenotype is affected by more than one gene. Numerous examples will be presented to illustrate a variety of heritable patterns observed in such situations.

Thus far, we have restricted our discussion to chromosomes other than the X and Y pair. By examining cases where genes are present on the X chromosome, illustrating **X-linkage**, we will see yet another modification of Mendelian ratios. Our discussion of modified ratios concludes with the consideration of sex-limited and sex-influenced inheritance, cases where the sex of the individual, but not necessarily the X chromosome, influences the phenotype. We conclude the chapter by showing how a given phenotype often varies depending on the overall environment in which a cell or an organism finds itself. This discussion points out that phenotypic expression depends on more than just the genotype of an organism.

4.1 Alleles Alter Phenotypes in Different Ways

Following the rediscovery of Mendel's work in the early 1900s, research focused on the many ways in which genes can influence an individual's phenotype. This course of investigation, stemming from Mendel's findings, is called neo-Mendelian genetics (*neo* from the Greek word meaning *since* or *new*).

Each type of inheritance described in this chapter was investigated when observations of genetic data did not precisely conform to the expected Mendelian ratios. Hypotheses that modified and extended the Mendelian principles were proposed and tested with specifically designed crosses. Explanations for these observations were in accord with the principle that a phenotype is under the influence of one or more genes located at specific loci on one or more pairs of homologous chromosomes.

To understand the various modes of inheritance, we must first examine the potential function of an allele. In some organisms where extensive populations have been studied, the allele that occurs most frequently in nature, the one that is arbitrarily designated as normal, is often referred to as the **wild-type allele**. This common allele is often, but not always, dominant. Wild-type alleles are responsible for the corresponding wild-type phenotype and serve as standards for comparison against other mutations occurring at a particular locus.

A mutant allele contains modified genetic information and often specifies an altered gene product. For example, in human populations, there are many known alleles of the gene encoding the β-chain of human hemoglobin. All such alleles store information necessary for the synthesis of the βchain polypeptide, but each allele specifies a slightly different form of the same molecule. Once manufactured, the product of an allele may or may not have its function altered.

The process of mutation is the source of alleles. For a new allele to be recognized when observing an organism, it must cause a change in the phenotype. A new phenotype results from a change in functional activity of the cellular product specified by that gene. Often, the mutation causes the dimunition or the loss of the specific wild-type function. For example, if a gene is responsible for the synthesis of a specific enzyme, a mutation in that gene may ultimately change the conformation of this enzyme and reduce or eliminate its affinity for the substrate. Such a case is designated as a **loss of function mutation**. If the loss is complete, the mutation has resulted in what is called a **null allele**.

Conversely, other mutations may enhance the function of the wild type product. Most often when this occurs, it is the result of increasing the quantity of the gene product. In such cases, the mutation is affecting the regulation of transcription of the gene under consideration. Such cases are designated **gain of function mutations**, which generally result in dominant alleles since one copy in a diploid organism is sufficient to alter the normal phenotype. Examples of gain in function mutations include the genetic conversion of protooncogenes, which regulate the cell cycle, to oncogenes, where regulation is overridden by excess gene product. The result is the creation of a cancerous cell.

Having examined gain or loss of function mutations, it is important to note that the possibility exists that a mutation will create an allele where no change in function can be detected. In this case, the mutation would not be immediately apparent since no phenotypic variation would be evident. However, such a mutation could be detetected if the DNA sequence of the gene was examined. It is also important to note that while a phenotypic trait may be affected by a single mutation, traits are often influenced by many gene products. In the case of enzymatic reactions, most are part of complex metabolic pathways. Therefore, phenotypic traits may be influenced by more than one gene and the allelic forms of each gene involved. In each of the many crosses discussed in the next few chapters, only one or a few gene pairs are involved. Keep in mind that in each cross, all genes that are not under consideration are assumed to have no effect on the inheritance patterns described.

How Do We Know?

How were early geneticists able to determine if mutant alleles were dominant or recessive to their wild-type counterparts?

4.2 Geneticists Use a Variety of Symbols for Alleles

In Chapter 3, we learned to symbolize alleles for very simple Mendelian traits. The initial letter of the name of a recessive trait, lowercased and italicized, denotes the recessive allele, and the same letter in uppercase refers to the dominant allele. Thus, for *tall* and *dwarf*, where *dwarf* is recessive, *D* and *d* represent the alleles responsible for these respective traits. Mendel used upper- and lowercase letters such as these to symbolize his unit factors.

Another useful system was developed in genetic studies of the fruit fly *Drosophila melanogaster* to discriminate between wild-type and mutant traits. This system uses the initial letter, or a combination of two or three letters, of the name of the mutant trait. If the trait is recessive, the lowercase form is used; if it is dominant, the uppercase form is used. The contrasting wild-type trait is denoted by the same letter, but with a superscript +.

For example, *ebony* is a recessive body color mutation in *Drosophila*. The normal wild-type body color is gray. According to the system just discussed, *ebony* is denoted by the symbol *e*, while gray is denoted by e^+. The responsible locus may be occupied by either the wild-type allele (e^+) or the mutant allele (e). A diploid fly may thus exhibit one of three possible genotypes.

e^+/e^+	gray homozygote (wild type)
e^+/e	gray heterozygote (wild type)
e/e	ebony homozygote (mutant)

The slash between the letters indicates that the two allele designations represent the same locus on two homologous chromosomes.

If we instead consider a mutant allele that is dominant to the normal wild-type allele, such as *Wrinkled* wing in *Drosophila*, we would use an intitial uppercase letter to designate dominance (*Wr*). Thus, the three possible genotypyes would be:

Wr/Wr	Wrinkled wings
Wr/Wr^+	Wrinkled wings
Wr^+/Wr^+	Normal, straight wings

It is because of dominance that both of the initial two genotypes express the mutant wrinkled-wing phenotype.

One advantage of the above system is that further abbreviation can be used when convenient: The wild-Type allele may simply be denoted by the + symbol. With *ebony* as an example, the designations of the three possible genotypes become

$+/+$	gray homozygote (wild type)
$+/e$	gray heterozygote (wild type)
e/e	ebony homozygote (mutant)

This system works nicely for other organisms as well, provided there is a distinct wild-type phenotype for the character under consideration. As we will see in Chapter 6, it is particularly useful when two or three genes linked together on the same chromosome are considered simultaneously. If no dominance exists, we may simply use uppercase letters and superscripts to denote alternative alleles (e.g., R^1 and R^2, L^M and L^N, and I^A and I^B). Their use will become apparent in ensuing sections of this chapter.

Two other points are important. First, although we have adopted a standard convention for assigning genetic symbols, there are many diverse systems of genetic nomenclature used to identify genes in various organisms. Usually, the symbol selected reflects the function of the gene or, occasionally, a disorder caused by a mutant gene. For example, the *CDK* gene from yeast refers to that encoding cyclin-dependent kinases. In bacteria, leu^- refers to a mutation that interrupts the biosynthesis of the amino acid leucine, where the wild-type gene is designated leu^+. The symbol *dnaA* represents a bacterial gene involved in DNA replication (and DnaA, without italics, designates the protein made by that gene). In humans, italicized capital letters are used to name genes: *BRCA1* represents one of the genes associated with susceptibility to *br*east *ca*ncer. Although these different systems may sometimes be confusing, they all represent different ways to symbolize genes.

4.3 In Incomplete Dominance, Neither Allele Is Dominant

Contrary to the Mendelian crosses reported in Chapter 3, a cross between parents with contrasting traits may sometimes generate offspring with an intermediate phenotype. For example, if plants such as four o'clocks or snapdragons with red flowers are crossed with white-flowered plants, the offspring have pink flowers. Because some red pigment is produced in the F_1 intermediate-colored pink flowers, neither the red nor white flower color is dominant. Such a situation is known as **incomplete dominance**.

If the phenotype is under the control of a single gene and two alleles, where neither is dominant, the results of the F_1 pink × pink cross can be predicted. The resulting F_2 generation shown in Figure 4–1 confirms the hypothesis that only one pair of alleles determines these phenotypes. The genotypic ratio (1:2:1) of the F_2 generation is identical to that of Mendel's monohybrid cross. However, because neither allele is dominant, the phenotypic ratio is identical to the genotypic ratio. Note that since neither of the alleles is recessive, we have chosen not to use upper- and lowercase letters as symbols. Instead, we denote the red and white alleles as R^1 and R^2. We could have chosen W^1 and W^2 or still other designations such as C^W and C^R, where *C* indicates color and the *W* and *R* superscripts indicate white and red, respectively.

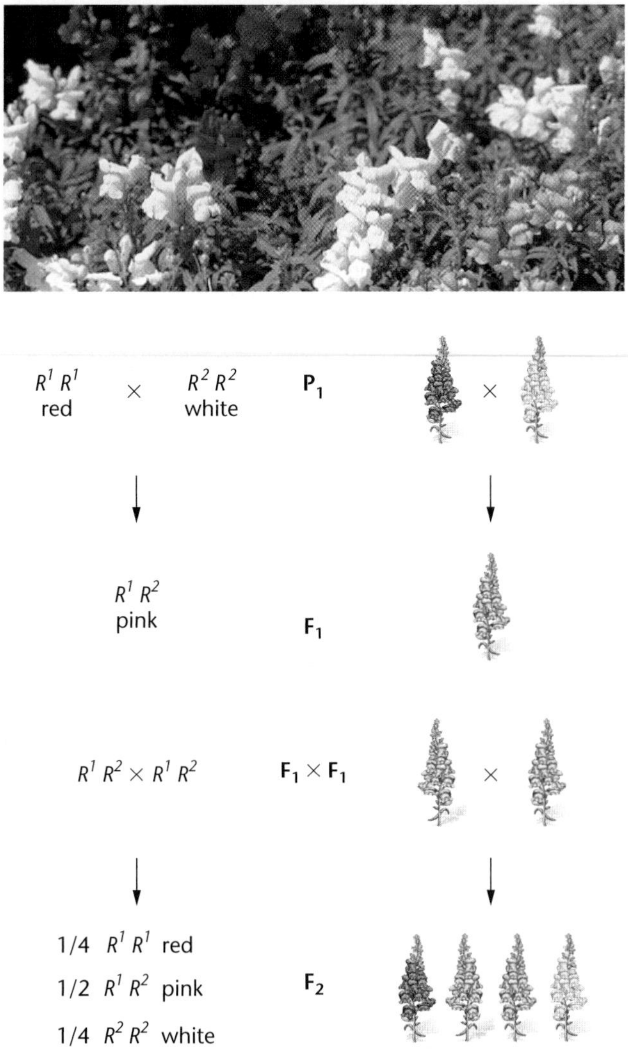

FIGURE 4–1 Incomplete dominance shown in the flower color of snapdragons.

How are we to interpret lack of dominance whereby an intermediate phenotype characterizes heterozygotes? The most accurate way is to consider gene expression in a quantitative way. In the case of flower color above, the mutation causing white flowers is most likely one where complete "loss of function" occurs. In this case, it is likely that the gene product of the wild type allele (R^1) is an enzyme that participates in a reaction leading to the synthesis of a red pigment. The mutant allele (R^2) produces an enzyme that cannot catalyze the reaction leading to pigment. The end result is that the heterozygote produces only about half the pigment of the red-flowered plant and the phenotype is pink.

Clear-cut cases of incomplete dominance are relatively rare. However, even when complete dominance seems apparent, careful examination of the gene product, rather than the phenotype, often reveals an intermediate level of gene expression. An example is the human biochemical disorder **Tay–Sachs disease**, in which homozygous recessive individuals are severely affected with a fatal lipid-storage disorder and neonates die during their first one to three years of life. In afflicted individuals, there is almost no activity of the responsible enzyme **hexosaminidase A**, which is normally involved in lipid metabolism. Heterozygotes, with only a single copy of the mutant gene, are phenotypically normal, but with only about 50 percent of the enzyme activity found in homozygous normal individuals. Fortunately, this level of enzyme activity is adequate to achieve normal biochemical function. This situation is not uncommon in enzyme disorders and illustrates the concept of the **threshold effect**, whereby normal phenotypic expression occurs anytime a certain level of gene product is attained. Most often, and in particular in Tay–Sachs disease, the threshold is less than 50 percent.

4.4 In Codominance, the Influence of Both Alleles in a Heterozygote Is Clearly Evident

If two alleles of a single gene are responsible for producing two distinct and detectable gene products, a situation different from incomplete dominance or dominance/recessiveness arises. In such a case, the joint expression of both alleles in a heterozygote is called **codominance**. The **MN blood group** in humans illustrates this phenomenon. Karl Landsteiner and Philip Levin discovered a glycoprotein molecule found on the surface of red blood cells that acts as a native antigen, providing biochemical and immunological identity to individuals. In the human population, two forms of this glycoprotein exist, designated M and N. An individual may exhibit either one or both of them.

The MN system is under the control of an autosomal locus found on chromosome 4, with two alleles designated L^M and L^N. Because humans are diploid, three combinations are possible, as follows, each resulting in a distinct blood type:

Genotype	Phenotype
L^ML^M	M
L^ML^N	MN
L^NL^N	N

As predicted, a mating between two heterozygous MN parents may produce children of all three blood types, as follows:

$$L^ML^N \times L^ML^N$$
$$\downarrow$$
$$1/4\ L^ML^M$$
$$1/2\ L^ML^N$$
$$1/4\ L^NL^N$$

The preceding example shows that in codominant inheritance, distinct expression of the gene products of both alleles can be detected. This characteristic distinguishes codominance from other modes of inheritance, such as incomplete dominance, where heterozygotes express an intermediate, or blended, phenotype. For codominance to be studied, both products must be phenotypically detectable. We shall see another example of codominance when we examine the ABO blood type system, which will be used in the following section to illustrate multiple alleles, another phenomenon that Mendel never encountered.

4.5 Multiple Alleles of a Gene May Exist in a Population

Because the information stored in any gene is extensive, mutations can modify the gene in many ways. Each change has the potential for producing a different allele. Therefore, for any gene, the number of alleles within members of a population of individuals is not necessarily restricted to two. When three or more alleles of the same gene are found, **multiple alleles** are said to be present, creating a characteristic mode of inheritance. It is important to realize that *multiple alleles can be studied only in populations*. Any individual diploid organism has, at most, two homologous gene loci that may be occupied by different alleles of the same gene. However, among members of a species, many alternative forms of the same gene can exist.

The ABO Blood Groups

The simplest case of multiple alleles is when three alternative alleles of one gene exist. This situation is illustrated in the inheritance of the **ABO blood groups** in humans, discovered by Karl Landsteiner in the early 1900s. The ABO system, like the MN blood types, is characterized by the presence of antigens on the surface of red blood cells. However, the A and B antigens are distinct from the MN antigens and are under the control of a different gene, located on chromosome 9. As in the MN system, one combination of alleles in the ABO system exhibits a codominant mode of inheritance.

The ABO phenotype of any individual is ascertained by mixing a blood sample with antiserum containing type A or type B antibodies. If the antigen is present on the surface of the person's red blood cells, it will react with the corresponding antibody and cause clumping, or agglutination, of the red blood cells. When an individual is tested in this way, one of four phenotypes may be revealed. Each individual has either the A antigen (A phenotype), the B antigen (B phenotype), the A and B antigens (AB phenotype), or neither antigen (O phenotype).

In 1924, it was hypothesized that these phenotypes were inherited as the result of three alleles of a single gene. This hypothesis was based on studies of the blood types of many different families. Although different designations can be used, we will use the symbols I^A, I^B, and I^O to distinguish these three alleles. The I designation stands for **isoagglutinogen**, another term for antigen. If we assume that the I^A and I^B alleles are responsible for the production of their respective A and B antigens and that I^O is an allele that does not produce any detectable A or B antigens, we can list the various genotypic possibilities and assign the appropriate phenotype to each:

Genotype	Antigen	Phenotype
I^AI^A	A	A
I^AI^O	A	
I^BI^B	B	B
I^BI^O	B	
I^AI^B	A, B	AB
I^OI^O	Neither	O

Note that in these assignments, the I^A and I^B alleles behave dominantly to the I^O allele, but codominantly to each other.

We can test the hypothesis that three alleles control ABO blood groups by examining potential offspring from the various combinations of matings, as shown in Table 4.1. If we assume heterozygosity wherever possible, we can predict which phenotypes can occur. These theoretical predictions have been upheld in numerous studies examining the blood types of children of parents with all possible phenotypic combinations. The hypothesis that three alleles control ABO blood types in the human population is now universally accepted.

Our knowledge of human blood types has several practical applications. One of the most important is testing the compatibility of blood transfusions. Another application involves cases of disputed parentage, where newborns are inadvertently mixed up in the hospital, or when it is uncertain whether a specific male is the father of a child. An examination of the ABO blood groups as well as other inherited antigens of the possible parents and the child may help to resolve the situation. For example, of all the matings shown in Table 4.1, the only one that

TABLE 4.1 POTENTIAL PHENOTYPES IN THE OFFSPRING OF PARENTS WITH ALL POSSIBLE ABO BLOOD GROUP COMBINATIONS, ASSUMING HETEROZYGOSITY WHENEVER POSSIBLE

Parents		Potential Offspring			
Phenotypes	Genotypes	A	B	AB	O
$A \times A$	$I^AI^O \times I^AI^O$	3/4	—	—	1/4
$B \times B$	$I^BI^O \times I^BI^O$	—	3/4	—	1/4
$O \times O$	$I^OI^O \times I^OI^O$	—	—	—	all
$A \times B$	$I^AI^O \times I^BI^O$	1/4	1/4	1/4	1/4
$A \times AB$	$I^AI^O \times I^AI^B$	1/2	1/4	1/4	—
$A \times O$	$I^AI^O \times I^OI^O$	1/2	—	—	1/2
$B \times AB$	$I^BI^O \times I^AI^B$	1/4	1/2	1/4	—
$B \times O$	$I^BI^O \times I^OI^O$	—	1/2	—	1/2
$AB \times O$	$I^AI^B \times I^OI^O$	1/2	1/2	—	—
$AB \times AB$	$I^AI^B \times I^AI^B$	1/4	1/2	1/2	—

can result in offspring with all four phenotypes is that between two heterozygous individuals, one showing the A phenotype and the other showing the B phenotype. On genetic grounds alone, a male or female may be unequivocally ruled out as the parent of a certain child. However, this type of genetic evidence never proves parenthood.

The A and B Antigens

The biochemical basis of the ABO blood type system has now been carefully worked out. The A and B antigens are actually carbohydrate groups (sugars) that are bound to lipid molecules (fatty acids) protruding from the membrane of the red blood cell. The specificity of the A and B antigens is based on the terminal sugar of the carbohydrate group.

Almost all individuals possess what is called the **H substance**, to which one or two terminal sugars are added. As shown in Figure 4–2, the H substance itself contains three sugar molecules, galactose (Gal), *N*-acetylglucosamine (AcGluNH), and fucose, chemically linked together. The I^A allele is responsible for an enzyme that can add the terminal sugar *N*-acetylgalactosamine (AcGalNH) to the H substance. The I^B allele is responsible for a modified enzyme that cannot add *N*-acetylgalactosamine, but instead can add a terminal galactose. Heterozygotes (I^AI^B) add either one or the other sugar at the many sites (substrates) available on the surface of the red blood cell, illustrating the biochemical basis of codominance in individuals of the AB blood type. Finally, persons of type O (I^OI^O) cannot add either terminal sugar, possessing only the H substance protruding from the surface of their red blood cells.

The molecular genetic basis of the mutations leading to the I^A, I^B, and I^O alleles has been clarified. We will return to this topic in Chapter 15 when we discuss mutation and mutagenesis.

The Bombay Phenotype

In 1952, a very unusual situation provided information concerning the genetic basis of the H substance. A woman in Bombay displayed a unique genetic history inconsistent with her blood type. In need of a transfusion, she was found to lack both the A and B antigens and was thus typed as O. However, as shown in the partial pedigree in Figure 4–3, one of her parents was type AB, and she was the obvious donor of an I^B allele to

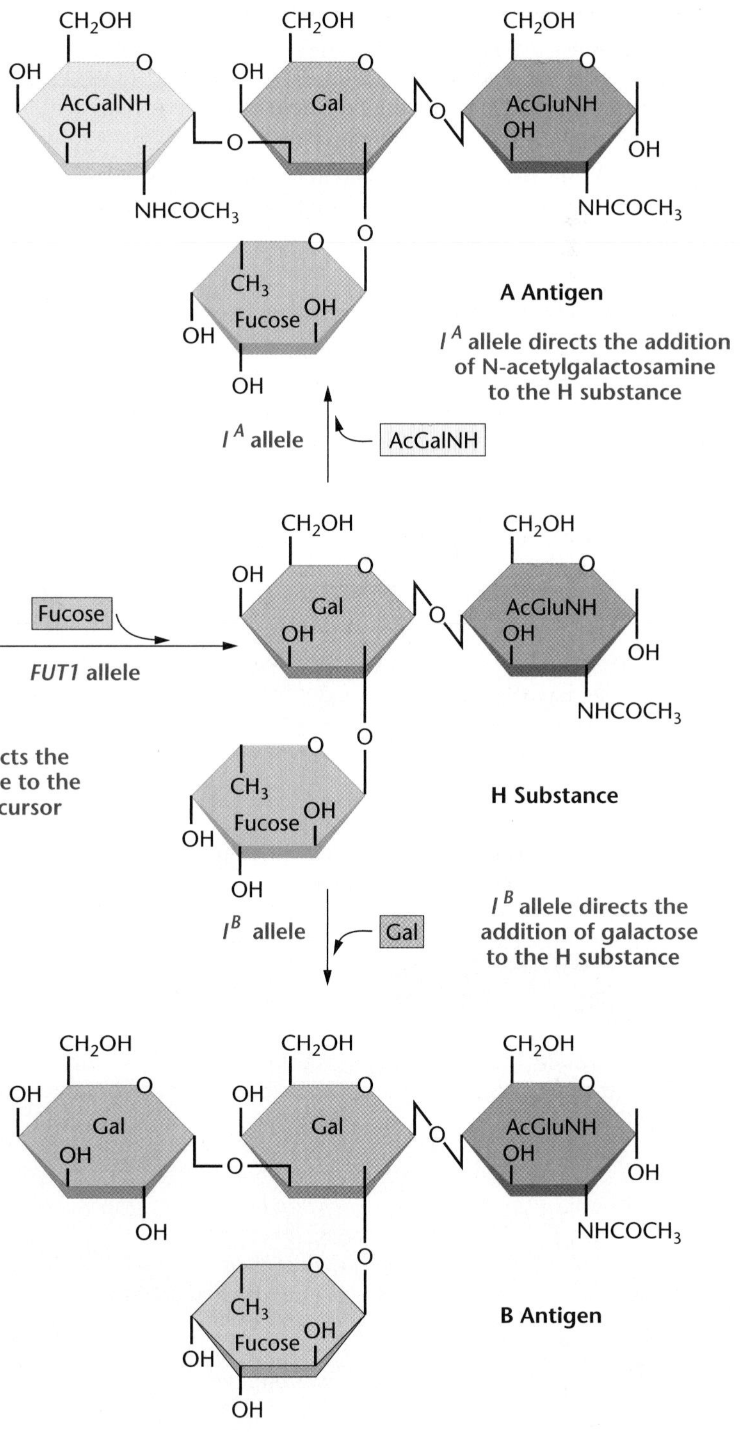

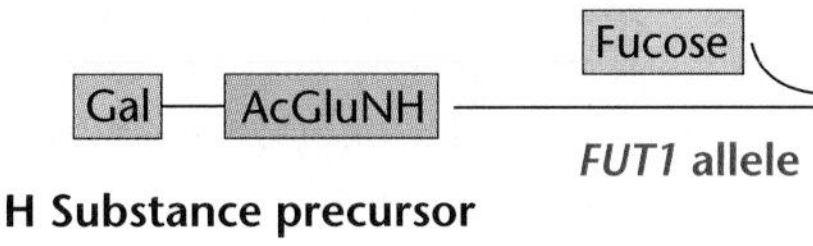

FUT1 allele directs the addition of fucose to the H substance precursor

FIGURE 4–2 The biochemical basis of the ABO blood groups. The *H* allele, present in almost all humans, directs the conversion of a precursor molecule to the H substance by adding a molecule of fucose to it. The I^A and I^B alleles are then able to direct the addition of terminal sugar residues to the H substance. The I^O allele is unable to direct either of these terminal additions. Gal: galactose; AcGluNH: *N*-acetylglucosamine; AcGalNH: *N*-acetylgalactosamine. Failure to produce the H substance results in the Bombay phenotype where individuals are type O, regardless of the presence of an I^A or I^B allele.

two of her offspring. Thus, she was genetically type B, but functionally type O!

This woman was subsequently shown to be homozygous for a rare recessive mutation in a gene designated *FUT1* (encoding an enzyme, fucosyl transferase), which prevented her from synthesizing the complete H substance. In this mutation, the terminal portion of the carbohydrate chain protruding from the red cell membrane lacks fucose, normally added by the enzyme. In the absence of fucose, the enzymes specified by the I^A and I^B alleles apparently are unable to recognize the incomplete H substance as a proper substrate. Thus, neither the terminal galactose nor *N*-acetylgalactosamine can be added, even though the appropriate enzymes capable of doing so are present and functional. As a result, the ABO system genotype cannot be expressed in individuals homozygous for the mutant form of the *FUT1* gene, and they are functionally type O. To distinguish them from the rest of the population, they are said to demonstrate the Bombay phenotype. The frequency of the mutant *FUT1* allele is exceedingly low. Hence, the vast majority of the human population can synthesize the H substance.

The *white* Locus in *Drosophila*

Many other phenotypes in both plants and animals are influenced by multiple allelic inheritance at a single locus. In *Drosophila*, for example, where the induction of mutations has been used extensively as a method of investigation, many alleles are present at practically every locus. The recessive eye mutation, *white*, discovered by Thomas H. Morgan and Calvin Bridges in 1912, represents only one of more than 100 alleles that can occupy this locus. In this allelic series, eye colors range from complete absence of pigment in the *white* allele, to deep ruby in the *white-satsuma* allele, to orange in the *white-apricot* allele, to a buff color in the *white-buff* allele. These alleles are designated w, w^{sat}, w^a, and w^{bf}, respectively (Table 4.2). In each case, the total amount of pigment in these mutant eyes is reduced to less than 20 percent of that found in the brick-red wild-type eye.

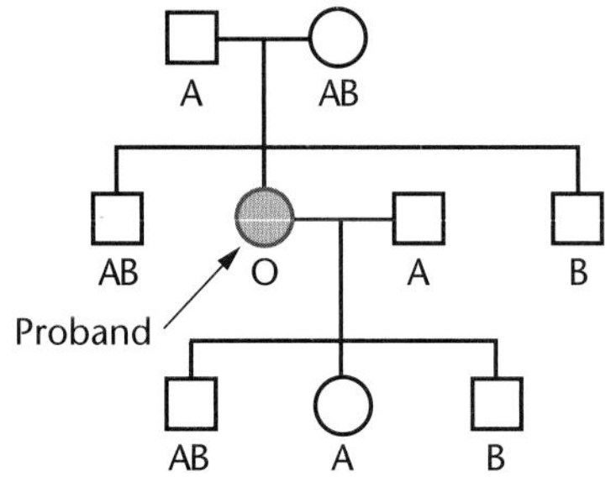

FIGURE 4–3 A partial pedigree of a woman displaying the Bombay phenotype. Functionally, her ABO blood group behaves as type O. Genetically, she is type B.

Now solve this

Problem 4.10 on page 93 involves a series of multiple alleles controlling coat color in rabbits.

Hint: Note particularly the hierachy of dominance of the various alleles. Remember also that even though there can be more than two alleles in a population, an individual can have at most two of these. Thus, the allelic distribution into gametes adheres to the principle of segregation.

4.6 Lethal Alleles Represent Essential Genes

Countless gene products are essential to an organism's survival. Loss of function mutations creating a nonfunctional gene product can sometimes be tolerated in the heterozygous state. In such a case, one wild-type allele may produce a sufficient quantity of the essential product for the organism to survive. However, such a mutation behaves as a **recessive lethal allele**, and homozygous recessive individuals will not survive. The time of death will depend on when the product is essential. In mammals, for example, this might occur during development, early childhood, or even during adulthood.

In some cases, the allele responsible for a lethal effect when it is homozygous may also result in a distinctive mutant phenotype when it is present heterozygously. *Such an allele is*

TABLE 4.2 SOME OF THE ALLELES PRESENT AT THE *WHITE* LOCUS OF *DROSOPHILA*

Allele	Name	Eye Color
w	*white*	pure white
w^a	*white-apricot*	yellowish orange
w^{bf}	*white-buff*	light buff
w^{bl}	*white-blood*	yellowish ruby
w^{cf}	*white-coffee*	deep ruby
w^e	*white-eosin*	yellowish pink
w^{mo}	*white-mottled orange*	light mottled orange
w^{sat}	*white-satsuma*	deep ruby
w^{sp}	*white spotted*	fine grain, yellow mottling
w^t	*white-tinged*	light pink

behaving as a recessive lethal, but is dominant with respect to the phenotype. For example, a mutation that causes yellow coat color in mice was discovered in the early part of the 20th century. The yellow coat varies from the normal agouti coat phenotype, as shown in Figure 4–4. Crosses between the various combinations of the two strains yield unusual results:

Crosses				
(A) agouti	×	agouti	⟶	all agouti
(B) yellow	×	yellow	⟶	2/3 yellow: 1/3 agouti
(C) agouti	×	yellow	⟶	1/2 yellow: 1/2 agouti

These results are explained on the basis of a single pair of alleles. With regard to coat color, the mutant yellow allele A^Y is dominant to the wild-type agouti allele A, so heterozygous mice will have yellow coats. However, the yellow allele also behaves as a homozygous recessive lethal. Mice of the genotype $A^Y A^Y$ die before birth, so no homozygous yellow mice are ever recovered. The genetic basis for these three crosses is provided in Figure 4–4.

Molecular analysis of the A gene in both normal agouti and mutant yellow mice has provided insight into how a mutation can be both dominant for one phenotypic effect (hair color) and recessive for another (embryonic development). The A^Y allele is a classic example of a "gain of function" mutation. Animals homozygous for the wild-type A allele have yellow pigment deposited as a band on the otherwise black hair shaft, resulting in the agouti phenotype. (See Figure 4–4.) Heterozygotes deposit yellow pignment along the entire length of hair shafts as a result of the deletion of the regulatory region preceding the DNA coding region of the A^Y allele. Without any means to regulate gene expression, one copy of the A^Y allele is always turned on in heterozygotes, resulting in the gain of function leading to the dominant effect.

The homozygous lethal effect has also been explained as a result of molecular analysis of the mutant gene. The extensive deletion of genetic material characterizing the A^Y allele actually extends into the coding region of an adjacent gene (*Merc*), rendering it nonfunctional. It is this gene that is critical to embryonic development. It is this "loss of function" in A^Y/A^Y homozygotes that causes lethality. Heterozygotes exceed the threshold level required of the wild type *Merc* gene product.

Many genes are known to exhibit similar properties in other organisms. In *Drosophila*, *Curly* wing (*Cy*), *Plum* eye (*Pm*), *Dichaete* wing (*D*), *Stubble* bristle (*Sb*), and *Lyra* wing (*Ly*) behave as homozygous lethals, but are dominant with respect to the expression of the mutant phenotype when heterozygous.

Dominant Lethal Mutations

In some instances, one copy of the wild-type gene is insufficient for normal development, in which case even the heterozygote will not survive. In such a circumstance, the mutation is behaving as a **dominant lethal allele**. When this occurs, the presence of only one normal allele encoding the gene product may be insufficient to achieve a critical threshold level of an essential gene product. Or, the presence of the mutant gene product may somehow override the normal function of the wild-type product.

One of the most tragic examples of a dominant lethal gene is that responsible for **Huntington disease** in humans (previously referred to as Huntington's chorea). Caused by the dominant allele H, the onset of the disease is usually delayed well into adulthood, typically at about age 40. Affected individuals then undergo gradual nervous and motor degeneration and eventually succumb to the disorder a number of years later. This lethal disorder is particularly tragic, because an affected individual may have produced a family before discovering the condition.

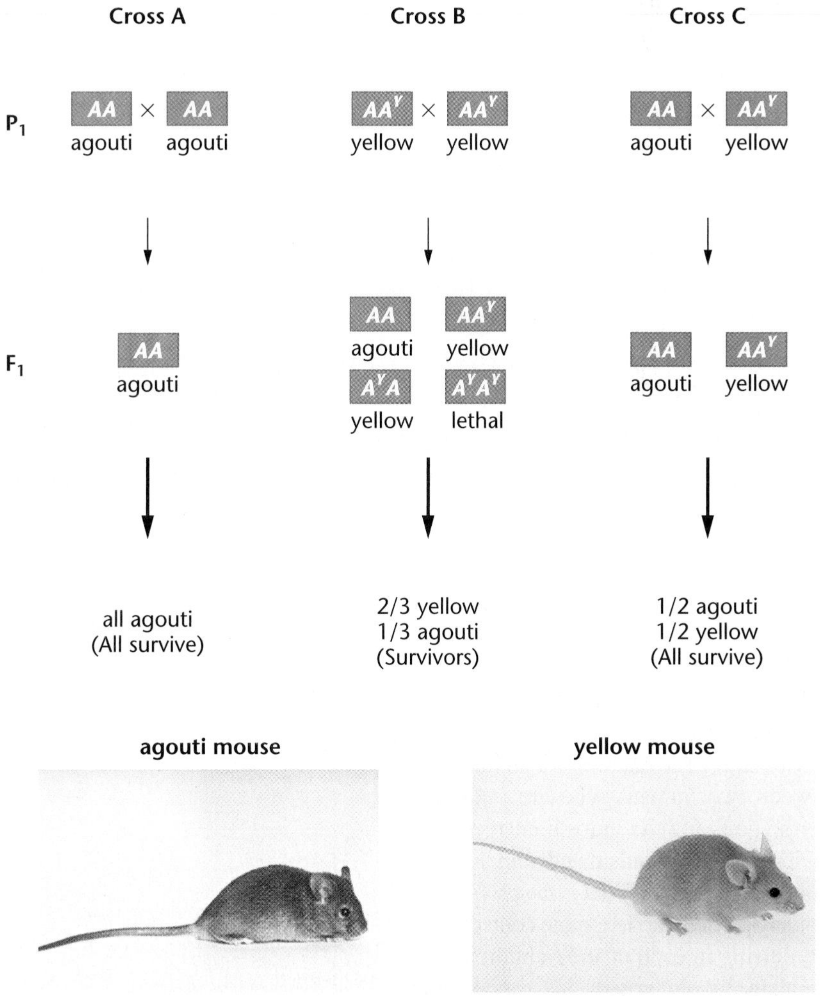

FIGURE 4–4 Inheritance patterns in three crosses involving the normal wild-type agouti allele (A) and the mutant yellow allele (A^Y) in the mouse. Note that the mutant allele behaves dominantly to the normal allele in controlling coat color, but it also behaves as a homozygous recessive lethal allele. The genotype A^YA^Y does not survive.

Affected individuals are heterozygous (*Hh*), having received the mutant allele from one of their parents. Thus, each of their offspring has a 50 percent probability of inheriting the lethal allele and developing the disease. The American folk singer and composer Woody Guthrie, father of Arlo Guthrie, died from this disease at age 39.

Dominant lethal alleles are very rare. For them to exist in a population, the affected individual must reproduce before dying, as can occur in Huntington disease. If all affected individuals die before reaching the reproductive age, the mutant allele will not be passed to future generations and will disappear from the population unless it arises again as a result of a new mutation.

HOW DO WE KNOW?

What experimental approach did early geneticists use to explain inheritance patterns that did not fit typical Mendelian ratios?

4.7 Combinations of Two Gene Pairs Involving Two Modes of Inheritance Modify the 9:3:3:1 Ratio

Each example discussed so far modifies Mendel's 3:1 monohybrid ratio. Therefore, combining any two of these modes of inheritance in a dihybrid cross will also modify the classic 9:3:3:1 ratio. Having established the foundation of the modes of inheritance of incomplete dominance, codominance, multiple alleles, and lethal alleles, we can now deal with the situation of two modes of inheritance occurring simultaneously. Mendel's principle of independent assortment applies to these situations, provided that the genes controlling each character are not linked on the same chromosome.

Consider, for example, a mating between two humans who are both heterozygous for the autosomal recessive allele causing albinism and who are both of blood type AB. What is the probability of any particular phenotypic combination occurring in each of their children? Albinism is inherited in the simple Mendelian fashion, and the blood types are determined by the series of three multiple alleles, I^A, I^B, and I^O. The solution to this problem is diagrammed in Figure 4–5.

Instead of the dihybrid cross yielding the four phenotypes in the classic 9:3:3:1 ratio, six phenotypes occur in a 3:6:3:1:2:1 ratio, establishing the expected probability for each phenotype. Figure 4–5 solves the problem using the forked-line method first described in Chapter 3. Recall that the forked-line method requires that the phenotypic ratios for each trait be computed individually (which can usually be done by inspection). All possible combinations can then be calculated. We could also solve the problem using the more conventional Punnett square.

This example is just one of many variants of modified ratios possible when different modes of inheritance are combined. We can deal in a similar way with any combination of two modes of inheritance. You will be asked to determine the phenotypes and their expected probabilities for many of these combinations in the problems at the end of the chapter. In each case, the final phenotypic ratio is a modification of the 9:3:3:1 dihybrid ratio.

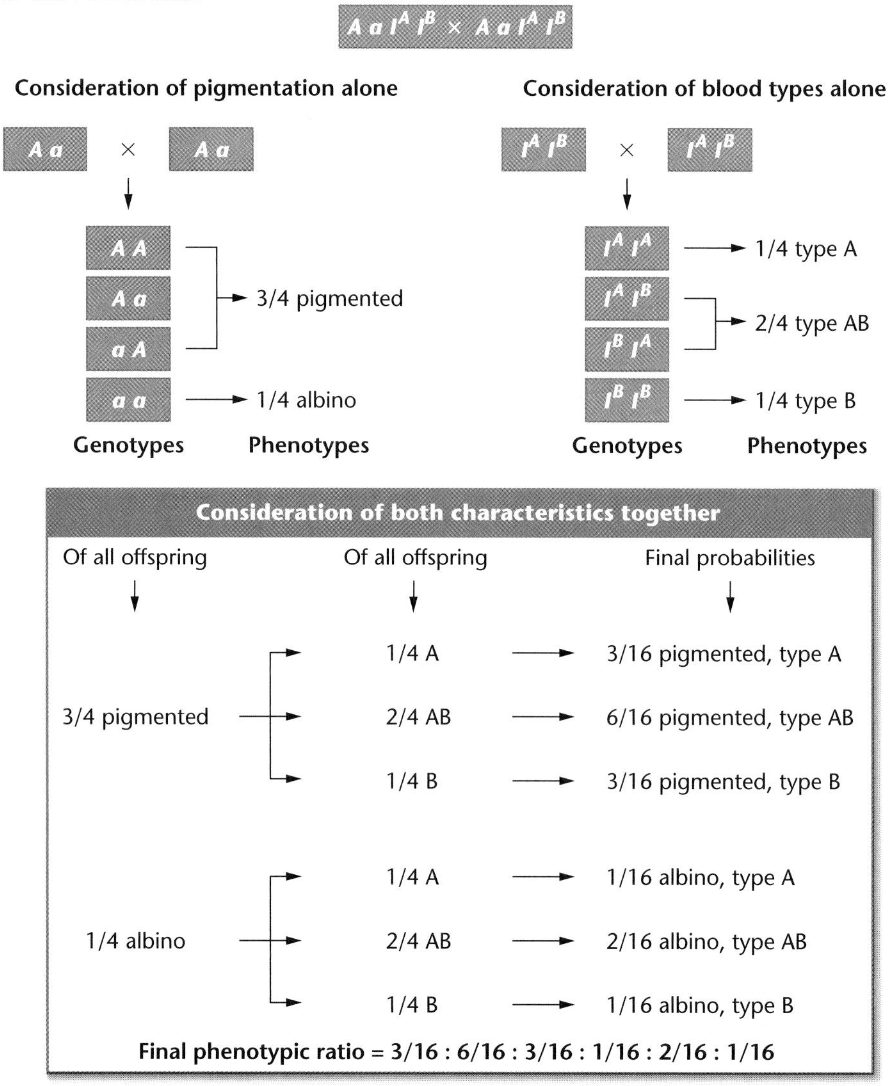

FIGURE 4–5 Calculation of the probabilities in a mating involving the ABO blood type and albinism in humans. The calculation is carried out by using the forked-line method.

4.8 Phenotypes Are Often Affected by More Than One Gene

Soon after Mendel's work was rediscovered, experimentation revealed that many traits characterized by discrete phenotypes are often affected by more than one gene. This discovery was significant because it revealed that genetic influence on the phenotype is much more complex than that which Mendel encountered in his crosses with the garden pea. Instead of single genes affecting the development of individual parts of a plant or animal body, it soon became clear that phenotypic characteristics, such as eye color, hair color, or fruit shape, are influenced by many genes and their resultant products.

The term **gene interaction** is often used to describe the idea that several genes influence a particular characteristic. This does not mean, however, that two or more genes or their products necessarily interact directly with one another to influence a particular phenotype. Rather, the cellular function of numerous gene products contributes to the development of a common phenotype. For example, the development of an organ such as the eye of an insect is exceedingly complex and leads to a structure with multiple phenotypic manifestations—most simply described as a specific size, shape, texture, and color. The development of the eye can best be envisioned as a complex cascade of developmental events. This process illustrates the developmental concept of **epigenesis**, whereby each ensuing step of development increases the complexity of the sensory organ and is under the control and influence of many genes. To clarify gene interaction, we will present numerous examples, several of which illustrate gene interaction at the biochemical level.

Epistasis

Some of the best examples of gene interaction are those showing the phenomenon of **epistasis**. Derived from the Greek word for *stoppage*, epistasis occurs when the effect of one gene or gene pair masks or modifies the effect of another gene or gene pair. The genes involved influence the same general phenotypic characteristic, sometimes in an antagonistic manner, as when masking occurs. In other cases, however, the genes involved exert their influence on one another in a complementary, or cooperative, fashion.

For example, the homozygous presence of a recessive allele may prevent or override the expression of other alleles at a second locus (or several other loci). In this case, the alleles at the initial locus are said to be **epistatic** to those at the second locus. The alleles at the second locus, which are masked, are described as being **hypostatic** to those at the first locus. As we will see, there are several variations on this theme. In another example of gene interaction, two gene pairs may **complement** one another such that at least one dominant allele at each locus is required to express a particular phenotype. If both are not present, an alternative phenotype results.

The Bombay phenotype discussed earlier is an example of the homozygous recessive condition at one locus masking the expression of a second locus. There, we established that the homozygous presence of the mutant form of the *FUT1* gene masks the expression of the I^A and I^B alleles. Only individuals containing at least one wild type *FUT1* allele can form the A or B antigen. As a result, individuals whose genotypes include the I^A or I^B allele and who lack a wild-type allele are of the type O phenotype, regardless of their potential to make either antigen. An example of the outcome of matings between individuals heterozygous at both loci is illustrated in Figure 4–6. If many such individuals have children, the phenotypic ratio of 3 A: 6 AB: 3 B: 4 O is expected in their offspring.

It is important to note two things when examining this cross and the predicted phenotypic ratio:

1. A key distinction exists in this cross compared to the modified dihybrid cross shown in Figure 4–5: *only one characteristic—blood type—is being followed.* In the modified dihybrid cross in Figure 4–5, blood type *and* skin pigmentation are followed as separate phenotypic characteristics.
2. Even though only a single character was followed, the phenotypic ratio was expressed in sixteenths. If we knew nothing about the H substance and the gene controlling it, we could still be confident that a second gene pair, other than that controlling the A and B antigens, was involved in the phenotypic expression. *When studying a single character, a ratio that is expressed in 16 parts (e.g., 3:6:3:4) suggests that two gene pairs are "interacting" during the expression of the phenotype under consideration.*

Unique Inheritance Patterns

The study of gene interaction has revealed a number of inheritance patterns that modify the Mendelian dihybrid F_2 ratio (9:3:3:1). In several of the subsequent examples, epistasis has the effect of combining one or more of the four phenotypic categories in various ways. The generation of these four groups is reviewed in Figure 4–7, along with several modified ratios.

As we discuss these and other examples (see Figure 4–8), we will make several assumptions and adopt certain conventions, as follows:

1. In each case, distinct phenotypic classes are produced, each clearly discernible from all others. Such traits illustrate discontinuous variation, where phenotypic categories are discrete and qualitatively different from one another.
2. The genes considered in each cross are not linked and therefore assort independently of one another during gamete formation. So that you may easily compare the results of different crosses, we designate alleles as *A*, *a* and *B*, *b* in each case.
3. When we assume that complete dominance exists between the alleles of any gene pair, such that *AA* and *Aa* or *BB* and *Bb* are equivalent in their genetic effects, we use the designations *A*– or *B*– for both combinations, where the dash (–) indicates that either allele may be present, without consequence to the phenotype.
4. All P_1 crosses involve homozygous individuals (e.g., *AABB* × *aabb*, *AAbb* × *aaBB*, or *aaBB* × *AAbb*). Therefore, each F_1 generation consists of only heterozygotes of genotype *AaBb*.

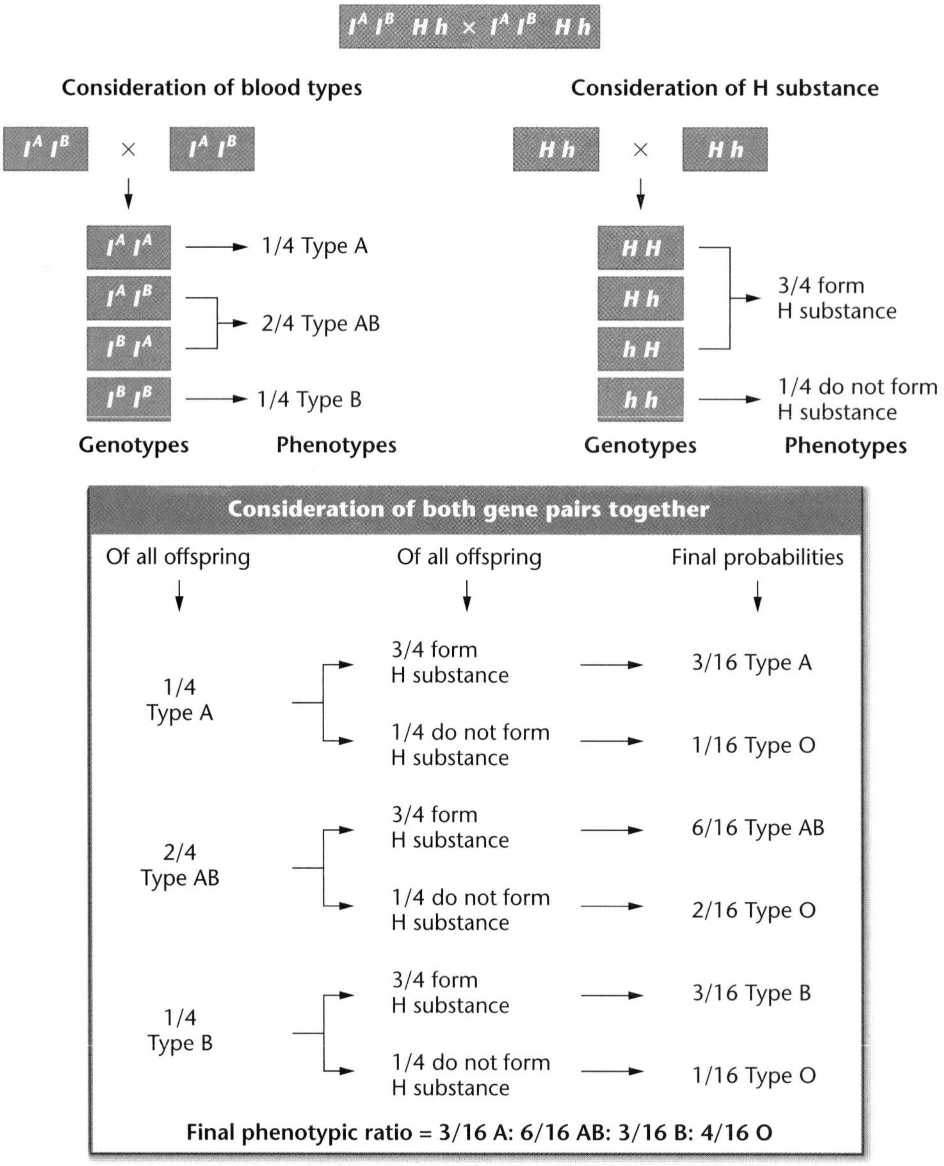

FIGURE 4–6 The outcome of a mating between individuals heterozygous at two genes that determine each individual's ABO blood type. Final phenotypes are calculated by considering both genes separately and then combining the results using the forked-line method.

5. In each example, the F_2 generation produced from these heterozygous parents is our main focus of analysis. When two genes are involved (Figure 4–7), the F_2 genotypes fall into four categories: 9/16 *A–B–*, 3/16 *A–bb*, 3/16 *aaB–*, and 1/16 *aabb*. Because of dominance, all genotypes in each category are equivalent in their effect on the phenotype.

The first case to be discussed is the inheritance of coat color in mice (Case 1 of Figure 4–8). As we saw in our discussion of lethal alleles, normal wild-type coat color is agouti, a grayish pattern formed by alternating bands of pigment on each hair. (See Figure 4–4.) Agouti is dominant to black (nonagouti) hair, which results from the homozygous expression of a recessive mutation, *a*. Thus, *A*– results in agouti, whereas *aa* yields black coat color. When homozygous, a recessive mutation, *b*, at a separate locus, eliminates pigmentation altogether, yielding albino mice (*bb*), regardless of the genotype at the *a* locus. The presence of at least one *B* allele allows pigmentation to occur in much the same way that the *FUT1* allele in humans allows the expression of the ABO blood types. In a cross between agouti (*AABB*) and albino (*aabb*), members of the F_1 are all *AaBb* and have agouti coat color. In the F_2 progeny of a cross between two F_1 double heterozygotes, the following genotypes and phenotypes are observed:

$$F_1: AaBb \times AaBb$$
$$\downarrow$$

F_2 Ratio	Genotype	Phenotype	Final Phenotypic Ratio
9/16	*A– B–*	agouti	9/16 agouti
3/16	*A– bb*	albino	4/16 albino
3/16	*aa B–*	black	3/16 black
1/16	*aa bb*	albino	

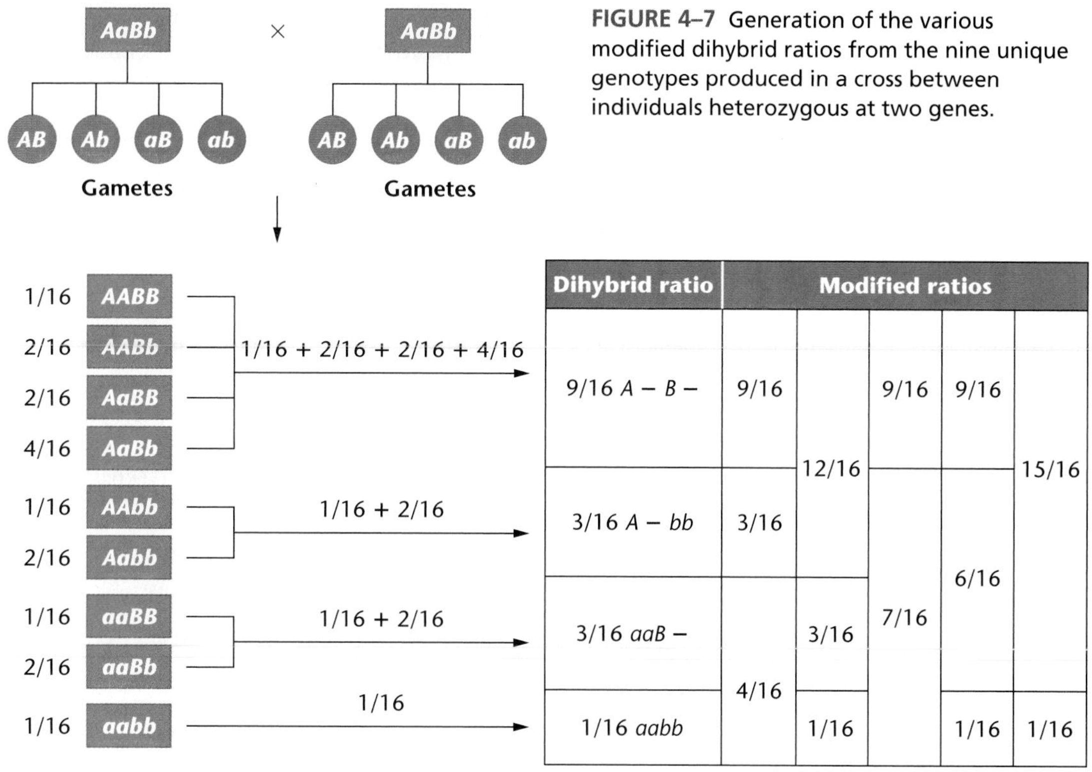

FIGURE 4–7 Generation of the various modified dihybrid ratios from the nine unique genotypes produced in a cross between individuals heterozygous at two genes.

We can envision gene interaction yielding the observed 9 : 3 : 4 F_2 ratio as a two-step process:

	Gene B		Gene A	
Precursor Molecule (colorless)	$\xrightarrow{B-}$	Black Pigment	$\xrightarrow{A-}$	Agouti Pattern

In the presence of a *B* allele, black pigment can be made from a colorless substance. In the presence of an *A* allele, the black pigment is deposited during the development of hair in a pattern producing the agouti phenotype. If the *aa* genotype occurs, all of the hair remains black. If the *bb* genotype occurs, no black pigment is produced, regardless of the presence of the *A* or *a* alleles, and the mouse is albino. Therefore, the homonzygous *bb* genotype masks or suppresses the expression of the *A* allele. As a result, this is referred to as *recessive epistasis.*

A variant of the above, called *dominant epistasis*, occurs when a dominant allele at one genetic locus masks the expression of the alleles of a second locus. For instance, Case 2 of Figure 4–8 deals with the inheritance of fruit color in summer squash. Here, the dominant allele *A* results in white fruit color regardless of the genotype at a second locus, *B*. In the absence of a dominant *A* allele (the *aa* genotype), *BB* or *Bb* results in yellow color, while *bb* results in green color. Therefore, if two white-colored double heterozygotes (*AaBb*) are crossed, an interesting phenotypic ratio occurs because of this type of epistasis:

F_1: *AaBb* × *AaBb*

↓

F_2 Ratio	Genotype	Phenotype	Final Phenotypic Ratio
9/16	A– B–	white	12/16 white
3/16	A– bb	white	
3/16	aa B–	yellow	3/16 yellow
1/16	aa bb	green	1/16 green

Of the offspring, 9/16 are *A–B–* and are thus white. The 3/16 bearing the genotypes *A–bb* are also white. Of the remaining squash, 3/16 are yellow (*aaB–*), while 1/16 are green (*aabb*). Thus, the modified phenotypic ratio of 12:3:1 occurs.

Our third example (Case 3 of Figure 4–8), first discovered by William Bateson and Reginald Punnett (of Punnett-square fame), is demonstrated in a cross between two strains of white-flowered sweet peas. Unexpectedly, the F_1 plants were all purple, and the F_2 occurred in a ratio of 9/16 purple to 7/16 white. The proposed explanation for these results suggests that the presence of at least one dominant allele of each of two gene pairs is essential in order for flowers to be purple. Thus, this cross represents a case of *complementary gene interaction.* All other genotype combinations yield white flowers because the homozygous condition of either recessive allele masks the expression of the dominant allele at the other locus.

The cross is shown as follows:

$$\begin{array}{ccc} P_1: \ AAbb & \times & aaBB \\ \text{white} & & \text{white} \\ & \downarrow & \\ & F_1: \text{All } AaBb \text{ (purple)} & \\ & \downarrow & \end{array}$$

F_2 Ratio	Genotype	Phenotype	Final Phenotypic Ratio
9/16	*A– B–*	purple	
3/16	*A– bb*	white	9/16 purple
3/16	*aa B–*	white	7/16 white
1/16	*aabb*	white	

We can now envision how two gene pairs might yield such results:

	Gene A		Gene B	
Precursor Substance (colorless)	$\xrightarrow[A-]{}$	Intermediate Product (colorless)	$\xrightarrow[B-]{}$	Final Product (purple)

At least one dominant allele from each pair of genes is necessary to ensure both biochemical conversions to the final product, yielding purple flowers. In the preceding cross, this will occur in 9/16 of the F_2 offspring. All other plants (7/16) have flowers that remain white.

These three examples illustrate in a simple way how the products of two genes interact to influence the development of a common phenotype. In other instances, more than two genes and their products are involved in controlling phenotypic expression.

Novel Phenotypes

Other cases of gene interaction yield novel, or new, phenotypes in the F_2 generation, in addition to producing modified dihybrid ratios. Case 4 in Figure 4–8 depicts the inheritance of fruit shape in the summer squash *Cucurbita pepo*. When plants with disc-shaped fruit (*AABB*) are crossed with plants with long fruit (*aabb*), the F_1 generation all have disc fruit. However, in the F_2 progeny, fruit with a novel shape—sphere—appear, as well as fruit exhibiting the parental phenotypes. These phenotypes are shown in Figure 4–9.

The F_2 generation, with a modified 9:6:1 ratio, is generated as follows:

$$\begin{array}{ccc} F_1: \ AaBb & \times & AaBb \\ \text{disc} & \downarrow & \text{disc} \end{array}$$

F_2 Ratio	Genotype	Phenotype	Final Phenotypic Ratio
9/16	*A– B–*	disc	9/16 disc
3/16	*A– bb*	sphere	6/16 sphere
3/16	*aa B–*	sphere	1/16 long
1/16	*aa bb*	long	

In this example of gene interaction, both gene pairs influence fruit shape equally. A dominant allele at either locus

<table>
<tr><th rowspan="2">Case</th><th rowspan="2">Organism</th><th rowspan="2">Character</th><th colspan="4">F_2 Phenotypes</th><th rowspan="2">Modified ratio</th></tr>
<tr><th>9/16</th><th>3/16</th><th>3/16</th><th>1/16</th></tr>
<tr><td>1</td><td>Mouse</td><td>Coat color</td><td>agouti</td><td>albino</td><td>black</td><td>albino</td><td>9:3:4</td></tr>
<tr><td>2</td><td>Squash</td><td>Color</td><td colspan="2">white</td><td>yellow</td><td>green</td><td>12:3:1</td></tr>
<tr><td>3</td><td>Pea</td><td>Flower color</td><td>purple</td><td colspan="3">white</td><td>9:7</td></tr>
<tr><td>4</td><td>Squash</td><td>Fruit shape</td><td>disc</td><td colspan="2">sphere</td><td>long</td><td>9:6:1</td></tr>
<tr><td>5</td><td>Chicken</td><td>Color</td><td colspan="2">white</td><td>colored</td><td>white</td><td>13:3</td></tr>
<tr><td>6</td><td>Mouse</td><td>Color</td><td>white-spotted</td><td>white</td><td>colored</td><td>white-spotted</td><td>10:3:3</td></tr>
<tr><td>7</td><td>Shepherd's purse</td><td>Seed capsule</td><td colspan="3">triangular</td><td>ovoid</td><td>15:1</td></tr>
<tr><td>8</td><td>Flour beetle</td><td>Color</td><td>6/16 sooty : 3/16 red</td><td>black</td><td>jet</td><td>black</td><td>6:3:3:4</td></tr>
</table>

FIGURE 4–8 The basis of modified dihybrid F_2 phenotypic ratios, resulting from crosses between doubly heterozygous F_1 individuals. The four groupings of the F_2 genotypes shown in Figure 4–7 and across the top of this figure are combined in various ways to produce these ratios.

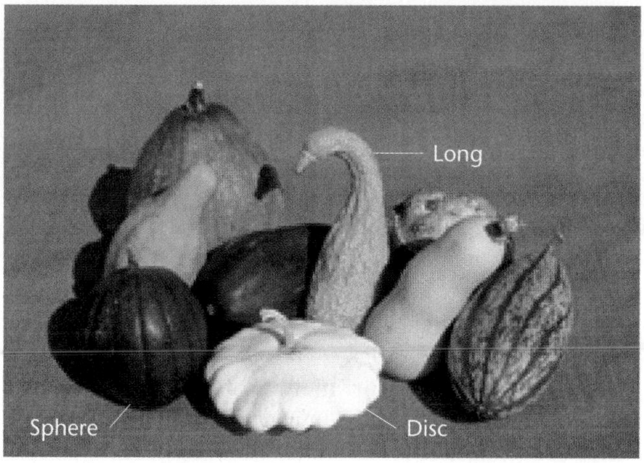

FIGURE 4–9 Summer squash exhibiting various fruit-shape phenotypes, where disc (white), long (orange gooseneck), and sphere (bottom left) are apparent.

ensures a sphere-shaped fruit. In the absence of dominant alleles, the fruit is long. However, if both dominant alleles (*A* and *B*) are present, the fruit is flattened into a disc shape.

Now solve this

Problem 4.17 on page 94 involves a single characteristic, flower color, that can take on one of three variations. You are asked to determine how many genes are involved in the inheritance of flower color, and what genotypes are responsible for what phenotypes.

Hint: The most important information is the data provided. You must analyze the raw data and convert the numbers to a meaningful ratio. This will guide you in determining how many gene pairs are involved. Then you can categorize the genotypic ratio in a way to match the phenotypic ratio.

Another interesting example of an unexpected phenotype arising in the F_2 generation is the inheritance of eye color in *Drosophila melanogaster*. The wild-type eye color is brick red. When two autosomal recessive mutants, *brown* and *scarlet*, are crossed, the F_1 generation consists of flies with wild-type eye color. In the F_2 generation, wild, scarlet, brown, and white-eyed flies are found in a 9:3:3:1 ratio. While this ratio is numerically the same as Mendel's dihybrid ratio, the *Drosophila* cross involves only one character: eye color. This is an important distinction to make as modified dihybrid ratios resulting from gene interaction are studied.

The *Drosophila* cross is an excellent example of gene interaction because the biochemical basis of eye color in this organism has been determined (Figure 4–10). *Drosophila*, as a typical arthropod, has compound eyes made up of hundreds of individual visual units called ommatidia. The wild-type eye color is due to the deposition and mixing of two separate pigment groups in each ommatidium—the bright-red pigments **drosopterins** and the brown pigment **xanthommatin**. Each pigment is produced by a separate biosynthetic pathway. Each step of each pathway is catalyzed by a separate enzyme and is thus under the control of a separate gene. As shown in Figure 4–10, the *brown* mutation, when homozygous, interrupts the pathway leading to the synthesis of the bright-red pigments. Because only xanthommatin pigments are present, the eye is brown. The *scarlet* mutation, affecting a gene located on a separate autosome, interrupts the pathway leading to the synthesis of the brown xanthommatins and renders the eye color bright red in homozygous mutant flies. Each mutation apparently causes the production of a nonfunctional enzyme. Flies that are double mutants and thus homozygous for both *brown* and *scarlet* lack both functional enzymes and can make neither of the pigments; they represent the novel white-eyed flies appearing in 1/16 of the F_2 generation. Note that the absence of pigment in these flies is not due to the X-linked *white* mutation, where pigments can be synthesized, but the necessary precursors cannot be transported into the cells making up the ommatidia.

Other Modified Dihybrid Ratios

The remaining cases (5–8) in Figure 4–8 illustrate additional modifications of the dihybrid ratio and provide still other examples of gene interactions. As you will note, ratios of 13:3, 10:3:3; 15:1, and 6:3:3:4 are illustrated. These cases, like the four preceding them, have two things in common. First, in arriving at a suitable explanation of the inheritance pattern of each one, we have not violated the principles of segregation and independent assortment. Therefore, the added complexity of inheritance in these examples does not detract from the validity of Mendel's conclusions. Second, the F_2 phenotypic ratio in each example has been expressed in sixteenths. When similar observations are made in crosses where the inheritance pattern is unknown, they suggest to geneticists that two gene pairs are controlling the observed phenotypes. You should make the same inference in your analysis of genetics problems. Other insights into solving genetics problems are provided in the "Insights and Solutions" section at the conclusion of this chapter.

HOW DO WE KNOW?

How did geneticists establish that inheritance of some phenotypic characteristics involves the interactions of two or more gene pairs? How did they determine how many gene pairs were involved?

4.9 Expression of a Single Gene May Have Multiple Effects

While the previous sections have focused on the effects of two or more genes on a single characteristic, the converse situation, where expression of a single gene has multiple phenotypic effects, is quite common. This phenomenon, apparent when careful examination of phenotypes is done, is referred to as **pleiotropy**. There are many excellent examples that can be drawn from human disorders, and we will review two such cases to illustrate this point.

The first disorder is **Marfan syndrome**, a human malady resulting from an autosomal dominant mutation in the gene encoding the connective tissue protein fibrillin. Because this protein is widespread in many tissues in the body, one would expect multiple effects of such a defect. In fact, fibrillin is important to the structural integrity of the lens of the eye, to the lining of vessels such as the aorta, and to bones, among others. As a result, the phenotype associated with Marfan syndrome includes lens dislocation, increased risk of aortic aneurism, and lengthened long bones in limbs. This disorder is of historical interest in that speculation abounds that Abraham Lincoln was afflicted.

A second example involves another human autosomal dominant disorder, **porphyria variegata**. Afflicted individuals cannot adequately metabolize the porphyrin component of hemoglobin when this respiratory pigment is broken down as red blood cells are replaced. The accumulation of excess porphyrins is immediately evident in the urine, which takes on a deep red color. However, this phenotypic characteristic is merely diagnostic. The severe aspects of the phenotype are due to the toxic nature of the buildup of porphyrins in the body, and particularly in the brain. Complete phenotypic characterization includes abdominal pain, muscular weakness, fever, a racing pulse, insomnia, headaches, vision problems (that can lead to blindness), delerium, and ultimately convulsions. As you can see, deciding which phenotypic trait best characterizes the disorder is impossible.

Like Marfan syndrome, porphyria variegata is also of historical significance. King George III, King of England during the American revolution, is believed to have suffered from episodes involving all of the above symptoms. He ultimately became blind and senile prior to his death.

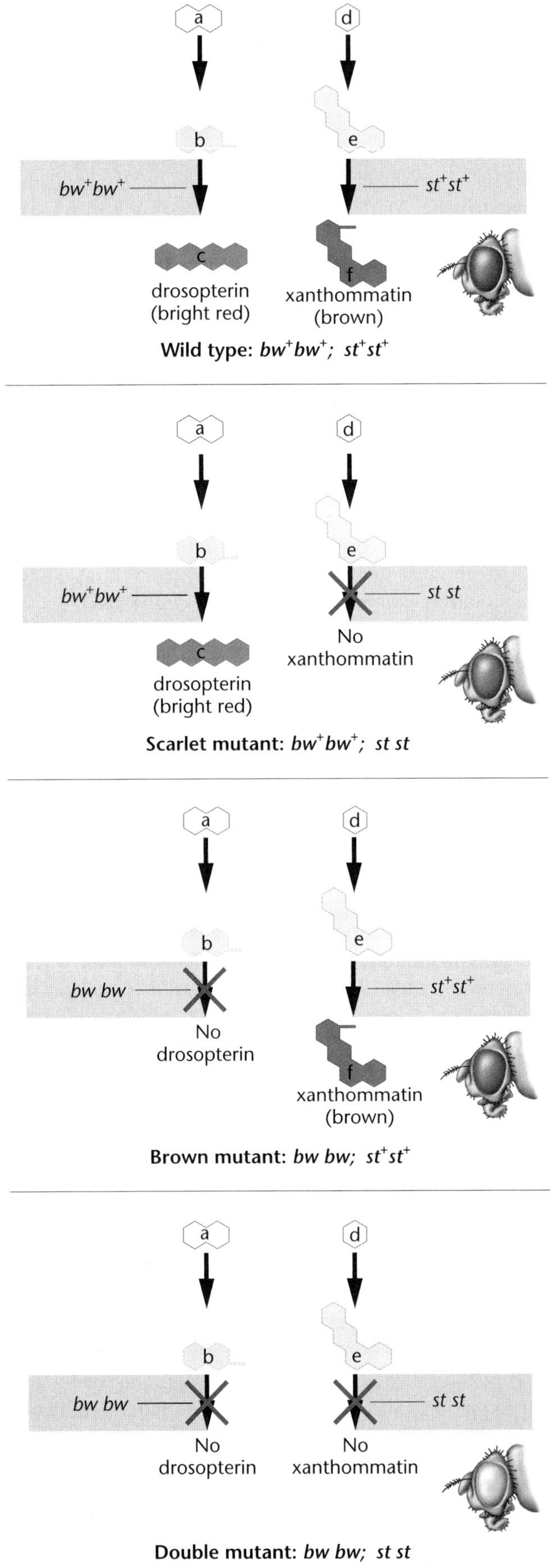

FIGURE 4–10 A theoretical explanation of the biochemical basis of the four eye-color phenotypes produced in a cross between *Drosophila* with brown eyes and scarlet eyes. In the presence of at least one wild-type bw^+ allele, an enzyme is produced that converts substance b to c, and the pigment drosopterin is synthesized. In the presence of at least one wild-type st^+ allele, substance e is converted to f, and the pigment xanthommatin is synthesized. The homozygous presence of the recessive *bw* and *st* mutant alleles blocks the synthesis of these respective pigment molecules. Either one, both, or neither of these pathways can be blocked, depending on the genotype.

There are many other examples that we could cite here to illustrate pleiotropy. Suffice it to say that if one looks carefully, most mutations display more than a single manifestation when expressed.

4.10 X-Linkage Describes Genes on the X Chromosome

In many animal and some plant species, one of the sexes contains a pair of unlike chromosomes that are involved in sex determination. In many cases, these are designated as the X and Y. For example, in both *Drosophila* and humans, males contain an X and a Y chromosome, whereas females contain two X chromosomes. While the Y chromosome must contain a region of pairing homology with the X chromosome if the two are to synapse and segregate during meiosis, a major portion of the Y chromosome in humans as well as other species is considered to be relatively inert genetically. While we now recognize a number of male-specific genes on the human Y chromosome, it lacks copies of most genes present on the X chromosome. As a result, genes present on the X chromosome exhibit unique patterns of inheritance in comparison with autosomal genes. The term **X-linkage** is used to describe such situations.

Below, we will focus on inheritance patterns resulting from genes present on the X, but absent from the Y chromosome. This situation results in a modification of Mendelian ratios, the central theme of this chapter.

X-Linkage in *Drosophila*

One of the first cases of X-linkage was documented in 1910 by Thomas H. Morgan during his studies of the *white* eye mutation in *Drosophila* (Figure 4–11). The normal wild-type red eye color is dominant to white eye color.

Morgan's work established that the inheritance pattern of the white-eye trait was clearly related to the sex of the parent carrying the mutant allele. Unlike the outcome of the typical Mendelian monohybrid cross where F_1 and F_2 data were very similar regardless of which P_1 parent exhibited the recessive mutant trait, reciprocal crosses between white-eyed and red-eyed flies did not yield identical results. Morgan's analysis led to the conclusion that the *white* locus is present on the X rather than on one of the autosomes. As such, both the gene and the trait are said to be X-linked.

Results of reciprocal crosses between white-eyed and red-eyed flies are shown in Figure 4–11. The obvious differences in phenotypic ratios in both the F_1 and F_2 generations are dependent on whether or not the P_1 white-eyed parent was male or female.

Morgan was able to correlate these observations with the difference found in the sex chromosome composition between male and female *Drosophila*. He hypothesized that in males with white eyes, the recessive allele for white eye is found on the X chromosome, but its corresponding locus is absent from the Y chromosome. Females thus have two available gene loci, one on each X chromosome, while males have only one available locus on their single X chromosome.

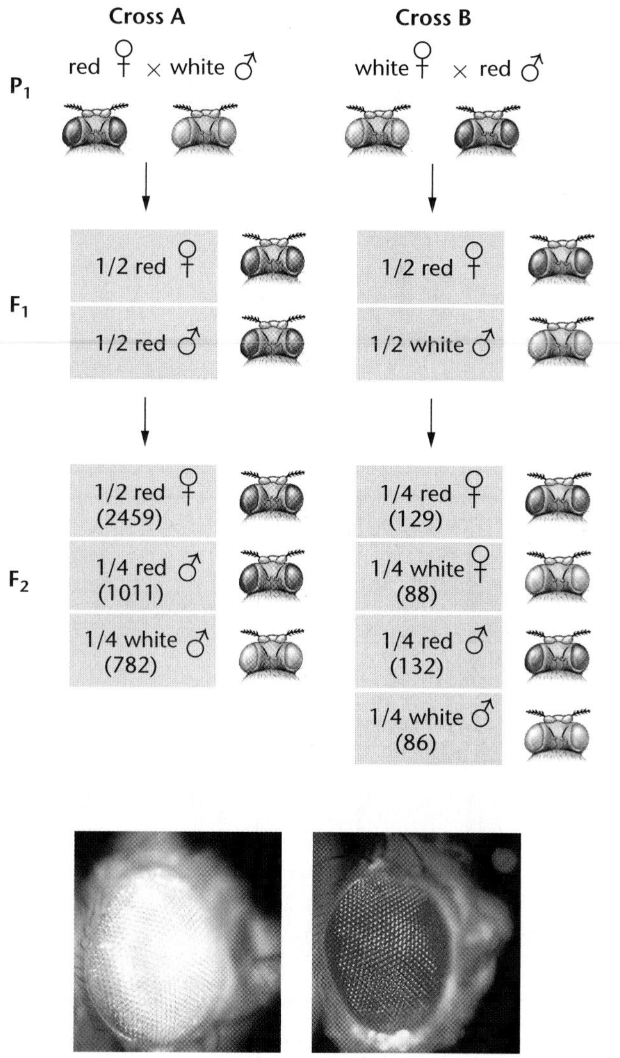

FIGURE 4–11 The F_1 and F_2 results of T. H. Morgan's reciprocal crosses involving the X-linked *white* mutation in *Drosophila melanogaster*. The actual data are shown in parentheses. The photographs show white eye and the brick-red wild-type eye color.

Morgan's interpretation of X-linked inheritance, shown in Figure 4–12, provides a suitable theoretical explanation for his results. Since the Y chromosome lacks homology with most genes on the X chromosome, whatever alleles are present on the X chromosome of the males will be directly expressed in the phenotype. Because males cannot be either homozygous or heterozygous for X-linked genes, this condition is referred to as **hemizygous**. In such cases, no alternative alleles are present, and the concept of dominance and recessiveness is irrelevant.

One result of X-linkage is the **crisscross pattern of inheritance**, whereby phenotypic traits controlled by recessive X-linked genes are passed from homozygous mothers to all sons. This pattern occurs because females exhibiting a recessive trait must contain the mutant allele on both X chromosomes. Because male offspring receive one of their mother's two X chromosomes and are hemizygous for all alleles present on that X, all sons will express the same

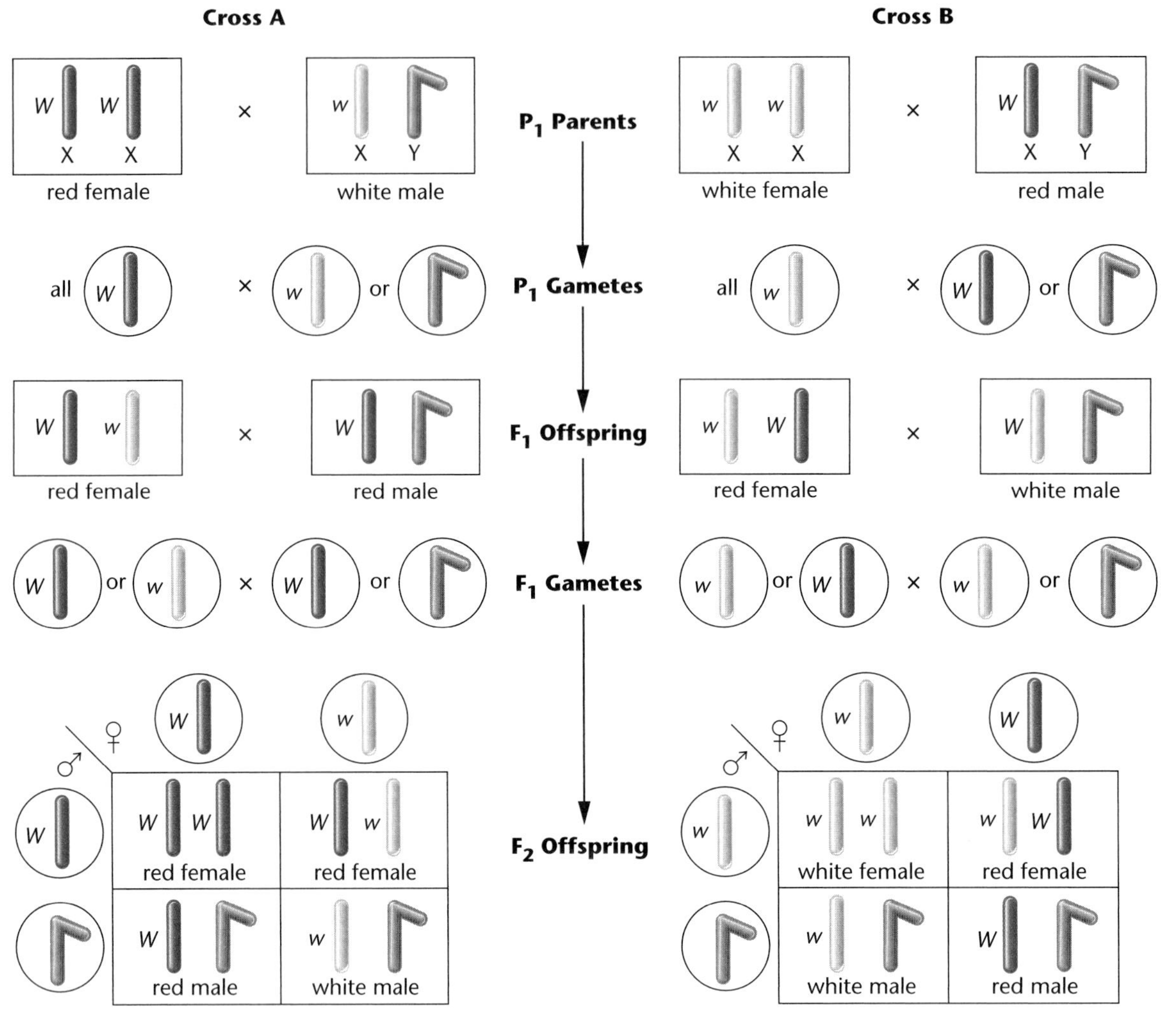

FIGURE 4–12 The chromosomal explanation of the results of the X-linked crosses shown in Figure 4–11.

recessive X-linked traits as their mother. This pattern of inheritance is apparent in the pedigree in Figure 4–13.

In addition to documenting the phenomenon of X-linkage, Morgan's work has taken on great historical significance. By 1910, the correlation between Mendel's work and the behavior of chromosomes during meiosis had provided the basis for the **chromosome theory of inheritance**, as postulated by Sutton and Boveri. (See Chapter 3.) Morgan's work, and subsequently that of his student Calvin Bridges, providing direct evidence that genes are transmitted on specific chromosomes, is considered the first solid experimental evidence in support of this theory. In the ensuing two decades, the outcome of research inspired by these findings provided indisputable evidence in support of this theory.

HOW DO WE KNOW?

How do we know that X-linkage exists and that specific genes are located on the sex-determining chromosomes?

X-Linkage in Humans

In humans, many genes and the traits controlled by them are recognized as being linked to the X chromosome. These X-linked traits can be easily identified in pedigrees, characterized by a criss-cross pattern of inheritance. A pedigree for one form of human color blindness is shown in Figure 4–13. The mother in generation I passes the trait on to all her sons, but to none of her daughters. If the offspring in generation II have children by normal individuals, the color-blind sons will produce all normal male and female offspring (III-1, 2, and 3); the normal-vision daughters will produce normal-vision female offspring (III-4, 6, and 7), as well as color-blind (III-8) and normal-vision (III-5) male offspring.

Many X-linked human genes have now been identified, as shown in Table 4.3. For example, the genes controlling two forms of hemophilia and two forms of muscular dystrophy are located on the X chromosome. In addition, numerous genes whose expression yields enzymes are X-linked. Glucose-6-phosphate dehydrogenase and hypoxanthine-guanine-phosphoribosyl transferase are two examples. In the latter case, the severe Lesch–Nyhan syn-

FIGURE 4–13 (a) A human pedigree of the X-linked color-blindness trait. (b) The most probable genotypes of each individual in the pedigree. The photograph is of an Ishihara color-blindness chart. Red-green color-blind individuals see a 3 rather than the 8 visualized by those with normal color vision.

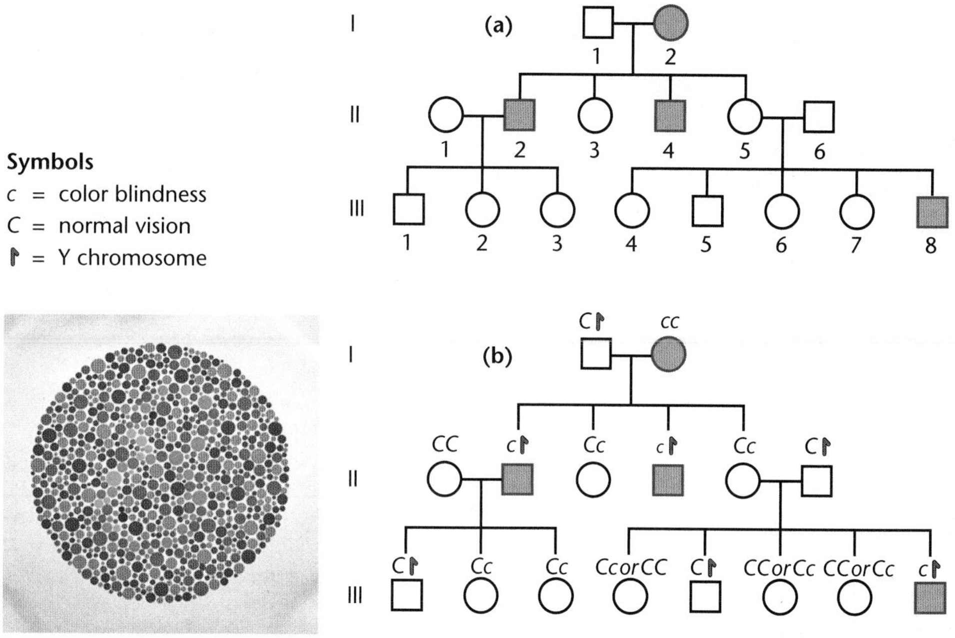

drome (discussed later in this chapter) results from the mutant form of the X-linked gene product.

Now Solve This

Problem 4.32 on page 96 asks you to determine if each of three pedigrees is consistent with X-linkage.

Hint: In X-linkage, because of hemizygosity, the genotype of males is immediately evident. Therefore, the key to solving this type of problem is to consider the possible genotypes of females that do not express the trait.

Because of the way in which X-linked genes are transmitted, unusual circumstances may be associated with recessive X-linked disorders in comparison to recessive autosomal disorders. For example, if an X-linked disorder debilitates or is lethal to the affected individual prior to reproductive maturation, the disorder occurs exclusively in males. This is because the only sources of the lethal allele in the population are heterozygous females who are "carriers" and do not express the disorder. They pass the allele to half of their sons, who develop the disorder because they are hemizygous but who rarely, if ever, reproduce. Heterozygous females also pass the allele to half of their daughters, who, like their mothers, become carriers but do not develop the disorder. Examples of such an X-linked disorder in humans include the Duchenne form of muscular dystrophy (DMD) and Lesch–Nyhan disease. (See the screened section on the next page.) DMD has an onset prior to age 6 and is often lethal around age 20. Affected males are unable to reproduce.

TABLE 4.3 HUMAN X-LINKED TRAITS

Condition	Characteristics
Color blindness, deutan type	Insensitivity to green light
Color blindness, protan type	Insensitivity to red light
Fabry's disease	Deficiency of galactosidase A; heart and kidney defects, early death
G-6-PD deficiency	Deficiency of glucose-6-phosphate dehydrogenase; severe anemic reaction following intake of primaquines in drugs and certain foods, including fava beans
Hemophilia A	Classical form of clotting deficiency; deficiency of clotting factor VIII
Hemophilia B	Christmas disease; deficiency of clotting factor IX
Hunter syndrome	Mucopolysaccharide storage disease resulting from iduronate sulfatase enzyme deficiency; short stature, clawlike fingers, coarse facial features, slow mental deterioration, and deafness
Ichthyosis	Deficiency of steroid sulfatase enzyme; scaly dry skin, particularly on extremities
Lesch–Nyhan syndrome	Deficiency of hypoxanthine-guanine phosphoribosyltransferase enzyme (HPRT) leading to motor and mental retardation, self-mutilation, and early death
Muscular dystrophy	Progressive, life-shortening disorder characterized by muscle degeneration and weakness; (Duchenne type) sometimes associated with mental retardation; deficiency of the protein dystrophin

Lesch–Nyhan Syndrome: The Molecular Basis of a Rare X-linked Recessive Disorder

Lesch–Nyhan syndrome (LNS) is a devastating disease that is first apparent in infants at age 3 to 6 months, when orange particles (sometimes referred to as orange sand) appear in the urine and discolor the affected infant's diaper. These urinary stones consist of urate crystals and they are a harbinger of many future difficulties that ultimately lead to premature death. LNS occurs only in males and is the result of the complete or nearly complete loss of activity of a critical enzyme, **hypoxanthine-guanine phoshoribosyltransferase (HPRT)**. This enzyme imparts the ability to metabolically recycle purines, one of the two major types of nitrogenous bases that make up nucleotides in DNA. While the purines adenine and guanine can be synthesized from basic chemical components, mammals have evolved the ability to extract them from DNA that is being degraded, recovering them in the form of the purine hypoxanthine. Under the direction of HPRT, hypoxanthine can be converted back to adenine and guanine-containing nucleotides. When this mechanism fails as a result of mutation, the excess hypoxanthine is converted to uric acid, which accumulates well beyond the body's ability to excrete it.

This so-called metabolic or biochemical disorder has numerous effects, the most severe being mental retardation, seizures, and aggressive, uncontrolled spastic movements (resembling cerebral palsy) that include self-mutilation of the fingers and lips. Patients require 24-hour care throughout their lives and almost always die prior to age 30, usually as a result of kidney failure. In February 2004, the oldest living LNS patient, Philip Barker, celebrated his thirty-third birthday in Bayview, New York.

The gene involved in LNS is located on the long arm of the X chromosome and consists of 44,000 base pairs (44 Kb). However, the HPRT gene product is only 218 amino acids long, thus requiring only 654 base pairs to encode it. Analysis of the cloned version of the gene reveals it to contain 9 exons and 8 introns. Mice have a nearly identical gene that is 95 percent homologous to its human counterpart.

In normal individuals the enzyme is ubiquitous in tissues throughout the body, but is present in greatest concentration in the basal ganglia of brain cells. No doubt this somehow relates to the behavioral phenotype characterizing LNS patients who lack enzyme activity in the brain and elsewhere in their bodies. In spite of extensive research efforts, there is no known cure. Because the responsible gene is recessive and X-linked and since affected males never reproduce, females, while they can be carriers of the mutant gene, never become homozygous and never develop LNS.

4.11 In Sex-Limited and Sex-Influenced Inheritance, an Individual's Sex Influences the Phenotype

In some cases, the expression of a specific phenotype is absolutely limited to one sex; in others, the sex of an individual influences the expression of a phenotype that is not limited to one sex or the other. This distinction differentiates **sex-limited inheritance** from **sex-influenced inheritance**.

In both types of inheritance, autosomal genes are responsible for the existence of contrasting phenotypes, but the expression of these genes is dependent on the hormone constitution of the individual. Thus, the heterozygous genotype may exhibit one phenotype in males and the contrasting one in females. In domestic fowl, for example, tail and neck plumage is often distinctly different in males and females (Figure 4–14), demonstrating sex-limited inheritance. Cock feathering is longer, more curved, and pointed, whereas hen feathering is shorter and more rounded. Inheritance of these feather phenotypes is controlled by a single pair of autosomal alleles whose expression is modified by the individual's sex hormones. As shown below, hen feathering is due to a dominant allele, *H*:

Genotype	Phenotype ♀	♂
HH	Hen feathered	Hen feathered
Hh	Hen feathered	Hen feathered
hh	Hen feathered	Cock feathered

FIGURE 4–14 Hen feathering (left) and cock feathering (right) in domestic fowl. The feathers in the hen are shorter and less curved.

However, regardless of the homozygous presence of the recessive *h* allele, all females remain hen-feathered. Only in males does the *hh* genotype result in cock feathering.

In certain breeds of fowl, the hen-feathering or cock-feathering allele has become fixed in the population. In the Leghorn breed, all individuals are of the *hh* genotype; as a result, male plumage differs from female plumage. Seabright bantams are all *HH*, showing no sexual distinction in feathering.

Now solve this

Problem 4.33 on page 96 involves the inheritance of colored spots in cattle, and you are asked to analyze the F_1 and F_2 ratios to determine the mode of inheritance.

Hint: Note particularly that the data are differentiated into male and female offspring and that the ratios in the F_2 vary according to sex (i.e., 3/8 of the males are mahogany while only 1/8 of the females are mahogany, etc.). This should immediately alert you to consider the possible influences that sex differences impart on the outcome of crosses. In this case, you should consider whether X-linkage, sex-limited, or sex-influenced inheritance might be involved.

Still another example of sex-limited inheritance involves the autosomal genes responsible for milk yield in dairy cattle. Regardless of the overall genotype that influences the quantity of milk production, those genes are obviously expressed only in females.

Cases of sex-influenced inheritance include pattern baldness in humans, horn formation in certain breeds of sheep (e.g., Dorsett Horn sheep), and certain coat patterns in cattle. In such cases, autosomal genes are responsible for the contrasting phenotypes displayed by both males and females, but the expression of these genes is dependent on the hormone constitution of the individual. Thus, the heterozygous genotype exhibits one phenotype in one sex and the contrasting one in the other. For example, pattern baldness in humans, where the hair is very thin or absent on the top of the head (Figure 4–15), is inherited in the following way:

Genotype	Phenotype ♀	Phenotype ♂
BB	Bald	Bald
Bb	Not bald	Bald
bb	Not bald	Not bald

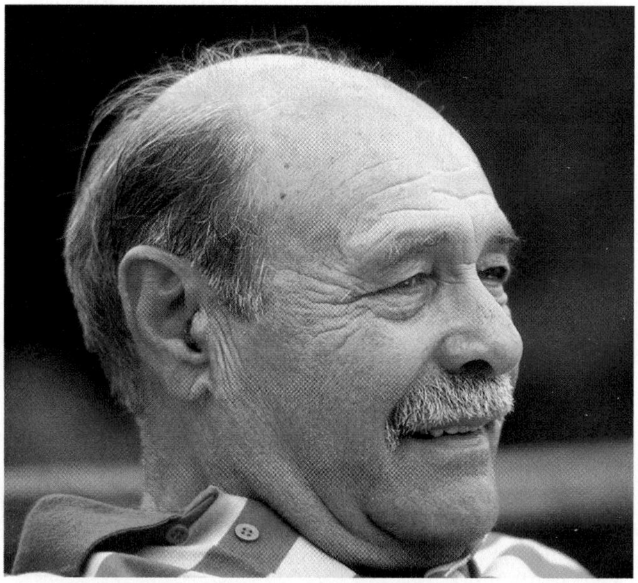

FIGURE 4–15 Pattern baldness, a sex-influenced autosomal trait in humans.

Even though females can display pattern baldness, this phenotype is much more prevalent in males. When females do inherit the *BB* genotype, the phenotype is much less pronounced than in males and is expressed later in life.

HOW DO WE KNOW?

How did we learn to distinguish between X-linkage and sex-limited inheritance?

4.12 Phenotypic Expression Is Not Always a Direct Reflection of the Genotype

We conclude this chapter with the consideration of *phenotypic expression*. In Chapters 2 and 3, we assumed that the genotype of an organism is always directly expressed in its phenotype. For example, peas homozygous for the recessive *d* allele (*dd*) will always be dwarf. We have discussed gene expression as though the genes operate in a closed "black-box" system in which the presence or absence of functional products directly determines the collective phenotype of an individual. The situation is actually much more complex. Most gene products function within the internal milieu of the cell, and cells interact with one another in various ways. Further, the organism must survive under diverse environmental influences. Thus, gene expression and the resultant phenotype are often modified through the interaction between an individual's particular genotype and the internal and external environment.

The degree of environmental influence can vary from inconsequential to subtle to very strong. Subtle interactions are the most difficult to detect and document, and they have led to "nature versus nurture" questions in which scientists debate the relative importance of genes versus environment—usually inconclusively. In this final section of this chapter, we will deal with some of the variables known to modify gene expression.

Penetrance and Expressivity

Some mutant genotypes are always expressed as a distinct phenotype, whereas others produce a proportion of individuals whose phenotypes cannot be distinguished from normal (wild type). The degree of expression of a particular trait may be studied quantitatively by determining the penetrance and expressivity of the genotype under investigation. The percentage of individuals that show at least some degree of expression of a mutant genotype defines the **penetrance** of the mutation. For example, the phenotypic expression of many mutant alleles in *Drosophila* is indistinguishable from wild type. If 15 percent of mutant flies show the wild-type appearance, the mutant gene is said to have a penetrance of 85 percent.

By contrast, **expressivity** reflects the *range of expression* of the mutant genotype. Flies homozygous for the recessive mutant gene *eyeless* yield phenotypes that range from the presence of normal eyes to a partial reduction in size to the complete absence of one or both eyes (Figure 4–16). Although

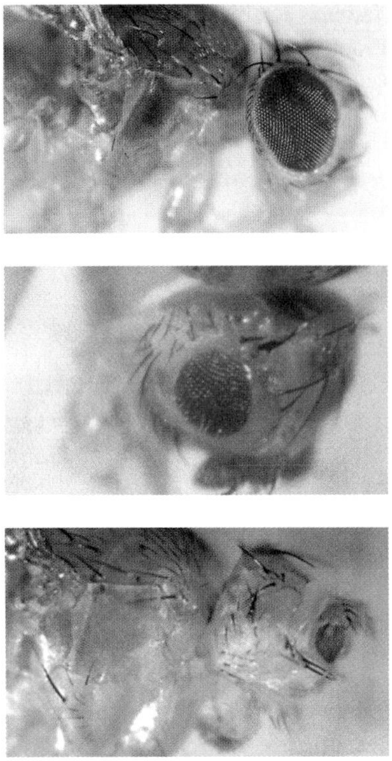

FIGURE 4–16 Variable expressivity as illustrated by the expression of the *eyeless* mutation in *Drosophila*. Gradations in phenotype range from wild type to partial reduction to eyeless.

the average reduction of eye size is one-fourth to one-half, expressivity ranges from complete loss of both eyes to completely normal eyes.

Examples such as the expression of the *eyeless* gene have provided the basis for experiments to determine the causes of phenotypic variation. If the laboratory environment is held constant and extensive variation is still observed, other genes may be influencing or modifying the *eyeless* phenotype. On the other hand, if the genetic background is not the cause of the phenotypic variation, environmental factors such as temperature, humidity, and nutrition may be involved. In the case of the *eyeless* phenotype, experiments have shown that both genetic background and environmental factors influence its expression.

Genetic Background: Suppression and Position Effects

Though it is difficult to assess the specific effect of the **genetic background** and the expression of a gene responsible for determining a potential phenotype, two effects of genetic background have been well characterized.

First, the expression of other genes throughout the genome may have an effect on the phenotype produced by the gene in question. The phenomenon of **genetic suppression** is an example. Mutant genes such as *suppressor of sable* (*su-s*), *suppressor of forked* (*su-f*), and *suppressor of Hairy-wing* (*su-Hw*) in *Drosophila* completely or partially restore the normal phenotype in an organism that is homozygous (or hemizygous) for the *sable*, *forked*, and *Hairy-wing* mutations, respectively. For example, flies hemizygous for both *forked* (a bristle mutation) and *su-f* have normal bristles. In each case, the suppressor gene causes the complete reversal of the expected phenotypic expression of the original mutation. Suppressor genes are excellent examples of the genetic background modifying primary gene effects.

Second, the physical location of a gene in relation to other genetic material may influence its expression. Such a situation is called a **position effect**. For example, if a region of a chromosome is relocated or rearranged (called a translocation or inversion event), normal expression of genes in that chromosomal region may be modified. This is particularly true if the gene is relocated to or near certain areas of the chromosome that are prematurely condensed and genetically inert, referred to as **heterochromatin**.

An example of a position effect involves female *Drosophila* heterozygous for the X-linked recessive eye-color mutant *white* (w). The w^+/w genotype normally results in a wild-type brick-red eye color. However, if the region of the X chromosome containing the wild-type w^+ allele is translocated so that it is close to a heterochromatic region, expression of the w^+ allele is modified. Instead of having a red color, the eyes are variegated, or mottled with red and white patches (Figure 4–17). Therefore, following translocation, the dominant effect of the normal w^+ allele is intermittent. A similar position effect is produced if a heterochromatic region is relocated next to the *white* locus on the X chromosome. Apparently, heterochromatic regions inhibit the expression of adjacent genes. Loci in many other organisms also exhibit position effects, providing proof that alteration of the normal arrangement of genetic information can modify its expression.

(a)

(b)

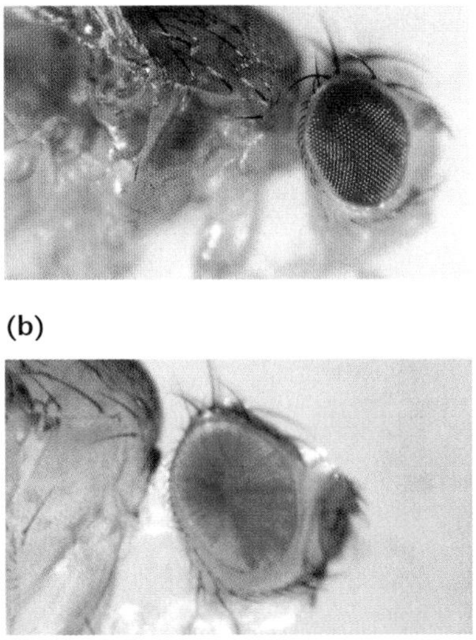

FIGURE 4–17 Position effect, as illustrated in the eye phenotype in two female *Drosophila* heterozygous for the gene *white*. (a) Normal dominant phenotype showing brick-red eye color. (b) Variegated color of an eye caused by rearrangement of the *white* gene to another location in the genome.

Temperature Effects

Because chemical activity depends on the kinetic energy of the reacting substances, which in turn depends on the surrounding temperature, we can expect temperature to influence phenotypes. One example is the evening primrose, which produces red flowers when grown at 23°C and white flowers when grown at 18°C. An even more striking example is seen in Siamese cats and Himalayan rabbits, which exhibit dark fur in certain body regions where the body temperature is slightly cooler, particularly the nose, ears, and paws (Figure 4–18). In these cases, it appears that the wild-type enzyme responsible for pigment production is functional at the lower temperatures present in the extremities, but it loses its catalytic function at the slightly higher temperatures found throughout the rest of the body.

Mutations that are affected by temperature are said to be **conditional** and are called **temperature sensitive**. Examples are known in viruses and a variety of organisms, including bacteria, fungi, and *Drosophila*. In extreme cases, an organism carrying a mutant allele may express a mutant phenotype when grown at one temperature, but express the wild-type phenotype when reared at another temperature. This type of temperature effect is useful in studying mutations that interrupt essential processes during development and are thus normally detrimental to the organism. For example, if bacterial viruses carrying certain temperature-sensitive mutations infect bacteria cultured at 42°C, known as the *restrictive condition*, infection progresses up to the point where the essential gene product is required (e.g., for viral assembly) and then arrests. If cultured under *permissive conditions* of 25°C, the gene product is functional, infection proceeds normally, and new viruses are produced. The use of temperature-sensitive mutations, which may be induced and isolated, has added immensely to the study of viral genetics.

Another case involves genes present in all organisms studied that are activated only when the organism finds itself under the stress of elevated environmental temperatures. First discovered in *Drosophila*, these are called **heat-shock genes**, which are responsible for producing a group of proteins believed to provide a protective function to the heat stress. The coordinated activation of these genes in eukaryotes is attributed to shared promotional elements involved in their transcriptional regulation.

Nutritional Effects

Another example where the phenotype is not a direct reflection of the organism's genotype involves **nutritional mutations**. In microorganisms, mutations that prevent synthesis of nutrient molecules are quite common, such as when an enzyme essential to a biosynthetic pathway becomes inactive. A microorganism bearing such a mutation is called an **auxotroph**. If the end product of a biochemical pathway can no longer be synthesized, and if that molecule is essential to normal growth and development, the mutation prevents growth and may be lethal. For example, if the bread mold *Neurospora* can no longer synthesize the amino acid leucine, proteins cannot be synthesized. If leucine is present in the growth medium, the detrimental effect is overcome. Nutritional mutants have been crucial to genetic studies in bacteria and also served as the basis for George Beadle and Edward Tatum's proposal, in the early 1940s, that one gene functions to produce one enzyme. (See Chapter 15.)

A slightly different set of circumstances exists in humans. The presence or absence of certain dietary substances, which normal individuals may consume without harm, can adversely affect individuals with abnormal genetic constitutions. Often, a mutation may prevent an individual from metabolizing some

(a)

(b)

FIGURE 4–18 (a) A Himalayan rabbit. (b) A Siamese cat. Both show dark fur color on the muzzle, ears, and paws. These patches are due to the effect of a temperature-sensitive allele responsible for pigment production at the lower temperatures of the extremities, which is inactive at slightly higher temperatures.

substance commonly found in normal diets. For example, those afflicted with the genetic disorder **phenylketonuria** cannot metabolize the amino acid phenylalanine. Those with **galactosemia** cannot metabolize galactose. However, if the dietary intake of the molecule is drastically reduced or eliminated, the associated phenotype may be ameliorated.

The fairly common case of **lactose intolerance**, in which individuals are intolerant of the milk sugar lactose, illustrates the general principles involved. Lactose is a disaccharide consisting of a molecule of glucose linked to a molecule of galactose, and makes up 7 percent of human milk and 4 percent of cow's milk. To metabolize lactose, humans require the enzyme **lactase**, which cleaves the disaccharide. Adequate amounts of lactase are produced during the first few years after birth. However, in many people, the level of this enzyme soon drops drastically. As adults, these individuals become intolerant of milk. The major phenotypic effect involves severe intestinal diarrhea, flatulence, and abdominal cramps. This condition is particularly prevalent in (although not limited to) Eskimos, Africans, Asians, and Americans with these heritages. In some of these cultures, milk is converted to cheese, butter, and yogurt, where the amount of lactose is reduced significantly and the adverse effects can be reduced. In the United States, milk low in lactose is commercially available, and to aid in the digestion of other lactose-containing foods, lactase is now a commercial product that can be ingested.

Onset of Genetic Expression

Not all genetic traits are expressed at the same time during an organism's life span. In most cases, the age at which a gene is expressed corresponds to the normal sequence of growth and development. In humans, the prenatal, infant, preadult, and adult phases require different genetic information. As a result, many severe inherited disorders are often not manifested until after birth. For example, **Tay–Sachs disease**, inherited as an autosomal recessive, is a lethal lipid-metabolism disease involving an abnormal enzyme, hexosaminidase A. Newborns appear phenotypically normal for the first few months. Then, developmental retardation, paralysis, and blindness ensue, and most affected children die by the age of three.

The **Lesch–Nyhan syndrome** (see page 84), inherited as an X-linked recessive disease, is characterized by abnormal nucleic acid metabolism (biochemical salvage of nitrogenous purine bases), leading to the accumulation of uric acid in blood and tissues, mental retardation, palsy, and self-mutilation of the lips and fingers. The disorder is due to a mutation in the gene encoding hypoxanthine-guanine phosphoribosyl transferase (HPRT). Newborns are normal for six to eight months prior to the onset of the first symptoms.

Still another example involves **Duchenne muscular dystrophy (DMD)**, an X-linked recessive disorder associated with progressive muscular wasting. It is not usually diagnosed until the age of 3 to 5 years. Even with modern medical intervention, the disease is often fatal in the early twenties.

Perhaps the most variable of all inherited human disorders regarding age of onset is **Huntington disease**. Inherited as an autosomal dominant, Huntington disease affects the frontal lobes of the cerebral cortex, where progressive cell death occurs over a period of more than a decade. Brain deterioration is accompanied by spastic uncontrolled movements, intellectual and emotional deterioration, and ultimately death. While onset has been reported at all ages, it most frequently occurs between ages 30 and 50, with a mean onset age of 38.

Conditions such as these support the concept that the critical expression of normal genes varies throughout the life cycle of organisms, including humans. Gene products may play more essential roles at certain times. Further, it is likely that the internal physiological environment of an organism changes with age.

Genetic Anticipation

Interest in studying the genetic onset of phenotypic expression has intensified with the discovery of heritable disorders that *exhibit a progressively earlier age of onset and an increased severity of the disorder in each successive generation.* This general phenomenon is referred to as **genetic anticipation**.

Myotonic dystrophy (DM), the most common type of adult muscular dystrophy, clearly illustrates genetic anticipation. Individuals afflicted with this autosomal dominant disorder exhibit extreme variation in the severity of symptoms. Mildly affected individuals develop cataracts as adults, but have little or no muscular weakness. Severely affected individuals demonstrate more extensive weakness, myotonia (muscle hyperexcitability), and may be mentally retarded. In its most extreme form, the disease is fatal just after birth. A great deal of excitement was generated in 1989, when C. J. Howeler and colleagues confirmed the correlation of increased severity with earlier onset. The researchers studied 61 parent–child pairs that expressed the disorder and, in 60 of the cases, age of onset was earlier or the severity greater in the child than in his or her parent.

In 1992, an explanation was put forward to explain both the molecular cause of the mutation responsible for DM as well as the basis of genetic anticipation. As we will see in Chapter 15, a short (3 nucleotide) DNA sequence of the DM gene is repeated a variable number of times and is unstable. Normal individuals have from 5 to 35 copies of this sequence, minimally affected individuals reveal about 150 copies, and severely affected individuals possess as many as 1500 copies. The most remarkable observation was that, in successive generations of DM individuals, the size of the repeated segment increases. Although it is not yet clear exactly how the expansion in size affects onset and phenotypic expression, the increased number of copies is now accepted as the basis of genetic anticipation in this disorder. Several other inherited human disorders, including the fragile-X syndrome, Kennedy disease, and Huntington disease, also reveal an association between the size of specific regions of the responsible gene and disease severity. We will return to this general topic when we discuss the molecular explanation in detail in Chapter 15.

Genomic (Parental) Imprinting

Our final example of modification of the laws of Mendelian inheritance involves the variation of phenotypic expression depending strictly on the parental origin of the chromosome carrying a particular gene, a phenomenon called **genomic** (or **parental**) **imprinting**. In some species, certain chromosomal regions and the genes contained within them somehow retain a memory, or an "imprint," of their parental origin that influences whether specific genes either are expressed or remain genetically silent—that is, are not expressed.

The imprinting step is thought to occur before or during gamete formation, leading to differentially marked genes (or chromosome regions) in sperm-forming versus egg-forming tissues. The process is clearly different from mutation because the imprint can be reversed in succeeding generations as genes pass from mother to son to granddaughter, and so on.

An example of imprinting involves the inactivation of one of the X chromosomes in mammalian females. In mice, prior to development of the embryo, imprinting occurs in tissues such that the X chromosome of paternal origin is genetically inactivated in all cells, while the genes on the maternal X chromosome remain genetically active. As embryonic development is subsequently initiated, the imprint is "released" and random inactivation of either the paternal *or* the maternal X chromosome occurs.

In 1991, more specific information established that three specific mouse genes undergo imprinting. One is the gene encoding insulinlike growth factor II (*Igf2*). A mouse that carries two nonmutant alleles of this gene is normal in size, whereas a mouse that carries two mutant alleles lacks a growth factor and is a dwarf. The size of a heterozygous mouse—one allele normal and one mutant (Figure 4–19)—depends on the parental origin of the normal allele. The mouse is normal in size if the normal allele came from the father, but dwarf if the normal allele came from the mother. From this, we can deduce that the normal *Igf2* gene is imprinted to function poorly during the course of egg production in females, but functions normally when it has passed through sperm-producing tissue in males.

Imprinting continues to depend on whether the gene passes through sperm-producing or egg-forming tissue leading to the next generation. For example, a heterozygous normal-sized male will, on average, donate to half his offspring a normal-functioning wild-type allele that will counteract a mutant allele received from the mother. In humans, two distinct genetic disorders are thought to be caused by differential imprinting of the same region of chromosome 15 (15q1). In both cases, the disorders *appear* to be due to an identical deletion of this region in one member of the chromosome 15 pair. The first disorder, **Prader–Willi syndrome (PWS)**, results when only an undeleted maternal chromosome remains. If only an undeleted paternal chromosome remains, an entirely different disorder, **Angelman syndrome (AS)**, results.

The two conditions are clearly different phenotypically. PWS entails mental retardation as well as a severe eating disorder marked by an uncontrollable appetite, obesity, diabetes, and growth retardation. Angelman syndrome involves distinct behavioral manifestations as well as mental retardation. We can conclude that the involved region of chromosome 15 is imprinted differently in male versus female gametes and that both a maternal and paternal region are required for normal development.

While numerous questions remain unanswered regarding genomic imprinting, it is now clear that many genes are subject to this process. More than 50 have been identified in mammals thus far. It appears that regions of chromosomes rather than specific genes are imprinted. The molecular mechanism of imprinting is still a matter for conjecture, but it seems certain that **DNA methylation** is involved. In vertebrates, methyl groups can be added to the carbon atom at position 5 in cytosine (see Chapter 10) as a result of the activity of the enzyme DNA methyltransferase. Methyl groups are added when the dinucleotide CpG or groups of CpG units (called CpG islands) are present along a DNA chain.

DNA methylation is a reasonable mechanism for establishing a molecular imprint, since there is evidence that a high level of methylation can inhibit gene activity and that active genes (or their regulatory sequences) are often undermethylated. Whatever the cause of this phenomenon, it is a fascinating topic and one that clearly establishes the requirement for epigenetic asymmetry between the maternal and paternal genome following fertilization. No doubt, the general phenomenon of genomic imprinting will remain an active area of research in the future.

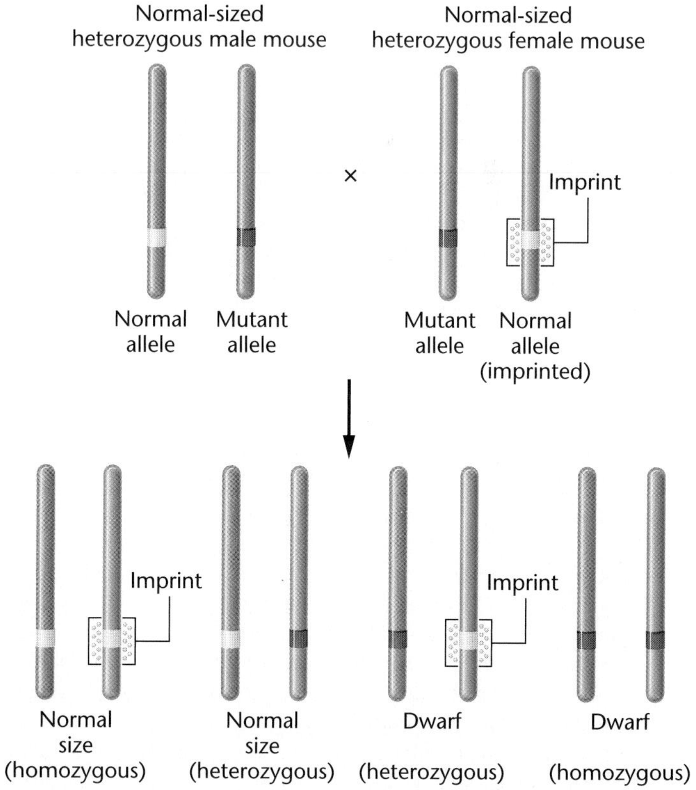

FIGURE 4–19 The effect of imprinting on the mouse *Igf2* gene, which produces dwarf mice in the homozygous condition. Heterozygous offspring that receive the normal allele from their father are normal in size. Heterozygotes that receive the normal allele from their mother, which has been imprinted, are dwarf.

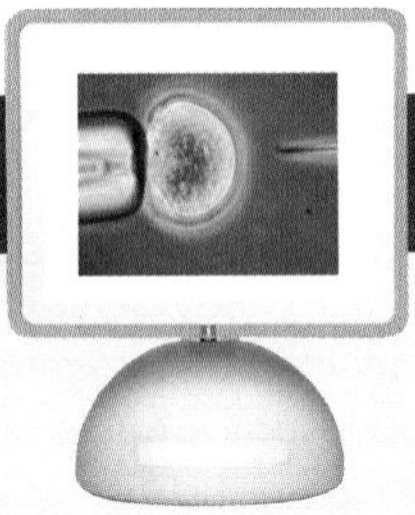

GENETICS, TECHNOLOGY, AND SOCIETY

Improving the Genetic Fate of Purebred Dogs

For dog lovers, nothing is quite so heartbreaking as watching a dog slowly go blind, standing by helplessly as he struggles to adapt to a life of perpetual darkness. That's what happens in progressive retinal atrophy (PRA), an inherited disorder first described in Gordon setters in 1911. Since then, PRA has been found in more than 60 other breeds of dogs, including Irish setters, border collies, Norwegian elkhounds, toy poodles, miniature schnauzers, cocker spaniels, and Siberian huskies.

The products of many genes are required for the development and maintenance of a healthy retina, and a defect in any one of them has the potential to cause retinal dysfunction. Decades of research have led to the identification of six such genes (*rcd1, rcd2, erd, Rho, Xpra,* and *prcd*), and more are likely to be discovered. Different genes are mutated in different breeds (e.g., *rcd1* in Irish setters, *rcd2* in collies, and *prcd* in most breeds). Each gene is associated with a different form of PRA that varies slightly in its clinical symptoms and rate of progression. PRA shows a recessive pattern of inheritance in most breeds, but is dominant in Mastiffs and X-linked in Siberian huskies.

Whichever mutation is responsible, PRA is almost ten times more common in certain purebred dogs than in mixed breeds. The development of distinct breeds of dogs has involved the intensive selection for desirable attributes, whether a particular size, shape, color, or behavior. Many desired characteristics are determined by recessive alleles. The fastest way to increase the homozygosity of these alleles and establish the characteristics in the population is to mate close relatives, which are likely to carry the same alleles. In the practice known as line breeding, for example, dogs may be mated to a cousin or a grandparent. Many breeders also produce hundreds of offspring from individual dogs that have won major dog shows, in an attempt to profit from impressive pedigrees. This "popular sire effect," as it has been termed, further increases the homozygosity of alleles in purebred dogs.

Unfortunately, the generations of inbreeding that have established favorable characteristics in purebreeds have also increased the homozygosity of certain harmful recessive alleles, such as PRA, resulting in other inherited diseases. Many breeds are plagued with inherited hip dysplasia, which is particularly prevalent in German shepherds. Deafness and kidney disorders are common genetic maladies in dalmatians. More than 300 such diseases have been characterized in purebred dogs, and many breeds have a predisposition to more than 20 of them. According to researchers at Cornell University, purebred dogs suffer the highest incidence of inherited disease of any animal: 25 percent of the 20 million purebred dogs in America are affected with one genetic ailment or another. While mutations are the cause of these genetic diseases, there is general agreement that breeding practices, if not executed properly, increase the frequency of disease.

Inbreeding is the likely explanation for the elevated frequency of PRA in certain breeds. Fortunately, advances in canine genetics are beginning to provide new tools for the breeding of healthy dogs. Since 1998, a genetic test has been available to identify mutations in the *prcd* gene on chromosome 9, which is responsible for progressive rod-cone degeneration (prcd), the most common form of PRA. This test is now being used to identify heterozygous carriers of *prcd* mutations—dogs that show no symptoms of PRA, but, if mated with other carriers, pass the trait on to about 25 percent of their offspring. Eliminating PRA carriers from breeding programs has almost completely eradicated this condition from Portuguese Water Dogs, and has greatly reduced its prevalence in many other breeds. OptiGen, a company devoted solely to testing and preventing inherited diseases in future generations of purebred dogs, now offers blood-based tests for all known forms of PRA and is developing tests for similar inherited disorders. The increased availability of genetic tests will soon allow for the diagnosis of diseases before symptoms develop, ensure that breeding animals are free from harmful recessive alleles, and, with hope, will offer a solution to the problems caused by the past century of inbreeding.

It took many years to identify *prcd* and other genes responsible for PRA. In the future, the isolation of genes underlying canine inherited disease should be faster, thanks to the Dog Genome Project, a collaborative effort involving scientists at the University of California, the University of Oregon, the Fred Hutchinson Cancer Research Center in Seattle, and other research centers. A genetic map of the 39 chromosomes in the dog was completed in 2003, and shows that the dog genome is smaller than the human genome but contains more than 18,000 homologs to human genes. Much to the delight of researchers and dog owners alike, information provided by the Dog Genome Project has already aided in combating retinal degenerative disease. Researchers at the University of Pennsylvania used gene therapy to treat Leber's Congenital Amaurosis (LCA), a disease similar to PRA. By injecting copies of normal genes into the retina of a congenitally blind puppy, researchers were able to partially restore the dog's vision. This marked the first time that congenital blindness was reversed in an animal larger than a mouse, and provides hope that the search for a treatment of PRA is not in vain.

The Dog Genome Project may have benefits for humans beyond the reduction of disease in their canine companions. This is because about 85 percent of the genes in the dog genome have equivalents in humans, and over 300 diseases affecting dogs also affect humans. In fact, the ten most common disorders of purebred dogs are all major human health concerns, including heart disease, epilepsy, allergies, and cancer. The identification of a disease-causing dog gene may be a shortcut to the isolation of the corresponding gene in humans. For example, the gene responsible for a common form of PRA in dogs (*prcd*) appears to be the canine version of the gene defect producing retinitis pigmentosa (RP) in humans, which afflicts about 1.5 million people worldwide. Despite much research, RP remains poorly understood, and current treatments only slow its progress. Understanding the genetic basis of PRA could lead to breakthroughs in the diagnosis and treatment of RP, potentially saving the sight of thousands of people every year. By contributing to the cure of human diseases, dogs may prove to be "man's best friend" in an entirely new way.

References

Kirkness, E.F. et al. 2003. The dog genome: survey sequencing and comparative analysis. *Science* 301:1898–1903.

Ray, K., Baldwin, V., Acland, G., and Aquirre, G. 1995. Molecular diagnostic tests for ascertainment of genotype at the rod cone dysplasia locus (*rcd1*) in Irish setters. *Curr. Eye Res.* 14:243.

Smith, C.A. 1994. New hope for overcoming canine inherited disease. *Am. J. Vet. Med. Assoc.* 204:41–46.

Chapter Summary

1. Since Mendel's work was rediscovered, the study of transmission genetics has expanded to include many alternative modes of inheritance. In many cases, phenotypes may be influenced by two or more genes in a variety of ways.
2. Incomplete, or partial, dominance is exhibited when intermediate phenotypic expression of a trait occurs in an organism that is heterozygous for two alleles.
3. Codominance is exhibited when distinctive effects of two alleles occur simultaneously in a heterozygous organism.
4. The concept of multiple allelism applies to populations, since a diploid organism may host only two alleles at any given locus. Within a population, however, many alternative alleles of the same gene occur.
5. Lethal mutations usually result in the impairment, inactivation, or the lack of synthesis of gene products essential during development. Such mutations may be recessive or dominant. The manifestation of some lethal genes, such as the gene causing Huntington disease, are not expressed until adulthood.
6. Phenotypes are often affected by more than one gene, leading to a variety of modifications of the Mendelian dihybrid and trihybrid ratios. Such cases are often classified under the general heading of gene interaction.
7. Epistasis may occur when two or more genes influence a single characteristic. Usually, the expression of one of the genes masks the expression of the other gene or genes.
8. Genes located on the X chromosome result in a characteristic mode of inheritance referred to as X-linkage. X-linked phenotypic ratios result because hemizygous individuals (those with an X and a Y chromosome) express all alleles present on their X chromosome.
9. Sex-limited and sex-influenced inheritance occur when the sex of the organism affects the phenotype controlled by a gene located on an autosome.
10. Phenotypic expression is not always the direct reflection of the genotype. Penetrance measures the percentage of organisms in a given population exhibiting evidence of the corresponding mutant phenotype. Expressivity, on the other hand, measures the range of phenotypic expression of a given genotype.
11. Phenotypic expression can be modified by genetic background, temperature, and nutrition. Position effects illustrate the existence of genetic background affecting phenotypic expression.
12. Genetic anticipation refers to the phenomenon where the onset of phenotypic expression occurs earlier and becomes more severe in each ensuing generation.
13. Genomic imprinting is a process whereby a region of either the paternal or maternal chromosome is modified (marked or imprinted), thereby affecting phenotypic expression. Expression therefore depends upon which parent contributes a mutant allele.

Insights and Solutions

Genetic problems take on added complexity if they involve two independent characters and multiple alleles, incomplete dominance, or epistasis. The most difficult types of problems are those that pioneering geneticists faced during laboratory or field studies. They had to determine the mode of inheritance by working backward from the observations of offspring to parents of unknown genotype.

1. Consider the problem of comb-shape inheritance in chickens, where walnut, rose, pea, and single are observed as distinct phenotypes. *How is comb shape inherited, and what are the genotypes of the* P_1 *generation of each cross?* Use the following data to answer these questions.

Cross 1:	single	× single	⟶	all single
Cross 2:	walnut	× walnut	⟶	all walnut
Cross 3:	rose	× pea	⟶	all walnut
Cross 4:	$F_1 \times F_1$ of Cross 3			
	walnut	× walnut	⟶	93 walnut
				28 rose
				32 pea
				10 single

Solution: At first glance, this problem appears quite difficult. However, as with other seemingly difficult problems, applying a systematic approach and breaking the analysis into steps usually saves the day. The approach involves two steps. First, analyze the data carefully for any useful information. Once you identify something that is clearly helpful, follow an empirical approach; that is, formulate an hypothesis and test it against the given data. Look for a pattern of inheritance consistent with all cases.

For example, this problem gives two immediately useful facts. First, in cross 1, P_1 singles breed true. Second, while P_1 walnut breeds true (cross 2), a walnut phenotype is also produced in crosses between rose and pea (cross 3). When these F_1 walnuts are mated (cross 4), all four comb shapes are produced in a ratio that approximates 9:3:3:1. This observation should immediately suggest a cross involving two gene pairs, because the resulting data closely resemble the ratio of Mendel's dihybrid crosses. Since only one character is involved and discontinuous phenotypes occur (comb shape), perhaps some form of epistasis is occurring. This may serve as a working hypothesis, and you must now propose how the two gene pairs interact to produce each phenotype.

If you call the allele pairs *A*, *a* and *B*, *b*, you might predict that, since walnut represents 9/16 in cross 4, *A–B–* will produce walnut. You might also hypothesize that in the case of cross 2, the genotypes were *AABB* × *AABB* where walnut was seen to breed true. (Recall that *A–* and *B–* mean *AA* or *Aa* and *BB* or *Bb*, respectively.) Because single is the phenotype representing 1/16 of the offspring of cross 4, we could predict that the phenotype is the result of the *aabb* genotype, which is consistent with cross 1.

Now we have only to determine the genotypes for rose and pea. A logical prediction would be that at least one dominant *A* or *B* allele, combined with the double recessive condition of the other allele pair, can account for these phenotypes; that is,

A–bb rose
and
$aa\ B$– pea

If, in cross 3, $AAbb$ (rose) were crossed with $aaBB$ (pea), all offspring would be $AaBb$ (walnut). This is consistent with the data, and we must now look at cross 4. We predict these walnut genotypes to be $AaBb$ (as before), and from the cross

$AaBb$ (walnut) $\times$ $AaBb$ (walnut)

we expect

9/16 A–B– (walnut)
3/16 A–bb (rose)
3/16 $aa\ B$– (pea)
1/16 $aa\ bb$ (single)

Our prediction is consistent with the data. The initial hypothesis of the interaction of two gene pairs proves consistent throughout, and the problem is solved.

This example demonstrates the need for a basic theoretical knowledge of transmission genetics. Then you must search for the appropriate clues, so that you can proceed in a stepwise fashion toward a solution. Mastering problem-solving requires practice, but will give you a great deal of satisfaction. Apply this general approach to the following problems.

2. In radishes, flower color may be red, purple, or white. The edible portion of the radish may be long or oval. When only flower color is studied, no dominance is evident, and red $\times$ white crosses yield all purple. If these F_1 purples are interbred, the F_2 generation consists of 1/4 red:1/2 purple:1/4 white. Regarding radish shape, long is dominant to oval in a normal Mendelian fashion.

(a) Determine the F_1 and F_2 phenotypes from a cross between a true-breeding red, long radish and one that is white and oval. Be sure to define all gene symbols initially.

Solution: This is a modified dihybrid cross where the gene pair controlling color exhibits incomplete dominance. Shape is controlled conventionally. First, establish gene symbols:

RR = red O– = long
Rr = purple oo = oval
rr = white

P_1: $RROO$ $\times$ $rroo$
(red long) (white oval)
F_1: all $RrOo$ (purple long)
$F_1 \times F_1$: $RrOo \times RrOo$

F_2:	1/4 RR	3/4 O–	3/16 $RR\ O$–	red long
		1/4 oo	1/16 $RR\ oo$	red oval
	2/4 Rr	3/4 O–	6/16 $Rr\ O$–	purple long
		1/4 oo	2/16 $Rr\ oo$	purple oval
	1/4 rr	3/4 O–	3/16 $rr\ O$–	white long
		1/4 oo	1/16 $rr\ oo$	white oval

Note that to generate the F_2 results, we have used the forked-line method. First, we consider the outcome of crossing F_1 parents for the color genes ($Rr \times Rr$). Then the outcome of shape is considered ($Oo \times Oo$).

(b) A red oval plant was crossed with a plant of unknown genotype and phenotype, yielding the following offspring:

103 red long : 101 red oval
98 purple long : 100 purple oval

Determine the genotype and phenotype of the unknown plant.

Solution: The two characters appear to be inherited independently, so consider them separately. The data indicate a 1/4 : 1/4 : 1/4 : 1/4 proportion. First, consider color:

P_1: red $\times$??? (unknown)
F_1: 204 red (1/2)
198 purple (1/2)

Because the red parent must be RR, the unknown must have a genotype of Rr to produce these results. Thus it is purple. Now, consider shape:

P_1: oval $\times$??? (unknown)
F_1: 201 long (1/2)
201 oval (1/2)

Since the oval plant must be oo, the unknown plant must have a genotype of Oo to produce these results. Thus it is long. The unknown plant is

$RrOo$ purple long

3. In humans, red–green color blindness is inherited as an X-linked recessive trait. A woman with normal vision whose father is color blind marries a male who has normal vision. Predict the color vision of their male and female offspring.

Solution: The female is heterozygous since she inherited an X chromosome with the mutant allele from her father. Her husband is normal. Therefore, the parental genotypes are

Cc $\times$ C ↾ (↾ represents the Y chromosome)

All female offspring are normal (CC or Cc). One half of the male children will be color blind (c ↾), and the other half will have normal vision (C ↾).

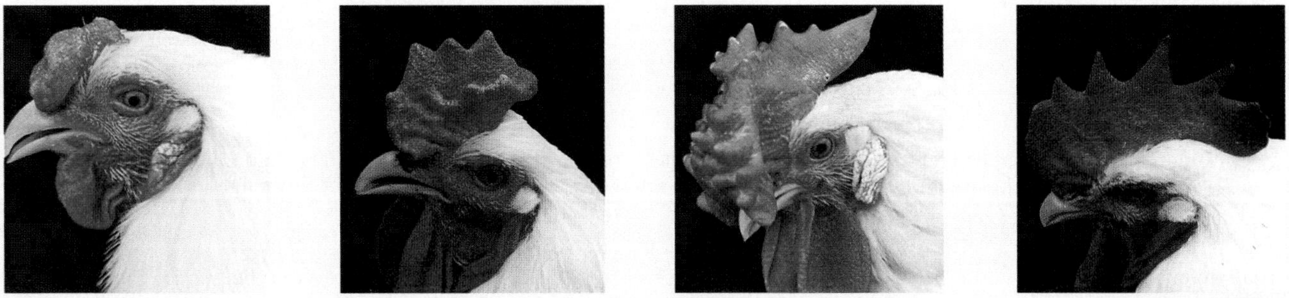

Walnut Pea Rose Single

Problems and Discussion Questions

1. In shorthorn cattle, coat color may be red, white, or roan. Roan is an intermediate phenotype expressed as a mixture of red and white hairs. The following data were obtained from various crosses:

red × red	⟶	all red
white × white	⟶	all white
red × white	⟶	all roan
roan × roan	⟶	1/4 red:1/2 roan:1/4 white

How is coat color inherited? What are the genotypes of parents and offspring for each cross?

2. Contrast incomplete dominance and codominance.
3. In foxes, two alleles of a single gene, P and p, may result in lethality (PP), platinum coat (Pp), or silver coat (pp). What ratio is obtained when platinum foxes are interbred? Is the P allele behaving dominantly or recessively in causing lethality? In causing platinum coat color?
4. In mice, a short-tailed mutant was discovered. When it was crossed to a normal long-tailed mouse, 4 offspring were short-tailed and 3 were long tailed. Two short-tailed mice from the F_1 generation were selected and crossed. They produced 6 short-tailed and 3 long-tailed mice. These genetic experiments were repeated three times with approximately the same results. What genetic ratios are illustrated? Hypothesize the mode of inheritance and diagram the crosses.
5. List all possible genotypes for the A, B, AB, and O phenotypes. Is the mode of inheritance of the ABO blood types representative of dominance? of recessiveness? of codominance?
6. With regard to the ABO blood types in humans, determine the genotype of the male parent and female parent shown here:

   Male parent: Blood type B; mother type O
   Female parent: Blood type A; father type B

   Predict the blood types of the offspring that this couple may have and the expected proportion of each.
7. In a disputed parentage case, the child is blood type O, while the mother is blood type A. What blood type would exclude a male from being the father? Would the other blood types prove that a particular male was the father?
8. The A and B antigens in humans may be found in water-soluble form in secretions, including saliva, of some individuals (*Se/Se* and *Se/se*) but not in others (*se/se*). The population thus contains "secretors" and "nonsecretors."
   (a) Determine the proportion of various phenotypes (blood type and ability to secrete) in matings between individuals that are blood type AB and type O, both of whom are *Se/se*.
   (b) How will the results of such matings change if both parents are heterozygous for the gene controlling the synthesis of the H substance (*Hh*)?
9. Describe what is meant by the phrase "gene interaction" as it pertains to the influence of genes on the phenotype.
10. In rabbits, a series of multiple alleles controls coat color in the following way: C is dominant to all other alleles and causes full color. The chinchilla phenotype is due to the c^{ch} allele, which is dominant to all alleles other than C. The c^h allele, dominant only to c^a (albino), results in the Himalayan coat color. Thus, the order of dominance is $C > c^{ch} > c^h > c^a$. For each of the following three cases, the phenotypes of the P_1 generations of two crosses are shown, as well as the phenotype of one member of the F_1 generation.

	P_1 Phenotypes		F_1 Phenotypes		
(a)	Himalayan × Himalayan	⟶	albino		
			×	⟶	??
	full color × albino	⟶	chinchilla		
(b)	albino × chinchilla	⟶	albino		
			×	⟶	??
	full color × albino	⟶	full color		
(c)	chinchilla × albino	⟶	Himalayan		
			×	⟶	??
	full color × albino	⟶	Himalayan		

For each case, determine the genotypes of the P_1 generation and the F_1 offspring, and predict the results of making each cross between F_1 individuals as previously shown.

Full color

Chinchilla

Himalayan

Albino

11. In the guinea pig, one locus involved in the control of coat color may be occupied by any of four alleles: C (full color), c^k (sepia), c^d (cream), or c^a (albino). Like coat color in rabbits (Problem 10), an order of dominance exists: $C > c^k > c^d > c^a$. In the following crosses, write the parental genotypes and predict the phenotypic ratios that would result:
    (a) sepia × cream where both guinea pigs had an albino parent
    (b) sepia × cream where the sepia guinea pig had an albino parent and the cream guinea pig had two sepia parents
    (c) sepia × cream where the sepia guinea pig had two full-color parents and the cream guinea pig had two sepia parents
    (d) sepia × cream, where the sepia guinea pig had a full-color parent and an albino parent and the cream guinea pig had two full-color parents
12. Three gene pairs located on separate autosomes determine flower color and shape as well as plant height. The first pair exhibits incomplete dominance, where the color can be red, pink (the heterozygote), or white. The second pair leads to personate (dominant) or peloric (recessive) flower shape, while the third gene pair produces either the dominant tall trait or the recessive dwarf trait. Homozygous plants that are red, personate, and tall are crossed to those that are white, peloric, and dwarf. Determine

the F_1 genotype(s) and phenotype(s). If the F_1 plants are interbred, what proportion of the offspring will exhibit the same phenotype as the F_1 plants?

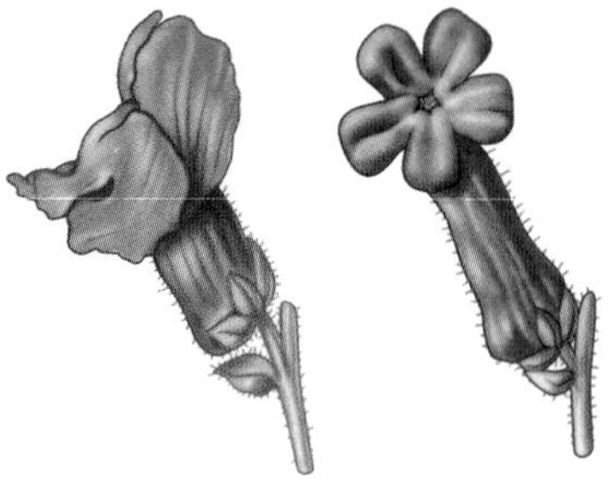

13. As in Problem 12, color may be red, white, or pink, and flower shape may be personate or peloric. For the following crosses, determine the P_1 and F_1 genotypes:

(a) red peloric × white personate F1: all pink personate
(b) red personate × white peloric F1: all pink personate
(c) pink personate × red peloric F1: 1/4 red peloric
1/4 red personate
1/4 pink peloric
1/4 pink personate
(d) pink personate × white peloric F1: 1/4 white personate
1/4 white peloric
1/4 pink personate
1/4 pink peloric

What phenotypic ratios would result from crossing the F_1 of (a) to the F_1 of (b)?

14. Horses can be cremello (a light cream color), chestnut (a brownish color), or palomino (a golden color with white in the horse's tail and mane). Of these phenotypes, only palominos never breed true.

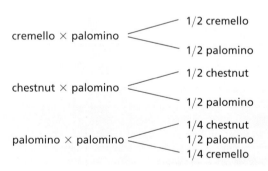

cremello × palomino → 1/2 cremello, 1/2 palomino

chestnut × palomino → 1/2 chestnut, 1/2 palomino

palomino × palomino → 1/4 chestnut, 1/2 palomino, 1/4 cremello

(a) From the results shown in the previous column, determine the mode of inheritance by assigning gene symbols and indicating which genotypes yield which phenotypes.
(b) Predict the F_1 and F_2 results of many initial matings between cremello and chestnut horses.

15. With reference to the eye-color phenotypes produced by the recessive, autosomal, unlinked *brown* and *scarlet* loci in *Drosophila* (see Figure 4–10), predict the F_1 and F_2 results of the following P_1 crosses. (Recall that when both the *brown* and *scarlet* alleles are homozygous, no pigment is produced, and the eyes are white.)
(a) wild type × white
(b) wild type × scarlet
(c) brown × white

16. Pigment in mouse fur is only produced when the *C* allele is present. Individuals of the *cc* genotype are white. If color is present, it may be determined by the *A*, *a* alleles. *AA* or *Aa* results in agouti color, while *aa* results in black coats.
(a) What F_1 and F_2 genotypic and phenotypic ratios are obtained from a cross between *AACC* and *aacc* mice?
(b) In three crosses between agouti females whose genotypes were unknown and males of the *aacc* genotype, the following phenotypic ratios were obtained:

(1)	(2)	(3)
8 agouti	9 agouti	4 agouti
8 white	10 black	5 black
		10 white

What are the genotypes of these female parents?

17. In some plants a red pigment, cyanidin, is synthesized from a colorless precursor. The addition of a hydroxyl group (OH^-) to the cyanidin molecule causes it to become purple. In a cross between two randomly selected purple plants, the following results were obtained:

94 purple
31 red
43 white

How many genes are involved in the determination of these flower colors? Which genotypic combinations produce which phenotypes? Diagram the purple × purple cross.

18. In rats, the following genotypes of two independently assorting autosomal genes determine coat color:

A–B–	(gray)
A–bb	(yellow)
aaB–	(black)
aabb	(cream)

A third gene pair on a separate autosome determines whether or not any color will be produced. The *CC* and *Cc* genotypes allow color according to the expression of the *A* and *B* alleles. However, the *cc* genotype results in albino rats regardless of the *A* and *B* alleles present. Determine the F_1 phenotypic ratio of the following crosses:

(a) *AAbbCC* × *aaBBcc*
(b) *AaBBCC* × *AABbcc*
(c) *AaBbCc* × *AaBbcc*
(d) *AaBBCc* × *AaBBCc*
(e) *AABbCc* × *AABbcc*

19. Given the inheritance pattern of coat color in rats, as described in Problem 18, predict the genotype and phenotype of the parents who produced the following offspring:

(a) 9/16 gray:3/16 yellow:3/16 black:1/16 cream
(b) 9/16 gray:3/16 yellow:4/16 albino
(c) 27/64 gray:16/64 albino:9/64 yellow:9/64 black: 3/64 cream
(d) 3/8 black:3/8 cream:2/8 albino
(e) 3/8 black:4/8 albino:1/8 cream

20. In a species of the cat family, eye color can be gray, blue, green, or brown, and each trait is true breeding. In separate crosses involving homozygous parents, the following data were obtained:

Cross	P_1	F_1	F_2
A	green × gray	all green	3/4 green:1/4 gray
B	green × brown	all green	3/4 green:1/4 brown
C	gray × brown	all green	9/16 green:3/16 brown 3/16 gray:1/16 blue

(a) Analyze the data. How many genes are involved? Define gene symbols and indicate which genotypes yield each phenotype.
(b) In a cross between a gray-eyed cat and one of unknown genotype and phenotype, the F_1 generation was not observed. However, the F_2 resulted in the same F_2 ratio as in cross C. Determine the genotypes and phenotypes of the unknown P_1 and F_1 cats.

21. In a plant, a tall variety was crossed with a dwarf variety. All F_1 plants were tall. When $F_1 \times F_1$ plants were interbred, 9/16 of the F_2 were tall and 7/16 were dwarf.
(a) Explain the inheritance of height by indicating the number of gene pairs involved and by designating which genotypes yield tall and which yield dwarf. (Use dashes where appropriate.)
(b) What proportion of the F_2 plants will be true breeding if self-fertilized? List these genotypes.

22. In a unique species of plants, flowers may be yellow, blue, red, or mauve. All colors may be true breeding. If plants with blue flowers are crossed to red-flowered plants, all F_1 plants have yellow flowers. When carried to an F_2 generation, the following ratio was observed:

9/16 yellow:3/16 blue:3/16 red:1/16 mauve

In still another cross using true-breeding parents, yellow-flowered plants are crossed with mauve-flowered plants. Again, all F_1 plants had yellow flowers and the F_2 showed a 9:3:3:1 ratio, as just shown.
(a) Describe the inheritance of flower color by defining gene symbols and designating which genotypes give rise to each of the four phenotypes.
(b) Determine the F_1 and F_2 results of a cross between true-breeding red and true-breeding mauve-flowered plants.

23. Five human matings (1–5), including both maternal and paternal phenotypes for ABO and MN blood-group antigen status, are shown in the following table:

Parental Phenotypes	Offspring
(1) A, M × A, N	(a) A, N
(2) B, M × B, M	(b) O, N
(3) O, N × B, N	(c) O, MN
(4) AB, M × O, N	(d) B, M
(5) AB, MN × AB, MN	(e) B, MN

Each mating resulted in one of the five offspring shown in the right-hand column (a–e). Match each offspring with one correct set of parents, using each parental set only once. Is there more than one set of correct answers?

24. A husband and wife have normal vision, although both of their fathers are red–green color blind, an inherited X-linked recessive condition. What is the probability that their first child will be (a) a normal son? (b) a normal daughter? (c) a color-blind son? (d) a color-blind daughter?

25. In humans, the ABO blood type is under the control of autosomal multiple alleles. Color blindness is a recessive X-linked trait. If two parents who are both type A and have normal vision produce a son who is color blind and is type O, what is the probability that their next child will be a female who has normal vision and is type O?

26. In *Drosophila*, an X-linked recessive mutation, scalloped (*sd*), causes irregular wing margins. Diagram the F_1 and F_2 results if (a) a scalloped female is crossed with a normal male; (b) a scalloped male is crossed with a normal female. Compare these results with those that would be obtained if scalloped were not X-linked.

27. Another recessive mutation in *Drosophila*, *ebony* (*e*), is on an autosome (chromosome 3) and causes darkening of the body compared with wild-type flies. What phenotypic F_1 and F_2 male and female ratios will result if a scalloped-winged female with normal body color is crossed with a normal-winged ebony male? Work this problem by both the Punnett square method and the forked-line method.

28. In *Drosophila*, the X-linked recessive mutation *vermilion* (*v*) causes bright red eyes, which is in contrast to brick-red eyes of wild type. A separate autosomal recessive mutation, *suppressor of vermilion* (*su-v*), causes flies homozygous or hemizygous for *v* to have wild-type eyes. In the absence of *vermilion* alleles, *su-v* has no effect on eye color. Determine the F_1 and F_2 phenotypic ratios from a cross between a female with wild-type alleles at the *vermilion* locus, but who is homozygous for *su-v*, with a *vermilion* male who has wild-type alleles at the *su-v* locus.

29. While *vermilion* is X-linked and brightens the eye color, *brown* is an autosomal recessive mutation that darkens the eye. Flies carrying both mutations, lose all pigmentation, and are white-eyed. Predict the F_1 and F_2 results of the following crosses:
(a) *vermilion* females × *brown* males
(b) *brown* females × *vermilion* males
(c) *white* females × wild-type males

30. In a cross in *Drosophila* involving the X-linked recessive eye mutation *white* and the autosomally linked recessive eye mutation *sepia* (resulting in a dark eye), predict the F_1 and F_2 results of crossing true-breeding parents of the following phenotypes:
(a) white females × sepia males;
(b) sepia females × white males.
Note that white is epistatic to the expression of sepia.

31. Consider the following three pedigrees all involving a single human trait:

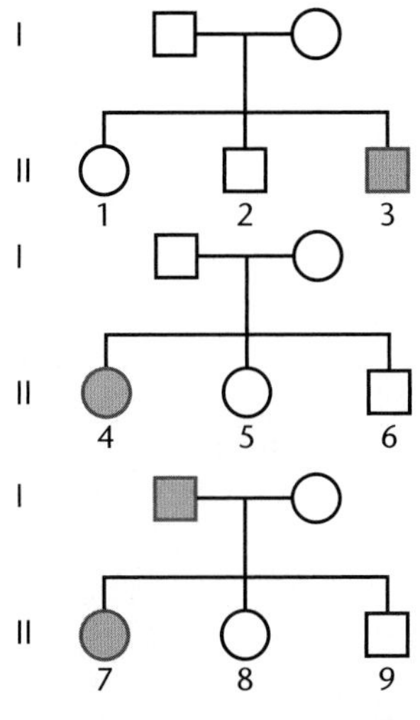

(a) Which conditions, if any, can be excluded?

Conditions: dominant and X-linked
dominant and autosomal
recessive and X-linked
recessive and autosomal

(b) For any condition that you excluded, indicate the single individual in generation II (e.g., II-1, II-2) that was most instrumental in your decision to exclude that condition. If none were excluded, answer "none apply."

(c) Given your conclusions in part (a), indicate the genotype of the individuals listed here.

II-1; II-6; II-9

If more than one possibility applies, list all possibilities. Use the symbols A and a for the genotypes.

32. Below are three pedigrees. For each, consider whether it could or could not be consistent with an X-linked recessive trait. In a sentence or two, indicate why or why not.

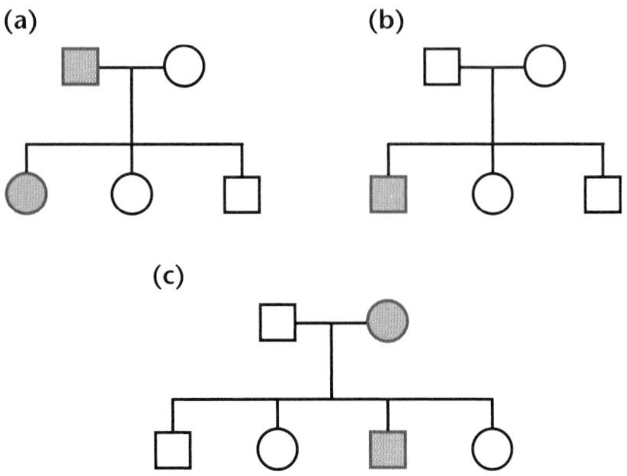

33. In spotted cattle, the colored regions may be mahogany or red. If a red female and a mahogany male, both derived from separate true-breeding lines, are mated and the cross carried to an F_2 generation, the following results are obtained:

F_1: 1/2 mahogany males
1/2 red females

F_2: 3/8 mahogany males
1/8 red males
3/8 red females
1/8 mahogany females

When the reciprocal of the initial cross is performed (mahogany female and red male), identical results are obtained. Explain these results by postulating how the color is genetically determined. Diagram the crosses.

34. Predict the F_1 and F_2 results of crossing a male fowl that is cock feathered with a true-breeding hen-feathered female fowl. Recall that these traits are sex limited.

35. Two mothers give birth to sons at the same time at a busy urban hospital. The son of mother 1 is afflicted with hemophilia, a disease caused by an X-linked recessive allele. Neither parent has the disease. Mother 2 has a normal son, despite the fact that the father has hemophilia. Several years later, couple 1 sues the hospital, claiming that these two newborns were swapped in the nursery following their birth. As a genetic counselor, you are called to testify. What information can you provide the jury concerning the allegation?

36. Discuss the topic of phenotypic expression and the many factors that impinge on it.

37. Contrast penetrance and expressivity as the terms relate to phenotypic expression.

38. Contrast the phenomena of genetic anticipation and genomic imprinting.

Extra-Spicy Problems

39. Labrador retrievers may be black, brown (chocolate), or golden (yellow) in color. While each color may breed true, many different outcomes occur if many litters are examined from a variety of matings, where the parents are not necessarily true breeding. Shown below are just some of the many possibilities.

(a) black	×	brown	⟶	all black
(b) black	×	brown	⟶	1/2 black 1/2 brown
(c) black	×	brown	⟶	3/4 black 1/4 golden
(d) black	×	golden	⟶	all black
(e) black	×	golden	⟶	4/8 golden 3/8 black 1/8 brown
(f) black	×	golden	⟶	2/4 golden 1/4 black 1/4 brown
(g) brown	×	brown	⟶	3/4 brown 1/4 golden
(h) black	×	black	⟶	9/16 black 4/16 golden 3/16 brown

Propose a mode of inheritance that is consistent with these data, and indicate the corresponding genotypes of the parents in each mating. Indicate as well the genotypes of dogs that breed true for each color.

40. A true-breeding purple-leafed plant isolated from one side of the rain forest in Puerto Rico (El Yunque) was crossed to a true-breeding white variety found on the other side of the rain forest. The F_1 offspring were all purple. A large number of $F_1 \times F_1$ crosses produced the following results:

purple: 4219; white: 5781 (Total = 10,000)

Propose an explanation for the inheritance of leaf color. As a geneticist, how might you go about testing your hypothesis? Describe the genetic experiments that you would conduct.

41. In Dexter and Kerry cattle, animals may be polled (hornless) or horned. The Dexter animals have short legs, whereas the Kerry animals have long legs. When many offspring were obtained from matings between polled Kerrys and horned Dexters, half were found to be polled Dexters and half polled Kerrys. When these two types of F_1 cattle were mated to one another, the following F_2 data were obtained:

3/8 polled Dexters
3/8 polled Kerrys
1/8 horned Dexters
1/8 horned Kerrys

A geneticist was puzzled by these data and interviewed farmers who had bred these cattle for decades. She learned that Kerrys were true breeding. Dexters, on the other hand, were not true breeding and never produced as many offspring as Kerrys. Provide a genetic explanation for these observations.

Kerry cow

Dexter bull

42. An alien geneticist escaping from a planet where genetic research had recently been prohibited brought with him to Earth two pure-breeding lines of pet frogs. One line croaked by uttering *rib-it rib-it* and had purple eyes. The other line croaked more softly by muttering *knee-deep knee-deep* and had green eyes. With a new-found freedom of inquiry, he mated the two types of frogs. In the F_1, all frogs had blue eyes and uttered *rib-it rib-it*. He proceeded to make many $F_1 \times F_1$ crosses, and when he fully analyzed the F_2 data, he realized they could be reduced to the following ratio:

27/64	blue-eyed, rib-it utterer
12/64	green-eyed, rib-it utterer
9/64	blue-eyed, knee-deep mutterer
9/64	purple-eyed, rib-it utterer
4/64	green-eyed, knee-deep mutterer
3/64	purple-eyed, knee-deep mutterer

(a) How many total gene pairs are involved in the inheritance of both traits? Support your answer.
(b) Of these, how many are controlling eye color? How can you tell? How many are controlling croaking?
(c) Assign gene symbols for all phenotypes and indicate the genotypes of the P_1 and F_1 frogs.
(d) Indicate the genotypes of the six F_2 phenotypes.
(e) After years of experiments, the geneticist isolated pure-breeding strains of all six F_2 phenotypes. Indicate the F_1 and F_2 phenotypic ratios of the following cross, using these pure-breeding strains:
blue-eyed, *knee-deep* mutterer × purple-eyed, *rib-it* utterer
(f) One set of crosses with his true-breeding lines initially caused the geneticist some confusion. When he crossed true-breeding purple-eyed, *knee-deep* mutterers with true-breeding green-eyed, *knee-deep* mutterers, he often got different results. In some matings, all offspring were blue-eyed, *knee-deep* mutterers, but in other matings all offspring were purple-eyed, *knee-deep* mutterers. In still a third mating, 1/2 blue-eyed, knee-deep mutterers and 1/2 purple-eyed, knee-deep mutterers were observed. Explain why the results differed.
(g) In another experiment, the geneticist crossed two purple-eyed, *rib-it* utterers together with the results shown here:

9/16 purple-eyed, *rib-it* utterer
3/16 purple-eyed, *knee-deep* mutterer
3/16 green-eyed, *rib-it* utterer
1/16 green-eyed, *knee-deep* mutterer

What were the genotypes of the two parents?

43. The pedigree below is characteristic of an inherited condition known as male precocious puberty, where affected males show signs of puberty by age 4:

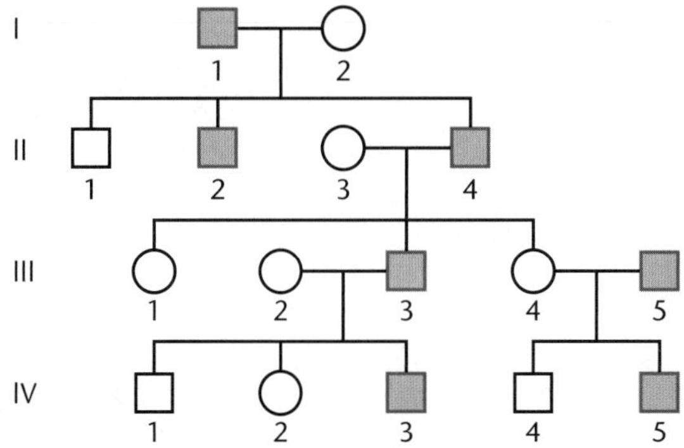

Propose a genetic explanation of this phenotype.

44. In birds, the male contains two identical sex chromosomes designated ZZ. The female contains one Z as well as a nearly blank chromosome, designated *W* (females are ZW). In parakeets, two genes control feather color. The presence of the dominant *Y* allele at the first gene results in the production of a yellow pigment. The dominant *B* allele at the second gene controls melanin production. When both genes are active, a green pigment results. If only the *Y* gene is active, the feathers are yellow. If only the *B* gene is active, a blue color is exhibited. If neither gene is active, the birds are albinos. Therefore, with our conventional designations, phenotypes are produced as follows:

Y–B–	green
Y–bb	yellow
yyB–	blue
yybb	albino

(a) A series of crosses established that one of the genes is autosomal and one is Z-linked. Based on the results of the following cross shown here, where both parents are true-breeding, determine which gene is Z-linked:

P_1:	green male × albino female
F_1:	1/2 green males, 1/2 green females
F_2:	6/16 green males
	2/16 yellow males
	3/16 green females
	1/16 yellow females
	3/16 blue females
	1/16 albino females

Support your answer by establishing the genotypes of the P_1 parents and working the cross through the F_2 generation.

(b) In a cross where the parental genotypes and whether the parents were true breeding were unknown, the offspring from repeated matings were recorded, as shown here:

13 green males
3 yellow males
11 blue females
5 albino females

Based on the results, determine the phenotypes and genotypes of the parents.

45. Students taking a genetics exam were asked to answer the following question, which centers around converting data to a "meaningful ratio", and then solving the problem. The instructor assumed that the final ratio would reflect two gene pairs, and most correct answers did. Here is the exam question.

"Flowers may be white, orange, or brown. When plants with white flowers are crossed with plants with brown flowers, all the F_1 flowers are white. For F_2 flowers, the following data were obtained:

48 white
12 orange
4 brown

Convert the F_2 data to a meaningful ratio that allows you to explain the inheritance of color. Determine the number of genes involved and the genotypes that yield each phenotype".

(a) Solve the problem for two gene pairs. What is the final F_2 ratio?
(b) A number of students failed to reduce the ratio and solved the problem using three gene pairs. When examined carefully, their solution was deemed a valid response. Solve the problem using three gene pairs.
(c) We now have a dilemma. The data are consistent with two alternative mechanisms of inheritance. Propose an experiment that executes crosses involving the original parents that would distinguish between the two solutions proposed by the students. Explain how this experiment would resolve the dilemma.

46. In four o'clock plants, many flower colors are observed. In a cross involving two true-breeding strains that were crimson and white, all of the F_1 generation were rose color. In the F_2, four new phenotypes appeared along with the P_1 and F_1 parental colors. The following ratio was obtained:

1/16 crimson	4/16 rose
2/16 orange	2/16 pale yellow
1/16 yellow	4/16 white
2/16 magneta	

Propose an explanation for the inheritance of the above flower colors.

47. Protooncogenes stimulate cells to progress through the cell cycle and begin mitosis. In cells that stop dividing, transcription of protooncogenes is inhibited by regulatory molecules. As is typical of all genes, protooncogenes contain a regulatory DNA region followed by a coding DNA region that specifies the amino acid sequence of the gene product. Consider two types of mutation in a protooncogene, one in the regulatory region that eliminates transcriptional control, and the other in the coding region that renders the gene product inactive. Characterize both of these mutant alleles as either gain-of-function or loss-of-function mutations and indicate whether each would be dominant or recessive.

48. In the human disorder sickle-cell anemia, many phenotypic traits are evident besides those where the red blood cells undergo sickling and clumping under low oxygen tension and have a shorter half-life than normal, causing anemia. The overall phenotype includes episodes of severe abdominal pain, weakness and fatigue, lengthened long bones, heart enlargement, kidney failure, and respiratory difficulties, including pneumonia. (a) What term describes cases of multiple phenotypic manifestations of a single gene? (b) Relate each of these phenotypic responses to the sickling and anemia. (c) Perform a Web search (or jump ahead to Chapter 14) and determine the molecular basis of the disorder.

SELECTED READINGS

Bartolomei, M.S., and Tilghman, S.M. 1997. Genomic imprinting in mammals. *Annu. Rev. Genet.* 31:493–525.

Brink, R.A., ed. 1967. *Heritage from Mendel*. Madison: University of Wisconsin Press.

Bultman, S.J., Michaud, E.J., and Woychik, R.P. 1992. Molecular characterization of the mouse *agouti* locus. *Cell* 71:1195–1204.

Carlson, E.A. 1987. *The gene: A critical history*, 2d ed. Philadelphia: Saunders.

Cattanach, B.M., and Jones, J. 1994. Genetic imprinting in the mouse: Implications for gene regulation. *J. Inherit. Metab. Dis.* 17:403–20.

Feil, R., and Kelsey, G. 1997. Genomic imprinting: A chromatin connection. *Am. J. Hum. Genet.* 61:1213–19.

Foster, H.L. et al., eds. 1981. The mouse in biomedical research, Vol. 1. *History, genetics, and wild mice*. Orlando, FL: Academic Press.

Grant, V. 1975. *Genetics of flowering plants*. New York: Columbia University Press.

Harper, P.S. et al. 1992. Anticipation in myotonic dystrophy: New light on an old problem. *Am. J. Hum. Genet.* 51:10–16.

Howeler, C.J. et al. 1989. Anticipation in myotonic dystrophy: Fact or fiction? *Brain* 112:779–97.

Morgan, T.H. 1910. Sex-limited inheritance in *Drosophila*. *Science* 32:120–22.

Nolte, D.J. 1959. The eye-pigmentary system of *Drosophila*. *Heredity* 13:233–41.

Pawelek, J.M., and Körner, A.M. 1982. The biosynthesis of mammalian melanin. *Am. Sci.* 70:136–45.

Peters, J.A., ed. 1959. *Classic papers in genetics*. Englewood Cliffs, NJ: Prentice-Hall.

Phillips, P.C. 1998. The language of gene interaction. *Genetics* 149:1167–71.

Race, R.R., and Sanger, R. 1975. *Blood groups in man*, 6th ed. Oxford, UK: Blackwell Scientific Publishers.

Sapienza, C. 1990. Parental imprinting of genes. *Sci. Am.* (Oct.) 363:52–60.

Siracusa, L.D. 1994. The *agouti* gene: Turned on to yellow. *Cell* 10:423–28.

Voeller, B.R., ed. 1968. *The chromosome theory of inheritance—Classic papers in development and heredity*. New York: Appleton-Century-Crofts.

Watkins, M.W. 1966. Blood group substances. *Science* 152:172–81.

Yoshida, A. 1982. Biochemical genetics of the human blood group ABO system. *Am. J. Hum. Genet.* 34:1–14.

Ziegler, I. 1961. Genetic aspects of ommochrome and pterin pigments. *Adv. Genet.* 10:349–403.

Chromosome Mapping in Eukaryotes

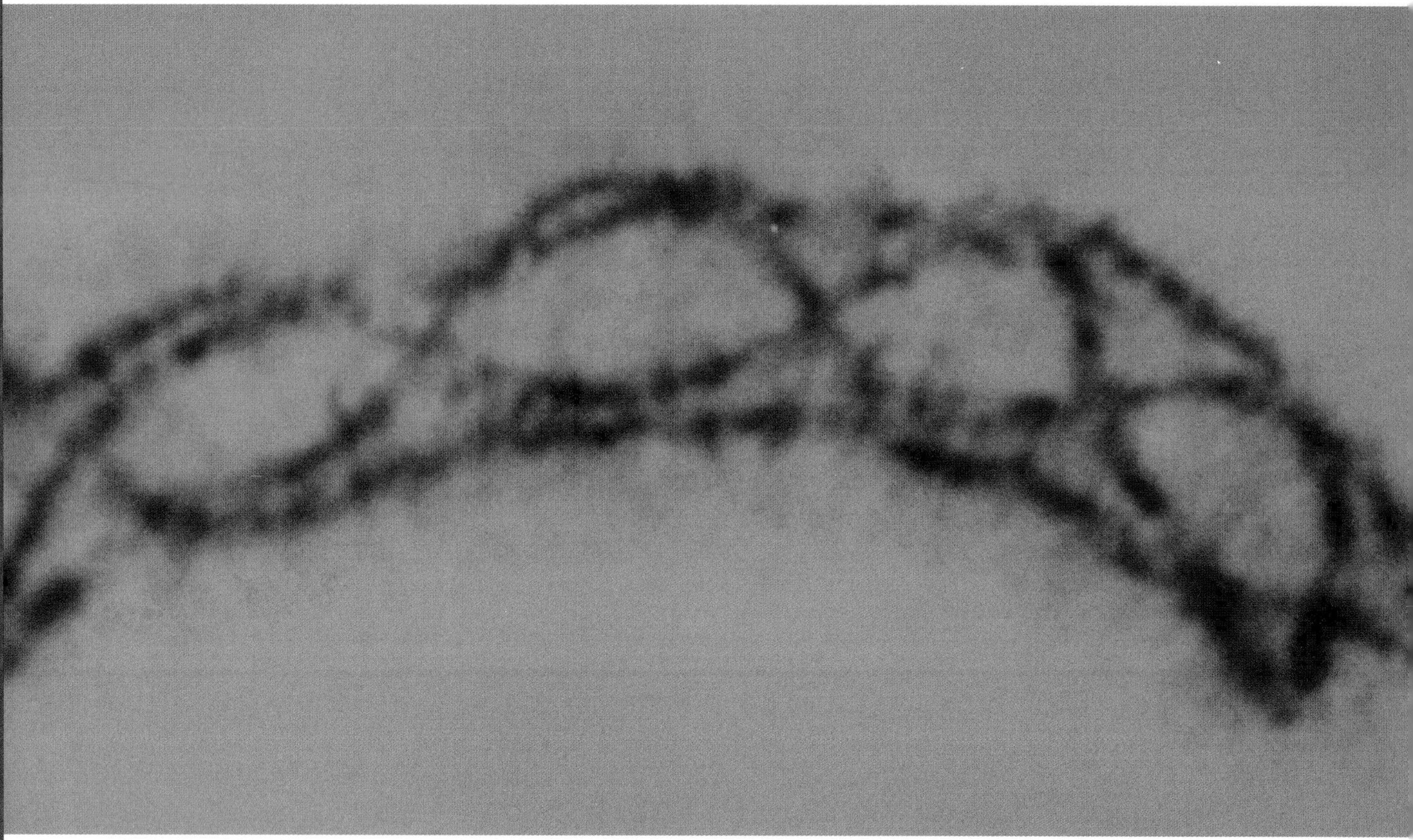

Chiasmata present between synapsed homologs during the first meiotic prophase.

CHAPTER CONCEPTS

- Chromosomes in eukaryotes contain many genes whose locations are fixed along the length of the chromosomes.
- Unless separated by crossing over, alleles present on a homolog segregate as a unit during gamete formation.
- Crossing over between homologs during meiosis creates recombinant gametes that enhance genetic variation.
- Crossing over between homologs serves as the basis for the construction of chromosome maps.
- Chromosome maps serve as an orderly characterization of genes in an organism.

Web Tutorial 5.1
Linkage and Recombination

Walter Sutton, along with Theodor Boveri, was instrumental in uniting the fields of cytology and genetics. As early as 1903, Sutton pointed out the likelihood that there must be many more "unit factors" than chromosomes in most organisms. Soon thereafter, genetic studies with several organisms revealed that certain genes were not transmitted according to the law of independent assortment; rather, these genes seemed to segregate as if they were somehow joined or linked together. Further investigations showed that such genes were part of the same chromosome, and they were indeed transmitted as a single unit.

We now know that most chromosomes consist of very large numbers of genes. Those that are part of the same chromosome are said to be *linked* and to demonstrate **linkage** in genetic crosses.

Because the chromosome, not the gene, is the unit of transmission during meiosis, linked genes are not free to undergo independent assortment. Instead, the alleles at all loci of one chromosome should, in theory, be transmitted as a unit during gamete formation. However, in many instances this does not occur. As we saw in Chapter 2, during the first meiotic prophase, when homologs are paired (or synapsed), a reciprocal exchange of chromosome segments may take place. This event, called **crossing over**, results in the reshuffling, or **recombination**, of the alleles between homologs.

Crossing over is currently viewed as an actual physical breaking and rejoining process that occurs during meiosis. You can see an example in the micrograph that opens this chapter. The exchange of chromosome segments provides an enormous potential for genetic variation in the gametes formed by any individual. This type of variation, in combination with that resulting from independent assortment, ensures that all offspring will contain a diverse mixture of maternal and paternal alleles.

The degree of crossing over between any two loci on a single chromosome is proportional to the distance between them, known as the **interlocus distance**. Thus, depending on which loci are being considered, the percentage of recombinant gametes varies. This correlation serves as the basis for the construction of **chromosome maps**, which indicate the relative locations of genes on the chromosomes.

In this chapter, we will discuss linkage, crossing over, and chromosome mapping. We will also consider a variety of other topics involving the exchange of genetic information, concluding the chapter with the rather intriguing question of why Mendel, who studied seven genes, did not encounter linkage. Or did he?

5.1 Genes Linked on the Same Chromosome Segregate Together

To provide a simplified overview of the major theme of this chapter, Figure 5–1 illustrates and contrasts the meiotic consequences of (a) independent assortment, (b) linkage *without* crossing over,

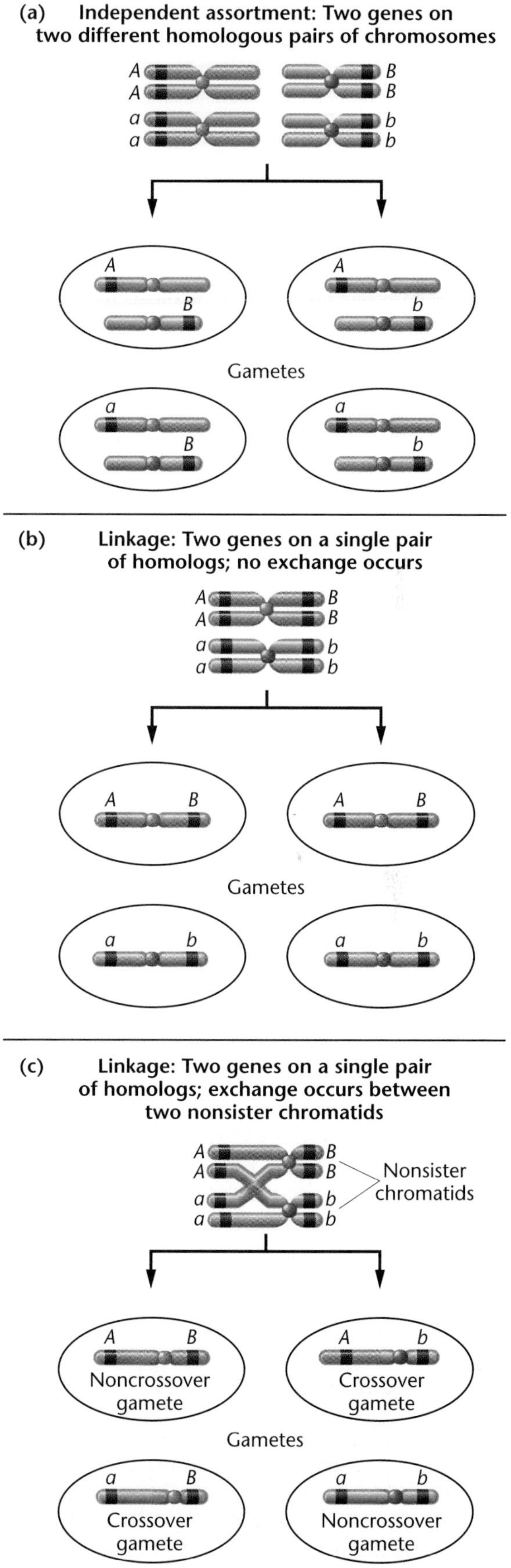
(a) Independent assortment: Two genes on two different homologous pairs of chromosomes
A
A
a
a
B
B
b
b
Gametes
(b) Linkage: Two genes on a single pair of homologs; no exchange occurs
Gametes
(c) Linkage: Two genes on a single pair of homologs; exchange occurs between two nonsister chromatids
Nonsister chromatids
Noncrossover gamete
Crossover gamete
Gametes
Crossover gamete
Noncrossover gamete

FIGURE 5–1 Results of gamete formation where two heterozygous genes are (a) on two different pairs of chromosomes; (b) on the same pair of homologs, but where no exchange occurs between them; and (c) on the same pair of homologs, where an exchange occurs between two nonsister chromatids.

and (c) linkage *with* crossing over. Figure 5–1(a) illustrates the results of independent assortment of two pairs of chromosomes, each containing one heterozygous gene pair. No linkage is exhibited. When a large number of meiotic events are observed, four genetically different gametes are formed in equal proportions. Each contains a different combination of alleles of the two genes.

We can compare these results with those that occur if the same genes are linked on the same chromosome. If no crossing over occurs between the two genes [Figure 5–1(b)], only two genetically different gametes are formed. Each gamete receives the alleles present on one homolog or the other, which has been transmitted intact as the result of segregation. This case illustrates **complete linkage**, which produces only **parental** or **noncrossover gametes**. The two parental gametes are formed in equal proportions. Though complete linkage between two genes seldom occurs, it is useful to consider the theoretical consequences of this concept when learning about crossing over.

Figure 5–1(c) illustrates the results of crossing over between two linked genes. As you can see, this crossover involves only two nonsister chromatids of the four chromatids present in the tetrad. This exchange generates two new allele combinations, called **recombinant** or **crossover gametes**. The two chromatids not involved in the exchange result in noncrossover gametes, like those in Figure 5–1(b).

The frequency with which crossing over occurs between any two linked genes is generally proportional to the distance separating the respective loci along the chromosome. In theory, two randomly selected genes can be so close to each other that crossover events are too infrequent to be easily detected. As shown in Figure 5–1(b), this circumstance, complete linkage, produces only parental gametes. On the other hand, if a small, but distinct distance separates two genes, few recombinant and many parental gametes will be formed. As the distance between the two genes increases, the proportion of recombinant gametes increases and that of the parental gametes decreases.

As we will discuss again later in this chapter, when the loci of two linked genes are far apart, the number of recombinant gametes approaches, but does not exceed, 50 percent. If 50 percent recombinants occurred, a 1:1:1:1 ratio of the four types (two parental and two recombinant gametes) would result. In such a case, transmission of two linked genes would be indistinguishable from that of two unlinked, independently assorting genes. That is, the proportion of the four possible genotypes would be identical, as shown in Figure 5–1(a) and (c).

Now solve this

Problem 5.9 on page 132 asks you to contrast the results of a testcross when two genes are unlinked versus linked, and when they are linked, if they are very far apart or relatively close together.

Hint: The results are indistinguishable when two genes are unlinked compared to the case where they are linked but so far apart that crossing over always intervenes between them during meiosis.

The Linkage Ratio

If complete linkage exists between two genes because of their close proximity and organisms heterozygous at both loci are mated, a unique F_2 phenotypic ratio results, which we shall designate the **linkage ratio**. To illustrate this ratio, we will consider a cross involving the closely linked recessive mutant genes *brown* (*bw*) eye and *heavy* (*hv*) wing vein in *Drosophila melanogaster* (Figure 5–2). The normal wild-type alleles bw^+ and hv^+ are both dominant and result in red eyes and thin wing veins, respectively.

In this cross, flies with mutant brown eyes and normal thin veins are mated to flies with normal red eyes and mutant heavy veins. To be more concise, we will discuss the flies simply in terms of their mutant phenotypes and say that brown-eyed flies are crossed with heavy-veined flies. Using the genetic symbols established in Chapter 4, we can represent linked genes by placing their allele designations above and below a single or double horizontal line. Those placed above the line are located at loci on one homolog and those below the line are at the homologous loci on the other homolog. Thus, we represent the P_1 generation as follows:

$$P_1: \frac{bw\ hv^+}{bw\ hv^+} \times \frac{bw^+\ hv}{bw^+\ hv}$$

brown, thin   red, heavy

Because the genes are located on an autosome, and not the X chromosome, no designation between males and females is necessary.

In the F_1 generation, each fly receives one chromosome of each pair from each parent; all flies are heterozygous for both gene pairs and exhibit the dominant traits of red eyes and thin veins:

$$F_1: \frac{bw\ \ hv^+}{bw^+\ hv}$$

red, thin

As shown in Figure 5–2, because of complete linkage, when the F_1 generation is interbred, each F_1 individual forms only parental gametes. Following fertilization, the F_2 generation will be produced in a 1:2:1 phenotypic and genotypic ratio. One fourth of this generation will show brown eyes and thin veins; half will show both wild-type traits, namely, red eyes and thin veins; and one-fourth will show red eyes and heavy veins. In more concise terms, the ratio is 1 brown:2 wild:1 heavy. Such a ratio is characteristic of complete linkage, which is observed only when two genes are very close together and the number of progeny is relatively small.

Figure 5–2 also demonstrates the results of a testcross with the F_1 flies. Such a cross produces a 1:1 ratio of brown, thin and red, heavy flies. Had the genes controlling these traits been incompletely linked or located on separate autosomes, the testcross would have produced four phenotypes, rather than two.

When we consider large numbers of mutant genes in any given species, genes located on the same chromosome will show evidence of linkage to one another. As a result, we can establish a **linkage group** for each chromosome. In theory, the number of linkage groups should correspond to the haploid number of chromosomes. In organisms with large numbers of mutant genes available for genetic study, this correlation has been confirmed.

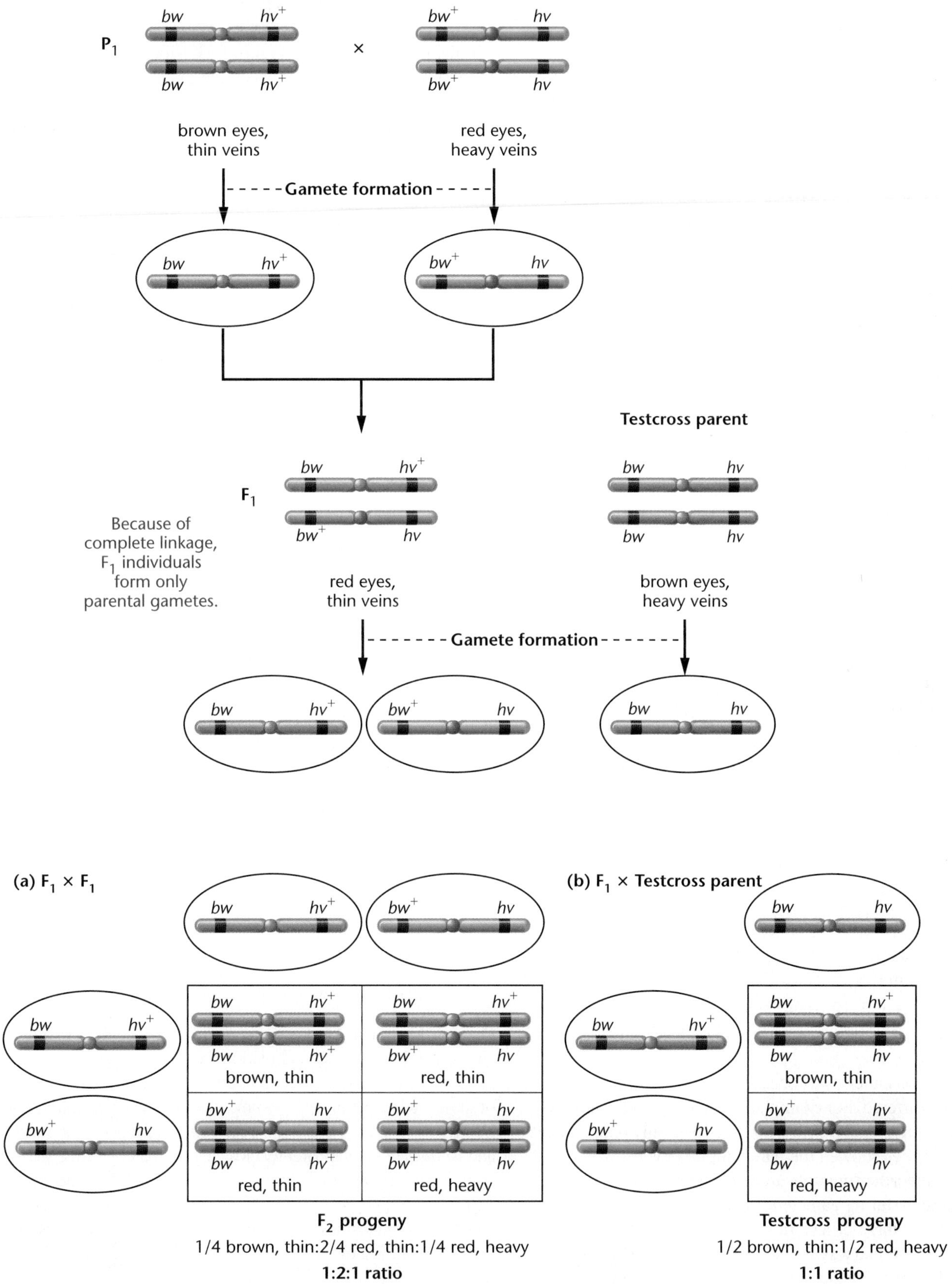

FIGURE 5–2 Results of a cross involving two genes located on the same chromosome where complete linkage is demonstrated. (a) The results of the cross. (b) The F_2 results of a testcross involving the F_1 progeny.

Web Tutorial 5.2
Linked Genes and Mapping

How Do We Know?

How did early geneticists establish that many genes were linked on single chromosomes and thus inherited as a unit, as opposed to being transmitted through gametes according to the Mendelian principle of independent assortment?

5.2 Crossing Over Serves as the Basis of Determining the Distance between Genes during Chromosome Mapping

It is highly improbable that two randomly selected genes linked on the same chromosome will be so close to one another along the chromosome that they demonstrate complete linkage. Instead, crosses involving two such genes will almost always produce a percentage of offspring resulting from recombinant gametes. The percentage is variable and depends upon the distance between the two genes along the chromosome. This phenomenon was first explained in 1911, by two *Drosophila* geneticists, Thomas H. Morgan and his undergraduate student, Alfred H. Sturtevant.

Morgan and Crossing Over

As you may recall from our discussion in Chapter 4, Morgan was the first to discover the phenomenon of X-linkage. In his studies, he investigated numerous *Drosophila* mutations located on the X chromosome. When he analyzed crosses involving only one trait, he was able to deduce the mode of X-linked inheritance. However, when he made crosses that simultaneously involved two X-linked genes, his results were puzzling at first. For example, as shown in cross A of Figure 5–3, he crossed mutant *yellow*-bodied (*y*) and *white*-eyed (*w*) females with wild-type males (gray body and red eyes). The F_1 females were wild type, whereas the F_1 males expressed both mutant traits. In the F_2 98.7 percent of the total offspring showed the parental phenotypes—yellow-bodied, white-eyed flies and wild-type flies (gray bodied, red eyed). The remaining 1.3 percent of the flies were either yellow bodied with red eyes or gray bodied with white eyes. It was as if the genes had somehow separated from each other during gamete formation in the F_1 flies.

When Morgan made crosses involving other X-linked genes, the results were even more puzzling (cross B of Figure 5–3). The same basic pattern was observed, but the proportion of F_2 phenotypes differed. For example, when he crossed *white*-eye, *miniature*-wing mutants with wild-type flies, only 62.8 percent of all the F_2 flies showed the parental phenotypes, while 37.2 percent of the offspring appeared as if the mutant genes had been separated during gamete formation.

As a result of these observations, Morgan was faced with two questions: (1) What was the source of gene separation? and (2) Why did the frequency of the apparent separation vary depending on the genes being studied? The answer Morgan proposed to the first question was based on his knowledge of earlier cytological observations made by F. A. Janssens and others. Janssens observed that synapsed homologous chromosomes in meiosis wrapped around each other, creating **chiasmata** (sing., **chiasma**) where points of overlap are evident. Morgan proposed that these chiasmata could represent points of genetic exchange.

In the crosses shown in Figure 5–3, Morgan postulated that if an exchange occurred between the mutant genes on the two X chromosomes of the F_1 females, it would lead to the percentages in the observed results. He suggested that such exchanges led to 1.3 percent recombinant gametes in the *yellow–white* cross and 37.2 percent in the *white–miniature* cross. On the basis of this and other experimentation, Morgan concluded that if linked genes exist in a linear order along the chromosome, then a variable amount of exchange occurs between any two genes.

As an answer to the second question, Morgan proposed that two genes located relatively close to each other along a chromosome are less likely to have a chiasma form between them than if the two genes are farther apart on the chromosome. Therefore, the closer two genes are, the less likely it is that a genetic exchange will occur between them. Morgan proposed the term **crossing over** to describe the physical exchange leading to recombination.

Sturtevant and Mapping

Morgan's student, Alfred H. Sturtevant, was the first to realize that his mentor's proposal could be used to map the sequence of linked genes. According to Sturtevant,

> "In a conversation with Morgan ... I suddenly realized that the variations in strength of linkage, already attributed by Morgan to differences in the spatial separation of the genes, offered the possibility of determining sequences in the linear dimension of a chromosome. I went home and spent most of the night (to the neglect of my undergraduate homework) in producing the first chromosomal map."

For example, Sturtevant compiled data on recombination between the genes represented by the *yellow*, *white*, and *miniature* mutants initially studied by Morgan and observed the following frequencies of crossing over between each pair of these three genes:

(1) *yellow, white*	0.5 percent
(2) *white, miniature*	34.5 percent
(3) *yellow, miniature*	35.4 percent

Because the sum of (1) and (2) is approximately equal to (3), Sturtevant argued that the recombination frequencies between linked genes are additive. On this assumption, he predicted that the order of the genes on the X chromosome was *yellow–white–miniature*. In arriving at this conclusion, he reasoned as follows: The *yellow* and *white* genes are apparently close to each other because the recombination frequency is low. However, both of these genes are quite far from *miniature*, since the *white, miniature* and *yellow, miniature* combinations show large recombination frequencies. Because *miniature* shows more recombination with *yellow* than with *white* (35.4 percent vs. 34.5 percent), it follows that *white* is between the other two genes, not outside of them.

Sturtevant knew from Morgan's work that the frequency of exchange could be taken as an estimate of the relative distance

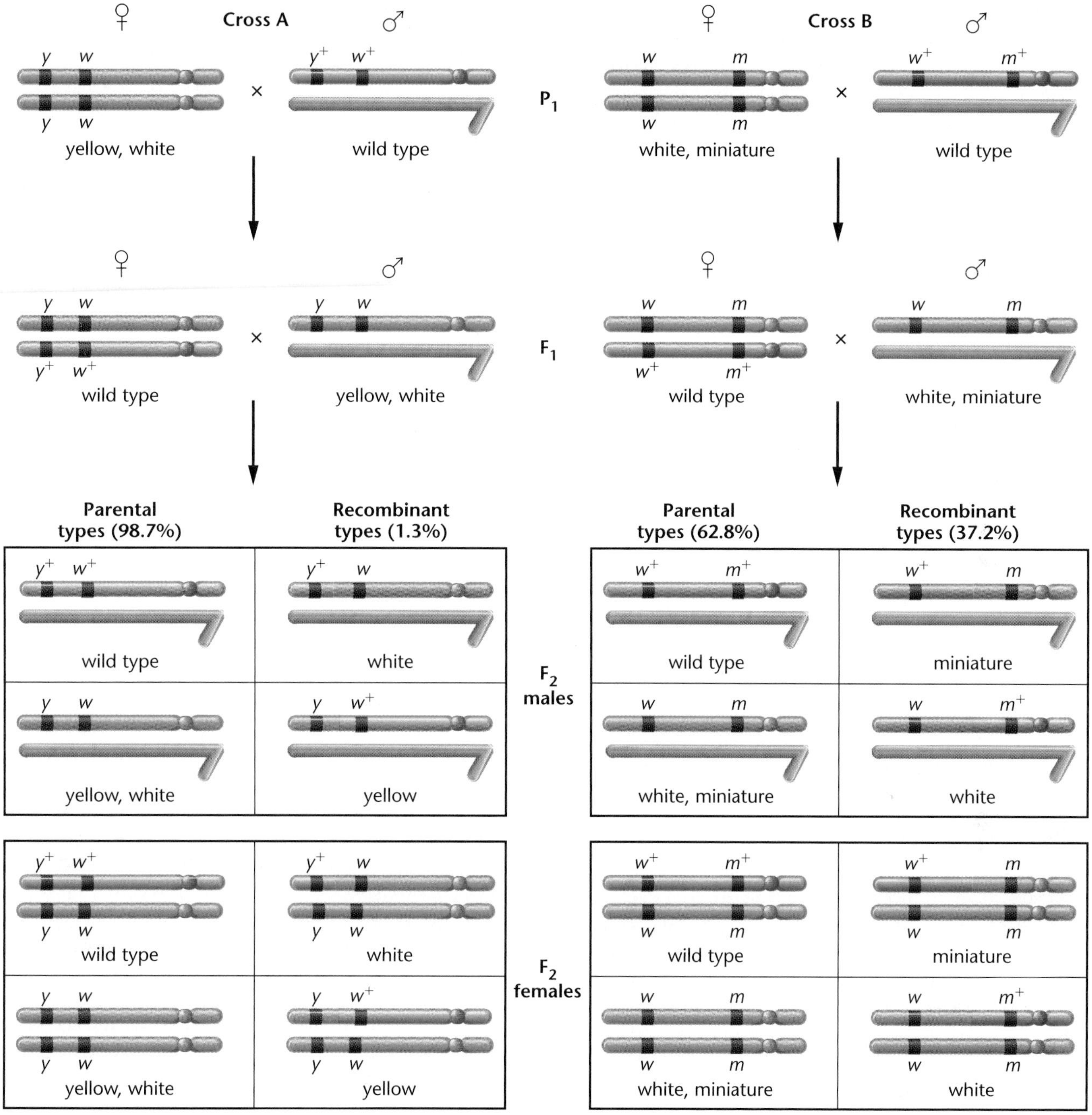

FIGURE 5–3 The F_1 and F_2 results of crosses involving the *yellow*-body, *white*-eye mutations and the *white*-eye, *miniature*-wing mutations. In cross A, 1.3 percent of the F_2 flies (males and females) demonstrate recombinant phenotypes, which express either *white* or *yellow*. In cross B, 37.2 percent of the F_2 flies (males and females) demonstrate recombinant phenotypes, which are either *miniature* or *white* mutants.

between two genes or loci along the chromosome. He constructed a map of the three genes on the X chromosome, with one map unit equated with 1 percent recombination between two genes.* In the preceding example, the distance between *yellow* and *white* would be 0.5 map unit (mu), and between *yellow* and *miniature* 35.4 mu. It follows that the distance between *white* and *miniature* should be (35.4 − 0.5), or 34.9 mu. This estimate is close to the actual frequency of recombination between *white* and *miniature* (34.5 mu). The simple map for these three genes is shown in Figure 5–4. It is important to keep in mind that map units such as those shown represent *relative* measurements of distance.

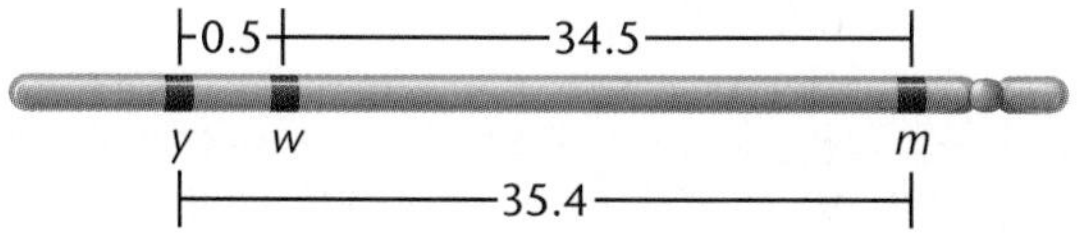

FIGURE 5–4 A map of the *yellow* (*y*), *white* (*w*), and *miniature* (*m*) genes on the X chromosome of *Drosophila melanogaster*. Each number represents the percentage of recombinant offspring produced in one of three crosses, each involving two different genes.

*In honor of Morgan's work, map units are often referred to as centimorgans (cM).

In addition to these three genes, Sturtevant considered two other genes on the X chromosome and produced a more extensive map, including all five genes. Soon thereafter, he and a colleague, Calvin Bridges, began a search for autosomal linkage in *Drosophila*. By 1923, they had clearly shown that linkage and crossing over were not restricted to X-linked genes, but could also be demonstrated with autosomes.

During their work, Sturtevant and Bridges made another interesting observation. In *Drosophila,* crossing over was shown to occur only in females. The fact that no crossing over occurs in males made the analysis of genetic mapping in *Drosophila* much less complex. In most other organisms, however, crossing over does occur in both sexes.

Although many refinements in chromosome mapping have developed since Sturtevant's initial work, his basic principles are accepted as correct and have been used to produce detailed chromosome maps of organisms for which large numbers of linked mutant genes are known. In addition to providing the basis for chromosome mapping, Sturtevant's findings were historically significant to the field of genetics. In 1910, the chromosomal theory of inheritance was still being widely disputed. Even Morgan was skeptical of this theory before he conducted the bulk of his experimentation. Research has now firmly established that chromosomes contain genes in a linear order and that these genes are the equivalent of Mendel's unit factors.

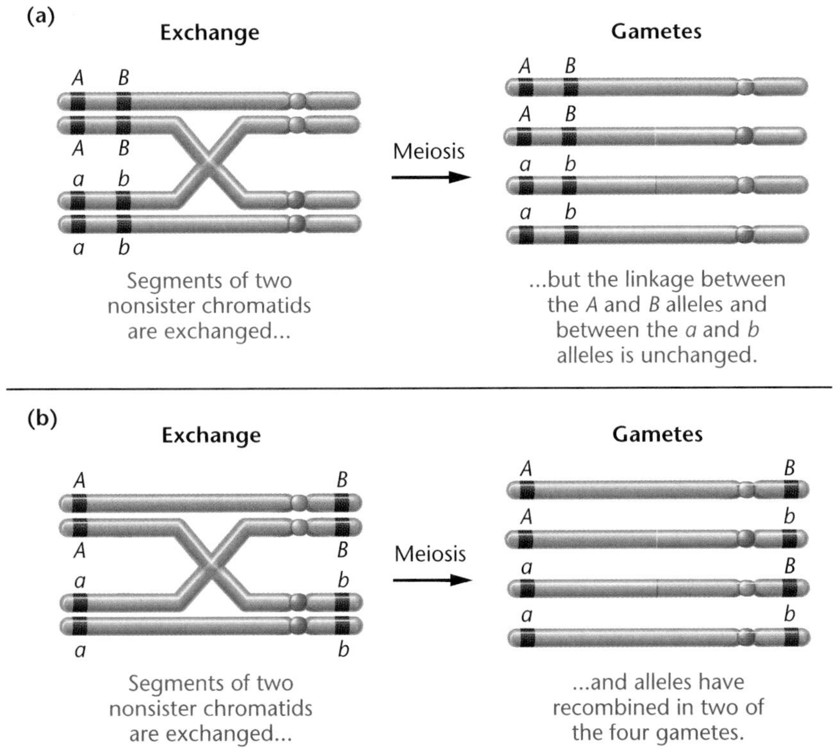
(a)
Exchange
A B
A B
a b
a b
Segments of two nonsister chromatids are exchanged...
Meiosis
Gametes
A B
A B
a b
a b
...but the linkage between the A and B alleles and between the a and b alleles is unchanged.
(b)
Exchange
A B
A B
a b
a b
Segments of two nonsister chromatids are exchanged...
Meiosis
Gametes
A B
A b
a B
a b
...and alleles have recombined in two of the four gametes.

FIGURE 5–5 Two examples of a single crossover between two nonsister chromatids and the gametes subsequently produced. In (a) the exchange does not alter the linkage arrangement between the alleles of the two genes, only parental gametes are formed, and the exchange goes undetected. In (b) the exchange separates the alleles, resulting in recombinant gametes, which are detectable.

Single Crossovers

Why should the relative distance between two loci influence the amount of recombination and crossing over observed between them? The basis for this variation is explained in the following analysis:

During meiosis, a limited number of crossover events occur in each tetrad. These recombinant events occur randomly along the length of the tetrad. Therefore, the closer that two loci reside along the axis of the chromosome, the less likely it is that any **single crossover** event will occur in between them. The same reasoning suggests that the farther apart two linked loci are, the more likely it is that a random crossover event will occur in between them.

In Figure 5–5(a), a single crossover occurs between two nonsister chromatids, but in an area of the chromosome that is not in between the two loci. Therefore, the crossover goes undetected because no recombinant gametes are produced. In Figure 5–5(b), where two loci are quite far apart, the crossover occurs in an area of the chromosome between them. This separates them, yielding recombinant gametes.

When a single crossover occurs between two nonsister chromatids, the other two chromatids of the tetrad are not involved in the exchange, and they enter the gamete unchanged. Following the completion of meiosis, they represent noncrossover gametes. In this case, even if a single crossover occurs 100 percent of the time between two linked genes, recombination will subsequently be observed in only 50 percent of the potential gametes formed. This concept is diagrammed in Figure 5–6. Theoretically, if we consider only single exchanges and observe 20 percent recombinant gametes, crossing over actually occurs between these two loci in 40 percent of the tetrads. The general rule is that under these conditions, the percentage of tetrads involved in an exchange between two genes is twice as great as the percentage of recombinant gametes produced. Therefore, the theoretical limit of recombination due to crossing over is 50 percent.

When two linked genes are more than 50 map units apart, a crossover can theoretically be expected to occur between them in 100 percent of the tetrads. If this prediction were achieved, each tetrad would yield equal proportions of the four gametes shown in Figure 5–6, just as if the genes were on different chromosomes and assorting independently. For a variety of reasons, this theoretical limit is seldom achieved.

HOW DO WE KNOW?

How was it established experimentally that the frequency of recombination (crossing over) between two genes is related to the distance between them?

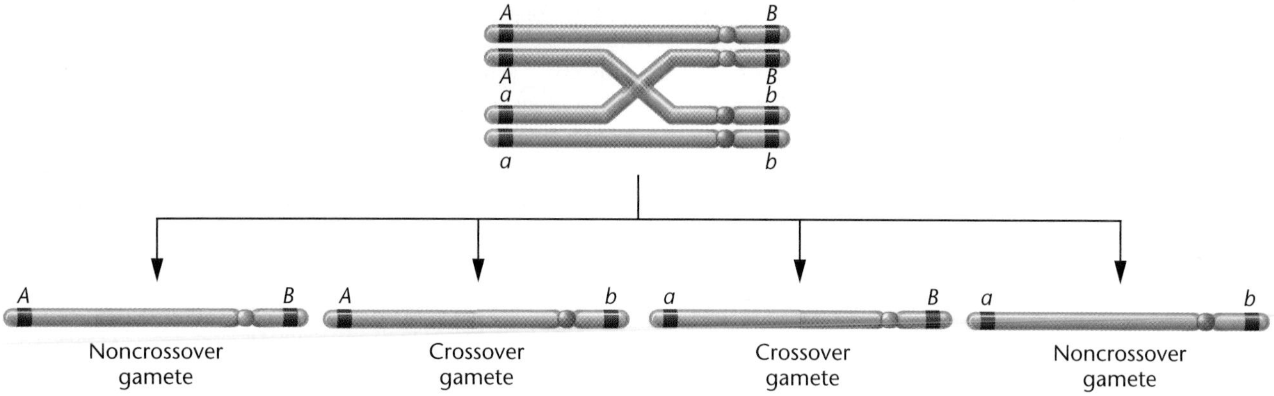

FIGURE 5–6 The consequences of a single exchange between two nonsister chromatids occurring in the tetrad stage. Two noncrossover (parental) and two crossover (recombinant) gametes are produced.

5.3 Determining the Gene Sequence during Mapping Relies on the Analysis of Multiple Crossovers

The study of single crossovers between two linked genes provides the basis of determining the *distance* between them. However, when many linked genes are studied, their *sequence* along the chromosome is more difficult to determine. Fortunately, the discovery that multiple exchanges occur between the chromatids of a tetrad has facilitated the process of producing more extensive chromosome maps. As we shall see next, when three or more linked genes are investigated simultaneously, it is possible to determine first the sequence of and then the distances between genes.

Multiple Exchanges

It is possible that in a single tetrad, two, three, or more exchanges will occur between nonsister chromatids as a result of several crossing over events. Double exchanges of genetic material result from **double crossovers (DCOs)**, as shown in Figure 5–7. To study a double exchange, three gene pairs must be investigated, each heterozygous for two alleles. Before we can determine the frequency of recombination among all three loci, we must review some simple probability calculations.

As we have seen, the probability of a single exchange occurring in between the *A* and *B* or the *B* and *C* genes is directly related to the physical distance separating the loci. The closer *A* is to *B* and *B* is to *C*, the less likely it is that a single exchange will occur in between either of the two sets of loci. In the case of a double crossover, two separate and independent events or exchanges must occur simultaneously. The mathematical probability of two independent events occurring simultaneously is equal to the product of the individual probabilities. This is the product law we introduced in Chapter 3.

Suppose that crossover gametes result from single exchanges between *A* and *B* 20 percent of the time ($p = 0.20$) and between *B* and *C* 30 percent of the time ($p = 0.30$). The probability of recovering a double-crossover gamete arising from two exchanges, between *A* and *B* and between *B* and *C*, is predicted to be $(0.20)(0.30) = 0.06$, or 6 percent. This calculation shows that the expected frequency of double-crossover gametes is always much lower than that of either single-crossover class of gametes alone.

If three genes are relatively close together along one chromosome, the expected frequency of double-crossover gametes is

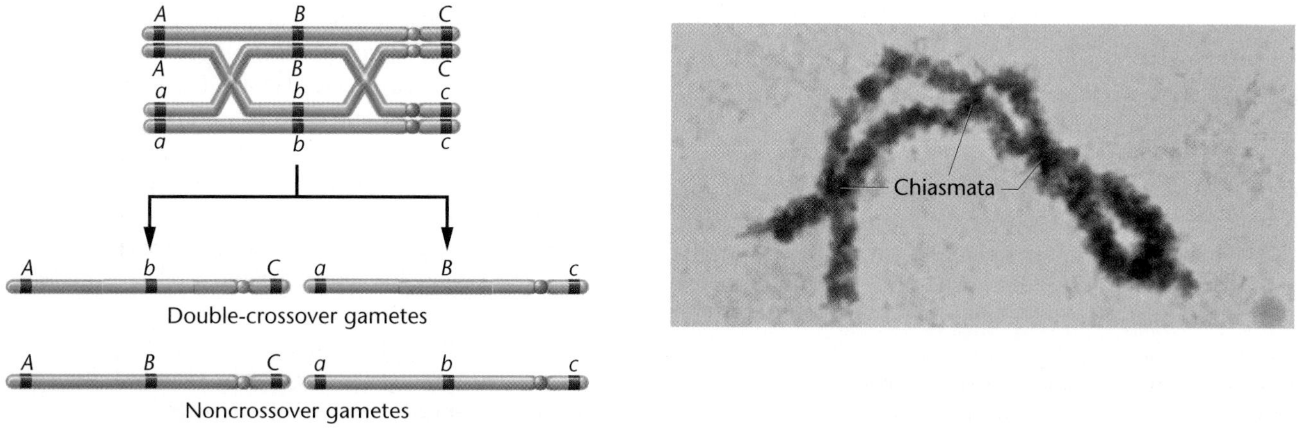

FIGURE 5–7 Consequences of a double exchange occurring between two nonsister chromatids. Because the exchanges involve only two chromatids, two noncrossover gametes and two double-crossover gametes are produced. The photograph illustrates several chiasmata found in a tetrad isolated during the first meiotic prophase stage.

extremely low. For example, assume the *A–B* distance in Figure 5–7 to be 3 mu and the *B–C* distance to be 2 mu. The expected double-crossover frequency would be $(0.03)(0.02) = 0.0006$, or 0.06 percent. This translates to only 6 events in 10,000. In such a mapping experiment, where closely linked genes are involved, very large numbers of offspring are required to detect double-crossover events. In this example, it would be unlikely that a double crossover would be observed even if 1000 offspring are examined. If we extend these probability considerations, it is evident that if four or five genes were being mapped, even fewer triple and quadruple crossovers could be expected to occur.

Now solve this

Problem 5.14 on page 132 asks you to contrast the results of crossing over when the arrangement of alleles along the homologs differs in an organism heterozygous for three genes.

Hint: Homologs enter noncrossover gametes unchanged and all crossover results must be derived from the noncrossover sequence of alleles.

Three-Point Mapping in *Drosophila*

The information presented in the previous section serves as the basis for mapping three or more linked genes in a single cross. To illustrate this, let us examine a situation involving three linked genes. Three criteria must be met for a successful mapping cross:

1. The genotype of the organism producing the crossover gametes must be heterozygous at all loci under consideration. If homozygosity occurred at any locus, all gametes produced would contain the same allele, precluding mapping analysis.
2. The cross must be constructed so that genotypes of all gametes can be accurately determined by observing the phenotypes of the resulting offspring. This approach is necessary because the gametes and their genotypes can never be observed directly. To overcome this problem, each phenotypic class must reflect the genotype of the gametes of the parents producing it.
3. A sufficient number of offspring must be produced in the mapping experiment to recover a representative sample of all crossover classes.

These criteria are met in the three-point mapping cross from *Drosophila melanogaster* shown in Figure 5–8. The cross involves three X-linked recessive mutant genes—*yellow* body color, *white* eye color, and *echinus* eye shape. To diagram the cross, we must assume some theoretical sequence, even though we do not yet know if it is correct. In Figure 5–8, we initially assume the sequence of the three genes to be *y–w–ec*. If this is incorrect, our analysis shall demonstrate it and reveal the correct sequence.

Web Tutorial 5.3 Three-Point Mapping

In the P_1 generation, males hemizygous for all three wild-type alleles are crossed to females that are homozygous for all three recessive mutant alleles. Therefore, the males are wild type with respect to body color, eye color, and eye shape, and they are said to have a wild-type phenotype. The females, on the other hand, exhibit the three mutant traits: *yellow* body color, *white* eyes, and *echinus* eye shape.

This cross produces an F_1 generation consisting of females heterozygous at all three loci and males that, because of the Y chromosome, are hemizygous for the three mutant alleles. Phenotypically, all F_1 females are wild type, while all F_1 males are *yellow*, *white*, and *echinus*. The genotype of the F_1 females fulfills the first criterion for constructing a map of the three linked genes; that is, it is heterozygous at the three loci and may serve as the source of recombinant gametes generated by crossing over. Note that, because of the genotypes of the P_1 parents, all three of the F_1 mutant alleles are on one homolog and all three wild-type alleles are on the other homolog. *Other arrangements are possible.* For example, the heterozygous F_1 female might have the *y* and *ec* mutant alleles on one homolog and the *w* allele on the other. This would occur if, in the P_1 cross, one parent was *yellow* and *echinus* and the other parent was *white*.

In our cross, the second criterion is met as a result of the gametes formed by the F_1 males. Every gamete will contain either an X chromosome bearing the three mutant alleles or a Y chromosome, which does not contain any of the three loci being considered. Whichever type participates in fertilization, the genotype of the gamete produced by the F_1 female will be expressed phenotypically in the F_2 female and male offspring. As a result, all noncrossover and crossover gametes produced by the F_1 female parent can be determined by observing the F_2 phenotypes.

With these two criteria met, we can construct a chromosome map from the crosses illustrated in Figure 5–8, but first, we must determine which F_2 phenotypes correspond to the various noncrossovers and crossover categories. Two of these can be determined immediately.

The **noncrossover** F_2 phenotypes are determined by the combination of alleles present in the parental gametes formed by the F_1 female. In this case, each gamete contains either three wild-type alleles *or* three mutant alleles, depending on which of the X chromosomes is unaffected by crossing over. As a result of segregation, approximately equal proportions of the two types of gametes, and subsequently the F_2 phenotypes, are produced. Since the F_2 phenotypes complement one another (i.e., one is wild type and the other is mutant for all three genes), they are called **reciprocal classes** of phenotypes.

The two noncrossover phenotypes are most easily recognized because *they occur in the greatest proportion of offspring.* Figure 5–8 shows that gametes (1) and (2) are present in the greatest numbers. Therefore, flies that are *yellow*, *white*, and *echinus* and those that are normal or wild type for all three characters constitute the noncrossover category and represent 94.44 percent of the F_2 offspring.

The second easily detected category is represented by the double-crossover phenotypes. Because of their probability of occurrence, *they must be present in the least numbers.* Remember that this group represents two independent, but simultaneous single-crossover events. Two reciprocal phenotypes can be identified: gamete (7), which shows the mutant traits *yellow* and *echinus*, but normal eye color; and gamete (8), which shows the mutant trait *white*, but normal body color and eye shape. Together these double-crossover phenotypes constitute only 0.06 percent of the F_2 offspring.

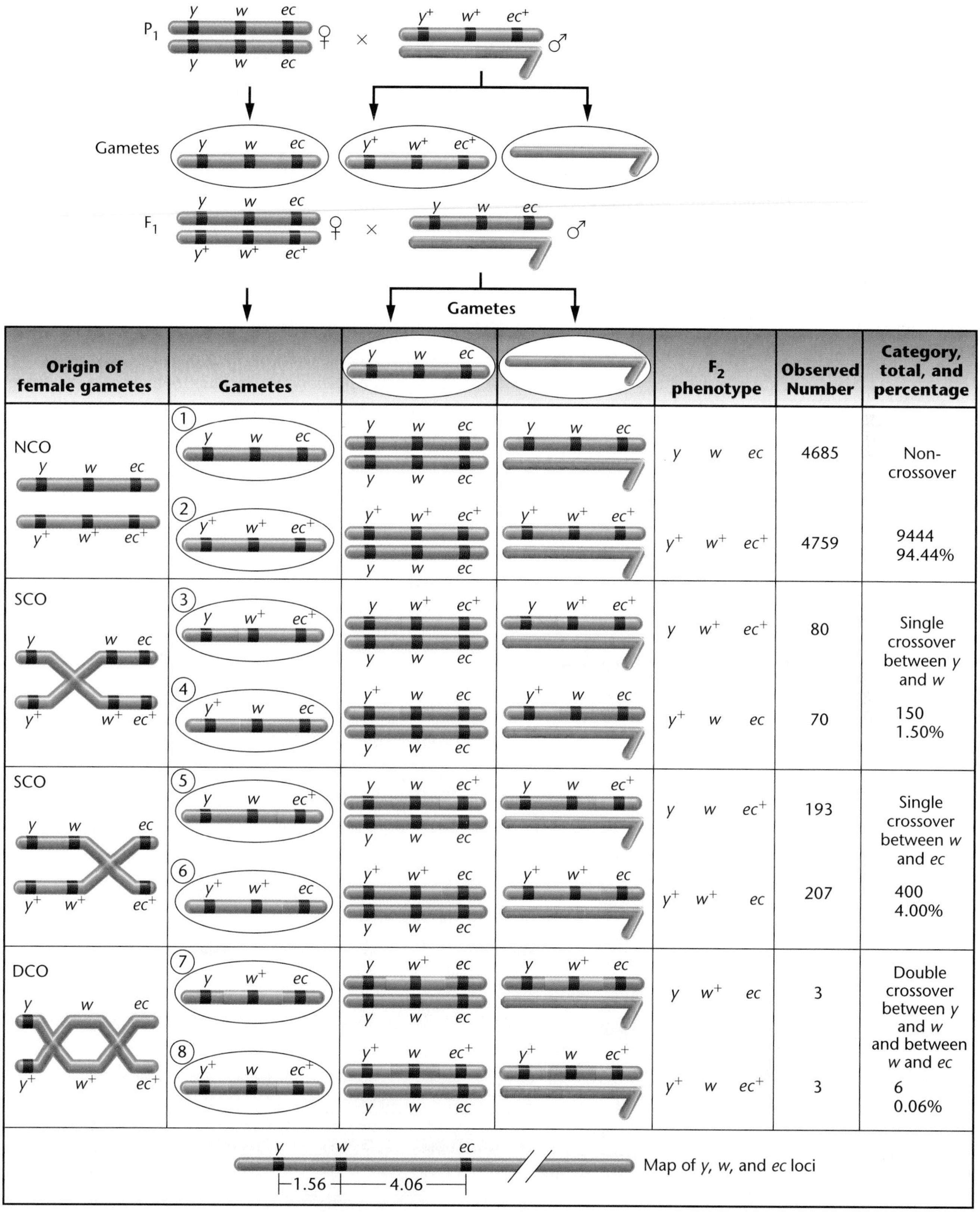

P1
Gametes
F1
Gametes
Origin of female gametes
Gametes
F2 phenotype
Observed Number
Category, total, and percentage
NCO
y w ec
4685
y+ w+ ec+
4759
Non-crossover
9444
94.44%
SCO
y w+ ec+
80
y+ w ec
70
Single crossover between y and w
150
1.50%
SCO
y w ec+
193
y+ w+ ec
207
Single crossover between w and ec
400
4.00%
DCO
y w+ ec
3
y+ w ec+
3
Double crossover between y and w and between w and ec
6
0.06%
1.56
4.06
Map of y, w, and ec loci

FIGURE 5–8 A three-point mapping cross involving the *yellow* (y or y^+), *white* (w or w^+), and *echinus* (ec or ec^+) genes in *Drosophila melanogaster*. NCO, SCO, and DCO refer to noncrossover, single-crossover, and double-crossover groups, respectively. Because of the complexity of this and several of the ensuing figures, centromeres have not been included on the chromosomes, and only two nonsister chromatids are initially shown in the left-hand column.

The remaining four phenotypic classes fall into two categories resulting from single crossovers. Gametes (3) and (4), reciprocal phenotypes produced by single-crossover events occurring between the *yellow* and *white* loci, are equal to 1.50 percent of the F_2 offspring. Gametes (5) and (6), constituting 4.00 percent of the F_2 offspring, represent the reciprocal phenotypes resulting from single-crossover events occurring between the *white* and *echinus* loci.

We can now calculate the map distances between the three loci. The distance between *y* and *w*, or between *w* and *ec*, is equal to the percentage of all detectable exchanges occurring between them. For any two genes under consideration, this will include all related single crossovers as well as all double crossovers. The latter are included because they represent two simultaneous single crossovers. For the *y* and *w* genes, this includes gametes (3), (4), (7), and (8), totaling 1.50 percent + 0.06 percent, or 1.56 mu. Similarly, the distance between *w* and *ec* is equal to the percentage of offspring resulting from an exchange between these two loci, including gametes (5), (6), (7), and (8), totaling 4.00 percent + 0.06 percent, or 4.06 mu. The map of these three loci on the X chromosome, based on these data, is shown at the bottom of Figure 5–8.

Determining the Gene Sequence

In the preceding example, we assumed the order, or sequence, of the three genes along the chromosome to be *y–w–ec*. Our analysis established that the sequence is consistent with the data. However, in most mapping experiments, the gene sequence is not known, and this constitutes another variable in the analysis. In our example, had the gene order been unknown, we could have used one of two methods (which we will study next) to determine it. In your own work, you should select one of these methods and use it consistently.

Method I This method is based on the fact that there are only three possible orders, each containing one of the three genes in between the other two:

(I)	*w–y–ec*	(*y* is in the middle)
(II)	*y–ec–w*	(*ec* is in the middle)
(III)	*y–w–ec*	(*w* is in the middle)

The following steps will allow you to determine which gene order is correct:

1. Assuming any of the three orders, first determine the *arrangement of alleles* along each homolog of the heterozygous parent giving rise to noncrossover and crossover gametes (the F_1 female in our example).
2. Determine whether a double-crossover event occurring within that arrangement will produce the *observed double-crossover phenotypes*. Remember that these phenotypes occur least frequently and can be identified easily.
3. If this order does not produce the correct phenotypes, try each of the other two orders. One of the three must work!

The next steps are shown in Figure 5–9, using the cross just discussed. The three possible orders are labeled I, II, and III, as shown previously. Either *y*, *ec*, or *w* must be in the middle:

1. If we assume order I with *y* between *w* and *ec*, the arrangement of alleles along the homologs of the heterozygote is

$$\frac{w \quad y \quad ec}{w^+ \quad y^+ \quad ec^+}$$

We know this because of the way in which the P_1 generation was crossed. The P_1 female contributed an X

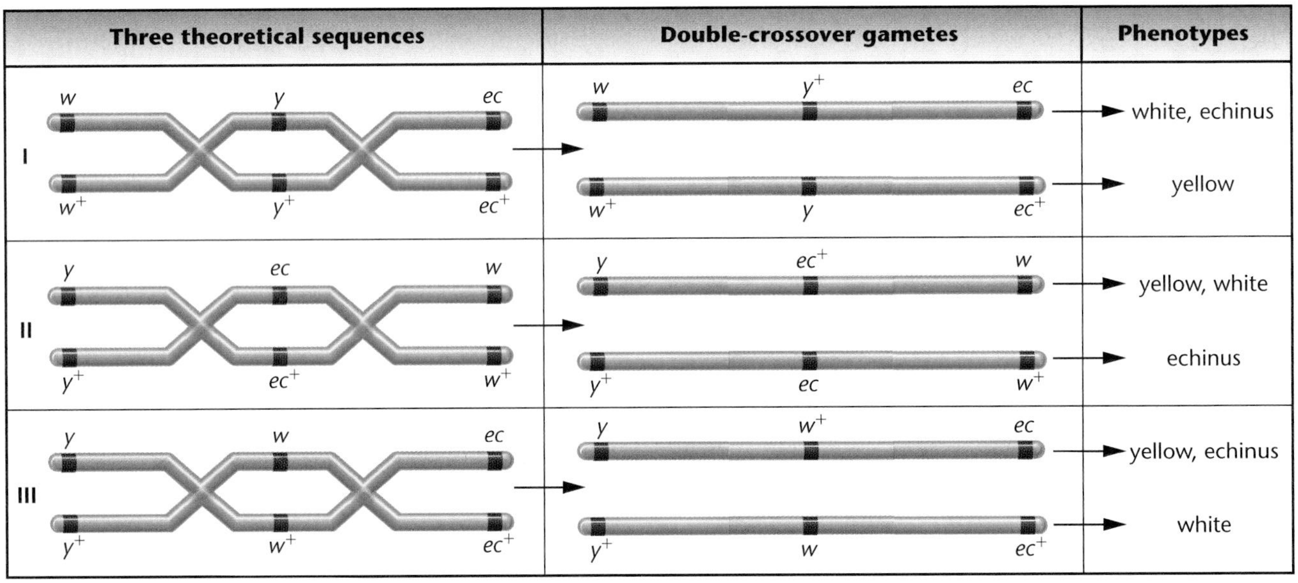

FIGURE 5–9 The three possible sequences of the *white*, *yellow*, and *echinus* genes, the results of a double crossover in each case, and the resulting phenotypes produced in a testcross. For simplicity, the two noncrossover chromatids of each tetrad are omitted.

chromosome bearing the *w*, *y*, and *ec* alleles, while the P_1 male contributed an X chromosome bearing the w^+, y^+, and ec^+ alleles.

2. A double crossover within the previous arrangement would yield the following gametes:

$$\underline{w \quad y^+ \quad ec} \quad \text{and} \quad \underline{w^+ \quad y \quad ec^+}$$

If *y* is in the middle, following fertilization, the F_2 double-crossover phenotypes will correspond to the foregoing gametic genotypes, yielding offspring that express the white, echinus phenotype and offspring that express the yellow phenotype. Determination of the actual double crossovers, however, reveals them to be yellow, echinus flies and white flies. Therefore, our assumed order is incorrect.

3. If we consider the other orders, one with ec/ec^+ alleles in the middle (II) or one with the w/w^+ alleles in the middle (III),

$$\text{(II)}\ \frac{y \quad ec \quad w}{y^+ \quad ec^+ \quad w^+} \quad \text{or} \quad \text{(III)}\ \frac{y \quad w \quad ec}{y^+ \quad w^+ \quad ec^+}$$

we see that arrangement II again provides *predicted* double-crossover phenotypes that do not correspond to the *actual* double-crossover phenotypes. The predicted phenotypes are yellow, white flies and echinus flies in the F_2 generation. Therefore, this order is also incorrect. However, arrangement III *will* produce the *observed* phenotypes—*yellow*, *echinus* flies and *white* flies. Therefore, this order, where the *w* gene is in the middle, is correct.

To summarize Method I, first, determine the arrangement of alleles on the homologs of the heterozygote yielding the crossover gametes. This is done by locating the reciprocal noncrossover phenotypes. Then test each of three possible orders to determine which one yields the observed (actual) double-crossover phenotypes. Whichever of the three does so represents the correct order. This method is summarized in Figure 5–9.

Method II Method II also begins by assuming the arrangement of alleles along each homolog of the heterozygous parent. In addition, it requires one further assumption:

Following a double-crossover event, the allele in the middle position will fall between the outside, or flanking, alleles that were present on the opposite parental homolog.

To illustrate, assume order (I), *w–y–ec*, in the following arrangement:

$$\frac{w \quad y \quad ec}{w^+ \quad y^+ \quad ec^+}$$

Following a double-crossover event, the *y* and y^+ alleles would be switched to this arrangement:

$$\frac{w \quad y^+ \quad ec}{w^+ \quad y \quad ec^+}$$

After segregation, two gametes would be formed:

$$\underline{w \quad y^+ \quad ec} \quad \text{and} \quad \underline{w^+ \quad y \quad ec^+}$$

Because the genotype of the gamete will be expressed directly in the phenotype following fertilization, the double-crossover phenotypes will be:

white, echinus flies, and yellow flies

Note that the *yellow* allele, assumed to be in the middle, is now associated with the two outside markers of the other homolog, w^+ and ec^+. However, these predicted phenotypes do not coincide with the observed double-crossover phenotypes. Therefore, the *yellow* gene is not in the middle.

This same reasoning can be applied to the assumption that the *echinus* gene or the *white* gene is in the middle. In the former case, we will reach a negative conclusion. If we assume that the *white* gene is in the middle, the *predicted* and *actual* double crossovers coincide. Therefore, we conclude that the *white* gene is located between the *yellow* and *echinus* genes.

To summarize Method II, determine the arrangement of alleles on the homologs of the heterozygote yielding crossover gametes. Then determine the actual double-crossover phenotypes. Simply select the single allele that has been switched so that it is now no longer associated with its original neighboring alleles.

In our example *y*, *ec*, and *w* are together in the F_1 heterozygote, as are y^+, ec^+, and w^+. In the F_2 double-crossover classes, it is *w* and w^+ that have been switched. The *w* allele is now associated with y^+ and ec^+, while the w^+ allele is now associated with the *y* and *ec* alleles. Therefore, the *white* gene is in the middle, and the *yellow* and *echinus* genes are the flanking markers.

? HOW DO WE KNOW?

What is the experimental basis for establishing the sequence of genes along a chromosome during a mapping experiment?

A Mapping Problem in Maize

Having established the basic principles of chromosome mapping, we will now consider a problem in maize (corn), in which the gene sequence and interlocus distances are unknown:

1. The previous mapping cross involved X-linked genes. Here, autosomal genes are considered.

2. In the previous discussion, we initially assumed a particular gene sequence, and we then introduced methods to determine whether it or a different sequence was correct. In this analysis of maize, the sequence is *initially* unknown.

3. In the discussion of this cross, we will use different symbols for alleles. Instead of using the symbols bm^+, v^+, and pr^+, we will simply use + to denote each wild-type allele. As first described in Chapter 4, this annotation of symbols is easier to manipulate, but it requires a strong understanding of mapping procedures.

This analysis differs from the preceding discussion in several ways and therefore will expand your knowledge of mapping procedures.

When we consider three autosomally linked genes in maize, the experimental cross must still meet the same three criteria established for the X-linked genes in *Drosophila:* (1) one parent must be heterozygous for all traits under consideration, (2) the gametic genotypes produced by the heterozygote must be apparent from observing the phenotypes of the offspring, and (3) a sufficient sample size must be available.

In maize, the recessive mutant genes *bm* (*brown* midrib), *v* (*virescent* seedling), and *pr* (*purple* aleurone) are linked on chromosome 5. Assume that a female plant is known to be heterozygous for all three traits. Nothing is known about the arrangement of the mutant alleles on the maternal and paternal homologs of this heterozygote, the sequence of genes, or the map distances between the genes. What genotype must the male plant have to allow successful mapping? To meet the second criterion, the male must be homozygous for all three recessive mutant alleles. Otherwise, offspring of this cross showing a given phenotype might represent more than one genotype, making accurate mapping impossible. Note that this is equivalent to performing a testcross.

Figure 5–10 diagrams this cross. As shown, we do not know either the arrangement of alleles or the sequence of loci in the

(a) Some possible allele arrangements and gene sequences in a heterozygous female

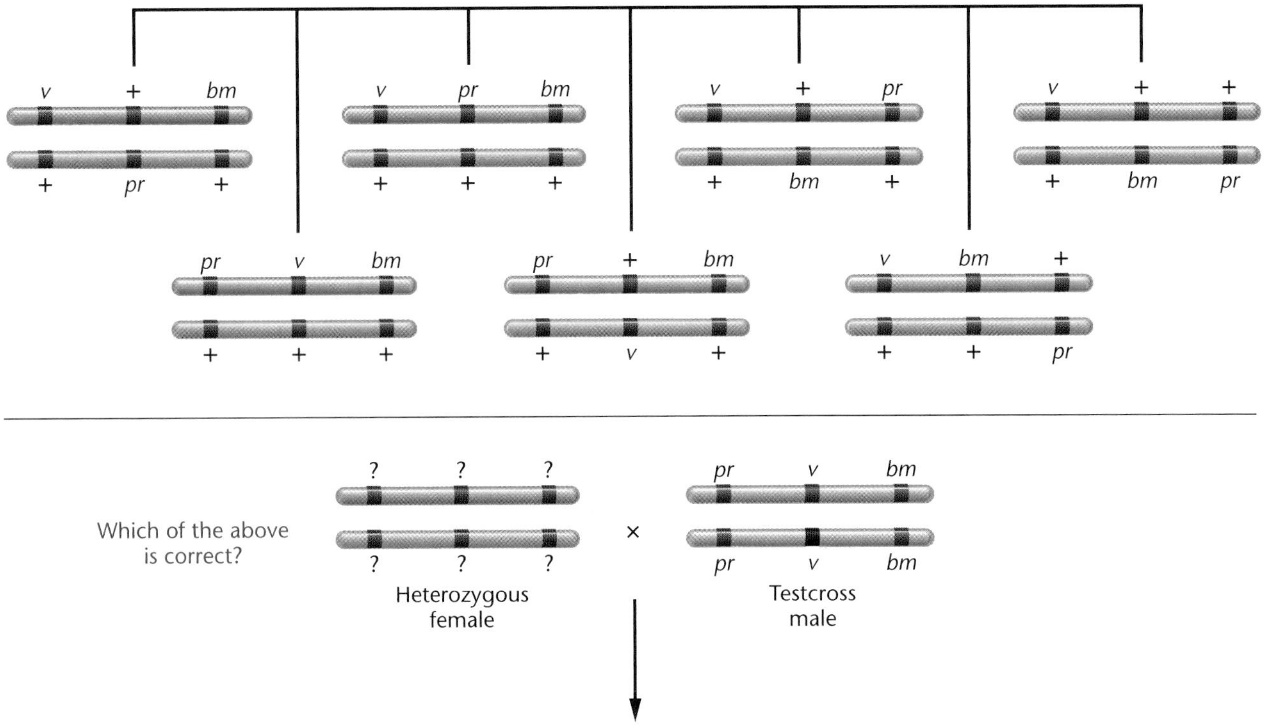

(b) Actual results of mapping cross*

Phenotypes of offspring	Number	Total and percentage	Exchange classification
+ v bm	230	467 42.1%	Noncrossover (NCO)
pr + +	237		
+ + bm	82	161 14.5%	Single crossover (SCO)
pr v +	79		
+ v +	200	395 35.6%	Single crossover (SCO)
pr + bm	195		
pr v bm	44	86 7.8%	Double crossover (DCO)
+ + +	42		

* The sequence *pr – v – bm* may or may not be correct

FIGURE 5–10 (a) Some possible allele arrangements and gene sequences in a heterozygous female. The data from a three-point mapping cross, depicted in (b), where the female is testcrossed, provide the basis for determining which combination of arrangement and sequence is correct. [See Figure 5–11(d).]

heterozygous female. Some of the possibilities are shown, but we have yet to determine which is correct. In the testcross male parent and the offspring, *we do not know the sequence*, and so we must designate it randomly. Note that we have chosen to place *v* in the middle. This may or may not be correct.

The offspring have been arranged in groups of two for each pair of reciprocal phenotypic classes. The two members of each reciprocal class are derived from either no crossing over (NCO), one of two possible single-crossover events (SCO), or a double-crossover event (DCO).

To solve this problem, it will help to refer to Figure 5–10 and Figure 5–11 as you consider the following questions:

1. *What is the correct heterozygous arrangement of alleles in the female parent?*

Determine the two noncrossover classes, those that occur with the highest frequency. In this case, they are + *v* *bm* and *pr* + +. Therefore, the alleles on the homologs of the female parent must be arranged as shown in Figure 5–11(a). These homologs segregate into gametes, unaffected by any recombination event. Any other arrangement of alleles could not yield the observed noncrossover classes. (Remember that + *v* *bm* is equivalent to pr^+ *v* *bm* and that *pr* + + is equivalent to *pr* v^+ bm^+).

Web Tutorial 5.4
Virtual Crossover Laboratory

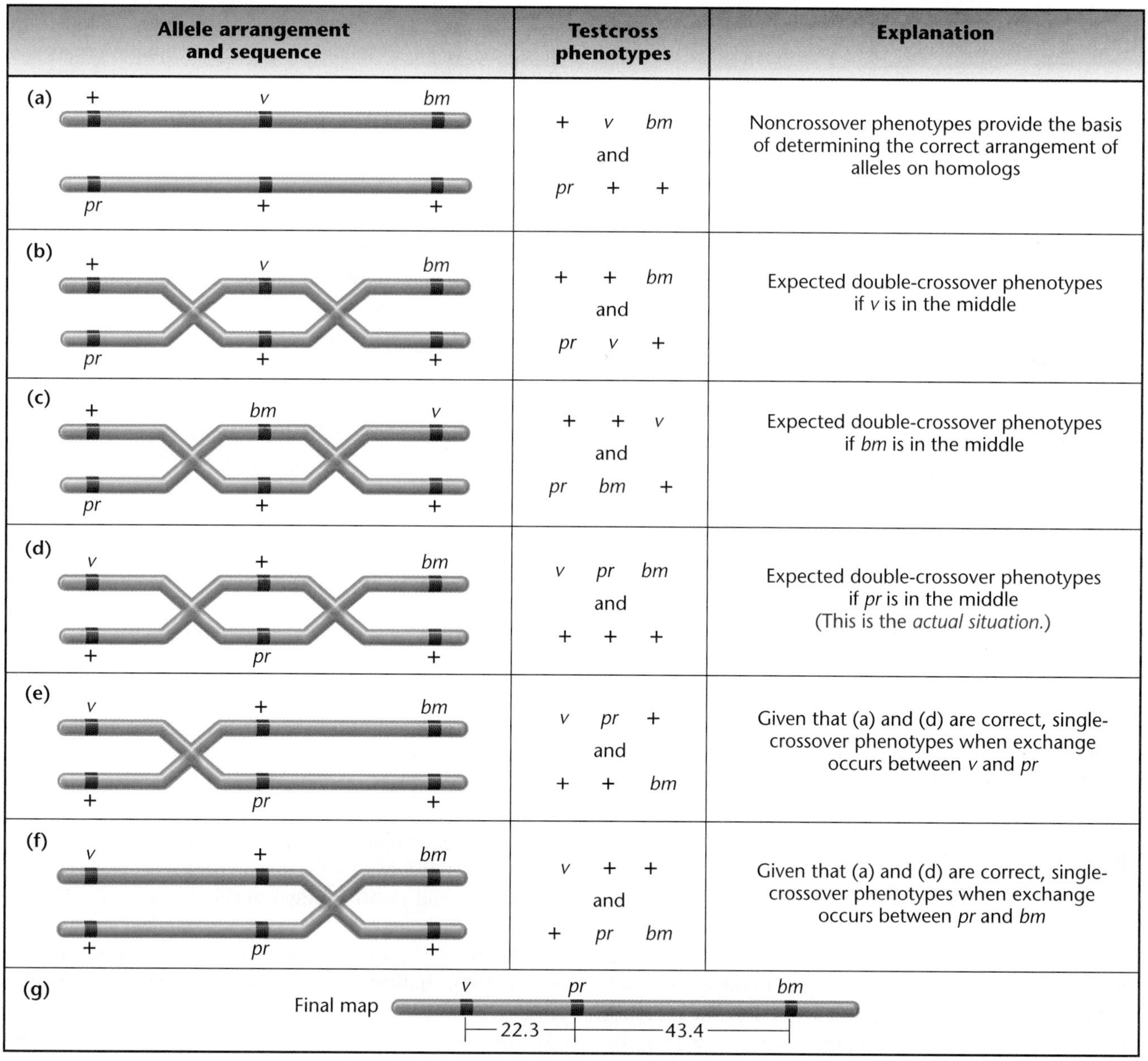

Allele arrangement and sequence
Testcross phenotypes
Explanation
(a) + v bm / pr + +
+ v bm and pr + +
Noncrossover phenotypes provide the basis of determining the correct arrangement of alleles on homologs
(b) + v bm / pr + +
+ + bm and pr v +
Expected double-crossover phenotypes if v is in the middle
(c) + bm v / pr + +
+ + v and pr bm +
Expected double-crossover phenotypes if bm is in the middle
(d) v + bm / + pr +
v pr bm and + + +
Expected double-crossover phenotypes if pr is in the middle (This is the actual situation.)
(e) v + bm / + pr +
v pr + and + + bm
Given that (a) and (d) are correct, single-crossover phenotypes when exchange occurs between v and pr
(f) v + bm / + pr +
v + + and + pr bm
Given that (a) and (d) are correct, single-crossover phenotypes when exchange occurs between pr and bm
(g) Final map
v pr bm
22.3
43.4

FIGURE 5–11 Producing a map of the three genes in the cross in Figure 5–10, where neither the arrangement of alleles nor the sequence of genes in the heterozygous female parent is known.

2. *What is the correct sequence of genes?*

To answer this question, we will first use the approach described in Method I. We know, based on the answer to Question 1, that the correct arrangement of alleles is

$$\frac{+ \quad v \quad bm}{pr \quad + \quad +}$$

But, is the assumed sequence of gene loci correct? If so, a double-crossover event will yield the observed double-crossover phenotypes following fertilization. Simple observation shows that it will not [Figure 5–11(b)]. Try the other two orders [Figure 5–11(c) and (d)], *keeping the same arrangement*:

$$\frac{+ \quad bm \quad v}{pr \quad + \quad +} \quad \text{or} \quad \frac{v \quad + \quad bm}{+ \quad pr \quad +}$$

Only the case on the right yields the observed double-crossover classes [Figure 5–11(d)]. Therefore, the *pr* gene is in the middle.

The same conclusion is reached if we used Method II to analyze the problem. In this case, no assumption of gene sequence is necessary. The arrangement of alleles in the heterozygous parent is

$$\frac{+ \quad v \quad bm}{pr \quad + \quad +}$$

The double-crossover gametes are also known:

$$pr \quad v \quad bm \quad \text{and} \quad + \quad + \quad +$$

We can see that it is the *pr* allele that has shifted, so as to be associated with *v* and *bm* following a double crossover. The latter two alleles (*v* and *bm*) were present together on one homolog, and they stayed together. Therefore, *pr* is the odd gene, so to speak, and it is in the middle. Thus, we arrive at the same arrangement and sequence as we did with Method I:

$$\frac{v \quad + \quad bm}{+ \quad pr \quad +}$$

3. *What is the distance between each pair of genes?*

Having established the correct sequence of loci as *v–pr–bm,* we can now determine the distance between *v* and *pr* and between *pr* and *bm*. Remember that the map distance between two genes is calculated on the basis of all detectable recombinational events occurring between them. This includes both the single- and double-crossover events involving the two genes being considered.

Figure 5–11(e) shows that the phenotypes *v pr +* and *+ + bm* result from single crossovers between *v* and *pr,* and Figure 5–10 shows that single crossovers account for 14.5 percent of the offspring. By adding the percentage of double crossovers (7.8 percent) to the number obtained for single crossovers, we calculate the total distance between *v* and *pr* to be 22.3 mu.

Figure 5–11(f) shows that the phenotypes *v + +* and *+ pr bm* result from single crossovers between the *pr* and *bm* loci, totaling 35.6 percent according to Figure 5–10. With the addition of the double-crossover classes (7.8 percent) the distance between *pr* and *bm* is 43.4 mu. The final map for all three genes in this example is shown in Figure 5–11(g).

5.4 Interference Affects the Recovery of Multiple Exchanges

Based on the previous discussion, the expected frequency of multiple exchanges, such as double crossovers, can be predicted once the distance between genes is established. For example, in the maize cross of the previous section, the distance between *v* and *pr* is 22.3 mu, and the distance between *pr* and *bm* is 43.4 mu. If the two single crossovers that make up a double crossover occur independently of one another, we can calculate the expected frequency of double crossovers (DCO_{exp}):

$$DCO_{exp} = (0.223) \times (0.434) = 0.097 = 9.7 \text{ percent}$$

Most often in mapping experiments, the observed DCO frequency is less than the expected number of DCOs. In the maize cross, for example, only 7.8 percent DCOs are observed when 9.7 percent are expected. This reduction is explained by the phenomenon called **interference**, which occurs when a crossover event in one region of the chromosome inhibits a second event in nearby regions.

To quantify the disparities that result from interference, we can calculate the **coefficient of coincidence (*C*)**:

$$C = \frac{\text{Observed DCO}}{\text{Expected DCO}}$$

In the maize cross, we have

$$C = \frac{0.078}{0.097} = 0.804$$

Once we have found *C*, we can quantify interference (*I*) by using this simple equation:

$$I = 1 - C$$

In the maize cross, we have

$$I = 1.000 - 0.804 = 0.196$$

If interference is complete and no double crossovers occur, then $I = 1.0$. If fewer DCOs than expected occur, *I* is a positive number and **positive interference** has occurred. If more DCOs than expected occur, *I* is a negative number and **negative interference** has occurred. In this example, *I* is a positive number (0.196), indicating that 19.6 percent fewer double crossovers occurred than expected.

Positive interference is most often observed in eukaryotic systems. In general, the closer genes are to one another along the chromosome, the more positive interference occurs. In fact, in

Drosophila, within a distance of 10 map units, interference is often complete (i.e., I = 0) and no multiple crossovers are recovered. Our observation suggests that interference may be explained by physical constraints preventing the formation of closely aligned chiasmata. The interpretation is consistent with the finding that interference decreases as the genes in question are located farther apart. In the maize cross illustrated in Figures 5–10 and 5–11, the three genes are relatively far apart, and 80 percent of the expected double crossovers are observed.

Now solve this

Problem 5.20 on page 133 asks you to solve a three-point mapping problem where only six phenotypic categories are observed, when eight categories are typical of such a cross.

Hint: If the distances between each pair of genes is relatively small, reciprocal pairs of single crossovers will be evident, but the sample size may be too small to recover the predicted number of double crossovers, excluding their appearance. You should write the missing gametes down, assigning them as double crossovers, and recording zeroes for their frequency of appearance.

5.5 As the Distance between Two Genes Increases, Mapping Experiments Become More Inaccurate

In theory, the frequency of crossing over between any two genes in a mapping experiment is expected to be directly proportional to the actual distance between two genes. However, in most cases, the experimentally derived mapping distance between two genes is an underestimate, and the farther apart the two genes are, the greater the inaccuracy. The discrepancy is due primarily to multiple exchanges that are predicted to occur between the two genes, but which are not recovered during experimental mapping. As we will explain next, the inaccuracy is the result of probability events that can be described using the **Poisson distribution**.

First, let us examine a mapping experiment that involves two exchanges between two genes that are far apart on a chromosome. As shown in Figure 5–12, there are three possible ways that two exchanges (equivalent to a double-crossover event) can occur between nonsister chromatids within a tetrad. A **two-strand double exchange** yields no recombinant chromatids, a **three-strand double exchange** yields 50 percent recombinant chromatids, and a **four-strand double exchange** yields 100 percent recombinant chromatids. In the aggregate, therefore,

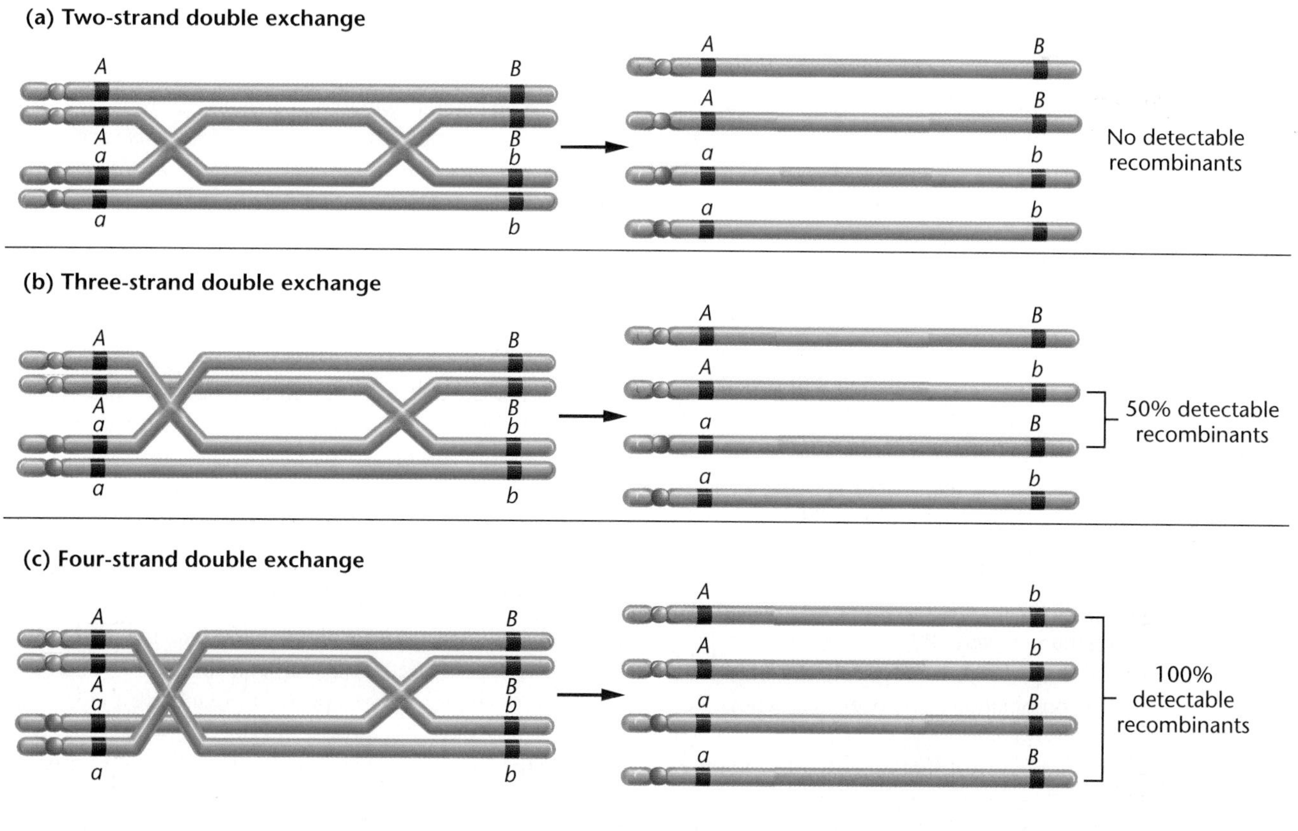

FIGURE 5–12 Three types of double exchanges that may occur between two genes. Two of them, (b) and (c), involve more than two chromatids. In each case, the detectable recombinant chromatids are bracketed.

these uncommon multiple events "even out" and two genes that are far apart on the chromosome theoretically yield the maximum of 50 percent recombination essential for *accurate* gene mapping.

In such a mapping experiment, double exchanges occuring between two genes (and, for that matter, all multiple exchanges) are relatively infrequent in comparison to the total number of single crossovers. As such, the *actual* occurrence of the infrequent events is subject to probability considerations based on the Poisson distribution. In our case, this distribution allows us to predict mathematically the frequency of samples that will *actually* undergo double exchanges. It is the failure of such exchanges to occur that leads to the underestimate of mapping distance.

The Poisson distribution is a mathematical function that assigns probabilities of observing various numbers of a specific event in a sample. To illustrate the effect of Poisson distribution, consider the analogy of an Easter egg hunt where 1000 children randomly search a large area for 1000 randomly hidden eggs. In one hour, all eggs are recovered. If all children are equally adept in the search, we can safely predict that many children will have one egg, but also that many will have either no eggs or more than one egg. The Poisson distribution allows us to predict mathematically the frequency (probability) of each outcome; that is, the frequency of children within the sample that recovered 0, 1, 2, 3, 4, ... eggs. Poisson distribution applies when the average number of events is small (a child finds an egg), while the total number of times the event that can occur within the sample is relatively large (1000 eggs can be found).

The Poisson terms used to calculate predicted distributions of events are

Distribution of Events	Probability
0	e^{-m}
1	me^{-m}
2	$(m^2/2)(e^{-m})$
3	$(m^3/6)(e^{-m})$
etc.	

where the mean number of independently occurring events is m and e represents the base of natural logarithms (e = about 2.7). For the Easter egg analogy, a calculation will reveal that over 300 children will fail to find an egg. Had we attempted to estimate the total number of youngsters in the hunt by tabulating the number who found at least one egg, we would have seriously underestimated the number of participants in the hunt.

In relation to chromosome mapping, we are interested in the cases where double exchanges may potentially occur between two genes, but because of the predictions based on Poisson distribution, no such exchanges actually occur within the data sample. Such an analysis creates what is called a **mapping function** that relates recombination (crossover) frequency (RF) to map distance.

We can now apply the Poisson distribution, assuming no interference occurs. Any class where m is one or more (one or more random crossovers) will yield, on average, 50 percent recombinant chromatids. Thus, we are interested in the zero term, which effectively reduces the number of recombinant chromatids. The proportion of meioses with one or more crossovers is equal to

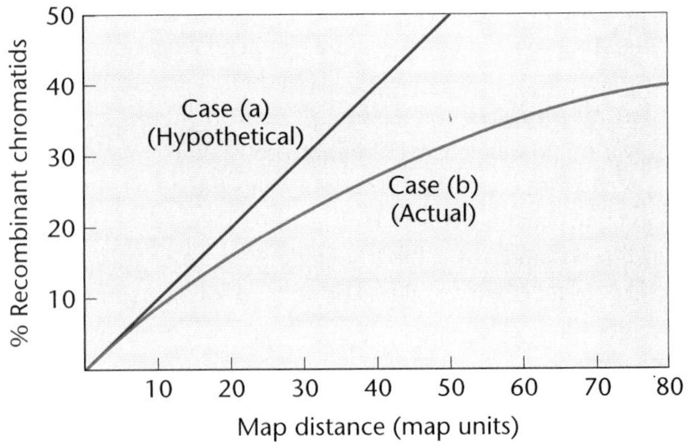

FIGURE 5–13 The relationship between the percentage of recombinant chromatids that occur and actual map distance when (a) Poisson distribution is used to predict the frequency of recombination in relation to map distance; and (b) there is a direct relationship between recombination and map distance.

one minus the fraction of zero crossovers $(1 - e^{-m})$, whereby 50 percent recombinant chromatids will occur. Therefore,

$$\text{percent observed recombination (RF)} = 0.5(1 - e^{-m}) \times 100$$

Solving this equation, we find that the curve (mapping function) shown in Case (a) of Figure 5–13 is generated. This is compared with the hypothetical Case (b) of Figure 5–13, where recombination is directly proportional to mapping distance—that is, where interference is complete and no multiple exchanges occur.

Careful examination of the graph reveals two important observations. When the actual map distance is low (i.e., 0–7 mu), the two lines coincide. *When two genes are close together, the accuracy of a mapping experiment is very high! However, as the distance between two genes increases, the accuracy of the experiment diminishes.* As predicted by the Poisson distribution, the absence of multiple exchanges has a very significant impact. For example, when 25 percent recombinant chromatids are detected, actual map distance is almost 35 mu! When just over 30 percent recombinants are detected, the true distance, discounting any interference, is 50 mu! Such inaccuracy has been well documented in a number of studies involving various organisms, including maize, *Drosophila*, and *Neurospora*.

5.6 *Drosophila* Genes Have Been Extensively Mapped

In organisms such as *Drosophila*, maize, and the mouse, where large numbers of mutants have been discovered and experimental crosses are easy to perform, extensive maps of each chromosome have been made. Figure 5–14 presents partial maps of the four chromosomes of *Drosophila*. Virtually every morphological feature of the fruit fly has been subjected to mutations. Each locus affected by mutation is first localized to one of the four chromosomes, or linkage groups, and then mapped in relation to other linked genes of that group. As you can see, the genetic map of the X chromosome is somewhat less

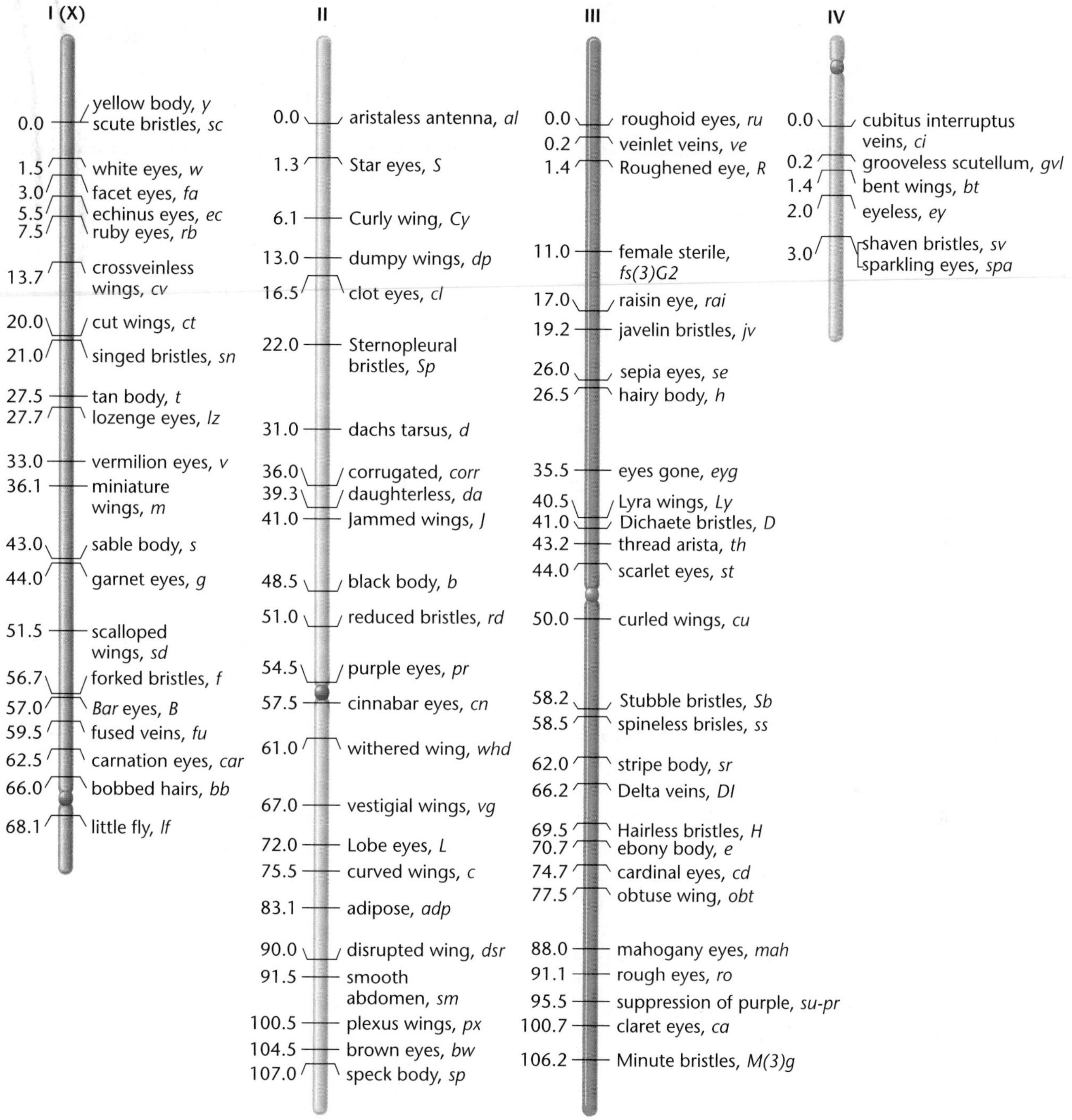

FIGURE 5–14 A partial genetic map of the four chromosomes of *Drosophila melanogaster*. The circle on each chromosome represents the position of the centromere.

extensive than that of autosome 2 or 3. In comparison to these three, autosome 4 is miniscule. Cytological evidence has shown that the relative lengths of the genetic maps correlate roughly with the relative physical lengths of these chromosomes.

5.7 Crossing Over Involves a Physical Exchange between Chromatids

Once genetic mapping was understood, information was sought concerning the relationship between chiasmata observed in meiotic prophase I and crossing over. For example, are chiasmata visible manifestations of crossover events? If so, then crossing over in higher organisms appears to be the result of an actual physical exchange between homologous chromosomes. That this is the case was demonstrated independently in the 1930s by Harriet Creighton and Barbara McClintock in *Zea mays* and by Curt Stern in *Drosophila*.

Because the experiments are similar, we will consider only one of them, the work with maize. Creighton and McClintock studied two linked genes on chromosome 9 of the maize plant. At one locus, the alleles *colorless* (*c*) and *colored* (*C*) control endosperm coloration. At the other locus, the alleles *starchy* (*Wx*) and *waxy* (*wx*) control the carbohydrate characteristics of

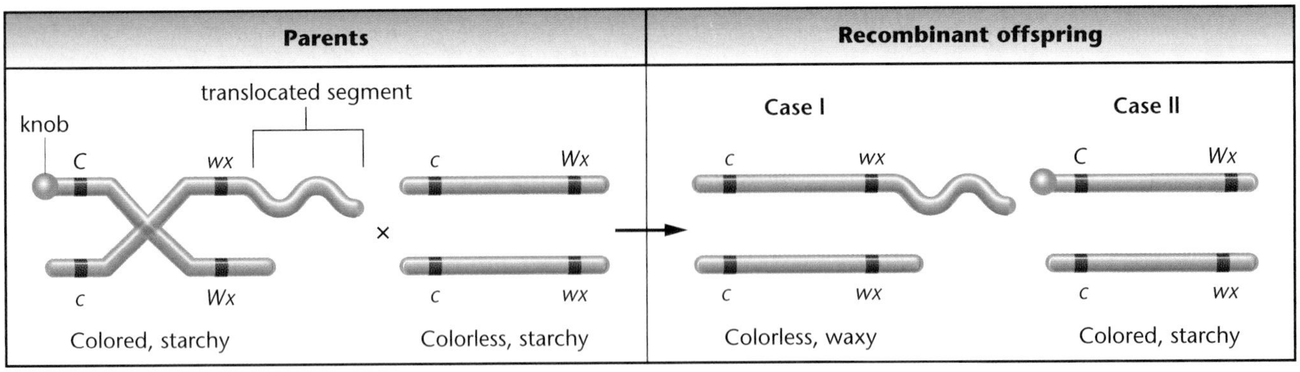

FIGURE 5–15 The phenotypes and chromosome compositions of parents and recombinant offspring in Creighton and McClintock's experiment in maize. The knob and translocated segment served as cytological markers which established that crossing over involves an actual exchange of chromosome arms.

the endosperm. The maize plant studied was heterozygous at both loci. That one of the homologs contained two unique cytological markers was the key to the experiment. The markers consisted of a densely stained knob at one end of the chromosome and a translocated piece of another chromosome (8) at the other end. The arrangements of alleles and cytological markers could be detected cytologically and are shown in Figure 5–15.

Creighton and McClintock crossed this plant to one homozygous for the color allele (*c*) and heterozygous for the endosperm alleles. They obtained a variety of different phenotypes in the offspring, but they were most interested in one that occurred as a result of crossing over involving the chromosome with the unique cytological markers. They examined the chromosomes of this plant, with the colorless, waxy phenotype (case I in Figure 5–15) for the presence of the cytological markers. If genetic crossing over is accompanied by a physical exchange between homologs, the translocated chromosome would still be present, but the knob would not. This was the case! In a second plant (case II), the phenotype colored, starchy should result from either nonrecombinant gametes or from crossing over. Some of the cases then ought to contain chromosomes with the dense knob, but not the translocated chromosome. This condition was also found, and the conclusion that a physical exchange had taken place was again supported. Along with Stern's findings with *Drosophila*, the work clearly established that crossing over has a cytological basis.

Once we have introduced the topics of the chemical structure and replication of DNA (Chapters 10 and 11), we will return to the topic of crossing over and look at how breakage and reunion occur between the strands of DNA making up chromatids. This discussion will provide a better understanding of genetic recombination.

How Do We Know?

What is the experimental basis for concluding that crossing over involves an actual physical exchange between homologs during meiosis?

5.8 Recombination Occurs between Mitotic Chromosomes

In 1936, Curt Stern considered whether exchanges similar to crossing over occur during mitosis. He was able to demonstrate that this indeed is the case in *Drosophila*. This finding, the first to demonstrate **mitotic recombination**, was considered unusual because homologs do not normally pair up during mitosis in most organisms. However, such synapsis appears to be the rule in *Drosophila*. Since Stern's discovery, genetic exchange during mitosis has also been shown to be a general event in certain fungi.

Stern observed small patches of mutant tissue in females heterozygous for the X-linked recessive mutations *yellow* body and *singed* bristles. Under normal circumstances, a heterozygous female is completely wild type (gray-bodied with straight, long bristles). He explained the appearance of the mutant patches by postulating that, during mitosis in certain cells during development, homologous exchanges could occur between the loci for *yellow* (*y*) and *singed* (*sn*) or between *singed* and the centromere. Figure 5–16 diagrams the nature of the proposed exchanges, in contrast to the case where no exchange occurs. The mutant patches that will occur are also depicted. When no exchange occurs, all tissue is wild type (gray body and gray, straight bristles). After the two types of exchanges, tissues are produced with either a *yellow* patch *or* with an adjacent *yellow* and *singed* patch (called a twin spot).

In 1958, George Pontecorvo and others described a similar phenomenon in the fungus *Aspergillus*. Although the vegetative stage is normally haploid, some cells fuse. The resultant diploid cells then divide mitotically. As in *Drosophila*, crossing over rarely occurs between linked genes during mitosis in this diploid stage, resulting in recombinant cells. Pontecorvo referred to these events that produce genetic variability as the **parasexual cycle**. On the basis of such exchanges, genes can be mapped by estimating the frequency of recombinant classes.

As a rule, if mitotic recombination occurs at all in an organism, it does so at a much lower frequency than meiotic crossing over. We assume that there is always at least one exchange per meiotic tetrad. By contrast, in organisms demonstrating mitotic exchange, it occurs in 1 percent or fewer of mitotic divisions.

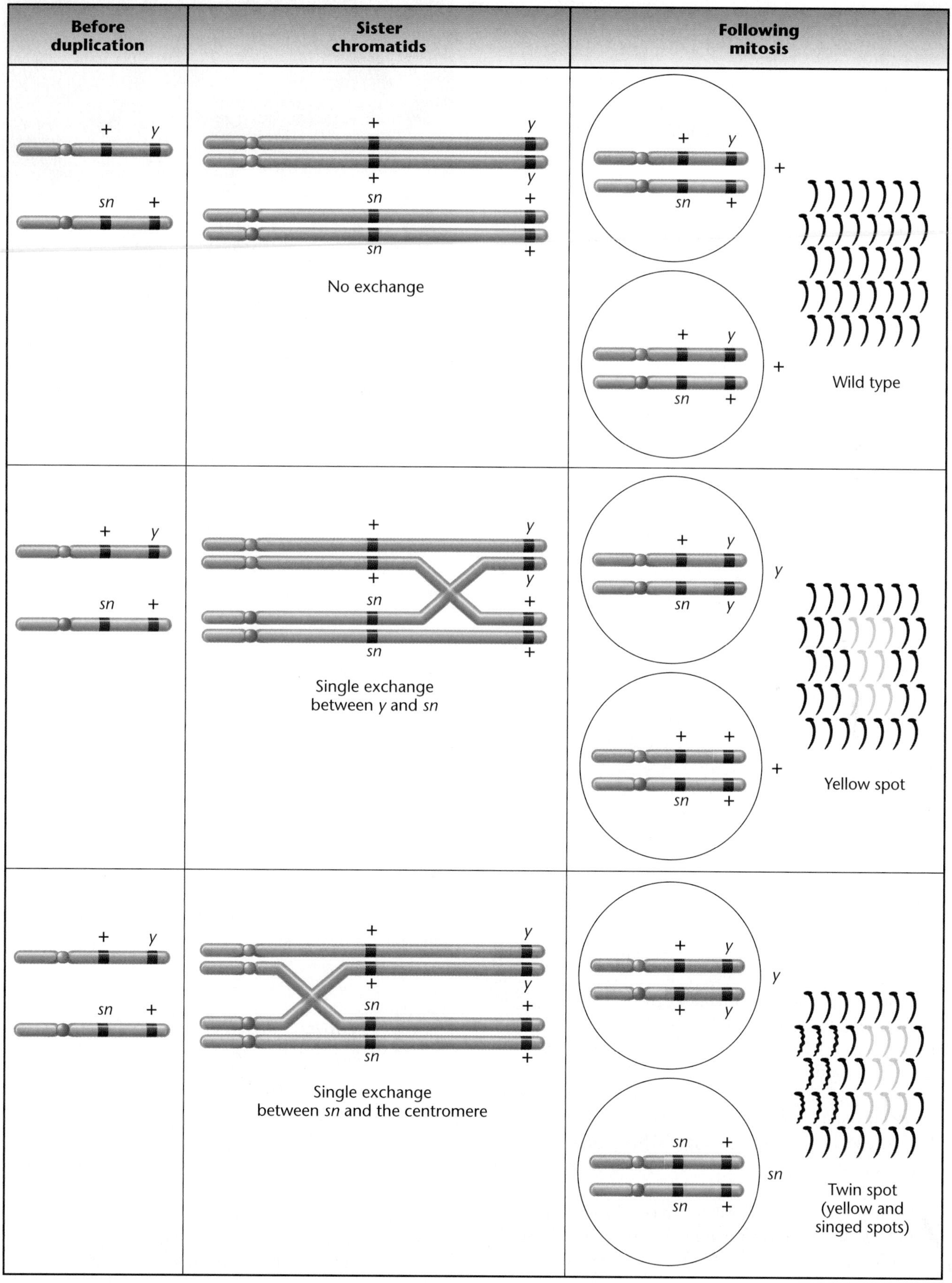

FIGURE 5–16 The production of mutant tissue in a female *Drosophila* heterozygous for the recessive *yellow* (*y*) and *singed* (*sn*) alleles as a result of mitotic recombination.

5.9 Exchanges Also Occur between Sister Chromatids

Knowing that crossing over occurs between synapsed homologs in meiosis, we might ask whether such a physical exchange occurs between homologs during mitosis. While homologous chromosomes do not usually pair up or synapse in somatic cells of diploid organisms (*Drosophila* is an exception), each individual chromosome in prophase and metaphase of mitosis consists of two identical sister chromatids, joined at a common centromere. Surprisingly, a number of experiments have demonstrated that reciprocal exchanges similar to crossing over occur even between sister chromatids. While these **sister chromatid exchanges (SCEs)** do not produce new allelic combinations, evidence is accumulating that attaches significance to these events.

Identification and study of SCEs are facilitated by several modern staining techniques. In one approach, cells are allowed to replicate for two generations in the presence of the thymidine analog bromodeoxyuridine (BUdR).* Following two rounds of replication, each pair of sister chromatids has one member with both strands labeled with BUdR. Using a differential stain, we find that chromatids with the analog in both strands stain less brightly than chromatids with it in only one strand. As a result, SCEs are readily detectable, if they occur. In Figure 5–17, numerous instances of SCE events are clearly evident. Because of the patchlike appearance, these sister chromatids are sometimes referred to as **harlequin chromosomes**.

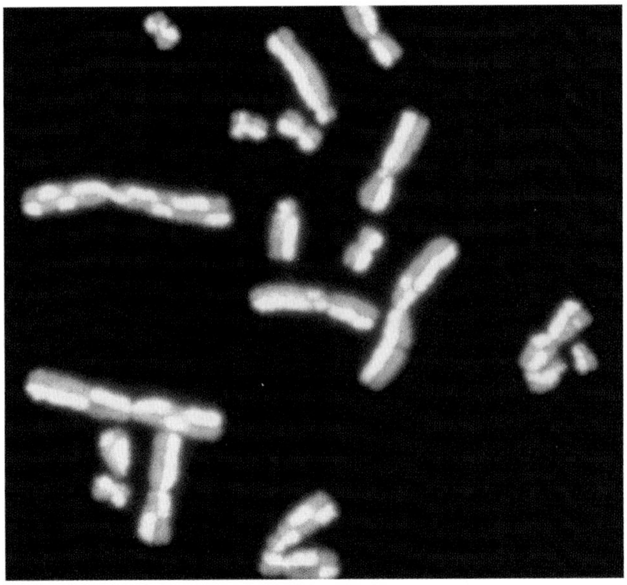

FIGURE 5–17 Demonstration of sister chromatid exchanges (SCEs) in mitotic chromosomes. Sometimes called harlequin chromosomes because of their patchlike appearance, chromatids containing the thymidine analog BUdR in both DNA strands fluoresce *less* brightly than do those with the analog in only one strand. These chromosomes were stained with 33258-Hoechst reagent and acridine orange and then viewed under fluorescence microscopy.

While the significance of SCEs is still uncertain, several observations have led to great interest in this phenomenon. We know, for example, that agents that induce chromosome damage (e.g., viruses, X rays, ultraviolet light, and certain chemical mutagens) also increase the frequency of SCEs. Further, an elevated frequency of SCEs is characteristic of **Bloom syndrome**, a human disorder caused by a mutation in the chromosome 15 *BLM* gene. This rare, recessively inherited disease is characterized by prenatal and postnatal retardation of growth, a great sensitivity of the facial skin to the sun, immune deficiency, a predisposition to malignant and benign tumors, and abnormal behavior patterns. The chromosomes from cultured leukocytes, bone marrow cells, and fibroblasts derived from homozygotes are very fragile and unstable when compared with those derived from homozygous and heterozygous normal individuals. Increased breaks and rearrangements between nonhomologous chromosomes are observed in addition to excessive amounts of sister chromatid exchanges. It is now evident that the *BLM* gene encodes an enzyme called **DNA helicase**, which is best known for its role in DNA replication. (See Chapter 11.)

The mechanisms of exchange between nonhomologous chromosomes and between sister chromatids may prove to share common features because the frequency of both events increases substantially in individuals with genetic disorders. These findings suggest that further study of sister chromatid exchange may contribute to an increased understanding of recombination mechanisms and the relative stability of normal and genetically abnormal chromosomes. We shall encounter still another demonstration of SCEs in Chapter 11 when we consider replication of DNA. (See Figure 11–5.)

*The abbreviation BrdU is also used to denote bromodeoxyuridine.

? How Do We Know?

How do we know that sister chromatids undergo recombination during mitosis?

5.10 Linkage Analysis and Mapping Can Be Performed in Haploid Organisms

We now turn to still another extension of our study of transmission genetics: linkage analysis and chromosome mapping in haploid eukaryotes. As we shall see, even though analysis of the location of genes relative to one another throughout the genome of haploid organisms may *seem* a bit more complex than in diploid organisms, the basic underlying principles are the same. In fact, many basic principles of inheritance were established during the study of haploid fungi.

While many single-celled eukaryotes are haploid during the vegetative stages of their life cycle, they also form reproduc-

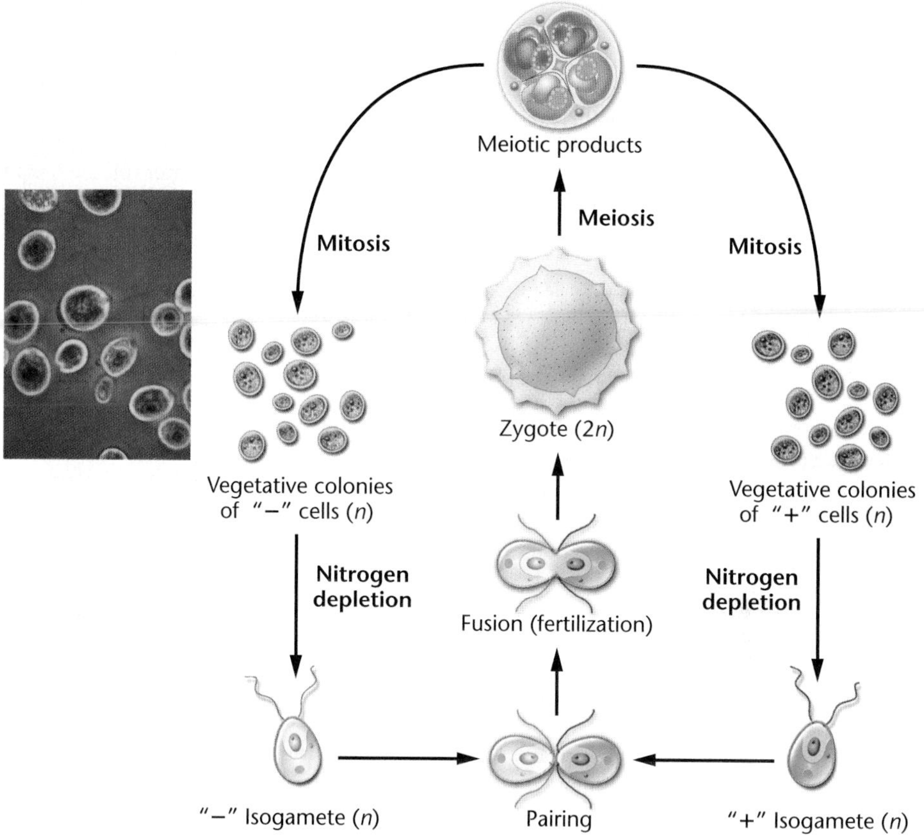

FIGURE 5–18 The life cycle of *Chlamydomonas*. The diploid zygote (in the center) undergoes meiosis, producing "+" or "−" haploid cells that undergo mitosis, yielding vegetative colonies. Unfavorable conditions stimulate them to form isogametes, which fuse in fertilization, producing a zygote that repeats the cycle. Vegetative colonies are illustrated photographically.

tive cells that fuse during fertilization, forming a diploid zygote. The zygote then undergoes meiosis and reestablishes haploidy. The haploid meiotic products are the progenitors of the subsequent members of the vegetative phase of the life cycle. Figure 5–18 illustrates this type of cycle in the green alga *Chlamydomonas*. Even though the haploid cells that fuse during fertilization *look* identical (and are thus called **isogametes**), a chemical identity that distinguishes two distinct types exists on their surface. As a result, all strains are either "+" or "−" and fertilization occurs only between unlike cells.

To perform genetic experiments with haploid organisms, genetic strains of different genotypes are isolated and crossed to one another. Following fertilization and meiosis, the meiotic products are retained together and can be analyzed. Such is the case in *Chlamydomonas* as well as in the fungus *Neurospora*, which we shall use as an example in the ensuing discussion. Following fertilization in *Neurospora* (Figure 5–19), meiosis occurs in a saclike structure called the **ascus**, within which the initial set of haploid products, called a **tetrad**, are retained. Tetrad has a quite different meaning here from its use to describe the four-stranded chromosome configuration characteristic of meiotic prophase I in diploids.

Following meiosis in *Neurospora*, each cell in the ascus divides mitotically, producing eight haploid **ascospores**. These can be dissected and examined morphologically or tested to determine their genotypes and phenotypes. Because the eight cells reflect the *sequence* of their formation following meiosis, the tetrad is "ordered" and we can do **ordered tetrad analysis**. This process is critical to our subsequent discussion.

Gene-to-Centromere Mapping

When a single gene (*a*) is analyzed in *Neurospora*, as diagrammed in Figure 5–20, the data can be used to calculate the map distance between that gene and the centromere. This process is sometimes referred to as **mapping the centromere**. It can be accomplished by experimentally determining the frequency of recombination using tetrad data. Remember that once the four meiotic products of the tetrad are formed, a mitotic division occurs, resulting in eight ordered products (ascospores).

If no crossover event occurs between the gene under study and the centromere, the pattern of ascospores (contained within an ascus) appears as shown in Figure 5–20(a) (*aaaa*++++). The pattern (++++*aaaa*) can also be formed, but it is indistinguishable from (*aaaa*++++). These patterns represent **first-division segregation**, because the two alleles are separated during the first meiotic division. However, crossover events will alter the pattern, as shown in Figure 5–20(b) (*aa*++*aa*++) and 5–20(c) (++*aaaa*++). Two other recombinant patterns also occur, depending on the chromatid orientation during the second meiotic division: (++*aa*++*aa*) and (*aa*++++*aa*). All four patterns, resulting from a crossover event between the *a* gene and the centromere, reflect **second-division segregation**, because the two alleles are not separated until the second meiotic division. Since the mitotic division simply replicates the patterns (from 4 to 8 ascospores), ordered tetrad data are usually

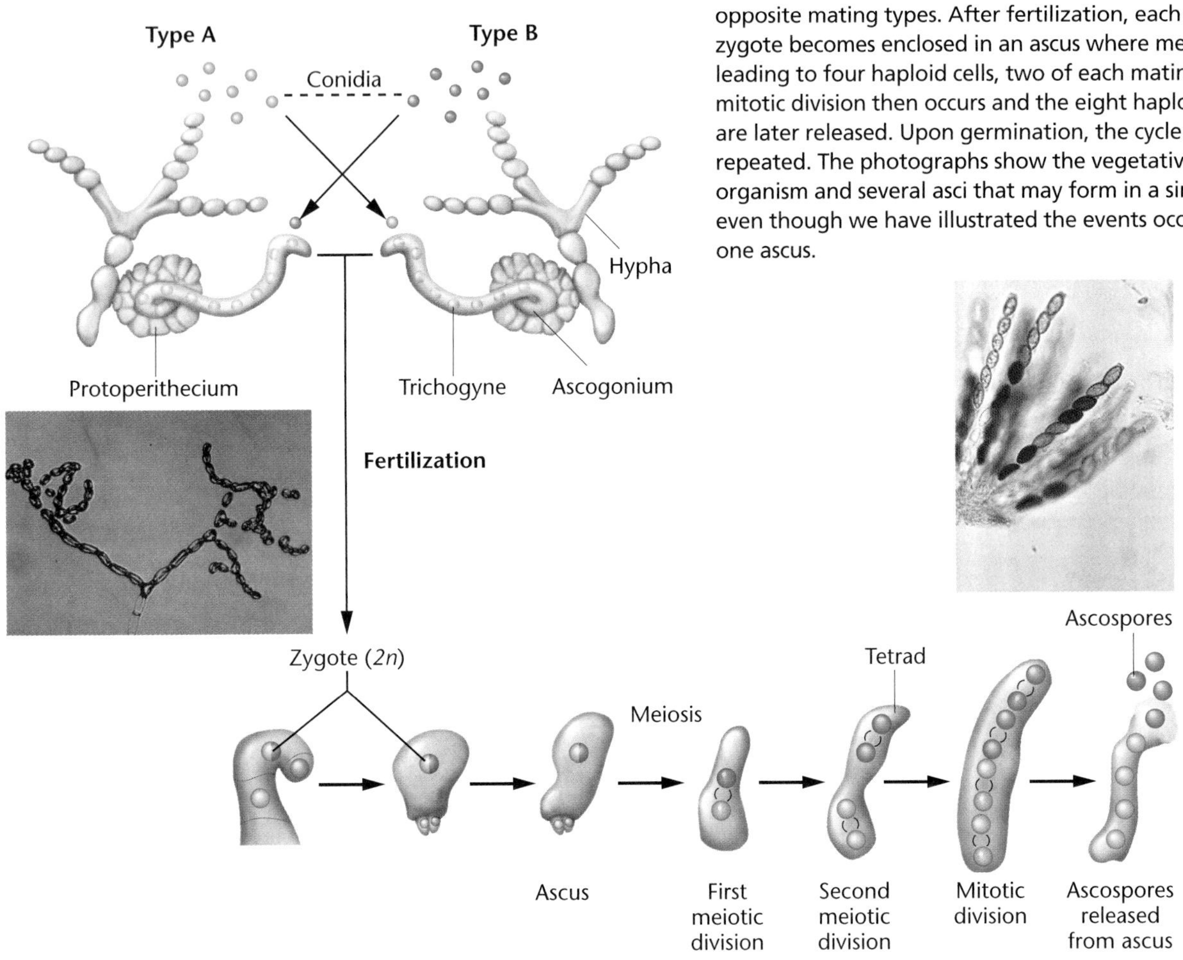

FIGURE 5–19 Sexual reproduction during the life cycle of *Neurospora* is initiated following fusion of conidia of opposite mating types. After fertilization, each diploid zygote becomes enclosed in an ascus where meiosis occurs, leading to four haploid cells, two of each mating type. A mitotic division then occurs and the eight haploid ascospores are later released. Upon germination, the cycle may be repeated. The photographs show the vegetative stage of the organism and several asci that may form in a single structure, even though we have illustrated the events occurring in only one ascus.

condensed to reflect the genotypes reflected in ascospore pairs that may be distinguished from one another. Six unique combinations are possible:

First-Division Segregation

(1)	*a*	*a*	+	+
(2)	+	+	*a*	*a*

Second-Division Segregation

(3)	*a*	+	*a*	+
(4)	+	*a*	+	*a*
(5)	+	*a*	*a*	+
(6)	*a*	+	+	*a*

To calculate the distance between the gene and the centromere, data must be tabulated from a large number of asci resulting from a controlled cross. We then use these data to calculate the distance (d):

$$d = \frac{1/2(\textit{second division segregant asci})}{\textit{total asci scored}} \times 100$$

The distance (d) reflects the percentage of recombination and is only half the number of second-division segregant asci. This is because crossing over in each occurs in only two of the four chromatids during meiosis.

To illustrate, assume that a represents albino and + represents wild type in *Neurospora.* In crosses between the two genetic types, suppose the following data are observed:

65 first-division segregants

70 second-division segregants

Thus, the distance between a and the centromere is

$$d = \frac{(1/2)(70)}{135} = 0.259 \times 100 = 25.9$$

or about 26 mu.

As the distance increases up to 50 units, in theory all asci should result from second-division segregation. However, numerous factors prevent this. As in diploid organisms, accuracy is greatest when the gene and the centromere are relatively close together. As we will discuss in the next section, we can also analyze haploid organisms in order to distinguish between linkage and independent assortment of two genes. Mapping distances between gene loci are calculated once linkage is established. As a result, detailed maps of organisms such as *Neurospora* and *Chlamydomonas* are now available.

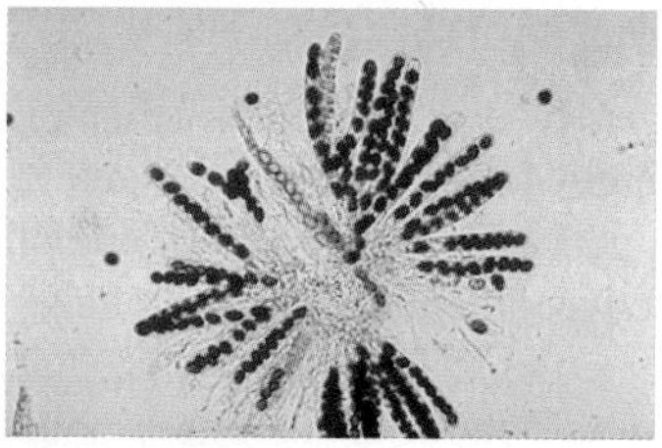

FIGURE 5–20 Three ways in which different ascospore patterns can be generated in *Neurospora*. Analysis of these patterns can serve as the basis of gene-to-centromere mapping. The photograph shows a variety of ascospore arrangements within *Neurospora* asci.

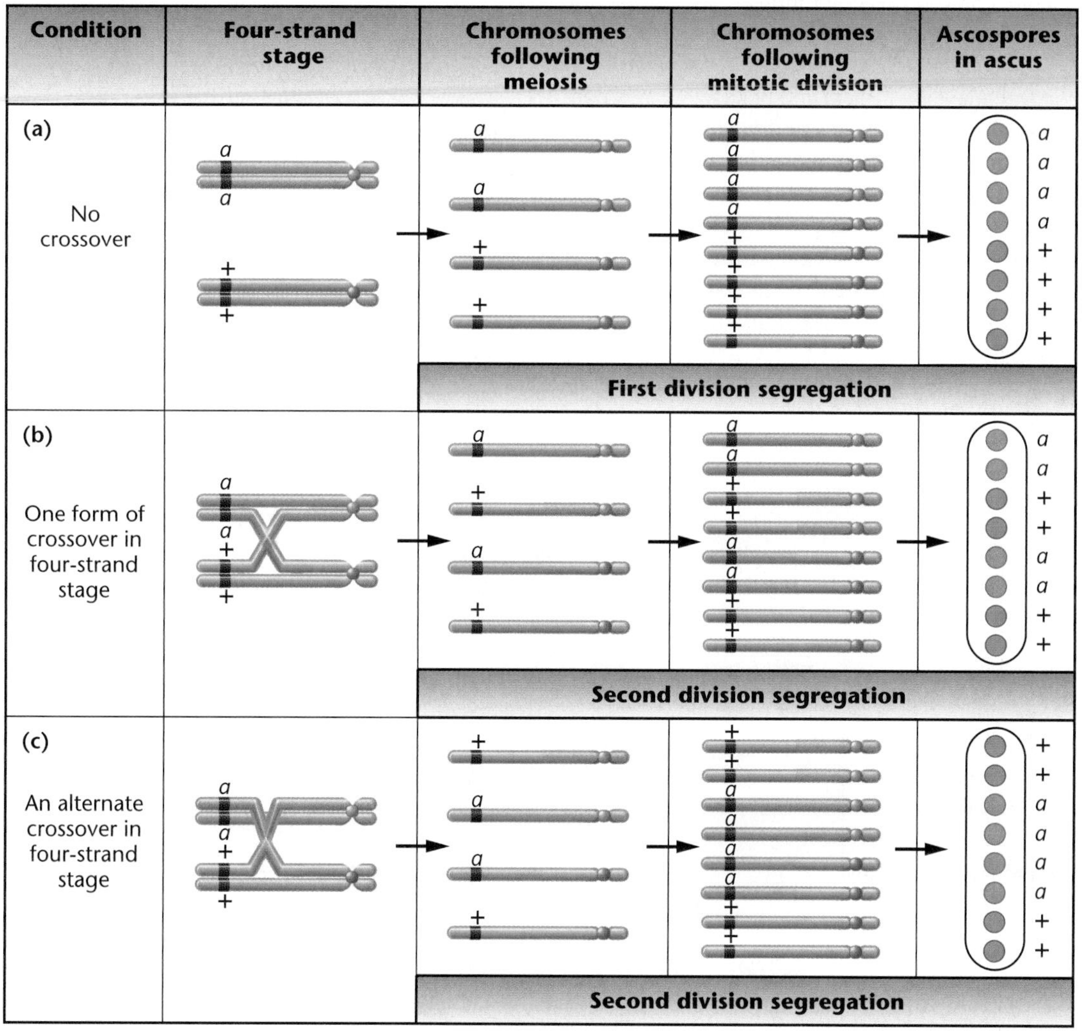

Ordered versus Unordered Tetrad Analysis

In our previous discussion, we assumed that the genotype of each ascospore and its position in the tetrad can be determined. To perform such an **ordered tetrad analysis**, individual asci must be dissected and each ascospore must be tracked as it germinates. This is a tedious process, but it is essential for two types of analysis:

1. To distinguish between first-division segregation and second-division segregation of alleles in meiosis.
2. To determine whether recombinational events are reciprocal or not. In the first case, such information is essential to "map the centromere," as we have just discussed. Thus, ordered tetrad analysis must be performed in order to map the distance between a gene and the centromere.

In the second case, ordered tetrad analysis has revealed that recombinational events are not always reciprocal, particularly when closely linked genes are studied in *Ascomycetes*. This observation has led to the investigation of the phenomenon called **gene conversion**. Because its discussion requires a background in DNA structure and analysis, we will return to this topic in Chapter 11.

It is much less tedious to isolate individual asci, allow them to mature, and then determine the genotypes of each ascospore, but not in any particular order. This approach is referred to as **unordered tetrad analysis**. As we shall see in the next section, such an analysis can be used to detemine whether or not two genes are linked on the same chromosome, and if so, to determine the map distance between them.

Linkage and Mapping

Analysis of genetic data derived from haploid organisms can be used to distinguish between linkage and independent assortment of two genes; it further allows mapping distances to be calculated between gene loci once linkage is established. We shall consider tetrad analysis in the alga *Chlamydomonas*. With the exception that the four meiotic products are not ordered and *do not* undergo a mitotic division following the completion of meiosis, the general principles discussed for *Neurospora* also apply to *Chlamydomonas*.

To compare independent assortment and linkage, we will consider two theoretical mutant alleles, *a* and *b*, representing two distinct loci in *Chlamydomonas*. Suppose that 100 tetrads derived from the cross *ab* × ++ yield the tetrad data shown in Table 5.1. As you can see, all tetrads produce one of three patterns. For example, all tetrads in category I produce two ++ cells and two *ab* cells and are designated as **parental ditypes (P)**. Category II tetrads produce two *a*+ cells and two +*b* cells and are called **nonparental ditypes (NP)**. Category III tetrads produce one cell of each of the four possible genotypes and are thus termed **tetratypes (T)**.

These data support the hypothesis that the genes represented by the *a* and *b* alleles are located on separate chromosomes. To understand why, you should refer to Figure 5–21. In parts

TABLE 5.1 TETRAD ANALYSIS IN *CHLAMYDOMONAS*

Category	I	II	III
Tetrad type	Parental (P)	Nonparental (NP)	Tetratypes (T)
Genotypes present	+ + + + *a b* *a b*	*a* + *a* + + *b* + *b*	+ + *a* + + *b* *a b*
Number of tetrads	43	43	14

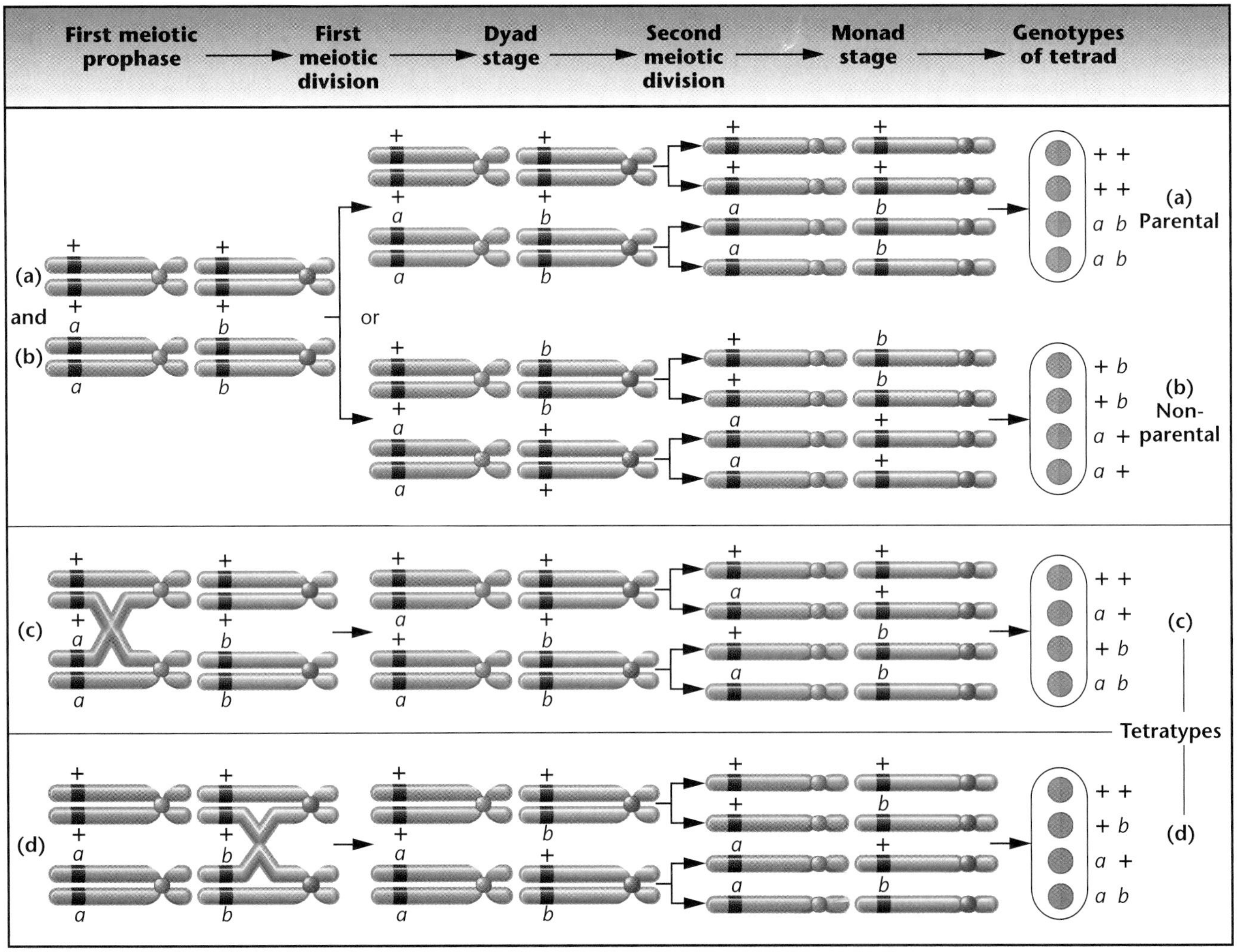

FIGURE 5–21 The origin of various genotypes found in tetrads in *Chlamydomonas* when two genes located on separate chromosomes are considered.

(a) and (b) of that figure, the origin of parental (P) and nonparental (NP) ditypes is demonstrated for two unlinked genes. According to the Mendelian principle of independent assortment of unlinked genes, approximately equal proportions of these tetrad types are predicted. Thus, when the parental ditypes are equal to the nonparental ditypes, then the two genes are not linked. The data in Table 5.1 confirm this prediction. Because independent assortment has occurred, it can be concluded that the two genes are located on separate chromosomes.

The origin of category III, the tetratypes, is diagrammed in Figure 5–21(c,d). The genotypes of tetrads in this category can be generated in two possible ways. Both involve a crossover event between one of the genes and the centromere. In Figure 5–21(c), the exchange involves one of the two chromosomes and occurs between gene *a* and the centromere; in Figure 5–21(d), the other chromosome is involved, and the exchange occurs between gene *b* and the centromere.

Production of tetratype tetrads does not alter the final ratio of the four genotypes present in all meiotic products. If the genotypes from 100 tetrads (which yield 400 cells) are computed, 100 of each genotype are found. This 1:1:1:1 ratio is predicted according to independent assortment.

Now consider the case where the genes *a* and *b* are linked (Figure 5–22). The same categories of tetrads will be produced. However, parental and nonparental ditypes will not necessarily occur in equal proportions; nor will the four genotypic combinations be found in equal numbers if the genotypes of all meiotic products are computed. For example, the following data might be encountered:

Category I	Category II	Category III
P	NP	T
64	6	30

Since the parental and nonparental categories are not produced in equal proportions, we can conclude that independent assortment is not in operation and that the two genes are linked. We can then proceed to determine the map distance between them.

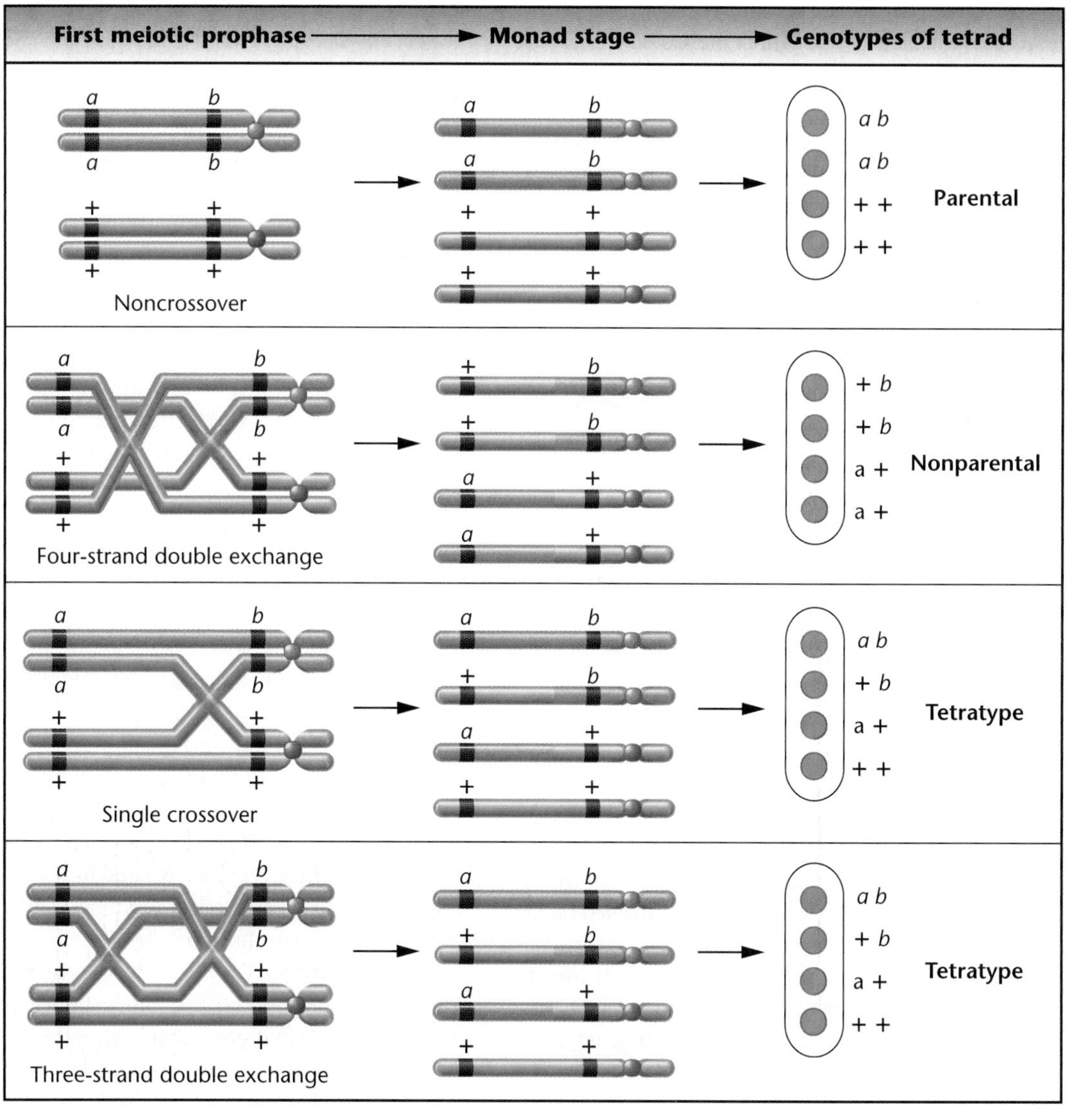

FIGURE 5–22 The various types of exchanges leading to the genotypes found in tetrads in *Chlamydomonas* when two genes located on the same chromosome are considered.

In the analysis of these data, we are concerned with the determination of which tetrad types represent genetic exchanges between the two genes. The parental ditype tetrads (P) arise only when no crossing over occurs between the two genes. The nonparental ditype tetrads (NP) arise only when a double exchange involving all four chromatids occurs between two genes. The tetratype tetrads (T) arise when either a single crossover occurs or when an alternative type of double exchange occurs between the two genes. The various types of exchanges described here are diagrammed in Figure 5–22.

When the proportion of the three tetrad types has been determined, it is possible to calculate the map distance between the two linked genes. The following formula computes the exchange frequency, which is proportional to the map distance between the two genes:

$$\text{exchange frequency (\%)} = \frac{\text{NP} + 1/2(\text{T})}{\text{total number of tetrads}} \times 100$$

In this formula, NP represents the nonparental tetrads; all meiotic products represent an exchange. The tetratype tetrads are represented by T; assuming only single exchanges, half of the meiotic products represents exchanges. The sum of the scored tetrads that fall into these categories is then divided by the total number of tetrads examined $(\text{P} + \text{NP} + \text{T})$. If this calculated number is multiplied by 100, it is converted to a percentage, which is directly equivalent to the map distance between the genes.

In our example, the calculation reveals that genes *a* and *b* are separated by 21 mu:

$$\frac{6 + 1/2(30)}{100} = \frac{6 + 15}{100} = \frac{21}{100} = 0.21 \times 100 = 21\%$$

Although we have considered linkage analysis and mapping of only two genes at a time, such studies often involve three or more genes. In these cases, both gene sequence and map distances can be determined.

5.11 Lod Score Analysis and Somatic Cell Hybridization Were Historically Important in Creating Human Chromosome Maps

For obvious reasons, our own species is not a good source of data for the types of extensive linkage analysis performed with experimental organisms. Thus, in humans, the earliest linkage studies had to rely on pedigree analysis. Attempts were made to establish whether a trait was X-linked or autosomal. As we established in Chapter 4, traits determined by genes located on the X chromosome result in characteristic pedigrees; thus, such genes were easier to identify. For autosomal traits, geneticists tried to distinguish clearly whether pairs of traits demonstrated linkage or independent assortment. When extensive pedigrees are available, it is possible to ascertain that the genes under consideration are closely linked (i.e., rarely separated by crossing over) from the fact that the two traits segregate together. For example, this approach established linkage between the genes encoding the **Rh antigens** and the gene responsible for the phenotype referred to as **elliptocytosis** (where the shape of erythrocytes is oval).

The difficulty arises, however, when two genes of interest are separated on a chromosome such that recombinant gametes are formed, obscuring linkage in a pedigree. In such cases, an approach relying on probability calculations, called the **lod score method**, helps to demonstrate linkage. First devised by J. B. S. Haldane and C. A. Smith in 1947 and refined by Newton Morton in 1955, the lod score (standing for *l*og of the *od*ds favoring linkage) assesses the probability that a particular pedigree involving two traits reflects linkage. First, the probability is calculated that family data (pedigrees) concerning two or more traits conform to the transmission of traits without linkage. Then the probability is calculated that the identical family data following these same traits result from linkage with a specified recombination frequency. The ratio of these probability values expresses the "odds" for, and against, linkage. The lod score method represented an important advance in assigning human genes to specific chromosomes and in constructing preliminary human chromosome maps. However, its accuracy is limited by the extent of the pedigree, and the initial results were discouraging because of these limitations and because of the relatively high haploid number of human chromosomes (23). By 1960, short of the assignment of some genes to the X chromosome, very little autosomal linkage or mapping information had become available.

In the 1960s, a newly developed technique, **somatic cell hybridization**, proved to be an immense aid in assigning human genes to their respective chromosomes. This technique, first discovered by Georges Barsky, relies on the fact that two cells in culture can be induced to fuse into a single hybrid cell. Barsky used two mouse cell lines, but it soon became evident that cells from different organisms will also fuse together. When fusion occurs, an initial cell type called a **heterokaryon** is produced. The hybrid cell contains two nuclei in a common cytoplasm. Using the proper techniques, it is possible to fuse human and mouse cells, for example, and isolate the hybrids from the parental cells.

As the heterokaryons are cultured *in vitro*, two interesting changes occur. Eventually, the nuclei fuse together, creating what is termed a **synkaryon**. Then, as culturing is continued for many generations, chromosomes from one of the two parental species are gradually lost. In the case of the human–mouse hybrid, human chromosomes are lost randomly until, eventually, the synkaryon has a full complement of mouse chromosomes and only a few human chromosomes. As we will see, it is the preferential loss of human chromosomes rather than mouse chromosomes that makes possible the assignment of human genes to the chromosomes upon which they reside.

The experimental rationale is straightforward. For example, if a specific human gene product is synthesized in a synkaryon containing three human chromosomes, then the gene responsible for that product must reside on one of the three human chromosomes remaining in the hybrid cell. Or, if the human gene product is absent, the responsible gene cannot be present on any of the remaining three human chromosomes. Ideally, a panel of 23 hybrid cell lines, each with only one unique human chromosome, would allow the immediate assignment to a particular chromosome of any human gene for which the product could be characterized.

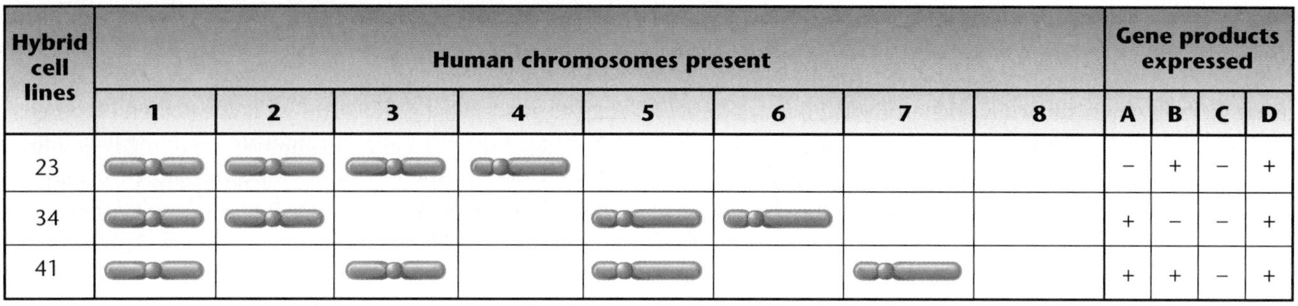

FIGURE 5–23 A hypothetical grid of data used in synteny testing to assign genes to their appropriate human chromosomes. Three somatic hybrid cell lines, designated 23, 34, and 41, have each been scored for the presence, or absence, of human chromosomes 1 through 8, as well as for their ability to produce the hypothetical human gene products *A*, *B*, *C*, and *D*.

In practice, a panel of cell lines, each with several remaining human chromosomes, is most often used. The correlation of the presence or absence of each chromosome with the presence or absence of each gene product is called **synteny testing**. Consider, for example, the hypothetical data provided in Figure 5–23, where four gene products (*A*, *B*, *C*, and *D*) are tested in relationship to eight human chromosomes. Let us carefully analyze the gene that produces product *A*:

1. Product *A* is not produced by cell line 23, but chromosomes 1, 2, 3, and 4 are present in cell line 23. Therefore, we can rule out the presence of gene *A* on those four chromosomes and conclude that it must be on chromosome 5, 6, 7, or 8.
2. Product *A* is produced by cell line 34, which contains chromosomes 5 and 6, but not 7 and 8. Therefore, gene *A* is on chromosome 5 or 6, but cannot be on 7 or 8 because they are absent, even though product A is produced.
3. Product *A* is also produced by cell line 41, which contains chromosome 5, but not chromosome 6. Therefore, gene *A* is on chromosome 5, according to this analysis.

Using a similar approach, we can assign gene *B* to chromosome 3. You should perform the same analysis to demonstrate for yourself that this is correct.

Gene *C* presents a unique situation. The data indicate that it is not present on any of the first seven chromosomes (1–7) because it is not produced by any of the three cell lines that collectively contain those chromosomes. While it might be on chromosome 8, no direct evidence supports this conclusion. Other panels are needed. We leave gene *D* for you to analyze. Upon what chromosome does it reside?

By using the approach just described, literally hundreds of human genes were assigned to specific chromosomes. Figure 5–24

FIGURE 5–24 Representative regional gene assignments for human chromosome 1 and the X chromosome. Many assignments were initially derived by using somatic cell hybridization techniques.

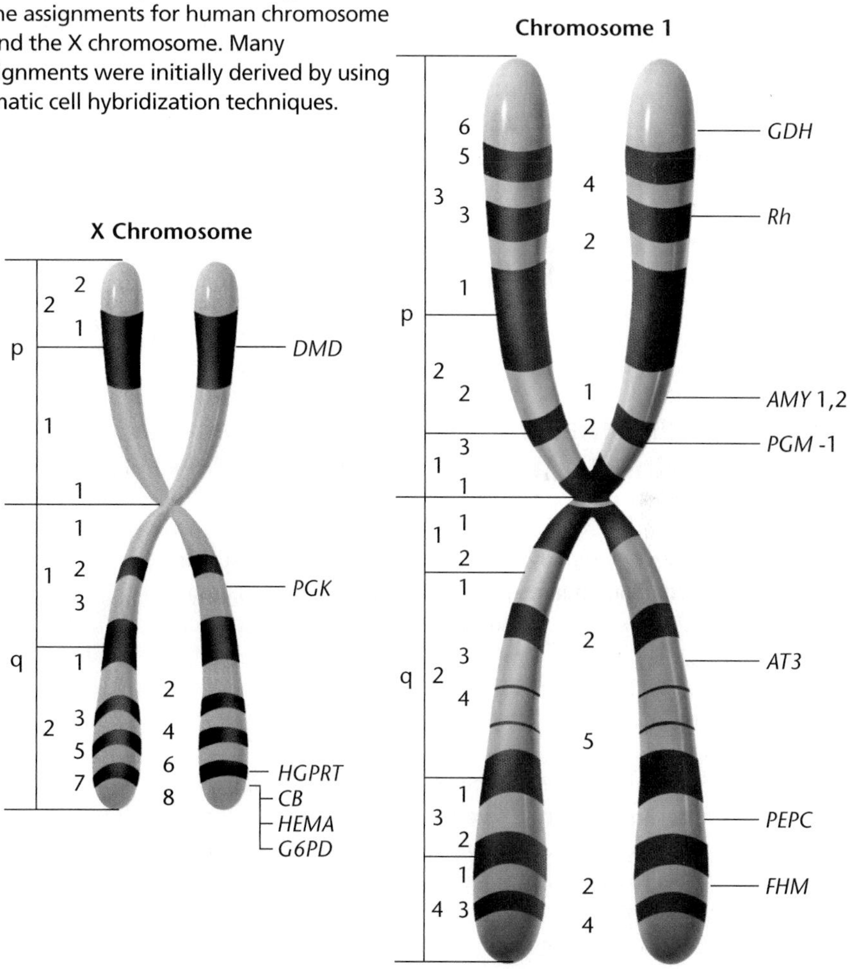

Key

- *AMY* Amylase (salivary and pancreatic)
- *AT3* Antithrombin (clotting factor IV)
- *CB* Color Blindness
- *DMD* Duchenne Muscular Dystrophy
- *FHM* Fumarate Hydratase (mitochondrial)
- *GDH* Glucose Dehydrogenase
- *G6PD* Glucose-6-Phosphate Dehydrogenase
- *HEMA* Hemophilia A (classic)
- *HGPRT* Hypoxanthine-Guanine-Phosphoribosyl Transferase (Lesch–Nyhan syndrome)
- *PEPC* Peptidase C
- *PGK* Phosphoglycerate Kinase
- *PGM* Phosphoglucomutase
- *Rh* Rhesus Blood Group (erythroblastosis fetalis)

illustrates some gene locations on two human chromosomes: X and 1. The gene assignments shown were either derived or confirmed with the use of somatic-cell hybridization techniques. To map genes for which the products have yet to be discovered, researchers have had to rely on other approaches. For example, by combining recombinant DNA technology with pedigree analysis, it has been possible to assign the genes responsible for Huntington disease, cystic fibrosis, and neurofibromatosis to their respective chromosomes 4, 7, and 17.

We conclude this discussion by addressing the next step in human gene mapping: assigning genes to different regions of a given chromosome. Sometimes in hybrid cell lines, fragments of a particular chromosome become transferred to another chromosome, resulting in a **translocation**. It is possible, using chromosome banding techniques, to identify the exact origin of the translocation and correlate the presence of a chromosomal segment in hybrid cells with specific gene expression. In that way, gene maps of human chromosomes may be compiled. The partial maps shown in Figure 5–24 illustrate this point. Although human chromosome maps are not as specific as the genetic map of *Drosophila*, these initial studies provided a great deal of information concerning the chromosome locations of a multitude of human genes. As we will see, modern technology involving recombinant DNA and the Human Genome Project have greatly extended our knowledge of gene locations within the human genome.

5.12 Gene Mapping Is Now Possible Using Molecular Analysis of DNA

While traditional methods based on recombinational analysis have produced detailed gene maps in several organisms, gene maps in other organisms that do not lend themselves to such studies, including humans, are greatly limited. Fortunately, the development of technology allowing direct analysis of DNA has greatly enhanced gene mapping in such organisms. We will address this topic using humans as an example.

Progress has initially relied on the discovery of **DNA markers** that have been identified during recombinant DNA and genomic studies. These markers can be followed in multigenerational pedigrees whereby they can be mapped on specific chromosomes. Then, human genes may be analyzed in relationship to these markers, establishing their positions along chromosomes, creating more detailed genetic maps. For example, the locations of DNA markers referred to as **microsatellites**, **minisatellites**, and **restriction fragment length polymorphisms (RFLPs)** can be determined (see Chapter 12). Then a gene, such as the one involved in **cystic fibrosis**, can be located in relation to such markers using the techniques of **chromosome walking** or **chromosome jumping** (Chapter 19).

Cystic fibrosis is an autosomal recessive exocrine disorder resulting in excessive, thick mucus that impedes the function of organs such as the lung and pancreas. It was first established that the gene causing this disorder is located on chromosome 7. Subsequently, using the chromosome walking technique, its exact location on the long arm of the chromosome was determined. Using this approach, much more detailed human gene maps have become available.

Gene Mapping Using Annotated Computer Databases

As we will see later in the text (Chapters 19 and 20), modern technology involving the Human Genome Project has had an even more profound impact on our knowledge of gene locations within the human genome (and in other animal and plant genomes). The Human Genome Project goal was first to obtain the sequence of the human genome, and then, by computer analysis, determine the location of the genes. These data are now available on several Web sites and are easily accessible through simple search engines.

A result of the many completed project databases is that it is now possible to determine the absolute location of any gene in a chromosome in base pair distance rather than recombination frequency. This distinguishes what is referred to as a **physical map** of the genome from the genetic maps described above. Distances can then be determined relative to other genes and relative to features such as the mini- and microsatellites as well as to markers designated as **single nucleotide polymorphisms (SNPs)**. The power of this approach is that it will soon be possible to construct chromosome maps for individuals that designate specific allele combinations at each gene site.

5.13 Did Mendel Encounter Linkage?

We conclude this chapter by examining a modern-day interpretation of the experiments forming the cornerstone of transmission genetics: Mendel's crosses with garden peas. Some observers believe that Mendel had extremely good fortune in his classic experiments with the garden pea. He did not encounter any *apparent* linkage relationships between the seven mutant characters in any of his crosses. Had Mendel obtained highly variable data characteristic of linkage and crossing over, these unorthodox observations might have hindered his successful analysis and interpretation.

The article by Stig Blixt, reprinted in its entirety in the box that follows, presents a modern-day interpretation of Mendel's experiments, demonstrating the inadequacy of this hypothesis. As we shall see, some of Mendel's genes were indeed linked. We shall leave it to Stig Blixt to enlighten you as to why Mendel did not detect linkage.

Why Didn't Gregor Mendel Find Linkage?

It is quite often said that Mendel was very fortunate not to run into the complication of linkage during his experiments. He used seven genes, and the pea has only seven chromosomes. Some have said that had he taken just one more, he would have had problems. This, however, is a gross oversimplification. The actual situation, most probably, is that Mendel worked with three genes in chromosome 4, two genes in chromosome 1, and one gene in each of chromosomes 5 and 7. (See Table 1.) It seems at first glance that, out of the 21 dihybrid combinations Mendel theoretically could have studied, no less than four (that is, *a–i*, *v–fa*, *v–le*, *fa–le*) ought to have resulted in linkages. As found, however, in hundreds of crosses and shown by the genetic map of the pea, *a* and *i* in chromosome 1 are so distantly located on the chromosome that no linkage is normally detected. The same is true for *v* and *le* on the one hand, and *fa* on the other, in chromosome 4. This leaves *v–le*, which ought to have shown linkage.

Mendel, however, seems not to have published this particular combination and thus, presumably, never made the appropriate cross to obtain both genes segregating simultaneously. It is therefore not so astonishing that Mendel did not run into the complication of linkage, although he did not avoid it by choosing one gene from each chromosome.

Stig Blixt
Weibullsholm Plant Breeding Institute, Landskrona, Sweden, and Centro de Energia Nuclear na Agricultura, Piracicaba, SP, Brazil.

TABLE 1 RELATIONSHIP BETWEEN MODERN GENETIC TERMINOLOGY AND CHARACTER PAIRS USED BY MENDEL

Character Pair Used by Mendel	Alleles in Modern Terminology	Located in Chromosome
Seed color, yellow–green	*I–i*	1
Seed coat and flowers, colored–white	*A–a*	1
Mature pods, smooth expanded–wrinkled indented	*V–v*	4
Inflorescences, from leaf axis–umbellate in top of plant	*Fa–fa*	4
Plant height, 0.5–1 m	*Le–le*	4
Unripe pods, green yellow	*Gp–gp*	5
Mature seeds, smooth wrinkled	*R–r*	7

CHAPTER SUMMARY

1. Genes located on the same chromosome are said to be linked. Alleles located on the same homolog, therefore, can be transmitted together during gamete formation. However, the mechanism of crossing over between homologs during meiosis results in the reshuffling of alleles, thereby contributing to genetic variability within gametes.
2. Early in this century, geneticists realized that crossing over provides an experimental basis for mapping the location of linked genes relative to one another along the chromosome.
3. Interference is a phenomenon describing the extent to which a crossover in one region of a chromosome influences the occurrence of a crossover in an adjacent region of the chromosome. The coefficient of coincidence (C) is a quantitative estimate of interference calculated by dividing the observed double crossovers by the expected double crossovers.
4. Due to statistical considerations described by the Poisson distribution, as the actual distance between two genes increases, experimentally determined mapping distances become more and more inaccurate (underestimated).
5. Extensive genetic maps have been created in organisms such as corn, mice, and *Drosophila*.
6. Cytological investigations of both maize and *Drosophila* reveal that crossing over involves a physical exchange of segments between nonsister chromatids.
7. In a few organisms, including *Drosophila* and *Aspergillus*, homologs pair during mitosis and crossing over occurs between them at a frequency far less than during meiosis.
8. An exchange of genetic material between sister chromatids can occur during mitosis as well. These events are referred to as sister chromatid exchanges (SCEs). An elevated frequency of such events is seen in the human disorder Bloom syndrome.
9. Linkage analysis and chromosome mapping are possible in haploid eukaryotes. Our discussion has included gene-to-centromere and gene-to-gene mapping as well as the consideration of how to distinguish between linkage and independent assortment.
10. Somatic cell hybridization techniques have made possible linkage and mapping analysis of human genes.
11. Evidence now suggests that several of the genes studied by Mendel are, in fact, linked. However, in such cases, the genes are sufficiently far apart to prevent the detection of linkage.

INSIGHTS AND SOLUTIONS

1. In a series of two-point map crosses involving three genes linked on chromosome 3 in *Drosophila*, the following distances were calculated:

cd–sr 13 mu

cd–ro 16 mu

(a) Determine the sequence and construct a map of these three genes.

Solution: It is impossible to do so. There are two possibilities based on these limited data,

Case 1: *cd* —13— *sr* —3— *ro*

or

Case 2: *ro* —16— *cd* —13— *sr*

(b) What mapping data will resolve this?

Solution: The map distance determined by crossing over between *ro* and *sr*. If case 1 is correct, it should be 3 mu, and if case 2 is correct, it should be 29 mu. In fact, this distance is 29 mu, demonstrating that case 2 is correct.

(c) Can we tell which of the sequences shown here is correct?

ro —16— *cd* —13— *sr*

or

sr —13— *cd* —16— *ro*

Solution: No; based on the mapping data, they are equivalent.

2. In *Drosophila, Lyra (Ly)* and *Stubble (Sb)* are dominant mutations located at loci 40 and 58, respectively, on chromosome 3. A recessive mutation with bright-red eyes was discovered and shown also to be on chromosome 3. A map was obtained by crossing a female who was heterozygous for all three mutations to a male homozygous for the bright-red mutation (which we refer to here as *br*). The following data are obtained:

Phenotype				Number
(1)	*Ly*	*Sb*	*br*	404
(2)	+	+	+	422
(3)	*Ly*	+	+	18
(4)	+	*Sb*	*br*	16
(5)	*Ly*	+	*br*	75
(6)	+	*Sb*	+	59
(7)	*Ly*	*Sb*	+	4
(8)	+	+	*br*	2
				Total = 1000

Determine the location of the *br* mutation on chromosome 3. By referring to Figure 5–14, predict what mutation has been discovered. How could you be sure?

Solution: First determine the *arrangement* of the alleles on the homologs of the heterozygous crossover parent (the female in this case) by locating the most frequent reciprocal phenotypes, which arise from the noncrossover gametes. These are phenotypes (1) and (2). Each one represents the arrangement of alleles on one of the homologs. Therefore, the arrangement is

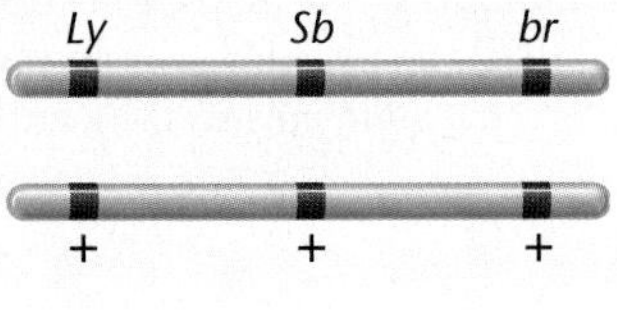

Second, determine the correct *sequence* of the three loci along the chromosome by determining which sequence will yield the *observed* double-crossover phenotypes that are the least frequent reciprocal phenotypes (7 and 8).

If the sequence is correct as written, then a double crossover depicted here,

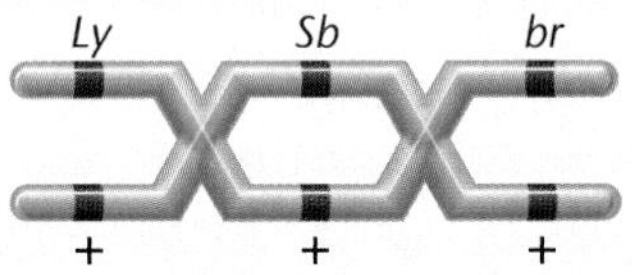

will yield *Ly* + *br* and + *Sb* + as phenotypes. Inspection shows that these categories (5 and 6) are actually single crossovers, not double crossovers. Therefore, the sequence, as written, is incorrect. There are only two other possible sequences. The *br* gene is either to the left of *Ly*, or it is between *Ly* and *Sb*:

Case A

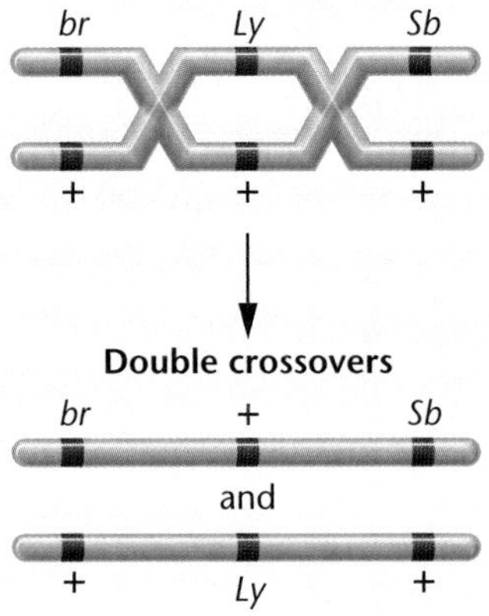

Case B

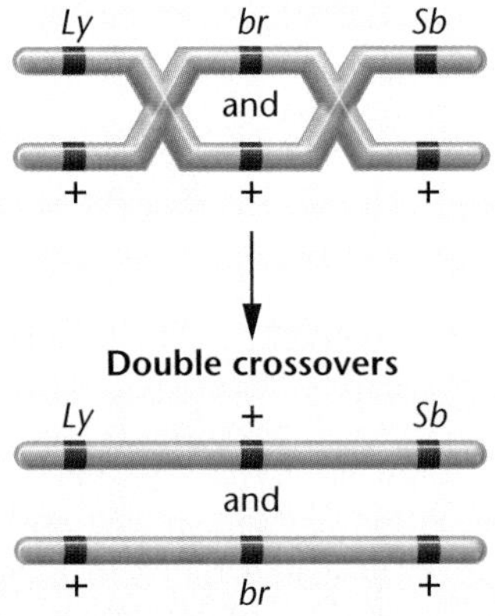

Comparison with the actual data shows that case B is correct. The double-crossover gametes (7) and (8) yield flies that express *Ly* and *Sb*, but not *br, or* express *br*, but not *Ly* and *Sb*. Therefore, the correct arrangement and sequence are as follows:

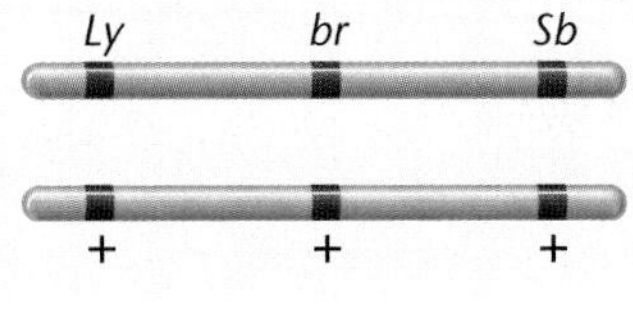

Once the correct arrangement and sequence are established, it is possible to determine the location of *br* relative to *Ly* and *Sb*. A single crossover between *Ly* and *br*, as shown here

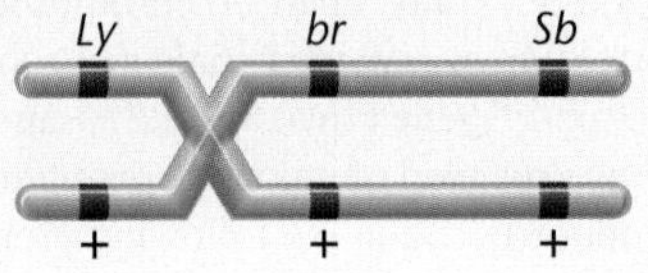

yields flies that are *Ly* + + and + *br Sb* (categories 3 and 4). Therefore, the distance between the *Ly* and *br* loci is equal to

$$\frac{18 + 16 + 4 + 2}{1000} = \frac{40}{1000} = 0.04 = 4 \text{ mu}$$

Remember that, because we need to know the frequency of all crossovers between *Ly* and *br*, we must add in the double crossovers, since they represent two single crossovers occurring simultaneously. Similarly, the distance between the *br* and *Sb* loci is derived mainly from single crossovers between them.

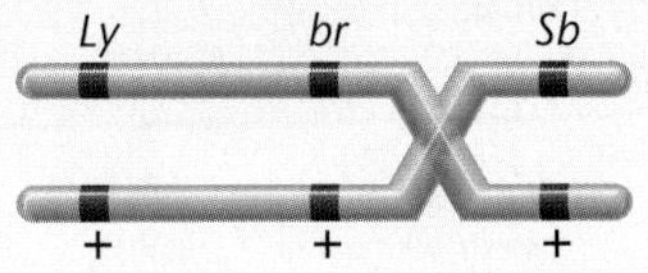

This event yields *Ly br* + and ++ *Sb* phenotypes (categories 5 and 6). Therefore, the distance equals

$$75 + 59 + 4 + 2/1000 = 140/1000 = 0.14 = 14 \text{ mu}$$

The final map shows that *br* is located at locus 44, since *Lyra* and *Stubble* are known:

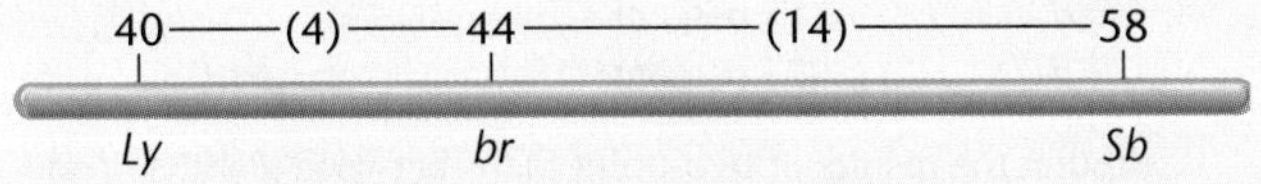

Inspection of Figure 5–14 reveals that the mutation *scarlet*, which has bright-red eyes, is known to exist at locus 44, so it is reasonable to hypothesize that the bright-red eye mutation is an allele of *scarlet*. To test this hypothesis, we could cross females of our bright-red mutant with known *scarlet* males. If the two mutations are alleles, no complementation will occur, and all progeny will reveal a bright-red mutant eye phenotype. If complementation occurs, all progeny will show normal brick-red (wild-type) eyes, since the bright red mutation and *scarlet* are at different loci. (They are probably very close together.) In such a case, all progeny will be heterozygous at both the bright eye and the *scarlet* loci and will not express either mutation because they are both recessive. This cross represents what is called an **allelism test**.

3. In rabbits, *black* (*B*) is dominant to *brown* (*b*), while *full color* (*C*) is dominant to *chinchilla* (c^{ch}). The genes controlling these traits are linked. Rabbits that are heterozygous for both traits and express *black, full color* were crossed to rabbits that express *brown, chinchilla*, with the following results:

31 brown, chinchilla

34 black, full color

16 brown, full color

19 black, chinchilla

Determine the arrangement of alleles in the heterozygous parents and the map distance between the two genes.

Solution: This is a two-point map problem, where the two reciprocal noncrossover phenotypes are recognized as those present in the highest numbers (*brown, chinchilla* and *black, full*). The less frequent reciprocal phenotypes (*brown, full* and *black, chinchilla*) arise from a single crossover. The arrangement of alleles is derived from the noncrossover phenotypes because they enter gametes intact. The cross is shown as follows:

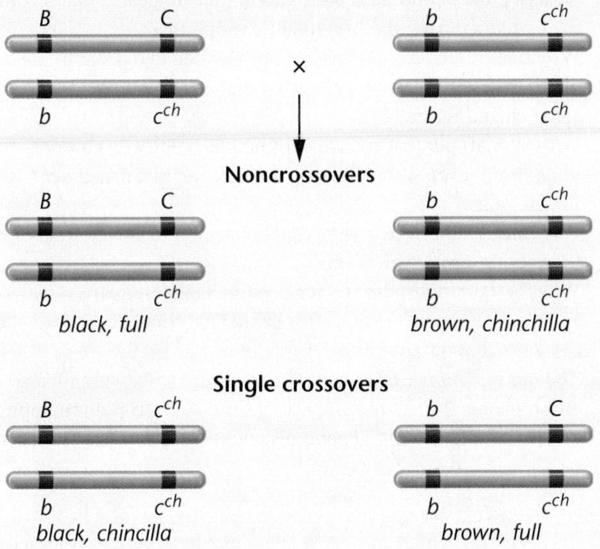

The single crossovers give rise to 35/100 offspring (35 percent). Therefore, the distance between the two genes is 35 mu.

4. In a cross in *Neurospora* where one parent expresses the mutant allele *a* and the other expresses a wild-type phenotype (+), the following data were obtained in the analysis of ascospores:

	Asci Types						
	1	2	3	4	5	6	
Sequence of ascospores in ascus	+	*a*	*a*	+	*a*	+	
	+	*a*	*a*	+	*a*	+	
	+	*a*	+	*a*	+	*a*	
	+	*a*	+	*a*	+	*a*	
	a	+	*a*	+	+	*a*	
	a	+	*a*	+	+	*a*	
	a	+	+	*a*	*a*	+	
	a	+	+	*a*	*a*	+	
	39	33	5	4	9	10	Total = 100

Calculate the gene-to-centromere distance.

Solution: Ascus types 1 and 2 represent first-division segregants (fds) where no crossing over occurred between the *a* locus and the centromere. All others (3–6) represent second-division segregation (sds). By applying the formula

$$\text{distance} = \frac{1/2 \text{ sds}}{\text{total asci}}$$

we obtain the following result:

$$\begin{aligned} d &= 1/2(5 + 4 + 9 + 10)/100 \\ &= 1/2(28)/100 \\ &= 0.14 \\ &= 14 \text{ mu} \end{aligned}$$

PROBLEMS AND DISCUSSION QUESTIONS

1. What is the significance of genetic recombination to the process of evolution?
2. Describe the cytological observation that suggests that crossing over occurs during the first meiotic prophase.
3. Why does more crossing over occur between two distantly linked genes than between two genes that are very close together on the same chromosome?
4. Why is a 50 percent recovery of single-crossover products the upper limit, even when crossing over *always* occurs between two linked genes?
5. Why are double-crossover events expected in lower frequency than single-crossover events?
6. What is the proposed basis for positive interference?
7. What two essential criteria must be met in order to execute a successful mapping cross?
8. The genes *dumpy* (*dp*), *clot* (*cl*), and *apterous* (*ap*) are linked on chromosome 2 of *Drosophila*. In a series of two-point mapping crosses, the following genetic distances were determined:

dp – *ap*	42
dp – *cl*	3
ap – *cl*	39

What is the sequence of the three genes?

9. Consider two hypothetical recessive autosomal genes *a* and *b*. Where a heterozygote is testcrossed to a double-homozygous mutant, predict the phenotypic ratios under the following conditions:
   (a) *a* and *b* are located on separate autosomes.
   (b) *a* and *b* are linked on the same autosome, but are so far apart that a crossover always occurs between them.
   (c) *a* and *b* are linked on the same autosome, but are so close together that a crossover almost never occurs.
   (d) *a* and *b* are linked on the same autosome about 10 mu apart.
10. In corn, colored aleurone (in the kernels) is due to the dominant allele *R*. The recessive allele *r*, when homozygous, produces colorless aleurone. The plant color (not the kernel color) is controlled by another gene with two alleles, *Y* and *y*. The dominant *Y* allele results in green color, whereas the homozygous presence of the recessive *y* allele causes the plant to appear yellow. In a testcross between a plant of unknown genotype and phenotype and a plant that is homozygous recessive for both traits, the following progeny were obtained:

Colored green	88
Colored yellow	12
Colorless green	8
Colorless yellow	92

Explain how these results were obtained by determining the exact genotype and phenotype of the unknown plant, including the precise association of the two genes on the homologs (i.e., the arrangement).

11. In the cross shown here, involving two linked genes, *ebony* (*e*) and *claret* (*ca*), in *Drosophila*, where crossing over does not occur in males, offspring were produced in a 2 + :1 *ca*:1 *e* phenotypic ratio:

♀		♂
$\dfrac{e \quad ca^+}{e^+ \quad ca}$	×	$\dfrac{e \quad ca^+}{e^+ \quad ca}$

These genes are 30 units apart on chromosome 3. What contribution did crossing over in the female make to these phenotypes?

12. With two pairs of genes involved (*P*/*p* and *Z*/*z*), a testcross (*ppzz*) with an organism of unknown genotype indicated that the gametes produced were in the following proportions:

PZ, 42.4 percent; *Pz*, 6.9 percent; *pZ*, 7.1 percent; and *pz*, 43.6 percent

Draw all possible conclusions from these data.

13. In a series of two-point map crosses involving five genes located on chromosome II in *Drosophila*, the following recombinant (single-crossover) frequencies were observed:

pr–adp	29
pr–vg	13
pr–c	21
pr–b	6
adp–b	35
adp–c	8
adp–vg	16
vg–b	19
vg–c	8
c–b	27

   (a) If *adp* gene is present near the end of chromosome II (locus 83), construct a map of these genes.
   (b) In another set of experiments, a sixth gene, *d*, was tested against *b* and *pr:*

d–b	17 percent
d–pr	23 percent

Predict the results of two-point maps between *d* and *c*, *d* and *vg*, and *d* and *adp*.

14. Two different female *Drosophila* were isolated, each heterozygous for the autosomally linked genes *b* (*black body*), *d* (*dachs tarsus*), and *c* (*curved wings*). These genes are in the order *d–b–c*, with *b* being closer to *d* than to *c*. Shown here is the genotypic arrangement for each female along with the various gametes formed by both:

Female A		**Female B**	
$\dfrac{d \; b \; +}{+ \; + \; c}$		$\dfrac{d \; + \; +}{+ \; b \; c}$	
↓	**Gamete formation**	↓	
(1) *d b c*	(5) *d*++	(1) d b +	(5) *d b c*
(2) + + +	(6) + *b c*	(2) + + *c*	(6) + + +
(3) + + *c*	(7) *d* + *c*	(3) *d* + *c*	(7) *d* + +
(4) *d b* +	(8) + *b* +	(4) + *b* +	(8) + *b c*

Identify which categories are noncrossovers (NCO), single crossovers (SCO), and double crossovers (DCO) in each case. Then, indicate the relative frequency in which each will be produced.

15. In *Drosophila*, a cross was made between females expressing the three X-linked recessive traits, *scute* bristles (*sc*), *sable* body (*s*), and *vermilion* eyes (*v*), and males that are wild type. In the F_1, all females were wild type, while all males expressed all three mutant traits. The cross was carried to the F_2 generation and 1000 offspring were counted, with the results shown here:

Phenotype	Offspring
sc s v	314
+ + +	280
+ s v	150
sc + +	156
sc + v	46
+ s +	30
sc s +	10
+ + v	14

No determination of sex was made in the data.

(a) Using proper nomenclature, determine the genotypes of the P_1 and F_1 parents.

(b) Determine the sequence of the three genes and the map distance between them.

(c) Are there more or fewer double crossovers than expected? Calculate the coefficient of coincidence. Does this represent positive or negative interference?

16. Another cross in *Drosophila* involved the recessive, X-linked genes *yellow* (*y*), *white* (*w*), and *cut* (*ct*). A yellow-bodied, white-eyed female with normal wings was crossed to a male whose eyes and body were normal, but whose wings were cut. The F_1 females were wild type for all three traits, while the F_1 males expressed the yellow-body, white-eye traits. The cross was carried to an F_2 progeny, and only male offspring were tallied. On the basis of the data shown here, a genetic map was constructed:

Phenotype	Male Offspring
y + ct	9
+ w +	6
y w ct	90
+ + +	95
+ + ct	424
y w +	376
y + +	0
+ w ct	0

(a) Diagram the genotypes of the F_1 parents.

(b) Assuming that *white* is at locus 1.5 on the X chromosome, construct a map.

(c) Were any double-crossover offspring expected?

(d) Could the F_2 female offspring be used to construct the map? Why or why not?

17. In *Drosophila, Dichaete* (*D*) is a mutation on chromosome 3 with a dominant effect on wing shape. It is lethal when homozygous. The genes *ebony* (*e*) and *pink* (*p*) are recessive mutations on chromosome 3 affecting the body and eye color, respectively. Flies from a *Dichaete* stock were crossed to homozygous *ebony, pink* flies, and the F_1 progeny, with a *Dichaete* phenotype, were backcrossed to the *ebony, pink* homozygotes. The following were the results of this backcross:

Phenotype	Number
Dichaete	401
ebony, pink	389
Dichaete, ebony	84
pink	96
Dichaete, pink	2
ebony	3
Dichaete, ebony, pink	12
wild type	13

(a) Diagram this cross, showing the genotypes of the parents and offspring of both crosses.

(b) What is the sequence and interlocus distance between these three genes?

18. *Drosophila* females homozygous for the third chromosomal genes *pink* and *ebony* (the same genes from Problem 17) were crossed with males homozygous for the second chromosomal gene *dumpy.* Because these genes are recessive, all offspring were wild type (normal). F_1 females were testcrossed to triply recessive males. If we assume that the two linked genes, *pink* and *ebony*, are 20 mu apart, predict the results of this cross. If the reciprocal cross were made (F_1 males—where no crossing over occurs—with triply recessive females), how would the results vary, if at all?

19. In *Drosophila*, two mutations, *Stubble* (*Sb*) and *curled* (*cu*), are linked on chromosome 3. *Stubble* is a dominant gene that is lethal in a homozygous state, and *curled* is a recessive gene. If a female of the genotype

$$\frac{Sb \quad cu}{+ \quad +}$$

is to be mated to detect recombinants among her offspring, what male genotype would you choose as a mate?

20. In *Drosophila*, a heterozygous female for the X-linked recessive traits *a*, *b*, and *c* was crossed to a male that phenotypically expressed *a*, *b*, and *c*. The offspring occurred in the following phenotypic ratios:

+	b	c	460
a	+	+	450
a	b	c	32
+	+	+	38
a	+	c	11
+	b	+	9

No other phenotypes were observed.

(a) What is the genotypic arrangement of the alleles of these genes on the X chromosome of the female?

(b) Determine the correct sequence and construct a map of these genes on the X chromosome.

(c) What progeny phenotypes are missing? Why?

21. Why did Stern observe more "twin spots" than *singed* spots in his study of somatic crossing over? If he had been studying *tan* body color (locus 27.5) and *forked* bristles (locus 56.7) on the X chromosome of heterozygous females, what relative frequencies of tan spots, forked spots, and "twin spots" would you predict might occur?

22. Are mitotic recombinations and sister chromatid exchanges effective in producing genetic variability in an individual? In the offspring of individuals?

23. What possible conclusions can be drawn from the observations that no synaptonemal complexes are observed and that no crossing over occurs in male *Drosophila* and female *Bombyx mori* (the commercial silkmoth)?

24. An organism of the genotype *AaBbCc* was testcrossed to a triply recessive organism (*aabbcc*). The genotypes of the progeny were as follows:

20	*AaBbCc*	20	*AaBbcc*
20	*aabbCc*	20	*aabbcc*
5	*AabbCc*	5	*Aabbcc*
5	*aaBbCc*	5	*aaBbcc*

(a) Assuming simple dominance and recessiveness in each gene pair, if these three genes were all assorting independently, how many genotypic and phenotypic classes would result in the offspring, and in what proportion?

(b) Answer the same question assuming the three genes are so tightly linked on a single chromosome that no crossover gametes were recovered in the sample of offspring.

(c) What can you conclude from the *actual* data about the location of the three genes in relation to one another?

25. Based on our discussion of the potential inaccuracy of mapping (see Figure 5–13), would you revise your answer to Problem 24? If so, how?

26. In a plant, fruit was either red or yellow and was either oval or long, where *red* and *oval* are the dominant traits. Two plants, both heterozygous for these traits, were testcrossed, with the following results:

Phenotype	Progeny Plant A	Progeny Plant B
red, long	46	4
yellow, oval	44	6
red, oval	5	43
yellow, long	5	47
	100	100

Determine the location of the genes relative to one another and the genotypes of the two parental plants.

27. In a plant heterozygous for two gene pairs (*Ab*/*aB*), where the two loci are linked and 25 mu apart, two such individuals were crossed. Assuming that crossing over occurs during the formation of both male and female gametes and that the *A* and *B* alleles are dominant, determine the phenotypic ratio of the offspring.

28. In a cross in *Neurospora* involving two alleles, *B* and *b*, the following tetrad patterns were observed:

Tetrad Pattern	Number
BBbb	36
bbBB	44
BbBb	4
bBbB	6
BbbB	3
bBBb	7

Calculate the distance between the gene and the centromere.

29. In *Neurospora*, the cross a+ × +*b* yielded only two types of ordered tetrads in approximately equal numbers:

	Spore Pair			
	1–2	3–4	5–6	7–8
Tetrad Type 1	a +	a +	+ b	+ b
Tetrad Type 2	+ +	+ +	a b	a b

What can be concluded?

30. Here are two sets of data derived from crosses in *Chlamydomonas*, involving three genes represented by the mutant alleles *a*, *b*, and *c*:

Genes	Cross	P	NP	T
1	a and b	36	36	28
2	*b* and *c*	79	3	18
3	*a* and *c*	?	?	?

Determine as much as you can concerning genetic arrangement of these three genes relative to one another. Assuming that *a* and *c* are linked and are 38 mu apart and that 100 tetrads are produced, describe the expected results of cross 3.

31. In *Chlamydomonas*, a cross *ab* × ++ yielded the following unordered tetrad data where *a* and *b* are linked:

Class	Tetrad	Number	Class	Tetrad	Number	Class	Tetrad	Number
(1)	+ + + + *a b* *a b*	38	(3)	*a* + *a* + + *b* + *b*	6	(5)	*a b* + + + *b* *a* +	2
(2)	+ + *a b* + + *a b*	5	(4)	*a b* *a* + + *b* + +	17	(6)	*a b* + *b* *a* + + +	3

(a) Identify the categories representing parental ditypes (P), nonparental ditypes (NP), and tetratypes (T).

(b) Explain the origin of category (2).

(c) Determine the map distance between *a* and *b*.

32. The following results are ordered tetrad pairs from a cross between strain *cd* and strain ++:

Tetrad Class						
1	2	3	4	5	6	7
c +	c +	c d	+ d	c +	c d	c +
c +	c d	c d	c +	++	++	+ d
+ d	++	++	c +	c d	c d	c d
+ d	+ d	++	+ d	+ d	++	++
1	17	41	1	5	3	1

They are summarized by tetrad classes.

(a) Name the ascus type of each class from 1 to 7 (P, NP, or T).

(b) The data support the conclusion that the *c* and *d* loci are linked. State the evidence in support of this conclusion.

(c) Calculate the gene–centromere distance for each locus.

(d) Calculate the distance between the two linked loci.

(e) Draw a chromosome map, including the centromere, and explain the discrepancy between the distances determined by the two different methods in parts (c) and (d).

(f) Describe the arrangement of crossovers needed to produce the ascus class 6.

33. In a cross in *Chlamydomonas*, *AB* × *ab*, 211 unordered asci were recovered:

10	*AB*,	*Ab*,	*aB*,	*ab*
102	*Ab*,	*aB*,	*Ab*,	*aB*
99	*AB*,	*AB*,	*ab*,	*ab*

(a) Correlate each of the three tetrad types in the problem with their appropriate tetrad designations (names).

(b) Are genes *A* and *B* linked?

(c) If they are linked, determine the map distance between the two genes. If they are unlinked, provide the maximum information you can about why you drew this conclusion.

34. A number of human–mouse somatic–cell hybrid clones were examined for the expression of specific human genes and the presence of human chromosomes. The results are summarized in the following table. Assign each gene to the chromosome upon which it is located.

	Hybrid cell clone					
	A	B	C	D	E	F
Genes expressed						
ENO1 (enolase-1)	−	+	−	+	+	−
MDH1 (malate dehydrogenase-1)	+	+	−	+	−	+
PEPS (peptidase S)	+	−	+	−	−	−
PGM1 (phospho-glucomutase-1)	−	+	−	+	+	−
Chromosomes (present or absent)						
1	−	+	−	+	+	−
2	+	+	−	+	−	+
3	+	+	−	−	+	−
4	+	−	+	−	−	−
5	−	+	+	+	+	+

Extra-Spicy Problems

35. A female of genotype

$$\frac{a \quad b \quad c}{+ \quad + \quad +}$$

produces 100 meiotic tetrads. Of these, 68 show no crossover events. Of the remaining 32, 20 show a crossover between *a* and *b*, 10 show a crossover between *b* and *c*, and 2 show a double crossover between *a* and *b* and between *b* and *c*. Of the 400 gametes produced, how many of each of the 8 different genotypes will be produced? Assuming the order *a*–*b*–*c* and the allele arrangement previously shown, what is the map distance between these loci?

36. In laboratory, a genetics student was assigned an unknown mutation in *Drosophila* that had a whitish eye. He crossed females from his true-breeding mutant stock to wild-type (brick-red–eyed) males, recovering all wild-type F_1 flies. In the F_2 generation, the following offspring were recovered in the following proportions:

wild type	5/8
bright red	1/8
brown eye	1/8
white eye	1/8

The student was stumped until the instructor suggested that perhaps the whitish eye in the original stock was the result of homozygosity for a mutation causing brown eyes *and* a mutation causing bright-red eyes, illustrating gene interaction. (See Chapter 4.) After much thought, the student was able to analyze the data, explain the results, and learn several things about the location of the two genes relative to one another. One key to his understanding was that crossing over occurs in *Drosophila* females, but not in males. Based on his analysis, what did the student learn about the two genes?

37. *Drosophila melanogaster* has one pair of sex chromosomes (XX or XY) and three pairs of autosomes, referred to as chromosomes 2, 3, and 4. A genetics student discovered a male fly with very short legs. Using this male, the student was able to establish a pure breeding stock of this mutant and found that it was recessive. She then incorporated the mutant into a stock containing the recessive gene *black* (body color located on chromosome 2) and the recessive gene *pink* (eye color located on chromosome 3). A female from the homozygous *black, pink, short* stock was then mated to a wild-type male. The F_1 males of this cross were all wild type and were then backcrossed to the homozygous *b p sh* females. The F_2 results appeared as shown in the table that follows. No other phenotypes were observed.

	Wild	Pink*	Black, Short	Black, Pink, Short
Females	63	58	55	69
Males	59	65	51	60

* *Pink* indicates that the other two traits are wild type, and so on.

(a) Based on these results, the student was able to assign *short* to a linkage group (a chromosome). Which one was it? Include a step-by-step reasoning.

(b) The student repeated the experiment, making the reciprocal cross, F_1 females backcrossed to homozygous *b p sh* males. She observed that 85 percent of the offspring fell into the given classes, but that 15 percent of the offspring were equally divided among *b* + *p*, *b* ++, + *sh p*, and + *sh* + phenotypic males and females. How can these results be explained and what information can be derived from the data?

38. In *Drosophila*, a female fly is heterozygous for three mutations, *Bar* eyes (*B*), *miniature* wings (*m*), and *ebony* body (*e*). Note that *Bar* is a dominant mutation. The fly is crossed to a male with normal eyes, miniature wings, and ebony body. The results of the cross are shown below:

111	miniature	101	Bar, ebony
29	wild type	31	Bar, miniature, ebony
117	Bar	35	ebony
26	Bar, miniature	115	miniature, ebony

Interpret the results of this cross. If you conclude that linkage is involved between any of the genes, determine the map distance(s) between them.

Selected Readings

Allen, G.E. 1978. *Thomas Hunt Morgan: The man and his science.* Princeton, NJ: Princeton University Press.

Chaganti, R., Schonberg, S., and German, J. 1974. A manyfold increase in sister chromatid exchange in Bloom's syndrome lymphocytes. *Proc. Natl. Acad. Sci. USA* 71:4508–12.

Creighton, H.S., and McClintock, B. 1931. A correlation of cytological and genetical crossing over in *Zea mays. Proc. Natl. Acad. Sci. USA* 17:492–97.

Douglas, L., and Novitski, E. 1977. What chance did Mendel's experiments give him of noticing linkage? *Heredity* 38:253–57.

Ellis, N.A. et al. 1995. The Bloom's syndrome gene product is homologous to RecQ helicases. *Cell* 83:655–66.

Ephrussi, B., and Weiss, M.C. 1969. Hybrid somatic cells. *Sci. Am.* (April) 220:26–35.

Lindsley, D.L., and Grell, E.H. 1972. *Genetic variations of Drosophila melanogaster.* Washington, DC: Carnegie Institute of Washington.

Morgan, T.H. 1911. An attempt to analyze the constitution of the chromosomes on the basis of sex-linked inheritance in *Drosophila. J. Exp. Zool.* 11:365–414.

Morton, N.E. 1955. Sequential test for the detection of linkage. *Am. J. Hum. Genet.* 7:277–318.

——— 1995. *LODs*—past and present. *Genetics* 140:7–12.

Neuffer, M.G., Jones, L., and Zober, M. 1968. *The mutants of maize.* Madison, WI: Crop Science Society of America.

Perkins, D. 1962. Crossing-over and interference in a multiply marked chromosome arm of *Neurospora. Genetics* 47:1253–74.

Ruddle, F.H., and Kucherlapati, R.S. 1974. Hybrid cells and human genes. *Sci. Am.* (July) 231:36–49.

Stahl, F.W. 1979. *Genetic recombination.* New York: W.H. Freeman.

Stern, H., and Hotta, Y. 1973. Biochemical controls in meiosis. *Annu. Rev. Genet.* 7:37–66.

——— 1974. DNA metabolism during pachytene in relation to crossing over. *Genetics* 78:227–35.

Sturtevant, A.H. 1913. The linear arrangement of six sex-linked factors in *Drosophila*, as shown by their mode of association. *J. Exp. Zool.* 14:43–59.

——— 1965. *A history of genetics.* New York: Harper & Row.

Voeller, B.R., ed. 1968. *The chromosome theory of inheritance: Classical papers in development and heredity.* New York: Appleton-Century-Crofts.

von Wettstein, D., Rasmussen, S.W., and Holm, P.B. 1984. The synaptonemal complex in genetic segregation. *Annu. Rev. Genet.* 18:331–414.

Wolff, S., ed. 1982. *Sister chromatid exchange.* New York: Wiley–Interscience.

Genetic Analysis and Mapping in Bacteria and Bacteriophages

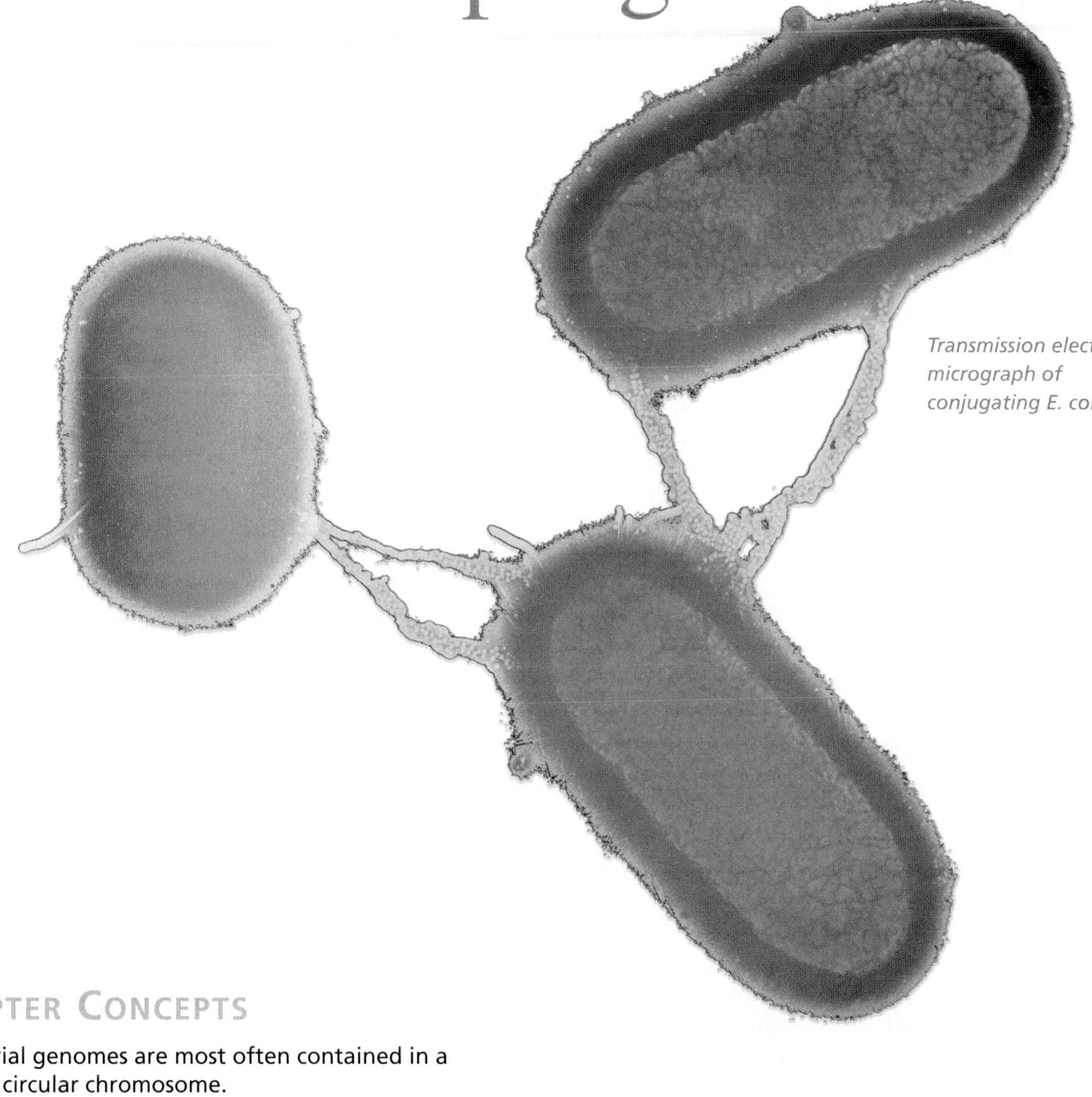

Transmission electron micrograph of conjugating E. coli.

CHAPTER CONCEPTS

- Bacterial genomes are most often contained in a single circular chromosome.
- Bacteria have developed numerous ways in which they can recombine genetic information between individual cells.
- The ability to undergo conjugation and to transfer the bacterial chromosome from one cell to another is governed by the presence of genetic information contained in the DNA of a "fertility," or F factor.
- The F factor can exist autonomously in the bacterial cytoplasm as a plasmid, or it can integrate into the bacterial chromosome, where it facilitates the transfer of the host chromosome to the recipient cell leading to genetic recombination.
- During conjugation, genetic recombination provides the basis for mapping bacterial genes.
- Bacteriophages are viruses that have bacteria as their hosts.
- During infection of the bacterial host, bacteriophage DNA is injected into the host cell, where it is replicated and directs the the reproduction of the bacteriophage.
- During bacteriophage infection, following replication, the phage DNA may undergo recombination, which may serve as the basis for intergenic and intragenic mapping.

We now shift our attention to a discussion of various genetic phenomena in **bacteria** (prokaryotes) and **bacteriophages**, viruses often referred to simply as phages. These viruses are so named because they have bacteria as their host. The study of bacteria and bacteriophages has been essential to the accumulation of knowledge in many areas of genetic study. For example, much of what we have previously discussed about molecular genetics initially was derived from experimental work with these organisms. Furthermore, as we shall see in Chapter 19, our knowledge of bacteria and their resident plasmids has served as the basis for their widespread use in DNA cloning and other recombinant DNA studies.

Bacteria and their viruses have been especially useful research organisms in genetics for a number of reasons. First, they have extremely short reproductive cycles. Literally hundreds of generations, giving rise to billions of genetically identical bacteria or phages, can be produced in short periods of time. They can also be studied in pure culture, whereby mutant strains of genetically unique bacteria or bacteriophages can be isolated and investigated independently.

A major focus in this chapter will be how one bacterial cell can acquire genetic information from another cell. This can occur in numerous ways, and once the information is taken into the recipient cell, it can recombine genetically with the recipient's chromosome, creating an altered bacterial genotype. Complex processes have evolved in these microorganisms that facilitate such genetic recombination. As we shall see, these processes are the basis for the chromosome mapping analysis in bacteria. We will also look at genetic recombination and mapping in bacteriophages.

6.1 Bacteria Mutate Spontaneously and Grow at an Exponential Rate

Genetic studies using bacteria depend upon our ability to study mutations in these organisms. Since the 1940s, it has been evident that genetically homogenous cultures of bacteria occasionally give rise to cells exhibiting heritable variation, particularly with respect to growth under unique environmental conditions. Prior to 1943, the source of this variation was hotly debated. The majority of bacteriologists believed that environmental factors induced changes in certain bacteria leading to their survival or adaptation to the new conditions. For example, strains of *E. coli* are known to be sensitive to infection by the bacteriophage T1. Infection by the bacteriophage T1 leads to the reproduction of the virus at the expense of the bacterial cell, which is lysed (destroyed) in the process. If a plate of *E. coli* is uniformly covered with T1, almost all cells are lysed. Rare *E. coli* cells, however, survive infection and are not lysed. If these cells are isolated and established in pure culture, all their descendants are resistant to T1 infection. The **adaptation hypothesis**, put forth to explain this type of observation, implies that the interaction of the phage and bacterium is essential to the acquisition of immunity. In other words, exposure to the phage "induces" resistance in the bacteria. On the other hand, the occurrence of **spontaneous mutations**, which occur in the presence or the absence of phage T1, suggested an alternative model to explain the origin of resistance in *E. coli*. In 1943, Salvador Luria and Max Delbruck presented convincing evidence that bacteria, like eukaryotic organisms, are capable of spontaneous mutation. This experiment, referred to as the **fluctuation test**, marks the initiation of modern bacterial genetic study. We will explore this discovery in Chapter 15. Spontaneous mutation is now considered the primary source of genetic variation in bacteria.

Mutant cells arising spontaneously in otherwise pure cultures can be isolated and established independently from the parent strain by using selection techniques. Selection is the growth of the organism under conditions where only the mutant you seek grows well, while the wild type does not grow. Using carefully designed selection, mutations for almost any desired characteristic can now be induced and isolated. Because bacteria and viruses usually contain only single chromosomes and are therefore haploid, all mutations are expressed directly in the descendants of mutant cells, adding to the ease with which these microorganisms can be studied.

Bacteria are grown in either a liquid culture medium or in a petri dish on a semisolid agar surface. If the nutrient components of the growth medium are very simple and consist only of an organic carbon source (such as glucose or a lactose) and a variety of inorganic ions, including Na^+, K^+, Mg^{++}, Ca^{++}, and NH_4^+ present as inorganic salts, it is called **minimal medium**. To grow on such a medium, a bacterium must be able to synthesize all essential organic compounds (e.g., amino acids, purines, pyrimidines, sugars, vitamins, and fatty acids). A bacterium that can accomplish this remarkable biosynthetic feat—one that we ourselves cannot duplicate—is termed a **prototroph**. It is said to be wild type for all growth requirements and can grow on minimal medium. On the other hand, if a bacterium loses, through mutation, the ability to synthesize one or more organic components, it is said to be an **auxotroph**. For example, a bacterium that loses the ability to make histidine is designated as a his^- auxotroph, in contrast to its prototrophic his^+ counterpart. For the his^- bacterium to grow, this amino acid must be added as a supplement to the minimal medium. Medium that has been extensively supplemented is called *complete medium*.

To study mutant bacteria in a quantitative fashion, an inoculum (a small amount such as 0.1 ml or 1.0 ml) of bacteria is placed in liquid culture medium. The bacteria exhibit a characteristic growth pattern, as illustrated in Figure 6–1. Initially, during the **lag phase**, growth is slow. Then, a period of rapid growth, called the **log phase**, ensues. During this phase, cells divide continually with a fixed time interval between cell divisions, resulting in exponential growth. When the bacteria reach a cell density of about 10^9 cells/ml, nutrients become limiting and cells enter the **stationary phase**. Because the doubling time during the log phase can be as short as 20 minutes, an initial inoculum of a few thousand cells added to the culture easily achieves maximum cell density during an overnight incubation.

Cells grown in liquid medium can be quantified by transferring them to semisolid medium in a petri dish. Following incubation and many divisions, each cell gives rise to a colony visible on the surface of the medium. By counting colonies, it is pos-

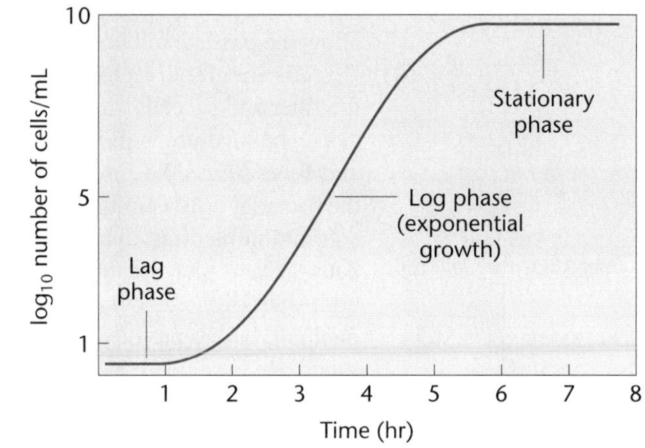

FIGURE 6–1 Typical bacterial population growth curve illustrating the initial lag phase, the subsequent log phase where exponential growth occurs, and the stationary phase that occurs when nutrients are exhausted.

sible to estimate the number of bacteria present in the original culture. If the number of colonies is too great to count, then successive dilutions (a technique called serial dilution) of the original liquid culture are made and plated, until the number of bacteria is reduced to the point where it can be counted. For example, assume that the three petri dishes in Figure 6–2 represent the plating of 1 ml of dilutions of 10^{-3}, 10^{-4}, and 10^{-5}, respectively (left to right).* We need only select the dish in which the number of colonies can be counted accurately. Because each colony presumably arose from a single bacterium, the number of colonies times the dilution factor represents the number of bacteria in the initial milliliter. In this case, the dish farthest to the right contains 15 colonies. Since it represents a dilution of 10^{-5}, we can estimate the initial concentration of bacteria to be 15×10^5 per milliliter.

6.2 Conjugation Is One Means of Genetic Recombination in Bacteria

Development of techniques that allowed the identification and study of bacterial mutations led to detailed investigations of the arrangement of genes on the bacterial chromosome. Such studies began in 1946, when Joshua Lederberg and Edward Tatum showed that bacteria undergo **conjugation**, a process by which genetic information from one bacterium is transferred to, and then recombined with, that of another bacterium. (See the photograph at the beginning of this chapter.) Like meiotic crossing over in eukaryotes, this process of **genetic recombination** in bacteria provided the basis for the development of chromosome mapping methodology. Note that the term *genetic recombination*, as applied to bacteria and bacteriophages, leads to the replacement of one or more genes present in one strain with those from a genetically distinct strain. While this is somewhat different from our use of the term *genetic recombination* in eukaryotes—where it describes crossing over resulting in reciprocal exchange events—the overall effect is the same: Genetic information on one chromosome is transferred to another, altering the genotype. Two other phenomena resulting in the transfer of genetic information from one bacterium to another, **transformation** and **transduction**, have also served as a basis for determining the arrangement of genes on the bacterial chromosome. We shall discuss these processes later in this chapter.

Lederberg and Tatum's initial experiments were performed with two multiple auxotroph strains (nutritional mutants) of *E. coli* strain K12. As shown in Figure 6–3, strain A required methionine (met) and biotin (bio) in order to grow, whereas strain B required threonine (thr), leucine (leu), and thiamine (thi). Neither strain would grow on minimal medium. The two strains were first grown separately in supplemented media, and then cells from both were mixed and grown together for several more generations. They were then plated on minimal medium. Any cells that grew on minimal medium were prototrophs. It is highly improbable that any of the cells containing two or three mutant genes would undergo spontaneous mutation simultaneously at two or three independent locations, leading to wild-type cells. Therefore, the researchers assumed that any prototrophs recovered must have arisen as a result of some form of genetic exchange and recombination between the two mutant strains.

In this experiment, prototrophs were recovered at a rate of one cell per 10^7 cells plated, a rate of 10^{-7}. The controls for this experiment involved separate plating of cells from strains A and B on minimal medium. No prototrophs were recovered. On the basis of these observations, Lederberg and Tatum proposed that, while the events were indeed rare, genetic recombination had occurred.

HOW DO WE KNOW?

How do geneticists detect and study mutant genes in bacteria?

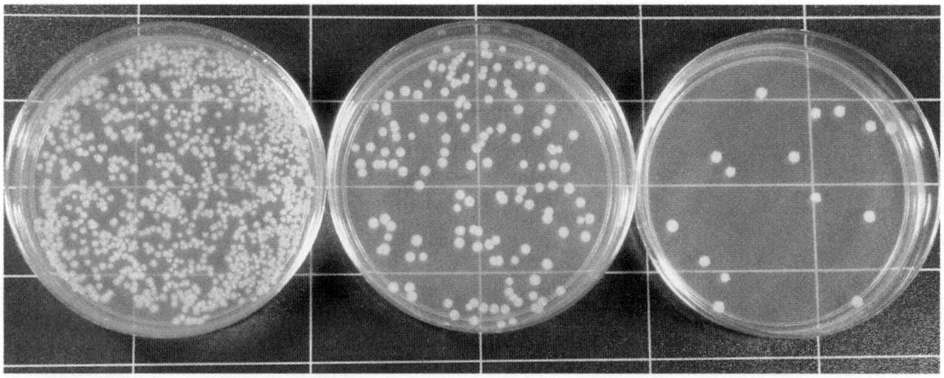

FIGURE 6–2 Results of the serial dilution technique and subsequent culture of bacteria. Each of the dilutions varies by a factor of 10. Each colony was derived from a single bacterial cell.

Web Tutorial 6.1 Bacterial Genetics

*10^{-5} represents a 1:100,000 dilution.

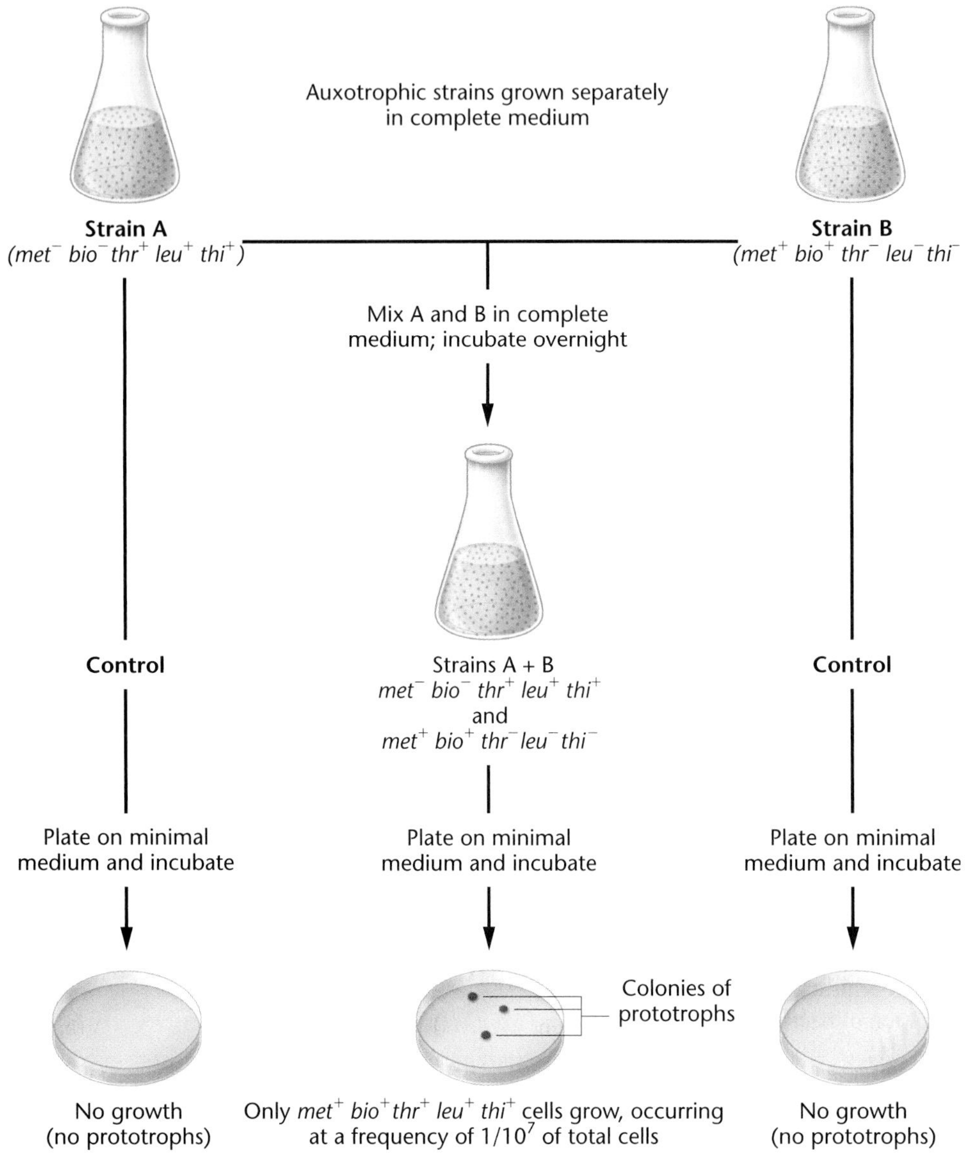

Auxotrophic strains grown separately in complete medium
Strain A
(met⁻ bio⁻ thr⁺ leu⁺ thi⁺)
Strain B
(met⁺ bio⁺ thr⁻ leu⁻ thi⁻)
Mix A and B in complete medium; incubate overnight
Control
Strains A + B
met⁻ bio⁻ thr⁺ leu⁺ thi⁺
and
met⁺ bio⁺ thr⁻ leu⁻ thi⁻
Control
Plate on minimal medium and incubate
Plate on minimal medium and incubate
Plate on minimal medium and incubate
Colonies of prototrophs
No growth
(no prototrophs)
Only met⁺ bio⁺ thr⁺ leu⁺ thi⁺ cells grow, occurring at a frequency of 1/10⁷ of total cells
No growth
(no prototrophs)

FIGURE 6–3 Genetic recombination involving two auxotrophic strains producing prototrophs. Neither auxotrophic strain will grow on minimal medium, but prototrophs do, suggesting that genetic recombination has occurred.

F^+ and F^- Bacteria

Lederberg and Tatum's findings were soon followed by numerous experiments designed to explain the physical nature and the genetic basis of conjugation. It quickly became evident that different strains of bacteria were involved in a unidirectional transfer of genetic material. When cells serve as donors of parts of their chromosomes, they are designated as **F^+ cells** (F for "fertility"). Recipient bacteria receive the donor chromosome material (now known to be DNA), and recombine it with part of their own chromosome. They are designated as **F^- cells.**

It was established subsequently that cell contact is essential for chromosome transfer. Support for this concept was provided by Bernard Davis, who designed a U-tube in which to grow F^+ and F^- cells (Figure 6–4). At the base of the tube is a sintered glass filter with a pore size that allows passage of the liquid medium, but that is too small to allow the passage of bacteria. The F^+ cells are placed on one side of the filter and F^- cells on the other side. The medium is moved back and forth across the filter so that the bacterial cells essentially share a common medium during incubation. Davis plated samples from both sides of the tube on minimal medium, but no prototrophs were found. He concluded that physical contact is essential to genetic recombination. We now know that this physical interaction is the initial stage of the process of conjugation and is mediated through a conjugation tube called the **F pilus** (or **sex pilus**). Bacteria often have many pili, which are microscopic tubelike extensions of the cell. Different types of pili perform different cellular functions, but all pili are involved in some way with adhesion. After contact has been initiated between mating pairs through F pili (Figure 6–5), transfer of the chromosome begins.

Later evidence established that F^+ cells contained a fertility factor (called the **F factor**) that confers the ability to donate part of their chromosome during conjugation. Experiments by Joshua and Esther Lederberg and by William Hayes and Luca Cavalli-Sforza showed that certain environmental conditions would eliminate the F factor from otherwise fertile cells. However, if these "infertile" cells were then grown with fertile donor cells, the F factor was regained.

The conclusion that the F factor is a mobile element was further supported by the observation that, after conjugation, recipient F^- cells always become F^+. Thus, in addition to the rare cases of transfer of genes from the bacterial chromosome (genetic recombination), the F factor itself is passed to all recipient cells. On this basis, the initial crosses of Lederberg and Tatum (Figure 6–3) may be designated as follows:

Strain A		Strain B
F^+ (DONOR)	×	F^- (RECIPIENT)

Characterization of the F factor confirmed these conclusions. Like the bacterial chromosome, though distinct from it, the F factor has been shown to consist of a circular, double-stranded DNA molecule, about 100,000 nucleotide pairs in length. The

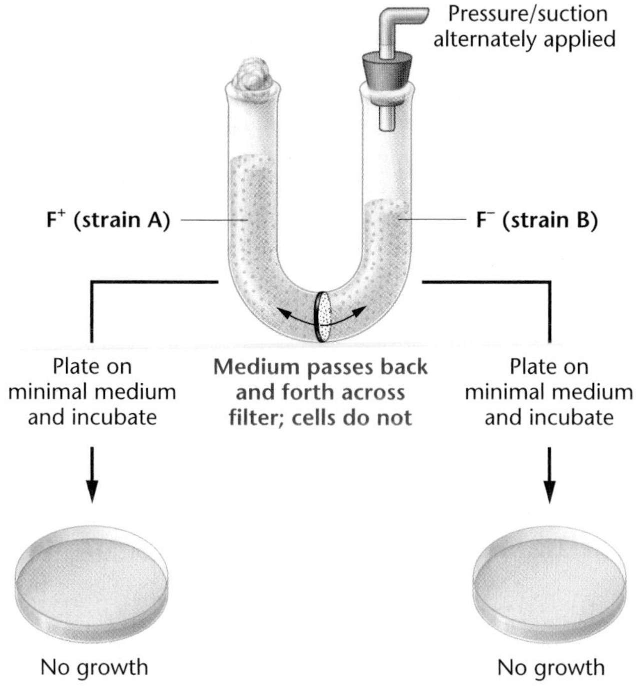

FIGURE 6–4 When strain A and strain B auxotrophs are grown in a common medium, but separated by a filter, no genetic recombination occurs and no prototrophs are produced. The apparatus shown is a Davis U-tube.

F factor contains, among others, more than 20 genes whose products are involved in the transfer of genetic information. Called *tra* genes, these include those genes essential to the formation of the sex pilus.

As we soon shall see, the F factor is in reality an autonomous genetic unit referred to as a **plasmid**. However, in our historical coverage of its discovery, we will continue in this chapter to refer to it as a "factor."

The behavior of the F factor during conjugation is diagrammed in a series of steps in Figure 6–6. It is believed that the transfer of the F factor involves separation of the two strands of its double helix and the movement of one of the two strands into the recipient cell (Step 2). The other strand remains in the donor cell. Both strands, one moving across the conjugation tube and one remaining in the donor cell, are replicated (Steps 3 and 4). The result is that both the donor and the recipient cells are F^+ at the completion of conjugation (Step 5).

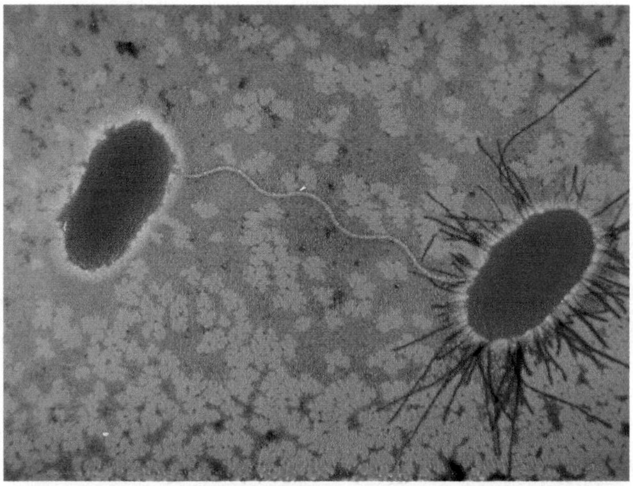

FIGURE 6–5 An electron micrograph of conjugation between an F^+ *E. coli* cell and an F^- cell. The sex pilus linking them is clearly visible.

In sum, *E. coli* cells may or may not contain the F factor. When it is present, the cell is able to form a sex pilus and potentially serve as a donor of genetic information. During conjugation, a copy of the F factor is almost always transferred from the F^+ cell to the F^- recipient, converting it to the F^+ state. The question remained as to exactly why such a low proportion of cells involved in these matings (10^{-7}) also had undergone genetic recombination. Also, it was unclear what the transfer of the F factor had to do with the transfer and recombination of a particular gene. The answers to these questions awaited further experimentation. Subsequent discoveries not only clarified how genetic recombination occurs, but also defined a mechanism by which the *E. coli* chromosome could be mapped. We shall first address chromosome mapping.

HOW DO WE KNOW?

How do we know that cell contact precedes genetic recombination in bacteria? How was it discovered that bacteria undergo genetic recombination, whereby genes may be transferred from the chromosome of one organism to that of another?

Hfr Bacteria and Chromosome Mapping

In 1950, Cavalli-Sforza treated an F^+ strain of *E. coli* K12 with nitrogen mustard, a potent chemical known to induce mutations. From these treated cells, he recovered a strain of donor bacteria that underwent recombination at a rate of $1/10^4$ (or 10^{-4}), 1000 times more frequently than the original F^+ strains. In 1953, William Hayes isolated another strain that demonstrated a similarly elevated frequency of recombination. Both strains were designated **Hfr**, for **high-frequency recombination**.

In addition to the higher frequency of recombination, another important difference was noted between Hfr strains and the original F^+ strains. If the donor is from an Hfr strain, recipient cells, while sometimes displaying genetic recombination, never become Hfr—that is, they remain F^-. In comparison, then,

$$F^+ \times F^- \longrightarrow \text{Recipient becomes } F^+$$

$$\text{Hfr} \times F^- \longrightarrow \text{Recipient remains } F^-$$

Perhaps the most significant characteristic of Hfr strains is the nature of recombination. In any given Hfr strain, certain genes are more frequently recombined than others, and some not at all. This nonrandom pattern of gene transfer was shown to vary from Hfr strain to Hfr strain. While these results were puzzling, Hayes interpreted them to mean that some physiological alteration of the F factor had occurred, resulting in the production of Hfr strains of *E. coli*.

In the mid-1950s, experimentation by Ellie Wollman and François Jacob explained the differences between cells that are Hfr and those that are F^+ and showed how Hfr strains allow genetic mapping of the *E. coli* chromosome. In Wollman's and

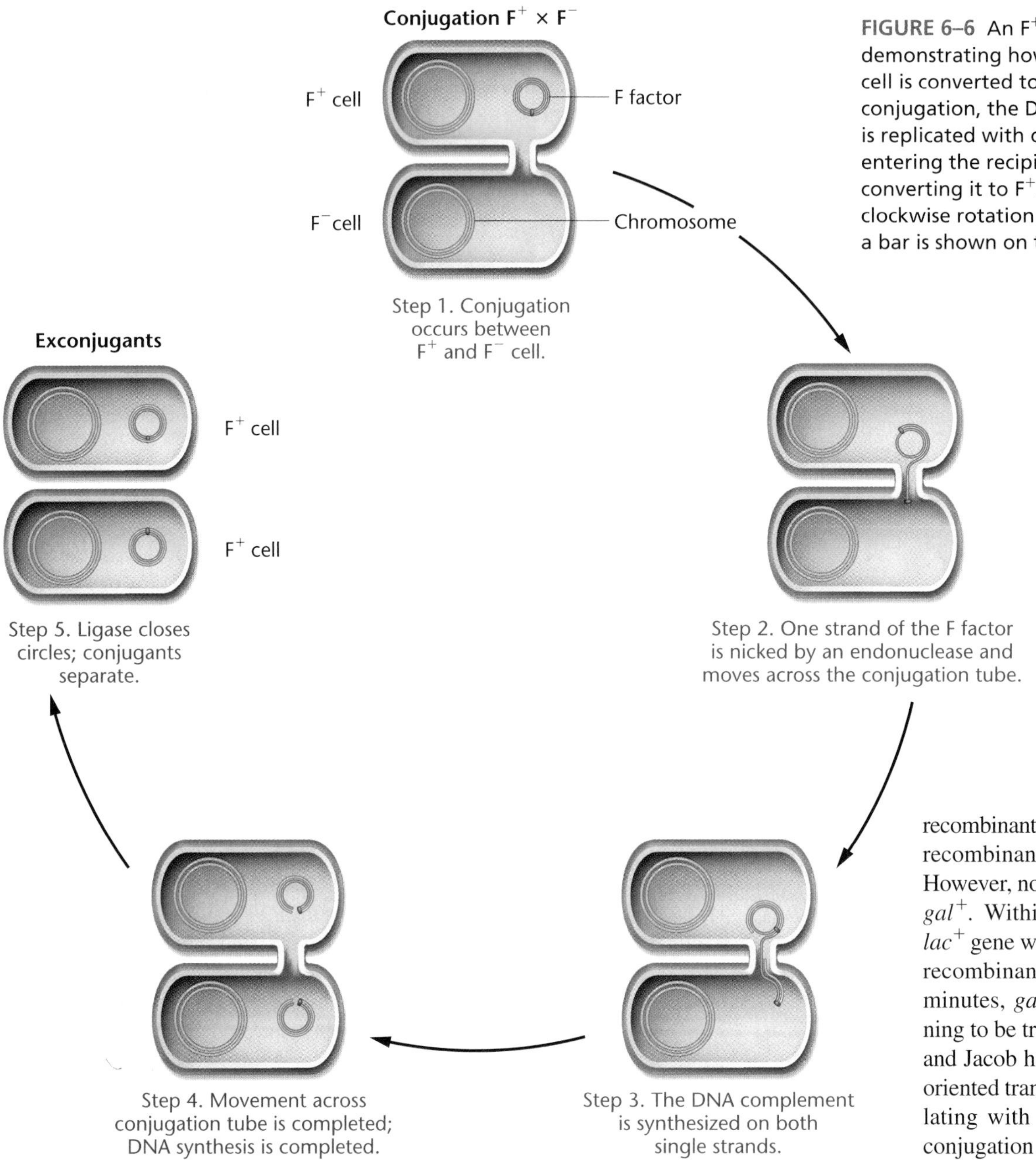
Conjugation F⁺ × F⁻
F⁺ cell
F factor
F⁻ cell
Chromosome
Step 1. Conjugation occurs between F⁺ and F⁻ cell.
Step 2. One strand of the F factor is nicked by an endonuclease and moves across the conjugation tube.
Step 3. The DNA complement is synthesized on both single strands.
Step 4. Movement across conjugation tube is completed; DNA synthesis is completed.
Step 5. Ligase closes circles; conjugants separate.
Exconjugants
F⁺ cell
F⁺ cell

FIGURE 6–6 An $F^+ \times F^-$ mating, demonstrating how the recipient F^- cell is converted to F^+. During conjugation, the DNA of the F factor is replicated with one new copy entering the recipient cell, converting it to F^+. To indicate the clockwise rotation during replication, a bar is shown on the F factor.

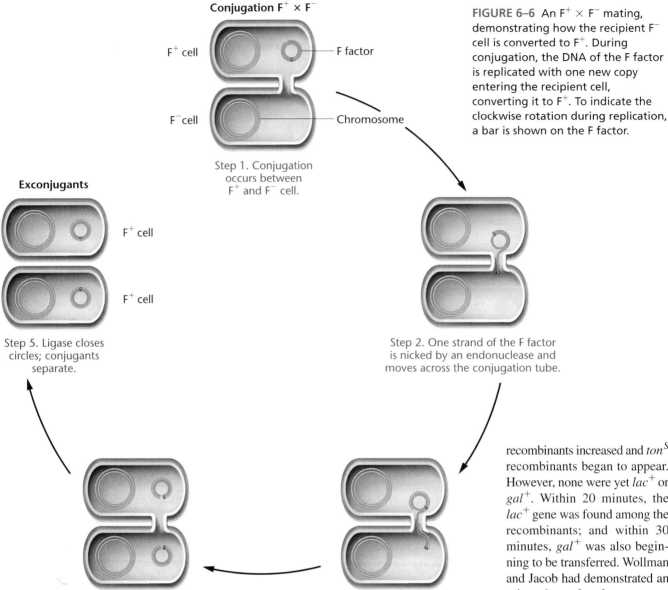

Jacob's experiments, Hfr and antibiotic-resistant F^- strains with suitable marker genes were mixed and recombination of specific genes assayed at different times. To accomplish this, a culture containing a mixture of an Hfr and an F^- strain was incubated, and samples were removed at various intervals and placed in a blender. The shear forces created in the blender separated conjugating bacteria so that the transfer of the chromosome was effectively terminated. To assay the cells for genetic recombination following the blender treatment, they were grown on medium containing the antibiotic so that only recipient cells were recovered.

Further experimentation using this process, called the **interrupted mating technique**, demonstrated that specific genes of a given Hfr strain were transferred and recombined sooner than others. Figure 6–7 illustrates this point. During the first 8 minutes after the two strains were mixed, no genetic recombination could be detected. At about 10 minutes, recombination of the azi^R gene could detected, but no transfer of the ton^S, lac^+, or gal^+ genes was noted. By 15 minutes, the percentage of azi^R recombinants increased and ton^S recombinants began to appear. However, none were yet lac^+ or gal^+. Within 20 minutes, the lac^+ gene was found among the recombinants; and within 30 minutes, gal^+ was also beginning to be transferred. Wollman and Jacob had demonstrated an oriented transfer of genes correlating with the length of time conjugation proceeded.

It appeared that the chromosome of the Hfr bacterium was transferred linearly and that the gene order and distance between genes, as measured in minutes, could be predicted from such experiments (Figure 6–8). This information served as the basis for the first genetic map of the *E. coli* chromosome. *Minutes* in bacterial mapping are equivalent to map units in eukaryotes.

Wollman and Jacob repeated the same type of experiment with other Hfr strains, obtaining similar results, but with one important difference. Although genes were always transferred linearly with time (as in their original experiment), which genes entered first and which followed later varied from Hfr strain to Hfr strain [Figure 6–9(a)]. When they reexamined the rate of entry of genes, and thus the different genetic maps for each strain, a definite pattern emerged. The major difference between each strain was simply the point of the origin and the direction in which entry proceeded from that point [Figure 6–9(b)].

To explain these results, Wollman and Jacob postulated that the *E. coli* chromosome is circular (a closed circle, with no free ends).

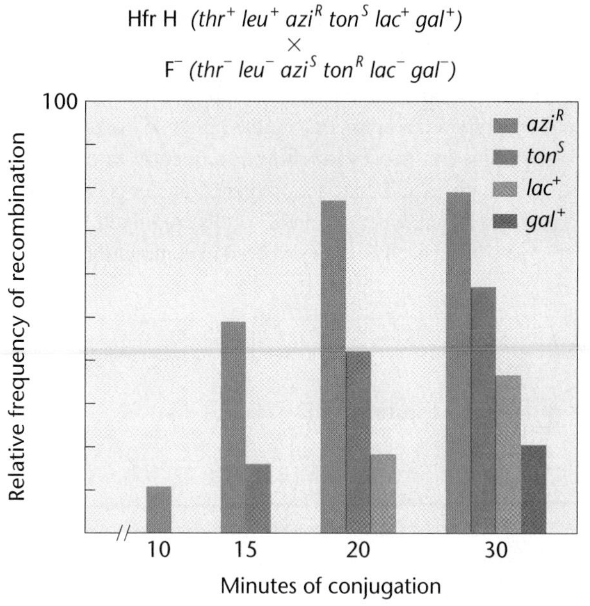

FIGURE 6–7 The progressive transfer during conjugation of various genes from a specific Hfr strain of *E. coli* to an F− strain. In this strain certain genes (*azi* and *ton*) are transferred sooner than others and recombine more frequently. Others (*lac* and *gal*) take longer to transfer and recombine with a lower frequency. Still others (*thr* and *leu*) are always transferred and are used in the initial screen for recombinants, but not shown in the histograms above.

If the point of origin (*O*) varied from strain to strain, a different sequence of genes would be transferred in each case. But what determines *O*? They proposed that, in various Hfr strains, the F factor integrated into the chromosome at different points and its position determined the *O* site. One such case of integration is shown during a number of steps in Figure 6–10. During conjugation between this Hfr and an F− cell, the position of the F factor determines the initial point of transfer (Steps 2 and 3). Those genes adjacent to *O* are transferred first. The F factor becomes the last part that can be transferred (Step 4). However, conjugation rarely, if ever, lasts long enough to allow the entire chromosome to pass across the conjugation tube (Step 5). This proposal explains why recipient cells, when mated with Hfr cells, remain F−.

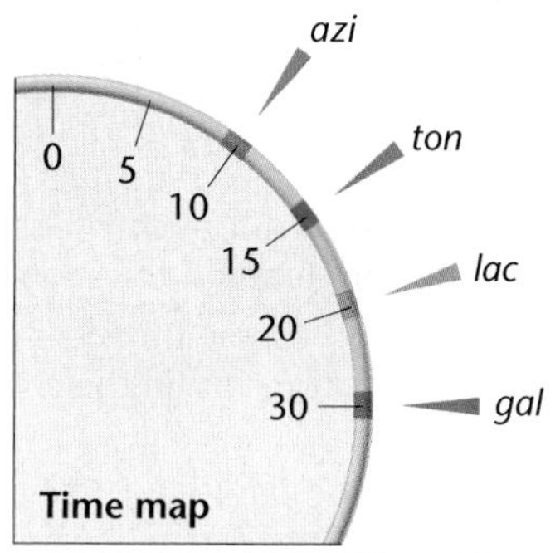

FIGURE 6–8 A time map of the genes studied in the experiment depicted in Figure 6–7.

Figure 6–10 also depicts how the two strands making up a DNA molecule unwind during transfer, allowing for the entry of one of the strands of DNA into the recipient (Step 3). Following replication, the entering DNA now has the potential to recombine with its homologous region of the host chromosome. The DNA strand that remains in the donor also undergoes replication.

The use of the interrupted mating technique with different Hfr strains allowed researchers to map the entire *E. coli* chromosome. Mapped in time units, strain K12 (or *E. coli* K12) was shown to be 100 minutes long. While modern genome analysis of the *E. coli* chromosome has now established the presence of just over 4000 protein-coding sequences, this original mapping procedure established the location of approximately 1,000 genes.

(a)

Hfr strain	Order of transfer (earliest)							(latest)
H	*thr* –	*leu* –	*azi* –	*ton* –	*pro* –	*lac* –	*gal* –	*thi*
1	*leu* –	*thr* –	*thi* –	*gal* –	*lac* –	*pro* –	*ton* –	*azi*
2	*pro* –	*ton* –	*azi* –	*leu* –	*thr* –	*thi* –	*gal* –	*lac*
7	*ton* –	*azi* –	*leu* –	*thr* –	*thi* –	*gal* –	*lac* –	*pro*

(b)

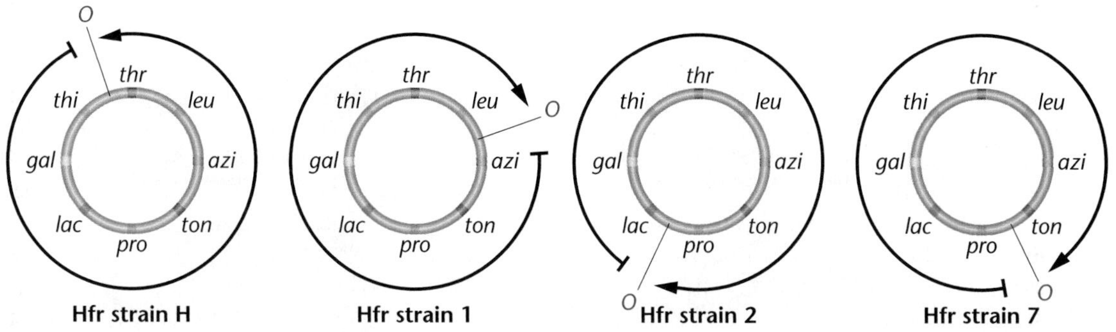

FIGURE 6–9 (a) The order of gene transfer in four Hfr strains, suggesting that the *E. coli* chromosome is circular. (b) The point where transfer originates (*O*) is identified in each strain. Note that transfer can proceed in either direction, depending on the strain. The origin is determined by the point of integration of the F factor into the chromosome, and the direction of transfer is determined by the orientation of the F factor as it integrates.

Now solve this

Problem 6.22 on page 162 involves an understanding of how the bacterial chromosome is transferred during conjugation, leading to recombination and mapping.

Hint: Chromosome transfer is strain-specific and depends on the position within the chromosome where the F factor is integrated.

Recombination in $F^+ \times F^-$ Matings: A Reexamination

The preceding model helped geneticists better understand how genetic recombination occurs during the $F^+ \times F^-$ matings. Recall that recombination occurs much less frequently in them than in Hfr $\times$ F^- matings and that random gene transfer is involved. The current belief is that, when F^+ and F^- cells are mixed, conjugation occurs readily and each F^- cell involved in conjugation with an F^+

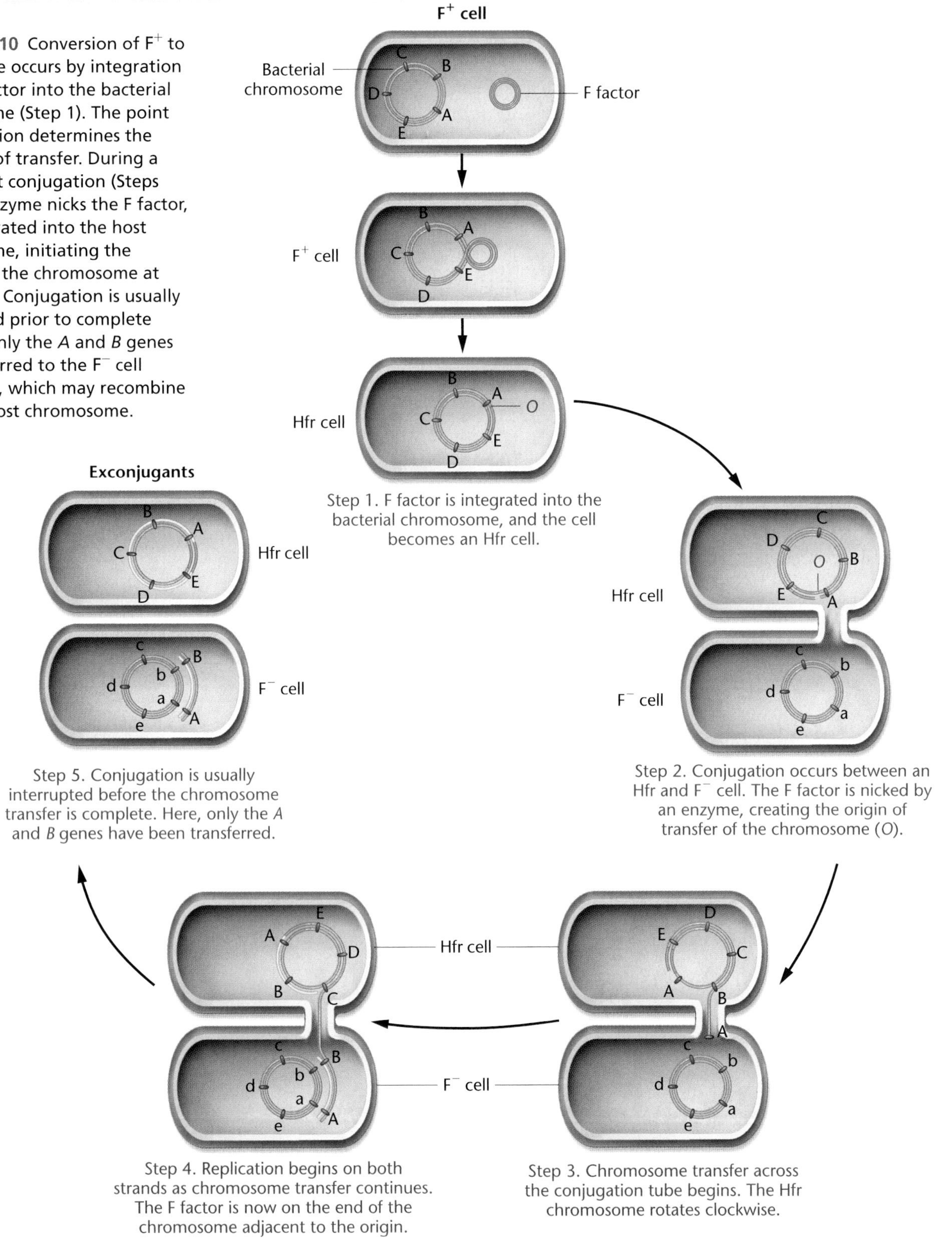
F^+ cell
Bacterial chromosome
F factor
F^+ cell
Hfr cell
Step 1. F factor is integrated into the bacterial chromosome, and the cell becomes an Hfr cell.
Hfr cell
F^- cell
Step 2. Conjugation occurs between an Hfr and F^- cell. The F factor is nicked by an enzyme, creating the origin of transfer of the chromosome (*O*).
Hfr cell
F^- cell
Step 3. Chromosome transfer across the conjugation tube begins. The Hfr chromosome rotates clockwise.
Step 4. Replication begins on both strands as chromosome transfer continues. The F factor is now on the end of the chromosome adjacent to the origin.
Exconjugants
Hfr cell
F^- cell
Step 5. Conjugation is usually interrupted before the chromosome transfer is complete. Here, only the *A* and *B* genes have been transferred.

FIGURE 6–10 Conversion of F^+ to an Hfr state occurs by integration of the F factor into the bacterial chromosome (Step 1). The point of integration determines the origin (*O*) of transfer. During a subsequent conjugation (Steps 2–4), an enzyme nicks the F factor, now integrated into the host chromosome, initiating the transfer of the chromosome at that point. Conjugation is usually interrupted prior to complete transfer. Only the *A* and *B* genes are transferred to the F^- cell (Steps 3–5), which may recombine with the host chromosome.

cell receives a copy of the F factor, but no genetic recombination occurs. However, at an extremely low frequency in a population of F^+ cells, the F factor integrates spontaneously into a random point in the bacterial chromosome, converting the F^+ cell to the Hfr state, as we saw in Figure 6–10. Therefore, in $F^+ \times F^-$ crosses, the extremely low frequency of genetic recombination (10^{-7}) is attributed to the rare, newly formed Hfr cells, which then undergo conjugation with F^- cells. Because the point of integration of the F factor is random, the genes transferred by any newly formed Hfr donor will also appear to be random within the larger F^+/F^- population. The recipient bacterium will appear as a recombinant cell, but will, in fact, remain F^-. If it subsequently undergoes conjugation with an F^+ cell, it will be converted to F^+.

The F′ State and Merozygotes

In 1959, during experiments with Hfr strains of *E. coli*, Edward Adelberg discovered that the F factor could lose its integrated status, causing the cell to revert to the F^+ state, as illustrated in a series of steps in Figure 6–11. When this occurs (Step 1), the F factor frequently carries several adjacent bacterial genes along with it (Step 2). Adelberg designated this condition **F′** to distinguish it from F^+ and Hfr. F′, like Hfr, is thus another special case of F^+. This conversion is described as one from Hfr to F ′.

The presence of bacterial genes within a cytoplasmic F factor creates an interesting situation. An F′ bacterium behaves like an F^+ cell, initiating conjugation with F^- cells (Step 3). When this occurs, the F factor, containing

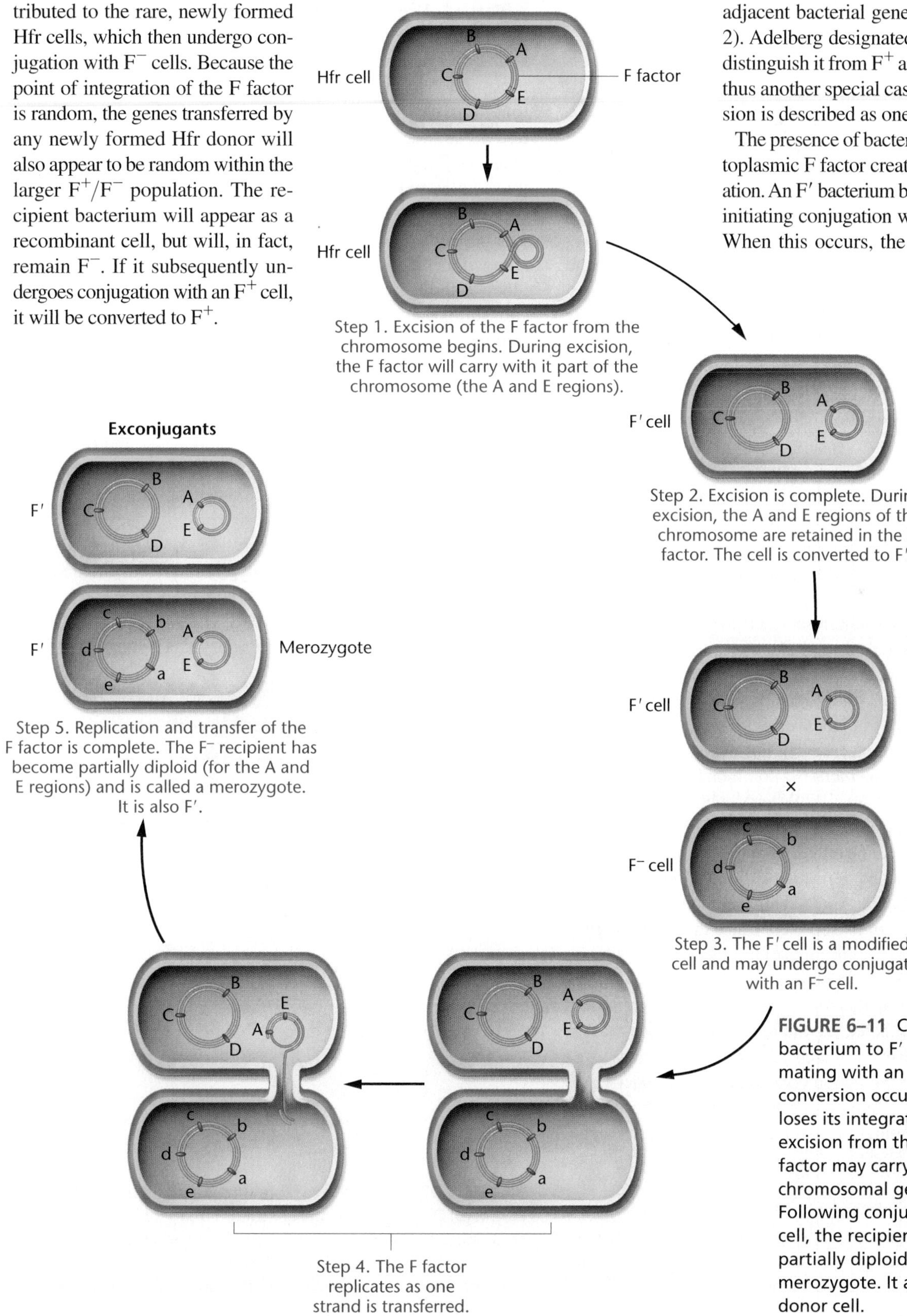
Hfr cell
F factor
Hfr cell
Step 1. Excision of the F factor from the chromosome begins. During excision, the F factor will carry with it part of the chromosome (the A and E regions).
F′ cell
Step 2. Excision is complete. During excision, the A and E regions of the chromosome are retained in the F factor. The cell is converted to F′.
F′ cell
×
F⁻ cell
Step 3. The F′ cell is a modified F⁺ cell and may undergo conjugation with an F⁻ cell.
Step 4. The F factor replicates as one strand is transferred.
Exconjugants
F′
F′
Merozygote
Step 5. Replication and transfer of the F factor is complete. The F⁻ recipient has become partially diploid (for the A and E regions) and is called a merozygote. It is also F′.

FIGURE 6–11 Conversion of an Hfr bacterium to F′ and its subsequent mating with an F^- cell. The conversion occurs when the F factor loses its integrated status. During excision from the chromosome, the F factor may carry with it one or more chromosomal genes (*A* and *E*). Following conjugation with an F^- cell, the recipient cell becomes partially diploid and is called a merozygote. It also behaves as an F^+ donor cell.

chromosomal genes, is transferred to the F^- cell (Step 4). As a result, whatever chromosomal genes are part of the F factor are now present in duplicate in the recipient cell (Step 5), because the recipient still has a complete chromosome. This creates a partially diploid cell called a **merozygote**. Pure cultures of F′ merozygotes can be established. They have been extremely useful in the study of genetic regulation in bacteria, as we will discuss in Chapter 16.

6.3 Mutational Analysis Led to the Discovery of the Rec Proteins Essential to Bacterial Recombination

Once researchers established that a unidirectional transfer of DNA occurs between bacteria, they were interested in determining how the actual recombination event occurs in the recipient cell. Just how does the donor DNA replace the comparable region in the recipient chromosome? As with many systems, the mechanism by which recombination occurs was deciphered through genetic studies. Major insights were gained as a result of the isolation of a group of mutations representing *rec* (for recombination) genes.

The first relevant observation in this case involved a series of mutant genes labeled *recA*, *recB*, *recC*, and *recD*. The first mutant gene, *recA*, was found to diminish genetic recombination in bacteria a thousandfold, nearly eliminating it altogether. The other *rec* mutations reduced recombination by about 100 times. Clearly, the normal wild-type products of these genes play some essential role in the process of recombination.

By looking for a functional gene product present in normal cells, but missing in *rec* mutant cells, researchers subsequently isolated several gene products and showed that they played a role in genetic recombination. The first is called the **RecA protein.*** This protein plays an important role in recombination involving either a single stranded DNA molecule or with the linear end of a double-stranded DNA molecule that has unwound. As it turns out, **single-strand displacement** is a common form of recombination in many bacterial species. When double-stranded DNA enters a recipient cell, one strand is often degraded, leaving the complementary strand as the only source of recombination. This strand must find its homologous region along the host chromosome, and once it does, RecA facilitates recombination.

The second is a more complex protein called the **RecBCD protein**, an enzyme consisting of polypeptide subunits encoded by three other *rec* genes. This protein is important when double-stranded DNA serves as the source of genetic recombination. RecBCD unwinds the helix, facilitating recombination that involves RecA. This genetic research has extended our knowledge of the process of recombination considerably and underscores the value of isolating mutations, establishing their phenotypes, and determining the biological role of the normal, wild-type gene.

*Note that the names of bacterial genes begin with lowercase letters and are italicized. The names of the corresponding gene products (proteins) begin with an uppercase letter and are not italicized. For example, the *recA* gene encodes the RecA protein.

How Do We Know?

How do we know that specific gene products such as those encoded by the *recA* genes are involved during genetic recombination?

6.4 F Factors Are Plasmids

In the preceding sections, we have examined the extrachromosomal heredity unit, the F factor. When it exists autonomously in the bacterial cytoplasm, it is composed of a double-stranded closed circle of DNA. These characteristics place the F factor in the more general category of genetic structures called **plasmids**. These structures [Figure 6–12(a)] may contain one or more genes—often, quite a few. Their replication depends on the same enzymes that replicate the chromosome of the host cell, and they are distributed to daughter cells along with the host chromosome during cell division.

Plasmids can be classified according to the genetic information specified by their DNA. The F factor confers fertility and contains genes essential for sex pilus formation, upon which conjugation and subsequent genetic recombination depend. Other examples of plasmids include the R and the Col plasmids.

Most **R plasmids** consist of two components: the **resistance transfer factor** (**RTF**) and one or more **r-determinants** [Figure 6–12(b)]. The RTF encodes genetic information essential to transfer of the plasmid between bacteria via conjugation, and the r-determinants are genes conferring resistance to antibiotics. While RTFs are quite similar in a variety of plasmids from different bacterial species, there is a wide variation in r-determinants, each of which is specific for resistance to one class of antibiotic. Sometimes, a bacterial cell contains r-determinant plasmids, but no RTF

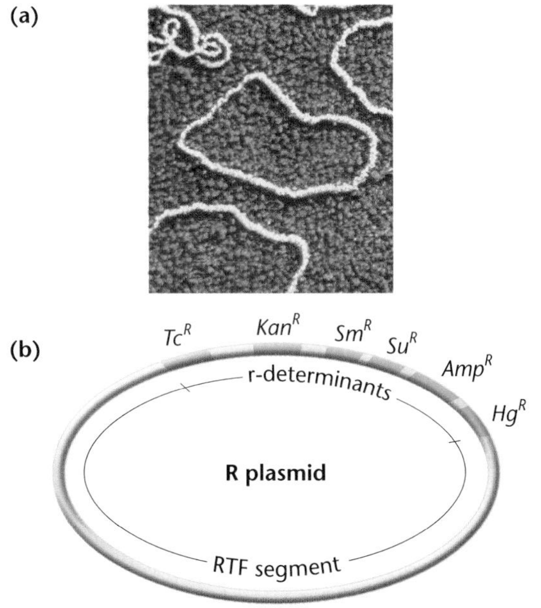

FIGURE 6–12 (a) Electron micrograph of a plasmid isolated from *E. coli*. (b) Diagrammatic representation of an R plasmid containing resistance transfer factors (RTFs) and multiple r-determinants (Tc, tetracycline; Kan, kanamycin; Sm, streptomycin; Su, sulfonamide; Amp, ampicillin; and Hg, mercury).

is present. Such a cell is resistant, but cannot transfer the genetic information for resistance to recipient cells. The most commonly studied plasmids, however, contain the RTF as well as one or more r-determinants, conferring resistance to tetracycline, streptomycin, ampicillin, sulfonamide, kanamycin, or chloramphenicol. Sometimes these occur in a single plasmid, conferring multiple resistance to several antibiotics [Figure 6–12(b)]. Bacteria bearing such plasmids are of great medical significance, not only because of their multiple resistance, but also because of the ease with which the plasmids may be transferred to other pathogenic bacteria, rendering them resistant to a wide range of antibiotics.

The **Col plasmid**, ColE1, derived from *E. coli*, is clearly distinct from R plasmids. It encodes one or more proteins that are highly toxic to bacterial strains that do not harbor the same plasmid. These proteins, called **colicins**, can kill neighboring bacteria. Bacteria that carry the plasmid are said to be colicinogenic. Present in 10 to 20 copies per cell, the Col plasmid also contains a gene encoding an immunity protein that protects the host cell from the toxin. Unlike an R plasmid, the Col plasmid is not usually transmissible by conjugation to other cells.

Interest in plasmids has increased dramatically because of their role in the genetic technology known as recombinant DNA research, to which we will return in Chapter 19. Specific genes from any source can be inserted into a plasmid, which may then be artificially inserted into a bacterial cell. As the altered cell replicates its DNA and undergoes division, the foreign gene is also replicated, thus cloning the genes.

6.5 Transformation Is Another Process Leading to Genetic Recombination in Bacteria

Transformation is a process that also provides a mechanism for the recombination of genetic information in some bacteria. In transformation, small pieces of extracellular DNA are taken up by a living bacterium, ultimately leading to a stable genetic change in the recipient cell. We are interested in transformation in this chapter because, in those bacterial species in which it occurs, the process can be used to map bacterial genes (although in a more limited way than conjugation). As we will see in Chapter 10, the process of transformation was also instrumental in establishing DNA as the genetic material.

The Transformation Process

Transformation leading to genetic recombination (illustrated in a number of steps in Figure 6–13) consists of several steps that achieve two basic outcomes: entry of foreign DNA into a recipient cell; and replacement by the donor DNA of its homologous region in the recipient chromosome. While completion of both outcomes leads to genetic recombination, the first step of transformation can occur without the second step, resulting in the addition of foreign DNA to the bacterial cytoplasm.

Let's begin with the first outcome. In a population of bacterial cells, only those in a particular physiological state, referred to as **competence**, take up DNA. Entry is thought to occur at a limited number of receptor sites on the surface of the bacterial cell (Step 1-2) and requires energy and specific transport molecules. This model is supported by the fact that substances that inhibit energy production or protein synthesis also inhibit transformation.

After entry into a competent cell, one of the two strands of the invading DNA molecule is digested by nucleases, leaving only a single strand to participate in recombination (Step 3). The surviving DNA strand aligns with its complementary region of the bacterial chromosome. In a process involving several enzymes, this segment of DNA replaces its counterpart in the chromosome, which is excised and degraded (Step 4).

For recombination to be detected, the transforming DNA must be derived from a different strain of bacteria, bearing some genetic variation. Once it is integrated into the chromosome, the recombinant region contains one host strand (present originally) and one mutant strand. Because these strands are from different sources, this region is referred to as a **heteroduplex**, and the two strands of DNA are not perfectly complementary in this region. Following one round of replication, one chromosome is restored to its original DNA sequence, identical to that of the recipient cell, and the other contains the mutant gene. Following cell division, one nonmutant (untransformed) cell and one mutant (transformed) cell are produced (Step 5).

Transformation and Linked Genes

For DNA to be effective in transformation, it must include between 10,000 to 20,000 nucleotide pairs, about 1/200 of the *E. coli* chromosome. This size is sufficient to encode several genes. Genes that are adjacent or very close to one another on the bacterial chromosome can be carried on a single segment of DNA of this size. Because of this fact, a single transfer event can result in the **cotransformation** of several genes simultaneously. Genes that are close enough to each other to be cotransformed are said to be *linked.* In contrast to the use of the term *linkage* in eukaryotes, which indicates all genes on a single chromosome, note that here linkage refers to the proximity of genes (i.e., genes next to, or close to, one another).

If two genes are not linked, simultaneous transformation can occur only as a result of two independent events involving two distinct segments of DNA. As in double crossing over in eukaryotes, the probability of two independent events occurring simultaneously is equal to the product of the individual probabilities. Thus, the frequency of two unlinked genes being transformed simultaneously is much lower than if they are linked.

In addition to establishing linkage relationships, relative mapping distances between linked genes can also be determined from the recombination data provided by transformation experiments. Though the analysis is more complex, such data are interpreted in a manner similar to chromosome mapping in eukaryotes.

Now solve this

Problem 6.8 on page 161 involves an understanding of how transformation can be used to determine if bacterial genes are "linked" in close proximity to one another.

Hint: Cotransformation occurs according to the laws of probability. Two "unlinked" genes are transformed as a

Web Tutorial 6.2
Phage Genetics

result of two separate events. In such a case, the probability of that occurrence is equal to the product of the individual probabilities.

6.6 Bacteriophages Are Bacterial Viruses

Bacteriophages, or **phages** as they are commonly known, are viruses that have bacteria as their hosts. During their reproduction, phages can be involved in still another mode of bacterial genetic recombination called transduction. To understand this process, we first must consider the genetics of bacteriophages, which themselves can undergo recombination. In this section, we will first examine the structure and life cycle of one type of bacteriophage. We will then discuss how these phages are studied during their infection of bacteria. Finally, we will contrast two possible modes of behavior once initial phage infection occurs. This information will serve as background for our subsequent discussion of transduction and bacteriophage recombination. A great deal of genetic research has been done using bacteriophages as a model system.

Phage T4: Structure and Life Cycle

Bacteriophage T4 is one of a group of related bacterial viruses referred to as T-even phages. It exhibits an intricate structure, as illustrated in Figure 6–14. Its genetic material, DNA, is contained within an icosahedral (a polyhedron with 20 faces) protein coat, together making up the head of the virus. The DNA is sufficient in quantity to encode more than 150 average-sized genes. The head is connected to a tail containing a collar and a contractile sheath surrounding a central core. Tail fibers, which protrude from the tail, contain binding sites in their tips that specifically recognize unique areas of the outer surface of the cell wall of the bacterial host, *E. coli*.

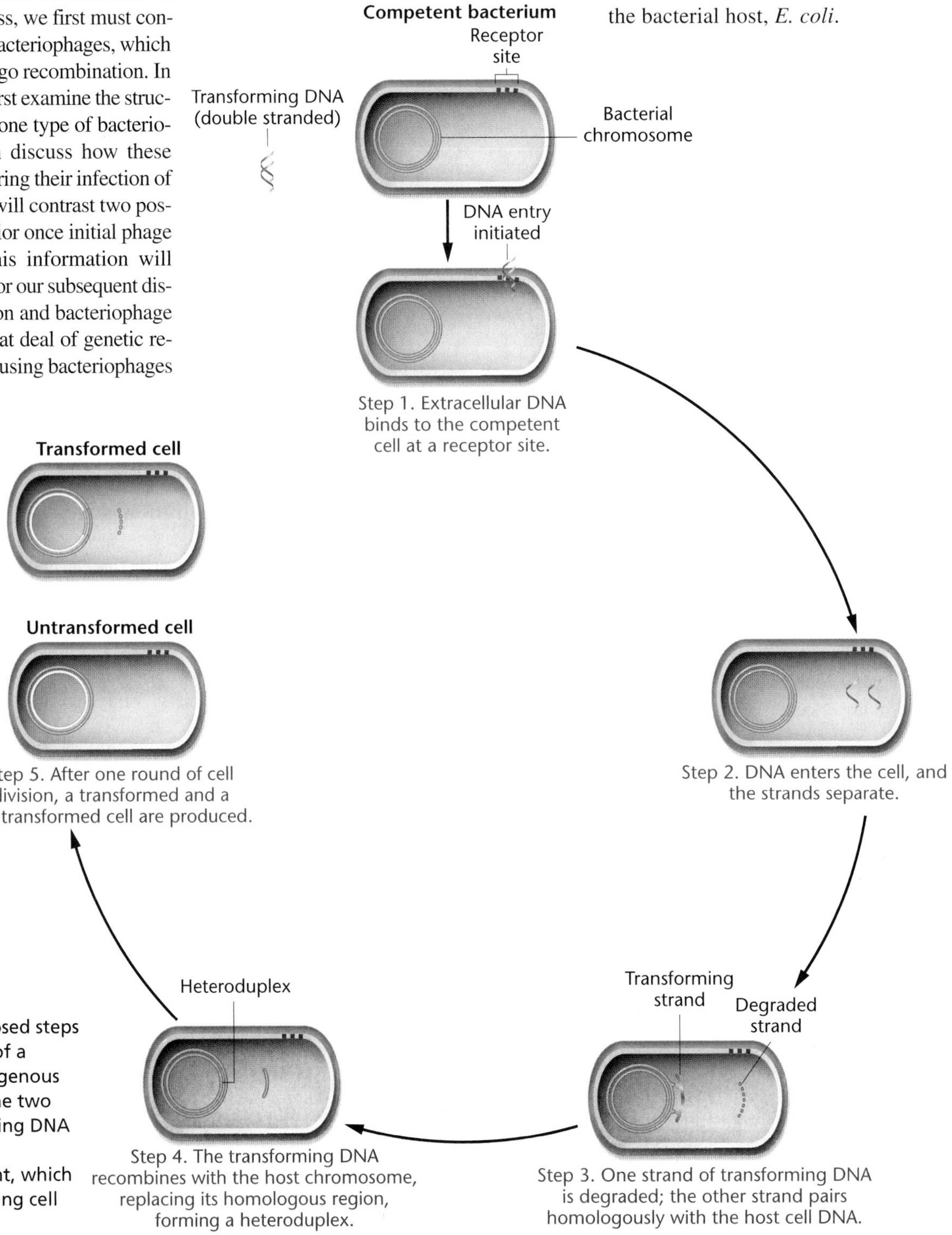
Competent bacterium
Receptor site
Transforming DNA (double stranded)
Bacterial chromosome
DNA entry initiated
Step 1. Extracellular DNA binds to the competent cell at a receptor site.
Step 2. DNA enters the cell, and the strands separate.
Transforming strand
Degraded strand
Step 3. One strand of transforming DNA is degraded; the other strand pairs homologously with the host cell DNA.
Heteroduplex
Step 4. The transforming DNA recombines with the host chromosome, replacing its homologous region, forming a heteroduplex.
Transformed cell
Untransformed cell
Step 5. After one round of cell division, a transformed and a nontransformed cell are produced.

FIGURE 6–13 Proposed steps for transformation of a bacterial cell by exogenous DNA. Only one of the two strands of the entering DNA is involved in the transformation event, which is completed following cell division.

The life cycle of phage T4 (Figure 6–15) is initiated when the virus binds by adsorption to the bacterial host cell. Then, an ATP-driven contraction of the tail sheath causes the central core to penetrate the cell wall. The DNA in the head is extruded, and it then moves across the cell membrane into the bacterial cytoplasm. Within minutes, all bacterial DNA, RNA, and protein synthesis is inhibited, and synthesis of viral molecules begins. At the same time, degradation of the host DNA is initiated.

A period of intensive viral gene activity characterizes infection. Initially, phage DNA replication occurs, leading to a pool of viral DNA molecules. Then, the components of the head, tail, and tail fibers are synthesized. The assembly of mature viruses is a complex process that has been well studied by William Wood, Robert Edgar, and others. Three sequential pathways occur: (1) DNA packaging as the viral heads are assembled, (2) tail assembly, and (3) tail-fiber assembly. Once DNA is packaged into the head, it combines with the tail components, to which tail fibers are added. Total construction is a combination of self-assembly and enzyme-directed processes.

When approximately 200 viruses are constructed, the bacterial cell is ruptured by the action of lysozyme (a phage gene product), and the mature phages are released from the host cell. The 200 new phages will infect other available bacterial cells, and the process will be repeated over and over again.

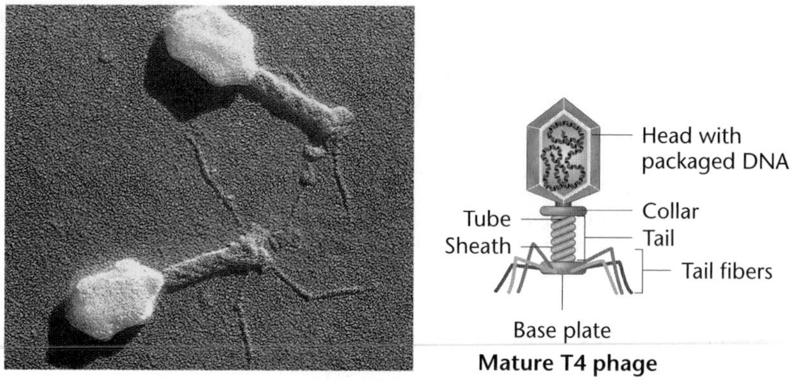

FIGURE 6–14 The structure of bacteriophage T4, including an icosahedral head filled with DNA, a tail consisting of a collar, tube, sheath, base plate, and tail fibers. During assembly, the tail components are added to the head and then tail fibers are added.

The Plaque Assay

The experimental study of bacteriophages and other viruses has played a critical role in our understanding of molecular genetics. During infection of bacteria, enormous quantities of bacteriophages may be obtained for investigation. Often, more than 10^{10} viruses per milliliter of culture medium are produced. Many genetic studies have relied on our ability to determine the number of phages produced following infection under specific culture conditions using a technique called the **plaque assay**.

This assay is illustrated in Figure 6–16, in which actual plaque morphology is also shown. A serial dilution is performed using a pure culture of bacteriophages derived from a virally infected bacterial culture. A 0.1-ml sample (called an aliquot) from one or more dilutions is added to a small volume of melted nutrient agar (about 3 ml) into which a few drops of a healthy bacterial culture have been mixed. The solution is then poured evenly over a base of solid nutrient agar in a Petri dish and allowed to solidify. Subsequently, during incubation, the bacteria reproduce

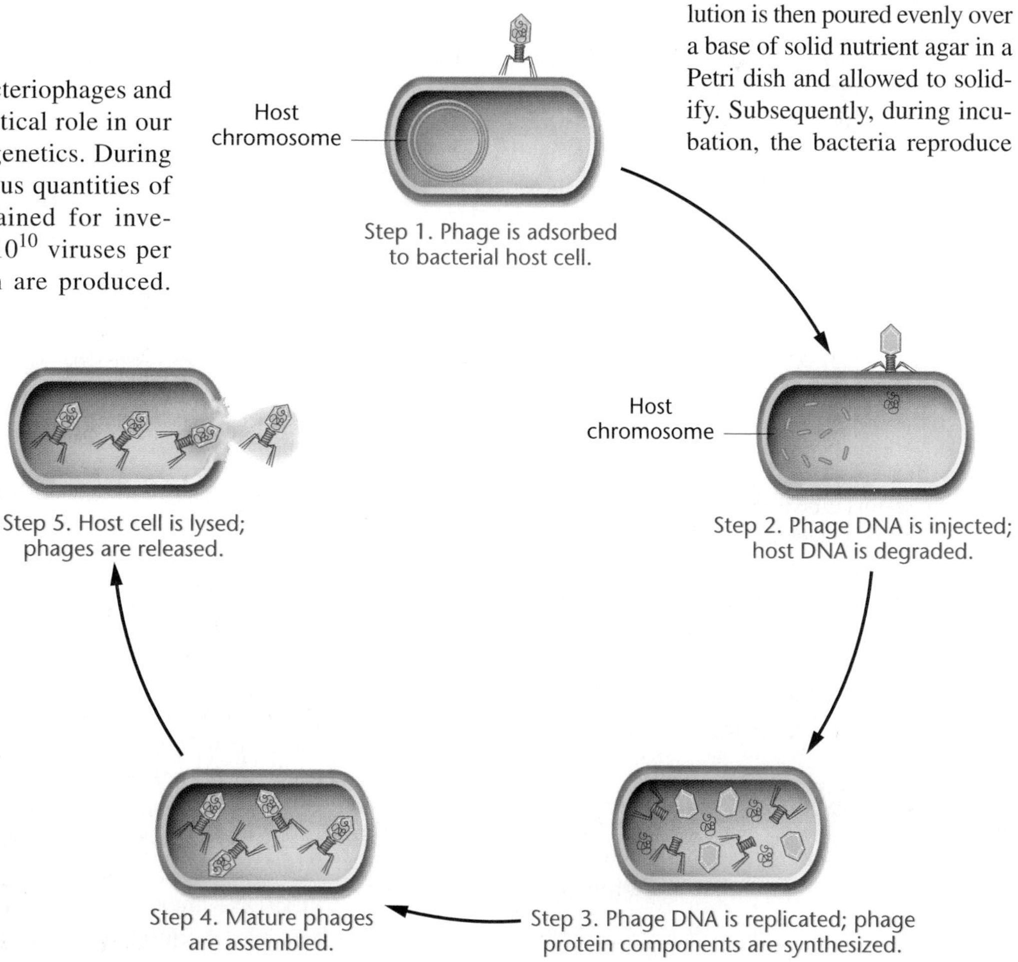

FIGURE 6–15 Life cycle of bacteriophage T4.

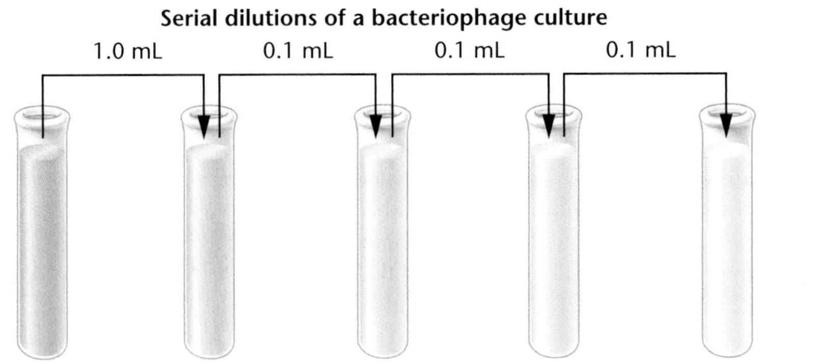

Total volume	10 mL	10 mL	10 mL	10 mL	10mL
Dilution	0	10^{-1}	10^{-3}	10^{-5}	10^{-7}
Dilution factor	0	10	10^3	10^5	10^7

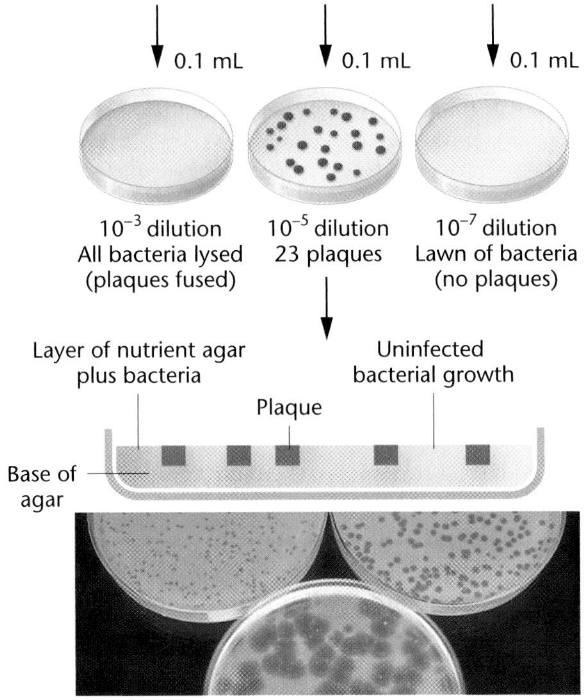

FIGURE 6–16 The plaque assay for bacteriophage analysis. Serial dilutions of a bacteriophage culture are first made. Then, three of the dilutions (10^{-3}, 10^{-5}, and 10^{-7}) are analyzed using the plaque assay technique. In each case, 0.1 ml of the diluted culture is used. Each plaque represents the initial infection of one bacterial cell by one bacteriophage. In the 10^{-3} dilution, so many phages are present that all bacteria are lysed. In the 10^{-5} dilution, 23 plaques are produced. In the 10^{-7} dilution, the dilution factor is so great that no phages are present in the 0.1 ml sample, and thus no plaques form.

and eventually cover the entire dish, forming what is called a lawn of cells. However, wherever phage infection occurs, a clear area, called a **plaque**, is formed. Plaques represent places where a single phage has initially infected one bacterium during incubation. As the phage reproduction cycle is repeated numerous times (newly produced phages invade adjacent bacterial cells), more bacterial cells are lysed and the plaque size increases. If the dilution factor is too low, many phages are present and the plaques are plentiful. In such cases, they may fuse, lysing the entire lawn of bacteria. This has occurred in the 10^{-3} dilution in Figure 6–16. On the other hand, if the dilution factor is increased, fewer phages are present in a given aliquot and plaques can be counted. From such data, the density of viruses in the initial culture can be estimated by counting plaques following serial dilution of an initial culture. For example, using the results shown in Figure 6–16, it is observed that there are 23 phage plaques derived from the 0.1-ml aliquot of the 10^{-5} dilution. This initial viral density in the undiluted sample, in which 23 plaques are observed from 0.1 ml of the 10^{-5} dilution, is calculated as

$$23 \text{ phage} \times 10^5/0.1 \text{ mL plated} = 23 \times 10^6 \text{ phage/mL}$$

While this figure is derived from the 10^{-5} dilution, it is interesting to consider the 10^{-7} dilution, where we can estimate that there will be only 0.23 phage per 0.1 ml. As a result, when 0.1 ml from this tube is assayed, it is predicted that most often, there will be no phage particles present. This possibility is borne out in Figure 6–16, in which an intact lawn of bacteria is depicted. The dilution factor is simply too great.

The use of the plaque assay has been invaluable in mutational and recombinational studies of bacteriophages. We will apply this technique more directly later in this chapter when we discuss Seymour Benzer's elegant genetic analysis of a single gene in phage T4.

How Do We Know?

When examining bacteria plated on medium in a petri dish, how do we know that bacteriophage infection has occurred?

Lysogeny

The relationship between virus and bacterium does not always result in viral reproduction and lysis. As early as the 1920s, it was known that some bacteriophages could enter a bacterial cell and establish a symbiotic relationship with it. The precise molecular basis of this symbiosis is now well understood. Upon entry, the viral DNA, instead of replicating in the bacterial cytoplasm, is integrated into the bacterial chromosome—a step characterizing the developmental stage referred to as **lysogeny**. Subsequently, each time the bacterial chromosome is replicated, the viral DNA is also replicated and passed to daughter cells following division. No new viruses are produced and no lysis of the bacterial cell occurs. However, in response to certain stimuli, such as chemical or ultraviolet-light treatment, the viral DNA may lose its integrated status and initiate replication, phage reproduction, and lysis of the bacterium.

Several terms are used to describe this relationship. The viral DNA integrated into the bacterial chromosome is called a **prophage**. Viruses that can either lyse the cell or behave as a prophage are called **temperate**. Those that can only lyse the cell are referred to as **virulent**. A bacterium harboring a prophage has been **lysogenized** and is said to be **lysogenic**; that is, it is capable of being lysed as a result of induced viral reproduction. The viral DNA, which can replicate either extrachromosomally or as part of the chromosome, is sometimes classified as an **episome**.

6.7 Transduction Is Virus-Mediated Bacterial DNA Transfer

In 1952, Norton Zinder and Joshua Lederberg were investigating possible recombination in the bacterium *Salmonella typhimurium*. Although they recovered prototrophs from mixed cultures of two different auxotrophic strains, subsequent investigations showed that recombination was not due to the presence of an F factor and conjugation, as in *E. coli*. What they discovered was a process of bacterial recombination mediated by bacteriophages and now called **transduction**.

The Lederberg–Zinder Experiment

Lederberg and Zinder mixed the *Salmonella* auxotrophic strains LA-22 and LA-2 together, and, when the mixture was plated on minimal medium, they recovered prototrophic cells. The LA-22 was unable to synthesize the amino acids phenylalanine and tryptophan ($phe^- trp^-$), and LA-2 could not synthesize the amino acids methionine and histidine ($met^- his^-$). Prototrophs ($phe^+ trp^+ met^+ his^+$) were recovered at a rate of about $1/10^5$ (or 10^{-5}).

Although these observations at first suggested that the recombination involved was the type observed earlier in conjugative strains of *E. coli*, using the Davis U-tube in an experiment showed otherwise (Figure 6–17). The two auxotrophic strains were separated by a sintered glass filter, thus preventing cell contact but allowing growth to occur in a common medium. Surprisingly, when samples were removed from both sides of the filter and plated independently on minimal medium, prototrophs were recovered only from the side of the tube containing LA-22 bacteria. Recall that if conjugation were responsible, using the Davis U-tube should prevent recombination altogether. (See Figure 6–4).

Since LA-2 cells appeared to be the source of the new genetic information (phe^+ and trp^+), how that information crossed the filter from the LA-2 cells to the LA-22 cells, allowing recombination to occur, was a mystery. It was postulated that an unknown source called a **filterable agent (FA)** was somehow responsible for transferring the information. Three subsequent observations were useful in identifying the FA:

1. The FA was produced by the LA-2 cells only when they were grown in association with LA-22 cells. If LA-2 cells were grown independently and that culture medium was then added to LA-22 cells, recombination did not occur. Therefore, LA-22 cells play some role in the production of FA by LA-2 cells and do so only when the two share common growth medium.
2. The addition of DNase, which enzymatically digests naked DNA, did not render the FA ineffective. Therefore, the FA is not DNA, ruling out transformation.
3. The FA could not pass across the filter of the Davis U-tube when the pore size was reduced below the size of bacteriophages.

Aided by these observations and aware of temperate phages that could lysogenize *Salmonella*, researchers proposed that the genetic recombination event was mediated by bacteriophage P22, present initially as a prophage in the chromosome of the LA-22 *Salmonella* cells. It was hypothesized that, rarely, P22 prophages might enter the lytic phase, reproduce, and be released by the LA-22 cells. Such phages, being much smaller than a bacterium, then cross the filter of the U-tube and subsequently infect and lyse some of the LA-2 cells. In the process of lysis of LA-2, the P22 phages occasionally packaged in their heads a region of the LA-2 chromosome. If this region contained the phe^+ and trp^+ genes, and if the phages subsequently passed back across the filter and infected LA-22 cells, these newly lysogenized cells would behave as prototrophs. The process of transduction, whereby bacterial recombination is mediated by a bacteriophage, is diagrammed in Figure 6–18.

HOW DO WE KNOW?

How do we know that cell contact is *not* essential to transduction?

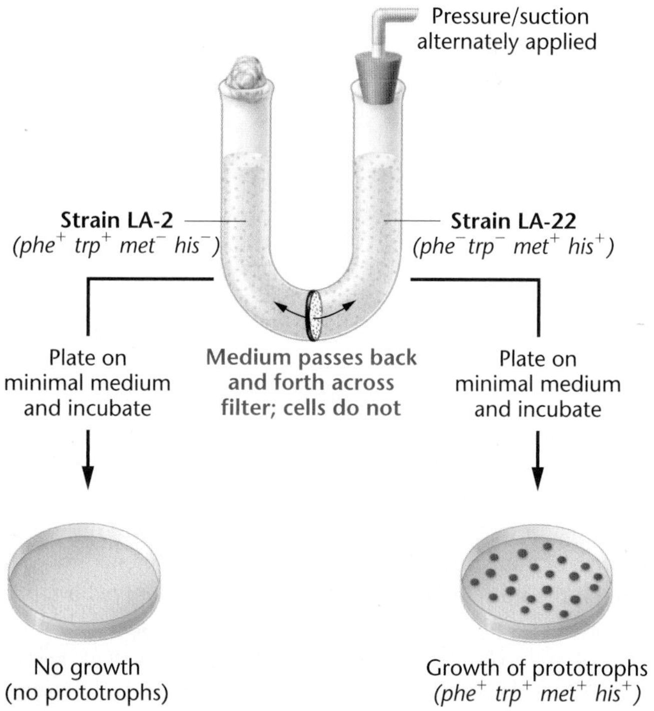

FIGURE 6–17 The Lederberg–Zinder experiment using *Salmonella*. After placing two auxotrophic strains on opposite sides of a Davis U-tube, Lederberg and Zinder recovered prototrophs from the side containing the LA-22 strain, but not from the side containing the LA-2 strain. These initial observations led to the discovery of the phenomenon called transduction.

FIGURE 6–18 Generalized transduction.

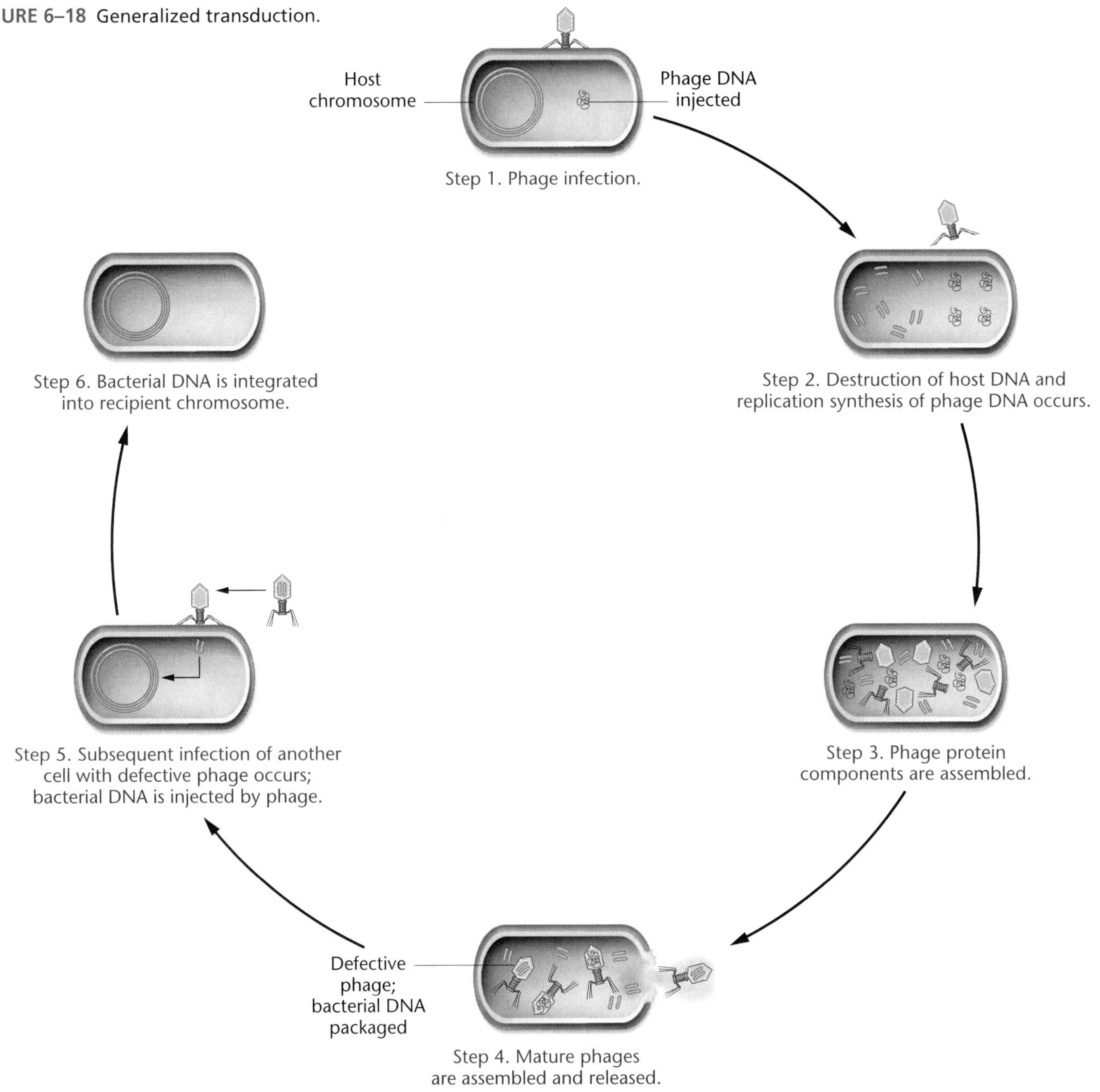

Host chromosome
Phage DNA injected
Step 1. Phage infection.
Step 2. Destruction of host DNA and replication synthesis of phage DNA occurs.
Step 3. Phage protein components are assembled.
Defective phage; bacterial DNA packaged
Step 4. Mature phages are assembled and released.
Step 5. Subsequent infection of another cell with defective phage occurs; bacterial DNA is injected by phage.
Step 6. Bacterial DNA is integrated into recipient chromosome.

The Nature of Transduction

Further studies revealed the existence of **transducing phages** in other species of bacteria. For example, *E. coli*, *Bacillus subtilis*, and *Pseudomonas aeruginosa* can be transduced by a number of different phages. The details of several different modes of transduction have also been established. Even though the initial discovery of transduction involved a temperate phage and a lysogenized bacterium, the same process can occur during the normal lytic cycle (infection and lysis). Sometimes, during what is called **specialized transduction**, a small piece of bacterial DNA is packaged along with the viral chromosome. In such cases, only a few specific bacterial genes are present in the transducing phage. In other cases, referred to as **generalized transduction**, the phage DNA is excluded completely and only bacterial DNA is packaged. Regions as large as 1 percent of the bacterial chromosome may become enclosed in the viral head. In both types of transduction, the ability to infect host cells is unaffected by the presence of foreign DNA.

During generalized transduction, when bacterial rather than viral DNA is injected into the bacterium, it either remains in the bacterial cytoplasm or recombines with the homologous region of the bacterial chromosome. If the bacterial DNA remains in the cytoplasm, it does not replicate, but is transmitted to one of the progeny cells following each division. When this happens, only a single cell, partially diploid for the transduced genes, is produced—a phenomenon called **abortive transduction**. If the bacterial DNA recombines with its homologous region of the bacterial chromosome, the transduced genes are replicated as part of the chromosome and passed to all daughter cells. This process is called **complete transduction**.

Both abortive and complete transduction are characterized by the random nature of DNA fragments and genes transduced. Each fragment of the bacterial chromosome has a finite, but small, chance of being packaged in the phage head. Most cases of generalized transduction are of the abortive type; some data suggest that complete transduction occurs 10 to 20 times less frequently.

Transduction and Mapping

Like transformation, generalized transduction was used in linkage and mapping studies of the bacterial chromosome. The fragment of bacterial DNA involved in a transduction event is large enough to include numerous genes. As a result, two genes that are close to one another along the bacterial chromosome (i.e., are linked) may be simultaneously transduced, a process called **cotransduction**. Two genes that are not close enough to one another along the chromosome to be included on a single DNA fragment require two independent events in order to be transduced into a single cell. Since this occurs with a much lower probability than cotransduction, linkage can be determined.

By concentrating on two or three linked genes, transduction studies can also determine the precise order of these genes. The closer linked genes are to each other, the greater the frequency of cotransduction. Mapping studies involving three closely aligned genes can be executed. The analysis of such an experiment involves the same assumptions as other mapping techniques.

6.8 Bacteriophages Undergo Intergenic Recombination

Around 1947, several research teams demonstrated that genetic recombination can be detected in bacteriophages. This led to the discovery that gene mapping can be performed in these viruses. Such studies relied on finding numerous phage mutations that could be visualized or assayed. As in bacteria and eukaryotes, these mutations allow genes to be identified and followed in mapping experiments. Before considering recombination and mapping in these bacterial viruses, we will briefly introduce several of the mutations that were studied.

Phage mutations often affect the morphology of the plaques formed following lysis of bacterial cells. For example, in 1946, Alfred Hershey observed unusual T2 plaques on plates of *E. coli* strain B. Where the normal T2 plaques are small and have a clear center surrounded by a diffuse (nearly invisible) halo, the unusual plaques were larger and possessed a more distinctive outer perimeter (Figure 6–19). When the viruses were isolated from these plaques and replated on *E. coli* B cells, the resulting plaque appearance was identical. Thus, the plaque phenotype was an inherited trait resulting from the reproduction of mutant phages. Hershey named the mutant *rapid lysis* (r) because the plaques were larger, apparently resulting from a more rapid or more efficient life cycle of the phage. It is now known that, in wild-type phages, reproduction is inhibited once a particular-sized plaque has been formed. The r mutant T2 phages are able to overcome this inhibition, producing larger plaques.

Salvador Luria discovered another bacteriophage mutation, *host range* (h). This mutation extends the range of bacterial hosts that the phage can infect. Although wild-type T2 phages can infect *E. coli* B, they normally cannot attach or be adsorbed to the surface

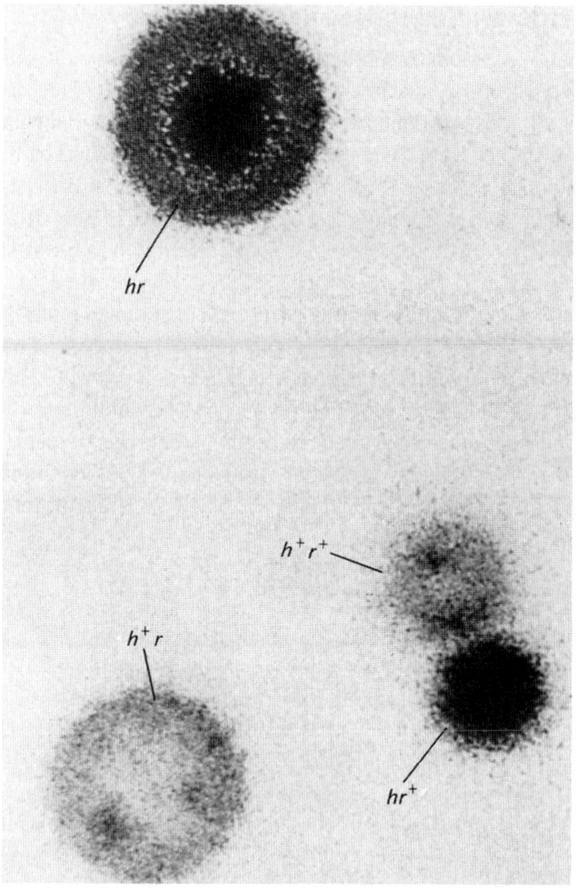

FIGURE 6–19 Plaque morphology phenotypes observed following simultaneous infection of *E. coli* by two strains of phage *T2*, h^+r and hr^+. In addition to the parental genotypes, recombinant plaques hr and h^+r^+ were also recovered.

of *E. coli* B-2. The h mutation, however, provides the basis for adsorption and subsequent infection of *E. coli* B-2. When grown on a mixture of *E. coli* B and B-2, the center of the h plaque appears much darker than the h^+ plaque (Figure 6–19).

Table 6.1 lists other types of mutations that have been isolated and studied in the T-even series of bacteriophages (e.g., T2, T4, T6). These mutations are important to the study of genetic phenomena in bacteriophages.

TABLE 6.1 SOME MUTANT TYPES OF T-EVEN PHAGES

Name	Description
Minute	Small plaques
Turbid	Turbid plaques on *E. coli* B
Star	Irregular plaques
UV-sensitive	Alters UV sensitivity
Acriflavin-resistant	Forms plaques on acriflavin agar
Osmotic shock	Withstands rapid dilution into distilled water
Lysozyme	Does not produce lysozyme
Amber	Grows in *E. coli* K12, but not B
Temperature-sensitive	Grows at 25°C, but not at 42°C

Mapping in Bacteriophages

Genetic recombination in bacteriophages was discovered during **mixed infection experiments**, in which two distinct mutant strains were allowed to simultaneously infect the same bacterial culture. These studies were designed so that the number of viral particles sufficiently exceeded the number of bacterial cells so as to ensure simultaneous infection of most cells by both viral strains. Because two loci are involved, recombination is referred to as **intergenic**. For example, in one study using the T2/*E. coli* system, the parental viruses were of either the h^+r (wild-type host range, rapid lysis) or the hr^+ (extended host range, normal lysis) genotype. If no recombination occurred, these two parental genotypes would be the only expected phage progeny. However, the recombinants h^+r^+ and hr were detected in addition to the parental genotypes. (See Figure 6–19). As with eukaryotes, the percentage of recombinant plaques divided by the total number of plaques reflects the relative distance between the genes:

$$(h^+r^+) + (hr)/\text{total plaques} \times 100 = \text{recombinational frequency}$$

Sample data for the *h* and *r* loci are shown in Table 6.2.

Similar recombinational studies have been performed with numerous mutant genes in a variety of bacteriophages. Data are analyzed in much the same way as they are in eukaryotic mapping experiments. Two- and three-point mapping crosses are possible, and the percentage of recombinants in the total number of phage progeny is calculated. This value is proportional to the relative distance between two genes along the DNA molecule constituting the chromosome.

How Do We Know?

How do we know that intergenic exchange occurs in bacteriophages?

Investigations into phage recombination support a model similar to that of eukaryotic crossing over—a breakage and reunion process between the viral chromosomes. A fairly clear picture of the dynamics of viral recombination is emerging. Following the early phase of infection, the chromosomes of the phages begin replication. As this stage progresses, a pool of chromosomes accumulates in the bacterial cytoplasm. If double infection by phages of two genotypes has occurred, then the pool of chromosomes initially consists of the two parental types. Genetic exchange between these two types will occur before, during, and after replication, producing recombinant chromosomes.

In the case of the $h^+r - hr^+$ example discussed here, recombinant h^+r^+ and hr chromosomes are produced. Each of these chromosomes may undergo replication, with new replicates undergoing exchange with each other and with parental chromosomes.

Furthermore, recombination is not restricted to exchange between two chromosomes—three or more may be involved simultaneously. As phage development progresses, chromosomes are randomly removed from the pool and packed into the phage head, forming mature phage particles. Thus, a variety of parental and recombinant genotypes are represented in progeny phages.

As we will see in the next section, powerful selection systems have made it possible to detect *intragenic* recombination in viruses, where exchanges occur at points within a single gene, as opposed to intergenic recombination, where exchanges occur between genes. Such studies have led to what has been called the fine-structure analysis of the gene.

TABLE 6.2 **Results of a Cross Involving the *h* and *r* Genes in Phage T2 ($hr^+ \times h^+r$)**

Genotype	Plaques	Designation
$h\ r^+$	42	Parental progeny
h^+r	34	76%
h^+r^+	12	Recombinants
$h\ r$	12	24%

Source: Data derived from Hershey and Rotman, 1949.

Now solve this

Problem 6.13 on page 162 involves an understanding of how intergenic mapping is performed in bacteriophages.

Hint: Mapping is based on recombination occuring during simultaneous infection of a bacterium by two or more strain-specific bacteriophages. The probability of recombination varies directly with the distance between two genes along the bacteriophage chromosome. Mapping in phages is similar to three-point mapping in eukaryotes. Just consider the bacterial cell as a "heterozygote" that harbors "homologous chromosomes."

6.9 Intragenic Recombination Occurs in Phage T4

We conclude this chapter with an account of an ingenious example of genetic analysis. In the early 1950s, Seymour Benzer undertook a detailed examination of a single locus, *rII*, in phage T4. Benzer successfully designed experiments to recover the extremely rare genetic recombinants arising as a result of intragenic exchange. Such recombination is equivalent to eukaryotic crossing over, but in this case, within a gene rather than between two genes. Benzer demonstrated that such recombination occurs between the DNA of individual bacteriophages during simultaneous infection of the host bacterium *E. coli*.

The end result of Benzer's work was the production of a detailed map of the *rII* locus. Because of the extremely detailed information provided from his analysis, and because these experiments occurred decades before DNA-sequencing techniques were developed, the insights concerning the internal structure of the gene were particularly noteworthy.

The *rII* Locus of Phage T4

The primary requirement in genetic analysis is the isolation of a large number of mutations in the gene being investigated. Mutants at the *rII* locus produce distinctive plaques when plated on *E. coli* strain B, allowing their easy identification. Figure 6–19 illustrates mutant *r* plaques compared to their wild-type r^+ counterparts in the related T2 phage. Benzer's approach was to isolate many independent *rII* mutants—he eventually obtained about 20,000—and to perform recombinational studies so as to produce a genetic map of this locus. Benzer assumed that most mutations, because they were randomly isolated, would represent different locations within the *rII* locus and would thus provide an ample basis for mapping studies.

The key to Benzer's analysis was that *rII* mutant phages, though capable of infecting and lysing *E. coli* B, could not successfully lyse a second related strain, *E. coli* K12(λ).* Wild-type phages, by contrast, could lyse both the B and the K12 strains. Benzer reasoned that these conditions provided the potential for a highly sensitive screening system. If phages from any two different mutant strains were allowed to simultaneously infect *E. coli* B, exchanges between the two mutant sites within the locus would produce rare wild-type recombinants (Figure 6–20). If the phage population, which contained more than 99.9 percent *rII* phages and less than 0.1 percent wild-type phages, were then allowed to infect strain K12, the wild-type recombinants would successfully reproduce and produce wild-type plaques. This is the critical step in recovering and quantifying rare recombinants.

By using serial dilution techniques, Benzer was able to determine the total number of mutant *rII* phages produced on *E. coli* B and the total number of recombinant wild-type phages that would lyse *E. coli* K12. These data provided the basis for calculating the frequency of recombination, a value proportional to the distance within the gene between the two mutations being studied. As we will see, this experimental design was extraordinarily sensitive. Remarkably, it was possible for Benzer to detect as few as one recombinant wild-type phage among 100 million mutant phages. When information from many such experiments is combined, a detailed map of the locus is possible.

Before we discuss this mapping, we need to describe an important discovery Benzer made during the early development of his screen—a discovery that led to the development of a technique used widely in genetics labs today, the **complementation assay**.

HOW DO WE KNOW?

How did Benzer establish that intragenic recombination occurred within the *rII* locus of bacteriophage T4?

Complementation by *rII* Mutations

Before Benzer was able to initiate intragenic recombination studies, he had to resolve a problem encountered during the early stages of his experimentation. While doing a control for his experiment in which K12 bacteria were simultaneously infected

*The inclusion of (λ) in the designation of K12 indicates that this bacterial strain is lysogenized by phage λ. This, in fact, is the reason that *rII* mutants cannot lyse such bacteria. In future discussions, this strain will simply be abbreviated as *E. coli* K12.

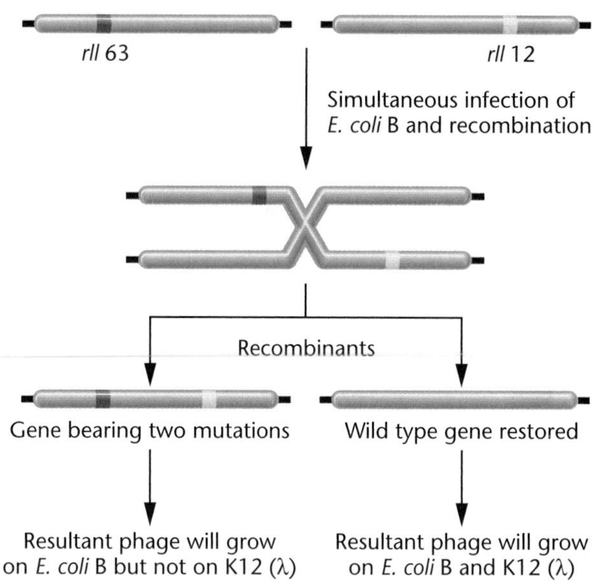

rII 63
rII 12
Simultaneous infection of E. coli B and recombination
Recombinants
Gene bearing two mutations
Wild type gene restored
Resultant phage will grow on E. coli B but not on K12 (λ)
Resultant phage will grow on E. coli B and K12 (λ)

FIGURE 6–20 Illustration of intragenic recombination between two mutations in the *rII* locus of phage T4. The result is the production of a wild-type phage that will grow on both *E. coli* B and K12 and a phage that has incorporated both mutations into the *rII* locus. It will grow on *E. coli* B, but not on *E. coli* K12.

with pairs of different *rII* mutant strains, Benzer sometimes found that certain pairs of the *rII* mutant strains lysed the K12 bacteria. This was initially quite puzzling, since only the wild-type *rII* was supposed to be capable of lysing K12 bacteria. How could two mutant strains of *rII*, each of which was thought to contain a defect in the same gene, show a wild-type function?

Benzer reasoned that, during simultaneous infection, each mutant strain provided something that the other lacked, thus restoring wild-type function. This phenomenon, which he called **complementation**, is illustrated in Figure 6–21(a). When many pairs of mutations were tested, each mutation fell into one of two possible **complementation groups**, A or B. Those that failed to complement one another were placed in the same complementation group, while those that did complement one another were each assigned to a different complementation group. Benzer coined the term **cistron**, which he defined as the smallest functional genetic unit, to describe a complementation group. In modern terminology, we know that a cistron represents a gene.

HOW DO WE KNOW?

How did Benzer determine that the *rII* locus was subdivided into two regions, or cistrons?

We now know that Benzer's A and B cistrons represent two separate genes in what we originally referred to as the *rII* locus. Complementation occurs when K12 bacteria are infected with two *rII* mutants, one with a mutation in the A gene and one with a mutation in the B gene. Therefore, there is a source of both wild-type gene products, since the A mutant provides wild-type B and the B mutant provides wild-type A. We can also explain why two strains that fail to complement, say two A cistron mutants, are actually mutations in the same gene. In

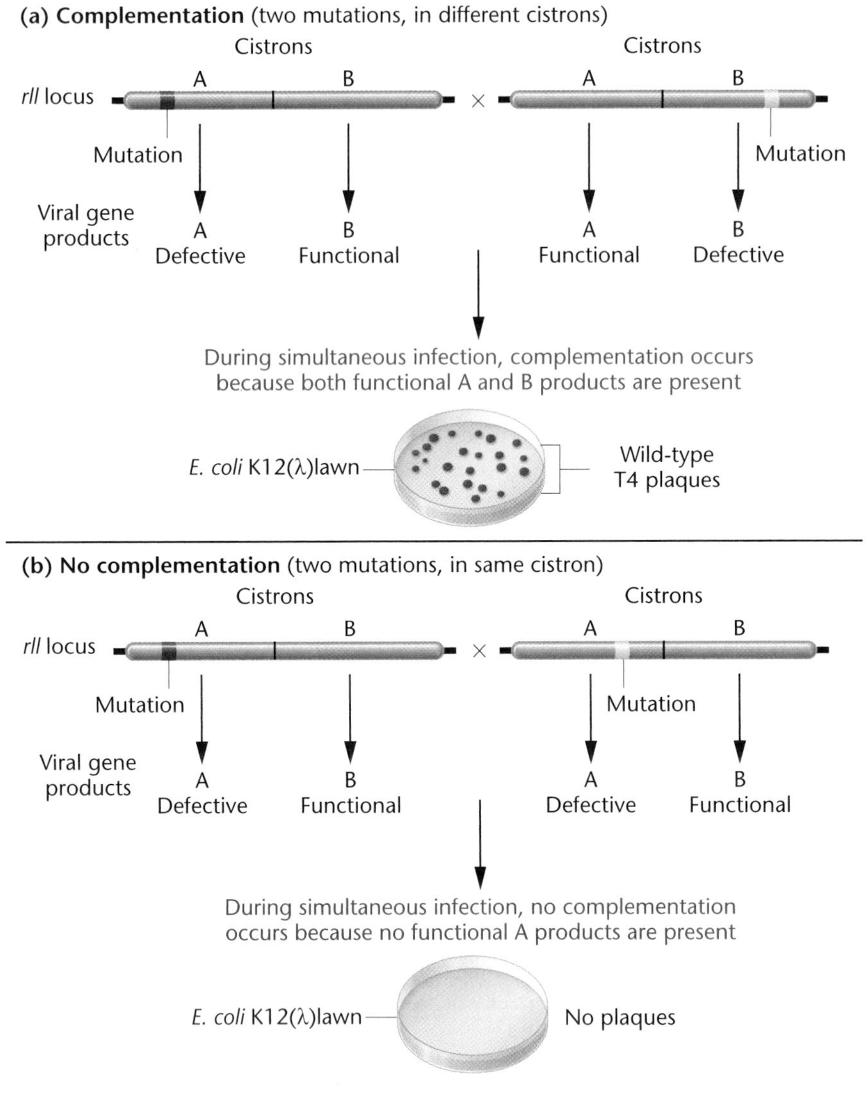

FIGURE 6–21 Comparison of two *rII* mutations that either (a) complement one another or (b) do not complement one another. Complementation occurs when each mutation is in a separate cistron. Failure to complement occurs when the two mutations are in the same cistron.

this case, if two A cistron mutants are combined, there will be a source of the wild-type B product, but no source of the wild-type A product [Figure 6–21(b)].

Once Benzer was able to place all *rII* mutations in either the A or the B cistron, he was set to return to his intragenic recombination studies, testing mutations in the A cistron against each other and testing mutations in the B cistron against each other.

Now solve this

Problem 6.18 on page 162 involves an understanding of why complementation occurs during simultaneous infection of a bacterial cell by two bacteriophage strains, each with a different mutation in the *rII* gene.

Hint: If each mutation alters a different genetic product, then each strain will provide the product that the other is missing, thus leading to complementation.

Recombinational Analysis

Of the approximately 20,000 *rII* mutations, roughly half fell into each cistron. Benzer set about mapping the mutations within each one. For example, focusing initially on the A cistron, if two *rII*A mutants were first allowed to infect *E. coli* B in a liquid culture, and if a recombination event occurred between the mutational sites in the A cistron, then wild-type progeny viruses would be produced at low frequency. If samples of the progeny viruses from such an experiment were then plated on *E. coli* K12, only the wild-type recombinants would lyse the bacteria and produce plaques. The total number of nonrecombinant progeny viruses was also determined by plating samples on *E. coli* B.

This experimental protocol is illustrated in Figure 6–22. The percentage of recombinants can be determined by counting the plaques at the appropriate dilution in each case. As in eukaryotic mapping experiments, the frequency of recombination is an estimate of the distance between the two mutations within the cistron. For example, if the number of recombinants is equal to 4×10^3 ml and the total number of progeny is 8×10^9 ml, then the frequency of recombination between the two mutants is

$$2\left(\frac{4 \times 10^3}{8 \times 10^9}\right) = 2(0.5 \times 10^{-6})$$
$$= 10^{-6}$$
$$= 0.000001$$

Multiplying by 2 is necessary because each recombinant event yields two reciprocal products, only one of which—the wild type—is detected.

Now solve this

Problem 6.20 on page 162 involves intragenic mapping within each cistron of the *rII* locus.

Hint: Recombination occurs within genes in the same way that it occurs between genes, but at a much reduced frequency.

Deletion Testing of the *rII* Locus

Although the system for assessing recombination frequencies described earlier allowed for mapping mutations within each cistron, testing 1000 mutants two at a time in all combinations would have required millions of experiments. Fortunately, Benzer was able to overcome this obstacle when he devised an analytical

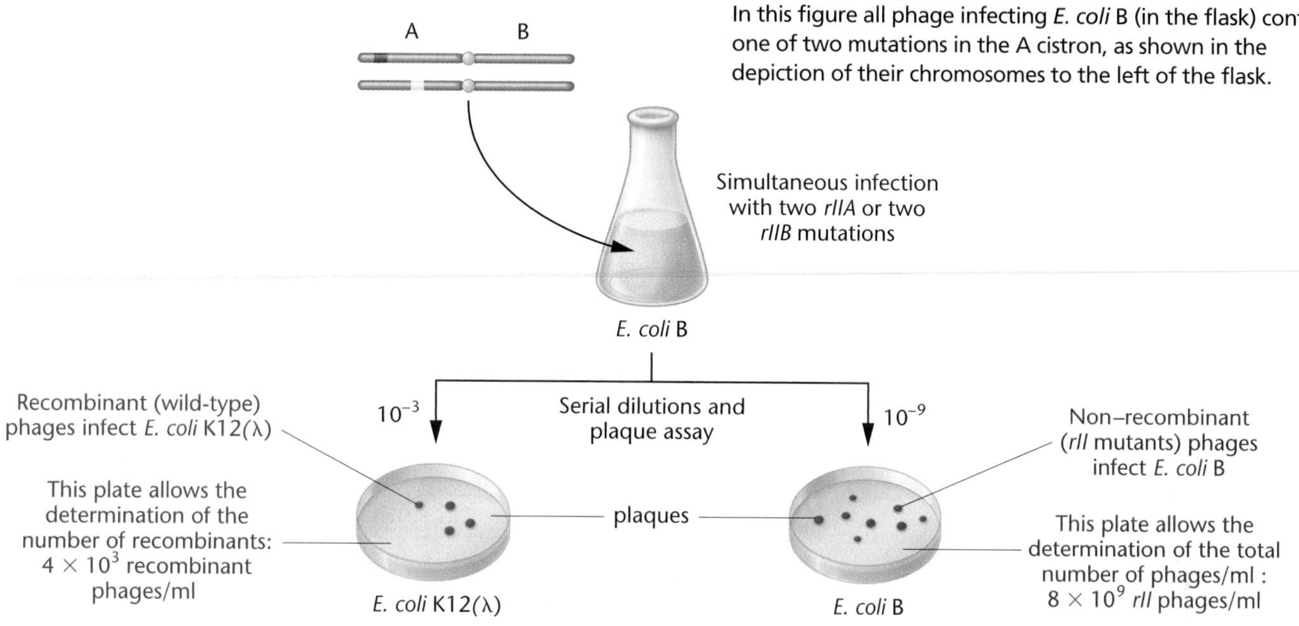
A
B
Simultaneous infection with two *rIIA* or two *rIIB* mutations
E. coli B
Serial dilutions and plaque assay
10^{-3}
10^{-9}
Recombinant (wild-type) phages infect *E. coli* K12(λ)
This plate allows the determination of the number of recombinants: 4×10^3 recombinant phages/ml
plaques
E. coli K12(λ)
E. coli B
Non–recombinant (*rII* mutants) phages infect *E. coli* B
This plate allows the determination of the total number of phages/ml : 8×10^9 *rII* phages/ml

FIGURE 6–22 The experimental protocol for recombinational studies between pairs of mutations in the same cistron. In this figure all phage infecting *E. coli* B (in the flask) contain one of two mutations in the A cistron, as shown in the depiction of their chromosomes to the left of the flask.

approach referred to as **deletion testing**. He discovered that some of the *rII* mutations were, in reality, deletions of small parts of each cistron. That is, the genetic changes giving rise to the *rII* properties were not a characteristic of point mutations. Most important, when a deletion mutation was tested during simultaneous infection with a point mutation located in the deleted part of the same cistron, it never yielded wild-type recombinants. The basis for the failure to do so is illustrated in Figure 6–23. Because the deleted area is lacking the area of DNA containing the mutation, no recombination is possible. Thus, a method was available that could roughly, but quickly, localize any mutation, provided it was contained within a region covered by a deletion.

The use of deletion testing could thus serve as the basis for the initial localization of each mutation. For example, as shown in Figure 6–24, seven overlapping deletions spanning various regions of the A and B cistrons were used for the initial screening of the point mutations. Depending on whether the viral chromosome bearing a point mutation does or does not undergo recombination with the chromosome bearing a deletion, each point mutation can be assigned to a specific area of the cistron. Further deletions within each of the seven areas can be used to localize, or map, each *rII* point mutation more specifically. Remember that, in each case, a point mutation is localized in the area of a deletion when it fails to give rise to any wild-type recombinants.

The *rII* Gene Map

After several years of work, Benzer produced a genetic map of the two cistrons composing the *rII* locus of phage T4 (Figure 6–25). Of the 20,000 mutations analyzed, 307 distinct sites within this locus were mapped in relation to one another. Areas containing many mutations, designated as **hot spots**, were apparently more susceptible to mutation than were areas in which only one or a few mutations were found. Additionally, Benzer discovered areas within the cistrons in which no mutations were localized. He estimated that as many as 200 recombinational units had not been localized by his studies.

The significance of Benzer's work is his application of genetic analysis to what had previously been considered an abstract unit—the gene. Benzer had demonstrated in 1955 that a gene is not an indivisible particle, but instead consists of mutational and recombinational units that are arranged in a specific order. Today, we know these are nucleotides composing DNA. His analysis, performed prior to the detailed molecular studies of the gene in the 1960s, is considered a classic example of genetic experimentation.

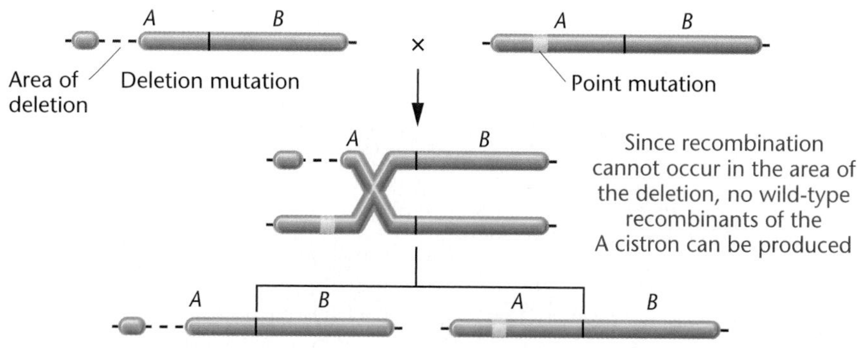
A
B
Area of deletion
Deletion mutation
×
Point mutation
Since recombination cannot occur in the area of the deletion, no wild-type recombinants of the A cistron can be produced

FIGURE 6–23 Demonstration that recombination between a phage chromosome with a deletion in the A cistron and another phage with a point mutation overlapped by that deletion cannot yield a chromosome with wild-type A and B cistrons.

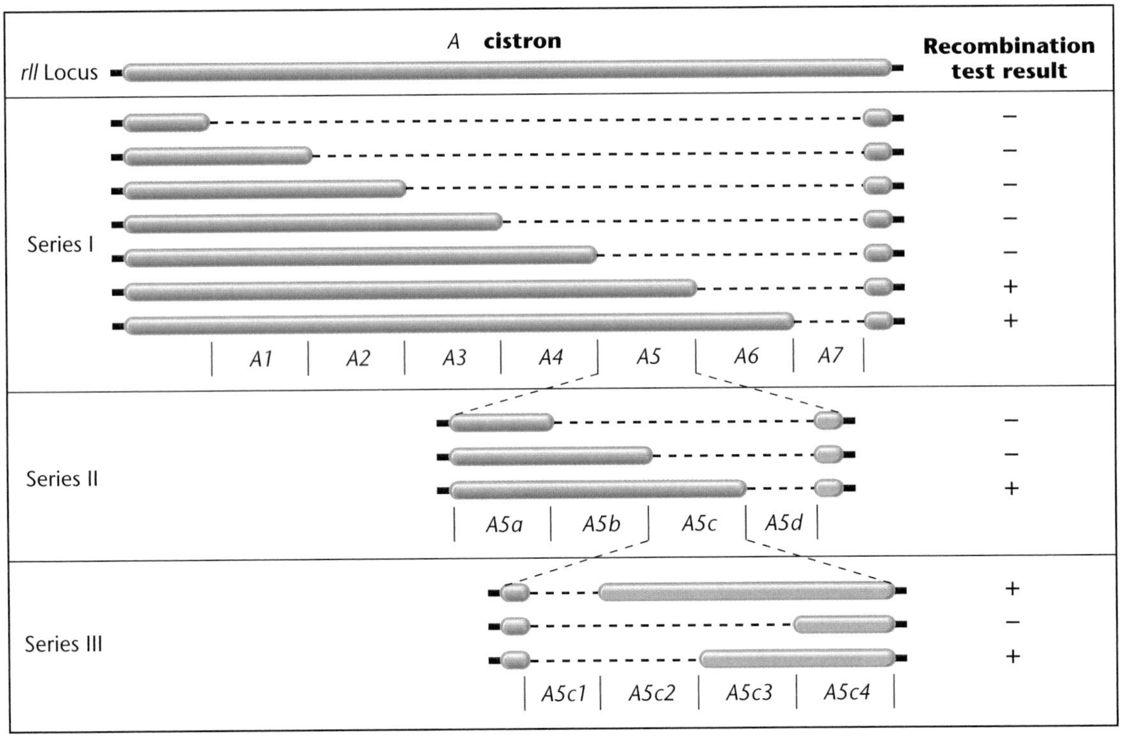

FIGURE 6–24 Three series of overlapping deletions in the A cistron of the *rII* locus used to localize the position of an unknown rII mutation. For example, if a mutant strain tested against each deletion (dashed areas) in Series I for the production of recombinant wild-type progeny shows the results at the right (+ or −), the mutation must be in segment A5. In Series II, the mutation is further narrowed to segment A5c, and in Series III to segment A5c3.

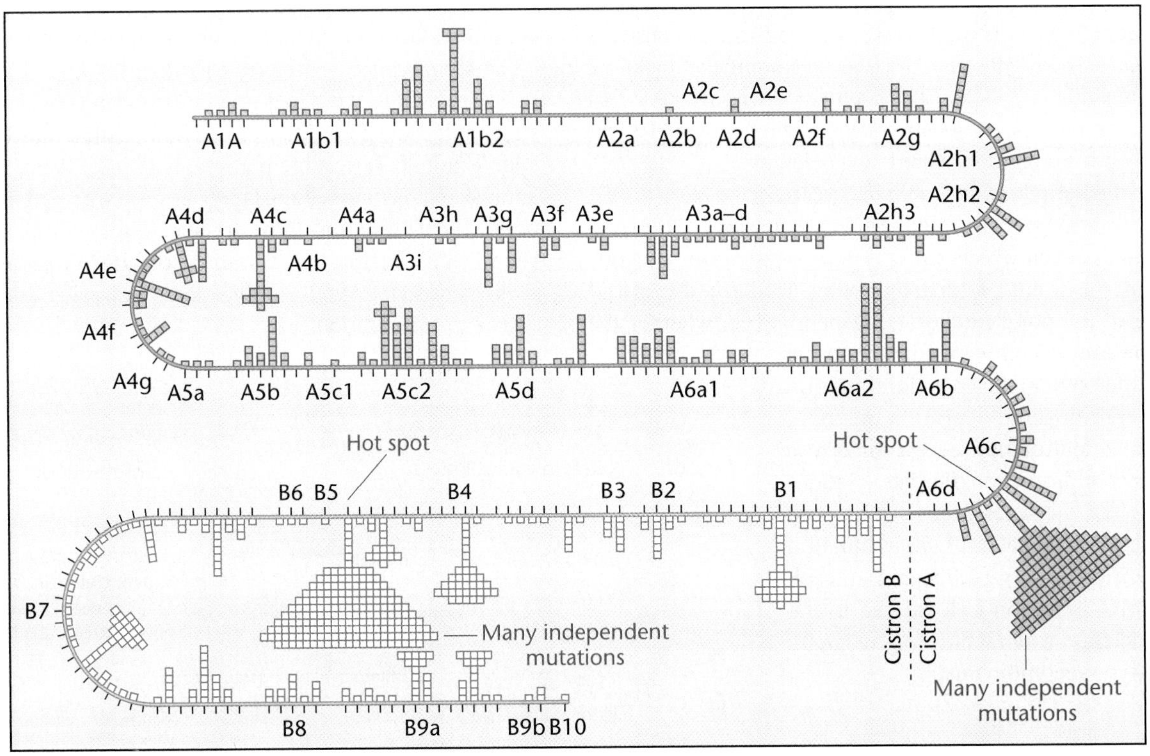

FIGURE 6–25 A partial map of mutations in the A and B cistrons of the *rII* locus of phage T4. Each square represents an independently isolated mutation. Note the two areas in which the largest number of mutations are present, referred to as "hot spots" (A6cd and B5).

Genetics, Technology, and Society

Bacterial Genes and Disease: From Gene Expression to Edible Vaccines

Using an expanding toolbox of molecular genetic tools, scientists are tackling some of the most serious bacterial diseases affecting our own species. As we will see, our newly acquired understanding of bacterial genes is leading directly to exciting new treatments based on edible vaccines. The story of cholera vaccines is a model for this research.

The causative agent of cholera is *Vibrio cholerae*, a curved, rod-shaped bacterium found mostly in rivers and oceans. Most genetic strains of *V. cholerae* are harmless; only a few are pathogenic. Infection occurs when a person drinks water or eats food contaminated with pathogenic *V. cholerae*. Once in the digestive system, these bacteria colonize the small intestine and produce proteins called enterotoxins. The cholera enterotoxin invades the mucosal cells lining the intestine, triggering a massive secretion of water and dissolved salts from these cells. This results in violent diarrhea, which, if untreated, is followed by severe dehydration, muscle cramps, lethargy, and often death. The enterotoxin consists of two polypeptides, called the A and B subunits, encoded by two separate genes. For the toxin to be active, one A subunit must be linked to five B subunits.

Cholera remains a leading cause of death of infants and children throughout the Third World, where basic sanitation is lacking and water supplies are often contaminated. For example, in July of 1994, 70,000 cases of cholera were reported among the Rwandans crowded into refugee camps in Goma, Zaire, leading to 12,000 fatalities. A cholera epidemic struck East Africa in the fall of 1997, killing nearly 3000. And after an absence of over 100 years, cholera reappeared in Latin America in 1991, spreading from Peru to Mexico and claiming more than 10,000 lives. In 2001, more than 40 cholera outbreaks in 28 countries were reported to the World Health Organization.

A new genetic technology is emerging to attack cholera. This technology centers on genetically engineered plants that act as vaccines. When a foreign gene is inserted into the plant genome, transcribed, and translated, a transgenic plant is produced that contains the foreign gene product. Immunization then requires eating these plants. In theory, the foreign protein in the plant acts as an antigen to stimulate the production of antibodies to protect against bacterial infection. Since it is the B subunit of the cholera enterotoxin that binds to intestinal cells, attention has focused on this polypeptide as the antigen, reasoning that antibodies against it will prevent toxin binding and render the bacteria harmless.

Leading the efforts to develop an edible vaccine has been Charles Arntzen and associates at Cornell University. To test the system, Arntzen is using the B subunit of an *E. coli* enterotoxin, which is similar in structure and immunological properties to the cholera protein. The first step was to obtain the DNA clone of the gene encoding the B subunit and to attach it to a promoter that would induce transcription in all tissues of the plant. Second, the hybrid gene was introduced into potato plants by means of *Agrobacterium*-mediated transformation. Arntzen and his colleagues chose the potato so they could assay the effectiveness of the antigen in the edible part of the plant, the tuber. Analysis showed that the engineered plants expressed their new gene and produced the enterotoxin B subunit. The third step was to feed mice a few grams of these genetically engineered tubers. Arntzen found that the mice produced specific antibodies against the B subunit and secreted these into the small intestine. And, most critically, mice later fed purified enterotoxin were protected from its effects and did not develop the symptoms of cholera. Clinical trials now involve human subjects to test the efficacy of the potato-produced vaccine. In those conducted in 1998, almost all of the volunteers developed an immune response, and none experienced adverse side effects.

The Arntzen group is also producing edible vaccines in bananas, which have several advantages over potatoes. Bananas can be grown almost anywhere throughout the tropical or subtropical developing countries of the world, exactly where vaccines are needed the most. And unlike potatoes, bananas are usually eaten raw, avoiding the potential inactivation of the antigenic proteins by cooking. Finally, bananas are well liked by infants and children, making this approach to immunization a more feasible one. If all goes as planned, it may someday be possible to immunize all Third-World children against cholera and other intestinal diseases, saving untold thousands of young lives.

Arntzen's experimentation has served as a model for other research efforts involving a variety of human diseases. A group from the John P. Robarts Institute in Ontario, Canada, has shown that potato-produced vaccines can prevent juvenile diabetes in mice. Meristem Therapeutics, based in France, is in clinical trials using corn geared toward alleviating the effects of cystic fibrosis. An Australian research team has successfully produced tobacco plants that contain a protein found in the measles virus. After demonstrating the induction of immunity by feeding mice extracts of the tobacco leaves, testing of the measles vaccine has begun on primates. Tobacco plants are also being used to produce the antiviral protein interleukin 10 to treat Crohn's disease. Similar efforts are underway to create vaccines against rabies, anthrax, tetanus, and AIDS.

Although edible vaccines appear to have a promising future, there are still many issues that must be addressed. One problem has been dosage. Some plants produce very little antigen, while others produce high levels that may induce tolerance instead of immunity. There are also concerns over potential environmental hazards associated with growing transgenic crops. In particular, measures must be taken to prevent transgenes from being introduced into native plant populations. In addition, some countries are morally opposed to genetically engineered foods. If these obstacles can be overcome, edible vaccines hold great promise for the amelioration of many human diseases.

References

Haq, T.A., Mason, H.S., Clements, J.D., and Arntzen, C.J. 1995. Oral immunization with a recombinant bacterial antigen produced in transgenic plants. *Science* 268:714–76.

Mason, H.S., Warzecha, H., Mor, T., and Arntzen, C.J. 2002. Edible plant vaccines: Applications for prophylactic and therapeutic molecular medicine. *Trends Mol. Med.* 8:324–329.

Webster, D.E., et al. 2002. Appetising solutions: An edible vaccine for measles. *Med. J. Australia* 176:434–437.

CHAPTER SUMMARY

1. Inherited phenotypic variation in bacteria can result from spontaneous mutation.
2. Genetic recombination in bacteria may result from three different modes of introduction of DNA into a recipient cell: conjugation, transformation, and transduction.
3. Conjugation is initiated by a bacterium housing a plasmid called the F factor. If the F factor is in the cytoplasm of a donor cell (F^+), the recipient F^- cell receives a copy of the F factor, converting it to the F^+ status.
4. If the F factor is integrated into the donor cell chromosome (Hfr), conjugation is initiated with the recipient cell, and the donor chromosome moves unidirectionally into the recipient, leading to recombination. Time mapping of the bacterial chromosome is based on the location of each gene relative the orientation and position of the F factor in the donor chromosome.
5. The products of a group designated as the *rec* genes are directly involved in the process of recombination between the invading DNA and the recipient bacterial chromosome.
6. Plasmids, such as the F factor, are autonomously replicating DNA molecules found in bacteria. Plasmids may contain many genes, including those conferring antibiotic resistance, as well as those necessary to achieve the plasmid's transfer during conjugation.
7. In the phenomenon of transformation, which does not require cell contact, exogenous DNA enters the host chromosome of a recipient bacterial cell. Linkage mapping of closely aligned genes is performed using this process.
8. Bacteriophages (viruses that infect bacteria) demonstrate a defined life cycle, during which they reproduce within the host cell. They can be studied using the plaque assay.
9. Bacteriophages may be lytic—that is, they infect, reproduce, and lyse the host cell—or they may lysogenize the host cell, where they infect it and integrate their DNA into the host chromosome.
10. Transduction is virus-mediated bacterial DNA recombination. When a lysogenized bacterium subsequently reenters the lytic cycle, the new bacteriophages serve as vehicles for the transfer of host (bacterial) DNA. In the process of generalized transduction, a random part of the bacterial chromosome is transferred. In specialized transduction, only specific genes adjacent to the point of insertion of the prophage are transferred.
11. Transduction is also used for bacterial linkage and mapping studies.
12. Various mutant phenotypes, including plaque morphology and host range, have been studied in bacteriophages. These have served as the basis for investigating genetic exchange and mapping between these viruses.
13. Genetic analysis of the *rII* locus in bacteriophage T4 allowed Seymour Benzer to study intragenic recombination. By isolating *rII* mutants, performing complementation analysis, recombinational studies, and deletion mapping, Benzer was able to locate and map more than 300 distinct sites within the two cistrons of the *rII* locus.

INSIGHTS AND SOLUTIONS

1. Time mapping using the interrupted mating technique was performed in a cross involving the genes *his*, *leu*, *mal*, and *xyl*. The recipient cells were auxotrophic for all four genes. After 25 minutes, mating was interrupted with the following results in recipient cells:

(a) 90 percent were xyl^+.
(b) 80 percent were mal^+.
(c) 20 percent were his^+.
(d) none were leu^+.

What are the positions of these genes relative to the origin (O) of the F factor and to one another?

Solution: Because the *xyl* gene was transferred most frequently, it is closest to O (very close). The *mal* gene is next and reasonably close to *xyl*, followed by the *his* gene. The *leu* gene is well beyond these three, since no recombinants are recovered that include it. The diagram here illustrates these relative locations along a piece of the circular chromosome:

2. Three strains of bacteria, each bearing a separate mutation, a^-, b^-, or c^-, were used as the sources of donor DNA in a transformation experiment. Recipient cells were wild type for those genes, but expressed the mutant d^-.

(a) Based on the data, and assuming that the location of the *d* gene precedes the *a*, *b*, and *c* genes, propose a linkage map for the four genes.

DNA Donor	Recipient	Transformants	Frequency of Transformants
a^-d^+	a^+d^-	a^+d^+	0.21
b^-d^+	b^+d^-	b^+d^+	0.18
c^-d^+	c^+d^-	c^+d^+	0.63

Solution: These data reflect the relative distances between each of the *a*, *b*, *c* genes and the *d* gene. The *a* and *b* genes are about the same distance away from the *d* gene and are thus tightly linked to one another. The *c* gene is more distant. Assuming that the *d* gene precedes the others, the map looks like this:

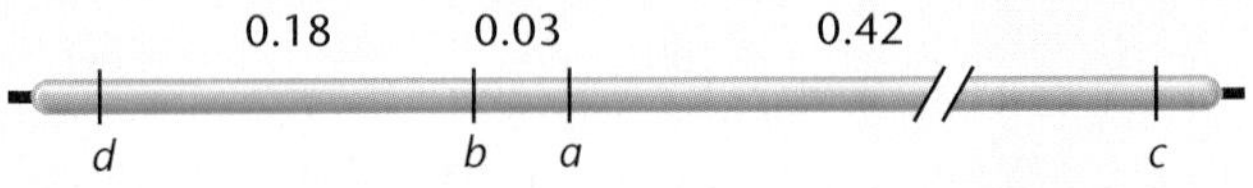

(b) If the donor DNA were wild type and the recipient cells were either a^-b^-, a^-c^-, or b^-c^-, in which case would wild-type transformants be expected most frequently?

Solution: Because the *a* and *b* genes are closely linked, they are most likely to be cotransformed in a single event. Thus, recipient cells of a^-b^- are most likely to be converted to wild type.

3. In four Hfr strains of bacteria, all derived from an original F^+ culture grown over several months, a group of hypothetical genes was studied and shown to be transferred in the following order:

Hfr Strain	Order of Transfer					
1	*E*	*R*	*I*	*U*	*M*	*B*
2	*U*	*M*	*B*	*A*	*C*	*T*
3	*C*	*T*	*E*	*R*	*I*	*U*
4	*R*	*E*	*T*	*C*	*A*	*B*

Assuming that *B* is the first gene along the chromosome, determine the sequence of all genes shown. One strain creates an apparent dilemma. Which one is it? Explain why the dilemma is only apparent and not real.

Solution: The solution is obtained by overlapping the genes in each strain in the sequence in which they were transferred, as follows:

```
         2  U M B A C T
Strain:  3          C T E R I U
         1              E R I U M B
```

Starting with *B*, the sequence of genes is *BACTERIUM*.

Strain 4 creates an apparent dilemma, which is resolved by realizing that the F factor is integrated in the opposite orientation; thus, the genes enter in the opposite sequence, starting with gene *R*:

MUIRETCAB
⟶

4. In Benzer's fine-structure analysis of the *rII* locus in phage T4, he was able to perform complementation testing of any pair of mutations once it was clear that the locus contained two cistrons. Complementation was assayed by simultaneously infecting *E. coli* K12 with two phage strains, each with an independent mutation, neither of which could alone lyse K12. From the data that follow, determine which mutations are in which cistron, assuming that mutation 1 (M-1) is in the A cistron and mutation 2 (M-2) is in the B cistron. Are there any cases where the mutation cannot be properly assigned?

Test Pair	Results*
1, 2	+
1, 3	–
1, 4	–
1, 5	+
2, 3	–
2, 4	+
2, 5	–

*+ or – indicates complementation or the failure of complementation, respectively.

Solution: M-1 and M-5 complement one another and, therefore, are not in the same cistron. Thus, M-5 must be in the B cistron. M-2 and M-4 complement one another. Using the same reasoning, M-4 is not with M-2 and, therefore, is in the A cistron. M-3 fails to complement either M-1 or M-2, and so it would seem to be in both cistrons. One explanation is that the physical basis of M-3 somehow overlaps both the A and the B cistrons. It might be a double mutation with one in each cistron. It might also be a deletion that overlaps both cistrons, making it impossible for it to complement either M-1 or M-2.

5. Another mutation, M-6, was tested with the results shown here:

Test Pair	Results
1, 6	+
2, 6	–
3, 6	–
4, 6	+
5, 6	–

Draw all possible conclusions about M-6.

Solution: These results are consistent with assigning M-6 to the B cistron.

6. Recombination testing was then performed for M-2, M-5, and M-6 so as to map the B cistron. Recombination analysis using both *E. coli* B and K12 showed that recombination occurred between M-2 and M-5 and between M-5 and M-6, but not between M-2 and M-6. Why not?

Solution: Either M-2 and M-6 represent identical mutations or one of them may be a deletion that overlaps the other, but not M-5. Furthermore, the data cannot rule out the possibility that both are deletions.

7. In recombination studies of *rII* locus in phage T4, what is the significance of the value determined by calculating phage growth on the K12 versus the B strains of *E. coli* following simultaneous infection in *E. coli* B? Which value is always greater?

Solution: By performing plaque analysis on *E. coli* B, where wild-type and mutant phages are both lytic, the total number of phages per milliliter can be determined. Because almost all cells are *rII* mutants of one type or another, this value is much larger. To avoid total lysis of the plate, extensive dilutions are necessary. On K12, *rII* mutations will not grow, but wild-type phages will. Because wild-type phages are the rare recombinants, there are relatively few of them and extensive dilution is not required.

PROBLEMS AND DISCUSSION QUESTIONS

1. Distinguish among the three modes of recombination in bacteria.
2. With respect to F^+ and F^- bacterial matings, answer the following questions:
   (a) How was it established that physical contact was necessary?
   (b) How was it established that chromosome transfer was unidirectional?
   (c) What is the basis of a bacterium being F^+?
3. List all major differences between (a) the $F^+ \times F^-$ and the Hfr × F^- bacterial crosses; and (b) the F^+, F^-, Hfr, and F′ bacteria.
4. Describe the basis for chromosome mapping in the Hfr × F^- crosses.
5. Why are the recombinants produced from an Hfr × F^- cross almost never F^+?
6. Describe the origin of F′ bacteria and merozygotes.
7. Describe what is known about the mechanism of the transformation process.
8. In a transformation experiment involving a recipient bacterial strain of genotype a^-b^-, the following results were obtained:

Transforming DNA	Transformants (%) a^+b^-	a^-b^+	a^+b^+
a^+b^+	3.1	1.2	0.04
a^+b^- and a^-b^+	2.4	1.4	0.03

What can you conclude about the location of the *a* and *b* genes relative to each other?

9. In a transformation experiment, donor DNA was obtained from a prototroph bacterial strain ($a^+b^+c^+$), and the recipient was a triple auxotroph ($a^-b^-c^-$). The following transformant classes were recovered:

$a^+b^-c^-$	180
$a^-b^+c^-$	150
$a^+b^+c^-$	210
$a^-b^-c^+$	179
$a^+b^-c^+$	2
$a^-b^+c^+$	1
$a^+b^+c^+$	3

What general conclusions can you draw about the linkage relationships among the three genes?

10. Explain the observations that led Zinder and Lederberg to conclude that the prototrophs recovered in their transduction experiments were not the result of F^+ mediated conjugation.

11. Define plaque, lysogeny, and prophage.

12. Differentiate between generalized and specialized transduction.

13. Two theoretical genetic strains of a virus ($a^-b^-c^-$ and $a^+b^+c^+$) are used to simultaneously infect a culture of host bacteria. Of 10,000 plaques scored, the following genotypes were observed:

$a^+b^+c^+$	4100	$a^-b^+c^-$	160
$a^-b^-c^-$	3990	$a^+b^-c^+$	140
$a^+b^-c^-$	740	$a^-b^-c^+$	90
$a^-b^+c^+$	670	$a^+b^+c^-$	110

Determine the genetic map of these three genes on the viral chromosome. Decide whether interference was positive or negative.

14. Describe the conditions under which genetic recombination may occur in bacteriophages.

15. The bacteriophage genome consists primarily of genes encoding proteins that make up the head, collar, tail, and tail fibers. When these genes are transcribed following phage infection, how are these proteins synthesized, since the phage genome lacks genes essential to ribosome structure?

16. If a single bacteriophage infects one *E. coli* cell present on a lawn of bacteria and, upon lysis, yields 200 viable viruses, how many phages will exist in a single plaque if only three more lytic cycles occur?

17. A culture of bacteriophages has been subjected to a series of dilutions, and a plaque assay was performed in each case, with the results shown below. What conclusion can be drawn in the case of each dilution, assuming that 0.1 mL was used in each plaque assay?

	Dilution Factor	Assay Results
(a)	10^4	All bacteria lysed
(b)	10^5	14 plaques
(c)	10^6	0 plaques

18. In complementation studies of the *rII* locus of phage T4, three groups of three different mutations were tested. For each group, only two combinations were tested. On the basis of each set of data (shown here), predict the results of the third experiment for each group.

Group A	Group B	Group C
$d \times e$—lysis	$g \times b$—no lysis	$j \times k$—lysis
$d \times f$—no lysis	$g \times i$—no lysis	$j \times l$—lysis
$e \times f$—?	$b \times i$—?	$k \times l$—?

19. In an analysis of other *rII* mutants, complementation testing yielded the following results:

Mutants	Results (+/− lysis)
1, 2	+
1, 3	+
1, 4	−
1, 5	−

(a) Predict the results of testing 2 and 3, 2 and 4, and 3 and 4 together.

(b) If further testing yielded the following results, what would you conclude about mutant 5?

Mutants	Results
2, 5	−
3, 5	−
4, 5	−

20. Using mutants 2 and 3 from the previous problem, following mixed infection on *E. coli* B, progeny viruses were plated in a series of dilutions on both *E. coli* B and K12 with the following results. What is the recombination frequency between the two mutants?

Strain Plated	Dilution	Plaques
E. coli B	10^{-5}	2
E. coli K12	10^{-1}	5

21. Another mutation, 6, was then tested in relation to mutations 1 through 5 from the previous problem. In initial testing, mutant 6 complemented mutants 2 and 3. In recombination testing with 1, 4, and 5, mutant 6 yielded recombinants with 1 and 5, but not with 4. What can you conclude about mutation 6?

22. When the interrupted mating technique was used with five different strains of Hfr bacteria, the following order of gene entry and recombination was observed:

Hfr Strain	Order				
1	*T*	*C*	*H*	*R*	*O*
2	*H*	*R*	*O*	*M*	*B*
3	*M*	*O*	*R*	*H*	*C*
4	*M*	*B*	*A*	*K*	*T*
5	*C*	*T*	*K*	*A*	*B*

On the basis of these data, draw a map of the bacterial chromosome. Do the data support the concept of circularity?

Extra-Spicy Problems

23. During the analysis of seven *rII* mutations in phage T4, mutants 1, 2, and 6 were in cistron A, while mutants 3, 4, and 5 were in cistron B. Of these, mutant 4 was a deletion overlapping mutant 5. The remainder were point mutations. Nothing was known about mutant 7. Predict the results of complementation (+ or −) between 1 and 2; 1 and 3; 2 and 4; and 4 and 5.
24. In recombination studies between 1 and 2 from the previous problem, the results shown below were obtained.
    (a) Calculate the recombination frequency.

Strain	Dilution	Plaques	Phenotypes
E. coli B	10^{-7}	4	*r*
E. coli K12	10^{-2}	8	+

    (b) When mutant 6 was tested for recombination with mutant 1, the data were the same for strain B, as shown above, but not for K12. The researcher lost the K12 data, but remembered that recombination was 10 times more frequent than when mutants 1 and 2 were tested. What were the lost values (dilution and colony numbers)?
    (c) Mutant 7 failed to complement any of the other mutants (1–6). Define the nature of mutant 7.
25. In *Bacillus subtilis*, linkage analysis of two mutant genes affecting the synthesis of the two amino acids, tryptophan (trp_2^-) and tyrosine (tyr_1^-), was performed using transformation. Examine the following data and draw all possible conclusions regarding linkage. What is the role of Part B of the experiment? [Reference: E. Nester, M. Schafer, and J. Lederberg (1963).]

Donor DNA	Recipient Cell	Transformants	No.
A. $trp_2^+\ tyr_1^+$	$trp_2^-\ tyr_1^-$	$trp^+\ tyr^-$	196
		$trp^-\ tyr^+$	328
		$trp^+\ tyr^+$	367
B. $trp_2^+\ tyr_1^-$ and $trp_2^-\ tyr_1^+$	$trp_2^-\ tyr_1^-$	$trp^+\ tyr^-$	190
		$trp^-\ tyr^+$	256
		$trp^+\ tyr^+$	2

26. An Hfr strain is used to map three genes in an interrupted mating experiment. The cross is $Hfr/a^+b^+c^+rif \times F^-/a^-b^-c^-rif^r$. (No map order is implied in the listing of the alleles; rif^r is resistance to the antibiotic rifampicin.) The a^+ gene is required for the biosynthesis of nutrient A, the b^+ gene for nutrient B, and c^+ for nutrient C. The minus alleles are auxotrophs for these nutrients. The cross is initiated at time = 0 and, at various times, the mating mixture is plated on three types of medium. Each plate contains minimal medium (MM) plus rifampicin plus specific supplements that are indicated in the following table. (The results for each time point are shown as the number of colonies growing on each plate.)

	Time of Interruption			
Supplements Added to MM	**5 min**	**10 min**	**15 min**	**20 min**
Nutrients A and B	0	0	4	21
Nutrients B and C	0	5	23	40
Nutrients A and C	4	25	60	82

    (a) What is the purpose of rifampicin in the experiment?
    (b) Based on these data, determine the approximate location on the chromosome of the *a*, *b*, and *c* genes relative to one another and to the F factor.
    (c) Can the location of the *rif* gene be determined in this experiment? If not, design an experiment to determine the location of *rif* relative to the F factor and to gene *b*.
27. A plaque assay is performed beginning with one mL of a solution containing bacteriophages. This solution is serially diluted three times by taking 0.1 mL and adding it to 9.9 mL of liquid medium. 0.1 mL of the final dilution is plated in the plaque assay and yields 17 plaques. What is the initial density of bacteriophages in original one mL?
28. In a cotransformation experiment, using various combinations of genes, two at a time, the following data were produced.

Successful Cotransformation	Unsuccessful Cotransformation
a and *d*; *b* and *c*; *b* and *f*	*a* and *b*; *a* and *c*; *a* and *f*; *d* and *b*; *d* and *c*; *d* and *f* *a* and *e*; *b* and *e*; *c* and *e* *d* and *e*; *f* and *e*

Determine which genes are "linked" and to whom.
29. Considering the data in Problem 28, another gene, *g*, was studied. It demonstrated positive cotransformation when tested with gene *f*. Predict the results of experiments when tested with genes *a*, *b*, *c*, *d*, and *e*.
30. Bacterial conjugation, mediated mainly by conjugative plasmids such as F, represents a potential health threat through the sharing of genes for pathogenicity or antibiotic resistance. Given that more than 400 different species of bacteria coinhabit a healthy human gut and more than 200 coinhabit human skin, Francisco Dionisio [*Genetics* (2002) 162:1525–1532] investigated the ability of plasmids to undergo between-species conjugal transfer. Data are presented below involving various species of the enterobacterial genus *Escherichia*. The data are presented as "log base 10" values where, for example, −2.0 would be equivalent to 10^{-2} as a rate of transfer. Assume that all differences between values presented are statistically significant.

	Donor			
Recipient	***E. chrysanthemi***	***E. blattae***	***E. fergusonii***	***E. coli***
E. chrysanthemi	−2.4	−4.7	−5.8	−3.7
E. blattae	−2.0	−3.4	−5.2	−3.4
E. fergusonii	−3.4	−5.0	−5.8	−4.2
E. coli	−1.7	−3.7	−5.3	−3.5

    (a) What general conclusion(s) can be drawn from these data?
    (b) In what species is within-species transfer most likely? In what species pair is between-species transfer most likely?
    (c) What is the significance of these findings in terms of human health?
31. A study was conducted in an attempt to determine which functional regions of a particular conjugative transfer gene (*traI*) are involved in the transfer of plasmid R27 in *Salmonella enterica*. The R27 plasmid is of significant clinical interest because it is capable of encoding multiple-antibiotic resistance to typhoid fever. To identify functional regions responsible for conjugal transfer, an analysis by Lawley et al. (2002.

J. Bacteriol. 184:2173–2180) was conducted whereby particular regions of the *tra1* gene were mutated and tested for their impact on conjugation. Shown below is a map of the regions tested and believed to be involved in conjugative transfer of the R27. Similar shading indicates related function. Numbers correspond to each functional region subjected to mutation analysis. Following the map is a table showing the effects of these mutations on R27 conjugation.

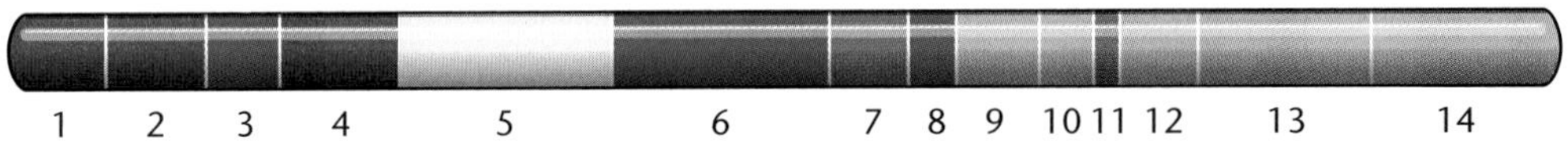

Effects of Mutations in Functional Regions of Transfer Region 1(*tra1*) on R27 Conjugation

R27 Mutation in Region	Conjugative Transfer	Relative Conjugation Frequency (%)*
1	+	100
2	+	100
3	–	0
4	+	100
5	–	0
6	–	0
7	+	12
8	–	0
9	–	0
10	–	0
11	+	13
12	–	0
13	–	0
14	–	0

(a) Given the preceding data, do all functional regions appear to influence conjugative transfer similarly?
(b) Which regions appear to have the most impact on conjugation?
(c) Which regions appear to have a limited impact on conjugation?
(d) What general conclusions might one draw from these data?

SELECTED READINGS

Benzer, S. 1962. The fine structure of the gene. *Sci. Am.* (Jan.) 206:70–86.

Brock, T. 1990. *The emergence of bacterial genetics.* Cold Spring Harbor, NY: Cold Spring Harbor Laboratory Press.

Cairns, J., Stent, G.S., and Watson, J.D., eds. 1966. *Phage and the origins of molecular biology.* Cold Spring Harbor, NY: Cold Spring Harbor Laboratory Press.

Campbell, A.M. 1976. How viruses insert their DNA into the DNA of the host cell. *Sci. Am.* (Dec.) 235:102–13.

Hayes, W. 1953. The mechanisms of genetic recombination in *Escherichia coli. Cold Spring Harbor Symp. Quant. Biol.* 18:75–93.

Hershey, A.D. 1946. Spontaneous mutations in a bacterial virus. *Cold Spring Harbor Symp. Quant. Biol.* 11:67–76.

Hershey, A.D., and Rotman, R. 1949. Genetic recombination between host range and plaque-type mutants of bacteriophage in single cells. *Genetics* 34:44–71.

Hotchkiss, R.D., and Marmur, J. 1954. Double marker transformations as evidence of linked factors in deoxyribonucleate transforming agents. *Proc. Natl. Acad. Sci. USA* 40:55–60.

Jacob, F., and Wollman, E.L. 1961. Viruses and genes. *Sci. Am.* (June) 204:92–106.

Lederberg, J. 1986. Forty years of genetic recombination in bacteria: A fortieth anniversary reminiscence. *Nature* 324:627–28.

Luria, S.E., and Delbruck, M. 1943. Mutations of bacteria from virus sensitivity to virus resistance. *Genetics* 28:491–511.

Lwoff, A. 1953. Lysogeny. *Bacteriol. Rev.* 17:269–336.

Miller, J.H. 1992. *A short course in bacterial genetics.* Cold Spring Harbor, NY: Cold Spring Harbor Press.

Morse, M.L., Lederberg, E.M., and Lederberg, J. 1956. Transduction in *Escherichia coli* K12. *Genetics* 41:141–56.

Nester, E., Schafer, M., and Lederberg, J. 1963. Gene linkage in DNA transfer: A cluster of genes in *Bacillus subtilis*. *Genetics* 48:529–51.

Novick, R.P. 1980. Plasmids. *Sci. Am.* (Dec.) 243:102–26.

Smith-Keary, P.F. 1989. *Molecular genetics of Escherichia coli.* New York: Guilford Press.

Stahl, F.W. 1987. Genetic recombination. *Sci. Am.* (Nov.) 256:91–101.

Stent, G.S. 1963. *Molecular biology of bacterial viruses.* New York: W.H. Freeman.

Tessman, I. 1965. Genetic ultrafine structure in the T4 *rII* region. *Genetics* 51:63–75.

Wollman, E.L., Jacob, F., and Hayes, W. 1956. Conjugation and genetic recombination in *Escherichia coli* K-12. *Cold Spring Harbor Symp. Quant. Biol.* 21:141–62.

Zinder, N.D. 1958. Transduction in bacteria. *Sci. Am.* (Nov.) 199:38–46.

______. 1992. Forty years ago: The discovery of bacterial transduction. *Genetics* 132:291–94.

Zinder, N.D., and Lederberg, J. 1952. Genetic exchange in *Salmonella*. *J. Bacteriol.* 64:679–99.

Sex Determination and Sex Chromosomes

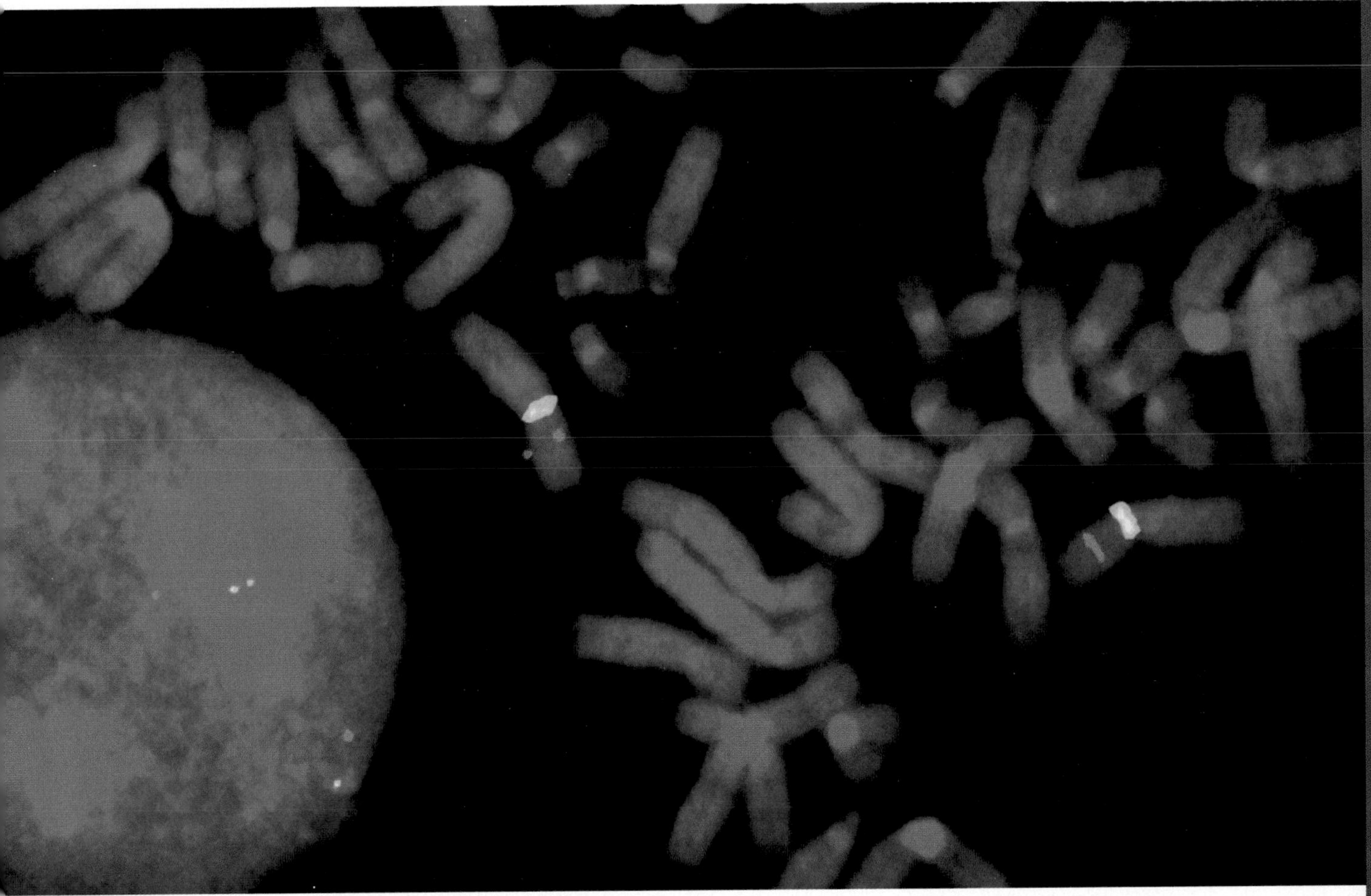

Human X chromosomes highlighted using fluorescence in situ hybridization (FISH), where specific probes bind to specific sequences of DNA. The green fluorescence is specific to the DNA of the X chromosome centromeres. The red fluorescence is specific to the DNA sequence of the Duchenne muscular dystrophy (DMD) gene.

Chapter Concepts

- Sexual reproduction, which greatly enhances genetic variation within species, requires mechanisms that result in sexual differentiation.
- A wide variety of genetic mechanisms have evolved in organisms leading to sexual dimorphism.
- Most often, specific genes, usually on a single chromosome, cause maleness or femaleness during development.
- In humans, the presence of extra X or Y chromosomes beyond the diploid number may be tolerated, but often lead to syndromes demonstrating distinctive phenotypes.
- While segregation of sex-determining chromosomes should theoretically lead to a one-to-one sex ratio of males to females, in humans, this ratio greatly favors males at conception.
- In mammals, females contain two X chromosomes compared to one in males, but the extra genetic information in females is compensated for by random inactivation of one of the X chromosomes early in development.
- In some reptilian species, temperature during incubation of eggs determines the sex of offspring.

In the biological world, a wide range of reproductive modes and life cycles are recognized. Organisms exist that display no evidence of sexual reproduction. Other species alternate between short periods of sexual reproduction and prolonged periods of asexual reproduction. In most diploid eukaryotes, however, sexual reproduction is the only natural mechanism resulting in new members of an existing species. Orderly transmission of genetic units from parents to offspring, and thus any phenotypic variability, relies on the processes of segregation and independent assortment occuring during meiosis. Meiosis produces haploid gametes so that, following fertilization, the resulting offspring maintain the diploid number of chromosomes characteristic of their species. Hence, meiosis ensures genetic constancy within members of the same species.

These events, which are involved in the perpetuation of all sexually reproducing organisms, depend ultimately on an efficient union of gametes during fertilization. In turn, successful mating between organisms, the basis for fertilization, depends on some form of sexual differentiation in organisms. Even though it is not overtly evident, this differentiation occurs for organisms as low on the evolutionary scale as bacteria and single-celled eukaryotic algae. In evolutionarily higher forms of life, the differentiation of the sexes is more evident as phenotypic dimorphism in the males and females of each species. The shield and spear ♂, the ancient symbol of iron and Mars, and the mirror ♀, the symbol of copper and Venus, represent the maleness and femaleness acquired by individuals.

Dissimilar, or **heteromorphic chromosomes**, such as the X–Y pair, often characterize one sex or the other, resulting in their label as **sex chromosomes**. Nevertheless, it is genes, rather than chromosomes, that ultimately serve as the underlying basis of **sex determination**. As we will see, some of these genes are present on sex chromosomes, but others are autosomal. Extensive investigation has revealed a wide variation in sex chromosome systems, even in closely related organisms, suggesting that mechanisms controlling sex determination have undergone rapid evolution in many instances.

In this chapter, we will first review several representative modes of sexual differentiation by examining the life cycles of three **model organisms** often studied in genetics: the green alga *Chlamydomonas*; the maize plant, *Zea mays*; and the nematode (roundworm), *Caenorhabditis elegans* (most often referred to as *C. elegans*). These will serve to contrast the different roles that sexual differentiation plays in the lives of diverse organisms. Then, we will delve more deeply into what is known about the genetic basis for the determination of sexual differences, with a particular emphasis on two other organisms: our own species, representative of mammals; and *Drosophila*, subject of pioneering sex-determining studies.

7.1 Sexual Differentiation and Life Cycles

In multicellular organisms, it is important to distinguish between **primary sexual differentiation**, which involves only the gonads where gametes are produced, and **secondary sexual differentiation**, which involves the overall appearance of the organism, including clear differences in such organs as mammary glands and external genitalia. In plants and animals, the terms **unisexual**, **dioecious**, and **gonochoric** are equivalent; they all refer to an individual containing only male *or* only female reproductive organs. Conversely, the terms **bisexual**, **monoecious**, and **hermaphroditic** refer to individuals containing both male *and* female reproductive organs, a common occurrence in both the plant and animal kingdoms. These organisms can produce fertile gametes of both sexes. The term **intersex** is usually reserved for individuals of intermediate sexual differentiation, who are most often sterile.

Chlamydomonas

The life cycle of the green alga *Chlamydomonas*, shown in Figure 7–1, is representative of organisms exhibiting only infrequent periods of sexual reproduction. Such organisms spend most of their life cycle in the haploid phase, asexually producing daughter cells by mitotic divisions. However, under unfavorable nutrient conditions, such as nitrogen depletion, certain daughter cells function as gametes. Following fertilization, a diploid zygote, which can withstand the unfavorable environment, is formed. When conditions become more suitable, meiosis ensues and haploid vegetative cells are again produced. In such species, there is little visible difference between the haploid vegetative cells that reproduce asexually and the haploid gametes that are involved in sexual reproduction. The two gametes that fuse together during mating are not usually morphologically distinguishable. Such gametes are called **isogametes**, and species producing them are said to be **isogamous**.

In 1954, Ruth Sager and Sam Granik demonstrated that gametes in *Chlamydomonas* could be subdivided into two **mating types**. Working with clones derived from single haploid cells, they showed that cells from a given clone would mate with cells from some, but not all other clones. When they tested mating abilities of large numbers of clones, all could be placed into one of two mating categories, either mt^+ or mt^- cells. "Plus" cells would only mate with "minus" cells, and vice versa, as represented in Figure 7–2. Following fertilization and meiosis, the four haploid cells (**zoospores**) produced were found to consist of two plus types and two minus types.

Further experimentation established that there is a chemical difference between plus and minus cells. When extracts were prepared from cloned *Chlamydomonas* cells (or their flagella) and then added to cells of the opposite mating type, clumping or agglutination occurred. No such agglutination occurred if the extract were added to cells of the mating type from which it was derived. These observations suggest that despite the morphological similarities between isogametes, a chemical differentiation has occurred between them. Therefore, in this alga, a primitive means of sex differentiation exists even though there is no morphological indication that such differentiation has occurred.

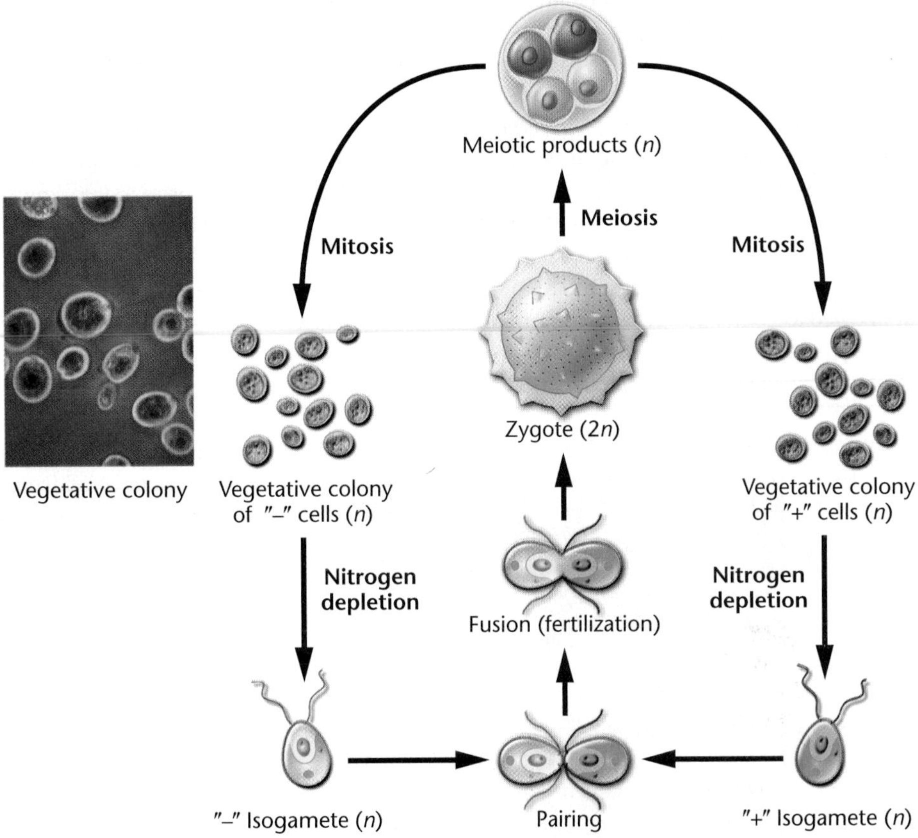

FIGURE 7–1 The life cycle of *Chlamydomonas*. Unfavorable conditions stimulate the formation of isogametes of opposite mating type that may fuse in fertilization. The resulting zygote undergoes meiosis, producing two haploid cells of each mating type. The photograph shows vegetative cells of this green alga.

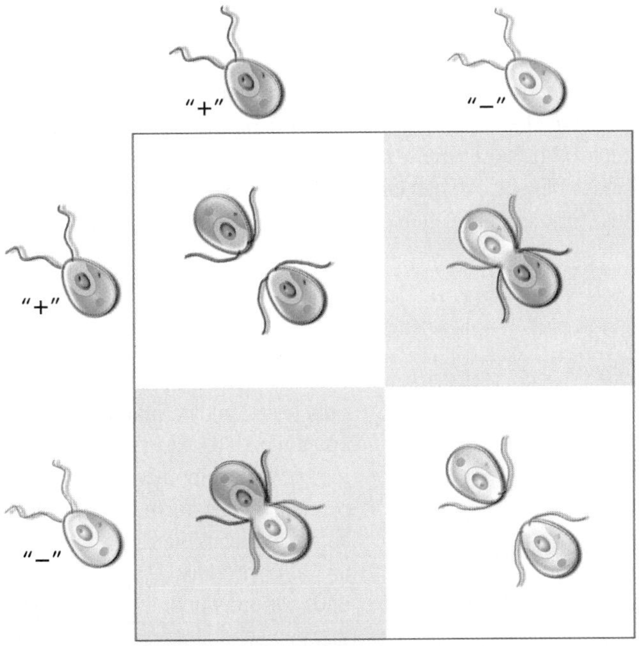

FIGURE 7–2 Illustration of mating types during fertilization in *Chlamydomonas*. Mating will occur only when plus (+) and minus (–) cells are together.

Zea mays

As we first discussed in Chapter 2 (see Figure 2–16), life cycles of many plants alternate between the haploid gametophyte stage and the diploid sporophyte stage. The processes of meiosis and fertilization link the two phases during the life cycle. The relative amount of time spent in the two phases varies between the major plant groups. In some nonseed plants, such as mosses, the haploid gametophyte phase and the morphological structures representing this stage predominate. The reverse is true in seed plants.

Maize (*Zea mays*), more popularly called corn, exemplifies a monoecious seed plant, where the sporophyte phase and the morphological structures representing this stage predominate during the life cycle. Both male and female structures are present on the adult plant. Thus, sex determination must occur differently in different tissues of the same organism, as illustrated in the life cycle of this familiar plant (Figure 7–3). The **stamens**, or **tassels**, produce diploid microspore mother cells, each of which undergoes meiosis and gives rise to four haploid microspores. Each haploid microspore in turn develops into a mature male microgametophyte—the pollen grain—which contains two sperm nuclei with identical genotypes.

Comparable female diploid cells, known as megaspore mother cells, exist in the **pistil** of the sporophyte. Following meiosis, only one of the four haploid megaspores survives. It usually divides mitotically three times, producing a total of eight genetically identical haploid nuclei enclosed in the embryo sac. Two of these nuclei unite near the center of the embryo sac, becoming the endosperm nuclei. At the end of the micropyle, the sac where the sperm enters, three nuclei remain: the oocyte nucleus and two synergids. The other three antipodal nuclei are clustered at the opposite end of the embryo sac.

Pollination occurs when pollen grains make contact with the silks (or stigma) of the pistil and develop extensive pollen tubes that grow toward the embryo sac. When contact is made at the micropyle, the two sperm nuclei enter the embryo sac. One sperm nucleus unites with the haploid oocyte nucleus and the other sperm nucleus unites with two endosperm nuclei. This process, known as *double fertilization*, results in the diploid zygote nucleus and the triploid endosperm nucleus, respectively. Each ear of corn may contain as many as 1000 of these structures, each of which develops into a single kernel. Each kernel, if allowed to germinate, gives rise to a new plant, the *sporophyte*.

The mechanism of sex determination and differentiation in a monoecious plant such as *Zea mays*, where the tissues forming both male and female gametes are of the same genetic constitution, was difficult to comprehend at first. However, the discovery of a large number of mutant genes that disrupt normal tassel and pistil formation supports the concept that normal products of these genes play an important role in sex determination by affecting the differentiation of male or female tissue in several ways.

For example, mutant genes that cause sex reversal provide valuable information. When homozygous, all mutations classified as *tassel seed* (*ts*) interfere with tassel production and induce the formation of female structures. Thus, it is possible for a single gene to cause a normally monoecious plant to become functionally only female. On the other hand, the recessive mutations *silkless* (*sk*) and *barren stalk* (*ba*) interfere with the development of the pistil, resulting in plants with only functional male reproductive organs.

Data gathered from studies of these and other mutants suggest that the products of many wild-type alleles of these genes interact in controlling sex determination. During development, certain cells are "determined" to become male or female structures. Following sexual differentiation into either male or female structures, male or female gametes are produced.

HOW DO WE KNOW?

How do we know that specific genes in maize play a role in sexual differentiation?

Caenorhabditis elegans

The nematode worm *Caenorhabditis elegans* [*C. elegans*, for short; Figure 7–4(a)] has become a popular organism in genetic studies, particularly during the investigation of the genetic control of development. Its usefulness is based on the fact that the hermaphroditic adult consists of exactly 959 cells, and the precise lineage of each cell can be traced back to specific embryonic origins. Among many interesting mutant phenotypes that have been studied, behavioral modifications have also been a favorite topic of inquiry.

There are two sexual phenotypes in these worms: males, which have only testes, and hermaphrodites that contain both testes and ovaries. During larval development of hermaphrodites, testes form that produce sperm, which is then stored. Ovaries are also produced, but oogenesis does not occur until the adult stage is reached several days later. The developed eggs are fertilized by the stored sperm during the process of self-fertilization.

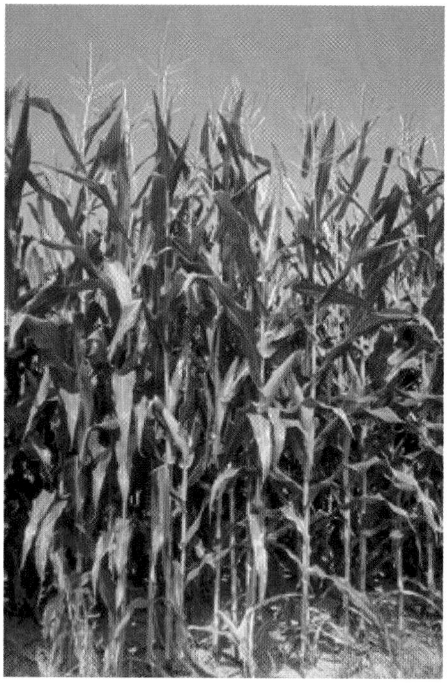

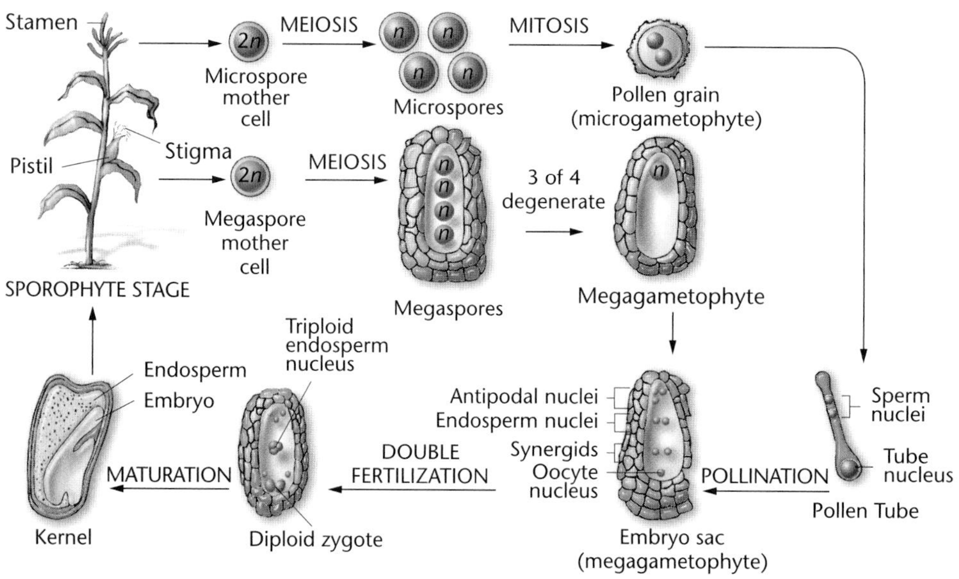

FIGURE 7–3 The life cycle of maize (*Zea mays*). The diploid sporophyte bears stamens and pistils that give rise to haploid microspores and megaspores, which develop into the pollen grain and the embryo sac that ultimately house the sperm and oocyte, respectively. Following fertilization, the embryo develops within the kernel and is nourished by the endosperm. Germination of the kernel gives rise to a new sporophyte (the mature corn plant), and the cycle repeats itself.

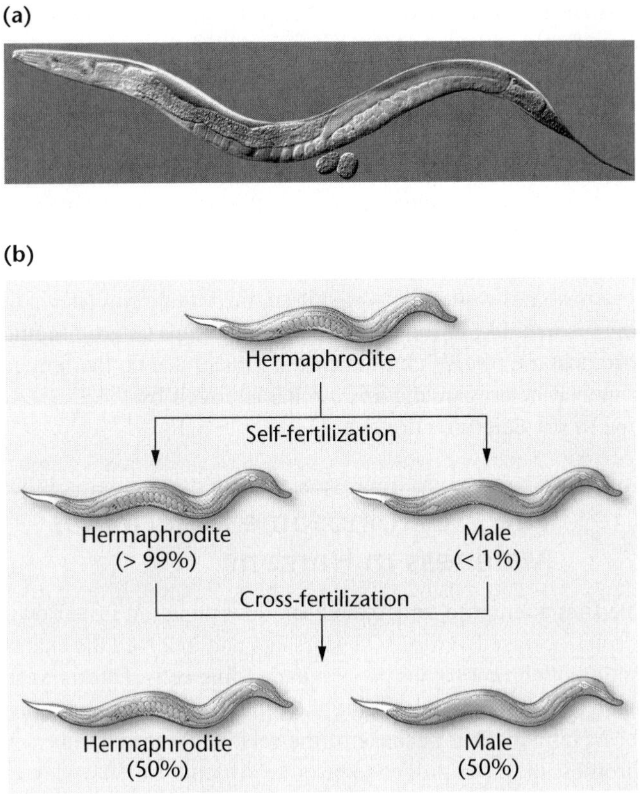

FIGURE 7–4 (a) Photomicrograph of a hermaphroditic nematode, *C. elegans*; (b) The outcomes of self-fertilization in a hermaphrodite and a mating of a hermaphrodite and a male worm.

The outcome of this process is quite interesting [Figure 7–4(b)]. The vast majority of organisms that result, like the parental worm, are hermaphrodites; less than 1 percent of the offspring are males. As adults, males can mate with hermaphrodites, producing about half male and half hermaphrodite offspring.

The genetic signal that determines maleness in contrast to hermaphroditic development is provided by genes located on both the X chromosome and autosomes. *C. elegans* lacks a Y chromosome altogether. Hermaphrodites have two X chromosomes, while males have only one X chromosome. It is believed that the ratio of X chromosomes to the number of sets of autosomes ultimately determines the sex of these worms. A ratio of 1.0 (two X chromosomes and two copies of each autosome) results in hermaphrodites and a ratio of 0.5 results in males. The absence of a heteromorphic Y chromosome is not uncommon in organisms.

Now solve this

Problem 7.22 on page 185 asks you to devise an experimental approach to elucidate the original findings regarding sex determination in a marine worm.

Hint: An obvious approach would be to attempt to isolate and devise experiments with the unknown factor affecting sex determination. An alternative approach, more in tune with genetic analysis, could involve the study of mutations that alter the normal outcomes.

7.2 X and Y Chromosomes Were First Linked to Sex Determination Early in the 20th Century

How sex is determined has long intrigued geneticists. In 1891, H. Henking identified a nuclear structure in the sperm of certain insects, which he labeled the *X-body*. Several years later, Clarance McClung showed that some grasshopper sperm contain an unusual genetic structure, which he called a *heterochromosome*, but the rest lack this structure. He mistakenly associated its presence with the production of male progeny. In 1906, Edmund B. Wilson clarified the findings of Henking and McClung when he demonstrated that female somatic cells in the insect *Protenor* contain 14 chromosomes, including 2 X chromosomes. During oogenesis, an even reduction occurs, producing gametes with 7 chromosomes, including one X. Male somatic cells, on the other hand, contain only 13 chromosomes, including a single X chromosome. During spermatogenesis, gametes are produced containing either 6 chromosomes, without an X, or 7 chromosomes, one of which is an X. Fertilization by X-bearing sperm results in female offspring, and fertilization by X-deficient sperm results in male offspring [Figure 7–5(a)].

The presence or absence of the X chromosome in male gametes provides an efficient mechanism for sex determination in this species and produces a 1:1 sex ratio in the resulting offspring. The mechanism, now called the **XX/XO** or ***Protenor*** **mode of sex determination**, depends on the random distribution of the X chromosome into half of the male gametes during segregation.

Wilson also experimented with the hemipteran insect *Lygaeus turicus*, in which both sexes have 14 chromosomes. Twelve of these are autosomes. In addition, the females have 2 X chromosomes, while the males have only a single X and a smaller heterochromosome labeled the **Y chromosome**. Females in this species produce only gametes of the (6A + X) constitution, but males produce two types of gametes in equal proportions: (6A + X) and (6A + Y). Therefore, following random fertilization, equal numbers of male and female progeny will be produced with distinct chromosome complements. This mode of sex determination is called the ***Lygaeus*** or **XX/XY** type [Figure 7–5(b)].

In *Protenor* and *Lygaeus* insects, males produce unlike gametes. As a result, they are described as the **heterogametic sex**, and in effect, their gametes ultimately determine the sex of the progeny in those species. In such cases, the female, who has like sex chromosomes, is the **homogametic sex**, producing uniform gametes with regard to chromosome numbers and types.

The male is not always the heterogametic sex. In other organisms, the female produces unlike gametes, exhibiting either the *Protenor* (XX/XO) or *Lygaeus* (XX/XY) mode of sex determination. Examples include moths and butterflies, most birds, some fish, reptiles, amphibians, and at least one species of plants (*Fragaria orientalis*). To immediately distinguish situations in which the female is the heterogametic sex, some

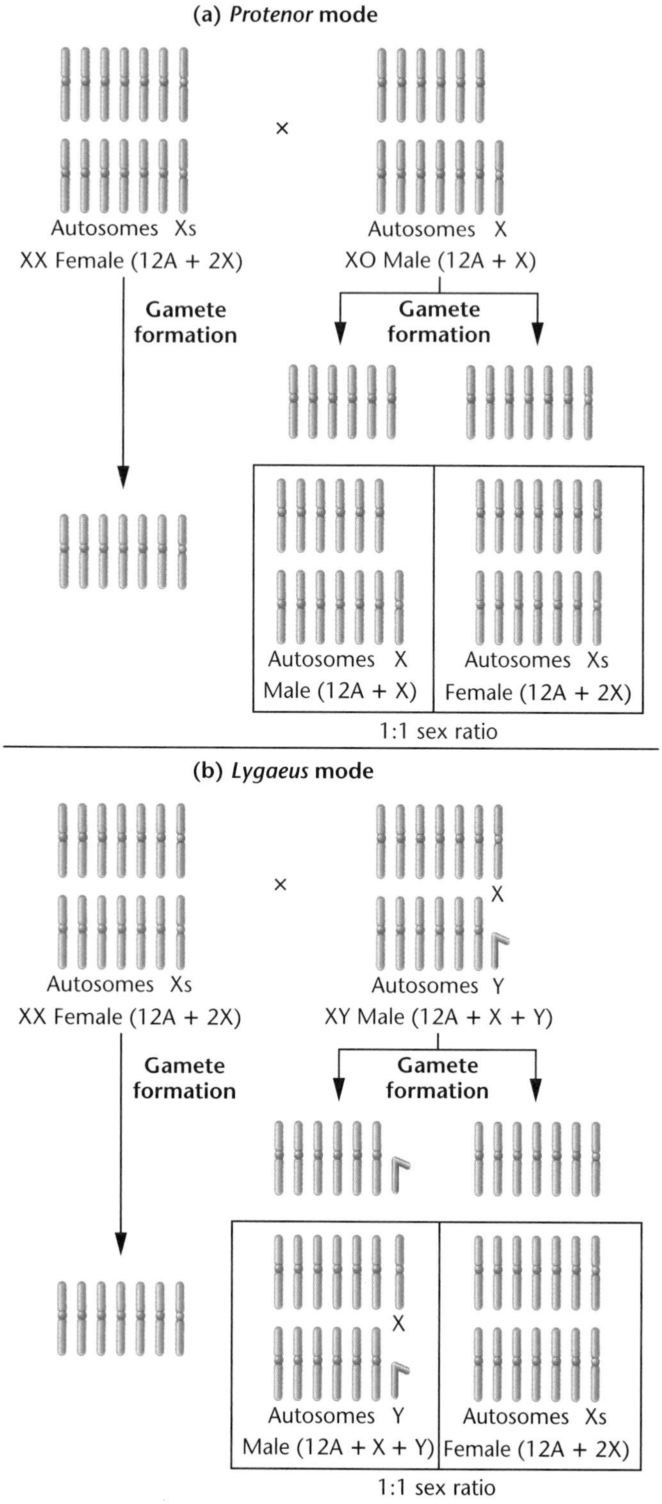

FIGURE 7–5 (a) The *Protenor* mode of sex determination where the heterogametic sex (the male in this example) is XO and produces gametes with or without the X chromosome; (b) The *Lygaeus* mode of sex determination, where the heterogametic sex (again, the male in this example) is XY and produces gametes with either an X or a Y chromosome. In both cases, the chromosome composition of the offspring determines its sex.

workers use the notation **ZZ/ZW**, where ZW is the heterogamous female, instead of the XX/XY notation.

The situation with fowl (chickens) illustrates the difficulty in establishing which sex is heterogametic and whether the *Protenor* or *Lygaeus* mode is operable. While genetic evidence supported the hypothesis that the female is the heterogametic sex, the cytological identification of the sex chromosome was not accomplished until 1961, because of the large number of chromosomes (78) characteristic of chickens. When the sex chromosomes were finally identified, the female was shown to contain an unlike chromosome pair, including a heteromorphic chromosome (the W chromosome). Thus, in fowl, the female is indeed heterogametic and is characterized by the *Lygaeus* type of sex determination.

7.3 The Y Chromosome Determines Maleness in Humans

The first attempt to understand sex determination in our own species occurred almost 100 years ago and involved the examination of chromosomes present in dividing cells. Efforts were made to accurately determine the diploid chromosome number of humans, but because of the relatively large number of chromosomes, this proved to be quite difficult. In 1912, H. von Winiwarter counted 47 chromosomes in a spermatogonial metaphase preparation. It was believed that the sex-determining mechanism in humans was based on the presence of an extra chromosome in females, who were thought to have 48 chromosomes. However, in the 1920s, Theophilus Painter observed between 45 and 48 chromosomes in cells of testicular tissue and also discovered the small Y chromosome, which is now known to occur only in males. In his original paper, Painter favored 46 as the diploid number in humans, but he later concluded incorrectly that 48 was the chromosome number in both males and females.

For 30 years, this number was accepted. Then, in 1956, Joe Hin Tjio and Albert Levan discovered a better way to prepare chromosomes. This improved technique led to a strikingly clear demonstration of metaphase stages showing that 46 was indeed the human diploid number. Later that same year, C. E. Ford and John L. Hamerton, also working with testicular tissue, confirmed this finding. The familiar karyotype of humans (Figure 7–6) is based on Tjio and Levan's technique.

Within the normal 23 pairs of human chromosomes, one pair was shown to vary in configuration in males and females. These two chromosomes were designated the X and Y sex chromosomes. The human female has two X chromosomes, and the human male has one X and one Y chromosome.

We might believe that this observation is sufficient to conclude that the Y chromosome determines maleness. However, several other interpretations are possible. The Y could play no role in sex determination; the presence of two X chromosomes could cause femaleness; or maleness could result from the lack of a second X chromosome. The evidence that clarified which explanation was correct awaited the study of variations in the human sex chromosome composition. As such investigations revealed, the Y chromosome does indeed determine maleness in humans.

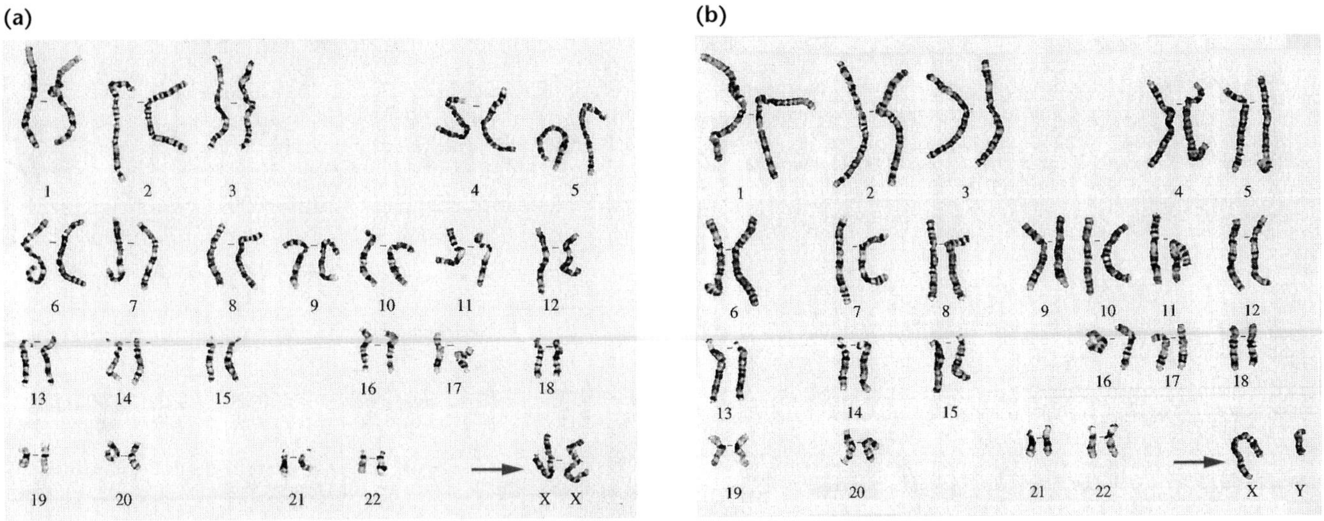

FIGURE 7–6 The traditional human karyotypes derived from a normal female and a normal male. Each contains 22 pairs of autosomes and two sex chromosomes. The female (a) contains two X chromosomes, while the male (b) contains one X and one Y chromosome (see arrows).

Klinefelter and Turner Syndromes

About 1940, scientists identified two human abnormalities characterized by aberrant sexual development, **Klinefelter syndrome** and **Turner syndrome**.* Individuals with Klinefelter syndrome have genitalia and internal ducts that are usually male, but their testes are rudimentary and fail to produce sperm. They are generally tall and have long arms and legs and large hands and feet.

Although some masculine development does occur, feminine sexual development is not entirely suppressed. Slight enlargement of the breasts (gynecomastia) is common, and the hips are often rounded. This ambiguous sexual development, referred to as intersexuality, may lead to abnormal social development. Intelligence is often below the normal range.

In Turner syndrome, the affected individual has female external genitalia and internal ducts, but the ovaries are rudimentary. Other characteristic abnormalities include short stature (usually under 5 feet), skin flaps on the back of the neck, and underdeveloped breasts. A broad, shieldlike chest is sometimes noted. Intelligence is often normal.

In 1959, the karyotypes of individuals with these syndromes were determined to be abnormal with respect to the sex chromosomes. Individuals with Klinefelter syndrome have more than one X chromosome. Most often they have an XXY complement in addition to 44 autosomes [Figure 7–7(a)]. People with this karyotype are designated **47,XXY**. Individuals with Turner syndrome are most often monosomic and have only 45 chromosomes, including just a single X chromosome. They are designated **45,X** [Figure 7–7(b)]. Note the convention used in designating the above chromosome compositions. The number indicates the total number of chromosomes present, and the information after the comma designates the relevant deviation from the normal diploid content. Both conditions result from nondisjunction, the failure of the X chromosomes to segregate properly during meiosis. (See Figures 2–17.)

These Klinefelter and Turner karyotypes and their corresponding sexual phenotypes allow us to conclude that the Y chromosome determines maleness in humans. In its absence, the sex of the individual is female, even if only a single X chromosome is present. The presence of the Y chromosome in the individual with Klinefelter syndrome is sufficient to determine maleness, even though male development is not complete. Similarly, in the absence of a Y chromosome, as in the case of individuals with Turner syndrome, no masculinization occurs.

Klinefelter syndrome occurs in about 2 of every 1000 male births. The karyotypes **48,XXXY**, **48,XXYY**, **49,XXXXY**, and **49,XXXYY** are similar phenotypically to 47,XXY, but manifestations are often more severe in individuals with a greater number of X chromosomes.

Turner syndrome can also result from karyotypes other than 45,X, including individuals called **mosaics** whose somatic cells display two different genetic cell lines, each exhibiting a different karyotype. Such cell lines result from a mitotic error during early development, the most common chromosome combinations being **45,X/46,XY** and **45,X/46,XX**. Thus, an embryo that began life with a normal karyotype can give rise to an individual whose cells show a mixture of karyotypes and who exhibits this syndrome.

Turner syndrome is observed in about 1 in 2000 female births, a frequency much lower than that for Klinefelter syndrome. One explanation for this difference is the observation that a substantial majority of 45,X fetuses die *in utero* and are aborted spontaneously. Thus, a similar frequency of the two syndromes may occur at conception.

---

*Although the possessive form of the names of most syndromes (eponyms) is sometimes used (e.g., Klinefelter's), the current preference is to use the nonpossessive form, which we have adopted for all human syndromes.

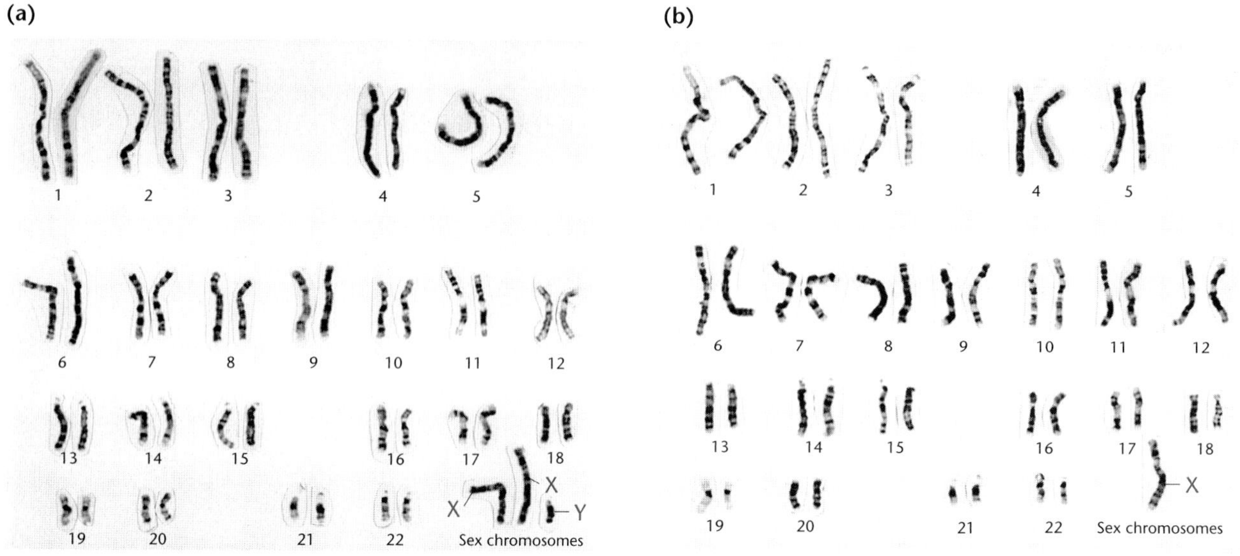

FIGURE 7–7 The karyotypes and phenotypic depictions of individuals with (a) Klinefelter syndrome (47,XXY) and (b) Turner syndrome (45,X).

How Do We Know?

What key information concerning the sex chromosome composition of Klinefelter and Turner syndrome individuals proves that X chromosomes play no role in human sex determination, while showing that the Y chromosome causes maleness and its absence causes femaleness in humans?

47,XXX Syndrome

The presence of three X chromosomes along with a normal set of autosomes (**47,XXX**) results in female differentiation. This syndrome, which is estimated to occur in about 1 of 1200 female births, is highly variable in expression. Frequently, 47,XXX women are perfectly normal. In other cases, underdeveloped secondary sex characteristics, sterility, and mental retardation may occur. In rare instances, **48,XXXX** and **49,XXXXX** karyotypes have been reported. The syndromes associated with these karyotypes are similar to but more pronounced than the 47,XXX. Thus, in many cases, the presence of additional X chromosomes appears to disrupt the delicate balance of genetic information essential to normal female development.

47,XYY Condition

Another human condition involving the sex chromosomes, **47,XYY**, has also been intensively investigated. Studies of this condition, where the only deviation from diploidy is the presence of an additional Y chromosome in an otherwise normal male karyotype, have led to an interesting controversy.

In 1965, Patricia Jacobs discovered 9 of 315 males in a Scottish maximum security prison to have the 47,XYY karyotype. These males were significantly above average in height and had been incarcerated as a result of antisocial (nonviolent) criminal acts. Of the nine males studied, seven were of subnormal intelligence, and all suffered personality disorders. Several other studies produced similar findings. The possible correlation between this chromosome composition and criminal behavior piqued considerable interest and extensive investigations of the phenotype and frequency of the 47,XYY condition in both criminal and noncriminal populations ensued. Above-average height (usually over 6 feet) and subnormal intelligence have been generally substantiated, and the frequency of males displaying this karyotype is indeed higher in penal and mental institutions compared with unincarcerated males. (See Table 7.1.) A particularly relevant question involves the characteristics displayed by XYY males who are not incarcerated. The only nearly constant association is that such individuals are over 6 feet tall!

A study addressing this issue was initiated to identify 47,XYY individuals at birth and to follow their behavioral patterns during preadult and adult development. By 1994, the two investigators, Stanley Walzer and Park Gerald, had identified about 20 XYY newborns in 15,000 births at Boston Hospital for Women. However, they soon came under great pressure to abandon their research. Those opposed to the study argued that the investigation could not be justified and might cause great harm to those individuals who displayed this karyotype. The opponents argued that (1) no association between the additional Y chromosome and abnormal behavior had been previously established in the population at large, and (2) "labeling" these individuals in the study might create a self-fulfilling prophecy. That is, as a result of participation in the study, parents, relatives, and friends might treat individuals identified as 47,XYY differently, ultimately producing the expected antisocial behavior. Despite the support of a government funding agency and the faculty at Harvard Medical School, Walzer and Gerald abandoned the investigation in 1995.

Since Walzer and Gerald's work, it has become apparent that many XYY males are present in the population who do not exhibit antisocial behavior and who lead normal lives. Therefore,

TABLE 7.1 FREQUENCY OF XYY INDIVIDUALS IN VARIOUS SETTINGS

Setting	Restriction	Number Studied	Number XYY	Frequency XYY
Control population	Newborns	28,366	29	0.10%
Mental–penal	No height restriction	4,239	82	1.93
Penal	No height restriction	5,805	26	0.44
Mental	No height restriction	2,562	8	0.31
Mental–penal	Height restriction	1,048	48	4.61
Penal	Height restriction	1,683	31	1.84
Mental	Height restriction	649	9	1.38

Source: Compiled from data presented in Hook, 1973, Tables 1–8. Copyright 1973 by the American Association for the Advancement of Science.

we must conclude that there is no consistent correlation between the extra Y chromosome and the predisposition of males to behavioral problems.

Sexual Differentiation in Humans

Once researchers had established that, in humans, it is the Y chromosome that houses genetic information necessary for maleness, they made efforts to pinpoint a specific gene or genes capable of providing the "signal" responsible for sex determination. Before we delve into this topic, it is useful to consider how sexual differentiation occurs in order to better comprehend how humans develop into sexually dimorphic males and females. During early development, every human embryo undergoes a period when it is potentially hermaphroditic. By the fifth week of gestation, gonadal primordia (the tissue that will form the gonad) arise as a pair of ridges associated with each embryonic kidney. Primordial germ cells migrate to these ridges, where an outer cortex and inner medulla form. The **cortex** is capable of developing into an ovary, while the inner **medulla** may develop into a testis. In addition, two sets of undifferentiated male (Wolffian) and female (Mullerian) ducts exist in each embryo.

If the cells of the genital ridge have the XY constitution, development of the medullary region into a testis is initiated around the seventh week. However, in the absence of the Y chromosome, no male development occurs, and the cortex of the genital ridge subsequently forms ovarian tissue. Parallel development of the appropriate male or female duct system then occurs, and the other duct system degenerates. A substantial amount of evidence indicates that in males, once testes differentiation is initiated, the embryonic testicular tissue secretes two hormones that are essential for continued male sexual differentiation.

In the absence of male development, as the 12th week of fetal development approaches, the oogonia within the ovaries begin meiosis and primary oocytes can be detected. By the 25th week of gestation, all oocytes become arrested in meiosis and remain dormant until puberty is reached some 10 to 15 years later. In males, on the other hand, primary spermatocytes are not produced until puberty is reached.

The Y Chromosome and Male Development

The human Y chromosome, unlike the X, has long been thought to be mostly blank genetically. It is now known that this is not true, even though the Y chromosome contains far fewer genes than does the X. Current analysis has revealed numerous genes and regions with potential genetic function, some with and some without homologous counterparts on the X chromosome. For example, present on both ends of the Y chromosome are the so-called **pseudoautosomal regions (PARs)** that share homology with regions on the X chromosome and which synapse and recombine with it during meiosis. The presence of such a pairing region is critical to segregation of the X and Y chromosomes during male gametogenesis. The remainder of the chromosome, about 95 percent of it, does not synapse or recombine with the X chromosome. As a result, it was originally referred to as the *nonrecombining region of the Y* (*NRY*). More recently, researchers have designated this region as the **male-specific region of the Y (MSY)**. As you will see, some portions of the MSY share homology with genes on the X chromosome, and some do not.

The human Y chromosome is diagrammed in Figure 7–8. The MSY is divided about equally between *euchromatic* regions containing functional genes and *heterochromatic* regions lacking genes. Within euchromatin, adjacent to the PAR of the short arm of the Y chromosome, is a critical gene that controls male sexual development, called the ***sex-determining region Y* (*SRY*)**. In humans, the absence of a Y chromosome almost always leads to female development, thus this gene is absent from the X chromosome. *SRY* encodes a gene product that somehow triggers the undifferentiated gonadal tissue of the embryo to form testes. This product is called the **testis-determining factor (TDF)**. *SRY* (or a closely related version) is present in all mammals thus far examined, which is indicative of its essential function throughout this diverse group of animals.

Our ability to identify the presence or absence of DNA sequences in rare individuals whose expected sex chromosome composition does not correspond to their sexual phenotype has provided evidence that *SRY* is the gene responsible for male sex determination. For example, there are human males who have two X and no Y chromosomes. Often, attached to one of their X

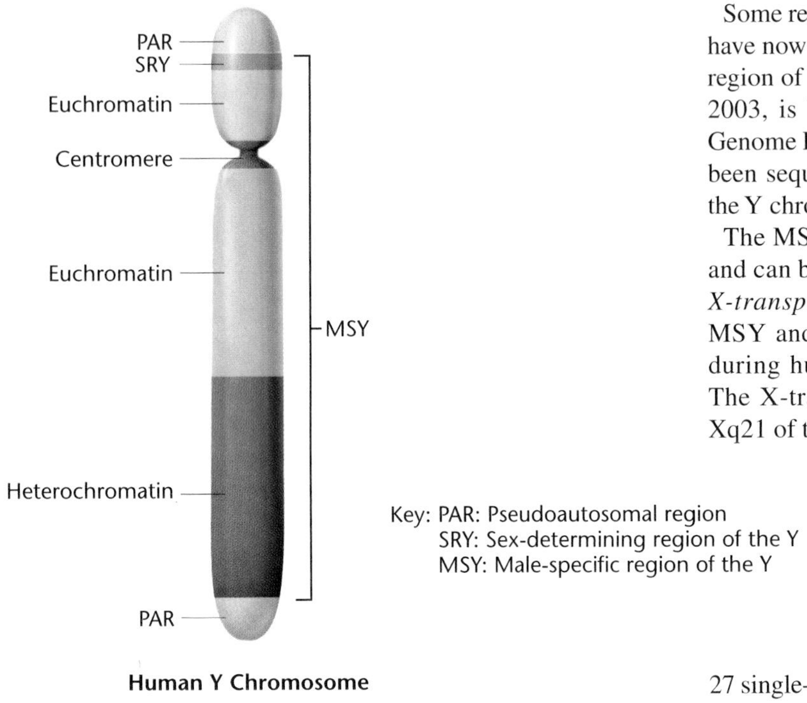

FIGURE 7–8 The various regions of the human Y chromosome.

chromosomes is the region of the Y that contains *SRY*. There are also females who have one X and one Y chromosome. Their Y is almost always missing the *SRY* gene. These observations argue strongly in favor of the role of *SRY* in providing the primary signal for male development.

Further support of this conclusion involves an experiment using **transgenic mice**. These animals are produced from fertilized eggs injected with foreign DNA that is subsequently incorporated into the genetic composition of the developing embryo. In normal mice, a chromosome region designated *Sry* has been identified that is comparable to *SRY* in humans. When mouse DNA containing *Sry* is injected into normal XX mouse eggs, most of the offspring develop into males.

The question of how the product of this gene triggers the embryonic gonadal tissue to develop into testes rather than ovaries is under intensive investigation. In humans, other autosomal genes are believed to be part of a cascade of genetic expression initiated by *SRY*. Examples include the *SOX9* gene and *WT1* (on chromosome 11), originally identified as an oncogene associated with Wilms tumor, which affects the kidney and gonads. Another, *SF1*, is involved in the regulation of enzymes affecting steroid metabolism. In mice, this gene is initially active in both the male and female bisexual genital ridge, persisting until the point in development when testis formation is apparent. At that time, its expression persists in males, but is extinguished in females. The link between these various genes and sex determination brings us closer to a complete understanding of how males and females arise in humans.

Some recent findings by David Page and his many colleagues have now provided a reasonably complete picture of the MSY region of the human Y chromosome. This work, completed in 2003, is based on information gained through the Human Genome Project, where the DNA of all chromosomes has now been sequenced. Page has spearheaded the detailed study of the Y chromosome for the past several decades.

The MSY consists of about 23 million base pairs (23 Mb) and can be divided into three regions. The first region is the *X-transposed region*. It contains about 15 percent of the MSY and was originally derived from the X chromosome during human evolution (about 3 to 4 million years ago). The X-transposed region is 99 percent identical to region Xq21 of the modern human X chromosome. Two genes, both with X-chromosome homologs, are present in this region.

The second area is designated the *X-degenerative region*. Containing about 20 percent of the MSY, this region contains DNA sequences that are even more distantly related to those present on the X chromosome. The X-degenerative region contains 27 single-copy genes, including numerous *pseudogenes*, whose sequences have degenerated sufficiently during evolution to render them nonfunctional. As with the genes present in the X-transposed region, all share some homology with counterparts on the X chromosome. These 27 genetic units include 14 that are capable of being transcribed, and each is present as a single copy. One of these is the *SRY* gene, discussed above. Other X-degenerative genes that encode protein products are expressed ubiquitously in all tissues in the body, but *SRY* is expressed only in the testes.

The third area, the *ampliconic region*, contains about 30 percent of the MSY, including most of the genes closely associated with testes development. These genes lack counterparts on the X chromosome and their expression is limited to the testes. There are 60 transcription units divided among 9 gene families in this region, most represented by multiple copies. Members of each family have nearly identical (>98 percent) DNA sequences. Each repeat unit is an **amplicon** and is contained within seven segments scattered across the euchromatic regions present on both the short and long arms of the Y chromosome. Genes in the ampliconic region encode proteins specific to the development and function of the testes, and the products of many of these genes are directly related to fertility in males. It is currently believed that a great deal of male sterility in our population can be linked to mutations in these genes.

This recent work has provided a comprehensive picture of the genetic information present on this unique chromosome. This information clearly refutes the so-called "wasteland" theory, prevalent only 20 years ago, that depicted the human Y chromosome as almost devoid of genetic information other than a gene or two that caused maleness. The knowledge we have gained provides the basis for a much clearer picture of how maleness is determined. Additionally, this information provides important clues as to the origin of the Y chromosome during human evolution.

Now solve this

Problem 7.28 on page 185 concerns itself with *SOX9*, a gene on an autosome, that when mutated appears to inhibit normal human male development.

Hint: Some genes are activated and produce their normal product as a result of expression of products of other genes found on different chromosomes—in this case, perhaps one that is on the Y chromosome.

7.4 The Ratio of Males to Females in Humans Is Not 1.0

The presence of heteromorphic sex chromosomes in one sex of a species but not the other provides a potential mechanism for producing equal proportions of male and female offspring. This potential is premised on the segregation of the X and Y (or Z and W) chromosomes during meiosis, such that half of the gametes of the heterogametic sex receive one of the chromosomes and half receive the other one. As we learned in the previous section, in humans, small pseudoautosomal regions of pairing homology do exist at both ends of the X and the Y chromosomes. Provided that both types of gametes are equally successful in fertilization and that the two sexes are equally viable during development, a one-to-one ratio of male and female offspring results.

Given the potential for the production of equal numbers of both sexes, the actual proportion of male to female offspring has been investigated and is referred to as the **sex ratio**. We can assess it in two ways. The **primary sex ratio** reflects the proportion of males to females conceived in a population. The **secondary sex ratio** reflects the proportion of each sex that is born. The secondary sex ratio is much easier to determine, but has the disadvantage of not accounting for any disproportionate embryonic or fetal mortality.

When the secondary sex ratio in the human population was determined in 1969 by using worldwide census data, it was found not to equal 1.0. For example, in the Caucasian population in the United States, the secondary ratio was a little less than 1.06, indicating that about 106 males were born for each 100 females. (In 1995, this ratio dropped to slightly less than 1.05.) In the African-American population in the United States, the ratio was 1.025. In other countries the excess of male births is even greater than reflected in these values. For example, in Korea, the secondary sex ratio was 1.15.

Despite these ratios, it is possible that the *primary sex ratio* is 1.0, and that it is altered between conception and birth. For the secondary ratio to exceed 1.0, then, prenatal female mortality would have to be greater than prenatal male mortality. However, this hypothesis has been examined and shown to be false. In fact, just the opposite occurs. In a Carnegie Institute study, reported in 1948, the sex of approximately 6000 embryos and fetuses recovered from miscarriages and abortions was determined, and fetal mortality was actually higher in males. On the basis of the data derived from that study, the primary sex ratio in U.S. Caucasians was estimated to be 1.079. It is now believed that this figure is much higher—between 1.20 and 1.60, suggesting that many more males than females are conceived in the human population.

It is not clear why such a radical departure from the expected primary sex ratio of 1.0 occurs. To come up with a suitable explanation, we must examine the assumptions upon which the theoretical ratio is based:

1. Because of segregation, males produce equal numbers of X- and Y-bearing sperm.
2. Each type of sperm has equivalent viability and motility in the female reproductive tract.
3. The egg surface is equally receptive to both X- and Y-bearing sperm.

While no direct experimental evidence contradicts any of these assumptions, the human Y chromosome is smaller than the X chromosome and therefore of less mass. Thus, it has been speculated that Y-bearing sperm are more motile than X-bearing sperm. If this is true, then the probability of a fertilization event leading to a male zygote is increased, providing one possible explanation for the observed primary ratio.

HOW DO WE KNOW?

What key experimental observations demonstrate that the primary sex ratio in humans strongly favors males at conception?

7.5 Dosage Compensation Prevents Excessive Expression of X-Linked Genes in Humans and Other Mammals

The presence of two X chromosomes in normal human females and only one X in normal human males is unique compared with the equal numbers of autosomes present in the cells of both sexes. On theoretical grounds alone, it is possible to speculate that this disparity should create a "genetic dosage" problem between males and females for all X-linked genes. Recall that in Chapter 4 we discussed the topic of X-linkage, the inheritance of traits under the control of genes located on one of the sex chromosomes. There, we saw that during meiosis, sex chromosomes, like autosomes, are subject to the laws of segregation and independent assortment during their distribution into gametes. Because females have two copies of the X chromosome and males only one, there is the potential for females to produce twice as much of each gene product for all X-linked genes. The additional X chromosomes in both males and females exhibiting the various syndromes discussed earlier in this chapter should compound this dosage problem even more. In this section, we will describe certain research findings regarding X-linked gene expression that demonstrate a genetic mechanism allowing for **dosage compensation**.

Barr Bodies

Murray L. Barr and Ewart G. Bertram's experiments with female cats, as well as Keith Moore and Barr's subsequent study with humans, demonstrate a genetic mechanism in mammals that compensates for X chromosome dosage disparities. Barr and Bertram observed a darkly staining body in interphase nerve cells of female cats that was absent in similar cells of males. In humans, this body can be easily demonstrated in female cells derived from the buccal mucosa (cheek cells) or in fibroblasts (undifferentiated connective tissue cells), but not in similar male cells (Figure 7–9). This highly condensed structure, about 1 μm in diameter, lies against the nuclear envelope of interphase cells. It stains positively in the Feulgen reaction, a cytochemical test for DNA.

Current experimental evidence demonstrates that this body, called a **sex chromatin body** or simply a **Barr body**, is an inactivated X chromosome. Susumo Ohno was the first to suggest that the Barr body arises from one of the two X chromosomes. This hypothesis is attractive because it provides a mechanism for dosage compensation. If one of the two X chromosomes is inactive in the cells of females, the dosage of genetic information that can be expressed in males and females is equivalent. Convincing but indirect evidence for this hypothesis comes from the study of the sex chromosome syndromes described earlier in this chapter. Regardless of how many X chromosomes exist, all but one of them appear to be inactivated and can be seen as Barr bodies. For example, no Barr body is seen in Turner 45,X females; one is seen in Klinefelter 47,XXY males; two in 47,XXX females; three in 48,XXXX females; and so on (Figure 7–10). Therefore, the number of Barr bodies follows an $N - 1$ rule where N is the total number of X chromosomes present.

Although this mechanism of inactivation of all but one X chromosome increases our understanding of dosage compensation, it further complicates our perception of other matters. Because one of the two X chromosomes is inactivated in normal human females, why then is the Turner 45,X individual not entirely normal? Why aren't females with the triplo-X and tetra-X karyotypes (47,XXX and 48,XXXX) completely unaffected by the additional X chromosome? Further, in Klinefelter syndrome (47,XXY), X chromosome inactivation effectively renders such individuals 46,XY. So why aren't these males unaffected by the extra X chromosome in their nuclei?

One possible explanation is that chromosome inactivation does not normally occur in the very early stages of development of those cells destined to form gonadal tissues. Another possible explanation is that not all of each X chromosome forming a Barr body is inactivated. If either hypothesis is correct, overexpression of certain X-linked genes might occur at critical times during development despite apparent inactivation of additional X chromosomes.

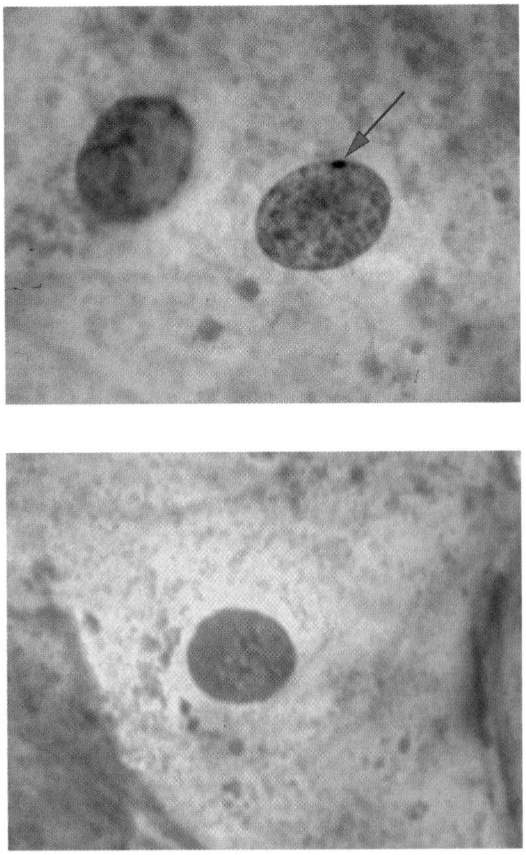

FIGURE 7–9 Photomicrographs comparing cheek epithelial cell nuclei from a male that fails to reveal Barr bodies (bottom) with a female that demonstrates Barr bodies (indicated by an arrow in the top image). This structure, also called a sex chromatin body, represents an inactivated X chromosome.

The Lyon Hypothesis

In mammalian females, one X chromosome is of maternal origin, and the other is of paternal origin. Which one is inactivated? Is the inactivation random? Is the same chromosome inactive in all somatic cells? In 1961, Mary Lyon and Liane Russell independently proposed a hypothesis that answers these questions. They postulated that the inactivation of X chromosomes occurs randomly in somatic cells at a point early in embryonic development. Further, once inacti-

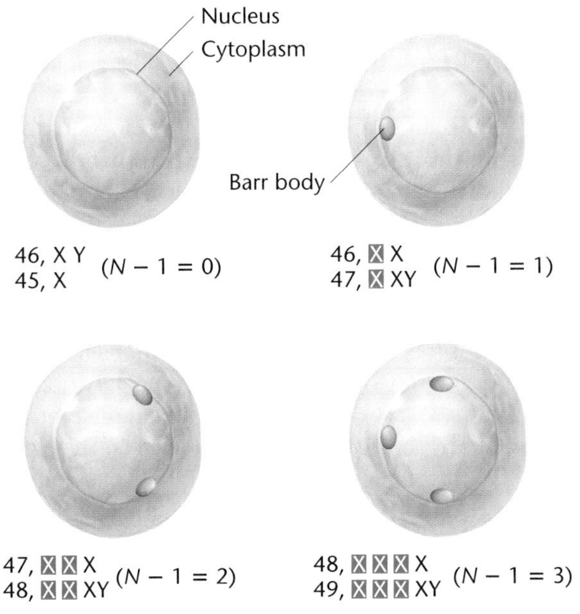

FIGURE 7–10 Barr body occurrence in various human karyotypes, where all X chromosomes except one ($N - 1$) are inactivated.

vation has occurred, all progeny cells have the same X chromosome inactivated.

This explanation, which has come to be called the **Lyon hypothesis**, was initially based on observations of female mice heterozygous for X-linked coat color genes. The pigmentation of these heterozygous females was mottled, with large patches expressing the color allele on one X and other patches expressing the allele on the other X. Indeed, such a phenotypic pattern would result if different X chromosomes were inactive in adjacent patches of cells. Similar mosaic patterns occur in the black and yellow-orange patches of female tortoiseshell and calico cats (Figure 7–11). Such X-linked coat color patterns do not occur in male cats because all their cells contain the single maternal X chromosome and are therefore hemizygous for only one X-linked coat color allele.

The most direct evidence in support of the Lyon hypothesis comes from studies of gene expression in clones of human fibroblast cells. Individual cells may be isolated following biopsy and cultured *in vitro*. If each culture is derived from a single cell, it is referred to as a **clone**. The synthesis of the enzyme **glucose-6-phosphate dehydrogenase (G6PD)** is controlled by an X-linked gene. Numerous mutant alleles of this gene have been detected, and their gene products can be differentiated from the wild-type enzyme by their migration pattern in an electrophoretic field.

Fibroblasts have been taken from females heterozygous for different allelic forms of G6PD and studied. The Lyon hypothesis predicts that if inactivation of an X chromosome occurs randomly early in development and is permanent in all progeny cells, such a female should show two types of clones, each showing only one electrophoretic form of G6PD, in approximately equal proportions.

In 1963, Ronald Davidson and colleagues performed an experiment involving 14 clones from a single heterozygous female. Seven showed only one form of the enzyme, and 7 showed only the other form. What was most important was that none of the 14 showed both forms of the enzyme. Studies of G6PD mutants thus provide strong support for the random permanent inactivation of either the maternal or paternal X chromosome.

How Do We Know?

What experimental evidence supports our belief that X chromosomal inactivation of either the paternal or maternal member is random in mammalian females?

The Lyon hypothesis is generally accepted as valid; in fact, the inactivation of an X chromosome into a Barr body is sometimes referred to as **lyonization**. One extension of the hypothesis is that mammalian females are mosaics for all heterozygous X-linked alleles—some areas of the body express only the maternally derived alleles, and others express only the paternally derived alleles. Two especially interesting examples involve **red-green color-blindness** and **anhidrotic ectodermal dysplasia**, both X-linked recessive disorders. In the former case, hemizygous males are fully color-blind in all retinal cells. However, heterozygous females display mosaic retinas with patches of defective color perception and surrounding areas with normal color perception. Males hemizygous for anhidrotic ectodermal dysplasia show absence of teeth, sparse hair growth, and lack of sweat glands. The skin of females heterozygous for this disorder reveals random patterns of tissue with and without sweat glands (Figure 7–12). In both examples, random inactivation of one or the other X chromosome early in the development of heterozygous females has led to these occurrences.

Now solve this

Problem 7.32 on page 186 is concerned with Carbon Copy, the first cloned cat, who was derived from a somatic nucleus of a calico cat.

Hint: The donor nucleus was from a differentiated ovarian cell of an adult female cat, which itself had inactivated one of its X chromosomes.

(a) (b)

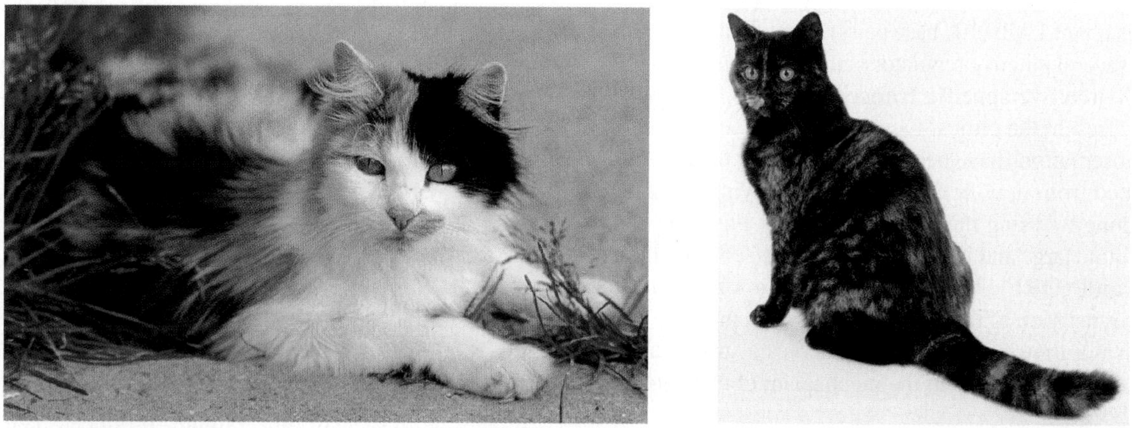

FIGURE 7–11 (a) A calico cat, where the random distribution of orange and black patches illustrates the Lyon hypothesis. The white patches are due to another gene; (b) A tortoiseshell cat, which lacks the white patches characterizing calicos.

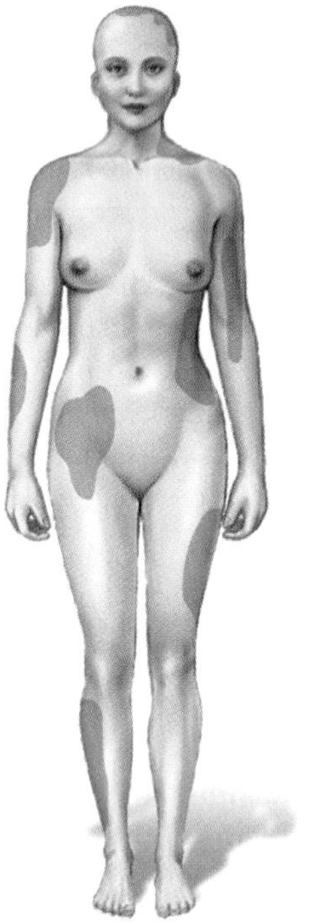

FIGURE 7–12 Depiction of the absence of sweat glands (shaded regions) in a female heterozygous for the X-linked condition anhidrotic ectodermal dysplasia. The locations vary from female to female, based on the random pattern of X chromosome inactivation during early development, resulting in unique mosaic distributions of sweat glands in heterozygotes.

The Mechanism of Inactivation

The least understood aspect of the Lyon hypothesis is the mechanism of chromosome inactivation in mammals. How are almost all genes of an entire chromosome inactivated? Recent investigations are beginning to clarify this issue. A single region of the mammalian X chromosome, called the **X-inactivation center (*Xic*)***, is the major control unit. Genetic expression of this region, located on the proximal end of the p arm, occurs only on the X chromosome that is inactivated. The constant association of expression of *Xic* and X chromosome inactivation supports the conclusion that this region is an important genetic component in the inactivation process.

The *Xic* is about 1 Mb (10^6 base pairs) in length and is known to contain several putative regulatory units and four genes. One of these, **X-inactive specific transcript (*Xist*)***, is now believed to represent the critical locus within the *Xic*. Several interesting observations have been made regarding the RNA that is transcribed from it, with much of the underlying work having been done by using the mouse *Xist* gene. First, the RNA product is quite large and lacks what is called an extended **open reading frame (ORF)**. An ORF includes the information necessary for translation of the RNA product into a protein. Thus, the RNA is not translated, but instead serves a structural role in the nucleus, presumably in the mechanism of chromosome inactivation. This finding has led to the belief that the RNA products of *Xist* spread over and coat the X chromosome bearing the gene that produced it, creating some sort of molecular "cage" that entraps it, leading to its inactivation. Inactivation is therefore said to be *cis*-acting.

Second, transcription of *Xist* occurs initially at low levels on all X chromosomes. As the inactivation process begins, however, transcription continues and is enhanced only on the X chromosome(s) that becomes inactivated.

In 1996, a research group led by Graeme Penny provided convincing evidence that transcription of *Xist* is the critical event in chromosome inactivation. These researchers were able to introduce a targeted deletion (7 kb) into this gene that destroyed its activity. As a result, the chromosome bearing the mutation lost its ability to become inactivated. Several interesting questions remain unanswered. First, in cells with more than two chromosomes, what sort of "counting" mechanism exists that designates all but one X chromosome to be inactivated? Second, what "blocks" the *Xic* of the active chromosome, preventing transcription of *Xist*? Third, how is inactivation of the same X chromosome or chromosomes maintained in progeny cells, as the Lyon hypothesis calls for? The inactivation signal must be stable as cells proceed through mitosis. Whatever the answers to these questions, we have taken an exciting step toward understanding how dosage compensation is accomplished in mammals.

7.6 The Ratio of X Chromosomes to Sets of Autosomes Determines Sex in *Drosophila*

Because males and females in *Drosophila melanogaster* (and other *Drosophila* species) have the same general sex chromosome composition as humans (males are XY and females are XX), we might assume that the Y chromosome also causes maleness in these flies. However, the elegant work of Calvin Bridges in 1916 showed this not to be true. He studied flies with quite varied chromosome compositions, leading him to the conclusion that the Y chromosome is not involved in sex determination in this organism. Instead, Bridges proposed that both the X chromosomes and autosomes together play a critical role in sex determination. Recall that in the nematode *C. elegans*, which lacks a Y chromosome, the sex chromosomes and autosomes are also critical to sex determination.

Bridges' work can be divided into two phases: (1) A study of offspring resulting from nondisjunction of the X chromosomes during meiosis in females and (2) subsequent work with progeny of females containing three copies of each chromosome, called triploid ($3n$) females. As we have seen previously in this chapter and earlier (see Figure 2–17), nondisjunction is the failure of paired chromosomes to segregate or separate during the anaphase stage of the first or second meiotic divisions. The result is the production of two types of abnormal gametes, one of which contains an extra chromosome ($n + 1$) and the other of which lacks a chromosome ($n - 1$). Fertilization of such gametes with a haploid gamete produces ($2n + 1$) or ($2n - 1$) zygotes. As in humans, if nondisjunction involves the X chromosome, in addition to the normal complement of

*On the human X chromosome, the inactivation center and the key gene are designated as *XIC* and *XIST*, respectively.

autosomes, both an XXY and an X0 sex chromosome composition may result. (The "0" signifies that there is neither a second X nor a Y chromosome present.) Contrary to what was later discovered in humans, Bridges found that the XXY flies were normal females, and the X0 flies were sterile males. The presence of the Y chromosome in the XXY flies did not cause maleness, and its absence in the X0 flies did not produce femaleness. From these data, he concluded that the Y chromosome in *Drosophila* lacks male-determining factors, but since the X0 males were sterile, it does contain genetic information essential to male fertility.

Bridges was able to clarify the mode of sex determination in *Drosophila* by studying the progeny of triploid females ($3n$), which have three copies each of the haploid complement of chromosomes. *Drosophila* has a haploid number of 4, thereby displaying three pairs of autosomes in addition to its pair of sex chromosomes. Triploid females apparently originate from rare diploid eggs fertilized by normal haploid sperm. Triploid females have heavy-set bodies, coarse bristles, and coarse eyes, and they may be fertile. Because of the odd number of each chromosome (3), during meiosis, a wide range of chromosome complements is distributed into gametes that give rise to offspring with a variety of abnormal chromosome constitutions. A correlation among the sexual morphology, chromosome composition, and Bridges' interpretation is shown in Figure 7–13.

Bridges realized that the critical factor in determining sex is the **ratio of X chromosomes to the number of haploid sets of autosomes (A)** present. Normal (2X:2A) and triploid (3X:3A) females each have a ratio equal to 1.0, and both are fertile. As the ratio exceeds unity (3X:2A, or 1.5, for example), what was originally called a superfemale is produced. Because this type of female is rather weak, infertile, and has lowered viability, it is now more appropriately called a **metafemale**.

Normal (XY:2A) and sterile (X0:2A) males each have a ratio of 1:2, or 0.5. When the ratio decreases to 1:3, or 0.33, as in the case of an XY:3A male, infertile **metamales** result. Other flies recovered by Bridges in these studies contained an X:A ratio intermediate between 0.5 and 1.0. These flies were generally larger, and they exhibited a variety of morphological abnormalities and rudimentary bisexual gonads and genitalia. They were invariably sterile and expressed both male and female morphology, thus being designated as **intersexes**.

Bridges' results indicate that in *Drosophila*, factors that cause a fly to develop into a male are not localized on the sex

FIGURE 7–13 Chromosome compositions, the ratios of X chromosomes to sets of autosomes, and the resultant sexual morphology in *Drosophila melanogaster*. The normal diploid male chromosome composition is shown as a reference on the left (XY/2A).

Normal diploid male

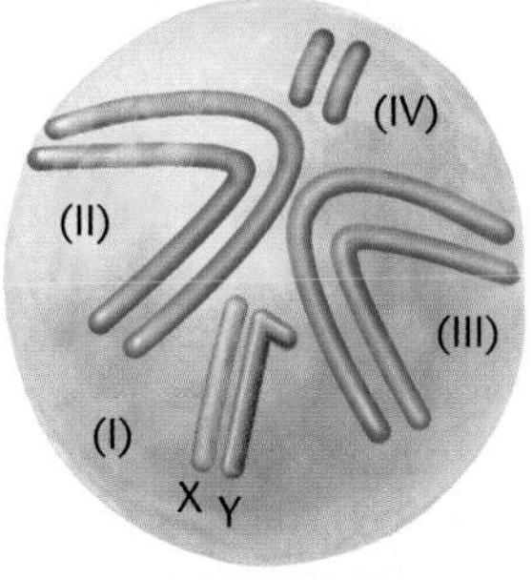

2 sets of autosomes
+
X Y

Chromosome composition	Chromosome formulation	Ratio of X chromosomes to autosome sets	Sexual morphology
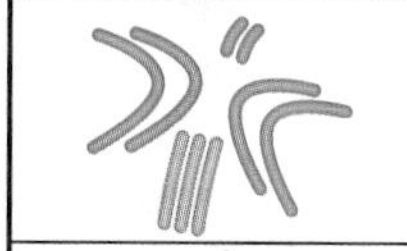	3X/2A	1.5	Metafemale
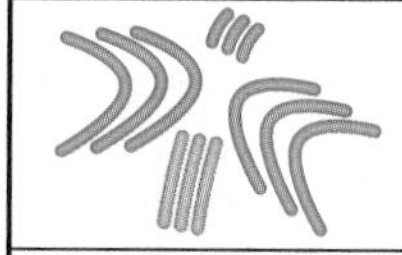	3X/3A	1.0	Female
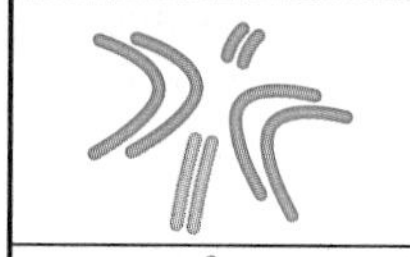	2X/2A	1.0	Female
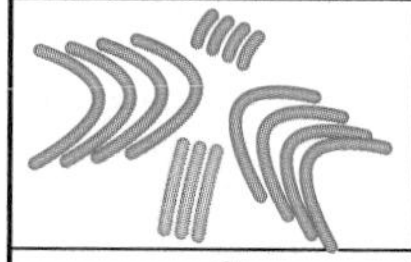	3X/4A	0.75	Intersex
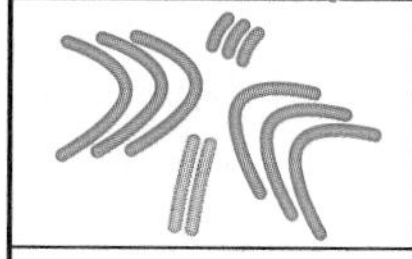	2X/3A	0.67	Intersex
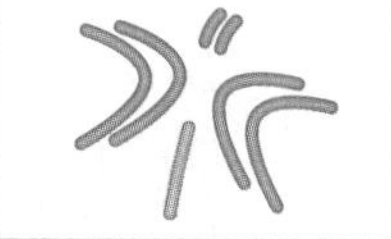	X/2A	0.50	Male
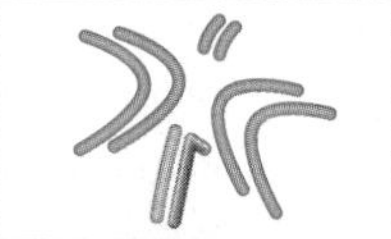	XY/2A	0.50	Male
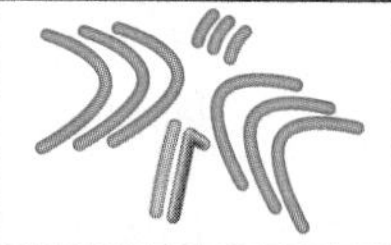	XY/3A	0.33	Metamale

chromosomes, but are instead found on the autosomes. Some female-determining factors, however, are localized on the X chromosomes. Thus, with respect to primary sex determination, male gametes containing one of each autosome plus a Y chromosome result in male offspring, not because of the presence of the Y, but because of the lack of a second X chromosome. This mode of sex determination is explained by the **genic balance theory**. Bridges proposed that a threshold for maleness is reached when the X:A ratio is 1:2 (X:2A), but that the presence of an additional X (XX:2A) alters the balance and results in female differentiation.

Numerous mutant genes have been identified that are involved in sex determination in *Drosophila*. The recessive autosomal gene *transformer* (*tra*), discovered over 50 years ago by Alfred H. Sturtevant, clearly demonstrated that a single autosomal gene could have a profound impact on sex determination. Females homozygous for *tra* are transformed into sterile males, but homozygous males are unaffected.

More recently, another gene, *Sex-lethal* (*Sxl*), has been shown to play a critical role, serving as a "master switch" in sex determination. Activation of the X-linked *Sxl* gene, which relies on a ratio of X chromosomes to sets of autosomes that equals 1.0, is essential to female development. In the absence of activation, resulting, for example, from an X:A ratio of 0.5, male development occurs. It is interesting to note that mutations that inactivate the *Sxl* gene, as originally studied in 1960 by Hermann J. Muller, kill female embryos, but have no effect on male embryos, consistent with the role of the gene, as described.

While it is not yet exactly clear how this ratio influences the *Sxl* locus, we do have some insights into the question. The *Sxl* locus is part of a hierarchy of gene expression and exerts control over still other genes, including *tra* (discussed in the previous paragraph) and *dsx* (*doublesex*), as well as others. Only in females is the wild-type allele of *tra* activated by the product of *Sxl* which, in turn, influences the expression of *dsx*. Depending on how the initial RNA transcript of *dsx* is processed (spliced), the resultant dsx protein activates either male- or female-specific genes required for sexual differentiation. Each step in this regulatory cascade requires a form of processing called **RNA splicing**, in which portions of the RNA are removed and the remaining fragments "spliced" back together prior to translation into a protein. In the case of the *Sxl* gene, its transcript may be spliced in several different ways, a phenomenon called **alternative splicing**. Two different RNA transcripts are produced in females and males. In potential females, the transcript is active and initiates a cascade of regulatory gene expression, ultimately leading to female differentiation. In potential males, the transcript is inactive, leading to a different pattern of gene activity, whereby male differentiation occurs. We will return to this topic in Chapter 17 when alternative splicing is again addressed as one of the mechanisms involved in the regulation of genetic expression in eukaryotes.

Dosage Compensation in *Drosophila*

Since *Drosophila* females contain two copies of X-linked genes whereas males contain only one copy, a dosage problem exists, as it does in mammals such as humans and mice. However, the mechanism of dosage compensation in *Drosophila* differs considerably from that in mammals, since X chromosome inactivation is not observed. Instead, male X-linked genes are transcribed at twice the level of the comparable genes in females. Interestingly, if groups of X-linked genes are moved (translocated) to autosomes, dosage compensation also affects them even when they are no longer part of the X chromosome.

As in mammals, considerable gains have been made recently in understanding the process of dosage compensation in *Drosophila*. At least four autosomal genes are known to be involved, under the same master-switch gene, *Sxl*, that induces female differentiation during sex determination. Mutations in any of these genes severely reduce the increased expression of X-linked genes in males, causing lethality.

Evidence supporting a mechanism of increased genetic activity in males is now available. The well-accepted model proposes that one of the autosomal genes, *mle* (*maleless*), encodes a protein that binds to numerous sites along the X chromosome, causing enhancement of genetic expression. The products of the other three autosomal genes also participate in and are required for *mle* binding.

This model predicts that the master-switch *Sxl* gene plays an important role during dosage compensation. In XY flies, *Sxl* is inactive; therefore, the autosomal genes are activated, causing enhanced X chromosome activity. On the other hand, *Sxl* is active in XX females and functions to inactivate one or more of the male-specific autosomal genes, perhaps *mle*. By dampening the activity of these autosomal genes, it ensures that they will not serve to double gene expression of X-linked genes in females, which would further compound the dosage problem.

Tom Cline has proposed that, before the aforementioned dosage compensation mechanism is activated, *Sxl* acts as a sensor for the expression of several other X-linked genes. In a sense, *Sxl* counts X chromosomes. When it registers the dose of their expression to be high, for example, as the result of two X chromosomes, the *Sxl* gene product is modified, and it dampens the expression of the autosomal genes. Although this model may yet be modified or refined, it is useful in guiding future research.

Clearly, an entirely different mechanism of dosage compensation exists in *Drosophila* (and probably many related organisms) than that in mammals. The development of elaborate mechanisms to equalize the expression of X-linked genes demonstrates the critical nature of gene expression. A delicate balance of gene products is necessary to maintain normal development of both males and females.

Drosophila Mosaics

Our knowledge of sex determination and X-linkage in *Drosophila* (Chapter 4) helps us to understand the unusual appearance of a unique fruit fly, shown in Figure 7–14. This fly was recovered from a stock where all other females were heterozygous for the X-linked genes *white* eye (*w*) and *miniature* wing (*m*). It is a **bilateral gynandromorph**, which means that one-half of its body (the left half) has developed as a male and the other half (the right half) as a female.

We can account for the presence of both sexes in a single fly in the following way. If a female zygote (heterozygous for *white*

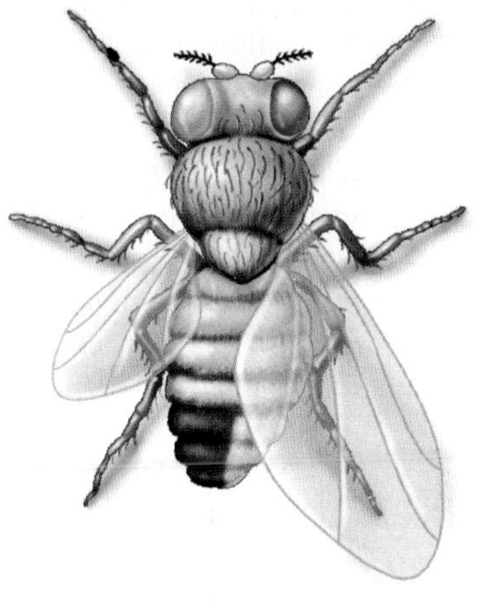

FIGURE 7–14 A bilateral gynandromorph of *Drosophila melanogaster* formed following the loss of one X chromosome in one of the two cells during the first mitotic division. The left side of the fly, composed of male cells containing a single X, expresses the mutant *white*-eye and *miniature*-wing alleles. The right side is composed of female cells containing two X chromosomes heterozygous for the two recessive alleles.

eye and *miniature* wing) were to lose one of its X chromosomes during the first mitotic division, the two cells would be of the XX and X0 constitution, respectively. Thus, one cell would be female and the other would be male. Each of these cells is responsible for producing all progeny cells that make up either the right half or the left half of the body during embryogenesis.

In the case of the bilateral gynandromorph, the original cell of X0 constitution apparently produced only identical progeny cells and gave rise to the left half of the fly, which, because of its chromosomal constitution, was male. Since the male half demonstrated the *white*, *miniature* phenotype, the X chromosome bearing the w^+, m^+ alleles was lost, while the *w*, *m*-bearing homolog was retained. All cells on the right side of the body were derived from the original XX cell, leading to female development. These cells, which remained heterozygous for both mutant genes, expressed the wild-type eye–wing phenotypes.

Depending on the orientation of the spindle during the first mitotic division, gynandromorphs can be produced where the "line" demarcating male versus female development occurs along or across any axis of the fly's body.

7.7 Temperature Variation Controls Sex Determination in Reptiles

We conclude this chapter by discussing several cases involving reptiles where the environment, specifically temperature, has a profound influence on sex determination. In contrast to **chromosomal** or **genotypic sex determination** (**CSD** or **GSD**), which describes all examples thus far presented, the cases that we will now discuss are categorized as **temperature-dependent sex determination** (**TSD**). As we shall see, the investigations leading to this information may well come closer to revealing the true nature of the primary basis of sex determination than any finding previously discussed.

In many species of reptiles, sex is predetermined at conception by sex chromosome composition, as is the case in many organisms already considered in this chapter. For example, in many snakes, including vipers, a ZZ/ZW mode is in effect, where the female is the heterogamous sex. However, in boas and pythons, it is impossible to distinguish one sex chromosome from the other in either sex. In lizards, both the XX/XY and ZZ/ZW systems are found, depending on the species. In other reptilian species, including all crocodiles, most turtles, and some lizards, sex determination is achieved according to the incubation temperature of eggs during a critical period of embryonic development. Since 1980, it has become clear that TSD is quite widespread among reptiles.

Interestingly, there are three distinct patterns of TSD, as illustrated in Figure 7–15. In the first two, low temperatures yield 100 percent females, while high temperatures yield 100 percent males (Case I), or just the opposite occurs (Case II). In the third pattern (Case III), low *and* high temperatures yield 100 percent females, while intermediate temperatures yield

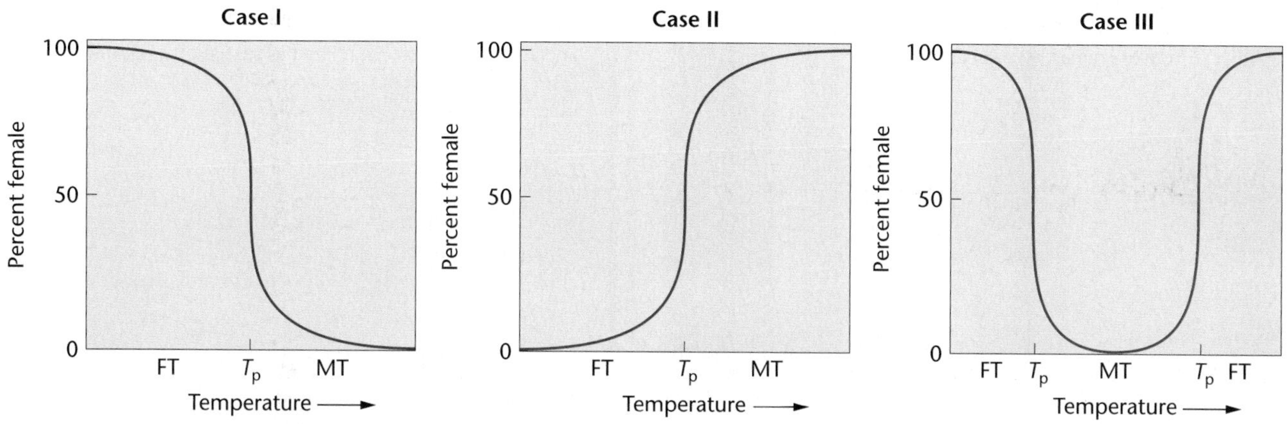

FIGURE 7–15 Three different patterns of temperature-dependent sex determination (TSD) in reptiles, as described in the text. The relative pivotal temperature T_P is crucial to sex determination during a critical point during embryonic development (FT = female-determining temperature; MT = male-determining temperature).

various proportions of males. The third pattern is seen in various species of crocodiles, turtles, and lizards, although other members of these groups are known to exhibit the other patterns. Two observations are noteworthy. First, under certain temperatures in all three patterns, both male and female offspring result. Secondly, the pivotal temperature (*P*) is fairly narrow, usually less than 5°C, and sometimes only 1°C.

The central question raised by these observations is what metabolic or physiological parameters that lead to the differentiation of one sex or the other are being affected by temperature? The answer is thought to involve steroids (mainly estrogens) and the enzymes involved in their synthesis. Studies have clearly demonstrated that the effects of temperature on estrogens, androgens, and inhibitors of the enzymes controlling their synthesis are involved in the sexual differentiation of ovaries and testes. One enzyme in particular, **aromatase**, converts androgens (male hormones such as testosterone) to estrogens (female hormones such as estradiol). The activity of this enzyme is correlated with the pathway followed during gonadal differentiation activity and is high in developing ovaries and low in developing testes. Researchers in this field, including Claude Pieau and colleagues, have proposed that a thermosensitive factor mediates the transcription of the reptilian aromatase gene which leads to temperature-dependent sex determination. Several other genes are likely to be involved in this mediation.

The involvement of sex steroids in gonadal differentiation has also been documented in birds, fishes, and amphibians. Thus, sex-determining mechanisms involving estrogens seem to be characteristic of nonmammalian vertebrates. The regulation of such a system, while temperature dependent in many reptiles, appears to be controlled by sex chromosomes (XX/XY or ZZ/ZW) in many of these other organisms. A final intriguing thought on this matter is that the product of *SRY*, a key component in mammalian sex determination, has been shown to bind *in vitro* to a regulatory portion of the aromatase gene, whereby it could act as a repressor of ovarian development.

Chapter Summary

1. In sexually reproducing organisms, meiosis, which both creates genetic variability and ensures genetic constancy, depends on fertilization. Fertilization ultimately relies on some form of sexual differentiation, which is achieved by a variety of sex-determining mechanisms.
2. The genetic basis of sexual differentiation is usually related to different chromosome compositions in the two sexes. The heterogametic sex either lacks one chromosome or contains a unique heteromorphic chromosome, usually referred to as the Y chromosome.
3. In humans, the study of individuals with altered sex chromosome compositions has established that the Y chromosome is responsible for male differentiation. The absence of the Y leads to female differentiation. Similar studies in *Drosophila* have excluded the Y in such a role, instead demonstrating that a balance between the number of X chromosomes and sets of autosomes is the critical factor.
4. The primary sex ratio in humans substantially favors males at conception. During embryonic and fetal development, male mortality is higher than that of females. The secondary sex ratio at birth still favors males by a small margin.
5. Dosage compensation mechanisms limit the expression of X-linked genes in females, who have two X chromosomes, as compared with males, who have only one X. In mammals, compensation is achieved by the inactivation of either the maternal or paternal X early in development. This process results in the formation of Barr bodies in female somatic cells. In *Drosophila*, compensation is achieved by stimulation of genes located on the single X chromosome to double their genetic expression.
6. The Lyon hypothesis states that, early in development, inactivation is random between the maternal and paternal X chromosomes. All subsequent progeny cells inactivate the same X as their progenitor cell. Mammalian females thus develop as genetic mosaics with respect to their expression of heterozygous X-linked alleles.
7. In many reptiles, the incubation temperature at a critical time during embryogenesis is often responsible for sex determination. Temperature influences the activity of enzymes involved in the metabolism of steroids related to sexual differentiation.

Genetics, Technology, and Society

A Question of Gender: Sex Selection in Humans

The desire to choose a baby's gender is as pervasive as human nature itself. Throughout time, people have resorted to varied and sometimes bizarre methods to achieve the preferred gender of their offspring. In medieval Europe, prospective parents would place a hammer under the bed to help them conceive a boy, or a pair of scissors to conceive a girl. Equally effective were practices based on the ancient belief that semen from the right testicle created male offspring and that from the left testicle created females. Men in ancient Greece would lie on their right side during intercourse in order to conceive a boy. Up until the 18th century, European men would tie off (or remove) their left testicle to increase the chances of getting a male heir.

In some cultures, efforts to control the sex of offspring has a darker side—female infanticide. In ancient Greece, the murder of female infants was so common that the male:female ratio in some areas approached 4 : 1. Some societies, even to present times, continue to practice female infanticide. In some parts of rural India, hundreds of families are reported to admitting to this practice, even as late as the 1990s. In 1997, the World Health Organization reported population data showing that about 50 million women were "missing" in China, likely due to institutionalized neglect of female children. The practice of female infanticide arises from poverty and age-old traditions. In these cultures, sons work and provide income and security, whereas daughters not only contribute no income but require large dowries when they marry. Under these conditions, it is easy to see why females are held in low esteem.

In recent times, sex-specific abortion has replaced much of the traditional female infanticide. Amniocentesis and ultrasound techniques have become lucrative businesses that provide prenatal sex determination. Studies in India estimate that hundreds of thousands of fetuses are aborted each year because they are female. As a result of sex-selective abortion, the female : male ratio in India was 927 : 1000 in 1991. In some Northern states, the ratio is as low as 600 : 1000. Although sex determination and selective abortion of female fetuses was outlawed in India and China in the mid-1990s, the practice is thought to continue.

In Western industrial countries, advances in genetics and reproductive technology offer parents ways to select their children's gender prior to implantation. Following *in vitro* fertilization, embryos can be biopsied and assessed for gender. Only sex-selected embryos are then implanted. The simplest and least invasive method for sex selection is preconception gender selection (PGS), which involves separating X- and Y-chromosome bearing spermatozoa. The only effective PGS method devised so far involves sorting the sperm based on their DNA content. Due to the different size of the X and Y chromosomes, X-bearing sperm contain 2.8–3.0 percent more DNA than Y-bearing sperm. Sperm samples are treated with a fluorescent DNA stain, then passed single file through a laser beam in a Fluorescence-Activated Cell Sorter (FACS) machine. The machine separates the sperm into two fractions based on the intensity of their DNA-fluorescence. Using this method, human sperm can be separated into X- and Y-chromosome fractions, with enrichments of about 85 percent and 75 percent, respectively. The sorted sperm are then used for standard intrauterine insemination. The Genetics and IVF Institute (Fairfax, Virginia) is presently using this PGS technique in an FDA-approved clinical trial. As of January 2002, 419 human pregnancies have resulted from the method. The company reports an approximately 80 to 90 percent success rate in producing the desired gender.

The emerging PGS methods raise a number of legal and ethical issues. Some people feel that prospective parents have the legal right to use sex-selection techniques as part of their fundamental procreative liberty. Others believe that this liberty does not extend to custom designing a child to the parents' specifications. Proponents state that the benefits far outweigh any dangers to offspring or society. The medical uses of PGS are a clear case for benefit. People at risk for transmitting X-linked diseases such as hemophilia or Duchenne muscular dystrophy can now enhance their chance of conceiving a female child who will not express the disease. As there are more than 500 known X-linked diseases and they are expressed in about 1 in 1000 live births, PGS could greatly reduce suffering for many families.

The greatest number of people undertake PGS for nonmedical reasons—i.e., to "balance" their families. It is possible that the ability to intentionally select the desired sex of an offspring may reduce overpopulation and economic burdens for families who would repeatedly reproduce to get the desired gender. In some cases, PGS may reduce the number of abortions of female fetuses. It is also possible that PGS may increase the happiness of both parents and children, as the child would be more "wanted." On the other hand, some argue that PGS serves neither the individual nor the common good. It is argued that PGS is inherently sexist, based on the concept of superiority of one sex over another, and leads to an increase in linking a child's worth to gender. Some fear that large-scale PGS will reinforce sex discrimination and lead to sex-ratio imbalances. Others feel that sexism and discrimination are not caused by sex ratios and would be better addressed through education and economic equality measures for men and women. The experience so far in Western countries suggest that sex-ratio imbalances would not result from PGS. More than half of couples in the United States who use PGS request female offspring. However, the consequences of widespread PGS in some Asian countries may be more problematic. Both India and China already have sex-ratio imbalances, which contribute to some socially undesirable side effects, such as millions of adult men being unable to marry.

Some critics of PGS argue that this technology may contribute to social and economic inequality, if it will be available only to those who can afford it. Other critics fear that social approval of PGS will open the door to accepting other genetic manipulations of children for characteristics such as skin or eye color. It is difficult to predict the full effects that PGS will bring to the world. But the gender-selection genie is now out of the bottle and will be unwilling to step back in.

References

Sills, E.S., Kirman, I., Thatcher, S.S. III, and Palermo, G.D. 1998. Sex-selection of human spermatozoa: Evolution of current techniques and applications. *Arch. Gynecol. Obstet.* 261:109–115.

Robertson, J.A. 2001. Preconception Gender Selection. *Am. J. Bioethics* 1:2–9.

Web Sites

Microsort technique, Genetics & IVF Institute, Fairfax, Virginia.

http://www.microsort.net

Female Infanticide, Gendercide Watch.

http://www.gendercide.org/case_infanticide.html

Insights and Solutions

1. In *Drosophila*, the X chromosomes may become attached to one another $(\widehat{XX})$ such that they always segregate together. Some flies contain both an attached X chromosome and a Y chromosome.

What sex would such a fly be? Explain why this is so.

Solution: The fly would be a female. The ratio of X chromosomes to sets of autosomes would be 1.0, leading to normal female development. The Y chromosome has no influence on sex determination in *Drosophila.*

Given this answer, predict the sex of the offspring that would occur in a cross between this fly and a normal one of the opposite sex.

Solution: All flies would have two sets of autosomes, but each would have one of the following sex chromosome compositions:

(1) $(\widehat{XX})$ X → a metafemale with 3 X's (called a trisomic)
(2) $(\widehat{XX})$ Y → a female like her mother
(3) XY → a normal male
(4) YY → no development occurs

If the offspring just mentioned were allowed to interbreed, what would be the outcome?

Solution: A stock would be created that would generate the attached X females generation after generation.

2. The *Xg* cell-surface antigen is coded for by a gene that is located on the X chromosome. There is no equivalent gene on the Y chromosome. Two codominant alleles of this gene have been identified: *Xg1* and *Xg2*. A woman of genotype *Xg2/Xg2* bears children with a man of genotype *Xg1/Y*, and they produce a son with Klinefelter syndrome of genotype *Xg1/Xg2Y*. Using proper genetic terminology, briefly explain how this individual was generated. In which parent and in which meiotic division did the mistake occur?

Solution: Because the son with Klinefelter syndrome is *Xg1/Xg2Y*, he must have received both the *Xg1* allele and the Y chromosome from his father. Therefore, nondisjunction must have occurred during meiosis I in the father.

Problems and Discussion Questions

1. As related to sex determination, what is meant by (a) homomorphic and heteromorphic chromosomes and (b) isogamous and heterogamous organisms?
2. Contrast the life cycle of a plant such as *Zea mays* with an animal such as *C. elegans.*
3. Discuss the role of sexual differentiation in the life cycles of *Chlamydomonas*, *Zea mays*, and *C. elegans*.
4. Distinguish between the concepts of sexual differentiation and sex determination.
5. Contrast the *Protenor* and *Lygaeus* modes of sex determination.
6. Describe the major difference between sex determination in *Drosophila* and in humans.
7. What specific observations (evidence) support the conclusions you have drawn about sex determination in *Drosophila* and humans?
8. Describe how nondisjunction in human female gametes can give rise to Klinefelter and Turner syndrome offspring following fertilization by a normal male gamete.
9. An insect species is discovered in which the heterogametic sex is unknown. An X-linked recessive mutation for *reduced wing* (*rw*) is discovered. Contrast the F_1 and F_2 generations from a cross between a female with reduced wings and a male with normal-sized wings when (a) the female is the heterogametic sex; (b) the male is the heterogametic sex.
10. Based on your answers in Problem 9, is it possible to distinguish between the *Protenor* and *Lygaeus* mode of sex determination based on the outcome of these crosses?
11. When cows have twin calves of unlike sex (fraternal twins), the female twin is usually sterile and has masculinized reproductive organs. This calf is referred to as a freemartin. In cows, twins may share a common placenta and thus fetal circulation. Predict why a freemartin develops.
12. An attached-X female fly, $\widehat{XX}Y$ (see the "Insights and Solutions" box), expresses the recessive X-linked *white* eye phenotype. It is crossed to a male fly that expresses the X-linked recessive miniature wing phenotype. Determine the outcome of this cross regarding the sex, eye color, and wing size of the offspring.
13. Assume that rarely, the attached X chromosomes in female gametes become unattached. Based on the parental phenotypes in Problem 12, what outcomes in the F_1 generation would indicate that this has occurred during female meiosis?
14. It has been suggested that any male-determining genes contained on the Y chromosome in humans cannot be located in the limited region that synapses with the X chromosome during meiosis. What might be the outcome if such genes were located in this region?
15. What is a Barr body, and where is it found in a cell?
16. Indicate the expected number of Barr bodies in interphase cells of the following individuals: Klinefelter syndrome; Turner syndrome; and karyotypes 47,XYY, 47,XXX, and 48,XXXX.
17. Define the Lyon hypothesis.
18. Can the Lyon hypothesis be tested in a human female who is homozygous for one allele of the X-linked *G6PD* gene? Why, or why not?
19. Predict the potential effect of the Lyon hypothesis on the retina of a human female heterozygous for the X-linked red-green color-blindness trait.
20. Cat breeders are aware that kittens expressing the X-linked calico coat pattern and tortoiseshell pattern are almost invariably females. Why?
21. What does the apparent need for dosage compensation mechanisms suggest about the expression of genetic information in normal diploid individuals?

22. The marine echiurid worm *Bonellia viridis* is an extreme example of the environment's influence on sex determination. Undifferentiated larvae either remain free-swimming and differentiate into females or they settle on the proboscis of an adult female and become males. If larvae that have been on a female proboscis for a short period are removed and placed in seawater, they develop as intersexes. If larvae are forced to develop in an aquarium where pieces of proboscises have been placed, they develop into males. Contrast this mode of sexual differentiation with that of mammals. Suggest further experimentation to elucidate the mechanism of sex determination in *B. viridis*.
23. How do we know that the primary sex ratio in humans is as high as 1.40 to 1.60?
24. Devise as many hypotheses as you can that might explain why so many more human male conceptions than human female conceptions occur.
25. In mice, the *Sry* gene (see Section 7.3) is located on the Y chromosome very close to one of the pseudoautosomal regions that pairs with the X chromosome during male meiosis. Given this information, propose a model to explain the generation of unusual males who have two X chromosomes (with an *Sry*-containing piece of the Y chromosome attached to one X chromosome).
26. The genes encoding the red and green color-detecting proteins of the human eye are located next to one another on the X chromosome and probably arose during evolution from a common ancestral pigment gene. The two proteins demonstrate 76 percent homology in their amino acid sequences. A normal-visioned woman with one copy of each gene on each of her two X chromosomes has a red color-blind son who was shown to contain one copy of the green-detecting gene and no copies of the red-detecting gene. Devise an explanation at the chromosomal level (during meiosis) that explains these observations.

Extra-Spicy Problems

27. The X-linked dominant mutation in the mouse, *Testicular feminization* (*Tfm*), eliminates the normal response to the testicular hormone testosterone during sexual differentiation. An XY mouse bearing the *Tfm* allele on the X chromosome develops testes, but no further male differentiation occurs—the external genitalia of such an animal are female. From this information, what might you conclude about the role of the *Tfm* gene product and the X and Y chromosomes in sex determination and sexual differentiation in mammals? Can you devise an experiment, assuming you can "genetically engineer" the chromosomes of mice, to test and confirm your explanation?
28. Campomelic dysplasia (CMD1) is a congenital human syndrome, featuring malformation of bone and cartilage. It is caused by an autosomal dominant mutation of a gene located on chromosome 17. Consider the following observations in sequence, and in each case, draw whatever appropriate conclusions are warranted.
    (a) Of those with the syndrome who are karyotypically 46,XY, approximately 75 percent are sex reversed, exhibiting a wide range of female characteristics.
    (b) The nonmutant form of the gene, called *SOX9*, is expressed in the developing gonad of the XY male, but not the XX female.
    (c) The *SOX9* gene shares 71 percent amino acid coding sequence homology with the Y-linked *SRY* gene.
    (d) CMD1 patients who exhibit a 46,XX karyotype develop as females, with no gonadal abnormalities.
29. In the wasp, *Bracon hebetor*, a form of parthenogenesis (where unfertilized eggs initiate development) resulting in haploid organisms is not uncommon. All haploids are males. When offspring arise from fertilization, females almost invariably result. P. W. Whiting has shown that an X-linked gene with nine multiple alleles (X_a, X_b, etc.) controls sex determination. Any homozygous or hemizygous condition results in males and any heterozygous condition results in females. If an X_a/X_b female mates with an X_a male and lays 50 percent fertilized and 50 percent unfertilized eggs, what proportion of male and female offspring will result?
30. Shown in the right column are two graphs that plot the percentage of males occurring against the atmospheric temperature during the early development of fertilized eggs in (a) snapping turtles and (b) most lizards. Interpret these data as they relate to the effect of temperature on sex determination.

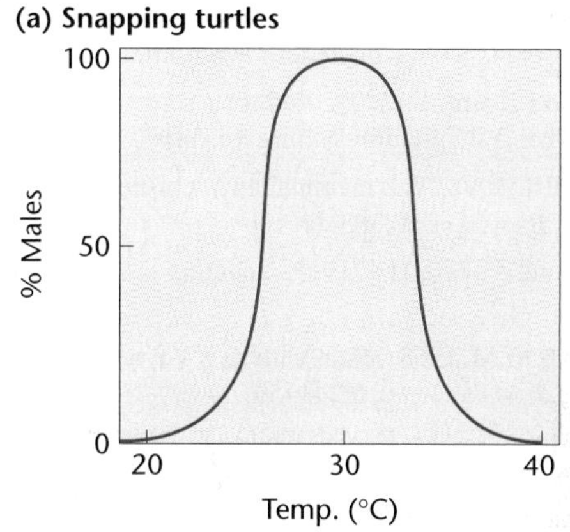

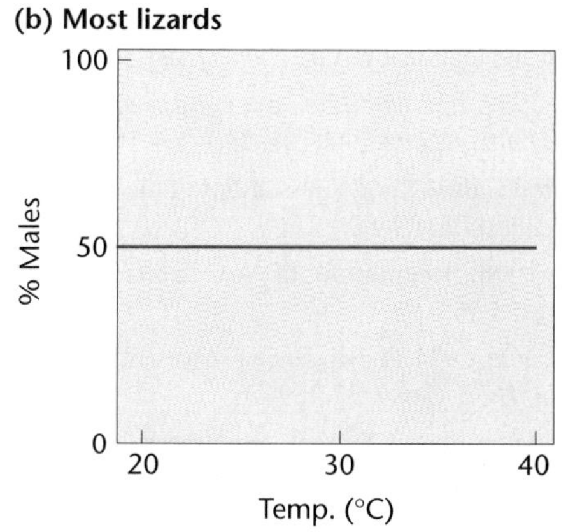

31. CC (Carbon Copy), the first cloned cat, was created from an ovarian cell taken from her genetic donor, Rainbow. The diploid nucleus from the cell was extracted and then injected into an enucleated egg. The resulting zygote was then allowed to develop in a petri dish and the cloned embryo was implanted in the uterus of a surrogate mother cat, who gave birth to CC. Rainbow is a calico cat. CC's surrogate mother is a tabby. Geneticists were very interested in the outcome of cloning a calico cat, because they were not certain if the cat would have patches of orange and black, just orange, or just black. Taking into account the Lyon hypothesis, explain the basis of the uncertainty.

Carbon Copy with her surrogate mother.

32. Let's assume hypothetically that Carbon Copy (see Problem 31) is indeed a calico with black and orange patches, along with the patches of white characterizing a calico cat. Would you expect CC to appear identical to Rainbow? Explain why or why not.
33. When Carbon Copy was born (see Problem 31), she had black patches and white patches, but completely lacked any orange patches. The knowledgeable students of genetics were not surprized at this outcome. Starting with the somatic ovarian cell used as the source of the nucleus in the cloning process, explain how this outcome occurred.

Selected Readings

Amory, J.K., et al. 2000. Klinefelter's syndrome. *Lancet* 356:333–35.

Avner, P., and Heard, E. 2001. X-chromosome inactivation: Counting, choice, and initiation. *Nature Reviews* 2:59–67.

Burgoyne, P.S. 1998. The mammalian Y chromosome: A new perspective. *Bioessays* 20:363–66.

Carrel, L., and Willard, H.F. 1998. Counting on Xist. *Nature Genetics* 19:211–12.

Court-Brown, W.M. 1968. Males with an XYY sex chromosome complement. *J. Med. Genet.* 5:341–59.

Davidson, R., Nitowski, H., and Childs, B. 1963. Demonstration of two populations of cells in human females heterozygous for glucose-6-phosphate dehydrogenase variants. *Proc. Natl. Acad. Sci. USA* 50:481–85.

Erickson, J.D. 1976. The secondary sex ratio of the United States, 1969–71: Association with race, parental ages, birth order, paternal education and legitimacy. *Ann. Hum. Genet.* (London) 40:205–12.

Hodgkin, J. 1990. Sex determination compared in *Drosophila* and *Caenorhabditis*. *Nature* 344:721–28.

Hook, E.B. 1973. Behavioral implications of the humans XYY genotype. *Science* 179:139–50.

Irish, E.E. 1996. Regulation of sex determination in maize. *BioEssays* 18:363–69.

Jacobs, P.A., et al. 1974. A cytogenetic survey of 11,680 newborn infants. *Ann. Hum. Genet.* 37:359–76.

Jegalian, K., and Lahn, B.T. 2001. Why the Y is so weird. *Sci. Am.* (Feb.) 284:56–61.

Koopman, P., et al. 1991. Male development of chromosomally female mice transgenic for *Sry*. *Nature* 351:117–21.

Lahn, B.T., and Page, D.C. 1997. Functional coherence of the human Y chromosome. *Science* 278:675–80.

Lucchesi, J. 1983. The relationship between gene dosage, gene expression, and sex in *Drosophila*. *Dev. Genet.* 3:275–82.

Lyon, M.F. 1972. X-chromosome inactivation and developmental patterns in mammals. *Biol. Rev.* 47:1–35.

———. 1988. X-chromosome inactivation and the location and expression of X-linked genes. *Am. J. Hum. Genet.* 42:8–16.

———. 1998. X-chromosome inactivation spreads itself: Effects in autosomes. *Am. J. Hum. Genet.* 63:17–19.

McMillen, M.M. 1979. Differential mortality by sex in fetal and neonatal deaths. *Science* 204:89–91.

Penny, G.D., et al. 1996. Requirement for Xist in X chromosome inactivation. *Nature* 379:131–37.

Pieau, C. 1996. Temperature variation and sex determination in reptiles. *BioEssays* 18:19–26.

Westergaard, M. 1958. The mechanism of sex determination in dioecious flowering plants. *Adv. Genet.* 9:217–81.

Whiting, P.W. 1939. Multiple alleles in sex determination in *Habrobracon*. *J. Morphology* 66:323–55.

Witkin, H.A., et al. 1996. Criminality in XYY and XXY men. *Science* 193:547–55.

Chromosome Mutations: Variation in Chromosome Number and Arrangement

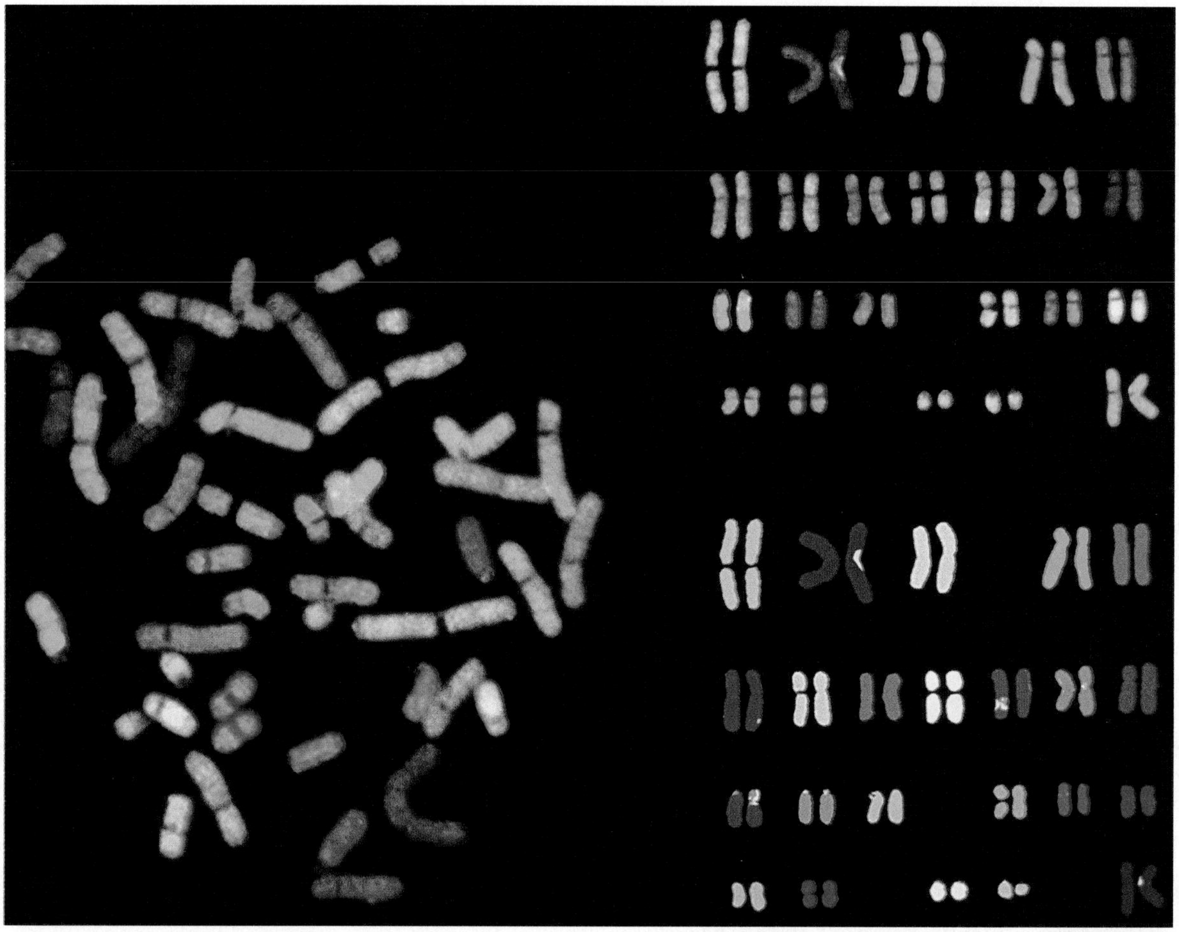

Spectral karyotyping of human chromosomes utilizing differentially labeled "painting" probes.

Chapter Concepts

- During meiosis, the failure of chromosomes to properly separate leads to variation in the chromosome content of gametes, and subsequently in offspring arising from such gametes.
- Plants often tolerate the abnormal content of genetic information, but manifest unique phenotypes. Such genetic variation has been an important factor in the evolution of plants.
- In animals, genetic information exists in a delicate equilibrium, whereby the gain or loss of a chromosome, or part of a chromosome, in an otherwise diploid organism often leads to lethality or to an abnormal phenotype.
- The rearrangement of genetic information within the genome of a diploid organism may be tolerated by that organism, but may affect the nature of gametes and the phenotypes of organisms arising from those gametes.
- Chromosomes are known to contain regions susceptible to breakage under specific conditions known as fragile sites, which lead to abnormal phenotypes.

Thus far, we have emphasized how mutations and the resulting alleles affect an organism's phenotype and how traits are passed from parents to offspring according to Mendelian principles. In this chapter, we shall look at phenotypic variation occuring as a result of changes that are more substantial than alterations of individual genes—modifications at the level of the chromosome.

Although most members of diploid species normally contain precisely two haploid chromosome sets, there are many known cases of variations from this pattern. Modifications include a change in the total number of chromosomes, the deletion or duplication of genes or segments of a chromosome, and rearrangements of the genetic material either within or among chromosomes. Taken together, such changes are called **chromosome mutations** or **chromosome aberrations**, to distinguish them from gene mutations. Because, according to Mendelian laws, the chromosome is the unit of genetic transmission, chromosome aberrations are passed on to offspring in a predictable manner, resulting in many unique genetic outcomes.

Since the genetic component of an organism is delicately balanced, even minor alterations of either the content or location of genetic information within the genome may result in some form of phenotypic variation. More substantial changes may be lethal, particularly in animals. Throughout the chapter, we shall consider the many types of chromosomal aberrations, the phenotypic consequences for the organism that harbors an aberration, and the impact of the aberration on offspring of the affected individual. We will also discuss the role of chromosome aberrations in the evolutionary process.

8.1 Specific Terminology Describes Variations in Chromosome Number

Variation in chromosome number ranges from the addition or loss of one or more chromosomes to the addition of one or more haploid sets of chromosomes. Before we embark on our discussion, it is useful to clarify the terminology that describes such changes. In the general condition known as **aneuploidy**, an organism gains or loses one or more chromosomes, but not a complete set. The loss of a single chromosome from an otherwise diploid genome is called *monosomy*. The gain of one chromosome results in *trisomy*. Such changes are contrasted with the condition of **euploidy**, where complete haploid sets of chromosomes are present. If more than two sets are present, the term **polyploidy** applies. Organisms with three sets are specifically *triploid*; those with four sets are *tetraploid*, and so on. Table 8.1 provides an organizational framework for you to follow as we discuss each of these categories of aneuploid and euploid variation and the subsets within them.

TABLE 8.1 TERMINOLOGY FOR VARIATION IN CHROMOSOME NUMBERS

Term	Explanation
Aneuploidy	$2n \pm x$ chromosomes
Monosomy	$2n - 1$
Trisomy	$2n + 1$
Tetrasomy, pentasomy, etc.	$2n + 2$, $2n + 3$, etc.
Euploidy	Multiples of n
Diploidy	$2n$
Polyploidy	$3n, 4n, 5n, \ldots$
Triploidy	$3n$
Tetraploidy, pentaploidy, etc.	$4n$, $5n$, etc.
Autopolyploidy	Multiples of the same genome
Allopolyploidy (Amphidiploidy)	Multiples of different genomes

8.2 Variation in the Number of Chromosomes Results from Nondisjunction

As we consider cases including the gain or loss of chromosomes, it is useful to examine how such aberrations originate. For instance, how do the syndromes arise where the number of sex-determining chromosomes in humans is altered? As you may recall from Chapter 7, in spite of a mechanism in somatic cells that inactivates all X chromosomes in excess of one, the gain (47,XXY), or the loss (45,X) of a sex-determining chromosome from an otherwise diploid genome alters the normal phenotype, resulting in **Klinefelter syndrome** or **Turner syndrome**, respectively. (See Figure 7–7.) Human females have also been known to have extra X chromosomes (e.g., 47,XXX, 48,XXXX), and males have had an extra Y chromosome (47,XYY).

Such chromosomal variation originates as a random error during the production of gametes. As first introduced in Chapter 2, **nondisjunction** is the failure of chromosomes or chromatids to disjoin and move to opposite poles during division. When this occurs in meiosis, the normal distribution of chromosomes into gametes is disrupted. The results of nondisjunction during meiosis I and meiosis II for a single chromosome of a diploid organism are shown in Figure 8–1. As you can see, for the affected chromosome, abnormal gametes can form containing either two members or none at all. Fertilizing these with a normal haploid gamete produces a zygote with either three members (trisomy) or only one member (monosomy) of this chromosome. Nondisjunction leads to a variety of autosomal aneuploid conditions in humans and other organisms.

Now solve this

Problem 8.15 on page 211 considers a female with Turner syndrome who expresses hemophilia, as did her father. You are asked which of her parents was responsible for the nondisjunction event leading to her syndrome.

Hint: The parent who contributed a gamete with an X chromosome underwent normal meiosis.

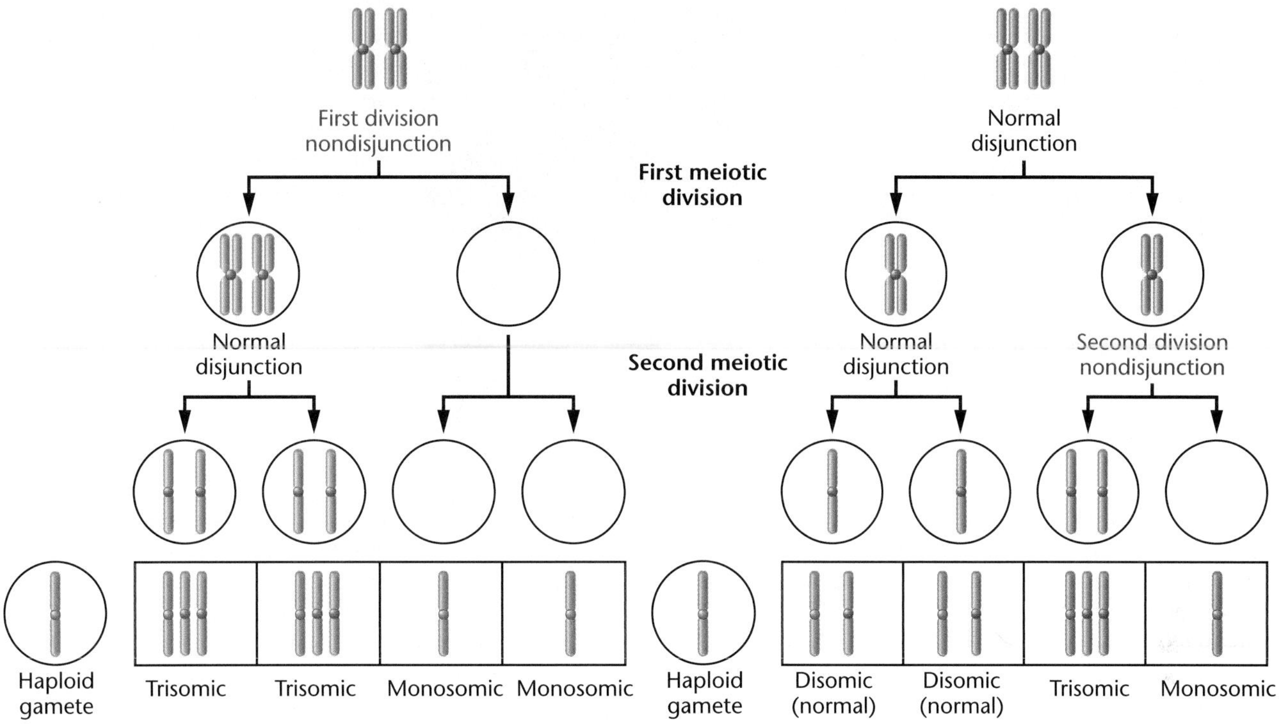

FIGURE 8–1 Nondisjunction during the first and second meiotic divisions. In both cases, some of the gametes formed either contain two members of a specific chromosome or lack that chromosome. Following fertilization by a gamete with a normal haploid content, monosomic, disomic (normal), or trisomic zygotes are produced.

8.3 Monosomy, the Loss of a Single Chromosome, May Have Severe Phenotypic Effects

We turn now to a consideration of variations in the number of autosomes and the genetic consequence of such changes. The most common examples of aneuploidy, where an organism has a chromosome number other than an exact multiple of the haploid set, are cases in which a single chromosome is either added to or lost from a normal diploid set. The loss of one chromosome produces a $2n - 1$ complement and is called **monosomy**.

Although monosomy for the X chromosome occurs in humans, as we have seen in 45,X Turner syndrome, monosomy for any of the autosomes is not usually tolerated in humans or other animals. In *Drosophila*, flies monosomic for the very small chromosome 4—a condition referred to as **Haplo-IV**—survive, but they develop more slowly, exhibit a reduced body size, and have impaired viability. Chromosome 4 contains no more than 5 percent of the genome of *Drosophila*. Monosomy for the larger chromosomes 2 and 3 is apparently lethal because such flies have never been recovered.

The failure of monosomic individuals to survive in many animal species is at first quite puzzling, since at least a single copy of every gene is present in the remaining homolog. One possible explanation involves the unmasking of recessive lethals that are tolerated in heterozygotes carrying the corresponding wild-type alleles. If an organism heterozygous for just one recessive lethal allele loses the homologous chromosome bearing the normal allele (which prevents lethality), the unpaired chromosome condition will lead to the death of the organism. Another possible explanation is that the expression of genetic information during early development is carefully regulated such that a delicate equilibrium of gene products is required to ensure normal development. While this is thought to be true during animal development, such a requirement does not appear to be so stringent in the plant kingdom, where aneuploidy is tolerated. Monosomy for autosomal chromosomes has been observed in maize, tobacco, the evening primrose *Oenothera*, and the Jimson weed *Datura*, among other plants. Nevertheless, such monosomic plants are almost always less viable than their diploid derivatives. Since both pollen grains and ovules must undergo extensive development after undergoing meiosis, but before participating in fertilization, they are particularly sensitive to the lack of one chromosome and are seldom viable.

Partial Monosomy in Humans: The Cri-du-Chat Syndrome

In humans, autosomal monosomy has not been reported beyond birth. Individuals with such chromosome complements are undoubtedly conceived, but none apparently survive embryonic and fetal development. There are, however, examples of survivors with **partial monosomy**, where only part of one chromosome is lost. These cases are also referred to as **segmental deletions**. One such case was first reported by Jerome LeJeune in 1963, when he described the clinical symptoms of the **cri-du-chat syndrome** ("cry of the cat"). This syndrome is associated with the loss of a small part of the short arm

Web Tutorial 8.1 Chromosome Separation in Meiosis: Disjunction

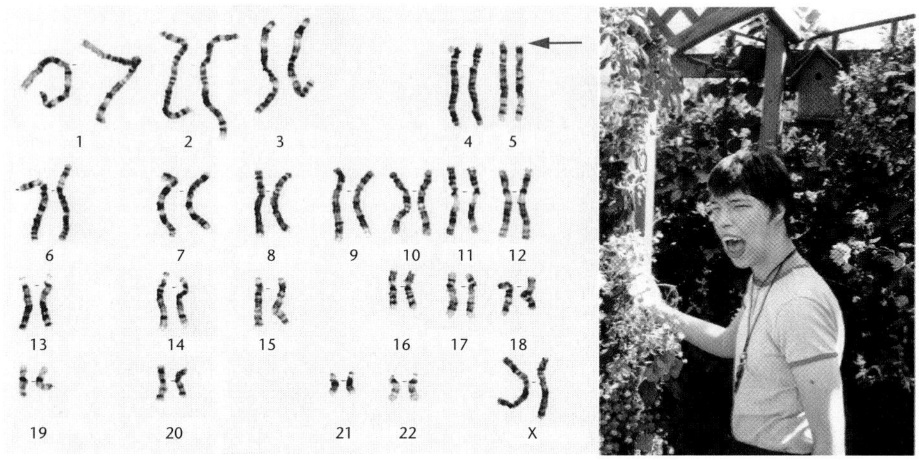

FIGURE 8–2 A representative karyotype and photograph of a child exhibiting cri-du-chat syndrome (46,−5p). In the karyotype, the arrow identifies the absence of a small part of the short arm of one member of the chromosome 5 homologs.

of chromosome 5 (Figure 8–2). Thus, the genetic constitution may be designated as **46, − 5p**, meaning that such an individual has all 46 chromosomes, but that some or all of the p arm (the petite or short arm) of one member of the chromosome 5 pair is missing.

Infants with this syndrome may exhibit anatomic malformations, including gastrointestinal and cardiac complications, and they are often mentally retarded. Abnormal development of the glottis and larynx is characteristic of this syndrome. As a result, the infant has a cry similar to that of the mewing of a cat, thus giving the syndrome its name.

Since 1963, hundreds of cases of cri-du-chat syndrome individuals have been reported worldwide. An incidence of 1 in 50,000 live births has been estimated. The length of the short arm that is deleted varies somewhat; longer deletions appear to have a greater impact on the physical, psychomotor, and mental skill levels of those children who survive. Although the effects of the syndrome are severe, many individuals achieve a level of social development in the trainable range. Those who receive home care and early special schooling are ambulatory, develop self-care skills, and learn to communicate verbally.

How Do We Know?

How do we determine the chromosome content in somatic nuclei of individuals?

8.4 Trisomy Involves the Addition of a Chromosome to a Diploid Genome

In general, the effects of trisomy $(2n + 1)$ parallel those of monosomy. However, the addition of an extra chromosome produces somewhat more viable individuals in both animal and plant species than does the loss of a chromosome. In animals, this is often true, provided that the chromosome involved is relatively small.

As in monosomy, the sex chromosome variation of the trisomic type has a less dramatic effect on the phenotype than does autosomal variation. Recall from our previous discussion that *Drosophila* females with three X chromosomes and a normal complement of two sets of autosomes (3X:2A) survive and reproduce, but are less viable than normal 2X:2A females. In humans, the addition of an extra X or Y chromosome to an otherwise normal male or female chromosome constitution (47,XXY, 47,XYY, and 47,XXX) leads to viable individuals exhibiting various syndromes. However, the addition of a large autosome to the diploid complement in both *Drosophila* and humans has severe effects and is usually lethal during development.

In plants, trisomic individuals are usually viable, but their phenotype may be altered. A classic example involves the Jimson weed *Datura*, a plant long known for its narcotic effect, whose diploid number is 24. Twelve different primary trisomic conditions are possible, and examples of each one have been recovered. Each trisomy alters the phenotype of the capsule of the fruit sufficiently to produce a unique phenotype (Figure 8–3). These capsule phenotypes were first thought to be caused by mutations in one or more genes.

Still another example is seen in the rice plant (*Oryza sativa*), which has a haploid number of 12. Trisomic strains for each chromosome have been isolated and studied. The plants of 11 of the strains can be distinguished from one another and from wild type. Trisomics for the longer chromosomes are the most distinctive and grow more slowly than the rest. This is in keeping with the belief that larger chromosomes cause greater genetic imbalance than smaller ones. In addition to growth rate, leaf structure, foliage, stems, grain morphology, and plant height vary between the various trisomies.

In plants as well as animals, trisomy may be detected during cytological observations of meiotic divisions. Since three copies of one of the chromosomes are present, pairing configurations are usually irregular. At any particular region along the chromosome length, only two of the three homologs may synapse, though different regions of the trio may be paired. When three copies of a chromosome are synapsed, the configuration is called a **trivalent**, which may be arranged on the spindle so that, during anaphase, one member moves to one pole and two go to the opposite pole (Figure 8–4). In some cases, one **bivalent** and one **univalent** (an unpaired chromosome) may be present instead of a trivalent prior to the first meiotic division. Meiosis thus produces gametes with a chromosome composition of $(n + 1)$, which can perpetuate the trisomic condition.

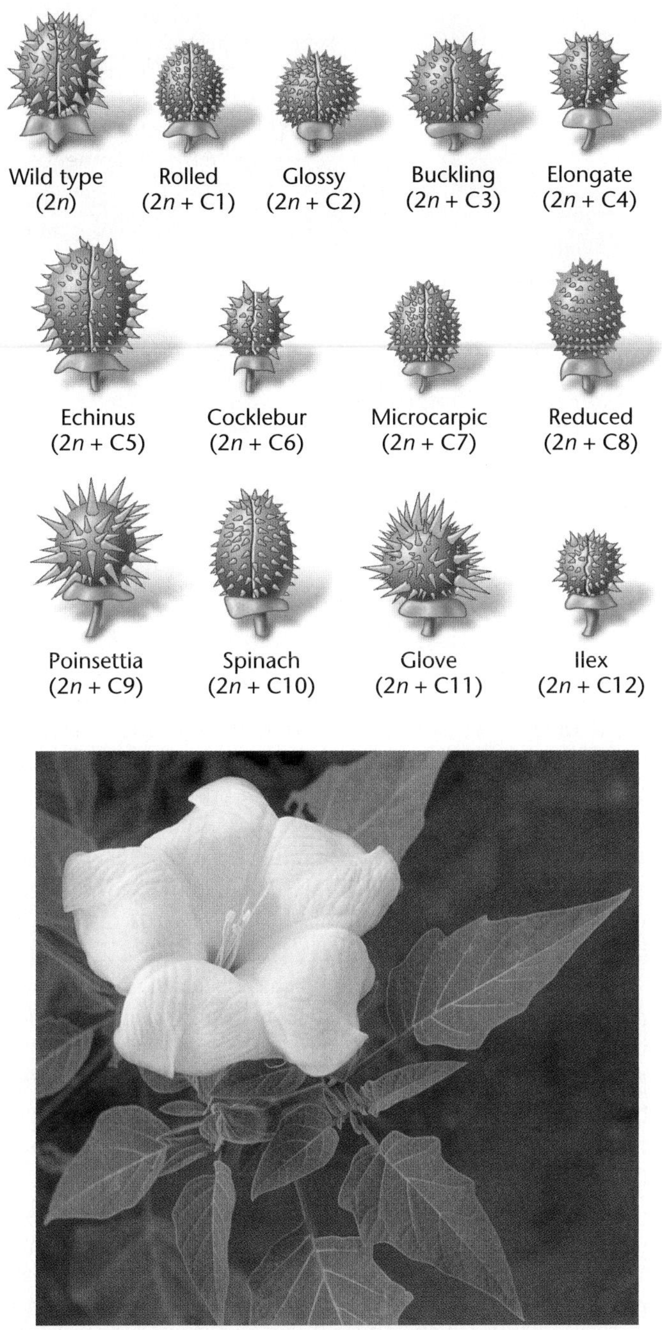

FIGURE 8–3 Drawings of capsule phenotypes of the fruits of the Jimson weed *Datura stramonium*. In comparison with wild type, each phenotype is the result of trisomy of 1 of the 12 chromosomes characteristic of the haploid genome. The photograph illustrates the plant itself.

Down Syndrome

The only human autosomal trisomy in which a significant number of individuals survive longer than a year past birth was discovered in 1866 by John Langdon Down. The condition is now known to result from trisomy of chromosome 21, one of the G group* (Figure 8–5), and is called **Down syndrome** or simply **trisomy 21** (designated **47, +21**). This trisomy is found in approximately one infant in every 800 live births.

*On the basis of size and centromere placement, human autosomal chromosomes are divided into seven groups: A (1–3), B (4–5), C (6–12), D (13–15), E (16–18), F (19–20), and G (21–22).

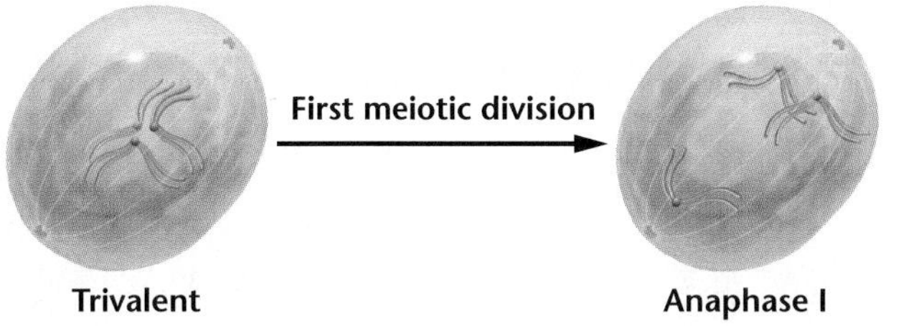

FIGURE 8–4 Diagrammatic representation of one possible pairing arrangement during meiosis I of three copies of a single chromosome, forming a trivalent configuration. During anaphase I, two chromosomes move toward one pole and one chromosome toward the other pole.

Typical of other conditions referred to as a syndrome, there are many phenotypic characteristics that may be present, but any single affected individual usually expresses only a subset of these. In the case of Down syndrome, there are 12 to 14 such characteristics, but each individual, on average, expresses 6 to 8 of them. Nevertheless, the outward appearance of these individuals is very similar, and they bear a striking resemblance to one another. This is, for the most part, due to a prominent epicanthic fold† in the corner of each eye and the typically flat face and round head. They are also characteristically short. They may also have protruding, furrowed tongues, which cause the mouth to remain partially open; and short, broad hands with fingers showing characteristic palm and fingerprint patterns. Physical, psychomotor, and mental development is retarded, and poor muscle tone is characteristic. Their life expectancy is shortened, although individuals are known to survive into their 50s.

Children afflicted with Down syndrome are prone to respiratory disease and heart malformations, and they show an incidence of leukemia approximately 20 times higher than that of the normal population. However, careful medical scrutiny and treatment throughout their lives has extended their survival significantly. A striking observation is that death of older Down syndrome adults is frequently due to Alzheimer's disease. The onset of this disease occurs at a much earlier age than it does in the normal population.

The origin of this trisomic condition is most often through nondisjunction of chromosome 21 during meiosis. Failure of paired homologs to disjoin during either anaphase I or II may lead to gametes with the $n + 1$ chromosome composition. About 75 percent of these errors leading to Down syndrome are attributed to nondisjunction during meiosis I. Following fertilization with a normal gamete, the trisomic condition is created.

†The epicanthic fold, or epicanthus, is a skin fold of the upper eyelid, extending from the nose to the inner side of the eyebrow. It covers the inner corner of the eye and appears to lower the inner corner of the eye, providing the eye with a slanted appearance. Besides Down syndrome, the epicanthus is prominent and a normal component of the eyes in many Asian groups.

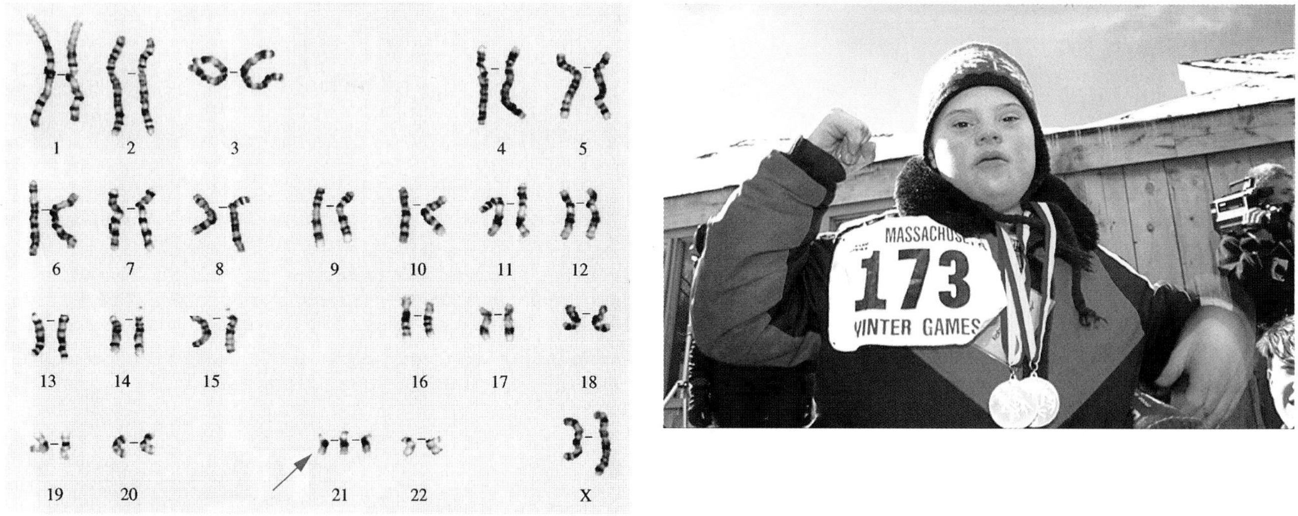

FIGURE 8–5 The karyotype and a photograph of a child with Down syndrome. In the karyotype, three members of the G-group chromosome 21 are present, creating the 47,+21 condition.

Chromosome analysis has shown that, while the additional chromosome may be derived from either the mother or father, the ovum is the source in about 95 percent of 47,+21 trisomy cases. Before the development of techniques involving polymorphic markers that clearly distinguish paternal from maternal homologs, this conclusion was supported by the more indirect evidence derived from studies of the age of mothers giving birth to infants afflicted with Down syndrome. Figure 8–6 shows the relationship between the incidence of Down syndrome births and maternal age, illustrating the dramatic increase as the age of the mother increases. While the frequency is about 1 in 1000 at maternal age 30, a tenfold increase to a frequency of 1 in 100 is noted at age 40. The frequency increases still further to about 1 in 50 at age 45. A very alarming statistic is that as the age of child-bearing women exceeds 45, the probability of a Down syndrome birth continues to increase substantially. In spite of this high probability, because the overwhelming proportion of pregnancies involve women under 35, more than half of Down syndrome births occur to women in this younger cohort.

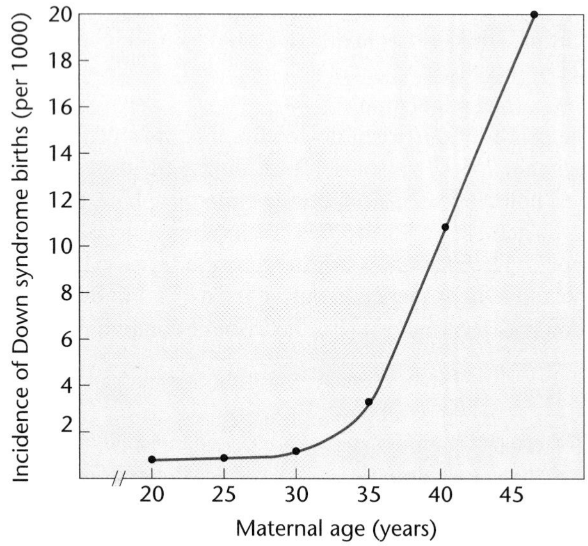

FIGURE 8–6 Incidence of Down syndrome births contrasted with maternal age.

While the nondisjunctional event that produces Down syndrome seems more likely to occur during oogenesis in women between the ages of 35 and 45, we do not know with certainty why this is so. However, one observation may be relevant. In human females, meiosis in all eggs is initiated during fetal development. Synapsis of homologs has occurred and recombination has been initiated. Then oocyte development is arrested in meiosis I. Thus, all primary oocytes have been formed by birth. Then, once ovulation begins at puberty, meiosis is reinitiated in one egg during each ovulatory cycle and continues into meiosis II. The process is once again arrested after ovulation and is not completed unless fertilization occurs.

The end result of this progression is that each succeeding ovum has been arrested in meiosis I for about a month longer than the one preceding it. As a result, women 30 or 40 years old produce ova that are significantly older and arrested longer than those they ovulated 10 or 20 years previously. However, no direct evidence proves that ovum age is the cause of the increased incidence of nondisjunction leading to Down syndrome.

These statistics obviously pose a serious problem for the woman who becomes pregnant late in her reproductive years. Genetic counseling early in such pregnancies is highly recommended. Such counseling informs prospective parents about the probability that their child will be affected and educates them about Down syndrome. Although some individuals with Down syndrome must be institutionalized, others benefit greatly from special education programs and may be cared for at home. Further, these children are noted for their affectionate, loving natures.

A genetic counselor may recommend a prenatal diagnostic technique, where fetal cells are isolated and cultured. **Amniocentesis** and **chorionic villus sampling (CVS)** are the two most familiar approaches, whereby fetal cells are obtained from the amniotic fluid or the chorion of the placenta, respectively. In a more current approach, fetal cells are derived directly from the maternal circulation. Once perfected, this approach will be preferable, because it is noninvasive, posing no risk to the fetus. Once fetal cells are obtained, the karyotype can then be determined by cytogenetic analysis. If the fetus is diagnosed as having Down syndrome, a therapeutic abortion is one option currently available to parents. Obviously, this is a difficult decision involving a number of religious and ethical issues.

Since Down syndrome is caused by a random error—nondisjunction of chromosome 21 during maternal or paternal meiosis—the occurrence of the disorder is *not* expected to be inherited. Nevertheless, Down syndrome occasionally runs in families. These instances, referred to as **familial Down syndrome**, involve a translocation of chromosome 21, another type of chromosomal aberration, which we will discuss later in the chapter.

HOW DO WE KNOW?

How do we know that the extra chromosome 21 present in Down syndrome children is almost always maternal in origin?

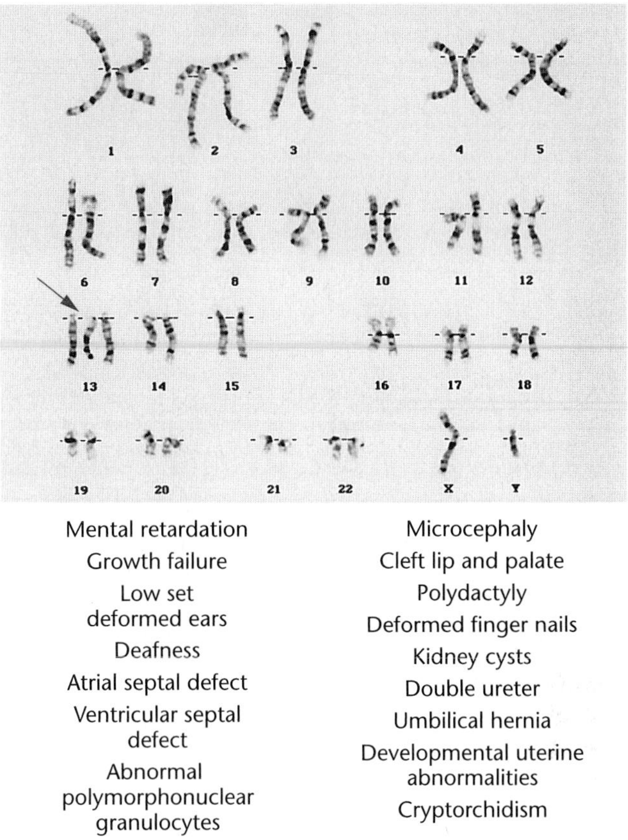

FIGURE 8–7 The karyotype and potential phenotypic characteristics associated with Patau syndrome, where three members of the D group chromosome 13 are present, creating the 47,+13 condition.

Patau Syndrome

In 1960, Klaus Patau and his associates observed an infant with severe developmental malformations and a karyotype of 47 chromosomes (Figure 8–7). The additional chromosome was medium sized, one of the acrocentric D group. It is now designated as chromosome 13. This **trisomy 13** condition has since been described in many newborns and is called **Patau syndrome (4, + 13)**. Affected infants are not mentally alert, are thought to be deaf, and characteristically have a harelip, cleft palate, and demonstrate polydactyly. Autopsies have revealed congenital malformation of most organ systems, a condition indicative of abnormal developmental events occurring as early as five to six weeks of gestation. The average survival of these infants is about three months.

The average maternal and paternal ages of parents of Patau infants are higher than the ages of parents of normal children, but they are not as high as the average maternal age in cases of Down syndrome. Both male and female parents average about 32 years of age when the affected child is born. Because the condition is so rare, occurring as infrequently as 1 in 19,000 live births, it is not known whether the origin of the extra chromosome is more often maternal, paternal, or whether it arises equally from either parent.

Edwards Syndrome

In 1960, John H. Edwards and his colleagues reported on an infant trisomic for a chromosome in the E group, now known to be chromosome 18 (Figure 8–8). Referred to as **trisomy 18 (47, + 18)**, this aberration is also named **Edwards syndrome** after its discoverer. The phenotype of this child, like that of individuals with Down and Patau syndromes, illustrates that the presence of an extra autosome produces congenital malformations and reduced life expectancy. These infants are smaller than the average newborn. Their skulls are elongated in an anterior–posterior direction, and their ears are set low and malformed. A webbed neck, congenital dislocation of the hips, and a receding chin are often characteristic of such individuals. Although the frequency of trisomy 18 is somewhat greater than that of trisomy 13, the average survival time is about the same, less than four months. Death is usually caused by pneumonia or heart failure.

Again, the average maternal age is high—34.7 years by one calculation. In contrast to Patau syndrome, the preponderance of Edwards syndrome infants are females. In one set of observations based on 143 cases, 80 percent were female. Overall, about 1 in 8000 live births exhibits this malady.

Viability in Human Aneuploidy

The reduced viability of individuals with recognized monosomic and trisomic conditions suggests that many other aneuploid conditions may arise, but the affected fetuses do not survive to term. This observation has been confirmed by karyotypic analysis of spontaneously aborted fetuses. In an extensive review of this subject in 1971 by David H. Carr, it was shown that a significant percentage of abortuses are trisomic for one or another of the autosomal chromosomes. Trisomies for every human chromosome were recovered. Monosomies,

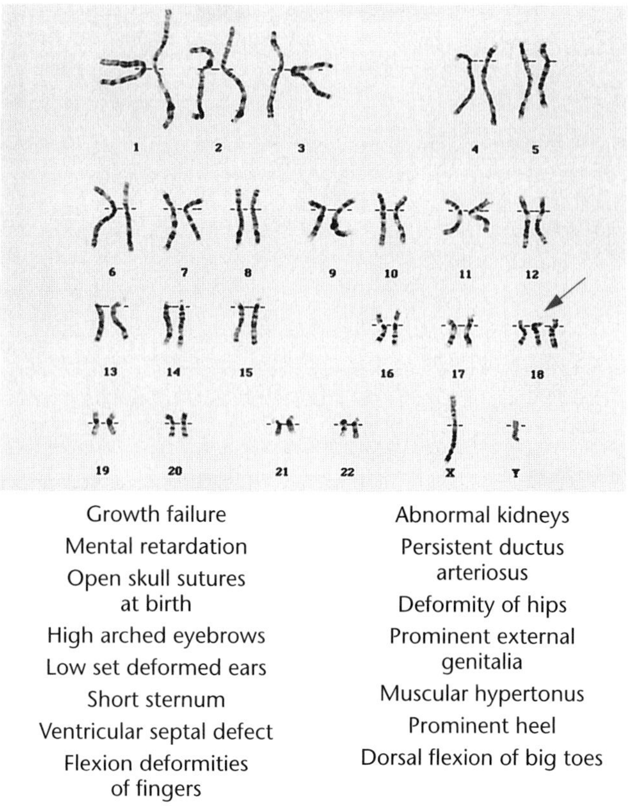

FIGURE 8–8 The karyotype and potential phenotypic characteristics associated with Edwards syndrome. Three members of the E group chromosome 18 are present, creating the 47,+18 condition.

however, were seldom found in the Carr study, even though nondisjunction should produce $n - 1$ gametes with a frequency equal to $n + 1$ gametes. This finding leads us to believe that gametes lacking a single chromosome are either so functionally impaired that they never participate in fertilization or that the monosomic embryo dies so early in its development that recovery occurs infrequently. Various forms of polyploidy and other miscellaneous chromosomal anomalies were also found in Carr's study.

Studies that included Carr's work revealed some striking statistics. It was estimated that about 15 to 20 percent of all human conceptions are terminated by spontaneous abortion and that about 30 percent of all resultant abortuses demonstrate some form of chromosomal anomaly. A calculation using these figures ($0.20 \times 0.30 = 0.06$) predicts that up to 6 percent of all pregnancies originate with an abnormal number of chromosomes. Most of these are terminated prior to birth. More recently derived figures are even more dramatic. It is now estimated that as many as 10 to 30 percent of all fertilized eggs in humans contain some error in chromosome number.

The largest percentage of chromosomal abnormalities are aneuploids. Surprisingly, an aneuploid with one of the highest incidence rates among abortuses is the 45,X condition, which produces an infant with Turner syndrome if the fetus survives to term. About 70 to 80 percent of aborted and liveborn 45,X conditions include the maternal X chromosome. Thus, the meiotic error leading to this syndrome occurs during spermatogenesis.

Collectively, these observations support the hypothesis that normal embryonic development requires a precise diploid complement of chromosomes maintaining a delicate equilibrium of expression of genetic information. The prenatal mortality of most aneuploids provides a barrier against the introduction of a variety of chromosome-based genetic anomalies into the human population.

HOW DO WE KNOW?

How do we know that human aneuploidy for each of the 22 autosomes occurs at conception, even though most often human aneuploids do not survive embryonic or fetal development and thus are never observed at birth?

8.5 Polyploidy, in Which More Than Two Haploid Sets of Chromosomes Are Present, Is Prevalent in Plants

The term *polyploidy* describes instances where more than two multiples of the haploid chromosome set are found. The naming of polyploids is based on the number of sets of chromosomes found: A **triploid** has $3n$ chromosomes; a **tetraploid** has $4n$; a **pentaploid**, $5n$; and so forth. Several general statements may be made about polyploidy. This condition is relatively infrequent in many animal species, but is well known in lizards, amphibians, and fish. It is much more common in plant species. Odd numbers of chromosome sets are not usually maintained reliably from generation to generation because a polyploid organism with an uneven number of homologs usually does not produce genetically balanced gametes. For this reason, triploids, pentaploids, and so on are not usually found in species depending solely upon sexual reproduction for propagation.

Polyploidy can originate in two ways: (1) The addition of one or more extra sets of chromosomes, identical to the normal haploid complement of the same species, results in **autopolyploidy**; and (2) the combination of chromosome sets from different species may occur as a consequence of interspecific matings resulting in **allopolyploidy** (from the Greek word *allo*, meaning other or different). The distinction between auto- and allopolyploidy is based on the genetic origin of the extra chromosome sets, as illustrated in Figure 8–9.

In our discussion of polyploidy, we will use certain symbols to clarify the origin of additional chromosome sets. For example, if A represents the haploid set of chromosomes of any organism, then

$$A = a_1 + a_2 + a_3 + a_4 + \cdots + a_n$$

Web Tutorial 8.2
Polyploidy

where a_1, a_2, and so on represent individual chromosomes, and where n is the haploid number. Using this nomenclature, a normal diploid organism would be represented simply as AA.

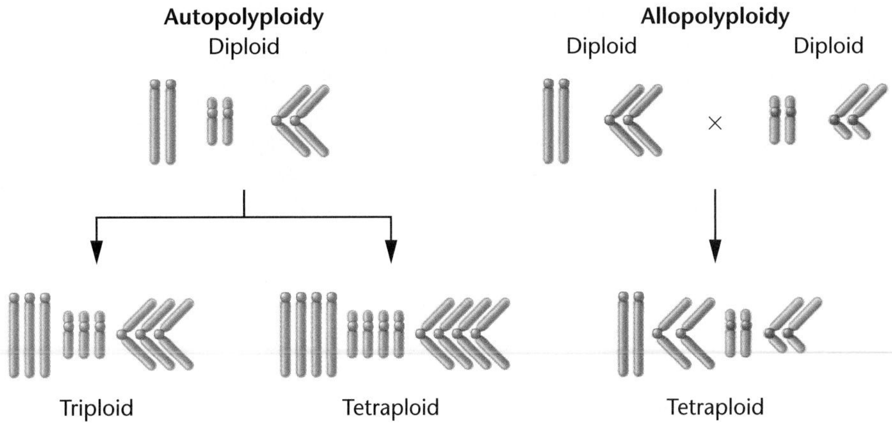

FIGURE 8–9 Contrasting chromosome origins of an autopolyploid with an allopolyploid karyotype.

Autopolyploidy

In autopolyploidy, each additional set of chromosomes is identical to the parent species. Therefore, triploids are represented as AAA, tetraploids are $AAAA$, and so forth.

Autotriploids arise in several ways. A failure of all chromosomes to segregate during meiotic divisions (first-division or second-division nondisjunction) can produce a diploid gamete. If such a gamete survives and is fertilized by a haploid gamete, a zygote with three sets of chromosomes is produced, or occasionally, two sperm may fertilize an ovum, resulting in a triploid zygote. Triploids can also be produced under experimental conditions by crossing diploids with tetraploids. Diploid organisms produce gametes with n chromosomes, whereas tetraploids produce $2n$ gametes. Upon fertilization, the desired triploid is produced.

Because they have an even number of chromosomes, **autotetraploids** ($4n$) are theoretically more likely to be found in nature than are autotriploids. Unlike triploids, which often produce genetically unbalanced gametes with odd numbers of chromosomes, tetraploids are more likely to produce balanced gametes when involved in sexual reproduction.

How polyploidy arises naturally is of great interest. In theory, if chromosomes have replicated, but the parent cell never divides and reenters interphase, the chromosome number will be doubled. That this very likely occurs is supported by the observation that tetraploid cells can be produced experimentally from diploid cells by applying cold or heat shock to meiotic cells, or by applying **colchicine** to somatic cells undergoing mitosis. Colchicine, an alkaloid derived from the autumn crocus, interferes with spindle formation, and thus, replicated chromosomes that cannot be separated at anaphase do not migrate to the poles. When colchicine is removed, the cell can reenter interphase. When the paired sister chromatids separate and uncoil, the nucleus will contain twice the diploid number of chromosomes and is therefore $4n$. This process is illustrated in Figure 8–10.

In general, autopolyploids are larger than their diploid relatives. Such an increase seems to be due to a larger cell size, rather than a greater number of cells. Although autopolyploids do not contain new or unique information compared with the diploid relative, the flower and fruit of plants are often increased in size, making such varieties of greater horticultural or commercial value. Economically important triploid plants include several potato species of the genus *Solanum*, Winesap apples, commercial bananas, seedless watermelons, and the cultivated tiger lily *Lilium tigrinum*. These plants are propagated asexually. Diploid bananas contain hard seeds, but the commercial, triploid, "seedless" variety has edible seeds. Tetraploid alfalfa, coffee, peanuts, and McIntosh apples are also of economic value, because they are either larger or grow more vigorously than do their diploid or triploid counterparts. The commercial strawberry is an octoploid.

We have long been curious how cells with increased ploidy values, where no new genes are present, express different phenotypes than their diploid counterparts. Our current ability to examine gene expression using modern biotechnology has provided some interesting insights. For example, Gerald Fink and his colleagues have been able to create strains of the yeast *Saccharomyces cerevisiae* with one, two, three, or four copies of the genome. Thus, each strain contains identical

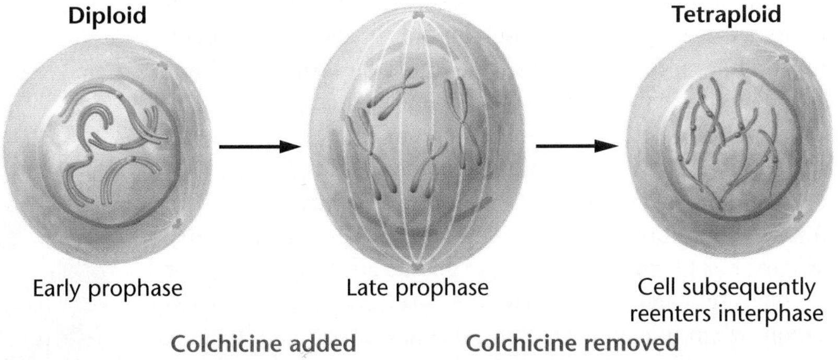

FIGURE 8–10 The potential involvement of colchicine in doubling the chromosome number, as occurs during the production of an autotetraploid. Two pairs of homologous chromosomes are followed. While each chromosome has replicated its DNA earlier during interphase, the chromosomes do not appear as double structures until late prophase. When anaphase fails to occur normally, the chromosome number doubles if the cell reenters interphase.

genes (they are said to be isogenic) but different ploidy values. They then proceeded to examine expression levels of all genes during the entire cell cycle of the organism. Using the rather stringent standards of a tenfold increase or decrease of gene expression, Fink and colleagues proceeded to identify ten cases where, as ploidy increased, gene expression was increased at least tenfold, and seven cases where it was reduced by a similar level.

One of these genes provides insights into how polyploid cells are larger than their haploid or diploid counterparts. In polyploid yeast, two **G1 cyclins**, Cln1 and Pc11 are repressed as ploidy increases, while the size of the yeast cells increases. This is explained based on the observation that G1 cyclins facilitate the cell's movement through G1, which is delayed when expression of these genes is repressed. The cell stays in G1 longer and, on average, grows to a larger size before it moves beyond the G1 stage of the cell cycle. Yeast cells also show different morphology as ploidy increases. Several of the other genes, repressed as ploidy increases, have been linked to cytoskeletal dynamics, which accounts for the morphological changes.

Allopolyploidy

Polyploidy can also result from hybridization of two closely related species. If a haploid ovum from a species with chromosome sets *AA* is fertilized by a haploid sperm from a species with sets *BB*, the resulting hybrid is *AB*, where $A = a_1, a_2, a_3, \ldots a_n$ and $B = b_1, b_2, b_3, \ldots b_n$. The hybrid plant may be sterile because of its inability to produce viable gametes. Most often, this occurs when some, or all, of the *a* and *b* chromosomes are not homologous and therefore cannot synapse in meiosis. As a result, unbalanced genetic conditions result. If, however, the new *AB* genetic combination undergoes a natural or induced chromosomal doubling, two copies of all *a* chromosomes and two copies of all *b* chromosomes are now present, and they will pair during meiosis. As a result, a fertile *AABB* tetraploid is produced. These events are illustrated in Figure 8–11. Since this polyploid contains the equivalent of four haploid genomes derived from two separate species, such an organism is called an **allotetraploid**. In such cases, when both original species are known, the equivalent term **amphidiploid** is preferred for describing the allotetraploid.

Amphidiploid plants are often found in nature. Their reproductive success is based on their potential for forming balanced gametes. Since two homologs of each specific chromosome are present, meiosis can occur normally (Figure 8–11), and fertilization can successfully propagate the plant sexually. This discussion assumes the simplest situation, where none of the chromosomes in set *A* are homologous to those in set *B*. In amphidiploids formed from closely related species, some homology between *a* and *b* chromosomes will no doubt exist. In this case, meiotic pairing is more complex. During synapsis, multivalents will be formed, resulting in the production of unbalanced gametes. In such cases, aneuploid varieties of amphidiploids may arise. Allopolyploids are rare in most animals because mating behavior is most often species specific; thus, the initial step in hybridization is unlikely to occur.

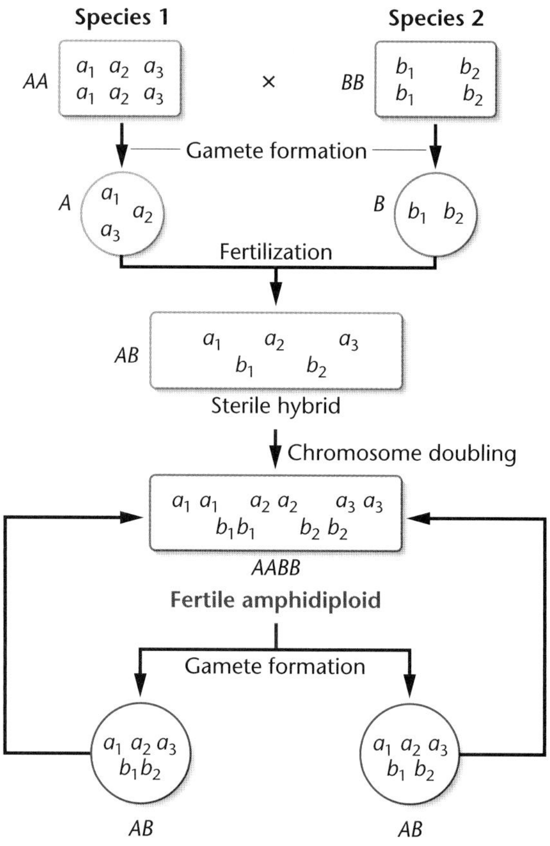

FIGURE 8–11 The origin and propagation of an amphidiploid. Species 1 contains genome *A* consisting of three distinct chromosomes, a_1, a_2, and a_3. Species 2 contains genome *B* consisting of two distinct chromosomes, b_1 and b_2. Following fertilization between members of the two species and chromosome doubling, a fertile amphidiploid containing two complete diploid genomes (*AABB*) is formed.

A classic example of amphidiploidy in plants is the cultivated species of American cotton, *Gossypium* (Figure 8–12). This species has 26 pairs of chromosomes: 13 are large and 13 are much smaller. When it was discovered that Old World cotton had only 13 pairs of large chromosomes, allopolyploidy was

FIGURE 8–12 The pods of the amphidiploid form of *Gossypium*, the cultivated cotton plant.

suspected. After an examination of wild American cotton revealed 13 pairs of small chromosomes, this speculation was strengthened. J. O. Beasley was able to reconstruct the origin of cultivated cotton experimentally. He crossed the Old World strain with the wild American strain and then treated the hybrid with colchicine to double the chromosome number. The result of these treatments was a fertile amphidiploid variety of cotton. It contained 26 pairs of chromosomes and characteristics similar to the cultivated variety.

Amphidiploids often exhibit traits of both parental species. An interesting example, but one with no practical economic importance, is that of the hybrid formed between the radish *Raphanus sativus* and the cabbage *Brassica oleracea*. Both species have a haploid number of 9. The initial hybrid consists of 9 *Raphanus* and 9 *Brassica* chromosomes (9R + 9B). While hybrids are almost always sterile, some fertile amphidiploids (18R + 18B) have been produced. Unfortunately, the root of this plant is more like the cabbage and its shoot more like the radish. Had the converse occurred, the hybrid might have been of economic importance.

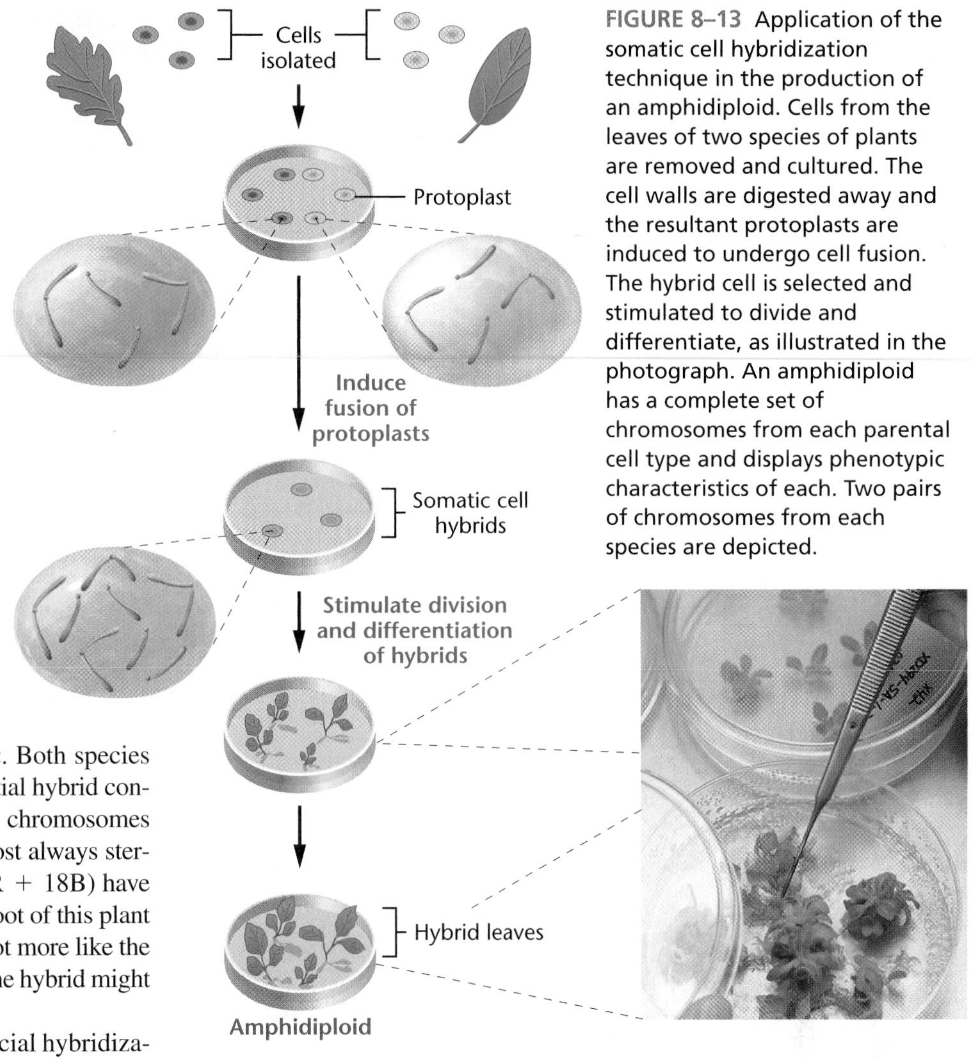

FIGURE 8–13 Application of the somatic cell hybridization technique in the production of an amphidiploid. Cells from the leaves of two species of plants are removed and cultured. The cell walls are digested away and the resultant protoplasts are induced to undergo cell fusion. The hybrid cell is selected and stimulated to divide and differentiate, as illustrated in the photograph. An amphidiploid has a complete set of chromosomes from each parental cell type and displays phenotypic characteristics of each. Two pairs of chromosomes from each species are depicted.

A much more successful commercial hybridization has been performed using the grasses wheat and rye. Wheat (genus *Triticum*) has a basic haploid genome of 7 chromosomes. Rye (genus *Secale*) also has a genome consisting of 7 chromosomes. In addition to normal diploids ($2n = 14$), cultivated allopolyploids exist, including tetraploid ($4n = 28$) and hexaploid ($6n = 42$) species. The only cultivated species is the diploid plant ($2n = 14$).

Using the technique outlined in Figure 8–11, geneticists have produced various hybrids. When tetraploid wheat is crossed with diploid rye, and the F_1 treated with colchicine, a hexaploid variety ($6n = 42$) is derived. The hybrid, designated *Triticale*, represents a new genus. Fertile hybrid varieties derived from various wheat and rye species can be crossed together or backcrossed. These crosses have created many variations of the genus *Triticale*. For example, a useful octaploid (an allooctaploid) variety with 56 chromosomes has been produced from a diploid rye and a hexaploid wheat plant.

The hybrid plants demonstrate characteristics of both wheat and rye. For example, certain hybrids combine the high protein content of wheat with the high content of the amino acid lysine in rye. The lysine content is low in wheat and thus is a limiting nutritional factor. Wheat is considered a high-yielding grain, whereas rye is noted for its versatility of growth in unfavorable environments. *Triticale* species, combining both traits, have the potential of significantly increasing grain production. Programs designed to improve crops through hybridization have long been underway in several underdeveloped countries of the world, as discussed in Chapter 1.

Recall that in a previous chapter, we discussed the use of **somatic cell hybridization** to map human genes. (See Chapter 6.) This technique has also been applied to the production of amphidiploid plants (Figure 8–13). Cells from the developing leaves of plants can be treated to remove their cell wall, resulting in **protoplasts**. These altered cells can be maintained in culture and stimulated to fuse with other protoplasts, producing somatic cell hybrids. If cells from different plant species are fused in this way, hybrid amphidiploid cells can be produced. Because protoplasts can be induced to divide and differentiate into stems that develop leaves, the potential for producing allopolyploids is available in the research laboratory. In some cases, entire plants can be derived from cultured protoplasts. If only stems and leaves are produced, these can be grafted onto the stem of another plant. If flowers are formed, fertilization may yield mature seeds, which, upon germination, yield an allopolyploid plant.

There are now many examples of allopolyploids that have been created commercially by using the aforementioned approach. Although most of them are not true amphidiploids (one or more chromosomes are missing), precise allotetraploids are sometimes produced.

Now solve this

In Problem 8.7 on page 211 you are asked to consider a hybrid plant derived from two different species that is more ornate than either of its parents, but is sterile.

Hint: Allopolyploid plants are often sterile when they contain an odd number of each chromosome, resulting in unbalanced gametes during meiosis.

Endopolyploidy

Endopolyploidy is the condition in which only certain cells in an otherwise diploid organism are polyploid. In such cells, replication and separation of chromosomes occur without nuclear division. Numerous examples of naturally occurring endopolyploidy have been observed. For example, vertebrate liver cell nuclei, including human ones, often contain $4n$, $8n$, or $16n$ chromosome sets. The stem and parenchymal tissue of apical regions of flowering plants are also often endopolyploid. Cells lining the gut of mosquito larvae attain a $16n$ ploidy, but during the pupal stages, such cells undergo very quick reduction divisions, giving rise to smaller diploid cells. In the water strider *Gerris*, wide variations in chromosome numbers are found in different tissues, with as many as 1024 to 2048 copies of each chromosome in the salivary gland cells. Since the diploid number in this organism is 22, the nuclei of these cells may contain more than 40,000 chromosomes.

Although the role of endopolyploidy is not clear, the proliferation of chromosome copies often occurs in cells where high levels of certain gene products are required. In fact, it is well established that certain genes whose product is in high demand in *every* cell exist naturally in multiple copies in the genome. Ribosomal and transfer RNA genes are examples of multiple-copy genes. In certain cells of organisms, where even this condition may not allow for a sufficient amount of a particular gene product, it may be necessary to replicate the entire genome, allowing an even greater rate of expression of that gene.

8.6 Variation Occurs in the Structure and Arrangement of Chromosomes

The second general class of chromosome aberrations includes structural changes that delete, add, or rearrange substantial portions of one or more chromosomes. Included in this broad category are deletions and duplications of genes or part of a chromosome and rearrangements of genetic material in which a chromosome segment is inverted, exchanged with a segment of a nonhomologous chromosome, or merely transferred to another chromosome. Exchanges and transfers are called *translocations*, in which the location of a gene is altered within the genome. These types of chromosome alterations are illustrated in Figure 8–14.

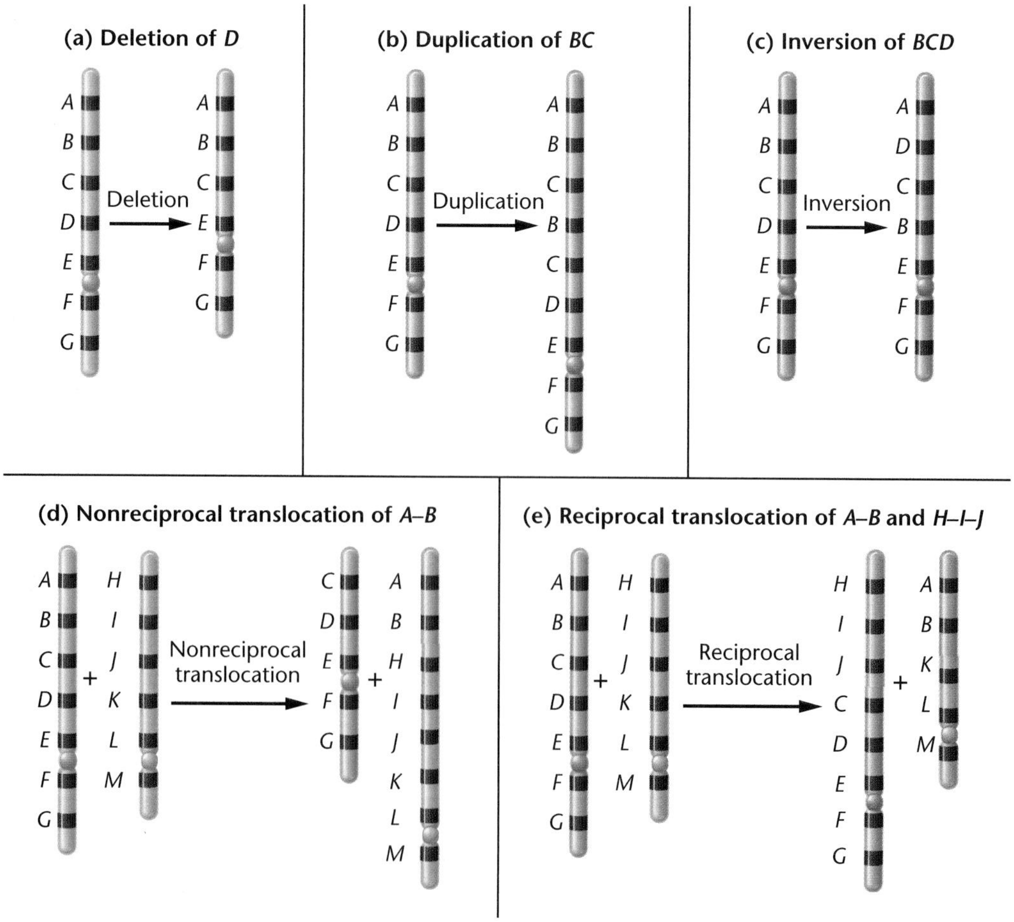

FIGURE 8–14 Overview of the five different types of rearrangement of chromosome segments.

Web Tutorial 8.3
Aberrations

In most instances, these structural changes are due to one or more breaks along the axis of a chromosome, followed by either the loss or rearrangement of genetic material. Chromosomes can break spontaneously, but the rate of breakage may increase in cells exposed to chemicals or radiation. Although the normal ends of chromosomes, known as telomeres, do not readily fuse with newly created ends of broken chromosomes or with other telomeres, the ends produced at points of breakage are described as being "sticky" and can rejoin other broken ends. If breakage and rejoining does not reestablish the original relationship, and if the alteration occurs in germ plasm, the gametes will contain the structural rearrangement, which is heritable.

If the aberration is found in one homolog, but not the other, the individual is said to be heterozygous for the aberration. In such cases, unusual but characteristic pairing configurations are formed during meiotic synapsis. These patterns are useful in identifying the type of change that has occurred. If no loss or gain of genetic material occurs, individuals bearing the aberration on one of the two homologs (described as being heterozygous for the aberration) are likely to be unaffected phenotypically. However, the unusual pairing arrangements often lead to gametes that are duplicated or deficient for some chromosomal regions. When this occurs, the offspring of "carriers" of certain aberrations often have an increased probability of demonstrating phenotypic manifestations.

8.7 A Deletion Is a Missing Region of a Chromosome

When a chromosome breaks in one or more places, and a portion of it is lost, the missing piece is referred to as a **deletion** (or a **deficiency**). The deletion can occur either near one end or from the interior of the chromosome. These are called **terminal** or **intercalary deletions**, respectively [Figure 8–15(a) and (b)]. The portion of the chromosome retaining the centromere region will usually be maintained when the cell divides, whereas the segment without the centromere will eventually be lost in progeny cells following mitosis or meiosis. For synapsis to occur between a chromosome with a large intercalary deficiency and a normal complete homolog, the unpaired region of the normal homolog must loop out of the linear structure into a **deletion** or **compensation loop** [Figure 8–15(c)].

As seen earlier in our discussion of the cri-du-chat syndrome, where only a small part of the short arm of chromosome 5 is lost, a deletion of a portion of a chromosome need not be very great before the effects become severe. If even more genetic information is lost as a result of a deletion, the aberration is often lethal. Such chromosome mutations never become available for study. Note that when a deletion involves a noticeable piece of a chromosome, as in cri-du-chat syndrome, we may also refer to the resulting condition as partial monosomy.

A final consequence of deletions can be noted in organisms heterozygous for a deficiency. Consider the mutant *Notch* phenotype in *Drosophila*. In these flies, the wings are notched on the posterior and lateral margins. Data from breeding studies indicate that the phenotype is controlled by an X-linked dominant

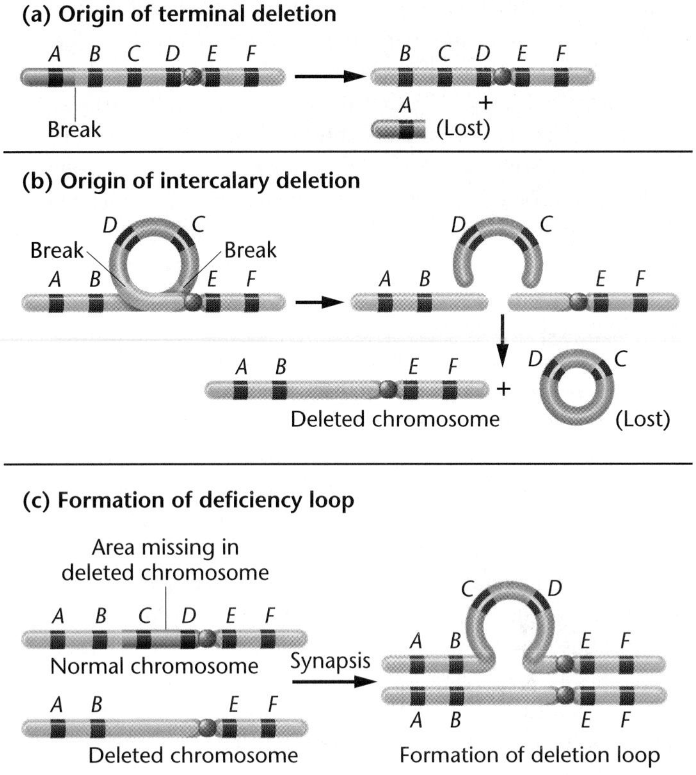

FIGURE 8–15 Origins of (a) a terminal and (b) intercalary deletion. In part (c), pairing occurs between a normal chromosome and one with an intercalary deletion by looping out the undeleted portion to form a deletion (or a compensation) loop.

mutation because heterozygous females have notched wings and transmit this allele to half of their female progeny. The mutation also appears to behave as a homozygous and hemizygous lethal because such females and males are never recovered. It has also been noted that if notched-winged females are also heterozygous for the closely linked recessive mutations *white*-eye, *facet*-eye, or *split*-bristle, they express these mutant phenotypes as well as *Notch*. Because these mutations are recessive, heterozygotes should express the normal, wild-type phenotypes. These genotypes and phenotypes are summarized in Table 8.2.

TABLE 8.2 NOTCH GENOTYPES AND PHENOTYPES

Genotype	Phenotype
$\frac{N^+}{N}$	*Notch* female
$\frac{N}{N}$	Lethal
$\frac{N}{\rightharpoondown}$	Lethal
$\frac{N^+ w}{N(w^+)*}$	*Notch, white* female
$\frac{N^+ fa\ spl}{N(fa^+ spl^+)*}$	*Notch, facet, split* female

* (deleted)

These observations have been explained through a cytological examination of the particularly large chromosomes characterizing certain larval cells of many insects. Called **polytene chromosomes** (see Chapter 12), they demonstrate a banding pattern specific to each region of each chromosome. In such cells of heterozygous *Notch* females, a deficiency loop was found along the X chromosome from the band designated 3C2 through band 3C11, as shown in Figure 8–16. These bands had previously been shown to include the loci for the *white, facet*, and *split* genes, among others. This region's deficiency in one of the two homologous X chromosomes has two distinct effects. First, it results in the *Notch* phenotype. Second, by deleting the loci for genes whose mutant alleles are present on the other X chromosome, the deficiency creates a partially hemizygous condition whereby the recessive *white, facet*, or *split* alleles are expressed whenever present. This type of phenotypic expression of recessive genes in association with a deletion is an example of the phenomenon called **pseudodominance**.

Many independently arising *Notch* phenotypes have been investigated. The common deficient band for all *Notch* phenotypes has now been designated as 3C7. In every case that *white* was also expressed pseudodominantly, the band 3C2 was also missing. The bands that cytologically distinguish the *Notch* locus from the *white* locus have been confirmed in this manner.

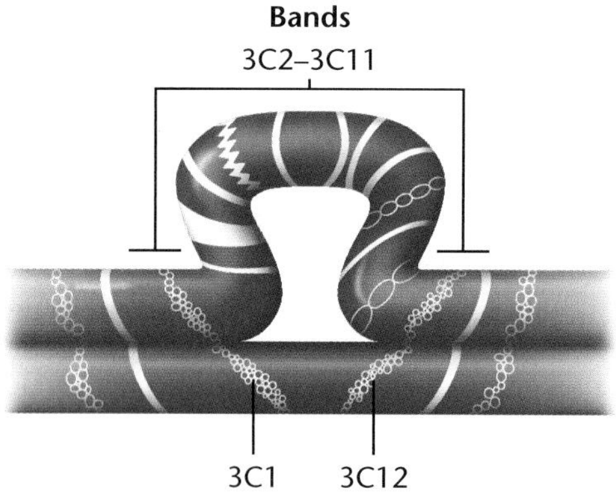

FIGURE 8–16 Deficiency loop formed in polytene chromosomes from the salivary glands of *Drosophila melanogaster* where the fly is heterozygous for a deletion. The deletion encompasses bands 3C2 through 3C11, corresponding to the region associated with the Notch phenotype.

8.8 A Duplication Is a Repeated Segment of the Genetic Material

When any part of the genetic material—a single locus or a large piece of a chromosome—is present more than once in the genome, it is called a **duplication**. As in deletions, pairing in heterozygotes may produce a compensation loop. Duplications can arise as the result of **unequal crossing over** between synapsed chromosomes during meiosis (Figure 8–17) or through a replication error prior to meiosis. In the former case, both a duplication and a deficiency are produced.

Next we'll consider three interesting aspects of duplications. First, they may result in gene redundancy. Second, as with deletions, duplications can produce phenotypic variation. Third, according to one convincing theory, duplications have also been an important source of genetic variability during evolution.

Gene Redundancy and Amplification: Ribosomal RNA Genes

Although many gene products are not needed in every cell of an organism, other gene products are known to be essential components of all cells. For example, ribosomal RNA must be present in abundance to support protein synthesis. The more metabolically active a cell is, the higher the demand for this molecule, which becomes a part of each ribosome. We might hypothesize that a single copy of a gene encoding rRNA is inadequate in many cells. Studies using the technique of molecular hybridization, which allows the determination of the percentage of the genome coding for specific RNA sequences, show that our hypothesis is correct. Indeed, it is the norm for organisms to have multiple copies of genes coding for rRNA. Collectively such DNA is called **rDNA**, and the general phenomenon is called *gene redundancy*. For example, in the common intestinal bacterium *Escherichia coli* (*E. coli*) about 0.4 percent of the haploid genome consists of rDNA. This is equivalent to 5 to 10 copies of the gene. In *Drosophila melanogaster*, 0.3 percent of the haploid genome, equivalent to 130 copies, consists of rDNA. Although the presence of multiple copies of the same gene is not restricted to those coding for rRNA, we focus on them in this section.

Studies of *Drosophila* have documented the critical importance of the extensive amounts of rRNA and ribosomes that

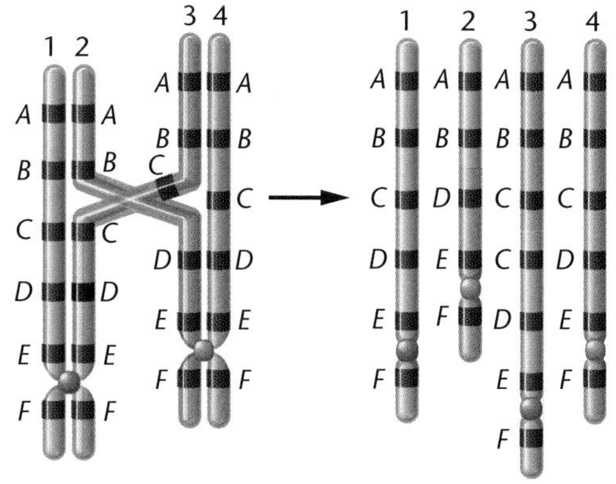

FIGURE 8–17 The origin of duplicated and deficient regions of chromosomes as a result of unequal crossing over. The tetrad at the left is mispaired during synapsis. A single crossover between chromatids 2 and 3 results in deficient and duplicated chromosomal regions. (See chromosomes 2 and 3, respectively, on the right.) The two chromosomes uninvolved in the crossover event remain normal in their gene sequence and content.

result from multiple copies of these genes. In *Drosophila*, the X-linked mutation *bobbed* has, in fact, been shown to be due to a deletion of a variable number of the 130 genes coding for rRNA. When the number of these genes is reduced and inadequate rRNA is produced, the result is mutant flies that have low viability, are underdeveloped, and have bristles reduced in size. Both general development and bristle formation, which occurs very rapidly during normal pupal development, apparently depend on a great number of ribosomes to support protein synthesis. Many *bobbed* alleles have been studied, and each has been shown to involve a deletion, often of a unique size. The reduction of both viability and bristle length correlates well with the relative number of rRNA genes deleted. We may conclude that the normal rDNA redundancy observed in wild-type *Drosophila* is near the minimum required for adequate ribosome production during normal development.

In some cells, particularly oocytes, even the normal redundancy of rDNA may be insufficient to provide adequate amounts of rRNA and ribosomes. Oocytes store abundant nutrients in the ooplasm for use by the embryo during early development. In fact, more ribosomes are included in the oocytes than in any other cell type. By considering how the amphibian *Xenopus laevis* (the South African clawed frog) acquires this abundance of ribosomes, we will see a second way in which the amount of rRNA is increased. This phenomenon is referred to as **gene amplification**.

The genes that code for rRNA are located in an area of the chromosome known as the **nucleolar organizer region (NOR)**. The NOR is intimately associated with the nucleolus, which is a processing center for ribosome production. Molecular hybridization analysis has shown that each NOR in *Xenopus* contains the equivalent of 400 redundant gene copies coding for rRNA. Even this number of genes is apparently inadequate to synthesize the vast amount of ribosomes that must accumulate in the amphibian oocyte to support development following fertilization. To further amplify the number of rRNA genes, the rDNA is selectively replicated, and each new set of genes is released from its template. Because each new copy is equivalent to a NOR, multiple small nucleoli are formed around each NOR in the oocyte. As many as 1500 of these "micronucleoli" have been observed in a single oocyte. If we multiply the number of micronucleoli (1500) by the number of gene copies in each NOR (400), we see that amplification in *Xenopus* oocytes can result in more than half a million gene copies. If each copy is transcribed only 20 times during the maturation of a single oocyte, in theory, sufficient copies of rRNA will be produced to result in more than one million ribosomes.

The *Bar*-Eye Mutation in *Drosophila*

Duplications can cause phenotypic variations that might at first appear to be caused by a simple gene mutation. The *Bar*-eye phenotype (Figure 8–18) in *Drosophila* is a classic example. Instead of the normal oval eye shape, *Bar*-eyed flies have narrow, slitlike eyes. This phenotype appears to be inherited as a dominant X-linked mutation. However, because both heterozygous females and hemizygous males exhibit the trait, but homozygous females show a more pronounced phenotype than either of these two cases, the inheritance more accurately illustrates the phenomenon of **semidominance**.

In the early 1920s, Alfred H. Sturtevant and Thomas H. Morgan discovered and investigated this apparent mutation. As illustrated in Figure 8–18(a), normal wild-type females (B^+/B^+) have about 800 facets in each eye. Heterozygous females (B^+/B) have about 350 facets, while homozygous females (B/B) average only about 70 facets. Females are occasionally recovered with even fewer facets and are designated as *double Bar* (B^D/B^+).

Some 10 years later, Calvin Bridges and Herman J. Muller compared the polytene X chromosome-banding pattern of the *Bar* fly with that of the wild-type fly. Recall from our discussion of deletions that such chromosomes contain specific banding patterns which characterize specific chromosomal regions. (See Figure 8–16.) The Bridges–Muller studies revealed that one copy of region 16A of the X chromosome was present in wild-type flies and that this region was duplicated in *Bar* flies and triplicated in *double-Bar* flies. These observations provided evidence that the *Bar* phenotype is not the result of a simple chemical change in the gene, but is instead a duplication. The *double-Bar* condition originates as a result of unequal crossing over, which produces the triplicated 16A region [Figure 8–18(b)].

Figure 8–18 also illustrates what is referred to as a **position effect**. You may recall from our introduction of this term in Chapter 4 that a position effect refers to altered gene expression resulting from new positioning of a gene within the genome. In this case, when the eye facet phenotypes of B/B and B^D/B^+ flies are compared, an average of 68 and 45 facets are found, respectively. In both cases, there are two extra 16A regions. However, when the repeated segments are distributed on the same homolog instead of being positioned on two homologs, the phenotype is more pronounced. Thus, the same amount of genetic information produces two distinct phenotypes, depending on the position of the genes.

How Do We Know?

How do we know that the *Bar*-eye mutation in *Drosophila* is due to a duplicated gene region rather than a change in the nucleotide sequence of a gene?

The Role of Gene Duplication in Evolution

During the study of evolution, it is intriguing to speculate on the possible mechanisms of genetic variation. The origin of unique gene products present in phylogenetically advanced organisms, but absent in less advanced ancestral forms is a topic of particular interest. In other words, how do "new" genes arise?

In 1970, Susumo Ohno published the provocative monograph *Evolution by Gene Duplication*, in which he elaborated on the

(a) Genotypes and Phenotypes

Genotype	Facet Number	Phenotype	= 16A segments
B^+/B^+	779		
B/B^+	358		
B/B	68		
B^D/B^+	45		

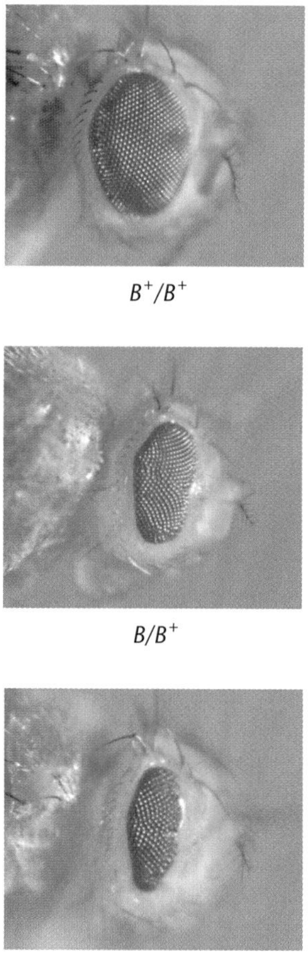

(b) Origin of B^D allele as a result of unequal crossing over

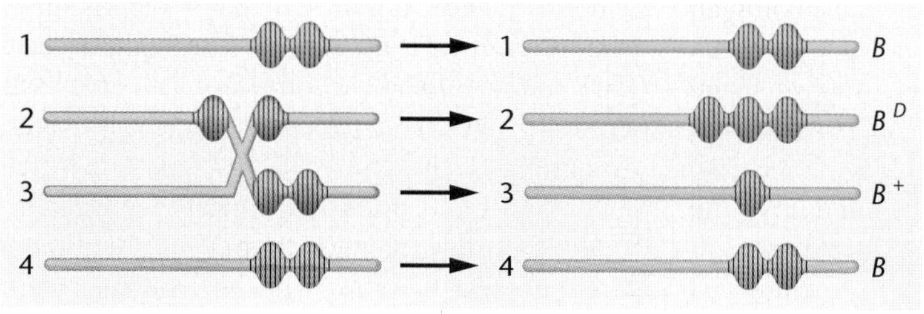

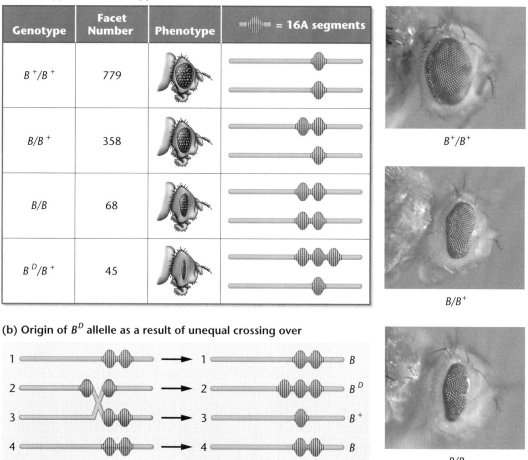

FIGURE 8–18 (a) The duplication genotypes and resultant *Bar*-eye phenotypes in *Drosophila*. Photographs show two *Bar*-eye phenotypes and the wild type (B^+/B^+). (b) The origin of the B^D (double-Bar) allele.

importance of gene duplication to the origin of new genes during evolution. While he was not the first to suggest that duplications might provide a "reservoir" from which new genes might arise, Ohno provided a detailed account of this idea. Ohno's thesis is based on the supposition that the gene products of essential genes, present as only a single copy in the genome, are indispensable to the survival of members of any species during evolution. Therefore, unique genes are not free to accumulate mutations that alter their primary function and potentially give rise to new genes.

However, if an essential gene were to become duplicated in the germ line, major mutational changes in this extra copy would be tolerated in future generations because the original gene provides the genetic information for its essential function. The duplicated copy would be free to acquire many mutational changes over extended periods of time. Over short intervals, the new genetic information may be of no practical advantage. However, over long evolutionary periods, the duplicated gene may change sufficiently so that its product assumes a divergent role in the cell. The new function may impart an "adaptive" advantage to organisms, enhancing their fitness. Ohno has outlined a mechanism through which sustained genetic variability may have originated.

Ohno's thesis is supported by the discovery of genes that have a substantial amount of their DNA sequence in common, but whose gene products are distinct. For example, the vertebrate digestive enzymes, trypsin and chymotrypsin, fit this description, as do the respiratory proteins, myoglobin and hemoglobin. The DNA sequence homology is great enough in these cases to conclude that members of each gene pair arose from a common ancestral gene through duplication. During evolution, the related genes diverged sufficiently so that their products became unique.

Other support includes the presence of **gene families**, regional groups of genes whose products perform the same general function. Again, members of a family show DNA sequence homology sufficient to conclude that they share a common origin. The various types of globin chains that are part of hemoglobin is one example. While the various globin chains function at different times during development, they are all a part of hemoglobin, functioning to transport oxygen. Other examples include the immunologically important T-cell receptors and antigens encoded by the major histocompatibility complex (MHC).

Many recent findings derived from our ability to sequence entire genomes have continued to support the idea that gene duplication has been a common feature of evolutionary

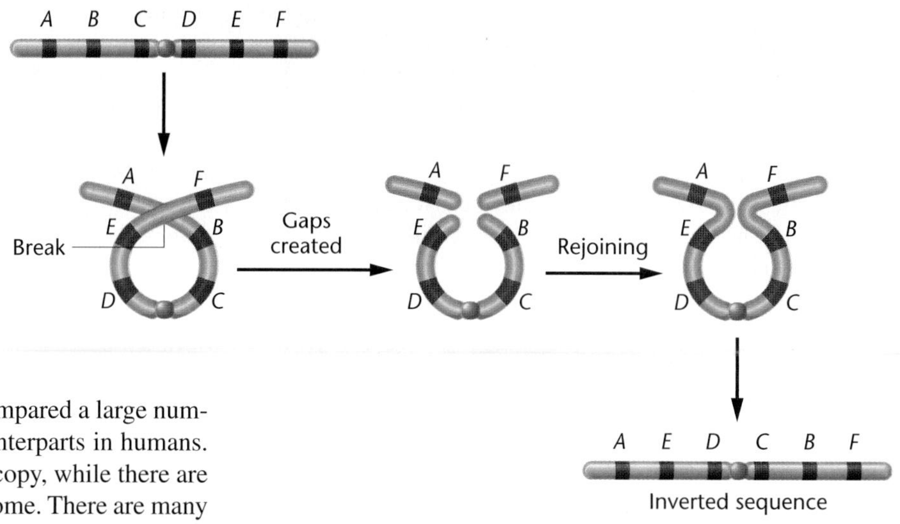

FIGURE 8–19 One possible origin of a pericentric inversion.

progression. For example, Jurg Spring compared a large number of genes in *Drosophila* and their counterparts in humans. In 50 genes studied, the fruit fly has one copy, while there are multiple copies present in the human genome. There are many other investigations that have provided similar findings. For example, of the approximately 5000 genes in yeast, there are 55 duplicated regions with 376 pairs of duplicated genes. In the mustard plant *Arabidopsis thaliana*, approximately 70 percent of the genome is duplicated. In humans, there are 1077 duplicated blocks of genes, with 781 of them containing five or more copies. Chromosome 18 and 20 contain large duplicated areas accounting for almost half of each chromosome.

An interesting debate is now occurring concerning a second aspect of Ohno's thesis—that major evolutionary jumps, such as the transition from invertebrates to vertebrates, may have involved the duplication of entire genomes. Ohno suggested that this might have occurred several times during the course of evolution. While it seems clear that genes, and even segments of chromosomes, have been duplicated, there is not yet any compelling evidence to convince evolutionary biologists that this was clearly the case. However, from the standpoint of gene expression, duplicating all genes proportionately seems more likely to be tolerated than duplicating just a portion of them. Whatever may be the case, it is interesting to examine evidence based on new technology that tests a hypothesis that was put forward more than 30 years ago, long before that technology was available.

8.9 Inversions Rearrange the Linear Gene Sequence

The **inversion**, another class of structural variation, is a type of chromosomal aberration in which a segment of a chromosome is turned around 180° within a chromosome.

An inversion does not involve a loss of genetic information, but simply rearranges the linear gene sequence. An inversion requires two breaks along the length of the chromosome and subsequent reinsertion of the inverted segment. Figure 8–19 illustrates how an inversion might arise. By forming a chromosomal loop prior to breakage, the newly created "sticky" ends are brought close together and rejoined. The inverted segment may be short or quite long and may or may not include the centromere. If the centromere *is* included in the inverted segment (as it is in Figure 8–19), the term **pericentric** is used to describe the inversion.

If the centromere is *not* part of the rearranged chromosome segment, the inversion is said to be **paracentric**. Although the gene sequence has been reversed in the paracentric inversion, the ratio of arm lengths extending from the centromere is unchanged (Figure 8–20). In contrast, some pericentric inversions create chromosomes with arms of different lengths from those of the noninverted chromosome, thereby changing the arm ratio. The change in arm length may sometimes be detected during the metaphase stage of mitotic or meiotic divisions.

Although inversions may seem to have a minimal impact on individuals bearing them, their consequences are of great interest to geneticists. Organisms heterozygous for inversions may produce aberrant gametes that have a major impact on their offspring. Inversions may also result in position effects and play an important role in the evolutionary process.

Consequences of Inversions during Gamete Formation

If only one member of a homologous pair of chromosomes has an inverted segment, normal linear synapsis during meiosis is not possible. Organisms with one inverted chromosome and

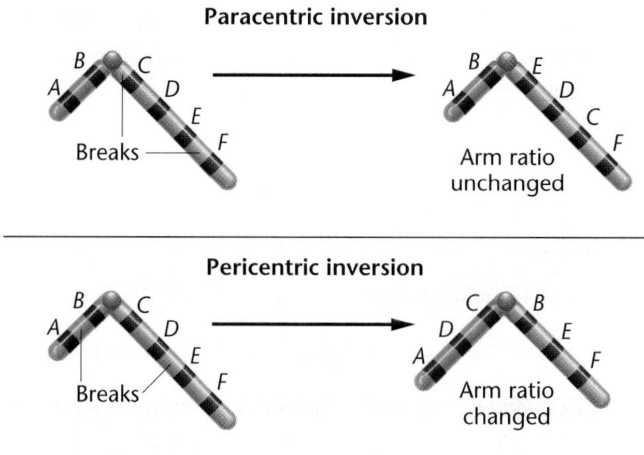

FIGURE 8–20 A comparison of the arm ratios of a submetacentric chromosome before and after the occurrence of a paracentric and pericentric inversion. Only the pericentric inversion results in an alteration of the original ratio.

one noninverted homolog are called **inversion heterozygotes**. Pairing between two such chromosomes in meiosis may be accomplished only if they form an **inversion loop** (Figure 8–21). In other cases, if no loop is formed, the homologs are seen to synapse everywhere but along the length of the inversion, where they appear separated.

If crossing over does not occur within the inverted segment of the inversion heterozygote, the homologs will segregate and result in two normal and two inverted chromatids that are distributed into gametes, and the inversion will be passed on to half of the offspring. However, if crossing over occurs within the inversion loop, abnormal chromatids are produced. For example, the effect of a single exchange event within a paracentric inversion is diagrammed in Figure 8–22(a). As in any meiotic tetrad, a single crossover between nonsister chromatids produces two parental chromatids and two recombinant chromatids. As shown, one recombinant chromatid is **dicentric** (two centromeres), and one recombinant chromatid is **acentric** (lacking a centromere). Both contain duplications and deletions of chromosome segments, as well. During anaphase, an acentric chromatid moves randomly to one pole or the other, or it may be lost, whereas a dicentric chromatid is pulled in two directions. This polarized movement produces a **dicentric bridge** that is cytologically recognizable. A dicentric chromatid will

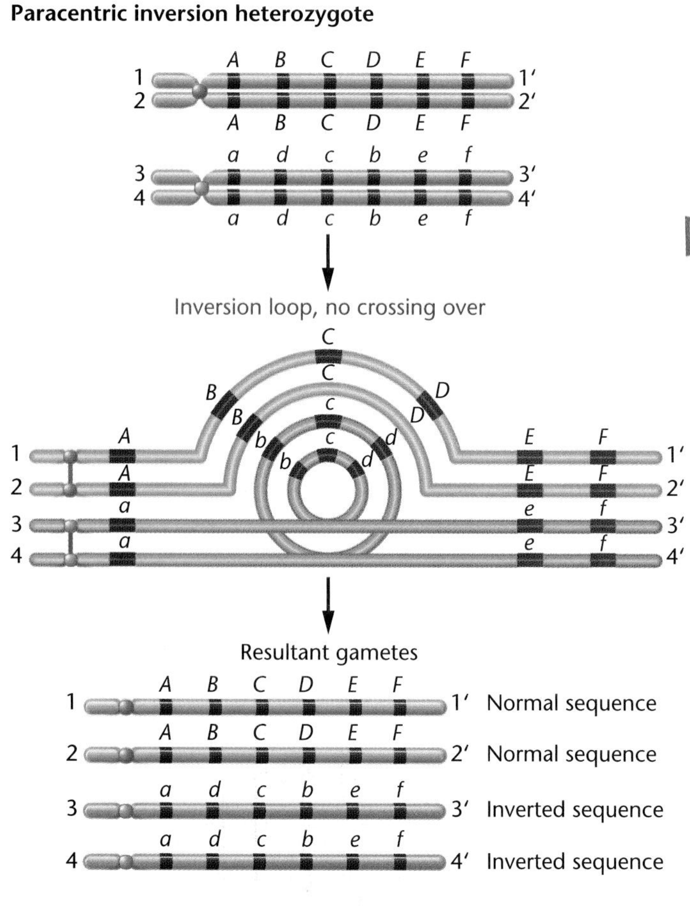

FIGURE 8–21 Illustration of how synapsis occurs in a paracentric inversion heterozygote by virtue of the formation of an inversion loop.

usually break at some point so that part of the chromatid goes into one gamete and part into another gamete during the reduction divisions. Therefore, gametes containing either broken chromatid are deficient in genetic material.

A similar chromosomal imbalance is produced as a result of a crossover event between a chromatid bearing a pericentric inversion and its noninverted homolog, as illustrated in Figure 8–22(b). The recombinant chromatids directly involved in the exchange have duplications and deletions. However, no acentric or dicentric chromatids are produced.

In plants, gametes receiving such aberrant chromatids usually fail to develop normally, leading to aborted pollen or ovules. Thus, lethality occurs prior to fertilization, and inviable seeds result. In animals, the gametes have developed prior to the meiotic error, so fertilization is more likely to occur in spite of the chromosome error. However, the end result is the production of inviable embryos following fertilization. In both cases, viability is reduced.

Because viable offspring do not result in either plants or animals, it appears as if the inversion suppresses crossing over, because offspring bearing crossover gametes are not recovered. However, in inversion heterozygotes, the inversion has the effect of *suppressing the recovery of crossover products* when chromosome exchange occurs within the inverted region. If crossing over always occurred within a paracentric or pericentric inversion, 50 percent of the gametes would be ineffective. The viability of the resulting zygotes is therefore greatly diminished. Furthermore, up to half of the viable gametes have the inverted chromosome, and the inversion will be perpetuated within the species. Moreover, the cycle will be repeated continuously during meiosis in future generations.

Now solve this

Problem 8.23 on page 212 considers a prospective male parent with a family history of stillbirths and malformed babies. He was shown to contain an inversion covering 70 percent of one member of chromosome 1. You are asked to explain the cause of the stillbirths.

Hint: Human chromosome 1 is metacentric, therefore an inversion covering 70 percent of the chromosome is no doubt a pericentric inversion.

Position Effects of Inversions

Another consequence of inversions involves the new positioning of genes relative to other genes and particularly to areas of the chromosome that do not contain genes, such as the centromere. If the expression of the gene is altered as a result of its relocation, a change in phenotype may result. Such a change is an example of what is called a position effect, as introduced in our earlier discussion of the *Bar* duplication.

We also introduced the general topic of position effect in Chapter 4, during our discussion of phenotypic expression. As we saw there, in *Drosophila* females heterozygous for the X-linked recessive mutation *white* eye (w^+/w), the X chromosome bearing the wild-type allele (w^+) may be inverted such that the *white* locus is moved to a point adjacent to the centromere. If the inversion is not present, a heterozygous female has wild-type red eyes, since the *white* allele is recessive. Females with the X chromosome inversion just described have eyes that are mottled or variegated, that is, with red and white patches. (See Figure 4–18.) Placement of the w^+ allele next to the centromere apparently inhibits wild-type gene expression, resulting in the loss of complete dominance over the *w* allele. Other genes, also located on the X chromosome, behave in the same manner when they are similarly relocated. Reversion to wild-type expression has sometimes been noted. When this has occurred, cytological examination has shown that the inversion has reestablished the normal sequence along the chromosome.

Evolutionary Advantages of Inversions

One major effect of an inversion is the maintenance of a set of specific alleles at a series of adjacent loci, provided that they are contained within the inversion. Because the recovery of crossover products is suppressed in inversion heterozygotes, a particular combination of alleles is preserved intact in the viable gametes. If the alleles of the involved genes provide a survival advantage to organisms maintaining them, the inversion is beneficial to the evolutionary survival of the species.

For example, if the set of alleles *ABcDef* is more adaptive than the sets *AbCdeF* or *abcdEF*, the favorable set of genes will not be disrupted by crossing over if it is maintained within a heterozygous inversion. As we will see in Chapter 26, there are documented examples where inversions are adaptive in this way. Specifically, Theodosius Dobzhansky has shown that the maintenance of different inversions on chromosome 3 of *Drosophila pseudoobscura* through many generations has been highly adaptive to this species. Certain inversions seem to be characteristic of enhanced survival under specific environmental conditions.

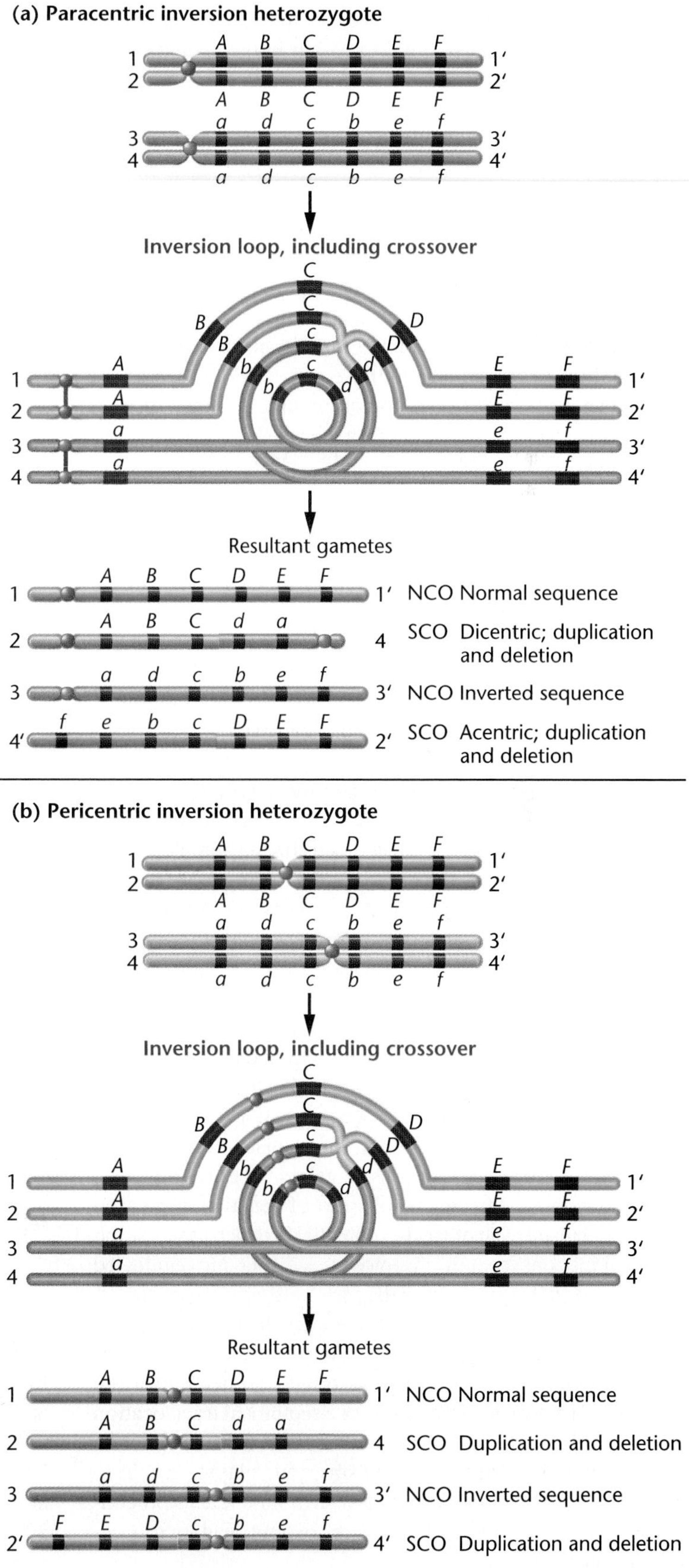

FIGURE 8–22 The effects of a single crossover within an inversion loop in cases involving (a) a paracentric inversion and (b) a pericentric inversion. In (a), two altered chromosomes are produced, one that is acentric and one that is dicentric. Both chromosomes also contain duplicated and deficient regions. In (b), two altered chromosomes are produced, both with duplicated and deficient regions. Ultimately, the number of viable gametes resulting from these events is reduced by about 50 percent because the abnormal outcomes are likely to be lethal.

8.10 Translocations Alter the Location of Chromosomal Segments in the Genome

Translocation, as the name implies, is the movement of a chromosomal segment to a new location in the genome. A **reciprocal translocation**, for example, involves the exchange of segments between two nonhomologous chromosomes. The least complex way for this event to occur is for two nonhomologous chromosome arms to come close to each other so that an exchange is facilitated. Figure 8–23(a) illustrates a simple reciprocal translocation in which only two breaks are required. If the exchange includes internal chromosome segments, four breaks are required, two on each chromosome.

The genetic consequences of reciprocal translocations are, in several instances, similar to those of inversions. For example, genetic information is not lost or gained; rather, there is only a rearrangement of genetic material. The presence of a translocation does not, therefore, directly alter the viability of individuals bearing it. Like an inversion, a translocation may also produce a position effect, because it may realign certain genes in relation to other genes. This exchange may create a new genetic linkage relationship that can be detected experimentally.

Homologs heterozygous for a reciprocal translocation undergo unorthodox synapsis during meiosis. As shown in Figure 8–23(b), pairing results in a crosslike configuration. As with inversions, genetically unbalanced gametes are also produced as a result of this unusual alignment during meiosis. In the case of translocations, however, aberrant gametes are not necessarily the result of crossing over. To see how unbalanced gametes are produced, focus on the homologous centromeres in Figure 8–23(b) and (c). According to the principle of independent assortment, the chromosome containing centromere 1 migrates randomly toward one pole of the spindle during the first meiotic anaphase; it travels with *either* the chromosome having centromere 3 *or* the chromosome having centromere 4. The chromosome with centromere 2 moves to the other pole, along with *either* the chromosome containing centromere 3 *or* centromere 4. This results in four potential meiotic products. The 1,4 combination contains chromosomes that are not involved in the translocation. The 2,3 combination, however, contains translocated chromosomes. These contain a complete complement of genetic information and are balanced. The other two potential products, the 1,3 and 2,4 combinations, contain chromosomes displaying duplicated and deleted (deficient) segments.

When incorporated into gametes, the resultant meiotic products are genetically unbalanced. If they participate in fertilization, lethality often results. As few as 50 percent of the progeny of parents heterozygous for a reciprocal translocation survive. This condition, called **semisterility**, has an impact on the reproductive fitness of organisms, thus playing a role in evolution. Furthermore, in humans, such an unbalanced condition results in partial monosomy or trisomy, leading to a variety of birth defects.

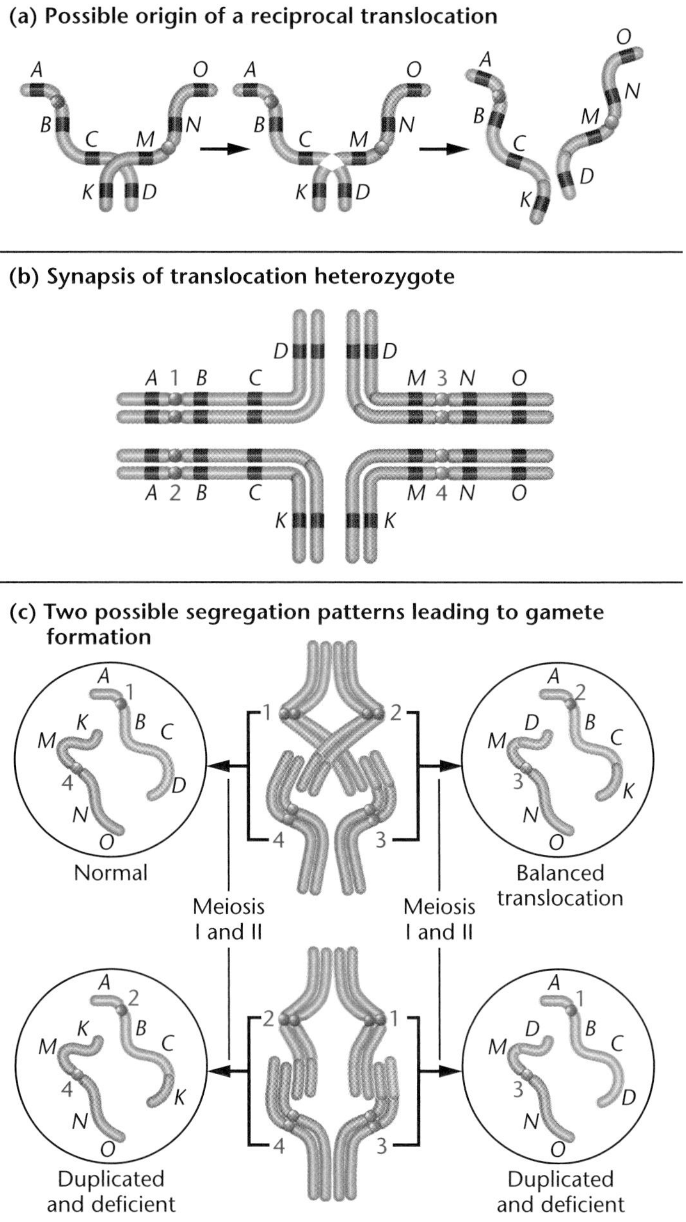

FIGURE 8–23 (a) The possible origin of a reciprocal translocation; (b) the synaptic configuration formed during meiosis in an individual that is heterozygous for the translocation. (c) Two possible segregation patterns, one of which leads to a normal and a balanced gamete (called **alternate segregation**) and one that leads to gametes containing duplications and deficiencies (called **adjacent segregation**).

Translocations in Humans: Familial Down Syndrome

Research conducted since 1959 has revealed numerous translocations in members of the human population. One common type, called a **Robertsonian translocation**, or **centric fusion**, involves breaks at the extreme ends of the short arms of two nonhomologous acrocentric chromosomes (13, 14, 15, 21, and 22). The small acentric fragments are lost, and the larger chromosomal segments fuse at their centromeric region, producing

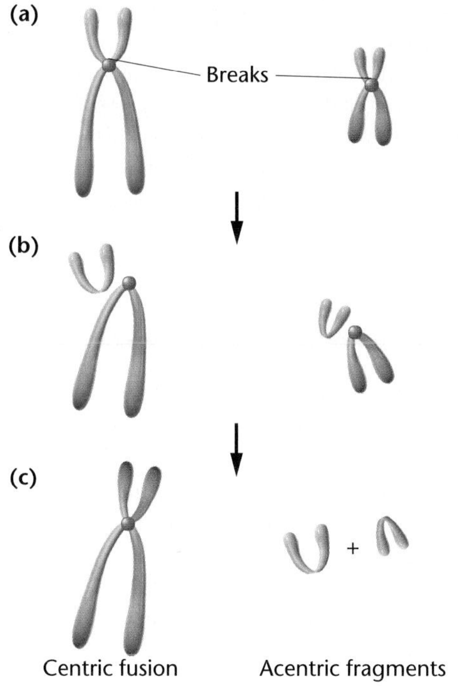

FIGURE 8–24 The possible origin of a Robertsonian translocation. Two independent breaks occur within the centromeric region on two nonhomologous chromosomes. Centric fusion of the long arms of the two acrocentric chromosomes creates the unique chromosome. Two acentric fragments remain.

a new, larger submetacentric or metacentric chromosome (Figure 8–24). These are the most common types of chromosome rearrangement in humans.

One such translocation accounts for cases in which Down syndrome is inherited or familial. Earlier in this chapter, we pointed out that most instances of Down syndrome are due to trisomy 21, resulting from nondisjunction during meiosis in one parent. Trisomy accounts for over 95 percent of all cases of Down syndrome. In such instances, the chance of the same parents producing a second afflicted child is extremely low. However, in the remaining families stricken with a Down child, the syndrome occurs in a much higher frequency over several generations.

Cytogenetic studies of the parents and their offspring from these unusual cases explain the cause of familial Down syndrome. Analysis reveals that one of the parents contains a **14/21 D/G translocation** (Figure 8–25). That is, one parent has the majority of the G-group chromosome 21 translocated to one end of the D-group chromosome 14. This individual is phenotypically normal even though he or she has only 45 chromosomes.

During meiosis, one fourth of the individual's gametes will have two copies of chromosome 21: a normal chromosome and most of a second copy translocated to chromosome 14. When such a gamete is fertilized by a standard haploid gamete, the resulting zygote has 46 chromosomes, but three copies of chromosome 21. These individuals exhibit Down syndrome. Other potential surviving offspring contain either the standard diploid genome (without a translocation) or the balanced translocation like the parent. Both cases result in phenotypically normal individuals. Knowledge of translocations has allowed geneticists to resolve the seeming paradox of an inherited trisomic phenotype in an individual with an apparent diploid number of chromosomes.

It is interesting to note that the carrier, who has 45 chromosomes and exhibits a normal phenotype, does not contain the *complete* diploid amount of genetic material. A small region is lost from both chromosomes 14 and 21 during the translocation event. This occurs because the ends of both chromosomes have broken off prior to their fusion. These specific regions are known to be two of many chromosomal locations housing multiple copies of the genes encoding rRNA, the major component of ribosomes. Despite the loss of up to 20 percent of these genes, the carrier is unaffected.

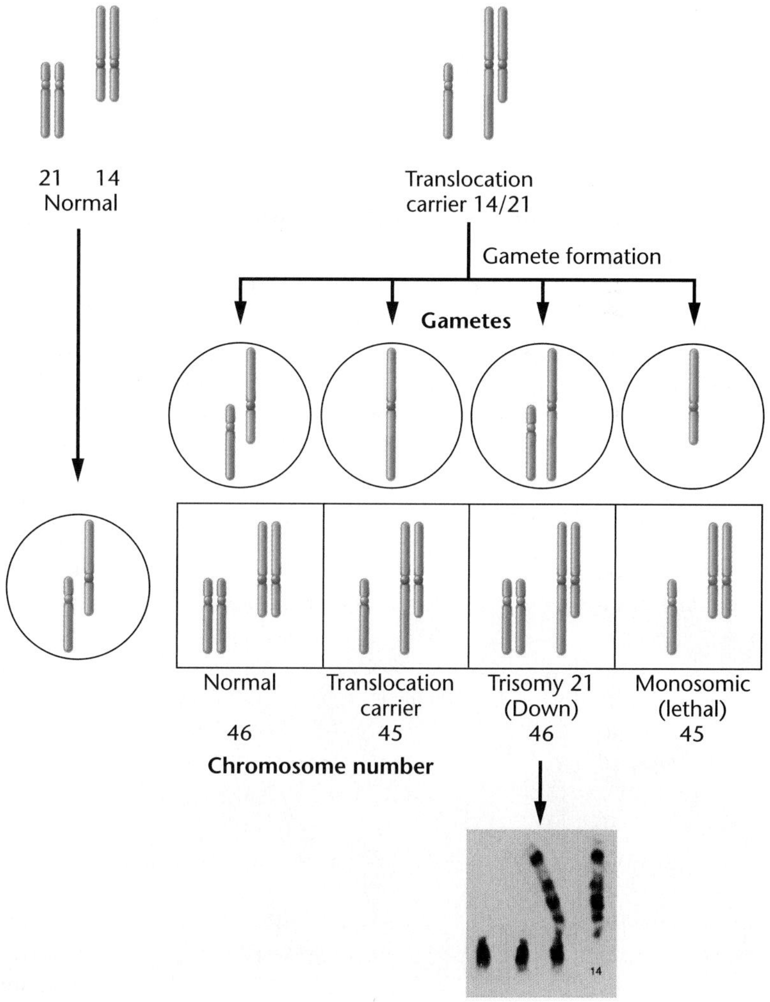

FIGURE 8–25 Chromosomal involvement in familial Down syndrome. The photograph illustrates the relevant chromosomes from a trisomy 21 offspring produced by a translocation carrier parent.

8.11 Fragile Sites in Humans Are Susceptible to Chromosome Breakage

We conclude this chapter with a brief discussion of the results of an interesting discovery made around 1970, during observations of metaphase chromosomes prepared following human cell culture. In cells derived from certain individuals, a specific area along one of the chromosomes failed to stain, giving the appearance of a gap. In other individuals whose chromosomes displayed such morphology, the gaps appeared at other places within the chromosome set. Such areas eventually became known as **fragile sites**, since they appeared to be susceptible to chromosome breakage when cultured in the absence of certain chemicals, such as folic acid, which is normally present in the culture medium. Fragile sites were at first considered curiosities, until a strong association was subsequently shown to exist between one of the sites and a form of mental retardation.

The cause of the fragility at these sites (and others) is unknown. Because they represent points along the chromosome that are susceptible to breakage, these sites may indicate regions where chromatin is not tightly coiled or compacted. Note that even though almost all studies of fragile sites have been carried out *in vitro*, using mitotically dividing cells, clear associations have been established between several of these sites and the corresponding altered phenotypes, including mental retardation and cancer.

Fragile X Syndrome (Martin–Bell Syndrome)

Many fragile sites do not appear to be associated with any clinical syndrome. However, individuals bearing a *folate-sensitive site* on the X chromosome (Figure 8–26) exhibit **fragile X syndrome** (or **Martin–Bell syndrome**), the most common form of inherited mental retardation. This syndrome affects about 1 in 4000 males and 1 in 8000 females. Because it is a dominant trait, females carrying only one fragile X chromosome can be mentally retarded. Fortunately, the trait is not fully penetrant, and many individuals with the genetic defect are not affected.

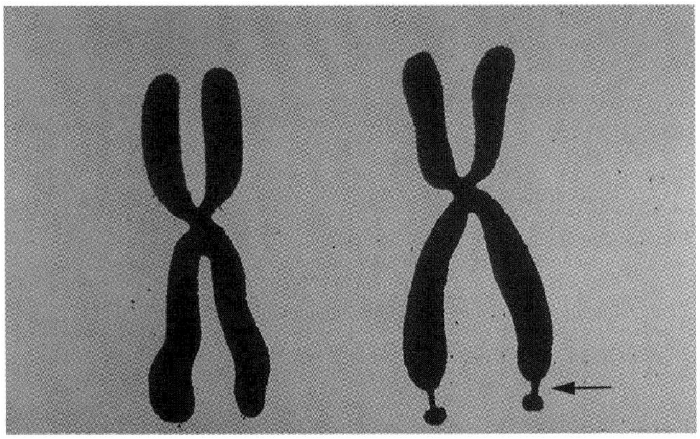

FIGURE 8–26 A normal human X chromosome (left) contrasted with a fragile X chromosome (right). The "gap" region (near the bottom of the chromosome) is associated with the fragile X syndrome.

In addition to mental deficiency, affected individuals often exhibit attention deficit disorder or autistic-like behavior. They also have characteristic long, narrow faces with protruding chins and enlarged ears, and males have increased testicular size.

A gene spanning the fragile site is responsible for this syndrome. This gene, known as *FMR1*, is one of a growing number of genes that have been discovered in which a sequence of three nucleotides is repeated many times, expanding the size of the gene. This phenomenon, called **trinucleotide repeats**, has been recognized in other human disorders, including Huntington disease. In *FMR1*, the trinucleotide sequence CGG is repeated in a transcribed, but an untranslated area called the 5′ UTR—the "untranslated region" that is upstream from the coding sequence of the gene. The number of repeats varies immensely within the human population, and a high number correlates directly with expression of fragile X syndrome. Normal individuals have between 6 and 54 repeats, whereas those with 55 to 230 repeats are considered to be "carriers" of the disorder. A level above 230 repeats leads to expression of the syndrome.

It is thought that when the number of repeats reaches this level, the CGG regions of the gene become chemically modified so that the bases within and around the repeat are methylated, causing inactivation of the gene. The normal product of the gene (FMRP—fragile X mental retardation protein) is an RNA-binding protein known to be expressed in the brain. Evidence is now accumulating directly linking the absence of the protein with the cognitive defects associated with the syndrome.

The protein is prominent in cells of the developing brain and normally shuttles between the nucleus and cytoplasm, transporting mRNAs to ribosomal complexes. It is believed that its absence in dendritic cells prevents the translation of other critical proteins during the development of the brain, presumably those produced from mRNAs that are transported by FMRP. Ultimately, the result is impairment of synaptic activity related to learning and memory. Molecular analysis has shown that FMRP is very selective, binding specifically to mRNAs that have a high guanine content. These RNAs form unique secondary structures called **G-quartets**, for which the protein has a strong binding affinity. That possession of such quartets is important to recognition by FMRP is evident by examining cases where the formation of these stacked quartets is inhibited (using Li^+). In such cases, FMRP binding to these RNAs is abolished.

The *Drosophila* homolog (*dfrx*) of the human *FRM1* gene has now been identified and used as a model for experimental investigation. The normal protein product of the fruit fly displays chemical properties that parallel its human counterpart. Mutations have been generated and studied that either overexpress the gene or eliminate its expression altogether. The latter case is designated a null mutation, where the gene has been "knocked out" using a molecular procedure now common in DNA biotechnology. The results of these studies show that the normal gene functions during the formation of synapses and that abnormal synaptic function accompanies the loss of *dfrx* expression. **Knockout mice** (mice that have had one or more genes experimentally excised from their genome; see Chapter 13) lacking this gene also display a similar phenotype. Ultimately, then, a closely related version of this gene is conserved in humans, *Drosophila*, and mice. In all three species, it has been shown to be essential to synaptic maturation and the subsequent establishment of proper cell communication in the brain. In humans, the neurological features of the syndrome are linked to the absence of its genetic expression.

From a genetic standpoint, one of the most interesting aspects of fragile X syndrome is the instability of the CGG repeats. An individual with 6 to 54 repeats transmits a gene containing the same number to his or her offspring. However, those with 55 to 230 repeats, while not at risk to develop the syndrome, may transmit to their offspring a gene with an increased number of repeats. The number of repeats continues to increase in future generations, illustrating the phenomenon known as **genetic anticipation**, first introduced in Chapter 4. Once the threshold of 200 is exceeded, expression of the malady becomes more severe in each successive generation as the number of trinucleotide repeats increases.

While the mechanism that leads to the trinucleotide expansion has not yet been established, several factors influencing the instability are known. Most significant is the observation that expansion from the carrier status (55 to 200 repeats) to the syndrome status (more than 200 repeats) occurs during the transmission of the gene by the maternal parent, but not by the paternal parent. Furthermore, several reports suggest that male offspring are more likely to receive the increased repeat size leading to the syndrome than are female offspring. Obviously, we have more to learn about the genetic basis of instability and expansion of DNA sequences.

Genetics, Technology, and Society

The Link between Fragile Sites and Cancer

While the study of the fragile X syndrome first brought unstable chromosome regions to the attention of geneticists, a second link between a fragile site and a human disorder was reported in 1996 by Carlo Croce, Kay Huebner, and their colleagues. They demonstrated an association between an autosomal fragile site and lung cancer. Many years of investigation led this research group to postulate that the defect associated with this fragile site contributes to the formation of a variety of different tumor types. Croce and Huebner first showed that the *FHIT* gene (standing for *f*ragile *hi*stidine *t*riad), located within the well-defined fragile site designated as *FRA3B* (on the p arm of chromosome 3), is often altered or missing in cells taken from tumors of individuals with lung cancer. Molecular analysis of numerous mutations showed that the DNA had been broken and incorrectly fused back together, resulting in deletions within the *FHIT* gene. In most cases, these mutations resulted in the loss of the normal protein product of this gene. More extensive studies have now revealed that this protein is absent in cells of many other cancers, including those of the esophagus, breast, cervix, liver, kidney, pancreas, colon, and stomach. Genes such as *FHIT* that are located within fragile regions undoubtedly exhibit an increased susceptibility to mutations and deletions.

We now have a better understanding of fragile sites, with more than 80 such regions identified in the Human Genome Data Base. Fragile sites are said to be recombinogenic, based on the observations that within these chromosomal regions, alterations such as translocations, sister-chromatid exchanges (SCEs), and other rearrangements between chromosomes frequently occur. It is now believed that breaks and gaps associated with fragile sites are induced by agents that inhibit DNA replication. Such gaps and breaks are therefore the result of DNA that has been incompletely replicated. Synthesis of DNA within these regions during the S phase of the cell cycle is delayed as compared to non fragile chromosomal regions. This hypothesis has been strengthened in the case of the *FRA3B* region, now shown to be characterized by delayed DNA replication. As a result, the DNA in this is region is not replicated by the time the cell enters the G_2 phase and subsequently initiates mitosis. The resulting metaphase chromosomes reveal distinct cytogenetic gaps and breaks within the fragile region. If these breaks are incorrectly repaired during cell cycle checkpoints, cancer-specific mutations may occur if the region contains genes involved in cell cycle control.

FHIT is a good example and has now been designated as a tumor-suppressor gene. The product of the normal gene is believed to recognize genetic damage in cells and to induce apoptosis, whereby a potentially malignant cell is targeted to undergo programmed cell death and is effectively destroyed. The failure to remove such cells as a result of mutations in the *FHIT* gene causes individuals to be particularly sensitive to carcinogen-induced damage, creating a susceptibility to cancer. For example, cigarette smokers who develop lung cancer demonstrate an increased expression of the FRA3B fragile region compared to the normal population. Still in question, however, is whether the alteration in the fragile site, leading to inactivation of the gene, causes cancer, *or* whether the behavior of malignant cells somehow induces breaks at fragile sites, subsequently inactivating or deleting this gene. Another important question under investigation is the extent of molecular polymorphism at the fragile site within the human population, causing some individuals to be more susceptible to the effects of carcinogens than others. Whatever the answers, these are significant and exciting studies that may prove valuable in the future for the prevention of common human cancers and the development of gene therapy.

Huebner, K & Croce, C. (2003). Cancer and the *FRA3B/FHIT* fragile locus: it's a HIT. *British J. Cancer*. 88:1501-1506.

Chapter Summary

1. Investigations into the uniqueness of each organism's chromosomal constitution have further enhanced our understanding of genetic variation. Alterations of the precise diploid content of chromosomes are referred to as chromosomal aberrations or chromosomal mutations.
2. Deviations from the expected chromosomal number, or mutations in the structure of the chromosome, are inherited in predictable Mendelian fashion. They often result in inviable organisms or substantial changes in the phenotype.
3. Aneuploidy is the gain or loss of one or more chromosomes from the diploid content, resulting in conditions of monosomy, trisomy, tetrasomy, etc. Studies of monosomic and trisomic disorders are increasing our understanding of the delicate genetic balance that enables normal development.
4. When complete sets of chromosomes are added to the diploid genome, polyploidy occurs. These sets can have identical or diverse genetic origin, creating either autopolyploidy or allopolyploidy, respectively.
5. Large segments of the chromosome can be modified by deletions or duplications. Deletions can produce serious conditions such as the cri-du-chat syndrome in humans, whereas duplications can be particularly important as a source of redundant or new genes.
6. Inversions and translocations, while altering the gene order along chromosomes, initially cause little or no loss of genetic information or deleterious effects. However, heterozygous combinations may cause genetically abnormal gametes following meiosis, with death a common result.
7. Fragile sites in human mitotic chromosomes have sparked research interest because one such site on the X chromosome is associated with the most common form of inherited mental retardation. Another fragile site, located on chromosome 3, has been linked to lung cancer.

Insights and Solutions

1. In a cross involving three linked genes, *a*, *b*, and *c* in maize, a heterozygote (*abc*/+++) was testcrossed (to *abc*/*abc*). Even though the three genes were separated along the chromosome, hence predicting that crossover gametes and the resultant phenotype should be observed, only two phenotypes were recovered: *abc* and +++. Additionally, the cross produced significantly fewer viable seeds (and thus plants) than expected. Can you propose why no other phenotypes were recovered and why viability was reduced?

Solution: One of the two chromosomes contains an inversion that overlaps all three genes, effectively precluding the recovery of any "crossover" offspring. If this is a paracentric inversion and the genes are clearly separated (assuring that a significant number of crossovers will occur between them), then numerous acentric and dicentric chromosomes will be formed, resulting in the observed reduction in viability.

2. A male *Drosophila* from a wild-type stock was discovered to have only seven chromosomes, whereas the normal $2n$ number is eight. Close examination revealed that one member of chromosome IV (the smallest chromosome) was attached to (translocated to) the distal end of chromosome II and was missing its centromere, consequently accounting for the reduction in chromosome number.

(a) Diagram all members of chromosomes II and IV during synapsis in Meiosis I.

Solution:

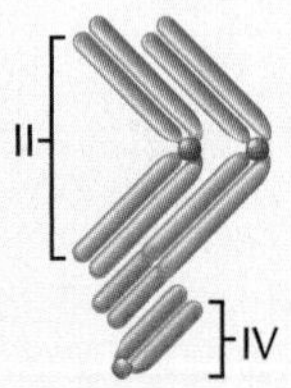

(b) If this male mates with a female with a normal chromosome composition who is homozygous for the recessive chromosome IV mutation *eyeless* (*ey*), what chromosome compositions will occur in the offspring regarding chromosomes II and IV?

Solution:

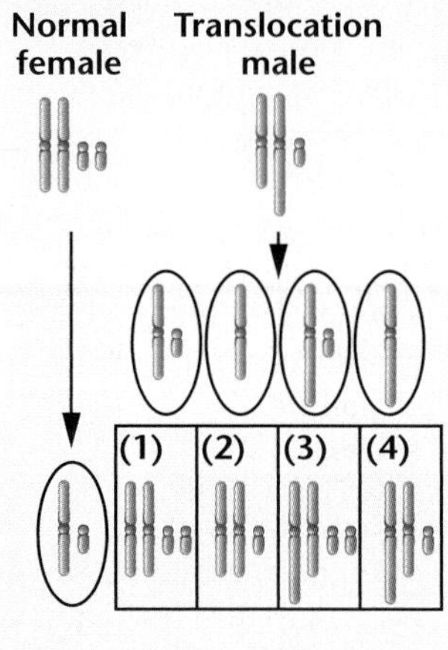

(c) With reference to the solution in (b), what phenotypic ratio will result regarding the presence of eyes, assuming all chromosome compositions survive?

Solution:

(1) normal (heterozygous)
(2) eyeless (monosomic, contains chromosome IV from mother)
(3) normal (heterozygous)
(4) normal (heterozygous)

The final ratio is 3/4 normal:1/4 eyeless.

3. If a Haplo-IV female *Drosophila* (containing only one chromosome 4, but an otherwise normal set of chromosomes) that has white eyes (an X-linked trait) and normal bristles is crossed with a male with a diploid set of chromosomes and normal red eye color, but who is homozygous for the recessive chromosome 4 bristle mutation, *shaven* (*sv*), what F_1 phenotypic ratio might be expected?

Solution: Let us first consider only the eye-color phenotypes. This is a straightforward X-linked cross. Offspring will appear as 1/2 red females:1/2 white males as shown here:

P_1:	ww	×	w^+/Y
	white female		red male
F_1:	$1/2\ ww^+$	:	$1/2\ w/Y$
	red female		white male

The bristle phenotypes will be governed by the fact that the normal-bristle P_1 female produces gametes, half of which contain a chromosome $4(sv^+)$ and half that have no chromosome 4 ("−" designates no chromosome). Following fertilization by sperm from the *shaven* male, half of the offspring will receive two members of chromosome 4 and be heterozygous for *sv*, expressing normal bristles. The other half will have only one copy of chromosome 4. Because its origin is from the male parent, where the chromosome bears the *sv* allele, these flies will express *shaven* because there is no wild-type allele present to mask this recessive allele:

P_1:	$sv^+/-$	×	sv/sv
	normal female		shaven male
F_1	$1/2\ sv^+/sv$ males and females		$= 1/2$ wild type
	$1/2\ sv/-$ males and females		$= 1/2$ shaven

Using the forked-line method, we can consider both eye color and bristle phenotypes together:

1/2 red-eyed females	→ 1/2 normal bristles	→ 1/4 red-eyed, normal-bristled females
	→ 1/2 shaven bristles	→ 1/4 red-eyed, shaven-bristled females
1/2 white-eyed males	→ 1/2 normal bristles	→ 1/4 white-eyed, normal-bristled males
	→ 1/2 shaven bristles	→ 1/4 white-eyed, shaven-bristled males

Problems and Discussion Questions

1. For a species with a diploid number of 18, indicate how many chromosomes will be present in the somatic nuclei of individuals who are haploid, triploid, tetraploid, trisomic, and monosomic.
2. Define and distinguish between the following pairs of terms:
   aneuploidy/euploidy
   monosomy/trisomy
   Patau syndrome/Edwards syndrome
   autopolyploidy/allopolyploidy
   autotetraploid/amphidiploid
   paracentric inversion/pericentric inversion
3. Contrast the relative survival times of individuals with Down syndrome, Patau syndrome, and Edwards syndrome. Speculate as to why such differences exist.
4. What evidence suggests that Down syndrome is more often the result of nondisjunction during oogenesis rather than during spermatogenesis?
5. What evidence exists that humans with aneuploid karyotypes occur at conception, but are usually inviable?
6. Contrast the fertility of an allotetraploid with an autotriploid and an autotetraploid.
7. When two plants belonging to the same genus, but different species were crossed together, the F_1 hybrid was more viable and had more ornate flowers. Unfortunately, this hybrid was sterile and could only be propagated by vegetative cuttings. Explain the sterility of the hybrid. How might a horticulturist attempt to reverse its sterility?
8. Describe the origin of cultivated American cotton.
9. Predict how the synaptic configurations of homologous pairs of chromosomes might appear when one member is normal and the other member has sustained a deletion or duplication.
10. Inversions are said to "suppress crossing over." Is this terminology technically correct? If not, restate the description accurately.
11. Contrast the genetic composition of gametes derived from tetrads of inversion heterozygotes where crossing over occurs within a paracentric and pericentric inversion.
12. Contrast the *Notch* locus with the *Bar* locus in *Drosophila*. What phenotypic ratios would be produced in a cross between *Notch* females and *Bar* males?
13. Discuss Ohno's hypothesis on the role of gene duplication in the process of evolution and the evidence in support of the hypothesis.
14. What roles have inversions and translocations played in the evolutionary process?
15. A human female with Turner syndrome (45,X) also expresses the X-linked trait hemophilia, as did her father. Which of her parents underwent nondisjunction during meiosis, giving rise to the gamete responsible for the syndrome?
16. The primrose, *Primula kewensis*, has 36 chromosomes that are similar in appearance to the chromosomes in two related species, *Primula floribunda* ($2n = 18$) and *Primula verticillata* ($2n = 18$). How could *P. kewensis* arise from these species? How would you describe *P. kewensis* in genetic terms?
17. Varieties of chrysanthemums are known that contain 18, 36, 54, 72, and 90 chromosomes; all are multiples of a basic set of 9 chromosomes. How would you describe these varieties genetically?

What feature is shared by the karyotypes of each variety? A variety with 27 chromosomes was discovered, but it was sterile. Why?

18. What is the effect of a rare double crossover within a pericentric inversion present heterozygously? What is the effect within a paracentric inversion present heterozygously?

19. *Drosophila* may be monosomic for chromosome 4 and remain fertile. Contrast the F_1 and F_2 results of the following crosses involving the recessive chromosome 4 trait, *bent bristles*:
(a) monosomic IV, bent bristles × diploid, normal bristles
(b) monosomic IV, normal bristles × diploid, bent bristles

20. *Drosophila* may also be trisomic for chromosome 4 and remain fertile. Predict the F_1 and F_2 results of crossing trisomic, bent bristles ($b/b/b$) × diploid, normal brisles (b^+/b^+).

21. Mendelian ratios are modified in crosses involving autotetraploids. Assume that one plant expresses the dominant trait green (seeds) and is homozygous (*WWWW*). This plant is crossed to one with white seeds that is also homozygous (*wwww*). If only one dominant allele is sufficient to produce green seeds, predict the F_1 and F_2 results of such a cross. Assume that synapsis between chromosome pairs is random during meiosis.

22. Having correctly established the F_2 ratio in Problem 21, predict the F_2 ratio of a "dihybrid" cross involving two independently assorting characteristics (e.g., $P_1 = WWWWAAAA \times wwwwaaaa$).

23. A couple, in looking ahead to planning a family, were aware that through the past three generations on the male's side a substantial number of stillbirths had occurred and several malformed babies were born who died early in childhood. The woman urged her husband to visit a genetic counseling clinic, where a complete karyotype-banding analysis was subsequently performed. Although it was found that he had a normal complement of 46 chromosomes, banding analysis revealed that one member of the chromosome 1 pair (in Group A) contained an inversion covering 70 percent of its length. The homolog of chromosome 1 and all other chromosomes showed the normal banding sequence.
(a) How would you explain the high incidence of past stillbirths?
(b) What would you predict about the probability of abnormality/normality of the couple's children?
(c) Would you advise the woman to "wait out" each pregnancy to term so as to determine whether the fetus was normal? If not, what would you suggest she do?

Extra-Spicy Problems

24. In a cross in *Drosophila*, a female heterozygous for the autosomally linked genes *a*, *b*, *c*, *d*, and *e* ($abcde/{+}{+}{+}{+}{+}$) was testcrossed to a male homozygous for all recessive alleles. Even though the distance between each of the these loci was at least 3 map units, only four phenotypes were recovered, yielding the following data:

Phenotype	No. of Flies
+ + + + +	440
a b c d e	460
+ + + + *e*	48
a b c d +	52
Total =	1000

Why are many expected crossover phenotypes missing? Can any of these loci be mapped from the data given here? If so, determine map distances.

25. A woman seeking genetic counseling was found to be heterozygous for a chromosomal rearrangement between the second and third chromosomes. Her chromosomes, compared with those in a normal karyotype, are diagrammed here:

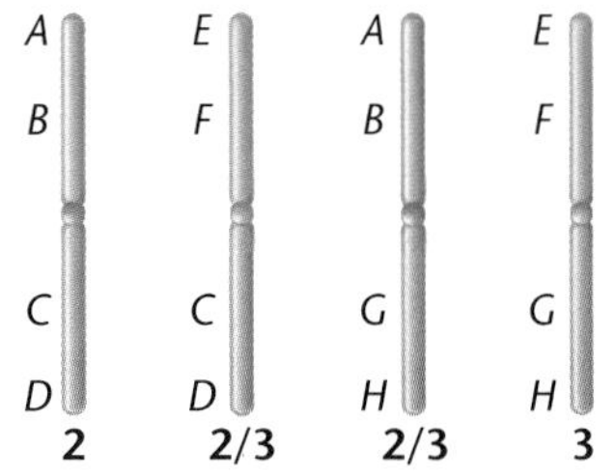

(a) What kind of chromosomal aberration is shown?
(b) Using a drawing, demonstrate how these chromosomes would pair during meiosis. Be sure to label the different segments of the chromosomes.
(c) This woman is phenotypically normal. Does that surprise you? Why or why not? Under what circumstances might you expect a phenotypic effect of such a rearrangement?
(d) This woman has had two miscarriages. She has come to you, an established genetic counselor, for advice. She raises the following questions: Is there a genetic explanation for her frequent miscarriages? Should she abandon her attempts to have a child of her own? If not, what is the chance that she could have a normal child? Provide an informed response to each of her concerns.

26. A boy with Klinefelter syndrome (47,XXY) is born to a mother who is phenotypically normal and a father who has the X-linked skin condition called anhidrotic ectodermal dysplasia. The mother's skin is completely normal with no signs of the skin abnormality. In contrast, her son has patches of normal skin and patches of abnormal skin.
(a) Which parent contributed the abnormal gamete?
(b) Using the appropriate genetic terminology, describe the meiotic mistake that occurred. Be sure to indicate which division the mistake occurred in.
(c) Using the appropriate genetic terminology, explain the son's skin phenotype.

27. To investigate the origin of nondisjunction, 200 human oocytes that had failed to be fertilized during *in vitro* fertilization procedures were examined (Angel, R. 1997. *Am. J. Hum. Genet.* 61:23–32). These oocytes had completed meiosis I and were arrested in metaphase II (MII). The majority (67 percent) had a normal MII-metaphase complement, showing 23 chromosomes, each consisting of two sister chromatids joined at a common centromere. The remaining oocytes all had abnormal chromosome compositions. Surprisingly, when trisomy is considered, none of the abnormal oocytes had 24 chromosomes.
(a) Interpret these results in regard to the origin of trisomy, as it relates to nondisjunction and when it occurs. Why are the results surprising?

A large number of the abnormal oocytes contained 22 1/2 chromosomes; that is, 22 chromosomes plus a single chromatid representing the 1/2 chromosome.
(b) What chomosome compositions will result in the zygote if such oocytes proceed through meiosis and are fertilized by normal sperm?

(c) How could the complement of 22 1/2 chromosomes arise? Answer this question with a drawing that includes several pairs of MII chromosomes.

(d) Do these findings support or dispute the generally accepted theory regarding nondisjunction and trisomy, as outlined in Figure 8–1?

28. In a human genetic study, a family was investigated with five phenotypically normal children. Two were "homozygous" for a Robertsonian translocation between chromosomes 19 and 20. (They contained two identical copies of the fused chromosome.) These children have only 44 chromosomes, but a complete genetic complement. Three of the children were "heterozygous" for the translocation, and contained 45 chromosomes, with one translocated chromosome, plus a normal copy of both chromosome 19 and 20. Two other pregnancies resulted in stillbirths. It was discovered that the parents were first cousins. Based on the above information, determine the chromosome compositions of the parents. What led to the stillbirths? Why was the discovery that the parents were first cousins a key piece of information in understanding the genetics of this family?

29. A 3-year-old child exhibited some early indication of Turner syndrome, which results from a 45,X chromosome composition. Karyotypic analysis demonstrated two cell types: 46,XX (normal) and 45,X. Propose a mechanism for the origin of this mosaicism.

30. A normal female is discovered with 45 chromosomes, one of which exhibits a Robertsonian translocation containing most of chromosomes 18 and 21. Discuss the possible outcomes of her offspring when her husband contains a normal karyotype.

31. In humans, cystic fibrosis (CF) is a recessive disorder characterized by abnormal mucous secretions that affect the lungs. Because it is a rare disorder, most affected children have two parents that are heterozygous carriers. However, an occasional CF child has only one parent that is a carrier. Propose an explanation for this phenomenon.

SELECTED READINGS

Antonarakis, S.E. 1998. Ten years of genomics, chromosome 21, and Down syndrome. *Genomics* 51:1–16.

Ashley-Koch, A.E., et al. 1997. Examination of factors associated with instability of the *FMR1* CGG repeat. *Am. J. Hum. Genet.* 63:776–85.

Beasley, J.O. 1942. Meiotic chromosome behavior in species, species hybrids, haploids, and induced polyploids of *Gossypium*. *Genetics* 27:25–54.

Blakeslee, A.F. 1934. New jimson weeds from old chromosomes. *J. Hered.* 25:80–108.

Boue, A. 1985. Cytogenetics of pregnancy wastage. *Adv. Hum. Genet.* 14:1–58.

Carr, D.H. 1971. Genetic basis of abortion. *Annu. Rev. Genet.* 5:65–80.

Croce, C.M., et al. 1996. The *FHIT* genKe at 3p14.2 is abnormal in lung cancer. *Cell* 85:17–26.

DeArce, M.A., and Kearns, A. 1984. The fragile X syndrome: The patients and their chromosomes. *J. Med. Genet.* 21:84–91.

Galitski, T., et al. 1999. Ploidy regulation of gene expression. *Science* 285:251–54.

Gersh, M., et al. 1995. Evidence for a distinct region causing a cat-like cry in patients with 5p deletions. *Am. J. Hum. Genet.* 56:1404–10.

Hassold, T.J., et al. 1980. Effect of maternal age on autosomal trisomies. *Ann. Hum. Genet.* (London) 44:29–36.

Hassold, T.J., and Hunt, P. 2001. To err (meiotically) is human: The genesis of human aneuploidy. *Nature Reviews Genetics* 2:280–91.

Hecht, F. 1988. Enigmatic fragile sites on human chromosomes. *Trends Genet.* 4:121–22.

Hulse, J.H., and Spurgeon, D. 1974. Triticale. *Sci. Am.* (Aug.) 231:72–81.

Kaiser, P. 1984. Pericentric inversions: Problems and significance for clinical genetics. *Hum. Genet.* 68:1–47.

Lewis, E.B. 1950. The phenomenon of position effect. *Adv. Genet.* 3:73–115.

Lewis, W.H., ed. 1980. *Polyploidy: Biological relevance*. New York: Plenum Press.

Lynch, M., and Conery, J.S. 2000. The evolutionary fate and consequences of duplicated genes. *Science* 290:1151–54.

Madan, K. 1995. Paracentric inversions: A review. *Hum. Genet.* 96:503–15.

Ohno, S. 1970. *Evolution by gene duplication*. New York: Springer-Verlag.

Oostra, B.A., and Verkerk, A.J. 1992. The fragile X syndrome: Isolation of the *FMR-1* gene and characterization of the fragile X mutation. *Chromosoma*. 101:381–87.

Patterson, D. 1987. The causes of Down syndrome. *Sci. Am.* (Aug.) 257:52–61.

Shepard, J.F. 1982. The regeneration of potato plants from protoplasts. *Sci. Am.* (May) 246:154–66.

Shepard, J.F., et al. 1983. Genetic transfer in plants through interspecific protoplast fusion. *Science* 219:683–88.

Sutherland, G. 1985. The enigma of the fragile X chromosome. *Trends Genet.* 1:108–11.

Taylor, A.I. 1968. Autosomal trisomy syndromes: A detailed study of 27 cases of Edwards syndrome and 27 cases of Patau syndrome. *J. Med. Genet.* 5:227–52.

Tjio, J.H., and Levan, A. 1956. The chromosome number of man. *Hereditas* 42:1–6.

Wharton, K.A., et al. 1985. *opa*: A novel family of transcribed repeats shared by the *Notch* locus and other developmentally regulated loci in *D. melanogaster. Cell* 40:55–62.

Wilkins, L.E., Brown, J.A., and Wolf, B. 1980. Psychomotor development in 65 home-reared children with cri-du-chat syndrome. *J. Pediatr.* 97:401–5.

Chapter Nine

Extranuclear Inheritance

The variegated leaves of the shrub Acanthopanax.

Chapter Concepts

- Many traits exhibit a pattern of inheritance that is not biparental, demonstrating what is called extranuclear inheritance.
- Extranuclear inheritance is often due to the expression of genetic information contained in mitochondria or chloroplasts.
- Extranuclear traits determined by mitochondrial DNA are most often transmitted through the maternal gamete.
- Extranuclear traits determined by chloroplast DNA may be transmitted uniparentally or biparentally.
- Expression of the maternal genotype during gametogenesis and during early development may have a strong influence on the phenotype of an organism.

Throughout the history of genetics, occasional reports have challenged the basic tenet of Mendelian transmission genetics—that the phenotype is transmitted by nuclear genes located on chromosomes of both parents. Observations have revealed inheritance patterns that fail to reflect Mendelian principles, and some indicate an apparent extranuclear influence on the phenotype. Before the role of DNA in genetics was firmly established, such observations were commonly regarded with skepticism. However, with the increasing knowledge of molecular genetics and the discovery of DNA in mitochondria and chloroplasts, **extranuclear inheritance** is now recognized as an important aspect of genetics.

There are several varieties of extranuclear inheritance. One major type, referred to above, is also described as **organelle heredity**. In such cases, DNA contained in mitochondria or chloroplasts determines certain phenotypic characteristics of the offspring. Examples are often recognized on the basis of the uniparental transmission of these organelles from the female parent through the egg to progeny. A second type involves **infectious heredity**, resulting from the symbiotic or parasitic association of a microorganism. In such cases, an inherited phenotype is affected by the presence of the microorganism in the cytoplasm of the host cells. The third variety involves the **maternal effect** on the phenotype, whereby nuclear gene products are stored in the egg and then transmitted through the ooplasm to offspring. These gene products are distributed to cells of the developing embryo and influence its phenotype.

The common element in all of these examples is the transmission of genetic information to offspring through the cytoplasm rather than through the nucleus, most often from only one of the parents. This shall constitute our definition of "extranuclear inheritance."

9.1 Organelle Heredity Involves DNA in Chloroplasts and Mitochondria

In this section, we will examine examples of inheritance patterns related to chloroplast and mitochondrial function. Before DNA was discovered in these organelles, however, the exact mechanism of transmission of the traits to be discussed next was not clear, except that their inheritance appeared to be linked to something in the cytoplasm rather than linked to genes in the nucleus. Transmission was most often from the maternal parent through the ooplasm, causing the results of reciprocal crosses to vary. Such a pattern is now more appropriately referred to as organelle heredity.

Analysis of the inheritance patterns resulting from mutant alleles in chloroplasts and mitochondria has been difficult for two reasons. First, the function of these organelles is dependent upon gene products from both nuclear and organelle DNA. Second, many organelles are contributed to each progeny. In such cases, if only one or a few of these contain a mutant gene, the corresponding mutant phenotype may not be revealed. Analysis is thus much more complex than for Mendelian characters.

First, we will discuss several of the classical examples that ultimately called attention to organelle heredity. Then, in the subsequent section, we will discuss information concerning DNA and the resultant genetic function in each organelle.

Chloroplasts: Variegation in Four O'Clock Plants

In 1908, Carl Correns (one of the rediscoverers of Mendel's work) provided the earliest example of inheritance linked to chloroplast transmission. Correns discovered a variant of the four o'clock plant, *Mirabilis jalapa*, that had branches with either white, green, or variegated leaves. As shown in Figure 9–1, inheritance in all possible combinations of crosses is strictly determined by the phenotype of the ovule source. For example, if the seeds (representing the progeny) were derived from ovules on branches with green leaves, all progeny plants bore only green leaves, regardless of the phenotype of the source of pollen. Correns concluded that inheritance was transmitted through the cytoplasm of the maternal parent because the pollen, which contributes little or no cytoplasm to the zygote, had no apparent influence on the progeny phenotypes. Since leaf coloration is related to the chloroplast, genetic information contained either in that organelle or somehow present in the cytoplasm and influencing the chloroplast must be responsible for this inheritance pattern.

	Location of Ovule		
Source of Pollen	White branch	Green branch	Variegated branch
White branch	White	Green	White, green, or variegated
Green branch	White	Green	White, green, or variegated
Variegated branch	White	Green	White, green, or variegated

FIGURE 9–1 Offspring from crosses between flowers from various branches of variegated four o'clock plants. The photograph shows the four o'clock plant.

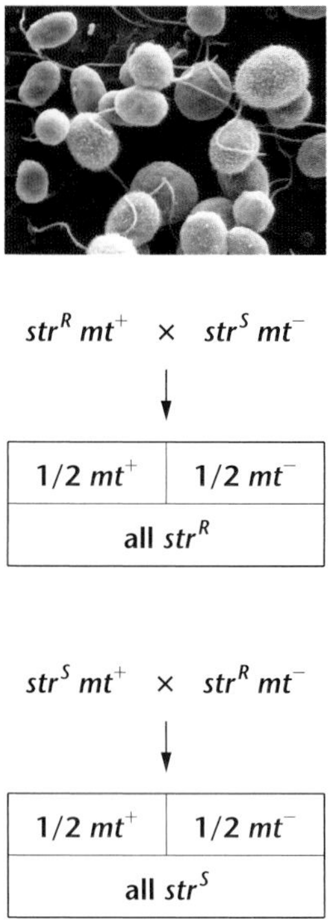

FIGURE 9–2 The results of reciprocal crosses between streptomycin-resistant (str^R) and streptomycin-sensitive (str^S) strains in the green alga *Chlamydomonas* (shown in the photograph).

Chloroplast Mutations in *Chlamydomonas*

The unicellular green alga *Chlamydomonas reinhardi* has provided an excellent system for the investigation of plastid inheritance. This haploid eukaryotic organism (Figure 9–2) has a single large chloroplast containing more than 50 copies of a circular double-stranded DNA molecule. Matings that reestablish diploidy are immediately followed by meiosis, and the various stages of the life cycle are easily studied in culture in the laboratory. The first cytoplasmic mutation, streptomycin resistance (str^R), was reported in 1954 by Ruth Sager. Although *Chlamydomonas*' two mating types, mt^+ and mt^-, appear to make equal cytoplasmic contributions to the zygote, Sager determined that the str^R phenotype is transmitted only through the mt^+ parent (Figure 9–2). Reciprocal crosses between sensitive and resistant strains yield different results depending on the genotype of the mt^+ parent, which is expressed in all offspring. As shown, one half of the offspring are mt^+ and one half of them are mt^-, indicating that mating type is controlled by a nuclear gene that segregates in a Mendelian fashion.

Since Sager's discovery, a number of other *Chlamydomonas* mutations (including resistance to, or dependence on, a variety of bacterial antibiotics) that show a similar uniparental inheritance pattern have been discovered. These mutations have all been linked to the transmission of the chloroplast, and their study has extended our knowledge of chloroplast inheritance.

Following fertilization, which involves the fusion of two cells of opposite mating type, the single chloroplasts of the two mating types fuse. After the resulting zygote has undergone meiosis and haploid cells are produced, it is apparent that the genetic information of the chloroplasts of progeny cells is derived only from the mt^+ parent. The genetic information present within the mt^- chloroplast has degenerated.

It is interesting to note that the inheritance of phenotypes affected by mitochondria is also uniparental in *Chlamydomonas*. However, studies of the transmission of several cases of antibiotic resistance have shown that it is the mt^- parent that transmits the genetic information to progeny cells. This is just the opposite of what occurs with chloroplast-derived phenotypes, such as str^R. The significance of inheriting one organelle from one parent and the other organelle from the other parent is not yet established.

How Do We Know?

How was it established that particular phenotypes are inherited as a result of genetic information present in the chloroplast rather than in the nucleus?

Now solve this

In Problem 9.8(b) on page 228 you are asked to consider the genetic outcome of crosses involving a mutation in *Chlamydomonas* and to propose how inheritance of the trait occurs.

Hint: Consider the two possibilities that inheritance of the trait is uniparental versus biparental.

Mitochondrial Mutations: The Case of *poky* in *Neurospora*

Mutations affecting mitochondrial function have been discovered and studied, revealing that mitochondria also contain a distinctive genetic system. As with chloroplasts, mitochondrial mutations are transmitted through the cytoplasm. In the current discussion, we will emphasize the link between mitochondrial mutations and the resultant extranuclear inheritance patterns.

In 1952, Mary B. Mitchell and Hershel K. Mitchell studied the bread mold *Neurospora crassa* (Figure 9–3). They discovered a slow-growing mutant strain and named it *poky*. (It is also designated *mi-1*, for *maternal inheritance*.) Slow growth is associated with impaired mitochondrial function, specifically caused by the absence of several cytochrome proteins essential for electron transport. In the absence of cytochromes, aerobic respiration leading to ATP synthesis is curtailed.

Results of genetic crosses between wild-type and *poky* strains suggest that the trait is maternally inherited. If the maternal parent is *poky* and the paternal parent is wild type, all progeny colonies are *poky*. The reciprocal cross produces normal wild-type colonies.

Studies with *poky* mutants show that occasionally hyphae from separate mycelia fuse with one another, giving rise to structures containing two or more nuclei in a common cytoplasm. If the hyphae contain nuclei of different genotypes, the structure is called a **heterokaryon**. The cytoplasm will contain mitochondria derived from both initial mycelia. A heterokaryon may give rise to haploid spores, or conidia, that produce new mycelia.

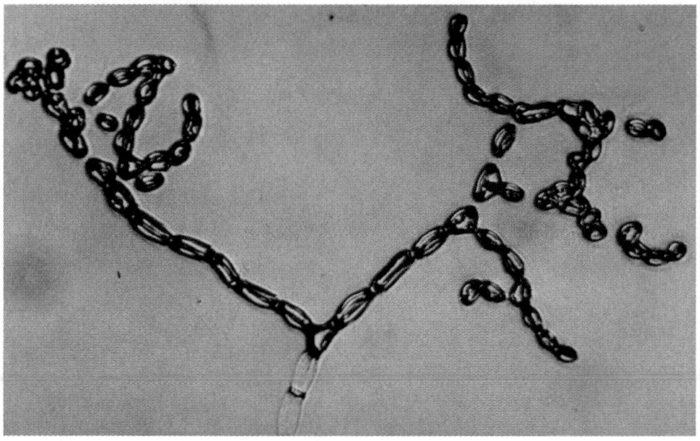

FIGURE 9–3 Micrograph illustrating the growth of the pink bread mold *Neurospora crassa*.

Heterokaryons produced by the fusion of *poky* and wild-type hyphae initially show normal rates of growth and respiration. However, mycelia produced through conidia formation become progressively more abnormal until they show the *poky* phenotype. This occurs despite the presumed presence of both wild-type and *poky* mitochondria in the cytoplasm of the hyphae.

To explain the initial growth and respiration pattern, we assume that the wild-type mitochondria support the respiratory needs of the hyphae. The subsequent expression of the *poky* phenotype suggests that the presence of the *poky* mitochondria may somehow prevent or depress the function of these wild-type mitochondria. Perhaps the *poky* mitochondria replicate more rapidly and "wash out" or dilute wild-type mitochondria numerically. Another possibility is that *poky* mitochondria produce a substance that inactivates the wild-type organelle or interferes with the replication of its DNA (mtDNA). This type of interaction makes *poky* an example of a broader category referred to as **suppressive mutations**, which may also include many other mitochondrial mutations of *Neurospora* and yeast.

Petites in *Saccharomyces*

Another extensive study of mitochondrial mutations has been performed with the yeast *Saccharomyces cerevisiae*. The first such mutation, described by Boris Ephrussi and his coworkers in 1956, was named *petite* because of the small size of the yeast colonies (Figure 9–4). Many independent petite mutations have since been discovered and studied, and all have a common characteristic: a deficiency in cellular respiration involving abnormal electron transport. Because this organism is a facultative anaerobe and can grow by fermenting glucose through glycolysis, it may survive the loss of mitochondrial function by generating energy anaerobically.

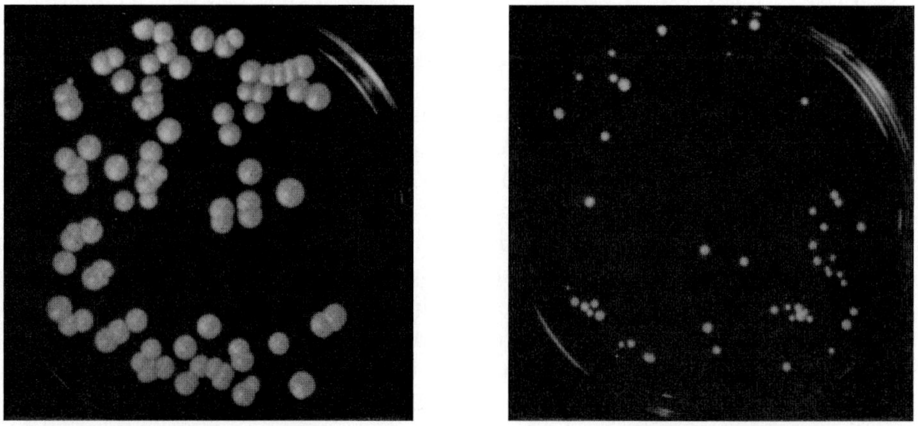

FIGURE 9–4 A comparison of normal versus petite colonies in the yeast *Saccharomyces cerevisiae*.

The complex genetics of petite mutations is diagrammed in Figure 9–5. A small proportion of these mutants are the result of nuclear mutations. They exhibit Mendelian inheritance and are thus called ***segregational petites***. The remainder demonstrate cytoplasmic transmission, producing one of two effects in matings. ***Neutral petites***, when crossed to wild type, produce meiotic products (called ascospores) that give rise only to wild-type, or normal, colonies. The same pattern continues if progeny of such cross are backcrossed to neutral petites. Since the majority of neutrals lack mtDNA completely or have lost a substantial portion of it, for offspring to be normal, they must also be inheriting normal mitochondria capable of aerobic respiration following reproduction. Thus, in yeast, mitochondria are inherited from both parental cells. The functional mitochondria are replicated in offspring and support normal mitochondrial function.

The third mutational type, the ***suppressive petite***, behaves like *poky* in *Neurospora*. Crosses between mutant and wild type give rise to mutant diploid zygotes, which after meiosis immediately yield haploid cells that are all mutant. Assuming that the offspring have received mitochondria from both parents, the *petite* mutation behaves "dominantly" and seems to suppress the function of the wild-type mitochondria. *Suppressive petites* also have deletions of mtDNA, but they are not nearly as extensive as deletions in the *neutral petites*.

Two major hypotheses that center around the organellar DNA have been advanced to explain suppressiveness. One explanation suggests that the mutant (or deleted) DNA in the mitochondria (mtDNA) replicates more rapidly, resulting in the mutant mitochondria "taking over" or dominating the phenotype by numbers alone. The second explanation suggests that recombination occurs between the mutant and wild-type mtDNA, introducing errors into or disrupting the normal mtDNA. It is not yet clear which one, if either, of these explanations is correct.

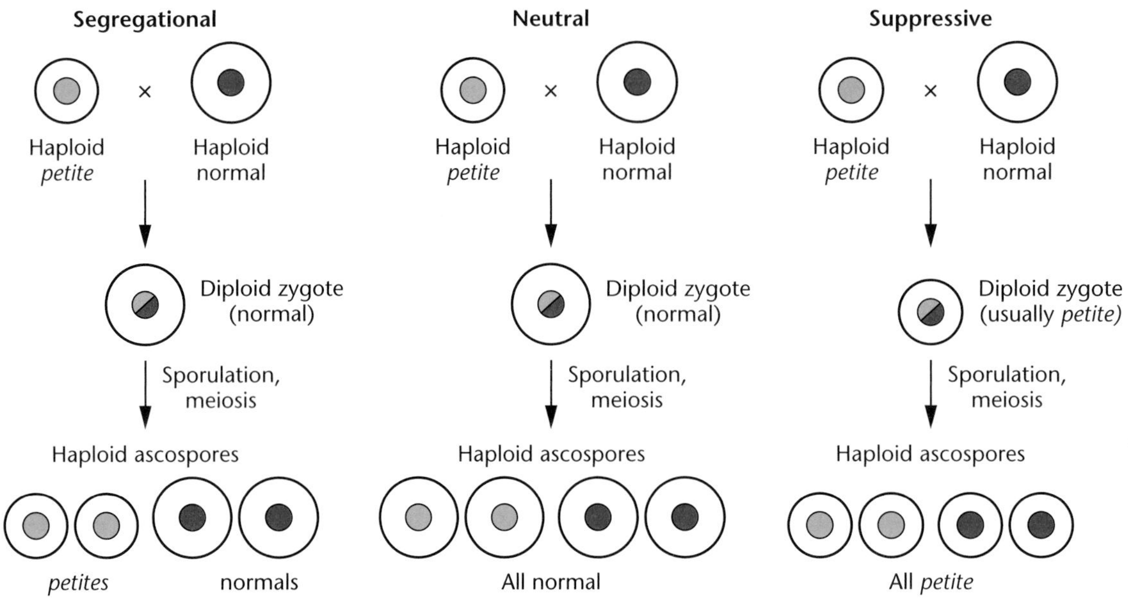

FIGURE 9–5 The outcome of crosses involving the three types of *petite* mutations affecting mitochondrial function in the yeast *Saccharomyces cerevisiae*.

HOW DO WE KNOW?

How did the discovery of three categories of *petite* mutations in yeast lead researchers to postulate extranuclear inheritance of colony size?

Now solve this

In Problem 9.4 on page 228 involving a newly isolated *petite* mutation crossed to a normal yeast strain, you are asked to draw conclusions about the petite mutation.

Hint: Remember that in yeast, inheritance of mitochondria is biparental.

9.2 Knowledge of Mitochondrial and Chloroplast DNA Helps Explain Organelle Heredity

That both mitochondria and chloroplasts contain their own DNA and a system for expressing genetic information was first suggested by the discovery of the mutations and the resultant inheritance patterns in plants, yeast, and other fungi, as discussed above. Because both mitochondria and chloroplasts are inherited through the maternal cytoplasm in most organisms, and since each of the earlier examples can be linked theoretically to the altered function of either chloroplasts or mitochondria, geneticists set out to look for more direct evidence of DNA in these organelles. Not only was unique DNA found to be a normal component of both mitochondria and chloroplasts, but careful examination of the nature of this genetic information was to provide essential clues as to the evolutionary origin of these organelles.

Organelle DNA and the Endosymbiotic Theory

Electron microscopists not only documented the presence of DNA in mitochondria and chloroplasts, but they also saw that this molecule exists in a form quite unlike that seen in the nucleus of the eukaryotic cells that house these organelles (Figures 9–6 and 9–7). This DNA looks remarkably similar to that seen in bacteria! This similarity, along with the observation of the presence of a unique genetic system capable of organelle-specific transcription and translation, led Lynn Margulis and others to the postulate known as the **endosymbiotic theory**. Basically, the theory states that mitochondria and chloroplasts arose independently about 2 billion years ago from free-living protobacteria that possessed the abililties now attributed to these organelles—aerobic respiration and photosynthesis, respectively. This idea proposes that the ancient bacteria were engulfed by larger primitive eukaryotic cells, which originally lacked the ability to respire aerobically or to capture energy from sunlight. A beneficial, symbiotic relationship subsequently developed, whereby the bacteria eventually lost their ability to function autonomously, while the eukaryotic host cells gained the ability to undergo either oxidative respiration or photosynthesis, as the case may be. Although some questions remain unanswered, evidence continues to accumulate in support of this theory, and its basic tenets are now widely accepted.

A brief examination of the modern-day mitochondria will help us better understand endosymbiotic theory. During the subsequent course of evolution following the invasion event, distinct branches of diverse eukaryotic organisms arose. As evolution progressed, the companion bacteria also underwent their own independent changes. The primary alteration was the transfer of many of the genes from the invading bacterium to the nucleus of the host. The products of these genes, while still functioning in the organelle, are nevertheless now encoded and

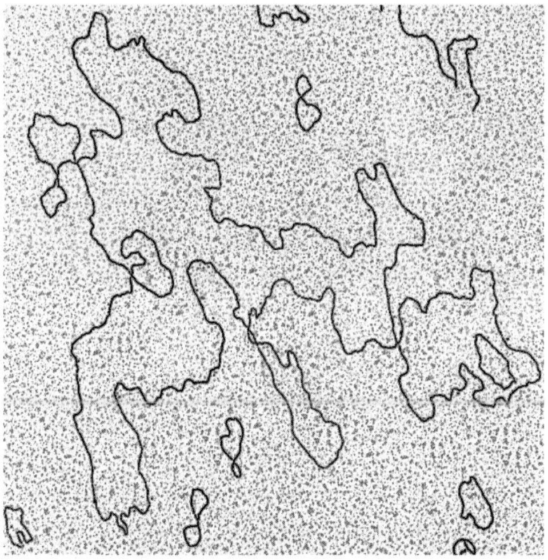

FIGURE 9–6 Electron micrograph of chloroplast DNA derived from lettuce.

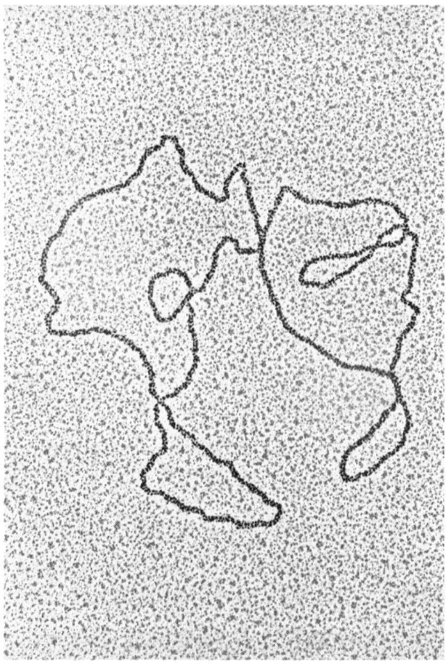

FIGURE 9–7 Electron micrograph of mitochondrial DNA derived from *Xenopus laevis.*

transcribed in the nucleus and translated in the cytoplasm prior to their transport into the organelle. What DNA remains today in the typical mitochondrial genome is miniscule compared with the free-living bacteria from which it was derived. The most gene-rich organelles now have fewer than 10 percent of the genes present in the smallest bacterium known.

Similar changes have characterized the evolution of chloroplasts. In the subsequent sections, we will explore in some detail what is known about modern-day chloroplasts and mitochondria.

HOW DO WE KNOW?

If we assume that the endosymbiotic theory is true, what experimental observations support this belief?

Molecular Organization and Gene Products of Chloroplast DNA

The details regarding the autonomous genetic system of chloroplasts have now been worked out, providing further support of the endosymbiotic theory. The chloroplast, responsible for photosynthesis, contains DNA as a source of genetic information and a complete protein-synthesizing apparatus. The molecular components of the chloroplast translation apparatus are jointly derived from both nuclear and organelle genetic information.

Chloroplast DNA (cpDNA), shown in Figure 9–6, is fairly uniform in size among different organisms, ranging between 100 and 225 kb in length. It shares many similarities to DNA found in prokaryotic cells. It is circular, double stranded, replicated semiconservatively, and free of the associated proteins characteristic of eukaryotic DNA. Compared with nuclear DNA from the same organism, it invariably shows a different buoyant density and base composition.

The size of cpDNA is much larger than that of mtDNA. To some extent, this can be accounted for by an increased number of genes. However, the biggest difference appears to be due to the presence in cpDNA of many long noncoding nucleotide sequences both between and within genes, the latter being introns. Duplications of many DNA sequences are also present. Since such noncoding sequences vary in different plants, these observations are indicative of the independent evolution occurring in chloroplasts following their initial invasion of a primitive eukaryoticlike cell.

In the green alga *Chlamydomonas*, there are about 75 copies of the chloroplast DNA molecule per organelle. Each copy consists of a length of DNA that contains 195,000 bp (195 kb). In higher plants, such as the sweet pea, multiple copies of the DNA molecule are also present in each organelle, but the molecule (134 kb) is considerably smaller than that in *Chlamydomonas*. Interestingly, genetic recombination between the multiple copies of DNA within chloroplasts has been documented in some organisms.

Numerous gene products encoded by chloroplast DNA function during translation within the organelle. For example, in a variety of higher plants (beans, lettuce, spinach, maize, and oats), two sets of the genes coding for the ribosomal RNAs—5*S*, 16*S*, and 23*S* rRNA—are present. Additionally, chloroplast DNA encodes numerous tRNAs, as well as many ribosomal proteins specific to the chloroplast ribosomes. For example, in the liverwort, whose cpDNA was the first to be sequenced, there are genes encoding 30 tRNAs, RNA polymerase, multiple rRNAs, and numerous ribosomal proteins. The variation of gene products encoded in cpDNA in different plants again attests to the independent evolution that occurred within chloroplasts.

Chloroplast ribosomes have a sedimentation coefficient slightly less than 70*S*, similar, but not identical, to bacterial ribosomes. Even though some chloroplast ribosomal proteins are encoded by chloroplast DNA and some by nuclear DNA, most, if not all, such proteins are distinct from their counterparts present in cytoplasmic ribosomes. Both observations provide direct support for the endosymbiotic theory.

Still other chloroplast genes specific to the photosynthetic function have been identified. For example, in the liverwort, there are 92 chloroplast genes encoding proteins that are part of the thylakoid membrane, the cellular component integral to the light-dependent reactions of photosynthesis. Mutations in these genes may inactivate photosynthesis. A typical distribution of genes between the nucleus and the chloroplast is illustrated by one of the major photosynthetic enzymes, ribulose-1-5-bisphosphate carboxylase (Rubisco). This enzyme has its small subunit encoded by a nuclear gene, whereas the large subunit is encoded by cpDNA.

Molecular Organization and Gene Products of Mitochondrial DNA

Extensive information is also available concerning the molecular aspects and gene products of mitochondrial DNA (mtDNA). In most eukaryotes, mtDNA exists as a double-stranded, closed circle (Figure 9–7) that replicates semiconservatively and is free of the chromosomal proteins characteristic of eukaryotic chromosomal DNA. An exception is found in some ciliated protozoans, in which the DNA is linear.

In size, mtDNA is much smaller than cpDNA and varies greatly among organisms, as demonstrated in Table 9.1. In a variety of animals, including humans, mtDNA consists of about 16,000 to 18,000 bp (16 to 18 kb). However, yeast (*Saccharomyces*) mtDNA consists of 75 kb. Plants typically exceed this amount—367 kb is present in mitochondria in the mustard plant, *Arabidopsis*. Vertebrates have 5 to 10 such DNA molecules per organelle, while plants have 20 to 40 copies per organelle.

We can say several other things about mtDNA. With only rare exceptions, introns (the intervening noncoding DNA sequences characteristic of nuclear DNA) are absent from mitochondrial genes and gene repetitions are seldom present. Nor is there usually much in the way of intergenic spacer DNA. This description is applicable particularly to species whose mtDNA is fairly small in size, such as humans. However, in *Saccharomyces*, with a much larger DNA molecule, much of the excess DNA is accounted for by introns and intergenic spacer DNA.

As will be discussed in Chapter 14, expression of mitochondrial genes uses several modifications of the otherwise standard genetic code. Replication is dependent on enzymes encoded by nuclear DNA. In humans, mtDNA encodes two ribosomal RNAs (rRNAs), 22 transfer RNAs (tRNAs), as well as 13 polypeptides essential to the oxidative respiratory functions of the organelle. In most cases, these polypeptides are part of multichain proteins, where the other subunits of each protein are often encoded in the nucleus, synthesized in the cytoplasm, and transported into the organelle. Thus, the protein-synthesizing apparatus and the molecular components for cellular respiration are jointly derived from nuclear and mitochondrial genes.

Another observation is that in vertebrate mtDNA, the two strands vary in density when examined by centrifugation. This serves as the basis to designate the strands as either heavy (H) or light (L). Interestingly, while most of the mitochondrial genes are encoded by the H strand, only a few of the genes are encoded by the L strand.

As might be predicted by the endosymbiont hypothesis, ribosomes found in the organelle differ from those present in the neighboring cytoplasm. Table 9.2 shows that mitochondrial ribosomes of different species vary considerably in their sedimentation coefficients, ranging from 55*S* to 80*S*, while cytoplasmic ribosomes are uniformly 80*S*.

Nuclear-coded gene products essential to biological activity in mitochondria are quite numerous. They include, for example, DNA and RNA polymerases, initiation and elongation factors essential for translation, ribosomal proteins, aminoacyl tRNA synthetases, and several tRNA species. These imported components are distinct from their cytoplasmic counterparts, even though both sets are coded by nuclear genes. For example, the synthetase enzymes essential for charging mitochondrial tRNA molecules (a process essential to translation) show a distinct affinity for the mitochondrial tRNA species as compared with the cytoplasmic tRNAs. Similar affinity has been shown for the initiation and elongation factors. Furthermore, while bacterial and nuclear RNA polymerases are known to be composed of numerous subunits, the mitochondrial variety consists of only one polypeptide chain. This polymerase is generally sensitive to antibiotics that inhibit bacterial RNA synthesis, but not to eukaryotic inhibitors. The various contributions of nuclear and mitochondrial gene products are contrasted in Figure 9–8.

TABLE 9.1 THE SIZE OF MTDNA IN DIFFERENT ORGANISMS

Organisms	Size (kb)
Human	16.6
Mouse	16.2
Xenopus (frog)	18.4
Drosophila (fruit fly)	18.4
Saccharomyces (yeast)	75.0
Pisum sativum (pea)	110.0
Arabidopsis (mustard plant)	367.0

TABLE 9.2 SEDIMENTATION COEFFICIENTS OF MITOCHONDRIAL RIBOSOMES

Kingdom	Examples	Sedimentation Coefficient (S)
Animals	Vertebrates	55–60
	Insects	60–71
Protists	*Euglena*	71
	Tetrahymena	80
Fungi	*Neurospora*	73–80
	Saccharomyces	72–80
Plants	Maize	77

Now solve this

In Problem 9.13 on page 229 you are asked to explain the number of different tRNA molecules encoded by human mitochorial DNA in contrast to the actual number required for translation of proteins within the organelle.

Hint: Mutations in certain nuclear genes have an impact on mitochondrial function.

9.3 Mutations in Mitochondrial DNA Cause Human Disorders

The DNA found in human mitochondria has been completely sequenced and contains 16,569 base pairs. As mentioned earlier, the mitochondrial gene products include the following:

13 proteins, required for aerobic cellular respiration;

22 transfer RNAs (tRNAs), required for translation;

2 ribosomal RNAs (rRNAs), required for translation.

Because a cell's energy supply is largely dependent on aerobic cellular respiration, disruption of any mitochondrial gene by mutation may potentially have a severe impact on that organism. We have seen this in our previous discussion of *petite* mutants in yeast, which would be a lethal mutation were it not for this organism's ability to respire anaerobically. In fact, mtDNA is particularly vulnerable to mutations. There are two possible reasons for this. First, the ability to repair mtDNA damage may not be equivalent to that of nuclear DNA. Second, the concentration of highly mutagenic free radicals generated by cell respiration that accumulate in a such a confined space very likely raises the mutation rate in mtDNA.

Fortunately, a zygote receives a large number of organelles through the egg, so if only one organelle or a few of them contains a mutation, its impact is greatly diluted by the many mitochondria that lack the mutation and function normally. During early development, cell division disperses the initial population of mitochondria present in the zygote, and in the newly formed cells, these organelles reproduce autonomously. Therefore, if a deleterious mutation arises or is present in the initial population of organelles, adults will have cells with a variable mixture of both normal and abnormal organelles. This condition is called **heteroplasmy**.

In order for a human disorder to be attributable to genetically altered mitochondria, several criteria must be met:

1. Inheritance must exhibit a maternal rather than a Mendelian pattern.
2. The disorder must reflect a deficiency in the bioenergetic function of the organelle.
3. There must be a specific genetic mutation in one or more of the mitochondrial genes.

Thus far, several disorders in humans are known to demonstrate these characteristics. For example, **myoclonic epilepsy and ragged red fiber disease (MERRF)** demonstrates a pattern of inheritance consistent with maternal inheritance. Only offspring of affected mothers inherit the disorder; the offspring of affected fathers are all normal. Individuals with this rare disorder express ataxia (lack of muscular coordination), deafness, dementia, and epileptic seizures. The disease is so named because of the presence of "ragged-red" skeletal muscle fibers that exhibit blotchy red patches resulting from the proliferation of aberrant mitochondria (Figure 9–9). Brain function, which also has a high energy demand, is also affected in this disorder, leading to the neurological symptoms.

Analysis of mtDNA from patients with MERFF has revealed a mutation in one of the 22 mitochondrial genes encoding a transfer RNA. Specifically, the gene encoding $tRNA^{Lys}$ contains an A-to-G transition within its sequence. This genetic alteration apparently interferes with the capacity for translation

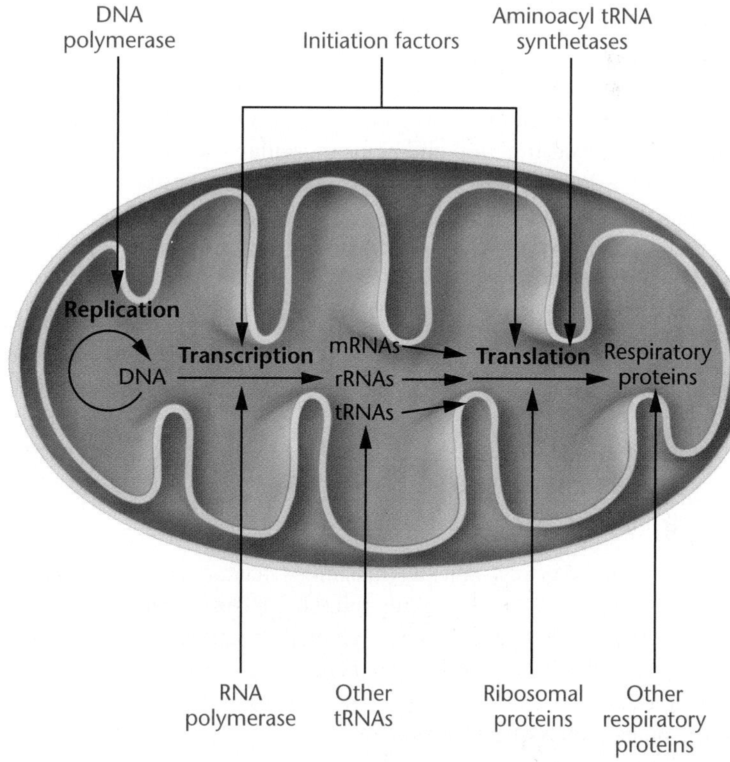

FIGURE 9–8 Gene products that are essential to mitochondrial function. Those shown entering the organelle are derived from the cytoplasm and encoded by nuclear genes.

within the organelle, which in turn leads to the various manifestations of the disorder.

The cells of affected individuals exhibit heteroplasmy, containing a mixture of normal and abnormal mitochondria. Different patients contain different proportions of the two, and even different tissues from the same patient exhibit various levels of abnormal mitochondria. Were it not for heteroplasmy, the mutation would very likely be lethal, testifying to the essential nature of mitochondrial function and its reliance on the genes encoded by mtDNA.

A second disorder, **Leber's hereditary optic neuropathy (LHON)**, also exhibits maternal inheritance as well as mtDNA lesions. The disorder is characterized by sudden bilateral blindness. The average age of vision loss is 27, but onset is quite variable. Four mutations have been identified, all of which disrupt normal oxidative phosphorylation. More than 50 percent of cases are due to a mutation at a specific position in the mitochondrial gene encoding a subunit of NADH dehydrogenase. This mutation is transmitted maternally to all offspring. It is interesting to note that in many instances of LHON, there is no family history; a significant number of cases appear to result from "new" mutations.

Individuals severely affected by a third disorder, **Kearns–Sayre syndrome (KSS)**, lose their vision, experience hearing loss, and display heart conditions. The genetic basis of KSS involves deletions at various positions within mtDNA. Many KSS patients are symptom-free as children, but display progressive symptoms as adults. The proportion of mtDNAs that reveal deletion mutations increases as the severity of symptoms increases.

The study of hereditary mitochondrial-based disorders provides insights into the critical importance of this organelle during normal development, as well as the relationship between mitochondrial function and neuromuscular and neurological disorders. Such study has also suggested a hypothesis for aging based on the progressive accumulation of mtDNA mutations and the accompanying loss of mitochondrial function.

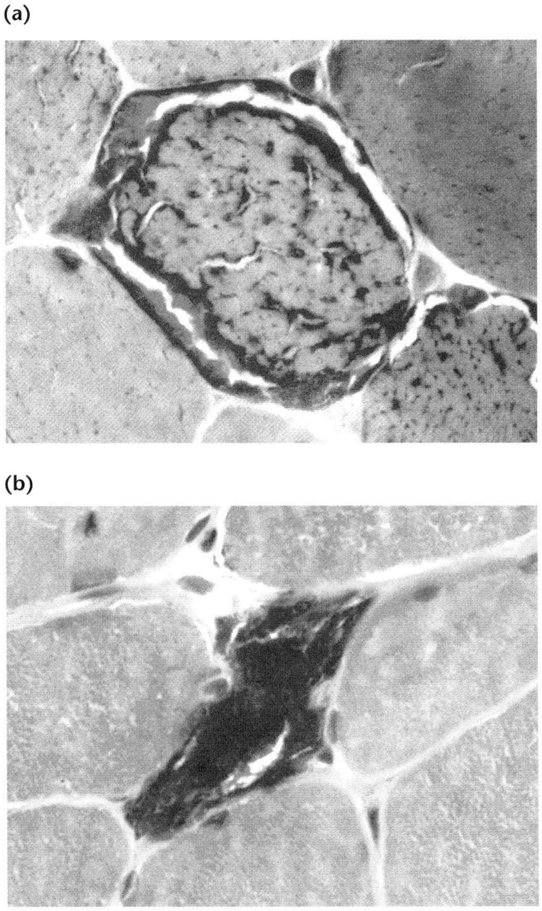

FIGURE 9–9 Ragged red fibers in skeletal muscle cells from patients with a mitochondrial disease. (a) The muscle fiber has mild proliferation. (See red rim and speckled cytoplasm.) (b) Marked proliferation where mitochondria have replaced most cellular structures.

9.4 Infectious Heredity Is Based on a Symbiotic Relationship between Host Organism and Invader

Examples abound of cytoplasmically transmitted phenotypes in eukaryotes that are due to an invading microorganism or particle. The foreign invader coexists in a symbiotic relationship, is usually passed through the maternal ooplasm to progeny cells or organisms, and confers a specific phenotype. We shall consider several examples of this phenomenon.

Kappa in *Paramecium*

First described by Tracy Sonneborn, certain strains of *Paramecium aurelia* are called **Killers** because they release a cytoplasmic substance called **paramecin** that is toxic and sometimes lethal to sensitive strains. This substance is produced by kappa particles that replicate in the Killer cytoplasm; one cell may contain 100 to 200 such particles. The kappa particles contain DNA and protein, and depend on a dominant nuclear gene, *K*, for their maintenance. To understand how the *K* gene and kappa particles are transmitted, we must look at the way *Paramecia* reproduce.

Paramecia are diploid protozoans that can undergo sexual exchange of genetic information through the process of conjugation, shown in Figure 9–10. Each *Paramecium* contains one macronucleus and two diploid micronuclei. Early in conjugation, both micronuclei in each mating pair undergo meiosis, resulting in eight haploid micronuclei. Seven of these degenerate, and the remaining one undergoes a single mitotic division. Each cell then donates one of the two haploid micronuclei to the other, recreating the diploid condition in both cells. As a result, exconjugates typically have identical genotypes. In some instances, conjugation may be accompanied by cytoplasmic exchange.

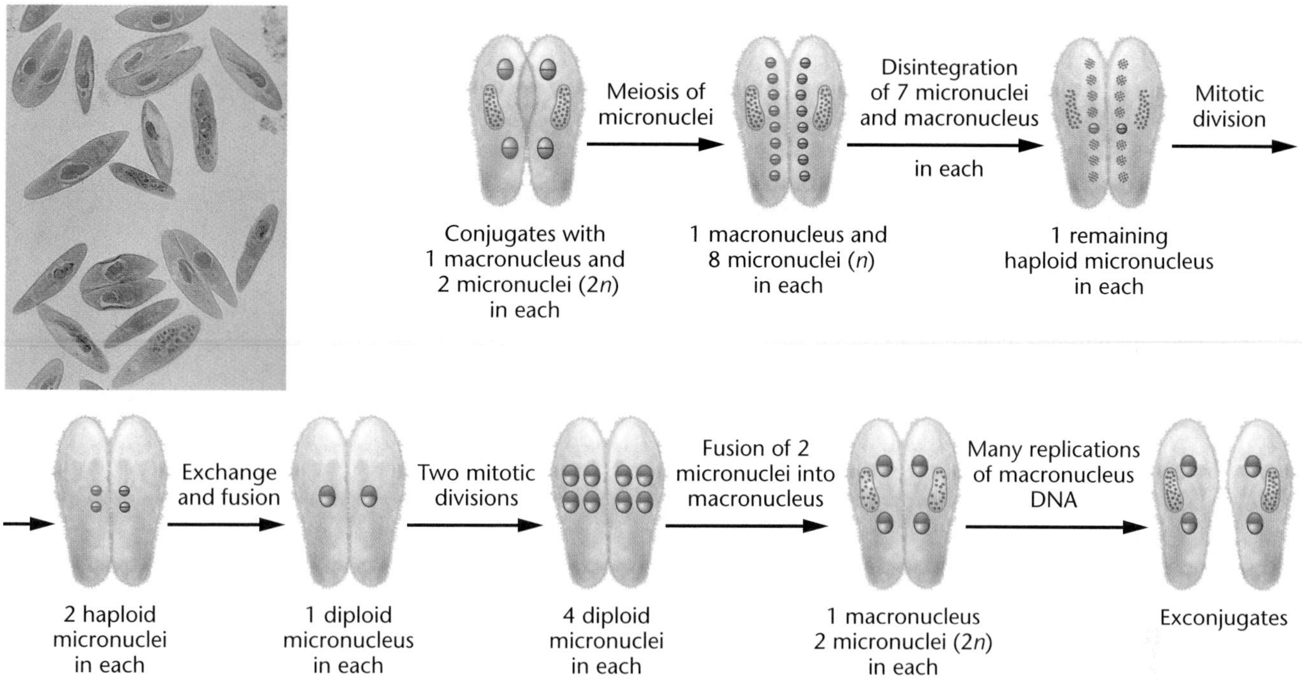

FIGURE 9–10 Genetic events occurring during conjugation in *Paramecium*. The photographic inset shows several pairs of organisms undergoing conjugation.

Autogamy is a similar process, involving only a single cell. Following meiosis of both micronuclei, seven products degenerate and one survives. This nucleus divides, and the resulting nuclei fuse to recreate the diploid condition. If the original cell was heterozygous, autogamy results in homozygosity, because the newly formed diploid nucleus is derived solely from a single haploid meiotic product. In a population of cells originally heterozygous, half of the new cells express one allele and half express the other allele.

Figure 9–11 illustrates the results of crosses between *KK* and *kk* cells, with and without cytoplasmic exchange. Sometimes, no cytoplasmic exchange occurs, even though the resultant cells may be *Kk* (or *KK*, following autogamy). When none occurs, no kappa particles are transmitted and cells remain sensitive. When exchange occurs, the cells become Killers, provided the kappa particles are supported by at least one dominant *K* allele.

Kappa particles are bacterialike and may contain temperate bacteriophages. One theory holds that these viruses of kappa may enter the vegetative state and proceed to reproduce. During this

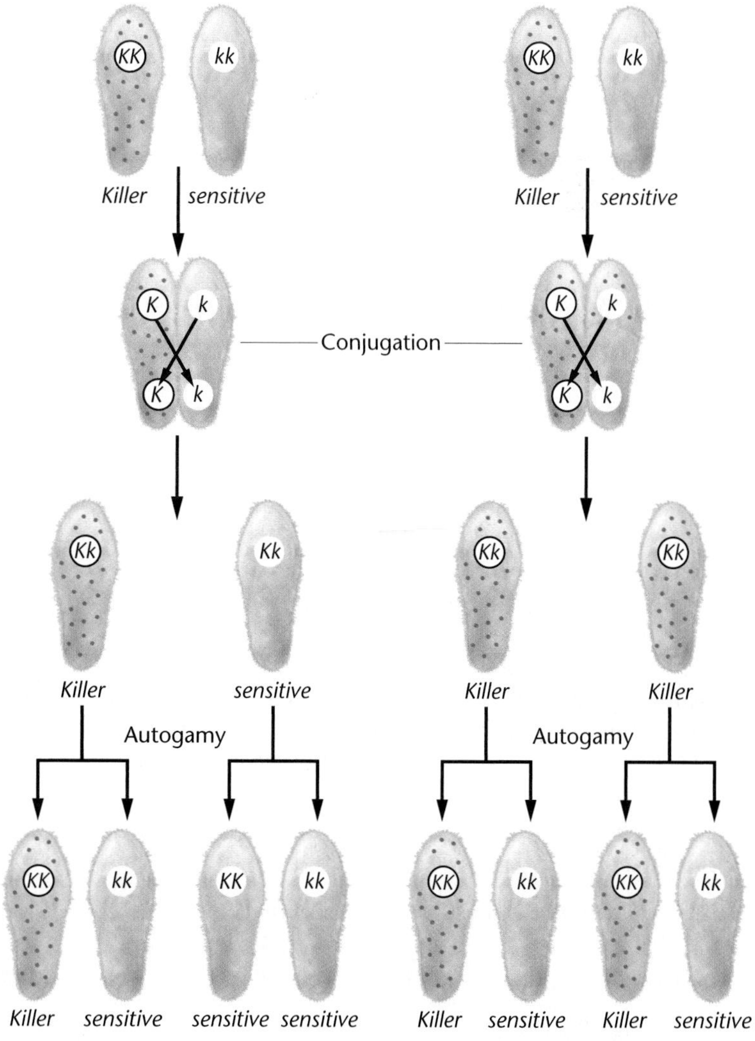

FIGURE 9–11 Results of crosses between Killer (*KK*) and sensitive (*kk*) strains of *Paramecium*, with and without cytoplasmic exchange during conjugation. The kappa particles (dots) are maintained only when a *K* allele is present.

multiplication, they produce the toxic products that are released and kill sensitive strains.

Infective Particles in *Drosophila*

Two examples of infectious heredity are known in *Drosophila*, **CO_2 sensitivity** and **sex ratio**. In the former, flies that would normally recover from carbon dioxide anesthetization instead become permanently paralyzed and are killed by CO_2. Sensitive mothers pass this trait to all offspring. Furthermore, extracts of sensitive flies induce the trait when injected into resistant flies. Phillip L'Heritier has postulated that sensitivity is due to the presence of a virus called **sigma**. The particle has been visualized under the electron microscope and is smaller than kappa. Attempts to transfer the virus to other insects have been unsuccessful, demonstrating that specific genes support the presence of sigma in *Drosophila*.

Our second example of infective particles comes from the study of *Drosophila bifasciata*. A small number of these flies produce predominantly female offspring if reared at 21°C or lower. This condition, designated sex ratio, is transmitted to daughters, but not to the low percentage of males produced. Such a phenomenon was subsequently investigated in *Drosophila willistoni*. In these flies, the injection of ooplasm from sex-ratio females into the abdomens of normal females induced the condition, suggesting that an extrachromosomal element is responsible for the sex ratio phenotype. The agent has now been isolated and shown to be a protozoan. While the protozoan has been found in both males and females, it is lethal primarily to developing male larvae. There is now some evidence that a virus harbored by the protozoan may be responsible for producing a male-lethal toxin.

9.5 In Maternal Effect, the Maternal Genotype Has a Strong Influence during Early Development

Maternal effect, also referred to as maternal influence, implies that an offspring's phenotype for a particular trait is under the control of nuclear gene products present in the egg. This is in contrast to biparental inheritance, where both parents transmit information on genes in the nucleus that determines the offspring's phenotype. In cases of maternal effect, the genetic information of the female gamete is transcribed, and the genetic products (either proteins or yet untranslated mRNAs) are present in the egg cytoplasm. Following fertilization, these products influence patterns or traits established during early development. Three examples will illustrate the influence of the maternal genome on particular traits.

Ephestia Pigmentation

A very straightforward illustration of a maternal effect is seen in the Mediterranean meal moth *Ephestia kuehniella*. The wild-type larva of this moth has a pigmented skin and brown eyes as

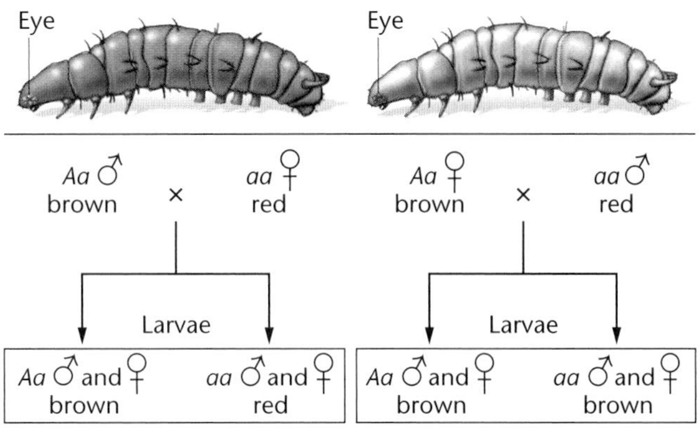

FIGURE 9–12 Maternal influence in the inheritance of eye pigment in the meal moth *Ephestia kuehniella*. Multiple light receptor structures (eyes) are present on each side of the anterior portion of larvae.

a result of the dominant gene *A*. The pigment is derived from a precursor molecule, kynurenine, which is in turn a derivative of the amino acid tryptophan. A mutation, *a*, interrupts the synthesis of kynurenine and, when homozygous, may result in red eyes and little pigmentation in larvae. However, as illustrated in Figure 9–12, results of the cross *Aa* × *aa*, depend on which parent carries the dominant gene. When the male is the heterozygous parent, a 1:1 brown–red-eyed ratio is observed in larvae, as predicted by Mendelian segregation. When the female is heterozygous for the *A* gene, however, all larvae are pigmented and have brown eyes, in spite of half of them being *aa*. As these larvae develop into adults, one-half of them gradually develop red eyes, reestablishing the 1:1 ratio.

One explanation for these results is that the *Aa* oocytes synthesize kynurenine or an enzyme necessary for its synthesis and accumulate it in the ooplasm prior to the completion of meiosis. Even in *aa* progeny whose mothers were *Aa*, this pigment is distributed in the cytoplasm of the cells of the developing larvae; thus, they develop pigmentation and brown eyes. Eventually, the pigment is diluted among many cells and depleted, resulting in the conversion to red eyes as adults. The *Ephestia* example demonstrates the maternal effect in which a cytoplasmically stored nuclear gene product influences the larval phenotype and, at least temporarily, overrides the genotype of the progeny.

Limnaea Coiling

Shell coiling in the snail *Limnaea peregra* is an excellent example of maternal effect on a permanent rather than a transitory phenotype. Some strains of this snail have left-handed, or sinistrally, coiled shells (*dd*), while others have right-handed, or dextrally, coiled shells (*DD* or *Dd*). These snails are hermaphroditic and may undergo either cross- or self-fertilization, providing a variety of types of matings.

Figure 9–13 illustrates the results of reciprocal crosses between true-breeding snails. As you can see, the crosses yield different outcomes, even though both are between sinistral and

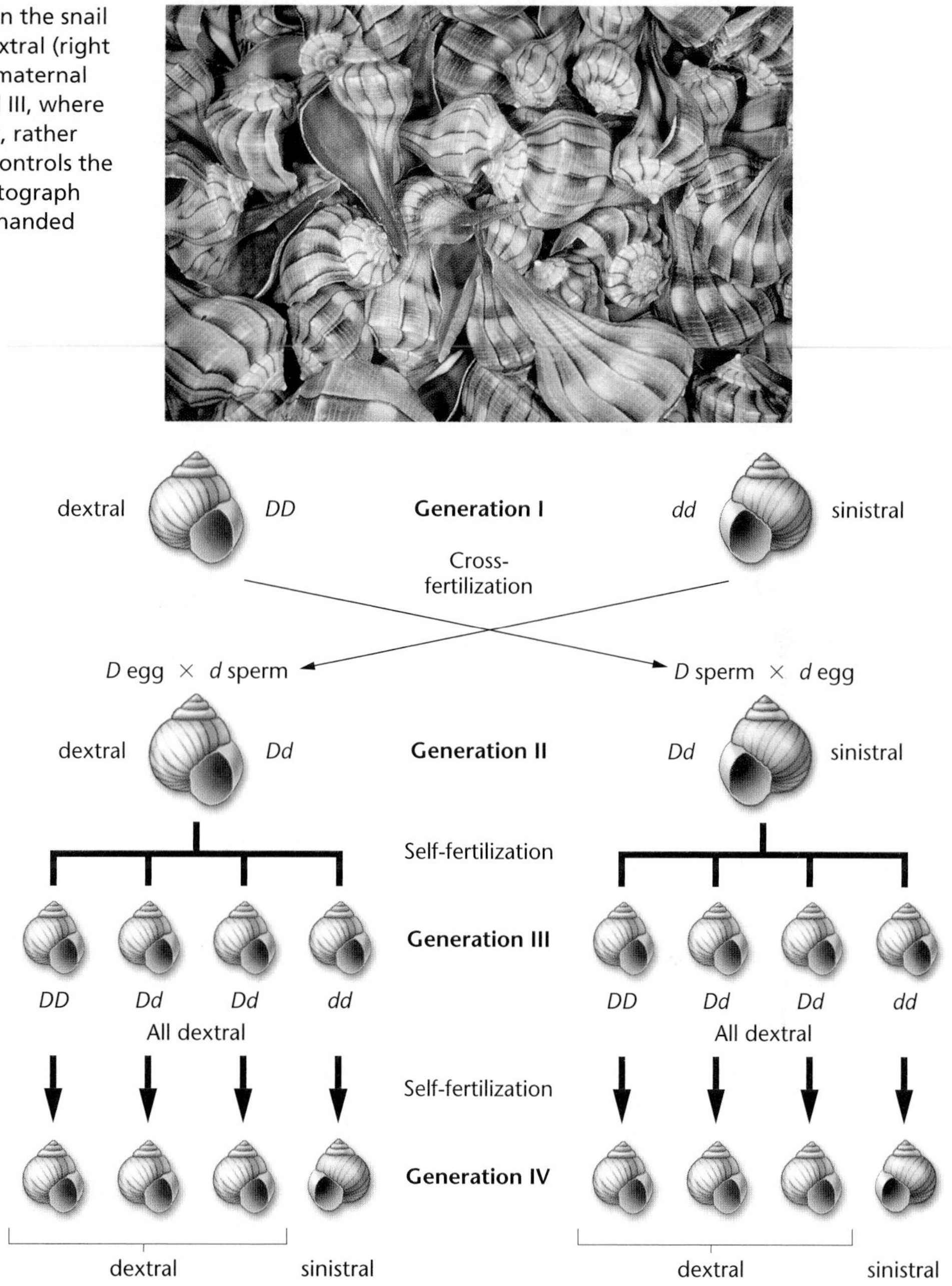

FIGURE 9–13 Inheritance of coiling in the snail *Limnaea peregra*. Coiling is either dextral (right handed) or sinistral (left handed). A maternal effect is evident in generations II and III, where the genotype of the maternal parent, rather than the offspring's own genotype, controls the phenotype of the offspring. The photograph illustrates a mixture of right- vs. left-handed coiled snails.

dextral organisms and produce all heterozygous offspring. Examination of the progeny reveals that their phenotypes depend on the genotypes of the ovum donor parents. If we adopt that conclusion as a working hypothesis, we can test it by examining the offspring in subsequent generations of self-fertilization events. In each case, the hypothesis is upheld. Ovum donors that are *DD* or *Dd* produce only dextrally coiled progeny. Ovum donors that are *dd* produce only sinistrally coiled progeny. The coiling pattern of the progeny snails is determined by the genotype of the parent producing the egg, regardless of the phenotype of that parent.

Investigation of the developmental events in *Limnaea* reveals that the orientation of the spindle in the first cleavage division after fertilization determines the direction of coiling. Spindle orientation appears to be controlled by maternal genes acting on the developing eggs in the ovary. The orientation of the spindle, in turn, influences cell divisions following fertilization and establishes the permanent adult coiling pattern. The dextral allele (*D*) produces an active gene product that causes right-handed coiling. If ooplasm from dextral eggs is injected into uncleaved sinistral eggs, they cleave in a dextral pattern. However, in the converse experiment, sinistral ooplasm has no effect when injected into dextral eggs. Apparently the sinistral allele is the result of a classic recessive mutation that encodes an inactive gene product.

We can conclude, then, that females that are either *DD* or *Dd* produce oocytes that synthesize the *D* gene product, which is stored in the ooplasm. Even if the oocyte contains only the *d* allele following meiosis and is fertilized by a *d*-bearing sperm, the resulting *dd* snail will be dextrally coiled (right handed).

HOW DO WE KNOW?

What key observations in crosses between dextrally and sinistrally coiled snails support the explanation that this phenotype is the result of "maternal effect" inheritance?

Embryonic Development in *Drosophila*

A more recently documented example of maternal effect involves various genes that control embryonic development in *Drosophila melanogaster*. The genetic control of embryonic development in *Drosophila*, discussed in greater detail in Chapter 23, is a fascinating story. The protein products of the maternal-effect genes function to activate other genes, which may in turn activate still other genes. This cascade of gene activity leads to a normal embryo whose subsequent development yields a normal adult fly. The extensive work by Edward B. Lewis, Christiane Nüsslein-Volhard, and Eric Wieschaus (who shared the 1995 Nobel Prize for Physiology or Medicine for their findings) has clarified how these and other genes function. Genes that illustrate maternal effect have products that are synthesized by the developing egg and stored in the oocyte prior to fertilization. Following fertilization, these products specify molecular gradients that determine spatial organization as development proceeds.

For example, the gene *bicoid* (bcd) plays an important role in specifying the development of the anterior portion of the fly. Embryos derived from mothers who are homozygous for this mutation (bcd^-/bcd^-) fail to develop anterior areas that normally give rise to the head and thorax of the adult fly. Embryos whose mothers contain at least one wild-type allele (bcd^+) develop normally, even if the genotype of the embryo is homozygous for the mutation. Consistent with the concept of *maternal effect*, the *genotype of the female parent*, not the *genotype of the embryo*, determines the phenotype of the offspring.

When we return to our discussion of this general topic in Chapter 23, we will see examples of other genes illustrating maternal effect, as well as many "zygotic" genes whose expression occurs during early development and that behave genetically in the conventional Mendelian fashion.

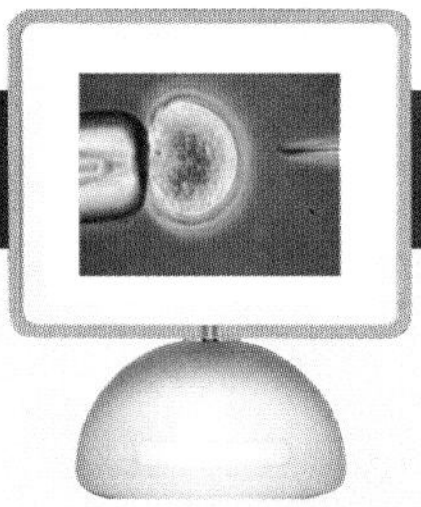

Genetics, Technology, and Society

Mitochondrial DNA and the Mystery of the Romanovs

By most accounts, Nicholas II, the last Tsar of Russia, was a substandard monarch. He was accused of bungling during the Russo-Japanese War of 1904–1905, and his regime was plagued by corruption and incompetence. Even so, he probably didn't deserve the fate that befell him and his family one summer night in 1918. As we shall see, a full understanding of that event has relied on, of all things, mitochondrial DNA.

After being forced to abdicate in 1917, ending 300 years of Romanov rule, Tsar Nicholas and the imperial family were banished to Ekaterinburg in western Siberia. There, it was believed, they would be out of reach of the fiercely anti-imperialist Bolsheviks, who were then fighting to gain control of the country. But the Bolsheviks eventually caught up with the Romanovs. During a July night in 1918, Tsar Nicholas, Tsarina Alexandra (granddaughter of Queen Victoria), their five children (Olga, 22; Tatiana, 21; Marie, 19; Anastasia, 17; and Alexei, 13), their family doctor, and three of their servants were awakened and brought to a downstairs room of the house where they were held prisoner. There they were made to form a double row against the wall, presumably so that a photograph could be taken. Instead, 11 men with revolvers burst into the room and opened fire. After exhausting their ammunition, they proceeded to bayonet the bodies and smash their faces in with rifle butts. The corpses were then hauled away and flung down a mineshaft, only to be pulled out two days later and dumped into a shallow grave, doused with sulfuric acid, and covered over. There they rested for more than 60 years.

An air of mystery soon developed around the demise of the Romanovs. Did all of the children of Nicholas and Alexandra die with their parents that bloody night, or did the youngest daughter, Anastasia, get away? Over the years, the possibility that Anastasia miraculously escaped execution has inspired countless books, a Hollywood movie, a ballet, a Broadway play, and, most recently, an animated movie. In all of these retellings, Anastasia reemerges to claim her birthright as the only surviving member of the Romanovs. Adding to the puzzle was one Anna Anderson. Two years after being dragged from a Berlin canal after a suicide attempt in 1920, she began to claim to be the Grand Duchess Anastasia. Despite a history of mental instability and a curious inability to speak Russian, she managed to convince many people.

The unraveling of the mystery began in 1979, when a Siberian geologist and a Moscow filmmaker discovered four skulls they believed to belong to the Tsar's family. It wasn't until the summer of 1991, after the establishment of glasnost, that exhumation began. Altogether, almost 1000 bone fragments were recovered, which were reassembled into nine skeletons—five females and four males. Based on measurements of the bones and computer-assisted superimposition of the skulls onto photographs, the remains were tentatively identified as belonging to the murdered Romanovs. But there were still two missing bodies, one of the daughters (believed to be Anastasia) and the boy Alexei.

The next step in authenticating the remains involved studies of DNA that were conducted by Pavel Ivanov, the leading Russian forensic DNA analyst, in collaboration with Peter Gill of the British Forensic Science Service. Their goals were to establish family relationships among the remains, and then to determine, by comparisons with living relatives, whether the family group was in fact the Romanovs. They froze bone fragments from the nine skeletons in liquid nitrogen, ground them into a fine powder, and extracted small amounts of DNA, which comprised both nuclear DNA and mitochondrial DNA (mtDNA). Genomic DNA typing of each skeleton confirmed the familial relationships and showed that, indeed, one of the princesses and Alexei were missing. Now final proof that the bones belonged to the Romanovs awaited the analysis of mtDNA.

Mitochondrial DNA is ideal for forensic studies for several reasons. First, since all the mitochondria in a human cell are descended from the mitochondria present in the egg, mtDNA is transmitted strictly from mother to offspring, never from father to offspring. Therefore, mtDNA sequences can be used to trace maternal lineages without the complicating effects of meiotic crossing over, which recombines maternal and paternal nuclear genes every generation. In addition, mtDNA is small (16,600 base pairs) and present in 500 to 1000 copies per cell, so it is much easier to recover intact than nuclear DNA.

Ivanov's group amplified two highly variable regions from the mtDNA isolated from all nine bone samples and determined the nucleotide sequences of these regions. By comparing these sequences with living relatives of the Romanovs, they hoped to establish the identity of the Ekaterinburg remains once and for all. The sequences from Tsarina Alexandra were an exact match with those from Prince Philip of England, who is her grandnephew, verifying her identity. However, authentication was more complicated for the Tsar.

The Tsar's sequences were compared with those from the only two living relatives who could be persuaded to participate in the study, Countess Xenia Cheremeteff-Sfiri (his great-grandniece) and James George Alexander Bannerman Carnegie, third Duke of Fife (a more distant relative, descended from a line of women stretching back to the Tsar's grandmother). These comparisons produced a surprise. At position 16169 of the mtDNA, the Tsar seemed to have either one or another base, a C or a T. The Countess and the Duke, in contrast, both had only T. The conclusion was that Tsar Nicholas had two different populations of mitochondria in his cells, each with a different base at position 16,169 of its DNA. This condition, called heteroplasmy, is now believed to occur in 10 to 20 percent of humans.

This ambiguity between the presumed Tsar and his two living maternal relatives cast doubt on the identification of the remains. Fortunately, the Russian government granted a request to analyze the remains of the Tsar's younger brother, Grand Duke Georgij Romanov, who died in 1899 of tuberculosis. The Grand Duke's mtDNA was found to have the same heteroplasmic variant at position 16,169, either a C or a T. It was concluded that the Ekaterinburg bones were the doomed imperial family. With years of controversy finally resolved, the now-authenticated remains of Tsar Nicholas II, Tsarina Alexandra, and three of their daughters were buried in the St. Peter and Paul Cathedral in St. Petersburg on July 17, 1998, 80 years to the day after they were murdered.

This DNA analysis did not solve the mystery of the fate of Anastasia, however. Did she die with her parents, sisters, and brother in 1918? Or is it possible that Anna Anderson was telling the truth, that she was the escaped duchess? In a separate study, Anna Anderson's nuclear and mtDNA was recovered from intestinal tissue preserved from an operation performed five years before her death in 1984. Analysis of this DNA proved that she was not Anastasia, but rather a Polish peasant named Franziska Schanzkowska.

If Anastasia's remains were not among those of her parents and sisters and if Anna Anderson was an imposter, what did happen to Anastasia? Most evidence suggests that Anastasia and her brother Alexei were not found with the others in the mass grave because their bodies were burned over the grave site two days after the killings and the ashes scattered, never to be found again. Not exactly a Hollywood ending.

References

Gibbons, A. 1998. Calibrating the mitochondrial clock. *Science* 279:28–29.

Ivanov, P. et al. 1996. Mitochondrial DNA sequence heteroplasmy in the Grand Duke of Russia Georgij Romanov establishes the authenticity of the remains of Tsar Nicholas II. *Nat. Genet.* 12:417–20.

Masse, R. 1996. *The Romanovs: The final chapter.* New York: Ballantine Books.

Stoneking, M. et al. 1995. Establishing the identity of Anna Anderson Manahan. *Nat. Genet.* 9:9–10.

Chapter Summary

1. Patterns of inheritance sometimes vary from that expected from the biparental transmission of nuclear genes. In such instances, phenotypes most often appear to result from genetic information transmitted through the egg.
2. Organelle heredity is based on the genotypes of chloroplast and mitochondrial DNA as these organelles are transmitted to offspring. Chloroplast mutations affect the photosynthetic capabilities of plants, whereas mitochondrial mutations affect cells highly dependent on ATP generated through cellular respiration. The resulting mutants display phenotypes related to the loss of function of these organelles.
3. Both chloroplasts and mitochondria originated in eukaryotic cells some 2 billion years ago as invading protobacteria. As evolution proceeded, many of the genes of the bacteria were transferred to the nucleus of the cell and a symbiotic relationship developed where these organelles enhanced the energetic capacity of the cell. Evidence in support of this endosymbiont hypothesis is extensive and centers around many observations involving the DNA and genetic machinery present in modern-day chloroplasts and mitochondria.
4. Another form of extranuclear inheritance is attributable to the transmission of infectious microorganisms. Kappa particles and CO_2-sensitivity and sex-ratio determinants are examples.
5. Maternal-effect patterns result when nuclear gene products controlled by the maternal genotype of the egg influence early development. *Ephestia* pigmentation, coiling in snails, and gene expression during early development in *Drosophila* are examples.

Insights and Solutions

1. Analyze the following hypothetical pedigree and determine the most consistent interpretation of how the trait is inherited and any inconsistencies:

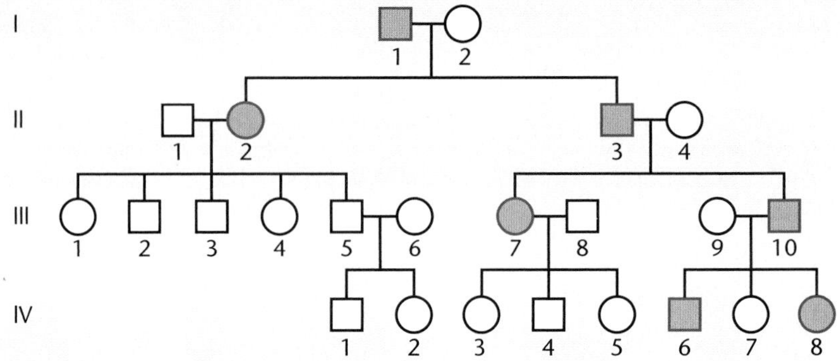

Solution: The trait is passed from all affected male parents to all but one offspring, but *never* passed maternally. Individual IV-7 (a female) is the only exception.

2. Can the explanation in Problem 1 be attributed to a gene on the Y chromosome? Defend your answer.

Solution: No, because male parents pass the trait to their daughters as well as to their sons.

3. Is the case in Problem 1 an example of a paternal effect or of paternal inheritance?

Solution: It has all the characteristics of paternal inheritance, because males pass the trait to almost all of their offspring. To assess whether the trait is due to a paternal effect (resulting from a nuclear gene in the male gamete), analysis of further matings would be needed.

Problems and Discussion Questions

1. What genetic criteria distinguish a case of extranuclear inheritance from a case of Mendelian autosomal inheritance? from a case of X-linked inheritance?
2. Streptomycin resistance in *Chlamydomonas* may result from a mutation in either a chloroplast gene or a nuclear gene. What phenotypic results would occur in a cross between a member of an mt^+ strain resistant in both genes and a member of an strain sensitive to the antibiotic? What results would occur in the reciprocal cross?
3. A plant may have green, white, or green-and-white (variegated) leaves on its branches, owing to a mutation in the chloroplast that prevents color from developing. Predict the results of the following crosses:

	Ovule Source		Pollen Source
(a)	Green branch	×	White branch
(b)	White branch	×	Green branch
(c)	Variegated branch	×	Green branch
(d)	Green branch	×	Variegated branch

4. In aerobically cultured yeast, a *petite* mutant is isolated. To determine the type of mutation causing this phenotype, the *petite* and wild-type strains are crossed. Shown here are three potential outcomes of such a cross. For each set of results, what conclusion about the type of *petite* mutation is justified?
   (a) all wild type
   (b) some *petite*: some wild type
   (c) all *petite*
5. In diploid yeast strains, sporulation and subsequent meiosis can produce haploid ascospores, which may fuse to reestablish diploid cells. When ascospores from a *segregational petite* strain fuse with those of a normal wild-type strain, the diploid zygotes are all normal. Following meiosis, ascospores are *petite* and normal. Is the *segregational petite* phenotype inherited as a dominant or a recessive trait?
6. Predict the results of a cross between ascospores from a *segregational petite* strain and a *neutral petite* strain. Indicate the phenotype of the zygote and the ascospores it may subsequently produce.
7. Described here are the results of three crosses between strains of *Paramecium*:

   (a) Killer × sensitive ⟶ 1/2 Killer : 1/2 sensitive
   (b) Killer × sensitive ⟶ all Killer
   (c) Killer × sensitive ⟶ 3/4 Killer : 1/4 sensitive

   Determine the genotypes of the parental strains.
8. *Chlamydomonas*, a eukaryotic green alga, is sensitive to the antibiotic erythromycin, which inhibits protein synthesis in prokaryotes.
   (a) Explain why.
   (b) There are two mating types in this alga, mt^+ and mt^-. If an mt^+ cell sensitive to the antibiotic is crossed with an mt^- cell that is resistant, all progeny cells are sensitive. The reciprocal cross (mt^+ resistant and mt^- sensitive) yields all resistant progeny cells. Assuming that the mutation for resistance is in the chloroplast DNA, what can you conclude from the results of these crosses?

9. In *Limnaea*, what results would you expect in a cross between a *Dd* dextrally coiled and a *Dd* sinistrally coiled snail, assuming cross-fertilization occurs as shown in Figure 9–13? What results would occur if the *Dd* dextral produced only eggs and the *Dd* sinistral produced only sperm?
10. In a cross of *Limnaea*, the snail contributing the eggs was dextral, but of unknown genotype. Both the genotype and the phenotype of the other snail are unknown. All F_1 offspring exhibited dextral coiling. Ten of the F_1 snails were allowed to undergo self-fertilization. One-half produced only dextrally coiled offspring, whereas the other half produced only sinistrally coiled offspring. What were the genotypes of the original parents?
11. In *Drosophila subobscura*, the presence of a recessive gene called *grandchildless* (*gs*) causes the offspring of homozygous females, but not those of homozygous males, to be sterile. Can you offer an explanation as to why females and not males are affected by the mutant gene?
12. A male mouse from a true-breeding strain of hyperactive animals is crossed with a female mouse from a true-breeding strain of lethargic animals. (These are both hypothetical strains.) All the progeny are lethargic. In the F_2 generation, all offspring are lethargic. What is the best genetic explanation for these observations? Propose a cross to test your explanation.
13. DNA in human mitochondria encode 22 different transfer RNA (tRNA) molecules. However, 32 different tRNA molecules are required for translation of proteins within mitochondria. How is this possible?
14. Consider the case where a mutation occurs that disrupts translation in a single human mitochondrion found in the oocyte participating in fertilization. What is the likely impact of this mutation on the offspring arising from this oocyte?

Extra-Spicy Problems

15. The specification of the anterior–posterior axis in *Drosophila* embryos is initially controlled by various gene products that are synthesized and stored in the mature egg following oogenesis. Mutations in these genes result in abnormalities of the axis during embryogenesis. These mutations illustrate *maternal effect*. How do such mutations vary from those involved in organelle heredity that illustrate *cytoplasmic inheritance?* Devise a set of parallel crosses and expected outcomes involving mutant genes that contrast maternal effect and cytoplasmic inheritance.
16. The maternal-effect mutation *bicoid* (*bcd*) is recessive. In the absence of the bicoid protein product, embryogenesis is not completed. Consider a cross between a female heterozygous for the bicoid (bcd^+/bcd^-) and a male homozygous for the mutation (bcd^-/bcd^-).
    (a) How is it possible for a male homozygous for the mutation to exist?
    (b) Predict the outcome (normal vs. failed embryogenesis) in the F_1 and F_2 generations of the cross described.
17. Shown here is a pedigree for a hypothetical human disorder:

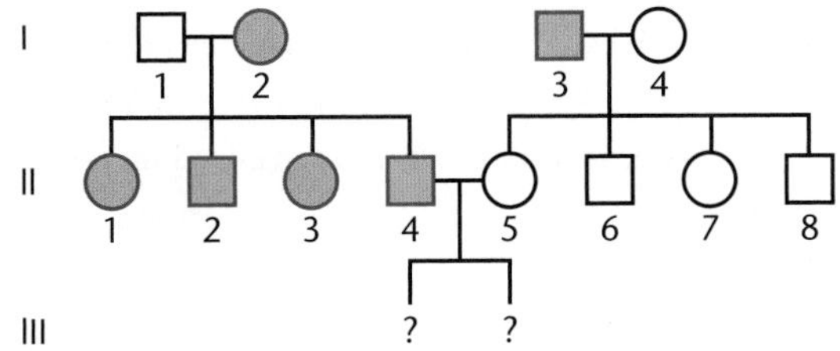

Analyze the pedigree and propose a genetic explanation for the nature of the disorder. Consistent with your explanation, predict the outcome of a mating between individuals II-4 and II-5.
18. In the late 1950s, Yuichiro Hiraizumi, a postdoctoral fellow at the University of Wisconsin, crossed a laboratory strain of *Drosophila melanogaster* that was homozygous for the recessive second chromosomal mutations *cinnabar* (*cn*) and *brown* (*bw*) with a wild-type strain collected in Madison, Wisconsin (Hiraizumi, Y. 1959. *Genetics* 44:232–50). The lab strain had white eyes, while the wild strain had red eyes and was subsequently designated *SD*. The resulting progeny, being heterozygous for *cn* and *bw*, had red eyes. When Hiraizumi backcrossed F_1 females with *cn bw*/*cn bw* males, 50 percent of the offspring had white eyes, as expected. When the reciprocal backcross was made (F_1 males with *cn bw*/*cn bw* females), less than 2 percent of the flies had white eyes.
    (a) Propose an explanation that is consistent with these results.
    (b) Design a genetic experiment to test your hypothesis.
    (c) *SD* stands for "segregation distortion." What is the significance of this description?
19. Extrachromosomally inherited traits are widespread among arthropods. In the two-spore ladybird beetle, *Adalia bipunctata*, a male-killing trait has been discovered whereby certain strains of females display a distorted sex ratio that favors female offspring (Werren, J., et al. 1994. *J. Bacteriol.* 176:388–94). Unaffected strains show a normal one-to-one sex ratio. Two key observations are that affected strains can be cured by antibiotics, and that in addition to their normal 18S and 28S rRNA, 16S rDNA can be detected by PCR (polymerase chain reaction) analysis. Of the modes of extranuclear inheritance described in the text (organelle heredity, infectious heredity, and maternal effects), which is most likely causing this altered sex ratio in *Adalia*?

20. The *abnormal oocyte* (*abo*) gene in *Drosophila melanogaster* causes a recessive maternal effect that reduces the viability of offspring. Lethality generally occurs during late embryogenesis, however some larvae do hatch. Interestingly, genetic manipulation of histone gene clusters in the mother's genome alters the severity of the *abo* maternal effect as shown in the following table (Berloco, M., et al. 2001. *Proc. Natl. Acad. Sci.* 98:12126–31).

Influence of Deficiencies of Histone Gene Clusters on the *abo* Maternal Effect

Maternal genotype		No. eggs	Progeny	Survival (adults/eggs)	Relative survival (E/C)
Df(2L)DS5, abo/abo	(E)	2905	969	0.33	
Df(2L)DS5, abo/+	(C)	2895	2400	0.83	0.40
Df(2L)DS6, abo/abo	(E)	1988	626	0.32	
Df(2L)DS6, abo/+	(C)	2000	1565	0.78	0.41
Df(2L)DS9, abo/abo	(E)	3450	570	0.17	
Df(2L)DS9, abo/+	(C)	3300	2970	0.90	0.19

Df(2)DS5 and *Df(2)DS6* are histone deficiencies; *Df(2)DS9* does not affect histone genes.
(E) = experimental; (C) = control

Analyze the information in the above table, and draw all possible conclusions regarding the the *abo* maternal effect and the role of the wild type *abo* gene.

Selected Readings

Adams, KL., et al. 2000. Repeated, recent and diverse transfers of a mitochondrial gene to the nucleus in flowering plants. *Nature* 408:354–57.

Choman, A. 1998. The myoclonic epilepsy and ragged-red fiber mutation provides new insights into human mitochondrial function and genetics. *Am. J. Hum. Genet.* 62:745–51.

Freeman, G., and Lundelius, J.W. 1982. The developmental genetics of dextrality and sinistrality in the gastropod *Lymnaea peregra. Wilhelm Roux Arch.* 191:69–83.

Gray, M.W., Burger, G., and Lang, B. 1999. Mitochondrial evolution. *Science* 283:1476–81.

Green, B.R., and Burton, H. 1970. Acetabularia chloroplast DNA: Electron microscopic visualization. *Science* 168:981–82.

Grivell, L.A. 1983. Mitochondrial DNA. *Sci. Am.* (Mar.) 248:78–89.

Larson, N.G., and Clayton, D.A. 1995. Molecular genetic aspects of human mitochondrial disorders. *Ann. Rev. Genet.* 29:151–78.

Margulis, L. 1970. *Origin of eukaryotic cells*. New Haven, CT: Yale University Press.

Mitchell, M.B., and Mitchell, H.K. 1952. A case of maternal inheritance in *Neurospora crassa. Proc. Natl. Acad. Sci. USA* 38:442–49.

Nüsslein-Volhard, C. 1996. Gradients that organize embryo development. *Sci. Am.* (Aug.) 275:54–61.

Preer, J.R. 1971. Extrachromosomal inheritance: Hereditary symbionts, mitochondria, chloroplasts. *Annu. Rev. Genet.* 5:361–406.

Sager, R. 1965. Genes outside the chromosomes. *Sci. Am.* (Jan.) 212:70–79.

______ 1985. Chloroplast genetics. *BioEssays* 3:180–84.

Schwartz, R.M., and Dayhoff, M.O. 1978. Origins of prokaryotes, eukaryotes, mitochondria and chloroplasts. *Science* 199:395–403.

Sonneborn, T.M. 1959. Kappa and related particles in *Paramecium. Adv. Virus Res.* 6:229–56.

Sturtevant, A.H. 1923. Inheritance of the direction of coiling in *Limnaea. Science* 58:269–70.

Wallace, D.C. 1997. Mitochondrial DNA in aging and disease. *Sci. Am.* Aug. 277:40–59.

______ 1999. Mitochondrial diseases in man and mouse. *Science* 283: 1482–88.

DNA Structure and Analysis

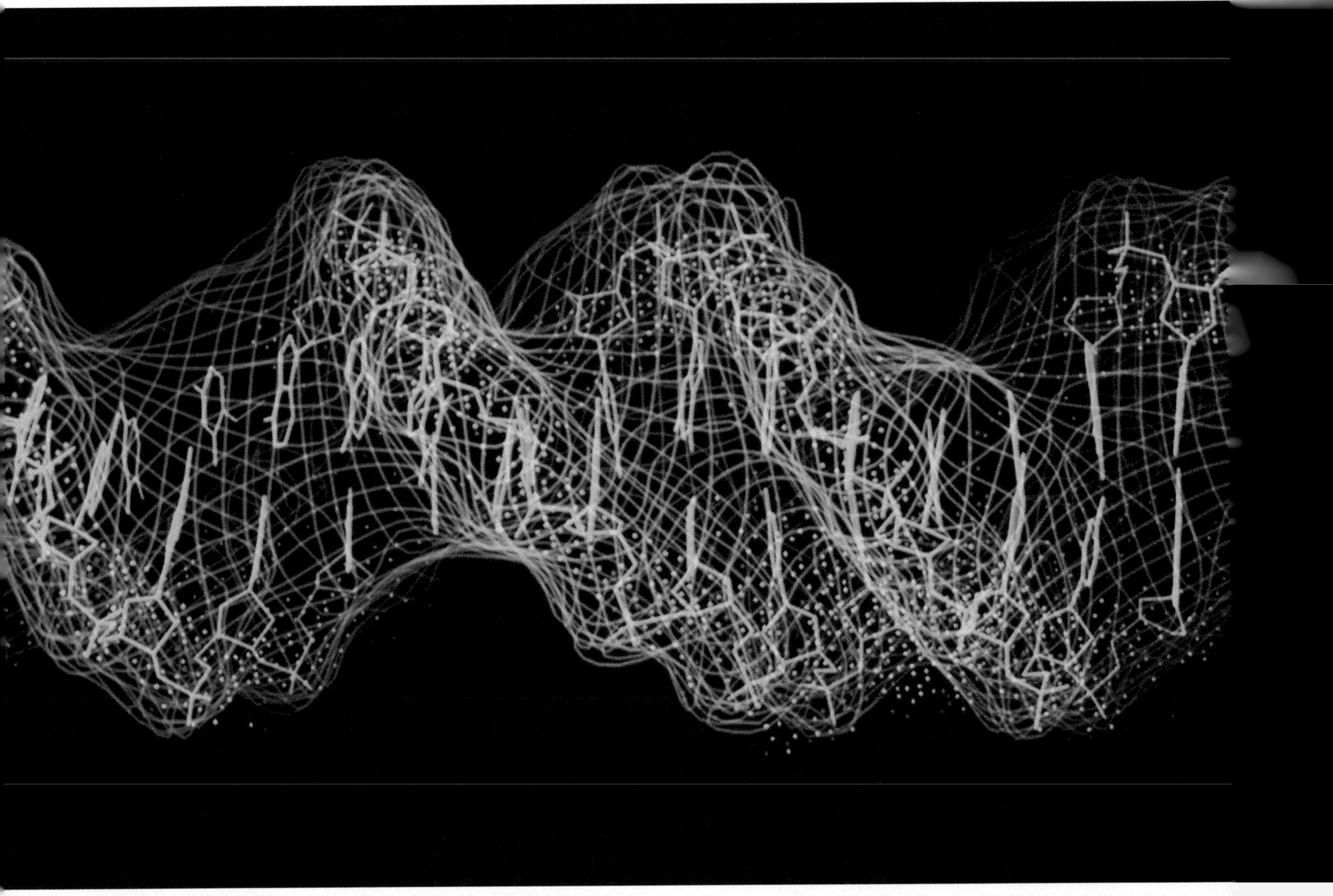

A computerized image of DNA.

Chapter Concepts

- With the exception of some viruses, DNA serves as the genetic material in all living organisms on earth.
- According to the Watson–Crick model, DNA exists in the form of right-handed double helix.
- The strands of the double helix are held together by hydrogen bonding between complementary nitrogenous base pairs.
- The structure of DNA provides the basis for storing and expressing genetic information.
- RNA has many similarities to DNA, but exists as a single-stranded molecule.
- In some viruses, RNA serves as the genetic material.

Earlier in the text, we discussed the existence of genes on chromosomes that control phenotypic traits and the way in which the chromosomes are transmitted through gametes to future offspring. Logically, some form of information must be contained in genes, which, when passed to a new generation, influences the form and characteristics of the offspring. We refer to this as **genetic information**. We might also conclude that this same information in some way directs the many complex processes leading to the adult form.

Until 1944, it was not clear what chemical component of the chromosome makes up genes and constitutes the genetic material. Because chromosomes were known to have both a nucleic acid and a protein component, both were considered candidates. In 1944, however, there emerged direct experimental evidence that the nucleic acid DNA serves as the informational basis for the process of heredity.

Once the importance of DNA in genetic processes was realized, work was intensified with the hope of discerning not only the structural basis of this molecule, but also the relationship of its structure to its function. Between 1944 and 1953, many scientists sought information that might answer one of the most significant and intriguing questions in the history of biology: How does DNA serve as the genetic basis for living processes? The answer was believed to depend strongly on the chemical structure of the DNA molecule, given the complex, but orderly functions ascribed to it.

These efforts were rewarded in 1953, when James Watson and Francis Crick set forth their hypothesis for the double-helical nature of DNA. The assumption that the molecule's functions would be easier to clarify once its general structure was determined proved to be correct. This chapter initially reviews the evidence that DNA is the genetic material and then discusses the elucidation of its structure.

10.1 The Genetic Material Must Exhibit Four Characteristics

For a molecule to serve the role of the genetic material, it must possess four major characteristics: **replication**, **storage of information**, **expression of information**, and **variation by mutation**. Replication of the genetic material is one facet of the cell cycle, a fundamental property of all living organisms. Once the genetic material of cells has been replicated, it must then be partitioned equally into daughter cells. During the formation of gametes, the genetic material is also replicated, but is partitioned so that each cell gets only one-half of the original amount of genetic material—the process of *meiosis*, discussed in Chapter 2. Although the products of mitosis and meiosis are different, these processes are both part of the more general phenomenon of cellular reproduction.

The characteristic of storage can be viewed as being able to serve as a repository of genetic information, whether or not it is expressed. It is clear that, while most cells contain a complete complement of DNA, at any given point they express only a part of this genetic potential. For example, bacteria "turn on" many genes in response to specific environmental conditions and turn them off when such conditions change. In vertebrates, skin cells may display active melanin genes, but never activate their hemoglobin genes; digestive cells activate many genes specific to their function, but do not activate their melanin genes.

Inherent in the concept of storage is the need for the genetic material to be able to encode the nearly infinite variety of gene products found among the countless forms of life present on our planet. The chemical language of the genetic material must be capable of this potential task as it stores information and as that information is transmitted to progeny cells and organisms.

Expression of the stored genetic information is a complex process and is the basis for the concept of **information flow** within the cell. Figure 10–1 shows a simplified illustration of this concept. The initial event is the transcription of DNA, resulting in the synthesis of three types of RNA molecules: messenger RNA (mRNA), transfer RNA (tRNA), and ribosomal RNA (rRNA). Of these, mRNAs are translated into proteins. Each type of mRNA is the product of a specific gene and leads to the synthesis of a different protein. Translation occurs in conjunction with rRNA-containing ribosomes and involves tRNA, which acts as an adaptor to convert the chemical information in mRNA to the amino acids that make up proteins. Collectively, these processes serve as the foundation for the **central dogma of molecular genetics**: *DNA makes RNA, which makes proteins.*

The genetic material must also serve as the basis of newly arising variability among organisms through the process of mutation. If a mutation—a change in the chemical composition of DNA—occurs, the alteration will be reflected during transcription and translation, often affecting the specified protein. If a mutation is present in gametes, it will be passed to future generations and, with time, may become distributed in the population. Genetic variation, which also includes rearrangements within and between chromosomes (as discussed in Chapter 8), provides the raw material for the process of evolution.

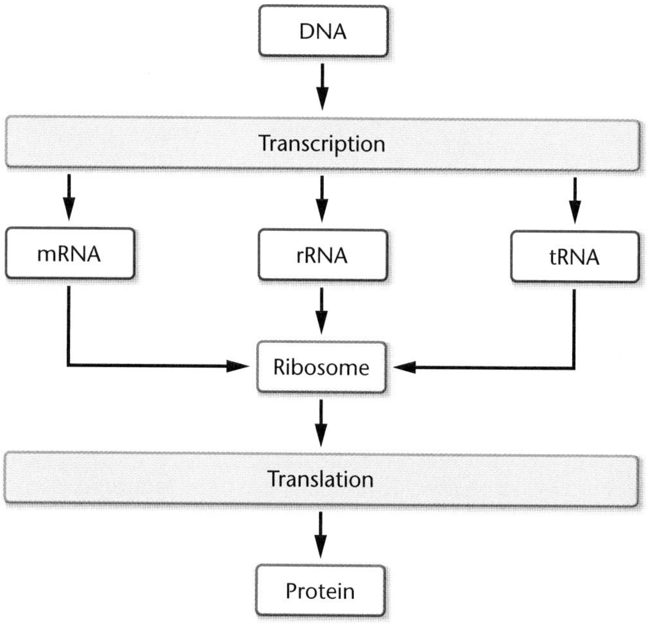

FIGURE 10–1 Simplified view of the central dogma describing information flow involving DNA, RNA, and proteins within cells.

10.2 Until 1944, Observations Favored Protein as the Genetic Material

The idea that genetic material is physically transmitted from parent to offspring has been accepted for as long as the concept of inheritance has existed. Beginning in the late 19th century, research into the structure of biomolecules progressed considerably, setting the stage for describing the genetic material in chemical terms. Although proteins and nucleic acid were both considered major candidates for the role of the genetic material, until the 1940s many geneticists favored proteins. Three factors contributed to this belief.

First, proteins are abundant in cells. Although the protein content may vary considerably, these molecules account for more than 50 percent of the dry weight of cells. Because cells contain such a large amount and variety of proteins, it is not surprising that early geneticists believed that some of this protein could function as the genetic material. The second factor was the accepted proposal for the chemical structure of nucleic acids during the early to mid-1900s. DNA was first studied in 1868 by a Swiss chemist, Friedrick Miescher. He was able to separate nuclei from the cytoplasm of cells and then isolate from these nuclei an acidic substance that he called **nuclein**. Miescher showed that nuclein contained large amounts of phosphorus and no sulfur, characteristics that differentiate it from proteins.

As analytical techniques were improved, nucleic acids, including DNA, were shown to be composed of four similar molecular building blocks called nucleotides. About 1910, Phoebus A. Levene proposed the **tetranucleotide hypothesis** to explain the chemical arrangement of these nucleotides in nucleic acids. He proposed that a very simple four-nucleotide unit, as shown in Figure 10–2, was repeated over and over in DNA. Levene based his proposal on studies of the composition of the four types of nucleotides. Although his actual data revealed proportions of the four nucleotides that varied considerably, he assumed a 1:1:1:1 ratio. The discrepancy was ascribed to inadequate analytical technique.

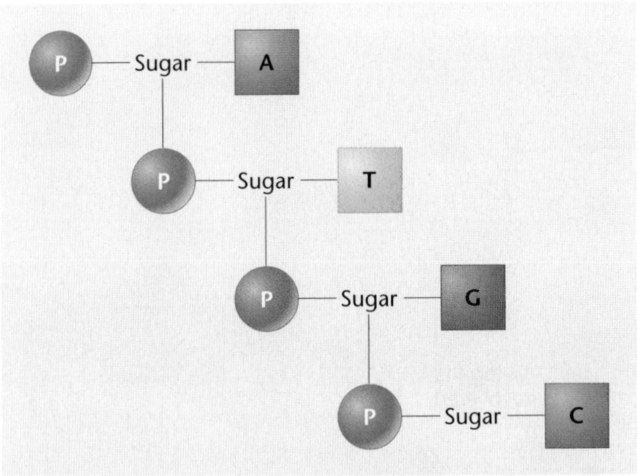

FIGURE 10–2 Diagrammatic depiction of Levene's proposed tetranucleotide containing one molecule each of the four nitrogenous bases: adenine (A), thymine (T), guanine (G), and cytosine (C). Each block represents a nitrogenous base. Other potential sequences of four nucleotides are possible.

Because a single covalently bonded tetranucleotide structure was relatively simple, geneticists believed nucleic acids could not provide the large amount of chemical variation expected for the genetic material. Proteins, on the other hand, contain 20 different amino acids, thus affording the basis for substantial variation. As a result, attention was directed away from nucleic acids, strengthening the speculation that proteins served as the genetic material.

The third contributing factor simply concerned the areas of most active research in genetics. Before 1940, most geneticists were engaged in the study of transmission genetics and mutation. The excitement generated in these areas undoubtedly diluted the concern for finding the precise molecule that serves as the genetic material. Thus, proteins were the most promising candidate and were accepted rather passively.

Between 1910 and 1930, other proposals for the structure of nucleic acids were advanced, but they were generally overturned in favor of the tetranucleotide hypothesis. It was not until the 1940s that the work of Erwin Chargaff led to the realization that Levene's hypothesis was incorrect. Chargaff showed that, for most organisms, the 1:1:1:1 ratio was inaccurate, disproving Levene's hypothesis.

10.3 Evidence Favoring DNA as the Genetic Material Was First Obtained during the Study of Bacteria and Bacteriophages

The 1944 publication by Oswald Avery, Colin MacLeod, and Maclyn McCarty concerning the chemical nature of a "transforming principle" in bacteria was the initial event leading to the acceptance of DNA as the genetic material. Their work, along with the subsequent findings of other research teams, constituted direct experimental proof that DNA, and not protein, is the biomolecule responsible for heredity. It marked the beginning of the era of molecular genetics, a period of discovery in biology that has served as the foundation for current studies in biotechnology and has moved us closer to an understanding of the basis of life. The impact of the initial findings on future research and thinking paralleled that of the publication of Darwin's theory of evolution and the subsequent rediscovery of Mendel's postulates of transmission genetics. Together, these events constituted the three great revolutions in biology.

Transformation: Early Studies

The research that provided the foundation for Avery, MacLeod, and McCarty's work was initiated in 1927 by Frederick Griffith, a medical officer in the British Ministry of Health. He performed experiments with several different strains of the bacterium *Diplococcus pneumoniae*.* Some were **virulent**

*This organism is now named *Streptococcus pneumoniae*.

(infectious) strains, which cause pneumonia in certain vertebrates (notably humans and mice), whereas others were **avirulent** (noninfectious) strains, which do not cause illness.

The difference in virulence is related to the polysaccharide capsule of the bacterium. Virulent strains have this capsule, whereas avirulent strains do not. The nonencapsulated bacteria are readily engulfed and destroyed by phagocytic cells in the animal's circulatory system. Virulent bacteria, which possess the polysaccharide coat, are not easily engulfed. Hence, they are able to multiply and cause pneumonia.

The presence or absence of the capsule is the basis for another characteristic difference between virulent and avirulent strains. Encapsulated bacteria form smooth, shiny-surfaced colonies (*S*) when grown on an agar culture plate; nonencapsulated strains produce rough colonies (*R*) (Figure 10–3). This feature allows virulent and avirulent strains to be distinguished easily by standard microbiological culture techniques.

Each strain of *Diplococcus* may be one of dozens of different types called **serotypes**. The specificity of the serotype is due to the detailed chemical structure of the polysaccharide constituent of the thick, slimy capsule. Serotypes are identified by immunological techniques and are usually designated by Roman numerals. In the United States, types I and II are most common in causing pneumonia. Griffith used types II and III in the critical experiments that led to new concepts about the genetic material. Table 10.1 summarizes the characteristics of Griffith's two strains.

Griffith knew from the work of others that only living virulent cells would produce pneumonia in mice. If heat-killed virulent bacteria are injected into mice, no pneumonia results, just as living avirulent bacteria fail to produce the disease. Griffith's

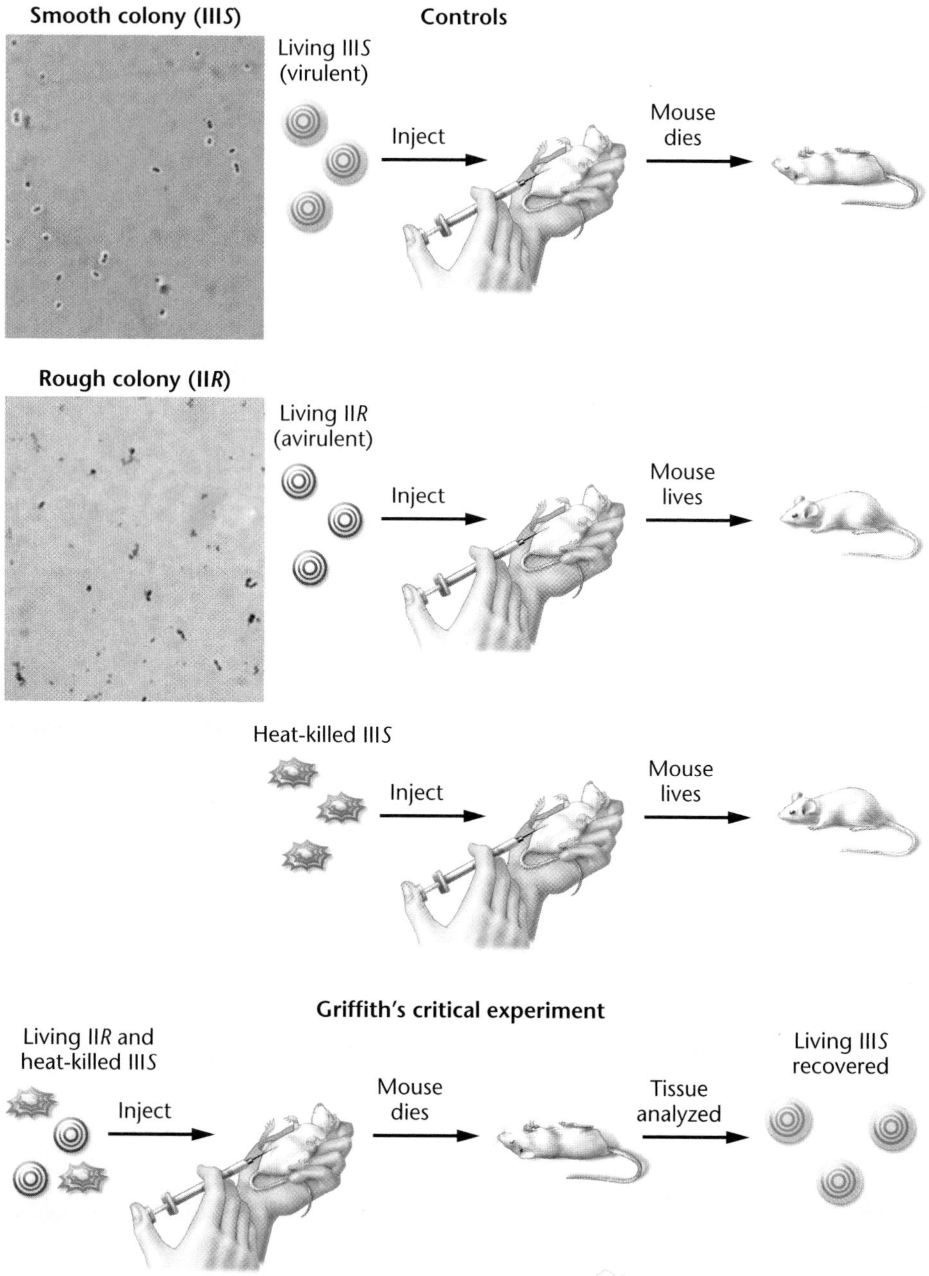

FIGURE 10–3 Griffith's transformation experiment. The photographs show bacterial colonies containing cells with capsules (type III*S*) and without capsules (type II*R*).

TABLE 10.1 **STRAINS OF *DIPLOCOCCUS PNEUMONIAE* USED BY FREDERICK GRIFFITH IN HIS ORIGINAL TRANSFORMATION EXPERIMENTS**

Serotype	Colony Morphology	Capsule	Virulence
II*R*	Rough	Absent	Avirulent
III*S*	Smooth	Present	Virulent

critical experiment (Figure 10–3) involved an injection into mice of living II*R* (avirulent) cells combined with heat-killed III*S* (virulent) cells. Since neither cell type caused death in mice when injected alone, Griffith expected that the double injection would not kill the mice. But, after five days, mice that received both types of cells were all dead. Analysis of their blood revealed a large number of living type III*S* (virulent) bacteria.

As far as could be determined regarding the capsular polysaccharide, these III*S* bacteria were identical to the III*S* strain from which the heat-killed cell preparation had been made. The control mice, injected only with living avirulent II*R* bacteria for this set of experiments, did not develop pneumonia and remained healthy. This finding ruled out the possibility that the avirulent II*R* cells had simply changed (or mutated) to virulent III*S* cells in the absence of the heat-killed III*S* fraction. Instead, some type of interaction was required between living II*R* and heat-killed III*S* cells.

Griffith concluded that the heat-killed III*S* bacteria were somehow responsible for converting live avirulent II*R* cells into virulent III*S* ones. Calling the phenomenon **transformation**, he suggested that the **transforming principle** might be some part of the polysaccharide capsule *or* some compound required for capsule synthesis, although the capsule alone did not cause pneumonia. To use Griffith's term, the transforming principle from the dead III*S* cells served as a "pabulum" for the II*R* cells.

Griffith's work led other physicians and bacteriologists to research the phenomenon of transformation. By 1931, Henry Dawson at the Rockefeller Institute had confirmed Griffith's observations and extended his work one step further. Dawson and his coworkers showed that transformation could occur *in vitro* (in a test tube). When heat-killed III*S* cells were incubated with living II*R* cells, living III*S* cells were recovered. Therefore, injection into mice was not necessary for transformation to occur. By 1933, Lionel J. Alloway had refined the *in vitro* system by using crude extracts of *S* cells and living *R* cells. The soluble filtrate from the heat-killed *S* cells was as effective in inducing transformation as were the intact cells. Alloway and others did not view transformation as a genetic event, but rather as a physiological modification of some sort. Nevertheless, the experimental evidence that a chemical substance was responsible for transformation was quite convincing.

Transformation: The Avery, MacLeod, and McCarty Experiment

The critical question, of course, was what molecule serves as the transforming principle? In 1944, after 10 years of work, Avery, MacLeod, and McCarty published their results in what is now regarded as a classic paper in the field of molecular genetics. They reported that they had obtained the transforming principle in a purified state, and that beyond reasonable doubt, the molecule responsible for transformation was DNA.

The details of their work, sometimes called the Avery, MacLeod, and McCarty experiment, are outlined in Figure 10–4. These researchers began their isolation procedure with large quantities (50–75 liters) of liquid cultures of type III*S* virulent cells. The cells were centrifuged, collected, and heat killed. Following homogenization and several extractions with the detergent deoxycholate (DOC), they obtained a soluble filtrate that, when tested, still contained the transforming principle. Protein was removed from the active filtrate by several chloroform extractions, and polysaccharides were enzymatically digested and removed. Finally, precipitation with ethanol yielded a fibrous mass that still retained the ability to induce transformation of type II*R* avirulent cells. From the original 75-liter sample, the procedure yielded 10 to 25 mg of the "active factor."

Further testing clearly established that the transforming principle was DNA. The fibrous mass was first analyzed for its nitrogen–phosphorus ratio, which was shown to coincide with the ratio of "sodium desoxyribonucleate," the chemical name then used to describe DNA. To solidify their findings, Avery, MacLeod, and McCarty sought to eliminate, to the greatest extent possible, all probable contaminants from their final product. Thus, it was treated with the proteolytic enzymes trypsin and chymotrypsin and then with an RNA-digesting enzyme, called **ribonuclease** (**RNase**). Such treatments destroyed any remaining activity of proteins and RNA. Nevertheless, transforming activity still remained. Chemical testing of the final product gave strong positive reactions for DNA. The final confirmation came with experiments using crude samples of the DNA-digesting enzyme **deoxyribonuclease** (**DNase**), which was isolated from dog and rabbit sera. Digestion with this enzyme was shown to destroy transforming activity. There could be little doubt that the active transforming principle was DNA.

The great amount of work, the confirmation and reconfirmation of the conclusions drawn, and the unambiguous logic of the experimental design involved in the research of these three scientists are truly impressive. Avery, MacLeod, and McCarty's conclusion in the 1944 publication was, however, very simply stated: "The evidence presented supports the belief that a nucleic acid of the desoxyribose* type is the fundamental unit of the transforming principle of *Pneumococcus* Type III."

Avery and his colleagues recognized the genetic and biochemical implications of their work when they observed that "nucleic acids of this type must be regarded not merely as structurally important but as functionally active in determining the biochemical activities and specific characteristics of pneumococcal cells." This suggested that the transforming principle interacts with the II*R* cell and gives rise to a coordinated series of enzymatic reactions culminating in the synthesis of the type III*S* capsular polysaccharide. Avery, MacLeod, and McCarty emphasized that, once transformation

*Desoxyribose is now spelled deoxyribose.

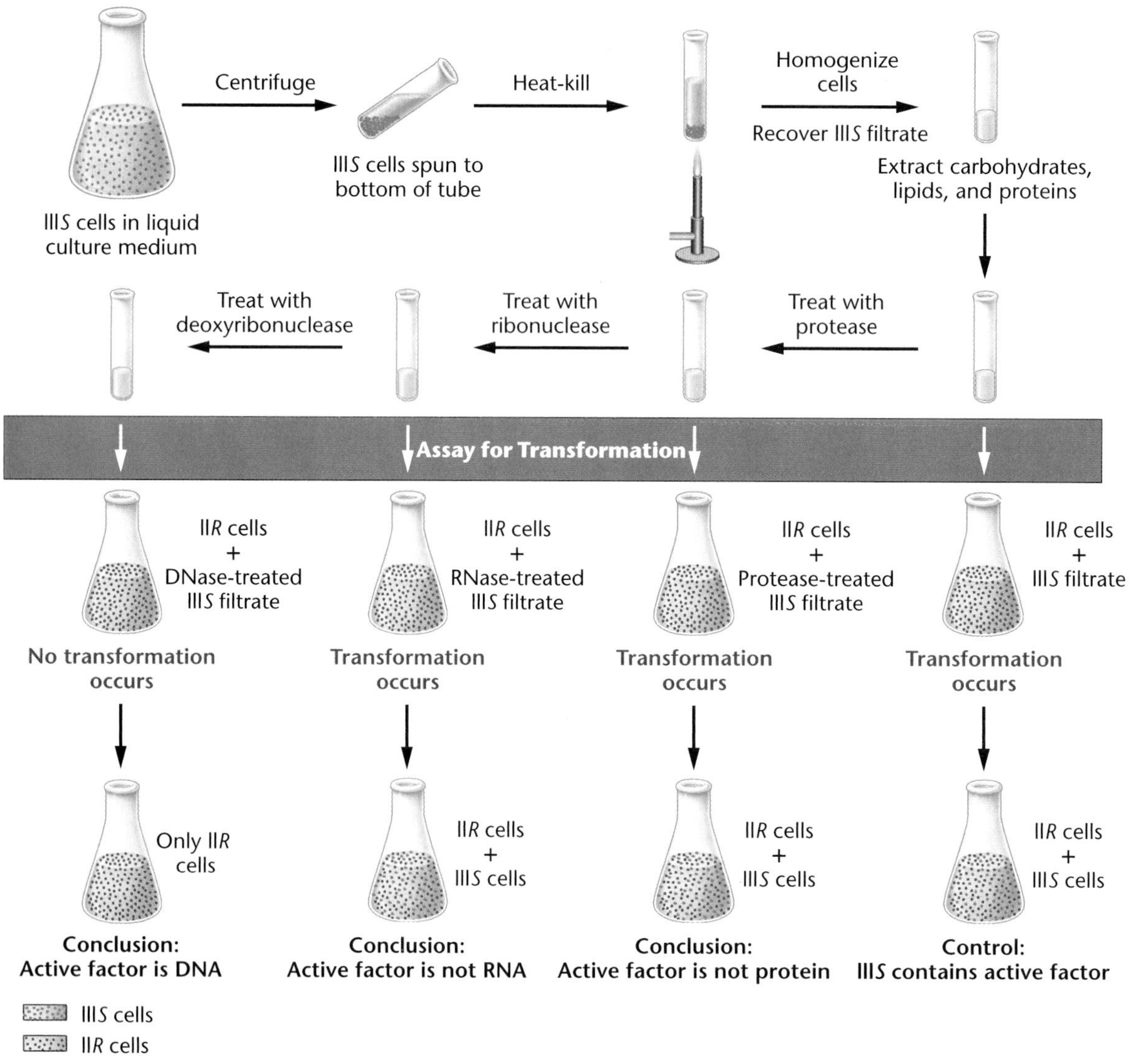

FIGURE 10–4 Summary of Avery, MacLeod, and McCarty's experiment, which demonstrated that DNA is the transforming principle.

occurs, the capsular polysaccharide is produced in successive generations. Transformation is therefore heritable, and the process affects the genetic material.

Immediately after the publication of the report, several investigators turned to, or intensified, their studies of transformation in order to clarify the role of DNA in genetic mechanisms. In particular, the work of Rollin Hotchkiss was instrumental in confirming that the critical factor in transformation was DNA and not protein. In 1949, in a separate study, Harriet Taylor isolated an **extremely rough (*ER*)** mutant strain from a rough (*R*) strain. This *ER* strain produced colonies that were more irregular than the *R* strain. The DNA from *R* accomplished the transformation of *ER* to *R*. Thus, the *R* strain, which served as the recipient in the Avery experiments, was shown to be able to serve as the DNA donor in transformation also.

Transformation has now been shown to occur in *Haemophilus influenzae, Bacillus subtilis, Shigella paradysenteriae*, and *Escherichia coli*, among many other microorganisms. Transformation of numerous genetic traits other than colony morphology has also been demonstrated, including ones involving resistance to antibiotics and the ability to metabolize various nutrients. These observations further strengthened the belief that transformation by DNA is primarily a genetic event, rather than simply a physiological change. This idea is pursued again in the "Insights and Solutions" section at the end of this chapter.

HOW DO WE KNOW?

What is the experimental basis for concluding that DNA, and not RNA or protein, is the transforming material?

The Hershey–Chase Experiment

The second major piece of evidence supporting DNA as the genetic material was provided during the study of the bacterium *Escherichia coli* and one of its infecting viruses, **bacteriophage T2**. Often referred to simply as a **phage**, the virus consists of a protein coat surrounding a core of DNA. Electron micrographs have revealed the phage's external structure to be composed of a hexagonal head plus a tail. The general aspects of the reproductive cycle of a T-even bacteriophage such as T2, as known in 1952, is shown in Figure 10–5. Briefly, the phage adsorbs to the bacterial cell and some component of the phage enters the bacterial cell. Following this infection step, the viral information "commandeers" the cellular machinery of the host and directs viral reproduction. In a reasonably short time, many new phages are constructed and the bacterial cell is lysed, releasing the progeny viruses. This process is referred to as the **lytic cycle**.

In 1952, Alfred Hershey and Martha Chase published the results of experiments designed to clarify the events leading to phage reproduction. Several of the experiments clearly established the independent functions of phage protein and nucleic acid in the reproduction process associated with the bacterial cell. Hershey and Chase knew from existing data that

1. T2 phages consist of approximately 50 percent protein and 50 percent DNA.
2. Infection is initiated by adsorption of the phage by its tail fibers to the bacterial cell.
3. The production of new viruses occurs within the bacterial cell.

It appeared that some molecular component of the phage—DNA or protein (or both)—enters the bacterial cell and directs viral reproduction. Which was it?

Hershey and Chase used the radioisotopes ^{32}P and ^{35}S to follow the molecular components of phages during infection (Figure 10–6). Because DNA contains phosphorus (P), but not sulfur, ^{32}P effectively labels DNA; and because proteins contain sulfur (S), but not phosphorus, ^{35}S labels protein. *This was a key point in the experiment.* If *E. coli* cells are first grown in the presence of ^{32}P or ^{35}S and then infected with T2 viruses, the progeny phages will have *either* a radioactively labeled DNA core *or* a radioactively labeled protein coat, respectively. These labeled phages can be isolated and used to infect unlabeled bacteria.

When labeled phages and unlabeled bacteria are mixed, an adsorption complex is formed as the phages attach their tail fibers to the bacterial wall. These complexes were isolated and subjected to a high shear force by placing them in a blender. This force stripped off the attached phages. When the mixture was centrifuged, the lighter phage particles separated from the heavier bacterial cells, allowing the isolation of both components (Figure 10–6). By tracing the radioisotopes, Hershey and Chase were able to demonstrate that most of the ^{32}P-labeled DNA had been transferred into the bacterial cell following adsorption; on the other hand, most of the ^{35}S-labeled protein remained outside the bacterial cell and was recovered in the phage "ghosts" (empty phage coats) after the blender treatment. Following this separation, the bacterial cells, which now contained viral DNA, were eventually lysed as new phages were produced. These progeny phages contained ^{32}P, but not ^{35}S.

Hershey and Chase interpreted these results to indicate that the protein of the phage coat remains outside the host cell and is not involved in directing the production of new phages. On the other

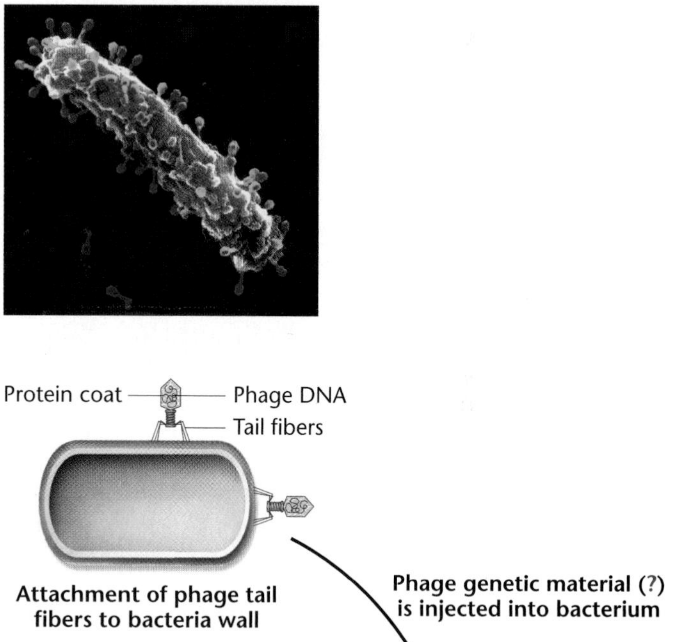
Protein coat
Phage DNA
Tail fibers
Attachment of phage tail fibers to bacteria wall
Phage genetic material (?) is injected into bacterium

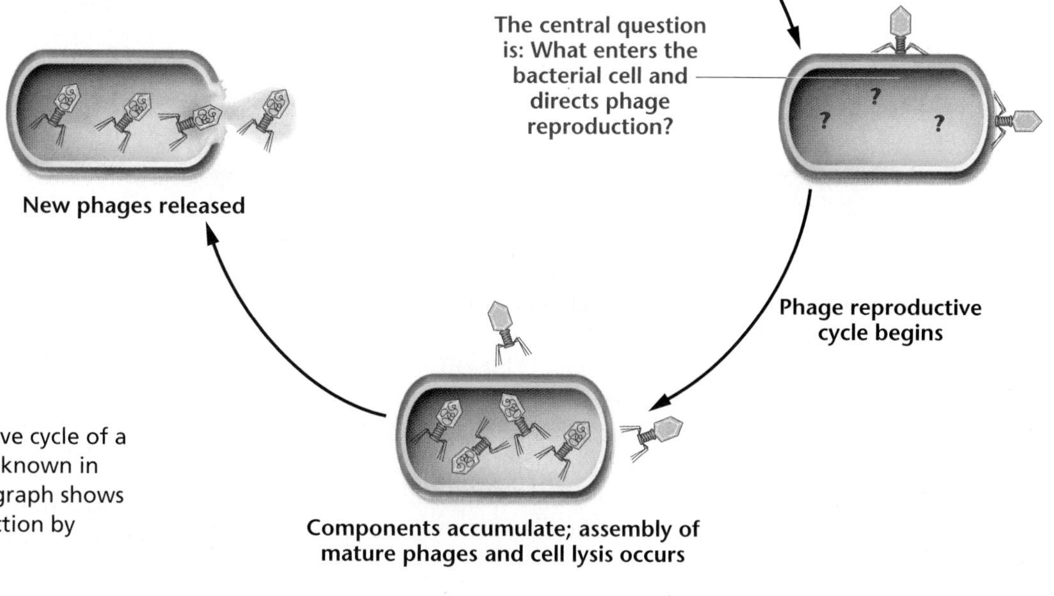
The central question is: What enters the bacterial cell and directs phage reproduction?
?
?
?
New phages released
Phage reproductive cycle begins
Components accumulate; assembly of mature phages and cell lysis occurs

FIGURE 10–5 Reproductive cycle of a T-even bacteriophage, as known in 1952. The electron micrograph shows an *E. coli* cell during infection by numerous phages.

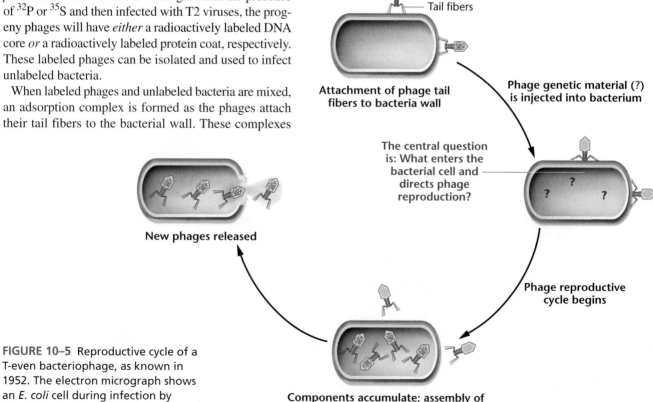

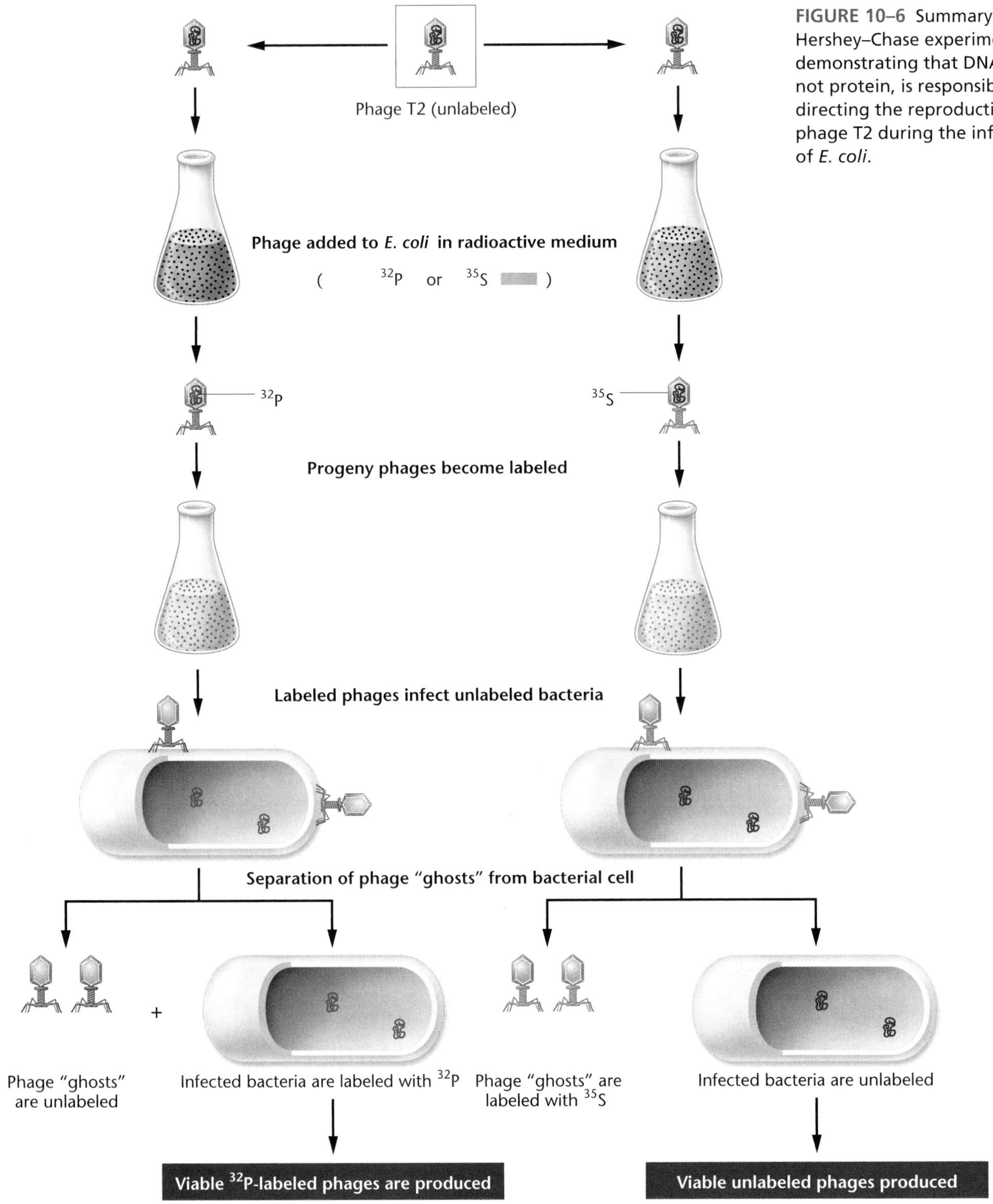

Phage T2 (unlabeled)
Phage added to E. coli in radioactive medium
(32P or 35S)
32P
35S
Progeny phages become labeled
Labeled phages infect unlabeled bacteria
Separation of phage "ghosts" from bacterial cell
+
Phage "ghosts" are unlabeled
Infected bacteria are labeled with 32P
Phage "ghosts" are labeled with 35S
Infected bacteria are unlabeled
Viable 32P-labeled phages are produced
Viable unlabeled phages produced

FIGURE 10–6 Summary of the Hershey–Chase experiment demonstrating that DNA, and not protein, is responsible for directing the reproduction of phage T2 during the infection of *E. coli*.

hand, and most important, phage DNA enters the host cell and directs phage reproduction. Hershey and Chase had demonstrated the genetic material in phage T2 is DNA, not protein.

These experiments, along with those of Avery and his colleagues, provided convincing evidence to most geneticists that DNA was the molecule responsible for heredity. Since then, many significant findings have been based on this supposition. These many findings, constituting the field of molecular genetics, are discussed in detail in subsequent chapters.

Now solve this

Problem 10.7 on page 260 asks about the possible success of applying the protocol of the Hershey–Chase experiment to the investigation of transformation.

Hint: As you attempt to apply the Hershey–Chase protocol, remember that in transformation, exogenous DNA enters the soon-to-be transformed cell and no cell-to-cell contact is involved in the process.

Transfection Experiments

During the eight years following the publication of the Hershey–Chase experiment, additional research using bacterial viruses provided even more solid proof that DNA is the genetic material. In 1957, several reports demonstrated that if *E. coli* was treated with the enzyme lysozyme, the outer wall of the cell can be removed without destroying the bacterium. Enzymatically treated cells are naked, so to speak, and contain only the cell membrane as their outer boundary. Such structures are called **protoplasts** (or **spheroplasts**). John Spizizen and Dean Fraser independently reported that by using spheroplasts, they were able to initiate phage reproduction with disrupted T2 particles. That is, provided that the cell wall is absent, a virus does not have to be intact for infection to occur. Thus, the outer protein coat structure may be essential to the movement of DNA through the intact cell wall, but it is not essential for infection when spheroplasts are used.

Similar, but more refined, experiments were reported in 1960 by George Guthrie and Robert Sinsheimer. DNA was purified from bacteriophage ϕX174, a small phage that contains a single-stranded circular DNA molecule of some 5386 nucleotides. When added to *E. coli* protoplasts, the purified DNA resulted in the production of complete ϕX174 bacteriophages. This process of infection by only the viral nucleic acid, called **transfection**, proved conclusively that ϕX174 DNA alone contains all the necessary information for production of mature viruses. Thus, the evidence supporting the conclusion that DNA serves as the genetic material was further strengthened, even though all direct evidence thus far had been obtained from bacterial and viral studies.

HOW DO WE KNOW?

What is the experimental basis for concluding that DNA, and not protein, is the molecule responsible for directing bacteriophage reproduction? Does this experiment discern between DNA and RNA as the responsible molecule?

10.4 Indirect and Direct Evidence Supports the Concept that DNA Is the Genetic Material in Eukaryotes

In 1950, eukaryotic organisms were not amenable to the types of experiments that in bacteria and viruses demonstrated that DNA is the genetic material. Nevertheless, it was generally assumed that the genetic material would be a universal substance and also serve this role in eukaryotes. Initially, support for this assumption relied on several circumstantial (indirect) observations that, taken together, inferred that DNA serves as the genetic material in eukaryotes. Subsequently, direct evidence established unequivocally the central role of DNA in genetic processes.

Indirect Evidence: Distribution of DNA

The genetic material should be found where it functions—in the nucleus as part of chromosomes. Both DNA and protein fit this criterion. However, protein is also abundant in the cytoplasm, whereas DNA is not. Both mitochondria and chloroplasts are known to perform genetic functions, and DNA is also present in these organelles. Thus, DNA is found only where primary genetic function occurs. Protein, on the other hand, is found everywhere in the cell. These observations are consistent with the interpretation favoring DNA over proteins as the genetic material.

TABLE 10.2 DNA CONTENT OF HAPLOID VERSUS DIPLOID CELLS OF VARIOUS SPECIES (IN PICOGRAMS)

Organism	*n*	2*n*
Human	3.25	7.30
Chicken	1.26	2.49
Trout	2.67	5.79
Carp	1.65	3.49
Shad	0.91	1.97

Sperm (*n*) and nucleated precursors to red blood cells (2*n*) were used to contrast ploidy levels.

Because it had earlier been established that chromosomes within the nucleus contain the genetic material, a correlation was expected to exist between the ploidy (*n*, 2*n*, etc.) of a cell and the quantity of the molecule that functions as the genetic material. Meaningful comparisons can be made between the amount of DNA and protein in gametes (sperm and eggs) and somatic or body cells. The latter are recognized as being diploid (2*n*) and containing twice the number of chromosomes as gametes, which are haploid (*n*).

Table 10.2 compares the amount of DNA found in haploid sperm and diploid nucleated precursors of red blood cells from a variety of organisms. The amount of DNA and the number of sets of chromosomes is closely correlated. No consistent correlation can be observed between gametes and diploid cells for proteins. These data thus provide further circumstantial evidence favoring DNA over proteins as the genetic material of eukaryotes.

Indirect Evidence: Mutagenesis

Ultraviolet (UV) light is one of a number of agents capable of inducing mutations in the genetic material. Bacteria and other organisms can be irradiated with various wavelengths of ultraviolet light and the effectiveness of each wavelength measured by the number of mutations it induces. When the data are plotted, an **action spectrum** of UV light as a mutagenic agent is obtained. This action spectrum can then be compared with the **absorption spectrum** of any molecule suspected to be the genetic material (Figure 10–7). *The molecule serving as the genetic material is expected to absorb at the wavelength(s) found to be mutagenic.*

UV light is most mutagenic at the wavelength λ of 260 nanometers (nm). Both DNA and RNA absorb UV light most strongly at 260 nm. On the other hand, protein absorbs most strongly at 280 nm, yet no significant mutagenic effects are observed at that wavelength. This indirect evidence supports the idea that a nucleic acid is the genetic material and tends to exclude protein.

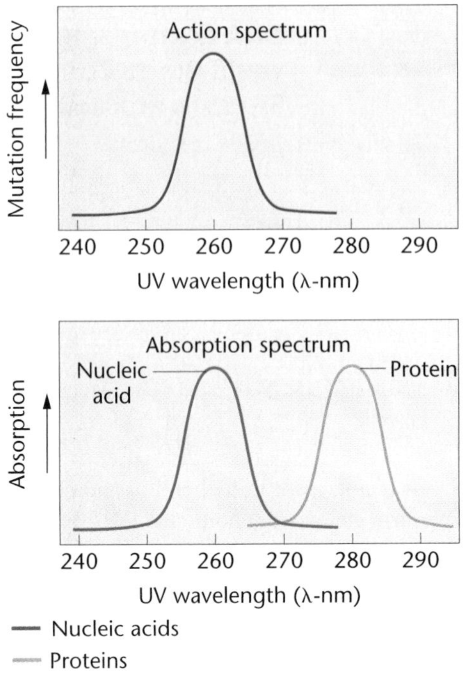

FIGURE 10–7 Comparison of the action spectrum (which determines the most effective mutagenic UV wavelength) and the absorption spectrum (which shows the range of wavelength where nucleic acids and proteins absorb UV light).

Direct Evidence: Recombinant DNA Studies

Although the circumstantial evidence just described does not constitute direct proof that DNA is the genetic material in eukaryotes, these observations spurred researchers to forge ahead, basing their work on this hypothesis. Today, there is no doubt of the validity of this conclusion; DNA *is* the genetic material in all eukaryotes.

The strongest evidence is provided by molecular analysis utilizing **recombinant DNA technology**. In this procedure, segments of eukaryotic DNA corresponding to specific genes are isolated and literally spliced into bacterial DNA. Such a complex can be inserted into a bacterial cell and its genetic expression monitored. If a eukaryotic gene is introduced, the presence of the corresponding eukaryotic protein product demonstrates directly that this DNA is not only present, but functional in the bacterial cell. This has been shown to be the case in countless instances. For example, the human gene products insulin and interferon are produced by bacteria following the insertion of the DNA representing human genes that encode these proteins. As the bacterium divides, the eukaryotic DNA is replicated along with the bacterial DNA and is distributed to the daughter cells, which also express the human genes by creating the the corresponding proteins.

The availability of vast amounts of DNA coding for specific genes, attainable as a result of recombinant DNA research, has led to other direct evidence that DNA serves as the genetic material. Work in the laboratory of Beatrice Mintz has demonstrated that DNA encoding the human b-*globin* gene, when microinjected into a fertilized mouse egg, is later found to be present and expressed in adult mouse tissue and transmitted to and expressed in that mouse's progeny. These mice are examples of what are called **transgenic animals**.

More recent work has introduced *rat* DNA encoding a growth hormone into fertilized *mouse* eggs. About one-third of the resultant mice grew to twice their normal size, indicating that foreign DNA was present and functional in the experimental mice. Subsequent generations of mice inherited this genetic information and also grew to a large size.

We will return to the topic of recombinant DNA later in the text. (See Chapters 19, 20, and 22.) The point to be made here is that in eukaryotes, DNA has been shown directly to meet the requirement of expression of genetic information. Later, we will see exactly how DNA is stored, replicated, expressed, and mutated.

10.5 RNA Serves as the Genetic Material in Some Viruses

Some viruses contain an RNA core rather than one composed of DNA. In these viruses, it would thus appear that RNA might serve as the genetic material—an exception to the general rule that DNA performs this function. In 1956, it was demonstrated that when purified RNA from **tobacco mosaic virus (TMV)** is spread on tobacco leaves, the characteristic lesions caused by viral infection subsequently appear on the leaves. It was concluded that RNA is the genetic material of this virus.

Soon afterwards, another type of experiment with TMV was reported by Heinz Fraenkel-Conrat and B. Singer, as illustrated in Figure 10–8. These scientists discovered that the RNA core and the protein coat from wild-type TMV and other viral strains could be isolated separately. In their work, RNA and coat proteins were separated and isolated from TMV and a second viral strain, **Holmes ribgrass (HR)**. Then, mixed viruses were reconstituted from the RNA of one strain and the protein of the other. When this "hybrid" virus was spread on tobacco leaves, the lesions that developed corresponded to the type of RNA in the reconstituted virus—that is, viruses with wild-type TMV RNA and HR protein coats produced TMV lesions and vice versa. Again, it was concluded that RNA serves as the genetic material in these viruses.

In 1965 and 1966, Norman R. Pace and Sol Spiegelman further demonstrated that RNA from the phage Qβ could be isolated and replicated *in vitro*. Replication was dependent on the enzyme **RNA replicase**, which was isolated from host *E. coli* cells following normal infection. When the RNA replicated *in vitro* was added to *E. coli* protoplasts, infection and viral multiplication occurred. Thus, RNA synthesized in a test tube serves as the genetic material in these phages by directing the production of all the components necessary for viral replication.

Finally, one other group of RNA-containing viruses bears mentioning. These are the **retroviruses**, which replicate in an unusual way. Following infection of the host cell, their RNA serves as a template for the synthesis of the complementary DNA molecule. The process, **reverse transcription**, occurs under the direction of an RNA-dependent DNA polymerase enzyme called **reverse transcriptase**. This DNA intermediate can be incorporated into the genome of the host cell, and when

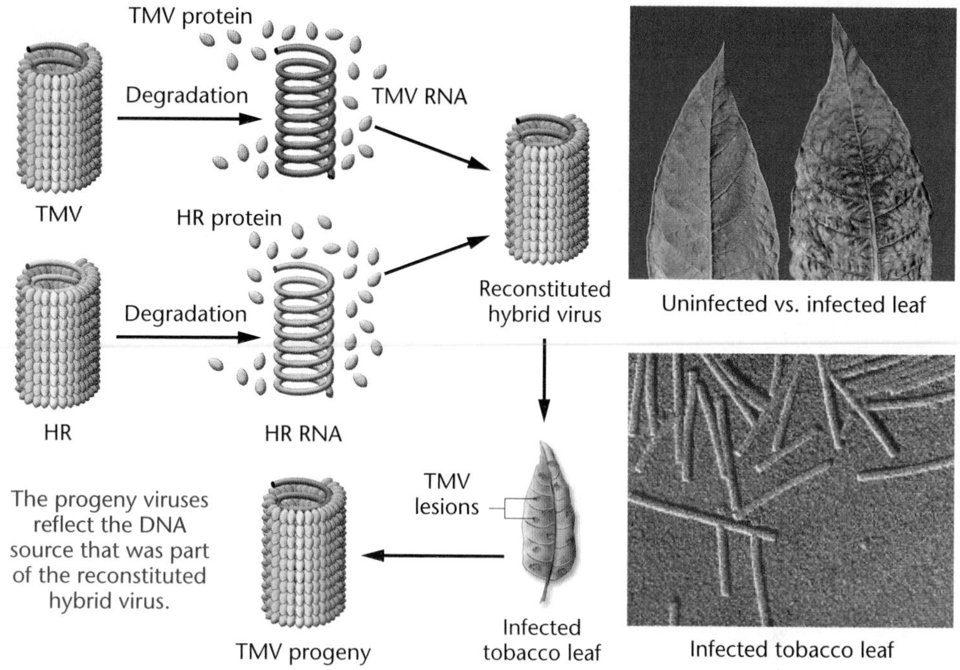

FIGURE 10–8 Reconstitution of hybrid tobacco mosaic viruses. In the hybrid, RNA is derived from the wild-type TMV virus, while the protein subunits are derived from the HR strain. Following infection, viruses are produced with protein subunits characteristic of the wild-type TMV strain and not those of the HR strain. The top photograph shows TMV lesions on a tobacco leaf compared with an uninfected leaf. At the bottom is an electron micrograph of mature viruses.

the host DNA is transcribed, copies of the original retroviral RNA chromosomes are produced. Retroviruses include the human immunodeficiency virus (HIV), which causes AIDS, as well as the RNA tumor viruses.

How Do We Know?

How was it experimentally determined that RNA serves as the genetic material in tobacco mosaic virus?

10.6 Knowledge of Nucleic Acid Chemistry Is Essential to the Understanding of DNA Structure

Having established thus far the critical importance of DNA and RNA in genetic processes, we will now take a brief look at the chemical basis of these molecules. As we shall see, the structural components of DNA and RNA are very similar. This chemical similarity is important in the coordinated functions played by these molecules during gene expression. Like the other major groups of organic biomolecules (proteins, carbohydrates, and lipids), nucleic acid chemistry is based on a variety of similar building blocks that are polymerized into chains of varying lengths.

Nucleotides: Building Blocks of Nucleic Acids

DNA is a nucleic acid, and **nucleotides** are the building blocks of all nucleic acid molecules. Sometimes called mononucleotides, these structural units consist of three essential components: a **nitrogenous base**, a **pentose sugar** (a 5-carbon sugar), and a **phosphate group**. There are two kinds of nitrogenous bases: the nine-member double-ring **purines** and the six-member single-ring **pyrimidines**. Two types of purines and three types of pyrimidines are commonly found in nucleic acids. The two purines are **adenine** and **guanine**, abbreviated **A** and **G**. The three pyrimidines are **cytosine**, **thymine**, and **uracil**, abbreviated **C**, **T**, and **U**. The chemical structures of A, G, C, T, and U are shown in Figure 10–9(a). Both DNA and RNA contain A, C, and G; only DNA contains the base T, whereas only RNA contains the base U. Each nitrogen or carbon atom of the ring structures of purines and pyrimidines is designated by an unprimed number. Note that corresponding atoms in the two rings are numbered differently in most cases.

The pentose sugars found in nucleic acids give them their names. Ribonucleic acids (RNA) contain **ribose**, while deoxyribonucleic acids (DNA) contain **deoxyribose**. Figure 10–9(b) shows the ring structures for these two pentose sugars. Each carbon atom is distinguished by a number with a prime sign (e.g., C-1′, C-2′). Compared with ribose, deoxyribose has a hydrogen atom at the C-2′ position rather than a hydroxyl group. The presence of a hydroxyl group at the C-2′ position thus distinguishes RNA from DNA.

If a molecule is composed of a purine or pyrimidine base and a ribose or deoxyribose sugar, the chemical unit is called a **nucleoside**. If a phosphate group is added to the nucleoside, the molecule is now called a **nucleotide**. Nucleosides and nucleotides are named according to the specific nitrogenous base (A, T, G, C, or U) that is part of the building block. The nomenclature and general structure of nucleosides and nucleotides are given in Figure 10–10.

The bonding among the components of a nucleotide is highly specific. The C-1′ atom of the sugar is involved in the chemical linkage to the nitrogenous base. If the base is a purine, the N-9 atom is covalently bonded to the sugar. If the base is a pyrimidine, the bonding involves the N-1 atom. Nucleotides may contain the phosphate group bonded to the C-2′, C-3′, or C-5′ atom of the sugar. However, the C-5′ phosphate configuration, shown in Figure 10–10, is the prevalent form in biological systems and the one found in DNA and RNA.

Web Tutorial 10.1
DNA Structure

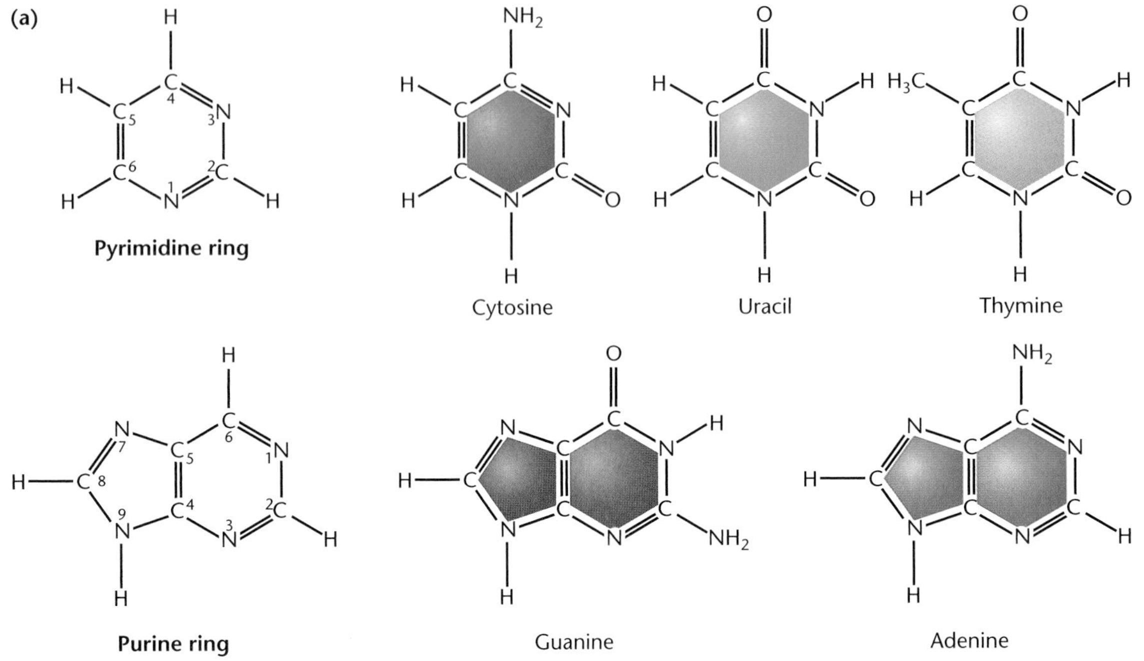

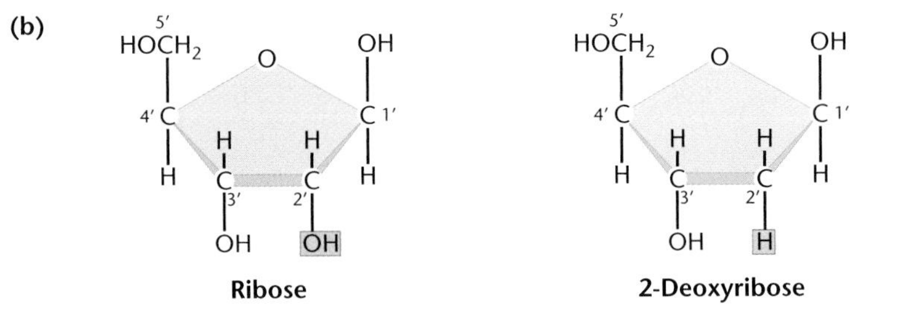

FIGURE 10–9 (a) Chemical structures of the pyrimidines and purines that serve as the nitrogenous bases in RNA and DNA. The nomenclature for numbering carbon and nitrogen atoms making up the two bases is shown within the structures that appear on the left.
(b) Chemical ring structures of ribose and 2-deoxyribose, which serve as the pentose sugars in RNA and DNA, respectively.

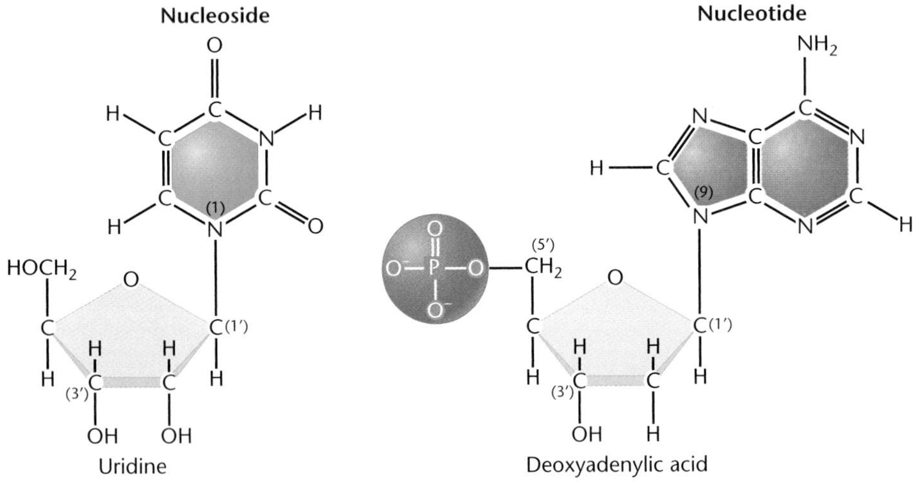

Ribonucleosides	Ribonucleotides
Adenosine Cytidine Guanosine Uridine	Adenylic acid Cytidylic acid Guanylic acid Uridylic acid
Deoxyribonucleosides	**Deoxyribonucleotides**
Deoxyadenosine Deoxycytidine Deoxyguanosine Deoxythymidine	Deoxyadenylic acid Deoxycytidylic acid Deoxyguanylic acid Deoxythymidylic acid

FIGURE 10–10 Structures and names of the nucleosides and nucleotides of RNA and DNA.

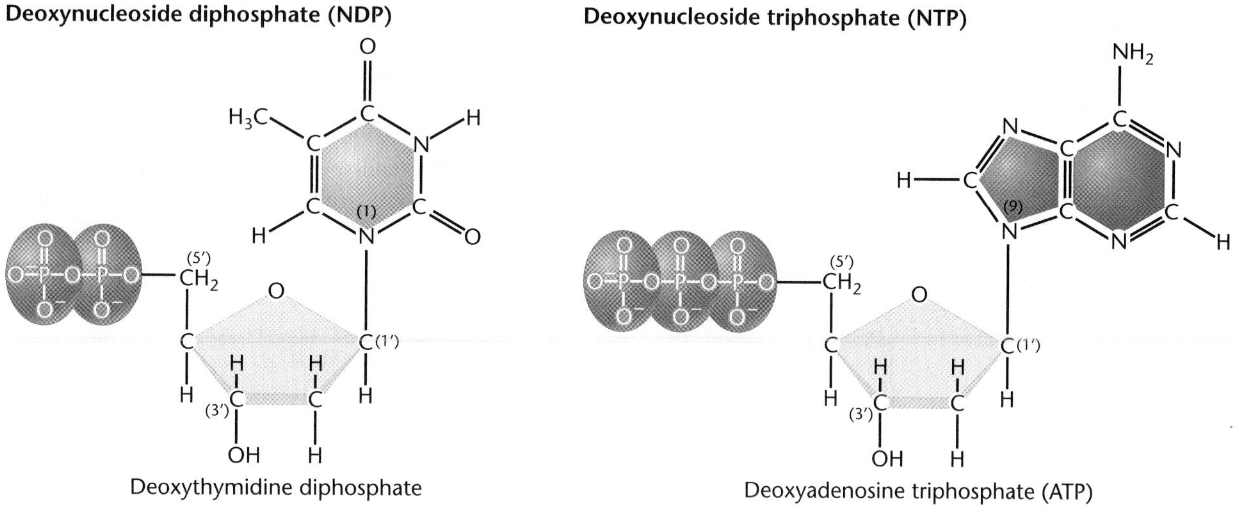

FIGURE 10–11 Basic structures of nucleoside diphosphates and triphosphates, as illustrated by deoxythymidine diphosphate and deoxyadenosine triphosphate.

Nucleoside Diphosphates and Triphosphates

Nucleotides are also described by the term **nucleoside monophosphate (NMP)**. The addition of one or two phosphate groups results in **nucleoside diphosphates (NDPs)** and **triphosphates (NTPs)**, as illustrated in Figure 10–11. The triphosphate form is significant because it serves as the precursor molecule during nucleic acid synthesis within the cell. (See Chapter 11.) Additionally, the triphosphates **adenosine triphosphate (ATP)** and **guanosine triphosphate (GTP)** are important in the cell's bioenergetics because of the large amount of energy involved in adding or removing the terminal phosphate group. The hydrolysis of ATP or GTP to ADP or GDP and inorganic phosphate P_i is accompanied by the release of a large amount of energy in the cell. When the chemical conversion of ATP or GTP is coupled to other reactions, the energy produced may be used to drive the reactions. As a result, both ATP and GTP are involved in many cellular activities, including numerous genetic events.

Polynucleotides

The linkage between two mononucleotides consists of a phosphate group linked to two sugars. A **phosphodiester bond** is formed, because phosphoric acid has been joined to two alcohols (the hydroxyl groups on the two sugars) by an ester linkage on both sides. Figure 10–12(a) shows the resultant phosphodiester bond in DNA. The same bond is found in RNA. Each structure has a **C-5′ end** and a **C-3′ end**. The joining of two nucleotides forms a **dinucleotide**; of three nucleotides, a **trinucleotide**; and so forth. Short chains consisting of approximately 20 nucleotides linked together are called **oligonucleotides**. Still longer chains are referred to as **polynucleotides**.

Because drawing polynucleotide structures, as shown in Figure 10–12(a) is time consuming and complex, a schematic shorthand method has been devised [Figure 10–12(b)]. The nearly vertical lines represent the pentose sugar; the nitrogenous base is attached at the top, or the C-1′ position. The diagonal line, with the P in the middle of it, is attached to the C-3′ position of one sugar and the C-5′ position of the neighboring sugar; it represents the phosphodiester bond. Several modifications of this shorthand method are in use, and they can be understood in terms of these guidelines.

Although Levene's tetranucleotide hypothesis (described earlier in this chapter) was generally accepted before 1940, research in subsequent decades revealed it to be incorrect. It was shown that DNA does not necessarily contain equimolar quantities of the four bases. Additionally, the molecular weight of DNA molecules was determined to be in the range of 10^6 to 10^9 daltons, far in excess of that of a tetranucleotide. The current view of DNA is that it consists of exceedingly long polynucleotide chains.

Long polynucleotide chains would account for the observed molecular weight and provide the basis for the most important property of DNA—storage of vast quantities of genetic information. If each nucleotide position in this long chain is occupied by any one of four nucleotides, extraordinary variation is possible. For example, a polynucleotide that is only 1000 nucleotides in length can be arranged 4^{1000} different ways, each one different from all other possible sequences. This potential variation in molecular structure is essential if DNA is to serve the function of storing the vast amounts of chemical information necessary to direct cellular activities.

10.7 The Structure of DNA Holds the Key to Understanding Its Function

The previous sections in this chapter have established that DNA is the genetic material in all organisms (with certain viruses being the exception) and have provided details as to the basic chemical components making up nucleic acids. What remained to be deciphered was the precise structure of DNA. That is, how are polynucleotide chains organized into DNA, which serves as the genetic material? Is DNA composed of a single

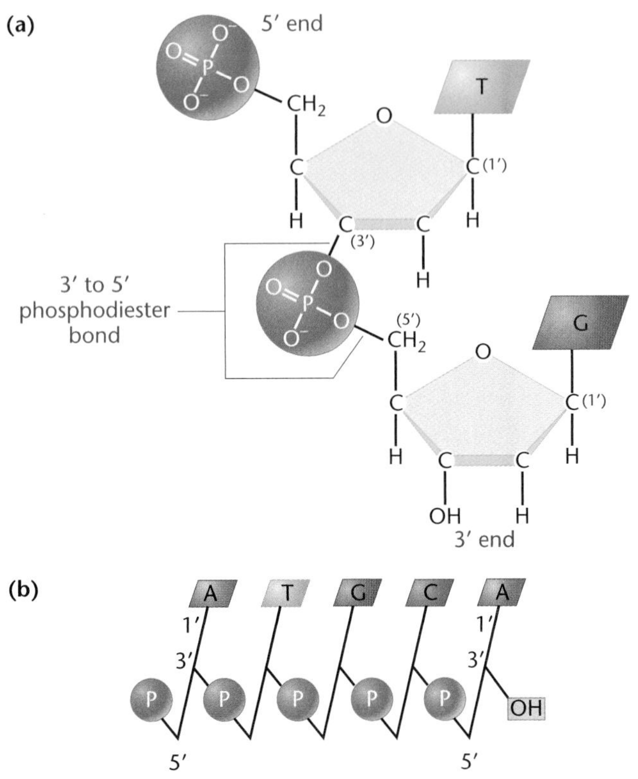

FIGURE 10–12 (a) Linkage of two nucleotides by the formation of a C-3′–C-5′ (3′–5′) phosphodiester bond, producing a dinucleotide. (b) Shorthand notation for a polynucleotide chain.

chain or more than one? If the latter is the case, how do the chains relate chemically to one another? Do the chains branch? And more important, how does the structure of this molecule relate to the various genetic functions served by DNA (i.e., storage, expression, replication, and mutation)?

From 1940 to 1953, many scientists were interested in solving the structure of DNA. Among others, Erwin Chargaff, Maurice Wilkins, Rosalind Franklin, Linus Pauling, Francis Crick, and James Watson sought information that might answer what many consider to be the most significant and intriguing question in the history of biology: How does DNA serve as the genetic basis for the living process? The answer was believed to depend strongly on the chemical structure and organization of the DNA molecule, given the complex but orderly functions ascribed to it.

In 1953, James Watson and Francis Crick proposed that the structure of DNA is in the form of a double helix. Their model was described in a short paper published in the journal *Nature*. (The article is reprinted in its entirety on page 250) In a sense, this publication constituted the finish line in a highly competitive scientific race to obtain what some considered to be the most significant finding in the history of biology. This "race," as recounted in Watson's book *The Double Helix* (see "Selected Readings"), demonstrates the many facets of human interactions that are part of a collaborative project that eventually led to the elucidation of DNA structure.

The data available to Watson and Crick, crucial to the development of their proposal, came primarily from two sources: (1) base composition analysis of hydrolyzed samples of DNA, and (2) X-ray diffraction studies of DNA. The analytical success of Watson and Crick can be attributed to model building that conformed to the existing data. If the correct solution to the structure of DNA is viewed as a puzzle, Watson and Crick, working at the Cavendish Laboratory in Cambridge, England, were the first to successfully put together all of the pieces.

As we pursue a discussion of this far-reaching discovery, you may find some background on James Watson and Francis Crick interesting. Watson began his undergraduate studies at the University of Chicago at age 15, and was originally interested in ornithology. He then pursued his Ph.D. at Indiana University, where he studied viruses. He was only 24 years old in 1953, when he and Crick proposed the double-helix theory. Crick, now considered one of the great theoretical biologists of our time, studied undergraduate physics at University College, London, and went on to perform military research during World War II. At the time of his collaboration with Watson, he was 35 years old and performing X-ray diffraction studies of polypeptides and proteins as a graduate student. The day of their major discovery, Crick is reputed to have walked into the Eagle Pub in Cambridge, where the two frequently lunched, and announced for all to hear, "We have discovered the secret of life." It turns out that more than 50 years later, many scientists would quite agree!

Base Composition Studies

Between 1949 and 1953, Erwin Chargaff and his colleagues used chromatographic methods to separate the four bases in DNA samples from various organisms. Quantitative methods were then used to determine the amounts of the four bases from each source. Table 10.3(a) provides some of Chargaff's original data. Parts (b) and (c) of the table show more recently derived base-composition information from various organisms that reinforce Chargaff's findings. As we shall see, Chargaff's data were critical to the successful model of DNA put forward by Watson and Crick.

On the basis of these data, which you should examine, the following conclusions may be drawn:

1. As shown in Table 10–3(b), the amount of adenine residues is proportional to the amount of thymine residues in the DNA of any species (columns 1, 2, and 5). Also, the amount of guanine residues is proportional to the amount of cytosine residues (columns 3, 4, and 6).

2. Based on the above proportionality, the sum of the purines (A + G) equals the sum of the pyrimidines (C + T), as shown in column 7.

3. The percentage of C + G does not necessarily equal the percentage of A + T. As we can see, the ratio between the two values varies greatly among species, as shown in column 8 and as is apparent in Table 10.3(c).

These conclusions indicate definite patterns of base composition of DNA molecules. The data provided the initial clue to "the puzzle." Additionally, they directly refute the tetranucleotide hypothesis, which stated that all four bases are present in equal amounts.

TABLE 10.3 **DNA BASE COMPOSITION DATA**

*(a) Chargaff's data**

	Molar proportions[a]			
	1	2	3	4
Organism's/Source	**A**	**T**	**G**	**C**
Ox thymus	26	25	21	16
Ox spleen	25	24	20	15
Yeast	24	25	14	13
Avian tubercle bacilli	12	11	28	26
Human sperm	29	31	18	18

(c) G + C content in several organisms

Organism	**%G + C**
Phage T2	36.0
Drosophila	45.0
Maize	49.1
Euglena	53.5
Neurospora	53.7

(b) Base compositions of DNAs from various sources

	Base composition				*Base ratio*		A + T/G + C *ratio*	
	1	2	3	4	5	6	7	8
Source	**A**	**T**	**G**	**C**	**A/T**	**G/C**	**(A + G)/(C + T)**	**(A + T)/(C + G)**
Human	30.9	29.4	19.9	19.8	1.05	1.00	1.04	1.52
Sea urchin	32.8	32.1	17.7	17.3	1.02	1.02	1.02	1.58
E. coli	24.7	23.6	26.0	25.7	1.04	1.01	1.03	0.93
Sarcina lutea	13.4	12.4	37.1	37.1	1.08	1.00	1.04	0.35
T7 bacteriophage	26.0	26.0	24.0	24.0	1.00	1.00	1.00	1.08

* *Source*: From Chargaff, 1950.
[a]Moles of nitrogenous constituent per mole of P. (Often, the recovery was less than 100 percent.)

X-Ray Diffraction Analysis

When fibers of a DNA molecule are subjected to X-ray bombardment, these rays are scattered according to the molecule's atomic structure. The pattern of scatter (diffraction) can be captured as spots on photographic film and analyzed, particularly for the overall shape of and regularities within the molecule. This process, **X-ray diffraction** analysis, was successfully applied to the study of protein structure by Linus Pauling and other chemists. The technique had been attempted on DNA as early as 1938 by William Astbury. By 1947, he had detected a periodicity within the structure of the molecule of 3.4 angstroms (Å)*, which suggested to him that the bases were stacked like coins on top of one another.

Between 1950 and 1953, Rosalind Franklin, working in the laboratory of Maurice Wilkins, obtained improved X-ray data from more purified samples of DNA (Figure 10–13). Her work confirmed the 3.4 Å periodicity seen by Astbury and suggested that the structure of DNA was some sort of helix. However, she did not propose a definitive model. Pauling had analyzed the work of Astbury and others and incorrectly proposed that DNA was a triple helix.

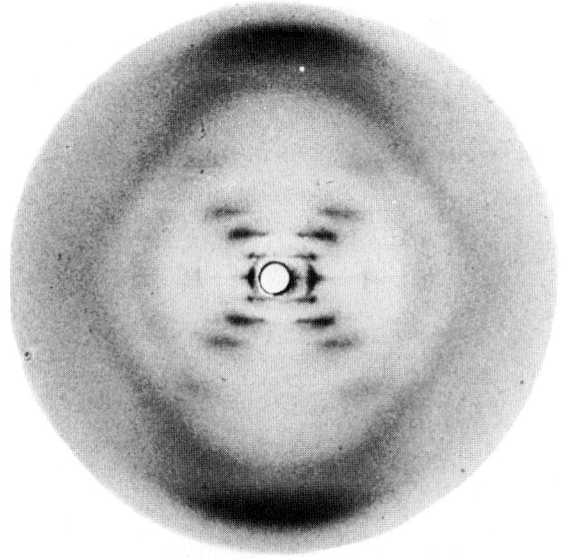

FIGURE 10–13 X-ray diffraction photograph of the B form of purified DNA fibers. The strong arcs on the periphery show closely spaced aspects of the molecule, providing an estimate of the periodicity of nitrogenous bases, which are 3.4 Å apart. The inner cross pattern of spots shows the grosser aspect of the molecule, indicating its helical nature.

The Watson–Crick Model

Watson and Crick published their analysis of DNA structure in 1953. By building models under the constraints of the information just discussed, they proposed the double-helical form of DNA, as shown in Figure 10–14(a). This model has the following major features:

*Today, measurement in nanometers (nm) is favored (1 nm = 10 Å).

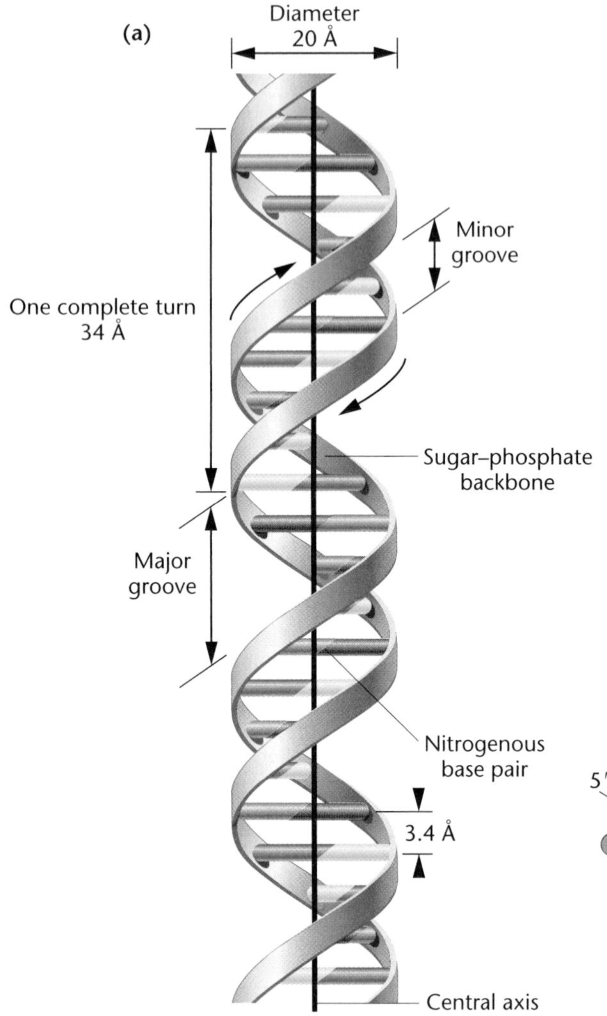

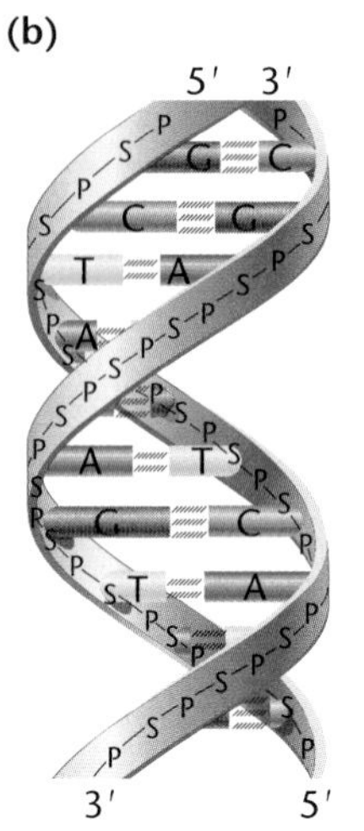

FIGURE 10–14 (a) The DNA double helix as proposed by Watson and Crick. The ribbonlike strands constitute the sugar-phosphate backbones, and the horizontal rungs constitute the nitrogenous base pairs, of which there are 10 per complete turn. The major and minor grooves are apparent. The solid vertical bar represents the central axis. (b) A detailed view depicting the bases, sugars, phosphates, and hydrogen bonds of the helix. (c) A demonstration of the antiparallel nature of the helix and the horizontal stacking of the bases.

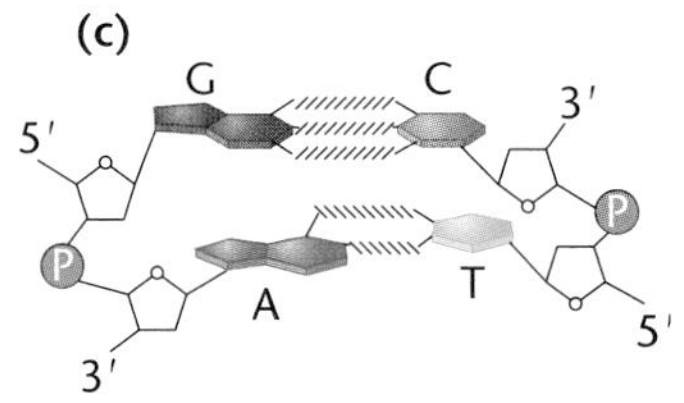

1. Two long polynucleotide chains are coiled around a central axis, forming a right-handed double helix.
2. The two chains are **antiparallel**; that is, their C-5′-to-C-3′ orientations run in opposite directions.
3. The bases of both chains are flat structures, lying perpendicular to the axis; they are "stacked" on one another, 3.4 Å (0.34 nm) apart, and are located on the inside of the helix.
4. The nitrogenous bases of opposite chains are paired to one another as the result of the formation of hydrogen bonds (described shortly); in DNA, only A═T and G≡C pairs are allowed.
5. Each complete turn of the helix is 34 Å (3.4 nm) long; hence, 10 bases exist per turn in each chain.
6. In any segment of the molecule, alternating larger **major grooves** and smaller **minor grooves** are apparent along the axis.
7. The double helix measures 20 Å (2.0 nm) in diameter.

The nature of base pairing (Point 4) is the most genetically significant feature of the model. Before discussing it in detail, several other important features warrant emphasis. First, the antiparallel nature of the two chains is a key part of the double-helix model. While one chain runs in the 5′-to-3′ orientation (what seems right-side up to us), the other chain is in the 3′-to-5′ orientation (and thus appears upside down). This is illustrated in Figure 10–14(c). Given the constraints of the bond angles of the various nucleotide components, the double helix could not be constructed easily if both chains ran parallel to one another.

Second, the right-handed nature of the helix is best appreciated by comparing such a structure to its left-handed counterpart, which is a mirror image, as shown in Figure 10–15.

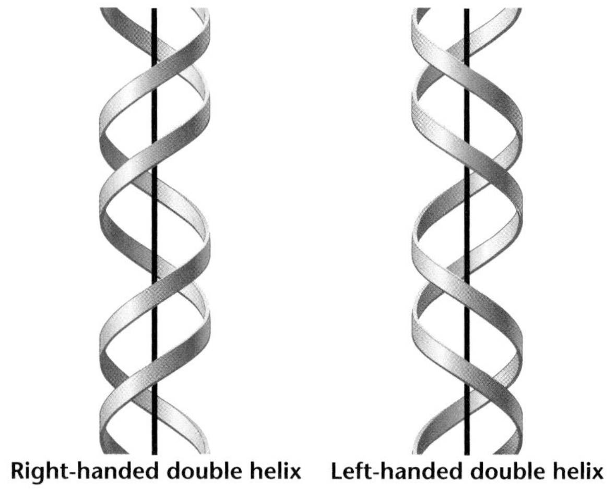

FIGURE 10–15 The right- and left-handed helical forms of DNA. Note that they are mirror images of one another.

The conformation in space of the right-handed helix is most consistent with the data that were available to Watson and Crick. As we shall see momentarily, an alternative form of DNA (Z-DNA) does exist as a left-handed helix.

The key to the model proposed by Watson and Crick is the specificity of base pairing. Chargaff's data had suggested that the amounts of A equaled T and that G equaled C. Watson and Crick realized that if A pairs with T and C pairs with G, thus accounting for these proportions, the members of each such base pair formed hydrogen bonds [Figure 10–14(b)], providing the chemical stability necessary to hold the two chains together. Arranged in this way, both major and minor grooves become apparent along the axis. Further, a purine (A or G) opposite a pyrimidine (T or C) on each "rung of the spiral staircase" of the proposed helix accounts for the 20 Å (2 nm) diameter suggested by X-ray diffraction studies.

The specific A=T and G≡C base pairing is the basis for the concept of **complementarity**, which describes the chemical affinity provided by the hydrogen bonds between the bases. As we will see, this concept is very important in the processes of DNA replication and gene expression.

Two questions are particularly worthy of discussion. First, why aren't other base pairs possible? Watson and Crick discounted the A=G and C≡T pairs because these represent purine–purine and pyrimidine–pyrimidine pairings, respectively. These pairings would lead to alternating diameters of more than, and less than, 20 Å because of the respective sizes of the purine and pyrimidine rings; in addition, the three-dimensional configurations formed by such pairings do not produce the proper alignment leading to sufficient hydrogen-bond formations. It is for this reason that A=C and G≡T pairings were also discounted, even though these pairs each consist of one purine and one pyrimidine.

The second question concerns hydrogen bonds. Just what is the nature of such a bond, and is it strong enough to stabilize the helix? A **hydrogen bond** is a very weak electrostatic attraction between a covalently bonded hydrogen atom and an atom with an unshared electron pair. The hydrogen atom assumes a partial positive charge, while the unshared electron pair—characteristic of covalently bonded oxygen and nitrogen atoms—assumes a partial negative charge. These opposite charges are responsible for the weak chemical attractions. As oriented in the double helix (Figure 10–16), adenine forms two hydrogen bonds with thymine, and guanine forms three hydrogen bonds with cytosine. Although two or three hydrogen bonds taken alone are energetically very weak, having as few as 20 base pairs with 40 to 60 hydrogen bonds imparts considerable

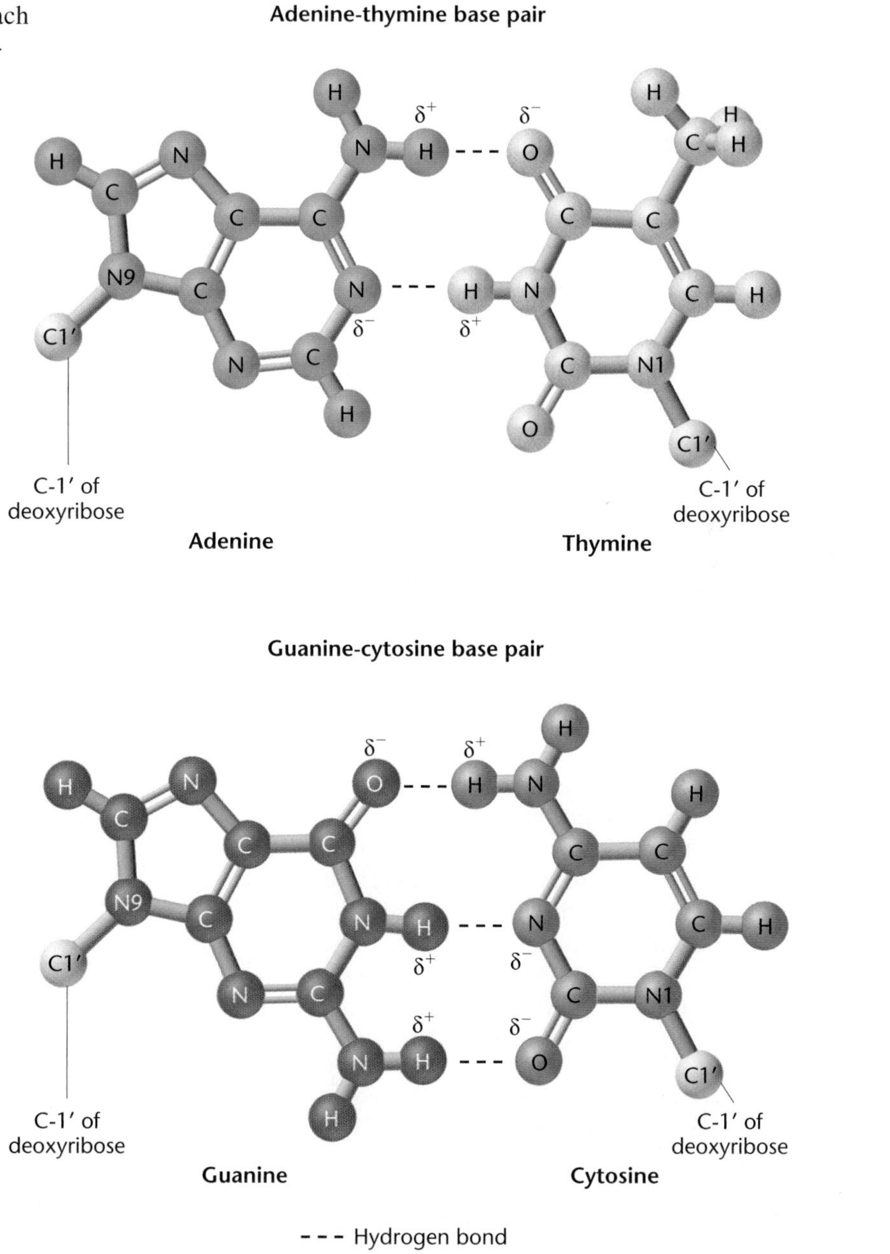

FIGURE 10–16 Ball-and-stick models of A=T and G≡C base pairs. The dashes (- - -) represent the hydrogen bonds that form between bases.

stability. Longer duplexes (two long polynucleotide chains) with thousands of bonds in tandem enhances this chemical stablility in a DNA double helix.

Still, another stabilizing factor is the arrangement of sugars and bases along the axis. In the Watson–Crick model, the hydrophilic sugar-phosphate backbone is on the outside of the axis, where both components may interact with water. The relatively hydrophobic or "water-fearing" nitrogenous bases are "stacked" almost horizontally on the interior of the axis, shielded from water. These molecular arrangements provide significant chemical stabilization to the helix.

A more recent and accurate analysis of the form of DNA that served as the basis for the Watson–Crick model has revealed a minor structural difference. A precise measurement of the number of base pairs (bp) per turn has demonstrated a value of 10.4, rather than the 10.0 predicted by Watson and Crick. Where, in the classic model, each base pair is rotated around the helical axis 36°, relative to the adjacent base pair, the new finding requires a rotation of 34.6°. This results in slightly more than 10 base pairs per 360° turn.

The Watson–Crick model had an immediate effect on the emerging discipline of molecular biology. Even in their initial 1953 article in *Nature*, the authors observed, "It has not escaped our notice that the specific pairing we have postulated immediately suggests a possible copying mechanism for the genetic material." Two months later, in a second article in *Nature*, Watson and Crick pursued this idea, suggesting a specific mode of replication of DNA: the semiconservative model. The second article also alluded to two new concepts: (1) the storage of genetic information in the sequence of the bases and (2) the mutations or genetic changes that would result from an alteration of bases. These ideas have received vast amounts of experimental support since 1953 and are now universally accepted.

Watson and Crick's "synthesis" of ideas was highly significant with regard to subsequent studies of genetics and biology. The nature of the gene and its role in genetic mechanisms could now be viewed and studied in biochemical terms. Recognition of their work, along with that of Wilkins, led to their receipt of the Nobel Prize in Physiology and Medicine in 1962. Unfortunately, Rosalind Franklin had died in 1958 at the age of 37, making her contributions ineligible for consideration since the award is not given posthumously. The Nobel Prize was to be one of many such awards bestowed for work in the field of molecular genetics.

How Do We Know?

How was it determined that the structure of DNA is a right-handed double helix with the two strands held together by hydrogen bonds formed between complementary nitrogenous bases?

10.8 Alternative Forms of DNA Exist

Under different conditions of isolation, several conformational forms of DNA have been recognized. At the time Watson and Crick performed their analysis, two forms—**A-DNA** and **B-DNA**—were known. Watson and Crick's analysis was based on Rosalind Franklin's X-ray studies of the B form, which is present under aqueous, low-salt conditions and is believed to be the biologically significant conformation.

While DNA studies around 1950 relied on the use of X-ray diffraction, more recent investigations have been performed using **single-crystal X-ray analysis**. The earlier studies achieved resolution of about 5 Å, but single crystals diffract X rays at about 1 Å, near atomic resolution. As a result, every atom is "visible" and much greater structural detail is available during analysis.

Using these modern techniques, A-DNA has now been scrutinized. It is prevalent under high-salt or dehydration conditions. In comparison to B-DNA (Figure 10–17), A-DNA is slightly more compact, with 9-base pairs in each complete turn of the helix, which is 23 Å in diameter. While it is also a right-handed helix, the orientation of the bases is somewhat different. They are tilted and displaced laterally in relation to the axis of the helix. As a result of these differences, the appearance of the major and minor grooves is modified compared with those in B-DNA. It seems doubtful that A-DNA occurs under biological conditions.

Three other right-handed forms of DNA helices have been discovered when investigated under laboratory conditions. These have been designated C-, D-, and E-DNA. **C-DNA** is found under even greater dehydration conditions than those observed during the isolation of A- and B-DNA. It has only 9.3 base pairs per turn and is, thus, less compact. Its helical diameter is 19 Å. Like A-DNA, C-DNA does not have its base pairs lying flat; rather, they are tilted relative to the axis of the helix. Two other forms, **D-DNA** and **E-DNA**, occur in helices lacking guanine in their base composition. They have even fewer base pairs per turn: 8 and 7, respectively.

Still another form of DNA, called **Z-DNA**, was discovered by Andrew Wang, Alexander Rich, and their colleagues in 1979, when they examined a small synthetic DNA oligonucleotide containing only C-G base pairs. Z-DNA takes on the rather remarkable configuration of a left-handed double helix (Figure 10–17). Like A- and B-DNA, Z-DNA consists of two antiparallel chains held together by Watson–Crick base pairs. Beyond these characteristics, Z-DNA is quite different. The left-handed helix is 18 Å (1.8 nm) in diameter, contains 12 base pairs per turn, and assumes a zigzag conformation (hence its name). The major groove present in B-DNA is nearly eliminated in Z-DNA.

More recent research by Jean-François Allemand and colleagues has shown that if DNA is artificially stretched, still another form of the molecule is assumed, called **P-DNA** (named for Linus Pauling). A model of P-DNA contrasted with the B form of DNA is quite interesting, since it is longer, more narrow, and the phosphate groups, found on the outside of B-DNA, are present inside the molecule. The nitrogenous bases, present inside the helix in B-DNA, are found closer to the external surface of the helix in P-DNA, and there are 2.62 bases per turn, in contrast to the 10.4 per turn in B-DNA.

The interest in alternative forms of DNA, such as Z and P, stems from the belief that DNA might have to assume a structure other than the B-form as it functions as the genetic material. During both replication and transcription (when its RNA complement is synthesized during gene expression), the strands of the helix must separate and become accessible to large enzymes, as well as a variety of other proteins involved in these processes. It is possible that changes in the shapes of the DNA facilitate these functions. Unique conformations might serve as points of molecular recognition for proteins. However, verification of the biological significance of alternative forms awaits the clear demonstration that they exist *in vivo*.

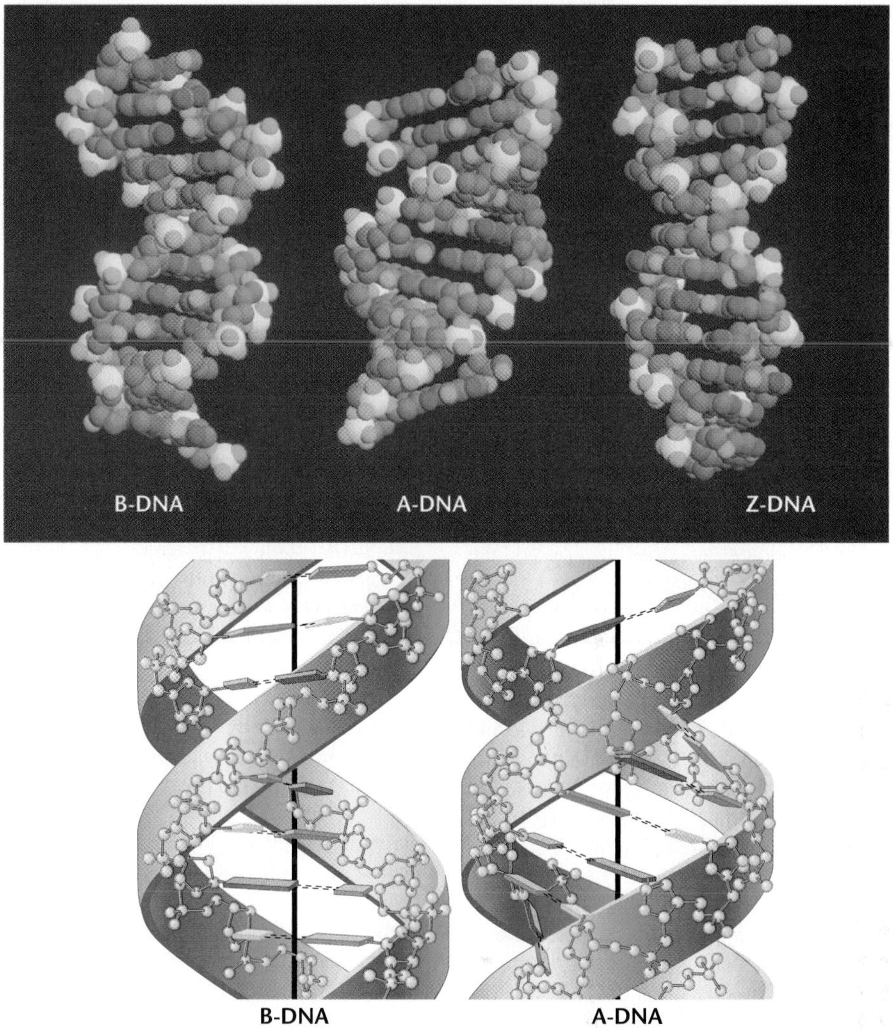

FIGURE 10–17 The top half of the figure shows computer-generated space-filling models of B-DNA (left), A-DNA (center), and Z-DNA (right). Below is an artist's depiction illustrating the orientation of the base pairs of B-DNA and A-DNA. (Note that in B-DNA the base pairs are perpendicular to the helix, while they are tilted and pulled away from the helix in A-DNA.)

Now solve this

Problem 10.36 on page 262 asks you to analyze DNA from a newly discovered source.

Hint: Knowing the nature and relative compositions of the nitrogenous bases of the unknown DNA will provide particulary important experimental information at the outset.

10.9 The Structure of RNA Is Chemically Similar to DNA, but Single Stranded

The second category of nucleic acids is ribonucleic acid, or RNA. The structure of these molecules is similar to DNA, with several important exceptions. Although RNA also has as its building blocks nucleotides linked into polynucleotide chains, the sugar ribose replaces deoxyribose and the nitrogenous base uracil replaces thymine. Another important difference is that most RNA is single stranded, although there are two important exceptions. First, RNA molecules sometimes fold back on themselves to form double-stranded regions of complementary base pairs. Second, some animal viruses that have RNA as their genetic material contain it in the form of double-stranded helices. Thus, there are several instances where RNA does not exist strictly as a linear single-stranded molecule.

Three major classes of cellular RNA molecules function during the expression of genetic information: **ribosomal RNA (rRNA)**, **messenger RNA (mRNA)**, and **transfer RNA (tRNA)**. These molecules all originate as complementary copies of one of the two strands of DNA segments during the process of transcription. That is, their nucleotide sequence is

Molecular Structure of Nucleic Acids: A Structure for Deoxyribose Nucleic Acid

We wish to suggest a structure for the salt of deoxyribose nucleic acid (D.N.A.). This structure has novel features which are of considerable biological interest. A structure for nucleic acid has already been proposed by Pauling and Corey.[1] They kindly made their manuscript available to us in advance of publication. Their model consists of three intertwined chains, with the phosphates near the fibre axis, and the bases on the outside. In our opinion, this structure is unsatisfactory for two reasons: (1) We believe that the material which gives the X-ray diagrams is the salt, not the free acid. Without the acidic hydrogen atoms it is not clear what forces would hold the structure together, especially as the negatively charged phosphates near the axis will repel each other. (2) Some of the van der Waals distances appear to be too small.

Another three-chain structure has also been suggested by Fraser (in the press). In his model the phosphates are on the outside and the bases on the inside, linked together by hydrogen bonds. This structure as described is rather ill-defined, and for this reason we shall not comment on it.

We wish to put forward a radically different structure for the salt of deoxyribose nucleic acid. This structure has two helical chains each coiled round the same axis. We have made the usual chemical assumptions, namely, that each chain consists of phosphate diester groups joining α-D-deoxyribofuranose residues with 3′, 5′ linkages. The two chains (but not their bases) are related by a dyad perpendicular to the fibre axis. Both chains follow right-handed helices, but owing to the dyad the sequences of the atoms in the two chains run in opposite directions. Each chain loosely resembles Furberg's[2] model No. 1; that is, the bases are on the inside of the helix and the phosphates on the outside. The configuration of the sugar and the atoms near it is close to Furberg's "standard configuration," the sugar being roughly perpendicular to the attached base. There is a residue on each chain every 3.4 Å in the *z*-direction. We have assumed an angle of 36° between adjacent residues in the same chain, so that the structure repeats after 10 residues on each chain, that is, after 34 Å the distance of a phosphorus atom from the fibre axis is 10 Å. As the phosphates are on the outside, cations have easy access to them.

The structure is an open one, and its water content is rather high. At lower water contents we would expect the bases to tilt so that the structure could become more compact.

The novel feature of the structure is the manner in which the two chains are held together by the purine and pyrimidine bases. The planes of the bases are perpendicular to the fibre axis. They are joined together in pairs, a single base from one chain being hydrogen-bonded to a single base from the other chain, so that the two lie side by side with identical *z*-coordinates. One of the pair must be a purine and the other a pyrimidine for bonding to occur. The hydrogen bonds are made as follows: purine position 1 to pyrimidine position 1; purine position 6 to pyrimidine position 6.

If it is assumed that the bases only occur in the structure in the most plausible tautomeric forms (that is, with the keto rather than the enol configurations) it is found that only specific pairs of bases can bond together. These pairs are: adenine (purine) with thymine (pyrimidine), and guanine (purine) with cytosine (pyrimidine).

In other words, if an adenine forms one member of a pair, on either chain, then on these assumptions the other member must be thymine; similarly for guanine and cytosine. The sequence of bases on a single chain does not appear to be restricted in any way. However, if only specific pairs of bases can be formed, it follows that if the sequence of bases on one chain is given, then the sequence on the other chain is automatically determined.

It has been found experimentally[3,4] that the ratio of the amounts of adenine to thymine, and the ratio of guanine to cytosine, are always very close to unity for deoxyribose nucleic acid.

It is probably impossible to build this structure with a ribose sugar in place of the deoxyribose, as the extra oxygen atom would make too close a van der Waals contact.

The previously published X-ray data[5,6] on deoxyribose nucleic acid are insufficient for a rigorous test of our structure. So far as we can tell, it is roughly compatible with the experimental data, but it must be regarded as unproved until it has been checked against more exact results. Some of these are given in the following communications. We were not aware of the details of the results presented there when we devised our structure, which rests mainly though not entirely on published experimental data and stereochemical arguments.

It has not escaped our notice that the specific pairing we have postulated immediately suggests a possible copying mechanism for the genetic material. Full details of the structure, including the conditions assumed in building it, together with a set of co-ordinates for the atoms, will be published elsewhere.

We are much indebted to Dr. Jerry Donohue for constant advice and criticism, especially on interatomic distances. We have also been stimulated by a knowledge of the general nature of the unpublished experimental results and ideas of Dr. M.H.F. Wilkins, Dr. R.E. Franklin and their co-workers at King's College, London. One of us (J.D.W.) has been aided by a fellowship from the National Foundation for Infantile Paralysis.

J.D. Watson
F.H.C. Crick

References

Medical Research Council Unit for the Study of the Molecular Structure of Biological Systems, Cavendish Laboratory, Cambridge, England.

[1] Pauling, L., and Corey, R.B., *Nature*, 171, 346 (1953); *Proc. U.S. Nat. Acad. Sci.*, 39, 84 (1953).

[2] Furberg, S., *Acta Chem. Scand.*, 6, 634 (1952).

[3] Chargaff, E., For references see Zamenhof, S., Brawerman, G., and Chargaff, E., *Biochim. et Biophys. Acta*, 9, 402 (1952).

[4] Wyatt, G.R., *J. Gen. Physiol.*, 36, 201 (1952).

[5] Astbury, W.T., *Symp. Soc. Exp. Biol. 1, Nucleic Acid*, 66 (Camb. Univ. Press, 1947).

[6] Wilkins, M.H.F., and Randall, J.T., *Biochim. et Biophys. Acta*, 10, 192 (1953).

TABLE 10.4 RNA CHARACTERIZATION

RNA Class	% Total RNA*	Components (Svedberg Coefficient)	Eukaryotic (E) or Prokaryotic (P)	Number of Nucleotides
Ribosomal (rRNA)	80	5*S*	P and E	120
		5.8*S*	E	160
		16*S*	P	1542
		18*S*	E	1874
		23*S*	P	2904
		28*S*	E	4718
Transfer (tRNA)	15	4*S*	P and E	75–90
Messenger (mRNA)	5	varies	P and E	100–10,000

*In *E. coli*

complementary to the deoxyribonucleotide sequence of DNA, which served as the template for their synthesis. Because uracil replaces thymine in RNA, uracil is complementary to adenine during transcription and during RNA base pairing.

Table 10.4 characterizes the major forms of RNA found in prokaryotic and eukaryotic cells. Different RNAs are distinguished according to their sedimentation behavior in a centrifugal field and their size, as measured by the number of nucleotides each contains. Sedimentation behavior depends on a molecule's density, mass, and shape; it is measured in units that designate a molecule's **Svedberg coefficient** (*S*). While higher *S* values almost always refer to molecules of greater molecular weight, the correlation is not direct; that is, a twofold increase in molecular weight does not lead to a twofold increase in *S*. This is because, in addition to a molecule's mass, the size and the shape of the molecule also affect its rate of sedimentation (*S*). As you can see, a wide variation exists in the size of the three classes of RNA.

Ribosomal RNA is generally the largest of these molecules (as is generally reflected in its *S* values) and usually constitutes about 80 percent of all RNA in an *E. coli* cell. Ribosomal RNAs are important structural components of **ribosomes**, which function as nonspecific workbenches during the synthesis of proteins during the process of translation. The various forms of rRNA found in prokaryotes and eukaryotes differ distinctly in size.

Messenger RNA molecules carry genetic information from the DNA of the gene to the ribosome, where translation occurs. They vary considerably in size, which is a reflection of the variation in the size of the protein encoded by the mRNA as well as the gene serving as the template for transcription of mRNA. While Table 10.4 shows that about 5 percent of RNA is mRNA in *E. coli*, this percentage varies from cell to cell and even at different times in the life of the same cell.

Transfer RNA, accounting for up to 15 percent of the RNA in a typical cell, is the smallest class of RNA molecules and carries amino acids to the ribosome during translation. Because more than one tRNA molecule interacts simultaneously with the ribosome, the molecule's smaller size facilitates these interactions.

We will discuss the functions of the three classes of RNA in much greater detail in Chapters 13 and 14. In addition, as we proceed through the text, we will encounter other unique RNAs that perform various genetic roles. For example, **small nuclear RNA (snRNA)** participates in processing mRNAs. **Telomerase RNA** is involved in DNA replication at the ends of chromosomes (Chapter 11), and **antisense RNA** and **short interfering RNA (siRNA)** are involved in gene regulation (Chapter 17). Our purpose in this section has been to contrast the structure of DNA, which stores genetic information, with that of RNA, which most often functions in the expression of that information.

10.10 Many Analytical Techniques Have Been Useful during the Investigation of DNA and RNA

Since 1953, the role of DNA as the genetic material and the role of RNA in transcription and translation have been clarified through detailed analysis of nucleic acids. We will consider several methods of analysis of these molecules in this chapter.

Absorption of Ultraviolet Light (UV)

Nucleic acids absorb ultraviolet (UV) light most strongly at wavelengths of 254 to 260 nm (Figure 10–7) due to the interaction between UV light and the ring systems of the purines and pyrimidines. Thus, any molecule containing nitrogenous bases (i.e., nucleosides, nucleotides, and polynucleotides) can be analyzed by using UV light. This technique is especially important in the localization, isolation, and characterization of nucleic acids.

Ultraviolet analysis is used in conjunction with many standard procedures that separate molecules. As we shall see in the next section, the use of UV absorption is critical to the isolation of nucleic acids following their separation.

Sedimentation Behavior

Nucleic acid mixtures can be separated by subjecting them to one of several possible centrifugation procedures (Figure 10–18). The mixture can be loaded on top of a solution prepared so that a concentration gradient has been formed from top to bottom. Then the entire mixture is centrifuged at high speeds in an ultracentrifuge. The mixture of molecules will migrate downward, with each component moving at a different rate. Centrifugation is stopped and the gradient eluted from the tube. Each fraction can then be measured spectrophotometrically for absorption at 260 nm. In this way, the previous position of a nucleic acid fraction along the gradient can be determined and the fraction isolated and studied further.

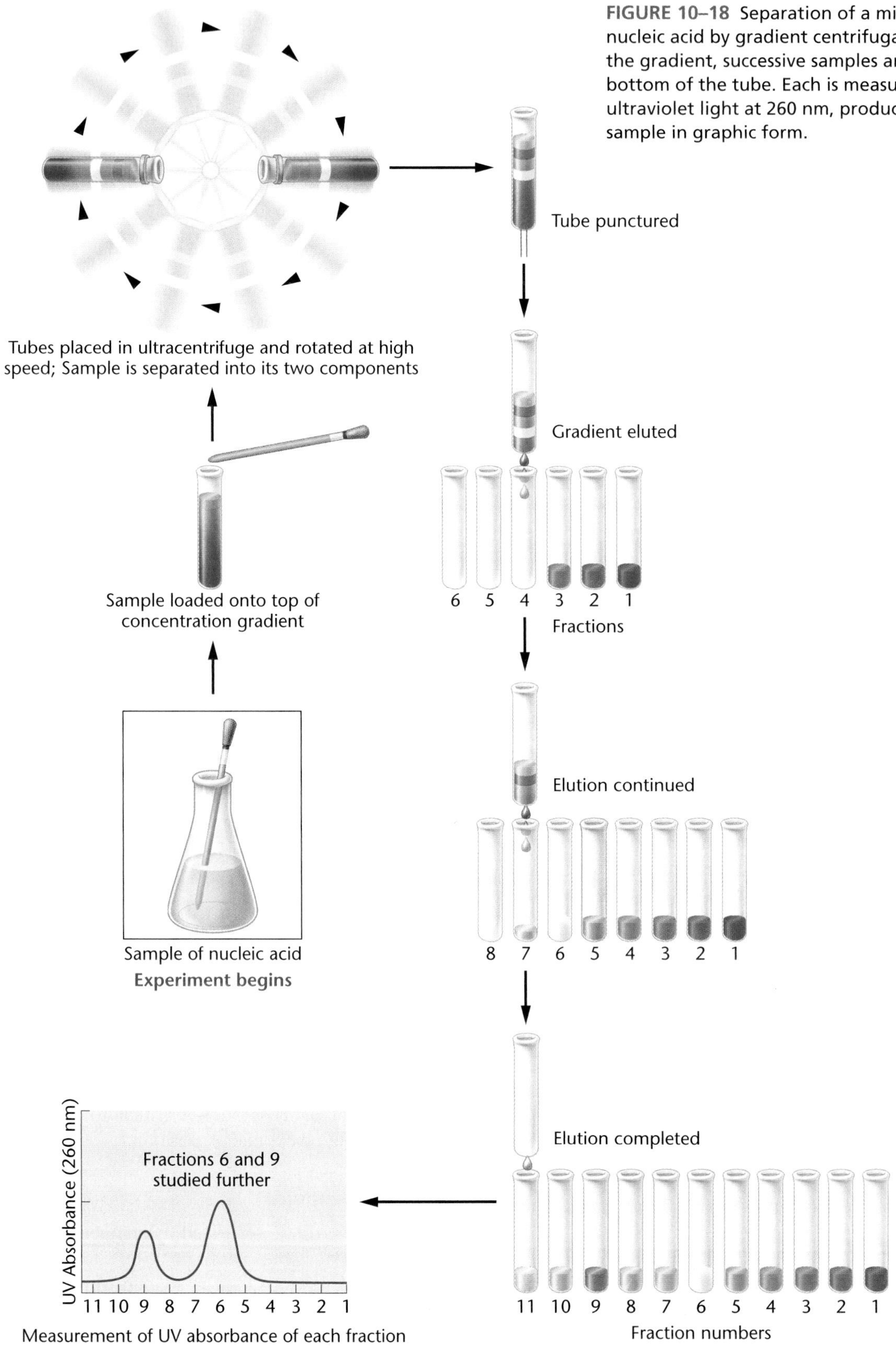

FIGURE 10–18 Separation of a mixture of two types of nucleic acid by gradient centrifugation. To fractionate the gradient, successive samples are eluted from the bottom of the tube. Each is measured for absorbance of ultraviolet light at 260 nm, producing a profile of the sample in graphic form.

The gradient centrifugations described rely on the sedimentation behavior of molecules in solution. Two major types of gradient centrifugation techniques are employed in the analysis of nucleic acids: sedimentation equilibrium and sedimentation velocity. Both require the use of high-speed centrifugation to create large centrifugal forces upon molecules in a gradient solution.

In **sedimentation equilibrium centrifugation** (sometimes called density gradient centrifugation), a density gradient is created that overlaps the densities of the individual components of a mixture of molecules. Usually, the gradient is made of a heavy metal salt, such as cesium chloride (CsCl). During centrifugation, the molecules migrate until they reach a point of neutral buoyant density. At this point, the centrifugal force on them is equal and opposite to the upward diffusion force, and no further migration occurs. If DNAs of different densities are used, they will separate as the molecules of each density reach equilibrium with the corresponding density of CsCl. The gradient may be fractionated and the components isolated (Figure 10–18). When properly executed, this technique provides high resolution in separating mixtures of molecules varying only slightly in density.

Sedimentation equilibrium centrifugation studies can also be used to generate data on the base composition of double-stranded DNA. The G≡C base pairs, compared with A=T pairs, are more compact and dense. As shown in Figure 10–19, the percentage of G≡C pairs in DNA is directly proportional to the molecule's buoyant density. By using this technique, we can make a useful molecular characterization of DNAs from different sources.

The second technique, **sedimentation velocity centrifugation**, employs an analytical centrifuge, which enables the migration of the molecules during centrifugation to be monitored with ultraviolet absorption optics. Thus, the "velocity of sedimentation" can be determined. As mentioned earlier, this velocity has been standardized in units called Svedberg coefficients (*S*).

In this technique, the molecules are loaded on top of the gradient, and the gravitational forces created by centrifugation drive them toward the bottom of the tube. Two forces work against this downward movement: (1) the viscosity of the solution creates a frictional resistance, and (2) part of the force of diffusion is directed upward. Under these conditions, the key variables are the mass and shape of the molecules being examined. In general, the greater the mass, the greater the sedimentation velocity. However, the molecule's shape affects the frictional resistance. Therefore, two molecules of equal mass, but different in shape, will sediment at different rates.

One use of the sedimentation velocity technique is the determination of **molecular weight (MW)**. If certain physical-chemical properties of a molecule under study are also known, the MW can be calculated based on the sedimentation velocity. The *S* values increase with molecular weight, but they are not directly proportional to it.

Denaturation and Renaturation of Nucleic Acids

When **denaturation** of double-stranded DNA occurs, the hydrogen bonds of the duplex structure break, the duplex unwinds, and the strands separate. However, no covalent bonds break. During strand separation, which can be induced by heat or chemical treatment, the viscosity of DNA decreases, and both the UV absorption and the buoyant density increase. Denaturation as a result of heating is sometimes referred to as melting. The increase in UV absorption of heated DNA in solution, called the **hyperchromic shift**, is easiest to measure. This effect is illustrated in Figure 10–20.

Because G≡C base pairs have one more hydrogen bond than do A=T pairs, they are more stable to heat treatment. Thus, DNA with a greater proportion of G≡C pairs than A=T pairs requires higher temperatures to denature completely. When absorption at 260 nm (OD_{260}) is monitored and plotted against temperature during heating, a **melting profile** of DNA is obtained. The midpoint of this profile, or curve, is called the **melting temperature** (T_m) and represents the point at which 50 percent of the strands are unwound or denatured

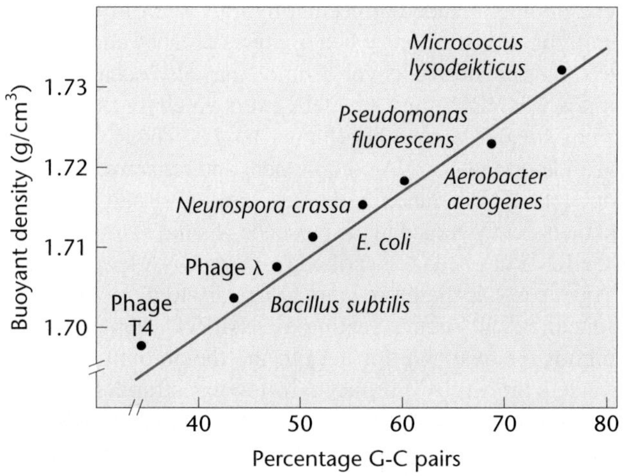

FIGURE 10–19 Percentage of guanine–cytosine G≡C base pairs in DNA, plotted against buoyant density for a variety of microorganisms.

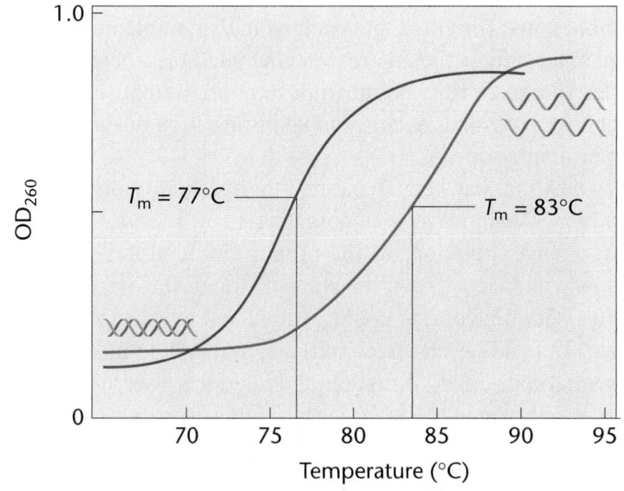

FIGURE 10–20 Increase in UV absorption vs. temperature (the hyperchromic effect) for two DNA molecules with different GC contents. The molecule with a melting point (T_m) of 83°C has a greater GC content than the molecule with a T_m of 77°C.

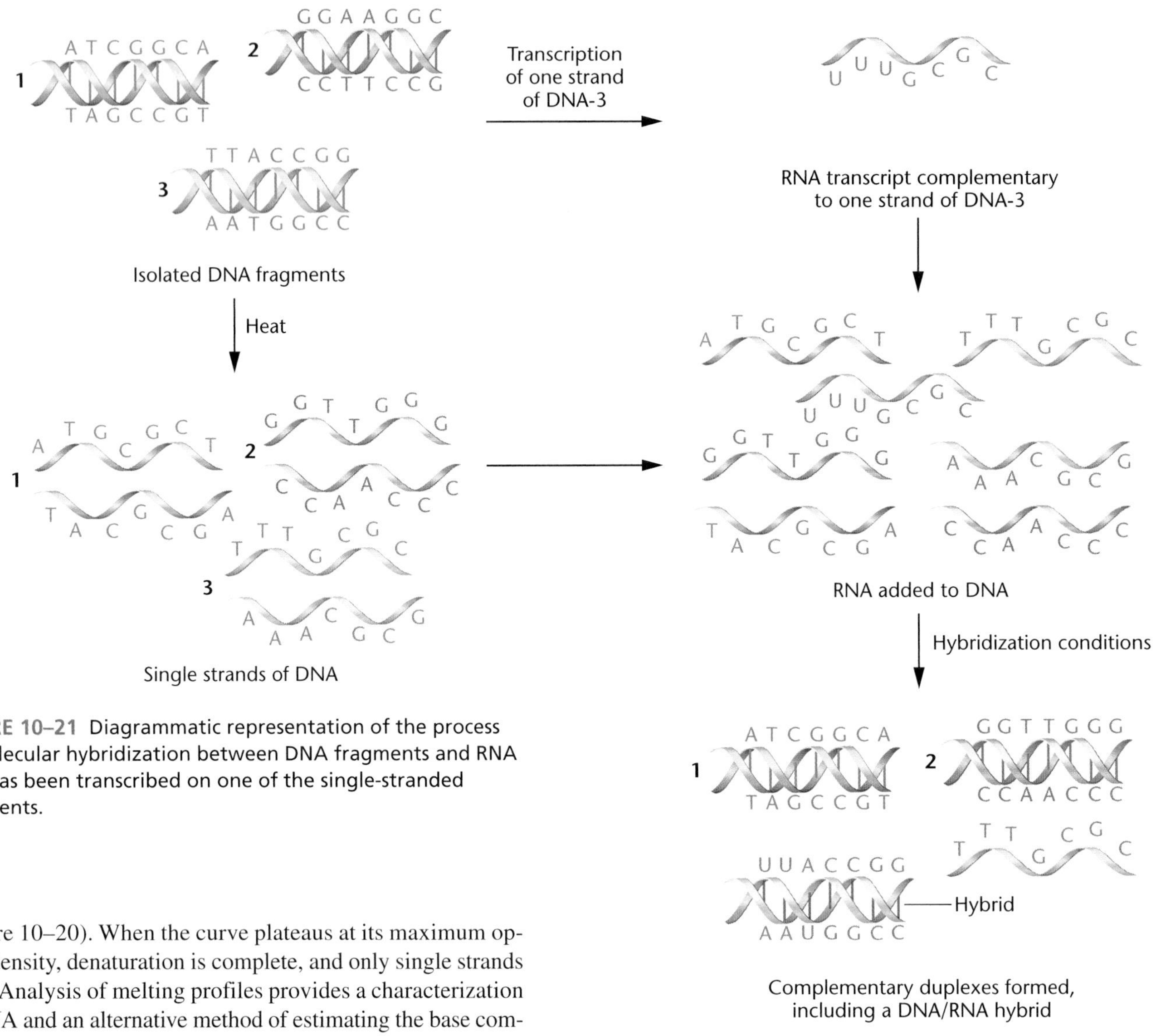

FIGURE 10–21 Diagrammatic representation of the process of molecular hybridization between DNA fragments and RNA that has been transcribed on one of the single-stranded fragments.

(Figure 10–20). When the curve plateaus at its maximum optical density, denaturation is complete, and only single strands exist. Analysis of melting profiles provides a characterization of DNA and an alternative method of estimating the base composition of DNA.

One might ask whether the denaturation process can be reversed; that is, can single strands of nucleic acids re-form a double helix, provided that each strand's complement is present? Not only is the answer yes, but such reassociation provides the basis for several important analytical techniques that have provided much valuable information during genetic experimentation.

If DNA that has been denatured thermally is cooled slowly, random collisions between complementary strands will result in their reassociation. At the proper temperature, hydrogen bonds will re-form, securing pairs of strands into duplex structures. With time during cooling, more and more duplexes will form. Depending on the conditions, a complete match is not essential for duplex formation, provided there are stretches of base pairing on at least two reassociating strands.

Molecular Hybridization

The property of denaturation–renaturation of nucleic acids is the basis for one of the most powerful and useful techniques in molecular genetics—**molecular hybridization**. This technique derives its name from the fact that renaturing single strands need not originate from the same nucleic acid source. For example, if DNA strands are isolated from two distinct organisms and some degree of base complementarity exists between them, double-stranded molecular hybrids will form during renaturation. Furthermore, when mixtures of DNA and RNA single strands are utilized, hybridization may also occur. A case in point is when RNA and the DNA from which it has been transcribed are present together (Figure 10–21). The RNA will find its single-stranded DNA complement and renature. In this example, the DNA strands are heated, causing strand separation, and then slowly cooled in the presence of single-stranded RNA. If the RNA has been transcribed from the DNA used in the experiment, and is therefore complementary to it, molecular hybridization will occur, creating a DNA:RNA duplex. Several methods are available for monitoring the amount of double-stranded molecules produced following strand separation. In early studies, radioisotopes were utilized to "tag" one of the strands and monitor its presence in hybrid duplexes that formed.

In the 1960s, molecular hybridization techniques contributed to our increased understanding of transcriptional events occurring at the gene level. Refinements of this process have occurred continually and have been the forerunners of work in studies of molecular evolution as well as the organization of DNA in chromosomes. Hybridization can occur in solution or when DNA is bound either to a gel or to a specialized binding filter. Such filters are used in a variety of **DNA blotting procedures**, whereby hybridization serves as a way to "probe" for complementary nucleic acid sequences. Blotting is used routinely in modern genomic analysis. Additionally, hybridization will occur even when DNA is part of tissue affixed to a slide, as in the FISH procedure (discussed in the next section), or when affixed to a glass chip, the basis of **DNA microarray analysis** (discussed in Chapter 22). Microarray analysis allows mass screening for a specific DNA sequence using thousands of cloned genes in a single assay.

Fluorescent *in situ* Hybridization (FISH)

A refinement using the technique of molecular hybridization has led to the use of DNA present in cytological preparations as the "target" for hybrid formation. When this approach is combined with the use of fluorescent probes to monitor hybridization, the technique is called **fluorescent *in situ* hybridization**, or simply the acronym **FISH**. In this procedure mitotic or interphase cells are fixed to slides and subjected to hybridization conditions. Single-stranded DNA or RNA is added, and hybridization is monitored. The nucleic acid serves as a "probe," since it will hybridize only with the specific chromosomal areas for which it is complementary. Before the use of fluorescent probes was refined, radioactive probes were used in these *in situ* procedures to allow detection on the slide. In this approach, the technique of autoradiography was utilized.

Fluorescent probes are prepared in a unique way. When DNA is used, it is first coupled to the small organic molecule biotin (creating biotinylated DNA). Once *in situ* hybridization is completed, another molecule (avidin or streptavidin) that has a high binding affinity for biotin is used. A fluorescent molecule such as fluorescein is linked to avidin (or streptavidin) and the complex is reacted with the cytological preparation. This procedure represents an extremely sensitive method for localizing the hybridized DNA.

Figure 10–22 illustrates the use of FISH in identifying the DNA specific to the centromeres of human chromosomes. The resolution of FISH is great enough to detect just a single gene within an entire set of chromosomes. The use of this technique in identifying chromosomal locations housing specific genetic information has been a valuable addition to the repertoire of experimental geneticists.

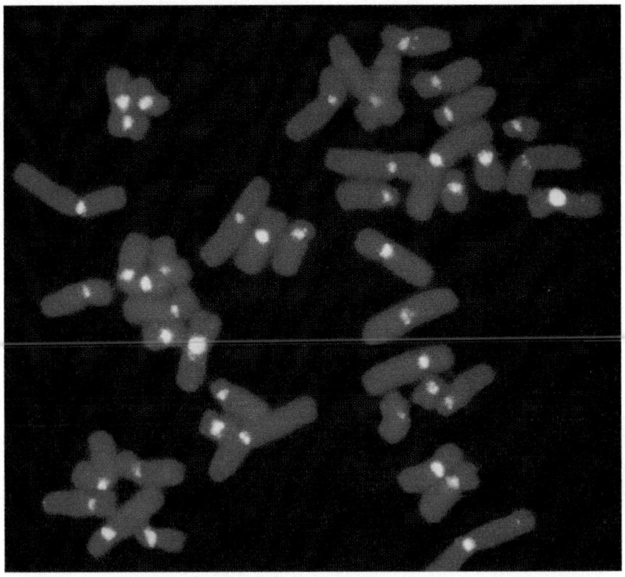

FIGURE 10–22 *In situ* hybridization of human metaphase chromosomes, using a fluorescent technique (FISH). The probe, specific to centromeric DNA, produces a yellow fluorescence signal indicating hybridization. The red fluorescence is produced by propidium iodide counterstaining of chromosomal DNA.

Reassociation Kinetics and Repetitive DNA

In one extension of molecular hybridization procedures, the *rate of reassociation* of complementary single DNA strands is analyzed. This technique, called **reassociation kinetics**, was first refined and studied by Roy Britten and David Kohne.

The DNA used in such studies is first fragmented into small pieces as a result of shearing forces introduced during isolation. The resultant DNA fragments cluster around a uniform average size of several hundred base pairs. These fragments of DNA are then dissociated into single strands by heating. Next, the temperature is lowered and reassociation is monitored. During reassociation, pieces of single-stranded DNA collide randomly. If they are complementary, a stable double strand is formed; if not, they separate and are free to encounter other DNA fragments. The process continues until all matches are made.

The results of such an experiment are presented in Figure 10–23. The percentage of reassociation of DNA fragments is plotted against a logarithmic scale of the product of C_0 (the initial concentration of DNA single strands in moles per liter of nucleotides), and t (the time, usually measured in minutes). The process of renaturation follows second-order rate kinetics according to the equation

$$\frac{C}{C_0} = \frac{1}{1 + kC_0t}$$

where C is the single-stranded DNA concentration remaining after time t has elapsed and k is the second-order rate constant. Initially, C equals C_0, and the fraction remaining single stranded is 100 percent.

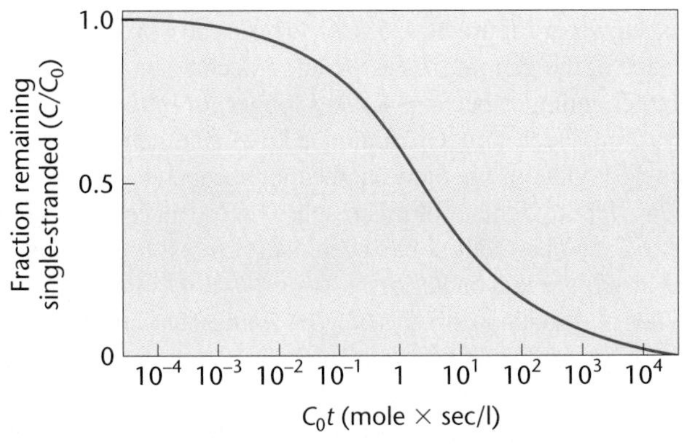

FIGURE 10–23 The ideal time course for reassociation of DNA (C/C_0) when, at time zero, all DNA consists of unique fragments of single-stranded complements. Note that the abscissa (C_0t) is plotted logarithmically.

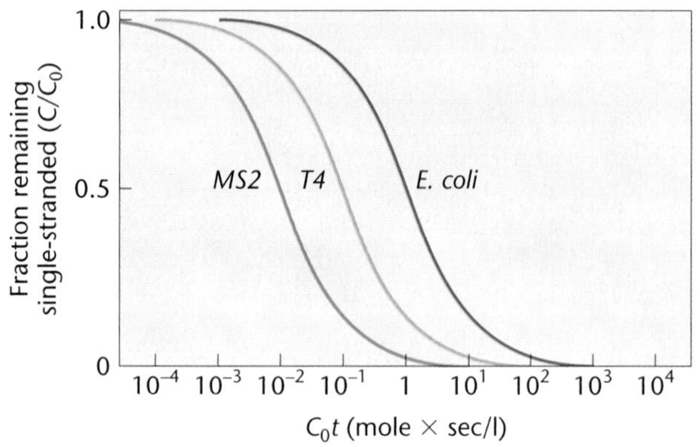

FIGURE 10–24 The reassociation rates (C/C_0) of DNA derived from phage MS2, phage T4, and *E. coli*. The genome of T4 is larger than MS2 and that of *E. coli* is larger than T4.

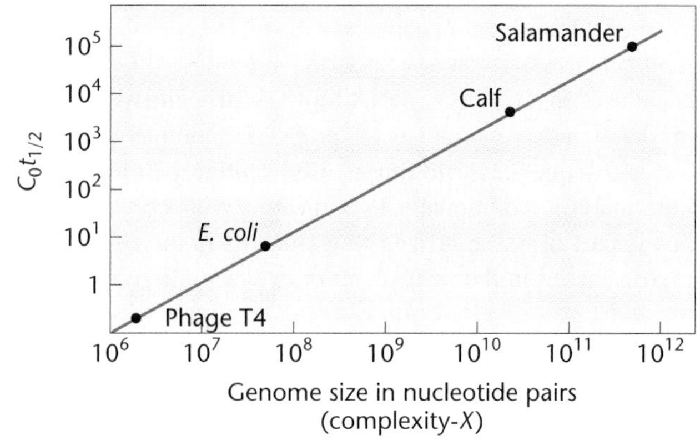

FIGURE 10–25 Comparison of $C_0t_{1/2}$ and genome size for phage T4, *E. coli*, calf, and salamander.

The initial shape of the curve reflects the fact that in a mixture of unique sequence fragments, each with one complement, initial matches take more time to make. Then, as many single strands are converted to duplexes, matches are made more quickly, reflecting an increase in the "slope" of the curve. Near the end of the reaction, the few remaining single strands require relatively greater time to make the final matches.

A great deal of information can be obtained from studies comparing the reassociation of DNA of different organisms. For example, we may compare the point in the reaction when one-half of the DNA is present as double-stranded fragments. This point is called the $C_0t_{1/2}$, or *half-reaction time*. Provided that all pairs of single-stranded DNA complements consist of unique nucleotide sequences and all are about the same size, $C_0t_{1/2}$ varies directly with the complexity of the DNA. Designated as X, complexity represents the length in nucleotide pairs of all unique DNA fragments laid end to end. If the DNA used in an experiment represents the entire genome, and if all DNA sequences are different from one another, then X is equal to the size of the haploid genome.

Figure 10–24 compares DNAs from two bacteriophages or one bacterial source, each with a different genome size. As can be seen, as genome size increases, the curves obtained are shifted farther and farther to the right, indicative of an extended reassociation time.

As shown in Figure 10–25, $C_0t_{1/2}$ is directly proportional to the size of the genome. Reassociation occurs at a reduced rate in larger genomes because it takes longer for initial matches if there are greater numbers of unique DNA fragments. This is so because collisions are random; the more sequences are present, the greater the number of mismatches before all correct matchings occur. The method has been useful in assessing genome size in viruses and bacteria.

When reassociation kinetics of DNA from eukaryotic organisms (whose genome sizes are much greater than phage or bacteria) were first studied, a surprising observation was made. Rather than exhibiting a reduced rate of reassociation, the data revealed that *some* DNA segments reassociate even more rapidly than those derived from *E. coli*. The remaining DNA, as expected because of its greater complexity, took longer to reassociate. For example, Britten and Kohne examined DNA derived from calf thymus tissue (Figure 10–26). Based on these observations, they hypothesized that the rapidly reassociating fraction must represent **repetitive DNA sequences** present many times in the calf genome. This interpretation would explain why these DNA segments reassociate so rapidly. Multiple copies of the same sequence are much more likely to make matches, thus reassociating more quickly than single copies. On the other hand, they hypothesized that the remaining DNA segments consist of unique nucleotide sequences present only once in the genome; because there are more of these unique sequences, increasing the DNA complexity in calf thymus (compared with *E. coli*), their reassociation takes longer. The *E. coli* curve has been added to Figure 10–26 for the sake of comparison.

It is now clear that repetitive DNA sequences are prevalent in the genome of eukaryotes and are key to our understanding of how genetic information is organized in chromosomes. Careful study has shown that various levels of repetition exist. In some cases, short DNA sequences are repeated over a million times. In other cases, longer sequences are repeated only a few times, or intermediate levels of sequence redundancy are present. We will return to this topic in Chapter 12, where we will

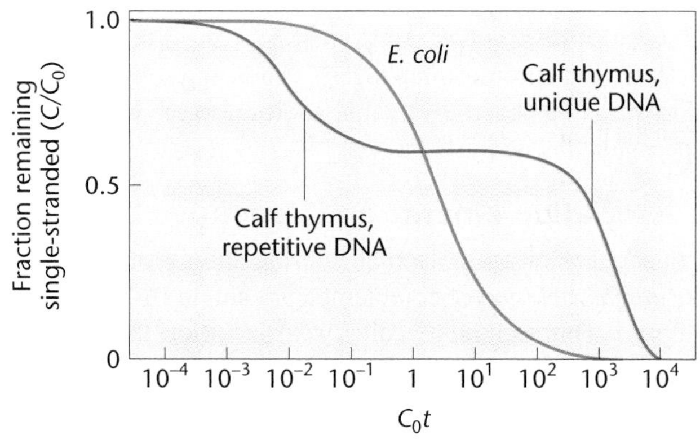

FIGURE 10–26 The C_0t curve of calf thymus DNA compared with *E. coli*. The repetitive fraction of calf DNA reassociates more quickly than that of *E. coli*, while the more complex unique calf DNA takes longer to reassociate than that of *E. coli*.

discuss the organization of DNA in genes and chromosomes. For now, we will conclude our discussion of repetitive DNA sequences by pointing out that the discovery of repetitive DNA was one of the first clues that much of the DNA in eukaryotes is not contained in genes that encode proteins. This concept will be developed and elaborated on as we expand our coverage of the molecular basis of heredity.

Now solve this

Problem 10.30 on page 261 asks you to to extrapolate information about C_0t analysis of DNA to the overall size of the DNA molecule.

Hint: Absolute C_0t values are directly proportional to the number of base pairs making up a DNA molecule.

Electrophoresis of Nucleic Acids

We conclude the chapter by considering an essential technique involved in the analysis of nucleic acids, **electrophoresis**. This technique separates different-sized fragments of DNA and RNA chains and is invaluable in current research investigations in molecular genetics.

In general, electrophoresis separates, *or resolves*, molecules in a mixture by causing them to migrate under the influence of an electric field. A sample is placed on a porous substance (a piece of filter paper or a semisolid gel), which is placed in a solution that conducts electricity. If two molecules have approximately the same shape *and* mass, the one with the greatest net charge will migrate more rapidly toward the electrode of opposite polarity.

As electrophoretic technology developed from its initial application to protein separation, researchers discovered that using gels of varying pore sizes significantly improved the resolution of this research technique. This advance is particularly useful for mixtures of molecules with a similar charge:mass ratio, but different sizes. For example, two polynucleotide chains of different *lengths* (e.g., 10 vs. 20 nucleotides) are both negatively charged based on the phosphate groups of the nucleotides. While they both move to the positively charged pole (the anode), the charge:mass ratio is the same for both chains, and separation based strictly on the electric field is minimal. However, using a porous medium such as **polyacrylamide gels** or **agarose gels**, which can be prepared with various pore sizes, allows these two molecules to be separated.

In such cases, *the smaller molecules migrate at a faster rate through the gel than the larger molecules* (Figure 10–27). The key to separation is based on the matrix (pores) of the gel, which restricts migration of larger molecules more than it restricts smaller molecules. The resolving power is so great that polynucleotides that vary by even one nucleotide in length are clearly separated. Once electrophoresis is complete, bands representing the variously sized molecules are identified either by autoradiography (if a component of the molecule is radioactive) or by the use of a fluorescent dye that binds to nucleic acids.

Electrophoretic separation of nucleic acids is at the heart of a variety of commonly used research techniques discussed later in the text (Chapters 19 and 22). Of particular note are the various "blotting" techniques (e.g., Southern blots and Northern blots), as well as DNA sequencing methods.

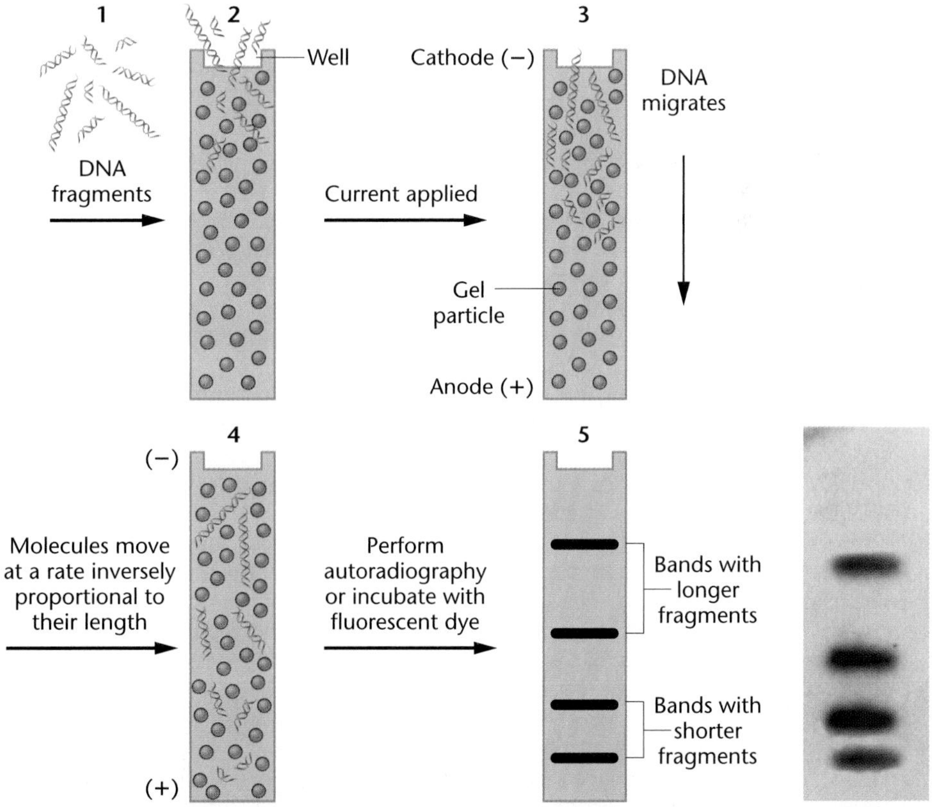

FIGURE 10–27 Electrophoretic separation of a mixture of DNA fragments that vary in length. The photograph shows an agarose gel with DNA bands corresponding to the diagram.

GENETICS, TECHNOLOGY, AND SOCIETY

The Twists and Turns of the Helical Revolution

Western civilization is frequently transformed by new scientific ideas that overturn our self-concepts and permanently alter our relationships with each other and the rest of the animate world. For 50 years, we have been in the midst of such a revolution—one as significant as those triggered by Darwin's theory of evolution or the Copernican rejection of Ptolemy's earth-centered universe.

The revolution began in April 1953 with Watson and Crick's discovery of the molecular structure of DNA. Their discovery that the DNA molecule consists of a twisted double helix, held together by weak bonds between specific pairs of bases, suddenly provided elegant solutions to age-old questions about the mechanisms of heredity, mutation, and evolution. Some of the greatest mysteries of life could be explained by the beauty and simplicity of a helix that replicates and shuffles the code of life.

After 1953, the double helix rapidly became the focus of modern science. Using knowledge of DNA's helical structure, molecular biologists quickly devised methods to purify, mutate, cut, and paste DNA in the test tube. They spliced DNA molecules from one organism into those of another, and then introduced these chimeric molecules into bacteria or cells in culture. They read the nucleotide sequences of genes and modified the traits of bacteria, fungi, fruit flies, and mice by removing and mutating their genes, or by introducing genes from other organisms. On the 50th anniversary of Watson and Crick's double-helical DNA model, the Human Genome Project announced the completion of the largest DNA project so far—sequencing the entire human genome.

In a mere 50 years, the helical revolution has touched the lives of millions of people. We can now test for simple genetic diseases, such as Tay–Sachs, cystic fibrosis, and sickle-cell anemia. We can manufacture large quantities of medically important proteins, such as insulin and growth hormone, using DNA technologies. DNA forensic tests help convict criminals, exonerate the innocent, and establish paternity. By following a trail of DNA sequences, anthropologists can now trace human origins back in time and place.

The helical revolution has profoundly altered our view of the living world. Although scientists dismiss the idea that humans are simply the products of their genes, popular culture endows DNA with almost magical powers. Genes are said to explain personality, career choice, criminality, intelligence—even fashion preferences and political attitudes. Advertisements hijack the language of genetics in order to grant inanimate objects a "genealogy" or "genetic advantage." Popular culture speaks of DNA as an immortal force, with the ability to affect morality and fate. The double helix is proclaimed as the essence of life, with the power to shape our future. Simple genetic explanations for our behavior appear to have more resonance for us than explanations involving social influences, economic factors, or free will. The beauty, symmetry, and biological significance of the double helix has insinuated itself into art, movies, advertising, and music. Paintings, sculpture, films—even video games and perfumes—use the language and imagery of genetics to confer upon the DNA molecule all the power and fears of modern technology.

But what of the future? Can we predict how the double helix and genetics will shape our world over the next 50 years? Although prophecy is certainly a risky business, some scientific developments seem assured. With the completion of the Human Genome Project, we will undoubtedly identify more and more of the genes that control normal and abnormal processes. In turn, this will enhance our ability to diagnose and predict genetic diseases. Over the next 50 years, we can look forward to biotechnologies as complex as gene therapies, prenatal diagnoses, and screening programs for susceptibilities to diseases as complicated as cancer and heart disease. We will continue to expand the applications of genetic engineering to agriculture as we manipulate plant and animal genes for enhanced productivity, disease resistance, and flavor.

The helical revolution will also continue to transform our concepts of ourselves and other creatures. As the human genome is compared to the genomes of other animals, it will become increasingly evident that we are closely related genetically to the rest of the animate world. The nucleotide sequence of the human genome differs only about 1 percent from that of chimpanzees, and some of our genes are virtually identical to homologous genes in plants, animals, and bacteria. As we realize the extent of our genetic kinship, extending over billions of years in a linear chain from the first life on earth, it is possible that this knowledge will alter our relationships with animals and with each other. When more genes are identified that contribute to phenotypic traits as simple as eye color and as complicated as intelligence or sexual orientation, it is possible that we will define ourselves even more as genetic beings and even less as creatures of free will or as the products of our environment.

Over the next 50 years, we will inevitably be faced with the practical and philosophical consequences of the DNA revolution. Will society harness DNA for everyone's benefit, or will this new genetic knowledge be used as a vehicle for discrimination? At the same time that modern genetics grants us more dominion over life, will it paradoxically increase our feelings of powerlessness? Will our new DNA-centered self-concepts increase our compassion for all life forms, or will it increase our perceived separation from the natural world? We will make our choices, and human history will proceed.

References

Dennis, C., and Campbell, P. 2003. The eternal molecule. (Introduction to a series of feature articles commemorating the 50th anniversary of the discovery of DNA structure.) *Nature* 421:396.

Web Site

A Revolution at 50. [*New York Times* article on the 50th anniversary of the discovery of DNA structure.]

http://www.nytimes.com/indexes/2003/02/25/health/genetics/index.html

CHAPTER SUMMARY

1. The existence of a genetic material capable of replication, storage, expression, and mutation is deducible from the observed patterns of inheritance in organisms.
2. Both proteins and nucleic acids were initially considered as the possible candidates for genetic material. Proteins are more diverse than nucleic acids (a requirement for the genetic material) and were favored owing to the advances being made in protein chemistry at the time. Additionally, Levene's tetranucleotide hypothesis had underestimated the magnitude of chemical diversity inherent in nucleic acids.
3. By 1952, transformation studies and experiments using bacteria infected with bacteriophages strongly suggested that DNA is the genetic material in bacteria and most viruses.
4. Initially, only circumstantial evidence supported the concept of DNA controlling inheritance in eukaryotes. These included DNA distribution in the cell, quantitative analysis of DNA, and UV-induced mutagenesis. More recent recombinant DNA techniques, as well as experiments with transgenic mice, have provided direct experimental evidence that the eukaryotic genetic material is DNA.
5. Numerous viruses provide an important exception to this general rule because many of them use RNA as their genetic material. These viruses include bacteriophages as well as some plant and animal viruses.
6. Establishment of DNA as the genetic material paved the way for the expansion of molecular genetics research and has served as the cornerstone for further important studies for nearly half a century.
7. During the late 1940s and early 1950s, a considerable effort was made to integrate accumulated information on the chemical structure of nucleic acids into a model of the molecular structure of DNA. The X-ray diffraction data of Franklin and Wilkins suggested that DNA was some sort of helix. In 1953, Watson and Crick were able to assemble a model of the double-helical DNA structure based on these X-ray diffraction studies as well as on Chargaff's analysis of base composition of DNA.
8. The DNA molecule exhibits antiparallel orientation and adenine–thymine and guanine–cytosine base-pairing complementarity along the polynucleotide chains. This model of DNA presents an obvious straightforward mechanism for its replication. The structure assumed by the helix appears to be a function of the nucleotide sequence and its chemical environment. Several alternative forms of the DNA helical structure exist. Watson and Crick described a B configuration, one of several right-handed helices. Wang and Rich discovered the left-handed Z-DNA currently being investigated for its physiological and genetic significance.
9. The second category of nucleic acids important in genetic function is RNA, which is similar to DNA with the exceptions that it is usually single stranded, the sugar ribose replaces the deoxyribose, and the pyrimidine uracil replaces thymine. Classes of RNA—ribosomal, transfer, and messenger—facilitate the flow of information from DNA to RNA to proteins, which are the end products of most genes.
10. The structure of DNA lends itself to various forms of analyses, which have in turn led to studies of the functional aspects of the genetic machinery. Absorption of UV light, sedimentation properties, denaturation–reassociation, and electrophoresis procedures are among the important tools for the study of nucleic acids. Reassociation kinetics analysis enabled geneticists to postulate the existence of repetitive DNA in eukaryotes, where certain nucleotide sequences are present many times in the genome.

INSIGHTS AND SOLUTIONS

The current chapter, in contrast to those preceding it, does not emphasize genetic problem solving. Instead, it recounts some of the initial experimental analyses that served as the cornerstone of modern genetics. Quite fittingly, then, our "Insights and Solutions" section shifts its emphasis to experimental rationale and analytical thinking, an approach that will continue through the remainder of the text whenever appropriate.

1. (a) Based strictly on the analysis of the transformation data of Avery, MacLeod, and McCarty, what objection might be made to the conclusion that DNA is the genetic material? What other conclusion might be considered? (b) What observations argue against this objection?

Solution: (a) Based solely on their results, it may be concluded that DNA is essential for transformation. However, DNA might have been a substance that caused capsular formation by *directly* converting nonencapsulated cells to ones with a capsule. That is, DNA may simply have played a catalytic role in capsular synthesis, leading to cells displaying smooth type III colonies.

Solution: (b) First, transformed cells pass the trait onto their progeny cells, thus supporting the conclusion that DNA is responsible for heredity, not for the direct production of polysaccharide coats. Second, subsequent transformation studies over a period of five years showed that other traits, such as antibiotic resistance, could be transformed. Therefore, the transforming factor has a broad general effect, not one specific to polysaccharide synthesis. This observation is more in keeping with the conclusion that DNA is the genetic material.

2. If RNA were the universal genetic material, how would it have affected the Avery experiment and the Hershey–Chase experiment?

Solution: In the Avery experiment, digestion of the soluble filtrate with RNase, rather than DNase would have eliminated transformation. Had this occurred, Avery and his colleagues would have concluded that RNA was the transforming factor. The Hershey and Chase results would not have changed, since ^{32}P would also label RNA, but not protein. Had they been using a bacteriophage with RNA as its nucleic acid, and had they known this, they would have concluded that RNA was responsible for directing the reproduction of their bacteriophage.

3. Sea urchin DNA, which is double stranded, was shown to contain 17.5 percent of its bases in the form of cytosine (C). What percentages of the other three bases are present in this DNA?

Solution: The amount of C≡G, so guanine is also present as 17.5 percent. The remaining bases, A and T, are present in equal amounts, and together they represent the rest of the bases $(100 - 35)$. Therefore, $A = T = 65/2 = 32.5$ percent.

4. The quest to isolate an important disease-causing organism was successful, and molecular biologists were hard at work. The organism contained as its genetic material a remarkable nucleic acid with a base composition of A = 21 percent, C = 29 percent, G = 29 percent, U = 21 percent. When heated, it showed a major hyperchromic effect, and when kinetics were studied, the nucleic acid of this organism provided the C_0t curve shown below, in contrast to that of phage T4 and *E. coli*. T4 contains 10^5 nucleotide pairs.

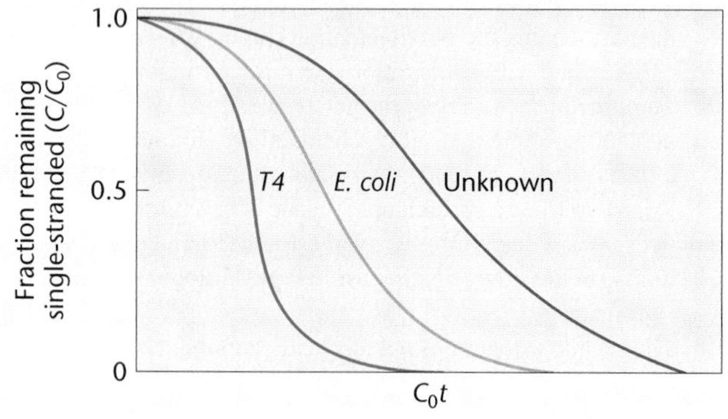

Analyze this information carefully, and draw *all* possible conclusions about the genetic material of this organism, based strictly on the preceding observations. As a test of your model, make one prediction that if upheld would strengthen your hypothesis about the nature of this molecule?

Solution: First of all, because of the presence of uracil (U), the molecule appears to be RNA. In contrast to normal RNA, this one has base ratios of A/U = G/C = 1, suggesting that the molecule may be a double helix. The hyperchromic shift and reassociation kinetics support this hypothesis. The kinetic study, based on the shape of the C_0t curve, reveals that there is no repetitive sequence DNA. Further, the complexity (X), or total length of unique sequence DNA, is greater than that of either phage T4 (10^5 nucleotide pairs) or *E. coli*. There *may* be a greater number of genes present compared to T4 or *E. coli*, but the excessive unique-sequence DNA may serve some other role, or simply play no genetic role. The prediction concerns the sugars. Our model suggests that ribose rather than deoxyribose should be present. If so, this observation would support the hypothesis that RNA is the genetic material in this organism.

Problems and Discussion Questions

1. The functions ascribed to the genetic material are replication, expression, storage, and mutation. What does each of these terms mean?
2. Discuss the reasons why proteins were generally favored over DNA as the genetic material before 1940. What was the role of the tetranucleotide hypothesis in this controversy?
3. Contrast the various contributions made to an understanding of transformation by Griffith, by Avery and his colleagues, and by Taylor.
4. When Avery and his colleagues had obtained what was concluded to be purified DNA from the IIIS virulent cells, they treated the fraction with proteases, RNase, and DNase, followed by the assay for retention or loss of transforming ability. What were the purpose and results of these experiments? What conclusions were drawn?
5. Why were ^{32}P and ^{35}S chosen for use in the Hershey–Chase experiment? Discuss the rationale and conclusions of this experiment.
6. Does the design of the Hershey–Chase experiment distinguish between DNA and RNA as the molecule serving as the genetic material? Why or why not?
7. Would an experiment similar to that performed by Hershey and Chase work if the basic design were applied to the phenomenon of transformation? Explain why or why not.
8. What observations are consistent with the conclusion that DNA serves as the genetic material in eukaryotes? List and discuss them.
9. What are the exceptions to the general rule that DNA is the genetic material in all organisms? What evidence supports these exceptions?
10. Draw the chemical structure of the three components of a nucleotide, and then link the three together. What atoms are removed from the structures when the linkages are formed?
11. How are the carbon and nitrogen atoms of the sugars, purines, and pyrimidines numbered?
12. Adenine may also be named 6-amino purine. How would you name the other four nitrogenous bases, using this alternative system? (O is oxy, and CH_3 is methyl.)
13. Draw the chemical structure of a dinucleotide composed of A and G. Opposite this structure, draw the dinucleotide composed of T and C in an antiparallel (or upside-down) fashion. Form the possible hydrogen bonds.
14. Describe the various characteristics of the Watson–Crick double-helix model for DNA.
15. What evidence did Watson and Crick have at their disposal in 1953? What was their approach in arriving at the structure of DNA?
16. What might Watson and Crick have concluded, had Chargaff's data from a single source indicated the following?

	A	T	G	C
%	29	19	21	31

Why would this conclusion be contradictory to Wilkins' and Franklin's data?

17. How do covalent bonds differ from hydrogen bonds? Define base complementarity.
18. List three main differences between DNA and RNA.
19. What are the three types of RNA molecules? How is each related to the concept of information flow?
20. What component of the nucleotide is responsible for the absorption of ultraviolet light? How is this technique important in the analysis of nucleic acids?
21. Distinguish between sedimentation velocity and sedimentation equilibrium centrifugation (density gradient centrifugation).
22. What is the basis for determining base composition, using density gradient centrifugation?
23. What is the physical state of DNA following denaturation?

24. Compare the following curves representing reassociation kinetics.

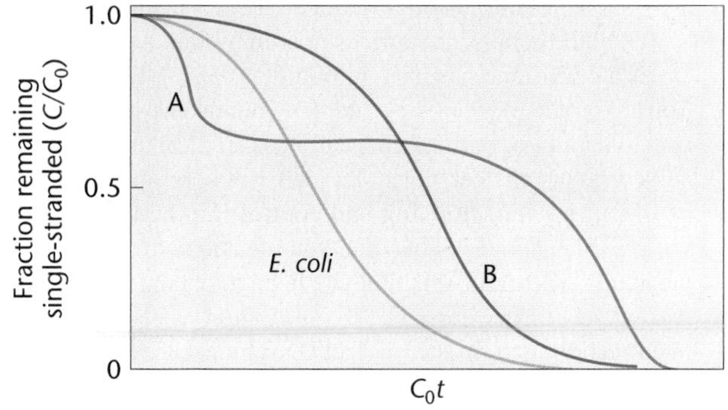

What can be said about the DNAs represented by each set of data compared with *E. coli?*

25. What is the hyperchromic effect? How is it measured? What does T_m imply?
26. Why is T_m related to base composition?
27. What is the chemical basis of molecular hybridization?
28. What did the Watson–Crick model suggest about the replication of DNA?
29. A genetics student was asked to draw the chemical structure of an adenine- and thymine-containing dinucleotide derived from DNA. His answer is shown here:

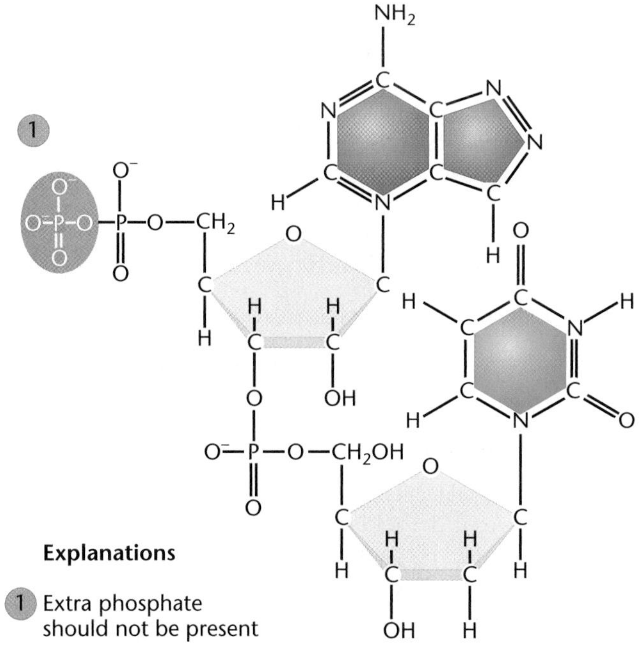

The student made more than six major errors. One of them is circled, numbered 1, and explained. Find five others. Circle them, number them 2 through 6, and briefly explain each by following the example given.

30. The DNA of the bacterial virus T4 produces a $C_0t_{1/2}$ of about 0.5 and contains 10^5 nucleotide pairs in its genome. How many nucleotide pairs are present in the genome of the virus *MS2* and the bacterium *E. coli*, whose respective DNAs produce $C_0t_{1/2}$ values of 0.001 and 10.0?
31. Considering the information in this chapter on B- and Z-DNA and right- and left-handed helices, carefully analyze the structures below, and draw conclusions about the helical nature of areas (a) and (b). Which is right handed and which is left handed?

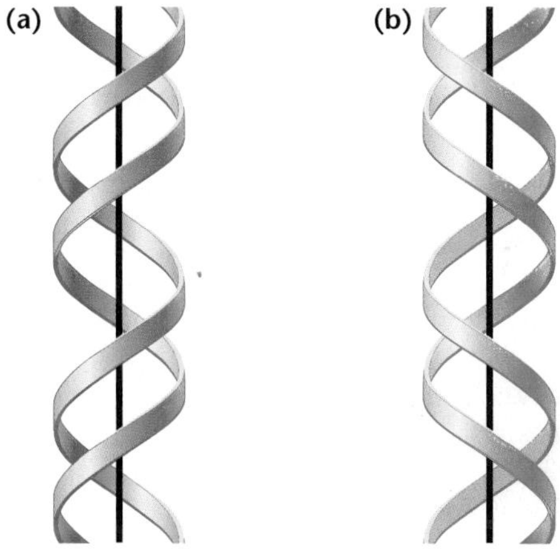

32. One of the most common spontaneous lesions that occurs in DNA under physiological conditions is the hydrolysis of the amino group of cytosine, converting it to uracil. What would be the effect on DNA structure of a uracil group replacing cytosine?
33. In some organisms, cytosine is methylated at carbon 5 of the pyrimidine ring after it is incorporated into DNA. If a 5-methyl cytosine is then hydrolyzed, as described in Problem 33, what base will be generated?

Extra-Spicy Problems

34. A primitive eukaryote was discovered that displayed a unique nucleic acid as its genetic material. Analysis revealed the following observations:
    (i) X-ray diffraction studies display a general pattern similar to DNA, but with somewhat different dimensions and more irregularity.
    (ii) A major hyperchromic shift is evident upon heating and monitoring UV absorption at 260 nm.
    (iii) Base composition analysis reveals four bases in the following proportions:

Adenine	=	8%
Guanine	=	37%
Xanthine	=	37%
Hypoxanthine	=	18%

(iv) About 75 percent of the sugars are deoxyribose, while 25 percent are ribose.

Attempt to solve the structure of this molecule by postulating a model that is consistent with the foregoing observations.

35. *Newsdate: March 1, 2015.* A unique creature has been discovered during exploration of outer space. Recently, its genetic material has been isolated and analyzed. This material is similar in some ways to DNA in its chemical makeup. It contains in abundance the 4-carbon sugar erythrose and a molar equivalent of phosphate groups. Additionally, it contains six nitrogenous bases: adenine (A), guanine (G), thymine (T), cytosine (C), hypoxanthine (H), and xanthine (X). These bases exist in the following relative proportions:

$$A = T = H \text{ and } C = G = X$$

X-ray diffraction studies have established a regularity in the molecule and a constant diameter of about 30 Å.
Together, these data have suggested a model for the structure of this molecule.
(a) Propose a general model of this molecule. Describe it briefly.
(b) What base-pairing properties must exist for H and for X in the model?
(c) Given the constant diameter of 30 Å, do you think that *either* (i) both H and X are purines or both pyrimidines, *or* (ii) one is a purine and one is a pyrimidine?

36. You are provided with DNA samples from two newly discovered bacterial viruses. Based on the various analytical techniques discussed in this chapter, construct a research protocol that would be useful in characterizing and contrasting the DNA of both viruses. For each technique that you include in the protocol, indicate the type of information you hope to obtain.

37. During gel electrophoresis, DNA molecules can easily be separated according to size because all DNA molecules have the same charge to mass ratio and the same shape (long rod). Would you expect RNA molecules to behave in the same manner as DNA during gel electrophoresis? Why or why not?

Selected Readings

Adleman, L.M. 1998. Computing with DNA. *Sci. Am.* (Aug.) 279:54–61.

Alloway, J.L. 1933. Further observations on the use of pneumococcus extracts in effecting transformation of type *in vitro*. *J. Exp. Med.* 57:265–78.

Avery, O.T., MacLeod, C.M., and McCarty, M. 1944. Studies on the chemical nature of the substance inducing transformation of pneumococcal types: Induction of transformation by a desoxyribonucleic acid fraction isolated from pneumococcus type III. *J. Exp. Med.* 79:137–58. (Reprinted in Taylor, J.H. 1965. *Selected papers in molecular genetics*. Orlando, FL: Academic Press.)

Britten, R.J., and Kohne, D.E. 1968. Repeated sequences in DNA. *Science* 161:529–40.

Chargaff, E. 1950. Chemical specificity of nucleic acids and mechanism for their enzymatic degradation. *Experientia* 6:201–9.

Dawson, M.H. 1930. The transformation of pneumococcal types: I. The interconvertibility of type-specific *S. pneumococci*. *J. Exp. Med.* 51:123–47.

Dickerson, R.E., et al. 1982. The anatomy of A-, B-, and Z-DNA. *Science* 216:475–85.

Dubos, R.J. 1976. *The professor, the Institute and DNA: Oswald T. Avery, his life and scientific achievements*. New York: Rockefeller University Press.

Felsenfeld, G. 1985. DNA. *Sci. Am.* (Oct.) 253:58–78.

Fraenkel-Conrat, H., and Singer, B. 1957. Virus reconstruction: II. Combination of protein and nucleic acid from different strains. *Biochim. Biophys. Acta* 24:530–48. (Reprinted in Taylor, J.H. 1965. *Selected papers in molecular genetics*, Orlando, FL: Academic Press.)

Franklin, R.E., and Gosling, R.G. 1953. Molecular configuration in sodium thymonucleate. *Nature* 171:740–41.

Griffith, F. 1928. The significance of pneumococcal types. *J. Hyg.* 27:113–59.

Guthrie, G.D., and Sinsheimer, R.L. 1960. Infection of protoplasts of *Escherichia coli* by subviral particles. *J. Mol. Biol.* 2:297–305.

Hershey, A.D., and Chase, M. 1952. Independent functions of viral protein and nucleic acid in growth of bacteriophage. *J. Gen. Phys.* 36:39–56. (Reprinted in Taylor, J.H. 1965. *Selected papers in molecular genetics*. Orlando, FL: Academic Press.)

Levene, P.A., and Simms, H.S. 1926. Nucleic acid structure as determined by electrometric titration data. *J. Biol. Chem.* 70:327–41.

McCarty, M.. 1985. *The transforming principle: Discovering that genes are made of DNA*. New York: W. W. Norton.

Olby, R. 1974. *The path to the double helix*. Seattle: University of Washington Press.

Pauling, L., and Corey, R.B. 1953. A proposed structure for the nucleic acids. *Proc. Natl. Acad. Sci. USA* 39:84–97.

Rich, A., Nordheim, A., and Wang, A.H.-J. 1984. The chemistry and biology of left-handed Z-DNA. *Annu. Rev. Biochem.* 53:791–846.

Spizizen, J. 1957. Infection of protoplasts by disrupted T2 viruses. *Proc. Natl. Acad. Sci. USA* 43:694–701.

Stent, G.S., ed. 1981. *The double helix: Text, commentary, review, and original papers*. New York: W. W. Norton.

Varmus, H. 1988. Retroviruses. *Science* 240:1427–35.

Watson, J.D. 1968. *The double helix*. New York: Atheneum.

Watson, J.D., and Crick, F.C. 1953a. Molecular structure of nucleic acids: A structure for deoxyribose nucleic acids. *Nature* 171:737–38.

———1953b. Genetic implications of the structure of deoxyribose nucleic acid. *Nature* 171:964.

Wilkins, M.H.F., Stokes, A.R., and Wilson, H.R. 1953. Molecular structure of desoxypentose nucleic acids. *Nature* 171:738–40.

DNA Replication and Recombination

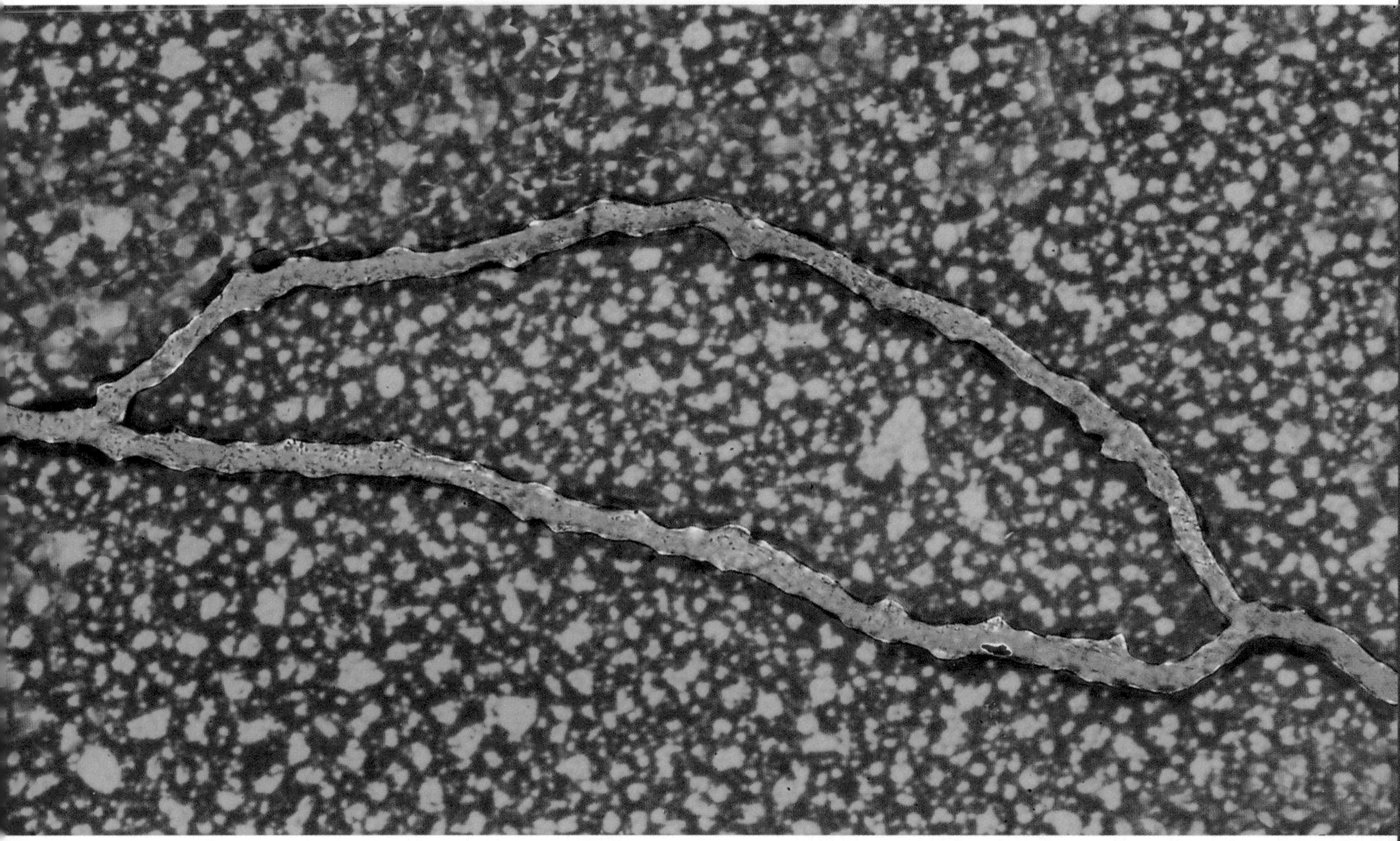

Transmission electron micrograph of human DNA from a HeLa cell, illustrating a replication fork characteristic of active DNA replication.

Chapter Concepts

- Genetic continuity between parental and progeny cells is maintained by semiconservative replication of DNA, as predicted by the Watson–Crick model.
- Semiconservative replication uses each strand of the parent double helix as a template, and each resulting double helix includes one "old" and one "new" strand of DNA.
- DNA synthesis is a complex but orderly process, occurring under the direction of a myriad of enzymes and other proteins.
- DNA synthesis involves the polymerization of nucleotides into polynucleotide chains.
- DNA synthesis is similar in prokaryotes and eukaryotes, but more complex in eukaryotes.
- Genetic recombination, an important process leading to the exchange of segments between DNA molecules, occurs under the direction of a group of enzymes.

Web Tutorial 11.1
DNA Replication

Following Watson and Crick's proposal for the structure of DNA, scientists focused their attention on how this molecule is replicated. Replication is an essential function of the genetic material and must be executed precisely if genetic continuity between cells is to be maintained following cell division. It is an enormous, complex task. Consider for a moment that more than 3×10^9 (3 billion) base pairs exist within the 23 chromosomes of the human genome. To duplicate faithfully the DNA of just one of these chromosomes requires a mechanism of extreme precision. Even an error rate of only 10^{-6} (one in a million) will still create 3000 errors (obviously an excessive number) during each replication cycle of the genome. While it is not error free, and much of evolution would not have occurred if it were, an extremely accurate system of DNA replication has evolved in all organisms.

As Watson and Crick noted in the concluding paragraph of their 1953 paper (reprinted on page 250), their proposed model of the double helix provided the initial insight into how replication occurs. Called *semiconservative replication*, this mode of DNA duplication was soon to receive strong support from numerous studies of viruses, prokaryotes, and eukaryotes. Once the general mode of replication was clarified, research to determine the precise details of DNA synthesis intensified. What has since been discovered is that numerous enzymes and other proteins are needed to copy a DNA helix. Because of the complexity of the chemical events during synthesis, this subject remains an extremely active area of research.

In this chapter, we will discuss the general mode of replication, as well as the specific details of DNA synthesis. The research leading to such knowledge is another link in our understanding of life processes at the molecular level.

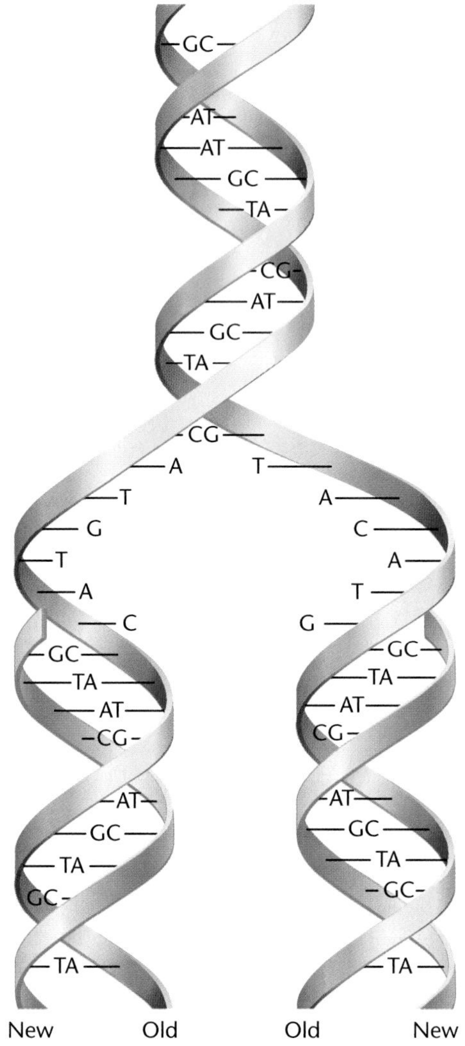

FIGURE 11–1 Generalized model of semiconservative replication of DNA. New synthesis is shown in teal.

11.1 DNA Is Reproduced by Semiconservative Replication

It was apparent to Watson and Crick that, because of the arrangement and nature of the nitrogenous bases, each strand of a DNA double helix could serve as a template for the synthesis of its complement (Figure 11–1). They proposed that, if the helix were unwound, each nucleotide along the two parent strands would have an affinity for its complementary nucleotide. As we learned in Chapter 10, the complementarity is due to the potential hydrogen bonds that can be formed. If thymidylic acid (T) were present, it would "attract" adenylic acid (A); if guanidylic acid (G) were present, it would attract cytidylic acid (C); likewise, A would attract T, and C would attract G. If these nucleotides were then covalently linked into polynucleotide chains along both templates, the result would be the production of two identical double strands of DNA. Each replicated DNA molecule would consist of one "old" and one "new" strand, hence the reason for the name **semiconservative replication**.

Two other theoretical modes of replication are possible that also rely on the parental strands as a template (Figure 11–2). In **conservative replication**, complementary polynucleotide chains are synthesized as described earlier. Following synthesis, however, the two newly created strands then come together and the parental strands reassociate. The original helix is thus "conserved."

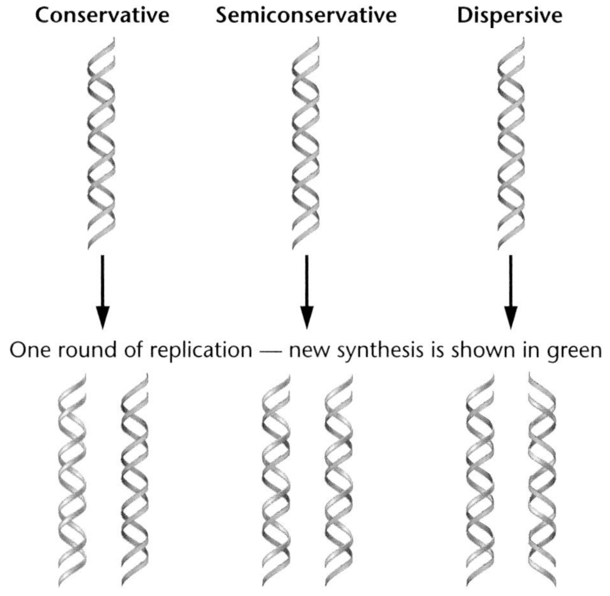

FIGURE 11–2 Results of one round of replication of DNA for each of the three possible modes by which replication could be accomplished.

In the second alternative mode, called **dispersive replication**, the parental strands are dispersed into two new double helices following replication. Hence, each strand consists of both old and new DNA. This mode would involve cleavage of the parental strands during replication. It is the most complex of the three possibilities and is therefore considered to be least likely to occur. It could not, however, be ruled out as an experimental model. Figure 11–2 shows the theoretical results of a single round of replication by each of the three different modes.

The Meselson–Stahl Experiment

In 1958, Matthew Meselson and Franklin Stahl published the results of an experiment providing strong evidence that semiconservative replication is the mode used by bacterial cells to produce new DNA molecules. They grew *E. coli* cells for many generations in a medium that had $^{15}NH_4Cl$ (ammonium chloride) as the only nitrogen source. A "heavy" isotope of nitrogen, ^{15}N contains one more neutron than the naturally occurring ^{14}N isotope; thus, molecules containing ^{15}N are more dense than those containing ^{14}N. Unlike radioactive isotopes, ^{15}N is stable. After many generations, all nitrogen-containing molecules, including the nitrogenous bases of DNA, contained the heavier isotope in the *E. coli* cells.

Critical to the success of this experiment, DNA containing ^{15}N can be distinguished from DNA containing ^{14}N. The experimental procedure involves the use of a technique referred to as **sedimentation equilibrium centrifugation**, or as it is also called, density gradient centrifugation (discussed in Chapter 10). Samples are forced by centrifugation through a density gradient of a heavy metal salt, such as cesium chloride. Molecules of DNA will reach equilibrium when their density equals the density of the gradient medium. In this case, ^{15}N-DNA will reach this point at a position closer to the bottom of the tube than will ^{14}N-DNA.

In this experiment (Figure 11–3), uniformly labeled ^{15}N cells were transferred to a medium containing only $^{14}NH_4Cl$. Thus, all "new" synthesis of DNA during replication contained only the "lighter" isotope of nitrogen. The time of transfer to the new medium was taken as time zero ($t = 0$). The *E. coli* cells were allowed to replicate over several generations, with cell samples removed after each replication cycle. DNA was isolated from each sample and subjected to sedimentation equilibrium centrifugation.

After one generation, the isolated DNA was present in only a single band of intermediate density—the expected result for semiconservative replication in which each replicated molecule was composed of one new ^{14}N-strand and one old ^{15}N-strand (Figure 11–4). This result was not consistent with the prediction of conservative replication, in which two distinct bands would occur, and thus this mode may be rejected.

After two cell divisions, DNA samples showed two density bands—one intermediate band and one lighter band corresponding to the ^{14}N position in the gradient. Similar results occurred after a third generation, except that the proportion of the band increased. This was again consistent with the interpretation that replication is semiconservative.

You may have realized that a molecule exhibiting intermediate density is also consistent with dispersive replication. However, Meselson and Stahl ruled out this mode of replication on the basis of two observations. First, after the first generation of replication in an ^{14}N-containing medium, they isolated the hybrid molecule and heat denatured it. Recall from Chapter 10 that heating will separate a duplex into single strands. When the densities of the single strands of the hybrid were determined, they exhibited *either* an ^{15}N profile *or* an ^{14}N profile, but *not* an intermediate density.

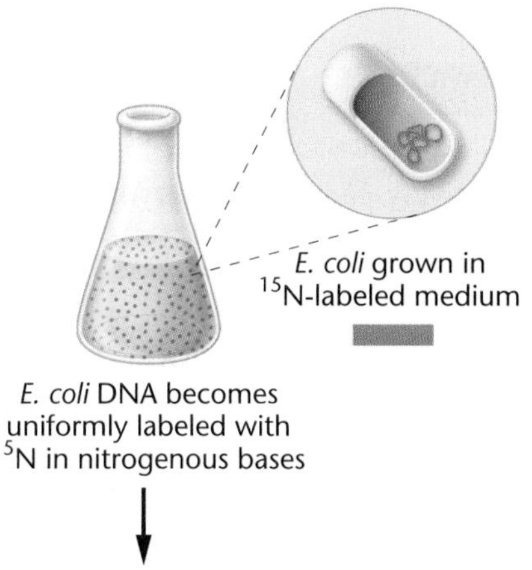

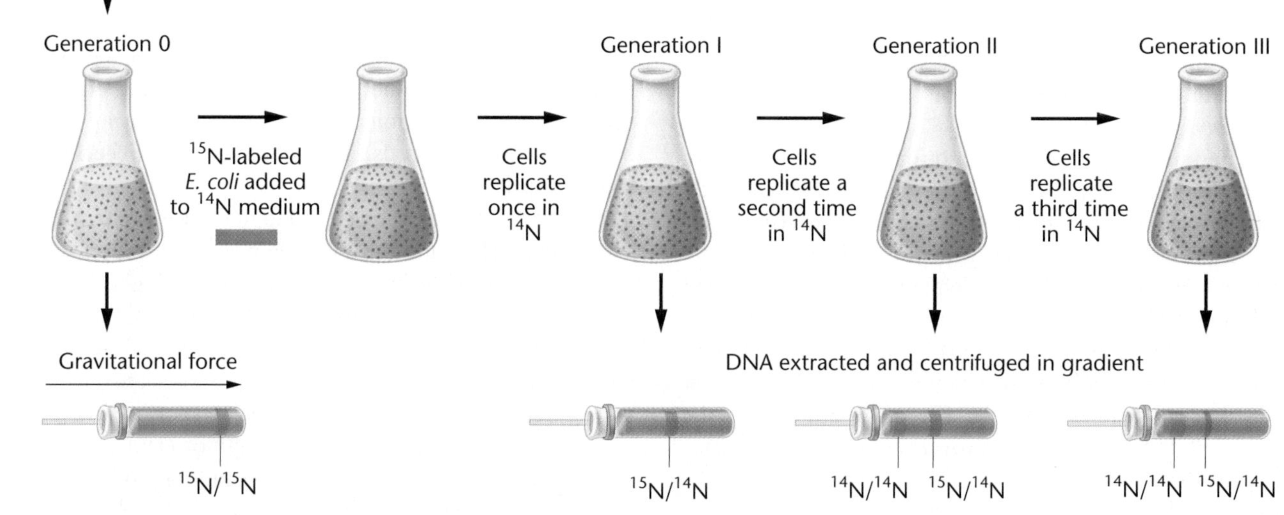

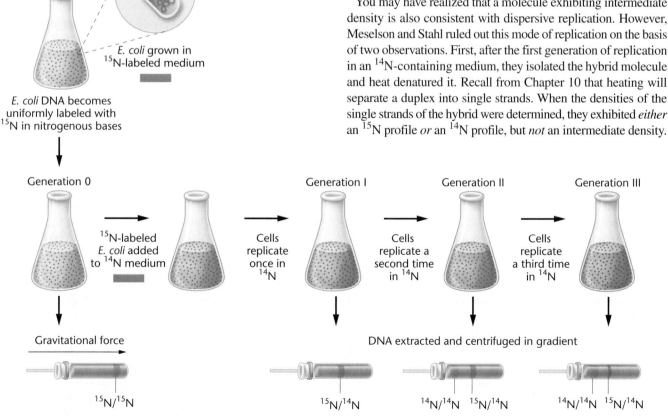

FIGURE 11–3 The Meselson–Stahl experiment.

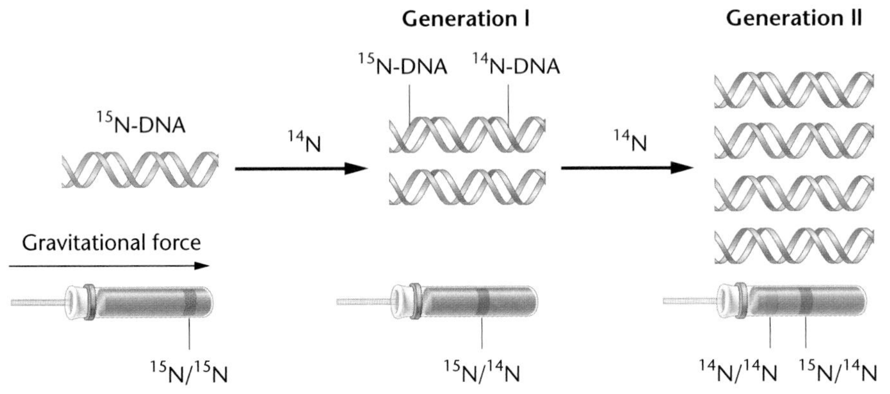

FIGURE 11–4 The expected results of two generations of semiconservative replication in the Meselson–Stahl experiment.

This observation is consistent with the semiconservative mode, but inconsistent with the dispersive mode.

Furthermore, if replication were dispersive, *all* generations after $t = 0$ would demonstrate DNA of an intermediate density. In each generation after the first, the ratio of $^{15}N/^{14}N$ would decrease, and the hybrid band would become lighter and lighter, eventually approaching the ^{14}N band. This result was not observed. The Meselson–Stahl experiment provided conclusive support for semiconservative replication in bacteria and tended to rule out both the conservative and dispersive modes.

Semiconservative Replication in Eukaryotes

In 1957, the year before the work of Meselson and his colleagues was published, J. Herbert Taylor, Philip Woods, and Walter Hughes presented evidence that semiconservative replication also occurs in eukaryotic organisms. They experimented with root tips of the broad bean *Vicia faba*, which are an excellent source of dividing cells. These researchers were able to monitor the process of replication by labeling DNA with ^{3}H-thymidine, a radioactive precursor of DNA, and performing autoradiography.

Autoradiography is a common technique that, when applied cytologically, pinpoints the location of a radioisotope in a cell. In this procedure, a photographic emulsion is placed over a histological preparation containing cellular material (root tips, in this experiment), and the preparation is stored in the dark. The slide is then developed, much as photographic film is processed. Because the radioisotope emits energy, following development the emulsion turns black at the approximate point of emission. The end result is the presence of dark spots or "grains" on the surface of the section, identifying the location of newly synthesized DNA within the cell.

Taylor and his colleagues grew root tips for approximately one generation in the presence of the radioisotope and then placed them in unlabeled medium in which cell division continued. At the conclusion of each generation, they arrested the cultures at metaphase by adding colchicine (a chemical derived from the crocus plant that poisons the spindle fibers) and then examined the chromosomes by autoradiography. They found radioactive thymidine only in association with chromatids that contained newly synthesized DNA. Figure 11–5 illustrates the replication of a single chromosome over two division cycles, including the distribution of grains.

These results are compatible with the semiconservative mode of replication. After the first replication cycle in the presence of the isotope, both sister chromatids show radioactivity, indicating that each chromatid contains one new radioactive DNA strand and one old unlabeled strand. After the second replication cycle, *which also takes place in unlabeled medium*, only one of the two sister chromatids of each chromosome should be radioactive, because half of the parent strands are unlabeled. With only the minor exceptions of sister chromatid exchanges (discussed in Chapter 6), this result was observed.

Together, the Meselson–Stahl experiment and the experiment by Taylor, Woods, and Hughes soon led to the general acceptance of the semiconservative mode of replication. Later studies with other organisms reached the same conclusion and also strongly supported Watson and Crick's proposal for the double-helix model of DNA.

HOW DO WE KNOW?

What is the experimental basis for concluding that DNA replicates semiconservatively in both prokaryotes and eukaryotes?

Origins, Forks, and Units of Replication

To enhance our understanding of semiconservative replication, let's briefly consider a number of relevant issues. The first concerns the **origin of replication**. Where along the chromosome is DNA replication initiated? Is there only a single origin, or does DNA synthesis begin at more than one point? Is any given point of origin random, or is it located at a specific region along the chromosome? Second, once replication begins, does it proceed in a single direction or in both directions away from the origin? In other words, is replication **unidirectional** or **bidirectional**?

To address these issues, we need to introduce two terms. First, at each point along the chromosome where replication is occurring, the strands of the helix are unwound, creating what is called a **replication fork**. Such a fork will initially appear at the point of origin of synthesis and then move along the DNA duplex as replication proceeds. If replication is bidirectional, two such forks will be present, migrating in opposite directions away from the origin. The second term refers to the length of DNA that is replicated following one initiation event at a single origin. This is a unit referred to as the **replicon**. In *E. coli*, the replicon consists of the entire genome of 4.2 Mb (4.2 million base pairs).

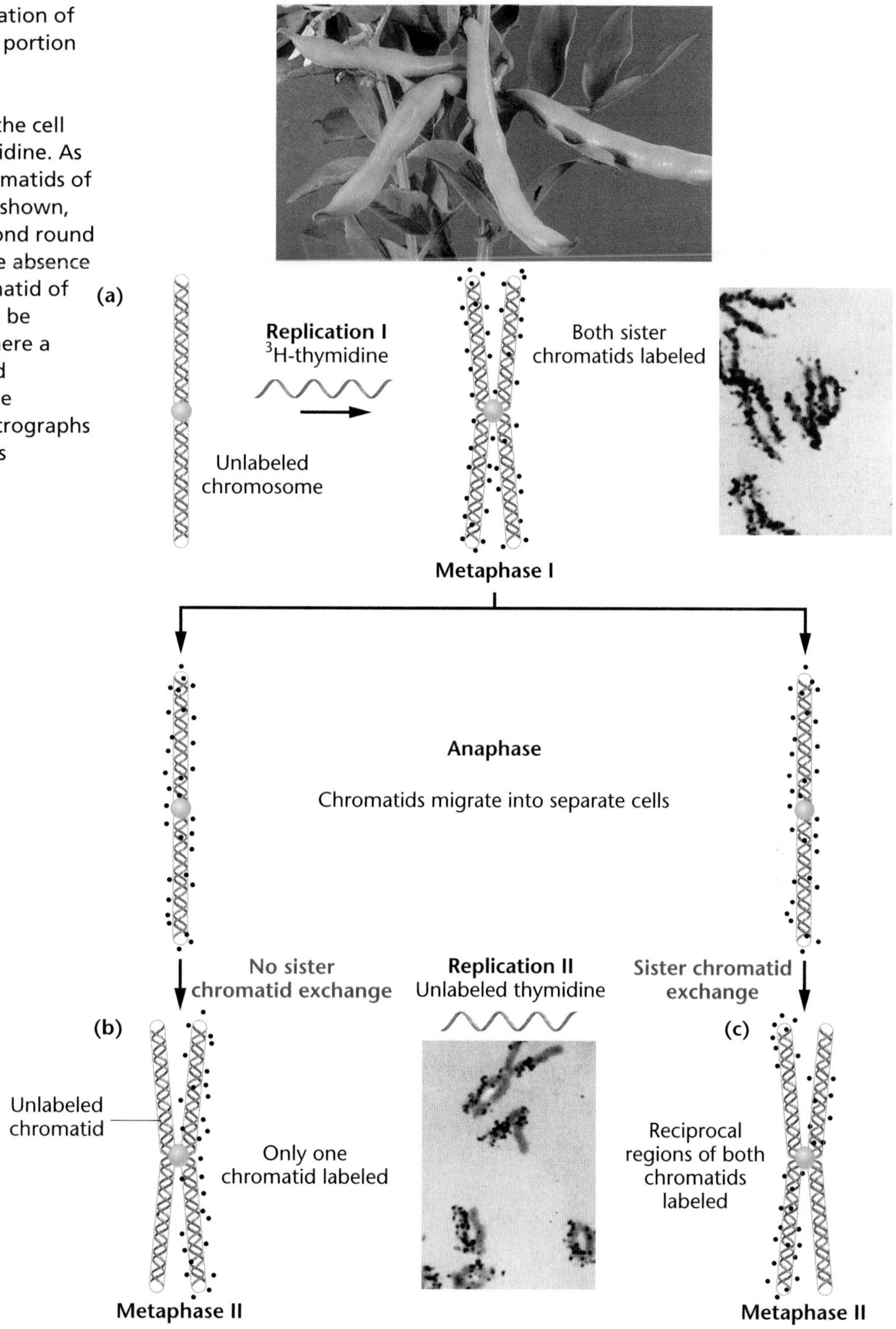

FIGURE 11–5 The Taylor–Woods–Hughes experiment, demonstrating the semiconservative mode of replication of DNA in root tips of *Vicia faba*. A portion of the plant is shown in the top photograph. (a) An unlabeled chromosome proceeds through the cell cycle in the presence of ^{3}H-thymidine. As it enters mitosis, both sister chromatids of the chromosome are labeled, as shown, by autoradiography. After a second round of replication (b), this time in the absence of ^{3}H-thymidine, only one chromatid of each chromosome is expected to be surrounded by grains. Except where a reciprocal exchange has occurred between sister chromatids (c), the expectation was upheld. The micrographs are of the actual autoradiograms obtained in the experiment.

The evidence is clear regarding the origin and direction of replication. John Cairns tracked replication in *E. coli*, using radioactive precursors of DNA synthesis and autoradiography. He was able to demonstrate that in *E. coli* there is only a single region, called *oriC*, where replication is initiated. Since DNA synthesis in bacteriophages and bacteria originates at a single point, the entire chromosome constitutes one replicon. The presence of only a single origin is characteristic of bacteria, which have only one circular chromosome.

Still other results, again relying on autoradiography, demonstrated that replication is bidirectional, moving away from *oriC* in both directions (Figure 11–6). This creates two replication forks that migrate farther and farther apart as replication proceeds. These forks eventually merge as semiconservative replication of the entire chromosome is completed at a termination region, called *ter*.

We will see later in this chapter that in eukaryotes, each chromosome contains multiple points of origin.

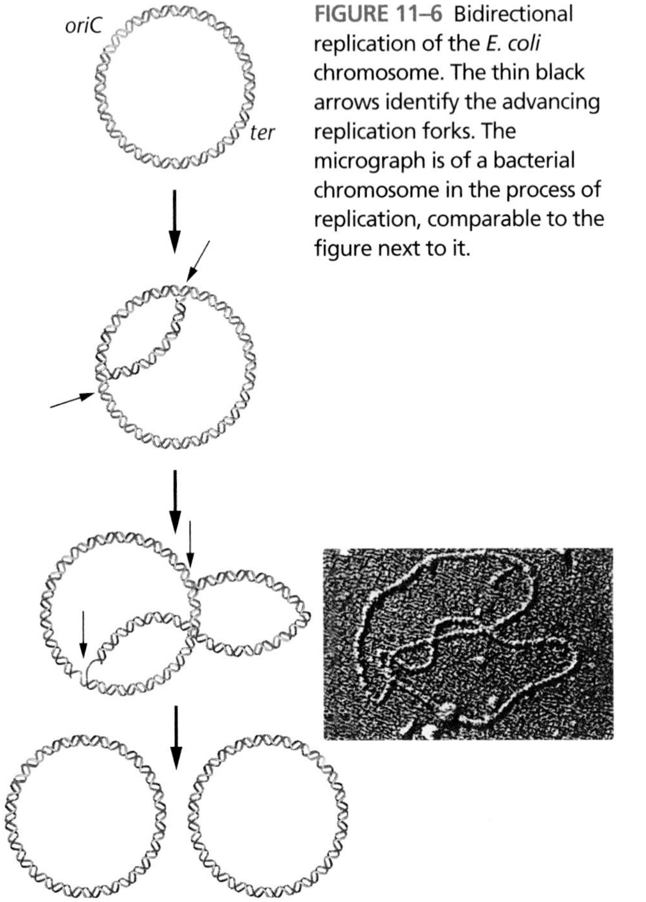

FIGURE 11–6 Bidirectional replication of the *E. coli* chromosome. The thin black arrows identify the advancing replication forks. The micrograph is of a bacterial chromosome in the process of replication, comparable to the figure next to it.

11.2 DNA Synthesis in Bacteria Involves Five Polymerases, as well as Other Enzymes

The determination that replication is semiconservative and bidirectional indicates only the *pattern* of DNA duplication and the association of finished strands with one another once synthesis is completed. A more complex issue is how the actual *synthesis* of long complementary polynucleotide chains occurs on a DNA template. As in most molecular biological studies, this question was first approached by using microorganisms. Research began about the same time as the Meselson–Stahl work, and the topic is still an active area of investigation. What is most apparent in this research is the tremendous complexity of the biological synthesis of DNA.

DNA Polymerase I

Studies of the enzymology of DNA replication were first reported by Arthur Kornberg and colleagues in 1957. They isolated an enzyme from *E. coli* that was able to direct DNA synthesis in a cell-free (*in vitro*) system. The enzyme is now called **DNA polymerase I**, as it was the first of several similar enzymes to be isolated.

Kornberg determined that there were two major requirements for *in vitro* DNA synthesis under the direction of DNA polymerase I: (1) all four deoxyribonucleoside triphosphates (dNTPs), and (2) template DNA.

If any one of the four deoxyribonucleoside triphosphates was omitted from the reaction, no measurable synthesis occurred. If derivatives of these precursor molecules other than the nucleoside triphosphate were used (nucleotides or nucleoside diphosphates), synthesis also did not occur. If no template DNA was added, synthesis of DNA occurred, but was reduced greatly.

Most of the synthesis directed by Kornberg's enzyme appeared to be exactly the type required for semiconservative replication. The reaction is summarized in Figure 11–7, which depicts the addition of a single nucleotide. The enzyme has since been shown to consist of a single polypeptide containing 928 amino acids.

The way in which each nucleotide is added to the growing chain is a function of the specificity of DNA polymerase I. As shown in Figure 11–8, the precursor dNTP contains the three phosphate groups attached to the 5′-carbon of deoxyribose. As the two terminal phosphates are cleaved during synthesis, the remaining phosphate attached to the 5′-carbon is covalently linked to the 3′-OH group of the deoxyribose to which it is added. Thus, **chain elongation** occurs in the **5′ to 3′ direction** by the addition of one nucleotide at a time to the growing 3′ end. Each step provides a newly exposed 3′-OH group that can participate in the next addition of a nucleotide as DNA synthesis proceeds.

Having shown how DNA was synthesized, Kornberg sought to demonstrate the accuracy, or fidelity, with which the enzyme replicated the DNA template. Because the technology to determine the nucleotide sequences of the template and the product was not yet available in 1957, he initially had to rely on several indirect methods.

One of Kornberg's approaches was to compare the nitrogenous base compositions of the DNA template with those of the recovered DNA product. Table 11.1 shows Kornberg's base composition analysis of three DNA templates. Within experi-

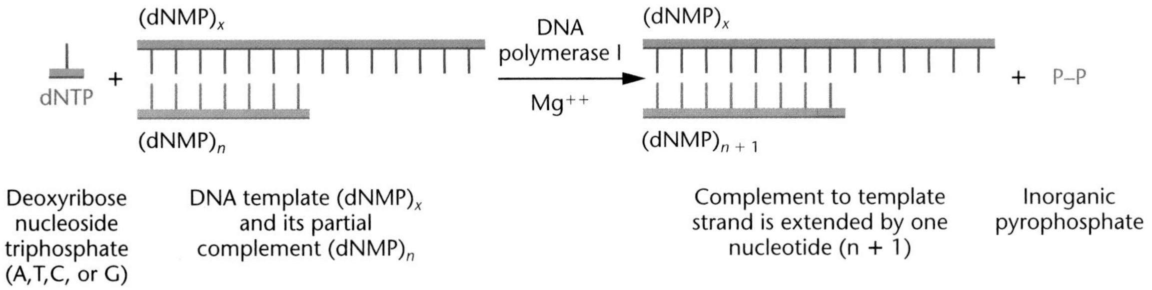

FIGURE 11–7 The chemical reaction catalyzed by DNA polymerase I. During each step, a single nucleotide is added to the growing complement of the DNA template, using a nucleoside triphosphate as the substrate. The release of inorganic pyrophosphate drives the reaction energetically.

FIGURE 11–8 Demonstration of 5′ to 3′ synthesis of DNA.

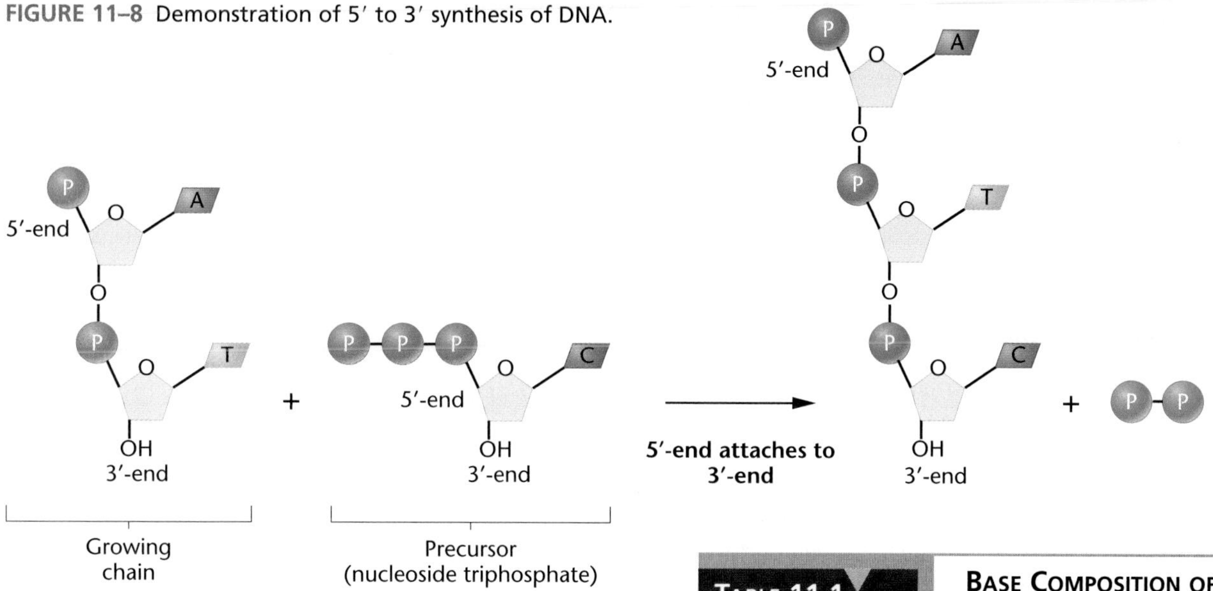

mental error, the base composition of each product agreed with the template DNAs used. These data, along with other types of comparisons of template and product, suggested that the templates were replicated faithfully.

Now solve this

Problem 11.30 on page 284 describes a procedure called nearest-neighbor analysis, an experimental approach which Kornberg used to test the fidelity of copying by DNA polymerase I.

Hint: As you analyze this procedure, remember that during synthesis of DNA, DNA polymerase I adds 5′-nucleotides to the 3′-OH group of the existing polynucleotide chain—that is, each nucleotide is added with the phosphate on the C-5′ of deoxyribose. However, the phosphodiesterase enzyme cleaves between the phosphate and the C-5′ atom, thereby producing 3′-nucleotides. As a result, the phosphate group is transferred to its "nearest neighbor," which is the key to your analysis.

Synthesis of Biologically Active DNA

Despite Kornberg's extensive work, not all researchers were convinced that DNA polymerase I was the enzyme that replicates DNA within cells (*in vivo*). Their reservations were that *in vitro* synthesis was much slower than the *in vivo* rate, that the enzyme was much more effective replicating single-stranded DNA than double-stranded DNA, and that the enzyme appeared to be able to *degrade* DNA as well as to *synthesize* it—that is, the enzyme exhibits exonuclease activity.

Uncertain of the true cellular function of DNA polymerase I, Kornberg pursued another approach. He reasoned that if the enzyme could be used to synthesize **biologically active DNA** *in vitro*, then DNA polymerase I must be the major catalyzing force for DNA synthesis within the cell. The term *biological activity* means that the DNA synthesized is capable of supporting metabolic activities and directs the reproduction of the organism from which it was originally duplicated.

TABLE 11.1 BASE COMPOSITION OF THE DNA TEMPLATE AND THE PRODUCT OF REPLICATION IN KORNBERG'S EARLY WORK

Organism	Template or Product	%A	%T	%G	%C
T2	Template	32.7	33.0	16.8	17.5
	Product	33.2	32.1	17.2	17.5
E. coli	Template	25.0	24.3	24.5	26.2
	Product	26.1	25.1	24.3	24.5
Calf	Template	28.9	26.7	22.8	21.6
	Product	28.7	27.7	21.8	21.8

Source: Kornberg (1960).

In 1967, Mehran Goulian, Robert Sinsheimer, and Kornberg experimented with the small bacteriophage ϕX174. This phage provides an ideal experimental system, because its genetic material is a very small (5386 nucleotides), circular single-stranded DNA molecule. In the normal course of ϕX174 infection, the circular single-stranded DNA, referred to as the (+) **strand**, enters the *E. coli* cell and serves as a template for the synthesis of the complementary (−) **strand**. The two strands (+ and −) remain together in a circular double helix, or duplex, called the **replicative form (RF)**. The RF serves as the template for its own replication, and subsequently, only (+) strands are produced. These strands are then packaged into viral coat proteins to form mature virus particles.

The experiment was carefully designed so that each newly synthesized strand could be distinguished and isolated from the template strand. The protocol of the experiment dictated that these (+) strands must have been synthesized *in vitro* under the direction of Kornberg's enzyme.

The critical test of biological activity was the process of transfection (Chapter 10), in which newly synthesized (+) strands were added to bacterial protoplasts (bacterial cells minus their cell wall). Following infection by the synthetic DNA, mature phages were produced—the synthetic DNA had successfully directed reproduction.

This demonstration of biological activity represented a precise assessment of faithful copying. If even a single error had occurred to alter the base sequence of any of the 5386 nucleotides constituting the ϕX174 chromosome, the change might easily have caused a mutation that would prohibit the production of viable phages.

DNA Polymerase II, III, IV, and V

Although DNA synthesized under the direction of polymerase I demonstrated biological activity, a more serious reservation about the enzyme's true biological role was raised in 1969. Paula DeLucia and John Cairns discovered a mutant strain of *E. coli* that was deficient in polymerase I activity. The mutation was designated *polA1*. In the absence of the functional enzyme, this mutant strain of *E. coli* still duplicated its DNA and successfully reproduced. However, the cells were highly deficient in their ability to repair DNA. For example, the mutant strain is highly sensitive to ultraviolet light (UV) and radiation, both of which damage DNA and are mutagenic. Nonmutant bacteria are able to repair a great deal of UV-induced damage.

These observations led to two conclusions:

1. At least one other enzyme that is responsible for replicating DNA *in vivo* is present in *E. coli* cells.

2. DNA polymerase I may serve a secondary function *in vivo*. This function is now believed by Kornberg and others to be critical to the *fidelity* of DNA synthesis, but the enzyme does not actually synthesize the entire complementary strand during replication.

To date, four other unique DNA polymerases have been isolated from cells lacking polymerase I activity and from normal cells that contain polymerase I. Table 11.2 contrasts several characteristics of DNA polymerase I with **DNA polymerase II** and **III.** While none of the three can *initiate* DNA synthesis on a template, all three can *elongate* an existing DNA strand, called a **primer**.

The DNA polymerase enzymes are all large proteins exhibiting a molecular weight in excess of 100,000 Daltons (Da). All three possess 3′ to 5′ exonuclease activity, which means that they have the potential to polymerize in one direction and then pause, reverse their direction, and excise nucleotides just added. As we will discuss later in the chapter, this activity provides a capacity to proofread newly synthesized DNA and to remove and replace incorrect nucleotides.

TABLE 11.2 PROPERTIES OF BACTERIAL DNA POLYMERASES I, II, AND III

Properties	I	II	III
Initiation of chain synthesis	−	−	−
5′-3′ polymerization	+	+	+
3′-5′ exonuclease activity	+	+	+
5′-3′ exonuclease activity	+	−	−
Molecules of polymerase/cell	400	?	15

DNA polymerase I also demonstrates 5′ to 3′ exonuclease activity. This activity allows the enzyme to excise nucleotides, starting at the end at which synthesis begins and proceeding in the same direction of synthesis. Two final observations probably explain why Kornberg isolated polymerase I and not polymerase III: polymerase I is present in greater amounts than is polymerase III, and it is also much more stable.

What then are the roles of the polymerases *in vivo*? Polymerase III is the enzyme responsible for the 5′ to 3′ polymerization essential to *in vivo* replication. Its 3′ to 5′ exonuclease activity also provides a proofreading function that is activated when it inserts an incorrect nucleotide. When this occurs, synthesis stalls and the polymerase "reverses course," excising the incorrect nucleotide. Then, it proceeds back in the 5′ to 3′ direction, synthesizing the complement of the template strand. Polymerase I is believed to be responsible for removing the primer, as well as for the synthesis that fills gaps produced during synthesis. Its exonuclease activities also allow for its participation in DNA repair. Polymerase II, as well as polymerase IV and V, are involved in various aspects of repair of DNA that has been damaged by external forces, such as ultraviolet light. Polymerase II is encoded by a gene activated by disruption of DNA synthesis at the replication fork.

We end this section by emphasizing the complexity of the DNA polymerase III molecule. Its active form, called a **holoenzyme**, consists of a dimer containing 10 different polypeptide subunits (Table 11.3) and has a molecular weight of 900,000 Da. The largest subunit, α, has a molecular weight of 140,000 Da and, along with subunits ϵ and θ, constitutes the core enzyme responsible for the polymerization activity. The α subunit is responsible for nucleotide polymerization on the template strands, whereas the ϵ subunit of the core enzyme possesses the 3′ to 5′ exonuclease activity.

A second group of five subunits (γ, δ, δ', χ, and ψ) forms what is called the γ complex, which is involved in "loading" the enzyme onto the template at the replication fork. This enzymatic function requires energy and is dependent on the hydrolysis of ATP. The β subunit serves as a "clamp" and prevents the core enzyme from falling off the template during polymerization. Finally, the τ subunit functions to dimerize two core polymerases facili-

TABLE 11.3 SUBUNITS OF THE DNA POLYMERASE III HOLOENZYME

Subunit	Function	Groupings
α	5′–3′ polymerization	Core enzyme: Elongates polynucleotide chain and proofreads
ϵ	3′–5′ exonuclease	
θ	core assembly	
γ, δ, δ', χ, ψ	Loads enzyme on template (Serves as clamp loader)	γ complex
β	Sliding clamp structure (processivity factor)	
τ	Dimerizes core complex	

tating simultaneous synthesis of both strands of the helix at the replication fork. The holoenzyme and several other proteins at the replication fork together form a huge complex (nearly as large as a ribosome) known as the **replisome**. We consider the function of DNA polymerase III in more detail later in this chapter.

How Do We Know?

What discoveries serve as the basis for concluding that DNA polymerase I (Kornberg's enzyme) is not the universal *in vivo* replicating enzyme in bacteria such as *E. coli*?

11.3 Many Complex Issues Must Be Resolved during DNA Replication

We have thus far established that in bacteria and viruses replication is semiconservative and bidirectional along a single replicon. We also know that synthesis is catalyzed by DNA polymerase III and occurs in the 5′ to 3′ direction. Bidirectional synthesis creates two replication forks that move in opposite directions away from the origin of synthesis. As we can see in the following points, many issues must still be resolved in order to provide a comprehensive understanding of DNA replication:

1. A mechanism must exist by which the helix undergoes localized unwinding and is stabilized in this "open" configuration so that synthesis may proceed along both strands.
2. As unwinding and subsequent DNA synthesis proceed, increased coiling creates tension further down the helix, which must be reduced.
3. A primer of some sort must be synthesized so that polymerization can commence under the direction of DNA polymerase III. Surprisingly, RNA, not DNA, serves as the primer.
4. Once the RNA primers have been synthesized, DNA polymerase III begins to synthesize the DNA complement of both strands of the parent molecule. Because the two strands are antiparallel to one another, continuous synthesis in the direction that the replication fork moves is possible along only one of the two strands. On the other strand, synthesis is discontinuous in the opposite direction.
5. The RNA primers must be removed prior to completion of replication. The gaps that are temporarily created must be filled with DNA complementary to the template at each location.
6. The newly synthesized DNA strand that fills each temporary gap must be joined to the adjacent strand of DNA.
7. While DNA polymerases accurately insert complementary bases during replication, they are not perfect and, occasionally, incorrect bases are added to the growing strand. A proofreading mechanism that also corrects errors is an integral process during DNA synthesis.

As we consider these points, examine Figures 11–9, 11–10, 11–11, and 11–12 to see how each issue is resolved. Figure 11–13 summarizes the model of DNA synthesis.

11.4 The DNA Helix Must Be Unwound

As discussed earlier, there is a single point of origin along the circular chromosome of most bacteria and viruses at which DNA synthesis is initiated. This region of the *E. coli* chromosome has been particularly well studied. Called ***oriC***, it consists of 245 base pairs characterized by repeating sequences of 9 and 13 bases (called **9mers** and **13mers**). As shown in Figure 11–9, one particular protein, called **DnaA** (because it is encoded by the gene called *dnaA*), is responsible for the initial step in unwinding the helix. A number of subunits of the DnaA protein bind to each of several 9mers. This step facilitates the

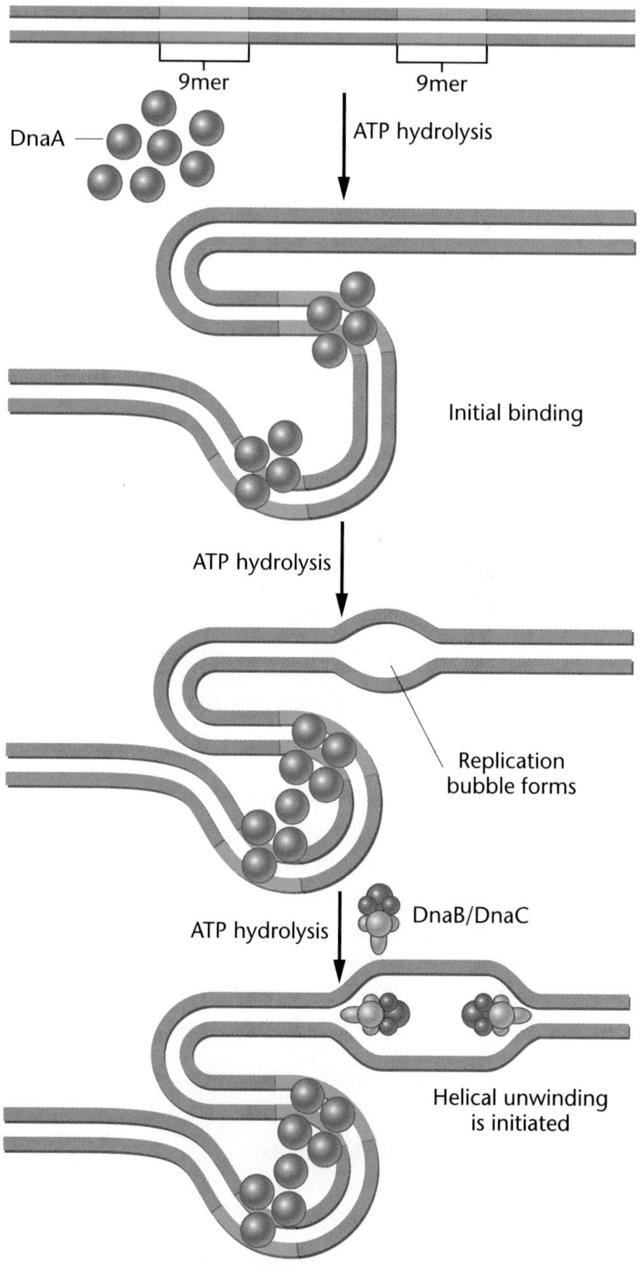

FIGURE 11–9 Helical unwinding of DNA during replication as accomplished by DnaA, DnaB, and DnaC proteins. Initial binding of many monomers of DnaA occurs at DNA sites containing repeating sequences of 9 nucleotides, called 9mers. Not illustrated are 13mers, which are also involved.

subsequent binding of **DnaB** and **DnaC** proteins that further open and destabilize the helix. Proteins such as these, which require the energy normally supplied by the hydrolysis of ATP in order to break hydrogen bonds and denature the double helix, are called **helicases**. Other proteins, called **single-stranded binding proteins (SSBPs)**, stabilize this open conformation.

As unwinding proceeds, a coiling tension is created ahead of the replication fork, often producing **supercoiling**. In circular molecules, supercoiling may take the form of added twists and turns of the DNA, much like the coiling you can create in a rubber band by stretching it out and then twisting one end. Such supercoiling can be relaxed by **DNA gyrase**, a member of a larger group of enzymes referred to as **DNA topoisomerases**. The gyrase makes either single- or double-stranded "cuts" and also catalyzes localized movements that have the effect of "undoing" the twists and knots created during supercoiling. The strands are then resealed. These various reactions are driven by the energy released during ATP hydrolysis.

Together, the DNA, the polymerase complex, and associated enzymes make up an array of molecules that participate in DNA synthesis and are part of what we have previously called the replisome.

11.5 Initiation of DNA Synthesis Requires an RNA Primer

Once a small portion of the helix is unwound, the initiation of synthesis may occur. As we have seen, DNA polymerase III requires a primer with a free 3′-hydroxyl group in order to elongate a polynucleotide chain. Since none is available in a circular chromosome, this prompted researchers to investigate how the first nucleotide could be added. It is now clear that RNA serves as the primer that initiates DNA synthesis.

A short segment of RNA (about 5 to 15 nucleotides long), complementary to DNA, is first synthesized on the DNA template. Synthesis of the RNA is directed by a form of RNA polymerase called **primase**, which does not require a free 3′ end to initiate synthesis. It is to this short segment of RNA that DNA polymerase III begins to add 5′-deoxyribonucleotides, initiating DNA synthesis. A conceptual diagram of initiation on a DNA template is shown in Figure 11–10. At a later point, the RNA primer must be clipped out and replaced with DNA. This is thought to occur under the direction of DNA polymerase I. Recognized in viruses, bacteria, and several eukaryotic organisms, RNA priming is a universal phenomenon during the initiation of DNA synthesis.

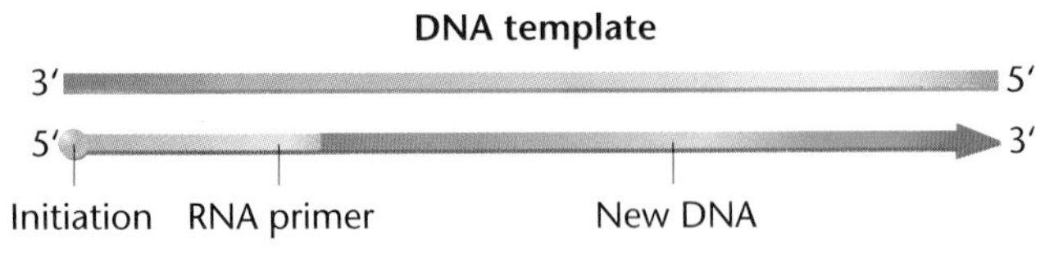

FIGURE 11–10 The initiation of DNA synthesis. A complementary RNA primer is first synthesized, to which DNA is added. All synthesis is in the 5′ to 3′ direction. Eventually, the RNA primer is replaced with DNA under the direction of DNA polymerase I.

11.6 Antiparallel Strands Require Continuous and Discontinuous DNA Synthesis

We must now reconsider the fact that the two strands of a double helix are **antiparallel** to each other—that is, one runs in the 5′–3′ direction, while the other has the opposite 3′–5′ polarity. Because DNA polymerase III synthesizes DNA in only the 5′–3′ direction, synthesis along an advancing replication fork occurs in one direction on one strand and in the opposite direction on the other.

As a result, as the strands unwind and the replication fork progresses down the helix (Figure 11–11), only one strand can serve as a template for **continuous DNA synthesis**. This newly synthesized DNA is called the **leading strand**. As the fork progresses, many points of initiation are necessary on the opposite DNA template, resulting in **discontinuous DNA synthesis** of the **lagging strand.***

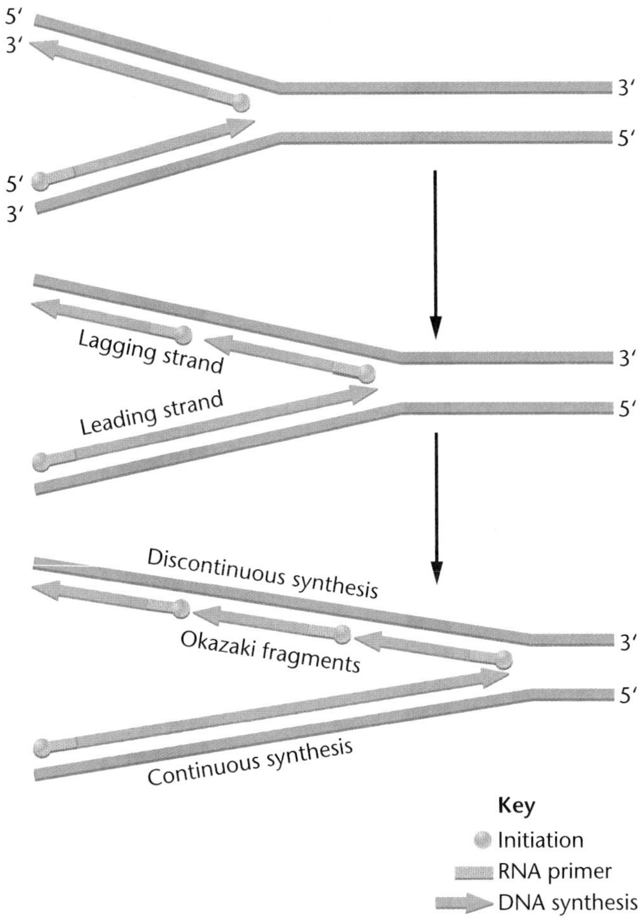

FIGURE 11–11 Opposite polarity of DNA synthesis along the two strands, necessary because the two strands of DNA run antiparallel to one another and DNA polymerase III synthesizes only in one direction (5′ to 3′). On the lagging strand, synthesis must be discontinuous, resulting in the production of Okazaki fragments. On the leading strand, synthesis is continuous. RNA primers are used to initiate synthesis on both strands.

*Because DNA synthesis is continuous on one strand and discontinuous on the other, the term **semidiscontinuous synthesis** is sometimes used to describe the overall process.

Evidence supporting the occurrence of discontinuous DNA synthesis was first provided by Reiji and Tuneko Okazaki. They discovered that when bacteriophage DNA is replicated in *E. coli*, some of the newly formed DNA that is hydrogen bonded to the template strand is present as small fragments containing 1000–2000 nucleotides. RNA primers are part of each such fragment. These pieces, now called **Okazaki fragments**, are converted into longer and longer DNA strands of higher molecular weight as synthesis proceeds.

Discontinuous synthesis of DNA requires enzymes that both remove the RNA primer and unite the Okazaki fragments into the lagging strand. As we have noted, DNA polymerase I removes the primer and replaces the missing nucleotides. Joining the fragments appears to be the work of **DNA ligase**, which is capable of catalyzing the formation of the phosphodiester bond that seals the nick between the discontinuously synthesized strands. The evidence that DNA ligase performs this function during DNA synthesis is strengthened by the observation of a ligase-deficient mutant strain (*lig*) of *E. coli*, in which a large number of unjoined Okazaki fragments accumulate.

How Do We Know?

How do we know that during replication, DNA synthesis is discontinuous on one of the two template strands?

11.7 Concurrent Synthesis Occurs on the Leading and Lagging Strands

Given the model just discussed, we might ask how DNA polymerase III synthesizes DNA on both the leading and lagging strands. Can both strands be replicated simultaneously at the same replication fork, or are the events distinct, involving two separate copies of the enzyme? Evidence suggests that both strands can be replicated simultaneously. As Figure 11–12 illustrates, if the lagging strand forms a loop, nucleotide polymerization can occur on both template strands under the direction of a dimer of the enzyme. After the synthesis of 100 to 200 base pairs, the monomer of the enzyme on the lagging strand will encounter a completed Okazaki fragment, at which point it releases the lagging strand. A new loop is then formed with the lagging template strand, and the process is repeated. Looping inverts the orientation of the template, but not the direction of actual synthesis on the lagging strand, which is always in the 5′ to 3′ direction.

Another important feature of the holoenzyme that facilitates synthesis at the replication fork is a dimer of the β subunit that forms a clamplike structure around the newly formed DNA duplex. This β-subunit clamp prevents the **core enzyme** (the α, ϵ, and θ subunits that are responsible for catalysis of nucleotide addition) from falling off the template as polymerization proceeds. Because the entire holoenzyme moves along the parent duplex, advancing the replication fork, the β-subunit dimer is often referred to as a sliding clamp.

Now solve this

In Problem 11.33 on page 284 a hypothetical organism is observed in which no Okasaki fragments are produced and a telomere problem exists. You are asked to suggest a DNA model consistent with these two features.

Hint: This observation suggests that DNA synthesis is continuous on both strands.

11.8 Proofreading and Error Correction Are an Integral Part of DNA Replication

The underpinning of DNA replication is the synthesis of a new strand that is precisely complementary to the template strand at each nucleotide position. Although the action of DNA polymerases is very accurate, synthesis is not perfect and a noncomplementary nucleotide is occasionally inserted erroneously. To compensate for such inaccuracies, the DNA polymerases all possess 3′ to 5′ exonuclease activity. This property imparts the potential for them to detect and excise a mismatched nucleotide (in the 3′–5′ direction). Once the mismatched nucleotide is removed, 5′ to 3′ synthesis can again proceed. This process, called **proofreading**, increases the fidelity of synthesis by a factor of about 100. In the case of the holoenzyme form of DNA polymerase III, the epsilon (ϵ) subunit is directly involved in the proofreading step. In strains of *E. coli* with a mutation that has rendered the ϵ subunit nonfunctional, the error rate (the mutation rate) during DNA synthesis is increased substantially.

11.9 A Coherent Model Summarizes DNA Replication

We can now combine the various aspects of DNA replication occurring at a single replication fork into a coherent model, as shown in Figure 11–13. At the advancing fork, a helicase is

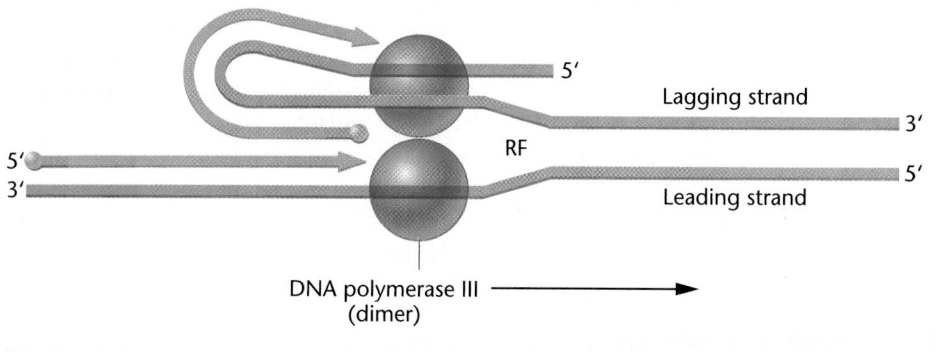

FIGURE 11–12 Illustration of how concurrent DNA synthesis may be achieved on both the leading and lagging strands at a single replication fork. The lagging template strand is "looped" in order to invert the physical direction of synthesis, but not the biochemical direction. The enzyme functions as a dimer, with each core enzyme achieving synthesis on one or the other strand.

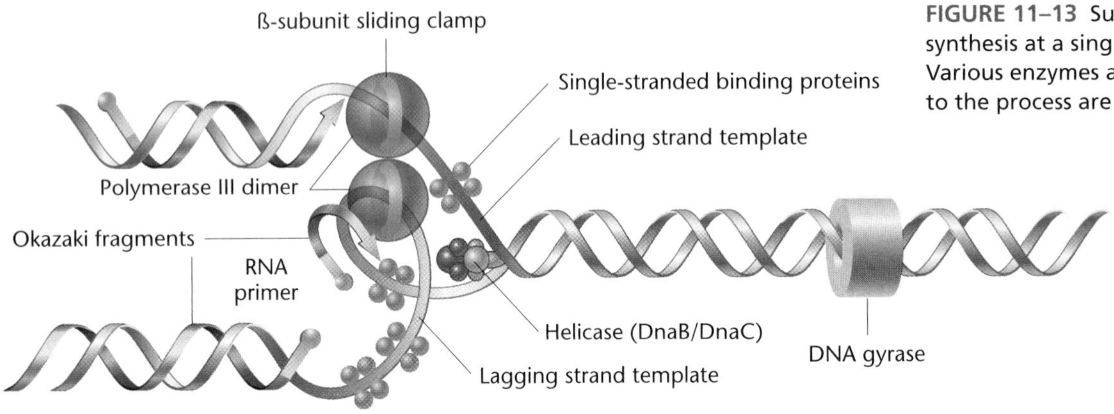

FIGURE 11–13 Summary of DNA synthesis at a single replication fork. Various enzymes and proteins essential to the process are shown.

unwinding the double helix. Once unwound, single-stranded binding proteins associate with the strands, preventing the reformation of the helix. In advance of the replication fork, DNA gyrase functions to diminish the tension created as the helix supercoils. Each half of the dimeric polymerase is a core enzyme bound to one of the template strands by a β-subunit sliding clamp. Continuous synthesis occurs on the leading strand, while the lagging strand must loop around in order for simultaneous synthesis to occur on both strands. Not shown in the figure, but essential to replication on the lagging strand, is the action of DNA polymerase I and DNA ligase, which together replace the RNA primer with DNA and join the Okazaki fragments, respectively.

Because the investigation of DNA synthesis is still an extremely active area of research, this model will no doubt be extended in the future. In the meantime, it provides a summary of DNA synthesis against which genetic phenomena can be interpreted.

11.10 Replication Is Controlled by a Variety of Genes

Much of what we know about DNA replication in viruses and bacteria is based on genetic analysis of the process. For example, we have already discussed studies involving the *polA1* mutation, which revealed that DNA polymerase I is not the major enzyme responsible for replication. Many other mutations interrupt or seriously impair some aspect of replication, such as the ligase-deficient and the proofreading-deficient mutations mentioned previously. Because such mutations are lethal ones, genetic analysis frequently uses **conditional mutations**, which are expressed under one condition, but not under a different condition. For example, a **temperature-sensitive mutation** may not be expressed at a particular *permissive* temperature. When mutant cells are grown at a *restrictive* temperature, the mutant phenotype is expressed, and can be studied. By examining the effect of the loss of function associated with the mutation, the investigation of such temperature-sensitive mutants can provide insight into the product and the associated function of the normal, nonmutant gene.

As shown in Table 11.4, a variety of genes in *E. coli* specify the subunits of the DNA polymerases and encode products involved in specification of the origin of synthesis, helix unwinding and stabilization, initiation and priming, relaxation of supercoiling, repair, and ligation. The discovery of such a large group of genes attests to the complexity of the process of replication, even in the relatively simple prokaryote. Given the enormous quantity of DNA that must be unerringly replicated in a very brief time, this level of complexity is not unexpected. As we will see in the next section, the process is even more involved and therefore more difficult to investigate in eukaryotes.

TABLE 11.4 SOME OF THE VARIOUS *E. COLI* MUTANT GENES AND THEIR PRODUCTS OR ROLE IN REPLICATION

Mutant Gene	Enzyme or Role
polA	DNA polymerase I
polB	DNA polymerase II
dnaE, N, Q, X, Z	DNA polymerase III subunits
dnaG	Primase
dnaA, I, P	Initiation
dnaB, C	Helicase at *oriC*
oriC	Origin of replication
gyrA, B	Gyrase subunits
lig	Ligase
rep	Helicase
ssb	Single-stranded binding proteins
rpoB	RNA polymerase subunit

Now solve this

Problem 11.25 on page 283 involves several temperature-sensitive mutations in *E. coli*, and you are asked to interpret the action of the gene that has mutated based on the phenotype that results.

Hint: Each mutation has disrupted one of the many steps essential to DNA synthesis. In each case, the mutant phenotype provides the clue as to which enzyme or function is affected.

11.11 Eukaryotic DNA Synthesis Is Similar to Synthesis in Prokaryotes, but More Complex

Research has shown that eukaryotic DNA is replicated in a manner similar to that of bacteria. In both systems, double-stranded DNA is unwound at replication origins, replication forks are formed, and bidirectional DNA synthesis creates leading and lagging strands from templates under the direction of DNA polymerase. Eukaryotic polymerases have the same fundamental requirements for DNA synthesis as do bacterial systems: four deoxyribonucleoside triphosphates, a template, and a primer. However, because eukaryotic cells contain much more DNA per cell, because eukaryotic chromosomes are linear rather than circular, and because this DNA is complexed with proteins, eukaryotes face many complexities not encountered by bacteria. Thus, the process of DNA synthesis is more complicated in eukaryotes and more difficult to study. However, a great deal is now known about the process.

Multiple Replication Origins

The most obvious difference between eukaryotic and prokaryotic DNA replication is that eukaryotic chromosomes contain multiple replication origins, in contrast to the single site that is part of the *E. coli* chromosome. Multiple origins, visible under the electron microscope as "replication bubbles" that form as the helix opens up (Figure 11–14), each provide two potential replication forks. Multiple origins are essential if replication of the entire genome of a typical eukaryote is to be completed in a reasonable time. Recall that (1) eukaryotes have much greater amounts of DNA than bacteria—e.g., yeast has four times as much DNA, and *Drosophila* has 100 times as much as *E. coli*—and (2) the rate of synthesis by eukaryotic DNA polymerase is much slower—only about 2000 nucleotides per minute, a rate 25 times less than the comparable bacterial enzyme. Under these conditions, replication from a single origin of a typical eukaryotic chromosome might take days to complete! However, replication of entire eukaryotic genomes is usually accomplished in a matter of hours.

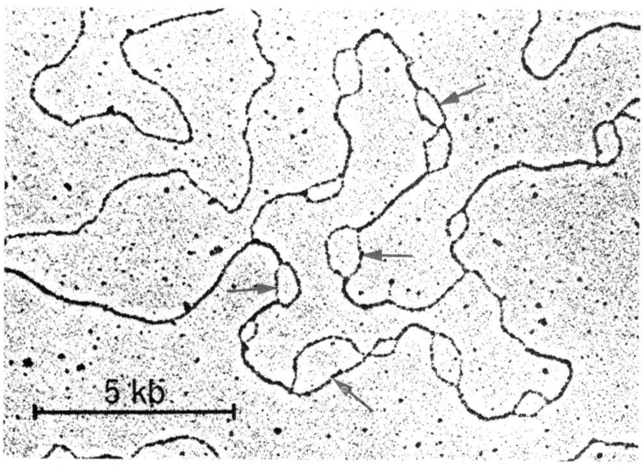

FIGURE 11–14 A demonstration of the multiple origins of replication along a eukaryotic chromosome. Each origin is apparent as a replication bubble along the axis of the chromosome. Arrows identify some of these replication bubbles.

Insights are now available concerning the molecular nature of the multiple origins and the initiation of DNA synthesis at these sites. Most information has originally been derived from the study of yeast (e.g., *Saccharomyces cerevisiae*), which has between 250–400 replicons per genome; subsequent studies have used mammalian cells, which have as many as 25,000 replicons. The origins in yeast have been isolated and are called **autonomously replicating sequences (ARSs)**. They consist of a unit containing a consensus sequence of 11 base pairs, flanked by other short sequences involved in efficient initiation. As we know from Chapter 2, DNA synthesis is restricted to the S phase of the eukaryotic cell cycle. Research has shown that the many origins are not all activated at once; instead, clusters of 20–80 adjacent replicons are activated sequentially throughout the S phase until all DNA is replicated.

How the polymerase finds the ARSs among so much DNA is an obvious recognition problem. The solution involves a mechanism that is initiated prior to the S phase. During the G1 phase of the cell cycle, all ARSs are initially bound by a group of specific proteins (six in yeast), forming what is called the **origin recognition complex (ORC)**. Mutations in either the ARSs or in any of the genes encoding these proteins of the ORC abolish or reduce initiation of DNA synthesis. Since these recognition complexes are formed in G1, but synthesis is not initiated at these sites until S, there must be still other proteins involved in the actual initiation signal. The most important of these proteins are specific kinases, key enzymes involved in phosphorylation, which are integral parts of cell cycle control. When bound along with ORC, a prereplication complex is formed that is accessible to DNA polymerase. After these kinases are activated, they serve to complete the initiation complex, directing localized unwinding and triggering DNA synthesis. Activation also inhibits reformation of the prereplication complexes once DNA synthesis has been completed at each replicon. This is an important mechanism, since it distinguishes segments of DNA that have completed replication from segments of unreplicated DNA, thus maintaining orderly and efficient replication. This ensures that replication only occurs once along each stretch of DNA during each cell cycle.

Eukaryotic DNA Polymerases

The most complex aspect of eukaryotic replication is the array of polymerases involved in directing DNA synthesis. As we will see, many different forms of the enzyme have been isolated and studied. However, only four are actually involved in replication of DNA, while the remainder are involved in repair processes. In order for the polymerases to have access to DNA, the topology of the helix must first be modified. As synthesis is triggered at each origin site, the double strands are opened up within an AT-rich region, allowing the entry of a helicase enzyme that proceeds to further unwind the double-stranded DNA. Before polymerases can begin synthesis, histone proteins complexed to the DNA (which form the characteristic nucleosomes of chromatin, as discussed in Chapter 12) also must be stripped away or otherwise modified. As DNA synthesis then proceeds, histones reassociate with the newly formed duplexes, reestablishing the characteristic nucleosome pattern (Figure 11–15). In eukaryotes, the synthesis of new histone proteins is tightly coupled to DNA synthesis during the S phase of the cell cycle.

The nomenclature and characteristics of six different polymerases are summarized in Table 11.5. Of these, three (Pol α, δ, and ϵ) are essential to nuclear DNA replication in eukaryotic cells. Two (Pol β and ζ) are thought to be involved in DNA repair (and still other repair forms have been discovered). The sixth form (Pol γ) is involved in the synthesis of mitochondrial DNA. Presumably, its replication function is limited to that organelle even though it is encoded by a nuclear gene. All but one of the six forms of the enzyme (the β form) consist of multiple subunits. Different subunits perform different functions during replication.

Pol α and δ are the major forms of the enzyme involved in initiation and elongation during nuclear DNA synthesis, so we concentrate our discussion on these. Two of the four subunits of the Pol α enzyme function in the synthesis of the RNA primers during the initiation of synthesis of both the leading and lagging strands. After the addition of about 10 ribonucleotides, another subunit functions to further elongate the RNA primer by adding 20–30 complementary deoxyribonucleotides. Pol α is said to possess low **processivity**, a term that essentially reflects the length of DNA that is synthesized by an enzyme before it dissociates from the template. Once the primer is in place, an event known as **polymerase switching** occurs, whereby Pol α dissociates from the template and is replaced by Pol δ. This form of the enzyme possesses high processivity and elongates the leading and lagging strands. It also possesses 3′ to 5′ exonuclease activity, which provides it with the potential to proofread. Pol ϵ, the third essential form, possesses the same general characteristics as Pol δ, but it is believed to operate under different cellular conditions, or to be restricted to synthesis of the lagging strand. In yeast, mutations that render Pol ϵ inactive are lethal, attesting to its essential function during replication.

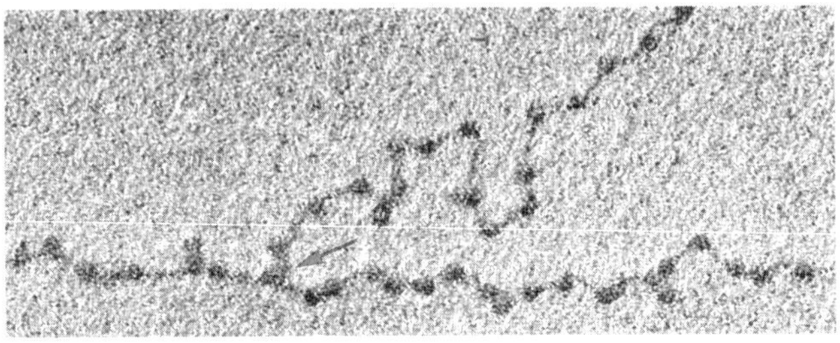

FIGURE 11–15 An electron micrograph of a eukaryotic-replicating fork demonstrating the presence of histone-protein-containing nucleosomes on both branches.

The process described applies to both leading and lagging strand synthesis. On both strands, RNA primers must be replaced with DNA. On the lagging strand, the Okazaki fragments, which are about 10 times smaller (100–150 nucleotides) in eukaryotes than in prokaryotes, must be ligated.

To accommodate the increased number of replicons, eukaryotic cells contain many more DNA polymerase molecules than do bacteria. While *E. coli* has about 15 copies of DNA polymerase III per cell, there may be up to 50,000 copies of the α form of DNA polymerase in animal cells. As has been pointed out, the presence of greater numbers of smaller replicons in eukaryotes compared with bacteria compensates for the slower rate of DNA synthesis in eukaryotes. *E. coli* requires 20 to 40 minutes to replicate its chromosome, while *Drosophila*, with 40 times more DNA, is known during embryonic cell divisions to accomplish the same task in only 3 minutes.

11.12 The Ends of Linear Chromosomes Are Problematic during Replication

A final difference that exists between prokaryotic and eukaryotic DNA synthesis involves the nature of the chromosomes. Unlike the closed, circular DNA of bacteria and most bacteriophages, eukaryotic chromosomes are linear. During replication, they face a special problem at the "ends" of these linear molecules, called the **telomeres**.

Before addressing this problem, we need to establish several things about telomeres. These structures consist of long stretches of a short repeating sequence of DNA bound to specific **telomere-associated proteins**. The unique quality of telomeres serves to preserve the integrity and stability of chro-

TABLE 11.5 PROPERTIES OF EUKARYOTIC DNA POLYMERASES

	Polymerase α	Polymerase β	Polymerase δ	Polymerase ϵ	Polymerase γ	Polymerase ζ
Location	**Nucleus**	**Nucleus**	**Nucleus**	**Nucleus**	**Mitochondria**	**Nucleus**
3′–5′ Exonuclease Activity	No	No	Yes	Yes	Yes	No
Essential to Nuclear Replication?	Yes	No	Yes	Yes	No	No

mosomes. Telomeres are essential because the double-stranded "ends" of DNA molecules at the termini of chromosomes potentially resemble what are called **double-stranded breaks (DSBs)** that can occur if the chromosome should be fragmented internally. In such case, the resulting double-stranded ends can fuse to other such ends and, if not, are vulnerable to degradation. Thus, telomeres are believed to protect the ends of intact eukaryotic chromosomes from such a fate.

Now, consider semiconservative replication near the end of a double-stranded DNA molecule. While synthesis of the leading strand proceeds normally to the end, a difficulty arises on the lagging strand once the RNA primer is removed (Figure 11–16). Normally, the newly created gap would be filled by addition of a nucleotide to the existing 3′-OH group provided during discontinuous synthesis (to the right of gap (b) in Figure 11–16). However, since it is the end of the DNA molecule, there is no strand present to provide the 3′-OH group. As a result, a gap remains on the lagging strand during each successive round of synthesis that will theoretically shorten the end of the chromosome by the length of the RNA primer. Because this is potentially such a significant problem, we can suppose that a molecular solution would have been developed early in evolution and subsequently be conserved and shared by almost all eukaryotes, which is indeed the case.

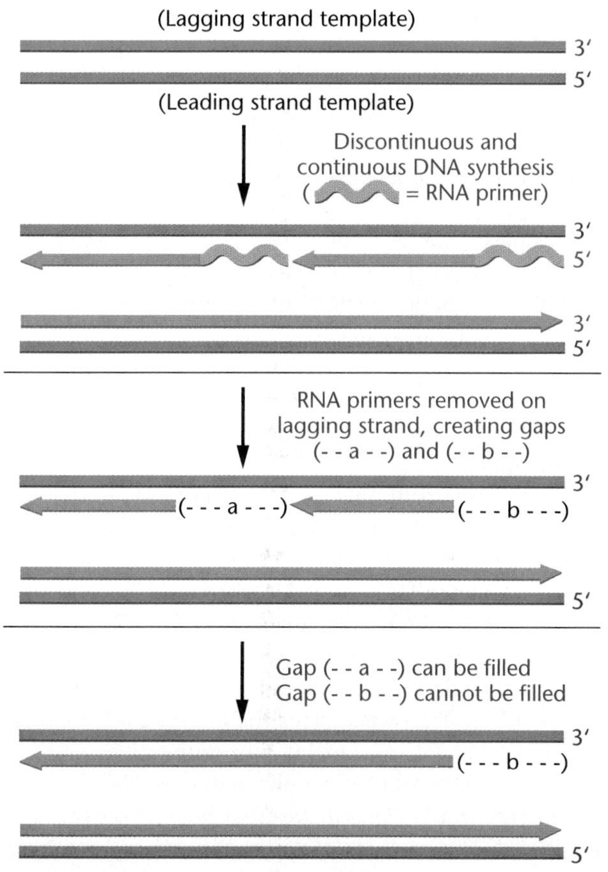

FIGURE 11–16 Diagram illustrating the difficulty encountered during the replication of the ends of linear chromosomes. A gap (- -b- -) is left following synthesis on the lagging strand.

A unique eukaryotic enzyme called **telomerase**, first discovered in the ciliated protozoan *Tetrahymena*, has helped us understand the solution. Before we look at how the enzyme actually works, let's look at what is accomplished. As pointed out above, telomeric DNA in eukaryotes is always found to consist of many short, repeated nucleotide sequences. For example, the template producing the lagging strand at the end of the *Tetrahymena* chromosome contains many repeats of the sequence 5′ TTGGGG-3′.

As illustrated in Figure 11–17, telomerase is capable of adding several repeats of this six-nucleotide sequence to the the 3′ end of the lagging strand (using 5′–3′ synthesis), thus preventing chromosome shortening. These repeats fold back on themselves forming a "hairpin loop" that is stabilized by unorthodox hydrogen bonding between opposite guanine residues (G═G), creating a free 3′-OH group on the end of the hairpin that can serve as a substrate for DNA polymerase I. This makes it possible to subsequently fill the gap that would otherwise shorten

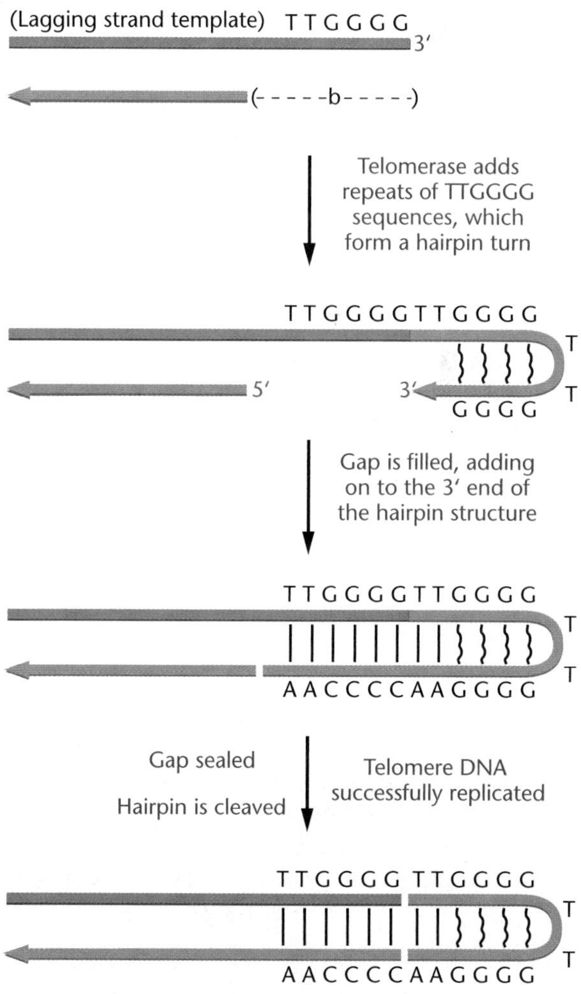

FIGURE 11–17 The predicted solution to the problem posed in Figure 11–16. The enzyme telomerase directs synthesis of the TTGGGG sequences, resulting in the formation of a hairpin structure. The gap can now be filled, and, following cleavage of the hairpin structure, the process averts the creation of a gap during replication of the ends of linear chromosomes.

the chromosome. The hairpin loop can then be cleaved off at its terminus, and the potential loss of DNA in each subsequent replication cycle is averted.

Detailed investigation by Elizabeth Blackburn and Carol Greider of how the *Tetrahymena* telomerase enzyme accomplishes the synthesis yielded an extraordinary finding. The enzyme is highly unusual in that it is a *ribonucleoprotein*, containing within its molecular structure a short piece of RNA that is essential to its catalytic activity. The RNA component serves as both a "guide" and a template for the synthesis of its DNA complement, a process referred to as *reverse transcription*. In *Tetrahymena*, the RNA contains several repeats of the sequence 5′-CCCCAA-3′, the complement of the repeating telomeric DNA sequence that must be synthesized (5′-TTGGGG-3′).

We can envision that part of the RNA sequence of the enzyme base pairs with the telomeric DNA, with the remainder of the RNA overlapping the end of the lagging strand. Reverse transcription of this overlapping sequence, which synthesizes DNA on an RNA template, will lead to the extension of the lagging strand, as depicted in Figure 11–17. It is believed that the enzyme is then translocated toward the end of the lagging strand and the sequence of events is repeated. In yeast, of which such synthesis has been investigated in detail, specific proteins are responsible for terminating synthesis, thus determining the number of repeating units added to the end of the lagging strand. Similar proteins no doubt function in the same way in other eukaryotes.

Analogous enzyme functions have now been found in almost all eukaryotes studied. In humans, the telomeric DNA sequence that creates the lagging strand and is repeated is 5′-TTAGGG-3′, differing from *Tetrahymena* by only one nucleotide.

As we shall see in Chapter 12, telomeric DNA sequences have been highly conserved throughout evolution, reflecting the critical function of telomeres. In the essay at the end of this chapter, we will see that telomere shortening has been linked to a molecular mechanism involved in cellular aging. In fact, in most eukaryotic somatic cells, telomerase is not active and thus, with each cell division, the telomeres of each chromosome shorten. After many divisions, the telomere is seriously eroded and the cell loses the capacity for further division. Malignant cells, on the other hand, maintain telomerase activity and are immortalized.

HOW DO WE KNOW?

How do we know experimentally that a telomere problem does, in fact, exist in eukaryotes during DNA replication?

11.13 DNA Recombination, Like DNA Replication, Is Directed by Specific Enzymes

We now turn to a topic previously discussed in Chapter 5—genetic recombination. There, we pointed out that the process of crossing over between homologs depends on breakage and rejoining of the DNA strands. Now that we have discussed the chemistry and replication of DNA, we can consider how recombination occurs at the molecular level. In general, the information that follows pertains to genetic exchange between any two homologous double-stranded DNA molecules, whether they be viral or bacterial chromosomes or eukaryotic homologs during meiosis. Genetic exchange at equivalent positions along two chromosomes with substantial DNA sequence homology is referred to as **general** or **homologous recombination**.

Several models attempt to explain homologous recombination, and they all share certain common features. First, all are based on proposals put forth independently by Robin Holliday and Harold L. K. Whitehouse in 1964. They also depend on the complementarity between DNA strands for the precision of exchange. Finally, each model relies on a series of enzymatic processes in order to accomplish genetic recombination.

One such model is shown in Figure 11–18. It begins with two paired DNA duplexes or homologs [Step (a)] in each of which, an endonuclease introduces a single-stranded nick at an identical position [Step (b)]. The ends of the strands produced by these cuts are then displaced and subsequently pair with their complements on the opposite duplex [Step (c)]. A ligase then seals the loose ends [Step (d)], creating hybrid duplexes called **heteroduplex DNA molecules**. The exchange creates a cross-bridged structure. The position of this cross bridge can then move down the chromosome by a process referred to as branch migration [Step (e)], which occurs as a result of a zipperlike action as hydrogen bonds are broken and then reformed between complementary bases of the displaced strands of each duplex. This migration yields an increased length of heteroduplex DNA on both homologs.

If the duplexes now separate [Step (f)] and the bottom portions rotate 180° [Step (g)], an intermediate planar structure called a *X* form—the characteristic **Holliday structure**—is created. If the two strands on opposite homologs previously uninvolved in the exchange are now nicked by an endonuclease [Step (h)] and ligation occurs [Step (i)] recombinant duplexes are created. Note that the arrangement of alleles is altered as a result of recombination.

Evidence supporting this model includes the electron microscopic visualization of *X*-form planar molecules from bacteria in which four duplex arms are joined at a single point of exchange [Figure 11–18(Step g)]. Further important evidence comes from the discovery of the **RecA protein** in *E. coli*. The molecule promotes the exchange of reciprocal single-stranded DNA molecules as occurs in Step (c) of the model. RecA also enhances the hydrogen-bond formation during strand displacement, thus initiating heteroduplex formation. Finally, many other enzymes essential to the nicking and ligation process have also been discovered and investigated. The products of the *recB*, *recC*, and *recD* genes are thought to be involved in the nicking and unwinding of DNA. Numerous mutations that prevent genetic recombination have been found in viruses and bacteria. These mutations represent genes whose products play an essential role in this process.

Web Tutorial 11.2
DNA Recombination

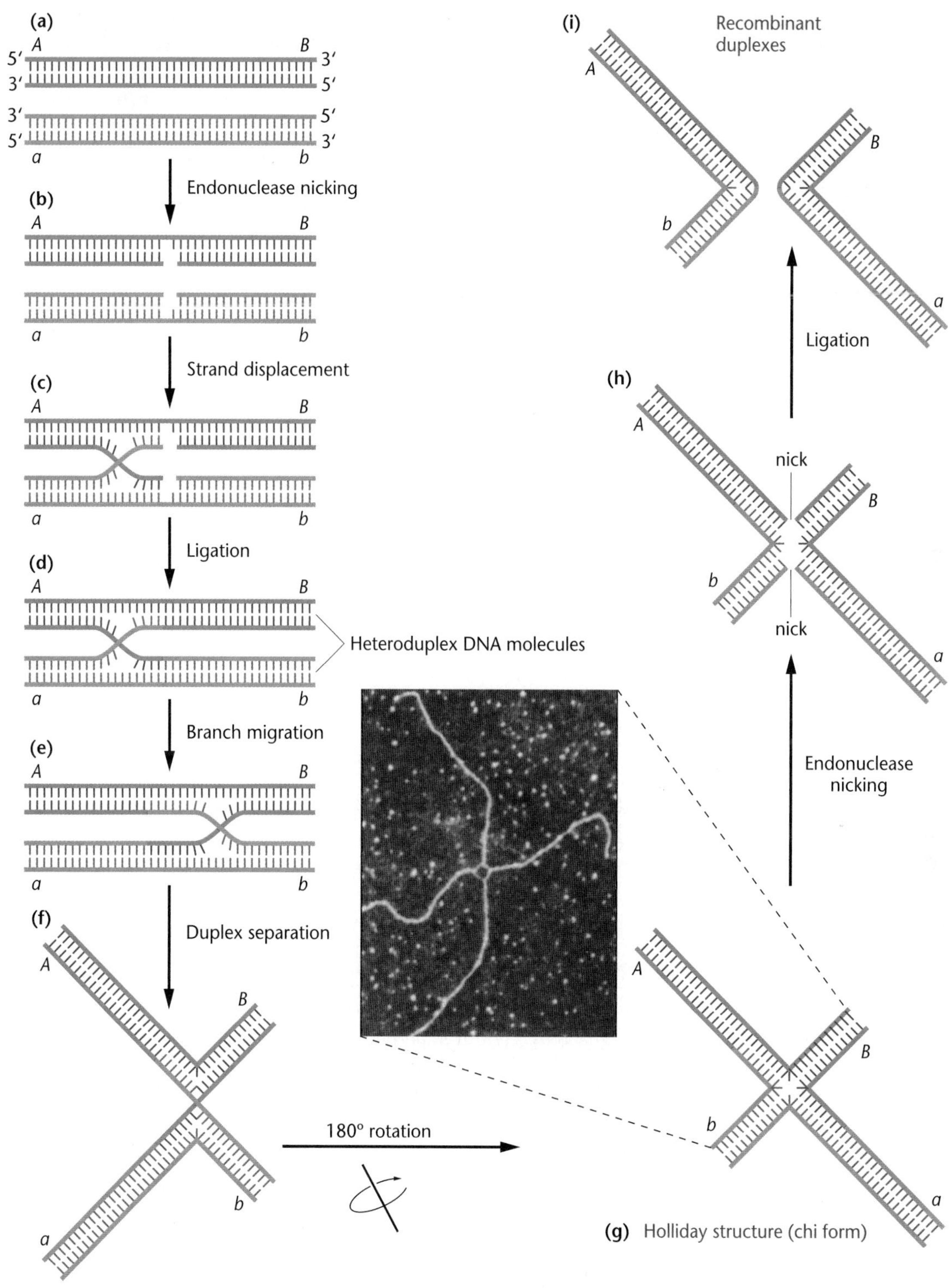

FIGURE 11–18 Model depicting how genetic recombination can occur as a result of the breakage and rejoining of heterologous DNA strands. Each stage is described in the text. The electron micrograph shows DNA in a χ-form structure similar to the diagram in (g); the DNA is an extended Holliday structure, derived from the *Col*E1 plasmid of *E. coli*. *David Dressler, Oxford University, England*

11.14 Gene Conversion Is a Consequence of DNA Recombination

A modification of the preceding model has helped us to better understand a unique genetic phenomenon known as **gene conversion**. Initially found in yeast by Carl Lindegren and in *Neurospora* by Mary Mitchell, gene conversion is characterized by a *nonreciprocal* genetic exchange between two closely linked genes. For example, if we were to cross two *Neurospora* strains, each bearing a separate mutation $(a+ \times +b)$, a *reciprocal* recombination between the genes would yield spore pairs of the $++$ and the *ab* genotypes. However, a nonreciprocal exhange yields one pair without the other. Working with pyridoxine mutants, Mitchell observed several asci-containing spore pairs with the ++ genotype, but not the reciprocal product (*ab*). Because the frequency of these events was higher than the predicted mutation rate and consequently could not be accounted for by mutation, they were called "gene conversions." They were so named because it appeared that one allele had somehow been "converted" to another in which genetic exchange had also occurred. Similar findings come from studies of other fungi as well.

Gene conversion is now considered to be a consequence of the process of DNA recombination. One possible explanation interprets conversion as a mismatch of base pairs during heteroduplex formation, as shown in Figure 11–19. Mismatched regions of hybrid strands can be repaired by the excision of one of the strands and the synthesis of the complement by using the remaining strand as a template. Excision may occur in either one of the strands, yielding two possible "corrections." One repairs the mismatched base pair and "converts" it to restore the original sequence. The other also corrects the mismatch, but does so by copying the altered strand, creating a base-pair substitution. Conversion may have the effect of creating identical alleles on the two homologs that were different initially.

In our example in Figure 11–19, suppose that the G≡C pair on one of the two homologs was responsible for the mutant allele, while the A=T pair was part of the wild-type gene sequence on the other homolog. Conversion of the G≡C pair to A=T would have the effect of changing the mutant allele to wild type, just as Mitchell originally observed.

Gene-conversion events have helped to explain other puzzling genetic phenomena in fungi. For example, when mutant and wild-type alleles of a single gene are crossed, asci should yield equal numbers of mutant and wild-type spores. However, exceptional asci with 3:1 or 1:3 ratios are sometimes observed. These ratios can be understood in terms of gene conversion. The phenomenon also has been detected during mitotic events in fungi, as well as during the study of unique compound chromosomes in *Drosophila*.

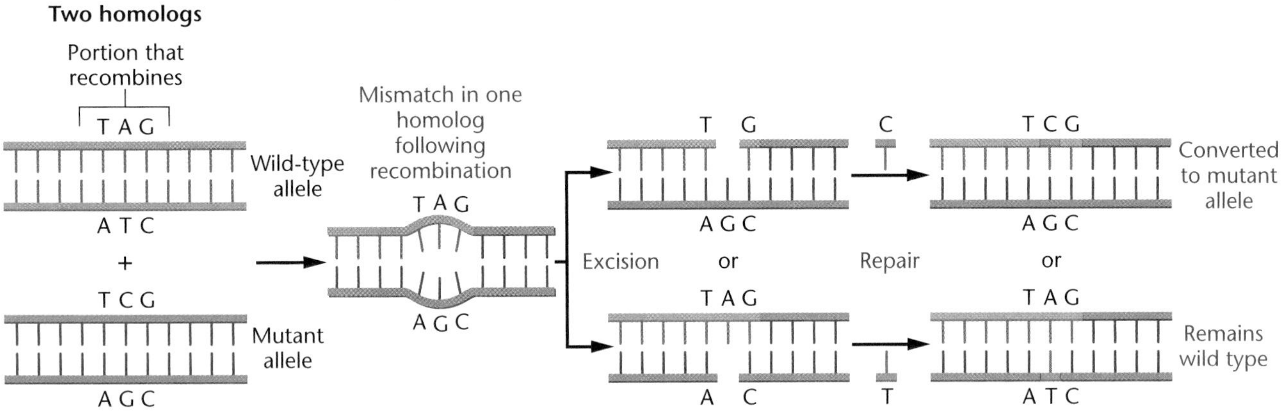

FIGURE 11–19 A proposed mechanism that accounts for the phenomenon of gene conversion. A base-pair mismatch occurs in one of the two homologs (bearing the mutant allele) during heteroduplex formation, which accompanies recombination in meiosis. During excision repair, one of the two mismatches is removed and the complement is synthesized. In one case, the mutant base pair is preserved. When it is subsequently included in a recombinant spore, the mutant genotype will be maintained. In the other case, the mutant base pair is converted to the wild-type sequence. When included in a recombinant spore, the wild-type genotype will be expressed, leading to a nonreciprocal exchange ratio.

GENETICS, TECHNOLOGY, AND SOCIETY

Telomerase: The Key to Immortality?

Humans, as do all multicellular organisms, grow old and die. As we age, our immune systems become less efficient, wound healing is impaired, and tissues and organs lose resilience. It has always been a mystery why we go through these age-related declines, and why each species has a characteristic finite life span. Why do we grow old? Can we reverse this march to mortality? Some recent discoveries suggest that the answers to these questions may lie at the ends of our chromosomes.

The study of human aging begins with a study of human cells growing in culture dishes. Like the organisms from which the cells are taken, cells in culture have a finite life span. This "replicative senescence" was noted more than 30 years ago by Hayflick. He reported that normal human fibroblasts lose their ability to grow and divide after about 50 cell divisions. These senescent cells remain metabolically active, but can no longer proliferate. Eventually, they die. Although we don't know whether cellular senescence directly causes organismal aging, the evidence is suggestive. For example, cells from young people go through more divisions in culture than do cells from older people; human fetal cells divide 60–80 times before undergoing senescence, whereas cells from older adults divide only 10–20 times. In addition, cells from species with short life spans stop growing after fewer divisions than do cells from species with longer life spans; mouse cells divide 11–15 times in culture, but tortoise cells undergo over 100 divisions. Moreover, cells from patients with genetic premature aging syndromes (such as Werner syndrome) undergo fewer divisions in culture than do cells from normal patients.

Another characteristic of aging cells is that their telomeres become shorter. *Telomeres* are the tips of linear chromosomes and consist of several thousand repeats of a short DNA sequence (TTAGGG in humans). Telomeres help preserve the structural integrity of chromosomes by protecting their ends from degrading or from fusing to other chromosomes. Telomeres are created and maintained by *telomerase*—a remarkable RNA-containing enzyme that adds telomeric DNA sequences onto the ends of linear chromosomes. Telomerase also solves the "end-replication" problem—which asserts that linear DNA molecules become shorter at each replication, because DNA polymerase cannot synthesize new DNA at the 3′ ends of each parent strand. By adding numerous telomeric repeat sequences onto the 3′ ends of chromosomes, telomerase prevents the chromosomes from shrinking into oblivion. Unfortunately, normal somatic cells contain little, if any, telomerase. As a result, telomere length decreases by about 100 base pairs every time a normal cell divides. Telomere shortening may act as a clock that counts cell division and instructs the cell to stop dividing.

Could we gain perpetual youth and vitality by increasing our telomere lengths? A recent study suggests that it may be possible to reverse senescence by artificially increasing the amount of telomerase in our cells. When the investigators introduced cloned telomerase genes into normal human cells in culture, telomeres lengthened by thousands of base pairs and the cells continued to grow long past their senescence point. These observations confirm that telomere length acts as a cellular clock. In addition, they suggest that some of the atrophy of tissues that accompanies old age may someday be reversed by activating telomerase genes. However, before we rush out to buy telomerase pills, we must consider a possible consequence of cellular immortality: cancer.

Although normal cells undergo senescence after a specific number of cell divisions, cancer cells do not. It is thought that cancers arise after several genetic mutations accumulate in a cell. These mutations disrupt the normal checks and balances that control cell growth and division. It therefore seems logical that cancer cells would also stop the normal aging clock. If their telomeres became shorter after each cell division, tumor cells would eventually succumb to aging and cease growth. However, if they synthesized telomerase, they would arrest the ticking of the senescence clock and become immortal. In keeping with this idea, more than 90 percent of human tumor cells contain telomerase activity and have stable telomeres. The correlation between uncontrolled tumor cell growth and the presence of telomerase activity is so good that telomerase assays are being developed as diagnostic markers for cancer. Although there is currently some debate about whether the presence of telomerase is a prerequisite for, or simply a consequence of, cell transformation, it is possible that acquiring telomerase activity may be an important step in the development of a cancer cell. In support of this hypothesis, recent studies show that the induction of telomerase activity, when combined with the inactivation of a tumor-suppressor gene ($p16^{INK4a}$), results in cell immortalization—an essential step toward tumor development. Therefore, any attempt to increase telomerase activity in normal cells carries the risk of enhancing the development of tumors.

An attractive possibility is that telomerase may be an ideal target for anticancer drugs. Drugs that inhibit telomerase might destroy cancer cells by allowing their telomeres to shorten, thereby forcing the cells into senescence. Because most normal human cells do not express telomerase, such a therapy might be specific for tumor cells and hence less toxic than most current anticancer drugs. Such antitelomerase anticancer drugs are currently under development by Geron Corporation. Although we don't yet know whether this approach will work in animals, it appears to work in cultured tumor cells. Tumor cells that are treated with an antitelomerase agent lose telomeric sequences and die after about 25 cell divisions.

Before antitelomerase drugs can be developed and used on humans, several questions must be answered. Is telomerase required by some normal human cells (such as lymphocytes and germ cells)? If so, antitelomerase drugs may be unacceptably toxic. Could some cancer cells compensate for the loss of telomerase by using other telomere-lengthening mechanisms (such as recombination)? If so, antitelomerase drugs may be doomed to failure. Even if we inhibit telomerase activity in tumor cells, could they undergo multiple divisions before reaching senescence and still damage the host?

Will telomerase allow us to both arrest cancers *and* reverse the descent into old age? Time will tell.

References

Bodnar, A.G., et al. 1998. Extension of life-span by introduction of telomerase into normal human cells. *Science* 279:349–52.

deLange, T. 1998. Telomeres and senescence: Ending the debate. *Science* 279:334–35.

Kiyono, T., et al. 1998. Both RB/$p16^{INK4a}$ inactivation and telomerase activity are required to immortalize human epithelial cells. *Nature* 396:84–88.

Chapter Summary

1. In theory, three modes of DNA replication are possible: semiconservative, conservative, and dispersive. Although all three rely on base complementarity, semiconservative replication is the most straightforward and was predicted by Watson and Crick.
2. In 1958, Meselson and Stahl resolved this question in favor of semiconservative replication in *E. coli*, showing that newly synthesized DNA consists of one old strand and one new strand. Taylor, Woods, and Hughes used root tips of the broad bean to demonstrate semiconservative replication in eukaryotes.
3. During the same period, Kornberg isolated the enzyme DNA polymerase I from *E. coli* and showed that it is capable of directing *in vitro* DNA synthesis, provided that a template and precursor nucleoside triphosphates are supplied.
4. The subsequent discovery of the *polA1* mutant strain of *E. coli*, capable of DNA replication despite its lack of polymerase I activity, cast doubt on the enzyme's *in vivo* replicative function. DNA polymerases II and III were then isolated. Polymerase III has been identified as the enzyme responsible for DNA replication *in vivo*.
5. During the process of DNA synthesis, the double helix unwinds, forming a replication fork at which synthesis begins. Proteins stabilize the unwound helix and assist in relaxing the coiling tension created ahead of the replication.
6. Synthesis is initiated at specific sites along each template strand by the enzyme primase, which results in a short segment of RNA that provides a suitable 3′ end upon which DNA polymerase III can begin polymerization.
7. Because of the antiparallel nature of the double helix, polymerase III synthesizes DNA continuously on the leading strand in a 5′–3′ direction. On the opposite strand, called the lagging strand, synthesis results in short Okazaki fragments that are later joined by DNA ligase.
8. DNA polymerase I removes and replaces the RNA primer with DNA, which is joined to the adjacent polynucleotide by DNA ligase.
9. The isolation of numerous phage and bacterial mutant genes affecting many of the molecules involved in the replication of DNA has helped to define the complex genetic control of the entire process.
10. DNA replication in eukaryotes is similar to, but more complex than, replication in prokaryotes. Multiple replication origins exist and multiple forms of DNA polymerase direct DNA synthesis.
11. Replication at the ends (telomeres) of linear molecules poses a special problem in eukaryotes that can be solved by a unique RNA-containing enzyme called telomerase.
12. Homologous recombination between genetic molecules relies on a series of enzymes that can cut, realign, and reseal DNA strands. The phenomenon of gene conversion may be best explained in terms of mismatch repair synthesis during these exchanges.

Insights and Solutions

1. Predict the theoretical results of conservative and dispersive replication of DNA under the conditions of the Meselson–Stahl experiment. Follow the results through two generations of replication after cells have been shifted to an ^{14}N-containing medium, using the following sedimentation pattern.

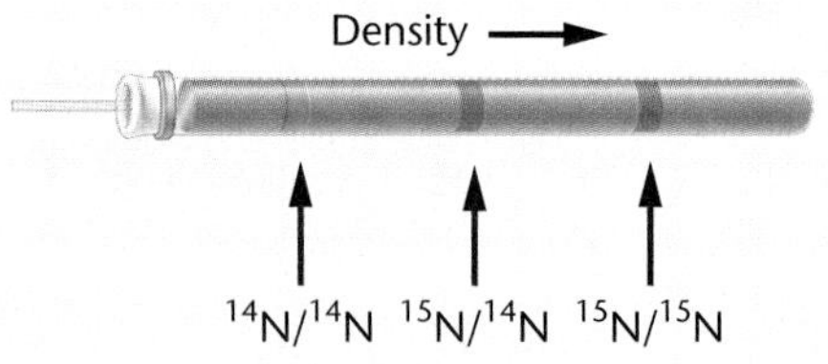

Solution:

Conservative replication

Generation I

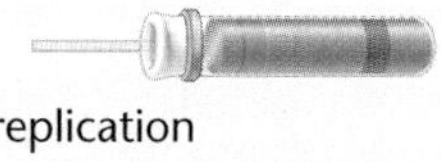

Generation II

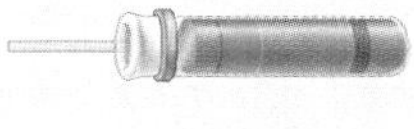

Dispersive replication

Generation I

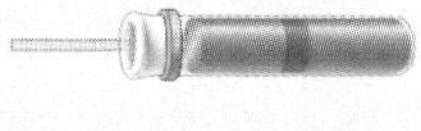

Generation II

2. Mutations in the *dnaA* gene of *E. coli* are lethal and can only be studied following the isolation of conditional, temperature-sensitive mutations. Such mutant strains grow nicely and replicate their DNA at the permissive temperature of 18°C, but they do not grow or replicate their DNA at the restrictive temperature of 37°C. Two observations were useful in determining the function of the DnaA protein product. First, *in vitro* studies using DNA templates that have unwound do not require the DnaA protein. Second, if intact cells are grown at 18°C and are then shifted to 37°C, DNA synthesis continues at this temperature until one round of replication is completed and then stops. What do these observations suggest about the role of the *dnaA* gene product?

Solution: At 18°C (the permissive temperature), the mutation is not expressed and DNA synthesis begins. Following the shift to the restrictive temperature, the already-initiated DNA synthesis continues, but no new synthesis can begin. Because the DnaA protein is not required for synthesis of unwound DNA, these observations suggest that, *in vivo*, the DnaA protein plays an essential role in DNA synthesis by interacting with the intact helix and somehow facilitating the localized denaturing necessary for synthesis to proceed.

3. DNA is allowed to replicate in moderately radioactive ^3H-thymidine for several minutes and is then switched to a highly radioactive medium for several more minutes. Synthesis is stopped and the DNA is subjected to autoradiography and electron microscopy. Interpret as much as you can regarding DNA replication from the drawing of the micrograph presented here.

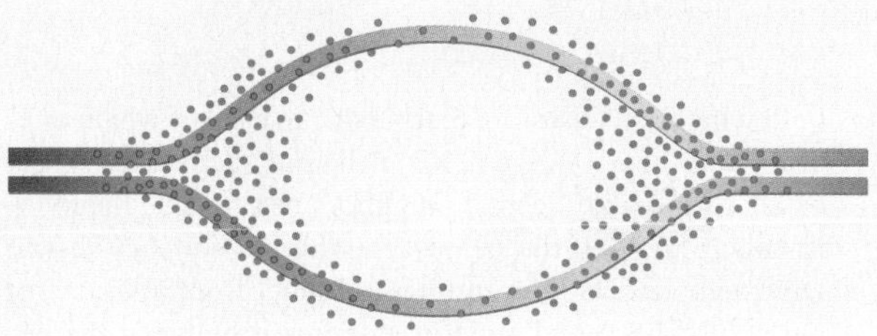

Solution: One interpretation is that there are two advancing replication forks proceeding in opposite directions and that replication is therefore bidirectional. The low density of grains represents synthesis that occurred during the first few minutes, beginning at the origin and proceeding outward in both directions. The higher density of grains represents the later minutes of synthesis when the level of radioactivity was increased.

PROBLEMS AND DISCUSSION QUESTIONS

1. Compare conservative, semiconservative, and dispersive modes of DNA replication.
2. Describe the role of ^{15}N in the Meselson–Stahl experiment.
3. In the Meselson–Stahl experiment, which of the three modes of replication could be ruled out after one round of replication? after two rounds?
4. Predict the results of the experiment by Taylor, Woods, and Hughes if replication were (a) conservative and (b) dispersive.
5. Reconsider Problem 35 in Chapter 10. In the model you proposed, could the molecule be replicated semiconservatively? Why? Would other modes of replication work?
6. What are the requirements for *in vitro* synthesis of DNA under the direction of DNA polymerase I?
7. In Kornberg's initial experiments, it was rumored that he grew *E. coli* in large Anheuser-Busch beer vats. (He was working at Washington University in St. Louis.) Why do you think this was helpful to his research effort involving polymerase I?
8. How did Kornberg test the fidelity of copying DNA by polymerase I?
9. Which of Kornberg's tests is the more stringent assay? Why?
10. Which characteristics of DNA polymerase I raised doubts that its *in vivo* function is the synthesis of DNA leading to complete replication?
11. Explain the theory of nearest neighbor frequency.
12. What is meant by "biologically active" DNA?
13. Why was the phage ϕX174 chosen for the experiment demonstrating biological activity?
14. Outline the experimental design of Kornberg's biological activity demonstration.
15. What was the significance of the *polA1* mutation?
16. Summarize and compare the properties of DNA polymerase I, II, and III.
17. List and describe the function of the 10 subunits constituting DNA polymerase III. Distinguish between the holoenzyme and the core enzyme.
18. Distinguish between (a) unidirectional and bidirectional synthesis and (b) continuous and discontinuous synthesis of DNA.
19. List the proteins that unwind DNA during *in vivo* DNA synthesis. How do they function?
20. Define and indicate the significance of (a) Okazaki fragments, (b) DNA ligase, and (c) primer RNA during DNA replication.
21. Outline the current model for DNA synthesis.
22. Why is DNA synthesis expected to be more complex in eukaryotes than in bacteria? How is DNA synthesis similar in the two types of organisms?
23. If the analysis of DNA from two different microorganisms demonstrated very similar base compositions, are the DNA sequences of the two organisms also nearly identical?
24. Suppose that *E. coli* synthesizes DNA at a rate of 100,000 nucleotides per minute and takes 40 minutes to replicate its chromosome.
    (a) How many base pairs are present in the entire *E. coli* chromosome?
    (b) What is the physical length of the chromosome in its helical configuration—that is, what is the circumference of the chromosome if it were opened into a circle?
25. Several temperature-sensitive mutant strains of *E. coli* display various characteristics. Predict what enzyme or function is being affected by each mutation.
    (a) Newly synthesized DNA contains many mismatched base pairs.
    (b) Okazaki fragments accumulate, and DNA synthesis is never completed.
    (c) No initiation occurs.
    (d) Synthesis is very slow.
    (e) Supercoiled strands are found to remain following replication, which is never completed.
26. Define gene conversion and describe how this phenomenon is related to genetic recombination.
27. Many of the gene products involved in DNA synthesis were initially defined by studying mutant *E. coli* strains that could not synthesize DNA.
    (a) The *dnaE* gene encodes the α subunit of DNA polymerase III. What effect is expected from a mutation in this gene? How could the mutant strain be maintained?
    (b) The *dnaQ* gene encodes the ϵ subunit of DNA polymerase. What effect is expected from a mutation in this gene?
28. In 1994, telomerase activity was discovered in human cancer cells in culture (*in vitro*). Although telomerase is not active in human somatic tissue, this discovery indicated that humans do contain the genes for telomerase proteins and telomerase RNA. Since inappropriate activation of telomerase can cause cancer, why do you think the genes coding for this enzyme have been maintained in the human genome throughout evolution? Are there any types of human body cells where telomerase activation would be advantageous or even necessary? Explain.

Extra-Spicy Problems

29. The genome of the fruit fly *Drosophila melanogaster* consists of approximately 1.6×10^8 base pairs. DNA synthesis occurs at a rate of 30 base pairs per second. In the early embryo, the entire genome is replicated in five minutes. How many *bidirectional origins of synthesis* are required to accomplish this feat?

30. To gauge the fidelity of DNA synthesis, Kornberg and colleagues used the nearest-neighbor frequency test, which determines the frequency with which any two bases occur adjacent to each other along the polynucleotide chain (*J. Biol. Chem.* 236:864–75). This test relies on the enzyme spleen phosphodiesterase. As we saw in Figure 11–8, during synthesis of DNA, 5′-nucleotides are inserted—that is, each nucleotide is added with the phosphate on the C-5′ of deoxyribose. However, as shown in the accompanying figure, the phosphodiesterase enzyme cleaves between the phosphate and the C-5′ atom, thereby producing 3′-nucleotides. The phosphates on only one of the four nucleotides (cytidylic acid, for example) are made radioactive with ^{32}P during DNA synthesis. Then, after enzymatic cleavage, the radioactive phosphate will be transferred to the base that is the "nearest neighbor" on the 5′ side of all cytidylic acid nucleotides.

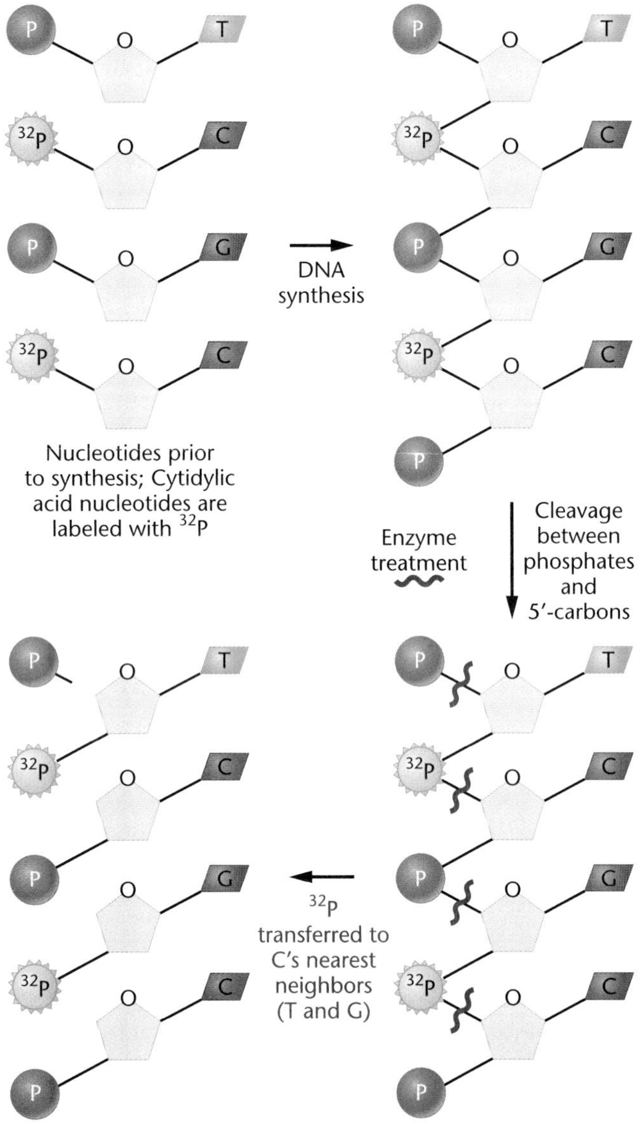

Following four separate experiments, in each of which only one of the four nucleotide types is made radioactive, the frequency of all 16 possible nearest neighbors can be calculated. When Kornberg applied the nearest-neighbor frequency test to the DNA template and the resultant product from a variety of experiments, he found general agreement between the nearest neighbor frequencies of the two.

Analysis of nearest neighbor data led Kornberg to conclude that the two strands of the double helix are in opposite polarity to one another. Demonstrate their approach by determining the outcome of such an analysis if the strands of DNA shown here are (a) antiparallel versus (b) parallel:

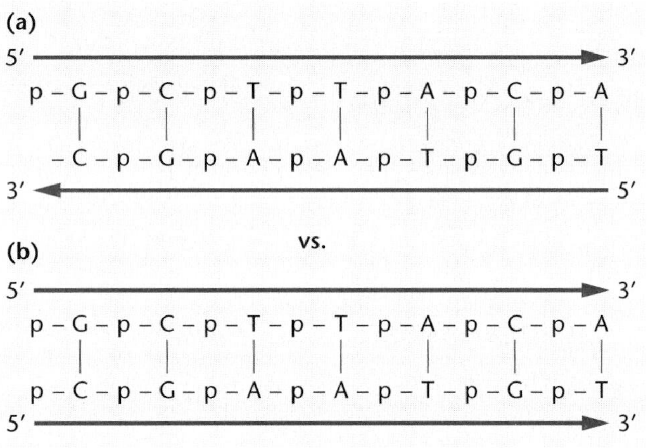

31. Assume a hypothetical organism in which DNA replication is conservative. Design an experiment similar to that of Taylor, Woods, and Hughes that will unequivocally establish this fact. Using the format established in Figure 11–5, draw sister chromatids and illustrate the expected results establishing this mode of replication.

32. DNA polymerases in all organisms only add 5′ nucleotides to the 3′ end of a growing DNA strand, never to the 5′ end. One possible reason for this is the fact that most DNA polymerases have a proofreading function that would not be *energetically* possible if DNA synthesis occurred in the 3′ to 5′ direction.

(a) Draw out the reaction that DNA polymerase would have to catalyze if DNA synthesis occurred in the 3′ to 5′ direction.

(b) Considering the information present in your drawing, speculate as to why proofreading would be problematic, while it isn't if synthesis occurs in the 5′ to 3′ direction.

33. An alien organism was investigated. It displayed characteristics of eukaryotes. When DNA replication was examined, two unique features were apparent: (1) no Okazaki fragments were observed; and (2) there was a telomere problem; i.e., telomeres shortened, but only on one end of the chromosome. Put forward a model of DNA that is consistent with both of these observations.

34. Assume that the sequence of bases given below is present on one nucleotide chain of a DNA duplex and that the chain has opened up at a replication fork. Synthesis of an RNA primer occurs on this template starting at the base that is underlined. (a) If the RNA primer consists of eight nucleotides, what is its base sequence? (b) In the intact RNA primer, which nucleotide has a free 3′-OH terminus?

3′.....GGCTACC<u>T</u>GGATTCA.....5′

35. Consider the diagram below. Assume that the phase G1 chromosome on the left underwent one round of replication in the presence of ^{3}H-thymidine and the metaphase chromosome on the right had both chromatids labeled. Which of the replicative models (conservative, dispersive, semiconservative) could be eliminated by this observation?

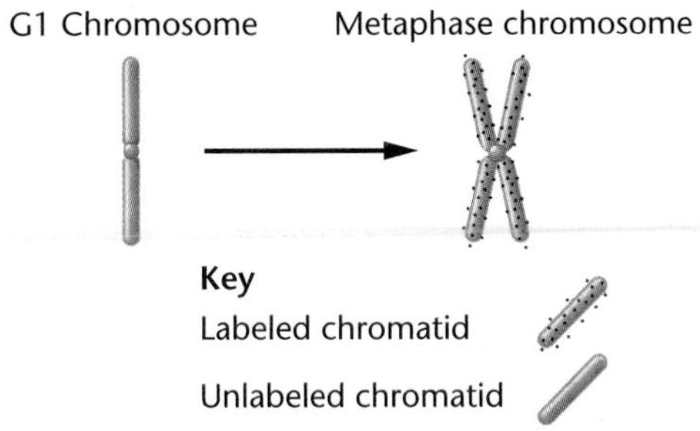

Selected Readings

Blackburn, E.H. 1991. Structure and function of telomeres. *Nature* 350:569–72.

Davidson, J.N. 1976. *The biochemistry of nucleic acids*, 8th ed. Orlando, FL: Academic Press.

DeLucia, P., and Cairns, J. 1969. Isolation of an *E. coli* strain with a mutation affecting DNA polymerase. *Nature* 224:1164–66.

Gilbert, D.M. 2001. Making sense of eukaryotic DNA replication origins. *Science* 294:96–100.

Greider, C.W. 1996. Telomeres, telomerase, and cancer. *Sci. Am.* (Feb.) 274:92–97.

______. 1998. Telomerase activity, cell proliferation, and cancer. *Proc. Natl. Acad. Sci. (USA)* 95:90–92.

Holliday, R. 1964. A mechanism for gene conversion in fungi. *Genet. Res.* 5:282–304.

Holmes, F.L. 2001. *Meselson, Stahl, and replication of DNA: A history of the "most beautiful experiment in biology."* New Haven: Yale U. Press.

Kim, J., Kaminker, P. and Campisi, J. 2002. Telomeres, aging and cancer: In search of a happy ending. *Oncogene* 21:503–11.

Kornberg, A. 1960. Biological synthesis of DNA. *Science* 131:1503–8.

______. 1974. *DNA synthesis.* New York: W.H. Freeman.

Kornberg, A., and Baker, T. 1992. *DNA replication*, 2d ed. New York: W.H. Freeman.

Meselson, M., and Stahl, F.W. 1958. The replication of DNA in *Escherichia coli. Proc. Natl. Acad. Sci. (USA)* 44:671–82.

Mitchell, M.B. 1955. Aberrant recombination of pyridoxine mutants of *Neurospora. Proc. Natl. Acad. Sci. (USA)* 41:215–20.

Okazaki, T., et al. 1979. Structure and metabolism of the RNA primer in the discontinuous replication of prokaryotic DNA. *Cold Spring Harbor Symp. Quant. Biol.* 43:203–22.

Radman, M., and Wagner, R. 1988. The high fidelity of DNA duplication. *Sci. Am.* (Aug.) 259:40–46.

______. 1987. Genetic recombination. *Sci. Am.* (Feb.) 256:90–101.

Taylor, J.H., Woods, P.S., and Hughes, W.C. 1957. The organization and duplication of chromosomes revealed by autoradiographic studies using tritium-labeled thymidine. *Proc. Natl. Acad. Sci. (USA)* 48:122–28.

Wang, J.C. 1982. DNA topoisomerases. *Sci. Am.* (July) 247:94–108.

______. 1987. Recent studies of DNA topoisomerases. *Biochim. Biophys. Acta* 909:1–9.

Whitehouse, H.L.K. 1982. *Genetic recombination: Understanding the mechanisms*. New York: Wiley.

CHAPTER TWELVE

DNA Organization in Chromosomes

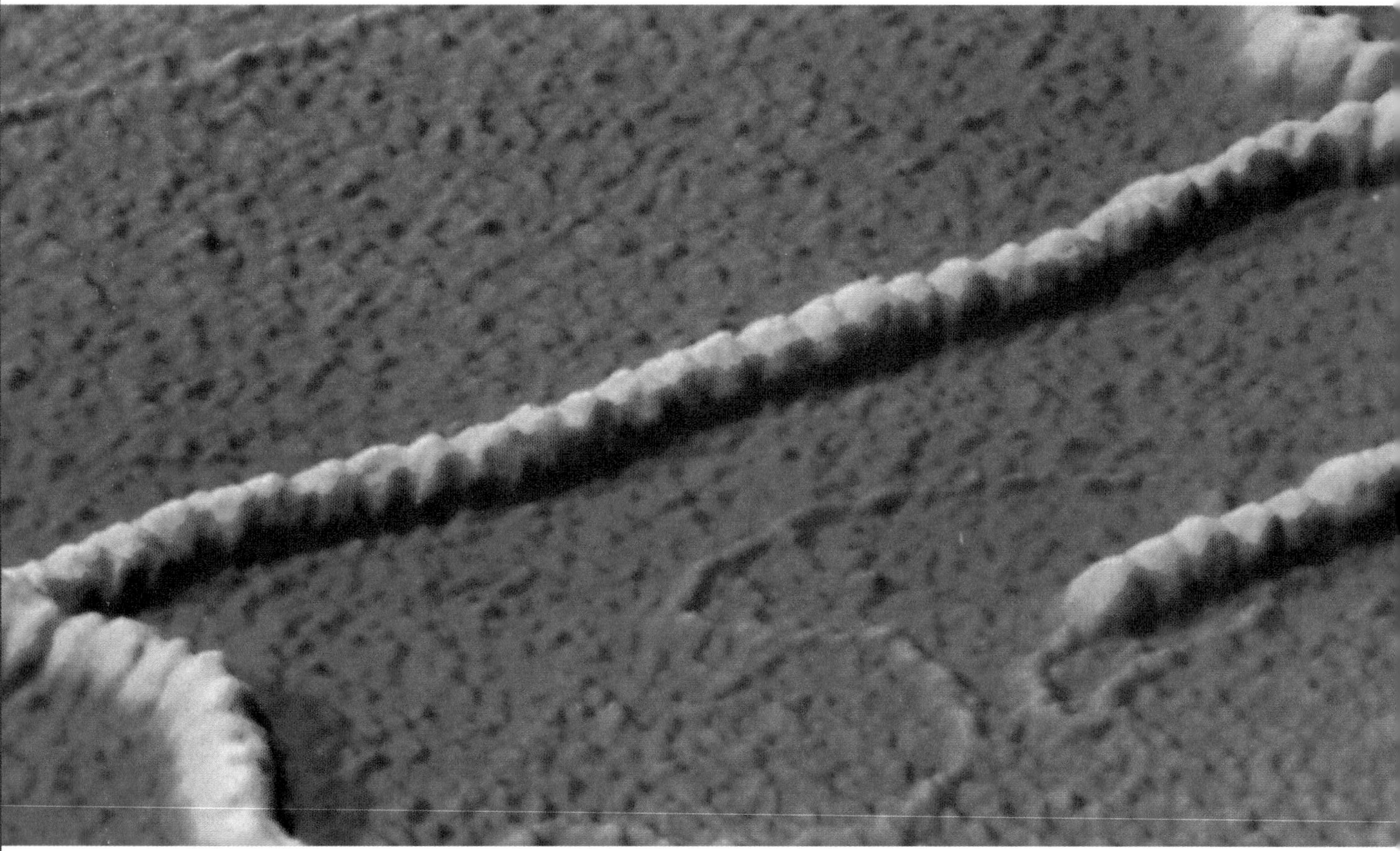

A chromatin fiber viewed using a scanning transmission microscope (STEM).

Chapter Concepts

- Genetic information in viruses, bacteria, mitochondria, and chloroplasts is most often contained in a short, circular DNA molecule, relatively free of associated proteins.
- Eukaryotic cells contain relatively large amounts of DNA organized into nucleosomes and present during most of the cell cycle as chromatin fibers.
- During division stages, chromatin fibers coil up into chromosomes.
- Eukaryotic genomes are characterized by both unique and repetitive DNA sequences.
- The vast majority of eukaryotic genomes consist of noncoding DNA sequences, some of which interrupt the coding regions of genes.

Once it was understood that DNA houses genetic information, it was of interest to determine how DNA is organized into genes and how these basic units of genetic function are organized into chromosomes. In short, the major question is how the genetic material is organized as it makes up the genome of organisms. There has been much interest in this question, because knowledge of the organization of the genetic material and associated molecules is important to the understanding of many other areas of genetics. For example, the way in which the genetic information is stored, expressed, and regulated must be related to the molecular organization of the genetic molecule, DNA. How genomic organization varies in different organisms—from viruses to bacteria to eukaryotes—will undoubtedly provide a better understanding of the evolution of organisms on Earth.

In this chapter, we focus on the various ways DNA is organized into chromosomes. The genetic material has been studied by using numerous approaches, including light microscopy and electron microscopy. More recently, molecular analysis has provided significant insights into chromosome organization. We will first survey what is known about chromosomes in viruses and bacteria. Then, we will examine the large specialized structures called polytene and lampbrush chromosomes. In the second half of the chapter, we will discuss some of the basic questions of eukaryotic genomic organization. For example, how is DNA complexed with proteins to form chromatin, and how are the chromatin fibers characteristic of interphase condensed into chromosome structures visible during mitosis and meiosis?

We will conclude the chapter by examining some aspects of DNA sequence organization that are characteristic of eukaryotic genomes. Later, in Chapter 20, we will turn to a discussion of the exciting studies that have clarified how genes are organized in chromosomes.

12.1 Viral and Bacterial Chromosomes Are Relatively Simple DNA Molecules

Compared with eukaryotes, the chromosomes of viruses and bacteria are much less complicated. They usually consist of a single nucleic acid molecule, largely devoid of associated proteins and containing relatively little genetic information in comparison to the multiple chromosomes constituting the genome of higher forms. These characteristics have greatly simplified analysis, providing a fairly comprehensive view of the structure of viral and bacterial chromosomes.

The chromosomes of viruses consist of a nucleic acid molecule—either DNA or RNA—that can be either single or double stranded. They can exist as circular structures (covalently closed circles), or they can take the form of linear molecules. The single-stranded DNA of the **ϕX174 bacteriophage** and the double-stranded DNA of the **polyoma virus** are closed circles housed within the protein coat of the mature viruses. The **bacteriophage lambda** (λ), on the other hand, possesses a linear double-stranded DNA molecule prior to infection, which closes to form a ring upon its infection of the host cell. Still other viruses, such as the **T-even series of bacteriophages**, have linear double-stranded chromosomes of DNA that do not form circles inside the bacterial host. Thus, circularity is not an absolute requirement for replication in some viruses.

Viral nucleic acid molecules have been visualized with the electron microscope. Figure 12–1 shows a mature bacteriophage λ with its double-stranded DNA molecule in the circular configuration. One constant feature shared by viruses, bacteria, and eukaryotic cells is the ability to package an exceedingly long DNA molecule into a relatively small volume. In λ, the DNA is 7 μm long and must fit into the phage head, which is less than 0.1 μm on any side.

(a) (b)

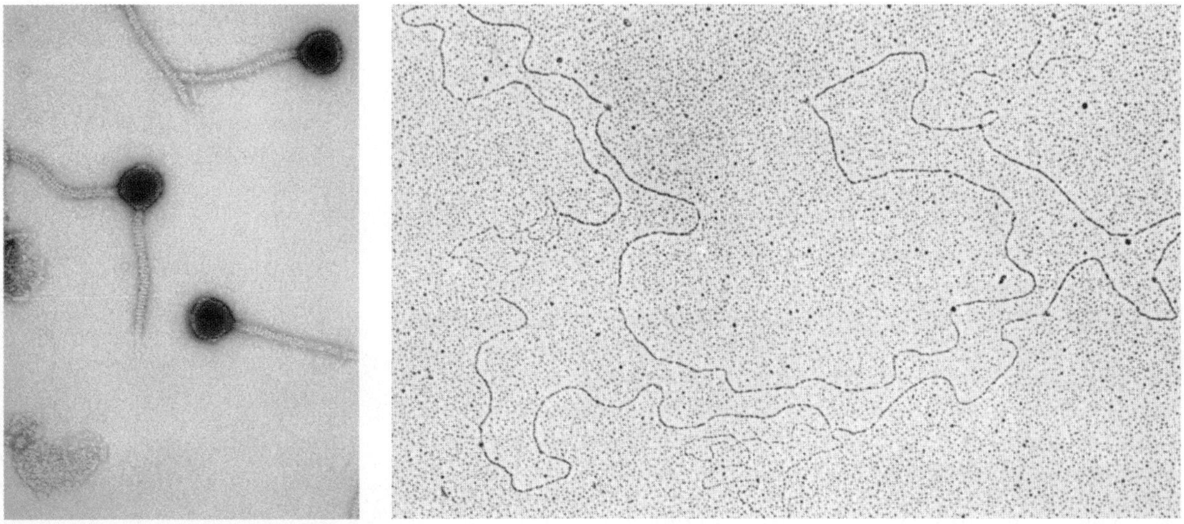

FIGURE 12–1 Electron micrographs of (a) phage λ and (b) the DNA isolated from it. The chromosome is 17 μm long. The phages are magnified about five times more than the DNA.

TABLE 12.1 THE GENETIC MATERIAL OF REPRESENTATIVE VIRUSES AND BACTERIA

	Organism	Nucleic Acid Type	SS or DS*	Length (μm)	Overall Size of Viral Head or Bacteria (μm)
Viruses	ϕX174	DNA	SS	2.0	0.025 × 0.025
	Tobacco mosaic virus	RNA	SS	3.3	0.30 × 0.02
	Lambda phage	DNA	DS	17.0	0.07 × 0.07
	T2 phage	DNA	DS	52.0	0.07 × 0.10
Bacteria	*Haemophilus influenzae*	DNA	DS	832.0	1.00 × 0.30
	Escherichia coli	DNA	DS	1200.0	2.00 × 0.50

*SS = single-stranded; DS = double-stranded

Table 12.1 compares the length of the chromosomes of several viruses with the size of their head structure. In each case, a similar packaging feat must be accomplished. The dimensions given for phage T2 may be compared with the micrograph of both the DNA and the viral particle shown in Figure 12–2. Seldom does the space available in the head of a virus exceed the chromosome volume by more than a factor of two. In many cases, almost all space is filled, indicating nearly perfect packing. Once packed within the head, the genetic material is functionally inert until it is released into a host cell.

Bacterial chromosomes are also relatively simple in form. They always consist of a double-stranded DNA molecule, compacted into a structure sometimes referred to as the **nucleoid**. *Escherichia coli*, the most extensively studied bacterium, has a large, circular chromosome measuring approximately 1200 μm (1.2 mm) in length. When the cell is gently lysed and the chromosome released, it can be visualized under the electron microscope (Figure 12–3).

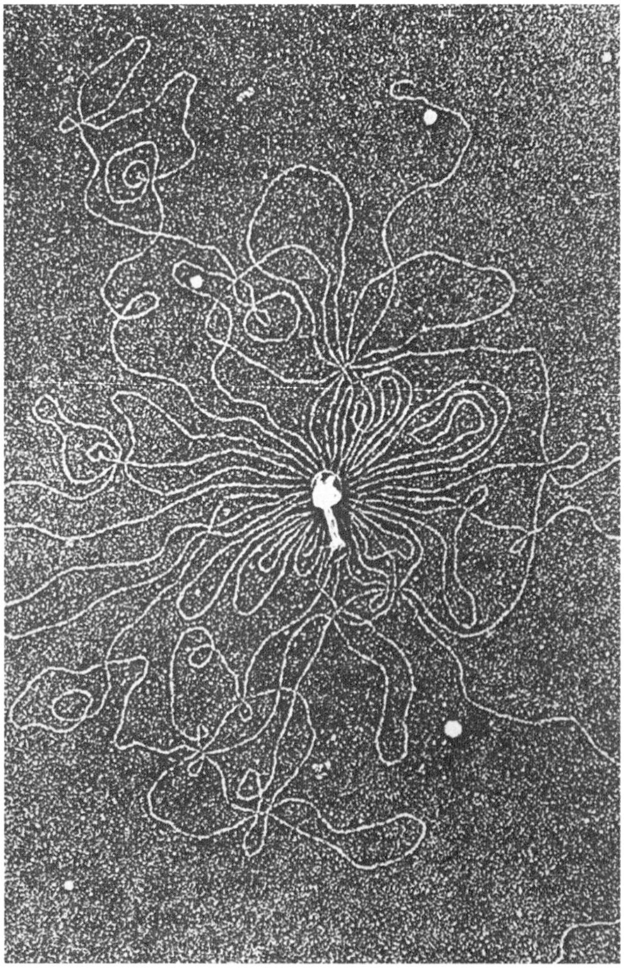

FIGURE 12–2 Electron micrograph of bacteriophage T2, which has had its DNA released by osmotic shock. The chromosome is 52 μm long.

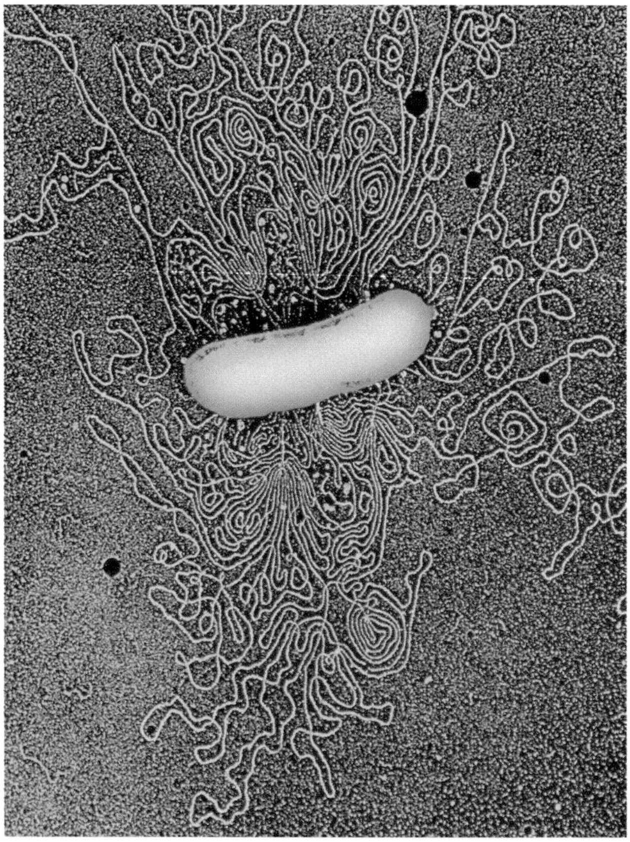

FIGURE 12–3 Electron micrograph of the bacterium *Escherichia coli*, which has had its DNA released by osmotic shock. The chromosome is 1200 μm long.

DNA in bacterial chromosomes is found to be associated with several types of **DNA-binding proteins**. Two, called **HU** and **H**, are small, but abundant in the cell, and contain a high percentage of positively charged amino acids that can bond ionically to the negative charges of the phosphate groups in DNA. These proteins are structurally similar to molecules, called **histones**, that are associated with eukaryotic DNA. (We will discuss histone organization later in this chapter.) Unlike the tightly packed chromosome present in the head of a virus, the bacterial chromosome is not functionally inert. Despite its somewhat compacted condition in the bacterial cell, the chromosome can be readily replicated and transcribed.

Now solve this

Problem 12.2 on page 302 involves the consideration of viral chromosomes that are linear in the bacteriophage, but circularize after they enter the bacterial host cell.

Hint: Recall from Chapter 11 that linear chromosomes face the "telomere problem" during their replication.

12.2 Supercoiling Is Common in the DNA of Viral and Bacterial Chromosomes

One major insight into the way in which DNA is organized and packaged in viral and bacterial chromosomes has come from the discovery of **supercoiled DNA**, which is particularly characteristic of closed circular molecules. Supercoiled DNA was first proposed as a result of a study of double-stranded DNA molecules derived from the polyoma virus, which causes tumors in mice. In 1963, it was observed that when such DNA was subjected to high-speed centrifugation, it was resolved into three distinct components, each of different density and compactness. The one that was least compact, and thus least dense, demonstrated a decreased sedimentation velocity; the other two fractions each showed greater velocities owing to their greater compaction and density. All three were of identical molecular weight.

In 1965, Jerome Vinograd proposed an explanation for these observations. He postulated that the two fractions of greatest sedimentation velocity both consisted of polyoma DNA molecules that are circular, whereas the fraction of lower sedimentation contained polyoma DNA molecules that are linear. Closed circular molecules are more compact and they sediment more rapidly than do linear molecules of the same length and molecular weight.

Vinograd proposed further that the more dense of the two fractions of circular molecules consisted of covalently closed DNA helices that are slightly *underwound* in comparison to the less dense circular molecules. Energetic forces stabilizing the double helix resist this underwinding, causing it to **supercoil** in order to retain normal base pairing. Vinograd proposed that it is the supercoiled shape that causes tighter packing and thus the increase in sedimentation velocity.

The transitions just described are illustrated in Figure 12–4. Consider a double-stranded linear molecule existing in the normal Watson–Crick right-handed helix [Figure 12–4(a)]. This helix contains 20 complete turns, which defines the **linking number** ($L = 20$) of this molecule. If the ends of the molecule are sealed, a closed circle is formed [Figure 12–4(b)], which is described as being *energetically relaxed.* Suppose, however, that the circle were cut open, underwound by several full turns, and resealed [Figure 12–4(c)]. Such a structure, where L is equal to 18, is *energetically strained,* and as a result, it will exist only temporarily in this form.

In order to assume a more energetically favorable conformation, the molecule can form supercoils in the opposite direction of the underwound helix. In our case [Figure 12–4(d)], two negative supercoils are introduced spontaneously, reestablishing the total number of original turns in the helix. The use of the term "negative" refers to the fact that, by definition, the supercoils are left-handed. The end result is the formation of a more compact structure with enhanced physical stability.

In most closed circular DNA molecules in bacteria and their phages, DNA is slightly underwound [as in Figure 12–4(c)]. For example, the virus SV40 contains 5200 base pairs. If the

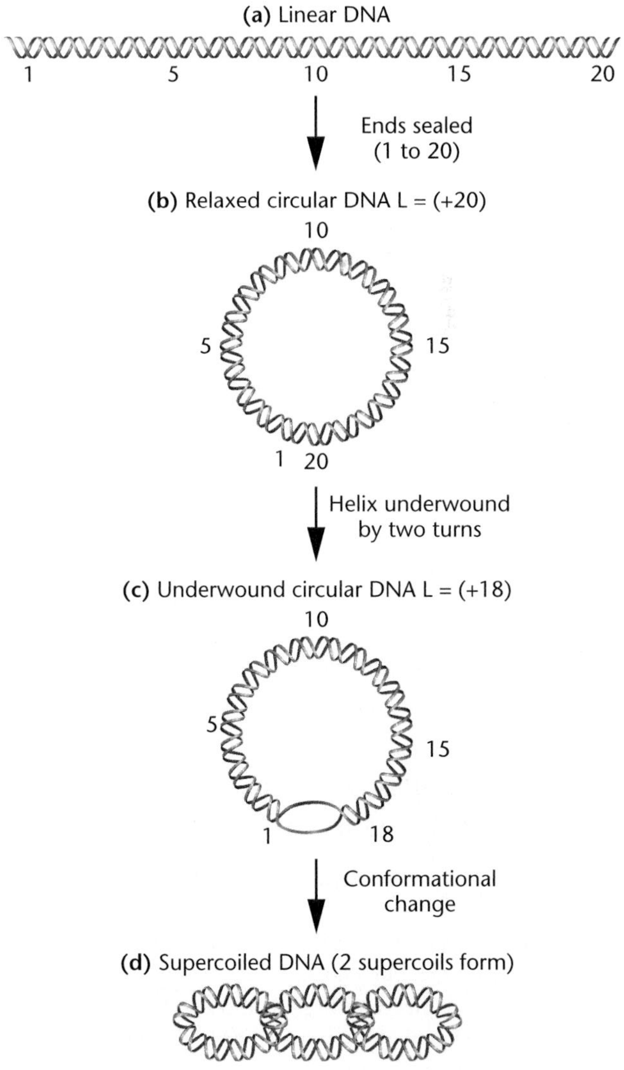

FIGURE 12–4 Depictions of the transformations leading to the supercoiling of circular DNA. *L* equals the linking number.

DNA is relaxed, 10.4 base pairs occupy each complete turn of the helix, and the linking number can be calculated as

$$L = \frac{5200}{10.4} = 500$$

When circular SV40 DNA is analyzed, it is underwound by 25 turns and L is equal to only 475. Predictably, 25 negative supercoils are observed. In *E. coli*, an even larger number of supercoils is observed, greatly facilitating chromosome condensation in the nucleoid region.

Two otherwise identical molecules that differ only in their linking number are said to be **topoisomers** of one another. But how can a molecule convert from one topoisomer to the other if there are no free ends, as is the case in closed circles of DNA? Biologically, this may be accomplished by any one of a group of enzymes that cut one or both of the strands and wind or unwind the helix before resealing the ends.

Appropriately, these enzymes are called **topoisomerases**. First discovered by Martin Gellert and James Wang, these catalytic molecules are known as either type I or type II, depending on whether they cleave one or both strands in the helix, respectively. In *E. coli*, topoisomerase I serves to reduce the number of negative supercoils in a closed-circular DNA molecule. Topoisomerase II introduces negative supercoils into DNA. This latter enzyme is thought to bind to DNA, twist it, cleave both strands, and then pass them through the loop that it has created. Once the phosphodiester bonds are reformed, the linking number is decreased and one or more supercoils form spontaneously.

Supercoiled DNA and topoisomerases are also found in eukaryotes. While the chromosomes in these organisms are not usually circular, supercoils can occur when areas of DNA are embedded in a lattice of proteins associated with the chromatin fibers. This association creates "anchored" ends providing the stability for the maintenance of supercoils once they are introduced by topoisomerases. As occurs in prokaryotes, DNA replication and transcription in eukaryotes creates supercoils downstream as the double helix unwinds and becomes accessible to the appropriate enzyme.

These enzymes may play still other genetic roles involving eukaryotic DNA conformational changes. Interestingly, topoisomerases are involved in separating (decatenating) the DNA of sister chromatids following replication.

12.3 Specialized Chromosomes Reveal Variations in Structure

We move next to the consideration of chromosomes present in eukaryotes. Before discussing these at the molecular level, we first introduce two cases of highly specialized chromosomes. Both types, polytene chromosomes and lampbrush chromosomes, are so large that their organization was discerned using light microscopy long before we understood that DNA is the genetic material. The study of these chromosomes provided many of our initial insights into the arrangement and function of the genetic information.

Polytene Chromosomes

Giant **polytene chromosomes** are found in various tissues (salivary, midgut, rectal, and malpighian excretory tubules) in the larvae of some flies, as well as in several species of protozoans and plants. Such structures were first observed by E. G. Balbiani in 1881. The large amount of information obtained from studies of these genetic structures provided a model system for subsequent investigations of chromosomes. What is particularly intriguing about polytene chromosomes is that they can be seen in the nuclei of interphase cells.

Each polytene chromosome observed under the light microscope reveals a linear series of alternating bands and interbands (Figure 12–5). The banding pattern is distinctive for each chromosome in any given species. Individual bands are sometimes called **chromomeres**, a more generalized term describing lateral condensations of material along the axis of a chromosome. Each polytene chromosome is 200 to 600 μm long.

Extensive study using electron microscopy and radioactive tracers led to an explanation for the unusual appearance of these chromosomes. First, polytene chromosomes represent paired homologs. This is highly unusual, since they are found in somatic cells, where, in most organisms, chromosomal material is normally dispersed as chromatin and homologs are not paired. Second, their large size and distinctiveness result from the many DNA strands that compose them. The DNA of these paired homologs undergoes many rounds of replication, but without strand separation or cytoplasmic division. As replication proceeds, chromosomes are created having 1000 to 5000 DNA strands that remain in parallel register (perfectly aligned) with one another. Apparently, it is the parallel register of so many DNA strands that gives rise to the distinctive band pattern along the axis of the chromosome.

The relationship between the structure of polytene chromosomes and the genes contained within them is intriguing. The presence of bands was initially interpreted as the visible manifestation of individual genes. The discovery that the strands

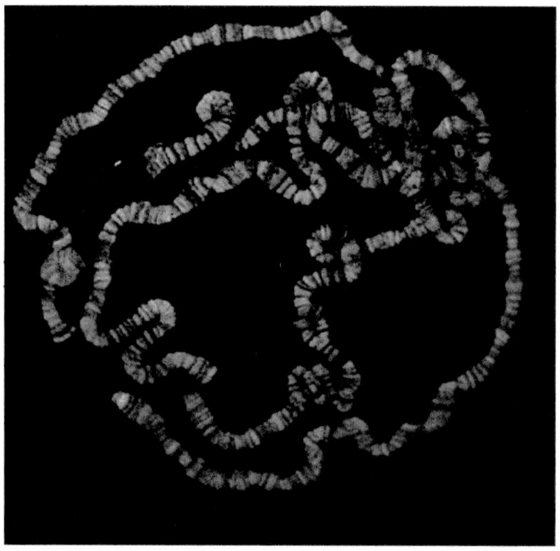

FIGURE 12–5 Polytene chromosomes derived from larval salivary gland cells of *Drosophila*.

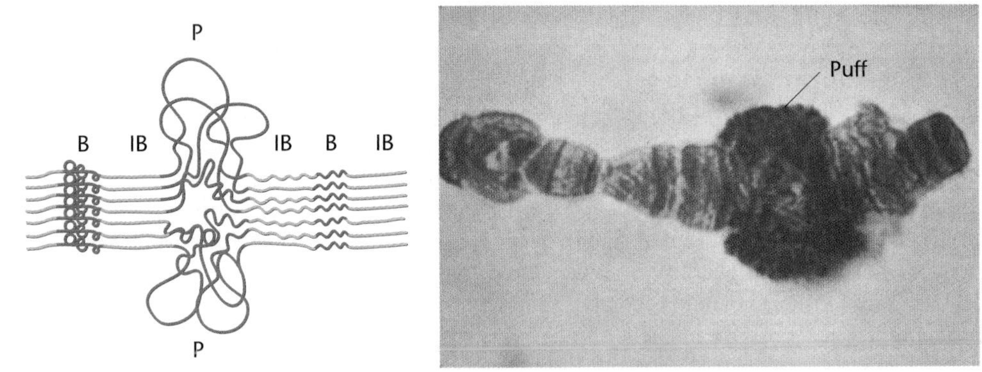

FIGURE 12–6 Photograph of a puff within a polytene chromosome. The diagram depicts the uncoiling of strands within a band (B) region to produce a puff (P) in polytene chromosomes. Interband (IB) regions are also labeled.

present in bands undergo localized uncoiling during genetic activity further strengthened this view. Each such uncoiling event results in what is called a **puff** because of its appearance (Figure 12–6). That puffs are visible manifestations of a high level of gene activity (transcription that produces RNA) is evidenced by their high rate of incorporation of radioactively labeled RNA precursors, as assayed by autoradiography. Bands that are not extended into puffs incorporate fewer radioactive precursors or none at all.

The study of bands during development in insects, such as *Drosophila* and the midge fly *Chironomus*, reveals differential gene activity. A characteristic pattern of band formation that is equated with gene activation is observed as development proceeds. Despite attempts to resolve the issue of the number of genes contained within each band, it is still not clear how many there are. It is known that a band may contain up to 10^7 base pairs of DNA, certainly enough DNA to encode 50 to 100 average-size genes.

Now solve this

Problem 12.4 on page 302 involves polytene chromosomes that are cultured in ^{3}H-thymidine and subjected to autoradiography.

Hint: ^{3}H-thymidine will only be incorporated during the synthesis of DNA.

Lampbrush Chromosomes

Another type of specialized chromosome that has provided insights into chromosomal structure is the **lampbrush chromosome**, so named because its appearance is similar to the brushes used to clean kerosene lamp chimneys in the 19th century. Lampbrush chromosomes were first discovered in 1892 in the oocytes of sharks and are now known to be characteristic of most vertebrate oocytes, as well as the spermatocytes of some insects. Therefore, they are meiotic chromosomes. Most of the experimental work has been done with material taken from amphibian oocytes.

These unique chromosomes are easily isolated from oocytes in the diplotene stage of the first prophase of meiosis, where they are active in directing the metabolic activities of the developing cell. The homologs are seen as synapsed pairs held together by chiasmata. However, instead of condensing, as most meiotic chromosomes do, lampbrush chromosomes are often extended to lengths of 500 to 800 μm. Later, in meiosis, they revert to their normal length of 15 to 20 μm. Based on these observations, lampbrush chromosomes are interpreted as extended, uncoiled versions of the normal meiotic chromosomes.

The two views of lampbrush chromosomes in Figure 12–7 provide significant insights into their morphology. In Figure 12–7(a), the meiotic configuration under the light microscope is shown. The linear axis of each structure contains a large number of condensed areas, referred to generally as chromomeres, that are repeated along the axis. Emanating from each chromomere is a pair of lateral loops, which give the chromosome its distinctive appearance. In Figure 12–7(b), the

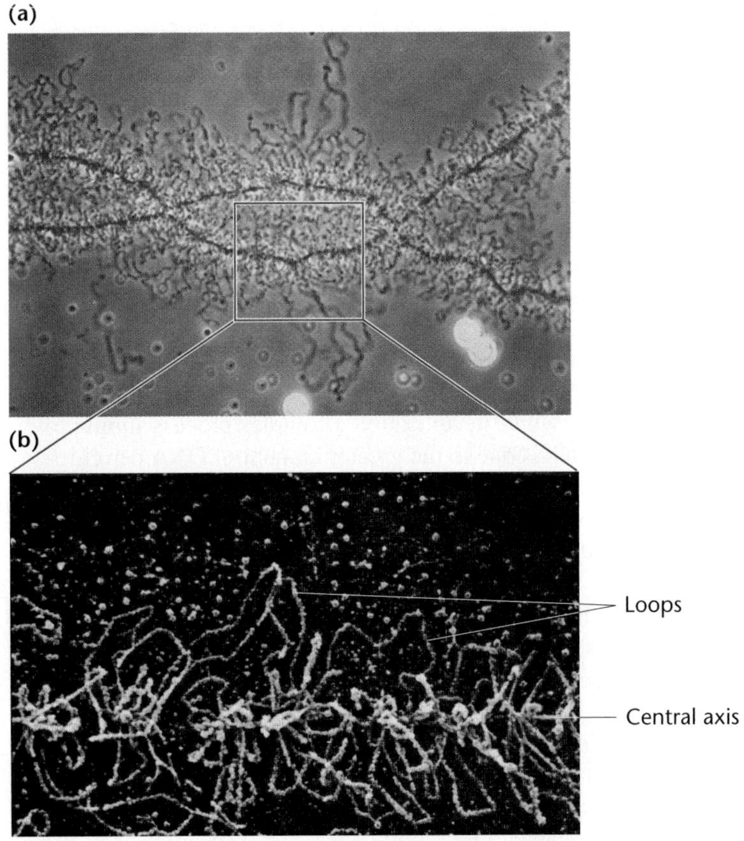

FIGURE 12–7 Lampbrush chromosomes derived from amphibian oocytes—(a) is a photomicrograph; (b) is a scanning electron micrograph.

scanning electron micrograph (SEM) is an enlargement that reveals adjacent loops present along one of the two axes of the chromosome. As with bands in polytene chromosomes, there is much more DNA present in each loop than is needed to encode a single gene. Such an SEM provides a clear view of the chromomeres and the chromosomal fibers emanating from them. Each chromosomal loop is thought to be composed of one DNA double helix, while the central axis is made up of two DNA helices. This hypothesis is consistent with the belief that each meiotic chromosome is composed of a pair of sister chromatids. Studies using radioactive RNA precursors reveal that the loops are active in the synthesis of RNA. The lampbrush loops, in a manner similar to puffs in polytene chromosomes, represent DNA that has been reeled out from the central chromomere axis during transcription.

How Do We Know?

What is the experimental basis for concluding that puffs in polytene chromosomes and loops in lampbrush chromosomes are areas of intense transcription of RNA?

12.4 DNA Is Organized into Chromatin in Eukaryotes

We now turn our attention to the way DNA is organized in eukaryotic chromosomes. Our focus will be on conventional eukaryotic cells, where DNA and proteins are complexed into a nucleoprotein structure referred to as **chromatin**. Chromosomes, visible only during mitosis, are composed of tightly coiled chromatin fibers. Following cell division, when cells enter the interphase stage of the cell cycle, the chromatin is uncoiled and chromosomes disappear. While in interphase, the chromatin is dispersed in the nucleus, and the DNA of each chromosome is replicated. As the cell cycle progresses, most cells reenter mitosis, whereupon chromatin coils into visible chromosomes once again. This condensation represents a contraction in length of some 10,000 times for each chromatin fiber.

The organization of DNA during the transitions just described is much more intricate and complex than in viruses or bacteria, which never exhibit a complex process similar to mitosis. This is due to the greater amount of DNA per chromosome, as well as the presence of a large number of proteins associated with eukaryotic DNA. For example, while DNA in the *E. coli* chromosome is 1200 μm long, the DNA in each human chromosome ranges from 19,000 to 73,000 μm in length. In a single human nucleus, all 46 chromosomes contain sufficient DNA to extend almost 2 meters. This genetic material, along with its associated proteins, is contained within a nucleus that usually measures about 5 to 10 μm in diameter.

Such intricacy parallels the structural and biochemical diversity of the many types of cells present in a multicellular eukaryotic organism. Different cells assume specific functions based on highly specific biochemical activity. While all cells carry a full genetic complement, different cells activate different sets of genes, so a highly ordered regulatory system governing the readout of the information must exist. Such a system must be imposed on or related to the molecular structure of the genetic material.

Because of the limitation of light microscopy, early studies of the structure of eukaryotic genetic material concentrated on intact chromosomes, preferably large ones, such as the polytene and lampbrush chromosomes. Subsequently, new techniques for biochemical analysis, as well as the examination of relatively intact eukaryotic chromatin and mitotic chromosomes under the electron microscope, have greatly enhanced our understanding of chromosome structure.

Chromatin Structure and Nucleosomes

As we have seen, the genetic material of viruses and bacteria consists of strands of DNA or RNA nearly devoid of proteins. In eukaryotic chromatin, a substantial amount of protein is associated with the chromosomal DNA in all phases of the eukaryotic cell cycle. The associated proteins are divided into basic, positively charged **histones** and less positively charged **nonhistones**. Of the proteins associated with DNA, the histones play the most essential structural role. Histones contain large amounts of the positively charged amino acids lysine and arginine, making it possible for them to bond electrostatically to the negatively charged phosphate groups of nucleotides. Recall that a similar interaction has been proposed for several bacterial proteins. There are five main types of histones (Table 12.2).

The general model for chromatin structure is based on the assumption that chromatin fibers, composed of DNA and protein, undergo extensive coiling and folding as they are condensed within the cell nucleus. X-ray diffraction studies confirm that histones play an important role in chromatin structure. Chromatin produces regularly spaced diffraction rings, suggesting that repeating structural units occur along the chromatin axis. If the histone molecules are chemically removed from chromatin, the regularity of this diffraction pattern is disrupted.

A basic model for chromatin structure was worked out in the mid-1970s. The following observations were particularly relevant to the development of this model:

Table 12.2 Categories and Properties of Histone Proteins

Histone Type	Lysine-Arginine Content	Molecular Weight (Da)
H1	Lysine-rich	23,000
H2A	Slightly lysine-rich	14,000
H2B	Slightly lysine-rich	13,800
H3	Arginine-rich	15,300
H4	Arginine-rich	11,300

1. Digestion of chromatin by certain endonucleases, such as micrococcal nuclease, yields DNA fragments that are approximately 200 base pairs in length, or multiples thereof. This demonstrates that enzymatic digestion is not random, for if it were, we would expect a wide range of fragment sizes. Hence, chromatin consists of some type of repeating unit, each of which is protected from enzymatic cleavage, except for the point at which any two units are joined. It is the area between units that is attacked and cleaved by the nuclease.
2. Electron microscopic observations of chromatin have revealed that chromatin fibers are composed of linear arrays of spherical particles (Figure 12–8). Discovered by Ada and Donald Olins, the particles occur regularly along the axis of a chromatin strand and resemble beads on a string. These particles, initially referred to as ν-bodies, are now called nucleosomes. These findings conform to the above observation that suggests the existence of repeating units.
3. Studies of precise interactions of histone molecules and DNA in the nucleosomes constituting chromatin show that histones H2A, H2B, H3, and H4 occur as two types of tetramers, $(H2A)_2 \cdot (H2B)_2$ and $(H3)_2 \cdot (H4)_2$. Roger Kornberg predicted that each repeating nucleosome unit consists of one of each tetramer in association with about 200 base pairs of DNA. Such a structure is consistent with previous observations and provides the basis for a model that explains the interaction of histones and DNA in chromatin.
4. When nuclease digestion time is extended, some of the 200 base pairs of DNA are removed from the nucleosome, creating what is called a **nucleosome core particle**, consisting of 147 base pairs. This number is consistent in all eukaryotes studied. The DNA lost in the prolonged digestion is responsible for linking nucleosomes together. This **linker DNA** is associated with the fifth histone, H1.
5. On the basis of this information, as well as on X-ray and neutron-scattering analyses of crystallized core particles by John T. Finch, Aaron Klug, and others, a detailed model of the nucleosome was put forward in 1984. In this model, the 147-bp DNA core is coiled around an octamer of histones in a left-handed superhelix, which completes about 1.7 turns per nucleosome.

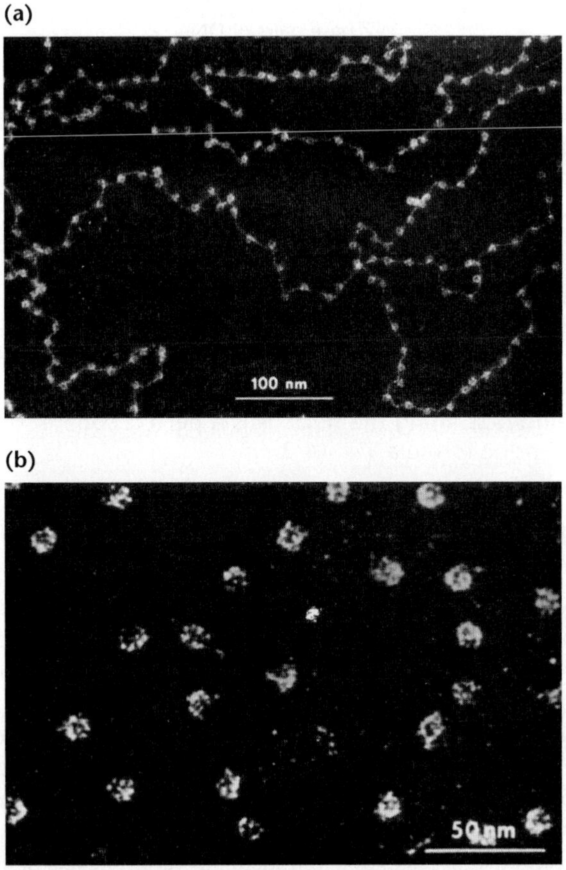

FIGURE 12–8 (a) Dark-field electron micrograph of nucleosomes present in chromatin derived from a chicken erythrocyte nucleus. (b) Dark-field electron micrograph of nucleosomes produced by micrococcal nuclease digestion.

The extensive investigation of nucleosomes now provides the basis for predicting how the chromatin fiber within the nucleus is formed and how it coils up into a mitotic chromosome. This model is illustrated in Figure 12–9. The 2-nm DNA molecule is initially coiled into a nucleosome that is about 11 nm in diameter [Figure 12–9(a)], consistent with the longer dimension of the ellipsoidal nucleosome. Significantly, the formation of the nucleosome represents the first level of DNA packing, whereby the helix is reduced to about one third of its original length by winding around the histones.

In the nucleus, the chromatin fiber seldom, if ever, exists in the extended form described in the previous paragraph. Instead, the 11-nm chromatin fiber is further packed into a thicker 30-nm structure that was initially called a solenoid [Figure 12–9(b)]. This larger fiber consists of numerous nucleosomes coiled closely together, creating the second level of packing. The exact details of the structure are not completely clear, but 30-nm chromatin fibers are characteristically seen under the electron microscope.

In the transition to the mitotic chromosome, still another level of packing occurs. The 30-nm structure forms a series of looped domains that further condense into the chromatin fiber, which is 300 nm in diameter [Figure 12–9(c)]. The fibers are then coiled into the chromosome arms that constitute a chromatid, which is part of the metaphase chromosome [Figure 12–9(d)]. While we show the chromatid arms to be 700 nm in diameter, the value undoubtedly varies among different organisms. At a value of 700 nm, a pair of sister chromatids making up a chromosome measures about 1400 nm.

The importance of the organization of DNA into chromatin and chromatin into mitotic chromosomes can be illustrated by considering a human cell that stores its genetic material in a nucleus about 5 to 10 μm in diameter. The haploid genome contains more than 3 billion base pairs of DNA distributed among 23 chromosomes. The diploid cell contains twice that amount. At 0.34 nm per base pair, this amounts to an enormous length of DNA (as stated earlier, almost 2 meters)! One estimate is that the DNA present in a typical human nucleus is organized into about 25×10^6 nucleosomes.

Web Tutorial 12.1
Chromosome Structure

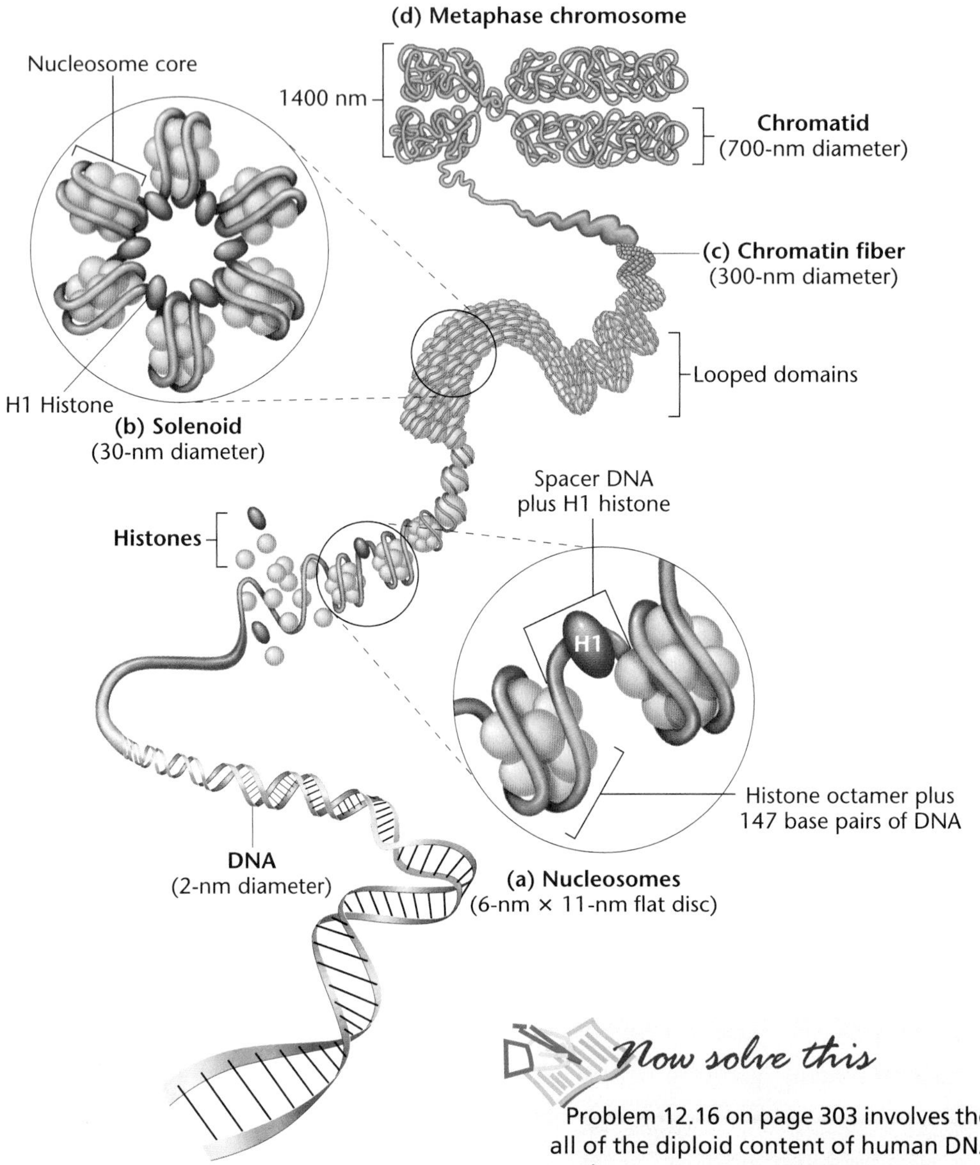

FIGURE 12–9 General model of the association of histones and DNA in the nucleosome, illustrating the way in which the chromatin fiber may be coiled into a more condensed structure, ultimately producing a mitotic chromosome.

In the overall transition from a fully extended DNA helix to the extremely condensed status of the mitotic chromosome, a packing ratio (the ratio of DNA length to the length of the structure containing it) of about 500 to 1 must be achieved. In fact, our model accounts for a ratio of only about 50 to 1. Obviously, the larger fiber can be further bent, coiled, and packed as even greater condensation occurs during the formation of a mitotic chromosome.

How Do We Know?

What experimental evidence supports the idea that eukaryotic chromatin exists in the form of repeating nucleosomes each consisting of about 200 base pairs and an octamer of histones?

Now solve this

Problem 12.16 on page 303 involves the extent to which all of the diploid content of human DNA can fit into the nucleus.

Hint: Assuming the nucleus is a perfect sphere, start with the formula $V = (4/3\ \pi r^3)$.

High Resolution Studies of the Nucleosome Core

As with many significant findings in genetics, the study of nucleosomes has answered some important questions, but at the same time led us to new ones. For example, in the preceding discussion, we have established that histone proteins play an important structural role in packaging DNA into the nucleosomes that make up chromatin. While helping to solve the structural problem of how to organize a huge amount of DNA within the eukaryotic nucleus, a new problem arises. The chromatin fiber, when complexed with histones and folded into various levels of compaction, makes the DNA inaccessible to interaction with important nonhistone proteins. For example, the variety of proteins that function in enzymatic and regulatory roles during the processes of replication and gene expression must interact directly with DNA. To accommodate these protein–DNA interactions, chromatin must be induced to change its structure, a process now referred to as **chromatin remodeling**. In the case of replication and gene expression, chromatin must relax its compact structure and expose regions of DNA to regulatory proteins, but be able to reverse the process during periods of inactivity.

Insights into how different states of chromatin structure may be achieved were forthcoming in 1997, when Timothy Richmond and members of his research team were able to significantly improve the level of resolution in X-ray diffraction studies of nucleosome crystals (from 7 Å in the 1984 studies to 2.8 Å in the 1997 studies). One model based on their work is shown in Figure 12–10. At this resolution, most atoms are visible, thus revealing the subtle twists and turns of the superhelix of DNA encircling the histones. Recall from our previous discussion that the double-helical ribbon represents 147 bp of DNA surrounding four pairs of histone proteins. This configuration is essentially repeated over and over in the chromatin fiber and is the principal packaging unit of DNA in the eukaryotic nucleus.

The work of Richmond and colleagues, extended to a resolution of 1.9 Å in 2003, has revealed the details of the location of each histone entity within the nucleosome. Of particular interest to the discussion of chromatin remodeling is the observation that there are unstructured histone tails that are not packed into the folded histone domains within the core of the nucleosome. For example, tails devoid of any secondary structure extending from histones H3 and H2B protrude through the minor groove channels of the DNA helix. The tails of histone H4 appear to make connection with adjacent nucleosomes. The significance of histone tails is that they provide potential targets for a variety of chemical interactions that may be linked to genetic functions along the chromatin fiber.

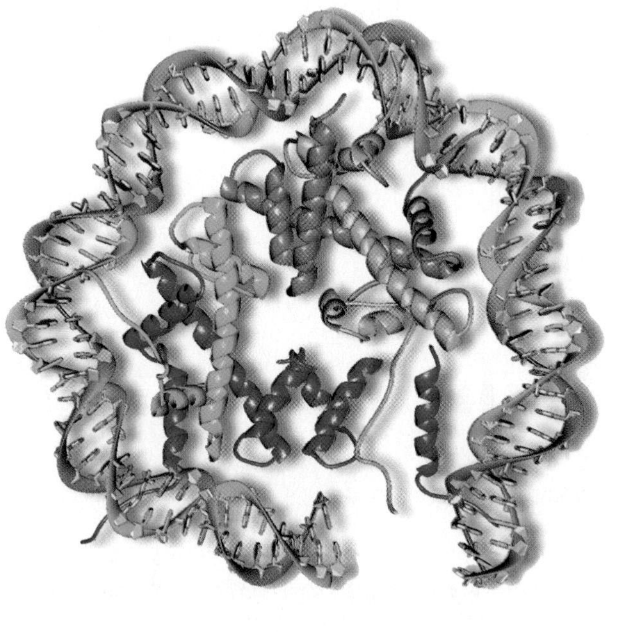

FIGURE 12–10 The nucleosome core particle derived from X-ray crystal analysis at 2.8 Å resolution. The double-helical DNA surrounds four pairs of histones.

Several of these potential chemical interactions are now recognized as important to genetic function. One of the most well-studied histone modifications involves **acetylation**, resulting from the action of histone acetyltransferase (HAT). This enzyme adds an acetyl group to the positively charged amino group present on the side chain of the amino acid lysine, effectively changing the net charge of the protein by neutralizing the positive charge. Lysine is in abundance in histones, and it has been known for some time that acetylation is linked to gene activation. It appears that high levels of acetylation open up the chromatin fiber, an effect that increases in regions of active genes and decreases in inactive regions. A well-known example involves the inactive X chromosome forming a Barr body in mammals (Chapter 7), in which histone H4 is known to be greatly underacetylated.

The two other important chemical modifications include the **methylation** and **phosphorylation** of amino acids that are part of histones. These chemical processes result from the action of enzymes called methyltransferases and kinases, respectively. Methyl groups can be added to both arginine and lysine of histones, and this change has been correlated with the activation of genes. Phosphate groups can be added to the hydroxyl groups of the amino acids serine and histidine, introducing a negative charge on the protein. During the cell cycle, increased phosphorylation, particularly of histone H3, is known to occur at characteristic times. Such chemical modification is believed to be related to the cycle of chromatin unfolding and condensation that occurs during and after DNA replication.

Interestingly, while methylation of histones in nucleosomes is positively correlated with gene activity in eukaryotes, methylation of the nitrogenous base cytosine within polynucleotide chains of DNA, forming **5-methyl cytosine**, is negatively correlated with gene activity. Methylation occurs most often when the nucleotide cytidylic acid is next to the nucleotide guanylic acid (forming what is called a **CpG island**).

The above discussion extends our knowledge of nucleosomes and chromatin organization and may serve as a general introduction to the concept of chromatin remodeling. A great deal more work must be done to elucidate the specific involvement of chromatin remodeling during genetic processes. In particular, the way in which the modifications are influenced by regulatory molecules within cells will provide important insights into our understanding of gene expression. What is clear is that the dynamic forms in which chromatin exists are vitally important to the way that all genetic processes directly involving DNA are executed. We will return to a more detailed discussion of the role of chromatin remodeling when we consider the regulation of eukaryotic gene expression in Chapter 17.

Heterochromatin

The discussion in the previous section implies that, along its entire length, each eukaryotic chromosome consists of one continuous double-helical DNA molecule present in repeating nucleosomes, which might lead to the belief that the whole chromosome is structurally uniform. However, in the early part of the 20th century, it was observed that, while most of the chromosome is present as uncoiled chromatin, some parts of the chromosome remain condensed and stain deeply during interphase. In 1928, the terms **euchromatin** and **heterochromatin** were coined to describe the parts of chromosomes that are uncoiled and those that remain condensed, respectively.

Subsequent investigation revealed a number of characteristics that distinguish heterochromatin from euchromatin. Heterochromatic areas are genetically inactive because they either lack genes or contain genes that are repressed. Also, heterochromatin replicates later during the S phase of the cell cycle than does euchromatin. The discovery of heterochromatin provided the first clues that parts of eukaryotic chromosomes do not always encode proteins.

Heterochromatin can be found in different regions of eukaryotic chromosomes. Early cytological studies showed that areas of the centromeres are composed of heterochromatin. The ends of chromosomes, called telomeres, are also heterochromatic. In some cases, in fact, whole chromosomes may be heterochromatic. Such is the case with a portion of the mammalian Y chromosome, much of which is genetically inert. And, as we discussed in Chapter 7, the inactivated X chromosome in mammalian females is condensed into an inert heterochromatic Barr body. In some species, such as mealy bugs, all chromosomes of one entire haploid set are heterochromatic.

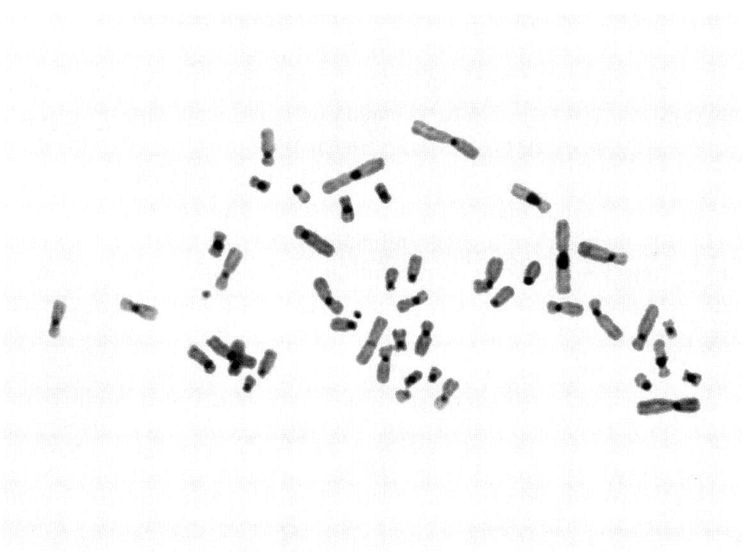

FIGURE 12–11 A human mitotic chromosome preparation processed to demonstrate C-banding. Only the centromeres stain.

When certain heterochromatic areas from one chromosome are translocated to a new site on the same or another nonhomologous chromosome, genetically active areas sometimes become genetically inert if they now lie adjacent to the translocated heterochromatin. As we saw in Chapter 4, this influence on existing euchromatin is one example of what is more generally referred to as a **position effect**. That is, the position of a gene or groups of genes relative to all other genetic material may affect their expression.

12.5 Chromosome Banding Differentiates Regions along the Mitotic Chromosome

Until about 1970, mitotic chromosomes viewed under the light microscope could be distinguished only by their relative sizes and the positions of their centromeres. Even in organisms with a low haploid number, two or more chromosomes are often indistinguishable from one another. However, new cytological procedures made possible differential staining along the longitudinal axis of mitotic chromosomes. Such methods are now referred to as chromosome-banding techniques, because the staining patterns resemble the bands of polytene chromosomes.

One of the first chromosome-banding techniques was devised by Mary Lou Pardue and Joe Gall. They found that if chromosome preparations from mice were heat denatured and then treated with Giemsa stain, a unique staining pattern emerged. Only the centromeric regions of mitotic chromosomes took up the stain! The staining pattern was thus referred to as **C-banding**. Relevant to our immediate discussion, this cytological technique identifies a specific area of the chromosome composed of heterochromatin. A micrograph of the human karyotype treated in this way is shown in Figure 12–11. Mouse chromosomes are all telocentric, thus localizing the stain at the end of each chromosome.

Other chromosome-banding techniques were developed about the same time. The most useful of these produces a staining pattern differentially along the length of each chromosome. This method, producing **G-bands** (Figure 12–12), involves the digestion of the mitotic chromosomes with the proteolytic enzyme trypsin, followed by Giemsa staining. The differential staining reactions reflect the heterogeneity and complexity of the chromosome along its length.

In 1971 a uniform nomenclature for human chromosome-banding patterns was established based on G-banding.

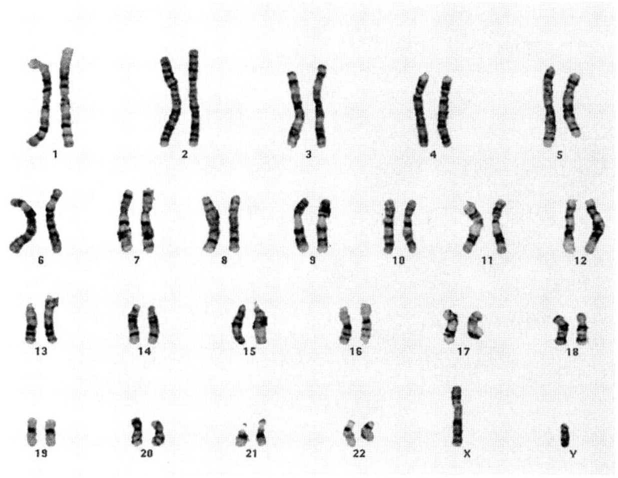

FIGURE 12–12 G-banded karyotype of a normal human male. Chromosomes were derived from cells in metaphase.

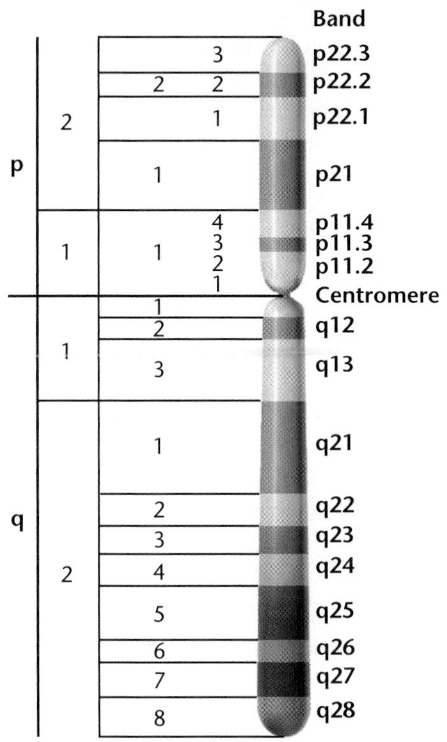

FIGURE 12–13 The regions of the human X chromosome distinguished by its banding pattern. The designations on the right identify specific bands.

Figure 12–13 illustrates the application of this nomenclature to the X chromosome. On the left of the chromosome are the various organizational levels of banding of the p and q arms that can be identified, while the designations on the right side identify specific regions within each specific region of the chromosome.

Although the molecular mechanisms involved in producing the various banding patterns are not well understood, the significance of banding has been great in cytogenetic analysis, particularly in humans. The pattern of banding of each chromosome is unique, allowing a distinction to be made even between those chromosomes that are identical in size and centromere placement (e.g., human chromosomes 4 and 5 and 21 and 22). So precise is the banding pattern of each chromosome that homologs can be distinguished from one another, and when a segment of one chromosome has been translocated to another chromosome, its origin can be determined with great precision.

12.6 Eukaryotic Chromosomes Demonstrate Complex Organization Characterized by Repetitive DNA

Thus far, we have examined how DNA is organized *into* chromosomes in bacteriophages, bacteria, and eukaryotes. We now begin an examination of what we know about the organization of particular sequences, ones that are repeated many times *within* eukaryotic chromosomes. These are referred to generally as **repetitive DNA**. As we shall see, there are many variations of this phenomenon in eukaryotic genomes. Subsequently in Chapter 20 we will focus on how genes themselves are organized within chromosomes.

We learned in Chapter 10 that, in addition to single copies of unique DNA sequences that make up genes, a great deal of the DNA sequences within eukaryotic chromosomes is repetitive in nature and that various levels of repetition occur within the genome of organisms. Many studies have now provided insights into repetitive DNA, demonstrating various classes of these sequences and their organization. Figure 12–14 outlines the various categories of repetitive DNA, which you can make reference to throughout the discussion in this section.

The figure also shows that some genes are present in more than one copy (referred to as multiple copy genes) and so are repetitive in nature. However, the majority of repetitive sequences is nongenic. In fact, the function, if any, of most repetitive sequences remains unknown. We will explore three main categories: (1) heterochromatin found associated with centromeres and making up telomeres; (2) tandem repeats of both short and longer DNA sequences; and (3) transposable sequences that are interspersed throughout the genome of eukaryotes.

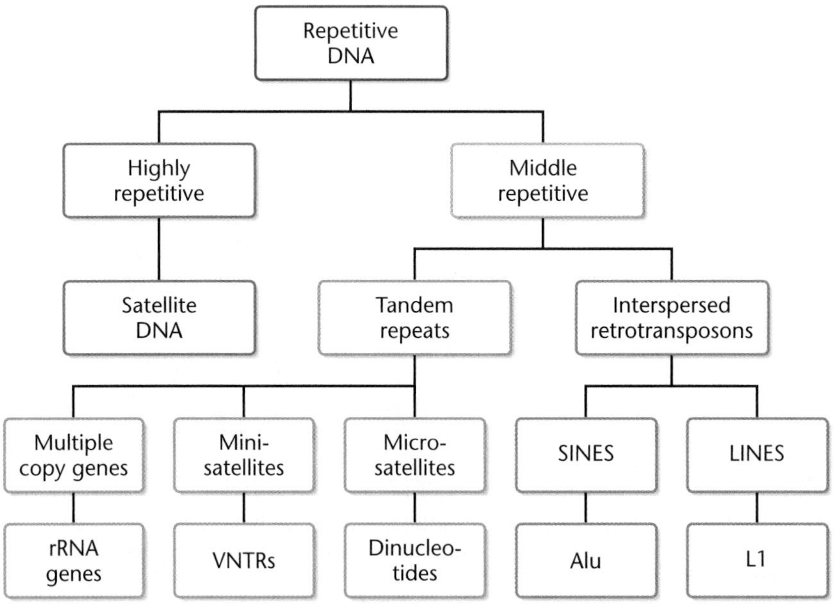

FIGURE 12–14 An overview of the various categories of repetitive DNA.

Repetitive DNA and Satellite DNA

The nucleotide composition of the DNA (e.g., the percentage of G≡C versus A=T pairs) of a particular species is reflected in its density, which can be measured with sedimentation equilibrium centrifugation (introduced in Chapter 10). When eukaryotic DNA is analyzed in this way, the majority of it is present as a single main peak or band of fairly uniform density. However, one or more additional peaks represent DNA that differs slightly in density. One such component, called **satellite DNA**, represents a variable proportion of the total DNA, depending on the species. For example, a profile of main-band and satellite DNA from the mouse is shown in Figure 12–15. By contrast, prokaryotes contain only main-band DNA and are thus devoid of satellite DNA.

The significance of satellite DNA remained an enigma until the mid-1960s, when Roy Britten and David Kohne developed a technique for measuring the reassociation kinetics of DNA that had previously been dissociated into single strands (Chapter 10). The researchers demonstrated that certain portions of DNA reannealed more rapidly than others and concluded that rapid reannealing was characteristic of multiple DNA fragments composed of identical or nearly identical nucleotide sequences—the basis for the descriptive term **repetitive DNA**. Recall that, in contrast, prokaryotic DNA is nearly devoid of anything other than unique, single-copy sequences.

When satellite DNA was subjected to analysis by reassociation kinetics, it fell into the category of *highly repetitive DNA* and is known to consist of short sequences repeated a large number of times. Further evidence suggested that these sequences are present as tandem repeats clustered in very specific chromosomal areas known to be heterochromatic—the regions flanking centromeres. This was discovered in 1969 when several researchers, including Mary Lou Pardue and Joe Gall, applied the technique of ***in situ* molecular hybridization** to the study of satellite DNA. The technique (see Appendix A) involves the molecular hybridization between an isolated fraction of radioactively labeled DNA or RNA probes and the DNA contained in the chromosomes of a cytological preparation. Following the hybridization procedure, autoradiography is performed to locate the chromosome areas complementary to the fraction of DNA or RNA.

In their work, Pardue and Gall demonstrated that molecular probes made from mouse satellite DNA hybridize with DNA of centromeric regions of mouse mitotic chromosomes (Figure 12–16). The following conclusions can be drawn: Satellite DNA differs from main-band DNA in its molecular composition, as established by buoyant density studies. It is also composed of short repetitive sequences. Finally, most satellite DNA is found in the heterochromatic centromeric regions of chromosomes.

HOW DO WE KNOW?

How do we know experimentally that satellite DNA consists of repetitive sequences and has been derived from regions of the centromere?

Centromeric DNA Sequences

Centromeres, described cytologically in the late 19th century as the *primary constrictions* along eukaryotic chromosomes, play several crucial roles during mitosis and meiosis. First, they are responsible for the maintenance of sister chromatid cohesion prior to the anaphase stage. It is at the point of the centromere along the chromosome that sister chromatids remain paired together during the early stages of mitosis and meiosis. Second, centromeres are the site of the formation of the kinetochore, the proteinaceous platform that is organized around the centromere and attaches to the microtubules of spindle fibers. Hence, centromeres mediate chromosome migration during the anaphase stage. This process is essential to the sep-

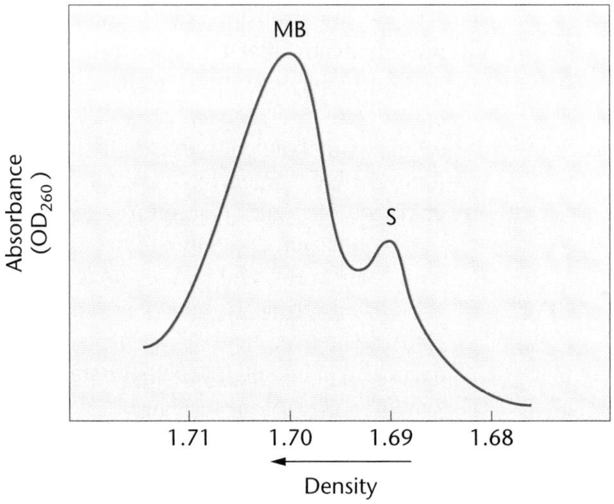

FIGURE 12–15 Separation of main-band (MB) and satellite (S) DNA from the mouse by using ultracentrifugation in a CsCl gradient.

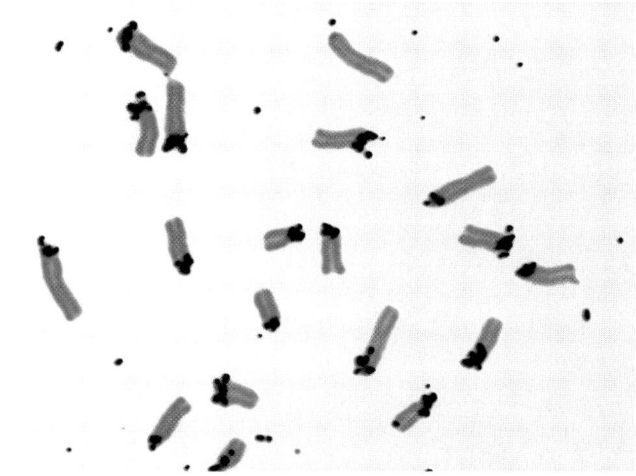

FIGURE 12–16 *In situ* hybridization between a radioactive probe representing mouse satellite DNA and mitotic chromosomes. The grains in the autoradiograph localize the chromosome regions (the centromeres) containing satellite DNA sequences.

aration of chromatids and thus the fidelity of chromosome distribution during cell division.

Most estimates of infidelity during mitosis are exceedingly low: 1×10^{-5} to 1×10^{-6} or 1 error per 100,000 to 1,000,000 cell divisions. As a result, it has been generally assumed that the analysis of the DNA sequence of centromeric regions will provide insights into the rather remarkable features of this chromosomal region. This DNA region is designated the **CEN**, which has now been defined and investigated in a number of organisms.

The analysis of the CEN regions of yeast *Saccharomyces cerevisiae* chromosomes provided the basis for a model system first described by John Carbon and Louis Clarke. Because each centromere serves an identical function, it is not surprising that all CENs were found to be remarkably similar in their organization. The CEN region of all 16 yeast chromosomes in *Saccaromyces cerevisiae* consists of about 125 bp, which can be divided into three regions (Figure 12–17). The first and third regions (I and III) are relatively short and highly conserved on yeast chromosomes, consisting of only 8 and 26 bp, respectively. Region II, which is larger (80–85 bp) and extremely A=T rich (up to 95 percent), varies in sequence among different chromosomes.

Mutational analysis suggests that regions I and II are less critical to centromere function than is region III. Mutations in the former regions are often tolerated, but mutations in region III can disrupt centromere function. While the DNA of this region appears to be essential to the eventual binding to the spindle fiber, DNA sequences are not unique to specific chromosomes. They can be experimentally exchanged between chromosomes without altering centromere function.

Based on the studies in yeast, it was assumed that similar findings would be forthcoming in the multicellular eukaryotes. However, to the surprise of researchers, the DNA sequence in organisms such as mammals, including humans, varies considerably and has not been highly conserved. This and other observations have led to the conclusion that, in these organisms, the CEN sequences themselves are not essential to centromere function.

The amount of DNA associated with the centromeric region in multicellular organisms is much more extensive than in yeast. Recall from our discussion that highly repetitive satellite DNA is localized in the centromere regions of mice. Such sequences, absent from yeast, but characteristic of most multicellular organisms, vary considerably in size. For example, the 10 bp sequence AATAACATAG is tandemly repeated many times in the centromeres of all four chromosomes of *Drosophila*. In humans, one of the most recognized satellite DNA sequences is the **alphoid family**. Found mainly in the centromere regions, a motif of alphoid DNA of 171 bp is present in tandem head-to-tail repeating arrays, totaling up to 3 million base pairs. While such a motif is present in other closely related primates, neither the sequence nor the number of repeats of the 171 bp sequence is conserved. While the precise role of this highly repetitive DNA in centromere function remains unclear, it is known that the sequences are not transcribed.

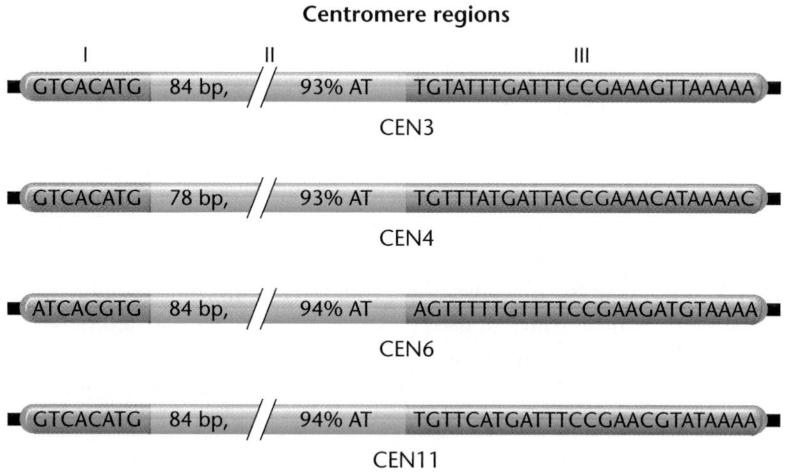

FIGURE 12–17 Nucleotide sequence information derived from DNA of the three major centromere regions of yeast chromosomes 3, 4, 6, and 11.

Telomeric DNA Sequences

Another important structure is the **telomere**, a component of chromosomes that we have discussed earlier in this chapter and in Chapter 11 when we considered problems associated with DNA replication at the telomere. Found at the ends of linear chromosomes, the function of telomeres is to provide stability to the chromosome by rendering chromosome ends generally inert in interactions with other chromosome ends. In contrast to broken chromosomes, whose ends may rejoin other such ends, telomere regions do not fuse with one another or with broken ends. It is thought that some aspect of the molecular structure of telomeres must be unique compared with most other chromosome regions.

As with centromeres, the analysis of telomeres was first approached by investigating the smaller chromosomes of simple eukaryotes, such as protozoans and yeast. The idea that all telomeres of all chromosomes in a given species might share a common nucleotide sequence has now been borne out.

Two types of telomere sequences have been discovered. The first type, simply called **telomeric DNA sequences**, consists of short tandem repeats. It is this group that contributes to the stability and integrity of the chromosome. In the ciliate *Tetrahymena*, more than 50 tandem repeats of the hexanucleotide sequence GGGGTT occur. In humans, the sequence GGGATT is repeated many times. The analysis of telomeric DNA sequences has shown them to be highly conserved throughout evolution, reflecting the critical role they play in maintaining the integrity of chromosomes.

The second type, **telomere-associated sequences**, also consists of repetitive sequences and is found both adjacent to and within the telomere. These sequences vary among organisms, and their significance remains unknown.

Interestingly, the telomeres of *Drosophila* are fundamentally different from those found thus far in all other organisms in that they are made up of transposable elements. The significance of this finding, however, is not yet clear.

As discussed in Chapter 11, replication of the telomere requires a unique RNA-containing enzyme, telomerase. In its absence, the DNA at the ends of chromosomes becomes shorter during each replication. Because single-celled eukaryotes are immortalized cells, telomerase is critical for the survival of such species. In multicellular organisms, such as humans, telomerase is active in germ-line cells, but is inactive in somatic cells. Chromosome shortening is considered part of the natural process of cell aging, serving as an internal clock. In human cancer cells, which have become immortalized, the transition to malignancy appears to require the activation of telomerase in order to overcome the normal senescence associated with chromosome shortening.

Middle Repetitive Sequences: VNTRs and Dinucleotide Repeats

A brief review of still another prominent category of repetitive DNA sheds more light on our understanding of the organization of the eukaryotic genome. In addition to highly repetitive DNA, which constitutes about 5 percent of the human genome (and 10 percent of the mouse genome), a second category, **middle** (or **moderately**) **repetitive DNA**, recognized by C_0t analysis, is fairly well characterized. Because we are learning a great deal about the human genome, we will use our own species to illustrate this category of DNA in genome organization.

While middle repetitive DNA does include some duplicated genes (such as those encoding ribosomal RNA), most prominent in this category are either noncoding tandemly repeated or interspersed sequences. No function has been ascribed to these components of the genome. An example includes those called **variable number tandem repeats (VNTRs)**. The repeating DNA sequence of VNTRs may be 15 to 100 bp long and is found within and between genes. Many such clusters are dispersed throughout the genome, and they are often referred to as **minisatellites**.

The number of tandem copies of each specific sequence at each location varies in individuals, creating localized regions of 1,000 to 20,000 bp (1–20 kb) in length. As we will see in Chapter 22, the variation in size (length) of these regions between individuals in humans was originally the basis for the forensic technique referred to as **DNA fingerprinting**.

Another group of tandemly repeated sequences consists of di-, tri-, tetra-, and pentanucleotides, also referred to as **microsatellites**. Like VNTRs, they are dispersed throughout the genome and vary among individuals in the number of repeats present at any site. For example, in humans, the most common microsatellite is the dinucleotide $(CA)_n$, where n equals the number of repeats. Most commonly, n is between 5 and 50. These clusters have served as useful molecular markers during genome analysis.

Repetitive Transposed Sequences: SINES and LINES

Still another category of repetitive DNA consists of sequences that are interspersed throughout the genome, rather than being tandemly repeated. They can be either short or long and many have the added distinction of being **transposable sequences**, which are mobile and can move to different locations within the genome. A large portion of eukaryotic genomes are composed of such sequences.

For example, **short interspersed elements**, called **SINES**, are less than 500 base pairs long and may be present 500,000 times or more in the human genome. The best characterized human SINE is a set of closely related sequences called the ***Alu* family** (the name is based on the presence of DNA sequences recognized by the restriction endonuclease *Alu*I). Members of this DNA family, also found in other mammals, are 200 to 300 base pairs long and are dispersed rather uniformly throughout the genome, both between and within genes. In humans, this family encompasses more than 5 percent of the entire genome.

Alu sequences are particularly interesting, although their function, if any, is yet undefined. Members of the *Alu* family are sometimes transcribed. The role of this RNA is not certain, but it is thought to be related to its mobility in the genome. In fact, *Alu* sequences are thought to have arisen from an RNA element whose DNA complement was dispersed throughout the genome as a result of the activity of reverse transcriptase (an enzyme that synthesizes DNA on an RNA template).

The group of **long interspersed elements (LINES)** represents still another category of repetitive transposable DNA sequences. In humans, the most prominent example is a family designated **L1**. Members of this sequence family are about 6400 base pairs long and are present up to 100,000 times, according to one estimate. Their 5′ end is highly variable, and their role within the genome has yet to be defined.

The basis for transposition of L1 elements is now clear. The L1 DNA sequence is first transcribed into an RNA molecule. The RNA then serves as the template for the synthesis of the DNA complement using the enzyme reverse transcriptase. This enzyme is encoded by a portion of the L1 sequence. The new L1 copy then integrates into the DNA of the chromosome at a new site. Because of the similarity of this transposition mechanism with that used by retroviruses, LINES are referred to as **retrotransposons**.

SINES and LINES represent a significant portion of human DNA. Both types of elements share the organizational feature of consisting of a mixture of about 70 percent unique and 30 percent repeating sequences within the DNA of each entity. Collectively, they constitute about 10 percent of the genome.

Middle Repetitive Multiple-Copy Genes

In some cases, middle repetitive DNA includes functional genes present tandemly in multiple copies. For example, many copies exist of the genes encoding ribosomal RNA. *Drosophila* has 120 copies per haploid genome. Single genetic units encode a large precursor molecule that is processed into the 5.8*S*, 18*S*, and 28*S* rRNA components. In humans, multiple copies of this gene are clustered on the p arm of the acrocentric chromosomes 13, 14, 15, 21, and 22. Multiple copies of the genes encoding 5*S* rRNA are transcribed separately from multiple clusters found together on the terminal portion of the p arm of chromosome 1.

12.7 The Vast Majority of a Eukaryotic Genome Does Not Encode Functional Genes

Given the preceding information involving various forms of repetitive DNA in eukaryotes, it is of interest to pose an important question: *What proportion of the eukaryotic genome actually encodes functional genes?* Taken together, the various forms of highly repetitive and moderately repetitive DNA constitute up to 40 percent of the human genome. Such an observation is not uncommon in eukaryotes. In addition to repetitive DNA, there is a large amount of single-copy DNA sequences as defined by C_0t analysis that appear to be noncoding. A small portion of them are called **pseudogenes**, which represent evolutionary vestiges of duplicated copies of genes that have undergone sufficient mutations to render them nonfunctional. In some cases, multiple copies of such genes exist.

While the proportion of the genome consisting of repetitive DNA varies among organisms, one feature seems to be shared: *Only a very small part of the genome actually codes for proteins.* For example, the 20,000 to 30,000 genes encoding proteins in sea urchin occupy less than 10 percent of the genome. In *Drosophila*, only 5 to 10 percent of the genome is occupied by genes coding for proteins. In humans, it appears that the estimated 20,000 to 25,000 functional genes occupy less than 5 percent of the genome.

The study of the various forms of repetitive DNA has significantly enhanced our understanding of genome organization. In the next chapter, we will explore the organization of genes within chromosomes.

Chapter Summary

1. The organization of the molecular components that form chromosomes is essential to understanding the function of the genetic material. Largely devoid of associated proteins, bacteriophage and bacterial chromosomes contain DNA molecules in a form equivalent to the Watson–Crick model.
2. Polytene and lampbrush chromosomes are examples of specialized structures that have extended our knowledge of genetic organization and function.
3. The eukaryotic chromatin fiber is a nucleoprotein organized into repeating units called nucleosomes. Composed of 200 base pairs of DNA, an octamer of four types of histones, plus one linker histone, the nucleosome is important in facilitating the conversion of the extensive chromatin fiber characteristic of interphase into the highly condensed chromosome seen in mitosis.
4. The structural heterogeneity of the chromosome axis has been established as a result of both biochemical and cytological investigation. Heterochromatin, prematurely condensed in interphase is, for the most part, genetically inert. The centromeric and telomeric regions, the Y chromosome, and the Barr body are examples.
5. DNA analysis has revealed unique nucleotide sequences in both the centromere and telomere regions of chromosomes that appear to be related to their respective roles.
6. Eukaryotic genomes demonstrate complex sequence organization characterized by numerous categories of repetitive DNA.
7. Repetitive DNA consists of either tandem repeats clustered in various regions of the genome or single sequences interspersed uniformly throughout the genome. In the former group, the size of each cluster varies among individuals, providing one form of biochemical identity. The latter group of sequences may be short or long, such as *Alu* and L1, respectively, and are transposable elements.
8. The vast majority of a eukaryotic genome does not encode functional genes. In humans, for example, less than 5 percent of the genome is used to encode the 20,000 to 25,000 genes found in our genome.

Insights and Solutions

A previously undiscovered single-celled organism was found living at a great depth on the ocean floor. Its nucleus contained only a single linear chromosome containing 7×10^6 nucleotide pairs of DNA coalesced with three types of histonelike proteins.

1. A short micrococcal nuclease digestion yielded DNA fractions consisting of 700, 1400, and 2100 base pairs. Predict what these fractions represent. What conclusions can be drawn?

Solution: The chromatin fiber may consist of a variation of nucleosomes containing 700 base pairs of DNA. The 1400- and 2100-bp fractions represent two and three nucleosomes, respectively, linked together. Enzymatic digestion may have been incomplete, leading to the latter two fractions.

2. The analysis of individual nucleosomes revealed that each unit contained one copy of each protein, and that the short linker DNA contained no protein bound to it. If the entire chromosome consists of nucleosomes (discounting any linker DNA), how many are there, and how many total proteins are needed to form them?

Solution: Since the chromosome contains 7×10^6 base pairs of DNA, the number of nucleosomes, each containing 7×10^2 base pairs, is equal to

$$7 \times 10^6/7 \times 10^2 = 10^4 \text{ nucleosomes}$$

The chromosome thus contains 10^4 copies of each of the three proteins, for a total of 3×10^4 molecules.

3. Analysis then revealed the organism's DNA to be a double helix similar to the Watson–Crick model, but containing 20 base pairs per complete turn of the right-handed helix. The physical size of the nucleosome was exactly double the volume occupied by that found in all other known eukaryotes, by virtue of increasing the distance along the fiber axis by a factor of two. Compare the degree of compaction (the number of turns per nucleosome) of this organism's nucleosome with that found in other eukaryotes.

Solution: The unique organism compacts a length of DNA consisting of 35 complete turns of the helix (700 base pairs per nucleosome/20 base pairs per turn) into each nucleosome. The normal eukaryote compacts a length of DNA consisting of 20 complete turns of the helix (200 base pairs per nucleosome/10 base pairs per turn) into a nucleosome half the volume of that in the unique organism. The degree of compaction is therefore less in the unique organism.

4. No further coiling or compaction of this unique chromosome occurs in the unique organism. Compare this situation with that of a eukaryotic chromosome. Do you think an interphase human chromosome 7×10^6 base pairs in length would be a shorter or longer chromatin fiber?

Solution: The eukaryotic chromosome contains still another level of condensation in the form of solenoids, which are dependent on the H1 histone molecule associated with linker DNA. Solenoids condense the eukaryotic fiber by still another factor of five. The length of the unique chromosome is compacted into 10^4 nucleosomes, each containing an axis length twice that of the eukaryotic fiber. The eukaryotic fiber consists of $7 \times 10^6/2 \times 10^2 = 3.5 \times 10^4$ nucleosomes, 3.5 more than the unique organism. However, they are compacted by the factor of five in each solenoid. Therefore, the chromosome of the unique organism is a longer chromatin fiber.

Problems and Discussion Questions

1. Contrast the size of the chromosome of bacteriophage λ and T2 with that of *E. coli*. How does this relate to the relative size and complexity of these phages and that bacterium?
2. Bacteriophages and bacteria almost always contain their DNA as circular (closed loops) chromosomes. Phage λ is an exception, maintaining its DNA in a linear chromosome within the viral particle. However, as soon as it is injected into a host cell, it circularizes before replication begins. Taking into account information in Chapter 11, what advantage exists in replicating circular DNA molecules compared to linear molecules?
3. Describe how giant polytene chromosomes are formed.
4. Salivary gland cells from *Drosophila* are isolated and cultured in the presence of radioactive thymidylic acid. Autoradiography is performed, revealing polytene chromosomes. Predict the distribution of the grains along the chromosomes.
5. What genetic process is occurring in a puff of a polytene chromosome?
6. Describe the structure of LINE sequences. Why are LINES referred to as retrotransposons?
7. During what genetic process are lampbrush chromosomes present in vertebrates?
8. Why might we predict that the organization of eukaryotic genetic material will be more complex than that of viruses or bacteria?
9. Describe the sequence of research findings that led to the development of the model of chromatin structure.
10. What is the molecular composition and arrangement of the components in the nucleosome?
11. Describe the transitions that occur as nucleosomes are coiled and folded, ultimately forming a chromatid.
12. Provide a comprehensive definition of heterochromatin, and list as many examples as you can.
13. Mammals contain a diploid genome consisting of at least 10^9 bp. If this amount of DNA is present as chromatin fibers, where each

group of 200 bp of DNA is combined with 9 histones into a nucleosome and each group of 6 nucleosomes is combined into a solenoid, achieving a final packing ratio of 50, determine (a) the total number of nucleosomes in all fibers, (b) the total number of histone molecules combined with DNA in the diploid genome, and (c) the combined length of all fibers.

14. Assume that a viral DNA molecule is a 50-μm-long circular strand of a uniform 20 Å diameter. If this molecule is contained in a viral head that is a 0.08-μm-diameter sphere, will the DNA molecule fit into the viral head, assuming complete flexibility of the molecule? Justify your answer mathematically.

15. How many base pairs are in a molecule of phage T2 DNA 52 μm long?

16. If a human nucleus is 10 μm in diameter, and it must hold as much as 2 m of DNA, which is complexed into nucleosomes that are 11 nm in diameter during full extension, what percentage of the volume of the nucleus is occupied by the genetic material?

Extra-Spicy Problems

17. Sun and others (2002. *Proc. Natl. Acad. Sci. (USA)* 99:8695–8700) studied *Drosophila* in which two normally active genes, w^+ (wild-type allele of the *white*-eye gene) and *hsp*26 (a heat-shock gene), were introduced (using a plasmid vector) into euchromatic and heterochromatic chromosomal regions. The relative activity of each gene was assessed, and an approximation of the data obtained is shown below. Considering three characteristics of heterochromatin, which is/are supported by the experimental data?

	Activity (relative percentage)	
Gene	Euchromatin	Heterochromatin
hsp26	100%	31%
w^+	100%	8%

18. Using molecular methods to "label" chromosomes with fluorescent dyes, Nagele and colleagues (1995. *Science* 270:1831–35) observed a precise nuclear positioning of chromosomes during an early stage of mitosis (prometaphase) in human fibroblast cells. Below is a sketch modified from this research that describes the relative positions of chromosomes 7, 8, 16, and X. Homologous chromosomes share the same color. Assuming that this pattern is consistent among other cells in humans, what conclusions can be drawn regarding the nuclear positions of the chromosomes during interphase and during the initial phases of mitosis? How could chromosomal localization influence gene function during interphase?

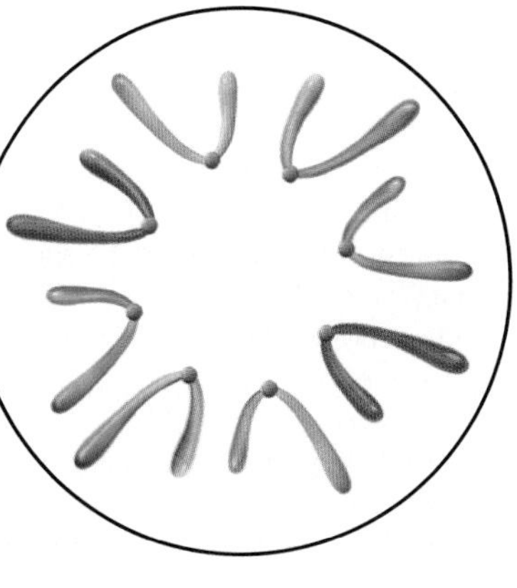

19. While much remains to be learned about the role of nucleosomes and chromatin structure and function, recent research indicates that *in vivo* chemical modification of histones is associated with changes in gene activity. For example, Bernstein and others (2000. *Proc. Natl. Acad. Sci. (USA)* 97:5340–45) determined that acetylation of H3 and H4 is associated with 21.1 percent and 13.8 percent increase in yeast gene activity, respectively, and that yeast heterochromatin is hypomethylated relative to the genome average. Speculate on the significance of these findings in terms of nucleosome–DNA interactions and gene activity.

20. In an article entitled "Nucleosome positioning at the replication fork," Lucchini and others (2002. *EMBO* 20:7294–302) state, "both the 'old' randomly segregated nucleosomes as well as the 'new' assembled histone octamers rapidly position themselves (within seconds) on the newly replicated DNA strands." Given this statement, how would one compare the distribution of nucleosomes and DNA in newly replicated chromatin? How could one experimentally test the distribution of nucleosomes on newly replicated chromosomes?

21. The human genome contains approximately 10^6 copies of an *Alu* sequence, one of the best-studied classes of short interspersed elements (SINES), per haploid genome. Individual *Alus* share a 282-nucleotide consensus sequence followed by a 3′-adenine-rich tail region (Schmid. 1998. *Nuc. Acids Res.* 26:4541–50). Given that there are approximately 3×10^9 bp per human haploid genome, about how many bp are spaced between each *Alu* sequence?

22. Below is a diagram of the general structure of the bacteriophage λ chromosome. Speculate on the mechanism by which it forms a closed ring upon infection of the host cell.

5′GGGCGGCGACCT—double-stranded region—3′
3′—double-stranded region—CCCGCCGCTGGA5′

23. Tandemly repeated DNA sequences with a repeat sequence of one to six base pairs for example, $(GACA)_n$, are called microsatellites and are common in eukaryotes. A particular subset, the trinucleotide repeats, is of great interest because of the role these repeats play in human neurodegenerative disorders (Huntington disease, myotonic dystrophy, spinal-bulbar muscular atrophy, spinocerebellar ataxia, and fragile X syndrome). Below are data modified from Toth and colleagues (originally derived in 2000, *Gen. Res.* 10:967–81) regarding the location of microsatellites within and between genes. What general conclusions can be drawn from these data?

Percentage of Microsatellite DNA Sequences within Genes and Between Genes

Taxonomic Group	Within Genes	Between Genes
Primates	7.4	92.6
Rodents	33.7	66.3
Arthropods	46.7	53.3
Yeasts	77.0	23.0
Other fungi	66.7	33.3

24. More information from the research effort in Problem 23 produced data regarding the pattern of the length of such repeats within genes. Each value in the following table represents the number of times a microsatellite of a particular sequence length, one to six bases long, is found in genes. For instance, in primates, a dinucleotide sequence (GC, for example) is found 10 times, while a trinucleotide is found 1126 times. In fungi, a repeat motif composed of 6 nucleotides (GACACC, for example) is found 219 times whereas a tetranucleotide repeat (GACA, for example) is found only 2 times. Analyze and interpret these data by indicating what general pattern is apparent regarding the distribution of various microsatellite lengths within genes. Of what significance might this general pattern display?

Distribution of Microsatellites by Unit Length within Genes

	Length of Repeated Motif (bp)					
Taxonomic Group	1	2	3	4	5	6
Primates	49	10	1126	29	57	244
Rodents	62	70	1557	63	116	620
Arthropods	12	34	1566	0	21	591
Yeasts	36	19	706	7	52	330
Other fungi	9	4	381	2	35	219

25. In spite of the considerable medical and biological significance of repetitive DNA sequences, the factors that determine their genesis and genomic distribution remain uncertain. Misalignment of repetitive DNA strands and DNA polymerase slippage have been described as mechanisms causing variation in the number of *existing* repeats. Until recently, there has been little information relating to the *initial* creation of a microsatellite genomic region. In 2001, Wilder and Hollocher (*Mol. Biol. Evol.* 18:384–92) sequenced DNA surrounding numerous tetranucleotide microsatellite regions in several strains within two species of *Drosophila* and observed the sequences shown below. (a) Identify the microsatellite tetranucleotide motif. Is it a perfect motif? Using Pu to represent a purine and Py to represent a pyrimidine, symbolize the tetranucleotide repeat in a form similar to $(CPuGPy)_n$. (b) What is the sequence of the nonmicrosatellite region? Is it a perfectly conserved region among all the species and strains listed?

Species (strain)	Base Sequence
D. nigrodunni-1	5′-TCGATATAGCCATGTCCGTCTGTCCGTCTGT
D. nigrodunni-2	5′-TCGATATAGCCATGTCCGTCTGTCCGTCTGT
D. nigrodunni-3	5′-TCGATATAGCAATGTCCGTCTGTCCGTCTGT
D. dunni-1	5′-TCGATATAGCAATGTCCGTCTGTCCGTCTGT
D. dunni-2	5′-TCGATATAGCCATGTCCGTCTGTCCGTCTGT

26. Regarding the findings and your analysis of data in Problem 25, what significance might there be to having a highly conserved nonmicrosatellite region flanking a specific microsatellite type?

27. Microsatellites are currently exploited as markers for paternity testing. A sample paternity test is shown below, where ten microsatellite markers were used to test samples from a mother, her child, and an alleged father. The name of the microsatellite locus is given in the left-hand column and the genotype of each individual is recorded as the number of repeats he or she carries at that locus. For example, at locus D9S302, the mother carries 30 repeats on one of her chromosomes and 31 on the other. In cases where an individual carries the same number of repeats on both chromosomes, only a single number is recorded. (Some of the numbers are followed by a decimal point, e.g., 20.2, and include a partial repeat as well as several complete repeats.) Assuming that these markers are inherited in a simple Mendelian fashion, can the alleged father be excluded as the source of the sperm that produced the child? Why or why not? Explain.

Microsatellite Locus-Chromosome Location	Mother	Child	Alleged Father
D9S302-9q31-q33	30	31	32
	31	32	33
D22S883-22pter-22qter	17	20.2	20.2
	22	22	
D18S535-18q12.2-q12.3	12	13	11
	14	14	13
D7SI 804-7pter-7qter	27	26	26
	30	30	27
D3S2387-3p24.2.3pter	23	24	20.2
	25.2	25.2	24
D4S2386-4pter-qter	12	12	12
			16
D5S1719-5pter-5qter	11	10.3	10
	11.3	11	10.3
CSF1PO-5q33.3.q34	11	11	10
		12	12
FESFPS-15q25-15qter	11	12	10
	12	13	13
TH01-11p15.5	7	7	7
			8

28. Recall from Chapter 10 (Figure 10–20) that when double-stranded DNA is heated, the optical density rises as the helix denatures and the strands separate. How might this technique be used to characterize different components and sequences of the eukaryotic genome?

29. If DNA is analyzed in *Drosophila melanogaster* using density gradient centrifugation and found to contain a "satellite" peak distinct from main-band DNA, describe a set of experiments to determine as much as you can about the nature of this DNA in contrast to main-band DNA.

30. At the end of the short arm of human chromosome 16 (16p), several genes associated with disease are present, including thalassemia and polycystic kidney disease. When that region of chromosome 16 was sequenced, gene-coding regions were found to be very close to the telomere-associated sequences. Could there be a possible link between the location of these genes and the presence of the telomere-associated sequences? What further information concerning the disease genes would be useful in your analysis?

SELECTED READINGS

Angelier, N., et al. 1984. Scanning electron microscopy of amphibian lampbrush chromosomes. *Chromosoma* 89:243–53.

Bauer, W.R., Crick, F.H.C., and White, J.H. 1980. Supercoiled DNA. *Sci. Am.* (July) 243:118–33.

Beerman, W., and Clever, U. 1964. Chromosome puffs. *Sci. Am.* (Apr.) 210:50–58.

Carbon, J. 1984. Yeast centromeres: Structure and function. *Cell* 37:352–53.

Chen, T.R., and Ruddle, F.H. 1971. Karyotype analysis utilizing differential stained constitutive heterochromatin of human and murine chromosomes. *Chromosoma* 34:51–72.

Corneo, G., et al. 1968. Isolation and characterization of mouse and guinea pig satellite DNA. *Biochemistry* 7:4373–79.

DuPraw, E.J. 1970. *DNA and chromosomes*. New York: Holt, Rinehart & Winston.

Gall, J.G. 1981. Chromosome structure and the *C*-value paradox. *J. Cell Biol.* 91:3s–14s.

Hewish, D.R., and Burgoyne, L. 1973. Chromatin substructure. The digestion of chromatin DNA at regularly spaced sites by a nuclear deoxyribonuclease. *Biochem. Biophys. Res. Comm.* 52:504–10.

Korenberg, J.R. and Rykowski, M.C. 1988. Human genome organization: *Alu*, LINES, and the molecular organization of metaphase chromosome bands. *Cell* 53:391–400.

Kornberg, R.D. 1975. Chromatin structure: A repeating unit of histones and DNA. *Science* 184:868–71.

Kornberg, R.D., and Klug, A. 1981. The nucleosome. *Sci. Am.* (Feb.) 244:52–64.

Lorch, Y., Zhang, N, and Kornberg, R.D. 1999. Histone octamer transfer by a chromatin-remodeling complex. *Cell*. 96:389–92.

Luger, K., et al. 1997. Crystal structure of the nucleosome core particle at 2.8 resolution. *Nature* 389:251–56.

Moyzis, R.K. 1991. The human telomere. *Sci. Am.* (Aug.) 265:48–55.

Olins, A.L., and Olins, D.E. 1974. Spheroid chromatin units (*n* bodies). *Science* 183:330–32.

———1978. Nucleosomes: The structural quantum in chromosomes. *Am. Sci.* 66:704–11.

Singer, M.F. 1982. SINES and LINES: Highly repeated short and long interspersed sequences in mammalian genomes. *Cell* 28:433–34.

Wolfe, A. 1998. *Chromatin: Structure and Function*. 3rd ed. San Diego: Academic Press.

Yunis, J.J., 1976. High resolution of human chromosomes. *Science* 191:1268–70.

Yunis, J.J., and Prakash, O. 1982. The origin of man: A chromosomal pictorial legacy. *Science* 215:1525–30.

Chapter Thirteen

The Genetic Code and Transcription

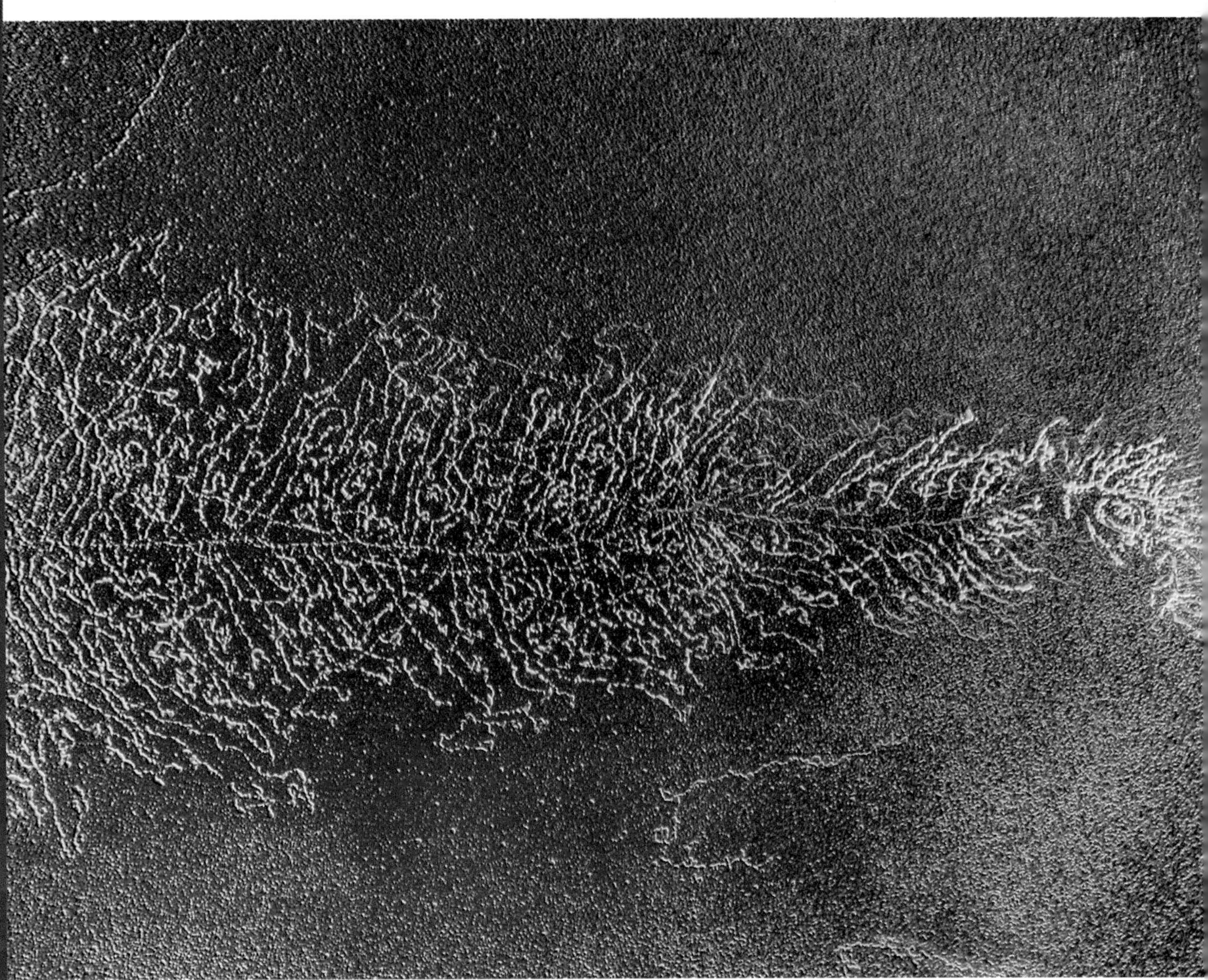

Electron micrograph visualizing the process of transcription.

Chapter Concepts

- Genetic information is stored in DNA using a triplet code that is nearly universal to all living things on Earth.
- The genetic code is initially transferred from DNA to RNA during the process of transcription.
- Once transferred to RNA, the genetic code exists as triplet codons, using the four ribonucleotides in RNA as the letters composing it.
- Using four different letters taken three at a time, 64 triplet sequences are possible to encode the 20 amino acids present in proteins, which serve as the end products of most genes.
- Several codons provide signals that initiate and terminate protein synthesis.
- The process of transcription is similar but more complex in eukaryotes compared to prokaryotes and bacteriophages that infect them.

As we saw in Chapter 10, the structure of DNA consists of a linear sequence of deoxyribonucleotides. This sequence ultimately dictates the components constituting proteins, the end product of most genes. A central issue is how such information stored as a nucleic acid can be decoded into a protein. Figure 13–1 provides a simple overview of how this transfer of information occurs. In the first step in gene expression, information present on one of the two strands of DNA (the template strand) is transferred into an RNA complement through the process of transcription. Once synthesized, this RNA acts as a "messenger" molecule, bearing the coded information—hence its name, **messenger RNA (mRNA)**. Such RNAs then associate with ribosomes, in which decoding into proteins occurs.

In this chapter, we will focus on the initial phases of gene expression by addressing two major questions. First, how is genetic information encoded? Second, how does the transfer from DNA to RNA occur, thus defining the process of transcription? As we will see, ingenious analytical research established that the genetic code is written in units of three letters—ribonucleotides present in mRNA that reflect the stored information in genes. Each triplet code word directs the incorporation of a specific amino acid into a protein as it is synthesized. As we can predict based on our prior discussion of the replication of DNA, transcription is also a complex process dependent on a major polymerase enzyme and a cast of supporting proteins. We will explore what is known about transcription in bacteria and then contrast this prokaryotic model with eukaryotes.

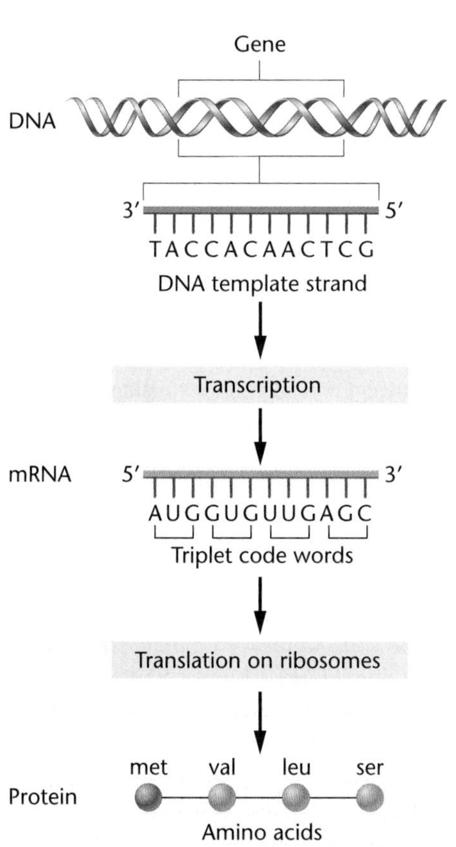

FIGURE 13–1 Flow of genetic information encoded in DNA to messenger RNA to protein.

In Chapter 14, we will continue our discussion of gene expression by addressing how translation occurs and then discussing the structure and function of proteins. Together, the information in these chapters provides a comprehensive picture of molecular genetics, which serves as the most basic foundation for the understanding of living organisms.

13.1 The Genetic Code Exhibits a Number of Characteristics

Before we consider the various analytical approaches that led to our current understanding of the genetic code, let's summarize the general features that characterize it:

1. The genetic code is written in linear form, using the ribonucleotide bases that compose mRNA molecules as "letters." The ribonucleotide sequence is derived from the complementary nucleotide bases in DNA.
2. Each "word" within the mRNA contains three ribonucleotide letters. Each group of *three* ribonucleotides, called a **codon**, specifies *one* amino acid.
3. The code is **unambiguous**, meaning that each codon specifies only a single amino acid.
4. The code is **degenerate**, meaning that a given amino acid can be specified by more than one codon. This is the case for 18 of the 20 amino acids.
5. The code contains "start" and "stop" signals, certain codons that are necessary to **initiate** and to **terminate** translation.
6. No internal punctuation ("commas") is used in the code. Thus, the code is said to be **commaless**. Once translation of mRNA begins, the codons are read one after the other with no breaks between them.
7. The code is **nonoverlapping**. After translation commences, any single ribonucleotide at a specific location within the mRNA is part of only one codon.
8. The code is nearly **universal**. With only minor exceptions, a single coding dictionary is used by almost all viruses, prokaryotes, archaea, and eukaryotes.

13.2 Early Studies Established the Basic Operational Patterns of the Code

Web Tutorial 13.1 Transcription

In the late 1950s, before it became clear that mRNA serves as an intermediate in transferring genetic information from DNA to proteins, it was considered that DNA itself might directly participate during the synthesis of proteins. Because ribosomes had already been identified, the initial thinking was that the DNA of a gene might somehow become associated with ribosomes, in which the stored information was decoded during the synthesis of proteins. Most early models, faced with the question of how four nucleotides could encode 20 amino acids, embraced the idea that a DNA code would be overlapping. That is, each nucleotide would be part of more than one contiguous code word. Such a concept soon became untenable as accumulating evidence demonstrated inconsistencies resulting from constraints placed by an overlapping code model on the resultant amino acid sequences. We will address some of this evidence shortly. Additionally, there was growing evidence that RNA might be involved in the process. In 1961, François Jacob and Jacques Monod postulated the existence of **messenger RNA (mRNA)**. Once mRNA was discovered, it soon became clear that, even though genetic information is stored in DNA, the code which is translated into proteins resides in RNA. The central question then was how only four letters—the four nucleotides—could specify 20 words, the amino acids.

The Triplet Nature of the Code

In the early 1960s, Sidney Brenner argued on theoretical grounds that the code must be a triplet, since three-letter words represent the minimal use of four letters to specify 20 amino acids. A code of four nucleotides, taken two at a time, for example, would provide only 16 unique code words (4^2). A triplet code would provide 64 words (4^3)—clearly more than the 20 needed—and would be much simpler than a four-letter code (4^4), which would specify 256 words.

The ingenious experimental work of Francis Crick, Leslie Barnett, Brenner, and R. J. Watts-Tobin presented the first solid evidence for the triplet nature of the code. These researchers induced insertion and deletion mutations in the *rII* locus of phage T4. (See Chapter 6.) Mutations in this locus cause rapid lysis and distinctive plaques. Such mutants will successfully infect strain B of *E. coli*, but cannot reproduce on a separate strain of *E. coli*, designated K12. Crick and his colleagues used the acridine dye proflavin to induce mutations. This mutagenic agent intercalates within the double helix of DNA, often causing the insertion or the deletion of one or more nucleotides during replication. As shown in Figure 13–2(a), an insertion of a single nucleotide causes the frame of reading to shift, whereby the specific sequence of all subsequent downstream codons (to the right of the insertion in the figure) are changed. Thus, they are referred to as **frameshift mutations**. Upon translation, the amino acid sequence of the encoded protein will be altered. When these mutations are present at the *rII* locus, T4 will not reproduce on *E. coli* K12.

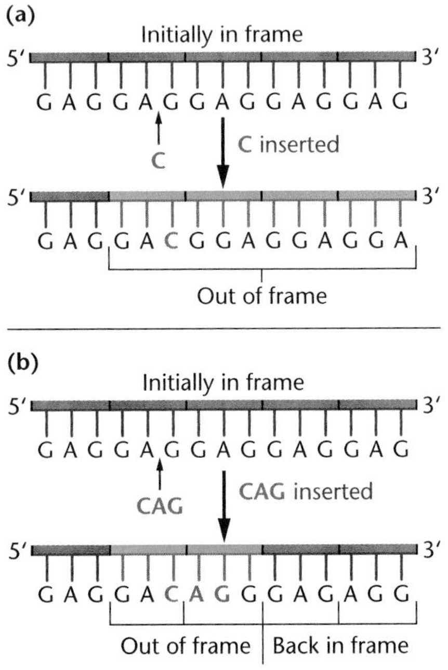

FIGURE 13–2 The effect of frameshift mutations on a DNA sequence repeating the triplet sequence GAG. (a) The insertion of a single nucleotide shifts all subsequent triplet reading frames. (b) The insertion of three nucleotides changes only two codons, but the frame of reading is then reestablished to the original sequence.

Crick and his colleagues reasoned that if phages with these induced mutations were treated again with proflavin, still other insertions or deletions would occur. A second change might result in a revertant phage, which would display wild-type behavior and successfully infect *E. coli* K12. For example, if the original mutant contained an insertion (+), a second event causing a deletion (−) close to the insertion would restore the original reading frame. In the same way, an event resulting in an insertion (+) might correct an original deletion (−).

In studying many mutations of this type, these researchers were able to compare various mutant combinations together within the same DNA sequence. They found that various combinations of one plus (+) and one minus (−) caused reversion to wild-type behavior. Still other observations shed light on the number of nucleotides constituting the genetic code. When two pluses were together or when two minuses were together, the correct reading frame *was not* reestablished. This argued against a doublet (two letter) code. However, when three plusses [Figure 13–2(b)] or three minuses were present together, the original frame *was* reestablished. These observations strongly supported the triplet nature of the code.

The Nonoverlapping Nature of the Code

Work by Sydney Brenner and others soon established that the code could not be overlapping. For example, given the assumption of a triplet code, Brenner considered restrictions that might be placed on it if it were overlapping. He considered theoretical nucleotide sequences encoding a protein consisting of three amino acids. In the nucleotide sequence GTACA, for example, parts of the central codon, TAC, are shared by the outer codons, GTA and ACA. Brenner reasoned that if this were the case, then only certain amino acids should be found adjacent to the one encoded by the central codon, leading him to conclude that if the code were overlapping, tripeptide sequences within proteins should be somewhat limited.

For example, when any given central amino acid is considered, only 16 combinations (2^4) of three-amino-acid sequences (a tripeptide) are theoretically possible. Looking at the available amino acid sequences of proteins that had been studied, he failed to find such restrictions in tripeptide sequences. For any central amino acid, he found many more than 16 different tripeptides. This observation led him to conclude that the code is not overlapping.

A second major argument against an overlapping code involved the effect of a single nucleotide change. With an overlapping code, two adjacent amino acids would be affected by such a point mutation. However, mutations in the genes coding for the protein coat of tobacco mosaic virus (TMV), human hemoglobin, and the bacterial enzyme tryptophan synthetase invariably revealed only single amino acid changes.

The third argument against an overlapping code was presented by Francis Crick in 1957, when he predicted that DNA would not serve as a direct template for the formation of proteins. Crick reasoned that any affinity between nucleotides and an amino acid would require hydrogen bonding. Chemically, however, such specific affinities seemed unlikely. Instead, Crick proposed that there must be an **adaptor molecule** that could covalently bind to the amino acid, yet also be capable of hydrogen bonding to a nucleotide sequence. Because various adaptors would somehow have to overlap one another at nucleotide sites during translation, Crick reasoned that physical constraints would make the process overly complex and perhaps even inefficient during translation. As we will see later in this chapter, Crick's prediction was correct; transfer RNA (tRNA) serves as the adaptor in protein synthesis, and the ribosome accommodates two tRNA molecules at a time during translation.

Crick's and Brenner's arguments, taken together, strongly suggested that, during translation, the genetic code is **nonoverlapping**. Without exception, this concept has been upheld.

? HOW DO WE KNOW?

On what basis did geneticists believe, even before direct experimentation was obtained, that the genetic code was a triplet and nonoverlapping?

The Commaless and Degenerate Nature of the Code

Between 1958 and 1960, information related to the genetic code continued to accumulate. In addition to his adaptor proposal, Crick hypothesized, on the basis of genetic evidence, that the code would be **commaless**—that is, he believed no internal punctuation would occur along the reading frame. Crick also speculated that only 20 of the 64 possible codons would specify an amino acid and that the remaining 44 would carry no coding assignment.

Was Crick wrong with respect to the 44 "blank" codes? That is, is the code **degenerate**, meaning that more than one codon specifies the same amino acid? Crick's frameshift studies suggested that, contrary to his earlier proposal, the code is degenerate. For the cases in which wild-type function is restored—that is, (+) and (−), (++) and (−−), (+++) and (−−−)—original frame of reading is also restored. However, there may be numerous codons between the various additions and deletions that would still be out of frame. If 44 of the 64 possible codons were blank (these are referred to as **nonsense codons**) and did not specify an amino acid, at least one of these would very likely occur in the length of nucleotides still out of frame. If a nonsense codon was encountered during protein synthesis, it was reasoned that the process would stop or be terminated at that point. If so, the product of the *rII* locus would not be made, and restoration would not occur. Because the various mutant combinations were able to reproduce on *E. coli* K12, Crick and his colleagues concluded that, in all likelihood, most if not all of the remaining 44 codes were not blank. It followed that the genetic code was **degenerate**. As we shall see, this reasoning proved to be correct.

13.3 Studies by Nirenberg, Matthaei, and Others Led to Deciphering of the Code

In 1961, Marshall Nirenberg and J. Heinrich Matthaei characterized the first specific coding sequences, which served as a cornerstone for the complete analysis of the genetic code. Their success, as well as that of others who made important contributions in deciphering the code, was dependent on the use of two experimental tools, an ***in vitro* (cell-free) protein-synthesizing system** and the enzyme **polynucleotide phosphorylase**, which allowed the production of synthetic mRNAs. These mRNAs served as templates for polypeptide synthesis in the cell-free system.

Synthesizing Polypeptides in a Cell-Free System

In the cell-free system, amino acids can be incorporated into polypeptide chains. This *in vitro* mixture, as might be expected, must contain the essential factors for protein synthesis in the cell: ribosomes, tRNAs, amino acids, and other molecules essential to translation. In order to follow (or trace) protein synthesis, one or more of the amino acids must be radioactive. Finally, an mRNA must be added, which serves as the template to be translated.

In 1961, mRNA had yet to be isolated. However, the use of the enzyme polynucleotide phosphorylase allowed the artificial synthesis of RNA templates, which could be added to the cell-free system. This enzyme, isolated from bacteria, catalyzes the reaction shown in Figure 13–3. Discovered in 1955 by Marianne Grunberg-Manago and Severo Ochoa, the enzyme functions metabolically in bacterial cells to degrade RNA. However, *in vitro*, with high concentrations of ribonucleoside diphosphates, the reaction can be "forced" in the opposite direction to synthesize RNA, as illustrated in the figure. In contrast to RNA polymerase, polynucleotide phosphorylase requires no DNA template. As a result, each addition of a ribonucleotide is random, based on the relative concentration of the four ribonucleoside diphosphates added to the reaction mixtures. The probability of the insertion of a specific ribonucleotide is proportional to the availability of that molecule, relative to other available ribonucleotides. *This point is absolutely critical to understanding the work of Nirenberg and others in the ensuing discussion.*

Taken together, the cell-free system for protein synthesis and the availability of synthetic mRNAs provided a means of deciphering the ribonucleotide composition of various codons encoding specific amino acids.

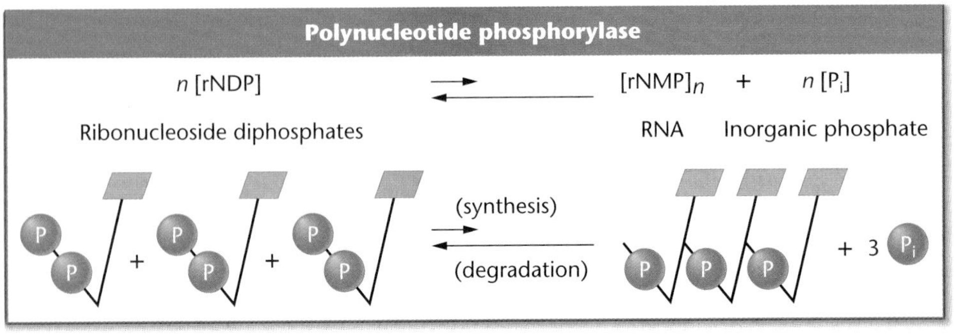

FIGURE 13–3 The reaction catalyzed by the enzyme polynucleotide phosphorylase. Note that the equilibrium of the reaction favors the degradation of RNA, but can be "forced" in the direction favoring synthesis.

Homopolymer Codes

In their initial experiments, Nirenberg and Matthaei synthesized **RNA homopolymers**, each consisting of only one type of ribonucleotide. Therefore, the mRNA added to the *in vitro* system was either UUUUUU . . . , AAAAAA . . . , CCCCCC . . . , or GGGGGG . . . . In testing each mRNA, they were able to determine which, if any, amino acids were incorporated into newly synthesized proteins. The researchers determined this by labeling one of the 20 amino acids added to the *in vitro* system and conducting a series of experiments, each with a different amino acid made radioactive.

For example, consider the experiments using ^{14}C-phenylalanine (Table 13.1). Nirenberg and Matthaei concluded that the message poly U (polyuridylic acid) directed the incorporation of only phenylalanine into the homopolymer polyphenylalanine. Assuming a triplet code, they had determined the first specific codon assignment: UUU codes for phenylalanine. In related experiments, they quickly found that AAA codes for lysine and CCC codes for proline. Poly G did not serve as an adequate template, probably because the molecule folds back on itself. Thus, the assignment for GGG had to await other approaches.

Note that the specific triplet codon assignments were possible only because homopolymers were used. The method yields only the composition of codons, but since three Us, Cs, or As can have only one possible sequence (i.e., UUU, CCC, and AAA), the actual codon was identified.

TABLE 13.1 INCORPORATION OF ^{14}C-PHENYLALANINE INTO PROTEIN

Artificial mRNA	Radioactivity (counts/min)
None	44
Poly U	39,800
Poly A	50
Poly C	38

Source: After Nirenberg and Matthaei (1961).

Mixed Copolymers

With these techniques in hand, Nirenberg and Matthaei, and Ochoa and coworkers turned to the use of **RNA heteropolymers**. In this technique, two or more different ribonucleoside diphosphates are added in combination to form the artificial message. The researchers reasoned that if they knew the relative proportion of each type of ribonucleoside diphosphate, they could predict the frequency of any particular triplet codon occurring in the synthetic mRNA. If they then added the mRNA to the cell-free system and ascertained the percentage of any particular amino acid present in the new protein, they could analyze the results and predict the *composition* of codons specifying specific amino acids.

This approach is illustrated in Figure 13–4. Suppose that A and C are added in a ratio of 1A:5C. The insertion of a ribonucleotide at any position along the RNA molecule during its synthesis is determined by the ratio of A:C. Therefore, there is a 1/6 possibility for an A and a 5/6 chance for a C to occupy each position. On this basis, we can calculate the frequency of any given codon appearing in the message.

For AAA, the frequency is $(1/6)^3$, or about 0.4 percent. For AAC, ACA, and CAA, the frequencies are identical—that is, $(1/6)^2(5/6)$, or about 2.3 percent for each. Together, all three 2A:1C codons account for 6.9 percent of the total three-letter sequences. In the same way, each of three 1A:2C codons accounts for $(1/6)(5/6)^2$, or 11.6 percent (or a total of 34.8 percent). CCC is represented by $(5/6)^3$, or 57.9 percent of the codons.

By examining the percentages of any given amino acid incorporated into the protein synthesized under the direction of this message, it is possible to propose probable base composition (Figure 13–4). Because proline appears 69 percent of the time, we could propose that proline is encoded by CCC (57.9 percent) and one codon consisting of 2C:1A (11.6 percent). Histidine, at 14 percent, is probably coded by one 2C:1A (11.6 percent) and one 1C:2A (2.3 percent). Threonine, at 12 percent, is likely coded by only one 2C:1A. Asparagine and glutamine each appear to be coded by one of the 1C:2A codons, and lysine appears to be coded by AAA.

Using as many as all four ribonucleotides to construct the mRNA, the researchers conducted many similar experiments. Although determination of the *composition* of triplet code words corresponding to all 20 amino acids represented a very significant breakthrough, the *specific sequences* of codons were still unknown. Their determination awaited still other approaches.

HOW DO WE KNOW?

What experimental evidence provided the initial insights into the compositions of codons encoding specific amino acids?

Possible compositions	Probability of occurrence of any triplet	Possible triplets	Final %
3A	$(1/6)^3 = 1/216 = 0.4\%$	AAA	0.4
1C:2A	$(5/6)(1/6)^2 = 5/216 = 2.3\%$	AAC ACA CAA	3 x 2.3 = 6.9
2C:1A	$(5/6)^2(1/6) = 25/216 = 11.6\%$	ACC CAC CCA	3 x 11.6 = 34.8
3C	$(5/6)^3 = 125/216 = 57.9\%$	CCC	57.9
			100.0

Chemical synthesis of message

RNA

C C C C C C C C A C C C C C C A A C C A C C C C C A C C C C C A C C C A A

Translation of message

Percentage of amino acids in protein		Probable base-composition assignments
Lysine	<1	AAA
Glutamine	2	1C:2A
Asparagine	2	1C:2A
Threonine	12	2C:1A
Histidine	14	2C:1A, 1C:2A
Proline	69	CCC, 2C:1A

FIGURE 13–4 Results and interpretation of a mixed copolymer experiment in which a ratio of 1A:5C is used (1/6A:5/6C).

Now solve this

Problem 13.24 on page 331 asks you to analyze a reciprocal pair of mixed copolymer experiments and to predict codon compositions to the amino acids they encode.

Hint: The data set generated in the reciprocal experiment is essential to solving the problem because analysis of the initial data set will provide you with more than one possible answer. However, there is only one answer consistent with both sets of data.

The Triplet Binding Assay

It was not long before more advanced techniques were developed. In 1964, Nirenberg and Philip Leder developed the **triplet binding assay**, leading to specific assignments of codons. The technique took advantage of the observation that ribosomes, when presented with an RNA sequence as short as three ribonucleotides, will bind to it and form a complex similar to what is found *in vivo*. The triplet acts like a codon in mRNA, attracting the complementary sequence within tRNA (Figure 13–5). Such a triplet sequence in tRNA that is complementary to a codon of mRNA is known as an **anticodon**.

Although it was not yet feasible to chemically synthesize long stretches of RNA, codons of known sequence could be synthesized in the laboratory to serve as templates. All that was needed was a method to determine which tRNA–amino acid was bound to the triplet RNA-ribosome complex. The test system devised was quite simple. The amino acid to be tested was made radioactive, and a charged tRNA was produced. Because code compositions were known, it was possible to narrow the decision as to which amino acids should be tested for each specific codon.

The radioactively charged tRNA, the RNA codon, and ribosomes are incubated together on a nitrocellulose filter, which will retain the larger ribosomes, but not the other smaller components, such as charged tRNA. If radioactivity is not retained on the filter, an incorrect amino acid has been tested. If radioactivity remains on the filter, it is retained because the charged tRNA has bound to the codon associated with the ribosome. In such a case, a specific codon assignment can be made.

Work proceeded in several laboratories, and in many cases clear-cut, unambiguous results were obtained. For example, Table 13.2 shows 26 codons assigned to nine amino acids. However, in some cases the degree of triplet binding was inefficient, and assignments were not possible. Eventually, about 50 of the 64 codons were assigned. These specific assignments of codons to amino acids led to two major conclusions. First, the genetic code is degenerate—that is, one amino acid may be specified by more than one codon. Second, the code is also unambiguous—that is, a single codon specifies only one amino acid. As we shall see later in this chapter, these conclusions have been upheld with only minor exceptions. The triplet binding technique was a major innovation in deciphering the genetic code.

How Do We Know?

How were the specific sequences of triplet codes determined experimentally?

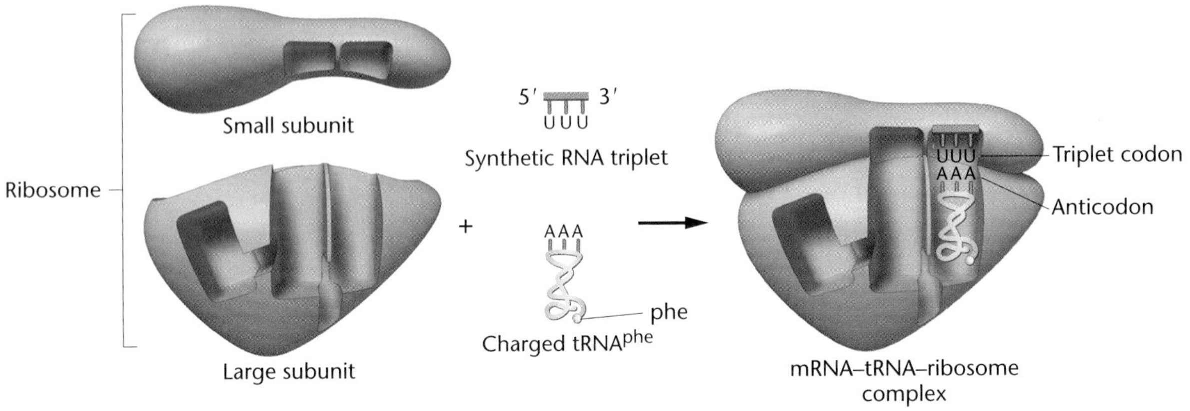

FIGURE 13–5 An example of the triplet-binding assay. The UUU triplet acts as a codon, attracting the complementary tRNA^{phe} anticodon AAA.

TABLE 13.2 AMINO ACID ASSIGNMENTS TO SPECIFIC TRINUCLEOTIDES DERIVED FROM THE TRIPLET-BINDING ASSAY

Trinucleotides	Amino Acid
UGU UGC	Cysteine
GAA GAG	Glutamic acid
AUU AUG AUA	Isoleucine
UUA UUG CUU	Leucine
CUC CUA CUG	Leucine
AAA AAG	Lysine
AUG	Methionine
UUU UUC	Phenylalanine
CCU CCC CCG CCA	Proline
UCU UCC UCA UCG	Serine

Repeating Copolymers

Yet another innovative technique used to decipher the genetic code was developed in the early 1960s by Gobind Khorana, who was able to chemically synthesize long RNA molecules consisting of short sequences repeated many times. First, he created shorter sequences (e.g., di-, tri-, and tetranucleotides), which were then replicated many times and finally joined enzymatically to form the long polynucleotides. As illustrated in Figure 13–6, a dinucleotide made in this way is converted to a message with two repeating codons. A trinucleotide is converted to a message with three potential codons, depending on the point at which initiation occurs, and a tetranucleotide creates four repeating codons.

When these synthetic mRNAs were added to a cell-free system, the predicted number of amino acids incorporated was upheld. Several examples are shown in Table 13.3. When such data were combined with those drawn from composition assignment and triplet binding, specific assignments were possible.

One example of specific assignments made from such data will illustrate the value of Khorana's approach. Consider the following three experiments in concert with one another: the repeating trinucleotide sequence UUCUUCUUC... produces three possible codons: UUC, UCU, and CUU, depending on the initiation point. When placed in a cell-free translation system, the polypeptides containing phenylalanine (phe), serine (ser), and leucine (leu) are produced. On the other hand, the repeating dinucleotide sequence UCUCUCUC... produces the codons UCU and CUC with the incorporation of leucine and serine into the polypeptide. The overlapping results indicate that the codons UCU and CUC

TABLE 13.3 AMINO ACIDS INCORPORATED USING REPEATED SYNTHETIC COPOLYMERS OF RNA

Repeating Copolymer	Codons Produced	Amino Acids in Polypeptides
UG	UGU	Cysteine
	GUG	Valine
AC	ACA	Threonine
	CAC	Histidine
UUC	UUC	Phenylalanine
	UCU	Serine
	CUU	Leucine
AUC	AUC	Isoleucine
	UCA	Serine
	CAU	Histidine
UAUC	UAU	Tyrosine
	CUA	Leucine
	UCU	Serine
	AUC	Isoleucine
GAUA	GAU	None
	AGA	None
	UAG	None
	AUA	None

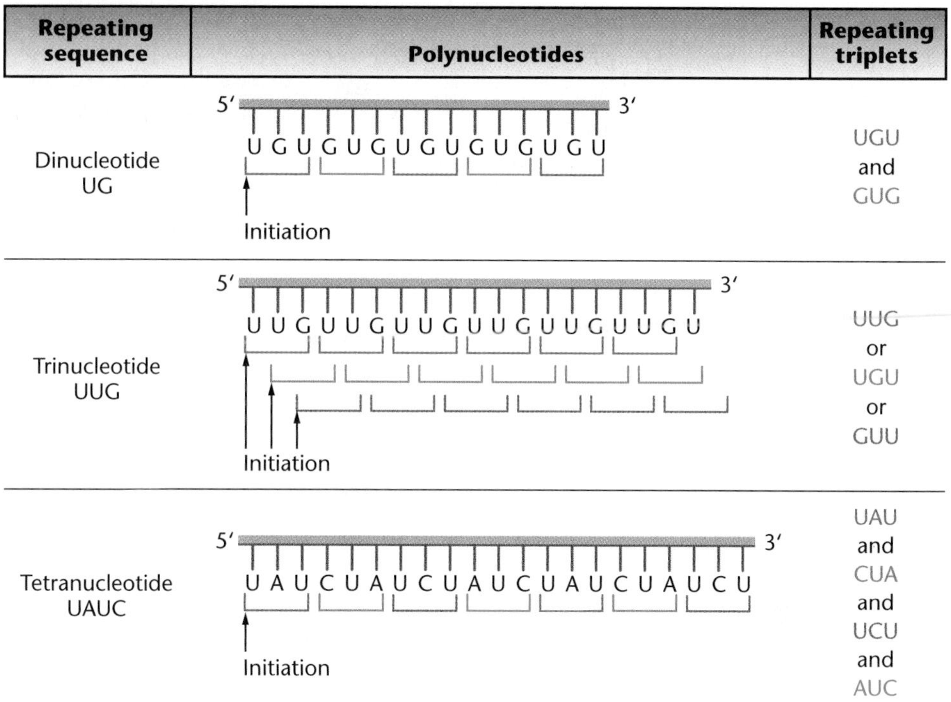
Repeating sequence
Polynucleotides
Repeating triplets
Dinucleotide UG
5′ U G U G U G U G U G U G U G U G 3′
Initiation
UGU and GUG
Trinucleotide UUG
5′ U U G U U G U U G U U G U U G U U G U 3′
Initiation
UUG or UGU or GUU
Tetranucleotide UAUC
5′ U A U C U A U C U A U C U A U C U A U C U 3′
Initiation
UAU and CUA and UCU and AUC

FIGURE 13–6 The conversion of di-, tri-, and tetranucleotides into repeating copolymers. The triplet codons produced in each case are shown.

specify leucine and serine, but not which codon specifies which amino acid. One can further conclude that *either* the CUU *or* the UUC codon also encodes leucine *or* serine, while the other encodes phenylalanine.

To derive more specific information, let's examine the results of using the repeating tetranucleotide sequence UUAC, which produces the codons UUA, UAC, ACU, and CUU. The CUU codon is one of the two in which we are interested. Three amino acids are incorporated: leucine, threonine, and tyrosine. Because CUU must specify only serine or leucine, and because, of these two, only leucine appears, we may conclude that CUU specifies leucine.

Once the latter is established, we can logically determine all other assignments. Of the two triplet pairs remaining (UUC and UCU from the first experiment *and* UCU and CUC from the second experiment), whichever codon is common to both must encode serine. This is UCU. By elimination, UUC is determined to encode phenylalanine and CUC is determined to encode leucine. While the logic must be carefully followed, four specific codons encoding three different amino acids have been assigned from these experiments.

From such interpretations, Khorana reaffirmed codons that were already deciphered and filled in gaps left from other approaches. For example, the use of two tetranucleotide sequences, GAUA and GUAA, suggested that at least two codons were termination signals. He reached this conclusion because neither of these sequences directed the incorporation of any amino acids into a polypeptide. Since there are no codons common to both messages, he predicted that each repeating sequence contained at least one codon that terminates protein synthesis. Table 13.3 lists the possible codons for the poly-(GAUA) sequence, of which UAG is a termination codon.

Now solve this

Problem 13.4 on page 330 asks you to consider several outcomes of repeating copolymer experiments.

Hint: In a repeating copolymer of RNA, initiation of translation can occur at any of several ribonucleotides. Determining the number of triplet codons produced by each possible initiation point is the key to solving this problem.

13.4 The Coding Dictionary Reveals Several Interesting Patterns among the 64 Codons

The various techniques applied to decipher the genetic code have yielded a dictionary of 61 triplet codon–amino acid assignments. The remaining three codons are termination signals, not specifying any amino acid. Figure 13–7 designates the assignments in a particularly illustrative form first suggested by Francis Crick.

Degeneracy and the Wobble Hypothesis

A general pattern of codon assignments becomes apparent when we look at the genetic coding dictionary. Most evident is that the code is **degenerate**, as the early researchers predicted. That is, almost all amino acids are specified by two, three, or four different codons. Three amino acids (serine, arginine, and leucine) are each encoded by six different codons. Only tryptophan and methionine are encoded by single codons.

Also evident is the pattern of degeneracy. Most often in a set of codons specifying the same amino acid, the first two letters are the same, with only the third differing. For example, as is

First position (5′-end)	Second position: U	Second position: C	Second position: A	Second position: G	Third position (3′-end)
U	UUU *phe*	UCU *ser*	UAU *tyr*	UGU *cys*	U
U	UUC *phe*	UCC *ser*	UAC *tyr*	UGC *cys*	C
U	UUA *leu*	UCA *ser*	UAA *Stop*	UGA *Stop*	A
U	UUG *leu*	UCG *ser*	UAG *Stop*	UGG *trp*	G
C	CUU *leu*	CCU *pro*	CAU *his*	CGU *arg*	U
C	CUC *leu*	CCC *pro*	CAC *his*	CGC *arg*	C
C	CUA *leu*	CCA *pro*	CAA *gln*	CGA *arg*	A
C	CUG *leu*	CCG *pro*	CAG *gln*	CGG *arg*	G
A	AUU *ile*	ACU *thr*	AAU *asn*	AGU *ser*	U
A	AUC *ile*	ACC *thr*	AAC *asn*	AGC *ser*	C
A	AUA *ile*	ACA *thr*	AAA *lys*	AGA *arg*	A
A	AUG *met*	ACG *thr*	AAG *lys*	AGG *arg*	G
G	GUU *val*	GCU *ala*	GAU *asp*	GGU *gly*	U
G	GUC *val*	GCC *ala*	GAC *asp*	GGC *gly*	C
G	GUA *val*	GCA *ala*	GAA *glu*	GGA *gly*	A
G	GUG *val*	GCG *ala*	GAG *glu*	GGG *gly*	G

Initiation Termination

FIGURE 13–7 The coding dictionary. AUG encodes methionine, which initiates most polypeptide chains. All other amino acids except tryptophan, which is encoded only by UGG, are represented by two to six codons. The codons UAA, UAG, and UGA are termination signals and do not encode any amino acids.

evident in Figure 13–7, the codons for phenylalanine (UUU and UUC in the top left corner of the coding table) differ only by their third letter. Either U or C in the third position specifies phenylalanine. Four codons specify valine (GUU, GUC, GUA, and GUG in the bottom left corner), and they differ only by their third letter. In this case, all four letters in the third position specify valine. Crick observed this pattern in the degeneracy throughout the code, and in 1966, he postulated the **wobble hypothesis**. Crick's hypothesis predicted that the initial two ribonucleotides of triplet codes are often more critical than the third member in attracting the correct tRNA. He postulated that hydrogen bonding at the third position of the codon–anticodon interaction would be *less* constrained and need not adhere as specifically to the established base-pairing rules.

More specifically, the wobble hypothesis proposes a more flexible set of base-pairing rules at the third position of the codon (Table 13.4). These modified base-pairing rules allow the anticodon of a single form of tRNA to pair with more than one codon in mRNA. Consistent with the wobble hypothesis and the degeneracy of the code, U at the first position (the 5′ end) of the tRNA anticodon may pair with A or G at the third position (the 3′ end) of the mRNA codon, and G may likewise pair with U or C. Inosine, one of the modified bases found in tRNA, may pair with C, U, or A. Applying these wobble rules, a minimum of about 30 different tRNA species are necessary to accommodate the 61 codons specifying an amino acid. If nothing more, wobble can be considered a potential economy measure, provided that the fidelity of translation is not compromised. Current estimates are that 30 to 40 tRNA species are present in bacteria and up to 50 tRNA species in animal and plant cells.

The Ordered Nature of the Code

Still another observation has become apparent in the pattern of codon sequences and their corresponding amino acids, leading to the description referred to as an **ordered genetic code**. Chemically similar amino acids often share one or two "middle" bases in the different triplets encoding them. For example, either U or C is often present in the second position of triplets that specify hydrophobic amino acids, including valine and alanine, among others. Two codons (AAA and AAG) specify the positively charged amino acids lysine. If only the middle letter of these codons is changed from A to G (AGA and AGG), the positively charged amino acid arginine is specified. Hydrophilic amino acids, such as serine or threonine are specified by triplet codons with G or C in the second position.

The chemical properties of amino acids will be discussed in more detail in Chapter 14. The end result of an "ordered" code is that it buffers the potential effect of mutation on protein function. While many mutations of the second base of triplet codons result in a change of one amino acid to another, the change is often to an amino acid with similar chemical properties. In such cases, protein function may not be noticeably altered.

Initiation, Termination, and Suppression

Initiation of protein synthesis is a highly specific process. In bacteria (in contrast to the *in vitro* experiments discussed earlier), the initial amino acid inserted into all polypeptide chains

TABLE 13.4 **ANTICODON–CODON BASE-PAIRING RULES**

Base at first position (5′ end) of tRNA	*Base at third position (3′ end) of mRNA*
A	U
C	G
G	C or U
U	A or G
I	A, U, or C

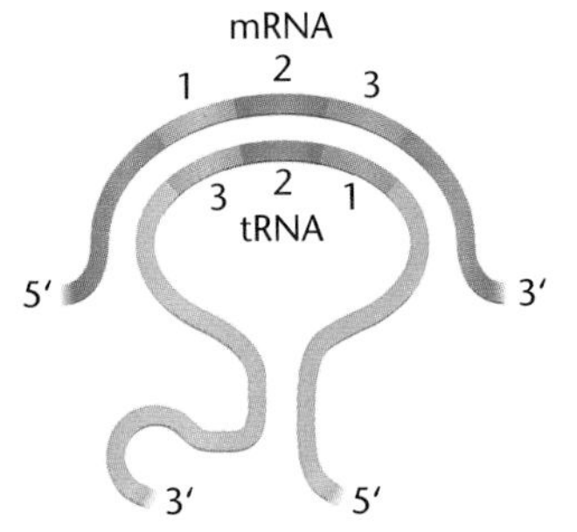

is a modified form of methionine—***N*-formylmethionine (fmet)**. Only one codon, AUG, codes for methionine, and it is sometimes called the **initiator codon**. However, when AUG appears internally in mRNA rather than at an initiating position, unformylated methionine is inserted into the polypeptide chain. Rarely, another codon, GUG, specifies methionine during initiation, though it is not clear why this happens, since GUG normally encodes valine.

In bacteria, either the formyl group is removed from the initial methionine upon the completion of synthesis of a protein, or the entire formylmethionine residue is removed. In eukaryotes, unformylated methionine is the initial amino acid during polypeptide synthesis. As in bacteria, it may be cleared from the polypeptide.

As mentioned in the preceding section, three other codons (UAG, UAA, and UGA)* serve as **termination codons**, punctuation signals that do not code for any amino acid. They are not recognized by a tRNA molecule, and translation terminates when they are encountered. Mutations that produce any of the three codons internally in a gene also result in termination. Consequently, only a partial polypeptide is synthesized, since it is prematurely released from the ribosome. When such a change occurs in the DNA, it is called a **nonsense mutation**.

Interestingly, a distinct mutation in a second gene may trigger suppression of premature termination resulting from a nonsense mutation (referred to as a **suppressor mutation**). These mutations cause the chain-termination signal to be read as a "sense" codon. The "correction" usually inserts an amino acid other than that found in the wild-type protein. But if the protein's structure is not altered drastically, it may function almost normally. Therefore, this second mutation has "suppressed" the mutant character resulting from the initial change to the termination codon.

Nonsense suppressor mutations occur in genes specifying tRNAs. If the mutation results in a change in the anticodon such that it becomes complementary to a termination code, there is the potential for insertion of an amino acid and suppression.

13.5 The Genetic Code Has Been Confirmed in Studies of Phage MS2

All aspects of the genetic code discussed so far yield a fairly complete picture. The code is triplet in nature, degenerate, unambiguous, and commaless, but it contains punctuation with respect to start and stop signals. These individual principles have been confirmed by the detailed analysis of the RNA-containing bacteriophage MS2 by Walter Fiers and his coworkers.

MS2 is a bacteriophage that infects *E. coli*. Its nucleic acid (RNA) contains only about 3500 ribonucleotides, making up only three genes. These genes specify a coat protein, an RNA-directed replicase, and a maturation protein (the A protein). This simple system of a small genome and few gene products allowed Fiers and his colleagues to sequence the genes and their products. The amino acid sequence of the coat protein was completed in 1970, and the nucleotide sequence of the gene and a number of nucleotides on each end of it were reported in 1972.

When the chemical nature of the gene and protein are compared, a colinear relationship is evident. That is, based on the coding dictionary, the linear sequence of nucleotides (and thus the sequence of triplet codons) corresponds precisely with the linear sequence of amino acids in the protein. Furthermore, the codon for the first amino acid is AUG, the common initiator codon; and the codon for the last amino acid is followed by two consecutive termination codons, UAA and UAG.

By 1976, the other two genes and their protein products were sequenced, providing similar confirmation. The analysis clearly shows that the genetic code, as established in bacterial systems, is identical in this virus. Other evidence suggests that the code is also identical in eukaryotes.

HOW DO WE KNOW?

How were the triplet codons verified experimentally using bacteriophage MS2?

13.6 The Genetic Code Is Nearly Universal

Between 1960 and 1978, it was generally assumed that the genetic code would be found to be universal, applying equally to viruses, bacteria, archaea, and eukaryotes. Certainly, the nature of mRNA and the translation machinery seemed to be very similar in these organisms. For example, cell-free systems derived from bacteria could translate eukaryotic mRNAs. Poly U was shown to stimulate translation of polyphenylalanine in cell-free systems when the components were derived from eukaryotes. Many recent studies involving recombinant DNA technology (see Chapter 19) reveal that eukaryotic genes can be inserted into bacterial cells and transcribed and translated. Within eukaryotes, mRNAs from mice and rabbits have been injected into amphibian eggs and efficiently translated. For the many eukaryotic genes that have been sequenced, notably those for hemoglobin molecules, the amino acid sequence of the encoded proteins adheres to the coding dictionary established from bacterial studies.

However, several 1979 reports on the coding properties of DNA derived from mitochondria of yeast and humans (mtDNA) undermined the principle of the universality of the genetic language. Since then, mtDNA has been examined in many other organisms.

Cloned mtDNA fragments were sequenced and compared with the amino acid sequences of various mitochondrial proteins, revealing several exceptions to the coding dictionary (Table 13.5). Most surprising was that the codon UGA, normally specifying termination, specifies the insertion of

*Historically, the terms *amber* (UAG), *ochre* (UAA), and *opal* (UGA) have been used to distinguish the three mutation possibilities.

TABLE 13.5 **EXCEPTIONS TO THE UNIVERSAL CODE**

Codon	Normal Code Word	Altered Code Word	Source
UGA	Termination	trp	Human and yeast mitochondria and *Mycoplasma*
CUA	Leu	thr	Yeast mitochondria
AUA	Ile	met	Human mitochondria
AGA	Arg	Termination	Human mitochondria
AGG	Arg	Termination	Human mitochondria
UAA	Termination	gln	*Paramecium*, *Tetrahymena*, and *Stylonychia*
UAG	Termination	gln	*Paramecium*

tryptophan during translation in yeast and human mitochondria. In human mitochondria, AUA, which normally specifies isoleucine, directs the internal insertion of methionine. In yeast mitochondria, threonine is inserted instead of leucine when CUA is encountered in mRNA.

In 1985, several other exceptions to the standard coding dictionary were discovered in the bacterium *Mycoplasma capricolum*, and in the nuclear genes of the protozoan ciliates *Paramecium*, *Tetrahymena*, and *Stylonychia*. For example, as shown in Table 13.5, one alteration converts the termination codons (UGA) to tryptophan. Several other code alterations convert a termination codon to glutamine (gln). These changes are significant because both a prokaryote and several eukaryotes are involved.

Note the apparent pattern in several of the altered codon assignments. The change in coding capacity involves only a shift in recognition of the third, or wobble, position. For example, AUA specifies isoleucine in the cytoplasm and methionine in the mitochondrion. In the cytoplasm, methionine is specified by AUG. In a similar way, UGA calls for termination in the cytoplasm, but calls for tryptophan in the mitochondrion. In the cytoplasm, tryptophan is specified by UGG. It has been suggested that such changes in codon recognition may represent an evolutionary trend toward reducing the number of tRNAs needed in mitochondria; only 22 tRNA species are encoded in human mitochondria, for example. However, until more examples are revealed, the differences must be considered to be exceptions to the previously established general coding rules.

13.7 Different Initiation Points Create Overlapping Genes

Earlier we stated that the genetic code is nonoverlapping—each ribonucleotide in an mRNA is part of only one codon. However, this characteristic of the code does not rule out the possibility that a single mRNA may have multiple initiation points for translation. If so, these points could theoretically create several different reading frames within the same mRNA, thus specifying more than one polypeptide. This concept of **overlapping genes** is illustrated in Figure 13–8(a).

That this might actually occur in some viruses was suspected when phage ϕX174 was carefully investigated. The circular DNA chromosome consists of 5386 nucleotides, which should encode a maximum of 1795 amino acids, sufficient for five or six proteins. However, this small virus in fact synthesizes 11 proteins consisting of more than 2300 amino acids. A comparison of the nucleotide sequence of the DNA and the amino acid sequences of the polypeptides synthesized has clarified the apparent paradox. At least four cases of multiple initiation have been discovered, creating overlapping genes [Figure 13–8(b)].

The sequences specifying the K and B polypeptides are initiated with separate reading frames within the sequence specifying the A polypeptide. The K sequence overlaps into the adjacent sequence specifying the C polypeptide. The E sequence is out of frame with, but initiated in that of the D polypeptide. Finally, the A′ sequence, while in frame, begins in the middle of the A sequence. They both terminate at the identical point. In all, seven different polypeptides are created from a DNA sequence that might otherwise have specified only three (A, C, and D).

A similar situation has been observed in other viruses, including phage G4 and the animal virus SV40. Like ϕX174, phage G4 contains a circular single-stranded DNA molecule.

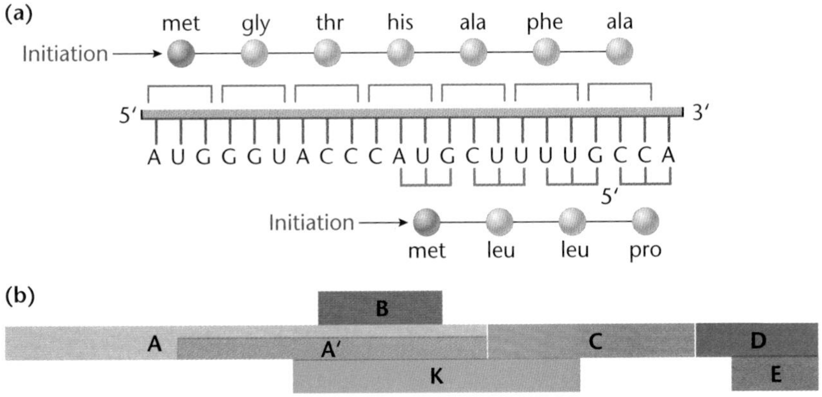

FIGURE 13–8 Illustration of the concept of overlapping genes. (a) An mRNA sequence initiated at two different AUG positions out of frame with one another will give rise to two distinct amino acid sequences. (b) The relative positions of the sequences encoding seven polypeptides of the phage ϕX174.

The use of overlapping reading frames optimizes the use of a limited amount of DNA present in these small viruses. However, such an approach to storing information has a distinct disadvantage in that a single mutation may affect more than one protein and thus increase the chances that the change will be deleterious or lethal. In the case we just discussed, a single mutation at the junction of genes *A* and *C* could affect three proteins (the A, C, and K proteins). It may be for this reason that overlapping genes are not common in other organisms.

13.8 Transcription Synthesizes RNA on a DNA Template

Even while the genetic code was being studied, it was quite clear that proteins were the end products of many genes. Hence, while some geneticists were attempting to elucidate the code, other research efforts were directed toward the nature of genetic expression. The central question was how DNA, a nucleic acid, is able to specify a protein composed of amino acids.

The complex, multistep process begins with the transfer of genetic information stored in DNA to RNA. The process by which RNA molecules are synthesized on a DNA template is called **transcription**. It results in an mRNA molecule complementary to the gene sequence of one of the two strands of the double helix. Each triplet codon in the mRNA is, in turn, complementary to the anticodon region of its corresponding tRNA as the amino acid is correctly inserted into the polypeptide chain during translation. The significance of the process of transcription is enormous, for it is the initial step in the process of information flow within the cell. The idea that RNA is involved as an intermediate molecule in the process of information flow between DNA and protein is suggested by the following observations:

1. DNA is, for the most part, associated with chromosomes in the nucleus of the eukaryotic cell. However, protein synthesis occurs in association with ribosomes located outside the nucleus in the cytoplasm. Therefore, DNA does not appear to participate directly in protein synthesis.
2. RNA is synthesized in the nucleus of eukaryotic cells, in which DNA is found, and is chemically similar to DNA.
3. Following its synthesis, most RNA migrates to the cytoplasm, in which protein synthesis (translation) occurs.
4. The amount of RNA is generally proportional to the amount of protein in a cell.

Like most new ideas in molecular genetics, the initial supporting experimental evidence for an RNA intermediate was based on studies of bacteria and their phages.

13.9 Studies with Bacteria and Phages Provided Evidence for the Existence of mRNA

In two papers published in 1956 and 1958, Elliot Volkin and his colleagues reported their analysis of RNA produced immediately after bacteriophage infection of *E. coli*. Using the isotope ^{32}P to follow newly synthesized RNA, they found that its base composition closely resembled that of the phage DNA, but was different from that of bacterial RNA (Table 13.6). This newly synthesized RNA was unstable, or short lived; however, its production was shown to precede the synthesis of new phage proteins. Thus, Volkin and his coworkers considered the possibility that synthesis of RNA is a preliminary step in the process of protein synthesis.

Although ribosomes were known to participate in protein synthesis, their role in this process was not clear. As we noted earlier, one possibility was that each ribosome is specific for the protein synthesized in association with it. That is, perhaps genetic information in DNA is transferred to the RNA of a ribosome during its synthesis so that different groups of ribosomes are restricted to the translation of particular proteins. The alternative hypothesis was that ribosomes are nonspecific "workbenches" for protein synthesis and that specific genetic information rests with a messenger RNA.

In an elegant experiment using the *E. coli*–phage system, the results of which were reported in 1961, Sidney Brenner, François Jacob, and Matthew Meselson clarified this question. They labeled uninfected *E. coli* ribosomes with heavy isotopes

TABLE 13.6 BASE COMPOSITIONS (IN MOLE PERCENTS) OF RNA PRODUCED IMMEDIATELY FOLLOWING INFECTION OF *E. COLI* BY THE BACTERIOPHAGES T2 AND T7 IN CONTRAST TO THE COMPOSITION OF RNA OF UNINFECTED *E. COLI*

	Adenine	Thymine	Uracil	Cystosine	Guanine
Postinfection RNA in T2-infected cells	33	—	32	18	18
T2 DNA	32	32	—	17*	18
Postinfection RNA in T7-infected cells	27	—	28	24	22
T7 DNA	26	26	—	24	22
E. coli RNA	23	—	22	18	17

* 5-hydroxymethyl cystosine

Source: From Volkin and Astrachan (1956); and Volkin, Astrachan, and Countryman (1958).

and then allowed phage infection to occur in the presence of radioactive RNA precursors. By following these components during translation, the researchers demonstrated that the synthesis of phage proteins (under the direction of newly synthesized RNA) occurred on bacterial ribosomes that were present prior to infection. The ribosomes appeared to be nonspecific, strengthening the case that another type of RNA serves as an intermediary in the process of protein synthesis.

That same year, Sol Spiegelman and his colleagues reached the same conclusion when they isolated ^{32}P-labeled RNA following the infection of bacteria and used it in molecular hybridization studies. They tried hybridizing this RNA to the DNA of both phages and bacteria in separate experiments. The RNA hybridized only with the phage DNA, showing that it was complementary in base sequence to the viral genetic information.

The results of these experiments agree with the concept of a messenger RNA (mRNA) being made on a DNA template and then directing the synthesis of specific proteins in association with ribosomes. This concept was formally proposed by François Jacob and Jacques Monod in 1961 as part of a model for gene regulation in bacteria. Since then, mRNA has been isolated and thoroughly studied. There is no longer any question about its role in genetic processes.

HOW DO WE KNOW?

What evidence do we have that the transfer of genetic information within a cell, stored in DNA, is initially transferred to an RNA intermediate?

13.10 RNA Polymerase Directs RNA Synthesis

To prove that RNA can be synthesized on a DNA template, it was necessary to demonstrate that there is an enzyme capable of directing this synthesis. By 1959, several investigators, including Samuel Weiss, had independently discovered such a molecule from rat liver. Called **RNA polymerase**, it has the same general substrate requirements as does DNA polymerase, the major exception being that the substrate nucleotides contain the ribose rather than the deoxyribose form of the sugar. Unlike DNA polymerase, no primer is required to initiate synthesis. The initial base remains as a nucleoside triphosphate (NTP). The overall reaction summarizing the synthesis of RNA on a DNA template can be expressed as

$$n(\text{NTP}) \xrightarrow[\text{enzyme}]{\text{DNA}} (\text{NMP})_n + n(\text{PP}_\text{i})$$

As the equation reveals, nucleoside triphosphates (NTPs) serve as substrates for the enzyme, which catalyzes the polymerization of nucleoside monophosphates (NMPs), or nucleotides, into a polynucleotide chain $(\text{NMP})_n$. Nucleotides are linked during synthesis by 5′ to 3′ phosphodiester bonds. (See Figure 10–8.) The energy created by cleaving the triphosphate precursor into the monophosphate form drives the reaction, and inorganic phosphates (PP_i) are produced.

A second equation summarizes the sequential addition of each ribonucleotide as the process of transcription progresses:

$$(\text{NMP})_n + \text{NTP} \xrightarrow[\text{enzyme}]{\text{DNA}} (\text{NMP})_{n+1} + \text{PP}_\text{i}$$

As this equation shows, each step of transcription involves the addition of one ribonucleotide (NMP) to the growing polyribonucleotide chain $(\text{NMP})_{n+1}$, using a nucleoside triphosphate (NTP) as the precursor.

RNA polymerase from *E. coli* has been extensively characterized and shown to consist of subunits designated α, β, β', ω, and σ. The active form of the enzyme, the **holoenzyme**, contains the subunits $\alpha_2\beta\beta'\sigma$ and has a molecular weight of almost 500,000 Da. Of these subunits, it is the β and β' polypeptides that provide the catalytic basis and active site for transcription. As we will see, the **σ subunit** [Figure 13–9(a)] plays a regulatory function involving the initiation of RNA transcription.

While there is but a single form of the enzyme in *E. coli*, there are several different σ factors, creating variations of the polymerase holoenzyme. On the other hand, eukaryotes display three distinct forms of RNA polymerase, each consisting of a greater number of polypeptide subunits than does the bacterial form of the enzyme. We shall return to a discussion of the eukaryotic form of the enzyme later in this chapter.

Now solve this

Problem 13.23 on page 331 asks you to consider the outcome of the transfer of complementary information from DNA to RNA and the amino acids encoded by this information.

Hint: In RNA, uracil is complementary to adenine, and while DNA stores genetic information in the cell, the code that is translated is contained in the RNA complementary to the template strand of DNA making up a gene.

Promoters, Template Binding, and the Sigma Subunit

Transcription results in the synthesis of a single-stranded RNA molecule complementary to a region along only one of the two strands of the DNA double helix. For the purpose of discussion, we will call the DNA strand that is transcribed the *template strand* and call its complement the *nontemplate strand*.

The initial step is referred to as **template binding** [Figure 13–9(b)]. In bacteria, the site of this initial binding is established when the σ (sigma) subunit of RNA polymerase recognizes specific DNA sequences called **promoters**. These regions are located in the 5′ (upstream) region, from the point of initial transcription of a gene. It is believed that the enzyme "explores" a length of DNA until it recognizes the promoter region and binds to about 60 nucleotide pairs of the helix, 40 of which are upstream from the point of initial transcription. Once this occurs, the helix is denatured or unwound locally, making the template strand of the DNA template accessible to the action of the enzyme. The point at which transcription actually begins is called the **transcription start site**.

The importance of promoter sequences cannot be overemphasized. They govern the efficiency of initiation of transcription. In bacteria, both strong promoters and weak promoters have been discovered, leading to a variation of initiation from once every 1 to 2 seconds to only once every 10 to 20 minutes. Mutations in promoter sequences may have the effect of reducing or enhancing the initiation of gene expression. Because the interaction of promoters with RNA polymerase governs transcription, the nature of the binding between them is at the heart of discussions concerning genetic regulation, the subject of Chapters 16 and 17. While we will pursue information involving promoter–enzyme interactions in greater detail in the aforementioned chapters, two points are appropriate here.

The first is the concept of **consensus sequences** of DNA. These are sequences that share homology in different genes of the same organism or in one or more genes of related organisms. Their conservation throughout evolution attests to the critical nature of their role in biological processes. Two such sequences have been found in bacterial promoters. One, TATAAT, is located 10 nucleotides upstream from the site of initial transcription (the -10 region, or **Pribnow box**). The other, TTGACA, is located 35 nucleotides upstream (the -35 region). These sequences are said to be ***cis*-acting elements**. The use of the term *cis* is drawn from organic chemistry nomenclature, meaning "next to" or on the same side as, in contrast to being "across from," or *trans*, to other functional groups. In molecular genetics, then, *cis*-elements are adjacent parts of the same DNA molecule in the gene itself. This is in contrast to ***trans*-acting factors**, molecules that bind to these DNA elements. As we will soon see, in most eukaryotic genes studied, a consensus sequence comparable to that in the -10 region has been recognized. Because it is rich in adenine and thymine residues, it is called the **TATA box**.

The second point is that the degree of RNA polymerase binding to different promoters varies greatly, which causes the variable gene expression mentioned earlier. Currently, this is attributed to sequence variation in the promoters.

A final general point to be made involves the sigma (σ) subunit in bacteria. The major form is designated as σ^{70}, based on its molecular weight of 70 kilodaltons (kDa). While the promoters of most bacterial genes recognize this form, there are several alternative forms of RNA polymerase in *E. coli* that have unique σ subunits associated with them (e.g., σ^{32}, σ^{54}, σ^{70}, σ^{S}, and σ^{E}). These recognize different promoter sequences and provide specificity to the initiation of transcription.

Initiation, Elongation, and Termination of RNA Synthesis

Once it has recognized and bound to the promoter, RNA polymerase catalyzes **initiation**, the insertion of the first 5′-ribonucleoside triphosphate, which is complementary to the first nucleotide at the start site of the DNA template strand. As we noted, no primer is required. Subsequent ribonucleotide complements are inserted and linked together by phosphodiester bonds as RNA polymerization proceeds. This process continues in the 5′ to 3′ direction, creating a temporary 8 bp DNA/RNA duplex whose chains run antiparallel to one another (Figure 13–9(b)).

After these ribonucleotides have been added to the growing RNA chain, the σ subunit usually dissociates from the holoenzyme, and **chain elongation** proceeds under the direction of the core enzyme [Figure 13–9(c)]. In *E. coli*, this process proceeds at the rate of about 50 nucleotides/second at 37°C.

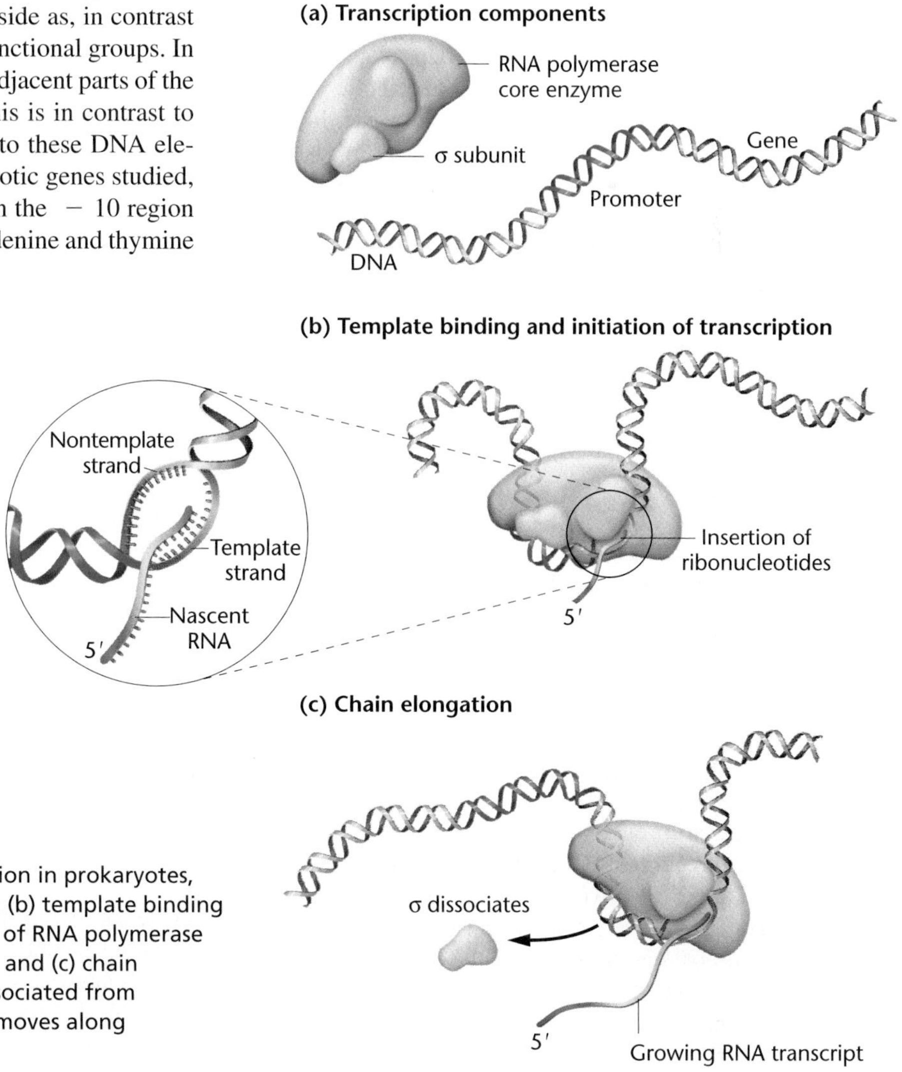

FIGURE 13–9 The early stages of transcription in prokaryotes, showing (a) the components of the process; (b) template binding at the -10 site involving the sigma subunit of RNA polymerase and subsequent initiation of RNA synthesis; and (c) chain elongation, after the sigma subunit has dissociated from the transcription complex and the enzyme moves along the DNA template.

Eventually, the enzyme traverses the entire gene until it encounters a specific nucleotide sequence that acts as a termination signal. Such termination sequences, about 40 base pairs in length, are extremely important in prokaryotes because of the close proximity of the end of one gene and the upstream sequences of the adjacent gene. An interesting aspect of termination in bacteria is that the sequence alluded to above is actually transcribed into RNA. The unique nature of this termination sequence causes the newly formed transcript to fold back on itself forming hydrogen bonds, and creating what is called a **hairpin secondary structure**. The hairpin is important to termination. In some cases, the termination of synthesis is also dependent upon the **termination factor, rho** (ρ). Rho is a large hexameric protein that physically interacts with the growing RNA transcript, facilitating termination of transcription.

When termination is achieved, the enzyme ceases to add ribonucleotides, the transcribed RNA molecule is released from the DNA template, and the core polymerase enzyme dissociates. The end result of transcription is the synthesis of an RNA molecule that is precisely complementary to the DNA sequence representing the template strand of a gene. Wherever an A, T, C, or G residue exists, a corresponding U, A, G, or C residue, respectively, is incorporated into the RNA molecule. Such RNA molecules ultimately provide the information leading to the synthesis of all proteins present in the cell.

It is important to note that in bacteria, groups of genes whose protein products are involved in the same metabolic pathway are often clustered together along the chromosome. In many such cases, the genes are contiguous and all but the last gene lack the encoded signals for termination of transcription. The result is that during transcription, a large mRNA is produced, encoding more than one protein. Since genes in phage and bacteria were historically referred to as cistrons, the RNA is called a **polycistronic mRNA**. Since the products of genes transcribed in this fashion are usually all needed at the same time, this is an efficient way to transcribe and subsequently translate the needed genetic information. In eukaryotes, **monocistronic** mRNAs are the rule.

13.11 Transcription in Eukaryotes Differs from Prokaryotic Transcription in Several Ways

Much of our knowledge of transcription has been derived from studies of prokaryotes. Most of the general aspects of the mechanics of these processes are similar in eukaryotes, but there are several notable differences:

1. Transcription in eukaryotes occurs within the nucleus under the direction of three separate forms of RNA polymerase. Unlike prokaryotes, in eukaryotes, the RNA transcript is not free to associate with ribosomes prior to the completion of transcription. For the mRNA to be translated, it must move out of the nucleus into the cytoplasm.

2. Initiation of transcription of eukaryotic genes requires that the compact chromatin fiber, characterized by nucleosome coiling (Chapter 12), must be uncoiled and the DNA made accessible to RNA polymerase and other regulatory proteins. This transition is referred to as **chromatin remodelling**, reflecting the dynamics involved in the conformational change that occurs as the DNA helix is opened.

3. Initiation and regulation of transcription involve a more extensive interaction between *cis*-acting upstream DNA sequences and *trans*-acting protein factors involved in stimulating and initiating transcription. In addition to promoters, other control units, called **enhancers**, may be located in the 5′ regulatory region upstream from the initiation point, but they have also been found within the gene or even in the 3′ downstream region beyond the coding sequence.

4. Maturation of eukaryotic mRNA from the primary RNA transcript involves many complex stages referred to generally as "processing." An initial processing step involves the addition of a 5′ cap and a 3′ tail to most transcripts destined to become mRNAs. Other extensive modifications occur to the internal nucleotide sequence of eukaryotic RNA transcripts that eventually serve as mRNAs. The initial (or primary) transcripts are most often much larger than those that are eventually translated. Sometimes called **pre-mRNAs**, they are part of a group of molecules found only in the nucleus—a group referred to collectively as **heterogeneous nuclear RNA (hnRNA)**. Such RNA molecules are of variable but large size and are complexed with proteins, forming **heterogeneous nuclear ribonucleoprotein particles (hnRNPs)**. Only about 25 percent of hnRNA molecules are converted to mRNA. In those that are converted, substantial amounts of the ribonucleotide sequence are excised and the remaining segments are spliced back together prior to nuclear export and translation. This phenomenon has given rise to the concepts of **split genes** and **splicing** in eukaryotes.

In the remainder of this chapter we will look at the basic details of transcription in eukaryotic cells. The process of transcription, is highly regulated, determining which DNA sequences are copied into RNA and when and how frequently they are transcribed. We will return to topics directly related to regulation of eukaryotic transcription in Chapter 17.

Initiation of Transcription in Eukaryotes

The transcription of selected, specific DNA regions by RNA polymerase is the basis of orderly genetic function in all cells. Eukaryotic RNA polymerase exists in three unique forms, each of which transcribes different types of genes, as indicated in Table 13.7. Each enzyme is larger and more complex than the single prokaryotic polymerase. For example, in yeast, the holoenzyme consists of two large subunits and 10 smaller subunits. In regard to the initial template binding step and promoter regions, most is known about RNA polymerase II (RNP II), which is responsible for the production of all mRNAs in eukaryotes.

The activity of RNP II is regulated by both the *cis*-acting elements in the gene itself, and by a number of *trans*-acting fac-

TABLE 13.7 **RNA POLYMERASES IN EUKARYOTES**

Form	Product	Location
I	rRNA	Nucleolus
II	mRNA, snRNA	Nucleoplasm
III	5S rRNA, tRNA	Nucleoplasm

tors that bind to these DNA elements. There are at least three *cis*-acting DNA elements that regulate the initiation of transcription by RNP II. The first of these sequences, called a **core-promoter element**, determines where RNP II binds to the DNA and where it begins copying the DNA into RNA. The other two types of regulatory DNA sequences, called **promoter** and **enhancer elements**, influence the efficiency or the rate of transcription by RNP II as the process proceeds from the core promoter element. Recall that in prokaryotes, the DNA sequence recognized by RNA polymerase is called the promoter. In eukaryotes, however, transcriptional initiation is controlled by these *cis*-acting DNA elements, that are *collectively* referred to as the promoter. Thus, in eukaryotes the term *promoter* refers to both the core-promoter sequence where RNP II binds, and to the promoter and enhancer elements in the DNA that influence RNP II activity.

An example of the *cis*-acting core promoter element is called the **Goldberg–Hogness** or **TATA box**, present in almost all eukaryotic genes. Located about 30 nucleotide pairs upstream (-30) from the start point of transcription, the TATA boxes share a consensus sequence that is a heptanucleotide consisting solely of A and T residues (TATAAAA). The sequence and function are analogous to that found in the -10 region of prokaryotic genes. The TATA box is the site where RNP II binds to the DNA template, and transcription typically begins approximately 25 to 30 bp downstream from the TATA box. In fact, the location of the TATA box determines the exact nucleotide position where transcription begins. As in prokaryotes, this position is called the *transcription start site*.

Another example of a *cis*-acting DNA sequence that is part of the eukaryotic promoter is the **CAAT box**. This specific DNA sequence, which contains the consensus sequence GGCCAATCT, is typically located upstream (in the 5′ region of the gene) within 100 nucleotides of the start of transcription. Still other types of promoter elements are found upstream of the transcription start site, and most genes contain one or more of them. Along with the CAAT box, they influence the efficiency of the promoter. The location of each element is based on studies of deletions of particular regions of the promoter, each of which reduces the efficiency of transcription.

The third type of *cis*-acting regulatory DNA elements are enhancers. Althought their locations can vary, enhancers are often found even farther upstream than the regions already mentioned, or even downstream or within the gene. Thus they have the potential to modulate transcription from a distance. Although they may not participate directly in RNP II binding to the core promoter, they are essential to highly efficient initiation of transcription.

Complementing the *cis*-acting regulatory sequences are various *trans*-acting factors that facilitate RNP II binding and, therefore, the initiation of transcription. These are proteins referred to as **transcription factors**. There are two broad categories of transcription factors: the **general transcription factors** that are absolutely required for all RNP II mediated transcription, and the **specific transcription factors** that influence the efficiency or the rate of RNP II transcription. We will consider the second category of transcription factors, the specific transcription factors, in more detail in Chapter 17. The general transcription factors, however, are essential because RNA polymerase II cannot bind directly to eukaryotic promoter sites and initiate transcription without their presence. The general transcription factors involved with human RNP II-binding are well characterized and designated **TFIIA**, **TFIIB**, and so on. One of these, **TFIID**, binds directly to the TATA-box sequence. TFIID consists of about 10 polypeptide subunits, one of which is sometimes called the **TATA-binding protein (TBP)**. Once initial binding to DNA occurs, at least seven other general transcription factors bind sequentially to TFIID, forming an extensive pre-initiation complex, which is then bound by RNA polymerase II.

Transcription factors with similar activity have been discovered in a variety of eukaryotes, including *Drosophila* and yeast. The nucleotide sequences in all organisms studied demonstrate a high degree of conservation. In Chapter 17, we will consider the role of transcription factors in eukaryotic gene regulation, as well as the various DNA-binding domains that characterize them.

Recent Discoveries Concerning RNA Polymerase Function

One approach to extend our understanding of the process of transcription has been to study the details of the structure and function of RNA. The ability to crystallize large nucleic acid–protein structures and perform X-ray diffraction analysis at a resolution below 5 Å has led to some remarkable observations. The work of Roger Kornberg and colleagues, using the enzyme isolated from yeast, has been particularly informative. It is useful to note here that achieving a resolution below 2.8 Å allows the visualization of each amino acid of every protein in the complex!

What Kornberg has discovered provides a highly detailed account of the most critical processes of transcription. RNA polymerase II in yeast contains two large subunits and ten smaller ones, forming a huge three-dimensional complex with a molecular weight of about 500 kDa. The promoter region of the DNA duplex that is to be transcribed enters a positively charged cleft formed between the two large subunits of the enzyme. The subunits form a structure resembling a pair of jaws that exist in two states. Prior to association with DNA, the cleft is open; once associated with DNA, the cleft partially closes, securing the duplex during the initiation of transcription. The critical region of the enzyme involved in the transition is about 50 kDa in size and is called the *clamp*.

Once secured by the clamp, a small duplex region of DNA separates at a position within the enzyme referred to as the *active center*, where complementary RNA synthesis is initiated on the DNA template strand. However, the entire complex remains unstable, and transcription most often terminates following the incorporation of only a few ribonucleotides. While it is not clear why, this so-called *abortive transcription* is repeated a number of times before a stable DNA:RNA hybrid including a transcript of 11 ribonucleotides is formed. Once this occurs, abortive transcription is overcome, a stable complex is achieved, and elongation of the RNA transcript proceeds in earnest. Transcription at this point is said to have achieved a level of highly processive RNA polymerization.

As transcription proceeds, the enzyme moves along the DNA, and at any given time, about 40 base pairs of DNA and 18 residues of the growing RNA chain are part of the enzyme complex. The RNA synthesized earliest runs through a groove in the enzyme and exits under a structure at the top and back designated as the *lid*. Another area, called the *pore*, has been identified at the bottom of the enzyme. It serves as the point through which RNA precursors gain entry into the complex.

Eventually, as transcription proceeds, the portion of the DNA that signals termination is encountered, and the complex once again becomes unstable, much as it was during the earlier state of abortive transcription. The clamp opens and both DNA and RNA are released from the enzyme as transcription is terminated. This completes a cycle that constitutes the paradigm of transcription. An unstable complex is formed during the initiation of transcription, stability is established once elongation manages to create a duplex of sufficient size, elongation proceeds, and then instability again characterizes termination of transcription.

These findings extend our knowledge of transcription considerably. It is significant that of the 10 subunits that are part of Kornberg's yeast model, all are very similar to their counterparts in the human enzyme. Nine of the 10 have also been conserved in the other forms of eukaryotic RNA polymerase (I and III). Based on the preceding description, try to mentally visualize the process of transcription from the time DNA associates with the enzyme until the transcript is released from the large molecular complex. If you have developed a clear set of images, you no doubt have acquired a clear understanding of transcription in eukaryotes, which is more complex than in prokaryotes.

Heterogeneous Nuclear RNA and Its Processing: Caps and Tails

We now proceed with the assumption that an initial transcript has been produced. The genetic code is written in the ribonucleotide sequence of mRNA. This information originated, of course, in the template strand of DNA, in which complementary sequences of deoxyribonucleotides exist. In bacteria, the relationship between DNA and RNA appears to be quite direct. The DNA base sequence is transcribed into an mRNA sequence, which is then immediately translated into an amino acid sequence according to the genetic code. In eukaryotes, by contrast, complex processing of mRNA occurs before it is transported to the cytoplasm to participate in translation.

By 1970, accumulating evidence showed that eukaryotic mRNA is transcribed initially as a precursor molecule much larger than that which is translated. This notion was based on the observation by James Darnell and his coworkers of the large heterogeneous nuclear RNA (hnRNA) in mammalian nuclei that contained nucleotide sequences common to the smaller mRNA molecules present in the cytoplasm. They proposed that the initial transcript of a gene results in a large RNA molecule that must first be processed in the nucleus before it appears in the cytoplasm as a mature mRNA molecule. The various processing steps, discussed in the sections that follow, are summarized in Figure 13–10.

The initial **posttranscriptional modification** of eukaryotic RNA transcripts destined to become mRNAs involves the 5′ end of these molecules, at which a **7-methylguanosine (7-mG) cap**

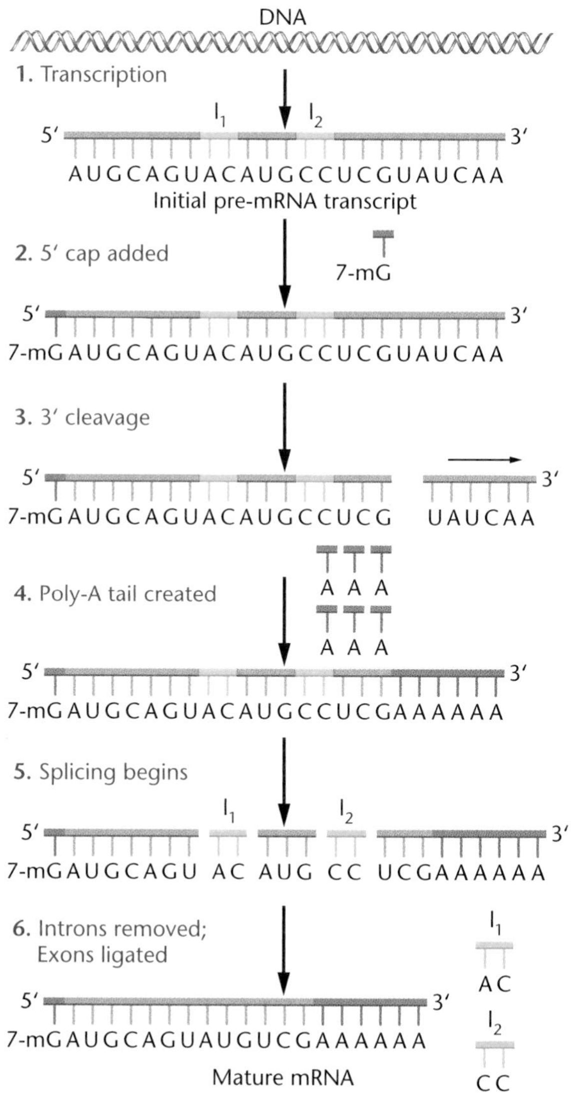

FIGURE 13–10 Posttranscriptional RNA processing in eukaryotes. Heterogeneous nuclear RNA (hnRNA) is converted to messenger (mRNA), which contains a 5′ cap and a 3′-poly-A tail, which then has introns spliced out.

is added (Figure 13–10, Step 2). The cap, which is added even before the initial transcript is complete, appears to be important to the subsequent processing within the nucleus, perhaps by protecting the 5′ end of the molecule from nuclease attack. Subsequently, this cap may be involved in the transport of mature mRNAs across the nuclear membrane into the cytoplasm and in the initiation of translation of the mRNA into protein. The cap is fairly complex and is distinguished by a unique 5′–5′ bonding between the cap and the initial ribonucleotide of the RNA. Some eukaryotes also contain a methyl group (CH_3) on the 2′-carbon of the ribose sugars of the first two ribonucleotides of the RNA.

Further insights into the processing of RNA transcripts during the maturation of mRNA came from the discovery that both hnRNAs and mRNAs contain at their 3′ end a stretch of as many as 250 adenylic acid residues. Such **poly-A sequences** are added after the 5′–7-mG cap has been added. First, the 3′ end of the initial transcript is cleaved enzymatically at a point some 10 to 35 ribonucleotides from a highly conserved AAUAAA sequence (Step 3). Then, polyadenylation occurs by the sequential addition of adenylic acid residues (Step 4). Poly A has now been found at the 3′ end of almost all mRNAs studied in a variety of eukaryotic organisms. In fact, poly-A tails have also been detected in some prokaryotic mRNAs. The exceptions in eukaryotes seem to be the RNAs that encode the histone proteins. While the AAUAAA sequence is not found on all eukaryotic transcripts, it appears to be essential to those that have it. If the sequence is changed as a result of a mutation, those transcripts that do have it cannot add the poly-A tail. In the absence of this tail, these RNA transcripts are rapidly degraded. Therefore, both the 5′ cap and the 3′ poly-A tail are critical if an RNA transcript is to be further processed and transported to the cytoplasm.

13.12 The Coding Regions of Eukaryotic Genes Are Interrupted by Intervening Sequences

One of the most exciting discoveries in the history of molecular genetics occurred in 1977, when Susan Berget, Philip Sharp, and Richard Roberts presented direct evidence that the genes of animal viruses contain *internal* nucleotide sequences that are not expressed in the amino acid sequence of the proteins they encode. These internal DNA sequences are present in initial RNA transcripts, but they are removed before the mature mRNA is translated (Figure 13–10, Steps 5 and 6). Such nucleotide segments are called **intervening sequences**, and the genes that contain them are known as **split genes**. Those DNA sequences that are not represented in the final mRNA product are also called **introns** ("int" for intervening), and those retained and expressed are called **exons** ("ex" for expressed). Splicing involves the removal of the ribonucleotide sequences present in introns as a result of an excision process and the rejoining of exons.

Similar discoveries were soon made in a variety of eukaryotes. Two approaches have been most fruitful. The first involves the molecular hybridization of purified, functionally mature mRNAs with DNA containing the genes specifying that message. Hybridization between nucleic acids that are not perfectly complementary results in **heteroduplexes**, in which introns present in the DNA, but absent in the mRNA loop out and remain unpaired. Such structures can be visualized with the electron microscope, as shown in Figure 13–11. The heteroduplex in the figure contains seven loops (A–G), representing seven introns whose sequences are present in DNA, but not in the final mRNA.

The second approach provides more specific information (Figure 13–12). It involves a comparison of nucleotide sequences of DNA with those of mRNA and the correlation with amino acid sequences. Such an approach allows the precise identification of all intervening sequences.

Thus far, most eukaryotic genes have been shown to contain introns. One of the first so identified was the **beta-globin gene** in mice and rabbits, studied independently by Philip Leder and Richard Flavell. The mouse gene contains an intron 550 nucleotides long, beginning immediately after the codon specifying the 104th amino acid. In the rabbit, there is an intron of 580 base pairs near the codon for the 110th amino acid. Additionally, another intron of about 120 nucleotides exists earlier in both genes. Similar introns have been found in the beta-globin gene in all mammals examined.

The **ovalbumin gene** of chickens has been extensively characterized by Bert O'Malley in the United States and Pierre

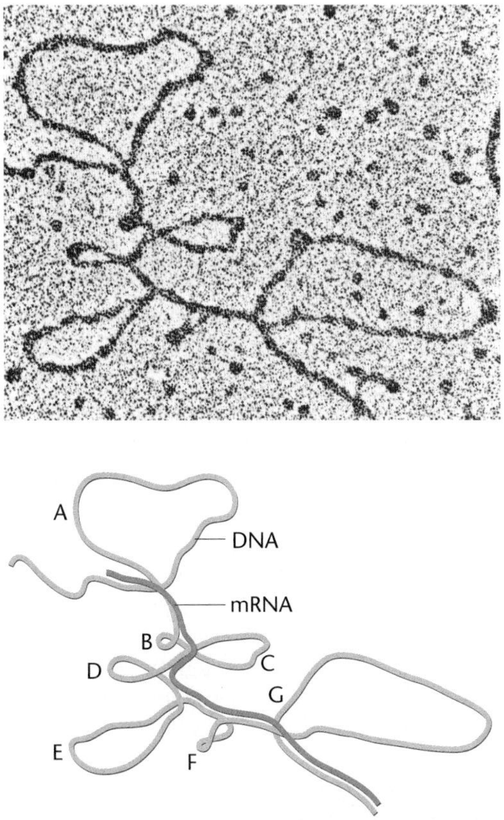

FIGURE 13–11 An electron micrograph and an interpretive drawing of the hybrid molecule (heteroduplex) formed between the template DNA strand of the chicken ovalbumin gene and the mature ovalbumin mRNA. Seven DNA introns, A–G, produce unpaired loops.

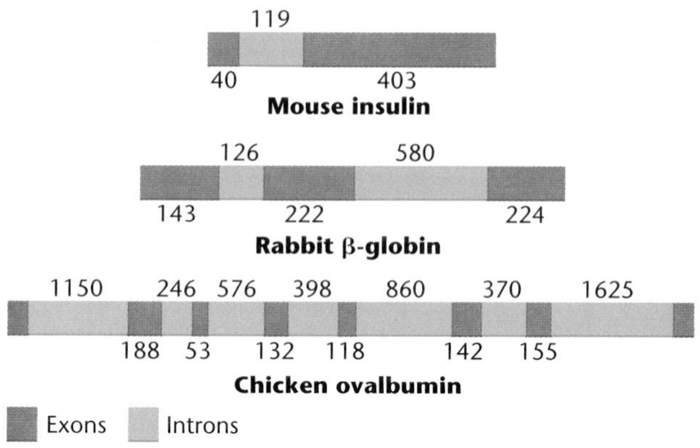

FIGURE 13–12 Intervening sequences in various eukaryotic genes. The numbers indicate the number of nucleotides present in various intron and exon regions.

Chambon in France. As shown in Figure 13–12, the gene contains seven introns, and the majority of the gene's DNA sequence is "silent," being composed of introns. The initial RNA transcript is nearly three times the length of the mature mRNA. Compare the ovalbumin gene in Figures 13–11 and 13–12. Can you match the unpaired loops in Figure 13–11 with the sequence of introns specified in Figure 13–12?

Few eukaryotic genes seem to be without introns. An extreme example of the number of introns in a single gene is that found in the gene coding for one of the subunits of collagen, the major connective tissue protein in vertebrates. The *pro-α-2(1) collagen* gene contains 50 introns. The precision of cutting and splicing that occurs must be extraordinary if errors are not to be introduced into the mature mRNA. What is equally noteworthy is a comparison of the size of genes with the size of the final mRNA once introns are removed. As shown in Table 13.8, only about 15 percent of the collagen gene consists of exons that finally appear in mRNA. For other proteins, an even more extreme picture emerges. Only about 8 percent of the albumin gene remains to be translated, and in the largest human gene known, dystrophin (which is the protein product absent in Duchenne muscular dystrophy), less than 1 percent of the gene sequence is retained in the mRNA. Several other human genes are also contrasted in Table 13.8.

While the vast majority of eukaryotic genes examined thus far contain introns, there are several exceptions. Notably, those coding for histones and for interferon appear to contain no introns. It is not clear why or how the genes encoding these molecules have been maintained throughout evolution without acquiring the extraneous information characteristic of almost all other genes.

HOW DO WE KNOW?

How do we know that eukaryotic genes contain noncoding sequences that must be removed from the RNA intermediate before its translation into proteins?

TABLE 13.8 CONTRASTING HUMAN GENE SIZE, MRNA SIZE, AND THE NUMBER OF INTRONS

Gene	Gene Size (kb)	mRNA Size (kb)	Number of Introns
Insulin	1.7	0.4	2
Collagen [*pro-α*-2(1)]	38.0	5.0	50
Albumin	25.0	2.1	14
Phenylalanine hydroxylase	90.0	2.4	12
Dystrophin	2000.0	17.0	50

Splicing Mechanisms: Autocatalytic RNAs

The discovery of split genes led to intensive attempts to elucidate the mechanism by which introns of RNA are excised and exons are spliced back together. A great deal of progress has already been made. Interestingly, it appears that somewhat different mechanisms exist for different types of RNA, as well as for RNAs produced in mitochondria and chloroplasts. Splicing has also been found during the maturation of tRNAs in prokaryotes. Let's take a closer look at eukaryotic splicing.

Introns can be categorized into several groups based on their splicing mechanisms. In group I—represented by introns that are part of the primary transcript of rRNAs—no additional components are required for intron excision; the intron itself is the source of the enzymatic activity necessary for its own removal. This amazing discovery, which contradicted expectations, was made in 1982 by Thomas Cech and his colleagues during a study of the ciliate protozoan *Tetrahymena*. Reflecting their autocatalytic properties, RNAs that are capable of splicing themselves are sometimes called **ribozymes**.

The **self-excision process** is illustrated in Figure 13–13. Chemically, two nucleophilic reactions (called transesterification reactions) occur, the first involving an interaction between guanosine—which acts as a cofactor in the reaction—and the primary transcript [Figure 13–13(a)]. The 3′-OH group of guanosine is transferred to the nucleotide adjacent to the 5′ end of the intron. The second reaction involves the interaction of the newly acquired 3′-OH group on the left-hand exon and the phosphate on the 3′ end of the right intron [Figure 13–13(b) and (c)]. The intron is spliced out and the two exon regions are ligated, leading to the mature RNA [Figure 13–13(d)].

Self-excision of group I introns, as described, is now known to apply to pre-rRNAs from other protozoans. Self-excision also seems to govern the removal of introns present in the primary mRNA and tRNA transcripts produced in mitochondria and chloroplasts. These are referred to as group II introns. Like group I molecules, splicing involves two autocatalytic reactions leading to the excision of introns. However, guanosine is not involved as a cofactor.

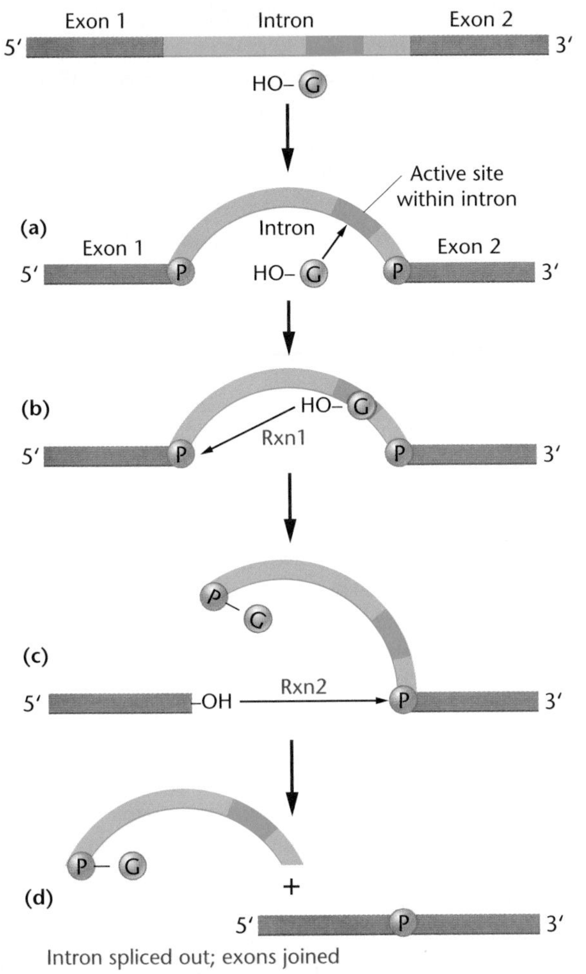

FIGURE 13–13 Splicing mechanism involved with group I introns removed from the primary transcript leading to rRNA. The process is one of self-excision involving two transesterification reactions.

Splicing Mechanisms: The Spliceosome

Introns are, of course, a major component of nuclear-derived pre-mRNA transcripts. Compared to the other RNAs we have been discussing, introns in nuclear-derived mRNA can be much larger—up to 20,000 nucleotides—and they are more plentiful. Their removal appears to require a much more complex mechanism.

The model diagrammed in Figure 13–14 illustrates the removal of one intron. First, the nucleotide sequences near the perimeters of this type of intron are often similar. Many begin at the 5′ end with a GU dinucleotide sequence, called the donor sequence, and terminate at the 3′ end with an AG dinucleotide, called the acceptor sequence. These, as well as other consensus sequences shared by introns, attract specific molecules that form a molecular complex essential to splicing. Such a complex, called a **spliceosome**, has been identified in extracts of yeast, as well as in mammalian cells. It is very large, being 40*S* in yeast and 60*S* in mammals. Perhaps the most essential component of spliceosomes is the unique set of **small nuclear RNAs (snRNAs)**. These RNAs are usually 100 to 200 nucleotides or less and are often complexed with proteins to form **small nuclear ribonucleoproteins** (**snRNPs**, or **snurps**). These are found only in the nucleus. Because they are rich in uridine residues, the snRNAs have been arbitrarily designated U1, U2, . . . U6.

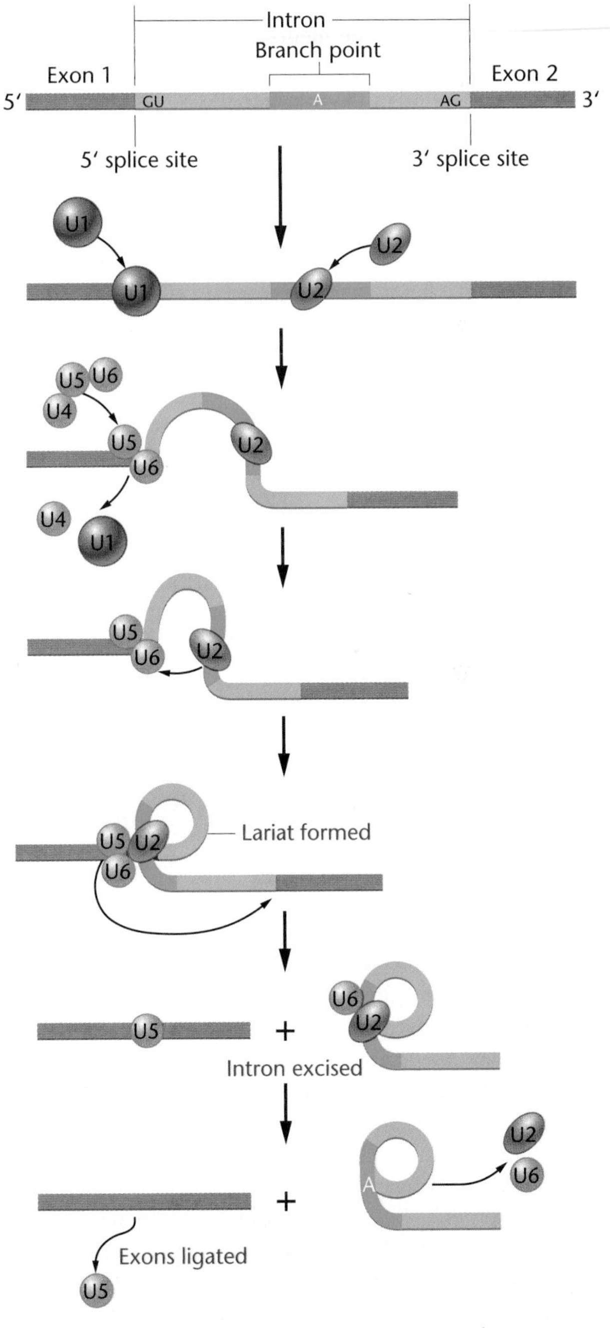

FIGURE 13–14 A model of the splicing mechanism involved with the removal of an intron from a pre-mRNA. Excision is dependent on various snRNAs (U1, U2, . . . , U6) that combine with proteins to form snRNPs (snurps), which function as part of a large structure referred to as the spliceosome. The lariat structure in the intermediate stage is characteristic of this mechanism.

The snRNA of U1 bears a nucleotide sequence complementary to the 5′ splice donor sequence end of the intron. Base pairing resulting from this complementarity promotes binding representing the initial step in the formation of the spliceosome. Following the ordered addition of the other snurps (U2, U4, U5, and U6), splicing commences. As with group I splicing, two transesterification reactions are involved. The first involves the interaction of the 2′-OH group from an adenine (A) residue present within the region of the intron called the **branch point**. The A residue attacks the 5′-splice site, cutting the RNA chain. In a subsequent step involving several other snurps, an intermediate structure is formed and the second reaction ensues, linking the cut 5′ end of the intron to the A. This results in the formation of a characteristic loop structure referred to as a lariat, which contains the excised intron. The exons are then ligated and the snurps are released.

The processing involved in splicing represents a potential regulatory step during gene expression. For example, several cases are known wherein introns present in pre-mRNAs *derived from the same gene* are spliced *in more than one way*, thereby yielding different collections of exons in the mature mRNA. This process, referred to as **alternative splicing**, yields a group of mRNAs that, upon translation, result in a series of related proteins called **isoforms**. A growing number of examples are now found in organisms ranging from viruses to *Drosophila* to humans. Alternative splicing of pre-mRNAs provides the basis for producing related proteins from a single gene. We shall return to this topic in our discussion of the regulation of gene expression in eukaryotes. (See Chapter 17.)

RNA Editing

In the late 1980s, still another unexpected form of posttranscriptional RNA processing was discovered in several organisms. In this form, referred to as **RNA editing**, the nucleotide sequence of a pre-mRNA is actually changed prior to translation. As a result, the ribonucleotide sequence of the mature RNA differs from the sequence encoded in the exons of the DNA from which the RNA was transcribed.

While there are numerous variations of RNA editing, the two main types studied initially include **substitution editing**, in which the identities of individual nucleotide bases are altered; and **insertion/deletion editing**, in which nucleotides are added to or subtracted from the total number of bases. Substitution editing is used in some nuclear-derived eukaryotic RNAs and is also very prevalent in mitochondrial and chloroplast RNAs transcribed in plants. *Physarum polycephalum*, a slime mold, uses both substitution and insertion/deletion editing for its mitochondrial mRNAs.

Trypanosoma, a parasite that causes African sleeping sickness, and its relatives use extensive insertion/deletion editing in mitochondrial RNAs. The number of uridines added to an individual transcript can make up more than 60 percent of the coding sequence, usually forming the initiation codon and placing the rest of the sequence into the proper reading frame. Insertion/deletion editing in trypanosomes is directed by **gRNA (guide RNA)** templates, which are also transcribed from the mitochondrial genome. These small RNAs are complementary to the edited region of the final, edited mRNAs. They base-pair with the preedited mRNAs to direct the editing machinery to make the correct changes.

The most well-studied examples of substitutional editing occur in mammalian nuclear-encoded mRNA transcripts. Apolipoprotein B (apo B) exists in both a long and a short form, though a single gene encodes both proteins. In human intestinal cells, apo B mRNA is edited by a single C to U change, which converts a CAA glutamine codon into a UAA stop codon and terminates the polypeptide at approximately half its genomically encoded length. The editing is performed by a complex of proteins that bind to a "mooring sequence" on the mRNA transcript just downstream of the editing site. In a second system, the synthesis of subunits constituting the glutamate receptor channels (GluR) in mammalian brain tissue is also affected by RNA editing. In this case, adenosine (A) to inosine (I) editing occurs in pre-mRNAs prior to their translation, during which I is read as guanosine (G). The enzymes produced by a family of three *ADAR* (*a*denosine *d*eaminase *a*cting on *R*NA) genes are believed to be responsible for the editing of various sites within the glutamate channel subunits. The double-stranded RNAs required for editing by the *ADAR* enzymes are provided by intron–exon pairing of the GluR mRNA transcripts. The editing changes alter the physiological parameters (solute permeability and desensitization response time) of the receptors containing the subunits.

The importance of RNA editing resulting from the action of *ADAR*s is most apparent when one examines situations in which these enzymes have lost their functional capacity as a result of mutation. In several investigations, the loss of function was shown to have a lethal impact in mice. In one study, embryos heterozygous for a defective *ADAR1* gene die during embryonic development as a result of a defective hematopoietic system. In another study, mice with two defective copies of *ADAR2* progress through development normally, but are prone to epileptic seizures and die while still in the weaning stage. Their tissues contain the unedited version of one of the GluR products. The defect leading to death is believed to be in the brain. Heterozygotes for the mutation are normal.

Findings such as these in mammals have established that RNA editing provides still another important mechanism of posttranscriptional modification, and that this process is not restricted to small or asexually reproducing genomes, such as mitochondria. Several new examples of RNA editing have been found each year since its discovery, and this trend is likely to continue. Further, the process has important implications for the regulation of genetic expression.

13.13 Transcription Has Been Visualized by Electron Microscopy

We conclude this chapter by presenting a striking visual demonstration of the transcription process based on the electron microscope studies of Oscar Miller, Jr., Barbara Hamkalo, and Charles Thomas. Their combined work has captured the transcription process in both prokaryotes and eukaryotes. Figure 13–15 shows micrographs and interpretive drawings from two organisms, the bacterium *E. coli* and the newt *Notophthalmus viridescens*. In both cases, multiple strands of RNA are seen to emanate from different points along a central DNA template. Many RNA strands result because numerous transcription events are occurring simultaneously along each gene. Progressively longer RNA strands are found farther downstream from the point of initiation of transcription, whereas the shortest strands are closest to the point of initiation.

An interesting picture emerges from the study of *E. coli* [Figure 13–15(a)]. Because prokaryotes lack nuclei, cytoplasmic ribosomes are not separated physically from the chromosome. As a result, ribosomes are free to attach to *partially* transcribed mRNA molecules and initiate translation. The longer RNA strands demonstrate the greatest number of ribosomes. In the case of the newt [Figure 13–15(b)], the segment of DNA is derived from oocytes that are known to produce an enormous amount of ribosomal (rRNA), one of the major components of the ribosome. To accomplish this synthesis, the genes specific for **ribosomal RNA (rDNA)** are replicated many times in the oocyte, a process called **gene amplification**. The micrograph shows several of these genes, each undergoing simultaneous transcription events. For each rRNA gene, longer and longer strands of incomplete rRNA molecules are produced as the enzymes move along the DNA strand. Visualization of transcription confirms our expectations based on the biochemical analysis of this process.

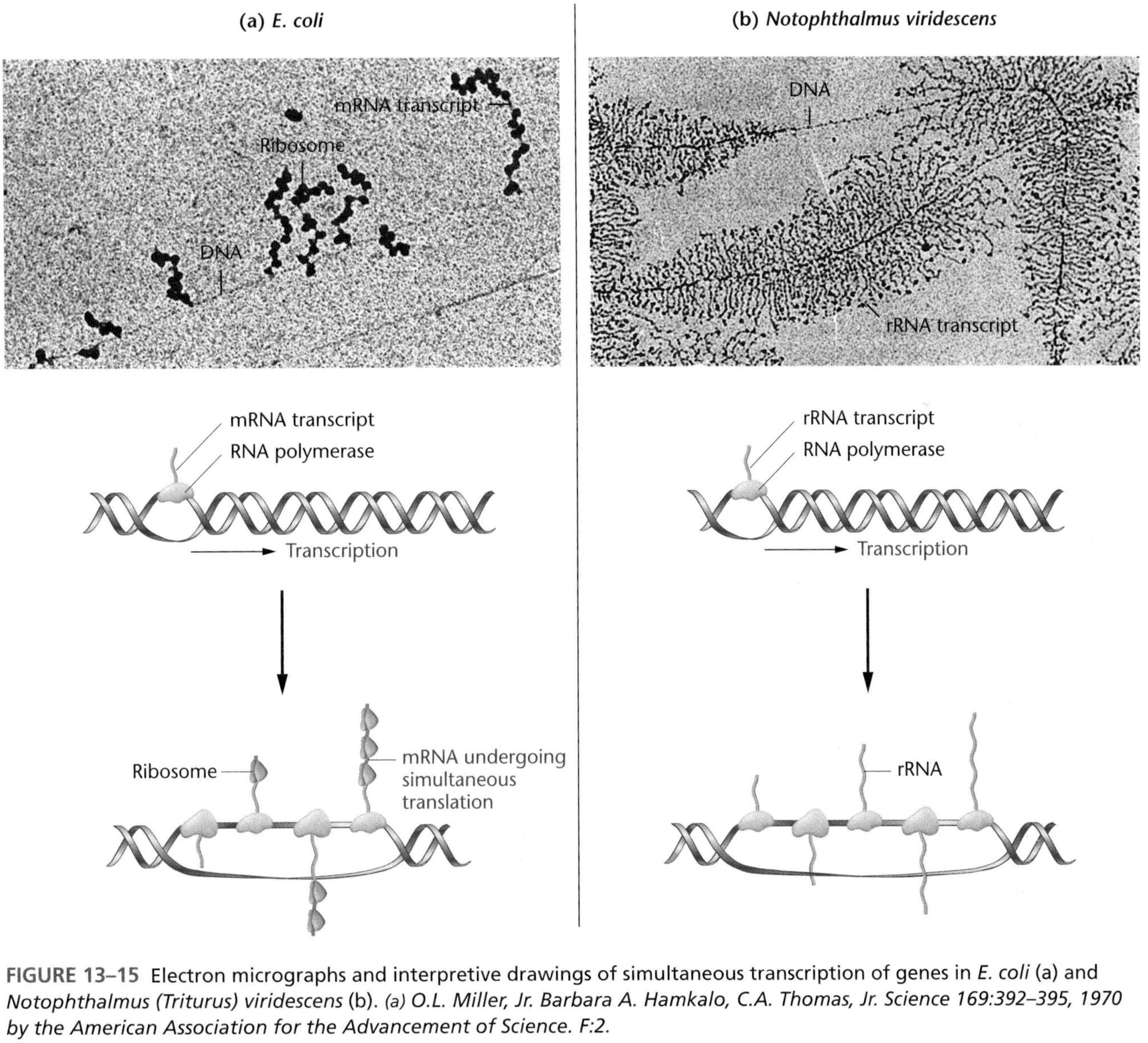

FIGURE 13–15 Electron micrographs and interpretive drawings of simultaneous transcription of genes in *E. coli* (a) and *Notophthalmus (Triturus) viridescens* (b). *(a) O.L. Miller, Jr. Barbara A. Hamkalo, C.A. Thomas, Jr. Science 169:392–395, 1970 by the American Association for the Advancement of Science. F:2.*

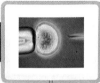

GENETICS, TECHNOLOGY, AND SOCIETY

Antisense Oligonucleotides: Attacking the Messenger

Standard chemotherapies for diseases such as cancer and AIDS are often accompanied by toxic side effects. Conventional therapeutic drugs target both normal and diseased cells, with diseased or infected cells being only slightly more susceptible than the patient's normal cells. Scientists have long wished for a magic bullet that could seek out and destroy the virus or cancer cell, leaving normal cells alive and healthy. Over the last decade, one particularly promising candidate for magic-bullet status has emerged—the **antisense oligonucleotide**.

Antisense therapies have arisen through an understanding of the molecular biology of gene expression. Gene expression is a two-step process. First, a single-stranded messenger RNA (mRNA) is copied from the template strand of the duplex DNA molecule. Second, the mRNA is complexed with ribosomes, and its coded information is translated into the amino acid sequence of a polypeptide. Normally, a gene is transcribed into RNA from only one strand of the DNA duplex. The resulting RNA is known as *sense RNA*. However, it is sometimes possible for the other DNA strand to be copied into RNA. The RNA produced by the transcription of the "wrong" strand of DNA is called *antisense RNA*. As with complementary strands of DNA molecules, complementary strands of RNA can form double-stranded molecules.

The formation of duplex structures between a sense and an antisense RNA may affect the sense RNA in at least two ways. Binding antisense RNA to sense RNA may physically block ribosome binding or elongation, hence inhibiting translation. Or binding antisense RNA to sense RNA may trigger the degradation of the sense RNA because double-stranded RNA molecules are attacked by intracellular ribonucleases. In both cases, gene expression is blocked.

What makes the antisense approach so exciting is its potential specificity. Scientists can design antisense RNA (or DNA) molecules of known nucleotide sequence and can then synthesize large amounts of these antisense nucleic acids *in vitro*. Usually, oligonucleotides of up to 20 nucleotides are used. It is theoretically possible to treat cells with these synthetic antisense oligonucleotides, have the oligonucleotides enter the cell and bind precise target mRNAs, and to turn off the synthesis of one specific protein. If that protein is necessary for virus reproduction or cancer cell growth (but is not necessary in normal cells), the antisense oligonucleotide should have only therapeutic effects.

In the past few years, laboratory tests of antisense drugs have been so promising that many clinical trials are in progress. In addition, one antisense drug (Vitravene©, from ISIS Pharmaceuticals and CIBA Vision Corp.) has been approved for sale in the United States. Vitravene is an antisense oligonucleotide used to treat cytomegalovirus-induced retinitis in AIDS patients. Cytomegalovirus (CMV) is a common virus that infects most people. Although it causes few problems in people with normal immune systems, it can cause serious symptoms in people with impaired immune systems (such as AIDS). Up to 40 percent of AIDS patients develop retinitis or blindness as a result of CMV infections of the eye. Clinical trials indicate that the introduction of antisense CMV oligonucleotides into the eyes of AIDS patients with CMV-induced retinitis significantly delays the disease progression compared to untreated groups. Laboratory studies suggest that antisense oligonucleotides may trigger the destruction of CMV mRNA and interfere with the adsorption of the virus to the surface of host cells.

Antisense oligonucleotides may also act as antiinflammatory drugs. Clinical trials are in progress to test antisense compounds in the treatment of asthma, rheumatoid arthritis, psoriasis, ulcerative colitis, and transplant rejection. One interesting antisense oligonucleotide (now in clinical trials conducted by ISIS and Boehringer Ingelheim) binds to the mRNA that encodes ICAM-1, a cell-surface glycoprotein that helps activate immune system and inflammatory cells. It is often overexpressed in body tissues that exhibit extreme inflammatory responses. Researchers hope that antisense oligonucleotides might temper inflammatory responses by reducing the expression of ICAM-1. The results of a recent clinical trial suggest that antisense ICAM-1 may be an effective treatment for the debilitating, chronic Crohn's disease, a form of inflammatory bowel disease that affects about 200,000 people in the United States. In one clinical trial, almost half of the Crohn's disease patients went into remission after treatment with antisense ICAM-1, whereas none of the placebo group did so. In this trial, the antisense ICAM-1 drugs appeared to be well-tolerated and safe.

Some interesting clinical trials are in progress to test antisense oligonucleotides as treatments for some types of cancer. These oligonucleotides are designed to reduce the synthesis of proteins that are either overexpressed in cancer cells or are present as mutant forms. These drugs will be evaluated as treatments for tumors of the ovary, prostate, breast, brain, colon, and lung. Other antisense drugs in the pipeline are designed to attack the hepatitis B, hepatitis C, human papilloma, and AIDS viruses. If antisense therapeutics pass the scrutiny of these clinical trials, we may indeed have acquired a molecular magic bullet to use in the battle against a variety of diseases.

More recently, another gene silencing technology, based on the antisense principle, has emerged. This technology, known as **RNA interference (RNAi)**, uses short double-stranded RNA molecules (~20 nucleotides long) called **short interfering RNA (siRNA)** to target specific mRNAs within cells. One strand of the short RNA recognizes a sequence within the mRNA, leading to the degradation or silencing of the mRNA. Although RNAi is highly efficient *in vitro* and is used in laboratory research, the application of RNAi to human disease is less advanced than the use of oligonucleotide-based antisense technologies. RNAi is discussed in more detail in Chapter 17 and 21.

References

Dykxhoorn, D.M., Novina, C.D., and Sharp, P.A. 2003. Killing the messenger: Short RNAs that silence gene expression. *Nature Revs. Molec. Cell Biol.* 4:457–67.

Nyce, J.W., and Metzger, W.J. 1997. DNA antisense therapy for asthma in an animal model. *Nature* 385:721–25.

Roush, W. 1997. Antisense aims for a renaissance. *Science* 276:1192–93.

Chapter Summary

1. The genetic code, stored in DNA, is copied to RNA, in which it is used to direct the synthesis of polypeptide chains. It is degenerate, unambiguous, nonoverlapping, and commaless.
2. The complete coding dictionary, determined using various experimental approaches, reveals that of the 64 possible codons, 61 encode the 20 amino acids found in proteins, while three codons terminate translation. One of these 61 is the initiation codon and specifies methionine.
3. The observed pattern of degeneracy often involves only the third letter of a triplet series and led Francis Crick to propose the wobble hypothesis.
4. Confirmation for the coding dictionary, including codons for initiation and termination, was obtained by comparing the complete nucleotide sequence of phage MS2 with the amino acid sequence of the corresponding proteins. Other findings support the belief that, with only minor exceptions, the code is universal for all organisms.
5. In some bacteriophages, multiple initiation points may occur during the transcription of RNA, resulting in multiple reading frames and overlapping genes.
6. Transcription—the initial step in gene expression—describes the synthesis, under the direction of RNA polymerase, of a strand of RNA complementary to a DNA template.
7. The processes of transcription, like DNA replication, can be subdivided into the stages of initiation, elongation, and termination. Also like DNA replication, the process relies on base-pairing affinities between complementary nucleotides.
8. Initiation of transcription is dependent on an upstream (5′) DNA region, called the promoter, that represents the initial binding site for RNA polymerase. Promoters contain specific DNA sequences, such as the TATA box, that are essential to polymerase binding.
9. Transcription is more complex in eukaryotes than in prokaryotes. The primary transcript is a pre-mRNA that must be modified in various ways before it can be efficiently translated. Processing, which produces a mature mRNA, includes the addition of a 7-mG cap and a poly-A tail, and the removal, through splicing, of intervening sequences, or introns. RNA editing of pre-mRNA prior to its translation also occurs in some systems.

Insights and Solutions

1. Calculate how many triplet codons would be possible had evolution seized on six bases (three complementary base pairs) rather than four bases within the structure of DNA. Would six bases be sufficient using a two-letter code, assuming 20 amino acids and start-and-stop codons?

Solution: Six things taken three at a time will produce $(6)^3$, or 216, triplet codes. If the code was a doublet, there would be $(6)^2$, or 36, two-letter codes, more than enough to accommodate 20 amino acids and start–stop punctuation.

2. In a heteropolymer experiment using 1/2C : 1/4A : 1/4G, how many different codons will occur in the synthetic RNA molecule? How often will the most frequent codon occur?

Solution: There will be $(3)^3$, or 27, codons produced. The most frequent will be CCC, present $(1/2)^3$, or 1/8, of the time.

3. In a regular copolymer experiment, in which UUAC is repeated over and over, how many different codons will occur in the synthetic RNA, and how many amino acids will occur in the polypeptide when this RNA is translated? Be sure to consult Figure 13–7.

Solution: The synthetic RNA will repeat four codons—UUA, UAC, ACU, and CUU—over and over. Because both UUA and CUU encode leucine, while ACU and UAC encode threonine and tyrosine, respectively, the polypeptides synthesized under the directions of such an RNA contain three amino acids in the repeating sequence leu-leu-thr-tyr.

4. Actinomycin D inhibits DNA-dependent RNA synthesis. This antibiotic is added to a bacterial culture in which a specific protein is being monitored. Compared to a control culture, into which no antibiotic is added, translation of the protein declines over a period of 20 minutes, until no further protein is made. Explain these results.

Solution: The mRNA, which is the basis for the translation of the protein, has a lifetime of about 20 minutes. When actinomycin D is added, transcription is inhibited, and no new mRNAs are made. Those already present support the translation of the protein for up to 20 minutes.

5. DNA and RNA base compositions were analyzed from a hypothetical bacterial species with the following results:

	(A + G)/(T + C)	(A + T)/(C + G)	(A + G)/(U + C)	(A + U)/(C + G)
DNA	1.0	1.2		
RNA			1.3	1.2

On the basis of these data, what can you conclude about the DNA and RNA of the organism? Are the data consistent with the Watson–Crick model of DNA? Is the RNA single stranded or double stranded, or can't we tell? If we assume that the entire length of DNA has been transcribed, do the data suggest that RNA has been derived from the transcription of one or both DNA strands, or can't we tell from these data?

Solution: This problem is a theoretical exercise designed to get you to look at the consequences of base complementarity as it affects the base composition of DNA and RNA. The base composition of DNA is consistent with the Watson–Crick double helix. In a double helix, we expect A + G to equal T + C. (The number of purines should equal the number of pyrimidines.) In this case, there is a preponderance of A=T base pairs (120 A=T pairs to every 100 G≡C pairs).

Given what we know about RNA, there is no reason to expect the RNA to be double stranded, but if it were double stranded, then we would expect that A=U and C≡G. If so, then (A + G)/(U + C) = 1. Since it doesn't equal unity, we can conclude that the RNA is not double stranded.

If all the DNA is transcribed, from either one or both strands, the ratio of (A + U)/(C + G) in RNA should be 1.2, and, as predicted, it is. Note that this ratio will not change, regardless of whether only one or both of the strands are transcribed. This is the case, because for every A═T pair in DNA, for example, transcription of RNA will yield one A *and* one U if both strands are transcribed. If just one strand is transcribed, transcription will yield one A *or* one U. In either case, the (A + U)/(C + G) ratio in RNA will reflect the (A + T)/(C + G) ratio in the DNA from which it was transcribed. To prove this to yourself, draw out a DNA molecule with 12 A═T pairs and 10 C≡G pairs and transcribe *both* strands. Then, transcribe *either* strand. Count the bases in the RNAs produced in both cases and calculate the ratios. Thus, we cannot determine whether just one or both strands are transcribed from the (A + U)/(C + G) ratio.

However, if both strands are transcribed, then the ratio of (A + G)/(U + C) should equal 1.0, and it doesn't. It equals 1.3. To verify this conclusion, examine the theoretical data you drew out on paper. One explanation for the observed ratio of 1.3 is that only one of the two strands is transcribed. If this is the case, then the (A + G)/(U + C) will reflect the proportion of (A + T) pairs that are A and the proportion of the G≡C pairs that are *on the DNA strand that is transcribed.* Another explanation is that transcription occurs only on one strand at any given point (e.g., for one gene), but on the other strand at other points (for other genes).

Problems and Discussion Questions

1. Early proposals regarding the genetic code considered the possibility that DNA served directly as the template for polypeptide synthesis. (See Gamow, 1954, in Selected Readings.) In eukaryotes, what difficulties would such a system pose? What observations and theoretical considerations argue against such a proposal?
2. In their studies of frameshift mutations, Crick, Barnett, Brenner, and Watts-Tobin found that either three pluses or three minuses restored the correct reading frame. If the code were a sextuplet (consisting of six nucleotides), would the reading frame be restored by either of the preceding combinations?
3. In a mixed copolymer experiment using polynucleotide phosphorylase, 3/4G:1/4C was added to form the synthetic message. The resulting amino acid composition of the ensuing protein was determined to be:

Glycine	36/64	(56 percent)
Alanine	12/64	(19 percent)
Arginine	12/64	(19 percent)
Proline	4/64	(6 percent)

From this information,

(a) indicate the percentage (or fraction) of the time each possible codon will occur in the message.

(b) determine one consistent base-composition assignment for the amino acids present.

(c) consider the wobble hypothesis, and predict as many specific codon assignments as possible.

4. When repeating copolymers are used to form synthetic mRNAs, dinucleotides produce a single type of polypeptide that contains only two different amino acids. On the other hand, using a trinucleotide sequence produces three different polypeptides, each consisting of only a single amino acid. Why? What will be produced when a repeating tetranucleotide is used?
5. The mRNA formed from the repeating tetranucleotide UUAC incorporates only three amino acids, but the use of UAUC incorporates four amino acids. Why?
6. In studies using repeating copolymers, AC . . . incorporates threonine and histidine, and CAACAA . . . incorporates glutamine, asparagine, and threonine. What triplet code can definitely be assigned to threonine?
7. In a coding experiment using repeating copolymers (as shown in Table 13.3), the following data were obtained:

Copolymer	Codons Produced	Amino Acids in Polypeptide
AG	AGA, GAG	Arg, Glu
AAG	AGA, AAG, GAA	Lys, Arg, Glu

AGG is known to code for arginine. Taking into account the wobble hypothesis, assign each of the four remaining different triplet codes to its correct amino acid.

8. In the triplet-binding technique, radioactivity remains on the filter when the amino acid corresponding to the codon is labeled. Explain the basis of this technique.
9. When the amino acid sequences of insulin isolated from different organisms were determined, some differences were noted. For example, alanine was substituted for threonine, serine was substituted for glycine, and valine was substituted for isoleucine at corresponding positions in the protein. List the single-base changes that could occur in codons of the genetic code to produce these amino acid changes.
10. In studies of the amino acid sequence of wild-type and mutant forms of tryptophan synthetase in *E. coli*, the following changes have been observed:

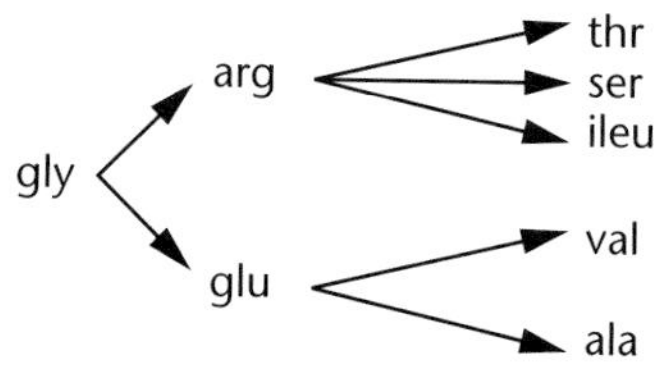

Determine a set of triplet codes in which only a single nucleotide change produces each amino acid change.

11. Why doesn't polynucleotide phosphorylase synthesize RNA *in vivo*?
12. Refer to Table 13.1. Can you hypothesize why a mixture of Poly U + Poly A would not stimulate incorporation of ^{14}C-phenylalanine into protein?

13. Predict the amino acid sequence produced during translation by the following short theoretical mRNA sequences (note that the second sequence was formed from the first by a deletion of only one nucleotide):

Sequence 1:	AUGCCGGAUUAUAGUUGA
Sequence 2:	AUGCCGGAUUAAGUUGA

What type of mutation gave rise to Sequence 2?

14. A short RNA molecule was isolated that demonstrated a hyperchromic shift indicating secondary structure. Its sequence was determined to be

AGGCGCCGACUCUACU

(a) Propose a two-dimensional model for this molecule.
(b) What DNA sequence would give rise to this RNA molecule through transcription?
(c) If the molecule were a tRNA fragment containing a CGA anticodon, what would the corresponding codon be?
(d) If the molecule were an internal part of a message, what amino acid sequence would result from it following translation? (Refer to the code chart in Figure 13–7.)

15. A glycine residue exists at position 210 of the tryptophan synthetase enzyme of wild-type *E. coli*. If the codon specifying glycine is GGA, how many single-base substitutions will result in an amino acid substitution at position 210? What are they? How many will result if the wild-type codon is GGU?

16. Refer to Figure 13–7 to respond to the following:
(a) Shown here is a theoretical viral mRNA sequence:

5′-AUGCAUACCUAUGAGACCCUUGGA-3′

Assuming that it could arise from overlapping genes, how many different polypeptide sequences can be produced? What are the sequences?

(b) A base substitution mutation that altered the sequence in (a) eliminated the synthesis of all but one polypeptide. The altered sequence is shown here:

5′-AUGCAUACCUAUGUGACCCUUGGA-3′

Determine why.

17. Most proteins have more leucine than histidine residues, but more histidine than tryptophan residues. Correlate the number of codons for these three amino acids with this information.
18. Define the process of transcription. Where does this process fit into the central dogma of molecular genetics (DNA makes RNA makes protein)?
19. What was the initial evidence for the existence of mRNA?
20. Describe the structure of RNA polymerase in bacteria. What is the core enzyme? What is the role of the sigma subunit?
21. In a written paragraph, describe the abbreviated chemical reactions that summarize RNA polymerase-directed transcription.
22. Messenger RNA molecules are very difficult to isolate in prokaryotes because they are rather quickly degraded in the cell. Can you suggest a reason why this occurs? Eukaryotic mRNAs are more stable and exist longer in the cell than do prokaryotic mRNAs. Is this an advantage or disadvantage for a pancreatic cell making large quantities of insulin?
23. The following represent deoxyribonucleotide sequences derived from the template strand of DNA:

Sequence 1:	5′-CTTTTTTGCCAT-3′
Sequence 2:	5′-ACATCAATAACT-3′
Sequence 3:	5′-TACAAGGGTTCT-3′

(a) For each strand, determine the mRNA sequence that would be derived from transcription.
(b) Using Figure 13–7, determine the amino acid sequence that is encoded by these mRNAs.
(c) For Sequence 1, what is the sequence of the nontemplate DNA strand?

Extra-Spicy Problems

24. In a mixed copolymer experiment, messengers were created with either 4/5C:1/5A or 4/5A:1/5C. These messages yielded proteins with the following amino acid compositions.

4/5C:1/5A		4/5A:1/5C	
Proline	63.0 percent	Proline	3.5 percent
Histidine	13.0 percent	Histidine	3.0 percent
Threonine	16.0 percent	Threonine	16.6 percent
Glutamine	3.0 percent	Glutamine	13.0 percent
Asparagine	3.0 percent	Asparagine	13.0 percent
Lysine	0.5 percent	Lysine	50.0 percent
	98.5 percent		99.1 percent

Using these data, predict the most specific coding composition for each amino acid.

25. Shown here are the amino acid sequences of the wild-type and three mutant forms of a short protein. Use this information to answer the following questions:

Wild type:	mer-trp-tyr-arg-gly-ser-pro-thr
Mutant 1:	met-trp
Mutant 2:	met-trp-his-arg-gly-ser-pro-thr
Mutant 3:	met-cys-ile-val-val-val-gln-his

(a) Using Figure 13–7, predict the type of mutation that occurred leading to each altered protein.
(b) For each mutant protein, determine the specific ribonucleotide change that led to its synthesis.
(c) The wild-type RNA consists of nine triplets. What is the role of the ninth triplet?
(d) For the first eight wild-type triplets, which, if any, can you determine specifically from an analysis of the mutant proteins? In each case, explain why or why not.
(e) Another mutation (Mutant 4) is isolated. Its amino acid sequence is unchanged, but mutant cells produce abnormally low amounts of the wild-type proteins. As specifically as you can, predict where in the gene this mutation exists.

26. The genetic code is degenerate. Amino acids are encoded by either 1, 2, 3, 4, or 6 triplet codons. (See Figure 13–7.) An interesting question is whether the frequency of triplet codes is in any way correlated with the frequency that amino acids appear in proteins? That is, is the genetic code optimized for its intended use? Some approximations of the frequency of appearance of nine amino acids in proteins in *E. coli* are

	Percentage
Met	2
Cys	2
Gln	5
Pro	5
Arg	5
Ile	6
Glu	7
Ala	8
Leu	10

(a) Determine how many triplets encode each amino acid.
(b) Devise a way to graphically express the two sets of information (data).
(c) Analyze your data to determine what, if any, correlations can be drawn between the relative frequency of amino acids making up proteins with the number of codons for each. Write a paragraph that states your specific and general conclusions.
(d) What would be the next steps in your analysis if you wanted to pursue this problem further?

27. As described in Chapter 12, *Alu* elements proliferate in the human genome by a process called retrotransposition, in which the *Alu* DNA sequence is transcribed into RNA, copied into double-stranded DNA, and then inserted back into the genome at a site distant from that of its "parent" *Alu*.

Clearly, this has been an extremely efficient process, since *Alus* have proliferated to about 10^6 copies in the human genome! This efficiency is largely due to the fact that *Alus*, like many small structural RNAs, carry their promoter sequences *within the transcribed region of the gene*, rather than 5′ to the transcription start site.

If *Alus* carried promoters upstream to the transcription site, similar to those of protein-coding genes, what would happen once they were retrotransposed? Would a retrotransposed *Alu* be able to proliferate? Explain.

28. M. Klemke and others (2001. *EMBO J.* 20:3849–60) discovered an interesting coding phenomenon in an exon within a neurologic hormone receptor gene in mammals, by which two protein entities (XLαs and ALEX) appear to be produced from the same exon. Below is the DNA sequence of the exon's 5′ end derived from a rat. The lowercase letters represent the initial coding portion for the XLαs protein, and the uppercase letters indicate the portion where the ALEX entity initiates. (For simplicity, and to correspond with the RNA coding dictionary, it is customary to represent the non-coding, nontemplate strand of the DNA segment.)

5′-gtcccaaccatgcccaccgatcttccgcctgcttctgaagATGCGGGCCCAG

(a) Convert the noncoding DNA sequence to the coding RNA sequence.
(b) Locate the initiator codon within the XLαs segment.
(c) Locate the initiator codon within the ALEX segment. Are the two initiator codons in frame?
(d) Provide the amino acid sequence for each coding sequence. In the region of overlap, are the two amino acid sequences the same?
(e) Are there any evolutionary advantages to having the same DNA sequence code for two protein products? Are there any disadvantages?

29. The concept of consensus sequences of DNA was introduced in this chapter as those sequences that are similar (homologous) in different genes of the same organism or in genes of different organisms. Examples were the Pribnow box and the −35 region in prokaryotes and the TATA-box region in eukaryotes. Work by Novitsky and colleagues (2002. *J. Virol.* 76:5435–51) indicates that among 73 isolates of HIV-Type 1C (a major contributor to the AIDS epidemic), a GGGNNNNNCC consensus sequence exists (where N equals any nitrogenous base) in the promoter–enhancer region of the NF-κB transcription factor, a *cis*-acting motif which is critical in initiating HIV transcription in human macrophages. The authors contend that finding this and other conserved sequences may be of value in designing an AIDS vaccine. What advantages would finding these consensus sequences confer? What disadvantages?

30. Theoretically, antisense oligodeoxynucleotides (relatively short, 7–20 nucleotides, single-stranded DNAs) can selectively block disease-causing genes. They do so by base-pairing with complementary regions of the RNA transcripts and inhibiting their function. The RNA/DNA hybrid is recognized by intracellular RNase H and the RNA portion of the hybrid is degraded. Cancer genes have often been chosen as potential targets for antisense drugs. Data from Cho and coworkers (2001. *Proc. Natl. Acad. Sci. [USA]* 98:9819–23), shown in the table below, established that in response to a single population of antisense oligodeoxynucleotides that target a specific kinase mRNA in a prostate cancer cell line, there was as much as a twentyfold (plus and minus) response differential of unrelated gene expression (as measured by mRNA production).

Gene Assayed	Change in Expression
Myosin light chain	+14.4X
G protein receptor	+3.7X
Collagen	−4.0X
Catalase	−6.7X

(a) Diagram your concept of the intracellular action of an antisense oligodeoxynucleotide.
(b) Diagram your concept of the action of RNase H.
(c) What might cause nonrelated gene expression to be decreased, as in the case of collagen and catalase?
(d) What might cause nonrelated gene expression to be increased, as in the case of myosin and G protein?
(e) Given the data presented here, what drawbacks might you expect in the use of antisense therapy for genetic diseases?

31. Recent observations indicate that alternative splicing is a common mechanism for eukaryotes to expand their repertoire of gene functions. Studies by Xu and colleagues (2002. *Nuc. Acids Res.* 30:3754–66) indicate that approximately 50 percent of human genes use alternative splicing, and approximately 15 percent of disease-causing mutations involve aberrant alternative splicing. Different tissues show remarkably different frequencies of alternative splicing, with the brain accounting for approximately 18 percent of such events. (a) What does alternative splicing mean, and what is an isoform? (b) What evolutionary strategy does alternative splicing offer, and why might some tissues engage in more alternative splicing than others?

SELECTED READINGS

Barrell, B.G., Air, G., and Hutchinson, C. 1976. Overlapping genes in bacteriophage ϕX174. *Nature* 264:34–40.

Barrell, B.G., Banker, A.T., and Drouin, J. 1979. A different genetic code in human mitochondria. *Nature* 282:189–94.

Bass, B.L., ed. 2000. *RNA Editing*. Oxford: Oxford University Press.

Brenner, S., Jacob, F., and Meselson, M. 1961. An unstable intermediate carrying information from genes to ribosomes for protein synthesis. *Nature* 190:575–80.

Brenner, S., Stretton, A.O.W., and Kaplan, D. 1965. Genetic code: The nonsense triplets for chain termination and their suppression. *Nature* 206:994–98.

Cattaneo, R. 1991. Different types of messenger RNA editing. *Annu. Rev. Genet.* 25:71–88.

Cech, T.R. 1986. RNA as an enzyme. *Sci. Am.* (Nov.) 255(5):64–75.

______. 1987. The chemistry of self-splicing RNA and RNA enzymes. *Science* 236:1532–39.

Chambon, P. 1981. Split genes. *Sci. Am.* (May) 244:60–71.

Cramer, P., et al. 2000. Architecture of RNA polymerase II and implications for the transcription mechanism. *Science* 288:640–49.

Crick, F.H.C. 1962. The genetic code. *Sci. Am.* (Oct.) 207:66–77.

______. 1966a. The genetic code: III. *Sci. Am.* (Oct.) 215:55–63.

______. 1966b. Codon–anticodon pairing: The wobble hypothesis. *J. Mol. Biol.* 19:548–55.

Crick, F.H.C., Barnett, L., Brenner, S., and Watts-Tobin, R.J. 1961. General nature of the genetic code for proteins. *Nature* 192:1227–32.

Darnell, J.E. 1983. The processing of RNA. *Sci. Am.* (Oct.) 249:90–100.

______. 1985. RNA. *Sci. Am.* (Oct.) 253:68–87.

Dickerson, R.E. 1983. The DNA helix and how it is read. *Sci. Am.* (Dec.) 249:94–111.

Dugaiczk, A., et al. 1978. The natural ovalbumin gene contains seven intervening sequences. *Nature* 274:328–33.

Fiers, W., et al. 1976. Complete nucleotide sequence of bacteriophage MS2 RNA: Primary and secondary structure of the replicase gene. *Nature* 260:500–507.

Gamow, G. 1954. Possible relation between DNA and protein structures. *Nature* 173:318.

Gnatt, A.L., et. al. 2001. Structural basis of transcription: An RNA polymerase II elongation complex at 3.3 Å resolution. *Science* 292:1876–82.

Hall, B.D., and Spiegelman, S. 1961. Sequence complementarity of T2-DNA and T2-specific RNA. *Proc. Natl. Acad. Sci. (USA)* 47:137–46.

Hamkalo, B. 1985. Visualizing transcription in chromosomes. *Trends Genet.* 1:255–60.

Jukes, T.H. 1963. The genetic code. *Am. Sci.* 51:227–45.

Khorana, H.G. 1967. Polynucleotide synthesis and the genetic code. *Harvey Lectures* 62:79–105.

Miller, O.L., Hamkalo, B., and Thomas, C. 1970. Visualization of bacterial genes in action. *Science* 169:392–95.

Nirenberg, M.W. 1963. The genetic code: II. *Sci. Am.* (March) 190:80–94.

O'Malley, B., et al. 1979. A comparison of the sequence organization of the chicken ovalbumin and ovomucoid genes. In *Eucaryotic gene regulation*, ed. R. Axel, et al., pp. 281–99. Orlando, FL: Academic Press.

Reed, R., and Maniatis, T. 1985. Intron sequences involved in lariat formation during pre-mRNA splicing. *Cell* 41:95–105.

Ross, J. 1989. The turnover of mRNA. *Sci. Am.* (April) 260:48–55.

Sharp, P.A. 1994. Nobel Lecture: Split genes and RNA splicing. *Cell* 77:805–15.

Steitz, J.A. 1988. Snurps. *Sci. Am.* (June) 258(6):56–63.

Volkin, E., and Astrachan, L. 1956. Phosphorus incorporation in *E. coli* ribonucleic acids after infection with bacteriophage T2. *Virology* 2:149–61.

Wang, Q., et. al. 2000. Requirement of the RNA editing deaminase *ADAR1* gene for embryonic erythropoiesis. *Science* 290:1765–68.

Watson, J.D. 1963. Involvement of RNA in the synthesis of proteins. *Science* 140:17–26.

Woychik, N.A. and Jampsey, M. 2002. The RNA polymerase II machinery: Structure illuminates function. *Cell* 108:453–64.

CHAPTER FOURTEEN

Translation and Proteins

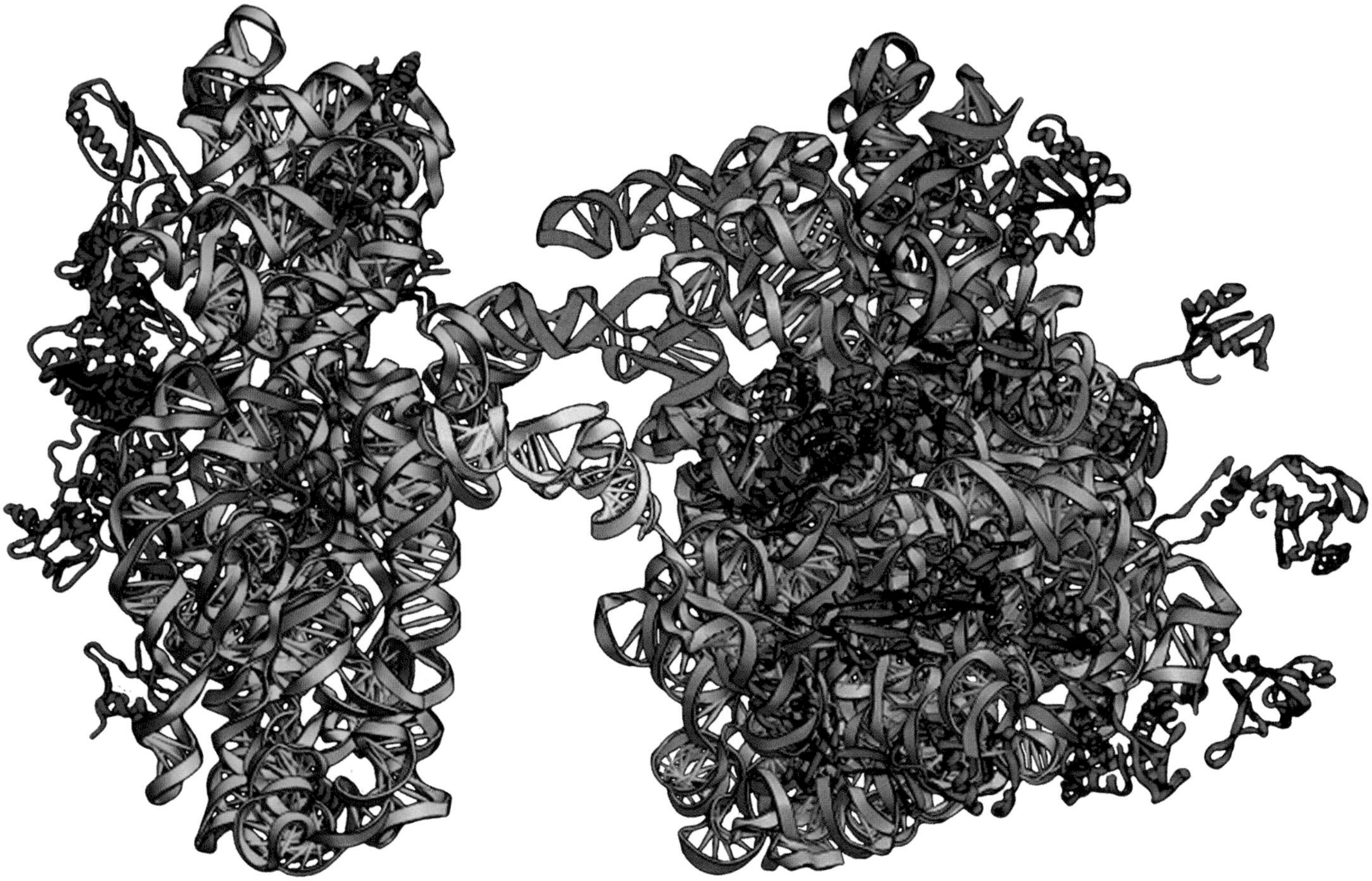

Crystal structure of a Thermus thermophilus 70S ribosome containing three bound transfer RNAs. Reprinted from the front cover of Science, Vol. 292, May 4, 2001 Crystal structure of a Thermus thermophilus 70S ribosome containing three bound transfer RNAs(top) and exploded views showing its different molecular components (middle and bottom). Image provided by Dr. Albion Baucom (baucom@biology.ucsc.edu). Copyright American Association for the Advancement of Science

CHAPTER CONCEPTS

- The ribonucleotide sequence of messenger RNA (mRNA) reflects genetic information stored in DNA that makes up genes and corresponds to the amino acid sequences in proteins encoded by those genes.
- The process of translation decodes the information in mRNA, leading to the synthesis of polypeptide chains.
- Translation involves the interactions of mRNA, tRNA, ribosomes, and a variety of translation factors essential to the initiation, elongation, and termination of the polypeptide chain.
- Proteins, the final product of most genes, achieve a three-dimensional conformation that is based on the primary amino acid sequences of the polypeptide chains making up each protein.
- The function of any protein is closely tied to its three-dimensional structure, which can be disrupted by mutation.

In Chapter 13, we established that a genetic code exists that stores information in the form of triplet codons in DNA, and that this information is initially expressed through the process of transcription into a messenger RNA complementary to the template strand of the DNA helix. However, the final product of gene expression, in most instances, is a polypeptide chain consisting of a linear series of amino acids whose sequence has been prescribed by the genetic code. In this chapter, we will examine how the information present in mRNA is translated to create polypeptides, which then fold into protein molecules. We will also review the evidence that confirmed that proteins are the end products of genes and discuss briefly the various levels of protein structure, diversity, and function. This information extends our understanding of gene expression and provides an important foundation for interpreting how the mutations that arise in DNA can result in the diverse phenotypic effects observed in organisms.

14.1 Translation of mRNA Depends on Ribosomes and Transfer RNAs

Translation of mRNA is the biological polymerization of amino acids into polypeptide chains. This process, alluded to in our earlier discussion of the genetic code, occurs only in association with ribosomes, which serve as nonspecific workbenches. The central question in translation is how triplet codons of mRNA direct specific amino acids into their correct position in the polypeptide. That question was answered once **transfer RNA (tRNA)** was discovered. This class of molecules adapts specific triplet codons in mRNA to their correct amino acids. The presence of an adaptor was postulated by Francis Crick in 1957.

In association with a ribosome, mRNA presents a codon that calls for a specific amino acid. A specific tRNA molecule contains within its nucleotide sequence three consecutive ribonucleotides complementary to the codon, called the **anticodon**, which can base-pair with the codon. Another region of this tRNA is covalently bonded to its corresponding amino acid.

Inside the ribosome, hydrogen bonding of tRNAs to mRNA holds the amino acids in proximity so that a peptide bond can be formed. The process occurs over and over as mRNA runs through the ribosome, and amino acids are polymerized into a polypeptide. Before we discuss the actual process of translation, we will first consider the structures of the ribosome and transfer RNA.

Ribosomal Structure

Because of its essential role in the expression of genetic information, the **ribosome** has been extensively analyzed. One bacterial cell contains about 10,000 of these structures, and a eukaryotic cell contains many times more. Electron microscopy has revealed that the bacterial ribosome is about 250 nm at its largest diameter and consists of two subunits, one large and one small. Both subunits consist of one or more molecules of rRNA and an array of **ribosomal proteins**. When the two subunits are associated with each other in a single ribosome, the structure is sometimes called a **monosome**.

The specific differences between prokaryotic and eukaryotic ribosomes are summarized in Figure 14–1. The subunit and rRNA components are most easily isolated and characterized on the basis of their sedimentation behavior in sucrose gradients (their rate of migration, abbreviated as *S*, as introduced in Chapter 10). In prokaryotes, the monosome is a 70*S* particle, and in eukaryotes it is approximately 80*S*. Sedimentation coefficients, which reflect the variable rate of migration of different-sized particles and molecules, are not additive. For example, the prokaryotic 70*S* monosome consists of a 50*S* and a 30*S* subunit, and the eukaryotic 80*S* monosome consists of a 60*S* and a 40*S* subunit.

The larger subunit in prokaryotes consists of a 23*S* RNA molecule, a 5*S* rRNA molecule, and 31 ribosomal proteins. In the eukaryotic equivalent, a 28*S* rRNA molecule is accompanied by a 5.8*S* and 5*S* rRNA molecule and 49 proteins. The smaller prokaryotic subunits consist of a 16*S* rRNA component and 21 proteins. In the eukaryotic equivalent, an 18*S* rRNA component and about 33 proteins are found. The approximate molecular weights and number of nucleotides of these components are shown in Figure 14–1.

Regarding the components of the ribosome, it is now clear that the RNA molecules provide the basis for all important catalytic functions associated with translation. The many proteins, whose functions were long a mystery, are thought to promote the binding of the various molecules involved in translation, and in general, to fine-tune the process. This conclusion is based on the observation that some of the catalytic functions in ribosomes still occur in experiments involving ribosomal protein-depleted ribosomes.

Molecular hybridization studies have established the degree of redundancy of the genes coding for the rRNA components. The *E. coli* genome contains seven copies of a single sequence that encodes all three components—23*S*, 16*S*, and 5*S*. The initial transcript of these genes produces a 30*S* RNA molecule that is enzymatically cleaved into these smaller components. Coupling of the genetic information encoding these three rRNA components ensures that, following multiple transcription events, equal quantities of all three will be present as ribosomes are assembled.

In eukaryotes, many more copies of a sequence encoding the 28*S*, 18*S*, and 5.8*S* components are present. In *Drosophila*, approximately 120 copies per haploid genome are each transcribed into a molecule of about 34*S*. This is processed into the 28*S*, 18*S*, and 5.8*S* rRNA species. In *X. laevis*, more than 500 copies per haploid genome are present. In mammalian cells, the initial transcript is 45*S*. The rRNA genes, called **rDNA**, are part of the moderately repetitive DNA fraction and are present in clusters at various chromosomal sites.

Each cluster in eukaryotes consists of **tandem repeats**, with each unit separated by a noncoding **spacer DNA** sequence. [See the micrograph in Figure 13.15(b).] In humans, these gene clusters have been localized near the ends of chromosomes 13, 14, 15, 21, and 22. The unique 5*S* rRNA component of eukaryotes is not part of this larger transcript. Instead, genes coding for this ribosomal component are distinct and located separately. In humans, a gene cluster encoding 5*S* rRNA has been located on chromosome 1.

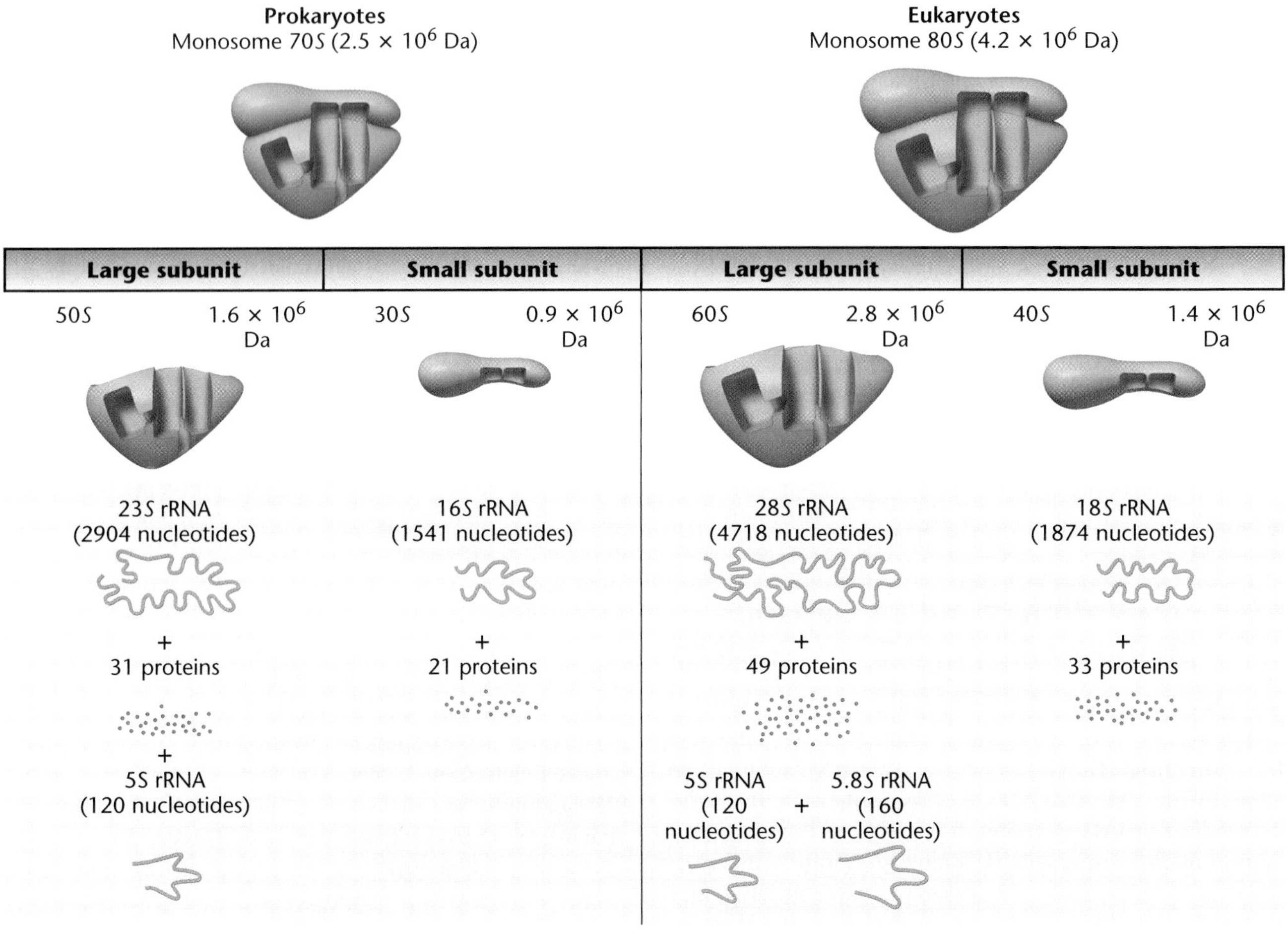

FIGURE 14–1 A comparison of the components in the prokaryotic and eukaryotic ribosome.

Despite the detailed knowledge available on the structure and genetic origin of the ribosomal components, a complete understanding of the function of these components has eluded geneticists. This is not surprising, since the ribosome is perhaps the most intricate of all cellular structures. In bacteria, the monosome has a combined molecular weight of 2.5 million Da!

tRNA Structure

Because of their small size and stability in the cell, transfer RNAs (tRNAs) have been investigated extensively and are the best characterized RNA molecules. They are composed of only 75 to 90 nucleotides, displaying a nearly identical structure in bacteria and eukaryotes. In both types of organisms, tRNAs are transcribed as larger precursors, which are cleaved into mature 4*S* tRNA molecules. In *E. coli*, for example, $tRNA^{tyr}$ (the superscript identifies the specific tRNA and the cognate amino acid that binds to it) is composed of 77 nucleotides, yet its precursor contains 126 nucleotides.

In 1965, Robert Holley and his colleagues reported the complete sequence of $tRNA^{ala}$ isolated from yeast. Of great interest was the finding that a number of nucleotides are unique to tRNA. As shown in Figure 14–2, each contains a modification of one of the four nitrogenous bases expected in RNA (G, C, A, and U). For example, inosinic acid is present, which contains the purine hypoxanthine. Ribothymidylic acid and pseudouridine are other examples. Variously referred to as *unusual, rare*, or *odd bases*, these modified structures are created *following* transcription, illustrating the more general concept of **posttranscriptional modification**. In this case, the unmodified base is inserted during transcription and, subsequently, enzymatic reactions catalyze the chemical modifications to the base.

Holley's analysis led him to propose the two-dimensional **cloverleaf model of tRNA**. It had been known that tRNA demonstrates a secondary structure due to base pairing. Holley discovered that he could arrange the linear sequence in such a way that several stretches of base pairing would result. Such an arrangement created a series of paired stems and unpaired loops resembling the shape of a cloverleaf. Loops consistently contained modified bases that did not generally form base pairs. Holley's model is shown in Figure 14–3.

Because the triplets GCU, GCC, and GCA specify alanine, Holley looked for an anticodon sequence complementary to one of these codons in his $tRNA^{ala}$ molecule. He found it in the form of CGI (the 3′ to 5′ direction), in one of the loops of the cloverleaf. The nitrogenous base I (inosinic acid) can form hydrogen bonds with U, C, or A, the third members of the triplets. Thus, the **anticodon loop** was established.

Studies of other tRNA species revealed many constant features. First, at the 3′ end, all tRNAs contain the sequence ...*pCCA*-3′, which are added post-transcriptionally. It is at this end of the molecule that the amino acid is joined cova-

Web Tutorial 14.1
Transfer RNAs

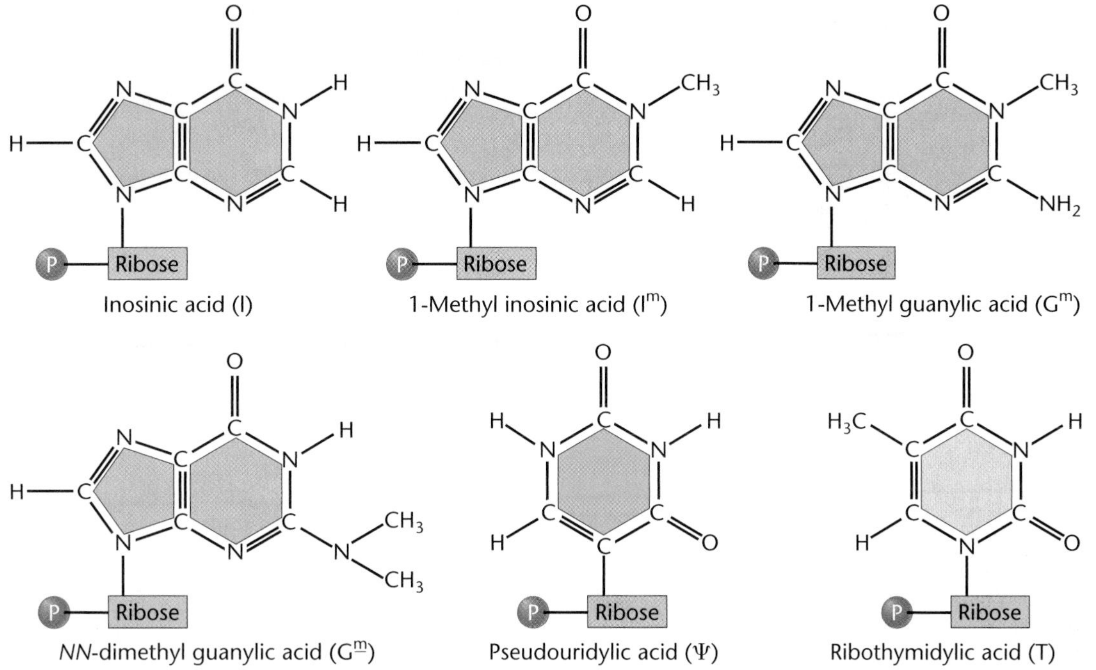

FIGURE 14–2 Unusual nitrogenous bases found in transfer RNA.

lently to the terminal adenosine residue. All tRNAs contain 5′-*G* . . . at the other end of the molecule.

In addition, the lengths of various stems and loops are very similar. Each tRNA examined also contains an anticodon complementary to the known amino acid codon for which it is specific, and all anticodon loops are present in the same position of the cloverleaf.

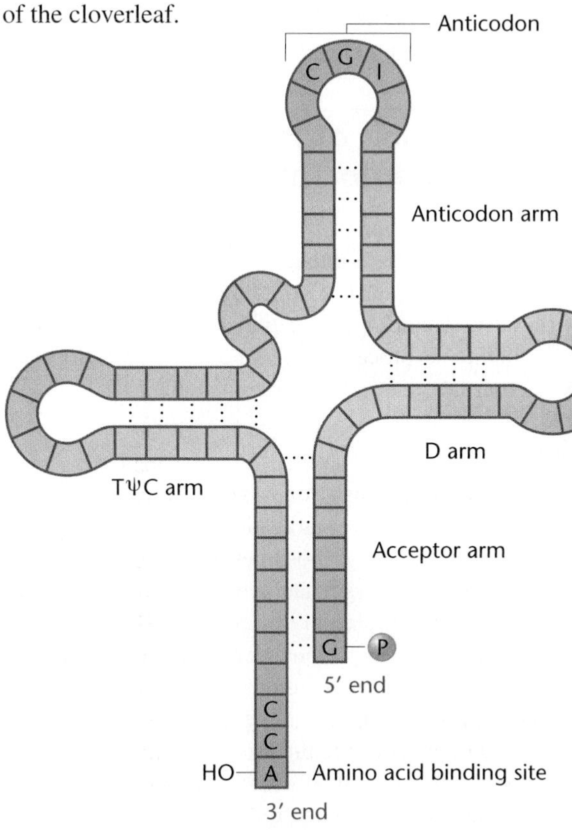

FIGURE 14–3 Holley's two-dimensional cloverleaf model of transfer RNA.

Because the cloverleaf model was predicted strictly on the basis of nucleotide sequence, there was great interest in the X-ray crystallographic examination of tRNA, which reveals a three-dimensional structure. By 1974, Alexander Rich and his colleagues in the United States, and J. Roberts, B. Clark, and Aaron Klug and their colleagues in England, had succeeded in crystallizing tRNA and performing X-ray crystallography at a resolution of 3 Å. At such resolution, the pattern formed by individual nucleotides is discernible.

As a result of these studies, a complete three-dimensional model of tRNA is now available (Figure 14–4). Both the anticodon loop and the 3′-acceptor region (to which the amino acid is covalently linked) have been located. It has been speculated that the shapes of the intervening loops and stems provide the three dimensional conformation that is recognized by the specific enzymes responsible for adding the amino acid to tRNA, a subject to which we now turn our attention.

HOW DO WE KNOW?

What experimentally derived information led to the two-dimensional cloverleaf model of tRNA, as proposed by Holley? What approach was used to locate the anticodon within tRNA?

Charging tRNA

Before translation can proceed, the tRNA molecules must be chemically linked to their respective amino acids. This activation process, called **charging**, or aminoacylation, occurs under the direction of enzymes called **aminoacyl tRNA synthetases**. Because there are 20 different amino acids, there must be at least 20 different tRNA molecules and as many different enzymes. In theory, because there are 61 triplet codes, there could

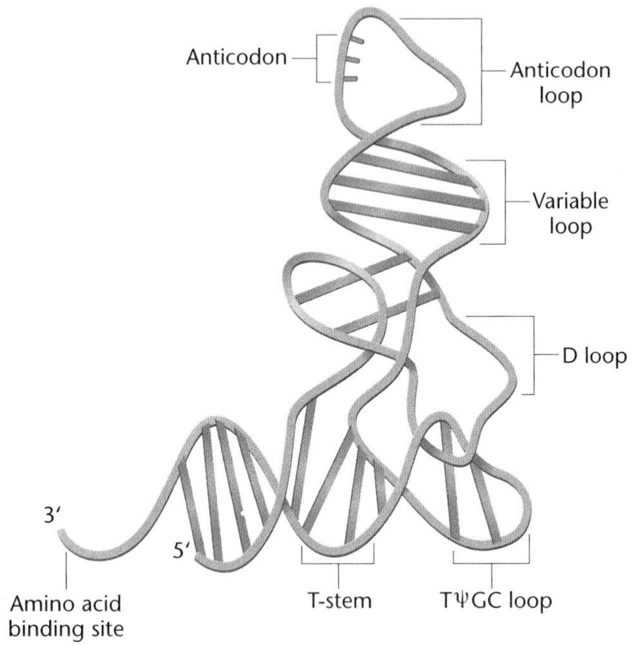

FIGURE 14–4 A three-dimensional model of transfer RNA.

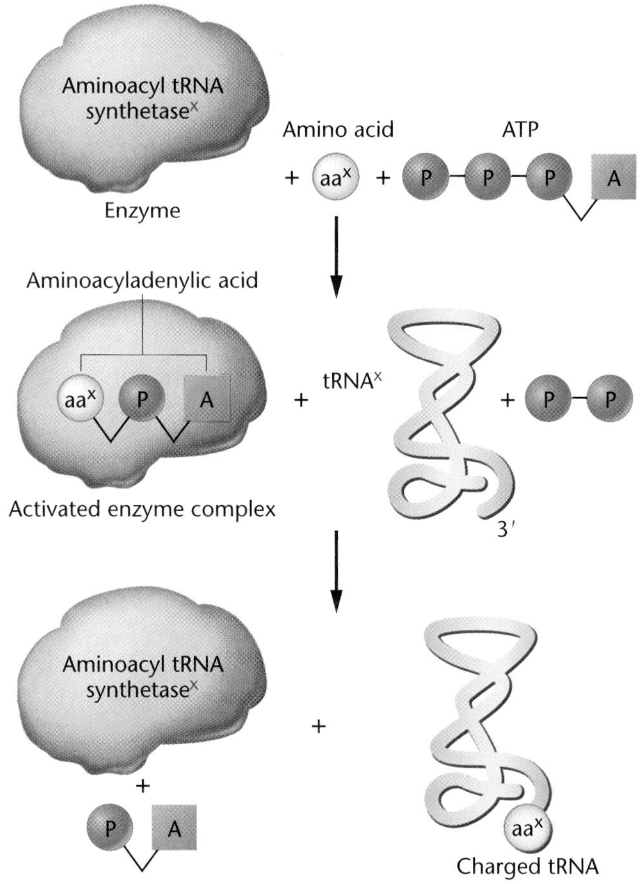

FIGURE 14–5 Steps involved in charging tRNA. The "x" denotes that for each amino acid only the corresponding specific tRNA and specific aminoacyl tRNA synthetase enzyme are involved in the charging process.

be the same number of specific tRNAs and enzymes. However, because of the ability of the third member of a triplet code to "wobble," it is now thought that there are at least 32 different tRNAs; it is also believed that there are only 20 synthetases, one for each amino acid, regardless of the greater number of corresponding tRNAs.

The charging process is outlined in Figure 14–5. In the initial step, the amino acid is converted to an activated form, reacting with ATP to create an **aminoacyladenylic acid**. A covalent linkage is formed between the 5′-phosphate group of ATP and the carboxyl end of the amino acid. This molecule remains associated with the synthetase enzyme, forming a complex that then reacts with a specific tRNA molecule. In the next step, the amino acid is transferred to the appropriate tRNA and bonded covalently to the adenine residue at the 3′ end. The charged tRNA may participate directly in protein synthesis. Aminoacyl tRNA synthetases are highly specific enzymes because they recognize only one amino acid and only the tRNAs corresponding to that amino acid, called **isoaccepting tRNAs**. This is a crucial point if fidelity of translation is to be maintained.

Now solve this

Problem 14.28 on page 358 is concerned with establishing whether tRNA or the amino acid added to the tRNA during charging is responsible for the response of charged tRNA to mRNA during translation.

Hint: In this experiment, when the triplet codon in mRNA calls for cysteine, alanine is inserted during translation, even though it is the "incorrect" amino acid.

14.2 Translation of mRNA Can Be Divided into Three Steps

In a way similar to transcription, the process of translation can be best described by breaking it into discrete phases. We will consider three such phases, each with its own set of illustrations (Figures 14–6, 14–7, and 14–8). Keep in mind that translation is a dynamic, continuous process. Many of the protein factors and their roles in translation are summarized in Table 14.1.

Initiation

Initiation of prokaryotic translation is depicted in Figure 14–6. Recall that the ribosome serves as a nonspecific workbench for the translation process. Many ribosomes, when they are not involved in translation, are dissociated into their large and small subunits. Initiation of translation in *E. coli* involves the small ribosomal subunit, an mRNA molecule, a specific charged initiator tRNA, GTP, Mg^{2+}, and a number of proteinaceous initiation factors (IFs). These are initially part of the small subunit and are required to enhance the binding affinity of the various translational components. Unlike ribosomal proteins, IFs are released from the ribosome once initiation is completed. In prokaryotes, the initiation codon of mRNA—AUG—calls for the modified amino acid N-**formylmethionine (f-met)**.

TABLE 14.1 **VARIOUS PROTEIN FACTORS INVOLVED DURING TRANSLATION IN *E. COLI***

Process	Factor	Role
Initiation of translation	IF1	Stabilizes 30*S* subunit
	IF2	Binds fmet-tRNA to 30*S*-mRNA complex; binds to GTP and stimulates hydrolysis
	IF3	Binds 30*S* subunit to mRNA; dissociates monosomes into subunits following termination
Elongation of polypeptide	EF-Tu	Binds GTP; brings aminoacyl-tRNA to the A site of ribosome
	EF-Ts	Generates active EF-Tu
	EF-G	Stimulates translocation; GTP-dependent
Termination of translation and release of polypeptide	RF1	Catalyzes release of the polypeptide chain from tRNA and dissociation of the translocation complex; specific for UAA and UAG termination codons
	RF2	Behaves like RF1; specific for UGA and UAA codons
	RF3	Stimulates RF1 and RF2

The small ribosomal subunit binds to several initiation factors, and this complex in turn binds to mRNA (Step 1). In bacteria, such binding involves a sequence of up to six ribonucleotides (AGGAGG, not shown in Figure 14–6), which *precedes* the initial AUG start codon of mRNA. This sequence (containing only purines and called the **Shine–Dalgarno sequence**) base-pairs with a region at the 3′ end of the 16*S* rRNA of the small ribosomal subunit, facilitating initiation.

Another initiation protein then enhances the binding of charged formylmethionyl tRNA to the small subunit in response to the AUG codon (Step 2). This step "sets" the reading frame so that all subsequent codons are translated accurately. The aggregate represents the **initiation complex**, to which the large ribosomal subunit binds. In this process, a molecule of GTP is hydrolyzed, providing the required energy, and the initiation factors are released (Step 3).

Elongation

The second phase of translation, elongation, is depicted in Figure 14–7. Once both subunits of the ribosome are assembled with the mRNA, binding sites for two charged tRNA molecules are formed. These are designated as the **P**, or **peptidyl**, and the **A**, or **aminoacyl**, **sites**. The charged initiator tRNA binds to the P site, provided that the AUG codon of mRNA is in the corresponding position of the small subunit.

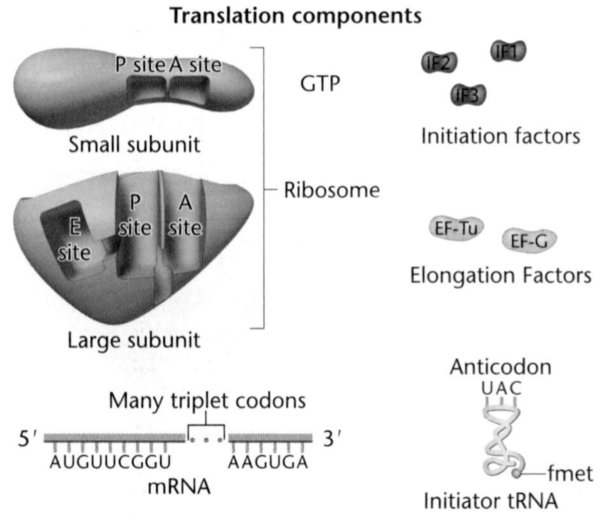

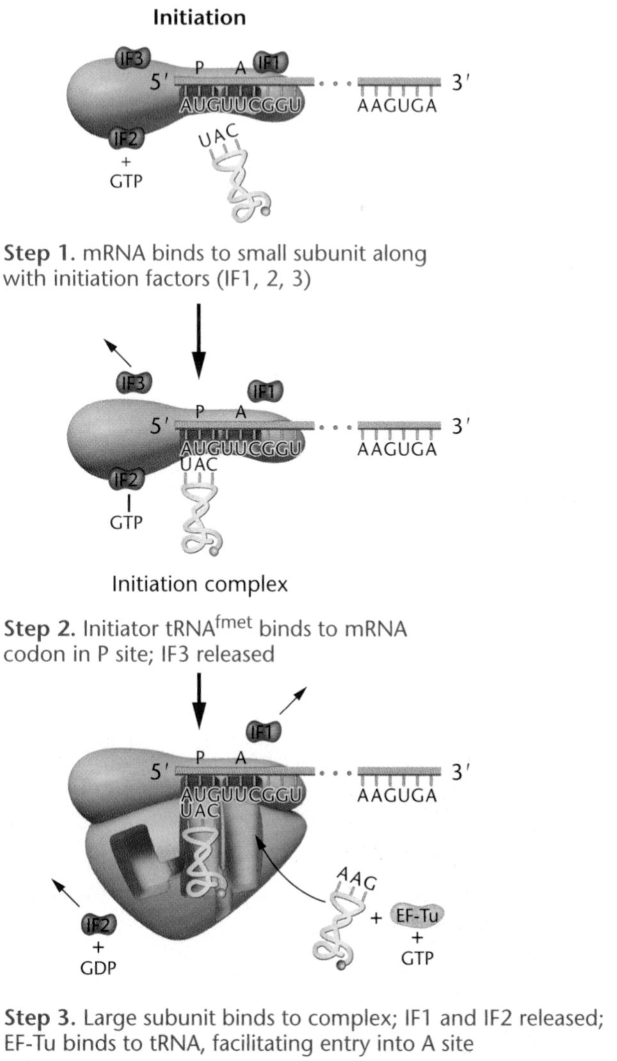

FIGURE 14–6 Initiation of translation. The components are depicted at the left.

Web Tutorial 14.2
Protein Translation

The lengthening of the growing polypeptide chain by one amino acid is called **elongation**. The sequence of the second codon in mRNA dictates which charged tRNA molecule will become positioned at the A site (Step 1). Once it is present, **peptidyl transferase** catalyzes the formation of the peptide bond that links the two amino acids together (Step 2). The catalytic activity of peptidyl transferase is a function of the 23*S* rRNA of the large subunit, not one of the ribosomal proteins. At the same time, the covalent bond between the amino acid and the tRNA occupying the P site is hydrolyzed (broken). The product of this reaction is a dipeptide, which is attached to the 3′ end of tRNA still residing in the A site.

Before elongation can be repeated, the tRNA attached to the P site, which is now uncharged, must be released from the large subunit. The uncharged tRNA moves transiently through a third site on the ribosome called the **E site** (E stands for exit). The entire **mRNA–tRNA–aa$_2$–aa$_1$** complex then shifts in the direction of the P site by a distance of three nucleotides (Step 3). This event requires several protein elongation factors (EFs) as well as the energy derived from hydrolysis of GTP. The result is that the third codon of mRNA is now in a position to accept its specific charged tRNA into the A site (Step 4). One simple way to distinguish the two sites in your mind is to remember that, *following the shift*, the P site contains a tRNA attached to a peptide chain (*P* for peptide), whereas the A site contains a tRNA with an amino acid attached (*A* for amino acid).

The sequence of elongation is repeated over and over (Steps 5 and 6). An additional amino acid is added to the growing polypeptide chain each time the mRNA advances through the ribosome. Once a polypeptide chain of reasonable size is assembled (about 30 amino acids), it begins to emerge from the base of the large subunit, as illustrated in Step 6. A tunnel exists within the large subunit, through which the elongating polypeptide emerges.

As we have seen, the role of the small subunit during elongation is one of "decoding" the codons present in mRNA, while the role of the large subunit is peptide-bond synthesis. The efficiency of the process is remarkably high: The observed error rate is only about 10^{-4}. An incorrect amino acid will occur only once in every 20 polypeptides of an average length of 500 amino acids! At 37°C, elongation in *E. coli* occurs at a rate of about 15 amino acids per second.

Termination

Termination, the third phase of translation, is depicted in Figure 14–8. Termination of protein synthesis is signaled by one or more of three triplet codons in the A site: UAG, UAA, or UGA. These codons do not specify an amino acid, nor do

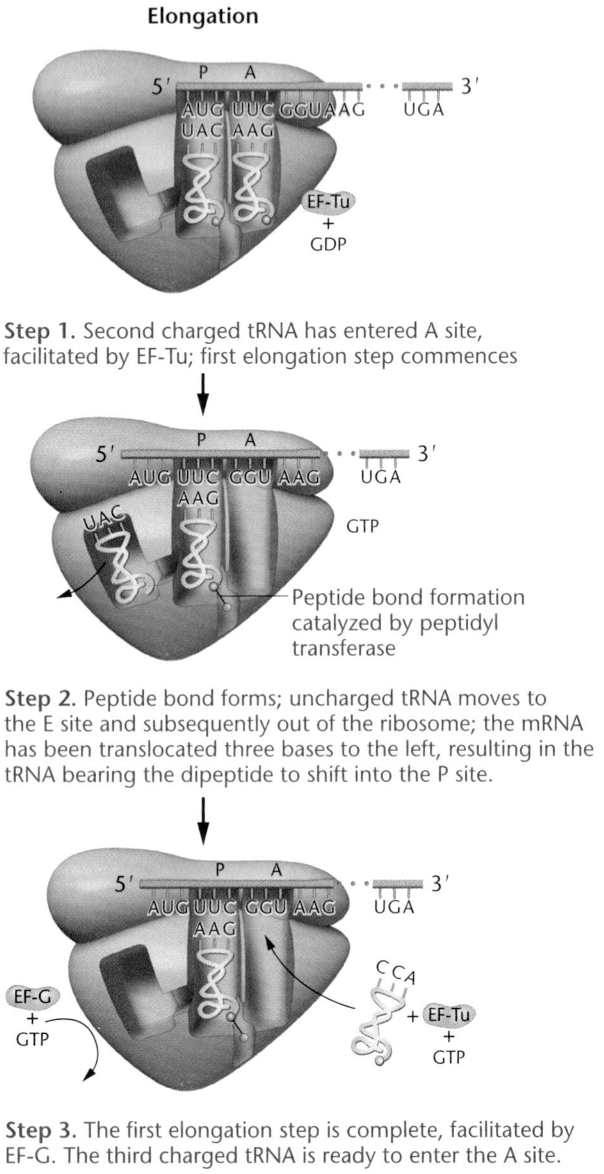
Elongation
P A
5′
3′
AUG UUC GGU AAG
UAC AAG
UGA
EF-Tu
+
GDP
Step 1. Second charged tRNA has entered A site, facilitated by EF-Tu; first elongation step commences
UAC
GTP
Peptide bond formation catalyzed by peptidyl transferase
Step 2. Peptide bond forms; uncharged tRNA moves to the E site and subsequently out of the ribosome; the mRNA has been translocated three bases to the left, resulting in the tRNA bearing the dipeptide to shift into the P site.
EF-G
+
GTP
CCA
+ EF-Tu
+
GTP
Step 3. The first elongation step is complete, facilitated by EF-G. The third charged tRNA is ready to enter the A site.

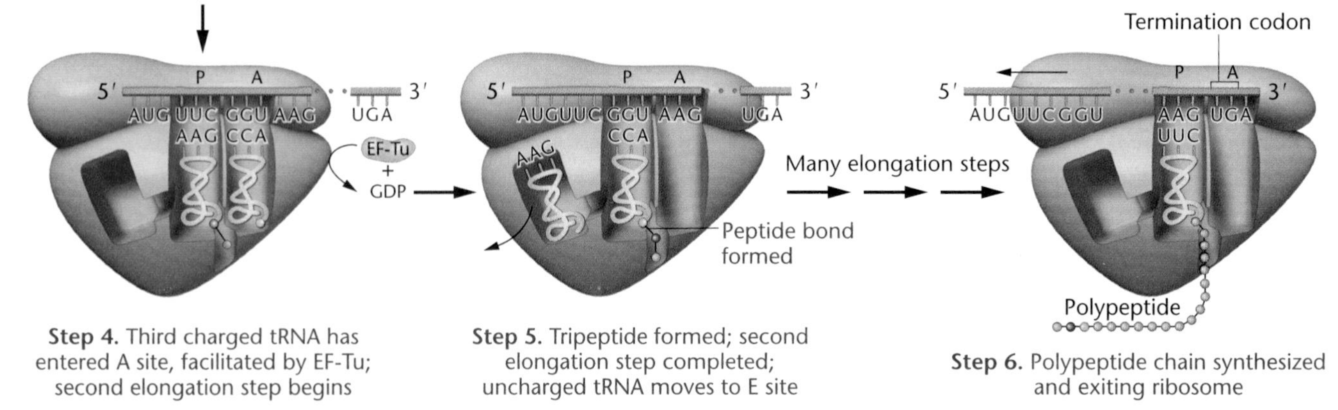
AAG CCA
EF-Tu
+
GDP
Peptide bond formed
Many elongation steps
Termination codon
AUGUUCGGU
AAG UGA
UUC
Polypeptide
Step 4. Third charged tRNA has entered A site, facilitated by EF-Tu; second elongation step begins
Step 5. Tripeptide formed; second elongation step completed; uncharged tRNA moves to E site
Step 6. Polypeptide chain synthesized and exiting ribosome

FIGURE 14–7 Elongation of the growing polypeptide chain during translation.

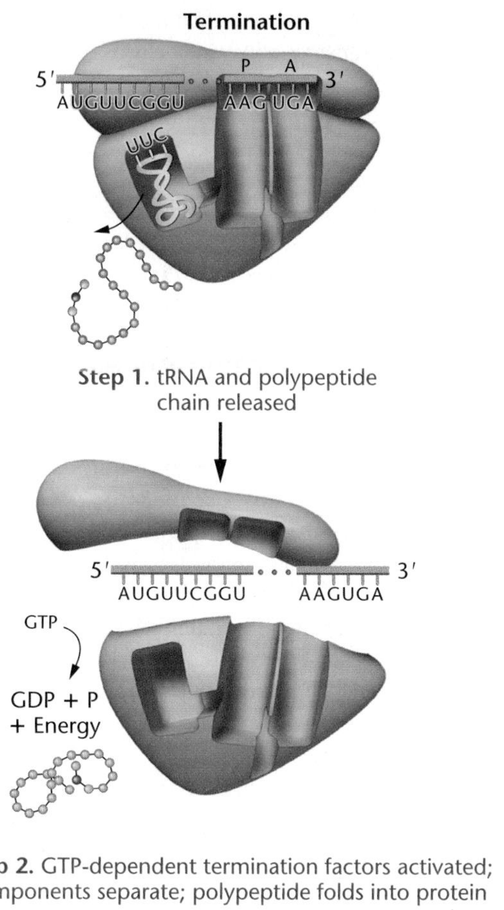

FIGURE 14–8 Termination of the process of translation.

they call for a tRNA in the A site. These codons are called **stop codons**, **termination codons**, or **nonsense codons**. The finished polypeptide is therefore still attached to the terminal tRNA at the P site, and the A site is empty. The termination codon signals the action of **GTP-dependent release factors**, which cleave the polypeptide chain from the terminal tRNA, releasing it from the translation complex (Step 1). Once this cleavage occurs, the tRNA is released from the ribosome, which then dissociates into its subunits (Step 2). If a termination codon should appear in the middle of an mRNA molecule as a result of mutation, the same process occurs, and the polypeptide chain is prematurely terminated.

How Do We Know?

What experimental information verifies that certain codons in mRNA specify chain termination during translation?

Polyribosomes

As elongation proceeds and the initial portion of mRNA has passed through the ribosome, the message is free to associate with another small subunit to form a second initiation complex. This process can be repeated several times with a single mRNA and results in what are called **polyribosomes** or just **polysomes**.

Polyribosomes can be isolated and analyzed following a gentle lysis of cells. Figures 14–9(a) and (b) illustrate these complexes as seen under the electron microscope. In Figures 14–9(a), you can see mRNA (the thin line) between the individual ribosomes. The micrograph in Figure 14–9(b) is even more remarkable, for it shows the polypeptide chains emerging from the ribosomes during translation. The formation of polysome complexes represents an efficient use of the components available for protein synthesis during a unit of time. Using the analogy of a tape and a tape recorder, in polysome complexes, one tape (mRNA) would be played simultaneously by several recorders (the ribosomes). But at any given moment, each recording (the polypeptide being synthesized in each ribosome) would be at a different point of completion.

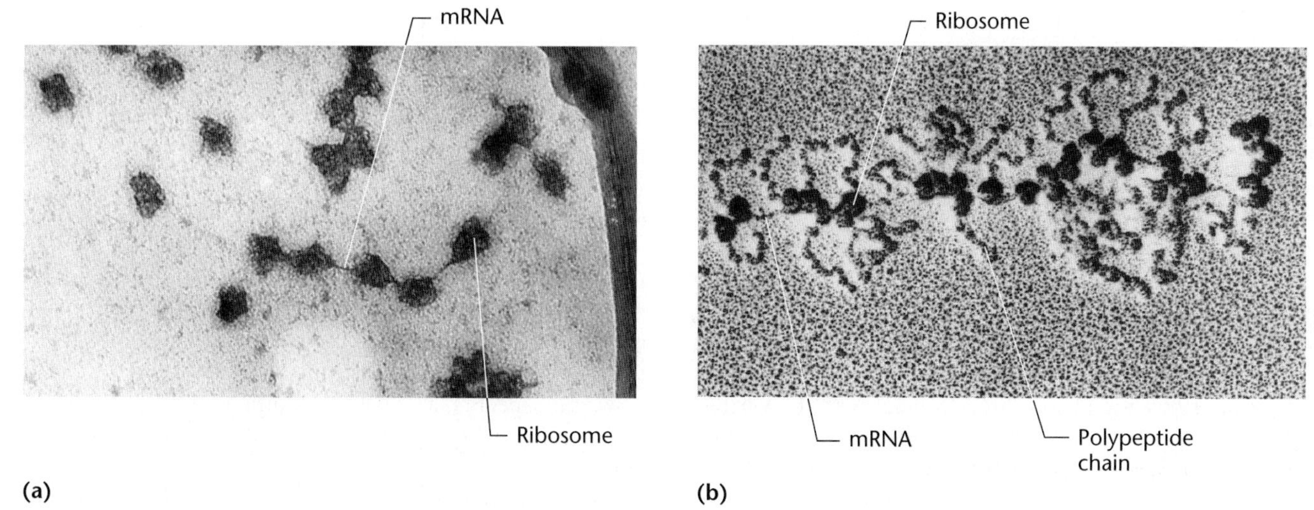

FIGURE 14–9 Polyribosomes visualized under the electron microscope. Those in (a) were derived from rabbit reticulocytes engaged in the translation of hemoglobin mRNA and in (b) from salivary gland cells of the midgefly, *Chironomus thummi*. In (b), the nascent polypeptide chain is apparent as it emerges from each ribosome. Its length increases as translation proceeds from left (5′) to right (3′) along the mRNA.

14.3 Crystallographic Analysis Has Revealed Many Details about the Functional Prokaryotic Ribosome

Our knowledge of the process of translation and the structure of the ribosome, as described in the previous sections, is based primarily on biochemical and genetic observations, in addition to the visualization of ribosomes under the electron microscope. Because of the tremendous size and complexity of the functional ribosome during active translation, obtaining crystals needed to perform X-ray diffraction studies has been extremely difficult. Nevertheless, great strides have been made in the past several years. First, the individual ribosomal subunits were crystallized and examined in several laboratories, most prominently that of V. Ramakrishnan. Then, in 2001, the crystal structure of the intact 70*S* ribosome, complete with associated mRNA and tRNAs, was examined by Harry Noller and colleagues. In essence, the entire translational complex was visualized at the atomic level. Both Ramakrishnan and Noller derived the ribosomes from the bacterium *Thermus thermophilus*.

Many noteworthy observations have been made from these investigations. A view of one of the models, based on Noller's findings, is shown as the opening photograph of this chapter. For example, the sizes and shapes of the subunits, measured at atomic dimensions, are in agreement with earlier estimates based on high-resolution electron microscopy. Further, the shape of the ribosome changes during different functional states, attesting to the dynamic nature of the process of translation. A great deal has also been learned about the prominence and location of the RNA components of the subunits. For example, about one third of the 16*S* RNA is responsible for producing a flat projection within the smaller 30*S* subunit referred to as the platform, which modulates movement of the mRNA–tRNA complex during translocation.

More information supports the concept that RNA is the real "player" in the ribosome during translation. The interface between the two subunits, considered to be the location in the ribosome where polymerization of amino acids occurs, is composed almost exclusively of RNA. In contrast, the numerous ribosomal proteins are found mostly on the periphery of the ribosome. These observations confirm what has been predicted on genetic grounds—the catalytic steps that join amino acids during translation occur under the direction of RNA, not proteins.

Another interesting finding involves the actual location of the three sites predicted to house tRNAs during translation. All three, the aminoacyl, the peptidyl, and the exit sites (designated the A, P, and E sites), have been identified, and in each case, the RNA of the ribosome makes direct contact with the various loops and domains of the tRNA molecule. This observation points to the importance of the different regions of tRNA and helps us understand why the specific three-dimensional conformation of all tRNA molecules has been preserved throughout evolution.

Still another observation is that the intervals between these A, P, and E sites are at least 20 Å, and perhaps as much as 50 Å, thus defining the atomic distance that the tRNA molecules must shift during each translocation event. This is considered a fairly large distance relative to the size of the tRNAs themselves. Analysis has led to the identification of molecular (RNA–protein) bridges existing between the three sites that are involved in the translocation events. Other such bridges are present at other key locations and have been related to ribosome function. These observations provide us with a much better picture of the dynamic changes that must occur within the ribosome during translation.

A final observation takes us back almost 50 years, when Francis Crick proposed the **wobble hypothesis**, as introduced in Chapter 13. The Ramakrishnan group has identified the precise location along the 16*S* rRNA of the 30*S* subunit involved in the decoding step between mRNA and tRNA. Two particular nucleotides of the 16*S* rRNA actually flip out and probe the codon:anticodon region and are believed to check for accuracy of base pairing during this interaction. Related to the wobble hypothesis, the stringency of this step is high for the first two base pairs, but less stringent for the third (or wobble) base pair.

While these findings represent landmark studies, numerous questions still remain about ribosome structure and function. In particular, the role of the many ribosomal proteins is yet to be clarified. Nevertheless, the models that are emerging based on the work of Noller, Ramakrishnan, and their many colleagues, provide us with a much better understanding of the mechanism of translation.

14.4 Translation Is More Complex in Eukaryotes

The general features of the model we just discussed were initially derived from investigations of the translation process in bacteria. As we saw, one of the main differences between translation in prokaryotes and eukaryotes is that, in the latter, translation occurs on ribosomes that are larger and whose rRNA and protein components are more complex than those of prokaryotes. (See Figure 14–1.) Another significant distinction is that while transcription and translation are coupled in prokaryotes, in eukaryotes these two processes are spatially and temporally separated. In eukaryotic cells transcription occurs in the nucleus and translation in the cytoplasm. This separation provides eukaryotic cells multiple opportunities to regulate genetic expression.

Several other differences are also important. Eukaryotic mRNAs are much longer lived than are their prokaryotic counterparts. Most exist for hours rather than minutes, remaining available much longer to orchestrate protein synthesis prior to their degradation by nucleases in the cell.

Several aspects that involve the initiation of translation are different in eukaryotes. First, as we discussed in our consideration of mRNA maturation, the 5′ end is capped with a 7-methylguanosine residue. The presence of this cap, absent in prokaryotes, is essential to efficient translation because RNAs lacking the cap are translated poorly. In addition, most eukaryotic mRNAs contain a short recognition sequence that surrounds the initiating AUG codon—5′-ACCAUGG. Named after its discoverer Marilyn Kozak, this sequence functions during initiation by creating the proper context next to the initiating AUG codon. If the Kozak sequence is missing, initator tRNA does not select the AUG codon and continues scanning the mRNA until it finds another AUG that is "in context" with the Kozak sequence. Thus, it is in a similar position but works differently than the Shine–Dalgarno sequence in prokaryotic mRNA. Nevertheless, both facilitate the initial binding of mRNA to the small subunit of the ribosome.

Another difference is that the amino acid formylmethionine is not required for the initiation of eukaryotic translation. However, as in prokaryotes, the AUG codon, which specifies methionine, is essential to the formation of the translational complex, and a unique transfer RNA ($tRNA_i^{met}$) is used during initiation.

Protein factors similar to those in prokaryotes guide the initiation, elongation, and termination of translation in eukaryotes. Many of these eukaryotic factors are clearly homologous to their counterparts in prokaryotes. However, a greater number of factors are usually required during each of these steps, and some are more complex than in prokaryotes.

Finally, recall that cytoplasmic ribosomes in eukaryotes are found either as "free-floating" cytosolic ribosomes, or as "membrane bound' ribosomes in association with the membranes that make up the endoplasmic reticulum (forming what is referred to as rough ER). The only difference between these two pools of ribosomes is the type of polypeptides they are translating. The association of ribosomes with the ER facilitates the transport of the newly synthesized proteins from the ribosomes directly into the channels of the endoplasmic reticulum. Recent studies using cryoelectron microscopy have established how this occurs. There is a tunnel in the large subunit of ribosomes that begins near the point at which the two subunits interface and exits near the back of the large subunit. It is believed that the location of the tunnel within the large subunit provides the conduit for the movement of the newly synthesized polypeptide chain out of the ribosome. In studies in yeast, newly synthesized polypeptides enter the ER through a membrane channel formed by a specific protein, Sec61. This channel is perfectly aligned with the exit point of the ribosomal tunnel. In prokaryotes, the polypeptides are released by the ribosome directly into the cytoplasm.

14.5 The Initial Insight That Proteins Are Important in Heredity Was Provided by the Study of Inborn Errors of Metabolism

Now, let's consider how we know that proteins are the end products of genetic expression. The first insight into the role of proteins in genetic processes was provided by observations made by Sir Archibald Garrod and William Bateson early in the 20th century. Garrod was born into an English family of medical scientists. His father was a physician with a strong interest in the chemical basis of rheumatoid arthritis, and his eldest brother was a leading zoologist in London. It is not surprising, then, that as a practicing physician, Garrod became interested in several human disorders that seemed to be inherited. Although he also studied albinism and cystinuria, we will describe his investigation of the disorder **alkaptonuria**. Individuals afflicted with this disorder cannot metabolize the alkapton 2,5-dihydroxyphenylacetic acid, also known as homogentisic acid. As a result, an important metabolic pathway (Figure 14–10) is blocked. Homogentisic acid accumulates in cells and tissues and is excreted in the urine. The molecule's

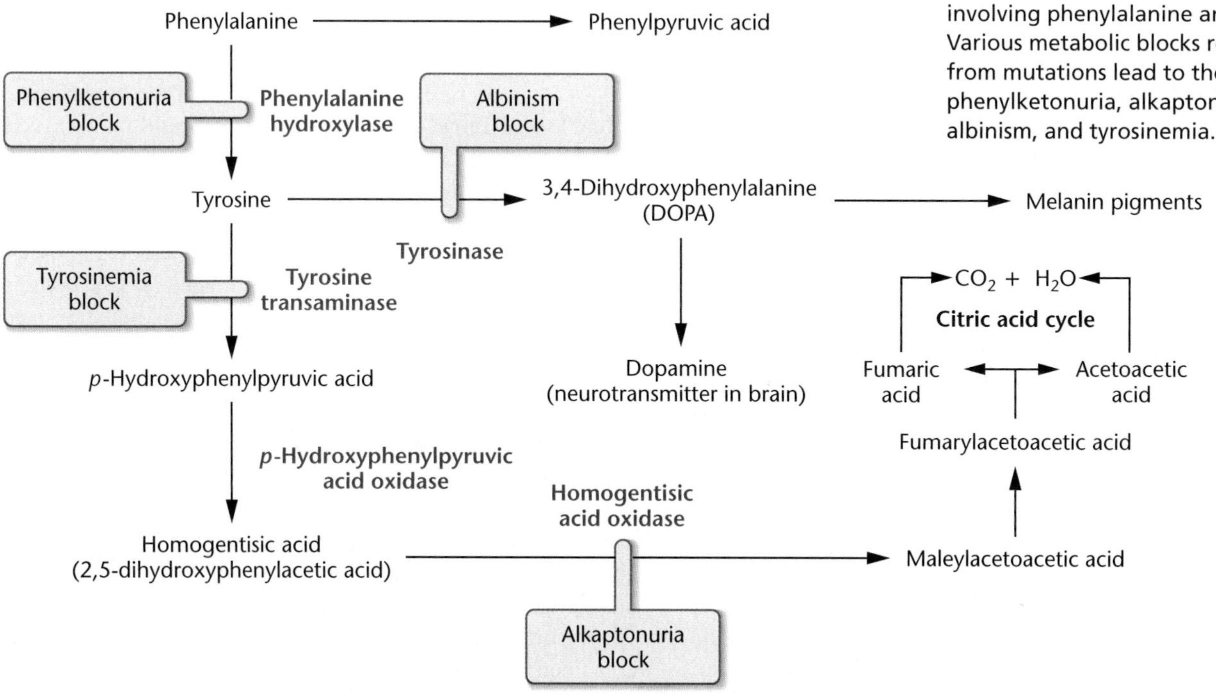

FIGURE 14–10 Metabolic pathway involving phenylalanine and tyrosine. Various metabolic blocks resulting from mutations lead to the disorders phenylketonuria, alkaptonuria, albinism, and tyrosinemia.

oxidation products are black and easily detectable in the diapers of newborns. The products tend to accumulate in cartilaginous areas, causing a darkening of the ears and nose. In joints, this deposition leads to a benign arthritic condition. Alkaptonuria is a rare, but not serious disease that persists throughout an individual's life.

Garrod studied alkaptonuria by increasing dietary protein or adding to the diet the amino acids phenylalanine or tyrosine, both of which are chemically related to homogentisic acid. Under such conditions, homogentisic acid levels increase in the urine of alkaptonurics, but not in unaffected individuals. Garrod concluded that normal individuals can break down, or catabolize, this alkapton, but that afflicted individuals cannot. By studying the disorder's pattern of inheritance, Garrod further concluded that alkaptonuria was inherited as a simple recessive trait.

On the basis of these conclusions, Garrod hypothesized that hereditary information controls chemical reactions in the body, and that the inherited disorders he studied are the result of alternative modes of metabolism. While *genes* and *enzymes* were not familiar terms during Garrod's time, he used the corresponding concepts of *unit factors* and *ferments*. Garrod published his initial observations in 1902.

Only a few geneticists, including Bateson, were familiar with, or referred to, Garrod's work. Garrod's ideas fit nicely with Bateson's belief that inherited conditions were caused by the lack of some critical substance. In 1909, Bateson published *Mendel's Principles of Heredity*, in which he linked ferments with heredity. However, for almost 30 years, most geneticists failed to see the relationship between genes and enzymes. Garrod and Bateson, like Mendel, were ahead of their time.

Phenylketonuria

The inherited human metabolic disorder, **phenylketonuria (PKU)**, results when another reaction in the pathway shown in Figure 14–10 is blocked. First described in 1934, this disorder can result in mental retardation and is transmitted as an autosomal recessive disease. Afflicted individuals are unable to convert the amino acid phenylalanine to the amino acid tyrosine. These molecules differ by only a single hydroxyl group (OH), present in tyrosine, but absent in phenylalanine. The reaction is catalyzed by the enzyme **phenylalanine hydroxylase**, which is inactive in affected individuals and active at a level of about 30 percent in heterozygotes. The enzyme functions in the liver. While the normal blood level of phenyalanine is about 1 mg/100 ml, phenylketonurics show a level as high as 50 mg/100 ml.

As phenylalanine accumulates, it may be converted to phenylpyruvic acid and, subsequently, to other derivatives. These are less efficiently resorbed by the kidney and tend to spill into the urine more quickly than phenylalanine. Both phenylalanine and its derivatives enter the cerebrospinal fluid, resulting in elevated levels in the brain. The presence of these substances during early development is thought to cause mental retardation.

As a result of early detection based on PKU screening of newborns, retardation can be ameliorated. When the condition is detected in the analysis of an infant's blood, a strict dietary regimen is instituted. A low-phenylalanine diet can reduce by-products such as phenylpyruvic acid, and the abnormalities characterizing the disease can be diminished. The screening of newborns occurs routinely in all states of the United States. Phenylketonuria occurs in approximately 1 in 11,000 births.

Knowledge of inherited metabolic disorders, such as alkaptonuria and phenylketonuria, has caused a revolution in medical thinking and practice. Human disease, once thought to be attributed solely to the action of invading microorganisms, viruses, or parasites, clearly can have a genetic basis. We know now that literally thousands of medical conditions are caused by errors in metabolism resulting from mutant genes. These human biochemical disorders include all classes of organic biomolecules.

14.6 Studies of *Neurospora* Led to the One-Gene:One-Enzyme Hypothesis

In two separate investigations that began in 1933, George Beadle was to provide the first convincing experimental evidence that genes are directly responsible for the synthesis of enzymes. The first investigation, conducted in collaboration with Boris Ephrussi, involved *Drosophila* eye pigments. Together, Beadle and Ephrussi confirmed that mutant genes that altered the eye color of fruit flies could be linked to biochemical errors that, in all likelihood, involved the loss of enzyme function. Encouraged by these findings, Beadle then joined with Edward Tatum to investigate nutritional mutations in the pink bread mold *Neurospora crassa*. This investigation led to the **one-gene:one-enzyme hypothesis**.

Analysis of *Neurospora* Mutants by Beadle and Tatum

In the early 1940s, Beadle and Tatum chose to work with *Neurospora* because much was known about its biochemistry, and mutations could be induced and isolated with relative ease. By inducing mutations, they produced strains that had genetic blocks of reactions essential to the growth of the organism.

Beadle and Tatum knew that this mold could manufacture nearly everything necessary for normal development. For example, using rudimentary carbon and nitrogen sources, the organism can synthesize nine water-soluble vitamins, 20 amino acids, numerous carotenoid pigments, and all essential purines and pyrimidines. Beadle and Tatum irradiated asexual conidia (spores) with X rays to increase the frequency of mutations and allowed them to be grown on "complete" medium containing all the necessary growth factors (e.g., vitamins, amino acids, etc.). Under such growth conditions, a mutant strain that would be unable to grow on minimal medium was able to grow by virtue of supplements present in the enriched complete medium. All the cultures were then transferred to minimal medium. If growth occurred on the minimal medium, the organisms were able to synthesize all the necessary growth factors themselves, and it was concluded that the culture did not contain a mutation. If no growth occurred, then it was concluded that the culture contained a nutritional mutation, and the only task remaining was to determine its type. Both cases are illustrated in Figure 14–11(a).

Many thousands of individual spores derived by this procedure were isolated and grown on complete medium. In subsequent tests on minimal medium, many cultures failed to grow, indicating that a nutritional mutation had been induced. To identify the mutant type, the mutant strains were tested on a series of different minimal media [Figure 14–11(b) and (c)], each containing groups of supplements, and subsequently on media containing single vitamins, amino acids, purines, or pyrimidines until one specific supplement that permitted growth was found. Beadle and Tatum reasoned that the supplement that restores growth is the molecule that the mutant strain could not synthesize.

The first mutant strain isolated required vitamin B_6 (pyridoxine) in the medium, and the second one required vitamin B_1 (thiamine). Using the same procedure, Beadle and Tatum eventually isolated and studied hundreds of mutants deficient in the ability to synthesize other vitamins, amino acids, nucleotides, or other substances.

The findings derived from testing more than 80,000 spores convinced Beadle and Tatum that genetics and biochemistry have much in common. It seemed likely that each nutritional mutation caused the loss of the enzymatic activity that facilitates an essential reaction in wild-type organisms. It also appeared that a mutation could be found for nearly any enzymatically controlled reaction. Beadle and Tatum had thus provided sound experimental evidence for the hypothesis that

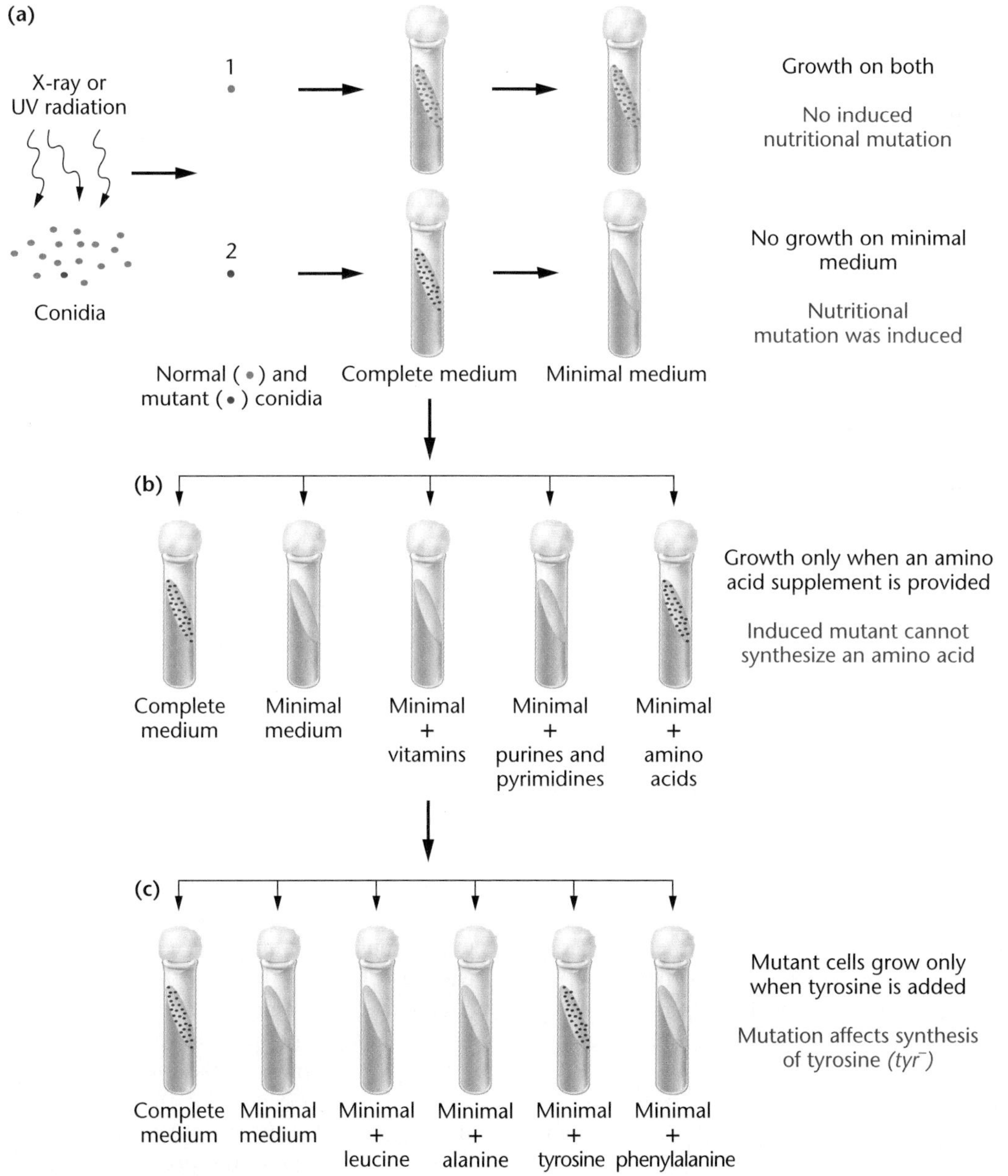

FIGURE 14–11 Induction, isolation, and characterization of a nutritional auxotrophic mutation in *Neurospora*. In (a), most conidia are not affected, but one conidium (shown in red) contains such a mutation. In (b) and (c), the precise nature of the mutation is determined to involve the biosynthesis of tyrosine.

one gene specifies one enzyme, an idea alluded to more than 30 years earlier by Garrod and Bateson. With modifications, this concept was to become a major principle of genetics.

HOW DO WE KNOW?

How do we know, when studying *Neurospora* nutritional mutations, that one gene specifies one enzyme?

Genes and Enzymes: Analysis of Biochemical Pathways

The one-gene:one-enzyme concept and its attendant methods have been used over the years to work out many details of metabolism in *Neurospora, Escherichia coli*, and a number of other microorganisms. One of the first metabolic pathways to be investigated in detail was that leading to the synthesis of the amino acid arginine in *Neurospora*. By studying seven mutant strains, each requiring arginine for growth (arg^-), Adrian Srb and Norman Horowitz were able to ascertain a partial biochemical pathway that leads to the synthesis of such a molecule. Their work illustrates how genetic analysis can be used to establish biochemical information.

Srb and Horowitz tested each mutant strain's ability to reestablish growth if either citrulline or ornithine, two compounds with close chemical similarity to arginine, was used as a supplement to minimal medium. If either was able to substitute for arginine, they reasoned that it must be involved in the biosynthetic pathway of arginine. The researchers found that both molecules could be substituted in one or more strains.

Of the seven mutant strains, four of them (*arg 4–7*) grew if supplied with either citrulline, ornithine, or arginine. Two of them (*arg 2* and *arg 3*) grew if supplied with citrulline or arginine. One strain (*arg 1*) would grow only if arginine were supplied; neither citrulline nor ornithine could substitute for it. From these experimental observations, the following pathway and metabolic blocks for each mutation were deduced:

$$\text{Precursor} \xrightarrow[\text{Enzyme A}]{arg\ 4\text{–}7} \text{Ornithine} \xrightarrow[\text{Enzyme B}]{arg\ 2\text{–}3} \text{Citrulline} \xrightarrow[\text{Enzyme C}]{arg\ 1} \text{Arginine}$$

The reasoning supporting these conclusions is based on the following logic: If mutants *arg 4* through *arg 7* can grow regardless of which of the three molecules is supplied as a supplement to minimal medium, the mutations preventing growth must cause a metabolic block that occurs *prior* to the involvement of ornithine, citrulline, or arginine in the pathway. When any one of these three molecules is added, its presence bypasses the block. As a result, it can be concluded that both citrulline and ornithine are involved in the biosynthesis of arginine. However, the sequence of their participation in the pathway cannot be determined on the basis of these data.

On the other hand, both the *arg 2* and the *arg 3* mutations grow if supplied citrulline, but not if they are supplied with only ornithine. Therefore, ornithine must occur in the pathway *prior* to the block. Its presence will not overcome the block. Citrulline, however, does overcome the block, so it must be involved beyond the point of blockage. Therefore, the conversion of ornithine to citrulline represents the correct sequence in the pathway.

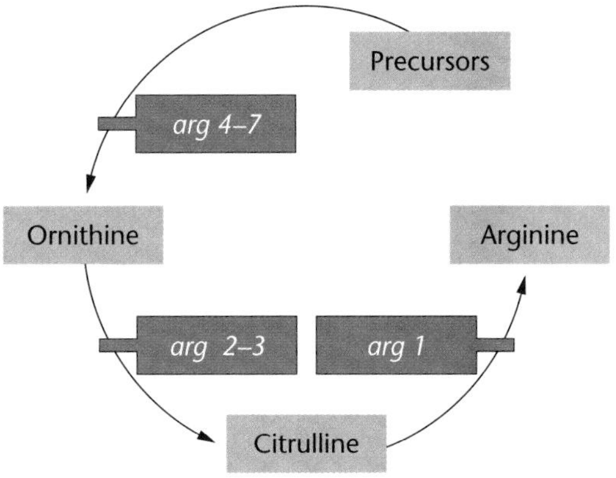

FIGURE 14–12 Abbreviated pathway resulting in the biosynthesis of arginine in *Neurospora*.

Finally, we can conclude that *arg 1* represents a mutation preventing the conversion of citrulline to arginine. Neither ornithine nor citrulline can overcome the metabolic block because both participate earlier in the pathway.

Taken together, these reasons support the sequence of biosynthesis outlined here. Since Srb and Horowitz's work in 1944, the detailed pathway has been worked out and the enzymes controlling each step characterized. The abbreviated metabolic pathway is shown in Figure 14–12.

Now solve this

Problem 14.12 on page 358 asks you to analyze data to establish a biochemical pathway in the bacterium *Salmonella*.

Hint: Apply the same principles and approach used to decipher biochemical pathways in *Neurospora*.

14.7 Studies of Human Hemoglobin Established That One Gene Encodes One Polypeptide

The one-gene:one-enzyme concept developed in the early 1940s was not immediately accepted by all geneticists. This is not surprising since it was not yet clear how mutant enzymes could cause variation in many phenotypic traits. For example, *Drosophila* mutants demonstrated altered eye size, wing shape, wing vein pattern, and so on. Plants exhibited mutant varieties of seed texture, height, and fruit size. How an inactive, mutant enzyme could result in such phenotypes was puzzling to many geneticists.

Two factors soon modified the one-gene:one-enzyme hypothesis. First, while nearly all enzymes are proteins, not all proteins are enzymes. As the study of biochemical genetics proceeded, it became clear that all proteins are specified by the information stored in genes, leading to the more accurate phraseology **one-gene:one-protein**. Second, proteins were often shown to have a subunit structure consisting of two or more polypeptide chains. This is the basis of the quarternary structure of proteins, which we will discuss later in the chapter. Because each distinct polypeptide chain is encoded by a

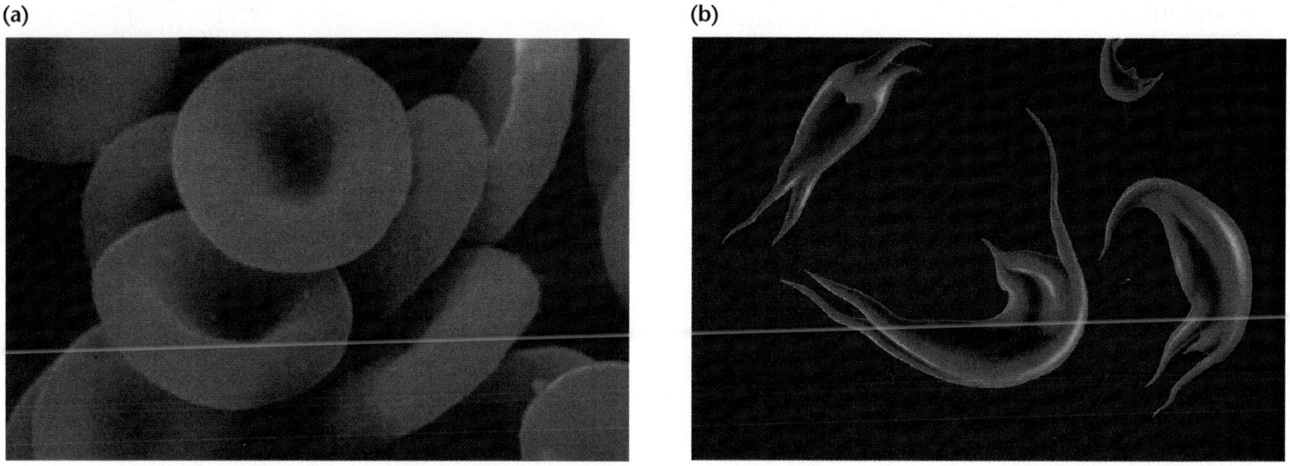

FIGURE 14–13 A comparison of erythrocytes from normal individuals (a) and from individuals with sickle-cell anemia (b).

separate gene, a more modern statement of Beadle and Tatum's basic principle is **one-gene:one-polypeptide chain**. These modifications of the original hypothesis became apparent during the analysis of hemoglobin structure in individuals afflicted with sickle-cell anemia.

Sickle-Cell Anemia

The first direct evidence that genes specify proteins other than enzymes came from the work on mutant hemoglobin molecules derived from humans afflicted with the disorder **sickle-cell anemia**. Affected individuals contain erythrocytes that, under low oxygen tension, become elongated and curved because of the polymerization of hemoglobin. The "sickle" shape of these erythrocytes is in contrast to the biconcave disc shape characteristic of those in normal individuals (Figure 14–13). Individuals with the disease suffer attacks when red blood cells aggregate in the venous side of capillary systems, where oxygen tension is very low. As a result, a variety of tissues may be deprived of oxygen and suffer severe damage. When this occurs, an individual is said to experience a sickle-cell crisis. If untreated, a crisis may be fatal. The kidneys, muscles, joints, brain, gastrointestinal tract, and lungs may be affected.

In addition to suffering crises, these individuals are anemic because their erythrocytes are destroyed more rapidly than are normal red blood cells. Compensatory physiological mechanisms include increased red-cell production by bone marrow and accentuated heart action. These mechanisms lead to abnormal bone size and shape as well as dilation of the heart.

In 1949, James Neel and E.A. Beet demonstrated that the disease is inherited as a Mendelian trait. Pedigree analysis revealed three genotypes and phenotypes controlled by a single pair of alleles, Hb^A and Hb^S. Normal and affected individuals result from the homozygous genotypes Hb^AHb^A and Hb^SHb^S, respectively. The red blood cells of heterozygotes, who exhibit the **sickle-cell trait**, but not the disease, undergo much less sickling because more than half of their hemoglobin is normal. Although largely unaffected, such heterozygotes are "carriers" of the defective gene, which is transmitted, on average, to 50 percent of their offspring.

In the same year, Linus Pauling and his coworkers provided the first insight into the molecular basis of sickle-cell anemia. They showed that hemoglobins isolated from diseased and normal individuals differ in their rates of electrophoretic migration. In this technique (described in Chapter 10 and Appendix A), charged molecules migrate in an electric field. If the net charge of two molecules is different, their rates of migration will be different. Hence, Pauling and his colleagues concluded that a chemical difference exists between normal and sickle-cell hemoglobin. The two molecules are now designated **HbA** and **HbS**, respectively.

Figure 14–14(a) illustrates the migration pattern of hemoglobin derived from individuals of all three possible genotypes when subjected to **starch gel electrophoresis**. The gel provides the supporting medium for the molecules during migration. In this experiment, samples are placed at a point of origin between the cathode ($-$) and the anode ($+$), and an electric field is applied. The migration pattern reveals that all molecules move toward the anode, indicating a net negative charge. However, HbA migrates farther than HbS, suggesting that its net negative charge is greater. The electrophoretic pattern of hemoglobin derived from carriers reveals the presence of both HbA and HbS, confirming their heterozygous genotype.

Pauling's findings suggested two possibilities. It was known that hemoglobin consists of four nonproteinaceous, iron-containing heme groups and a globin portion containing four polypeptide chains. Theoretically, the alteration in net charge in HbS could be due to a chemical change in either component.

Work carried out between 1954 and 1957 by Vernon Ingram resolved the question. He demonstrated that the chemical change occurs in the primary structure of the globin portion of the hemoglobin molecule. Using the **fingerprinting technique**, Ingram showed that HbS differs from HbA in amino acid composition. Human adult hemoglobin contains two identical alpha (α) chains of 141 amino acids and two identical beta (β) chains of 146 amino acids in its quaternary structure.

The fingerprinting technique involves the enzymatic digestion of the protein into peptide fragments. The mixture is then placed on absorbent paper and exposed to an electric field,

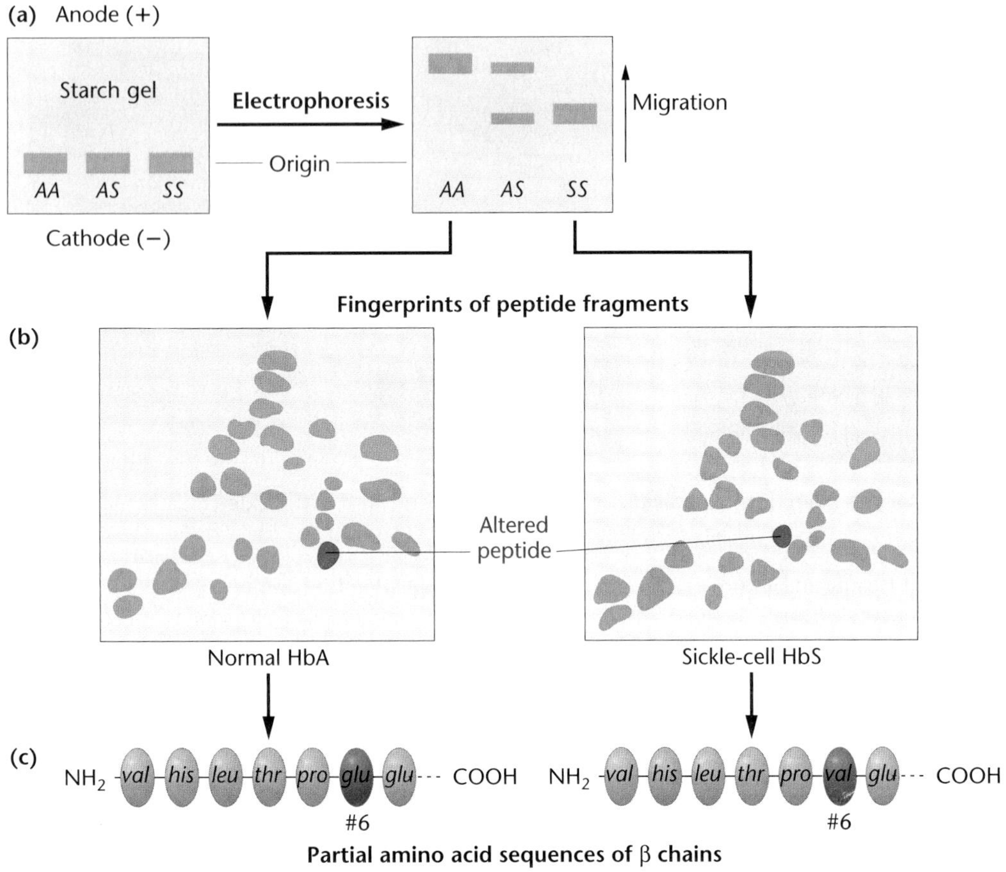

FIGURE 14–14 Investigation of hemoglobin derived from Hb^AHb^A and Hb^SHb^S individuals using electrophoresis, fingerprinting, and amino acid analysis. Hemoglobin from individuals with sickle-cell anemia (Hb^SHb^S) (a) migrates differently in an electrophoretic field, (b) shows an altered peptide in fingerprint analysis, and (c) shows an altered amino acid, valine, at the sixth position in the β chain. During electrophoresis, heterozygotes (Hb^AHb^S) reveal both forms of hemoglobin.

where migration occurs according to net charge. The paper is then turned at a right angle and placed in a solvent, in which chromatographic action causes the migration of the peptides in the second direction. The end result is a two-dimensional separation of the peptide fragments into a distinctive pattern of spots or a "fingerprint." Ingram's work revealed that HbS and HbA differed by only a single peptide fragment [Figure 14–14(b)]. Further analysis then revealed a single amino acid change: Valine was substituted for glutamic acid at the sixth position of the β chain, accounting for the peptide difference [Figure 14–14(c)].

The significance of this discovery has been multifaceted. It clearly establishes that a single gene provides the genetic information for a single polypeptide chain. Studies of HbS also demonstrate that a mutation can affect the phenotype by directing a single amino acid substitution. Also, by providing the explanation for sickle-cell anemia, the concept of inherited **molecular disease** was firmly established. Finally, this work led to a thorough study of human hemoglobins, which has provided valuable genetic insights.

In the United States, sickle-cell anemia is found almost exclusively in the African-American population. It affects about one in every 625 African-American infants. Currently, about 50,000 to 75,000 individuals are afflicted. In about 1 of every 145 African-American married couples, both partners are heterozygous carriers. In these cases, each of their children has a 25 percent chance of having the disease.

Human Hemoglobins

Having introduced human hemoglobins in an historical context, it may be useful to extend our discussion and provide an update of what is currently known about these molecules in our species. Molecular analysis reveals that a variety of hemoglobin molecules are produced in humans at different stages of the life cycle. All are tetramers consisting of numerous combinations of seven distinct polypeptide chains, each encoded by a separate gene. The expression of these various genes is developmentally regulated.

Almost all adult hemoglobin consists of **HbA**, which contains two α and two β **chains**. Recall that the mutation in sickle-cell anemia involves the β chain. HbA represents about 98 percent of all hemoglobin found in an individual's erythrocytes after the age of six months. The remaining 2 percent consists of **HbA$_2$**, a minor adult component. This molecule contains two α chains and two **delta (δ) chains**. The δ chain is very similar to the β chain, consisting of 146 amino acids.

During embryonic and fetal development, quite a different set of hemoglobins is found. The earliest set to develop is called **Gower 1**, containing two **zeta (ζ) chains**, which are most similar to α chains, and two **epsilon (ϵ) chains**, which are most similar to β chains. By eight weeks gestation, the embryonic form is gradually replaced by still another hemoglobin molecule with still different chains. This molecule is called **HbF**, or **fetal hemoglobin**, and consists of two α chains and two **gamma (γ) chains**. There are two types of γ chains designated

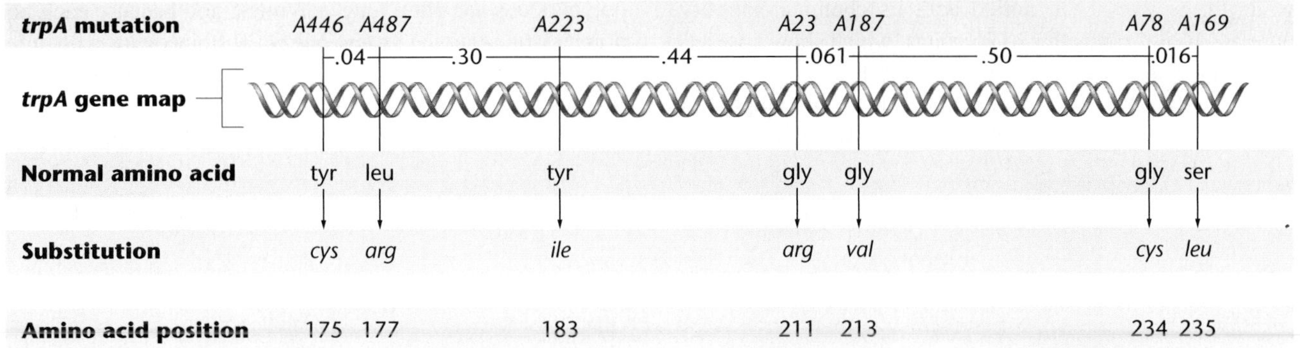

FIGURE 14–15 Demonstration of colinearity between the genetic map of various *trpA* mutations in *E. coli* and the affected amino acids in the protein product. The numbers shown between mutations represent linkage distances.

TABLE 14.2 CHAIN COMPOSITIONS OF HUMAN HEMOGLOBINS FROM CONCEPTION TO ADULTHOOD

Hemoglobin Type	Chain Composition
Embryonic-Gower 1	$\zeta_2\epsilon_2$
Fetal-HbF	$\alpha_2{}^{G}\gamma_2$
	$\alpha_2{}^{A}\gamma_2$
Adult-HbA	$\alpha_2\beta_2$
Minor adult-HbA_2	$\alpha_2\delta_2$

${}^{G}\gamma$ and ${}^{A}\gamma$. Both are most similar to β chains and differ from each other by only a single amino acid. The nomenclature and sequence of appearance of the five tetramers described are summarized in Table 14.2.

14.8 The Nucleotide Sequence of a Gene and the Amino Acid Sequence of the Corresponding Protein Exhibit Colinearity

Once it was established that genes specify the synthesis of polypeptide chains, the next logical question was how the genetic information contained in the nucleotide sequence of a gene can be transferred to the amino acid sequence of a polypeptide chain. It seemed most likely that a colinear relationship (the concept of **colinearity**) would exist between the two molecules. That is, the order of nucleotides in the DNA of a gene would correlate directly with the order of amino acids in the corresponding polypeptide.

The initial experimental evidence in support of this concept was derived from studies by Charles Yanofsky of the *trpA* gene that encodes the A subunit of the enzyme **tryptophan synthetase** in *E. coli*. Yanofsky isolated many independent mutants that had lost the activity of the enzyme, mapped them, and established their location with respect to one another within the gene. He then determined where the amino acid substitution had occurred in each mutant protein. When the two sets of data were compared, the colinear relationship was apparent. The location of each mutation in the *trpA* gene correlates with the position of the altered amino acid in the A polypeptide of tryptophan synthetase. This comparison is illustrated in Figure 14–15.

HOW DO WE KNOW?

What experimental information supports the concept of colinearity, by which there is a direct correspondence between the nucleotide sequence making up a gene and the amino acid sequence making up a polypeptide chain?

14.9 Protein Structure Is the Basis of Biological Diversity

Having established that the genetic information is stored in DNA and influences cellular activities through the proteins it encodes, we turn now to a brief discussion of protein structure. How can these molecules play such a critical role in determining the complexity of cellular activities? As we will see, the fundamental aspects of the structure of proteins provide the basis for incredible complexity and diversity. At the outset, we should differentiate between the terms **polypeptides** and **proteins**. Both describe molecules composed of amino acids. The molecules differ, however, in their state of assembly and functional capacity. *Polypeptides* are the precursors of proteins, as assembled on the ribosome during translation. When released from the ribosome following translation, a polypeptide folds up and assumes a higher order of structure. When this occurs, a three-dimensional conformation in space emerges, and in many cases, several polypeptides interact to produce such a conformation. When the final conformation is achieved, the molecule, now fully functional, is appropriately called a *protein*. It is the three-dimensional conformation that is essential to the function of the molecule.

The polypeptide chains of proteins, like nucleic acids, are linear nonbranched polymers. There are 20 amino acids that serve as the subunits (the building blocks) of proteins. Each amino acid has a **carboxyl group**, an **amino group**, and a **radical (R) group** (a side chain) bound covalently to a **central carbon atom**.

The R group gives each amino acid its chemical identity. Figure 14–16 illustrates the 20 R groups, which show a variety of configurations and can be divided into four main classes: (1) **nonpolar (hydrophobic)**, (2) **polar (hydrophilic)**, (3) **negatively charged**, and (4) **positively charged**. Because polypeptides are often long polymers, and because each position may be occupied by any one of 20 amino acids with unique chemical properties, enormous variation in chemical conformation and activity is possible. For example, if an average polypeptide is composed of 200 amino acids (a molecular weight of

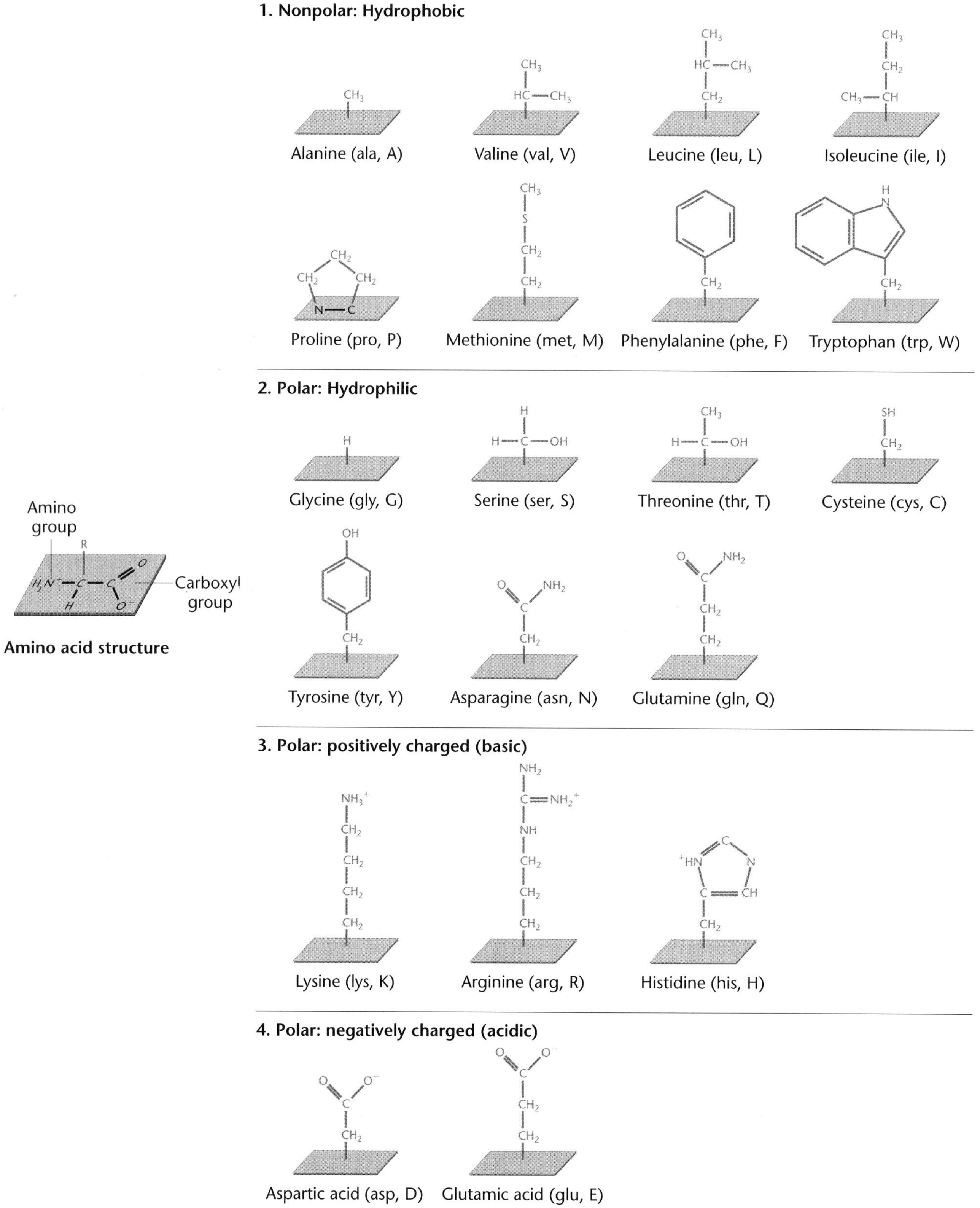

FIGURE 14–16 Chemical structures and designations of the 20 amino acids found in living organisms, divided into four major categories.

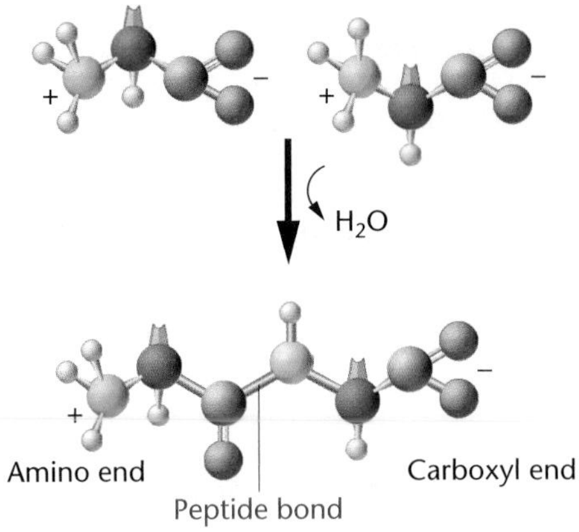

FIGURE 14–17 Peptide bond formation between two amino acids, resulting from a dehydration reaction.

about 20,000 Da), 20^{200} different molecules, each with a unique sequence, can be created by using 20 different building blocks.

Around 1900, German chemist Emil Fischer determined the manner in which the amino acids are bonded together. He showed that the amino group of one amino acid can react with the carboxyl group of another amino acid during a dehydration (condensation) reaction, releasing a molecule of H_2O. The resulting covalent bond is known as a *peptide bond* (Figure 14–17). Two amino acids linked together constitute a dipeptide, three a tripeptide, and so on. Once 10 or more amino acids are linked by peptide bonds, the chain may be referred to as a *polypeptide*. Generally, no matter how long a polypeptide is, it will contain an amino group at one end (the N-terminus) and a carboxyl group at the other end (the C-terminus).

Four levels of protein structure are recognized: primary, secondary, tertiary, and quaternary. The sequence of amino acids in the linear backbone of the polypeptide constitutes its **primary structure**. This sequence is specified by the sequence of deoxyribonucleotides in DNA through an mRNA intermediate. The primary structure of a polypeptide helps determine the specific characteristics of the higher orders of organization as a protein is formed.

The **secondary structure** refers to a regular or repeating configuration in space assumed by amino acids aligned closely to one another in the polypeptide chain. In 1951, Linus Pauling and Robert Corey predicted, on theoretical grounds, an ***α* helix** as one type of secondary structure. The α-helix model [Figure 14–18(a)] has since been confirmed by X-ray crystallographic studies. It is rodlike and has the greatest possible theoretical stability. The helix is composed of a spiral chain of amino acids stabilized by hydrogen bonds.

The side chains (the R groups) of amino acids extend outward from the helix, and each amino acid residue occupies a vertical distance of 1.5 Å in the helix. There are 3.6 residues per turn. While left-handed helices are theoretically possible, all proteins demonstrating an α helix are right handed.

Also in 1951, Pauling and Corey proposed a second structure, the ***β*-pleated-sheet** configuration. In this model, a single polypeptide chain folds back on itself, or several chains run in either parallel or antiparallel fashion next to one another. Each such structure is stabilized by hydrogen bonds formed between atoms present on adjacent chains [Figure 14–18(b)]. A single zigzagging plane is formed in space with adjacent amino acids 3.5 Å apart.

As a general rule, most proteins demonstrate a mixture of α-helix and β-pleated-sheet structures. Globular proteins, most of which are round in shape and water soluble, usually contain a core of β-pleated-sheet structure, as well as many areas demonstrating helical structures. The more rigid structural proteins, many of which are water insoluble, rely on more extensive β-pleated-sheet regions for their rigidity. For example, **fibroin**, the protein made by the silk moth, depends extensively on this form of secondary structure.

While the secondary structure describes the arrangement of amino acids within certain areas of a polypeptide chain, **tertiary protein structure** defines the three-dimensional conformation of the entire chain in space. Each polypeptide twists and turns and loops around itself in a very specific fashion, characteristic of the specific protein. Three aspects of this level

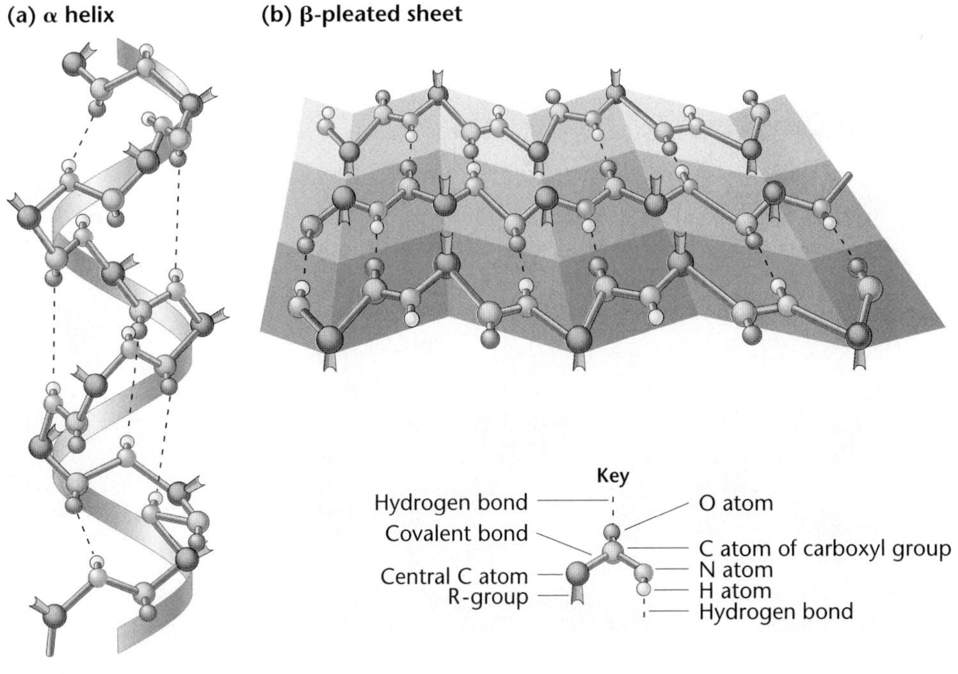

FIGURE 14–18 (a) The right-handed α-helix which represents one form of secondary structure of a polypeptide chain. (b) The β-pleated-sheet configuration, an alternative form of secondary structure of polypeptide chains. For the sake of clarity, some atoms are not shown.

of structure are most important in determining the conformation and in stabilizing the molecule:

1. Covalent disulfide bonds form between closely aligned cysteine residues to form the unique amino acid cystine.
2. Nearly all of the polar hydrophilic R groups are located on the surface, where they can interact with water.
3. The nonpolar hydrophobic R groups are usually located on the inside of the molecule, where they interact with one another, avoiding interaction with water.

It is important to emphasize that the three-dimensional conformation achieved by any protein is a product of the primary structure of the polypeptide. Thus, the genetic code need only specify the sequence of amino acids in order to encode information that leads ultimately to the final assembly of proteins. The three stabilizing factors depend on the location of each amino acid relative to all others in the chain. As folding occurs, the most thermodynamically stable conformation possible results.

A model of the three-dimensional tertiary structure of the respiratory pigment **myoglobin** is shown in Figure 14–19. This level of organization is extremely important because the specific function of any protein is directly related to its three-dimensional conformation.

The **quaternary level of organization** applies only to proteins composed of more than one polypeptide chain, indicating the conformation of the various chains in relation to one another. This type of protein is called oligomeric, and each chain is called a protomer, or, less formally, a subunit. The individual protomers have conformations that fit together with other subunits in a specific complementary fashion. Hemoglobin, an oligomeric protein consisting of four polypeptide chains (two α and two β chains), has been studied in great detail. Its quaternary structure is shown in Figure 14–20. Most enzymes, including DNA and RNA polymerase, demonstrate quaternary structure.

Now solve this

Problem 14.31 on page 359 asks you to consider the potential impact of several amino acid substitutions that result due to mutations in one of the genes encoding one of the chains making up human hemoglobin.

Hint: When considering the three amino acids (glutamic acid, lysine, and valine), consider the net charge of each R group.

Posttranslational Modification

Before turning to a discussion of protein function, it is important to point out that polypeptide chains, like RNA transcripts, are often modified once they have been synthesized. This additional processing is broadly described as **posttranslational modification**. Although many of these alterations are detailed biochemical transformations and beyond the scope of our discussion, you should be aware that they occur and that they are critical to the functional capability of the final protein product. Several examples of posttranslational modification are presented here:

1. *The N-terminus amino acid is usually removed or modified.* For example, either the formyl group or the entire formylmethionine residue in bacterial polypeptides is usually removed enzymatically. In eukaryotic polypeptide chains, the amino group of the initial methionine residue is often removed, and the amino group of the N-terminal residue is chemically modified (acetylated).
2. *Individual amino acid residues are sometimes modified.* For example, phosphates may be added to the hydroxyl groups of certain amino acids, such as tyrosine. Modifications such as these create negatively charged residues that may form an ionic bond with other molecules. The process of phosphorylation is extremely important in regulating many cellular activities and is a result of the action

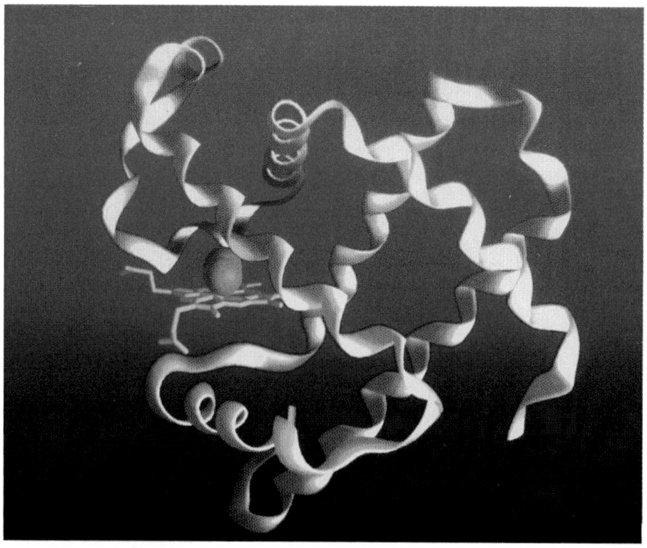

FIGURE 14–19 The tertiary level of protein structure in the respiratory pigment myoglobin. The bound oxygen atom is shown in red.

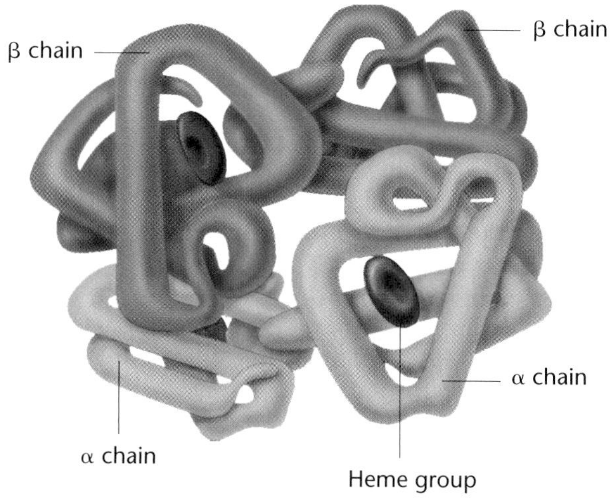

FIGURE 14–20 The quaternary level of protein structure as seen in hemoglobin. Four chains (two α and two β) interact with four heme groups to form the functional molecule.

of enzymes called **kinases**. At other amino acid residues, methyl groups or acetytl groups may be added enzymatically, which can similarly affect the function of the modified polypeptide chain.

3. *Carbohydrate side chains are sometimes attached.* These are added covalently, producing **glycoproteins**, an important category of extracellular molecules, such as those specifying the antigens in the ABO blood-type system in humans.

4. *Polypeptide chains may be trimmed.* For example, insulin is first translated into a longer molecule that is enzymatically trimmed to its final 51-amino-acid form.

5. *Signal sequences are removed.* At the N-terminal end of some proteins, a sequence of up to 30 amino acids is found that plays an important role in directing the protein to the location in the cell in which it functions. This is called a **signal sequence**, and it determines the final destination of a protein in the cell. The process is called **protein targeting**. For example, proteins whose fate involves secretion or ones that are to become part of the plasma membrane are dependent on specific sequences for their initial transport into the lumen of the endoplasmic reticulum. While the signal sequence of various proteins with a common destination might differ in their primary amino acid sequence, they share many chemical properties. For example, those destined for secretion all contain a string of up to 15 hydrophobic amino acids preceded by a positively charged amino acid at the N-terminus of the signal sequence. Once the polypeptides are transported, but prior to achieving their functional status as proteins, the signal sequence is enzymatically removed from these polypeptides.

6. *Polypeptide chains are often complexed with metals.* The tertiary and quaternary levels of protein structure often include and are dependent on metal atoms. The function of the protein is thus dependent on the molecular complex that includes both polypeptide chains and metal atoms. Hemoglobin, containing four iron atoms along with four polypeptide chains, is a good example.

These types of posttranslational modifications are obviously important in achieving the functional status specific to any given protein. Because the final three-dimensional structure of the molecule is intimately related to its specific function, how polypeptide chains ultimately fold into their final conformations is also an important topic. For many years, it was thought that protein folding was a spontaneous process whereby the molecule achieved maximum thermodynamic stability, based largely on the combined chemical properties inherent in the amino acid sequence of the polypeptide chain(s) composing the protein. However, numerous studies have shown that, for many proteins, folding is dependent upon members of a family of still other, ubiquitous proteins called **chaperones**. Chaperone proteins (sometimes called *molecular chaperones* or *chaperonins*) function to facilitate the folding of other proteins. While the mechanism by which chaperones function is not yet clear, like enzymes, they do not become part of the final product. Initially discovered in *Drosophila*, in which they are called heat-shock proteins, chaperones have been discovered in a variety of organisms, including bacteria, animals, and plants. Ultimately, protein folding is a critically important process, not only because misfolded proteins may be nonfunctional, but also because improperly folded proteins can be dangerous. It is becoming clear that some disorders, such as **Alzheimer disease** in humans and the spongiform encephalopathies (**mad cow disease** in cattle and **Creutzfeldt–Jakob disease** in humans) are caused by the presence of incorrectly folded neural proteins. Currently, many laboratories are focused on trying to understand how protein folding occurs normally, as well as how the presence of misfolded polypeptides "poisons" the folding of normal polypeptides and ultimately causes cell death and disease.

14.10 Protein Function Is Directly Related to the Structure of the Molecule

The essence of life on Earth rests at the level of diverse cellular function. One can argue that DNA and RNA simply serve as vehicles to store and express genetic information. However, proteins are at the heart of cellular function. And it is the capability of cells to assume diverse structures and functions that distinguishes most eukaryotes from less evolutionarily advanced organisms, such as bacteria. Therefore, an introductory understanding of protein function is critical to a complete view of genetic processes.

Proteins are the most abundant macromolecules found in cells. As the end products of genes, they play many diverse roles. For example, the respiratory pigments **hemoglobin** and **myoglobin** bind to oxygen. Hemoglobin transports the oxygen, which is essential for cellular metabolism. **Collagen** and **keratin** are structural proteins associated with the skin, connective tissue, and hair of organisms. **Actin** and **myosin** are contractile proteins, found in abundance in muscle tissue. Still other examples are the **immunoglobulins**, which function in the immune system of vertebrates; **transport proteins**, involved in movement of molecules across membranes; some of the **hormones** and their **receptors**, which regulate various types of chemical activity; and **histones**, which bind to DNA in eukaryotic organisms.

The largest group of proteins with a related function are the **enzymes**. Since we have referred to these molecules throughout this chapter, it may be useful to extend our discussion and include a more detailed description of their biological role. These molecules specialize in catalyzing chemical reactions within living cells. Enzymes increase the rate at which a chemical reaction reaches equilibrium, but they do not alter the end point of the chemical equilibrium. Their remarkable, highly

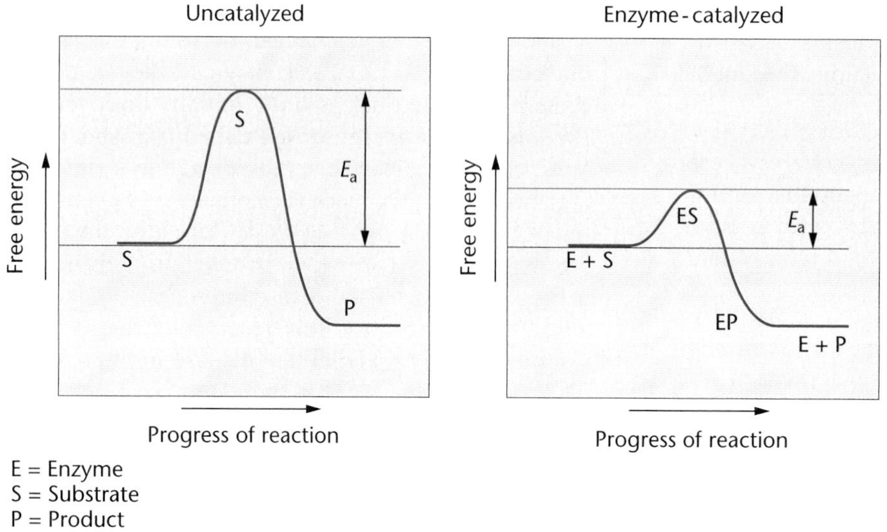

FIGURE 14–21 Energy requirements of an uncatalyzed versus an enzymatically catalyzed chemical reaction. The energy of activation (E_a) necessary to initiate the reaction is substantially lower as a result of catalysis.

specific catalytic properties largely determine the metabolic capacity of any cell type. The specific functions of many enzymes involved in the genetic and cellular processes of cells are described throughout the text.

Biological catalysis is a process whereby the **energy of activation** for a given reaction is lowered (Figure 14–21). The energy of activation is the increased kinetic energy state that molecules must usually reach before they react with one another. While this state can be attained as a result of elevated temperatures, enzymes allow biological reactions to occur at lower physiological temperatures. Thus, enzymes make life as we know it possible.

The catalytic properties and specificity of an enzyme are determined by the chemical configuration of the molecule's **active site**. This site is associated with a crevice, a cleft, or a pit on the surface of the enzyme, which binds the reactants, or substrates, facilitating their interaction. Enzymatically catalyzed reactions control metabolic activities in the cell. Each reaction is either **catabolic** or **anabolic**. Catabolism is the degradation of large molecules into smaller, simpler ones with the release of chemical energy. Anabolism is the synthetic phase of metabolism, yielding the various components that make up nucleic acids, proteins, lipids, and carbohydrates.

14.11 Proteins Are Made Up of One or More Functional Domains

We conclude this chapter by briefly discussing the finding that regions made up of specific amino acid sequences are associated with specific functions in protein molecules. Such sequences, usually between 50 and 300 amino acids, constitute what are called **protein domains** and are represented by modular portions of the protein that fold independently of the rest of the molecule into stable, unique conformations. Different domains impart different functional capabilities. Some proteins contain only a single domain, while others contain two or more.

The significance of domains rests at the tertiary structure level of proteins. Each such modular unit can be a mixture of secondary structures, including both α helices and β-pleated-sheets. The unique conformation that is assumed in a single domain imparts a specific function to the protein. For example, a domain may serve as the catalytic site of an enzyme, or it may impart the capability to bind to a specific ligand. Thus, in the study of proteins, you will hear of *catalytic domains*, *DNA-binding domains*, and so on. The result is that a protein must be looked at as being composed of a series of structural and functional modules. Obviously, the presence of multiple domains in a single protein increases the versatility of each molecule and adds to its functional complexity.

Exon Shuffling and the Origin of Protein Domains

An interesting proposal to explain the genetic origin of protein domains was put forward by Walter Gilbert in 1977. Gilbert suggested that the amino-acid-coding regions of genes in higher organisms consist of collections of exons originally present in ancestral genes, brought together through recombination during the course of evolution. Referring to the process as **exon shuffling**, Gilbert proposed that exons are modular in the sense that each might encode a single protein domain. Serving as the basis of the useful part of a protein, a single type of domain might function in a variety of proteins. In Gilbert's proposal, during evolution, many exons could be "mixed and matched" to form unique genes in eukaryotes.

Several observations lend support to this proposal. Most exons are fairly small, averaging about 150 base pairs and encoding about 50 amino acids. Such a size is consistent with the production of many functional domains in proteins. Second, recombinational events that lead to exon shuffling would be expected to occur within areas of genes represented by introns. Because introns are free to accumulate mutations without harm to the organism, recombinational events would tend to further randomize their nucleotide sequences. Over extended evolutionary period, diverse sequences would tend to accumulate. This is, in fact, what is observed: introns range from 50 to 20,000 bases and exhibit fairly random base sequences.

Since 1977, a serious research effort has been directed toward the analysis of gene structure. In 1985, more direct evidence in favor of Gilbert's proposal of exon modules was presented. For example, the human gene encoding the membrane receptor for low-density lipoproteins (LDL) was isolated and sequenced.

The LDL receptor protein is essential to the transport of plasma cholesterol into the cell. It mediates endocytosis and is expected to have numerous functional domains. These include the capability of this protein to bind specifically to the LDL substrates and to interact with other proteins at different levels of the membrane during transport across it. In addition, the receptor molecule is modified posttranslationally by the addition of a carbohydrate. Thus a domain where the carbohydrate is added must also be present. Given these functional constraints, one would predict that the LDL receptor polypeptide would contain several distinct domains.

Detailed analysis of the gene encoding the LDL receptor supports the concept of exon modules and their shuffling during evolution. The gene is quite large—45,000 base pairs—and contains 18 exons, which represent only slightly less than 2600 nucleotides. These exons are related to the functional domains of the protein *and* appear to have been recruited from other genes during evolution.

Figure 14–22 illustrates these relationships. The first exon encodes a signal sequence that is removed from the protein before the LDL receptor becomes part of the membrane. The next five exons represent the domain specifying the binding site for cholesterol. This domain is made up of a 40-amino-acid sequence repeated seven times. The next domain consists of a sequence of 400 amino acids bearing a striking homology to the mouse peptide hormone epidermal growth factor (EGF). This region is encoded by eight exons and contains three repetitive sequences of 40 amino acids. A similar sequence is also found in many other polypeptides, including three blood-clotting proteins. The 15th exon specifies the domain for the posttranslational addition of the carbohydrate, while the remaining 2 exons specify regions of the protein that are integrated into the membrane, anchoring the receptor to specific sites called *coated pits* on the cell surface.

These observations concerning the LDL exons are fairly compelling in support of the theory of exon shuffling during evolution. Certainly, there is no disagreement concerning the concept of protein domains being responsible for specific molecular interactions.

What remains controversial and evocative in the exon-shuffling theory is the question of when introns first appeared on the evolutionary scene. In 1978, W. Ford Doolittle proposed that these intervening sequences were part of the genome of the most primitive ancestors of modern-day eukaryotes. In support of this "intron-early" idea, Gilbert has argued that if intron similarities in a DNA sequence are found in identical positions within genes shared by totally unrelated eukaryotes (such as humans, chickens, and corn), they must have also been present in primitive ancestral genomes.

If Doolittle's proposal is correct, why are introns absent in most prokaryotes and infrequent in yeast? Gilbert argues that they were present at one point during evolution, but as the genome of these primitive organisms evolved, they were lost. This resulted from strong selection pressure to streamline chromosomes to minimize the energy expenditure supporting replication and gene expression. Further, streamlining led to more error-free mRNA production. However, supporters of the opposing "intron-late" school, including Jeffrey Palmer, argue that introns' first appearance came much later during evolution, when they became a part of a single group of eukaryotes that are ancestral to modern-day members, but not to prokaryotes.

The advent of large scale genome sequencing has served to increase the controversy. We now have the ability to decipher the complete nucleotide sequence of all the DNA in specific organisms. By comparing the amino-acid coding and the non-coding DNA sequences from evolutionarily spaced species, geneticists hope to gain insight into how these sequences evolved. Currently, no indisputable evidence in support of either the "intron-early" or the "intron-late" theory has been found, and the question has remained difficult to resolve.

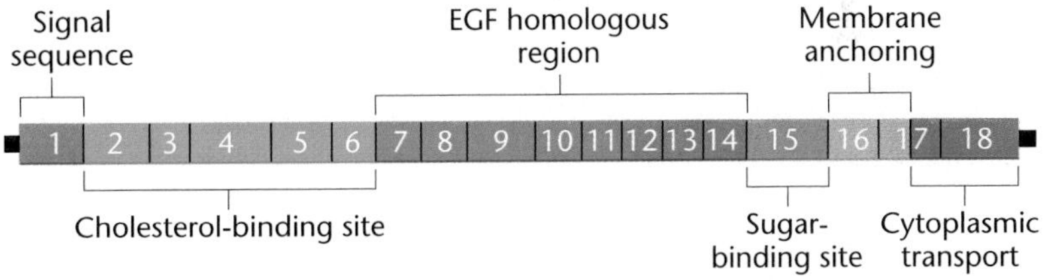

FIGURE 14–22 A comparison of the 18 exons making up the gene encoding the LDL receptor protein. The exons are organized into five functional domains and one signal sequence.

GENETICS, TECHNOLOGY, AND SOCIETY

Mad Cow Disease: The Prion Story

In March 1996, the British government announced that a new brain disease had killed 10 young Britons, and that the victims might have caught the disease by eating affected beef. Bovine spongiform encephalopathy (BSE), popularly known as "mad cow disease," slowly destroys brain cells and is always fatal. Recent studies confirm that BSE and the human disease, new variant Creutzfeldt–Jakob disease (nvCJD), are so similar at the molecular and pathological levels that they are very likely to be the same disease. Mad cow disease triggered political turmoil in Europe, a worldwide ban on British beef, and the near-collapse of the \$8.9 billion British beef industry. The European Union demanded the slaughter and incineration of 4.7 million British cattle, a campaign that cost the government more than \$12 billion in compensation to farmers, milk import, and the purchase of stock for new herds. Most European countries, as well as the Middle East, Japan, and Canada have discovered BSE in at least one cow. Although most nvCJD cases have occurred in Britain, cases have also appeared in France, Italy, Ireland, South America, Canada, and the United States. More than 125 people have now died of nvCJD, and epidemiologists estimate that many more cases will appear in the next two decades.

BSE, nvCJD, and Creutzfeldt–Jakob disease (CJD) are all members of a group of neurological diseases known as transmissible spongiform encephalopathies (TSEs), which affect animals and humans. In this group of diseases, the affected brain tissue eventually resembles a sponge (hence, spongiform) and is riddled with proteinaceous deposits. Victims of the disease lose motor function, become demented, and ultimately die. CJD and nvCJD differ symptomatically in that victims of nvCJD display psychological symptoms such as depression or anxiety prior to developing neurological symptoms such as shaking or paralysis. Also, patients with nvCJD are usually young (16–40 years), whereas CJD normally affects people over 55. A number of CJD cases arise spontaneously and randomly, at a rate of one per million per year worldwide, but CJD can also be inherited as an autosomal dominant condition.

CJD can be transmitted through corneal or nervous tissue grafts or by injection of a growth hormone derived from human pituitary glands. Kuru, a CJD-like disease of the Fore people of New Guinea, was transmitted from person to person through ritualistic cannibalism. Transmissible spongiform encephalopathies in animals include scrapie (sheep and goats), chronic wasting disease (deer and elk), and BSE. Like Kuru, BSE is passed from animal to animal by ingestion of diseased animal remains, particularly neural tissue. The epidemic of BSE in Britain occurred because diseased cows and sheep were processed and fed to cattle as a protein supplement. In 1998, the British government banned the use of cows and sheep as feed for other cows and sheep, and the epidemic has subsided. The European Union has now banned the use of feeds containing animal products for all livestock. In contrast, the United States and Canada still allow nonruminant animals to consume feed containing ruminants, and ruminant animals to consume feed containing nonruminants, as well as certain ruminant by-products, including blood, gelatin, and fat. However, these regulations may change as a result of the discovery of a BSE case in Canada in 2003. In addition, recent studies suggest that BSE may be transmitted via blood or tallow.

For many years, the analysis of spongiform encephalopathies defied the best efforts of scientists. The diseases are difficult to study for a number of reasons. First, they require injection of affected brain material into the brains of experimental animals, and the diseases take months or years to develop. In addition, the infectious agent is apparently not a virus or bacterium, and affected animals do not develop antibodies against these mysterious agents.

There are no treatments for the diseases, and the only way to make a firm diagnosis is to examine brain tissue after death. The infectious material is unaffected by radiation or nucleases that damage nucleic acids; however, it is destroyed by some reagents that hydrolyze or modify proteins. In the early 1980s, American scientist Stanley Prusiner purified the infectious agent and concluded that it consists of only protein. He proposed that scrapie is spread by an infectious protein particle that he called a *prion*. His hypothesis was dismissed by most scientists, as the idea of an infectious agent with no DNA or RNA as genetic material was heretical. However, Prusiner and others presented evidence supporting the prion hypothesis, and the notion that the disease can be transmitted by an infectious particle that contains no genetic material has gained acceptance.

If prions are composed of protein only, how do they cause disease? The answer may be as strange as the disease itself. The protein that makes up a prion (PrP) is a version of a normal protein that is synthesized in neurons and found in the brains of all adult animals. The difference between normal PrP and prion PrP lies in their secondary protein structures. Normal, noninfectious PrP folds into α helices, whereas infectious prion PrP folds into β-pleated sheets. When a normal PrP molecule contacts a prion PrP molecule, the normal protein is somehow unfolded and refolded into the abnormal PrP conformation. Once the normal PrP molecule has been transformed into an abnormal PrP molecule, it spreads its lethal conformation to neighboring normal PrP molecules, and the process takes off in a chain reaction. Normal PrP is a soluble protein that is easily destroyed by heat or enzymes that digest proteins. However, abnormal infectious PrP is insoluble in detergents, resists both heat and protease digestion, and is nearly indestructible. Hence, spongiform encephalopathies can be considered diseases of secondary protein structure.

Many urgent questions need to be addressed. How extensive is BSE contamination of the world's food supply? How many humans are affected with nvCJD but don't show symptoms? Can humans or animals act as asymptomatic carriers of prion diseases? Can prions exist in other parts of the body besides the brain and spinal cord, and if so, can nvCJD be spread through blood transfusions, from mother to fetus, or by sterilized surgical instruments? Can we develop diagnostic tests and therapies for BSE and nvCJD? Are we near the end of the BSE story, or is it just beginning?

References

Balter, M. 2000. Tracking the human fallout from "mad cow disease." *Science* 289:1452–54.

Belay, E.D. 1999. Transmissible spongiform encephalopathies in humans. *Annu. Rev. Microbiol.* 53:283–314.

Spencer, C.A. 2004. *Mad Cows and Cannibals: A Guide to the Transmissible Spongiform Encephalopathies*. Upper Saddle River: Pearson/Prentice-Hall.

Web Sites

Online compilation of news reports about mad cow disease, *http://organicconsumers.org/madcow.htm*

The Official Mad Cow Disease Home Page (a compilation of news reports), *http://www.mad-cow.org*.

CHAPTER SUMMARY

1. Translation describes the synthesis in cells of polypeptide chains, under the direction of mRNA and in association with ribosomes. This process ultimately converts the information stored in the genetic code of the DNA making up a gene into the corresponding sequence of amino acids making up the polypeptide.
2. Translation is a complex energy-requiring process that also depends on charged tRNA molecules and numerous protein factors. Transfer RNA (tRNA) serves as the adaptor molecule between an mRNA codon and the appropriate amino acid.
3. The processes of translation, like transcription, can be subdivided into the stages of initiation, elongation, and termination. The process relies on base-pairing affinities between complementary nucleotides and is more complex in eukaryotes than in prokaryotes.
4. The first insight that proteins are the end products of genes was provided by Garrod through his study of inherited metabolic disorders in humans early in the 20th century. Basic to his studies were inborn errors of metabolism leading to cystinuria, albinism, and alkaptonuria.
5. The investigation of nutritional requirements in *Neurospora* by Beadle and colleagues made it clear that mutations cause the loss of enzyme activity. Their work led to the concept of the one-gene:one-enzyme hypothesis.
6. The one-gene:one-enzyme hypothesis was later revised. Pauling and Ingram's investigations of hemoglobins from patients with sickle-cell anemia led to the discovery that one gene directs the synthesis of only one polypeptide chain.
7. Thorough investigations have revealed the existence of several major types of human hemoglobin molecules found in the embryo, fetus, and adult. Specific genes control each polypeptide chain constituting these various hemoglobin molecules.
8. The proposal suggesting that a gene's nucleotide sequence specifies in a colinear way the sequence of amino acids in a polypeptide chain was confirmed by experiments involving mutations in the tryptophan synthetase gene in *E. coli*.
9. Proteins, the end products of genes, demonstrate four levels of structural organization that together provide the chemical basis for their three-dimensional conformation, which is the basis of the molecule's function.
10. Of the myriad functions performed by proteins, the most influential role is assumed by enzymes. These highly specific, cellular catalysts play a central role in the production of all classes of molecules in living systems.
11. Proteins consist of one or more functional domains, which are shared by many different molecules. The origin of these domains may be the result of exon shuffling during evolution.

INSIGHTS AND SOLUTIONS

1. The growth responses that follow were obtained by using four mutant strains of *Neurospora* and the related compounds A, B, C, and D. None of the mutations grow on minimal medium. Draw all possible conclusions.

	Growth Product			
Mutation	**C**	**D**	**B**	**A**
1	–	–	–	–
2	–	+	+	+
3	–	–	+	+
4	–	–	+	–

Solution: First, nothing can be concluded about mutation 1, except that it is lacking some essential growth factor, perhaps even unrelated to the biochemical pathway represented by mutations 2–4; nor can anything be concluded about compound C. If it is involved in the pathway, it is a product synthesized prior to the synthesis of A, B, and D. We must now analyze these three compounds and the control of their synthesis by the enzymes encoded by genes 2, 3, and 4. Because product B allows growth in all three cases, it may be considered the "end product." It bypasses the block in all three instances. Using similar reasoning, product A precedes B in the pathway, since it allows a bypass in two of the three steps. Product D precedes B, yielding the more complete solution:

$$C(?) \rightarrow D \rightarrow A \rightarrow B$$

Now, determine which mutations control which steps. Since mutation 2 can be alleviated by products D, B, and A, it must control a step prior to all three products, perhaps the direct conversion to D, although we cannot be certain. Mutation 3 is alleviated by B and A, so its effect must precede them in the pathway. Thus, we will assign it as controlling the conversion of D to A. Likewise, we can assign mutation 4 to the conversion of A to B, leading to the more complete solution

$$C(?) \xrightarrow{2(?)} D \xrightarrow{3} A \xrightarrow{4} B$$

PROBLEMS AND DISCUSSION QUESTIONS

1. List and describe the role of all of the molecular constituents present in a functional polyribosome.
2. Contrast the roles of tRNA and mRNA during translation and list all enzymes that participate in the transcription and translation process.
3. Francis Crick proposed the "adaptor hypothesis" for the function of tRNA. Why did he choose that description?
4. During translation, what molecule bears the codon? the anticodon?
5. The α chain of eukaryotic hemoglobin is composed of 141 amino acids. What is the minimum number of nucleotides in an mRNA

coding for this polypeptide chain? Assuming that each nucleotide is 0.34 nm long in the mRNA, how many triplet codes can, occupy at one time the space in a ribosome that is 20 nm in diameter?

6. Summarize the steps involved in charging tRNAs with their appropriate amino acids.
7. For it to carry out its role, each transfer RNA requires at least four specific recognition sites that must be inherent in its tertiary structure. What are they?
8. Discuss the potential difficulties involved in designing a diet to alleviate the symptoms of phenylketonuria.
9. Phenylketonurics cannot convert phenylalanine to tyrosine. Why don't these individuals exhibit a deficiency of tyrosine?
10. Phenylketonurics are often more lightly pigmented than are normal individuals. Can you suggest a reason why this is so?
11. The synthesis of flower pigments is known to be dependent on enzymatically controlled biosynthetic pathways. In the crosses shown here, postulate the role of mutant genes and their products in producing the observed phenotypes:

(a) P_1: white strain A × white strain B
F_1: all purple
F_2: 9/16 purple : 7/16 white

(b) P_1: white × pink
F_1: all purple
F_2: 9/16 purple : 3/16 pink : 4/16 white

12. A series of mutations in the bacterium *Salmonella typhimurium* results in the requirement of either tryptophan or some related molecule in order for growth to occur. From the data shown here, suggest a biosynthetic pathway for tryptophan:

	Growth Supplement				
Mutation	**Minimal Medium**	**Anthranilic Acid**	**Indole Glycerol Phosphate**	**Indole**	**Tryptophan**
trp-8	−	+	+	+	+
trp-2	−	−	+	+	+
trp-3	−	−	−	+	+
trp-1	−	−	−	−	+

13. The study of biochemical mutants in organisms such as *Neurospora* has demonstrated that some pathways are branched. The data shown here illustrate the branched nature of the pathway resulting in the synthesis of thiamine:

	Growth Supplement			
Mutation	**Minimal Medium**	**Pyrimidine**	**Thiazole**	**Thiamine**
thi-1	−	−	+	+
thi-2	−	+	−	+
thi-3	−	−	−	+

Why don't the data support a linear pathway? Can you postulate a pathway for the synthesis of thiamine in *Neurospora?*

14. Explain why the one-gene:one-enzyme concept is not considered totally accurate today.
15. Why is an alteration of electrophoretic mobility interpreted as a change in the primary structure of the protein under study?
16. Contrast the polypeptide-chain components of each of the hemoglobin molecules found in humans.
17. Using sickle-cell anemia as a basis, describe what is meant by a molecular or genetic disease. What are the similarities and dissimilarities between this type of a disorder and a disease caused by an invading microorganism?
18. Contrast the contributions of Pauling and Ingram to our understanding of the genetic basis for sickle-cell anemia.
19. Hemoglobins from two individuals are compared by electrophoresis and by fingerprinting. Electrophoresis reveals no difference in migration, but fingerprinting shows an amino acid difference. How is this possible?
20. Describe what colinearity means. Of what significance is the concept of colinearity in the study of genetics?
21. Certain mutations called *amber* in bacteria and viruses result in premature termination of polypeptide chains during translation. Many *amber* mutations have been detected at different points along the gene coding for a head protein in phage T4. How might this system be further investigated to demonstrate and support the concept of colinearity?
22. Define and compare the four levels of protein organization.
23. List as many different categories of protein functions as you can. Wherever possible, give an example of each category.
24. How does an enzyme function? Why are enzymes essential for living organisms on Earth?
25. Does Fiers's work with phage MS2, discussed in Chapter 13, constitute more direct evidence in support of colinearity than Yanofsky's work with the *trpA* locus in *E. coli* (discussed in this chapter)? Explain.
26. Shown here are several amino acid substitutions in the α and β chains of human hemoglobin:

Hb Type	**Normal Amino Acid**	**Substituted Amino Acid**
Hb Toronto	ala	asp (α-5)
HbJ Oxford	gly	asp (α-15)
Hb Mexico	gln	glu (α-54)
Hb Bethesda	tyr	his (β-145)
Hb Sydney	val	ala (β-67)
HbM Saskatoon	his	tyr (β-63)

Using the code table (Figure 13–7), determine how many of them can occur as a result of a single nucleotide change.

27. Early detection and adherence to a strict dietary regime has relieved much of the mental retardation previously occurring in those afflicted with phenylketonuria (PKU). Affected individuals now often lead normal lives and have families. For various reasons, such individuals adhere less rigorously to their diet as they get older. Predict the effect of such a situation on newborns of mothers with PKU who neglect their diets.

Extra-Spicy Problems

28. In 1962, F. Chapeville and others reported an experiment in which they isolated radioactive ^{14}C-cysteinyl-tRNA^{cys} (charged tRNA^{cys} + cysteine). They then removed the sulfur group from the cysteine, creating alanyl-tRNA^{cys} (charged tRNA^{cys} + alanine). When alanyl-tRNA^{cys} was added to a synthetic mRNA calling for cysteine, but not alanine, a polypeptide chain was synthesized con-

taining alanine. What can you conclude from this experiment? (Chapeville, et al. 1962. *Proc. Natl. Acad. Sci.* [*USA*] 48:1086–93.)

29. Three independently assorting genes are known to control the biochemical pathway here that provides the basis for flower color in a hypothetical plant:

$$\text{Colorless} \xrightarrow{A-} \text{yellow} \xrightarrow{B-} \text{green} \xrightarrow{C-} \text{speckled}$$

Homozygous recessive mutations, which interrupt each step, are known. Determine the phenotypic results in the F_1 and F_2 generations resulting from the P_1 crosses given, involving true-breeding plants:

(a) speckled (*AABBCC*)	×	yellow (*AabbCC*)
(b) yellow (*AabbCC*)	×	green (*AABBcc*)
(c) colorless (*aaBBCC*)	×	green (*AABBcc*)

30. How would the results vary in cross (a) of Problem 29 if genes A and B were linked with no crossing over between them? How would the results of cross (a) vary if genes A and B were linked and 20 mu apart?

31. HbS results from the amino acid change of glutamic acid to valine at the number 6 position in the B chain of human hemoglobin. HbC is the result of a change at the same position in the B chain, but lysine replaces glutamic acid. Return to the genetic code table (Figure 13–7) and determine whether single nucleotide changes can account for these mutations. Then view Figure 14–16 and examine the R groups in the amino acids glutamic acid, valine, and lysine. Describe the chemical differences between the three amino acids. Predict how the changes might alter the structure of the molecule and lead to altered hemoglobin function. HbS results in anemia and resistance to malaria, whereas in those with HbA, the parasite *Plasmodium falciparum* invades red blood cells and causes the disease. Predict whether those with HbC are likely to be anemic and whether they would be resistant to malaria.

32. Most wood lily plants have orange flowers and are true breeding. A genetics student discovered both red and yellow true-breeding variants and proceeded to hybridize the three strains. Being knowledgeable about transmission genetics, the student surmised that two gene pairs were involved. On this basis, she proposed that these genes account for the various colors as shown here:

P_1	F_1	F_2
(1) orange × red	all orange	3/4 orange 1/4 red
(2) orange × yellow	all orange	3/4 orange 1/4 yellow
(3) red × yellow	all orange	9/16 orange 4/16 yellow 3/16 red

(a) What outcome(s) of the crosses suggested that two gene pairs were at work?

(b) Assuming that she was correct, and using the mutant-gene symbols *y* for yellow and *r* for red, propose which genotypes give rise to which phenotypes.

(c) The student then devised several possible biochemical pathways for the production of these pigments, as well as which steps are enzymatically controlled by which genes. These are listed here.

I. white precursor I $\xrightarrow{y\text{ gene}}$ red pigment
white precursor II $\xrightarrow{r\text{ gene}}$ yellow pigment

II. white precursor I $\xrightarrow{r\text{ gene}}$ red pigment
white precursor II $\xrightarrow{y\text{ gene}}$ yellow pigment

III. white precursor $\longrightarrow$ red $\xrightarrow{y\text{ gene}}$ yellow $\xrightarrow{r\text{ gene}}$ orange

IV. white precursor $\longrightarrow$ red $\xrightarrow{r\text{ gene}}$ yellow $\xrightarrow{y\text{ gene}}$ orange

V. white precursor $\longrightarrow$ yellow $\xrightarrow{y\text{ gene}}$ red $\xrightarrow{r\text{ gene}}$ orange

VI. white precursor $\longrightarrow$ yellow $\xrightarrow{r\text{ gene}}$ red $\xrightarrow{y\text{ gene}}$ orange

Which of these (one or more) is (are) consistent with your proposal in (b); which was based on the original data? Assume that a mixture of red and yellow pigment results in orange flowers. Defend your answer.

33. Deep in a previously unexplored South American rain forest, a species of plants was discovered with true-breeding varieties whose flowers were either pink, rose, orange, or purple. A very astute plant geneticist made a single cross, carried to the F_2 generation, as shown here:

P_1:	purple × pink
F_1:	all purple
F_2:	27/64 purple 16/64 pink 12/64 rose 9/64 orange

Based solely on these data, he was able to propose both a mode of inheritance for flower pigmentation and a biochemical pathway for the synthesis of these pigments.

Carefully study the data. Create a hypothesis of your own to explain the mode of inheritance. Then propose a biochemical pathway consistent with your hypothesis. How could you test the hypothesis by making other crosses?

34. The emergence of antibiotic-resistant strains of *Enterococci* and transfer of resistant genes to other bacterial pathogens have highlighted the need for new generations of antibiotics to combat serious infections. To grasp the range of potential sites for the action of existing antibiotics, sketch the components of the translation machinery (e.g., see Step 3 of Figure 14–6), and using a series of numbered pointers, indicate the specific location for the action of the antibiotics shown in the following table.

Antibiotic	Action
1. Streptomycin	Binds to 30*S* ribosomal subunit
2. Chloramphenicol	Inhibits peptidyl transferase of 70*S* ribosome
3. Tetracycline	Inhibits binding of charged tRNA to ribosome
4. Erythromycin	Binds to free 50*S* particle and prevents formation of 70*S* ribosome
5. Kasugamycin	Inhibits binding of $tRNA^{fmet}$
6. Thiostrepton	Prevents translocation by inhibiting EF-G

35. Development of antibiotic resistance by pathogenic bacteria represents a major health concern. One potential new antibiotic is evernimicin, which was isolated from *Micromonospora carbonaceae*. Evernimicin is an oligosaccharide antibiotic with activity against a broad range of gram-positive pathogenic bacteria. To determine the mode of action of this drug, Adrian and others (2000. *Antimicrob. Ag. and Chemo.* 44:3101–06) analyzed 23*S* ribosomal DNA mutants that showed reduced sensitivity to evernimicin. They and others discovered two classes of mutants that conferred resistance: 23*S* rRNA nucleotides 2475–2483 and ribosomal protein L16. This suggests that these two ribosomal components are structurally and functionally linked. It turns out that the tRNA anticodon stem-loop appears to bind to the A site of the ribosome at rRNA bases 2465–2485. This finding conforms to the proposed function of L16, which appears to be involved in attracting the aminoacyl stem of the tRNA to the ribosome at its A site. Using your sketch of the translation machinery from Problem 34 along with this information, designate where the proposed antibacterial action of evernimicin is likely to occur.

36. The flow of genetic information from DNA to protein is mediated by messenger RNA. If you introduce short DNA strands (called antisense oligonucleotides) that are complementary to mRNAs, hydrogen bonding may occur and "label" the DNA/RNA hybrid for ribonuclease-H degradation of the RNA. Lloyd and others (2001. *Nuc. Acids Res.* 29:3664–73) compared the effect of different length antisense oligonucleotides upon ribonuclease-H–mediated degradation of tumor necrosis factor (*TNFα*) mRNA. *TNFα* exhibits antitumor and proinflammatory activities. The graph below indicates the efficacy of various-sized antisense oligonucleotides in causing ribonuclease-H cleavage. (a) Describe how antisense oligonucleotides interrupt the flow of genetic information in a cell. (b) What general conclusion is apparent in the graph? (c) What factors other than oligonucleotide length are likely to influence antisense efficacy *in vivo*?

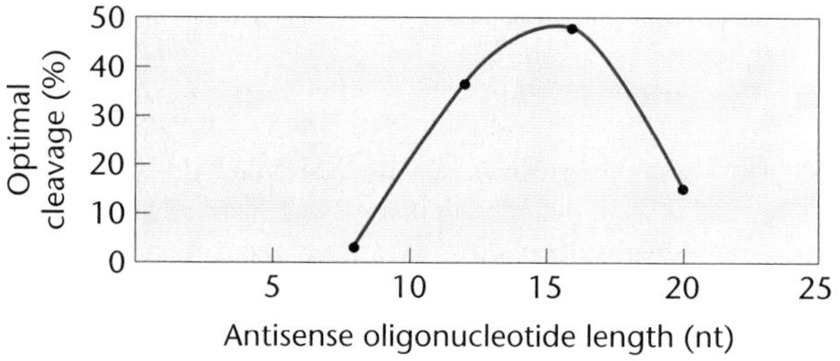

SELECTED READINGS

Anfinsen, C.B. 1973. Principles that govern the folding of protein chains. *Science* 181:223–30.

Bartholome, K. 1979. Genetics and biochemistry of phenylketonuria—Present state. *Hum. Genet.* 51:241–45.

Beadle, G.W., and Tatum, E.L. 1941. Genetic control of biochemical reactions in *Neurospora*. *Proc. Natl. Acad. Sci. USA* 27:499–506.

Beet, E.A. 1949. The genetics of the sickle-cell trait in a Bantu tribe. *Ann. Eugenics* 14:279–84.

Brenner, S. 1955. Tryptophan biosynthesis in *Salmonella typhimurium*. *Proc. Natl. Acad. Sci. USA* 41:862–63.

Doolittle, R.F. 1985. Proteins. *Sci. Am.* (Oct.) 253:88–99.

Frank, J. 1998. How the ribosome works. *Amer. Scient.* 86:428–39.

Garrod, A.E. 1902. The incidence of alkaptonuria: A study in chemical individuality. *Lancet* 2:1616–20.

______. 1909. *Inborn errors of metabolism*. London: Oxford University Press. (Reprinted 1963, Oxford University Press, London.)

Garrod, S.C. 1989. Family influences on A.E. Garrod's thinking. *J. Inher. Metab. Dis.* 12:2–8.

Ingram, V.M. 1957. Gene mutations in human hemoglobin: The chemical difference between normal and sickle-cell hemoglobin. *Nature* 180:326–28.

Koshland, D.E. 1973. Protein shape and control. *Sci. Am.* (Oct.) 229:52–64.

Lake, J.A. 1981. The ribosome. *Sci. Am.* (Aug.) 245:84–97.

Maniatis, T., et al. 1980. The molecular genetics of human hemoglobins. *Annu. Rev. Genet.* 14:145–78.

Murayama, M. 1966. Molecular mechanism of red cell sickling. *Science* 153:145–49.

Neel, J.V. 1949. The inheritance of sickle-cell anemia. *Science* 110:64–66.

Nirenberg, M.W., and Leder, P. 1964. RNA codewords and protein synthesis. *Science* 145:1399–1407.

Nomura, M. 1984. The control of ribosome synthesis. *Sci. Am.* (Jan.) 250:102–14.

Pauling, L., Itano, H.A., Singer, S.J., and Wells, I.C. 1949. Sickle-cell anemia: A molecular disease. *Science* 110:543–48.

Ramakrishnan, V. 2002. Ribosome structure and the mechanism of translation. *Cell* 108:557–72.

Rich, A., and Houkim, S. 1978. The three-dimensional structure of transfer RNA. *Sci. Am.* (Jan.) 238:52–62.

Rich, A., Warner, J.R., and Goodman, H.M. 1963. The structure and function of polyribosomes. *Cold Spring Harbor Symp. Quant. Biol.* 28:269–85.

Richards, F.M. 1991. The protein-folding problem. *Sci. Am.* (Jan.) 264:54–63.

Ross, J. 1989. The turnover of mRNA. *Sci. Am.* (Apr.) 260:48–55.

Rould, M.A., et al. 1989. Structure of *E. coli* glutaminyl-tRNA synthetase complexed with $tRNA^{gln}$ and ATP at 2.8 resolution. *Science* 246:1135–42.

Scott-Moncrieff, R. 1936. A biochemical survey of some Mendelian factors for flower colour. *J. Genet.* 32:117–70.

Scriver, C.R., and Clow, C.L. 1980. Phenylketonuria and other phenylalanine hydroxylation mutants in man. *Annu. Rev. Genet.* 14:179–202.

Srb, A.M., and Horowitz, N.H. 1944. The ornithine cycle in *Neurospora* and its genetic control. *J. Biol. Chem.* 154:129–39.

Warner, J., and Rich, A. 1964. The number of soluble RNA molecules on reticulocyte polyribosomes. *Proc. Natl. Acad. Sci. USA* 51:1134–41.

Wimberly, B.T., et al. 2000. Structure of the 30*S* ribosomal subunit. *Nature* 407:327–33.

Yanofsky, C., Drapeau, G., Guest, J., and Carlton, B. 1967. The complete amino acid sequence of the tryptophan synthetase A protein and its colinear relationship with the genetic map of the *A* gene. *Proc. Natl. Acad. Sci. USA* 57:296–98.

Yusupov, M.M., et.al. 2001. Crystal structure of the ribosome at 5.5A resolution. *Science* 292:883–96.

Gene Mutation, DNA Repair, and Transposition

Mutant erythrocytes derived from an individual with sickle cell anemia.

Chapter Concepts

- Mutations are the source of genetic variation and the basis for natural selection. They are also the source of genetic damage that contributes to cell death, genetic diseases, and cancer.
- Mutations have a wide range of effects on organisms depending upon the type and location of the nucleotide change within the gene and the genome.
- Mutations can occur spontaneously as a result of natural biological and chemical processes, or can be induced by external factors, such as chemicals or radiation.
- The rates of spontaneous mutation vary between organisms and between genes within an organism.
- Organisms invoke a number of DNA repair mechanisms to counteract mutations. These mechanisms range from proofreading of replication errors to base excision and homologous recombination repair.
- Mutations in genes whose products control DNA repair lead to genome hypermutability and human DNA repair diseases, such as xeroderma pigmentosum.
- Transposable elements create mutations by moving into and out of chromosomes. They induce mutations within coding regions and in gene regulatory regions, and cause chromosome breaks.

The ability of DNA molecules to store, replicate, transmit, and decode information is the basis of genetic function. But equally important is the capacity of DNA to make mistakes. Without the variation that arises from changes in DNA sequences, there would be no phenotypic variability, no adaptation to environmental changes, and no evolution. Gene mutations are the source of most new alleles and are the origin of genetic variation within populations. On the downside, they are also the source of genetic changes that can lead to cell death, genetic diseases, and cancer.

Mutations provide the basis for genetic analysis. The phenotypic variability resulting from mutations allows geneticists to identify and study the genes responsible for the modified trait. In genetic investigations, mutations act as identifying "markers" for genes so that they can be followed during their transmission from parents to offspring. Without phenotypic variability, genetic analysis would be impossible. For example, if all pea plants displayed a uniform phenotype, Mendel would have had no foundation for his research. Certain organisms lend themselves to the induction of mutations that are readily detected and studied. These model organisms include viruses, bacteria, fungi, fruit flies, certain plants, nematodes, and mice. In Chapter 21, we deal at length with how geneticists carry out mutational analysis in these model organisms to dissect gene function.

In Chapter 8, we examined mutations in large regions of chromosomes—chromosomal mutations. In this chapter, we will explore mutations primarily at the level of changes in the base-pair sequence of DNA within individual genes—**gene mutations**. We will also describe how the cell defends itself from mutations using various mechanisms of **DNA repair**. The chapter will conclude with a discussion of **transposable elements**, sometimes called "jumping genes" or **transposons**, which can disrupt gene function and carry genetic information within and between DNA molecules.

15.1 Mutations Are Classified in Various Ways

A mutation can be defined as an alteration in DNA sequence. Any base-pair change in any part of a DNA molecule can be considered a mutation. A mutation may comprise a single base pair substitution, a deletion or insertion of one or more base pairs, or may constitute a major alteration in the structure of a chromosome.

Mutations may or may not occur within regions of a gene that code for protein or in regions outside the gene that affect how the gene is expressed. (See Chapters 13, 14, 16, and 17 for descriptions of gene expression and gene regulation.) Because of this, mutations may or may not bring about a detectable change in phenotype. The extent to which a mutation changes the characteristics of an organism depends upon where the mutation occurs and the degree to which the mutation alters the gene.

Mutations can occur within somatic cells or within germ cells. Those that occur in germ cells are heritable and are the basis for genetic diversity and evolution, as well as genetic diseases. Those that occur in somatic cells may lead to localized cell death, altered cellular function, or tumors.

Because of the wide range of types and effects of mutations, geneticists classify mutations according to several different schemes. These organizational schemes are not mutually exclusive. In this section, we outline some of the ways in which gene mutations are classified.

Spontaneous, Induced, and Adaptive Mutations

All mutations are either spontaneous or induced, although these two categories overlap to some degree. **Spontaneous mutations** are those that happen naturally. No specific agents are associated with their occurrence, and they are generally assumed to be random changes in the nucleotide sequences of genes. Most such mutations are linked to normal biological or chemical processes in the organism that alter the structure of nitrogenous bases. Often, spontaneous mutations occur during the enzymatic process of DNA replication, an idea that we will discuss later in this chapter. Once an error is present in the genetic code, it may be reflected in the amino acid composition of the specified protein. If the amino acid change is present in a part of the protein critical to its structure or biochemical activity, a phenotypic change can result.

In contrast to spontaneous mutations, mutations that result from the influence of an extraneous factor are considered to be **induced mutations**. Induced mutations may be the result of either natural or artificial agents. For example, radiation from cosmic and mineral sources and ultraviolet radiation from the sun are energy sources to which most organisms are exposed and, as such, may be factors that cause induced mutations. The earliest demonstration of the artificial induction of mutations occurred in 1927, when Hermann J. Muller reported that X rays could cause mutations in *Drosophila*. In 1928, Lewis J. Stadler reported that X rays had the same effect on barley. In addition to various forms of radiation, numerous natural and man-made chemical agents are also mutagenic, as we will see later in the chapter.

The concept of **adaptive mutation** revolves around the controversial idea that organisms may "select" or "direct" the nature of gene mutation in order to adapt to a particular environmental pressure. In 1943, Salvador Luria and Max Delbrück presented the first direct evidence that mutations are not adaptive, but occur spontaneously. This experiment marked the beginning of modern bacterial genetic study. Their experiment, known as the **Luria–Delbrück fluctuation test**, is an example of exquisite analytical and theoretical work.

Luria and Delbrück carried out their experiments with the *E. coli*–T1 system. The T1 bacteriophage is a bacterial virus that infects *E. coli* cells, and lyses the infected bacteria. A similar bacteriophage, T4, is described in Chapter 6. Luria and Delbrück grew many small individual liquid cultures of phage-sensitive *E. coli* and then added numerous aliquots of each culture to petri dishes of agar medium containing T1 bacteriophages. To obtain precise quantitative data, they also determined the total number of bacteria added to each plate prior to incubation. Following incubation, each plate was scored for the number of phage-resistant bacterial colonies that grew in the presence of the bacteriophage. This was easy to ascertain because only mutant cells were not lysed and thus survived to be counted.

The experimental rationale for distinguishing between the two hypotheses (mutations occur spontaneously versus mutations occur as a result of adaptation) was as follows.

Hypothesis 1: Adaptive Mutation. In this scenario, every bacterium has a small, but constant, probability of acquiring T1 resistance as a result of contact with the phages in the petri dish. In this case, the bacteria are mutating to resistance as a result of their incubation with bacteriophage. Therefore, the number of resistant cells will depend only on the number of bacteria and phages added to each plate. The final results should be independent of all other experimental conditions. Therefore, the adaptation hypothesis predicts that if a constant number of bacteria and phages is present on each plate, and if the incubation time is constant, there should be little fluctuation in the number of resistant cells from plate to plate and from experiment to experiment.

Hypothesis 2: Spontaneous Mutation. On the other hand, if resistance is acquired as a result of mutations that occur randomly, without the stimulus of incubation with bacteriophage, resistance will occur at a low rate during the incubation in liquid medium *prior to plating*—that is, before any contact with the phage occurs. When mutations occur *early* during incubation, the subsequent reproduction of the mutant bacteria will produce a relatively large number of resistant cells. When mutations occur *later* during incubation, far fewer resistant cells will be produced. The random mutation hypothesis therefore predicts that the number of resistant cells will fluctuate significantly from experiment to experiment, and from tube to tube, reflecting the varying times at which most spontaneous mutations occurred in liquid culture.

Table 15.1 shows a representative set of data from the Luria–Delbrück experiments. The middle column shows the number of mutants recovered from a series of aliquots derived from one large individual liquid culture. In the large culture, the mutants are evenly distributed because the culture is constantly mixed. As a result, these data serve as a control because the number in each aliquot should be nearly identical. As predicted, little fluctuation is observed. In contrast, the right-hand column shows the number of resistant mutants recovered from each of 10 independently incubated liquid cultures. The amount of fluctuation in the data will support only one of the two alternative hypotheses. For this reason, the experiment has been designated the **fluctuation test**. Fluctuation is measured by the amount of variance, a statistical calculation.

As seen in the right-hand column, a great fluctuation *was* observed among independently incubated cultures in the Luria–Delbrück experiment, thus supporting the hypothesis that mutations arise randomly, even in the absence of selective pressure, and are inherited in a stable fashion.

Although the concept of spontaneous mutation in viruses, bacteria, and higher organisms has been accepted for some time, the possibility that organisms might also be capable of selecting a specific set of mutations as a result of environmental pressures has long intrigued geneticists. Some recent and controversial research has suggested that under some stressful nutritional conditions such as starvation, bacteria may be capable of activating mechanisms that create a hypermutable state in genes that would, when mutated, enhance survival. The conclusions from these studies are still a source of debate, but keep alive the interest in the possibility of adaptive mutation.

TABLE 15.1 THE LURIA–DELBRÜCK EXPERIMENT DEMONSTRATING THAT SPONTANEOUS MUTATIONS ARE THE SOURCE OF PHAGE-RESISTANT BACTERIA

	Number of T1-Resistant Bacteria	
Sample No.	Same Culture (Control)	Different Cultures
1	14	6
2	15	5
3	13	10
4	21	8
5	15	24
6	14	13
7	26	165
8	16	15
9	20	6
10	13	10
Mean	16.7	26.2
Variance	15.0	2178.0

Source: After Luria and Delbrück (1943).

Classification Based on Location of Mutation

Mutations may be classified according to the cell type or chromosomal locations in which they occur. **Somatic mutations** may occur in any cell in the body except germ cells. **Germ-line mutations** occur in gametes. **Autosomal mutations** occur within genes located on the autosomes, whereas **X-linked mutations** occur within genes located on the X chromosome.

Mutations arising in somatic cells are not transmitted to future generations. When a **recessive autosomal mutation** occurs in a somatic cell of a diploid organism, it is unlikely to result in a detectable phenotype. The expression of most such mutations is likely to be masked by the wild-type allele. Somatic mutations will have a greater impact if they are dominant or, in males, if they are X-linked, since such mutations are most likely to be immediately expressed. Similarly, the impact of dominant or X-linked somatic mutations will be more noticeable if they occur early in development, when a small number of undifferentiated cells replicate to give rise to several differentiated tissues or organs. Dominant mutations that occur in cells of adult tissues are often masked by the thousands upon thousands of nonmutant cells in the same tissue that perform the normal function.

Mutations in gametes are of greater significance because they are transmitted to offspring as part of the germ line. They have the potential of being expressed in all cells of an offspring. Inherited **dominant autosomal mutations** will be expressed phenotypically in the first generation. **X-linked**

recessive mutations arising in the gametes of a homogametic female may be expressed in hemizygous male offspring. This will occur provided that the male offspring receives the affected X chromosome. Because of heterozygosity, the occurrence of an autosomal recessive mutation in the gametes of either males or females (even one resulting in a lethal allele) may go unnoticed for many generations, until the resultant allele has become widespread in the population. Usually, the new allele will become evident only when a chance mating brings two copies of it together into the homozygous condition. In some cases a recessive mutant allele can bring about a detectable phenotype when in a heterozygous or hemizygous state. This occurs in a situation known as **haploinsufficiency**, in which the remaining wild-type allele cannot synthesize sufficient levels of normal gene product to carry out the gene product's function. Some human diseases are the result of haploinsufficiency in genes that encode transcription factors.

Classification Based on Type of Molecular Change

Geneticists often classify gene mutations in terms of the nucleotide changes that create the mutation. A change of one base pair to another in a DNA molecule is known as a **point mutation**, or **base substitution** (Figure 15–1). A change of one nucleotide of a triplet within a protein-coding portion of a gene may result in the creation of a new triplet that codes for a different amino acid in the protein product. If this occurs, the mutation is known as a **missense mutation**. A second possible outcome is that the triplet will be changed into a stop codon, resulting in the termination of translation of the protein. This is known as a **nonsense mutation**. Nonsense mutations are also discussed in Chapter 13. If the point mutation alters a codon but does not result in a change in the amino acid at that position in the protein (due to degeneracy of the genetic code), it can be considered a **silent mutation**.

You will often see two other terms used to describe base substitutions. If a pyrimidine replaces a pyrimidine or a purine replaces a purine, a **transition** has occurred. If a purine and a pyrimidine are interchanged, a **transversion** has occurred.

Another type of change is the insertion or deletion of one or more nucleotides at any point within the gene. As illustrated in Figure 15–1, the loss or addition of a single letter causes all of the subsequent three-letter words to be changed. These are called **frameshift mutations** because the frame of triplet reading during translation is altered. Frameshift mutations are also discussed in Chapter 13. A frameshift mutation will occur when any number of bases are added or deleted, except multiples of three, which would reestablish the initial frame of reading. The analogy in Figure 15–1 demonstrates that insertions and deletions have the potential to change all the subsequent triplets in a gene. It is possible that one of the many altered triplets will be UAA, UAG, or UGA, the termination codons. When one of these triplets is encountered during translation, polypeptide synthesis is terminated at that point. Obviously, the results of frameshift mutations can be very severe.

Classification Based on Phenotypic Effects

Depending upon their type and location, mutations can have a wide range of phenotypic effects, from silent mutations to dominant lethals.

As discussed in Chapter 4, a **loss-of-function mutation** is one that eliminates the function of the gene product. Any type of mutation, from a point mutation to deletion of the entire gene, may lead to a loss of function. These mutations are also known as **null mutations** or **gene knockouts**. It is possible for a loss-of-function mutation to be either dominant or recessive. A dominant loss-of-function mutation may result from the presence of a defective protein product that binds to, or inhibits the action of, the normal gene product, which is also present in the same organism. A **gain-of-function** mutation results in a gene product with a new function. This may be due to a change in the amino acid sequence of the protein that confers a new activity, or it may result from a mutation in the regulatory region of the gene, leading to expression of the gene at an abnormal level, time, or place. Most gain-of-function mutations are dominant.

The most easily observed mutations are those affecting a **morphological trait**. These mutations are also known as visible mutations and are recognized by their ability to alter a normal or wild-type phenotype. For example, all of Mendel's pea characters and many genetic variations encountered in *Drosophila* fit this designation, since they cause obvious changes to the morphology of the organism.

Web Tutorial 15.1
Mutations at the DNA Level

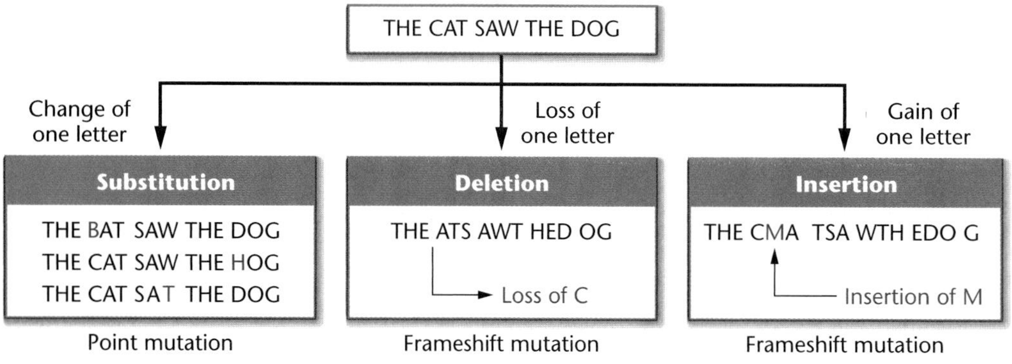

FIGURE 15–1 Analogy of the effects of substitution, deletion, and insertion of one letter in a sentence composed of three-letter words, demonstrating point and frameshift mutations.

A second broad category of mutations includes those that exhibit **nutritional** or **biochemical** effects. In bacteria and fungi, a typical nutritional mutation results in a loss of ability to synthesize an amino acid or vitamin. In humans, sickle-cell anemia and hemophilia are examples of biochemical mutations. While such mutations in these organisms do not always affect morphological characters, they can have an effect on the well-being and survival of the affected individual.

A third category consists of mutations that affect behavior patterns of an organism. For example, the mating behavior or circadian rhythms of animals can be altered. The primary effect of **behavioral mutations** is often difficult to analyze. For example, the mating behavior of a fruit fly may be impaired if it cannot beat its wings. However, the defect may be in the flight muscles, the nerves leading to them, or the brain, where the nerve impulses that initiate wing movements originate.

Still another type of mutation may affect the regulation of genes. For example, as we will see in the *lac* operon discussed in Chapter 16, a regulatory gene can produce a product that controls the transcription of other genes. In other instances, a region of DNA either close to, or far away from, a gene may also modulate its activity. In either case, a mutation in a regulatory gene or a gene control region can disrupt normal regulatory processes and permanently activate or inactivate a gene. Our knowledge of genetic regulation has been dependent on the study of such **regulatory mutations**.

It is also possible that a mutation may interrupt a process that is essential to the survival of the organism. In this case, it is referred to as a **lethal mutation**. For example, a mutant bacterium that has lost the ability to synthesize an essential amino acid will cease to grow and eventually will die when placed in a medium lacking that amino acid. Various inherited human biochemical disorders are also good examples of lethal mutations. For example, Tay–Sachs disease and Huntington disease are caused by mutations that result in lethality, but at different points in the life cycle of humans.

Another interesting aspect of mutations is when their expression depends on the environment in which the organism finds itself. Such mutations are called **conditional mutations**, because the mutation is present in the genome of an organism, but can be detected only under certain conditions. Among the best examples of conditional mutations are **temperature-sensitive mutations**. At a "permissive" temperature, a mutant gene product functions normally, but loses its function at a different, "restrictive" temperature. Therefore, when the organism is shifted from the permissive to the restrictive temperature, the impact of the mutation becomes apparent. Temperature sensitive mutations are also discussed in Chapter 4 (see Figure 4–18). The study of conditional mutations has been important in experimental genetics, particularly in understanding the function of genes essential to cell division. The use of temperature-sensitive mutants in dissecting cell cycle genes is discussed in Chapter 21.

HOW DO WE KNOW?

How do we know that mutations occur spontaneously, in the absence of selective pressure?

15.2 The Spontaneous Mutation Rate Varies Greatly among Organisms

Examination of the spontaneous **mutation rate** in a variety of organisms reveals some interesting points (Table 15.2). First, the rate of spontaneous mutation is exceedingly low for all organisms studied. Second, the rate varies considerably between different organisms. Third, even within the same species, the spontaneous mutation rate varies from gene to gene.

Viral and bacterial genes undergo spontaneous mutation on an average of about 1 in 100 million (10^{-8}) cell divisions. *Neurospora* exhibits a similar rate, but maize, *Drosophila*, and humans demonstrate a rate several orders of magnitude higher. The genes studied in these groups average between 1/1,000,000 and 1/100,000 (10^{-6} and 10^{-5}) mutations per gamete formed. Mouse genes are another order of magnitude higher in their spontaneous mutation rate—1/100,000 to 1/10,000 (10^{-5} to 10^{-4}). It is not clear why such a large variation occurs in mutation rates. The variation between organisms might reflect the relative efficiency of their DNA proofreading and repair systems. We will discuss these systems later in the chapter.

Deleterious Mutations in Humans

The vast majority of mutations are likely to occur in the large portions of the genome that do not contain genes. These are considered to be **neutral mutations**, because they do not usually affect gene products. While this information is of great interest to geneticists, perhaps a more significant piece of information regarding spontaneous mutations is the rate at which those mutations that are deleterious occur. Of greatest interest is the rate of occurrence of these mutations in our own species. For decades, the rate has been difficult to determine at the nucleotide level. However, recent molecular techniques, including DNA sequencing, have allowed accurate estimates to be made. The rate of deleterious mutations in humans is surprisingly high, at least 1.6 deleterious genetic changes per individual per generation.

James Crow clearly and insightfully discusses the possible consequences of deleterious mutations to our species and the methods used to estimate deleterious mutations in a short article that was published in the journal *Nature* in 1999. Dr. Crow is one of the foremost researchers and pioneers in mutation research. His article "The Odds of Losing at Genetic Roulette" is reprinted on page 367.

HOW DO WE KNOW?

How do geneticists know what the deleterious mutation rates are for a particular organism or a particular gene within an organism?

TABLE 15.2 RATES OF SPONTANEOUS MUTATIONS AT VARIOUS LOCI IN DIFFERENT ORGANISMS

Organism	Character	Gene	Rate	Units
Bacteriophage T2	Lysis inhibition	$r \rightarrow r^+$	1×10^{-8}	Per gene replication
	Host range	$h^+ \rightarrow h$	3×10^{-9}	
	Lactose fermentation	$lac^- \rightarrow lac^+$	2×10^{-7}	
	Lactose fermentation	$lac^+ \rightarrow lac^-$	2×10^{-6}	
	Phage T1 resistance	$Tl\text{-}s \rightarrow Tl\text{-}r$	2×10^{-8}	
	Histidine requirement	$his^+ \rightarrow his^-$	2×10^{-6}	
	Histidine independence	$his^- \rightarrow his^+$	4×10^{-8}	
E. coli	Streptomycin dependence	$str\text{-}s \rightarrow str\text{-}d$	1×10^{-9}	Per cell division
	Streptomycin sensitivity	$str\text{-}d \rightarrow str\text{-}s$	1×10^{-8}	
	Radiation resistance	$rad\text{-}s \rightarrow rad\text{-}r$	1×10^{-5}	
	Leucine independence	$leu^- \rightarrow leu^+$	7×10^{-10}	
	Arginine independence	$arg^- \rightarrow arg^+$	4×10^{-9}	
	Tryptophan independence	$trp^- \rightarrow trp^+$	6×10^{-8}	
Salmonella typhimurium	Tryptophan independence	$trp^- \rightarrow trp^+$	5×10^{-8}	Per cell division
Diplococcus pneumoniae	Penicillin resistance	$pen^s \rightarrow pen^r$	1×10^{-7}	Per cell division
Chlamydomonas reinhardi	Streptomycin sensitivity	$str^r \rightarrow str^s$	1×10^{-6}	Per cell division
Neurospora crassa	Inositol requirement	$inos^+ \rightarrow inos^-$	8×10^{-8}	Mutant frequency among asexual spores
	Adenine independence	$ade^- \rightarrow ade^+$	2×10^{-8}	
Zea mays	Shrunken seeds	$sh^+ \rightarrow sh^-$	1×10^{-6}	Per gamete per generation
	Purple	$pr^+ \rightarrow pr^-$	1×10^{-5}	
	Colorless	$c^+ \rightarrow c^-$	2×10^{-6}	
	Sugary	$su^+ \rightarrow su^-$	2×10^{-6}	
Drosophila melanogaster	Yellow body	$y^+ \rightarrow y$	1.2×10^{-6}	Per gamete per generation
	White eye	$w^+ \rightarrow w$	4×10^{-5}	
	Brown eye	$bw^+ \rightarrow bw$	3×10^{-5}	
	Ebony body	$e^+ \rightarrow e$	2×10^{-5}	
	Eyeless	$ey^+ \rightarrow ey$	6×10^{-5}	
Mus musculus	Piebald coat	$s^+ \rightarrow s$	3×10^{-5}	Per gamete per generation
	Dilute coat color	$d^+ \rightarrow d$	3×10^{-5}	
	Brown coat	$b^+ \rightarrow b$	8.5×10^{-4}	
	Pink eye	$p^+ \rightarrow p$	8.5×10^{-4}	
Homo sapiens	Hemophilia	$h^+ \rightarrow h$	2×10^{-5}	Per gamete per generation
	Huntington disease	$Hu^+ \rightarrow Hu$	5×10^{-6}	
	Retinoblastoma	$R^+ \rightarrow R$	2×10^{-5}	
	Epiloia	$Ep^+ \rightarrow Ep$	1×10^{-5}	
	Aniridia	$An^+ \rightarrow An$	5×10^{-6}	
	Achondroplasia	$A^+ \rightarrow A$	5×10^{-5}	

15.3 Spontaneous Mutations Arise from Replication Errors and Base Modifications

In this section, we will outline the processes that lead to spontaneous mutations. It is useful to keep in mind, though, that many of the base modifications that occur during spontaneous mutagenesis also occur, at a higher rate, during induced mutagenesis.

DNA Replication Errors

As we learned in Chapter 11, the process of DNA replication is imperfect. Occasionally, DNA polymerases insert incorrect nucleotides in a replicated strand of DNA. Although DNA polymerases can correct most of these replication errors using their inherent 3′ to 5′ exonuclease proofreading capacity, misincorporated nucleotides may persist after replication. If these errors are not detected and corrected by DNA repair mechanisms (discussed later in this chapter), they may lead to mutations. The fact that bases can take several forms, known as **tautomers**, also increases the chance of mispairing during DNA replication. Tautomeric shifts are discussed below. Replication errors due to mispairing predominantly lead to point mutations.

Replication Slippage

In addition to point mutations, DNA replication can lead to the introduction of small insertions or deletions. These mutations can occur when one strand of the DNA template loops out and

The Odds of Losing at Genetic Roulette

The rate at which deleterious mutations occur in a genome is clearly a quantity of interest—not least when the genome is our own. It becomes particularly important if, as some have argued[1], the rate in some species is high, at several new mutations each generation. Yet the deleterious mutation rate has been notoriously difficult to measure, and no convincing estimates exist for any vertebrate. On page 344 of this issue[2], Eyre-Walker and Keightley give the first such estimates for ourselves, chimpanzees, and gorillas.

Getting an estimate of the total mutation rate is relatively simple. Neutral mutations—those which neither enhance nor impair the organism carrying them—accumulate through the generations at a rate equal to the mutation rate[3]. The mutation rate can be determined by the rate of change of presumed neutral regions: areas of the genome, such as introns and pseudogenes, which are not translated into proteins. For mammals, these rates, extrapolated to the whole genome, lead to enormous numbers, on the order of 100 new mutations per individual[1]. Of course these can't all be deleterious, but no one knows what the proportion is. This proportion could be determined in principle by comparing the slower rate of evolutionary change of the genome as a whole with the faster rate expected under neutral assumptions; but this involves statistical uncertainties and extensive sequencing[4].

Eyre-Walker and Keightley[2] have made the analysis feasible by concentrating on protein-coding regions. They measured the amino-acid changes in 46 proteins in the human ancestral line after its divergence from the chimpanzee. Among 41,471 nucleotides, they found 143 non-synonymous substitutions—mutations where swapping one DNA base for another changes an amino acid, and therefore the final protein made by that gene. If these had evolved at the neutral rate, 231 would be expected. The difference, 88 (38 percent), is an estimate of the number of deleterious mutations that have been eliminated by natural selection and have therefore made no contribution to contemporary populations.

Translating these numbers into mutation rates gave a total rate of 4.2 mutations per person per generation, and a deleterious rate of 1.6. The rates of chimpanzees and gorillas were very similar, the deleterious rates being 1.7 and 1.2, respectively. The authors took 60,000 as the gene number and 25 years as the generation length. The number 1.6 is probably an underestimate, for various reasons. For instance, mutations outside the coding region are not counted and some of these regions—such as those controlling gene expression—are expected to be subject to natural selection. The gene number may also be an underestimate. If there have been mutations that increase fitness, they would also cause the number of deleterious mutations to be underestimated. A less conservative, and probably more realistic, estimate doubles the value, giving three new deleterious mutations per person per generation.

What's the significance? Every deleterious mutation must eventually be eliminated from the population by premature death or reduced relative reproductive success, a "genetic death"[5]. That implies three genetic deaths per person! Why aren't we extinct? If harmful mutations were eliminated independently, as in an asexual species, it has been estimated that this would lower population fitness to a fraction e^{-3}, or 5 percent, of the mutation-free value[6], leading to the inevitable extinction of species with limited reproductive capacity. A way out is for mutations to be eliminated in bunches. This happens if selection operates such that individuals with the most mutations are preferentially eliminated, for example if harmful mutations interact. But such a process can only work in sexual species, where mutations are shuffled each generation by genetic recombination[1,7]. The existence of a high deleterious mutation rate strengthens the argument that a major advantage of sex is that it is an efficient way to eliminate harmful mutations[1]. It also raises again the possibility of fitness decline or even extinction in rare species from too many harmful mutations[8].

Presumably, we humans have profited in the past by sexual reproduction's ability to reduce the effect of a high mutation rate on fitness. In the recent past the intensity of natural selection has been greatly reduced, especially where a high standard of living means that most infants reach reproductive age. From this it would seem that natural selection will weed out mutations more slowly than they accumulate. This effect may be accentuated by trends for males to start or continue reproducing later in life, because the sperm of older men contains more base-substitution mutations[8]. In a time of rapid environmental improvement, how this genetic decline will affect our health can only be guessed at.

Eyre-Walker and Keightley[2] noticed that the proportion of harmful mutations in the 46 genes in their study is greater in humans than in the equivalent genes of rodents. Their preferred explanation is that slightly deleterious mutations have become fixed in the population, by a process known as random genetic drift, during periods of human history when the breeding population size was low—especially during genetic "bottlenecks." This would increase whatever effect the accumulated mutations are having on current human welfare. Are some of our headaches, stomach upsets, weak eyesight, and other ailments the result of mutation accumulation? Probably, but in our present state of knowledge, we can only speculate.

James F. Crow, Department of Genetics University of Wisconsin Madison, WI 93706

[1]Kondrashov, A.S. *J. Theor. Biol.* 175:583–94 (1995).

[2]Eyre-Walker, A. & Keightley, P.D. *Nature* 397:344–47 (1999).

[3]Kimura, M. *The Neutral Theory of Molecular Evolution* (Cambridge University Press, 1983).

[4]Kondrashov, A.S. & Crow, J.F. *Hum. Mutat.* 2:229–34 (1993).

[5]Muller, H.J. *Am. J. Hum. Genet.* 2:111–76 (1950).

[6]Kimura, M. & Maruyama, T. *Genetics* 54:1337–51 (1966).

[7]Crow, J.F. *Proc. Natl. Acad. Sci. USA* 94:8380–86 (1997).

[8]Lande, R. *Evolution* 48:1460–69 (1994).

becomes displaced during replication, or when DNA polymerase slips or stutters during replication. If a loop occurs in the template strand during replication, DNA polymerase may miss the looped-out nucleotides and a small deletion in the new strand will be introduced. If DNA polymerase repeatedly introduces nucleotides that are not present in the template strand, an insertion of one or more nucleotides will be introduced, creating an unpaired loop on the newly synthesized strand. Insertions and deletions may lead to frameshift mutations or amino acid additions or deletions in the gene product.

Replication slippage can occur anywhere in the DNA, but appears to have a distinct preference for regions containing repeated sequences. Repeat sequences are hot-spots for DNA mutation, and in some cases contribute to hereditary diseases (discussed in Section 15.5). The hypermutability of repeat sequences in noncoding regions of the genome also forms the basis for current methods of forensic DNA analysis. (See Chapter 22.)

Tautomeric Shifts

In 1953, immediately after they had proposed the molecular structure of DNA, Watson and Crick published a paper in which they discussed the genetic implications of this structure. They recognized that the purines and pyrimidines found in DNA could exist in **tautomeric** forms—that is, nitrogenous bases can exist in alternate chemical forms called structural isomers, each differing by only a single proton shift in the molecule. The biologically important tautomers involve the keto–enol forms of thymine and guanine, and the amino–imino forms of cytosine and adenine. Because these shifts change the bonding structure of the molecule, Watson and Crick suggested that **tautomeric shifts** could result in base pair changes or mutations.

The most stable tautomers of nitrogenous bases take part in the standard base-pairings that serve as the basis of the double-helix model of DNA. The less frequently occurring, transient tautomers are capable of hydrogen bonding with noncomplementary bases. However, the pairing is always between a pyrimidine and a purine. Figure 15–2 compares the normal base-pairing relationships with the rare unorthodox pairings. Anomalous T≡G and C=A pairs, among others, may be formed.

A mutation occurs during DNA replication when a rare tautomer in the template strand pairs with a noncomplementary base. In the next round of replication, the "mismatched" members of the base pair are separated, and each becomes the template for its normal complementary base. The end result is a point mutation (Figure 15–3).

Depurination and Deamination

Some of the most common causes of spontaneous mutations are due to DNA base damage. **Depurination** involves the loss of one of the nitrogenous bases in an intact double-helical DNA molecule. Most frequently, such an event involves purines—either guanine or adenine. These bases may be lost if the glycosidic bond linking the 1′-C of the deoxyribose and the number 9 position of the purine ring is broken. This leads to the creation of an **apurinic (AP) site** on one strand of the DNA. Geneticists estimate that thousands of such spontaneous lesions are formed daily in the DNA of mammalian cells in culture. If AP sites are not repaired, there will be

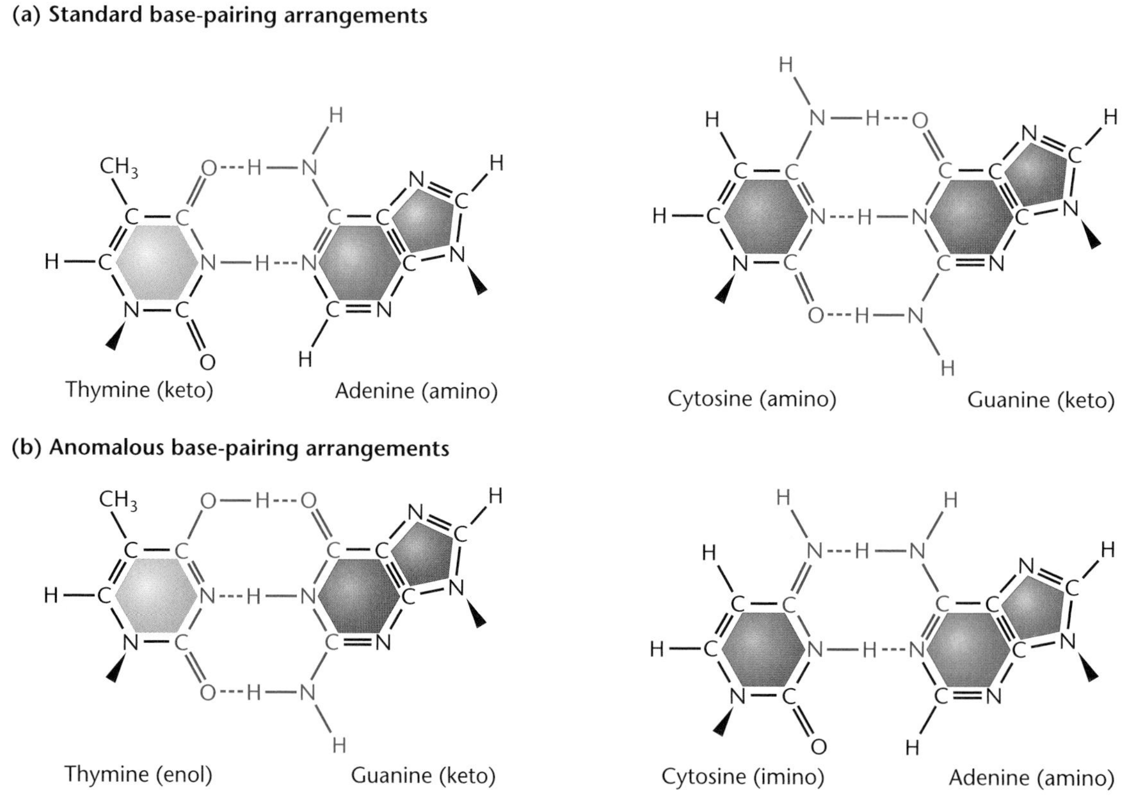

FIGURE 15–2 Standard base-pairing relationships (a), compared with anomalous base-pairing that occurs as a result of tautomeric shifts (b). The long triangle indicates the point at which the base bonds to the pentose sugar.

no base at that position to act as a template during DNA replication. As a result, DNA polymerase may introduce a nucleotide at random at that site.

In the process of **deamination**, an amino group is converted to a keto group in cytosine and adenine (Figure 15–4). In these two cases, cytosine is converted to uracil and adenine is changed to hypoxanthine. The major effect of these changes is to alter the base-pairing specificities of these two bases during DNA replication. For example, cytosine normally pairs with guanine. Following its conversion to uracil, which pairs with adenine, the original G≡C pair is converted to an A=U pair and then, following an additional replication, is converted to an A=T pair. When adenine is deaminated, the original A=T pair is converted to a G≡C pair because hypoxanthine pairs naturally with cytosine. Deamination may occur spontaneously or as a result of treatment with chemical mutagens such as nitrous acid (HNO_2).

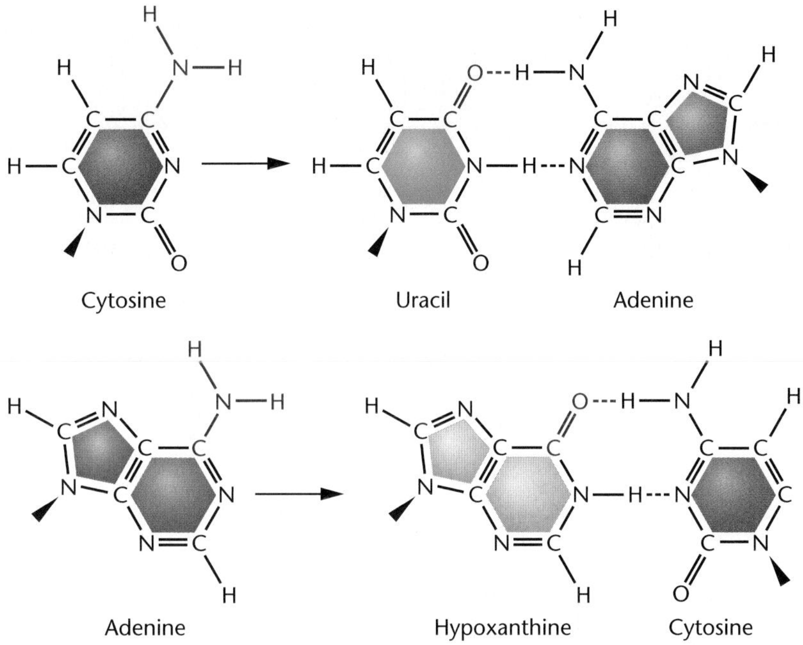

FIGURE 15–4 Deamination of cytosine and adenine, leading to new base pairing and mutation. Cytosine is converted to uracil, which base pairs with adenine. Adenine is converted to hypoxanthine, which base pairs with cytosine.

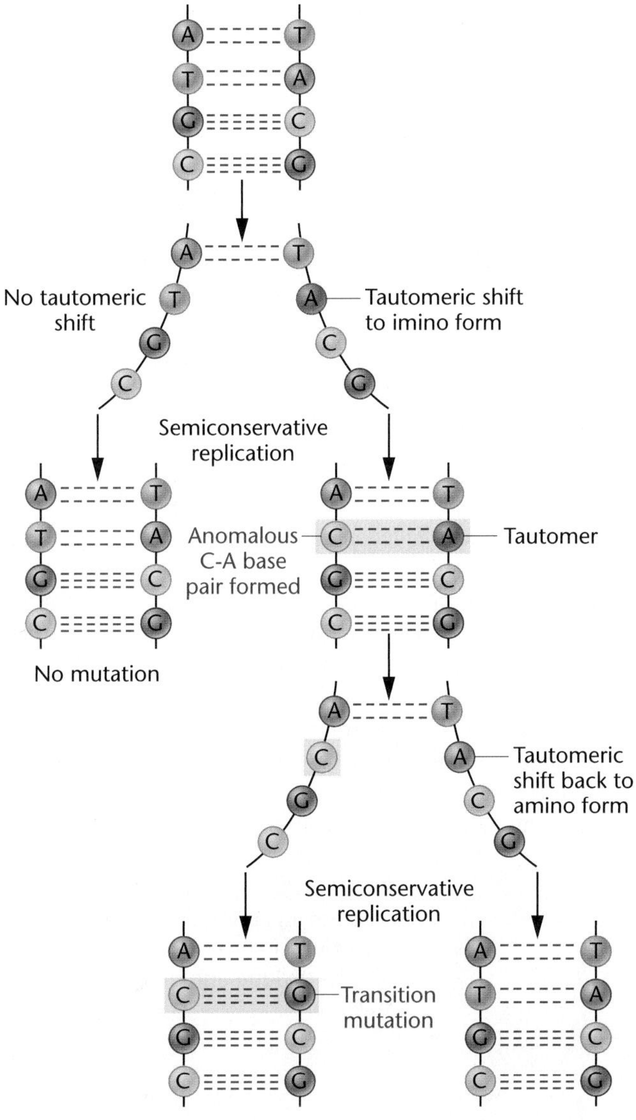

FIGURE 15–3 Formation of a T=A to C≡G transition mutation as a result of a tautomeric shift in adenine.

Oxidative Damage

DNA may also suffer damage from the by-products of normal cellular processes. These by-products include reactive oxygen species (electrophilic oxidants) that are generated during normal aerobic respiration. For example, superoxides ($O_2^{\cdot -}$), hydroxyl radicals (·OH), and hydrogen peroxide (H_2O_2) are created during cellular metabolism and are constant threats to the integrity of DNA. Reactive oxidants, also generated by exposure to high-energy radiation, can produce more than 100 different types of chemical modifications in DNA, including modifications to bases that lead to mispairing during replication.

Transposons

Transposable genetic elements (transposons) are also agents of spontaneous mutations in both eukaryotes and prokaryotes. Transposons will be discussed later in this chapter (Section 15.8).

15.4 Induced Mutations Arise from DNA Damage Caused by Chemicals and Radiation

All cells on Earth are exposed to a plethora of agents that have the potential to damage DNA and cause mutations (**mutagens**). Some of these agents, such as some fungal toxins, cosmic rays, and ultraviolet light, are natural components of our environment. Others, including some industrial pollutants, medical X rays, and chemicals within tobacco smoke, can be considered as unnatural or man-made additions to our

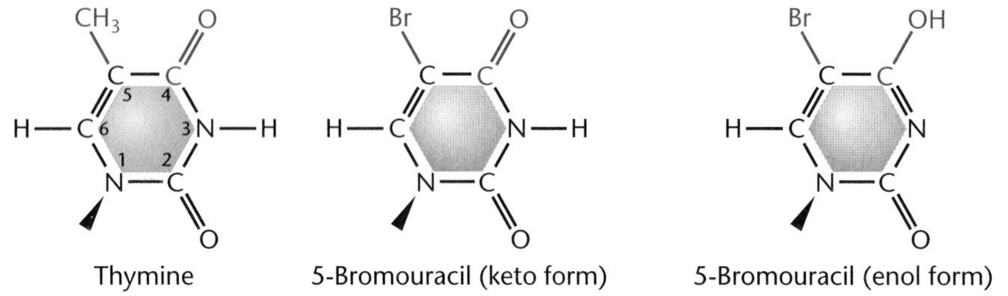

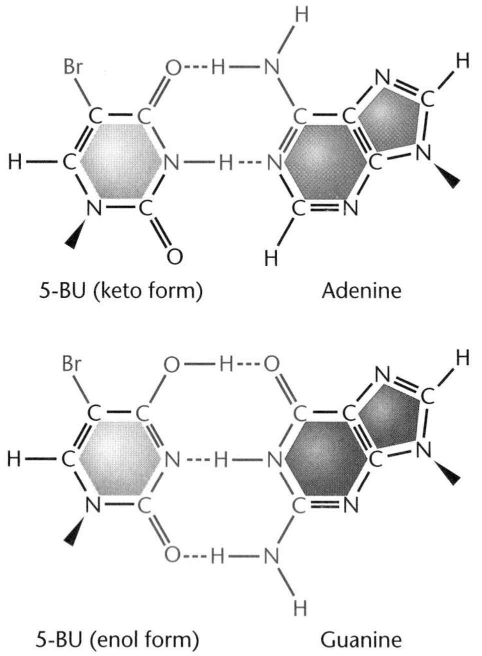

FIGURE 15–5 Similarity of 5-bromouracil (5-BU) structure to thymine structure. In the common keto form, 5-BU base pairs normally with adenine, behaving as an analog. In the rare enol form, it pairs anomalously with guanine.

modern world. On the positive side, geneticists harness some mutagens in order to analyze genes and gene functions. (See Chapter 21.) The mechanisms by which some of these natural and unnatural agents lead to mutations are outlined below.

Base Analogs

These mutagenic chemicals can substitute for purines or pyrimidines during nucleic acid biosynthesis. For example, **5-bromouracil (5-BU)** a derivative of uracil, behaves as a thymine analog and is halogenated at the number 5 position of the pyrimidine ring. If 5-BU is chemically linked to deoxyribose, the nucleoside analog **bromodeoxyuridine (BrdU)** is formed. Figure 15–5 compares the structure of this analog with that of thymine. The presence of the bromine atom in place of the methyl group increases the probability that a tautomeric shift will occur. If 5-BU is incorporated into DNA in place of thymine and a tautomeric shift to the enol form occurs, 5-BU base pairs with guanine. After one round of replication, an A═T to G≡C transition results. Furthermore, the presence of 5-BU within DNA increases the sensitivity of the molecule to ultraviolet (UV) light, which itself is mutagenic, as discussed below.

There are other base analogs that are mutagenic. For example, **2-amino purine (2-AP)** can act as an analog of adenine. In addition to its base-pairing affinity with thymine, 2-AP can also base pair with cytosine, leading to possible transitions from A═T to G≡C following replication.

Alkylating Agents

The sulfur-containing mustard gases, discovered during World War I, were some of the first groups of chemical mutagens identified in chemical warfare studies. Mustard gases are **alkylating agents**—that is, they donate an alkyl group, such as CH_3 or CH_3CH_2, to amino or keto groups in nucleotides. **Ethylmethane sulfonate (EMS)**, for example, alkylates the keto groups in the number 6 position of guanine and in the number 4 position of thymine. As with base analogs, base-pairing affinities are altered and transition mutations result. For example, **6-ethyl guanine** acts as an analog of adenine and pairs with thymine (Figure 15–6). Table 15.3 lists the chemical names and structures of several frequently used alkylating agents known to be mutagenic.

Acridine Dyes and Frameshift Mutations

Chemical mutagens called acridine dyes cause frameshift mutations. As illustrated in Figure 15–1, frameshift mutations result from the addition or removal of one or more base pairs in the polynucleotide sequence of the gene. **Proflavin**, the most widely studied acridine mutagen, and **acridine orange** are illustrated in Figure 15–7. Acridine dyes are about the same dimensions as nitrogenous base pairs and **intercalate**, or wedge, between the purines and pyrimidines of intact DNA. Intercalation introduces contortions in the DNA helix and causes deletions and insertions that create frameshift mutations.

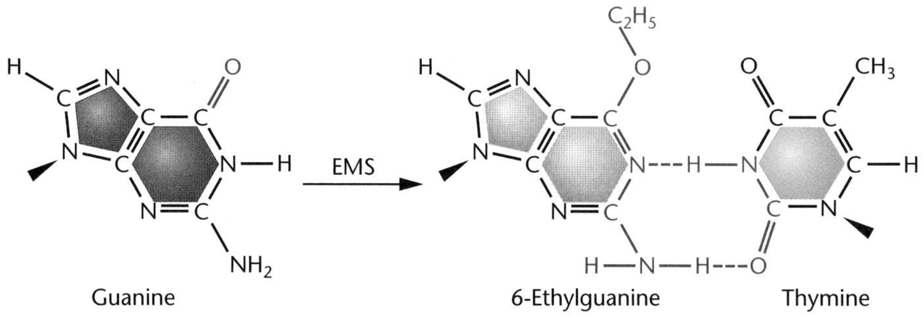

FIGURE 15–6 Conversion of guanine to 6-ethylguanine by the alkylating agent ethylmethane sulfonate (EMS). The 6-ethylguanine base pairs with thymine.

TABLE 15.3 ALKYLATING AGENTS

Common Name or Symbol	Chemical Name	Chemical Structure
Mustard gas (sulfur)	Di-(2-chloroethyl) sulfide	$Cl-CH_2-CH_2-S-CH_2-CH_2-Cl$
EMS	Ethylmethane sulfonate	$CH_3-CH_2-O-\overset{\overset{O}{\Vert}}{\underset{\underset{O}{\Vert}}{S}}-CH_3$
EES	Ethylethane sulfonate	$CH_3-CH_2-O-\overset{\overset{O}{\Vert}}{\underset{\underset{O}{\Vert}}{S}}-CH_2-CH_3$

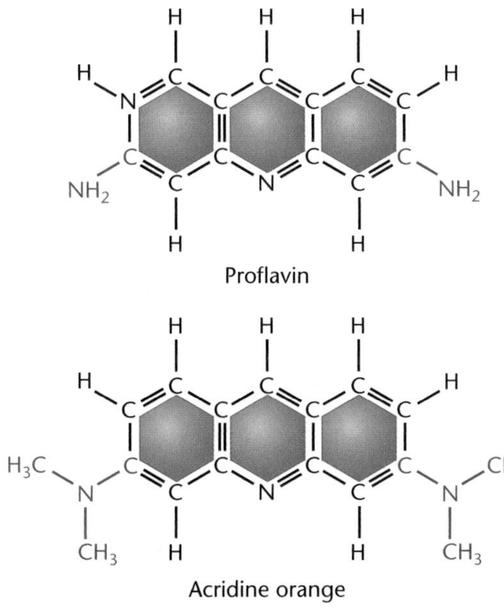

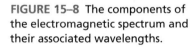

FIGURE 15–7 Chemical structures of proflavin and acridine orange, which intercalate into DNA and cause frameshift mutations.

One model suggests that the resultant frameshift mutations are generated at gaps produced in DNA during replication, repair, or recombination. During these events, there is the possibility of slippage and improper base pairing of one strand with the other. The model suggests that the intercalation of acridine into an improperly base-paired region can extend the existence of these slippage structures. If so, the probability increases that the mispaired configuration will exist when DNA synthesis or rejoining occur, thereby resulting in an addition or deletion of one or more bases in one of the strands.

Ultraviolet Light and Thymine Dimers

All energy on Earth consists of a series of electromagnetic components of varying wavelength (Figure 15–8). Referred to as the **electromagnetic spectrum**, the energy associated with the various components varies inversely with wavelength. Everything longer than, and including, visible light is benign when it interacts with most organic molecules. However, everything of shorter wavelength than visible light, being inherently more energetic, has a disruptive impact on organic molecules, including those comprising living tissue. For example, in Chapter 10, we

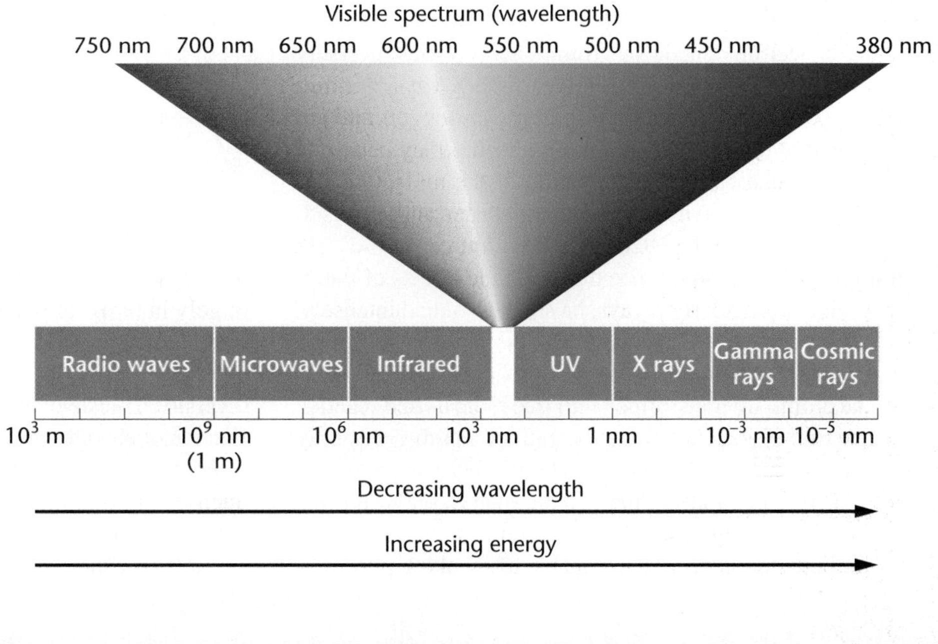

FIGURE 15–8 The components of the electromagnetic spectrum and their associated wavelengths.

Web Tutorial 15.2
DNA Repair: Photoreactivation and Excision

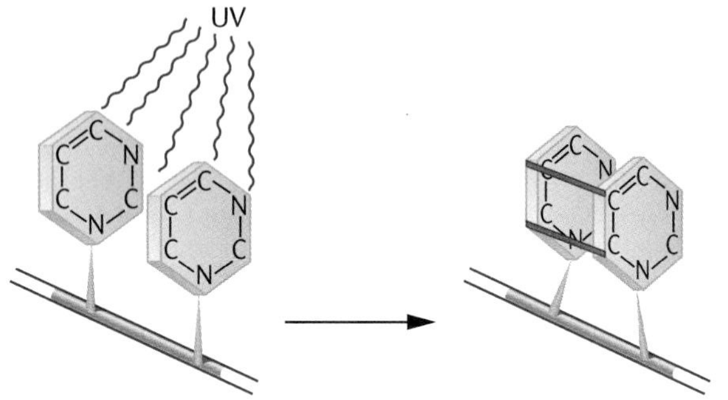

FIGURE 15–9 Induction of a thymine dimer by UV radiation, leading to distortion of the DNA. The covalent crosslinks occur between the atoms of the pyrimidine ring.

emphasized the fact that purines and pyrimidines absorb **ultraviolet (UV) radiation** most intensely at a wavelength of about 260 nm, a property that is useful in the detection and analysis of nucleic acids. In 1934, as a result of studies involving *Drosophila* eggs, it was discovered that UV radiation is mutagenic. The major effect of UV radiation is the creation of **pyrimidine dimers**, particularly those involving two thymine residues (Figure 15–9). While cytosine–cytosine and thymine–cytosine dimers may also be formed, they are less prevalent than thymine dimers. The dimers distort the DNA conformation and inhibit normal replication. As a result, errors can be introduced in the base sequence of DNA during replication. When UV-induced dimerization is extensive, it is responsible (at least in part) for the killing effects of UV radiation on cells.

As we will see later, several mechanisms have evolved to correct UV-induced lesions. We will illustrate the essential nature of such repair mechanisms in humans by discussing the inherited disorder xeroderma pigmentosum, in which any of seven different genes controlling DNA repair are mutated. Individuals with this genetic condition have been described as "children of the night," since they cannot risk exposure to the ultraviolet component of sunlight without the risk of epidermal malignancy.

Ionizing Radiation

Within the electromagnetic spectrum, energy varies inversely with wavelength. As shown in Figure 15–8, **X rays**, **gamma rays**, and **cosmic rays** have even shorter wavelengths and are more energetic than UV radiation. As a result, they penetrate deeply into tissues, causing ionization of the molecules encountered along the way. Hermann J. Muller and Lewis J. Stadler established in the 1920s that these sources of **ionizing radiation** are mutagenic. Since that time, the effects of ionizing radiation, particularly X rays, have been studied intensely.

As X rays penetrate cells, electrons are ejected from the atoms of molecules encountered by the radiation. Thus, stable molecules and atoms are transformed into free radicals and reactive ions. The trail of ions left along the path of a high-energy ray can initiate a variety of chemical reactions. These reactions can directly or indirectly affect the genetic material, altering the purines and pyrimidines in DNA and resulting in point mutations. Ionizing radiation is also capable of breaking phosphodiester bonds, disrupting the integrity of chromosomes, and producing a variety of chromosomal aberrations, such as deletions, translocations, and chromosomal fragmentation.

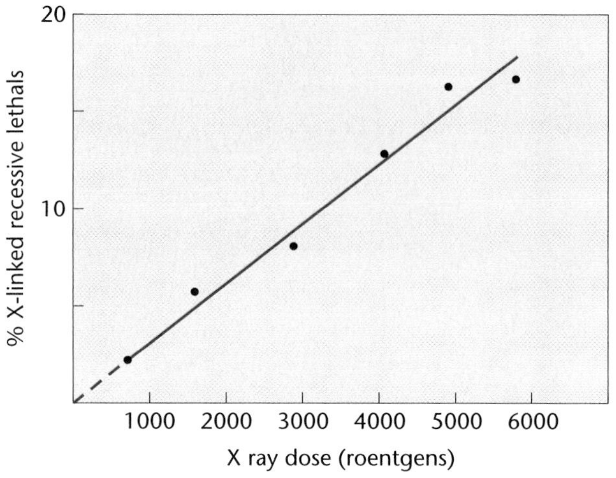

FIGURE 15–10 Plot of the percentage of X-linked recessive mutations induced by increasing doses of X rays. If extrapolated, the graph intersects the zero axis as shown by the dashed line.

Figure 15–10 shows a graph of the percentage of induced X-linked recessive lethal mutations versus the dose of X rays administered. There is a linear relationship between X ray dose and the induction of mutation; for each doubling of the dose, twice as many mutations are induced. Because the line intersects near the zero axis, this graph suggests that even very small doses of radiation are mutagenic.

Now solve this

Problem 15.23 on page 389 asks you to describe ways in which the chemotherapy drug melphalan kills cancer cells, and how these cells can repair the DNA damage caused by this drug.

Hint: To answer this question, review how alkylating agents alter bases in DNA. Think about how changes in base-pairing lead to mutations and the effects that mutations have in replicating cells. In Section 15.8, you will learn about the ways in which cells repair the types of mutations introduced by alkylating agents.

15.5 Genomics and Gene Sequencing Have Enhanced Our Understanding of Mutations in Humans

So far, we have discussed the molecular basis of mutation largely in terms of nucleic acid chemistry. Historically, scientists learned about mutations by analyzing the amino acid sequences of proteins within populations that show substantial diversity. This diversity, which arises during evolution, is a reflection of changes in the triplet codons following substitution, insertion, or deletion of one or more nucleotides in the DNA sequences of protein-coding genes.

As our ability to analyze DNA more directly increases, we are able to examine the actual nucleotide sequence of genes and to gain greater insights into the nature of mutations. Several techniques capable of accurate, rapid sequencing of DNA (see Chapter 19) have greatly extended our knowledge of mutations. In this section, we examine the results of a number of studies that have investigated the actual gene sequences of various mutations that affect human health.

ABO Blood Types

The ABO system is based on a series of antigenic determinants found on erythrocytes and other cells, particularly epithelial cells. As we discussed in Chapter 4, there are three alleles of a single gene that encodes the glycosyltransferase enzyme. The H substance is modified to either the A or B antigen, as a result of the product of the I^A or I^B allele, respectively. Failure to modify the H substance results from the null I^O allele. (See Figure 4–2.)

The glycosyltransferase gene has been sequenced in 14 people with different ABO status. When the DNAs of the I^A and I^B alleles are compared, four consistent nucleotide substitutions are found. It is assumed that the resulting changes in the amino acid sequence of the glycosyltransferase gene product lead to the different modifications of the H substance.

The I^O allele situation is unique and interesting. Individuals homozygous for this allele have type O blood, lack glycosyltransferase activity, and fail to modify the H substance. Analysis of the DNA of this allele shows one consistent change that is unique compared with the sequences of the other alleles—the deletion of a single nucleotide early in the coding sequence, causing a frameshift mutation. A complete messenger RNA is transcribed, but at translation, the reading frame shifts at the point of the deletion and continues out of frame for about 100 nucleotides before a stop codon is encountered. At this point, the polypeptide chain terminates prematurely, resulting in a nonfunctional product.

These findings provide a direct molecular explanation of the ABO allele system and the basis for the biosynthesis of the corresponding antigens. The molecular basis for the antigenic phenotypes is clearly the result of mutations within the gene encoding the glycosyltransferase enzyme.

Muscular Dystrophy

Muscular dystrophies are a group of genetic diseases characterized by progressive muscle weakness and degeneration. There are many types of muscular dystrophy which differ in their severities, onsets and genetic causes. Two related forms of muscular dystrophy—**Duchenne muscular dystrophy (DMD)** and **Becker muscular dystrophy (BMD)**—are recessive, X-linked conditions. DMD is the more severe of the two diseases, with a rapid progression of muscle degeneration and involvement of the heart and lungs. Males with DMD usually lose the ability to walk by the age of 12 and may die in their early 20s. Because the condition is recessive and X-linked, and affected males usually die before they can reproduce, females are rarely affected by the disorder. The incidence of 1 in 5000 live male births makes DMD one of the most common life-shortening hereditary diseases. In contrast, BMD does not involve the heart or lungs and progresses slowly, from adolescence to the age of fifty or more.

The gene responsible for DMD and BMD—the *dystrophin* gene—is unusually large, consisting of about 2.5 million base pairs. In normal (unaffected) individuals, transcription and subsequent processing of the *dystrophin* initial transcript results in a messenger RNA containing only about 14,000 bases (14 kb). It is translated into the protein **dystrophin**, which consists of 3685 amino acids. In most cases of the less severe BMD, the dystrophin protein can be detected in the sarcolemma of smooth, striated, and cardiac myofibers. However, it is rarely detectable in the muscles of DMD patients. These observations led to the hypothesis that mutations causing BMD may not alter the reading frame of the gene, but mutations causing DMD may change the reading frame early in the gene, resulting in premature termination of dystrophin translation. This hypothesis would be consistent with the observed differences in severity of these two forms of the disorder.

In an extensive study performed in 1989, J. T. Den Dunnen and associates analyzed the DNA of 194 patients (160 DMD and 34 BMD). In most cases, the results were consistent with the "reading frame" hypothesis. With few exceptions, DMD mutations change the reading frame of the dystrophin gene, whereas BMD mutations usually do not. Subsequent studies have shown that about two thirds of mutations in the dystrophin gene that lead to DMD and BMD are deletions and duplications. Only one third are point mutations, and of these, the majority are small insertions or deletions. The majority of DMD mutations, both rearrangements and point mutations, lead to premature termination of translation. This, in turn, leads to degradation of the improperly translated dystrophin transcript and low levels of dystrophin protein. In contrast, the majority of BMD gene rearrangements and point mutations may alter the internal sequence of the dystrophin transcript and protein, but do not alter the translation reading frame.

These observations reflect the fact that a mutation caused by a random single-nucleotide substitution is more likely to be tolerated without a devastating phenotypic effect than the addition or loss of nucleotides that may alter the translational reading frame. There are three reasons for this:

1. A nucleotide substitution may not change the encoded amino acid, since the genetic code is degenerate.
2. If an amino acid substitution does result, the change may not be present at a location within the protein that is critical to its function.
3. Even if the altered amino acid is present at a critical region, it may still have little or no effect on the function of the protein. For example, an amino acid might be changed to another with nearly identical chemical properties or to one with very similar physical properties, such as shape.

Trinucleotide Repeats in Fragile X Syndrome, Myotonic Dystrophy, and Huntington Disease

Beginning about 1990, molecular analysis of the genes responsible for a number of inherited human disorders provided a remarkable set of observations. Researchers discovered that some mutant genes contained expansions of **trinucleotide repeat sequences**, usually from fewer than 15 copies in normal individuals to a large number in affected individuals.

For example, the genes responsible for fragile X syndrome, myotonic dystrophy, and Huntington disease contain a specific trinucleotide DNA sequence repeated many times. While these types of repeated sequences are also present in the nonmutant (normal) allele of each gene, the mutations found in individuals with these disorders consist of significant increases in the number of times the trinucleotide is repeated. Table 15.4 summarizes the cases discussed in this section, with particular emphasis on the size of the repeats in normal and diseased individuals.

In Chapter 4 we discussed the onset of the expression of various phenotypes. In several cases, a correlation exists between the number of repeats and the age of manifestation of mutant phenotypes. The greater the number of repeats, the earlier disease onset occurs. Further, in affected individuals, the number of repeats may increase in each subsequent generation. This general phenomenon is known as **genetic anticipation**.

We begin with a short discussion of **fragile X syndrome**, described in detail in Chapter 8. The responsible gene, *FMR-1*, may have several hundred to several thousand copies of the trinucleotide sequence CGG, located in the 5′ untranslated region of the gene. Individuals with up to 54 copies are normal and do not display the mental retardation associated with the syndrome. Individuals with 54 to 230 copies are considered carriers. Although they are normal, their offspring may contain even more copies and express the syndrome. The large regions of CGG repeats in the gene's regulatory region result in loss of expression of the FMRP protein, thought to be an RNA-binding protein affecting brain cell function.

Myotonic dystrophy, or DM (short for its original name, dystrophia myotonica), is the most common form of adult muscular dystrophy, affecting 1 in 8000 individuals. The dominantly inherited disorder is not as severe as DMD and it is highly variable both in symptoms and age of onset. Mild myotonia (atrophy and weakness) of the musculature of the face and extremities is most common. Cataracts, reduced cognitive ability, and cutaneous and intestinal tumors are also part of the syndrome.

The affected gene, *MDPK*, is located on the long arm of chromosome 19. It encodes a serine–threonine protein kinase, MDPK. This protein is the product of 15 exons of the gene, the last of which encodes the 3′-untranslated RNA sequence of the mRNA. It is this sequence that houses the multiple copies of the trinucleotide CTG. Individuals with 5 to 37 copies of the CTG repeat are normal, and the number of copies is stable from generation to generation. Individuals with more than 37 copies exhibit symptoms ranging from mild to severe, with onset occurring anywhere between birth and age 60. Both the severity and onset are directly related to the size of the repeated sequence. Minimally affected patients have up to 150 repeats, while severely affected patients have up to 1500 copies of the CTG triplet. Genetic anticipation is exhibited in the offspring. The mechanism by which these repeated regions cause DM is still uncertain.

Huntington disease (HD) is a fatal neurodegenerative disease, inherited as an autosomal dominant. (Also see Chapter 4.) The gene responsible for HD is located on chromosome 4. The gene contains the trinucleotide CAG sequence, repeated 10 to 35 times in normal individuals. The CAG repeat is located within the coding region of the gene, and encodes a polyglutamine tract. The CAG repeat sequence exists in significantly increased numbers (up to 120) in diseased individuals. Much earlier onset occurs when the number of copies is closer to the upper range. Interestingly, in still another disorder, **spinobulbar muscular atrophy** (Kennedy disease), the involved gene (different from that in Huntington disease) also contains repeated copies of the CAG triplet. However, only 35 to 60 copies of the sequence cause individuals to be affected.

The role of such repeated sequences in normal and mutant genes remains a mystery. Their locations within the gene vary in each case. In Huntington and Kennedy diseases, the repeats lie within the coding portion of the gene. In the case of Huntington disease, this causes the mutant **huntingtin** protein to contain an excess of glutamine residues. Such is not the case in the other two disorders, however. In the gene responsible for fragile X syndrome, the repeat is upstream (the 5′ end) of the gene's coding region, in an area that is most often involved in regulating gene expression. In the case of myotonic dystrophy, the repeat is downstream of the coding region (the 3′ end).

The mechanism by which the repeated sequence expands from generation to generation is of great interest. Present-day thinking is that expansion can result from both errors during replication as well as in mechanisms responsible for repairing damaged DNA. Whatever the cause may be, this general instability of these short repeats seems to be more prevalent in humans than in many other organisms.

Now solve this

Problem 15.24 on page 389 asks you to speculate on why the *dystrophin* gene appears to suffer a large number of mutations.

Hint: In answering this question, you may want to think about the structure and function of this gene. Consider what features of *dystrophin* gene structure might make it susceptible to the accumulation of many mutations.

TABLE 15.4 SUMMARY OF TRINUCLEOTIDE-REPEAT DISORDERS

	Trinucleotide Repeat	Number in Normal Individuals	Number in Affected Individuals
Huntington disease	CAG	6–35	36–120
Myotonic dystrophy	CTG	5–37	37–1500
Fragile X syndrome	CGG	6–230	>230
Spinobulbar muscular atrophy	CAG	10–35	35–60

15.6 Genetic Techniques, Cell Cultures, and Pedigree Analysis Are All Used to Detect Mutations

Before geneticists can directly study the mutational process or obtain mutant organisms for genetic investigations, they must be able to detect mutations. The ease and efficiency of detecting mutations has generally determined the organism's usefulness in genetic studies. In Chapter 21, we examine how *Saccharomyces cerevisiae*, *Drosophila,* and mice are used as model organisms in genetic analysis. In this section, we briefly examine several other examples to illustrate how mutations are detected.

Detection in Bacteria and Fungi

Detection of mutations is most efficient in such haploid microorganisms as bacteria and fungi. Detection often depends on a selection system by which mutant cells can be isolated easily from nonmutant cells. The general principles are similar in bacteria and fungi. To illustrate, we describe how nutritional mutations in the fungus *Neurospora crassa* are detected.

Neurospora is a pink mold that normally grows on bread. It can also be cultured in the laboratory. This eukaryotic mold is haploid in the vegetative phase of its life cycle. Thus, mutations can be detected without the complications generated by heterozygosity in diploid organisms. Wild-type *Neurospora* grows on a **minimal culture medium** of glucose, a few inorganic acids and salts, a nitrogen source such as ammonium nitrate, and the vitamin biotin. Induced nutritional mutants will not grow on minimal medium, but will grow on a supplemented or **complete medium** that also contains numerous amino acids, vitamins, nucleic acid derivatives, and so forth. Microorganisms that are nutritional wild types (requiring only minimal medium) are called **prototrophs**, while those mutants that require a specific supplement to the minimal medium are called **auxotrophs**.

Nutritional mutants can be detected and isolated by their failure to grow on minimal medium and their ability to grow on complete medium. The mutant cells cannot synthesize some essential compound that is absent in minimal medium, but is present in complete medium. Once a nutritional mutant is detected and isolated, the missing compound is determined by attempts to grow the mutant strain in a series of tubes, each containing minimal medium supplemented with a single compound. The auxotrophic mutation can be defined in this way. Beadle and Tatum used this detection technique during their "one-gene:one-enzyme" studies.

Detection in Plants

Genetic variation in plants is extensive. Mendel's peas were the basis for his discovery of the fundamental postulates of transmission genetics, and subsequent studies of plants have enhanced our understanding of gene interaction, polygenic inheritance, linkage, sex determination, chromosome rearrangements, and polyploidy. Many genetic variants are detected simply by visual observation, but techniques also exist for detecting biochemical mutations in plants.

One technique involves the analysis of a plant's biochemical composition. For example, biochemical analysis of the maize mutant, *opaque 2*, led to the discovery of its high lysine content. These analyses involved the isolation of proteins from maize endosperm, hydrolysis of the proteins, and determination of the amino acid composition. Because maize protein is usually low in lysine, this mutation significantly improves the nutritional value of the plant. Once the mutant was discovered, plant geneticists and other specialists analyzed the amino acid compositions of various strains of other grain crops, including rice, wheat, barley, and millet. The results of such analyses have been useful in combating malnutrition diseases resulting from inadequate protein or the lack of essential amino acids in the diet.

Other detection techniques involve the culturing of plant cell lines in defined medium. Cultured plant cells are treated like microorganisms, and their resistance to herbicides or disease toxins can be determined by adding these compounds to the culture medium. Techniques for isolating and analyzing conditional lethal mutants can be used on plant cells in tissue culture and then applied to the genetics of higher plants, providing a means of detection that would not be possible in an intact plant.

Detection in Humans

Because humans are obviously not suitable experimental organisms, techniques developed for the detection of mutations in organisms such as bacteria and plants are not available to human geneticists. To determine the mutational basis for any human characteristic or disorder, geneticists may first analyze a pedigree that traces the family history as far back as possible. If a trait is shown to be inherited, the pedigree may be useful in determining whether the mutant allele is behaving as a dominant or a recessive and whether it is X-linked or autosomal.

Dominant mutations are the simplest to detect. If they are present on the X chromosome, affected fathers pass the trait to all of their daughters. If dominant mutations are autosomal, approximately 50 percent of the offspring of an affected

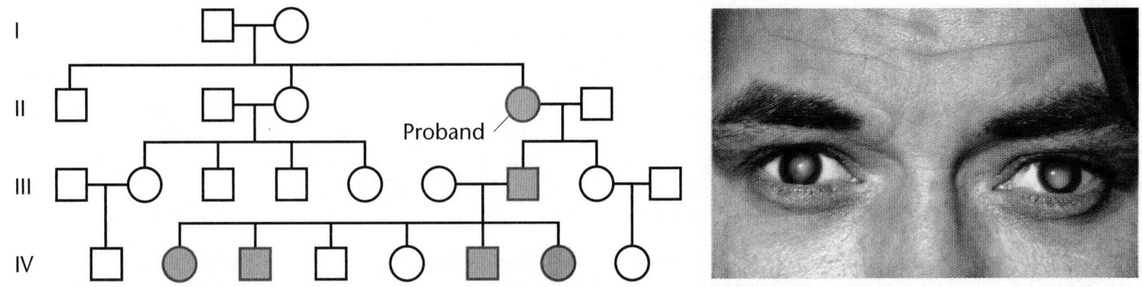

FIGURE 15–11 A hypothetical pedigree of inherited cataracts of the eye in humans.

heterozygous individual are expected to show the trait. Figure 15–11 shows a pedigree illustrating the initial occurrence of an autosomal dominant mutation for cataracts of the eye. The parents in Generation I were unaffected, but one of their three offspring (Generation II) developed cataracts, so the original mutation presumably occurred in a gamete of one of the parents. The affected female, the proband, produced two children, of which the male child was affected (Generation III). Of his six offspring (Generation IV), four were also affected. These observations are consistent with, but do not prove, an autosomal dominant mode of inheritance. However, the high percentage of affected offspring in Generation IV favors this conclusion. Also, the presence of an unaffected daughter in this generation (Generation IV) argues against X-linkage, because she received her X chromosome from her affected father. Hence, the conclusion is sound, provided that the mutant allele is completely penetrant.

X-linked recessive mutations can also be detected by pedigree analysis, as discussed in Chapter 4. The most famous case of an X-linked recessive mutation in humans is that of **hemophilia**, which was found in the descendants of Britain's Queen Victoria. Inspection of the pedigree (Figure 15–12) leaves little doubt that Victoria was heterozygous for the trait. Her father was not affected and there is no evidence to show that her mother was a carrier, as was Victoria. The recessive mutation for hemophilia has occurred many times in human populations, but the political consequences of the mutation that occurred in the British royal family were sweeping. Two of Queen Victoria's daughters were carriers and they passed hemophilia into the Russian and Spanish royal families. It is possible that Tsarina Alexandra's attempts to deal with her son's hemophilia resulted in Rasputin's presence in the royal household and may have contributed to the start of the Russian revolution.

It is also possible to determine the recessive nature of alleles by pedigree analysis. Because this type of mutation is "hidden" when heterozygous, it is not unusual for the trait to appear only intermittently through a number of generations. A mating between an affected individual and a homozygous normal individual will produce unaffected heterozygous carrier children. Matings between two carriers will produce, on average, one-fourth affected offspring.

In addition to pedigree analysis, human cells may now be routinely cultured *in vitro*. This procedure has allowed the study of many more mutations than any other form of analysis. Analysis of enzyme activity, protein migration in electrophoretic fields, and direct sequencing of DNA and proteins are among the techniques that have identified mutations and demonstrated the wide genetic variation among individuals in human populations. As we will see later in the chapter, initial studies of the human disorder xeroderma pigmentosum relied on the *in vitro* cell culture technique.

More recently, genomics and reverse genetic techniques have expanded the techniques available for geneticists to study human mutations. Some of these new methods are described in Chapter 21.

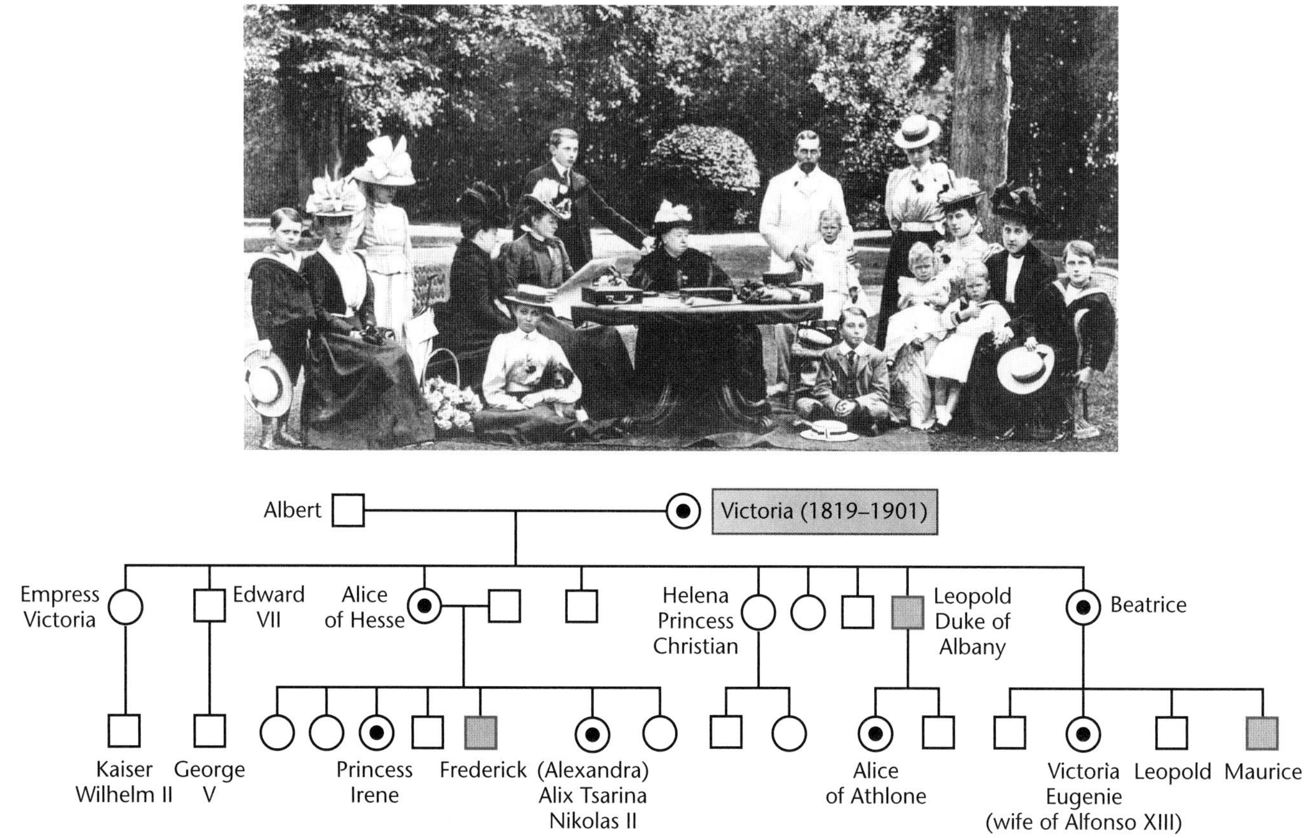

FIGURE 15–12 A partial pedigree of hemophilia in the British royal family descended from Queen Victoria. The pedigree is typical of the transmission of X-linked recessive traits. Circles with a dot in them indicate presumed female carriers heterozygous for the trait. The photograph shows Queen Victoria (seated at the table, center) and some of her immediate family.

Now solve this

Problem 15.26 on page 389 asks you to explain how X-linked hemophilia could have arisen in the British royal family.

Hint: To answer this question, you might consider the ways in which mutations occur, in what types of cells they can occur, and how they are inherited.

15.7 The Ames Test Is Used to Assess the Mutagenicity of Compounds

There is great concern about the possible mutagenic properties of any chemical that enters the human body, whether through the skin, the digestive system, or the respiratory tract. For example, great attention has been given to materials in air and water pollution, food preservatives and additives, artificial sweeteners, herbicides, pesticides, and pharmaceutical products. As we have seen, mutagenicity may be tested in various organisms, including fungi, plants, and cultured mammalian cells. The most common test, which involves bacteria, was devised by Bruce Ames.

The **Ames test** (Figure 15–13) uses any of four strains of the bacterium *Salmonella typhimurium* that have been selected for their sensitivity to specific types of mutagenesis. For example, one strain is used to detect base pair substitutions, and another three strains detect various frameshift mutations. Each mutant strain is unable to synthesize histidine, and therefore requires histidine for growth (his^- strains). The assay measures the frequency of reverse mutation, which yields wild-type bacteria (his^+ revertants). These *Salmonella* strains also have a high sensitivity to mutagens due to the presence of mutations in genes involved in excision repair and synthesis of the lipopolysaccharide barrier that coats bacteria and protects them from external substances.

It is interesting to note that many substances entering the human body are relatively innocuous until activated metabolically, usually in the liver, to a more chemically reactive product. Thus, the Ames test includes a step in which the test compound is incubated *in vitro* in the presence of a mammalian liver extract. Alternatively, test compounds may be injected into a mouse in which they are modified by liver enzymes and then recovered for use in the Ames test.

In the initial use of Ames testing in the 1970s, a large number of known **carcinogens** (cancer-causing agents) were examined and more than 80 percent of these were shown to be strong mutagens. This is not surprising, as the transformation of cells to the malignant state occurs as a result of mutations. (See Chapter 18.) Although a positive response in the Ames test does not prove that a compound is carcinogenic, the Ames test is useful as a preliminary screening device. The Ames test is used extensively during the development of industrial and pharmaceutical chemical compounds.

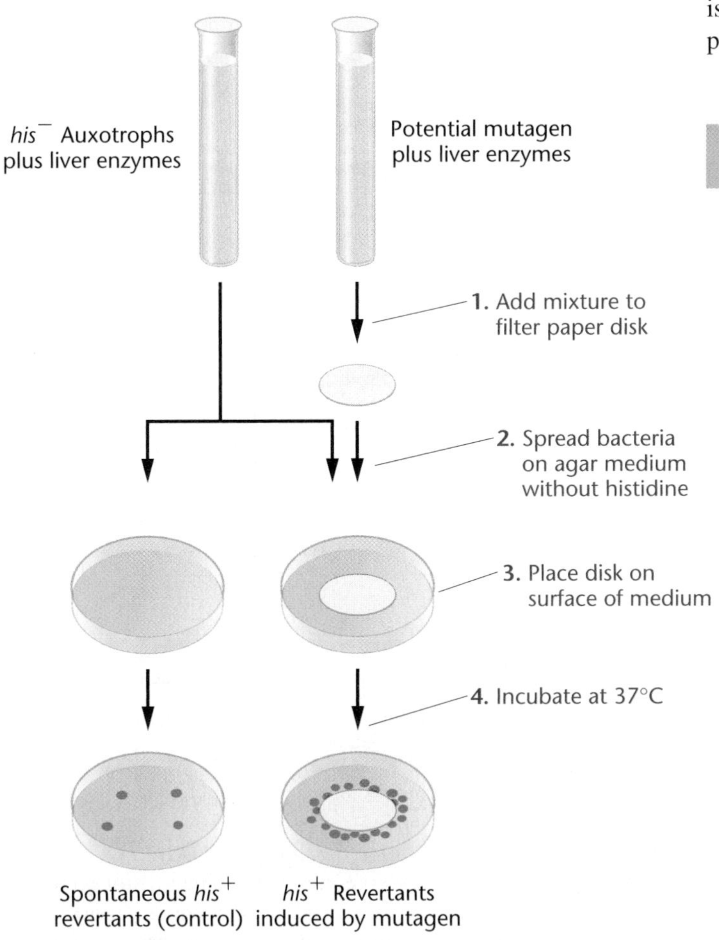
his⁻ Auxotrophs plus liver enzymes
Potential mutagen plus liver enzymes
1. Add mixture to filter paper disk
2. Spread bacteria on agar medium without histidine
3. Place disk on surface of medium
4. Incubate at 37°C
Spontaneous his⁺ revertants (control)
his⁺ Revertants induced by mutagen

FIGURE 15–13 The Ames test, which screens compounds for potential mutagenicity.

15.8 Organisms Use DNA Repair Systems to Counteract Mutations

Living systems have evolved a variety of elaborate repair systems that are able to counteract many of the forms of DNA damage resulting from internal and external agents. The extensive variety and complex nature of DNA repair systems attest to the critical importance of DNA repair. As we will see, such repair systems are absolutely essential to the maintenance of the genetic integrity of organisms, and as such, to the survival of organisms on Earth. Of foremost concern in humans is the potential to counteract genetic damage that results in genetic diseases and cancer. The link between defective DNA repair and cancer susceptibility is described in Chapter 18.

We now embark on a review of some systems of DNA repair. Since the field is expanding rapidly, our goal here is to survey the major approaches that organisms use to counteract genetic damage.

Proofreading and Mismatch Repair

Some of the most common types of mutations in DNA arise during DNA replication when an incorrect nucleotide is inserted by DNA polymerase. The enzyme in bacteria (DNA polymerase III) makes an error approximately once every 100,000 insertions, leading to an error rate of 10^{-5}. Fortunately, the enzyme polices its own synthesis by **proofreading**

Web Tutorial 15.3
Chemical Mutagenesis: Ames Test

each step, catching 99 percent of those errors. During polymerization, when an incorrect nucleotide is inserted, the enzyme has the potential to recognize the error and "reverse" its direction and behave as an exonuclease, cutting out the incorrect nucleotide and then replacing it with the correct one. This improves the efficiency of replication one hundredfold, creating only $1/10^7$ mismatches immediately following DNA replication, for a final error rate of 10^{-7}.

To cope with those errors that remain after proofreading, still another mechanism, called **mismatch repair**, may be activated. Robin Holliday proposed this mechanism more than 20 years ago, and the molecular basis of the process is now well established. As in repair of other DNA lesions, the alterations or mismatches are detected, the incorrect nucleotides are removed, and the incorrect nucleotides replaced with the correct ones. But a special problem exists during the correction of a mismatch. How does the repair system recognize which strand is correct (the template) and which contains the incorrect base (the newly synthesized strand)? If the mismatch is recognized, but no discrimination occurs, the excision would be random, and the strand bearing the correct base would be clipped out 50 percent of the time. Hence, strand discrimination by a repair enzyme is a critical step.

The process of strand discrimination has been elucidated at least in some bacteria including *E. coli*, and is based on **DNA methylation**. These bacteria contain an enzyme, adenine methylase, which recognizes the DNA sequence

$$5' \ldots \text{GATC} \ldots 3'$$
$$3' \ldots \text{CTAG} \ldots 5'$$

as a substrate during DNA replication, adding a methyl group to each of the adenine residues. This modification is stable throughout the cell cycle.

Following a round of replication, the newly synthesized strand remains temporarily unmethylated, as the methylase lags behind the DNA polymerase. It is at this point that the repair enzyme recognizes the mismatch and preferentially binds to the unmethylated (newly synthesized) DNA strand. A nick is made by an endonuclease enzyme, either 5′ or 3′ to the mismatch on the unmethylated strand. The nicked DNA strand is then unwound and degraded by an exonuclease, until the region of the mismatch is reached. Finally, DNA polymerase fills in the gap created by the exonuclease, using the correct DNA strand as a template. DNA ligase then seals the gap.

A series of *E. coli* gene products, Mut H, L, and S, are involved in mismatch repair. Mutations in any of these genes result in bacterial strains deficient in mismatch repair. Mut S recognizes the initial mismatch, and in fact, can also recognize small insertion and deletion loops created during DNA synthesis. The Mut L and H proteins align and nick the strand bearing the mutation. The mismatched region is digested by an exonuclease, the gap is filled in by DNA polymerase, and finally sealed by DNA ligase. In bacteria, the repair process is remarkably efficient, reducing the error rate one thousandfold (99.9 percent of the mismatches are corrected). While the preceding mechanism is based on studies of *E. coli*, similar mechanisms involving homologous proteins exist in yeast and in mammals.

Postreplication Repair and the SOS Repair System

Many other types of repair have been discovered, illustrating the diversity of mechanisms that have evolved to overcome DNA damage. One system, called **postreplication repair** responds *after* damaged DNA has escaped repair and failed to be completely replicated, thus its name. As illustrated in Figure 15–14, when DNA bearing a lesion of some sort (such as a pyrimidine dimer) is being replicated, DNA polymerase may stall at the lesion and then skip over it, leaving a gap on the newly synthesized strand. To correct the gap, the RecA protein directs a recombinational exchange with the corresponding region on the undamaged parental strand of the same polarity (the "donor" strand). When the undamaged segment of DNA replaces the damaged segment, this transfers the gap to the donor strand. This gap can be filled by repair synthesis as replication proceeds. Because a recombinational event is involved in this

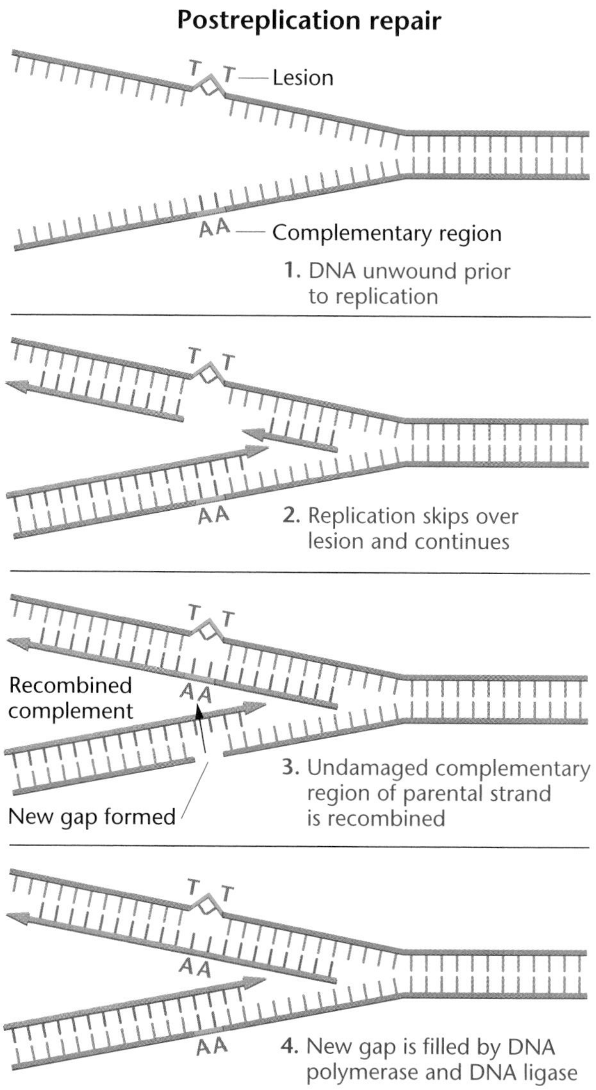
Postreplication repair
T T
Lesion
AA
Complementary region
1. DNA unwound prior to replication
T T
AA
2. Replication skips over lesion and continues
T T
Recombined complement
AA
New gap formed
3. Undamaged complementary region of parental strand is recombined
T T
AA
AA
4. New gap is filled by DNA polymerase and DNA ligase

FIGURE 15–14 Postreplication repair occurs if DNA replication has skipped over a lesion such as a thymine dimer. Through the process of recombination, the correct complementary sequence is recruited from the parental strand and inserted into the gap opposite the lesion. The new gap is filled by DNA polymerase and DNA ligase.

type of DNA repair, postreplication repair is also referred to as **homologous recombination repair**.

Still another repair pathway in *E. coli* is called the **SOS repair system**, which also responds to damaged DNA, but in a different way. In the presence of DNA mismatches and gaps during replication, bacteria can induce the expression of about 20 genes (including *lexA*, *recA*, and *uvr*) whose products allow DNA replication to occur even in the presence of these types of lesions. As this type of repair is a last resort to DNA damage, it is known as SOS repair. During SOS repair, DNA synthesis becomes error-prone, inserting random and possibly incorrect nucleotides in places that would normally stall DNA replication. As a result, SOS repair itself becomes mutagenic—although it may allow the cell to survive DNA damage that might otherwise kill it.

Photoreactivation Repair: Reversal of UV Damage in Prokaryotes

As illustrated in Figure 15–9, UV light is mutagenic as a result of the creation of pyrimidine dimers. The study of UV-induced mutagenicity paved the way for the discovery of many forms of natural repair of DNA damage. The first relevant discovery concerning repair of UV damage in bacteria was made in 1949, when Albert Kelner observed the phenomenon of **photoreactivation repair**. He showed that UV-induced damage to *E. coli* DNA could be partially reversed if, following irradiation, the cells were exposed briefly to light in the blue range of the visible spectrum. The photoreactivation repair process is also temperature dependent, suggesting that the light-induced mechanism involves an enzymatically controlled chemical reaction.

Further studies of photoreactivation have revealed that the process is dependent on the activity of a protein called the **photoreactivation enzyme (PRE)**. The enzyme's mode of action is to cleave the bonds between thymine dimers, thus directly reversing the effect of UV radiation on DNA (Figure 15–15). Although the enzyme will associate with a dimer in the dark, it must absorb a photon of light to cleave the dimer. In spite of the potential for diminishing UV-induced mutations, photoreactivation repair is not absolutely essential in *E. coli*, since a mutation creating a null allele in the gene coding for the PRE is not lethal. In addition, the enzyme cannot be detected in humans and other eukaryotes, who must rely on other repair mechanisms to reverse the effects of UV radiation.

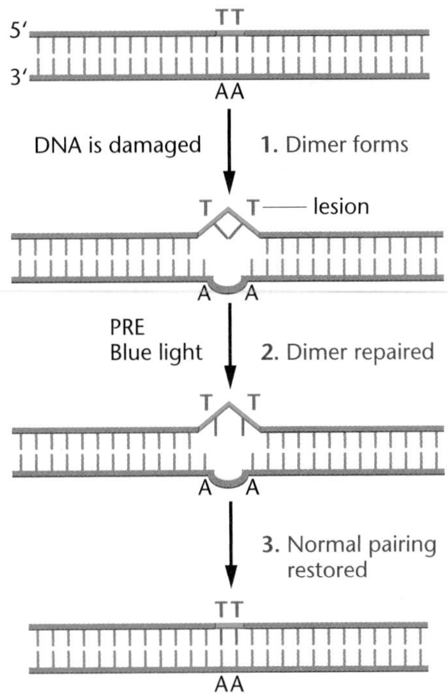

FIGURE 15–15 Damaged DNA repaired by photoreactivation repair. The bond creating the thymine dimer is cleaved by the photoreactivation enzyme (PRE), which must be activated by blue light in the visible spectrum.

Base and Nucleotide Excision Repair

In addition to PRE, light-independent repair systems exist in all prokaryotes and eukaryotes. The basic mechanisms involved in these types of repair—referred to as **excision repair**—consist of the following three steps and can be described generally as "cut-and-paste" systems.

1. The distortion or error present on one of the two strands of the DNA helix is recognized and enzymatically clipped out by a nuclease. Excisions in the phosphodiester backbone usually include a number of nucleotides adjacent to the error as well, leaving a gap on one strand of the helix.
2. A DNA polymerase fills in the gap by inserting deoxyribonucleotides complementary to those on the intact strand, using it as a replicative template. The enzyme adds these bases to the free 3′-OH end of the clipped DNA. In *E. coli*, this is usually performed by DNA polymerase I.
3. DNA ligase seals the final "nick" that remains at the 3′-OH end of the last base inserted, closing the gap.

There are two types of excision repair: base excision repair and nucleotide excision repair. **Base excision repair (BER)** corrects damage to nitrogenous bases created by spontaneous hydrolysis or agents that chemically alter them.

The first step in the BER pathway in *E. coli* involves the recognition of the chemically altered base by **DNA glycosylases**, which are specific to different types of DNA damage (Figure 15–16). For example, the enzyme uracil–DNA glycosylase recognizes the presence of uracil in DNA. The enzyme first cuts the glycosidic bond between the base and the sugar, creating an apyrimidinic (AP) site. Such a sugar with a missing base is then recognized by an enzyme called **AP endonuclease**. The endonuclease makes a cut in the phosphodiester backbone at the AP site. This creates a distortion in the DNA helix that is recognized by the excision-repair system, which is then activated, leading to the correction of the error.

Although much has been learned about DNA glycosylases in *E. coli*, much less is known about the DNA glycosylases in eukaryotes. In addition, unlike the nucleotide excision pathway we are going to discuss next, there is no human or animal disease model available that has a defective BER pathway, so it is difficult to assess the importance of BER in eukaryotes.

While BER recognizes and replaces modified bases in DNA, the **nucleotide excision repair (NER)** pathway repairs "bulky"

Web Tutorial 15.4 DNA Repair: Proofreading, Mismatch, and SOS

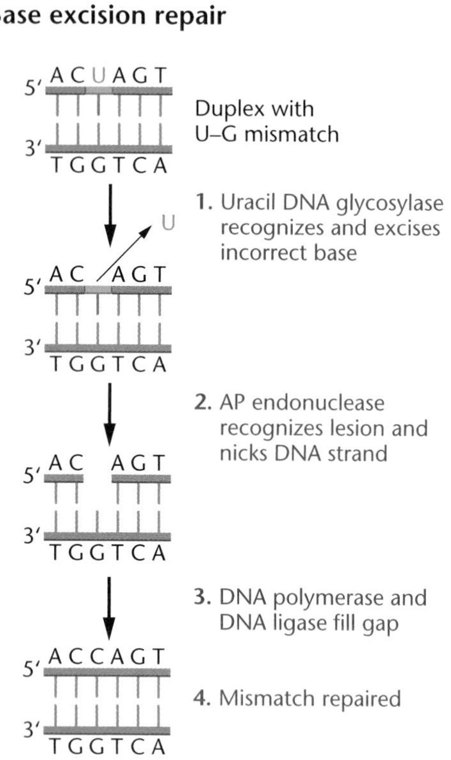

FIGURE 15–16 Base excision repair (BER) accomplished by uracil DNA glycosylase, AP endonuclease, DNA polymerase, and DNA ligase. Uracil is recognized as a noncomplementary base, excised, and replaced with the complementary base (C).

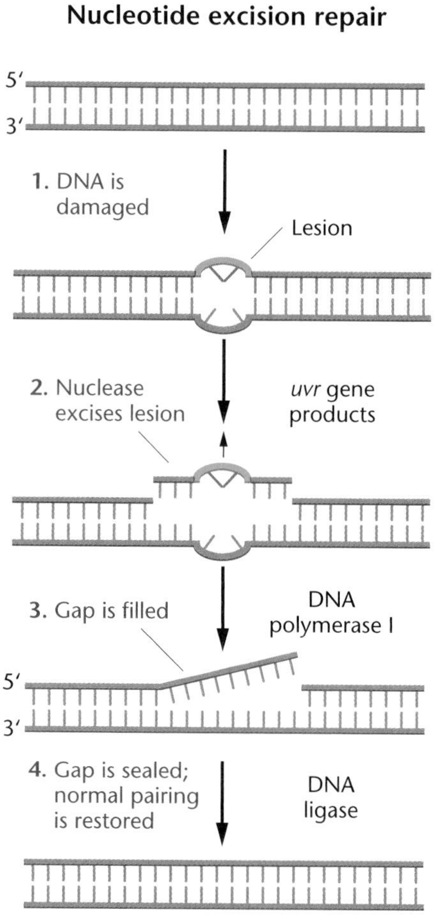

FIGURE 15–17 Nucleotide excision repair (NER) of a UV-induced thymine dimer. During repair, 13 bases are excised in prokaryotes and 28 bases are excised in eukaryotes.

lesions in DNA that alter or distort the double helix, such as the UV-induced pyrimidine dimers discussed previously. However, all types of bulky lesions leading to distortions of the helix are subject to NER.

The NER pathway (Figure 15–17) was first discovered in *E. coli* by Paul Howard-Flanders and coworkers, who isolated several independent mutants that are sensitive to UV radiation. One group of genes was designated *uvr* (ultraviolet repair) and included the *uvrA*, *uvrB*, and *uvrC* mutations. In the NER pathway, the *uvr* gene products are involved in recognizing and clipping out lesions in the DNA. Usually, a very specific number of nucleotides is clipped out around both sides of the lesion. In *E. coli*, there are usually a total of 13 nucleotides removed, which include the lesion. The repair is then completed by DNA polymerase I and DNA ligase, in a manner similar to that occurring in BER. The undamaged strand opposite the lesion is used as a template for the replication resulting in repair.

Xeroderma Pigmentosum and Nucleotide Excision Repair in Humans

The mechanism of nucleotide excision repair (NER) in eukaryotes is much more complicated than that in prokaryotes, and involves many more proteins. Much of what is known about the system in humans has come from detailed studies of individuals with **xeroderma pigmentosum (XP)**, a rare recessive genetic disorder that predisposes individuals to severe skin abnormalities. These individuals have lost their ability to undergo NER; as a result, individuals suffering from XP who are exposed to the UV radiation in sunlight exhibit reactions that range from initial freckling and skin ulceration to the development of skin cancer.

The condition is severe and may be lethal, although early detection and protection from sunlight can arrest it. Figure 15–18 compares two XP individuals, one of whom has been detected early and protected from sunlight. Because sunlight contains UV radiation, a causal relationship was predicted between thymine dimer production and XP. To test this prediction, the repair of UV-induced lesions was investigated in human fibroblast cultures derived from XP and normal individuals. (Fibroblasts are undifferentiated connective tissue cells.) The results of these studies suggested that the XP phenotype was caused by more than one mutant gene.

In 1968, James Cleaver showed that cells from XP patients were deficient in **unscheduled DNA synthesis** (DNA synthesis other than that occurring during chromosome replication), which is elicited in normal cells by UV radiation. Because this type of synthesis is thought to represent the activity of DNA polymerization during excision repair, the lack of unscheduled DNA synthesis in XP patients suggested that XP may be a deficiency in excision repair.

The link between XP and excision repair was further strengthened by studies using **somatic cell hybridization** (introduced in Chapter 5). Fibroblast cells from any two unrelated XP patients, when grown together in tissue culture, can fuse together, forming a heterokaryon. A heterokaryon is a single cell with two nuclei but

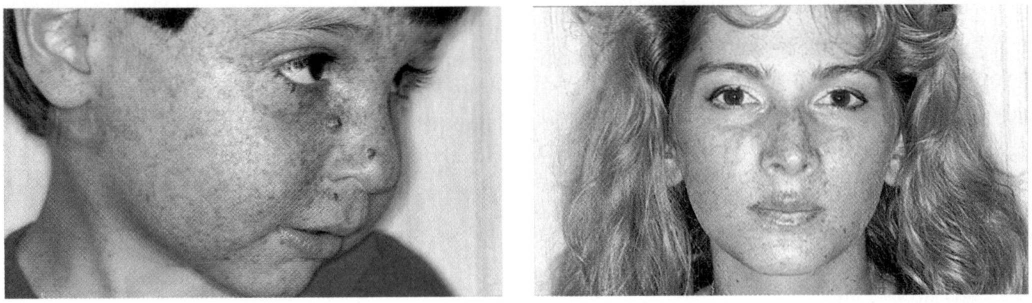

FIGURE 15–18 Two individuals with xeroderma pigmentosum (XP). The 4-year-old boy on the left shows marked skin lesions induced by sunlight. Mottled redness (erythema) and irregular pigment changes in response to cellular injury are apparent. Two nodular cancers are present on his nose. The 18-year-old girl on the right has been carefully protected from sunlight since her diagnosis of xeroderma pigmentosum in infancy. Several cancers have been removed and she has worked as a successful model.

a common cytoplasm. After fusion of two XP cells, excision repair, as assayed by unscheduled DNA synthesis, can be measured. If the mutation in each of the two XP cells occurs in the same gene, the heterokaryon will still be unable to undergo excision repair. This is because there is no normal copy of the relevant gene present in the heterokaryon. However, if excision repair occurs in the heterokaryon, the mutations in the two XP cells must have been present in two different genes. Hence, the two mutants are said to demonstrate **complementation**. (Complementation is also described in Chapters 6 and 21.) Complementation occurs because the heterokaryon has at least one normal copy of each gene in the fused cell. By fusing XP cells from a large number of XP patients, it was possible to determine how many genes contribute to the XP phenotype.

Based on many studies, XP patients have been divided into seven **complementation groups**, suggesting that at least seven different genes are involved in excision repair in humans. These seven human genes, *XPA* to *XPG* (*X*eroderma *P*igmentosum gene *A* to *G*), have now been identified, and a homologous gene for each has been identified in yeast. Approximately 20 percent of XP patients do not fall into any of the seven complementation groups. These patients manifest similar symptoms, but their fibroblasts do not demonstrate defective excision repair. There is some evidence that these XP cells are less efficient in normal DNA replication.

As a result of the study of defective genes in xeroderma pigmentosum, a great deal is now known about how NER counteracts DNA damage in normal cells. The first step in humans is the recognition of the damaged DNA by proteins encoded by the *XPC*, *XPE*, and *XPA* genes. These proteins then recruit the remainder of the repair proteins to the site of DNA damage. The *XPB* and *XPD* genes encode helicases and are also components of the basal transcription factor, TFIIH. The *XPF* and *XPG* genes encode nucleases. The excision repair complex containing these and other factors excises a 28-nucleotide-long fragment from the DNA strand that contains the lesion.

Another defect in the human NER pathway has been found in **Cockayne syndrome (CS)**, which is characterized by impairment of both physical and neurological development. While CS patients, like XP patients, are highly photosensitive, no predisposition to tumor formation characterizes the disease. Three of the five genes involved in CS are also involved in XP (*XPB*, *XPD*, and *XPG*). Products of the five CS complementation groups are involved in **transcription-coupled repair**, a type of NER in which only the transcribed strand of actively transcribed genes is repaired.

Double-Strand Break Repair in Eukaryotes

Thus far, we have discussed repair pathways that deal with damage within one strand of DNA. We conclude our discussion of DNA repair by considering what happens when both strands of the DNA helix are cleaved—as a result of exposure to ionizing radiation, for example. While such damage occurs in bacteria as well, we will focus our discussion on eukaryotic cells.

In these cases, a specialized form of DNA repair, the **DNA double-strand break repair (DSB repair)** pathway, is activated and is responsible for reattaching the two DNA strands. Recently, interest has grown in DSB repair because defects in this pathway are associated with X-ray hypersensitivity and immune deficiency. Such defects may also underlie familial disposition to breast and ovarian cancer.

Another pathway involved in double-strand break repair is referred to as **homologous recombinational repair**. In a process similar to that of postreplication repair, damaged DNA is recombined and replaced with homologous undamaged DNA. In this case, the first step in the process involves the activity of an enzyme that recognizes the double-strand break, then digests back the 5′ ends of the broken DNA helix, leaving overhanging 3′ ends. These overhanging ends then interact with a region of an undamaged sister chromatid, aligning the homologous sequences and allowing DNA polymerase to copy the undamaged DNA sequence into the damaged DNA strand. This interaction of two sister chromatids is necessary because, when both strands of one helix are broken, there is no undamaged parental DNA strand available to use as the source of the complementary template DNA sequence during repair. Therefore, the genetic information present in the homologous region of the sister chromatid is "recruited" to replace the damaged double-stranded break. The process usually occurs during the late S/G2 phase of the cell cycle, after DNA replication, a time when sister chromatids are available to be used as repair templates. In yeast, the system depends upon a complex of at least five proteins, called the RAD52 complex.

A second pathway, called **nonhomologous recombinational repair**, or **end joining**, also repairs double-strand breaks.

However, as the name implies, the mechanism does not recruit a homologous region of DNA during repair. This system is activated in G1, prior to DNA replication. End joining involves a complex of three proteins including DNA-dependent protein kinase. These proteins bind to the free ends of the broken DNA, resulting in the ends being trimmed and ligated back together. Because some nucleotide sequence is lost in the process of end joining, it is an error-prone repair system.

HOW DO WE KNOW?

How do we know that cells repair most of the spontaneous DNA damage that afflicts them?

15.9 Transposable Elements Move within the Genome and May Disrupt Genetic Function

We conclude this chapter by discussing **transposable elements**, also known as **transposons** or "jumping genes." These DNA elements can move or transpose within the genome, inserting themselves into various locations within and between chromosomes. Transposable elements were first discovered in maize by Barbara McClintock more than 50 years ago. However, the idea that some genetic information may not be fixed within the genome was slow to find acceptance. The concept of moveable genes was quite alien to the classical understanding of genes as discrete loci that can be mapped to specific chromosomes. It was not until transposable elements were discovered in other organisms, and their molecular basis revealed, that the significance of transposable elements was confirmed.

Transposons are present in the genomes of all organisms from bacteria to humans. Not only are they ubiquitous, but they also comprise large portions of some eukaryotic genomes. For example, recent genomic sequencing has revealed that almost 50 percent of the human genome is derived from transposable elements. Some organisms with unusually large genomes, such as salamanders and barley, contain hundreds of thousands of copies of various types of transposable elements. The function of these elements is unknown, and it is remarkable that eukaryotic genomes tolerate such a load of apparently useless or "junk" DNA. Data from human genome sequencing suggest that some genes may have evolved from transposons, and that the activity of transposons helps to modify and reshape the genome. In this sense, some transposable elements may confer benefits upon their hosts.

Transposable elements have also proved to be valuable tools in genetic research. As will be described in Chapter 21, geneticists have harnessed transposons as mutagens, as cloning tags, and as vehicles for introducing foreign DNA into model organisms.

In this chapter, we will discuss transposable elements as naturally occurring mutagens. The movement of transposons from one place in the genome to another has the capacity to disrupt genes and cause mutations, as well as to create chromosomal damage such as double-strand breaks.

Insertion Sequences

There are two types of transposable elements in bacteria: insertion sequences and bacterial transposons. **Insertion sequences (IS elements)** were first characterized at the molecular level in the 1970s. They can move from one location to another and, if they insert into a gene or gene regulatory region, may cause mutations.

IS elements were first identified during analyses of mutations in the *gal* operon of *E.coli*. Researchers discovered that certain mutations in this operon were due to the presence of several hundred base pairs of extra DNA inserted into the beginning of the operon. Surprisingly, the segment of mutagenic DNA could spontaneously excise from this location, restoring wild-type function to the *gal* operon.

Subsequent research revealed that several other DNA elements could behave in a similar fashion, inserting into bacterial chromosomes and affecting gene function. These DNA elements are relatively short, not exceeding 2000 bp (2 kb). The first insertion sequence to be characterized in *E. coli*, IS1, is about 800 bp long. Other IS elements such as IS2, 3, 4, and 5 are about 1250 to 1400 bp in length. IS elements are present in multiple copies in bacterial genomes. For example, the *E.coli* chromosome contains 5 to 8 copies of IS1, 5 copies each of IS2 and IS3, as well as copies of IS elements on plasmids such as F factors.

All IS elements contain two features that are essential for their movement. First, they contain a gene that encodes an enzyme called **transposase**. This enzyme is responsible for making staggered cuts in chromosomal DNA, into which, or out of which, the IS element can insert. Second, the ends of IS elements contain inverted terminal repeats (ITRs). ITRs are short segments of DNA that have the same nucleotide sequence as each other, but are oriented in the opposite direction (Figure 15–19). Although Figure 15–19 shows the ITRs to consist of only a few nucleotides, many more are actually present. For example, IS1 ITRs contain about 20 nucleotide pairs, those of IS2 and IS3 contain about 40 nucleotide pairs, and that of IS4 contains about 18 nucleotide pairs. ITRs are essential for transposition and act as recognition sites for the binding of the transposase enzyme.

Bacterial Transposons

Bacterial transposons (**Tn elements**) are larger than IS elements and contain protein-coding genes that are unrelated to their transposition. Some Tn elements, such as Tn10, are comprised of a drug-resistance gene flanked by two IS elements present in opposite orientations. The IS elements encode the

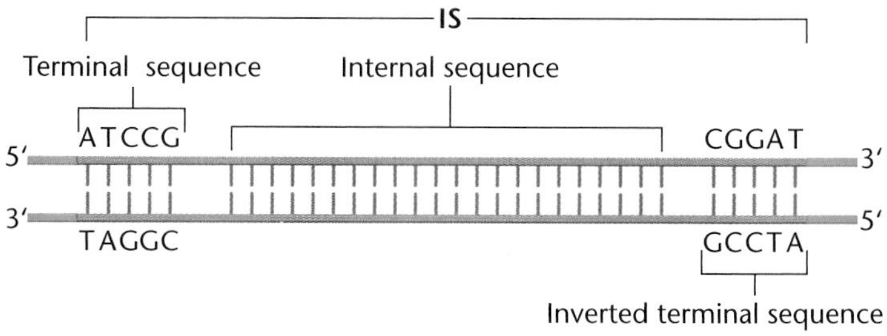

FIGURE 15–19 An insertion sequence (IS), shown in purple. The terminal sequences are perfect inverted repeats of one another.

transposase enzyme that is necessary for transposition of the Tn element. Other types of Tn elements, such as Tn3, have shorter inverted repeat sequences at their ends and encode their transposase enzyme from a transposase gene located in the middle of the Tn element. Like IS units, Tn elements are mobile in both bacterial chromosomes and in plasmids, and can cause mutations if they insert into genes or gene regulatory regions.

The presence of inverted repeats at the 5′ and 3′ termini of Tn elements has been revealed by electron microscopic analysis (Figure 15–20). When double-stranded DNA from a plasmid containing a Tn element is separated into single strands and each strand is allowed to reanneal separately, the ITRs, being complementary, form a heteroduplex. All other areas of the Tn and plasmid remain single-stranded and form loops on either end of the double-stranded stem.

Tn elements are currently of particular interest because they can introduce multiple drug resistance onto bacterial plasmids. These plasmids, called R factors, may contain many Tn elements conferring simultaneous resistance to heavy metals, antibiotics, and other drugs. These elements can move from plasmids onto bacterial chromosomes and can spread multiple drug resistance between different strains of bacteria.

The *Ac–Ds* System in Maize

About 20 years before the discovery of transposons in bacteria, Barbara McClintock discovered mobile genetic elements in corn plants (maize). She did this by analyzing the genetic behavior of two mutations, ***Dissociation* (*Ds*)** and ***Activator* (*Ac*)** expressed in either the endosperm or aleurone layers (Figure 15–21). She then correlated her genetic observations with cytological examinations of the maize chromosomes. Initially, McClintock determined that *Ds* was located on chromosome 9. If *Ac* was also present in the genome, *Ds* induced breakage at a point on the chromosome adjacent to its own location. If chromosome breakage occurred in somatic cells during their development, progeny cells often lost part of the broken chromosome, causing a variety of phenotypic effects.

Subsequent analysis suggested to McClintock that both *Ds* and *Ac* elements sometimes moved to new chromosomal locations. While *Ds* moved only if *Ac* was also present, *Ac* was capable of autonomous movement. Where *Ds* came to reside determined its genetic effects—that is, it might cause chromosome breakage or it might inhibit expression of a certain gene. In cells in which *Ds* caused a gene mutation, *Ds* might move again, restoring the gene mutation to wild-type.

Figure 15–22 illustrates the types of movements and effects brought about by *Ds* and *Ac* elements. In McClintock's original observation, pigment synthesis was restored in cells in which the *Ds* element jumped out of chromosome 9. McClintock concluded that the *Ds* and *Ac* genes were **mobile controlling elements**. We now commonly refer to them as transposable elements, a term coined by another great maize geneticist, Alexander Brink.

Several *Ac* and *Ds* elements have now been analyzed, and the relationship between the two elements has been clarified (Figure 15–23). The *Ac* element is 4563 nucleotides long and its structure is strikingly similar to that of bacterial transposons.

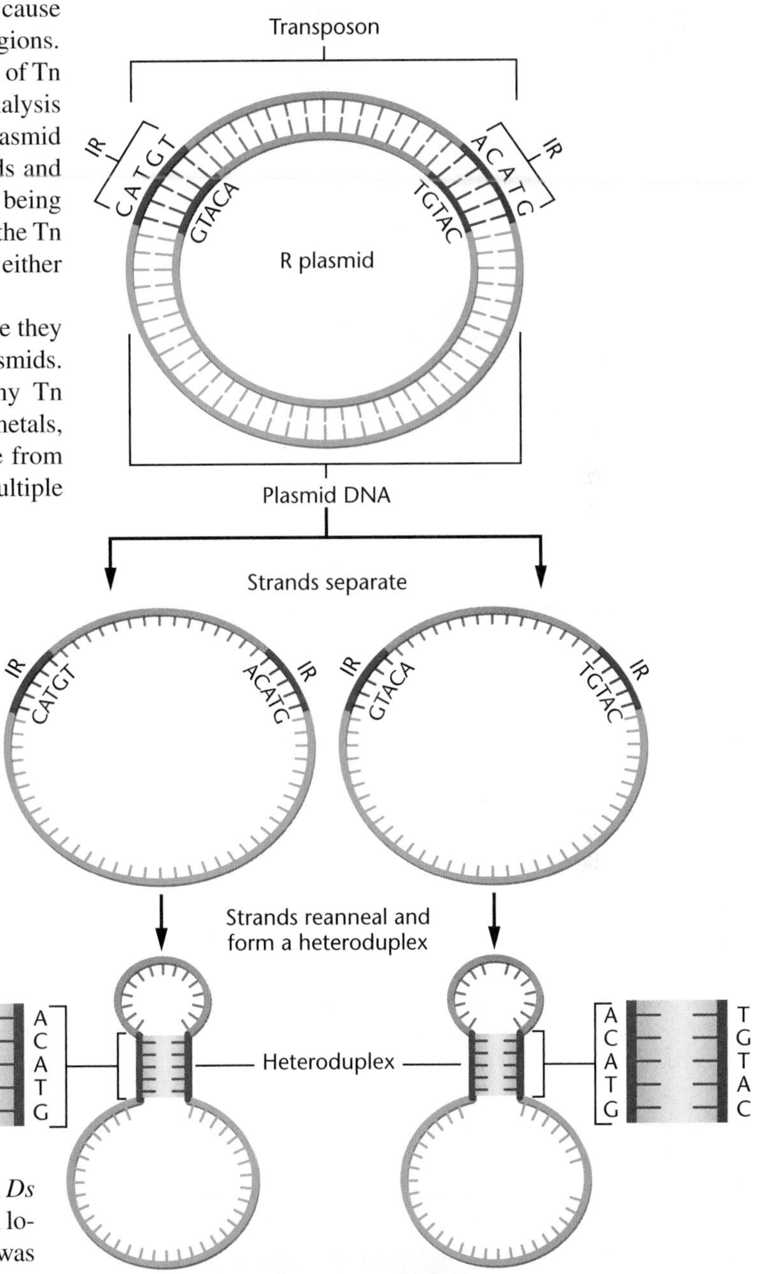

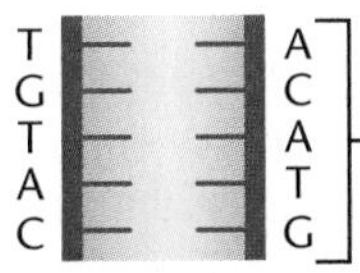

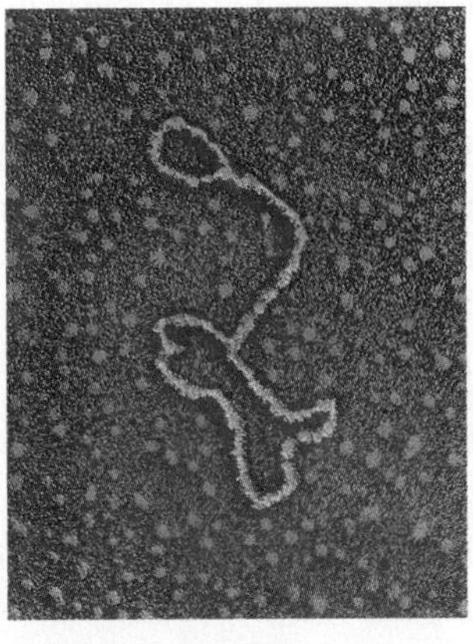

FIGURE 15–20 Heteroduplex formation as a result of inverted repeat sequences within a transposon inserted into a bacterial plasmid. The electron micrograph illustrates the final heteroduplex.

(a)

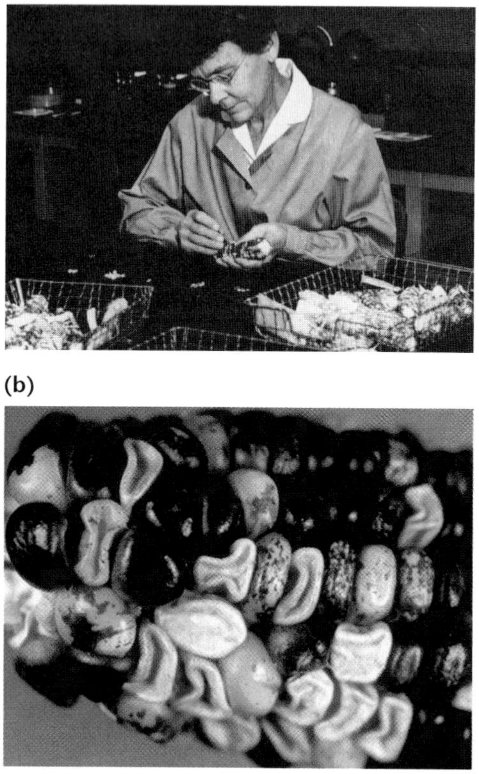

(b)

FIGURE 15–21 (a) Barbara McClintock analyzing corn variants in her laboratory. McClintock earned the Nobel Prize in Physiology or Medicine in 1983, at the age of 81, for her pioneering research on mobile genetic elements. (b) Ear of corn, showing a range of pigment variations between and within kernels, due to transposition of the *Ds* element.

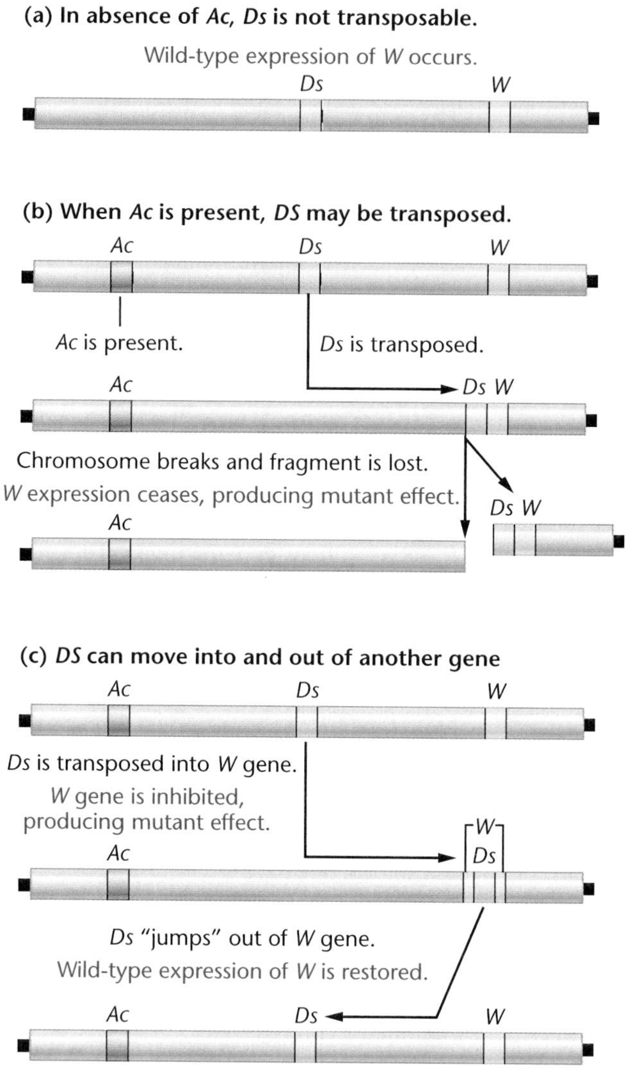

FIGURE 15–22 Effects of *Ac* and *Ds* elements on gene expression. (a) If *Ds* is present in the absence of *Ac*, there is normal expression of a distantly located hypothetical gene *W*. (b) In the presence of *Ac*, *Ds* transposes to a region adjacent to *W*. *Ds* can induce chromosome breakage, which may lead to loss of a chromosome fragment bearing the *W* gene. (c) In the presence of *Ac*, *Ds* may transpose into the *W* gene, disrupting *W* gene expression. If *Ds* subsequently transposes out of the *W* gene, *W* gene expression may return to normal.

The *Ac* sequence contains two 11-base-pair imperfect ITRs, two open reading frames (ORFs), and three noncoding regions. One of the two ORFs encodes the *Ac* transposase enzyme. The first *Ds* element studied (*Ds-a*) is nearly identical to *Ac* except for a 194-bp deletion within the transposase gene. The deletion of part of the transposase gene in the *Ds-a* element explains its dependence on the *Ac* element for transposition. Several other *Ds* elements have also been sequenced, and each contains an even larger deletion within the transposase gene. In each case, however, the ITRs are retained.

Although the significance of Barbara McClintock's mobile controlling elements was not fully appreciated following her initial observations, molecular analysis has since verified her conclusions. She was awarded the Nobel Prize in Physiology or Medicine in 1983.

Mobile Genetic Elements and Wrinkled Peas: Mendel Revisited

Recent work on transposable elements in plants has led us back to Gregor Mendel and his observations of the inheritance of round and wrinkled peas. The two phenotypes are controlled by alleles of a single gene, *rugosus*. It is now known that the wrinkled phenotype is caused by the absence of an enzyme, **starch-branching enzyme (SBEI)**, that controls the formation of branch points in starch molecules. The lack of starch synthesis in wrinkled peas leads to the accumulation of sucrose and a higher water content and osmotic pressure in the developing seeds. As the seeds mature, those that are wrinkled (genotype *rr*) lose more water than do the smooth seeds (*RR* or *Rr*), producing the wrinkled phenotype.

The structural gene for SBEI has been cloned and characterized in both wild-type and mutant genotypes. In the *rr* genotype, the SBEI protein is nonfunctional because the *SBEI* gene is interrupted by a 0.8-kb insertion, resulting in the production of an abnormal RNA transcript. The inserted DNA has 12-bp inverted repeats at each end that are highly homologous to the ITRs of the transposable element *Ac* from maize and to other *Ac*-like elements from snapdragons and parsley.

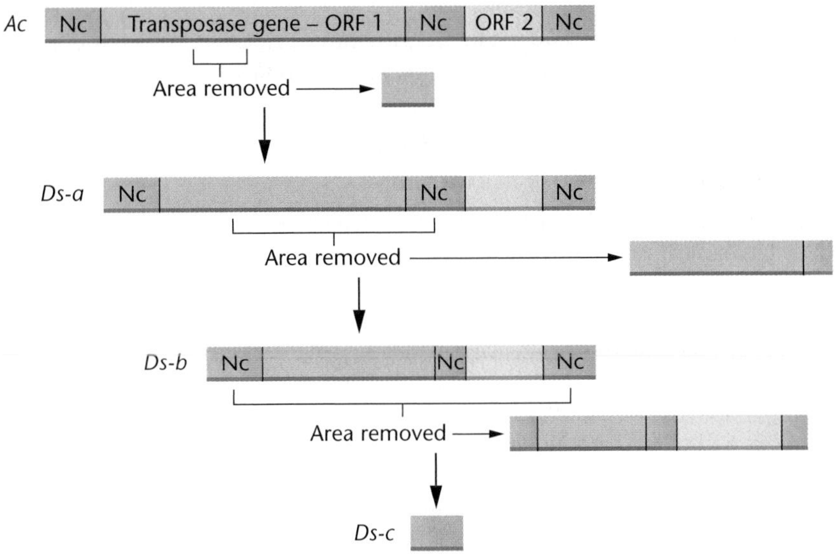

FIGURE 15–23 A comparison of the structure of an *Ac* element and three *Ds* elements. The imperfect inverted repeats are at the ends of the *Ac* element. The transposase gene is in the open reading frame, ORF 1. No function has yet been assigned to ORF 2. Noncoding regions are designated Nc. As this illustration shows, *Ds-a* appears to be an Ac element containing a small deletion in the gene encoding the transposase enzyme.

Copia Elements in *Drosophila*

In 1975, David Hogness and his colleagues David Finnegan, Gerald Rubin, and Michael Young identified a class of DNA elements in *Drosophila melanogaster* that they designated as ***copia***. These elements are transcribed into "copious" amounts of RNA (hence their name). *Copia* elements are present in 10 to 100 copies in the genomes of *Drosophila* cells. Mapping studies show that they are transposable to different chromosomal locations and are dispersed throughout the genome.

Each *copia* element consists of approximately 5000 to 8000 bp of DNA, including a long **direct terminal repeat (DTR)** sequence of 276 bp at each end. Within each DTR is an inverted terminal repeat (ITR) of 17 bp (Figure 15–24). The short ITR sequences are characteristic of *copia* elements. The DTR sequences are found in other transposons in other organisms, but they are not universal.

Insertion of *copia* is dependent on the presence of the ITR sequences and seems to occur preferentially at specific target sites in the genome. The *copia*-like elements demonstrate regulatory effects at the point of their insertion in the chromosome. Certain mutations, including those affecting eye color and segment formation, are due to *copia* insertions within genes. For example, the eye-color mutation *white–apricot* (w^a), an allele of the *white* (w) gene, contains a *copia* element within the gene. Transposition of the copia element out of the w^a allele can restore the allele to wild-type.

Copia elements are only one of approximately 30 families of transposable elements in *Drosophila*, each of which is present in up to 20 to 50 copies in the genome. Together, these families constitute about 5 percent of the *Drosophila* genome and over half of the middle repetitive DNA of this organism. One study suggests that 50 percent of all visible mutations in *Drosophila* are the result of the insertion of transposons into otherwise wild-type genes.

P Element Transposons in *Drosophila*

Perhaps the most significant *Drosophila* transposable elements are the ***P* elements**. These were discovered while studying the phenomenon of hybrid dysgenesis, a condition characterized by sterility, elevated mutation rates, and chromosome rearrangements in the offspring of crosses between certain strains of fruit flies. Hybrid dysgenesis is caused by high rates of *P* element transposition in the germ line, in which transposons insert themselves into or near genes, thereby causing mutations. *P* elements range from 0.5 to 2.9 kb long, with 31-bp ITRs. Full-length *P* elements encode at least two proteins, one of which is the transposase enzyme that is required for transposition, and another is a repressor protein that inhibits transposition. The transposase gene is expressed only in the germ line, accounting for the tissue specificity of *P* element transposition. Strains of flies that contain full-length *P* elements inserted into their genomes are resistant to further transpositions due to the presence of the repressor protein encoded by the *P* elements.

Mutations can arise from several kinds of insertional events. If a *P* element inserts into the coding region of a gene, it can terminate transcription of the gene and destroy normal gene expression. If it inserts into the promoter region of a gene, it can affect the level of expression of the gene. Insertions into introns can affect splicing or cause the premature termination of transcription.

As described in Chapter 21, geneticists have harnessed *P* elements as tools for genetic analysis. One of the most useful applications of *P* elements is as vectors to introduce transgenes into *Drosophila*—a technique known as **germ-line transformation**. *P* elements are also used to generate mutations and to clone mutant genes. In addition, researchers are perfecting methods to target *P* element insertions to precise single-chromosomal sites, which should increase the precision of germ-line transformation in the analysis of gene activity.

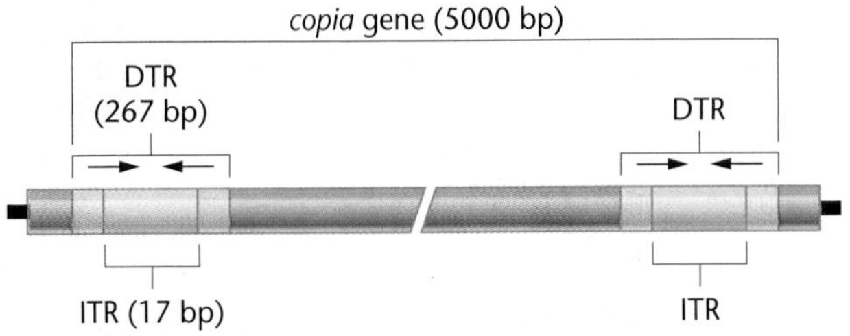

FIGURE 15–24 Structural organization of a copia transposable element in *Drosophila melanogaster*, showing the terminal repeats.

How Do We Know?

How do we know that transposons move within the genome?

Transposable Elements in Humans

The human genome, like that of other eukaryotes, is riddled with DNA derived from transposons. Recent genomic sequencing data reveal that approximately half of the human genome is comprised of transposable element DNA.

As we saw in Chapter 12, the major families of human transposons are the long interspersed elements and short interspersed elements (**LINES** and **SINES**). LINES consist of DNA elements of about 6 kilobase pairs in length, present in up to 850,000 copies. In all, LINES account for 21 percent of human genomic DNA. SINES are about 100 to 500 bp long with about 1.5 million copies present in human cells. SINES comprise about 13 percent of human genomic DNA. Other families of transposable elements account for a further 11 percent of the human genome. As coding sequences comprise only about 5 percent of the human genome, there is about tenfold more transposable element DNA in the human genome than DNA in functional genes.

Although most human transposons appear to be inactive, the potential mobility and mutagenic effects of transposable elements have far-reaching implications for human genetics, as can be seen in a recent example of a transposon "caught in the act." The case involves a male child with hemophilia. One cause of hemophilia is a defect in blood-clotting factor VIII, the product of an X-linked gene. Haig Kazazian and his colleagues found LINES inserted at two points within the gene. Researchers were interested in determining if one of the mother's X chromosomes also contained this specific LINE. If so, the unaffected mother would be heterozygous and pass the LINE-containing chromosome to her son. The surprising finding was that the LINE sequence was *not* present on either of her X chromosomes, but *was* detected on chromosome 22 of both parents. This suggests that this mobile element may have transposed from one chromosome to another in the gamete-forming cells of the mother, prior to being transmitted to the son.

LINE insertions into the human *dystrophin* gene have resulted in at least two separate cases of Duchenne muscular dystrophy. In one case, a transposon inserted into exon 48 and in another case, a transposon inserted into exon 44, both leading to frameshift mutations and premature termination of translation of the dystrophin protein. There are also reports that LINES have inserted into the *APC* and *c-myc* genes, leading to mutations that may have contributed to the development of some colon and breast cancers. In the latter cases, the transposition had occurred within one or a few somatic cells.

SINE insertions are also responsible for a number of human disease cases. In one case, an ***Alu* element** integrated into the BRCA2 gene, inactivating this tumor suppressor gene and leading to a familial case of breast cancer. Other genes that have been mutated by *Alu* integrations are the factor IX gene (leading to hemophilia B), the *ChE* gene (leading to acholinesterasemia) and the *NF1* gene (leading to neurofibromatosis).

Chapter Summary

1. Mutations provide the foundation for genetic variation, evolution, and genetic analysis.
2. A mutation is defined as an alteration in DNA sequence. A mutation may or may not create a detectable phenotype.
3. Mutations may be either spontaneous or induced, somatic or germ-line, autosomal or sex-linked.
4. Mutations can have many different effects on gene function depending upon the type of nucleotide changes that comprise the mutation. Point mutations within coding regions of a gene may lead to missense mutations or nonsense mutations. Insertions or deletions within coding regions may lead to frameshift mutations and premature termination of translation.
5. The phenotypic effects of mutations may be varied, from loss-of-function mutations and gene knockouts, to gain-of-function mutations that confer a new activity to a gene product.
6. Spontaneous mutation rates vary considerably between organisms and between different genes in the same organism.
7. Spontaneous mutations arise from many mechanisms, from errors during DNA replication to damage caused to DNA bases as a result of normal cellular metabolism.
8. Mutations can be induced by many types of chemicals and radiation. These agents can damage both bases and the sugar-phosphate backbone of DNA molecules.
9. The human ABO blood groups owe their variation to point mutations or frameshift mutations within the gene encoding the enzyme that modifies the H substance.
10. The majority of mutations that lead to muscular dystrophies are insertion and deletion mutations that create changes in the translation reading frame of the *dystrophin* gene.
11. Mutations in repeat regions of some genes lead to human diseases such as the fragile X syndrome, myotonic dystrophy, and Huntington disease.
12. Although a range of genetic and biochemical techniques are used to analyze mutations in model organisms, characterization of human mutations usually begins with pedigree analysis.
13. The Ames test allows scientists to estimate the mutagenicity and cancer-causing potential of chemical agents by following the rate of mutation in specific strains of bacteria.
14. Organisms counteract both spontaneous and induced mutations using DNA repair systems. Errors in DNA synthesis can be repaired by proofreading during replication, mismatch repair, and postreplication repair.
15. Other types of DNA damage can be repaired by photoreactivation repair, SOS repair, base excision and nucleotide excision repair, and double-strand break repair.
16. Mutations in genes controlling nucleotide excision repair in humans can lead to diseases such as xeroderma pigmentosum and Cockayne syndrome.
17. Transposable elements can move within a genome and create mutations if they insert within or near genes. They have also been harnessed as genetic dissection tools to create mutations, clone genes, and introduce foreign genes into model organisms.

GENETICS, TECHNOLOGY, AND SOCIETY

Chernobyl's Legacy

On April 26, 1986, Reactor 4 of the Chernobyl Nuclear Power Station exploded and ejected massive amounts of radioactive material into the surrounding countryside and throughout the Northern Hemisphere. The accident killed 31 emergency workers and caused acute radiation sickness in more than 200 others. In the nine days following the initial explosion, as the reactor's temperature approached meltdown, radioactive fission products including radioactive iodine, xenon, strontium, and cesium, were released into the atmosphere. Fallout traveled through central Europe, reaching Finland and Sweden three days after the initial explosion and the United Kingdom and North America within a week. Millions of people were exposed to measurable amounts of radioactivity. People living within 30 km (about 14 miles) of Chernobyl were exposed to high levels of radioactivity prior to their evacuation 36 hours after the accident. Equally high were the exposures of some of the 600,000 military and civilian workers who were sent to Chernobyl to decontaminate the area and encase the shattered reactor in a sarcophagus.

The Chernobyl incident was the world's largest accidental release of radioactive material. The question that remains is whether Chernobyl's radioactive pollution directly threatens the long-term health of millions of people.

More is known about the effects of ionizing radiation on human health than any other toxic agent (except perhaps cigarette smoke). Ionizing radiation (such as X rays and gamma rays) is a subset of radiation that possesses sufficient energy to eject electrons from atoms. Ionizing radiation can damage any cellular component, alter nucleotides, and induce double-strand breaks in DNA. These DNA lesions can lead to mutations or chromosomal translocations.

High doses of ionizing radiation increase the risk of developing certain cancers. The survivors of the atomic-bomb blasts of Hiroshima and Nagasaki showed an increased incidence of leukemia within two years of the bombing. Breast cancers increased 10 years after exposure, as did cancers of the lung, thyroid, colon, ovary, stomach, and nervous system. Since ionizing radiation induces DNA damage, the offspring of the survivors were expected to show increases in birth defects, yet these increases were not detected.

The problem in extrapolating from Hiroshima to Chernobyl is the difference in dose. In atomic-bomb survivors, the cancer rate increased among people exposed to at least 200 mSv (mSv = millisievert, a unit dose of absorbed radiation). It is estimated that people living in the most contaminated areas close to Chernobyl may have received about 50 mSv, and some cleanup workers were exposed to approximately 250 mSv. Outside the Chernobyl area, radiation doses were estimated at 0.4 to 0.9 mSv in Germany and Finland, 0.01 mSv in the United Kingdom, and 0.0006 mSv in the United States.

To put these numbers in perspective, the average dose for medical diagnostic procedures (such as chest and dental X rays) is 0.39 mSv per year. A person's radiation exposure from natural background sources (cosmic rays, rocks, and radon gas) is about 2 to 3 mSv per year. Smokers expose themselves to an average dose of about 2.8 mSv per year from the intake of naturally occurring radioactive materials in tobacco smoke.

It appears that mutation rates among plants and animals exposed to Chernobyl's radioactive waste may be two- to tenfold higher than normal. Also, Chernobyl cleanup workers exhibit a 25 percent higher mutation frequency at the *HPRT* locus. Mutation rates at minisatellites among children born in polluted areas are two fold higher than those from controls in the United Kingdom. But do these increased mutation rates translate into health effects?

Estimates of 26 leukemia cases higher than the 25 to 30 that would occur spontaneously were projected in the 115,000 people evacuated from Chernobyl. It was also estimated that up to 17,000 additional cancers might occur among Europeans, above the 123 million that would occur normally. At present, however, there have been no detectable increases in leukemias or solid tumors in contaminated areas of the former Soviet Union, Finland, or Sweden, or in the 600,000 Chernobyl cleanup workers.

Despite the lack of detectable increases in leukemias and solid tumors, one type of cancer has increased in the Chernobyl area. The rate of childhood thyroid cancer has reached more than 100 cases per million children per year, whereas normal rates are expected to be between 0.5 and 3 cases per million children per year. Although epidemiologists debate whether these increases are entirely due to radiation or to increased reporting, the scale of the increase and the fact that radioactive isotopes of iodine made up a significant portion of the Chernobyl fallout make the link between thyroid cancer and Chernobyl's fallout a plausible one.

The most profound immediate effects of Chernobyl have been psychological. Chernobyl cleanup workers have demonstrated a 50 percent increase in suicides and a detectable increase in smoking- and alcohol-related disease. This mirrors other studies showing that 45 percent of people living within 300 km of Chernobyl believe that they have a radiation-induced illness. Health effects such as depression, sleep disturbance, hypertension, and altered perception have been documented. Post-traumatic stress may be a greater threat to health than the actual radiation exposure from the accident. People feel that they live in constant danger and are simply awaiting the results of a cancer "lottery." Even if cancer rates and genetic defects do not increase dramatically, the indirect health effects from the Chernobyl Nuclear Power Station explosion have been and continue to be immense.

References

Ginzberg, H.M. 1993. The psychological consequences of the Chernobyl accident—Findings from the International Atomic Energy Agency study. *Publ. Health Rep*. 108:184–92.

Jacob, P. et al. 2000. Thyroid cancer risk in Belarus after the Chernobyl accident: Comparison with external exposures. *Rad. and Env. Biophys*. 39:25–31.

Rahu, M. et al. 1997. The Estonian study of Chernobyl cleanup workers: II. Incidence of cancer and mortality. *Radiation Res*. 147:653–57.

Williams, D. 1994. Chernobyl, eight years on. *Nature* 371:556.

Web Sites

United Nations Scientific Committee on the Effects of Atomic Radiation, 3rd International Conference, June 2001.
http://www.unscear.org/pdffiles/kievconlus.pdf

INSIGHTS AND SOLUTIONS

1. The base analog 2-amino purine (2-AP) substitutes for adenine during DNA replication, but it may base pair with cytosine. The base analog 5-bromouracil (5-BU) substitutes for thymidine, but it may base pair with guanine. Follow the double-stranded trinucleotide sequence shown here through three rounds of replication, assuming that, in the first round, both analogs are present and become incorporated wherever possible. In the second and third round of replication, they are removed. What final sequences occur?

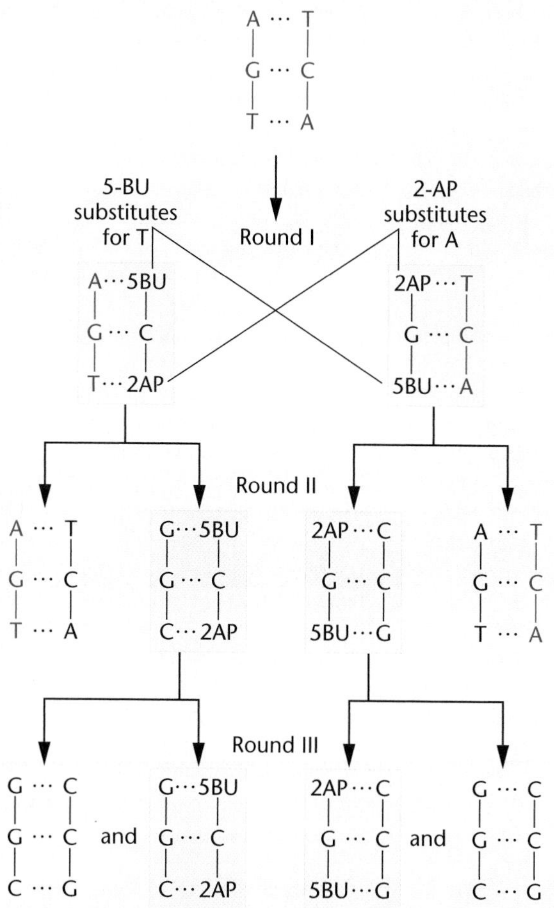

2. A rare dominant mutation expressed at birth was studied in humans. Records showed that six cases were discovered in 40,000 live births. Family histories revealed that in two cases, the mutation was already present in one of the parents. Calculate the spontaneous mutation rate for this mutation. What are some underlying assumptions that may affect our conclusions?

Solution: Only four cases represent a new mutation. Because each live birth represents two gametes, the sample size is from 80,000 meiotic events. The rate is equal to

$$4/80{,}000 = 1/20{,}000 = 5 \times 10^{-5}$$

We have assumed that the mutant gene is fully penetrant and is expressed in each individual bearing it. If it is not fully penetrant, our calculation may be an underestimate because one or more mutations may have gone undetected. We have also assumed that the screening was 100 percent accurate. One or more mutant individuals may have been "missed," again leading to an underestimate. Finally, we assumed that the viability of the mutant and nonmutant individuals is equivalent and that they survive equally *in utero*. Therefore, our assumption is that the number of mutant individuals at birth is equal to the number at conception. If this were not true, our calculation would again be an underestimate.

3. Consider the following estimates:

(a) There are 5.5×10^9 humans living on this planet.

(b) Each individual has about 30,000 (0.3×10^5) genes.

(c) The average mutation rate at each locus is 10^{-5}.

How many spontaneous mutations are currently present in the human population? Assuming that these mutations are equally distributed among all genes, how many new mutations have arisen at each gene in the human population?

Solution: First, since each individual is diploid, there are two copies of each gene per person, each arising from a separate gamete. Therefore, the total number of spontaneous mutations is

$$\begin{aligned}&(2 \times 0.3 \times 10^5 \text{ genes/individual}) \times (5.5 \times 10^9 \text{ individuals}) \\ &\quad \times (10^{-5} \text{ mutations/gene}) \\ &= (0.6 \times 10^5) \times (5.5 \times 10^9) \times (10^{-5}) \text{ mutations} \\ &= 3.3 \times 10^9 \text{ mutations in the population}\end{aligned}$$

$$\begin{aligned}&3.3 \times 10^9 \text{ mutations}/0.6 \times 10^5 \text{ genes} \\ &= 5.5 \times 10^4 \text{ mutations per gene in the population}\end{aligned}$$

PROBLEMS AND DISCUSSION QUESTIONS

1. Discuss the importance of mutations in genetic studies.
2. Describe a technique for the detection of nutritional mutants in *Neurospora*.
3. Most mutations are thought to be deleterious. Why, then, is it reasonable to state that mutations are essential to the evolutionary process?
4. Why is a random mutation more likely to be deleterious than beneficial?
5. Most mutations in a diploid organism are recessive. Why?
6. What is meant by a conditional mutation?
7. Describe a tautomeric shift and how it may lead to mutation.
8. Contrast and compare the mutagenic effects of deaminating agents, alkylating agents, and base analogs.
9. Acridine dyes induce frameshift mutations. Why are frameshift mutations likely to be more detrimental than point mutations, in which a single pyrimidine or purine has been substituted?
10. Why are X rays more potent mutagens than UV radiation?
11. Compare the mechanisms by which UV radiation and X rays cause mutations.
12. Contrast the various types of DNA repair mechanisms known to counteract the effects of UV radiation. What is the role of visible light in repairing UV-induced mutations?
13. Mammography is an accurate screening technique for the early detection of breast cancer in humans. Because this technique uses X rays diagnostically, it has been highly controversial. Can you explain why? What reasons justify the use of X rays for such a type of medical diagnosis?
14. Explain the molecular basis of fragile X syndrome, myotonic dystrophy, and Huntington disease and how the types of mutations involved in each relate to the severity of the disorders, as well as to the phenomenon called genetic anticipation.
15. Describe how the Ames test screens for potential environmental mutagens. Why is it thought that a compound that tests positively in the Ames test may also be carcinogenic?
16. What genetic defects result in the disorder xeroderma pigmentosum (XP) in humans? How do these defects create the phenotypes associated with the disorder?
17. In a bacterial culture in which all cells are unable to synthesize leucine (leu^-), a potent mutagen is added, and the cells are allowed to undergo one round of replication. At that point, samples are taken, a series of dilutions is made, and the cells are plated on either minimal medium or minimal medium containing leucine. The first culture condition (minimal medium) allows the growth of only leu^+ cells, while the second culture condition (minimum medium with leucine added) allows the growth of all cells. The results of the experiment are as follows:

Culture Condition	Dilution	Colonies
minimal medium	10^{-1}	18
minimal + leucine	10^{-7}	6

What is the rate of mutation at the locus involved with leucine biosynthesis?
18. Compare the various transposable genetic elements in bacteria, maize, *Drosophila*, and humans. What properties do they share?
19. The initial discovery of IS elements in bacteria involved their presence upstream (5′) of three genes controlling galactose metabolism. All three genes were affected simultaneously although there was only one IS insertion. Offer an explanation as to why this might occur.
20. *Ty*, a transposable element in yeast, contains an open reading frame (ORF) that encodes the enzyme reverse transcriptase. This enzyme synthesizes DNA from an RNA template. Speculate on the role of this gene product in the transposition of *Ty* within the yeast genome.
21. It has been noted that most transposons in humans and other organisms are located in noncoding regions of the genome—regions such as introns, pseudogenes, and stretches of particular types of repetitive DNA. There are several ways to interpret this observation. Describe two possible interpretations. Which interpretation do you favor? Why?
22. Speculate on how improved living conditions and medical care in the developed nations might affect human mutation rates, both neutral and deleterious.
23. The cancer drug melphalan is an alkylating agent of the mustard gas family. It acts in two ways: by causing alkylation of guanine bases and by crosslinking DNA strands together. Describe two ways in which melphalan might kill cancer cells. What are two ways in which cancer cells could repair the DNA damaging effects of melphalan?
24. Muscular dystrophies are some of the most common inherited diseases in humans, resulting from a large number of different mutations in the *dystrophin* gene. Speculate on why this gene appears to suffer so many mutations.
25. Describe how the process of DNA replication can lead to expansion of trinucleotide repeat regions in the gene responsible for Huntington disease.
26. The origin of the mutation that led to hemophilia in Queen Victoria's family is controversial. Her father did not have X-linked hemophilia and there is no evidence that any members of her mother's family had the condition. What are some possibilities to explain how the mutation arose? What types of mutations could lead to the disease?

Extra-Spicy Problems

27. Presented here are hypothetical findings from studies of heterokaryons formed from seven human xeroderma pigmentosum cell strains:

	XP1	XP2	XP3	XP4	XP5	XP6	XP7
XP1	–						
XP2	–	–					
XP3	–	–	–				
XP4	+	+	+	–			
XP5	+	+	+	+	–		
XP6	+	+	+	+	–	–	
XP7	+	+	+	+	–	–	–

Note: + = complementing; – = noncomplementing

These data are measurements of the occurrence of unscheduled DNA synthesis in the fused heterokaryon. None of the strains alone shows any unscheduled DNA synthesis. What does unscheduled DNA synthesis represent? Which strains fall into the same complementation groups? How many different groups are revealed based on these data? What can we conclude about the genetic basis of XP from these data?

28. Imagine yourself as one of the team of geneticists who launched the study of the genetic effects of high-energy radiation on the surviving Japanese population immediately following the atomic bomb attacks at Hiroshima and Nagasaki in 1945. Demonstrate your insights into both chromosomal and gene mutation by outlining a comprehensive short-term and long-term study that addresses the topic of this study. Be sure to include strategies for considering the effects on both somatic and germ-line tissues.

29. Cystic fibrosis (CF) is a severe autosomal recessive disorder in humans that results from a chloride ion–channel defect in epithelial cells. More than 500 mutations have been identified in the 24 exons of the responsible gene (*CFTR*, for cystic fibrosis transmembrane regulator), including dozens of different missense mutations and frameshift mutations, as well as numerous splice-site defects. Although all affected CF individuals demonstrate chronic obstructive lung disease, there is variation in pancreatic enzyme insufficiency (PI). Speculate which types of observed mutations are likely to give rise to less severe symptoms of CF, including only minor PI. Some of the 300 sequence alterations that have been detected within the exon regions of the *CFTR* gene do not give rise to cystic fibrosis. Taking into account your accumulated knowledge of the genetic code, gene expression, protein function, and mutation, describe why this might be so, as if you were explaining it to a freshman biology major.

30. Electrophilic oxidants are known to create modified bases in DNA named 7,8-dihydro-8-oxoguanine (oxoG). Whereas guanine base pairs with cytosine, oxoG base pairs with either cytosine or adenine.
    (a) What are the sources of reactive oxidants within cells that cause this type of base alteration?
    (b) Drawing on your knowledge of nucleotide chemistry, draw the structure of oxoG, and, below it, draw guanine. Opposite guanine, draw cytosine, including the hydrogen bonds that allow these two molecules to base pair. Does the structure of oxoG, in contrast to guanine, provide any hint as to why it base pairs with adenine?
    (c) Assume that an unrepaired oxoG lesion is present in the helix of DNA opposite cytosine. Following several rounds of replication, predict the type of mutation that will occur.
    (d) Which DNA repair mechanisms might work to counteract an oxoG lesion? Which of these is likely to be most effective? (Reference: Bruner, S.D., et al. 2000. *Nature* 403:859–62.)

31. Among Betazoids in the world of *Star Trek* ®, the ability to read minds is under the control of a gene called "mindreader" (abbreviated *mr*). Most Betazoids can read minds, but rare recessive mutations in the *mr* gene result in two alternative phenotypes: *delayed-receivers* and *insensitives*. Delayed-receivers have some mind-reading ability but perform the task much more slowly than normal Betazoids. Insensitives cannot read minds at all. Betazoid genes do not have introns, so the gene only contains coding DNA. It is 3332 nucleotides in length and Betazoids use a *four letter* genetic code.
The table below shows some data from five unrelated *mr* mutations.

Description of Mutation		Phenotype
mr-1	Nonsense mutation in codon 829	delayed-receiver
mr-2	Missense mutation in codon 52	delayed-receiver
mr-3	Deletion of nucleotides 83–150	delayed-receiver
mr-4	Missense mutation in codon 192	insensitive
mr-5	Deletion of nucleotides 83–93	insensitive

For each mutation, provide a plausible explanation for why it gives rise to its associated phenotype and not to the other phenotype. For example, hypothesize why the *mr-1* nonsense mutation in codon 829 gives rise to the milder delayed-receiver phenotype rather than the more severe insensitive phenotype. Then repeat this type of analysis for the other mutations. (More than one explanation is possible, so be creative within *plausible* bounds!)

32. Skin cancer carries a lifetime risk nearly equal to all other cancers combined. Below is a graph (modified from Kraemer. 1997. *Proc. Natl. Acad. Sci. (USA)* 94:11–14), depicting the age of onset of skin cancers in patients with or without XP, where cumulative percentage of skin cancer is plotted against age. The non-XP curve is based on 29,757 cancers surveyed by the National Cancer Institute, and the curve representing those with XP is based on 63 skin cancers from the Xeroderma Pigmentosum Registry. (a) Provide an overview of the information contained in the graph. (b) Explain why individuals with XP show such an early age of onset.

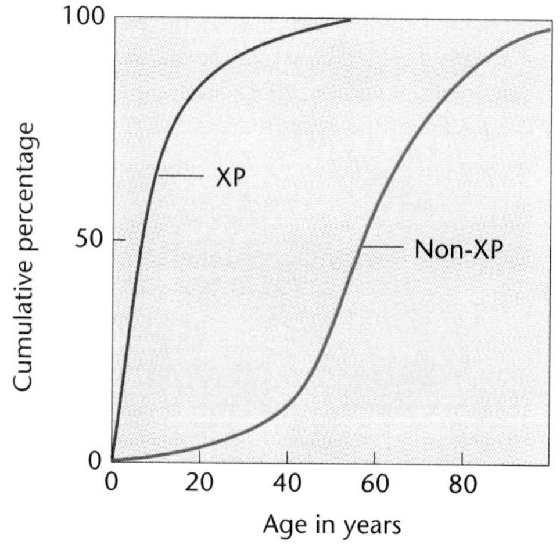

SELECTED READINGS

Ames, B.N., McCann, J., and Yamasaki, E. 1975. Method for detecting carcinogens and mutagens with the *Salmonella*/mammalian microsome mutagenicity test. *Mut. Res.* 31:347–64.

Bates, G., and Leharch, H. 1994. Trinucleotide repeat expansions and human genetic disease. *BioEssays* 16:277–84.

Becker, M.M., and Wang, Z. 1989. Origin of ultraviolet damage in DNA. *J. Mol. Biol.* 210:429–38.

Cairns, J., Overbaugh, J., and Miller, S. 1988. The origin of mutants. *Nature* 335:142–45.

Cleaver, J.E. 1990. Do we know the cause of xeroderma pigmentosum? *Carcinogenesis* 11:875–82.

Cohen, S.N., and Shapiro, J.A. 1980. Transposable genetic elements. *Sci. Am.* (Feb.) 242:40–49.

Comfort, N.C. 2001. *The tangled field: Barbara McClintock's search for the patterns of genetic control*. Cambridge, MA: Harvard University Press.

Crow, J.F., and Denniston, C. 1985. Mutation in human populations. *Adv. Hum. Genet.* 14:59–216.

Den Dunnen, J.T., et al. 1989. Topography of the Duchenne muscular dystrophy (DMD) gene. *Am. J. Hum. Genet.* 45:835–47.

Friedberg, E.C., Walker, G.C., and Siede, W. 1995. *DNA repair and mutagenesis*. Washington, DC: ASM Press.

Hall, B.G. 1990. Spontaneous point mutations that occur more often when advantageous than when neutral. *Genetics* 126:5–16.

Jiricny, J. 1998. Eukaryotic mismatch repair: An update. *Mutation Research* 409:107–21.

Knudson, A.G. 1979. Our load of mutations and its burden of disease. *Am. J. Hum. Genet.* 31:401–13.

Landers, E.S. et al. 2001. Initial sequencing and analysis of the human genome. *Nature* 409:860–921.

Little, J.W., and Mount, D.W. 1982. The SOS regulatory system of *E. coli. Cell* 29:11–22.

Massie, R., and Massie, S. 1975. *Journey*. New York: Knopf.

MacDonald, M.E., et al. 1993. A novel gene containing a trinucleotide repeat that is expanded and unstable in Huntington's disease chromosome. *Cell* 72:971–80.

McClintock, B. 1956. Controlling elements and the gene. *Cold Spring Harbor Symp. Quant. Biol.* 21:197–216.

Miki, Y. 1998. Retrotransposal integration of mobile genetic elements in human disease. *J. Human Genet.* 43:77–84.

Nickoloff, J.A. and Hoekstra, M.F., eds. 2001. *DNA damage and repair*, Vol. III: *Advances from phage to humans*. Totowa, NJ: Humana Press.

O'Hare, K. 1985. The mechanism and control of *P* element transposition in *Drosophila. Trends Genet.* 1:250–54.

Radman, M., and Wagner, R. 1988. The high fidelity of DNA duplication. *Sci. Am.* (Aug.) 259:40–46.

Topal, M.D., and Fresco, J.R. 1976. Complementary base pairing and the origin of substitution mutations. *Nature* 263:285–89.

Wells, R.D. 1994. Molecular basis of genetic instability of triplet repeats. *J. Biol. Chem.* 271:2875–78.

Yamamoto, F., et al. 1990. Molecular genetic basis of the histo-blood group ABO system. *Nature* 345:229–33.

Regulation of Gene Expression in Prokaryotes

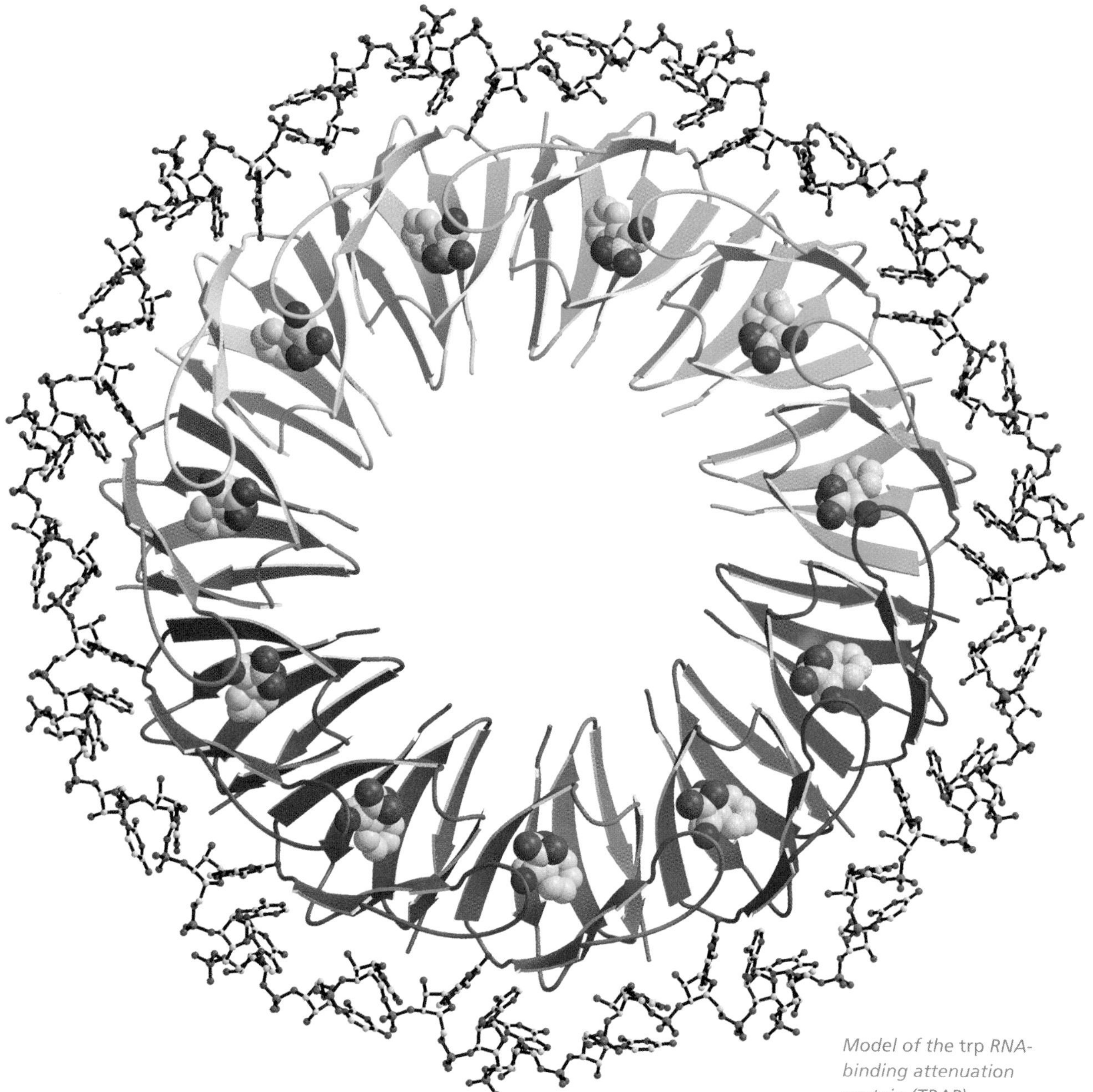

Model of the trp *RNA-binding attenuation protein (TRAP).*

CHAPTER CONCEPTS

- In bacteria, regulation of gene expression is often linked to the metabolic needs of the cell.
- Efficient expression of genetic information in bacteria is dependent on intricate regulatory mechanisms that exert control over transcription.
- Mechanisms that regulate transcription are categorized as exerting either positive or negative control of gene expression.
- Genes that encode enzymes with related functions tend to be organized in clusters and are often under the coordinated control of a single regulatory unit called an operon.
- Transcription of genes within operons is either inducible or repressible.
- The metabolic end product of biosynthetic pathways often serves as the inducer or repressor of gene expression.

We have previously established how DNA is organized into genes, how genes store genetic information, and how this information is expressed. We now consider one of the most fundamental issues in molecular genetics: *How is genetic expression regulated?* Evidence in support of the idea that genes can be turned on and off is very convincing. Detailed analysis of proteins in *E. coli*, for example, has shown that concentrations of the 4000 or so polypeptide chains encoded by the genome vary widely. Some proteins may be present in as few as 5 to 10 molecules per cell, whereas others, such as ribosomal proteins and the many proteins involved in the glycolytic pathway, are present in as many as 100,000 copies per cell. Though most prokaryote gene products exist continuously at a basal level (a few copies), the level can be increased dramatically. Clearly, fundamental regulatory mechanisms must exist to control the expression of the genetic information.

In this chapter, we will explore some of what is known about the regulation of genetic expression in bacteria. As we have seen in a number of previous chapters, these organisms served as excellent research organisms during many seminal investigations in molecular genetics. Bacteria have been especially useful research organisms in genetics for a number of reasons. First, they have extremely short reproductive cycles. Literally hundreds of generations, giving rise to billions of genetically identical bacteria or phages, can be produced in overnight cultures. They can also be studied in "pure culture," whereby mutant strains of genetically unique bacteria can be isolated and investigated independently.

Relevant to our current topic, bacteria also served as an excellent model system for studies involving the induction of genetic transcription in response to changes in environmental conditions. Our focus will be on regulation at the level of the gene. Keep in mind that posttranscriptional regulation also occurs in bacteria. However, we will defer discussion of this level of regulation to a subsequent chapter when we consider eukaryotic regulation.

16.1 Prokaryotes Exhibit Efficient Genetic Mechanisms to Respond to Environmental Conditions

Regulation of gene expression has been extensively studied in prokaryotes, particularly in *E. coli*. We have learned that highly efficient genetic mechanisms have evolved to turn genes on and off, depending on the cell's metabolic need for the respective gene products. Not only do bacteria respond to changes in their environment, but they also regulate gene activity involved in a variety of normal cellular responses (including the replication, recombination, and repair of their DNA) and in cell division and development.

The idea that microorganisms regulate the synthesis of gene products is not a new one. As early as 1900, it was shown that when lactose (a galactose and glucose-containing disaccharide) is present in the growth medium of yeast, the organisms produce enzymes specific to lactose metabolism. When lactose is absent, the enzymes are not manufactured. Soon thereafter, investigators were able to generalize that bacteria adapt to their environment, producing certain enzymes only when specific chemical substrates are present. Such enzymes were thus referred to as **adaptive** (sometimes also called facultative). In contrast, enzymes that are produced continuously, regardless of the chemical makeup of the environment, were called **constitutive**. Since then, the term *adaptive* has been replaced with the more accurate term **inducible**, reflecting the role of the substrate, which serves as the **inducer** in enzyme production.

More recent investigation has revealed a contrasting system, whereby the presence of a specific molecule inhibits genetic expression. This is usually true for molecules that are end products of anabolic biosynthetic pathways. For example, the amino acid tryptophan can be synthesized by bacterial cells. If a sufficient supply of tryptophan is present in the environment or culture medium, it is energetically inefficient for the organism to synthesize the enzymes necessary for tryptophan production. A mechanism has evolved whereby tryptophan plays a role in repressing transcription of mRNA essential to the production of the appropriate biosynthetic enzymes. In contrast to the inducible system controlling lactose metabolism, the system governing tryptophan expression is said to be **repressible**.

Regulation, whether it is inducible or repressible, may be under either **negative** or **positive control**. Under negative control, genetic expression occurs *unless it is shut off by some form of a regulator molecule*. In contrast, under positive control, transcription occurs *only if a regulator molecule directly stimulates RNA production*. In theory, either type of control can govern inducible or repressible systems. Our discussion in the ensuing sections of this chapter will help clarify these contrasting systems of regulation. For the enzymes involved in lactose and tryptophan, negative control is operative.

16.2 Lactose Metabolism in *E. coli* Is Regulated by an Inducible System

Beginning in 1946 with the studies of Jacques Monod and continuing through the next decade with significant contributions by Joshua Lederberg, François Jacob, and Andre L'woff, genetic and biochemical evidence involving lactose metabolism was amassed. Insights were provided into the way in which the gene activity is repressed when lactose is absent, but induced when it is available. In the presence of lactose, the concentration of the enzymes responsible for its metabolism increases rapidly from a few molecules to thousands per cell. The enzymes responsible for lactose metabolism are thus *inducible*, and lactose serves as the *inducer*.

In prokaryotes, genes that code for enzymes with related functions (e.g., genes involved with lactose metabolism) tend to be organized in clusters, and they are often under the coordinated genetic control of a single regulatory unit. The location of this unit is almost always upstream to the gene cluster it controls and, as we learned earlier in our discussion of transcription (Chapter 13), we refer to it as a ***cis*-acting site**. Interactions at the site involve binding molecules that control transcription of the gene cluster. Such molecules are called ***trans*-acting**

elements. Actions at the regulatory site determine whether the genes are expressed and thus whether the corresponding enzymes or other protein products are present. Binding a *trans*-acting element at a *cis*-acting site can regulate the gene cluster either negatively (by turning the genes off) or positively (by turning genes in the cluster on). In this section, we discuss how such bacterial gene clusters are coordinately regulated.

The discovery of a regulatory gene and a regulatory site that are part of the gene cluster was paramount to the understanding of how gene expression is controlled in the system. Neither of these regulatory elements encodes enzymes necessary for lactose metabolism—that is, the function of the three genes in the cluster. As illustrated in Figure 16–1, the three structural genes and the adjacent regulatory site constitute the **lactose**, or ***lac*, operon**. Together, the entire gene cluster functions in an integrated fashion to provide a rapid response to the presence or absence of lactose.

HOW DO WE KNOW?

What evidence established that lactose serves as the inducer of a gene whose product is related to lactose metabolism?

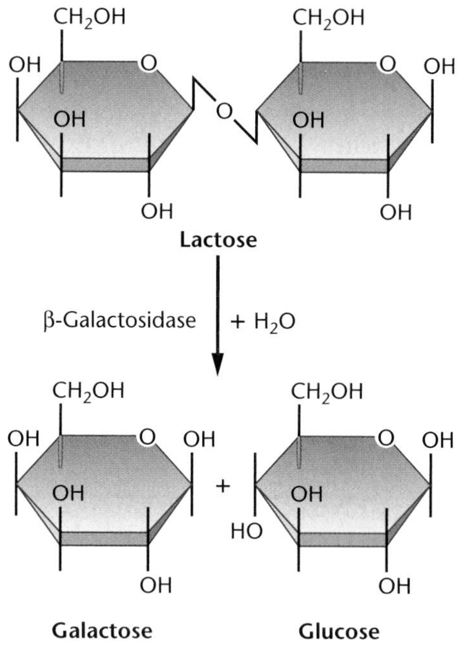

FIGURE 16–2 The catabolic conversion of the disaccharide lactose into its monosaccharide units, galactose and glucose.

Structural Genes

Genes coding for the primary structure of the enzymes are called **structural genes**. There are three structural genes in the *lac* operon. The *lacZ* gene encodes ***β*-galactosidase** an enzyme whose primary role is to convert the disaccharide lactose to the monosaccharides glucose and galactose (Figure 16–2). This conversion is essential if lactose is to serve as the primary energy source in glycolysis. The second gene, *lacY*, specifies the primary structure of **permease**, an enzyme that facilitates the entry of lactose into the bacterial cell. The third gene, *lacA*, codes for the enzyme **transacetylase**. While its physiological role is still not completely clear, it may be involved in the removal of toxic by-products of lactose digestion from the cell.

To study the genes coding for these three enzymes, researchers isolated numerous mutations in order to eliminate the function of one or the other enzyme. Such lac^- mutants were first isolated and studied by Joshua Lederberg. Mutant cells that fail to produce active β-galactosidase ($lacZ^-$) or permease ($lacY^-$) are unable to use lactose as an energy source. Mutations were also found in the transacetylase gene. Mapping studies by Lederberg established that all three genes are closely linked or contiguous to one another in the order *Z–Y–A*. (See Figure 16–1).

Another observation is relevant to what became known about the structural genes. Knowledge of their close linkage led to the discovery that all three genes are transcribed as a single unit, resulting in a polycistronic mRNA (Figure 16–3). This results in the coordinate regulation of all three genes, since a single message serves as the basis for translation of all three gene products.

The Discovery of Regulatory Mutations

How does lactose activate structural genes and induce the synthesis of the related enzymes? A partial answer comes from the discovery and study of **gratuitous inducers**, chemical analogues of lactose such as the sulfur analogue **isopropylthiogalactoside (IPTG)**, shown in Figure 16–4. Gratuitous inducers behave like natural inducers, but they do not serve as substrates for the enzymes that are subsequently synthesized. Their discovery provides strong evidence that the primary induction event does *not* depend on the interaction between the inducer and the enzyme.

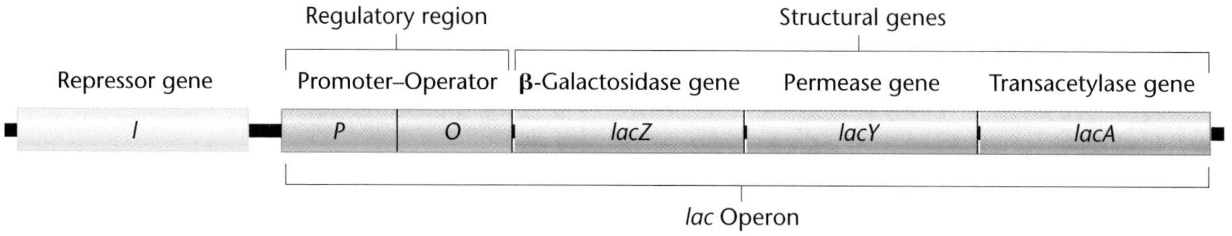

FIGURE 16–1 A simplified overview of the genes and regulatory units involved in the control of lactose metabolism. (This region of DNA is not drawn to scale.) A more detailed model will be developed later in this chapter. (See Figure 16–10.)

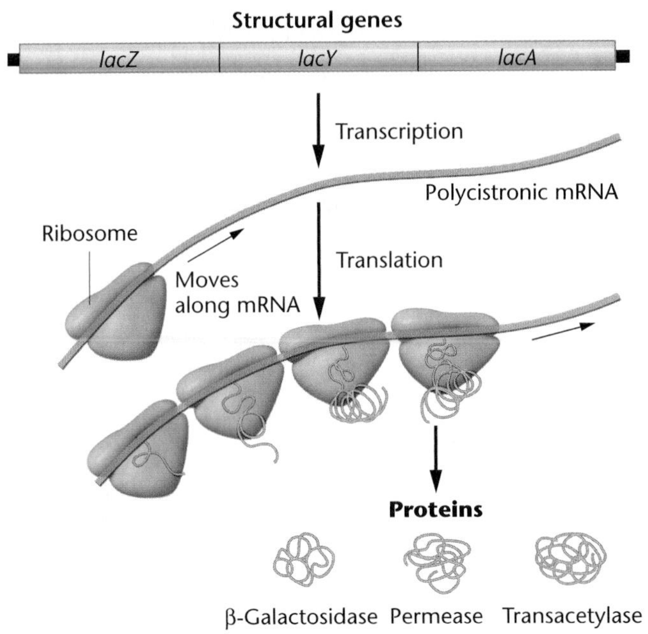

FIGURE 16–3 The structural genes of the *lac* operon are transcribed into a single polycistronic mRNA, which is translated simultaneously by several ribosomes into the three enzymes encoded by the operon.

What, then, is the role of lactose in induction? The answer to this question required the study of another class of mutations called **constitutive mutants**. In this type of mutation, the enzymes are produced regardless of the presence or absence of lactose. Maps of the first type of constitutive mutation, $lacI^-$, showed that it is located at a site on the DNA close to, but distinct from, the structural genes. The *lacI* gene is appropriately called a **repressor gene**. A second set of constitutive mutations producing identical effects was found in a region immediately adjacent to the structural genes. This class of mutations, designated $lacO^C$, identifies the **operator region** of the operon. Because inducibility has been eliminated in both types of constitutive mutation (the enzymes are continually produced), regulation has been disrupted by genetic changes.

The Operon Model: Negative Control

Around 1960, Jacob and Monod proposed a scheme involving negative control called the **operon model**, whereby a group of genes is regulated and expressed together as a unit. As we saw in Figure 16–1, the *lac* operon they proposed consists of the *Z*, *Y*, and *A* structural genes, as well as the adjacent sequences of DNA referred to as the *operator region*. They argued that the *lacI* gene regulates the transcription of the structural genes by producing a **repressor molecule**, and that the repressor is **allosteric**, meaning that the molecule reversibly interacts with another molecule, causing both a conformational change in three-dimensional shape and a change in chemical activity. Figure 16–5 illustrates the components of the *lac* operon as well as the action of the *lac* repressor in the presence and absence of lactose.

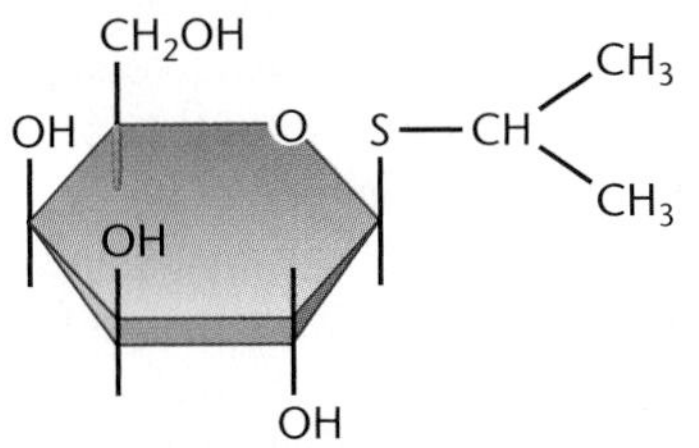

FIGURE 16–4 The gratuitous inducer isopropylthiogalactoside (IPTG).

Jacob and Monod suggested that the repressor normally interacts with the DNA sequence of the operator region. When it does so, it inhibits the action of RNA polymerase, effectively repressing the transcription of the structural genes [Figure 16–5(b)]. However, when lactose is present, this sugar binds to the repressor and causes an allosteric conformational change. This change alters the binding site of the repressor, rendering it incapable of interacting with operator DNA [Figure 16–5(c)]. In the absence of the repressor–operator interaction, RNA polymerase transcribes the structural genes, and the enzymes necessary for lactose metabolism are produced. Because transcription occurs only when the repressor *fails* to bind to the operator region, regulation is said to be under *negative control*.

The operon model uses these potential molecular interactions to explain the efficient regulation of the structural genes. In the absence of lactose, the enzymes encoded by the genes are not needed and are repressed. When lactose is present, it indirectly induces the activation of the genes by binding with the repressor.* If all lactose is metabolized, none is available to bind to the repressor, which is again free to bind to operator DNA and repress transcription.

Both the I^- and O^C constitutive mutations interfere with these molecular interactions, allowing continuous transcription of the structural genes. In the case of the I^- mutant, seen in Figure 16–6(a), the repressor protein is altered and cannot bind to the operator region, so the structural genes are always turned on. In the case of the O^C mutant [Figure 16–6(b)], the nucleotide sequence of the operator DNA is altered and will not bind with a normal repressor molecule. The result is the same: structural genes are always transcribed.

HOW DO WE KNOW?

What experimental observations underlie the prediction that a repressor molecule is produced that regulates the *lac* operon?

Genetic Proof of the Operon Model

The operon model is a good one because it leads to three major predictions that can be tested to determine its validity. The major predictions to be tested are that (1) the *I* gene produces a diffusible cellular product; (2) the *O* region is involved in regulation, but does not produce a product; and (3) the *O* region must be adjacent to the structural genes in order to regulate transcription.

*Technically, allolactose, an isomer of lactose, is the inducer. When lactose enters the bacterial cell, some of it is converted to allolactose by the β-galactosidase enzyme.

Web Tutorial 16.1
Regulation of Gene Expression: Prokaryotes

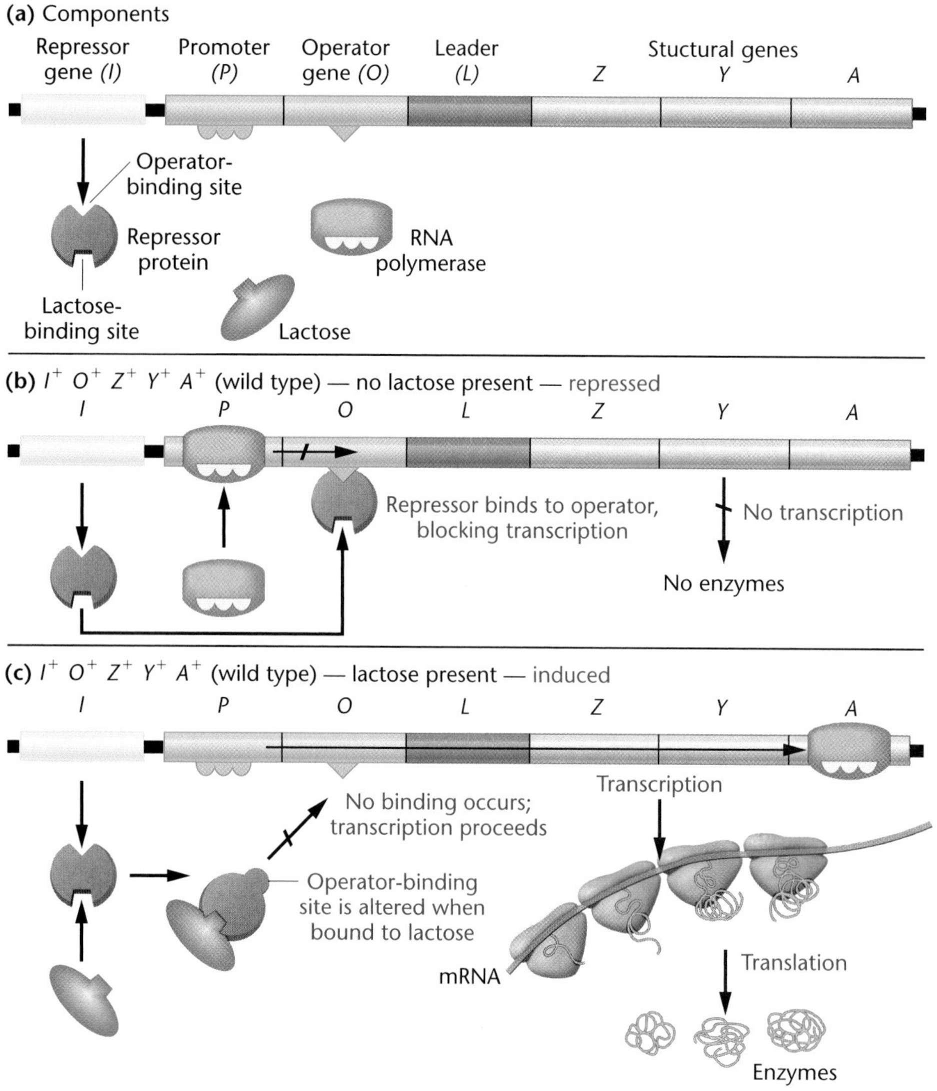

FIGURE 16–5 The components of the wild-type *lac* operon and the response in the absence and the presence of lactose.

The construction of partially diploid bacteria allows us to assess these assumptions, particularly those that predict *trans*-acting regulatory elements. For example, as introduced in Chapter 6, the F plasmid may contain chromosomal genes, in which case it is designated F′. When an F^+ cell acquires such a plasmid, it contains its own chromosome plus one or more additional genes present in the plasmid. This creates a host cell, called a **merozygote**, that is diploid for those genes. The use of such a plasmid makes it possible, for example, to introduce an I^+ gene into a host cell whose genotype is I^-, or to introduce an O^+ region into a host cell of genotype O^C. The Jacob–Monod operon model predicts how regulation should be affected in such cells. Adding an I^+ gene to an I^- cell should restore inducibility, because the normal wild-type repressor, which is a *trans*-acting factor, would be produced by the inserted I^+ gene. Adding an O^+ region to an O^C cell should have no effect on constitutive enzyme production, since regulation depends on an O^+ region immediately adjacent to the structural genes—that is, O^+ is a *cis*-acting regulator.

Results of these experiments are shown in Table 16.1, where Z represents the structural genes. The inserted genes are listed after the designation F′. In both cases described here, the Jacob–Monod model is upheld (part B of Table 16.1). Part C shows the reverse experiments, where either an I^- gene or an O^C region is added to cells of normal inducible genotypes. As the model predicts, inducibility is maintained in these partial diploids.

Another prediction of the operon model is that certain mutations in the *I* gene should have the opposite effect of I^-. That is, instead of being constitutive because the repressor can't bind the operator, mutant repressor molecules should be produced that cannot interact with the inducer, lactose. As a result, the repressor would always bind to the operator sequence, and the structural genes would be permanently repressed (Figure 16–7). If this were the case, the presence of an additional I^+ gene would have little or no effect on repression.

In fact, such a mutation, I^S, was discovered wherein the operon is "superrepressed," as shown in part D of Table 16.1. An additional I^+ gene does not effectively relieve repression of gene activity. These observations are consistent with the idea that the repressor contains separate allosteric sites involved in regulation.

TABLE 16.1 A COMPARISON OF GENE ACTIVITY (+ OR −) IN THE PRESENCE OR ABSENCE OF LACTOSE FOR VARIOUS *E. COLI* GENOTYPES

	Genotype	Presence of β-Galactosidase Activity: Lactose Present	Lactose Absent
	$I^+O^+Z^+$	+	−
A.	$I^+O^+Z^-$	−	−
	$I^-O^+Z^+$	+	+
	$I^+O^cZ^+$	+	+
B.	$I^-O^+Z^+/F'I^+$	+	−
	$I^+O^cZ^+/F'O^+$	+	+
C.	$I^+O^+Z^+/F'I^-$	+	−
	$I^+O^+Z^+/F'O^c$	+	−
D.	$I^sO^+Z^+$	−	−
	$I^sO^+Z^+/F'I^+$	−	−

Note: In parts B to D, most genotypes are partially diploid, containing an F factor plus attached genes (F′).

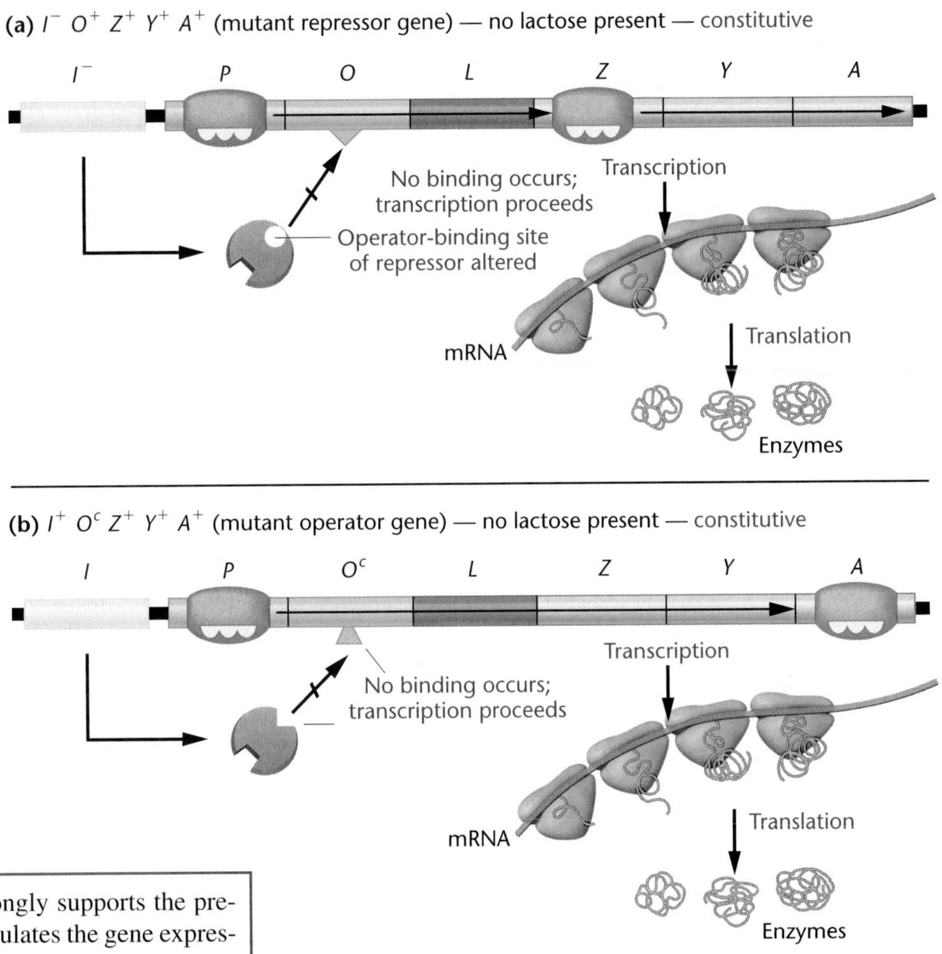

FIGURE 16–6 The response of the *lac* operon in the absence of lactose when a cell bears either the I^- or the O^C mutation.

HOW DO WE KNOW?

What experimental evidence strongly supports the prediction that a repressor molecule regulates the gene expression in the *lac* operon?

Now solve this

Problem 16.6 on page 408 asks you to predict the outcome of gene expression under the conditions of varying genotypes and cellular conditions.

Hint: Determine initially whether the repressor is active or inactive based on whether the gene encoding it is wild type or mutant. Then, consider the impact of the presence or absence of lactose.

Isolation of the Repressor

Although Jacob and Monod's operon theory succeeded in explaining many aspects of genetic regulation in prokaryotes, the nature of the repressor molecule was not known when their landmark paper was published in 1961. While they had assumed that the allosteric repressor was a protein, RNA was also a candidate because activity of the molecule required the ability to bind to DNA. Despite many attempts to isolate and characterize the hypothetical repressor molecule, no direct chemical evidence was immediately forthcoming. A single *E. coli* cell contains no more than 10 or so copies of the *lac*

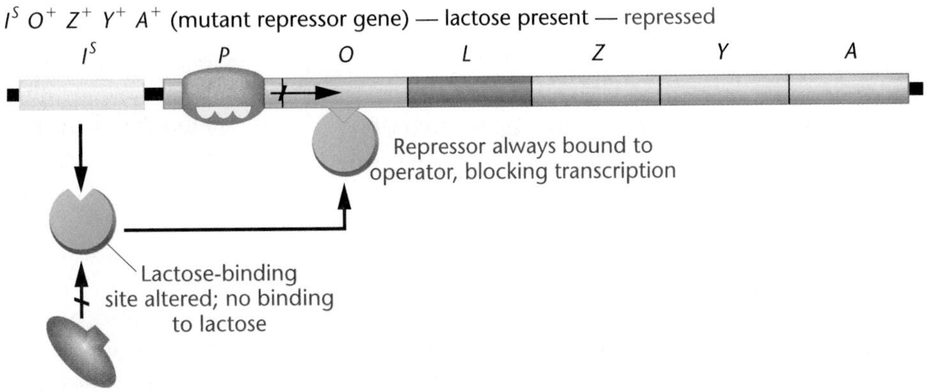

FIGURE 16–7 The response of the *lac* operon in the presence of lactose in a cell bearing the I^S mutation.

repressor; direct chemical identification of 10 molecules in a population of millions of proteins and RNAs in a single cell presented a tremendous challenge.

In 1966, Walter Gilbert and Benno Müller-Hill reported the isolation of the *lac* repressor in partially purified form. To achieve the isolation, they used a *regulator quantity* (I^q) mutant strain that contains about 10 times as much repressor as do wild-type *E. coli* cells. Also instrumental in their success was the use of the gratuitous inducer IPTG, which binds to the repressor, and the technique of **equilibrium dialysis**. In this technique, extracts of I^q cells were placed in a dialysis bag and allowed to attain equilibrium with an external solution of radioactive IPTG, which is small enough to diffuse freely in and out of the bag. At equilibrium, the concentration of IPTG was higher inside the bag than in the external solution, indicating that an IPTG-binding material was present in the cell extract and that this material was too large to diffuse across the wall of the bag.

Ultimately, the IPTG-binding material was purified and shown to have various characteristics of a protein. In contrast, extracts of I^- constitutive cells having no *lac* repressor *activity* did not exhibit IPTG-binding, strongly suggesting that the isolated protein was the repressor molecule.

To confirm this thinking, Gilbert and Müller-Hill grew *E. coli* cells in a medium containing radioactive sulfur and then isolated the IPTG-binding protein, which was labeled in its sulfur-containing amino acids. This protein was mixed with DNA from a strain of phage lambda (λ), carrying the $lacO^+$ gene. The DNA sedimented at 40*S*, while the IPTG-binding protein sedimented at 7*S*. The DNA and protein were mixed and sedimented in a gradient, using ultracentrifugation. The radioactive protein sedimented at the same rate as did DNA, indicating that the protein binds to the DNA. Further experiments showed that the IPTG-binding, or repressor, protein binds only to DNA containing the *lac* region and does not bind to *lac* DNA containing an operator-constitutive (O^C) mutation.

Now solve this

Problem 16.8 on page 408 asks you to describe how the *lac* repressor was isolated.

Hint: You must first be clear that equilibrium dialysis allows small molecules such as lactose or the gratuitous inducer IPTG to move freely back and forth across the membrane, while large molecules such as the repressor cannot and remain in the same position throughout the experiment.

16.3 The Catabolite-Activating Protein (CAP) Exerts Positive Control over the *lac* Operon

As is apparent from the preceding discussion of the *lac* operon, the role of β-galactosidase is to cleave lactose into its components glucose and galactose. Then, for galactose to be used by the cell, it is converted to glucose. What if the cell found itself in an environment that contained an ample amount of lactose *and* glucose? It would not be energetically efficient for a cell to be "induced" by lactose to make β-galactosidase, since what it really needs, glucose, is already present. As we shall see next, still another molecular component, called the **catabolite-activating protein (CAP)**, is involved in effectively repressing the expression of the *lac* operon when glucose is present. This inhibition, called **catabolite repression**, reflects the greater simplicity with which glucose may be metabolized in comparison to lactose. The cell "prefers" glucose, and if it is available, the *lac* operon is not activated, even when lactose is present.

To understand CAP and its role in regulation, let's backtrack for a moment. When the *lac* repressor is bound to the inducer, the *lac* operon is activated and RNA polymerase transcribes the structural genes. As stated in Chapter 13, transcription is initiated as a result of the binding that occurs between RNA polymerase and the nucleotide sequence of the **promoter region**, found upstream (5′) from the initial coding sequences. Within the *lac* operon, the promoter is found between the *I* gene and the operator region (*O*). (See Figure 16–1.) Careful examination has revealed that polymerase binding is never very efficient unless CAP is also present to facilitate the process.

The mechanism is summarized in Figure 16–8. In the absence of glucose and under inducible conditions, CAP exerts **positive control** by binding to the CAP site, facilitating RNA polymerase binding at the promoter, and thus transcription. Therefore, for maximal transcription, the repressor must be bound by lactose (so as not to repress operon expression), *and* CAP must be bound to the CAP-binding site.

This leads to the central question about CAP. What role does glucose play in inhibiting CAP binding when it is present? The answer involves still another molecule, **cyclic adenosine monophosphate (cAMP)**, upon which CAP binding is dependent. In order to bind to the promoter, CAP must be bound to cAMP. The level of cAMP is itself dependent on an enzyme, **adenyl cyclase**, which catalyzes the conversion of ATP to cAMP.* (See Figure 16–9.)

The role of glucose in catabolite repression is now clear. It inhibits the activity of adenyl cyclase, causing a decline in the level of cAMP in the cell. Under this condition, CAP cannot form the CAP–cAMP complex essential to the positive control of transcription of the *lac* operon.

Like the *lac* repressor, CAP and cAMP–CAP have been examined by using X-ray crystallography. CAP is a dimer that inserts into adjacent regions of a specific nucleotide sequence of the DNA making up the promoter. The cAMP–CAP complex, when bound to DNA, bends it, causing it to assume a new conformation.

Binding studies in solution further clarify the mechanism of gene activation. Alone, neither cAMP–CAP nor RNA polymerase has a strong affinity to bind to *lac* promoter DNA, nor does either molecule have a strong affinity to bind to the other. However, when both are together in the presence of the *lac* promoter DNA, a tightly bound complex is formed, an example of what is called **cooperative binding** in biochemical terms. In the case of cAMP–CAP and

*Because of its involvement with cAMP, CAP is also called cyclic AMP receptor protein (CRP), and the gene encoding the protein is named *crp*. Since the protein was first named CAP, we will adhere to the initial nomenclature.

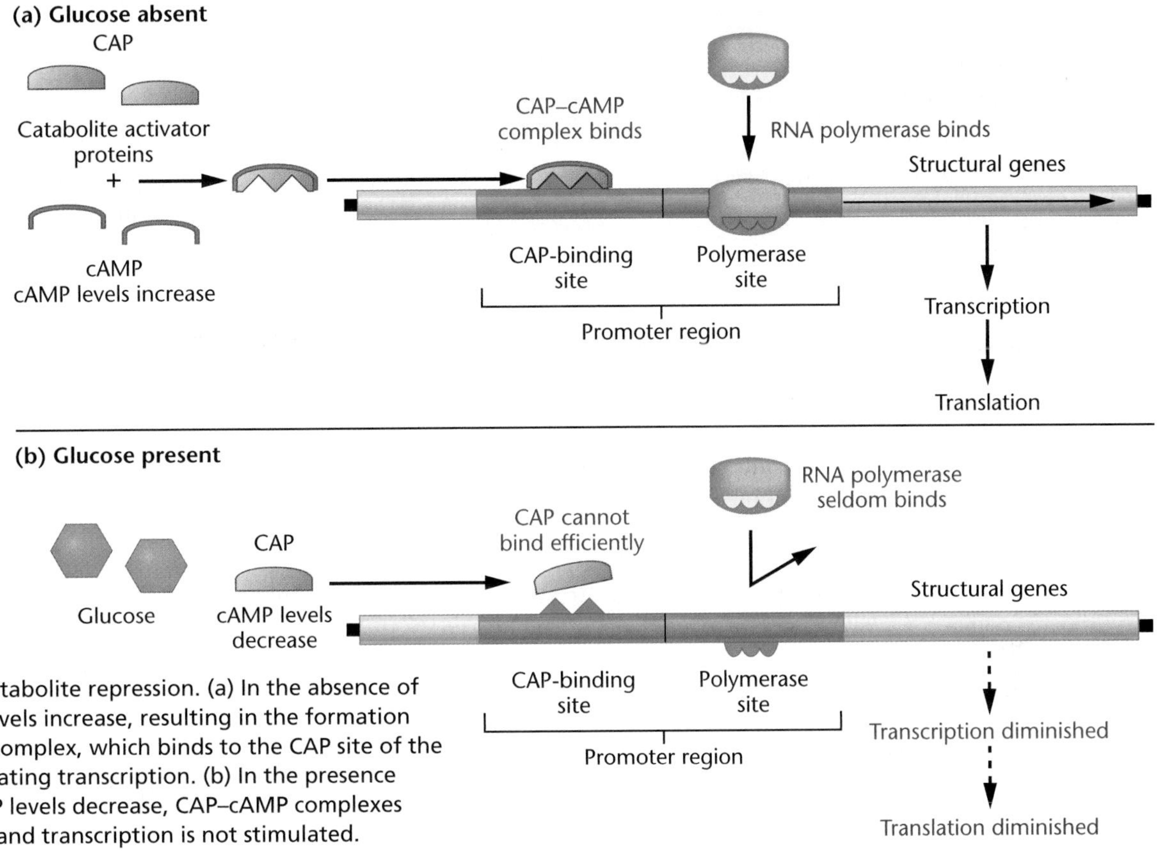

FIGURE 16–8 Catabolite repression. (a) In the absence of glucose, cAMP levels increase, resulting in the formation of a CAP–cAMP complex, which binds to the CAP site of the promoter, stimulating transcription. (b) In the presence of glucose, cAMP levels decrease, CAP–cAMP complexes are not formed, and transcription is not stimulated.

the *lac* operon, the phenomenon illustrates the high degree of specificity that is involved in the genetic regulation of just one small group of genes.

Regulation of the *lac* operon by catabolite repression results in efficient energy use, because the presence of glucose will override the need for the metabolism of lactose, should it also be available to the cell. Catabolite repression involving CAP has also been observed for other inducible operons, including those controlling the metabolism of galactose and arabinose.

Now solve this

Problem 16.10 on page 408 asks you to assess the level of gene activity in the *lac* operon when considering various combinations of the presence or absence of both lactose and glucose.

Hint: You must keep in mind that regulation involving lactose is a negative control system, while regulation involving glucose is a positive control system.

16.4 Crystal Structure Analysis of Repressor Complexes Has Confirmed the Operon Model

We now have thorough knowledge of the biochemical nature of the regulatory region of the *lac* operon, identifying the precise locations of its various components relative to one another (Figure 16–10). In 1996, Mitchell Lewis, Ponzy Lu, and their colleagues succeeded in determining the crystal structure of the *lac* repressor, as well as the structure of the repressor bound to the inducer and to operator DNA. As a result, previous information that was based on genetic and biochemical data has now been complemented with the

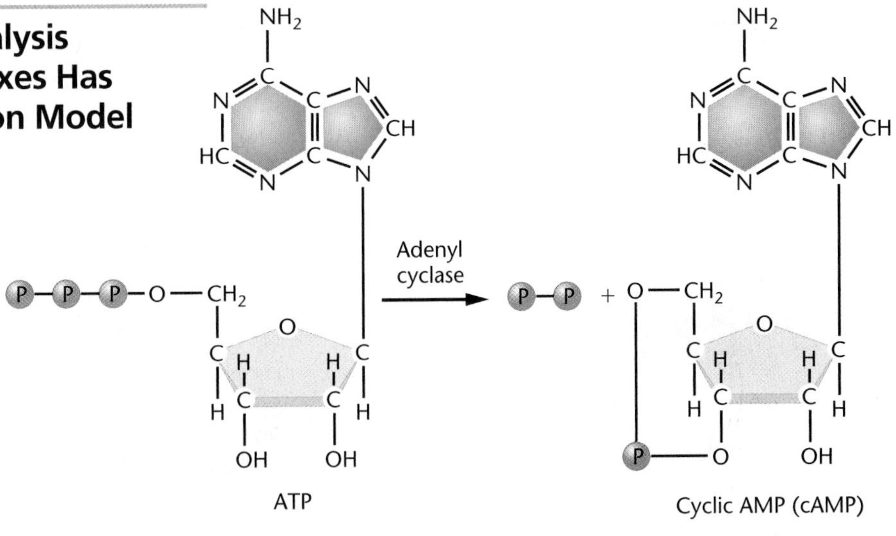

FIGURE 16–9 The formation of cAMP from ATP, catalyzed by adenyl cyclase.

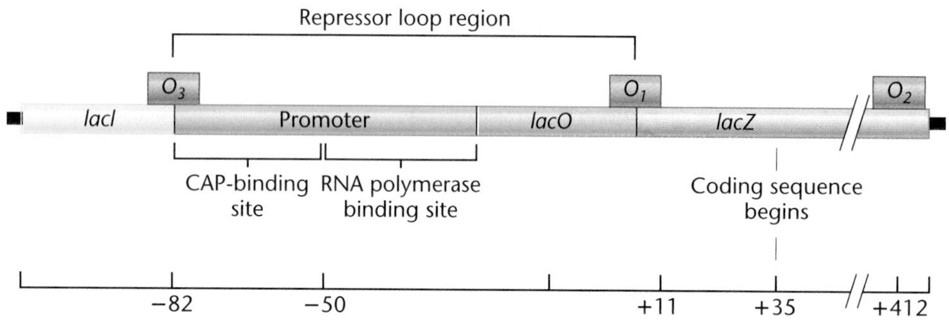

FIGURE 16–10 A detailed depiction of various regulatory regions involved in the control of genetic expression of the *lac* operon, as described in the text. The numbers on the bottom scale represent nucleotide sites upstream and downstream from the initiation of transcription.

missing structural interpretation. Together, a nearly complete picture of the regulation of the operon has emerged.

The repressor, as the gene product of the *I* gene, is a monomer consisting of 360 amino acids. Within the monomer, the region of inducer binding has been identified [Figure 16–11(a)]. While dimers also bind, the functional repressor contains four such subunits, creating a homotetramer. The tetramer can be cleaved with a protease under controlled conditions and yields five fragments. Four are derived from the N-terminal ends of the tetramer, and they bind, to operator DNA. The fifth fragment is the remaining core of the tetramer, derived from the COOH-terminus ends; it binds to lactose and gratuitous inducers such as IPTG. Analysis has revealed that, at any single time, each tetramer can bind to two symmetrical operator DNA helices [Figure 16–11(b)].

The operator DNA that was previously defined by mutational studies ($lacO^C$) and confirmed by DNA-sequencing analysis is located just upstream from the beginning of the actual coding sequence of the *lacZ* gene. Crystallographic studies show that the actual region of repressor binding of this primary operator O_1 consists of 21 base pairs. Two other auxiliary operator regions have been identified, as shown in Figure 16–10. One, O_2, is 401 base pairs downstream from the primary operator within the *lacZ* gene. The other, O_3, is 93 base pairs upstream from O_1, just above the CAP site. *In vivo*, all three operators must be bound for maximum repression.

Binding by the repressor at two operator sites distorts the conformation of DNA, causing it to bend away from the repressor. When a model is created involving dual binding of operators O_1 and O_3 [Figure 16–11(c)], the 93 base pairs of DNA that intervene must jut out, forming what is called a **repression loop**. This model positions the promoter region that binds RNA polymerase on the inside of the loop, which prevents access during repression. In addition, the repression loop positions the CAP-binding site in a way to facilitate CAP interaction with RNA polymerase upon subsequent induction. The finding of DNA looping during repression is

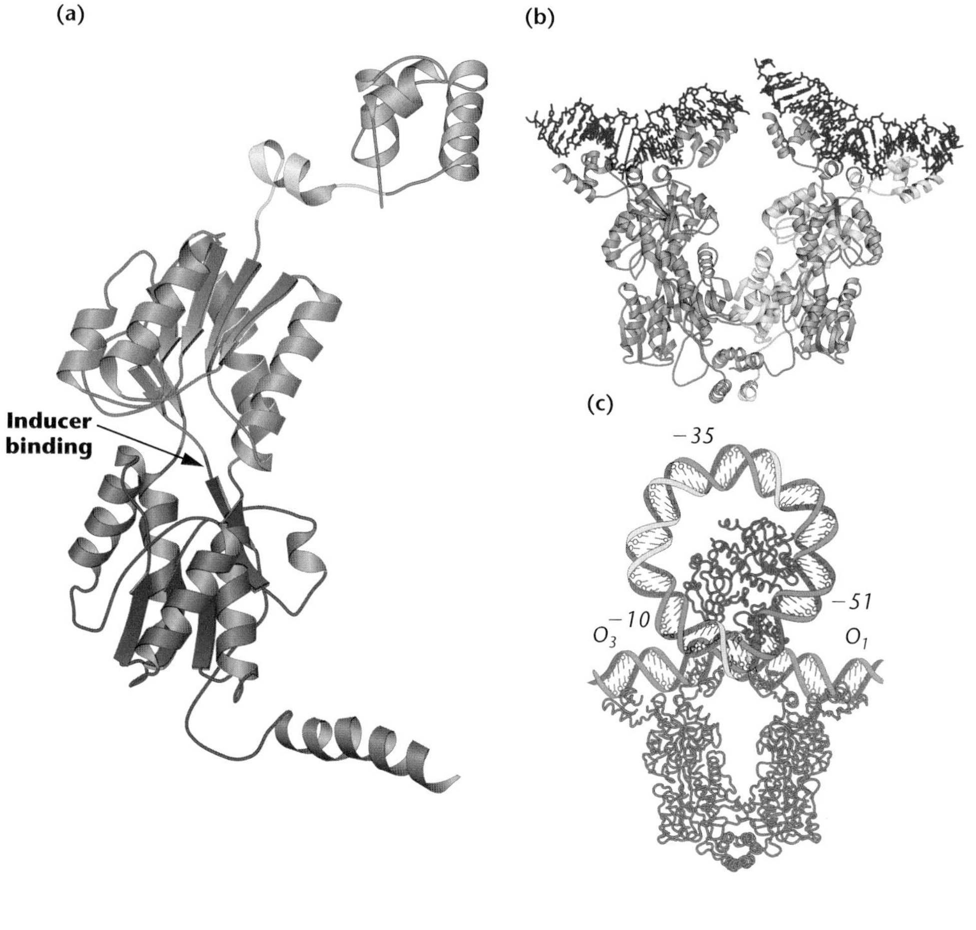

FIGURE 16–11 Models of the *lac* repressor and its binding to operator sites in DNA, as generated from crystal structure analysis. (a) The repressor monomer. The arrow points to the inducer-binding site. The DNA-binding region is shown in red. (b) The repressor dimer bound to two 21-base-pair segments of operator DNA (shown in blue). (c) The repressor and CAP (shown in dark blue) bound to the *lac* DNA. Binding to operator regions O_1 and O_3 creates a 93-base-pair repression loop of promoter DNA.

similar to transitions that are predicted to occur in eukaryotic systems. (See Chapter 17.)

Studies have also defined the three-dimensional conformational changes that accompany the allosteric transitions occuring during the interactions with the inducer molecules. Taken together, the crystallographic studies bring us to a new level of understanding of the regulatory process occurring within the *lac* operon, confirming the findings and predictions of Jacob and Monod in their model set forth over 40 years ago and based strictly on genetic grounds.

16.5 The tryptophan (*trp*) Operon in *E. coli* Is a Repressible Gene System

Although the process of induction had been known for some time, it was not until 1953 that Monod and colleagues discovered a repressible operon. Wild-type *E. coli* are capable of producing the enzymes necessary for the biosynthesis of amino acids as well as other essential macromolecules. Focusing his studies on the amino acid tryptophan and the enzyme **tryptophan synthetase**, Monod discovered that if tryptophan is present in sufficient quantity in the growth medium, the enzymes necessary for its synthesis are not produced. Energetically, repression of the genes involved in the production of these enzymes is highly economical for the cell when ample tryptophan is present.

Further investigation showed that a series of enzymes encoded by five contiguous genes on the *E. coli* chromosome is involved in tryptophan synthesis. These genes are part of an operon, and in the presence of tryptophan, all are coordinately repressed, and none of the enzymes are produced. Because of the great similarity between this repression and the induction of enzymes for lactose metabolism, Jacob and Monod proposed a model of gene regulation analogous to the *lac* system (Figure 16–12).

To account for repression, they suggested the presence of a *normally inactive repressor* that alone cannot interact with the operator region of the operon. However, the repressor is an allosteric molecule that can bind to tryptophan. When this amino acid is present, the resultant complex of repressor and tryptophan attains a new conformation that binds to the operator, repressing transcription. Thus, when tryptophan, the end product of this anabolic pathway, is present, the system is repressed and enzymes are not made. Since the regulatory complex inhibits transcription of the operon, this repressible system is under negative control. And, as tryptophan participates in repression, it is referred to as a **corepressor** in this regulatory scheme.

Evidence for the *trp* Operon

Support for the concept of a repressible operon was soon forthcoming, based primarily on the isolation of two distinct categories of constitutive mutations. The first class, *trpR*$^-$, maps at a

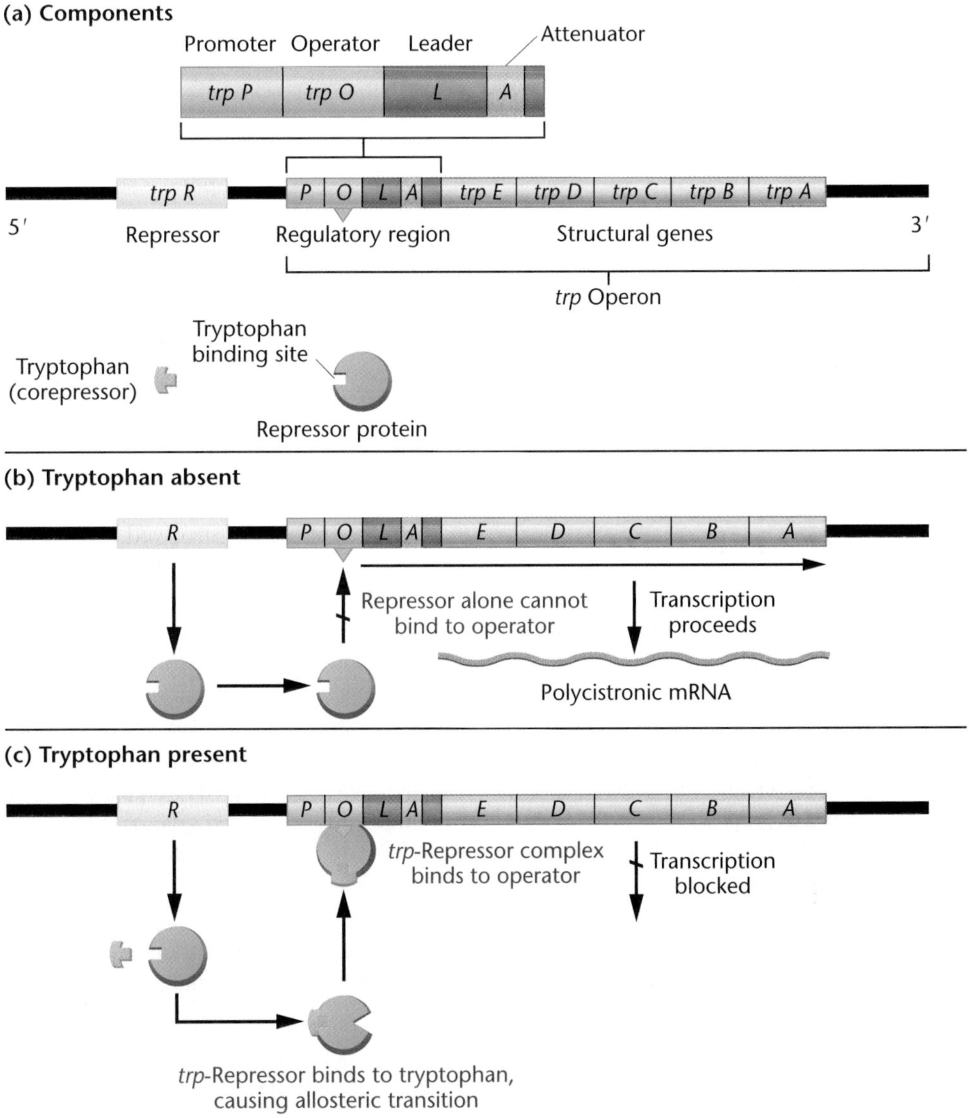

FIGURE 16–12 (a) The components involved in the regulation of the tryptophan operon. (b) Regulatory conditions are depicted that involve either activation or (c) repression of the structural genes. In the absence of tryptophan, an inactive repressor is made that cannot bind to the operator (*O*), thus allowing transcription to proceed. In the presence of tryptophan, it binds to the repressor, causing an allosteric transition to occur. This complex binds to the operator region, leading to repression of the operon.

considerable distance from the structural genes. This locus represents the gene coding for the repressor. Presumably, the mutation either inhibits the interaction of the repressor with tryptophan or inhibits repressor formation entirely. Whichever the case, no repression ever occurs in cells with the *trpR*$^-$ mutation. As expected, if the *trpR*$^+$ gene encodes a functional repressor molecule, the presence of a copy of this gene will restore repressibility.

The second constitutive mutant is analogous to that of the operator of the lactose operon, because it maps immediately adjacent to the structural genes. Furthermore, the addition of a wild-type operator gene into mutant cells (as an external element) does not restore enzyme repression. This is predictable if the mutant operator no longer interacts with the repressor–tryptophan complex.

The entire *trp* operon has now been well defined, as shown in Figure 16–12. Five contiguous structural genes (*trp E, D, C, B,* and *A*) are transcribed as a polycistronic message directing translation of the enzymes that catalyze the biosynthesis of tryptophan. As in the *lac* operon, a promoter region (*trpP*) represents the binding site for RNA polymerase, and an operator region (*trpO*) binds the repressor. In the absence of binding, transcription is initiated within the overlapping *trpP–trpO* region and proceeds along a **leader sequence** 162 nucleotides prior to the first structural gene (*trpE*). Within that leader sequence, still another regulatory site has been demonstrated, called an *attenuator*—the subject of the next section of this chapter. As we shall see, this regulatory unit is an integral part of the control mechanism of the operon.

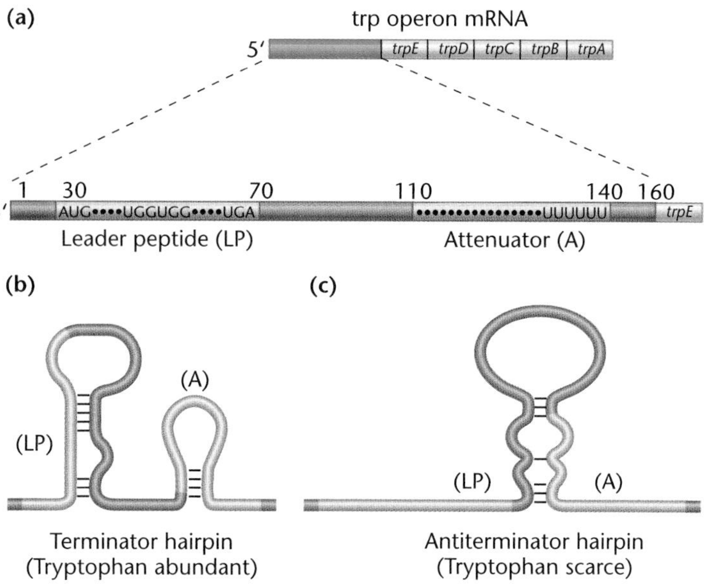

FIGURE 16–13 Diagram of the involvement of the leader sequence of the mRNA transcript of the *trp* operon of *E. coli* during attenuation. (a) The 5′-leader sequence of the *trp* operon is expanded to identify the portions encoding the leader peptide (LP) and the attenuator region (A). Critical nucleotide sequences, including the UGG triplets that encode tryptophan, are also identified; (b) the terminator hairpin, which forms when the ribosome proceeds during translation of the *trp* codons; and (c) the antiterminator hairpin, which forms when the ribosome stalls at the *trp* codons because tryptophan is scarce.

HOW DO WE KNOW?

How do we know that the *trp* operon is a repressible control system, in contrast to the *lac* operon, which is an inducible control system?

16.6 Attenuation Is a Critical Process during the Regulation of the *trp* Operon in *E. coli*

Charles Yanofsky, his coworker Kevin Bertrand, and their colleagues observed that, even when tryptophan is present and the *trp* operon is repressed, initiation of transcription can still occur, leading to the initial portion of the mRNA (the 5′-leader sequence). Hence, while the activated repressor binds to the operator region, it does not strongly inhibit the *initial expression* of the operon, suggesting that there must be a subsequent mechanism by which tryptophan somehow inhibits transcription and thus enzyme synthesis. Yanofsky discovered that, following initiation of transcription, *in the presence of high concentrations of tryptophan*, mRNA synthesis is usually terminated at a point about 140 nucleotides along the transcript.

This process is called **attenuation**,* indicative of the effect of diminishing genetic expression of the operon.

However, when tryptophan is absent, or present in very low concentrations, transcription is initiated and *not* subsequently terminated, instead continuing beyond the leader sequence along the DNA encoding the structural genes, starting with the *trpE* gene. As a result, a polycistronic mRNA is produced and the enzymes essential to the biosynthesis of tryptophan are subsequently translated. The *trp* operon and the genetic components involved in the attenuation process are diagrammed in Figure 16–13(a).

Identification of the site involved in attenuation was made possible by the isolation of various deletion mutations in the region 115 to 140 nucleotides into the leader sequence. Such mutations abolish attenuation. This site is referred to as the **attenuator**. An explanation of how attenuation occurs and how it is overcome, put forward by Yanofsky and colleagues, is summarized in Figure 16–13. The initial DNA sequence that is transcribed gives rise to an mRNA molecule that has the potential to fold into two mutually exclusive stem-loop structures referred to as "hairpins." In the presence of excess tryptophan, the hairpin that is formed behaves as a **terminator** structure and transcription is almost always abolished prematurely. On the other hand, if tryptophan is scarce, the alternative structure referred to as the **antiterminator hairpin** is formed. Transcription is allowed to proceed past the involved DNA sequence, and the entire mRNA is subsequently produced. These hairpins are illustrated in Figure 16–13(b) and (c).

---

*"Attenuation" is derived from the verb *attenuate*, meaning "to reduce in strength, weaken, or impair."

The question is how the absence (or a low concentration) of tryptophan allows attenuation to be bypassed. A key point in Yanofsky's model is that the leader transcript must be translated in order for the antiterminator hairpin to form. He discovered that the leader transcript includes two triplets (UGG) that encode tryptophan preceded upstream by an initial AUG sequence that prompts the initiation of translation by ribosomes. When adequate tryptophan is present, charged $tRNA^{trp}$ is also present. As a result, translation proceeds past these triplets, and the *terminator hairpin* is formed, as illustrated in Figure 16–13(b). If cells are starved of tryptophan, charged $tRNA^{trp}$ is unavailable. The ribosome then "stalls" during the translation of triplets as charged $tRNA^{trp}$ is called for, but is unavailable because of the lack of tryptophan. This event induces the formation of the antiterminator hairpin within the transcript, as shown in Figure 16–13(c). As a result, attenuation is overcome, and transcription proceeds, leading to expression of the entire set of structural genes.

Details of the secondary structures of the transcript have since been worked out and described. They satisfy conditions leading either to continued transcription or termination (attenuation). Additionally, other mutations in the leader sequence have been isolated that were predicted to alter the secondary structure of the transcript and its impact on attenuation. In each case, the predicted result has been upheld. Yanofsky's model, while complex, has significantly extended our knowledge of genetic regulation. Furthermore, the experimental evidence supporting it represents an integrated approach involving genetic and biochemical analysis, which is becoming commonplace in molecular genetic research.

Attenuation appears to be a mechanism common to other bacterial operons in *E. coli* that regulate the enzymes essential to the biosynthesis of amino acids. In addition to tryptophan, operons involved in threonine, histidine, leucine, and phenylalanine metabolism display attenuators in their leader sequences. As with the *trp* operon, each is known to contain multiple codons calling for the amino acid being regulated, which prompt "stalling" of translation if the appropriate amino acid is missing. For example, the leader sequence in the histidine operon codes for seven histidine residues in a row. The threonine operon leader sequence calls for eight threonine residues. As with tryptophan, when the amino acid is present, stalling does not occur, a terminator hairpin structure is formed, and attenuation occurs.

Given that attenuation involves the synchronous nature of transcription and translation in space and time, the phenomenon is unique to prokaryotes. Such a regulatory mechanism is impossible in eukaryotes since transcription occurs in the nucleus and translation occurs in the cytoplasm.

16.7 TRAP and AT Proteins Govern Attenuation in *B. subtilis*

It is useful to point out that not all organisms, even those of the same type, solve problems such as gene regulation in exactly the same way. Thus, it is not unusual for us to discover new strategies arising during evolution. Often, the new approach is a variation on the same theme. Such is the case regarding the regulation of the *trp* operon in bacteria.

As we saw previously, *E. coli*, a Gram negative bacterium, uses charged $tRNA^{trp}$ and a terminator hairpin in the leader sequence of the transcript as the underlying basis for attenuating transcription of its *trp* operon. The Gram positive bacterium *Bacillus subtilis* also uses attenuation and hairpins to regulate its *trp* operon. In fact, *B. subtilis* relies on attenuation as the sole mechanism for regulation, lacking a mechanism that represses transcription entirely in this operon, as is present in *E. coli*.

However, the molecular signals that cause attenuation in *B. subtilis* do not invoke the process of translation and stalling, as does *E. coli*, to induce the hairpin that terminates transcription. Instead, a specific protein, isolated in the 1990s by Charles Yanofsky and coworkers, either binds or does not bind to the attenuator leader sequence, thereby inducing the alternative terminator or antiterminator configurations, respectively.

The way attenuation is accomplished in *B. subtilis* is unique. The protein, ***trp* RNA-binding attenuation protein (TRAP)**, binds to tryptophan if it is present in the cell. TRAP consists of 11 subunits, forming a symmetrical quaternary protein structure. Each subunit can bind one molecule of tryptophan, incorporating it into a deep pocket within the protein [Figure 16–14(a)]. When fully saturated with tryptophan, this protein can bind to the 5′-leader sequence of the RNA transcript, which contains 11 triplet repeats of either GAG or UAG, each separated by several spacer nucleotides. Each triplet is linked to one of the subunits, which contains a binding pocket for the triplet. The binding [Figure 16–14(b)] forms an RNA belt around

(a)

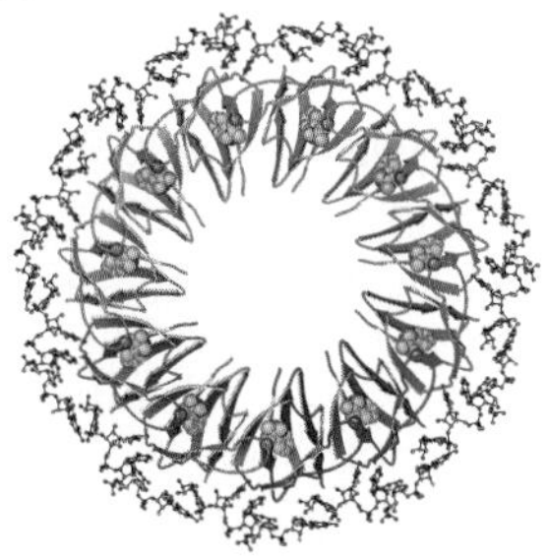

(b)

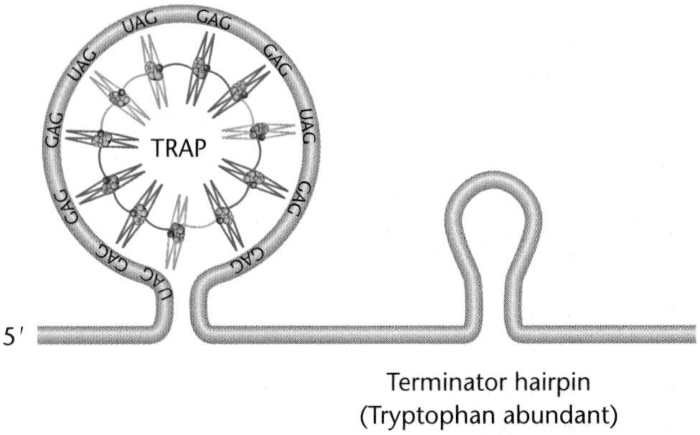

FIGURE 16–14 (a) Model of the *trp* RNA-binding attenuation protein (TRAP). The symmetrical molecule consists of 11 subunits, each capable of binding to one molecule of tryptophan that is embedded within the subunit. (b) The interaction of a tryptophan-bound TRAP molecule with the leader sequence of the *trp* operon of *B. subtilis*. This interaction induces the formation of the terminator hairpin, attenuating expression of the *trp* operon.

TRAP, preventing the antiterminator hairpin from forming. This results in the formation of the terminator configuration, thus leading to premature termination of transcription and, consequently, attenuation of expression of the operon.

The preceding attenuation strategy involves an interesting mechanism of regulation. There are two observations that suggested to Yanofsky and his colleagues that the attenuation strategy in *B. subtilis* may be even more complex than it might appear. First, regulation of the *trp* operon in *B. subtilis* is extremely sensitive to a wide range of tryptophan concentrations. Such a fine tuning suggests that something more than a simple on–off mechanism attributed to TRAP may be at work. Second, a mutation in the gene encoding tryptophanyl-tRNA synthetase leads to the overexpression of the *trp* operon, even in the presence of excess tryptophan. This enzyme is responsible for charging tRNA^{trp}, and as uncharged *t*RNA^{trp} molecules accumulate, regulation is interrupted. The interruption of regulation would suggest that uncharged tRNA^{trp} plays some essential role in attenuation, leading Yanofsky and his coworker Angela Valbuzzi to ask how tRNA^{trp} might be involved with TRAP. What they found extends our knowledge of the system of regulation.

They discovered that there exists still another protein, **anti-TRAP (AT)**, which provides a metabolic signal that tRNA^{trp} is uncharged, thereby indicating that tryptophan is very scarce in the cell. Yanofsky and Valbuzzi theorized that uncharged tRNA^{trp} induces a separate operon to express the AT gene. The AT protein then associates with TRAP, specifically when it is in the tryptophan-activated state, inhibiting binding to its target leader RNA sequence.

Such a finding not only adds still another dimension to our understanding of this system of gene regulation, but it also explains the original observation of the mutation in the tRNA^{trp} synthetase gene that leads to the overexpression of the *trp* operon. The mutation prevents charging of tRNA^{trp}, even in the presence of tryptophan. As uncharged tRNA^{trp} accumulates, it induces AT, which binds to TRAP. This prevents TRAP from binding to the leader RNA sequence, thus overinducing the *trp* operon.

The preceding description provides insights into the complexity of regulatory mechanisms in bacteria. That such intricate strategies have resulted during the evolutionary process attests to the critical importance of carefully regulating gene expression in bacteria.

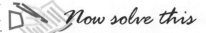

Now solve this

Problem 16.20 on page 409 considers the regulation of the *B. subtilis trp* operon, which involves the TRAP protein. You are asked to predict whether the structural genes are expressed or not under various conditions.

Hint: Unlike *E. coli*, which uses tryptophan as a repressor as well as the process of attenuation in the regulation of the *trp* operon, in *B. subtilis*, regulation relies solely on the process of attenuation and the TRAP protein.

16.8 The *ara* Operon Is Controlled by a Regulator Protein That Exerts Both Positive and Negative Control

We conclude this chapter with a brief discussion of the **arabinose (*ara*) operon** as studied in *E. coli*. This inducible operon is unique because the same regulatory protein is capable of exerting both positive or negative control, and as a result, either induces or represses gene expression. The various panels in Figure 16–15 accompany the following description of the conditions in which the operon is defined and shown to be either active or inactive:

1. The metabolism of the sugar arabinose is under the direction of the enzymatic products of three structural genes, *ara B*, *A*, and *D*. Their transcription is controlled by the regulatory protein AraC, encoded by the *araC* gene, that interacts with two regulatory regions, *araI* and *araO*$_2$.* [Figure 16–15(a)]. These sites can be bound individually or coordinately by the AraC protein.

2. The *I* region bears that designation because when only it is bound by AraC, the system is induced [Figure 16–15(b)]. For this to occur, both arabinose and cAMP must be present. Thus, like induction of the *lac* operon, a CAP-binding site is present in the promoter region that modulates catabolite repression in the presence of glucose.

3. In the absence of both arabinose and cAMP, the AraC protein binds coordinately to both the *I* site *and* the O_2 site (so

---

*An additional operator region (O_1) is also present, but is not involved in the regulation of the *ara* structural gene.

named because it was the second operator region to be discovered in this operon). When both *I* and O_2 are bound by AraC, a conformational change occurs in the DNA, whereby a tight loop is formed [Figure 16–15(c)] and the structural genes are repressed.

4. The O_2 region is found about 200 nucleotides upstream from the *I* region. Binding at either regulatory region involves a dimer of AraC. When both regions are bound, the dimers interact leading to the loop that causes repression. Presumably, this loop inhibits the access of RNA polymerase to the promoter region. The region of DNA located between *I* and O_2. which loops out, is critical to the formation of the repression complex. Genetic alteration through insertion or deletion of even a few nucleotides is sufficient to interfere with loop formation and, therefore, repression.

These findings involving the *ara* operon serve to illustrate the degree of complexity that exists in the regulation of a group of related genes. As we alluded to at the beginning of this chapter, the development of regulatory mechanisms has provided evolutionary advantages to bacterial systems that allow them to adjust to a variety of natural environments. Without question, these systems are keenly equipped genetically not only to survive under varying physiological conditions, but to do so with great biochemical efficiency.

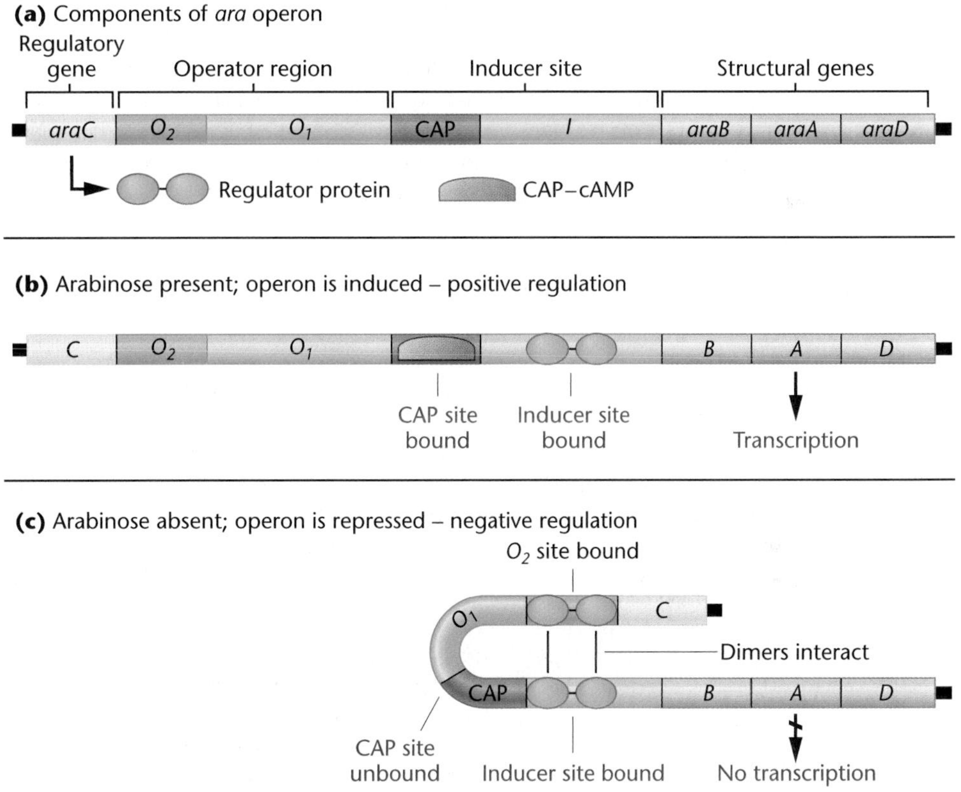

FIGURE 16–15 Genetic regulation of the *ara* operon. The regulatory protein of the *araC* gene acts as either an inducer (in the presence of arabinose) or a repressor (in the absence of arabinose).

GENETICS, TECHNOLOGY, AND SOCIETY

Quorum Sensing: How Bacteria Talk to One Another

For decades, scientists have regarded bacteria as independent microbes, incapable of cell-to-cell communication. However, recent research has shown that many bacteria can regulate gene expression and coordinate group behavior through a form of bacterial communication termed ***quorum sensing***. Through this process, bacteria send and receive chemical signals called autoinducers that relay information about population size. When the population size reaches a "quorum," defined in the business world as the minimum number of members of an organization that must be present to conduct business, the autoinducers regulate gene expression accordingly. Quorum sensing has been described in more than 70 species of bacteria, and its applications are widespread. Some marine bacteria use autoinducers to control bioluminescence, while other pathogenic bacteria use similar signals to regulate the expression of toxic virulence factors. Elucidating the complex mechanisms of quorum sensing has caused us to reevaluate our understanding of prokaryotic gene regulation, and has already led to promising applications in the development of alternatives to antibiotic drugs.

The field of quorum sensing began in the 1960 when researchers noticed that during the day, the squid lowers the concentration of the luminescent marine bacterium *Vibrio fischeri* in its light organ and the bacteria no longer glow. *V. fischeri* has a symbiotic relationship with the squid, *Euprymna scolopes*. While the squid hunts for food at night, it uses light emitted by the *V. fischeri* present in its light organ to illuminate the ocean floor, thus perfectly countering the shadows created by moonlight that normally act as a beacon for the squid's predators. This symbiotic relationship provides the bacteria with a protected, nutrient-rich environment in the squid's light organ.

What turns the bacteria's luminescent (*lux*) genes on in response to high cell density and off in response to low cell density? The chemical "language" of bacteria involved in quorum sensing is comprised of two main classes of autoinducer molecules: homoserine lactones (HSLs), used by Gram-negative bacteria, and oligopeptides, used by Gram-positive bacteria. In *V. fischeri*, the responsible autoinducer is an HSL. At a critical population size, or quorum, the HSL regulates the lux operon by binding directly to transcription factors that regulate gene expression. Oligopeptides also activate transcription factors, but do so by initiating phosphorylation cascades.

Because most bacterial species use unique autoinducers that vary in their chemical structure and properties, quorum sensing was thought to mediate communication only among members of the same species. Then, in 1994, Bonnie Bassler and her colleagues at Princeton University made headlines when they discovered an autoinducer molecule in the marine bacterium *Vibrio harveyi* that was also present in many diverse types of bacteria. This molecule, autoinducer-2 (AI-2), has the potential to mediate "quorum sensing cross talk" and serve as a universal language for bacterial communication between species. Because the accumulation of AI-2 is proportional to cell number, and since the structure of AI-2 may vary slightly between different species, the current hypothesis is that AI-2 can transmit information about both cell density and species composition of a bacterial community. Such a communication system would be highly advantageous in nature, where populations of bacteria commonly contain hundreds of different species. Although presently only the AI-2 structure of *V. harveyi* is known, researchers are currently studying other bacterial species to determine if slight modifications of this molecule represent different "words" in the AI-2 bacterial language.

Pathogenic bacteria also use quorum sensing to regulate gene expression. Instead of controlling bioluminescence, these bacteria communicate to coordinate the production of harmful virulence substances or to avoid detection by the immune system. For example, *Vibrio cholerae*, the causative agent of cholera, uses AI-2 and an additional species-specific autoinducer to activate the genes controlling the production of cholera toxin. *Pseudomonas aeruginosa*, the Gram-negative bacterium that often affects cystic fibrosis patients, uses quorum sensing to regulate the production of elastase, a protease that disrupts the respiratory epithelium and interferes with ciliary function. *P. aeruginosa* also uses autoinducers to control the production of biofilms, tough protective shells that resist host defenses and make treatment with antibiotics nearly impossible. Other bacteria determine cell density through quorum sensing to delay the production of toxic substances until the colony is large enough to overpower the host's immune system and establish an infection.

Each year, millions of hospital patients acquire bacterial infections. Although some can be successfully treated with antibiotics, antibiotic-resistent strains of bacteria are evolving faster than scientists can develop alternative treatments. *Staphylococci* infections alone affect 500,000 patients with implanted medical devices such as catheters and artificial heart valves, resulting in approximately 90,000 deaths each year. Because many bacteria rely on quorum sensing to regulate disease-causing genes, treatments that block quorum sensing may aid in combating infections. In 1999, Bonnie Bassler formed a company called Quorex, with the goal of using quorum-sensing research to develop new antibacterial agents. In June 2004, researchers showed that implanted medical devices coated with a quorum-sensing inhibitor called RIP can reduce the prevalence of *Staphylococcus* infections in rats. Similarly, research is also underway to develop drugs that block quorum sensing in *Bacillus anthracis*, the bacterium more commonly known as anthrax, and *Pseudomonas aeruginosa*, the bacterium that may be lethal to those with cystic fibrosis or to others with compromised immune systems. Similar drugs that inhibit mechanisms of bacterial communication are being tested for their ability to inhibit the formation of biofilms, restore the potency of antibiotics, and limit the development of new antibiotic-resistent strains of bacteria. Thus, what began as a fascinating observation in the glowing squid has launched an exciting era of research in bacterial genetics that may one day prove of great clinical significance.

References

Camara, M., Hardman, A., Williams, P., and Milton, D. 2002. Quorum sensing in *Vibrio cholerae*. *Nature Genetics*. 32:217–218.

Federle, M.J., and Bassler, B.L. 2003. Interspecies communication in bacteria. *J. Clin. Invest.* 112:1291–1299.

Suga, H., and Smith, K.M. 2003. Molecular mechanisms of bacterial quorum sensing as a new drug target. *Current Opinion in Chemical Biology*. 7:586–591.

Chapter Summary

1. A system of genetic regulation must exist if the complete genome is not to be continuously active in transcription throughout the life of every cell of all species. Highly refined mechanisms have evolved that regulate transcription, optimizing genetic efficiency.
2. Genetic analysis of bacteria has been pursued successfully since the 1940s. The ease of obtaining large quantities of pure cultures of mutant strains of bacteria as experimental material has made bacteria an organism of choice in numerous types of genetic studies.
3. The discovery and study of the *lac* operon in *E. coli* pioneered the study of gene regulation in bacteria. Genes involved in the metabolism of lactose are coordinately regulated by a negative control system, whereby gene expression is induced by lactose. In its absence, the system is shut down.
4. The catabolite-activating protein (CAP) is essential to the binding of RNA polymerase to the promoter and subsequent transcription in the *lac* and other operons. Therefore, CAP exerts positive control over gene expression. For CAP to stimulate the *lac* operon, it must be bound to cyclic AMP.
5. When glucose is present, cyclic AMP levels are low, CAP binding does not occur, and the structural genes are not expressed. This phenomenon is called catabolite repression. Such a mechanism is efficient energetically, since glucose is preferred, even in the presence of lactose.
6. The *lac* repressor has been isolated and studied. Crystal structure analysis has demonstrated how it interacts with the DNA of the operon as well as with inducers. These studies have demonstrated conformational changes in DNA leading to the formation of a repression loop that inhibits binding between RNA polymerase and the promoter region of the operon.
7. The biosynthesis of tryptophan in *E. coli* involves a number of enzymes whose presence or absence is controlled by another distinct operon. In contrast to the inducible operon controlling lactose metabolism, the *trp* operon is repressible. In the presence of tryptophan, an active repressor is made that shuts off the expression of the structural genes. Like the *lac* operon, the *trp* operon functions under negative control.
8. An additional regulatory step, referred to as attenuation, has been studied in the *trp* operon. The mechanism varies between *E. coli* and *B. subtilis*, but both depend on alternative hairpin structures that form under different condition in the leader sequence of the mRNA.
9. The *ara* operon is unique in that the regulator protein exerts both positive and negative control over expression of the genes specifying the enzymes that metabolize the sugar arabinose.

Insights and Solutions

1. A theoretical operon (*theo*) in *E. coli* contains several structural genes encoding enzymes that are involved sequentially in the biosynthesis of an amino acid. Unlike the *lac* operon, in which the repressor gene is separate from the operon, the gene encoding the regulator molecule is contained within the *theo* operon. When the end product (the amino acid) is present, it combines with the regulator molecule, and this complex binds to the operator, repressing the operon. In the absence of the amino acid, the regulatory molecule fails to bind to the operator, and transcription proceeds.

Characterize this operon, then consider the following mutations, as well as the situation in which the wild-type gene is present along with the mutant gene in partially diploid cells (F′):

(a) Mutation in the operator region.

(b) Mutation in the promoter region.

(c) Mutation in the regulator gene.

In each case, will the operon be active or inactive in transcription, assuming that the mutation affects the regulation of the *theo* operon? Compare each response with the equivalent situation of the *lac* operon.

Solution: The *theo* operon is repressible and under negative control. When there is no amino acid present in the medium (or the environment), the product of the regulatory gene cannot bind to the operator region, and transcription proceeds under the direction of RNA polymerase. The enzymes necessary for the synthesis of the amino acid are produced, as is the regulator molecule. If the amino acid *is* present, or is present after sufficient synthesis occurs, the amino acid binds to the regulator, forming a complex that interacts with the operator region, causing repression of transcription of the genes within the operon.

The *theo* operon is similar to the tryptophan system, except that the regulator gene is within the operon, rather than separate from it. Therefore, in the *theo* operon, the regulator gene is itself regulated by the presence or absence of the amino acid.

(a) As in the *lac* operon, a mutation in the *theo* operator gene inhibits binding with the repressor complex, and transcription occurs constitutively. The presence of an F′ plasmid bearing the wild-type allele would have no effect, since it is not adjacent to the structural genes.

(b) A mutation in the *theo* promoter region would no doubt inhibit binding to RNA polymerase and therefore inhibit transcription. This would also happen in the *lac* operon. A wild-type allele present in an F′ plasmid would have no effect.

(c) A mutation in the *theo* regulator gene, as in the *lac* system, may inhibit either its binding to the repressor or its binding to the operator gene. In both cases, transcription will be constitutive, because the *theo* system is repressible. Both cases result in the failure of the regulator to bind to the operator, allowing transcription to proceed. In the *lac* system, failure to bind the corepressor lactose would permanently repress the system. The addition of a wild-type allele would restore repressibility, provided that this gene was transcribed constitutively.

Problems and Discussion Questions

1. Contrast the need for the enzymes involved in lactose and tryptophan metabolism in bacteria when lactose and tryptophan, respectively, are (a) present and (b) absent.
2. Contrast positive versus negative control of gene expression.
3. Contrast the role of the repressor in an inducible system and in a repressible system.
4. Even though the *lac Z, Y*, and *A* structural genes are transcribed as a single polycistronic mRNA, each gene contains the appropriate initiation and termination signals essential for translation. Predict what will happen when a cell growing in the presence of lactose contains a deletion of one nucleotide (a) early in the *Z* gene and (b) early in the *A* gene.
5. For the *lac* genotypes shown in the accompanying table, predict whether the structural genes (*Z*) are constitutive, permanently repressed, or inducible in the presence of lactose.

Genotype	Constitutive	Repressed	Inducible
$I^+O^+Z^+$			x
$I^-O^+Z^+$			
$I^+O^cZ^+$			
$I^-O^+Z^+/F'I^+$			
$I^+O^cZ^+/F'O^+$			
$I^sO^+Z^+$			
$I^sO^+Z^+/F'I^+$			

6. For the genotypes and condition (lactose present or absent) shown in the accompanying table, predict whether functional enzymes, nonfunctional enzymes, or no enzymes are made.

Genotype	Condition	Functional Enzyme Made	Nonfunctional Enzyme Made	No Enzyme Made
$I^+O^+Z^+$	No lactose			X
$I^+O^cZ^+$	Lactose			
$I^-O^+Z^-$	No lactose			
$I^-O^+Z^-$	Lactose			
$I^-O^+Z^+/F'I^+$	No lactose			
$I^+O^cZ^+/F'O^+$	Lactose			
$I^+O^+Z^-/F'I^+O^+Z^+$	Lactose			
$I^-O^+Z^-/F'I^+O^+Z^+$	No lactose			
$I^sO^+Z^+/F'O^+$	No lactose			
$I^+O^cZ^+/F'O^+Z^+$	Lactose			

7. The locations of numerous $lacI^-$ and $lacI^S$ mutations have been determined within the DNA sequence of the $lacI^+$ gene. Among these, $lacI^-$ mutations were found to occur in the 5′-upstream region of the gene, while $lacI^S$ mutations were found to occur farther downstream in the gene. Is the location of the two types of mutations within the gene consistent with what is known about the function of the repressor that is the product of the $lacI^+$ gene?
8. Describe the experimental rationale developed that allowed the *lac* repressor to be isolated.
9. What properties demonstrate the *lac* repressor to be a protein? Describe the evidence that it indeed serves as a repressor within the operon scheme.
10. Predict the level of genetic activity of the *lac* operon as well as the status of the *lac* repressor and the CAP protein under the cellular conditions listed in the accompanying table.

	Lactose	Glucose
(a)	−	−
(b)	+	−
(c)	−	+
(d)	+	+

11. Predict the effect on the inducibility of the *lac* operon of a mutation that disrupts the function of (a) the *crp* gene, which encodes the CAP protein, and (b) the CAP-binding site within the promoter.
12. Describe the role of attenuation in the regulation of tryptophan biosynthesis.
13. Attenuation of the *trp* operon was viewed as a relatively inefficient way to achieve genetic regulation when it was first discovered in the 1970s. Since then, however, attenuation has been found to be a relatively common regulatory strategy. Assuming that attenuation is a relatively inefficient way to achieve genetic regulation, what might explain its widespread use?
14. Neelaredoxin is a protein of 15 kDa that is a gene product common in anaerobic prokaryotes. It has superoxide-scavenging activity and it is *constitutively expressed*. In addition, its expression is not further *induced* during its exposure to O_2 or H_2O_2 (Silva, G., et al. 2001. *J. Bacteriol.* 183:4413–20). What do the terms *constitutively expressed* and *induced* mean in terms of neelaredoxin synthesis?
15. Milk products such as cheeses and yogurts are dependent on the conversion by a variety of anaerobic bacteria, including several *Lactobacillus,* of lactose to glucose and galactose, ultimately producing lactic acid. These conversions are dependent on both permease and β-galactosidase as part of the *lac* operon. After selection for rapid fermentation for the production of yogurt, one *Lactobacillus* subspecies lost its ability to regulate *lac* operon expression (Lapierre, L., et al. 2002. *J. Bacteriol.* 184:928–35). Would you consider it likely that in this subspecies the *lac* operon is "on" or "off"? What genetic events would likely contribute to the loss of regulation as described above?

Extra-Spicy Problems

16. Bacterial strategies to evade various natural or human-imposed antibiotics are varied and include membrane-bound efflux pumps that export antibiotics from the cell. In a review of efflux pumps (Grkovic, S., et al. 2002. *Microb. and Mol. Biol. Rev.* 66:671–701), it is stated that because energy is required to drive the efflux pumps, there is a selective disadvantage to activate the pumps in the absence of the antibiotic. It is also stated that a given antibiotic may play a role in the regulation by interacting with either an activator protein or a repressor protein, depending on the particular system involved. How might such systems be categorized in terms of *negative control* (*inducible* or *repressible*) or *positive control* (*inducible* or *repressible*)?

17. In a theoretical operon, genes *A*, *B*, *C*, and *D* represent the repressor gene, the promoter sequence, the operator gene, and the structural gene, *but not necessarily in that order*. This operon is concerned with the metabolism of a theoretic molecule (tm). From the data provided in the accompanying table, first decide whether the operon is inducible or repressible. Then, assign *A*, *B*, *C*, and *D* to the four parts of the operon. Explain your rationale. (AE = active enzyme; IE = inactive enzyme; NE = no enzyme)

Genotype	tm Present	tm Absent
$A^+B^+C^+D^+$	AE	NE
$A^-B^+C^+D^+$	AE	AE
$A^+B^-C^+D^+$	NE	NE
$A^+B^+C^-D^+$	IE	NE
$A^+B^+C^+D^-$	AE	AE
$A^-B^+C^+D^+/F'A^+B^+C^+D^+$	AE	AE
$A^+B^-C^+D^+/F'A^+B^+C^+D^+$	AE	NE
$A^+B^+C^-D^+/F'A^+B^+C^+D^+$	AE + IE	NE
$A^+B^+C^-D^+/F'A^+B^+C^+D^+$	AE	NE

18. A bacterial operon is responsible for the production of the biosynthetic enzymes needed to make the theoretical amino acid tisophane (tis). The operon is regulated by a separate gene, *R*, deletion of which causes the loss of enzyme synthesis. In the wild-type condition, when tis is present, no enzymes are made; in the absence of tis, the enzymes are made. Mutations in the operator gene (O^-) result in repression regardless of the presence of tis.
Is the operon under positive or negative control? Propose a model for (a) repression of the genes in the presence of tis in wild-type cells and (b) the mutations.

19. A marine bacterium is isolated and shown to contain an inducible operon whose genetic products metabolize oil when it is encountered in the environment. Investigation demonstrates that the operon is under positive control and that there is a *reg* gene whose product interacts with an operator region (*o*) to regulate the structural genes designated *sg*.

In an attempt to understand how the operon functions, a constitutive mutant strain and several partial diploid strains were isolated and tested with the results shown here:

Host Chromosome	F′ Factor	Phenotype
wild type	none	inducible
wild type	*reg* gene from mutant strain	inducible
wild type	operon from mutant strain	constitutive
mutant strain	*reg* gene from wild type	constitutive

Draw all possible conclusions about the mutation as well as the nature of regulation of the operon. Is the constitutive mutation in the *trans*-acting *reg* element or in the *cis*-acting *o* operator element?

20. The SOS repair genes in *E. coli* (discussed in Chapter 13) are negatively regulated by the *lexA* gene product, called the LexA repressor. When a cell sustains extensive damage to its DNA, the LexA repressor is inactivated by the *recA* gene product (RecA), and transcription of the SOS genes is increased dramatically.

One of the SOS genes is the *uvrA* gene. You are studying the function of the *uvrA* gene product in DNA repair. You isolate a mutant strain that shows constitutive expression of the UvrA protein. You name this mutant strain $uvrA^C$. The following simple diagram shows the *lexA* and *uvrA* operons:

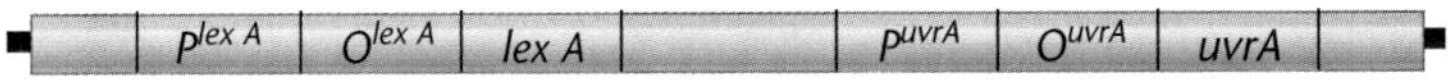

(a) Describe two different mutations that would result in a *uvrA* constitutive phenotype. Indicate the actual genotypes involved.
(b) Outline a series of genetic experiments, using partial diploid strains, that would allow you to determine which of the two possible mutations you have isolated.

21. A fellow student considers the issues in Problem 20 and argues that there is a more straightforward, nongenetic experiment that could differentiate between the two types of mutations. While the experiment requires no fancy genetics, you must be able to easily assay the products of the other SOS genes. Propose such an experiment.

22. In Figure 16–13, numerous critical regions of the leader sequence of mRNA that play important roles during the process of attenuation in the *trp* operon are depicted. Shown below (running across the entire page), beginning at about position 30 downstream from the 5′-end, is the sequence of ribonucleotides that make up the leader sequence:

AUGAAAGCAAUUUUCGUACUGAAAGGUUGGUGGCGCACUUCCUGAAACGGGCAGUGUAUUCACCAUGCGUAAAGCAAUCAGAUACCCAGCCCGCCUAAUGAGCGGGCUUUUUUUU

Within this molecule are the sequences that serve as the basis of the formation of the secondary structures. These structures are critical in the formation of the alternative hairpins, as well as the successive triplets that encode tryptophan, wherein stalling during translation occurs. Obtain a large piece of paper (such as manila wrapping paper) and, along with several other students from your genetics class, work through the base sequence, attempting to identify the *trp* codons and which parts of the molecule represent the base-pairing regions that form the terminator and antiterminator hairpins shown in Figure 16–13.

23. Consider the regulation of the *trp* operon in *B. subtilis*. For each of the following cases, indicate whether the structural genes in the operon are being expressed or not expressed:

    (a) TRAP present, $tRNA^{trp}$ abundant, tryptophan abundant.
    (b) TRAP present, $tRNA^{trp}$ abundant, tryptophan scarce.
    (c) TRAP present, $tRNA^{trp}$ scarce, tryptophan abundant.
    (d) TRAP absent, $tRNA^{trp}$ abundant, tryptophan abundant.

    In each case, also indicate whether AT is present. Describe the role of TRAP and AT in attenuation.

SELECTED READINGS

Antson, A.A., et al. 1999. Structure of the trp RNA-binding attenuation protein, TRAP, bound to RNA. *Nature* 401:235–42.

Beckwith, J.R., and Zipser, D., eds. 1970. *The Lactose Operon*. Cold Spring Harbor, NY: Cold Spring Harbor Laboratory Press.

Bertrand, K., et al. 1975. New features of the regulation of the tryptophan operon. *Science* 189:22–26.

Gilbert, W., and Müller-Hill, B. 1966. Isolation of the *lac* repressor. *Proc. Natl. Acad. Sci. USA* 56:1891–98.

______. 1967. The *lac* operator is DNA. *Proc. Natl. Acad. Sci. USA* 58:2415–21.

Jacob, F., and Monod, J. 1961. Genetic regulatory mechanisms in the synthesis of proteins. *J. Mol. Biol.* 3:318–56.

Lee, D., and Schleif, R. 1989. *In vivo* DNA loops in *araCBAD:* Size limits and helical repeats. *Proc. Natl. Acad. Sci. USA* 86:476–80.

Lewis, M., et al. 1996. Crystal structure of the lactose operon repressor and its complexes with DNA and inducer. *Science* 271:1247–54.

Stroynowski, I., and Yanofsky, C. 1982. Transcript secondary structures regulate transcription termination at the attenuator of *S. marcescens* tryptophan operon. *Nature* 298:34–38.

Valbuzzi, A., and Yanofsky, C. 2001. Inhibition of the *B. subtilis* regulatory protein TRAP by the TRAP-inhibitory protein, AT. *Science* 293:2057–61.

Yanofsky, C. 1981. Attenuation in the control of expression of bacterial operons. *Nature* 289:751–58.

Yanofsky, C., and Kolter, R. 1982. Attenuation in amino acid biosynthetic operons. *Annu. Rev. Genet.* 16:113–34.

Regulation of Gene Expression in Eukaryotes

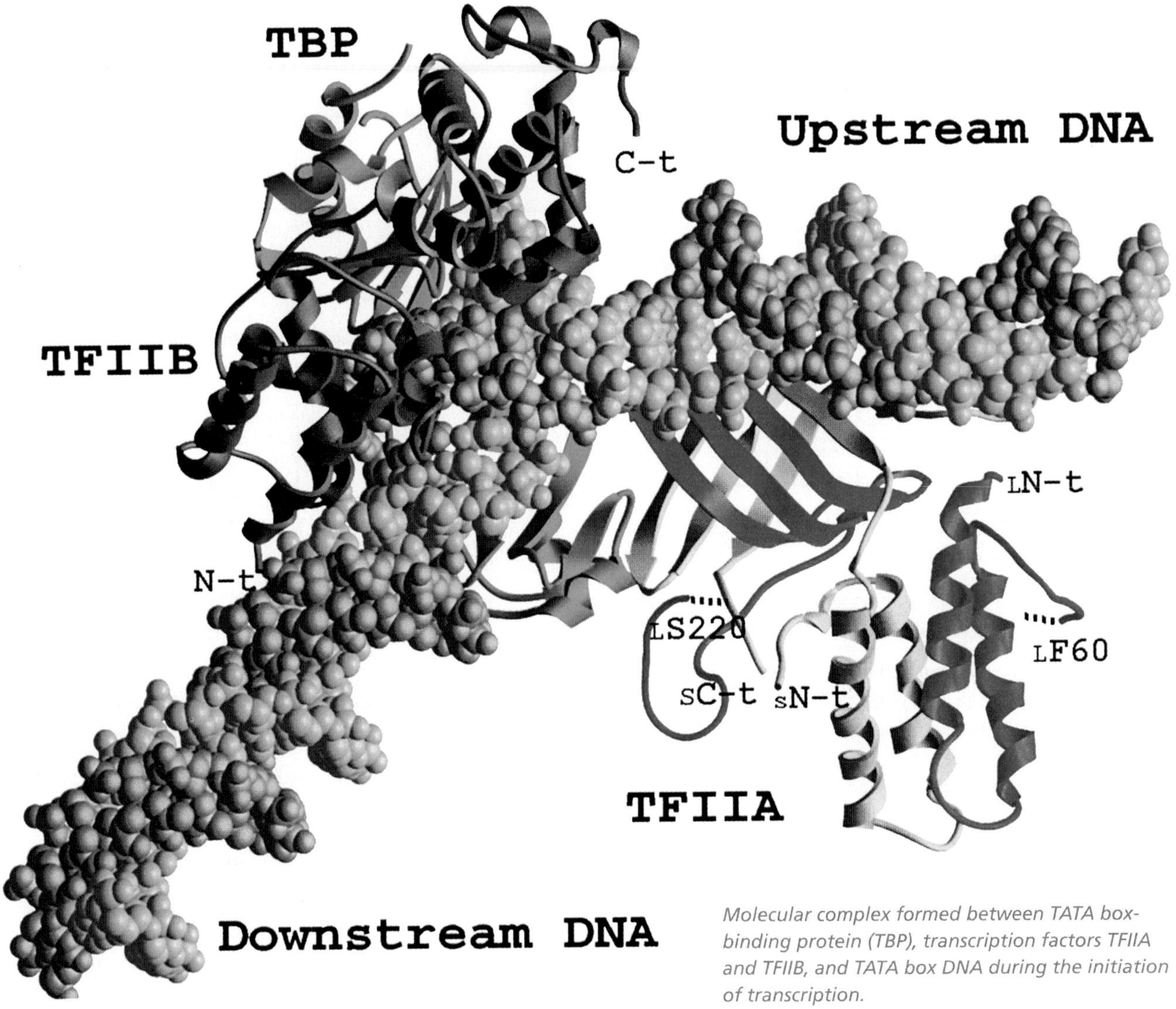

Molecular complex formed between TATA box-binding protein (TBP), transcription factors TFIIA and TFIIB, and TATA box DNA during the initiation of transcription.

Chapter Concepts

- Eukaryotic gene regulation is very different from prokaryotic gene regulation.
- Organization of chromatin in the nucleus plays an important part in regulating gene expression in eukaryotes. Chromatin must be remodeled to provide access to sites of transcription.
- Assembly of the components of transcription occurs at promoter sites. Proteins bound to enhancer sites affect the rate of transcription initiation.
- Factors such as DNA methylation may be important in regulating gene expression.
- Posttranscriptional regulation includes alternative splicing of pre-mRNA.
- Control of mRNA stability is one way to regulate posttranslational gene expression.

Web Tutorial 17.1 Regulation of Gene Expression: Eukaryotes

In multicellular eukaryotes, differential gene expression is at the heart of embryonic development and maintenance of the adult state. Cells of the pancreas, for example, do not make retinal pigment, nor do retinal cells make insulin. So how does an organism express a subset of genes in one cell type and a different subset of genes in another cell type? At the cellular level, this regulation is *not* accomplished by eliminating unused genetic information; instead, mechanisms activate specific portions of the genome and repress the expression of other genes. Regulation of gene expression occurs by positive (activation of transcription) and negative (repression of transcription) mechanisms. The activation and repression of selected loci represent a delicate balancing act for an organism; expression of a gene at the wrong time, in the wrong cell type, or in abnormal amounts can lead to a deleterious phenotype—cancer or cell death—even when the gene itself is normal.

In this chapter, we will review some of the general features of eukaryotic gene regulation, outline the components needed to initiate transcription, and discuss how these components interact. We will also consider the role of posttranscriptional mechanisms in regulating gene expression in eukaryotes. When appropriate, we will compare the more complex approaches exhibited by eukaryotes to the less complex counterparts in prokaryotes.

17.1 Eukaryotic Gene Regulation Differs from Regulation in Prokaryotes

There are several reasons why gene regulation is more complex in eukaryotes than in prokaryotes:

1. Eukaryotic cells contain a much greater amount of genetic information than do prokaryotic cells, and this DNA is complexed with histones and other proteins to form chromatin. As we shall see, the structure of chromatin—whether it is open (decondensed) and available to be transcribed, or closed (condensed) and not available—is the major on/off switch for gene regulation. (Remember that in prokaryotes, DNA sequences—including operators and activators—serve as the on/off switch.)
2. Genetic information in eukaryotes is carried on many chromosomes (rather than just on one), and these chromosomes are enclosed within a double-membrane-bound nucleus.
3. Since the genetic information in eukaryotes is segregated from the cytoplasm, transcription is spatially and temporally separated from translation; transcription occurs in the nucleus and translation occurs later in the cytoplasm.
4. The transcripts of eukaryotic genes are processed before transport to the cytoplasm.
5. Eukaryote mRNA has a much longer half-life ($t_{1/2}$) than does prokaryotic mRNA. Most prokaryotes are single-celled organisms and must be able to rapidly respond to environmental changes, and rapid decay of mRNA is an adaptive mechanism allowing rapid response.
6. Because mRNA is much more stable, eukaryotes extensively utilize translational controls.

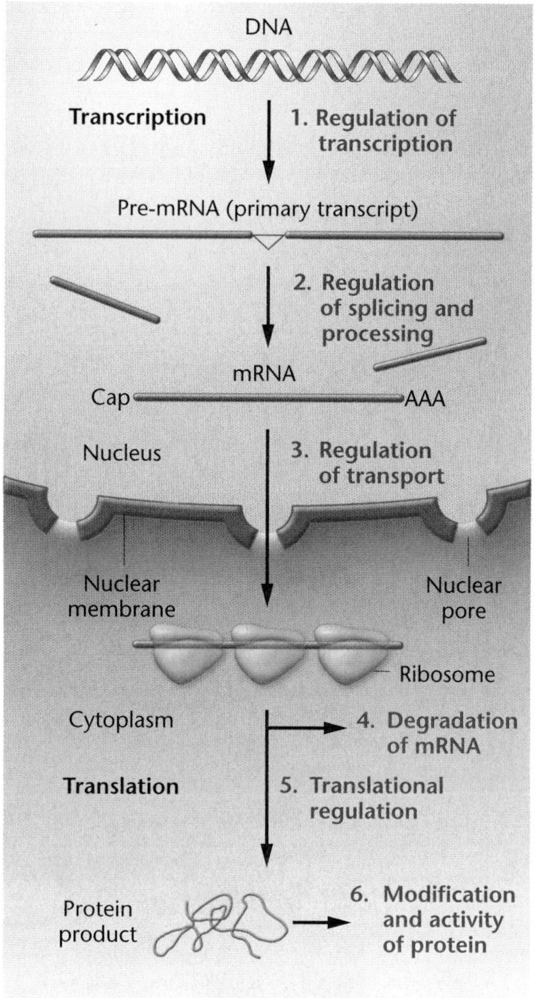

FIGURE 17–1 Various levels of regulation that are possible during the expression of the genetic material.

7. Most eukaryotes are multicellular with differentiated cell types. These different cell types typically use different sets of genes to make different proteins, even though each cell contains a complete set of genes.

Hence, it is evident that the regulation of eukaryotic gene expression can potentially occur at many levels (Figure 17–1) that include (1) transcriptional control, (2) posttranscriptional control (i.e., processing of the pre-mRNA), (3) transport to the cytoplasm, (4) stability of the mRNA, (5) translational control (i.e., selecting which mRNAs are translated), and (6) post-translational modification of the protein product. As most eukaryotic genes are regulated in part at the transcriptional level, throughout the chapter we will emphasize transcriptional control, though we will discuss other levels of control as well.

17.2 Chromosome Organization in the Nucleus Influences Gene Expression

During interphase of the cell cycle, chromosomes are unwound and cannot be seen as intact structures by light microscopy. The development of chromosome-painting techniques has

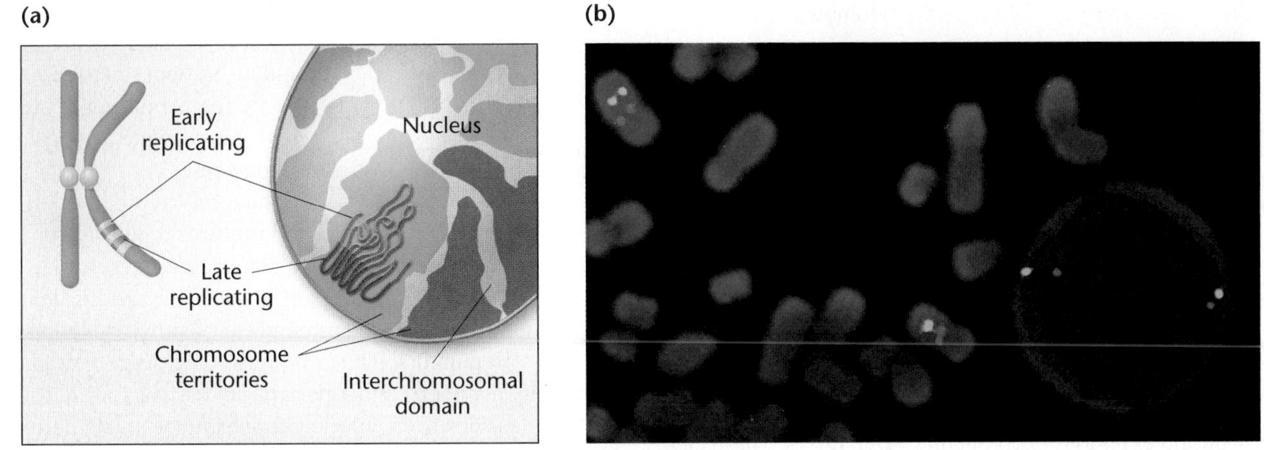

FIGURE 17–2 (a) In the nucleus, each chromosome occupies a discrete territory, and is separated from other chromosomes by an interchromosomal domain, where mRNA transcription and processing is thought to occur. (b) FISH probes hybridized to human chromsome 7. At left is hybridization to metaphase chromosomes. In the nucleus (right) the probes reveal the location of the territories occupied by chromosome 7.

revealed that the interphase nucleus is not a bag of tangled chromosome arms, but is, instead, a highly organized structure. Further, it is becoming clear that nuclear organization of chromosomes and dynamic changes in chromatin structure are key elements in controlling gene expression in eukaryotic cells. Before we examine events at the molecular level that initiate gene expression, we will first examine the structural mechanisms that determine whether or not a gene will be transcribed.

In the interphase nucleus, each chromosome occupies a discrete domain called a **chromosome territory** that separates it from other chromosomes (Figure 17–2). Chromosomes are organized in the nucleus according to their size and gene density, with gene-poor chromosomes located peripherially, and gene-dense chromosomes more internal. Channels between chromosomes are called **interchromosomal compartments**. Each compartment is a space between chromosomes that contains little or no DNA.

Chromosome structure is continuously rearranged so that transcriptionally active genes are cycled to the edge of chromosome territories at the border of the interchromosomal domain channels (Figure 17–2). Although evidence suggests that transcription in many, if not most, genes occurs when they are in direct contact with the border between the chromosome and the interchromosomal compartment, other evidence suggests that transcription can also take place within chromosomal territories.

Once a gene has been moved to the edge of a chromosome territory, initiation of gene expression requires two steps: (1) the remodeling and activation of chromatin by enzymes that alter nucleosome structure, making promoter sites accessible to the transcription machinery; and (2) the recruitment of coactivators that assemble the proteins necessary for transcription, including general transcription factors and RNA polymerase II. Each of these mechanisms will be described in detail in later sections of the chapter. In the next sections, we will describe the components required for transcription and the events leading to the production of mRNA molecules from a transcribed gene.

17.3 Transcription Initiation Is a Major Form of Gene Regulation

The structure of eukaryotic genes was first introduced in Chapters 12 and 15. Recall that these genes have three types of *cis*-regulatory sequences that control transcription: promoters, silencers, and enhancers. Figure 17–3 shows these and other elements that control expression of a eukaryotic gene. Transcription of DNA into an mRNA molecule is a complex, highly regulated process involving several different types of DNA sequences, the interactions of well over 100 proteins, chromatin remodeling and the bending and looping of DNA sequences. We will begin by discussing some of the DNA sequences regulating transcription.

Promoters Have a Modular Organization

Promoters are nucleotide sequences that serve as recognition sites for the transcriptional machinery. They represent the region necessary to *initiate* transcription at a basal level. Promoters are located immediately adjacent to the genes they regulate. Promoter regions are usually several hundred nucleotides in length and specify the site at which transcription begins and the direction of transcription along the DNA.

The promoter region of most genes contains several elements, including TATA, CAAT, and GC boxes. The region to which RNA polymerase II binds is a sequence called the **TATA box**, also referred to as the **core promoter**. Located about 25 to 30 bases upstream from the transcription start site (designated as -25 to -30), this box consists of a 7- to 8-bp consensus sequence (a sequence conserved in most genes studied) composed of AT base pairs often flanked on either side by GC-rich regions. Genetic analysis of TATA sequences show that mutations within TATA sequences reduce transcription, and that deletions may alter the initiation point of transcription (Figure 17–4).

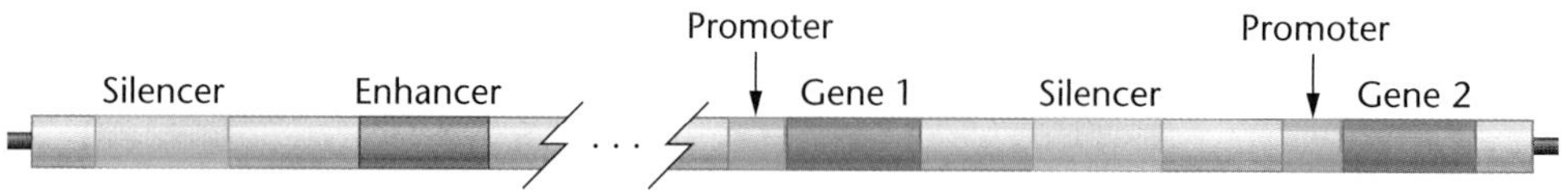

FIGURE 17–3 Expression of eukaryotic genes is controlled by regulatory elements directly adjacent to the gene, such as promoters, and by sequences that can be far from the transcriptional unit, such as enhancers and silencers.

How Do We Know?

What evidence confirms that promoter regions control the initiation of transcription?

Many promoter regions also contain **CAAT boxes**. These elements have the consensus sequence CAAT or CCAAT. The CAAT box frequently appears 70 to 80 bp upstream from the start site. Mutational analysis suggests that CAAT boxes are sites where proteins bind to DNA and are critical to the promoter's ability to facilitate transcription. Mutations on either side of this element have no effect on transcription, whereas mutations within the CAAT sequence dramatically lower the rate of transcription (Figure 17–4). Another element often seen in some promoter regions, called the **GC box**, has the consensus sequence GGGCGG and is often found at about position −110. The CAAT and GC elements bind transcription factors and function more like enhancers, which we will cover in the next section.

To sum up, promoter sequences in eukaryotic genes are modules that vary in location and organization (Figure 17–5). There are no universal components of promoter regions; genes differ in the type, number, spacing, and orientation of promoter elements.

Enhancers Control the Rate of Transcription

Transcription of eukaryotic genes is regulated not only by the promoter region, but also by DNA sequences called **enhancers**. Enhancers can be on either side of a gene, at some distance from the gene, or even within the gene. They are called *cis* regulators because they are found adjacent to the structural genes they regulate, as opposed to *trans* regulators (e.g., binding proteins), which can regulate a gene on any chromosome.

Enhancers typically interact with multiple regulatory proteins and transcription factors, and can increase the efficiency of transcription initiation or activate the promoter. Within enhancer sites, binding sites are often found for positive as well as negative gene regulators. Thus, there is some degree of analogy between enhancers and operator and activator regions in prokaryotes. However, enhancers appear to be much more complex in both structure and function. Enhancers greatly stimulate the transcriptional activity of a promoter and can be distinguished from promoters by the following characteristics:

1. The position of an enhancer is not fixed; it can be upstream, downstream, or within the gene it regulates.
2. Its orientation can be inverted without significant effect on its action.
3. If an enhancer is experimentally moved to another location in the genome, or if an unrelated gene is placed near an enhancer, the transcription of the adjacent gene is enhanced.

How Do We Know?

How do we know that the actions of enhancers are not necessarily gene-specific?

An example of an enhancer located *within* the gene it regulates is found in the immunoglobulin heavy-chain gene, in which an enhancer is located in an intron between two coding regions. This enhancer is active only in cells expressing the immunoglobulin genes, indicating that tissue-specific gene expression can be modulated through enhancers. Internal enhancers have been discovered in other eukaryotic genes, including the immunoglobulin light-chain gene.

Downstream enhancers are found in the human β-globin gene and the chicken thymidine kinase gene. In chickens, an enhancer located between the β-globin gene and the ε-globin genes works in one direction to control transcription of the ε-globin gene during embryonic development and in the opposite direction to regulate expression of the β-globin gene during adult life. In yeast, regulatory sequences similar to enhancers,

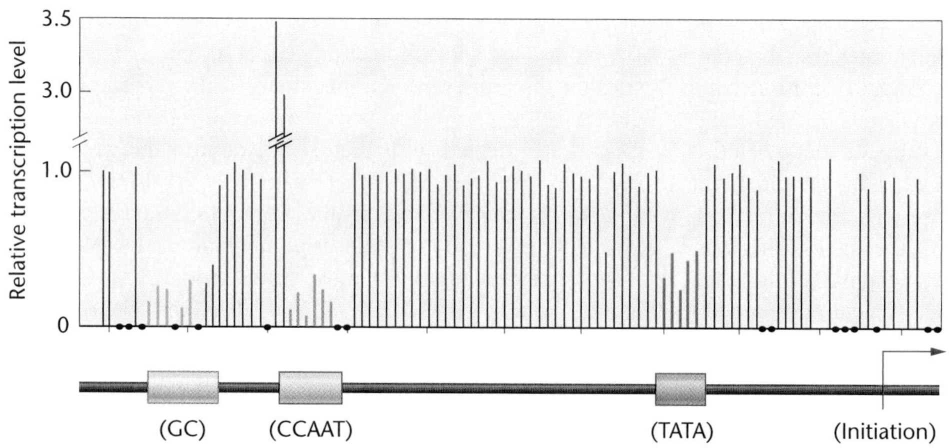

FIGURE 17–4 Summary of the effects of point mutations in the promoter region on transcription of the β-globin gene. Each line represents the level of transcription produced by a single nucleotide mutation (relative to wild type) in a separate experiment. Dots represent nucleotides in which no mutation was obtained. Note that mutations within the specific elements of the promoter have the greatest effect on the level of transcription.

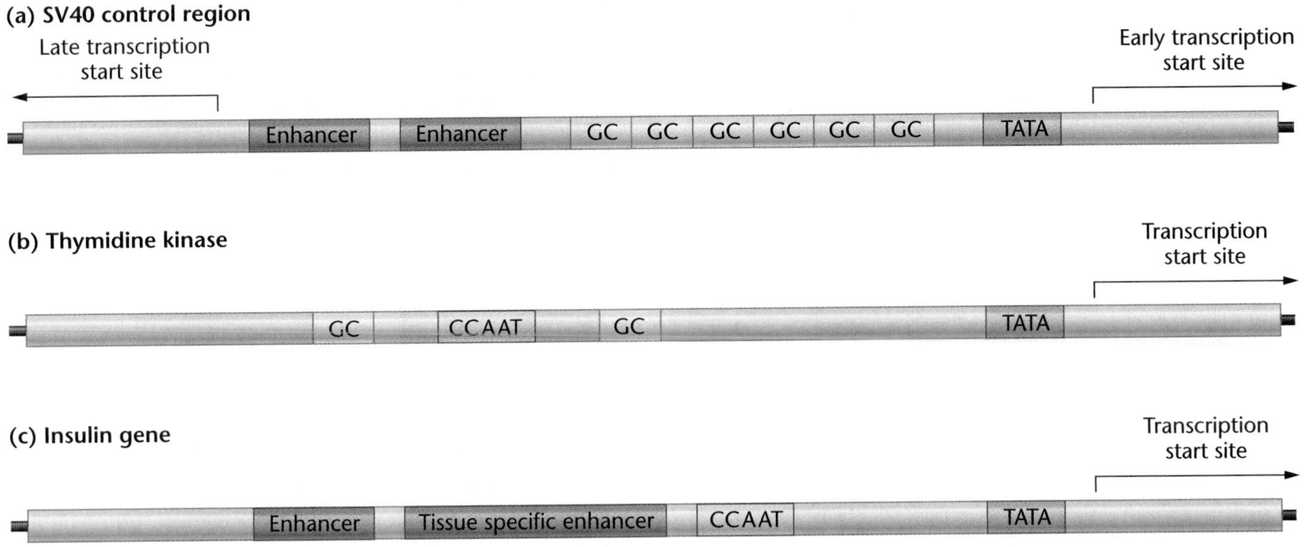

FIGURE 17–5 Organization of the promoter regions in several genes expressed in eukaryotic cells, illustrating the variable nature, number, and arrangement of controlling elements.

called **upstream activator sequences (UAS)**, can function upstream at variable distances and in either orientation. They differ from enhancers in that they cannot function downstream of the transcription start point.

Like promoters, enhancers are modular, and contain several different short DNA sequences. The enhancer of the SV40 virus (which is transcribed inside a eukaryotic cell) has a complex structure consisting of two adjacent 72-bp sequences located some 200 bp upstream from a transcriptional start point (Figure 17–6). Each of the two 72-bp regions contains five sequences that contribute to the maximum rates of transcription. If one or the other of these regions is deleted, there is no effect on transcription; but if both are deleted, *in vivo* transcription is greatly reduced.

Enhancers differ from promoters in a number of significant ways. While promoter sequences are essential for basal-level transcription, enhancers are necessary for the full level of transcription. In addition, enhancers are responsible for time- and tissue-specific gene expression. Since enhancers are able to stimulate levels of transcription at a distance, an intriguing question is how are they able to do this? As we will see, they perform at least two functions. First, transcription factors bind to enhancers and alter the configuration of chromatin; second, by bending or looping the DNA, they bring distant enhancers and their promoters into close proximity to form complexes with transcription factors and polymerases. In the new configuration, transcription is stimulated to a higher level, increasing the overall rate of RNA synthesis.

17.4 Transcription in Eukaryotes Requires Several Steps

As outlined earlier, activation of transcription in eukaryotes proceeds in a series of steps. First in this series is the remodeling of chromatin to open the DNA and make it available for transcription. After the DNA has been opened, transcription factors bind to promoter regions in the DNA along with RNA polymerase, forming a transcription initiation complex. This section describes the events of chromatin remodeling, and the next section details the molecular events in the initiation of transcription.

Transcription Requires Chromatin Remodeling

The DNA in eukaryotic chromosomes is combined with histones and nonhistone proteins to form chromatin. Some chromosomal regions are highly condensed in the interphase nucleus and form transcriptionally inert heterochromatin. Investigators have shown that genes in transcriptionally inert

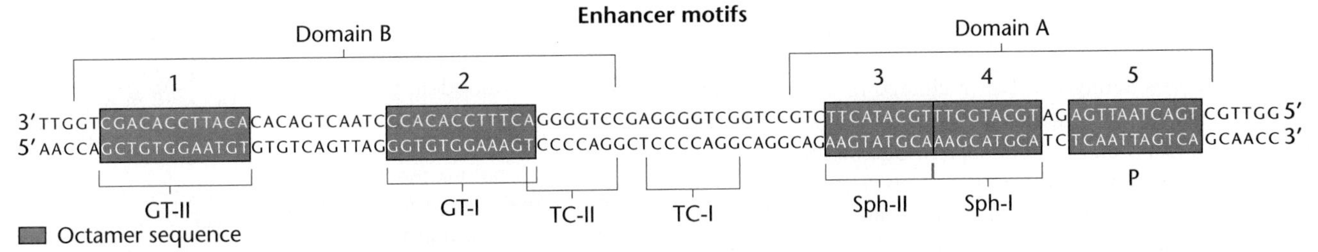

FIGURE 17–6 DNA sequence for the SV40 enhancer. Boxed sequences are those required for maximum enhancer effect. The brackets below the sequence name the various sequence motifs within this region. The two domains of the enhancer (A and B) are indicated.

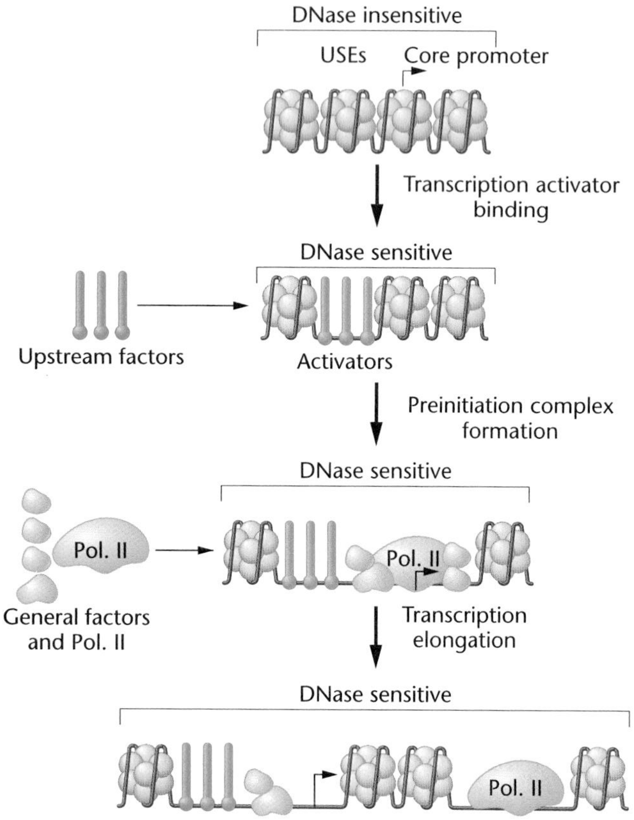

FIGURE 17–7 Nucleosomes can inhibit multiple steps required for gene transcription. The binding of enhancer transcription factors requires accessing nucleosomal DNA and may result in displacement of the nucleosome. Similarly, the formation of the basal transcription complex at the TATA box, other upstream elements (USEs) and the core promoter site is also suppressed by nucleosomes. Finally, RNA polymerase II is slowed by the presence of nucleosomes.

regions of the genome are relatively resistant to *in vitro* digestion with DNAse I. Euchromatic regions on the other hand, have an open chromatin configuration, and are sensitive to DNAse I digestion. Genes in these regions can be transcribed.

These levels of chromatin organization are a reflection of the higher order structure of nucleosomes, the basic repeating unit of chromatin. (Nucleosomes and chromosome structure are discussed in Chapter 12.) Changes in chromatin organization, referred to as **chromatin remodeling**, are essential for many processes, including the binding of polymerases and the initiation of transcription (Figure 17–7), as well as DNA replication, repair, and recombination.

Chromatin remodeling involves a change in the interaction between DNA and histones in nucleosomes. When remodeling is completed, promoter sequences are freed from histones and accessible to proteins that initiate transcription. Chromatin remodeling is carried out by a diverse group of protein complexes that all have ATP-ase activity. One of the best-studied remodeling complexes is the **SWI/SNF** complex. This is a large 11-subunit complex first described in yeast and subsequently found widely distributed in other eukaryotes, including humans. One of the subunits contains a domain that allows nonspecific DNA binding, and another is an ATP-ase. The proteins in this complex were originally identified as transcriptional activators, since their action leads to promoter activation.

Nucleosome remodeling complexes can be targeted to specific sites in several ways [Figure 17–8(a)]. Transcription activators, including those containing domains called leucine zippers (described below) can direct binding of the SWI/SNF complex. Other possible ways of targeting these complexes involve binding and stabilization by acetylated histones [Figure 17–8(b)], or by binding to methylated DNA [Figure 17–8(c)]. (DNA methylation is discussed later in this chapter.)

Nucleosome remodeling complexes may alter nucleosome structure by several different mechanisms. As seen in Figure 17–9(a), this can entail altering the contacts between the DNA and the nucleosome histone proteins, causing the nucleosome to slide farther down the DNA molecule. Alternatively, the path of the DNA around the nucleosome core particle may be altered, pulling the DNA off the nucleosome, as seen in Figure 17–9(b). Another possibility is that the structure of the nucleosome core particle itself may be altered, producing a nucleosome dimer, as seen in Figure 17–9(c).

Histone Modification Is Part of Chromatin Remodeling

A second mechanism of chromatin alteration is histone modification. The chemical alteration of the histone component of nucleosomes is catalyzed by **histone acetyltransferase enzymes (HAT)**. When an acetate group is added to specific basic amino acids on the histone tails, the attraction between the basic histone protein and acidic DNA is lessened (Figure 17–10). As in the remodeling complexes, HATs are targeted to genes by specific transcription factors. The remodeling of a chromatin region containing a gene involves interaction between a remodeling complex and a HAT. Of course, what can be opened can also be closed. In this case, histone **deacetylases (HDACs)** can remove acetate groups from histone tails. Typically, from the focal point of an enhancer, remodeling spreads toward the promoter of the gene and then into the transcription unit. **Insulator elements**, short DNA

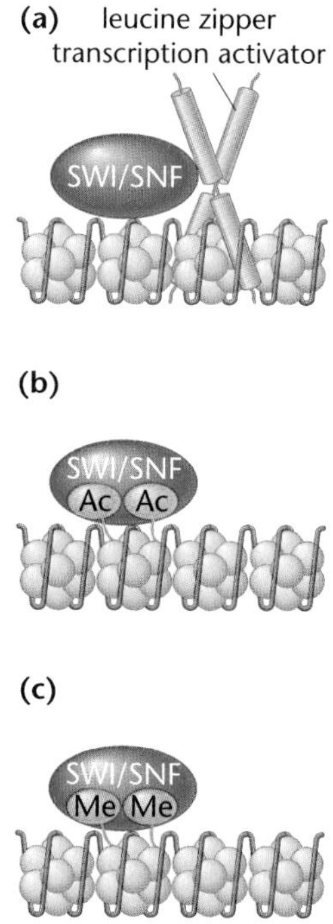

FIGURE 17–8 The SWI/SNF nucleosome remodeling complexes can be directed to specific DNA sites in several ways. (a) Transcription factors, including those with leucine zipper domains can target binding. (b) Histone components of nucleosomes modified by acetylation can serve as SWI/SNF targets. (c) Methlyated DNA regions can also be target sites for nucleosome remodeling complexes.

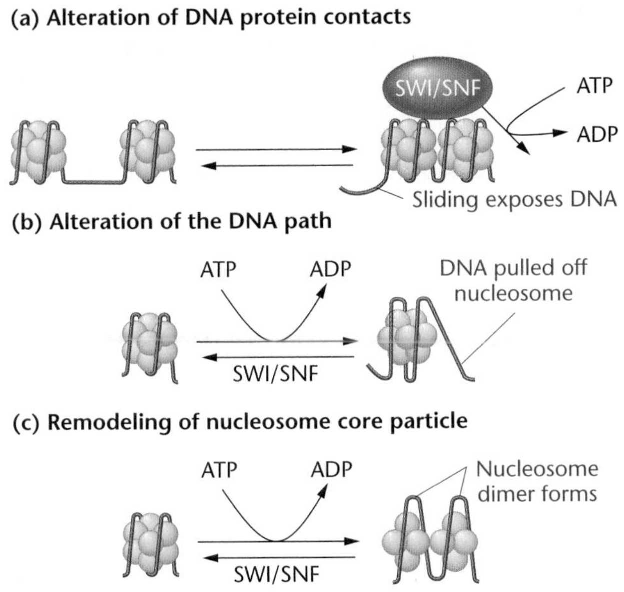

FIGURE 17–9 Three mechanisms that might be used to alter nucleosome structure by the ATP hydrolysis-dependent remodeling complex SWI/SNF. (a) DNA–histone contacts may be loosened. (b) The path of the DNA around an unaltered nucleosome core particle may be altered. (c) The conformation of the nucleosome core particle may be altered.

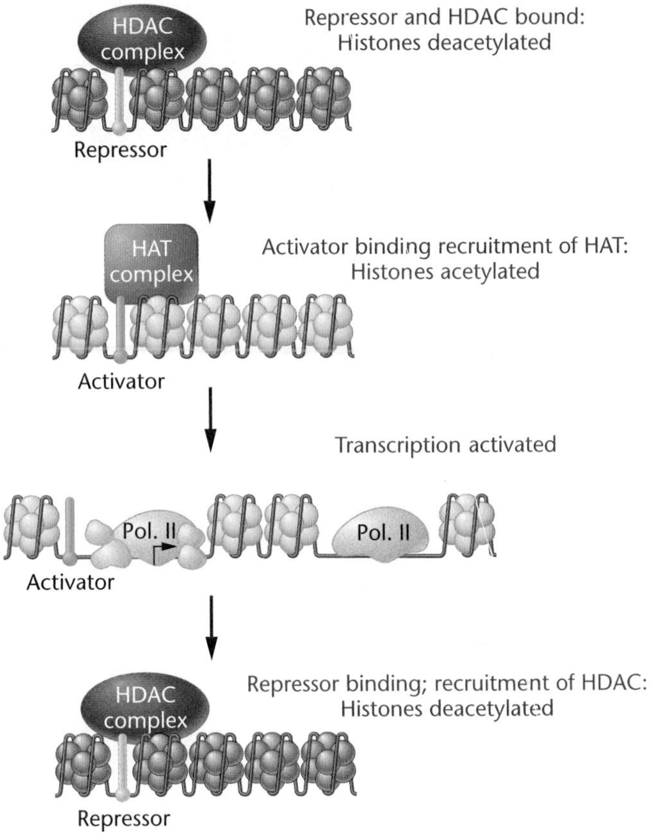

FIGURE 17–10 Proposed model of the action of HAT and HD complexes. Transcription factors recruit the complex to the gene, which either adds or removes acetyl groups, aiding in either opening or closing the chromatin structure.

sequences that bind specific proteins, can act as barriers to prevent the spread of remodeling into neighboring genes. In addition to acetylation, histones can be modified by phosphorylation and methylation. These structural alterations occur at specific amino acid residues in histones. This observation led to the proposal that specific, reversible patterns of covalent histone modification lead to gene activation or gene silencing. According to this idea, these patterns of enzyme-mediated modification constitute a series of signals known as the **histone code**, a regulatory mechanism that alters chromatin conformation and in turn, gene activity. Some evidence is accumulating that specific histone modifications serve as binding sites for proteins involved in gene regulation.

Our emphasis here is on remodeling that opens chromatin and makes promoter regions available for binding of transcription factors that initiate the chain of events associated with transcription.

17.5 Assembly of the Basal Transcription Complex Occurs at the Promoter

Transcriptional control in eukaryotes involves interaction between DNA sequences adjacent to promoter regions and DNA-binding proteins. The regulatory sequences adjacent to a wide range of genes have been identified, mapped, and sequenced. DNAse-protection and gel-retardation assays have identified DNA-binding proteins in cell extracts, and affinity chromatography has been used to isolate transcriptional factors present in very low concentrations. The result has been a virtual explosion of information about eukaryotic transcription factors. This section will review some of this information, including the characteristics of transcription factors, their interaction with DNA sequences, the involvement of other regulatory factors, and the initiation of transcription by RNA polymerases.

RNA Polymerases and Transcription

Eukaryotic genes use three RNA polymerases for transcription. (Prokaryotic genes are all transcribed by a single RNA polymerase.) These three are classified as type I (ribosomal RNAs), type II (mRNAs and snRNAs), and type III (tRNAs, 5*S* rRNA, and several other small cellular RNAs). The promoter for each type of polymerase has a different nucleotide sequence and binds different transcription factors. Our discussion here will be confined to events associated with genes transcribed by RNA polymerase II.

Formation of the Transcription Initiation Complex

A series of proteins called **basal** or **general transcription factors** act in *trans* to control initiation of transcription. These proteins are not part of the RNA polymerase II molecule, but assemble at the promoter in a specific order, forming a transcriptional pre-initiation complex (PIC) that provides a platform for the polymerase to recognize and bind to the promoter (Figure 17–11). To initiate formation of the transcription

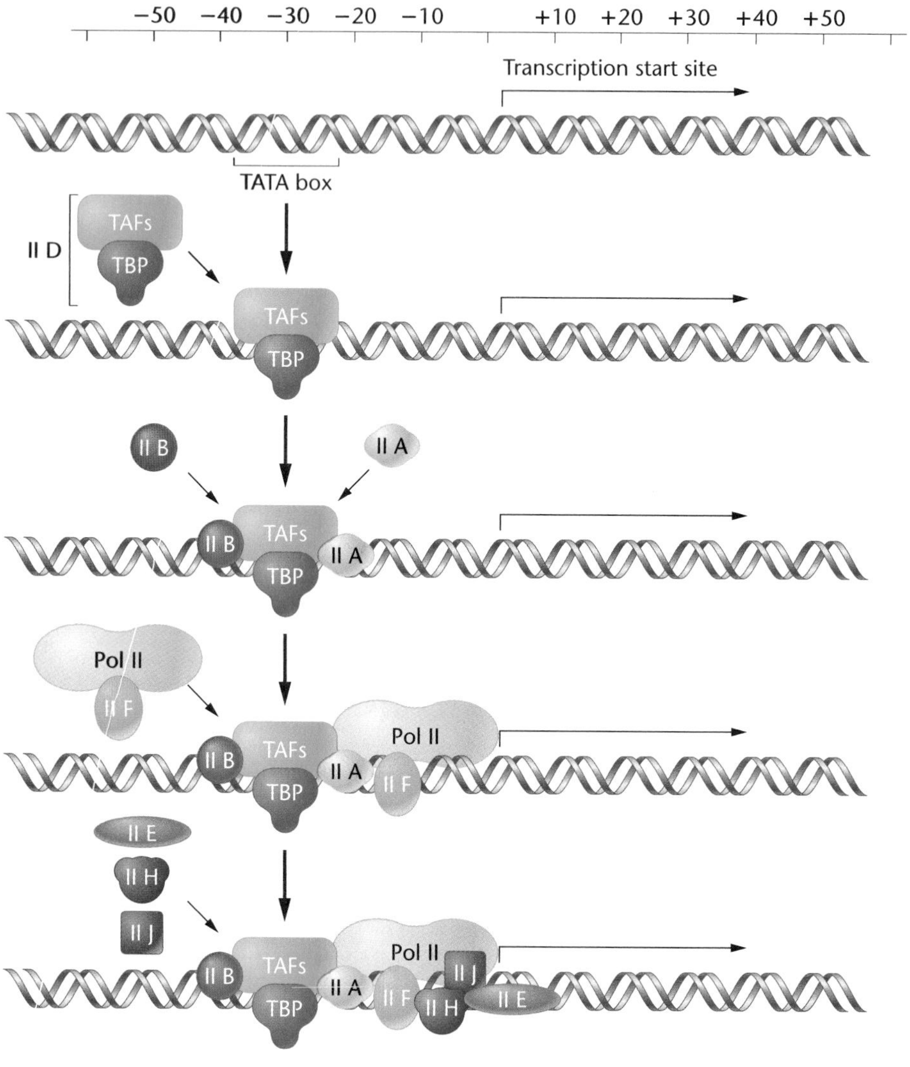

FIGURE 17–11 The assembly of transcription factors required for the initiation of transcription by RNA polymerase II.

complex, a complex called TFIID binds to the TATA box through one of its subunits, called TBP (*T*ATA *B*inding *P*rotein) (Figure 17–12). The TFIID complex is composed of TBP plus approximately 13 proteins called TAFs (TATA Associated Factors). About 20 base pairs of DNA are involved in binding the TBP and the other subunits in the TFIID transcription factor.

TFIID responds to contact with activator proteins by conformational changes that expedite binding of additional factors, such as TFIIB. TFIIB is a protein that interacts with TBP and DNA sequences upstream of TATA. TFIIA, RNA polymerase II, and some additional factors such as TFIIF then bind, followed by TFIIE, TFIIH, and TFIIJ. In the final stage, RNA polymerase leaves the TATA box, and the transcription of the adjacent gene ensues at a *basal* level.

In addition to the general transcription factors that form a platform for the binding of RNA polymerase and the initiation of transcription at a basal level, other transcription factors bind at enhancer sites and affect the rate of transcription. These proteins may act as *positive factors* to increase transcription (activators) or as *negative factors* (repressors) to decrease transcription. These proteins control where and when genes are expressed and the rate of expression. Activators bind to the enhancer and interact with the transcription complex at the promoter and can increase the rate of transcription initiation by as much as a hundredfold.

Activators Bind to Enhancers and Change the Rate of Transcription Initiation

Activators are modular proteins that bind to enhancer DNA sequences, forming a complex known as an **enhanceosome** (Figure 17–13). Once formed, proteins in the enhanceosome interact with proteins in the transcription complex. To do this, activators have two functional domains (clusters of amino acids that carry out a specific function). One domain binds to DNA sequences present in the enhancer **(DNA-binding domain)**, and another activates transcription through protein–protein interaction **(*trans*-activating domain)**. The *trans*-activating domain binds to RNA polymerase or to other transcription factors at the promoter. The existence of separate domains in eukaryotic transcription factors was first demonstrated by the genetic analysis of mutants in yeast. The domains of eukaryotic transcription factors take on several forms. The DNA-binding domains have characteristic three-dimensional structural patterns or **motifs**. There are several classes of these motifs, including **helix–turn–helix (HTH)**, zinc finger, and basic leucine zipper (bZIP). This list is not

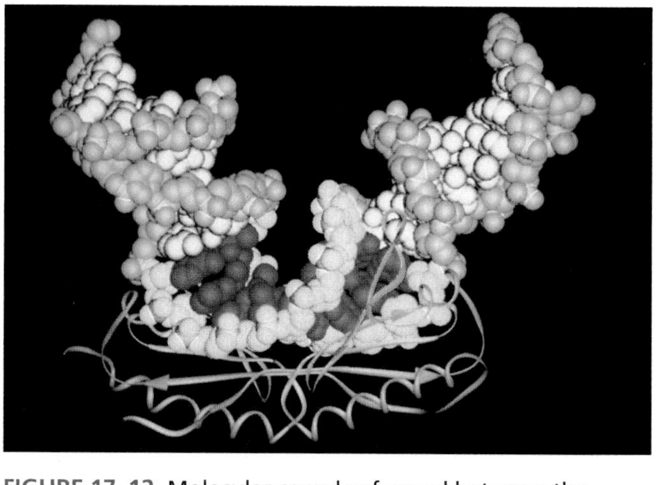

FIGURE 17–12 Molecular complex formed between the TATA-box binding protein (green) and the 8-bp TATA-box nucleotide sequence (red) and its phosphodiester backbone (yellow).

exhaustive, and other new groups will undoubtedly be established as new factors are characterized.

The first DNA-binding domain to be discovered was the helix–turn–helix (HTH) motif. This motif, also present in prokaryotes, is characterized by its geometric conformation, rather than a distinctive amino acid sequence (Figure 17–14). Two adjacent α-helices separated by a "turn" of several amino acids enables the protein to bind to DNA (hence the name of the motif). Unlike several of the other DNA-binding motifs, the HTH pattern cannot fold or function alone, but is always part of a larger DNA-binding domain. Amino acid residues outside the HTH motif are important in regulating DNA recognition and binding.

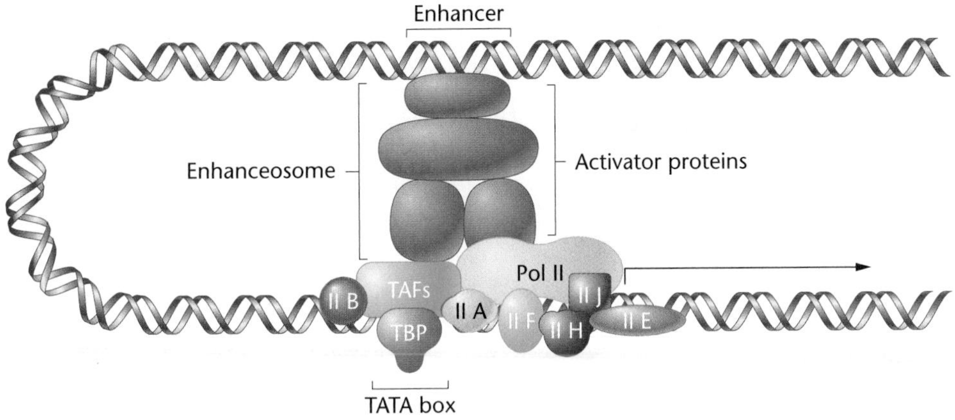

FIGURE 17–13 Formation of DNA loops allows factors that bind to enhancers at a distance from the promoter to interact with regulatory proteins in the transcription complex and to maximize transcription.

How Do We Know?

How have we become aware that transcription factors have multiple binding domains?

The potential for forming HTH geometry has been recognized in distinct regions of a large number of eukaryotic genes known to regulate developmental processes. Present almost universally in eukaryotic organisms is the **homeobox**, a stretch of 180 bp specifying a 60-amino-acid **homeodomain** sequence that can form an HTH structure. Of the 60 amino acids, many are basic (arginine and lysine), and a conserved sequence is found among these genes. Because of their significance to animal developmental processes, we will discuss homeobox-containing genes in Chapter 23.

The **zinc-finger** motif was originally discovered in the TFIIIA transcription factor of the frog, *Xenopus laevis*. This structural motif has now been identified in protooncogenes (Chapter 18), in genes that regulate development in *Drosophila* (the *Krüppel* gene, see Chapter 23), in proteins whose synthesis is induced by growth factors and differentiation signals, and in transcription factors. Zinc-finger proteins, representing one of the major families of eukaryotic transcription factors, are involved in many aspects of gene regulation. There are several types of zinc-finger proteins, each with a distinctive structural pattern.

A typical zinc-finger protein contains clusters of two cysteines and two histidines at repeating intervals (Figure 17–15). The consensus amino acid repeat is $CysN_{2–4}CysN_{12–14}HisN_3His$ (where N is any amino acid). The interspersed cysteine and histidine residues covalently bind zinc atoms, folding the amino acids into loops known as zinc fingers. Each finger consists of approximately 23 amino acids (with a loop of 12 to 14 amino acids between the Cys and His residues) and a linker between loops consisting of 7 or 8 amino acids. The amino acids in the loop interact with, and bind to, specific DNA sequences. Studies have shown that zinc fingers bind in the major groove of the DNA helix and wrap at least partway around the DNA. Within the major groove, the zinc finger contacts, and may form, hydrogen bonds with a set of bases, especially in G-rich strands. The number of fingers in a zinc-finger transcription factor ranges from 2 to 13, as does the length of the DNA-binding sequence.

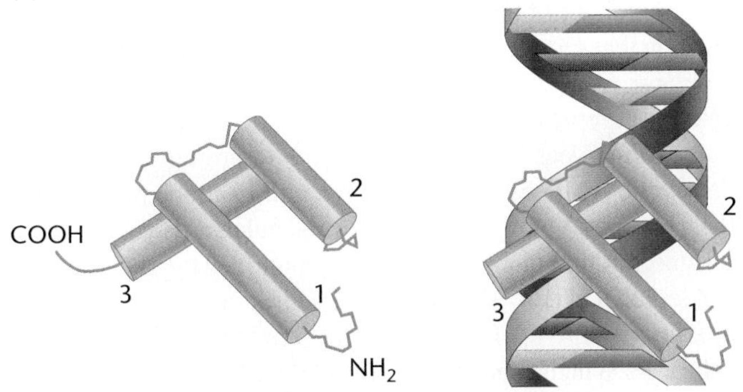

FIGURE 17–14 A *helix–turn–helix* or *homeodomain* in which (a) three planes of the α-helix of the protein are established, and (b) these domains bind in the grooves of the DNA molecule.

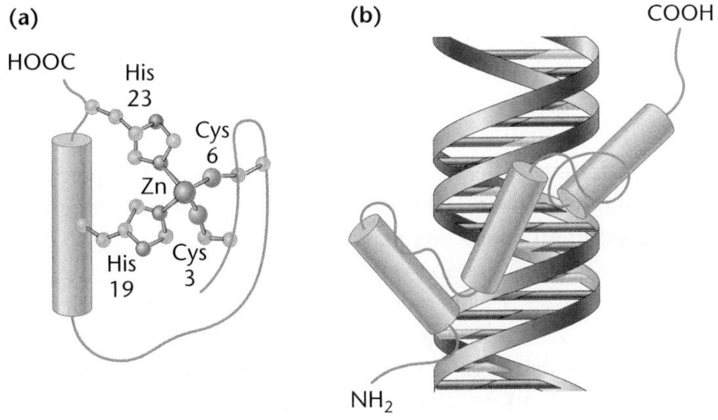

FIGURE 17–15 (a) A zinc finger in which cysteine and histidine residues bind to a Zn^{++} atom. (b) This loops the amino acid chain out into a fingerlike configuration.

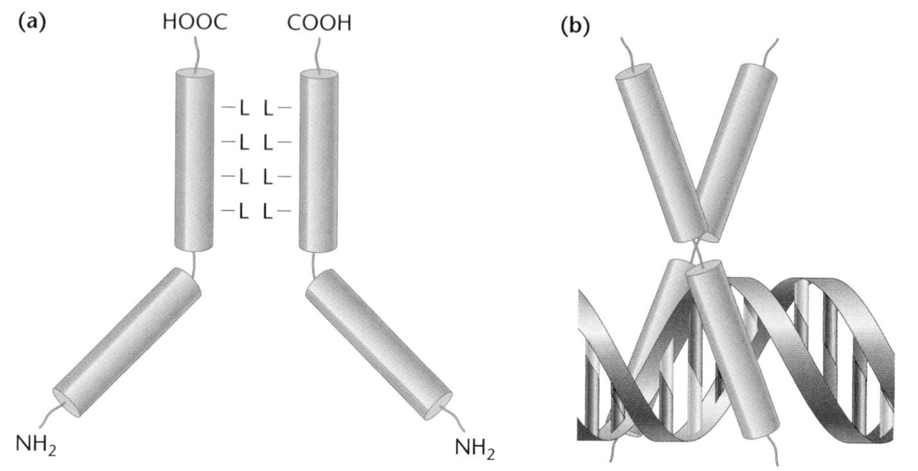

FIGURE 17–16 (a) A leucine zipper is the result of dimers from leucine residue at every other turn of the α-helix in facing stretches of two polypeptide chains. (b) When the α-helical regions form a leucine zipper, the regions beyond the zipper form a Y-shaped region that grips the DNA in a scissorlike configuration.

A third type of DNA-binding domain is represented by the **basic leucine zipper (bZIP)**. This DNA-binding domain is adjacent to the **leucine zipper**, a region that allows protein–protein dimerization. The leucine zipper contains four leucine residues spaced 7 amino acids apart and flanked by basic amino acids. This region forms a helix with leucine residues protruding at every other turn. When two such molecules dimerize, the leucine residues "zip" together (Figure 17–16). The dimer contains two basic α-helical regions adjacent to the zipper that bind to phosphate residues and specific bases in DNA, making the dimer look like a pair of scissors.

Transcription factors also contain domains that interact with proteins in the basal transcription complex and control the level of transcription initiation. These *trans*-activating domains, distinct from the DNA-binding domains, can occupy from 30 to 100 amino acids. They interact with other transcription factors (such as those that bind to the promoter) or directly with RNA polymerase. (See Figure 17–13.) In addition, many transcription factors also contain domains that bind **coactivators**, such as hormones or small metabolites that regulate their activity.

Overall, the picture of transcriptional regulation in eukaryotes is complex, but several important generalizations can be drawn. First, the structural organization of chromatin and alterations in chromatin structure allowing binding of transcription factors are considered to be the primary levels of regulation. Second, the regulation of transcription by transcription factors is largely positive, although transcription repressors are now being recognized as important components of gene regulation.

In addition, different transcription factors may compete for binding to a given DNA sequence or to two overlapping sequences, or the same site may bind different factors in two different tissues. Transcription factor concentration and the strength with which each factor binds to the DNA will dictate which factor binds. Finally, because of the variability in the location and modular nature of enhancers, and because of the variety of transcription factors, an initiation complex at the promoter can interact with multiple enhancers, with individual proteins contacting the basal-transcription complex at the promoter in a number of ways, all involving protein–protein interaction.

Now solve this

Question 17.13 on page 431 asks you to diagram how supercoiling affects enhancers.

Hint: Remember that changes in chromatin conformation and positioning of elements associated with the action of enhancers are important factors.

17.6 Gene Regulation in a Model Organism: Positive Induction and Catabolite Repression in the *gal* Genes of Yeast

One of the first model systems used to study eukaryotic genetic regulation was the genes in yeast that encode enzymes for the breakdown of galactose. These *gal* genes serve a function similar to the *lac* operon and *ara* operon in *E. coli*. Expression of the *gal* genes is **inducible**—that is, they are regulated by the presence or absence of their substrate, galactose. In the absence of galactose, the genes are not transcribed. If galactose is added to the growth medium, transcription begins immediately, and the mRNA concentration of the transcripts increases a thousandfold. However, transcription is activated only if the concentration of glucose is low. So, like the *lac* and *ara* operons, the *gal* genes are under a second level of control, **catabolite repression**. (See Chapter 16.) A mutation in the regulator of the *gal* genes, *GAL4*, prevents activation, indicating that transcription is under **positive control**—that is, the regulator must be present to turn on gene transcription. Yeast catabolite repression differs from that in bacteria. This phenomenon in yeast involves an enzyme (a protein kinase), and not cyclic AMP, as in bacteria. We will examine how two of the *gal* genes, *GAL1* and *GAL10*, are regulated (Figure 17–17).

Transcription of these two genes is controlled by a central control region, called **UAS$_G$** (upstream activating sequence), of approximately 170 bp. Recall that UASs are functionally similar to enhancers in higher eukaryotes. The chromatin structure of a UAS is constitutively open, or **DNase hypersensitive**, meaning that it is free of nucleosomes. This open chromatin structure requires action of the SWI/SNF remodeling complex. Within the UAS are

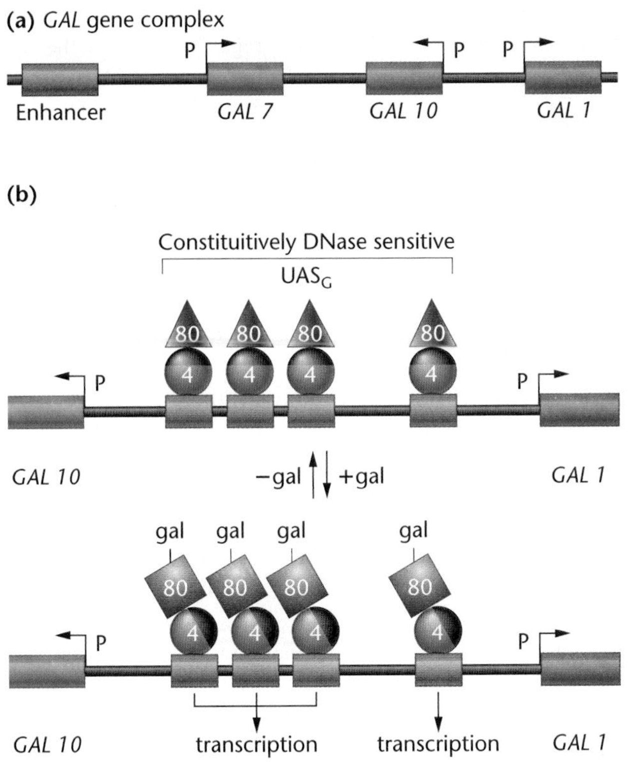

FIGURE 17–17 Model of *GAL1* and *GAL10* UAS structures showing the binding sites for the Gal4p positive regulator and their interaction with the Gal80p negative regulator. Induction is shown as the structure of the Gal80p becomes altered, exposing the activation domain of Gal4p.

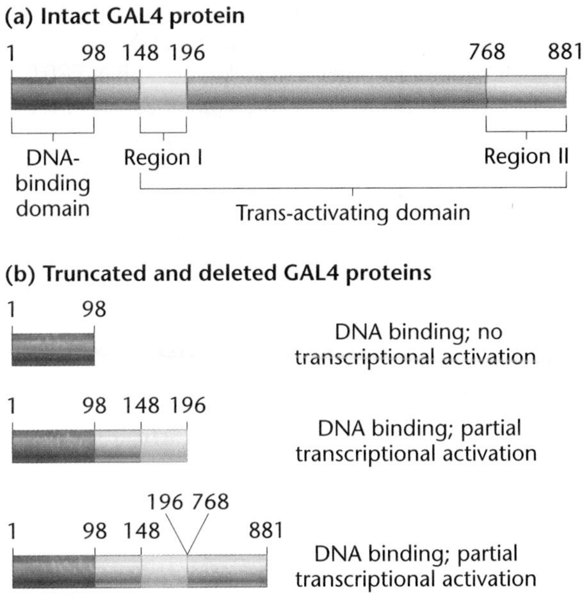

FIGURE 17–18 The Gal4p protein has three domains that participate in the activation of the *GAL1* and *GAL10* loci. Amino acids 1–98 bind to DNA-recognition sites in UAS_G. Amino acids 148–196 and 768–881 are required for transcriptional activation. Gal80p binding activity resides in the amino acid region 851–881.

four binding sites for the Gal4 protein (Gal4p). These sites are permanently occupied by Gal4p, whether or not the *gal* genes have been activated. Gal4p is, in turn, negatively regulated by Gal80p, another *gal* gene regulator. Gal80p is always bound to Gal4p, covering its activation domain, shown in Figure 17–17 as a dark patch. Induction occurs when the inducer, a phosphorylated galactose, binds to Gal80p or Gal4p (or both) and causes a structural alteration, exposing the Gal4p activation domain. As in *E. coli*, these genes are under a second level of control—catabolite repression—and thus cannot be turned on in the presence of glucose. In this case, however, the catabolite does not operate through cAMP, but rather through a complex system involving a protein kinase. Through activation, additional factors are recruited to further remodel the chromatin, encompassing the two promoters exposing the TATA boxes to the TBP.

Gal4p is a protein consisting of 881 amino acids. It includes a DNA-binding domain that recognizes and binds to sequences in the UAS_G and a *trans*-activating domain that activates transcription [Figure 17–18(a)]. These functional domains were identified by cloning and expressing truncated *GAL4* genes and by assaying the gene products for their ability to bind DNA and to activate transcription. Because the protein contains two functional domains, deleting one of these regions allows the other to remain intact and functional.

Mutational analysis identified the DNA-binding domain as a region at the N-terminus of the protein. This region recognizes and binds to the nucleotide sequences in UAS_G. Proteins with deletions from amino acid 98 to the C-terminus (amino acid 881) retain the ability to bind to the UAS_G region. Similar experiments also identified two regions of the protein involved in transcriptional activation: region I (amino acids 148–196) and region II (amino acids 768–881). Constructs containing the DNA-binding domain (amino acids 1–96) and region I (amino acids 148–196), or the DNA-binding domain and region II (amino acids 768–881), have reduced transcriptional activity [Figure 17–18(b)].

To explore the nature of the activating regions, researchers started with a protein containing the DNA-binding region and the activating region I (GAL4 1–238), and they isolated point mutations in region I that result in increased activation. Most single mutations that increase activation result from amino acid substitutions that increase the negative charge of region I and cause a threefold increase in transcription. A multiple mutant with four more negative charges than the starting protein causes a ninefold increase in transcriptional activation and has almost 80 percent of the activity of the intact Gal4p. However, negative charges are not the only factor in activation; some mutants with decreased transcriptional activity have about the same number of negative charges as the parental GAL4 1–238 molecule.

These results strongly suggest that activation directs contact between the activating domain of the transcription factor and other proteins. How does this protein–protein interaction activate transcription? There are several possibilities, including chromatin remodeling, stabilization of binding between the promoter DNA and RNA polymerase, increasing the rate of unwinding the double-stranded DNA within the transcribed region, accelerating release of RNA polymerase from the promoter, or attracting and stabilizing other factors that bind to the promoter or to the RNA polymerase.

An attractive candidate for an activator target is TFIID—a complex of proteins that contains the TATA binding protein (TBP)—and about 13 other TAFs. The idea that TFIID is a target for activation is supported by the finding that some *GAL4* constructs stimulate transcription in mammalian cell extracts and change the conformation of DNA-bound TFIID. In addition, other studies show that although TBP alone can bind to the promoter, TAFs must be present for the complex to respond to activators (as shown in Figure 17–13).

17.7 DNA Methylation and Regulation of Gene Expression

Alteration of chromatin conformation is one way of regulating eukaryotic gene expression. In this section, we will consider another mechanism that plays a role in gene regulation: the chemical modification of DNA by adding or removing methyl groups from the DNA bases.

After replication, the DNA of most eukaryotic organisms is modified by the enzyme-mediated addition of methyl groups to bases and sugars. Base methylation most often involves adding methyl groups to cytosine. In the genome of any given eukaryotic species, approximately 5 percent of the cytosine residues are methylated. However, the extent of methylation can be tissue specific and can vary from less than 2 percent to more than 7 percent. The ability of base methylation to alter gene expression is known from studies on the *lac* operon in *E. coli*. Methylating DNA in the operator region, even at a single cytosine residue, can cause a marked change in the affinity of the repressor for the operator. Methylation occurs at position 5 of the cytosine base, causing the methyl group to protrude into the major groove of the DNA helix where it alters the binding of proteins to the DNA. Methylation occurs most often in the cytosine of CG doublets in DNA, usually in both strands:

$$5'\text{—}{}^{m}\text{CpG—}3'$$
$$3'\text{—GpC}^{m}\text{—}5'$$

Whether or not DNA is methylated can be determined by restriction enzyme analysis. The enzyme *Hpa*II cleaves at the recognition sequence CCGG; however, if the second cytosine is methylated, the enzyme will not cut the DNA. The enzyme *Msp*I cuts at the same CCGG site, regardless of whether the second cytosine is methylated. If a segment of DNA is unmethylated, both enzymes produce the same restriction pattern of bands. As shown in Figure 17–19, if one site is methylated, digestion with *Hpa*II produces an altered pattern of fragments. Using this method in eukaryotes to analyze the methylation status of a given gene in different tissues shows that if a gene is expressed, it is usually not methylated or has only a low level of methylation.

Evidence for the role of methylation in eukaryotic gene expression is somewhat indirect and is based on a number of observations. First, an inverse relationship exists between the degree of methylation and the degree of expression. That is, low amounts of methylation are associated with high levels of gene expression, and high levels of methylation are associated with low levels of gene expression. In mammalian females, the inactivated X chromosome, which is almost totally inactive, has a higher level of methylation than does the active X chromosome. Within the inactive X, regions that escape inactivation have much lower levels of methylation than those seen in adjacent inactive regions.

Second, methylation patterns are tissue specific and, once established, are heritable for all cells of that tissue. Perhaps the strongest evidence for the role of methylation in gene expression comes from studies using base analogs. The nucleotide 5-azacytidine is incorporated into DNA in place of cytidine and cannot be methylated (Figure 17–20), causing the undermethylation of the sites where it is incorporated. The incorporation of 5-azacytidine causes changes in the pattern of gene expression and can stimulate expression of alleles on inactivated X chromosomes.

The analog 5-azacytidine is being used in clinical trials for the treatment of sickle-cell anemia. In this autosomal recessive condition, a mutant hemoglobin protein with a defect in β-globin causes changes in shape of red blood cells that, in turn, generate a cascade of clinical symptoms. During embryogenesis, the ε- and γ^{g}-globin genes are expressed, but normally become transcriptionally inactive after birth, when β-globin synthesis begins. Treating affected individuals with 5-azacytidine causes a reduction in the amount of methylation in

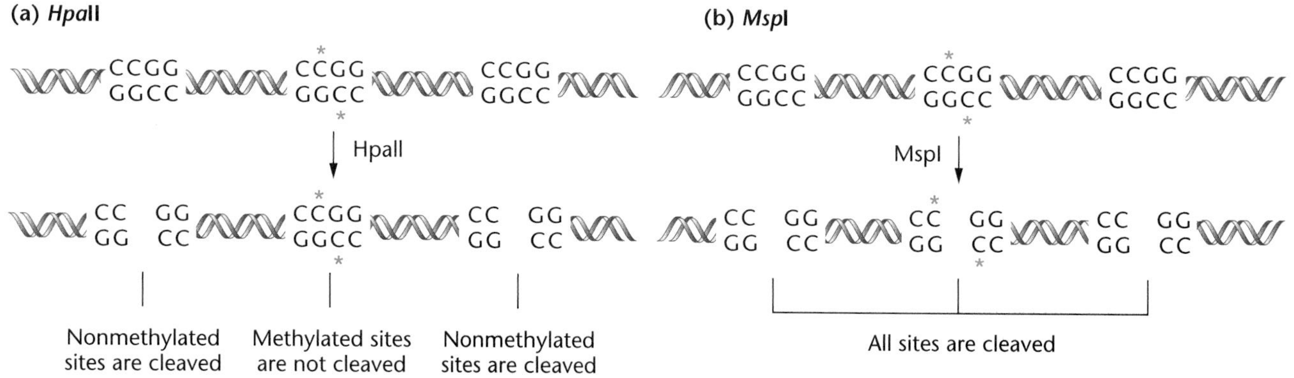

FIGURE 17–19 The restriction enzymes *Hpa*II and *Msp*I recognize and cut at CCGG sequences. (a) If the second cytosine is methylated (indicated by an asterisk), *Hpa*II will not cut. (b) The enzyme *Msp*I cuts at all CCGG sites, whether or not the second cytosine is methylated. Thus, the state of methylation of a given gene in a given tissue can be determined by cutting DNA extracted from that tissue with *Hpa*II and *Msp*I.

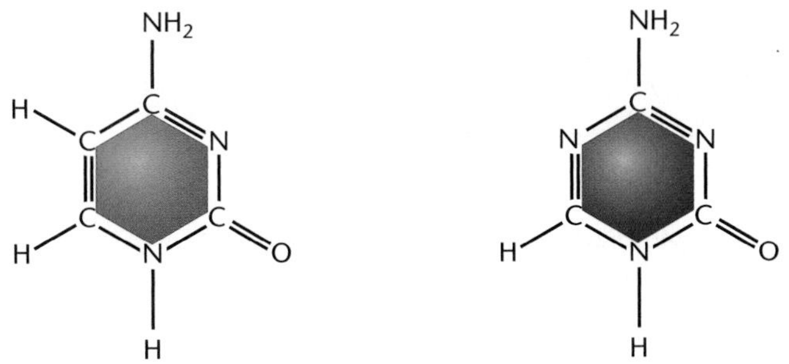

Cytosine 5-Azacytosine

FIGURE 17–20 The base 5-azacytosine, which has a nitrogen at the 5 position and can be incorporated into DNA in place of deoxycytidine during DNA synthesis. The base 5-azacytosine cannot be methylated, causing undermethylation of the CpG dinucleotide wherever it has been incorporated.

the ε-globin and γ^{g}-globin genes and initiates reexpression of these embryonic and fetal genes. The ε- and γ-proteins replace β-globin in hemoglobin molecules, bringing about a reduction in the amount of sickling in the red blood cells.

Though the available evidence indicates that the absence of methyl groups in DNA is related to increases in gene expression, methylation cannot be regarded as a general mechanism for gene regulation because methylation is not a general phenomenon in eukaryotes. In *Drosophila*, for example, there is no methylation of DNA. Thus, methylation may represent only one of a number of ways in which gene expression can be regulated by genomic changes, and it seems likely that similar mechanisms remain to be discovered.

How might methylation affect gene regulation? One possibility comes from the observation that certain proteins bind to 5-methyl cytosine without regard to the DNA sequence. These proteins could recruit corepressors or histone deacetylases (or both) to remodel the chromatin, changing it from an open structure to a closed structure, or may recruit nucleosome remodeling complexes.

Now solve this

Question 17.14 on page 432 asks you to interpret the effect of methylation of DNA on the expression of a gene.

Hint: Remember that the location of various regulatory sequences outside the gene will affect the results.

17.8 Posttranscriptional Regulation of Gene Expression

As we have seen, regulation of genetic expression can occur at many points along the pathway from DNA to protein. Although transcriptional control is perhaps the major type of regulation in eukaryotes, **posttranscriptional regulation** also occurs in many organisms. Eukaryotic nuclear RNA transcripts are modified prior to translation, noncoding introns are removed, the remaining exons are precisely spliced together, and the mRNA is modified by the addition of a cap at the 5′ end and a poly-A tail at the 3′ end. The message is then exported to the cytoplasm. Each of these processing steps offers several possibilities for regulation. We will examine two that are especially important in eukaryotes—alternative splicing of a single mRNA transcript to give multiple mRNAs and regulation of the stability of mRNA itself.

Alternative Splicing Pathways for mRNA

Alternative splicing can generate different forms of mRNA from a single pre-mRNA molecule, so that expression of one gene can give rise to a family of proteins, with similar or different functions. Changes in splicing patterns can have many different effects on the translated protein. Small changes can alter enzymatic activity, receptor binding capacity, or protein localization in the cell. Changes in splicing patterns are important events in development, apoptosis, axon-to-axon connection in the nervous system, and many other processes. Mutations that affect regulation of splicing are the basis of several genetic disorders.

Alternative splicing increases the number of proteins that can be made from each gene. As a result, the number of proteins that a cell can make (its **proteome**) is not directly related to the number of genes in the genome, and protein diversity can exceed gene number by an order of magnitude. Alternative splicing is found in all metazoans, but is especially common in vertebrates, including humans. Now that the coding portion of the human genome has been sequenced, the task in the postgenomic era is to identify and catalog all the proteins and RNAs produced from this genome, and understand their functions. It has been estimated that 30 to 60 percent of the genes in the human genome use alternative splicing. Thus, humans can produce several hundred thousand different proteins (or perhaps more) from the 25,000–30,000 or so genes in the haploid genome.

Figure 17–21 illustrates an example of alternative splicing in the pre-mRNA transcribed from a gene containing twelve exons. This RNA has multiple sites of alternative splicing and can produce many different mRNAs. In the cell, different copies of the pre-mRNA can be spliced in different combinations, simultaneously producing many different mRNAs and proteins in a single gene. (Recall from Chapter 14 that splicing takes place in the spliceosome, a molecular complex that catalyzes splicing events.)

Alternative Splicing and Cell Function

To understand the significance of alternative splicing, let's examine the role it plays in the function of cochlear hair cells and heaving. To hear sounds in the world around us, our ears detect sound waves across a thousandfold range of frequencies. Inside the cochlea of the inner ear, the basilar membrane carries four rows of hair cells (Figure 17–22). Each cell responds to a different and narrow range of frequencies. Ensuring that hair cells are tuned to receive different frequencies is controlled in part by alternative splicing of pre-mRNA transcripts of the *SLO* gene, which encodes a calcium-regulated potassium channel. There are at least eight sites of alternative splicing in *SLO* pre-mRNA. At most sites, there is an alternative exon that can be included or excluded, or an exon with alternative splice sites. If each site is regulated independently, more than 500 different mRNAs can be produced from the transcript of the *SLO* gene.

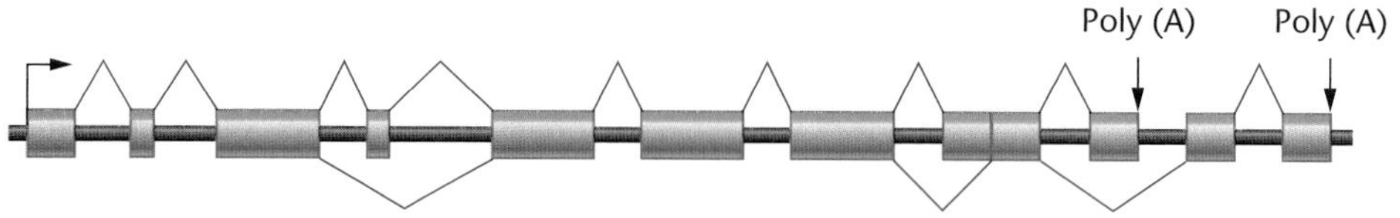

FIGURE 17–21 Patterns of alternative splicing in a eukaryotic mRNA. Exons are represented by cylinders. The normal splicing pattern is shown above the exons; alternative patterns of splicing are shown below the exons. Note that alternative splicing can add exons and a new poly (A) site. Alternative splice sites are often used in multiple combinations, resulting in many different mRNAs from a single transcript.

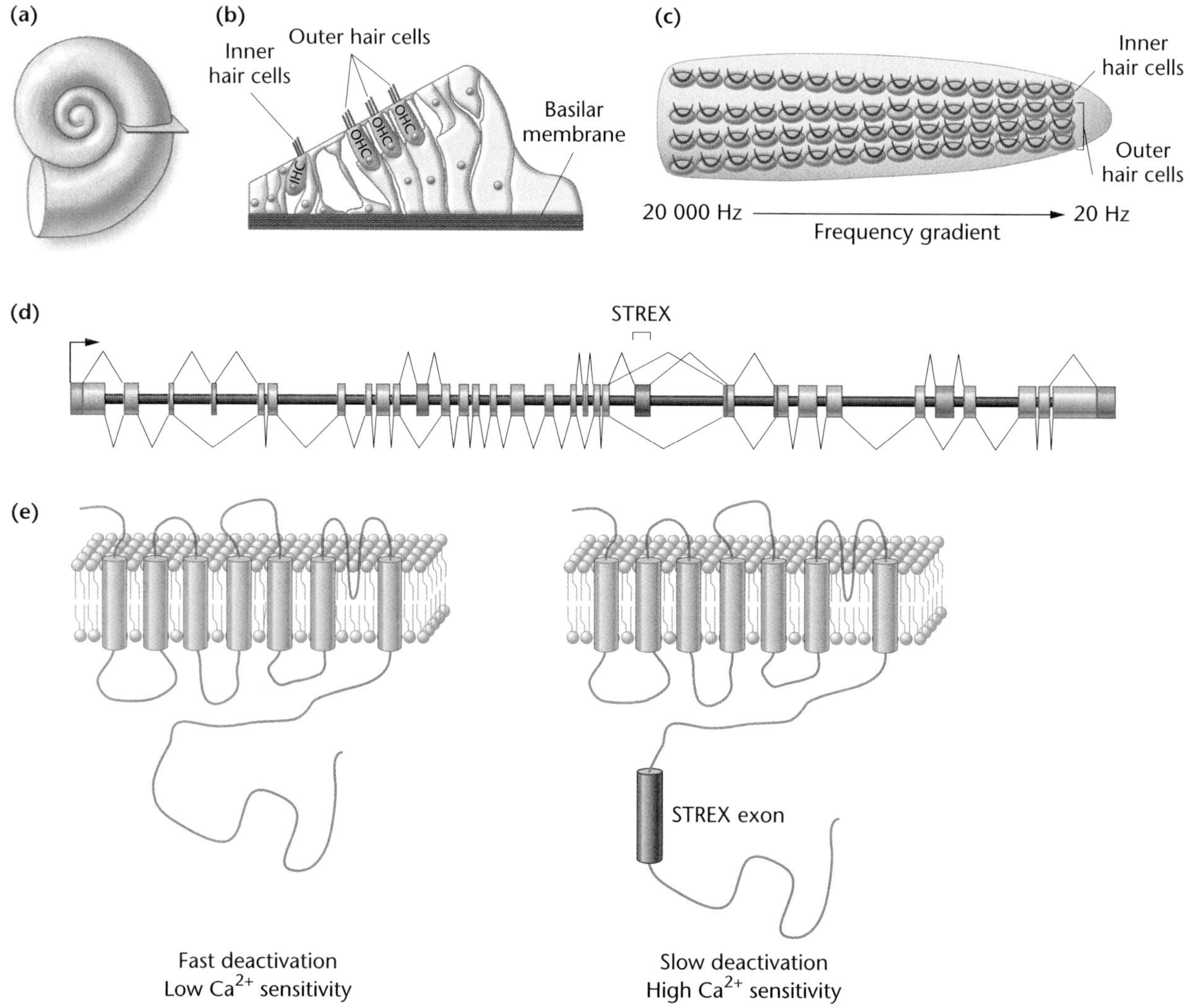

FIGURE 17–22 Alternative splicing in the human *SLO* gene. (a) The cochlea of the inner ear, which contains a basilar membrane carrying hair cells. (b) A cross-sectional view of the basilar membrane, showing the single row of inner cells, and three rows of outer cells. (c) The basilar membrane laid out to show the arrangement of hair cells and the frequency gradient of sound received by hairs along the membrane. (d) Exon–intron organization of the *SLO* gene. Normal splicing events are shown above the exons; alternative sites are shown below the exons. Constitutive exons are silver, alternative exons are purple, and the STREX exon is magenta. (e) The presence or absence of the STREX exon affects the function of the calcium-sensitive potassium channel found in the plasma membrane of the inner hair cells.

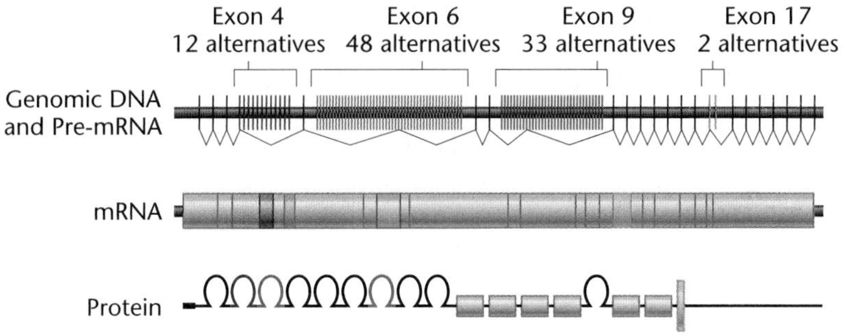

FIGURE 17–23 (Top) Organization of the *Dscam* gene in *Drosophila melanogaster* and the transcribed pre-mRNA. Each mRNA will contain one of the 12 possible exons for exon 4 (red), one of the 48 possible exons for exon 6 (blue), one of 33 for exon 9 (green), and one of 2 for exon 17 (yellow). If all possible combinations of these exons are used, the *Dscam* gene can encode 38,016 different versions of the DSCAM protein.

Some forms of the *SLO* proteins have different physiological properties that may play a role in ensuring we can hear sounds across a broad range of frequencies.

Alternative Splicing Amplifies the Number of Proteins Produced by a Genome

Given the existence of alternate splicing, how many different polypeptides can be derived from the same pre-mRNA? One answer to that question comes from work on a gene in *Drosophila*. During development, cells of the nervous system must accurately connect with each other. Even in *Drosophila*, with only about 250,000 neurons, this is a formidable task. Axons are cellular processes that form connections with other nerve cells. The *Drosophila Dscam* gene encodes a protein that guides axon growth, ensuring that neurons are correctly wired together. In *Dscam* pre-mRNA, exons 4, 6, 9, and 17 each have an array of possible exons (Figure 17–23). These are spliced into the mRNA in an exclusive fashion, so that only one of each of the possible exons is represented. There are 12 alternatives for exon 4; 48 alternatives for exon 6; 33 alternatives for exon 9; and 2 alternatives for exon 17. If all combinations of these exons are used in alternative splicing, the *Dscam* gene can produce 38,016 different proteins.

In a more extreme example, *para*, another gene expressed in the nervous system of *Drosophila*, has 13 alternative exons, but also undergoes another form of posttranscriptional modification called editing at 11 positions. RNA editing involves base substitutions made after transcription and splicing. Taking into account both alternative splicing and editing, the *para* gene can theoretically produce more than 1 million different transcripts.

The *Drosophila* genome carries about 13,000 genes, but the *Dscam* gene alone can produce 2.5 times that many proteins, and while the *para* gene may be an extreme example, it should be obvious that the *Drosophila* proteome is much more complex than its genome. Because alternative splicing is far more common in vertebrates, the combinations of proteins that can be produced from the human genome may be astronomical.

RNA Silencing of Gene Expression

In the last several years, the discovery that regulatory RNA molecules play an important role in controlling gene expression has given rise to a new field of research. First discovered in plants, short RNA molecules, ~21 nucleotides long, are now known to regulate gene expression in the cytoplasm by repressing translation of mRNAs and degrading mRNAs. More recently, similar RNAs have been shown to act in the nucleus to alter chromatin structure and bring about **gene silencing**.

The best-studied form of RNA silencing is called **RNA interference (RNAi)** in animals, and posttranscriptional gene silencing (PTGS) in plants. This process begins with a double stranded RNA (about 70 nucleotides long) that is processed by a protein (called Dicer) with double stranded RNAse activity. Dicer has two catalytic domains, and functions as a dimeric enzyme [Figure 17–24(a)]. One of the catalytic domains in each monomer is inactive, and alignment and

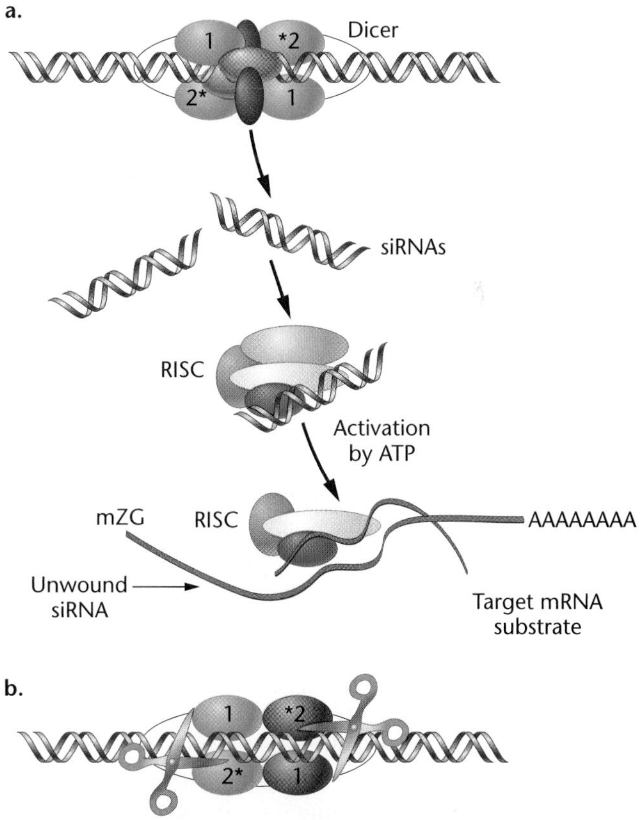

FIGURE 17–24 (a) The action of Dicer and RISC (RNA-induced silencing complex). Dicer binds to double-stranded RNA molecules and cleaves them into ~21 nucleotide molecules called small interfering RNAs (siRNAs). These bind to a multiprotein RISC complex and are unwound to form single-stranded molecules that target mRNAs with complementary sequences, marking them for degradation. (b) Binding of the catalytic domains of Dicer monomers to RNA. Domains marked with an asterisk are inactive. Cutting by the active domains produces fragments ~21 nucleotides long.

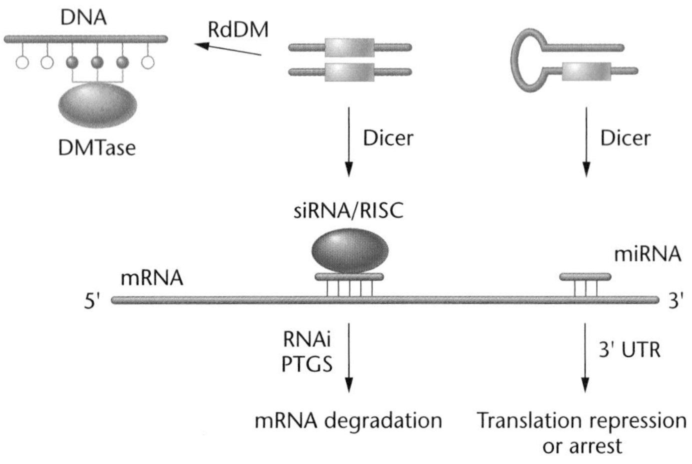

FIGURE 17–25 Mechanisms of gene regulation by RNA gene silencing. In the cytoplasm, two systems operate to silence genes. (Middle) In siRNA mediated silencing, a precursor RNA molecule is processed by Dicer, a protein with RNAse activity to form an antisense single stranded RNA that combines with a protein complex with endonuclease activity. siRNA/RISC (RNA-induced silencing complex) binds to mRNAs with complementary sequences, and cuts the mRNA into fragments that are degraded. This process is called RNAi in animal cells, and posttranslational gene silencing (PTGS) in plants. (Right) A partially double-stranded precursor is processed by Dicer to yield microRNA (miRNA) that binds to complementary 3′-untranslated regions (UTRs) of mRNA, inhibiting translation. In plants, miRNAs cause arrest of translation. (Left) Small RNAs, processed by Dicer, play a role in RNA-directed DNA methylation (RdDM). These RNAs combine with DNA methyl transferases (DMTases) to methylate cytosine residues in promoter regions (purple circles), silencing genes.

cutting by the active domains [Figure 17–24(b)] results in cleavage at about 21 nucleotide intervals. The product is a short (~21 nucleotide) RNA called **short interfering RNA (siRNA)**. The siRNA unwinds into sense and antisense single strands. The antisense strand combines with a protein complex, **RISC** (*R*NA-*i*nduced *s*ilencing *c*omplex) that recognizes, binds to, and cleaves mRNA containing sequences complementary to the antisense strand of siRNA (Figure 17–25). RISC cuts the mRNA at or near the middle of the region paired with the siRNA, and the mRNA fragments are degraded. Within RISC, members of the Argonaute family of proteins (Ago-1, Ago-2) may play important roles in mRNA cleavage.

RNAi plays an important role in cellular defense against invading viruses, and in silencing transposons. However, preliminary evidence indicates that these sequences may also play roles in control of development.

A second form of RNA silencing is mediated by short RNA molecules called **microRNAs (miRNAs)**. miRNA is derived from one regions of a ~70 to 130 nucleotide imperfectly paired hairpin precursor by cleavage with Dicer. In animal cells, the product, a ~19–24 nucleotide RNA pairs with the 3′-untranslated regions (UTRs) of mature mRNA molecules and blocks translation (Figure 17–25). In plants, miRNAs can arrest translation, or initiate mRNA degradation.

In *C. elegans* and in vertebrates, miRNA genes may constitute up to 0.2 to 0.5 percent of the genome, and encompass several hundred genes. The expression pattern for many miRNA genes is consistent with a role for these RNAs in controlling development. Most of the targets of *Arabidposis* miRNAs are transcription factors that are important in determining cell fate in the developing plant.

More recently, it has been discovered that short RNAs can target specific regions of the genome for chromatin modification. siRNAs and miRNAs depend on RNA–RNA sequence recognition and binding. However, RNA can also base pair with DNA, and short RNAs are involved in several forms of genome modification that regulate gene expression. These include **RNA-directed DNA methylation (RdDM)** of cytosine (Figure 17–25). RdDM is a highly specific process, and is limited to the region of RNA–DNA pairing. In RdDM, CG dinucleotides and other C residues in promoter regions are methylated, leading to gene silencing.

Short RNA molecules have been shown to regulate gene expression by two cytoplasmic mechanisms (mRNA degradation and inhibition or arrest of translation), and in the nucleus by the methylation of cytosine residues in DNA, leading to gene silencing. In addition, several other types of small, endogenous RNAs have been discovered and have yet to be classified, making it seem likely that there are many more functionally distinct classes of small RNAs to be discovered that play important roles in regulating genome structure and function.

17.9 Alternative Splicing and mRNA Stability Can Regulate Gene Expression

The demonstration that alternative splicing is common and can produce many different proteins from a single gene raises several questions. First, are all the protein variants produced by alternative splicing functional? Perhaps pre-mRNA splicing is an inherently inefficient process, and only a subset of splicing variants produces functional proteins. A second related question is whether splicing is a random process, or one that is directed and regulated in some way. If splicing is regulated, then perhaps it is important at specific stages of the life cycle, or in cells with specific functions.

Although many examples of alternative splicing are known, only a small number have been conclusively shown to be regulated and have functional significance. The best known of these is the process of sex determination in *Drosophila*, in which alternative transcript splicing of five genes sets the embryo on the pathway to male or female sexual development.

Sex Determination in *Drosophila*: A Model for Regulation of Alternative Splicing

As outlined in Chapter 7, sex in *Drosophila* is determined by the ratio of X chromosomes to sets of autosomes (X:A). When the ratio is 0.5 (1X:2A), males are produced, even when no Y chromosome is present; when the ratio is 1.0 (2X:2A), females are produced. Intermediate ratios (2X:3A) produce intersexes. Chromosomal ratios are interpreted by a small number of genes that initiate a cascade of splicing events resulting in the pro-

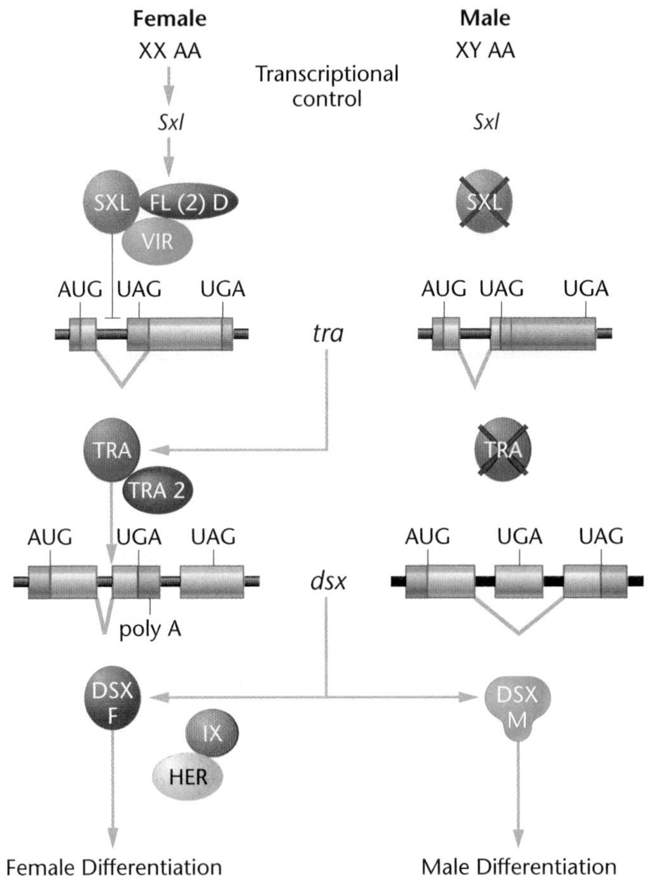

FIGURE 17–26 Hierarchy of gene regulation for sex determination in *Drosophila*. In females, the X:A ratio activates transcription of the *Sxl* gene. The product of this gene binds to premRNA of the *tra* gene and directs its splicing in a female-specific fashion. The female-specific TRA protein, in combination with the TRA-2 protein, directs female-specific splicing of *dsx* pre-mRNA, resulting in a female-specific protein (DSX-F). This protein, in combination with the IX protein, suppresses the pathway of male sexual development and activates the female pathway. In males, the X:A ratio does not activate *Sxl*. The result is a male-specific processing of *tra* pre-mRNA, resulting in no functional TRA protein, which in turn leads to male-specific processing of *dsx* transcripts, which results in a male-specific (DSX-M) protein that activates the pathway for male sexual development.

duction of male or female somatic cells and the corresponding male or female phenotypes. Three major genes in this pathway are S*ex lethal* (*Sxl*), *transformer* (*tra*), and *doublesex* (*dsx*). We will review some of the key steps in this process.

The regulatory gene at the beginning of this cascade (Figure 17–26) is the gene *sex lethal (Sxl)*, which encodes an RNA binding protein. The SXL protein is expressed only in female embryos. In the presence of SXL, female splicing patterns are expressed, and male splicing patterns are repressed.

One of the targets of SXL is the set of transcripts from the *transformer* (*tra*) gene. When SXL is present, *tra* pre-mRNAs are spliced to produce a functional protein. If SXL is absent, *tra* splicing results in a truncated, nonfunctional protein. How does this work? The *tra* pre-mRNA includes a termination codon in exon 2. In females, SXL binds to the pre-mRNA and splicing removes exon 2 from the mRNA. In males (where no SXL is present), splicing incorporates the stop codon into the mature mRNA. Translation in females produces a female-specific, functional protein that controls later steps in sexual differentiation, but in males, translation is prematurely terminated by the stop codon, resulting in an inactive gene product.

The next gene in the cascade, *dsx*, is a critical control point in the development of sexual phenotype. It produces a functional mRNA and protein in both females and males. However, the pre-mRNA is processed in a sex-specific manner to produce different transcripts. In females, the functional TRA protein is a splicing factor that binds to the *dsx* pre-mRNA and directs splicing in a female-specific pattern. In males, no TRA protein is present, and splicing of the *dsx* pre-mRNA results in a male-specific mRNA and protein. The female DSX protein (DSX-F) and the male protein (DSX-M) are both transcription factors, but have opposite effects. DSX-F works in a coordinated way with the protein encoded by *intersex* (*ix*) to repress genes that lead to male sexual development. DSX-M works independently of the IX protein to express genes in the male pathway of development, and to repress genes in the female pathway.

In sum, the S*xl* gene acts as a switch that selects the pathway of sexual development by controlling splicing of the *dsx* transcript in a female-specific fashion. The SXL protein is produced only in embryos with an X:A ratio of 1. Failure to control the splicing of the *dsx* transcript in a female mode results in the default splicing of the transcript in a male mode, leading to the production of a male phenotype.

Controlling mRNA Stability

After mRNA precursors are processed and transported, they enter the population of cytoplasmic mRNA molecules, from which messages are recruited for translation. All mRNA molecules have a characteristic life span (called a half-life, or $t_{1/2}$); they are degraded some time after they are synthesized, usually in the cytoplasm. The lifetimes of different mRNA molecules vary widely. Some are degraded within minutes after synthesis, whereas others last hours, or even months and years (in the case of mRNAs stored in oocytes). The mRNA stability, and thus its *turnover rate*, is intrinsic to the mRNA sequence. Stability of some mRNAs may be controlled by endogenous small RNAs as described above.

Another way to control mRNA stability is through **translation level** control—translation of the message controls its stability. One of the best studied examples of translational regulation is the synthesis of α- and β-tubulins, the subunit components of eukaryotic microtubules. Treatment of a cell with the drug colchicine leads to a rapid disassembly of microtubules and an increase in the concentration of the α- and β-subunits. Under these conditions, the synthesis of α- and β-tubulins drops dramatically. However, when cells are treated with vinblastine, a drug that also causes microtubule disassembly, the synthesis of tubulins is increased. Though the two drugs both cause microtubule disassembly, vinblastine precipitates the subunits, lowering the concentration of free α- and β-subunits. At low concentrations, the synthesis of tubulins is stimulated, whereas at higher concentrations, synthesis is inhibited. This type of translational regulation is known as **autoregulation**.

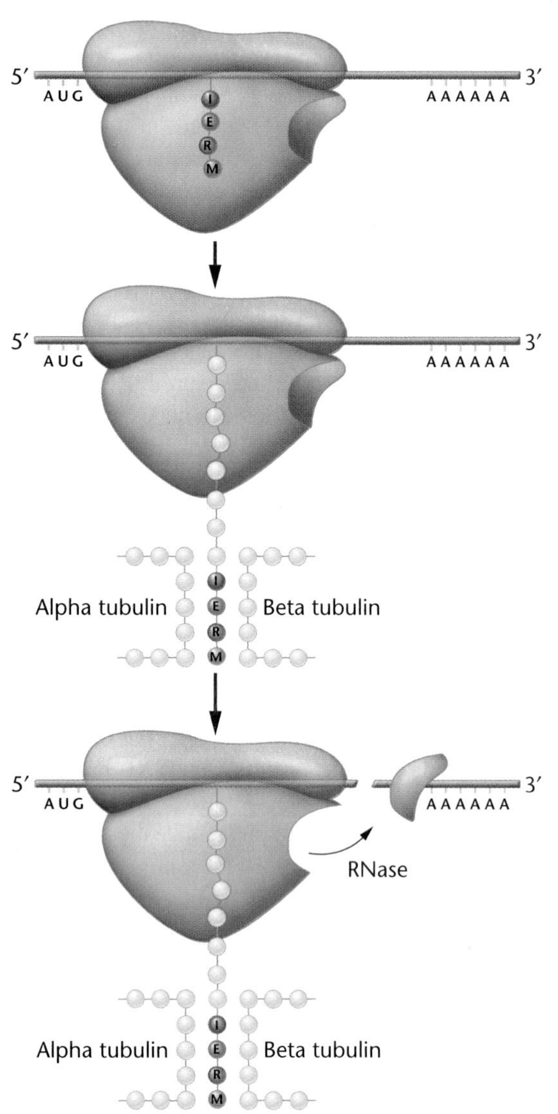

FIGURE 17–27 Model of posttranslational regulation of tubulin synthesis.

Work by Don Cleveland and his colleagues has described how the stability of tubulin mRNA is regulated. Fusion of tubulin gene fragments to a cloned thymidine kinase gene shows that the first 13 nucleotides at the 5′ end of the mRNA following the transcription start site (encoding the amino terminal region of the protein) cause the hybrid tubulin–thymidine kinase mRNA to be regulated as efficiently as intact tubulin mRNA. These first 13 nucleotides encode the amino acids Met-Arg-Glu-Ile (abbreviated as MREI). Deletions, translocations, or point mutations in this 13-nucleotide segment abolish regulation. The examination of the cytoplasmic distribution of mRNAs demonstrates that only tubulin mRNAs bound to ribosomes are depleted by drug treatment. Copies of tubulin mRNAs not associated with ribosomes are protected from degradation. Also, translation must proceed to at least codon 41 of the mRNA in order for regulation to occur.

A model, shown in Figure 17–27, has been proposed to explain these observations. In this model, regulation occurs after the process of translation has begun. The first four amino acids (MREI) of the tubulin gene product (in this case, β-tubulin) constitute a recognition element to which the regulatory factors bind. The concentration of α- and β-tubulin subunits in the cytoplasm may serve this regulatory function. The protein–protein interaction invokes the action of an RNase that may be a ribosomal component or a nonspecific cytoplasmic RNase. Action by the RNase degrades the tubulin mRNA in the act of translation, shutting down tubulin biosynthesis. As an alternative, the binding of the regulatory element may cause the ribosome to stall in its translocation along the mRNA, leaving it exposed to the action of RNase. Translation must proceed to at least codon 41 because the first 30 to 40 amino acids translated occupy a tunnel within the large ribosomal subunit, and only the translation of this number of amino acids will make the MREI recognition sequence accessible for binding.

Translation-coupled mRNA turnover has been proposed as a regulatory mechanism for other genes, including histones, some transcription factors, lymphokines, and cytokines. This suggests the existence of a regulatory mechanism that acts on mRNAs in the act of translation, probably by protein–protein interaction between a regulatory element and the nascent polypeptide chain, resulting in the activation of RNase, which breaks down the mRNA.

HOW DO WE KNOW?

How are we aware that tubulin mRNA stability is regulated by the concentration of tubulin subunits?

CHAPTER SUMMARY

1. Mechanisms controlling gene regulation in eukaryotes are governed by the properties of expanded genome size, the spatial and temporal separation between transcription and translation, and, in most organisms, multicellularity.
2. The roles played by chromatin structure, transcription factors, and signal transduction, as well as the expression of specific gene sets, are studied as models for gene regulation in higher organisms.
3. Transcription in eukaryotes is controlled by the interaction between promoters and enhancers. Other regulatory sequences, such as the CAAT box, are elements of promoter regions found near the promoter itself. Enhancer elements, which appear to control the degree of transcription, can be located before, after, or within the gene expressed.
4. Transcription factors are proteins that bind to DNA-recognition sequences within the promoters and enhancers and activate transcription through protein–protein interactions.
5. A gene may have a different methylation patterns in different tissues, which are often correlated with different regulatory patterns.
6. Several types of posttranscriptional control of gene expression are possible in eukaryotes. One such mechanism is alternative processing of a single class of pre-mRNAs to generate different mRNA species. Other mechanisms, governed by small RNAs, work in the nucleus and cytoplasm to regulate gene expression.

GENETICS, TECHNOLOGY, AND SOCIETY

Human Genetic Diseases and Loss of Gene Regulation

When we think of mutant genes that cause human genetic disorders, we usually think of mutations that change a single nucleotide in the coding portion of the gene and produce an altered protein, as in sickle cell anemia, or a deletion that knocks out a codon, as in the *Δ508* mutation in cystic fibrosis. However, gene expression is a complex process requiring a number of sequential steps and the action of large, multiprotein complexes.

As we learn more about the process of gene regulation and about the molecular mechanisms of human genetic disorders, it is apparent that mutations affecting all steps in the process of gene regulation can result in genetic disorders.

One of the first steps in eukaryotic gene expression involves remodeling chromatin to make it accessible for transcription. Remodeling involves the action of the ATP-dependent SWI/SNF protein complex. Mutations of genes encoding the enzymes in the SWI/SNF complex cause diseases with a wide range of phenotypes. These include Williams syndrome, an autosomal dominant disorder with heart and lung defects, and growth deficiencies, alpha-thalassemia and X-linked mental retardation. This gene encodes the Daxx cotranscription factor that targets chromatin remodeling to specific promoters, and may act by causing aberrant methylation of repetitive DNA sequences.

Histone acetylation and other modifications are important steps in regulating gene expression. Histone acetyltransferases (HATs) and histone deacetylases (HDACs) cooperate with the SWI/SNF factors to remodel chromatin as a first step in transcription. Mutation in a histone acetylase gene (*CREBBP, cAMP-responsive element binding protein, also called the CREB binding protein*) is responsible for Rubinstein-Taybi syndrome, an autosomal dominant disorder with growth retardation, facial and hand abnormalities, and mental retardation. In general, developmental disorders with a broad range of phenotypes usually result from inactivation of histone acetylases or deacetylases. Abnormal activation of HATs or HDACs is associated with several forms of cancer.

Alzheimer disease (AD) is a devastating neurodegenerative disorder of the elderly, affecting over 4.5 million in the United States, a number expected to triple in the next 20 years. Affected individuals suffer from a gradual loss of memory, judgement and personality changes and the ability to function in social settings leading to death 8 to 10 years after onset of the disease. Brains of those with AD have characteristic lesions caused by the accumulation of extracelluar deposits of an improperly processed polypeptide called amyloid β-protein. About 25 percent of all cases have a genetic component. One of these, familial AD type 3 is an early onset, rapidly progressing, and aggressive form of the disease caused by a mutation in the *Presenelin 1* (*PS1*) gene. The PS1 protein has several functions; it controls the rate of amyloid β-protein production, and the expression of several genes. In mutant form, it leads to overproduction of the amyloid protein and overexpression of genes that increase the rate of neurodegeneration, events that accelerate the progress of type 3 AD.

Other diseases result from transcriptional derepression, rather than abnormalities in transcription initiation. Facioscapulohumeral dystrophy (FSHD) is an autosomal dominant disorder associated with progressive weakening of muscles in the shoulders, abdomen, and pelvis. In FSHD, there is no mutant gene causing the disorder. Instead, the disorder is caused by alterations in the organization of the genome at the tip of the long arm of chromosome 4, called the 4q35 region. Normally, a set of 3.3 kb repetitive DNA sequences, called D4Z4 repeats, represses transcription of 4q35 genes by forming a multiprotein DNA-binding complex. In people with FSHD, many or most of the D4Z4 repeats are deleted. This deletion abolishes the repression of 4q35 genes, and expression of these genes causes FSHD.

From what we are learning about the molecular mechanisms of human genetic disorders, it is becoming clear that mutations in genes associated with chromatin remodeling or gene expression result in complex, multisystem phenotypes. Work with model organisms is revealing how genetic and epigenetic factors contribute to the phenotypes of complex diseases, and coupled with information from the Human Genome Project, we are entering a new and exciting era in human genetics research.

References

Cho, K.S., Elizondo, L.I., and Boerkoel, C.F. 2004. Advances in chromatin remodeling and human disease. *Curr. Opin. Genet. Develop.* 14:308–315.

Gabellini, D., Green, M.R., and Tupler, R. 2004. When enough is enough: genetic diseases associated with transcriptional derepression. *Curr. Opin. Genet. Develop.* 14:301–307.

Insights and Solutions

1. Regulatory sites for eukaryotic genes are usually located within a few hundred nucleotides of the transcriptional starting site, but they can be located up to several kilobases away. DNA sequence-specific binding assays have been used to detect and isolate protein factors present at low concentrations in nuclear extracts. In these experiments, DNA-binding sequences are bound to material on a column, and nuclear extracts are passed over the column. The idea is that if proteins that specifically bind to the sequence represented on the column are present, they will bind to the DNA, and they can be recovered from the column after all other nonbinding material has been washed away. Once a DNA-binding protein has been identified and isolated, the problem is to devise a general method for screening cloned libraries for the genes encoding these and other DNA-binding factors. Determining the amino acid sequence of the protein and constructing synthetic oligonucleotide probes is time consuming and useful only for one factor at a time. Knowing the strong affinity for binding between the protein and its DNA-recognition sequence, how would you screen for genes encoding binding factors?

Solution: Several general strategies have been developed, and one of the most promising has been devised by Steve McKnight's laboratory at the Fred Hutchinson Cancer Center. The cDNA isolated from cells expressing the binding factor is cloned into the lambda vector, gt11. Plaques of this library, containing proteins derived from expression of cDNA inserts, are adsorbed onto nitrocellulose filters and probed with double-stranded radioactive DNA corresponding to the binding site. If a fusion protein corresponding to the binding factor is present, it will bind to the DNA probe. After the unbound probe is washed off, the filter is subjected to autoradiography and the plaques corresponding to the DNA-binding signals can be identified. An added advantage of this strategy is filter recycling by washing the bound DNA from the filters for reuse. Such an ingenious procedure is similar to the colony-hybridization and plaque-hybridization procedures described for library screening in Chapter 18, and it provides a general method for isolating genes encoding DNA-binding factors.

2. Compartmentalization of eukaryotic cellular contents theoretically provides an opportunity for gene regulation. One focus for determining regulation is the import and export of proteins and nucleic acids through nuclear pores. Molecular cargo appears to be transported primarily by two classes of proteins, importin and exportin, as well as other proteins. (Macara, I. 2001. *Microbiol. and Mol. Biol. Rev.* 65:570–94.) Suggest a role for such transport proteins in eukaryotic gene regulation.

Solution: If nucleocytoplasmic transport mechanisms can recognize different RNA species and either deny, facilitate, or redirect such species, then posttranscriptional regulation could result. For instance, even if two different cells generate similar concentrations of a given RNA, differential treatment by the transport machinery will bring about regulation. In addition, if certain gene activating or repressing molecules are selectively imported or exported from the nucleus, differential gene activity could be the result. If such selective transport exists, then an additional level of genetic regulation will be available to eukaryotes.

Problems and Discussion Questions

1. Why is gene regulation assumed to be more complex in a multicellular eukaryote than in a prokaryote? Why is the study of this phenomenon in eukaryotes more difficult?
2. List and define the levels of gene regulation discussed in this chapter.
3. Distinguish between the regulatory elements referred to as promoters and enhancers.
4. Is the binding of a transcription factor to its DNA recognition sequence necessary and sufficient for an initiation of transcription at a regulated gene? What else plays a role in this process?
5. Contrast transcription factors designated as true activators with those designated as antirepressors.
6. Contrast and compare the regulation of the *gal* genes in yeast with the *lac* genes in *E. coli*.
7. Write an essay comparing the control of gene regulation in eukaryotes and prokaryotes at the level of initiation of transcription. How do the regulatory mechanisms work? What are the similarities and differences in these two types of organisms regarding the specific components of the regulatory mechanisms? Address how the differences or similarities relate to the biological context of the control of gene expression.
8. In the autoregulation of tubulin synthesis, two models for the mechanism were proposed: (1) the tubulin subunits bind to the mRNA, and (2) the subunits interact with the nascent tubulin polypeptides. To distinguish between these two models, Cleveland and colleagues introduced mutations into the 13-base regulatory element of the β-tubulin gene. Some of the mutations resulted in amino acid substitution, while others did not. In addition, they shifted the reading frame of the intact 13-base sequence. Presented here are the results of a mutagenesis study of the mRNA:

	met	arg	glu	lys	Autoregulation
Wild type	AUG	AGG	GAA	ATC	+
Second codon mutations	____	UGG	____	____	–
	____	GGG	____	____	–
	____	CGG	____	____	+
	____	AGA	____	____	+
	____	AGC	____	____	–
Third codon mutations	____	____	GAC	____	+
	____	____	AAC	____	–
	____	____	UAU	____	–
	____	____	UAC	____	–

Which of the two models is supported by the results? What experiments would you do to confirm this?

9. You are interested in studying transcription factors and have developed an *in vitro* transcription system by using a defined segment of DNA that is transcribed under the control of a eukaryotic promoter. The transcription of this DNA occurs when you add purified RNA polymerase II, TFIID (the TATA binding factor), and TFIIB and TFIIE (which bind to RNA polymerase). You perform a series of experiments that compare the efficiency of transcription in the "defined system" with the efficiency of transcription in a crude nuclear extract. You test the two systems with your template DNA and with various deletion templates that you have generated. The following are the results of your study:

(a) Why is there no transcription from the −11 deletion template?
(b) How do the results for the nuclear extract and the defined system differ from the *undeleted* template? How would you interpret these results?
(c) For the various deleted templates, compare the results with the nuclear extract and the purified system. How would you interpret the results with the deleted templates? Be very specific about what you can conclude from these data.

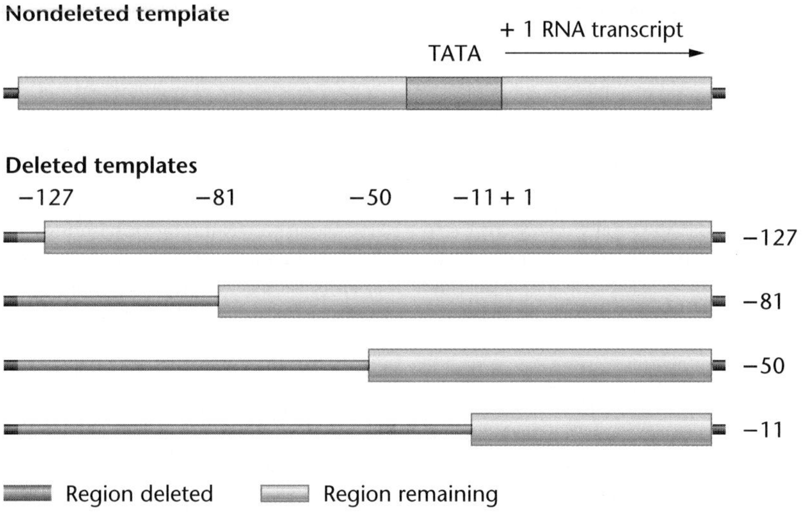

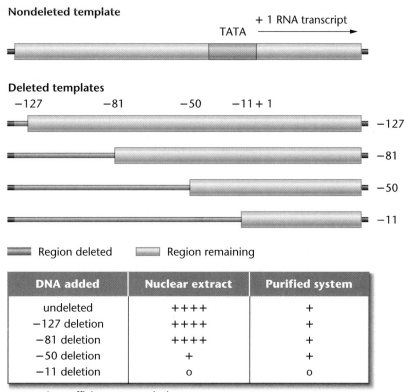

DNA added	Nuclear extract	Purified system
undeleted	++++	+
−127 deletion	++++	+
−81 deletion	++++	+
−50 deletion	+	+
−11 deletion	o	o

+ Low efficiency transcription
++++ High efficiency transcription
o No transcription

10. While it is customary to consider transcriptional regulation in eukaryotes as resulting from the positive or negative influence of different factors binding to DNA, a more complex picture is emerging. For instance, Ducret and others (1999. *Mol. and Cell. Biol.* 19:7076–87) described the action of a transcriptional repressor (Net) that is regulated by nuclear export. Under neutral conditions, Net inhibits transcription of target genes; however, when phosphorylated, Net stimulates transcription of target genes. When stress conditions exist in a cell (ultraviolet light or heat shock), Net is excluded from the nucleus, and target genes are transcribed. Devise a model that includes diagrams that provide a consistent explanation of these three conditions.

11. Given that alternative splicing can lead to different populations of RNAs from a single primary transcript, there has been considerable appeal to regard selective nuclear transport of RNAs as a form of genetic regulation in eukaryotes. The discovery of what appears to be an array of nuclear fibrous elements adds to that appeal. Pederson (2000. *Molecular Biology of the Cell* 11:799–805) reviewed literature from the late 1990s, which reported on various labeling experiments designed to determine whether gene regulation in eukaryotes is related to nuclear sorting of pre-mRNAs. The vast majority of the results indicated that nuclear poly-A RNA moves by diffusion. What influence would these results have on models that relate gene regulation to nuclear RNA transport?

12. Noncoding DNA sequences in eukaryotes are viewed more by what is absent from the DNA (promoters, exons, genes, terminators) than by what possible functions might exist. Nevertheless, it is possible that noncoding DNA may be involved in cell-specific genetic regulation by generating particular chromatin folding patterns that open or conceal certain genetic regions for transcription. What is an inherent weakness in this suggestion, and how does this weakness apply to other suggested models for regulation in eukaryotes?

13. DNA supercoiling occurs when coiling tension is generated ahead of the replication fork where it is relieved by DNA gyrase. Supercoiling may also be involved in genetic regulation. Liu and colleagues (2001. *Pro. Natl. Acad. Sci. [USA]* 98:14,883–88) discovered that transcriptional enhancers operating over a long distance (2500 base pairs) are dependent on DNA supercoiling while enhancers operating over shorter distances (110 base pairs) are not so dependent. Using a diagram, suggest a way in which supercoiling may positively influence enhancer activity over long distances.

14. DNA methylation is commonly associated with reduction of transcription. Irvine and colleagues (2002. *Mol. and Cell. Biol.* 22:6689–96) studied the impact of the location of DNA methylation relative to gene activity in human cells. The data below describe the relative expression of a reporter gene (luciferase) with variable DNA methylation outside and within the transcription region. What general conclusions can be drawn from these data?

DNA Segment	Patch size of Methylation (kb)	Number of Methylated CpGs	Relative Luciferase Expression
Outside transcription unit	0.0	0	490X
(0–7.6 kb away)	2.0	100	290X
	3.1	102	250X
	12.1	593	2X
Inside transcription unit	0.0	0	490X
	1.9	108	80X
	2.4	134	5X
	12.1	593	2X

15. In some organisms, including mammals, there is an inverse relationship between the presence of 5-methylcytosine (m^5C) in CpG sequences and gene activity. In addition, m^5C may be involved in the recruitment of proteins that repress chromatin thus segregating the genome into transcriptionally active and inactive regions. Overall, genomic DNA is relatively poor in CpG sequences due to the accumulation of m^5C compared to thymine transitions; however, unmethylated CpG islands are often associated with genes. Oakes and colleagues (2003. *Proc. Natl. Acad. Sci.* 100:1775–80) have determined that patterns of DNA methylation in spermatozoa vary with age in rats and suggest that such age-related alterations in DNA methylation may be one mechanism underlying age-related abnormalities in mammals. Consider the above information and provide an explanation which relates paternal age-related alterations in DNA methylation and birth abnormalities.

16. Genome sequencing has provided an opportunity to examine long-range similarities among a variety of organisms and allow estimations of the frequency of gene homologies. Comparisons of mouse and human genomes indicate that less than 1 percent of the mouse genes have no apparent homolog in the human genome thereby indicating considerable sequence similarity. But another issue surfaces in terms of the expression of such genomes. Are patterns of alternative splicing also conserved? Thanaraj, Clark, and Muilu (2003. *Nucl. Acids Res.* 31:2544–52) determined that approximately 68 percent of the alternative splicing patterns in the human are conserved in the mouse. What criteria would one use to arrive at such a percentage?

Extra-Spicy Problems

17. Mouse quaking is caused by a recessive dysmyelination mutation (*qk*), which appears to involve defects in alternative splicing of an evolutionarily conserved signal transduction protein. Homozygotes (*qk*/*qk*) suffer tremors during exertion. Mature mice may experience seizures and remain motionless for many seconds. Below is a figure that describes the splicing patterns in various genotypes of quaking and normal mice. Portions of brain MAG (myelin associated glycoprotein) RNA (including exons 11, 12, and 13) from young (14 day) and adult (2 month) mice are represented. The relative concentration of each RNA is indicated in the gel diagram. (Figure and data are modified from Wu, J., et al. 2002. *Proc. Natl. Acad. Sci.* 99:4233–38.) Given this information, describe the alternative splicing patterns with regards to exons included/excluded as a function of age and genotype.

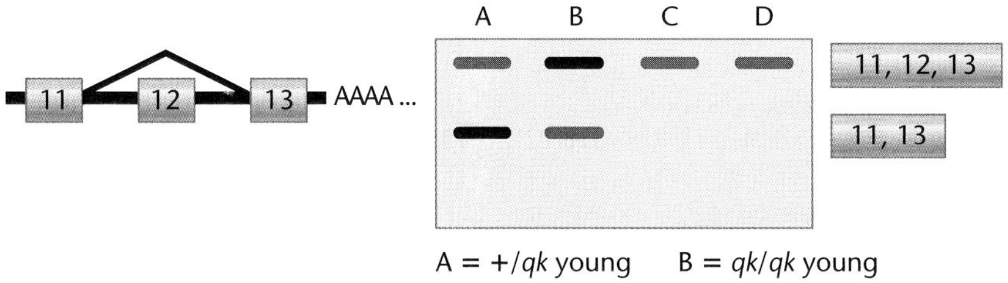

A = +/*qk* young B = *qk*/*qk* young
C = +/*qk* adult D = *qk*/*qk* adult

18. It has been estimated that approximately half of human genes produce alternatively spliced mRNA isoforms. In some cases incorrectly spliced RNAs lead to human pathologies. Xu and Lee (2003. *Nucl. Acids Res.* 31:5635–43) examined human cancers for splice-specific changes and found that many of the changes disrupted tumor suppressor function. In general what types of RNAs would be expected when there is a loss of splicing specificity? How might loss of splicing specificity be associated with cancer?

19. Philips and Cooper (2000. *Cell. Mol. Life Sci.* 57:235–49) have estimated that approximately 15 percent of disease-causing mutations involve errors in alternative splicing. However, there is an interesting case where an exon deletion appears to be a phenomenon that enhances dystrophin production in muscle cells of DMD (Duchenne muscular dystrophy) patients. It turns out that a deletion of exon 45 is the most frequent DMD-causing mutation. But some individuals with Becker muscular dystrophy (BMD), a milder form of muscular dystrophy, have deletions in exons 45 *and* 46, a condition that was experimentally verified by van Deutekom and van Ommen (2003. *Nat. Rev. Genetics* 4:774–83). Given that deletions often cause frameshift mutations, provide a possible explanation for the enhanced production of dystrophin in the presence of exon 45 *and* 46 deletions in BMD patients.

SELECTED READINGS

Becker, P.B., and Hörz, W. 2002. ATP-dependent nucleosome remodeling. *Annu. Rev. Biochem.* 71:247–73.

Black, D.L. 2000. Protein diversity from alternative splicing: a challenge for bioinformatics and postgenomic biology. *Cell* 103:367–70.

______. 2003. Mechanisms of alternative pre-mRNA splicing. *Annu. Rev. Biochem.* 72:291–336.

Butler, J.E.F., and Kadonaga, J.T. 2002. The RNA polymerase II core promoter: a key component in the regulation of gene expression. *Genes Dev.* 16:2583–92.

Dillon, N., and Sabbattini, P. 2000. Functional gene expression domains: Defining the functional unit of eukaryotic gene regulation. *Bioessays* 22:657–65.

Graveley, B.R. 2001. Alternative splicing: increasing diversity in the proteomic world. *Trends Genet.* 17:100–107.

Gregory, P.D., Wagner, K., and Hürz, W. 2001. Histone acetylation and chromatin remodeling. *Exp. Cell Res.* 265:195–202.

Jackson, D.A. 2003. The anatomy of transcription sites. *Curr. Opin. Cell Biol.* 15:311–17.

Jenuwein, T., and Allis, C.D. 2001. Translating the histone code. *Science* 293:1074–80.

Lee, T.I., and Young, R.A. 2000. Transcription of eukaryotic protein-coding genes. *Ann. Rev. Genet.* 34:77–137.

Lusser, A., and Kadonaga, J.T. 2003. Chromatin remodeling by ATP-dependent molecular machines. *BioEssays* 25:1192–1200.

Mahy, N.L., Perry, P.E., Gilchrist, S., Baldock, R.A., and Bickmore, W.A. 2002. Spatial organization of active and inactive genes and noncoding DNA within chromosome territories. *J. Cell Biol.* 157:579–89.

Meister, G., and Tuschl, T. 2004. Mechanisms of gene silencing by double-standard RNA. *Nature* 431:343–349.

Mello, C. C., and Conte, D., Jr. 2004. Revealing the world of RNA interference. *Nature* 431:338–342.

Peterson, C.L., and Logie, C. 2000. Recruitment of chromatin-remodeling machines. *J. Cell. Biochem.* 78:179–185.

Spector, D. 2003. The dynamics of chromosome organization and gene regulation. *Annu. Rev. Biochem.* 72:573–608.

Sudarsanam, P., and Winston, F. 2000. The SWI/SNF family of nucleosome-remodeling complexes and transcriptional control. *Trends in Genet.* 16:345–51.

Turner, B.M. 2000. Histone acetylation and an epigenetic code. *Bioessays* 22:836–45.

Vershure, P.J., van der Kraan, I., Enserink, J.M., Moné, M.J., Manders, E.M.M., and van Driel, R. 2002. Large-scale chromatin organization and the localization of proteins involved in gene expression in human cells. *The J. of Histochem. Cytochem.* 50:1303–10.

CHAPTER EIGHTEEN

Cell Cycle Regulation and Cancer

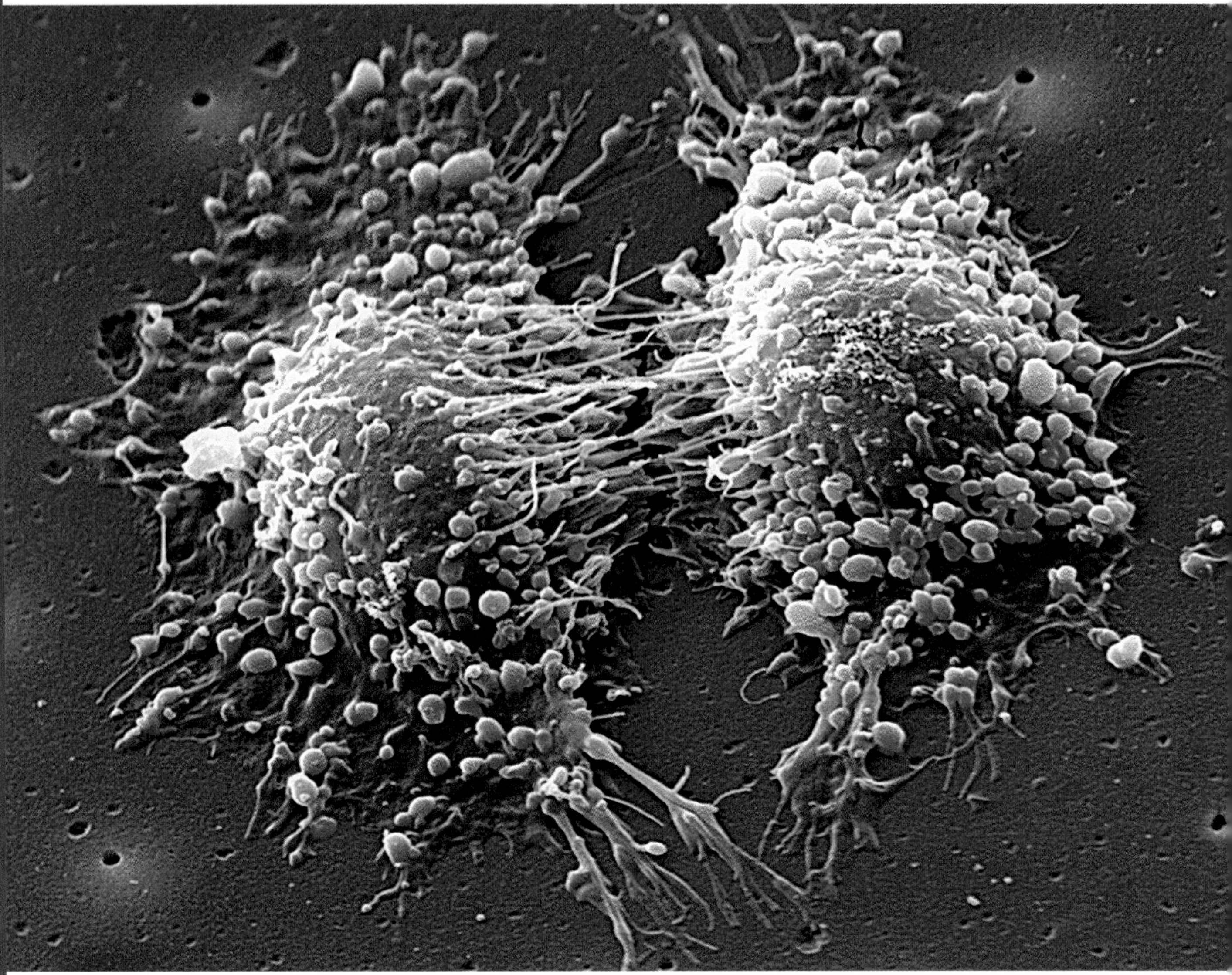

Colored scanning electron micrograph of two prostate cancer cells in the final stages of cell division (cytokinesis). The cells are still joined by strands of cytoplasm.

CHAPTER CONCEPTS

- Cancer is a group of genetic diseases affecting fundamental aspects of cellular function, including DNA repair, the cell cycle, apoptosis, differentiation, and cell–cell contact.
- Most cancer-causing mutations occur in somatic cells; only about 1 percent of cancers have a hereditary component.
- Mutations in cancer-related genes lead to abnormal proliferation and loss of control over how cells spread and invade surrounding tissues.
- The development of cancer is a multistep process requiring mutations in genes controlling many aspects of cell proliferation and metastasis.
- Cancer cells show high levels of genomic instability, leading to the accumulation of multiple mutations in cancer-related genes.
- Mutations in proto-oncogenes and tumor suppressor genes contribute to the development of cancers.
- Oncogenic viruses introduce oncogenes into infected cells and stimulate cell proliferation.
- Environmental agents contribute to cancer by damaging DNA.

Cancer is the second leading cause of death in Western countries, surpassed only by heart disease. It strikes people of all ages, and one out of three people will experience a cancer diagnosis sometime in his or her lifetime. Each year, more than 1 million cases of cancer are diagnosed in the United States and more than 500,000 people die from the disease (Table 18.1). Lung cancer is the biggest cancer killer, followed by cancers of the breast, prostate, and colon.

Given its capacity to cause immense human suffering, our society has directed significant resources toward understanding the causes of cancer and developing effective treatments. Over the last 30 years, scientists have discovered that cancer is a genetic disease, characterized by an interplay of oncogenes and tumor suppressor genes leading to the uncontrolled growth and spread of cancer cells. In addition, completion of the human genome project is opening the door to a wealth of new information about the mutations that trigger a cell to become cancerous. This new understanding of cancer genetics is also leading to new gene-specific treatments, some of which are now entering clinical trials. Some scientists predict that gene therapies will replace chemotherapies within the next 25 years.

The goal of this chapter is to highlight our current understanding of the nature and causes of cancer. As we will see, cancer is a genetic disease that arises from mutations in genes controlling many basic aspects of cellular function. We will examine the relationship between genes and cancer, and consider how mutations, chromosomal changes, and environmental agents play roles in the development of cancer.

18.1 Cancer Is a Genetic Disease

Perhaps the most significant development in understanding the causes of cancer is the realization that cancer is a genetic disease. Genomic alterations that are associated with cancer range from single-nucleotide substitutions to large-scale chromosome rearrangements, amplifications, and deletions (Figure 18–1). However, unlike other genetic diseases, cancer is caused by mutations that occur predominantly in somatic cells. Only about 1 percent of cancers are associated with germ-line mutations that increase a person's susceptibility to certain types of cancer. Another important difference between cancer and other genetic

TABLE 18.1 CANCER PROBABILITIES IN THE UNITED STATES

Cancer Site	Gender	Birth to 39	Age 40–59	Age 60–79	Birth to Death
All sites	Male	1 in 62	1 in 12	1 in 3	1 in 2
	Female	1 in 52	1 in 11	1 in 4	1 in 3
Breast	Female	1 in 235	1 in 25	1 in 15	1 in 8
Prostate	Male	< 1 in 10,000	1 in 53	1 in 7	1 in 6
Lung, bronchus	Male	1 in 3300	1 in 92	1 in 17	1 in 13
	Female	1 in 3180	1 in 120	1 in 25	1 in 17
Colon, rectum	Male	1 in 1500	1 in 124	1 in 29	1 in 18
	Female	1 in 1900	1 in 149	1 in 33	1 in 18

Source: American Cancer Society

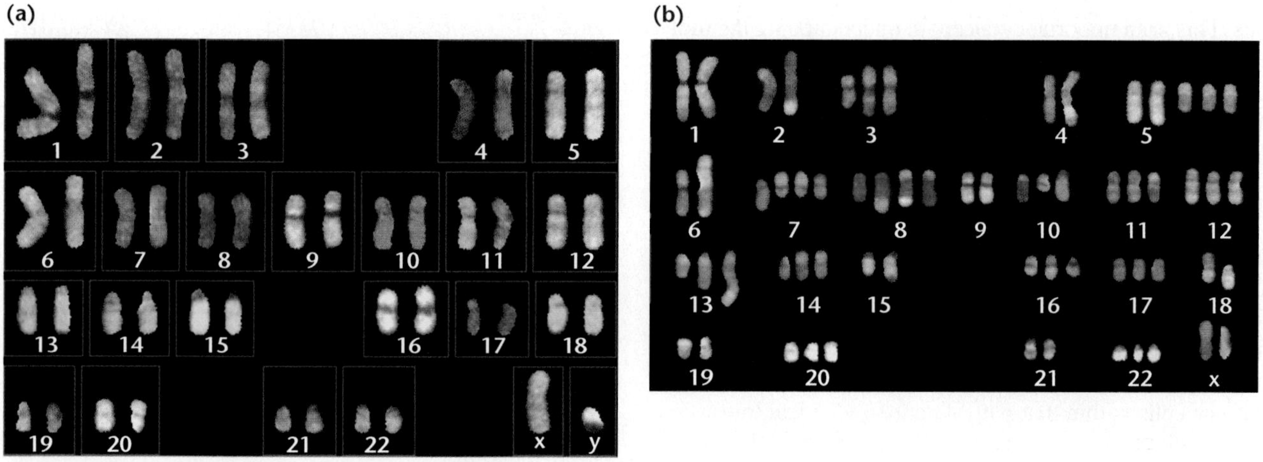

FIGURE 18–1 (a) Spectral karyotype of a normal cell. (b) Karyotype of a cancer cell showing translocations, deletions, and aneuploidy—characteristic features of cancer cells.

diseases is that cancers rarely arise from a single mutation, but from the accumulation of many mutations—as many as six to ten. The mutations that lead to cancer affect multiple cellular functions, including repair of DNA damage, cell division, apoptosis, cellular differentiation, and cell–cell contact.

What Is Cancer?

Clinically, cancer is defined as a large number of complex diseases, up to a hundred, that behave differently depending upon the cell types from which they originate. Cancers vary in their ages of onset, growth rates, invasiveness, prognoses, and responsiveness to treatments. However, at the molecular level, all cancers exhibit common characteristics that unite them as a family.

All cancer cells share two fundamental properties: (1) abnormal cell growth and division (**cell proliferation**), and (2) abnormalities in the normal restraints that keep cells from spreading and invading other parts of the body (**metastasis**). In normal cells, these functions are tightly controlled by genes that are expressed appropriately in time and place. In cancer cells, these genes are either mutated or are expressed inappropriately.

It is this combination of uncontrolled cell proliferation and metastatic spread that makes cancer cells dangerous. When a cell simply loses genetic control over cell growth, it may grow into a multicellular mass, a **benign tumor**. Such a tumor can often be removed by surgery and usually causes no serious harm. However, if cells in the tumor also acquire the ability to break loose, enter the bloodstream, invade other tissues, and form secondary tumors (metastases), they become malignant. **Malignant tumors** are difficult to treat, and may become life threatening. As we will see later in the chapter, there are multiple steps and genetic mutations that convert a benign tumor into a dangerous malignant tumor.

The Clonal Origin of Cancer Cells

Although benign and malignant tumors may contain billions of cells, and metastatic tumors may invade and grow in numerous parts of the body, all cancer cells in the primary and secondary tumors are clonal, meaning that they originated from a common ancestral cell that accumulated numerous specific mutations. This is an important concept in understanding the molecular causes of cancer, and has implications for its diagnosis.

Numerous data support the concept of cancer clonality. For example, reciprocal chromosomal translocations are characteristic of many cancers, including leukemias and lymphomas, two cancers involving white blood cells. Cancer cells from patients with **Burkitt's lymphoma** show reciprocal translocations between chromosome 8 (with translocation breakpoints at or near the *c-myc* gene) and chromosomes 2, 14, or 22 (with translocation breakpoints at or near one of the immunoglobulin genes). Each Burkitt's lymphoma patient exhibits unique breakpoints in their *c-myc* and immunoglobulin gene DNA sequences; however, all lymphoma cells within that patient contain identical translocation breakpoints. This demonstrates that all cancer cells in each case of Burkitt's lymphoma arise from a single cell, and this cell passes on its genetic aberrations to its progeny.

Another demonstration that cancer cells are clonal is their pattern of X-chromosome inactivation. As explained in Chapter 7, female humans are mosaic, with some cells containing an inactivated paternal X-chromosome and other cells containing an inactivated maternal X-chromosome. X-chromosome inactivation occurs early in development and occurs at random. All cancer cells within a tumor, and within all primary and metastatic tumors in one female individual, contain the same inactivated X-chromosome. This supports the concept that all the cancer cells in that patient arose from a common ancestor cell.

Cancer As a Multistep Process, Requiring Multiple Mutations

Although we know that cancer is a genetic disease initiated by mutations that lead to uncontrolled cell proliferation and metastasis, a single mutation is not sufficient to transform a normal cell into a tumor-forming (tumorigenic), malignant cell. If it were sufficient, then cancer would be far more prevalent than it is. In humans, mutations occur spontaneously at a rate of about 10^{-6} mutations per gene, per cell division, mainly due to the intrinsic error rates of DNA replication. As there are approximately 10^{16} cell divisions in a human body during a lifetime, a person might suffer up to 10^{10} mutations per gene somewhere in the body, during his or her lifetime. However, only about one person in three will suffer from cancer.

The phenomenon of age-related cancer is another indication that cancer develops from the accumulation of several mutagenic events in a single cell. The incidence of most cancers rises exponentially with age (Figure 18–2). If only a single mutation were sufficient to convert a normal cell to a malignant one, then cancer incidence would appear to be independent of age. The age-related incidence of cancer suggests that up to ten independent mutations, occurring randomly and with a low probability, are necessary before a cell is transformed into a malignant cancer cell. Another indication that cancer is a multistep process is the delay that occurs between exposure to **carcinogens** (cancer-causing agents) and the appearance of the cancer. For example, an incubation period of 5 to 8 years separated exposure of people

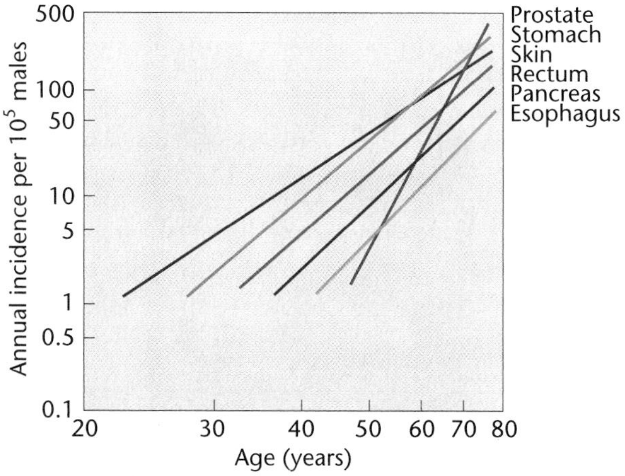

FIGURE 18–2 The incidence of most cancers rises exponentially with age. This graph shows that the logarithmic plot of the incidence rate has a linear relationship with the logarithmic plot of the patient's age.

to the radiation of the atomic explosions at Hiroshima and Nagasaki and the onset of leukemias. Similarly, from the 1930s to 1950s tuberculosis patients were often treated with X rays. Some of those patients developed breast cancer, but only after a lag of about 15 years after their treatments.

The multistep nature of cancer development is supported by the observation that cancers often develop in progressive steps, from mildly aberrant cells to increasingly tumorigenic and malignant. This progressive nature of cancer is illustrated by the development of cervical cancers (Figure 18–3). In a normal cervix, cells in the basal layer divide and differentiate into **quiescent** (nondividing) cells that are eventually shed from the surface of the cervix. Sometimes, basal layer cells acquire mutations that allow them to grow abnormally, and these cells appear to lose part of their ability to differentiate and become quiescent. These areas of abnormally dividing cells are known as **dysplasia**. If areas of dysplasia progress over a period of years, with cells further proliferating, acquiring mutations and failing to differentiate, they become known as **carcinomas *in situ***. Up until this stage, it is relatively easy to treat and cure a cervical carcinoma. However, in about 20 to 30 percent of cases, carcinomas *in situ* may gradually progress to full malignancy, characterized by cells that disengage from the tumor, cross the basal lamina, and invade surrounding tissues. At these later stages, an invasive cervical cancer is more difficult to treat.

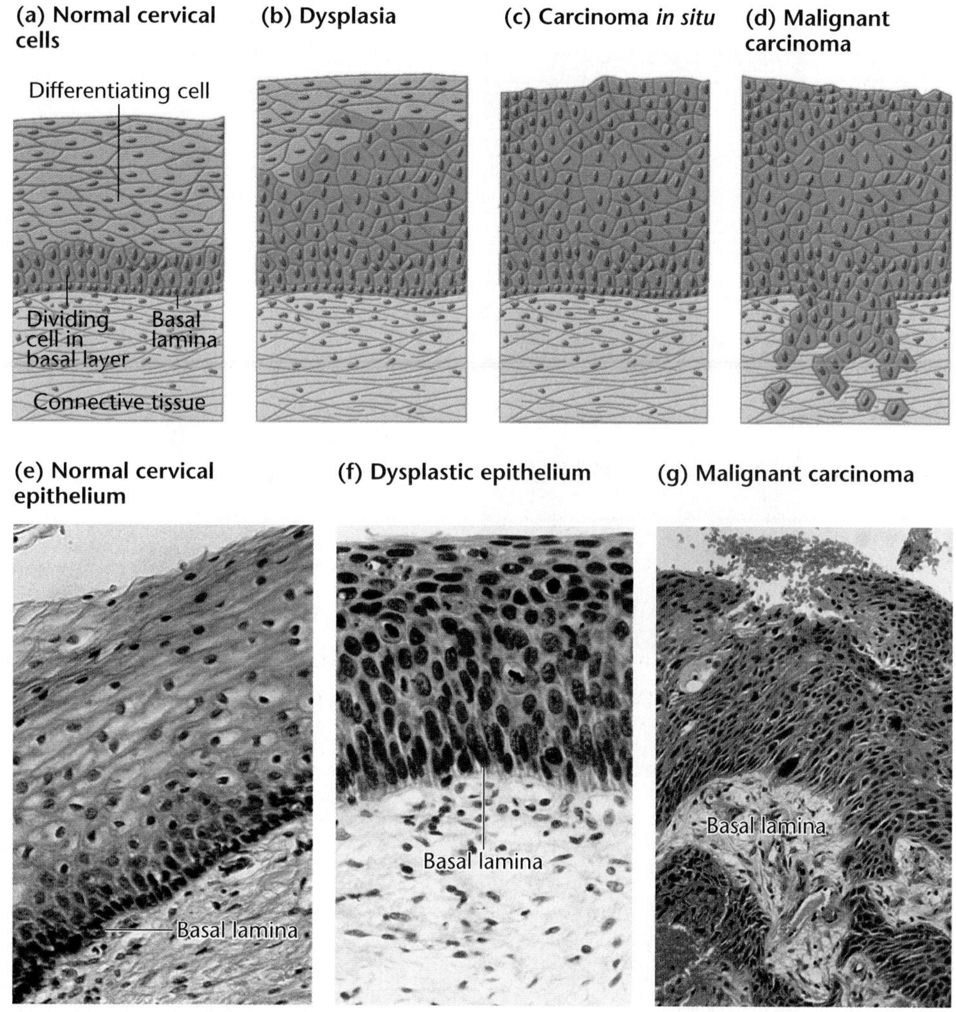

FIGURE 18–3 Stages in the development of cervical cancer. (a) to (d) Steps in progression from normal cervical epithelium to a malignant carcinoma. (e) to (g) Light microscope photos of cervical biopsy samples. Note the invasive cells in (g) that have crossed the basal lamina (light area in center) and entered connective tissue below.

Each step in **tumorigenesis** (the development of a malignant tumor) appears to be the result of one or more genetic alterations that release the cells progressively from the controls that normally operate on proliferation and malignancy. As we will discover in subsequent sections of this chapter, the genes that undergo mutations leading to cancer (called oncogenes and tumor suppressor genes) are those that control DNA damage repair, the cell cycle, cell–cell contact, and programmed cell death.

We will now investigate each of these fundamental processes, the genes that control them and how mutations in these genes may lead to cancer.

How Do We Know?

How did we come to realize that cancer is a multistep genetic disease?

18.2 Cancer Cells Contain Genetic Defects Affecting Genomic Stability and DNA Repair

Cancer cells show higher than normal rates of mutation, chromosomal abnormalities, and genomic instability. In fact, many researchers believe that the fundamental defect in cancer cells is a derangement of the cells' normal ability to repair DNA damage. This loss of genomic integrity leads to mutations in specific genes that control various aspects of cell proliferation, programmed cell death, and cell–cell contact. In turn, the accumulation of mutations in genes controlling these processes leads to cancer. The high level of genomic instability seen in cancer cells is known as the **mutator phenotype**.

The genomic instability in cancer cells manifests itself in the presence of gross defects such as translocations, aneuploidy, chromosome loss, DNA amplification, and chromosome deletions (Figures 18–1 and 18–4). Cancer cells that are grown in

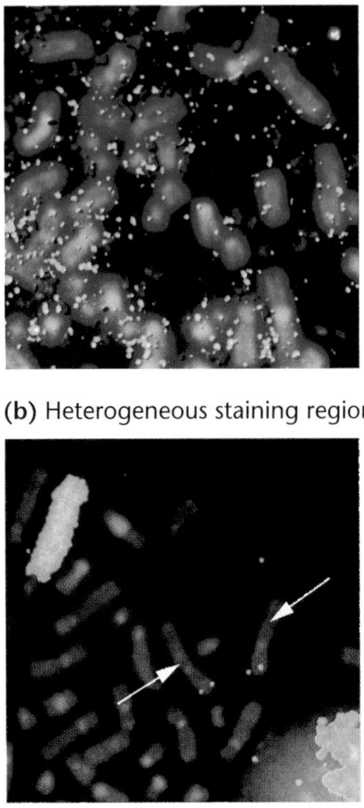

FIGURE 18–4 DNA amplifications in neuroblastoma cells. (a) Two cancer genes (*MYCN* in red and *MDM2* in green) are amplified as small DNA fragments that remain separate from chromosomal DNA within the nucleus. These units of amplified DNA are known as double minute chromosomes. Normal chromosomes are stained blue. (b) Multiple copies of the *MYCN* gene are amplified within one large region called a heterogeneous staining region (green). Single copies of the *MYCN* gene are visible on the normal parental chromosomes (white arrows). Normal chromosomes are stained red.

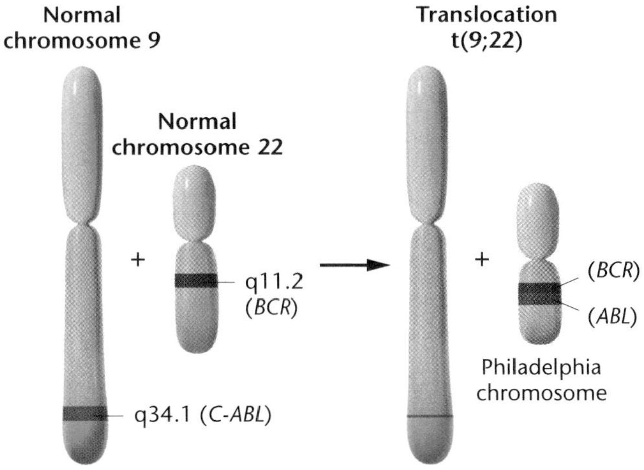

FIGURE 18–5 A reciprocal translocation involving the long arms of chromosomes 9 and 22 results in the formation of a characteristic chromosome, the Philadelphia chromosome, which is associated with chronic myelogenous leukemia (CML). The t(9;22) translocation results in the fusion of the *C-ABL* proto-oncogene on chromosome 9 with the *BCR* gene on chromosome 22. The fusion protein is a powerful hybrid molecule that allows cells to escape control of the cell cycle, contributing to the development of CML.

cultures in the lab also show a great deal of genomic instability—duplicating, losing and translocating chromosomes or parts of chromosomes. Often cancer cells show specific chromosomal defects that are used to diagnose the type and stage of the cancer. For example, leukemic white blood cells from patients with **chronic myelogenous leukemia (CML)** bear a specific translocation, in which the *C-ABL* gene on chromosome 9 is translocated into the *BCR* gene on chromosome 22. This translocation creates a structure known as the **Philadelphia chromosome** (Figure 18–5). The *BCR-ABL* fusion gene codes for a chimeric BCR-ABL protein. The normal ABL protein is a **protein kinase** that acts within signal transduction pathways, transferring growth factor signals from the external environment to the nucleus. The BCR-ABL protein is an abnormal signal transduction molecule in CML cells, constantly stimulating these cells to proliferate.

In keeping with the concept of the cancer mutator phenotype, a number of inherited cancers are caused by defects in genes that control DNA repair. For example, xeroderma pigmentosum (XP) is a rare hereditary disorder that is characterized by extreme sensitivity to ultraviolet light and other carcinogens. Patients with XP often develop malignant skin cancer. Cells from patients with XP are defective in nucleotide excision repair, with mutations appearing in any one of seven genes whose products are necessary to carry out DNA repair. XP cells are impaired in their ability to repair DNA lesions such as thymine dimers induced by UV light. The relationship between XP and genes controlling nucleotide excision repair is described in Chapter 15.

Another hereditary cancer, **hereditary nonpolyposis colorectal cancer (HNPCC)**, is also caused by mutations in genes controlling DNA repair. HNPCC is an autosomal dominant syndrome, affecting about one in every 200 people (Figure 18–6). Patients affected by HNPCC have an increased risk of developing colon, ovary, uterine, and kidney cancers. Cells from patients with HNPCC show higher than normal mutation rates and genomic instability. At least eight genes are associated with HNPCC, and four of these genes are involved in DNA mismatch repair. Inactivation of any of these four genes—*MSH2*, *MSH6*, *MLH1*, and *MLH3*—causes a rapid accumulation of genome-wide mutations and subsequent development of colorectal and other cancers.

The observation that hereditary defects in genes controlling nucleotide excision repair and DNA mismatch repair lead to high rates of cancer lends support to the idea that the mutator phenotype is a significant contributor to the development of cancer.

Now solve this

Problem 18.17 on page 454 asks how the BCR-ABL hybrid fusion protein, found in CML leukemic white blood cells, could be used as a target for cancer therapy.

Hint: Most cancer therapies, including radiation and chemotherapies, aim to kill cells that are constantly dividing. However, many normal cells in the body also divide and are killed by cancer therapies, leading to side

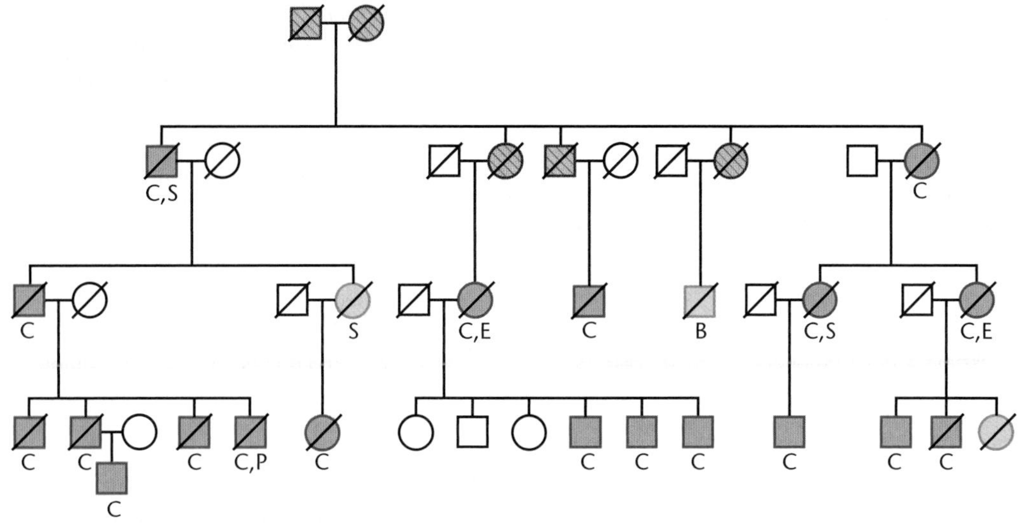

FIGURE 18–6 Pedigree of a family with HNPCC. Families with HNPCC are defined as those in which at least three relatives in two generations have been diagnosed with colon cancer, with one relative diagnosed at less than 50 years of age. Colon cancer: C; stomach cancer: S; endometrial cancer: E; pancreatic cancer: P; bladder/urinary cancer: B. Blue symbols indicate family members with colon cancer; diagonal stripes mean that diagnosis is uncertain; orange symbols indicate other tumors. Symbols with slashes indicate deceased individuals. *Reprinted with permission from Aaltonen et.al. Clues to the pathogenesis of familial colorectal cancer. Science 260:812-816, Figure 1. Copyright 1993 AAAS.*

effects. The fact that the BCR-ABL fusion protein is found only in CML white blood cells provides a potential cancer-specific target for cancer chemotherapies. This feature has indeed been used as a starting point for developing a new cancer therapy. To learn about this new CML therapy, called Gleevec, see *http://www.nci.nih.gov/newscenter/qandagleevec.*

18.3 Cancer Cells Contain Genetic Defects Affecting Cell Cycle Regulation

One of the fundamental aberrations in all cancer cells is a loss of control over cell proliferation. Cell proliferation is the process of cell growth and division that is essential for all development and tissue repair in multicellular organisms. Although some cells, such as epidermal cells of the skin or blood-forming cells in the bone marrow, continue to grow and divide throughout the organism's lifetime, most cells in adult multicellular organisms are in a nondividing, quiescent, and differentiated state. Differentiated cells are those that are specialized for a specific function, such as photoreceptor cells of the retina or muscle cells of the heart. The most extreme examples of nonproliferating cells are nerve cells, which do not grow or divide, even to replace damaged tissue. In contrast, many differentiated cells, such as those in the liver and kidney, are able to grow and divide when stimulated by extracellular signals and growth factors. In this way, multicellular organisms are able to replace dead and damaged tissue. However, the growth and differentiation of cells must be strictly regulated; otherwise, the integrity of organs and tissues would be compromised by inappropriate types and quantities of cells. Normal regulation over cell proliferation involves a large number of gene products that control steps in the cell cycle, programmed cell death, and the response of cells to external growth signals. In cancer cells, many of the genes that control these functions are mutated or aberrantly expressed, leading to uncontrolled cell proliferation.

In this section, we will review steps in the cell cycle, some of the genes that control the cell cycle, and how these genes, when mutated, lead to cancer.

The Cell Cycle and Signal Transduction

As described in Chapter 2, the cellular events that occur from one cell division to the next comprise the cell cycle (Figure 2–5). The interphase stage of the cell cycle is the interval between mitotic divisions. During this time, the cell grows and replicates its DNA. During G1, the cell prepares for DNA synthesis by accumulating the enzymes and molecules required for DNA replication. G1 is followed by S phase, during which the cell's chromosomal DNA is replicated. During G2, the cell continues to grow and prepare for division. During M phase, the duplicated chromosomes condense, sister chromosomes separate to opposite poles, and the cell divides in two.

In early-to mid-G1, the cell makes a decision either to enter the next cell cycle or to withdraw from the cell cycle into quiescence. Continuously dividing cells do not exit the cell cycle, but proceed through G1, S, G2, and M phases until they receive signals to stop growing. If the cell stops proliferating, it enters the G0 phase of the cell cycle. During G0, the cell remains metabolically active but does not grow or divide. Most differentiated cells in multicellular organisms can remain in this G0 phase indefinitely. Some, such as neurons, never reenter the cell cycle.

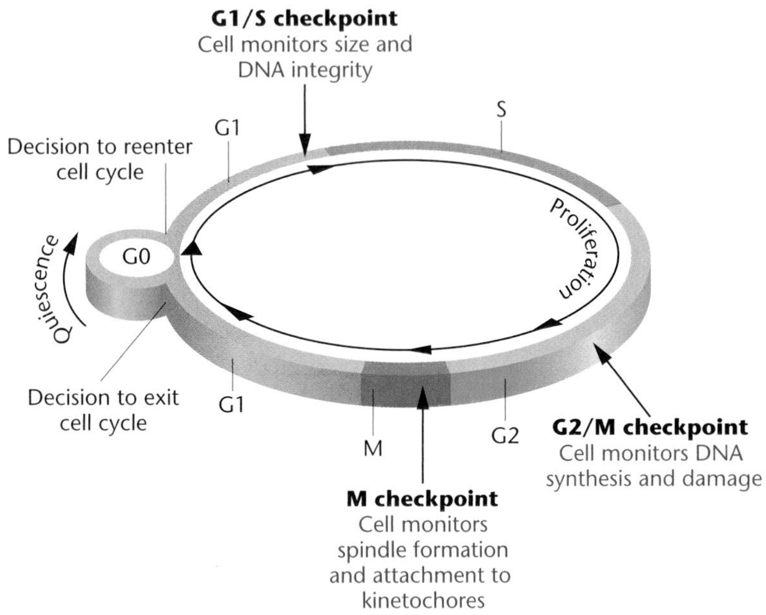

FIGURE 18–7 Checkpoints and proliferation decision points monitor the progress of the cell through the cell cycle.

In contrast, cancer cells are unable to enter G0, and continuously cycle. Their rate of proliferation is not necessarily any greater than a normal proliferating cell; however, they are not able to become quiescent at the appropriate time or place.

Cells in G0 can often be stimulated to reenter the cell cycle by external growth signals. These signals are delivered to the cell by molecules such as growth factors and hormones that bind to cell surface receptors, which then relay the signal from the plasma membrane to signal transduction molecules located in the cytoplasm. The transfer of signals through the cytoplasm usually involves a cascade of protein phosphorylations and configuration changes that send the signal down a multiprotein pathway leading to the nucleus. Ultimately, signal transduction initiates a program of gene expression that propels the cell out of G0 back into the cell cycle. Cancer cells often have defects in signal transduction pathways. Sometimes, abnormal cell surface receptors or cytoplasmic signal transduction molecules send continuous growth signals to the nucleus even in the absence of external growth signals. In addition, malignant cells may not respond to external signals from surrounding cells—signals that would normally inhibit cell proliferation within a mature tissue.

Cell Cycle Control and Checkpoints

In normal cells, progress through the cell cycle is tightly regulated and the completion of each step is necessary prior to initiating the next step. There are at least three distinct points in the cell cycle at which the cell monitors its internal equilibrium before proceeding to the next stage of the cell cycle. These are the **G1/S**, the **G2/M**, and **M checkpoints** (Figure 18–7). At the G1/S checkpoint, the cell monitors its size and determines if its DNA has been damaged. If the cell has not achieved an adequate size, or if the DNA has been damaged, further progress through the cell cycle is halted until these conditions are corrected. If cell size and DNA integrity are normal, the G1/S checkpoint is traversed, and the cell proceeds to S phase. The second important checkpoint is the G2/M checkpoint, where physiological conditions in the cell are monitored prior to entering mitosis. If DNA replication or repair to any DNA damage has not been completed, the cell cycle arrests until these processes are complete. The third major checkpoint occurs during mitosis and is called the M checkpoint. At this checkpoint, both the successful formation of the spindle-fiber system and the attachment of spindle fibers to the kinetochores associated with the centromeres are monitored. If spindle fibers are not properly formed or attachment is inadequate, mitosis is arrested.

In addition to regulation at cell cycle checkpoints, regulation of cell cycle progress is mediated by two classes of proteins: **cyclins** and **cyclin-dependent kinases (CDKs)**. The cell synthesizes and destroys cyclin proteins in a precise pattern during the cell cycle (Figure 18–8). When a cyclin is present, it binds to a specific CDK, triggering activity of the CDK/cyclin complex. The CDK/cyclin complexes then selectively phosphorylate and activate other proteins that in turn bring about the changes necessary to advance the cell through the cell cycle. For example, in G1 phase, CDK4/cyclin D complexes activate proteins that trigger transcription of sets of genes whose products (such as DNA polymerase δ and DNA ligase) are required for DNA replication during S phase. Another CDK/cyclin complex, CDK1/cyclinB, phosphorylates a number of proteins that trigger the events of early mitosis, such as nuclear membrane breakdown, chromosome condensation and cytoskeletal reorganization (Figure 18–9). Mitosis can only be completed, however,

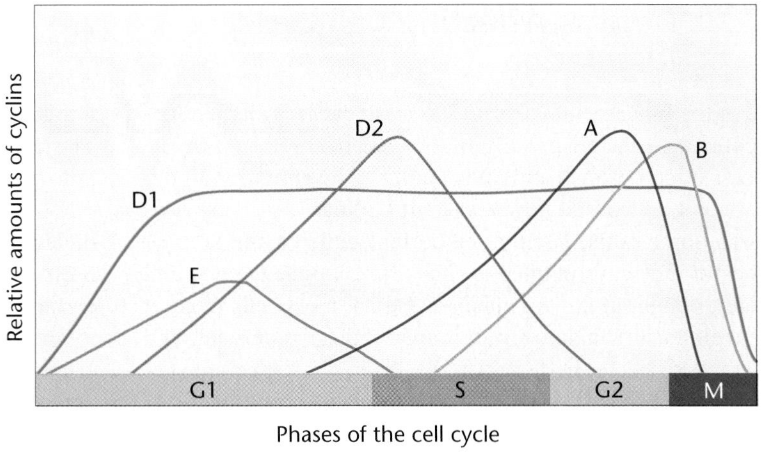

FIGURE 18–8 Relative expression times and amounts of cyclins during the cell cycle. Cyclin D1 accumulates early in G1 and is expressed at a constant level through most of the cycle. Cyclin E accumulates in G1, reaches a peak, and declines by mid S phase. Cyclin D2 begins accumulating in the last half of G1, reaches a peak just after the beginning of S, and then declines by early G2. Cyclin A appears in late G1, accumulates through S phase, peaks at the G2/M transition, and is rapidly degraded. Cyclin B peaks at the G2/M transition and declines rapidly in M phase.

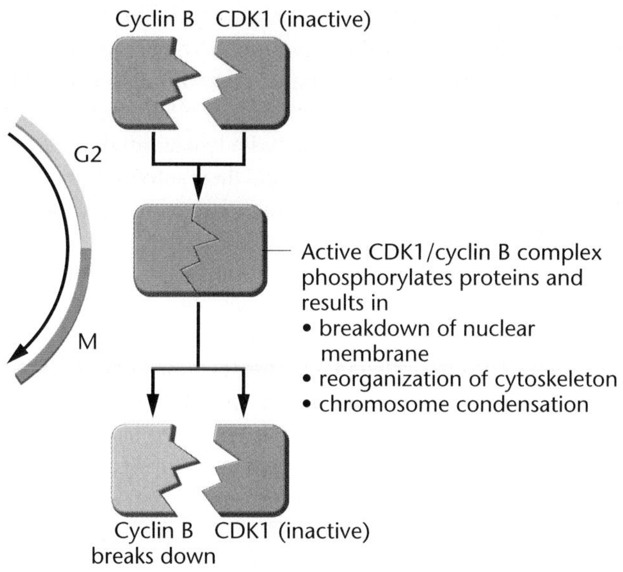

FIGURE 18–9 Transition from G2 to M phase is controlled by CDK1 and cyclin B. These molecules interact to form a complex that adds phosphate groups to cellular components. These in turn bring about the structural and biochemical changes necessary for mitosis (M phase).

when cyclin B is degraded and the protein phosphorylations characteristic of M phase are reversed. Although a large number of different protein kinases exist in cells, only a few are involved in cell cycle regulation.

Each of the cell cycle checkpoints, as well as the activities of cell cycle control molecules, are genetically regulated. In general, the cell cycle is regulated by an interplay of genes whose products either promote or suppress cell division. Mutation or mis-expression of any of the genes controlling the cell cycle contributes to the development of cancer in several ways. For example, if genes that control the G1/S or G2/M checkpoints are defective, the cell may continue to cycle before repairing DNA damage. This may lead to the accumulation of more mutations in genes that lead to uncontrolled proliferation or metastasis. Similarly, if genes that control progress through the cell cycle, such as those that encode the cyclins, are expressed inappropriately, the cell may be primed for continuous cycling and may be unable to exit the cell cycle into G0.

As described above, if DNA replication, repair or chromosome assembly are aberrant, the cell halts its progress through the cell cycle until the condition is corrected. This reduces the number of mutations and chromosomal abnormalities that accumulate in proliferating cells. However, if DNA or chromosomal damage is so severe that repair is impossible, the cell may initiate a second line of defense—a process called **apoptosis**, or **programmed cell death**. Apoptosis is a genetically controlled process whereby the cell commits suicide. Apoptosis is also initiated during normal multicellular development in order to eliminate certain cells that do not contribute to the final adult organism. Apoptosis differs from other types of cell death, such as necrosis, in that the host's immune system is not activated by the dying cell. The steps in apoptosis are the same for damaged cells and for cells eliminated during development: Nuclear DNA becomes fragmented, internal cellular structures are disrupted, and the cell dissolves into small spherical structures known as apoptotic bodies (Figure 18–10). In the final step, the apoptotic bodies are engulfed by the immune system's phagocytic cells. A series of proteases called **caspases** are responsible for initiating apoptosis and for digesting intracellular components. Apoptosis is genetically controlled, in that

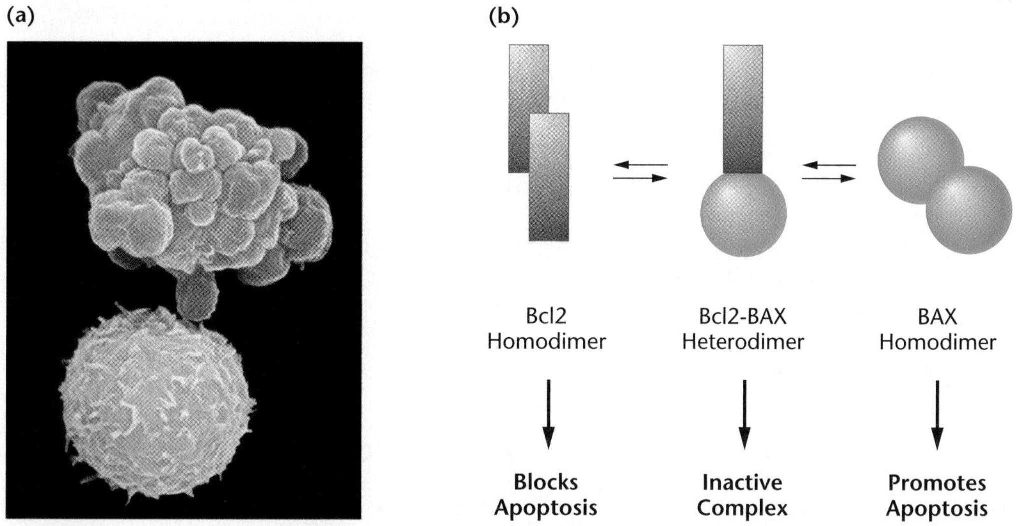

FIGURE 18–10 (a) A normal white blood cell (bottom) and a white blood cell undergoing apoptosis (top). Apoptotic bodies appear as grape-like clusters on the cell surface. (b) The relative concentrations of the Bcl2 and BAX proteins regulate apoptosis. A normal cell contains a balance of Bcl2 and BAX, which form inactive heterodimers. A relative excess of Bcl2 results in Bcl2 homodimers, which prevent apoptosis. Cancer cells with Bcl2 overexpression are resistant to chemotherapies and radiation therapies. A relative excess of BAX results in BAX homodimers, which induce apoptosis. In normal cells, activated p53 protein induces transcription of BAX and inhibits transcription of Bcl2, leading to cell death. In many cancer cells, p53 is defective, preventing the apoptotic pathway from removing the cancer cells.

regulation of specific gene products such as the Bcl2 and BAX proteins can trigger or prevent apoptosis. By removing damaged cells, programmed cell death reduces the number of mutations that are propagated to the next generation, including those in cancer-causing genes. The same genes that regulate cell cycle checkpoints can trigger apoptosis. As we will see, these genes are mutated in a wide range of cancers.

HOW DO WE KNOW?

How do we know that DNA damage is an important contributor to the development of cancer?

18.4 Many Cancer-Causing Genes Disrupt Control of the Cell Cycle

Two general categories of genes are mutated or mis-expressed in cancer cells—the proto-oncogenes and the tumor suppressor genes (Table 18.2). **Proto-oncogenes** are genes whose products promote cell growth and division. They do this by encoding transcription factors that stimulate expression of other genes, signal tranduction molecules that stimulate cell division, or cell cycle regulators that move the cell through the cell cycle. Proto-oncogene products may be located in the plasma membrane, cytoplasm, or nucleus, and their activities are controlled in various ways, including regulation at the transcriptional, translational, and protein-modification levels. When cells become quiescent and cease division, they repress the expression of most proto-oncogene products. In cancer cells, one or more proto-oncogenes are altered in such a way that their activities cannot be controlled in a normal fashion. This is sometimes due to a mutation in the proto-oncogene resulting in a protein product that acts abnormally. In other cases, proto-oncogenes may encode normal protein products, but the genes are overexpressed or cannot be transcriptionally repressed at the correct time. In these cases, the proto-oncogene product is continually in an "on" state, which may constantly stimulate the cell to divide. When a proto-oncogene is mutated or aberrantly expressed, and contributes to the development of cancer, it is

TABLE 18.2 SOME PROTO-ONCOGENES AND TUMOR SUPPRESSOR GENES

Proto-oncogene	Normal Function	Alteration in Cancer	Associated Cancers
Ha-ras	Signal transduction molecule, binds GTP/GDP	Point mutations	Colorectal, bladder, many types
c-erbB	Transmembrane growth factor receptor	Gene amplification, point mutations	Glioblastomas, breast cancer, cervix
c-myc	Transcription factor, regulates cell cycle, differentiation, apoptosis	Translocation, amplification, point mutations	Lymphomas, leukemias, lung cancer, many types
c-fos	Transcription factor, responds to growth factors	Overexpression	Osteosarcomas, many types
c-kit	Tyrosine kinase, signal transduction	Mutation	Sarcomas
c-raf	Cytoplasmic serine-threonine kinase, signal transduction	Gene rearrangements	Stomach cancer
RARα	Hormone-dependent transcription factor, differentiation	Chromosomal translocations with PML gene, fusion product	Acute promyelocytic leukemia
E6	Human papillomavirus encoded oncogene, inactivates p53	HPV infection	Cervical cancer
MDM2	Binds and inactivates p53, abrogates cell cycle checkpoints	Gene amplification, overexpression	Osteosarcomas, liposarcomas
Cyclins	Bind to CDKs, regulate cell cycle	Gene amplification, overexpression	Lung, esophagus, many types
CDK2, 4	Cyclin-dependent kinases, regulate cell cycle phases	Overexpression, mutation	Bladder, breast, many types

Tumor Suppressor	Normal Function	Alteration in Cancer	Associated Cancers
p53	Cell cycle checkpoints, apoptosis	Mutation, inactivation by viral oncogene products	Brain, lung, colorectal, breast, many types
RB1	Cell cycle checkpoints, binds E2F	Mutation, deletion, inactivation by viral oncogene products	Retinoblastoma, osteosarcoma, many types
APC	Cell–cell interaction	Mutation	Colorectal cancers, brain, thyroid
Bcl2	Apoptosis regulation	Overexpression blocks apoptosis	Lymphomas, leukemias
XPA–XPG	Nucleotide excision repair	Mutation	Xeroderma pigmentosum, skin cancers
BRCA2	DNA repair	Point mutations	Breast, ovarian, prostate cancers

known as an **oncogene** (cancer-causing gene). Oncogenes are those that have experienced a gain-of-function alteration. As a result, only one allele of a proto-oncogene needs to be mutated or mis-expressed in order to trigger uncontrolled growth. Hence, oncogenes confer a dominant cancer phenotype.

Tumor suppressor genes are those whose products normally regulate cell cycle checkpoints and initiate the process of apoptosis. In normal cells, proteins encoded by tumor suppressor genes halt progress through the cell cycle in response to DNA damage or growth-suppression signals from the extracellular environment. When tumor suppressor genes are mutated or inactivated, cells are unable to respond normally to cell cycle checkpoints, or are unable to undergo programmed cell death if DNA damage is extensive. This leads to a further increase in mutations, and the inability to leave the cell cycle when the cell should become quiescent. When both alleles of a tumor suppressor gene are inactivated, and other changes in the cell keep it growing and dividing, cells may become tumorigenic.

Below are some examples of proto-oncogenes and tumor suppressor genes that contribute to cancer when mutated. There are more than 200 oncogenes and tumor suppressor genes now known, and more will likely be discovered as cancer research continues.

The *Cyclin D1* and *Cyclin E* Proto-oncogenes

As we learned earlier in this chapter, cyclins are synthesized and degraded throughout the cell cycle. They form complexes with CDK molecules, and regulate their activities. These CDK/cyclin complexes are important regulators of each phase of the cell cycle. Several cyclin genes are known to be associated with the development of cancer. For example, the gene encoding cyclin D1 is amplified in a number of cancers, including those of the breast, bladder, lung, and esophagus. DNA amplification creates multiple copies of the *cyclin D1* gene and hence may lead to higher than normal levels of the cyclin D1 protein in these cancer cells. High levels of cyclin D1 protein may contribute to uncontrolled entry into S phase. In other cancers, *cyclin D1* is overexpressed even in the absence of gene amplification. In certain parathyroid tumors and B-cell lymphomas, the *cyclin D1* gene is involved in chromosomal aberrations such as translocations. These alterations to the *cyclin D1* gene may also cause it to be expressed abnormally. Similarly, the *cyclin E* gene is amplified or overexpressed in some leukemias and cancers of the breast and colon. It is possible that overexpression of these key cell cycle regulators, or loss of their periodic degradation, keeps cells primed for continuous cell cycling, and they can no longer leave the cell cycle, enter the quiescent G0 phase, or undergo cell differentiation.

The *ras* Proto-oncogenes

Some of the most frequently mutated genes in human tumors are those of the *ras* gene family. These genes are mutated in more than 40 percent of human tumors. The *ras* gene family encodes signal transduction molecules that are associated with the cell membrane and regulate cell growth and division. Ras proteins normally transmit signals from the cell membrane to the nucleus, stimulating the cell to divide in response to external growth factors. Ras proteins cycle between an inactive (switched off) and an active (switched on) state by binding either GDP or GTP. When a cell encounters a growth factor (such as platelet derived growth factor or epidermal growth factor), growth factor receptors on the cell membrane bind to the growth factor, resulting in autophosphorylation of the cytoplasmic portion of the growth factor receptor. This causes recruitment of proteins known as nucleotide exchange factors to the plasma membrane. These nucleotide exchange factors cause Ras to release GDP and bind GTP, thereby activating Ras. The active, GTP-bound form of Ras then sends its signals through cascades of protein phosphorylations in the cytoplasm (Figure 18–11). The end-point of these cascades is activation of nuclear transcription factors that stimulate expression of genes whose products drive the cell from quiescence into the cell cycle. Once Ras has sent its signals to the nucleus, it hydrolyzes GTP to GDP and becomes inactive. Mutations that convert the proto-oncogene *ras* to an oncogene prevent the Ras protein from hydrolyzing GTP to GDP and hence freeze the Ras protein into its "on" conformation, constantly stimulating the cell to divide. A comparison of amino acid sequences of Ras proteins from normal and cancer cells reveals that

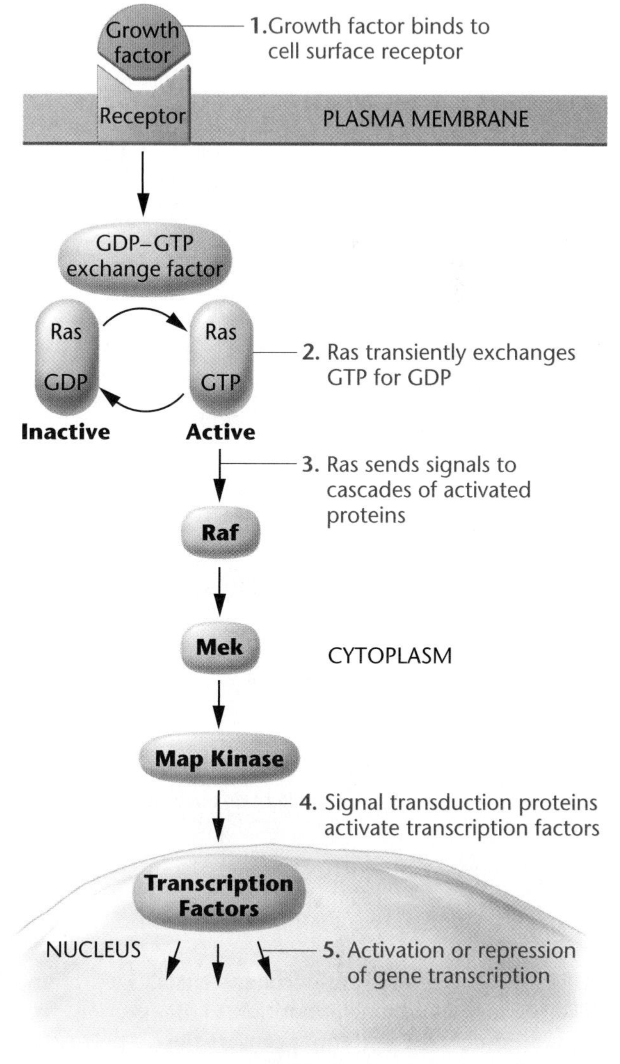

FIGURE 18–11 A signal transduction pathway mediated by Ras.

oncogenic Ras proteins have single amino acid substitutions at either position 12 or 61 (Figure 18–12). Each of these amino acid changes can be created by a single nucleotide substitution in the *ras* gene (Figure 18–13).

The *p53* Tumor Suppressor Gene

The most frequently mutated gene in human cancers—occurring in more than 50 percent of all cancers—is the ***p53*** gene. The *p53* gene encodes a nuclear protein that acts as a transcription factor that represses or stimulates transcription of more than 50 different genes (Figure 18–14).

Normally, the p53 protein is continuously synthesized but is rapidly degraded and therefore is present in cells at a low level. In addition, the p53 protein is normally bound to another protein called Mdm2, which prevents the phosphorylations and acetylations that convert the p53 protein from an inactive to an active form. Several types of events cause a rapid increase in the nuclear levels of activated p53 protein. These include chemical damage to DNA, double-stranded breaks in DNA induced by ionizing radiation, or the presence of DNA-repair intermediates generated by exposure of cells to ultraviolet light. The increase in activated p53 protein results from increases in protein phosphorylation, acetylation, and an increase in p53 protein stability.

The p53 protein initiates two different responses to DNA damage: arrest of the cell cycle followed by DNA repair, or apoptosis and cell death if DNA cannot be repaired. Each of these responses is accomplished by p53 acting as a transcription factor that stimulates or represses the expression of genes involved in each of these responses.

In normal cells, p53 can arrest the cell cycle at several phases. To arrest the cell cycle at the G1/S checkpoint, activated p53 protein stimulates transcription of a gene encoding the p21 protein. The p21 protein inhibits the CDK4/cyclin D1 complex, hence preventing the cell from moving from the G1 phase into S phase. Activated p53 protein also regulates expression of genes that retard the progress of DNA replication, thus allowing time for DNA damage to be repaired during S phase. By regulating expression of other genes, activated p53 can block cells at the G2/M checkpoint, if DNA damage occurs during S phase.

Activated p53 can also instruct a damaged cell to commit suicide by apoptosis. It does so by activating the transcription of the *Bax* gene and repressing transcription of the *Bcl2* gene. In normal cells, the BAX protein is present in a heterodimer with

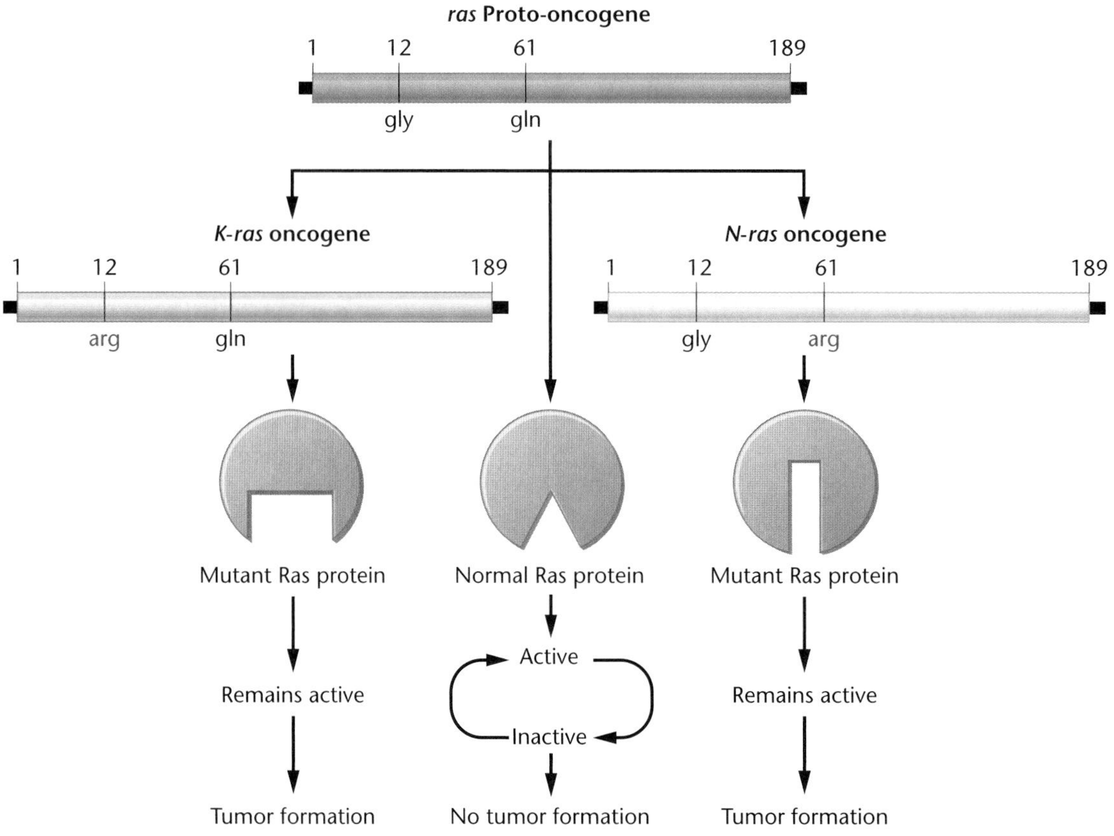

FIGURE 18–12 The *ras* proto-oncogene encodes a protein of 189 amino acids. In the normal protein, glycine is encoded at position 12, and glutamine at position 61. Ras proteins from several tumors contain single amino acid substitutions at one of these positions. These mutations convert the *ras* proto-oncogene into a tumor-promoting oncogene. *K-ras* and *N-ras* are mutant alleles of the *ras* proto-oncogene.

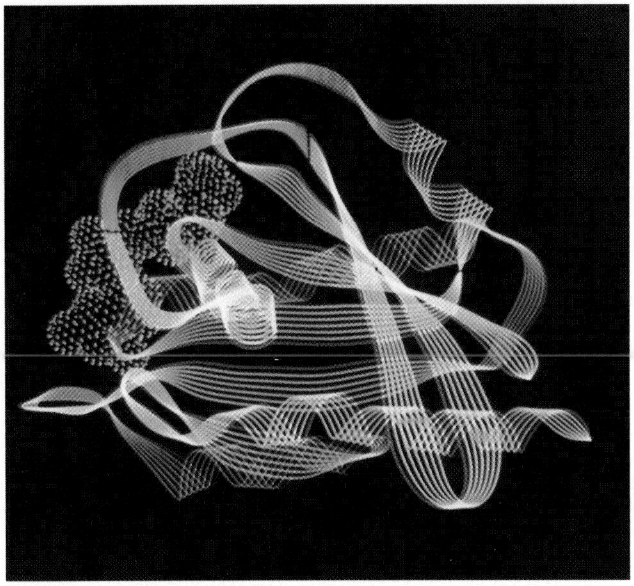

FIGURE 18–13 A three-dimensional computer-generated image of Ras proteins in two different conformations. Normal Ras proteins act as molecular switches controlling cell growth and differentiation. The switch is on (conformation shown in blue) when GTP binds to the protein, and off (conformation shown in yellow) when GTP is hydrolyzed to GDP. Switching the protein between states alters the conformation of the protein in the two regions (blue and yellow). Oncogenic mutant Ras proteins are stuck in the on state, and continuously signal for cell growth.

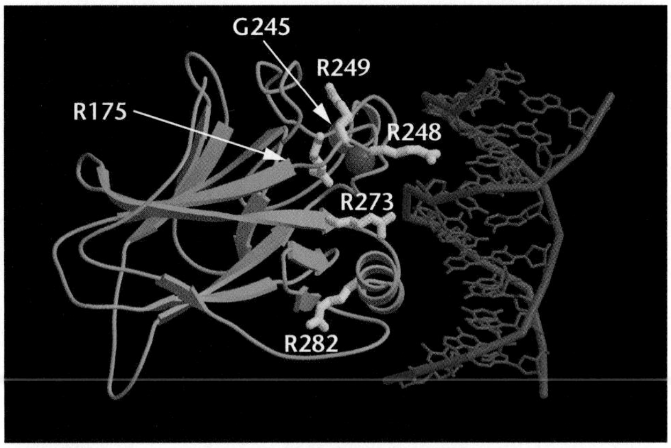

FIGURE 18–14 A computer-generated image of the p53 protein (left), showing it complexed with DNA (right), with the six most frequently mutated regions of the protein shown. R is the abbreviation for arginine, G is the abbreviation for glycine, and the numbers represent the amino acid positions of the mutations.

the Bcl2 protein and the cell remains viable. When the levels of BAX protein increase in response to p53 stimulation of *Bax* gene transcription, BAX homodimers are formed and these homodimers activate the cellular changes that lead to cellular self-destruction (Figure 18–10). In cancer cells that lack functional p53, BAX protein levels do not increase in response to cellular damage and apoptosis may not occur.

Hence, cells lacking functional p53 are unable to arrest at cell cycle checkpoints or to enter apoptosis in response to DNA damage. As a result, they move unchecked through the cell cycle, regardless of the condition of the cell's DNA. Cells lacking p53 have high mutation rates and accumulate the types of mutations and chromosomal aberrations that lead to cancer. Because of the importance of the *p53* gene to genomic integrity, it is often referred to as the "guardian of the genome."

The *RB1* Tumor Suppressor Gene

The loss or mutation of the ***RB1* (retinoblastoma 1)** tumor suppressor gene contributes to the development of many cancers, including those of the breast, bone, lung, and bladder. The *RB1* gene was originally identified as a result of studies on retinoblastoma, an inherited disorder in which tumors develop in the eyes of young children. Retinoblastoma occurs with a frequency of about 1 in 14,000 to 20,000 individuals. In the familial form of the disease, individuals inherit one mutated allele of the *RB1* gene and have an 85 percent chance of developing retinoblastomas as well as an increased chance of developing other cancers. All somatic cells of patients with hereditary retinoblastoma contain one mutated allele of the *RB1* gene. However, it is only when the second normal allele of the *RB1* gene is lost or mutated in certain retinal cells that retinoblastoma develops. The wild-type allele can be inactivated by a single base-pair mutation, a deletion, or other type of chromosomal aberration. In individuals that do not have this hereditary condition, retinoblastoma is extremely rare, as it requires two separate somatic mutations in a retinal cell in order to inactivate both copies of the *RB1* gene (Figure 18–15).

The **retinoblastoma protein (pRB)**, like the p53 protein, is a tumor suppressor protein that controls the G1/S cell cycle checkpoint. The pRB protein is found in the nuclei of all cell types at all stages of the cell cycle. However, its activity varies throughout the cell cycle, depending upon its phosphorylation state. When cells are in the G0 phase of the cell cycle, the pRB protein is nonphosphorylated and binds to transcription factors such as E2F, inactivating them (Figure 18–16). When the cell is stimulated by growth factors, it enters G1 and approaches S phase. Throughout the G1 phase, the pRB protein becomes phosphorylated by the CDK4/cyclin D1 complex. Phosphorylated pRB is inactive and releases its bound regulatory proteins. When E2F and other regulators are released by pRB, they are free to induce the expression of over 30 genes whose products are required for the transition from G1 into S phase. After cells traverse S, G2, and M phases, pRB reverts to a nonphosphorylated state, binds to regulatory proteins such as E2F and keeps them sequestered until required for the next cell cycle. In normal quiescent cells, the pRB protein is active and prevents passage into S phase. In many cancer cells, including retinoblastoma cells, both copies of the *RB1* gene are defective, inactive, or absent, and progression through the cell cycle is not regulated.

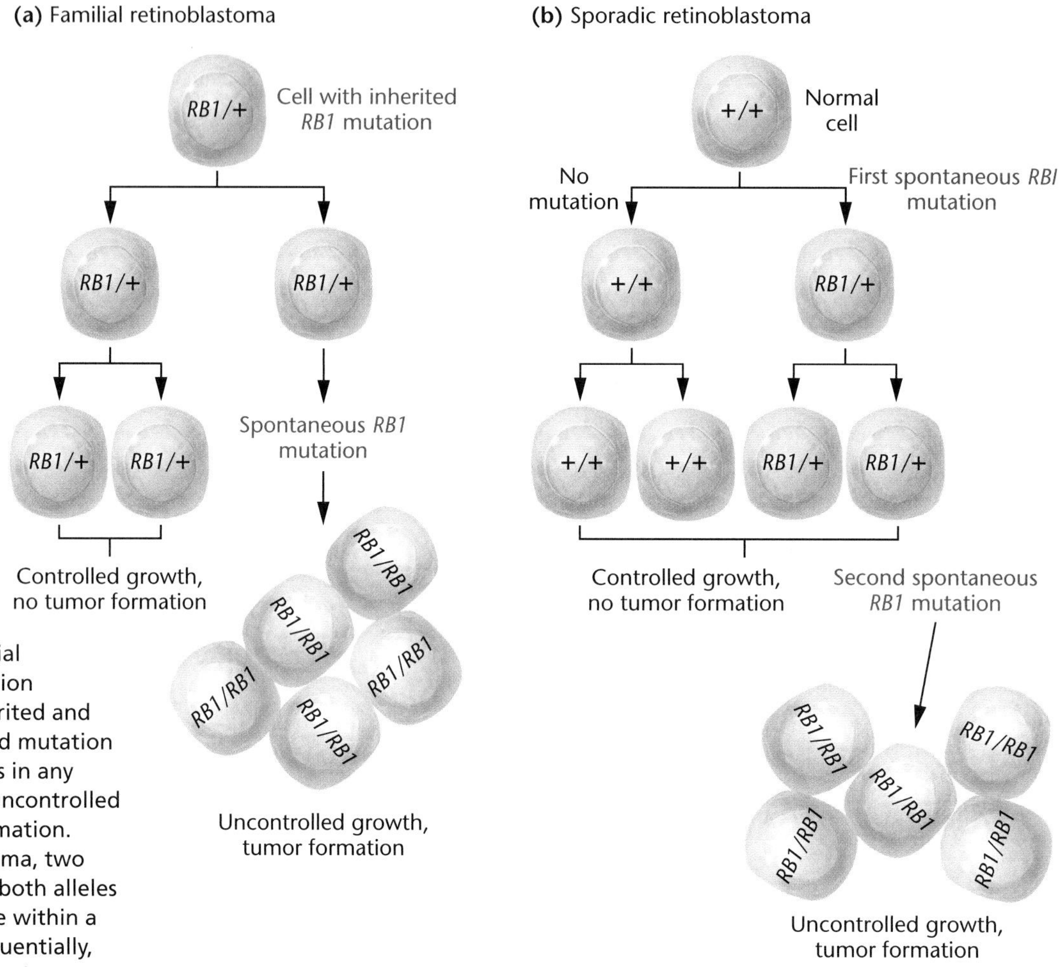

FIGURE 18–15 (a) In familial retinoblastoma, one mutation (designated as *RB1*) is inherited and present in all cells. A second mutation at the retinoblastoma locus in any retinal cell contributes to uncontrolled cell growth and tumor formation. (b) In sporadic retinoblastoma, two independent mutations in both alleles of the retinoblastoma gene within a single cell are acquired sequentially, also leading to tumor formation.

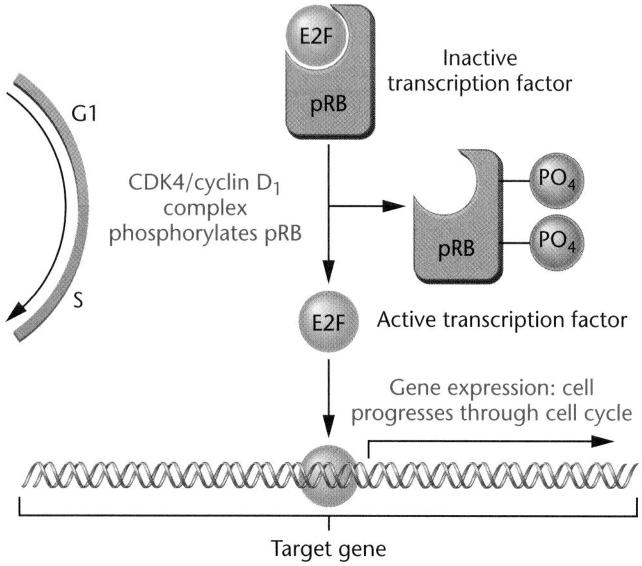

FIGURE 18–16 In the nucleus during G1, pRB interacts with and inactivates transcription factor E2F. As the cell moves from G1 to S phase, a CDK4/cyclinD1 complex forms and adds phosphate groups to pRB. As pRB becomes phosphorylated, E2F is released and becomes transcriptionally active, allowing the cell to pass through S phase. Phosphorylation of pRB is transitory; as CDK/cyclin complexes are degraded and the cell moves through the cell cycle to early G1, pRB phosphorylation declines.

Now solve this

Problem 18.23 on page 455 asks you to explain how in the inherited Li–Fraumeni syndrome, mutations in one allele of the *p53* gene can give rise to a wide variety of different cancers.

Hint: To answer this question, you might review the cellular functions regulated by the normal *p53* tumor suppressor gene. Consider how each of these functions could be affected if the *p53* gene product is either absent or defective. To understand how mutations in one allele of a tumor suppressor gene result in tumors that contain aberrations in both alleles, read about loss of heterozygosity in Section 18.6.

18.5 Cancer Is a Genetic Disorder Affecting Cell–Cell Contact

As discussed at the beginning of this chapter, uncontrolled growth alone is insufficient to create a malignant and life-threatening cancer. Cancer cells must also acquire the features of metastasis, which include the ability to disengage from the original tumor site, to enter the blood or lymphatic system, to invade surrounding tissues and to develop into secondary tumors. In order to leave the site of the primary tumor and invade other tissues, tumor cells must digest components of the **extracellular matrix** and **basal lamina**, which normally contain and separate tissues (Figure 18–17). The extracellular matrix and basal lamina are composed of proteins and carbohydrates. They form the scaffold for tissue growth and normally inhibit the migration of cells.

The ability to invade the extracellular matrix is also a property of some normal cell types. For example, implantation of the embryo in the uterine wall during pregnancy requires cell migration across the extracellular matrix. In addition, white blood cells reach the site of infection by penetrating capillary walls. The mechanisms of invasion are probably similar in these normal cells and in cancer cells. The difference is that in normal cells, the invasive ability is tightly regulated, whereas in tumor cells, this regulation has been lost.

Although less is known about the genes that control metastasis than about those controlling the cell cycle, it is likely that metastasis is controlled by a large number of genes, including those that encode cell adhesion molecules and proteolytic enzymes. For example, epithelial tumors have a lower than normal level of the E-cadherin glycoprotein, which is responsible for cell–cell adhesion in normal tissues. Also, proteolytic enzymes such as **metalloproteinases** are present at higher than normal levels in highly malignant tumors, and are not susceptible to the normal controls conferred by regulatory molecules such as **tissue inhibitors of metalloproteinases (TIMPs)**. It has been shown that the level of aggressiveness of a tumor correlates positively with the levels of proteolytic enzymes expressed by the tumor. Hence, inappropriately expressed cell adhesion and proteinase enzymes may assist malignant tumor cells by loosening the normal constraints on cell location and creating holes through which the tumor cells can pass into and out of the circulatory system.

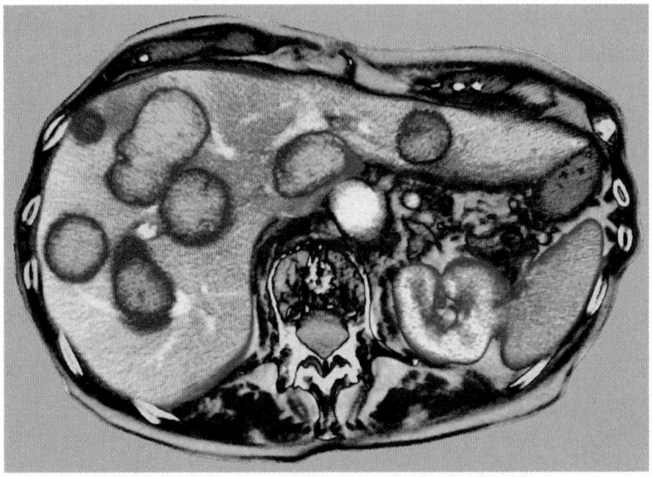

FIGURE 18–17 Colored computer tomography (CT) scan of a patient with a number of metastatic tumors in the liver. Tumors appear as light blue round structures surrounded by red; the liver as yellow. The secondary tumors spread from a primary tumor located elsewhere in the body. The backbone appears as a light blue structure in the lower center of the image. Parts of the rib cage appear as light blue ovals located around the outside edges of the image.

Now solve this

Problem 18.25 on page 455 describes the isolation of a gene that is mutated in metastatic tumors. The gene appears to be a member of the serine protease family. You are asked to conjecture how this gene might contribute to the development of highly invasive cancers.

Hint: As you learned in this section, metastatic cancer cells have the ability to escape their adhesions to other cells, travel through the circulatory system, and invade distant tissues. You might think about what types of metastatic processes are facilitated by having a protease molecule that lacks normal regulation. Also consider what types of mutations in this gene could lead to increased metastasis.

18.6 Predisposition to Some Cancers Can Be Inherited

Although the vast majority of human cancers are sporadic, a small fraction (1–2 percent of all cancers) have an hereditary or familial component. At present, about 50 forms of hereditary cancer are known (Table 18.3).

Most inherited cancer-susceptibility genes, although transmitted in a Mendelian dominant fashion, are not sufficient in themselves to trigger development of a cancer. At least one other somatic mutation in the other copy of the gene must occur in order to drive a cell toward tumorigenesis. In addition, mutations in other genes are usually necessary to fully express the cancer phenotype. As mentioned above, inherited mutations in the *RB1* gene predispose individuals to developing various cancers. Although the normal somatic cells of these patients are heterozygous for the *RB1* mutation, their tumors contain mutations in both copies of the gene. The second somatic mutation may be a point mutation, deletion, or gene disruption caused by a chromosomal aberration. The phenomenon whereby the second, wild-type, allele is mutated in a tumor is known as **loss of heterozygosity**. Although loss of heterozygosity is an essential first step in

TABLE 18.3 INHERITED PREDISPOSITIONS TO CANCER

Tumor Predisposition Syndromes	Chromosome
Early-onset familial breast cancer	17q
Familial adenomatous polyposis	5q
Familial melanoma	9p
Gorlin syndrome	9q
Hereditary nonpolyposis colon cancer	2p
Li–Fraumeni syndrome	17p
Multiple endocrine neoplasia, type 1	11q
Multiple endocrine neoplasia, type 2	22q
Neurofibromatosis, type 1	17q
Neurofibromatosis, type 2	22q
Retinoblastoma	13q
Von Hippel–Lindau syndrome	3p
Wilms tumor	11p

expression of these inherited cancers, further mutations in other proto-oncogenes and tumor suppressor genes are necessary for the passage of tumor cells to full malignancy.

The development of hereditary colon cancer illustrates how inherited mutations in one allele of a gene contribute only one step in the multistep pathway leading to malignancy.

About 1 percent of colon cancer cases result from a genetic predisposition to cancer known as **familial adenomatous polyposis (FAP)**. In FAP, individuals inherit one mutant copy of the *APC* (adenomatous polyposis) gene located on the long arm of chromosome 5. Mutations include deletions, frameshift, and point mutations. The normal function of the *APC* gene product is to act as a tumor suppressor controlling cell–cell contact and growth inhibition by interacting with the β-catenin protein. The presence of a heterozygous *APC* mutation causes the epithelial cells of the colon to partially escape cell cycle control, and the cell divides to form a small cluster of cells called a **polyp** or adenoma. People who are heterozygous for this condition develop hundreds to thousands of colon and rectal polyps early in life (Figure 18–18). In FAP, each polyp is a clone of cells, all of which carry an *APC* gene mutation, and the development of polyps is a dominant trait. Although it is not necessary for the second allele of the *APC* gene to be mutated in polyps at this stage, in the majority of cases, the second *APC* allele becomes mutant in a later stage of cancer development. Over time, some polyps acquire a series of mutations that convert these benign polyps into malignant tumors. The relative order of mutations in the development of FAP is shown in Figure 18–19.

The second mutation in polyp cells that contain an *APC* gene mutation occurs in the *ras* proto-oncogene. The *ras* gene product stimulates cell growth and division by transmitting growth signals from the cell surface to the nucleus. The combined *APC* and *ras* gene mutations trigger the development of intermediate adenomas. These adenomas have defects in normal cell differentiation and will grow in culture, in the absence of contact with other cells and hence are described as **transformed**. The third step toward malignancy requires loss of function of both alleles of the *DCC* (deleted in colon cancer) gene. The *DCC* gene product is thought to be involved with cell adhesion and differentiation. Mutations in both *DCC* alleles result in the formation of late stage adenomas with a number of finger-like outgrowths (villi). In order for late adenomas to progress to cancerous adenomas, they suffer loss of functional *p53* genes. As discussed earlier in this chapter, the p53 gene product is essential for arresting the cell cycle in response to DNA damage, and loss of p53 results in high mutation rates throughout the genome. The final steps toward malignancy involve mutation in an unknown number of genes associated with metastasis (Figure 18–19).

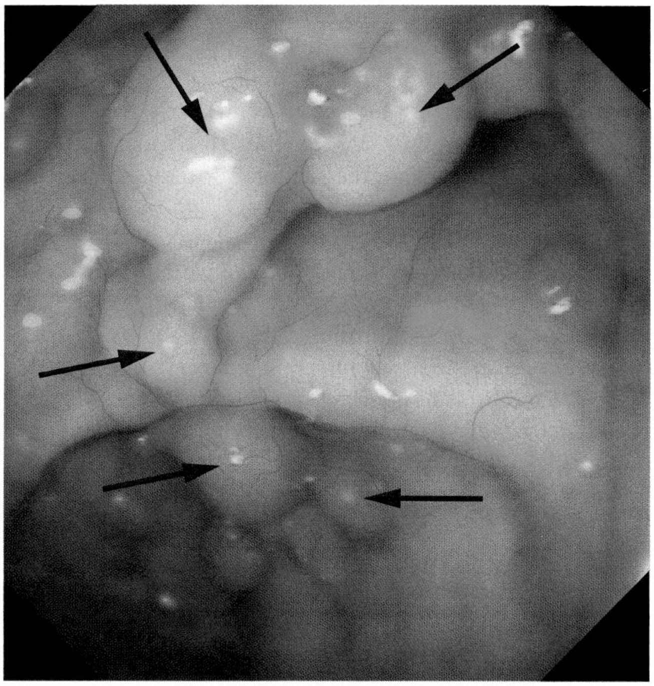

FIGURE 18–18 Endoscopic view of a colon from a patient with familial polyposis. Up to hundreds of polyps may grow on the lining of the colon, some developing into cancers. Arrows indicate polyps.

Now solve this

Problem 18.7 on page 454 asks you to explain why some tobacco smokers and some people with inherited mutations in cancer-related genes never develop cancer.

Hint: When considering the reasons why cancer is not an inevitable consequence of genetic or environmen-

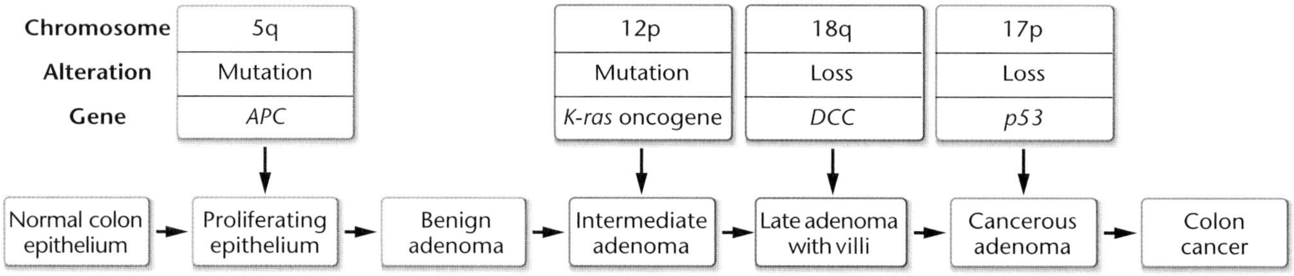

FIGURE 18–19 A model for the multistep development of colon cancer. The first step is the loss or inactivation of one allele of the *APC* gene on chromosome 5. In FAP cases, one mutant *APC* allele is inherited. Subsequent mutations involving genes on chromosomes 12, 17, and 18 in cells of benign adenomas can lead to a malignant transformation that results in colon cancer. Although the mutations on chromosomes 12, 17, and 18 usually occur at a later stage than those involving chromosome 5, the sum of changes is more important than the order in which they occur.

Web Tutorial 18.1 The Multistep Nature of Cancer

tal conditions, you might think about the multistep nature of cancer and the types of genetically controlled functions that are abnormal in cancer cells. It might also be useful to consider how each individual's genetic background may affect mutation rates and differences in DNA repair functions.

18.7 Viruses Contribute to Cancer in Both Humans and Animals

Viruses that cause cancer in animals have played a significant role in the search for knowledge about the genetics of human cancer. Most of these animal viruses are RNA viruses known as **retroviruses**. Because they transform cells into cancer cells, they are known as acute transforming retroviruses. The first of these **acute transforming retroviruses** was discovered in 1910 by Francis Peyton Rous. Rous was studying sarcomas (solid tumors of muscle, bone, or fat) in chickens, and he observed that extracts from these tumors caused the formation of new sarcomas when they were injected into tumor-free chickens. Several decades later, the agent within the extract that caused the sarcomas was identified as a retrovirus, and named the **Rous sarcoma virus (RSV)**. In 1966, Rous was honored with the Nobel Prize in Physiology or Medicine for his work, which established the link between viruses and cancer.

To understand how retroviruses cause cancer in animals, it is necessary to know how these viruses replicate in cells. When a retrovirus infects a cell, its RNA genome is copied into DNA by the **reverse transcriptase** enzyme, which is brought into the cell with the infecting virus. The DNA copy then enters the nucleus of the infected cell, where it integrates at random into the host cell's genome. The integrated DNA copy of the retroviral RNA is called a **provirus**. The proviral DNA contains powerful enhancer and promoter elements in its U5 and U3 sequences at the ends of the provirus (Figure 18–20). The U5 promoter uses the host cell's transcription proteins, directing transcription of the viral genes (*gag*, *pol*, and *env*). The products of these genes are the proteins and RNA genomes that make up the new retroviral particles. Because the provirus is integrated into the host genome, it is replicated along with the host's DNA during the cell's normal cell cycle. Retroviruses may not kill a cell, but may continue to use the cell as a factory to replicate more viruses that will then infect surrounding cells.

A retrovirus may cause cancer in two different ways. First, the proviral DNA may integrate by chance near one of the cell's normal proto-oncogenes. The strong promoters and enhancers in the provirus then stimulate high levels or inappropriate timing of transcription of the proto-oncogene, leading to stimulation of host cell proliferation. Second, a retrovirus may pick up a copy of a host proto-oncogene and integrate it into its genome (Figure 18–20). The new viral oncogene may be mutated during the process of transfer into the virus, or may be expressed at abnormal levels because it is now under control of viral promoters. An oncogene carried by retroviruses is called a *v-onc*; the normal cellular version of the gene is called a *c-onc*. Retroviruses that carry these *v-onc* genes can infect and transform normal cells into tumor cells. In the case of RSV, the oncogene that was captured from the chicken genome was the *c-src* gene. Through the study of many acute transforming viruses of animals, scientists have identified dozens of proto-oncogenes. Some of these are listed in Table 18.4. The oncogenes from which the *v-onc* genes were derived can cause cancer by other mechanisms besides being present as *v-onc* genes in retroviruses. These include mutation or abnormal expression, as described previously in this chapter. Similarly, scientists have identified many proto-oncogenes by cloning and characterizing genes that are abnormally expressed as a result of retroviral integration.

So far, no acute transforming retroviruses have been identified in humans. However, RNA and DNA viruses contribute to the development of human cancers in a variety of ways. It is thought that worldwide, about 15 percent of cancers are associated with viruses, making virus infection the second greatest risk factor for cancer, next to tobacco smoking.

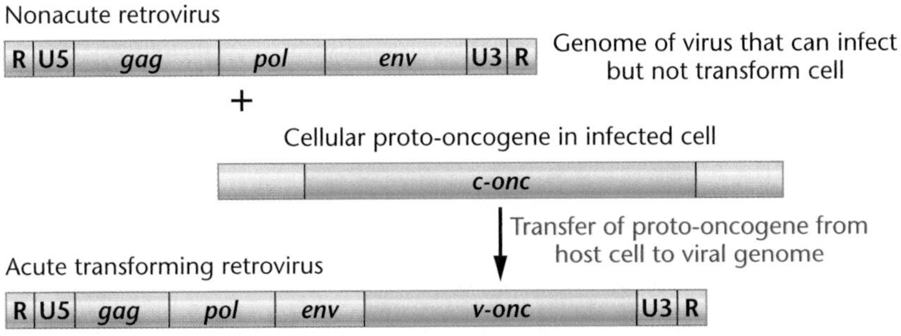

FIGURE 18–20 The genome of a typical retrovirus is shown at the top of the diagram. The genome contains repeats at the termini (R), the U5 and U3 regions that contain promoter and enhancer elements, and the three major genes that encode viral structural proteins (*gag* and *env*) and the viral reverse transcriptase (*pol*). RNA transcripts of the entire viral genome comprise the new viral genomes. If the retrovirus acquires all or part of a host cell proto-oncogene (*c-onc*), this gene (now known as a *v-onc*) is expressed along with the viral genes, leading to overexpression or inappropriate expression of the *v-onc* gene. The *v-onc* gene may also acquire mutations that enhance its transforming ability.

The most significant contributors to virus-induced cancers are the papillomaviruses (HPV 16 and 18), human T-cell leukemia virus (HTLV-1), hepatitis B virus, and Epstein–Barr virus (Table 18.5). Like other risk factors for cancer, including hereditary predisposition to certain cancers, virus infection alone is not sufficient to trigger human cancers. Other factors, including DNA damage or the accumulation of mutations in one or more of a cell's oncogenes and tumor suppressor genes, are required to move a cell down the multistep pathway to cancer.

Because viruses are comprised solely of a nucleic acid genome surrounded by a proteinaceous coat, viruses must utilize the host cell's biosynthetic machinery in order to reproduce themselves. As most viruses need the host cell to be in an actively growing state in

TABLE 18.4 **REPRESENTATIVE VIRAL ONCOGENES**

Oncogene	Origin	Species	Cellular function
v-abl	Abelson murine leukemia virus	Mouse	Tyrosine kinase
v-erbB	Avian erythroblastosis virus	Chicken	EGF receptor
v-fos	FBJ osteosarcoma virus	Mouse	Transcription factor
v-myc	Avian myelocytomatosis virus	Chicken	Transcription factor
v-Ha-ras	Harvey murine sarcoma virus	Mouse	GTP-binding protein
v-sis	Simian sarcoma virus	Monkey	Platelet-derived growth factor
v-src	Rous sarcoma virus	Chicken	Tyrosine kinase signal protein

order to access the host's DNA synthetic enzymes, these viruses often contain genes encoding products that stimulate the cell cycle. If the virus does not kill the host cell, the potential exists for a loss of cell cycle control and the beginning of tumorigenesis.

HPV 16 and 18 are thought to contribute to the development of some anogenital cancers, especially cervical cancer (Figure 18–21). Up to 90 percent of cervical cancers contain HPV DNA; however, only a small fraction of HPV-infected women develop cervical cancer, and only after long latency periods. HPV 16 and 18 encode two genes, *E6* and *E7*, both of which can be considered virus-encoded oncogenes. The E6 protein binds to a cell's p53 protein, leading to its inactivation. As discussed previously, the p53 protein is an essential component of cell cycle checkpoints, and loss of p53 function leads to genomic instability. E6 may also inactivate apoptosis by interacting with an apoptosis protein known as BAK. The E7 protein binds to cellular pRB protein, causing pRB to release the transcription factor E2F, which in turn triggers expression of genes whose products are involved in DNA replication. E7 may also stimulate S phase progression by activating cyclins A and E and by blocking CDK inhibitor proteins.

The HTLV-1 retrovirus also contains genes that stimulate cell proliferation, leading to the development of adult T-cell leukemia. HTLV-1 infection is a worldwide epidemic, affecting up to 20 million people. In about 5 percent of these cases, virus infection may lead to leukemia, after long latency periods. Most of the transforming capacity of HTLV-1 is due to the product of its *pX* gene, called Tax protein. Tax is essential for HTLV-1 replication and is able to indepen-

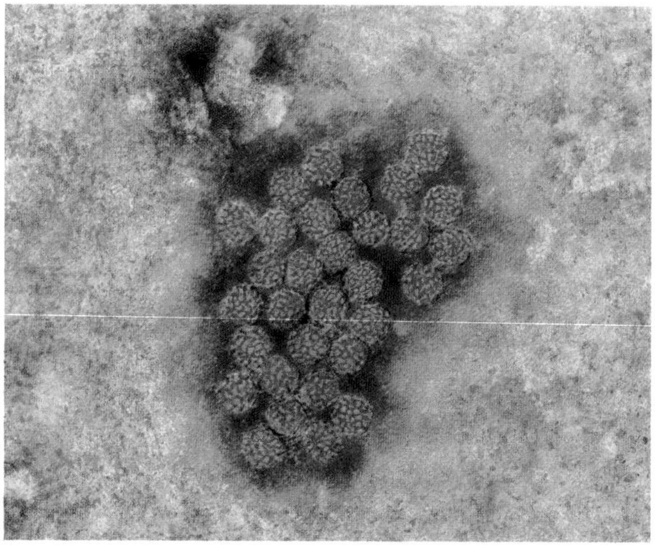

FIGURE 18–21 Colored electron micrograph of a cluster of human papillomaviruses (HPV). These DNA viruses cause warts on hands, feet, and mucous membranes. HPV 16 and 18 are implicated in the development of cervical cancer.

TABLE 18.5 **HUMAN VIRUSES ASSOCIATED WITH CANCER**

Virus	Cancer	Oncogenes	Mechanism
Human papillomavirus 16, 18	Cervical cancer	*E6, E7*	Inhibit p53 and pRB tumor suppressors
Hepatitis B virus	Liver cancer	*HBx*	Signal transduction, stimulates cell cycle
Epstein–Barr virus	Burkitt's lymphoma, naso-pharyngeal cancer	Unknown	Unknown
Human herpesvirus 8	AIDS-related Kaposi's sarcoma	Several possible	Unknown
Human T-cell leukemia virus	Adult T-cell leukemia	*pX*	Stimulates cell cycle

dently transform cells in culture and induce tumors in transgenic mice that express the protein. Tax acts as a transcription factor, binding to the viral promoter region and stimulating transcription of the viral genome. In addition, Tax stimulates transcription of a number of cellular genes, including the proto-oncogenes *c-fos* and *c-jun*, growth factor genes and growth factor receptor genes. This leads to stimulation of quiescent cells and their entry into the cell cycle. In addition, Tax represses transcription of some tumor suppressor genes such as the *p18INK4c* gene. This tumor suppressor gene encodes a protein that normally inhibits CDK4 and arrests the cell cycle in G1. By repressing transcription of *p18INK4c*, Tax increases the activity of CDK4, leading to phosphorylation of pRB, release of E2F, and expression of genes required for the transition from G1 to S phase. In this way, Tax helps drive the cell through the cell cycle. In addition to its transcriptional effects, Tax also inhibits p53 activity, leading to abrogation of cell cycle checkpoints and increases in mutation rates.

How Do We Know?

How do we know that some viruses contain genes that contribute to the development of cancer?

18.8 Environmental Agents Contribute to Human Cancers

Any substance or event that damages DNA has the potential to be carcinogenic. Unrepaired or inaccurately repaired DNA introduces mutations which, if they occur in proto-oncogenes or tumor suppressor genes, can lead to disregulation of the cell cycle or disruption of controls over cell contact and invasion.

Our environment, both natural and man-made, contains abundant carcinogens. These include chemicals, radiation, some viruses, and chronic infections. Perhaps the most significant carcinogen in our environment is tobacco smoke. Epidemiologists estimate that about 30 percent of human cancer deaths are associated with cigarette smoking. Tobacco smoke contains a number of cancer-causing chemicals, some of which preferentially mutate proto-oncogenes such as *ras* and tumor suppressor genes such as *p53*.

Diet is often implicated in the development of cancer. Consumption of red meat and animal fat is associated with some cancers, such as colon, prostate, or breast cancer. The mechanisms by which these substances may contribute to carcinogenesis are not clear, but may involve stimulation of cell division through hormones or creation of carcinogenic chemicals during cooking. Alcohol may cause inflammation of the liver and contribute to liver cancer.

Although most people perceive the man-made, industrial environment to be a highly significant contributor to cancer, it may account for only a small percentage of total cancers, and often in only specialized situations. Some of the most mutagenic agents, and hence potentially the most carcinogenic, are natural substances and natural processes. For example, aflatoxin, a component of a mold that grows on peanuts and corn, is one of the most carcinogenic chemicals known. Most chemical carcinogens are of the alkylating, aralkylating, or arylhydroxylamine classes, all of which react with DNA. Although some of these substances, such as nitrosamines, are components of man-made substances, many are naturally occurring. For example, natural pesticides and antibiotics found in plants may be carcinogenic, and the human body itself creates alkylating agents in the acidic environment of the gut. Nevertheless, these observations do not diminish the serious cancer risks to specific populations who are exposed to man-made carcinogens, such as synthetic pesticides or asbestos.

DNA lesions brought about by natural radiation (X rays, ultraviolet light), natural dietary substances, and the external environment contribute the majority of environmentally caused mutations that lead to cancer. Normal metabolism creates oxidative end-products that can damage DNA, proteins, and lipids. It is estimated that the human body suffers about 10,000 damaging DNA lesions per day due to the actions of oxygen free radicals. DNA repair enzymes deal successfully with most of this damage; however, some damage may lead to permanent mutations. The process of DNA replication itself is mutagenic. Hence, substances like growth factors or hormones that simulate cell division are ultimately mutagenic and perhaps carcinogenic. Chronic inflammation due to infection also stimulates tissue repair and cell division, resulting in DNA lesions accumulating during replication. These mutations may persist, particularly if cell cycle checkpoints are compromised due to mutations or inactivation of tumor suppressor genes such as *p53* or *RB1*.

Both ultraviolet (UV) light and ionizing radiation (such as X rays and gamma rays) induce DNA damage. (See Chapter 15.) UV damage in sunlight is well accepted as an inducer of skin cancers. Ionizing radiation has clearly demonstrated itself as a carcinogen in studies of populations exposed to neutron and gamma radiation from atomic blasts such as those in Hiroshima and Nagasaki. Another significant environmental component, radon gas, may be responsible for up to 50 percent of the ionizing radiation exposure for the U.S. population and could contribute to lung cancers in some populations.

Now solve this

Problem 18.16 on page 454 asks you to evaluate what percentage of money spent on cancer research should be devoted to research and education on preventing cancer and what percentage should be devoted to research on cancer cures.

Hint: In answering this question, think about the relative rates of environmentally induced and spontaneous cancers. Second, consider the proportion of environmentally induced cancers that are due to lifestyle choices. (An interesting source of information on this topic can be found in Ames, B.N. et al. 1995. The causes and prevention of cancer. *Proc. Natl. Acad. Sci. USA* 92:5258–65.)

GENETICS, TECHNOLOGY, AND SOCIETY

Breast Cancer: The Double-Edged Sword of Genetic Testing

These are exhilarating times for genetics and biotechnology. Close on the heels of the completion of the Human Genome Project has come a rush of optimism about future applications of genetics. Scientists and the media predict that gene technologies will soon diagnose and cure diseases as diverse as diabetes, asthma, heart disease, and Parkinson disease.

The prospect of using genetics to prevent and cure a whole range of diseases is exciting. However, in our enthusiasm, we often forget that these new technologies have significant limitations and profound ethical concerns. The story of genetic testing for breast cancer illustrates how we must temper our high expectations with respect for uncertainty.

Breast cancer is the most common cancer among women and the second leading cause of all cancer deaths (after lung cancer). Each year, more than 190,000 new cases are diagnosed in the United States. Breast cancer is not limited to women; about 1400 men are also diagnosed with the disease each year. A woman's lifetime risk of developing breast cancer is about 12 percent, and the risk increases with age.

Approximately 5 to 10 percent of breast cancers are familial, defined by the appearance of several cases of breast or ovarian cancer among near blood relatives and the early onset of these diseases. In 1994, two genes were identified that show linkage to familial breast cancers. Germline mutations in these genes (*BRCA1* and *BRCA2*) are associated with the majority of familial breast cancers. The molecular functions of *BRCA1* and *BRCA2* are still uncertain, although they appear to be involved in repairing damaged DNA. Mutations in these genes act as autosomal dominants with variable penetrance. Women who bear mutations in *BRCA1* or *BRCA2* have a 36 to 85 percent lifetime risk of developing breast cancer and a 16 to 60 percent risk of developing ovarian cancer. Men with germline mutations in *BRCA2* have a 6 percent lifetime breast cancer risk—a hundredfold increase over the general male population.

BRCA1 and *BRCA2* genetic tests detect any of the over 2000 different mutations that are known to occur within the coding regions of these genes, but the tests have limitations. They do not detect mutations in regulatory regions outside the coding region—mutations that could cause aberrant expression of these genes. Also, little is known about how any particular mutation manifests itself in terms of cancer risk, and the effects of each mutation may be modified by environmental factors and by interactions with other susceptibility genes.

Many patients at risk for familial breast cancer opt to undergo genetic testing. These patients feel that test results will help them to prevent breast or ovarian cancers, will guide them in childbearing decisions, and allow them to inform family members at risk. But none of these benefits is clear-cut.

A woman whose *BRCA* test results are negative may be relieved and feel that she is not subject to familial breast cancer. However, her risk of developing breast cancer is still 12 percent (the population risk), and she should continue to monitor for the disease. Also, a negative *BRCA* genetic test does not eliminate the possibility that she bears an inherited mutation in another gene that increases breast cancer risk or that *BRCA1* or *BRCA2* gene mutations exist in regions of the genes that are inaccessible to current genetic tests.

A woman whose test results are positive faces difficult choices. Her treatment options are poor, consisting of close monitoring, prophylactic mastectomy or oophorectomy (removal of breasts and ovaries respectively), and taking drugs such as tamoxifen. Prophylactic surgery reduces her risks, but does not eliminate them, as cancers can still occur in tissues that remain after surgery. Drugs such as tamoxifen reduce her risks, but have serious side effects. Genetic tests not only affect the patient but also affect the patient's entire family. People often experience fear, anxiety, and guilt on learning that they are carriers of a genetic disease. Studies show that people who refuse genetic test results often suffer from even more anxiety than those who opt to learn the results. Confidentiality is also a major concern. Patients fear that their genetic test results may be leaked to insurance companies or employers, jeopardizing their prospects for jobs or affordable health and life insurance. One study shows that a quarter of eligible patients refuse *BRCA* gene testing because of concerns about cost, confidentiality, and potential discrimination.

Genetic testing is such a new development that the health system has lagged behind the science. Because genetic testing has both psychological and medical ambiguities, genetic counseling is imperative for patients and their families. However, there are insufficient numbers of genetic counselors with experience in genetic testing, and even in the most qualified hands, issues are complex and difficult. Physicians often have limited knowledge of human clinical genetics and feel inadequate to advise their patients. The federal government and the insurance industries have yet to develop comprehensive policies concerning genetic tests and genetic information. Given the unclear interpretation of *BRCA* genetic tests, the relatively ineffective treatment options, and the potential for psychological and societal side effects, it is not surprising that only about 60 percent of familial breast cancer patients and their families decide to take the genetic tests.

The unanswered questions about *BRCA1* and *BRCA2* genetic testing are many and important. What cancer risks are associated with which mutations? Should all people have access to *BRCA* tests, or only those at high risk? How can we ensure that the high costs of genetic tests and counseling do not limit this new technology to only a portion of the population? As we develop genetic tests for more and more diseases over the next few decades our struggle with these issues will continue to grow.

References

Surbone, A. 2001. Ethical implications of genetic testing for breast cancer susceptibility. *Crit. Rev. in Onc./Hem.* 40:149–57.

Web Sites

Genetic Testing for BRCA1 and BRCA2: It's Your Choice [online]. National Institutes of Health. *http://cis.nci.nih.gov/fact/3_62.htm*

Chapter Summary

1. Cancer is a genetic disease, predominantly of somatic cells. About 1 percent of cancers are associated with germline mutations that increase the susceptibility to certain cancers.
2. Cancer cells show two basic properties: abnormal cell proliferation and a propensity to spread and invade other parts of the body (metastasis). Genes controlling these aspects of cellular function are either mutated or expressed inappropriately in cancer cells.
3. Cancers are clonal, meaning that all cells within a tumor originate from a single cell that contains a number of mutations.
4. The development of cancer is a multistep process, requiring mutations in several cancer-related genes.
5. Cancer cells show high rates of mutation, chromosomal abnormalities, and genomic instability. This leads to the accumulation of mutations in specific genes that control aspects of cell proliferation, apoptosis, and cell–cell contact.
6. Cancer cells have defects in the regulation of cell cycle progression, cell cycle checkpoints, and signal transduction pathways.
7. Proto-oncogenes are normal genes that promote cell growth and division. When proto-oncogenes are mutated or mis-expressed in cancer cells, they are known as oncogenes.
8. Tumor suppressor genes normally regulate cell cycle checkpoints and apoptosis. When tumor suppressor genes are mutated or inactivated, cells cannot correct DNA damage. This leads to accumulation of mutations that may cause cancer.
9. Inherited mutations in cancer-susceptibility genes are not sufficient to trigger cancer. A second somatic mutation, in the other copy of the gene, is necessary to trigger tumorigenesis. In addition, mutations in other cancer-related genes are necessary for the development of hereditary cancers.
10. RNA and DNA tumor viruses contribute to cancers by stimulating cells to proliferate. They do this by introducing new oncogenes, interfering with the cell's normal tumor suppressor gene products or stimulating the expression of a cell's proto-oncogenes.
11. Environmental agents such as chemicals, radiation, viruses, and chronic infections contribute to the development of cancer. The most significant environmental factors that affect human cancers are tobacco smoke, diet, and natural radiation.

Insights and Solutions

1. In disorders such as retinoblastoma, a mutation in one allele of the *RB1* gene can be inherited from the germline, causing an autosomal dominant predisposition to the development of eye tumors. To develop tumors, a somatic mutation in the second copy of the *RB1* gene is necessary, indicating that the mutation itself acts as a recessive trait. Given that the first mutation can be inherited, in what ways can a second mutational event occur?

Solution: In considering how this second mutation arises, we must look at several types of mutational events, including changes in nucleotide sequence and events that involve whole chromosomes or chromosome parts. Retinoblastoma results when both copies of the *RB1* locus are lost or inactivated. With this in mind, you must first list the phenomena that can result in a mutational loss or the inactivation of a gene.

One way the second *RB1* mutation can occur is by a nucleotide alteration that converts the remaining normal *RB1* allele to a mutant form. This alteration can occur through a nucleotide substitution or by a frameshift mutation caused by the insertion or deletion of nucleotides during replication. A second mechanism involves the loss of the chromosome carrying the normal allele. This event would take place during mitosis, resulting in chromosome 13 monosomy, and leaving the mutant copy of the gene as the only *RB1* allele. This mechanism does not necessarily involve loss of the entire chromosome; deletion of the long arm (*RB1* is on 13q) or an interstitial deletion involving the *RB1* locus and some surrounding material would have the same result. Alternatively, a chromosome aberration involving loss of the normal copy of the *RB1* gene might be followed by duplication of the chromosome carrying the mutant allele. Two copies of chromosome 13 would be restored to the cell, but the normal *RB1* allele would not be present. Finally, a recombination event followed by chromosome segregation could produce a homozygous combination of mutant *RB1* alleles.

2. Proto-oncogenes can be converted to oncogenes in a number of different ways. In some cases, the proto-oncogene itself becomes amplified up to hundreds of times in a cancer cell. This is the case for the *cyclin D1* gene, which is amplified in some cancers. In other cases, the proto-oncogene may be mutated in a limited number of specific ways leading to alterations in the gene product's structure. The *ras* gene is an example of a proto-oncogene that becomes oncogenic after suffering point mutations in specific regions of the gene. Explain why these two proto-oncogenes (*cyclin D1* and *ras*) undergo such different alterations in order to convert them into oncogenes.

Solution: The first step to solving this question is to understand the normal functions of these proto-oncogenes, and to think about how either amplification or mutation would affect each of these functions.

The cyclin D1 protein regulates progression of the cell cycle from G1 into S phase, by binding to CDK4 and activating this kinase. The cyclin D1/CDK4 complex phosphorylates a number of proteins including pRB that in turn activates other proteins in a cascade that results in transcription of genes whose products are necessary for DNA replication in S phase. The simplest way to increase the activity of cyclin D1 would be to increase the number of cyclin D1 molecules available for binding to the cell's endogenous CDK4 molecules. This can be accomplished by several mechanisms, including amplification of the *cyclin D1* gene. A point mutation in the *cyclin D1* gene would most likely interfere with the ability of the cyclin D1 protein to bind to CDK4; hence, mutations in the gene would probably repress cell cycle progression rather than stimulate it.

The *ras* gene product is a signal transduction protein that operates as an on/off switch in response to external stimulation by growth factors. It does so by binding either GTP (the on state) or GDP (the off state). Oncogenic mutations in the *ras* gene occur in specific regions that alter the ability of the Ras protein to exchange GDP for GTP. Oncogenic Ras proteins are locked in the on conformation, bound to GTP. In this way, they constantly stimulate the cell to divide. An amplification of the *ras* gene would simply provide more molecules of normal Ras protein, which would still be capable of on/off regulation. Hence, simple amplification of *ras* would be less likely to be oncogenic.

3. Explain why many oncogenic viruses contain genes whose products interact with tumor suppressor proteins.

Solution: In order to answer this question, it is useful to consider what viruses try to accomplish in cells, and what the roles of tumor suppressors are in normal cells.

The goal of oncogenic viruses is not to cause cancer, but to maximize the potential for viral replication. Viruses are relatively simple entities, consisting of only a nucleic acid genome—either RNA or DNA—and a relatively small number of structural and enzymatic proteins. They depend upon their host cells for much of the biosynthetic machinery and structural components necessary to replicate their genomes and assemble new virus particles. For example, many viruses require the host cell's RNA polymerase II enzyme, transcription factors, and ribonucleotide precursors in order to transcribe their viral genes. They also require components of the host cell's DNA replication machinery to replicate viral genomes. Hence, the ideal host cell for a virus infection is one that is within the cell cycle, preferably in late G1 to early S phase.

Most cells in a higher eukaryote such as humans are quiescent (in G0 phase). In order to stimulate the infected cell to enter the cell cycle and become primed for DNA replication, many viruses contain genes that encode growth-stimulating proteins. As we learned in this chapter, tumor suppressor proteins are those involved in either restraining the cell cycle at checkpoints, or in triggering the process of programmed cell death. Both of these functions are inhibitory to the goals of a typical virus; therefore, many viruses have evolved methods to inactivate tumor suppressors. One of the ways in which viral proteins can inactivate tumor suppressors is to bind to them and inhibit their functions. The tumor suppressors p53 and pRB are common targets of viral regulatory proteins, such as the E6 and E7 proteins of HPV 16 and 18. By inactivating tumor suppressors, these viruses are able to maintain the cell within the cell cycle. For the host, however, this growth stimulation in the absence of functional cell cycle checkpoints can lead to increased mutation accumulation and possible tumorigenesis.

Problems and Discussion Questions

1. As a genetic counselor, you are asked to assess the risk for a couple with a family history of retinoblastoma who are thinking about having children. Both the husband and wife are phenotypically normal, but the husband has a sister with familial retinoblastoma in both eyes. What is the probability that this couple will have a child with retinoblastoma? Are there any tests that you could recommend to help in this assessment?
2. What events occur in each phase of the cell cycle? Which phase is most variable in length?
3. Where are the major regulatory points in the cell cycle?
4. List the functions of kinases and cyclins, and describe how they interact to cause cells to move through the cell cycle.
5. (a) How does pRb function to keep cells at the G1 checkpoint? (b) How do cells get past the G1 checkpoint to move into S phase?
6. What is the difference between saying that cancer is inherited and saying that the predisposition to cancer is inherited?
7. Although tobacco smoking is responsible for a large number of human cancers, not all smokers develop cancer. Similarly, some people who inherit mutations in the tumor suppressor genes *p53* or *RB1* never develop cancer. Describe some reasons for these observations.
8. What is apoptosis and under what circumstances do cells undergo this process?
9. Define tumor suppressor genes. Why is a mutation in a single copy of a tumor suppressor gene expected to behave as a recessive gene?
10. In the Rous sarcoma virus (RSV) genome, the host-cell proto-oncogene is converted into an oncogene. How does this conversion occur?
11. Part of the Ras protein is associated with the plasma membrane, and part extends into the cytoplasm. How does the Ras protein transmit a signal from outside the cell into the cytoplasm? What happens in cases where the *ras* gene is mutated?
12. If a cell suffers damage to its DNA while in S phase, how can this damage be repaired before the cell enters mitosis?
13. Distinguish between oncogenes and proto-oncogenes. In what ways can proto-oncogenes be converted to oncogenes?
14. Of the two classes of genes associated with cancer, tumor suppressor genes and oncogenes, mutations in which group can be considered gain-of-function mutations? In which group are the loss-of-function mutations? Explain.
15. How do translocations such as the Philadelphia chromosome contribute to cancer?
16. Given that cancers can be environmentally induced and that some environmental factors are the result of lifestyle choices such as smoking, sun exposure, and diet, what percentage of the money spent on cancer research do you think should be devoted to research and education on preventing cancer rather than on finding cancer cures?
17. In CML, leukemic blood cells can be distinguished from other cells of the body by the presence of a functional BCR-ABL hybrid protein. Explain how this characteristic provides an opportunity to develop a therapeutic approach to a treatment for CML.
18. How do normal cells protect themselves from accumulating mutations in genes that could lead to cancer? How do cancer cells differ from normal cells in these processes?

19. Describe the difference between an acute transforming virus and a virus that does not cause tumors.
20. Explain how environmental agents such as chemicals and radiation cause cancer.
21. Radiotherapy (treatment with ionizing radiation) is one of the most effective current cancer treatments. It works by damaging DNA and other cellular components. In which ways could radiotherapy control or cure cancer, and why does radiotherapy often have significant side effects?
22. Assume that a young woman in a suspected breast cancer family takes the *BRCA1* and *BRCA2* genetic tests and receives negative results. That is, she does not test positive for the mutant alleles of *BRCA1* or *BRCA2*. Can she consider herself free of risk for breast cancer?
23. People with a genetic condition known as Li–Fraumeni syndrome inherit one mutant copy of the *p53* gene. These people have a high risk of developing a number of different cancers, such as breast cancer, leukemias, bone cancers, adrenocortical tumors, and brain tumors. Explain how mutations in one cancer-related gene can give rise to such a diverse range of tumors.
24. Explain the differences between a benign and malignant tumor.
25. As part of a cancer research project, you have discovered a gene that is mutated in many metastatic tumors. After determining the DNA sequence of this gene, you compare the sequence with those of other genes in the human genome sequence database. Your gene appears to code for an amino acid sequence that resembles sequences found in some serine proteases. Conjecture how your new gene might contribute to the development of highly invasive cancers.

Extra-Spicy Problems

26. A study by Bose and colleagues (1998. *Blood* 92:3362–67) and a previous study by Biernaux and others (1996. *Bone Marrow Transplant* 17: (Suppl. 3) S45–S47) showed that *BCR-ABL* fusion gene transcripts can be detected in 25 to 30 percent of healthy adults who do not develop chronic myelogenous leukemia (CML). Explain how these individuals can carry a fusion gene that is transcriptionally active yet do not develop CML.
27. Those who inherit a mutant allele of the *RB1* gene are at risk for developing a bone cancer called osteosarcoma. You suspect that in these cases, osteosarcoma requires a mutation in the second *RB1* allele and have cultured some osteosarcoma cells and obtained a cDNA clone of a normal human *RB1* gene. A colleague sends you a research paper revealing that a strain of cancer-prone mice develop malignant tumors when injected with osteosarcoma cells, and you obtain these mice. Using these three resources, what experiments would you perform to determine (a) whether osteosarcoma cells carry two *RB1* mutations, (b) whether osteosarcoma cells produce any pRB protein, and (c) if the addition of a normal *RB1* gene will change the cancer-causing potential of osteosarcoma cells?
28. The compound benzo[*a*]pyrene is found in cigarette smoke. This compound chemically modifies guanine bases in DNA. Such abnormal bases are usually removed by an enzyme that hydrolyzes the base, leaving an apurinic site. If such a site is left unrepaired, an adenine is preferentially inserted across from the apurinic site. In a study of lung cancer patients (Harris, A. 1991. *Nature* 350:377–78), tumor cells from 15 out of 25 patients had a G to T transversion in the *p53* gene, which has a known role in cancer formation. You are testifying as an expert witness in a court case in which the widow of a man who was a lifelong smoker and died of lung cancer is suing a company for manufacturing the tobacco products that killed her husband. What would you tell the jury? (Science only please—no personal expositions on lawyers or the legal system!)
29. Table 18.6 (p. 456) summarizes some of the data that have been collected on *BRCA1* mutations in families with a high incidence of both early-onset breast and ovarian cancer. Table 18.7 shows neutral polymorphisms found in control families (with no increased frequency of breast and ovarian cancer). (a) Note the coding effect of the mutation found in kindred group 2082 in Table 18.6. This results from a single base-pair substitution. Draw the normal double-stranded DNA sequence for this codon (with the 5′ and 3′ ends labeled), and show the sequence of events that generated this mutation, assuming that it resulted from an uncorrected mismatch event during DNA replication. (b) Examine the types of mutations that are listed in Table 18.6, and determine if the *BRCA1* gene is likely to be a tumor suppressor gene or an oncogene. (c) Although the mutations in Table 18.6 are clearly deleterious and cause breast cancer in women at very young ages, each of the kindred groups had at least one woman who carried the mutation but lived until age 80 without developing cancer. Name at least two different mechanisms (or variables) that could underlie variation in the expression of a mutant phenotype and propose an explanation for the incomplete penetrance of this mutation. How do these mechanisms or variables relate to this explanation?
30. Examine Table 18.7 (p. 456). (a) What is meant by a neutral polymorphism? (b) What is the significance of this table in the context of examining a family or population for *BRCA1* mutations that predispose an individual to cancer? (c) Is the PM2 polymorphism likely to result in a neutral missense mutation or a silent mutation? (d) Answer part (c) for the PM3 polymorphism.

TABLE 18.6 PREDISPOSING MUTATIONS IN *BRCA1*

		Mutation		
Kindred	Codon	Nucleotide Change	Coding Effect	Frequency in Control Chromosomes
1901	24	−11 bp	Frameshift or splice	0/180
2082	1313	C → T	Gln → Stop	0/170
1910	1756	Extra C	Frameshift	0/162
2099	1775	T → G	Met → Arg	0/120
2035	NA*	?	Loss of transcript	NA*

Source: 1994. *Science* 266:66–71.

*NA indicates not applicable, as the regulatory mutation is inferred, and the position has not been identified.

TABLE 18.7 NEUTRAL POLYMORPHISMS IN *BRCA1*

			*Frequency in Control Chromosomes**			
Name	Codon Location	Base in Codon†	A	C	G	T
PM1	317	2	152	0	10	0
PM6	878	2	0	55	0	100
PM7	1190	2	109	0	53	0
PM2	1443	3	0	115	0	58
PM3	1619	1	116	0	52	0

*The number of chromosomes with a particular base at the indicated polymorphic site (A, C, G, or T) is shown.

† Position 1, 2, or 3 of the codon.

SELECTED READINGS

Ames, B.N., Gold, L.S., and Willett, W.C. 1995. The causes and prevention of cancer. *Proc. Natl. Acad. Sci. USA* 92:5258–65.

Bernards, R., and Weinberg, R.A. 2002. A progression puzzle. *Nature* 418:823.

Brown, M.A. 1997. Tumor suppressor genes and human cancer. *Adv. Genet.* 36:45–135.

Cavenee, W.K., and White, R.L. 1995. The genetic basis of cancer. *Sci. Am.* (Mar.) 272:72–79.

Compagni, A., and Christofori, G. 2000. Recent advances in research on multistage tumorigenesis. *Brit. J. Cancer* 83:1–5.

Cornelis, J.F., et al. 1998. Metastasis. *Am. Scient.* 86:130–41.

Fearon, E.R. 1997. Human cancer syndromes: Clues to the origin and nature of cancer. *Science* 278:1043–50.

Futreal, P.A., et al. 2004. A census of human cancer genes. *Nature Reviews Cancer* 4:177–83.

Hartwell, L.H., and Kastan, M.B. 1994. Cell cycle control and cancer. *Science* 266:1821–27.

Lengauer, C., Kinzler, K.W., and Vogelstein, B. 1997. Genetic instability in colorectal cancer. *Nature* 386:623–27.

Nurse, P. 1997. Checkpoint pathways come of age. *Cell* 91:865–67.

Raff, M. 1998. Cell suicide for beginners. *Nature* 396:119–22.

Sherr, C. J. 1996. Cancer cell cycles. *Science* 274:1672–77.

Vogelstein, B., and Kinzler, K.W. 1993. The multistep nature of cancer. *Trends in Genetics* 9:138–41.

Weinberg, R.A. 1995. The retinoblastoma protein and cell cycle control. *Cell* 81:323–30.

Weinberg, R.A. 1996. How cancer arises. *Sci. Am.* (Sept.) 275:62–70.

Recombinant DNA Technology

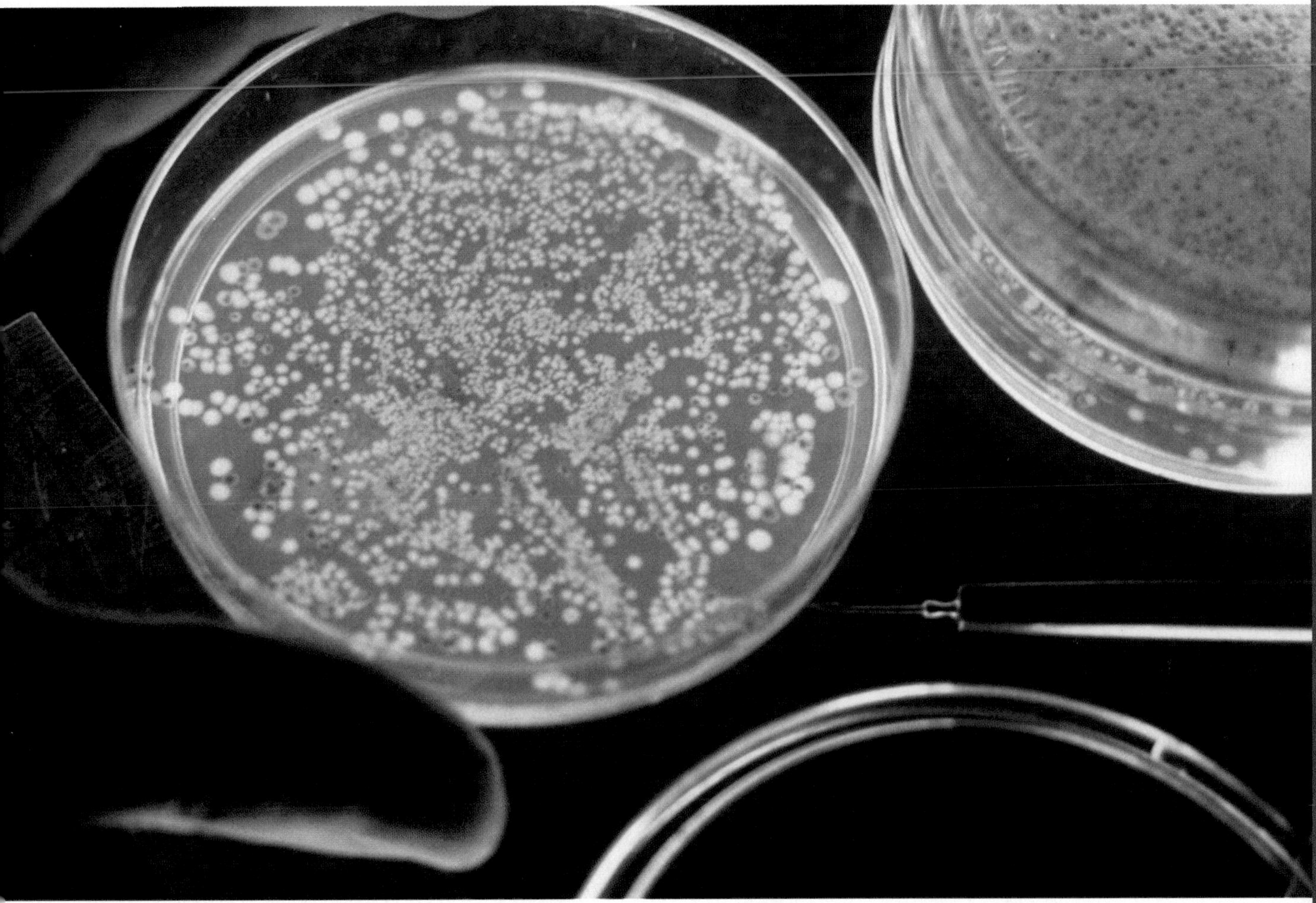

A Petri dish revealing the growth of host cells after uptake of recombinant plasmids.

Chapter Concepts

- Recombinant DNA technology creates artificial combinations of DNA molecules.
- Recombinant DNA technology depends in part on the ability to cleave and rejoin DNA segments at specific base sequences.
- Recombinant DNA molecules usually consist of DNA from two different sources, most often representing DNA from different species.
- The most useful application of recombinant DNA technology is in cloning a DNA segment of interest.
- Cloning inserts specific DNA segments into vectors (such as plasmids), which are transferred into host cells (such as bacterial cells), where the recombinant molecules replicate as the host cells divide.
- Recombinant molecules can be cloned in prokaryotic or eukaryotic host cells.
- DNA segments can be quickly and efficiently amplified thousands of times using the Polymerase Chain Reaction (PCR).
- Cloned DNA segments are characterized in several ways, the most specific being DNA sequencing.
- Recombinant DNA technology has revolutionized our ability to investigate the genomes of diverse species.

In 1971, a paper published by Kathleen Danna and Daniel Nathans marked the beginning of the recombinant DNA era. The paper described the isolation of an enzyme from a bacterial strain and the use of the enzyme to cleave viral DNA at specific nucleotide sequences. It contained the first published photograph of DNA cut with one of these proteins, called a restriction enzyme. Using restriction enzymes and a number of other resources, techniques to create, replicate, and analyze recombinant DNA molecules were developed in the mid to late 1970s. This set of methods, called recombinant DNA technology, was a major advance in research, and allowed scientists to isolate and study specific DNA sequences. For their contributions to the development of this technology, Nathans, Hamilton Smith, and Werner Arber were awarded the 1978 Nobel Prize for Physiology or Medicine.

Recombinant DNA technology makes it possible to identify and isolate a single gene from the thousands or tens of thousands present in a genome and to produce large quantities of this gene in the form of cloned DNA molecules. In addition, the cloned gene can be transferred into cells that will synthesize its encoded gene product. This product can then be recovered and purified for use in research, medicine, or industry.

Clones are identical organisms, cells, or molecules descended from a single ancestor. Cloning a gene produces many identical copies that can be used for numerous purposes, including research into its structure and organization or the commercial production of its encoded protein. In this chapter, we review the basic methods of recombinant DNA technology used to isolate, replicate, and analyze genes. In Chapter 22, we will discuss some applications of this technology to research, medicine, the legal system, agriculture, and industry.

19.1 Recombinant DNA Technology Combines Several Experimental Techniques

The term **recombinant DNA** refers to a combination of DNA molecules that are not found together in nature. Although genetic processes such as crossing over produce recombined DNA molecules, the term *recombinant DNA* is generally reserved for molecules produced by joining DNA obtained from different biological sources.

Recombinant DNA technology uses methods derived from nucleic acid biochemistry, coupled with genetic techniques originally developed for the study of bacteria and viruses. This technology is a powerful tool for producing potentially unlimited quantities of a gene. Although several methods are available, the basic procedure involves the following steps:

1. DNA to be cloned is purified from cells or tissues.
2. Proteins called **restriction enzymes** are used to generate specific DNA fragments. These enzymes recognize and cut DNA molecules at specific nucleotide sequences.
3. The fragments produced by restriction enzymes are joined to other DNA molecules that serve as vectors, or carrier molecules. A vector joined to a DNA fragment is a **recombinant DNA molecule**.
4. The recombinant DNA molecule is transferred to a host cell. Within the host cell, the recombinant molecule replicates, producing dozens of identical copies, or clones, of the recombinant molecule.
5. As host cells replicate, the recombinant DNA molecules within them are passed on to all their progeny, creating a population of host cells, each of which carries copies of the cloned DNA sequence.
6. The cloned DNA can be recovered from host cells, purified, and analyzed.
7. The cloned DNA can then be transcribed, its mRNA translated, and the encoded gene product isolated and used for research or sold commercially.

19.2 Recombinant DNA Technology Is the Foundation of Genome Analysis

Recombinant DNA and gene-cloning technology provides scientists with methods for isolating large quantities of specific genes or other DNA sequences from large and complex genomes. For example, the human genome contains more than 3 billion nucleotides and 25,000 to 30,000 genes. Restriction enzymes cut the genome into smaller fragments that can be manipulated, separated, copied, and studied individually. This allows researchers to investigate many aspects of gene organization and function, factors that regulate gene expression as well as the nature and functions of encoded proteins. We will begin our discussion of this technology by considering two important components used to construct recombinant DNA molecules.

19.3 Restriction Enzymes Cut DNA at Specific Recognition Sequences

Restriction enzymes are produced by bacteria as a defense mechanism against infection by viruses. More than 200 restriction enzymes have been identified, and about 100 of these are commonly used by researchers. Each restriction enzyme binds to DNA (Figure 19–1) and recognizes a specific nucleotide sequence called a **recognition sequence**. The enzyme cuts both strands of the DNA within the recognition sequence in a specific **cleavage pattern**. Restriction enzymes' usefulness in cloning derives from their ability to accurately and reproducibly cut genomic DNA into fragments called **restriction fragments**. The size of restriction fragments is determined by how often a given restriction enzyme cuts the DNA. Enzymes with a four base recognition sequence such as *Alu*I (AGCT) will cut, on average, every 256 base pairs ($4^n = 4^4 = 256$) if all four nucleotides are present in equal proportions, producing many small fragments. Enzymes like *Not*I have an eight base recognition sequence (GCGGCCGC) and cut the DNA on average every 65,500 base pairs ($4^n = 4^8$), producing fewer, larger fragments. The actual fragment sizes produced by cutting DNA with a given restriction enzyme vary because the number and location of recognition sequences are not always distributed randomly in DNA.

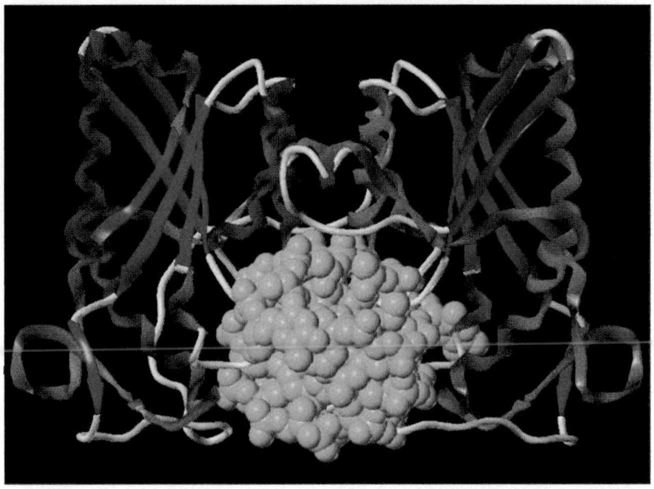

FIGURE 19–1 The restriction enzyme *Bam*H1 bound to a DNA molecule (green). Restriction enzymes cut DNA at specific sites.

How Do We Know?

How do we know whether a restriction enzyme will cut DNA frequently or infrequently when all bases are equally represented?

Most recognition sequences contain a form of symmetry in which the nucleotide sequence reads the same on both strands of the DNA when read in the 5′ to 3′ direction (a **palindrome**). Each restriction enzyme binds to DNA at a specific recognition sequence, and cuts the DNA in a characteristic cleavage pattern. Many restriction enzymes cut the DNA strands in an offset way, producing fragments with single-stranded tails, or cut both strands across a line of symmetry in the palindromic recognition sequence to produce blunt ends. Some restriction enzymes and their recognition sequences are shown in Figure 19–2.

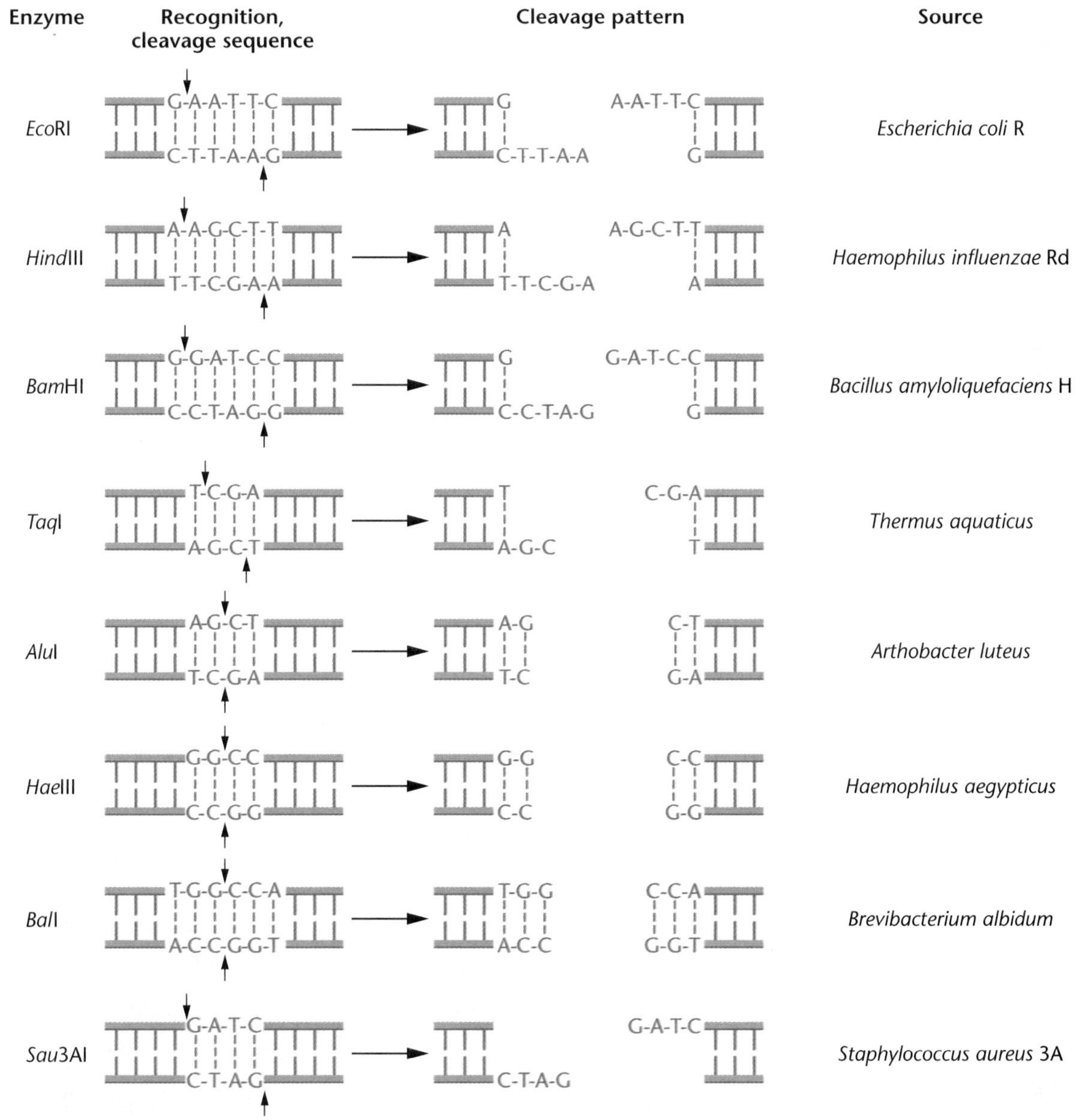

FIGURE 19–2 Some common restriction enzymes, with their recognition sequences, cleavage patterns, and sources.

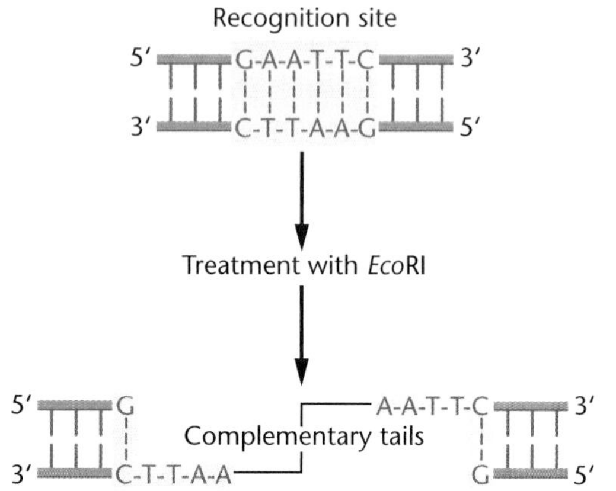

FIGURE 19–3 The restriction enzyme *Eco*RI recognizes and binds to the palindromic nucleotide sequence GAATTC. Cleavage of the DNA at this sequence produces complementary single-stranded tails. These single-stranded tails can anneal with single-stranded tails from other DNA fragments to form recombinant DNA molecules.

One of the first restriction enzymes to be identified was isolated from *Escherichia coli* strain R and is designated *Eco*RI (pronounced echo-r-one). Its six-nucleotide recognition sequence and cleavage pattern are shown in Figure 19–3. DNA fragments produced by *Eco*RI digestion have overhanging single-stranded tails ("sticky ends") that can form hydrogen bonds with complementary single-stranded tails on DNA fragments from any other source. If they are mixed under the proper conditions, DNA fragments from two sources form recombinant molecules by hydrogen bonding of their sticky ends. The fragments can be covalently linked to form recombinant DNA molecules by the enzyme **DNA ligase** (Figure 19–4).

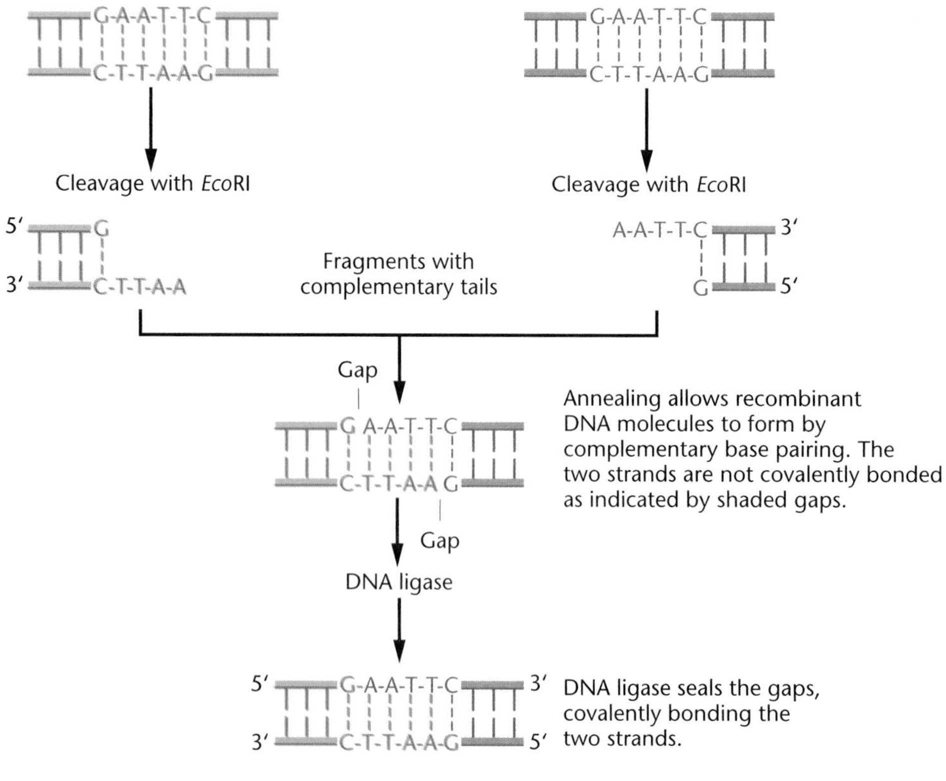

FIGURE 19–4 DNA from different sources is cleaved with *Eco*RI and mixed to allow annealing to form recombinant molecules. The enzyme DNA ligase then chemically bonds these annealed fragments into an intact recombinant DNA molecule.

19.4 Vectors Carry DNA Molecules to Be Cloned

DNA restriction fragments cannot directly enter host cells to be copied. However, when a restriction fragment is joined to another DNA molecule called a vector, it can gain entry to a host cell, where it can be replicated or cloned into many copies. Vectors are, in essence, carrier DNA molecules that transfer and replicate (clone) inserted DNA fragments. Many different vectors are available for cloning, with different host cell specificities, differences in the size of inserts that can be carried, and differences in other properties, such as the number of copies produced, number of recognition sequences available for cloning, and the number and type of marker genes.

To serve as a vector, a DNA molecule must be able to independently replicate itself and any DNA fragment it carries. The vector should also contain several recognition sequences to allow insertion of DNA fragments to be cloned. To insert a DNA fragment, the vector is cut with a restriction enzyme and mixed with a collection of DNA fragments produced by cutting with the same enzyme. Vectors carrying an inserted fragment are called **recombinant vectors**, and each is an example of a recombinant DNA molecule, produced by joining DNA from two different sources.

To distinguish host cells that have taken up vectors from host cells that do not contain vectors, the vector should carry a selectable marker gene (usually antibiotic resistance genes or genes for enzymes absent from the host cell). Finally, the vector and its inserted DNA fragment should be easy to recover from the host cell.

Plasmid Vectors

Genetically modified plasmids were the first vectors developed and are still widely used for cloning. These plasmid vectors were derived from naturally occurring, extrachromosomal, double-stranded DNA molecules that replicate autonomously within bacterial cells (Figure 19–5). The genetics of plasmids and their host bacterial cells have been discussed in Chapters 6 and 16. In this section, we will emphasize the use of plasmids as vectors for cloning DNA. Plasmids have been extensively modified by genetic engineering to serve as cloning vectors. Many genetically engineered plasmid vectors are now available with a range of useful features. For example, although only a single plasmid

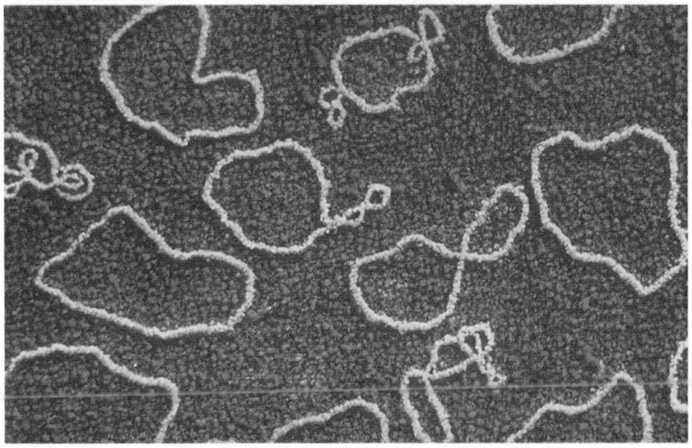

FIGURE 19–5 A color-enhanced electron micrograph of circular plasmid molecules isolated from *E. coli*. Genetically engineered plasmids are used as vectors for cloning DNA.

generally enters a bacterial host cell, once inside, some plasmids can increase their copy number so that several hundred copies are present. As vectors, these plasmids allow many copies of cloned DNA to be produced. For added convenience, these vectors have also been genetically engineered to contain a number of convenient restriction enzyme recognition sequences and marker genes that reveal their presence in host cells.

One such plasmid is **pUC18** (Figure 19–6), which has several useful features as a vector.

1. It is small (2686 bp), so it can carry relatively large DNA inserts.

2. It has an origin of replication and can produce up to 500 copies of inserted DNA fragments per cell.

3. A large number of restriction enzyme recognition sequences have been engineered into pUC18, conveniently clustered in one region called a **polylinker**.

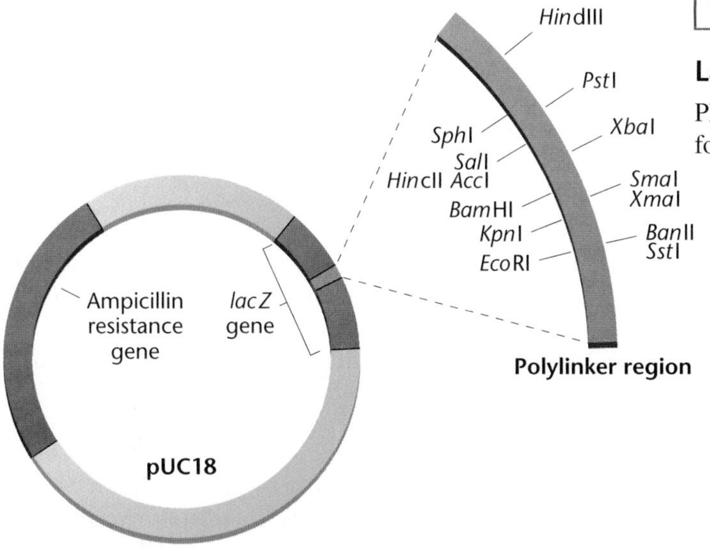

FIGURE 19–6 A diagram of the plasmid pUC18 showing the polylinker region, located within a *lacZ* gene. DNA inserted into the polylinker region disrupts the *lacZ* gene, resulting in white colonies that allow direct identification of bacterial colonies carrying cloned DNA inserts.

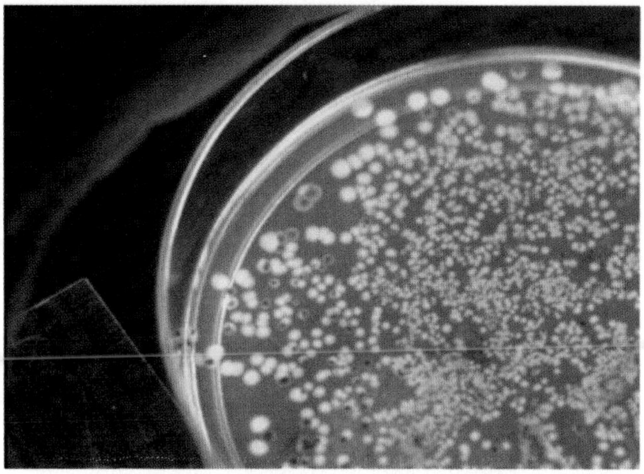

FIGURE 19–7 A Petri dish showing the growth of bacterial cells after uptake of recombinant plasmids. The medium on the plate contains a compound called Xgal. DNA inserts into the pUC18 vector disrupt the gene responsible for the formation of blue colonies. Cells in the blue colonies do not carry any cloned DNA inserts, whereas the white colonies contain vectors carrying DNA inserts.

4. It allows recombinant plasmids to be easily identified. For example, pUC18 carries a fragment of the bacterial *lacZ* gene, and the polylinker is inserted into this fragment. Expression of *lacZ* causes bacterial host cells carrying pUC18 to produce blue colonies when grown on medium containing a compound known as Xgal. If a DNA fragment is inserted into the polylinker, the *lacZ* gene is inactivated and a bacterial cell carrying pUC18 with an inserted DNA fragment forms white colonies, making them easy to identify (Figure 19–7).

HOW DO WE KNOW?

How do we know when DNA fragments have been successfully incorporated into plasmid vectors?

Lambda (λ) Phage Vectors

Plasmid vectors generally carry up to 10 kb of inserted DNA, but for many experiments, larger pieces of DNA are necessary. For these purposes, genetically modified strains of **λ phage** are used as vectors (Figure 19–8). The genome of λ phage has been completely mapped and sequenced. For use as a vector, the central third of its chromosome can be replaced with foreign DNA without affecting the phage's ability to infect cells and form plaques (Figure 19–9).

To clone DNA using this vector, the phage DNA is purified and cut with a restriction enzyme such as *Eco*RI, producing three chromosomal fragments: the left arm, the right arm, and the dispensable central region. The arms are isolated and mixed with DNA from another source that also has been cut with *Eco*RI. Ligation with DNA ligase produces recombinant vectors. Recombinant λ vectors are packaged into phage protein heads *in vitro* and introduced into bacterial host cells. Inside the bacteria, the vectors replicate and form many copies of infective phage, each of which carries a DNA insert. As they reproduce, they lyse their bacterial

Web Tutorial 19.2
Recombinant DNA Technology: Vectors

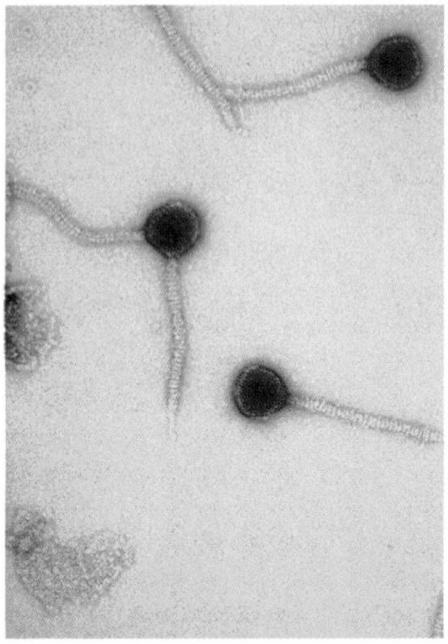

FIGURE 19–8 A colorized electron micrograph of phage λ, widely used as a vector in recombinant DNA work.

host cells, forming clear spots—known as plaques—on Petri plates, from which the cloned DNA can be recovered.

Phage vectors can carry inserts of about 20 kb, more than twice as long as DNA inserts in plasmid vectors. This is an important advantage when cloning large genes or small genomes. In addition, some phage vectors accept only inserts of a minimum size; they do not carry relatively useless small inserts only a few dozen or a few hundred nucleotides in length.

Cosmid Vectors

Cosmids are hybrid vectors created by combining parts of the lambda chromosome with parts of plasmids. Cosmids contain the *cos* sequence of phage lambda, necessary for packing phage DNA into phage protein coats and have plasmid sequences necessary for replication. They also contain a plasmid-derived antibiotic resistance gene (Figure 19–10) to help identify host cells carrying recombinant cosmids. After insertion of DNA fragments, recombinant cosmids are packaged into lambda protein heads, forming infective phage particles. Once inside a bacterial host cell, the cosmid replicates as a plasmid. Because only a very small part of the lambda genome has been retained, cosmids can carry DNA inserts that are much larger than those carried by lambda vectors. Cosmids can carry almost 50 kb of inserted DNA, whereas phage vectors can accommodate DNA inserts 10–15 kb in length.

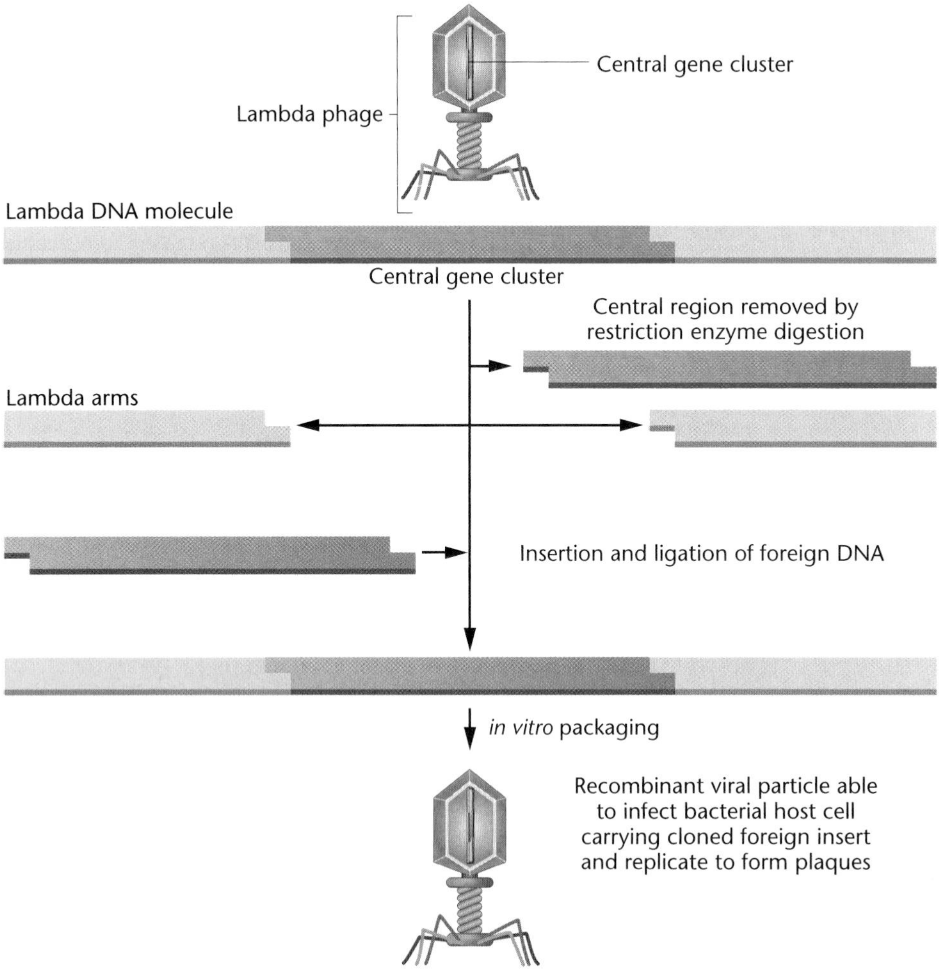

FIGURE 19–9 λ phage as a vector. DNA is extracted from the phage, the central gene cluster is removed, and the DNA to be cloned is ligated into the arms of the λ phage chromosome. The recombinant chromosome is then packaged into phage proteins to form a recombinant virus.

Other hybrid vectors with origins of replication derived from different sources (e.g., animal viruses such as SV40) can replicate in more than one type of host cell, and are called shuttle vectors. These vectors usually contain genetic markers that are selectable in both types of host cells and are used to shuttle DNA inserts between *E. coli* and another type of host cell, such as yeast. Often such vectors are employed in studying gene expression.

Bacterial Artificial Chromosomes

Mapping and analyzing large complex eukaryotic genomes require cloning vectors that can carry very large DNA fragments. In addition, since some human genes are 1000 kb to more than 2000 kb long, vectors with large cloning capacities are necessary to carry these genes.

A vector with a large cloning capacity is based on the fertility plasmid (F factor) of bacteria and is called a **bacterial artificial chromosome (BAC)**. Recall from Chapter 6 that F factors are independently replicating plasmids that transfer genetic information during bacterial conjugation. Because F factors can carry fragments

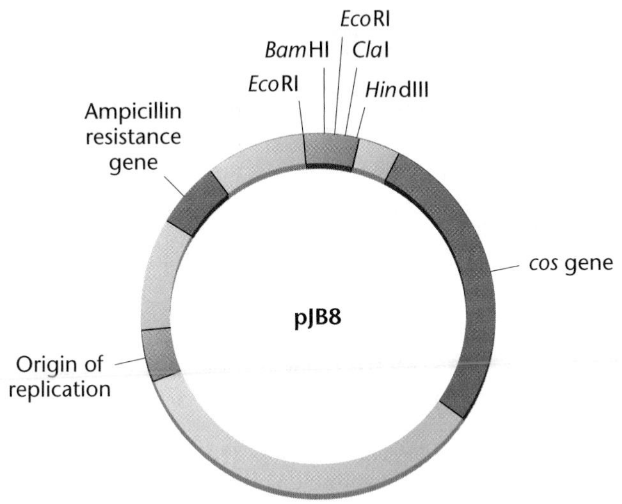

FIGURE 19–10 The cosmid pJB8 contains a bacterial origin of replication (*ori*), a single cos sequence (*cos*), an ampicillin resistance gene (*amp*, for selection of colonies that have taken up the cosmid), and a region containing four sites for cloning (*Bam*HI, *Eco*RI, *Cla*I, and *Hind*III). Because the vector is small (5.4 kb long), it can accept foreign DNA segments between 33 and 46 kb in length. The *cos* sequence allows cosmids carrying large inserts to be packaged into lambda viral coat proteins as though they were viral chromosomes. The viral coats carrying the cosmid can be used to infect a suitable bacterial host, and the vector, carrying a DNA insert, will be transferred into the host cell. Once inside, the *ori* sequence allows the cosmid to replicate as a bacterial plasmid.

of the bacterial chromosome up to 1 Mb in length, they have been engineered to act as vectors for eukaryotic DNA and can carry inserts of about 300 kb (Figure 19–11). BAC vectors carry F factor genes for replication and copy number, and have at least one antibiotic resistance marker as well as a polylinker containing a number of clustered restriction enzyme recognition sequences for inserting foreign DNA. In addition, the polylinker is flanked by promoter sequences that can be used to generate RNA molecules for the expression of the cloned gene, for use as probes in chromosome walking, and for sequencing the cloned insert.

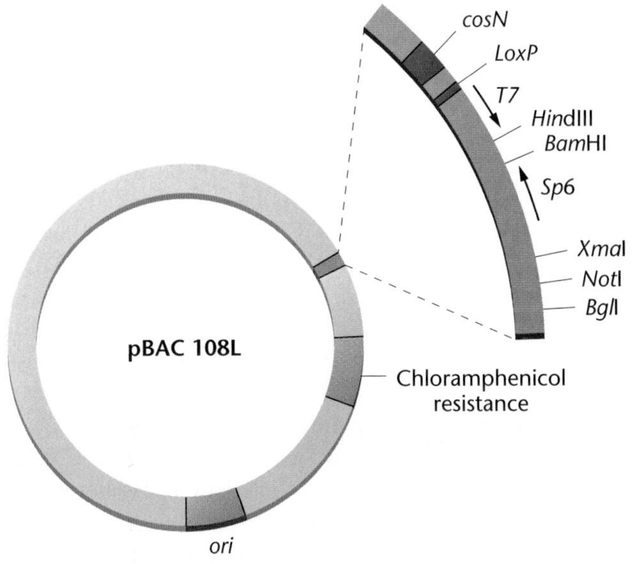

FIGURE 19–11 A bacterial artificial chromosome (BAC). The polylinker carries a number of unique sites for the insertion of foreign DNA. The arrows labeled T7 and Sp6 are promoter regions that allow expression of genes cloned between these regions.

Expression Vectors

Expression vectors are engineered to produce many copies of a selected protein in a host cell. Expression vectors are available for both prokaryotic and eukaryotic host cells. An expression vector for use in an *E. coli* host cell is pET (Figure 19–12). To use this vector, the gene to be expressed is cloned into a recognition sequence in the polylinker, placing the gene adjacent to a T7 viral promoter and the bacterial *lac* operator. The host cell genome has been modified to carry the gene for the T7 viral polymerase, the *lac* promoter, and the *lac* operator. After recombinant plasmids are inserted into host cells, expression is induced by adding the lactose analog IPTG to the medium. IPTG displaces repressor from the *lac* operator, activating the T7 polymerase gene on the bacterial chromosome *and* the gene in the polylinker. T7 polymerase binds to the T7 promoter and transcribes the gene in the polylinker, producing large quantities of the encoded protein.

19.5 DNA Was First Cloned in Prokaryotic Host Cells

As discussed earlier, scientists use recombinant DNA technology to construct *and* replicate recombinant DNA molecules to make cloned copies of specific sequences. Replication takes place after transfer of recombinant molecules into host cells. This cloning method, known as cell-based cloning, was the first method developed, and is widely used to make cloned DNA.

One of the most commonly used prokaryotic hosts is a laboratory strain of the bacterium *E. coli* known as K12. *E. coli* strains such as K12 are genetically well characterized, and can host a wide range of vectors. The following steps are required to create recombinant DNA molecules and transfer them to an *E. coli* host cell (Figure 19–13) where they are cloned:

1. The DNA to be cloned is isolated and treated with a restriction enzyme to create fragments ending in a specific sequence.
2. The fragments are ligated to plasmid molecules that have been cut with the same restriction enzyme, creating a recombinant vector.
3. The recombinant vector is transferred into an *E. coli* host cell, where the recombinant plasmid replicates to form dozens of copies.
4. The bacteria are plated on nutrient medium, where they form colonies, and are screened to identify those that have taken up the recombinant plasmids.

Because the cells in each colony are derived from a single ancestral cell, all the cells in the colony and the plasmids they contain are genetically identical clones. Similarly, phages containing foreign DNA are used to infect *E. coli* host cells, and when plated on solid medium, each resulting plaque represents a cloned descendant of a single ancestral bacteriophage.

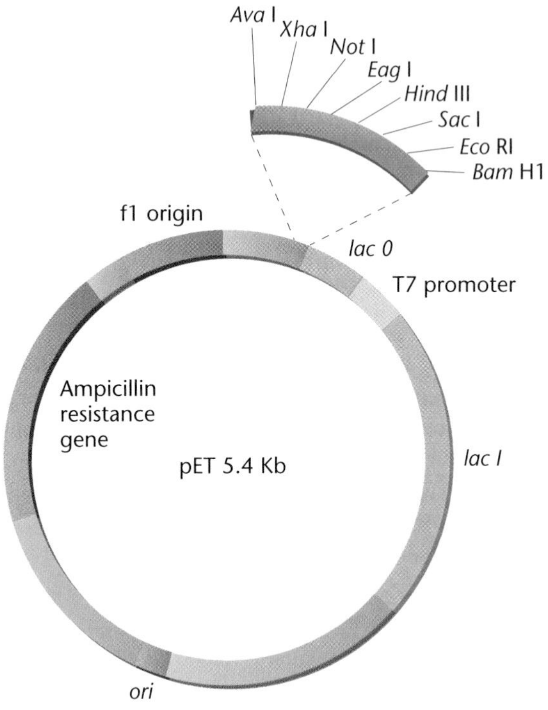

FIGURE 19–12 A pET expression vector. This system uses a genetically engineered host cell. The host cell carries the viral T7 RNA polymerase gene under the control of a *lac* promoter and operator, making it inducible by the lactose analog IPTG. For expression of a target gene, pET vectors carrying the target gene are inserted into host cells. Growth of the host on IPTG derepresses the T7 RNA polymerase gene and the target gene in the pET vector (which is under *lac O* control), leading to expression of the target gene. This system combines a strong promoter with tight regulation, and expression only occurs in the presence of IPTG.

Now solve this

Question 19 on page 481 relates to preparing a library in a vector carrying two antibiotic resistance genes.

Hint: Inserting foreign DNA into the vector disrupts one of the resistance genes in the plasmid. Bacteria taking up plasmids will be able to grow on medium containing only one of the antibiotics.

19.6 Yeast Cells Are Used as Eukaryotic Hosts for Cloning

The yeast *Saccharomyces cerevisiae* is widely used as a host cell for the cloning and expression of eukaryotic genes for five reasons. (1) Although yeast is a eukaryotic organism, it can be grown and manipulated in much the same way as bacterial cells. (2) The genetics of yeast has been intensively studied, providing a large catalog of mutations and a highly developed genetic map. (3) The entire yeast genome has been sequenced, and most genes in the organism have been identified. (4) To study the function of some eukaryotic proteins, it is necessary to use a host cell that can posttranslationally modify the protein so it folds into a functional form. Bacterial host cells cannot carry out these modifications, and rapidly degrade misfolded proteins. (5) Yeast has been used for centuries in the baking and brewing industries and is considered to be a safe organism for producing proteins for vaccines and therapeutic agents. Table 19.1 lists some of the products of cloning in yeast.

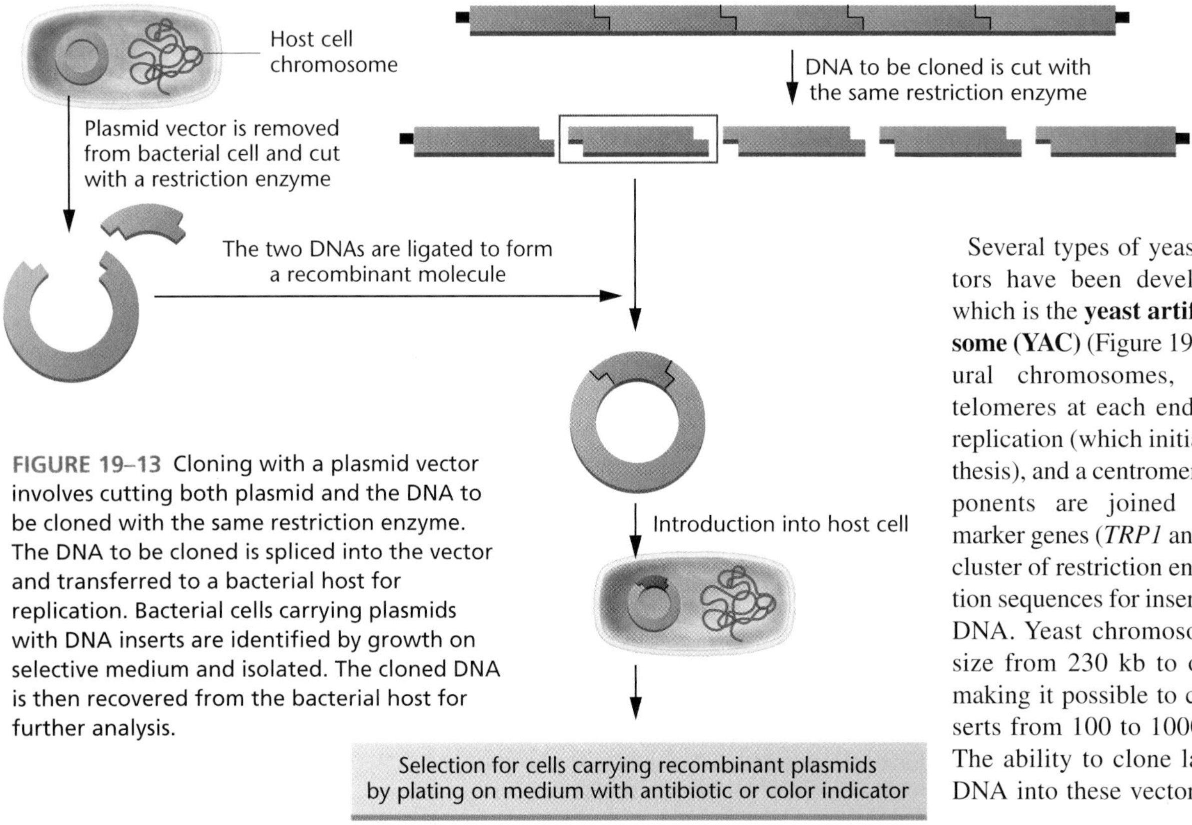

FIGURE 19–13 Cloning with a plasmid vector involves cutting both plasmid and the DNA to be cloned with the same restriction enzyme. The DNA to be cloned is spliced into the vector and transferred to a bacterial host for replication. Bacterial cells carrying plasmids with DNA inserts are identified by growth on selective medium and isolated. The cloned DNA is then recovered from the bacterial host for further analysis.

Several types of yeast cloning vectors have been developed, one of which is the **yeast artificial chromosome (YAC)** (Figure 19–14). Like natural chromosomes, a YAC has telomeres at each end, an origin of replication (which initiates DNA synthesis), and a centromere. These components are joined to selectable marker genes (*TRP1* and *URA3*) and a cluster of restriction enzyme recognition sequences for insertion of foreign DNA. Yeast chromosomes range in size from 230 kb to over 1900 kb, making it possible to clone DNA inserts from 100 to 1000 kb in YACs. The ability to clone large pieces of DNA into these vectors makes them

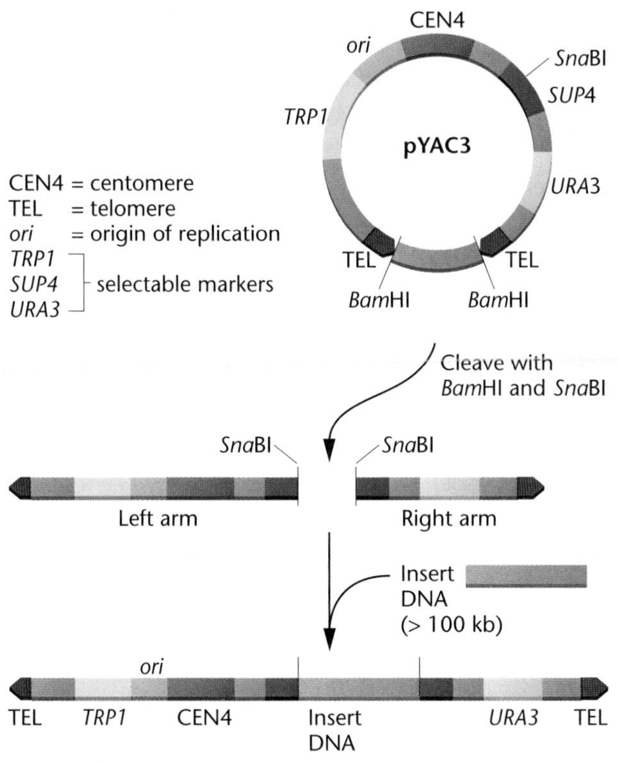

FIGURE 19–14 The yeast artificial chromosome pYAC3 contains telomere sequences (TEL), a centromere (CEN4) derived from yeast chromosome 4, and an origin of replication (*ori*). These elements give the cloning vector the properties of a chromosome. *TRP1* and *URA3* are yeast genes that are selectable markers for the left and right arms of the chromosome. Within the *SUP4* gene is a restriction enzyme recognition sequence for the enzyme *Sna*B1. Two *Bam*H1 recognition sequences flank a spacer segment. Cleavage with *Sna*B1 and *Bam*H1 breaks the artificial chromosome into two arms. The DNA to be cloned is treated with *Sna*B1 producing a collection of fragments. The arms and fragments are ligated together, and the artificial chromosome is inserted into yeast host cells. Because yeast chromosomes are large, the artificial chromosome accepts inserts in the million base-pair range.

an important tool in genome projects, including the Human Genome Project. (See Chapter 20.)

Although cloning in yeast vectors and host cells is currently the most advanced eukaryotic system used, other systems, including human artificial chromosome vectors using mammalian cells as hosts, are being developed.

TABLE 19.1 **RECOMBINANT PROTEINS SYNTHESIZED IN YEAST CELLS**

Hepatitis B virus surface protein
Malaria parasite protein
Epidermal growth factor
Platelet-derived growth factor
α_1-antitrypsin
Clotting factor XIIIA

19.7 Genes Can Be Transferred to Eukaryotic Cells

In addition to yeast, both plant and animal cells can take up DNA from their environment. Moreover, several different types of vectors, including YACs, can be used to transfer DNA into eukaryotic cells. When the vector is a plasmid, DNA transfer is referred to as **transformation**. When the vector is a virus, the term **transfection** is used to describe uptake.

Plant Cell Hosts

Gene transfer into higher plants uses bacterial plasmid vectors. The soil bacterium *Agrobacterium tumifaciens* infects plant cells and produces tumors (called plant galls) in many species of plants. Tumor formation is associated with the presence of a tumor-inducing (Ti) plasmid carried in the bacteria (Figure 19–15). When Ti plasmid-carrying bacteria infect plant cells, a segment of the Ti plasmid, known as *T-DNA*, is transferred into the genome of the host plant cell. Genes in the T-DNA segment control tumor formation and the synthesis of compounds required for growth of the infecting bacteria. Foreign genes can be inserted into the T-DNA segment and the recombinant plasmid transferred into plant cells by infection with *A. tumifaciens*. Once inside the cell, the foreign DNA is inserted into the plant genome when the T-DNA integrates into a host cell chromosome. Plant cells carrying a recombinant Ti plasmid can be grown in tissue culture to form a cell mass called a *callus*. By changing the culture medium, the callus can be induced to form roots and shoots, and eventually a mature plant carrying a foreign gene.

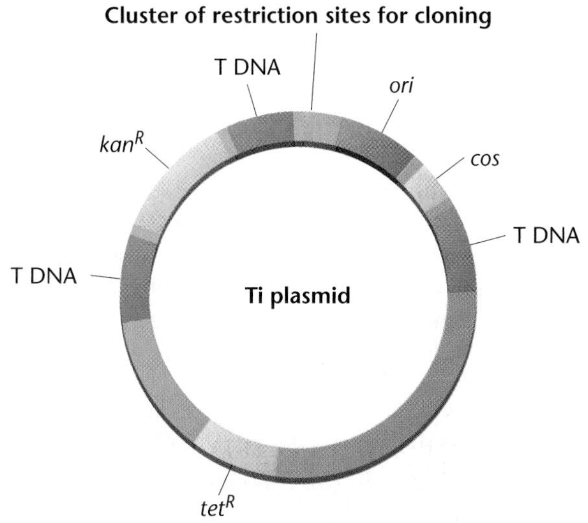

FIGURE 19–15 A Ti plasmid designed for cloning in plants. Segments of T DNA, including those necessary for integration, are combined with bacterial segments that incorporate cloning sites and antibiotic resistance genes (kan^R and tet^R). The vector also contains an origin of replication (*ori*) and a lambda *cos* sequence that permits recovery of cloned inserts from the host plant cell.

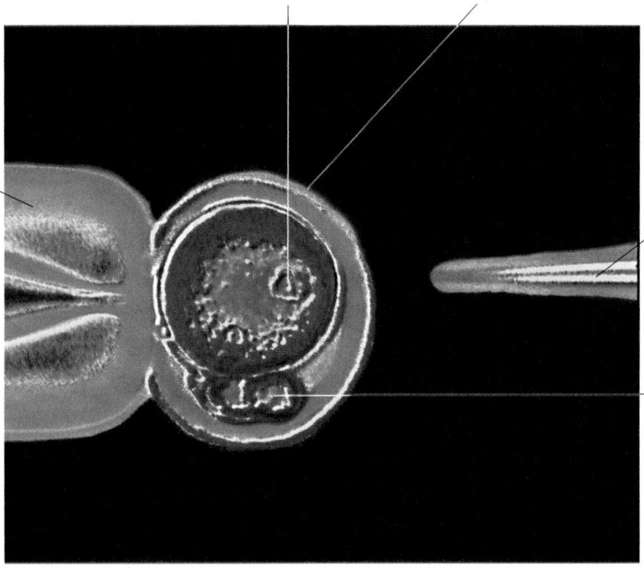

FIGURE 19–16 Cloned DNA can be transferred in mammals by direct injection into the oocytes.

Plants (or animals) carrying a foreign gene are called **transgenic** organisms. In Chapter 22, we will see how gene transfer has been used to alter crop plants.

Mammalian Cell Hosts

DNA can be transferred into mammalian cells by several methods, including endocytosis, or encapsulation of DNA into artificial membranes (liposomes), followed by fusion with cell membranes. DNA can also be transferred using YACs and vectors based on retroviruses. DNA introduced into a mammalian cell by any of these methods is usually integrated into the host genome. Genes transferred into fertilized eggs are used to produce transgenic animals. These same methods are also used to replicate cloned genes using mammalian cells as hosts.

YACs are used as vectors for several reasons, one of which is to increase the efficiency of gene transfer into the germ line of mice. The first step in producing transgenic mice involves transferring a recombinant YAC into the nucleus of an appropriate mouse cell, such as a fertilized egg or an embryonic stem cell, followed by the integration of the DNA into a chromosome.

YACs are transferred to mice in several ways. The first uses microinjection of purified YAC DNA into the nucleus of a mouse oocyte (Figure 19–16). Transgenic zygotes are then implanted in foster mothers for development. YACs are also transferred into mouse embryonic stem (ES) cells. This can be done by fusing a yeast cell carrying a YAC with a mouse stem cell, transferring the YAC and all or most of the yeast genome into the stem cell. These transgenic ES cells are injected into early stage mouse embryos, where they participate in the formation of adult tissues, including those that form germ cells. The ability to transfer large DNA segments into mice has applications in many areas of research. Some of these are described in Chapter 22.

Other vectors for mammalian cells are based on genetically engineered avian and mouse retroviruses. These retroviruses have single-stranded RNA molecules as their genomes. After infection of the host cell, the RNA is transcribed by reverse transcriptase into a double-stranded DNA (dsDNA) molecule. The dsDNA integrates into the host genome and is passed on to daughter cells during cell division. The retroviral genome can be engineered to remove a number of viral genes, creating vectors that accept foreign DNA, including human genes. These vectors are used to treat genetic disorders by gene therapy, a topic that is also discussed in Chapter 22.

19.8 The Polymerase Chain Reaction Makes DNA Copies Without Host Cells

Recombinant DNA techniques were developed in the early 1970s and revolutionized research in genetics and molecular biology. These methods also gave birth to the booming biotechnology industry. However, cloning DNA using vectors and host cells is often labor intensive and time-consuming. In 1986, another technique, called the **polymerase chain reaction (PCR)**, was developed. This advance again revolutionized recombinant DNA methodology and further accelerated the pace of biological research. The significance of this method was underscored by the awarding of the 1993 Nobel Prize in Chemistry to Kary Mullis for developing the PCR technique.

PCR is a rapid method of DNA cloning that extends the power of recombinant DNA research and eliminates the need for host cells in DNA cloning. Although cell-based cloning is still widely used, PCR is the method of choice in many applications, including molecular biology, human genetics, evolution, development, conservation, and forensics.

PCR copies a specific DNA sequence through a series of *in vitro* reactions and can amplify target DNA sequences present in infinitesimally small quantities in a population of other DNA molecules. As a prerequisite for PCR, some information about the nucleotide sequence of the DNA to be cloned is required. The sequence information is used to synthesize two oligonucleotide primers: one for the 5′ end and one for the 3′ end of the DNA sequence that will be cloned. The primers are added to a sample of DNA that has been converted into single strands. The primers bind to complementary nucleotides flanking the sequence to be cloned. A heat-stable DNA polymerase is added after hybridization has occurred, and it synthesizes a second strand of DNA (Figure 19–17). Repeating these steps makes more copies of the DNA.

In practice, the PCR reaction involves three steps. The amount of amplified DNA produced is theoretically limited only by the number of times these steps are repeated.

1. The DNA to be cloned is *denatured* into single strands. The DNA can come from many sources, including genomic DNA, mummified remains, fossils, or forensic samples such as dried blood or semen, single hairs, or dried samples from medical records. Heating to 90–95°C denatures the double-stranded DNA, which dissociates into single strands (usually in about 5 minutes).
2. The temperature of the reaction is lowered to somewhere between 50°C and 70°C, and at this *annealing* temperature, the primers bind to the single-stranded DNA. As described above, the primers are synthetic oligonucleotides (15–30 nucleotides long) complementary to sequences flanking the target DNA to be copied. The primers serve as starting points for synthesizing new DNA strands complementary to the target DNA.
3. A heat-stable form of DNA polymerase (*Taq* polymerase) is added to the reaction mixture. DNA synthesis is carried out at temperatures between 70°C and 75°C. The *Taq* polymerase *extends* the primers by adding nucleotides in the 5′ to 3′ direction, making a double-stranded copy of the target DNA.

Each set of three steps—**denaturation** of the double-stranded DNA, **primer annealing**, and **extension** by polymerase—is a cycle. PCR is a chain reaction because the number of new DNA strands is doubled in each cycle, and the new strands, along with the old strands, serve as templates in the next cycle. Each cycle, which takes about 5 minutes, can be repeated, and in less than 3 hours, 25–30 cycles result in a more than millionfold increase in the amount of DNA (Figure 19–17). This process is automated by machines called *thermocyclers* that can be programmed to carry out a predetermined number of cycles, yielding large amounts of a specific DNA sequence that can be used for many purposes, including cloning into plasmid vectors, DNA sequencing, clinical diagnosis, and genetic screening.

PCR-based DNA cloning has several advantages over cell-based cloning. PCR is rapid and can be carried out in a few hours, rather than the days required for cell-based cloning. In addition, the design of PCR primers is done automatically with computer software, and the commercial synthesis of the oligonucleotides is also fast and economical.

PCR is also very sensitive and amplifies specific DNA sequences from vanishingly small DNA samples, including the DNA in a single cell. This feature of PCR is invaluable in several areas, including genetic testing, forensics, and molecular paleontology. DNA samples that are partly degraded, mixed with other materials, or embedded in a medium (such as amber) can be used when conventional cloning would be difficult or impossible.

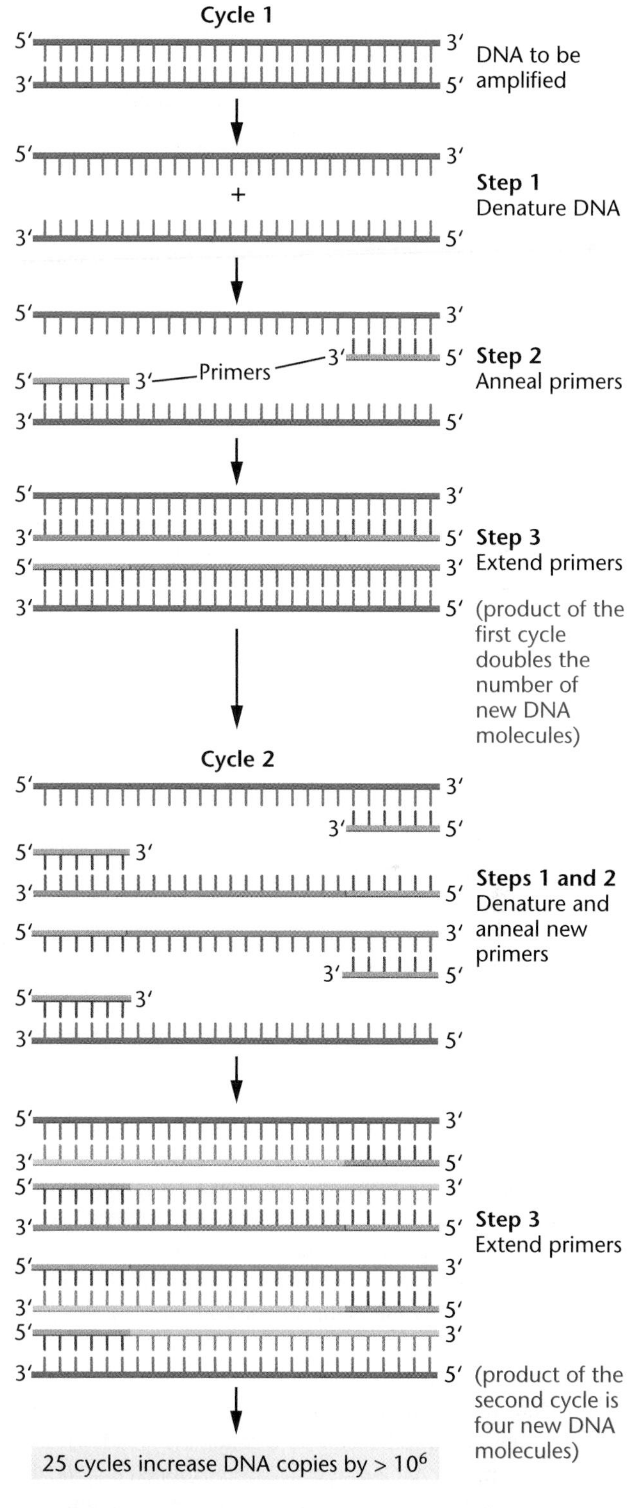

FIGURE 19–17 In the polymerase chain reaction (PCR), the target DNA is denatured into single strands; each strand is then annealed to a short, complementary primer. DNA polymerase and nucleotides extend the primers in the 3′ direction, using the single-stranded DNA as a template. The result is a newly synthesized double-stranded DNA molecule with the primers incorporated into it. Repeated cycles of PCR can amplify the original DNA sequence by more than a millionfold.

Web Tutorial 19.3 Polymerase Chain Reaction

Limitations of PCR

Although PCR is a valuable technique, it does have limitations: Some information about the nucleotide sequence of the target DNA must be known, and even minor contamination of the sample with DNA from other sources can cause problems. For example, cells shed from the skin of a researcher can contaminate samples gathered from a crime scene or taken from fossils, making it difficult to obtain accurate results. PCR reactions must always be performed in parallel with carefully designed and appropriate controls.

Other Applications of PCR

PCR DNA cloning is now one of the most widely used techniques in genetics and molecular biology. PCR and its variations have many other applications. It quickly identifies restriction enzyme recognition sequence variants as well as variations in tandemly repeated DNA sequences that can be used as genetic markers in gene mapping studies. Gene-specific primers provide a way of screening for mutations in genomic DNA, allowing the nature of the mutation to be determined quickly. Primers can be designed to distinguish between target sequences differing only by a single nucleotide. This makes it possible to synthesize allele-specific probes for genetic testing. Random primers indiscriminately amplify DNA, and are particularly advantageous for exploring uncharacterized DNA regions adjacent to known regions. PCR has been used to enforce the worldwide ban on the sale of certain whale products and to settle arguments about the pedigree background of purebred dogs. In short, PCR is one of the most versatile techniques in modern genetics.

19.9 Libraries Are Collections of Cloned Sequences

Only relatively small DNA segments result from cloning, and these may represent only a single gene or even a portion of a gene. As a result, a large collection of clones is needed to explore even a small fraction of an organism's genome. A set of DNA clones derived from a single individual is a cloned *library*. These libraries can represent an entire genome, a single chromosome, or a set of genes that are expressed in a single cell type.

Genomic Libraries

Ideally, a **genomic library** contains at least one copy of all the sequences in an organism's genome. Genomic libraries are constructed using host cell cloning methods, since PCR-cloned DNA fragments are relatively small. In making a genomic library, DNA is extracted from cells or tissues, cut with restriction enzymes, and the fragments ligated into vectors. Since some vectors (such as plasmids) carry only a few thousand base pairs of inserted DNA, selecting the vector so that the library contains the whole genome in the smallest number of clones is an important consideration.

How big does a genomic library have to be to have a 95 or 99 percent chance of containing all the sequences in a genome? The number of clones required to contain a genome depends on several factors, including the average size of the cloned inserts, the size of the genome to be cloned, and the level of probability desired. The number of clones in a library can be calculated as

$$N = \frac{\ln(1 - P)}{\ln(1 - f)}$$

where N is the number of required clones, P is the probability of recovering a given sequence, and f is the fraction of the genome in each clone.

Suppose we wish to prepare a human genome library large enough to have a 99 percent chance of containing all the sequences in the genome. Because the human genome is so large, the choice of vector is a primary consideration in making this library. If we construct the library using a plasmid vector with an average insert size of 5 kb, then more than 2.4 million clones would be required for a 99 percent probability of recovering any given sequence from the genome. Because of its size, this library would be difficult to screen efficiently. If a phage vector with an average insert size of 17 kb is used as a vector, then about 800,000 clones would be required for a 99 percent probability of finding any given human sequence. While it is much smaller than a plasmid library, screening a phage library of this size would still be a labor-intensive chore. However, if the library was constructed in a YAC vector with an average insert size of 1 Mb, then the library would contain only about 14,000 YACs, making it relatively easy to screen. Vectors with large cloning capacities such as YACs are an essential part of the Human Genome Project.

Chromosome-Specific Libraries

A library made from a subgenomic fraction such as a single chromosome can be of great value in cloning specific genes and in the study of chromosome organization. In *Drosophila*, DNA from a small segment of the X chromosome about 50 polytene bands long was isolated by microdissection. The DNA in this chromosomal fragment was purified, cut with a restriction endonuclease, and cloned into a lambda vector. This X-chromosomal region contains the genes *white*, *zeste*, and *Notch*, as well as an insertion site for a transposable element that can translocate a chromosomal segment to more than 100 other loci scattered throughout the genome. Although technically difficult, this procedure produced a library that contains only the genes of interest and their adjacent sequences, saving the time and effort that would have been needed to screen a genomic library to recover all these clones.

Cloned libraries prepared from individual human chromosomes are made by using a technique known as flow cytometry. To isolate individual chromosomes, mitotic cells are collected, and the metaphase chromosomes are stained with two fluorescent dyes, one that binds to AT pairs, the other to GC pairs. The stained chromosomes flow past a laser beam that stimulates them to fluoresce, and a photometer sorts and fractionates the chromosomes by differences in dye binding and light scattering (Figure 19–18). Once the chromosomes are isolated, DNA is extracted, cut with a restriction enzyme, and the DNA fragments are cloned into a vector. Cloned libraries for each human chromosome are available, and these libraries played an important role in the Human Genome Project (to be discussed in Chapter 20).

Individual chromosomes have been isolated for library construction in other ways. A version of gel electrophoresis known as **pulsed-field gel electrophoresis** was used to isolate yeast chromosomes for the construction of chromosome-specific libraries (Figure 19–19). A cloned library of yeast chromosome III (315 kb) was the starting point for the Yeast Genome Project, a consortium of laboratories that sequenced this chromosome and the rest of the yeast genome.

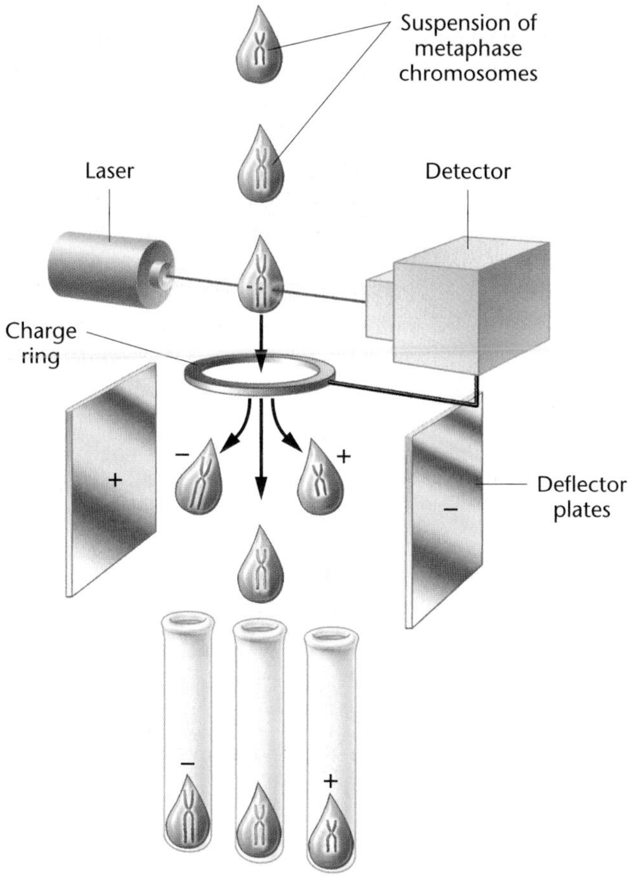

FIGURE 19–18 In chromosome sorting, metaphase chromosomes are stained with two fluorescent dyes, one that stains AT base pairs, and another that stains GC base pairs. Microdrops containing the stained chromosomes flow past a laser that stimulates the dyes to fluoresce. Fluorescence of the two dyes combines to produce a unique signal for each chromosome that is read by a detector. As drops flow through a ring, an electrical charge may be applied to the drop depending on the chromosome it carries. The drops fall past a deflector plate that directs the drops into small tubes, producing a collection of individual chromosomes that can be used as sources of DNA to make chromosome-specific libraries.

One of the unexpected results of sequencing chromosome III was the discovery that about half of all the genes on this chromosome were previously unknown. It was difficult for many geneticists to accept the fact that the time-tested methods of mutagenesis and gene mapping used for decades was so inefficient. However, this finding was confirmed and extended when the sequence of yeast chromosome XI (664 kb) was published and when the sequencing of the entire yeast genome was completed in 1996.

Single chromosome libraries are valuable in gaining access to genetic loci when conventional methods such as mutagenesis and genetic analysis have been unsuccessful and where other probes, such as mRNA or gene products, are unavailable or unknown. In addition, chromosome-specific libraries provide a means for studying the molecular organization and even the nucleotide sequence in a defined region of the genome.

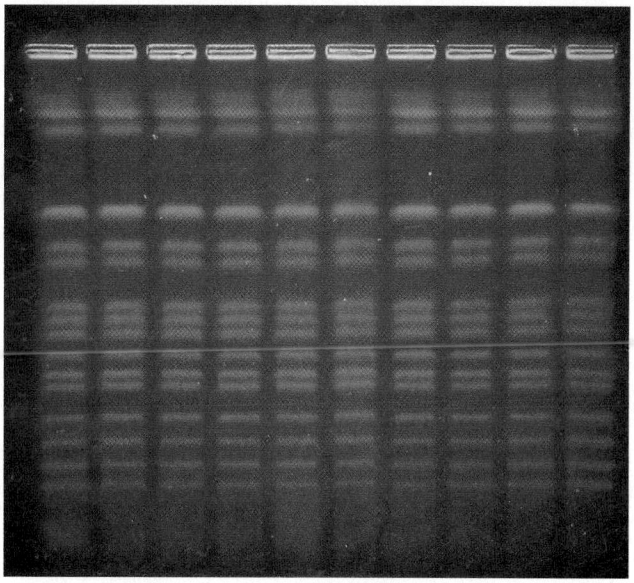

FIGURE 19–19 Intact yeast chromosomes separated using a method of electrophoresis employing pulsed field gel electrophoresis. In each lane, 15 of the 16 yeast chromosomes are visible, separated by size, with the largest chromosomes at the top.

cDNA Libraries

To study specific events in development, cell death, cancer, and other biological processes, a library of the subset of the genome that is expressed in a given cell type at a given time can be a valuable tool. Genomic libraries and chromosome libraries contain all the genes in a genome or on a chromosome, but these collections cannot be directly used to find genes that are transcriptionally active in a cell.

A **cDNA library** contains DNA copies made from the mRNA molecules present in a cell population at a given time, and represents the genes transcriptionally active in the cell at the time the library is made. It is called a cDNA library because the DNA is complementary to the nucleotide sequence of the mRNA.

Clones in a cDNA library are not the same as the clones in a genomic library. Eukaryotic mRNA is processed from pre-mRNA transcripts, and intron sequences are removed during processing. In addition, an mRNA molecule does not include the sequences adjacent to the gene that regulate its activity.

A cDNA library is prepared by isolating mRNA from a population of cells. This is possible because almost all eukaryotic mRNA molecules contain a poly-A tail at their 3′ ends. mRNA with poly-A tails are isolated and used as a template for the synthesis of **complementary DNA (cDNA)** molecules. The cDNA molecules are subsequently cloned into vectors to produce a cDNA library that is a snapshot of genes that were transcriptionally active at a given time.

To make a cDNA library, the first step is mixing mRNAs with poly-A tails with oligo-dT primers, which anneal to the poly-A, forming a partially double-stranded product (Figure 19–20). The enzyme **reverse transcriptase** extends the primer and synthesizes a complementary DNA copy of the mRNA sequence. The product of this reaction is an mRNA–DNA double-stranded hybrid molecule. Action of the enzyme **RNAse H** introduces

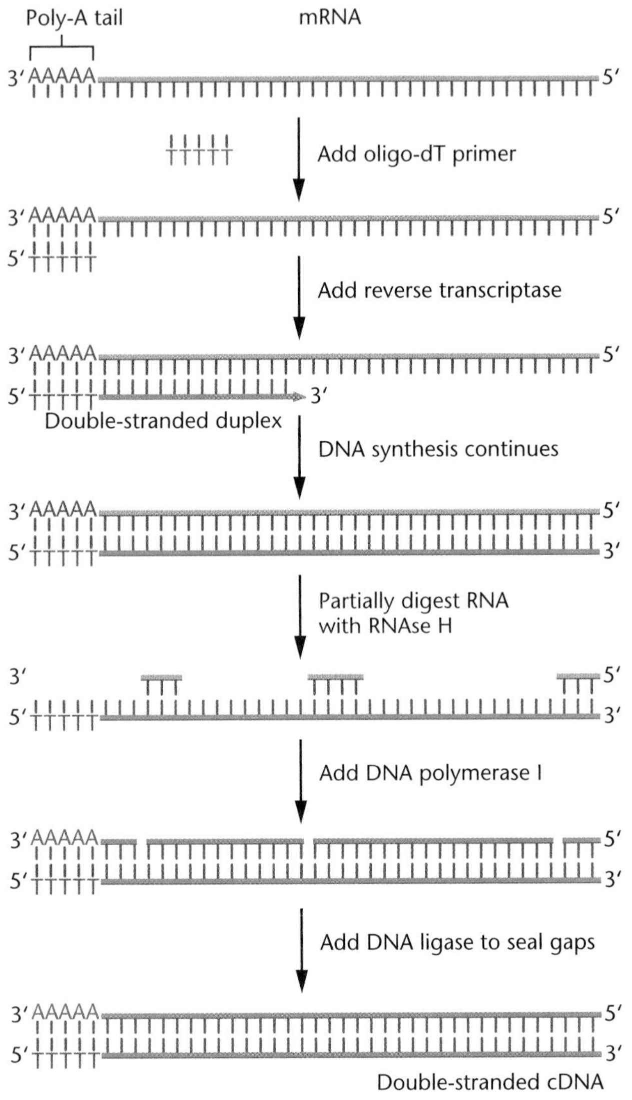

FIGURE 19–20 Producing cDNA from mRNA. Because many eukaryotic mRNAs have a polyadenylated tail (A) of variable length at their 3′ end, a short oligo-dT annealed to this tail serves as a primer for the enzyme reverse transcriptase. Reverse transcriptase uses the mRNA as a template to synthesize a complementary DNA strand (cDNA) and forms an mRNA/cDNA double-stranded duplex. The mRNA is digested with the enzyme RNAse H, producing gaps in the RNA strand. The 3′ ends of the remaining RNA serves as a primer for DNA polymerase, which synthesizes a second DNA strand. The result is a double-stranded cDNA molecule that can be cloned into a suitable vector, or used directly as a probe for library screening.

nicks in the RNA strand by partially digesting the RNA. The remaining RNA fragments serve as primers for the enzyme DNA polymerase I. (This situation is similar to synthesis of the lagging strand of DNA in prokaryotes.) DNA polymerase I synthesizes a second DNA strand and removes the RNA primers, producing double-stranded cDNA.

The cDNA can be cloned into a plasmid or phage vector by attaching linker sequences to the ends of the cDNA. Linker sequences are short double-stranded oligonucleotides containing a restriction enzyme recognition sequence (e.g., *Eco*RI). After attachment to the cDNAs, the linkers are cut with *Eco*RI and ligated to vectors treated with the same enzyme. Many cDNA libraries are available from cells and tissues in specific stages of development, or different organs such as brain, muscle, kidney, and such. These libraries provide an instant catalog of all the genes active in a cell at a specific time.

A cDNA library can also be prepared using a variation of PCR called **reverse transcriptase PCR (RT-PCR)**. In this procedure, reverse transcriptase is used to generate single-stranded cDNA copies of mRNA molecules as described earlier. This reaction is followed by PCR to copy the single-stranded DNA into double-stranded molecules, and then amplify these into many copies. Taq polymerase and random DNA primers (instead of primers specific for a given gene) are added to the single-stranded cDNA, and after primer binding, Taq polymerase extends the primers, making double-stranded cDNA. Additional cycles of PCR make many copies of the cDNA. The amplified cDNA is cloned into plasmid vectors to produce a cDNA library. RT-PCR is more sensitive than conventional cDNA preparation, and is a powerful tool for identifying mRNAs that may be present in only one or two copies per cell.

Now solve this

Question 19.13 on page 481 involves calculating how many clones it takes to make a *Drosophila* genomic library using a plasmid vector.

Hint: Remember there are three parameters in this calculation: the size of the genome, the average size of the cloned inserts, and the probability of having a gene included in the library.

19.10 Specific Clones Can Be Recovered from a Library

A genomic library often consists of several hundred thousand clones. To find a specific gene, we need to identify and isolate only the clone or clones containing that gene. We must also determine whether a given clone contains all or only part of the gene we are studying. Several methods allow us to sort through a library to recover clones of interest. The choice of method often depends on the circumstances and available information about the gene being sought.

Probes Identify Specific Clones

Probes are used to screen a library to recover clones of a specific gene. A **probe** is any DNA or RNA sequence that has been labeled in some way, and is complementary to some part of a cloned sequence present in the library.

When used in a hybridization reaction, the probe binds to any complementary DNA sequences present in one or more clones. Probes can be labeled with radioactive isotopes, or with compounds that undergo chemical or color reactions to indicate the location of a specific clone in a library.

Probes are derived from a variety of sources—even related genes isolated from other species can be used if enough of the DNA sequence is conserved. For example, extrachromosomal copies of the ribosomal RNA genes of the African clawed frog

Xenopus laevis can be isolated by centrifugation and cloned into plasmid vectors. Because ribosomal gene sequences are highly conserved, clones carrying human ribosomal genes can be recovered from a genomic library using cloned fragments of *Xenopus* ribosomal DNA as probes.

If the gene to be selected from a genomic library is expressed in certain cell types, a cDNA probe can be used. This technique is particularly helpful when purified or enriched mRNA for a gene product can be obtained. For example, β-globin mRNA is present in high concentrations in certain stages of red blood cell development. The mRNA purified from these cells can be copied by reverse transcriptase into a cDNA molecule for use as a probe. In fact, a cDNA probe was originally used to recover the structural gene for human β-globin from a cloned genomic library.

Screening a Library

To screen a *plasmid library*, clones from the library are grown on nutrient agar plates, where they form hundreds or thousands of colonies (Figure 19–21). A replica of the colonies is made by gently pressing a nylon filter onto the plate's surface. This transfers bacterial colonies from the plate to the filter. The filter is processed to lyse the bacterial cells, denature the double-stranded DNA released from the cells into single strands, and bind these strands to the filter.

The DNA on the filter is screened by incubation with a labeled nucleic acid probe. The probe is heated and quickly cooled to form single-stranded molecules, and added to a solution containing the filter. If the nucleotide sequence of any of the DNA on the filter is complementary to a probe, a double-stranded hybrid molecule will form (one strand from the probe and the other from the cloned DNA on the filter). After incubation of the probe and the filter, unbound and/or excess probe molecules are washed away, and the filter is assayed to detect the hybrid molecules. If a radioactive probe has been used, the filter is overlaid with a piece of X-ray film. Radioactive decay in the probe molecules bound to DNA on the filter will expose the film, producing dark spots when the film is developed. These spots represent colonies on the plate containing the cloned gene of interest (Figure 19–21). Using the position of spots on the film as a guide, the corresponding colony on the plate is identified and recovered. The cloned DNA it contains can be used in further experiments. With some nonradioactive probes, a chemical reaction emits photons of light (chemiluminescence) to expose the photographic film and reveal the location of colonies carrying the gene of interest.

HOW DO WE KNOW?

How do we know which clone in a library contains a gene of interest?

To screen a phage library, a slightly different method, called **plaque hybridization**, is used. A solution of phage carrying DNA inserts is spread over a lawn of bacteria growing on a plate. The phages infect the bacterial cells and form plaques as they replicate. Each plaque, which appears as a clear spot on the plate, represents the progeny of a single phage and is a clone. The plaques are transferred to a nylon membrane. The phages are disrupted and the DNA on the filter is denatured into single strands and screened with a labeled probe. Phage plaques are much smaller than plasmid colonies, and many plaques can be screened on a single filter, making this method more efficient for screening large genomic libraries.

Now solve this

Question 19.18 on page 481 involves selecting a cloned gene from a cDNA library.

Hint: cDNA clones do not have all the sequences of a genomic library, but do have the coding sequences, and can be selected with the proper probe.

19.11 Cloned Sequences Can Be Characterized in Several Ways

The recovery and identification of genes and other DNA sequences by cloning or by PCR is a powerful tool for analyzing genomic structure and function. In fact, much of the Human Genome Project is based on such techniques. In the following sections, we consider some of these methods, which are used to provide information about the organization and function of cloned sequences.

Restriction Mapping

One of the first steps in characterizing a DNA clone is the construction of a **restriction map**. A restriction map establishes the number, order, and distance between restriction enzyme cleavage sites along a cloned segment of DNA. Restriction maps for different cloned DNAs are usually different enough to serve as an identity tag for that clone. Recall that restriction map units are expressed in **base pairs (bp)** or, for longer lengths, **kilobase (kb)** pairs. Restriction maps provide information about the length of a cloned insert and the location of restriction enzyme cleavage sites within the clone. These data can be used to reclone fragments of a gene or compare its internal organization with that of other cloned sequences.

Fragments generated by cutting DNA with restriction enzymes can be separated by gel electrophoresis, a method that separates fragments by size, with the smallest pieces moving farthest. The fragments appear as a series of bands that can be visualized by staining the DNA with ethidium bromide and viewing under ultraviolet illumination (Figure 19–22).

Figure 19–23 shows the construction of a restriction map from a cloned DNA segment. For this map, let's begin with a cloned DNA segment 7.0 kb in length. Three samples of the cloned DNA are digested with restriction enzymes—one with *Hin*dIII, one with *Sal*I, and one with both *Hin*dIII and *Sal*I. The fragments are separated by gel electrophoresis (discussed in Chapter 10), and stained with ethidium bromide, producing a series of bands on the gel. These bands are photographed or scanned for analysis. The molecular weights of the fragments are measured by comparing their location on the gel to a set of molecular weight standards run in an adjacent lane. The restriction map is constructed by analyzing the number and length of the fragments. When the DNA is cut with *Hin*dIII, two

Web Tutorial 19.4
Cloning Libraries

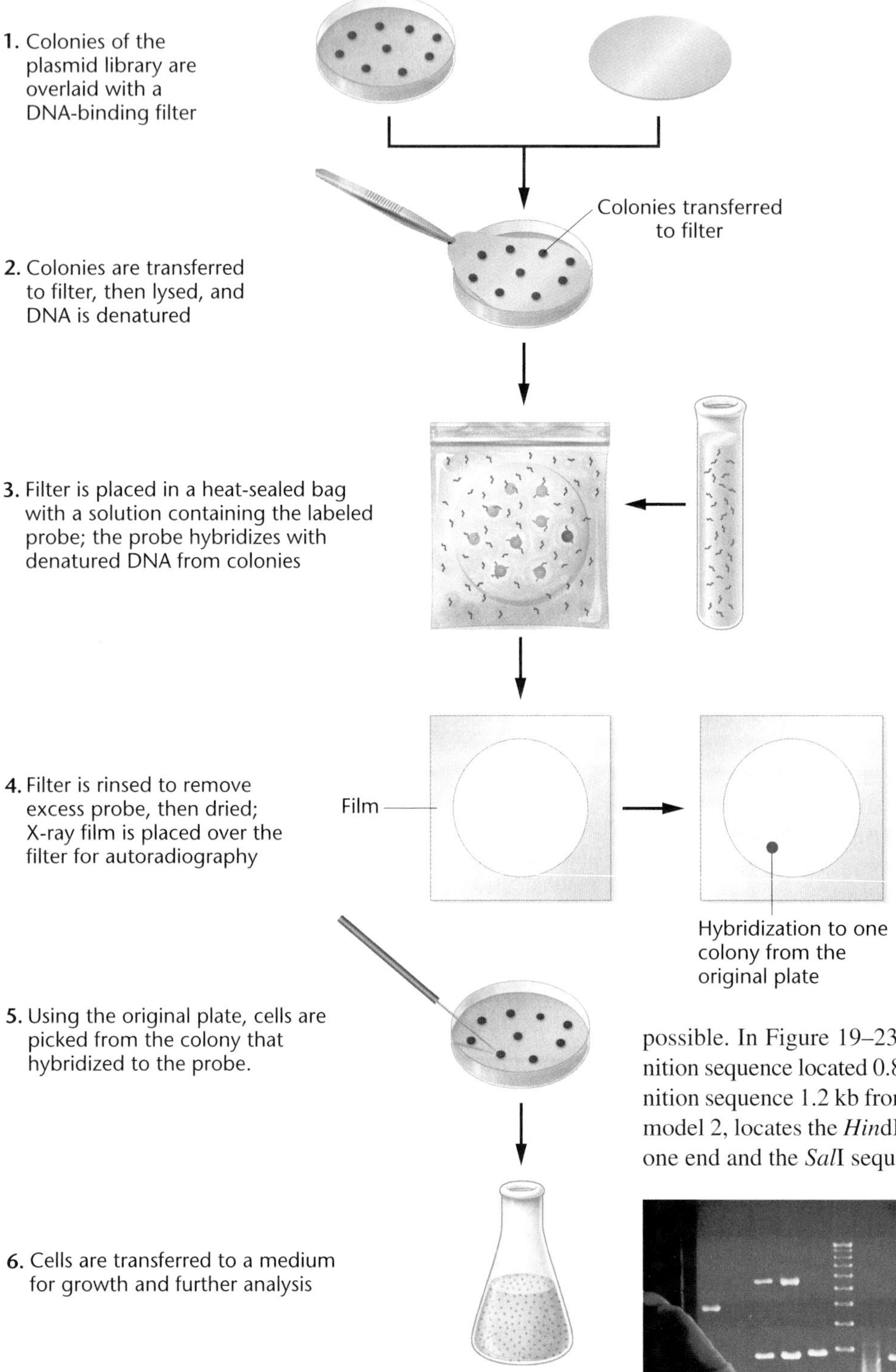
1. Colonies of the plasmid library are overlaid with a DNA-binding filter
2. Colonies are transferred to filter, then lysed, and DNA is denatured
Colonies transferred to filter
3. Filter is placed in a heat-sealed bag with a solution containing the labeled probe; the probe hybridizes with denatured DNA from colonies
4. Filter is rinsed to remove excess probe, then dried; X-ray film is placed over the filter for autoradiography
Film
Hybridization to one colony from the original plate
5. Using the original plate, cells are picked from the colony that hybridized to the probe.
6. Cells are transferred to a medium for growth and further analysis

FIGURE 19–21 Screening a plasmid library to recover a cloned gene. The library, present in bacteria on Petri plates, is overlaid with a DNA binding filter, and colonies are transferred to the filter. Colonies on the filter are lysed, and the DNA denatured to single strands. The filter is placed in a hybridization bag along with buffer and a labeled single-stranded DNA probe. During incubation, the probe forms a double-stranded hybrid with complementary sequences on the filter. The filter is removed from the bag and washed to remove excess probe. Hybrids are detected by placing a piece of X-ray film over the filter and exposing it for a short time. The film is developed, and hybridization events are visualized as spots on the film. Colonies containing the insert that hybridized to the probe are identified from the orientation of the spots. Cells are picked from this colony for growth and further analysis.

fragments (0.8 and 6.2 kb) are produced, confirming that the cloned insert is 7.0 kb in length and contains only one recognition sequence for this enzyme (located 0.8 kb from one end). When the DNA is cut with *Sal*I, two fragments (1.2 and 5.8 kb) result, indicating that the insert also has only one recognition sequence for this enzyme, located 1.2 kb from one end of the cloned DNA segment.

These results show that the DNA contains one restriction enzyme recognition sequence for each enzyme, but the relative arrangement of the two restriction enzyme equences is unknown. From the information available, two different maps are possible. In Figure 19–23, model 1 shows the *Hin*dIII recognition sequence located 0.8 kb from one end and the *Sal*I recognition sequence 1.2 kb from the same end. The alternative map, model 2, locates the *Hin*dIII recognition sequence 0.8 kb from one end and the *Sal*I sequence 1.2 kb from the other end.

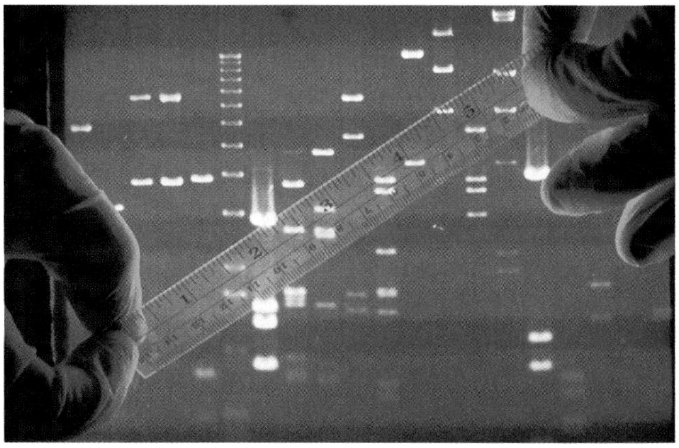

FIGURE 19–22 An agarose gel containing separated DNA fragments stained with a dye (ethidium bromide) and visualized under ultraviolet light. Smaller fragments migrate faster and farther than do larger fragments, resulting in the distribution shown.

The correct model is determined by analyzing the results from the sample digested with *Hin*dIII *and Sal*I. Model 1 predicts that digestion with both enzymes will generate three fragments: 0.4, 0.8, and 5.8 kb in length; model 2 predicts that there will be three fragments of 0.8, 5.0, and 1.2 kb. The pattern and molecular weights seen on the gel after digestion with both enzymes indicates that model 1 is correct. (See Figure 19–23.)

Restriction maps are an important way of characterizing cloned DNA and can be constructed in the absence of any other information about the DNA, including its coding capacity or function. In conjunction with other techniques, restriction mapping can define the boundaries of a gene, dissect the internal organization of a gene and its flanking regions, and locate mutations within genes.

Restriction digestion of clones plays an important role in mapping genes to specific human chromosomes and to defined regions of individual chromosomes. In addition, if a restriction enzyme recognition sequence is closely linked to a mutant allele, this sequence can be used as a marker in genetic testing to identify carriers of recessively inherited disorders, or to prenatally diagnose a fetal genotype. This topic will be discussed in Chapter 22.

Nucleic Acid Blotting

Many of the techniques described in this chapter rely on hybridization between complementary nucleic acid (DNA or RNA) molecules. One of the most widely used methods for detecting such hybrids is called Southern blotting (after Edward Southern, who devised it). The **Southern blot** method can be used to identify which clones in a library contain a given DNA sequence (such as ribosomal DNA, a globin gene, etc.), and to characterize the size of the fragments, thus deriving a restriction map of the cloned DNA. Southern blots can also be used to determine whether a clone contains all or only part of a gene and to ascertain the overall size and sequence organization of a gene or DNA sequence of interest. Fragments of genomic

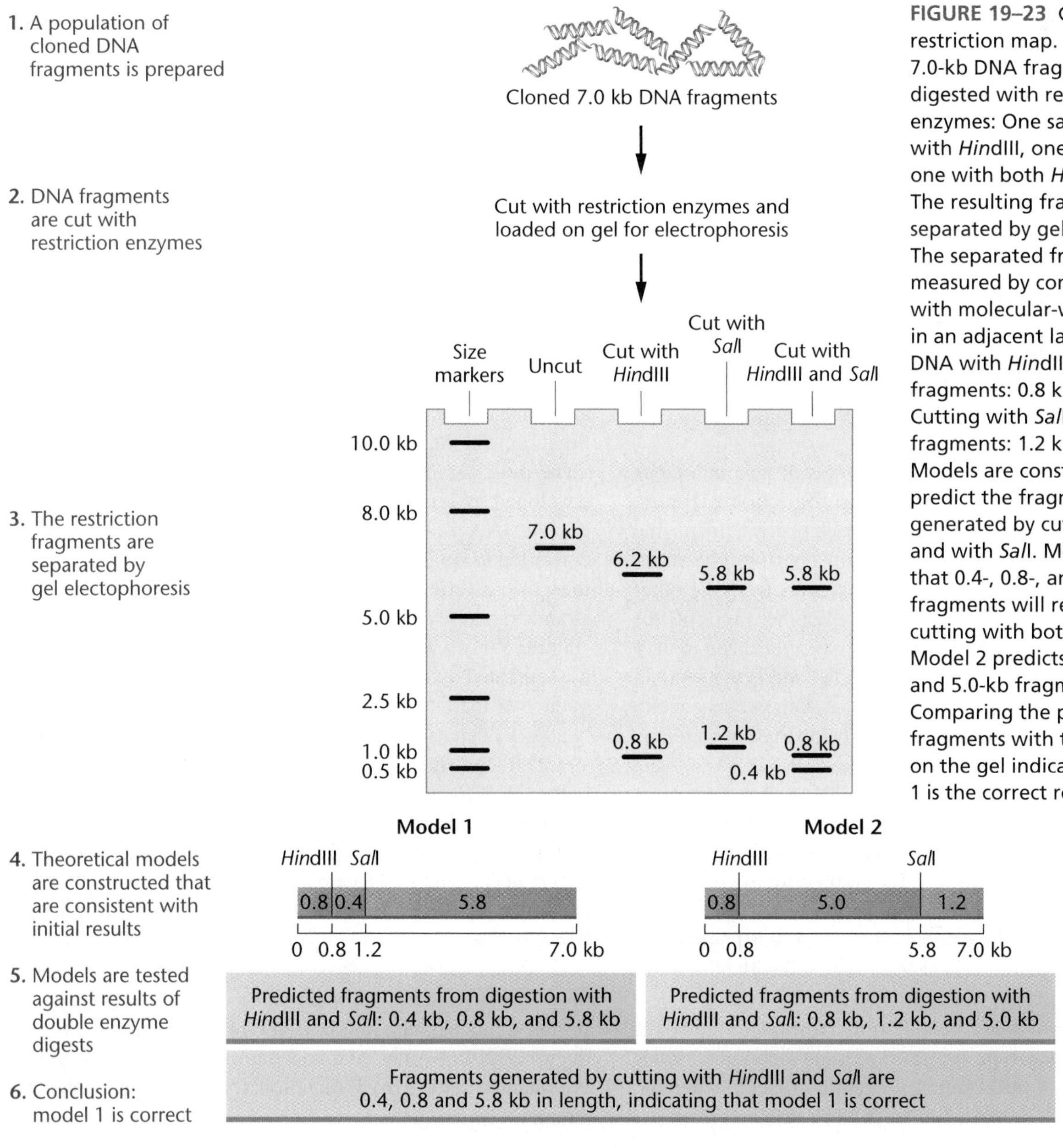
1. A population of cloned DNA fragments is prepared
Cloned 7.0 kb DNA fragments
2. DNA fragments are cut with restriction enzymes
Cut with restriction enzymes and loaded on gel for electrophoresis
3. The restriction fragments are separated by gel electophoresis
Size markers
Uncut
Cut with HindIII
Cut with SalI
Cut with HindIII and SalI
10.0 kb
8.0 kb
5.0 kb
2.5 kb
1.0 kb
0.5 kb
7.0 kb
6.2 kb
5.8 kb
5.8 kb
0.8 kb
1.2 kb
0.8 kb
0.4 kb
4. Theoretical models are constructed that are consistent with initial results
Model 1
HindIII SalI
0.8 0.4 5.8
0 0.8 1.2 7.0 kb
Model 2
HindIII SalI
0.8 5.0 1.2
0 0.8 5.8 7.0 kb
5. Models are tested against results of double enzyme digests
Predicted fragments from digestion with HindIII and SalI: 0.4 kb, 0.8 kb, and 5.8 kb
Predicted fragments from digestion with HindIII and SalI: 0.8 kb, 1.2 kb, and 5.0 kb
6. Conclusion: model 1 is correct
Fragments generated by cutting with HindIII and SalI are 0.4, 0.8 and 5.8 kb in length, indicating that model 1 is correct

FIGURE 19–23 Constructing a restriction map. Samples of the 7.0-kb DNA fragments are digested with restriction enzymes: One sample is digested with *Hin*dIII, one with *Sal*I, and one with both *Hin*dIII and *Sal*I. The resulting fragments are separated by gel electrophoresis. The separated fragments are measured by comparing them with molecular-weight standards in an adjacent lane. Cutting the DNA with *Hin*dIII generates two fragments: 0.8 kb and 6.2 kb. Cutting with *Sal*I produces two fragments: 1.2 kb and 5.8 kb. Models are constructed to predict the fragment sizes generated by cutting with *Hin*dIII and with *Sal*I. Model 1 predicts that 0.4-, 0.8-, and 5.8-kb fragments will result from cutting with both enzymes. Model 2 predicts that 0.8-, 1.2-, and 5.0-kb fragments will result. Comparing the predicted fragments with those observed on the gel indicates that model 1 is the correct restriction map.

Web Tutorial 19.5 Restriction Mapping

clones isolated by Southern blots can be isolated and recloned, providing a way to isolate parts of a gene, including its adjacent control regions, or coding regions.

This technique has two components: separation of DNA fragments by gel electrophoresis and hybridization of the fragments using labeled probes. Gel electrophoresis characterizes the number of fragments produced by restriction digestion of DNA and their molecular weights. Hybridization characterizes the DNA sequences present in the fragments. The DNA to be characterized by Southern blot hybridization can come from several sources, including clones selected from a library or genomic DNA. Our discussion will use examples from both sources to show how Southern blots are used to characterize the number, size, organization, and sequence content of DNA fragments.

To make a Southern blot, DNA is cut into fragments with one or more restriction enzymes, and the fragments are separated by gel electrophoresis (Figure 19–24), producing a series of bands. The DNA in the gel is stained and photographed or scanned to reveal the number and molecular weights of the restriction fragments. In preparation for hybridization, DNA in the gel is denatured with alkaline treatment to form single-stranded fragments. The gel is then overlaid with a DNA-binding membrane, usually nitrocellulose or a nylon filter. The DNA fragments in the gel are transferred to the membrane by placing the membrane and gel on a wick (often a sponge) sitting in a buffer solution. Layers of paper towels or blotting paper are placed on top of the filter and held in place with a weight. Capillary action draws buffer up through the gel, transferring the DNA fragments to the membrane.

The filter is placed in a heat-sealed bag with a labeled, single-stranded DNA probe for hybridization. DNA fragments on the filter that are complementary to the probe's nucleotide sequence form double-stranded hybrids. Excess probe is washed away, and the hybridized fragments are visualized on a piece of film (Figures 19–24 and 19–25).

In Figure 19–25, researchers cut samples of genomic DNA from two strains of *E. coli* with several restriction enzymes. The pattern of fragments obtained for each restriction enzyme is shown in Figure 19–25(a). The fragments from one strain are shown in lanes a, c, and e; and fragments from the other strain are shown in lanes b, d, and f. A Southern blot of this gel is shown in Figure 19–25(b). The probe hybridized to only one band in each lane, and the size of the band is the same in the two strains. These results indicate that both strains contain the DNA sequence of interest and that the restriction pattern is very similar for the two strains.

In addition to characterizing cloned DNAs, Southern blots can be used to create restriction maps within and near a gene and to identify DNA fragments carrying all or parts of a single gene in a mixture of fragments. By comparing the pattern of bands in normal cells to those from patients with genetic disorders or cancer, Southern blots also detect rearrangements, deletions, and duplications in genes associated with human genetic disorders and cancers.

To determine whether a gene is transcriptionally active in a given cell or tissue type, a related blotting technique probes for the presence of mRNA complementary to a cloned gene. To do this, mRNA is extracted from a specific cell or tissue type and separated by gel electrophoresis. The resulting pattern of RNA bands is transferred to a membrane, as in Southern blotting. The membrane is then hybridized to a labeled single-stranded DNA probe derived from a cloned copy of the gene. If mRNA complementary to the DNA probe is present, it will be detected as a band on the film. Because the original procedure (DNA bound to a filter) is known as a Southern blot, this procedure (RNA bound to a filter) is called a **northern blot**. (Following this somewhat perverse logic, another procedure involving proteins bound to a filter is known as a **western blot**.)

Northern blots provide information about the expression of specific genes and are used to study patterns of gene expression in embryonic tissues, cancer, and genetic disorders. Northern blots also detect alternatively spliced mRNAs (multiple types of transcripts derived from a single gene) and are used to derive other information about transcribed mRNAs. If marker RNAs of known size are run as controls, northern blots can be used to measure the size of a gene's mRNA transcripts. In addition, the amount of transcribed RNA present in a cell type is related to the density of the RNA band on the film. Measuring band density gives the relative transcriptional activity. Thus, northern blots characterize and quantify the transcriptional activity of genes in different cells, tissues, and organisms.

19.12 DNA Sequencing Is the Ultimate Way to Characterize a Clone

In a sense, a cloned DNA molecule or any DNA, from a clone to a genome, is completely characterized only when its nucleotide sequence is known. The ability to sequence DNA has greatly enhanced our understanding of genome organization and enhanced our knowledge about genes including structure, function, and the mechanisms of regulation.

The most commonly used method of DNA sequencing was developed by James Sanger and his colleagues. In this procedure, a DNA molecule whose sequence is to be determined is converted to single strands that are used as a template for synthesizing a series of complementary strands. Each of these strands randomly terminates at a different, specific nucleotide (Figure 19–26). This produces a series of DNA fragments that are separated by electrophoresis and analyzed to reveal the sequence of the DNA. In the first step of this reaction, DNA is heated to denature it and form single strands. The single-stranded DNA is mixed with primers that anneal to the 3′ end of the DNA. Samples of the primer-bound single-stranded DNA are distributed into four tubes. In the next step, DNA polymerase and the four deoxyribonucleotide triphosphates (dATP, dCTP, dGTP, and dTTP) are added to each tube. In addition, each tube receives a small amount of one modified deoxyribonucleotide, called a dideoxynucleotide (e.g., ddATP, ddCTP, ddGTP, and ddTTP). Dideoxynucleotides have a 3′-H instead of a 3′-OH group. One of the deoxyribonucleotides or the primer is labeled with radioactivity for later analysis of the sequence. DNA polymerase is added to each tube, and the primer is elongated in the 5′ to 3′ direction, forming a complementary strand to the template.

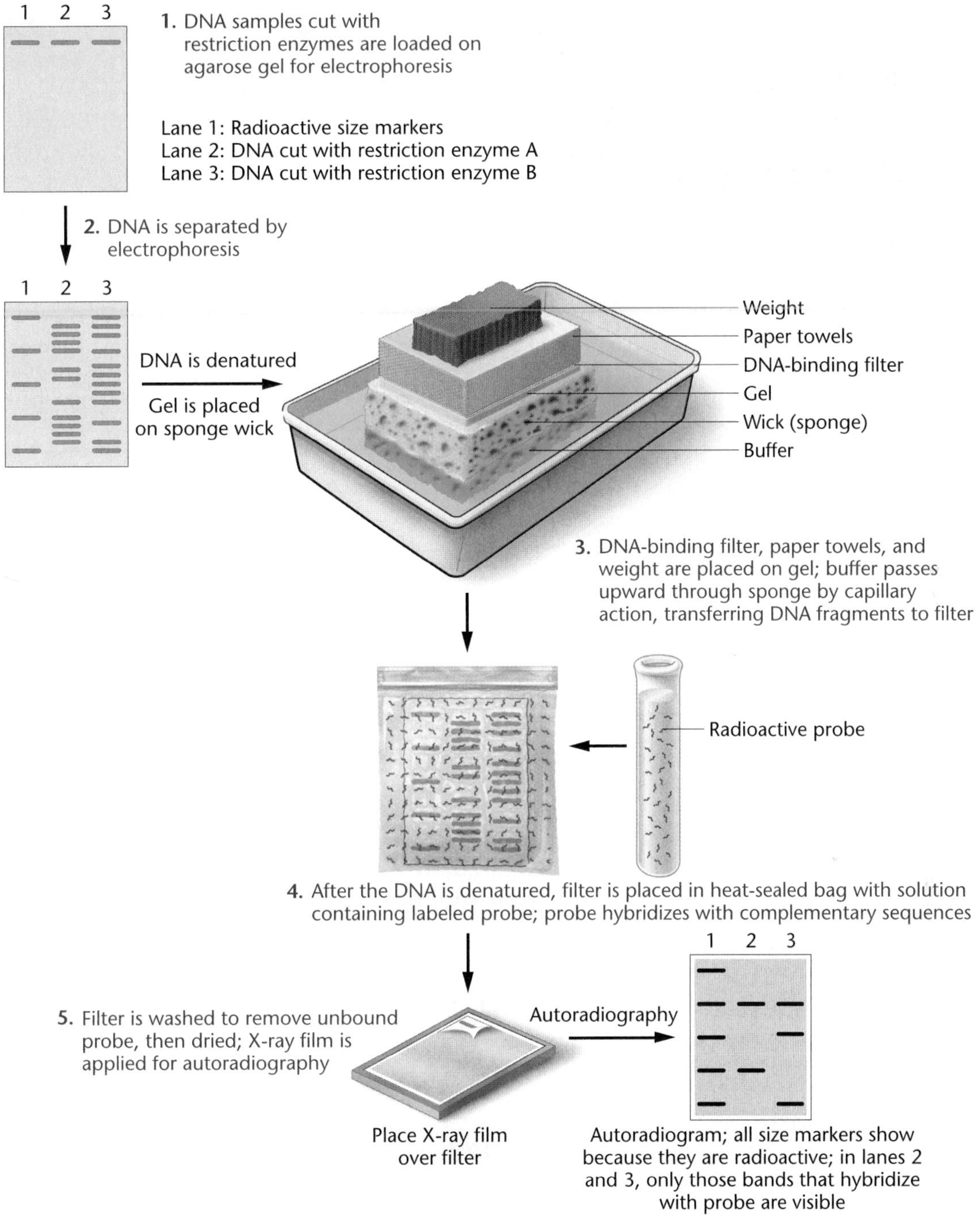
1 2 3
1. DNA samples cut with restriction enzymes are loaded on agarose gel for electrophoresis
Lane 1: Radioactive size markers
Lane 2: DNA cut with restriction enzyme A
Lane 3: DNA cut with restriction enzyme B
2. DNA is separated by electrophoresis
1 2 3
DNA is denatured
Gel is placed on sponge wick
Weight
Paper towels
DNA-binding filter
Gel
Wick (sponge)
Buffer
3. DNA-binding filter, paper towels, and weight are placed on gel; buffer passes upward through sponge by capillary action, transferring DNA fragments to filter
Radioactive probe
4. After the DNA is denatured, filter is placed in heat-sealed bag with solution containing labeled probe; probe hybridizes with complementary sequences
1 2 3
5. Filter is washed to remove unbound probe, then dried; X-ray film is applied for autoradiography
Autoradiography
Place X-ray film over filter
Autoradiogram; all size markers show because they are radioactive; in lanes 2 and 3, only those bands that hybridize with probe are visible

FIGURE 19–24 In the Southern blotting technique, samples of the DNA to be probed are cut with restriction enzymes and the fragments separated by gel electrophoresis. The pattern of fragments is visualized and photographed under ultraviolet illumination by staining the gel with ethidium bromide. The gel is then placed on a sponge wick in contact with a buffer solution and covered with a DNA-binding filter. Layers of paper towels or blotting paper are placed on top of the filter and held in place with a weight. Capillary action draws the buffer through the gel, transferring the pattern of DNA fragments from the gel to the filter. The DNA fragments on the filter are then denatured into single strands and hybridized with a labeled DNA probe. The filter is washed to remove excess probe and overlaid with a piece of X-ray film for autoradiography. The hybridized fragments show up as bands on the X-ray film.

As DNA synthesis takes place, the polymerase occasionally inserts a dideoxynucleotide instead of a deoxyribonucleotide into a growing DNA strand. Since the dideoxynucleotide has no 3′-OH group, it cannot form a 3′ bond with another nucleotide, and DNA synthesis terminates. For example, in the tube with added ddATP, the polymerase inserts ddATP instead of dATP, causing termination of chain elongation (Figure 19–26). As the reaction proceeds, the tube with ddATP will accumulate DNA molecules that terminate at all positions containing A. In the other tubes, reactions terminate at

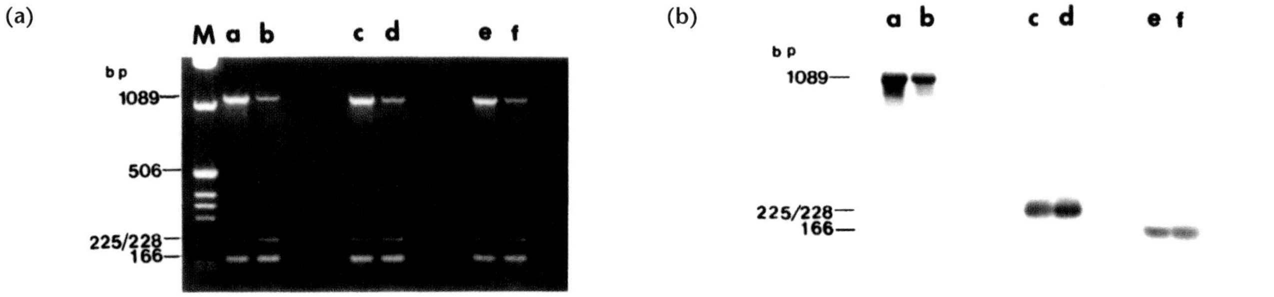

FIGURE 19–25 (a) Agarose gel stained with ethidium bromide to show DNA fragments. Lane M contains the size markers. Lanes a, c, and e contain DNA from one bacterial strain of *E. coli*, and lanes b, d, and f contain DNA from another strain. (b) A Southern blot prepared from the gel in part (a). Only those bands containing DNA sequences complementary to the probe show hybridization. Both strains contain the probed sequence, and the two strains cannot be distinguished from one another on this basis. Size markers were not radioactive and are not visible on the Southern blot.

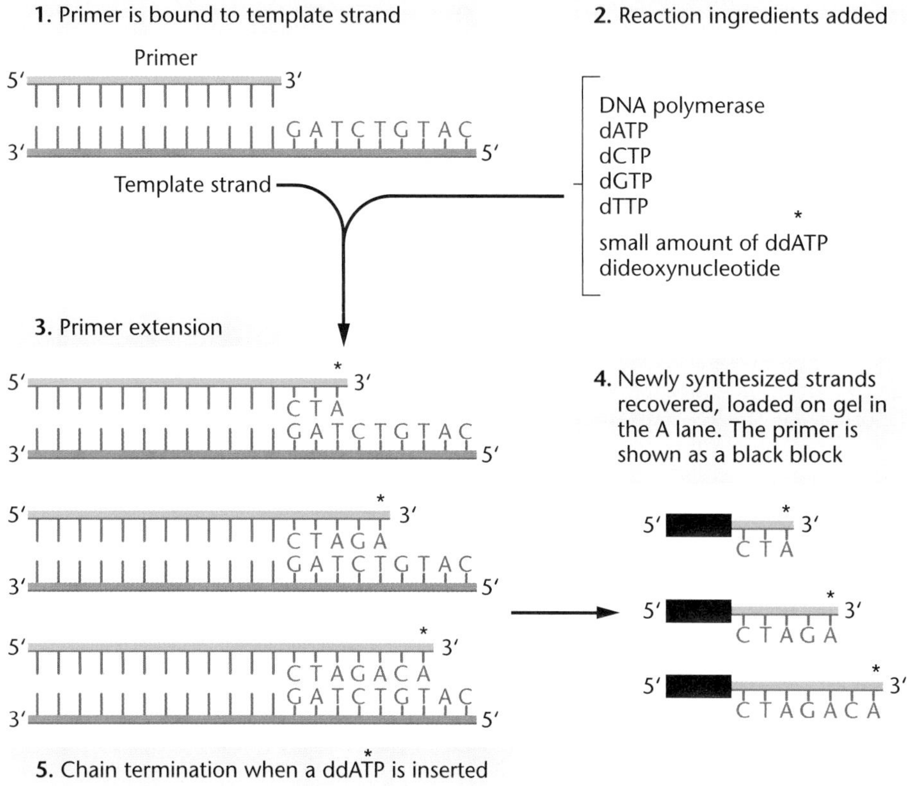

FIGURE 19–26 DNA sequencing using the chain termination method. (1) A primer is annealed to a sequence adjacent to the DNA being sequenced (usually at the insertion site of a cloning vector). (2) A reaction mixture is added to the primer–template combination. This includes DNA polymerase, the four dNTPs (one of which is radioactively labeled) and a small amount of one dideoxynucleotide. Four tubes are used, each containing a different dideoxynucleotide (ddATP, ddCTP, etc.). (3) During primer extension, the polymerase occasionally inserts a ddNTP instead of a dNTP, terminating the synthesis of the chain, because the ddNTP does not have the 3′-OH group needed to attach the next nucleotide. In the figure, ddATP and the A inserted from this dideoxynucleotide are indicated with an asterisk. Over the course of the reaction, all possible termination sites will have a ddNTP inserted. (4) The newly synthesized strands are removed from the template, and the mixture is placed on a gel. DNA fragments from the reaction tube containing ddATP and terminating with A are loaded in the A lane, those ending in C are loaded in the C lane, and so forth.

G, C, and T, respectively. The DNA fragments from each reaction tube (one for each dideoxynucleotide) are separated in adjacent lanes by gel electrophoresis. The result is a series of bands forming a ladderlike pattern that is visualized by developing film exposed to the gel (Figure 19–27). The nucleotide sequence is read directly from bottom to top, corresponding to the 5′ to 3′ sequence of the DNA strand complementary to the template.

DNA sequencing in large-scale genome sequencing projects is automated, and uses machines that can sequence several hundred thousand nucleotides per day. In this procedure, each of the four dideoxynucleotide analogs are labeled with a different colored fluorescent dye (Figure 19–28) so that chains terminating in adenosine are labeled with one color, those ending in cytosine with another color, and so forth. All four labeled dideoxynucleotides are added to a single tube, and after primer extension by DNA polymerase, the reaction products are loaded into one lane on a gel. The gel is scanned with a laser, causing each band to fluoresce a different color. A detector in the sequencing machine reads the color of each band and determines whether it represents an A, T, C, or G. The data are represented as a series of colored peaks, each corresponding to one nucleotide in the sequence (Figure 19–29).

DNA Sequencing and Genome Projects

DNA sequencing is the heart of genome projects. Using a combination of recombinant DNA techniques and nucleotide sequencing, the genomes of more than 100 species of prokaryotes have been sequenced, with hundreds more projects underway. The Human Genome Project sponsored by the U.S. Department of Energy and the National Institutes of Health and a private project sponsored by the biotechnology company Celera used a combination of genomic and chromosome-specific libraries to finish sequencing the coding portion of the human genome in 2003. Genome projects for many other eukaryotic species have also been completed with the use of DNA sequencing technology, and dozens more are underway.

Web Tutorial 19.6
DNA Sequencing

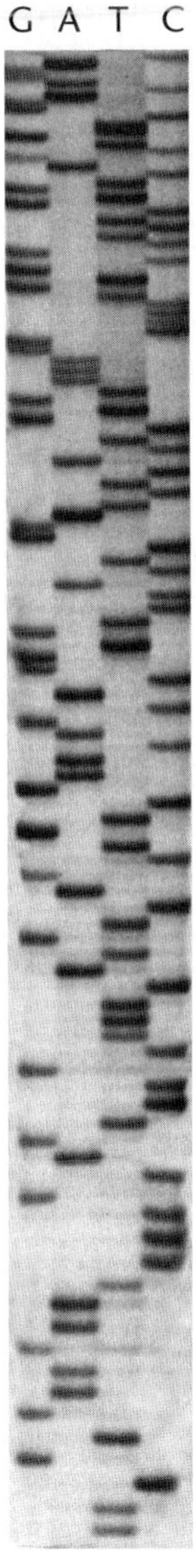

FIGURE 19–27 DNA sequencing gel showing the separation of newly synthesized fragments in the four sequencing reactions (one per lane). To obtain the base sequence of the DNA fragment, the gel is read from the bottom, beginning with the lowest band in any lane, then the next lowest, then the next, and so on. For example, the sequence of the DNA on this gel begins with 5′-TT at the very bottom of the gel, and proceeds upwards as 5′-TTCGTGAAGAA and so forth.

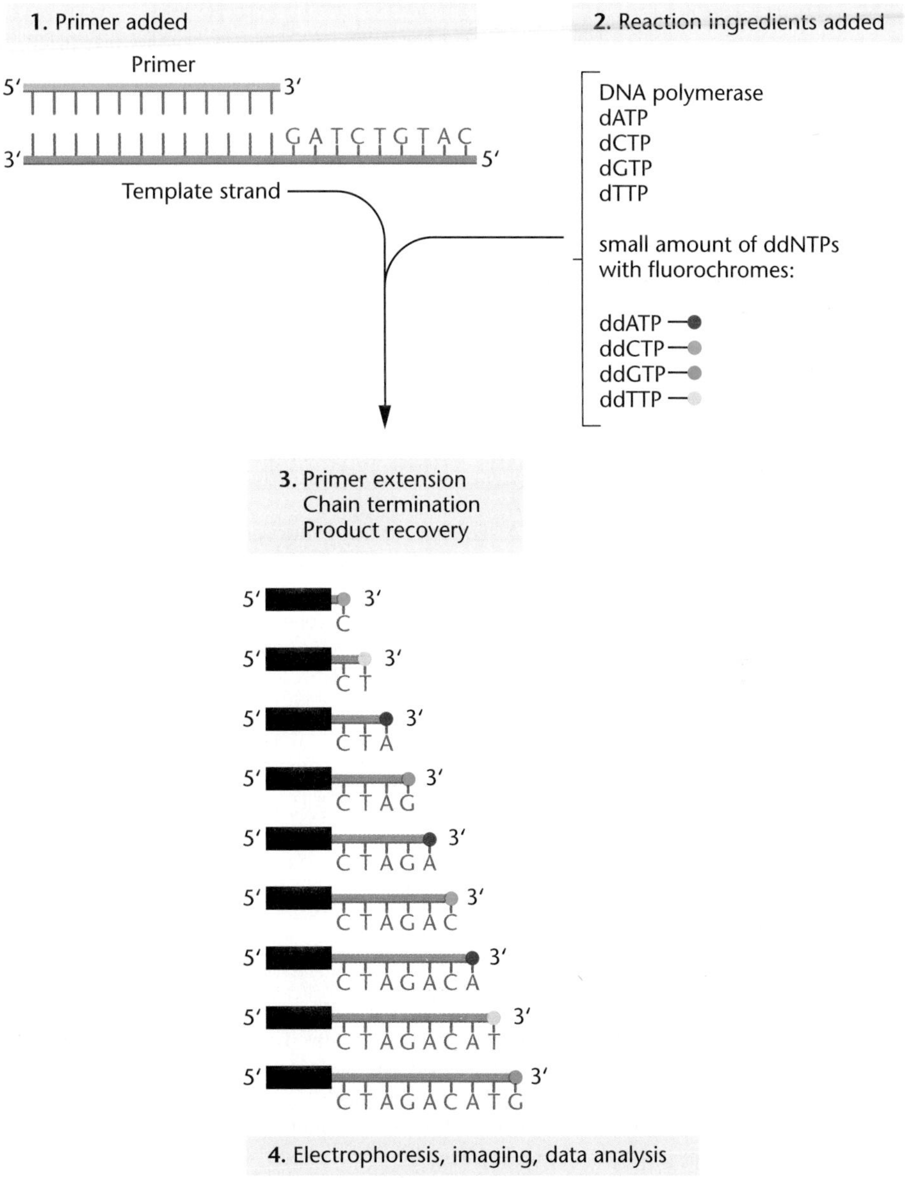

FIGURE 19–28 In DNA sequencing using dideoxynucleotides labeled with fluorescent dyes, all four ddNTPs are added to the same tube, and during primer extension, all combinations of chains are produced. The products of the reaction are added to a single lane on a gel, and the bands are read by a detector and imaging system. This process is now automated, and robotic machines, such as those used in the Human Genome Project, sequence several hundred thousand nucleotides in a 24-hour period and then store and analyze the data automatically. The sequence obtained is by extension of the primer, and is read from the newly synthesized strand, not the template strand. Thus, the sequence obtained begins with 5′-CTAGACATG.

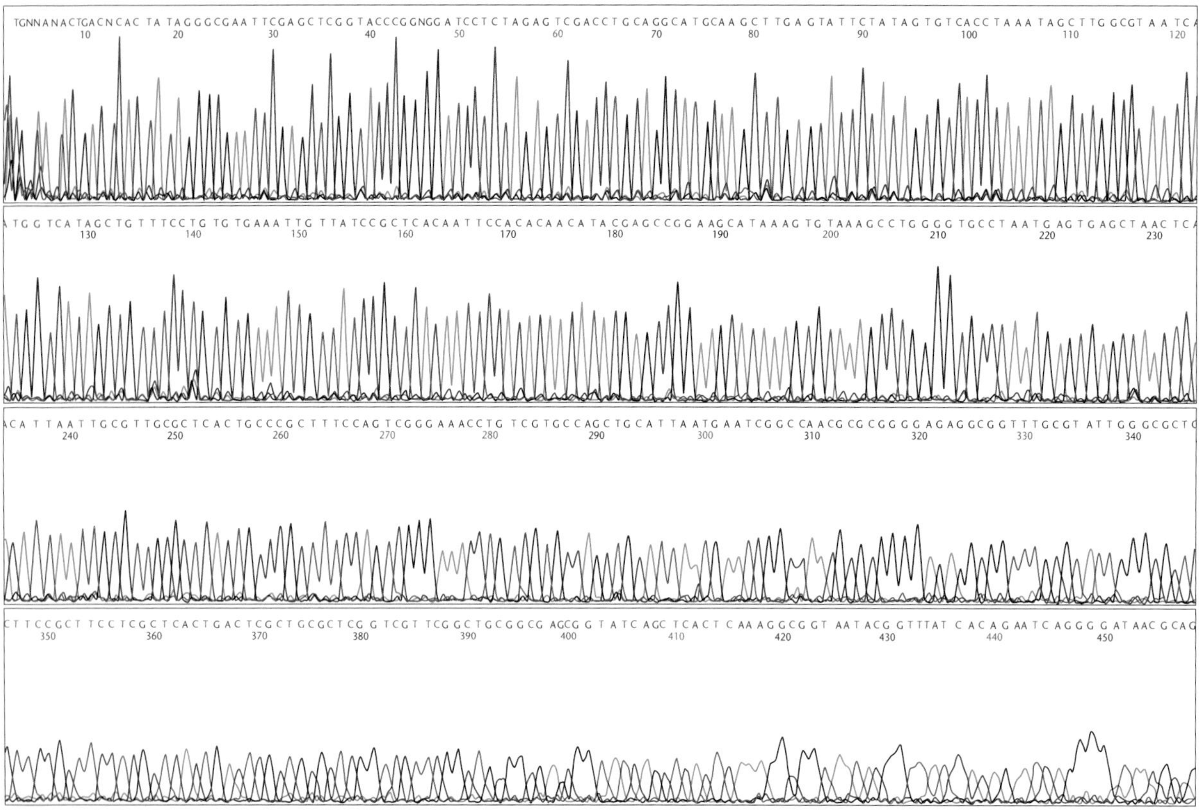

FIGURE 19–29 Automated DNA sequencing using fluorescent dyes, one for each base. Each peak represents the correct nucleotide in the sequence. The primer sequence is located to the left of the bases at the upper left of the figure, and are not shown here. The bases labeled as N are ambiguous, and cannot be identified with certainty. These ambiguous base readings are more likely to occur near the primer because the quality of sequence determination deteriorates the closer the sequence is to the primer. The separated bases are read in order along the axis from left to right. Thus, the sequence begins as 5′-TGNNANACTGACNCAC. Numbers below the bases indicate length of the sequence in base pairs.

Chapter Summary

1. The cornerstone of recombinant DNA technology is a class of enzymes called restriction enzymes, which cut DNA at specific recognition sequences. The fragments thus produced are joined with DNA vectors to form recombinant DNA molecules.
2. Vectors replicate autonomously in host cells and facilitate the manipulation of the newly created recombinant DNA molecules. Vectors are constructed from many sources, including bacterial plasmids and phages.
3. Recombinant DNA molecules are transferred into a host, and cloned copies are produced during host-cell replication. A variety of host cells may be used for replication, including bacteria, yeast, and mammalian cells. Cloned copies of foreign DNA sequences are recovered, purified, and analyzed.
4. The polymerase chain reaction (PCR) is a method for amplifying a specific DNA sequence that is present in a collection of DNA sequences, such as genomic DNA. The PCR method allows DNA to be cloned without host cells and is a rapid, sensitive method with wide-ranging applications.
5. Once cloned, DNA sequences are analyzed through a variety of methods, including restriction mapping and DNA sequencing. Other methods, such as Southern blotting, use hybridization to identify genes and flanking regulatory regions within the cloned sequences.

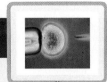

Genetics, Technology, and Society

DNA Fingerprints and Forensics: The Case of the Telltale Palo Verde

The use of DNA analysis in forensics is making it harder and harder to get away with many crimes. It's now a simple matter to isolate DNA from tissue left at a crime scene, a splattering of blood, some skin left under a victim's fingernails, or even the cells clinging to the base of a hair shaft. A variety of techniques can be used to determine the likelihood that the sample came from a suspect in the case. In recent years, the PCR technique has been increasingly used in forensics, both because it is fast (taking only hours) and because it requires only a small amount of DNA (about a nanogram). One variation of the PCR method that may be especially valuable is called the random amplified polymorphic DNA (RAPD, pronounced "rapid") procedure. Under the right conditions, the RAPD procedure generates a DNA profile that is unique to each individual and that can thus be used for personal identification.

The RAPD procedure involves the amplification of a set of DNA fragments of unknown sequence. First, an amplification primer is mixed with a sample of DNA. The primer will anneal to all sites in the DNA sample that have a matching base sequence. For primers 10 nucleotides in length, binding will occur at a few thousand sites scattered randomly throughout the genome. When primers happen to bind to DNA sites that are not too far apart (within about 2000 nucleotides), DNA amplification can occur between these sites, eventually producing millions of copies of the DNA sequence lying between the binding sites. When the products of such a reaction are separated by electrophoresis on an agarose gel, each amplified DNA segment will appear as a distinct band. The locations of the primer binding sites in the genome of any two individuals in a population are likely to differ, due to minor DNA sequence differences (e.g., base changes, insertions, and deletions). Since these affect the initial binding to the primers, the number and locations of the DNA bands on the gel will vary from one individual to another. The resulting DNA profile is referred to as a DNA fingerprint. It represents a characteristic "snapshot" of the genome of an individual, allowing it to be distinguished from those of other individuals in the population.

When a DNA profile from tissue found at a crime scene matches that of a suspect, it does not prove that the tissue belongs to the suspect; instead, it excludes all those who have a different pattern. The strategy, therefore, is to generate five or more DNA profiles from the same sample, using different amplification primers. The more profiles that match between the sample and the suspect, the more unlikely it is that the sample at the crime scene came from someone other than the suspect. The likelihood that a complete match of all the banding patterns occurs simply by chance can be calculated, giving, for example, a 1-in-100,000 to a 1-in-1,000,000 probability of a "random match."

One of the most interesting uses of DNA fingerprinting in a criminal case did not involve the suspect's own DNA, but rather the DNA of plants growing at the crime scene. On the night of May 2, 1992, a Phoenix woman was strangled and her body dumped near an abandoned factory in the surrounding desert. Investigators discovered a pager near the body, making its owner a prime suspect. When questioned, this man admitted being with the woman the day of the murder, but claimed he had not been near the factory and suggested that the woman must have stolen his pager from his pickup truck. A search of the pickup provided the crucial clue, placing the suspect at the factory: In its bed were two seed pods from a palo verde tree.

The palo verde tree (*Cercidium floridum*), native to the desert of the Southwest, is a member of the bean family. In an adaptation to the hot, dry climate, it puts out very small leaves. To compensate for this, its stems become green and carry out photosynthesis. This gives the plant its name; palo verde is Spanish for "green stem." The tree may not resemble a bean plant in some respects, but, like a typical bean plant, it produces seeds in pods that drop to the ground when mature.

The homicide investigators assigned to the case wondered whether it could be proved that the seed pods found in the bed of the suspect's truck had fallen from a palo verde tree near the place where the body was found. If so, it would be a key piece of evidence placing the suspect at the scene. But how could this be demonstrated? The investigators turned to Dr. Timothy Helentjaris, then at the University of Arizona in nearby Tucson. Helentjaris proceeded to determine whether the DNA profile of the seed pods from the pickup truck matched that of any of the palo verde trees near the scene. He understood that for a match between a seed pod and a tree to have significance, he would first have to show that palo verdes differ genetically from one tree to the next. Otherwise, a pod could never be unambiguously assigned to one particular tree. Fortunately, he found considerable variation in DNA profiles among the palo verde trees.

Helentjaris was then given the two seed pods from the suspect's truck along with pods collected from 12 palo verde trees from the vicinity of the factory. The investigators knew which of the 12 trees was the key one at the crime scene, but they did not tell Helentjaris. He extracted DNA from seeds from each pod and performed RAPD reactions using an amplification primer that he had earlier found to reproducibly yield profiles of 10–15 distinct bands. The result of this controlled experiment was unmistakable: The DNA profile from one of the pods found in the truck exactly matched that of only 1 of the 12 trees, the one nearest to where the body was found. In an important additional test, Helentjaris showed that this DNA profile differed from that of pods collected from 18 other trees located at random sites around Phoenix. Collectively, Helentjaris's analysis led to the estimate of the likelihood of a random match at a little less than 1 in 1,000,000.

Helentjaris's analysis was admitted as evidence in the trial, placing the suspect at the crime scene. At the completion of the five week trial, the suspect was found guilty of first-degree murder. His conviction was upheld upon appeal, and he is currently serving a life sentence without parole.

References

Ayala, F.J., and Black. B. 1993. Science and the courts. *Am. Sci.* 81:230–39.

Krawczak, M., and Schmidtke, J. 1994. *DNA fingerprinting*. Oxford: BIOS Scientific.

Marx, L. 1988. DNA fingerprinting takes the witness stand. *Science* 240:1616–18.

Insights and Solutions

1. The recognition sequence for the restriction enzyme *Sau*3A is GATC (Figure 19–2); the recognition sequence for the enzyme *Bam*HI is GGATCC, where the four internal bases are identical to the *Sau*3A sequence. Therefore, the single-stranded ends produced by the two enzymes are identical. Suppose you have a cloning vector that contains a *Bam*HI recognition sequence and foreign DNA that you have cut with *Sau*3A. (a) Can this DNA be ligated into the *Bam*HI site of the vector, and if so, why? (b) Can the DNA segment cloned into this sequence be cut from the vector with *Sau*3A? With *Bam*HI? What potential problems do you see with the use of *Bam*HI?

Solution: (a) DNA cut with *Sau*3A can be ligated into the vector's *Bam*HI cutting site because the single-stranded ends generated by the two enzymes are identical. (b) The DNA can be cut from the vector with *Sau*3A because the recognition sequence for this enzyme (GATC) is maintained on each side of the insert. Recovering the cloned insert with *Bam*HI is more problematic. In the ligated vector, the conserved sequences are GGATC (left) and GATCC (right). The correct base will *follow* the conserved sequence (to produce GGATCC on the left) only about 25 percent of the time, and the correct base will *precede* the conserved sequence (and produce GGATCC on the right) about 25 percent of the time as well. Thus, *Bam*HI will be able to cut the insert from the vector ($0.25 \times 0.25 = 0.0625$), or only about 6 percent of the time.

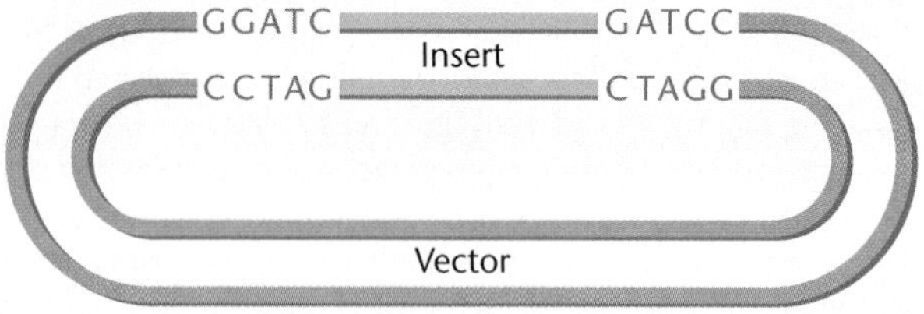

2. In setting up a PCR reaction, which set of primers would you choose to amplify the sequence shown as a series of asterisks?

5′-TTAAGATCCGTTACGTATGC * * * * * * AACCCGTTCCTACGAACCTT-3′

3′-AATTCTAGGCAATGCATACG * * * * * * TTGGGCAAGGATGCTTGGAA-5′

Primers:

Set 1:	5′-TTAAGATCCGTT-3′	5′-CGTTCCTACGAA-3′
Set 2:	5′-GATCCGTTACGT-3′	5′-TTCGTAGGAACG-3′
Set 3:	5′-CGTATGCATTGC-3′	5′-TTCCAAGCATCC-3′

Solution: A PCR reaction requires two primers, one for each strand. The primers are complementary to sequences near the 3′ end of each single stranded DNA. In the first step of PCR, the DNA to be copied is heated to separate the molecule into single strands. Then the primers are annealed to the single strands, and elongated in the $5' \rightarrow 3'$ direction by the addition of Taq polymerase. The only set of primers that will anneal properly is Set 2. One primer has a complementary sequence 4 base pairs in from the 3′ end of the lower strand of DNA, and the other primer has a complementary sequence 4 base pairs in from the 3′ end of the upper strand.

Problems and Discussion Questions

1. What roles do restriction enzymes, vectors, and host cells play in recombinant DNA studies?
2. Why is oligo-dT an effective primer for reverse transcriptase?
3. The human insulin gene contains a number of introns. In spite of the fact that bacterial cells do not excise introns from mRNA, explain how a gene like this can be cloned into a bacterial cell and produce insulin.
4. Restriction enzymes recognize palindromic sequences in intact DNA molecules and cleave the double-stranded helix at these sequences. Inasmuch as the bases are internal in a DNA double helix, how is this recognition accomplished?
5. Although the potential benefits of cloning in higher plants are obvious, the development of this field has lagged behind cloning in bacteria, yeast, and mammalian cells. Can you think of a reason for this?
6. Using DNA sequencing on a cloned DNA segment, you recover the nucleotide sequence shown below. Does this segment contain a palindromic recognition sequence for a restriction enzyme? What is the double-stranded sequence of the palindrome? What enzyme would cut at this sequence? (Consult Figure 19–2 for a list of restriction enzyme recognition sequences.)

   CAGTATCCTAGGCAT

7. Restriction enzyme sequences are palindromic; that is, they read the same in the 5′ to 3′ direction on each strand of DNA. What is the advantage of having restriction recognition sequences organized in this way?
8. List the advantages and disadvantages of using plasmids and YACs as cloning vectors.
9. Listed below are several restriction enzymes with four and six base recognition sequences. For each of these enzymes, if we assume random distribution and equal amounts of each nucleotide in the DNA to be cut, how far apart are each of these recognition sequences on average?

*Taq*I	TCGA	*Hin*dIII	AAGCTT
*Alu*I	AGCT	*Bal*I	TGGCCA
*Hae*III	GGCC	*Bam*HI	GGATCC

10. Some restriction enzymes have recognition sequences that are specific and unambiguous, while other recognition sequences are ambiguous, so that any purine, pyrimidine, or nucleotide can occupy a given position in the cutting site. In the following examples, *Not*I has an unambiguous recognition sequence, *Hin*fI has an ambiguous sequence (N = any nucleotide), and *Xho*II also has an ambiguous recognition sequence (Pu = any purine and Py = any pyrimidine). Assuming random distribution and equal amounts of each nucleotide in the DNA to be cut, how far apart are each of these restriction recognition sequences on average?

*Not*I	GCGGCCGC
*Hin*fI	GANTC
*Xho*II	PuGATCPy

11. What are the advantages of using a restriction enzyme with relatively few recognition sequences? When would you use such enzymes?

12. An ampicillin-resistant, tetracycline-resistant plasmid, pBR322, is cleaved with *Pst*I, which cleaves within the ampicillin gene. The cut plasmid is ligated with *Pst*I-digested *Drosophila* DNA to prepare a genomic library, and the mixture is used to transform *E. coli* K12. (a) Which antibiotic should be added to the medium to select cells that have incorporated a plasmid? (b) What growth pattern should be selected to obtain plasmids containing *Drosophila* inserts? (c) How can you explain the presence of colonies that are resistant to both antibiotics?

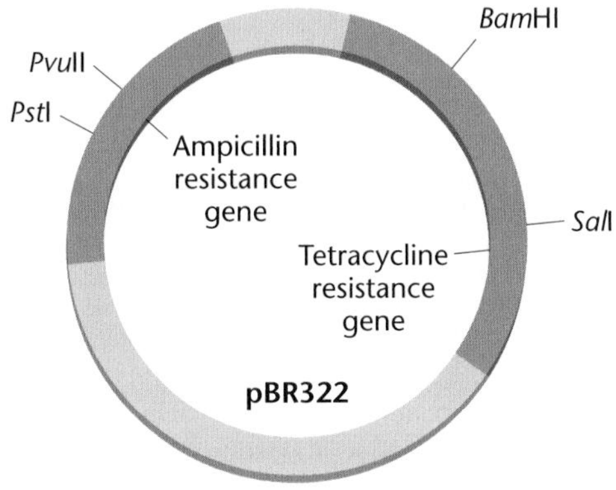

13. Plasmids isolated from the clones in Problem 12 are found to have an average insert length of 5 kb. Given that the *Drosophila* genome is 1.5×10^5 kb long, how many clones would be necessary to give a 99 percent probability that this library contains all genomic sequences?

14. In a control experiment, a plasmid containing a *Hin*dIII recognition sequence within a kanamycin resistance gene is cut with *Hin*dIII, religated, and used to transform *E. coli* K12 cells. Kanamycin-resistant colonies are selected, and plasmid DNA from these colonies is subjected to electrophoresis. Most of the colonies contain plasmids that produce single bands that migrate at the same rate as the original intact plasmid. A few colonies, however, produce two bands, one of original size and one that migrates much higher in the gel. Diagram the origin of this slow band as a product of ligation.

15. You have just created the world's first genomic library from the African okapi, a relative of the giraffe. No genes from this genome have been previously isolated or described. You wish to isolate the gene encoding the oxygen-transporting protein β-globin from the okapi library. This gene has been isolated from humans, and its nucleotide sequence and amino acid sequence are available in databases. Using the information available about the human β-globin gene, what two strategies can you use to isolate this gene from the okapi library?

16. When making cDNA, the single-stranded DNA produced by reverse transcriptase can be made double stranded by treatment with DNA polymerase I. However, no primer is required with the DNA polymerase. Why is this?

17. What should you consider in deciding which vector to use in constructing a genomic library of eukaryotic DNA?

18. You are given a cDNA library of human genes prepared in a bacterial plasmid vector. You are also given the cloned yeast gene that encodes EF-1a, a protein that is highly conserved among eukaryotes. Outline how you would use these resources to identify the human cDNA clone encoding EF-1a.

19. Once you have isolated the human cDNA clone for EF-1a in Problem 18, you sequence the clone and find that it is 1384 nucleotide pairs long. Using this cDNA clone as a probe, you isolate the DNA encoding EF-1a from a human genomic library. The genomic clone is sequenced and found to be 5282 nucleotides long. What accounts for the difference in length observed between the cDNA clone and the genomic clone?

20. You have recovered a cloned DNA segment and determine that the insert is 1300 bp in length. To characterize this cloned segment, you isolate the insert and decide to construct a restriction map. Using enzyme I and enzyme II, followed by gel electrophoresis, you determine the number and size of the fragments produced by enzymes I and II alone and in combination shown in the table below. Construct a restriction map from these data, showing the positions of the restriction enzyme cutting sites relative to one another and the distance between them in units of base pairs.

Enzymes	Restriction Fragment Sizes (bp)
I	350, 950
II	200, 1100
I and II	150, 200, 950

21. To create a cDNA library, cDNA can be cloned into vectors. In analyzing cDNA clones, it is often difficult to find clones that are full length—that is, extending to the 5′ end of the mRNA. Why is this so?

22. Although the capture and trading of great apes has been banned in 112 countries since 1973, it is estimated that about 1000 chimpanzees are removed annually from Africa and smuggled into Europe, the United States, and Japan. This illegal trade is often disguised by simulating births in captivity. Until recently, genetic identity tests to uncover these illegal activities were not used because of the lack of highly polymorphic markers and the difficulties of obtaining chimpanzee blood samples. Recently, a study was reported in which DNA samples were extracted from freshly plucked chimpanzee hair roots and used as templates for PCR. The primers used in these studies flank highly polymorphic sites in human DNA that result from variable numbers of tandem nucleotide repeats. Several offspring and their putative parents were tested to determine whether the offspring are "legitimate" or the product of illegal trading. The data are shown in the Southern blot below.

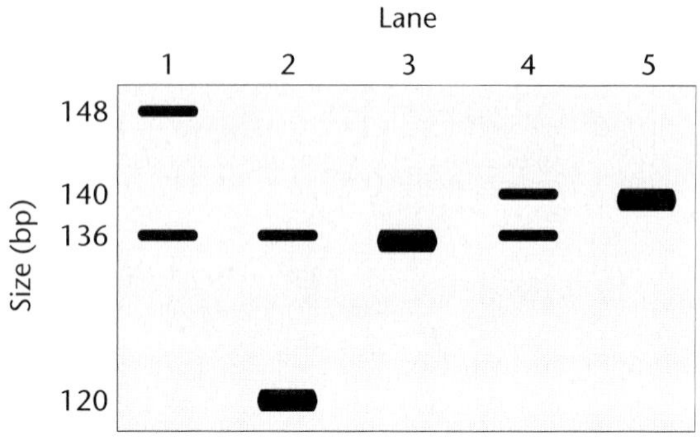

Lane 1: father chimpanzee
Lane 2: mother chimpanzee
Lanes 3–5: putative offspring A, B, C

Examine the data carefully, and choose the best conclusion.

(a) None of the offspring is legitimate.

(b) Offspring B and C are not the products of these parents and were probably purchased on the illegal market. The data are consistent with offspring A being legitimate.

(c) Offspring A and B are products of the parents shown, but C is not and was therefore probably purchased on the illegal market.

(d) There are not enough data to draw any conclusions. Additional polymorphic sites should be examined.

(e) No conclusion can be drawn because "human" primers were used.

23. The accompanying partial restriction map shows a recombinant plasmid, pBIO220, formed by cloning a piece of *Drosophila* DNA (striped box), including the gene *rosy* into the vector pBR322, which also contains the penicillin resistance gene, *pen*. The vector part of the plasmid contains only the two E recognition sequences shown, and no A or B sequences. The gel shows several restriction digests of pBIO220.

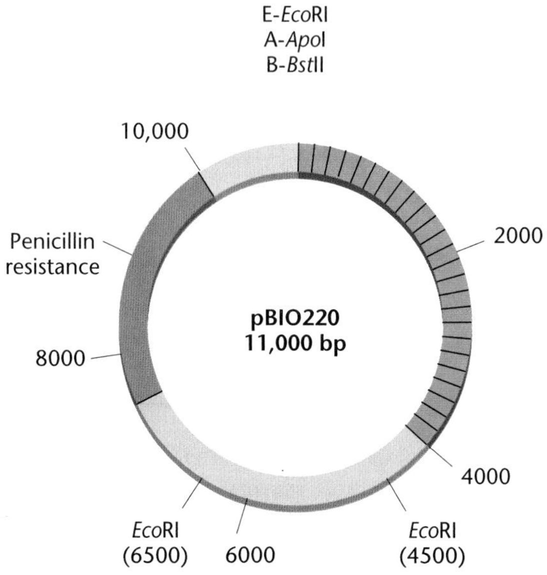

(a) Use the stained gel pattern to deduce where restriction enzyme recognition sequences are located in the cloned fragment.

(b) A PCR-amplified copy of the entire 2000-bp *rosy* gene was used to probe a Southern blot of the same gel. Use the Southern-blot results to deduce the location of *rosy* in the cloned fragment. Redraw the map showing the location of the *rosy* gene.

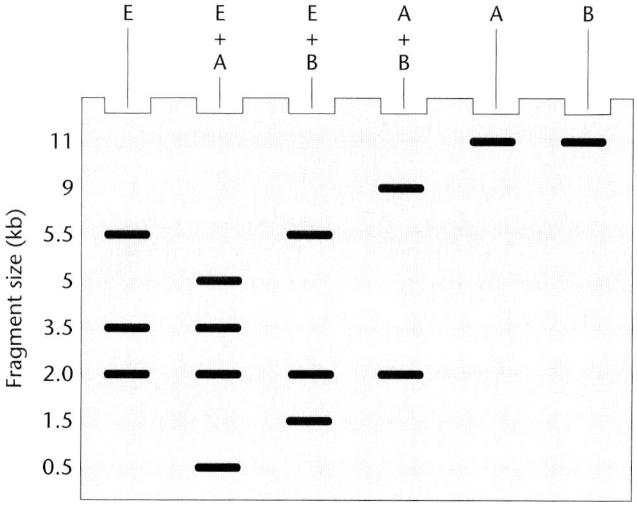

Stained gel

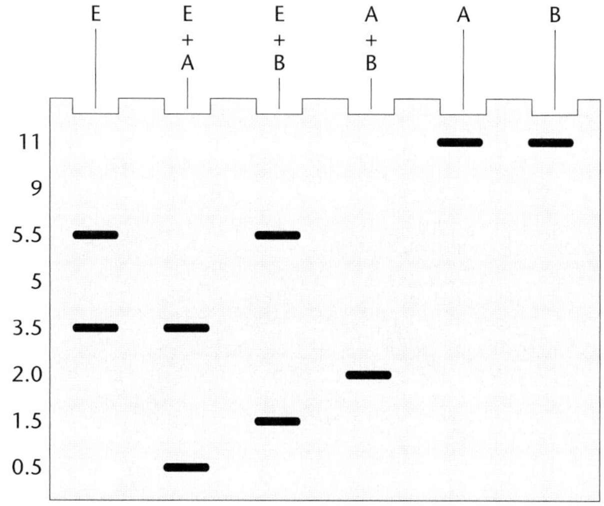

Southern blot

24. One of the six restriction maps shown on the next page is consistent with the pattern of bands shown in the accompanying figure in the gel after digestion with several restriction endonucleases. The enzymes that were used are shown.

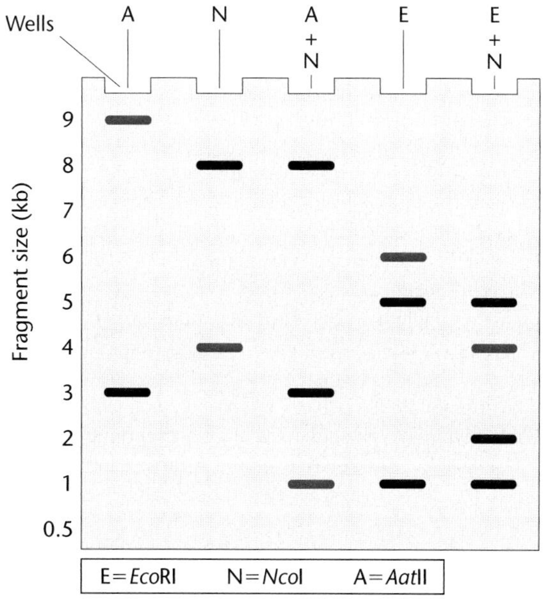

(a) From your analysis of the pattern of bands on the gel, select the correct map and explain your reasoning.

(b) In a Southern blot prepared from this gel, the highlighted bands (pink) hybridized with the gene *pep*. Where is the *pep* gene located?

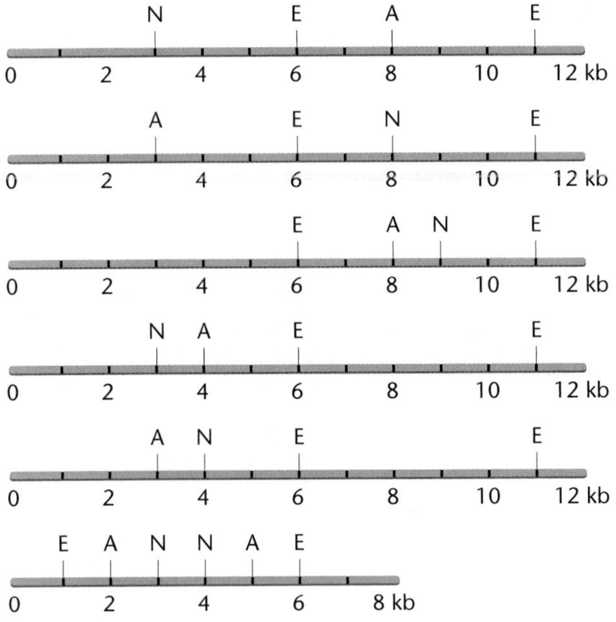

25. List the steps involved in screening a genomic library. What must be known before starting such a procedure? What are the potential problems with such a procedure, and how can they be overcome or minimized?

26. To estimate the number of cleavage sites in a particular piece of DNA with a known size, you can apply the formula $N/4^n$, where N is the number of base pairs in the target DNA, and n is the number of bases in the recognition sequence of the restriction enzyme. If the recognition sequence for *Bam*HI is GGATCC and the λ phage DNA contains approximately 48,500 bp, how many cleavage sites would you expect?

27. In a typical PCR reaction, what phenomena are occurring at temperature ranges (a) 90–95°C, (b) 50–70°C, and (c) 70–75°C?

28. A widely used method for calculating the annealing temperature for a primer used in PCR is 5 degrees below the T_m(°C), which is computed by the equation $81.5 + 0.41(\%GC) - (675/N)$, where %GC is the percentage of GC nucleotides in the oligonucleotide, and N is the length of the oligonucleotide. Notice from the formula that both the GC content *and* the length of the oligonucleotide are variables. Assuming you have the oligonucleotide shown below as a primer, compute the annealing temperature for PCR. What is the relationship between T_m(°C) and %GC? Why? (*Note:* In reality, this computation provides only a starting point for empirical determination of the most useful annealing temperature.)

5′-TTGAAAATATTTCCCATTGCC-3′

29. We usually think of enzymes as being most active at around 37°C, yet in PCR the DNA polymerase is subjected to multiple exposures of relatively high temperatures and seems to function appropriately at 70–75°C. What is special about the DNA polymerizing enzymes typically used in PCR?

30. How are dideoxynucleotides (ddNTPs) used in the chain termination method of DNA sequencing?

31. Assume you have conducted a standard DNA sequencing using the chain-termination method. You performed all the steps correctly and electrophoresed the resulting DNA fragments correctly, but when you looked at the sequencing gel, many of the bands were duplicated (in terms of length) in other lanes. What might have happened?

SELECTED READINGS

Burger, S.L., and Kimmel, A.R. 1987. *Guide to Molecular Cloning Techniques. Methods in Enzymology*. Vol. 152. San Diego: Academic Press.

Cohen, S.N., Chang, A.C., Boyer, H.W., and Helling, R.B. 1973. Construction of biologically functional plasmids *in vitro*. *Proc. Nat. Acad. Sci.* 70:3240–3244.

Colosimo, A., Goncz, K.K., Holmes, A.R., et al. 2000. Transfer and expression of foreign genes in mammalian cells. *BioTech.* 29:314–31.

Danna, K., and Nathans, D. 1971. Specific cleavage of simian virus 40 DNA by restriction endonuclease of *Hemophilus influenzae*. *Proc. Nat. Acad. Sci.* 68:2913–2917.

Mullis, K.B. 1990. The unusual origin of the polymerase chain reaction. *Sci. Am.* (Apr.) 262:56–65.

Old, R.W., and Primrose, S.B. 1994. *Principles of Genetic Manipulation: An Introduction to Genetic Engineering*, 5th ed. Palo Alto: Blackwell Scientific.

Oste, C. 1988. Polymerase chain reaction. *BioTech.* 6:162–67.

Sambrook, J. and Russell, D.W. 2001. *Molecular Cloning: A Laboratory Manual*, 3rd ed. Cold Spring Harbor: Cold Spring Harbor Press.

Sanger, F., and Couslon, A.R. 1975. A rapid method for determining sequences in DNA by primed synthesis with DNA polymerase. *J. Mol. Biol.* 94:441–448.

Southern, E. 1975. Detection of specific sequences among DNA fragments separated by gel electrophoresis. *J. Mol. Biol.* 98:503–07.

Genomics and Proteomics

Arabidopsis thaliana, one of the model organisms used to study genetics.

Chapter Concepts

- Genomics uses methods of sequencing and analysis to study the complete genome of organisms.
- Genomics has several subfields, including structural genomics, functional genomics, and comparative genomics.
- The analysis of genomic sequence data is refining our ideas about the size and organization of prokaryotic genomes.
- The human genome contains 25,000 to 30,000 genes organized in clusters and separated by large spans of intergenic spacer DNA.
- Comparative genomics is useful in studying the evolutionary relationships among genes, gene families, and organisms.
- Proteomics is used to provide information about the protein content, structure, and function of subcellular components.

The term **genome** was coined in 1920, at a time when geneticists began to turn from the study of individual genes to focus on the larger picture: the set of genes carried by an organism (the definition of a genome). To systematically identify and map all the genes in an organism's genome, geneticists developed a two-part approach: (1) Identify spontaneous mutations or collect mutants by using chemical or physical agents, and (2) Generate genetic maps by linkage analysis using the mutant strains. Although a wide range of organisms was originally studied, emphasis was gradually placed on a few, including *Drosophila*, maize, mice, bacteria, and yeast. These organisms are widely studied in genetics, and became model systems for invertebrates (*Drosophila*), crop plants (maize), mammals including humans (mice), prokaryotes (*E. coli*), and lower eukaryotes including fungi (yeast). (See Chapter 1 for a discussion of model organisms.) In some organisms, such as *Drosophila*, genes were mapped to specific sites on chromosomes, creating physical maps.

This approach, developed about 90 years ago, was the backbone of genetic analysis, and is still widely used in genetics. Chapter 21 considers these methods of genetic analysis in detail, as well as modern methods used to dissect gene function. The limitation of mutational analysis is that at least one mutation for each gene is required to identify all genes in a genome, and obtaining mutations can be a time-consuming endeavor. In addition, mutations often have a lethal phenotype or no clear phenotype, making it difficult or impossible to map the mutated gene.

Beginning in the 1980s, human geneticists began using recombinant DNA technology as a second approach to genetic analysis by mapping cloned DNA sequences to specific chromosomes. Most of these sequences were not genes, but markers such as restriction fragment length polymorphisms (RFLPs), single nucleotide polymorphisms (SNPs), and other molecular markers. Once assigned to chromosomes, these markers were used in pedigree analysis to establish linkage between the markers and genetic disorders. This method, called positional cloning (described in detail in Chapter 22), was used to map, isolate, clone, and sequence the genes for cystic fibrosis, neurofibromatosis, and dozens of other disorders. Although positional cloning identified one gene at a time, by the mid-1980s, more than 3500 genes and markers had been assigned to human chromosomes, putting geneticists on the threshold of producing high resolution maps of all human chromosomes.

In 1977, while recombinant DNA-based advances in genetic mapping were being developed, Fred Sanger and colleagues began the field of **genomics** (the study of genomes) by using a newly developed method of DNA sequencing to sequence the 5400-nucleotide genome of the virus ϕX174. Other viral genomes were sequenced in short order, but the process was slow and labor-intensive, limiting its use to small genomes. Over the next decade, the development of newer, automated sequencing methods made it possible to consider sequencing the larger and more complex genomes of eukaryotes, including the 3.2 billion nucleotides that make up the haploid human genome, which is some 600,000 times larger than the ϕX174 genome.

With several hundred genomes from both prokaryotes and eukaryotes now sequenced and more than a thousand projects underway, genomics has grown into a major field with wide-ranging goals (Table 20.1). Genomics encompasses several subfields, including **structural genomics**, **functional genomics**, and **comparative genomics**. Structural genomics includes the construction of genomic sequence data, gene discovery, and localization and the construction of gene maps. Functional genomics studies the biological function of genes and their regulation and products, and comparative genomics compares gene or protein sequences from different genomes to elucidate functional and evolutionary relationships. Comparative genomics is also important in developing model organisms for human diseases.

An outgrowth of genomics is **proteomics**, the study of the proteins present in a cell at a given time under a given set of conditions. Proteomics includes (1) the identification of all proteins expressed in a cell under a given set of conditions, (2) the nature and extent of any posttranslational modification of proteins, (3) the nature and extent of protein–protein interactions in the cell, and (4) the subcellular location of proteins in various cellular compartments.

Gathering and analyzing genome information is a large-scale, labor-intensive endeavor requiring the coordinated effort of many laboratories. For this reason, geneticists often form collaborations, called genome projects. One of the largest and best-known of these is the **Human Genome Project (HGP)**.

The Human Genome Project is a coordinated international effort to determine the sequence of the 3.2 billion base pairs in the haploid human genome and identify all the genes in the genome. Under the umbrella of the HGP, genome projects for a number of model organisms were established, including bacteria (*Escherichia coli*), yeast (*Saccharomyces cerevisiae*), a nematode (*Caenorhabditis elegans*), the fruit fly (*Drosophila melanogaster*), and the mouse (*Mus musculus*).

In this chapter, we describe the technology of genomics and review the significant findings in structural genomics from a variety of genome projects using prokaryotic and eukaryotic organisms. This review includes insights into the organization of selected genomes and how genes are organized in human chromosomes. We will also discuss some applications of comparative

TABLE 20.1 **GOALS OF GENOMICS**

- Compiling the genomic sequences of organisms
- Establishing the location of all genes in a genome
- Annotating the gene set in a genome
- Establishing the function of all genes in a genome
- Generating gene expression profiles for cells under differing conditions
- Identifying all the proteins that can be produced from a given genome
- Comparing genes and proteins between organisms to establish evolutionary relationships

genomics, including efforts to determine the minimum number of genes necessary for life, and the development of model systems for human disease. Finally, we will examine how the methods and applications of proteomics provide insight into the structure and function of organelles.

20.1 Genomics: Sequencing Is the Basis for Identifying and Mapping All Genes in a Genome

Geneticists use two different methods for sequencing genomes. The **clone-by-clone method** was the first to be developed [Figure 20–1(a)]. It begins with the construction of cloned libraries of large fragments that include all the DNA in an organism's genome. Next, the clones are assembled into genetic and physical maps encompassing the entire genome. The nucleotide sequence is determined clone by clone until the entire genome is sequenced. The clone-by-clone method was chosen for the publicly funded Human Genome Project sponsored by the NIH and the Department of Energy. This method relies on the construction of restriction maps and large genomic clones to provide the foundation for genome sequencing. Making large clones and generating restriction maps are time-limiting steps in this approach, and genome projects with this approach have long timelines for completion.

The **shotgun method** [Figure 20–1(b)] was made possible by advances in sequencing technology and development of software to assemble sequences from many short overlapping clones into a continuous sequence. In this method, two or three preparations of genomic DNA are made. One is sheared into short fragments of about 2 kb, another is sheared into longer fragments of 15–20 kb, and another preparation of 200–300 kb fragments might be made. A genomic library is constructed from each preparation. Clones are selected at random from each library and sequenced. Software is used to assemble long stretches of sequence from overlapping short fragments from the libraries, using the sequences from the larger clones as a framework.

This method, developed by Craig Venter and his colleagues at The Institute for Genome Research (TIGR), was used to sequence the genome of the bacterium *Haemophilus influenzae* in 1995. This was the first complete genome sequence from a free-living organism. After refining the method and using it to sequence the genomes of other prokaryotes, the shotgun method was used to sequence eukaryotic genomes, including *Drosophila*, dog, and humans. In a later section of this chapter, we will discuss the Human Genome Project in more detail and review some of the properties of the human genome discovered with these sequencing methods.

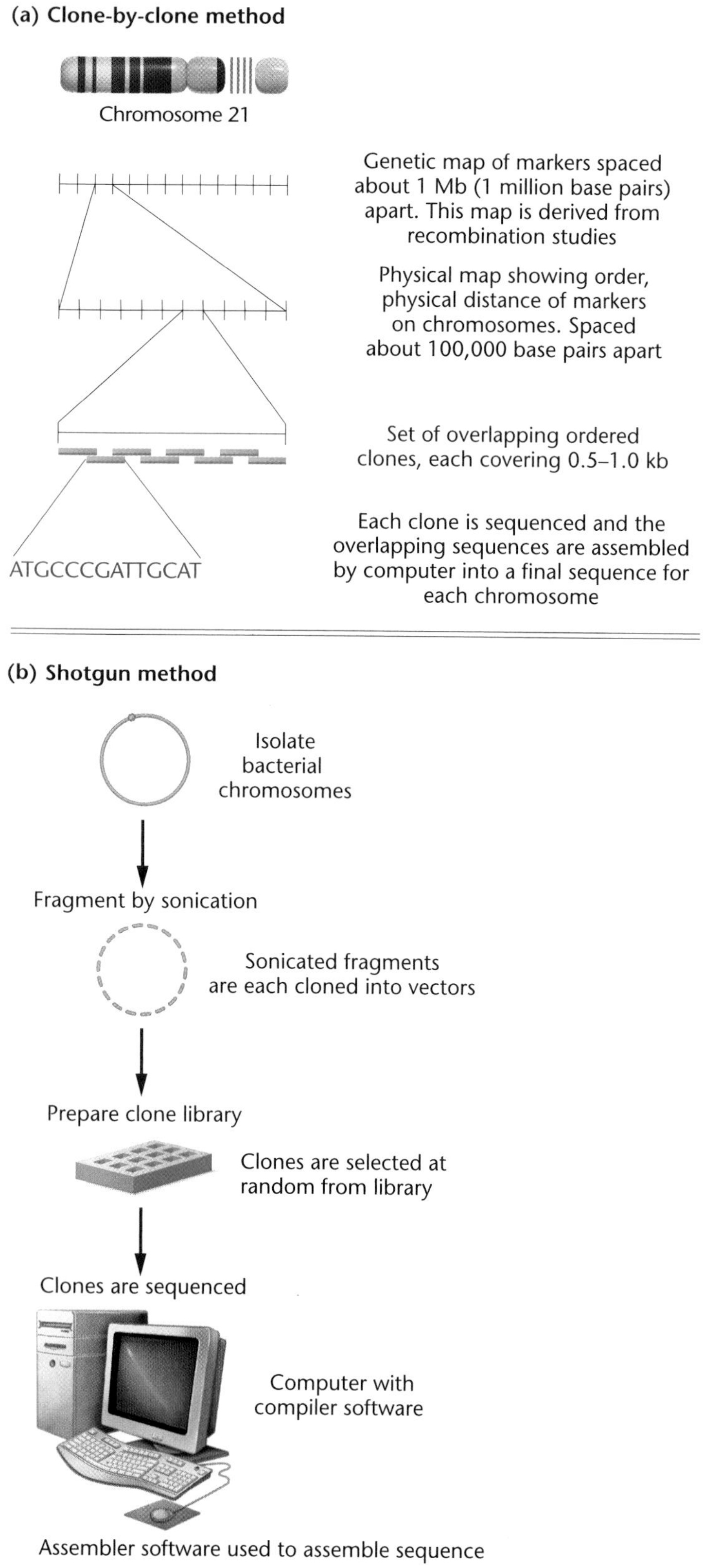
(a) Clone-by-clone method
Chromosome 21
Genetic map of markers spaced about 1 Mb (1 million base pairs) apart. This map is derived from recombination studies
Physical map showing order, physical distance of markers on chromosomes. Spaced about 100,000 base pairs apart
Set of overlapping ordered clones, each covering 0.5–1.0 kb
ATGCCCGATTGCAT
Each clone is sequenced and the overlapping sequences are assembled by computer into a final sequence for each chromosome
(b) Shotgun method
Isolate bacterial chromosomes
Fragment by sonication
Sonicated fragments are each cloned into vectors
Prepare clone library
Clones are selected at random from library
Clones are sequenced
Computer with compiler software
Assembler software used to assemble sequence

FIGURE 20–1 (a) In the clone-by-clone method, clones from a genomic library are organized into genetic and physical maps of each chromosome. After the clones are arranged into physical maps, they are broken into smaller, overlapping clones that cover each chromosome. Each smaller clone is sequenced, and the genomic sequence is assembled by stringing together the nucleotide sequence of the clones. (b) In the shotgun method, a genomic library is constructed from fragments of genomic DNA. Clones are selected from the library at random and sequenced. The sequence is assembled by looking for sequence overlaps between clones from different libraries. This is done with a computer, using assembler software designed for genomic analysis.

20.2 An Overview of Genomic Analysis

Genome projects generate large amounts of DNA sequence information. These data are useful only when they have been analyzed and interpreted. Several types of analyses using software and databases are necessary to identify and characterize gene-coding regions of anonymous DNA sequences. Some of the goals of these analyses include:

1. Ensuring the sequence is complete and accurate.
2. Identifying genes by finding open reading frames (ORFs).
3. Finding promoter sites, transcription initiation sites, and translation initiation sites to confirm gene identity.
4. In eukaryotic genomes, finding splice sites, introns, and exons.
5. Translating the DNA sequence into a protein sequence, searching all six reading frames.
6. Comparing the DNA sequence to known protein sequences to verify that a gene has been located.

The predicted protein sequences are analyzed to predict function. The function can often be inferred if a sequence of one or more homologous proteins with known function can be found in a database. The ultimate goal of sequence analysis is to derive a complete functional description of all genes in the genome of an organism.

Compiling the Sequence

To ensure that the nucleotide sequence of a genome is complete and error-free, the genome is sequenced more than once. This process is known as **compiling**. For example, using the shotgun method on the genome of the bacterium *Pseudomonas aeruginosa*, researchers sequenced the 6.3 million nucleotides seven times to ensure that the sequence was accurate. Even with this level of redundancy, the assembler software recognized 1604 regions requiring further clarification. These regions were reanalyzed and resequenced to improve accuracy. Finally, the shotgun sequence was compared with the sequence of two widely separated genomic regions obtained by conventional cloning. The sequence of the 81,843 nucleotides cloned and sequenced by the clone-by-clone method was in perfect agreement with the sequence obtained by the shotgun method. This level of care is not unusual; similar precautions are taken in all genome projects.

The HGP sequenced the euchromatic portion of the human genome 12 times. The privately run shotgun-cloning project based at Celera, a biotechnology company, examined this part of the genome more than 35 times. A draft of the genome was completed in 2001, with several tasks unfinished. These included obtaining the remining 10 percent of the sequence, correcting errors (proofreading the genome), and filling sequence gaps (totaling less than 150 Mb). Updates to the sequence were published in 2003 and 2004. Sometime in the future, researchers will begin sequencing the 15 percent of the genome that contains heterochromatin. A final version of the euchromatic portion of the genome was published in 2003. Heterochromatic regions of the genome were originally excluded by design, as they contain long stretches of repetitive DNA sequences initially thought to contain no genes. However, while sequencing the *Drosophila* genome, researchers discovered that heterochromatic regions contain a small number of genes (about 50 in *Drosophila*). As a result, heterochromatic regions of the human genome must now be sequenced to ensure that all genes are identified. Once the human genome or any other genome is sequenced, compiled, and proofread, the next stage—annotation—begins.

HOW DO WE KNOW?

How can the accuracy and completeness of a genome be verified? Is there a way to detect missing sequences?

Annotating the Sequence

After a genome has been sequenced, compiled, and checked for accuracy, the next task is to find all the genes that encode products (proteins and RNA). This is the first step in **annotation**—a process that identifies genes, their regulatory sequences, and their function(s). Annotation also identifies non-protein coding genes (including ribosomal RNA, transfer RNA, and small nuclear RNAs) and finds and characterizes the mobile genetic elements and repetitive-sequence families that may be present in the genome.

Locating protein-coding genes is done by analyzing the sequence using software. Sequences that encode genes have several identifying features. Protein-coding genes are composed of **open reading frames (ORFs)**, nucleotide triplets that can be translated into the amino acid sequence of a protein. ORFs begin with an initiation sequence (usually ATG) and end with a termination sequence (TAA, TAG, or TGA).

Because genetic information is encoded in groups of three nucleotides, it is not clear from raw sequence data whether to begin the analysis at the first, second, or third nucleotide of the sequence. Analysis therefore begins with the first nucleotide and examines the sequence three nucleotides at a time, searching for ORFs. Further rounds of analysis begin with the second and third nucleotide in the sequence, respectively, looking for ORFs three nucleotides at a time. Thus, there are three possible reading frames in a DNA sequence.

However, searching a DNA sequence for all three reading frames does not finish the job. The data provide only the nucleotide sequence of a single DNA strand. This strand may *not* be the one that encodes a gene; that information might be on the complementary strand. This means that ORF searches must examine the sequenced strand *and* its complementary strand to determine if either contains a protein-coding gene. For this analysis, the sequenced strand is converted to its complementary strand (usually by a software program). As with the sequenced strand, the ORF search on the complementary strand analyzes all three possible reading frames. This means that six possible reading frames must be examined in searching for ORFs in an unidentified DNA sequence.

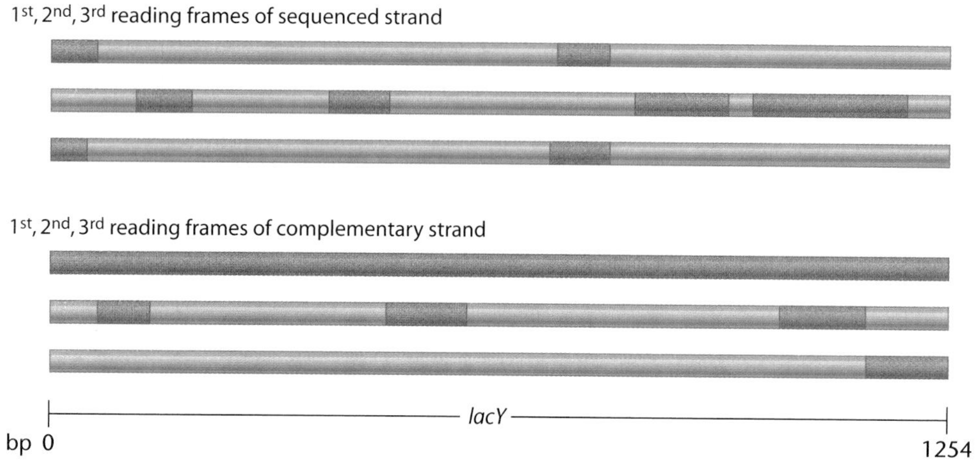

FIGURE 20–2 ORF analysis of the nucleotide sequence of the *E. coli lacY* gene. ORFs for each of the six possible reading frames, three on each DNA strand, are shown as dark regions. Only ORFs longer than 50 codons are marked. Because most genes are much longer than 50 codons, ORF length is one indication of an actual gene. In this case, only one ORF covers the entire sequence and shows the location and orientation of the *lacY* gene.

Figure 20–2 shows the results of a computer-based search for ORFs on both strands and each reading frame for the nucleotide sequence of the *lacY* gene in *E. coli*. Notice that each reading frame identifies ORFs, but only one ORF covers the entire sequence. Incorrect reading frames usually specify very short peptide sequences.

If ORF sequences have been identified in the sequence before analysis begins, the number of reading frames to be scanned can be reduced. Searching for ORFs starting with an ATG followed at some distance by a termination sequence is one strategy for finding genes.

Now solve this

Problem 20.7 on page 513 involves searching a genomic sequence for ORFs.

Hint: In setting limits for the size of ORFs that may represent genes, be aware of the average gene length in the organism's genome.

The organization of genes in eukaryotic genomes (including the human genome) makes direct searching for ORFs more difficult. First, many eukaryotic genes have introns, noncoding regions that interrupt coding regions (exons). As a result, many, if not most, eukaryotic genes are not organized as continuous ORFs; instead, the sequence consists of ORFs interspersed with sequences representing introns. Early versions of ORF-scanning software often interpreted exons as separate genes because termination codons are common in introns. Second, genes in humans and other eukaryotes are often widely spaced, increasing the chances of finding false ORFs in the regions between gene clusters.

Figure 20–3(a) shows a portion of the human genome sequence. From a casual inspection, it is not clear whether this sequence contains one or more genes, and if so, where each begins and ends. In analyzing this sequence, identifiable footprints of genes provide clues to what the sequence contains. Control regions at the beginning of genes are marked by a TATA sequence. Splice sites between exons and introns have a predictable sequence (most begin with GT and end with AG), as does the end of the gene where a poly-A tail is added to the transcript. In addition, if a DNA sequence encodes a protein, the sequence contains an open reading frame. Knowledge derived from sequencing cDNA and annotated genes from other organisms also provide clues that can be used in deciphering a sequence.

Computer software is used to scan sequences, searching for these and other features of genes. Analysis of the sequence in Figure 20–3(b) shows it contains a control region and three exons. The two unshaded regions between the exons represent introns, regions removed as the mRNA is processed. Using this sequence to search genomic sequence databases shows this is the sequence of the human β-globin gene.

Newer versions of ORF-scanning software designed for ORF analysis of eukaryotic genomes are highly efficient. These programs search for features mentioned above, and other characteristics of genes such as codon bias and in some genomes, CpG islands.

Codon bias is the selective use of one or two codons to encode amino acids that can be specified by a number of different codons. For example, alanine can be encoded by GCA, GCT, GCC, and GCG. If the codons were used randomly, each would be used about 25 percent of the time. Yet in the human genome, GCC is used 41 percent of the time, and GCG only 11 percent of the time. Codon bias is present in exons but should not be present in introns or intergenic spacers. **CpG islands** are DNA regions containing many copies of this nucleotide doublet. They are found in gene-rich regions of the mammalian genome, often adjacent to genes. However, searching for codon bias and other features of eukaryotic genes such as CpG islands is not foolproof, and scanning by eye is a common backup technique.

20.3 Functional Genomics Classifies Genes and Identifies Their Functions

After a genomic sequence is annotated, the next task is to assign functions to all the genes in the sequence. Some of the genes have had functions assigned by the classic method of mutagenesis and linkage mapping (described in Chapter 21), but many other genes have no clear-cut function assigned. One approach to assigning

(a)

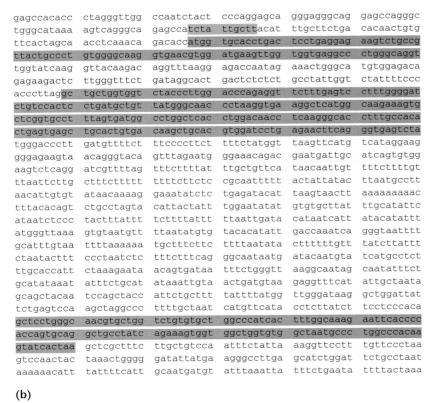

```
gagccacacc  ctagggttgg  ccaatctact  cccaggagca  gggagggcag  gagccagggc
tgggcataaa  agtcagggca  gagccatcta  ttgcttacat  ttgcttctga  cacaactgtg
ttcactagca  acctcaaaca  gacaccatgg  tgcacctgac  tcctgaggag  aagtctgccg
ttactgccct  gtggggcaag  gtgaacgtgg  atgaagttgg  tggtgaggcc  ctgggcaggt
tggtatcaag  gttacaagac  aggtttaagg  agaccaatag  aaactgggca  tgtggagaca
gagaagactc  ttgggtttct  gataggcact  gactctctct  gcctattggt  ctattttccc
acccttaggc  tgctggtggt  ctacccttgg  acccagaggt  tctttgagtc  ctttggggat
ctgtccactc  ctgatgctgt  tatgggcaac  cctaaggtga  aggctcatgg  caagaaagtg
ctcggtgcct  ttagtgatgg  cctggctcac  ctggacaacc  tcaagggcac  ctttgccaca
ctgagtgagc  tgcactgtga  caagctgcac  gtggatcctg  agaacttcag  ggtgagtcta
tgggaccctt  gatgttttct  ttccccttct  tttctatggt  taagttcatg  tcataggaag
gggagaagta  acagggtaca  gtttagaatg  ggaaacagac  gaatgattgc  atcagtgtgg
aagtctcagg  atcgttttag  tttcttttat  ttgctgttca  taacaattgt  tttcttttgt
ttaattcttg  ctttcttttt  ttttcttctc  cgcaattttt  actattatac  ttaatgcctt
aacattgtgt  ataacaaaag  gaaatatctc  tgagatacat  taagtaactt  aaaaaaaaac
tttacacagt  ctgcctagta  cattactatt  tggaatatat  gtgtgcttat  ttgcatattc
ataatctccc  tactttattt  tcttttattt  ttaattgata  cataatcatt  atacatattt
atgggttaaa  gtgtaatgtt  ttaatatgtg  tacacatatt  gaccaaatca  gggtaatttt
gcatttgtaa  ttttaaaaaa  tgctttcttc  ttttaatata  ctttttttgtt  tatcttattt
ctaatacttt  ccctaatctc  tttctttcag  ggcaataatg  atacaatgta  tcatgcctct
ttgcaccatt  ctaaagaata  acagtgataa  tttctgggtt  aaggcaatag  caatatttct
gcatataaat  atttctgcat  ataaattgta  actgatgtaa  gaggtttcat  attgctaata
gcagctacaa  tccagctacc  attctgcttt  tattttatgg  ttgggataag  gctggattat
tctgagtcca  agctaggccc  ttttgctaat  catgttcata  cctcttatct  tcctcccaca
gctcctgggc  aacgtgctgg  tctgtgtgct  ggcccatcac  tttggcaaag  aattcacccc
accagtgcag  gctgcctatc  agaaagtggt  ggctggtgtg  gctaatgccc  tggcccacaa
gtatcactaa  gctcgctttc  ttgctgtcca  atttctatta  aaggttcctt  tgttccctaa
gtccaactac  taaactgggg  gatattatga  agggccttga  gcatctggat  tctgcctaat
aaaaaacatt  tattttcatt  gcaatgatgt  atttaaatta  tttctgaata  ttttactaaa
```

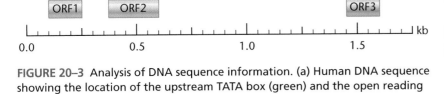

(b)

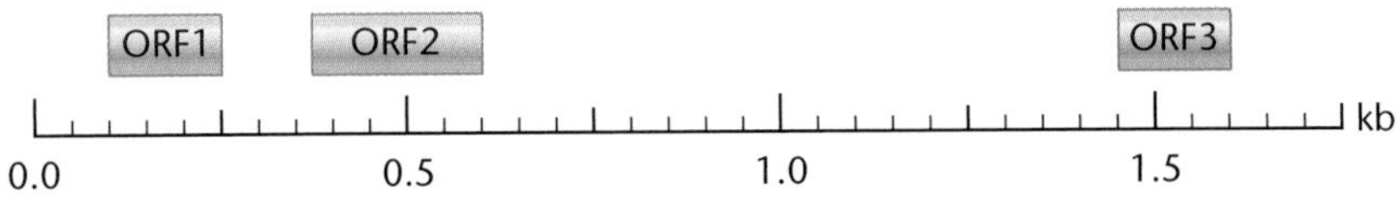

FIGURE 20–3 Analysis of DNA sequence information. (a) Human DNA sequence showing the location of the upstream TATA box (green) and the open reading frames (ORFs) for three exons (blue) of the human β globin gene.
(b) A diagram showing the size and location of open reading frames for the three exons of the human β-globin gene encoded by the sequence in (a).

functions to these genes is the use of homology searches. This analysis has several components:

1. Search databases such as GenBank to find similar genes isolated from other organisms.
2. Compare the sequence of an ORF with that of a well-characterized gene from another organism.
3. Scan the ORF for functional motifs, regions of DNA that encode protein domains such as ion channels, DNA-binding regions, or secretion/export signals, etc.

Let's examine how this strategy is used in the analysis of a prokaryotic genome.

Functional Genomics of a Bacterial Genome

Table 20.2 shows the results of an ORF analysis in the *P. aeruginosa* genome. In this analysis, as in all functional genomic analyses, ORFs are classified on a scale according to confidence levels. Level 1 genes are those whose function is already known from conventional genetic analysis or are genes otherwise known to have a demonstrated function. Level 2 genes have a strong homology to genes with known functions in other organisms. Level 3 genes have a proposed function based on encoded motifs or limited homology to genes in other organisms. Level 4 genes have no known or proposed function. In a functional analysis of the *P. aeruginosa* genome (Table 20.3), about half the genes have no known or proposed function. This situation is not unique to *P. aeruginosa*; in almost all organisms whose genomes have been sequenced to date, about half of their genes have no known function.

The goal in the analysis of the *P. aeruginosa* genome, as in all genomic analyses, is to move all of the genes up to Level 1 or 2 and to understand how these genes and their products interact in the biology of the organism. The levels used in functional genomics do not reflect the importance of a gene to an organism's survival or reproduction but are assignments based on our knowledge of a gene's function.

Strategies for Functional Assignments of Unknown Genes

Yeast (*S. cerevisiae*) is one of the model eukaryotic organisms in genetic analysis. Despite decades of study using classical genetic analysis, only about one third of the 6200 genes revealed by genomic sequencing had been fully characterized (assigned as Level 1 genes). In the years following publication of the yeast genome, the number of characterized genes has risen to over 3700, and an additional 500 or so genes have been assigned as Level 2 or 3 genes based on homology. This leaves about 1900–2000 genes that encode proteins of no known function.

Because an array of genetic, biochemical, and molecular techniques are well established for studying yeast, it is instructive to see how these are being applied in functional analyses of the yeast genome. The use of these techniques has changed—they are now being integrated and applied to carry out functional analysis on a genomic scale, instead of studying just one gene or protein at a time. In Chapter 21 we explore some current methods geneticists employ to analyze gene function in model organisms. Functional genomics now uses an extensive array of experimental approaches, from classical mutagenesis screens to reverse genetics, a method that begins with a genome sequence or a purified protein and progresses to functional analyses.

TABLE 20.2 ANALYSIS OF THE *PSEUDOMONAS AERUGINOSA* GENOME

General Features

Genome size (bp)	6,264,403
G + C content	66.6%
Coding regions	89.4%

Coding Sequences

Confidence level	ORFs	(%)	Definition
1	372	6.7	*P. aeruginosa* genes with demonstrated function
2	1059	19.0	Strong homology to genes with demonstrated function from other organisms
3	1590	28.5	Genes with proposed function based on motif searches or limited homology
4a	769	13.8	Homologs of reported genes of unknown function
4b	1780	32.0	No homology to any reported sequences
Total	5570	100	

Source: Stover, *et al.* 2000. Complete genome sequence of *Pseudomonas aeruginosa* PA01, an opportunistic pathogen. *Nature* 406:959–64. Table 1, p. 961.

TABLE 20.3 FUNCTIONAL CLASSES OF PREDICTED GENES IN *PSEUDOMONAS*

	ORFs	
Functional Class	**Number**	**%**
Adaptation, protection (e.g., cold shock proteins)	60	1.1
Amino acid biosynthesis and metabolism	150	2.7
Antibiotic resistance and susceptibility	19	0.3
Biosynthesis of cofactors, prosthetic groups, and carriers	119	2.1
Carbon compound catabolism	130	2.3
Cell division	26	0.5
Cell wall	83	1.5
Central intermediary metabolism	64	1.1
Chaperones and heat shock proteins	52	0.9
Chemotaxis	43	0.8
DNA replication, recombination, modification, and repair	81	1.5
Energy metabolism	166	3.0
Fatty acid and phospholipid metabolism	56	1.0
Membrane proteins	7	0.1
Motility and attachment	65	1.2
Nucleotide biosynthesis and metabolism	60	1.1
Protein secretion/export apparatus	83	1.5
Putative enzymes	409	7.3
Related to phage, transposon, or plasmid	38	0.7
Secreted factors (toxins, enzymes, alginate)	58	1.0
Transcriptional regulators	403	7.2
Transcription, RNA processing, and degradation	45	0.8
Translation, posttranslational modification, and degradation	149	2.7
Transport of small molecules	555	10.0
Two-component regulatory systems	118	2.1
Hypothetical	1774	31.8
Unknown (conserved hypothetical)	757	13.6
Total	5570	100

Source: Stover, *et al.* 2000. Complete genome sequence of *Pseudomonas aeruginosa* PA01, an opportunistic pathogen. *Nature* 406:959–64.

How Do We Know?

How is the function of proteins encoded in a genome determined? What if the gene encoding the protein is unique and has never been seen in any other organism?

20.4 Prokaryotic Genomes Have Some Unexpected Features

Sequencing of more than one hundred prokaryotic and eukaryotic genomes has been completed. The results indicate that there are significant differences in genome organization between prokaryotes and eukaryotes. Therefore, we will consider the organization of these genomes separately. Since most prokaryotes have small genomes amenable to shotgun cloning and sequencing, most of the completed projects have focused on two types of prokaryotes, Eubacteria and Archaea (Table 20.4), and there are more than 200 additional projects under way. Many of the prokaryotic genomes already sequenced are from organisms that cause human diseases, such as cholera, tuberculosis, and leprosy.

Table 20.4 Genome Size and Gene Number in Selected Prokaryotes

	Genome Size (Mb)	Number of Genes
Archaea		
Archaeoglobus fulgidis	2.17	2493
Methanococcus jammaschii	1.66	1738
Thermoplasma acidophilium	1.56	1509
Eubacteria		
E. coli	4.64	4397
Bacillus subtilis	4.21	4212
H. influenzae	1.83	1791
Aquifex aeolicus	1.55	1552
Rickettsia prowazekii	1.11	834
Mycoplasma pneumoniae	0.82	710
Mycoplasma genitalium	0.58	503

Size Range of Eubacterial Genomes

Based on genome project results, we can make a number of generalizations concerning the size and organization of the genomes of Eubacteria. Traditionally, the bacterial genome has been thought of as relatively small (less than 5 Mb) and contained within a single circular DNA molecule. The flood of genomic information now available has challenged all parts of this viewpoint. Although most prokaryotic genomes are small, genome sizes vary across a surprisingly wide range. In fact, there is some overlap in size between larger bacterial genomes (30 Mb in *Bacillus megaterium*) and smaller eukaryotic genomes (12.1 Mb in yeast).

Linear Chromosomes and Multiple Chromosomes in Bacteria

While most prokaryotic genomes sequenced to date *are* circular DNA molecules, an increasing number of genomes composed of linear DNA molecules are being identified, including the genome of *Borrelia burgdorferi*, the organism that causes Lyme disease. Perhaps more important, genome data are blurring the distinction between plasmids and bacterial chromosomes and redefining our view of the bacterial genome as a single DNA molecule (Table 20.5).

Plasmids are small, usually circular DNA molecules containing genes that are not essential to their bacterial host cell. Plasmids replicate using the same enzymes used to copy the chromosome of the host cell, and are distributed to daughter cells along with the host chromosome during cell division. Because plasmids carry nonessential genes and can be transferred from one cell to another, and because the same plasmid is often present in bacteria from different species, plasmid genes are not included as part of a bacterial genome. However, *B. burgdorferi* carries approximately 17 plasmids, which contain at least 430 genes. Some of these genes are essential to the bacterium, including genes for purine biosynthesis and membrane proteins, which in most species are carried on the bacterial chromosome.

Table 20.5 Bacterial Genome Organization

Bacterial species	Chromosome Number and Organization	Plasmid Number
Agrobacterium tumefaciens	one linear + one circular	two circular
Bacillus subtilis	one circular	
Bacillus thuringiensis	one circular	six
Borrella burgdorferi	one linear	>17 circular + linear
Buchnera sp.	one circular	two circular
Deinococcus radiodurans	two circular	two circular
Escherichia coli	one circular	
Vibrio cholerae	two circular	

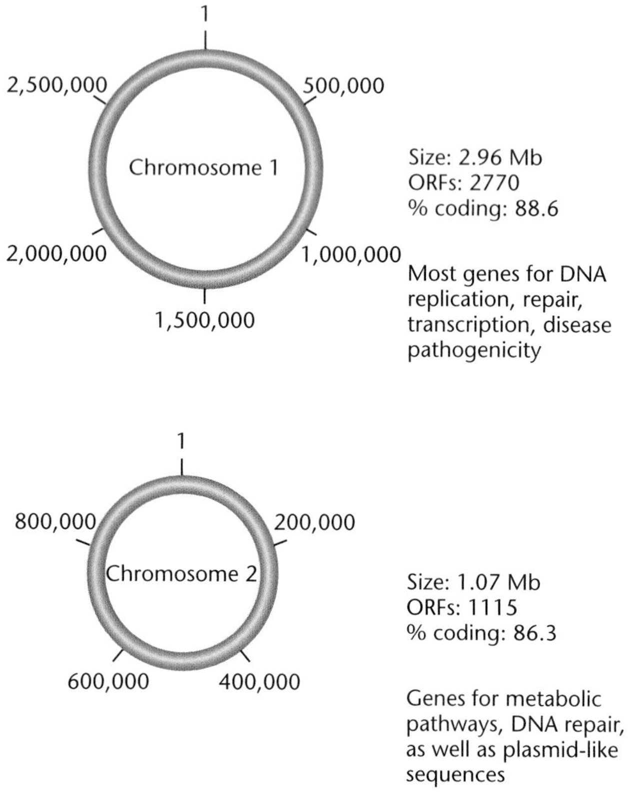

FIGURE 20–4 The *Vibrio cholerae* genome is contained in two chromosomes. The larger chromosome (chromosome 1) contains most of the genes for essential cellular functions and infectivity. Most of the genes on chromosome 2 (52 percent of 1115) are of unknown function. The bias in gene content and the presence of plasmid-like sequences on chromosome 2 suggest that this chromosome was a megaplasmid captured by an ancestral *Vibrio* species.

Recently, genome sequencing of *Vibrio cholerae*, the organism responsible for cholera, revealed the presence of two circular chromosomes (Figure 20–4). Chromosome 1 (2.96 Mb) contains 2770 ORFs, and chromosome 2 (1.07 Mb) contains 1115 ORFs, some of which encode essential genes, such as ribosomal proteins. Based on its nucleotide sequence, chromosome 2 is apparently derived from a plasmid captured by an ancestral species, or it could represent a plasmid with many inserted chromosomal genes.

Two chromosomes are also present in other species of *Vibrio*. The existence of prokaryotic genomes containing multiple chromosomes raises an important question. What is the difference between a chromosome and a large plasmid? Most researchers agree that bacterial chromosomes have some definable properties. First is gene content; genes essential to survival (e.g., genes for DNA replication, transcription, metabolic genes, etc.) are found on chromosomes. Second is size; chromosomes should constitute a significant portion of the genome. Third, chromosomes should be found in all strains of a species, while plasmids are not. In *Vibrio cholerae*, chromosome 2 contains essential genes, represents 40 percent of the genome, and all strains examined to date contain this chromosome. Based on these criteria, it is a chromosome, even though it contains plasmid-derived sequences. Other questions about the origin of this chromosome and its evolutionary relationship to plasmids remain to be investigated, and may shed light on its ability to cause cholera.

Other bacteria also have genomes with two chromosomes, including *Agrobacterium tumefaciens*, *Rhodobacter sphaeroides*, and *Rhizobium meliloti* (Table 20.5). The finding that some bacterial species have two chromosomes also raises the question of what regulatory mechanisms control gene expression and metabolism in a multichromosome prokaryotic genome. Answers to this and other questions raised by genomic discoveries will redefine some ideas about prokaryotic genomes and the nature of plasmids and may provide clues about the evolution of multichromosome eukaryotic genomes.

20.5 Genomes of Eubacteria

We can make two generalizations about the organization of protein-coding genes in the chromosomes of eubacteria. First, gene density is very high, averaging about one gene per kilobase of DNA. For example, *E. coli* has a 4.6 Mb genome containing 4288 protein-coding genes, a density very close to one gene per kilobase. *Mycoplasma genitalium* has a small genome (0.58 Mb), with 503 genes, and its gene density is also close to one gene per kilobase pair. This close packing of genes in prokaryotic genomes results in a very high proportion of the DNA (approximately 85–90 percent) serving as coding DNA. In *M. genitalium*, there are only about 110–125 bp of DNA between genes. Typically, only a small amount of a bacterial genome is noncoding DNA, usually in the form of transposable elements that can move from one place to another in the genome. Introns are extremely rare in bacterial genomes.

A second generalization we can make is that bacterial genomes are characterized by the presence of operons. The organization of a small segment of the *E. coli* genome is shown in Figure 20–5. The lines above the genes show how those genes are organized into transcription units with promoters that locate the start of transcription and the di-

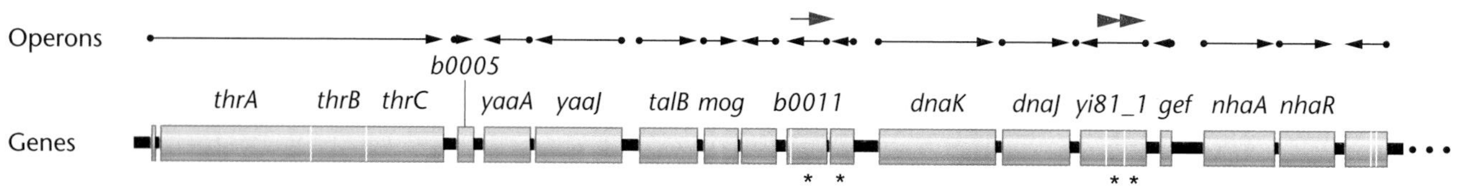

FIGURE 20–5 A portion of the *E. coli* chromosome showing genes and operons. A dot indicates the promoter for each gene or operon. Arrows indicate the direction of transcription. Overlapping genes are identified by asterisks.

rection of transcription for each unit. Some of these units are organized as operons. In *E. coli*, 27 percent of the predicted transcription units are contained in operons (almost 600 operons). In other bacterial genomes, the organization of genes into transcriptional units is challenging our ideas about the nature of operons. In *Aquifex aeolicus*, most genes are parts of polycistronic transcription units we would usually call operons. However, our working definition that operons contain genes whose products are part of a single biochemical pathway does not hold true in this organism. In *A. aeolicus* (Figure 20–6), one operon contains six genes: two for DNA recombination, one for lipid synthesis, one for nucleic acid synthesis, one for protein synthesis, and one that encodes a protein for cell motility. Other operons in this species also contain genes with several different functions. This finding, combined with similar results from other genome projects, raises questions about the nature of operons and their role in biochemical processes in bacterial cells and may cause us to redefine our concept of the operon.

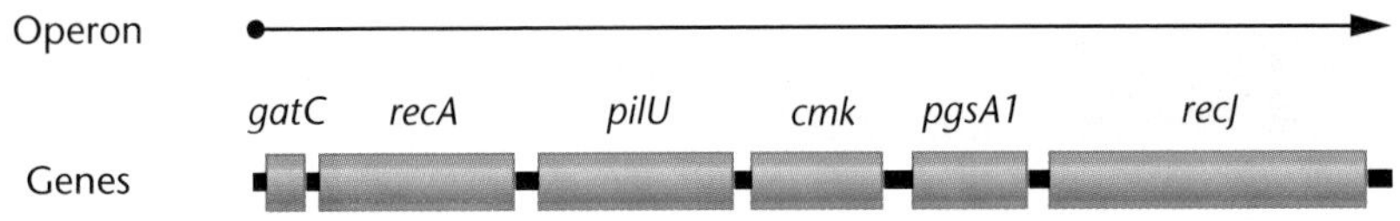

FIGURE 20–6 An operon from the *A. aeolicus* genome contains genes for protein synthesis (*gatC*), for DNA recombination (*recA* and *recJ*), for a motility protein (*pilU*), for nucleotide biosynthesis (*cmk*), and for lipid biosynthesis (*pgsA1*). This organization challenges the conventional idea that genes in an operon encode products that control a common biochemical pathway.

Note that there are several cases of overlapping genes in *E. coli* (Figure 20–5), where one gene is nested inside another on the same DNA strand (shown in green). Overlapping genes are common in viruses (where space is at a premium), but are infrequent in bacteria and most other organisms.

Genomes of Archaea

Archaea (formerly known as Archaebacteria) are one of the three major domains of living organisms. The other two are the Eubacteria (true bacteria, without a membrane-bound nucleus, such as *E. coli* and *Bacillus*) and Eukarya (eukarotes, organisms with a membrane-bound nucleus). Archaea, like Eubacteria, are prokaryotes; that is, they have no nucleus. They have only recently been recognized as a separate domain of life.

Typically, Archaea are extremophiles; they live in extreme environments with very high temperature, high salt content, high pressure, or extreme pH. With the completion of several genome projects from the Archaea, we can examine their relationship to organisms in the other domains. The archaeon *Methanococcus jannaschii* (Figure 20–7) was originally isolated from a high-temperature deep-sea vent 2600 meters below sea level, where it lives at a pressure of 270 atmospheres and a temperature near 300°C. It has a circular, double-stranded DNA genome of 1.66 Mb with 1738 protein-coding genes organized into three chromosomes: a large circular chromosome of 166 Mb and two small circular chromosomes of 58.4 and 16.5 kb. Most of the genes in this organism (58 percent) do not match any other known genes, but the majority of genes involved in energy production, cell division, and general metabolism closely resemble those of Eubacteria. Gene organization also resembles Eubacteria; genes are densely packed, have operons, and do not have introns.

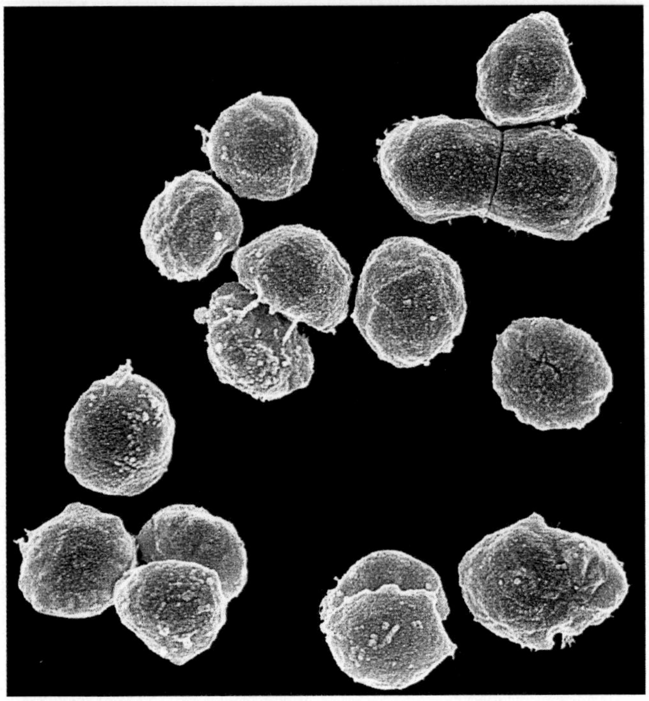

FIGURE 20–7 A colorized scanning electron micrograph showing cells of the archeon, *Methanococcus jannashcii*, a prokaryote with three chromosomes.

On the other hand, Archaea genes also have significant similarities to eukaryotic genes. The genes involved in RNA synthesis, protein synthesis, and DNA synthesis closely resemble those in eukaryotes. Most surprising is the presence of histone chromosomal proteins and evidence that chromosomal DNA is organized into chromatin. Although they have no introns in their protein-coding genes, Archaea do have introns in their tRNA genes, as do eukaryotes.

20.6 Eukaryotic Genomes Have Several Organizational Patterns

Although the basic features of the eukaryotic genome are similar in different species, genome size is highly variable (Table 20.6). Genome sizes range from about 10 Mb in fungi to over 100,000 Mb in some flowering plants (a ten-thousandfold range), the number of chromosomes per genome ranges from two into the hundreds (about a hundredfold range), but the number of genes varies much less than either genome size or chromosome number.

TABLE 20.6 GENOME SIZE AND GENE NUMBER IN SELECTED EUKARYOTES

Organism	Genome Size (Mb)	Number of Genes
S. cerevisiae (yeast)	12.1	6548
Plasmodium falciparum (malaria)	30	≈6500
C. elegans (nematode)	97	>20,000
Arabidopsis thalania (mustard plant)	120	≈20,000
D. melanogaster (fruit fly)	170	≈16,000
Oryza sativa (rice)	415	≈20,000
Zea mays (maize)	2500	≈20,000
Homo sapiens (human)	3300	≈25,000
Hordeum vulgare (barley)	5300	≈20,000

General Features of Eukaryotic Genomes

Compared with prokaryotes, eukaryotes have a relatively low gene density. Figure 20–8 compares gene density in genome regions of *E. coli* with those of yeast, humans, and maize. In general, more complex eukaryotes have less compact genomes with lower levels of gene density. If we compare regions of different eukaryotic genomes, several features become apparent:

- **Gene density.** The 50-kb region of the yeast genome from chromosome III in Figure 20–8(b) contains more than 20 genes, while the 50-kb segment of the human genome from chromosome 16 in Figure 20–8(c) contains only 6 genes.
- **Introns.** None of the yeast genes shown in Figure 20–8(b) contain introns, whereas most human genes contain introns. In fact, the entire yeast genome has only 239 introns, while just a single gene in the human genome can contain more than 100 introns.
- **Repetitive sequences.** The presence of introns and the existence of repetitive sequences are two major reasons for the wide range of genome sizes in eukaryotes. In some plants, such as maize, repetitive sequences are the dominant feature of the genome. The maize genome has about 2500 Mb of DNA, more than two thirds of which is composed of repetitive DNA, resulting in very low gene density in the genome of this species [Figure 20–8(d)].

Transcriptional Units in the *C. elegans* Genome

Although eukaryotic genomes share many features, specific genomic features vary widely. The nematode *C. elegans* has a 97 Mb genome organized into six chromosomes, with a total of about 20,000 genes. Gene density is much lower than in yeast—3 times more genes and 8 times more DNA, with an average density of about 1 gene per 5 kb. About half of the *C. elegans* genome is composed of intergenic spacer DNA, and a significant fraction of the genome is highly repetitive simple-sequence DNA (sequences such as ATAT repeated tens of thousands of times).

The organization of genes in the *C. elegans* genome differs dramatically from most other eukaryotes. Approximately 25 percent of the genes in *C. elegans* genes are organized into polycistronic transcription units, or operons, like those in bacteria (Figure 20–9). In addition, compared with yeast, introns are far more prevalent in *C. elegans* genes, whose average gene has five introns. In fact, 26 percent of the *C. elegans* genome is introns; in addition, many of its genes are found within the introns of other genes (Figure 20–10).

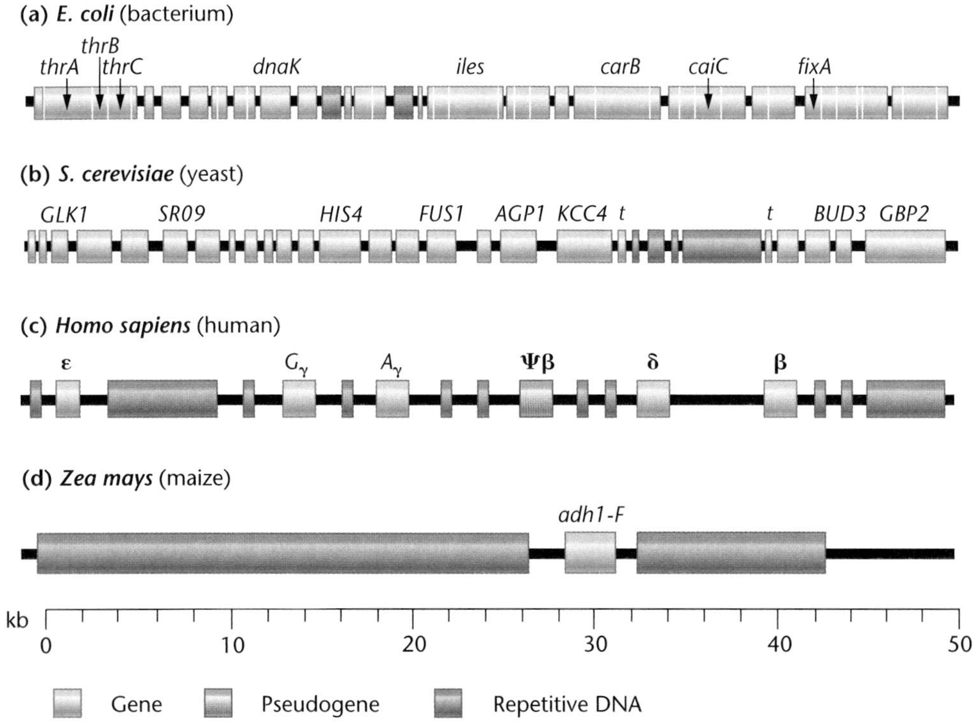

FIGURE 20–8 (a) In *E. coli*, gene density is high, and there are very few repetitive sequences. (b) A 50-kb region from chromosome 3 of yeast contains over 20 genes and little repetitive DNA. (c) A 50-kb region from human chromosome 11 contains 5 genes, 1 pseudogene, and stretches of repetitive DNA. (d) 50 kb of the maize genome surrounding the *Adh* locus. This gene is surrounded by long stretches of repetitive DNA. In eukaryotes (b–d), gene density is lower, and portions of the genome are occupied by repetitive DNA sequences.

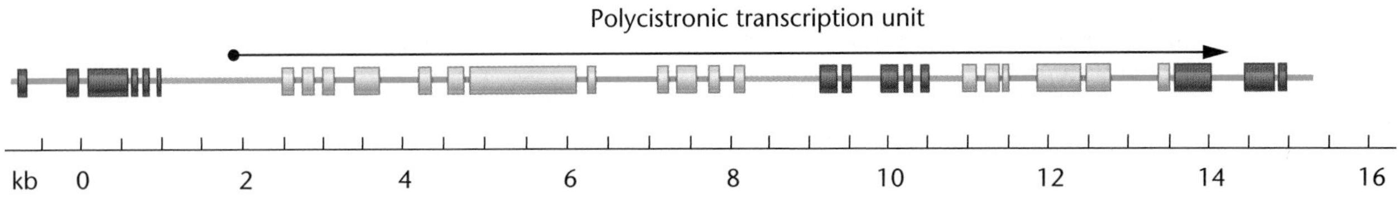

FIGURE 20–9 A region on chromosome 3 of *C. elegans* showing the location and organization of five genes with introns for each gene indicated by a different color. The central three genes are part of an operon and are transcribed as a set to form a polycistronic mRNA. Operons are common in bacterial genomes but rare in most eukaryotic genomes. In *C. elegans*, however, about 25 percent of all genes are part of operons.

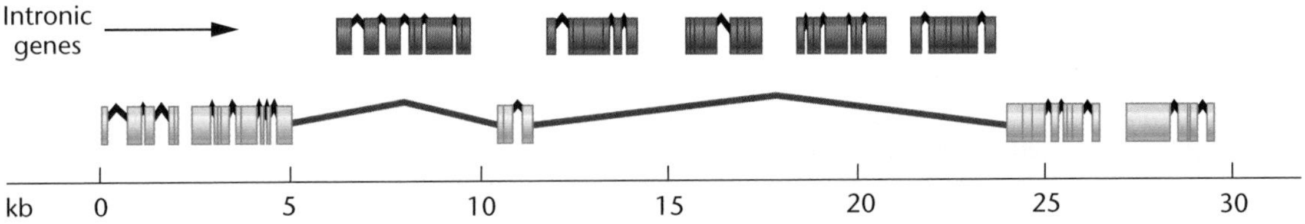

FIGURE 20–10 A 30-kb section of the *C. elegans* genome showing the location and organization of several genes. The first gene encodes a lipase, spans just over 25 kb, and contains two large introns. Five other genes are located within the introns of the lipase gene.

Genomes of Higher Plants

To geneticists, the small flowering plant *Arabidopsis thaliana* is the *Drosophila* of the plant world (Figure 20–11). Its small genome, 120 Mb distributed in 5 chromosomes, is comparable in size to those of *C. elegans* and *Drosophila*. It contains an estimated 20,000 genes with a gene density of 1 gene per 5 kb, which also resembles that of *C. elegans* and *Drosophila*. At least half of its genes are closely related to genes found in bacteria and humans. Because of its compact genome, *Arabidopsis* is a model organism for studying other plants with larger genomes, such as the economically important agricultural crop plants such as rice, maize, and barley.

Analysis of the nucleotide sequence of chromosomes 2 and 4 of *Arabidopsis* reveals a dynamic genome. Both chromosomes contain many tandem gene duplications (239 duplications on chromosome 2, involving 539 genes) as well as larger duplications involving 4 blocks of DNA sequences, spanning 2.5 Mb. There is some interchromosomal gene duplication as well; genes on chromosome 4 are also present on chromosome 5. Genomic and gene duplications have played a large role in genome evolution, as will be discussed in a later section.

Other plants, such as maize and barley, have genomes more than an order of magnitude larger than *Arabidopsis*, but have about the same number of genes (see Table 20.6). Genes in these large-genome plants are clustered in stretches of DNA separated by long stretches of intergenic DNA (Figure 20–12). Collectively, the gene clusters (known as the gene space) occupy only 12–24 percent of the genome. In maize, the intergenic DNA is composed mainly of transposons. Because plants closely related to *Arabidopsis* have larger genomes but similar numbers of genes, the small genome of *Arabidopsis* is thought to derive from a genomic contraction, in which almost all of the gene-empty regions disappeared, along

FIGURE 20–11 *Arabidopsis thaliana*, a small plant that is used as a genetic model to study genome organization and development of flowering plants.

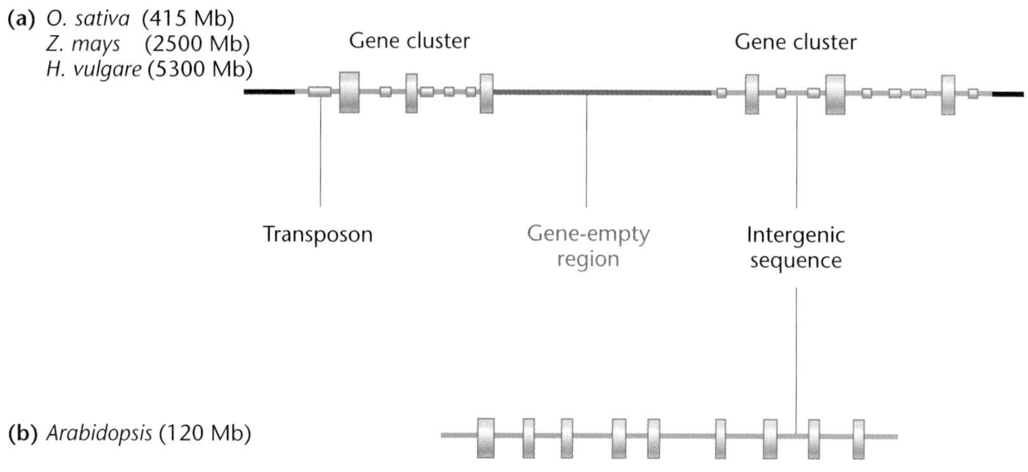

FIGURE 20–12 Genome organization in several large-genome plants and the compact genome of *Arabidopsis*. (a) The vertical boxes are genes located in clusters, separated by long, gene-empty spaces of repetitive DNA sequences. Within the gene clusters, the intergenic spaces (green) contain many transposons (small horizontal boxes). (b) In the *Arabidopsis* genome, gene-empty regions have been lost, and transposable elements have been lost or reduced. The result is a much smaller genome with genes at a much higher density throughout the genome.

with most of the transposon sequences. In addition, there may have been a reduction in the size and number of introns, accounting for the compact genome size in this plant.

20.7 The Human Genome: The Human Genome Project (HGP)

The Human Genome Project (HGP) extends the tradition of genetic mapping started by geneticists in the early years of the twentieth century. The goal of mutational analysis is the identification and mapping of all the genes in an organism's genome. This goal is shared by the HGP and all other genome projects, using recombinant DNA technology and DNA sequencing instead of mutational analysis.

The HGP has produced a plethora of information, much of which is still being analyzed and interpreted. What is clear so far is that genomic information from a wide range of organisms demonstrates that humans and all other species share a common set of genes essential for cellular function and reproduction, confirming that all living organisms arose from a common ancestor. These evolutionary relationships facilitate the analysis of the human genome and the development of model organisms to study inherited human diseases.

Origins of the Human Genome Project

In the United States, a project to sequence the human genome was proposed in 1986, and in 1988, the National Institutes of Health (NIH) and the United States Department of Energy created a joint committee to develop a plan for the project. The Human Genome Project (HGP) began in 1990 under the direction of James Watson, the codiscoverer of the double helix structure of DNA.

Instead of finding and mapping markers and disease genes one at a time, the HGP set out to sequence all the DNA in the human genome, identify and map the thousands of genes to the 24 chromosomes, and establish the function of all genes. To deal with the impact that genetic information would have on society, the HGP set up the ELSI program (Ethical, Legal and Social Impact) to ensure that genetic information would be safeguarded and not used in discriminatory ways. Once established in the United States, the HGP quickly grew into an international effort involving programs in other countries, especially Japan, Germany, and England, all coordinated by the Human Genome Organization (HUGO).

Although not reflected in the name, the Human Genome Project also included projects to sequence the genomes of model organisms used in experimental genetics. These include *E. coli*, *S. cerevisiae*, *C. elegans*, *D. melanogaster*, and *M. musculus*.

Genome projects have grown in number since the start of the HGP, and have been completed or are underway for literally hundreds of prokaryotic and eukaryotic species, many of which are of agricultural importance, or cause diseases in plants, animals, and humans. To date, the genomes for more than 125 species have been completely sequenced, and hundreds more are being sequenced.

Major Features of the Human Genome

In June 2000, the public and private genome projects met at the White House and jointly announced the completion of a draft sequence of the human genome (Figure 20–13). In February 2001, they each published an analysis covering about 96 percent of the euchromatic region of the genome. The remaining work of completing the sequence by filling in the gaps was completed in 2003, when the finished sequence was published.

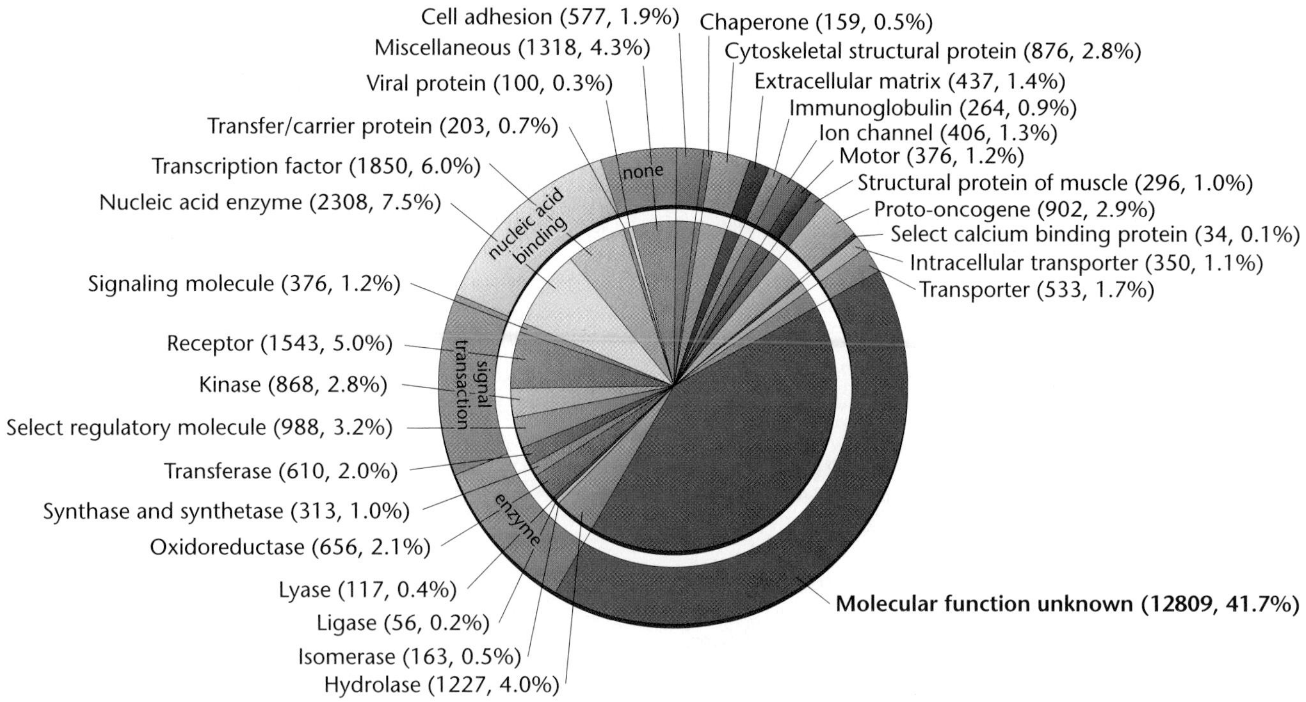

FIGURE 20–13 A preliminary list of assigned functions for 26,588 genes in the human genome based on similarity to proteins of known function. Among the most common genes are those involved in nucleic acid metabolism (7.5 percent of all genes identified), transcription factors (6.0 percent), receptors (5 percent), protein kinases (2.8 percent), and cytoskeletal structural proteins (2.8 percent). A total of 12,809 predicted proteins (41 percent) have unknown functions, reflecting the work that is still needed to fully decipher our genome.

Attention is now directed at analyzing and interpreting the vast amount of data gathered in this project. The major features of the human genome are presented in Table 20.7.

The Unfinished Tasks in Human Genome Sequencing

As of the end of 2004, more than 99 percent of the euchromatic portion of the human genome sequence has been finished. Two types of gaps remain in the sequence. There are about 324 gaps in the euchromatic portion of the genome, spanning about 24 Mb of sequence. Most of these are associated with duplicated regions of the genome that are difficult to assemble. Major gaps in the heterochromatic regions of the genome, including regions around centromeres, the ribosomal DNA genes, and telomeres cover more than 200 Mb of sequence. The heterochromatic regions of the human genome were excluded from the genome projects, and these gaps were anticipated. Heterochromatic regions contain extensive sets of tandemly repeated DNA sequences with few if any genes. Although the short arms of chromosomes 13 to 15 and 21 to 22 are heterochromatic and contain large blocks of repetitive sequences, a few genes have been localized to these regions. Gaps of both

TABLE 20.7 **MAJOR FEATURES OF THE HUMAN GENOME**

- The human genome contains more than 3 billion nucleotides, but protein-coding sequences make up only about 5 percent of the genome.
- At least 50 percent of the genome is derived from transposable elements, such as LINE and *Alu* sequences.
- The human genome contains between 25,000 and 30,000 genes, far fewer than the predicted number of 50,000–100,000 genes.
- More than 40 percent of the genes identified have no known molecular function.
- Genes are not uniformly distributed on the 24 human chromosomes. Gene-rich clusters are separated by gene-poor 'deserts' that account for 20 percent of the genome. These deserts correlate with G bands seen in stained chromosomes. Chromosome 19 has the highest gene density, and chromosomes 13 and the Y chromosome have the lowest gene density.
- Human genes are larger and contain more and larger introns than genes in invertebrate genomes such as *Drosophila*. The largest known human gene encodes dystrophin, a muscle protein. This gene, associated in mutant form with muscular dystrophy, is 2.5 Mb in length, larger than many bacterial chromosomes. Most of this gene is composed of introns.
- The number of introns in human genes ranges from 0 (histone genes) to 234 (in the gene for *titin*, a muscle protein).

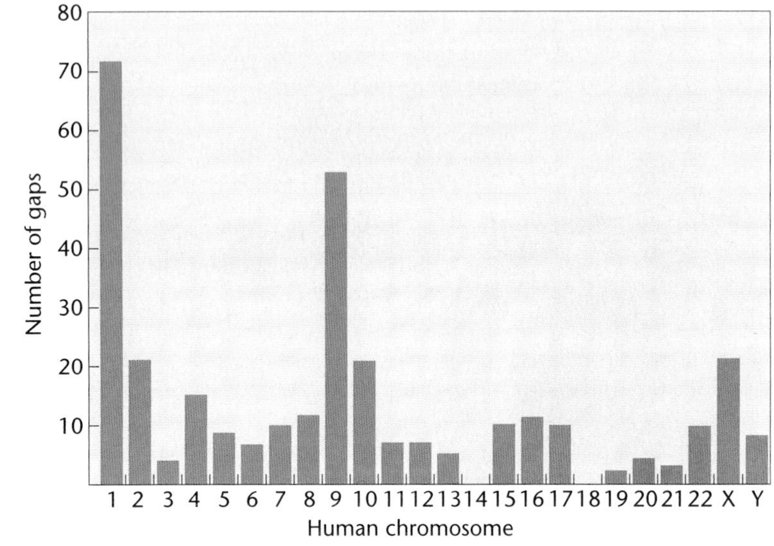

FIGURE 20–14 Distribution of sequence gaps across all human chromosomes. In 2004, there were just over 300 gaps remaining to be closed, with only chromosomes 14 and 18 free from any gaps. Researchers are using several methods to clone and sequence the DNA spanning these gaps. *From Eichler E., Clark R.A. She, X. 2004. An assessment of the sequence gaps: unfinished business in a finished human genome. Nature Reviews Genetics 5(5):345-54 copyright 2004 Macmillan Publishers Ltd.*

types are nonrandomly distributed in the genome (Figure 20–14), and because of differences in size and sequence content, several different strategies may be needed to fill in all the gaps.

Chromosomal Organization of Human Genes

The organization of genes on the long arms of human chromosomes 21 and 22 reveals some interesting features about the genetic landscape of human chromosomes (Figure 20–15). One of the most intriguing findings is the difference in gene density on these two chromosomes. The amount of sequence reported for these chromosomes is similar (33.65 Mb for chromosome 21 and 34.65 Mb for chromosome 22), but chromosome 21 has only 225 genes (about 1 gene per 150 kb of DNA), while chromosome 22 has 541 genes (about 1 gene per 64 kb of DNA).

Closer examination of gene distribution reveals that the genes on these chromosomes are not evenly spaced (Figure 20–16). On chromosome 21, the region of the long arm close to the centromere is gene-poor and averages 1 gene per 304 kb of DNA. The region of the long arm close to the telomere has a much higher gene density, with a gene every 95 kb of DNA. Several regions of chromosome 21 are almost devoid of genes; 1 region spanning 7 Mb contains only 1 gene, and 3 other regions of 1 Mb each contain no genes. Together, these gene-poor regions add up to 10 Mb, or about one third the length of the long arm. Chromosome 22 has a 2.5-Mb region near the telomere and 2 smaller regions (about 1 Mb each) located elsewhere that lack genes. This organization helps to explain the fact that most of the human genome does not contain protein-coding genes.

Chromosomes 21 and 22 both contain duplicated regions. On chromosome 21, a 220-kb region is duplicated near both ends of the long arm, and another 10-kb region is duplicated near the centromere. On chromosome 22, a 60-kb segment is duplicated. A study of chromosome breakpoints associated with translocations and deletions on chromosome 21 indicates that breakpoints cluster within and near duplicated regions, suggesting that these regions may mediate events involved in these chromosomal aberrations. Future analysis of the duplicated regions and their associated repetitive sequences may provide a molecular explanation for chromosome breakage.

Our Genome and the Chimpanzee Genome

The nucleotide sequence of chimpanzee chromosome 22 was completed in 2004. Comparisons with chromosome 21, the human equivalent, offers some interesting insights into what makes some primates humans and makes others chimpanzees.

Humans and chimpanzees last shared a common ancestor about 6 million years ago, and the two genomes have been evolving separately since then. The chimpanzee chromosome and its human ortholog (genes or chromosomes descended from a common ancestor are orthologs) have accumulated nucleotide substitutions that total 1.44 percent of the sequence. The most surprising difference is the discovery of 68,000 insertions or dele-

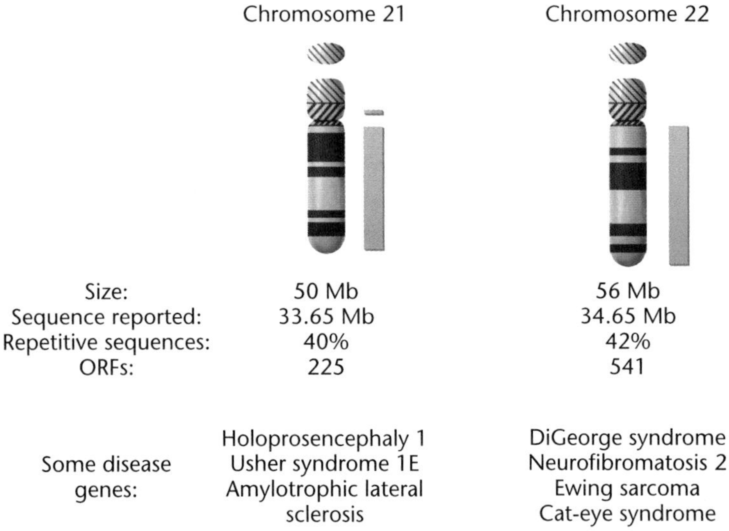

FIGURE 20–15 Human chromosome 21 (left) and chimpanzee chromosome 22 (right). The two chromosomes are similar in size (33.12 Mb for human chromosome 21 and 32.7 Mb for chimpanzee chromosome 22) and gene content (284 for human chromosome 21 and 272 for chimpanzee chromosome 22). The chromosomes share 179 genes with coding sequences of identical length and similar intron-exon boundaries. These shared genes have an average nucleotide and amino acid identity of 99.29% and 99.18% respectively, providing evidence for the close evolutionary relationship between the two species.

tions (called **indels**) in one chromosome or the other, a frequency of 1 indel every 470 bases. Many of these are *Alu* insertions in human chromosome 21. Although the overall chromosomal sequence differences are small, only 17 percent of the genes analyzed are identical in both chromosomes; the other 83 percent encode genes with one or more amino acid differences.

The physical differences between the chromosomes (Table 20.8) and their genes are only part of the story. Using DNA microarrays, researchers compared tissue expression patterns of 202 human chromosome 21 genes in human and chimp brain and liver cells. They found expression of 60 genes in brain and 40 in liver in one or the other species. Of these, 21 showed significant differences in levels of expression. Some of these genes play important roles in the development of the heart, brain, and nervous system. If chromosome 21 represents about 1 percent of the human genome, then there are potentially thousands of genes with differences in amino acid sequence or expression patterns between the two species. Some of these changes may be more crucial than others, and comparison of the whole genome sequences between humans and chimpanzees and between these species and other primates may help identify the genes that help make us human.

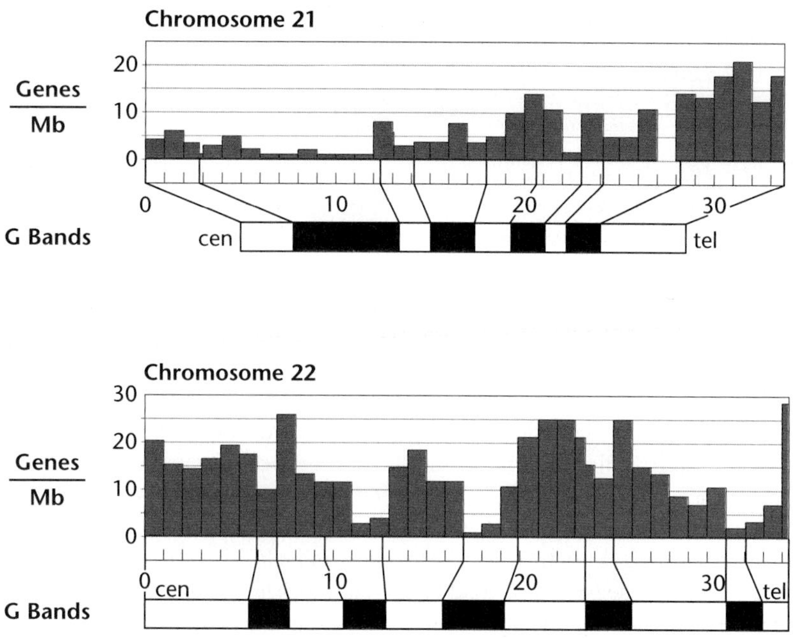

FIGURE 20–16 Gene density in the long arms of human chromosomes 21 and 22 (genes per Mb of DNA). In chromosome 21, the region in the dark staining band is relatively gene-poor, and the region near the distal tip is gene-rich. On chromosome 22, there is a small region near the chromosome tip that is gene-poor and two other regions, closer to the centromere that are relatively gene poor.

TABLE 20.8 Comparisons between Human Chromosome 21 and Chimpanzee Chromosome 22

	Human 21	Chimpanzee 22
Size (bp)	33,127,944	32,799,845
G + C Content	40.94	41.01
CpG Islands	950	885
SINES (*Alu* elements)	15,137	15,048
Genes	284	272
Pseudogenes	98	89

20.8 Comparative Genomics Is a Versatile Tool

Comparative genomics is a new and rapidly developing field of biology that directly compares the genetic information of one organism with that of another. It is a field with many research and practical applications, including gene discovery; the development of model organisms for human diseases; the elucidation of evolutionary history and relationships between genes, genomes, and species; and the relationship between organisms and their environment. Comparative genomics uses a wide range of techniques and resources, including the construction and use of databases containing nucleic acid and amino acid sequences, cytogenetic techniques of gene mapping, such as fluorescent *in situ* hybridization (FISH), as well as experimental methods, such as mutagenesis. Comparative genomics uses these resources to identify genetic differences and similarities between organisms, to determine how these differences contribute to differences in phenotype, life cycle, or other attributes, and to ascertain the evolutionary history of these genetic differences.

In the following sections, we will discuss several applications of comparative genomics at several levels (genes, chromosomes, and genomes), as well as its role in model-system selection, evolution, and drug discovery.

Finding New Genes Using Comparative Genomics

The completion of the euchromatic human genome sequence has facilitated the identification and cataloging of our genes. However, since 95 percent of the genome is not protein-coding, vast stretches of sequence must be searched to identify functional genes. Comparative genome analysis using the human genome and other genomes can be useful in searching the human genome sequence for previously unidentified genes. We'll examine one example of many that have emerged from comparisons of the human and mouse genomes.

The long arm of human chromosome 11 contains a four-gene cluster (*APOAI-APOAIV*) that encodes apolipoproteins. These proteins control blood levels of lipids and triglycerides. Mutations in these genes dramatically alter blood lipid levels, and are a major factor in cardiovascular disease. To investigate whether this region of chromosome 11 might contain additional apolipoprotein genes, Edward Rubin and his colleagues compared mouse and human genome sequences in a 200-kb region near the human *APOA* gene cluster. This comparison identified an open reading frame in the human sequence encoding a 366 amino acid

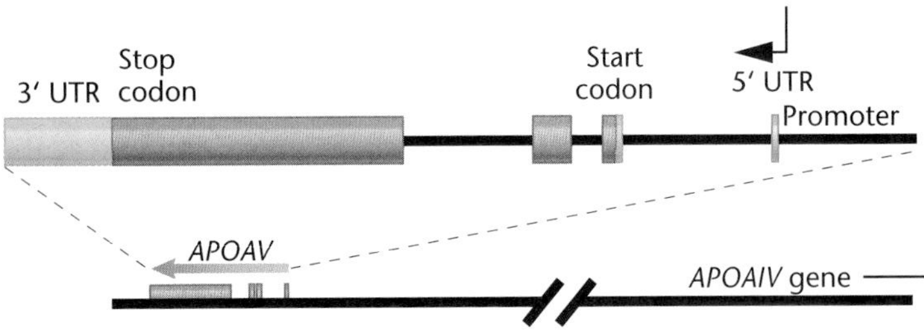

FIGURE 20–17 Organization of a human apolipoprotein gene (*APOAV*) that was identified by comparing ~ 200kb of DNA sequence from the long arm of human chromosome 11 with DNA sequence from the mouse genome. This region on chromosome 11 contains several other apolipoprotein genes, but *APOAV* was identified only by comparing mouse and human sequence, emphasizing the importance of comparative genome analysis. *Reprinted with permission from Pennacchio L.A., et al., An Apolipoprotein influencing triglycerides in humans and mice revealed by comparative sequencing. Science 294:169-173, Figure 1. Copyright 2001 AAAS.*

protein with high sequence homology to the mouse *Apoav* gene and the human *APOAIV* gene (Figure 20–17).

Analysis of the predicted protein structure revealed domains with lipid-binding properties. This region of the human genome was cloned and used in Southern blots to identify mRNA transcripts in human tissues, confirming this sequence is an actively transcribed gene. Transgenic mice carrying this new human gene (named *APOAV*) had blood lipid levels only about one third as high as controls, while knockout mice lacking the mouse *Apoav* gene had blood lipid levels four times higher than controls. These experimental results demonstrate the important role of this newly identified human gene in controlling levels of lipids and triglycerides, an important risk factor in cardiovascular disease. This discovery also illustrates the power of comparative genomics to identify previously unknown genes with functional significance.

Comparative genomic studies using a database of cloned mouse cDNA sequences have identified several human genes associated with genetic disorders. Table 20.9 lists human disease genes previously mapped to chromosomal loci, but not isolated or cloned. By computer mapping mouse cDNA sequences onto the human genome sequence, candidate human genes for these disorders were identified. These candidate genes can be confirmed by analysis of mutations in affected families and by pedigree studies, expanding the catalog of cloned genes associated with human genetic disorders.

TABLE 20.9 HUMAN DISEASE GENES IDENTIFIED BY MOUSE COMPARATIVE GENOMICS

Disease	Phenotype
Meckel syndrome	Defects in central nervous system
Neural tube defect	Spinal neural tube defects
Wolf–Hirschhorn syndrome	Lack of physical growth and mental retardation
Spinulosa decalvans	Thickening of skin, hair loss
Benign family convulsions	Seizures beginning at 6 months of age
Cerebellar ataxia, infantile onset	Progressive loss of balance, sudden deafness
Vacuolar neuromyopathy	Progressive muscle weakness
Tylosis with esophageal cancer	Thickening of skin, adult development of esophageal cancer

Comparative Genomics and Model Organisms

Analysis of the growing number of genome sequences confirms that all living organisms are related and descended from a common ancestor. Similar gene sets are used in all organisms for basic cellular functions, such as DNA replication, transcription, and translation. These genetic relationships provide the basis for the development and use of model organisms to study inherited human disorders and the interaction of genes and environment in complex diseases, such as cardiovascular disease and behavioral disorders. As mentioned earlier, these model organisms include yeast, *Drosophila*, *C. elegans*, and the mouse. Recently, the dog genome has been sequenced and analyzed (Figure 20–18), providing another useful model with which to study our own genome.

The dog offers several advantages for studying heritable human diseases. Dogs share many genetic disorders with humans, including over 400 single gene disorders, sex chromosome aneuploidies, multifactorial diseases including epilepsy, and genetic predispositions to cancer.

Comparative chromosome painting using fluorescently labeled probes from one species hybridized to chromosomes of another species is used to visualize homologies between the genomes of dif-

FIGURE 20–18 Shadow, J. Craig Venter's poodle, whose DNA was used as the source of the dog genome sequence assembled by Venter and his colleagues.

ferent species. About 90 conserved blocks of the dog genome can be mapped to human chromosomes by comparative FISH (Figure 20–19). These blocks contain DNA sequences with a high degree of similarity between the dog and the human genomes, reflecting the evolutionary relationship between dogs and our species.

Analysis of genome sequences indicates that at least 60 percent of inherited diseases in dogs have similar or identical molecular causes, such as point mutations, deletions, etc., as in humans. In addition, at least 50 percent of the genetic diseases in dogs are breed-specific, so the mutant allele segregates in relatively homogeneous genetic backgrounds. Dog breeds resemble isolated human populations in having a small number of founders and a long period of relative genetic isolation. These properties facilitate the use of individual breeds as models of human genetic disorders.

The availability of the dog genome sequence, the large number of genetic disorders in dogs, and their molecular similarity to mutations in homologous human genes makes the dog an important model for human gene therapy. Gene therapy (discussed in Chapter 22) treats an inherited disorder by transferring normal genes into target cells where they are expressed to correct an abnormal phenotype. Gene therapy for single gene disorders has been successful in dogs and this procedure can be used to develop and refine the modified viruses used as vectors for transferring copies of a normal gene into mutant cells before attempting gene therapy in humans.

Comparative Analysis of Nuclear Receptors and Drug Development

Analysis of genome sequences provides an opportunity to study genes and their encoded products that mediate the action of prescription drugs. **Nuclear receptors (NRs)** are a superfamily of transcription factors that regulate transcription of target gene sets in response to steroid hormones. More than 10 percent of commonly prescribed drugs work by binding to NRs.

Figure 20–20 (a) shows the domains found in typical NR proteins. A variable N-terminal group, AF1, is responsible for cell and target gene specificity. A highly conserved DNA-binding domain (DBD) with zinc finger elements is centrally located. Near the C-terminus is a ligand-binding domain (LBD) that binds to steroid hormones. In conjuction with coactivator molecules, the activated NR remodels chromatin to facilitate binding of transcription complexes as a first step in activating gene expression.

Comparisons of NR-encoding genes in human and nonhuman genomes have helped catalog amino acid sequences required for ligand recognition. This provides information on the three-dimensional structure required for a drug to bind to and activate an NR. Comparative genomics also identifies model organisms with NRs similar to those of humans. These organisms will have the same response to specific drugs as humans and can be used to develop new drugs to treat human diseases.

Farnesoid X-activated receptor (FXR) is an indispensable regulator of lipid metabolism. The human *FXR* gene was identified in 1995. Analysis of genome sequences revealed the unexpected presence of an *FXR* pseudogene in the human genome. Pseudogenes are nonfunctional versions of genes that resemble other gene sequences, but contain significant nucleotide substitutions, deletions, and duplications that prevent their expression.

More surprising was the discovery that other mammalian genomes contain a functional second *FXR* gene (*FXRβ*). This gene is coexpressed with *FXR* in embryonic and adult tissues of the mouse, but binds to a different set of ligands and activates a separate set of genes in response to binding. This means that there may be significant differences between humans and other mammals in the metabolism of cholesterol and other lipids,

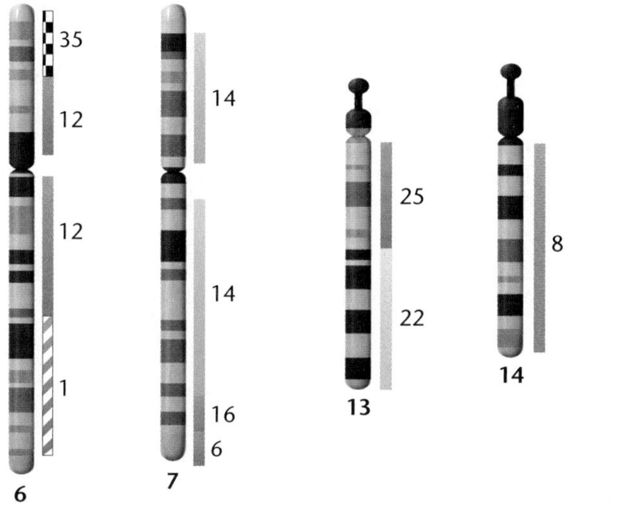

FIGURE 20–19 Chromosomal painting, a variation of fluorescent *in situ* hybridization (FISH) has been used to find homologous segments on human and dog chromosomes. These homologies can be used to compare dog and human DNA sequences in the search for genes in both the dog and in the human genome. Human chromosome 6 has sequences found on three dog chromosomes (1, 12, and 35). Human chromosome 7 is composed of sequences found on dog chromosomes 6, 14 and 16. The acrocentric human chromosomes 13 and 14 contain sequences found on dog chromosomes 22 and 25, and on dog chromosome 8 respectively.

(a)

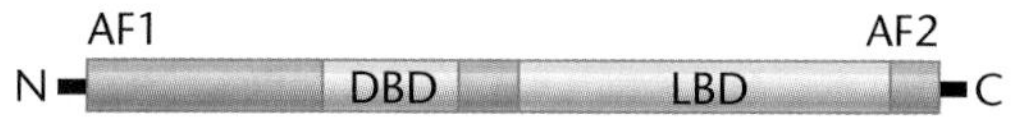

(b)

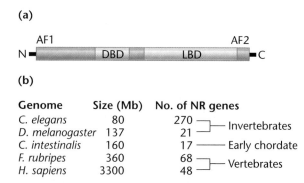

Genome	Size (Mb)	No. of NR genes	
C. elegans	80	270	Invertebrates
D. melanogaster	137	21	
C. intestinalis	160	17	Early chordate
F. rubripes	360	68	Vertebrates
H. sapiens	3300	48	

FIGURE 20–20 (a) Organization of the motifs and domains encoded by nuclear receptor genes. AF1 is a sequence that is critical for cell and target gene specificity. DBD is a higly conserved DNA binding domain, containing two zinc-finger regions. The DBD region binds to specific genomic sequences and determines which genes are activated. The LBD, ligand-binding domain activates the nuclear receptor when it binds a signal molecule (a ligand). The AF2 motif plays an important role in activation of the nuclear receptor protein. (b) Analysis of nuclear receptor genes from a variety of organisms provides important information on the size, number and amino acid sequence of nuclear receptor genes. Highly conserved regions in these genes are being investigated as drug targets.

and that these organisms may be poor models for the development of new drugs to control lipid levels in humans.

In addition, cloning and sequencing of the *FXRα* gene in hamsters led to the identification of several splice variants with altered activation domains. Sequence data from the hamster variants were used to identify human variants that are differentially expressed in embryonic and adult tissues and have different levels of response to certain ligands. This finding indicates that regulation of lipid metabolism in humans is more complex than previously thought and that comparative genome analysis has become an important tool in drug discovery and development.

The Minimum Genome for Living Cells

What is the minimum number of genes necessary to support life? We cannot fully answer this question until we define what we mean by a living organism, and until we know the functions of each gene in the genomes that have been sequenced. Many scientists would define life as a free-living cell—that is, one that can exist outside another organism. Although we do not know the minimum number of genes to support independent life, we can use the small genomes of obligate parasites to speculate on the minimum number of genes required to maintain life. To do this, we can compare sequence information from the bacterial genomes of *Mycoplasma genitalium* and *M. pneumoniae,* both of which are human parasitic pathogens. These two closely related organisms are among the simplest self-replicating prokaryotes known, and can serve as model systems to understand the essential functions of a self-replicating cell. *M. genitalium* has a genome of 580 kb, while that of *M. pneumoniae* is 816 kb. They are members of a group of bacteria that lack a cell wall and invade other organisms, often causing disease in a wide range of hosts, including insects, plants, and humans. (In humans, these bacteria cause genital and respiratory infections.)

The *M. genitalium* genome has 480 protein-coding genes, the smallest bacterial genome among nearly 100 genomes sequenced to date. The genome of a related species, *M. pneumoniae*, has those 480 genes plus an additional 197, for a total of 677 protein-coding genes. In contrast, the 1.8 Mb genome of *Haemophilus influenzae* (the first bacterial genome sequenced) has 1703 genes. Table 20.10 summarizes the functions of some genes in these bacteria, and compares them to *E. coli.*

The availability of genome sequences allows us to ask whether the 480 genes carried by *M. genitalium* (and shared with *M. pneumoniae*) are close to the minimum gene set needed for life. In other words, can we define life in terms of a number of specific genes? A combination of comparative and experimental methods can be used to answer this question. The comparative approach is based on the premise that genes shared by distantly related organisms are likely to be essential for life. By comparing the gene sets shared by different organisms, it should be possible to catalog those that are shared and develop a list of genes needed for life.

M. genitalium and *H. influenzae* last shared a common ancestor about 1.5 billion years ago. The genes that are still shared between these species should represent those essential for life. By comparing the nucleotide sequences of the 480 *M. genitalium* genes with the 1703 *H. influenzae* genes, researchers identified 240 **orthologous** genes between the two species. **Orthologs** are genes descended from a common ancestral gene that have the same function in different species. In addition to the 240 shared genes, 16 genes were identified that have different sequences, but perform identical functions. These represent essential functions that are performed by nonorthologous genes. Thus, from comparative genomics alone, it appears that 256 genes may represent the minimum gene set needed for life.

Craig Venter and his colleagues used an experimental approach to determine how many of the 480 *M. genitalium* genes are essential for life. They used transposons to selectively mutate genes in *M. genitalium*. Mutations in essential genes produced a lethal phenotype, but if the gene was nonessential, mutating the gene did not affect viability. In essence, they mutated genes one at a time by transposon insertion, and if the phenotype was lethal, they classified the gene as essential. They found that many of the 480 genes were nonessential, and that the minimum gene set for *M. genitalium* is about 265–300 genes. This is close to the value of 256 derived from comparative genome analysis.

TABLE 20.10 GENE FUNCTION IN *M. GENITALIUM*, *H. INFLUENZAE*, AND *E. COLI*

Functional Class	*E. coli*	*M. genitalium*	*H. influenzae*
Protein-coding genes	4288	470	1,727
DNA replication, repair	115	32	87
Transcription	55	12	27
Translation	182	101	141
Regulatory proteins	178	7	64
Amino acid biosynthesis	131	1	68
Nucleic acid biosynthesis	58	19	53
Lipid metabolism	48	6	25
Energy metabolism	243	31	112
Uptake, transport proteins	427	34	123

Source: Blattner, F., et al. 1997. The complete genome sequence of *Escherichia coli* K-12. *Science* 277:1453–62. Table 4, p. 1458; Fraser, C.M., et al. 1995. The minimal gene complement of *Mycoplasma genitalium*. *Science* 270:397–403. Table 2, p. 400.

Now solve this

Problem 20.22 on page 514 involves differential gene loss in members of the prokaryotic genus *Buchneria*.

Hint: Remember that symbiotic organisms coevolve with their hosts.

The search for the minimum gene set needed for life is complicated by the fact that cells often activate different gene sets when exposed to different conditions. Consequently, there may be different minimum gene sets needed for parasitic organisms, symbiotic organisms, and free-living organisms. To define life at the molecular level in a more direct way, Venter has started a project to develop and test a synthetic genome. This project has several formidable hurdles to overcome. The first is selecting which genes should be included in the genome. Next is synthesizing a genome consisting of several hundred thousand nucleotides. Once synthesized, the team plans on destroying the DNA in an *M. genitalium* cell and inserting the synthetic genome. Whether the cytoplasmic components will recognize and activate the synthetic genome is unknown, but if successful, Venter and his colleagues will have defined life at the molecular level, raising ethical questions about creating life forms.

20.9 Comparative Genomics: Multigene Families Diversify Gene Function

Genes belonging to multigene families share similar, but not identical DNA sequences as a result of descent with mutation from a single ancestral gene. Their gene products frequently have similar functions, and the genes are often, but not always found together in a single location along a chromosome. To see how analysis of multigene families provides insight into eukaryotic genome organization and evolution, we first examine cases where groups of genes encode very similar but not identical polypeptide chains that become part of proteins with closely related functions. Multiple proteins that arise from single-gene duplications are known as **paralogs**. The globin gene family responsible for encoding the various polypeptides in hemoglobin molecules exemplifies a paralogous multigene family that arose by duplication and dispersal to different chromosomal sites.

Gene Duplications

The almost uninterrupted flow of sequence data from genome projects is providing evidence that multigene families are present in many, if not all, genomes. In addition to genome-wide duplications, small blocks of genes and single genes can be duplicated by several mechanisms. Unequal crossing over is a recombination event between members of a homologous pair of chromosomes in which a DNA segment is duplicated in one of the recombination products (Chapter 8).

Once generated, members of multigene families may remain linked on a single chromosome or they may disperse to other parts of the genome. Several mechanisms drive this process, including inversion, translocation, and transposition by mobile elements (Chapter 13).

Molecular phylogenetics has traced the ancestry and relationships among members of gene families. One of the best-studied examples of divergence is the globin gene superfamily (Figure 20–21). In this family, duplication of an ancestral gene encoding an oxygen transport protein occurred some 800 million years ago. This split produced two sister genes, one of which evolved into the modern-day myoglobin gene. (Myoglobin is an oxygen-carrying protein found in muscle.) The other gene became the ancestral globin gene. About 500 million years ago, the ancestral globin gene duplicated to form prototypes of the α- and β-globin gene subfamilies. The **α-globin** and **β-globin** genes encode the proteins found in hemoglobin, the oxygen-carrying molecule in red blood cells. Additional duplications within these genes occurred within the last 200 million years. Subsequent events dispersed members of this superfamily, and

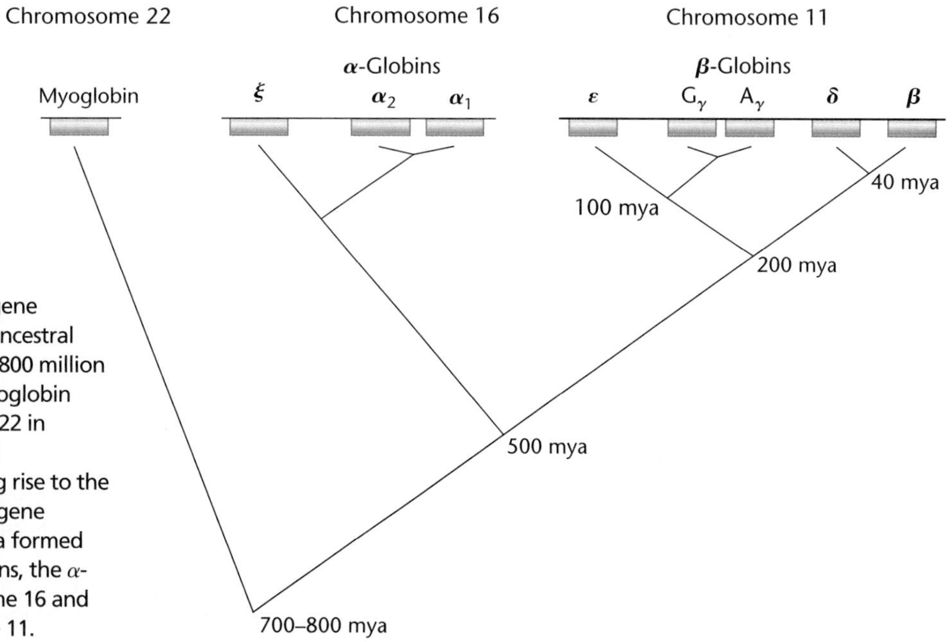

FIGURE 20–21
The evolutionary history of the globin gene superfamily. A duplication event in an ancestral gene gave rise to 2 lineages about 700–800 million years ago (mya). One line led to the myoglobin gene, which is located on chromosome 22 in humans; the other underwent a second duplication event about 500 mya, giving rise to the ancestors of the α-globin and β-globin gene subfamilies. Duplications about 200 mya formed the β-globin gene subfamilies. In humans, the α-globin genes are located on chromosome 16 and the β-globin genes are on chromosome 11.

each subfamily now resides on a separate chromosome.

Similar patterns of evolution are observed in other gene families, including the trypsin–chymotrypsin family of proteases, the homeotic selector genes of animals, and the rhodopsin family of visual pigments.

Evolution of Gene Families: The Globin Genes

The human α-globin and β-globin genes are two of the most intensively studied regions of the human genome. Proteins encoded by these genes are components of hemoglobin. Adult hemoglobin is a tetramer, containing two α- and two β-polypeptides (Figure 20–22). Each polypeptide incorporates a heme group that reversibly binds oxygen. There is a cluster of 3 α-globin genes on the short arm of chromosome 16, and another cluster with 5 β-globin genes on the short arm of chromosome 11. Members of both subfamilies share nucleotide sequence similarity, but members of the same subfamily have the greatest amount of sequence similarity (Figure 20–23).

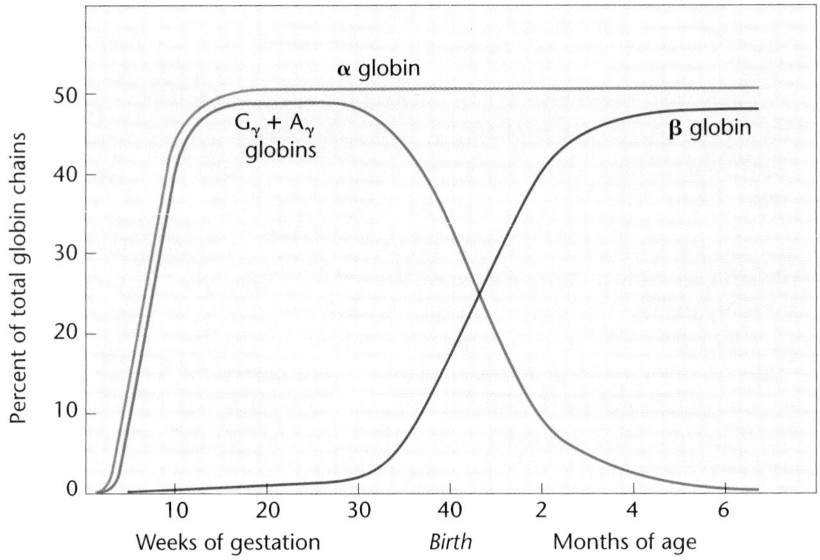

FIGURE 20–24 Changes in globin gene expression during human development. The alpha genes are switched on early in development, and continue throughout life. The Gγ and Aγ members of the beta family are expressed during fetal development, and are switched off just before birth. The β-globin gene is switched on just before birth and is expressed throughout life.

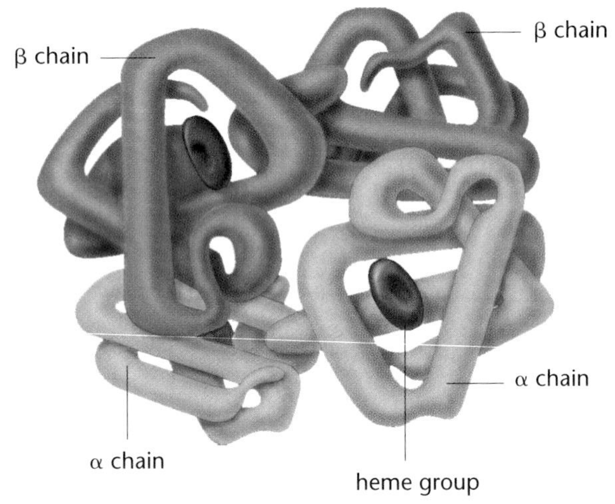

FIGURE 20–22 Each functional hemoglobin molecule consists of two alpha and two beta chains, each of which carries a heme group.

Within each subfamily, genes are coordinately turned on and off during embryonic, fetal, and adult stages of development (Figure 20–24). For both the α- and β-globin gene subfamilies, this expression occurs in the same order in which the genes are arranged on the chromosome.

The α-globin gene subfamily [Figure 20–25(a)] spans more than 30 kb and contains 3 genes: the ζ (zeta) gene, expressed only in early embryogenesis, and 2 copies of the α gene, expressed during the fetal (α_1) and adult stages (α_2). In addition, two pseudogenes (ζ and α_1) are present in the cluster. Pseudogenes are designated by the prefix ψ (psi), followed by the symbol of the gene they most resemble. Thus, the designation $\psi\alpha_1$ indicates a pseudogene of the adult α_1 gene.

The organization of the α-globin subfamily members and the location of their introns and exons reveal several features. First, as is common in eukaryotes, the DNA encoding the three functional α genes occupies only a small portion of the region containing the subfamily. Most of the DNA in this region is intergenic spacer. Second, each functional gene in

α-globin V – L S P A D K T N V K A A W G K V G A H A G E Y G A E A L E R M F L S F P T T K T Y F P H F – D L S H
β-globin V H L T P E E K S A V T A L W G K V – – N V D E V G G E A L G R L L V V Y P W T Q R F F E S F G D L S T

α-globin – – – G S A Q V K G H G K K V A – D A L T N A V A H V D D – M P N A L S A L S D L H A H K L R V D P V N
β-globin A V M G N P K V K A H G K K V L – G A F S D G L A H L D N – L K G T F A T L S E L H C D K L H V D P E N

α-globin L L S H C L L V T L A A H L P A E F T P A V H A S L D K F L A S V S T V L T S K Y R – 141
β-globin L L G N V L V C V L A H H F G K E F T P P V Q A A Y Q K V V A G V A N A L A H K Y H – 146

FIGURE 20–23 The amino acid sequence of the alpha (α) and beta (β) globin genes using the single letter abbreviations for the amino acids (See figure 14–16). Shaded areas indicate identical amino acids. The two genes are descended from a common ancestor, and diverged from each other about 500 million years ago.

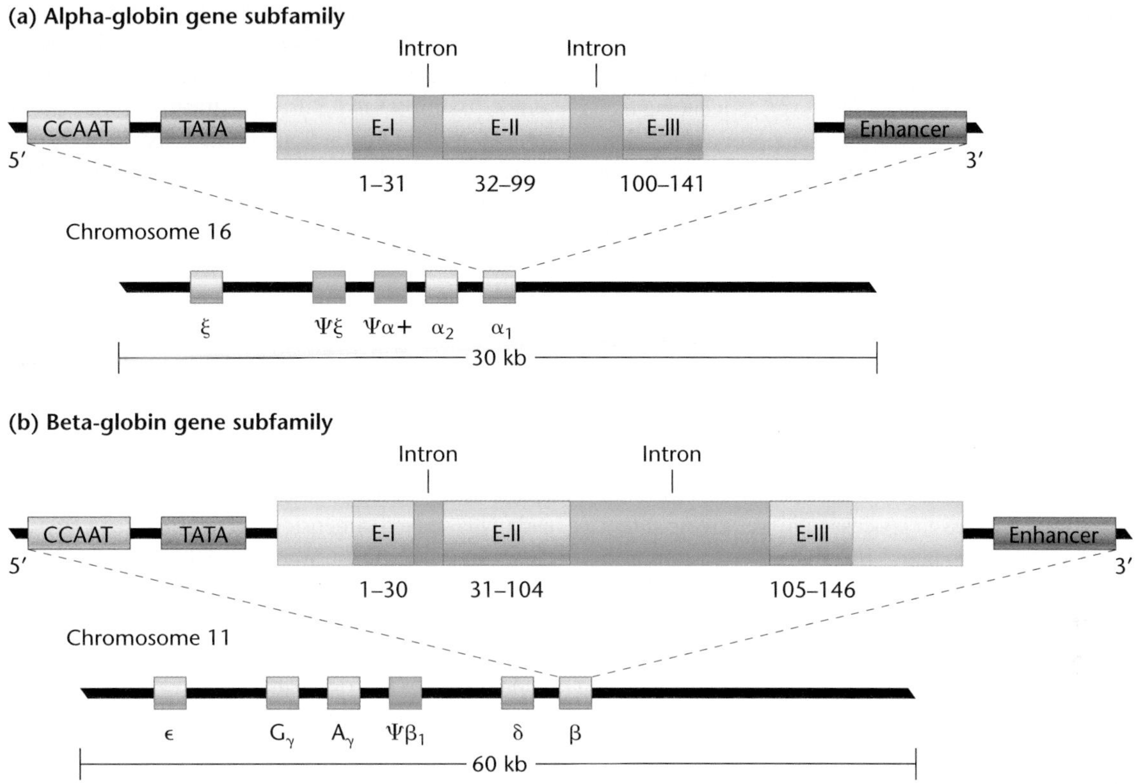

FIGURE 20–25 (a) Organization of the α-globin gene subfamily on chromosome 16, and (b) the β-globin gene subfamily on chromosome 11. Also shown is the internal organization of the α_1 gene and the β gene. Each gene contains three exons (E-I, E-II, E-III) and two introns. The numbers below the exons indicate the amino acids in the gene product encoded by each exon.

this subfamily contains two introns at precisely the same positions. Third, the nucleotide sequences within corresponding exons are nearly identical in the ζ and α genes. Both of these genes encode polypeptide chains of 141 amino acids. However, their intron sequences are highly divergent, even though they are about the same size. Note that much of the nucleotide sequence of each gene is contained in these noncoding introns.

The human β-globin gene cluster is longer than the α-globin cluster and contains five genes spaced over 60 kb of DNA [Figure 20–25(b)]. As with the α-globin gene subfamily, the order of genes on the chromosome parallels their order of expression during development. Of the five genes, three are expressed prior to birth. The ε (epsilon) gene is expressed only during embryogenesis, while the two nearly identical γ genes (G_γ and A_γ) are expressed only during fetal development. The polypeptide products of the two γ genes differ only by a single amino acid. The two remaining genes, δ and β, are expressed after birth. Finally, a single pseudogene $\psi\beta_1$ is present within the subfamily. All five functional genes in this cluster encode proteins with 146 amino acids and have two similarly sized introns at exactly the same positions. The second intron in the β-globin genes is significantly larger than its counterpart in the functional α-globin genes. These similarities reflect the evolutionary history of each subfamily and the events such as gene duplication, nucleotide substitution, and chromosome translocations that produced the present-day globin superfamily.

20.10 Proteomics Identifies and Analyzes the Proteins in a Cell

Proteome is a relatively new term that defines the complete set of proteins encoded by a genome. In a narrower sense, it can be used to describe the set of proteins expressed in a cell at a given time. As genome projects provide information about the number and kinds of genes present in prokaryotic and eukaryotic genomes, the question of protein function is becoming a central issue in biology. As stated earlier, in most of the genomes sequenced to date, many newly discovered genes have no known function. Others have only presumed functions assigned by analogy with known genes made by sequence comparisons from databases, not experimental evidence from laboratory experiments. For example, in *E. coli* and *S. cerevisiae*, more than half of the genes encoded by their genomes have no known function. In the human genome, about 41 percent of the genes are of unknown function.

Reconciling the Number of Genes and the Number of Proteins

As more genomic sequence data become available for eukaryotes, it is apparent that the complexity of an organism is not necessarily related to the number of genes in its genome. For example, how is it that *Drosophila* has fewer genes than the less complex nematode *C. elegans*? Or, to look at it another way, how can *C. elegans* have almost as many genes as humans? The answer to these questions involves an understanding of gene function. In the pregenomic era, gene function was equated with identifying an encoded gene product. However, sequencing has revealed that the link between gene and gene product is often complex. Genes can have multiple transcription start sites that produce several different types of transcripts. Alternative splicing and editing of pre-mRNA molecules can generate dozens of different proteins from a single gene. Altogether, it is estimated that 40–60 percent of human genes produce more than one protein by alternative splicing.

Once made, many gene products are modified by cleavage of end groups (such as signal sequences, propeptides, or initiator methionine residues), by the addition of chemical groups (e.g., methyl, acetyl, phosphoryl), or by linkage to sugars and lipids. Proteins are internally and externally cross-linked and in some rare and interesting cases, even processed by removing internal amino acid sequences (called **inteins**). Over a hundred mechanisms of posttranslational modification are known. Analysis of protein function is complicated by the fact that many proteins work via protein–protein interactions or as part of a large molecular complex. Thus, the human genome, which has 25,000–30,000 protein-coding genes, has the capacity to produce several hundred thousand different gene products. Proteomics is used to reconcile the differences between the number of genes in a genome and the number of proteins observed in cells. Proteomics provides information about a protein's function, structure, posttranslational modifications, protein–protein interactions, cellular localization, variants, and relationships (shared domains, evolutionary history) to other proteins—for every protein encoded in a genome.

Proteomics Technology

Proteomics uses techniques for separating and identifying proteins isolated from cells. The most commonly used combination of techniques involves two-dimensional gel electrophoresis (2DGE) and mass spectrometry (MS). In 2DGE (Figure 20–26), proteins extracted from a cell are loaded onto a polyacrylamide gel and in the first dimension, proteins are separated according to their electrical charge. When completed, the gel is rotated 90°, and in the second dimension, proteins are separated according to their molecular weight. When the gels are stained, proteins are revealed as spots; typical gels show from 200 to 10,000 spots (Figure 20–27). The proteins isolated by 2DGE must then be identified. In the past, proteins were identified by their binding to an antibody, or by fragmentation and step-wise analysis to produce the amino acid sequence. These time-consuming and labor-intensive methods are not useful in analyzing the hundreds or thousands of proteins that can be isolated from a cell.

To speed up and automate the process, new techniques of mass spectrometry were developed, allowing the mass of large molecules to be accurately determined (Figure 20–28). One method is called **matrix-assisted laser desorption and ionization (MALDI)**. The procedure has several steps. First, a protein isolated from a 2D gel is digested with the enzyme trypsin producing a characteristic set of peptide fragments. Each fragment is analyzed by MALDI, producing a peptide mass fingerprint. Second, a database search algorithm (e.g., MS-BLAST) is used to search protein databases that contain the amino acid sequence of known proteins. In searching the database, the algorithim breaks each known protein into a series of predicted peptide fragments (a process known as virtual trypsin digests) and computes the mass

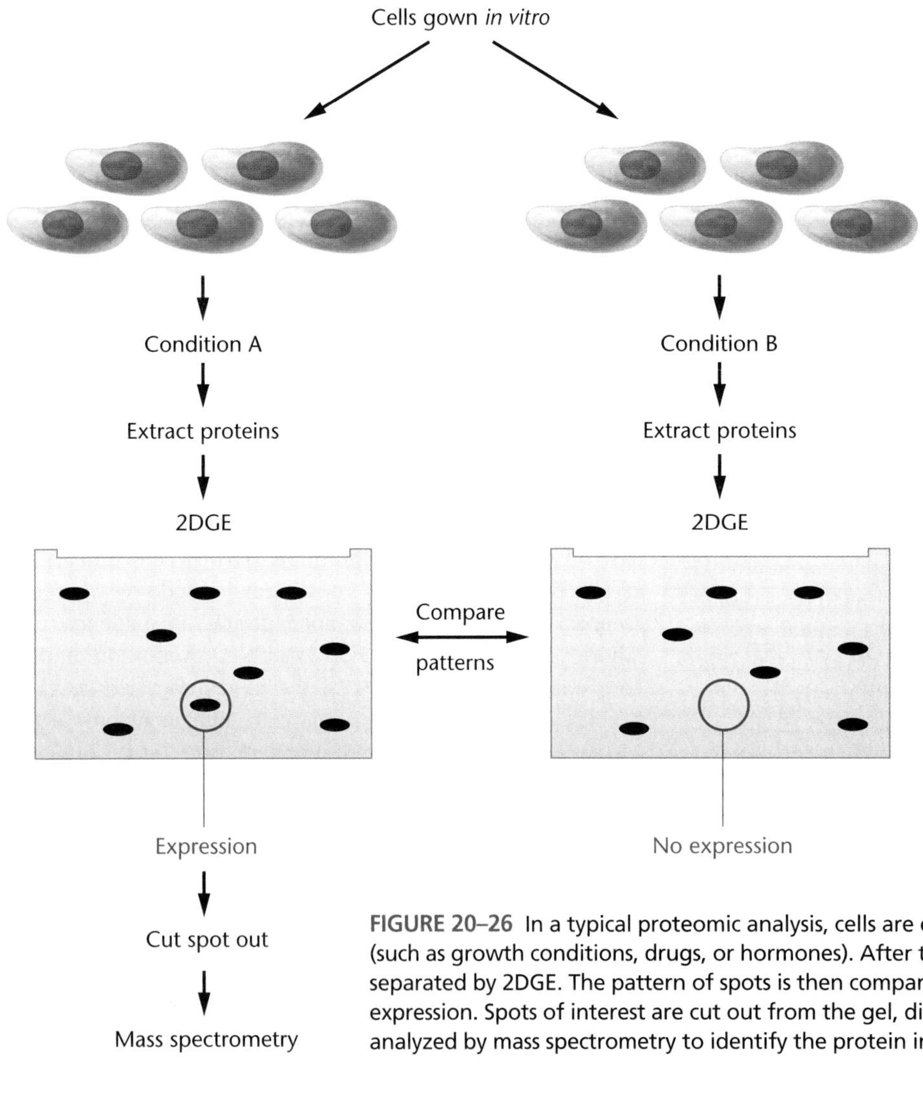

FIGURE 20–26 In a typical proteomic analysis, cells are exposed to two different conditions (such as growth conditions, drugs, or hormones). After treatment, proteins are extracted and separated by 2DGE. The pattern of spots is then compared for evidence of differential gene expression. Spots of interest are cut out from the gel, digested into peptide fragments, and analyzed by mass spectrometry to identify the protein in the spot.

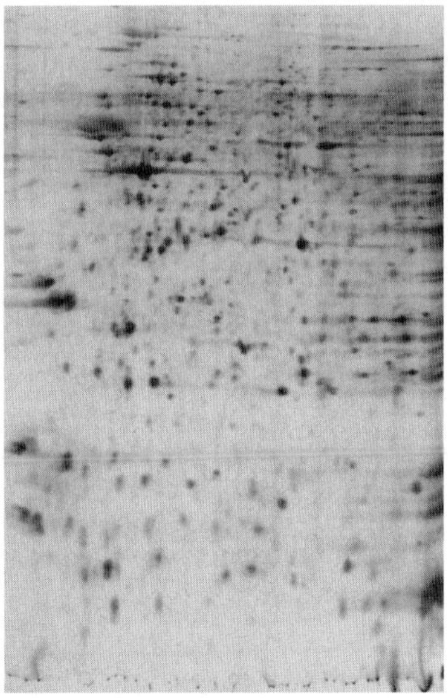

FIGURE 20–27 A two-dimensional protein gel, showing the separated proteins as spots. Several thousand proteins can be displayed on such gels.

of each fragment. The masses of the predicted fragments are compared to the ones in the protein sample derived by MALDI. After searching the database, the search program produces a series of matches between the proteins in the database and the protein being analyzed. The matches are listed from highest to lowest. A perfect match identifies a protein with no ambiguity. However, because of alternative splicing or posttranslational modification, there may not be a perfect match, and the identity of the protein may have to be confirmed by other mass spectrometry techniques. High-performance instruments can identify hundreds of proteins per day, and banks of spectrometers can process thousands of samples in a single day. New instruments with faster sample processing times and increased sensitivity and accuracy are under development and will allow proteomics to have a significant impact on many areas of biology. Collectively, these techniques and approaches are referred to as **high throughput technology**.

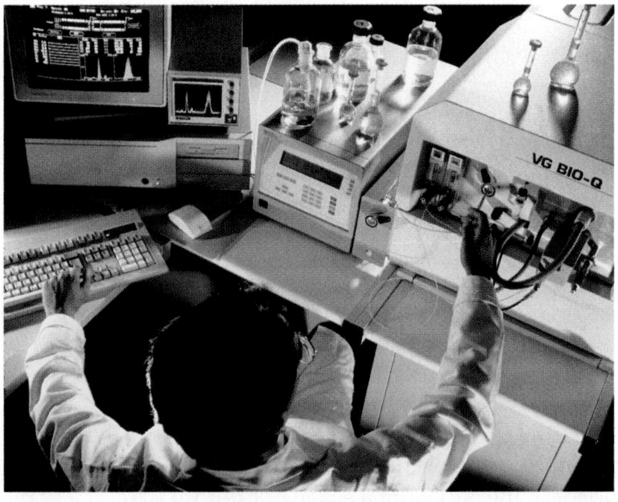

FIGURE 20–28 Proteins are digested and analyzed by mass spectrometry to identify the gene products present in a cell at a given time.

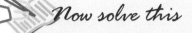

Now solve this

In Problem 20.27 on page 514 a protein domain is used to query a database and recover proteins with a given sequence. You are asked to explain why the function of the query sequence is not related to the function of the other proteins.

Hint: Remember that protein domains may have related functions, and that proteins can contain several different domains.

The Bacterial Proteome Changes with Alterations in the Environment

As outlined earlier, *M. genitalium*, with a genome of 480 genes, represents one of the simplest, independently living organisms known. Valerie Wasinger and her colleagues used proteomics to provide a snapshot in which genes are expressed under two different growth conditions in *M. genitalium*: exponential growth and the stationary phase following rapid growth.

Using 2DGE, Wasinger's group identified 427 protein spots in exponentially growing cells. Of these, 201 were analyzed and identified by peptide digestion, mass spectrometry, and comparison to known proteins. The analysis uncovered 158 known proteins (33 percent of the proteome) and 17 unknown proteins. The remaining spots included fragments derived from larger proteins, different forms of the same protein (isoforms), and posttranslationally modified products. The identified proteins included enzymes involved in energy metabolism, DNA replication, transcription, translation, and transport of materials across the cell membrane.

During the transition from exponential growth to the stationary phase, there was a 42 percent reduction in the number of proteins synthesized. In addition, some new proteins appeared, while other proteins underwent dramatic changes in abundance. These changes are apparently a consequence of nutrient depletion, increased acidity of the growth medium, and other adaptations to environmental changes.

Wasinger's analysis helps establish the minimum number of expressed genes required for maintenance of the living state and the changes in gene expression that accompany the transition to the stationary phase.

This study also points out one of the limitations of current proteomics technology: Only the most abundantly expressed proteins can be detected with 2DGE. In this study, most of the proteome was unexpressed or undetected because the proteins were present in very low amounts (too low to be detected on gels) or were not solubilized and recovered by the extraction methods used in Wasinger's study. In spite of these limitations, proteomic analysis provides a wide range of information that cannot be obtained by genome sequencing.

Proteome Analysis of an Organelle: The Nucleolus

Proteome analysis faces several inherent problems because the proteome is a very dynamic system. Cellular proteins have a vast range of concentrations (more than a millionfold) and hundreds of posttranslational and alternative splicing forms, many of which may be difficult to separate from one another. In addition, there are cell-cycle specific variations in abundance and many protein–protein interactions to deal with. Another problem is technical: For several reasons, some of which were outlined earlier, proteins displayed by 2DGE usually represent only a fraction of those present in a cell.

Isolation of a portion of the cell's proteome is one way to reduce the impact of these limitations. This approach reduces the complexity of the sample and improves the separation and quantitative analysis of the sample's peptide fragments. One way to do this is by isolating subcellular organelles and components. Subproteome analysis has been used to successfully study the proteomes of nuclear pores, cell surfaces, the nucleolus, and other cellular components.

The **nucleolus** is usually the largest and most prominent organelle in the eukaryotic nucleus and is a dynamic structure. The nucleolus forms early in the G1 phase of the cell cycle at the chromosomal sites of the rDNA genes [Figure 20–29(a)], and is disassembled just before mitosis. Three distinct subcompartments of the nucleolus have been described: the fibrillar center (FC), the dense fibrillar compartment (DFC), and the granular compartment (GC) [Figure 20–29(b)]. Traditionally, the nucleolus has been regarded solely as a ribosomal processing center, in which the ribosomal RNAs are transcribed in the FC, processed in the DFC and, along with 5*S* RNA and ribosomal proteins, assembled into ribosomal subunits in the GC.

Recent studies have suggested that the nucleolus may have other important functions, including assembly of spliceosomes, telomerase, and other small nuclear RNA-protein complexes. An analysis of the nucleolar proteome will help confirm these new functions and define others.

To study the nucleolar proteome, Jens Andersen and his colleagues isolated nucleoli from HeLa cells, a human cancer cell line, and analyzed the sample purity using several methods, including electron microscopy [Figure 20–30(a)]. Nucleolar proteins were separated using 2DGE. A variety of methods was used to prepare the sample prior to performing MS, thus enabling the investigators to identify more than 400 nucleolar proteins [Figure 20–30(c)].

This study identified many proteins already known to be associated with the nucleolus, as well as many nuclear proteins not known to exist in the nucleolus. The role of the nucleolus in assembly of spliceosomes and other RNA-protein complexes was confirmed. More important, discovery of more than 100 previously unknown proteins, making up 32 percent of the nucleolar proteome, indicates that the list of nucleolar functions will probably grow much larger as these proteins are characterized and as functional roles are assigned to them.

Using fluorescent tagging of one of the newly discovered nucleolar proteins (Paraspeckle Protein 1, or PSP1), Andersen's group was able to show that this protein accumulates in a previously unknown compartment of the nucleus, called **paraspeckles**, along with at least two other newly identified proteins [Figure 20–30(b)]. All human cells examined to date contain 10–20 nuclear paraspeckles in the space between chromosomes, and these are associated with RNA splicing components. The studies show that PSP1 and the other two proteins shuttle between the nucleolus and the nucleus in a transcription-dependent fashion. When transcription is inhibited, these three proteins move from the nucleus into the nucleolus and localize in a cap at the periphery of the nucleolus [Figure 20–30(b)]. This study emphasizes that we have much to learn about the dynamic interactions between the nucleus and nucleolus and that the nucleolus is much more than a ribosome factory.

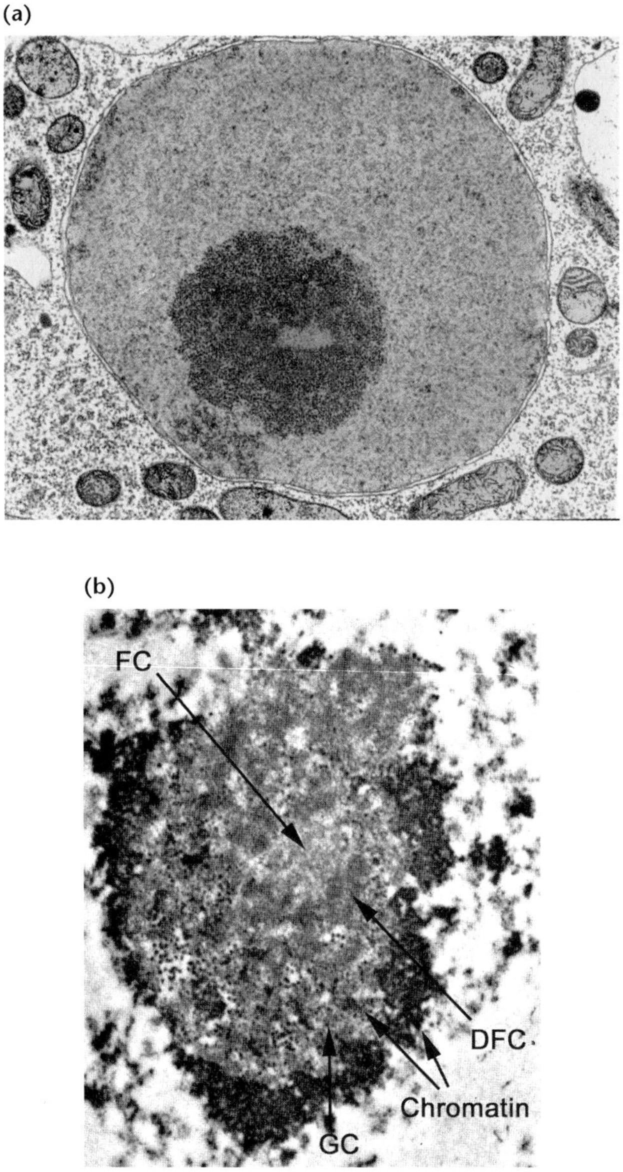

FIGURE 20–29 (a) A transmission electron micrograph showing nucleoli in the nucleus of a eukaryotic cell. (b) A transmission electron micrograph of the nucleolus showing the nucleolar compartments. Key: FC: fibrillar center; DFC: dense fibrillar compartment; and GC: granular compartment.

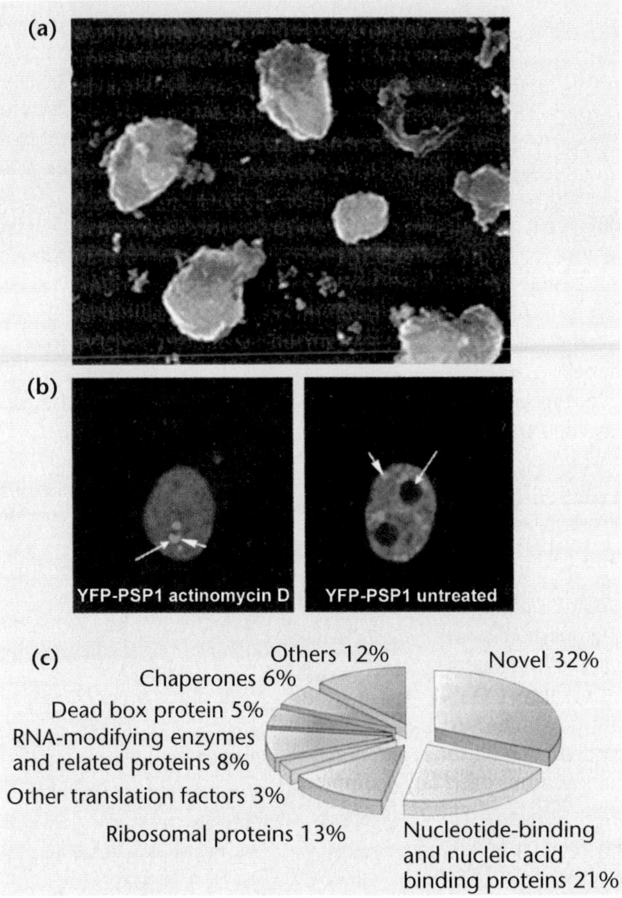

FIGURE 20–30 (a) Scanning electron micrograph showing purified human nucleoli. Proteins were solubilized, fractionated, and analyzed by mass spectrometry. (b) At right, localization of one newly identified nucleolar protein (PSP1), within nuclear structures called paraspeckles. At left, inhibition of transcription by treating the cells with actinomycin D causes the nucleolus to reorganize as well as the formation of cap structures containing PSP1. (c) Functional assignments of the more than 400 proteins identified in this proteomic analysis. Dead box proteins are enzymes that participate in folding RNA, including spliceosome formation. Note that 32 percent of the proteins identified were previously unknown.

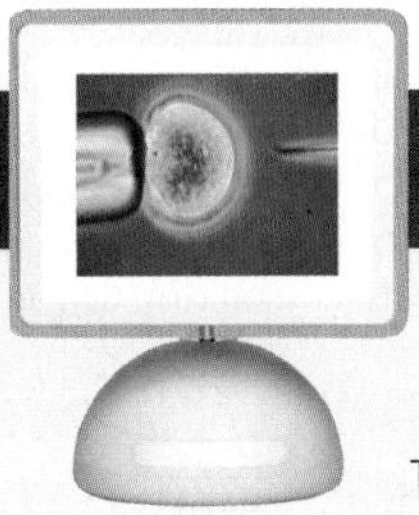

GENETICS, TECHNOLOGY, AND SOCIETY

Beyond Dolly: The Cloning of Humans

The birth of Dolly, a healthy white-faced lamb, took the world by surprise. Until February 1997, the idea that an animal could be cloned from the cells of another adult animal was science fiction—something from *Brave New World* or *The Boys from Brazil.* For many people, the notion that animals could be cloned conjured up scenes of multiple replicated Adolf Hitlers or creations gone astray, as in Shelley's *Frankenstein.* For decades, the public had been comforted by scientists who asserted that it would be impossible to clone a new mammal from cells of an adult. It was thought that cells from adult animals could not be reprogrammed in such a way as to form a new, entire organism. But then Dolly appeared.

Dolly, the first mammal cloned from adult cells, was brought into being by a group of embryologists led by Ian Wilmut and Keith Campbell at the Roslin Institute in Scotland. Their reason for developing methods to clone farm animals was to provide identical transgenic animals that secrete pharmaceutical products, such as blood clotting factors or insulin, into their milk. In this way, animals could be used as bioreactors to synthesize proteins that are difficult or expensive to synthesize in bacteria or in the test tube.

The cloning method that Wilmut and Campbell used to create Dolly—a procedure called *nuclear transfer*—was first suggested by embryologist Hans Spemann in 1938. The idea is simply to replace the nucleus of a newly fertilized (or unfertilized) egg with the nucleus from an adult cell, thereby creating a "reconstructed" zygote containing the nucleus from one animal and the egg cytoplasm from another. In theory, the genetic information in the donor nucleus should direct all further embryonic development, and the new organism should be a genetic replica of the adult that donated the nucleus. Although this procedure sounds simple in theory, it proved to be extremely difficult in practice. Even when the technical procedures such as collecting eggs, removing nuclei, and transplanting donor nuclei were perfected, the cloning procedure would work only if the donor nucleus came from an embryo. Because adult nuclei come from specialized structures such as the skin, liver, and kidney, they are "differentiated" and express only a small subset of all the genes in the nuclear genome. As all the genes in the genome must be actively expressed in an embryo, adult nuclei proved to be inadequate for the job.

Wilmut and Campbell accomplished what was perceived to be the impossible by genetically reprogramming the adult nuclei before transferring them into recipient eggs. They removed donor cells from the udder of an adult ewe and starved the cells so that they became quiescent and entered the G0 phase of the cell cycle. For unknown reasons, this starvation procedure allowed the silent genes within the udder cell nucleus to be turned on after the nucleus was transplanted into the cytoplasm of a sheep's egg. In addition, the researchers passed an electric current through the egg to facilitate nuclear transfer and to stimulate the new "zygote" to begin dividing. To create Dolly, over 200 udder-cell nuclei were transferred. Of these nuclei, only 29 developed into embryos, and 13 were transferred into surrogate mother ewes. One pregnancy resulted, which culminated in the birth of Dolly.

Although Dolly was the first mammal to be cloned from adult cells, the approach has since been used to create cloned mice from adult mouse cells. In addition, transgenic sheep and cattle have been cloned by similar procedures, except that fetal-cell nuclei were used as donors, rather than adult nuclei. One of these transgenic sheep ("Polly") bears the human gene for blood-clotting factor IX and secretes factor IX into her milk, paving the way for the production of pharmaceuticals from cloned farm animals.

Besides the benefits for drug production, cloning promises other advantages. For example, cloning might allow scientists to preserve and replicate endangered species, cure genetic diseases in farm animals, and provide animal models of human diseases for which there are presently no research models.

Despite the obvious advantages for agriculture and medicine, the cloning of mammals has led to outcry and concern. The idea that humans could be cloned has been termed immoral, repugnant, and ethically wrong. Within days of the announcement about Dolly, bills were introduced into the U.S. Congress to prohibit research into human cloning, and worldwide bans were called for. Frightening scenarios were proposed: rich and powerful people cloning themselves for reasons of vanity; people with serious illnesses cloning replicas to act as organ donors; and legions of human clones suffering loss of autonomy, individuality, and kinship ties. But is it really possible to clone humans? And if so, should human cloning ever be done?

The answer to the first question is simple: The same technology used to create Dolly could be used to clone a human. Most of the necessary technical procedures are being used now for *in vitro* fertilization. And it seems likely that adult human nuclei could be reprogrammed similarly to the adult sheep nuclei that created Dolly.

The answer to the second question is not as simple. Both scientific and ethical issues cloud our judgment about human cloning. Present technology cannot guarantee the health of any cloned individual, human or animal. It is possible that clones could exhibit a higher risk of genetic disease or cancer, because of the accumulation of mutations in the donated adult nucleus. Clones might also show accelerated aging, due to shortened telomeres that are present in the donor's chromosomes. If the donor's nucleus is not completely reprogrammed prior to cloning, the clone could undergo abnormal development as well. Until the safety issues are resolved, it may be prudent to suspend any attempts to clone humans.

The ethical issues are even more problematic. It is possible to foresee positive aspects to human cloning. Infertile couples or couples who suffer from genetic disease on one side of the family could choose to clone one of the partners in order to raise a child who is biologically related. Cloning a person's cells *in vitro* could provide a source of cells or tissues to treat a number of serious diseases. On the other hand, it is equally possible to foresee negative aspects. Would cloning seriously alter what it means to be a unique human being? What would be the fate of clones created for organ transplantation? Could the technique be misused by the unscrupulous for social or political goals? It is important for society to consider these issues now in order to ensure responsible and beneficial outcomes of this new technology.

References

Cibelli, J.B., et al. 1998. Cloned transgenic calves produced from nonquiescent fetal fibroblasts. *Science* 280:1256–58.

Pennisi, E. 1998. After Dolly, a pharming frenzy. *Science* 279:646–48.

Robertson, J.A. 1998. Human cloning and the challenge of regulation. *N. Eng. J. Med.* 339:118–25.

Solter, D. 1998. Dolly as a clone—and no longer alone. *Nature* 394:315–16.

Wilmut, I. 1998. Cloning, for medicine. *Sci. Am.* (Dec.) 58–63.

Wilmut, I. et al. 1997. Viable offspring derived from fetal and adult mammalian cells. *Nature* 385:810–13.

CHAPTER SUMMARY

1. Genome sequences are analyzed in several steps to ensure that the sequence is accurate, to identify all encoded genes, and to classify known genes into functional categories. They are then deposited into searchable databases.
2. Bacterial genomes have very high gene density, averaging one gene per kilobase pair of DNA. Typically, as much as 90 percent of the chromosome encodes genes. Many genes are organized into polycistronic transcription units that do not contain introns. Archaea are a prokaryotic domain. In some ways, their genes, proteins, and metabolism resemble those of eukaryotes.
3. Eukaryotic genomes are organized into two or more chromosomes, each containing a linear double-stranded DNA molecule. The gene density is much lower than that in bacteria. Genes typically are not organized into operons; rather, each is a separate transcription unit. However, the nematode *C. elegans* has many genes organized into operons. Eukaryotic genes are often interrupted with introns.
4. Complex multicellular eukaryotes differ from the less complex yeast in a number of ways. These eukaryotes have more genes and much more DNA. This results in gene densities falling to 1 gene per 5 kb or even 1 gene per 10–20 kb or more. A higher proportion of eukaryotic genes have introns. The number of introns per gene increases, and the size of introns increases as complexity increases from yeast to *C. elegans* to humans. Some plants, such as *Arabidopsis*, have a gene structure and organization that is indistinguishable from animals. Other plants, such as maize, have a different organization, with vast blocks of transposable elements separating islands of genes.
5. In the human genome, large differences exist in gene density on different chromosomes, with gene-rich regions alternating with

gene-poor regions. Duplicated segments are a common feature of the chromosomes sequenced to date.

6. Many eukaryotic genes have undergone duplication followed by sequence divergence, leading to multigene families. The globin gene clusters are prime examples of this phenomenon.

7. Proteomics is used to study the expression of genes in bacterial cells under different growth conditions and is providing insight into the gene sets cells use during growth. These techniques have also been used to study the architecture of subcellular elements, such as the nucleolus.

Insights and Solutions

1. How are gene duplications generated, and how do they contribute to genome evolution?

Solution: Gene duplications can be generated by unequal crossing over and DNA slippage. In addition, chromosomal aberrations such as inversions and translocations can lead to gene duplication. It is possible to duplicate genes through horizontal transfer involving transposons, plasmids, and other transferred DNAs. Gene duplication contributes to evolution by providing mutational opportunities without compromising the functionality of the genome. Gene duplications allow for mutational experimentation and therefore evolution.

2. Recent sequencing of the heterochromatic regions (repeat-rich sequences concentrated in centric and telomeric areas) of the *Drosophila* genome indicates that within 20.7 Mb, there are 297 protein-coding genes (Bergman, et al 2002. *genomebiology3 (12)@genomebiology.com/2002/3/12/RESEARCH/0086*). Given that the euchromatic regions of the genome contain 13,379 protein-coding genes in 116.8 Mb, what general conclusion is apparent?

Solution: Gene density in euchromatic regions of the *Drosophila* genome is about one gene per 8,730 base pairs, while gene density in heterochromatic regions is one gene per 69,696.9 bases (20.7 Mb/297). Clearly, a given region of heterochromatin is much less likely to contain a gene than the same sized region in euchromatin.

Problems and Discussion Questions

1. Used in gene annotation, gene prediction programs allow researchers to identify likely coding regions in DNA sequences. Annotation is complicated when genes are complex, contain multiple initiation sites, or contain numerous exons. For example, Pavy and others (1999. *Bioinformatics* 15:887–99) determined that even for the most well-studied organisms, such programs can predict correct exon boundaries only about 80 percent of the time. Given this percentage, what is the likelihood of determining the correct exon boundaries in a gene with five exons?

2. Recent genome-sequencing efforts have provided considerable insight to the molecular nature of living systems. However, it has become increasingly apparent that to fully comprehend the genome, reannotation, manual verification, and more advanced techniques will be needed. Haas and colleagues (2002. *genomebiology3(6)@genomebiology.com/2002/3/6/RESEARCH/0029*) recently reannotated the *Arabidopsis* genome and found 240 new genes, 92 of which are homologous to known proteins. In addition, they identified a new class of exons, called micro-exons, which vary in length from 3 to 25 base pairs. Based on this information, how would you qualify the statement that "to know the sequence of DNA is to know the blueprint of life"?

3. In a recent draft annotation and overview of the human genome sequence, Wright and others (2001. *genomebiology2(7)@genomebiology.com/2001/2/7/RESEARCH/0025*) presented a graph similar to the one shown in the next column. The graph provides the approximate number of embryo-specific genes for each chromosome. Review earlier information in the text on human chromosomal aneuploids, and correlate that information with the graph. Does this graph provide insight as to why some aneuploids occur and others do not?

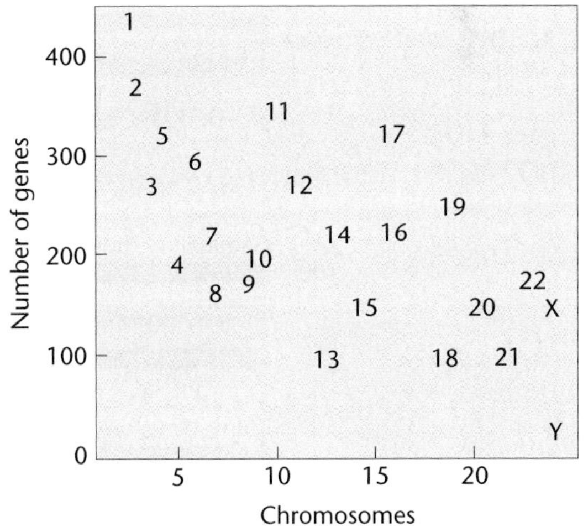

4. The process of annotating a sequenced genome is continual. In March 2000, the first annotated sequence of the *Drosophila* genome was released, which predicted 13,601 protein-coding genes within the euchromatic region of the genome. Shown below are selected data from Release 2 and Release 3. (Modified from Misra et al. 2002. *genomebiology3(12)@genomebiology.com/2002/3/12/RESEARCH/0083.*) (a) Assuming a uniform distribution for Release 3, approximately how many base pairs of DNA are between protein-coding genes in *Drosophila*? (b) On average, approximately how many exons are there per gene for Release 3? (c) Approximately how many introns are there per gene? (d) What appears to be the most significant difference between Release 2 and Release 3?

Criteria	Release 2	Release 3
Total length of euchromatin	116.2 Mb	116.8 Mb
Total protein-coding genes	13,474	13,379
Protein-coding exons	50,667	54,934
Introns	48,381	48,257
Genes with alternative transcripts	689	2729

5. The data given in Problem 4 indicate that the more closely researchers examine genome sequences, the more complex the interpretations of those data will become. Misra and colleagues (2002) found that nested and overlapping genes are common in *Drosophila*. They determined that approximately 7.5 percent of all Release 3 genes were included within the introns of other genes and the majority are transcribed from the opposite strand of the including gene. In addition, they found that about 15 percent of the annotated genes involve the overlap of mRNAs on opposite strands. What impact will this information have on genome annotation, and what clinical significance might it have?

6. In April 2003, scientists announced that the Human Genome Project had finished its mission and that the human genome was completely known. One of the leaders of the project was quoted as saying, "We have before us the instruction set that carries each of us from the one-cell egg through adulthood to the grave." However, from the beginning, the HGP excluded the heterochromatic regions at the tips of chromosomes and the regions surrounding the centromeres. The human genome is about 3000 Mb, and heterochromatic regions comprise about 15 percent of this total. If gene density in human heterochromatin is the same as it is in *Drosophila*, how many genes remain to be discovered in humans? Would you be comfortable stating that all human genes have been identified? See question 2 in the Insights and Solutions section.

7. One of the main problems in annotation is deciding how long an ORF must be before it is accepted as a gene. Shown below are three different ORF scans of the *E. coli* genome region containing the *lacY* gene. Regions shaded in brown indicate ORFs. The scans have been set to accept ORFs of 50, 100, and 300 nucleotides as genes. How many putative genes are detected in each scan? The longest ORF covers 1254 bp; the next longest covers 234 bp; and the shortest covers 54 bp. How can you decide how many genes are actually in this region? In this type of ORF scan, is it more likely that the number of genes in the genome will be overestimated or underestimated? Why?

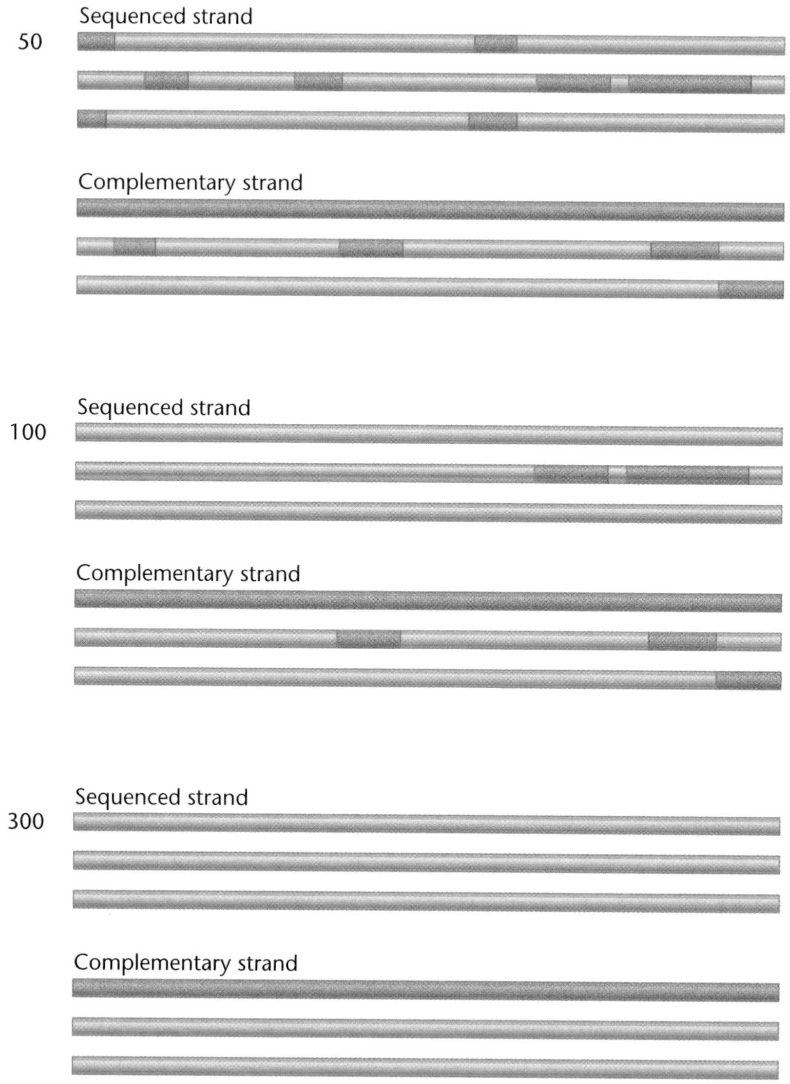

8. To deal with the problems of correctly annotating microbial genomes, Marie Skovgaard and her colleagues (2001. *Trends Genet.* 17:425–28) compared the annotated number of genes in genomes derived from sequence analysis to the number of known proteins in each organism as reported in a protein database. The results of their study are summarized in the graph below. The errors range from a few percent for *M. genitalium* to almost 100 percent for *A. pernix*. The general trend shown in the graph is that the error rate increases as the GC content of the genome increases. What explanation might account for this? What precautions should be taken in annotating the genomes with high GC content?

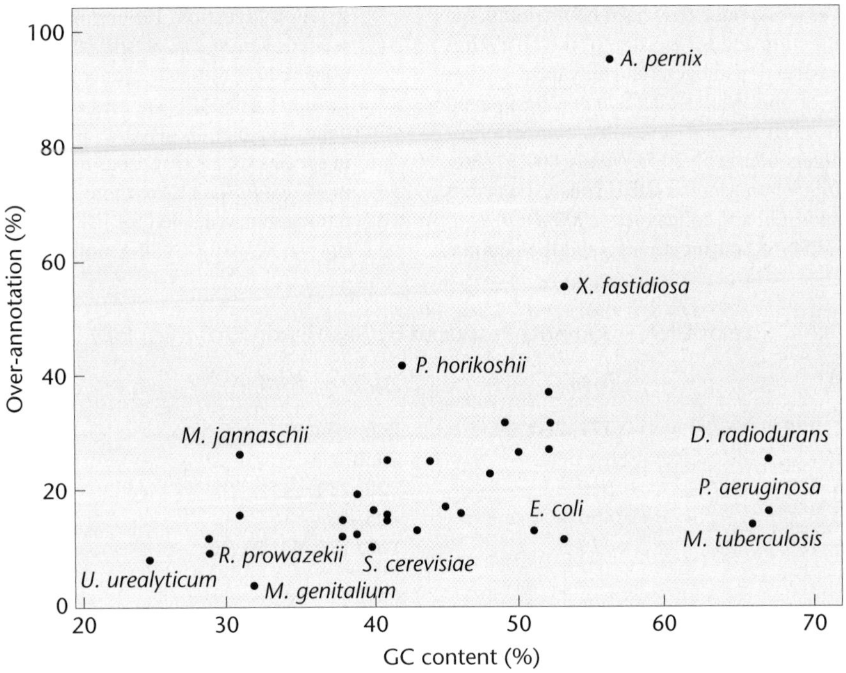

9. What are CpG islands, and how are they used in analysis of genomes?
10. What is functional genomics? How does it differ from comparative genomics?
11. What features do Archaea genomes share with eukaryotic genomes?
12. What is the definition of an operon? How does information from the *Aquifex aeolicus* genome alter our ideas about operons? How do findings about transcription units in the *C. elegans* genome fit into the classic definition of operons?
13. Plasmids can be transferred between species of bacteria, and most carry nonessential genes. For these and other reasons, plasmid genes have not been included as part of the genomes of bacterial species. The bacterium *B. burgdorferi* contains 17 plasmids carrying 430 genes, some of which are essential for life. Should plasmids carrying essential genes be considered as part of an organism's genome? What about other plasmids that do not carry such genes? In other words, how do we define an organism's genome in these cases?
14. Compare and contrast the chemical nature, size, and form assumed by the genetic material of Eubacteria and yeast. Do the same with Archaea and yeast. Which group has more differences and which has more similarities to yeast?
15. Why might we predict that the organization of eukaryotic genetic material is more complex than that of viruses or bacteria?
16. Compare the organization of bacterial genes to that of eukaryotic genes. What are the major differences?
17. *C. elegans* is a eukaryotic organism with a genome of 97 Mb and about 20,000 genes. What organizational features of this genome are unusual when compared to the genomes of other eukaryotes, such as yeast and *Drosophila*?
18. Based on the completion of the genome sequence of *Arabidopsis* in 2000, Simillion and others (2002. *Proc. Nat. Acad. Sci. [USA]* 99:13627–32) have estimated that the *Arabidopsis* genome has undergone three rounds of genome duplication or polyploidization events in the last 100 million years. Presently, of the approximately 25,000 genes in *Arabidopsis*, these researchers estimate that about 80 percent are in duplicated regions of the genome. Examination of Table 20.6 reveals that *Arabidopsis*, rice, maize, and barley have approximately the same number of genes, but that *Arabidopsis* has much less DNA. What type of genomic organization appears to account for the vast differences in DNA content with similar gene numbers in these species?
19. Annotation of the human genome sequence reveals that our genome contains 25,000–30,000 genes. Proteomic analysis indicates that human cells are capable of synthesizing more than 300,000 different proteins. How can this discrepancy be reconciled?
20. List some of the following general features of the human genome: size, how much codes for proteins, how much is composed of repetitive sequences, where genes are distributed on chromosomes, and how many genes it contains.
21. The discovery that *M. genitalium* has a genome of 0.58 Mb and only 470 protein-coding genes has sparked interest in determining the minimum number of genes needed for a living cell. In the search for organisms with smaller and smaller genomes, a new species of Archaea, *Nanoarchaeum equitans*, was discovered in a high-temperature vent on the ocean floor (Huber et al. 2002. *Nature* 417:63–67). This prokaryote has one of the smallest cell sizes ever discovered, and its genome is only about 0.5 Mb. However, organisms such as *M. genitalium*, *N. equitans*,

and other microbes with very small genomes are either parasites or symbionts. How does this affect the search for a minimum genome? Should the definition of the minimum genome size for a living cell be redefined?

22. In the search for the smallest bacterial genome and the minimum number of genes necessary for life, attention has turned to species of *Buchnera*, which live as intracellular symbionts in aphid intestinal cells. As symbionts, they need not maintain the genes necessary for infection and for evasion of the host's immune system as do parasites or pathogens and may have smaller genomes. The genome of one species of *Buchnera*, designated APS, has been sequenced. It has 564 genes in a circular chromosome of 640 kb (Shigenobou, et al. 2000. *Nature* 407:81–86). To determine whether *Buchnera* genome size is conserved across different groups of aphids, Gil and colleagues (2002. *Proc. Nat. Acad. Sci. [USA]* 99:4454–58) physically mapped the genome sizes of nine *Buchnera* genomes that were isolated from five aphid families. The genomes were sized by digestion with restriction enzymes, and separation of the resulting fragments was done by gel electrophoresis. The data for some *Buchnera* species are given in the following table. Although there are some discrepancies in sizes, the sum of the fragments corresponds to the size of the chromosome that appeared on the gel without restriction digestion. From your analysis of the data, is genome reduction in *Buchnera* still occurring? How do the genome sizes obtained for these species compare with the genome of *M. genitalium*? The APS species of Buchnera contains 564 coding genes in a 641 kb genome. How many genes should be present in species CCE? How does this compare to the number of genes in *M. genitalium*? Are there other ways to determine the minimum genome needed for life without searching for other bacterial species with small genomes?

Size of DNA Fragments Produced by Restriction Enzymes, (kb)

Buchnera	*ApaI*	*RsrII*	*ApaI* + *RsrII*	Total DNA Length kb
APS	286, 226, 73, 52, 3.4	277, 264, 99	240, 104, 99, 73, 52, 45, 24, 3.4	640±
THS	545	545	320, 234	544±
CCE	440	450	405, 46	448±
CCU	265, 135, 50, 25	475	200, 136, 64, 50, 30	476±

Extra-Spicy Problems

23. In addition to comparisons of nucleotide sequences for determining phylogenetic relationships among organisms, studies of gene order have become an informative source for genome studies. In addition to providing evidence of evolutionary relationships, gene-order data have been used to predict gene function and functional interactions of proteins. Tamames (2001) (*Genome Biology 2: RESEARCH0020.* E pub 2001 Jun 01) determined that in prokaryotes, loss of gene order occurs when phylogenetic distance increases, but contrary to expectation, significant conservation is maintained in distant groups. What factors might you expect to contribute to the conservation of gene order among distantly related species?

24. Genomic sequencing has opened doors to numerous studies that help us understand the evolutionary forces shaping the genetic makeup of organisms. Using databases containing the sequences of 25 genomes, Kreil and Ouzounis (2001) (*Nucl. Acids Res.* 29:1608–15) examined the relationship between GC content and global amino acid composition. They found that it is possible to identify thermophilic species on the basis of their amino acid compositions alone, which suggests that evolution in a hot environment selects for a certain whole-organism amino acid composition. In what way might evolution in extreme environments influence genome and amino acid composition? How might evolution in extreme environments influence the interpretation of genome sequence data?

25. What are pseudogenes, and how are they produced?

26. The β-globin gene family consists of 60 kb of DNA, yet only 5 percent of the DNA encodes gene products. Account for as much of the remaining 95 percent of the DNA as you can.

27. Annotation of the proteome attempts to relate each protein to function in time and space. Traditionally, protein annotation depended on an amino acid sequence comparison between a query protein and a protein with known function. If two proteins shared a considerable portion of their sequence, the query would inherit the function of the annotated protein. Below is a representation of the "look-the-same" method of protein annotation involving a query sequence and three different human proteins (modified from Rigoutsos, et al. 2002. *Nucl. Acids Res.* 30:3901–16). Note that the query sequence aligns to common domains within the three other proteins. What argument might you present to suggest that the function of the query is not related to the function of the other three proteins?

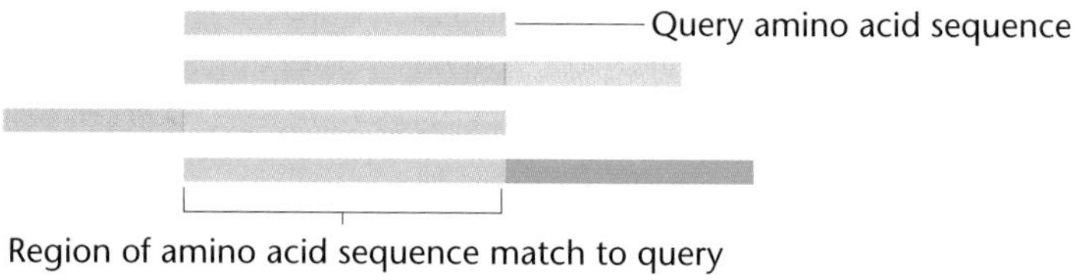

SELECTED READINGS

Aebersold, R., and Mann, M. 2003. Mass-spectrometry-based proteomics. *Nature* 422:198–207.

Andersen, J. et al. 2002. Directed proteomic analysis of the human nucleolus. *Curr. Biol.* 12:1–11.

Brett, D., et al. 2002. Alternative splicing and genome complexity. *Nature Genetics* 30:29–30.

Harrison, P.M., et al. 2002. A question of size: the eukaryotic proteome and the problem in defining it. *Nucl. Acids Res.* 30:1083–90.

International Chimpanzee Chromosome 22 Consortium. 2004. DNA sequence and comparative analysis of chimpanzee chromosome 22. *Nature* 429:383–88.

International Human Genome Sequencing Consortium. 2001. Initial sequencing and analysis of the human genome. *Nature* 409:860–921.

Kiyosawa, H., et al. 2004. Systematic genome-wide approach to positional candidate cloning for identification of novel human disease genes. *Internal Med. J.* 34:79–90.

Nobrega, M.A., and Pennachio, L.A. 2003. Comparative genomic analysis as a tool for biological discovery. *J. Physiol.* 554:31–39.

Schriml, L.M., et al. 2003. Human disease genes and their cloned mouse orthologs: exploration of the FANTOM2 cDNA sequence data set. *Genome Res.* 13:1496–1500.

Switnoski, M., Szczerbal, I., and Nowacka, J. 2004. The dog genome map and its use in mammalian comparative genomics. *J. Appl. Physiol.* 45:195–214.

Tyers, M., and Mann, M. 2003. From genomics to proteomics. *Nature* 422:193–97.

Venter, J.C., et al. 2001. The sequence of the human genome. *Science* 291:1304–51.

CHAPTER TWENTYONE

Dissection of Gene Function: Mutational Analysis in Model Organisms

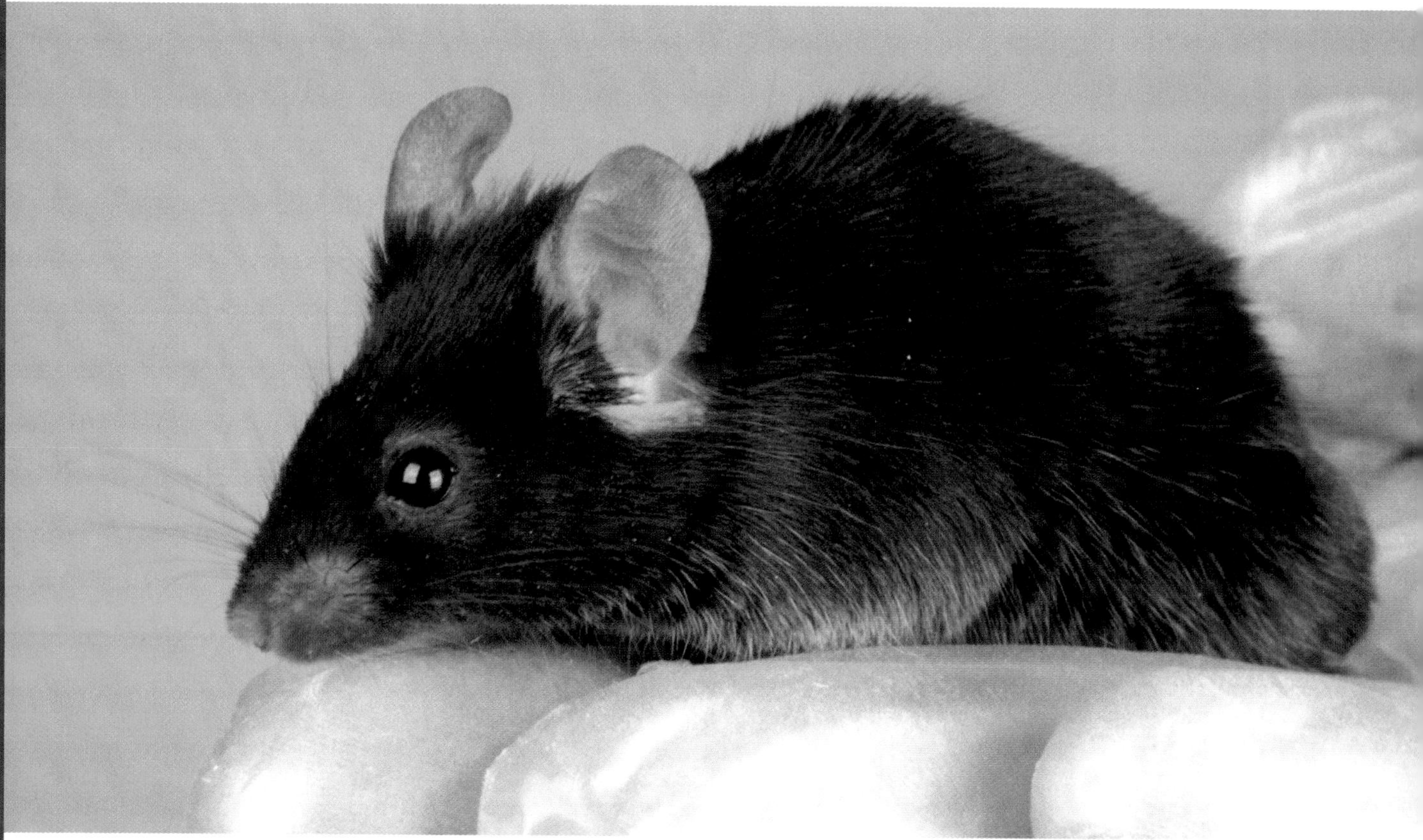

A mouse bearing mutations in the dystrophin *gene and used as a model for research into human muscular dystrophy. The mouse was treated with gene therapy, by introducing the normal gene into its muscle cells.*

CHAPTER CONCEPTS

- Geneticists explore the relationships between genotype and phenotype by using a wide variety of experimental tools. The gene mutant is the most powerful of these experimental tools.
- A good model organism for genetic analysis must be easy to grow, have a short generation time, have a small genome, produce abundant progeny, and be readily mutagenized and crossed.
- Forward genetic analysis begins with the isolation of mutants and is followed by defining genetic pathways, cloning the gene, and creating more mutants to identify genetic interactions.
- Reverse genetic analysis begins with a cloned wild-type gene or a purified protein and progresses to site-directed mutagenesis and phenotypic analysis—the opposite order from that used in forward genetic analysis.
- Functional genomics and high-throughput technologies allow geneticists to dissect the interactions of thousands of gene products simultaneously, supporting the new sciences of genomics, proteomics, and transcriptomics.

A major goal of genetics is to understand what genes are and how they work. As genes control every aspect of biological activity, knowledge of which genes are involved in each process, and how the products of these genes control phenotype and influence function, are central questions in all biological research. Genetic analysis has dramatically extended our understanding of biological processes as simple as biochemical reactions in a single cell or as complex as the developmental steps that lead to the creation of multicellular organisms.

But how do geneticists dissect gene function? How do they discover which genes are involved in which biological processes and how these genes control phenotype?

Since the emergence of genetics as a science in the late 19th century, geneticists have attempted to answer these questions using a wide variety of experimental tools. The most powerful of these experimental tools is the gene mutant, as studied in one of several model organisms.

In classical genetic analysis, geneticists attempt to answer genetic questions by first collecting a number of individuals that display mutations affecting the phenotype of interest. From there, they determine whether the phenotype is controlled by one or more genes, which genes are responsible for which steps in the pathway leading to the phenotype, and how each gene product controls phenotype at the biochemical level. Starting from a series of mutants, geneticists identify, clone, and characterize the function of genes.

Modern genomics and molecular biology have vastly expanded the tools available for genetic analysis. It is now possible to begin a genetic analysis with a cloned gene of unknown function, create specific mutations within the cloned gene, and test the phenotype and function of these mutant genes within model organisms—so-called "reverse genetics." In addition, the modern tools of genomics, such as DNA microarrays, automated sequencers, and computerized bioinformatics allow geneticists to explore overall patterns of gene expression affecting specific phenotypes or biological processes.

In this chapter, we explore some of the methods by which geneticists dissect gene function. We will learn why genetic analysis requires the use of model organisms and how such organisms allow geneticists to answer specific biological questions. We will examine forward and reverse genetic techniques, as well as the new molecular methods of functional genomics, transgenics, and gene knockouts.

21.1 Geneticists Use Model Organisms That Are Genetically Tractable

In order to dissect the genes and processes that regulate biological functions, geneticists undertake their experiments using **model organisms**. Not all organisms are ideal candidates for genetic analysis. For example, the research goal may be to understand the genetics of cancer in humans or the genetic control of development in an important agricultural plant. However, geneticists cannot effectively perform mutagenesis, do rapid controlled matings, or clone genes in most complex, slow-growing life forms such as our own species. In addition, investigators must take into account ethical considerations when undertaking genetic research in humans and many other animals.

A good model organism for genetic analysis must be easy to grow, have a short generation time, produce abundant progeny, and be readily mutagenized and crossed. In addition, the organism must carry out the biological process to be studied. One might question, though, whether knowledge gained about gene functions in simple laboratory organisms is relevant to understanding gene function in higher plants and animals.

Fortunately, genetics, molecular biology, and genome sequencing reveal that many of the genes and molecular processes governing biological functions are shared across evolution. Many gene sequences are conserved from yeast to higher vertebrates, and these genes often carry out similar functions in a wide range of organisms. For example, the genes that regulate cell cycle checkpoints in yeast have homologs in humans. The human versions of these genes regulate the cell cycle and act as tumor suppressor genes. Approximately 200 of the 300 currently known human disease genes have sequence similarities to genes in fruit flies and nematodes. Approximately 100 of these genes are also similar to genes in yeast. Over half of the known human cancer genes have homologs in *Drosophila*. In addition, many early developmental processes, and the genes that control them, are conserved between flies, nematodes, mice, and humans. Geneticists take advantage of these genetic similarities to study the fundamentals of metabolism, development, and disease in simple laboratory organisms, then apply this knowledge to more complex eukaryotes.

Features of Genetic Model Organisms

The most extensively used eukaryotic organisms in genetic research are *Saccharomyces cerevisiae* (budding yeast), *Schizosaccharomyces pombe* (fission yeast), *Drosophila melanogaster* (fruit fly), *Caenorhabditis elegans* (nematode), *Arabidopsis thaliana* (mustard plant), and *Mus musculus* (mouse). Each of these model organisms benefits from a large storehouse of genetic knowledge acquired over decades of genetic research, including databases of their entire DNA sequences and collections of strains bearing specific deletions and mutations, making genetic analysis rapid and efficient. In addition, each organism has features making it the model of choice for a particular study. In this chapter, we will focus on genetic dissection techniques in three model organisms: yeast, *Drosophila*, and mouse.

How Do We Know?

What experimental information is now available that makes organisms such as yeast or *Drosophila* appropriate models in which to study genes that affect human biological processes?

Yeast as a Genetic Model Organism

The yeast *Saccharomyces cerevisiae*, also known as budding yeast or brewer's yeast, is one of the most popular model organisms for genetic research. Geneticists often refer to the "awesome

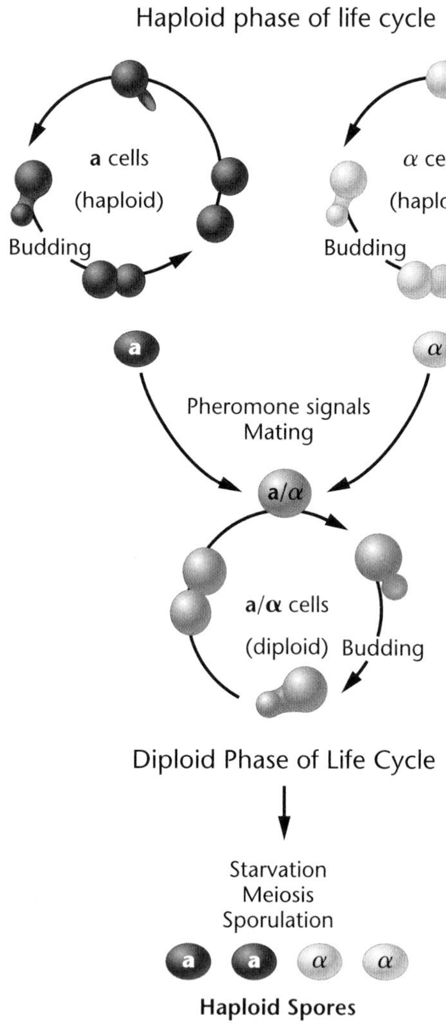

FIGURE 21–1 Life cycle of the budding yeast *S. cerevisiae*. Yeast can grow as either haploid or diploid cells. Both haploid and diploid cells divide by budding. Haploid cells, when stimulated by pheromones, fuse to form diploids. In response to starvation, diploid cells undergo meiosis and sporulation, yielding four haploid spores.

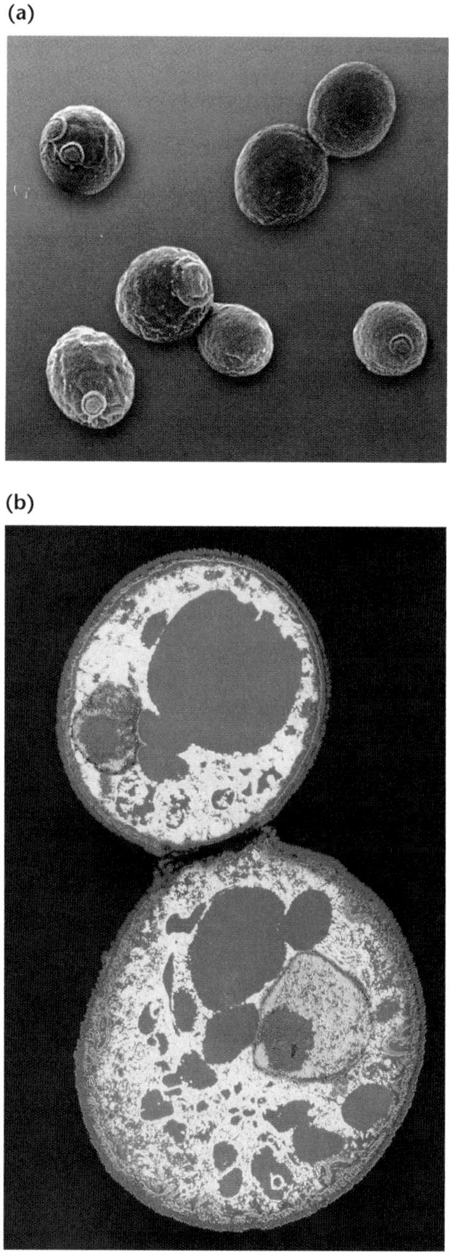

FIGURE 21–2 *Saccharomyces cerevisiae*. (a) Yeast cells in various stages of the cell cycle. Scars on the surface of cells indicate sites of previous budding events.
(b) Photomicrograph of budding yeast cell. Cell wall is shown in red, cytoplasm in blue/yellow, and nucleus in red/green.

power of yeast genetics" because of the ease with which genes can be manipulated and characterized in this organism.

Yeast cells undergo both haploid (1n) and diploid (2n) phases in their life cycle (Figure 21–1). During these phases, cell division occurs by budding, a mitotic process by which a smaller genetically identical daughter cell buds from the surface of the mother cell, grows, and eventually splits from the mother (Figure 21–2). Haploid cells occur in two mating types, a and α. Both a and α cells grow mitotically by budding until a chemical signal known as a pheromone stimulates them to mate. The fusion of one a with one α cell is followed by fusion of their nuclei and formation of a diploid yeast cell. The diploid cell either continues to bud or undergoes meiosis and sporulates, depending upon the availability of nutrients. The process of meiosis results in genetic reshuffling and formation of four haploid spores.

The alternating haploid and diploid nature of the yeast life cycle is particularly useful for genetic analysis. It is easy to detect recessive mutations in haploid cells, as the recessive allele is not masked by a wild-type allele. The diploid phase makes further studies, such as complementation analysis, possible. Also, recessive lethal mutations can be maintained in a diploid yeast strain that carries the mutation on one chromosome and the wild-type allele on the other chromosome.

Another major advantage of yeast as a genetic model is the availability of DNA sequence information and collections of mutants and deletion strains. *S. cerevisiae* was the first eukaryote to have its genome sequenced, the project being completed in 1996. Bioinformatic analysis of the yeast genome reveals that there are approximately 6000 yeast genes. About 4000 of these genes have a known function or an anticipated function based on sequence similarities to other known genes. Yeast geneticists also have available a library of yeast strains containing deletions in each open reading frame (ORF) of the genome, allowing immediate access to null strains for each yeast gene.

These features, combined with the ease with which yeast can be grown, mutagenized, transformed by plasmids, and genetically manipulated, make yeast a popular model for genetic dissection. Molecular techniques for deleting genes or replacing them with genes that have been mutagenized *in vitro* also lend power to yeast genetic analysis. The major disadvantage of yeast is its weakness for studies of multicellular communication and development. These types of studies require the use of model organisms such as *Drosophila, C. elegans, Arabidopsis*, or mouse.

Drosophila as a Genetic Model Organism

Like yeast, *Drosophila* is amenable to genetic analysis in part because of the ease with which it can be grown and the size of its genome—about 13,000 genes on four chromosomes (Figure 21–3). *Drosophila* has a short generation time of about 10 days from fertilized egg to adult. Each female fly produces about 3000 offspring in her lifetime. The stages of the *Drosophila* life cycle and some of the genes controlling development are described in Chapter 23.

Perhaps the greatest strength of *Drosophila* for genetic analysis is its easily observed body plan development in embryonic, larval, and adult stages. The fly's outer skeleton presents scientists with abundant features, such as eye colors, wing shapes, bristles, and segment organization that can be easily identified using a light microscope. Changes in these features reflect mutations in genes controlling differentiation and developmental processes.

An important feature in *Drosophila* genetics is the absence of crossing over in males and only moderate amounts of crossing over in females. The absence of male crossing over means that it is possible to retain the chromosome linkage relationships of genes that are inherited through the male parent. This feature simplifies several aspects of genetic analysis, including the use of sophisticated **genetic screen** techniques, as described later.

Unlike yeast, *Drosophila* does not have a haploid phase in its life cycle. Therefore, geneticists have devised ingenious experimental tools to examine recessive mutations and to maintain stocks of mutant organisms bearing recessive lethal mutations. One of the most important of these tools is the **balancer chromosome**. Researchers created these chromosomes by bombarding *Drosophila* with X rays, leading to multiple overlapping chromosomal inversions. The presence of multiple inversions prevents the recovery of crossover products (as described in Chapter 8). As a result, a balancer chromosome and its homologous normal chromosome remain as intact entities within the population with no recombinant chromosomes passing to the progeny. Balancer chromosomes also bear a dominant marker gene, such as a gene for eye color or wing shape. This allows geneticists to visually identify the presence of a balancer chromosome in individual flies during crosses. In addition, balancer chromosomes contain a recessive lethal gene, that prevents homozygous *balancer/balancer* flies from surviving. Therefore, the only offspring that survive matings of heterozygotes with recessive lethal mutations are heterozygotes and wild-type

(a)

(b)

FIGURE 21–3 Growing and manipulating *Drosophila* cultures. (a) Flies feed and lay their eggs in agar medium poured into milk bottles. Larvae feed within the medium, then crawl up the sides of the jar to pupate. (b) Geneticists observe anaesthetized *Drosophila* under a dissecting microscope and sort flies into groups using a small brush.

homozygotes. Balancer chromosomes exist for each of the three autosomal chromosomes, as well as for the X chromosome. The use of a balancer chromosome to recover X-linked recessive lethal mutations is described in the next section. An innovative use of balancer chromosomes to recover, maintain, and analyze recessive lethal mutations during mutant screens is described in the case study at the end of this chapter, in Section 21.5.

Geneticists also exploit *P* element transposons as powerful genetic tools in *Drosophila*. As described in Chapter 15, *P* elements are mobile transposable elements that can move into and out of the *Drosophila* genome. Transposition occurs in the presence of the *P* element transposase enzyme that recognizes and acts upon the 31 bp inverted repeats at each end of the *P* element DNA. *Drosophila* geneticists have harnessed *P* elements as vectors to introduce cloned genes into the *Drosophila* genome. They first insert a cloned gene of interest into the middle of a *P* element, which also contains a gene for a visible characteristic such as eye color. Next, they inject the recombinant *P* element DNA into eggs, along with a helper plasmid that encodes the transposase gene (Figure 21–4). The transposase gene is transcribed and translated in the germ cells of the early embryo, enabling the recombinant *P* element bearing the gene of interest to insert itself into the embryo's germ cell DNA. Because the helper plasmid does not integrate into the genome or persist during development, further transposition does not occur. *P* element-mediated transformation is one of the most efficient methods for introducing cloned genes into higher eukaryotes.

Drosophila geneticists also harness *P* elements as mutagens and as tools to assist gene cloning. To create *P* element insertion mutations, they cross two strains of flies. One strain contains a *P* element that encodes the transposase enzyme but lacks the *P* element ends that are necessary for transposition. The other strain contains a *P* element with intact ends, as well as a gene that encodes a visible marker such as eye color. This second *P* element also contains a gene for antibiotic resistance and ORI sequences from bacterial plasmid DNA (Figure 21–5). The germ cells of F_1 progeny of a cross between these two strains synthesize the transposase enzyme and carry the mobile *P* element. In the F_1 germ cells, the mobile *P* element will excise and insert itself at random throughout the genome. Some of these insertions will be within and around important genes, disrupting their function. Researchers then screen those F_2 offspring that show the visible marker (e.g., eye color) for mutant phenotypes that are due to the insertion of *P* elements into new positions in the *Drosophila* genome.

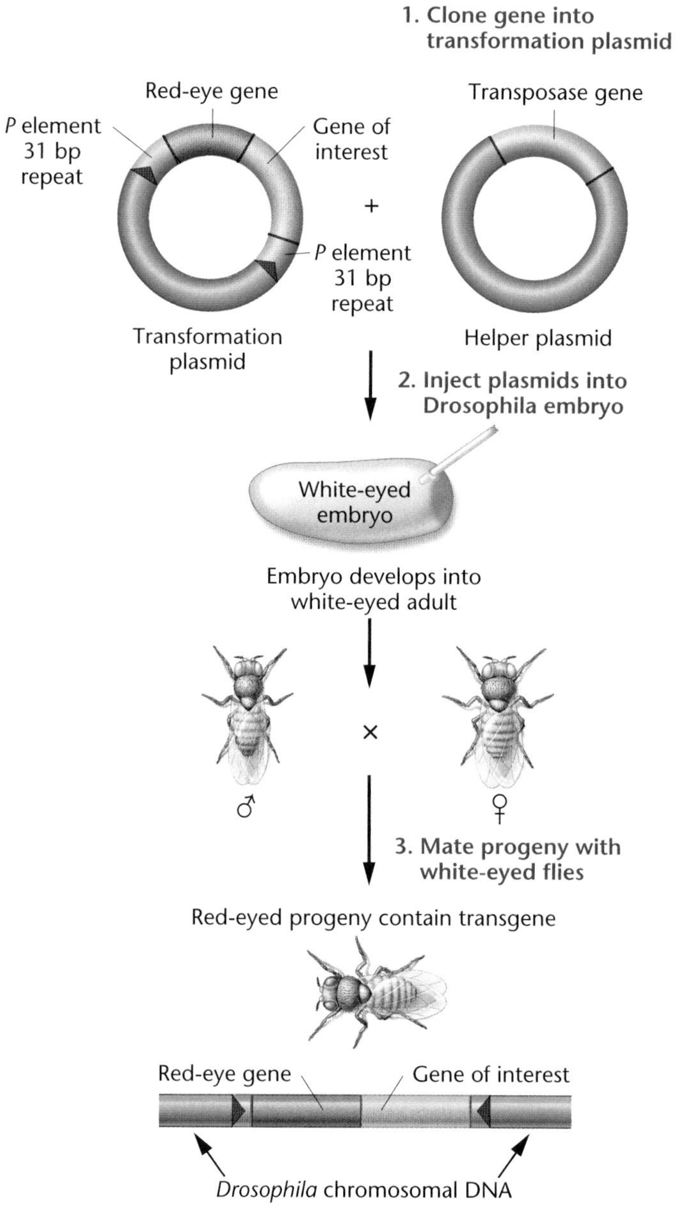

FIGURE 21–4 *P* element-mediated transformation of *Drosophila*. The gene of interest is cloned into the transformation plasmid, which also contains a dominant gene for eye color (e.g., wild-type red eye). The genes are flanked by *P* element ends (31 bp inverted repeats). The transformation plasmid and the helper plasmid containing the transposase gene are co-injected into fertilized eggs from a homozygous recessive mutant white-eye female. The *P* element sequences along with the two transgenes insert into the DNA of the embryo's germ cells. Progeny of this fly will express the eye color marker (red), if the transformation plasmid inserted into the parent's germ cells.

The next step is to clone the unknown gene into which the *P* element has been inserted. As shown in Figure 21–5, researchers purify genomic DNA from the new mutant strain of flies, then digest the DNA with a restriction enzyme that cuts once within the *P* element and at various positions outside the *P* element in or near the unknown disrupted gene. This digestion releases a fragment of DNA containing part of the *P* element and part of the unknown gene. They then treat the digested DNA with DNA ligase, which causes the fragments of digested DNA to form closed circles. The resulting closed circular DNA molecules are then transformed into bacteria. Only those bacteria that take up the closed circles bearing the antibiotic resistance gene (along with the attached unknown gene) will grow as colonies on medium containing the antibiotic. After isolating the plasmid DNA from the surviving bacteria, the unknown gene is sequenced and identified.

Currently, the Berkeley *Drosophila* Genome Project has collected strains of flies bearing *P* element insertions in about 25 percent of *Drosophila's* 13,000 genes. The goal of this project is to generate null mutations in each essential gene, thus producing a "gene knockout kit."

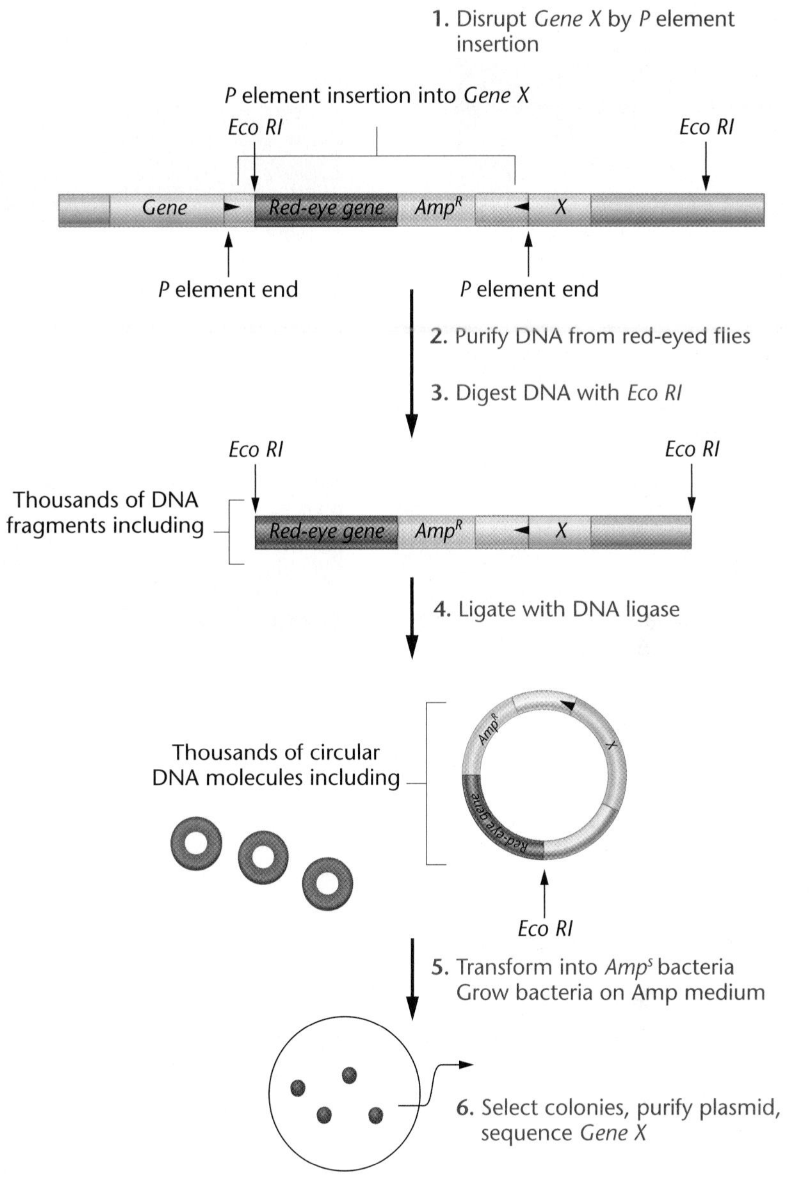

FIGURE 21–5 *P* element-mediated mutagenesis and cloning. *Gene X* is disrupted by the insertion of a *P* element containing a gene for red eyes, bacterial plasmid sequences, and the ampicillin resistance gene. DNA is purified from red-eyed flies and the DNA is digested with a restriction enzyme (e.g., *EcoRI*) that cuts once within the *P* element and at thousands of sites throughout the *Drosophila* genome. The digested DNA is ligated into circles. DNA is transformed into ampicillin-sensitive bacteria. Surviving bacteria contain the plasmid DNA which also contains *P* element and *Gene X* sequences.

The Mouse as a Genetic Model Organism

Of all the model organisms used in genetic research, the mouse is perhaps the most relevant, accessible model for human disease studies. Mice have relatively short generation times of eight to nine weeks. They can be easily kept in laboratory situations and have eight or more offspring per mating. In addition, mice and humans have similar body plans and undergo similar steps in development.

Mouse and human genomes are approximately the same size (about 3 billion base pairs of DNA) with similar numbers of chromosomes (20 pairs in mice). Most human genes have homologs in mice. Interestingly, genes that are linked on the same chromosome in humans are often also linked in the mouse. Geneticists use this similarity in gene organization to identify and map genes in one species once the genes have been mapped in the other species.

Genes that have sequence similarity in humans and mice often, but not always, control the same biological processes. As exon sequences are usually well conserved between mouse and humans, probes prepared from cloned genes in one species can often be used to detect and clone the corresponding genes in the other species.

Despite the many advantages of using mice experimentally, they are more difficult to grow, cross, and mutagenize than lower eukaryotes such as yeast or *Drosophila*. Large scale genetic screens to identify genes of interest cannot be readily performed in mice. Nevertheless, specific genes can be inserted, deleted, or subjected to gene targeting in the mouse (as described in Section 21.3). This makes the mouse a useful system in which to determine the function and regulation of specific genes.

One of the most important genetic dissection techniques in mice is the creation of transgenic organisms. A mouse with a foreign piece of DNA (usually a gene or gene regulatory region) introduced into its genome is referred to as a transgenic mouse. The transgene may be a wild-type gene from another mouse, animal, or plant, or may be a gene that has been mutated *in vitro*.

The method of creating a transgenic mouse is conceptually very simple (Figure 21–6). Researchers isolate newly fertilized eggs from a female mouse and inject purified transgenic DNA into the nucleus of the egg. The eggs are then replaced into the oviduct of a pseudopregnant female mouse. In 20 to 50 percent of the injected eggs, the transgenic DNA inserts into a chromosome by recombination, due to the actions of naturally occurring DNA repair enzymes. Researchers screen the transgenic mice by obtaining a

Web Tutorial 21.1 Creating a Mouse

(a)

(b)

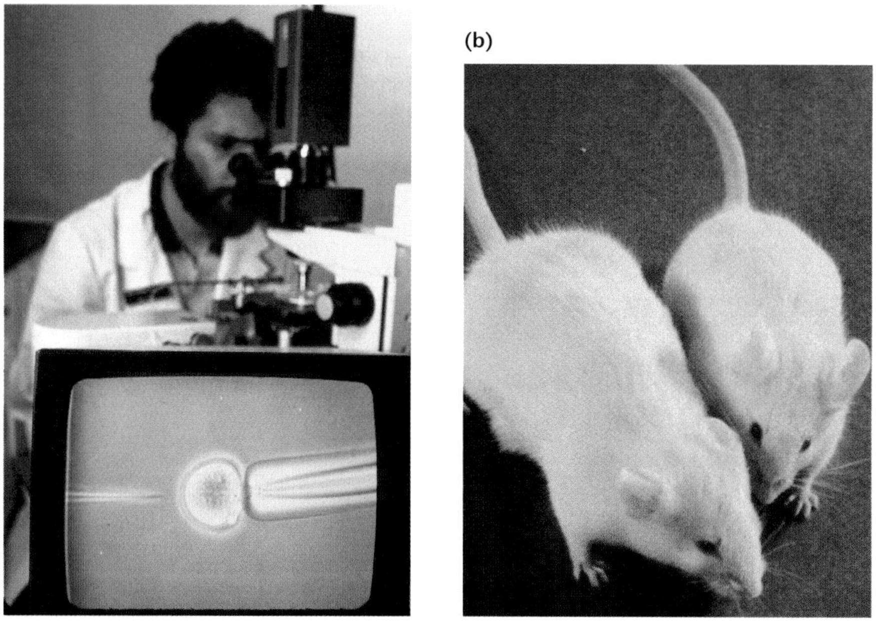

FIGURE 21–6 (a) Scientist microinjecting cloned DNA into a fertilized egg. The injections are performed by manipulating the egg and microinjection needle under a light microscope, seen in the background. The injection procedure is displayed on the screen in the foreground. The egg is held by a suction pipette (on the right-hand side of the egg). (b) A transgenic mouse with its nontransgenic sibling. The mouse on the left is transgenic for a rat growth hormone gene, cloned downstream from a mouse metallothionein promoter. When the transgenic mouse was fed zinc, the metallothionein promoter induced the transcription of the growth hormone gene, stimulating the growth of the transgenic mouse.

sample of tail tissue, purifying the DNA and performing either a Southern blot or a polymerase chain reaction (PCR) assay, to verify that the transgene is present in the animal's genome. As long as the integrated DNA is present in the germ-line cells, the transgene will be inherited in all of the mouse's offspring. Sibling matings can then generate homozygous transgenic lines.

One of the most common uses of mouse transgenics is to determine the function of human genes that have previously been cloned. For example, as described in Chapter 7, geneticists verified the function of the human *SRY* gene by injecting the mouse *Sry* gene into normal XX mouse eggs. The transgenic XX *Sry* mice developed as males, demonstrating that *SRY* plays an important role in bringing about male development. An example of how transgenic mice are being used to devise therapies for Lou Gehrig disease is described in Section 21.5.

One way in which researchers can dissect the normal function of a gene is to clone it next to a regulatory region that will cause the gene to be expressed at abnormal times, in abnormal places, or at higher than normal levels. This strategy was used to determine the role that the human *c-myc* oncogene plays in the development of various cancers.

Prior to transgenic studies, evidence for the link between the *c-myc* gene and human cancer was mostly circumstantial. Most normal cells contain low levels of the Myc protein, but many cancer cells contain high levels, or levels that are not regulated normally during the cell cycle. Also, in cancers such as Burkitt's lymphoma, the *c-myc* gene is disrupted by specific reciprocal translocations that place the *c-myc* gene next to immunoglobulin gene promoters and enhancers. Despite this evidence for a link between *c-myc* expression and cancer, researchers still questioned whether abnormal expression of the *c-myc* gene was directly responsible for cancer formation in white blood cells and in other tissues (such as breast tissue). In one study, researchers cloned the *c-myc* gene next to an immunoglobulin gene enhancer and introduced this composite transgene into mice. The resulting transgenic mice expressed high levels of *c-myc* mRNA in blood cells, and these mice had higher than normal rates of lymphoma and leukemia. In another study, researchers cloned the *c-myc* gene next to the promoter of the mouse mammary tumor virus and created transgenic mice bearing this transgene. The transgene was expressed at high levels in breast and other tissues and about 50 percent of the transgenic mice developed cancers by the age of 14 months. The results of these studies confirmed that the *c-myc* gene is an important determinant in the development of a range of different cancers. However, as not all tissues that expressed the *c-myc* gene developed cancers, researchers concluded that inappropriate expression of *c-myc* is not sufficient for tumor development. These studies also illustrated that several genetic changes must occur in a cell in order to develop full malignancy (as discussed in Chapter 18).

A second important genetic dissection technique in mice is the gene knockout and its sister technique, targeted gene replacement. Together, they are known as **gene targeting**. A knockout mouse is a true-breeding mouse strain lacking the function of a gene because the gene has been replaced with either a null allele or a specific mutagenized allele. Geneticists use knockout mice as models for some human genetic disorders. To create a mouse model of a human genetic disease, researchers clone the mouse gene that is homologous to the gene that causes the human disorder, subject it to site-directed mutagenesis, and then replace the normal mouse gene with the mutagenized allele. Mouse models of cystic fibrosis and Duchenne muscular dystrophy have been created using this technology. Geneticists also use mouse knockouts to study the genetic control of early development and behavior. The technology for creating gene knockouts and gene replacements in mice is described in Section 21.3.

Now solve this

Problem 21.2 on page 546 asks you to select an appropriate model organism in which to study a number of genetic processes and human diseases, from cell division to cystic fibrosis.

Hint: To answer this question, you might consider the limitations and advantages of using each model organism discussed in this chapter. Then, think about how these features would affect research into each of these research topics.

21.2 Geneticists Dissect Gene Function Using Mutations and Forward Genetics

In this section, we explore the methods used in classical genetic analysis, sometimes called **forward genetics**. Forward genetic analysis begins with the isolation of mutants that show differences in phenotype for the process of interest. Mutant isolation is followed by defining **gene pathways**, cloning of the gene, and creating more mutants in order to understand the biological pathway. In this way, mutants define the normal function of the gene.

Generating Mutants with Radiation, Chemicals, and Transposon Insertion

The first step in forward genetic analysis is to define an experimental question—for example, which genes control the early development of an organ system? The next step is to predict the types of phenotypes that would emerge if the genes involved were mutated. Geneticists then begin to collect mutants displaying those phenotypes.

Geneticists sometimes examine spontaneous, naturally occurring mutations; however, as we saw in Chapter 15, mutations are generally rare in nature. For example, mice experience one mutation per 100,000 genes per generation. Researchers would need to screen millions of fruit flies, yeast cells, or mice in order to detect one relevant mutant.

The process of mutant hunting is greatly enhanced by mutagenizing the model organism before screening for mutations. The goal is to create one mutation at random in the genome of each organism in the population so that only one gene product is disrupted in each organism, leaving the rest of the organism wild type. When mutagenesis is thorough enough that each gene is mutagenized at least once in the treated population, the mutagenesis is said to be **saturated**.

As we saw in Chapter 15, any substance that alters a DNA sequence is mutagenic. In genetic analysis, certain mutagens are preferred because of the types of mutations they trigger. For example, ionizing radiation causes chromosome breaks, deletions, translocations, and other major rearrangements. Mutations created by radiation are likely to be null mutations, which may have severe effects on the phenotype. In contrast, ultraviolet light and certain chemicals such as ethyl methane sulfonate (EMS) and nitrosoguanidine cause single base pair changes or small deletions and insertions. With these mutagenic agents, a range of mild to severe mutant phenotypes may emerge, depending upon where in the gene the lesion occurs. It is more likely that single base pair mutations will result in the creation of conditional mutations, such as temperature-sensitive mutations, which are particularly useful for the study of essential gene functions.

Geneticists also use transposons to create mutations. The random insertion of a transposon, such as a *Drosophila P* element, into the genome can cause major disruptions of gene function. These large insertions usually create null mutations if they transpose into a gene's ORF. They may create a number of different effects on gene expression if they transpose into a gene's regulatory sequences, either in flanking regions around the gene or within the gene's introns.

Screening for Mutants

How are mutants detected, grown, and maintained once the organisms are mutagenized? Most mutations result in a loss of gene function. In addition, most loss-of-function mutations are recessive, and in diploid organisms the wild-type allele on the other chromosome usually directs the synthesis of sufficient wild-type gene product to bring about a normal phenotype. To further complicate the situation, many loss-of-function recessive mutations are lethal, as most gene products are essential for proper development and biological function.

In some cases, a mutation will be dominant. In this case, in a diploid organism, the wild-type gene on the other chromosome either cannot produce sufficient normal gene product to overcome the effects of the mutant gene or the mutant gene product interferes with the function of the normal product. If the mutation is dominant, the mutant phenotype will be immediately observed, even in a diploid organism.

The most straightforward method that geneticists use to detect mutants is a **genetic screen**. Frequently, a genetic screen involves the visual examination of large numbers of mutagenized organisms. For example, the screen used to select cell cycle mutants in yeast (Section 21.5) initially required time-lapse photomicroscopic analysis of thousands of yeast colonies growing at two different temperatures. Researchers detected cells with mutations in cell cycle regulatory genes by examining the ratio of yeast bud size to mother cell size, indicative of growth arrest at specific points in the cell cycle. Similarly, the screen for *Drosophila* mutations affecting development and segmentation (Section 21.5) required visual inspection of tens of thousands of *Drosophila* strains in order to select those that lacked white-eyed offspring. The investigators then followed up the screen by microscopically inspecting thousands of larvae for the presence of abnormal body segment patterns.

It is easy to see how dominant mutations can be detected during a genetic screen. As long as the dominant mutation is not lethal, the phenotype will be visible immediately in any organism that bears a dominant mutation in the relevant gene. The mutant can then be crossed and heterozygous or homozygous stocks maintained. If the dominant mutation is lethal, it will not usually be detected.

However, it is more likely that mutations will be recessive. Haploid organisms such as yeast have a significant advantage for

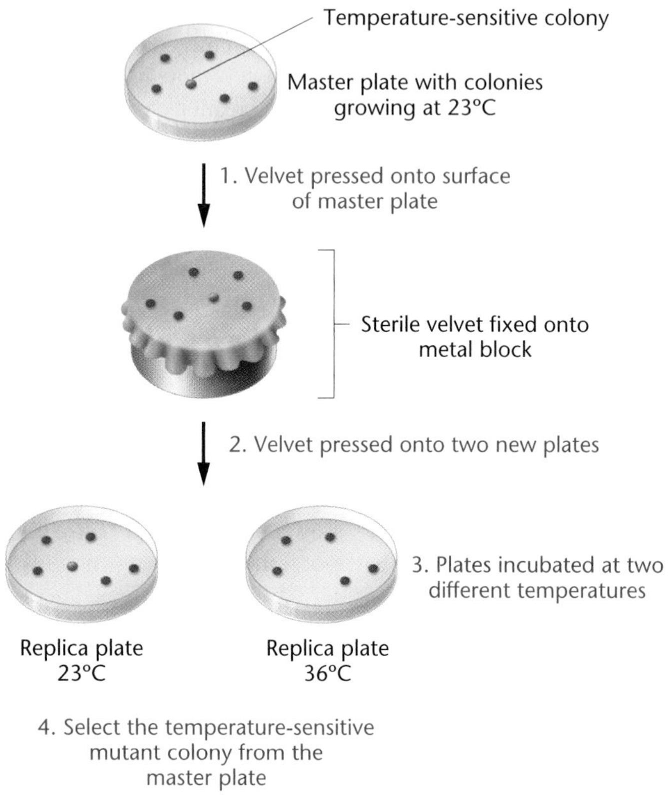
Temperature-sensitive colony
Master plate with colonies growing at 23°C
1. Velvet pressed onto surface of master plate
Sterile velvet fixed onto metal block
2. Velvet pressed onto two new plates
3. Plates incubated at two different temperatures
Replica plate 23°C
Replica plate 36°C
4. Select the temperature-sensitive mutant colony from the master plate

FIGURE 21–7 Replica plating technique. After mutagenesis, yeast cells are spread onto a plate containing growth medium. Some cells from the colonies that grow from each single cell are transferred to sterile velvet. Cells are transferred from velvet onto two new plates. The replica plates are incubated at either the permissive (23°C) temperature or the restrictive (36°C) temperature. Colonies that do not grow at the restrictive temperature are selected. These colonies have temperature-sensitive mutations.

the detection of recessive mutations, as the mutant phenotype will be immediately evident in the mutated organism. Detecting recessive mutations in diploid organisms such as *Drosophila* or mice requires the mutated organism to be mated and the F_1 progeny crossed with each other to reveal the one-quarter F_2 that will be homozygous for the recessive allele.

But how do geneticists detect recessive mutations that occur in essential genes?

Recessive lethal mutations will not be detectable in haploid organisms, such as yeast. To overcome this limitation, yeast geneticists isolate conditional mutants. A **conditional mutation** is one that allows the mutant gene product to function normally under the **permissive condition** (usually a normal growth temperature), but to function abnormally under the **restrictive condition** (usually a higher temperature). Temperature sensitivity occurs because the protein encoded by the mutant gene misfolds and becomes nonfunctional at higher temperatures. The strategy for selecting temperature-sensitive lethal recessive mutations is simple (Figure 21–7). After mutagenesis, the yeast cells are allowed to grow into colonies on growth medium in a petri dish, at 23°C. A piece of sterile velvet is then pressed onto the colonies, and some yeast cells from each colony transfer onto the velvet. The velvet is then pressed onto two new plates containing sterile growth medium. The yeast cells transfer from the velvet onto the replica plates. One replica plate is incubated at the permissive temperature and one is incubated at the restrictive temperature. Yeast colonies that do not grow at the restrictive temperature may contain temperature-sensitive recessive lethal mutations. Researchers then return to the original plate, growing at the permissive temperature, and select the colony for further analysis. Hartwell and coworkers isolated 150 cell cycle mutants using this strategy. (See Section 21.5.)

Diploid organisms that are heterozygous for a recessive lethal allele will not show the mutant phenotype and those that are homozygous for the lethal allele will die. So how is it possible to detect recessive lethal mutations in a population of diploid organisms?

Geneticists have devised some intricate strategies to detect and recover recessive lethal mutations in diploid organisms. An example of a method to detect recessive lethal mutations is the ***ClB* technique**. Hermann Muller devised this technique in the 1920s in order to demonstrate that X rays cause mutations in *Drosophila*. The *ClB* technique detects recessive lethal mutations on the X chromosome (Figure 21–8). To carry out this technique, geneticists treat wild-type males with a mutagen such as radiation, and then mate these males to untreated heterozygous *ClB* females. These *ClB* female flies carry a balancer X chromosome with three important features. First, the balancer chromosome contains inversions that prevent crossing over (designated by the *C*). Second, it contains a recessive lethal allele (*l*) which eliminates any offspring that are homozygous for the balancer. Third, the balancer chromosome bears a dominant gene for *Bar* eye (*B*), making it possible to visually inspect offspring for the presence of the *ClB* chromosome.

Next, the researchers select individual F_1 females with *Bar* eyes. These flies have one X chromosome from their mother (the *ClB* chromosome) and one X chromosome from their father (which may or may not contain a newly induced recessive lethal mutation). Researchers then backcross each *Bar*-eyed female to a wild-type male. If a recessive lethal mutation is present on the paternal X chromosome, there will be no viable males in the backcross progeny. Half of them will die because they are hemizygous for the *ClB* chromosome (the *l* gene is lethal), and the other half will die because they are hemizygous for the newly induced lethal allele. The absence of males signals the presence of a recessive lethal mutation on the X chromosome. The mutation can be maintained in the population by selecting the wild-type (red-eyed) females and backcrossing them.

Geneticists have also devised techniques to detect recessive lethal mutations on autosomes, as described in Section 21.5.

HOW DO WE KNOW?

How do geneticists know whether a mutation in one of the four *Drosophila* chromosomes is a recessive lethal mutation, a dominant mutation, or a recessive nonlethal mutation?

Selecting for Mutants

In a genetic screen, each organism must be individually examined for the phenotype of interest. Often tens of thousands of individuals must be screened to find the required number of mutants. Although effective, this strategy is obviously slow and labor-intensive. In some cases, a different method, **selection**, can be employed to reduce the amount of time and labor. The goal of selection is to create conditions that remove the irrelevant wild-type or mutant organisms from the population, leaving only the mutants that are sought. Usually, this is accomplished by killing or inhibiting the growth of organisms that do not display the relevant phenotype.

Selection is simplest if the mutant phenotype enhances survival under certain conditions. For example, a mutant gene that confers resistance to a drug can be easily selected by growing the organisms on a medium containing the drug. Also, a mutation that corrects a defect in a metabolic pathway, such as an inability to metabolize galactose, can be selected if the organism is grown in a medium that only contains galactose as a carbon source.

Although some mutations lend themselves to direct selection, most do not because the majority of mutations are loss-of-function mutations. However, geneticists can use selection to study **suppressor mutations** (mutations that restore normal function to an existing loss-of-function mutant). Suppressor mutations help define other genes in multistep pathways leading to the phenotype of interest, as described below.

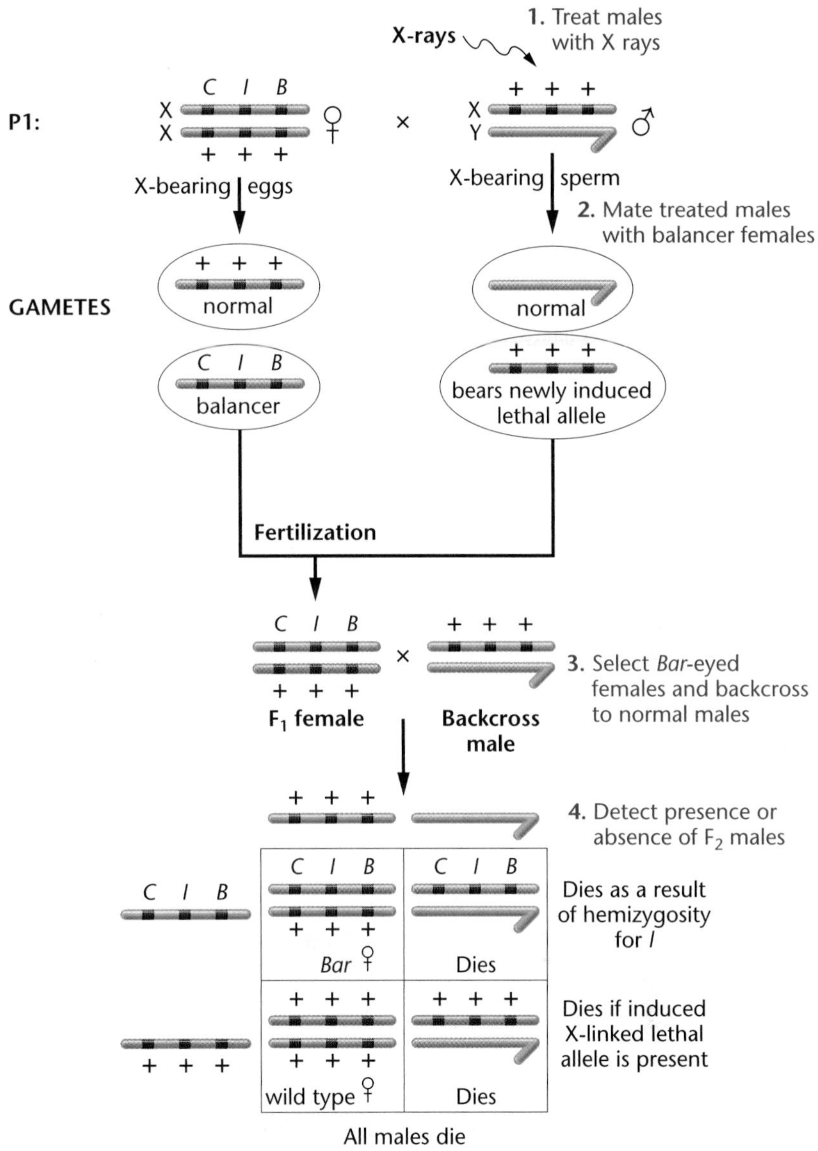

FIGURE 21–8 The *CIB* technique for the detection of induced, X-linked recessive lethal mutations in *Drosophila*.

Defining the Genes

The goal of genetic dissection is to discover all the genes that affect a phenotype and to determine how the genes function. Hence, it is important to define the number of genes that lead to a specific phenotype.

As we have seen in previous chapters, a single gene can have several different alleles—weak alleles, null alleles, or variants with different phenotypes. For example, the ABO blood groups represent three different alleles at the same locus that can lead to six different genotypes and four different phenotypes. (See Chapter 4.) The I^A and I^B alleles direct the synthesis of the A and B antigens, and the I^O allele is a null, nonfunctional allele. The I^A and I^B alleles are codominant, and the I^O allele is recessive to the other two. It is also possible for a gene to act within a multigene pathway leading to a particular phenotype. A mutation in any one of the genes within a pathway may result in the same phenotype, as each mutant gene product prevents the pathway from functioning. For example, the gene represented by the I^A, I^B, and I^O alleles acts within the same pathway as the gene represented by the *H* allele. As we saw in Chapter 4, people with a homozygous null mutation in the *H* gene show the same phenotype as those with a homozygous I^O genotype.

How can geneticists determine whether the mutations in the collection of mutant organisms represent alleles of the same gene, or whether they represent mutations in a number of different genes within a pathway?

To do this, geneticists perform either complementation or recombination analysis. **Complementation analysis** allows investigators to determine whether two mutations are in the same gene—that is, whether they are alleles—or whether they represent mutations in separate genes. To perform complementation analysis, researchers simply cross two homozygous mutant strains together and examine the F_1 progeny. There are two alternative outcomes and interpretations of such a cross, as illustrated in Figure 21–9. One possible outcome is that all the progeny are wild type. In this case, complementation has occurred; therefore, the two mutations must lie within separate genes. Complementation occurs when the F_1 progeny are heterozygous for each gene and the wild-type allele at each locus directs the synthesis of a wild-type protein that is required for normal phenotype. The other possible outcome is that all the F_1 progeny are mutant. In this case, complementation has not occurred; therefore, the two mutations lie within the same gene. Complementation does not occur when both alleles of the gene are mutant and no wild-type gene product is synthesized.

Complementation analysis may be used to screen any number of individual mutations that result in the same phenotype. All mutations that are present in any single gene are said to fall into the same **complementation group**, and they will complement mutations in all other groups. If large numbers of mutations affecting the same trait are available and studied using complementation analysis, it is possible to predict the total number of genes involved in determining that trait. In Chapter 15, we discussed how complementation analysis was used to study the genes involved in the inherited human disorder xeroderma pigmentosum. Results of these studies revealed that mutations in any one of seven complementation groups (genes) can lead to the disorder. Later in this chapter, we will learn how Leland Hartwell and his colleagues used complementation analysis to show that 32 genes, representing 150 independently isolated mutants, regulate steps in the yeast cell cycle. Drs. Nüsslein-Volhard and Wieschaus performed complementation tests on their collection of 580 embryonic pattern mutants. These mutants fell into 139 complementations groups, defining the number of *Drosophila* genes that affect embryonic development.

Dominant mutations cannot be tested by complementation, but they can be tested by **recombination analysis**. Recombination analysis provides an estimate of genetic linkage by calculating the amount of recombination (crossing over) that occurs between mutant loci when these loci are present in the same organism. If two mutations occur in the same gene, they will be tightly linked. If present in different genes, they will most likely be unlinked. Methods used to calculate linkage by crossover and recombination in yeast, *Drosophila,* and other organisms were described in Chapter 5.

Dissecting Genetic Networks: Epistasis and Pathways

Products of several different genes may act in a pathway leading to a particular phenotype. If complementation or recombination analysis reveals that two or more genes contribute to the same phenotype, they may be acting in one pathway. How, then, do geneticists determine where these genes operate within a single pathway? One way to do this is to examine how two mutant genes interact during epistasis analysis. As we saw in Chapter 4,

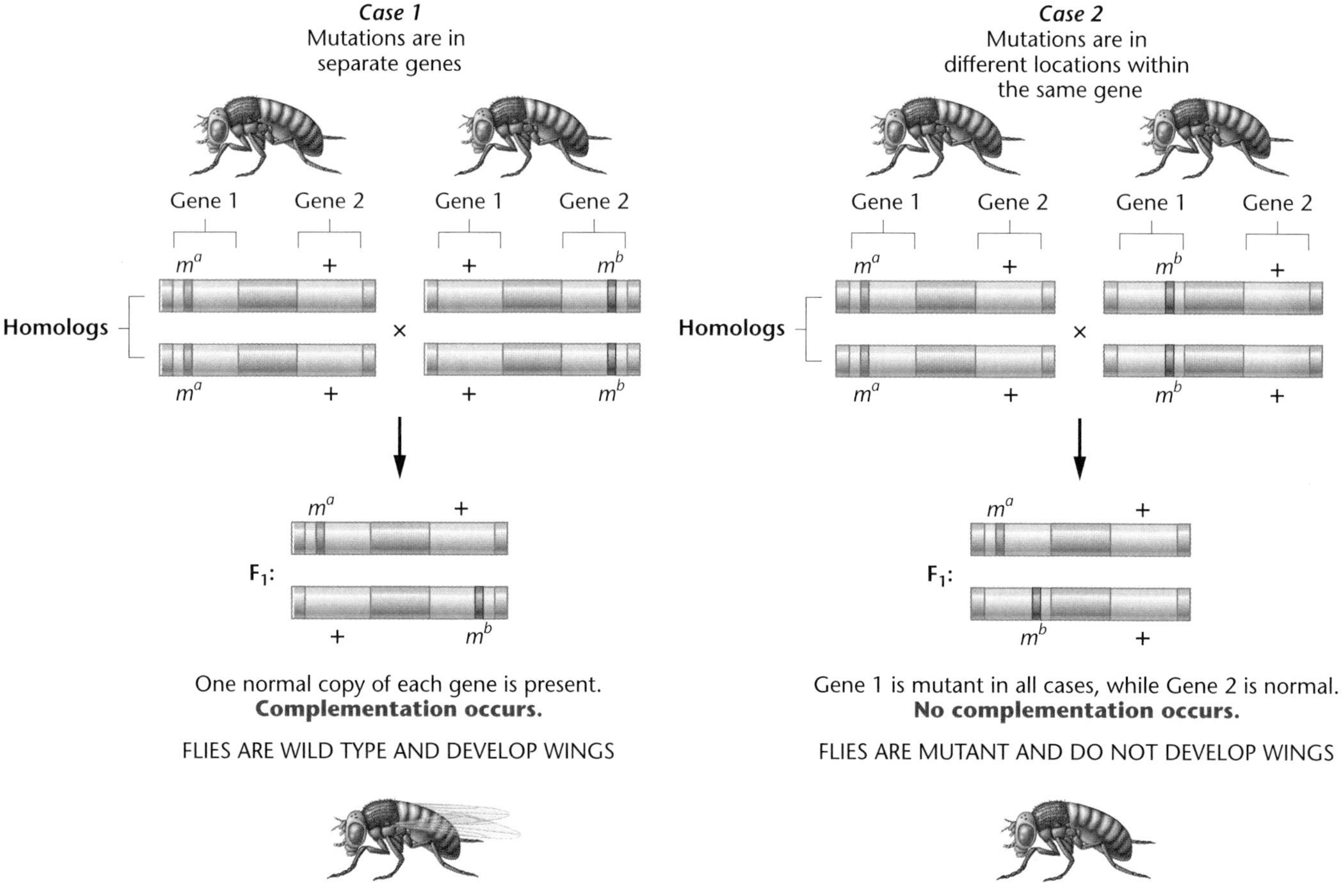

FIGURE 21–9 Complementation analysis to determine whether two *wingless* mutations in *Drosophila* are in separate genes or are within the same gene.

epistasis occurs when the effect of one gene masks or modifies the effect of another gene. If one gene is epistatic to another, the mutation in the gene whose effects are manifest earliest in the pathway will be the one whose phenotype is evident in a double mutant. In order for epistasis analysis to work, each mutant must show a slightly different phenotype.

To do an epistasis analysis, two homozygous recessive mutants with discernable phenotypes are crossed and the F_2 progeny are examined. The phenotypic ratios of the F_2 progeny reveal which of the two genes controls the earlier or later steps in the pathway. (See Chapter 4.) Alternatively, a double mutant organism, homozygous for both mutations, can be created and the phenotype of this double mutant determined. If the phenotype of the double mutant is the same as one of the single mutants, that single mutation likely represents the earlier step in the pathway (Figure 21–10). However, one must keep in mind that other types of gene interactions can be epistatic to each other, although they may not operate within the same biochemical pathway. Hence, other types of genetic and biochemical tests need to be completed before a full genetic pathway can be dissected.

In their genetic analysis of the cell cycle in yeast, Hartwell and colleagues constructed double mutants of each of their single mutants. (See Section 21.5.) Their analysis showed that a mutant with a defect in a gene controlling the first of six stages in the cell cycle could not complete the remaining five stages. Similarly, mutants with a defect in a gene controlling the second of these stages arrested at the second stage and could not complete the remaining four stages. A double mutant containing both mutations arrested at the first stage, confirming that the first mutant contained a mutation in the gene controlling a step prior to the second mutant. Using this strategy, the investigators ordered all 32 cell cycle mutations into six stages of the yeast cell cycle pathway.

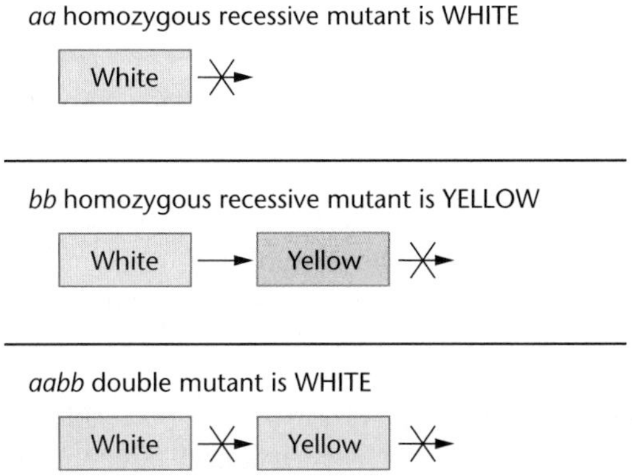

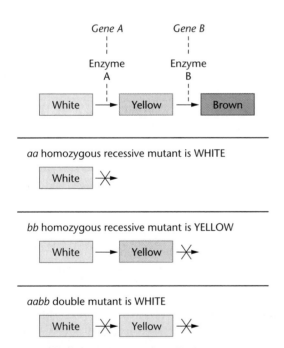

FIGURE 21–10 Example of epistasis analysis for two genes that control steps in a biochemical pathway that converts a colorless coat pigment to yellow, and then to brown. Wild-type coat color is brown. The double mutant (aabb) displays the same phenotype as the single mutant (aa). This shows that *Gene A* controls a step prior to the step controlled by *Gene B*.

Extending the Analysis: Suppressors and Enhancers

A forward genetic analysis may identify only some of the genes within a pathway. It is possible that the original mutagenesis procedure did not mutagenize each of the genes in the pathway, or that not all genes in the pathway were recovered in the screen. How can geneticists find the other genes in the pathway, if only some have been mutated?

One way to identify other genes in a pathway is to carry out a second mutagenesis screen. The goal of the second screen is to identify genes that dominantly enhance or suppress the first mutant phenotype. To perform the second screen, geneticists subject the first mutant strain to a second round of mutagenesis, and screen for more severe or milder phenotypes.

A suppressor mutation is a second mutation that rescues the original mutant phenotype. For example, a yeast temperature-sensitive mutant that cannot replicate its genome at the restrictive temperature may suffer a second mutation that returns the mutant to wild type in terms of DNA replication. There are a number of mechanisms by which suppressor mutations work (Figure 21–11). A mutation may occur in a gene encoding a protein that would normally interact with the protein encoded by the first mutant gene. In the original mutant organism, the two proteins would not interact because the first protein was abnormal, perhaps misfolding so that it could not bind to its protein partner. However, in the presence of the suppressor, the two abnormal proteins interact, restoring function. Another type of suppressor mutation may occur in a different pathway, but the pathway can, when mutated, substitute for or bypass the abnormal pathway in the first mutant. A third type of suppressor is the high copy suppressor. These suppressors rescue the first mutant by supplying high levels of another gene product that may compensate for or stabilize the abnormal mutated gene product. Geneticists use these types of suppressors to define other genes involved in the genetic pathway.

Not all suppressor mutations are useful for identifying other genes in a pathway. Sometimes, the suppressor mutation occurs in the same gene as the first, correcting the defect. This could be a direct reversion of the original mutation or a second mutation in the same gene that corrects a defect. One example is a mutation that reestablishes the correct reading frame. Also, a suppressor may be a nonsense suppressor mutation in a tRNA gene that may allow a termination codon in the mutant gene to be recognized as an amino acid codon, allowing a functional protein to be translated from the mutant gene.

Enhancer mutations are the opposite of suppressor mutations; they increase the intensity of the first mutant phenotype.

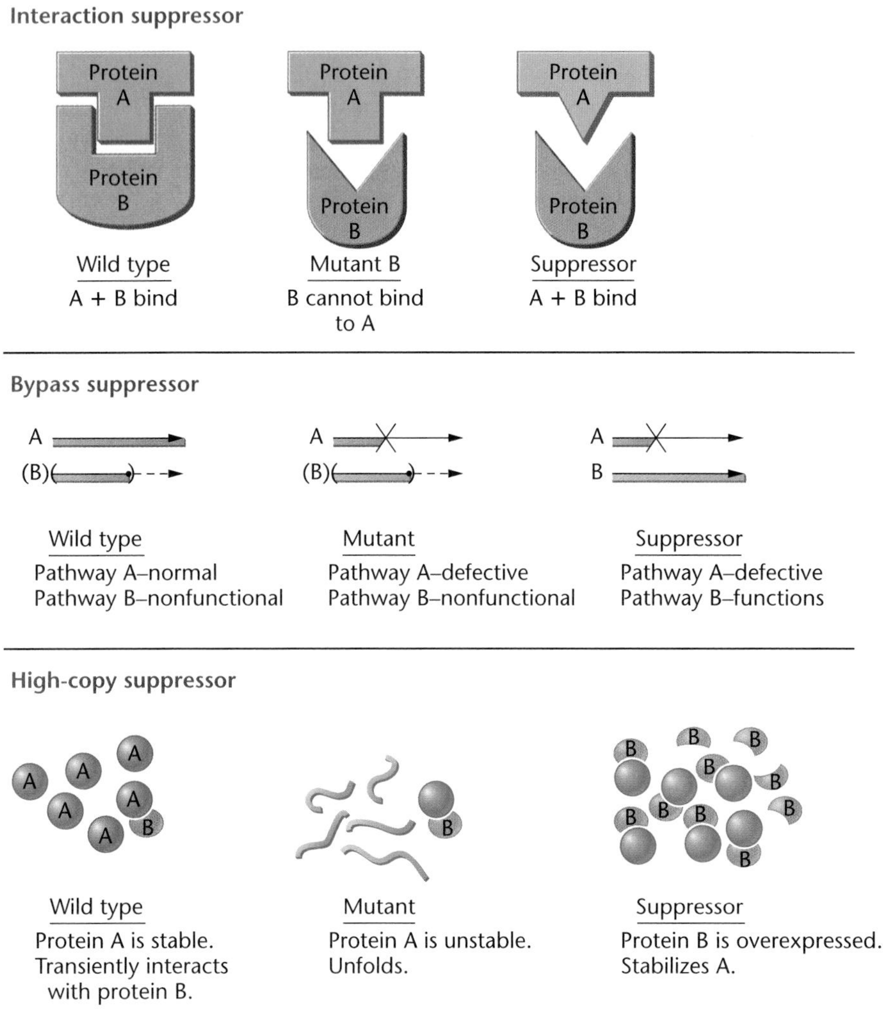

FIGURE 21–11 Three types of suppressor mutations.

Geneticists use enhancer mutations in similar ways to suppressor mutations, as they point the way to other genes in a genetic network.

How Do We Know?

How do we know how many genes contribute to a particular phenotype in a model organism?

Extending the Analysis: Cloning the Genes

Once geneticists generate mutants and define genetic networks, the genes involved are identified by cloning. Now that the genomes of many model organisms are fully sequenced and their genomes' ORFs identified, the job of cloning is greatly simplified.

In yeast, it is possible to clone genes by **functional complementation** with a yeast cDNA library. In this method, plasmids encoding all the wild-type genes from yeast are methodically transformed into the mutant strain, until one cDNA clone restores the mutant to wild type. The cDNA clone that restored the mutant to wild type is then purified from the transformed strain of yeast, and the DNA sequence of the cDNA is determined. The complementing gene is then identified by searching the yeast genomic sequence against the unknown plasmid sequence. Once the wild-type gene corresponding to the mutated gene has been identified, this cloned wild-type gene can be used to isolate the mutant gene from the mutant organism. Geneticists can design PCR primers based on the DNA sequence of the wild-type gene. They then use these primers to amplify all or part of the mutated gene in the mutant organism. The amplified DNA fragments can be directly sequenced to identify the nature of the mutation that led to the mutant phenotype.

Although other model organisms, such as *Drosophila* and mice, can be transformed with cDNA or genomic DNA clones, the process is not efficient enough in these organisms to allow cloning by functional complementation. How then, can geneticists clone a gene based on a mutant phenotype in these organisms?

The first step in cloning genes from higher organisms is for the researcher to map the position of the gene by linkage analysis. The gene can be mapped to a chromosome and then to a general chromosome region using standard linkage analysis, such as that described in Chapter 5. When the approximate location of the gene is known, the candidate gene can be identified by a technique called **chromosome walking**. To perform a chromosome walk, investigators determine the DNA sequence of a fragment of genomic DNA and use that fragment as a probe to isolate the clone representing the next segment. The genomic DNA within this next clone is then sequenced and used as a probe for the next segment. This is repeated until the entire region has been cloned and sequenced. Geneticists can identify the presence of ORFs within the DNA sequence using computer search programs. To verify which of the ORFs within a region of DNA encode the gene of interest, geneticists may specifically mutate or knock out the ORFs and test these mutations in model organisms, as described later in this chapter.

In *Drosophila*, mutations generated by *P* element insertion can be readily cloned, as described previously (Figure 21–5). Special *P* elements with plasmid sequences can be recovered with the mutated gene attached, or the *P* element itself can be used as a tag with which to select the corrrect DNA from a genomic library created from the mutant organism.

Extending the Analysis: Biochemical Functions

After the gene is cloned and its DNA sequence determined, genetic dissection enters a new and equally challenging phase, that of defining the functions of the gene product. This phase of the analysis uses a wide range of bioinformatic, genetic, and biochemical technologies.

Geneticists may begin the analysis by comparing the gene's DNA sequence to sequences of other genes in the same or different organisms. If a gene of similar sequence exists in another species and the function of that gene is known, it is possible to use this information to deduce a possible function for the unknown gene product. Investigators may analyze the gene's DNA sequence for the presence of amino acid **sequence motifs** in the gene product that may provide clues as to the gene product's function. For example, the gene may contain a sequence that encodes a kinase motif, suggesting that the gene encodes an enzyme that phosphorylates other proteins. Or the gene may contain a sequence that encodes a DNA binding region found in certain types of transcription factors, suggesting that the unknown gene regulates expression of other genes. Hypotheses such as these can then be tested by further biochemical or genetic tests.

Researchers apply a wide range of **molecular genetic tools** to extend the analysis of candidate genes. These tools include techniques to analyze gene expression in specific tissues, during specific times in development, or in regions of the cell during the cell cycle. Investigators can also examine whether other proteins interact with the protein encoded by the gene, using a variety of biochemical and genetic tests. A range of powerful tools that are now used to define gene function are site-directed mutagenesis, gene replacement, gene knockouts, and examination of phenotypes in transgenic organisms. These methods are described in the next section.

Now solve this

Problem 21.25 on page 547 involves planning a genetic screen for genes that affect a visible phenotype; in this case, the tanning and hardening of adult *Drosophila* cuticle.

Hint: Although the phenotype is a visible one, you must also consider the possibility of lethality at any stage of the fly's life cycle. In Section 21.2, you learned about methods used to recover lethal mutations in *Drosophila.* The last part of the problem deals with a negative outcome of the screen, as no mutants were isolated. To suggest reasons for this outcome, you might consider aspects such as the mutagenesis itself or how the screening was done. Think about the fact that experiments do not always proceed according to plan.

21.3 Geneticists Dissect Gene Function Using Genomics and Reverse Genetics

Forward genetic analysis begins with a collection of mutants and progresses to defining genetic pathways, cloning the wild-type gene, and analyzing biochemical functions of the gene product. Although this approach continues to be a powerful one with which to dissect gene function, modern genomics and molecular biology are challenging forward genetics as the dissection tool of choice. Although researchers debate whether forward genetics will become obsolete as a research tool, most genomics and molecular-based technologies still require mutation analysis of candidate genes, as this is the ultimate demonstration of biological function.

Reverse genetics is a somewhat broad term describing a number of gene analysis methods that begin with a cloned wild-type gene or purified protein and progress to site-directed mutagenesis and phenotypic analysis—the opposite order from that used in forward genetics.

Reverse genetics is particularly relevant today, after completion of genome sequencing from a number of organisms including yeast, *Drosophila,* and humans. Genomic data tell us that there are approximately 6200 genes in yeast and 25,000 to 30,000 genes in humans; however, only about half of yeast genes and less than 10 percent of human genes have been assigned a function. Researchers can now begin with a novel gene sequence in a genomics database and probe the gene's function. They can also analyze genome-wide patterns of gene expression prior to knowing the function of individual genes being analyzed. In some model organisms, such as yeast and *Drosophila*, investigators are systematically deleting each of the genome's ORFs and studying the resulting phenotypes. These gene-specific and genome-wide approaches comprise the field known as **functional genomics**.

Reverse genetics is particularly suited to identifying human disease genes. As humans cannot be mutagenized and mated in controlled situations, it is not possible to use classical forward genetics to identify disease genes. Using molecular biological techniques, geneticists clone and study human disease genes beginning only with a disease phenotype or in some cases a purified mutant protein. Functional genomics adds to this capability by allowing investigators to identify large numbers of genes whose expression levels are modified in certain disease states, and by facilitating the identification of these genes.

In this section, we explore how geneticists use reverse genetic approaches to dissect gene function. In the next section we will describe some of the newer methods of functional genomics.

Genetic Analysis Beginning with a Purified Protein

One starting point for a reverse genetic analysis is a protein suspected to be involved in a phenotype of interest. If sufficient quantities of the candidate protein can be purified, geneticists can determine the amino acid sequence of the protein. This is done by sequentially cleaving the N-terminal amino acids from the protein and identifying the cleaved amino acids in an automated amino acid sequencer. Investigators then examine the amino acid sequence and deduce the gene's DNA sequence corresponding to the protein's amino acid sequence. Because the DNA code is degenerate, several possible DNA sequences could be responsible for coding a particular amino acid sequence. Therefore, researchers synthesize various combinations of oligonucleotides that will correspond to all possible codon combinations (Figure 21–12). They then label the mixture of synthetic DNA oligonucleotides with radioactivity or indicator dyes and use this mixture to probe a genomic or cDNA library. If one or more oligonucleotides hybridize to a

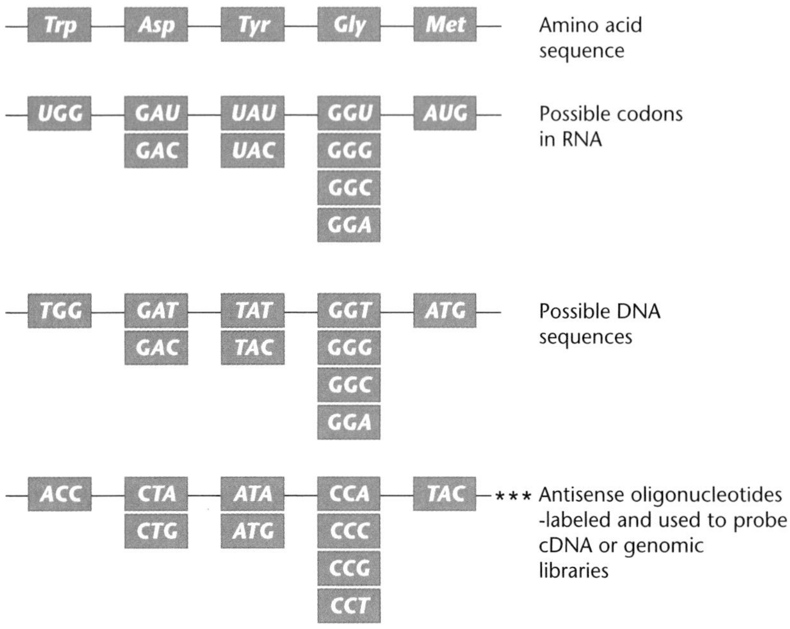

FIGURE 21–12 Strategy used to clone the *Factor VIII* gene from a porcine cDNA library. The purified Factor VIII protein was cleaved into peptides and the amino acid sequence of each peptide was determined. All possible codon combinations that could code for a five amino acid sequence were calculated, as well as the corresponding DNA sequences. A mixture of antisense oligonucleotides was synthesized for each possible DNA sequence that could encode the five amino acid peptide. This mixture was labeled (***) and used to probe a porcine cDNA library in order to select the porcine *Factor VIII* gene. *From Gitschier, J. et al. 1984. Characterization of the human factor VIII gene. Nature 312:326-30 copyright 1984 Macmillan Publishers Ltd.*

clone in the library, the clone is selected for further analysis. (See Figure 19–13).

Researchers used this protein-first strategy to clone the gene responsible for hemophilia A, a disease causing defects in blood clotting. From pedigree studies and linkage analysis, geneticists discovered that hemophilia A is an X-linked recessive trait, likely controlled by a single mutated gene. A reasonable hypothesis was that the gene responsible for the defect encoded a blood clotting factor. Biochemical analysis of clotting factors in the blood of normal individuals and hemophiliacs revealed that one protein, Factor VIII, was absent in hemophiliacs. Because Factor VIII is present in very low quantities even in normal blood, investigators purified large quantities of the protein from pig blood. They then determined the amino acid sequence of a portion of the Factor VIII protein and predicted the DNA sequence that would encode the protein. They synthesized DNA oligonucleotides complementary to this DNA sequence, probed a porcine genomic DNA library and selected the Factor VIII gene from the library. They then used the porcine genomic clone to probe a human genomic library. By comparing the DNA sequences of the Factor VIII gene from normal and hemophiliac individuals, they determined that the Factor VIII gene was mutated in people with hemophilia, confirming that the Factor VIII gene is responsible for the disease.

An alternative approach to cloning a gene beginning with the purified protein involves the use of antibodies. To generate an antibody to a purified protein, investigators inject the protein into laboratory animals, such as rabbits or mice. These animals will synthesize antibodies that recognize the injected protein, and the antibodies are purified from blood samples. Investigators then use the purified antibodies as probes. They first label the antibodies with radioactivity or fluorescent dyes and hybridize the antibodies to cDNA expression libraries. Expression libraries contain cDNA clones inserted into special vectors containing a promoter that will drive the expression of the cDNA within the bacterial host cells (Figure 21–13). Colonies of bacteria containing cDNA expression clones will produce small quantities of the proteins encoded by the cDNAs. Antibodies specific for the purified protein will specifically recognize the protein produced in the bacteria from the cDNA vector, allowing investigators to select the relevant clone. The gene encoding the enzyme tyrosinase, which is responsible for albinism, was cloned in this way, using tyrosinase-specific antibodies.

Genetic Analysis Beginning with a Mutant Model Organism

In some cases, it is possible to clone a gene by functional complementation. In the previous section, we learned that geneticists use functional complementation in mutant yeast strains to clone wild-type yeast genes. Functional complementation can also involve transforming a mutant organism with cloned genes from another organism in order to identify the gene controlling the mutant phenotype in the other organism.

An example of functional complementation is the use of mutant yeast strains to clone homologous genes from humans. Many genes and gene products are highly conserved during evolution, both in their sequences and their functions. Hence, it is sometimes possible to functionally complement a yeast mutation with a wild-type human gene. Researchers have used this approach to identify human genes that encode transcription factors, purine and pyrimidine biosynthesis enzymes, and cell cycle proteins. One of the first successful functional complementations between yeast and human genes was the identification of the human *CDC2* gene. The *CDC2* gene encodes a kinase protein that regulates several stages of the cell cycle. In 1987, Paul Nurse and colleagues cloned the human *CDC2* gene by tranforming a yeast temperature-sensitive *cdc2* mutant strain with clones from a human cDNA library. One of the human cDNA clones functionally complemented the yeast mutation, returning the yeast strain to wild type. They recovered the complementing plasmid from the yeast strain, sequenced it, and discovered 63 percent sequence identity between the yeast and human Cdc2 proteins. In both organisms, the *CDC2* gene controls key steps in the cell division cycle.

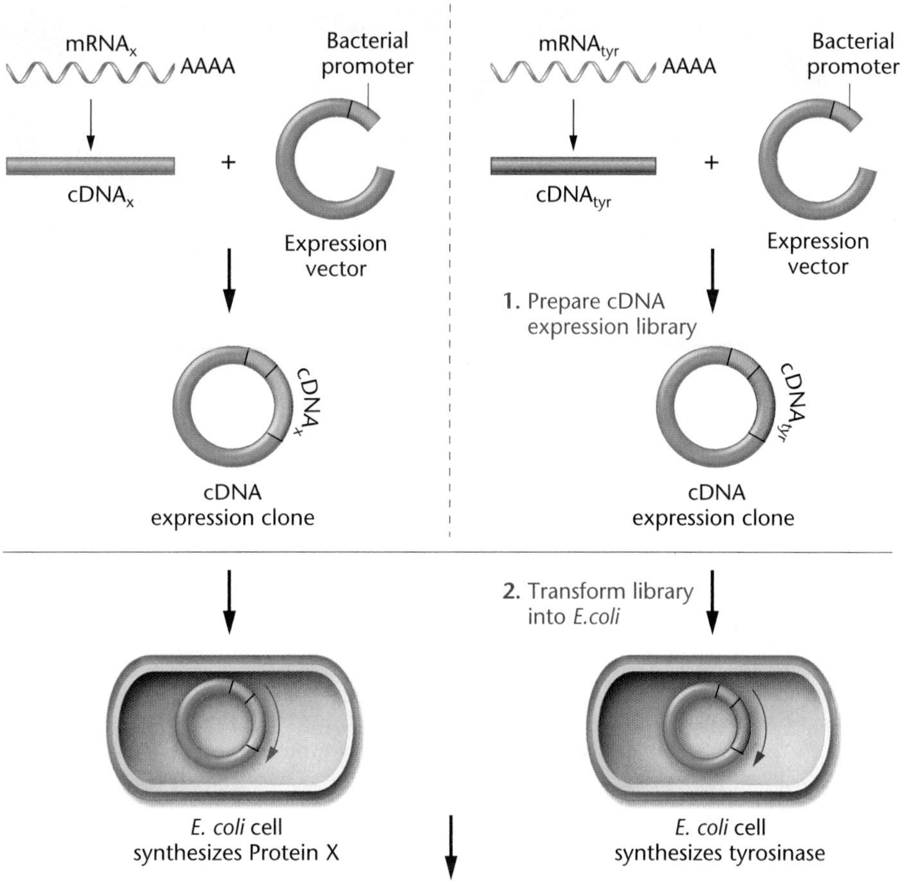

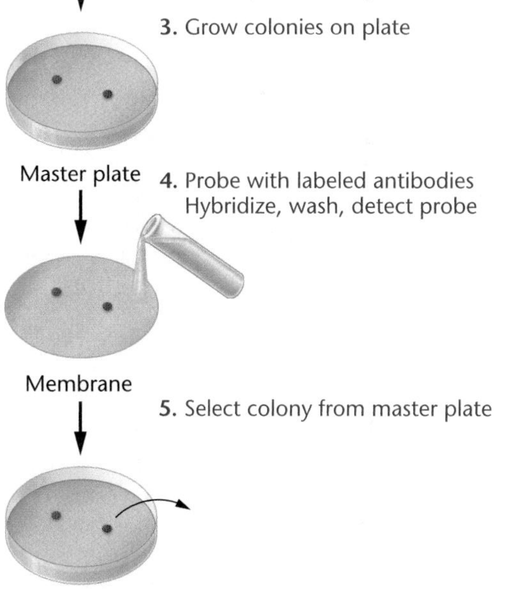

FIGURE 21–13 Cloning the tyrosinase gene. A cDNA library is prepared by inserting the cDNAs into vectors that contain a bacterial promoter. After transformation into bacterial cells, the cDNA inserts are transcribed from the bacterial promoter and the mRNAs are translated into proteins. The protein-containing bacteria are grown on plates. Cells from the colonies are transferred to a membrane, and the membrane is probed with antibodies that recognize the tyrosinase protein. Colonies that bind antibody are selected from the master plate.

Geneticists also use functional complementation in cultured cells to clone human genes. For example, the Chinese hamster cell line EM9 is defective for DNA repair, making it hypersensitive to ionizing radiation. Functional complementation of cloned wild-type human genes into this cell line led to the identification of the human DNA repair gene, *XRCC1*. The introduction of *XRCC1* into the EM9 cell line corrected the cell line's hypersensitivity to ionizing radiation. In addition, some human tumor suppressor genes have been identified and cloned by transforming wild-type human genes into tumor cell lines. Genes that suppress the growth of the tumor cells were selected, sequenced, and further studied to verify their roles as tumor suppressors.

Genetic Analysis Beginning with the Cloned Gene

Once a candidate gene is identified—from a forward mutational analysis, a protein-first approach, a functional complementation assay, or simply selected from a random collection of clones—it must be subjected to further genetic and biochemical analysis to verify its function.

Often the first step that geneticists use to characterize a candidate gene is to compare the gene's DNA sequence to sequences available in **DNA databases**, such as GenBank. Investigators employ computerized sequence comparison programs to reveal regions of sequence similarity and to set up **sequence alignments** between the candidate gene and other genes in the database. If a similar gene is present in another organism, its function may be known, providing hints about the function of the candidate gene. Once researchers know the gene's DNA sequence, they translate the DNA sequence into the amino acid sequence of the protein product and compare the amino acid sequence to that of other proteins, also available in public sequence databases, such as SwissProt.

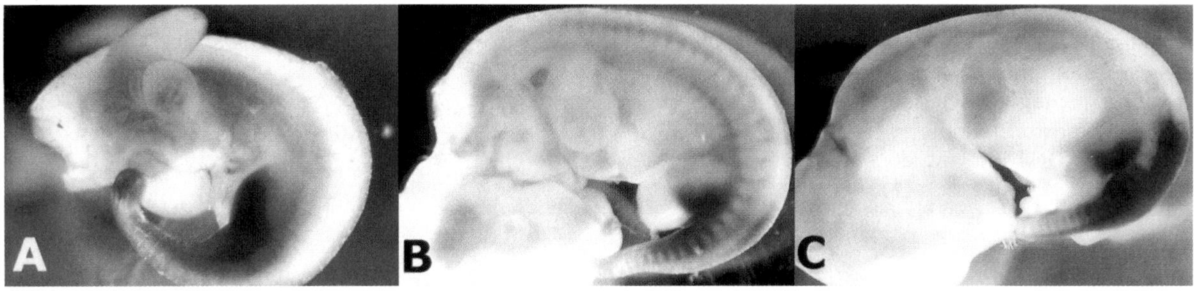

FIGURE 21–14 *In situ* hybridization of whole mouse embryos, showing distribution of the *Hoxc11* mRNA during development. RNA is detectable as dark blue staining, near the posterior end of the embryo. Head of the embryo is at left; tail at right. Mouse embryos at 10.5 (A), 11.5 (B) and 12.5 (C) days of gestation, showing *Hoxc11* mRNA concentrated in hindlimbs, vertebrae, and cells that will later form kidney and reproductive organs. These data suggest that the *Hoxc11* gene product is involved in the early development of these structures.

Researchers use computer programs such as BLAST to search protein sequences for regions of similarly to known amino acid motifs in other proteins. If the candidate protein contains a motif such as a DNA binding region, a kinase region, a membrane-spanning motif, or a secretion signal, it may suggest the protein's functions. These putative functions can then be tested further by molecular biological or biochemical techniques.

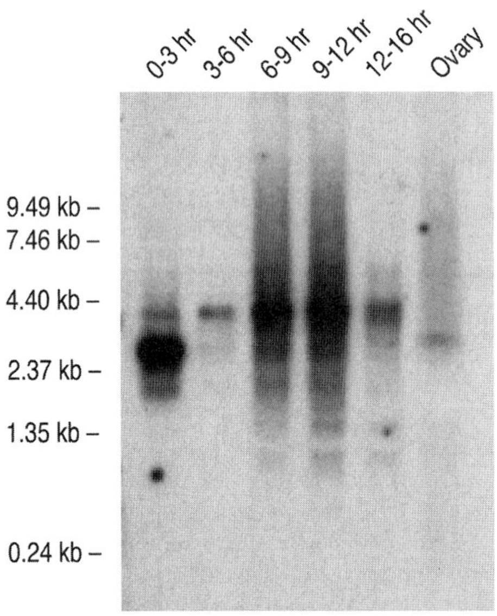

FIGURE 21–15 Northern blot analysis of *dfmr1* gene expression in *Drosophila* ovaries and embryos. An approximately 2.8 kilobase (kb) *dfmr1* transcript is present in ovaries and 0 to 3 hr old embryos. The *dfmr1* transcript peaks in abundance between 9 and 12 hrs of development, and measures 4.0 kb. These data suggest that *dfmr1* gene expression may be regulated at the levels of transcription or transcript processing during embryogenesis. The *dfmr1* gene is a homolog of the human *FMR1* gene. Loss-of-function mutations in *FMR1* result in human fragile X mental retardation.

A second approach to aid in determining the function of a candidate gene is to analyze its patterns of **gene expression**. Investigators can determine the tissue-specific or temporal-specific patterns of gene expression using a number of techniques. An RNA ***in situ* hybridization** may reveal the presence of the mRNA in one or more tissues in a multicellular organism, or in one or more structures within a single cell (Figure 21–14). To perform an *in situ* hybridization, investigators label a cDNA clone with fluorescent or visible dyes, then hybridize the labeled probe to a thin section of tissue. After washing off the unhybridized probe, they observe the section under a light or fluorescent microscope. The presence of stain defines where the gene is expressed at the mRNA level and may suggest its function. For example, if the mRNA is present only in the liver of embryonic mice, the gene may play a critical role in early development of the liver. Alternatively, if the mRNA is present in all cells at all stages of development, the gene may encode a ubiquitous housekeeping protein.

Another way to determine the time and place of a gene's expression is to perform a northern blot analysis (described in Chapter 19). The technique involves purifying mRNA from the tissue of interest, subjecting the mRNA to electrophoresis, transferring the separated mRNAs onto a filter, and probing the filter with a labeled DNA probe (Figure 21–15). Northern blots can be more labor-intensive than *in situ* hybridization, but are more useful for quantitating the mRNA.

Although the presence of mRNA often reflects the tissue and temporal expression of a candidate gene, the expression pattern of the protein may be more revealing, especially if the gene's regulation occurs at posttranscriptional stages. If the expression of the candidate gene occurs at the level of translation or protein stability, the protein and mRNA profiles may not be identical. To verify the tissue and temporal expression of the protein, investigators assay the protein's expression profiles by **immunofluorescence staining**. Immunofluorescence staining uses an antibody that specifically binds to the protein of interest. To do the staining, the antibodies are either directly labeled with a fluorescent tag, or are bound to a second antibody that recognizes the first antibody. The second antibody is conjugated to the fluorescent tag. As in *in situ* hybridization, the labeled probe is hybridized to a thin section of tissue, the excess is washed away, and the section is observed using a fluorescence microscope (Figure 21–16).

The search for a candidate gene's function often involves a large number of biochemical and molecular biological assays, whose descriptions are beyond the scope of this chapter. For example, if the gene is suspected to encode a protein kinase, investigators may assay the ability of the purified gene product to phosphorylate various substrates *in vitro*. They may examine the interactions of the gene product with other cellular proteins by precipitating specific complexes from living cells and identifying the proteins within the complex. If the gene product is suspected to be a transcription factor or other DNA binding protein, tools exist that allow researchers to identify the DNA sequences that bind to these factors, and ultimately the genes that are controlled by them (as described in Section 21.4).

(a)

(b)

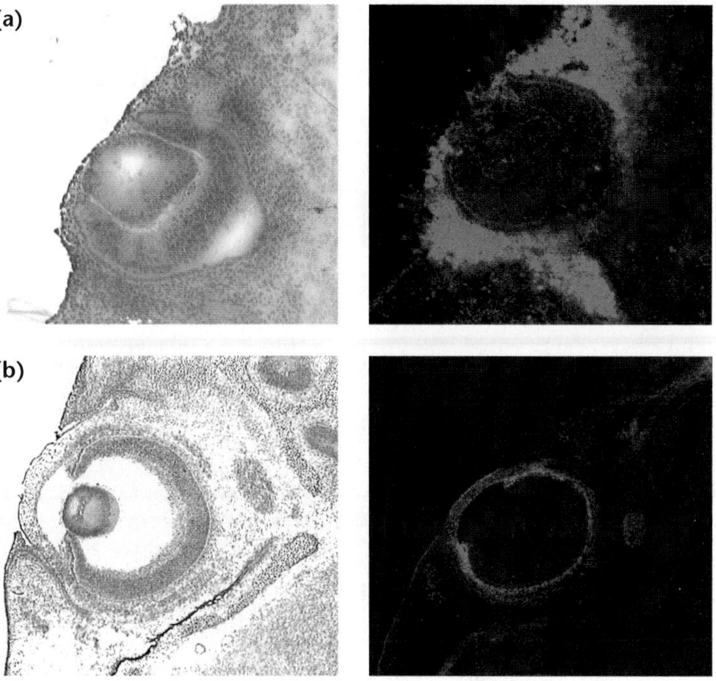

FIGURE 21–16 Immunofluorescence staining of the developing mouse eye, showing the location of the Pitx2 protein. The two right panels show mouse embryonic eyes, stained with a DNA-specific stain that denotes the nucleus of each cell. The two left panels are the corresponding immunofluorescence stained samples, viewed with a fluorescence microscope. Embryonic stages are 11.5 days (a) and 15.5 days (b). The pattern of Pitx2 protein expression is consistent with the Pitx2 gene's proposed role in early eye development. C: cornea, i: iris, pm: periocular mesenchyme.

Genetic Analysis Using Gene Targeting Technologies

The ultimate test of a gene's function comes from the study of *in vivo* phenotypes of mutant organisms. In the reverse genetic approach, the creation of mutants follows the isolation of the gene. In addition, the type of mutation is under precise control of the investigator. Researchers can make specific mutations in the cloned candidate gene, introduce the mutated gene into a model organism, and examine the phenotype. Specific genes can be altered, deleted, and moved from one organism to another.

Two powerful techniques that geneticists employ to understand gene function are gene knockouts and gene replacements following site-directed mutagenesis, together known as gene targeting.

Targeted gene knockouts involve the deletion or disruption of a specific gene. This can be achieved in most model organisms, including yeast, *Drosophila,* and mice. In yeast, gene replacement is particularly efficient due to high levels of **mitotic recombination** in this organism. One method is illustrated in Figure 21–17. In this example, researchers insert the *KanMX* gene (which confers resistance to the antibiotic G418) into the middle of the cloned gene of interest, replacing a portion of the gene. This replacement retains DNA sequences on either side of the *KanMX* gene that are identical to the gene of interest. Linear pieces of DNA containing the *KanMX* gene and surrounding sequences are transformed into diploid yeast cells. The DNA fragments undergo homologous recombination with the yeast chromosome, replacing the wild-type gene with the *KanMX* cassette. The cells undergo sporulation, the spores are grown in the presence of G418, and the survivors selected. In the resulting haploid yeast, these G418-resistant cells will be null for the gene of interest.

In mice, gene knockouts are more complicated to engineer (Figure 21–18). The first requirement is a culture of mouse **embryonic stem (ES) cells**. ES cells are removed from early mouse embryos at the blastocyst stage and grown in tissue culture. Under the right conditions, these cells will grow and divide as single cells without differentiating. The second requirement is a disrupted version of the gene of interest. Researchers disrupt the cloned gene by splicing in a piece of DNA containing an antibiotic-resistance gene, such as neo^r. This fragment then becomes part of a larger recombinant DNA molecule that also contains the herpes simplex virus *tk* gene. The neo^r gene confers resistance to the antibiotic neomycin. The viral *tk* gene encodes an enzyme, thymidine kinase, that selectively phosphorylates nucleoside analogues such as ganciclovir. If a cell expressing the *tk* gene is grown in the presence of ganciclovir, the phosphorylated nucleoside analog becomes a toxic inhibitor of cellular DNA replication. The vector containing the neo^r and *tk* genes is known as the **targeting vector**.

When the targeting vector is transformed into the ES cells in culture, two possible recombination events take place. First, the targeting vector may undergo homologous

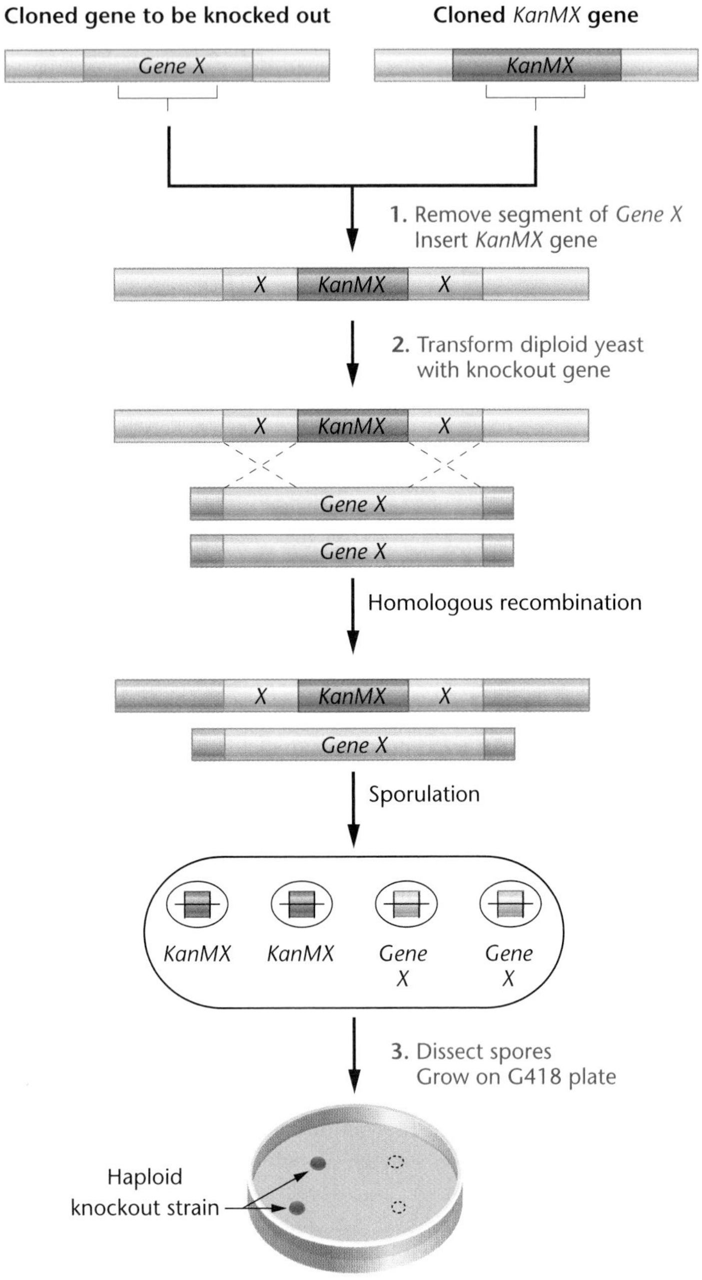

FIGURE 21–17 Targeted gene knockout in yeast. The *KanMX* gene is inserted into a cloned copy of the gene to be disrupted (*Gene X*). The *KanMX*-disrupted *Gene X* is transformed into diploid yeast cells. Homologous recombination occurs between the cloned knockout gene and one chromosome of the diploid yeast cell, yielding a heterozygous knockout yeast strain. After sporulation, the four haploid spores are dissected from the ascus and grown on a plate containing medium and the antibiotic G418. Any spores that grow contain the *KanMX*-disrupted *Gene X*.

recombination with the ES cell gene of interest, in the regions of sequence identity that flank the *neo*^{r} gene. Second, the targeting vector may undergo random integration into the ES cell genome. To select ES cells that have undergone homologous recombination, geneticists grow the transformed cells in medium containing neomycin and ganciclovir. The only cells that survive the presence of neomycin have incorporated the targeting vector that contains the *neo*^{r} gene. The only neomycin-resistant cells that survive in the presence of ganciclovir have undergone homologous recombination with the targeting vector, and hence have not incorporated the *tk* gene. After this drug selection, the surviving ES cells are tested by Southern blot or PCR analysis to verify that the gene of interest in the ES cells has been disrupted with the targeting vector DNA.

Once the ES cells with the targeted disruption are in hand, they are used to create the knockout mouse. Researchers obtain blastocysts from a mouse that has a different coat color from the mouse that donated the ES cells. For example, the ES cells may have been derived from a mouse with a dominant coat color such as brown (agouti), whereas the blastocyst may have been donated by a recessive black mouse. The ES cells are injected into the blastocyst, and these cells become part of the growing embryo, sometimes including the germ line. The resulting mouse will be a chimera, as it is derived from two different sources of cells. The chimeric mice are easy to identify, as they have patches of brown and black fur. Chimeric mice are mated to black mice. If the knockout gene is present in the germ line of a chimeric mouse, half of its progeny will be heterozygous for agouti coat color as well as the knockout gene. After testing the DNA of the chimeric mouse's progeny, progeny containing the knockout gene are mated to establish homozygous knockout mice.

This elaborate procedure is used to establish knockout mice as models for human diseases, as well as vehicles to test the function of cloned genes. For example, this approach was used to verify that mutations in the *CFTR* gene are responsible for the disease cystic fibrosis.

Targeted gene replacement is an equally valuable tool with which to dissect the function of a cloned gene. Gene replacement usually involves substituting a gene containing specifically designed mutations for the wild-type version of the gene of interest. But how do geneticists introduce specific mutations in a cloned gene?

To do this, they undertake site-directed mutagenesis. Site-directed mutagenesis allows researchers to introduce a specific mutation at a precise site within a cloned gene. Geneticists use site-directed mutagenesis to delete, insert, or change one or more nucleotides in a DNA clone. The technique was first developed by Michael Smith at the University of British Columbia in the 1980s. Dr. Smith received the Nobel Prize in Chemistry in 1993, reflecting the significance of this technique to modern genetic analysis. Several variations of this site-directed mutagenesis technique now exist, including those based on the use of PCR. Here, we will discuss the classic method of site-directed mutagenesis based on Dr. Smith's technique.

The first step in the classic method of site-directed mutagenesis is to clone the gene of interest and determine its DNA sequence. The DNA sequence is examined and the site

FIGURE 21–18 Creating a knockout mouse. (a) Inserting the knockout gene into embryonic stem (ES) cells. ES cells are derived from an agouti (brown) mouse. (b) Creating the knockout mouse strain from knockout ES cells.

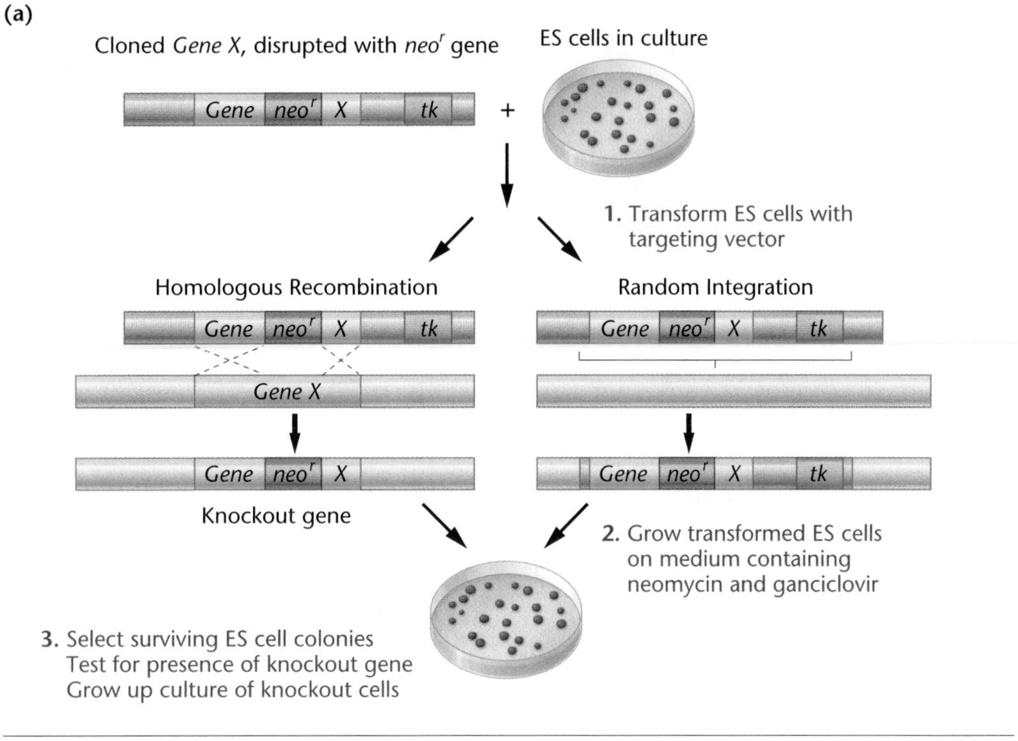
(a)
Cloned Gene X, disrupted with neo^r gene
ES cells in culture
Gene neo^r X tk
+
1. Transform ES cells with targeting vector
Homologous Recombination
Random Integration
Gene neo^r X tk
Gene X
Gene neo^r X
Knockout gene
Gene neo^r X tk
2. Grow transformed ES cells on medium containing neomycin and ganciclovir
3. Select surviving ES cell colonies
Test for presence of knockout gene
Grow up culture of knockout cells

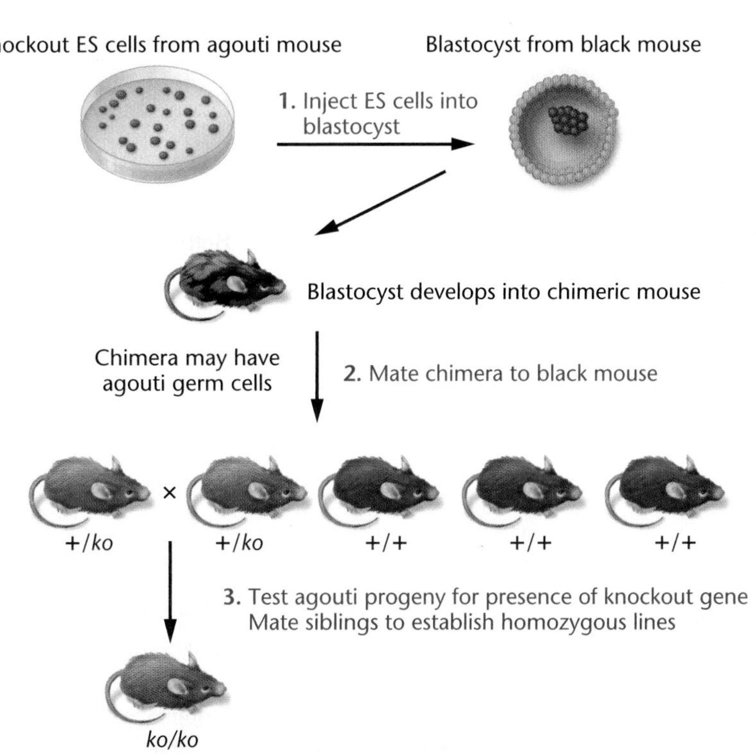
(b)
Knockout ES cells from agouti mouse
Blastocyst from black mouse
1. Inject ES cells into blastocyst
Blastocyst develops into chimeric mouse
Chimera may have agouti germ cells
2. Mate chimera to black mouse
+/ko
×
+/ko
+/+
+/+
+/+
3. Test agouti progeny for presence of knockout gene
Mate siblings to establish homozygous lines
ko/ko

of the mutation and its nature are chosen (Figure 21–19). The gene, or a fragment from the gene, is cloned into the M13 bacteriophage (described in Chapter 19). M13 exists in two forms: the double-stranded DNA form known as replicative form and a single-stranded DNA form that is extruded from the bacterial cell. Once the gene of interest is cloned into M13, the single-stranded form of the M13 genome is harvested and used *in vitro* as the template on which to create the site-directed mutation.

The second step in site-directed mutagenesis is to design and synthesize an oligonucleotide that is complementary to a region of the gene of interest, but contains the desired mutation. The oligonucleotide may be as short as 18 nucleotides. The mutation may be a change in one or more nucleotides, or may be a small deletion or insertion. The synthetic oligonucleotide is then hybridized to the single-stranded M13 DNA molecule so that it anneals to the region of complementarity between the oligonucleotide and the cloned gene in M13. The region that is not complementary (i.e., the mutation) does not hybridize with the gene of interest in the single-stranded M13 clone. A purified DNA polymerase is added to the reaction, along with a mixture of nucleotides (dATP, dCTP, dTTP, and dGTP). The DNA polymerase extends the 3′ end of the oligonucleotide to create a double-stranded M13 DNA molecule. DNA ligase is added to seal the newly synthesized DNA strand. The resulting double-stranded M13 molecule is then transformed into *E. coli* bacteria. The bacteria replicate the double-stranded M13 DNA. Because DNA

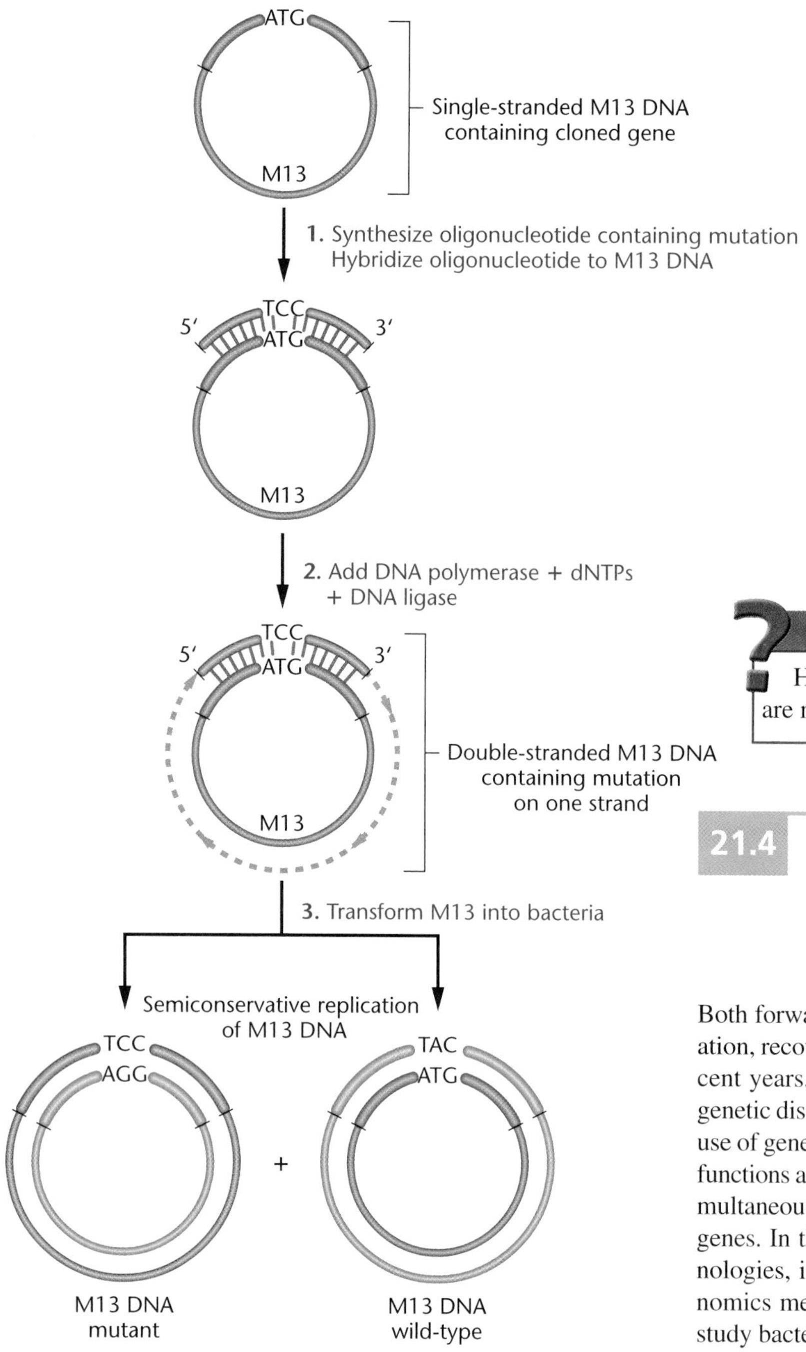

FIGURE 21–19 Oligonucleotide-mediated site-directed mutagenesis.

replication is semiconservative, half of the replicated DNA molecules will bear the mutation and half will bear the wild-type version of the cloned gene. The mutated versions of the M13 molecules are then selected using a number of molecular screening procedures.

After the mutated version of the gene is synthesized, it may be subcloned from the M13 vector into any number of different vectors and introduced into cultured cells or model organisms to test the mutant gene function. Geneticists can introduce the newly mutated gene into an organism using transgenic technologies. If the organism is null for the gene of interest, the transgene will provide the only copy of the gene, and its phenotype can be assessed. This is the strategy currently being used to investigate mutations in the human *CFTR* gene, which is responsible for the disease cystic fibrosis. The mutated *CFTR* genes have been introduced into *CFTR* knockout mouse models to determine how each mutation controls the range of symptoms that occur in cystic fibrosis.

Sometimes it is desirable to directly replace an endogenous gene in the model organism with a site-directed mutagenized copy. This is done in yeast and mice following similar procedures used to create gene knockouts in those organisms. Instead of replacing the wild-type gene with a null gene containing an antibiotic resistance gene, the wild-type gene is replaced by the *in vitro* mutagenized copy.

HOW DO WE KNOW?

How do we know that mutations in the *CFTR* gene are responsible for the human disease cystic fibrosis?

21.4 Geneticists Dissect Gene Function Using Functional Genomic and RNAi Technologies

Both forward and reverse genetic techniques center on the creation, recovery, and analysis of gene mutations. However, in recent years, geneticists have devised alternative approaches to genetic dissection. These methods do not necessarily require the use of gene mutants, but involve nongenetic inactivation of gene functions and various high-throughput technologies that allow simultaneous examination of expression from a large number of genes. In this section, we introduce several of these new technologies, including RNAi and high-throughput functional genomics methods. Some functional genomics methods used to study bacterial and yeast genomes are discussed in Chapter 20.

RNAi: Genetics without Mutations

Both forward and reverse genetic analyses center on the creation and examination of gene mutants. These approaches, although highly effective, are labor-intensive and cannot be used in organisms in which gene targeting is not possible. Recently, researchers have devised a new and potentially powerful dissection tool, **RNA interference (RNAi)**. This new technology allows investigators to specifically create single-gene defects without resorting to the creation of heritable mutations. These RNAi-mediated **gene silencing** technologies are relatively inexpensive and allow rapid analysis of gene function. Researchers hope to use RNAi technology to systematically knock out the function of each gene in model organisms. The technique also may be useful for gene therapy, in which specific disease genes can be silenced.

In nature, cells employ RNAi to defend themselves from viruses or invading transposons. (See Chapter 17). The double-stranded RNA molecules produced by these agents are recognized by enzymes within the cell that degrade double-stranded RNAs. When the RNA is degraded, no protein is translated and gene expression is blocked. Another way in which RNAi works is to inhibit transcription of viral or transposon genes. In this case, the antisense strands of double-stranded RNA molecules bind to the complementary regions of a gene, leading to recruitment of proteins that modify DNA and inhibit transcription. Cells also use RNAi to regulate expression of their endogenous genes. Short fragments of antisense RNA complementary to a sense mRNA can bind to the mRNA and inhibit translation.

To use RNAi as a research tool, investigators introduce short synthetic double-stranded RNA molecules into cells (Figure 21–20). These molecules trigger the same RNA-degradation pathway that is triggered by viral or transposon double-stranded RNAs. The antisense strand from the double-stranded RNA hybridizes to the complementary mRNA in the cells, targeting the mRNA for degradation.

Alternatively, investigators can transform cells with vectors that express RNA hairpin structures which are partially double-stranded. If the vectors that express the double-stranded RNA molecules become integrated into a cell's genome, all the progeny of that cell will express the RNAi molecules and one specific gene will be silenced in those cells. Silencing may result from RNA degradation or inhibition of transcription or translation. Another variation on the RNAi technique is to drive the expression of the RNAi gene from an inducible promoter. For example, if the *Drosophila* heat shock gene promoter is cloned next to the gene encoding the RNAi, and the RNAi vector is then transformed into *Drosophila* by *P* element-mediated transformation, the resulting flies can be gene silenced by simply raising the temperature. This approach is particularly useful for silencing genes that are essential for viability during development.

RNAi methods have been used successfully to investigate gene functions in organisms that are usually difficult to manipulate genetically. These include mosquitoes, trypanosomes, and mammalian cells. RNAi has also been used to silence the expression of genes that are aberrantly expressed in cancer cells. When these cancer-related genes are silenced, the cancer cells lose their ability to form tumors when introduced into mice. Although RNAi technologies, sometimes referred to as "genetics without mutations," are just beginning to be developed, they hold great promise for both investigative and therapeutic uses.

High-Throughput Functional Genomics Techniques

The major strength of classical genetic analysis is that investigators can examine the effects of altering one gene at a time, against a background of normal gene function in the rest of the genome. However, since the advent of modern genomics, various **high-throughput technologies** have been developed that allow investigators to probe the genetic interactions of thousands of genes simultaneously. These technologies include DNA and protein expression **microarrays**, automated methods for the isolation and dissection of large protein complexes and genome-wide searches for protein-DNA interaction sites. Proteomics, the study of global patterns of protein expression, has been discussed in Chapter 20. As these techniques are based on the wealth of data provided by completion of genome sequencing projects, they are often referred to as functional genomics techniques.

These methods provide a more global picture of gene expression than that revealed by conventional genetic analysis, and may help dissect aspects of genetic pathways and gene interactions that a mutational approach might miss.

Gene Expression Microarrays

DNA microarrays (DNA chips) are used to examine the expression of thousands of genes simultaneously. DNA microarrays are simply pieces of glass (chips) onto which DNA samples are applied in an ordered pattern—i.e., an array. (See Chapter 22.) Often these microarrays are manufactured by automated machines that place microscopic droplets of specific DNA samples in specific positions on the chip. The DNA samples that are applied may be any type of cloned DNA, but are often short oligonucleotides that are synthesized *in vitro*. More than 30,000 different DNA samples may be applied to a DNA microarray, representing thousands of different genes. Theoretically, all the genes in an organism's genome can be present on a microarray, allowing analysis of gene expression of the entire genome.

DNA microarrays may be used to compare the patterns of gene expression in two or more different tissues, in tissues at two or more different times during development, or in cells from normal or diseased tissues.

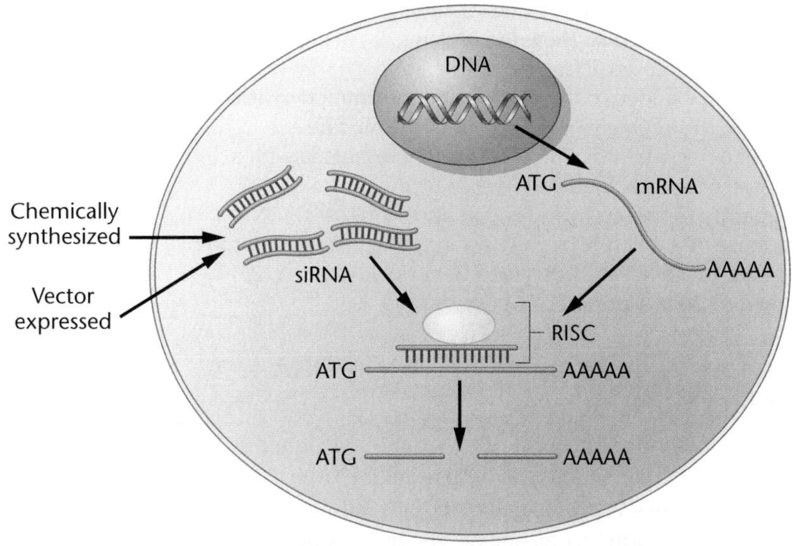

FIGURE 21–20 RNA interference (RNAi). Short double-stranded RNA molecules called short interfering RNAs (siRNAs) are introduced into a cell. In the cell, they become part of the RNA-induced silencing complex (RISC). The antisense strand of the siRNA binds to the complementary mRNA within the cell, targeting the cellular mRNA for cleavage by enzymes within the RISC. The cleaved mRNA is subsequently degraded, and is not translated.

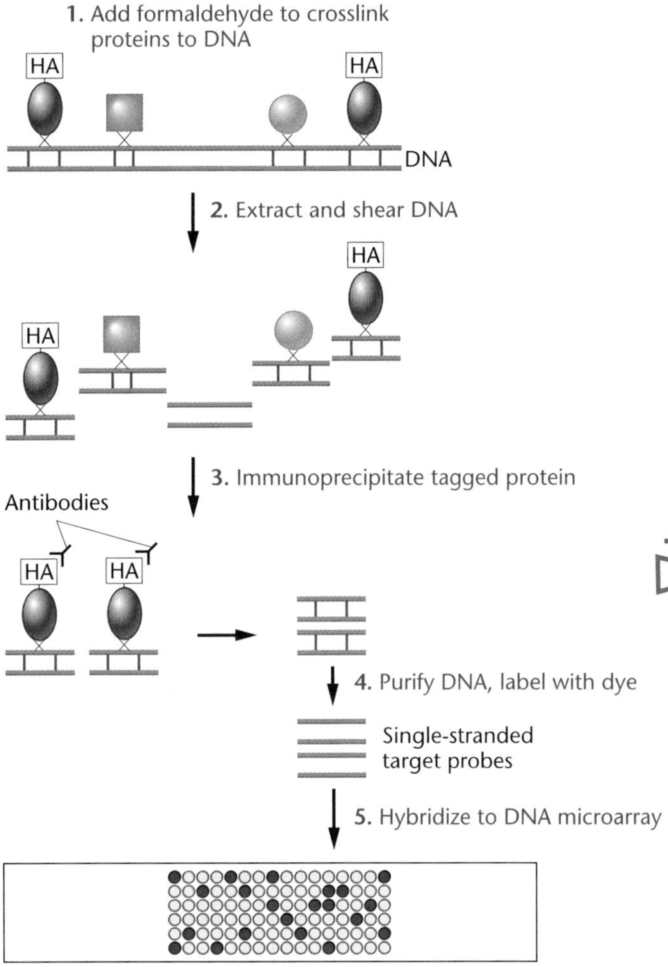
1. Add formaldehyde to crosslink proteins to DNA
HA
HA
DNA
2. Extract and shear DNA
HA
HA
3. Immunoprecipitate tagged protein
Antibodies
HA
HA
4. Purify DNA, label with dye
Single-stranded target probes
5. Hybridize to DNA microarray

FIGURE 21–21 Genome-wide screen for transcription factor binding sites. The top line shows various proteins bound to their DNA binding sites. One protein is tagged with a protein fragment (HA) that can be recognized by a specific antibody. Formaldehyde is added to tissues or cultured cells to crosslink these proteins to DNA. Then, the DNA is extracted from the cells and sheared into small fragments. The antibody that recognizes the HA tag is added to the mixture and the antibody, with its protein/DNA fragment, is pulled out of the mixture (immunoprecipitation). The DNA is purified, labeled with a fluorescent dye. Labeled DNA is then hybridized to a DNA microarray. A positive hybridization signal indicates a DNA sequence that is bound by the HA-tagged protein. *Reprinted by permission from Nature Reviews Genetics 2(4):302-312, copyright 2001 Macmillan Publishers Ltd.*

Genome-Wide Mapping of Protein–DNA Binding Sites

Another functional genomic technique is designed to map protein–DNA interactions, and is particularly useful for identifying genes that are regulated by DNA-binding transcription factors. To perform this technique, researchers treat tissues or single cells such as yeast with formaldehyde (Figure 21–21). Any proteins that are tightly bound to DNA will be crosslinked to the DNA by the formaldehyde. The researchers then extract DNA from the cells and shear it into small fragments. To isolate DNA fragments that are bound to a specific protein, they add an antibody that recognizes the protein. The antibody is then precipitated from the mixture, along with the protein it recognizes and any DNA fragments that are crosslinked to the protein. The investigators then purify the immunoprecipitated DNA fragments, amplify these DNA fragments by PCR, and label the fragments with a fluorescent tag. The labeled DNA fragments are hybridized to a DNA microarray containing cloned DNAs representing the organism's entire genome. Any spot on the microarray that hybridizes to the labeled DNA represents a DNA sequence that bound the protein. As each spot on the DNA microarray is known, the identity of each positive signal can be determined. In one application of this genome-wide method, more than 200 previously unknown targets of a transcription activator protein that functions at the G1/S cell cycle interface were identified.

Now solve this

Problem 21.15 on page 546 asks you to think critically about the limitations of one of the new high-throughput functional genomics methods—DNA microarrays (DNA chips)—and how to overcome them.

Hint: When thinking about limitations, you might consider potential technical problems due to the nature of the material on the chips. You might also think about interpreting the data that come from these types of experiments. Also think about the number of steps involved and what could go wrong at each step.

21.5 Geneticists Advance Our Understanding of Molecular Processes by Undertaking Genetic Research in Model Organisms: Three Case Studies

In this chapter, we have outlined many of the diverse approaches to genetic dissection—from forward and reverse genetics to high-throughput functional genomics techniques. The power of these methods has depended upon their application within appropriate model systems. For example, geneticists have revealed the basic molecular mechanisms of DNA replication, recombination, transcription, and translation by manipulating gene mutations in the single-celled prokaryote, *E. coli*. Similarly, some of the most important insights into control of the cell cycle and gene expression have arisen from analyses of gene function in the single-celled eukaryote, *S. cerevisiae*. Mutagenesis and gene knockouts in the simple plant, *Arabidopsis thaliana*, are increasing our knowledge about plant genomics, gene regulation, and gene–environment interactions. Pioneering research on multicellular development was made possible through the creative application of genetic techniques in simple eukaryotes such as *C. elegans* and *Drosophila*.

The following three case studies illustrate the power of genetic analysis in model organisms. They also provide a narrative on how geneticists carry out their research. The knowledge gained about gene control of the cell cycle, development, and human disease genes have far-reaching implications not only for understanding the basic processes of life, but also for potential therapies to arrest or cure cancer and human genetic diseases.

Leland H. Hartwell
Fred Hutchinson Cancer Research Center, Seattle, WA, USA

R. Timothy (Tim) Hunt
Imperial Cancer Research Fund, London, United Kingdom

Sir Paul M. Nurse
Imperial Cancer Research Fund, London, United Kingdom

FIGURE 21–22 Drs. Hartwell, Hunt, and Nurse, recipients of the 2001 Nobel Prize in Physiology or Medicine "for their discoveries of key regulators of the cell cycle."

Yeast: Cell Cycle Genes

On October 8, 2001, Drs. Leland Hartwell, Timothy Hunt, and Paul Nurse jointly accepted the Nobel Prize in Physiology or Medicine. Working independently, and using different model organisms, these three investigators had made important discoveries about the genes that control cell growth and division (Figure 21–22). This case study tells the story of how Dr. Hartwell and his colleagues undertook a series of elegant genetic experiments using the budding yeast *S. cerevisiae*. These experiments led to the identification of dozens of genes that regulate cell division. More recent work reveals that homologous genes operate in higher organisms, with implications for understanding and treating human cancers.

As he described in his Nobel Lecture, Dr. Hartwell began his research career with an interest in understanding cancer. Although he had begun his research looking at control of DNA synthesis in cultured mammalian cells, he soon switched to *Saccharomyces cerevisiae* as a model system. This organism had several major advantages for Hartwell's research into cell division. Yeast can be grown easily in the lab as single cells and are easy to manipulate genetically. Yeast cells have both a haploid and diploid growth phase, allowing rapid identification of recessive mutations in the haploid stage and analysis by complementation in the diploid stage. In addition, Hartwell and his colleagues realized that they could identify the stage of the yeast cell cycle by observing cellular morphology (Figure 21–23). At the beginning of S phase of the cell cycle, yeast cells develop a bud (the daughter cell) that extends from the surface of the mother cell. As the cell cycle progresses, the bud increases in size, the nucleus migrates to the region between the two cells, and the nucleus and the two cells divide in two. Hence, the relative size of the bud and mother cell, as well as the nuclear morphology, indicate the stage of the cell cycle. In a population of normal growing yeast cells, cells are in all stages of the cell cycle.

Hartwell's genetic screen for cell cycle mutations began by treating a yeast culture with the mutagenic chemical nitrosoguanidine. Hartwell and his colleagues then spread the cells onto growth plates so that each cell would form a single colony. Once colonies formed, they replica plated the colonies on each plate onto two other plates—one plate to be incubated at room temperature (23°C) and the other plate at the restrictive temperature (36°C). All colonies grew at 23°C, whereas colonies from cells that had suffered a temperature-sensitive mutation did not grow at 36°C (Figure 21–7). Using this approach Hartwell's team isolated more than one thousand temperature-sensitive mutant strains.

Although his research into cell cycle regulation seems, in retrospect, to have been a logical progression, at the beginning

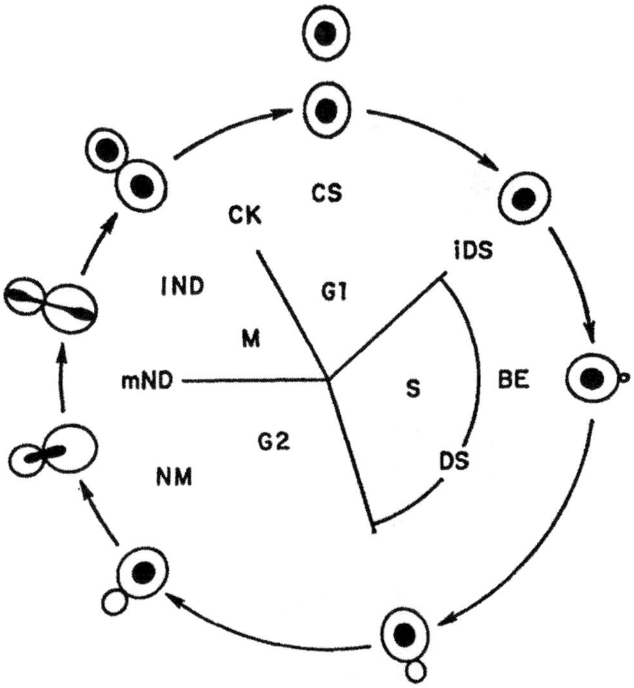

FIGURE 21–23 Yeast cell morphology throughout the cell cycle. Diagram showing stages of the cell cycle (G1, S, G2, and M) as well as steps in the cell cycle defined by the mutants in Hartwell's study. iDS: initiation of DNA synthesis, BE: bud emergence, DS: DNA synthesis, NM: nuclear migration, mND: medial nuclear division, lND: late nuclear division, CK: cytokinesis, CS: cell separation. This figure is reproduced from the 1974 *Science* paper by Hartwell, Culotti, Pringle and Reid, describing genetic control of the cell division cycle in yeast.

of his research Dr. Hartwell had no idea that some temperature-sensitive mutants had defects in the cell cycle. His first observation was that some temperature-sensitive mutants developed odd shapes at the restrictive temperature, but these cells returned to normal shapes at the permissive temperature. Hartwell asked an undergraduate student, Brian Reid, to determine how these cells changed from odd shapes to normal shapes. To do this, Reid decided to use time-lapse photography to examine yeast cells as they changed from normal to odd and back to normal. He set up a microscope in a warm room (98°F), and spent his summer photographing yeast cells. At the end of the project, Reid had discovered that the odd shaped cells were arrested at particular points within the cell cycle.

Reid's conclusions provided the technical insights that allowed Hartwell's group to screen their temperature-sensitive mutants for cell cycle defects. They observed each mutant by time-lapse photomicroscopy after a shift to 36°C. Yeast cells with a temperature-sensitive defect in a gene controlling one point in the cell cycle arrested with a uniform phenotype characteristic of that point in the cell cycle (Figure 21–24). For example, at 36°C, all cells blocked at the G2/M interphase appeared to have a large bud with a nucleus at the isthmus between the mother and daughter cell [Figure 21–24(b)].

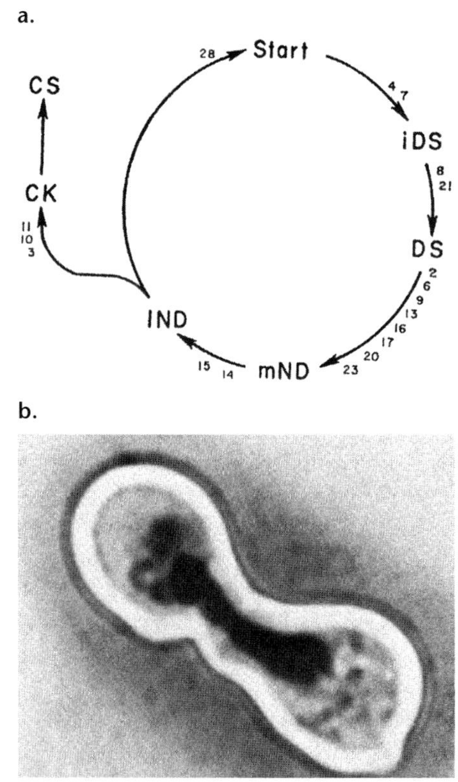

FIGURE 21–24 Yeast cell cycle mutants, ordered into steps in the cell cycle. (a) Hartwell's model of cell cycle steps, with 18 of his cell division cycle (*cdc*) mutants marked at the step affected in the mutant. Abbreviations are the same as those in Figure 21–23. (b) Microscopic view of a mutant yeast cell, growing at the restrictive temperature. The cell is arrested at G2/M with a characteristic morphology. Image (a) is from Dr. Hartwell's 1974 *Science* paper "Genetic control of the cell division cycle in yeast".

Using these methods, the investigators identified about 150 temperature-sensitive mutants that displayed cell division cycle (cdc) defects. The next question was whether the 150 mutations had all occurred in separate genes, or whether there were multiple hits within only a few genes. To do this, the researchers performed complementation analysis. Two haploid yeast strains, each with a *cdc* temperature-sensitive mutation, were crossed to form diploids. If the resulting diploid strains were wild type, the two mutations were in separate genes; if the diploid strains were mutant, the two mutations were in the same gene. In this way, they determined that these 150 *cdc* temperature-sensitive mutations occurred in 32 different genes.

The next question was whether these 32 genes regulated 32 separate steps within the cell cycle, or whether some affected the same steps as others. In order to position each mutant relative to the other, Hartwell and colleagues performed epistasis analysis. They compared the cell cycle phenotypes of single mutants that arrest cells at two different stages, with the cell cycle phenotypes of double mutants. Whichever mutant showed epistasis over the other mutant represented a mutation in an earlier stage of the cell cycle. Using these methods, they ordered the mutations into six cell cycle events, each controlled by one or more genes [Figure 21–24(a)]. This analysis also confirmed that each step in the cell cycle is dependent upon the successful completion of the previous step.

One particular mutant, *cdc28*, revealed significant properties. The investigators determined that the product of the *CDC28* gene controlled a step in late G1, and successful completion of this step appeared to be necessary for all subsequent events in the cell cycle. Once cells had passed the *CDC28* step, they were committed to finish the cell cycle before they could arrest growth in response to starvation or undergo mating. Hartwell and his colleagues called this decision point "start."

Since Hartwell and his coinvestigators published the results of their study in 1974, the importance of the *CDC28* gene has been confirmed. Homologs of this gene are found in virtually all eukaryotic organisms, and the functions of these homologs are similar in each. The *S. cerevisiae CDC28* gene and the homologous *cdc2* gene from *Schizosaccharomyces pombe* (fission yeast) functionally complement each other, as do the *CDC28* gene and the *Xenopus laevis* (frog) gene encoding the kinase component of MPF (maturation promoting factor). This evolutionary conservation of cell cycle control genes was used to identify and clone the human *CDC2* gene by complementing an *S. pombe cdc2* mutant strain with a human cDNA library. The product of the *CDC2/28* gene has now been identified as a cyclin-dependent kinase (CDK). CDKs control the cell cycle by interacting with cyclin molecules, and by phosphorylating proteins in such a way as to activate or inactivate them. The precise and elegant cyclic phosphorylations that orchestrate each stage of the eukaryotic cell cycle are described in Chapter 18. The realization that cell cycle control is a key component in normal and cancer cell function, and that CDKs are central to these functions, have validated Dr. Hartwell's choice of yeast as a relevant model organism for research into the development of cancer.

FIGURE 21–25 Christiane Nüsslein-Volhard and Eric Wieschaus.

Drosophila: The Heidelberg Screens

When Drs. Christiane Nüsslein-Volhard and Eric Wieschaus (Figure 21–25) finished their graduate research training in the late 1970s, they decided to combine their forces to take on a formidable challenge—unlocking the secrets of how a single-celled egg develops into a complex, multicellular organism. Their common interest in early development led them to their first independent research jobs at the European Molecular Biology Laboratory (EMBL) in Heidelberg, where they shared the same small research lab for three years. Although they, like other researchers in their field, believed that genes must control the body patterns that arise during embryogenesis, it seemed almost impossible to determine which genes were responsible. It was estimated that more than 10,000 gene products (RNAs and proteins) were present in a developing embryo, and the relative levels of these gene products changed constantly throughout development. Within this morass of fluctuating molecular regulation, how could anyone identify the relevant genes?

Drs. Nüsslein-Volhard and Wieschaus decided to plunge into the seemingly impossible task of identifying all the genes that control early development. No scientist had undertaken such an ambitious task and it was not at all clear whether they would succeed.

They chose the fruit fly *Drosophila melanogaster* as the model organism for these studies. The *Drosophila* egg develops rapidly—about 20 hours from fertilization to hatching into a larva and about 9 days until emergence as an adult fly. Between fertilization and hatching, one cell divides and differentiates into about 40,000 cells at which time the embryonic segment patterns are fully established. In addition, *Drosophila* genetics is well developed, with established methods to mutagenize and maintain mutant strains. The two investigators decided to restrict themselves to examining early embryonic and larval body patterns and to observing these patterns in the larval cuticle. This external skeleton provides visual landmarks that reflect the underlying body plan. This decision was important, as it allowed them to observe developmental features in fruit flies that bore recessive lethal mutations. Embryos and larvae that were destined to die could be observed and scored for defects prior to, or at, death.

Not knowing how many genes they were looking for, Nüsslein-Volhard and Wieschaus decided to establish a large number of mutant strains of flies. As many of the mutations they generated would be recessive lethal mutations, they also had to devise a method that would allow them to maintain these mutations within heterozygous stocks of flies, but still generate homozygous mutant flies that they could observe and screen for defects. To do this, they developed the following strategy. They first fed male flies with the mutagenic chemical ethyl methane sulphonate (EMS), which causes a high frequency of mutations, particularly single-base changes. They then crossed the mutagenized males to females carrying a balancer chromosome for the chromosome that they wanted to screen.

Figure 21–26 shows their strategy for screening mutations on the second chromosome. The male is homozygous recessive for two mutations (*cn* and *bw*) on the second chromosome, that results in a white-eye phenotype. The second chromosomes in the female balancer strain bear a dominant gene for curly wings (*CyO*) on one homolog and a dominant temperature-sensitive lethal gene on the other homolog (*TS*). Of the first generation offspring, males with curly wings and red eyes were individually selected. These males had one second chromosome from their male parent (likely bearing a mutation) and one second chromosome from the balancer female, bearing the curly wing marker. These first generation males were then backcrossed to the same balancer strain of females. The second generation was then grown at a restrictive temperature that killed any flies that inherited the dominant temperature-sensitive *TS* gene. In addition, any flies that inherited two copies of the curly marker would not live, as the homozygous *CyO* chromosome is lethal. Hence, the only viable second generation offspring contained one copy of the *CyO* balancer chromosome and one copy of the mutagenized white-eye marked chromosome. This second generation was crossed within itself to produce the third generation. As the *CyO* homozygotes were not viable, there were only three possible progeny in the third generation. The researchers knew that if any of the third generation progeny displayed white eyes, then the homozygous white *cn bw* chromosome did not contain a recessive lethal mutation. However, if a recessive lethal allele was present on the *cn bw* marked chromosome, there would be no white-eyed offspring. In this case, the strain could be maintained indefinitely by mating siblings. The only survivors in these balanced stocks would be red-eyed, curly-winged heterozygotes, with the recessive lethal gene on one homolog of the second chromosome. This same procedure was repeated for each of the *Drosophila* chromosomes to be tested.

In the Heidelberg screens, Nüsslein-Volhard and Wieschaus examined the third generations of each strain for those that contained no viable homozygotes (no white-eyed flies). They then screened these strains to determine whether the lethality for the homozygous *cn bw* chromosome occurred in late embryogenesis or early larval phases. For over a year, the two investigators

FIGURE 21–26 Strategy for screening recessive lethal mutations on the *Drosophila* second chromosome.

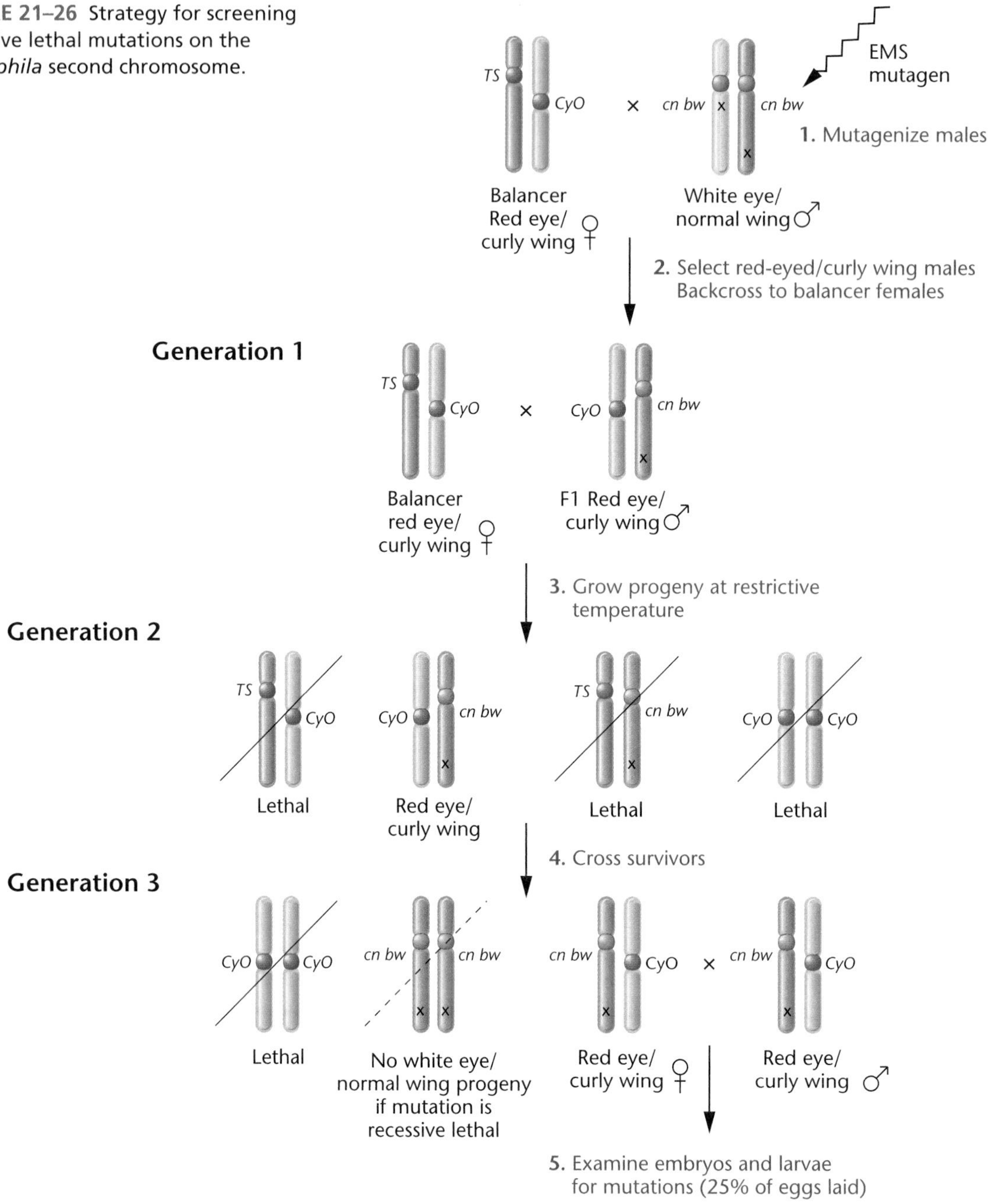

worked together, using a two-person microscope, examining thousands of late embryos for abnormalities in body organization and segmentation (Figure 21–27). Between 1979 and 1980, they established about 27,000 individual lines of *Drosophila*. They determined that about 18,000 of these lines had recessive lethal mutations, and of these, about 4000 mutations were lethal during embryogenesis. Of these embryonic lethals, about 580 lines showed visible defects in the external cuticle. They next addressed the question of how many genes were represented by these 580 embryonic lethal mutations. Complementation tests revealed that these 580 mutations lay within only 139 genes.

There were several surprising outcomes of the Heidelberg screens. One was the small number of genes that appeared to regulate embryonic pattern formation, despite the apparent complexity of the process. The second was the fact that these genes could be classified into groups, as many genes appeared to affect the same developmental step. The genes that Nüsslein-Volhard and Wieschaus identified could be classified according to the time at which they act during development, as well as the parts of the embryo that are programmed by these genes. For example, the 30 segmentation genes fell into three groups. Mutations in the gap genes (such as *Krüppel* and *knirps*) all show large contiguous deletions of segments. Mutations in the pair rule genes (such as *even-skipped*) show deletions in every other segment. Mutations in segment polarity genes (such as *engrailed* and *wingless*) show defects in every segment. This suggested to the investigators that segmentation requires three levels of organization in a specific temporal pattern. (Details of the Nüsslein-Volhard and Wieschaus models of embryonic development are presented in Chapter 23.)

Drs. Nüsslein-Volhard and Wieschaus published the results of their study in the journal *Nature* in 1980. Their work was im-

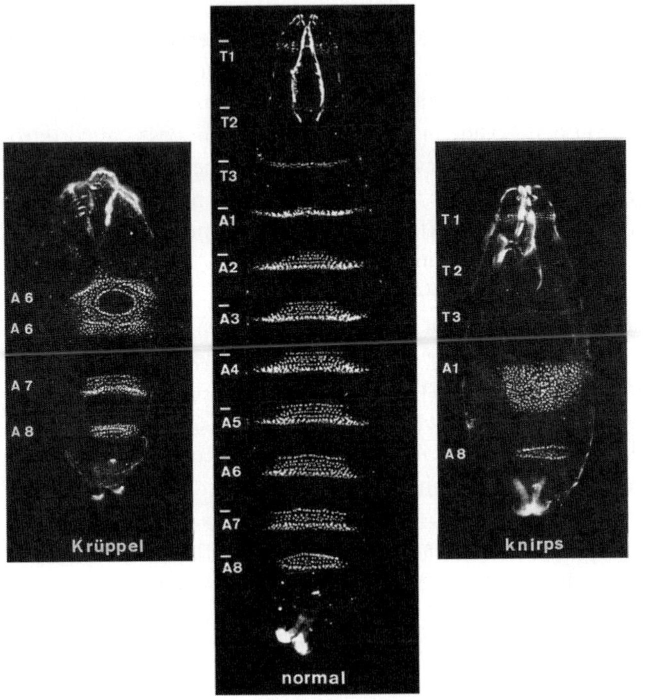

FIGURE 21–27 Cuticle preparations of late *Drosophila* embryos. Features on the cuticle indicate location, number, and orientation of segments. The normal embryo (center) shows all three thoracic (T) segments and all eight abdominal (A) segments. The two gap gene mutants, *Krüppel* and *knirps*, lack a number of contiguous thoracic or abdominal segments, as indicated. Photos are from Nüsslein-Volhard's and Wieschaus's 1980 *Nature* paper.

mediately recognized for its significance. When scientists began to look in other organisms for developmental genes, they discovered remarkable similarities to *Drosophila*. Many *Drosophila* developmental genes have homologs in higher vertebrates including humans, and these vertebrate genes similarly affect early embryonic pattern formation. Most of the genes identified by Nüsslein-Volhard and Wieschaus have now been cloned. Most encode transcription factors and cell signaling molecules. Their patterns of expression reflect the regions affected in the corresponding mutant.

In recognition of the significance of their research, Drs. Nüsslein-Volhard and Wieschaus received the Nobel Prize in Physiology or Medicine in 1995.

The Mouse: A Model for ALS Gene Therapy

Amyotrophic lateral sclerosis (ALS) is an incurable neurodegenerative disease that leads to the progressive destruction of motor neurons in the brain and spinal cord. When motor neurons die, the brain no longer controls movement and the muscles weaken and atrophy. This leads to difficulties with speech, breathing, and swallowing, and eventually to full paralysis. Despite this, ALS does not affect intellectual ability, hearing, sight, or other senses. ALS sufferers usually have normal bowel, bladder, and sexual function. Every year, about 1 to 2 people per 100,000 are diagnosed with ALS throughout the world. The causes of ALS are unknown. There are no drugs that prevent or cure ALS. Most therapies merely alleviate the severity of the symptoms. ALS is also called Lou Gehrig's disease, after the famous NY Yankees baseball player who succumbed to the illness in 1941. The accomplished physicist Stephen Hawking has suffered from ALS for most of his life and shows remarkable courage and stamina in dealing with his condition. (See Figure 21–28.)

Approximately 10 percent of ALS cases are familial, inherited in an autosomal dominant fashion. About 20 percent of these familial cases are associated with dominant mutations in the Cu/Zn superoxide dismutase gene, *SOD1*. The *SOD1* gene encodes an antioxidant enzyme (SOD) that converts superoxide radicals to hydrogen peroxide and molecular oxygen. By destroying oxidative free radicals, SOD protects cells from the damage caused by these by-products of aerobic metabolism.

As ALS is difficult to study in human patients, researchers needed an animal model in which to find the causes and devise treatments for the disease. One year after the identification of ALS-related mutations in the human *SOD1* gene, geneticists generated the first mouse model for the disease. They did this by creating transgenic mice bearing a human *SOD1* gene found in many familial ALS cases. Because ALS is inherited as an autosomal dominant trait, it was hoped that transgenic mice bearing one copy of a mutant *SOD1* gene would exhibit symptoms resembling ALS. The defect in these mutant genes is an exon 4 mutation that results in a substitution of alanine for glycine at amino acid 93 in the SOD protein (the G93A mutant). Using standard molecular genetic techniques, the investigators cloned the 14.5 kilobase pair human genomic DNA fragment containing the G93A gene along with the gene's promoter region. They then injected this DNA into mouse eggs and returned the eggs to a surrogate mother. To determine which of the resulting mice were transgenic, the investigators performed PCR analysis on DNA isolated from small samples of the mouse tails. Those DNA samples that showed positive PCR signals were examined by Southern blot to verify that the G93A gene was present in the mouse genome. A number of transgenic *SOD* mice were generated and these transmitted the transgene to 50 percent of their progeny.

The transgenic *SOD* mice develop symptoms of ALS similar to those seen in humans. By four months of age, these mice show weakness in their hind limbs and show muscle wasting. Over a two-week period, they become progressively

FIGURE 21–28 Lou Gehrig (left) and Stephen Hawking (right), two famous sufferers from amyotrophic lateral sclerosis (ALS).

paralyzed, and they die by five months. Post mortem examination reveals the loss of spinal motor neurons and the presence of fibrous material in the neurons—also a feature of the human disease.

What have investigators learned about the causes of ALS from the transgenic mouse models? First, the G93A mutation does not appear to affect the activity of the SOD enzyme. In addition, mice carrying homozygous null mutations in the *SOD1* gene are normal. Therefore, loss of SOD enzyme activity is unlikely to be the cause of ALS. Second, the disease is not likely to be caused by simple overexpression of *SOD1*. Control transgenic mice that express the normal human *SOD1* gene produce large amounts of SOD enzyme, but do not develop ALS-like symptoms. Together, these studies indicate that there is significance to the presence of mutations in *SOD1*, despite the fact that they do not affect enzyme activity. Researchers now think that the *SOD1* mutations in ALS may cause the SOD protein to become toxic. The nature of this toxicity is currently being studied.

SOD transgenic mice are also providing insights into potential therapies for ALS. These mice are being used to test potential drug therapies, such as selenium, zinc, vitamin E, and nerve growth factors. Most of these drugs have had little or no effect on the symptoms of the transgenic *SOD* mice, and in some cases, these results have been confirmed in human clinical trials. The only promising drug treatment so far appears to be creatine, which improves motor activity and survival in G93A *SOD* mice.

In 2003 an exciting development in ALS treatment was reported. Previously, nerve growth factors, such as ciliary neurotrophic factors and insulin-like growth factor (IGF-1), showed little effect when tested in clinical trials. These trials involved injecting the growth factors under the skin or directly into spinal fluid, but it was not clear whether the factors were reaching the relevant motor neurons. In the August 8, 2003 issue of *Science*, B. K. Kaspar and colleagues reported a significant increase in survival of the G93A transgenic *SOD* mouse by delivering the *IGF-1* gene to neurons via gene therapy.

The researchers cloned the *IGF-1* gene into the DNA of a specially engineered adeno-associated virus (AAV). This virus had been genetically altered so that it could not reproduce; however, it retained an important characteristic of the parent virus—the ability to be taken up by nerve endings in the muscle and to be transported through the nerve axon into the motor neuron cell bodies in the spinal cord (a process called retrograde transfer). After the virus arrives at the motor neuron cell body, it remains intact and its DNA does not integrate into the neuronal genome. After retrograde transport, the AAV–IGF vector directed the transcription, translation, and secretion of biologically active IGF-1 (Figure 21–29).

In one experiment, Kaspar and colleagues injected the AAV–IGF viruses, or control AAV, into the muscles of G93A *SOD* mice at 60 days of age (prior to development of ALS-like symptoms). At about 90 days, the control mice developed the disease; however, the AAV–IGF treated mice remained disease-free until day 120. The maximum lifespan of the control animals was 140 days, whereas that of the AAV–IGF-treated animals was 265 days. Most significantly, the benefits also occurred in G93A *SOD* mice who were already symptomatic. The results of this gene therapy study in transgenic mice are promising for human ALS patients. Clinical trials of this gene therapy method are now being designed.

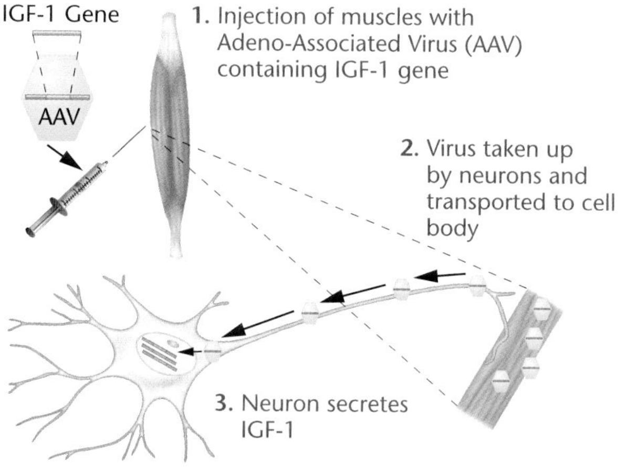

FIGURE 21–29 Gene therapy technique in the transgenic *SOD* mouse.

Chapter Summary

1. Geneticists dissect gene function in model organisms that display the biological process to be studied. These model organisms are chosen because they produce abundant progeny in a short period of time, are easily mutagenized, and can be subject to controlled crosses.
2. It is possible to use model organisms such as yeast or *Drosophila* for genetic studies because many of the genes and molecular processes governing biological functions are shared across evolution. Many gene sequences are conserved from yeast to higher vertebrates, and these genes often carry out similar functions in a wide range of organisms.
3. Forward genetic analysis begins with the isolation of mutants that show differences in phenotype for the process of interest. Mutant isolation is followed by defining gene pathways, cloning of the gene, and creating more mutants to understand the biological pathway. In this way, mutants define the normal function of the gene.
4. In forward genetic screens, the goal is to disrupt one gene at a time, and to disrupt each of the genes involved in a specific genetic pathway. In this way, each relevant gene will be mutated at least once in a population of mutagenized organisms. In order to create enough mutations to disrupt every gene that is involved in a genetic pathway, geneticists must subject the model organism to mutagenesis, usually by treating the organisms with chemicals or radiation.
5. In order to extend a forward genetic analysis, geneticists may mutagenize the first mutant and screen for second mutations that suppress or enhance the original mutant phenotype.

6. Reverse genetic analysis begins with a cloned wild-type gene or purified protein and progresses to site-directed mutagenesis and phenotypic analysis—the opposite order from that used in forward genetics.
7. Once a gene is cloned—by any method, including selection from a genome sequencing project or from functional complementation—it can be studied using a wide array of genetic and biochemical assays. These include computerized homology searches, gene expression assays, gene targeting techniques, and site-directed mutagenesis.
8. RNAi techniques allow geneticists to create gene knockouts at the expression level, in the absence of any genetic mutation to the gene to be disrupted.
9. DNA microarrays and other high-throughput functional genomics techniques provide insight into the interactions of thousands of gene products simultaneously.

INSIGHTS AND SOLUTIONS

1. You have isolated eight stains of *Drosophila*, each bearing a recessive lethal mutation somewhere on chromosome 3. You have designated these strains as el^1, el^2, el^3, etc., as the mutations cause embryonic lethality (*el*) accompanied by missing thoracic segments. Each strain is maintained in balanced stocks. You want to determine whether these mutations are allelic or occur in separate genes on chromosome 3. You do this by performing complementation tests, by crossing all combinations of stocks and examining the embryos for the presence of 25 percent lethality with segmentation defects. You obtain the following data. The genotypes crossed are indicated in the top row and left column. Viable progeny are indicated by a (+) and 25 percent embryonic lethality with segmentation defects is indicated by a (−).

	el^1	el^2	el^3	el^4	el^5	el^6	el^7	el^8
el^1	−	−	−	+	+	−	−	+
el^2		−	−	+	+	−	−	+
el^3			−	+	+	−	−	+
el^4				−	−	+	+	+
el^5					−	+	+	+
el^6						−	−	+
el^7							−	+
el^8								−

How many genes are represented by these mutations? Which complementation group does each mutant fall into?

Solution: One way to analyze complementation data is to systematically examine each cross. Obviously, each mutant when crossed with itself leads to the lethal segmentation phenotype. However, el^1, when crossed to el^2, el^3, el^6 and el^7, also leads to the lethal segmentation phenotype. Therefore, all of these mutants (el^1, el^2, el^3, el^6, and el^7) belong to one complementation group. That leaves el^4, el^5 and el^8 to be analyzed. When the el^4 and el^5 mutants are crossed, they produce the mutant embryonic lethal with segmentation defects phenotype. Therefore, these two mutants (el^4 and el^5) belong to one complementation group, and this complementation group is distinct from the first one. That leaves el^8, which complements all the other mutants and is therefore in a complementation group of its own. Hence, there are three complementation groups represented in these data.

2. In forward genetic dissection, the goal is to conduct a saturation genetic screen. In this way, each gene that might affect a phenotype should be mutated at least once in a population of mutagenized organisms. However, some genes may not appear to suffer mutations, even in a saturation screen. Can you suggest two reasons why a gene might never be revealed in a genetic screen?

Solution: The principle behind mutagenesis screens is that mutations in relevant genes will affect a detectable phenotype. If the phenotype cannot be detected, a mutation may be missed. For example, several genes may independently contribute to the same phenotype. This is known as gene redundancy. Mutating one of these genes will not affect the phenotype; therefore, the gene will not be revealed by a forward genetic screen. Another reason that a mutation may be missed is that most or all mutations in the gene lead to dominant lethality. In haploid organisms such as yeast, the mutagenized organisms would die and may not be screened for a phenotype. Similarly, in diploid organisms such as *Drosophila*, any F_1 progeny of the mutagenized flies that inherit the dominant lethal mutated gene would die and would not be detected, unless the lethality was accompanied by an observable phenotype prior to death.

3. You have isolated a temperature-sensitive mutant strain of *S. cerevisiae* that is null for the *CDC53* gene. This gene functions during the yeast cell cycle, allowing cells to progress from mitosis into the G1 phase of the cell cycle. At the restrictive temperature, $cdc53^{ts}$ cells arrest in mid-mitosis, with a characteristic phenotype.

(a) What would be the phenotype of haploid $cdc53^{ts}$ cells after 36 hours growth at the restrictive temperature?

(b) How would you use this mutant strain of yeast to identify genes whose products interact with the CDC53 protein, allowing CDC53 to function properly at the M/G1 transition?

(c) How would you use this mutant strain to clone the *CDC53* gene from human cells?

Solution (a): To answer this question, you need to know the phenotypes of *S. cerevisiae* cells during the cell cycle (Figure 21–23). Cells that cannot progress from mitosis into G1 remain joined, although the replicated DNA may migrate into the daughter cell. Therefore, at the restrictive temperature, all the cells in the $cdc53^{ts}$ culture will have this phenotype.

Solution (b): One way to do this is to undertake a suppressor screen. In a suppressor screen, the $cdc53^{ts}$ strain is subjected to a second round of mutagenesis, and cells are screened for a normal cell cycle at the restrictive temperature. These suppressor mutants would then be examined to determine whether the second mutation occurred in the same gene (a reversion) or at other sites (a second-site suppressor). Second-site-suppressor mutations may occur in genes whose products interact with the CDC53 protein. A second approach would be to use an antibody that recognizes CDC53 to immunoprecipitate CDC53 from extracts of wild-type yeast cells. Under the right conditions, any interacting proteins will immunoprecipitate along with the CDC53 protein. These proteins could be purified, their amino acid sequences determined, and this sequence used to undertake a reverse genetic screen to clone their genes.

Solution (c): It may be possible to use the *cdc53*^{ts} strain in a functional complementation assay. In some cases, a human gene will substitute for the loss of function in a yeast mutant, returning the yeast mutant to wild type. To do functional complementation, yeast cells are systematically transformed with clones from a human cDNA library. The transformed *cdc53*^{ts} cells are then screened for normal cell division at the restrictive temperature. If any of these cDNA clones revert the *cdc53*^{ts} strain to wild type, they may contain a functionally complementary human clone. The transformed, reverted yeast strain can be grown, the cDNA clone recovered from the yeast strain, and the identity of the cDNA insert determined by DNA sequencing.

PROBLEMS AND DISCUSSION QUESTIONS

1. What is a suppressor mutation and how do geneticists use these mutations to characterize gene function?
2. Which model organism—yeast, *Drosophila*, mouse, or human—would you choose to study the following genetic conditions and processes? Explain why.
   (a) Control of kidney development
   (b) Cancer
   (c) Cystic fibrosis
   (d) Purine metabolism
   (e) Immune function
   (f) Cell division
3. You are trying to create a knockout mouse bearing a homozygous null mutation in the proto-oncogene *c-myc*. You begin by tranforming ES cells from a normal mouse strain with your cloned DNA, which bears a neomycin-resistance gene inserted into the middle of the second exon of the murine *c-myc* gene. You now have 20 colonies of ES cells that grow in the presence of medium containing neomycin. Describe two methods you could use to verify that these ES cells are *c-myc* knockouts.
4. Describe the method of oligonucleotide-mediated site-directed mutagenesis.
5. Geneticists often employ suppressor mutations to define gene function. Suppressor mutants are generated in the same way as other mutants—by exposing the organism to the effects of a mutagen. In one experiment, *Drosophila* that have a recessive mutation in an eye color gene (pink eye) are subjected to EMS mutagenesis. The progeny of the mutagenized flies are screened for eye color. Twenty-three individual flies with wild-type (red) eyes appear in the mutagenesis screen.
   (a) How would you determine whether the suppressor mutations were due to EMS mutagenesis, or were spontaneous mutations?
   (b) How would you determine whether the suppressor mutation was a reversion of the first mutation or was due to a second-site mutation?
   (c) How would you locate the site of the suppressor mutation?
   (d) One of the suppressor mutations occurred in the same gene as the original mutation; however, it was not a simple reversion of the original mutation. What other possibilities could explain the suppression?
6. What are *Drosophila* balancer chromosomes, what features do they have, and what do geneticists use them for?
7. How do transposons generate mutations? What sorts of mutations do these mobile elements create?
8. How do geneticists use mouse coat color genes as part of the method to make knockout mice? In the absence of coat color markers, is there another way that the method could work?
9. Yeast with mutations in the *pro-1* gene are unable to synthesize the amino acid proline, which is required for normal growth. The wild-type strain of yeast is capable of synthesizing proline. How would you select suppressor mutants of this yeast strain?
10. In the *ClB* technique for detection of mutations in *Drosophila*,
    (a) What type of mutation may be detected?
    (b) What is the importance of the *l* gene?
    (c) Why is it necessary to have the crossover suppressor (*C*) on the X chromosome?
    (d) What is *C*?
11. Why are geneticists so enthusiastic about yeast as a model organism? What is the advantage of having a haploid life cycle?
12. In order to study the cell cycle, geneticists generate recessive temperature-sensitive mutants in *S. cerevisiae*. If you were designing an experiment to collect a number of temperature-sensitive recessive mutations in the cell division cycle in yeast, how would you select these mutants? How could you distinguish between *cdc* mutations and other types of temperature-sensitive mutations?
13. As part of your research into how the cell walls of budding yeast are formed, you purify a protein that is synthesized as the cells divide and is incorporated into the new cell wall as the cells grow. Your next task is to clone the gene.
    (a) What strategy would you use to clone the cell wall protein gene?
    (b) Once you have the gene cloned, what would you do to determine whether this protein was important for growth of the yeast cell wall?
14. You have selected an interesting-looking gene from the mouse genome sequence. You think that this gene might be involved in mouse development, as DNA in one part of the gene encodes a classic homeobox domain. You want to find out where and when this gene is expressed. What experiments would you do to determine the temporal and tissue specific expression of your gene?
15. Although modern high-throughput genomics methods such as gene expression arrays are generally lauded as technologies of the future, many scientists are skeptical about their value. Can you think of two limitations in using these DNA arrays? How might these limitations be overcome?
16. Describe the steps involved in *P* element-mediated transformation in *Drosophila*. What are the differences between this mode of transformation and the methods used to generate transgenic mice?
17. The Berkeley *Drosophila* Genome Project is creating a collection of *Drosophila* strains, each strain bearing a *P*-element insertion in a different gene. The goal is to generate null mutations in each essential gene of the *Drosophila* genome. How will this collection of stocks be useful for genetic analysis? Describe one experiment that would use the Berkeley collection.

18. One complication of mouse transgenics is that the transgene may integrate at random into the coding region, or the regulatory region, of an endogenous mouse gene. What might be the consequences of such random integrations? How might this complicate genetic analysis of the transgene?
19. Is it possible to undertake a forward genetic analysis in *Drosophila* without using a mutagen? If so, explain how this might be done.
20. Describe the difference between selecting for mutants and screening for mutants.
21. Over a period of several years, you have purified a small amount of protein from human blood, a protein that you suspect is involved in the blood clotting cascade. Because there is so little protein, you cannot generate antibodies to use as probes on a cDNA expression library. As an alternative, you obtain a bit of the amino acid sequence from the amino terminal end of your protein. The amino acid sequence is:

    Met-Pro-Lys-Glu-Ala-Ile-Gly-

    (a) How would you use this information to clone the gene?
    (b) Once you have cloned the gene, how could you verify that it is the gene that encodes your protein?
22. What is the difference between genetic complementation analysis and functional complementation?
23. What is RNAi and how does it work? Describe two ways that investigators can introduce RNAi molecules into cells.
24. A DNA microarray analysis reveals that 12 genes are expressed at higher than normal levels in a mouse liver tumor than in the normal liver. An examination of the DNA sequences of these 12 genes shows that they have all been sequenced and the sequences are available in the mouse genome database; however, there is no further information about them except their size, numbers of exons and introns, etc. You choose to study 2 of these 12 genes. What would be the first three steps you would take toward determining the identity and function of these two genes?

Extra-Spicy Problems

25. Your Ph.D. thesis project is to identify all the genes involved in a biosynthetic pathway in *Drosophila*. The pathway leads to the tanning and hardening of the adult cuticle after the fly emerges from the pupa.
    (a) What type of mutagen would you choose and why?
    (b) Would you expect that your mutations would be lethal? Why? How could you tell whether mutations in this pathway were lethal or not?
    (c) How would you screen for the mutants if you didn't know whether they would be lethal or not?
    (d) After you perform numerous mutational screens using thousands of flies, you obtain no mutations that affect the tanning and hardening of the adult cuticle. What possibilities could explain this? What would you do to test your hypotheses?
26. You are interested in the genetic basis of a human disease characterized by neural degeneration and abnormal protein deposition in the brain. A candidate gene has been cloned from humans, and it is possible that this gene is involved in causing the disease. However, it is difficult to analyze this gene in humans. You are working in a research lab that uses *S. cerevisiae* as a model organism for metabolic studies. Would you consider using yeast as a model organism to study this human disease? Explain why you would or would not, and what experiments you would do before making your final decision.
27. As part of a genomics project, you obtain the DNA sequence of a fragment of genomic DNA from a wild-type strain of yeast. After examining the sequence, you see that it contains a long ORF preceded by a promoter and has a polyadenylation sequence at the 3′ end of the ORF. Hence, this fragment of DNA likely contains a gene. You are interested in determining what gene it is and what function the gene performs. Your first step is to create a mutant yeast strain that is null for this gene. How could you use your genomic clone of this gene to create a mutant? Provide details of each step you would use in this procedure.
28. While analyzing some new genomic information from the mouse genome project, you notice a DNA sequence that would code for an interesting stretch of amino acids. This amino acid sequence forms a motif known as a helix-loop-helix domain. These domains are found in dozens of proteins in organisms as diverse as flies, *Arabidopsis,* and humans. How would you verify that this DNA sequence was part of a gene and was not simply a random piece of "junk" DNA? Assuming that the DNA sequence was part of a *bona fide* gene, how would you go about investigating the identity and potential function of this mouse gene?
29. A number of mouse models for human cystic fibrosis (CF) exist. Each mouse strain is transgenic, bearing a specific *CFTR* gene mutation. These mutations are the same as those seen in humans with variations of CF. These transgenic CF mice are being used to study the range of different phenotypes that characterize CF in humans. They are also used as models to test potential CF drugs. Unfortunately, most transgenic mouse CF strains do not show one of the most characteristic symptoms of human CF, that of lung congestion. Can you think of a reason why mouse CF strains do not display this symptom of human CF?
30. A famous research group wanted to find the tissue expression profile of a new homeobox (*HOX*) gene that they had discovered during a DNA sequence search of the *Drosophila* genome database. In their first set of experiments, they labeled a cDNA probe for their new *HOX* gene and hybridized the probe to three-day-old *Drosophila* larvae. The cDNA probe hybridized to the malpighian tubules and the nervous system. Delighted with their new data, and in the midst of writing up their scientific paper for publication, they decided to verify their result by performing an immunofluorescence assay, using the HOX protein antibody. To their horror, the antibody bound exclusively to larval testes. Explain why there might be such a difference between the *in situ* hybridization and immunofluorescence results. What should they do next?

Selected Readings

Adams, J.M. et al. 1985. The *c-myc* oncogene driven by immunoglobulin enhancers induces lymphoid malignancy in transgenic mice. *Nature* 318:533–38.

Capecchi, M.R. 1989. Altering the genome by homologous recombination. *Science* 244:1288–92.

Dykxhoorn, D.M. et al. 2003. Killing the messenger: short RNAs that silence gene expression. *Nature Reviews Molecular Cell Biology* 4:457–67.

Forsburg, S.L. 2001. The art and design of genetic screens: Yeast. *Nature Reviews Genetics* 2:659–68.

Fraser, A. 2004. RNA interference: Human genes hit the big screen. *Nature* 428:375–78.

Gitschier, J. et al. 1984. Characterization of the human factor VIII gene. *Nature* 312:326–30.

Hartwell, L.H. et al. 1974. Genetic control of the cell division cycle in yeast. *Science* 183:46–51.

Hartwell, L.H. et al. 1970. Genetic control of the cell division cycle in yeast, I. Detection of mutants. *Proc. Natl. Acad. Sci.* 66:352–59.

Kaspar, B.K. et al. 2003. Retrograde viral delivery of IGF-1 prolongs survival in a mouse ALS model. *Science* 301:839–42.

Kwon, B.S, et al. 1987. Isolation and sequence of a cDNA clone for human tyrosinase that maps at the mouse *c-albino* locus. *Proc. Natl. Acad. Sci. USA* 84:7473–77.

Leder, A. et al. 1986. Consequences of widespread deregulation of the *c-myc* gene in transgenic mice: Multiple neoplasms and normal development. *Cell* 45:485–95.

Nüsslein-Volhard, C. and Wieschaus, E. 1980. Mutations affecting segment number and polarity in *Drosophila*. *Nature* 287:795–801.

Oliver, S.G. 1997. From gene to screen with yeast. *Curr. Opin. Genetics and Development* 7:405–09.

Probst, F.J. et al. 1998. Correction of deafness in shaker-2 mice by an unconventional myosin in a BAC transgene. *Science* 280:1444–47.

Steinmetz, L.M. and Davis, R.W. 2004. Maximizing the potential of functional genomics. *Nature Reviews Genetics* 5:190–201.

St. Johnston, D. 2002. The art and design of genetic screens: *Drosophila melanogaster*. *Nature Reviews Genetics* 3:176–88.

Wieschaus, E. 1995. From molecular patterns to morphogenesis: The lessons from *Drosophila*. Nobel lecture, December 8, 1995. Published in Nobel E-museum: *http://www.nobel.se/medicine/laureates/1995/press.html*.

Applications and Ethics of Biotechnology

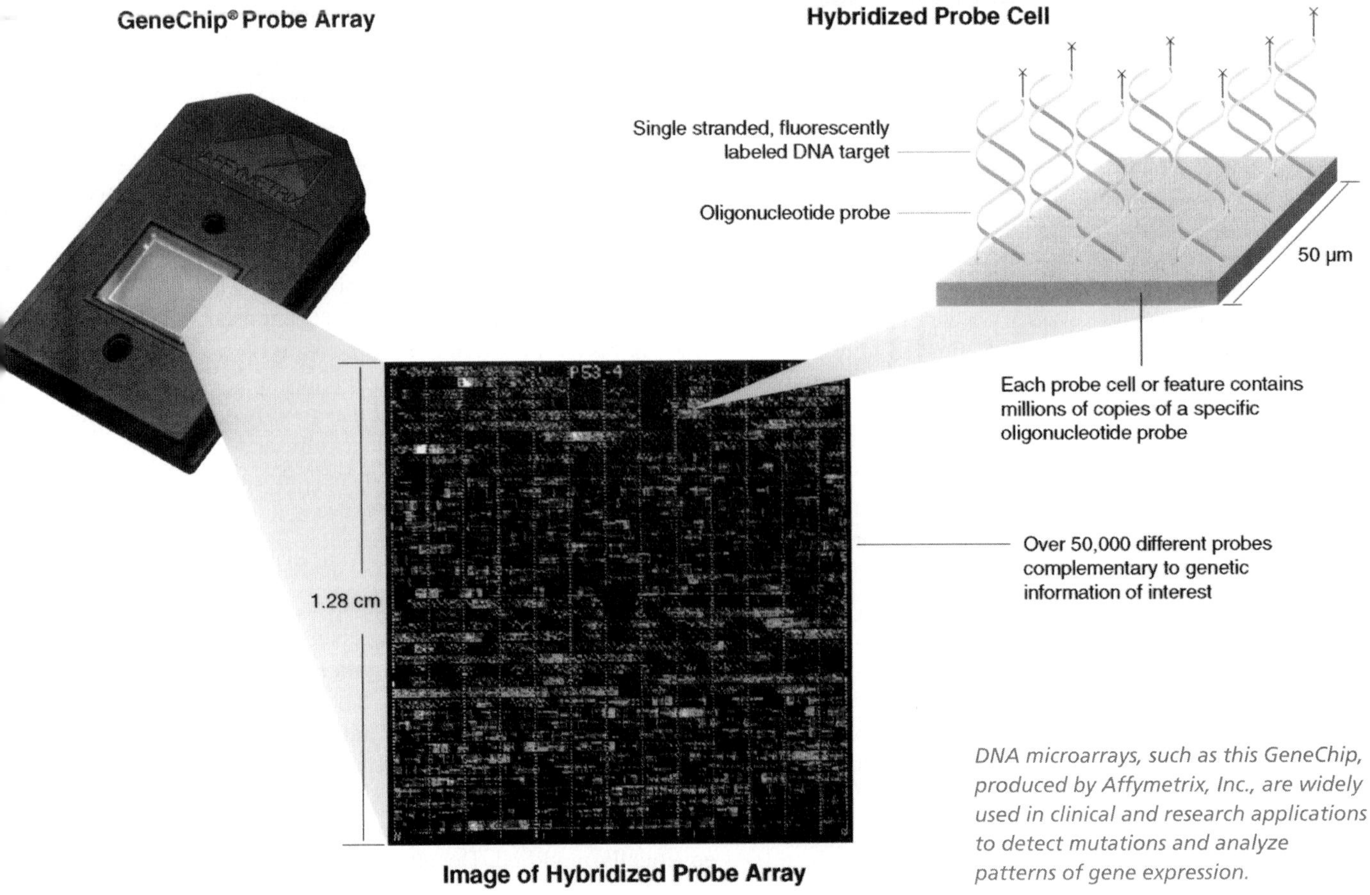

DNA microarrays, such as this GeneChip, produced by Affymetrix, Inc., are widely used in clinical and research applications to detect mutations and analyze patterns of gene expression.

Chapter Concepts

- Biotechnology has revolutionized agriculture through the development and introduction of genetically modified crop plants that are resistant to herbicides and pests, and have improved nutritional value.
- Genetically modified plants and animals can produce human proteins used to treat disease.
- Recombinant DNA technology is used to diagnose genetic disorders, detect heterozygotes, and scan the human genome to detect diseases that may not develop for decades. Use of this technology presents unresolved ethical problems.
- Gene therapy by transfer of cloned copies of functional alleles into target tissues is used to treat genetic disorders with mixed results. Future developments in vector technology are needed.
- Gene therapy and information derived from the Human Genome Project pose ethical problems that are being addressed by changes in public policy and laws.
- Major advances in mapping human genes and identifying gene products and functions have come about through use of recombinant DNA techniques.
- DNA fingerprinting can identify specific individuals in a large population and is widely used in forensics.

In 1982, the Food and Drug Administration approved the marketing of human insulin produced by genetically engineered bacterial cells. Previously, insulin was chemically extracted from the pancreas glands of cows and pigs obtained from slaughterhouses. The production of insulin by recombinant DNA technology ensures a steady supply of a purified gene product to treat disease, and was a milestone in the development of the biotechnology industry. The spread of recombinant DNA beyond the research laboratory, exemplified by the production of human insulin, marked the beginning of significant changes in our everyday lives. More than twenty years after the appearance of recombinant insulin, we are now surrounded by the products of biotechnology. Biotechnology is used to diagnose disease and produce medicines, industrial chemicals, and even milk and other foods found in the supermarket.

In Chapter 19, we described the methods used to create and analyze recombinant DNA molecules, and in Chapter 20 we examined the use of this technology in genomics. This chapter describes how these tools are being used in commercial and scientific applications in the biotechnology industry and the ethical problems posed by their use. We will consider a cross-section of applications that illustrate the power of biotechnology in (1) generating new varieties of plants and animals, (2) producing food, (3) diagnosing and treating diseases, (4) forensics, and (5) mapping human genes. Biotechnology is also being used to produce therapeutic proteins in genetically altered plants and animals. We begin by explaining how biotechnology has changed basic methods of food production and generated economic, social, and environmental controversies. Next, we examine the impact of recombinant DNA on medicine, from the prenatal analysis of genotypes, genome scanning, and the treatment of genetic disorders by gene therapy. In addition, we will consider the impact of biotechnology on forensics. Finally, we will discuss some of the ways biotechnology is being used in basic research, using human **gene mapping** as an example.

22.1 Biotechnology Has Revolutionized Agriculture

Although recombinant DNA techniques were originally developed to facilitate basic research on gene organization and regulation of expression, gene transfer methods are also the foundation of the biotechnology industry. Biotechnology is used to manufacture a wide range of products, including hormones, clotting factors, herbicide-resistant plants, enzymes for food production, and vaccines. In the last decade, this industry has grown into a multibillion-dollar segment of the economy. In this section, we will review some of the current uses of biotechnology in agriculture and the issues raised by this application of recombinant DNA technology.

Transgenic Crops and Herbicide Resistance

Damage from weed infestation destroys about 10 percent of crops worldwide. To combat this problem, farmers use more than 100 different herbicides at an annual cost of more than 10 billion dollars. Some of these herbicides also kill crop plants, while others remain in the environment or are carried by water runoff to contaminate local water supplies. Creating herbicide-resistant plants is one way to improve crop yields, while at the same time reducing the environmental impact of herbicides. To do this, vectors carrying genes for herbicide resistance are transferred to crop plants.

We will examine how resistance to one herbicide was transferred to crops such as soybeans and maize. The herbicide **glyphosate** is effective at very low concentrations, is not toxic to humans, and is rapidly degraded by soil microorganisms. Glyphosate kills plants by inhibiting the action of a chloroplast enzyme called EPSP synthase. This enzyme is important in amino acid biosynthesis in both bacteria and plants. Without the ability to synthesize vital amino acids, plants wither and die.

The first step in producing a glyphosate-resistant crop plant was the isolation and cloning of an EPSP synthase gene from a glyphosate-resistant strain of *E. coli*. This EPSP gene was cloned into a vector and placed adjacent to plant virus promoter sequences and upstream from plant transcription-termination sequences. The recombinant vector was transferred into the bacterium *Agrobacterium tumifaciens* (Figure 22–1). Plasmid-carrying bacteria were, in turn, used to infect cells in discs cut from plant leaves. Calluses formed after infection with *A. tumifaciens* were tested for their ability to grow in the presence of glyphosate. Glyphosate-resistant calluses were grown into transgenic plants and sprayed with glyphosate at concentrations four times higher than that needed to kill wild-type plants. Transgenic plants that overproduced the EPSP synthase grew and developed, while the control plants withered and died.

Glyphosate-resistant corn and soybeans are now on the market in the United States and other countries. Since their introduction in 1996, genetically engineered crops that tolerate herbicides and resist insect pests have been planted in several countries (Figure 22–2). In the United States, the three major genetically engineered crops are maize, soybeans, and cotton. In 2000, 26 percent of the maize, 68 percent of the soybeans, and 69 percent of the cotton planted in the United States were genetically modified. Developing countries have also engineered genetically modified crops that have been widely adopted by farmers. In China, for example, more than 30 percent of the cotton crop planted in 2000 was made up of strains genetically modified to resist insect pests.

Nutritional Enhancement of Crop Plants

In the last 50 to 100 years, genetic improvement of crop plants through the traditional methods of artificial selection and genetic crosses has resulted in dramatic improvements in productivity and nutritional enhancement. For example, maize yields have increased fourfold over the last 60 years, and more than half of this increase is due to genetic improvement by artificial selection. Broccoli contains gluconsinolates, compounds thought to have a role in cancer prevention by activation of the anticancer enzyme quinolone reductase. New hybrid varieties produced by conventional crosses have a hundredfold increase in the ability to induce enzyme levels in mouse cells.

Gene transfer by recombinant DNA techniques offers a new way to enhance the nutritional value of plants. Many crop plants

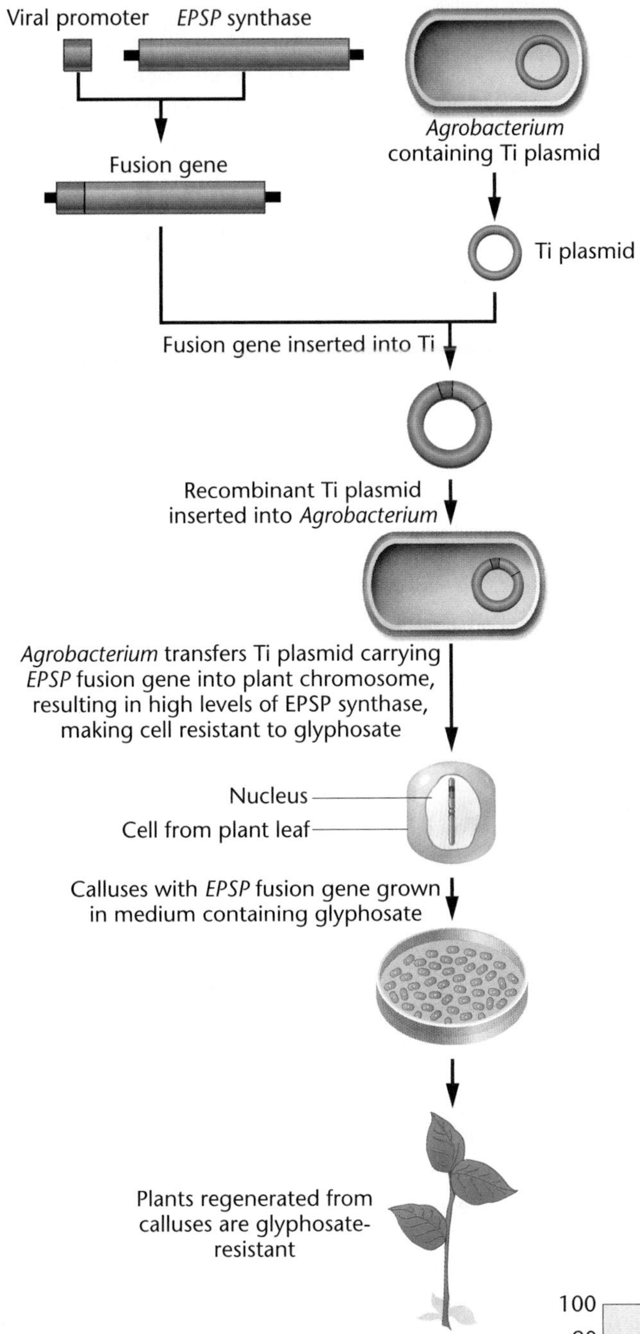

FIGURE 22–1 In gene transfer of glyphosate resistance, the *EPSP* synthase gene is fused to a promoter from the cauliflower mosaic virus. This fusion gene is then transferred to a Ti plasmid vector, and the recombinant vector is inserted into an *Agrobacterium* host. *Agrobacterium* infection of cultured plant cells transfers the *EPSP* synthase fusion gene into a plant-cell chromosome. Cells that acquire the gene are able to synthesize large quantities of EPSP synthase, making them resistant to the herbicide glyphosate. Resistant cells are selected by growth in herbicide-containing medium. Plants regenerated from these cells are herbicide-resistant.

are deficient in some of the nutrients required in the human diet, and biotechnology is being used to produce crops that meet these dietary requirements. One major example of this is the production of "golden rice" with enhanced levels of β-carotene, a precursor to vitamin A (Figure 22–3). In this case, three genes encoding enzymes in the biosynthetic pathway leading to carotenoid synthesis were transferred to the rice genome using methods of recombinant DNA technology. Two of these genes came from the daffodil, and one from a bacterium. Vitamin A deficiency is prevalent in many areas of Asia and Africa, and more than 500,000 children a year become permanently blind as a result of this deficiency.

Golden rice is now being field tested and should be available for planting by farmers in 2006. Other work is directed at enhancing the levels of key fatty acids, antioxidants, and other vitamins and minerals already present in crop plants. These efforts are directed at addressing the dietary lack of nutrients affecting more than 40 percent of the world's population. More important, the idea that crops can be grown for health as well as food is changing our view of agriculture. As discussed later, this idea extends to the production of plant-based human proteins, antibodies, and vaccines.

Concerns about Genetically Modified Organisms

Most genetically modified food products contain an introduced gene encoding a protein that confers a desired trait (herbicide resistance, insect resistance, etc.). Much of the concern over genetically modified plants centers on issues of safety and environmental consequences. Are genetically modified plants containing a new protein safe to eat? In general, if the proteins are not toxic or allergenic and do not have any other negative physiological effects, they are not considered to be a significant hazard to health. In the case of herbicide-resistant EPSP-containing food plants, the protein is readily degraded in digestive fluids, is nontoxic to mice at doses thousands of times

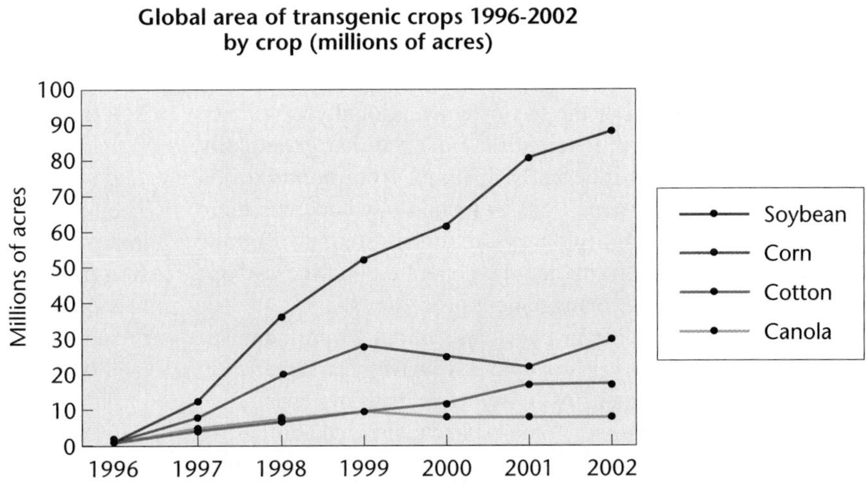

FIGURE 22–2 Since their introduction in 1996, genetically modified soybean, corn, and cotton crops have been planted on millions of acres annually in several countries. The United States accounts for two thirds of the world production of genetically modified crops.

FIGURE 22–3 A photograph of golden rice, a strain genetically modified to produce β-carotene, a precursor to vitamin A. Many children in countries where rice is a dietary staple have diets deficient in vitamin A and develop blindness.

higher than any potential human exposure, and has no amino acid sequence similarity to known protein toxins and allergens. Regulatory oversight by appropriate government agencies using standardized methods for the evaluation of proteins in genetically modified foods is being developed in the United States and in Europe. In Europe, labeling of food containing genetically modified ingredients is mandatory, but in the United States such labeling is not required at the present time.

What about the environmental risks associated with genetically modified plants? Environmental risks include gene transfer by cross breeding with wild plants, toxicity, and invasiveness of the modified plant, resulting in loss of natural species (loss of biodiversity). In fact, these problems are the same problems facing the use of conventional crop plants, and there is no current scientific evidence that genetically modified crops are inherently different from nonmodified crops. However, as gene transfer technology becomes more sophisticated and multiple traits are transferred, novel traits or combinations of traits may be generated and these crops may require specific management procedures.

Agriculturally important plants and animals were domesticated nearly 10,000 years ago, and we have been genetically modifying these organisms by selective breeding ever since, producing the diversity of domesticated plants and animals we have today. Biotechnology has changed the rate at which new plants and animals can be developed and, by enabling gene transfer between species, has altered the types of changes that can be made. However, unlike selective-breeding programs, which have introduced thousands of genetically altered strains, biotechnology has generated concerns about the release of genetically modified organisms into the environment and about the safety of eating such products. If biotechnology is to achieve a new green revolution, these concerns need to be addressed through prudent research and education of the public.

22.2 Pharmaceutical Products Are Synthesized in Genetically Altered Organisms

The first human gene product manufactured by using recombinant DNA and licensed for therapeutic use was human insulin, which became available in 1982. Insulin is a protein hormone that regulates glucose metabolism. Individuals who cannot produce insulin have diabetes, a disease that, in its more severe form (type I), affects more than 2 million individuals in the United States.

Clusters of cells embedded in the pancreas synthesize a precursor peptide known as preproinsulin. As this polypeptide is secreted from the cell, amino acids are cleaved from the end and the middle of the chain. These cleavages produce the mature insulin molecule, which contains two polypeptide chains (the *A* and *B* chains), joined by disulfide bonds. Insulin circulating in the blood regulates the uptake of glucose. Diabetics are unable to produce sufficient insulin and must take insulin as a medication.

Insulin Production in Bacteria

Although synthetic human insulin is now produced by another process, a look at the original method is instructive, as it shows both the promise and the difficulty of applying recombinant DNA technology. A functional insulin molecule contains two polypeptide chains, *A* and *B*. The *A* subunit has 21 amino acids, and the *B* subunit has 30. In the original bacterial process, synthetic genes for the *A* and *B* subunits were constructed by oligonucleotide synthesis (63 nucleotides for the *A* polypeptide and 90 nucleotides for the *B* polypeptide). Each synthetic oligonucleotides was inserted into a vector adjacent to the *lacZ* a gene encoding the bacterial form of the enzyme, β-galactosidase. When transferred to a bacterial host, the *lacZ* gene and the adjacent synthetic oligonucleotide were transcribed and translated as a unit. The product, a **fusion polypeptide**, contained the amino acid sequence for β-galactosidase attached to the amino acid sequence for one of the insulin subunits (Figure 22–4). The fusion proteins were purified from bacterial extracts and treated with cyanogen bromide, a chemical that cleaves the fusion protein from the β-galactosidase. Each insulin subunit was produced and purified by this process. When they were mixed, the two subunits spontaneously united, forming an intact, active insulin molecule. The purified insulin was then packaged for use by diabetics.

Several genetically engineered proteins for therapeutic use have been produced by similar methods (Table 22.1). In most cases, cloning a human gene into a plasmid and inserting the recombinant vector into a bacterial host produces the proteins. After ensuring that the transferred gene is expressed, large quantities of the transformed bacteria are produced, and the human protein is recovered and purified.

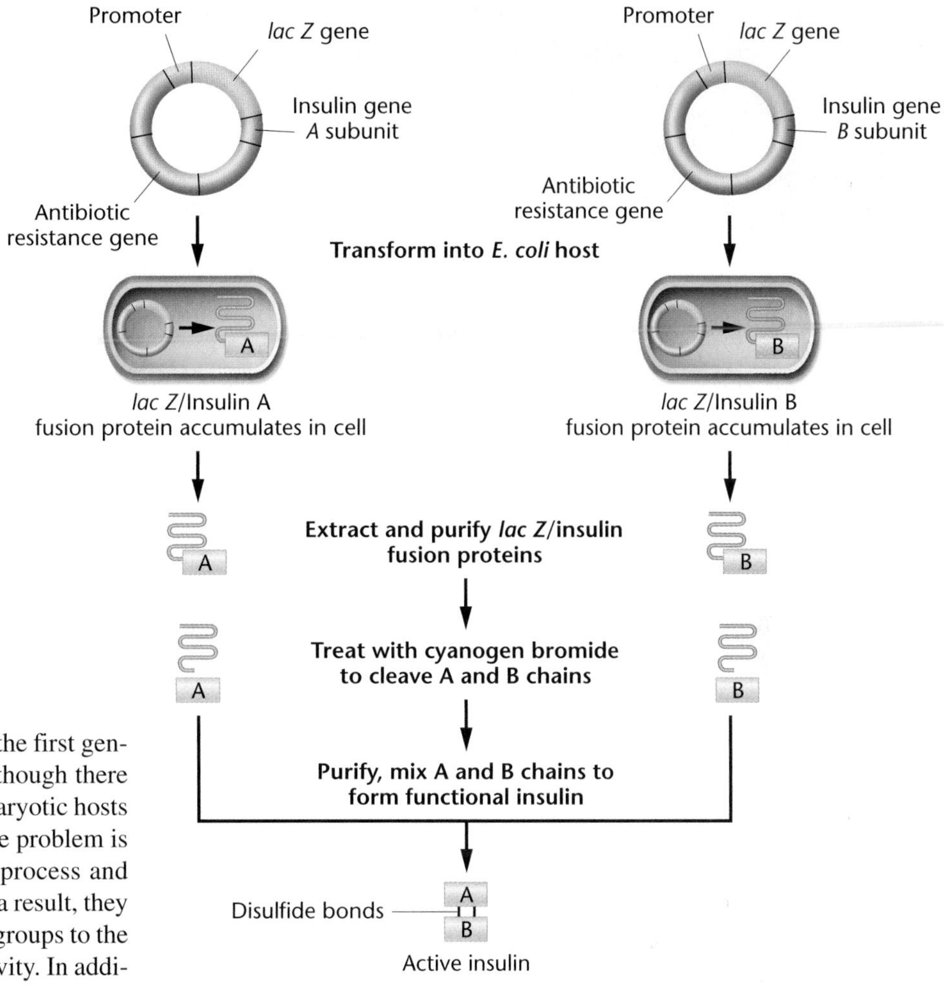

FIGURE 22–4 To synthesize recombinant human insulin, synthetic oligonucleotides encoding the insulin *A* and *B* chains were inserted at the tail end of a cloned *E. coli lacZ* gene. The recombinant plasmids were transferred to *E. coli* hosts, where the β-gal/insulin fusion protein was synthesized and accumulated in the host cells. Fusion proteins were then extracted from the host cells and purified. Insulin chains were released from the β-galactosidase by treatment with cyanogen bromide. The insulin subunits were purified and mixed to produce a functional insulin molecule.

Transgenic Animal Hosts and Pharmaceutical Products

Bacterial hosts were used to produce the first generation of therapeutic proteins, even though there are some disadvantages in using prokaryotic hosts to synthesize eukaryotic proteins. One problem is that bacterial cells cannot correctly process and modify many eukaryotic proteins. As a result, they cannot add the sugars and phosphate groups to the protein needed for full biological activity. In addition, eukaryotic proteins produced in prokaryotic cells often don't fold into the proper three-dimensional configuration and, as a result, are inactive. To overcome these difficulties and increase yields, second-generation methods use eukaryotic hosts. Rather than being produced by host cells grown in tissue culture, human proteins such as α_1-antitrypsin are produced in the milk of livestock.

A deficiency of the enzyme α_1-antitrypsin is associated with the heritable form of emphysema, a progressive and fatal respiratory disorder common among people of European ancestry. To produce α_1-antitrypsin for use in treating this disease, the human gene was cloned into a vector at a site adjacent to a sheep promoter sequence that activates transcription in milk-producing cells. Genes placed next to this promoter are expressed only in mammary tissue. This fusion gene was microinjected into sheep zygotes fertilized *in vitro* (Figure 22–5). The fertilized zygotes were transferred to surrogate mothers. The resulting transgenic sheep developed normally and after mating, produced milk containing high concentrations of functional human α_1-antitrypsin. This human protein is present in concentrations of up to 35 grams per liter of milk, and can be easily extracted and purified. A small herd of lactating transgenic sheep can easily provide an adequate supply of this protein. Herds of other transgenic animals acting as biofactories are becoming part of the pharmaceutical industry. In fact, the famous sheep, Dolly, was cloned to facilitate the establishment of a flock of sheep that would consistently produce high levels of human proteins.

Human proteins produced in transgenic animals undergo clinical testing as a first step in the therapeutic use of recombinant human proteins. A recombinant human enzyme, α-glucosidase,

TABLE 22.1 GENETICALLY ENGINEERED PHARMACEUTICAL PRODUCTS NOW AVAILABLE OR IN CLINICAL TRIALS

Gene Product	Condition Treated
Atrial natriuretic factor	Heart failure, hypertension
Epidermal growth factor	Burns, skin transplants
Erythropoietin	Anemia
Factor VIII	Hemophilia
Gamma Interferon	Cancer
Granulocyte colony-stimulating factor	Cancer
Hepatitis B vaccine	Hepatitis
Human growth hormone	Dwarfism
Insulin	Diabetes
Interleukin-2	Cancer
Superoxide dismutase	Transplants
Tissue plasminogen activator	Heart attack

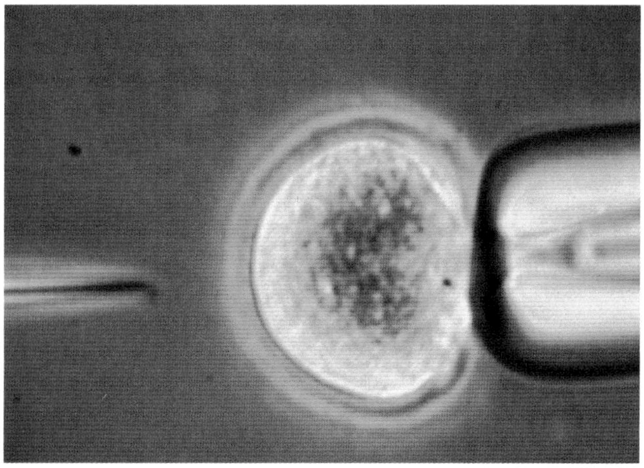

FIGURE 22–5 A micropipette is used to transfer cloned genes into the nucleus of a mammalian zygote. The injected zygote will then be transferred to the uterus of a surrogate mother for development.

produced in rabbit milk, is now in clinical tests for treating children with **Pompe disease**. This progressive and fatal metabolic disorder is caused by a lack of the enzyme α-glucosidase and is inherited as an autosomal recessive condition. In early onset Pompe disease, infants have poor muscle tone, and may never sit or stand. Most die before 2 years of age from respiratory or cardiac complications. In one of the first tests, the recombinant enzyme was given to affected children weekly, with no significant side effects. All of the children showed normal enzyme activity in the tissues analyzed, and all showed improvements in their symptoms. If large-scale trials are successful, recombinant α-glucosidase from transgenic animals will become the preferred method of treatment for this disorder.

Transgenic Plants and Edible Vaccines

One of the most beneficial applications of biotechnology may be the production of vaccines. Vaccines stimulate the immune system to produce antibodies against disease-causing organisms and thereby confer immunity against specific diseases. Two types of vaccines are commonly used; inactivated vaccines, prepared from killed samples of the infectious virus or bacteria; and attenuated vaccines, which are live viruses or bacteria that can no longer reproduce, but can cause a usually mild form of the disease. Biotechnology is being used to produce a new type of vaccine called a subunit vaccine, which would consist of one or more surface proteins of the virus or bacterium. This surface protein is an antigen that stimulates the immune system to make antibodies that act against the virus or bacterium. One of the first subunit vaccines was made against hepatitis B, a virus that causes liver damage and cancer. The gene for the hepatitis B surface protein was cloned into a yeast-expression vector and produced in yeast host cells. The protein is extracted and purified from the host cells and packaged for use as a vaccine.

Subunit vaccines produced by biotechnology provide a source of pure vaccine manufactured under controlled conditions and are widely used. However, vaccination programs in developing countries are faced with serious problems of manufacturing, transportation, and storage. Most vaccines need refrigeration, and must be injected under sterile conditions. In the rural areas of many countries, refrigeration and facilities for sterilization of instruments are not available. To overcome these problems, biotechnology is used to develop inexpensive vaccines synthesized in edible food plants. Such vaccines are inexpensive to produce, do not require refrigeration, and do not have to be given under sterile conditions by trained medical personnel.

As a model system, the gene encoding antigenic subunit of hepatitis B vaccine has been transferred to the tobacco plant and expressed in its leaves (Figure 22–6). For use as a source of vaccine, the gene would be inserted into food plants such as grains or vegetables. Other edible vaccines are in clinical trials. A vaccine against a bacterium that causes diarrhea has been produced in genetically engineered potatoes (Figure 22–7) and used to successfully vaccinate human volunteers who ate small quantities (50–100 g) of the potatoes. (See the Genetics, Technology, and Society essay in Chapter 6.) In trials with another vaccine, transgenic spinach expressing rabies viral antigens was fed to volunteers. Eight of 14 volunteers showed significant increases in rabies-specific antibodies. Tests using vaccine-producing bananas are currently under way. These proof-of-principle trials establish that edible vaccines are feasible. The success of these tests means that genetically engineered edible plants will soon be available to vaccinate infants, children, and adults against many infectious diseases.

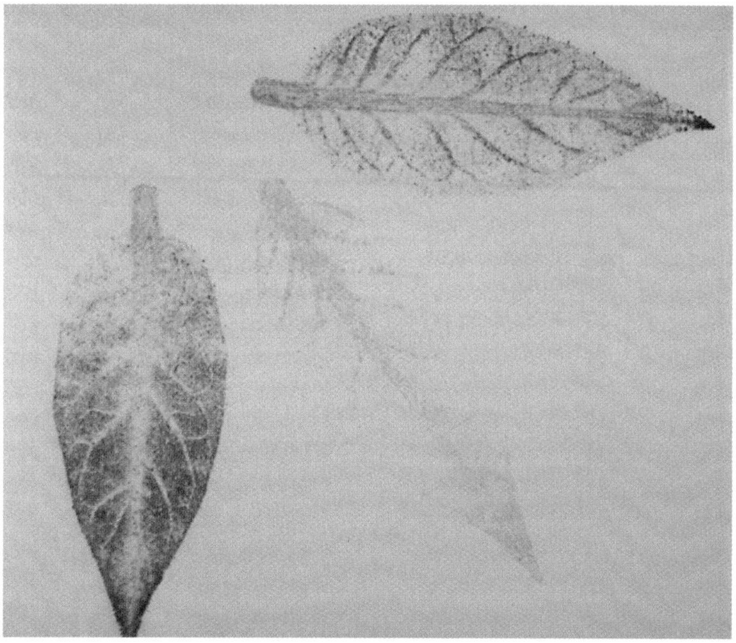

FIGURE 22–6 Transgenic tobacco plants carrying an antigenic subunit of the hepatitis B virus were generated. Leaves from transgenic plants were treated with antibodies against hepatitis B antigen, and stained to show that the plants produced the antigen. The central leaf in the photograph is a leaf from a normal tobacco plant and is unstained. Tobacco plants are used in these experiments because they are a common experimental organism in plant molecular biology. For use as a vaccine, the antigenic subunit would be transferred to an edible crop plant.

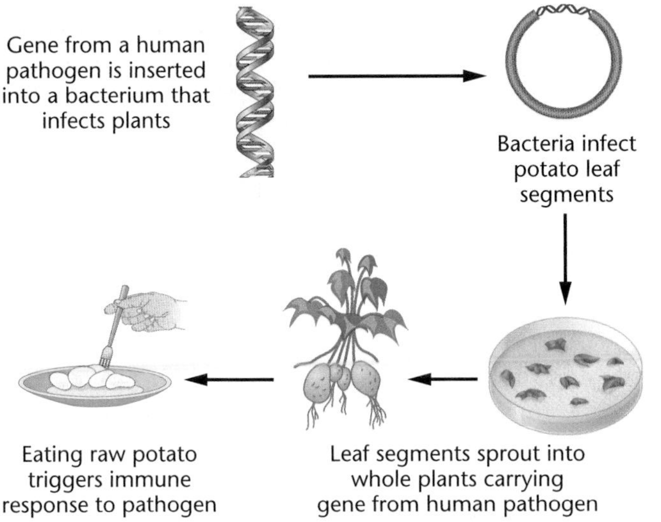
Gene from a human pathogen is inserted into a bacterium that infects plants
Bacteria infect potato leaf segments
Leaf segments sprout into whole plants carrying gene from human pathogen
Eating raw potato triggers immune response to pathogen

FIGURE 22–7 In making an edible vaccine, a gene from a pathogen (a disease-causing agent, such as a virus or bacterium) is tranferred into a vector and then into a bacterial host that infects plant cells. Infection of potato plant leaves transfers the vector and the pathogen's gene into the nuclei of potato leaf cells. The leaf segments are grown into mature potato plants that express the pathogenic gene. Eating the raw potato triggers an immune response to the protein encoded by the pathogen's gene, conferring immunity to infection by this pathogen.

Now solve this

Problem 22.4 on page 571 asks you to interpret data on the development of an edible vaccine.

Hint: Antibody development by the smallest possible portion of a protein is important to ensure vaccine specificity.

22.3 Biotechnology Is Used to Diagnose and Screen Genetic Disorders

Many genetic disorders can be prenatally diagnosed using biotechnology. The most widely used methods of obtaining samples for prenatal diagnosis are amniocentesis and chorionic villus sampling (CVS). In amniocentesis, a needle is used to withdraw amniotic fluid from the uterus (Figure 22–8), and the fluid and the cells it contains are analyzed for chromosomal or genetic disorders. In CVS, a catheter is inserted into the uterus, and a small tissue sample of the fetal chorion is retrieved. Cytogenetic, biochemical, and recombinant DNA-based testing are then performed on the tissue.

When coupled with these sample-recovery methods, biotechnology is a highly sensitive and accurate tool for the prenatal detection of genetic disorders. In prenatal testing, the fetal genotype can be examined directly, rather than using the few tests available for normal or mutant gene products. This capability is particularly important because frequently, gene products cannot be detected before birth, even when tests are available. For example, defects in the adult β-globin protein cannot be detected prenatally because the β-globin gene is not expressed until a few days after birth.

Prenatal Diagnosis of Sickle-Cell Anemia

Sickle-cell anemia is an autosomal recessive condition common in people with family origins in areas of West Africa, the Mediterranean basin, and parts of the Middle East and India. Sickle-cell anemia is caused by a single amino acid substitution in the β-globin gene. This change is brought about by a single-nucleotide substitution that eliminates a cutting site for the restriction enzymes *Mst*II and *Cvn*I. As a result, the mutation alters the pattern of restriction fragments seen on Southern blots. These differences in restriction cutting sites are used to prenatally diagnose sickle-cell anemia and to establish the parental genotypes and the genotypes of other family members who may be heterozygous carriers of this condition. For prenatal diagnosis, fetal cells are obtained by amniocentesis or CVS; for analysis of family members, a blood sample is collected. DNA is extracted from the samples and digested with *Mst*II. This enzyme cuts three times in the region of the normal β-globin gene, producing two small DNA fragments. In the mutant sickle-cell allele, the middle *Mst*II

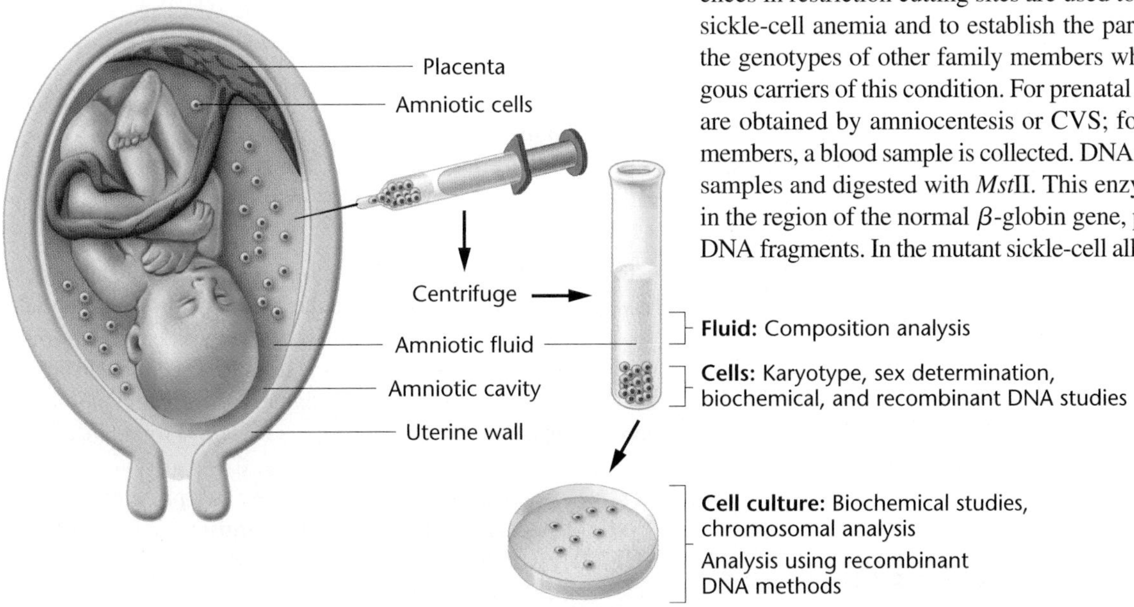
Placenta
Amniotic cells
Centrifuge
Amniotic fluid
Amniotic cavity
Uterine wall
Fluid: Composition analysis
Cells: Karyotype, sex determination, biochemical, and recombinant DNA studies
Cell culture: Biochemical studies, chromosomal analysis
Analysis using recombinant DNA methods

FIGURE 22–8 For amniocentesis, the position of the fetus is first determined by ultrasound, and a needle is then inserted through the abdominal and uterine walls to recover fluid and fetal cells for cytogenetic or biochemical analysis.

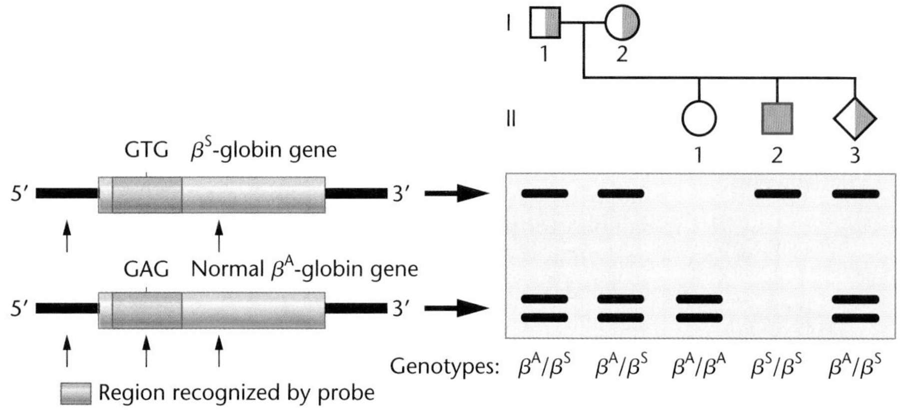

FIGURE 22–9 A Southern blot diagnosis of sickle-cell anemia, with arrows representing the location of restriction enzyme recognition sites. In the mutant β-globin allele (β^S), a point mutation (GAG→GTG) has destroyed a restriction enzyme cutting site, resulting in a single large fragment on a Southern blot. In the pedigree, the family has one unaffected homozygous normal daughter (II-1), an affected son (II-2), and an unaffected carrier fetus (II-3). The genotype of each family member can be read directly from the blot and is shown below each lane.

site is destroyed by the mutation, and one large restriction fragment is produced by *Mst*II digestion (Figure 22–9). The restriction-digested DNA fragments are separated by gel electrophoresis, transferred to a nylon membrane, and visualized by Southern blot hybridization.

Figure 22–9 shows the results of genetic testing for sickle-cell anemia in one family. The parents (I-1 and I-2) are both heterozygous carriers of the mutation. *Mst*II digestion of the parent's DNA produced a large band (the mutant allele) and two smaller bands (the normal allele) in each case. The parents' first child (II-1) is homozygous normal because she has only the two smaller bands. The second child (II-2) has sickle-cell anemia; he has only one large band and is homozygous for the mutant allele. The fetus (II-3) has a large band and two small bands and is therefore heterozygous for sickle-cell anemia. He or she will be unaffected, but will be a carrier.

Now solve this

Problem 22.24 on page 572 uses RFLP analysis to determine whether sisters in a family are carriers of hemophilia.

Hint: Differences in the number and location of restriction enzyme sites creates RFLPs that can be used to determine genotypes. In this problem, these differences are key factors in analyzing genotypes.

Only about 5–10 percent of all point mutations can be detected by restriction enzyme analysis. However, if a mutant gene has been well characterized and the mutated region has been sequenced, synthetic oligonucleotides can be used as probes to detect mutant alleles.

Single-Nucleotide Polymorphisms and Genetic Screening

Synthetic probes known as **allele-specific oligonucleotides (ASO)** can identify alleles that differ by as little as a single nucleotide. In contrast to restriction enzyme analysis, which is limited to cases for which a mutation changes a restriction site, ASOs detect single-nucleotide changes of all types, including those that do not affect restriction enzyme cutting sites. As a result, this method offers increased resolution and wider application. Under proper conditions, an ASO will hybridize only with its complementary sequence and not with other sequences, even those that vary by as little as a single nucleotide. A method using ASOs and PCR analysis is now available to screen for many disorders, including sickle-cell anemia.

In this procedure, DNA is extracted, denatured into single strands and used to amplify a region of the β-globin gene by PCR. A small amount of the amplified DNA is spotted onto filters, and each filter is hybridized to an ASO synthesized from a normal or mutant copy of the β-globin gene (Figure 22–10). After visualization, the genotype can be read directly from the filters. With an ASO for the normal sequence (Figure 22–10a), the homozygous normal (*AA*) genotype produces a dark spot (two copies of the normal allele), and the heterozygous genotype (*AS*) produces a light spot (one copy of the normal allele). The homozygous recessive sickle-cell genotype will not bind the probe, so no spot will be visible; with a probe used for the mutant allele (Figure 22–10b), the pattern is reversed. This rapid, inexpensive, and highly accurate technique is used to diagnose a wide range of genetic disorders caused by point mutations.

Now solve this

Problem 22.23 on page 572 asks whether DNA sequences from normal and sickle cell alleles of the β-globin gene will bind to a given ASO.

Hint: ASO analysis is done under conditions that allow only identical nucleotide sequences to hybridize to the ASO on the filter. Mismatches as small as a single nucleotide will prevent hybridization.

In cases where the nucleotide sequence of the normal allele is known and where the molecular nature of the mutant allele has been identified, ASOs are directly synthesized from normal and mutant copies of the gene and used to screen for disorders that involve deletions instead of single nucleotide mutations. In **cystic fibrosis (CF)**, a deletion called $\Delta 508$ is found in 70 percent of all mutant copies of the gene. CF is an autosomal recessive disorder associated with a defect in a protein called the **cystic fibrosis transmembrane conductance regulator (CFTR)**, which regulates chloride ion transport across the plasma membrane. To detect heterozygous carriers of the $\Delta 508$ mutation, allele-specific oligonucleotides are made by PCR from cloned samples of the normal allele and the mutant allele. DNA extracted from white blood cells of the individuals to be tested is spotted on a nylon filter and hybridized to each

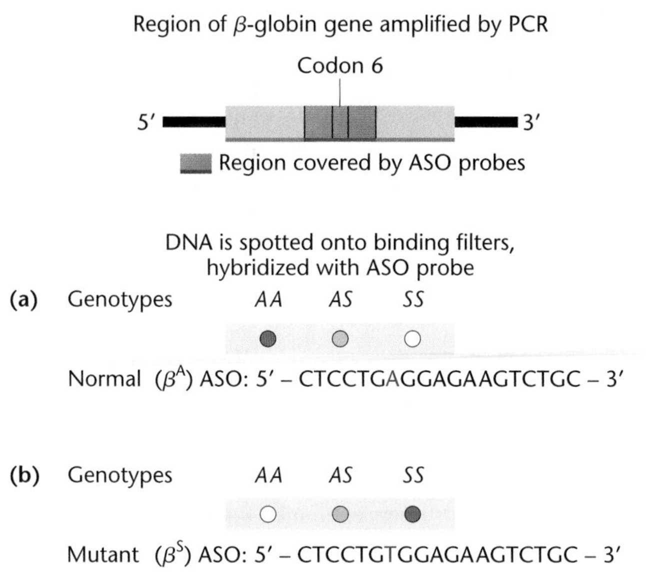

FIGURE 22–10 To determine a genotype with allele-specific oligonucleotides (ASOs), the β-globin gene is amplified by PCR, using DNA extracted from white blood cells. The amplified DNA is denatured and spotted onto strips of DNA-binding filters. Each strip is hybridized to a specific ASO and visualized on X-ray film after hybridization and exposure. If all three genotypes are hybridized to an ASO from the normal β-globin allele, the pattern in (a) will be observed: *AA*-homozygous individuals have normal hemoglobin that has two copies of the normal β-globin gene and will show heavy hybridization; *AS*-heterozygous individuals carry one normal β-globin allele and one mutant allele and will show weaker hybridization; *SS*-homozygous sickle-cell individuals carry no normal copy of the β-globin gene and will show no hybridization to the ASO probe for the normal β-globin allele. (b) The same genotypes hybridized to the probe for the sickle-cell β-globin allele will show the reverse pattern: no hybridization by the *AA* genotype, weak hybridization by the heterozygote (*AS*), and strong hybridization by the homozygous sickle-cell genotype (*SS*).

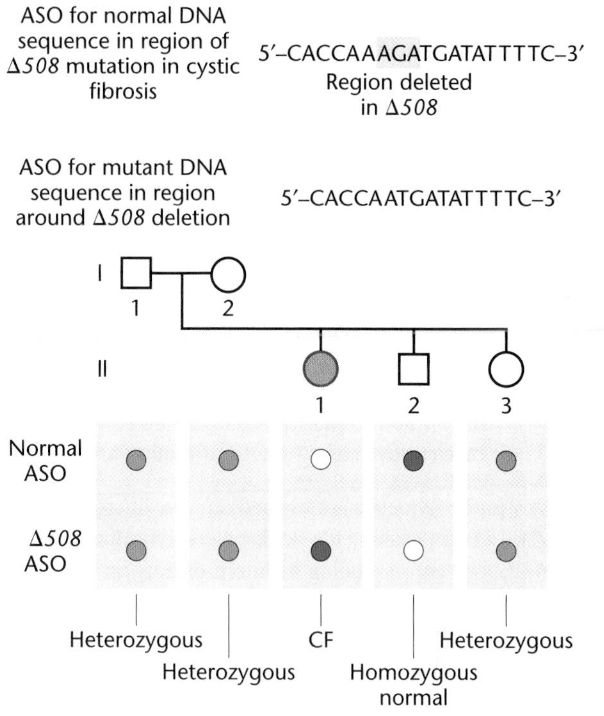

FIGURE 22–11 ASOs for the region spanning the most common mutation in CF (the Δ*508* allele), are prepared from cloned copies of the normal allele and a Δ*508* allele and then spotted on a DNA-binding membrane. In screening, the CF alleles carried by an individual are amplified by PCR, labeled, and hybridized to the membrane. The genotype of each family member can then be read directly from the filter. DNA from the parents (I-1 and I-2) hybridizes to both ASOs, indicating that they each carry one normal allele and one mutant allele and are therefore heterozygous. DNA from II-1 hybridizes only to the Δ*508* ASO, indicating that this family member is homozygous for the mutation and has cystic fibrosis. DNA from II-2 hybridizes only to the ASO from the normal CF allele, indicating that this individual carries two normal alleles. DNA from II-3 shows two hybridization spots and is thus heterozygous.

ASO (Figure 22–11). In affected individuals, only the ASO made from the mutant allele hybridizes; in heterozygotes, both ASOs hybridize; and in normal homozygotes, only the ASO from the normal allele hybridizes.

CF affects approximately 1 in 2000 individuals of northern European descent, and screening for CF can be used in these populations to detect carriers and counsel people about their genetic status with respect to CF. However, not all of the known mutations for this gene (more than 1000 mutations have been identified) can be screened, so a negative result does not eliminate someone as a heterozygous carrier—and it is likely that more CF mutations remain to be identified. Consequently, CF screening is not widespread, but will no doubt become commonplace when tests can cover 98–99 percent of all possible CF mutations.

DNA Microarrays

The use of allele-specific nucleotides has been coupled with the technology of the semiconductor industry to produce **DNA microarrays** (also called DNA chips) that can be used to test hundreds or thousands of genes in a single assay. Microarrays are prepared on small glass plates divided into very small areas called fields (small squares). Each field can be as small as half the width of a human hair. In each field, copies of a specific synthetic DNA probe about 20 nucleotides in length are attached to the glass (Figure 22–12). Microarray fields can carry different nucleotide combinations as well as normal and different mutant alleles of specific genes. Along a row of fields, the sequence of the synthetic probe differs by one nucleotide from field to field. Thus, a set of four fields (one for each nucleotide) is necessary to test the nucleotide content of a given position in a DNA molecule. The current generation of DNA chips can hold between 280,000 and 560,000 fields, but chips with several million fields are now being tested.

To use a microarray in genetic testing, DNA is extracted from an individual and selected genes are amplified by PCR. The PCR products are tagged with one or more fluorescent dyes, denatured into single strands, and pumped into the microarray. PCR fragments with a nucleotide sequence that exactly match a probe sequence on the microarray will bind, and those that do not perfectly

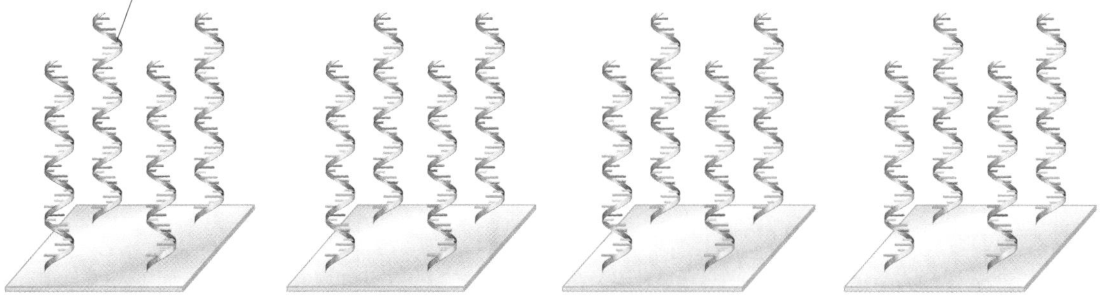

FIGURE 22–12 In making a DNA microarray, short, 15–30 base-pair, single-stranded DNA molecules of known sequence are attached to a glass substrate. Each cluster of identical molecules occupies an area known as a field on the microarray. Each field is about half the width of a human hair.

match any probe sequence are washed off. The microarray is then scanned by a laser, causing the fields where hybridization has occurred to fluoresce, producing a pattern of dots on the microarray (Figure 22–13). Linked software analyzes the pattern of hybridization, and the data can be presented in several forms.

DNA microarrays are used to scan for mutations in the *p53* gene, present in 60 percent of all cancers, and to screen for mutations in the *BRCA1* gene, which predispose women to breast cancer. In addition to testing for mutations in single genes, DNA chips can be made to screen thousands of genes thought to be involved in disorders exhibiting multifactorial inheritance or that arise following several mutational events. For example, microarrays are used to analyze gene expression patterns in cancer diagnosis and in complex disorders such as bipolar illness and schizophrenia.

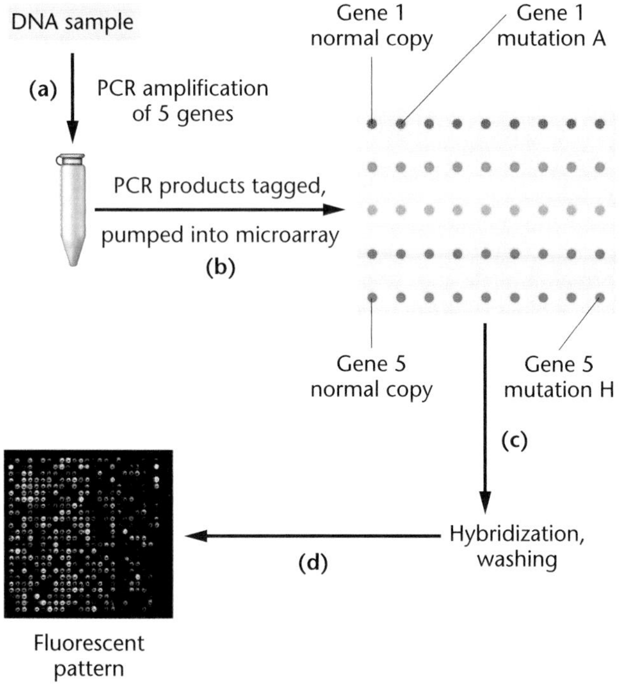

FIGURE 22–13 In a DNA microarray, DNA extracted from a blood sample is amplified by PCR. (a) In this example, primers for five genes are used. (b) The microarray contains single-stranded probes for the normal allele of each of the genes (column 1) and eight mutant alleles for each of the five genes 1 row = 1 gene. (c) The single-stranded PCR products are tagged with fluorescent probes and pumped into the microarray. (d) The resulting hybridization is revealed by the pattern and color of the spots on the microarray. The mutations in each gene have been arbitrarily labeled as A through H.

Drug Development

Many forms of cancer show a distinctive pattern of gene expression that differs from that of normal cells and from other forms of cancer. In a typical microarray analysis, mRNA is isolated from normal cells and cancer cells arising from the same cell type. The mRNA populations represent all the genes expressed in the normal cells and the cancer cells. Using reverse transcriptase, the mRNA from each cell type is converted to cDNA and labeled with a fluorescent dye. cDNA from normal cells is linked to a fluorescent dye (e.g., green); the cDNA from the cancer cells is tagged with a second fluorescent dye (e.g., red). The labeled cDNAs are mixed together and pumped through a DNA microarray carrying probes representing genes expressed in both normal and cancerous cells. The cDNAs bind to complementary single-stranded probe DNAs on the chip, but not to other DNAs. After washing off nonbinding cDNAs, a laser scans the microarray, and the pattern of hybridization shows up as a series of colored dots, with each dot corresponding to a field on the microarray.

Green dots on the microarray represent genes expressed only in normal cells. Red dots represent genes expressed only in cancer cells. Probes that bind cDNAs from both cell types equally will produce yellow dots (a combination of red and green). In practice, some probes bind to cDNA from one cell type, some bind to others, and some bind to both, producing dots with colors that range from red to green (Figure 22–14). Colors other than red or green represent intermediate amounts of binding by the two cDNAs and indicate different levels of gene expression in the normal cell and in the cancer cell.

Genes expressed only in cancer cells represent targets for the development of drugs that interfere with the action of the encoded proteins as well as the development of drugs that inhibit transcription of specific cancer genes in these cells. Because different forms of the same cancer often have different and characteristic patterns of gene expression, microarray analysis can also be used to plan an individualized program of cancer therapy.

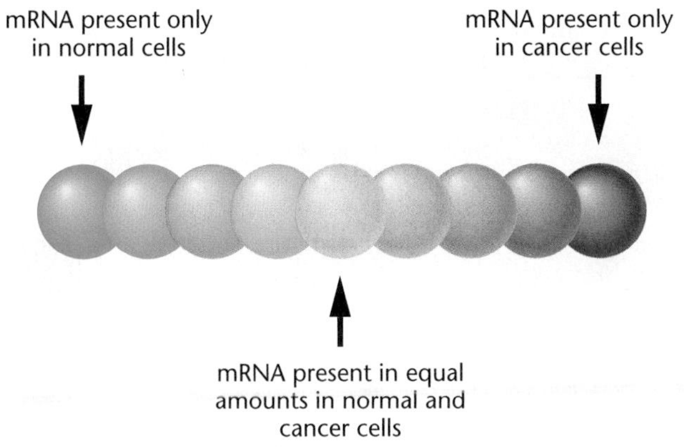

FIGURE 22–14 The color of dots on a DNA microarray represent levels of gene expression. Green dots represent genes expressed in one cell type (e.g., normal cells), and red dots represent genes expressed in another cell type (e.g., cancer cells). Intermediate colors represent different levels of expression of the same gene in the two cell types.

Disease Diagnosis

Using a microarray carrying 18,000 probes representing genes expressed during the development of normal and cancerous white blood cells, Ash Allzadeh and colleagues used cDNA to analyze gene expression in one type of **non-Hodgkin's lymphoma** (a cancer of white blood cells). About 40 percent of patients have an aggressive form of this cancer called **diffuse large B-cell lymphoma (DLBCL)**. About 40 percent of patients with DLBCL respond well to chemotherapy and have long survival times. The other 60 percent do not respond to chemotherapy and die after a short time. Based on almost 1.8 million measurements of gene expression patterns of the 18,000 genes from normal and cancerous cells, Allzadeh and his team discovered that there are two different cell types in DLCBL, with almost inverse patterns of expression (Figure 22–15). One cell type, called GC B-like, has an expression pattern similar to developing B cells, found in lymph glands. The second cell type, called activated B-like cells, has an expression pattern similar to mature B cells that have left the lymph glands and moved into the circulatory system.

A graph of patient survival plotted against gene expression patterns reveals that patients with the activated B-like pattern had much lower survival rates (Figure 22–16). They concluded that DLCBL is actually two different diseases with different outcomes. The differences in gene expression patterns are being used to diagnose each form of the disease, and to develop a program of chemotherapy specific for each form of cancer. The use of microarrays has revolutionized both the diagnosis and treatment of these diseases.

Genome Scanning

DNA microarrays that carry all the genes in the human genome have also been developed (Figure 22–17). Use of these microarrays, called **genome scanning**, makes it possible to analyze someone's DNA for dozens or hundreds of disease alleles, including those that predispose the person to heart attacks, diabetes, Alzheimer disease, and other genetically defined disease subtypes. Genome scanning makes it possible

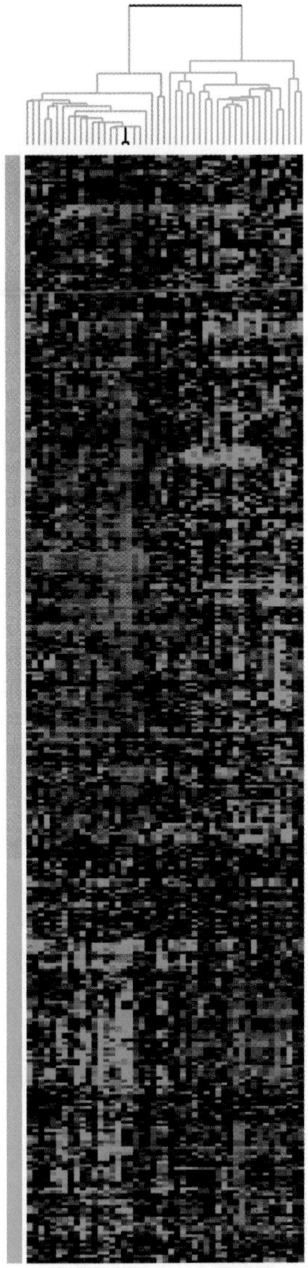

FIGURE 22–15 This microarray analysis profiled 18,000 genes expressed in normal and cancerous lymphocytes. The analysis was repeated to cover 1.8 million individual assays of gene expression. The malignant cells fall into two clusters. The orange cluster contains cells with expression profiles for GC B-like DLBCL cells. The blue cluster contains cells with expression profiles for activated B-like DLBCL cells. The cells within each cluster (shown across the top of the figure) are grouped by how closely their expression profiles resemble each other. The more closely the profiles resemble each other, the closer they are. The colors represent ratios of relative gene expression compared to normal control cells. Red represents expression greater than the mean level in controls, green represents expression lower than the mean level in controls, and the color intensity represents the magnitude of the difference from the mean.

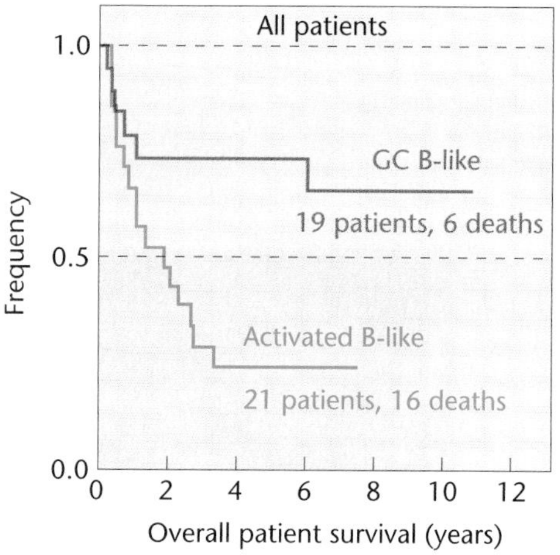

FIGURE 22–16 This graph sorts patients by gene expression profile and survival probability. Those with activated B-like profiles have a much higher rate of death (16 in 21) than those with GC B-like profiles (6 in 19).

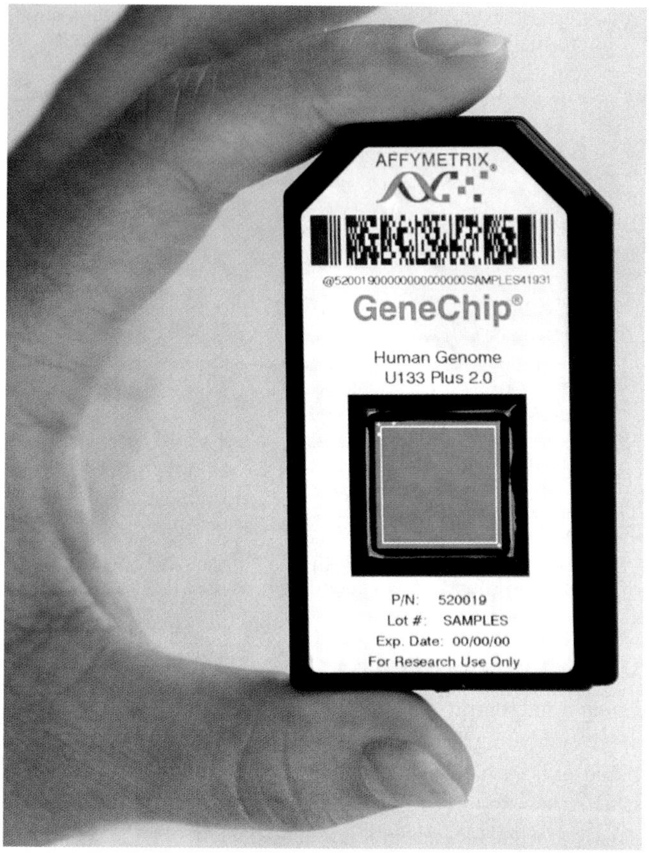

FIGURE 22–17 A first generation microarray containing 50,000 human candidate genes, called a GeneChip, is being marketed by Affymetrix, Inc. Use of these microarrays allows the entire human genome to be scanned at once, cataloging the normal and variant alleles carried by an individual.

to scan an individual's genome and define risks for specific diseases years or decades before such conditions appear. Although genome scanning is expensive, costs are expected to drop rapidly, and in a few years, genome scanning is expected to become widely available and to help change medical diagnosis. Perhaps in decades to come, all newborns will be scanned to determine their lifetime risks for genetic disorders.

Genetic Testing and Ethical Dilemmas

We have considered examples of prenatal diagnosis, heterozygote screening of adults for single recessive disorders, and genome scanning. Use of all these technologies present varying degrees of ethical problems. Genome scanning can predict someone's lifetime risk of disease, thereby identifying people who are presently healthy but at high risk of contracting a genetic disease in the future, including many that may not develop for years. These advances will affect our health, reproductive patterns, and medical care in fundamental ways.

The use of this technology also raises legal, social, and ethical issues that will not be easy to resolve. For example, what should people know before deciding to have a genome scan or a genetic test for a single disorder? How can we protect the information revealed by such tests? How can we define and prevent genetic discrimination? We know that heterozygotes for sickle-cell anemia are more resistant to malaria than people who are homozygous for the normal allele; this type of protection may be true of other genetic disorders as well. How do we keep the beneficial aspects of some mutations as we strive to eliminate their destructive aspects? Some mutations have horrific consequences; others gave rise to our very existence as humans. We know a great deal about the molecular nature of many different human mutations, but our legal and ethical understanding and consensus lags far behind our scientific understanding. Thoughtful and wide-ranging public debate is essential as we explore the use of biotechnology.

Many of the potential risks and benefits of genetic testing are still unknown. We can test for many genetic diseases for which there are no effective treatments to cure or mitigate the clinical consequences. Should we test people for these disorders? With present technology, a negative result does not necessarily rule out future development of a disease; nor does a positive result always mean that an individual will get the disease. How can we effectively communicate the results of testing and the risks to those being tested? Public policy and laws on genetic testing are being formulated more slowly than the technology and use of genetic testing. Lawmakers and other groups made up of scientists, health-care professionals, ethicists, and consumers are debating these issues and formulating policy options.

22.4 Genetic Disorders Can Be Treated by Gene Therapy

For more than two decades, *gene products* such as insulin have been used for therapeutic treatment of genetic disorders. Methods for transferring specific *genes* into mammalian cells, originally developed as research tools, are now being used to treat genetic disorders, a process known as **gene therapy**. In theory, gene therapy transfers a normal allele into a somatic cell that carries one or more mutant alleles. Expression of the normal allele results in a functional gene product whose action produces a normal phenotype. Delivery of these structural genes and their regulatory sequences is accomplished by using a vector or gene transfer system.

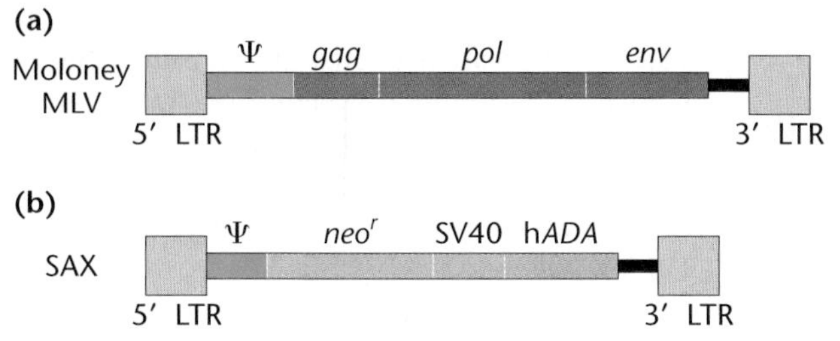
(a)
Moloney MLV
Ψ
gag
pol
env
5′ LTR
3′ LTR
(b)
SAX
Ψ
neo^r
SV40 hADA
5′ LTR
3′ LTR

FIGURE 22–18 The native Moloney MLV genome contains a sequence required for encapsulation, as well as genes that encode viral coat proteins (*gag*), an RNA-dependent DNA polymerase (*pol*), and surface glycoproteins (*env*). At each end, the genome is flanked by long terminal repeat (LTR) sequences that control transcription and integration into the host genome. The SAX vector retains the LTR and ψ sequences and includes a bacterial neomycin resistance (*neo^r*) gene that can be used as a selective marker. As shown, the vector carries a cloned human adenosine deaminase (h*ADA*) gene, which is fused to an SV40 early region promoter–enhancer. The SAX construct is typical of retroviral vectors that are used in human gene therapy.

FIGURE 22–19 Ashanti De Silva, the first person to be treated by gene therapy.

In the first generation of gene-therapy trials, the most common method of gene transfer used genetically modified retroviruses as vectors. These vectors were based on a mouse virus called Moloney murine leukemia virus (MLV) (Figure 22–18). This vector was created by removing a cluster of three genes from the virus, and inserting a cloned human gene. After packaging in a viral protein coat, the recombinant vector is used to infect cells. Once inside, the virus cannot replicate itself because of the missing viral genes. In the cell, the recombinant virus with the inserted human gene moves to the nucleus and integrates into a site on a chromosome, and becomes part of the genome. If the gene is expressed, it produces a normal gene product that can correct a mutation carried by the affected individual. In initial attempts at gene therapy, several heritable disorders, including severe combined immunodeficiency (SCID), familial hypercholesterolemia, and cystic fibrosis, were treated. Let's examine the first attempt to use gene therapy to treat a young girl with SCID and then review the trials that are currently under way using a new generation of viral vectors.

Gene Therapy for Severe Combined Immunodeficiency (SCID)

Gene therapy began in 1990 with the treatment of a young girl named Ashanti DeSilva (Figure 22–19), who has a heritable disorder called severe combined immunodeficiency (SCID). Affected individuals have no functional immune system and usually die from what would normally be minor infections. Ashanti has an autosomal form of SCID caused by a mutation in the gene encoding the enzyme **adenosine deaminase (ADA)**. Her gene therapy began when clinicians isolated some of her white blood cells, called T cells (Figure 22–20). These cells, which are part of the immune system, were mixed with a retroviral vector carrying an inserted copy of the normal *ADA* gene. The virus infected many of the T cells, and a normal copy of the *ADA* gene was inserted into the genome of some T

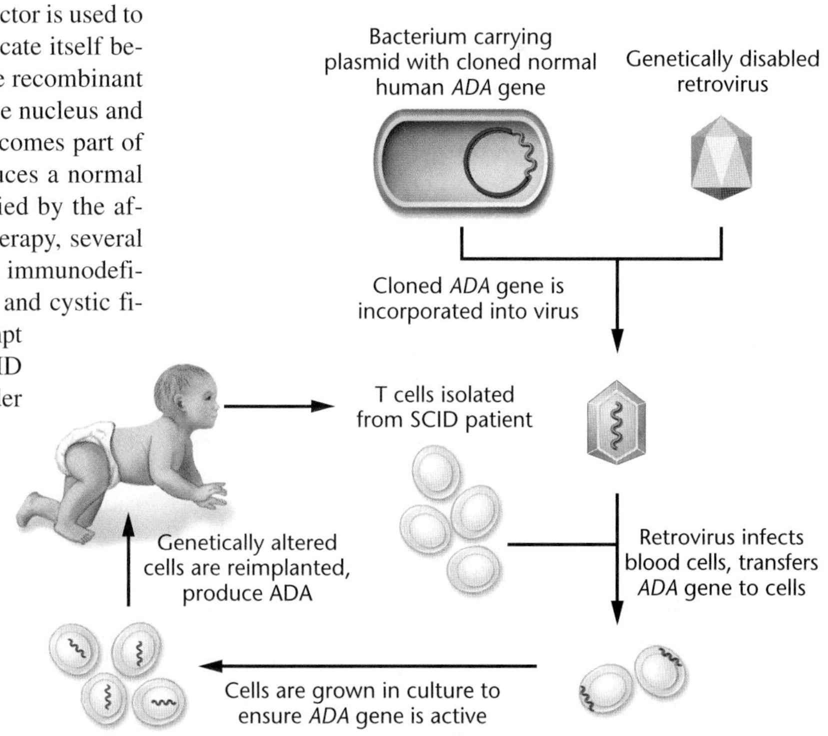
Bacterium carrying plasmid with cloned normal human ADA gene
Genetically disabled retrovirus
Cloned ADA gene is incorporated into virus
T cells isolated from SCID patient
Retrovirus infects blood cells, transfers ADA gene to cells
Genetically altered cells are reimplanted, produce ADA
Cells are grown in culture to ensure ADA gene is active

FIGURE 22–20 To treat SCID using gene therapy, a cloned human *ADA* gene is transferred into a viral vector, which is then used to infect white blood cells removed from the patient. The transferred *ADA* gene is incorporated into a chromosome and becomes active. After growth to enhance their numbers, the cells are reimplanted into the patient, where they produce ADA, allowing the development of an immune response.

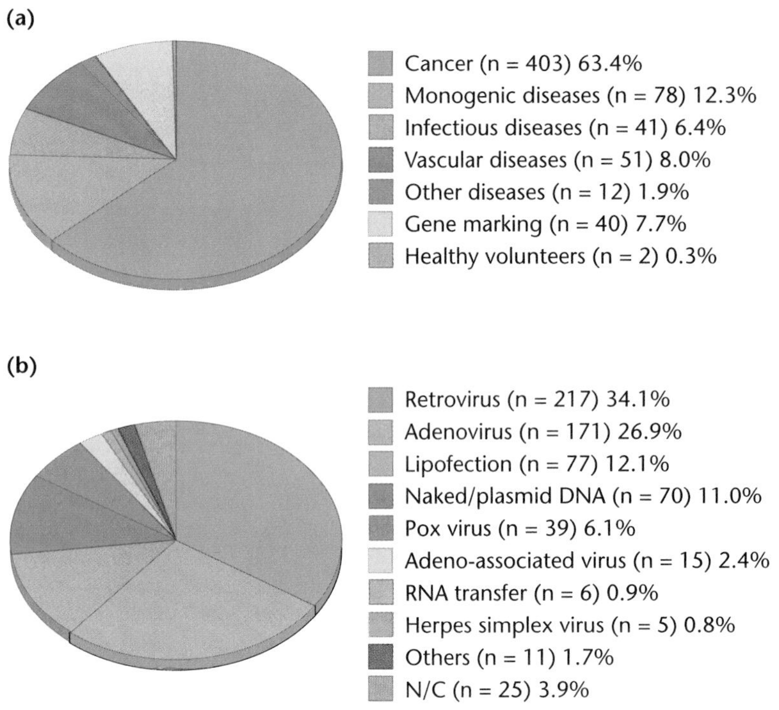

FIGURE 22–21 (a) Pie chart summarizing, by disease, more than 630 gene therapy trials under way worldwide. Most trials involve cancer treatment. (b) Gene therapy trials ranked by the vectors used. Retroviruses are the most widely used vectors, accounting for 34 percent of the total. Newer vectors, such as adeno-associated virus account for only a small percentage of vectors. N/C = not classified

cells. After mixing with the vector and infection, the T cells were grown in the laboratory and analyzed to make sure that the transferred *ADA* gene was expressed. The final step in gene therapy was injection of a billion or so genetically altered T cells into her bloodstream. Some of these T cells migrated to Ashanti's bone marrow and began dividing, and producing ADA. She has ADA protein expression in 25–30 percent of her T cells, which is enough to allow her to lead a normal life.

Unfortunately, a second child treated a short time later expressed the normal *ADA* gene in less than 1 percent of her white blood cells after treatment—a level not high enough to be effective. In later trials, attempts were made to transfer the *ADA* gene into the bone marrow cells that form T cells, but these attempts were mostly unsuccessful. Although gene therapy was originally developed as a treatment for single-gene (monogenic) inherited diseases, this technique was rapidly adapted for the treatment of acquired diseases such as cancer, neurodegenerative diseases, cardiovascular disease, and infectious diseases, such as HIV. As a result, most gene therapy treatments and trials involve these disorders (Figure 22–21a). In fact, gene therapy is used to treat cancer more often than any other condition.

HOW DO WE KNOW?

How do we know that gene therapy delivers a functional normal gene to target cells?

In addition to retroviral vectors, other viruses and other methods are being used to transfer genes into human cells (Figure 22–21b). These methods include the use of viral vectors, chemically assisted transfer of genes across cell membranes, and fusion of cells with artificial vesicles containing cloned DNA sequences.

Problems and Failures in Gene Therapy

Over a 10-year period, from 1990 to 1999, more than 4000 people underwent gene transfer for a variety of genetic disorders. These trials often failed and thus led to a loss of confidence in gene therapy. Hopes for gene therapy plummeted even further in September 1999 when a teenager, Jesse Gelsinger, died during gene therapy. His death was triggered by a massive inflammatory response to the vector, a modified adenovirus. (Adenoviruses cause colds and respiratory infections.)

The outlook for gene therapy brightened in 2000, when a group of French researchers reported the first large-scale success in gene therapy. Three patients with a fatal X-linked form of SCID developed immune systems after being treated with a retroviral vector carrying a normal gene. This success was repeated in subsequent treatments of several other patients. However, two patients later developed leukemia-like disorders. An analysis of cancer cells from the two patients showed that the retroviral vector had inserted near or into a gene called *LMO2*. This insertion activated *LMO2*, causing uncontrolled white blood cell proliferation and development of the leukemia-like disorder. Once again the ideas and methods used in the whole field of gene therapy underwent analysis to identify problems and develop new strategies.

The Future of Gene Therapy

Most problems with gene therapy have been traced to the vectors. First-generation vectors, such as MLV and adenovirus, have several serious drawbacks. First, integration of the retroviral genome (including the cloned human gene) into the host cell's genome occurs only if the host cells are replicating their DNA. In the body, only a very small number of cells in the highly differentiated cell types that are targets for gene therapy are synthesizing DNA at any given time. Second, most of these viral vectors eventually cause an immune response in the patient, as happened in Jesse Gelsinger's case. Third, insertion of viral genomes into the host chromosome can inactivate or mutate an indispensable gene, as in the case of the two French patients. Fourth, retroviruses have a low cloning capacity and cannot carry inserted sequences much larger than 8 kb—many human genes, even discounting introns, exceed this size. Finally, there is the possibility of producing an infectious virus if a recombination event takes place between the vector and other retroviral genomes already present in the host cell.

TABLE 22.2 NEW VECTORS FOR GENE THERAPY

Vector	Cell Targets	Cloning Capacity	Advantages	Disadvantages
Adenovirus	Lung, respiratory tract	7.5 kb	Efficient transfection	Strong immune response
Adeno-associated virus	Fibroblasts, T cells, others	4.5 kb	Transfect many cell types	Small insert size
Retroviruses	Proliferating cells	8 kb	Prolonged expression	Low transfection efficiency
Lentiviruses	Stem cells, proliferating cells	8 kb	Efficient transfection	Related to HIV

The disappointing and deadly results from gene therapy trials using first-generation vectors led to a widespread crisis of confidence in gene therapy in the medical and other scientific communities. To overcome these problems, new viral vectors and strategies for targeting cells are being developed. The properties of current viral vectors and those under development are summarized in Table 22.2. In general, these vectors fall into two categories: those that integrate into the host-cell genome (retroviruses and others) and those that remain in the nucleus but do not integrate into the chromosomes (adeno-associated viruses and others). Researchers hope that the use of new vectors will circumvent several of the problems encountered with earlier vectors and will also have new design features to allow regulation of insertion sites and the levels of gene product.

In one application of new strategies for gene therapy, scientists reported the successful production of the blood hormone erythropoeitin in rhesus monkeys and mice whose muscle cells had been injected with a viral vector carrying an erythropoeitin gene. These animals produced the hormone only when they received the antibiotic rapamycin. These results are particularly encouraging because the stimulated levels of the hormone are quite high, the presence of the foreign DNA has not triggered an immune response, and the gene can be repeatedly activated.

This trial used one of the most promising new viral vectors—adeno-associated virus (AAV). This virus enters cells that are not synthesizing DNA, does not integrate into the host-cell genome, and is a very small virus, so it does not typically cause an immune response. In the trials with erythropoeitin, the vector contained a rapamycin-responsive promoter that is used to switch on the erythropoietin gene. The gene is expressed only when cells carrying the vector are exposed to rapamycin. This therapy is being considered for patients who have low numbers of blood cells, such as dialysis patients who currently receive regular injections of the erythropoietin hormone.

22.5 Gene Therapy Raises Many Ethical Concerns

Gene therapy raises several ethical concerns, and many forms of therapy are still sources of intense debate. At present, all gene therapy trials are restricted to using somatic cells as targets for gene transfer. This form of gene therapy is called **somatic gene therapy**; only one individual is affected, and the therapy is done with the permission and informed consent of the patient. The ethical guidelines for gene therapy, as it is currently performed, have been revised and strengthened in the wake of Jesse Gelsinger's death. This sort of therapy can be initiated only after careful review by several levels of administrators, and the trials are monitored to protect the interests of the patient.

Two other forms of gene therapy have *not* been approved, primarily because of the unresolved ethical issues surrounding them. The first is called **germ-line therapy**, whereby germ cells (the cells that give rise to the gametes—i.e., sperms and eggs) or mature gametes are used as targets for gene transfer. In this approach, the transferred gene is incorporated into all the cells of the body, including the germ cells. This means that individuals in future generations will also be affected, without their consent. Is this procedure ethical? Do we have the right to make this decision for future generations? Thus far, the concerns have outweighed the potential benefits, and such research is prohibited.

The second unapproved form of gene therapy—which raises an even greater ethical dilemma—is termed **enhancement gene therapy**, whereby people are enhanced for some desired trait. This use of gene therapy is extremely controversial and is strongly opposed by many people. Should genetic technology be used to enhance human potential? For example, should it be permissible to use gene therapy to increase height, enhance athletic ability, or extend intellectual potential? Presently, the consensus is that enhancement therapy, like germ-line therapy, is an unacceptable use of gene therapy. However, there is an ongoing debate and many issues are still unresolved. For example, the U.S. Food and Drug Administration now permits growth hormone produced by recombinant DNA technology to be used as a growth enhancer, in addition to its current medical use for the treatment of growth-associated genetic disorders. Critics charge that the use of a gene product for enhancement will lead to the use of transferred genes for the same purpose. The outcome of these debates may affect not only the fate of individuals but the direction of our society as well.

22.6 Ethical Issues Are an Outgrowth of the Human Genome Project

Geneticists now use recombinant DNA technology to identify genes, diagnose genetic disorders, screen populations for heterozygous carriers, and treat disorders by gene therapy. Knowledge gained by sequencing the human genome will greatly advance our understanding of human genetics, and will have a great impact on biomedical research and health care. However, applications of the knowledge gained from the project raise ethical, social, and legal issues that must be identified, debated, and resolved. Resolutions often take the form of laws or public policy. The ethical debate surrounding gene therapy discussed in the previous section represents a subset of the broader ethical issues raised by knowledge gained as an outcome of the Human Genome Project.

The Ethical, Legal, and Social Implications (ELSI) Program

When the **Human Genome Project (HGP)** was first discussed, scientists and the general public raised concerns about how genome information would be used and how the interests of both individuals and society can be protected. To address these concerns, the **Ethical, Legal, and Social Implications (ELSI) Program** was established as an adjunct to the Human Genome Project. The ELSI program considers a number of issues, including the impact of genetic information on individuals, the privacy and confidentiality of genetic information, implications for medical practice, genetic counseling, and reproductive decision making. Through research grants, workshops, and public forums, ELSI is formulating policy options to address these issues.

ELSI focuses on four areas: (1) privacy and fairness in the use and interpretation of genetic information, (2) ways of transferring genetic knowledge from the research laboratory to clinical practice, (3) ways to ensure that participants in genetic research know and understand the potential risks and benefits of their participation and give informed consent, and (4) public and professional education. It is hoped that, as the HGP moves from generating information about the genetic basis of disease to improving treatments, promoting prevention, and developing cures, these and other ethical issues will be identified and an international consensus will be developed on appropriate policies and laws.

22.7 Finding and Mapping Genes in the Human Genome with Recombinant DNA Technology

When the first human genetic disorders were mapped, the phenotype was usually associated with a mutant gene product that could be identified in affected individuals. By using the mutant protein as a marker, pedigree analysis was used to establish the pattern of inheritance and in a small number of cases, to establish the chromosomal locus of the gene. However, aside from genes on the X chromosome, progress in mapping human genes was slow because of the lack of markers. Real progress in mapping genes and identifying the encoded protein has come only in the last two decades. Now, with our increasingly detailed knowledge of the genome, we can map a gene without knowing anything about its product.

Web Tutorial 22.1
Restriction Fragment Length Polymorphisms

RFLPs as Genetic Markers

Variations in nucleotide sequences occur throughout the human genome (mostly in noncoding regions) about once in every 200 nucleotides. These nucleotide changes occur at specific sites and are created by substitutions, deletions, or insertions of one or more nucleotides. Such variations can sometimes create or destroy restriction enzyme cutting sites. If a restriction enzyme site created by nucleotide variation is present on one chromosome but absent on its homolog, the two chromosomes can be distinguished from one another by their pattern of restriction fragments on a Southern blot (Figure 22–22). The region of chromosome A shown in the figure contains three *Bam*HI sites;

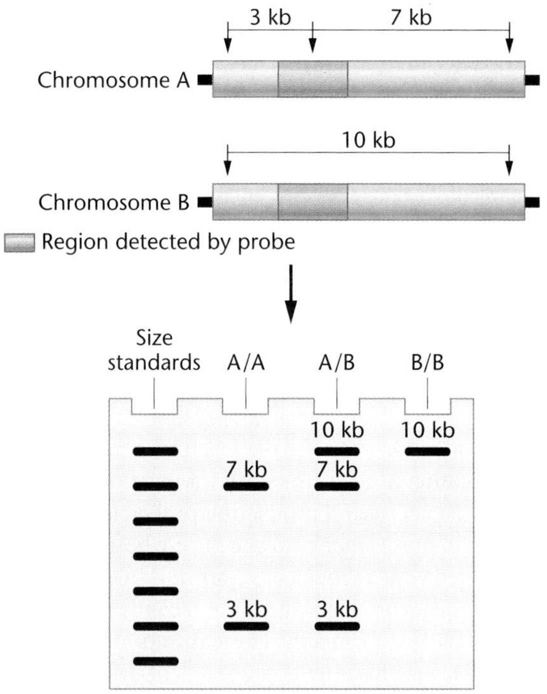

Genotypes	Fragment sizes
Homozygous for chromosome A (A/A)	3, 7 kb
Heterozygous (A/B)	3, 7, 10 kb
Homozygous for chromosome B (B/B)	10 kb

FIGURE 22–22 The alleles on chromosomes A and B represent DNA segments from homologous chromosomes. The region that hybridizes to a probe is shown in green; arrows indicate the location of restriction enzyme cutting sites that define the alleles. On chromosome A, three cutting sites generate fragments of 7 kb and 3 kb. On chromosome B, only two cutting sites are present, generating a single fragment that is 10 kb in length. The absence of the cutting site in chromosome B could be the result of a single base mutation within the enzyme recognition or cutting site. Because these differences in restriction cutting sites are inherited in a codominant fashion, there are three possible genotypes: *AA*, *AB*, and *BB*. The allele combination carried by any individual can be detected by restriction digestion of genomic DNA (obtained, for example, from a blood sample or skin fibroblasts), followed by gel electrophoresis, transfer to a DNA binding filter, and hybridization to the appropriate probe. The fragment patterns for the three possible genotypes are shown as they would appear on a Southern blot.

its homolog, chromosome B, contains two such sites. When chromosome A is cut with *Bam*HI, 3-kb and 7-kb fragments are generated, whereas only a single 10-kb fragment is generated when chromosome B is cut. These fragments can be visualized on a Southern blot by using a probe from this chromosomal region.

Variations in DNA fragment length generated by restriction enzymes are called **restriction fragment-length polymorphisms (RFLPs)**. RFLPs are quite common; thousands have been identified in the human genome, and most have been assigned to specific chromosome regions. These variations can be detected on Southern blots as codominant alleles and can be used as markers to follow the inheritance of genetic disorders from generation to generation in an affected family.

RFLPs can be used to map the chromosomal locus of a genetic disorder if the RFLP cosegregates with the genetic disorder in a multigenerational family (three generations or more). Selecting which RFLP to use in the family being studied is a matter of trial and error. Many different RFLPs may have to be tested to find a set for which most individuals of the family are heterozygous. If family members are heterozygous for an RFLP assigned to a specific chromosome, each member of the chromosome pair being tested can be identified as it passes from generation to generation. This provides a way to establish linkage between a specific chromosome (identified by the RFLP marker) and the phenotype of the disease.

Linkage Analysis Using RFLPs

To map a genetic disorder, the inheritance of a given RFLP and the disorder are traced through a family (Figure 22–23). If loci for the RFLP and the disorder are near each other on the same chromosome, they will show linkage and be inherited together. However, most RFLPs will not show linkage to the disease phenotype because the RFLP and the disease loci are on different chromosomes or there is frequent recombination between the RFLP and the disease loci. These data however, are not useless—they help to identify chromosomes and chromosome regions that do *not* carry the disease locus.

Many genes have been mapped in humans by using RFLP analysis, and by compiling studies from many families, the location of many markers have been determined and genetic maps for all human chromosomes have been constructed (Figure 22–24). The unit for linkage is the **centimorgan (cM)**, named after the geneticist T. H. Morgan. One centimorgan is equal to a recombination frequency of 1 percent between two loci.

Positional Cloning: The Gene for Neurofibromatosis

Using RFLPs to map genes is a departure from previous methods of mapping, which worked from an identified gene product to the gene locus. The search for the chromosomal locus for the gene causing type 1 neurofibromatosis (NF1) was one of the first to use RFLP analysis in gene mapping. NF1 is inherited as an autosomal dominant condition affecting 1 in 3000 individuals, and is associated with nervous system defects, including benign tumors and a high incidence of learning disorders.

The *NF1* gene was mapped in 1987 by RFLP analysis in several steps. First, a group of research laboratories around the world compared the inheritance of NF1 to the inheritance of dozens of RFLP markers in families with NF1. Each RFLP marker had been assigned to a specific human chromosome or chromosome

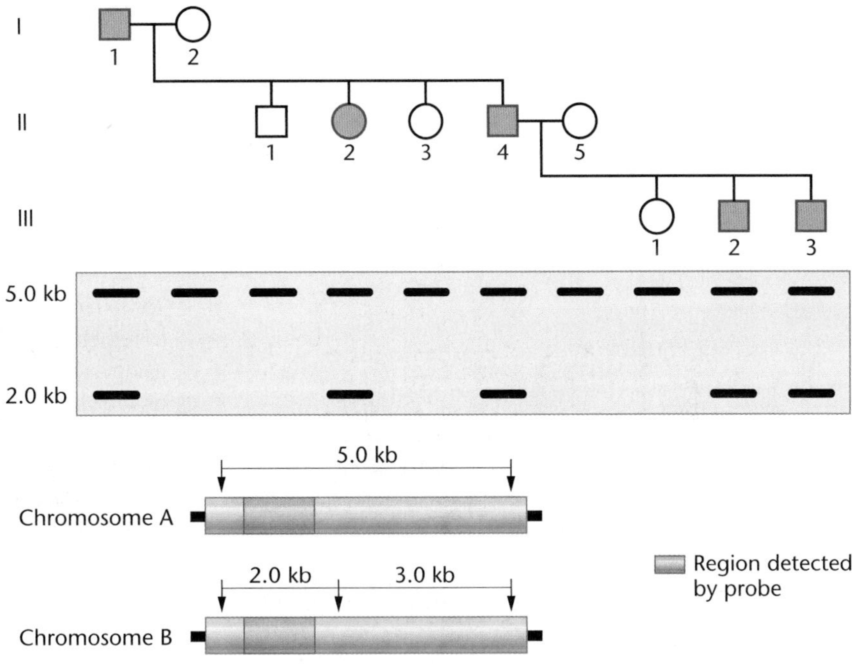

FIGURE 22–23 This pedigree shows a family with members affected by a dominant trait (filled symbols). Family members also carry two alleles of an RFLP locus assigned to chromosome 17. A 5.0-kb allele (allele *A*) is present on one homolog; the *B* allele on the other homolog consists of two fragments (2.0 kb and 3.0 kb). The probe used in the Southern blot detects the 5-kb *A* allele and the 2.0-kb portion of the *B* allele. The RFLP pattern for each family member is shown below the appropriate pedigree symbol. In generation II, both II-2 and II-4 are affected, and received a *B* allele from their affected father. Individual III-1, who is unaffected, received an *A* allele from each parent (II-4 and II-5). Individual III-2 is affected and received an *A* allele from his mother and a *B* allele from his affected father, as did the youngest son (III-3), who is affected. Not all members of a population are affected by the disorder, and some copies of the chromosome carrying the *B* allele will not carry the mutant allele responsible for the disorder. Analysis of the pedigree and the Southern blot suggests that in this family, the mutant allele for the dominant trait and the RFLP *B* allele are on the same chromosome and are therefore linked on chromosome 17. Assigning a mutant allele to a chromosome by RFLP analysis is the first step in mapping a gene.

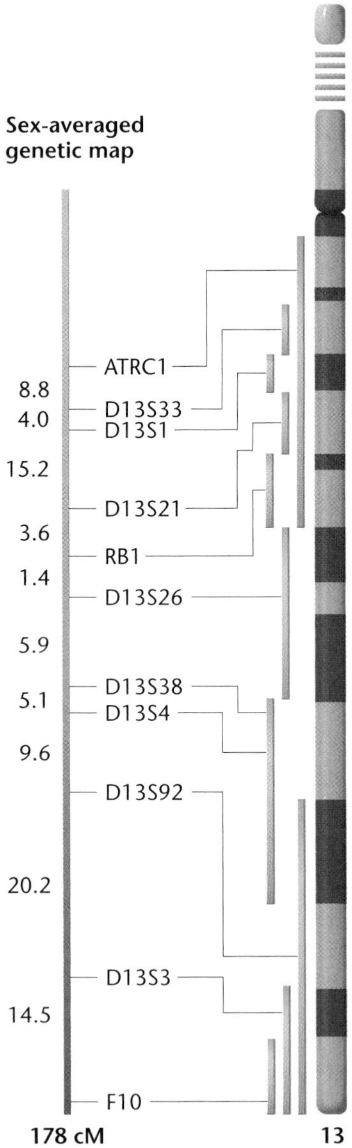

FIGURE 22–24 A genetic map (vertical bar at the left) and a physical map (right) of human chromosome 13. The genetic map shows the order and distance (in centimorgans [cM]) of the markers. At right is the physical map showing the chromosomal location of markers. The vertical bars next to the chromosome represent the region where the marker is located. The genetic map for females is 203 cM, and that for males is 158 cM, reflecting the difference in recombination frequencies between females and males. When the two maps are averaged together, the result is the sex-averaged map of 178 cM shown on the left.

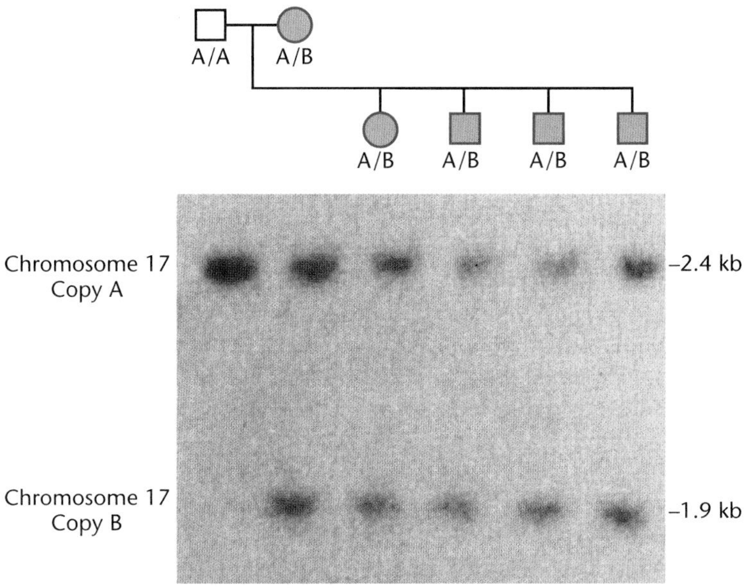

FIGURE 22–25 Segregation of a 1.9-kb RFLP allele with type 1 neurofibromatosis (NF1) in each of four affected offspring and their mother. This RFLP is detected by probe pA10-41, which is a DNA segment from a region near the centromere of human chromosome 17. On the basis of this result as well as results from tests using other probes, the locus for NF1 was assigned to chromosome 17. *From Barker, D., et al. 1987. "Gene for von Recklinghausen Neurofibromatosis is in the Pericentromeric Region of Chromosome 17". Science 236:1100–1102, Fig. 1. © 1987 by the American Association for the Advancement of Science*

region. The first round of analysis did not identify the chromosome carrying the *NF1* gene, but did produce an **exclusion map**, indicating which RFLPs were not linked to the disease and therefore which chromosomes do *not* carry the *NF1* locus. This analysis also produced evidence that there might be linkage between the *NF1* gene and RFLPs on chromosomes 5, 10, and 17. The second step used RFLPs from these chromosomes to study the inheritance of *NF1* in families, and produced conclusive evidence that the disorder is closely linked to an RFLP near the centromere of chromosome 17 (Figure 22–25). In a third step, more than 30 RFLP markers from this region were used to analyze 13,000 individuals from NF1 families, and the gene was mapped to region 17q11.2. Finally, the locus for NF1 was identified by using a collection of genomic clones that spanned a small region of 17q11.2. The gene's identity was confirmed when mutant versions of the gene were found in individuals with NF1.

Once the gene was identified, the amino acid sequence of the gene product was reconstructed from the DNA sequence. Databases with protein-sequence information were scanned to identify similar proteins in other organisms. Researchers determined that the NF1 gene product is similar to signal transduction proteins. Analysis of the NF1 protein, **neurofibromin**, confirmed that it is involved in the transduction of intracellular signals and the down-regulation of a gene that controls cell division. Mutation in the *NF1* gene leads to a loss of control of cell growth, causing the production of the small tumors characteristic of this disorder.

Researchers began the project on mapping the NF1 gene with no direct knowledge of the nature of the gene product or the mutational events that produce the NF1 phenotype. The mapping, cloning and sequencing of this gene and the identification of the gene product took a little over three years, and is an example of **positional cloning**, a method of gene mapping that requires no prior knowledge of the gene product. Using this strategy, an ever-increasing number of human genes are being mapped, isolated, and cloned.

Fluorescent *in situ* Hybridization (FISH) Gene Mapping

In addition to methods such as RFLP mapping, genes can be mapped directly to metaphase chromosomes in a process called *in situ* hybridization. (See Chapter 10 for a discussion of DNA

(a)

(b)

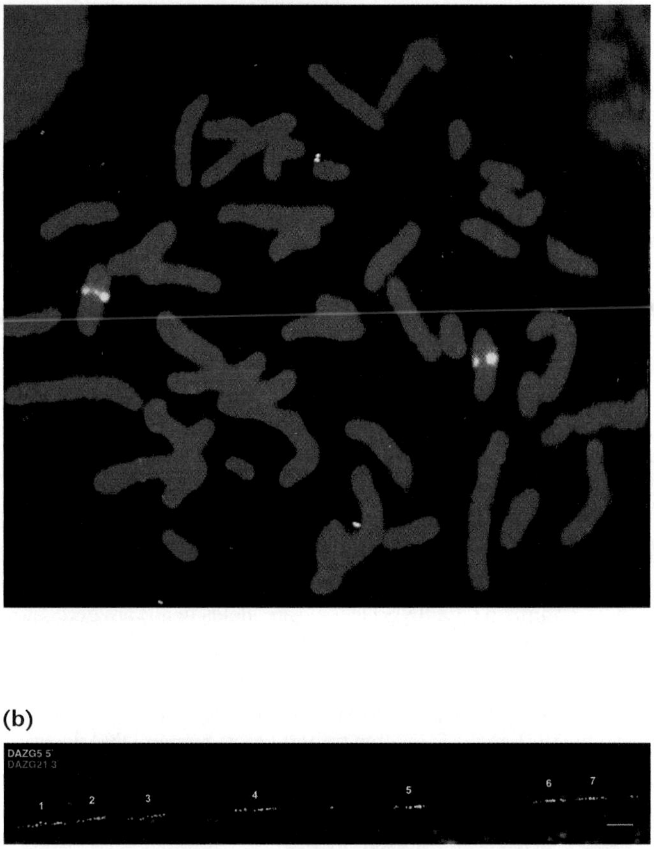

FIGURE 22–26 (a) Mapping a gene by fluorescent *in situ* hybridization (FISH) to metaphase chromosomes. The sites of hybridization appear as yellow dots on the sister chromatids of the chromosome pair. (b) Localization of four clusters of two different DNA sequences (DAZG5 and DAZG2) along a chromosome fiber.

hybridization.) Metaphase chromosomes spread on microscope slides are denatured, converting the double-stranded chromosomal DNA into single-stranded molecules. A labeled probe, containing all or part of a cloned gene, is hybridized to the chromosome preparation. The probe can be labeled with a radioactive isotope, antibodies, or fluorescent dyes. In contrast to the use of gels in Southern blotting, the resulting hybrids are examined and detected under a microscope.

When fluorescent dyes are used to label the probes, a fluorescence microscope is used to examine the chromosome preparation and to determine the site of hybridization. This technique, known as **fluorescent *in situ* hybridization (FISH)** can be used to directly assign a cloned gene to a chromosomal locus (Figure 22–26a). The technique has a resolution of about 1Mb and is now used routinely to map human genes.

A tenfold higher level of resolution (around 100 kb) can be obtained by using fiber FISH, a variation of the FISH technique. In fiber FISH, chromosome fibers are stretched along the glass slide during chromosome preparation. (See Chapter 12 for a discussion of chromosome organization.) Fluorescent probes are hybridized to the fibers (Figure 22–26b). This method is used to directly establish the order and distance between two or more genes on a chromosome.

22.8 DNA Fingerprints Can Identify Individuals

Web Tutorial 22.2 DNA Fingerprinting

As discussed in the previous section, the presence or absence of restriction sites in the human genome can be used as genetic markers. Another genetic marker, discovered in the mid-1980s, is based on variations in the length of repetitive DNA sequence clusters. These polymorphisms in human DNA serve as the basis for DNA fingerprinting. DNA fingerprinting is used for a wide variety of applications, from identifying paternity to forensics to conservation biology.

Minisatellites (VNTRs) and Microsatellites (STRs)

Minisatellites are repeating clusters of 10–100 nucleotides. For example, the sequence

5′-**GACTGCCTGCTAAGAT**GACTGCCTGCTAAGAT**GAC**
TGCCTGCTAAGATGACTGCCTGCTAAGAT**GACTGCCT**
GCTAAGATGACTGCCTGCTAAGAT**GACTGCCTGCTAA**
GATGACTGCCTGCTAAGAT**GACTGCCTGCTAAGAT**-3′

is composed of nine tandem repeats of the 16 base-pair sequence GACTGCCTGCTAAGAT. Clusters of such sequences are widely dispersed in the human genome. The number of repeats at each locus ranges from 2 to more than 100. These loci, known as **variable-number tandem repeats (VNTRs)** were introduced in Chapter 12 as examples of middle repetitive DNA. The number of repeats at a given locus is variable, and each variation represents a VNTR allele. Many loci have dozens of alleles each; as a result, heterozygosity is common.

A pattern of bands is produced when DNA is cut with restriction enzymes and visualized by Southern blotting using VNTR sequences as probes. This pattern was called a DNA fingerprint (Figure 22–27) because the pattern of bands is always the same for a given individual, no matter what tissue is used as the source of the DNA; but the pattern varies from individual to individual as do real fingerprints. In fact, there is so much variation in the band pattern from individual to individual that, theoretically, if enough VNTRs are analyzed, each person's pattern is unique. This technique was developed during the 1980s by Alec Jeffries and was first used to solve the murders of two schoolgirls in a high-profile case in Great Britain.

An important limitation of DNA fingerprint analysis with VNTRs is that it requires a relatively large sample of DNA (10,000 cells or about 50 μg)—more than is usually found at a typical crime scene—and the DNA must be relatively intact (nondegraded). As a result, DNA fingerprinting is more useful in paternity testing than in criminal cases. In paternity testing, blood drawn from the child, mother, and alleged father provide an unlimited source of fresh, intact cells for DNA extraction and analysis.

To overcome the problems in DNA sample size and conditions necessary for analysis with VNTRs, a different, but related set of sequences are used as markers, and are analyzed by PCR rather than RFLP analysis. These sequences, called **short tandem repeats (STRs)**, are very similar to VNTRs, but the repeated motif is shorter—between two and nine base pairs. Thirteen tetrameric (four base-pair repeat) STRs have been developed into a marker panel (called the Combined DNA Index System or CODIS panel) used by the FBI and other law enforcement agencies to do DNA

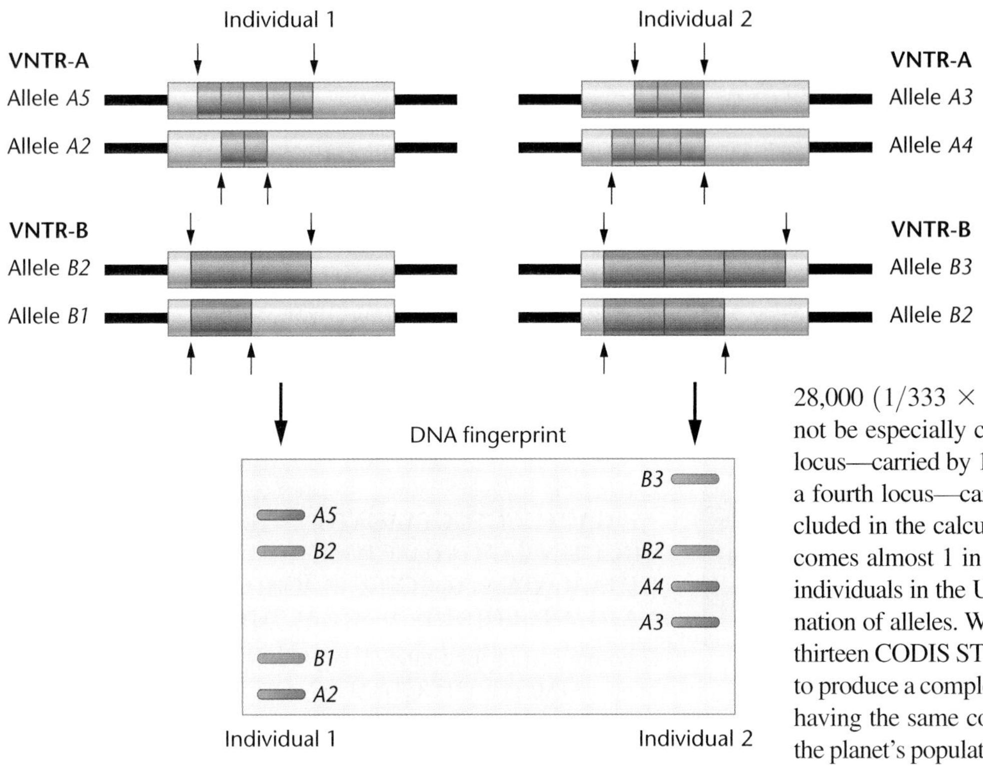

FIGURE 22–27 VNTR alleles at two loci (*A* and *B*) are shown for each individual. (Arrows mark restriction enzyme cutting sites flanking the VNTRs.) Restriction enzyme digestion produces a series of fragments that can be detected as bands on a Southern blot (bottom). The number of repeats at each locus is variable, so the overall pattern of bands is distinct for each individual, even though one band is shared (the *B2* allele band). The pattern is a DNA fingerprint.

typing of crime suspects (Figure 22–28) and to create a database containing DNA profiles of convicted felons.

STR analysis by PCR is used routinely in forensic casework to generate DNA profiles from trace samples (e.g., single hairs, saliva left on a cigarette butt or toothbrush) and from samples that are old and/or degraded (e.g., a skull found in a field, ancient Egyptian mummies). As a result, STRs have replaced VNTRs in most forensic laboratories. In addition, STR typing is less expensive, less labor-intensive, and much faster than RFLP analysis, so it is rapidly replacing VNTR typing in most paternity laboratories as well.

The results of STR analysis are analyzed and interpreted using statistics, probability, and population genetics. The population frequency of each STR allele in the standard set has been measured in many groups of people across the United States and the world. Using this information, the probability of having any combination of these 13 alleles can be calculated. For example, if an allele of locus 1 is carried by 1 in 333 individuals, and an allele of locus 2 is carried by 1 in 83 individuals, the probability that someone will carry both alleles is equal to the product of their individual frequencies, or nearly 1 in 28,000 ($1/333 \times 1/83$). This overall probability may not be especially convincing, but if an allele of a third locus—carried by 1 in 100 individuals—and an allele of a fourth locus—carried by 1 in 25 individuals—are included in the calculations, the combined frequency becomes almost 1 in 70 million. That is, only about four individuals in the U.S. population will carry this combination of alleles. When the genotype of the alleles of all thirteen CODIS STR loci (26 alleles in all) are generated to produce a complete STR profile, the chance of anyone having the same combination is 1 in 100 trillion. Since the planet's population is only about 6 billion, it is easy to see why STR analysis is often referred to as human DNA identification testing.

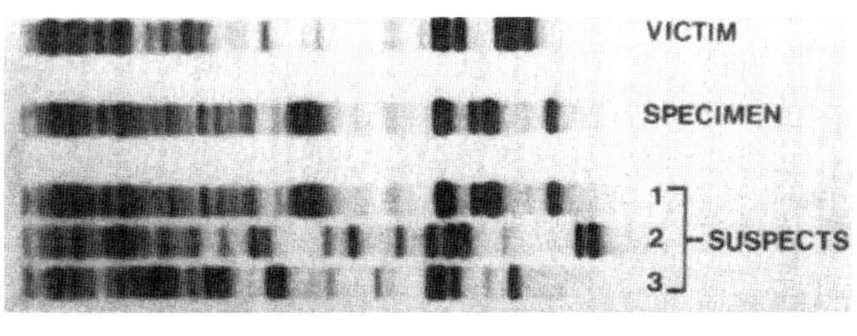

FIGURE 22–28 DNA profiles prepared using STRs reveal patterns from the victim, the evidence and three suspects in a criminal case.

HOW DO WE KNOW?

How do we know for sure that any individual aside from an identical twin can be identified by DNA profiling?

Forensic Applications

In a criminal case, if a suspect's DNA profile does not match that of the evidence, the suspect can be excluded as the criminal (exclusions occur in about 30 percent of all cases). When a match *is* made between DNA obtained at a crime scene and the DNA of a suspect, there are two possible interpretations: (1) the DNA fingerprint came from the suspect, or (2) it is from someone else with the same pattern of bands. Thus, the DNA fingerprint evidence does not of itself prove the suspect is guilty; it is one piece of evidence to be considered along with the other facts in the case.

DNA evidence was first used in a criminal trial in England in 1986. In the United States, it was first used in a Florida rape case in 1987. Since then, thousands of criminal trials have used DNA profiles. In 1998, the FBI started a national database cataloging the DNA profiles of convicted felons as well as DNA samples recovered from crime scenes. This database now contains more than 1.7 million genetic profiles that are used to match evidence against suspects.

Some states also have established DNA profile databanks. One of the largest is in Virgina, which has more than 200,000 samples. As the database has grown, it has become an important tool in solving criminal cases where there was no arrest and no suspects. By matching DNA from these crime scenes with DNA profiles in the database, more than 1600 crimes were solved using only DNA evidence between 2000 and 2003.

GENETICS, TECHNOLOGY, AND SOCIETY

Gene Therapy—Two Steps Forward or Two Steps Back?

In September 1999, 18-year-old Jesse Gelsinger received his first dose of gene therapy. Large numbers of adenovirus vectors bearing the ornithine transcarbamylase (*OTC*) gene were injected into his hepatic artery. The virus vectors were expected to lodge in his liver, enter liver cells, and trigger the production of OTC protein. In turn, the OTC protein might correct his genetic defect and perhaps cure him of his liver disease. However, within hours, a massive immune reaction surged through Jesse's body. He developed a high fever, his lungs filled with fluid, multiple organs shut down, and he died four days later of acute respiratory failure. In the aftermath of the tragedy, several government and scientific inquiries were initiated. Investigators learned that clinical trial scientists had not reported other adverse reactions to gene therapy, and that some of the scientists were affiliated with private companies that could benefit financially from the trials. They found that serious side effects in animal studies were not explained to patients during informed-consent discussions and some clinical trials were proceeding too quickly in the face of data that suggested caution. The U.S. Food and Drug Administration (FDA) scrutinized gene therapy trials across the country, halted a number of trials, and shut down several gene therapy programs. Other research groups voluntarily suspended their gene therapy studies.

Jesse's death dealt a severe blow to the struggling field of gene therapy—a blow from which it was still reeling when a second tragedy hit.

In April 2000, a French group announced that gene therapy for X-linked severe combined immunodeficiency (X-SCID) had succeeded. It was the first unequivocal success in the gene therapy field. In this study, the young patients' bone marrow cells were removed, treated with a retrovirus bearing the γc transmembrane protein gene, and transplanted back to the patients. Nine of 11 patients were cured of their immune deficiency and were able to lead normal lives. Published reports of the study were greeted with enthusiasm by the gene therapy community. But elation turned to despair in 2003, when it became clear that two of the nine children who had been cured of X-SCID had developed leukemia as a direct result of their therapy. The FDA immediately halted 27 similar gene therapy clinical trials and, once again, gene therapy underwent a profound reassessment.

Up until the apparent success of the French X-SCID clinical trials, gene therapy had suffered not only from the scandals and scrutiny that emerged from Jesse Gelsinger's death but also from the skepticism of scientists and the general public about the feasibility of this much-promoted therapeutic technique.

Since the first clinical trial for gene therapy in 1990, almost 500 gene therapy clinical trials involving over 4000 patients have been initiated. These trials aimed to cure cancers, inherited diseases such as hemophilia and cystic fibrosis, and infectious diseases such as AIDS. Despite high expectations and intense publicity, therapeutic benefits have been unclear at best, and more frequently, absent.

The most significant positive outcome of gene therapy came from the first clinical trial in 1990. Ashanti De Silva, who had received retroviral-transduced T cells for severe combined immunodeficiency (SCID), now leads a normal life, but the reasons for her success are not completely resolved. She had been given a new drug treatment that replaced her missing ADA enzyme prior to and after gene therapy. Hence, it is still not known how much of her cure is due to gene therapy and how much is due to drug treatment.

As of late 2004, no human gene-therapy product has been approved for sale. Critics of gene therapy continue to criticize research groups for undue haste, conflicts of interest, sloppy clinical trial management, and for promising much but delivering little. In the mid-1990s, a National Institutes of Health review committee concluded that significant problems remain in all basic aspects of gene therapy, including those aspects that led to Jesse Gelsinger's death and the X-SCID leukemias—adverse immune reactions to viral vectors and the side effects of retroviral vector integration into the host genome.

The question remains whether gene therapy can ever recover from these setbacks and fulfill its promise as a cure for genetic diseases. At present, about 200 gene therapy clinical trials are under way in the United States—most are for cancer, and are in phase I trials (which examine safety and dosage, but not efficacy). Tighter restrictions on clinical trial protocols are designed to correct some of the procedural problems that emerged from the Gelsinger case. In addition, basic science is proceeding with development of safe, effective vectors, such as those that insert vector sequences into specific regions of the genome (reducing the possibility of cancer or vector-gene silencing) and those bearing receptors on their surfaces that allow them to infect specific cell types. However, the steps leading to successful gene therapy are many. There is a need to optimize tissue-specific expression of therapeutic genes, to efficiently transduce blood cells in culture, to predict and control immune reactions to vectors, and to develop better animal models in which to test gene therapies prior to clinical trials.

Many scientists feel that we should continue gene therapy research and clinical trials despite the setbacks. However, they have a more sober view of its progress. Clinical trials for any new therapy are potentially dangerous, and often animal studies will not accurately reflect the reaction of individual humans to a new drug or procedure. Inevitably, more adverse reactions to gene therapy will emerge in the clinical trials as methods become more effective. Those in the field now believe that the road ahead will be longer and more difficult than first imagined, but not impossible. Perhaps we should view gene therapy as we have antibiotics, organ transplants, and manned space travel. There will be setbacks and even tragedies, but step by small step, we will move toward a technology that could—someday—provide cures for many severe genetic diseases.

References

Thomas, C. E., Ehrhardt, A., and Kay, M. A. 2003. Progress and problems with the use of viral vectors for gene therapy. *Nat. Rev. Gen.* 4:346–58.

Web Sites

Thompson, L. Sept./Oct. 2000. *Human gene therapy: Harsh lessons, high hopes [on-line]. FDA Consumer Magazine.*
http://www.fda.gov/fdac/features/2000/500_gene.html

Chapter Summary

1. The biotechnology industry uses recombinant DNA methods to improve crop plants by transferring herbicide and insect resistance and enhancing the nutritional value of plants. Biotechnology is also used to produce human gene products in a variety of hosts, ranging from bacteria to farm animals. In the near future, it may be possible to use food plants to vaccinate people against infectious disease.
2. Recombinant DNA technology offers a new approach to genetic analysis. Instead of relying on the isolation and mapping of mutant genes, large cloned segments of the genome are manipulated to make genetic and physical maps that use molecular markers rather than phenotypes visible at the level of the organism.
3. Recombinant DNA methods are used in the prenatal diagnosis of human genetic disorders, allowing direct examination of the genotype, whereas previous methods relied on gene expression and the identification of the gene product. It can also identify carriers of genetic disorders, the basis for proposals to screen the population for a number of genetic disorders, including sickle-cell anemia and cystic fibrosis.
4. The availability of cloned human genes has led to their use in gene therapy. In somatic gene therapy, a cloned normal copy of a gene is transferred into a vector; the vector then transfers the gene to a target tissue that takes up and expresses the cloned copy of the gene, thereby altering the mutant phenotype. The development of new and more effective vector systems means that gene therapy probably will become a standard method for treatment of genetic disorders.
5. Gene therapy and the Human Genome Project have raised many ethical issues that are yet to be resolved. The development and application of biotechnology in medicine have moved faster than a consensus about how to use and interpret this technology. Research and education are needed to make informed decisions.
6. Cloned DNA is finding a wide range of applications, including gene mapping and the identification and isolation of the genes responsible for genetic disorders. Positional cloning, which is based on recombinant DNA technology, allows a gene to be mapped and identified with no knowledge of the nature or function of the gene product.
7. Recombinant DNA techniques that detect allelic variants of variable tandem nucleotide repeats (DNA fingerprints) have found applications in forensics, paternity testing, and a wide range of other fields, including archaeology, conservation biology, and public health.

Insights and Solutions

1. Probes for DNA fingerprinting can be derived from a single locus or multiple loci. Two multiple-loci probes have been widely employed in both criminal and civil cases and derive from minisatellite loci on chromosome 1 (1cen-q24) and chromosome 7 (7q31.3). These probes, which are used because they produce a highly individual fingerprint, have determined paternity in thousands of cases over the last few years. The results of DNA fingerprinting of a mother (M), putative father (F), and child (C), using the aforementioned probes, are shown in the accompanying figure. The child has 6 maternal bands, 11 paternal bands, and 5 bands shared between the mother and the alleged father. Based on this fingerprint, can you conclude that this man is the father of the child?

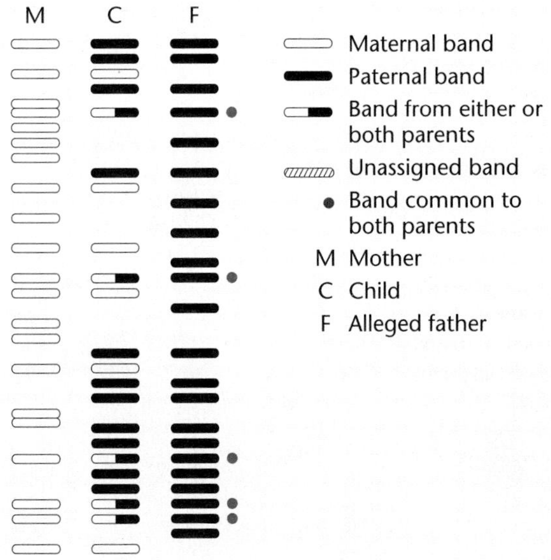

Solution: All bands present in the child can be assigned as coming from either the mother or the father. In other words, all the bands in the child's DNA fingerprint that are not maternal are present in the father. Since the father and the child share 11 bands and the child has no unassigned bands, paternity can be assigned with confidence. In fact, the chance that this man is not the father is on the order of 10^{-13}.

2. The DNA fingerprints of a mother, a child, and the alleged father in a second case are shown in the figure below. The child has 8 maternal bands, 15 paternal bands, 6 bands that are common to both the mother and the alleged father, and 1 band that is not present in either the mother or the alleged father. What are the possible explanations for the presence of the last band? Based on your analysis of the band pattern, which explanation is most likely?

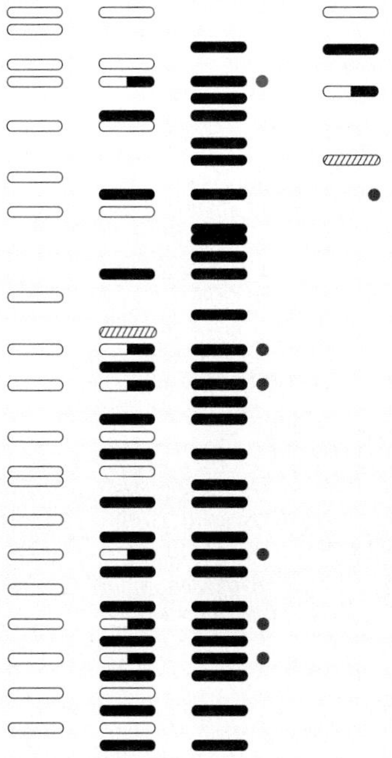

Solution: In this case, one band in the child cannot be assigned to either parent. Two possible explanations are that the child is mutant for one band or that the man tested is not the father. To estimate the probability of paternity, the mean number of resolved bands (n) is determined and the mean probability (x) that a band in individual A matches that in a second, unrelated individual B is calculated. In this case, because the child and the father share 15 bands in common, the probability that the man tested is not the father is very low (probably 10^{-7} or lower). As a result, the most likely explanation is that the child is mutant for a single band. In fact, in 1419 cases of genuine paternity resolved by the minisatellite probes on chromosomes 1 and 7, single bands in the children were recorded in 399 cases, accounting for 28 percent of all cases.

3. Infection by HIV-1 (human immunodeficiency virus) is responsible for the destruction of cells in the immune system and results in the symptoms of AIDS (acquired immunodeficiency syndrome). HIV infects and kills cells of the immune system that carry a cell-surface receptor known as CD4. An HIV surface protein known as gp120 binds to the CD4 receptor and allows the virus to enter into the cell. The gene encoding the CD4 protein has been cloned. How might this clone be used along with recombinant DNA techniques to combat HIV infection?

Solution: Several methods that use the CD4 gene are being explored to combat HIV infection. First, because infection depends on an interaction between the viral gp120 protein and the CD4 protein, the cloned CD4 gene has been modified to produce a soluble form of the protein (sCD4). The idea is that HIV can be prevented from infecting cells if the gp120 protein of the virus is bound up with the soluble form of the CD4 protein. Thus it is unable to bind to CD4 proteins on the surface of immune system cells. Studies in cell culture systems indicate that the presence of sCD4 effectively prevents HIV infection of tissue culture cells. However, studies in HIV-positive humans have been somewhat disappointing, mainly because the strains of HIV used in the laboratory are different from those found in infected individuals. Since HIV-infected cells carry the viral gp120 protein on their surface, the CD4 gene has been fused with genes encoding bacterial toxins to kill them. The resulting fusion protein contains CD4 regions that bind to gp120 and toxin regions that should kill the infected cell. In tissue culture experiments, cells infected with HIV are killed by the fusion protein, whereas uninfected cells survive. Researchers hope that the targeted delivery of drugs and toxins can be used in therapeutic applications to treat HIV infection.

PROBLEMS AND DISCUSSION QUESTIONS

1. In attempting to vaccinate people against diseases by having them eat antigens (such as the cholera toxin), the antigen must reach the cells of the small intestine. What are some potential problems of this method? Why don't absorbed food molecules stimulate the immune system and make you allergic to the food you eat?
2. One of the main safety issues associated with genetically modified crops is the potential for allergenicity caused by introduction of an allergen or caused by changing the level of expression of a host allergen. Since common allergenic proteins often have identical stretches of a few (6 or 7) amino acids in common, Kleter and Peijnenburg (2002. *BMC Struct. Biol.* 2:8.) developed a method for screening transgenic crops to evaluate potential allergenic properties. How do you think they accomplished this?
3. As of the end of 2004, there are more now more than 1000 cloned farm animals in the United States. Within the next two years, milk from cloned cows and their offspring (born naturally) may be available in supermarkets. These animals are clones, but have not been transgenically modified, and they are no different than are identical twins. Should milk from such animals and their naturally born offspring be labeled as coming from cloned cows or their descendants? Why?
4. One of the major causes of sickness, death, and economic loss in the cattle industry is *Mannheimia haemolytica*, which causes bovine pasteurellosis (shipping fever). Noninvasive delivery of vaccine using transgenic plants expressing immunogens would reduce labor costs and trauma to livestock. An early step toward developing an edible vaccine is to determine whether an injected version of an antigen (usually a derivative of the pathogen) is capable of stimulating the development of antibodies in a test organism. The table below assesses the ability of a transgenic portion of a toxin (Lkt) of *M. haemolytica* cloned into white clover to stimulate development of specific antibodies in rabbits. (Table modified from Lee et al. 2001. *Infect. and Immunity* 69:5786–93.) (a) What general conclusion can you draw from the data? (b) With regards to development of a usable edible vaccine, what work remains to be done?

Immunogen Injected	Antibody Production in Serum
Lkt50*—saline extract	+
Lkt50—column extract	+
Mock injection	–
Pre-injection	–

*Lkt50 is a smaller derivative of Lkt that lacks all hydrophobic regions.
+ indicates at least 50% neutralization of toxicity of Lkt; – indicates no neutralization activity.

5. Outline the steps involved in transferring glyphosate resistance to a crop plant. Do you envision that this trait can escape from the crop plant and make weeds glyphosate-resistant? Why or why not?
6. Now that enhancement therapy using one gene product has been approved, is this justification for using enhancement gene therapy?
7. Recombinant adenoviruses have been used in a number of preclinical studies to determine the efficacy of gene therapy for rheumatoid arthritis and osteoarthritis. Genes can be delivered by injection to the tissues that need them. Christopher Evans and colleagues (2001. *Arthritis Res.* 3:142–46), estimated that approximately 20 percent of all human gene therapy trials have used adenoviruses for gene delivery. The death of a patient in 1999 after infusion of adenoviral vectors has caused concern. As you

consider the use of viral vectors as therapy-delivery vehicles for human pathologies, what factors seem of paramount concern?

8. Define somatic gene therapy, germ-line therapy, and enhancement gene therapy. Which of these is currently in use?

9. *Transductional targeting* is a preferred route for the delivery of therapeutics for human diseases. It involves the development of tissue-specific interactions between the viral vector and a specific tissue. A genetic approach used by Ponnazhagan and colleagues (2002. *J. Virol.* 76:12,900–07) involves engineering the capsid of an adeno-associated virus (type 2) vector to target specific human cell types. Nongenetic approaches are also possible. Speculate on problems associated with the genetic approach of capsid alteration and problems that might be associated with nongenetic approaches to transductional targeting.

10. The development of safe vectors for human gene therapy has been a goal since 1990. Of the variety of problems associated with viral-based vectors, many such viruses (i.e., SV40) have transformation properties thought to be mediated by binding and inactivating gene products such as p53, retinoblastoma protein (pRB), and others. Mark Cooper and colleagues (1997. *Proc. Nat. Acad. Sci. (USA)* 94:6450–55) developed SV40-based vectors that are deficient in binding the p53, pRB, and other proteins. Why would you specifically want to avoid inactivating p53, pRB, and related proteins?

11. Gene therapy for human genetic disorders involves transferring a copy of the normal human gene into a vector and using the vector to transfer the cloned human gene into target tissues. Presumably, the gene enters the target tissue and becomes active, and the gene product relieves the symptoms. (a) Why are disorders such as muscular dystrophy difficult to treat by gene therapy? (b) What are the potential problems of using retroviruses as vectors? (c) Should gene therapy involve germ-line tissue instead of somatic tissue? What are some of the potential ethical problems associated with the former approach?

12. Sequencing the human genome and the development of microarray technology promises to improve our understanding of normal and abnormal cell behavior. In an article on the applications of microarray technology in breast cancer research (2001. *Breast Can. Res.* 3:158–75), Cooper stated that "data from such studies could revolutionize cancer diagnosis." (a) What is microarray technology? (b) In what way might this technology revolutionize our understanding of cancer?

13. What are the advantages of using STRs instead of VNTRs for DNA identification?

14. In producing physical maps of markers and cloned sequences, what advantage does *Drosophila* offer that other organisms, including humans, do not?

15. (a) Outline the steps involved in identifying a gene by positional cloning. What steps may cause difficulty in this process? (b) Once a region on a chromosome has been identified as containing a given gene, what kind of mutations would speed the process of identifying the locus?

16. Why is positional cloning a useful strategy in identifying and mapping a mutant gene? For what kind of genetic disorders would positional cloning be most appropriate?

17. What is an exclusion map and why is it useful?

18. The phenotype of many behavioral traits, such as manic depression and schizophrenia, may be controlled by several genes, each located on a different chromosome or chromosome segment. Can positional cloning be used to map and isolate such genes? What if a trait is controlled by six genes, each equally contributing to the phenotype in an additive way? Can positional cloning be used in this case? Why or why not?

19. Suppose you develop a screening method for cystic fibrosis that allows you to identify the predominant mutation *Δ508* and the next six most prevalent mutations. What must you consider before using this method to screen a population for this disorder?

20. A couple with European ancestry seeks genetic counseling before having children, because of a history of cystic fibrosis (CF) in the husband's family. ASO testing for CF reveals that the husband is heterozygous for the *Δ508* mutation and the wife is heterozygous for the *R117* mutation. You are the couple's genetic counselor. At a meeting with you, they express their conviction that they are not at risk for having an affected child because they each carry different mutations and cannot have a child who is homozygous for either mutation. What would you say to them?

21. Dominant mutations can be categorized according to whether they increase or decrease the overall activity of a gene or gene product. Although a **loss-of-function mutation** (a mutation that inactivates the gene product) is usually recessive, for some genes, one dose of the gene product is not sufficient to produce a normal phenotype. In this case, a loss-of-function mutation in the gene will be dominant, and the gene is said to be *haploinsufficient*. A second category of dominant mutations is **gain-of-function mutations**, which result in increased activity or expression of the gene or gene product. The phenotype of such a mutation results from too much gene product. The gene therapy technique currently used in clinical trials involves the "addition" to somatic cells of a normal copy of a gene. In other words, a normal copy of the gene is inserted into the genome of the mutant somatic cell, but the mutated copy of the gene is not removed or replaced. Will this strategy work for either of the two aforementioned types of dominant mutation?

22. Why are most recombinant human proteins produced in animal or plant hosts instead of bacterial host cells?

23. The DNA sequence surrounding the site of the sickle-cell mutation in the β-globin for normal and mutant genes is shown below.

5′– GACTCCTGAGGAGAAGT – 3′
3′– CTGAGGACTCCTCTTCA – 5′
Normal DNA

5′– GACTCCTGTGGAGAAGT – 3′
3′– CTGAGGACACCTCTTCA – 5′
Sickle-cell DNA

Each type of DNA is denatured into single strands and applied to a filter. The paper containing the two spots is hybridized to an ASO of the sequence

5′-GACTCCTGAGGAGAAGT-3′

Which spot, if either, will hybridize to this probe?

24. One form of hemophilia, an X-linked disorder of blood clotting, is caused by a mutation in clotting factor VIII. Many single-nucleotide mutations of this gene have been described, making the detection of mutant genes by Southern blots inefficient. There is, however, an RFLP for the enzyme *Hin*dIII contained in an intron of the factor VIII gene that can often be used in screening, as shown in the figure below.

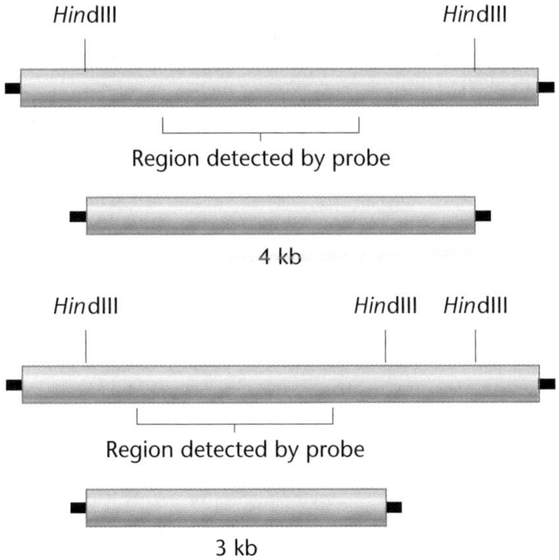

A female whose brother has hemophilia has a 50 percent risk of being a carrier of this disorder. To test her status, DNA is obtained from her white blood cells and those of family members, cut with *Hin*dIII, and the fragments are probed and visualized by Southern blotting. Using the results below, determine whether either of the females in generation II is a carrier for hemophilia.

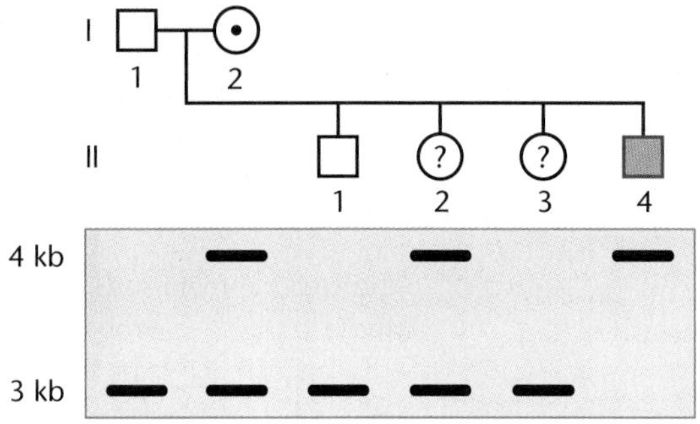

25. The human insulin gene contains introns. Since bacterial cells will not excise introns from mRNA, how can a gene like this be cloned into a bacterial cell and produce insulin?
26. In mice transfected with the rabbit β-globin gene, the rabbit gene is active in several tissues, including the spleen, brain, and kidney. In addition, some mice suffer from thalassemia (a form of anemia) caused by an imbalance in the coordinate production of α- and β-globins. Which problems associated with gene therapy are illustrated by these findings?
27. Genome scanning to detect susceptibility to diseases is already in use. As this technology becomes used more widely, medical records will contain the results of such testing. Who should have access to this information? Should employers, potential employers, or insurance companies be allowed to have this information? Would you favor or oppose having the government establish and maintain a central database containing the results of genome scanning on members of the population?
28. What limits the use of differences in restriction enzyme sites as a way of detecting point mutations in human genes?

Extra-Spicy Problems

29. You are asked to assist with a prenatal genetic test for a couple, each of whom is found to be a carrier for a deletion in the β-globin gene that produces β-thalassemia when homozygous. The couple already has one child who is unaffected and is not a carrier. The woman is pregnant, and the couple wants to know the status of the fetus. You receive DNA samples obtained from the fetus by amniocentesis, and from the rest of the family by extraction from white blood cells. Using a probe for the deletion, you obtain the blot shown below. Is the fetus affected? What is its genotype for the β-globin gene?

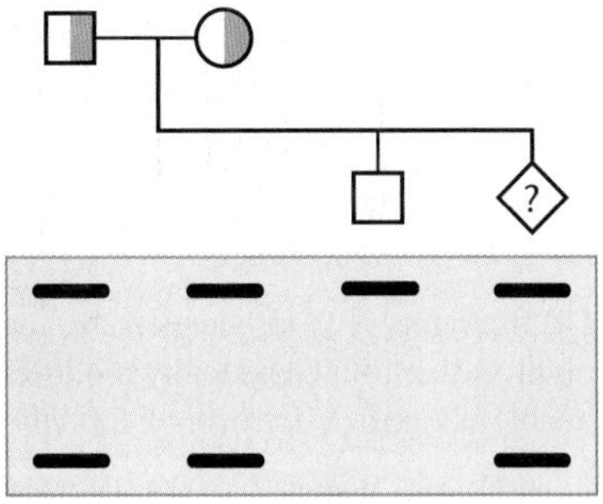

30. Shown below is a pedigree tracking the inheritance of a rare disease. (a) Which mode or modes of inheritance are excluded by or consistent with this pedigree?

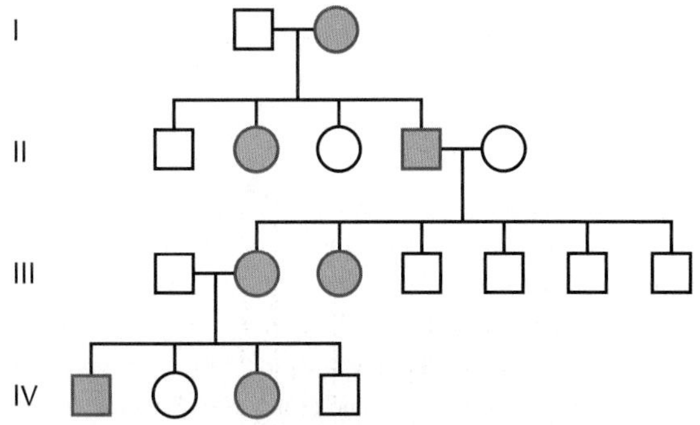

DNA samples from generations II and III above are obtained and subjected to RFLP linkage analysis. One RFLP is found on chromosome 10 (identified by probe *A*), and the other is found on chromosome 21 (identified by probe *B*). The results of the analysis are shown on the next page. Assume that additional data were gathered on this family and that they were consistent with the data shown and statistically significant. (b) On which chromosome is the disease gene located? (c) Individual III-1 is married to a normal man whose RFLP genotype is *A1A2* and *B1B1*. What kind of prenatal diagnostic test can be done to determine whether the child this couple is expecting will be normal? Describe what you can conclude regarding the result, and indicate the accuracy of the test. (d) Individual III-4 marries a woman of genotype *B1B1*. This couple has a child who is *B1B1*. The father assumes that the child is illegitimate, but the mother, who has taken a genetics course, argues that the child could be the result of an event that occurred in the father's germ line and cites two possibilities. What are they?

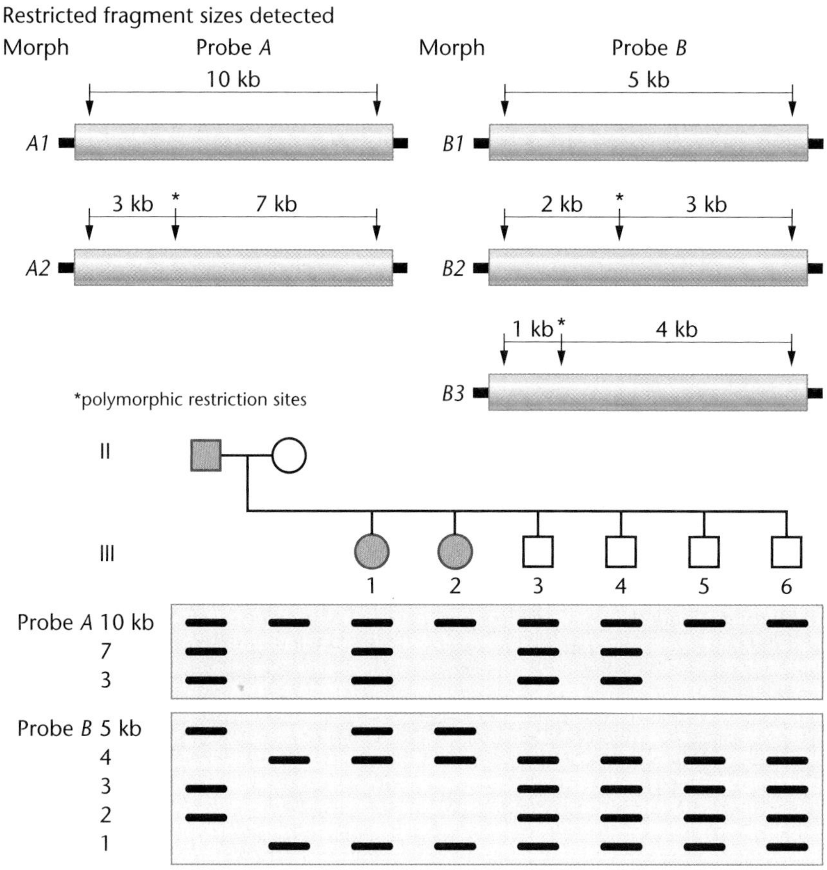

SELECTED READINGS

Anderson, W.F. 2000. The best of times, the worst of times. *Science* 288:627–28.

Atherton, K.T. 2002. Safety assessment of genetically modified crops. *Toxicol.* 81/182:421–26.

Bleck, O., McGrath, J.A., and South, A.P. 2001. Searching for candidate genes in the new millennium. *Clin. Exp. Dermatol.* 26:279–83.

Cavanna-Calvo, M. et al. 2000. Gene therapy of severe combined immunodeficiency (SCID)-XI disease. *Science* 288:669–72.

Dale, P.J., Clarke, B., and Fontes, E.M. G. 2002. Potential for the environmental impact of transgenic crops. *Nat. Biotechnol.* 20:567–74.

Daniell, H., Streatfield, S.J., and Wycoff, K. 2001. Medical molecular farming: Production of antibodies *Trends in Plant Sci.* 6:219–226.

Engler, O.B. et al. 2001. Peptide vaccines against hepatitis B virus: From animal model to human studies. *Mol. Immunol.* 38:457–65.

Gaugitsch, H. 2002. Experience with environmental issues in GM crop production and the likely future scenarios. *Toxicol. Letters* 127:351–57.

Kmiec, E.B. 1999. Gene therapy. *Amer. Scient.* 87:240–47.

Pray, C.E., Huang, J., and Rozelle, S. 2002. Five years of Bt cotton in China: the benefits continue. *The Plant J.* 31:423–30.

Riley, J.H. et al. 2000. The use of single nucleotide polymorphisms in the isolation of common disease genes. *Pharmacogen.* 1:39–47.

Schillberg, S., Fischer, T., and Emans, N. 2003. Molecular farming of recombinant antibodies in plants. *Cell Mol. Life Sci.* 60:433–45.

Thomas, C.E., Ehrhardt, A., and Kay, M.A. 2003. Progress and problems with the use of viral vectors for gene therapy. *Nat. Rev. Genet.* 4:346–58.

Walmsley, A.M., and Artzen, C.J. 2003. Plant cell factories and mucosal vaccines. *Curr. Opin. Biotechnol.* 14:145–50.

Wisniewski, J-P., Frangne, N., Massonneau, A., and Dumas, C. 2002. Between myth and reality: Genetically modified maize, an example of a sizeable scientific controversy. *Biochimie* 84:1095–1103.

Yang. X., Tian, X.C., and Wang, B. 2000. Transgenic farm animals: applications in agriculture and biomedicine. *Biotechnol. Annu. Rev.* 5:269–92.

Developmental Genetics of Model Organisms

This unusual four-winged Drosophila has developed an extra set of wings as a result of a homeotic mutation.

Chapter Concepts

- Gene action in development is based on differential transcription of selected genes and not on retention and loss of genes in the genome.
- Multicellular animals use a small number of signaling systems and regulatory networks to construct adult body forms from the zygote, making it possible to use animal models to study human development.
- Differentiation is controlled by cascades of gene action that follow the establishment of developmental fate.
- Plants independently evolved developmental mechanisms paralleling those of animals.
- Cell–cell signaling programs the developmental fate of adjacent and distant cells.
- Apoptosis, programmed cell death, is a normal part of development.

In multicellular plants and animals, a fertilized egg undergoes a cycle of developmental events that ultimately give rise to an adult member of the species from which the egg and sperm were derived. Thousands, millions, or even billions of cells are organized into a cohesive and coordinated unit that we perceive as a living organism (Figure 23–1). The series of events through which organisms attain their final adult forms is studied by developmental biologists. This area of study is perhaps the most intriguing in biology. Not only does the comprehension of developmental processes require the knowledge of many different biological disciplines—such as molecular, cellular, and organismal biology—but also an understanding of how biological processes change over time and ultimately transform the organism.

In the past hundred years, investigations in embryology, genetics, biochemistry, molecular biology, cell physiology, biophysics, and evolution have all contributed to the study of development. The findings have largely pointed out the tremendous complexity of developmental processes. Unfortunately, a description of *what* happens does not answer the "why" and "how" of development.

Over the last two decades, however, genetic analysis and molecular biology have shown that in spite of wide diversity in the size and shape of the adult stage, all multicellular organisms share many genes, genetic pathways, and molecular signaling mechanisms in the processes leading from the zygote to the adult. Development is marked by two important events: **determination**, the time when a specific developmental fate for a cell becomes fixed, and **differentiation**, the process by which a cell achieves its final form and function. In addition, genomics has shown that close relationships exist among higher organisms. Genetic analysis has identified genes that regulate developmental processes, and we are beginning to understand how the action and interaction of these genes control basic developmental processes in members of the animal kingdom.

In this chapter, the primary emphasis will be on how genetics has been used to study development. This area, called developmental genetics, has contributed tremendously to our understanding of developmental processes, because genetic information is required for the molecular and cellular functions mediating developmental events and contributes to the final phenotype of the newly formed organism.

(a)

(b)

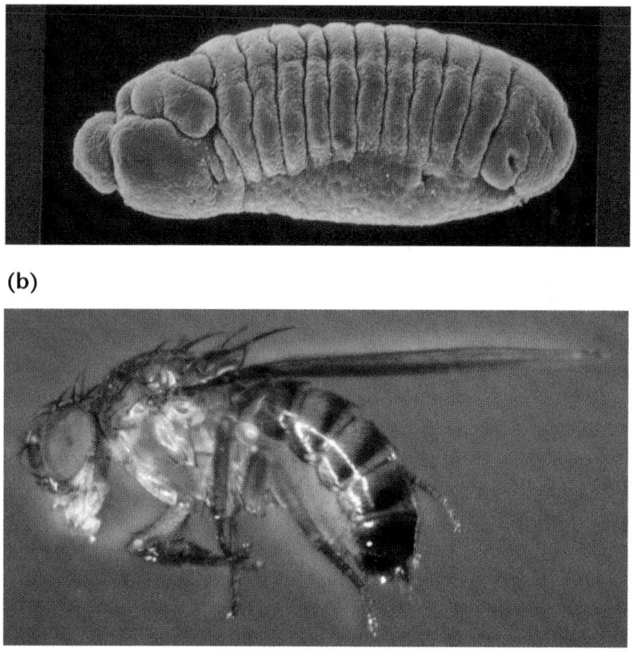

FIGURE 23–1 (a) A *Drosophila* embryo and (b) the adult fly that develops from it.

23.1 Developmental Genetics Seeks to Explain How a Differentiated State Develops from an Organism's Genome

Developmental geneticists use mutations to ask important questions about development: What genes are expressed? When are they expressed? In what cells are they expressed? and At what level does expression take place? These questions about gene expression are the foundation for exploring the molecular basis for developmental processes such as determination, induction, cell–cell communication, and cellular differentiation. The goal is to use genetic analysis to establish a causal relationship between the presence or absence of inducers, receptors, transcriptional events, and cell and tissue interactions, and the observable morphological events that accompany development.

A useful way to define **development** is to say that it is the *attainment of a differentiated state* by all the cells in an organism. For example, a cell in a blastula-stage embryo (when the embryo is just a ball of uniform-looking cells) is undifferentiated, while a red blood cell synthesizing hemoglobin in the adult body is differentiated. How do cells get from the undifferentiated to the differentiated state? The process involves the progressive activation of different gene sets in different cells of the embryo. From a genetics perspective, one way of defining the collection of different cell types that form during development in multicellular organisms is to catalog the genes that are active in each cell type. In other words, development depends on patterns of differential gene expression.

The idea that differentiation is accomplished by activating and inactivating genes at different times and in different cell types is called the **variable gene activity hypothesis**. Its underlying assumptions are, first, that each cell contains an entire genome, and second, that differential transcription of selected genes controls the development and differentiation of each cell. In multicellular organisms, evolution has conserved the genes involved in development, the patterns of differential transcription, and the ensuing developmental mechanisms, allowing the use of genetically well-characterized model organisms to dissect development.

23.2 Conservation of Developmental Mechanisms and the Use of Model Organisms

Genetic analysis of development in a wide range of organisms has demonstrated that there are only a small number of developmental mechanisms and signaling systems used in all multicellular organisms. For example, eye formation in *Drosophila* and eye formation in humans is controlled by the same gene sets working through the same regulatory network. In fact, mouse eye genes inserted into flies regulate eye formation in *Drosophila*! As another example, most of the differences in size and shape between zebras and zebrafish are controlled by different patterns of expression in one gene set, not by different genes. Genome sequencing projects have confirmed that genes from a wide range of organisms are conserved.

Although many developmental mechanisms are similar among all multicellular animals, evolution has also generated new and different approaches to transform a zygote into an adult. This has involved several mechanisms, including mutation, gene duplication and divergence, the assignment of new functions to old genes, and the recruitment of genes to new developmental pathways. For example, the turtle shell, a structure unique among reptiles, is formed by adapting a signaling pathway to a new function. The shell forms by outgrowth of the dorsal body wall under control of the fibroblast growth factor 10 (*FGF-10*) gene, a component of a signal pathway that controls development of other outgrowths: limbs and lungs. The emphasis in this chapter, however, will be on the similarities among species.

Model Organisms in the Study of Development

Historically, geneticists have used a handful of organisms to study development. These include yeast (*Saccharomyces cerevisiae*), the fruit fly (*Drosophila melanogaster*), the nematode (*Caenorhabditis elegans*), zebrafish (*Danio rerio*), the mouse (*Mus musculus*), and the small flowering plant *Arabdopsis thaliana*. The genetics of these species has been well characterized, and the genomes of each have been sequenced. A large catalog of mutants affecting important steps in development is available for all these species, and there are well-established procedures for making and maintaining new mutant strains. With the exception of plants, all these species share gene regulatory pathways and developmental mechanisms with humans. The use of these organisms has helped identify genes that control disorders of human development, and provided insights into gene–environment interactions that affect the outcome of key steps in development.

Analysis of Developmental Mechanisms

In the space of this chapter, we cannot survey all aspects of development, nor explore the genetic analysis of all developmental mechanisms triggered by the fusion of sperm and egg. Instead, we will focus on three general processes in development: specification of the body axis, the program of gene expression that turns undifferentiated cells into differentiated cells, and the role of cell–cell communication in development. Interwoven with the discussion of these mechanisms will be several other topics: the role of master switch genes in controlling developmental pathways, cytoplasmic components in the oocyte that control early transcriptional patterns, how cells become determined to form adult structures, and patterns of the genetics of signaling systems that alter patterns of gene expression.

We will use three model systems: *D. melanogaster*, *C. elegans*, and *A. thalania* to illustrate these three developmental processes and the related topics. Programming cells for a specific developmental pathway often involves action of a master gene that works as a binary switch. We will introduce this topic before examining the role of differential gene expression in the progressive restriction of developmental options leading to the formation of the body axis. We will then expand the discussion to include the selection of pathways that result in differentiated cells in plants and animals, and consider the role of cell–cell communication in development. Finally, the chapter will conclude with a discussion of the most dramatic restriction of cell fate during development—namely, programmed cell death (apoptosis).

Basic Concepts in Developmental Genetics

In higher eukaryotes, development begins with the formation of the zygote, a cell generated by the fusion of the sperm and oocyte. The oocyte is a cell with a heterogeneous cytoplasm and a nonuniform distribution of its components. Following fertilization and early cell divisions, the nuclei of progeny cells find themselves in different environments as the maternal cytoplasm is distributed into the new cells. Evidence suggests that in different cells, the cytoplasm exerts different influences on the genetic material, causing differential transcription at specific points during development. Hence, cytoplasmic localization, or the inheritance of particular cytoplasmic components by individual embryonic cells, plays a major regulatory role during development. Early gene products synthesized by the zygote's genome further alter the cytoplasm of each cell, producing a still different cellular environment that, in turn, leads to the activation of other genes, and so on. As the number of cells in the embryo increases, they influence one another by cell–cell interaction. The environmental factors acting on the genetic material, therefore, now include the cytoplasm of individual cells, as well as signals from other cells. Although early in embryogenesis most cells show no evidence of structural or functional specialization, the combination of their localized cytoplasmic components and their position in the developing embryo determine the ultimate structure and functions they will assume. It is as if their fate has been programmed prior to the actual events leading to specialization.

As embryonic cells respond to their continually changing external and internal environments, they embark on a developmental pathway. Selection of a specific developmental pathway for a cell often involves action by a single master gene that initiates a cascade of expression by other genes.

23.3 Master Switch Genes Program Genomic Expression

As part of the progressive restriction of transcriptional activity that accompanies development, certain genes act as switches, decreasing the number of alternative developmental pathways that a cell can follow. Each decision point is usually binary—that is, there are two alternative developmental fates for a cell at a given time—and the action of a switch gene programs the cell to follow only one of these pathways. These genes are called **master regulatory genes** or **binary switch genes**. They are defined by their ability to initiate the complete development of an organ or a tissue type. We will briefly describe how a binary switch gene controls the formation of the eye and how this regulatory pathway is used in all organisms with eyes.

The Control of Eye Formation

Drosophila adults have compound eyes that develop in the preadult stage [Figure 23–2(a)]. In flies homozygous for the recessive allele *eyeless*, eye development is abnormal, and adults have no eyes [Figure 23–2(b)]. The protein encoded by the wild-type allele acts very early in development and is expressed in all cells that will give rise to the eye. In loss of function mutants, cells normally destined to become part of the eye degenerate late in development and undergo apoptosis (discussed at the end of this chapter), and no eyes are formed. The *eyeless* gene is one of seven genes controlling eye formation in *Drosophila*. Loss of function mutations in any of these genes produce adults without eyes. Because *eyeless* is the first gene in this set to be expressed during development, this gene has been dubbed the master regulator of eye development. The eyeless gene encodes a transcription factor, and is regarded as a master regulator or binary switch gene because it operates at the beginning of a developmental pathway and it activates the expression of other genes that also encode transcription factors. Activation of these genes leads to the expression of a large gene set that controls the development and differentiation of the adult eye.

Expression of *eyeless* in other organs such as legs, wings, and antennae produces eyes in other locations [Figure 23–3(a)]. This indicates that *eyeless* initiates the program of gene expression for eye formation in cells that do not normally form eyes and is able to override the program of determination and differentiation present in these cells.

The *eyeless* gene is part of a network of seven genes [Figure 23–3(b)]. This network is regarded as the master regulator of eye formation. Some of the genes in the network are required for the expression of other genes, while others reinforce the transcription of network members, and others are responsible for growth of the eye.

The eyeless gene and the other genes in this network are important because this gene set has been conserved through evolution and is used by all animals, including humans, to make eyes. The human and fly genes are compared in Table 23.1. The discovery that *eyeless* directs the formation of eyes in vertebrates has forced reevaluation of the long-held belief that the compound eye of insects and the single-lens eye of vertebrates evolved separately. Such an assumption was based on the observation that the compound insect eye and the vertebrate camera eye have different embryonic origins, develop by different pathways, and are structurally very different.

(a) (b)

FIGURE 23–2 (a) The normal compound eyes of adult *Drosophila*. (b) In flies homozygous for the *eyeless (ey)* mutant, eye development is abnormal, and adults have no eyes. The *ey* gene is a master regulator of eye development in all animals.

TABLE 23.1

Vertebrate Gene	*Drosophila* Homolog	Expression in Vertebrate Eye	Loss of Function
Pax6	*eyeless, twin of eyeless*	Lens placode, optic vesicle	Aniridia (human), *small eye* (mouse)
Bmp4	*dpp*	Optic vesicle, head ectoderm	No lens (mouse)
Bmp7	*60A*	Optic vesicle, head ectoderm	No lens (mouse)
Eyal	*eyes absent*	Perioptic mesenchyme, weak lens expression	No eye phenotype (human) or (mouse), some human mutations lead to cataracts and anterior defects
Six3	*sine oculis*	Lens placode, optic vesicle	Very small eyes (human)
Optx2	*optix*	Optic vesicle	No eyes (human)
Dach1	*dachshund*	Optic vesicle	*n.d.

*n.d., not determined

Source: Wawersik, S. and Maas, R.L. 2000. Vertebrate eye development as modelled in *Drosophila*. *Hum. Mol. Genet.* 9:917–925, Table 1, p. 921.

In 1995, Walter Gehring and his colleagues generated transgenic flies that carried copies of the mouse homologue of *eyeless*. The results were striking and led to two inescapable conclusions. First, this single switch gene was capable of triggering the formation of extra eyes on the wings, antennae, and legs of flies [Figure 23–3(a)], and second, because the mouse gene works in flies, the invertebrate and vertebrate eye *are* in fact homologous at the molecular level. Therefore, these eyes are related evolutionarily. The downstream targets of these transcription factors are also conserved, indicating that steps in the genetic regulation of eye development are shared between species that diverged over half a billion years ago (from a common ancestor without eyes). This evolutionary conservation makes it possible to use genetic analysis in *Drosophila* to study the development of eyes in humans and to understand the molecular basis for inherited eye defects.

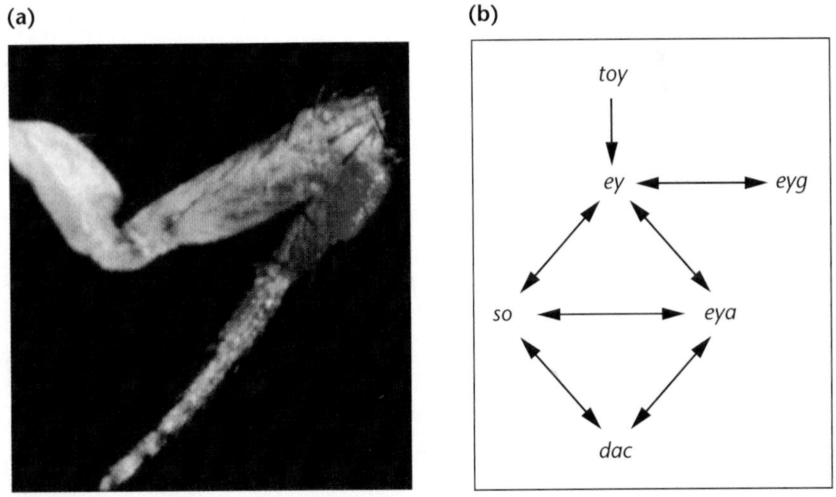

FIGURE 23–3 (a) Eye formation in transgenic flies carrying the mouse homolog of eyeless. Complete eyes are formed at ectopic locations, including the leg. (b) The network of genes that regulate eye formation in *Drosophila*.

HOW DO WE KNOW?

What evidence links eye formation in humans and other vertebrates to eye formation in invertebrates such as *Drosophila*?

23.4 Genetics of Embryonic Development in *Drosophila*: Specification of the Body Axis

Why a certain cell turns specific genes on or off at specific stages of development is a central question in developmental biology. At present, there is no simple answer to this question. However, information derived from the study of model organisms gives us a starting point. In particular, the genetic and molecular analysis of embryonic development in *Drosophila* highlights the key role of molecular components placed in the oocyte cytoplasm during oogenesis in controlling gene expression.

Overview of *Drosophila* Development

Beginning with the fertilized egg, *Drosophila* passes through a preadult period of development with five distinct phases: the embryo, three larval stages, and the pupal stage. The adult fly emerges from the pupal case about 10 days after fertilization (Figure 23–4). Externally, the *Drosophila* egg has a number of structures that delineate the anterior, posterior, dorsal, and ventral regions (Figure 23–5). The anterior end of the egg contains the micropyle, a specialized conical structure for the entrance of sperm into the egg, while the posterior end is rounded and marked by a series of aeropyles (openings that allow gas exchange during development). The dorsal side of the egg is flattened and contains the chorionic appendages, while the ventral side is curved. Internally, the egg cytoplasm

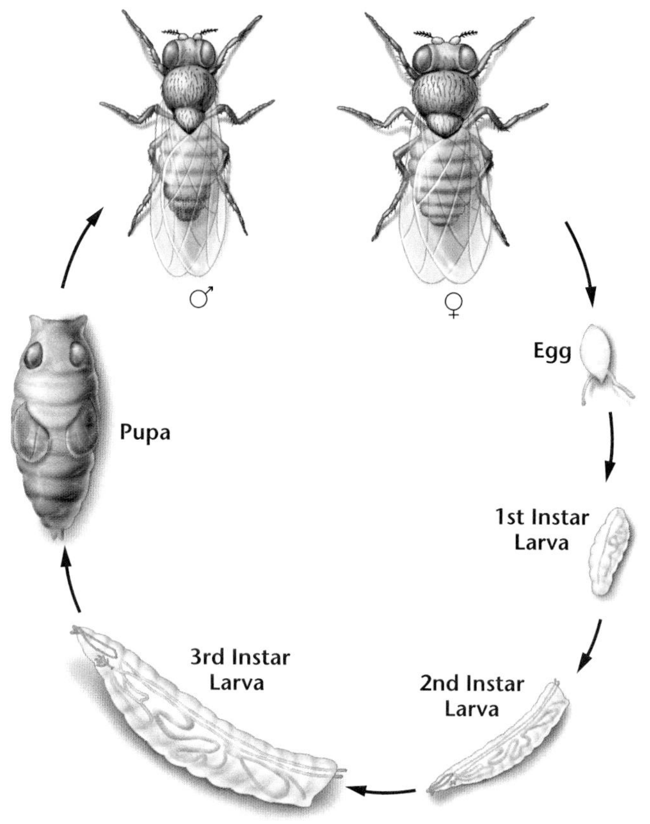
♂
♀
Egg
Pupa
1st Instar
Larva
3rd Instar
Larva
2nd Instar
Larva

FIGURE 23–4 *Drosophila* life cycle.

is organized into a series of maternally derived molecular gradients. These gradients play a key role in establishing the developmental fates of nuclei that migrate into specific regions of the embryo.

Immediately after fertilization, the zygote nucleus undergoes a series of divisions without cytokinesis [Figure 23–6(a)

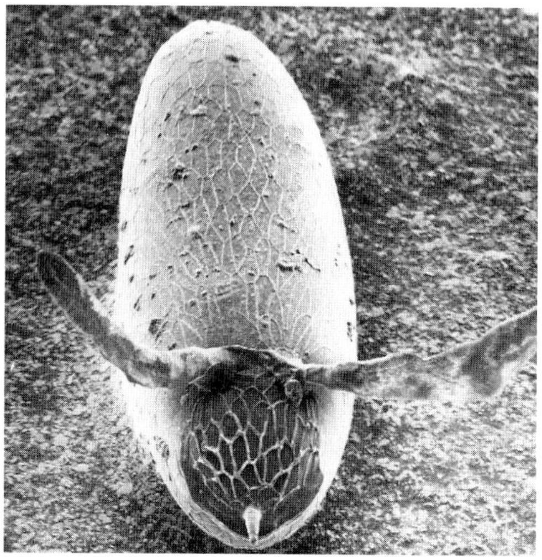

FIGURE 23–5 Scanning electron micrograph of a newly oviposited *Drosophila* egg. The micropyle is the small, white projection at the anterior tip (at the bottom). The dorsal surface (with the two chorionic appendages) is facing up.

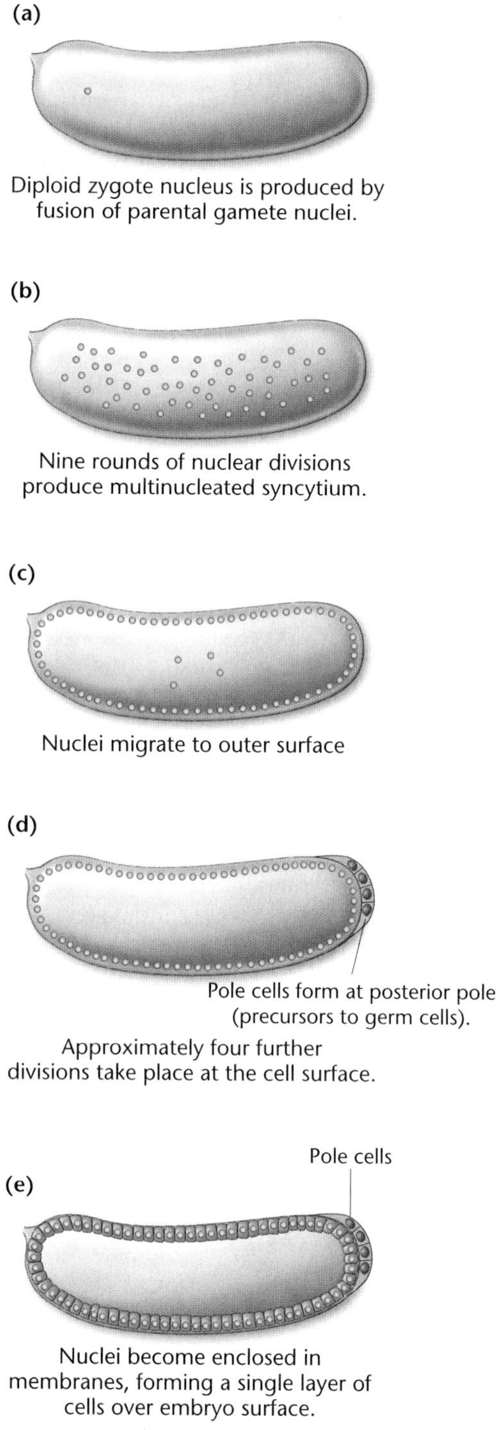
(a)
Diploid zygote nucleus is produced by fusion of parental gamete nuclei.
(b)
Nine rounds of nuclear divisions produce multinucleated syncytium.
(c)
Nuclei migrate to outer surface
(d)
Pole cells form at posterior pole (precursors to germ cells).
Approximately four further divisions take place at the cell surface.
Pole cells
(e)
Nuclei become enclosed in membranes, forming a single layer of cells over embryo surface.

FIGURE 23–6 Early stages of embryonic development in *Drosophila*. (a) Fertilized egg with zygotic nucleus, about 30 minutes after fertilization. (b) Nuclear divisions occur about every 10 minutes, producing a multinucleate cell, the syncytial blastoderm. (c) After approximately nine divisions (512 nuclei), the nuclei migrate to the outer surface or cortex of the egg. (d) At the surface, four additional rounds of nuclear division occur. A small cluster of cells, the pole cells, form at the posterior pole about 2.5 hours after fertilization. These cells will form the germ cells of the adult. (e) About 3 hours after fertilization, the nuclei become enclosed in membranes, forming a single layer of cells over the embryo surface, creating the cellular blastoderm.

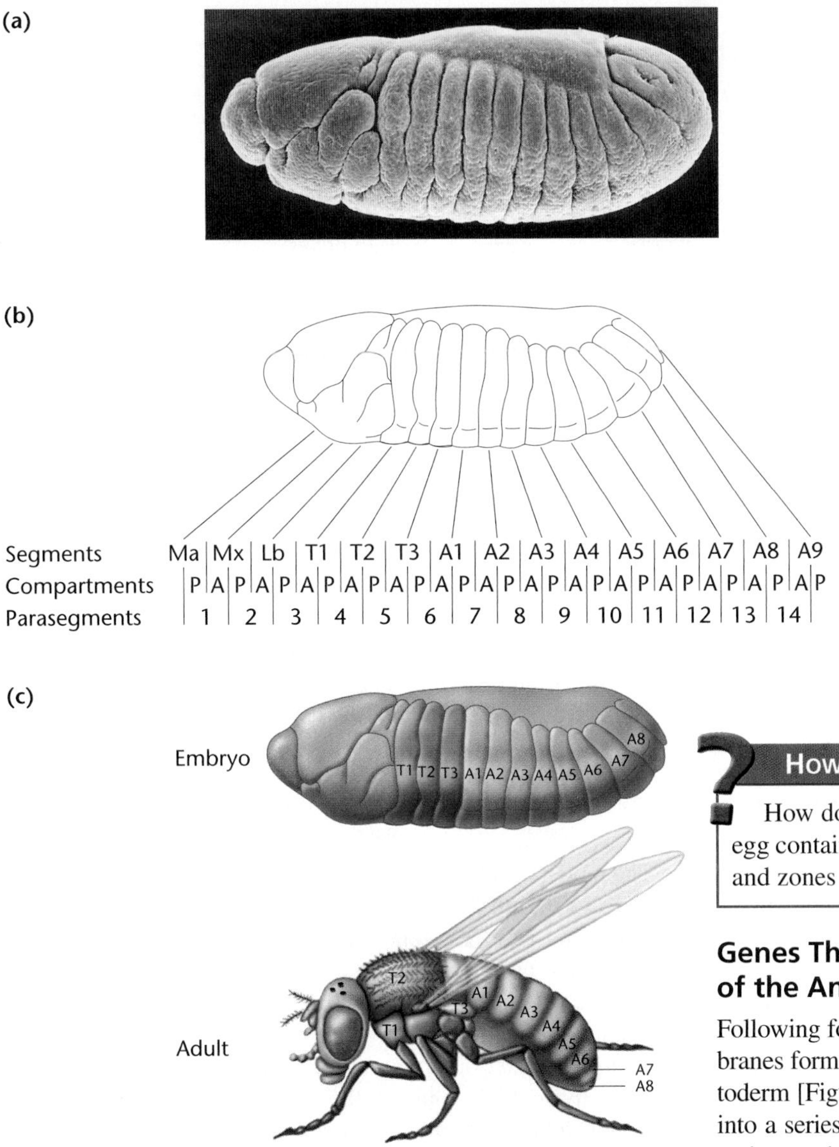

FIGURE 23–7 Segmentation in *Drosophila*. (a) Scanning electron micrograph of a *Drosophila* embryo at about 10 hours after fertilization. At this stage, the segmentation pattern of the body is clearly established. (b) The segments, compartments, and parasegments of the *Drosophila* embryo. Ma, Mx, and Lb represent the segments that will form head structures. T1–T3 are thoracic segments, and A1–A9 are abdominal segments. Each segment is divided into anterior (A) and posterior (P) compartments. Parasegments represent an early pattern specification that is later refined into the segmental plan of the body. Note that all the parasegments are shifted forward by one compartment to form the segments. (c) The segmented embryo and the adult structures that will form each segment.

and (b)]. After nine rounds of division, the zygote cytoplasm contains approximately 512 nuclei in a single cell. The nuclei migrate to the egg's outer surface, or cortex, where further divisions take place [Figure 23–6(c) and (d)], forming the syncytial blastoderm (a syncytium is any cell with more than one nucleus.) Arranged around the periphery of the egg, the nuclei are surrounded by cytoplasm containing localized gradients of maternally derived transcripts and proteins. A program of gene expression is initiated in the nuclei, under the control of these cytoplasmic components, leading to different developmental programs in different cells.

The formation of germ cells at the posterior pole of the embryo demonstrates the regulatory role of these localized cytoplasmic components [Figure 23–6(d) and (e)]. Experiments have shown that any nucleus placed in the posterior cytoplasm (the pole plasm) will form germ cells. Hence, the cytoplasm of the posterior pole contains maternal components that direct nuclei to form germ cells. The transcriptional programs triggered in the rest of the migrating nuclei form the embryo's anterior–posterior and dorsal–ventral axes of symmetry.

? HOW DO WE KNOW?

How do we know that the cytoplasm of the *Drosophila* egg contains maternal components organized into gradients and zones of unequal concentration?

Genes That Regulate Formation of the Anterior–Posterior Body Axis

Following formation of the syncytial blastoderm, plasma membranes form around individual nuclei, creating the cellular blastoderm [Figure 23–6(e)]. Later, the embryo becomes organized into a series of segments defined by the expression patterns of various embryonic genes (Figure 23–7). Still later, these give rise to the adult segments, though they are shifted slightly in position. Within the segments, cells first become determined to form either the anterior or posterior compartment of the segment. In later stages, a cell's developmental options become further restricted, so that all the cells in a compartment are eventually programmed to form a single structure in the adult body.

The adult body plan of *Drosophila* derives its overall organization from the larval body plan and is composed of head, thoracic, and abdominal segments. Much of the larval body is broken down during the pupal stage, and many adult body structures are constructed from small cell clusters called **imaginal discs**, formed during larval stages. There are 12 bilaterally paired discs—eye-antennal discs, leg discs, wing discs, and so forth—and one genital disc. Figure 23–8 shows which imaginal discs will give rise to which adult structures.

Genetic Analysis of Embryogenesis

Knowledge of anatomical development in *Drosophila* has been useful, but genetic analysis has provided a wealth of information about the events in embryogenesis. Genes controlling

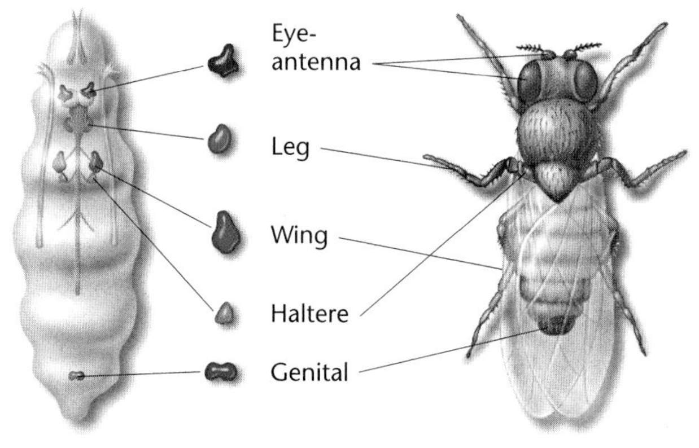

FIGURE 23–8 Imaginal discs of *Drosophila* larva and the adult structures derived from them.

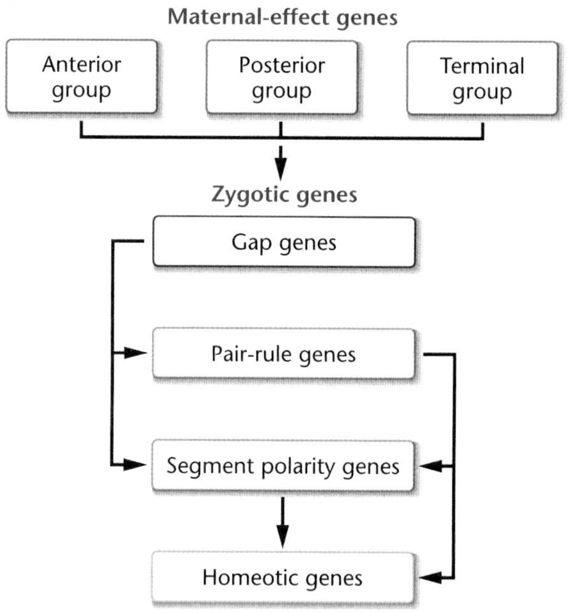

FIGURE 23–9 The hierarchy of genes involved in establishing the segmented body plan in *Drosophila*. Gene products from the maternal genes regulate the expression of the first three groups of zygotic genes (gap, pair-rule, and segment polarity, collectively called the segmentation genes), which in turn control expression of the homeotic genes.

embryonic development are of two types: maternal-effect genes and zygotic genes. Products of maternal-effect genes (mRNA and/or proteins) are deposited in the developing egg during oogenesis. Many of these products are distributed in a gradient or concentrated in specific regions of the egg cytoplasm. Female flies carrying deleterious mutations in maternal-effect genes are sterile since none of the embryos of females homozygous for a recessive mutation receive wild-type gene products from their mother, and therefore develop abnormally. In *Drosophila*, these maternal-effect genes encode transcription factors, receptors, and proteins that regulate translation. During embryonic development, these gene products activate or repress expression of the zygotic genome in a temporal and spatial sequence.

Zygotic genes are expressed in the developing embryo, and are therefore transcribed after fertilization. The phenotype of flies in this class with deleterious mutations exhibit embryonic lethality. In a cross between two flies heterozygous for a recessive zygotic mutation, one fourth of the embryos (the homozygotes) fail to develop normally and die. In *Drosophila*, many zygotic genes are transcribed in specific regions of the embryo that depend on the distribution of maternal-effect proteins.

Much of what is known about the zygotic genes that regulate development comes from a mutagenic analysis performed by Christiane Nüsslein-Volhard and Eric Wieschaus, who systematically screened for mutations that affect embryonic development. They examined thousands of dead offspring of mutagenized flies, looking for recessive embryonic lethal mutations with defects in external structures. The parents were thus identified as carriers of these mutations, which they grouped into three classes: *gap*, *pair-rule*, and *segment polarity* genes. In a paper published in 1980, these two scientists proposed a model in which embryonic development is initiated by gradients of maternal-effect gene products. The positional information laid down by these molecular gradients along the anterior–posterior axis of the embryo is interpreted by two sets of zygotic genes: (1) those identified in their screen—gap, pair-rule, and segment polarity genes—collectively called **segmentation genes** (these genes divide the embryo into a series of stripes or segments and define the number, size, and polarity of each segment), and (2) homeotic genes, which specify the identity or fate of each segment (Figure 23–9).

The model developed by Nüsslein-Volhard and Wieschaus is shown in Figure 23–10. The developmental process begins when maternal-effect gene products responsible for forming the anterior–posterior axis, which were placed in the egg during oogenesis, are activated immediately after fertilization [Figure 23–10(a)]. Their activity restricts cells to forming either anterior or posterior structures. Gene products from this gradient activate transcription of the gap genes that divide the embryo into a limited number of broad regions [Figure 23–10(b)]. Gap proteins are transcription factors that activate pair-rule genes, whose products divide the embryo into smaller regions about two segments wide [Figure 23–10(c)]. The combined action of all gap genes defines segment borders. The pair-rule genes in turn activate the segment polarity genes, which then divide the segments into anterior and posterior compartments [Figure 23–10(d)]. The collective action of the maternal genes that form the anterior–posterior axis and the segmentation genes define the fields of action for the homeotic (*Hox*) genes [Figure 23–10(e)].

In a second screening, Eric Wieschaus and Trudi Schüpbach examined thousands of flies for maternal-effect mutations that affected external structures of the embryo. They estimated that about 40 maternal-effect genes and 50 to 60 zygotic genes regulate embryonic development in *Drosophila*. This means that only about 100 genes control normal embryogenesis, a surprisingly low number given the complexity of the structures formed from the fertilized egg. For their work on the genetic control of development in *Drosophila*, Nüsslein-Volhard, Wieschaus, and E. B. Lewis—the geneticist who initially identified and studied many of these genes in the 1970s—were awarded the 1995 Nobel Prize for Physiology or Medicine.

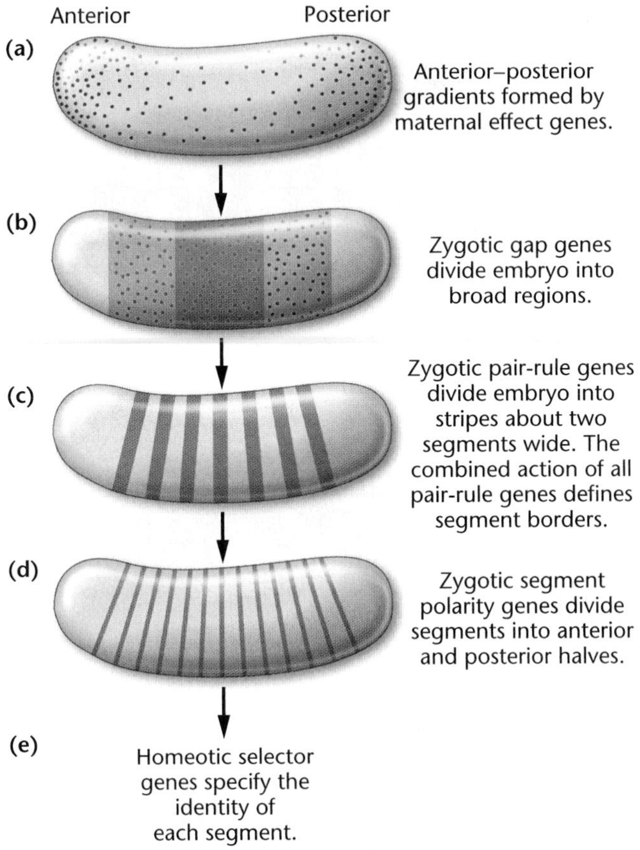
Anterior
Posterior
(a)
Anterior–posterior gradients formed by maternal effect genes.
(b)
Zygotic gap genes divide embryo into broad regions.
(c)
Zygotic pair-rule genes divide embryo into stripes about two segments wide. The combined action of all pair-rule genes defines segment borders.
(d)
Zygotic segment polarity genes divide segments into anterior and posterior halves.
(e)
Homeotic selector genes specify the identity of each segment.

FIGURE 23–10 (a) Progressive restriction of cell fate during development in *Drosophila*. Gradients of maternal proteins are established along the anterior–posterior axis of the embryo. (b), (c), and (d) Three groups of segmentation genes progressively define the body segments. (e) Individual segments are given identity by the homeotic genes.

Now solve this

Problem 23.7 on page 597 involves screening for mutants that affect external structures of the embryo.

Hint: In reconciling the results of your study with the conclusions of Wieschaus and Schüpbach, remember the differences between genes and alleles.

23.5 Zygotic Genes Program Segment Formation in *Drosophila*

Zygotic genes are activated or repressed in a positional gradient by the maternal-effect gene products. The expression of three subsets of these genes divides the embryo into a series of segments along the anterior–posterior axis. These segmentation genes are normally transcribed in the developing embryo, and their mutations have embryonic lethal phenotypes.

Over 20 segmentation loci have been identified (Table 23.2). They are classified on the basis of their mutant phenotypes: (1) gap genes delete a group of adjacent segments, (2) pair-rule genes affect every other segment and eliminate a specific part of each affected segment, and (3) segment polarity genes cause defects in homologous portions of each segment. We'll examine each group in greater detail.

TABLE 23.2 SEGMENTATION GENES IN DROSOPHILA

Gap Genes	Pair-Rule Genes	Segment Polarity Genes
Krüppel	*hairy*	*engrailed*
knirps	*even-skipped*	*wingless*
hunchback	*runt*	*cubitis interruptus^D*
giant	*fushi-tarazu*	*hedgehog*
tailless	*odd-paired*	*fused*
buckebein	*odd-skipped*	*armadillo*
	sloppy-paired	*patched*
		gooseberry
		paired
		naked
		disheveled

Gap Genes

Transcription of gap genes is activated or inactivated by gene products previously expressed along the anterior–posterior axis and by other genes of the maternal gradient system. When mutated, these genes produce large gaps in the embryo's segmentation pattern. *hunchback* mutants lose head and thorax structures, *Krüppel* mutants lose thoracic and abdominal structures, and *knirps* mutants lose most abdominal structures. Transcription of gap genes divides the embryo into a series of broad regions (the head, thorax, and abdomen). Within these regions, different combinations of gene activity eventually specify both the type of segment that forms and the proper order of segments in the body of the larva, pupa, and adult. To date, all gap genes that have been cloned encode transcription factors with zinc finger DNA-binding motifs. The expression of the gap genes correlates roughly with their mutant phenotypes: *hunchback* at the anterior, *Krüppel* in the middle, and *knirps* at the posterior (Figure 23–11). Gap genes control the transcription of pair-rule genes.

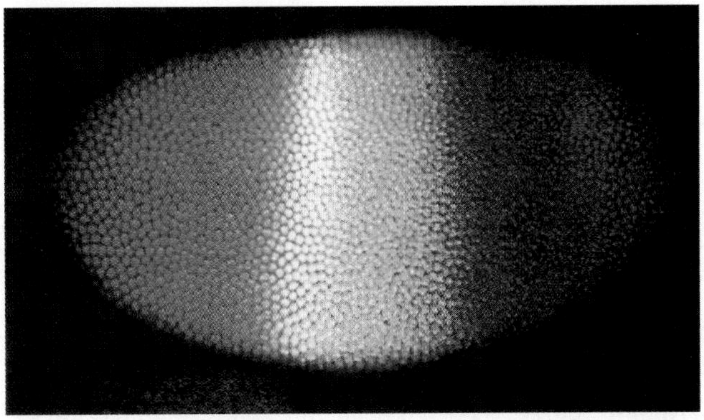

FIGURE 23–11 Expression of several gap genes in a *Drosophila* embryo. The hunchback protein is shown in orange and Krüppel is indicated in green. The yellow stripe is created when cells contain both hunchback and Krüppel proteins.

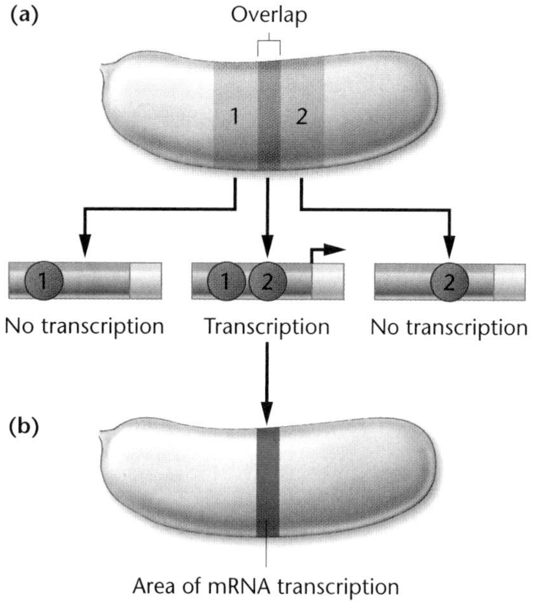

FIGURE 23–12 New patterns of gene expression can be generated by overlapping regions containing two different gene products. (a) Transcription factors 1 and 2 are present in an overlapping region of expression. If both transcription factors must bind to the promoter of a target gene to trigger expression, the gene will be active only in cells containing both factors (most likely in the zone of overlap). (b) The expression of the target gene in the restricted region of the embryo.

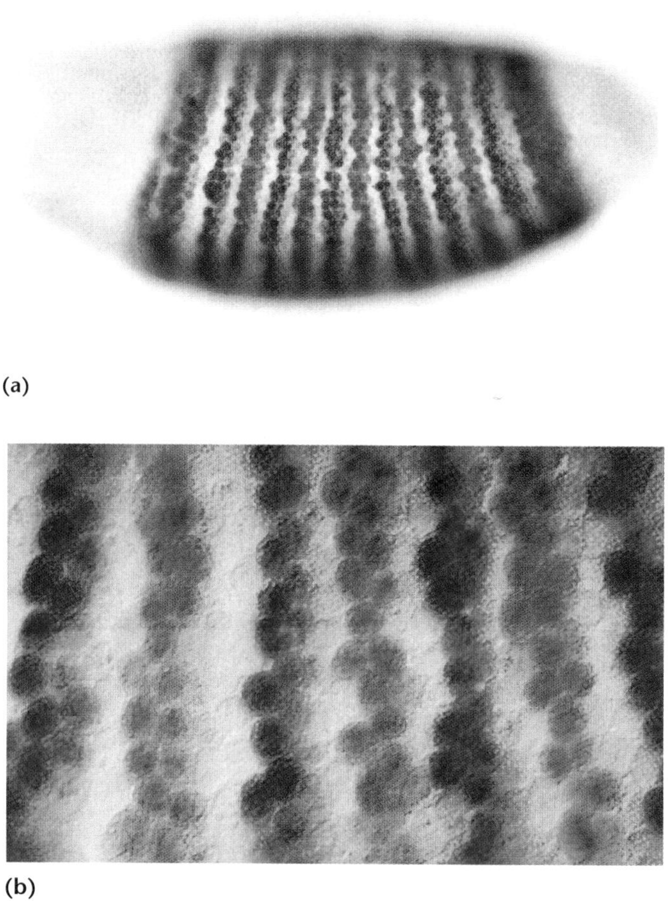

FIGURE 23–13 Stripe pattern of pair-rule gene expression in *Drosophila* embryo. This embryo is stained to show patterns of expression of the genes *even-skipped* and *fushi-tarazu*; (a) low-power view, and (b) high-power view of the same embryo.

Pair-Rule Genes

Pair-rule genes divide the broad regions established by gap genes into sections about one segment wide. Mutations in pair-rule genes eliminate segment-size sections at every other segment. The pair-rule genes are expressed in narrow bands or stripes of nuclei that extend around the circumference of the embryo. The expression of this gene set first establishes the boundaries of segments and then establishes the developmental fate of the cells within each segment by controlling the segment polarity genes. At least eight pair-rule genes act to divide the embryo into a series of stripes. However, the boundaries of these stripes overlap, meaning that cells in the stripes express different combinations of pair-rule genes in an overlapping fashion (Figure 23–12). Many pair-rule genes encode transcription factors containing helix–turn–helix homeodomains. The transcription of the pair-rule genes is mediated by the action of gap gene products, but the pattern is resolved into highly delineated stripes by the interaction among the gene products of the pair-rule genes themselves (Figure 23–13).

Segment Polarity Genes

Expression of segment polarity genes is controlled by transcription factors encoded by pair-rule genes. Within each segment, segment polarity genes become active in a single band of cells that extends around the embryo's circumference (Figure 23–14). This divides the embryo into 14 segments and the products of the segment polarity genes control the cellular identity within each segment. Some segment polarity genes, including *engrailed*, encode transcription factors. Rather than activating transcription, however, the engrailed protein competitively inhibits activation by other homeodomain proteins, resulting in the establishment of a segment border.

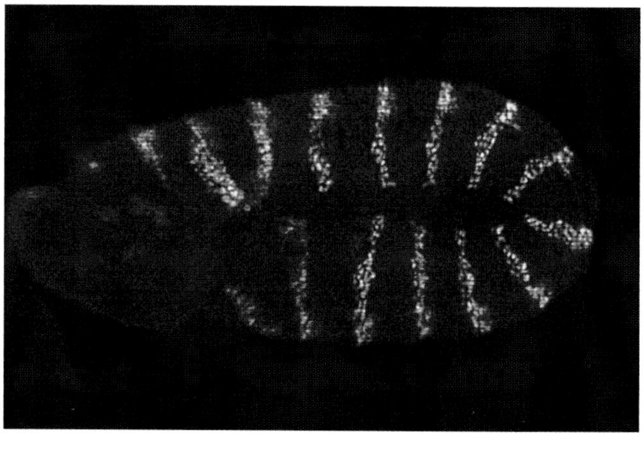

FIGURE 23–14 The 14 stripes of expression of the segment polarity gene *engrailed* in a *Drosophila* embryo.

23.6 Homeotic Genes Control the Developmental Fate of Segments along the Anterior–Posterior Axis

As segment boundaries are established by action of the segmentation genes, the homeotic (from the Greek word for "same") genes are activated as targets of the zygotic genes. Expression of homeotic genes determines which adult structures will be formed by each body segment. In *Drosophila,* this includes the antennae, mouth parts, legs, wings, thorax, and abdomen. Mutants of these genes are called **homeotic mutants**, because the structure formed by one segment is transformed so it is the same as that of a neighboring segment. For example, the wild-type allele of *Antennapedia (Antp)* specifies formation of a leg on the second segment of the thorax. Dominant gain-of-function *Antp* mutations cause this gene to be expressed in the head as well, and mutant flies have a leg on their head in place of an antenna (Figure 23–15).

TABLE 23.3 ***Hox* GENES OF *DROSOPHILA***

Antennapedia Complex	*Bithorax* Complex
labial	*Ultrabithorax*
Antennapedia	*abdominal A*
Sex combs reduced	*Abdominal B*
Deformed	
proboscipedia	

Hox Genes in *Drosophila*

The *Drosophila* genome contains two clusters of *Hox* genes on chromosome 3 (Table 23.3). One cluster, the *Antennapedia (Antp-C)* complex, contains five genes that specify structures in the head and first two thoracic segments (Figure 23–16).

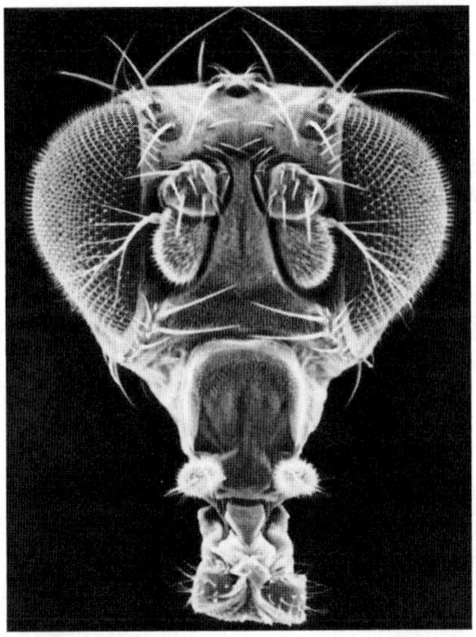

(a)

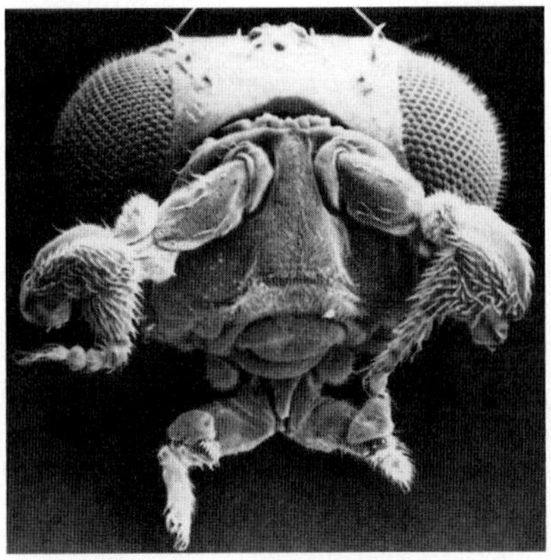

(b)

FIGURE 23–15 *Antennapedia (Antp)* mutation in *Drosophila*. (a) Head from wild-type *Drosophila*, showing the antenna and other head parts. (b) Head from an *Antp* mutant, showing the replacement of normal antenna structures with legs. This is caused by activation of the *Antp* gene in the head region.

(a)

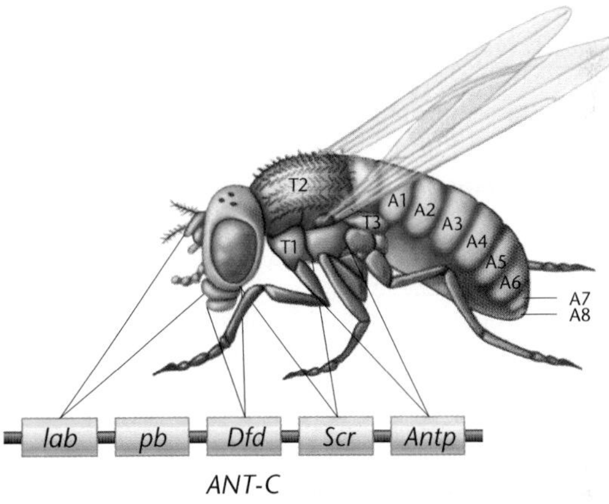

(b)

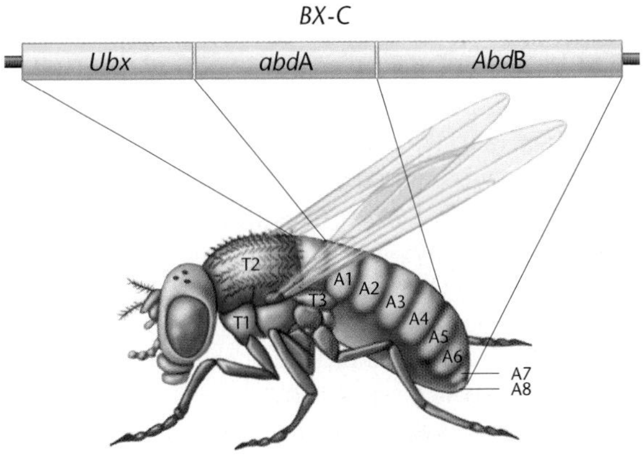

FIGURE 23–16 Genes of the *Antennapedia* complex and the adult structures they specify. (a) In the *ANT-C* complex, the *labial* (*lab*) and *Deformed* (*Dfd*) genes control the formation of head segments. The *Sex comb reduced* (*Scr*) and *Antennapedia* (*Ant*) genes specify the identity of the first two thoracic segments. The remaining gene in the complex, *proboscipedia* (*pb*), may not act during embryogenesis, but may be required to maintain the differentiated state in adults. In mutants, the labial palps are transformed into legs. (b) In the *BX-C* complex, *Ultrabithorax* (*Ubx*) controls formation of structures in the posterior compartment of T2 and structures in T3. The two other genes, *abdominal A* (*abdA*) and *Abdominal B* (*AbdB*), specify the segmental identities of the eight abdominal segments (A1–A8).

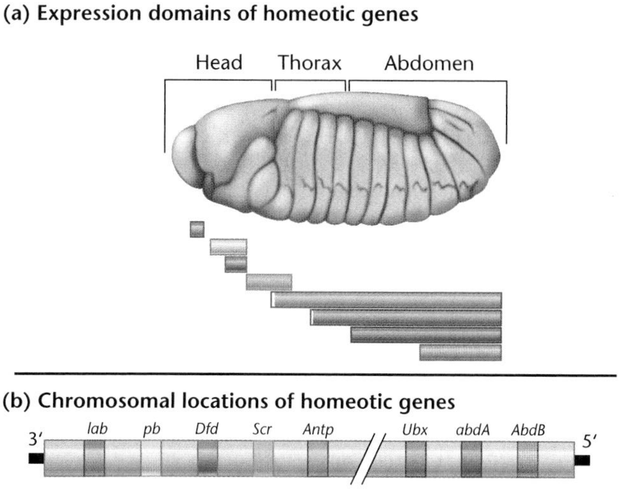

FIGURE 23–17 The colinear relationship between the spatial pattern of expression and chromosomal locations of homeotic genes in *Drosophila*. (a) *Drosophila* embryo and the domains of homeotic gene expression in the embryonic epidermis and central nervous system. (b) Chromosomal location of homeotic genes. Note that the order of genes on the chromosome correlates with the sequential anterior borders of their expression domains.

The second cluster, the *bithorax (BX-C)* complex contains three genes that specify structures in the posterior portion of the second thoracic segment, the entire third thoracic segment, and abdominal segments (Figure 23–16).

The *Hox* genes have two properties in common. First, each *Hox* gene (listed in Table 23.3) encodes a transcription factor that includes a 180 bp DNA-binding domain known as a **homeobox**. (*Hox* is a contraction of homeobox.) The homeobox encodes a sequence of 60 amino acids known as a **homeodomain**. Second, expression of the genes is colinear. Genes at the 3′-end of a cluster are expressed at the anterior end of the embryo, those in the middle are expressed in the middle of the embryo, and genes at the 5′-end of a cluster are expressed at the embryo's posterior region (Figure 23–17). Although *Hox* genes were first identified in *Drosophila*, they are found in the genomes of most eukaryotes with segmented body plans, including zebrafish, *Xenopus*, chickens, mice, and humans (Figure 23–18).

To summarize, genes that control development in *Drosophila* act in a temporally and spatially ordered cascade, beginning with the genes that establish the anterior–posterior (and dorsal–ventral) axis of the egg and early embryo. Gradients of maternal mRNAs and proteins along the anterior–posterior axis activate the gap genes, which subdivide the embryo into broad bands. Gap genes in turn activate the pair-rule genes, which divide the embryo into segments. The final group of segmentation genes, the segment polarity genes, divides each segment into anterior and posterior regions arranged linearly along the anterior–posterior axis. The segments are then given identity by the *Hox* genes. Therefore, this progressive restriction of developmental potential of the *Drosophila* embryo's cells (all of which occurs during the first third of embryogenesis), involves a cascade of gene action, with regulatory proteins acting at both the transcriptional and translational levels.

Hox Genes and Human Genetic Disorders

Although first described in *Drosophila*, *Hox* genes are found in the genomes of all multicellular animals where they play a funda-

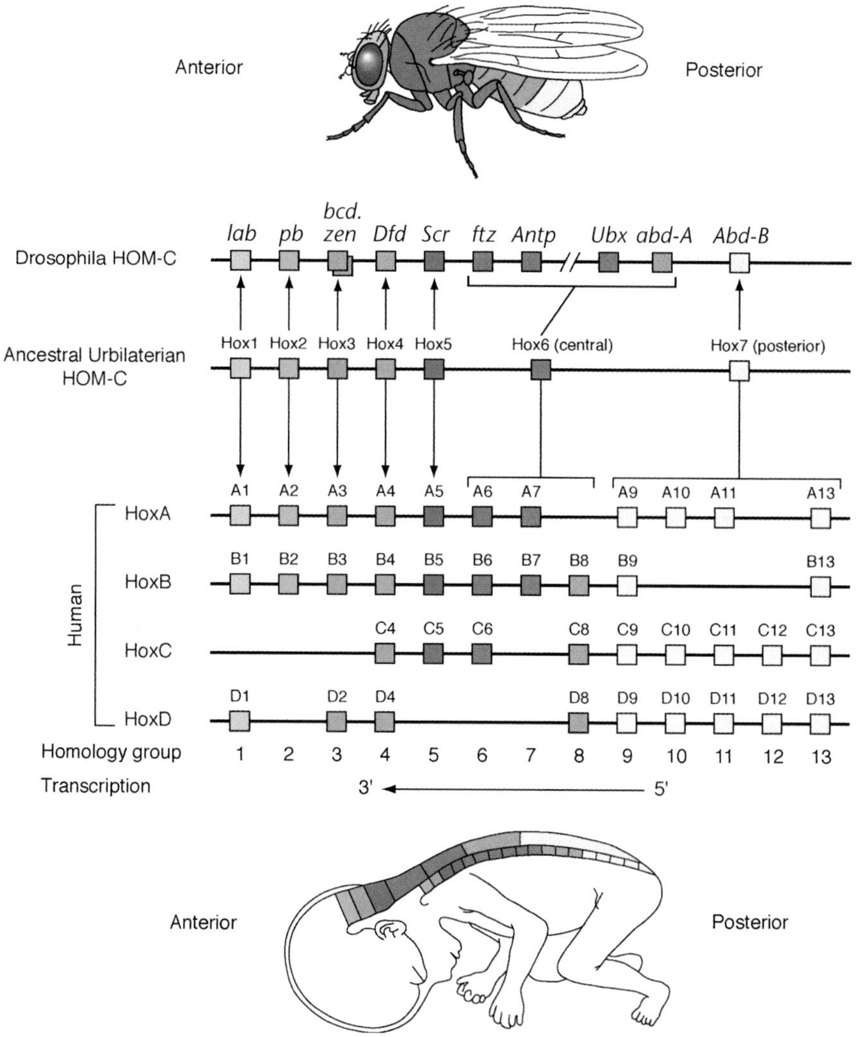

FIGURE 23–18 Conservation of organization and patterns of expression in *Hox* genes. (Top) The structures formed in adult *Drosophila* are shown, with the colors corresponding to members of the *Hox* cluster that control their formation. (Middle) The reconstructed *Hox* cluster of the common ancestor to all bilateral organisms contains seven genes. (Bottom) The arrangement and expression pattern of the four clusters of Hox genes in an early human embryo. Note that some of the posterior genes are expressed in the limbs. The expression pattern is inferred from that observed in mice. As in *Drosophila*, genes at the left end of the cluster form anterior structures, and genes at the right end of the cluster form posterior structures. Because of duplications, genes with equal homology to an ancestral sequence are indicated by brackets.

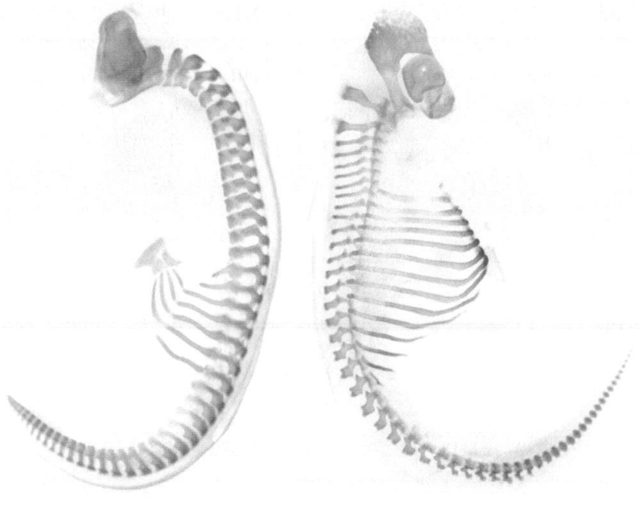

FIGURE 23–19 Patterns of *Hox* gene expression control the formation of structures along the anterior-posterior axis of bilaterally symmetrical animals in a species-specific manner. In the chick (left) and the mouse (right) expression of the same set of *Hox* genes is differentially programmed in time and space to produce different body forms.

mental role in shaping the body and its appendages. Humans and most vertebrates have four clusters of *Hox* genes (*HOXA*, *HOXB*, *HOXC*, and *HOXD*) containing 39 genes. The conservation of sequence, the order of genes in the *Hox* clusters, and their pattern of expression in vertebrates suggests that these gene sets function in flies and humans to control the pattern of structures along the anterior–posterior axis (Figure 23–19). Evidence for this comes from mutational studies in chicks and mice showing that *HOXD* genes near the 5′ end of the cluster play critical roles in limb development. In addition, this role for *HOXD* genes in humans was confirmed by the discovery that a number of inherited limb malformations are caused by specific mutations in *HOXD* genes. For example, mutations in *HOXD13* cause synpolydactlyly (SPD), a malformation characterized by extra fingers and toes, and abnormalities in bones of the hands and feet (Figure 23–20).

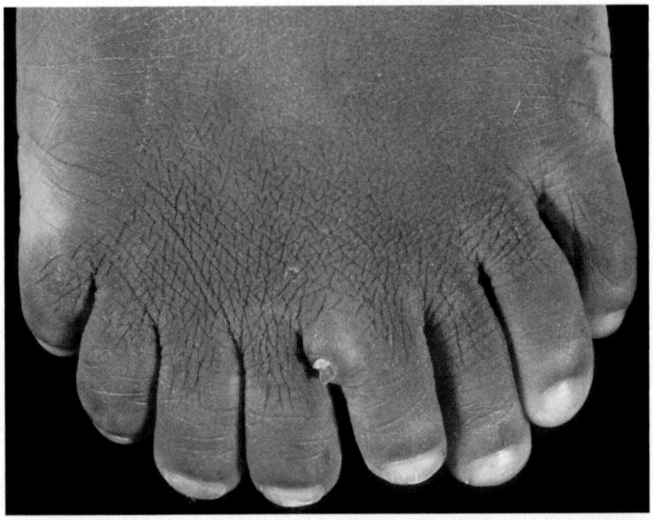

FIGURE 23–20 Mutations in posterior *Hox* genes (*HoxD13* in this case) in humans result in malformations of the limbs, shown here as extra toes. This condition is known as synpolydactlyly. Mutations in *HoxD13* are also associated with abnormalities of the bones in the hands and feet.

Control of *Hox* Gene Expression

In *Drosophila*, a number of genes controlling *Hox* gene expression have been identified, including *extra sex combs (esc)*, *Polycomb (Pc)*, *supersex combs (sxc)*, and *trithorax (trx)*. In the mutant *extra sex combs (esc)*, some of the head and all of the thoracic and abdominal segments develop as posterior segments (Figure 23–21), indicating that this gene normally controls the expression of *BX-C* genes in all body segments. The mutation does not affect either the number or the polarity of the segments. It does affect their developmental fate, indicating that the esc^+ gene product stored in the egg by the maternal genome may be required to interpret the information gradient correctly in the egg cortex.

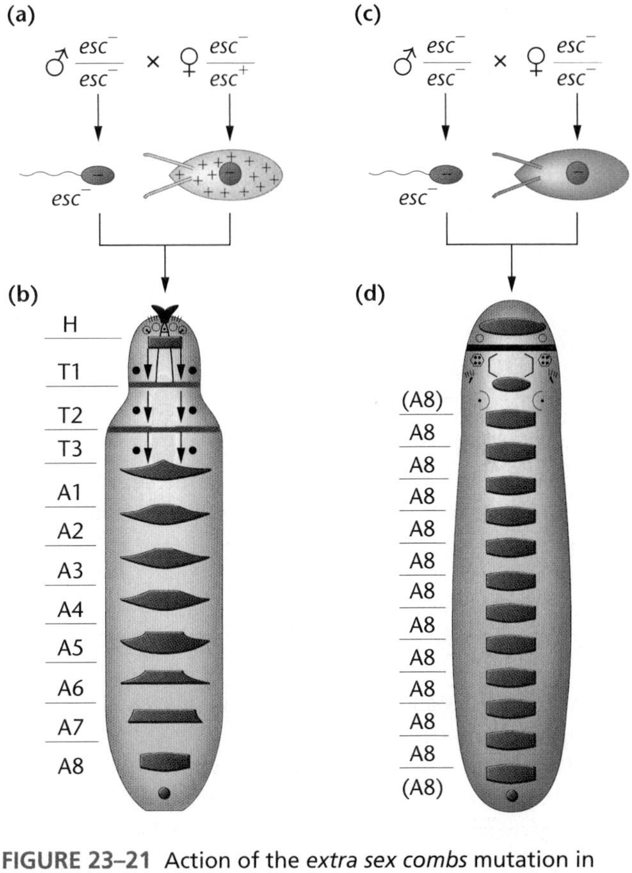

FIGURE 23–21 Action of the *extra sex combs* mutation in *Drosophila*. (a) Heterozygous females form wild-type *esc* gene product and store it in the oocyte. (b) When the esc^- egg formed at meiosis by the heterozygous female is fertilized by esc^- sperm, it produces wild-type larva with a normal segmentation pattern (maternal rescue). Borderlines between the head and thorax and between the thorax and abdomen are marked with arrows. (c) Homozygous esc^- females produce defective eggs, which, when fertilized by esc^- sperm, (d) produce a larva in which most of the segments of the head, thorax, and abdomen are transformed into the eighth abdominal segment. Maternal rescue demonstrates that the *esc* gene product is produced by the maternal genome and is stored in the oocyte for use in the embryo. (H, head; T1–T3, thoracic segments; A1–A8, abdominal segments.)

The proteins encoded by members of the *Polycomb* family control expression of *Hox* genes by altering chromatin conformation, blocking the binding of transcription factors. These proteins assemble at the site of a gene and form multiprotein complexes that modify chromatin and promote gene silencing. The *trithorax* proteins reverse the chromatin-based inactivation of genes. These proteins act as transcriptional activators and reverse the blocks put in place by inactivating proteins such as *Polycomb*-encoded products.

Other than the genes identified in *Drosophila*, little is known about the genetic control of *Hox* genes in other organisms, including humans. Results to date indicate that the regulatory circuits that operate upstream of *Hox* genes are not widely conserved in animal systems. If this is the case, the study of model organisms may not be useful in understanding how *Hox* genes are regulated in humans.

Now solve this

Problem 23.17 on page 597 involves analysis of gene expression in embryos with different genetic backgrounds.

Hint: It is this genetic background that provides the clues to the timing of expression and regulatory patterns of the two genes in question.

23.7 Cascades of Gene Action Control Differentiation

In addition to homeobox genes in the *Hox* gene cluster, there is a large and diverse family of other homeobox genes in eukaryotic genomes. In *Drosophila*, one of these, *Distal-less (Dll)*, plays an important role in the development of appendages, including the antennae, mouthparts, legs, and wings. This gene, which maps to chromosome 2 and encodes a homeobox transcription factor, is the earliest known gene expressed in appendage formation. Mutations in *Dll* produce a wide range of phenotypes, including transformation of the antennae into legs and the formation of shortened legs that are missing distal structures.

In the antennae, expression of the wild-type *Dll* allele has two roles: It directs cells to form an antenna instead of a leg and controls the proximal (closest to the body) to distal (farthest from the body) specification of antennal structures.

The antenna is the ear and nose of the fly and has several components, including the arista and three antennal segments [Figure 23–22(a) and (b)]. The arista vibrates in response to sound waves; these vibrations initiate signals that are transferred through the second antennal segment, to the first antennal segment, and to the antennal nerve to the brain. The third antennal segment is covered with olfactory receptors. These are stimulated by odors that generate signals then transferred to the brain via the first two antennal segments and a nerve connected to the brain.

Expression of *Dll* in the early pupal stage [Figure 23–22(b)] is restricted to the second and third antennal segments and the arista, and is the first step in a cascade of gene expression

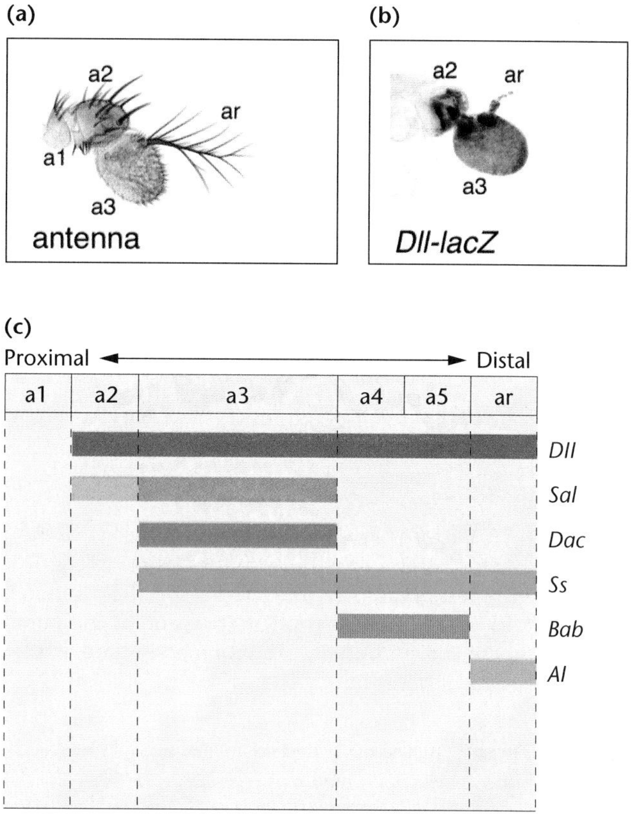

FIGURE 23–22 Action of the *Dll* gene in *Drosophila* produces (a) the wild-type antenna, divided into the arista (ar) and three antennal segments (a1, a2, a3). (b) *Dll* expression (shown in blue) in a late pupal antenna. Expression at this stage is limited to the arista and a3. (c) Activation of *Dll* target genes in antennal segments during the third larval instar just before the pupal stage. At this stage, *Dll* is expressed in all segments except a1. Each of these target genes is a transcription factor, which in turn activates a cascade of other genes.

associated with differentiation. At the larval-pupal transformation, five genes are activated by *Dll* and each is expressed in a segment-specific pattern [Figure 23–22(c)]. Four of these genes (*spalt*, *spineless*, *bric a brac*, and *aristaless*) encode transcription factors. Each of these genes in turn, controls multiple target genes. The fifth gene, *dachshund*, encodes a protein of unknown function that is confined to the nucleus. The genes activated by *Dll* expression initiate a cascade of segment-specific gene expression causing differentiation of the antenna and arista, but the details of how this is accomplished are still being studied.

In humans, there are six genes (*Dlx1–Dlx6*) in the *Distal-less* gene family with a homeobox sequence closely related to the *Drosophila Dll* gene. The *Drosophila Dll* gene is expressed in the head and appendages of the developing fly. The human *Dlx3* gene is expressed in the developing head, and mutations in this gene are responsible for an inherited condition called trichodentoosseous syndrome (TDO). This autosomal dominant disorder causes deficiencies in the calcification of bones in the skull and enamel defects in teeth.

The discovery that TDO is caused by a mutation in a homeobox gene began with mutational analysis of *Drosophila*

FIGURE 23–23 The flowering plant *Arabidopsis thalania*, used as a model organism in plant genetics.

development and illustrates the valuable role that model organisms play in understanding human genetic disorders. Human - *Distal-less* genes were discovered by screening a human cDNA library with a cloned probe containing the homeobox sequence of the *Drosophila Dll* gene in 1994 and the gene for *Dlx3* was mapped to chromosome 17 using fluorescent *in situ* hybridization (FISH) in 1995. TDO was originally described in 1966, but the nature of the gene and its location were unknown. In 1997, geneticists found that TDO was linked to markers on chromosome 17. Further work mapped TDO to the same locus as *Dlx3*, and in 1998, analysis of mutations in affected individuals confirmed that TDO is caused by a 4bp deletion in *Dlx3*.

23.8 Plants Have Evolved Systems That Parallel the *Hox* Genes of Animals

Flower development in *Arabidopsis thalania* (Figure 23–23), a small plant in the mustard family, has been used to study pattern formation in plants. A cluster of undifferentiated cells, called the *floral meristem*, gives rise to flowers (Figure 23–24). Each flower consists of four organs—sepals, petals, stamens, and carpels—that develop from concentric rings of cells within the meristem [Figure 23–25(a)]. Each organ develops from a different whorl.

FIGURE 23–24 (a) Parts of the *Arabidopsis* flower. The floral organs are arranged concentrically. The sepals form the outermost ring, followed by petals and stamens, with carpels on the inside. (b) View of the flower from above.

Homeotic Genes in *Arabidopsis*

Three classes of floral homeotic genes control the development of these organs: Class A genes specify sepals, class A and class B genes specify petals, and class B and class C genes control stamen formation. Class C genes alone specify carpels [Figure 23–25(b)]. The genes in each class are listed in Table 23.4. Class A genes are active in the two outermost whorls (sepals and petals), class B genes are expressed in the second and third whorls (petals and stamens) and class C genes act in the third and fourth whorls (stamens and carpels). The organ formed depends

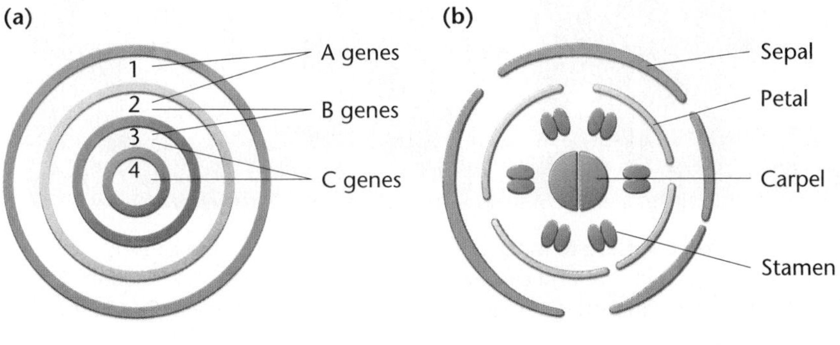

FIGURE 23–25 Cell arrangement in the floral meristem. (a) The four concentric rings, or whorls, labeled 1–4, give rise to (b) Arrangement of the sepals, petals, stamens, and carpels, respectively in the mature flower.

TABLE 23.4 HOMEOTIC SELECTOR GENES IN ARABIDOPSIS

Class A	*APETALA1 (AP1)**
	APETALA2 (AP2)
Class B	*APETALA3 (AP3)*
	PISTILLATA (P1)
Class C	*AGAMOUS (AG)*

* By convention, wild-type genes in *Arabidopsis* use capital letters.

on the expression pattern of the three gene classes. If only class A genes are expressed, sepals form. If class A *and* class B genes are expressed, petals form. Activity of class B *and* class C genes leads to stamen formation. If only class C genes are expressed, carpels form.

As in *Drosophila*, mutations in homeotic genes cause organs to form in abnormal locations. For example, in *AP2* mutants (lacking activity of an A class gene), the order of organs is carpel, stamen, stamen, and carpel instead of the normal order, sepal, petal, stamen, and carpel [Figure 23–26(a) and (b)]. If both class A and class B gene activity are eliminated by mutation, then all whorls will have only class C genes active (which will spread to the first and second whorls, because class A is absent), and the order of organs will be sepal, petal, petal, and sepal [Figure 23–26(d)].

Evolutionary Divergence in Homeotic Genes

Plants and animals diverged from a common ancestor about 1.6 billion years ago, after the origin of eukaryotes and probably before the rise of multicellular organisms. From genetic analysis of development in both *Drosophila* and *Arabidopsis*, it is clear that each organism uses a set of master regulatory genes to establish the body axis and specify the identity of structures along the axis. In *Drosophila*, this task is accomplished in part by the *Hox* genes, which encode a set of transcription factors sharing a homeobox domain. In *Arabidopsis*, however, the floral homeotic genes are members of a different family of transcription factors, called the MADS-box proteins. Each member of this family contains a common sequence of 58 amino acids with no similarity in amino acid sequence or protein structure with the *Hox* genes. Both gene sets encode transcription factors, both sets are master regulators of development expressed in a pattern of overlapping domains, and both specify identity of structures.

Reflecting their common evolutionary origins, the genomes of both organisms contain members of the homeobox and MADS-box genes, but these genes have been adapted for different uses in the plant and animal kingdoms, indicating that developmental mechanisms evolved independently in each group.

In both plants and animals, the action of transcription factors depends on changes in chromatin structure that makes genes available for expression. Mechanisms of transcription initiation are conserved in plants and animals, and this is reflected in homology of genes in *Drosophila* and *Arabidopsis* that maintain patterns of expression initiated by regulatory gene sets. Action of the floral homeotic genes is controlled by a gene called *CURLY LEAF*. This gene shares significant homology with members of the *Drosophila Polycomb* gene family, a group of genes that regulate homeobox genes during development in the fruit fly. Both these genes encode proteins that alter chromatin conformation and shut off gene expression. Thus, although different genes are used to control development, both plants and animals use an evolutionarily conserved mechanism to regulate expression of these gene sets.

23.9 Cell–Cell Interactions in *C. elegans* Development

During development in multicellular organisms, cell–cell interactions influence the transcriptional programs and developmental fate of neighboring cells. Cell–cell interaction is an important process in the embryonic development of most eukaryotic organisms, including *Drosophila*, as well as vertebrates such as *Xenopus*, mice, and humans.

Signaling Systems in Development

In early development, metazoans use five signaling systems; after organogenesis begins, another five signal systems are added to those already in use. These systems act both independently and in coordinated networks to send and receive developmental

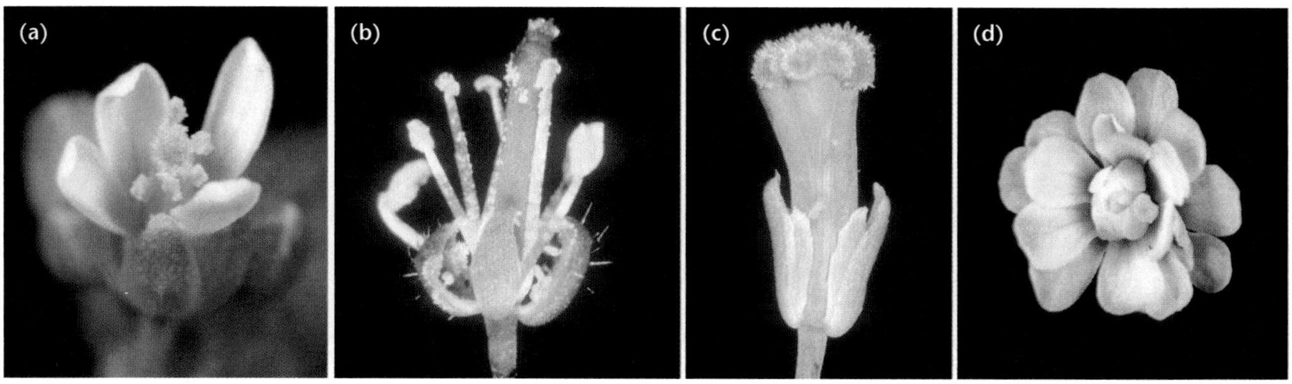

FIGURE 23–26 (a) Wild-type flowers of *Arabidopsis* have (from outside to inside) sepals, petals, stamens, and carpels. (b) Homeotic *APETALA2* mutant flower, with carpels, stamens, stamens, and carpels. (c) *PISTILLATA* mutants have sepals, sepals, carpels, and carpels. (d) *AGAMOUS* mutants have petals and sepals at places where stamens and carpels should form.

TABLE 23.5 SIGNALING SYSTEMS USED IN EARLY EMBRYONIC DEVELOPMENT

Wnt Pathway
Dorsalization of body
Female reproductive development
Dorsal–ventral differences
TGF-β Pathway
Mesoderm induction
Left–right asymmetry
Bone development
Hedgehog Pathway
Notochord induction
Somitogenesis
Gut/visceral mesoderm
Receptor Tyrosine Kinase Pathway
Mesoderm maintenance
Notch/Delta Pathway
Blood cell development
Neurogenesis
Retina development

Taken from Gerhart, J. 1999. 1998 Warkany lecture: Signaling pathways in development. *Teratology* 60:226–39.

signals that elicit specific transcriptional responses. The signal networks establish anterior–posterior polarity and body axes, coordinate pattern formation, and direct the differentiation of tissues and organs. The signaling pathways used in early development and some of the developmental processes they control are listed in Table 23.5. After an introduction to the components and interactions of one of these systems—the Notch signaling pathway—we will briefly examine its role in the development of the vulva in the nematode, *Caenorhabditis elegans*.

The Notch Signaling Pathway

The genes in the Notch pathway are named after the *Drosophila* mutants in which the parts of the pathway were discovered. Notch is a short-range signaling system that works through direct cell–cell contact to control the developmental fate of the interacting cells. The *Notch* gene (and the equivalent gene in other organisms) encodes a transmembrane signal receptor (Figure 23–27). The signal is another transmembrane protein encoded by the *Delta* gene (and its equivalents). Because both the signal and receptor are membrane bound, the Notch signal system works only between adjacent cells. When the Delta protein binds to the Notch receptor, the cytoplasmic tail of the Notch protein cleaves off and binds to a cytoplasmic protein encoded by the *Su(H)* (suppressor of Hairy) gene. This protein complex moves into the nucleus and binds to transcriptional cofactors activating transcription of a gene set that controls a specific developmental pathway (Figure 23–27).

Variations of this pathway control a number of different developmental processes in *Drosophila*, establishing the dorsal–ventral boundary in the wing disc and directing the fate of cells in the nervous system, muscle, and gut. One of the main roles of the Notch signal system is specifying the fate of equivalent cells in a population. In its simplest form, this interaction involves two neighboring cells that are developmentally equivalent. In humans, four members of the Notch family (*NOTCH1–NOTCH4*) have been identified. Mutations in these genes and other genes in the Notch pathway are responsible for a number of human developmental disorders, including Alagille syndrome (AGS), and spondylocostal dysotosis (SD). We will explore the role of the Notch signaling system in development of the vulva in *C. elegans*, after a brief introduction to nematode embryogenesis.

Overview of *C. elegans* Development

The nematode *C. elegans* is widely used to study the genetic control of development. This organism has several advantages for such studies: (1) the genetics of the organism are well known, (2) the genome sequence is available, and (3) adults are formed from a small number of cells that follow a developmental program that is unchanged from individual to individual. Adult nematodes are about 1 mm long and mature from a fertilized egg in about 2 days (Figure 23–28). The life cycle consists of an embryonic stage (about 16 hours), four larval stages (L1 through L4), and the adult stage. Adults are of two

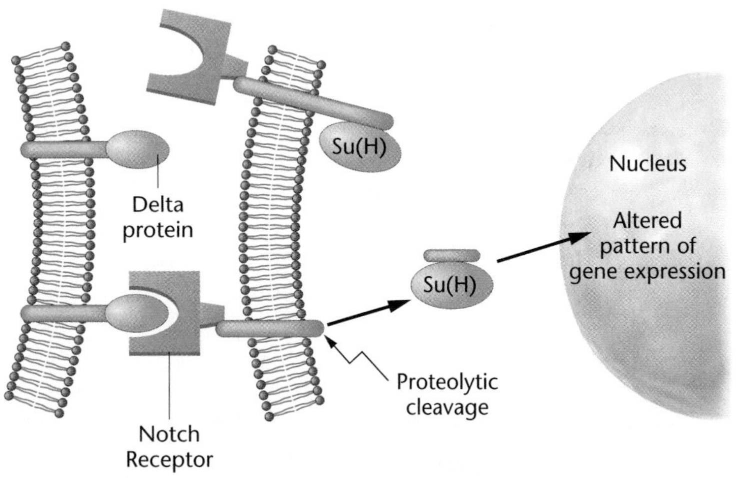

FIGURE 23–27 Components of the Notch signaling system. The cell carrying the Delta transmembrane protein is the sending cell; the cell carrying the transmembrane Notch protein receives the signal. Binding of Delta to Notch triggers a proteolytic-mediated activation of transcription. The fragment cleaved from the cytoplasmic side of the Notch protein combines with the Su(H) protein, and moves to the nucleus where it activates a program of gene transcription.

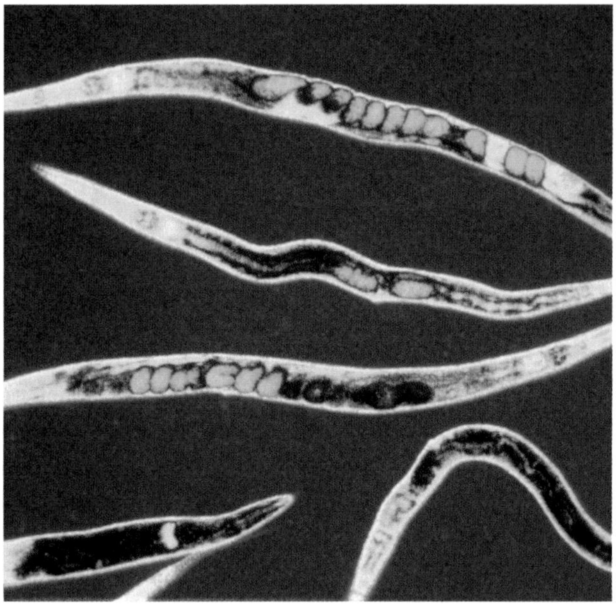

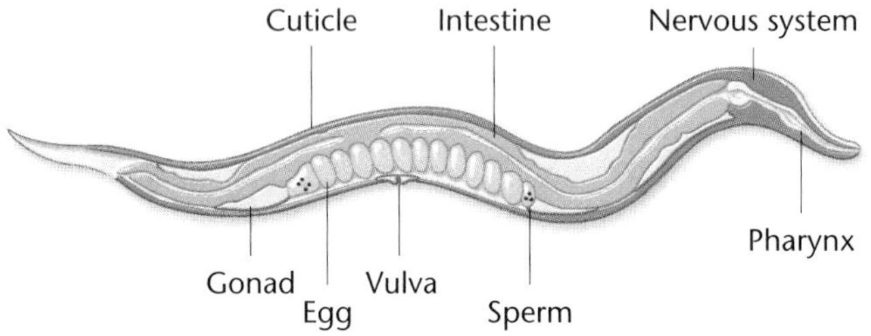

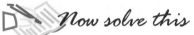

FIGURE 23–28 An adult *Caenorhabditis elegans*. This nematode, about 1 mm in length, consists of 959 cells and has been used to study many aspects of the genetic control of development.

sexes: XX self-fertilizing hermaphrodites that can make both eggs and sperm and XO males. Self-crossing of mutagen-treated hermaphrodites quickly results in homozygous stocks of mutant strains, and hundreds of mutants have been generated, catalogued, and mapped.

Adult hermaphrodites have 959 somatic cells (and about 2000 germ cells). The exact cell lineage from fertilized egg to adult has been mapped (Figure 23–29) and is invariant from individual to individual. Knowing the lineage of each cell, we can easily follow events resulting from mutations that alter cell fate or from killing of cells with laser microbeams or ultraviolet irradiation. In *C. elegans* hermaphrodites, the fate of cells in the development of the reproductive system is determined by cell–cell interaction giving us insight into how gene expression and cell–cell interaction work together to specify developmental outcomes.

Now solve this

Problem 23.28 on page 598 involves two genes that control sex determination in *C. elegans*.

Hint: In solving this problem, remember to consider the action of gene products and the effect of loss-of-function mutations on expression of other genes or the action of other proteins.

Genetic Analysis of Vulva Formation

C. elegans adult hermaphrodites lay eggs through the vulva, an opening located about midbody (Figure 23–28). The vulva is formed in stages during larval development, and the process involves several rounds of cell–cell interactions.

During development in *C. elegans*, two developmentally equivalent neighboring cells, Z1.ppp and Z4.aaa, interact with each other so that one becomes the gonadal anchor cell and the other becomes a precursor to the ventral uterus. The determination of which becomes which occurs during the second larval stage (L2) and is controlled by the Notch receptor gene, *lin-12*. In recessive *lin-12(0)* mutants (a loss-of-function mutant), both cells become anchor cells. The dominant mutation *lin-12(d)* (a gain-of-function mutation) causes both to become uterine precursors. Thus, it appears that the expression of the *lin-12* gene causes the selection of the uterine pathway, since in the absence of the LIN-12 (Notch) receptor, both cells become anchor cells.

However, as shown in Figure 23–30, the situation is more complex than it first appears. Initially, the neighboring cells are

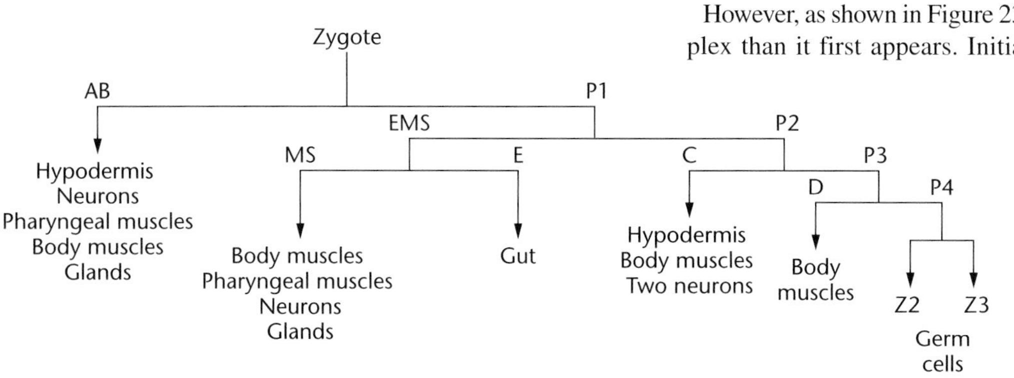

FIGURE 23–29 A truncated cell lineage chart for *C. elegans*, showing early divisions and the tissues and organs. Each vertical line represents a cell division, and horizontal lines connect the two cells produced. For example, the first cell division creates two new cells from the zygote, AB and P1. The cells in this chart refer to those present in the first-stage larva L1. During subsequent larval stages, further cell divisions will produce the 959 somatic cells of the adult hermaphrodite worm.

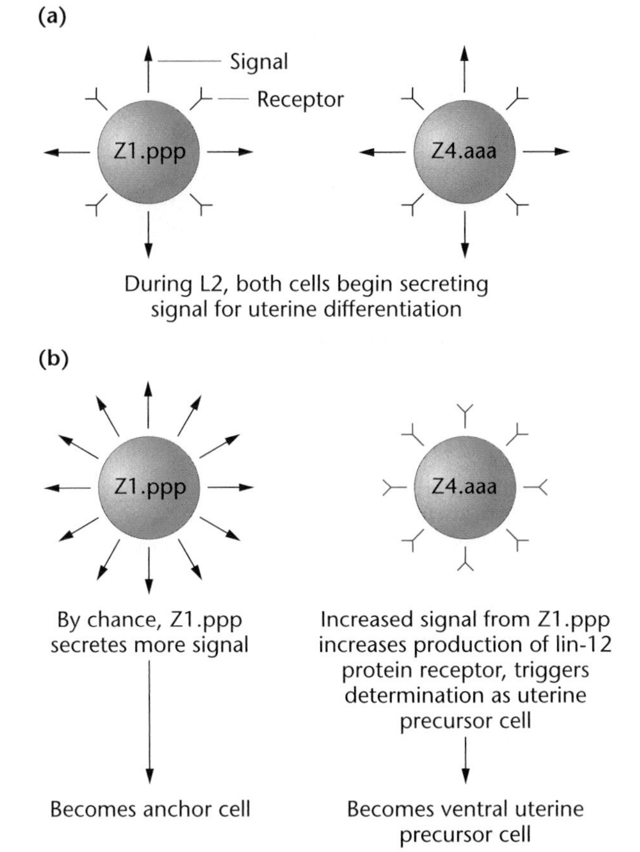

FIGURE 23–30 Cell–cell interaction in anchor cell determination. (a) During L2, two neighboring cells begin the secretion of chemical signals for the induction of uterine differentiation. (b) By chance, cell Z1.ppp secretes more of these signals, causing cell Z4.aaa to increase production of the receptor for signals. The action of increased signals causes Z4.aaa to become the ventral uterine precursor cell and allows Z1.ppp to become the anchor cell.

developmentally equivalent. Each cell displays low levels of the Notch signal (encoded by the *lag-2* gene), and low levels of the Notch receptor. This situation is unstable, and both cells cannot continue sending and receiving developmental signals. By chance, the cell secreting more of the LAG-2 (Delta) signal causes the neighboring cell to increase production of the LIN-12 receptor. As a result, the cell producing more receptor becomes the ventral uterine precursor cell, and the other cell, with more signal, becomes the anchor cell. The critical factor in this first round of cell–cell interaction is the balance between the LAG-2 (Delta) gene product and the LIN-12 (Notch) gene product.

A second round of cell–cell interactions in larval development involves the anchor cell (located in the gonad) and six precursor cells (located in the skin, or hypodermis) adjacent to the gonad. The precursor cells are named P3.p to P8.p and collectively are called Pn.p cells. The fate of each Pn.p cell is specified by its position relative to the anchor cell. Figure 23–31 shows the developmental pathway described in the following paragraphs.

Sometime in larval stage 3, the *lin-3* gene is expressed in the anchor cell. The gene product, LIN-3, is a signal protein related to vertebrate epidermal growth factor (EGF). All six Pn.p cells express a receptor encoded by *let-23*, a gene homologous to the vertebrate EGF receptor. The binding of LIN-3 to the LET-23 receptor triggers an intracellular cascade of events that determines whether the precursor cells will form the primary vulval precursor cell or secondary vulval cells. In other words, *let-23* establishes the primary and secondary fates of precursor cells. Recessive loss-of-function *let-23* mutations cause all Pn.p cells to act as if they have not received any signal, and as a result, no vulva forms.

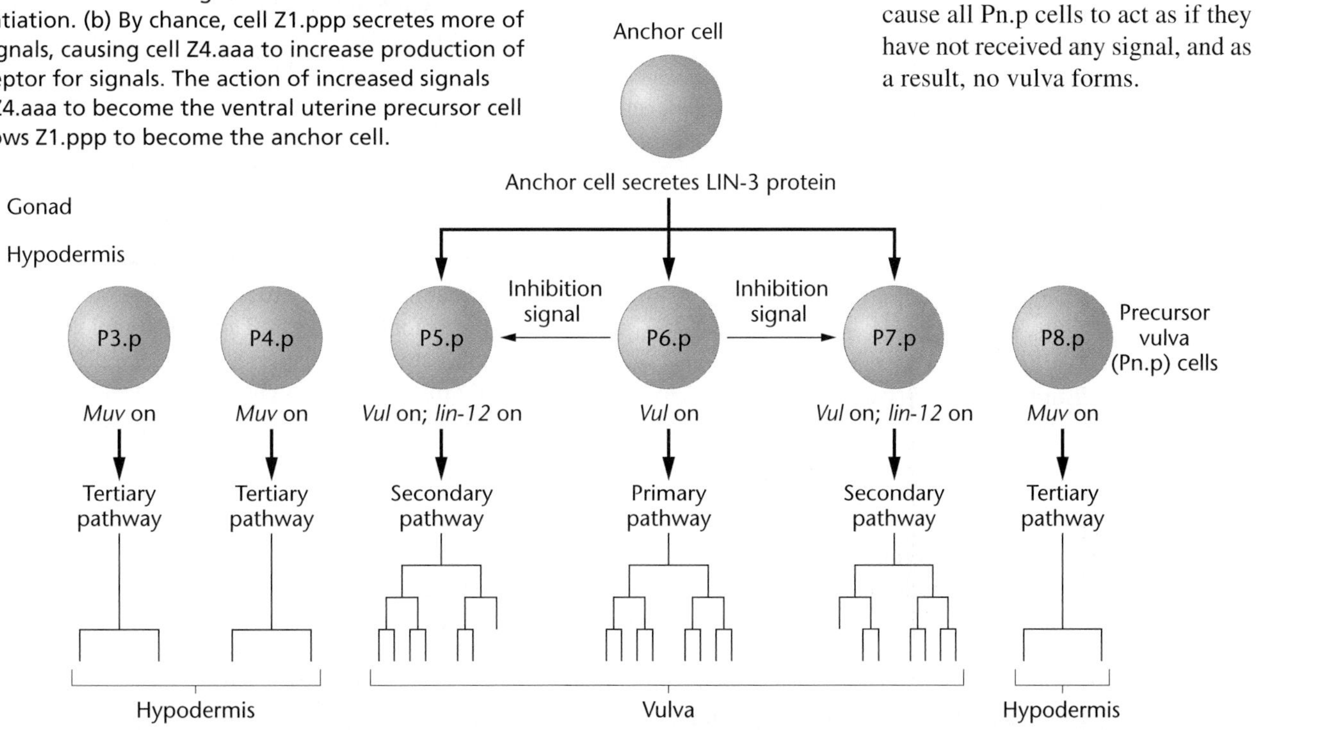

FIGURE 23–31 Cell lineage determination in *C. elegans* vulva formation. A signal from the anchor cell in the form of LIN-3 protein is received by three precursor vulval cells (Pn.p cells). The cells closest to the anchor cell become primary vulval precursor cells, and adjacent cells become secondary precursor cells. Primary cells secrete a signal that activates the *lin-12* gene in secondary cells, preventing them from becoming primary cells. Flanking precursor cells, which receive no signal from the anchor cell, increase activity of the *Muv* gene and become skin (hypodermis) cells, instead of vulval cells.

Web Tutorial 23.1 Programmed Cell Death

The signals transmitted from anchor cells to the Pn.p cells also involve the *let-60* gene. Recessive mutations in *let-60* cause precursor cells to develop as though they have not received a signal from the anchor cell. Dominant *let-60* alleles have the opposite phenotype, causing all precursor cells to respond, thus forming multiple vulvas. The *C. elegans let-60* is a homolog of the *ras* gene, a human protooncogene. The *let-60* dominant gain-of-function mutant that causes multiple vulva formation has a Gly–Glu mutation at amino acid 13, the same mutation that converts *c-ras* to an oncogene.

How Do We Know?

How do we know whether a signaling system in vulva development works only on adjacent cells or uses signals that can affect more distant cells?

Normally, the cell closest to the anchor cell (P6.p) receives the strongest signal initiated by LIN-3 binding to LET-23. This signal activates expression of the *Vulvaless (Vul)* gene (the gene is named for its mutant phenotype) in P6.p, and the cell adopts the primary pathway of differentiation—it divides three times to produce vulva cells. The two neighboring cells (P5.p and P7.p) receive a lower amount of signal and initiate a secondary fate. These cells divide asymmetrically to form additional vulva cells.

To reinforce these developmental pathways, a third level of cell–cell interaction is used. The P6.p cell activates *lin-12* in the two neighboring cells (P5.p and P7.p). This signal prevents P5.p and P7.p from adopting the division pattern of the primary cell. In other words, cells in which both *Vul* and *lin-12* are active cannot become primary vulva cells. The three remaining precursor cells (P3.p, P4.p, and P8.p) receive no signal from the anchor cell. In these cells the *Multivulva (Muv)* gene is expressed, *Muv* represses *Vul*, and the three cells develop as skin cells.

Thus, three levels of cell–cell interactions are used in the developmental pathway leading to vulva formation in *C. elegans*. First, two neighboring cells interact to establish the identity of the anchor cell. Second, the anchor cell interacts with three vulval precursor cells to establish the identity of the primary vulval precursor cell (usually P6.p) and two secondary cells (P5.p and P7.p). Third, the primary vulval cell interacts with the secondary cells to suppress their ability to adopt the pathway of the primary cell. Each interaction is accompanied by secretion of molecular signals and by reception and processing of these signals in neighboring cells.

This theme of cell–cell interactions acting in a spatial and temporal cascade to specify the developmental fates of individual cells is a developmental theme repeated over and over in organisms from prokaryotes to humans.

In humans, there are four Notch genes encoding receptors that play important roles in generating cells of the immune system and bone marrow stem cells. Mutations in the Notch signaling pathway are responsible for several inherited disorders, including Alagille syndrome. This autosomal dominant disorder produces developmental abnormalities of the liver, the heart and circulatory system, and the skeleton.

23.10 Programmed Cell Death Is Required for Normal Development

Programmed cell death, or **apoptosis**, is a genetically controlled developmental program that shapes and molds tissues and organs. One well-known example of programmed cell death is the formation of digits in the vertebrate limb. This process requires the death of the cells between the digits (Figure 23–32). The genes that control apoptosis were first identified in *C. elegans*. In *C. elegans*, normal development relies on programmed cell death. The number of cells that die during the worm's development is always the same: 131 of 1090 in hermaphrodites and 147 of 1178 in males. In addition, the time in development at which a given cell dies and the identity of the cells that die are always the same.

Mutational analysis indicates that, although programmed cell death occurs in cells in different lineages, all cells use the same genetic pathway. In *C. elegans*, 15 genes are involved in cell death. These genes control four processes: (1) decisions about cell death, (2) implementation of decisions, (3) engulfment of dying cells, and (4) degradation of cell debris within the engulfing cells.

Expression of *ced-3* and *ced-4* are necessary for execution of the cell death program; mutations that inactivate either of these genes result in survival of cells that normally die. Expression of *ced-3* and *ced-4* is controlled by *ced-9*. Gain-of-function mutations that cause constitutive expression or over expression of *ced-9* prevent cell death. Conversely, loss-of-function mutants that inactivate *ced-9* cause embryonic lethality, meaning that *ced-9* works by inactivating *ced-3* and *ced-4* in surviving cells (Figure 23–32). In other words, *ced-9* is a *binary switch gene* for apoptosis. Cells that express *ced-9* survive and those that do not, die.

The *ced-9* gene of *C. elegans* has a human homolog, *bcl-2*, a protooncogene that controls apoptosis. Mutations that overexpress *bcl-2* prevent death in cells that would normally die. In humans, overexpression of *bcl-2* is found in follicular lymphoma, a form of cancer. Transfer of a cloned human *bcl-2* gene into *ced-9* null mutant *C. elegans* embryos prevents apoptosis, indicating that nematodes and mammals share a common pathway for this process. Thus, the homology in molecules and mechanisms among species across the phylogenetic tree, illustrated throughout this chapter, extends to the molecules and mechanisms required for cells to die.

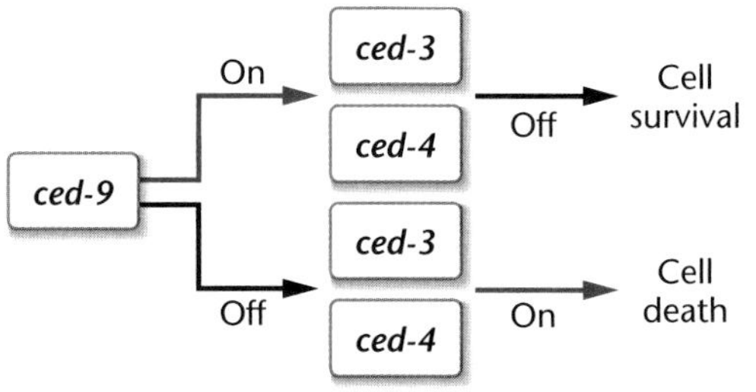

FIGURE 23–32 In the genetic pathway controlling cell death, the gene *ced-9* acts as a binary switch. If *ced-9* is active, it represses the expression of *ced-3* and *ced-4*, and the cell lives. If *ced-9* is inactive, *ced-3* and *ced-4* are expressed, and the cell dies.

GENETICS, TECHNOLOGY, AND SOCIETY

Stem Cell Wars

Stem cell research is at the center of a battle fought by scientists, politicians, advocacy groups, religious leaders, and ethicists. Proponents fight for the right to carry out stem cell research, claiming that it is revolutionary, if not miraculous—with the potential to cure diabetes, Parkinson disease, and spinal cord injuries, and also improve the quality of life for millions. Critics lobby for an end to stem cell research, warning that it will propel us down the slippery slope toward disregard for human life. Although stem cell research is the focus of presidential proclamations, highly publicized media campaigns, and legislative bans, few of us understand it sufficiently to evaluate its pros and cons. What is stem cell research, and why does it spark such intense controversy?

Stem cells are primitive cells that replicate indefinitely and have the unique capacity to differentiate into cells with specialized functions, such as those found in the heart, brain, liver, and muscle. Stem cells are the origin of all the cells that make up the approximately 200 distinct types of tissues in our bodies. In contrast to stem cells, mature, fully differentiated cells do not replicate or undergo transformations into different cell types. Some types of stem cells are defined as *totipotent*, meaning that they have the ability to differentiate into any mature cell type in the body. Other types of stem cells are pluripotent, and are able to differentiate into only one of several mature cell types.

In the last few years, several research teams have isolated and cultured human pluripotent stem cells. These cells remain undifferentiated and grow indefinitely in culture dishes. When treated with growth factors or hormones, these pluripotent stem cells differentiate into cells that have characteristics of neural, bone, kidney, liver, heart, or pancreatic cells.

The fact that pluripotent stem cells grow prolifically in culture and differentiate into more specialized cells has created great excitement. Some foresee a day when stem cells may be a cornucopia from which to harvest unlimited numbers of specialized cells to replace cells in damaged and diseased tissues. Hence, stem cells could be used to treat Parkinson disease, type 1 diabetes, chronic heart disease, kidney and liver failure, Alzheimer disease, Duchenne muscular dystrophy, and spinal cord injuries. Some predict that stem cells will be genetically modified to eliminate transplant rejection or to deliver specific gene products, thereby correcting genetic defects or treating cancers. The excitement about stem cell therapies has been fueled by reports of dramatically successful experiments in animals. For example, mice with spinal cord injuries regained their mobility, and bowel and bladder control after they were injected with human stem cells. Both proponents and critics of stem cell research agree that stem cell therapies could be revolutionary. Why, then, should stem cell research be so contentious?

The answer to that question lies in the source of pluripotent stem cells. To date, all pluripotent stem cell lines have been derived from five-day embryonic blastocysts. Blastocysts at this stage consist of about 2110 cells, most of which will develop into placental and supporting tissues for the early embryo. The inner cell mass of the blastocyst consists of about 30–40 pluripotent stem cells that develop into all tissues of the embryo. *In vitro* fertilization clinics grow fertilized eggs to the five-day blastocyst stage prior to uterine transfer. Embryonic stem (ES) cell lines are created by dissecting out the inner cell mass of five-day blastocysts and growing the undifferentiated cells in culture dishes. All human ES cell lines have been derived from unused five-day blastocysts that were discarded by *in vitro* fertilization clinics.

The fact that early embryos are destroyed in the process of establishing human ES cell lines disturbs people who believe that preimplantation embryos are persons with rights; however, it does not disturb people who believe that these embryos are too primitive to have an inherent moral status. Both sides in the debate put forth lengthy arguments revolving around the fundamental question of what constitutes a human being.

Critics of ES cell research argue that we may be able to benefit from stem cell therapies without resorting to the use of ES cells. This argument is based on recent reports about the plasticity of adult stem cells. Adult stem cells are undifferentiated cells that are present in differentiated tissues such as blood and brain. They divide within the differentiated tissue and differentiate into mature cells that make up the tissue in which they are found. Adult stem cells have been found in bone marrow, blood, the retina, the brain, skeletal muscle, the liver, skin, and the pancreas. The best-known adult stem cells are hematopoietic stem cells (HSCs) which are found in bone marrow, peripheral blood, and umbilical cords. HSCs differentiate into mature blood cell types such as red blood cells, lymphocytes, and macrophages. HSCs have been used clinically for many years, as transplant material to reconstitute the immune systems of patients undergoing treatment for cancer and autoimmune diseases. Interestingly, recent studies suggest that adult stem cells may have the capacity to differentiate into other cell types.

Although these reports are intriguing, it is still too early to know whether adult stem cells will hold the same pluripotent promise as ES cells. Adult stem cells are rare, difficult to identify and isolate, and grow poorly, if at all, in culture. However, if these obstacles can be overcome, adult stem cells may provide an ethical alternative to ES cell and calm the raging debate. On the other hand, new philosophical dilemmas could be created. If adult stem cells are found to exhibit the same pluripotency as ES cells, they could have the same potential to create a human embryo—dragging critics and proponents of stem cell research back into the same moral quagmire. At the present time, it is impossible to predict whether either adult or embryonic stem cells will be as miraculous as predicted by scientists and the popular press. But if stem cell research progresses at its current rapid pace, we won't have long to wait.

References

Robertson. J. A. 2001. Human embryonic stem cell research: ethical and legal issues. *Nat. Rev. Gen.* 2:74–78.

Freed, C. R. 2002. Will embryonic stem cells be a useful source of dopamine neurons for transplant into patients with Parkinson's disease? *Proc. Nad. Acad. Sci. (USA)* 99:1755–57.

Web Sites

National Institutes of Health. 2001. "Stem Cells: Scientific Progress and Future Research Directions."

http://stemcells.nih.gov/info/scireport/ (or for the .pdf version: http://stemcells.nih.gov/info/scireport/PDFs/fullrptstem.pdf)

National Institutes of Health. 2004. "Stem Cell Information."

http://stemcells.nih.gov/index.asp

CHAPTER SUMMARY

1. The role of genetic information during development and differentiation is one of the major questions in biology and has been studied extensively. Geneticists are exploring this question by isolating developmental mutations and identifying the genes involved in controlling developmental processes.
2. Determination is the regulatory event whereby cell fate becomes fixed during early development. Determination precedes the actual differentiation or specialization of distinctive cell types.
3. During embryogenesis, specific gene activity appears to be affected by the internal environment of the cell, or localized cytoplasmic components. The regulation of early events is mediated by the maternal cytoplasm, which then influences zygotic gene expression. As development proceeds, both the cell's internal environment and its external environment become further altered by the presence of early gene products and communication with other cells.
4. In *Drosophila*, both genetic and molecular studies have confirmed that the egg contains information that specifies the body plan of the larva and adult and that interactions of embryonic nuclei with the maternal cytoplasm initiate transcriptional programs characteristic of specific developmental pathways.
5. Extensive genetic analysis of embryonic development in *Drosophila* has led to the identification of maternal-effect genes that lay down the anterior–posterior and dorsal–ventral axes of the embryo. In addition, these maternal-effect genes activate sets of zygotic segmentation genes, initiating a cascade of gene regulation that ends with the determination of segment identity by the selector genes.
6. In *C. elegans*, the stereotyped lineage of all cells allows developmental biologists to study the cell–cell signaling required for organogenesis and to determine which genes are required for the normal process of programmed cell death.

INSIGHTS AND SOLUTIONS

1. In the slime mold *Dictyostelium*, experimental evidence suggests that cyclic AMP (cAMP) plays a central role in the developmental program leading to spore formation. The genes encoding the cAMP cell-surface receptor have been cloned, and the amino acid sequence of the protein components is known. To form reproductive structures, free-living individual cells aggregate together and then differentiate into one of two cell types, prespore cells or prestalk cells. Aggregating cells secrete waves or oscillations of cAMP to foster the aggregation of cells, and then continuously secrete cAMP to activate genes in the aggregated cells at later stages of development. It has been proposed that cAMP controls cell–cell interaction and gene expression. It is important to test this hypothesis by using several experimental techniques. What different approaches can you devise to test this hypothesis, and what specific experimental systems would you employ to test them?

Solution: Two of the most powerful forms of analysis in biology involve the use of biochemical analogues (or inhibitors) to block gene transcription or the action of gene products in a predictable way, and the use of mutations to alter the gene and its products. These two approaches can be used to study the role of cAMP in the developmental program of *Dictyostelium*. First, compounds chemically related to cAMP, such as GTP and GDP, can be used to test whether they have any effect on the processes controlled by cAMP. In fact, both GTP and GDP lower the affinity of cell-surface receptors for cAMP, effectively blocking the action of cAMP. To inhibit the synthesis of the cAMP receptor, it is possible to construct a vector containing a DNA sequence that transcribes an antisense RNA (a molecule that has a base sequence complementary to the mRNA). Antisense RNA forms a double-stranded structure with the mRNA, preventing it from being transcribed. If normal cells are transformed with a vector that expresses antisense RNA, no cAMP receptors will be produced. It is possible to predict that such cells will fail to respond to a gradient of cAMP and, consequently, will not migrate to an aggregation center. In fact, that is what happens. Such cells remain dispersed and nonmigratory in the presence of cAMP. Similarly, it is possible to determine whether this response to cAMP is necessary to trigger changes in the transcriptional program by assaying for the expression of developmentally regulated genes in cells expressing this antisense RNA.

Mutational analysis can be used to dissect components of the cAMP receptor system. One approach is to use transformation with wild-type genes to restore mutant function. Similarly, because the genes for the receptor proteins have been cloned, it is possible to construct mutants with known alterations in the component proteins and transform them into cells to assess their effects.

2. In the sea urchin, early development up to gastrulation may occur even in the presence of actinomycin D, which inhibits RNA synthesis. However, if actinomycin D is present early in development but removed at the end of blastula formation, gastrulation does not proceed. In fact, if actinomycin D is present only between the sixth and eleventh hours of development, gastrulation (normally occurring at the fifteenth hour) is arrested. What conclusions can be drawn concerning the role of gene transcription between hours six and fifteen?

Solution: Maternal mRNAs are present in the fertilized sea urchin egg. Thus, a considerable amount of development can take place without transcription of the embryo's genome. Because gastrulation is inhibited by prior treatment with actinomycin D, it appears that transcripts from the embryo's genome are required to initiate or maintain gastrulation. This transcription must take place between the sixth and fifteenth hours of development.

3. If it were possible to introduce one of the homeotic genes from *Drosophila* into an *Arabidopsis* embryo homozygous for a homeotic flowering gene, would you expect any of the *Drosophila* genes to negate (rescue) the *Arabidopsis* mutant phenotype? Why or why not?

Solution: The *Drosophila* homeotic genes belong to the *Hox* gene family, while *Arabidopsis* homeotic genes belong to the MADS-box protein family. Both gene families are present in *Drosophila* and *Arabidopsis*, but they have evolved different functions in the animal and the plant kingdom. As a result, it is unlikely that a transferred *Drosophila Hox* gene would rescue the phenotype of a MADS-box mutant, but only an actual experiment would confirm this.

Problems and Discussion Questions

1. Carefully distinguish between the terms *differentiation* and *determination*. Which phenomenon occurs initially during development?
2. Nuclei from almost any source may be injected into *Xenopus* oocytes. Studies have shown that these nuclei remain active in transcription and translation. How can such an experimental system be useful in developmental genetic studies?
3. The homunculus doctrine postulated that miniature adult entities are contained within the egg and merely unfold and grow to give rise to a mature organism. What sorts of isolated evidence presented in this chapter might have led to this doctrine? Why is the epigenetic theory held as correct today?
4. (a) What are the imaginal discs of *Drosophila*? (b) When do they form, how many are there, and what structures do they form in the adult?
5. Distinguish between the syncytial blastoderm stage and the cellular blastoderm stage in *Drosophila* embryogenesis.
6. (a) What are maternal-effect genes? (b) When are gene products from these genes made, and where are they located? (c) What aspects of development do maternal-effect genes control? (d) What is the phenotype of maternal-effect mutations?
7. Suppose you initiate a screen for maternal-effect mutations in *Drosophila* affecting external structures of the embryo and your screen identifies more than 100 mutations that affect external structures. Weischaus and Schüpbach estimated from their screening that there are about 40 maternal-effect genes. How do you reconcile these different results?
8. (a) What are zygotic genes, and when are their gene products made? (b) What is the phenotype associated with zygotic gene mutations? Does the maternal genotype contain zygotic genes?
9. List the main classes of zygotic genes. What is the function of each class of these genes?
10. Experiments have shown that any nuclei placed in the polar cytoplasm at the posterior pole of the *Drosophila* egg will differentiate into germ cells. If polar cytoplasm is transplanted into the anterior end of the egg just after fertilization, what will happen to nuclei that migrate into this cytoplasm at the anterior pole?
11. How can you determine whether a particular gene is being transcribed in different cell types?
12. You observe that a particular gene is being transcribed during development. How can you tell whether the expression of this gene is under transcriptional or translational control?
13. What are *Hox* genes? What properties do they have in common? Are all homeotic genes *Hox* genes?
14. The homeotic mutation *Antennapedia* causes mutant *Drosophila* to have legs in place of antennae and is a dominant gain-of-function mutation. What are the properties of such mutations? How does the *Antennapedia* gene change antennae into legs?
15. The *Drosophila* homeotic mutation *spineless aristapedia* (ss^a) results in the formation of a miniature tarsal structure (normally part of the leg) on the end of the antenna. From your knowledge of imaginal discs, what insight is provided by ss^a concerning the role of genes during determination?
16. Embryogenesis and oncogenesis (generation of cancer) share a number of features including cell proliferation, apoptosis, cell migration and invasion, formation of new blood vessels, and differential gene activity. Embryonic cells are relatively undifferentiated and cancer cells appear to be undifferentiated or dedifferentiated. Homeotic gene expression directs early development, and mutant expression leads to loss of the differentiated state or an alternative cell identity. M.T. Lewis (2000. *Breast Can. Res.* 2:158–69) suggested that breast cancer may be caused by the altered expression of homeotic genes. When he examined 11 such genes in cancers, eight were underexpressed while three were overexpressed compared with controls. Given what you know about homeotic genes, what is the likelihood that they are involved in oncogenesis?
17. In *Drosophila*, both *fushi tarazu* (*ftz*) and *engrailed* genes encode homeobox transcription factors and are capable of eliciting the expression of other genes. Both genes work at about the same time during development and in the same region to specify cell fate in body segments. To discover if *ftz* regulates the expression of *engrailed*, if *engrailed* regulates *ftz*; or if both are regulated by another gene, you perform a mutant analysis. In ftz^- embryos (*ftz*/*ftz*), engrailed protein is absent; in $engrailed^-$ embryos (*eng*/*eng*), *ftz* expression is normal. What does this tell you about the regulation of these two genes—does the *engrailed* gene regulate *ftz*, or does the *ftz* gene regulate *engrailed*?
18. Early development depends on the temporal and spatial interplay between maternally supplied material and mRNA and the onset of zygotic gene expression. Maternally encoded mRNAs must be produced, positioned, and degraded (Surdej and Jacobs-Lorena, 1998. *Mol. Cell Biol.* 18:2892–2900). For example, transcription of the *bicoid* gene that determines anterior–posterior polarity in *Drosophila* is maternal. The mRNA is synthesized in the ovary by nurse cells and then transported to the oocyte, where it localizes to the anterior ends of oocytes. After egg deposition, *bicoid* mRNA is translated and unstable bicoid protein forms a decreasing concentration gradient from the anterior end of the embryo, where *gap* genes along the anterior half of the embryo are activated. At the start of gastrulation, *bicoid* mRNA has been degraded. Consider two models to explain the degradation of *bicoid* mRNA: (1) degradation may result from signals within the mRNA (intrinsic model), or (2) degradation may result from the mRNA's position within the egg (extrinsic model). Experimentally, how could one distinguish between these two models?
19. Formation of germ cells in *Drosophila* and many other embryos is dependent on their position in the embryo and their exposure to localized cytoplasmic determinants. Nuclei exposed to cytoplasm in the posterior end of *Drosophila* eggs (the pole plasm) form cells that develop into germ cells under the direction of maternally derived components. R. Amikura et al. (2001. *Proc. Nat. Acad. Sci. (USA)* 98:9133–38) consistently found mitochondria-type ribosomes outside mitochondria in the germ plasma of *Drosophila* embryos and postulated that they are intimately related to germ-cell specification. If you were studying this phenomenon, what would you want to know about the activity of these ribosomes?
20. One of the most interesting aspects of early development is the remodeling of the cell cycle from rapid cell divisions, apparently lacking G1 and G2 phases, to slower cell cycles with measurable G1 and G2 phases and checkpoints. During this remodeling, maternal mRNAs that specify cyclins are deadenylated and zygotic genes are activated to produce cyclins. Audic et al. (2001. *Mol. and Cell. Biol.* 21:1662–71) suggest that deadenylation requires transcription of zygotic genes. Present a diagram that captures the significant features of these findings.

Extra-Spicy Problems

21. In studying gene action during development, it is desirable to be able to position genes in a hierarchy or pathway of action to establish which genes are primary and in what order genes act. There are several ways of doing this. One is to make double mutants and study the outcome. The gene *fushi-tarazu* (*ftz*) is expressed in early embryos at the seven-stripe stage. All of the genes involved in forming the anterior–posterior pattern affect the expression of this gene, as do the *gap* genes. However, expression of segment-polarity genes is affected by *ftz*. What is the location of *ftz* in this hierarchy?
22. A number of genes that control expression of *Hox* genes in *Drosophila* have been identified. One of these homozygous mutants is *extra sex combs*, where some of the head and all of the thorax and abdominal segments develop as the last abdominal segment. In other words, all affected segments develop as posterior segments. What does this phenotype tell you about which set of *Hox* genes is controlled by the *extra sex combs* gene?
23. The *apterous* gene in *Drosophila* encodes a protein required for wing patterning and growth. It is also known to function in nerve development, fertility, and viability. When human and mouse genes whose protein products closely resemble *apterous* were used to generate transgenic *Drosophila* (Rincon-Limas et al. 1999. *Proc. Nat. Acad. Sci. [USA]* 96:2165–70), the apterous mutant phenotype was *rescued*. In addition, the whole-body expression patterns in the transgenic *Drosophila* were similar to normal *apterous*. (a) What is meant by the term *rescued* in this context? (b) What do these results indicate about the molecular nature of development?
24. In *Arabidopsis*, flower development is controlled by sets of homeotic genes. How many classes of these genes are there, and what structures are formed by their individual and combined expression?
25. The floral homeotic genes of *Arabidopsis* are MADS-box proteins, while in *Drosophila*, they are *Hox* genes, belonging to the homeobox gene family. In both *Arabidopsis* and *Drosophila*, members of the *Polycomb* gene family control expression of these divergent homeotic genes. How do *Polycomb* genes control expression of two very different sets of homeotic genes?
26. Vulval development in *C. elegans* begins when two neighboring cells (Z1.ppp and Z4.aaa) interact with each other by cell–cell signaling involving two components: a membrane-bound signal molecule and a membrane-bound receptor. By chance, one cell produces more signal and causes its neighbor to produce more receptor. The signal-producing cell becomes the anchor cell and the receptor-producing cell becomes the uterine precursor. This form of cell–cell interaction is called the Notch/Delta signaling system. Although it is a widely used signaling mechanism in metazoans, this pathway works only in adjacent cells. Why is this so, and what are the advantages and disadvantages of such a system?
27. The identification and characterization of genes that control sex determination has been another focus of investigators working with *C. elegans*. As with *Drosophila*, sex in this organism is determined by the ratio of X chromosomes to sets of autosomes. A diploid wild-type male has one X chromosome, and a diploid wild-type hermaphrodite has two X chromosomes. Many different mutations have been identified that affect sex determination. Loss-of-function mutations in a gene called *her-1* cause an XO nematode to develop into a hermaphrodite and have no effect on XX development. (That is, XX nematodes are normal hermaphrodites.) In contrast, loss-of-function mutations in a gene called *tra-1* cause an XX nematode to develop into a male. Deduce the roles of these genes in wild-type sex determination from this information.
28. Based on the information in Problem 27 and the analysis of the phenotypes of single- and double-mutant strains, a model for sex determination in *C. elegans* has been generated. This model proposes that the *her-1* gene controls sex determination by establishing the level of activity of the *tra-1* gene, which in turn, controls the expression of genes involved in generating the various sexually dimorphic tissues. Given this information, (a) does the *her-1* gene product have a negative or a positive effect on the activity of the *tra-1* gene? (b) What would be the phenotype of a *tra-1*, *her-1* double mutant?

Selected Readings

Baker, N.E. 2001. Master regulatory genes: telling them what to do. *BioEssays* 23:763–66.

Goodman, F. 2002. Limb malformations and the human *HOX* genes. *Am. J. Med. Genet.* 112:256–65.

Gridley, T. 2003. Notch signaling and inherited human diseases. *Hum. Mol. Genet.* 12:R9–R13.

Halder, G., Callaerts, P., and Gehring, W.J. 1995. Induction of ectopic eyes by targeted expression of the *eyeless* gene in *Drosophila. Science* 267:1788–92.

Kumar, J.P., and Moses, K. 2001. EGF receptor and Notch signaling act upstream of eyeless/PAX6 to control eye specification. *Cell* 104:687–97.

Lall, S., and Patel, N.H. 2001. Conservation and divergence in molecular mechanisms of axis formation. *Annu. Rev. Genet.* 35:407–37.

Nüsslein-Volhard, C., and Weischaus, E. 1980. Mutations affecting segment number and polarity in *Drosophila. Nature* 287:795–801.

Pires-daSilva, A., and Sommer, R.J. 2002. The evolution of signaling pathways in animal development. *Nature Rev. Genet.* 4:39–49.

Rudel, D., and Sommer, R.J. 2003. The evolution of developmental mechanisms. *Dev. Biol.* 264:15–37.

Verakasa, A., Del Campo, M., and McGinnis, W. 2000. Developmental patterning genes and their conserved functions: from model organisms to humans. *Mol. Genet. Metabol.* 69:85–100.

Wang, M., and Sternberg, P.W. 2001. Pattern formation during *C. elegans* vulval induction. *Curr. Top. Dev. Biol.* 51:189–220.

Quantitative Genetics and Multifactorial Traits

A field of pumpkins, where size is under polygenic control.

Chapter Concepts

- Quantitative inheritance results in a range of measurable phenotypes for a polygenic trait.
- Not all polygenic traits show continuous variation.
- Quantitative traits can be explained in Mendelian terms.
- The study of polygenic traits relies on statistical analysis.
- Heritability estimates the genetic contribution to phenotypic variability.
- Twin studies allow an estimation of heritability in humans.
- Quantitative trait loci can be mapped.

We have looked at examples of phenotypic variation that can be classified into distinct and separate categories: pea plants were tall or dwarf; squash fruit shape was spherical, disc shaped, or elongated; and fruit fly eye color was red or white. (See Chapter 4.) Traits such as these with a small number of discrete phenotypes show **discontinuous variation**. Typically in these traits a genotype will produce a single identifiable phenotype, although phenomena such as variable penetrance and expressivity, pleiotropy, and epistasis can obscure the relationship between genotype and phenotype, even for simple discontinuous traits. Other traits, however, including many of medical or agricultural importance, are more complex. They show much more variation with a continuous range of phenotypes that cannot be easily classified into distinct categories. Examples of such traits showing **continuous variation** include human height or weight, milk or meat production in cattle, crop yield, and seed protein content. Continuous variation across a range of phenotypes is measured and described in quantitative terms, so this genetic phenomenon is known as **quantitative inheritance**. Because the varying phenotypes result from the input of genes at multiple loci, quantitative traits are sometimes referred to as **polygenic** (literally "many genes"). For traits showing continuous variation, the genotype generated at fertilization establishes the quantitative range within which a particular individual can fall. However the final phenotype is often also influenced by environmental factors to which that individual is exposed. Human height, for example, is partly genetically determined, but is also affected by environmental factors such as nutrition. The terms **complex** and **multifactorial** are used to describe traits where a range of phenotypes results from gene action and environmental influences.

In this chapter, we will examine examples of quantitative inheritance and some of the statistical techniques used to study complex traits. We will also consider how geneticists assess the relative importance of genetic versus environmental factors contributing to continuous phenotypic variation, and we will discuss approaches to mapping and characterizing quantitative genes within the genome.

24.1 Not All Polygenic Traits Show Continuous Variation

When a continuous trait is examined in a population, individual measurements form a graduation of phenotypes with no clear categories. It is possible for a single-gene trait to show a graduation of phenotypes within a population, if for example there are different degrees of gene penetrance among individuals. Single-gene traits can also be multifactorial if the interaction of alleles with the environment produces a range of different phenotypes. More often, however, a continuous quantitative trait is the result of polygenic inheritance, and polygenic traits are frequently multifactorial, with environmental factors contributing to the range of phenotypes observed.

In addition to continuous quantitative traits where phenotypic variation can fall at any point along a scale of measurement, there are two other classes of polygenic traits. **Meristic** traits are those in which the phenotypes are recorded by counting whole numbers. Examples of meristic traits include the number of seeds in a pod or the number of eggs laid by a chicken in a year. These are quantitative traits but they do not have an infinite range of phenotypes: For example, a pod may contain 2, 4, or 6 seeds but not 5.75. **Threshold traits** are polygenic (and frequently environmental factors affect the phenotypes, so they are also multifactorial), but they are distinguished from continuous and meristic traits by having a small number of discrete phenotypic classes. Threshold traits are of interest to human geneticists as an increasing number of diseases are now suspected to show this pattern of polygenic inheritance. One example is Type II diabetes, also known as adult-onset diabetes because it typically affects individuals who are middle aged or older. A population can be divided into just two phenotypic classes for this trait—individuals who have Type II diabetes and those who do not—so at first glance this may appear to more closely resemble a simple monogenic trait. However, no single adult-onset diabetes gene has been identified. Instead, the combination of alleles present at multiple contributing loci gives an individual a greater or lesser likelihood of developing the disease. These varying levels of liability form a continuous range: At one extreme are those at very low risk for Type II diabetes, while at the other end of the distribution are those whose genotypes make it highly likely they will develop the disease (Figure 24–1). As with many threshold traits, environmental factors also play a role in determining the final phenotype, with diet and lifestyle having significant impact on whether an individual with moderate to high genetic liability will actually develop Type II diabetes.

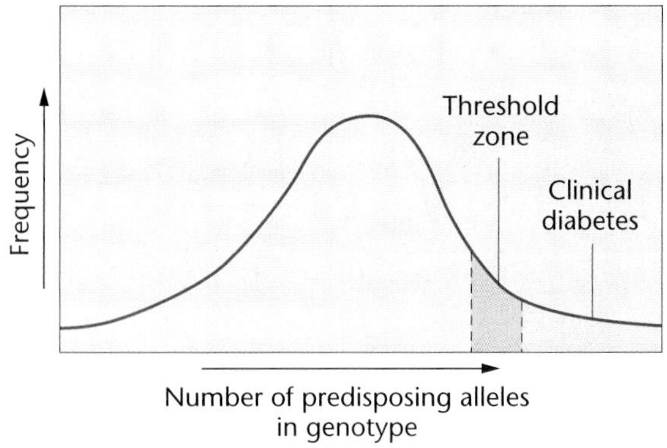

FIGURE 24–1 Threshold trait—Type II diabetes.

How Do We Know?

How do we know that threshold traits are actually polygenic even though they may have as few as two discrete phenotypic classes?

24.2 Quantitative Traits Can Be Explained in Mendelian Terms

The issue of whether continuous phenotypic variation could be explained in Mendelian terms caused considerable controversy in the early 1900s. Some scientists argued that although

Mendel's unit factors or genes explained patterns of discontinuous segregation with discrete phenotypic classes, they could not also account for the range of phenotypes seen in quantitative patterns of inheritance. However, geneticists William Bateson and Gudny Yule, adhering to a Mendelian explanation, proposed the **multiple-factor** or **multiple-gene hypothesis** in which many genes, each individually behaving in a Mendelian fashion, contribute to the phenotype in a *cumulative* or *quantitative* way.

The Multiple Gene Hypothesis for Quantitative Inheritance

The multiple gene hypothesis was based largely on a key set of experimental results published by Hermann Nilsson-Ehle in 1909. Nilsson-Ehle used grain color in wheat to test the concept that the cumulative effects of alleles at multiple loci produce the range of phenotypes seen in quantitative traits. In one set of experiments, wheat with red grain was crossed to wheat with white grain (Figure 24–2). The F_1 generation demonstrated an intermediate pink color, which at first sight suggested incomplete dominance of two alleles at a single locus. However, in the F_2, Nilsson-Ehle did not observe the 3:1 segregation typical of a monhybrid cross. Instead, approximately 15/16 of the plants showed some degree of red grain, while 1/16 of the plants showed white grain. Careful examination of the F_2 revealed that grain with color could be classified into four different shades of red. Because the F_2 ratio occurred in sixteenths, it appears that two genes, each with two alleles, control the phenotype and that they segregate independently from one another in a Mendelian fashion.

If each gene has one potential **additive allele** that contributes to the red grain color and one potential **nonadditive allele** that fails to produce any red pigment, we can see how the multiple-factor hypothesis could account for the various grain color phenotypes. In the P_1, both parents were homozygous; the red parent contains only additive alleles (AABB in Figure 24–2), while the white parent contains only nonadditive alleles (aabb). The F_1 is heterozygous AaBb, contains two additive (A and B) and two nonadditive (a and b) alleles, and expresses the intermediate pink phenotype. In the F_2, each offspring has 4, 3, 2, 1, or 0 additive alleles. F_2 plants with no additive alleles are white (aabb) like one of the P_1 parents, while F_2 plants with 4 additive alleles are red (AABB) like the other P_1 parent. Plants with 3, 2, or 1 additive alleles constitute the other three categories of red color observed in the F_2. The greater the number of additive alleles in the genotype the more intense the red color expressed in the phenotype, as each additive allele present contributes equally to the cumulative amount of pigment produced in the grain.

Nilsson-Ehle's results showed how continuous variation could still be explained in a Mendelian fashion, with additive alleles at multiple loci influencing the phenotype in a quantitative manner, but each individual allele segregating according to Mendelian rules. As we saw in Nilsson-Ehle's initial cross, if two loci, each with two alleles, were involved, then five F_2 phenotypic categories in a 1:4:6:4:1 ratio would be expected. However, there is no reason why three, four, or more loci cannot function in a similar fashion in controlling various quantitative phenotypes. As more quantitative loci become involved, greater and greater numbers of classes appear in the F_2 in more complex ratios. The number of phenotypes and the expected F_2 ratios for crosses involving up to five gene pairs are illustrated in Figure 24–3.

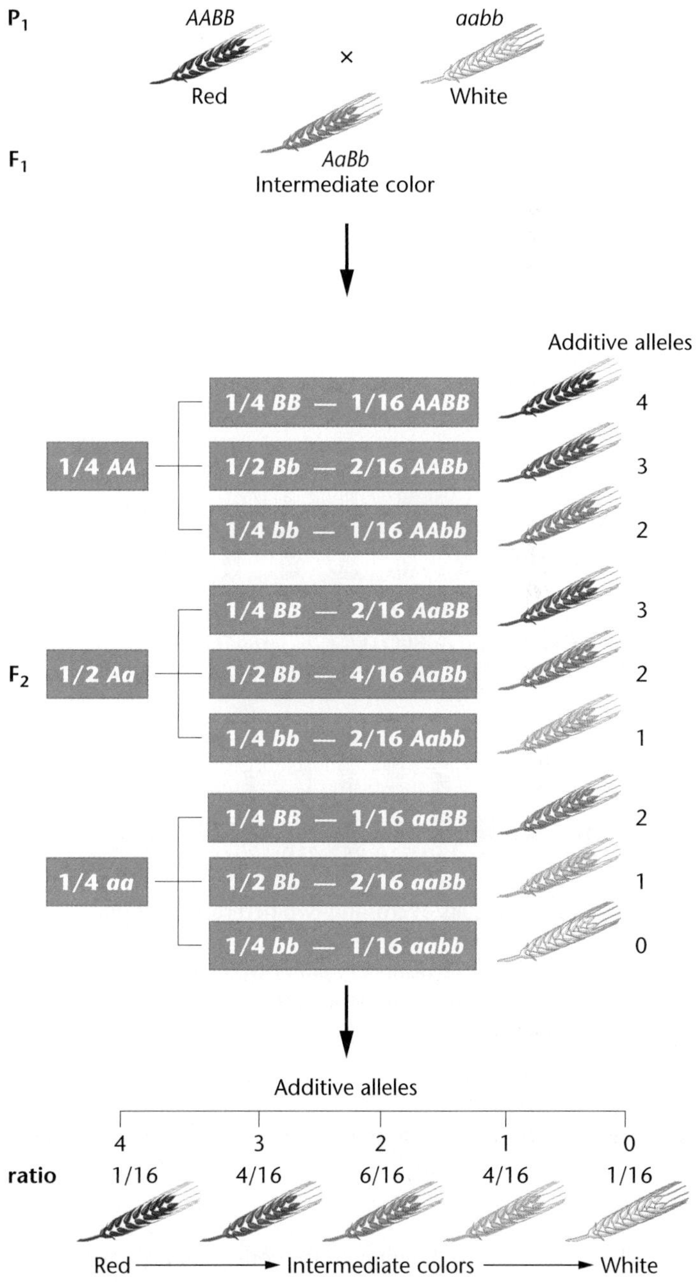

FIGURE 24–2 How the multiple-factor hypothesis counts for the 1:4:6:4:1 phenotypic ratio of grain color when all alleles designated by an uppercase letter are additive and contribute an equal amount of pigment to the phenotype.

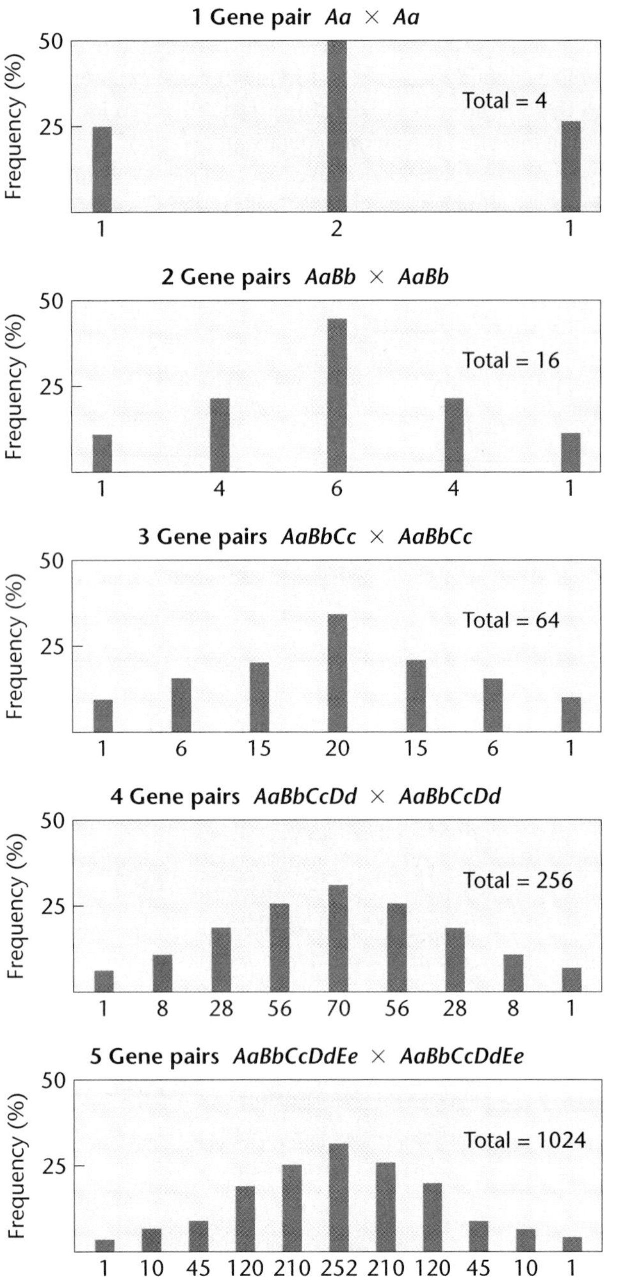

FIGURE 24–3 The results of crossing two heterozygotes when polygenic inheritance is in operation with 1–5 gene pairs. Each histogram bar indicates a distinct F_2 phenotypic class from one extreme (left end) to the other extreme (right end). Each phenotype results from a different number of additive alleles.

Additive Alleles: The Basis of Continuous Variation

The multiple-gene hypothesis embodies the following major points:

1. Phenotypic traits showing continuous variation can be quantified by measuring, weighing, counting, and so on.
2. Two or more gene loci, often scattered throughout the genome, account for the hereditary influence on the phenotype in an *additive way.* Because many genes may be involved, inheritance of this type is called *polygenic.*
3. Each gene locus may be occupied by either an additive allele, which contributes a constant amount to the phenotype, or by a nonadditive allele, which does not contribute quantitatively to the phenotype.
4. The contribution to the phenotype of each additive allele, while small, is approximately equal.
5. Together, the additive alleles contributing to a single quantitative character produce substantial phenotypic variation.

Calculating the Number of Polygenes

Various formulae have been developed for estimating the number of polygenes contributing to a quantitative trait. For example, if the ratio of F_2 individuals resembling *either* of the two extreme P_1 phenotypes can be determined, the number of polygenes involved (n) may be calculated as follows:

$$1/4^n = \text{ratio of } F_2 \text{ individuals expressing either extreme phenotype}$$

In the example of the red and white wheat grain color summarized in Figure 24–2, 1/16 of the progeny are either red *or* white like the P_1 phenotypes. This ratio can be substituted on the right side of the equation to solve for n:

$$\frac{1}{4^n} = \frac{1}{16}$$

$$\frac{1}{4^2} = \frac{1}{16}$$

$$n = 2$$

Table 24.1 lists the ratio and the number of F_2 phenotypic classes produced in crosses involving up to five gene pairs.

For low numbers of polygenes, it is sometimes easier to use the $(2n + 1)$ rule. If n equals the number of additive loci involved in the trait, then $2n + 1$ will determine the total number of possible phenotypes. For example, if $n = 2$, $2n + 1 = 5$ and each phenotype is the result of 4, 3, 2, 1, or 0 additive alleles. If $n = 3$, $2n + 1 = 7$ and each phenotype is the result of 6, 5, 4, 3, 2, 1, or 0 additive alleles.

It should be noted, however, that both these simple methods for estimating the number of polygenes involved in a quantitative trait assume that all the relevant alleles contribute equally and

TABLE 24.1 DETERMINATION OF THE NUMBER OF POLYGENES (n) INVOLVED IN A QUANTITATIVE TRAIT

n	Individuals Expressing either Extreme Phenotype	Distinct Phenotypic Classes
1	1/4	3
2	1/16	5
3	1/64	7
4	1/256	9
5	1/1024	11

additively, and also that phenotypic expression in the F_2 is not affected significantly by environmental factors. As we will see later, for many quantitative traits, these assumptions may not be true.

HOW DO WE KNOW?

How can the number of polygenes involved in a quantitative trait be determined?

Now solve this

Problem 24.3 on page 613 gives F_1 and F_2 ranges for a quantitative trait and asks you to calculate the number of polygenes involved.

Hint: Remember the ratio (not the number) of parental phenotypes reappearing in the F_2 is the key.

24.3 The Study of Polygenic Traits Relies on Statistical Analysis

One of the most important challenges facing genetic researchers working with quantitative traits is determining how much of the phenotypic variation observed in a population is due to genotypic differences among individuals and how much is due to environmental factors. Ways of partitioning variance are discussed in more detail in the next section of this chapter, but first we need to consider the basic statistical tools used with data derived from measuring quantitative phenotypic variation. It is not usually feasible to measure expression of a polygenic trait in every individual in a population, so a random subset of individuals is usually selected for measurement to provide a **sample**. It is important to remember that the accuracy of the information obtained depends on whether the sample is truly random and representative of the population from which it was drawn. Suppose, for example, that a student wants to determine the average height of the 100 students in his genetics class and for his sample he measures the two students sitting next to him, both of whom happen to be on the college basketball team. It is unlikely that this sample will provide a good estimate of the average height of the class for two reasons: first, it is too small; second, it is not a representative subset of the class (unless all 100 students are on the basketball team).

If the sample measured for expression of a quantitative trait is sufficiently large and also representative of the population from which it is drawn, we often find that the data form a **normal distribution** with a characteristic bell-shaped curve when plotted as a frequency histogram (Figure 24–4). Several statistical methods are useful in the analysis of traits that exhibit a normal distribution, including the mean, variance, standard deviation, standard error of the mean, and covariance.

The Mean

The mean provides information about where the central point lies along a range of measurements for a quantitative trait. We can see from Figure 24–5 that the distributions of two sets of phenotypic measurements graphed cluster around a central

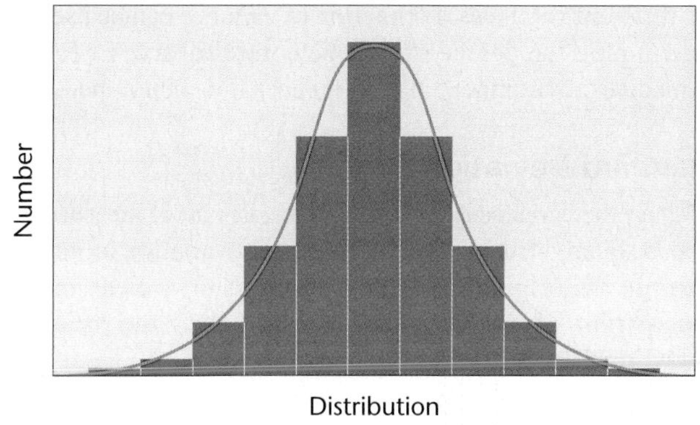

FIGURE 24–4 Normal frequency distribution characterized by a bell-shaped curve.

value. This clustering is called a **central tendency** and the central point is the **mean** $\overline{X}$.

The mean is the arithmetic average of a set of measurements and is calculated as

$$\overline{X} = \frac{\Sigma X_i}{n}$$

where $\overline{X}$ is the mean, ΣX_i represents the sum of all individual values in the sample, and n is the number of individual values.

The mean provides a useful descriptive summary of the sample, but it tells us nothing about the range or spread of the data. As illustrated in Figure 24–5, a symmetrical distribution of values in the sample may, in one case, be clustered near the mean. Or, a set of measurements may have the same mean, but be distributed more widely around it. A second statistic, the **variance**, provides information about the spread of data around the mean.

Variance

The variance for a sample is the average squared distance of all measurements from the mean. It is calculated as

$$s^2 = \frac{\Sigma(X_i - \overline{X})^2}{n - 1}$$

where the sum (Σ) of the squared differences between each measured value (X_i) and the mean $(\overline{X})$ is divided by one less than the total sample size $n - 1$.

As Figure 24–5 shows, it is possible for two sets of sample measurements for a quantitative trait to have the same mean

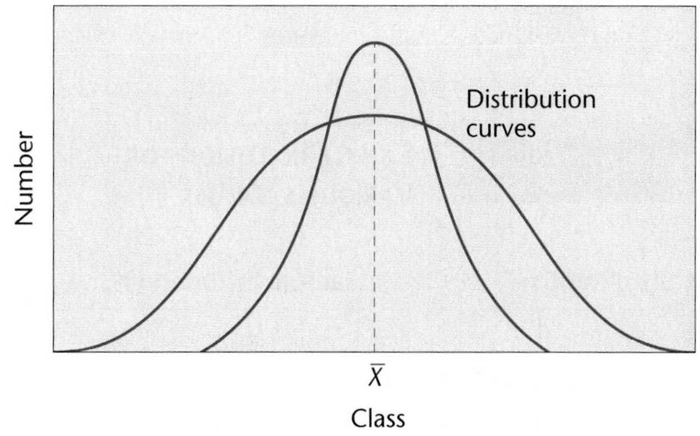

FIGURE 24–5 Two normal frequency distributions with the same mean but different amounts of variation.

but different variances. Estimation of variance can be useful in determining the degree of genetic control of traits when the immediate environment also influences the phenotype.

Standard Deviation

Because the variance is a squared value, its unit of measurement is also squared (m^2, g^2, etc.). To express variation around the mean in the original units of measurement, we can use the square root of the variance, a term called the **standard deviation** (s):

$$s = \sqrt{s^2}$$

Table 24.2 shows the percentage of individual values within a normal distribution that fall within different multiples of the standard deviation. One standard deviation to either side of the mean includes 68 percent of all values in the sample. More than 95 percent of all values are found within two standard deviations to either side of the mean. This means that the standard deviation s can also be interpreted as a probability. For example, a sample measurement picked at random has a 68 percent probability of falling within the range of one standard deviation on either side of the mean.

Standard Error of the Mean

If multiple samples are taken from a population and measured for the same quantitative trait, we might find that their means varied. Theoretically, larger truly random samples will represent the population more accurately and their means will be closer to each other. To measure the accuracy of the sample mean we use the **standard error of the mean** ($S_{\overline{X}}$), calculated as

$$S_{\overline{X}} = \frac{s}{\sqrt{n}}$$

where s is the standard deviation and $\sqrt{n}$ is the square root of the sample size. Because the standard error of the mean is computed by dividing s by $\sqrt{n}$, it is always a smaller value than the standard deviation.

Covariance

Often geneticists working with quantitative traits find they have to consider two phenotypic characters simultaneously. For example, a poultry breeder might find **covariance** between body weight and egg production in hens: heavier birds tend to lay more eggs. The covariance statistic measures how much variation is common to both quantitative traits. It is calculated by taking the deviations from the mean for each trait (just as we did for estimating variance) for each individual in the sample. This gives a pair of values that is multiplied together; the sum of all these individual products is then divided by one fewer than the number in the sample. Thus the covariance $\mathbf{cov}_{xy}$ of two sets of trait measurements, x and y is calculated as

$$\text{cov}_{XY} = \frac{\Sigma[(X_i - \overline{X})(Y_i - \overline{Y})]}{n - 1}$$

The covariance can be then be standardized as a further statistic, the **correlation coefficient (*r*)**. The calculation is

$$r = \text{cov}_{XY}/S_{\overline{X}}S_{\overline{Y}}$$

where S_x is the standard deviation of the first set of quantitative measurements x and S_y is the standard deviation of the second set of quantitative measurements y. Values for the correlation coefficient r can range from -1 to $+1$. Positive values mean that an increase in measurement for one trait tends to be associated with an increase in measurement for the other, while negative r values mean that increases in one trait are associated with decreases in the other. Therefore, if heavier hens do tend to lay more eggs, a positive r value can be expected. A negative r value, on the other hand, suggests that greater egg production is more likely from less heavy birds. One important point to note about correlation coefficients is that even high r values do not imply that a cause-and-effect relationship exists between two traits. Correlation simply tells us the extent to which variation in one quantitative trait is associated with variation in another, not what causes that variation.

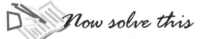

Now solve this

Problem 24.15 on page 614 gives data for two quantitative traits in a flock of sheep. Are the traits correlated?

Hint: First calculate the standard deviation for each trait, then the covariance.

TABLE 24.2 SAMPLE INCLUSION FOR VARIOUS *s* VALUES

Multiples of *s*	Sample Included (%)
$\overline{X} \pm 1s$	68.3
$\overline{X} \pm 1.96s$	95.0
$\overline{X} \pm 2s$	95.5
$\overline{X} \pm 3s$	99.7

Analysis of a Quantitative Character

One homozygous tomato variety produces fruit averaging 18 oz in weight while fruit from a different homozygous variety average 6 oz. The F_1 obtained by crossing these two varieties has fruit weights ranging from 10 to 14 oz. The F_2 population contains individuals that produce fruit ranging from 6 to 18 oz. The results characterizing both generations are shown in Table 24.3.

The mean value for the fruit weight in the F_1 generation can be calculated as

$$\overline{X} = \frac{\Sigma X_i}{n} = \frac{626}{52} = 12.04$$

Similarly, the mean value for fruit weight in the F_2 generation is calculated as

$$\overline{X} = \frac{\Sigma X_i}{n} = \frac{872}{72} = 12.11$$

TABLE 24.3 DISTRIBUTION OF F_1 AND F_2 PROGENY

		Weight (oz.)												
		6	7	8	9	10	11	12	13	14	15	16	17	18
Number of	F_1:					4	14	16	12	6				
Individuals	F_2:	1	1	2	0	9	13	17	14	7	4	3	0	1

Although these mean values are similar, the frequency distributions in Table 24.3 show more variation in the F_2 generation. This variation can be quantified as the sample variance s^2, calculated as the sum of the squared differences between each value and the mean, divided by one less than the total number of observations. So

$$s^2 = \frac{\Sigma(X_i - \overline{X})^2}{n - 1}$$

The variance is 1.29 for the F_1 generation and 4.27 for the F_2 generation. When converted to the standard deviation $\left(s = \sqrt{s^2}\right)$, the values become 1.13 and 2.06, respectively. Therefore, the distribution of tomato weight in the F_1 generation can be described as 12.04 ± 1.13, and in the F_2 generation it can be described as 12.11 ± 2.06.

Assuming both tomato varieties are homozygous at the relevant loci and that the alleles controlling fruit weight act additively, we can estimate the number of polygenes involved in this trait. Since $1/72$ of the F_2 offspring have a phenotype that overlaps one of the parental strains (72 total F_2 offspring; one weighs 6 oz, one weighs 18 oz; see Table 24.3), the use of the formula $1/4^n = 1/72$ indicates that n is between 3 and 4, indicative of the number of genes that control fruit weight in these tomato strains.

24.4 Heritability Estimates the Genetic Contribution to Phenotypic Variability

The question most often asked by geneticists working with multifactorial traits is how much of the observed phenotypic variation in a population is due to genotypic differences among individuals and how much is due to environment. The term **heritability** is used to describe the proportion of total phenotypic variation in a population due to genetic factors. For a multifactorial trait in a given population, a high heritability estimate indicates that much of the variation can be attributed to genetic factors, with the environment having less impact on expression of the trait. With a low heritability estimate, environmental factors are likely to have a greater impact on phenotypic variation within the population.

The concept of heritability is frequently misunderstood and misused. It should be emphasized that heritability does not indicate how much of a trait is genetically determined, or the extent to which an individual's phenotype is due to genotype. In recent years, such misinterpretations of heritability for human quantitative traits has led to controversy, notably in relation to measurements of "intelligence quotient" or IQ. Variation in heritability estimates for IQ among different racial groups tested led to incorrect suggestions that unalterable genetic factors control differences in intelligence levels among human races. Such suggestions misrepresented the meaning of heritability and ignored the contribution of genotype-by-environment interaction to phenotypic variation in a population. Moreover, heritability is not fixed for a trait. For example, a heritability estimate for egg production in a flock of chickens kept in individual cages might be high, indicating that differences in egg output among individual birds is largely due to genetic differences as they all have very similar environments. For a different flock kept outdoors, heritability for egg production might be much lower, as variation among different birds may also reflect differences in their individual environments. Such differences could include how much food each bird manages to find and whether it competes successfully for a good roosting spot at night. Thus a heritability estimate tells us the proportion of phenotypic variation that can be attributed to genetic variation *within a certain population in a particular environment*. If we measure heritability for the same trait among different populations in a range of environments, we frequently find the calculated heritability values have large standard errors. This is an important point to remember when considering heritability estimates for traits in human populations. A mean heritability estimate of 0.65 for human height does not mean that your height is 65 percent due to your genes, but rather that in the populations sampled, on average, 65 percent of the overall variation in stature could be explained by genotypic differences among individuals.

With this subtle, but important distinction in mind, we will now consider how geneticists divide the phenotypic variation observed in a population into genetic and environmental components. As we saw in the previous section, this variation can be quantified as a sample variance: taking measurements of the trait in question from a representative sample of the population and determining the extent of the spread of those measurements around the sample mean. This gives us an estimate of the total phenotypic variance in the population (V_P). Heritability estimates are obtained by using different experimental and statistical techniques to partition V_P into **genotypic variance** (V_G) and **environmental variance** (V_E) components.

A third factor contributing to overall levels of phenotypic variation is the extent to which individual genotypes affect the phenotype differently among varying environments. For example, wheat variety A may yield an average of 20 bushels an acre on poor soil, while variety B yields an average of 17 bushels. On good soil, variety A yields 22 bushels, while variety B averages

25 bushels an acre. There are differences in yield between the two genotypically distinct varieties, so variation in wheat yield has a genetic component. Both varieties yield more on good soil, so yield is also affected by environment. However, we also see that the two varieties do not respond to better soil conditions equally: The genotype of wheat variety B achieves a greater increase in yield on good soil than does variety A. Thus, we have differences in the interaction of genotype with environment contributing to variation for yield in populations of wheat plants. This third component of phenotypic variation is **genotype-by-environment interaction variance** ($V_{G \times E}$) (Figure 24–6).

We can now summarize all the components of total phenotypic variance V_P using the following equation:

$$V_P = V_G + V_E + V_{G \times E}$$

In other words, total phenotypic variance can be subdivided into genotypic variance, environmental variance, and genotype-by-environment interaction variance. When obtaining heritability estimates for a multifactorial trait, researchers often assume that the genotype-by-environment interaction variance is small enough to ignore or combine it with the environmental variance. However, it is worth remembering that this kind of approximation is another reason why heritability values are estimates for a given population, not a fixed attribute for a trait.

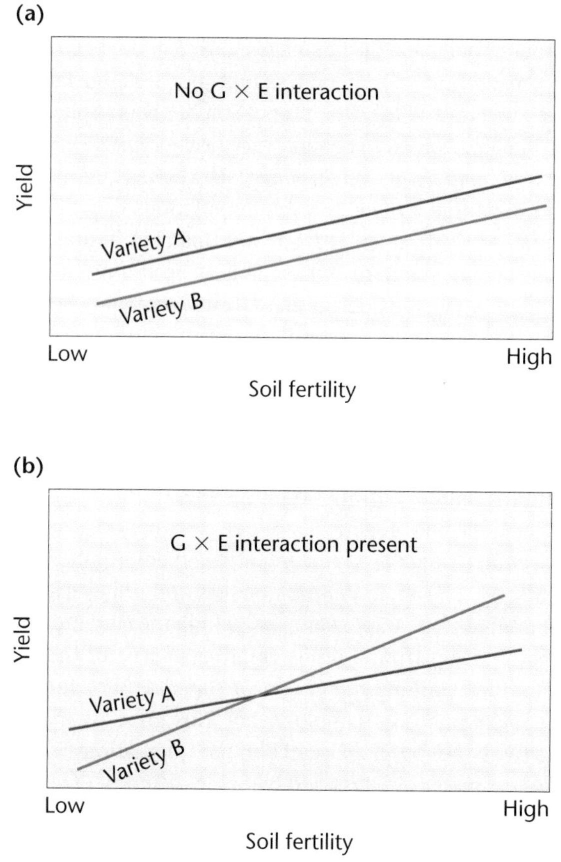

FIGURE 24–6 Differences in yield between two wheat varieties at different soil fertility levels. (a) No G × E interaction: The varieties show genetic differences in yield but respond equally to increasing soil fertility. (b) G × E interaction present: Variety A outyields B at low soil fertility but B yields more than A at high fertility levels.

Animal and plant breeders use a range of experimental techniques to estimate heritabilities by partitioning measurements of phenotypic variance into genotypic and environmental components. One approach uses inbred strains containing genetically homogenous individuals with highly homozygous genotypes. Experiments are then designed to test the effects of a range of environmental conditions on phenotypic variability. Variation *between* different inbred strains reared in a constant environment is due predominantly to genetic factors. Variation *among* members of the same inbred strain reared under different conditions is more likely to be due to environmental factors. Other approaches involve analysis of variance for a quantitative trait among offspring from different crosses, or comparing expression of a trait among offspring and parents reared in the same environment. More details of these experimental techniques can be found in some of the Selected Readings at the end of this chapter.

Broad-Sense Heritability

Broad-sense heritability (represented by the term H^2) measures the contribution of the genotypic variance to the total phenotypic variance. It is estimated as a proportion:

$$H^2 = \frac{V_G}{V_P}$$

Heritability values for a trait in a population range from 0.0 to 1.0. A value approaching 1.0 indicates that the environmental conditions have little impact on phenotypic variance, which is therefore largely due to genotypic differences among individuals in the population. Low values close to 0.0 indicate that environmental factors, not genotypic differences, are largely responsible for the observed phenotypic variation within the population studied. Few quantitative traits have very high or very low heritability estimates, suggesting that both genetics and environment play a part in the expression of most phenotypes in the range.

The genotypic variance component V_G used in broad-sense heritability estimates includes all types of genetic variation in the population: It does not distinguish between quantitative trait loci with alleles acting additively as opposed to those with epistatic or dominance effects. Broad-sense heritability estimates also assume that the genotype-by-environment variance component is negligible. While broad-sense heritability estimates for a trait are of general genetic interest, these limitations mean this kind of heritability is not very useful in breeding programs. Animal or plant breeders wishing to develop improved strains of livestock or higher yielding crop varieties need more precise heritability estimates for the traits they wish to manipulate in a population. Therefore, another type of estimate, narrow-sense heritability, has been devised that is of more practical use.

Narrow-Sense Heritability

Narrow-sense heritability (h^2) is the proportion of phenotypic variance due to additive genotypic variance alone. Genotypic variance can be divided into subcomponents representing the different modes of action of alleles at quantitative trait loci. As not all the genes involved in a quantitative trait affect the phenotype in the same way, this partitioning distinguishes between three different kinds of gene action contributing to genotypic variance. **Additive variance**, V_A, is the genotypic variance due to the additive action of alleles at quantitative trait loci. **Dominance variance**, V_D, is the deviation from the additive components that results when phenotypic expression in heterozygotes is not precisely intermediate between the two homozygotes. **Interactive variance**, V_I, is the deviation from the additive components that occurs when two or more loci behave epistatically. The amount of interactive variance is often negligible and so this component may be excluded from calculations of total genotypic variance.

The partitioning of the total genotypic variance V_G is summarized in the equation

$$V_G = V_A + V_D + V_I$$

and a narrow-sense heritability estimate based only on that portion of the genotypic variance due to additive gene action becomes

$$h^2 = \frac{V_A}{V_P}$$

Omitting V_I and separating V_P into genotypic and environmental variance components we obtain

$$h^2 = \frac{V_A}{V_E + V_A + V_D}$$

Heritability estimates are used in animal and plant breeding to indicate the potential response of a population to artificial selection for a quantitative trait. Narrow-sense heritability, h^2, provides a more accurate prediction of selection response than broad-sense heritability, H^2, and therefore h^2 is more widely used by breeders.

Now solve this

Problem 24.12 on page 614 gives data for two quantitative traits in a herd of hogs and asks you to choose the one the farmer should select for.

Hint: Consider which variance component's proportion is the best indicator of potential response to selection.

Artificial Selection

Artificial selection is the process of choosing specific individuals with preferred phenotypes from an initially heterogeneous population for future breeding purposes. Theoretically, if artificial selection based on the same trait preferences is repeated over multiple generations, a population can be developed containing a high frequency of individuals with the desired characteristics. If selection is for a simple trait controlled by just one or two genes subject to little environmental influence, generating the desired population of plants or animals is relatively fast and easy. However, many traits of economic importance in crops and livestock, such as grain yield in plants, weight gain or milk yield in cattle, and speed or stamina in horses, are polygenic and frequently multifactorial. Artificial selection for such traits is slower and more complex. Narrow sense heritability estimates are valuable to the plant or animal breeder because—as we have just seen—they estimate the proportion of total phenotypic variance for the trait that is due to additive genetic variance. Quantitative trait alleles with additive action are those most easily manipulated by the breeder. Alleles at quantitative trait loci that generate dominance effects or interact epistatically (and therefore contribute to V_D or V_I) are less responsive to artificial selection. Thus narrow-sense heritability, h^2, can be used to predict the impact of selection. The higher the estimated value for h^2 in a population, the greater the change in phenotypic range for the trait that the breeder will see in the next generation.

Partitioning the genetic variance components to calculate h^2 and predict response to selection is a complex task requiring careful experimental design and analysis. The simplest approach is to select individuals with superior phenotypes for the desired quantitative trait from a heterogenous population and breed offspring from those individuals. The mean scores for the trait can then be compared to (1) the original population (M), (2) the selected individuals used as parents (M1), and (3) the offspring resulting from interbreeding the selected parents (M2). The relationship between these three means and h^2 is

$$h^2 = \frac{M2 - M}{M1 - M}$$

This equation can be further simplified by defining M2 − M as the **response** (R) and M1 − M as the **selection differential** (S), so h^2 reflects the ratio of the response observed to the total response possible. Thus,

$$h^2 = \frac{R}{S}$$

A narrow-sense heritability value obtained in this way by selective breeding and measuring the response in the offspring is referred to as an estimate of **realized heritability**.

As an example of a realized heritability estimate, suppose that we measure the diameter of corn kernels in a population where the mean diameter M was 20 mm. From this population, we select a group with the smallest diameters, for which the mean M1 equals 10 mm. The selected plants are interbred, and the mean diameter M2 of the progeny kernels is 13 mm. We can calculate the realized heritability h^2 to estimate the potential for artificial selection on kernel size:

$$h^2 = \frac{M2 - M}{M1 - M}$$

$$h^2 = \frac{13 - 20}{10 - 20}$$

$$= \frac{-7}{-10}$$

$$= 0.70$$

This value for narrow-sense heritability indicates the selection potential for kernel size is relatively high.

The longest running artificial selection experiment known is still being conducted at the State Agricultural Laboratory in Illinois. Since 1896, corn has been selected for both high and low oil content. After 76 generations, selection continues to result in increased oil content (Figure 24–7). With each cycle of successful selection, more of the corn plants accumulate a higher percentage of additive alleles involved in oil production. Consequently, the narrow-sense heritability h^2 of increased oil content in succeeding generations has declined (see parenthetical values at generations 9, 25, 52, and 76 in Figure 24–7) as artificial selection comes closer and closer to optimizing the genetic potential for oil production. Theoretically, the process will continue until all individuals in the population possess a uniform genotype that includes all the additive alleles responsible for high oil content. At that point, h^2 will be reduced to zero, and response to artificial selection will cease. The decrease in response to selection for low oil content shows that heritability for the trait is approaching this point.

Table 24.4 lists narrow-sense heritability estimates expressed as percentage values for a variety of quantitative traits in different organisms. As you can see, these h^2 values vary, but heritability tends to be low for quantitative traits that are essential to an organism's survival. Remember, this does not mean that there is no genetic contribution to the observed phenotypes for such traits. Instead, the low h^2 values show that natural selection has already largely optimized the genetic component of these traits during evolution. Egg production, litter size, and conception rate are examples where such physiological limitations on selection have already been established. Traits that are less critical to survival, such as body weight, tail length, and wing length, have higher heritabilities as more genotypic variation for such traits is still present in the population. Remember also, that any single heritability estimate provides information only about one population in a specific environment. Therefore, narrow-sense heritability estimates are more valuable as predictors of response to selection when they are based on data collected in many populations and environments and where a clear trend is established.

TABLE 24.4 ESTIMATES OF HERITABILITY FOR TRAITS IN DIFFERENT ORGANISMS

Trait	Heritability (h^2)
Mice	
Tail length	60%
Body weight	37
Litter size	15
Chickens	
Body weight	50
Egg production	20
Egg hatchability	15
Cattle	
Birth weight	45
Milk yield	44
Conception rate	3

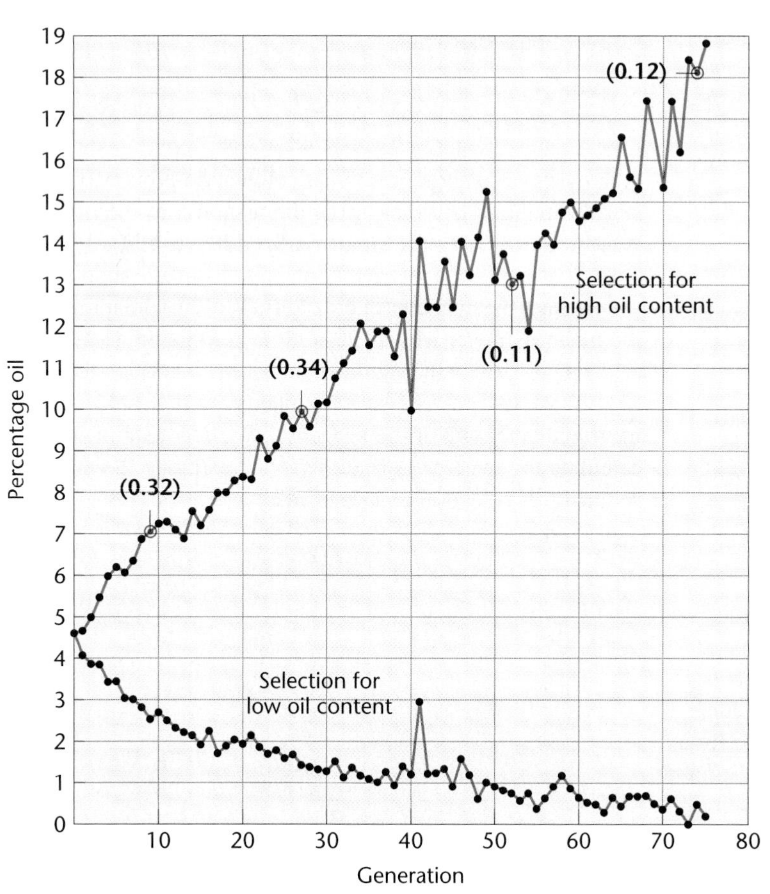

FIGURE 24–7 Response of corn selected for high and low oil content over 76 generations. The numbers in parentheses at generations 9, 25, 52, and 76 for the "high oil" line indicate the calculation of heritability at these points in the continuing experiment.

24.5 Twin Studies Allow an Estimation of Heritability in Humans

For obvious reasons, traditional heritability studies are not possible in humans. However, human twins can be useful subjects for examining how much variance for a multifactorial trait is genotypic as opposed to environmental. **Monozygotic (MZ)** or **identical twins**, derived from the division and splitting of a single egg following fertilization, are genotypically identical. **Dizygotic (DZ)** or **fraternal twins**, on the other hand, originate from two separate fertilization events and are as genetically similar as any other two siblings, with about 50 percent of their genes in common. For a given trait, therefore, phenotypic differences between pairs of identical twins will be equivalent to the environmental variance V_e (because the genotypic variance is zero). Phenotypic differences between dizygotic twins, however, represent both environmental variance V_e and approximately half the genotypic variance V_g. Comparison of phenotypic variances for the same trait in monozygotic and dizygotic sets of twins provides an estimate of broad sense heritability for the trait.

Another approach used in twin studies has been to compare identical twin pairs reared together with pairs that were separated and raised in different settings. For any particular trait, average similarities or differences can be investigated. Twins are said to be **concordant** for a given trait if both express it or neither expresses it. If one expresses the trait and the other does not, the pair is said to be **discordant**. Comparison of concordance values of MZ versus DZ twins reared together illustrates the potential value for heritability assessment. (See "Now Solve This" below.)

Such data must be examined very carefully before any conclusions are drawn. If the concordance value approaches 90 to 100 percent in monozygotic twins, we might be inclined to interpret that value as indicating a large genetic contribution to the expression of the trait. In some cases—for example, blood types and eye color—we know that this is indeed true. In the case of measles, however, a high concordance value merely indicates that the trait is almost always induced by a factor in the environment—in this case, a virus.

It is more meaningful to compare the *difference* between the concordance values of monozygotic and dizygotic twins. If these values are significantly higher for monozygotic twins than for dizygotic twins, we suspect that there is a strong genetic component involved in the determination of the trait. We reach this conclusion because monozygotic twins, with identical genotypes, would be expected to show a greater concordance than genetically related, but not genetically identical, dizygotic twins. In the case of measles, where concordance is high in both types of twins, the environment is assumed to contribute significantly. Such an analysis is useful because phenotypic characteristics that remain similar in different environments are likely to have a strong genetic component.

Interesting as they are, human twin studies contain some unavoidable sources of error. For example, identical twins are often treated more similarly by parents and teachers than are nonidentical twins, especially where the nonidentical pairs are of different sex. This may inflate the environmental variance for nonidentical twin pairs. Another possible error source is genotype: environment interaction, which can increase the total phenotypic variance for nonidentical twins compared to identical twins raised in the same environment. Heritability estimates for human traits based on twin studies should therefore be considered approximations and examined very carefully before any conclusions are drawn.

Now solve this

Problem 24.24 on page 615 lists the percentages of monozygotic (MZ) and dizygotic (DZ) twins expressing the same phenotype for different traits and asks you to evaluate the relative importance of genetic and environmental factors.

Hint: Consider how MZ twins differ genetically from DZ twins. If each pair of twins was raised in the same environment, consider how that would affect expression of genetically determined vs. environmentally determined phenotypes.

24.6 Quantitative Trait Loci Can Be Mapped

The kind of pedigree analysis we looked at earlier in the text (see Chapters 3 and 4) is of little use in identifying the multiple genes involved in quantitative traits. Environmental effects, interaction among segregating alleles, and the number of genes contributing to a polygenic phenotype make it difficult to isolate the effect of any one individual gene. However, because many quantitative traits are of economic or medical importance, it is useful to identify the genes involved and determine their location in the genome. Multiple genes contributing to a quantitative trait are known as **quantitative trait loci (QTLs)**. Identifying these loci provides information on how many genes are involved in a quantitative trait and whether they all contribute equally to the trait or some genes influence the phenotype more strongly than others. Mapping reveals whether the QTLs associated with a trait are grouped on a single chromosome or scattered throughout the genome.

To find and map QTLs, researchers look for associations between particular DNA sequences within the genome and phenotypes falling within a certain range of the quantitative phenotype. One way to do this is to cross homozygous inbred lines that have different phenotypic extremes for the trait of interest to create an F_1 generation whose members will be heterozygous at most of the loci contributing to the trait. Additional crosses, either among F_1 individuals or between the F_1 and the inbred parent lines, result in a F_2 generation with a high degree of segregation for different QTL genotypes and associated phenotypes. This segregating F_2 is known as the **QTL mapping population**. Researchers measure expression of the trait by individuals in the mapping population and identify different genotypes using DNA markers such as RFLPs and microsatellites. (See Chapter 22 for a more detailed description of marker technology.) Computer-based statistical analysis is then used to examine the correlation between marker genotypes and phenotypic variation for the trait. If a DNA marker is not linked to a QTL, then the phenotypic mean score for the trait will not vary among individuals with different genotypes at that marker locus. However, if a DNA marker is linked to a QTL, then genotypes differing at that marker locus will also differ in their expression of the trait. When this occurs, the marker locus and the QTL are said to *cosegregate*. Consistent cosegregation establishes the presence of a QTL at or near the DNA marker along the chromosome: The marker and QTL are linked. When numerous QTLs for a given trait have been located, a genetic map is created giving the positions of the genes involved on the different chromosomes.

DNA markers are now available for many organisms of agricultural importance, making possible systematic mapping of QTLs. For example, hundreds of RFLP markers have been located in the tomato. QTL analysis is performed by crossing plants with extreme, but opposite, phenotypes and following the crosses through several generations. Many loci have been identified that are responsible for quantitative traits such as fruit size, shape, soluble solid content, and acidity. These loci are distributed on all 12 chromosomes representing the haploid genome of this plant. Several chromosomes contain loci for multiple traits.

Mapping and characterizing QTLs in the tomato has been the focus of a highly successful research effort conducted by Steven Tanksley and his colleagues at Cornell University. Their research shows that approximately 30 QTLs contribute to tomato fruit size and shape. However, these QTLs are not equal in their effects: 10 genes scattered among seven chromosomes account for most of the observed phenotypic variation for these traits. Different alleles at just one of these loci, *fw2.2* on chromosome 2, can change fruit size by up to 30 percent. *fw2.2* has been isolated, cloned, and transferred between plants, with interesting results. While the cultivated tomato can weigh up to 1000 grams, fruit from the related wild species thought to be the progenitor of the modern tomato weighs only a few grams. Two distinct alleles of *fw2.2*, identified as a result of RFLP mapping studies, appear to have played a major part in the development of the domesticated tomato. One allele is present in all wild small-fruited varieties of tomatoes investigated. The other allele is present in all domesticated large-fruited varieties. When the cloned *fw2.2* allele from small-fruited varieties is transferred to a plant that normally produces large tomatoes, the transformed plant produces fruits that are greatly reduced in weight (Figure 24–8). In the varieties studied by Tanksley's group, the reduction averaged 17 grams, a significant phenotypic change caused by the action of a single QTL. Further analysis of *fw2.2* revealed that the large-fruited allele at this locus codes for a protein acting as a negative repressor of cell division, resulting in increased mitosis in the cortical tissue that forms the fruit. The small-fruited *fw2.2* allele codes for the same protein but has mutations in the promoter region that reduce transcription levels, resulting in lower rates of cell division and much smaller fruit.

FIGURE 24–8 Phenotypic effect of the *fw2.2* transgene in tomato. When the allele causing small fruit is transferred to a plant that normally procedures large fruit (+), the fruit is reduced in size. A control fruit (−), which is normally larger, is shown for comparison.

Marker-based mapping can identify chromosomal regions where QTLs are found, but identifying individual quantitative genes usually requires additional techniques. One approach is to develop DNA markers that are tightly linked to a gene of interest and then use positional cloning (see Chapter 22) to identify the gene itself. Clearly, finding and characterizing all the genes involved in a quantitative trait is a long and complex task. *fw2.2* is at present one of few quantitative genes that have been isolated, cloned, and studied in detail. However, mapping QTLs and defining the function of genes present in these regions in agriculturally important plants will enhance programs designed to improve crop yields. These studies also pave the way for similar research on animal genes, including our own species.

Chapter Summary

1. Quantitative inheritance results in a range of phenotypes for a trait due to the action of multiple genes combined with environmental factors.
2. Many polygenic quantitative traits show continuous variation that cannot be classified into discrete phenotypic classes. Some threshold traits are also polygenic although they have a small number of distinct phenotypic classes.
3. Statistical methods are used to analyze quantitative traits, including the mean, variance, standard deviation, standard error, and covariance.
4. Heritability estimates the relative contribution of genetic versus environmental factors to the range of phenotypic variation seen in a quantitative trait. Twin studies are used to estimate heritabilities for polygenic traits in humans.
5. Multiple genes contributing to a quantitative trait are known as quantitative trait loci or QTLs. Mapping techniques using DNA markers to can be used to locate and characterize QTLs.

Genetics, Technology, and Society

The Green Revolution Revisited

Of the greater than 6 billion people now living on Earth, about 750 million don't have enough to eat. And despite efforts to limit population growth, an additional 1 million people are expected to go hungry each year for the next several decades.

Will we be able to solve this problem? The past gives us some reasons to be optimistic. In the 1950s and 1960s, in the face of looming population increases, plant scientists around the world set about to increase the production of crop plants, including the three most important grains, wheat, rice, and maize. These efforts became known as the Green Revolution. The approach was three pronged: (1) to increase the use of fertilizers, pesticides, and irrigation water; (2) to bring more land under cultivation; and (3) to develop improved varieties of crop plants by intensive plant breeding. While highly successful in recent years, the rate of increase in grain yields has slowed. If food production is to keep pace with the projected increase in the world's population, plant breeders will have to depend more and more on the genetic improvement of crop plants to provide higher yields. But is this possible? Are we approaching the theoretical limits of yield in important crop plants? Recent work with rice suggests that the answer to this question is a resounding no.

Rice ranks third in worldwide production, just behind wheat and maize. About 2 billion people, fully one third of Earth's population, depend on rice for their basic nourishment. The majority of the world's rice crop is grown and consumed in Asia, but it is also a dietary staple in Africa and Central America. The Green Revolution for rice began in 1960, with the establishment of the International Rice Research Institute (IRRI), headquartered at Los Baños, Philippines. The goal was to breed rice with improved disease resistance and higher yield. Breeders were almost too successful: The first high-yield varieties were so top heavy with grain that they tended to fall over. (Plant breeders call this "lodging.") To reduce lodging, IRRI breeders crossed a high-yield line with a dwarf native variety to create semidwarf lines, which were introduced to farmers in 1966. Due in large part to the adoption of the semidwarf lines, the world production of rice doubled in the next 25 years.

Rice breeders cannot afford to rest on their laurels, however, as the yield of modern rice varieties has not improved much in recent years. Predictions suggest that a 70 percent increase in the annual rice harvest may be necessary to keep pace with anticipated population growth during the next 30 years. Breeders are now looking to wild rice varieties for further crop improvement. Leading the way are Susan McCouch, Steven Tanksley, and their coworkers at Cornell University. To test the hypothesis that wild-rice species carry genes that will improve the yield of cultivated varieties, they crossed cultivated rice (*Oryza sativa*) with a low-yield wild ancestor species (*Oryza rufipogon*) and then successively backcrossed the interspecific hybrid to cultivated rice for three generations. In theory, this would create lines whose genomes were about 95 percent from *O. sativa* and 5 percent from *O. rufipogon*. When testing these backcrossed lines for grain yield, they found that several of them outproduced cultivated rice by as much as 30 percent. These results demonstrated strikingly that even though wild-rice relatives have low yields and appear to be inferior to cultivated rice, they still carry genes that will increase the yield of elite rice varieties. It will now be up to breeders to exploit the wild-rice relatives.

But introducing favorable genes from a wild relative into a cultivated variety by conventional breeding is a long and involved process, often requiring a decade or more of crossing, selection, backcrossing, and more selection. Future improvements in cultivated species must be quicker if crop yields are to keep up with population growth. Fortunately, modern gene-mapping techniques are now leading to the identification of QTLs that control complex traits such as yield and disease resistance. This makes possible a more direct approach to crop improvement, which has been termed the *advanced backcross QTL method*.

First, a cultivated variety is crossed with a wild relative, just as *O. sativa* was interbred with *O. rufipogon*. The hybrid is then backcrossed to the cultivated variety to generate lines that contain only a small fraction of the "wild" genome. The backcross lines with the best qualities (e.g., highest yield and most disease resistance) are selected, and the "wild QTLs" responsible for the superior performance are identified, using a detailed molecular linkage map. Once the beneficial QTLs are discovered by this method, they may be introduced into other cultivated varieties. In order for this strategy to succeed, it is essential that wild crop relatives be preserved as a storehouse of potentially useful genes. Efforts were begun in the 1970s to protect the existing crop relatives of many plants in their natural habitats and to preserve them in seed banks. As the work of McCouch and Tanksley and others has shown, it is not possible to predict which wild varieties may be needed decades or even centuries from now to contribute their beneficial alleles to cultivated varieties. To prevent the loss of superior genes, the widest possible spectrum of wild species must be preserved, even those that have no obvious favorable characteristics.

Almost 60 years ago, the great Russian plant geneticist N. I. Vavilov suggested that wild relatives of crop plants could be the source of genes to improve agriculture. In the coming century, Vavilov's vision may finally be realized, as genes from long-neglected wild crop relatives, identified by new molecular methods, spark a revitalized Green Revolution.

References

Mann, C. 1997. Reseeding the green revolution. *Science* 277:1038–43.

Ronald, P.C. 1997. Making rice disease-resistant. *Sci. Am.* (Nov.) 277:98–105.

Tanksley, S.D., and McCouch, S.R. 1997. Seed banks and molecular maps: Unlocking genetic potential from the wild. *Science* 277:1063–66.

Xiao, J. et al. 1996. Genes from wild rice improve yield. *Nature* 384:223–24.

Insights and Solutions

1. In a plant, height varies from 6 to 36 cm. When 6-cm and 36-cm plants were crossed, all plants were 21 cm. In the F_2 generation, a continuous range of heights was observed. Most were around 21 cm, and 3 of 200 were as short as the 6-cm P_1 parent.

(a) What mode of inheritance is illustrated, and how many gene pairs are involved?

Solution: Polygenic inheritance is illustrated where a continuous trait is involved and where alleles contribute additively to the phenotype. The 3/200 ratio of F_2 plants is the key to determining the number of gene pairs. This reduces to a ratio of 1/66.7, very close to 1/64. Using the formula $1/4^n = 1/64$ (where 1/64 is equal to the proportion of F_2 phenotypes as extreme as either P_1 parent), $n = 3$. Therefore, three gene pairs are involved.

(b) How much does each additive allele contribute to height?

Solution: The variation between the two extreme phenotypes is

$$36 - 6 = 30 \text{ cm.}$$

Because there are six potential additive alleles (*AABBCC*), each contributes

$$30/6 = 5 \text{ cm}$$

to the base height of 6 cm, which results when no additive alleles (*aabbcc*) are part of the genotype.

(c) List all genotypes that give rise to plants that are 31 cm.

Solution: All genotypes that include 5 additive alleles will be 31 cm (5 alleles × 5 cm + 6 cm base height = 31 cm). Therefore, *AABBCc*, *AABbCC*, and *AaBBCC* are the genotypes that will result in plants that are 31 cm.

(d) In a cross separate from the earlier mentioned F_1 a plant of unknown phenotype and genotype was testcrossed with the following results:

1/4 11 cm
2/4 16 cm
1/4 21 cm

An astute genetics student realized that the unknown plant could be only one phenotype, but could be any of three genotypes. What were they?

Solution: When testcrossed (with *aabbcc*), the unknown plant must be able to contribute either one, two, or three additive alleles in its gametes in order to yield the three phenotypes in the offspring. Since no 6-cm offspring are observed, the unknown plant never contributes all nonadditive alleles (*abc*). Only plants that are homozygous at one locus and heterozygous at the other two loci will meet these criteria. Therefore, the unknown parent can be any of three genotypes, all of which have a phenotype of 26 cm:

AABbCc
AaBbCC
AaBBCc

For example, in the first genotype (*AABbCc*),

$$AABbCc \times aabbcc$$

yields

1/4 *AaBbCc* 21 cm
1/4 *AaBbcc* 16 cm
1/4 *AabbCc* 16 cm
1/4 *Aabbcc* 11 cm

which is the ratio of phenotypes observed.

2. The results shown in the following table (at the top of pp. 613) were recorded for ear length in corn:

For each of the parental strains and the F_1 generation, calculate the mean values for ear length.

Solution: The mean values can be calculated as follows:

$$\overline{X} = \frac{\Sigma X_i}{n} \quad P_A: \overline{X} = \frac{\Sigma X_i}{n} = \frac{378}{57} = 6.63$$

$$P_B: \overline{X} = \frac{\Sigma X_i}{n} = \frac{1697}{101} = 16.80$$

$$F_1: \overline{X} = \frac{\Sigma X_i}{n} = \frac{836}{69} = 12.11.$$

3. For the corn plant described in Problem 2, compare the mean of the F_1 with that of each parental strain. What does this tell you about the type of gene action involved?

Solution: The F_1 mean (12.11) is almost midway between the parental means of 6.63 and 16.80. This indicates that the genes in question may be additive in effect.

4. The mean and variance of corolla length in two highly inbred strains of *Nicotiana* and their progeny are shown here:

One parent (P_1) has a short corolla and the other parent (P_2) has a long corolla. Calculate the broad-sense heritability (H^2) of corolla length in this plant.

Strain	Mean (mm)	Variance (mm)
P_1 short	40.47	3.12
P_2 long	93.75	3.87
F_1 ($P_1 \times P_2$)	63.90	4.74
F_2 ($F_1 \times F_1$)	68.72	47.70

Solution: The formula for estimating heritability is $H^2 = V_G/V_P$, where V_G and V_P are the genetic and phenotypic components of variation, respectively. The main issue in this problem is obtaining some estimate of two components of phenotypic variation: genetic and environmental factors. V_P is the combination of genetic and environmental variance. Because the two parental strains are true breeding, they are assumed to be homozygous, and the variance of 3.12 and 3.87 is considered to be the result of environmental influences. The average of these two values is 3.50. The F_1 is also genetically homogeneous and gives us an additional estimate of the impact of environmental factors. By averaging this value along with that of the parents,

$$\frac{4.74 + 3.50}{2} = 4.12$$

we obtain a relatively good idea of environmental impact on the phenotype. The phenotypic variance in the F_2 is the sum of the genetic P_1 and environmental (V_E) components. We have estimated the environmental input as 4.12, so 47.70 minus 4.12 gives us an estimate of V_G, which is 43.58. Heritability then becomes 43.58/47.70 or 0.91. This value, when viewed in percentage form, indicates that about 91 percent of the variation in corolla length is due to genetic influences.

	Length of Ear in cm																
	5	6	7	8	9	10	11	12	13	14	15	16	17	18	19	20	21
Parent A	4	21	24	8													
Parent B									3	11	12	15	26	15	10	7	2
F_1					1	12	12	14	17	9	4						

PROBLEMS AND DISCUSSION QUESTIONS

1. What is the difference between continuous and discontinuous variation? Which of the two is most likely to be the result of polygenic inheritance?
2. Define the following: (a) polygene, (b) additive alleles, (c) correlation, (d) monozygotic and dizygotic twins, (e) heritability, and (f) QTL.
3. A homozygous plant with 20 cm diameter flowers is crossed with a homozygous plant of the same species that has 40 cm diameter flowers. The F_1 plants all have flowers 30 cm in diameter. In the F_2 generation of 512 plants, 2 plants have flowers 20 cm in diameter, 2 plants have flowers 40 cm in diameter and the remaining 510 plants have flowers of a range of sizes in between.
   (a) Assuming that all alleles involved act additively, how many genes control flower size in this plant?
   (b) What frequency distribution of flower diameter would you expect to see in the progeny of a backcross between an F_1 plant and the large-flowered parent?
   (c) Calculate the mean, variance, and standard deviation for flower diameter in the backcross progeny from (b).
4. A dark-red strain and a white strain of wheat are crossed and produce an intermediate, medium-red F_1. When the F_1 plants are interbred, an F_2 generation is produced in a ratio of 1 dark-red: 4 medium-dark-red: 6 medium-red: 4 light-red: 1 white. Further crosses reveal that the dark-red and white F_2 plants are true breeding.
   (a) Based on the ratio of offspring in the F_2, how many genes are involved in the production of color?
   (b) How many additive alleles are needed to produce each possible phenotype?
   (c) Assign symbols to these alleles and list possible genotypes that give rise to the medium-red and the light-red phenotypes.
   (d) Predict the outcome of the F_1 and F_2 generations in a cross between a true-breeding medium-red plant and a white plant.
5. Height in humans depends on the additive action of genes. Assume that this trait is controlled by the four loci R, S, T, and U and that environmental effects are negligible. Instead of additive versus nonadditive alleles, assume that additive and partially additive alleles exist. Additive alleles contribute two units, and partially additive alleles contribute one unit to height.
   (a) Can two individuals of moderate height produce offspring that are much taller or shorter than either parent? If so, how?
   (b) If an individual with the minimum height specified by these genes marries an individual of intermediate or moderate height, will any of their children be taller than the tall parent? Why or why not?
6. An inbred strain of plants has a mean height of 24 cm. A second strain of the same species from a different geographical region also has a mean height of 24 cm. When plants from the two strains are crossed together, the F_1 plants are the same height as the parent plants. However, the F_2 generation shows a wide range of heights; the majority are like the P_1 and F_1 plants, but approximately 4 of 1000 are only 12 cm high, and about 4 of 1000 are 36 cm high.
   (a) What mode of inheritance is occurring here?
   (b) How many gene pairs are involved?
   (c) How much does each gene contribute to plant height?
   (d) Indicate one possible set of genotypes for the original P_1 parents and the F_1 plants that could account for these results.
   (e) Indicate three possible genotypes that could account for F_2 plants that are 18 cm high and three that account for F_2 plants that are 33 cm high.
7. Erma and Harvey were a compatible barnyard pair, but a curious sight. Harvey's tail was only 6 cm, while Erma's was 30 cm. Their F_1 piglet offspring all grew tails that were 18 cm. When inbred, an F_2 generation resulted in many piglets (Erma and Harvey's grandpigs), whose tails ranged in 4-cm intervals from 6 to 30 cm (6, 10, 14, 18, 22, 26, and 30). Most had 18-cm tails, while 1/64 had 6-cm tails and 1/64 had 30-cm tails.
   (a) Explain how tail length is inherited by describing the mode of inheritance, indicating how many gene pairs are at work and designating the genotypes of Harvey, Erma, and their 18 cm offspring.
   (b) If one of the 18 cm F_1 pigs is mated with one of the 6 cm F_2 pigs, what phenotypic ratio would be predicted if many offspring resulted? Diagram the cross.
8. In the following table, average differences of height, weight, and fingerprint ridge count between monozygotic twins (reared together and apart), dizygotic twins, and nontwin siblings are compared:

Trait	MZ Reared Together	MZ Reared Apart	DZ Reared Together	Sibs Reared Together
Height (cm)	1.7	1.8	4.4	4.5
Weight (kg)	1.9	4.5	4.5	4.7
Ridge count	0.7	0.6	2.4	2.7

   Based on the data in this table, which of these quantitative traits have the highest heritability values?
9. What kind of heritability estimates (broad sense or narrow sense) are obtained from human twin studies?
10. List as many human traits as you can that are likely to be under the control of a polygenic mode of inheritance.

11. Corn plants from a test plot are measured, and the distribution of heights at 10-cm intervals is recorded in the following table:

Height (cm)	Plants (no.)
100	20
110	60
120	90
130	130
140	180
150	120
160	70
170	50
180	40

Calculate (a) the mean height, (b) the variance, (c) the standard deviation, and (d) the standard error of the mean. Plot a rough graph of plant height against frequency. Do the values represent a normal distribution? Based on your calculations, how would you assess the variation within this population?

12. The following variances were calculated for two traits in a herd of hogs.

Trait	V_P	V_G	V_A
Back fat	30.6	12.2	8.44
Body length	52.4	26.4	11.7

(a) Calculate broad sense (H^2) and narrow sense (h^2) heritabilities for each trait in this herd.

(b) Which of the two traits will respond best to selection by a breeder? Why?

13. The mean and variance of plant height of two highly inbred strains (P_1 and P_2) and their progeny (F_1 and F_2) are shown here:

Strain	Mean (cm)	Variance
P_1	34.2	4.2
P_2	55.3	3.8
F_1	44.2	5.6
F_2	46.3	10.3

Calculate the broad-sense heritability (H^2) of plant height in this species.

14. A hypothetical study investigated the vitamin A content and the cholesterol content of eggs from a large population of chickens. The variances (V) were calculated, as shown here:

	Trait	
Variance	Vitamin A	Cholesterol
V_P	123.5	862.0
V_E	96.2	484.6
V_A	12.0	192.1
V_D	15.3	185.3

(a) Calculate the narrow-sense heritability (h^2) for both traits.

(b) Which trait, if either, is likely to respond to selection?

15. The following table shows measurements for fiber lengths and fleece weight in a small flock of eight sheep.

	Sheep Fiber Length (cm)	Fleece Weight (kg)
1	9.7	7.9
2	5.6	4.5
3	10.7	8.3
4	6.8	5.4
5	11.0	9.1
6	4.5	4.9
7	7.4	6.0
8	5.9	5.1

(a) What is the mean, variance, and standard deviation for each trait in this flock?

(b) What is the covariance of the two traits?

(c) What is the correlation coefficient for fiber length and fleece weight?

(d) Do you think greater fleece weight is caused by an increase in fiber length? Why or why not?

16. In a herd of dairy cows the narrow-sense heritability for milk protein content is 0.76 and for milk butterfat it is 0.82. The correlation coefficient between milk protein content and butterfat is 0.91. If the farmer selects for cows producing more butterfat in their milk, what will be the most likely effect on milk protein content in the next generation?

17. In an assessment of learning in *Drosophila,* flies may be trained to avoid certain olfactory cues. In one population, a mean of 8.5 trials was required. A subgroup of this parental population that was trained most quickly (mean = 6.0) was interbred and their progeny examined. These flies demonstrated a mean training value of 7.5. Calculate realized heritability for olfactory learning in *Drosophila*.

18. Suppose you want to develop a population of *Drosophila* that would rapidly learn to avoid certain substances the flies could detect by smell. Based on the heritability estimate you obtained in Problem 17, would it be worth doing this by artificial selection? Why or why not?

19. In a population of tomato plants, mean fruit weight is 60 g and h^2 is 0.3. Predict the mean weight of the progeny if tomato plants whose fruit averaged 80 g were selected from the original population and interbred.

20. In a population of 100 inbred genotypically identical rice plants, variance for grain yield is 4.67. What is the heritability for yield? Would you advise a rice breeder to improve yield in this strain of rice plants by selection?

Extra-Spicy Problems

21. A mutant strain of *Drosophila* was isolated and shown to be resistant to an experimental insecticide, whereas normal (wild-type) flies were sensitive to the chemical. Following a cross between resistant flies and sensitive flies, isolated populations were derived that had various combinations of chromosomes from the two strains. Each was tested for resistance, as shown in the figure at the top of the next page. Analyze the data and draw any appropriate conclusion about which chromosome(s) contain a gene responsible for inheritance of resistance to the insecticide.

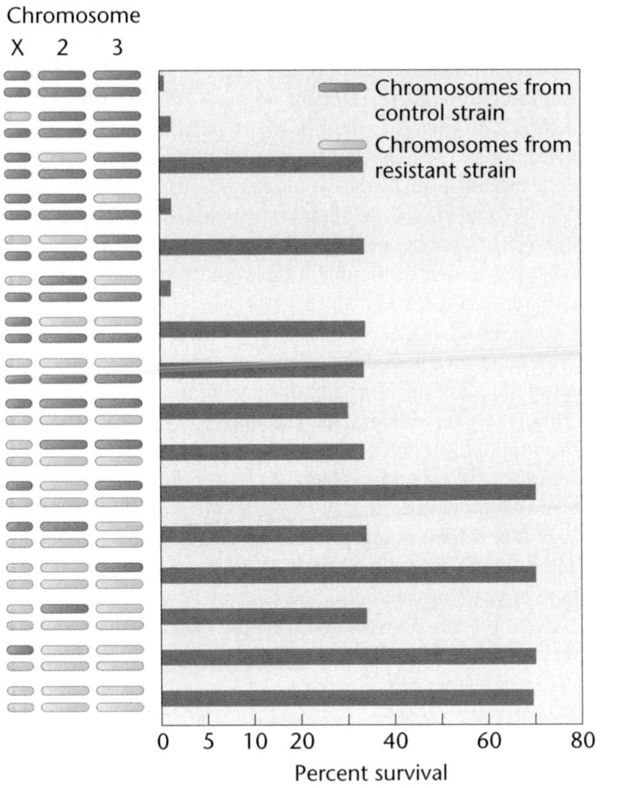

22. In 1988, Horst Wilkens investigated blind cavefish, comparing them with members of a sibling species with normal vision that are found in a lake. (We will call them cavefish and lakefish.) Wilkens found that cavefish eyes are about seven times smaller than lakefish eyes. F_1 hybrids have eyes of intermediate size. These data as well as the $F_1 \times F_1$ cross and those from backcrosses ($F_1 \times$ cavefish and $F_1 \times$ lakefish) are depicted here:

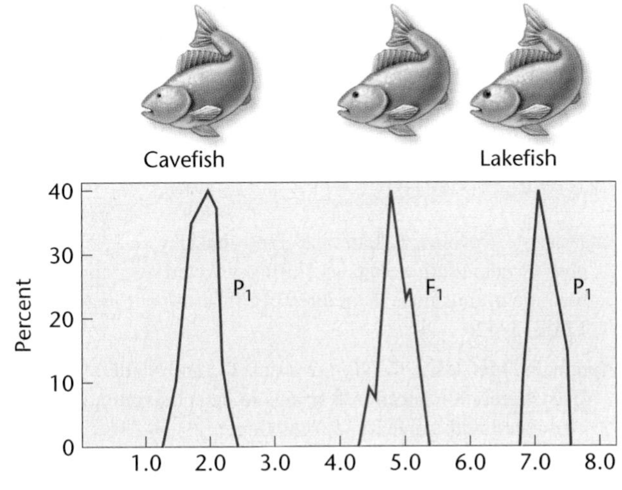

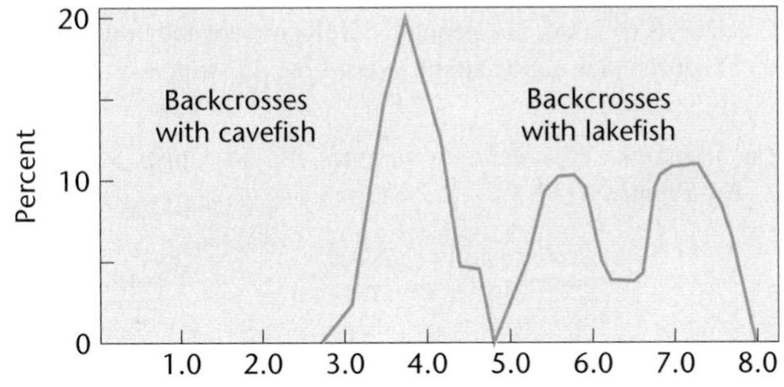

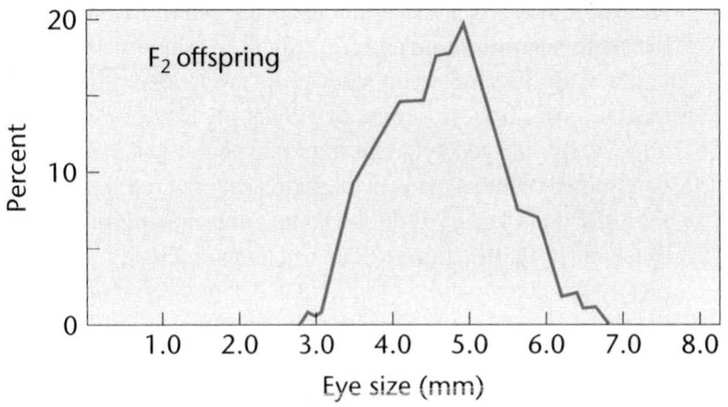

Examine Wilkens' results and respond to the following questions:

(a) Based strictly on the F_1 and F_2 results of Wilkens' initial crosses, what possible explanation concerning the inheritance of eye size seems most feasible?

(b) Based on the results of the F_1 backcross with cavefish, is your explanation supported? Explain.

(c) Based on the results of the F_1 backcross with lakefish, is your explanation supported? Explain.

(d) Wilkens examined about 1000 F_2 progeny and estimated that 6–7 genes are involved in determining eye size. Is the sample size adequate to justify this conclusion? Propose an experimental protocol to test the hypothesis.

(e) A comparison of the embryonic eye in cavefish and lakefish revealed that both reach approximately 4 mm in diameter. However, lakefish eyes continue to grow, while cavefish eye size is greatly reduced. Speculate on the role of the genes involved in this problem. [*Reference:* Wilkens, H. (1988). *Ecol. Biol.* 23:271–367.]

23. A 3″ plant was crossed with a 15″ plant and all F_1 plants were 9″. In the F_2, plants exhibited a "normal distribution" with heights of 3, 4, 5, 6, 7, 8, 9, 10, 11, 12, 13, 14, and 15″.

(a) What ratio will constitute the "normal distribution" in the F_2?

(b) What outcome will occur if the F_1 plants are test crossed with plants that are homozygous for all nonadditive alleles?

24. The table below gives the percentage of twin pairs studied in which both twins expressed the same phenotype for a trait (concordance). Percentages listed are for different traits in monozygotic (MZ) and dizygotic (DZ) twins. Assuming that both twins in each pair were raised together in the same environment, what do you conclude about the relative importance of genetic versus environmental factors for each trait?

Trait	MZ%	DZ%
Blood Types	100	66
Eye Color	99	28
Mental retardation	97	37
Measles	95	87
Hair color	89	22
Handedness	79	77
Idiopathic epilepsy	72	15
Schizophrenia	69	10
Diabetes	65	18
Identical allergy	59	5
Cleft lip	42	5
Club foot	32	3
Mammary cancer	6	3

25. In a cross between a strain of large guinea pigs and a strain of small guinea pigs, the F_1 are phenotypically uniform with an average size about intermediate between that of the two parental

strains. Among 1014 individuals, 3 are about the same size as the small parental strain and 5 are about the same size as the large parental strain. How many gene pairs are involved in the inheritance of size in these strains of guinea pigs?

26. Type A1B brachydactyly (short middle phalanges) is a genetically determined trait that maps to the short arm of chromosome 5 in humans. If you classify individuals as either having or not having brachydactyly, the trait appears to follow a single locus, incompletely dominant pattern of inheritance. However, if one examines the fingers and toes of affected individuals, one sees a range of expression from extremely short to only slightly short. What might cause such variation in the expression of brachydactyly?

27. In a series of crosses between two true-breeding strains of peaches, the F_1 generation was uniform, producing 30 g peaches. The F_2 fruit mass ranges from 38 to 22 g at intervals of 2 g. (a) Using these data, determine the number of polygenic loci involved in the inheritance of peach mass. (b) Using gene symbols of your choice, give the genotypes of the parents and the F_1.

28. Students in a genetics laboratory began an experiment in an attempt to increase heat tolerance in two strains of *Drosophila melanogaster*. One strain was trapped from the wild six weeks before the experiment was to begin; the other was obtained from a *Drosophila* repository at a university laboratory. In which strain would you expect to see the most rapid and extensive response to heat-tolerance selection?

29. Consider a true-breeding plant, *AABBCC*, crossed with another true-breeding plant, *aabbcc*, whose resulting offspring are *AaBbCc*. If you cross the F_1s where independent assortment is operational, the expected fraction of offspring in each phenotypic class is given by the expression $N!/[M!(N - M!)]$, where N is the total number of alleles (six in this example) and M is the number of uppercase alleles. In a cross of $AaBbCc \times AaBbCc$, what proportion of the offspring would be expected to contain two uppercase alleles?

30. Canine hip dysplasia is a quantitative trait that continues to affect most large breeds of dogs in spite of approximately 40 years of effort to reduce the impact of this condition. Breeders and veterinarians rely on radiographic and universal registries to enhance the development of breeding schemes to reduce its incidence. Recent data (Wood and Lakhani. 2003. *Vet. Rec.* 152:69–72) indicate that there is a "month-of-birth" effect on hip dysplasia in Labrador retrievers and Gordon setters, whereby the frequency and extent of expression of this disorder varies depending on the time of year dogs are born. Speculate on how breeders attempt to "select" out this disorder and what the month-of-birth phenomenon indicates about the expression of polygenic traits?

SELECTED READINGS

Brink, R., ed. 1967. *Heritage from Mendel.* Madison: University of Wisconsin Press.

Browman, K.W. 2001. Review of statistical methods of QTL mapping in experimental crosses. *Lab Animal* 30:44–52.

Crow, J.F. 1993. Francis Galton: Count and measure, measure and count. *Genetics* 135:1.

Dudley, J.W. 1977. 76 generations of selection for oil and protein percentage in maize. In *Proc. Intern. Conf. on Quant. Genet.* E. Pollack, O. Kempthorne, and T. Bailey eds. pp. 459–73. Ames: Iowa State University Press.

Falconer, D.S., and Mackay, F.C. 1996. *Introduction to quantitative genetics,* 4th ed. Essex, England: Longman.

Farber, S. 1980. *Identical twins reared apart.* New York: Basic Books.

Feldman, M.W., and Lewontin, R.C. 1975. The heritability hangup. *Science* 190:1163–66.

Frary, A., et al. 2000. *fw2.2*: A quantitative trait locus key to the evolution of tomato fruit size. *Science* 289:85–88.

Haley, C. 1991. Use of DNA fingerprints for the detection of major genes for quantitative traits in domestic species. *Anim. Genet.* 22:259–77.

______ 1996. Livestock QTLs: Bringing home the bacon. *Trends Genet.* 11:488–90.

Lander, E., and Botstein, D. 1989. Mapping Mendelian factors underlying quantitative traits using RFLP linkage maps. *Genetics* 121:185–99.

Lander, E., and Schork, N. 1994. Genetic dissection of complex traits. *Science* 265:2037–48.

Lewontin, R.C. 1974. The analysis of variance and the analysis of causes. *Am. J. Hum. Genet.* 26:400–11.

Lynch, M., and Walsh, B. 1998. *Genetics and analysis of quantitative traits.* Sunderland, MA: Sinauer Associates.

Macy, T.F.C. 2001. Quantitative trait loci in *Drosophila*. *Nature Reviews Genetics* 2:11–19.

Newman, H.H., Freeman, F.N., and Holzinger, K.T. 1937. *Twins: A study of heredity and environment.* Chicago: University of Chicago Press.

Paterson, A., Deverna, J., Lanini, B., and Tanksley, S. 1990. Fine mapping of quantitative traits loci using selected overlapping recombinant chromosomes in an interspecific cross of tomato. *Genetics* 124:735–42.

Plomin, R., McClearn, G., Gora-Maslak, G., and Neiderhiser, J. 1991. Use of recombinant inbred strains to detect quantitative trait loci associated with behavior. *Behav. Genet.* 21:99–116.

Stuber, C.W. 1996. Mapping and manipulating quantitative traits in maize. *Trends Genet.* 11:477–87.

Tanksley, S.D. 2004. The genetic, developmental and molecular bases of fruit size and shape variation in tomato. *Plant Cell* 16:S181–S189.

Zar, J.H. 1999. *Biostatistical analysis*, 4th ed. Upper Saddle River, NJ: Prentice Hall.

Population Genetics

These lady-bird beetles, from the Chiricahua Mountains in Arizona, show considerable phenotypic variation.

CHAPTER CONCEPTS

- Allele frequencies in population gene pools can vary in space and time.
- The relationship between allele frequencies and genotype frequencies in an ideal population is described by the Hardy–Weinberg law.
- Natural selection is a major force driving allele frequency change and population evolution.
- Migration and drift can also cause changes in allele frequency.
- Nonrandom mating changes population genotype frequency but not allele frequency.
- Mutation creates new alleles in a population gene pool.

When Alfred Russel Wallace and Charles Darwin first identified natural selection as the mechanism of adaptive evolution in the mid 19th century, there was no accurate model of the mechanisms responsible for variation and inheritance. Gregor Mendel published his work on the inheritance of traits in 1866, but it received little notice at the time. The rediscovery of Mendel's work in 1900 began a 30-year effort to reconcile Mendel's concept of genes and alleles with the theory of evolution. In a key insight, biologists realized that the frequency at which phenotypic traits occur in a population is linked to the relative abundance of the alleles influencing those traits. This laid the foundation for the emergence of **population genetics**, the study of genetic variation in populations and how it changes over time. Population geneticists investigate patterns of genetic variation or **genetic structure** within and among groups of interbreeding individuals. As changes in genetic structure form the basis for evolution of a population, population genetics has become an important subdiscipline of evolutionary biology.

Early in the 20th century a number of workers, including G. Udny Yule, William Castle, Godfrey Hardy, and Wilhelm Weinberg, formulated the basic principles of population genetics. Theoretical geneticists Sewall Wright, Ronald Fisher, and J. B. S. Haldane developed mathematical models to describe the genetic structure of populations. More recently, field researchers have been able to test these models using biochemical and molecular techniques that measure variation directly at the protein and DNA levels. These experiments examine allele frequencies and the forces that act on them, such as selection, mutation, migration, and genetic drift. In this chapter, we will consider general aspects of population genetics and discuss other areas of genetics that relate to evolution.

How Do We Know?

Population geneticists study changes in allele frequency in population gene pools over time, the nature and amount of genetic diversity in populations, and patterns of distribution of different genotypes. As you read this chapter think about these questions: How do geneticists detect the presence of different alleles in a population? How are different genotypes identified? Are differences in morphological phenotype (the physical appearance of an individual) the only way that genetically distinct individuals in a population can be identified?

25.1 Allele Frequencies in Population Gene Pools Vary in Space and Time

A **population** is a group of individuals sharing a common set of genes that lives in the same geographic area and actually or potentially interbreeds. All the alleles shared by these individuals constitute the **gene pool** for the population. Often when we examine a single genetic locus in this population, we find that distribution of the alleles at this locus results in individuals with different genotypes. A key element of population genetics is the computation of frequencies of various alleles in the gene pool, frequencies of different genotypes in the population, and how these frequencies change from one generation to the next. Population geneticists use these calculations to address questions such as: How much genetic variation is present in a population? Are genotypes randomly distributed in time and space or do discernable patterns exist? What processes affect the composition of a population's gene pool? Do these processes produce genetic divergence among populations?

Populations are dynamic; they expand and contract through changes in birth and death rates, migration, or contact with other populations. Often some individuals within a population will produce more offspring than others, contributing a disproportionate amount of their alleles to the next generation. Thus the dynamic nature of a population can, over time, lead to changes in the population's gene pool.

25.2 The Hardy–Weinberg Law Describes the Relationship between Allele Frequencies and Genotype Frequencies in an Ideal Population

The theoretical relationship between the relative proportions of alleles in the gene pool and the frequencies of different genotypes in the population was elegantly described in the early 1900s in a mathematical model developed independently by the British mathematician Godfrey H. Hardy and the German physician Wilhelm Weinberg. This model, called the **Hardy–Weinberg law**, describes what happens to alleles and genotypes in an "ideal" population that is infinitely large and random mating, and that is not subject to any evolutionary forces such as mutation, migration, or selection. Under these conditions the Hardy–Weinberg model makes two predictions:

1. The frequencies of the alleles in the gene pool do not change over time.
2. If two alleles at a locus, *A* and *a*, are considered, then after one generation of random mating the frequencies of genotypes *AA*:*Aa*:*aa* in the population can be calculated as

$$p^2 + 2pq + q^2 = 1$$

where $p =$ frequency of allele *A* and $q =$ frequency of allele *a*.

A population that meets these criteria, and in which the frequencies p and q of two alleles at a locus do result in the predicted genotypic frequencies, is said to be in Hardy–Weinberg equilibrium. As we will see later in this chapter, it is rare for a real population to conform totally to the Hardy–Weinberg model, and for all allele and genotype frequencies to remain unchanged for generation after generation.

The Hardy–Weinberg model uses Mendelian principles of segregation and simple probability to explain the relationship between allele and genotype frequencies in a population. We can demonstrate how this works by considering the example of a single locus with two alleles, *A* and *a*, in a population where

Web Tutorial 25.1 Population Genetics

the frequency of A is 0.7, and the frequency of a is 0.3. Note that $0.7 + 0.3 = 1$, indicating that all the alleles for gene A present in the gene pool are accounted for. We assume that individuals mate randomly, following Hardy–Weinberg requirements, so for any one zygote, the probability that the female gamete will contain A is 0.7, and the probability that male gamete will contain A is also 0.7. The probability that *both* gametes will contain A is $0.7 \times 0.7 = 0.49$. Thus we predict that genotype AA will occur 49 percent of the time. The probability that a zygote will be formed from a female gamete carrying A and a male gamete carrying a is $0.7 \times 0.3 = 0.21$, and the probability of a female gamete carrying a being fertilized by a male gamete carrying A is $0.3 \times 0.7 = 0.21$, so the frequency of genotype Aa is $0.21 + 0.21 = 0.42 = 42$ percent. Finally, the probability that a zygote will be formed from two gametes carrying a is $0.3 \times 0.3 = 0.09$ so the frequency of genotype aa is 9 percent. As a check on our calculations, note that $0.49 + 0.42 + 0.09 = 1.0$, confirming that we have accounted for all of the zygotes. These calculations are summarized in Figure 25–1.

We started with the frequency of a particular allele in a specific gene pool and calculated the probability that certain genotypes would be produced from this pool. When the zygotes develop into adults and reproduce, what will be the frequency distribution of alleles in the new gene pool? Under the Hardy–Weinberg law, we assume that all genotypes have equal rates of survival and reproduction. This means that in the next generation, all genotypes contribute equally to the new gene pool. The AA individuals constitute 49 percent of the population, and we can predict that the gametes they produce will constitute 49 percent of the gene pool. These gametes all carry allele A. Likewise, Aa individuals constitute 42 percent of the population, so we predict that their gametes will constitute 42 percent of the new gene pool. Half (0.5) of these gametes will carry allele A. Thus, the frequency of allele A in the gene pool is $0.49 + (0.5)0.42 = 0.7$. The other half of the gametes produced by Aa individuals will carry allele a. The aa individuals constitute 9 percent of the population, so their gametes will constitute 9 percent of the new gene pool. These gametes all carry allele a. Thus, we can predict that the allele a in the new gene pool is $(0.5)0.42 + 0.09 = 0.3$. As a check on our calculation, note that $0.7 + 0.3 = 1.0$ accounting for all of the gametes in the gene pool of the new generation.

We have arrived back where we began, with a gene pool in which the frequency of allele A is 0.7 and the frequency of allele a is 0.3. For the general case of the Hardy–Weinberg law, we use variables instead of numerical values for the allele frequencies. Imagine a gene pool in which the frequency of allele A is p, and the frequency of allele a is q, such that $p + q = 1$. If we randomly draw male and female gametes from the gene pool and pair them to make a zygote, the probability that both will carry allele A is $p \times p$. Thus, the frequency of genotype AA among the zygotes is p^2. The probability that the female gamete carries A and the male gamete carries a is $p \times q$, and that the female gamete carries a and the male gamete carries A is $q \times p$. Thus, the frequency of genotype Aa among the zygotes is $2pq$. Finally, the probability that both gametes carry a is $q \times q$, making the frequency of genotype aa among the zygotes q^2. Therefore, the distribution of genotypes among the zygotes is

$$p^2 + 2pq + q^2 = 1$$

This is summarized in Figure 25–2.

These calculations demonstrate the two main predictions of the Hardy–Weinberg law: Allele frequencies in our population do not change from one generation to the next, and genotype frequencies after one generation of random mating can be predicted from the allele frequencies. In other words, this population does not change or evolve with respect to the locus we have examined. Remember, however, the assumptions about the theoretical population described by the Hardy–Weinberg model:

1. Individuals of all genotypes have equal rates of survival and equal reproductive success—that is, there is no selection.

Eggs \ Sperm	fr(A) = 0.7	fr(a) = 0.3
fr(A) = 0.7	fr(AA) = 0.7×0.7 = 0.49	fr(Aa) = 0.7×0.3 = 0.21
fr(a) = 0.3	fr(aA) = 0.3×0.7 = 0.21	fr(aa) = 0.3×0.3 = 0.09

FIGURE 25–1 Calculating genotype frequencies from allele frequencies. Gametes represent withdrawals from the gene pool to form the genotypes of the next generation. In this population, the frequency of the A allele is 0.7, and the frequency of the a allele is 0.3. The frequencies of the genotypes in the next generation are calculated as 0.49 for AA, 0.42 for Aa, and 0.09 for aa. Under the Hardy–Weinberg law, the frequencies of A and a remain constant from generation to generation.

Eggs \ Sperm	fr(A) = p	fr(a) = q
fr(A) = p	fr(AA) = p^2	fr(Aa) = pq
fr(a) = q	fr(aA) = qp	fr(aa) = q^2

FIGURE 25–2 The general case of allele and genotype frequencies under Hardy–Weinberg assumptions. The frequency of allele A is p and the frequency of allele a is q. After mating, the three genotypes AA, Aa, and aa have the frequencies p^2, $2pq$, and q^2, respectively.

2. No new alleles are created or converted from one allele into another by mutation.
3. Individuals do not migrate into or out of the population.
4. The population is infinitely large, which in practical terms means that the population is large enough that sampling errors and other random effects are negligible.
5. Individuals in the population mate randomly.

These assumptions are what make the Hardy–Weinberg law so useful in population genetic research. By specifying the conditions under which the population cannot evolve, the Hardy–Weinberg model identifies the real-world forces that cause allele frequencies to change. In other words, by holding certain conditions constant, Hardy–Weinberg isolates the forces of evolution and allows them to be quantified. Application of the Hardy–Weinberg model can also reveal "neutral genes" in a population gene pool—those not being operated on by the forces of evolution.

There are three additional important consequences of the Hardy–Weinberg law. First, it shows that dominant traits do not necessarily increase from one generation to the next. Second, it demonstrates that **genetic variability** can be maintained in a population since, once established in an ideal population, allele frequencies remain unchanged. Third, if we invoke Hardy–Weinberg assumptions, then knowing the frequency of just one genotype enables us to calculate the frequencies of all other genotypes at that locus. This is particularly useful in human genetics because we can calculate the frequency of heterozygous carriers for recessive genetic disorders even when all we know is the frequency of affected individuals. (See section 25.4.)

Now solve this

Problem 25.1 on page 638 asks you to calculate frequencies of a dominant and a recessive allele in a population where you know the phenotype frequencies.

Hint: Determine which allele frequency (p or q) you must estimate first when homozygous dominant and heterozygous genotypes have the same phenotype.

25.3 The Hardy–Weinberg Law Can Be Applied to Human Populations

To show how allele frequencies are measured in a real population, and to demonstrate the kind of questions asked by population genetics researchers, we will consider a human population example drawn from research into genetic factors that influence an individual's susceptibility to infection by HIV-1, the virus responsible for the bulk of AIDS cases worldwide. Of particular interest to AIDS researchers is the small number of individuals who make high-risk choices (such as unprotected sex with HIV-positive partners), yet remain uninfected. In 1996, Rong Liu and colleagues discovered that two exposed, but uninfected, individuals were homozygous for a mutant allele of a gene called *CC-CKR5*.

The *CC-CKR-5* gene, located on chromosome 3, encodes a protein called the C–C chemokine receptor-5, often abbreviated CCR5. Chemokines are signaling molecules associated with the immune system. When white blood cells with CCR5 receptor proteins bind to chemokines, the cells respond by moving into inflamed tissues to fight infection.

The CCR5 protein is also a receptor for strains of HIV-1. To gain entry into cells, a protein called Env (short for envelope protein) on the surface of HIV-1 binds to the CD4 protein on the surface of the host cell. Binding to CD4 causes Env to change shape and form a second binding site. This second site binds to CCR5, which, in turn, initiates the fusion of the viral protein coat with the host cell membrane. The merging of viral envelope with cell membrane transports the HIV viral core into the host cell's cytoplasm.

The mutant allele of the *CCR5* gene contains a 32-bp deletion in one of its coding regions. As a result, the protein encoded by the mutant allele is shortened and made nonfunctional. The protein never makes it to the cell membrane, so HIV-1 cannot enter these cells. The gene's normal allele is called *CCR51* (we will designate this as *1*), and its allele with the 32-bp deletion is called *CCR5-Δ32* (we will designate this as *Δ32*). The two uninfected individuals described by Liu both had the genotype *Δ32/Δ32*. As a result, they had no CCR5 receptors on the surface of their cells and were resistant to infection by strains of HIV-1 that require CCR5 as a coreceptor.

At least three *Δ32/Δ32* individuals who are infected with HIV-1 have been found. Curiously, researchers have not discovered any adverse effects associated with this genotype. Heterozygotes with genotype 1/*Δ32* are susceptible to HIV-1 infection, but evidence suggests that they progress more slowly to AIDS. Table 25.1 summarizes the genotypes possible at the *CCR5* locus and the phenotypes associated with each.

The discovery of the *CCR5-Δ32* allele, and the fact that it provides some protection against AIDS, generates two important questions: Which human populations harbor the *Δ32* allele, and how common is it? To address these questions, several teams of researchers surveyed a large number of people from a variety of populations. Genotypes were determined by direct analysis of DNA (Figure 25–3). We'll look at one sample population of 100 French individuals from a survey in Brittany in which 79 individuals have genotype *1/1*, 20 have genotype and 1/*Δ32*, and 1 has genotype *Δ32/Δ32*. This sample has 158 *1* alleles (carried

Table 25.1 *CCR5* Genotypes and Phenotypes

Genotype	Phenotype
1/1	Susceptible to sexually transmitted strains of HIV-1
1/Δ32	Susceptible, but may progress to AIDS slowly
Δ32/Δ32	Resistant to most sexually transmitted strains of HIV-1

TABLE 25.2 **METHODS OF DETERMINING ALLELE FREQUENCIES FROM DATA ON GENOTYPES**

A. Counting Alleles

Genotype	*1/1*	*1/Δ32*	*Δ32/32*	Total
Number of individuals	79	20	1	100
Number of 1 alleles	158	20	0	178
Number of *Δ32* alleles	0	20	2	22
Total number of alleles	158	40	2	200

Frequency of *CCR5I* in sample: 178/200 = 0.89 = 89%
Frequency of *CCR5-Δ32* in sample: 22/200 = 0.11 = 11%

B. From Genotype Frequencies

Genotype	*1/1*	*1/Δ32*	*Δ32/Δ32*	Total
Number of individuals	79	20	1	100
Genotype frequency	79/100 = 0.79	20/100 = 0.20	1/100 = 0.01	1.00

Frequency of *CCR51* in sample: 0.79 + (0.5)0.20 = 0.89
Frequency of *CCR5-Δ32* in sample: (0.5)0.20 + 0.01 = 0.11

by the *1/1* individuals) plus 20 *1* alleles carried by $1/\Delta 32$ individuals, for a total of 178. The frequency of the *CCR51* allele in the sample population is thus 178/200 = 0.89 = 89 percent. The *CCR5-Δ32* allele count shows 20 carried by $1/\Delta 32$ individuals, plus 2 carried by the $\Delta 32/\Delta 32$ individual, for a total of 22. The frequency of the *CCR5-Δ32* allele is thus 22/200 = 0.11 = 11 percent. Notice that $p + q = 1$ which confirms that we have accounted for the entire gene pool. Table 25.2 shows two methods for computing the frequencies of the *1* and *Δ32* alleles in the Brittany sample.

Figure 25–4 shows the frequency of the *CCR5-Δ32* allele in the 18 European populations surveyed. The studies show that populations in northern Europe around the Baltic Sea have the highest frequencies of the *Δ32* allele. There is a sharp gradient in allele frequency from north to south, and populations in Sardinia and Greece have very low frequencies. In populations without European ancestry, the *Δ32* allele is essentially absent. The highly patterned global distribution of the *Δ32* allele presents an evolutionary puzzle that we'll return to later in the chapter.

Can we expect the *CCR5-Δ32* allele to increase in human populations in which it is currently rare? From what we now know about the Hardy–Weinberg law, we can say that if (1) individuals of all genotypes have equal rates of survival and reproduction, (2) there is no mutation, (3) no differential migration into or out of the population, (4) the population is extremely large, and (5) the individuals in the population choose their mates randomly with respect to the locus of interest, then the frequency of the *Δ32* allele in human populations will not change. However, as

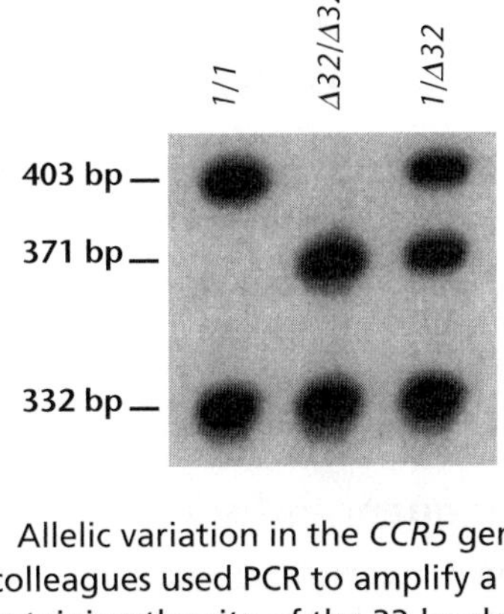

FIGURE 25–3 Allelic variation in the *CCR5* gene. Michel Samson and colleagues used PCR to amplify a part of the *CCR5* gene containing the site of the 32-bp deletion, cut the resulting DNA fragments with a restriction enzyme, and ran the fragments on an electrophoresis gel. Each lane reveals the genotype of a single individual. The *1* allele produces a 332-bp fragment and a 403-bp fragment; the *Δ32* allele produces a 332-bp fragment and a 371-bp fragment. Heterozygotes produce three bands.

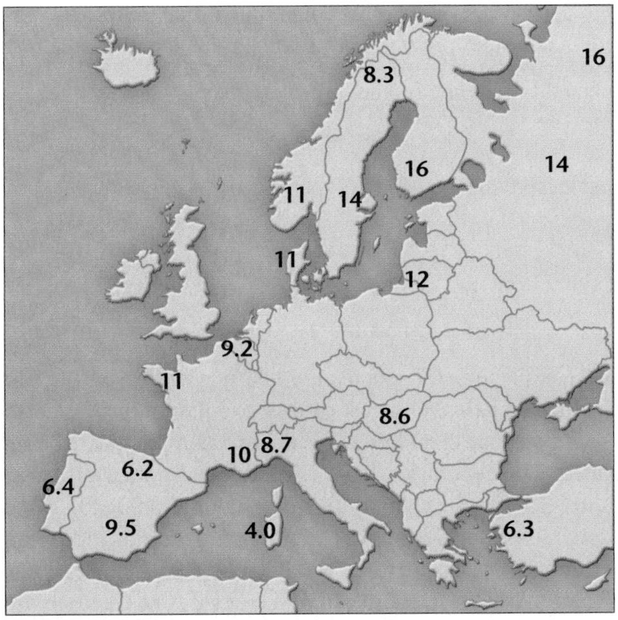

FIGURE 25–4 The frequency (percentage) of the *CCR5-Δ32* allele in 18 European populations.

we shall see in the last half of the chapter, when the assumptions of the Hardy–Weinberg law are not met—because of natural selection, mutation, migration, or genetic drift—the allele frequencies in a population may change from one generation to the next. Nonrandom mating does not, by itself, alter *allele* frequencies, but by altering *genotype* frequencies, it indirectly affects the course of evolution. The Hardy–Weinberg law tells geneticists where to look to find the causes of evolution in populations. We return to the *CCR5-Δ32* allele later in the chapter to see how a population genetics perspective has generated fruitful hypotheses for further research.

Testing for Hardy–Weinberg Equilibrium

One way we establish whether one or more of the Hardy–Weinberg assumptions do not hold in a given population is by determining whether the population's genotypes are in equilibrium. To do this, we first determine the frequencies of the genotypes, either directly from the phenotypes (if heterozygotes are recognizable) or by analyzing proteins or DNA sequences. We then calculate the allele frequencies from the genotype frequencies, as demonstrated earlier. Finally, we use the allele frequencies in the parental generation to predict the offspring's genotype frequencies. According to the Hardy–Weinberg law, the genotype frequencies are predicted to fit the $p^2 + 2pq + q^2 = 1$ relationship. If they do not, then one or more of the assumptions are invalid for the population in question.

We will use the *CCR5* genotypes of a population in Britain to demonstrate the Hardy–Weinberg law. The population includes 283 individuals, of which 223 have genotype *1/1*; 57 have genotype $1/\Delta 32$; and 3 have genotype $\Delta 32/\Delta 32$. These numbers represent genotype frequencies of $223/283 = 0.788$, $57/283 = 0.201$, and $3/283 = 0.011$, respectively. From the genotype frequencies, we compute the *CCR51* allele frequency as 0.89 and the frequency of the *CCR5-Δ32* allele as 0.11. From these allele frequencies, we can use the Hardy–Weinberg law to determine whether this population is in equilibrium. The allele frequencies predict the genotype frequencies as follows:

Expected Frequency of genotype

$$1/1 = p^2 = (0.89)^2 = 0.792$$

Expected Frequency of genotype

$$1/\Delta 32 = 2pq = 2(0.89)(0.11) = 0.196$$

Expected Frequency of genotype

$$\Delta 32/\Delta 32 = q^2 = (0.11)^2 = 0.012$$

These expected frequencies are nearly identical to the observed frequencies. Our test of this population has failed to provide evidence that Hardy–Weinberg assumptions are being violated. The conclusion is confirmed by a χ^2 analysis. (See Chapter 3.) The χ^2 value in this case is tiny: 0.00023. To be statistically significant at even the most generous, accepted level, $p = 0.05$, the χ^2 value would have to be 3.84. (In a test for Hardy–Weinberg equilibrium, the degrees of freedom are given by $k - 1 - m$, where k is the number of genotypes and m is the number of independent allele frequencies estimated from the data. Here, $k = 3$ and $m = 1$ since calculating only one allele frequency allows us to determine the other by subtraction. Thus, we have $3 - 1 - 1 = 1$ (degrees of freedom.)

On the other hand, if the Hardy–Weinberg test had demonstrated that the population is not in equilibrium, it would indicate that one or more assumptions are not being met. Therese Markow and colleagues documented a real human population that is not in Hardy–Weinberg equilibrium. These researchers studied 122 Havasupai, a population of Native Americans in Arizona. They determined the genotype of each Havasupai individual at two loci in the major histocompatibility complex (MHC). These genes, *HLA-A* and *HLA-B*, encode proteins that are involved in the immune system's discrimination between self and nonself. The immune systems of individuals heterozygous at MHC loci appear to recognize a greater diversity of foreign invaders and thus may be better able to fight disease. Markow and colleagues observed significantly more individuals who are heterozgous at both loci and significantly fewer homozygous individuals than would be expected under the Hardy–Weinberg law. Violation of either, or both, of two Hardy–Weinberg assumptions could explain the excess of heterozygotes among the Havasupai. First, Havasupai fetuses, children, and adults who are heterozygous for *HLA-A* and *HLA-B* may have higher rates of survival than individuals who are homozygous (i.e., selection is occurring). Second, rather than choosing their mates randomly, Havasupai people may somehow prefer mates whose MHC genotypes differ from their own.

Now solve this

Problem 25.6 on page 637 asks you to determine whether two populations are in Hardy–Weinberg equilibrium, based on their genotype frequencies.

Hint: Start by determining the allele frequencies based on these data.

25.4 The Hardy–Weinberg Law Can Be Used for Multiple Alleles, X-Linked Traits, and Estimating Heterozygote Frequencies

Having established the foundation of the Hardy–Weinberg law, we can now consider how it may be used in estimating allele frequencies within populations. We will examine calculations as applied to multiple alleles, X-linked traits, and heterozygote frequencies.

Calculating Frequencies for Multiple Alleles

We commonly find several alleles of a single locus in a population. The ABO blood group in humans (discussed in Chapter 4) is such an example. The locus I (isoagglutinin) has three alleles I^A, I^B, *and* I^O, yielding six possible genotypic combinations (I^AI^A, I^BI^B, I^OI^O, I^AI^B, I^AI^O, I^BI^O) Remember that in this case I^A and I^B are codominant alleles, and both of these are dominant to I^O. The result is that homozygous I^AI^A and heterozygous I^AI^O individuals are phenotypically identical, as are I^BI^B and I^BI^O individuals, so we can distinguish only four phenotypic combinations.

By adding another variable to the Hardy–Weinberg equation, we can calculate both the genotype and allele frequencies for the situation involving three alleles. Let p, q, and r represent the frequencies of alleles A, B, and O, respectively. Note that because there are three alleles

$$p + q + r = 1$$

Under Hardy–Weinberg assumptions, the frequencies of the genotypes are given by

$$(p + q + r)^2 = p^2 + q^2 + r^2 + 2pq + 2pr + 2qr = 1$$

If we know the frequencies of blood types for a population, we can then estimate the frequencies for the three alleles of the ABO system. For example, in one population sampled, the following blood-type frequencies are observed: A = 0.53, B = 0.13, O = 0.26. Because the I^O allele is recessive, the population's frequency of type O blood equals the proportion of the recessive genotype r^2. Thus,

$$r^2 = 0.26$$
$$r = \sqrt{0.26}$$
$$r = 0.51$$

Using r, we can estimate the allele frequencies for the I^A and I^B alleles. The I^A allele is present in two genotypes, I^AI^A *and* I^AI^O alleles. The frequency of the I^AI^A genotype is represented by p^2 and the I^AI^O genotype by $2pr$. Therefore, the combined frequency of type A blood and type O blood is given by

$$p^2 + 2pr = r^2 = 0.53 + 0.26$$

If we factor the left side of the equation and take the sum of the terms on the right, we get

$$(p + r)^2 = 0.79$$
$$p + r = \sqrt{0.79}$$
$$p = 0.89 - r$$
$$p = 0.89 - 0.51 = 0.38$$

Having estimated p and r, the frequencies of allele I^A and allele I^O, we can now estimate the frequency for the I^B allele:

$$p + q + r = 1$$
$$q = 1 - p - r$$
$$= 1 - 0.38 - 0.51$$
$$= 0.11$$

The phenotypic frequencies and genotypic frequencies for this population are summarized in Table 25.3.

Calculating Frequencies for X-linked Traits

The Hardy–Weinberg law can be used to calculate allele and genotype frequencies for X-linked traits, but we need to remember that in an XY sex-determination system, the homogametic (XX) sex will have two copies of an X-linked allele while the heterogametic sex (XY) only has one copy. Thus, for mammals where the female is XX and the male is XY, the frequency of the X-linked allele in the gene pool and the frequency of males expressing the X-linked trait will be the same. This is because each male only has one X chromosome, and the probability of any individual male receiving an X chromosome with the allele in question must be equal to the frequency of the allele. The probability of any individual female having the allele in question on both X chromosomes will be q^2 where q is the frequency of the allele.

To illustrate this for a recessive X-linked trait, consider the example of red-green color blindness, which affects 8 percent of human males. The frequency of the color blindness allele is therefore 0.08; in other words, 8 percent of X chromosomes carry it. The other 92 percent of X chromosomes carry the dominant allele for normal red-green color vision. If we assign p as the frequency of the normal allele and q as the frequency of the color blindness allele, then $p = 0.92$ and $q = 0.08$. The frequency of color blind females (with two affected X chromosomes) is $q^2 = (0.08)^2 = 0.0064$ and the frequency of carrier females (one normal and one affected X chromosome) is $2pq = 2(0.08)(0.92) = 0.147$. In other words, 14.7% of females carry the allele for red-green color blindness and can pass it to their children, although they themselves have normal color vision.

An important consequence of the difference in allele frequency for X-linked genes between male and female gametes is that for a rare recessive allele, the trait will be expressed at

TABLE 25.3 CALCULATING GENOTYPE FREQUENCIES FOR MULTIPLE ALLELES WHERE THE FREQUENCY OF ALLELE $I^A = 0.38$, ALLELE $I^B = 0.11$, AND ALLELE $I^O = 0.51$

Genotype	Genotype Frequency	Phenotype	Phenotype Frequency
I^AI^A	$p^2 = (0.38)^2 = 0.14$	A	0.53
I^AI^O	$2pr = 2(0.38)(0.51) = 0.39$		
I^BI^B	$q^2 = (0.11)^2 = 0.01$	B	0.12
I^BI^O	$2qr = 2(0.11)(0.51) = 0.11$		
I^AI^B	$2pq = 2(0.38)(0.11) = 0.084$	AB	0.08
I^OI^O	$r^2 = (0.51)^2 = 0.26$	O	0.26

a much higher frequency among XY individuals than those who are XX. So, for example, diseases such as hemophilia and Duchenne muscular dystrophy in humans, both caused by recessive mutations on the X chromosome, are much more common in boys who need only inherit a single copy of the mutated allele to suffer from the disease. Girls who inherit two affected X chromosomes will also have the disease, but with a rare allele the probability of this occurring is small. Females have a greater probability of being carriers of recessive X-linked traits, as we saw previously in the example of color blindness.

Now solve this

Problem 25.18 on page 637 asks you to calculate the number of heterozygous carriers for an X-linked trait in a human population.

Hint: First determine the genotype of carriers. What proportion of the population can potentially be a carrier?

Calculating Heterozygote Frequency

In another application, the Hardy–Weinberg law allows us to estimate the frequency of heterozygotes in a population. The frequency of a recessive trait can usually be determined by counting such individuals in a sample of the population. With this information and the Hardy–Weinberg law, we can then calculate the allele and genotype frequencies.

Cystic fibrosis, an autosomal recessive trait, has an incidence of about $1/2500 = 0.0004$ in people of northern European ancestry. Individuals with cystic fibrosis are easily distinguished from the population at large by such symptoms as salty sweat, excess amounts of thick mucus in the lungs, and susceptibility to bacterial infections. Because this is a recessive trait, individuals with cystic fibrosis must be homozygous. Their frequency in a population is represented by q^2 provided that mating has been random in the previous generation. The frequency of the recessive allele therefore is

$$q = \sqrt{q^2} = \sqrt{0.0004} = 0.02$$

Since $p + q = 1$, then the frequency of p is

$$p = 1 - q = 1 - 0.02 = 0.98$$

In the Hardy–Weinberg equation, the frequency of heterozygotes is $2pq$. Thus,

$$2pq = 2(0.98)(0.02)$$
$$= 0.04 \text{ or 4 percent, or } 1/25$$

Thus, heterozygotes for cystic fibrosis are rather common in the population (4 percent), even though the incidence of homozygous recessives is only 1/2500, or 0.04 percent.

In general, the frequencies of all three genotypes can be estimated once the frequency of either allele is known and Hardy–Weinberg assumptions are invoked. The relationship between genotype and allele frequency is shown in Figure 25–5. It is important to note that heterozygotes increase rapidly in a population as the values of p and q move from 0 or 1. This observation confirms our conclusion that when a recessive trait such as cystic fibrosis is rare, the majority of those carrying the allele are heterozygotes. In populations in which the frequencies of p and q are between 0.33 and 0.67, heterozygotes occur at higher frequency than either homozygote.

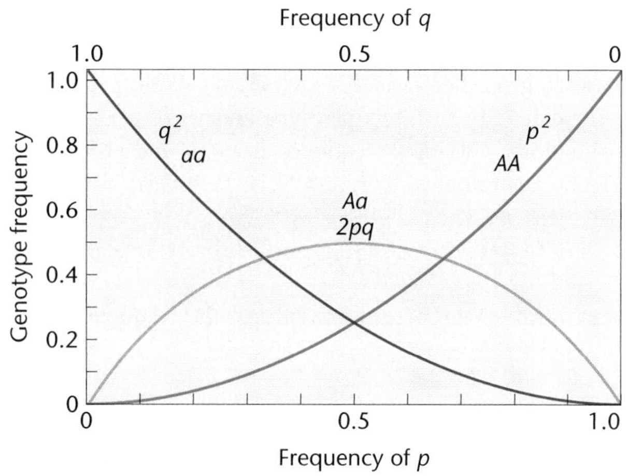

FIGURE 25–5 The relationship between genotype and allele frequencies derived from the Hardy–Weinberg equation.

Now solve this

Problem 25.17 on page 637 asks you to calculate the frequency of heterozygous carriers of the recessive trait albinism in a human population.

Hint: First determine how you calculate the frequency of the albinism allele in this population.

25.5 Natural Selection Is a Major Force Driving Allele Frequency Change

We have noted that the Hardy–Weinberg law establishes an ideal population that allows us to estimate allele and genotype frequencies at a given locus in populations in which the assumptions of random mating, absence of selection, absence of mutation, equal viability, and fertility hold. Obviously, it is difficult to find natural populations in which all these assumptions hold for all loci. In nature, populations are dynamic, and changes in size and gene pool are common. The Hardy–Weinberg law allows us to investigate populations that vary from the ideal. In this and the following sections, we discuss factors that prevent populations from reaching Hardy–Weinberg equilibrium, or that drive populations toward a different equilibrium, and the relative contribution of these factors to evolutionary change.

Natural Selection

The first assumption of the Hardy–Weinberg law is that individuals of all genotypes have equal rates of survival and equal reproductive success. If this assumption does not hold, allele frequencies may change from one generation to the next. To see why, let's imagine a population of 100 individuals in which the frequency of allele A is 0.5 and that of allele a is 0.5.

Assuming the previous generation mated randomly, we find that the genotype frequencies in the present generation are $(0.5)^2 = 0.25$ for *AA*, $2(0.5)(0.5) = 0.5$ for *Aa*, and $(0.5)^2 = 0.25$ for *aa*. Because our population contains 100 individuals, we have 25 *AA* individuals, 50 *Aa* individuals, and 25 *aa* individuals. Now suppose that individuals with different genotypes have different rates of survival: All 25 *AA* individuals survive to reproduce, 90 percent or 45 of the *Aa* individuals survive to reproduce, and 80 percent or 20 of the *aa* individuals survive to reproduce. When the survivors reproduce, each contributes two gametes to the new gene pool, giving us $2(25) + 2(45) + 2(20) = 180$ gametes. What are the frequencies of the two alleles in the surviving population? We have 50 *A* gametes from *AA* individuals, plus 45 *A* gametes from *Aa* individuals, so the frequency of allele *A* is $(50 + 45)/180 = 0.53$. We have 45 *a* gametes from *Aa* individuals, plus 40 *a* gametes from *aa* individuals, so the frequency of allele *a* is $(45 + 40)/180 = 0.47$.

These differ from the frequencies we started with. Allele *A* has increased, while allele *a* has declined. A difference among individuals in survival or reproduction rate (or both) is called **natural selection**. Natural selection is the principal force that shifts allele frequencies within large populations and is one of the most important factors in evolutionary change.

Fitness and Selection

Selection occurs whenever individuals with a particular genotype enjoy an advantage in survival or reproduction over other genotypes. However, selection may vary from less than 1 percent to 100 percent for a lethal gene. In the previous example, selection was strong. Weak selection might involve just a fraction of a percent difference in the survival rates of different genotypes. Advantages in survival and reproduction ultimately translate into increased genetic contribution to future generations. An individual's genetic contribution to future generations is called its **fitness**. Thus, genotypes associated with high rates of reproductive success are said to have high fitness, whereas genotypes associated with low reproductive success are said to have low fitness.

Hardy–Weinberg analysis also allows us to examine fitness. By convention, population geneticists use the letter *w* to represent fitness. Thus, w_{AA} represents the relative fitness of genotype *AA*, w_{Aa} the relative fitness of genotype *Aa*, and w_{aa} the relative fitness of genotype *aa*. Assigning the values $w_{AA} = 1$, $w_{Aa} = 0.9$, and $w_{aa} = 0.8$ could mean, for example, that all *AA* individuals survive, 90 percent of the *Aa* individuals survive, and 80 percent of the *aa* individuals survive, as in the previous example.

Let's consider selection against deleterious alleles. Fitness values $w_{AA} = 1$, $w_{Aa} = 1$, and $w_{aa} = 0$ describe a situation in which *a* is a homozygous lethal allele. As homozygous recessive individuals die without leaving offspring, the frequency of allele *a* will decline. The decline in the frequency of allele *a* is described by the equation

$$q_g = \frac{q_0}{1 + gq_0}$$

where q_g is the frequency of allele *a* in generation *g*, q_0 is the starting frequency of *a* (i.e., the frequency of *a* in generation zero), and *g* is the number of generations that have passed.

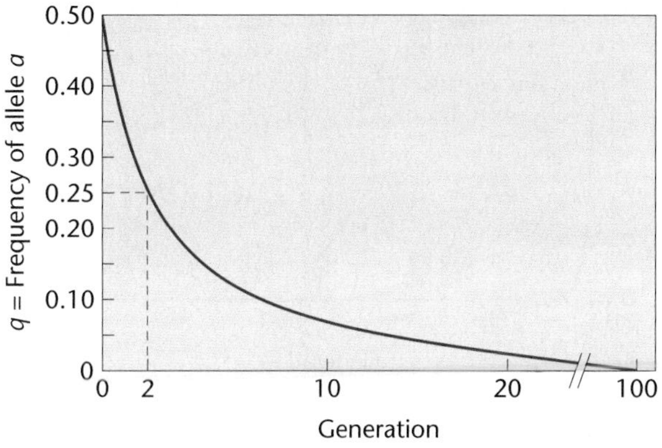

Generation	p	q	p^2	$2pq$	q^2
0	0.50	0.50	0.25	0.50	0.25
1	0.67	0.33	0.44	0.44	0.12
2	0.75	0.25	0.56	0.38	0.06
3	0.80	0.20	0.64	0.32	0.04
4	0.83	0.17	0.69	0.28	0.03
5	0.86	0.14	0.73	0.25	0.02
6	0.88	0.12	0.77	0.21	0.01
10	0.91	0.09	0.84	0.15	0.01
20	0.95	0.05	0.91	0.09	< 0.01
40	0.98	0.02	0.95	0.05	< 0.01
70	0.99	0.01	0.98	0.02	< 0.01
100	0.99	0.01	0.98	0.02	< 0.01

FIGURE 25–6 Change in the frequency of a lethal recessive allele, *a*. The frequency of *a* is halved in two generations, and halved again by the sixth generation. Subsequent reductions occur slowly because the majority of *a* alleles are carried by heterozygotes.

Figure 25–6 shows what happens to a lethal recessive allele with an initial frequency of 0.5. At first, because of the high percentage of *aa* genotypes, the frequency of allele *a* declines rapidly. The frequency of *a* is halved in only two generations. By the sixth generation, the frequency is halved again. By now, however, the majority of *a* alleles are carried by heterozygotes. Because *a* is recessive, these heterozygotes are not selected against. This means that as time continues to pass, the frequency of allele *a* declines ever more slowly. As long as heterozygotes continue to mate, it is difficult for selection to completely eliminate a recessive allele from a population.

Of course, a deleterious allele need not be recessive; many other scenarios are possible. A deleterious allele may be codominant, so that heterozygotes have intermediate fitness. Or an allele may be deleterious in the homozygous state, but beneficial in the heterozygous state, like the allele for sickle-cell anemia. Figure 25–7 shows examples in which a deleterious allele is codominant, but the intensity of selection varies from strong to weak. In each case, the frequency of the deleterious allele, *a*, starts at 0.99 and declines over time. However, the rate of decline depends heavily on the strength of selection. When only

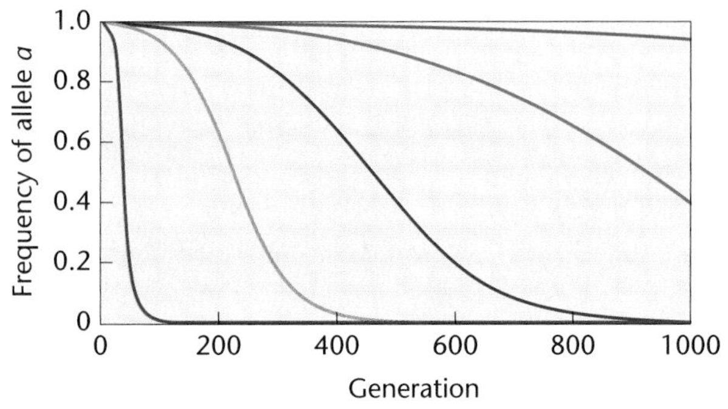

Selection Against Allele *a*					
	Strong ←				→ Weak
w_{AA}	1.0	1.0	1.0	1.0	1.0
w_{Aa}	0.90	0.98	0.99	0.995	0.998
w_{aa}	0.80	0.96	0.98	0.99	0.996

FIGURE 25–7 The effect of selection on allele frequency. The rate at which a deleterious allele is removed from a population depends heavily on the strength of selection.

90 percent of the heterozygotes and 80 percent of the *aa* homozygotes survive (red curve), the frequency of allele *a* drops from 0.99 to less than 0.01 in 85 generations. However, when 99.8 percent of the heterozygotes and 99.6 percent of the *aa* homozygotes survive (blue curve), it takes 1000 generations for the frequency of allele *a* to drop from 0.99 to 0.93. Two important conclusions can be drawn from this graph. First, given thousands of generations, even weak selection can cause substantial changes in allele frequencies; because evolution generally occurs over a large number of generations, selection is a powerful force of evolution. Second, for selection to produce rapid changes in allele frequencies, changes measurable within a human life span, the differences in fitness among genotypes must be large, or generation time must be short.

The manner in which selection affects allele frequencies allows us to make some inferences about the *CCR5-Δ32* allele that we discussed earlier. Because individuals with genotype *Δ32/Δ32* are resistant to most sexually transmitted strains of HIV-1, while individuals with genotypes *1/1* and *1/Δ32* are susceptible, we might expect AIDS to act as a selective force causing the frequency of the *Δ32* allele to increase over time. Indeed it probably will, but the increase in frequency is likely to be slow in human terms.

We can do a rough calculation as follows: Imagine a population in which the current frequency of the *Δ32* allele is 0.10. Under Hardy–Weinberg assumptions, the genotype frequencies in this population are 0.81 for *1/1*, 0.18 for *1/Δ32* and 0.01 for *Δ32/Δ32*. Imagine also that 1 percent of the *1/1* and *1/Δ32* individuals in this population will contract HIV and die of AIDS.

Based on our assumptions, we can assign fitness levels to the genotypes as follows: $w_{1/1} = 0.99$; $w_{1/\Delta32} = 0.99$; $w_{\Delta32/\Delta32} = 1.0$. Given the assigned fitness, we can predict that the frequency of the *CCR5-Δ32* allele in the next generation will be 0.100091. In fact, it will take about 100 generations (about 2000 years) for the frequency of the *Δ32* allele to reach just 0.11 (Figure 25–8). In other words, the frequency of the *Δ32* allele will probably not change much over the next few generations in most populations that currently harbor it. A population genetic perspective sheds light on the *CCR5-Δ32* story in other ways as well. Two research groups have analyzed genetic variation at marker loci closely linked to the *CCR5* gene. Both groups concluded that most, if not all, present-day copies of the *Δ32* allele are descended from a single ancestral copy that appeared in northeastern Europe at most a few thousand years ago. In fact, one group estimates that the common ancestor of all *Δ32* alleles existed just 700 years ago. How could a new allele rise from a frequency of virtually zero to as high as 20 percent in roughly 30 generations?

It seems that there must have been strong selection in favor of the *Δ32* allele, most likely in the form of an infectious disease. The agent of selection cannot have been HIV-1, because HIV-1 moved from chimpanzees to humans too recently. Because selection occurred about 700 years ago, J. C. Stephens suggests that the agent of selection was bubonic plague. During the Black Death of 1346–1352, between a quarter and a third of all Europeans died from plague. Bubonic plague is caused by the bacterium *Yersinia pestis*. This bacterium manufactures a protein that kills some kinds of white blood cells. Stephens hypothesizes that the process through which the bacterial protein kills white cells involves the *CCR5* gene product. If true, some mechanism makes individuals homozygous for the *Δ32* allele more likely to survive plague epidemics.

Based on their study of the *CCR5* locus in Jewish and European populations, William Klitz and his colleagues have suggested that the *Δ32* allele may have arisen as early as the eighth century. They also suggest that smallpox may be the selective agent responsible for the rapid increase in the frequency of this allele. Both HIV and variola, the virus responsible for smallpox, use the CCR5 receptor for infecting cells, and with a fatality rate of 25 percent, waves of smallpox would be a powerful selective agent.

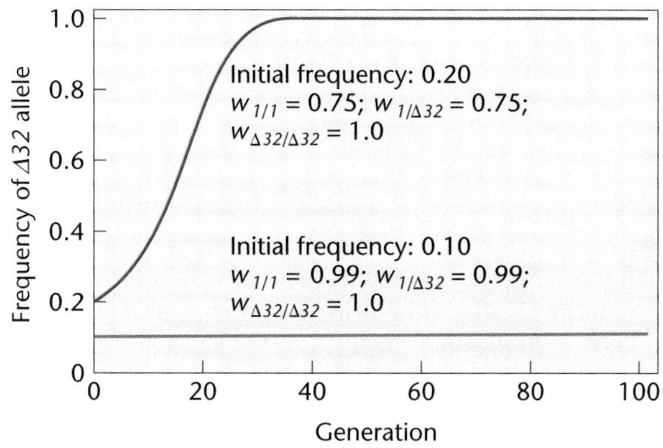

FIGURE 25–8 The rate at which the frequency of the *CCR5-Δ32* allele changes in hypothetical populations with different initial frequencies and different fitnesses.

If past epidemics are responsible for the high frequency of the *CCR5-Δ32* allele in European populations, then the virtual absence of the allele in non-European populations is at first somewhat puzzling. Perhaps plague has been more common in Europe than elsewhere. Alternatively, perhaps other populations have different alleles of the *CCR5* gene that also confer protection against the plague. Teams of researchers looking for other alleles of the *CCR5* gene in various populations have found a total of 20 mutant alleles, including *Δ32*. Sixteen of these alleles encode proteins different in structure from that encoded by the *CCR51* allele. Some, like *Δ32*, are loss-of-function alleles. Some of the alleles appear confined to Asian populations, others to African populations. Their frequencies are as high as 3 to 4 percent. Together, these discoveries are consistent with the hypothesis that alteration or loss of the CCR5 protein protects against an as-yet-unidentified infectious disease or diseases.

Now solve this

Problem 25.8 on page 637 asks you to calculate the effect of varying fitness levels for different genotypes on allele frequencies.

Hint: First calculate the genotype frequencies in the next generation.

Selection in Natural Populations

Geneticists have done a great deal of research on the effect of natural selection on allele frequencies in both laboratory and natural populations. Among the most detailed studies of natural populations are those that involve insects exposed to pesticides. Christine Chevillon and colleagues, for example, studied the effect of the insecticide chlorpyrifos on allele frequencies in populations of house mosquitoes (Figure 25–9). Chlorpyrifos kills mosquitoes by interfering with the function of the enzyme acetylcholinesterase (ACE), which under normal circumstances breaks down the neurotransmitter acetylcholine. An allele of ACE called Ace^R encodes a slightly altered version of ACE that is immune to interference by chlorpyrifos.

FIGURE 25–9 Pesticides used to control insects act as selective agents, changing allele frequencies of resistance genes in the populations exposed to the pesticide.

Chevillon measured the frequency of the Ace^R allele in nine populations. In the first four locations, chlorpyrifos had been used to control mosquitoes for 22 years; in the last five locations, chlorpyrifos had never been used. Chevillon predicted that the frequency of Ace^R would be higher in the exposed populations. The researchers also predicted that the frequencies of alleles for enzymes unrelated to the physiological effects of chlorpyrifos would show no such pattern. Among the control enzymes studied was aspartate amino transferase 1. The results appear in Figure 25–10. As the researchers predicted, the frequency of the Ace^R allele was significantly higher in the exposed populations. Also as predicted, the frequencies of the most common alleles of the control enzyme gene showed no such trends. The explanation is that during the 22 years of

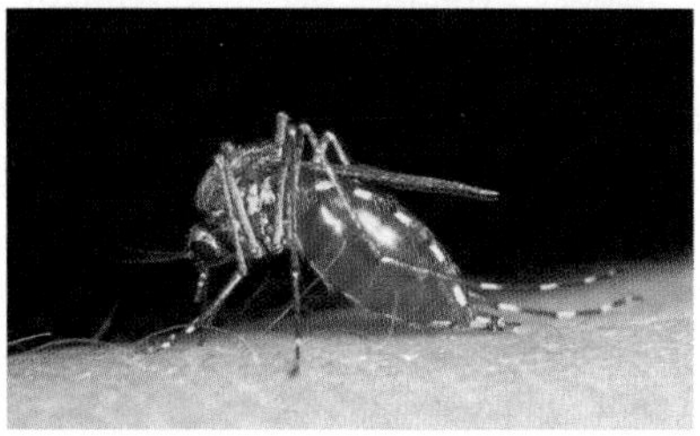

House mosquito, *Culex pipiens.*

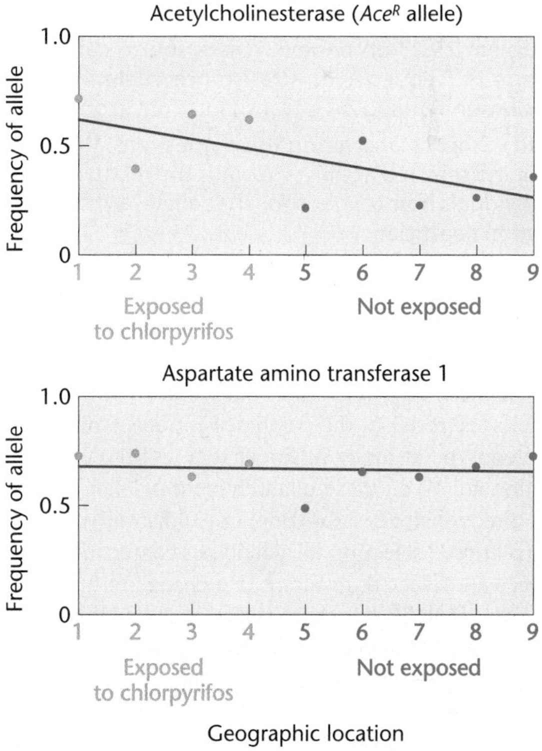

FIGURE 25–10 The effect of selection on allele frequencies in natural populations. (a) The frequency of the Ace^R allele, which confers resistance to the insecticide chlorpyrifos, is higher in house mosquito populations exposed to chlorpyrifos. (b) The frequency of an allele for an enzyme unrelated to chlorpyrifos metabolism (aspartate amino transferase 1) shows no such pattern.

exposure to the pesticide, mosquitoes had higher rates of survival if they carried the Ace^R allele. In other words, the Ace^R allele had been favored by natural selection.

In the case of mosquito populations exposed to chlorpyrifos, rate of change in allele frequencies at the ACE locus will depend on the intensity of selection against the mosquitoes that do not have the Ace^R allele. If selection is operating against an allele in the population with an initial frequency q, the new frequency q' after one generation can be calculated using the following equation:

$$q' = \frac{q(1 - sq)}{1 - sq^2}$$

where s is the selection coefficient against individuals homozygous for q. Thus, if the relative fitness of mosquitoes that do not have the Ace^R allele and are not resistant to chlorpyrifos is 0.75 and the initial frequency of the normal unmutated version of the ACE allele is 0.95, then after one generation of exposure to the insecticide the new frequency q' of the normal allele will be:

$$q' = \frac{0.95(1 - 0.25 \times 0.95)}{1 - 0.25 \times 0.95^2} = 0.935$$

This shows that even with a relatively strong selection in its favor, a new allele occurring at a low frequency will require multiple generations to produce a significant change in the gene pool.

Now solve this

Problem 25.23 on page 637 asks you to calculate the change in frequency of a resistance allele in an insect population feeding on a transgenic corn crop.

Hint: Start by answering these questions: Which allele is being selected against? What is the relative fitness of individuals homozygous for that allele? What is the selection coefficient?

Natural Selection and Quantitative Traits

As we saw in Chapter 24, many phenotypic traits are controlled not by alleles at a single locus, but are quantitative: The phenotype is the result of the combined influence of the individual's genotype at many different loci and the environment. Selection acting on these quantitative traits can be classified as (1) directional, (2) stabilizing, or (3) disruptive.

In **directional selection** the genotypes conferring phenotypic extremes are selected, resulting in a change in the population mean over time. This form of selection is widely practiced in plant and animal breeding. If the trait is polygenic, the most extreme phenotypes that the genotype can express will appear in the population only after prolonged selection. An example of a long-running experiment in directional artificial selection is the study in high and low oil content in corn conducted at the State Agricultural Laboratory in Illinois, described in Chapter 24.

In nature, directional selection can occur when one of the phenotypic extremes becomes selected for or against, usually as a result of changes in the environment. A carefully documented example comes from research by Peter and Rosemary Grant and their colleagues, who used the medium ground finches (*Geospiza fortis*) of Daphne Major Island in the Galapagos Islands. The beak size of these birds varies enormously. In 1976, for example, some birds in the population had beaks less than 7 mm deep, while others had beaks more than 12 mm deep. Beak size is heritable, which means that large-beaked parents tend to have large-beaked offspring, and small-beaked parents tend to have small-beaked offspring. In 1977, a severe drought on Daphne killed some 80 percent of the finches. Big-beaked birds survived at higher rates than small-beaked birds, because when food became scarce, the big-beaked birds were able to eat a greater variety of seeds. When the drought ended in 1978 and the survivors paired off and bred, the offspring inherited their parents' big beaks. Between 1976 and 1978, the beak depth of the average finch in the Daphne Major population increased by just over 0.5 mm, shifting the average beak size toward one phenotypic extreme.

Stabilizing selection, in contrast, tends to favor intermediate types, with both extreme phenotypes being selected against. Over time, this will reduce the population variance, but without a significant shift in the mean. One of the clearest demonstrations of stabilizing selection is provided by the data of Mary Karn and Sheldon Penrose on human birth weight and survival for 13,730 children born over an 11-year period. Figure 25–11 shows the distribution of birth weight and the percentage of mortality at 4 weeks of age. Infant mortality increases on either side of the optimal birth weight of 7.5 pounds and quite dramatically so at the low end. At the genetic level, stabilizing selection acts to keep a population well adapted to its environment. In this situation, individuals closer to the average for a given trait will have higher fitness.

Disruptive selection is selection against intermediates and for both phenotypic extremes. It can be viewed as the opposite of stabilizing selection because the intermediate types are selected against. This will result in a population with an increasingly bimodal distribution for the trait, as we can see in Figure 25–12. In one set of experiments, John Thoday applied disruptive selection to a population of *Drosophila* on the basis of bristle number. In every generation, he allowed only the flies with high- or low-bristle numbers to breed. After several generations, most of the flies could be easily placed in a low- or high-bristle category (Figure 25–13). In natural populations, such a situation might exist for a population in a heterogeneous environment.

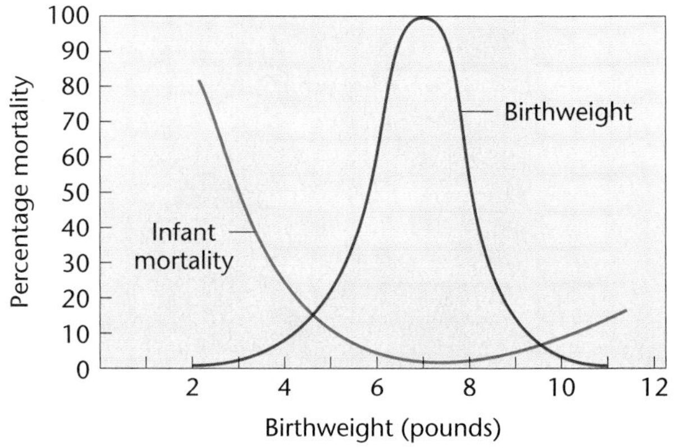

FIGURE 25–11 Relationship between birth weight and mortality in humans.

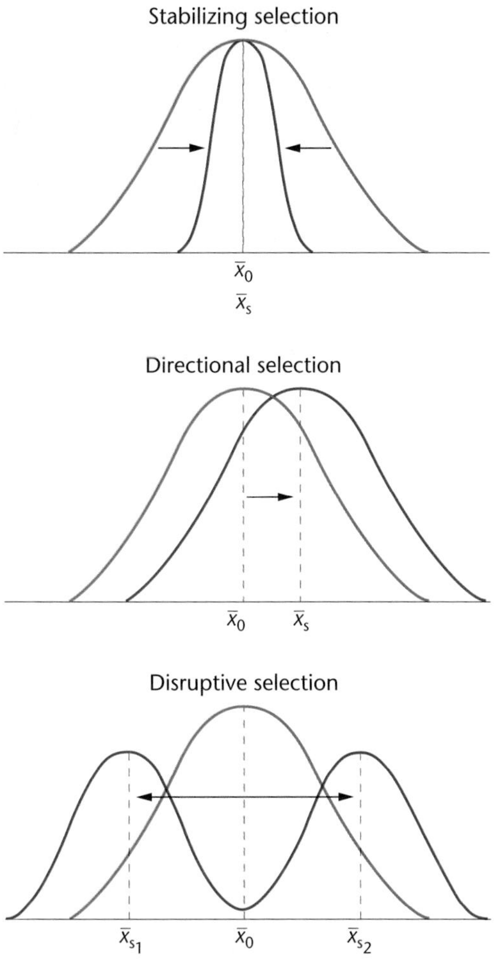

FIGURE 25–12 The impact of stabilizing, directional, and disruptive selection. In each case, the mean of an original population (green) and the mean of the population following selection (red) is shown.

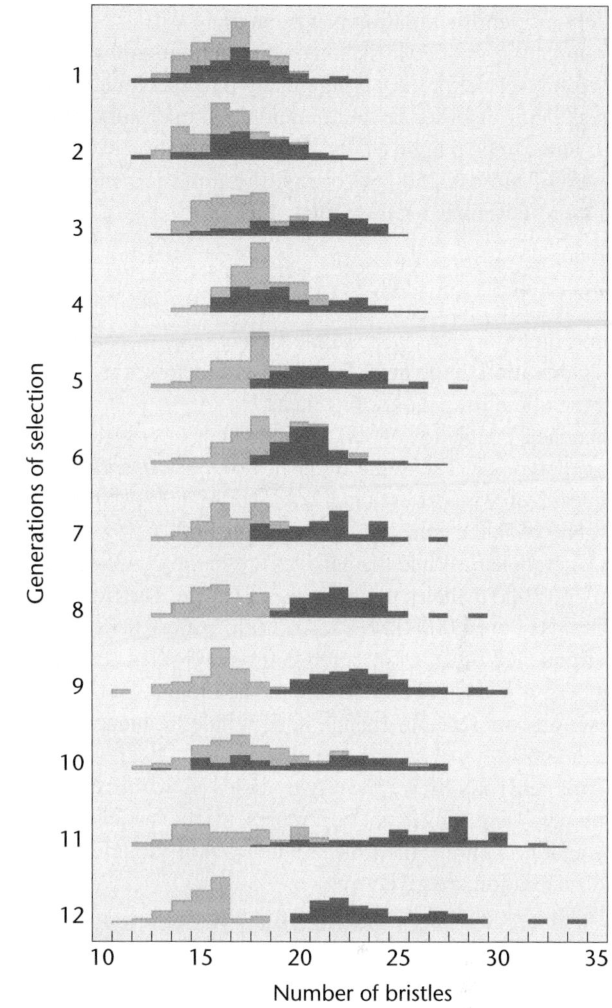

FIGURE 25–13 The effect of disruptive selection on bristle number in *Drosophila*. When individuals with the highest and lowest bristle number were selected, the population showed a nonoverlapping divergence in only 12 generations.

25.6 Mutation Creates New Alleles in a Gene Pool

Within a population, the gene pool is reshuffled each generation to produce new genotypes in the offspring. Because the number of possible genotypic combinations is so large, population members alive at any given time represent only a fraction of all possible genotypes. The enormous genetic reserve present in the gene pool allows Mendelian assortment and recombination to produce new genotypic combinations continuously. But assortment and recombination do not produce new alleles. **Mutation** alone acts to create new alleles. It is important to keep in mind that mutational events occur at random—that is, without regard for any possible benefit or disadvantage to the organism. In this section, we consider whether mutation is, by itself, a significant factor in causing allele frequencies to change.

To determine whether mutation is a significant force in changing allele frequencies, we measure the rate at which mutations are produced. As most mutations are recessive, it is difficult to observe mutation rates directly in diploid organisms. Indirect methods that use probability and statistics or large-scale screening programs are employed. For certain dominant mutations, however, a direct method of measurement can be used. To ensure accuracy, several conditions must be met:

1. The allele must produce a distinctive phenotype that can be distinguished from similar phenotypes produced by recessive alleles.
2. The trait must be fully expressed or completely penetrant so that mutant individuals can be identified.
3. An identical phenotype must never be produced by nongenetic agents such as drugs or chemicals.

Mutation rates can be stated as the number of new mutant alleles per given number of gametes. Suppose that for a given gene that undergoes mutation to a dominant allele, 2 out of 100,000 births exhibit a mutant phenotype. In these two cases, the parents are phenotypically normal. Because the zygotes that produced these births each carry two copies of the gene, we have actually surveyed 200,000 copies of the gene (or 200,000 gametes). If we assume that the affected births are each heterozygous, we have uncovered two mutant alleles out of 200,000. Thus, the mutation rate is 2/200,000 or 1/100,000, which in scientific notation is written as 1×10^{-5} In humans, a dominant form of dwarfism known as **achondroplasia** fulfills the requirements for measuring mutation rates. Individuals with this skeletal disorder have an enlarged skull, short arms and legs, and can be diagnosed by X-ray examination at birth. In a survey of almost 250,000 births, the mutation rate (μ) for achondroplasia has been calculated as

$$\mu = 1.4 \times 10^{-5} \pm 0.5 \times 10^{-5}$$

Knowing the rate of mutation, we can estimate the extent to which mutation can cause allele frequencies to change from one generation to the next. We represent the normal allele as *d* and the allele for achondroplasia as *D*.

Imagine a population of 500,000 individuals in which everyone has genotype *dd*. The initial frequency of *d* is 1.0, and the initial frequency of *D* is 0. If each individual contributes two gametes to the gene pool, the gene pool will contain 1,000,000 gametes, all carrying allele *d*. While the gametes are in the gene pool, 1.4 of every 100,000 *d* alleles mutates into a *D* allele. The frequency of allele *d* is now $(1{,}000{,}000 - 14)/1{,}000{,}000 = 0.999986$, and the frequency of *D* is $14/1{,}000{,}000 = 0.000014$. From these numbers, it will clearly be a long time before mutation, by itself, causes any appreciable change in the allele frequencies in this population.

More generally, if we have two alleles, *A* with frequency *p* and *a* with frequency *q*, and if μ represents the rate of mutations converting *A* into *a*, then the frequencies of the alleles in the next generation are given by

$$p_{g+1} = p_g - \mu p_g \text{ and } q_{g+1} = q_g + \mu p_g$$

where p_{g+1} and q_{g+1} represent the allele frequencies in the next generation, and p_g and q_g represent the allele frequencies in the present generation.

Figure 25–14 shows the replacement rate (change over time) in allele *A* for a population in which the initial frequency of *A* is 1.0 and the rate of mutation (μ) converting *A* into *a* is 1.0×10^{-5}. At this mutation rate, it will take about 70,000 generations to reduce the frequency of *A* to 0.5. Even if the rate of mutation increases through exposure to higher levels of radioactivity or chemical mutagens, the impact of mutation on allele frequencies will be extremely weak. The ultimate source of the genetic variability, mutation provides the raw material for evolution, but *by itself* plays a relatively insignificant role in changing allele frequencies. Instead, the fate of alleles created by mutation is more likely to be determined by natural selection (discussed previously) and genetic drift (discussed later).

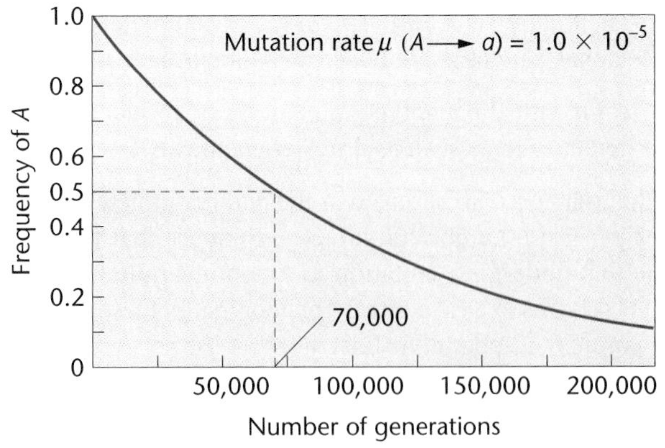

FIGURE 25–14 Replacement rate of an allele by mutation alone, assuming an average mutation rate of 1.0×10^{-5}.

An evolutionary perspective on mutation can lead to medical discoveries. The autosomal recessive disease cystic fibrosis, which we discussed previously in relation to calculating heterozygote frequencies, is caused by a loss-of-function mutation in the gene for a cell-surface protein called the cystic fibrosis transmembrane conductance regulator (CFTR).

The frequency for the mutant alleles causing cystic fibrosis is about 2 percent in European populations. Until recently, most individuals with two mutant alleles died before reproducing, meaning that selection against homozygous recessive individuals was rather strong. This creates a puzzle: In the face of selection against them, what has maintained the mutant alleles at an overall frequency of 2 percent?

The mutation-selection balance hypothesis posits that mutation is constantly creating new alleles to replace the ones being eliminated by selection. However, for this scenario to work, the rate of mutations creating new alleles would have to be rather high, on the order of 5×10^{-4} to counteract the effect of selection. Many evolutionary geneticists instead prefer an alternative explanation, the heterozygote superiority hypothesis. According to the heterozygote superiority hypothesis, selection against homozygous mutant individuals is counterbalanced by selection in favor of heterozygotes. The most popular agent of selection in favor of heterozygotes is resistance to an as-yet-unidentified disease.

Recent work suggests that cystic fibrosis heterozygotes may have enhanced resistance to typhoid fever. Typhoid fever is caused by the bacterium *Salmonella typhi*, which infiltrates cells of the intestinal lining. In laboratory studies, mouse intestinal cells that were heterozygous for *CFTR-Δ508*, the analog of the most common cystic fibrosis mutation in humans, acquired 86 percent fewer bacteria than did cells homozygous for the wild-type allele. Whether humans heterozygous for *CFTR-Δ508* also enjoy resistance to typhoid fever remains to be established. If they do, then cystic fibrosis will join sickle-cell anemia as an example of heterozygote superiority.

25.7 Migration and Gene Flow Can Alter Allele Frequencies

Occasionally, a species divides into populations that are separated geographically. Various evolutionary forces, including selection, can establish different allele frequencies in such populations. **Migration** occurs when individuals move between the populations. Imagine a species in which a single locus has two alleles, *A* and *a*. There are two populations of this species, one on a mainland and one on an island. The frequency of *A* on the mainland is represented by p_m, and the frequency of *A* on the island is p_i. Under the influence of migration from the mainland to the island, the frequency of *A* in the next generation on the island ($P_{i'}$) is given by

$$P_{i'} = (1 - m)p_i + mp_m$$

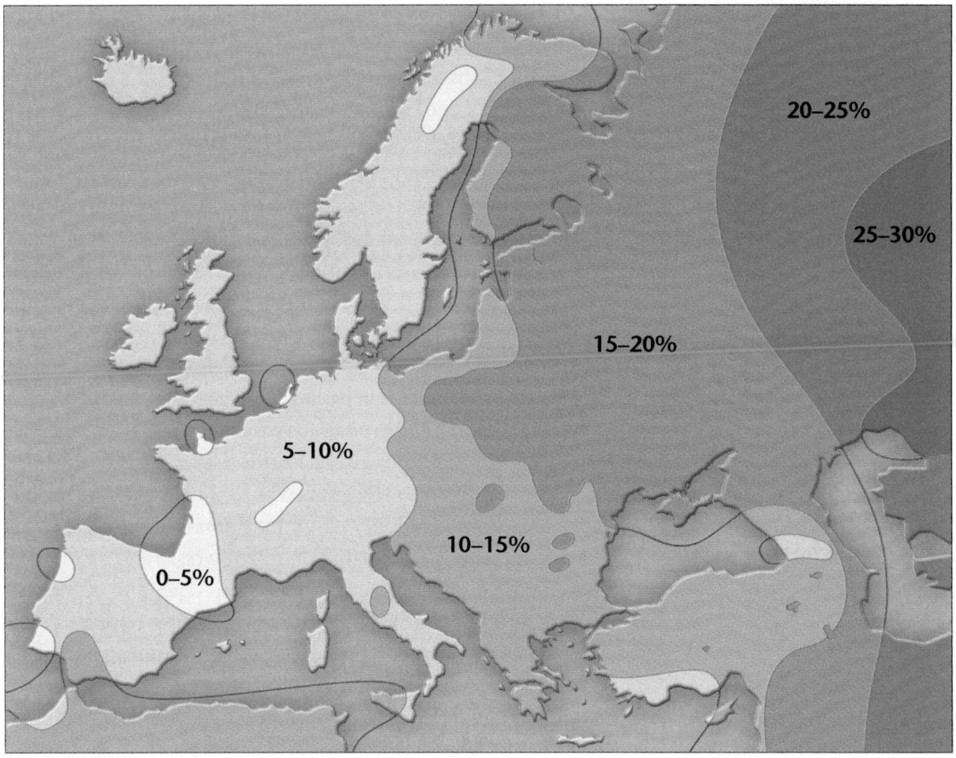

FIGURE 25–15 Migration as a force in evolution. The *B* allele of the *ABO* locus is present in a gradient from east to west. This allele shows the highest frequency in Central Asia and the lowest is in northeastern Spain. The gradient parallels the waves of Mongol migration into Europe following the fall of the Roman Empire and is a genetic relic of human history.

where m represents migrants from the mainland to the island.

Under these conditions, the frequency of A in the next generation on the island ($P_{i'}$) will be affected by migration. For example, assume that $p_i = 0.4$ and $p_m = 0.6$ and that 10 percent of the parents of the next generation are migrants from the mainland, so that $m = 0.1$. In the next generation, the frequency of allele A on the island will be

$$p_{i'} = [(1 - 0.1) \times 0.4] + (0.1 \times 0.6)$$
$$= 0.36 + 0.06$$
$$= 0.42$$

In this case, migration from the mainland has changed the frequency of A on the island from 0.40 to 0.42 in a single generation.

If either m is large or p_m is very different from p_i, then a rather large change in the frequency of A can occur in a single generation. If migration is the only force acting to change the allele frequency on the island, then an equilibrium will be attained only when $p_i = p_m$. These calculations reveal that the change in allele frequency attributable to migration is proportional to the differences in allele frequency between the donor and recipient populations and to the rate of migration. As m can have a wide range of values, the effect of migration can substantially alter allele frequencies in populations, as shown for the B allele of the ABO blood group in Figure 25–15. Although migration can be difficult to quantify, it can often be estimated.

Migration can also be regarded as the flow of genes between populations that were once, but are no longer, geographically isolated. Esteban Parra and colleagues measured allele frequencies for several different DNA sequence polymorphisms in African-American and European-American populations and in African and European populations representative of the ancestral populations from which the two American populations are descended. One locus they studied, a restriction site polymorphism called *FY-NULL*, has two alleles, *FY-NULL*1* and *FY-NULL*2*. Figure 25–16 shows the frequency of *FY-NULL*1* in each population. The frequency of this allele is 0 in the three African populations and 1.0 in the three European populations, but lies between these extremes in all African-American and European-American populations. The simplest explanation for these data is that genes have mixed between American populations with predominantly African ancestry and American populations with predominantly European ancestry. Based on *FY-NULL* and several other loci, researchers estimate that African-American populations derive between 11.6 and

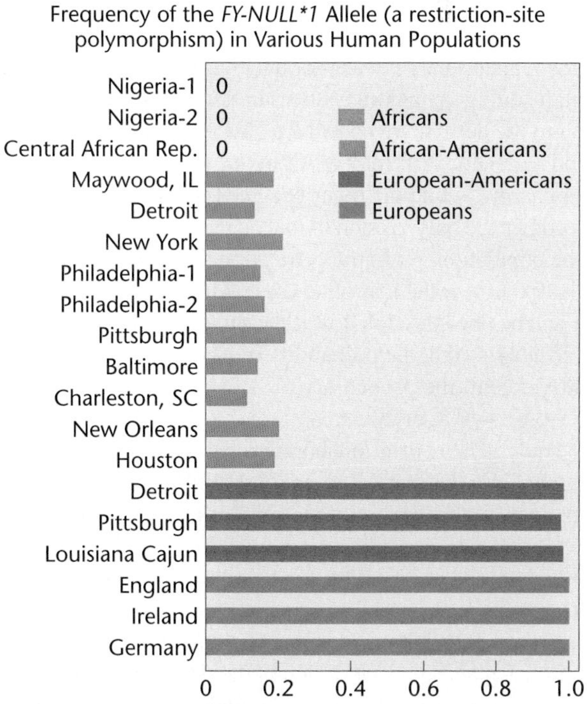

FIGURE 25–16 Frequency histogram of a restriction site polymorphism allele in several populations. These and other data demonstrate that African-American and European-American populations are of mixed ancestry.

22.5 percent of their ancestry from Europeans and that European-American populations derive between 0.5 and 1.2 percent of their ancestry from Africans.

Now solve this

Problem 25.25 on page 638 asks you to calculate the change in frequency of a mutant allele in a population of bighorn sheep that has been augmented by artificially introducing individuals from another population.

Hint: If the introduced sheep came from a population that lacked the mutation, would you expect the allele frequency in the augmented population to increase or decrease?

25.8 Genetic Drift Causes Random Changes in Allele Frequency in Small Populations

Genetic drift occurs when the number of reproducing individuals in a population is too small to ensure that all the alleles in the gene pool will be passed on to the next generation in their existing frequencies. Drift is defined as a change in allele frequency resulting from random gamete sampling. To visualize this, imagine a sexually reproducing diploid population consisting of 25 males and 25 females. Even if all individuals in the population reproduce as 25 separate mating pairs (an unlikely scenario in real life) to generate 25 offspring, the number of gametes forming the gene pool of the next generation will be just 50. It is unlikely that such a small number of gametes will accurately reflect the genetic structure of the parent population. Consequently, individual alleles may be over- or underrepresented in the gamete pool, resulting in random changes in frequency from one generation to the next. If we consider a single locus with two alleles, *A* and *a*, genetic drift may result in one of the alleles eventually disappearing, while the other becomes fixed— in other words, it becomes the only version of that gene present in the gene pool of the population. A simple correlation describes the likelihood of the fixation or the loss of an allele. The probability that an allele will be fixed through drift is the same as its initial frequency. So, if $p(A) = 0.8$, the probability of *A*'s becoming fixed is 0.8, or 80 percent; the probability of *A* being lost through drift is $(1 - 0.8) = 0.2$, or 20 percent.

To study genetic drift in laboratory populations of *Drosophila melanogaster*, Warwick Kerr and Sewall Wright set up over 100 lines, with four males and four females as the parents for each line. Within each line, the frequency of the sex-linked bristle mutant *forked* (f) and its wild-type allele (f^+) was 0.5. In each generation, four males and four females were chosen at random to parent the next generation. After 16 generations, the complete loss of one allele and the fixation of the other had occurred in 70 lines—29 in which only the *forked* allele was present and 41 in which the wild-type allele had become fixed. The remaining lines were still segregating the two alleles or had gone extinct. If fixation had occurred randomly, then an equal number of lines should have become fixed for each allele. In fact, the experimental results do not differ statistically from the expected ratio of 35:35, demonstrating that alleles can spread through a population and eliminate other alleles by chance alone.

How are small populations created in nature? A disaster such as an epidemic might occur, leaving a small number of survivors to breed. Or a small group might emigrate from the larger population and become founders in a new environment, such as a volcanically created island. Among endangered plants and animals, habitat loss often results in small isolated populations where loss of genetic diversity through drift can threaten the long-term survival of the species. (This is discussed in more detail in Chapter 27.)

Allele frequencies in certain isolated human populations demonstrate the role of drift as an evolutionary force in natural populations. The Pingelap atoll in the western Pacific Ocean (lat. 6° N, long. 160° E) has in the past been devastated by typhoons and famine, and around 1780, there were only about nine surviving males. Today there are fewer than 2000 inhabitants, all of whose ancestry can be traced to the typhoon survivors. About 4 to 10 percent of the current population is colorblind from infancy. These people are affected by an autosomal recessive disorder, **achromatopsia**, which causes ocular disturbances, a form of color blindness, and cataract formation. The disorder is extremely rare in the human population as a whole. However, the mutant allele is present at a relatively high frequency in the Pingelap population. Genealogical reconstruction shows that one of the original survivors (about 30 individuals) was heterozygous for the condition. If we assume that he was the only carrier in the founding population, the initial gene frequency was $1/60$, or 0.016. On average, about 7 percent of the current population is affected (a homozygous recessive genotype), so the frequency of the allele in the present population has increased to 0.26. The story of the inhabitants of Pingelap is told by Oliver Sacks, in his book *The Island of the Colorblind*.

Another example of genetic drift involves the Dunkers, a small, isolated religious community who emigrated from the German Rhineland to Pennsylvania. Because their religious beliefs do not permit marriage with outsiders, the population has grown only through marriage within the group. When the frequencies of the ABO and MN blood group alleles are compared, significant differences are found between the Dunkers, the German population, and the U.S. population. The frequency of blood group A in the Dunkers is about 60 percent, in contrast to 45 percent in the U.S. and German populations. The I^B allele is nearly absent in the Dunkers. Type M blood is found in about 45 percent of the Dunkers, compared with about 30 percent in the U.S. and German populations. As there is no evidence for a selective advantage of these alleles, it is apparent that the observed frequencies are the result of chance events in a relatively small isolated population.

25.9 Nonrandom Mating Changes Genotype Frequency but Not Allele Frequency

We have explored how violations of the first four assumptions of the Hardy–Weinberg law, in the form of selection, mutation, migration, and genetic drift, can cause allele frequencies to change.

The fifth assumption is that the members of a population mate at random; in other words, any one genotype has an equal probability of mating with any other genotype in the population. Nonrandom mating can change the frequencies of genotypes in a population. Subsequent selection for or against certain genotypes has the potential to affect the overall frequencies of the alleles they contain, but it is important to note that nonrandom mating *does not itself directly change allele frequencies.*

Nonrandom mating can take one of several forms. In **positive assortive mating** similar genotypes are more likely to mate than dissimilar ones. This often occurs in humans: A number of studies have indicated that many people are more attracted to individuals who physically resemble them (and are therefore more likely to be genetically similar as well). **Negative assortive mating** occurs when dissimilar genotypes are more likely to mate; some plant species have inbuilt pollen/stigma recognition systems that prevent fertilization between individuals with the same alleles at key loci. However, the form of nonrandom mating most commonly found to affect genotype frequencies in population genetics is **inbreeding.**

P_1: Self-fertilization *Aa*

	AA	*Aa*	*aa*
F_1	0.250	0.500	0.250
F_2	0.375	0.250	0.375
F_3	0.437	0.125	0.437
F_4	0.468	0.063	0.468
F_n	$\frac{1-\frac{1}{2^n}}{2}$	$\frac{1}{2^n}$	$\frac{1-\frac{1}{2^n}}{2}$

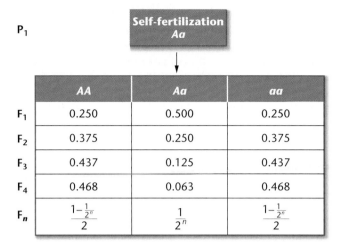

FIGURE 25–17 Reduction in heterozygote frequency brought about by self-fertilization. After *n* generations, the frequencies of the genotypes can be calculated according to the formulas in the bottom row.

Inbreeding

Inbreeding occurs when mating individuals are more closely related than any two individuals drawn from the population at random; loosely defined, inbreeding is mating among relatives. For a given allele, inbreeding increases the proportion of homozygotes in the population. A completely inbred population will theoretically consist only of homozygous genotypes. To demonstrate this, let us consider the most extreme form of inbreeding, **self-fertilization**, which although rare in animals is widespread in plant species. Figure 25–17 shows the results of four generations of self-fertilization, starting with a single individual heterozygous for one pair of alleles. By the fourth generation, only about 6 percent of the individuals are still heterozygous, and 94 percent of the population is homozygous. Note, however, that the frequencies of alleles *A* and *a* remain unchanged at 50 percent.

Not all inbreeding in populations occurs through self-fertilization, and different degrees of inbreeding exist. To describe the intensity of inbreeding in a population, geneticist Sewall Wright devised the **coefficient of inbreeding** (***F***). *F* quantifies the probability that the two alleles of a given gene in an individual are identical *because they are descended from the same single copy of the allele in an ancestor.* If $F = 1$, all individuals in the population are homozygous, and both alleles in every individual are derived from the same ancestral copy. If $F = 0$, no individual has two alleles derived from a common ancestral copy.

One simple method of estimating *F* for a population is based on the inverse relationship between inbreeding and the frequency of heterozygotes: As the level of inbreeding increases, the proportion of heterozygotes declines. Therefore, *F* can be calculated as

$$F = \frac{H_e - H_o}{H_e}$$

where H_e is the expected heterozygosity based on the Hardy–Weinberg law and H_o the observed heterozygosity in a population. Note that in a completely random mating population the expected and observed levels of heterozygosity will be equal and $F = 0$.

A different method can be used to estimate *F* for an individual. Figure 25–18 shows a pedigree of a first-cousin marriage. The fourth-generation female (shaded pink) is the daughter of first cousins (yellow). Suppose her great-grandmother (green) was a carrier of a recessive lethal allele, *a*. What is the probability that the fourth-generation female will inherit two copies of her great-grandmother's lethal allele? For this to happen, (1) the great-grandmother had to pass a copy of the allele to her son, (2) her son had to pass it to his daughter, and (3) his daughter has to pass it to her daughter (the pink female). Also, (4) the great-grandmother had to pass a copy of the allele to her daughter, (5) her daughter had to pass it to her son, and (6) her son has to pass it to his daughter (the pink female). Each of the six necessary events has an individual probability of 1/2, and they *all* have to happen, so the probability that the pink female will

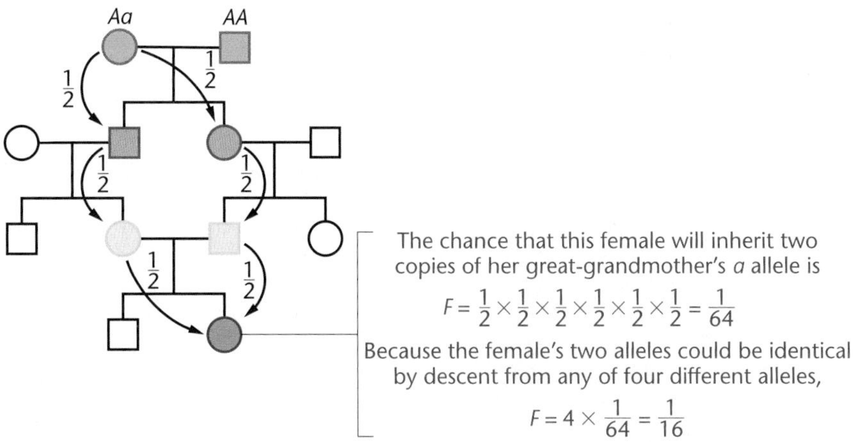
Aa
AA
1/2
The chance that this female will inherit two copies of her great-grandmother's a allele is
$F = \frac{1}{2} \times \frac{1}{2} \times \frac{1}{2} \times \frac{1}{2} \times \frac{1}{2} \times \frac{1}{2} = \frac{1}{64}$
Because the female's two alleles could be identical by descent from any of four different alleles,
$F = 4 \times \frac{1}{64} = \frac{1}{16}$

FIGURE 25–18 Calculating the coefficient of inbreeding *(F)* for the offspring of a first-cousin marriage.

inherit two copies of her great-grandmother's lethal allele is $(1/2)^6 = 1/64$. To calculate an overall value of F for the pink female as a child of a first-cousin marriage, remember that she could also inherit two copies of any of the other three alleles present in her great-grandparents. Because any of four possibilities would give the pink female two alleles identical by descent from an ancestral copy,

$$F = 4 \times (1/64) = 1/16.$$

Now solve this

Problem 25.22 on page 637 asks you to calculate the inbreeding coefficient of a population based on DNA marker data.

Hint: First, work out the frequencies of the DNA marker alleles. If this population is in Hardy–Weinberg equilibrium, how many heterozygotes would you expect to see?

Genetic Effects of Inbreeding

Inbreeding results in the production of individuals homozygous for recessive alleles that were previously concealed in heterozygotes. Because many recessive alleles are deleterious when homozygous, one consequence of inbreeding is an increased chance that an individual will be homozygous for a recessive deleterious allele. Inbred populations often have a lowered mean fitness. **Inbreeding depression** is a measure of the loss of fitness caused by inbreeding. In domesticated plants and animals, inbreeding and selection have been used for thousands of years, and these organisms already have a high degree of homozygosity at many loci. Further inbreeding will usually produce only a small loss of fitness. However, inbreeding among individuals from large, randomly mating populations can produce high levels of inbreeding depression. This effect can be seen by examining the mortality rates in offspring of inbred animals in zoo populations (Table 25.4). The potential impact of inbreeding on populations of threatened and endangered species is discussed further in Chapter 27.

TABLE 25.4 MORTALITY IN OFFSPRING OF INBRED ZOO ANIMALS

Species	n		Noninbred	Inbred	Inbreeding Coefficient
Zebra	32	Lived:	20	3	0.250
		Died:	7	2	
Eld's deer	24	Lived:	13	0	0.250
		Died:	4	7	
Giraffe	19	Lived:	11	2	0.250
		Died:	3	3	
Oryx	42	Lived:	35	0	0.250
		Died:	2	5	
Dorcas gazelle	92	Lived:	36	17	0.269
		Died:	14	25	

In humans, inbreeding increases the risk of spontaneous abortions, neonatal deaths, congenital deformities, and recessive genetic disorders. Although less common than in the past, inbreeding occurs in many regions of the world where social customs favor marriage between first cousins. Alan Bittles and James Neel analyzed data from numerous studies on different cultures. They found that the rate of child mortality (i.e., death in the first several years of life) varies dramatically from culture to culture. No matter what the baseline mortality rate for children of unrelated parents, however, children of first cousins virtually always have a higher death rate—typically by about 4.5 percentage points.

It is important to note that inbreeding is not always harmful. Indeed, inbreeding has long been recognized as a useful tool for breeders of domesticated plants and animals. When an inbreeding program is initiated, homozygosity increases, and some breeding stocks become fixed for favorable alleles and others for unfavorable alleles. By selecting the more viable and vigorous plants or animals, the proportion of individuals carrying desirable traits can be increased.

If members of two inbred lines are mated, hybrid offspring are often more vigorous in desirable traits than either of the parental lines. This phenomenon is called **hybrid vigor**. When such an approach was used in breeding programs established for maize, crop yields increased tremendously. However, hybrid vigor is highest in the F_1 generation and typically declines in subsequent generations due to additional allelic segregation and recombination. Consequently, the F_1 hybrids must be regenerated each time by crossing the original inbred parental lines.

Hybrid vigor has been explained in two ways. The first theory, the **dominance hypothesis**, incorporates the obvious reversal of inbreeding depression, which inevitably must occur in outcrossing. Consider a cross between two strains of maize with the following genotypes:

Strain A		Strain B		F_1
aaBBCCddee	$\times$	*AAbbccDDEE*	$\rightarrow$	*AaBbCcDdEe*

The F_1 hybrids are heterozygotes at all loci shown. Any deleterious recessive alleles present in the homozygous form in the parents are masked by the more favorable dominant alleles in the hybrids. Such masking is thought to cause hybrid vigor.

The second theory, **overdominance**, holds that in many cases the heterozygote is superior to either homozygote. This may relate to the fact that in the heterozygote two forms of a gene product may be present, providing a form of biochemical diversity. Thus, the cumulative effect of heterozygosity at many loci accounts for the hybrid vigor. Most likely, hybrid vigor results from a combination of phenomena explained by both hypotheses.

We have seen that nonrandom mating can drive genotype frequencies in a population away from their expected values under the Hardy–Weinberg law. This can indirectly affect the course of evolution. As in stocks purposely inbred by animal and plant breeders, inbreeding in a natural population may increase the frequency of homozygotes for a deleterious recessive allele. With domestic stocks, this increases the efficiency with which selection removes the deleterious allele from the population. Consequently, once deleterious genes are removed, inbreeding no longer causes problems.

GENETICS, TECHNOLOGY, AND SOCIETY

Tracking Our Genetic Footprints out of Africa

Where did we come from? Are we one human family with minor differences, or are we separated into races with profound and ancient roots? For millennia, our efforts to answer these questions invoked legends, mythologies, and the creation stories of our many religions. Over the last century, paleoanthropologists have applied a variety of sophisticated scientific tools to explore our origins and human kinships. The evolving story of our beginnings, based on modern genetics, is as fascinating and controversial as any creation myth.

Based on the physical traits and distribution of hominid fossils, most paleoanthropologists agree that a large-brained, tool-using hominid named *Homo erectus* appeared in east Africa about 2 million years ago. This species used simple stone tools, hunted but did not fish, did not build houses or fireplaces, and lacked ritual burial practices. About 1.7 million years ago, *H. erectus* spread into Eurasia and south Asia. Most scientists also agree that *H. erectus* likely developed into several hominid types including Neanderthals (in Europe) and Peking man or Java man (in Asia). These hominids were anatomically robust, with large heavy skeletons and skulls. Neanderthals and other *H. erectus* groups disappeared 50,000 to 30,000 years ago—around the same time that anatomically modern humans (*H. sapiens*) appeared all over the world.

It is at this point in our history—when ancient hominids gave way to lighter-skeletoned, anatomically modern humans—that controversy arises.

At present, there are two dominant hypotheses to explain our origins: The out-of-Africa and multiregional hypotheses. The multiregional hypothesis is based primarily on archaeological and fossil evidence. It proposes that *H. sapiens* developed gradually and simultaneously all over the world from existing *H. erectus* groups, including Neanderthals. Interbreeding between these groups eventually made *H. sapiens* a genetically homogeneous species. Natural selection over 1.5 million years then created the regional variants (races) that we see today. In the multiregional view, our genetic makeup should include contributions from Neanderthals and other *H. erectus* groups. In contrast, the out-of-Africa hypothesis, based primarily on genetic analyses of modern human populations, contends that *H. sapiens* evolved from the descendants of *H. erectus* in sub-Saharan Africa about 200,000 to 400,000 years ago. A small band of *H. sapiens* (fewer than 10,000) then left Africa, expanded, and migrated into Europe and Asia around 100,000 years ago. By about 60,000 years ago, populations of *H. sapiens* reached Australia and later migrated into North America. In the out-of-Africa model, *H. sapiens* replaced all the preexisting *H. erectus* types, without interbreeding. In this way, *H. sapiens* became the only species in the genus by about 30,000 years ago.

Although still contentious, most genetic evidence appears to support the out-of-Africa hypothesis. Humans all over the globe are remarkably similar genetically. DNA sequences from any two people chosen at random are 99.9% identical. There is more genetic identity between two persons chosen at random from a human population than there is between two chimpanzees chosen at random from a chimpanzee population. Interestingly, about 90% of the genetic differences that do exist occur between individuals, rather than between populations. This unusually high degree of genetic relatedness in all humans around the world supports the idea that our species arose recently from a small founding group of humans.

Studies of mitochondrial DNA sequences from current human populations reveal that the highest levels of genetic variation occur within African populations. Africans show twice the mitochondrial DNA-sequence diversity of non-Africans. This implies that the earliest branches of *H. sapiens* diverged in Africa and had a longer time to accumulate mitochondrial DNA mutations, which are thought to accumulate at a constant rate over time.

DNA sequences from mitochondrial, Y-chromosome, and chromosome-21 markers support the idea that our roots are in east Africa and the migration out of Africa occurred through Ethiopia, along the coast of the Arabian peninsula, and outward to Eurasia and Southeast Asia. Recent data based on DNA-sequence diversity in nuclear microsatellites further support the notion that humans migrated out of Africa and dispersed throughout the world from a small founding population. Sub-Saharan African populations show the highest levels of microsatellite heterozygosity, followed by those in the Middle East, Europe, East Asia, Oceania, and the Americas—in that order. Native American populations show about 15% less microsatellite heterozygosity than that seen in Africans.

By comparing DNA-sequence differences between populations around the world and by extrapolating back to a time when all sequences would have been the same, paleoanthropologists propose that modern *H. sapiens* developed from a small group in Africa between 200,000 and 400,000 years ago. The time of the out-of-Africa migration is calculated to be 50,000 to 100,000 years ago.

The recent sequencing of Neanderthal mitochondrial DNA shows that it is so different from ours that Neanderthals were likely a separate species and that Neanderthals and *H. sapiens* diverged about 600,000 years ago. Hence, it appears unlikely that Neanderthals and perhaps other *H. erectus* groups such as Peking man contributed significantly to the *H. sapiens* gene pool.

So, if all people on Earth are so similar genetically, how did we come to have such a range of physical differences, which some describe as racial differences? Many geneticists believe that the genetic changes responsible for these characteristics such as skin color and facial features could accumulate over short periods of time, especially if these characteristics are adaptive to particular climatic and geographic conditions.

As with any explanation of human origins, the out-of-Africa hypothesis is actively debated and may undergo mutation—or even extinction—over time. As methods to sequence DNA from ancient fossils improve, it may be possible to fill the gaps in our genetic pathway leading out of Africa and help us to resolve those age-old questions about our origins.

References

Cavalli-Sforza, L.L., and Feldman M.W., 2003. The application of molecular genetic approaches to the study of human evolution. *Nat. Gen. (Suppl.)* 33: 266–75.

Web Sites

Johanson, D. 2001. *Origins of modem humans: Multiregional or out of Africa?* [on-line.]

http://www.actionbioscience.orglevolutionlj ohanson.html

Chapter Summary

1. Populations evolve as a result of changes in an allele frequency at a number of loci over a period of time. Population genetics studies the factors driving change in allele frequencies and the amount and distribution of genetic variation in populations.
2. The Hardy–Weinberg law provides a simple mathematical model describing the relationship between allele frequency and genotype frequency in a population. This allows prediction of allele or genotype frequencies at a given locus in a population under a set of simple assumptions. If there is no selection, mutation, or migration, if the population is large, and if individuals mate at random, then allele frequencies will not change from one generation to the next.
3. The Hardy–Weinberg formula can be used to investigate whether or not a population is in evolutionary equilibrium at a given locus and to estimate the frequency of heterozygotes in a population from the frequency of homozygous recessives.
4. By specifying the conditions under which allele frequencies will not change, the Hardy–Weinberg law identifies the forces that can drive evolution in a population. Selection, mutation, migration, and genetic drift can cause change in allele frequencies.
5. Nonrandom mating alters genotype frequency but not allele frequency in a population. Inbreeding, or mating between relatives, is the form of nonrandom mating with the most significant impact: It increases the frequency of homozygotes in a population and decreases the frequency of heterozygotes.
6. Natural selection is the most powerful of the forces affecting allele frequency. The rate of change under natural selection depends on initial allele frequencies, selection intensity, and the relative fitness of different genotypes.
7. Mutation and migration introduce new alleles into a population, but usually have only small effects on allele frequencies. Persistence of new alleles in a population depends on the fitness they confer and the action of selection.
8. Genetic drift produces random change in allele frequencies as a result of gamete sampling. It can have a major impact in small populations.

Insights and Solutions

1. Tay–Sachs disease is caused by loss-of-function mutations in a gene on chromosome 15 that encodes a lysosomal enzyme. Tay–Sachs is inherited as an autosomal recessive condition. Among Ashkenazi Jews of central European ancestry, about 1 in 3600 children is born with the disease. What fraction of the individuals in this population are carriers?

Solution: If we let p represent the frequency of the wild-type enzyme allele and q the total frequency of recessive loss-of-function alleles, and if we assume that the population is in Hardy–Weinberg equilibrium, then the frequencies of the genotypes are given by p^2 for homozygous normal, $2pq$ for carriers, and q^2 for individuals with Tay–Sachs. The frequency of Tay–Sachs alleles is thus

$$q = \sqrt{q^2} = \sqrt{\frac{1}{3600}} = 0.017$$

Since $p + q = 1$, we have

$$p = 1 - q = 1 - 0.017 = 0.983$$

Therefore, we can estimate that the frequency of carriers is

$$2pq = 2(0.983)(0.017) = 0.033 \text{ or } 1 \text{ in } 30.$$

2. *Eugenics* is the term employed for the selective breeding of humans to bring about improvements in populations. As a eugenic measure, it has been suggested that individuals suffering from serious genetic disorders should be prevented (sometimes by force of law) from reproducing (by sterilization, if necessary) in order to reduce the frequency of the disorder in future generations. Suppose that such a recessive trait were present in the population at a frequency of 1 in 40,000 and that affected individuals did not reproduce. In 10 generations, or about 250 years, what would be the frequency of the condition? Are the eugenic measures effective in this case?

Solution: Let q represent the frequency of the recessive allele responsible for the disorder. Because the disorder is recessive, we can estimate that

$$q = \sqrt{\frac{1}{40{,}000}} = 0.005$$

If all affected individuals are prevented from reproducing, then in an evolutionary sense, the disorder is lethal: Affected individuals have zero fitness. This means that we can predict the frequency of the recessive allele 10 generations in the future by using the following equation:

$$q_g = q_0/(1 + gq_0)$$

Here, $q_0 = 0.005$, and $g = 10$, so we have

$$q_{10} = \frac{(0.005)}{[1 + (10 \times 0.005)]}$$
$$= 0.0048$$

If $q_{10} = 0.0048$, then the frequency of homozygous recessive individuals will be roughly

$$(q_{10})^2 = (0.0048)^2 = 0.000023 = \frac{1}{43{,}500}$$

The frequency of the genetic disorder has been reduced from 1 in 40,000 to 1 in 43,500 in 10 generations, indicating that this eugenic measure has limited effectiveness.

PROBLEMS AND DISCUSSION QUESTIONS

1. The ability to taste the compound PTC is controlled by a dominant allele *T*, while individuals homozygous for the recessive allele *t* are unable to taste PTC. In a genetics class of 125 students, 88 can taste PTC and 37 cannot. Calculate the frequency of the *T* and *t* alleles in this population and the frequency of the genotypes.
2. Calculate the frequencies of the *AA*, *Aa*, and *aa* genotypes after one generation if the initial population consists of 0.2 *AA*, 0.6 *Aa*, and 0.2 *aa* genotypes and meets the requirements of the Hardy–Weinberg relationship. What genotype frequencies will occur after a second generation?
3. Consider rare disorders in a population caused by an autosomal recessive mutation. From the frequencies of the disorder in the population given, calculate the percentage of heterozygous carriers:
   (a) 0.0064 (b) 0.000081 (c) 0.09
   (d) 0.01 (e) 0.10
4. What must be assumed in order to validate the answers in Problem 3?
5. In a population where only the total number of individuals with the dominant phenotype is known, how can you calculate the percentage of carriers and homozygous recessives?
6. Determine whether the following two sets of data represent populations that are in Hardy–Weinberg equilibrium (use χ^2 analysis if necessary):
   (a) *CCR5* genotypes: *1/1*, 60 percent; *1/Δ32*, 35.1 percent; *Δ32/Δ32*, 4.9 percent
   (b) Sickle-cell hemoglobin: *AA*, 75.6 percent; *AS*, 24.2 percent; *SS*, 0.2 percent
7. If 4 percent of a population in equilibrium expresses a recessive trait, what is the probability that the offspring of two individuals who do not express the trait will express it?
8. Consider a population in which the frequency of allele *A* is $p = 0.7$ and the frequency of allele *a* is $q = 0.3$, and where the alleles are codominant. What will be the allele frequencies after one generation if the following occurs?
   (a) $w_{AA} = 1$, $w_{Aa} = 0.9$, and $w_{aa} = 0.8$
   (b) $w_{AA} = 1$, $w_{Aa} = 0.95$, $w_{aa} = 0.9$
   (c) $w_{AA} = 1$, $w_{Aa} = 0.99$, $w_{aa} = 0.98$
   (d) $w_{AA} = 0.8$, $w_{Aa} = 1$, $w_{aa} = 0.8$
9. If the initial allele frequencies are $p = 0.5$ and $q = 0.5$ and allele *a* is a lethal recessive, what will be the frequencies after 1, 5, 10, 25, 100, and 1000 generations?
10. Under what circumstances might a lethal dominant allele persist in a population?
11. Determine the frequency of allele *A* in an island population after one generation of migration from the mainland under the following conditions:
    (a) $p_i = 0.6$; $p_m = 0.1$; $m = 0.2$
    (b) $p_i = 0.2$; $p_m = 0.7$; $m = 0.3$
    (c) $p_i = 0.1$; $p_m = 0.2$; $m = 0.1$
12. Assume that a recessive autosomal disorder occurs in 1 of 10,000 individuals (0.0001) in the general population and that in this population about 2 percent (0.02) of the individuals are carriers for the disorder. Estimate the probability of this disorder occurring in the offspring of a marriage between first cousins. Compare this probability to the population at large.
13. What is the basis of inbreeding depression?
14. Describe how inbreeding can be used in the domestication of plants and animals. Discuss the theories underlying these techniques.
15. Evaluate the following statement: Inbreeding increases the frequency of recessive alleles in a population.
16. In a breeding program to improve crop plants, which of the following mating systems should be employed to produce a homozygous line in the shortest possible time?
    (a) self-fertilization
    (b) brother–sister matings
    (c) first-cousin matings
    (d) random matings
    Illustrate your choice with pedigree diagrams.
17. If the albino phenotype occurs in 1/10,000 individuals in a population at equilibrium, and albinism is caused by an autosomal recessive allele *a*, calculate the frequency of
    (a) the recessive mutant allele
    (b) the normal dominant allele
    (c) heterozygotes in the population
    (d) matings between heterozygotes.
18. In a human population of 4,000 there are two males diagnosed with hemophilia. Assuming this population is 50 percent male and 50 percent female, how many female carriers of hemophilia will there be?
19. One of the first Mendelian traits identified in humans was a dominant condition known as *brachydactyly*. This gene causes an abnormal shortening of the fingers or toes (or both). At the time, it was thought by some that the dominant trait would spread until 75 percent of the population would be affected (because the phenotypic ratio of dominant to recessive is 3:1). Show that the reasoning is incorrect.
20. Achondroplasia is a dominant trait that causes a characteristic form of dwarfism. In a survey of 50,000 births, five infants with achondroplasia were identified. Three of the affected infants had affected parents, while two had normal parents. Calculate the mutation rate for achondroplasia and express the rate as the number of mutant genes per given number of gametes.
21. A prospective groom, who is normal, has a sister with cystic fibrosis (CF), an autosomal recessive disease. Their parents are normal. The brother plans to marry a woman who has no history of CF in her family. What is the probability that they will produce a CF child? They are both Caucasian and the overall frequency of CF in the Caucasian population is 1/2500—that is, 1 affected child per 2500. (Assume the population meets the Hardy–Weinberg assumptions.)
22. A botanist studying waterlilies in an isolated pond observed three leaf shapes in the population: round, arrowhead, and scalloped. Marker analysis of DNA from 125 individuals showed the round-leafed plants to be homozygous for allele *r1* while the plants with arrowhead leaves were homozygous for a different allele at the same locus, *r2*. Plants with scalloped leaves showed DNA profiles with both the *r1* and *r2* markers. Frequency of the *r1* marker was estimated at 0.81.

    If the botanist counted 20 plants with scalloped leaves in the pond, what is the inbreeding coefficient *F* for this population?
23. A farmer plants transgenic Bt corn that is genetically modified to produce its own insecticide. Of the corn borer larvae feeding on these Bt corn plants, only 10 percent survive unless they have at

least one copy of the dominant resistant allele B that confers resistance to the Bt insecticide. When the farmer first plants Bt corn, the frequency of the B resistance allele in the corn borer population is 0.02. What will be the frequency of the resistance allele after one generation of corn borers fed on Bt corn?

24. In an isolated population of 50 desert bighorn sheep, a mutant recessive allele *c* has been found to cause curled coats in both males and females. The normal dominant allele *C* produces straight coats. A biologist studying these sheep counts four with curled coats, and takes blood samples for DNA marker analysis, which reveals that 17 of the straight-coated sheep are carriers of the *c* allele. What is the inbreeding coefficient *F* for this population?

25. To increase genetic diversity in the bighorn sheep population described in Problem 24, 10 sheep are introduced from a population in a different region where the *c* mutation is absent. Assuming that random mating occurs between the original and the introduced sheep, and that the *c* allele is selectively neutral, what will be the frequency of *c* in the next generation?

Extra-Spicy Problems

26. A form of dwarfism known as Ellis–van Creveld syndrome was first discovered in the late 1930s, when Richard Ellis and Simon van Creveld shared a train compartment on the way to a pediatrics meeting. In the course of conversation, they discovered that they each had a patient with this syndrome. They published a description of the syndrome in 1940. Affected individuals have a short-limbed form of dwarfism and often have defects of the lips and teeth, and polydactyly (extra fingers). The largest pedigree for the condition was reported in an Old Order Amish population in eastern Pennsylvania by Victor McKusick and his colleagues (1964). In that community, about 5 per 1000 births are affected, and in the population of 8000, the observed frequency is 2 per 1000. All affected individuals have unaffected parents, and all affected cases can trace their ancestry to Samuel King and his wife, who arrived in the area in 1774. It is known that neither King nor his wife were affected with the disorder. There are no cases of the disorder in other Amish communities, such as those in Ohio or Indiana.

(a) From the information provided, derive the most likely mode of inheritance of this disorder. Using the Hardy–Weinberg law, calculate the frequency of the mutant allele in the population and the frequency of heterozygotes, assuming Hardy–Weinberg conditions.

(b) What is the most likely explanation for the high frequency of the disorder in the Pennsylvania Amish community and its absence in other Amish communities?

27. The graph below shows the variation in the frequency of a particular allele (*A*) that occurred over time in two relatively small, independent populations exposed to very similar environmental conditions. A student has analyzed the graph and concluded that the best explanation for these data is that the selection for allele *A* is occurring in Population 1. Is the student's conclusion correct? Explain.

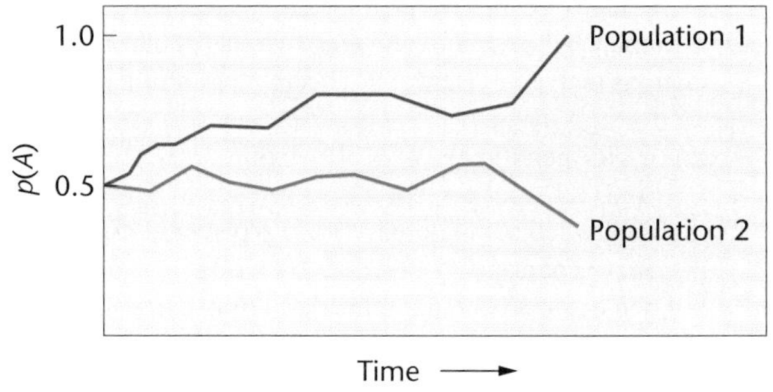

Selected Readings

Ansari-Lari, M.A., et al. 1997. The extent of genetic variation in the *CCR5* gene. *Nature Genetics* 16:221–22.

Ballou, J., and Ralls, K. 1982. Inbreeding and juvenile mortality in small populations of ungulates: A detailed analysis. *Biol. Conserv.* 24:239–72.

Barton, N.H., and Turelli, M. 1989. Evolutionary quantitative genetics: How little do we know? *Annu. Rev. Genet.* 23:337–70.

Bittles, A.H., and Neel, J.V. 1994. The costs of human inbreeding and their implications for variations at the DNA level. *Nature Genet.* 8:117–21.

Carrington, M., and Kissner, T., et al. 1997. Novel alleles of the chemokine-receptor gene *CCR5*. *Am. J. Hum. Genet.* 61:1261–67.

Chevillon, C., et al. 1995. Population structure and dynamics of selected genes in the mosquito *Culex pipiens*. *Evolution* 49: 997–1007.

Crow, J.F. 1986. *Basic concepts in population, quantitative and evolutionary genetics*. New York: W. H. Freeman.

Dudley, J.W. 1977. 76 generations of selection for oil and protein percentage in maize. In *Proceedings of the International Conference on Quantitative Genetics*, ed. E. Pollack, et al., pp. 459–73. Ames, IA: Iowa State University Press.

Fisher, R.A. 1930. *The genetical theory of natural selection*. Oxford, U.K.: Clarendon Press. (Reprinted Dover Press, 1958)

Freeman, S., and Herron, J.C. 2001. *Evolutionary analysis*. Upper Saddle River, NJ: Prentice Hall.

Freire-Maia, N. 1990. Five landmarks in inbreeding studies. *Am. J. Med. Genet.* 35:118–20.

Gayle, J.S. 1990. *Theoretical population genetics*. Boston, MA: Unwin-Hyman.

Grant, P.R. 1986. *Ecology and evolution of Darwin's finches*. Princeton, NJ: Princeton University Press.

Hartl, D.L. and Clark, A.G. 1997. *Principles of population genetics*. 3d ed. Sunderland, MA: Sinauer Associates.

Jones, J.S. 1981. How different are human races? *Nature* 293:188–90.

Karn, M.N., and Penrose, L.S. 1951. Birth weight and gestation time in relation to maternal age, parity and infant survival. *Ann. Eugen.* 16:147–64.

Kerr, W.E., and Wright, S. 1954. Experimental studies of the distribution of gene frequencies in very small populations of *Drosophila melanogaster. Evolution* 8:172–77.

Khoury, M.J., and Flanders, W.D. 1989. On the measurement of susceptibility to genetic factors. *Genet. Epidemiol.* 6:699-701.

Laikre, L., Ryman, N., and Thompson, E.A. 1993. Hereditary blindness in a captive wolf (*Canis lupus*) population: Frequency reduction of a deleterious allele in relation to gene conservation. *Conservation Biol.* 7:592–601.

Leibert, F., et al. 1998. The *DCCR5* mutation conferring protection against HIV-1 in Caucasian populations has a single and recent origin in northeastern Europe. *Hum. Molec. Genet.* 7:399–406.

Liu, R., et al. 1996. Homozygous defect in HIV-1 coreceptor accounts for resistance in some multiply-exposed individuals to HIV-1 infection. *Cell* 86:367–77.

Lucotte, G., and Mercier, G. 1998. Distribution of the *CCR5* gene 32-bp deletion in Europe. *J. Acquired Immune Deficiency Syndrome and Human Retrovirology* 19:174–77.

Markow, T., et al. 1993. HLA polymorphism in the Havasupai: Evidence for balancing selection. *Am. J. Hum. Genet.* 53:943–52.

Martinson, J.J., et al. 1997. Global distribution of the *CCR5* gene 32-bp deletion. *Nature Genet.* 16:100–103.

Mettler, L.E., Gregg, T., and Schaffer, H.E. 1988. *Population genetics and evolution.* 2d ed. Englewood Cliffs, NJ: Prentice Hall.

Parra, E.J., et al. 1998. Estimating African American admixture proportions by use of population-specific alleles. *Am. J. Hum. Genet.* 63:1839–51.

Pier, G.B., et al. 1998. *Salmonella typhi* uses CFTR to enter intestinal epithelial cells. *Nature* 393:79–82.

Quillent, C., et al. 1998. HIV-1-resistance phenotype conferred by combination of two separate inherited mutations of *CCR5* gene. *The Lancet* 351:14–18.

Renfrew, C. 1994. World linguistic diversity. *Sci. Am.* (Jan.) 270:116–23.

Roberts, D.F. 1988. Migration and genetic change. *Hum. Biol.* 60:521–39.

Samson, M., et al. 1996. Resistance to HIV-1 infection in Caucasian individuals bearing mutant alleles of the *CCR-5* chemokine receptor gene. *Nature* 382:722–25.

Spiess, E.B. 1989. *Genes in populations.* 2d ed. New York: Wiley.

Stephens, J.C., et al. 1998. Dating the origin of the *CCR5-Δ32* AIDS-resistance allele by the coalescence of haplotypes. *Am. J. Hum. Genet.* 62:1507–15.

Wallace, B. 1989. One selectionist's perspective. *Quart. Rev. Biol.* 64:127–45.

Woodworth, C.M., Leng, E.R., and Jugenheimer, R.W. 1952. Fifty generations of selection for protein and oil in corn. *Agron. J.* 44:60–66.

Yudin, N.S., et al. 1998. Distribution of *CCR5-Δ32* gene deletion across the Russian part of Eurasia. *Hum. Genet.* 102:695–98.

Evolutionary Genetics

Light and dark forms of the peppered moth Biston betularia on light tree bark. These moths are often cited as examples of rapid evolutionary change resulting from selective pressures during the industrial revolution in Great Britain.

Chapter Concepts

- Speciation can occur by transformation or by splitting gene pools.
- Most populations and species harbor considerable genetic variation.
- The genetic structure of populations changes across space and time.
- The definition of species is a great challenge for evolutionary biology.
- A reduction in gene flow between populations, accompanied by divergent selection or genetic drift, can lead to speciation.
- Genetic differences among populations or species are used to reconstruct evolutionary history.

As we embark on a discussion of genetics and evolution, consider the following scenarios:

(1) In late 1986, a Florida dentist tested positive for HIV. Several months later, he was diagnosed with AIDS. He continued to practice general dentistry for two more years, until one of his patients, a woman with no known risk factors, discovered that she, too, was infected with HIV. When the dentist publicly urged his other patients to have themselves tested, several more were found to be HIV positive. Did this dentist transmit HIV to his patients, or did the patients become infected by some other means?

(2) Fossil evidence indicates that Neanderthals (*Homo neanderthalensis*) lived in Europe and western Asia from 300,000 to 30,000 years ago. For some of that time, Neanderthals coexisted with anatomically modern humans (*Homo sapiens*). Because they lived in the same places at the same time, did Neanderthals and modern humans interbreed, so that descendants of the Neanderthals are alive today, or did the Neanderthals die off, so that their lineage is now extinct?

(3) The mitochondria that inhabit our cells have their own DNA. The organization and function of the mitochondrial genome bear many similarities to the organization and function of bacterial genomes. Did our mitochondria originate as free-living bacteria that took up residence inside other cells, or is the mitochondrial genome ultimately derived from nuclear chromosomes?

These scenarios pose three apparently unrelated questions. However, these questions in fact have something in common: They all can be addressed by using genetic data to reconstruct evolution. The acquisition and analysis of such data is the focus of the current chapter. We will use the methods of genetic analysis and the reconstruction of evolutionary history to address the above questions at the end of the chapter.

Evolution results from two processes: the transformation and splitting of lineages. Together, they produce a diversity of populations and species. In Chapter 25, we described the evolution of populations in terms of changes in allele frequencies and we outlined the forces that can cause such frequencies to change. In this chapter, we see how the population genetic processes of microevolution can be extended to macroevolutionary events. Mutation, migration, selection, and drift, individually and collectively, alter allele frequencies and bring about evolutionary divergence that eventually may result in the formation of species. The process of speciation is aided by environmental or ecological diversity. If a population is spread over a geographic range encompassing a number of ecologically distinct subenvironments with different selection pressures, the populations occupying these areas may gradually adapt and become genetically differentiated. Differentiated populations are dynamic. They may remain in existence, become extinct, reunite with each other, or continue to diverge until they form new species.

Because isolated populations and species accumulate genetic diversity, we can use patterns of genetic differences in various groups to reconstruct evolutionary history. After exploring the genetic structure of populations, their divergence across space and time, and the process of speciation, we will discuss how genetic data can be used to answer questions that have an evolutionary context, such as those in the three scenarios at the beginning of the chapter.

26.1 Speciation Can Occur by Transformation or by Splitting Gene Pools

Figure 26–1 shows an evolutionary tree, or **phylogeny**, that describes the history of several hypothetical lizard species. The passage of time is plotted horizontally as changes in the phenotype of the lizards occur and the lineages diverge or separate.

The history of these lizards begins with species 1. For some time, that species experiences evolutionary **stasis**; that is, it does not change. Species 1 then undergoes a process of steady transformation, called **phyletic evolution**, or **anagenesis**, and becomes species 2. During this process, there is only one species present at all times, and identification of the precise time at which species 1 becomes 2 is difficult. After species 2 forms, it undergoes **cladogenesis**, whereby it gives rise to two distinct and independent daughter species. Further transformation of these daughter species produces the species we see today, species 3 and species 4. These extant species are in many cases experimentally verifiable. That is, they may be tested to determine their reproductive compatibility (one species) or incompatibility (two species).

In his 1859 book, *On the Origin of Species*, Charles Darwin amassed evidence that all species derive from a single common ancestor by transformation and speciation:

> All living things have much in common, in their chemical composition, their germinal vesicles, their cellular structure, and their laws of growth and reproduction.... Therefore I should infer...that probably all the organic beings which have ever lived on this earth have descended from some one primordial form.

Everything biologists have since learned supports Darwin's conclusion that only one tree of life exists. To understand evolution, we must understand the mechanisms that transform one species into another and that split one species into two or more.

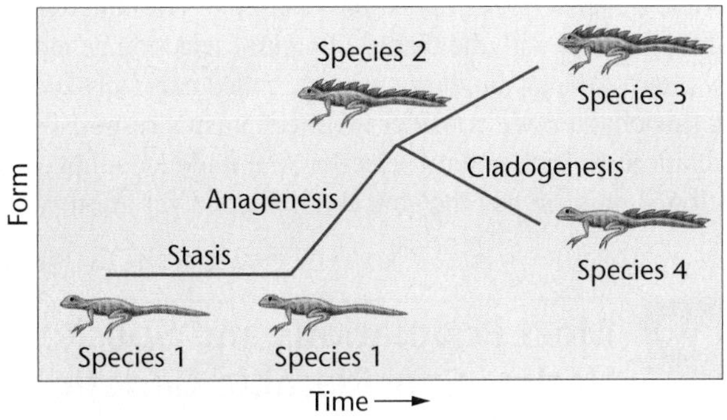

FIGURE 26–1 In phyletic evolution, or anagenesis, one species is transformed over time into another species. At all times, only one species exists. In cladogenesis, one species splits into two or more species.

Chapter 25 covered the mechanisms responsible for the transformation of species. Chief among them is natural selection, discovered independently by Darwin and by Alfred Russel Wallace. The Wallace–Darwin concept of natural selection can be summarized as follows:

1. Individuals of a species exhibit variations in phenotype—for example, showing differences in size, agility, coloration, defenses against enemies, ability to obtain food, courtship behaviors, flowering times, etc.
2. Many of these variations, even small and seemingly insignificant ones, are heritable and passed on to offspring.
3. Organisms tend to reproduce in an exponential fashion. More offspring are produced than can survive. This causes members of a species to engage in a struggle for survival, competing with other members of the community for scarce resources. Offspring also must avoid predators, and in sexually reproducing species, adults must compete for mates.
4. In the struggle for survival, individuals with particular phenotypes will be more successful than individuals with others, allowing the former to survive and reproduce at higher rates.

As a consequence of natural selection, species change. The phenotypes that confer improved ability to survive and reproduce become more common, and the phenotypes that confer poor prospects for survival and reproduction may eventually disappear. Under certain conditions populations that at one time could interbreed may lose that capability, thus segregating their adaptations into particular niches. If this adaptive selection is imposed on the hybrids produced by parents from different populations, selection favors the formation of two species where there was once only one.

Although Wallace and Darwin proposed that natural selection explains how evolution occurs, they could not explain either the origin of the variations that provided the raw material for evolution or how such variations are passed from parents to offspring. In the 20th century, as biologists applied the principles of Mendelian genetics to populations, the source of variation (mutation) and the mechanism of inheritance (segregation of alleles) were both explained. We now view evolution as due to changes in allele frequencies in populations over time. This union of population genetics with the theory of natural selection generated a new view of the evolutionary process, called *neo-Darwinism*.

In this chapter, we consider the mechanisms responsible for speciation. As with natural selection, our understanding of speciation is built on key insights about genetic variation.

26.2 Most Populations and Species Harbor Considerable Genetic Variation

At first glance, it seems that members of a well-adapted population should be highly homozygous because the most favorable allele at each locus has become fixed. Certainly, an examination of most populations of plants and animals reveals many phenotypic similarities among individuals. However, a large body of evidence indicates that most populations contain a high degree of heterozygosity. This built-in genetic diversity is concealed, so to speak, because it is not necessarily apparent in the phenotype; hence, detecting it is not a simple task. Nevertheless, with the use of the techniques discussed next, such investigation has been successful.

Artificial Selection

One way to determine whether there is genetic variation affecting a phenotypic character is to impose artificial selection on the character. A phenotype with no genetic variation will not respond to selection; if there is genetic variation, the phenotype will change over a few generations. A dramatic example of this is the domestic dog. The broad array of sizes, shapes, colors, and behaviors seen in different breeds of dogs all arose from selection on the genetic variation present in wild wolves, from which all domestic dogs are descended. On a shorter time scale, laboratory selection experiments on the fruit fly *Drosophila melanogaster* have caused significant changes over a few generations in almost every phenotype imaginable, including size, shape, developmental rate, fecundity, and behavior.

Protein Polymorphisms

Gel electrophoresis separates protein molecules on the basis of differences in size and electrical charge. If a nucleotide variation in a structural gene results in the substitution of a charged amino acid, such as glutamic acid, for an uncharged amino acid, such as glycine, the net electrical charge on the protein will be altered. This difference in charge can be detected as a change in the rate at which proteins migrate through an electrical field. In the mid-1960s, John Hubby and Richard Lewontin used gel electrophoresis to measure protein variation in natural populations of *Drosophila*, and Harry Harris used the same techniques to measure human variation. In subsequent years, researchers have used the technique to study genetic variation in a wide range of organisms (Table 26.1), although subsequently developed DNA techniques have in many cases replaced the use of enzyme electrophoresis.

The electrophoretically distinct forms of a protein produced by different alleles are called *allozymes*. As shown in the table, a surprisingly large percentage of loci examined from diverse species produce distinct allozymes. Of the populations listed, approximately 30 loci per species were examined, and about 30 percent of the loci were polymorphic, with an average of 10 percent allozyme heterozygosity per diploid genome. These estimates clearly support the notion that organismal genotypes harbor vast amounts of DNA-based variability.

These values apply only to genetic variation detectable by altered protein migration in an electric field. Electrophoresis probably detects only about 30 percent of the actual variation due to amino acid substitutions, because many substitutions do not change the net electric charge on the molecule. Richard Lewontin estimated that about two thirds of all loci in a population are polymorphic. In any individual within the population, about one third of the loci exhibit genetic variation in the

TABLE 26.1 ALLOZYME HETEROZYGOSITY AT THE PROTEIN LEVEL

Species	Populations Studied	Loci Examined	Polymorphic Loci* per Population (%)	Heterozygotes per Locus (%)
Homo sapiens (humans)	1	71	28	6.7
Mus musculus (mouse)	4	41	29	9.1
Drosophila pseudoobscura (fruit fly)	10	24	43	12.8
Limulus polyphemus (horseshoe crab)	4	25	25	6.1

* A polymorphic locus is one for which a population harbors more than one allele.

Source: From Lewontin, 1974, p. 117.

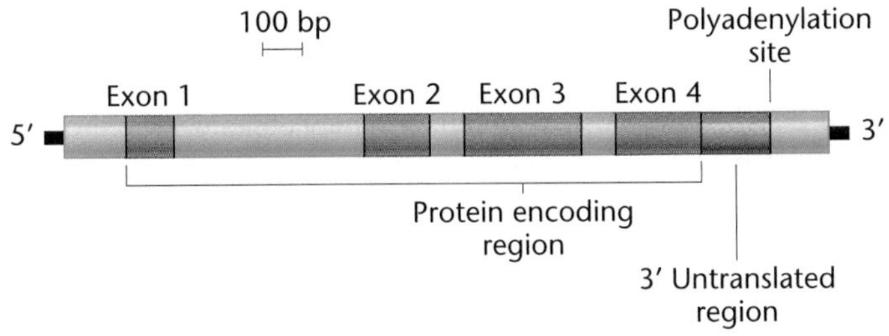

FIGURE 26–2 Organization of the *Adh* locus of *Drosophila melanogaster*.

form of heterozygosity. The significance of genetic variation detected by electrophoresis is controversial. Some argue that allozymes are functionally equivalent and therefore do not play any role in adaptive evolution. We address this argument later in this section.

Variations in Nucleotide Sequence

The most direct way to estimate genetic variation is to compare the nucleotide sequences of genes carried by individuals in a population. With the development of techniques for cloning and sequencing DNA, and with the proliferation of genome projects, nucleotide sequence variations have been catalogued for an increasing number of genes and genomes. In an early study, Alec Jeffreys used restriction enzymes to detect polymorphisms in 60 unrelated individuals. His aim was to estimate the total number of DNA sequence variants in humans. His results showed that within the genes of the β-globin cluster, 1 in 100 base pairs showed polymorphic variation. If that region is representative of the genome, this study indicates that at least 3×10^7 nucleotide variants per genome are possible.

In another study, Martin Kreitman examined the *alcohol dehydrogenase* locus (*Adh*) in *Drosophila melanogaster* (Figure 26–2). This locus encodes two allozymic variants: the *Adh-f* and the *Adh-s* alleles. These differ by only a single amino acid (thr versus lys at codon 192). To determine whether the amount of genetic variation detectable at the protein level (one amino acid difference) corresponds to the variation at the nucleotide level, Kreitman cloned and sequenced *Adh* loci from five natural populations of *Drosophila*.

The 11 cloned loci contained 43 nucleotide variations from the consensus *Adh* sequence of 2721 base pairs. These variations are distributed throughout the gene: 14 in the exon coding regions, 18 in the introns, and 11 in the untranslated and flanking regions. Of the 14 variations in coding regions, only one leads to an amino acid replacement—the one in codon 192, producing the two observed electrophoretic variants. The other 13 coding region nucleotide substitutions do not lead to amino acid replacements. Are the differences in the number of allozyme and nucleotide variants the result of natural selection? If so, of what significance is this fact? Let's examine these questions.

Among the most intensively studied loci to date is the locus encoding the cystic fibrosis transmembrane conductance regulator (CFTR). Recessive loss-of-function mutations in the *CFTR* locus cause **cystic fibrosis**, a disease with symptoms including salty skin and the production of an excess of thick mucus in the lungs, which leads to susceptibility to bacterial infections. Geneticists have examined the *CFTR* locus in some 30,000 chromosomes from individuals with cystic fibrosis and have found more than 500 different mutations that can cause the disease. Among these mutations are missense mutations, amino acid deletions, nonsense mutations, frameshifts, and splice defects.

Figure 26–3 shows a map of the 27 exons in the *CFTR* locus, with most exons identified by function. The histogram above the map shows the locations of some of the disease causing mutations and the number of copies of each that have been found. A single mutation, a 3-bp deletion in exon 10 called $\Delta F508$ accounts for 67 percent of all mutant cystic fibrosis alleles, but several other mutations are found in at least 100 of the chromosomes surveyed. In populations of European ancestry, between 1 in 44 and 1 in 20 individuals are heterozygous carriers of mutant alleles. Note that Figure 26–3 includes only the sequence variants that alter the function of the CFTR protein. There are undoubtedly many more *CFTR* alleles with silent sequence variants that do not change the structure of the protein and that do not affect its function.

Studies of other organisms, including the rat, the mouse, and the mustard plant *Arabadopsis thaliana* have produced similar estimates of nucleotide diversity in various genes. These studies

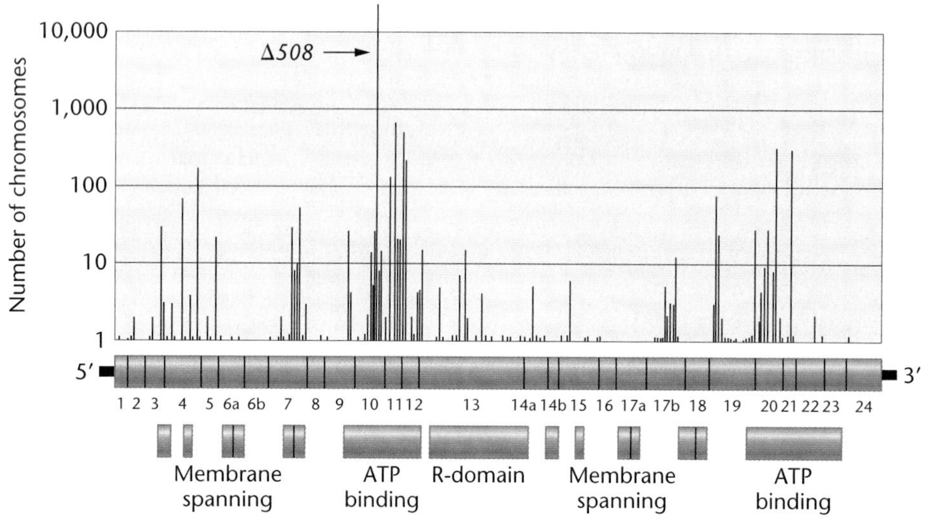

FIGURE 26–3 The locations of disease-causing mutations in the cystic fibrosis gene. The histogram shows the number of copies of each mutation geneticists have found. (The vertical axis is on a logarithmic scale.) The genetic map below the histogram shows the locations and relative sizes of the 27 exons of the *CFTR* locus. The boxes at the bottom indicate the functions of different domains of the CFTR protein.

indicate that there is an enormous reservoir of genetic variability within most populations and that, at the level of DNA, most, and perhaps all, genes exhibit diversity from individual to individual.

Now solve this

Problem 26.11 on page 676 asks what types of nucleotide substitutions will not be detected by electrophoretic studies of proteins.

Hint: Detection of enzymes in gels relies on their conversion of their natural substrate. Allozymes that migrate at different speeds in the gel matrix do so because of their variable charge.

HOW DO WE KNOW?

How do we determine how much genetic variation exists in a population?

Explaining the High Level of Genetic Variation in Populations

As mentioned earlier, the finding that populations harbor considerable genetic variation at the amino acid and nucleotide levels came as a surprise to many evolutionary biologists. The early consensus was that selection would favor a single optimal (wild-type) allele at each locus and that, as a result, populations would have high levels of homozygosity. This expectation was obviously wrong, and considerable research and argument ensued concerning the forces that maintain genetic variation.

One view about the high degree of genetic variation argues that it reflects primarily the action of mutation and genetic drift. The **neutral theory** of molecular evolution, proposed by Motoo Kimura, argues that mutations leading to amino acid substitutions are usually detrimental, with a very small fraction that are favorable. Some mutations are neutral or functionally equivalent to the allele that is replaced. Those polymorphisms that are favorable or detrimental are preserved or removed from the population, respectively, by natural selection. However, the frequency of the neutral alleles in a population will be determined by mutation rates and random genetic drift. Some neutral mutations will drift to fixation in the population; other neutral mutations will be lost. At any given time, the population may contain several neutral alleles at any particular locus. The diversity of alleles at most polymorphic loci does not, however, reflect the action of natural selection, but instead is a function of population size (larger populations have more variation) and the fractions of mutations that are neutral.

The alternative explanation for variation is **natural selection**. There are several examples in which enzyme or protein polymorphisms are maintained by adaptation to certain environmental conditions. The well-known advantage of sickle-cell anemia heterozygotes when infected by malarial parasites is such an example, and another was discussed in Chapter 25, when we considered evidence that polymorphism at the *CFTR* locus may reflect the superior fitness of heterozygotes in areas where typhoid fever is common.

Fitness differences of a fraction of a percent would be sufficient to maintain a polymorphism, but this would be difficult to measure. Current data are therefore insufficient to determine what fraction of molecular genetic variation is neutral and what fraction is subject to selection. The neutral theory nonetheless serves a crucial function: By pointing out that some genetic variation is expected simply as a result of mutation and drift, the neutral theory provides a working hypothesis for studies of molecular evolution. In other words, biologists must find positive evidence that selection is acting on allele frequencies at a particular locus before they can reject the simpler assumption that only mutation and drift are at work.

26.3 The Genetic Structure of Populations Changes across Space and Time

As population geneticists discovered that most populations harbor considerable genetic diversity, they also found that the genetic structure of populations varies across space and time. To illustrate, we consider studies on *Drosophila pseudoobscura* conducted by Theodosius Dobzhansky and his colleagues. This

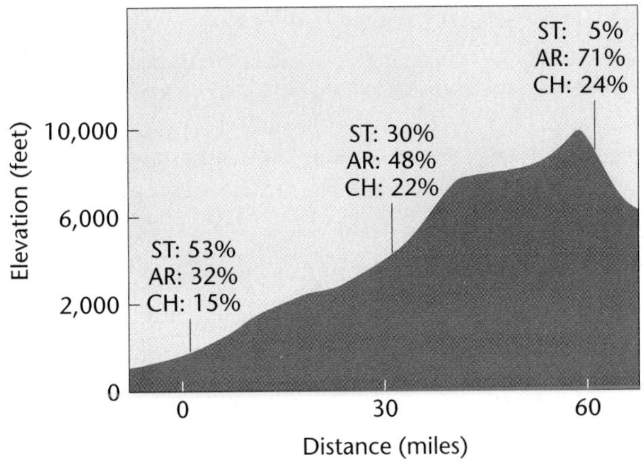

FIGURE 26–4 Inversions in chromosome 3 of *D. pseudoobscura* at different elevations in the Sierra Nevada range near Yosemite National Park.

species is found over a wide range of environmental habitats, including the western and southwestern United States. Although the flies throughout this range are morphologically similar, Dobzhansky's team discovered that populations from different locations vary in the arrangement of genes on chromosome 3. They found several different inversions in this chromosome that can be detected by loop formations in larval polytene chromosomes. Each inversion sequence is named after the locale in which it was first discovered (e.g., AR = Arrowhead, British Columbia, and CH = Chiricahua Mountains, Arizona). The inversion sequences were compared with one standard sequence, arbitrarily designated ST.

Figure 26–4 compares the frequencies of three arrangements at different elevations in the Sierra Nevada Mountains in California. The ST arrangement is most common at low elevations; at 8000 feet, AR is the most common and ST least common. In the populations studied, the frequency of the CH arrangement gradually increases with elevation, a phenomenon that is probably the result of natural selection and that parallels the gradual environmental changes occurring at ascending elevations, such as decreasing air temperature.

Dobzhansky's team also found that, for populations collected at a single site throughout the year, inversion frequencies change as well. That is, cyclic variations in chromosome arrangements occur as the seasons change, as shown in Figure 26–5. Such variation was consistently observed over a period of several years. The frequency of ST always declines during the spring, and that of CH increases in the spring.

To test the hypothesis that this cyclic change is a response to natural selection, Dobzhansky and his group devised a laboratory experiment. They constructed large population cages from which samples of *D. pseudoobscura* could be removed periodically and studied. They began with a population that was 88 percent CH and 12 percent ST. The flies were maintained at 25°C and sampled over a 1-year period. As shown in Figure 26–6, the frequency of ST increased gradually until it was present at a level of 70 percent. At that point, an equilibrium between ST and CH was reached. When the same experiment was performed at 16°C, no change in inversion frequency occurred. The researchers concluded that the equilibrium reached at 25°C was in response to the elevated temperature, the only variable in the experiment.

The results of the study indicate that a balance in the frequency of the two inversions and the gene arrangements they contain is superior to either inversion by itself. The equilibrium reached in the experiment presumably represents the highest mean fitness in the population under controlled laboratory conditions. This interpretation of the experiment suggests that natural selection is the driving force maintaining the diversity in chromosome 3 inversions.

In a more extensive study, Dobzhansky and his colleagues sampled *D. pseudoobscura* populations over a broad geographic range. They found 22 different chromosome arrangements in populations from 12 locations. In Figure 26–7, the frequencies of five of these inversions are shown according to geographic location. The differences are largely quantitative, with most populations differing only in the relative frequencies of inversions. Collectively, Dobzhansky's data show that the genetic structure of *D. pseudoobscura* populations changes from place to place and from one time to another. At least some of this variation in population genetic structure is the result of natural selection.

Another example of how the genetic structure of a species varies among populations is provided by the work of Dennis A. Powers and Patricia Schulte on the mummichog (*Fundulus heteroclitus*), a small fish (5 to 10 cm long) that lives in inlets,

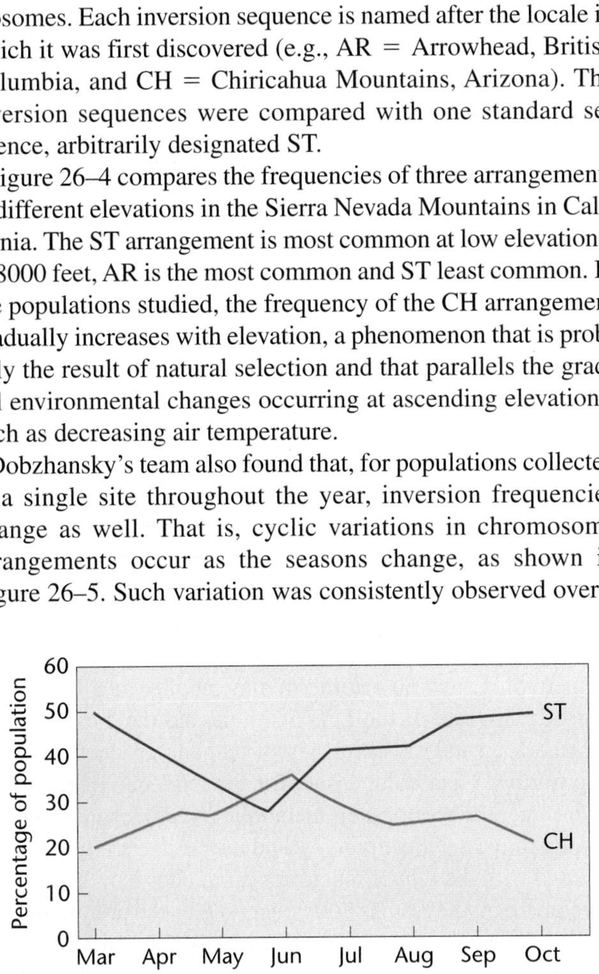

FIGURE 26–5 Changes in the ST and CH arrangements in *D. pseudoobscura* throughout the year.

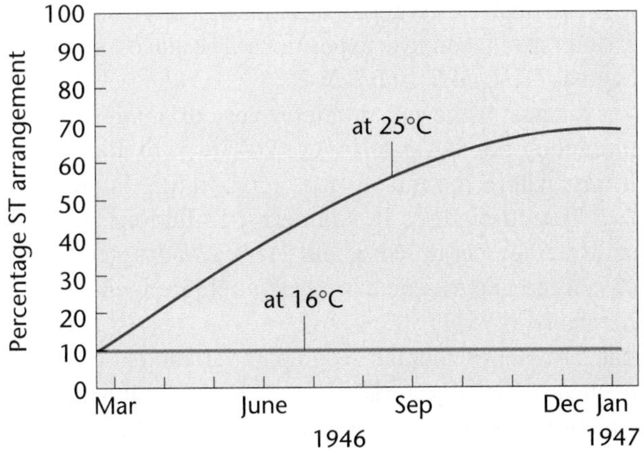

FIGURE 26–6 Increase in the ST arrangement of *D. pseudoobscura* in population cages under laboratory conditions.

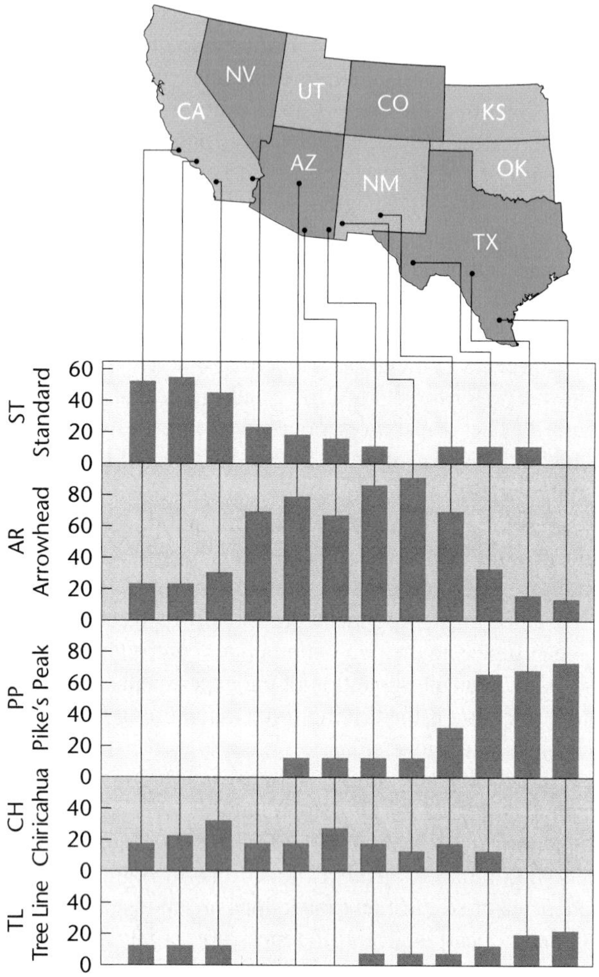

FIGURE 26–7 Relative frequencies (percentages) of five chromosomal inversions in *D. pseudoobscura* in different geographic regions.

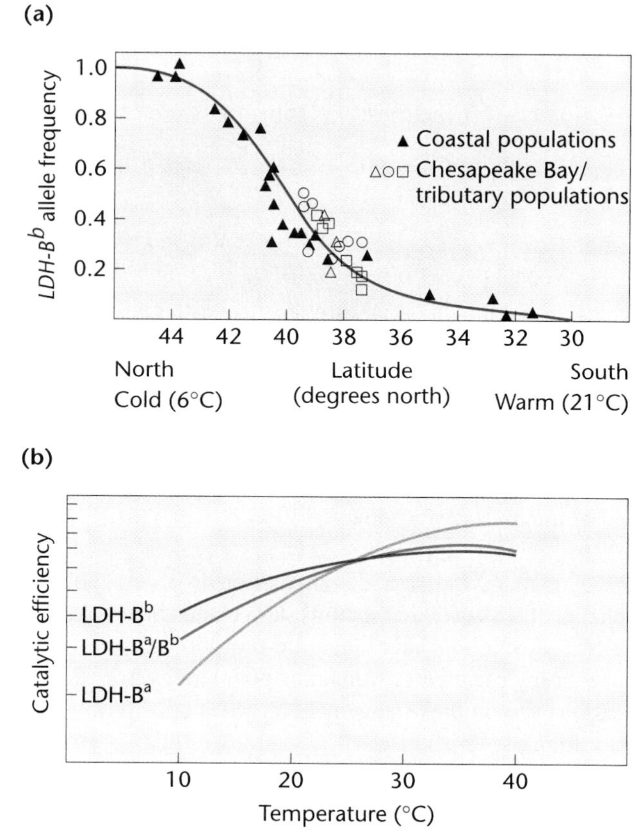

FIGURE 26–8 Variation in genetic structure among mummichog populations. (a) Frequencies of the B^b allele in populations along the Atlantic coast of North America. (b) Catalytic efficiency of LDH-B allozymes as a function of temperature.

bays, and estuaries along the Atlantic coast of North America from Florida to Newfoundland. These workers measured allele frequencies at the locus encoding the enzyme lactate dehydrogenase-B (LDH-B), which is made in the liver, heart, and red skeletal muscle. LDH-B converts lactate to pyruvate and is thus pivotal in both the manufacture of glucose and aerobic metabolism. Two allozymes of LDH-B are seen on gels; they differ at two amino acid positions. The alleles encoding the allozymes are *Ldh-*B^a and *Ldh-*B^b.

Frequencies of the *Ldh-B* alleles vary dramatically among mummichog populations [Figure 26–8(a)]. In northern populations, where the mean water temperature is about 6°C, *Ldh-*B^b predominates. In southern populations, where the mean water temperature is about 21°C, *Ldh-*B^a predominates. Between the geographic extremes, allele frequencies are intermediate.

To determine whether the geographic variation in *Ldh-B* allele frequencies is due to natural selection, Powers and Schulte studied the biochemical properties of the LDH-B allozymes. They found that the enzyme encoded by *Ldh-*B^b has higher catalytic efficiency at low temperatures, whereas the gene product of *Ldh-*B^a is more efficient at high temperatures [Figure 26–8(b)]. A mixture of the two forms has intermediate efficiency at all temperatures.

In addition to the functional differences between the LDH-B allozymes, the *Ldh-B* alleles have nucleotide sequence differences in their regulatory regions. One such difference results in the *Ldh-*B^b allele's rate of transcription being more than twice that of the *Ldh-*B^a allele. As a result, northern fish have higher concentrations of the LDH-B enzyme in their cells.

Differences in the catalytic efficiency and transciption rate of the two *Ldh-B* alleles are consistent with the hypothesis that mummichog populations are adapted to the temperatures at which they live. The fish are ectotherms: Their body temperature is determined by their environment. In general, low body temperatures slow an ectotherm's metabolic rate. The higher transcription rate of the *Ldh-*B^b allele and the superior low-temperature catalytic efficiency of its gene product appear to help northern fish compensate for the tendency of their cold environment to reduce their metabolic rate. The superior high-temperature catalytic efficiency and lower transcription rate associated with the *Ldh-*B^a allele appear to allow southern fish to economize on the resources devoted to glucose production and aerobic metabolism. On the basis of this and other evidence, Powers and Schulte suggest that differences among mummichog populations in *Ldh-B* allele frequencies are the result of natural selection.

How Do We Know?

What evidence has been obtained from laboratory studies to indicate that natural selection is the cause of genetic differences among natural populations of a species?

26.4 The Definition of Species Is a Great Challenge for Evolutionary Biology

The genetic processes that work on populations to change them across space and time can result in the transformation and splitting of lineages, ultimately leading to speciation. However, it is difficult to determine the exact time when a new species forms, partly because of our difficulty in defining what a species is. Charles Darwin, who was unaware of the underlying genetic basis for the difference between species, relied on a morphological approach to species definition, in which closely related species are separated by distinct gaps in the values of morphological characters. This approach is useful with both extant species and those that we know only from the fossil record, but it also poses several problems. Even within recognized species there are often large gaps in character values and, conversely, there are many examples of morphologically very similar species that nonetheless can not interbreed and so appear to be distinct.

Problems with the morphological approach and our growing understanding of population genetics and evolution led to a new conception of species in the mid-twentieth century, called the biological species concept. In this view, a species is defined as *a group of interbreeding or potentially interbreeding populations that is reproductively isolated in nature from all other such groups*. As such, species are distinct gene pools, so that speciation occurs when one gene pool divides into two or more separate gene pools. Changes in morphology, physiology, and adaptation to an ecological niche may also occur, but are not considered necessary components of speciation, which can take place gradually or within a few generations.

Zoologists in particular favor the biological species concept since it is easily applied to animals, most of whom have distinct sexes and exchange genes through sexual reproduction. However, this interpretation also has problems. Many species, such as most prokaryotes, do not reproduce sexually and so individuals within the species do not interbreed. Yet clearly, these individuals are not all different species! Also, about 20 percent of plants and 10 percent of birds and butterflies easily hybridize yet are recognized as distinct species, and many plants regularly self-pollinate. The presence in genomes of genes resulting from horizontal gene transfer also suggests that hybridization is not uncommon, and this argues against defining species solely by reproductive isolation. Also, it is impossible to use the biological species concept for extinct species since we are not able to determine the presence of reproductive isolation in fossils. Even for extant species it may be impossible to test for reproductive isolation in a suite of closely related putative species.

In practice, many taxonomists (those scientists that describe species and study their evolutionary relationships) rely on the type of morphological characteristics that Darwin emphasized. With the advent of modern genetics, a great wealth of new molecular and sequence information can be added to these data sets to enhance our ability to identify distinct species via physical characteristics. This has produced a third type of species concepts based on phylogeny, which argue that a species consists of those organisms that are monophyletic and share one or more uniquely derived characteristics. This approach has its problems as well, because we cannot be certain that we have determined the correct relationships between all the possible relatives in a phylogeny in order to identify monophyletic clades.

Finally, some biologists use an ecological species concept, in which species are defined as groups of individuals that live in the same type of environment and have shared ecological requirements, filling the same niche. The main problem here is that there are examples in which populations that seem to be the same species (i.e., they readily interbreed with one another) have very distinct niches.

The scientific community is engaged in an ongoing discussion about the definition of a species. It is one of the most important and challenging questions in evolutionary biology, but may have no clear answer. As it stands now, biologists use the definition of species that is most sensible and useful for their own field of scientific inquiry. The biological species concept is most closely allied to the population genetic mechanisms of evolution. Its emphasis on reproductive isolation not only helps define species, but also provides a focus for thinking about the process of speciation. Whether or not complete reproductive isolation is a requirement for separating one species from another taxonomically, it is clear that a reduction in gene flow between populations is a powerful evolutionary force.

Now solve this

In Problem 26.4 on page 660 you are asked how phenotypically similar species can be identified as separate species, rather than as the same species with some phenotypic variation.

Hint: The species concept that you rely upon will influence the answer to this question.

26.5 A Reduction in Gene Flow between Populations, Accompanied by Divergent Selection or Genetic Drift, Can Lead to Speciation

We have learned that most populations harbor considerable genetic variation and that different populations within a species may have different alleles or allele frequencies at a variety of loci. The genetic divergence of these populations can be caused by natural selection, genetic drift, or both. In Chapter 25, we saw that the migration of individuals between populations, together with the gene flow that accompanies that migration, tends to homogenize allele frequencies among populations. In other words, migration counteracts the tendency of populations to diverge.

When gene flow between populations is reduced or absent, the populations may diverge to the point that members of one population are no longer able to interbreed successfully with members of the other. When populations reach the point where they are reproductively isolated from one another, they have become different species, according to the biological species concept.

The biological barriers that prevent or reduce interbreeding between populations are called **reproductive isolating mechanisms**, classified in Table 26.2. These mechanisms may be ecological, physiological, behavioral, seasonal or temporal, or mechanical.

Prezygotic isolating mechanisms prevent individuals from mating in the first place. Individuals from different populations may not find each other at the right time, may not recognize each other as suitable mates, or may try to mate, but find that they are unable to do so.

Postzygotic isolating mechanisms create reproductive isolation even when the members of two populations are willing and able to mate with each other. For example, genetic divergence may have reached the stage where the viability or fertility of hybrids is reduced. Hybrid zygotes may be formed, but all or most may be inviable. Alternatively, the hybrids may be viable, but be sterile or suffer from reduced fertility. Yet again, the hybrids themselves may be fertile, but their progeny may have lowered viability or fertility. These postzygotic mechanisms act at or beyond the level of the zygote and are generated by genetic divergence.

In some models of speciation, postzygotic isolation evolves first and is followed by prezygotic isolation. Postzygotic isolating mechanisms waste gametes and zygotes and lower the reproductive fitness of hybrid survivors. Selection will therefore favor the spread of alleles that reduce the formation of hybrids, leading to the development of prezygotic isolating mechanisms, which in turn prevent interbreeding and the formation of hybrid zygotes and offspring. In animal evolution, the most effective prezygotic mechanism is behavioral isolation, involving courtship behavior.

Genes that control development of reproductive organs may play a vital role in the evolution of major new lineages, even beyond speciation. Recent research on *Arabidopsis thaliana* and other taxa indicates that evolutionary changes in MADS-box homeotic gene regulation and function have been a driving force in the evolution of floral form in the flowering plants. This is key to reproductive isolation and divergence, since successful pollination and therefore mating within a plant species depends on quite specific floral anatomy. Research on *A. thaliana* in the laboratories of Vivian Irish and her colleagues has focused on APETALA3 (AP3) and PISTILLATA (PI), floral homeotic genes that encode MADS-domain containing transcription factors used to determine stamen and petal identities during floral whorl development. They have shown that these genes are expressed in different amounts in the perianths (petals/sepals) of groups of angiosperms with divergent floral anatomy. This finding suggests that changes simply in the expression of developmental genes may change floral morphology evolutionarily. The Irish group also has investigated evolutionary changes in the function of these genes. They grew transgenic *A. thaliana* plants constructed from an *AP3* mutant with its carboxy-terminus replaced by an ancestral 'paleoAP3' carboxy-terminal motif obtained from an evolutionarily older plant lineage that does not produce petals in its flowers. (They had shown that the carboxy-terminal motif is required for the protein's function.) The transgenic plants made stamens in their flowers, but not petals, suggesting that the function of the developmental gene product changed when the lineage that led to *A. thaliana* diverged from the ancestral lineage. Such so-called evo-devo collaborations between evolutionary biologists and developmental geneticists are leading to exciting new

TABLE 26.2 REPRODUCTIVE ISOLATING MECHANISMS

Prezygotic Mechanisms (prevent fertilization and zygote formation)

1. **Geographic or ecological:** The populations live in the same regions but occupy different habitats
2. **Seasonal or temporal:** The populations live in the same regions but are sexually mature at different times
3. **Behavioral (only in animals):** The populations are isolated by different and incompatible behavior before mating
4. **Mechanical:** Cross-fertilization is prevented or restricted by differences in reproductive structures (genitalia in animals, flowers in plants)
5. **Physiological:** Gametes fail to survive in alien reproductive tracts

Postzygotic Mechanisms (fertilization takes place and hybrid zygotes are formed, but these are nonviable or give rise to weak or sterile hybrids)

1. **Hybrid nonviability or weakness**
2. **Developmental hybrid sterility:** Hybrids are sterile because gonads develop abnormally or meiosis breaks down before completion
3. **Segregational hybrid sterility:** Hybrids are sterile because of abnormal segregation into gametes of whole chromosomes, chromosome segments, or combinations of genes
4. **F_2 breakdown:** F_1 hybrids are normal, vigorous, and fertile, but the F_2 contains many weak or sterile individuals

Source: From G. Ledyard Stebbins, *Processes of organic evolution*, 3rd ed., copyright 1977, p. 143. Reprinted by permission of Prentice Hall, Upper Saddle River, NJ.

discoveries about the pivotal role of molecular evolution in developmental genes during major evolutionary events such as the origin of new lineages and species.

Examples of Speciation

In this section, we consider two examples of speciation, one from a laboratory study and the other from a field study. Diane Dodd and her colleagues studied the evolution of digestive physiology in *Drosophila pseudoobscura*. They collected flies from a wild population and established several separate laboratory populations. Some of the laboratory populations were raised on starch-based medium and others on maltose-based medium. Both food sources were stressful for the flies. It was only after several months of evolution by natural selection that the populations became adapted to their artificial diets and began to thrive.

Dodd's team wanted to know whether the starch-adapted populations and the maltose-adapted populations, which diverged under strong selection and in the absence of gene flow, had become different species. Roughly a year after the populations were established, a series of mating trials were performed. For each trial, 48 flies were placed in a bottle: 12 males and 12 females from a starch-adapted population and 12 males and 12 females from a maltose-adapted population. The researchers then noted which flies mated.

Dodd's group predicted that if the populations adapted to different media had speciated, then the flies would prefer to mate with members of their own population. If the populations had not speciated, then the flies would mate at random. The results appear in Table 26.3. Roughly 600 of the 900 matings observed were between males and females from the same population. In other words, the differently adapted fly populations showed partial premating isolation. Dodd and her colleagues concluded that the populations had begun to speciate, but had not yet completed the process.

The Isthmus of Panama, which created a land bridge connecting North and South America and simultaneously separated the Caribbean Sea from the Pacific Ocean, formed roughly 3 million

TABLE 26.3 **INCIPIENT SPECIATION IN LABORATORY POPULATIONS OF *DROSOPHILA PSEUDOOBSCURA* (NUMBER OF MATINGS INVOLVING EACH KIND OF MALE–FEMALE PAIR)**

	Female	
	Starch-adapted	Maltose-adapted
Male		
Starch-adapted	290	153
Maltose-adapted	149	312

Source: Compiled from Dodd, D.M.B. 1989. Reproductive isolation as a consequence of adaptive divergence in *Drosophila pseudoobscura*. *Evolution* 43:1308–11.

FIGURE 26–9 A snapping shrimp (genus *Alpheus*).

years ago. Nancy Knowlton and colleagues took advantage of a natural experiment that the formation of the Isthmus of Panama had performed on several species of snapping shrimps (Figure 26–9). After identifying seven Caribbean species of snapping shrimp, they matched each one with a Pacific species to form a pair. Members of each pair were more similar to each other in structure and appearance than either was to any other species in its own ocean. Analysis of allozyme allele frequencies and mitochondrial DNA sequences confirmed that the members of each pair were one another's closest genetic relatives.

The Knowlton team's interpretation of these data is that, prior to the formation of the isthmus, the ancestors of each pair were a single species. When the isthmus closed, the seven ancestral species were each divided into two separate populations, one in the Caribbean and the other in the Pacific.

Meeting in a dish in Knowlton's lab for the first time in 3 million years, would Caribbean and Pacific members of a species pair recognize each other as suitable mates? Knowlton placed males and females of a species pair together and noted their behavior toward each other. She then calculated the relative inclination of Caribbean–Pacific couples to mate versus that of Caribbean–Caribbean or Pacific–Pacific couples. For three of the seven species pairs, transoceanic couples refused to mate altogether. For the other four species pairs, transoceanic couples were 33, 45, 67, and 86 percent as likely to mate with each other as were same-ocean pairs. Of the same-ocean couples that mated, 60 percent produced viable clutches of eggs. Of the transoceanic couples that mated, only 1 percent produced viable clutches. We can conclude from these results that 3 million years of separation has resulted in complete or nearly complete speciation, involving strong pre- and postzygotic isolating mechanisms for all seven species pairs.

The Minimum Genetic Divergence Required for Speciation

How much genetic separation is required between two populations before they become different species? We shall consider two examples, one from an insect and the other from a plant, which demonstrate that in some cases the answer is "not very much."

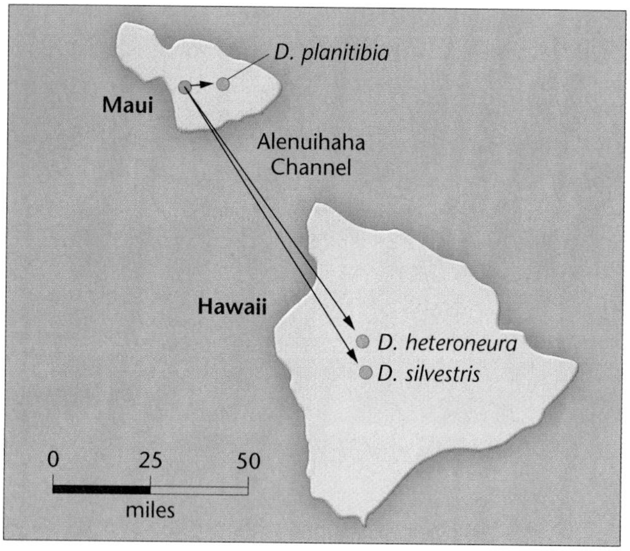

FIGURE 26–10 Proposed pathway for Hawaii's colonization by members of the *D. planitibia* species. The purple circle represents a population ancestral to the three present-day species.

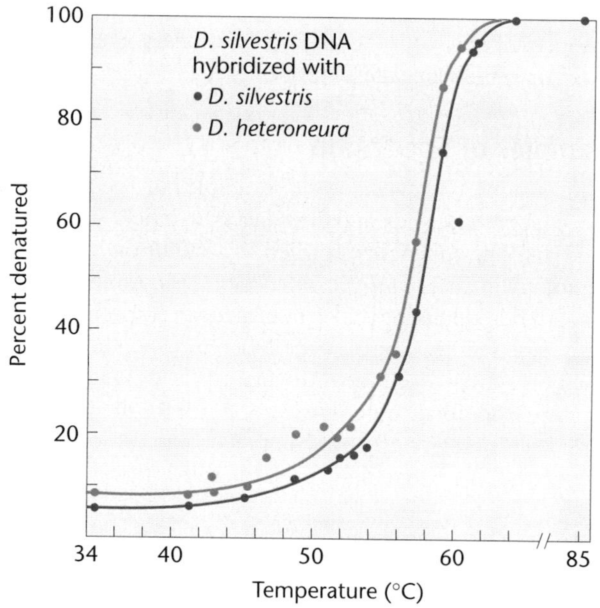

FIGURE 26–12 Nucleotide sequence diversity in the *Drosophila planitibia* species complex. The shift to the left by the heterologous hybrids indicates the degree of nucleotide sequence divergence.

Researchers estimate that *Drosophila heteroneura* and *D. silvestris*, found only on the island of Hawaii, diverged from a common ancestral species only about 300,000 years ago. The two species are thought to be descended from *D. planitibia* colonists from the older island of Maui (Figure 26–10). The two species are clearly separated from each other by different and incompatible courtship and mating behaviors (a prezygotic isolating mechanism), by morphology, and by body and wing pigmentation (Figure 26–11).

In spite of their morphological and behavioral divergence, demonstrating significant differences between these species in chromosomal inversion patterns or protein polymorphisms is difficult. DNA hybridization studies carried out on the two species indicate that the sequence diversity between them is only about 0.55 percent (Figure 26–12). Thus, nucleotide sequence diversity may precede the development of protein or chromosomal polymorphisms.

Genetic evidence suggests that the differences between *D. heteroneura* and *D. silvestris* are controlled by a relatively small number of genes. For example, as few as 15 to 19 major loci may be responsible for the morphological differences between the species, demonstrating that the process of speciation likely involves only a small number of genes.

Studies using two closely related species of the monkey flower, a plant that grows in the Rocky Mountains and areas west, confirm that species can be separated by only a few genetic differences. One species, *Mimulus cardinalis*, is fertilized by hummingbirds and does not interbreed with *Mimulus lewisii*, which is fertilized by bumblebees. H. D. Bradshaw and his colleagues studied genetic differences related to reproduction in the two species—namely, flower shape, size, and color and nectar production. For each trait, a difference in a single gene provided at least 25 percent of the variation observed among laboratory-created hybrids. *M. cardinalis* makes 80 times more nectar than does *M. lewisii*, and a single gene is responsible for at least half the difference. A single gene also controls a large part of the differences in flower color between the two species (Figure 26–13). In this case, as in the Hawaiian *Drosophila*, species differences can be traced to a relatively small number of genes.

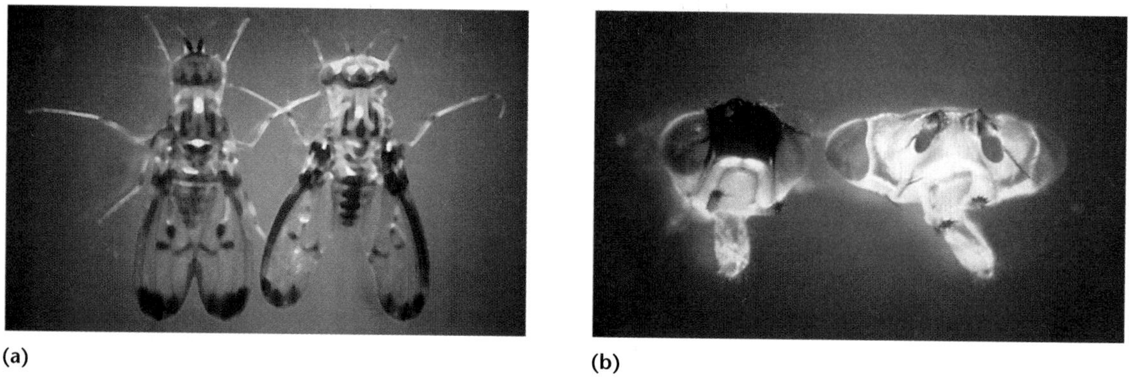

(a) (b)

FIGURE 26–11 (a) Differences in pigmentation patterns in *D. sylvestris* (left) and *D. heteroneura* (right). (b) Head morphology in *D. sylvestris* (left) and *D. heteroneura* (right).

FIGURE 26–13 Flowers of two closely related species of monkey flowers, (a) *Mimulus cardinalis* and (b) *Mimulus lewisii.*

HOW DO WE KNOW?

How can the minimum genetic divergence between two species be determined?

In at Least Some Instances, Speciation Is Rapid

How much time is required for speciation? In many cases, speciation takes place slowly over a long period. In other cases, however, speciation can be surprisingly rapid.

The Rift Valley lakes of East Africa support hundreds of species of cichlid fish. Lake Victoria (Figure 26–14), for example, has more than 400 species. Cichlids have some morphological diversity, but, more dramatically, they are highly specialized for different niches (Figure 26–15). Some eat algae floating on the water's surface, whereas others are bottom feeders, insect feeders, mollusk eaters, and predators on other fish species. Lake Tanganyika has a similar array of species. Genetic analyses indicate that the species in a given lake are all more closely related to each other than to species from other lakes. The implication is that most or all the species in, say, Lake Tanganyika are descended from a single common ancestor and that they evolved within their home lake.

Lake Victoria is between 250,000 and 750,000 years old, and there is evidence that the lake may have dried out nearly completely less than 14,000 years ago. Is it possible that the 400 species in the lake today evolved from a common ancestral species in less than 14,000 years?

In a study of cichlid origins in Lake Tanganyika, Norihiro Okada and colleagues examined the insertion of a novel family of DNA sequences called short interspersed elements (SINES) (discussed in Chapter 12) into the genomes of cichlid species in Lake Tanganyika. SINES are repetitive DNA sequences inserted into the genome at random. They are a type of retroposon, and integration of a SINE at a locus is an irreversible event. If a SINE is present at the same locus in the genome of all species examined, this is strong evidence that all of those species descend from a common ancestor. Using a SINE called AFC, Okada's team screened 33 species of cichlids belonging to four groups (called species tribes). In each tribe, the SINE was present at all the sites tested, indicating that the species in each tribe are descended from a single ancestral species (Figure 26–16). If further research using SINES and other molecular markers produces similar results with the Lake Victoria species, it means that the 400 cichlid species in this lake evolved in less than 14,000 years. If confirmed, this finding would represent the fastest evolutionary radiation ever documented in vertebrates.

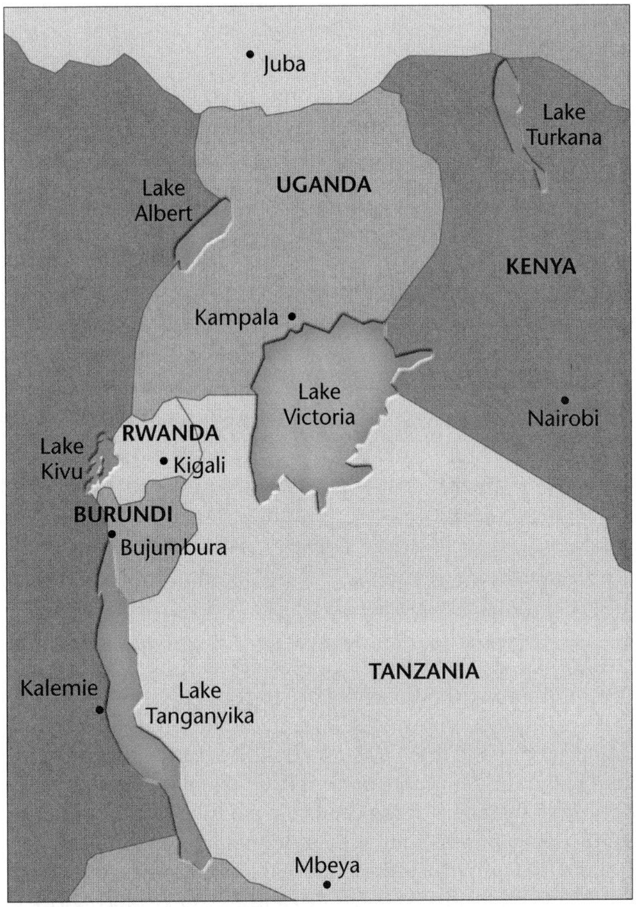

FIGURE 26–14 Lake Victoria in the Rift Valley of East Africa is home to more than 400 species of cichlids.

Even faster speciation is possible through the mechanism of polyploidy. The formation of animal species by

FIGURE 26–15 Cichlids occupy a diverse array of niches, and each species is specialized for a distinct food source.

polyploidy is rare, but polyploidy is an important factor in plant evolution. It is estimated that one-half of all flowering plants have evolved by this mechanism. One form of polyploidy is allopolyploidy (see Chapter 8), produced by doubling the chromosome number in a hybrid formed by crossing two species.

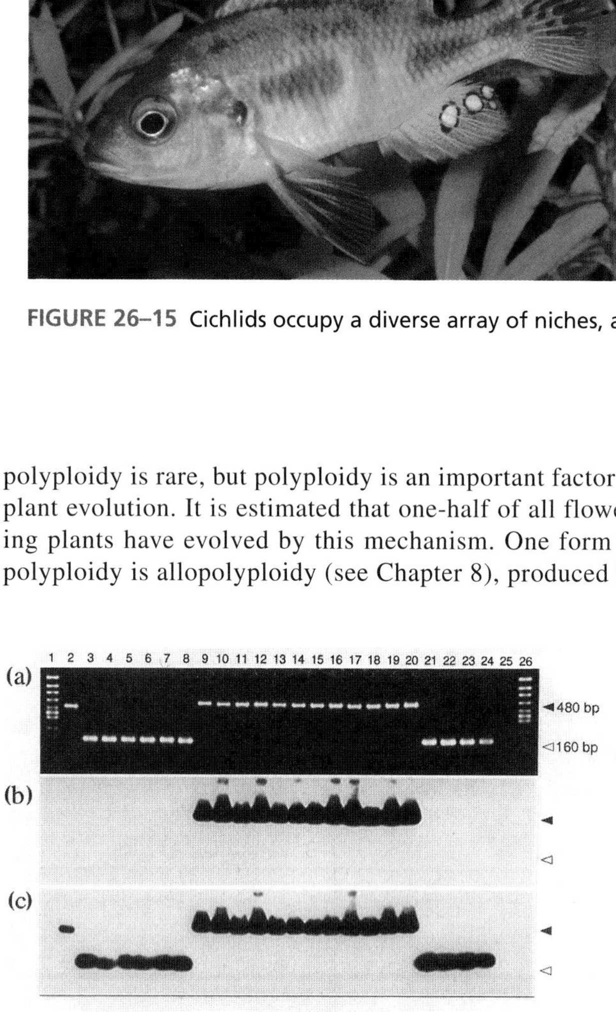

FIGURE 26–16 Using repetitive DNA elements to trace species relationships in cichlids from Lake Tanganyika. (a) Photograph of an agarose gel showing PCR fragments generated from primers that flank the AFC family of SINES. Large fragments containing this family are present in all samples of genomic DNA from members of the Lamprologini tribe of cichlids (lanes 9–20). DNA from the species in lane 2 has a similar, but shorter fragment, which may represent another repetitive sequence. DNA from other species (lanes 3–8 and 21–24) produce short, non-SINE-containing fragments. (b) A Southern blot of the gel from (a) probed with DNA from the AFC family of SINES. AFC is present in DNA from all species in the Lamprologini tribe (lanes 9–20), but not in DNA from other species (lanes 2–8 and 21–24). (c) A second Southern blot of the gel from (a) probed with the genomic sequence at which the AFC SINE inserts. All species examined (lanes 2–24) contain the insertion sequence. The larger fragments in Lamprologini DNA (lanes 9–20) correspond to those in the previous blot, showing that the SINE is inserted at the same site in all tribal species. In sum, these data show that the AFC SINE is present only in members of this tribe, is present in all member species, and is inserted at the same site in all cases. These results are interpreted as showing a common origin for the species of the tribe in question.

If two species of related plants have the genetic constitution SS and TT (where S and T represent the haploid set of chromosomes in each species), the hybrid would have the chromosome constitution ST. Normally, such a plant would be sterile, because there are few or no homologous chromosome pairs and aberrations arise during meiosis. However, if the hybrid undergoes a spontaneous doubling of its chromosome number, a tetraploid SSTT plant would be produced. This chromosomal aberration might occur during mitosis in somatic tissue, giving rise to a partially tetraploid plant that produces some tetraploid flowers. Alternatively, aberrant meiotic events may produce ST gametes that, when fertilized, would yield SSTT zygotes. The SSTT plants would be fertile because they would possess homologous chromosomes producing viable ST gametes. This new, true-breeding tetraploid would have a combination of characteristics derived from the parental species and would be reproductively isolated from them because hybrids would be triploids and consequently sterile.

The tobacco plant *Nicotiana tabacum* ($2n = 48$) is the result of a doubling of the chromosome number in the hybrid between *N. otophora* ($2n = 24$) and *N. silvestris* ($2n = 24$) (Figure 26–17). The origin of *N. tabacum* is an example of virtually instantaneous speciation.

Now solve this

In Problem 26.26 on page 661 you are asked to interpret whether neutral sequence polymorphisms found among the cichlid species in African lakes help to explain their evolution.

Hint: If a sequence is neutral, then it should not be subject to any form of selection through time, and should only be subject to random mutation and genetic drift.

FIGURE 26–17 The cultivated tobacco plant *Nicotiana tabacum* is the result of hybridization between two other species.

26.6 We Can Use Genetic Differences among Populations or Species to Reconstruct Evolutionary History

Early in this chapter, we noted that the course of evolutionary history involves the transformation of one species into another and the splitting of species into two or more new species. Our examples have demonstrated that speciation is associated with changes in the genetic structure of populations and with genetic divergence. Therefore, we should be able to use genetic differences among species to reconstruct their evolutionary histories.

In an important early example of phylogeny reconstruction, W. M. Fitch and E. Margoliash assembled data on the amino acid sequence for cytochrome c in a variety of organisms. **Cytochrome c** is a respiratory pigment found in the mitochondria of eukaryotes, and its amino acid sequence has evolved very slowly. For example, the amino acid sequence in humans and chimpanzees is identical, and humans and rhesus monkeys show only one amino acid difference. We should expect little evolution in a protein whose proper function is essential for fitness, but such utter similarity is remarkable, considering that the fossil record indicates that the lines leading to humans and monkeys diverged from a common ancestral species approximately 20 million years ago.

Column (a) of Table 26.4 shows the number of amino acid differences between cytochrome c in humans and a variety of other species. The table is broadly consistent with our intuitions about how closely related we are to these other species. For example, we are more closely related to other mammals than we are to insects, and we are more closely related to insects than we are to fungi. Likewise, our cytochrome c differs in 10 amino acids from that of dogs, in 24 amino acids from that of moths, and in 38 amino acids from that of yeast.

However, more than one nucleotide change may be required to change a given amino acid. When the nucleotide changes necessary for all amino acid differences observed in a protein are totaled, the **minimal mutational distance** between the genes of any two species is established. Column (b) in Table 26.4 shows such an analysis of the genes encoding cytochrome c. As expected, these values are larger than the corresponding number of amino acids separating humans from the other nine organisms listed.

Fitch used data on the minimal mutational distances between the cytochrome c genes of 19 organisms to reconstruct their evolutionary history. The result is an estimate of the evolutionary tree, or phylogeny, that unites the species (Figure 26–18). The black dots on the tips of the twigs represent the extant species, which are connected to the inferred common ancestors represented by red dots. The ancestral species evolved and diverged to produce the modern species. The common ancestors are connected to still earlier common ancestors, culminating in a single common ancestor for all the species on the tree, represented by the red dot on the extreme left.

A Method for Estimating Evolutionary Trees from Genetic Data

Many methods use data on genetic differences to estimate phylogenies. It is beyond the scope of this chapter to review all of them. Instead, we shall present just one method, called the *u*nweighted *p*air *g*roup *m*ethod using *a*rithmetic averages, or UPGMA. This method is not the most powerful one for

TABLE 26.4 AMINO ACID DIFFERENCES AND THE MINIMAL MUTATIONAL DISTANCES BETWEEN CYTOCHROME C IN HUMANS AND OTHER ORGANISMS

Organism	(a) Amino Acid Differences	(b) Minimal Mutational Distance
Human	0	0
Chimpanzee	0	0
Rhesus monkey	1	1
Rabbit	9	12
Pig	10	13
Dog	10	13
Horse	12	17
Penguin	11	18
Moth	24	36
Yeast	38	56

Source: From W.M. Fitch and E. Margoliash, Construction of phylogenetic trees, *Science* 155:279–84, 20 January 1967. Copyright 1967 by the American Association for the Advancement of Science.

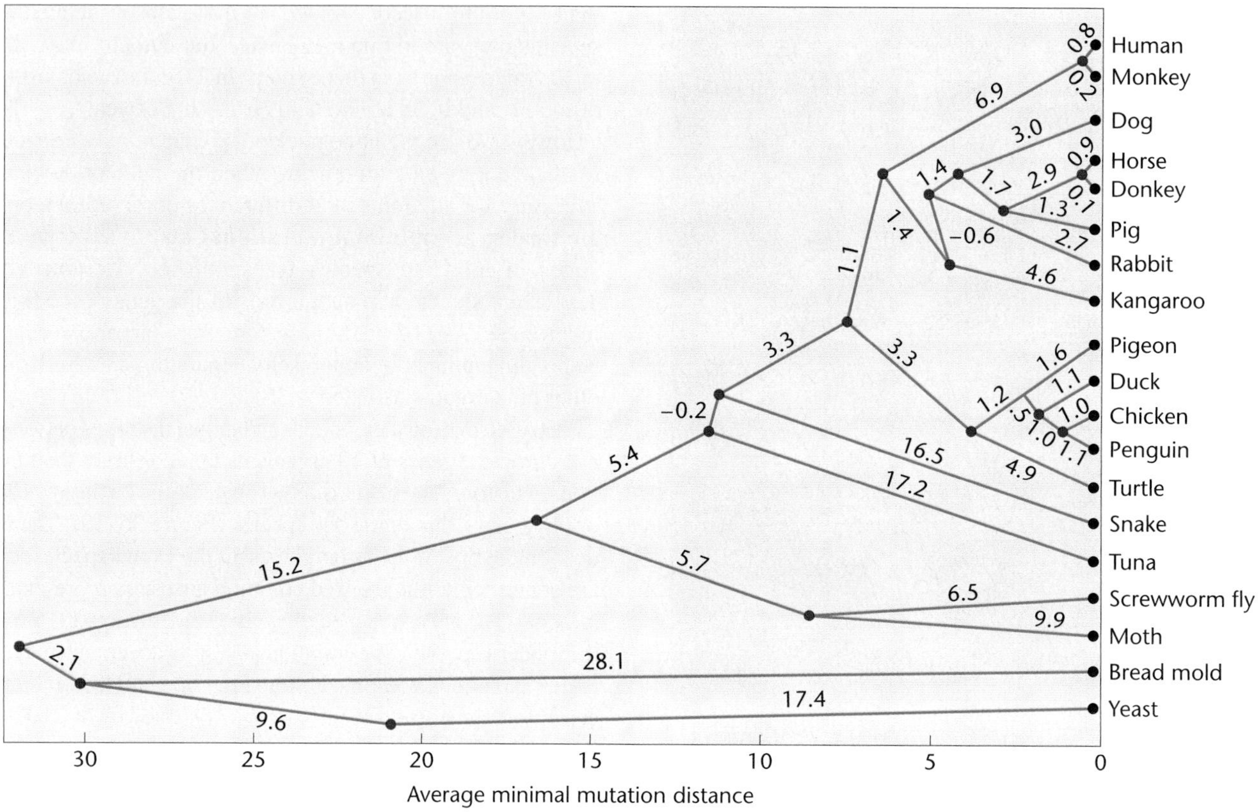

FIGURE 26–18 Phylogeny constructed by comparing homologies in cytochrome c amino acid sequences. *Reprinted with permission from Fitch, W.M., and Margoliash E. 1967. Construction of phylogenetic trees. Science 279:279-84. Figure 2. Copyright 1967 AAAS.*

estimating phylogenies from genetic data, but, in spite of its lengthy name, it is intuitively straightforward. Furthermore, UPGMA works reasonably well under many circumstances.

The starting point for UPGMA is a table of genetic distances among a group of species [Figure 26–19(a)]. To demonstrate the technique, we will use data from DNA hybridization studies by Charles Sibley and Jon Alquist, who computed genetic distances between humans and four species of apes: the common chimpanzee, the gorilla, the siamang, and the common gibbon.

The method uses the following steps:

1. Search for the smallest genetic distance between any pair of species. In Figure 26–19(a), this distance is 1.628, the distance between human and chimpanzee. Once identified, the corresponding species pair is placed on neighboring branches of an evolutionary tree [Figure 26–19(b)]. The length of each branch is half the genetic distance between the species (1.628/2), so the branches connecting human and chimpanzee to their common ancestor are each about 0.81.

2. Recalculate the genetic distances between this species pair and all other species [Figure 26–19(c)]. The genetic distance between the human–chimpanzee (Hu–Ch) cluster* and the other species is the average of the distances between each member of the cluster and the other species. For example, the genetic distance between the human–chimp cluster and the gorilla is the average of the human–gorilla distance and the chimp–gorilla distance, or

$$(2.267 + 2.21)/2 = 2.2385.$$

3. Repeat steps 1 and 2 until all the species have been added to the tree.

The smallest distance in the recalculated table is 1.95, the distance between siamang and gibbon [Figure 26–19(c)]. To build the next section of the tree, siamang and gibbon are placed on neighboring branches, with lengths equal to half of 1.95, or 0.98. [Figure 26–19(d)].

Recalculating the table, we now find that there are two clusters plus the gorilla [Figure 26–19(e)]. The genetic distance between the human–chimp cluster and the siamang–gibbon cluster is the average of four distances: human–siamang, human–gibbon, chimp–siamang, and chimp–gibbon.

Now the smallest genetic distance is 2.239, the distance between the human–chimp cluster and the gorilla. We therefore add a gorilla branch to the tree and connect it to the common ancestor of the human–chimp cluster [Figure 26–19(f)]. The branches are drawn so that the distance between the tips of any two branches in the human–chimp–gorilla cluster is 2.239.

Recalculating the table for the final time [Figure 26–19(g)], we find that the genetic distance between the human–chimp–gorilla cluster and the siamang–gibbon cluster is the average of six genetic distances: human–siamang, human–gibbon, chimp–siamang,

---

*We say "cluster" even though what we really have is a pair; the reason is that we can (and will) form larger groups and it is convenient to call them by the same, more inclusive name—ergo *cluster*.

(a)

	Human	Chimp	Gorilla	Siamang	Gibbon
Human	–				
Chimp	**1.628**	–			
Gorilla	2.267	2.21	–		
Siamang	4.7	5.133	4.543	–	
Gibbon	4.779	4.76	4.753	1.95	–

(c)

	Hu-Ch	Gorilla	Siamang	Gibbon
Hu-Ch	–			
Gorilla	2.2385	–		
Siamang	4.9165	4.543	–	
Gibbon	4.7695	4.753	**1.95**	–

(e)

	Hu-Ch	Gorilla	Si-Gi
Hu-Ch	–		
Gorilla	**2.239**	–	
Si-Gi	4.843	4.648	–

(g)

	Hu-Ch-Go	Si-Gi
Hu-Ch-Go	–	
Si-Gi	**4.778**	–

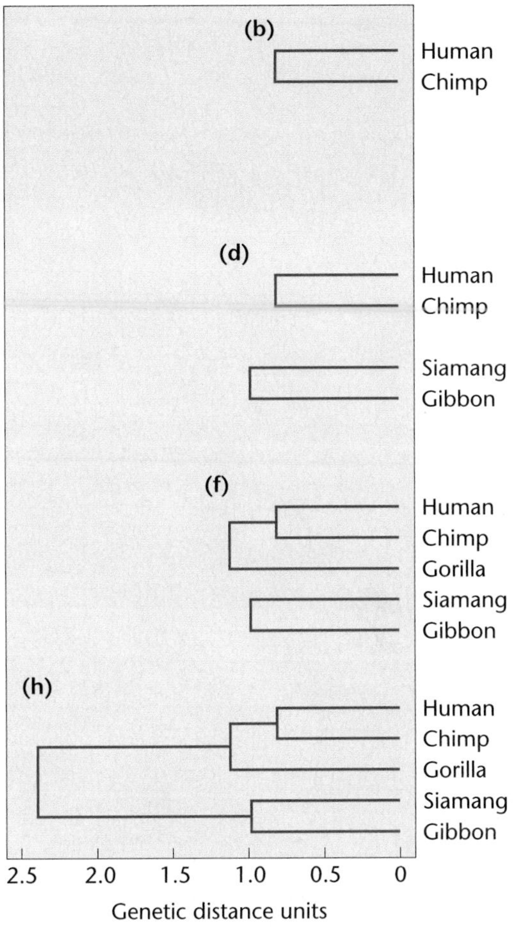

FIGURE 26–19 Phylogeny reconstruction by UPGMA.

chimp–gibbon, gorilla–siamang, and gorilla–gibbon. This distance is 4.778, which allows us to complete our evolutionary tree [Figure 26–19(h)]. The tree indicates that humans and chimpanzees are one another's closest relatives. That is not to say that humans evolved from chimpanzees; rather, humans and chimpanzees share a more recent common ancestor than either shares with other species on the tree.

The chief shortcoming of UPGMA is that it provides no means of determining how well its trees fit the data, compared with other possible trees. For example, how much better does the tree we just created fit our data than a tree that shows chimpanzees and gorillas as closest relatives?

Evolutionary geneticists have developed several techniques for searching the set of all possible trees connecting a group of species and calculating the relative performance of each. One set of techniques, called **parsimony** methods, compares trees by using the minimum number of evolutionary changes each requires and then selects the simplest possible tree. Often, however, there is a set of equally parsimonious trees, so these methods often must be combined with other criteria to develop the most accurate phylogeny. Another set of analytical tools, called **maximum-likelihood** methods, starts with a model of the evolutionary process, calculates how likely it is that evolution will produce each possible tree under the model, and selects the most likely tree. It is important to keep in mind that a phylogeny produced from data on genetic distances is not necessarily the way species actually evolved, but instead is a reasonable estimate of their evolutionary histories, based on the methods used. Phylogenies based on combined data sets from several loci are usually more reliable than those based on a single locus.

Molecular Clocks

In many cases, we would like to estimate not only which members of a set of species are most closely related, but also when their common ancestors lived. Sometimes we can do so, thanks to **molecular clocks**—amino acid sequences or nucleotide sequences in which evolutionary changes accumulate at a constant rate over time.

Research by Walter M. Fitch and colleagues on the influenza A virus shows how molecular clocks are used. Fitch and his associates sequenced part of the hemagglutinin gene from flu viruses that had been isolated at different times over a 20-year period. They calculated the number of nucleotide differences among the various viruses and constructed an evolutionary tree [Figure 26–20(b)]. Most strains have gone extinct, leaving no descendants among the more recently isolated viruses. Fitch's group then plotted the number of nucleotide substitutions between the first virus and each subsequent virus against the year in which the virus was isolated [Figure 26–20(a)]. The points all fall very close to a straight line, indicating that nucleotide substitutions in this gene have accumulated at a steady rate.

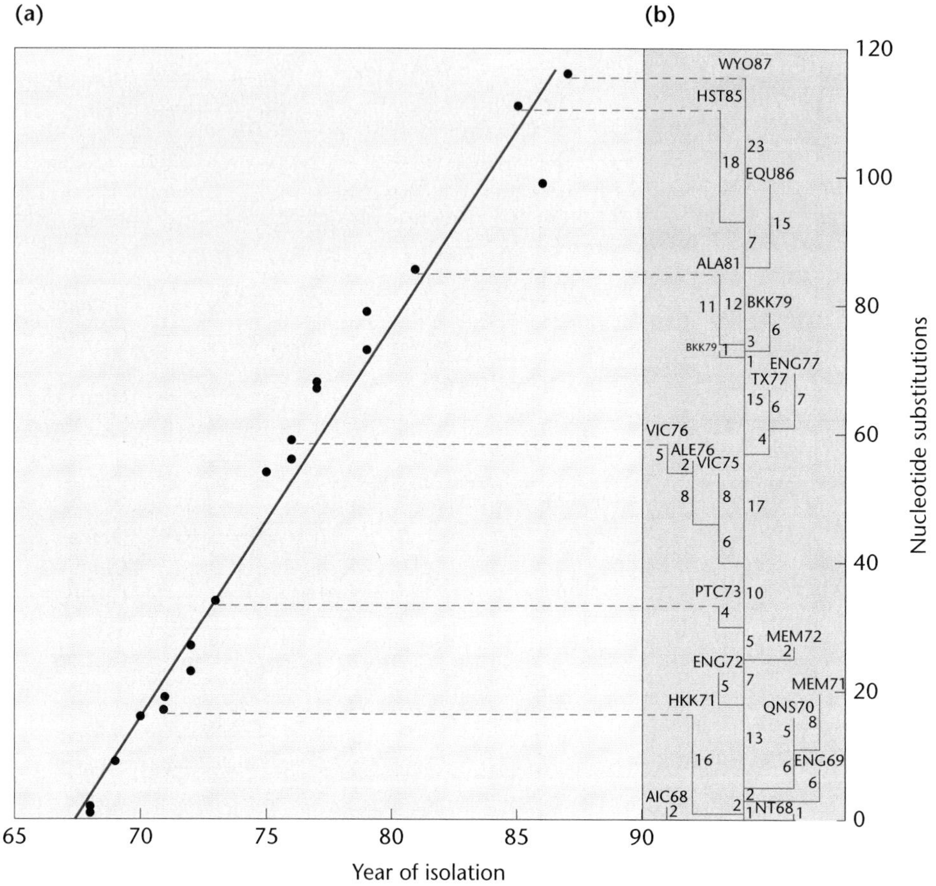

FIGURE 26–20 Molecular clock in the influenza A hemagglutinin gene. (a) Number of nucleotide differences between the first isolate and each subsequent isolate as a function of year of isolation. (b) Estimate of the phylogeny of the isolates.

The hemagglutinin gene thus serves as a molecular clock. Molecular clocks are used to compare the sequences of new flu viruses as they appear each year and to estimate the time that has passed since each diverged from a common ancestor.

Molecular clocks must be carefully calibrated and used with caution. For example, Fitch's data indicate that strains of influenza A that jump from birds to humans have evolved much more rapidly than strains that have remained in birds. Hence, a molecular clock calibrated from human strains of the virus would be highly misleading if applied to bird strains.

How Do We Know?

How can we determine the last common ancestor shared by two divergent species?

26.7 Reconstructing Evolutionary History Allows Us to Answer a Variety of Questions

Evolutionary analysis addresses a wide range of questions, some of which relate to evolution directly, while others deal with contemporary issues and even criminal activity. To conclude this chapter, we provide answers to the questions posed in the scenarios set forth in the beginning.

Transmission of HIV from a Dentist to His Patients

In late 1986, a Florida dentist tested positive for HIV. Several months later, he was diagnosed with AIDS. He continued to practice general dentistry for two more years, until one of his patients, a young woman with no known risk factors, discovered that she, too, was infected with HIV. When the dentist publicly urged his other patients to have themselves tested, several more were found to be HIV positive. Did this dentist transmit HIV to his patients, or did the patients become infected by some other means?

At first glance, this situation appears to be a long way from a discussion of evolution; however, the movement of a virus from one individual to another is similar to the founding of a new island population by a small number of migrants. Gene flow between the ancestral population and the new population is nonexistent, and the populations are free to diverge, as a result of genetic drift, adaptation to different environments, or both.

If the dentist passed his HIV infection to his patients, then the viral strains isolated from these patients should be more closely related to each other and to the dentist's strain than to strains from any other individuals living in the same area. Following this line of reasoning, Chin-Yih Ou and colleagues sequenced portions of the gene for the HIV envelope protein from viruses collected from the dentist, 10 of his patients, and several other HIV-infected individuals from the area who served as local controls.

An evolutionary tree for the HIV strains produced from these sequence data is shown in Figure 26–21. The viruses isolated from the participants in the study are at the tips of the branches; branch points join these to the inferred common ancestors. The HIV samples from patients A, B, C, E, G, and I all share a more recent common ancestor with each other and with the dentist's strains than with any of the HIV samples from any of the other local individuals. The evolutionary relationship among these HIV strains indicates that the dentist did, indeed, transmit his infection to those patients. In contrast, the viruses taken from patients D, F, H, and J are all more closely related to strains from local controls (LC) than they are to the dentist's strains. These patients got their infection from someone other than the dentist. Patient J, in fact, seems to have acquired his infection from two different sources. The Centers for Disease Control, which tracks HIV cases, has not been able to determine how the dentist-to-patient transmission took place. There have been no other documented cases of provider-to-patient transmission in a healthcare setting.

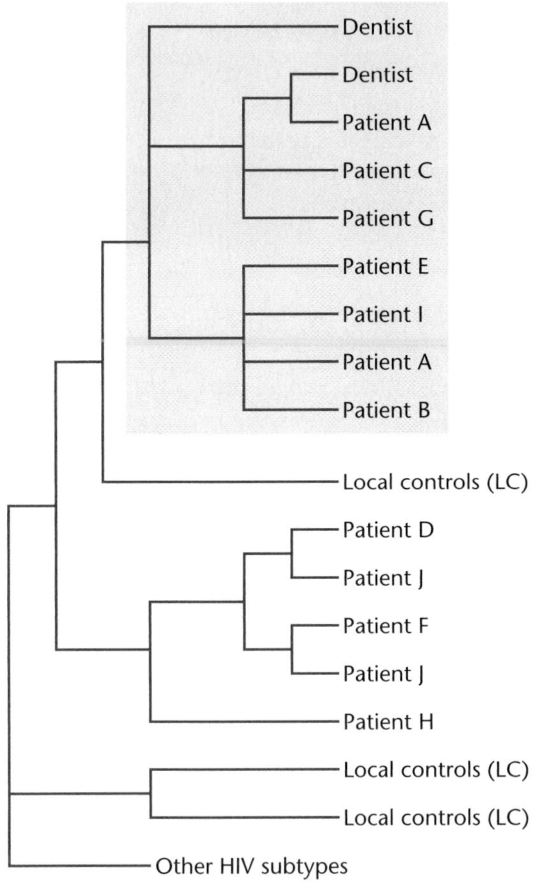

FIGURE 26–21 A phylogeny of HIV strains taken from a dentist, his patients, and several local controls (LC). The group of viral strains in the shaded portion of the phylogeny are all derived from a common ancestral strain.

The Relationship of Neanderthals to Modern Humans

Paleontological evidence indicates that the Neanderthals, *Homo neanderthalensis*, lived in Europe and western Asia from some 300,000 to 30,000 years ago. For at least 30,000 years, Neanderthals coexisted with anatomically modern humans (*H. sapiens*) in several areas. Questions about Neanderthals and modern humans remain unanswered: (1) Can Neanderthals be regarded as direct ancestors of modern humans? (2) Did Neanderthals and *H. sapiens* interbreed, so that descendants of the Neanderthals are alive today? Or did the Neanderthals die off and become extinct?

To resolve these issues, several groups have focused their attention on the recovery and analysis of DNA extracted from Neanderthal bones. In 1997, Svante Pääbo, Matthias Krings, and their colleagues extracted fragments of mitochondrial DNA from a Neanderthal skeleton found in Feldhofer cave near Düsseldorf, Germany. After confirming that they had indeed isolated Neanderthal gene fragments, the researchers placed the Neanderthal sequences on a phylogenetic tree together with sequences from more than 2000 modern humans [Figure 26–22(a)]. On the basis of Pääbo and Krings's analysis, Neanderthals appear to be a distant relative of modern humans. Using a molecular clock calibrated with chimpanzees and humans, the researchers calculated that the last common ancestor of Neanderthals and modern humans lived roughly 600,000 years ago, four times as long ago as the last common ancestor of all modern humans. However, considerable caution is required when drawing conclusions based on a single Neanderthal specimen.

More recently, Igor Ovchinnikov and his colleagues analyzed mitochondrial DNA recovered from Neanderthal remains discovered in Mezmaiskaya cave in the Caucasus Mountains east of the Black Sea. Although the two Neanderthal sequences (i.e., Pääbo and Krings's, on the one hand, and Ovchinnikov's, on the other) are from individuals from different geographic regions more than 1000 miles apart, they vary by only about 3.5 percent, indicating that they derive from a single gene pool. Further, the amount of variation between the two Neanderthal sequences is comparable to that seen among modern humans. Phylogenetic analysis places the two Neanderthals in a group that is distinct from modern humans [Figure 26–22(b)]; the conclusion to be drawn from the two studies is that although Neanderthals and humans have a common ancestor, the Neanderthals were a separate hominid line and did not contribute mitochondrial genes to *H. sapiens*. Taken together, the studies suggest that when the Neanderthals disappeared, their lineage went extinct with them.

The Origin of Mitochondria

Mitochondria are cellular organelles that contain their own genome in the form of a circular DNA molecule. In humans, the mitochondrial genome encodes 13 proteins required for oxidative phosphorylation and ATP production, 22 tRNA molecules, and 2 rDNA genes. The organization and function of the mitochondrial genome bear many similarities to those of bacterial genomes. Therefore, mitochondria are thought to derive from bacteria; that is, they are descended from free-living prokaryotic organisms that became intracellular symbionts in nucleated host cells. Once incorporated into a nucleated cell, the genome of the mitochondrion became smaller over time, and the reduced

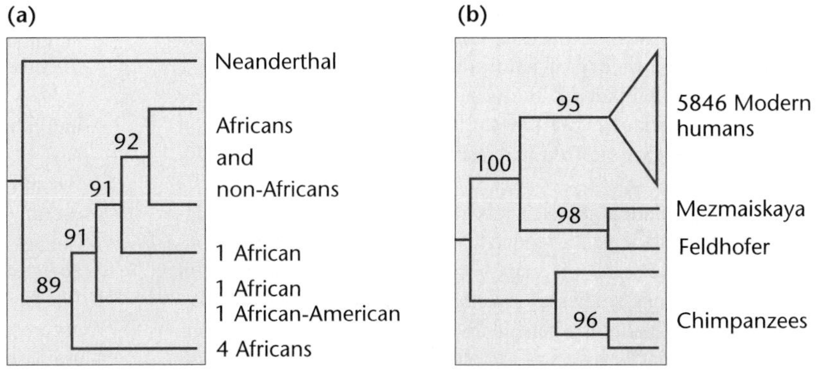

FIGURE 26–22 (a) A phylogeny estimated from mitochondrial DNA sequences of one Neanderthal and over 2000 modern humans. *(a) From Krings, M et al., 1997. Cell 90: 19-30. Figure 7A, p.26, copyright 1997 with permission from Elsevier; (b) From Ovchinnikov, I.V. et al., Molecular analysis of Neanderthal DNA from the northern Caucasus Nature 404: 490-93 copyright 2000 Macmillan Publishers Ltd.*

genome we now see in mitochondria is thought to be the product of gene loss and gene transfer to the nucleus of the host cell. However, many questions about the origin and evolution of mitochondria remain. Among them is one relating to the bacterial origins of mitochondria: Which group of bacteria are they descended from?

Researchers investigating this question have used information derived from nucleotide sequencing of the mitochondrial genes for small-subunit ribosomal RNAs (SSU rRNA). These sequences are under strong functional constraint and so are highly conserved. Phylogenetic analysis of the sequences identified a group of bacteria known as the α-proteobacteria (purple bacteria) as the closest living relatives of mitochondria. More recent work has divided this group into two subclasses, one of which is called the rickettsial subdivision.

Based on the SSU phylogeny, the rickettsia have been identified as the bacteria most closely related to mitochondria. Interestingly, rickettsia, like mitochondria, live only inside eukaryotic cells.

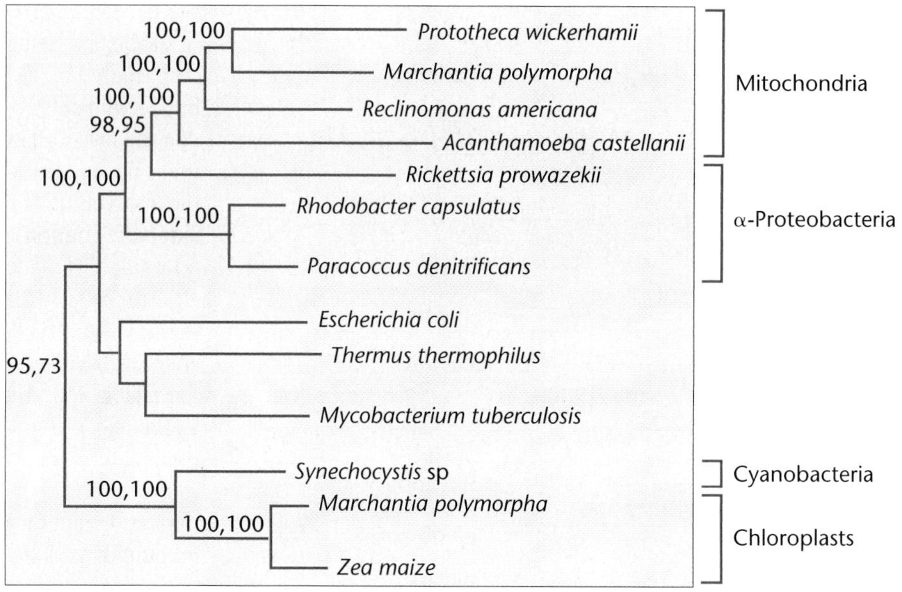

FIGURE 26–23 The evolutionary relationships of mitochondria, α-proteobacteria, cyanobacteria, and chloroplasts. Note that, although *R. prowazekii* is an α-proteobacterium it is more closely related to mitochondria than to other α-proteobacteria. *From Andersson, S.G.E. et al., The genome sequence of Rickettsia prowazekii and the origin of mitochondria. Nature 396: 133-43, copyright 1998 Macmillan Publishers Ltd.*

Unlike mitochondria, however, rickettsia are disease-causing parasites, responsible for typhus, one of the most serious diseases in the history of our species. Genomic sequencing of one rickettsial species, *Rickettsia prowazekii*, provides additional insight into the relationship between these bacteria and mitochondria. Using the genome sequence information for genes involved in ATP synthesis derived from *R. prowazekii*, other bacteria, and mitochondria from several sources, Siv Andersson and colleagues constructed a phylogenetic tree (Figure 26–23). Their tree indicates a close evolutionary relationship between *R. prowazekii* and mitochondria. Evidence from SSU analysis results in a similar tree. Thus, two lines of evidence, from SSUs and proteins involved in ATP synthesis, provide strong evidence that the mitochondria we carry in our cells derive from an ancient ancestor of the rickettsia.

Chapter Summary

1. Today's species are the products of an evolutionary history that included the transformation, splitting, and divergence of lineages. Alfred Russel Wallace and Charles Darwin formulated the theory of natural selection, which provides a major mechanism for the transformation of lineages. The genetic basis of evolution and the role of natural selection in changing allele frequencies were discovered in the 20th century.
2. When population geneticists began studying the genetic structure of populations, they found that most populations harbor considerable genetic diversity, which becomes apparent in electrophoretic studies of proteins and at the level of DNA sequences. Whether the genetic diversity of populations is maintained primarily by mutation plus genetic drift or by natural selection is a matter of some debate.
3. The geographic ranges of most species encompass a degree of environmental diversity. As a result of both adaptation to different environments and genetic drift, different populations within a species may have different alleles or allele frequencies at many loci.
4. Gene flow among populations tends to homogenize their genetic composition. When gene flow is reduced, genetic drift and adaptation to different environments can cause populations to diverge. Eventually, populations may become so different that the individuals in one population either will not or cannot mate with the individuals in the other. At this point, the divergent populations have become different species.
5. Because speciation is associated with genetic divergence, we can use the genetic differences among species to infer their evolutionary history. By comparing amino acid or nucleotide sequences, we can determine the genetic distances among species. We can then use the genetic distances to reconstruct evolutionary trees. The simplest methods for reconstructing phylogenies are based on the assumption that the least divergent species are one another's closest relatives.
6. The reconstruction of evolutionary trees is a key technique in answering a diversity of interesting questions. Examples include tracing the route of transmission of a disease in an epidemic, deciphering recent events in human evolution, and determining the evolutionary relationships among all organisms.

GENETICS, TECHNOLOGY, AND SOCIETY

What Can We Learn from the Failure of the Eugenics Movement?

The eugenics movement had its origins in the ideas of the English scientist Francis Galton, who became convinced from his study of the appearance of geniuses within families (including his own) that intelligence is inherited. Galton concluded in his 1869 book *Hereditary Genius* that it would be "quite practicable to produce a highly gifted race of men by judicious marriages during several consecutive generations." The term *eugenics*, coined by Galton in 1883, refers to the improvement of the human species by such selective mating. Once Mendel's principles were rediscovered in 1900, the *eugenics* movement flourished.

The eugenicists believed that a wide range of human attributes were inherited as Mendelian traits, including many aspects of behavior, intelligence, and moral character. Their overriding concern was that the presumed genetically "feebleminded" and immoral in the population were reproducing faster than the genetically superior, and that this differential birthrate would result in the progressive deterioration of the intellectual capacity and moral fiber of the human race. Several remedies were proposed. Positive eugenics called for the encouragement of especially "fit" parents to have more children. More central to the goals of the eugenicists, however, was the negative eugenics approach aimed at discouraging the reproduction of the genetically inferior or, better yet, eliminating it altogether.

In the United States, the eugenics movement enjoyed wide popular support for a time and had a significant impact on public policy. Partially at the urging of prominent eugenicists, 30 states passed laws compelling the sterilization of criminals, epileptics, and inmates in mental institutions; most states enacted laws invalidating marriages between the "feebleminded" and others considered eugenically unfit. The crowning legislative achievement of the eugenics movement, however, was the passage of the Immigration Restriction Act of 1924, which severely limited the entry of immigrants from eastern and southern Europe due to their perceived mental inferiority.

Throughout the first two decades of the century, most geneticists passively accepted the views of eugenicists, but by the 1930s critics recognized that the goals of the eugenics movement were determined more by racism, class prejudice, and anti-immigrant sentiment than by sound genetics. Increasingly, prominent geneticists began to speak out against the eugenics movement, among them William Castle, Thomas Hunt Morgan, and Hermann Muller. When the horrific extremes to which the Nazis took eugenics became known, a strong reaction developed that all but ended the eugenics movement.

Paradoxically, the eugenics movement arose at the same time basic Mendelian principles were being developed, principles that eventually undermined the theoretical foundation of eugenics. Today, any student who has completed an introductory course in genetics should be able to identify several fundamental mistakes the eugenicists made.

1. They assumed that complex human traits such as intelligence and personality were strictly inherited, completely disregarding any environmental contribution to the phenotype. Their reasoning was that because certain traits ran in families, they must be genetically determined.

2. They assumed that these complex traits were determined by single genes with dominant and recessive alleles. This belief persisted despite research showing that multiple genes contribute to many phenotypes.

3. They assumed that a single ideal genotype existed in humans. Presumably, such a genotype would be highly homozygous in order to be sustained. This precept runs counter to current evidence suggesting that a high level of homozygosity is often deleterious, supporting the superiority of the heterozygote.

4. They assumed that the frequency of recessively inherited defects in the population could be significantly lowered by preventing homozygotes from reproducing. In fact, for recessive traits that are relatively rare, most of the recessive alleles in the population are carried by asymptomatic heterozygotes who are spared from such selection. Negative eugenic practices, no matter how harsh, are relatively ineffective at eliminating such traits.

5. They thought that those deemed genetically unfit in the population might outreproduce those thought to be genetically fit. This is the exact reverse of the Darwinian concept of fitness, which equates reproductive success with fitness. (Galton should have understood this, being Darwin's first cousin!)

More than seven decades have passed since the eugenics movement was in full bloom. We now have a much more sophisticated understanding of genetics, as well as a greater awareness of its potential misuses. But the application of current genetic technologies makes possible a "new eugenics" of a scope and power that Francis Galton could not have imagined. In particular, prenatal genetic screening and *in vitro* fertilization enable the selection of children according to their genotype, a power that will dramatically increase as more and more genes are associated with inherited diseases and perhaps even behaviors.

As we move into this new genetic age, we must not forget the mistakes made by the early eugenicists. We must remember that the phenotype is a complex interaction between the genotype and the environment, and not lapse into an approach that treats a person as only a collection of genes. We must keep in mind that many genes may contribute to a particular phenotype, whether a disease or a behavior, and that the alleles of these genes may interact in unpredictable ways. We must not fall prey to the assumption that there is an ideal genotype. The success of all populations in nature is believed to be enhanced by genetic diversity. And most of all, we must not use genetic information to advance ideological goals. We may find that there is a fine line between the legitimate uses of genetic technologies, such as having healthy children, and other eugenic practices. It will be up to us to decide exactly where the line falls.

References

Allen, G.E., Jacoby, R., and Glauberman, N. (eds.). 1995. Eugenics and American social history, 1880–1950. *Genome* 31:885–89.

Hartl, D.L. 1988. A primer of population genetics, 2nd ed. Sunderland, MA: Sinauer.

Kevles, D.J. 1985. In the name of eugenics: Genetics and the uses of human heredity. Berkeley: Univ. of California Press.

Insights and Solutions

1. Sequence analysis of DNA can be accomplished by a number of techniques. Protein sequencing, on the other hand, is made more complex by the fact that 20 different subunits, as opposed to just the four nucleotides of DNA, need to be unambiguously identified and enumerated. Because of their unique properties, the N-terminal and C-terminal amino acids in a protein are easy to identify, but the array in between offers a difficult challenge, because many proteins contain hundreds of amino acids. How is it that protein sequencing is accomplished?

Solution: The strategy for protein sequencing is the same as for DNA sequencing: Divide and conquer. To accomplish this, specific enzymes are used that reproducibly cleave proteins between certain amino acids. The use of different enzymes produces overlapping fragments, each of which is isolated and its amino acid sequence determined by chemical means. Sequences from overlapping fragments are then assembled to give a sequence for the entire protein. Alternatively, researchers can first sequence the gene that encodes the protein and then use the genetic code to infer the protein's amino acid sequence.

2. A single plant twice the size of others in the same population suddenly appears. Normally, plants of that species reproduce by self-fertilization and by cross-fertilization. Is this new giant plant simply a variant, or could it be a new species? How would you determine which it is?

Solution: One of the most widespread mechanisms of speciation in higher plants is polyploidy, the multiplication of entire sets of chromosomes. The result of polyploidy is usually a larger plant with larger flowers and seeds. There are two ways of testing the new variant to determine whether it is a new species. First, the giant plant should be crossed with a normal-sized plant to see whether the former produces viable, fertile offspring. If it does not, then the two different types of plants would appear to be reproductively isolated. Second, the giant plant should be cytogenetically screened to examine its chromosome complement. If it has twice the number of its normal-sized neighbors, it is a tetraploid that may have arisen spontaneously. If the chromosome number differs by a factor of two and the new plant is reproductively isolated from its normal-sized neighbors, it is a new species.

Problems and Discussion Questions

1. What is the neo-Darwinian definition of evolution?
2. Define speciation. How does this differ from evolution?
3. Wallace and Darwin identified natural selection as a force acting on phenotypic variation. What two processes related to phenotypic variation were they unable to explain?
4. Two closely related species may have very few phenotypic differences. How can we justify classifying them as two different species instead of one species with a small range of phenotypic variation?
5. Natural selection can lead to speciation. Is this the only way species can form?
6. The maintenance of allozymic diversity in natural populations remains a controversial topic. Nevo (2001. *Proc. Natl. Acad. Sci. [USA]* 98:6233–40) examined polymorphisms among 111 above-ground and 132 subterranean mammalian species and found that subterranean mammals are less polymorphic than those living above ground. Provide a possible explanation.
7. Discuss the rationale behind the statement that inversions in chromosome 3 of *D. pseudoobscura* represent genetic variation.
8. Dobzhansky's studies on populations of *D. pseudoobscura* showed changes in the frequency of the ST and CH chromosome arrangements throughout the year. Why hasn't one of these arrangements been eliminated over a long period of time?
9. Describe how populations with substantial genetic differences can form. What is the role of natural selection?
10. Price et al. (1999. *J. Bacteriol.* 181:2358–62) conducted a genetic study of the toxin transport protein (PA) of *Bacillus anthracis*, the bacterium that causes anthrax in humans. Within the 2294-nucleotide gene in 26 strains they identified 5 point mutations—2 missense and 3 synonyms—among different isolates. Necropsy samples from an anthrax outbreak in 1979 revealed a novel missense mutation and 5 unique nucleotide changes among 10 victims. The authors concluded that these data indicate little or no horizontal transfer between different *B. anthracis* strains. (a) Which types of nucleotide changes (missense or synonyms) cause amino acid changes? (b) What is meant by horizontal transfer? (c) On what basis did the authors conclude that evidence of horizontal transfer is absent from their data?
11. What types of nucleotide substitutions will not be detected by electrophoretic studies of a gene's protein product?
12. A recent study examining the mutation rates of 5669 mammalian genes (17,208 sequences) indicates that, contrary to popular belief, mutation rates among lineages with vastly different generation lengths and physiological attributes are remarkably constant (Kumar, S., and Subramanian S. 2002. *Proc. Natl. Acad. Sci. [USA]* 99:803–08). The average rate is estimated at 12.2×10^{-9} per bp per year. What is the significance of this finding in terms of mammalian evolution?
13. Discuss the arguments supporting the neutral theory of molecular evolution. What counterarguments are proposed by the selectionists? Of what value is the debate concerning the neutralist theory?
14. What genetic changes take place during speciation?
15. The mummichog shows *Ldh* allele frequency differences from cold northern waters to warm southern waters. (a) Is there a correlation between allele type and allele frequency and water temperature? (b) Why are differences in catalytic efficiency *and* differences in transcription rates important in this adaptive strategy?
16. List the barriers that prevent interbreeding, and give an example of each.
17. What are the two groups of reproductive isolating mechanisms? Which of these is regarded as more efficient, and why?
18. In the tobacco plant *N. tabacum*, formed as a hybrid between *N. otophora* and *N. silvestris*, would you expect to find higher levels of heterozygosity in the polyploid species than in the diploid parental species? Is this an advantage or disadvantage?
19. Why are many species of flowering plants polyploid, but only a few animal species are polyploid?
20. Some critics have warned that the use of gene therapy to correct genetic disorders will affect the course of human evolution. Evaluate this criticism in light of what you know about population genetics and evolution, distinguishing between somatic gene therapy and germline gene therapy.

Extra-Spicy Problems

21. Shown below are two homologous lengths of the α and β chains of human hemoglobin. Consult the genetic code dictionary (Figure 13–7) and determine how many amino acid substitutions may have occurred as a result of a single nucleotide substitution. For any that cannot occur as the result of a single change, determine the minimal mutational distance.

α:	Ala	Val	Ala	His	Val	Asp	Asp	Met	Pro
β:	Gly	Leu	Ala	His	Leu	Asp	Asn	Leu	Lys

22. Determine the minimal mutational distances between these amino acid sequences of cytochrome c from various organisms. Compare the distance between humans and each organism.

Human:	Lys	Glu	Glu	Arg	Ala	Asp
Horse:	Lys	Thr	Glu	Arg	Glu	Asp
Pig:	Lys	Gly	Glu	Arg	Glu	Asp
Dog:	Thr	Gly	Glu	Arg	Glu	Asp
Chicken:	Lys	Ser	Glu	Arg	Val	Asp
Bullfrog:	Lys	Gly	Glu	Arg	Glu	Asp
Fungus:	Ala	Lys	Asp	Arg	Asn	Asp

23. The data below represent short DNA sequences from five species: human, chimpanzee, gorilla, orangutan, and baboon. The sequences are each 50 bp long. They represent a short piece of a gene for testis-specific protein Y, located on the Y chromosome. The complete human sequence is given; the sequences for the other four species are shown only where they differ from the human sequence. Calculate the genetic difference between each pair of species. (For example, chimps and gorillas differ in 2 out of 50 bases, or 4 percent; thus, the genetic difference between chimps and gorillas is 0.04.) Then use UPGMA to reconstruct the phylogeny for these five species. Is your phylogeny consistent with those shown in Figures 26–19?

24. The genetic difference between two *Drosophila* species *D. heteroneura* and *D. sylvestris*, as measured by nucleotide diversity, is about 1.8 percent. The difference between chimpanzees (*P. troglodytes*) and humans (*H. sapiens*) is about the same, yet the latter species are classified in different genera. In your opinion, is this valid? Explain why.

25. The use of nucleotide sequence data to measure genetic variability is complicated by the fact that the genes of higher eukaryotes are complex in organization and contain 5' and 3' flanking regions as well as introns. Researchers have compared the nucleotide sequence of two cloned alleles of the γ-*globin* gene from a single individual and found a variation of 1 percent. Those differences include 13 substitutions of one nucleotide for another, and 3 short DNA segments that have been inserted in one allele or deleted in the other. None of the changes takes place in the gene's exons (coding regions). Why do you think this is so, and should it change our concept of genetic variation?

26. In a recent study of cichlid fish inhabiting Lake Victoria in Africa, Nagl et al. (1998. *Proc. Natl. Acad. Sci. [USA]* 95:14,238–43) examined suspected neutral sequence polymorphisms in noncoding genomic loci in 12 lake species and their putative riverine ancestors. At all loci, the same polymorphism was found in nearly all of the tested species from Lake Victoria, both lacustrine and riverine. Different polymorphisms at these loci were found in cichlids at other African lakes. (a) Why would you suspect neutral sequences to be located in noncoding genomic regions? (b) What conclusions can be drawn from these polymorphism data in terms of cichlid ancestry in these lakes?

27. Given that there are approximately 400 cichlid species in Lake Victoria and that it dried up almost completely about 14,000 years ago, what evidence indicates that extremely rapid evolutionary adaptation rather than extensive immigration occurred?

```
Human:       AGAGGTTTTTCAGTGAATGAAGCTATTTTTAAGGGAGTGTGATTGCTGCC
Chimpanzee:                    C
Gorilla:                T      C                            C
Orangutan:    T        GT   C  C                 C          C
Baboon:       C C  G  G T      C        G       GC      C   C G
```

28. Comparisons of Neanderthal mitochondrial DNA with that of modern humans indicate they are not related to modern humans and did not contribute to our mitochondrial heritage. However, because Neanderthals and modern humans are separated by at least 25,000 years, this does not rule out some forms of interbreeding causing the modern European gene pool being derived from both Neanderthals and early humans (called Cro-Magnons). To resolve this question, Caramelli et al. (2003. *Proc. Natl. Acad. Sci. [USA]* 100:6593–97) analyzed mitochondrial DNA sequences from 25,000 year old Cro-Magnon remains and compared them to four Neanderthal specimens and a large data set derived from modern humans. The results are shown in the graph at the right. The x-axis represents age of the specimens in thousands of years; the y-axis represents the average genetic distance. Modern humans are indicated by filled squares; Cro-Magnons, open squares; and Neanderthals, diamonds. (a) What can you conclude about the relationship between Cro-Magnons and modern Europeans? What about the relationship between Cro-Magnons and Neanderthals? (b) From these data, does it seem likely that Neanderthals made any contributions to the Cro-Magnon gene pool or the modern European gene pool?

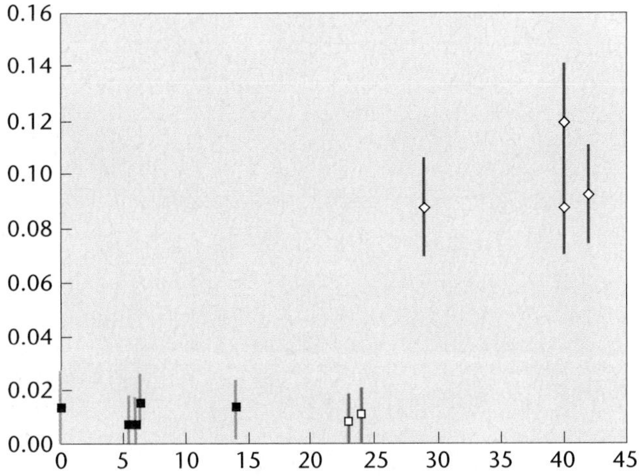

SELECTED READINGS

Adcock, G.J., Dennis, E.S., Easteal, S., Huttley, G.A., Jermiin, L.S., Peacock, W.J., and Thorne, A. 2001. Mitochondrial DNA sequences in ancient Australians: Implications for modern human origins. *Proc. Nat. Acad. Sci.* 98:537–42.

Anderson, W., Dobzhansky, T., Pavlovsky, O., Powell, J., and Yardley, D. 1975. Genetics of natural populations: XLII. Three decades of genetic change in *Drosophila pseudoobscura. Evolution* 29:24–36.

Ayala, F.J., 1984. Molecular polymorphism: How much is there, and why is there so much? *Dev. Genet.* 4:379–91.

Bradshaw, H.D., Jr., Otto, K.G., and Frewen, B.E. 1998. Quantitative trait loci affecting differences in floral morphology between two species of monkey flower (*Mimulus*). *Genetics* 149:367–82.

Dobzhansky, T. 1947. Adaptive changes induced by natural selection in wild populations of *Drosophila. Evolution* 1:1–16.

Freeman, S., and Herron, J.C. 2001. *Evolutionary analysis*. Upper Saddle River, NJ: Prentice Hall.

Gould, S.J. 1982. Darwinism and the expansion of evolutionary theory. *Science* 216:380–87.

Hardison, R. 1999. The evolution of hemoglobin. *Amer. Scient.* 87:126–37.

Ke, Y., et al. 2001. African origin of modern humans in East Asia: A tale of 12,000 Y chromosomes. *Science* 292:1151–53.

Kimura, M. 1989. The neutral theory of molecular evolution and the world view of the neutralists. *Genome* 31:24–31.

King, M.C., and Wilson, A.C. 1975. Evolution at two levels: Molecular similarities and biological differences between humans and chimpanzees. *Science* 188:107–16.

Knowlton, N., Weigt, L.A., Solorzano, L.A., Mills, D.K., and Bermingham, E. 1993. Divergence in proteins, mitochondrial DNA, and reproductive compatibility across the Isthmus of Panama. *Science* 260:1629–32.

Kramer, E.M., and Irish, V.F. 1999. Evolution of genetic mechanisms controlling petal development. *Nature* 399:144–48.

Kreitman, M. 1983. Nucleotide polymorphism at the alcohol dehydrogenase locus of *Drosophila melanogaster. Nature* 304:412–17.

Lamb, R.S., and Irish, V.F. 2003. Functional divergence within the APETALA3/PISTILLATA floral homeotic gene lineages. *Proc. Natl. Acad. Sci.* 100:6558–63.

Lewontin, R.C., and Hubby, J.L. 1966. A molecular approach to the study of genic heterozygosity in natural populations: II. Amount of variation and degree of heterozygosity in natural populations of *Drosophila pseudoobscura. Genetics* 54:595–609.

Ou, C.-Y., et al. 1992. Molecular epidemiology of HIV transmission in a dental practice. *Science* 256:1165–71.

Powers, D.A., and Schulte, P.M. 1998. Evolutionary adaptations of gene structure and expression in natural populations in relation to a changing environment: A multidisciplinary approach to address the million-year saga of a small fish. *J. Exper. Zool.* 282:71–94.

Relethford, J.H. 2001. Ancient DNA and the origin of modern humans. *Proc. Nat. Acad. Sci.* 98:390–91.

Stiassny, M.L.J., and Meyer, A. 1999. Cichlids of the Rift Lakes. *Sci. Am.* (Feb.) 280:64–69.

Takahashi, K., 1998. A novel family of short interspersed repetitive elements (SINES) from cichlids: The pattern of insertion of SINES at orthologous loci support the proposed monophyly of four major groups of cichlid fishes in Lake Tanganyika. *Mol. Biol. Evol.* 15:391–407.

Conservation Genetics

Cryogenically preserved seeds of rare crop varieties at the U.S. Department of Agriculture National Seed Storage Laboratory.

Chapter Concepts

- Species require genetic diversity for long-term survival and adaptation.
- Small, isolated populations are particularly vulnerable to genetic effects.
- Endangered populations that recover in size may not redevelop genetic diversity.
- Conservation and breeding efforts focus on retaining genetic diversity for long-term species survival.

As the twenty-first century progresses, the diversity of life on Earth will be under increasing pressure from the direct and indirect effects of explosive human population growth. Approximately 10 million *Homo sapiens* lived on the planet 10,000 years ago. This number grew to 100 million 2000 years ago and to 2.5 billion by 1950. Within the span of a single lifetime, the world's human population more than doubled to 5.5 billion in 1993 and is projected to reach as high as 19 billion by 2100 (Figure 27–1).

The effect on other species of accelerating human population growth has been dramatic. Data from the World Conservation Union (IUCN) show that, globally, 25 percent of all mammal species, 11 percent of birds, 20 percent of reptiles, 25 percent of amphibians, and 34 percent of fish species are vulnerable or endangered. The situation for plants is no better. Based on IUCN surveys, 12 percent of all vascular plant species worldwide—some 34,000 species—are threatened. Not just wild species are at risk. Genetic diversity in domesticated plants and animals is also being lost as many traditional crop varieties and livestock breeds disappear. The Food and Agriculture Organization (FAO) estimates that since 1900, 75 percent of the genetic diversity in agricultural crops has been lost. Out of approximately 5000 different breeds of domesticated farm animals worldwide, one-third are at risk of being lost.

Why should we be concerned about losing **biodiversity**—that is, the biological variation represented by these different plants and animals? As the fossil record shows, unrelated to human influences, many different plants and animals that once inhabited the planet have become extinct over millions of years, and others have taken their places. Biologists are concerned, however, at the accelerated rate of species extinctions we are witnessing today, all of which can be ascribed to direct or indirect human impacts. Deliberate hunting or harvesting of plants and animals by humans, habitat destruction through human development activities, and the indirect effects of global climate change are making it increasingly difficult for many species to survive.

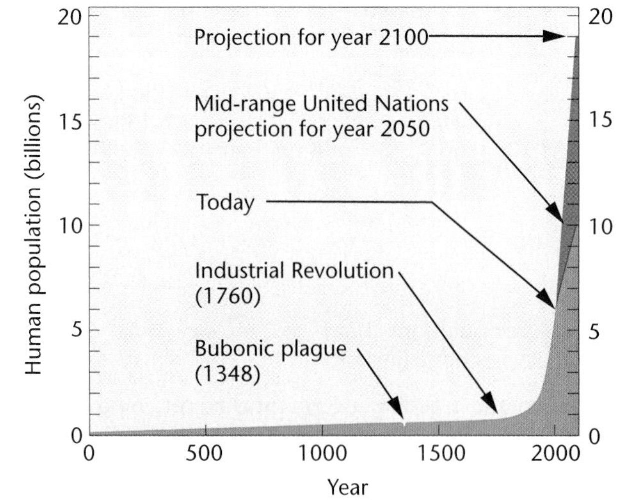

FIGURE 27–1 Growth in human population over the past 2000 years and projected through 2100.

Some biologists fear that **ecosystems**—the complex webs of interdependent, but diverse plants and animals found together in the same environment—may collapse if key sustaining species are lost. This portends possible consequences for our own long-term survival. Other scientists have pointed out that we may be losing the unknown economic potential or other benefits of unexploited plants and animals if we allow them to become extinct. A much-publicized, recent example is the Pacific yew (*Taxus brevifolia*), a rare tree found in coastal forests of the western United States. This slow-growing species was ignored because of its small size and poor timber qualities and was frequently destroyed during logging operations as a "trash tree." In 1991, it was found to be a source of taxol, a compound now widely used as a powerful drug for treating cancer. Quite apart from practical benefits, scientists and nonscientists have made the case that human life will be diminished both spiritually and aesthetically if we do not strive to maintain the fascinating and often beautiful variety of living organisms with which we share the planet.

Conservation biologists work to understand and maintain biodiversity, studying the factors that lead to species decline and the ways in which species can be preserved. The new field of **conservation genetics** has emerged in the last 20 years as scientists have begun to recognize that genetics will be an important tool in maintaining and restoring population viability. The applications of genetics to conservation biology are multifaceted; in this chapter, we will explore but a few of them. Underlying the increasingly important role of genetics in conservation biology is the recognition that biodiversity depends on genetic diversity and that maintaining biodiversity in the long term is unlikely if genetic diversity is lost.

How Do We Know?

In this chapter we will focus on the genetic approach used during conservation efforts. Conservation geneticists assess genetic diversity and work to maintain population numbers for long-term species survival. As you study this topic, you should try to answer several fundamental questions:

- How do we determine the level of genetic diversity in a species?
- How do we know that diminished genetic diversity is detrimental to species survival?
- How do we determine the most effective approach to countering decreased population size?
- How do we attempt to "conserve" existing genetic diversity?

27.1 Genetic Diversity Is at the Heart of Conservation Genetics

While biodiversity encompasses the variation represented by all existing species of plants and animals on this planet at any given time, **genetic diversity** is not as easy to define and study. Genetic diversity can be considered on two levels: **interspecific diversity** and **intraspecific diversity**.

FIGURE 27–2 (a) Tropical rain forest is an ecosystem with high interspecific diversity. (b) A coastal marsh in North Carolina exemplifies an ecosystem with low interspecific diversity.

Diversity between species, or *interspecific diversity*, is reflected in the number of different plant and animal species present in an ecosystem. Some ecosystems have a very high level of species diversity, such as a tropical rainforest in which hundreds of different plant and animal species may be found within a few square meters [Figure 27–2(a)]. Other ecosystems, especially where plants and animals must adapt to a harsh environment, may have much lower levels of interspecific diversity [Figure 27–2(b)]. Lists of the different plant and animal species found in a particular environment are used to compile species inventories. The inventories identify diversity hot spots: geographic areas with especially high levels of interspecific diversity where conservation efforts can be focused. Conservation biologists working at the ecosystem level are interested in preventing species from being lost and in restoring species that were once part of the system but are no longer present. The reestablishment of the gray wolf (*Canis lupus*) in Yellowstone National Park and the release of captive-bred California condors (*Gymnogyps californianus*) into the mountain ranges they previously occupied in the southwest United States are examples of current attempts by conservation biologists to restore missing species to their ecosystems.

Intraspecific diversity—the diversity within a species—is reflected in the level of genetic variation occurring between individuals within a single population of a given species (*intrapopulation diversity*) or between different populations of the same species (*interpopulation diversity*). Genetic variation within populations can be measured as the frequency of individuals in the population that are heterozygous at a given locus or as the number of different alleles at a locus that are present in the population gene pool. When DNA-profiling techniques are used, the percentage of polymorphic loci—those represented by varying bands on the DNA profile in different individuals—can be calculated to indicate the extent of genetic diversity in a population. In outbreeding species, most intraspecific genetic diversity is found at the intrapopulation level.

Significant interpopulation diversity can occur if populations are separated geographically and there is no migration or exchange of gametes between them. On the other hand, predominantly inbreeding species (such as self-fertilizing plants) tend to have greater levels of interpopulation than intrapopulation diversity. There is a limited number of genotypes dominating an individual population, but greater variation between different populations. Understanding the breeding system and the distribution of genetic variation in an endangered species is important to its conservation, not only to ensure continued production of offspring but also to determine the best strategy for maintaining intraspecific diversity. For example, would it be more effective to preserve a few large populations or many small, distinct ones? This information can then be used to guide conservation or restoration efforts.

Loss of Genetic Diversity

Loss of genetic diversity in nondomesticated species is usually associated with a reduction in population size. This may be due to excessive hunting or harvesting. For example, biologists blame commercial overfishing for the collapse in the early 1990s of the deep-sea cod populations off the Newfoundland coast—once one of the most productive fisheries in the world. Habitat loss is also a major cause of population decline. As the global human population increases, more land is developed for housing and transport systems or is put into agricultural production, reducing or eliminating areas that were once home to wild plants and animals. The shrinking available habitat reduces populations of wild species and often also isolates them from each other as individual populations become trapped in pockets of undeveloped land surrounded by areas taken over for agriculture, urban development, or other human uses. This process is known as **population fragmentation**. When populations are no longer in contact with each other, gene flow through migration or gamete exchange between them ceases, and an important mechanism for maintaining genetic variation is lost.

In domesticated species, loss of genetic diversity is not usually the result of habitat loss or collapsing population numbers; there is little risk that cows or corn as species will become extinct any time soon. Reduction in diversity within domesticated species can instead be traced to changes in agricultural practice and consumer demand. Modern farming techniques have greatly increased production levels, but they have also led to greater genetic uniformity. As farmers switch to new crop varieties or improved livestock strains on a large scale, they abandon cultivation of many older local types, which may then disappear if efforts are not made to preserve them.

For example, in 1900, more than 100 different kinds of potatoes were available in the United States. Today, three quarters of commercial potato production in this country depends on just nine varieties. A single type, Russet Burbank, makes up 43 percent of the total acreage planted. Modern crop varieties or livestock strains may be better adapted to meet the demands of modern agricultural production, but older types often contain useful genes that can still play a vital role in survival functions, such as resistance to disease, cold, or drought. When the Russian wheat aphid (*Diuraphis noxia*), an insect that causes serious damage to cereal crops, invaded the United States in 1986, plant breeders eventually found genes conferring resistance to this pest—not in modern American wheat lines, but in old traditional varieties from the former Soviet Union, that fortunately had been collected and preserved. The old Russian wheats were crossed with United States wheat plants to transfer the resistance genes, creating new commercial wheat varieties that resist Russian wheat aphid attack and saving wheat growers from crop losses in the millions of dollars.

Identifying Genetic Diversity

For many years, population geneticists based estimates of intraspecific diversity on phenotypic differences between individuals, such as different colors of seeds or flowers or variation in markings (Figure 27–3). DNA profiling is a more direct molecular approach used to detect and quantify genetic differences between individuals drawn from a population. This technique has become an important tool for detecting and assessing intra- and interpopulation genetic variation. Nuclear, mitochondrial, and chloroplast DNA can all be analyzed to determine levels of genetic variation. We have already described applications of DNA profiling techniques using either restriction enzymes or PCR in Chapter 22. These techniques have also been used with DNA from species of concern to conservation biologists to examine levels and distribution of genetic variation and guide conservation efforts.

In one recent study, DNA marker analysis was performed on the only remaining population of a critically endangered plant species (*Limonium cavanillesii*) growing on the Mediterranean coast of Spain. Researchers found very low levels of diversity but did detect genetically distinct individual plants within the population from which seeds could be collected. In another recent study, DNA marker analysis of the southwestern willow flycatcher, an endangered migratory bird species that nests in the southwestern United States, found significant levels of genetic diversity within populations at different breeding sites, but little interpopulation diversity. This indicated to the scientists studying the flycatcher that migration of individual birds between breeding populations is important in maintaining diversity in the species and that conservation efforts should be directed toward preserving nesting sites that allow such migration to continue.

FIGURE 27–3 Phenotypic variation in seed color and markings in the common bean (*Phaseolus vulgaris*) reveals high levels of intraspecific diversity.

Another application of DNA markers in conservation is **DNA fingerprinting**. The use of DNA-profiling techniques in forensic science was discussed in Chapter 22; similar techniques can be used to uncover illegal trade in endangered plant and animal species that are protected by law. For example, since 1986, an international moratorium on commercial whaling has been in place, with only limited hunting of a few species permitted. In 1994, scientists based in New Zealand and Hawaii used PCR to amplify mitochondrial DNA (mtDNA) sequences extracted from meat on sale in markets in Japan, where whale meat is prized as a delicacy. The DNA fingerprints revealed several meat samples from humpback and fin whales, which are protected species. Two meat samples were not from whales at all, but were dolphin meat! This information assisted Japanese customs officials in tracking down sources of illegal whale meat imports.

Unlawful trade in animal products also occurs in the United States. In a recent study by scientists at the University of Florida, amplification of mtDNA sequences from turtle meat sold in Louisiana and Florida revealed that one quarter of the samples were actually alligator, not turtle. Most of the rest were from small freshwater turtles, indicating that populations of freshwater turtle species were declining through overharvesting and thus were in need of protection. Further refinements to DNA extraction and profiling techniques now allow accurate identification of turtle species from meat already spiced, cooked, and served in restaurants. We now have a tool that hopefully will help to eliminate the illegal harvesting of these endangered reptiles.

27.2 Population Size Has a Major Impact on Species Survival

Some species have never been numerous, especially those that are adapted to survive in unusual habitats. Biologists refer to such species as *naturally rare*. *Newly rare* species, on the other hand, are those whose numbers are in decline because of pressures such as habitat loss. Populations of such species may not only be small in number but also fragmented and isolated from other populations. Both decreased population size and increased isolation have important genetic consequences, leading to an increased risk of loss of diversity.

How small must a population be before it is considered endangered? It varies somewhat with species, but small populations can quickly become vulnerable to genetic phenomena that increase the risk of extinction. In general, a population of fewer than 100 individuals is considered extremely sensitive to these problems, which include genetic drift, inbreeding, and reduction in gene flow. The effects of such problems on species survival are substantial. Studies of bighorn sheep, for example, have shown that populations of fewer than 50 are highly likely to become extinct within 50 years. Projections based on computer models show that for all species, populations of fewer than 10,000 are likely to be limited in adaptive genetic variation and at least 100,000 individuals must be present if a population is to show long-term sustainability.

Determining the number of individuals a population must contain in order to have long-term sustainability is complicated by the fact that not all members of a population are equally likely to produce offspring: some will be infertile, too young, or too old. The *effective population size* (N_e) is defined as the number of individuals in a population having an equal probability of contributing gametes to the next generation. N_e is almost always smaller than the *absolute population size* (N). The effective population size can be calculated in different ways, depending on the factors that are preventing all individuals in a population from contributing equally to the next generation. In a sexually reproducing population that contains different numbers of males and females, for example, the effective population size is calculated as

$$N_e = \frac{4(N_m N_f)}{N_m + N_f}$$

where N_m is the number of males and N_f the number of females in the population. Hence a population of 100 males and 100 females would have an effective size of $4(100 \times 100)/(100 + 100) = 200$. In contrast, if there were 180 males and only 20 females, the effective population size would be $4(180 \times 20)/(180 + 20) = 72$.

Effective population size is also influenced by fluctuations in absolute population size from one generation to the next. Here, the effective population size is the harmonic mean of the numbers in each generation, so

$$N_e = 1 \Big/ \frac{1}{t}\left(\frac{1}{N_1} + \frac{1}{N_2} \cdots + \frac{1}{N_t}\right)$$

where t is the total number of generations being considered. For example, if a population went through a temporary reduction in size in generation 2, so that $N_1 = 100$, $N_2 = 10$, and $N_3 = 100$, then

$$N_e = 1 \Big/ \frac{1}{3}\left(\frac{1}{100} + \frac{1}{10} + \frac{1}{100}\right) = \frac{1}{0.04} = 25$$

In this case, although the mean actual number of individuals in the population over three generations was 70, the effective population size during that time was only 25. A severe temporary reduction in size such as this is known as a **population bottleneck**. Bottlenecks occur when a population or species is reduced to a few reproducing individuals whose offspring then increase in numbers over subsequent generations to reestablish the population. Although the number of individuals may be restored to healthier levels, genetic diversity in the newly expanded population is often severely reduced because gametes from the handful of surviving individuals functioning as parents do not represent all of the different allelic frequencies present in the original gene pool.

Now solve this

Question 27.13 on page 676 asks you to calculate effective population size N_e for an endangered gorilla population.

Hint: Consider what unusual feature of the social structure of the gorilla population will affect N_e.

Captive-breeding programs, in which a few surviving individuals from an endangered species are removed from the wild and their offspring raised in a protected environment to rebuild the population, inevitably create population bottlenecks. Bottlenecks also occur naturally when a small number of individuals from one population migrate to establish a new population elsewhere. When a new population derived from a small subset of individuals has significantly less genetic diversity than the original population, it exhibits the **founder effect**. Reduced levels of genetic diversity due to a founder effect can persist for many generations, as shown by studies of two species of Antarctic fur seal (*Arctocephalus gazella* and *Arctocephalus tropicalis*). Seal hunters in the 18th and 19th centuries had severely reduced populations of these species, eliminating them from parts of their natural range in the Southern Ocean.* Although the number of Antarctic fur seals has now rebounded and the two species have recolonized much of their original habitat, mtDNA fingerprinting reveals that a founder effect can still be detected in *A. gazella*, with reduced genetic variation in current populations descended from a handful of surviving individuals.

FIGURE 27–4 The cheetah (*Acinonyx jubatus*), a species with reduced genetic variation following population bottlenecks.

The cheetah (*Acinonyx jubatus*), shown in Figure 27–4, is another well-studied example of a species with reduced genetic variation resulting from at least one severe population bottleneck that occurred in its recent history. Allozyme studies of South African cheetah populations have shown levels of genetic variation that are less than 10 percent of those found in other mammals. The abnormal spermatozoa and poor reproductive rates commonly observed in cheetahs are thought to be linked to the lack of genetic diversity, although when and how the population bottleneck occurred in this species is still unclear.

Now solve this

Question 27.22 on page 676 ask you to compare allozyme data for cheetah populations with similar data from other cat species.

Hint: Consider what genetic factors potentially reduce the levels of genetic variation reflected in the data. Which species will be most affected by these factors?

*In 2000, this fifth world ocean was delimited from the southern portions of the Atlantic, Indian, and Pacific Oceans.

27.3 Genetic Effects Are More Pronounced in Small, Isolated Populations

Small isolated populations, such as those found in threatened and endangered species, are especially vulnerable to genetic drift, inbreeding, and reduction in gene flow. These phenomena act on the gene pool in different ways, but ultimately have similar effects in that they all can further reduce genetic diversity and long-term species viability.

Genetic Drift

If the number of breeding individuals in a population is small, fewer gametes will form the next generation. The alleles carried by these gametes may not be a representative sample of all those present in the population; purely by chance, some alleles may be underrepresented or not present at all, which will cause changes in allele frequency over time, resulting in **genetic drift**. This phenomenon has already been described in Chapter 25. A serious result of genetic drift in populations with a small effective population size is the loss of genetic variation. Genetic drift is a random process, so both deleterious and advantageous alleles can become fixed within a small population. This means a useful allele can be lost even if it has the potential to increase fitness or long-term adaptability.

Inbreeding

In small populations, the chance of **inbreeding** (matings between closely related individuals) is greater. As described in Chapter 25, inbreeding increases the proportion of homozygotes in a population, thus increasing the possibility that an individual may be homozygous for a deleterious allele. We have already seen in Chapter 25 how the extent of inbreeding in a population can be quantified as the **inbreeding coefficient (*F*)**, which measures the probability that two alleles of a given gene are derived from a common ancestral allele. Remember that the inbreeding coefficient is inversely related to the frequency of heterozygotes in the population, and can be calculated as

$$F = \frac{2pq - H}{2pq}$$

where $2pq$ is the expected frequency of heterozygotes based on the Hardy–Weinberg law, and H is the actual frequency of heterozygotes in the population.

In a declining population that has become small enough for drift to occur, heterozygosity (H) will decrease with each generation. The smaller the effective population size, the more rapid the decrease in H and the resulting increase in F, as can be seen from the equation

$$H_t/H_0 = \left(1 - 1/2N_e\right)^t$$

where H_0 is the initial frequency of heterozygotes, H_t is the frequency of heterozygotes after t generations, and N_e is the effective population size. Figure 27–5 compares rates of increase in F for different effective population sizes.

Web Tutorial 27.1 Genetic Drift

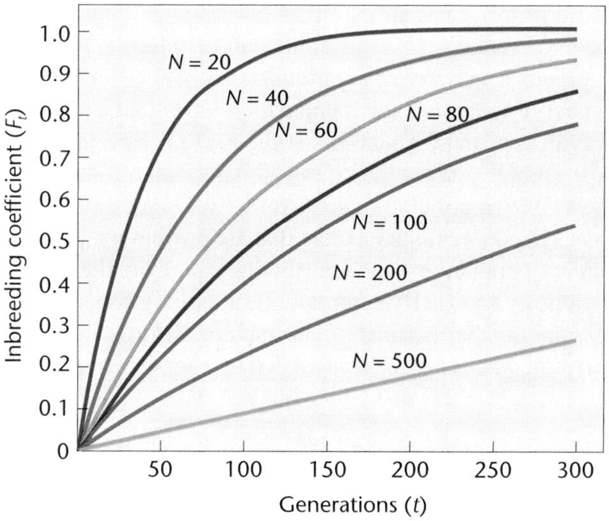

FIGURE 27–5 Increase in inbreeding coefficient (F) in theoretical populations with different effective population size, or N_e.

Now solve this

Question 27.1 on page 675 asks you to calculate effective population sizes, heterozygote frequencies, and an inbreeding coefficient for a jackal population that crashed and then recovered.

Hint: Calculate N_e for each generation first, and remember that heterozygote frequency and inbreeding coefficient are inversely related.

What effect does inbreeding have on the long-term survival of a population? In some cases, inbreeding may not immediately reduce the amount of genetic variation present in the overall gene pool of a species in which numbers of individuals remain high. Self-pollinating plants, for example, often show high levels of homozygosity and relatively little genetic variation within a single population. However, they tend to have considerable variation between different populations, each of which has adapted to slightly different local environmental conditions. On the other hand, in most outbreeding species, including all mammals, inbreeding is associated with reduced fitness and lower survival rates among offspring. This **inbreeding depression** can result from increased homozyosity for deleterious alleles. The number of deleterious alleles present in the gene pool of a population is called the **genetic load** (or genetic burden).

In some species, inbreeding accompanied by selection against less fit individuals homozygous for deleterious alleles has resulted in the elimination of these alleles from the gene pool, a process known as *purging the genetic load*. Species that have successfully purged their genetic load do not show continued reduction in fitness, even after many generations of inbreeding. This is true of numerous domesticated species, especially self-pollinating plants such as wheat. However, computer simulation experiments have shown that it may take 50 generations or more to complete the purging process, during which time inbreeding depression will still occur.

Alternatively, inbreeding depression can result from heterozygous individuals having a higher level of fitness than either of the corresponding homozygotes. In this case, the long-term survival of the population requires that inbreeding be avoided and that the levels of all alleles in the gene pool be maintained. Both of these requirements can be difficult in a species that has already suffered a significant reduction in population size.

The effects of inbreeding depression and loss of genetic variation in a small isolated population have been documented in the case of the Isle Royale gray wolves (Figure 27–6). Around 1950, a pair of gray wolves apparently crossed an ice bridge from the Canadian mainland to Isle Royale in Lake Superior. The island had no other wolves and had an abundance of moose, which became the wolves' main food source. By 1980, the Isle Royale wolf population had increased to over 50 individuals. Over the next decade, however, wolf numbers declined to fewer than a dozen with no new litters being born, despite plentiful food and no apparent sign of disease. The genetic variation of the remaining wolves was examined by mtDNA analysis and nuclear DNA fingerprinting. It was found that the Isle Royale wolves had levels of homozygosity that were twice as high as wolves in an adjacent mainland population. Furthermore, the wolves all possessed the same mtDNA genotype, consistent with descent from the same female. Hence, the degree of relatedness between individual Isle Royale wolves was equivalent to that of full siblings, suggesting that the wolves' reproductive failure was due to inbreeding depression, a phenomenon that has also been seen in captive-wolf populations.

Reduction in Gene Flow

Gene flow, the gradual exchange of alleles between two populations, is brought about by the dispersal of gametes or the migration of individuals. It is an important mechanism for introducing new alleles into a gene pool and increasing genetic variation. Migration is the main route for gene flow in animals; we examined the effect on population allele frequencies of migration and gene flow in Chapter 25. In plants, gene flow occurs not through movement of individuals but as a result of cross-pollination between different populations and, to some extent, through seed dispersal. Isolation and fragmentation of populations in rare and declining species significantly reduces gene flow and the potential for maintaining genetic diversity. As we have already discussed, habitat loss is a major threat to species survival. It is not unusual for a threatened or endangered species to be restricted to small separate pockets of the remaining habitat. This isolates and fragments the surviving populations so that movement of individuals can no longer occur between them, thus preventing gene flow.

The term **metapopulation** is used to describe a population consisting of spatially separated subpopulations with limited gene flow, especially if local extinctions and replacements of some of the subpopulations occur over time. One well-studied metapopulation is that of the endangered red-cockaded woodpecker (*Picoides borealis*) (Figure 27–7), which was once common in pinewoods throughout the southeastern United States. Habitat loss because of logging has reduced the remaining populations of this bird to small scattered sites isolated from each other, with virtually no migration between them. Studies using allozymes and DNA profiling show that the smallest surviving woodpecker populations (where $N = <100$) have suffered the greatest loss of genetic diversity and are at most risk of inbreeding depression, compared with larger populations where $N = >100$. Management of the red-cockaded woodpecker now includes efforts to increase the genetic diversity of the smallest populations by introducing birds from different larger populations to artificially recreate the gene flow by migration that would have occurred in the original unfragmented distribution of the species.

FIGURE 27–6 Isle Royale gray wolf (*Canis lupus*).

FIGURE 27–7 The red-cockaded woodpecker (*Picoides borealis*).

27.4 Genetic Erosion Diminishes Genetic Diversity

The loss of previously existing genetic diversity from a population or species is referred to as **genetic erosion**. Why does it matter if a population loses genetic diversity, especially if the numbers of individuals remain high? Genetic erosion has two important effects on a population. First, it can result in the loss of potentially useful alleles from the gene pool, thus reducing the ability of the population to adapt to changing environmental conditions and increasing its risk of extinction. Several decades before the dawn of modern genetics, Charles Darwin recognized the importance of diversity to long-term species survival and evolutionary success. In *On the Origin of Species* (1859), Darwin wrote,

> The more diversified the descendants of any one species ... by so much will they be better enabled to seize on many widely diversified places in the polity of nature, and so enabled to increase in numbers.

The story of the peppered moth *Biston betularia* in the 19th century illustrates the importance of maintaining allelic diversity in the gene pool. The allele producing the darker melanic phenotype did not confer any obvious advantage to the species until the moth's environment in parts of Great Britain was significantly altered as a result of increasing industrialization and the accompanying pollution. The continued presence of the melanic allele in the moth's gene pool, however, enabled it to adapt to the new environmental conditions and survive, just as the persistence of the nonmelanic allele in the population now allows the moth to readapt as its environment changes once again. What might have happened to the peppered moth if the melanic allele had been lost from its gene pool before the industrial revolution?

The second important effect of genetic erosion is a reduction in levels of heterozygosity. At the population level, reduced heterozygosity will be seen as an increase in the number of individuals homozygous at a given locus. At the individual level, a decrease in the number of heterozygous loci within the genotype of a particular plant or animal will occur. As we have seen, loss of heterozygosity is a common consequence of reduced population size. Alleles may be lost through genetic drift or because individuals carrying them die without reproducing. Smaller populations also increase the likelihood of inbreeding, which inevitably increases homozygosity.

Obviously, once an allele is lost from a gene pool, the potential for heterozygosity is greatly reduced, or it is completely eliminated if there are only two alleles at the locus in question and one is now fixed. The level of homozygosity that can be tolerated varies with species. Studies of populations showing higher-than-normal levels of homozygosity have documented a range of deleterious effects, including reduced sperm viability and reproductive abnormalities in African lions, increased offspring mortality in elephant seals, and reduced nesting success in woodpeckers.

As yet, no evidence has emerged from field studies conclusively linking the extinction of a wild population to genetic erosion. However, laboratory studies using *Drosophila melanogaster* to model evolutionary events show that loss of genetic variation does reduce the ability of a population to adapt to changing environmental conditions. Fruit flies, with their small size and rapid generation time of only 10 to 14 days, are a useful model organism for studying evolutionary events, especially because large populations can be readily developed and maintained through multiple generations.

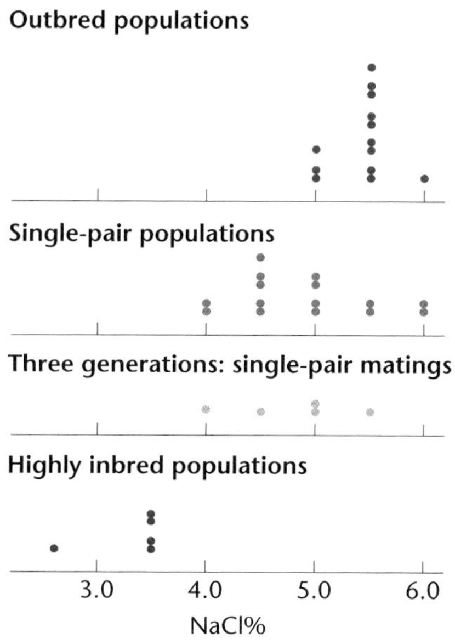

FIGURE 27–8 Effects of bottlenecks in various populations on evolutionary potential in *Drosophila*, as shown by distributions of NaCl concentrations at extinction.

In one set of experiments carried out by scientists at Macquarie University in Sydney, Australia, *Drosophila* populations were reduced to a single pair for up to three generations to simulate a population bottleneck and were then allowed to increase in number. The capacity of the bottlenecked populations to tolerate increasing levels of sodium chloride was compared with that of normal outbred populations. Researchers found that the bottlenecked populations became extinct at lower salt concentrations (Figure 27–8).

Other experiments comparing inbred and outbred *Drosophila* populations exposed to environmental stresses, such as high temperatures and ethanol presence, have shown that populations with even a low level of inbreeding have a much greater probability of extinction at lower levels of environmental stress than outbred populations. Experimental results such as these indicate that genetic erosion does indeed reduce the long-term viability of a population by reducing its capacity to adapt to changing environmental conditions.

27.5 Conservation of Genetic Diversity Is Essential to Species Survival

Scientists working to maintain biological diversity face several dilemmas. Should they focus on preserving individual populations, or should they take a broader approach by trying to conserve not just one species but all the interdependent plants and animals in an ecosystem? How can genetic diversity be maintained in a species whose numbers are declining? Can genetic diversity lost from a population be restored? Early conservation efforts often focused only on the population size of an endangered species. Biologists now recognize that a complex interplay of different factors must be considered during conservation efforts, including the need to examine the habitat and role of a species within an ecosystem, as well as the importance of genetic variation for long-term survival.

Ex Situ Conservation: Captive Breeding

Ex situ (Latin for off-site) **conservation** involves removing plants or animals from their original habitat to an artificially maintained location such as a zoo or botanic garden. This living collection can then form the basis of a captive-breeding program.

These programs, where a few surviving individuals from an endangered species are removed from the wild and their offspring raised in a protected environment to rebuild the population, have been instrumental in bringing a number of species back from the brink of extinction. However, such programs can have undesirable genetic consequences that potentially jeopardize the long-term survival of the species even after population numbers have been restored. A captive-breeding program is rarely initiated until very few individuals are left in the wild, when the original genetic diversity of the species is already depleted. The breeding program is frequently based on a small number of captured individuals, so genetic diversity is further reduced by the founder effect. Since captive-breeding facilities often accommodate only a small number of individuals in the breeding group, inbreeding is difficult to avoid, especially for animal species that form harems (breeding groups dominated by a single male), so that N_e is considerably smaller than N. Finally, unintended selection for genotypes more suited to captive-breeding conditions over time can reduce the overall capacity of the recovered population to adapt and survive in the wild.

How can this type of program be managed to minimize these genetic effects? As we have already seen, loss of genetic diversity measured as heterozygosity over t generations can be expressed as

$$H_t/H_0 = (1 - 1/2N_e)^t$$

This equation indicates that the loss of genetic diversity H_t/H_0 will be greater with a smaller effective population size N_e and a larger number of generations t. We can expand this equation for a captive-breeding population as follows:

$$H_t/H_0 = [1 - (1/2N_{fo})]\{1 - 1/[2N(N_e/N)]\}^{t-1}$$

where N_{fo} is the effective size of the founding population, N is the mean population size, and N_e is the mean effective population size of the group over t generations. Examination of this equation shows that maximum genetic diversity in a captive-breeding group will be maintained by (1) using the largest possible number of founding individuals to maximize N_{fo}, (2) maximizing N_e/N so that as many individuals as possible produce offspring each generation, and (3) minimizing the number of generations in captivity to reduce t. The importance of maximizing N_{fo}, to increase allelic diversity in the founding

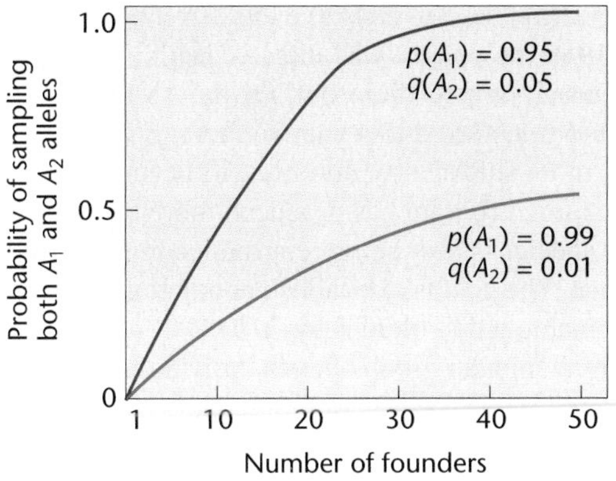

FIGURE 27–9 Effect of captive-population founder number on probability of sampling both *A* and *a* alleles at a locus.

population can also be seen in Figure 27–9. Successful programs of this sort for endangered species are managed with these three goals in mind. In addition, keeping good pedigree records to avoid matings between relatives and exchanging individuals between different breeding programs when possible will reduce inbreeding.

Now solve this

Question 27.14 on page 676 asks you to estimate the potential for genetic erosion in a captive population of rare red pandas.

Hint: If N_e/N is maintained at 0.42 and actual population size at 50, you can calculate N_e. Then use the equation given above.

Captive Breeding: The Black-Footed Ferret

The black-footed ferret, *Mustela nigripes* (Figure 27–10), is an excellent example of a species that has been successfully subjected to captive breeding. This ferret was once widespread throughout the plains of the western United States, but by the 1970s they were considered to be extinct. Decades of trapping and poisoning of both the ferret and its main prey, the prairie dog, had decimated their populations. However, in 1981 a small surviving colony of black-footed ferrets was discovered on a ranch near Meeteetse, Wyoming. Conservation

FIGURE 27–10 The black-footed ferret (*Mustela nigripes*).

biologists first tried to conserve this ferret population *in situ*, but in 1985 the colony was infected with canine distemper, which nearly wiped them out. Of the 18 ferrets that were saved and transferred to a captive facility, only 8 were considered to be sufficiently unrelated to become the founders. The breeding program has produced more than 3000 ferrets, and the species is now being reintroduced to parts of its original range. The goal is to establish populations of 1500 ferrets in sustainable wild colonies by 2010.

The small founder group of eight individuals caused a severe bottleneck for this species. Additional loss of genetic diversity in the captive-breeding program occurred because of drift, limited reproduction from several of the founder females, and a high rate of breeding by one founder male. The risk of inbreeding and further genetic erosion in this type of program is still a problem.

Genetic management strategies, such as using DNA markers to identify the most genetically varied individuals, maintaining careful pedigree records to avoid mating closely related animals, and developing techniques for artificial insemination and sperm cryopreservation have helped to conserve the genetic diversity that remained in the species. Conservation geneticists working on the recovery project estimate that all existing black-footed ferrets share about 12 percent of their genome. This is roughly the equivalent of being full cousins. The effects of this degree of genetic similarity in the expanding ferret population are unclear. Occasional abnormalities such as webbed feet and kinked or short tails have been observed in the captive population, but without a noninbred population available for comparison, it is unclear whether this is evidence that inbreeding is increasing the homozygosity for deleterious alleles.

Scientists disagree as to whether the long-term future of the black-footed ferret is jeopardized by the severe bottleneck and subsequent loss of genetic diversity the species has experienced. Some geneticists suggest that because the one surviving Meeteetse population was so isolated, it may have already become sufficiently inbred to purge any deleterious alleles. Other researchers point out that studies of black-footed ferret DNA extracted from museum specimens show that the ferrets alive today have lost significant genetic diversity compared with earlier prebottleneck populations, and they suggest that loss of fitness because of inbreeding will inevitably be seen over time.

Ex Situ Conservation and Gene Banks

Another form of *ex situ* conservation is provided by establishing **gene banks**. In contrast to housing entire animals or plants, these collections instead provide long-term storage and preservation for reproductive components, such as sperm, ova, and frozen embryos in the case of animals, and seeds, pollen, and cultured tissue in the case of plants. Many more individual genotypes can be preserved for longer periods in a gene bank than in a living collection. Cryopreserved gametes or seeds can be used to reconstitute lost or endangered animals or plants after many years in storage. Because they are expensive to construct and maintain, most gene banks are used to conserve these components of domesticated species having economic value.

Gene banks have been established in many countries to help preserve genetic material of agricultural importance, such as traditional crop varieties that are no longer grown or old livestock breeds that are becoming rare. One of the most important *ex situ* collections in the United States is the National Center for Genetic Resources Preservation, a Department of Agriculture (USDA) facility in Fort Collins, Colorado, which maintains more than 300,000 different accessions of crop varieties and related wild species. Some of the accessions are stored as seeds and others as cryogenically preserved tissue from which whole plants can be regenerated. (See this chapter's opening photograph.) Animal genetic resources, including frozen semen and embryos from endangered livestock breeds, are also preserved at this facility.

Ex situ conservation using gene banks, while often vital, has several disadvantages. A major problem with gene banks is that even large collections cannot contain all the genetic variation that is present in a species. Conservation geneticists attempt to address this problem by identifying a **core collection** for a species. The core collection is a subset of individual genotypes that represents as much as possible of the genetic variation within a species; preserving the core collection takes priority over randomly collecting and preserving large numbers of genotypes. Another disadvantage of *ex situ* conservation is that the artificial conditions under which a species is preserved in a living collection or gene bank often create their own selection pressures. When seeds of a rare plant species are maintained in cold storage, for example, selection may occur for those genotypes better adapted to withstand the lower temperatures, and genotypes may be lost that would actually have greater fitness in the plant's natural environment. Yet another problem posed by *ex situ* conservation is that while the greatest biological diversity in both domesticated and nondomesticated species is frequently found in underdeveloped countries, most *ex situ* collections are situated in developed countries that have the resources to establish and maintain them. This leads to conflict over who owns and has access to the potentially valuable genetic resources maintained in such collections.

In Situ Conservation

In situ (Latin for on-site) **conservation** attempts to preserve the population size and biological diversity of a species while it is maintained in its original habitat. The use of species inventories to identify diversity hot spots is an important tool for determining the best places to establish parks and reserves where plants and animals can be protected from hunting or collecting and where their habitat can be preserved.

For domesticated species, there is increasing interest in "on-farm" preservation, whereby farmers are encouraged with additional resources and financial incentives to maintain traditional crop varieties and livestock breeds. Nondomesticated species with economic potential have also been targeted for *in situ* preservation. In 1998, the USDA established its first *in situ* conservation sites for a wild plant to protect populations of the native rock grape (*Vitis rupestris*) in several eastern states. The rock grape is prized by winegrowers not for its fruit, but for its roots; grape vines grafted onto wild rock grape rootstock are resistant to phylloxera, a serious pest of wine grapes.

The advantage of *in situ* conservation for the rock grape, as with other species, is that larger populations with greater genetic diversity can be maintained. Species conserved *in situ* will also continue to live and reproduce in the environments to which they are adapted, reducing the likelihood that novel selection pressures will produce undesirable changes in allele frequency. However, as the global human population continues to rise, setting aside suitable areas for *in situ* conservation becomes a greater challenge. For example, in the case of the North American brown bear, even in large preserves like Yellowstone National Park, a species' migration and gene flow may be eliminated with the consequent loss of genetic diversity. This problem is even more acute in smaller, more fragmented areas of protected habitat.

Population Augmentation

What genetic considerations should accompany efforts to restore populations or species that are in decline? As we have already seen, populations that go through bottlenecks continue to suffer from low levels of genetic diversity, even after their numbers have recovered. We have also seen that inbreeding and drift contribute to genetic erosion in small populations, and fragmentation interrupts migration and gene flow, further reducing diversity. Captive-breeding programs designed to restore a critically endangered species from a few surviving individuals risk genetic erosion in the renewed population from founder effect and inbreeding, as in the case of the black-footed ferret.

An alternative strategy used by conservation biologists is **population augmentation**—boosting the numbers of a declining population by transplanting and releasing individuals of the same species captured or collected from more numerous populations elsewhere. Attempts to reestablish gene flow in severely fragmented populations of the endangered red-cockaded woodpecker by augmenting the smallest populations were described earlier. Other population augmentation projects in the United States have involved bighorn sheep and grizzly bears in the Rocky Mountains. This method has also been employed with the Florida panther, *Felis coryi* (Figure 27–11), using an isolated population of less than 50 animals confined to the area around the Big Cypress Swamp and Everglades National Park in south Florida. As we will discuss in the Genetics, Technology, and Society essay at the end of this chapter, DNA-profiling patterns show high levels of inbreeding in the Florida panther, with reduced fitness because of severe reproductive abnormalities and increased susceptibility to parasite infections. In this effort, seven unrelated animals from a captive population of South and North American panthers were released into the Everglades in the 1960s and allowed to interbreed with the Florida population, thus producing more genetically diverse family groups. However, further population augmentation in this species has been controversial, as some biologists argue that the unique features that allow the Florida panther to be classified as a separate subspecies will be lost if mating with other panthers takes place.

FIGURE 27–11 The Florida panther (*Felis coryi*).

Despite such controversies, population augmentation appears to be a valuable restoration tool. This strategy increases population numbers, as well as genetic diversity if the transplanted individuals are unrelated to those in the population to which they are introduced. However, one potential problem with population augmentation is **genetic swamping**. This occurs when the gene pool of the original population is overwhelmed by different genotypes from the transplanted individuals, thus altering allele frequencies.

Another difficulty with population augmentation can be caused by **outbreeding depression**, where reduced fitness occurs in the progeny from matings between genetically diverse individuals. Outbreeding depression occurring in the F_1 generation is thought to be due to the offspring being less well adapted to local environmental conditions than the parents.

This phenomenon has been documented in some plant species in which seeds of the same species—but from a different location—were used to revegetate a damaged area. Outbreeding depression that occurs in the F_2 and later generations is thought to be due to the disruption of **coadapted gene complexes**—groups of alleles that have evolved to work together to produce the best level of fitness in an individual. This type of outbreeding depression has been documented in F_2 hybrid offspring from matings between fish from different salmon populations in Alaska. These studies suggest that restoring the most beneficial type and amount of genetic diversity in a population is more complicated than previously thought, reinforcing the argument that the best long-term strategy for species survival is to prevent the loss of diversity in the first place.

GENETICS, TECHNOLOGY, AND SOCIETY

Gene Pools and Endangered Species: The Plight of the Florida Panther

In the last 400 years, more than 700 of Earth's animal and plant species have become extinct. In the United States alone, at least 30 species have suffered extinction in the last decade. In addition, hundreds of genetically distinct plant and animal species are now endangered. This dramatic loss of biological diversity is a direct consequence of human activity. As humans increase in number and spread over Earth's surface, we harness more and more of Earth's resources for our use. As a consequence of these activities, we have cleared the forests, polluted the water, and permanently altered Earth's natural balance.

Although the destruction continues, we are beginning to understand the implications of our actions and to make efforts to save some of the more endangered life forms. However, despite our best efforts to save threatened species, we often intercede too late—after population numbers have suffered severe declines and after important ecosystems have been permanently compromised. As a result, much of the genetic diversity that existed in these species is lost and the populations must be rebuilt from reduced gene pools. This genetic uniformity can reduce fitness, as well as expose genetic diseases. Reduced fitness may further deplete the numbers of the threatened plant or animal, and the population spirals downward to extinction.

The story of the Florida panther provides a dramatic example of an animal brought to the verge of extinction and the challenges that must be overcome to restore it to a healthy place in the ecosystem. The Florida panther is one of 30 subspecies of cougar and is one of the most endangered mammals in the world. The panthers once roamed the southeastern corner of North America, from South Carolina and Arkansas to the southern tip of Florida. As people settled in the panthers' habitats, the panthers were considered to be a threat to livestock and humans. So the panthers were killed by hunting, poisoning, highway collisions, and loss of the habitat that supports their prey—primarily wild deer and hogs. By 1967, Florida panthers were listed as endangered, and only about 30 Florida panthers were left. Population estimates predicted that the Florida panther would be extinct by the year 2055.

Because of about 20 generations of geographical isolation and inbreeding, Florida panthers had the lowest levels of genetic heterozygosity of any subspecies of cougar. The loss of genetic diversity manifested itself in the appearance of some severe genetic defects. For example, almost 80 percent of panther males born after 1989 in the Big Cypress region showed a rare, heritable (autosomal dominant or X-linked recessive) condition known as cryptorchidism—failure of one or both testicles to descend. This defect is associated with low testosterone levels and reduced sperm count. Life-threatening congenital heart defects also appeared in the Florida panther population, possibly because of an autosomal dominant gene defect. In addition, some immune deficiencies emerged, making the small panther population more susceptible to diseases and further contributing to the population's decline. Other less serious genetic features appeared, such as a kink in the tail and a whorl of fur on the back.

Over the last two decades, a faint glimmer of hope has appeared for the Florida panther. Federal and state agencies, as well as private individuals, have implemented a Florida Panther Recovery Program. The program's goal is to exceed 500 breeding animals by the year 2010. If successful, the panther species would be granted a 95 percent probability of survival, while retaining up to 90 percent of its genetic diversity. The plan includes a captive-breeding program, strict protection, increasing and improving the panther's habitat, and educating the public and private landowners. Wildlife underpasses have been constructed on highways in panther territory, and these have significantly reduced panther highway fatalities (which account for about half of panther deaths). By 2002, Florida panther numbers had increased to about 100 animals.

To retard the detrimental effects of inbreeding, eight wild female Texas panthers (a related subspecies from western Texas) were released into Florida panther territory in 1995. Three of the females died prior to breeding; however, the remaining Texas females gave birth to litters of healthy kittens. It is estimated that 40 to 70 of the 100 panthers in South Florida are now hybrids of Texas panthers and Florida panthers. None of the hybrids appears to have the kinked tail or other traits characteristic of the inbred Florida panthers. The hybrid cats are more genetically diverse than purebred Florida panthers, with up to 20 to 30 percent of their genetic material being contributed by the Texas panthers. This is of concern to some biologists, who worry that the Texas cats may genetically swamp the distinct Florida panther population.

Paradoxically, the success of the restoration program has become a problem. Now that the population has reached about 100 animals, the Florida panther has almost exceeded the capacity of its existing habitat. Biologists estimate that the population must reach at least 250 individuals in order to be self-sustaining. To reach this number, the panthers will need to expand their territory, again putting them in direct competition with human expansion in Florida. Biologists are now evaluating potential new territories for the growing population of Florida panthers—including parts of Louisiana, Arkansas, and South Carolina.

Despite the recent success of the restoration program, the survival of the Florida panther is far from certain. The panther's comeback will require years of monitoring and frequent intervention. In addition, people must be willing to share their land with wild creatures that are dangerous and do not directly further human interests. However, public support for the return of the Florida panther has been strong, so there may be hope for this unique, impressive animal.

References

Maehr, D.S., and Lacy, R.C. 2002. Avoiding the lurking pitfalls in Florida panther recovery. *Wildlife Soc. Bull.* 30(3):971–78.

Mansfield, K.G., and Land, E.D. 2002. Cryptorchidism in Florida panthers: Prevalence, features and influence of genetic restoration. *J. of Wildlife Diseases* 38(4):693–98.

Web Sites

Derr, M. 2002. Florida panther's great leap hits a wall [online]. *The New York Times*, 15 October 2002.

http://www.nytimes.com/2002/10/15/science/life/15PANT.html

Florida Panther Net [online].

http://www.panther.state.fl.us/

Chapter Summary

1. Biodiversity is lost as increasing numbers of plants and animal species are threatened with extinction. Conservation genetics applies principles of population genetics to the preservation and restoration of threatened species. A major concern of conservation geneticists is the maintenance of genetic diversity.
2. Genetic diversity includes interspecific diversity, which is reflected by the number of different species present in an ecosystem, and intraspecific diversity, which is reflected by genetic variation within a population or between different populations of the same species. Genetic diversity can be measured by examining different phenotypes in the population or, at the molecular level, by using allozyme analysis or DNA-profiling techniques.
3. Major declines in a species' population numbers reduce genetic diversity and contribute to their risk of extinction as a result of genetic drift, inbreeding, or loss of gene flow. Populations that suffer severe reductions in effective size and then recover have passed through a population bottleneck and often show reduced genetic diversity.
4. Loss of genetic diversity reduces the capacity of a population to adapt to changing environmental conditions because useful alleles may disappear from the gene pool. Reduced genetic diversity also results in greater levels of homozygosity in a population, often leading to an accumulation of deleterious alleles and inbreeding depression.
5. Conservation of genetic diversity depends on *ex situ* methods, such as living collections, captive-breeding programs, and gene banks, as well as *in situ* approaches, such as the establishment of parks and preserves.
6. Population augmentation, in which individuals are transplanted into a declining population from a more numerous population of the same species located elsewhere, can be used to increase numbers and genetic diversity. However, the risk of outbreeding depression and genetic swamping accompanies this process.

Insights and Solutions

1. Is a rare species found as several fragmented subpopulations more vulnerable to extinction than an equally rare species found as one larger population? What factors should be considered when managing fragmented populations of a rare species?

Solution: A rare species in which the remaining individuals are divided among smaller isolated sub-populations can appear to be less vulnerable. If one subpopulation becomes extinct through local causes, such as disease or habitat loss, then the remaining subpopulations may still survive. However, genetic drift will cause smaller populations to experience more rapidly increasing homozygosity over time compared with larger populations. Even with random mating, the change in heterozygosity from one generation to the next because of drift can be calculated as

$$H_1 = H_0(1 - 1/2N)$$

where H_0 is the frequency of heterozygotes in the present generation, H_1 is the frequency of heterozygotes in the next generation, and N is the number of individuals in the population. Thus, in a small population of 50 individuals with an initial heterozygote frequency of 0.5, heterozygosity will decline to $0.5(1 - 1/100) = 0.495$, a loss of 0.5 percent in just one generation. In a larger population of 500 individuals and the same initial heterozygote frequency, after one generation, heterozygosity will be $0.5(1 - 1/1000) = 0.4995$, a loss of only 0.05 percent.

Therefore, smaller populations are likely to show the effects of homozygosity for deleterious alleles sooner than larger populations, even with random mating. If populations are fragmented so that movement of individuals or gametes between them is prevented, management options could include transplanting individuals from one subpopulation to another to enable gene flow to occur. Establishment of "wildlife corridors" of undisturbed habitat that connect fragmented populations could be considered. In captive populations, exchange of breeding adults (or their gametes through shipment of preserved semen or pollen) can be undertaken. Management for increased population numbers, however, is vital to prevent further genetic erosion through drift.

Problems and Discussion Questions

1. A wildlife biologist studied four generations of a population of rare Ethiopian jackals. When the study began, there were 47 jackals in the population and analysis of microsatellite loci from these animals showed a heterozygote frequency of 0.55. In the second generation, an outbreak of distemper occurred in the population and only 17 animals survived to adulthood. These jackals produced 20 surviving offspring, which in turn gave rise to 35 progeny in the fourth generation. (a) What was the effective population size for the four generations of this study? (b) Based on this effective population size, what is the heterozygote frequency of the jackal population in generation 4? (c) What is the inbreeding coefficient in generation 4, assuming an inbreeding coefficient of $F = 0$ at the beginning of the study, no change in microsatellite allele frequencies in the gene pool, and random mating in all generations?
2. Chondrodystrophy, a lethal form of dwarfism, has recently been reported in captive populations of the California condor and has killed embryos in 5 out of 169 fertile eggs. Chondrodystrophy in condors appears to be caused by an autosomal recessive allele with an estimated frequency of 0.09 in the gene pool of this

species. (a) How do you think California condor populations should be managed in the future to minimize the effect of this lethal allele? (b) What are the advantages and disadvantages of attempting to eliminate it from the gene pool?

3. A geneticist is studying three loci, each with one dominant and one recessive allele, in a small population of rare plants. She estimates the frequencies of the alleles at each of these loci as $A = 0.75$, $a = 0.25$; $B = 0.80$, $b = 0.20$; $C = 0.95$, $c = 0.05$. What is the probability that all the recessive alleles will be lost from the population through genetic drift?
4. How are genetic drift and inbreeding similar in their effects on a population? How are they different?
5. You are the manager of a game park in Africa with a native herd of just 16 black rhinos, an endangered species worldwide. Describe how you would manage this herd to establish a viable population of black rhinos in the park. What genetic factors would you take into consideration in your management plan?
6. Compare the causes and effects of inbreeding depression and outbreeding depression.
7. Cloning, using the techniques similar to those pioneered by the Scottish scientists who produced Dolly the sheep, has been proposed as a way to increase the numbers of some highly endangered mammalian species. Discuss the advantages and disadvantages of using such an approach to aid long-term species survival.
8. In a population of wild poppies found in a remote region of the mountains of eastern Mexico, almost all of the members have pale yellow flowers, but breeding experiments show that pale yellow is recessive to deep orange. Using the tools of the conservation geneticists described in this chapter, how could you experimentally determine whether the prevalence of the recessive phenotype among the eastern Mexican poppy population is due to natural selection or simply due to the effects of genetic drift and/or inbreeding?
9. Contrast *ex situ* conservation techniques with *in situ* conservation techniques.
10. Describe how the captive-breeding *ex situ* conservation approach is applied to a severely endangered species.
11. Explain why a low amount of genetic diversity in a species is a detriment to the survival of that species.
12. Contrast allozyme analysis with RFLP analysis as measures of genetic diversity.
13. A population of endangered lowland gorillas is studied by conservation biologists in the wild. The biologists count 15 gorillas but observe that the population consists of two harems, each dominated by a different single male. One harem contains eight females and the other has five. What is the effective population size, N_e, of the gorilla population?
14. Twenty endangered red pandas are taken into captivity to found a breeding group. (a) What is the probability that at least one of the captured pandas has the genotype B_1/B_2 if $p(B_1) = 0.99$? (b) Careful management keeps the N_e/N ratio for the red panda breeding population at 0.42. If the overall size of the captive population is maintained at 50 individuals, what proportion of the heterozygosity present in the founding population will still be present after 5 generations in captivity?
15. Use your analysis of the red panda in Problem 14 to make suggestions for managing the captive-breeding population of red pandas to maintain as much genetic diversity as possible.

Extra-Spicy Problems

16. Przewalski's horse (*Equus przewalskii*), thought by some biologists to represent the ancestral species from which modern horses were domesticated, is classified as a separate species from the domestic horse although members of the two species can mate and produce fertile offspring. Przewalski's horse was hunted to extinction in its native habitat in the Asian steppes by the 1920s. The few hundred Przewalski's horses alive today are descended from 12 surviving individuals that had been taken into captivity. The founder breeding group also included a domestic mare. Currently, there is interest in reintroducing Przewalski's horse to its original range. (a) What genetic factors associated with the animals available for reintroduction do you think should be considered before a decision is reached? (b)What advice would you give to the conservation biologists managing the reintroduction project?
17. DNA profiles based on different kinds of molecular markers are increasingly used to measure levels of genetic diversity in populations of threatened and endangered species. Do you think these marker-based estimates of genetic diversity are reliable indicators of a population's potential for survival and adaptation in a natural environment? Why, or why not?
18. Seed banks provide managed protection for the conservation of species of economic and noneconomic importance. Schoen and colleagues (*Proc. Natl. Acad. Sci. [USA]* 1998. 95:394–99) quantified considerable fitness decay in seeds maintained in long-term storage. When germination falls below 65–85 percent, regeneration (planting and seed collection) of a finite sample of the stored seed type is recommended. What genetic consequences might you expect to accompany the conservation practice of long-term seed storage and regeneration?
19. According to Holmes (*Proc. Natl. Acad. Sci. [USA] 2001* 98:5072–77), "One of the first questions a resource manager asks about threatened and endangered species is: How bad is it?" More formally, the question seeks a population viability analysis that includes estimates of extinction risk. What factors would you consider significant in providing an estimate of extinction risk for a species?
20. Microsatellite loci are short (2-to-5 base pair) tandem repeats, which are abundantly and somewhat randomly distributed in all eukaryotic chromosomes. They show high mutability such that new alleles are produced more frequently than in traditional protein-coding genes. It is likely that most microsatellites are selectively neutral and highly heterozygous in natural populations. Knowing that cheetahs underwent a population bottleneck approximately 12,000 years ago, North American pumas 10,000 years ago, and Gir Forest lions 1,000 years ago, in which of these species would you expect to see the highest degree of microsatellite polymorphism? the lowest?
21. Considering the behavior and evolution of microsatellites described in Problem 20, provide a graph that relates microsatellite polymorphism (variance) with the number of years since a species' bottleneck. Select variances from 1.0 to 9.0, where increasing variance represents increasing polymorphism, and set the range of years since the bottleneck from 10,000 to 50,000.
22. Allozymes are electrophoretically distinct forms of a particular protein, while microsatellites and minisatellites are repetitive DNA sequences that have been found in all eukaryotes studied to date. Such repetitive sequences are rarely associated with coding sequences of DNA. The data shown here represent the per-

cent heterozygosity of allozymes and microsatellite and minisatellite DNAs in nuclear genomes of four species of felines (modified from Driscoll et al. 2002. *Genome Res.* 12:414–23). (a) Which species appears to contain the greatest genetic variability? Why might this species be so variable? (b) Which species appears to have the least genetic variability? (c) Why are allozymes less variable than minisatellite or microsatellite DNAs?

	% Heterozygosity			
Marker	**Cheetah**	**Lion**	**Puma**	**Domesticated Cat**
Allozyme	1.4	0.0	1.8	8.2
Minisatellite	43.3	2.9	10.3	44.9
Microsatellite	46.7	7.9	14.7	68.1

23. If we assume that one of the species in Problem 22 had undergone a recent population crisis in which the number of effective breeders reached critically low levels, which species do you think it would be?

24. For many conservation efforts, scientists lack sufficient data to make definitive conservation decisions. Yet the allocation of habitats for conservation cannot be delayed until such data are available. In these cases, conservation efforts are often directed toward three classes of species: flagships (high-profile species), umbrellas (species requiring large areas for habitat), and biodiversity indicators (species representing diverse, especially productive habitats) (Andelman and Fagan, 2000. *Proc. Natl. Acad. Sci. [USA]* 97:5954–59). In terms of protecting threatened species, what advantages and disadvantages might accompany investing scarce talent and resources in each of these classes?

Selected Readings

Baker, C.S., and Palumbi, S.R. 1994. Which whales are hunted? A molecular genetic approach to monitoring whaling. *Science* 265:1538–39.

Bonnell, M.L., and Selander, R.K. 1974. Elephant seals: Genetic variation and near extinction. *Science* 184:908–09.

Daniels, S.J., and Walters, J.R. 2000. Inbreeding depression and its effects on natal dispersal in red-cockaded woodpeckers. *Condor* 102:482–91.

Dobson, A., and Lyles, A. 2000. Black-footed ferret recovery. *Science* 288:985.

Frankham, R. 1995. Conservation genetics. *Ann. Rev. Genet.* 29:305–27.

Gharrett, A.J., and Smoker, W.W. 1991. Two generations of hybrids between even-year and odd-year pink salmon (*Oncorhynchus gorbuscha*): A test for outbreeding depression? *Canadian J. Fish. Aquat. Sci.* 48:426–38.

Hedrick, P.W. 2001. Conservation genetics: Where are we now? *Trends Ecol. Evol.* 16:629–36.

Lacy, R.C. 1997. Importance of genetic variation to the viability of mammalian populations. *J. Mammalogy* 78:320–35.

Paetkau, D., et al. 1998. Variation in genetic diversity across the range of North American brown bears. *Conserv. Biol.* 12:418–29.

Palacios, C., and Gonzalez-Candelas, F. 1999. AFLP analysis of the critically endangered *Limonium cavanillesii. J. of Heredity* 90:485–89.

Ralls, K., et al. 2000. Genetic management of chondrodystrophy in California condors. *Animal Conserv.* 3:145–53.

Roman J., and Bowen, B.W. 2000. The mock turtle syndrome: Genetic identification of turtle meat purchased in the southeastern United States of America. *Animal Conserv.* 3:61–65.

Wayne, R.K., et al. 1991. Conservation genetics of the endangered Isle Royale gray wolf. *Conserv. Biol.* 5:41–51.

Wilson, E.O., (ed.). 1988. *Biodiversity*. Washington, DC: National Academy of Sciences.

Wynen, L.P., et al. 2000. Postsealing genetic variation and population structure of two species of fur seal. *Mol. Ecol.* 9:299–314.

Appendix A

Glossary

abortive transduction An event in which transducing DNA fails to be incorporated into the recipient chromosome. See *transduction.*

acentric chromosome Chromosome or chromosome fragment with no centromere.

acquired immunodeficiency syndrome (AIDS) An infectious disease caused by a retrovirus named the human immunodeficiency virus (HIV). The disease is characterized by a gradual depletion of T lymphocytes, recurring fever, weight loss, multiple opportunistic infections, and rare forms of pneumonia and cancer associated with collapse of the immune system.

acridine dyes A class of organic compounds that bind to DNA and intercalate into the double-stranded structure, producing local disruptions of base pairing. These disruptions result in nucleotide additions or deletions in the next round of replication.

acrocentric chromosome Chromosome with the centromere located very close to one end. Human chromosomes 13, 14, 15, 21, and 22 are acrocentric.

active site That portion of a protein, usually an enzyme, whose structural integrity is required for function (e.g., the substrate binding site of an enzyme).

adaptation A heritable component of the phenotype that confers an advantage in survival and reproductive success. The process by which organisms adapt to the current environmental conditions.

additive genes See *polygenic inheritance.*

additive variance Genetic variance that is attributed to the substitution of one allele for another at a given locus. This variance can be used to predict the rate of response to phenotypic selection in quantitative traits.

A-DNA An alternative form of the right-handed double-helical structure of DNA in which the helix is more tightly coiled, with 11 base pairs per full turn of the helix. In the A form, the bases in the helix are displaced laterally and tilted in relation to the longitudinal axis. It is not yet clear whether this form has biological significance.

albinism A condition caused by the lack of melanin production in the iris, hair, and skin. It is most often inherited as an autosomal recessive trait in humans.

aleurone layer In seeds, the outer layer of the endosperm.

alkaptonuria An autosomal recessive condition in humans caused by the lack of the enzyme homogentisic acid oxidase. Urine of homozygous individuals turns dark upon standing because of oxidation of excreted homogentisic acid. The cartilage of homozygous adults blackens from deposition of a pigment derived from homogentisic acid. Affected individuals often develop arthritic conditions.

allele One of the possible mutational states of a gene, distinguished from other alleles by phenotypic effects.

allele frequency Measurement of the proportion of individuals in a population carrying a particular allele.

allele-specific oligonucleotide (ASO) Synthetic nucleotides, usually 15–20 bp in length, that under carefully controlled conditions will hybridize only to a perfectly matching complementary sequence. Under these conditions, ASOs with a one-nucleotide mismatch will not hybridize.

allelic exclusion In a plasma cell heterozygous for an immunoglobulin gene, the selective action of only one allele.

allelism test See *complementation test.*

allolactose A lactose derivative that acts as the inducer for the *lac* operon.

allopatric speciation Process of speciation associated with geographic isolation.

allopolyploid Polyploid condition formed by the union of two or more distinct chromosome sets with a subsequent doubling of chromosome number.

allosteric effect Conformational change in the active site of a protein brought about by interaction with an effector molecule.

allotetraploid Diploid for two genomes derived from different species.

allozyme An allelic form of a protein that can be distinguished from other forms by electrophoresis.

alpha fetoprotein (AFP) A 70-kDa glycoprotein synthesized during embryonic development by the yolk sac. High levels of this protein in the amniotic fluid are associated with neural tube defects such as spina bifida; lower-than-normal levels may be associated with Down syndrome.

alternative splicing Generation of different protein molecules from the same pre-mRNA by changing the number and order of exons in the mRNA product.

***Alu* sequence** An interspersed DNA sequence of approximately 300 bp found in the genome of primates that is cleaved by the restriction enzyme *Alu*I. These sequences are composed of a head-to-tail dimer. The first monomer is approximately 140 bp and the second, approximately 170 bp. In humans, they are dispersed throughout the genome and are present in 300,000–600,000 copies, constituting some 3–6 percent of the genome. See *short interspersed elements.*

amber codon The codon UAG, which does not code for an amino acid but for chain termination.

Ames test An assay developed by Bruce Ames to detect mutagenic and carcinogenic compounds, using reversion to histidine independence in the bacterium *Salmonella typhimurium.*

amino acid Any of the subunit building blocks that are covalently linked to form proteins.

aminoacyl tRNA Covalently linked combination of an amino acid and a tRNA molecule.

amniocentesis A procedure used to test for fetal defects in which fluid and fetal cells are withdrawn from the amniotic layer surrounding the fetus.
amphidiploid See *allotetraploid*.
anabolism The metabolic synthesis of complex molecules from less complex precursors.
analog A chemical compound structurally similar to another, but differing by a single functional group (e.g., 5-bromodeoxyuridine is an analog of thymidine).
anaphase Stage of cell division in which chromosomes begin moving to opposite poles of the cell.
anaphase I The stage in the first meiotic division during which members of homologous pairs of chromosomes separate from one another.
aneuploidy A condition in which the chromosome number is not an exact multiple of the haploid set.
angstrom (Å) Unit of length equal to 10^{-10} meters.
annotation Analysis of genomic nucleotide sequence data to identify the protein-coding genes, the nonprotein-coding genes, their regulatory sequences, and their function(s).
antibody Protein (immunoglobulin) produced in response to an antigenic stimulus with the capacity to bind specifically to an antigen.
anticipation A phenomenon first observed in myotonic dystrophy, where the severity of the symptoms increases from generation to generation and the age of onset decreases from generation to generation. This phenomenon is caused by the expansion of trinucleotide repeats within or near a gene.
anticodon The nucleotide triplet in a tRNA molecule that is complementary to and binds with the codon triplet in an mRNA molecule.
antigen A molecule, often a cell-surface protein, that is capable of eliciting the formation of antibodies.
antiparallel Describing molecules in parallel alignment, but running in opposite directions. Most commonly used to describe the opposite orientations of the two strands of a DNA molecule.
apoptosis A genetically controlled program of cell death, activated as part of normal development or as a result of cell damage.
ascospore A meiotic spore produced in certain fungi.
ascus In fungi, the sac enclosing the four or eight ascospores.
asexual reproduction Production of offspring in the absence of any sexual process.
assortative mating Nonrandom mating between males and females of a species. Selection of mates with the same genotype is positive; selection of mates with opposite genotypes is negative.
ATP Adenosine triphosphate. A nucleotide that is the main energy source in cells.
attached-X chromosome Two conjoined X chromosomes that share a single centromere.
attenuator A nucleotide sequence between the promoter and the structural gene of some operons that regulates the transit of RNA polymerase, reducing transcription of the related structural gene.
autogamy A process of self-fertilization resulting in homozygosis.
autoimmune disease The production of antibodies that results from an immune response to one's own molecules, cells, or tissues. Such a response results from the inability of the immune system to distinguish self from nonself. Diseases such as arthritis, scleroderma, systemic lupus erythematosus, and juvenile-onset diabetes are examples of autoimmune diseases.
autonomously replicating sequences (ARS) Origins of replication, about 100 nucleotides in length, found in yeast chromosomes. ARS elements are also present in organelle DNA.
autopolyploidy Polyploid condition resulting from the duplication of one diploid set of chromosomes.
autoradiography Production of a photographic image by radioactive decay. Used to localize radioactively labeled compounds within cells and tissues.
autosomes Chromosomes other than the sex chromosomes. In humans, there are 22 pairs of autosomes.
autotetraploid An autopolyploid condition composed of four similar genomes. In this situation, genes with two alleles (*A* and *a*) can have five genotypic classes: *AAAA* (quadraplex), *AAAa* (triplex), *AAaa* (duplex), *Aaaa* (simplex), and *aaaa* (nulliplex).
auxotroph A mutant microorganism or cell line that requires a substance for growth that can be synthesized by wild-type strains.
backcross A cross involving an F_1 heterozygote and one of the P_1 parents (or an organism with a genotype identical to one of the parents).
bacteriophage A virus that infects bacteria (also, *phage*).
bacteriophage λ A member of the lambdoid family of viruses that attach to, infect, and replicate within bacterial cells, destroying the host cell in the process. Genetically modified lambda phages are used as vectors in recombinant DNA research.
bacteriophage μ A group of phages whose genetic material behaves as an insertion sequence that can inactivate host genes and rearrange host chromosomes.
balanced lethals Recessive, nonallelic lethal genes, each carried on different homologous chromosomes. When organisms carrying balanced lethal genes are interbred, only organisms with genotypes identical to the parents (heterozygotes) survive.
balanced polymorphism Genetic polymorphism maintained in a population by natural selection.
Barr body Densely staining nuclear mass seen in the somatic nuclei of mammalian females. Discovered by Murray Barr, this body represents an inactivated X chromosome.
base analog See *analog*.
base substitution A single base change in a DNA molecule that produces a mutation. There are two types of substitutions: *transitions*, in which a purine is substituted for a purine or a pyrimidine for a pyrimidine; and *transversions*, in which a purine is substituted for a pyrimidine or vice versa.
β-galactosidase A bacterial enzyme encoded by the *lacZ* gene that converts lactose into galactose and glucose.
bidirectional replication A mechanism of DNA replication in which two replication forks move in opposite directions from a common origin of replication.
biodiversity The genetic diversity present in populations and species of plants and animals.
biometry The application of statistics and statistical methods to biological problems.
biotechnology Commercial and/or industrial processes that utilize biological organisms or products.
bivalents Synapsed homologous chromosomes in the first prophase of meiosis.
Bombay phenotype A rare variant of the ABO system in which affected individuals do not have A or B antigens and thus appear as blood type O, even though their genotype may carry unexpressed alleles for the A and/or B antigens.
bottleneck Fluctuation in allele frequency that occurs when a population undergoes a temporary reduction in size.
BSE (bovine spongiform encephalopathy) A fatal, degenerative brain disease of cattle caused by prion infection. This infection is transmissible to humans and other animals. Also know as mad cow disease.
BrdU (5-bromodeoxyuridine) A mutagenically active analog of thymidine in which the methyl group at the 5′ position in thymine is replaced by bromine; also abbreviated BUdR.
buoyant density A property of particles (and molecules) that depends upon their actual density, as determined by partial specific volume and degree of hydration. Provides the basis for density gradient separation of molecules or particles.

CAAT box A highly conserved DNA sequence found in the untranslated promoter region of eukaryotic genes. This sequence is recognized by transcription factors.

CAP Catabolite activator protein; a protein that binds cAMP and regulates the activation of inducible operons.

carcinogen A physical or chemical agent that causes cancer.

carrier An individual heterozygous for a recessive trait.

catabolism A metabolic reaction in which complex molecules are broken down into simpler forms, often accompanied by the release of energy.

catabolite activator protein See *CAP*.

catabolite repression The selective inactivation of an operon by a metabolic product of the enzymes encoded by the operon.

***cdc* mutation** A class of cell division cycle mutations in yeasts that affect the timing and progression through the cell cycle.

cDNA DNA synthesized from an RNA template by the enzyme reverse transcriptase.

cDNA library A collection of cloned cDNA sequences.

cell cycle Sum of the phases of growth of an individual cell type; divided into G1 (gap 1), S (DNA synthesis), G2 (gap 2), and M (mitosis).

cell-free extract A preparation of the soluble fraction of cells, made by lysing cells and removing the particulate matter, such as nuclei, membranes, and organelles. Often used to carry out the synthesis of proteins by the addition of specific, exogenous mRNA molecules.

CEN In yeasts, fragments of chromosomal DNA, about 120 bp in length, that when inserted into plasmids confer the ability to segregate during mitosis. These segments contain at least three types of sequence elements associated with centromere function.

centimeter (cm) A unit of length equal to 10^{-2} meter.

centimorgan (cM) A unit of distance between genes on chromosomes. One centimorgan represents a value of 1 percent crossing over between two genes.

central dogma The concept that information flow progresses from DNA to RNA to proteins. Although exceptions are known, this idea is central to an understanding of gene function.

centric fusion See *Robertsonian translocation*.

centriole A cytoplasmic organelle composed of nine groups of microtubules, generally arranged in triplets. Centrioles function in the generation of cilia and flagella and serve as foci for the spindles in cell division.

centromere Specialized region of a chromosome to which sister chromatids remain attached after replication and the site to which spindle fibers attach during cell division. Location of the centromere determines the shape of the chromosome during the anaphase portion of cell division. Also known as the primary constriction.

centrosome Region of the cytoplasm containing a pair of centrioles.

chaperone A protein that regulates the folding of a polypeptide into a functional three-dimensional shape.

character An observable phenotypic attribute of an organism.

charon phages A group of genetically modified lambda phages designed to be used as vectors (carriers) for cloning foreign DNA. Named after the ferryman in Greek mythology who carried the souls of the dead across the River Styx.

chemotaxis Negative or positive response to a chemical gradient.

chiasma (pl., **chiasmata**) The crossed strands of nonsister chromatids seen in diplotene of the first meiotic division. Regarded as the cytological evidence for exchange of chromosomal material, or crossing over.

chi-square (χ^2) analysis Statistical test to determine if an observed set of data fits a theoretical expectation.

chloroplast A cytoplasmic self-replicating organelle containing chlorophyll. The site of photosynthesis.

chorionic villus sampling (CVS) A technique of prenatal diagnosis that intravaginally retrieves chorionic fetal cells and uses them to detect cytogenetic and biochemical defects in the embryo.

chromatid One of the longitudinal subunits of a replicated chromosome; it is joined to its sister chromatid at the centromere.

chromatin The complex of DNA, RNA, histones, and nonhistone proteins that make up uncoiled chromosomes characteristic of the eukaryotic interphase nucleus.

chromatography Technique for the separation of a mixture of solubilized molecules by their differential migration over a substrate.

chromocenter An aggregation of centromeres and heterochromatic elements of polytene chromosomes.

chromomere A coiled, beadlike region of a chromosome most easily visualized during cell division. The aligned chromomeres of polytene chromosomes are responsible for their distinctive banding pattern.

chromosomal aberration Any change resulting in the duplication, deletion, or rearrangement of chromosomal material.

chromosomal mutation See *chromosomal aberration*.

chromosomal polymorphism Alternative structures or arrangements of a chromosome that are carried by members of a population.

chromosome In prokaryotes, one or more DNA molecules containing the genome; in eukaryotes, a DNA molecule complexed with RNA and proteins to form a threadlike structure containing genetic information arranged in a linear sequence and visible during mitosis and meiosis.

chromosome banding Technique for the differential staining of mitotic or meiotic chromosomes to produce a characteristic banding pattern, or selective staining of certain chromosomal regions such as centromeres, the nucleolus organizer regions, and GC- or AT-rich regions—not to be confused with the banding pattern present in polytene chromosomes, which is produced by the alignment of chromomeres.

chromosome map A diagram showing the location of genes on chromosomes.

chromosome puff A localized uncoiling and swelling in a polytene chromosome, usually regarded as a sign of active transcription.

chromosome theory of inheritance The idea put forward independently by Walter Sutton and Theodore Boveri that chromosomes are the carriers of genes and the basis for the Mendelian mechanisms of segregation and independent assortment.

chromosome walking A method for analyzing long stretches of DNA, in which the end of a cloned segment of DNA is subcloned and used as a probe to identify other clones that overlap the first clone.

***cis* configuration** The arrangement of two genes or two mutant sites within a gene on the same homolog, such as

$$\frac{a^1 \quad a^2}{+ \quad +}$$

contrasts with a *trans* arrangement, where the mutant alleles are located on opposite homologs.

***cis–trans* test** A genetic test to determine whether two mutations are located within the same cistron.

cistron That portion of a DNA molecule coding for a single polypeptide chain; defined by a genetic test as a region within which two mutations cannot complement each other.

cline A gradient of genotype or phenotype distributed over a geographic range.

clonal selection Theory of the immune system that proposes that antibody diversity precedes exposure to the antigen and that the antigen functions to select the cells containing its specific antibody to undergo proliferation.

clone Identical molecules, cells, or organisms derived from a single ancestor by asexual or parasexual methods, such as a DNA segment that has been enzymatically inserted into a plasmid or chromosome of a phage or a bacterium and replicated to form many copies.

cloned DNA library A collection of cloned DNA molecules representing all or part of an individual's genome.
code See *genetic code.*
codominance Condition in which the phenotypic effects of a gene's alleles are fully and simultaneously expressed in the heterozygote.
codon A triplet of nucleotides that specifies or encodes the information for a single amino acid. Sixty-one codons specify the amino acids used in proteins, and three codons called stop codons signal termination of growth of the polypeptide chain.
coefficient of coincidence A ratio of the observed number of double crossovers divided by the expected number of such crossovers.
coefficient of inbreeding The probability that two alleles present in a zygote are descended from a common ancestor.
coefficient of selection (*s*) A measurement of the reproductive disadvantage of a given genotype in a population. For example, if for genotype *aa* only 99 of 100 individuals reproduce, then the selection coefficient is 0.1.
colchicine An alkaloid compound that inhibits spindle formation during cell division. Used in the preparation of karyotypes to collect a large population of cells inhibited at the metaphase stage of mitosis.
colinearity The linear relationship between the nucleotide sequence in a gene (or the RNA transcribed from it) and the order of amino acids in the polypeptide chain specified by the gene.
competence In bacteria, the transient state or condition during which the cell can bind and internalize exogenous DNA molecules, making transformation possible.
complementarity Chemical affinity between nitrogenous bases as a result of hydrogen bonding. Responsible for the base-pairing between the strands of the DNA double helix.
complementation test A genetic test to determine whether two mutations occur within the same gene. If two mutations are introduced into a cell simultaneously and produce a wild-type phenotype (i.e., they complement each other), they are often nonallelic. If a mutant phenotype is produced, the mutations are noncomplementing and are often allelic.
complete linkage A condition in which two genes are located so close to each other that no recombination occurs between them.
complexity The total number of nucleotides or nucleotide pairs in a population of nucleic acid molecules as determined by reassociation kinetics.
complex locus A gene within which a set of functionally related pseudoalleles can be identified by recombinational analysis (e.g., the *bithorax* locus in *Drosophila*).
complex trait A trait whose phenotype is determined by the interaction of multiple genes and environmental factors.
concatemer A chain or linear series of subunits linked together. The process of forming a concatemer is called concatenation (e.g., multiple units of a phage genome produced during replication).
concordance Pairs or groups of individuals identical in their phenotype. In twin studies, a condition in which both twins exhibit or fail to exhibit a trait under investigation.
conditional mutation A mutation that expresses a wild-type phenotype under certain (permissive) conditions and a mutant phenotype under other (restrictive) conditions.
conjugation Temporary fusion of two single-celled organisms for the sexual transfer of genetic material.
consanguineous Related by a common ancestor within the previous few generations.
consensus sequence A basically common, although not necessarily identical, sequence of nucleotides in DNA or amino acids in proteins.
conservation genetics The branch of genetics concerned with the preservation and maintenance of wild species of plants and animals in their natural environments.
continuous variation Phenotype variation exhibited by quantitative traits distributed from one phenotypic extreme to another in an overlapping or continuous fashion.
cosmid A vector designed to allow cloning of large segments of foreign DNA. Cosmids are composed of the *cos* sites of phage λ inserted into a plasmid. In cloning, the recombinant DNA molecules are packaged into phage protein coats, and after infection of bacterial cells, the recombinant molecule replicates and can be maintained as a plasmid.
coupling conformation See *cis configuration.*
covalent bond A nonionic chemical bond formed by the sharing of electrons.
Creutzfeldt–Jakob disease (CJD) A progressive degenerative and fatal disease of the brain and nervous system caused by mutations in the prion protein gene on chromosome 20 that produce aberrant forms of the encoded protein. CDJ is inherited as an autosomal dominant trait.
cri-du-chat syndrome A clinical syndrome in humans produced by a deletion of a portion of the short arm of chromosome 5. Afflicted infants have a distinctive cry that sounds like a cat.
crossing over The exchange of chromosomal material (parts of chromosomal arms) between homologous chromosomes by breakage and reunion. The exchange of material between nonsister chromatids during meiosis is the basis of genetic recombination.
cross-reacting material (CRM) Nonfunctional form of an enzyme, produced by a mutant gene, that is recognized by antibodies made against the normal enzyme.
C-terminal amino acid The terminal amino acid in a peptide chain that carries a free carboxyl group.
C terminus The end of a polypeptide that carries a free carboxyl group of the last amino acid. By convention, the structural formula of polypeptides is written with the C terminus at the right.
***C* value** The haploid amount of DNA present in a genome.
***C* value paradox** The apparent paradox that there is no relationship between the size of the genome and the evolutionary complexity of species. For example, the *C* value (haploid genome size) of amphibians varies by a factor of 100.
cyclic adenosine monophosphate (cAMP) An important regulatory molecule in both prokaryotic and eukaryotic organisms.
cyclins A class of proteins found in eukaryotic cells that are synthesized and degraded in synchrony with the cell cycle and regulate passage through stages of the cycle.
cytogenetics A branch of biology in which the techniques of both cytology and genetics are used to study heredity.
cytokinesis The division or separation of the cytoplasm during mitosis or meiosis.
cytological map A diagram showing the location of genes at particular chromosomal sites.
cytoplasmic inheritance Non-Mendelian form of inheritance involving genetic information transmitted by self-replicating cytoplasmic organelles such as mitochondria, chloroplasts, etc.
cytoskeleton An internal array of microtubules, microfilaments, and intermediate filaments that confers shape and the ability to move on a eukaryotic cell.
dalton (Da) A unit of mass equal to that of the hydrogen atom, which is 1.67×10^{-24} gram. A unit used in designating molecular weights.
Darwinian fitness See *fitness.*
deficiency A chromosomal mutation involving the loss or deletion of chromosomal material.
degenerate code The genetic code, where a given amino acid may be represented by more than one codon. For example, some amino acids (leucine) have six codons, while others (isoleucine) have three.
deletion See *deficiency.*
deme A local interbreeding population.

denatured DNA DNA molecules that have been separated into single strands.

de novo Newly arising; synthesized from less complex precursors rather than having been produced by modification of an existing molecule.

density gradient centrifugation A method of separating macromolecular mixtures by the use of centrifugal force and solutions of varying density. In buoyant density gradient centrifugation using cesium chloride, the cesium solution establishes a gradient under the influence of the centrifugal field and a mixture of macromolecules such as DNA sediment in the gradient until the density of the cesium chloride solution equals their own, separating them by differences in density.

deoxyribonuclease A class of enzymes that breaks down DNA into oligonucleotide fragments by introducing single-stranded breaks into the double helix.

deoxyribonucleic acid (DNA) A macromolecule usually consisting of antiparallel polynucleotide chains held together by hydrogen bonds, in which the sugar residues are deoxyribose. The primary carrier of genetic information.

deoxyribose The five-carbon sugar associated with the deoxyribonucleotides found in DNA.

dermatoglyphics The study of the surface ridges of the skin, especially of the hands and feet.

determination A regulatory event that establishes a specific pattern of future gene activity and developmental fate for a given cell.

diakinesis The final stage of meiotic prophase I in which the chromosomes become tightly coiled and compacted and move toward the periphery of the nucleus.

dicentric chromosome A chromosome having two centromeres.

dideoxynucleotide A nucleotide containing a dexoyribose sugar lacking 3′a hydroxyl group. Stops further chain elongation when incorporated into a growing polynucleotide; used in the Sanger method of DNA sequencing.

differentiation The process of complex changes by which cells and tissues attain their adult structure and functional capacity.

dihybrid cross A genetic cross involving two characters in which the parents possess different forms of each character (e.g., yellow, round × green, wrinkled peas).

diploid A condition in which each chromosome exists in pairs; having two of each chromosome.

diplotene A stage of meiotic prophase immediately after pachytene. In diplotene, one pair of sister chromatids begins separating from the other, and chiasmata become visible. These overlaps move laterally toward the ends of the chromatids (terminalization).

directional selection A selective force that changes the frequency of an allele in a given direction, either toward fixation or toward elimination.

discontinuous replication of DNA The synthesis of DNA in discontinuous fragments on the lagging strand of the replication fork. The fragments, known as Okazaki fragments, are joined by DNA ligase to form a continuous strand.

discontinuous variation Phenotypic data that fall into two or more distinct, nonoverlapping classes.

discordance In twin studies, a situation where one twin expresses a trait but the other does not.

disjunction The separation of chromosomes at the anaphase stage of cell division.

disruptive selection Simultaneous selection for phenotypic extremes in a population, usually resulting in the production of two phenotypically discontinuous strains.

dizygotic twins Twins produced from separate fertilization events; two ova fertilized independently. Also known as fraternal twins.

DNA See *deoxyribonucleic acid.*

DNA fingerprinting A molecular method for identifying an individual member of a population or species. The pattern of DNA fragments is obtained by restriction enzyme digestion, followed by Southern blot hybridization using minisatellite probes. See also *STR sequences.*

DNA footprinting See *footprinting*.

DNA gyrase One of the DNA topoisomerases that acts during DNA replication to reduce molecular tension caused by supercoiling. DNA gyrase produces, then seals double-stranded breaks.

DNA ligase An enzyme that forms a covalent bond between the 5′ end of one polynucleotide chain and the 3′ end of another polynucleotide chain. It is also called polynucleotide-joining enzyme.

DNA polymerase An enzyme that catalyzes the synthesis of DNA from deoxyribonucleotides and a template DNA molecule.

DNase Deoxyribonucleosidase; an enzyme that degrades or breaks down DNA into fragments or constitutive nucleotides.

dominance The expression of a trait in the heterozygous condition.

dominant suppression A form of epistasis in which a dominant allele at one locus suppresses the effect of a dominant allele at another locus, resulting in a 13:3 phenotypic ratio.

dosage compensation A genetic mechanism that regulates the levels of gene products at certain loci on the X chromosome in mammals such that males and females have equal amounts of a gene product. In mammals, this is accomplished by random inactivation of one X chromosome.

double crossover Two separate events of chromosome breakage and exchange occurring within the same tetrad.

double helix The model for DNA structure proposed by James Watson and Francis Crick, involving two antiparallel hydrogen-bonded polynucleotide chains wound into a right-handed helical configuration, with 10 base pairs per full turn of the double helix. Often called B-DNA.

Duchenne muscular dystrophy An X-linked recessive genetic disorder caused by a mutation in the gene for dystrophin, a protein found in muscle cells. Affected males show a progressive weakness and wasting of muscle tissue. Death ensues by about age 20 caused by respiratory infection or cardiac failure.

duplication A chromosomal aberration in which a segment of the chromosome is repeated.

dyad The products of tetrad separation or disjunction at the first meiotic prophase. Consists of two sister chromatids joined at the centromere.

dystrophin A protein that attaches to the inside of the muscle cell plasma membrane and stabilizes the membrane during muscle contraction. Mutations in the gene encoding dystrophin cause Duchenne and Becker muscular dystrophy. See *Duchenne muscular dystrophy.*

effective population size The number of individuals in a population with an equal probability of contributing gametes to the next generation.

effector molecule Small, biologically active molecule that regulates the activity of a protein by binding to a specific receptor site on the protein.

electrophoresis A technique that separates a mixture of molecules by their differential migration through a stationary medium (such as a gel) in an electrical field.

endocytosis The uptake by a cell of fluids, macromolecules, or particles by pinocytosis, phagocytosis, or receptor-mediated endocytosis.

endomitosis Chromosomal replication that is not accompanied by either nuclear or cytoplasmic division.

endonuclease An enzyme that hydrolyzes internal phosphodiester bonds in a polynucleotide chain or nucleic acid molecule.

endoplasmic reticulum A membranous organelle system in the cytoplasm of eukaryotic cells. The outer surface of the membranes may be ribosome-studded (rough ER) or smooth ER.

endopolyploidy The increase in chromosome sets that results from endomitotic replication within somatic nuclei.

endosymbiont theory The proposal that self-replicating cellular organelles such as mitochondria and chloroplasts were originally free-living organisms that entered into a symbiotic relationship with nucleated cells.

enhancer Originally identified as a 72-bp sequence in the genome of a virus, SV40, that increases the transcriptional activity of nearby structural genes. Similar sequences that enhance transcription have been identified in the genomes of eukaryotic cells. Enhancers can act over a distance of thousands of base pairs and can be located upstream, downstream, 5', 3', or internal to the gene they affect, and thus are different from promoters.

environment The complex of geographic, climatic, and biotic factors within which an organism lives.

enzyme A protein or complex of proteins that catalyzes a specific biochemical reaction.

epigenesis The idea that an organism develops by the appearance and growth of new structures. Opposed to preformationism, which holds that development is the growth of structures already present in the egg.

episome A circular genetic element in bacterial cells that can replicate independently of the bacterial chromosome or integrate and replicate as part of the chromosome.

epistasis Nonreciprocal interaction between genes such that one gene interferes with or prevents the expression of another gene. In *Drosophila* for example, the recessive gene *eyeless*, when homozygous, prevents the expression of eye color genes present in the genome.

epitope That portion of a macromolecule or cell that acts to elicit an antibody response; an antigenic determinant. A complex molecule or cell can contain several such sites.

equational division A division of each chromosome into longitudinal halves that are distributed into two daughter nuclei. Chromosome division in mitosis is an example of equational division.

equatorial plate See *metaphase plate*.

euchromatin Chromatin or chromosomal regions that are lightly staining and are relatively uncoiled during the interphase portion of the cell cycle. Euchromatic regions contain most of the structural genes.

eugenics The improvement of the human species by selective breeding. Positive eugenics refers to the promotion of breeding of people with favorable genes, and negative eugenics refers to the discouragement of breeding among those with undesirable traits.

eukaryotes Organisms having true nuclei and membranous organelles and whose cells demonstrate mitosis and meiosis.

euphenics Medical or genetic intervention to reduce the impact of defective genotypes.

euploid Polyploid with a chromosome number that is an exact multiple of a basic chromosome set.

evolution The origin of plants and animals from preexisting types. Descent with modifications.

excision repair Removal of damaged DNA segments followed by repair. Excision can include the removal of individual bases (base repair) or a stretch of damaged nucleotides (nucleotide repair). The gap created by excision is filled by DNA polymerase, and the ends are ligated to form an intact molecule.

exon (extron) The DNA segments of a gene that are transcribed and translated into proteins.

exonuclease An enzyme that breaks down nucleic acid molecules by breaking the phosphodiester bonds at the 3′- or 5′-terminal nucleotides.

expressed sequence tags (ESTs) All or part of the nucleotide sequence of cDNA clones. Used as markers in construction of genetic maps.

expression vector Plasmids or phages carrying promoter regions designed to cause expression of inserted DNA sequences.

expressivity The degree or range in which a phenotype for a given trait is expressed.

extranuclear inheritance Transmission of traits by genetic information contained in cytoplasmic organelles such as mitochondria and chloroplasts.

F^- cell A bacterial cell that does not contain a fertility factor, and that acts as a recipient in bacterial conjugation.

F^+ cell A bacterial cell containing a fertility factor, and that acts as a donor in bacterial conjugation.

F factor An episome in bacterial cells that confers the ability to act as a donor in conjugation (also, *fertility factor*).

F′ factor A fertility factor that contains a portion of the bacterial chromosome.

F_1 generation First filial generation; the progeny resulting from the first cross in a series.

F_2 generation Second filial generation; the progeny resulting from a cross of the F_1 generation.

F pilus See *pilus*.

familial trait A trait transmitted through and expressed by members of a family.

fate map A diagram of an embryo showing the location of cells whose development fate is known.

fertility factor See *F factor*.

filial generations See F_1, F_2 *generations*.

fingerprint The unique pattern of ridges and whorls on the tip of a human finger. Also, the pattern obtained by enzymatically cleaving a protein or nucleic acid and subjecting the digest to two-dimensional chromatography or electrophoresis. See also *DNA fingerprinting*.

FISH See fluorescence *in situ* hybridization.

fitness A measure of the relative survival and reproductive success of a given individual or genotype.

fixation In population genetics, a condition in which all members of a population are homozygous for a given allele.

fluctuation test A statistical test developed by Salvadore Luria and Max Delbrück to determine whether bacterial mutations arise spontaneously or are produced in response to selective agents.

fluorescence *in situ* hybridization (FISH) A method of *in situ* hybridization that utilizes probes labeled with a fluorescent tag, causing the site of hybridization to fluoresce when viewed in ultraviolet light under a microscope.

flush–crash cycle A period of rapid population growth followed by a drastic reduction in population size.

f-met See *formylmethionine*.

folded-fiber model A model of eukaryotic chromosome organization in which each sister chromatid consists of a single fiber, composed of double-stranded DNA and protein, which is wound like a tightly coiled skein of yarn.

footprinting A technique for identifying a DNA sequence that binds to a particular protein, based on the idea that the phosphodiester bonds in the region covered by the protein are protected from digestion by deoxyribonucleases. See also *DNA footprinting*.

formylmethionine (f-met) A molecule derived from the amino acid methionine by attachment of a formyl group to its terminal amino group. This is the first amino acid inserted in all bacterial polypeptides. Also known as *N*-formyl methionine.

founder effect A form of genetic drift. The establishment of a population by a small number of individuals whose genotypes carry only a fraction of the different kinds of alleles in the parental population.

fragile site A heritable gap or nonstaining region of a chromosome that can be induced to generate chromosome breaks.

fragile X syndrome A genetic disorder caused by the expansion of a CGG trinucleotide repeat and a fragile site at Xq27.3 within the *FMR-1* gene. Fragile X syndrome is the most common form of mental retardation. Affected males have distinctive facial features and are mentally retarded. Carrier females lack physical symptoms but as a group have higher rates of mental retardation than normal individuals.

frameshift mutation A mutational event leading to the insertion of one or more base pairs in a gene, shifting the codon reading frame in all codons that follow the mutational site.

fraternal twins See *dizygotic twins*.

G1 checkpoint A point in the G1 phase of the cell cycle when a cell becomes committed to initiate DNA synthesis and continue the cycle or withdraw into the G0 resting stage.

G0 A point in the G1 phase where cells withdraw from the cell cycle and enter a nondividing but metabolically active state.

gamete A specialized reproductive cell with a haploid number of chromosomes.

gap genes Genes expressed in contiguous domains along the anterior–posterior axis of the *Drosophila* embryo that regulate the process of segmentation in each domain.

gene The fundamental physical unit of heredity whose existence can be confirmed by allelic variants and which occupies a specific chromosomal locus. A DNA sequence coding for a single polypeptide.

gene amplification The process by which gene sequences are selected and differentially replicated either extrachromosomally or intrachromosomally.

gene conversion The process of nonreciprocal recombination by which one allele in a heterozygote is converted into the corresponding allele.

gene duplication An event in replication leading to the production of a tandem repeat of a gene sequence.

gene flow The gradual exchange of genes between two populations; brought about by the dispersal of gametes or the migration of individuals.

gene frequency The percentage of alleles of a given type in a population.

gene interaction Production of novel phenotypes by the interaction of alleles of different genes.

gene mutation See *point mutation*.

gene pool The total of all alleles possessed by reproductive members of a population.

generalized transduction The transduction of any gene in the bacterial genome by a phage.

genetically modified organism (GMO) A plant or animal that carries a gene from another species transferred to its genome using recombinant DNA technology, and where the gene is expressed to produce a gene product.

genetic anticipation The phenomenon of a progressively earlier age of onset and increasing severity of symptoms in successive generations for a genetic disorder. See *anticipation*.

genetic background All genes carried in the genome other than the one being studied.

genetic code The nucleotide triplets that code for the 20 amino acids or for chain initiation or termination.

genetic counseling Analysis of risk for genetic defects in a family and the presentation of options available to avoid or ameliorate possible risks.

genetic drift Random variation in allele frequency from generation to generation, most often observed in small populations.

genetic engineering The technique of altering the genetic constitution of cells or individuals by the selective removal, insertion, or modification of individual genes or gene sets.

genetic equilibrium Maintenance of allele frequencies at the same value in successive generations. A condition in which allele frequencies are neither increasing nor decreasing.

genetic erosion The loss of genetic diversity from a population or a species.

genetic fine structure Intragenic recombinational analysis that provides mapping information at the level of individual nucleotides.

genetic load Average number of recessive lethal genes carried in the heterozygous condition by an individual in a population.

genetic polymorphism The stable coexistence of two or more discontinuous genotypes in a population. When the frequencies of two alleles are carried to an equilibrium, the condition is called balanced polymorphism.

genetics The branch of biology that deals with heredity and the expression of inherited traits.

genome The set of genes carried by an individual.

genomic imprinting A condition where the expression of a gene depends on whether the gene has been inherited from a male or a female parent.

genomics The study of genomes, including nucleotide sequence, gene content, organization, and gene number.

genotype The specific allelic or genetic constitution of an organism; often, the allelic composition of one or a limited number of genes under investigation.

germ line An embryonic cell lineage that forms the reproductive cells (eggs and sperm).

germ plasm Hereditary material transmitted from generation to generation.

Goldberg–Hogness box A short nucleotide sequence 20–30 bp upstream from the initiation site of eukaryotic genes to which RNA polymerase II binds. The consensus sequence is TATAAAA. Also known as TATA box.

graft-versus-host disease (GVHD) In transplants, reaction by immunologically competent cells of the donor against the antigens present on the cells of the host. Often a fatal condition in human bone marrow transplants.

green revolution A program that resulted in a two- to threefold increase in crop yields generated by the development of new varieties of cereal plants with shorter stems and increased disease resistance.

gynandromorph An individual composed of cells with both male and female genotypes.

gyrase One of a class of enzymes known as topoisomerases. Gyrase converts closed circular DNA to a negatively supercoiled form prior to replication, transcription, or recombination. See *DNA gyrase*.

haploid A cell or organism having a single set of unpaired chromosomes. The gametic chromosome number.

haplotype The set of alleles from closely linked loci carried by an individual and usually inherited as a unit.

Hardy–Weinberg law The principle that both gene and genotype frequencies will remain in equilibrium in an infinitely large population in the absence of mutation, migration, selection, and nonrandom mating.

heat shock A transient response following exposure of cells or organisms to elevated temperatures. The response involves activation of a small number of loci, inactivation of some previously active loci, and selective translation of heat shock mRNA. Appears to be a nearly universal phenomenon observed in organisms ranging from bacteria to humans.

helicase An enzyme that participates in DNA replication by unwinding the double helix near the replication fork.

helix–turn–helix motif The structure of a region of DNA-binding proteins in which a turn of four amino acids holds two α helices at right angles to each other.

hemizygous Conditions where a gene is present in a single dose in an otherwise diploid cell. Usually applied to genes on the X chromosome in heterogametic males.

hemoglobin (Hb) An iron-containing, oxygen-carrying protein occurring chiefly in the red blood cells of vertebrates.

hemophilia An X-linked trait in humans associated with defective blood-clotting mechanisms.

heredity Transmission of traits from one generation to another.

heritability A measure of the degree to which observed phenotypic differences for a trait are genetic.

heterochromatin The heavily staining, late-replicating regions of chromosomes that are condensed in interphase. Thought to be devoid of structural genes.

heteroduplex A double-stranded nucleic acid molecule in which each polynucleotide chain has a different origin. These structures may be produced as intermediates in a recombinational event or by the *in vitro* reannealing of single-stranded, complementary molecules.

heterogametic sex The sex that produces gametes containing unlike sex chromosomes.

heterogeneous nuclear RNA (hnRNA) The collection of RNA transcripts in the nucleus, representing precursors and processing intermediates to rRNA, mRNA, and tRNA. Also represents RNA transcripts that will not be transported to the cytoplasm, such as snRNA.

heterokaryon A somatic cell containing nuclei from two different sources.

heterozygote An individual with different alleles at one or more loci. Such individuals will produce unlike gametes and therefore will not breed true.

Hfr A strain of bacteria exhibiting a high frequency of recombination. These strains have a chromosomally integrated F factor that is able to mobilize and transfer part of the chromosome to a recipient F^- cell.

histocompatibility antigens See *HLA*.

histones Proteins complexed with DNA in the nucleus. They are rich in the basic amino acids arginine and lysine, and function in coiling DNA to form nucleosomes.

HLA Cell-surface proteins, produced by histocompatibility loci, involved in the acceptance or rejection of tissue and organ grafts and transplants.

hnRNA See *heterogeneous nuclear RNA*.

Holliday structure An intermediate in bidirectional DNA recombination seen in the transmission electron microscope as an X-shaped structure showing four single-stranded DNA regions.

homeobox A sequence of about 180 nucleotides that encodes a sequence of 60 amino acids called a *homeodomain*, which is part of a DNA-binding protein that acts as a transcription factor.

homeotic mutation A mutation that causes a tissue normally determined to form a specific organ or body part to alter its differentiation and form another structure.

homogametic sex The sex that produces gametes that do not differ with respect to sex chromosome content; in mammals, the female is homogametic.

homologous chromosomes Chromosomes that synapse or pair during meiosis. Chromosomes that are identical with respect to their genetic loci and centromere placement.

homozygote An individual with identical alleles at one or more loci. These individuals will produce identical gametes and will therefore breed true.

homunculus The miniature individual imagined by preformationists to be contained within the sperm or egg.

H substance The carbohydrate group present on the surface of red blood cells. When unmodified, it results in blood type O; when modified by the addition of monosaccharides, it results in types A, B, and AB.

human immunodeficiency virus (HIV) A human retrovirus associated with the onset and progression of AIDS. See *acquired immunodeficiency syndrome*.

hybrid An individual produced by crossing two parents of different genotypes.

hybrid vigor The superiority of a heterozygote over either homozygote for a given trait.

hydrogen bond An electrostatic attraction between a hydrogen atom bonded to a strongly electronegative atom such as oxygen or nitrogen and another atom that is electronegative or contains an unshared electron pair.

hypervariable regions The regions of antibody molecules that attach to antigens. These regions have a high degree of diversity in amino acid content.

identical twins See *monozygotic twins*.

Ig See *immunoglobulin*.

imaginal disc Discrete groups of cells set aside during embryogenesis in holometabolous insects, which are determined to form the external body parts of the adult.

immunoglobulin (Ig) The class of serum proteins having the properties of antibodies.

inborn error of metabolism A genetically controlled biochemical disorder; usually an enzyme defect that produces a clinical syndrome.

inbreeding Mating between closely related organisms.

inbreeding depression A decrease in viability, vigor, or growth in progeny after several generations of inbreeding.

incomplete dominance Expression of heterozygous phenotype that is distinct from and often intermediate to that of either parent.

incomplete linkage Occasional separation of two genes on the same chromosome by a recombinational event.

independent assortment The independent behavior of each pair of homologous chromosomes during their segregation in meiosis I. The random distribution of maternal and paternal homologs into gametes.

inducer An effector molecule that activates transcription.

inducible enzyme system An enzyme system under the control of a regulatory molecule, or inducer, which acts to block a repressor and allow transcription.

initiation codon The triplet of nucleotides (AUG) in an mRNA molecule that codes for the insertion of the amino acid methionine as the first amino acid in a polypeptide chain.

insertion sequence See *IS element*.

***in situ* hybridization** A technique for the cytological localization of DNA sequences complementary to a given nucleic acid or polynucleotide.

intercalating agent A compound that inserts between bases in a DNA molecule, disrupting the alignments and pairing of bases in the complementary strands (e.g., acridine dyes).

interference A measure of the degree to which one crossover affects the incidence of another crossover in an adjacent region of the same chromatid. Negative interference increases the chances of another crossover; positive interference reduces the probability of a second crossover event.

interphase That portion of the cell cycle between divisions.

intervening sequence See *intron*.

intron A portion of DNA between coding regions in a gene that is transcribed but does not appear in the mRNA product.

inversion A chromosomal aberration in which the order of a chromosomal segment has been reversed.

inversion loop The chromosomal configuration resulting from the synapsis of homologous chromosomes, one of which carries an inversion.

in vitro Literally, *in glass*; outside the living organism; occurring in an artificial environment.

in vivo Literally, *in the living*; occurring within the living body of an organism.

IS element A mobile DNA segment that is transposable to any of a number of sites in the genome.

isoagglutinogen An antigenic factor or substance present on the surface of cells that is capable of inducing the formation of an antibody.

isochromosome An aberrant chromosome with two identical arms and homologous loci.
isolating mechanism Any barrier to the exchange of genes between different populations of a group of organisms. In general, isolation can be classified as spatial, environmental, or reproductive.
isotopes Forms of chemical elements that have the same number of protons and electrons but differ in the number of neutrons contained in the atomic nucleus.
isozyme Any of two or more distinct forms of an enzyme that have identical or nearly identical chemical properties but differ in some property such as net electrical charge, pH optima, number and type of subunits, or substrate concentration.
κ particles DNA-containing cytoplasmic particles found in certain strains of *Paramecium aurelia*. When these self-reproducing particles are transferred into the growth medium, they release a toxin, paramecin, that kills other sensitive strains. A nuclear gene, *K*, is responsible for maintaining kappa particles in the cytoplasm.
karyokinesis The process of nuclear division.
karyotype The chromosome complement of a cell or an individual. Often used to refer to the arrangement of metaphase chromosomes in a sequence according to length and position of the centromere.
kilobase (kb) A unit of length consisting of 1000 nucleotides.
kinetochore A fibrous structure with a size of about 400 nm, located within the centromere. It appears to be the site of microtubule attachment during division.
Klinefelter syndrome A genetic disorder in human males caused by the presence of an extra X chromosome. Klinefelter males are XXY instead of XY. This syndrome is associated with enlarged breasts, small testes, sterility, and occasionally, mild mental retardation.
knockout mice In producing knockout mice, a cloned normal gene is inactivated by the insertion of a marker, such as an antibiotic resistance gene. The altered gene is transferred to embryonic stem cells, where the altered gene will replace the normal gene (in some cells). These cells are injected into a blastomere embryo, producing a mouse that is bred to yield mice homozygous for the mutated gene.
***lac* repressor protein** A protein that binds to the operator in the *lac* operon and blocks transcription.
lagging strand In DNA replication, the strand synthesized in a discontinuous fashion, 5′ to 3′ away from the replication fork. Each short piece of DNA synthesized in this fashion is called an Okazaki fragment.
lampbrush chromosomes Meiotic chromosomes characterized by extended lateral loops, which reach maximum extension during diplotene. Although most intensively studied in amphibians, these structures occur in meiotic cells of organisms ranging from insects to humans.
lariat structure A structure formed by an intron via a 5′ to 3′ bond during processing and removal of that intron from an mRNA molecule.
leader sequence That portion of an mRNA molecule from the 5′ end to the beginning codon; may contain regulatory or ribosome binding sites.
leading strand During DNA replication, the strand synthesized continuously 5′ to 3′ from the origin of replication toward the replication fork.
leptotene The initial stage of meiotic prophase I, during which the chromosomes become visible and are often arranged in a bouquet configuration, with one or both ends of the chromosomes gathered at one spot on the inner nuclear membrane.
lethal gene A gene whose expression results in death.
leucine zipper A structural motif in a DNA-binding protein that is characterized by a stretch of leucine residues spaced at every seventh amino acid residue, with adjacent regions of positively charged amino acids. Leucine zippers on two polypeptides may interact to form a dimer that binds to DNA.
linkage Condition in which two or more nonallelic genes tend to be inherited together. Linked genes have their loci along the same chromosome; they do not assort independently, but can be separated by crossing over.
linkage group A group of genes that have their loci on the same chromosome.
linking number The number of times that two strands of a closed, circular DNA duplex cross over each other.
locus (pl., **loci**) The site or place on a chromosome where a particular gene is located.
lod score A statistical method used to determine whether two loci are linked or unlinked. A lod (log of the odds) score of 4 indicates that linkage is 10,000 times more likely than nonlinkage. By convention, lod scores of 3–4 are signs of linkage.
long interspersed elements (LINES) Repetitive sequences found in the genomes of higher organisms, such as the 6-kb L1 sequences found in primate genomes.
long terminal repeat (LTR) Sequence of several hundred base pairs found at the ends of retroviral DNAs.
Lyon hypothesis The random inactivation of the maternal or paternal X chromosome in somatic cells of mammalian females early in development. All daughter cells will have the same X chromosome inactivated, producing a mosaic pattern of expression of genes on the X chromosome.
lysis The disintegration of a cell brought about by the rupture of its membrane.
lysogenic bacterium A bacterial cell carrying the DNA of a temperate bacteriophage integrated into its chromosome.
lysogeny The process by which the DNA of an infecting phage becomes repressed and integrated into the chromosome of the bacterial cell it infects.
lytic phase The condition in which a temperate bacteriophage loses its integrated status in the host chromosome (becomes induced), replicates, and lyses the bacterial cell.
major histocompatibility (MHC) loci In humans, the HLA complex; and in mice, the H2 complex.
mapping functions Map distance estimates from recombination when the recombination frequency in a region exceeds 15–20 percent, and double crossovers are undetectable.
map unit A measure of the genetic distance between two genes, corresponding to a recombination frequency of 1 percent. See *centimorgan*.
maternal effect Phenotypic effects on the offspring produced by the maternal genome. Factors transmitted through the egg cytoplasm that produces a phenotypic effect in the progeny.
maternal influence See *maternal effect*.
maternal inheritance The transmission of traits via cytoplasmic genetic factors such as mitochondria or chloroplasts.
mean The arithmetic average.
median The value in a group of numbers below and above which there is an equal number of data points or measurements.
meiosis The process in gametogenesis or sporogenesis during which one replication of the chromosomes is followed by two nuclear divisions to produce four haploid cells.
melting profile (T_m) The temperature at which a population of double-stranded nucleic acid molecules is half-dissociated into single strands. This is taken to be the melting temperature for that species of nucleic acid.
merozygote A partially diploid bacterial cell containing, in addition to its own chromosome, a chromosome fragment introduced into the cell by transformation, transduction, or conjugation.
messenger RNA (mRNA) An RNA molecule transcribed from DNA and translated into the amino acid sequence of a polypeptide.

metabolism The sum of chemical changes in living organisms by which energy is generated and used.
metacentric chromosome A chromosome with a centrally located centromere, producing chromosome arms of equal lengths.
metafemale In *Drosophila*, a poorly developed female of low viability in which the ratio of X chromosomes to sets of autosomes exceeds 1.0. Previously called a superfemale.
metamale In *Drosophila*, a poorly developed male of low viability in which the ratio of X chromosomes to sets of autosomes is less than 0.5. Previously called a supermale.
metaphase The stage of cell division in which condensed chromosomes lie in a central plane between the two poles of the cell and during which the chromosomes become attached to the spindle fibers.
metaphase plate The arrangement of mitotic or meiotic chromosomes at the equator of the cell during metaphase.
methylation Enzymatic transfer of methyl groups from S-adenosylmethionine to biological molecules including phospholipids, proteins, RNA, and DNA. Methylation of DNA is associated with reduction in gene expression and with epigenetic phenomena such as imprinting.
MHC See *major histocompatibility loci.*
micrometer (μm) A unit of length equal to 1×10^{-6} meter. Previously called a micron.
micron See *micrometer.*
migration coefficient An expression of the proportion of migrant genes entering the population per generation.
millimeter (mm) A unit of length equal to 1×10^{-3} meter.
minimal medium A medium containing only those nutrients that will support the growth and reproduction of wild-type strains of an organism.
minisatellite Short tandem repeats of 10–100 nucleotides widely dispersed in the genome of eukaryotes. The number of repeats at each locus is variable; these loci are known as variable number tandem repeats (VNTRs). Each variation represents a VNTR allele, and many loci have dozens of alleles. VNTRs are used in DNA fingerprinting. See also *STR sequences.*
mismatch repair A process of excision repair, during which an unpaired base or bases is excised, followed by the synthesis of a new segment, using the complementary strand as a template.
missense mutation A mutation that alters a codon to that of another amino acid, causing an altered translation product to be made.
mitochondrial DNA (mtDNA) Double-stranded, self-replicating circular DNA found in mitochondria. mtDNA encodes mitochondrial ribosomal RNAs, transfer RNAs, and proteins used in oxidative respiratory functions of the organelle.
mitochondrion Found in the cells of eukaryotes, a cytoplasmic, self-reproducing organelle that is the site of ATP synthesis.
mitogen A substance that stimulates mitosis in nondividing cells (e.g., phytohemagglutinin).
mitosis A form of cell division resulting in the production of two cells, each with the same chromosome and genetic complement as the parent cell.
mode In a set of data, the value occurring in the greatest frequency.
monohybrid cross A genetic cross between two individuals involving only one character (e.g., $AA \times aa$).
monosomic An aneuploid condition in which one member of a chromosome pair is missing; having a chromosome number of $2n - 1$.
monozygotic twins Twins produced from a single fertilization event; the first division of the zygote produces two cells, each of which develops into an embryo. Also known as identical twins.
mRNA See *messenger RNA.*
mtDNA See *mitochondrial DNA.*
multigene family A gene set descended from a common ancestor by duplication and subsequent divergence from a common ancestor. The globin genes are an example of a multigene family.
multiple alleles Three or more alleles of the same gene.
multiple-factor inheritance See *polygenic inheritance.*
multiple infection Simultaneous infection of a bacterial cell by more than one bacteriophage, often of different genotypes.
mu (μ) phage A phage group in which the genetic material behaves like an insertion sequence; capable of insertion, excision, transposition, inactivation of host genes, and induction of chromosomal rearrangements. See *bacteriophage μ.*
mutagen Any agent that causes an increase in the rate of mutation.
mutant A cell or organism carrying an altered or mutant gene.
mutation The process that produces an alteration in DNA or chromosome structure; the source of most alleles.
mutation rate The frequency with which mutations take place at a given locus or in a population.
muton The smallest unit of mutation in a gene, corresponding to a single base change.
nanometer (nm) A unit of length equal to 1×10^{-9} meter.
natural selection Differential reproduction of some members of a species resulting from variable fitness conferred by genotypic differences.
neutral mutation A mutation with no immediate adaptive significance or phenotypic effect.
nonautonomous transposon A transposable element that lacks a functional transposase gene.
noncrossover gamete A gamete that contains no chromosomes that have undergone genetic recombination.
nondisjunction An error during cell division in which homologous chromosomes (in meiosis) or the sister chromatids (in mitosis) fail to separate and migrate to opposite poles; responsible for defects such as monosomy and trisomy.
nonsense codon The nucleotide triplet in an mRNA molecule that signals the termination of translation. Three such termination codons are known: UGA, UAG, and UAA.
nonsense mutation A mutation that changes an amino acid codon into a termination codon: UAG, UAA, or UGA. Leads to premature termination during translation of mRNA.
NOR See *nucleolar organizer region.*
normal distribution A probability function that approximates the distribution of random variables. The normal curve, also known as a Gaussian or bell-shaped curve, is the graphic display of the normal distribution.
northern blot A technique in which RNA molecules are separated by electrophoresis and transferred by capillary action to a nylon or nitrocellulose membrane. Specific RNA molecules can be identified by hybridization to a labeled nucleic acid probe.
N-terminal amino acid The terminal amino acid in a peptide chain that carries a free amino group.
N terminus The free amino group of the first amino acid in a polypeptide. By convention, the structural formula of polypeptides is written with the N terminus at the left.
nuclease An enzyme that breaks bonds in nucleic acid molecules.
nucleoid The DNA-containing region within the cytoplasm in prokaryotic cells.
nucleolar organizer region (NOR) A chromosomal region containing the genes for rRNA; most often found in physical association with the nucleolus.
nucleolus A nuclear organelle that is the site of ribosome biosynthesis; usually associated with or formed in association with the NOR.
nucleoside A purine or pyrimidine base covalently linked to a ribose or deoxyribose sugar molecule.

nucleosome A complex of four histone molecules, each present in duplicate, wrapped by two turns of a DNA molecule. One of the basic units of eukaryotic chromosome structure.
nucleotide A nucleoside covalently linked to a phosphate group. Nucleotides are the basic building blocks of nucleic acids. The nucleotides commonly found in DNA are deoxyadenylic acid, deoxycytidylic acid, deoxyguanylic acid, and deoxythymidylic acid. The nucleotides in RNA are adenylic acid, cytidylic acid, guanylic acid, and uridylic acid.
nucleotide pair The pair of nucleotides (A and T or G and C) in opposite strands of a DNA molecule that are hydrogen-bonded to each other.
nucleus The membrane-bound cytoplasmic organelle of eukaryotic cells that contains the chromosomes and nucleolus.
null allele A mutant allele that produces no functional gene product. Usually inherited as a recessive trait.
null hypothesis Used in statistical tests, it states that there is no difference between the observed and expected data sets. Statistical methods such as chi-square analysis are used to test the probability of this hypothesis.
nullisomic Describes an individual with a chromosomal aberration in which both members of a chromosome pair are missing.
Okazaki fragment The small, discontinuous strands of DNA produced on the lagging strand during DNA synthesis.
oligonucleotide A linear sequence of about 10–20 nucleotides connected by 5′–3′ phosphodiester bonds.
oncogene A gene whose activity promotes uncontrolled proliferation in eukaryotic cells.
open reading frame (ORF) A nucleotide sequence organized as triplets that encodes the amino acids of a protein. Located between an initiation codon and a termination codon within a gene.
operator region A region of a DNA molecule that interacts with a specific repressor protein to control the expression of an adjacent gene or gene set.
operon A genetic unit consisting of one or more structural genes encoding polypeptides, and an adjacent operator gene that controls the transcriptional activity of the structural gene or genes.
origin of replication (ori) Sites along the length of a chromosome where DNA replication begins.
outbreeding depression Reduction in fitness in the offspring produced by mating genetically diverse parents. It is thought to result from a lowered adaptation to local environmental conditions.
overdominance The phenomenon where heterozygotes have a phenotype that is more extreme than either homozygous genotype.
overlapping code A genetic code first proposed by George Gamow in which any given nucleotide is shared by three adjacent codons.
pachytene The stage in meiotic prophase I when the synapsed homologous chromosomes split longitudinally (except at the centromere), producing a group of four chromatids called a tetrad.
pair-rule genes Genes expressed as stripes around the blastoderm embryo during development of the *Drosophila* embryo.
palindrome A word, number, verse, or sentence that reads the same backward or forward (e.g., *tis Ivan on a visit*). In nucleic acids, a sequence in which the base pairs read the same on complementary strands in the 5′ to 3′ direction. For example:

5′-GAATTC-3′
3′-CTTAAG-5′

These often occur as sites for restriction endonuclease recognition and cutting.
pangenesis A discarded theory of development that postulated the existence of pangenes, small particles from all parts of the body that concentrated in the gametes, passing traits from generation to generation, blending the traits of the parents in the offspring.
paracentric inversion A chromosomal inversion that does not include the centromere.
parental gamete See *noncrossover gamete*.
parthenogenesis Development of an egg without fertilization.
partial diploids See *merozygote*.
partial dominance See *incomplete dominance*.
patroclinous inheritance A form of genetic transmission in which the offspring have the phenotype of the father.
pedigree In human genetics, a diagram showing the ancestral relationships and transmission of genetic traits over several generations in a family.
P element Transposable DNA element found in *Drosophila* responsible for hybrid dysgenesis.
penetrance The frequency, expressed as a percentage, with which individuals of a given genotype manifest at least some degree of a specific mutant phenotype associated with a trait.
peptide bond The covalent bond between the amino group of one amino acid and the carboxyl group of another amino acid.
pericentric inversion A chromosomal inversion that involves both arms of the chromosome and thus involves the centromere.
phage See *bacteriophage*.
phenotype The observable properties of an organism that are genetically controlled.
phenylketonuria (PKU) A hereditary condition in humans associated with the inability to metabolize the amino acid phenylalanine. The most common form is caused by the lack of the enzyme phenylalanine hydroxylase.
Philadelphia chromosome The product of a reciprocal translocation that contains the short arm of chromosome 9 carrying the *C-ABL* oncogene and the long arm of chromosome 22 carrying the *BCR* gene.
phosphodiester bond In nucleic acids, the covalent bond between a phosphate group and adjacent nucleotides, extending from the 5′ carbon of one pentose (ribose or deoxyribose) to the 3′ carbon of the pentose in the neighboring nucleotide. Phosphodiester bonds form the backbone of nucleic acid molecules.
photoreactivation enzyme (PRE) An exonuclease that catalyzes the light-activated excision of ultraviolet-induced thymine dimers from DNA.
photoreactivation repair Light-induced repair of damage caused by exposure to ultraviolet light. Associated with an intracellular enzyme system.
phyletic evolution The gradual transformation of one species into another over time; vertical evolution.
pilus A filamentlike projection from the surface of a bacterial cell. Often associated with cells possessing F factors.
plaque A clear area on an otherwise opaque bacterial lawn, caused by the growth and reproduction of phages.
plasmid An extrachromosomal, circular DNA molecule (often carrying genetic information) that replicates independently of the host chromosome.
pleiotropy Condition in which a single mutation simultaneously affects several characters.
ploidy Term referring to the basic chromosome set or to multiples of that set.
point mutation A mutation that can be mapped to a single locus. At the molecular level, a mutation that results in the substitution of one nucleotide for another.
polar body A cell produced in females at either the first or second meiotic division, which contains almost no cytoplasm as a result of an unequal cytokinesis.
polycistronic mRNA A messenger RNA molecule that encodes the amino acid sequence of two or more polypeptide chains in adjacent structural genes.

polygenic inheritance The transmission of a phenotypic trait whose expression depends on the additive effect of a number of genes.

polylinker A segment of DNA that has been engineered to contain multiple sites for restriction enzyme digestion. Polylinkers are usually found in engineered vectors such as plasmids.

polymerase chain reaction (PCR) A method for amplifying DNA segments that uses cycles of denaturation, annealing to primers, and DNA polymerase-directed DNA synthesis.

polymerases Enzymes that catalyze the formation of DNA and RNA from deoxynucleotides and ribonucleotides, respectively.

polymorphism The existence of two or more discontinuous, segregating phenotypes in a population.

polynucleotide A linear sequence of more than 20 nucleotides, joined by 5′–3′ phosphodiester bonds. See also *oligonucleotide*.

polypeptide A molecule made up of amino acids joined by covalent peptide bonds. This term is used to denote the amino acid chain before it assumes its functional three-dimensional configuration.

polyploid A cell or individual having more than two sets of chromosomes.

polyribosome See *polysome*.

polysome A structure composed of two or more ribosomes associated with mRNA, engaged in translation. Formerly called polyribosome.

polytene chromosome A chromosome that has undergone several rounds of DNA replication without separation of the replicated chromosomes, thus forming a giant, thick chromosome with aligned chromomeres producing a characteristic banding pattern.

population A local group of individuals belonging to the same species, which are actually or potentially interbreeding.

position effect Change in expression of a gene associated with a change in the gene's location within the genome.

postzygotic isolation mechanism A factor that prevents or reduces inbreeding by acting after fertilization to produce nonviable, sterile hybrids or hybrids of lowered fitness.

preadaptive mutation A mutational event that later becomes of adaptive significance.

preformationism The discredited idea that an organism develops by growth of structures already present in the egg or sperm.

prezygotic isolation mechanism A factor that reduces inbreeding by preventing courtship, mating, or fertilization.

Pribnow box A 6-bp sequence upstream from the beginning of transcription in prokaryotic genes, to which the σ subunit of RNA polymerase binds. The consensus sequence for this box is TATAAT.

primary protein structure The sequence of amino acids in a polypeptide chain.

primary sex ratio Ratio of males to females at fertilization.

primer In nucleic acids, a short length of RNA or single-stranded DNA required for the functioning of polymerases.

prion An infectious pathogenic agent devoid of nucleic acid and composed of a protein, PrP, with a molecular weight of 27,000–30,000 Da. Prions are known to cause scrapie, a degenerative neurological disease in sheep; bovine spongiform encephalopathy (BSE or mad cow disease) in cattle; and similar diseases in humans, including kuru and Creutzfeldt–Jakob disease.

probability Ratio of the frequency of a given event to the frequency of all possible events.

proband An individual in whom a genetically determined trait of interest is first detected. Formerly known as a propositus.

probe A macromolecule such as DNA or RNA that has been labeled and can be detected by an assay such as autoradiography or fluorescence microscopy. Probes are used to identify target molecules, genes, or gene products.

product law In statistics, the law that holds that the probability of two independent events occurring simultaneously is the product of their independent probabilities.

progeny The offspring produced from a mating.

prokaryotes Organisms lacking nuclear membranes, meiosis, and mitosis. Bacteria and blue–green algae are examples of prokaryotic organisms.

promoter Region having a regulatory function and to which RNA polymerase binds prior to the initiation of transcription.

proofreading A molecular mechanism for correcting errors in replication, transcription, or translation. Also known as editing.

prophage A phage genome integrated into a bacterial chromosome. Bacterial cells carrying prophages are said to be lysogenic.

propositus (female, **proposita**) See *proband*.

protein A molecule composed of one or more polypeptides, each composed of amino acids covalently linked together.

proteomics The study of the expressed proteins present in a cell at a given time.

protooncogene A gene that normally functions to initiate or maintain cell division. Protooncogenes can be converted to oncogenes by alterations in structure or expression.

protoplast A bacterial or plant cell with the cell wall removed. Sometimes called a spheroplast.

prototroph A strain (usually microorganisms) that is capable of growth on a defined, minimal medium. Wild-type strains are usually regarded as prototrophs.

pseudoalleles Genes that behave as alleles to one another by complementation, but can be separated from one another by recombination.

pseudoautosomal inheritance Inheritance of alleles located within the regions of the Y chromosome that are homologous to the X chromosome. Because these alleles are located on both the X and Y chromosome, their pattern of inheritance is indistinguishable from that of autosomal inheritance.

pseudodominance The expression of a recessive allele on one homolog caused by the deletion of the dominant allele on the other homolog.

pseudogene A nonfunctional gene with sequence homology to a known structural gene present elsewhere in the genome. They differ from their functional relatives by insertions or deletions and by the presence of flanking direct repeat sequences of 10–20 nucleotides.

puff See *chromosome puff.*

punctuated equilibrium A pattern in the fossil record of long periods of species stability, punctuated with brief periods of species divergence.

quantitative inheritance See *polygenic inheritance*.

quantitative trait loci (QTL) Two or more genes that act on a single polygenic trait.

quantum speciation Formation of a new species within a single or a few generations by a combination of selection and drift.

quaternary protein structure Types and modes of interaction between two or more polypeptide chains within a protein molecule.

race A genotypically or geographically distinct subgroup within a species.

rad A unit of absorbed dose of radiation with an energy equal to 100 ergs per gram of irradiated tissue.

radioactive isotope One of the forms of an element, differing in atomic weight and possessing an unstable nucleus that emits ionizing radiation during decay.

random amplified polymorphic DNA (RAPD) A PCR method that uses random primers about 10 nucleotides in length to amplify unknown DNA sequences.

random mating Mating between individuals without regard to genotype.

reading frame Linear sequence of codons in a nucleic acid.
reannealing Formation of double-stranded DNA molecules from dissociated single strands.
recessive A term describing an allele that is not expressed in the heterozygous condition.
reciprocal cross A paired cross in which the genotype of the female in the first cross is present as the genotype of the male in the second cross, and vice versa.
reciprocal translocation A chromosomal aberration in which nonhomologous chromosomes exchange parts.
recombinant DNA A DNA molecule formed by joining two heterologous molecules. Usually applied to DNA molecules produced by *in vitro* ligation of DNA from two different organisms.
recombinant gamete A gamete containing a new combination of genes produced by crossing over during meiosis.
recombination The process that leads to the formation of new gene combinations on chromosomes.
recon A term coined by Seymour Benzer to denote the smallest genetic units between which recombination can occur.
reductional division The chromosome division that halves the diploid chromosome number. The first division of meiosis is a reductional division.
redundant genes Gene sequences present in more than one copy per haploid genome (e.g., ribosomal genes).
regulatory site A DNA sequence that is involved in the control of expression of other genes, usually involving an interaction with another molecule.
rem Radiation equivalent in man; the dosage of radiation that will cause the same biological effect as one roentgen of X-rays.
renaturation The process by which a denatured protein or nucleic acid returns to its normal three-dimensional structure.
repetitive DNA sequences DNA sequences present in many copies in the haploid genome.
replicating form (RF) Double-stranded nucleic acid molecules present as an intermediate during the reproduction of certain viruses.
replication The process of DNA synthesis.
replication fork The Y-shaped region of a chromosome associated with the site of replication.
replicon A chromosomal region or free genetic element containing the DNA sequences necessary for the initiation of DNA replication.
replisome The term used to describe the complex of proteins, including DNA polymerase, that assembles at the bacterial replication fork to synthesize DNA.
repressible enzyme system An enzyme or group of enzymes whose synthesis is regulated by the intracellular concentration of certain metabolites.
repressor A protein that binds to a regulatory sequence adjacent to a gene and blocks transcription of the gene.
reproductive isolation Absence of interbreeding between populations, subspecies, or species. Reproductive isolation can be brought about by extrinsic factors, such as behavior, and intrinsic barriers, such as hybrid inviability.
resistance transfer factor (RTF) A component of R plasmids that confers the ability for cell–cell transfer of the R plasmid by conjugation.
resolution In an optical system, the shortest distance between two points or lines at which they can be perceived to be two points or lines.
restriction endonuclease Nuclease that recognizes specific nucleotide sequences in a DNA molecule and cleaves or nicks the DNA at those sites. Derived from a variety of microorganisms, those enzymes that cleave both strands of the DNA are used in the construction of recombinant DNA molecules.
restriction fragment length polymorphism (RFLP) Variation in the length of DNA fragments generated by a restriction endonuclease. These variations are caused by mutations that create or abolish cutting sites for restriction enzymes. RFLPs are inherited in a codominant fashion and can be used as genetic markers.
restrictive transduction See *specialized transduction.*
retrovirus Viruses with RNA as genetic material that utilize the enzyme reverse transcriptase during their life cycle.
reverse transcriptase A polymerase that uses RNA as a template to transcribe a single-stranded DNA molecule as a product.
reversion A mutation that restores the wild-type phenotype.
R factor (R plasmid) Bacterial plasmids that carry antibiotic resistance genes. Most R plasmids have two components: an r-determinant, which carries the antibiotic resistance genes, and the resistance transfer factor (RTF).
RFLP See *restriction fragment length polymorphism.*
Rh factor An antigenic system first described in the rhesus monkey. Recessive r/r individuals produce no Rh antigens and are Rh negative, while R/R and R/r individuals have Rh antigens on the surface of their red blood cells and are classified as Rh positive.
ribonucleic acid (RNA) A nucleic acid characterized by the sugar ribose and the pyrimidine uracil, usually a single-stranded polynucleotide. Several forms are recognized, including ribosomal RNA, messenger RNA, transfer RNA, and heterogeneous nuclear RNA.
ribose The five-carbon sugar associated with the ribonucleotides found in RNA.
ribosomal RNA (rRNA) The RNA molecules that are the structural components of the ribosomal subunits. In prokaryotes, these are the 16*S*, 23*S*, and 5*S* molecules; in eukaryotes, they are the 18*S*, 28*S*, and 5*S* molecules.
ribosome A ribonucleoprotein organelle consisting of two subunits, each containing RNA and protein. Ribosomes are the site of translation of mRNA codons into the amino acid sequence of a polypeptide chain.
RNA See *ribonucleic acid.*
RNA editing Alteration of the nucleotide sequence of an mRNA molecule after transcription and before translation. There are two main types of editing: substitution editing, which changes individual nucleotides, and insertion/deletion editing, in which individual nucleotides are added or deleted.
RNA polymerase An enzyme that catalyzes the formation of an RNA polynucleotide strand using the base sequence of a DNA molecule as a template.
RNase A class of enzymes that hydrolyzes RNA.
Robertsonian translocation A form of chromosomal aberration that involves the fusion of the long arms of acrocentric chromosomes at the centromere.
roentgen (R) A unit of measure of the amount of radiation corresponding to the generation of 2.083×10^9 ion pairs in one cubic centimeter of air at 0°C at an atmospheric pressure of 760 mm of mercury.
rolling circle model A model of DNA replication in which the growing point or replication fork rolls around a circular template strand; in each pass around the circle, the newly synthesized strand displaces the strand from the previous replication, producing a series of contiguous copies of the template strand.
R point The point during the G1 stage of the cell cycle when either a commitment is made to DNA synthesis and another cell cycle or the cell withdraws from the cycle and becomes quiescent. Also known as the restriction point.
rRNA See *ribosomal RNA.*
RTF See *resistance transfer factor.*
S_1 nuclease A deoxyribonuclease that cuts and degrades single-stranded molecules of DNA.
satellite DNA DNA that forms a minor band when genomic DNA is centrifuged in a cesium salt gradient. This DNA usually consists of short sequences repeated many times in the genome.

SCE See *sister chromatid exchange.*

secondary protein structure The α-helical or β-pleated-sheet form of a protein molecule brought about by the formation of hydrogen bonds between amino acids.

secondary sex ratio The ratio of males to females at birth.

secretor An individual having soluble forms of the blood group antigens A and/or B present in saliva and other body fluids. This condition is caused by a dominant, autosomal gene unlinked to the *ABO* locus (*I* locus).

sedimentation coefficient See *Svedberg coefficient unit.*

segment polarity genes Genes that regulate the spatial pattern of differentiation within each segment of the developing *Drosophila* embryo.

segregation The separation of homologous chromosomes into different gametes during meiosis.

selection The force that brings about changes in the frequency of alleles and genotypes in populations through differential reproduction.

selection coefficient (*s*) A quantitative measure of the relative fitness of one genotype compared with another. See *coefficient of selection.*

selfing In plant genetics, the fertilization of ovules of a plant by pollen produced by the same plant. Reproduction by self-fertilization.

semiconservative replication A model of DNA replication in which a double-stranded molecule replicates in such a way that the daughter molecules are composed of one parental (old) and one newly synthesized strand.

semisterility A condition in which a percentage of all zygotes is inviable.

sex chromatin body See *Barr body.*

sex chromosome A chromosome, such as the X or Y in humans, which is involved in sex determination.

sexduction Transmission of chromosomal genes from a donor bacterium to a recipient cell by the F factor.

sex-influenced inheritance Phenotypic expression that is conditioned by the sex of the individual. A heterozygote may express one phenotype in one sex and the alternate phenotype in the other sex.

sex-limited inheritance A trait that is expressed in only one sex even though the trait may not be X-linked.

sex ratio See *primary and secondary sex ratio.*

sexual reproduction Reproduction through the fusion of gametes, which are the haploid products of meiosis.

Shine–Dalgarno sequence The nucleotides AGGAGG present in the leader sequence of prokaryotic genes that serve as a ribosome binding site. The 16*S* RNA of the small ribosomal subunit contains a complementary sequence to which the mRNA binds.

short interspersed elements (SINES) Repetitive sequences found in the genomes of higher organisms, such as the 300-bp *Alu* sequence.

shotgun experiment The cloning of random fragments of genomic DNA into a vehicle such as a plasmid or phage, usually to produce a library from which clones of specific interest can be selected.

sibling species Species that are morphologically almost identical, but which are reproductively isolated from one another.

sickle-cell anemia A genetic disease in humans caused by an autosomal recessive gene, fatal in the homozygous condition if untreated. Caused by an alteration in the amino acid sequence of the β chain of globin.

sickle-cell trait The phenotype exhibited by individuals heterozygous for the sickle-cell gene.

sigma (σ) factor A polypeptide subunit of the RNA polymerase that recognizes the binding site for the initiation of transcription.

single-stranded binding proteins (SSBs) In DNA replication, proteins that bind to and stabilize single-stranded regions of DNA that result from the action of unwinding proteins.

sister chromatid exchange (SCE) A crossing over event that can occur in meiotic and mitotic cells; involves the reciprocal exchange of chromosomal material between sister chromatids joined by a common centromere. Such exchanges can be detected cytologically after BrdU incorporation into the replicating chromosomes.

site-directed mutagenesis A process that uses a synthetic oligonucleotide containing a mutant base or sequence as a primer for inducing a mutation at a specific site in a cloned gene.

small nuclear RNA (snRNA) Species of RNA molecules ranging in size from 90 to 400 nucleotides. The abundant snRNAs are present in 1×10^4 to 1×10^6 copies per cell, associated with proteins, and form RNP particles known as snRNPs or *snurps.* Six uridine-rich snRNAs known as U1–U6 are located in the nucleoplasm, and their complete nucleotide sequence is known. snRNAs have been implicated in the processing of pre-mRNA and may have a range of cleavage and ligation functions.

snurps See *small nuclear RNA.*

solenoid structure A level of eukaryotic chromosome structure generated by the supercoiling of nucleosomes.

somatic cell genetics The use of cultured somatic cells to investigate genetic phenomena by parasexual techniques involving the fusion of cells from different organisms.

somatic cells All cells other than the germ cells or gametes in an organism.

somatic mutation A nonheritable mutational event occurring in a somatic cell.

somatic pairing The pairing of homologous chromosomes in somatic cells.

SOS response The induction of enzymes to repair damaged DNA in *Escherichia coli.* The response involves activation of an enzyme that cleaves a repressor, activating a series of genes involved in DNA repair.

Southern blot A technique developed by Edward Southern in which DNA fragments produced by restriction enzyme digestion are separated by electrophoresis and transferred by capillary action to a nylon or nitrocellulose membrane. Specific DNA fragments can be identified by hybridization to a labeled nucleic acid probe.

spacer DNA DNA sequences found between genes, usually repetitive DNA segments.

specialized transduction Genetic transfer of only specific host genes by transducing phages.

speciation The process by which new species of plants and animals arise.

species A group of actually or potentially interbreeding individuals that is reproductively isolated from other such groups.

spheroplast See *protoplast.*

spindle fibers Cytoplasmic fibrils formed during cell division that are involved with the separation of chromatids at anaphase and their movement toward opposite poles in the cell.

spliceosome The nuclear macromolecule complex within which splicing reactions occur to remove introns from pre-mRNAs.

spontaneous mutation A mutation that is not induced by a mutagenic agent.

spore A unicellular body or cell encased in a protective coat that is produced by some bacteria, plants, and invertebrates; it is capable of survival in unfavorable environmental conditions; and it can give rise to a new individual upon germination. In plants, spores are the haploid products of meiosis.

SRY The sex-determining region of the Y gene found near the pseudoautosomal boundary of the Y chromosome. Accumulated evidence indicates that this gene is the testis-determining factor (TDF).

stabilizing selection Preferential reproduction of those individuals having genotypes close to the mean for the population. A selective elimination of genotypes at both extremes.

standard deviation A quantitative measure of the amount of variation in a sample of measurements from a population.
standard error A quantitative measure of the amount of variation in a sample of measurements from a population.
sterile The condition of being unable to reproduce. Also, free from contaminating microorganisms.
strain A group with common ancestry that has physiological or morphological characteristics of interest for genetic study or domestication.
STR sequences Short tandem repeats of 2–9 base pairs found within minisatellite sequences. These sequences are used to prepare DNA profiles in forensics, paternity identification, and other applications.
structural gene A gene that encodes the amino acid sequence of a polypeptide chain.
sublethal gene A mutation causing lowered viability, with death before maturity in less than 50 percent of the individuals carrying the gene.
submetacentric chromosome A chromosome with the centromere placed so that one arm of the chromosome is slightly longer than the other.
subspecies A morphologically or geographically distinct interbreeding population of a species.
sum law The law that holds that the probability of one or the other of two mutually exclusive events occurring is the sum of their individual probabilities.
supercoiled DNA A form of DNA structure in which the helix is coiled upon itself. Such structures can exist in stable forms only when the ends of the DNA are not free, as in a covalently closed circular DNA molecule.
superfemale See *metafemale*.
supermale See *metamale*.
suppressor mutation A mutation that acts to completely or partially restore the function lost by a previous mutation at another site.
Svedberg coefficient unit (*S*) A unit of measure for the rate at which particles (molecules) sediment in a centrifugal field. This unit is a function of several physicochemical properties, including size and shape. A sedimentation value of 1×10^{-13} sec is defined as one Svedberg coefficient unit.
symbiont An organism coexisting in a mutually beneficial relationship with another organism.
sympatric speciation Process of speciation involving populations that inhabit, at least in part, the same geographic range.
synapsis The pairing of homologous chromosomes at meiosis.
synaptonemal complex (SC) An organelle consisting of a tripartite nucleoprotein ribbon that forms between the paired homologous chromosomes in the pachytene stage of the first meiotic division.
syndrome A group of signs or symptoms that occur together and characterize a disease or abnormality.
synkaryon The nucleus of a zygote that results from the fusion of two gametic nuclei. Also used in somatic cell genetics to describe the product of nuclear fusion.
syntenic test In somatic cell genetics, a method for determining whether two genes are on the same chromosome.
TATA box See *Goldberg–Hogness box*.
tautomeric shift A reversible isomerization in a molecule, brought about by a shift in the localization of a hydrogen atom. In nucleic acids, tautomeric shifts in the bases of nucleotides can cause changes in other bases at replication and are a source of mutations.
TDF (testis-determining factor) The product of the *SRY* gene on the Y chromosome that controls the developmental switch point for the development of the indifferent gonad into a testis.
telocentric chromosome A chromosome in which the centromere is located at the end of the chromosome.
telomerase The enzyme that adds short, tandemly repeated DNA sequences to the ends of eukaryotic chromosomes.
telomere The terminal chromomere of a chromosome.
telophase The stage of cell division in which the daughter chromosomes reach the opposite poles of the cell and re-form nuclei. Telophase ends with the completion of cytokinesis.
telophase I In the first meiotic division, when duplicated chromosomes reach the poles of the dividing cell.
temperate phage A bacteriophage that can become a prophage and confer lysogeny upon the host bacterial cell.
temperature-sensitive mutation A conditional mutation that produces a mutant phenotype at one temperature range and a wild-type phenotype at another temperature range.
template The single-stranded DNA or RNA molecule that specifies the nucleotide sequence of a strand synthesized by a polymerase molecule.
terminalization The movement of chiasmata toward the ends of chromosomes during the diplotene stage of the first meiotic division.
tertiary protein structure The three-dimensional structure of a polypeptide chain brought about by folding upon itself.
test cross A cross between an individual whose genotype at one or more loci may be unknown and an individual who is homozygous recessive for the genes in question.
tetrad The four chromatids that make up paired homologs in the prophase of the first meiotic division. The four haploid cells produced by a single meiotic division.
tetrad analysis Method for the analysis of gene linkage and recombination, using the four haploid cells produced in a single meiotic division.
tetranucleotide hypothesis An early theory of DNA structure proposing that the molecule was composed of repeating units, each consisting of the four nucleotides containing adenine, thymine, cytosine, and guanine.
theta (θ) structure An intermediate in the bidirectional replication of circular DNA molecules. At about midway through the cycle of replication, the intermediate resembles the Greek letter theta.
thymine dimer A pair of adjacent thymine bases in a single polynucleotide strand that have chemical bonds formed between carbon atoms 5 and 6. This lesion, usually caused by exposure to ultraviolet light, inhibits DNA replication unless repaired by the appropriate enzymes.
T_m See *melting profile*.
topoisomerase A class of enzymes that convert DNA from one topological form to another. During DNA replication, these enzymes facilitate the unwinding of the double-helical structure of DNA.
totipotent The ability of a cell or embryo part to give rise to all adult structures. This capacity is usually progressively restricted during development.
trait Any detectable phenotypic variation of a particular inherited character.
***trans* configuration** The arrangement of two mutant sites on opposite homologs, such as

$$\frac{a^1 \quad +}{+ \quad a^2}$$

contrasts with a *cis* arrangement, where the sites are located on the same homolog.
transcription Transfer of genetic information from DNA by the synthesis of an RNA molecule copied from a DNA template.
transcriptome The set of mRNA molecules present in a cell at any given time.
transdetermination Change in developmental fate of a cell or group of cells.
transduction Virally mediated genes transfer from one bacterium to another or the transfer of eukaryotic genes mediated by retrovirus.

transfer RNA (tRNA) A small ribonucleic acid molecule that contains a three-base segment (anticodon) that recognizes a codon in mRNA, a binding site for a specific amino acid, and recognition sites for interaction with the ribosomes and the enzyme that links it to its specific amino acid.
transformation Heritable change in a cell or an organism brought about by exogenous DNA.
transgenic organism An organism whose genome has been modified by the introduction of external DNA sequences into the germline.
transition A mutational event in which one purine is replaced by another or one pyrimidine is replaced by another.
translation The derivation of the amino acid sequence of a polypeptide from the base sequence of an mRNA molecule in association with a ribosome.
translocation A chromosomal mutation associated with the transfer of a chromosomal segment from one chromosome to another. Also used to denote the movement of mRNA through the ribosome during translation.
transmission genetics The field of genetics concerned with the mechanisms by which genes are transferred from parent to offspring.
transposable element A DNA segment that translocates to other sites in the genome, essentially independent of sequence homology. Usually such elements are flanked by short inverted repeats of 20–40 base pairs at each end. Insertion into a structural gene can produce a mutant phenotype. Insertion and excision of transposable elements depends on two enzymes, transposase and resolvase. Such elements have been identified in both prokaryotes and eukaryotes.
transversion A mutational event in which a purine is replaced by a pyrimidine or a pyrimidine is replaced by a purine.
trinucleotide repeat A tandemly repeated cluster of three nucleotides (such as CTG) in or near a gene, which undergoes an expansion in copy number, resulting in a disease phenotype.
triploidy The condition in which a cell or organism possesses three haploid sets of chromosomes.
trisomy The condition in which a cell or organism possesses two copies of each chromosome, except for one, which is present in three copies. The general form for trisomy is therefore $2n + 1$.
tRNA See *transfer RNA*.
tumor-suppressor gene A gene that encodes a gene product that normally functions to suppress cell division. Mutations in tumor-suppressor genes result in the activation of cell division and tumor formation.
Turner syndrome A genetic condition in human females caused by a 45, X genotype. Such individuals are phenotypically female but are sterile because of undeveloped ovaries.
unequal crossing over A crossover between two improperly aligned homologs, producing one homolog with three copies of a region and the other with one copy of that region.
unique DNA DNA sequences that are present only once per genome.
universal code The assumption that the genetic code is used by all life forms. In general, this is true; some exceptions are found in mitochondria, ciliates, and mycoplasmas.
unwinding proteins Nuclear proteins that act during DNA replication to destabilize and unwind the DNA helix ahead of the replicating fork.
variable number tandem repeats (VNTRs) Short, repeated DNA sequences (2–20 nucleotides) present as tandem repeats between two restriction enzyme sites. Variations in the number of repeats creates DNA fragments of differing lengths following restriction enzyme digestion.
variable region Portion of an immunoglobulin molecule that exhibits many amino acid sequence differences between antibodies of differing specificities.
variance A statistical measure of the variation of values from a central value, calculated as the square of the standard deviation.
variegation Patches of differing phenotypes, such as color, in a tissue.
vector In recombinant DNA, an agent such as a phage or plasmid into which a foreign DNA segment will be inserted.
viability The measure of the number of individuals in a given phenotypic class that survive, relative to another class (usually wild type).
virulent phage A bacteriophage that infects and lyses the host bacterial cell.
VNTRs See *variable number tandem repeats*.
western blot A technique in which proteins are separated by gel electrophoresis and transferred by capillary action to a nylon membrane or nitrocellulose sheet. A specific protein can be identified through hybridization to a labeled antibody.
wild type The most commonly observed phenotype or genotype, designated as the norm or standard.
wobble hypothesis An idea proposed by Francis Crick, stating that the third base in an anticodon can align in several ways to allow it to recognize more than one base in the codons of mRNA.
writhing number The number of times that the axis of a DNA duplex crosses itself by supercoiling.
W, Z chromosomes Sex chromosomes in species where the female is the heterogametic sex (WZ).
X chromosome Sex chromosome present in species where females are the homogametic sex (XX).
X inactivation In mammalian females, the random cessation of transcriptional activity of one X chromosome. This event, which occurs early in development, is a mechanism of dosage compensation. Molecular basis of inactivation is unknown, but involves a region called the X-inactivation center (XIC) on the proximal end of the p arm. Some loci on the tip of the short arm of the X can escape inactivation. See also *Barr body*, *Lyon hypothesis*.
XIST A locus in the X-chromosome inactivation center that may control inactivation of the X chromosome in mammalian females.
X-linkage The pattern of inheritance resulting from genes located on the X chromosome.
X-ray crystallography A technique to determine the three-dimensional structure of molecules through diffraction patterns produced by X-ray scattering by crystals of the molecule under study.
YAC A cloning vector in the form of a yeast artificial chromosome, constructed using chromosomal elements including telomeres (from a ciliate), centromeres, origin of replication, and marker genes from yeast. YACs are used to clone long stretches of eukaryotic DNA.
Y chromosome Sex chromosome in species where the male is heterogametic (XY).
Y-linkage Mode of inheritance shown by genes located on the Y chromosome.
Z-DNA An alternative structure of DNA in which the two antiparallel polynucleotide chains form a left-handed double helix. Z-DNA has been shown to be present in chromosomes and may have a role in regulation of gene expression.
zein Principal storage protein of maize endosperm, consisting of two major proteins, with molecular weights of 19,000 and 21,000 Da.
zinc finger A DNA-binding domain of a protein that has a characteristic pattern of cysteine and histidine residues that complex with zinc ions, throwing intermediate amino acid residues into a series of loops or fingers.
zygote The diploid cell produced by the fusion of haploid gametic nuclei.
zygotene A stage of meiotic prophase I in which the homologous chromosomes synapse and pair along their entire length, forming bivalents. The synaptonemal complex forms at this stage.

Appendix B

Answers to Selected Problems

Chapter 1

2. Based on the parallels between Mendel's model of heredity and the behavior of chromosomes, the chromosome theory of inheritance emerged. It states that inherited traits are controlled by genes residing on chromosomes that are transmitted by gametes.
4. A gene variant is called an allele. There can be many such variants in a population, but for a diploid organism, only two such alleles can exist in any given individual.
6. *Genes*, linear sequence of nucleotides, usually exert their influence by producing proteins through the process of transcription and translation. Genes are the functional units of heredity. They associate, sometimes with proteins, to form *chromosomes*.
8. The central dogma of molecular genetics refers to the relationships among DNA, RNA, and protein. The processes of *transcription* and *translation* are integral to understanding these relationships.
10. Restriction enzymes (endonucleases) cut double-stranded DNA at particular base sequences. When a vector is cleaved with the same enzyme, complementary ends are created such that ends, regardless of their origin, can be combined and ligated to form intact double-stranded structures. Such recombinant forms are often useful for industrial, research, and/or pharmaceutical efforts.
12. Supporters of organismic patenting argue that it is needed to encourage innovation and allow the costs of discovery to be recovered. Capital investors assume that there is a likely chance that their investments will yield positive returns. Others argue that natural substances should not be privately owned and that once owned by a small number of companies, free enterprise will be stifled. Individuals and companies needing vital, but patented products, may have limited access.
14. Model organisms are not only useful, but necessary, for understanding genes that influence human diseases. Given that genetic/molecular systems are highly conserved across broad phylogenetical lines, what is learned in one organism is usually applied to all organisms. Most model organisms have peculiarities, such as ease of growth, genetic understanding, or abundant offspring, which make them straightforward and especially informative in genetic studies.

Chapter 2

2. Chromosomes that are homologous share many properties including:

   Overall length
   Position of the centromere
   Banding patterns
   Type and location of genes
   Autoradiographic pattern

   Diploidy is a term often used in conjunction with the symbol $2n$. It means that both members of a homologous pair of chromosomes are present. The change from a diploid ($2n$) to haploid (n) occurs during *reduction division* when tetrads become dyads during meiosis I.
6. Centromere placement may be medial (metacentric), off center (submetacentric), near one end (acrocentric), or at one end (telocentric). Notice the different anaphase shapes of chromosomes as they move to the poles: metacentric (a), submetacentric (b), acrocentric (c), telocentric (d).

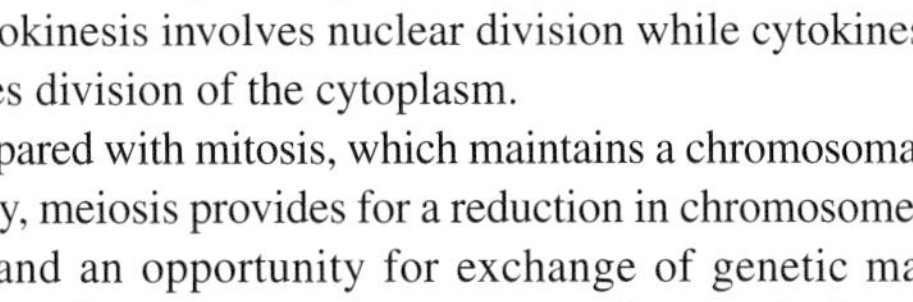

8. Carefully read the section on mitosis and cell division in the text. Major divisions of the cell cycle include interphase and mitosis. Interphase is composed of four phases: G1, G0, S, and G2. During the S phase, chromosomal DNA doubles. Karyokinesis involves nuclear division while cytokinesis involves division of the cytoplasm.
10. Compared with mitosis, which maintains a chromosomal constancy, meiosis provides for a reduction in chromosome number, and an opportunity for exchange of genetic material between homologous chromosomes. In mitosis there is no change in chromosome number or kind in the two daughter cells whereas in meiosis numerous potentially different haploid (n) cells are produced.
12. Sister chromatids are genetically identical, except where mutations may have occurred during DNA replication. Nonsister chromatids are genetically similar if on homologous chromosomes or genetically dissimilar if on nonhomologous chromosomes. If crossing over occurs, then chromatids attached to the same centromere will no longer be identical.

14. Not necessarily. If crossing over occurred in meiosis I, then the chromatids in the secondary oocyte are not identical. Once they separate during meiosis II, unlike chromatids reside in the ootid and the second polar body.

16. Through independent assortment of chromosomes at anaphase I of meiosis, daughter cells (secondary spermatocytes and secondary oocytes) may contain different sets of maternally and paternally derived chromosomes. Crossing over, which happens at a much higher frequency in meiotic cells as compared to mitotic cells, allows maternally and paternally derived chromosomes to exchange segments thereby increasing the likelihood that daughter cells are variable.

18. If there are eight combinations possible for part (c) in the previous problem, there would be 16 combinations with the addition of another chromosome pair.

20. One half of each tetrad will have a maternal homolog: $(1/2)^{10}$.

22. In angiosperms, meiosis results in the formation of microspores (male) and megaspores (female), which give rise to the haploid male and female gametophyte stage. Micro- and megagametophytes produce the pollen and the ovules respectively. Following fertilization, the sporophyte is formed.

24. The folded-fiber model is based on each chromatid consisting of a single fiber wound like a skein of yarn. Each fiber consists of DNA and protein. A coiling process occurs during the transition of interphase chromatin to more condensed chromosomes during prophase of mitosis or meiosis. Such condensation leads to a 5000-fold contraction in the length of the DNA within each chromatid.

26. Duplicated chromosomes A^m, A^p, B^m, B^p, C^m, and C^p will align at metaphase, with the centromeres dividing and sister chromatids going to opposite poles at anaphase.

28. As long as you have accounted for eight possible combinations in the previous problem, there would be no new ones added in this problem.

30. See the products of nondisjunction of chromosome *C* at the end of meiosis I as follows.

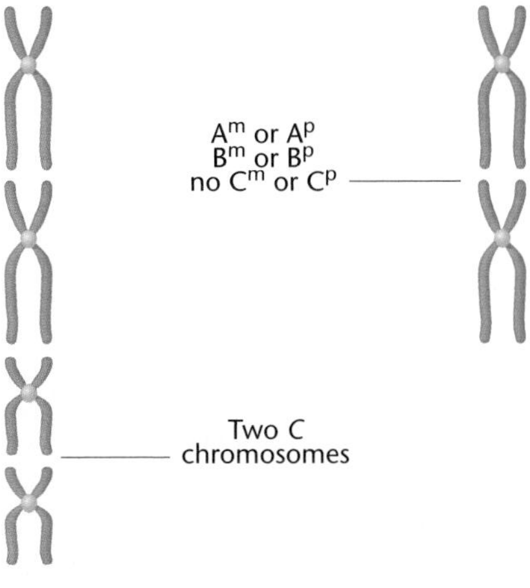

At the end of meiosis II, assuming that, as the problem states, the *C* chromosomes separate as dyads instead of monads during meiosis II, you would have monads for the *A* and *B* chromosomes, and dyads (from the cell on the left) for both *C* chromosomes as one possibility.

Chapter 3

2. Since albinism is inherited as a recessive trait, genotypes *AA* and *Aa* should produce the normal phenotype, while *aa* will give albinism.

(a) The parents must both be heterozygous (*Aa*).

(b) The female must be *aa*. Since all the children are normal, one would consider the male to be *AA* instead of *Aa*. However, the male could be *Aa*. Under that circumstance, the likelihood of having six normal children is 1/64.

(c) The female must be *aa*. The fact that half of the children are normal and half are albino indicates a typical "test cross" in which the *Aa* male is mated to the *aa* female.

(d)

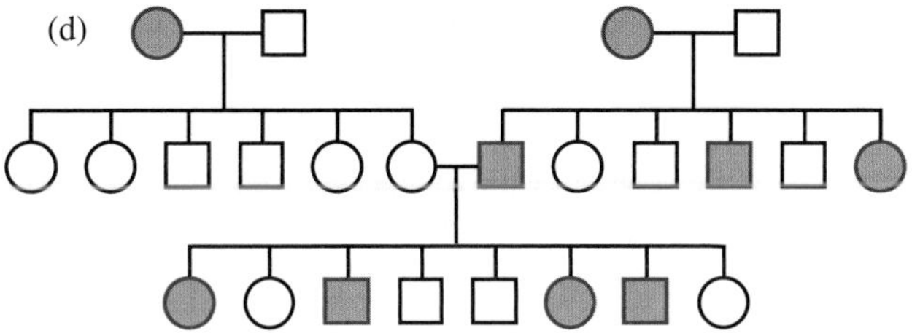

The 1:1 ratio of albino to normal in the last generation theoretically results because the mother is *Aa* and the father is *aa*.

4. First, organisms contained **unit factors** for various traits. Second, if these **factors occurred in pairs**, there existed the possibility that some organisms would "breed true" if homozygous, while others would not (heterozygotes). If one "factor" of a pair had a **dominant influence** over the other, then he could explain how two organisms, looking the same, could be genetically different (homozygous or heterozygous). Third, if the **paired elements separate (segregate)** from each other during gamete formation and if gametes combine at random, he could account for the 3:1 ratios in the monohybrid crosses. The fourth postulate, independent assortment, cannot be demonstrated by a monohybrid cross because two gene pairs must be involved to do so.

6. P = checkered; p = plain. Checkered is tentatively assigned the dominant function because in a casual examination of the data, especially cross (b), we see that checkered types are more likely to be produced than plain types.

Cross (a): *PP* × *PP* or *PP* × *Pp*
Cross (b): *PP* × *pp*
Cross (c): Because all the offspring from this cross are plain, there is no doubt that the genotype of both parents is *pp*.

Genotypes of all individuals:

	F₁ Progeny	
P₁ Cross	***Checkered***	***Plain***
(a) *PP* × *PP*	*PP*	
(b) *PP* × *pp*	*Pp*	
(c) *pp* × *pp*		*pp*
(d) *PP* × *pp*	*Pp*	
(e) *Pp* × *pp*	*Pp*	*pp*
(f) *Pp* × *Pp*	*PP*, *Pp*	*pp*
(g) *PP* × *Pp*	*PP*, *Pp*	

8. $WWgg = 1/16$

10. In Problem #9, (d) fits this description.

12. Mendel's four postulates are related to the diagram below.

(1) Factors occur in pairs. Notice *A* and *a*.

(2) Some genes have dominant and recessive alleles. Notice *A* and *a*.

(3) Alleles segregate from each other during gamete formation. When homologous chromosomes separate from each other at anaphase I, alleles will go to opposite poles of the meiotic apparatus.

(4) One gene pair separates independently from other gene pairs. Different gene pairs on the same homologous pair of chromosomes (if far apart) or on nonhomologous chromosomes will separate independently from each other during meiosis.

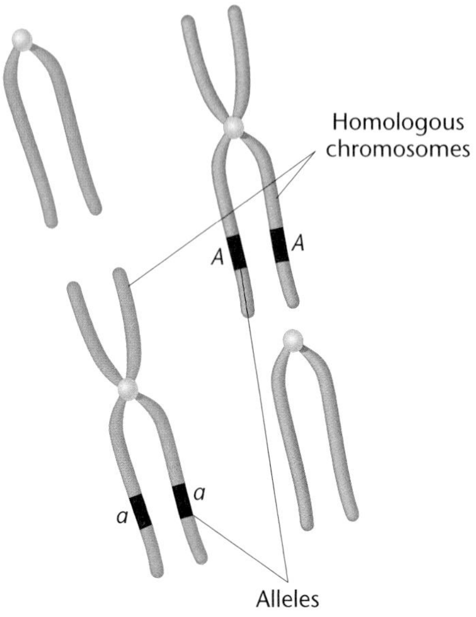

14. Homozygosity refers to a condition where both genes of a pair are the same (*i.e.*, *AA* or *GG* or *hh*), whereas heterozygosity refers to the condition where members of a gene pair are different (*i.e.*, *Aa* or *Gg or Bb*).
16. The general formula for determining the number of kinds of gametes produced by an organism is 2^n where n = number of *heterozygous* gene pairs.
 (a) 4: *AB*, *Ab*, *aB*, *ab*
 (b) 2: *AB*, *aB*
 (c) 8: *ABC*, *ABc*, *AbC*, *Abc*, *aBC*, *aBc*, *abC*, *abc*
 (d) 2: *ABc*, *aBc*
 (e) 4: *ABc*, *Abc*, *aBc*, *abc*
 (f) $2^5 = 32$
18. *G* = yellow seeds; *g* = green seeds.

	Phenotypes	**Genotypes**
P_1:	Yellow × green	*GG* × *gg*
F_1:	all yellow	*Gg*
F_2:	6022 yellow	1/4 *GG*; 2/4 *Gg*
	2001 green	1/4 *gg*

Notice that of the *yellow* F_2 *offspring*, 1/3 are *GG* and 2/3 are *Gg*. If you selfed the 1/3 *GG* types then all the offspring (the 166) would breed true, whereas the others (353 which are *Gg*) should produce offspring in a 3:1 ratio when selfed.
GG × *GG* = all *GG*
Gg × *Gg* = 1/4 *GG*; 2/4 *Gg*; 1/4 *gg*

20. Symbols:

Seed shape	**Seed color**
W = round	*G* = yellow
w = wrinkled	*g* = green

P_1: *WWgg* × *wwGG*
F_1: *WwGg* cross to *wwgg*
(which is a typical testcross)

The offspring will occur in a typical 1:1:1:1 as
1/4 *WwGg* (round, yellow)
1/4 *Wwgg* (round, green)
1/4 *wwGg* (wrinkled, yellow)
1/4 *wwgg* (wrinkled, green)

22. (a) $\chi^2 = .064$
 probability (*p*) value between 0.9 and 0.5. We would therefore say that there is a "good fit" between the observed and expected values.
 (b) $\chi^2 = 0.39$
 The *p* value in the table for 1 degree of freedom is still between 0.9 and 0.5. The deviation in each case can be attributed to chance.
24. For the test of a 3:1 ratio, the χ^2 value is 33.3 with an associated *p* value of less than 0.01 for 1 degree of freedom. For the test of a 1:1 ratio, the χ^2 value is 25.0, again with an associated *p* value of less than 0.01 for 1 degree of freedom. Based on these probability values, both null hypotheses should be rejected.
26. If a gene is dominant, it will not skip generations nor will it be passed to offspring unless the parents have the gene. On the other hand, genes that are recessive can skip generations and exist in a carrier state in parents. Notice that II-4 and II-5 produce a female child (III-4) with the affected phenotype. On these criteria alone, the gene must be viewed as being recessive. Note: if a gene is recessive and X-linked (to be discussed later), the pattern will often be from affected male to carrier female to affected male.
 I-1 (*Aa*), I-2 (*aa*), I-3 (*Aa*), I-4(*Aa*)
 II-1 (*aa*), II-2 (*Aa*), II-3 (*aa*), II-4 (*Aa*), II-5 (*Aa*), II-6 (*aa*), II-7 (*AA* or *Aa*), II-8 (*AA* or *Aa*)
 III-1 (*AA* or *Aa*), III-2 (*AA* or *Aa*), III-3 (*AA* or *Aa*), III-4 (*aa*), III-5 (probably *AA*), III-6 (*aa*)
 IV-1 through IV-7 all *Aa*.
28. 1/8
30. 3/4
32. 4/9
34. 1/9
36. $P = [8!(3/4)^6(1/4)^2]/6!2!$
38. **(a)**

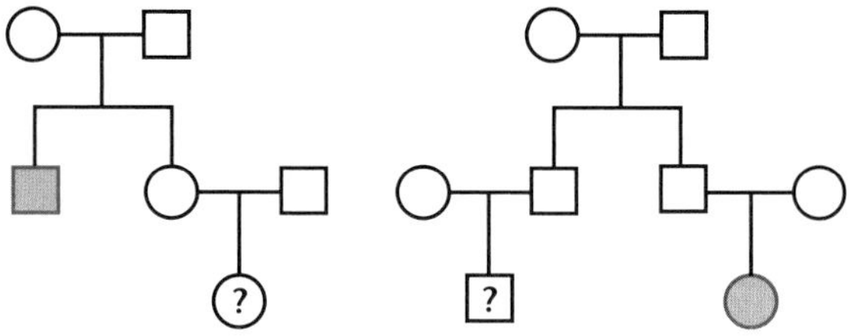

 (b) 1/12
 (c) 6/12
 (d) 5/12
40. $P = [5!\,(3/4)^3(1/4)^2]/3!2!$
 $= 135/512$
42. (a) For Set I, the $\chi^2 = 2.15$ with *p* being between 0.2 and 0.05. So one would accept the null hypothesis of no significant difference between the expected and observed values.
 For Set II, the $\chi^2 = 21.43$ and $p < 0.001$. One would reject the null hypothesis and assume a significant difference between the observed and expected values.
 (b) In most cases, more confidence is gained as the sample size increases, however, depending on the organism or experiment, there are practical limits on sample size.

Chapter 4

2. *Incomplete dominance* can be viewed more as a quantitative phenomenon where the heterozygote is intermediate (approximately) between the limits set by the homozygotes. *Codominance* can be viewed in a more qualitative manner where both of the alleles in the heterozygote are expressed.

4. Cross 1:

short tail × normal long tail ⟶

approximately 1/2 short, 1/2 long

This tells you that one type is heterozygous, the other homozygous.

Cross 2:

short tail × short tail ⟶

6 short tail, 3 long tail (2/3 short, 1/3 long)

At this point one would consider that the 2/3 *short* are heterozygotes, and *long* is the homozygous class. Also, short is dominant to long. Since these ratios were repeated and verified, one can conclude that a 2:1 ratio is not a statistical artifact and that the following genotypic model would hold. Because long is the "normal" does not mean that it is dominant.

Symbolism:

S = short, s = long

Cross 1:

$Ss \times ss$ ⟶

1/2 *Ss* (short), 1/2 *ss* (long)

Cross 2:

$Ss \times Ss$ ⟶

1/4 *SS* (lethal), 2/4 *Ss* (short), 1/4 *ss* (long)

6. $$I^AI^O \times I^BI^O$$

	I^B	I^O
I^A	$I^A I^B$(AB)	$I^A I^O$(A)
I^O	$I^B I^O$(B)	$I^O I^O$(O)

The ratio would be

1(A):1(B):1(AB):1(O).

8. Symbolism:

Se = secretor *se* = nonsecretor

I^AI^B *Sese* × I^OI^O *Sese*

(a)

	I^O*Se*	I^O*se*
I^A *Se*	A, secretor	A, secretor
I^A *se*	A, secretor	A, nonsecretor
I^B *Se*	B, secretor	B, secretor
I^B *se*	B, secretor	B, nonsecretor

Overall ratio: 3/8 A, secretor
1/8 A, nonsecretor
3/8 B, secretor
1/8 B, nonsecretor

(b) 1/4 of all individuals will have blood type O.

10. (a) Phenotypes:

Himalayan × Himalayan ⇒ albino

Genotypes: c^hc^a c^hc^a c^ac^a

The Himalayan parents must both be heterozygous to produce an albino offspring.

Phenotypes:

full color × albino ⇒ chinchilla

Genotypes: Cc^{ch} c^ac^a $c^{ch}c^a$

Therefore the cross of albino with chinchilla would be as follows:

$c^ac^a \times c^{ch}c^a$ ⟶

1/2 chinchilla; 1/2 albino

(b) Phenotypes:

albino × chinchilla ⇒ albino

Genotypes: c^ac^a $c^{ch}c^a$ c^ac^a

Phenotypes:

full color × albino ⇒ full color

Genotypes: $C\_$ c^ac^a Cc^a

It is impossible to determine the complete genotype of the full color parent but the full color offspring must be as indicated, Cc^a.

Therefore the cross of the albino with full color would be as follows:

$c^ac^a \times Cc^a$ ⟶

1/2 full color; 1/2 albino

(c) Phenotypes:

chinchilla × albino ⇒ Himalayan

Genotypes: $c^{ch}c^h$ c^ac^a c^hc^a

The chinchilla parent must be heterozygous for Himalayan because of the Himalayan offspring.

Phenotypes:

full color × albino ⇒ Himalayan

Genotypes: Cc^h c^ac^a c^hc^a

Therefore a cross between the two Himalayan types would produce the following offpsring:

$c^hc^a \times c^hc^a$

⇓

3/4 Himalayan; 1/4 albino

12. 18/64

14. (a) $C^{ch}C^{ch}$ = chestnut
C^cC^c = cremello
$C^{ch}C^c$ = palomino

(b) The F_1 resulting from matings between cremello and chestnut horses would be expected to be all palomino. The F_2 would be expected to fall in a 1:2:1 ratio as in the third cross in part (a) above.

16. (a) In a cross of

AACC × *aacc*,

the offpsring are all *AaCc* (agouti) because the *C* allele allows pigment to be deposited in the hair, and when it is it will be agouti. F_2 offspring would have the following "simplified" genotypes with the corresponding phenotypes:

*A_C_* = 9/16 (agouti)

A_cc = 3/16

(colorless because *cc* is epistatic to *A*)

*aaC_* = 3/16 (black)

aacc = 1/16

(colorless because *cc* is epistatic to *aa*)

The two colorless classes are phenotyically indistinguishable, therefore the final ratio is 9:3:4.

(b) Results of crosses of female agouti

(*A_C_*) × *aacc* (males)

are given in three groups:

(1) To produce an even number of agouti and colorless offspring, the female parent must have been *AACc* so that half of the offspring are able to deposit pigment because of *C*, and when they do, they are all agouti (having received only *A* from the female parent).

(2) To produce an even number of agouti and black offspring the mother must have been *Aa* and so that no colorless offspring were produced, the female must have been *CC*. Her genotype must have been *AaCC*.

(3) Notice that half of the offspring are colorless, therefore the female must have been *Cc*. Half of the pigmented offspring are black and half are agouti, therefore the female must have been *Aa*. Overall, the *AaCc* genotype seems appropriate.

18. (a) *AaBbCc* ⇒ gray (*C* allows pigment)
(b) *A_B_Cc* ⇒ gray (*C* allows pigment)
(c) 16/32 albino;
9/32 gray;
3/32 yellow;
3/32 black;
1/32 cream
(d) 9/16 (gray);
3/16 (black);
4/16 (albino)
(e) 3/8 (gray);
1/8 (yellow);
4/8 (albino)

20. (a) Cross A:
P_1: *AABB* × *aaBB*
F_1: *AaBB*
F_2: 3/4 *A_BB*: 1/4 *aaBB*

Cross B:
P_1: *AABB* × *AAbb*
F_1: *AABb*
F_2: 3/4 *AAB_*: 1/4 *AAbb*

Cross C:
P_1: *aaBB* × *AAbb*
F_1: *AaBb*
F_2: 9/16 *A_B_*: 3/16 *A_bb*:
3/16 *aaB_*: 1/16 *aabb*:

(b) The genotype of the unknown P_1 individual would be *AAbb* (brown) while the F_1 would be *AaBb* (green).

22. (a) Assign the phenotypes as given, then see if patterns emerge.

*A_B_* = 9/16 (yellow)

A_bb = 9/16 (blue)

*aaB_* = 9/16 (red)

aabb = 9/16 (mauve)

See that each type can exist as a full homozygote. If plants with blue flowers (homozygotes) are crossed to red-flowered homozygotes, the F_1 plants would have yellow flowers. If yellow-flowered plants are crossed with mauve-flowered plants, the F_1 plants are yellow and the F_2 will occur in a 9:3:3:1 ratio. All of the observations fit the model as proposed.

(b) If one crosses a true-breeding red plant (*aaBB*) with a mauve plant (*aabb*), the F_1 should be red (*aaBb*). The F_2 would be as follows:

aaBb × *aaBb* →

3/4 *aaB_* (red): 1/4 *aabb* (mauve)

24. (a) 1/4
(b) 1/2
(c) 1/4
(d) zero

26. Symbolism: Normal wing margins = sd^+; scalloped = sd

(a) P_1: $X^{sd}X^{sd} \times X^+/Y$ →

F_1: 1/2 X^+X^{sd} (female, normal)
1/2 X^{sd}/Y (male, scalloped)
F_2: 1/4 X^+X^{sd} (female, normal)
1/4 $X^{sd}X^{sd}$ (female, scalloped)
1/4 X^+/Y (male, normal)
1/4 X^{sd}/Y (male, scalloped)

(b) P_1: $X^+/X^+ \times X^{sd}/Y$ →

F_1: 1/2 X^+X^{sd} (female, normal)
1/2 X^+/Y (male, normal)
F_2: 1/4 X^+X^+ (female, normal)
1/4 X^+X^{sd} (female, normal)
1/4 X^+/Y (male, normal)
1/4 X^{sd}/Y (male, scalloped)

If the *scalloped* gene were not X-linked, then all of the F_1 offspring would be wild (phenotypically) and a 3:1 ratio of normal to scalloped would occur in the F_2.

28. P_1: X^+X^+; *su-v/su-v* × X^v/Y; $su\text{-}v^+/su\text{-}v^+$
F_1: 1/2 X^+X^v; $su\text{-}v^+/su\text{-}v$ (female, normal)
1/2 X^+/Y; $su\text{-}v^+/su\text{-}v$ (male, normal)

F_2: 2/4 females, $X^+/\_$ < 3/4 $su\text{-}v^+/\_$; 1/4 *su-v/su-v*
1/4 males, X^+/Y < 3/4 $su\text{-}v^+/\_$; 1/4 *su-v/su-v*
1/4 males, X^v/Y < 3/4 $su\text{-}v^+/\_$; 1/4 *su-v/su-v*

8/16 wild type females; (none of the females are homozygous for the *vermilion* gene)

5/16 wild type males; (4/16 because they have no *vermilion* gene and 1/16 because the X-linked, hemizygous *vermilion* gene is suppressed by *su-v/su-v*)

3/16 vermilion males; (no suppression of the *vermilion* gene)

30. (a) w/w; $se^+/se^+ \times w^+/Y$; se/se
⇓
F_1: w^+/w; se^+/se = wild females
w/Y; se^+/se = white-eyed males
F_2: 3/16 males wild
4/16 males white
1/16 males sepia
3/16 females wild
4/16 females white
1/16 females sepia

(b) w^+/w^+; $se/se \times w/Y$; se^+/se^+
⇓
F_1: w^+/w; se^+/se = wild females
w^+/Y; se^+/se = wild males
F_2: 3/16 males wild
4/16 males white
1/16 males sepia
6/16 females wild
2/16 females sepia

32. Let *a* represent the mutant gene and *A* represent its normal allele.

(a) This pedigree is consistent with an X-linked recessive trait because the male would contribute an X chromosome carrying the *a* mutation to the *aa* daughter. The mother would have to be heterozygous *Aa*.

(b) This pedigree is consistent with an X-linked recessive trait because the mother could be *Aa* and transmit her *a* allele to her one son (*a*/Y) and her *A* allele to her other son.

(c) This pedigree is not consistent with an X-linked mode of inheritance because the *aa* mother has an *A*/Y son.

34. F_1: all hen-feathering
F_2: All of the offspring would be hen-feathered except for 1/8 males, which are cock-feathered.

36. Phenotypic expression is dependent on the genome of the organism, the immediate molecular and cellular environment of the genome, and numerous interactions between a genome, the organism, and the environment.

38. *Anticipation* occurs when a heritable disorder exhibits a progressively earlier age of onset and an increased severity in successive generations. *Imprinting* occurs when phenotypic expression is influenced by the parental origin of the chromosome carrying a particular gene.

40. A first glance would seem to favor a 9:7 ratio; however, the phenotypes would have to be reversed for such a result to fit. Therefore, one must consider an alternative explanation. A 27:9:9:9:3:3:3:1 ratio fits very well with A_B_C_ being purple and any homozygous recessive combination giving white. Thus a 27(purple):37(white) ratio fits well. To test this hypothesis, one might take the purple F_1's and cross them to the pure breeding (*aabbcc*) white type. Such a cross should give a 1(purple):7(white).

42. (a) Because the denominator in the ratios is 64, one would begin to consider that there are three independently assorting gene pairs operating in this problem. However, because there are only two characteristics (eye color and croaking), one might hypothesize that two gene pairs are involved in the inheritance of one trait while one gene pair is involved in the other.

(b) Notice that there is a 48:16 (or 3:1) ratio of rib-it to knee-deep and a 36:16:12 (or 9:4:3) ratio of blue to green to purple eye color. Because of these relationships one would conclude that croaking is due to one (dominant/recessive) gene pair while eye color is due to two gene pairs. Because there is a (9:4:3) ratio regarding eye color, some gene interaction (epistasis) is indicated.

(c, d) Symbolism:
Croaking: *R_* = rib-it; *rr* = knee-deep
Eye color:
Since the most frequent phenotype is blue eye, let *A_B_* represent the genotypes. For the purple class, "a 3/16 group" use the *A_bb* genotypes. The "4/16" class (green) would be the *aaB_* and the *aabb* groups.

(e) The cross involving a blue-eyed, knee-deep frog and a purple-eyed, rib-it frog would have the genotypes:

$$AABBrr \times AAbbRR$$

which would produce an F_1 of *AABbRr*, which would be blue-eyed and rib-it. The F_2 will follow a pattern of a 9:3:3:1 ratio because of homozygosity for the *A* locus and heterozygosity for both the *B* and *R* loci.

9/16	*AAB_R_*	= blue-eyed, rib-it
3/16	*AAB_rr*	= blue-eyed, knee-deep
3/16	*AAbbR_*	= purple-eyed, rib-it
1/16	*AAbbrr*	= purple-eyed, knee-deep

(f) The different results can arise because of the genetic variety possible in producing the green-eyed frogs. Since there is no dependence on the *B* locus, the following genotypes can define the green phenotype:

aaBB, aaBb, aabb

(g) Notice that the ratio of purple-eyed to green-eyed frogs is 3:1, therefore expect the parents to be heterozygous for the *A* locus. Because the ratio of rib-it to knee-deep is also 3:1, expect both parents to be heterozygous at the *R* locus. The *B* locus would have the *bb* genotype because both parents are purple-eyed as given in the problem. Both parents would therefore be *AabbRr*.

44. (a) P_1: *YYBB* × *y***W***bb*
F_1: *YyBb* and *Y***W***Bb*
Crossing these F_1's gives the observed ratios in the F_2.

(b) Given a blue male with the genotype *yyBb* and a green female with the genotype *Y***W***Bb*, the offspring are as given in part (b) of the question.

46. Given the following genotypes of the parents:

aabb = crimson
AABB = white

the F_1 consist of *AaBb* genotypes with a rose phenotype.

In the F_2, the following genotypes correspond to the given phenotypes:

*AAB_*	= white	4/16
AaBB	= magenta	2/16
AaBb	= rose	4/16
Aabb	= orange	2/16
aaBB	= yellow	1/16
aaBb	= pale yellow	2/16
aabb	= crimson	1/16

48. (a) The term *pleiotropy* is used when there are multiple phenotypic manifestations of a single gene. (b) Each phenotypic response is a result of the inability of the red blood cells to adequately supply tissues. Sickle-cell hemoglobin fails to adequately carry oxygen and the sickle shape of the red blood cells blocks their passage to various tissues. As a result of a vaso-occlusive crisis, tissue necrosis occurs and a variety of health problems result. (c) The molecular basis of sickle-cell anemia is a substitution of valine for glutamic acid at the sixth position of the β chain.

Chapter 5

2. First, in order for chromosomes to engage in crossing over, they must be in proximity. It is likely that the side-by-side pairing that occurs during synapsis is the earliest time during the cell cycle that chromosomes achieve that necessary proximity. Second, chiasmata are visible during prophase I of meiosis and it is likely that these structures are intimately associated with the genetic event of crossing over.

4. Because crossing over occurs at the four-strand stage of the cell cycle (that is, after S phase) each single crossover involves only two of the four chromatids.

6. Interference is often explained by a physical rigidity of chromatids such that they are unlikely to make sufficiently sharp bends to allow crossovers to be close together.

8.
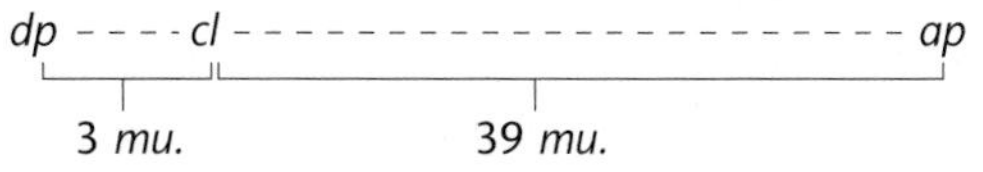

10. The most frequent phenotypes in the offspring, the parentals, are colored, green (88) and colorless, yellow (92). This indicates that the heterozygous parent in the test cross is coupled

$$RY/ry \times ry/ry$$

with the two dominant genes on one chromosome and the two recessives on the homolog. Seeing that there are 20 crossover progeny among the 200, or 20/200, the map distance would be 10 map units (20/200 × 100 to convert to percentages) between the *R* and *Y* loci.

12. The most frequent classes are *PZ* and *pz*. These classes represent the parental (noncrossover) groups, which indicates that the original parental arrangement in the test cross was

$$PZ/pz \times pz/pz$$

Adding the crossover percentages together (6.9 + 7.1) gives 14%, which would be the map distance between the two genes.

14.

	female A:	*female* B:	*Frequency:*
NCO	3, 4	7, 8	first
SCO	1, 2	3, 4	second
SCO	7, 8	5, 6	third
DCO	5, 6	1, 2	fourth

The single crossover classes, which represent crossovers between the genes that are closer together (*d-b*) would occur less frequently than the classes of crossovers between more distant genes (*b-c*).

16. (a) *y w*+/++ *ct* × *y w* +/Y

(b) *y* - - - - *w* - *ct*
0.0 1.5 20.0

(c) There were

$$.185 \times .015 \times 1000 = 2.775$$

double crossovers expected.

(d) Because the cross to the F_1 males included the normal (wild type) gene for *cut wings* it would not be possible to unequivocally determine the genotypes from the F_2 phenotypes for all classes.

18. Provide the genotypes of the parents in the original cross and the reciprocal. Use a semicolon to indicate that two different chromosome pairs are involved.

P_1: females: +/+; *p e*/*p e*
×
males: *dp*/*dp*; ++/++

F_1: females: +/*dp*; ++/*p e*
×
males: *dp*/*dp*; *p e*/*p e*

0.20 wild type
0.05 ebony
0.05 pink
0.20 pink, ebony
0.20 dumpy
0.05 dumpy, ebony
0.05 dumpy, pink
0.20 dumpy, pink, ebony

For the reciprocal cross:

.25 wild type
.25 pink, ebony
.25 dumpy
.25 dumpy, pink, ebony

The results would change because of no crossing over in males.

20. (a,b) +*bc*/*a*++

$$a - b = \frac{32 + 38 + 0 + 0}{1000} \times 100$$
$$= 7 \text{ map units}$$

$$b - c = \frac{11 + 9 + 0 + 0}{1000} \times 100$$
$$= 2 \text{ map units}$$

(c) The progeny phenotypes that are missing are ++*c* and *a b* +, which, of 1000 offspring, 1.4 (.07 × .02 × 1000) would be expected. Perhaps by chance or some other unknown selective factor, they were not observed.

22. Because sister chromatids are genetically identical (with the exception of rare new mutations) crossing over between sisters provides no increase in genetic variability. Individual genetic variability could be generated by somatic crossing over because certain patches on the individual would be genetically different from other regions. This variability would be of only minor consequence in all likelihood. Somatic crossing over would have no influence on the offspring produced.

24. (a) There would be $2^n = 8$ genotypic and phenotypic classes and they would occur in a 1:1:1:1:1:1:1:1 ratio.

(b) There would be two classes and they would occur in a 1:1 ratio.

(c) There are 20 map units between the *A* and *B* loci and locus *C* assorts independently from both *A* and *B* loci.

26. Assign the following symbols, for example:

R = Red *r* = yellow
O = Oval *o* = long

Progeny A: *Ro*/*rO* × *rroo* = 10 map units
Progeny B: *RO*/*ro* × *rroo* = 10 map units

28. The percentage of second division segregation is 20/100 or 20%. Dividing by 2 (because only two of the four chromatids are involved in any single crossover event) gives 10 map units.

30. For Cross 1:

$$\frac{36 + 14}{100} = 50 \text{ map units}$$

Because there are 50 map units between genes *a* and *b*, they are not linked.

For Cross 2:

$$\frac{3 + 9}{100} = 12 \text{ map units}$$

Because genes *a* and *b* are not linked they could be on nonhomologous chromosomes or far apart (50 map units or more) on the same chromosome. Because genes *c* and *b* are linked and therefore on the same chromosome, it is also possible that genes *a* and *c* are on different chromosome pairs. Under that condition, the NP and P (parental ditypes) would be equal; however, there is a possibility that the following arrangement occurs and that genes *a* and *c* are linked.

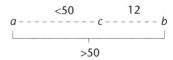

<50 12
a - - - - - - - - - - - *c* - - - - - - - *b*
>50

32. (a)

Tetrad in Problem	*Class*
1	NP
2	T
3	P
4	NP
5	T
6	P
7	T

(b) If P > NP, then the genes are linked. If P = NP they are independently assorting. In the problem given here, P = 44 and NP = 2. Therefore the genes are linked.

(c) For the centromere to *c* distance: = 7.2 map units
For the centromere to *d* distance: = 15.9 map units

(d) 20 map units

(e) 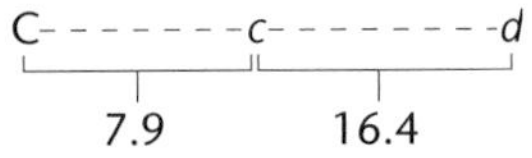

The discrepancy between the two mapping systems is caused by the manner in which first and second division segregation products are scored. For instance, in tetrad arrangement #1 there are actually two crossovers between the *d* gene and the centromere, but it is still scored as a first division segregation. In tetrad arrangement #4, three crossovers occur between the *d* gene and the centromere, but they are scored as one. If one draws out all the crossovers needed to produce the tetrad arrangements in this problem it would become clear that there are many crossovers between the *d* gene and the centromere, which go undetected in the scoring of the arrangements of the *d* gene itself. This will cause one to underestimate the distance and give the discrepancy noted.

One could account for these additional crossover classes to make the map more accurate.

(f) 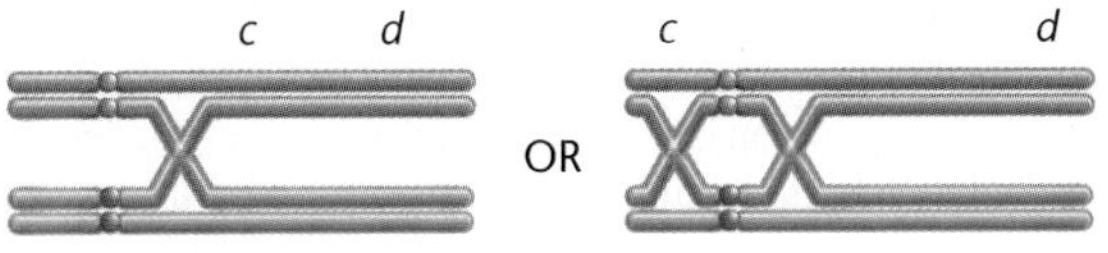

34. *MDH1*: chromosome 2
PEPS: chromosome 4
PMG1: chromosome 1

36. There is no crossing over in males, and if two genes are on the same chromosome, there will be complete linkage of the genes in the male gametes. In females, crossing over will produce parental and crossover gametes. What you will have is the following gametes from the females (left) and males (right):

$bw^+\ st^+$	1/4	$bw^+\ st^+$	1/2
$bw^+\ st$	1/4	$bw\ st$	1/2
$bw\ st^+$	1/4		
$bw\ st$	1/4		

Combining these gametes will give the ratio presented in the table of results.

38. The cross would be as follows:
$B^+B\ m^+m\ e^+e \times B^+?\ m?\ e?$
(The *?* is used at this point to indicate that we have no information allowing us to decide whether any of the alleles in the male are X-linked.) There are approximately as many ebony offspring (282) as those with wild body color (283). Therefore, we can conclude that the *ebony* locus is not linked to *B* or *m*. The most frequent offspring regarding eye shape and wing size are wild-miniature and Bar-wild. This suggests that the arrangement is "trans" or "repulsion" as indicated below:
$B\ m^+/B^+m;\ e^+/e$

At this point and without prior knowledge, we still don't know whether any of the genes are X-linked; however, it is of no consequence to the solution of the problem. (In actuality, both *B* and *m* loci are X-linked).

Mapping the distance between *B* and *m* would be as follows:

$$(57 + 64)/(226 + 218 + 57 + 64) \times 100 = 121/565 \times 100 = 21.4 \text{ map units.}$$

We would conclude that the *ebony* locus is either far away from *B* and *m* (50 map units or more) or it is on a different chromosome. In fact, *ebony* is on a different chromosome.

Chapter 6

2. (a) The requirement for physical contact between bacterial cells during conjugation was established by placing a filter in a U-tube such that the medium can be exchanged but the bacteria cannot come in contact.

(b) By treating cells with streptomycin, an antibiotic, it was shown that recombination would not occur if one of the two bacterial strains was inactivated. However, if the other was similarly treated, recombination would occur.

(c) An F^+ bacterium contains a circular, double-stranded, structurally independent, DNA molecule that can direct recombination.

4. Mapping the chromosome in an Hfr $\times$ F^- cross takes advantage of the oriented transfer of the bacterial chromosome through the conjugation tube. For each F type, the point of insertion and the direction of transfer are fixed, therefore breaking the conjugation tube at different times produces partial diploids with corresponding portions of the donor chromosome being transferred. The length of the chromosome being transferred is contingent on the duration of conjugation, thus mapping of genes is based on time.

6. The F^+ element can enter the host bacterial chromosome and upon returning to its independent state, it may pick up a piece of a bacterial chromosome. When combined with a bacterium with a complete chromosome, a partial diploid, or merozygote, is formed.

8. In the first data set, the transformation of each locus, a^+ or b^+, occurs at a frequency of .031 and .012 respectively. To determine if there is linkage one would determine whether the frequency of double transformants a^+b^+ is greater than that expected by a multiplication of the two independent events. Multiplying .031 $\times$ .012 gives .00037 or approximately 0.04%. From this information, one would consider no linkage between these two loci. Notice that this frequency is approximately the same as the frequency in the second experiment, where the loci are transformed independently.

10. In their experiment a filter was placed between the two auxotrophic strains, which would not allow contact. F-mediated conjugation requires contact, and without that contact, such conjugation cannot occur. The treatment with DNase showed that the filterable agent was not naked DNA.

12. In *generalized transduction* virtually any genetic element from a host strain may be included in the phage coat and thereby be transduced. In *specialized (restricted) transduction* only those genetic elements of the host that are closely linked to the insertion point of the phage can be transduced. Specialized transduction involves the process of lysogeny.

14. Viral recombination occurs when there is a sufficiently high number of infecting viruses so that there is a high likelihood that more than one type of phage will infect a given bacterium. Under this condition, phage chromosomes can recombine by crossing over.

16. Starting with a single bacteriophage, one lytic cycle produces 200 progeny phage, three more lytic cycles would produce $(200)^4$ or 1,600,000,000 phage.

18. For Group A, *d* and *f* are in the same complementation group (gene) while *e* is in a different one. Therefore

$$e \times f = +.$$

For Group B, all three mutations are in the same gene, hence

$$h \times i = -.$$

In Group C, *j* and *k* are in different complementation groups as are *j* and *l*. It would be impossible to determine whether *l* and *k* are in the same or different complementation group if the *r*II region had more than two cistrons. However, because only two complementation regions exist, and both are not in the same one as *j*, *k* and *l* must both be in the other.

20. 5×10^{-4}

22. T C H R O M B A K

24. (a) 4×10^{-5}

(b) The dilution would be 10^{-3} and the colony number would be 8×10^3.

(c) Mutant 7 might well be a deletion spanning parts of both A and B cistrons.

26. (a) Rifampicin eliminates the donor strain, which is *rif*^{s}.

(b) *b* *a* *c* F

(c) To determine the location of the *rif* gene one could use a donor strain, which was *rif*^{r} but sensitive to another antibiotic (ampicillin, for example). The interrupted mating experiment is conducted as usual on an ampicillin-containing medium, but the recombinants must be replated on a rifampicin medium to determine which ones are sensitive.

28. If two genes are cotransforming at a relatively high rate they are said to be "linked" in a sense that they are closer together than two genes that do not cotransform. The data indicate that *a* and *d* are linked, *b* and *c* are linked, and since *f* cotransforms with *b*, then *b*, *c*, and *f* are likely to be linked. However if the arrangement is *c* *b* *f* or the reverse, there is a possibility that whereas both *c* and *f* are "linked" to *b*, *c* and *f* may not be linked strongly enough to cotransform. Gene *e* does not cotransform with any gene so it must be independent of the other linkage groups.

30. (a) Some strains, *E. fergusonii*, for example, undergo relatively low transfer as a donor strain, while others, such as *E. chrysanthemi*, undergo relatively frequent transfer as a donor strain. Within-species transfer is not necessarily more frequent than between-species transfer. The direction of transfer (which is the donor and recipient strain) in some cases influences the frequency of transfer; for example, notice the frequencies of transfer when *E. chrysanthemi* is the donor and *E. coli* is the recipient (-1.7), compared to when *E. coli* is the donor and *E. chrysanthemi* is the recipient (-3.7).

(b) *E. chrysanthemi* (-2.4); *Ecoli-E. chrysanthem* (-1.7).

(c) Conjugative plasmids can share genes when bacteria are in proximity, and since such plasmids may contain either pathological genes or genes that compromise the use of antibiotics, any harmful variant that develops in one species may be spread to others. While a particular gene may be harmless in one bacterium, it may confer pathogenicity or drug resistance to a different species.

Chapter 7

4. Sexual differentiation is the response of cells, tissues, and organs to signals provided by the genetic mechanisms of sex determination. In other words, genes are present that signal developmental pathways whereby the sexes are generated. Sexual differentiation is the complex set of responses to those genetic signals.

6. In *Drosophila* it is the balance between the number of X chromosomes and the number of haploid sets of autosomes that determines sex. In humans there is a small region on the Y chromosome that determines maleness.

8. In *primary* nondisjunction half of the gametes contain two X chromosomes while the complementary gametes contain no X chromosomes. Fertilization, by a Y-bearing sperm cell, of those female gametes with two X chromosomes would produce the XXY Klinefelter syndrome. Fertilization of the "no-X" female gamete with a normal X-bearing sperm will produce Turner syndrome.

10. No. Since the Y chromosome cannot be detected in these crosses, there is no way to distinguish the two modes of sex determination.

12. Because attached-X chromosomes have a mother-to-daughter inheritance and the father's X is transferred to the son, one would see daughters with the white-eye phenotype and sons with the miniature-wing phenotype.

14. Because synapsis of chromosomes in meiotic tissue is often accompanied by crossing over, it would be detrimental to sex-determining mechanisms to have sex-determining loci on the Y chromosome transferred, through crossing over, to the X chromosome.

16. Klinefelter syndrome (XXY) = 1
Turner syndrome (XO) = 0
47, XYY = 0
47, XXX = 2
48, XXXX = 3

18. Unless other markers, cytological or molecular, are available, one cannot test the Lyon hypothesis with homozygous X-linked genes. The test requires identification of allelic alternatives to see differences in X chromosome activity.

20. Dosage compensation and the formation of Barr bodies occurs only when there are two or more X chromosomes. Males normally have only one X chromosome, therefore such mosaicism cannot occur. Females normally have two X chromosomes. There are cases of male calico cats that are XXY.

22. In mammals, the scheme of sex determination is dependent on the presence of a piece of the Y chromosome. If present, a male is produced. In *Bonellia viridis*, the female proboscis produces some substance that triggers a morphological, physiological, and behavioral developmental pattern that produces males. To elucidate the mechanism, one could attempt to isolate and characterize the active substance by testing different chemical fractions of the proboscis. Second, mutant analysis usually provides critical approaches into developmental processes. Depending on characteristics of the organism, one could attempt to isolate mutants that lead to changes in male or female development. Third, by using microtissue transplantations, one could attempt to determine which anatomical "centers" of the embryo respond to the chemical cues of the female.

24. One could account for the significant departures from a 1:1 ratio of males to females by suggesting that at anaphase I of meiosis, the Y chromosome more often goes to the pole that produces the more viable sperm cells. One could also speculate that the Y-bearing sperm has a higher likelihood of surviving in the female reproductive tract, or that the egg surface is more receptive to Y-bearing sperm. At this time the mechanism is unclear.

26. Because of the homology between the *red* and *green* genes, there exists the possibility for an irregular synapsis (see the figure below) that, following crossing over, would give a chromosome with only one (*green*) of the duplicated genes. When this X chromosome combines with the normal Y chromosome, the son's phenotype can be explained.

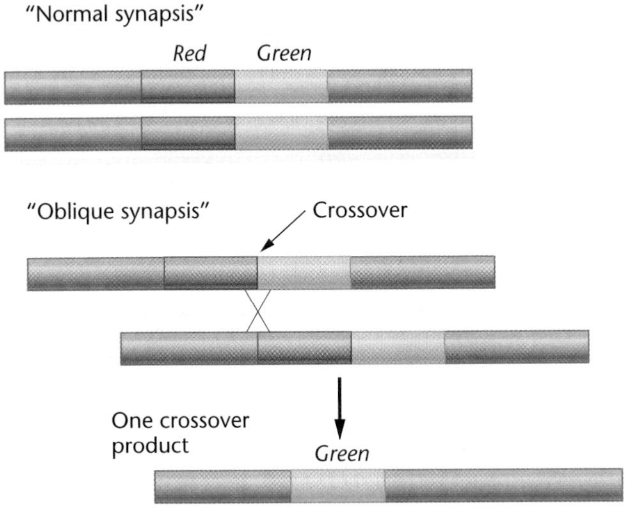

28. (a) Something is missing from the male-determining system of sex determination either at the level of the genes, gene products, or receptors, etc.
(b) The *SOX9* gene or its product is probably involved in male development. Perhaps it is activated by *SRY*.
(c) There is probably some evolutionary relationship between the *SOX9* gene and *SRY*. There is considerable evidence that many other genes and pseudogenes are also homologous to *SRY*.
(d) Normal female sexual development does not require the *SOX9* gene or gene product(s).

30. In snapping turtles, sex determination is strongly influenced by temperature such that males are favored in the 26–34°C range. Lizards, on the other hand, appear to have their sex determined by factors other than temperature in the 20–40°C range.

32. The white patches of CC are due to an autosomal gene *S* for white spotting that prevents pigment formation in the cell lineages in which it is expressed. Homozygous *SS* cats have more white than heterozygous *Ss* cats and there is no absolute pattern of patches due to the *S* allele. So the distribution of white patches would be expected to be different from Rainbow. In addition, since X chromosome inactivation is random, CC would have a different patch pattern from her genetic mother on the random X inactivation basis alone.

Chapter 8

4. The fact that there is a significant maternal age effect associated with Down syndrome indicates that nondisjunction in older females contributes disproportionately to the number of Down syndrome individuals. In addition, certain genetic and cytogenetic marker data indicate the influence of female nondisjunction.

6. Because an allotetraploid has a possibility of producing bivalents at meiosis I, it would be considered the most fertile of the three. Having an even number of chromosomes to match up at the metaphase I plate, autotetraploids would be considered to be more fertile than autotriploids.

8. American-cultivated cotton has 26 pairs of chromosomes—13 large, 13 small. Old world cotton has 13 pairs of large chromosomes and American wild cotton has 13 pairs of small chromosomes. It is likely that an interspecific hybridization occurred followed by chromosome doubling. These events probably produced a fertile amphidiploid (allotetraploid). Experiments have been conducted to reconstruct the origin of American cultivated cotton.

10. While there is the appearance that crossing over is suppressed in inversion "heterozygotes," the phenomenon extends from the fact that the crossover chromatids end up being abnormal in genetic content. As such they fail to produce viable (or competitive) gametes or lead to zygotic or embryonic death.

12. The mutant *Notch* in *Drosophila* produces flies with abnormal wings. It is a sex-linked dominant gene (a deletion) that also behaves as a recessive lethal. A deficiency (compensation) loop indicates that bands 3C2 through 3C11 are involved. Loci near *Notch* display pseudodominance. On the other hand the *Bar* gene results from a duplication of a sex-linked region (16A) and results in abnormal eye shape.

Females: N^+/N × males: B/Y

1/4	N^+/B	females; Bar
1/4	N/B	females; Notch, Bar
1/4	N^+/Y	males; wild
1/4	N/Y	**lethal**

The final phenotypic ratio would be 1:1:1 for the phenotypes shown above.

14. It is likely that when certain combinations of genes are of selective advantage in a specific and stable environment, it would be beneficial to the organism to protect that gene combination from disruption through crossing over. By having the genes in an inversion, crossover chromatids are not recovered and therefore are not passed on to future generations. Translocations offer an opportunity for new gene combinations by associations of genes from nonhomologous chromosomes. Under certain conditions such new combinations may be of selective advantage and meiotic conditions have evolved so that segregation of translocated chromosomes yields a relatively uniform set of gametes.

16. The primrose, *Primula kewensis*, with its 36 chromosomes, is likely to have formed from the hybridization and subsequent chromosome doubling of a cross between the two other species, each with 18 chromosomes.

18. The rare double crossovers in the boundaries of a paracentric or pericentric inversion produce only minor departures from the standard chromosomal arrangement as long as the crossovers involve the same two chromatids. With two-strand double crossovers, the second crossover negates the first. However, three-strand and four-strand double crossovers have consequences that lead to anaphase bridges as well as a high degree of genetically unbalanced gametes.

20. In the trisomic, segregation will be "2 × 1" as illustrated below:

P_1:

$b/b/b$ × b^+/b^+

gametes: (bb)(b) (b^+)

F_1:

$b^+/b^1/b^2$ × b^+/b
(normal bristles) (normal bristles)

Notice that there are several segregation patterns created by the trivalent at anaphase I.

gametes:

b^+ b^1b^2 b^+b^1 b^+b^2 b^2 b^1	b^+ b

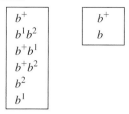

F_2:
b^+b^+ = normal bristles
$b^+b^1b^2$ = normal bristles
$b^+b^+b^1$ = normal bristles
$b^+b^+b^2$ = normal bristles
b^+b^2 = normal bristles
b^+b^1 = normal bristles
b^+b = normal bristles
bb^1b^2 = bent bristles
b^+bb^1 = normal bristles
b^+bb^2 = normal bristles
bb^2 = bent bristles
bb^1 = bent bristles

22. Given some of the information in the above problem the expression would be as follows:

$(35/36W{:}1/36w)(35/36A{:}1/36a)$

$(35/36)^2$ W——— A———

$35/(36)^2$ W——— $aaaa$

$35/(36)^2$ $wwwwA$———

$1/(36)^2$ $wwwwaaaa$

24. Considering that there are at least three map units between each of the loci, and that only four phenotypes are observed, it is likely that genes *a b c d* are included in an inversion and that crossovers that do occur among these genes are not recovered because of their genetically unbalanced nature. In a sense, the minimum distance between loci *d* and *e* can be estimated as 10 map units

$(48 + 52/1000)$;

However, this is actually the distance from the *e* locus to the breakpoint that includes the inversion.

The "map" is therefore as drawn below:

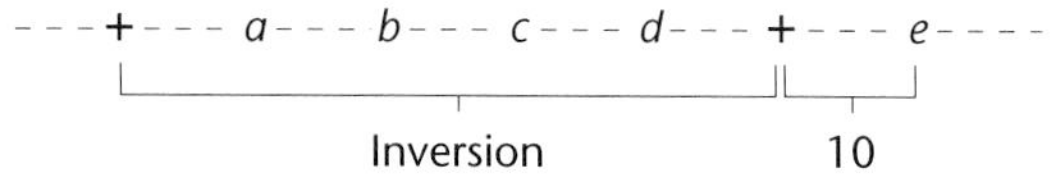

- - - + - - - a - - - b - - - c - - - d - - - + - - - e - - - -
Inversion
10

26. (a) The father must have contributed the abnormal X-linked gene.

(b) Since the son is XXY and heterozygous for anhidrotic dysplasia, he must have received both the defective gene and the Y chromosome from his father. Thus nondisjunction must have occurred during meiosis I.

(c) This son's mosaic phenotype is caused by X-chromosome inactivation, a form of dosage compensation in mammals.

28. Below is a description of breakage/reunion events that illustrate a translocation in relatively small, similarly sized, chromosomes 19 (metacentric) and 20 (metacentric/submetacentric). The case described here is shown occurring before S phase duplication. The same phenomenon is shown in the text as occurring after S phase. Since the likelihood of such a translocation is fairly small in a general population, inbreeding played a significant role in allowing the translocation to "meet itself."

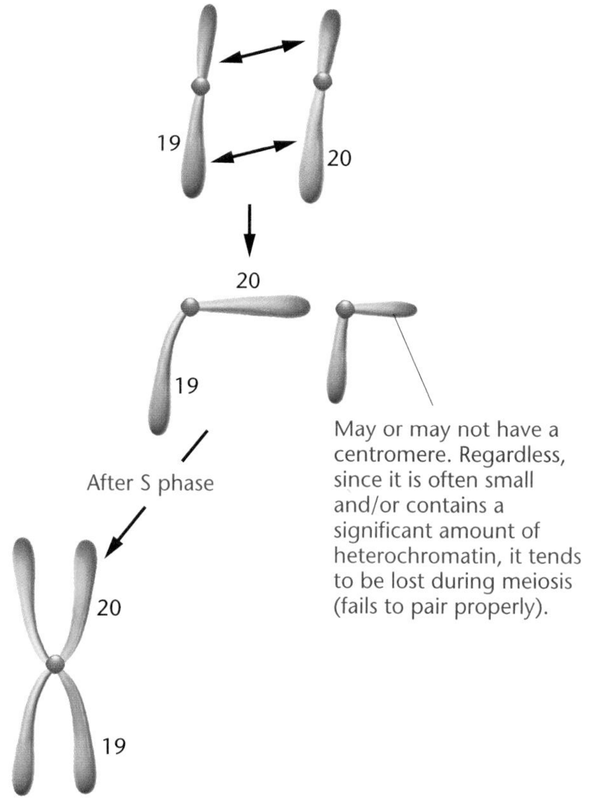

19
20
20
19
May or may not have a centromere. Regardless, since it is often small and/or contains a significant amount of heterochromatin, it tends to be lost during meiosis (fails to pair properly).
After S phase
20
19

30. This female will produce meiotic products of the following types:

normal: 18 + 21
translocated: 18/21
translocated plus 21: 18/21 + 21
deficient: 18 only

Fertilization with a normal 18 + 21 sperm cell will produce the following offspring:

normal: 46 chromosomes
translocation carrier: 45 chromosomes 18/21 + 18 + 21
trisomy 21: 46 chromosomes 18/21 + 21 + 21
monosomic: 45 chromosomes 18 + 18 + 21, lethal

Chapter 9

2. The mt^+ strain (resistant for the nuclear and chloroplast genes) contributes the "cytoplasmic" component of streptomycin resistance, which would negate any contribution from the mt^- strain. Therefore, all the offspring will have the streptomycin resistance phenotype. In the reciprocal cross, with the mt^+ strain being streptomycin sensitive, all the offspring will be sensitive.

4. (a) neutral
(b) segregational (nuclear mutations)
(c) suppressive

6. The inheritance patterns for the two, *segregational* and *neutral*, are quite different. The segregational mode is dependent on nuclear genes, while that of the neutral type is dependent on cytoplasmic influences namely mitochondria. If the two are crossed as stated in the problem, then one would expect, in the diploid zygote, the *segregational* allele to be "covered" by normal alleles from the neutral strain. On the other hand, as the nuclear genes are again "exposed" in the haploid state of the ascospores, one would expect a 1:1 ratio of normals to petites. The petite phenoytpe is caused by the nuclear, *segregational* gene.

8. (a) It is likely that mitochondria and chloroplasts evolved from bacteria in a symbiotic relationship, therefore it is not surprising that certain antibiotics that influence bacteria will also influence all mitochondria and chloroplasts.
(b) Since the inheritance of resistance or sensitivity is dependent on the status of the mt^+ gene, clearly, the mt^+ strain is the donor of the *cp*DNA.

10. The fact that all of the offspring (F_1) showed a dextral coiling pattern indicates that one of the parents (maternal parent) contains the *D* allele. Taking these offspring and seeing that their progeny (call these F_2) occur in a 1:1 ratio indicates that half of the offspring (F_1) are *dd*. In order to have these results, one of the original parents must have been *Dd* while the other must have been *dd*.
Parents: $Dd \times dd$
Offspring (F_1): 1/2 *Dd*, 1/2 *dd*
(all dextral because of the maternal genotype)
Progeny (F_2):
All those from *Dd* parents will be dextral while all those from *dd* parents will be sinistral.

12. Since there is no evidence for segregation patterns typical of chromosomal genes and Mendelian traits, some form of extranuclear inheritance seems possible. If the *lethargic* gene is dominant, then a maternal effect may be involved. In that case, some of the F_2 progeny would be hyperactive because maternal effects are only temporary, affecting only the immediate progeny. If the lethargic condition is caused by some infective agent, then perhaps injection experiments could be used. If caused by a mitochondrial defect, then the condition would persist in all offspring of lethargic mothers, through more than one generation.

14. Since an initial mutation does not involve all copies of mtDNA within a mitochondrion, the original state is heteroplasmic and the mutated mtDNA is rare. If the new mutation confers no selective advantage to the host cell, the frequency of the mutation is likely to diminish. However, if the mutation confers a selective advantage to the cell, it is likely to gain in frequency. Depending on a number of factors, including chance as well as the extent of the selective advantage, an mtDNA mutation may become prominent and eventually establish homoplasmy.

16. (a) The presence of bcd^-/bcd^- males can be explained by the maternal effect: mothers were bcd^+/bcd^-.
(b) The cross

female bcd^+/bcd^- × male bcd^-/bcd^-

will produce an F_1 with normal embryogenesis because of the maternal effect. In the F_2, any cross having bcd^+/bcd^- mothers will have phenotypically normal embryos. Offspring of any cross involving homozygous bcd^-/bcd^- mothers will have problems with embryogenesis.

18. (a) A locus, *Segregation Distortion* (*SD*), is present on the wild type chromosome. Some aspect of *SD* causes a shift in the segregation ratio by allowing sperm to carry the *SD* chromosome at the expense of the homolog.
(b) One could use this *SD* chromosome in a variety of crosses and determine that the abnormal segregation is based on a particular chromosomal element. One could even map the *SD* locus on the second chromosome (as has been done).
(c) Segregation Distortion describes a condition in which typical Mendelian segregation is distorted from the 50:50.

20. Deficiencies that remove histone genes contribute to increased survival of progeny of *abo*/*abo* mothers. Since deficiencies in histone genes reduce the severity of the maternal effect it is likely that an overabundance of histones is involved. The observation that the addition of heterochromatin also reduces the severity of the maternal effect may be due to a sequestering of the histone overload by the heterochromatin. One might therefore speculate that the *abo* gene is a regulator of histone production. The normal allele specifies a negative regulator of histone genes while the mutant *abo* gene fails to exert such negative control.

Chapter 10

2. Prior to 1940 most of the interest in genetics centered on the transmission of similarity and variation from parents to offspring (transmission genetics). While some experiments examined the possible nature of the hereditary material, abundant knowledge of the structural and enzymatic properties of proteins generated a bias that worked to favor proteins as the hereditary substance. In addition, proteins were composed of as many as twenty different subunits (amino acids), thereby providing ample structural and functional variation for the multiple tasks that must be accomplished by the genetic material. The tetranucleotide hypothesis (structure) provided insufficient variability to account for the diverse roles of the genetic material.

4. Specific degradative enzymes, proteases, RNase, and DNase were used to selectively eliminate components of the extract and, if transformation is concomitantly eliminated, then the eliminated fraction is the transforming principle. DNase eliminates DNA and transformation, therefore it must be the transforming principle.

6. Actually, phosphorus is found in approximately equal amounts in DNA and RNA; therefore, labeling with ^{32}P would "tag" both RNA and DNA. However, the T2 phage, in its mature state, contains very little if any RNA, therefore DNA would be interpreted as being the genetic material in T2 phage.

8. By comparing DNA content in various cell types (sperm and somatic cells) and observing that the *action* and *absorption* spectra of ultraviolet light were correlated, DNA was considered to be the genetic material. This suggestion was supported by the fact that DNA was shown to be the genetic material in bacteria and some phage. Direct evidence for DNA being the genetic material comes from a variety of observations including gene transfer that has been facilitated by recombinant DNA techniques.

10. Linkages among the three components require the removal of water (H_2O).

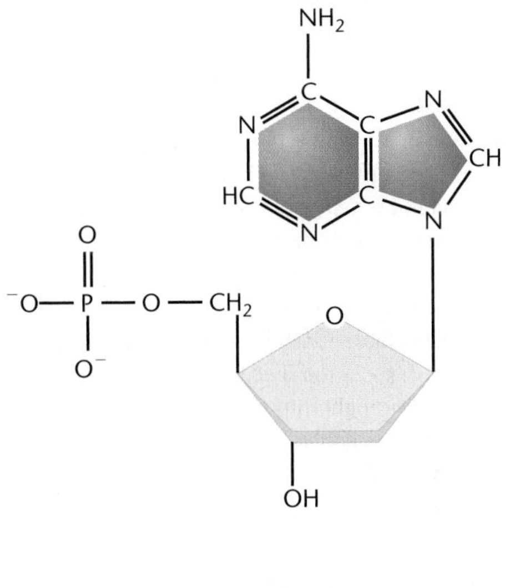

12. Guanine: 2-amino-6-oxypurine
Cytosine: 2-oxy-4-aminopyrimidine
Thymine: 2,4-dioxy-5-methylpyrimidine
Uracil: 2,4-dioxypyrimidine

16. Because in double-stranded DNA, A=T and G=C (within limits of experimental error), the data presented would have indicated a lack of pairing of these bases in favor of a single-stranded structure or some other nonhydrogen-bonded structure.

Alternatively, from the data it would appear that A=C and T=G, which would negate the chance for typical hydrogen bonding since opposite charge relationships do not exist. Therefore, it is quite unlikely that a tight helical structure would form at all. In conclusion, Watson and Crick might have concluded that hydrogen bonding is not a significant factor in maintaining a double-stranded structure.

18. Three main differences between RNA and DNA are the following:
(1) uracil in RNA replaces thymine in DNA,
(2) ribose in RNA replaces deoxyribose in DNA, and
(3) RNA often occurs as both single- and partially double-stranded forms whereas DNA most often occurs in a double-stranded form.

20. The nitrogenous bases of nucleic acids (nucleosides, nucleotides, and single- and double-stranded polynucleotides) absorb UV light maximally at wavelengths 254 to 260 nm. Using this phenomenon, one can often determine the presence and concentration of nucleic acids in a mixture. Since proteins absorb UV light maximally at 280 nm, this is a relatively simple way of dealing with mixtures of biologically important molecules.

UV absorption is greater in single-stranded molecules (hyperchromic shift) as compared to double-stranded structures, therefore one can easily determine, by applying denaturing conditions, whether a nucleic acid is in the single- or double-stranded form. In addition, A—T rich DNA denatures more readily than G—C rich DNA, therefore one can estimate base content by denaturation kinetics.

22. Guanine and cytosine are held together by three hydrogen bonds whereas adenine and thymine are held together by two. Because G—C base pairs are more compact, they are more dense than A—T pairs. The percentage of G—C pairs in DNA is thus proportional to the buoyant density of the molecule.

24. For curve A in the problem, there is evidence for a rapidly renaturing species (repetitive) and a slowly renaturing species (unique). The fraction that reassociates faster than the *E. coli* DNA is highly repetitive and the last fraction (with the highest $C_ot_{1/2}$ value) contains primarily unique sequences. Fraction B contains mostly unique, relatively complex DNA.

26. Because G—C base pairs are formed with three hydrogen bonds while A—T base pairs by two such bonds, it takes more energy (higher temperature) to separate G—C pairs.

28. In one sentence of Watson and Crick's paper in *Nature*, they state,

"It has not escaped our notice that the specific pairing we have postulated immediately suggests a possible copying mechanism for the genetic material."

The model itself indicates that unwinding of the helix and separation of the double-stranded structure into two single strands immediately exposes the specific hydrogen bonds through which new bases are brought into place.

30. MS-2 = 200 base pairs
E. coli = 2×10^6 base pairs

32. Since cytosine pairs with guanine and uracil pairs with adenine, the result would be a base substitution of G:C to A:T after rounds of replication.

34. (i) The X-ray diffraction studies would indicate a helical structure, for it is on the basis of such data that a helical pattern is suggested. The fact that it is irregular may indicate different diameters (base pairings), additional strands in the helix, kinking, or bending.
(ii) The hyperchromic shift would indicate considerable hydrogen bonding, possibly caused by base pairing.
(iii) Such data may suggest irregular base pairing in which purines bind purines (all the bases presented are purines) thus giving the atypical dimensions.
(iv) Because of the presence of ribose, the molecule may show more flexibility, kinking, and/or folding.

While there are several situations possible for this model, the phosphates are still likely to be far apart (on the outside) because of their strong like charges. Hydrogen bonding probably exists on the inside of the molecule and there is probably considerable flexibility, kinking, and/or bending.

36. Heat application would yield a hyperchromic shift if the DNA is double-stranded. One could also get a rough estimation of the GC content from the kinetics of denaturation and the degree of sequence complexity from comparative renaturation studies. Determination of base content by hydrolysis and chromatography could be used for comparative purposes and could also provide evidence as to the strandedness of the DNA. Antibodies for Z-DNA could be used to determine the degree of left-handed structures, if present. Sequencing the DNA from both viruses would indicate sequence homology. In addition, through various electronic searches readily available on the Internet (Web site: *blast@ncbi.nlm.nih.gov*, for example) one could determine whether similar sequences exist in other viruses or in other organisms.

Chapter 11

2. By labeling the pool of nitrogenous bases of the DNA of *E. coli* with a heavy isotope ^{15}N, it would be possible to "follow" the "old" DNA.

4. (a) Under a conservative scheme all of the newly labeled DNA will go to one sister chromatid, while the other sister chromatid will remain unlabeled.
(b) Under a dispersive scheme all of the newly labeled DNA will be interspersed with unlabeled DNA.

6. The *in vitro* replication requires a DNA template, a divalent cation (Mg^{++}), and all four of the deoxyribonucleoside triphosphates: dATP, dCTP, dTTP, and dGTP. The lower case "d" refers to the deoxyribose sugar.

8. Two general analytical approaches showed that the products of DNA polymerase I were probably copies of the template DNA. Because *base composition* can be similar without reflecting sequence similarity, the least stringent test was the comparison of base composition. By comparing *nearest neighbor frequencies*, Kornberg determined that there is a very high likelihood that the product was of the same base sequence as the template.

10. The *in vitro* rate of DNA synthesis using DNA polymerase I is slow, being more effective at replicating single-stranded DNA than double-stranded DNA. In addition, it is capable of degrading as well as synthesizing DNA. Such degradation suggested that it functioned as a repair enzyme. In addition, DeLucia and Cairns discovered a strain of *E. coli* (*pol*A1) that still replicated its DNA but was deficient in DNA polymerase I activity.

12. Biologically active DNA implies that the DNA is capable of supporting typical metabolic activities of the cell or organism and is capable of faithful reproduction.

14. DNA polymerase I and DNA ligase are used to synthesize and label an RF duplex. DNase is used to nick one of the two strands. The heavy (^{32}P/BU-containing) DNA is isolated by denaturation and centrifugation and DNA polymerase and DNA ligase are used to make a synthetic complementary strand. When isolated, this synthetic strand is capable of transfecting *E. coli* protoplasts, from which new ϕX174 phages are produced.

16. All three enzymes share several common properties. None can *initiate* DNA synthesis on a template but all can *elongate* an existing DNA strand, assuming there is a template strand as shown in the figure below. Polymerization of nucleotides occurs in the 5′ to 3′ direction where each 5′ phosphate is added to the 3′ end of the growing polynucleotide.

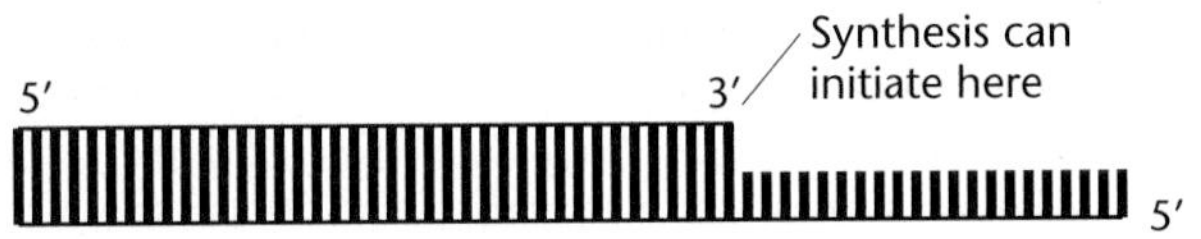

All three enzymes are large complex proteins with a molecular weight in excess of 100,000 daltons and each has 3' to 5' exonuclease activity.

DNA polymerase I:
- exonuclease activity
- present in large amounts
- relatively stable
- removal of RNA primer

DNA polymerase II:
- possibly involved in repair function

DNA polymerase III:
- exonuclease activity
- essential for replication
- complex molecule

18. Given a stretch of double-stranded DNA, one could initiate synthesis at a given point and either replicate strands in one direction only (unidirectional) or in both directions (bidirectional). Synthesis of complementary strands occurs in a *continuous* 5′ > 3′ mode on the leading strand in the direction of the replication fork, and in a *discontinuous* 5′ > 3′ mode on the lagging strand opposite the direction of the replication fork.

20. *Okazaki fragments* are relatively short (1000 to 2000 bases in prokaryotes) DNA fragments that are synthesized in a discontinuous fashion on the lagging strand during DNA replication. Such fragments appear to be necessary because template DNA is not available for 5′ > 3′ synthesis until some degree of continuous DNA synthesis occurs on the leading strand in the direction of the replication fork. The isolation of such fragments provides support for the scheme of replication. DNA *ligase* is required to form phosphodiester linkages in gaps that are generated when DNA polymerase I removes RNA primer and meets newly synthesized DNA ahead of it. The discontinuous DNA strands are ligated together into a single continuous strand. *Primer* RNA is formed by RNA primase to serve as an initiation point for the production of DNA strands on a DNA template. None of the DNA polymerases is capable of initiating synthesis without a free 3′ hydroxyl group. The primer RNA provides that group and thus can be used by DNA polymerase III.

22. Because there is a much greater amount of DNA to be replicated and DNA replication is slower, there are multiple initiation sites for replication in eukaryotes (and increased DNA polymerase per cell) in contrast to the single replication origin in prokaryotes. Replication occurs at different sites during different intervals of the S phase. The proposed functions of four DNA polymerases are described in the text.

24. (a) In *E. coli*, 100 kb are added to each growing chain per minute. Therefore the chain should be about 4,000,000 bp.
(b) Given
$(4 \times 10^6 \text{ bp}) \times 0.34 \text{ nm/bp} = 1.36 \times 10^6 \text{ nm or } 1.3 \text{ mm}$

26. *Gene conversion* is likely to be a consequence of genetic recombination in which nonreciprocal recombination yields products where one allele is "converted" to another. Gene conversion is now considered a result of heteroduplex formation that is accompanied by mismatched bases. When these mismatches are corrected, the "conversion" occurs.

28. Telomerase activity is present in germ-line tissue to maintain telomere length from one generation to the next. In other words, telomeres cannot shorten indefinitely without eventually eroding genetic information.

30.

Initial Labeled Base	***Labeled Base after Spleen Phosphodiesterase Digestion***	
	ANTIPARALLEL	**PARALLEL**
G	A,T	C,T
C	G,A,G	G,A
T	C,T, G	C,T,A,G
A	T,C,A,T	T,C,A,G

One can determine which model occurs in nature by comparing the pattern in which the labeled phosphate is shifted following spleen phosphodiesterase digestion. Focus your attention on the antiparallel model and notice that the frequency that "C" (for example) is the 5′ neighbor of "G" is not necessarily the same as the frequency that "G" is the 5′ neighbor of "C." However, in the parallel model (b), the frequency that "C" is the 5′ neighbor of "G" is the same as the frequency that "G" is the 5′ neighbor of "C." By examining such "digestion frequencies" it can be determined that DNA exists in the opposite polarity.

32. (a) DNA polymerase would catalyze a bond between the 5′ end of the last nucleotide added and the 3′ end of the incoming nucleotide. In this reaction, the energy would be provided by the cleavage of the gamma- and beta-phosphates of the last nucleotide added to the chain rather than of the incoming nucleotide.
(b) If DNA polymerase removed a base, it would not be able to add any more bases to the chain because the penultimate base would have a monophosphate rather than a triphosphate and there would be no source of energy for the polymerization reaction.

34. (a) 5′ACCUAAGU
(b) U

Chapter 12

2. By having a circular chromosome there are no free ends to present the problem of linear chromosomes—namely, complete replication of terminal sequences.

4. Since eukaryotic chromosomes are "multirepliconic" in that there are multiple replication forks along their lengths, one would expect to see multiple clusters of radioactivity.
6. Long interspersed elements (LINES) are repetitive transposable DNA sequences in humans. The most prominent family, designated **L1**, is about 6.4 kb each and is represented about 100,000 times. LINES are often referred to as retrotransposons because their mechanism of transposition resembles that used by retroviruses.
8. Because of the diverse cell types of multicellular eukaryotes, a variety of gene products is required, which may be related to the increase in DNA content per cell. In addition, the advantage of diploidy automatically increases DNA content per cell. However, seeing the question in another way, it is likely that a much higher *percentage* of the genome of a prokaryote is actually involved in phenotype production than in a eukaryote.

   Eukaryotes have evolved the capacity to obtain and maintain what appears to be large amounts of "extra," perhaps "junk," DNA. Prokaryotes on the other hand, with their relatively short life cycle, are extremely efficient in their accumulation and use of their genome.

   Given the larger amount of DNA per cell and the requirement that the DNA be partitioned in an orderly fashion to daughter cells during cell division, certain mechanisms and structures (mitosis, nucleosomes, centromeres, etc.) have evolved for packaging and distributing the DNA. In addition, the genome is divided into separate entities (chromosomes) perhaps to facilitate the partitioning process in mitosis and meiosis.
10. Nucleosomes are octomeric structures of two molecules of each histone (H2A, H2B, H3, and H4) except H1. Between the nucleosomes and complexed with linker DNA is histone H1. A 146-base pair sequence of DNA wraps around the nucleosome.
12. *Heterochromatin* is chromosomal material that stains deeply and remains condensed when other parts of chromosomes, euchromatin, are otherwise pale and decondensed. Heterochromatic regions replicate late in S phase and are relatively inactive in a genetic sense because there are few genes present or if they are present, they are repressed. Telomeres and the areas adjacent to centromeres are composed of heterochromatin.
14. Volume of DNA:

   $$3.14 \times 10 \text{ Å} \times 10 \text{ Å} \times (50 \times 10^4 \text{ Å}) = 1.57 \times 10^8 \text{ Å}^3$$

   Volume of capsid:

   $$4/3\,(3.14 \times 400 \text{ Å} \times 400 \text{ Å} \times 400 \text{ Å}) = 2.67 \times 10^8 \text{ Å}^3$$

   Because the capsid head has a greater volume than the volume of DNA, the DNA will fit into the capsid.
16. About 36.3%
18. Chromosomes are not randomly distributed within nuclei. Homologous chromosomes tend to distribute themselves opposite each other and in an antiparallel manner, meaning that their positions are in reverse order on opposite sides of the nucleus. Assuming that such patterns are maintained throughout the entire cell cycle, it is possible that chromosomal positions may influence gene function and/or chromosomal behavior during mitosis and/or meiosis. If gene function is influenced not only by gene position in a chromosome but also by gene position in a nucleus, then an alternative explanation for position effect exists.
20. Nucleosomes follow a *dispersive* pattern with each daughter chromatid containing a mixture of old and original nucleosomes. One could test the distribution of nucleosomes by conducting an autoradiographic experiment similar to Taylor–Woods–Hughes, but instead of labeling the DNA with ^{3}H-thymidine, one would label some or all the histones H2A, H2B, H3, and H4 in nucleosomes.
22. Bacteriophage lambda is composed of a double-stranded, linear DNA molecule of about 48,000 base pairs. It is capable of forming a closed, double-stranded circular molecule because of a 12-base pair, single-stranded, complementary "overhanging" sequence at the 5′ end of each single strand.
24. The general frequency and pattern of various trinucleotide repeat motifs are similar in all taxonomic groups. Within-gene trinucleotide repeats are the most frequent repeat motif in all taxonomic groups followed by hexanucleotide repeats. One explanation might be that various microsatellite types (mono, di, tri, etc.) are generated at different rates in different genomic regions (within and between genes). A second possibility is that selection acts differentially depending on the type and location of a repeat. The correlation between the high frequency of tri- and hexanucleotide repeats within genes and a triplet code specifying particular amino acids within genes may not be coincidental.
26. If microsatellites in general are flanked by a conserved sequence, those conserved sequences may be involved in the generation and/or maintenance of the microsatellite. Alternatively, the microsatellite may generate the nonmicrosatellite region. Any hypothesis presented is in need of additional investigation before definitive statements can be made.
28. Generally, the higher the AT content of a DNA strand, the lower the temperature of melting or denaturation. Since one can monitor the degree of strand separation with absorption of ultraviolet light (optical density) one can get an estimate of the AT/GC content in a given stretch of DNA. For relatively short strands of DNA the temperature of melting is also related to strand length. Other factors being equal, the shorter the strand, the lower the T_m.
30. Since both genes mentioned in the problem are located near the end of chromosome 16 it is possible that erosion of the end of the chromosome is related to each disease. Examination of the gene by *in situ* hybridization and molecular cloning indicates that thalassemia involves a terminal deletion in distal portion of 16p. To learn more about such conditions, visit *http://www.ncbi.nlm.nih.gov/* and follow the OMIM link.

Chapter 13

2. Given a sextuplet code, restoration of the reading frames would only occur with the addition or loss of 6 nucleotides.
4. Because of a triplet code, a trinucleotide sequence will, once initiated, remain in the same reading frame and produce the same code all along the sequence regardless of the initiation site. If a tetranucleotide is used there are four different initiation sites for reading a triplet code, therefore there will be four different sequences.
6. From the repeating polymer ACACA . . . one can say that threonine is either CAC or ACA. From the polymer CAACAA . . . with ACACA . . . , ACA is the only codon in common. Therefore, threonine would have the codon ACA.
8. The basis of the technique is that if a trinucleotide contains bases (a codon), which are complementary to the anticodon of a charged tRNA, a relatively large complex is formed that contains the ribosome, the tRNA, and the trinucleotide. This complex is trapped in the filter whereas the components by themselves are not trapped. If the amino acid on a charged, trapped tRNA is radioactive, then the filter becomes radioactive.

10. Apply the most conservative pathway of change.

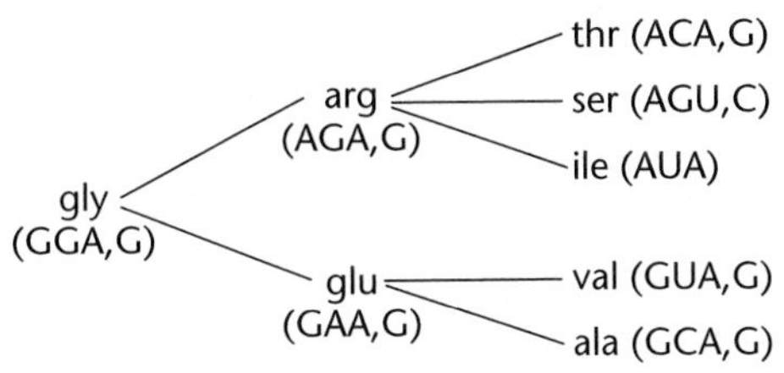

12. Because Poly U is complementary to Poly A, double-stranded structures will be formed. In order for an RNA to serve as a messenger RNA it must be single-stranded, thereby exposing the bases for interaction with ribosomal subunits and tRNAs.

14. (a)

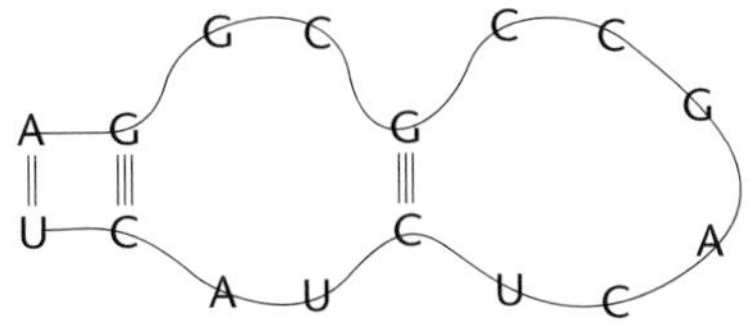

(b) TCCGCGGCTGAGATGA (use complementary bases, substituting T for U)

(c) GCU

(d) Assuming that the AGG . . . is the 5′ end of the mRNA, then the sequence would be

arg-arg-arg-leu-tyr

16. (a) met-his-thr-tyr-glu-thr-leu-gly
met-arg-pro-leu-asp (or glu)

(b) In the shorter of the two reading sequences (the one using the internal AUG triplet), a UGA triplet was introduced at the second codon. While not in the reading frames of the longer polypeptide (using the first AUG codon) the UGA triplet eliminates the product starting at the second initiation codon.

18. The central dogma of molecular genetics and to some extent, all of biology, states that DNA produces, through transcription, RNA, which is "decoded" (during translation) to produce proteins.

20. RNA polymerase from *E. coli* is a complex, large (almost 500,000 daltons) molecule composed of subunits $(\alpha, \beta, \beta', \sigma)$ in the proportion $\alpha 2, \beta, \beta', \sigma$ for the holoenzyme. The β subunit provides catalytic function while the sigma (σ) subunit is involved in recognition of specific promoters. The core enzyme is the protein without the sigma.

22. While some folding (from complementary base pairing) may occur with mRNA molecules, they generally exist as single-stranded structures that are quite labile. Eukaryotic mRNAs are generally processed such that the 5′ end is "capped" and the 3′ end has a considerable string of adenine bases. It is thought that these features protect the mRNAs from degradation. Such stability of eukaryotic mRNAs probably evolved with the differentiation of nuclear and cytoplasmic functions. Because prokaryotic cells exist in a more unstable environment (nutritionally and physically, for example) than many cells of multicellular organisms, rapid genetic response to environmental change is likely to be adaptive. To accomplish such rapid responses, a labile gene product (mRNA) is advantageous. A pancreatic cell, which is developmentally stable and existing in a relatively stable environment, could produce more insulin on stable mRNAs for a given transcriptional rate.

24. Proline: C_3, and one of the C_2A triplets
Histidine: one of the C_2A triplets
Threonine: one C_2A triplet, and one A_2C triplet
Glutamine: one of the A_2C triplets
Asparagine: one of the A_2C triplets
Lysine: A_3

26. (a, b) Use the code table to determine the number of triplets that code each amino acid, then construct a graph and plot such as this one below:

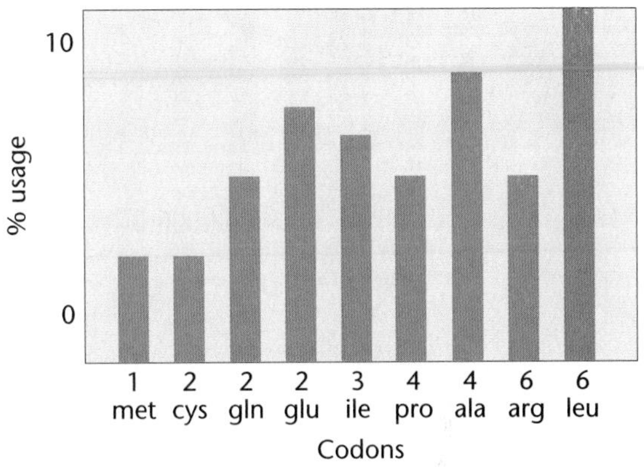

(c) There appears to be a weak correlation between the relative frequency of amino acid usage and the number of triplets for each.

(d) To continue to investigate this issue, one might examine additional amino acids in a similar manner. In addition, different phylogenetic groups use code synonyms differently. It may be possible to find situations in which the relationships are more extreme. One might also examine more proteins to determine whether such a weak correlation is stronger with different proteins.

28. (a) gucccaaccaugcccaccgaucuuccgccugcuucugaagAUGCGGGCCCAG

(b) 5′ gtc cca acc **atg** ccc acc gat ctt ccg cct gct tct gaa gAT GCG GGC CCA G

(c) 5′ gtcccaaccatgcccaccgatcttccgcctgcttctgaag **ATG** CGG GCC CAG
The two initiator codons are not in phase.

(d) 5′ gtc cca acc **atg** ccc acc gat ctt ccg cct gct tct gaa gAT GCG GGC CCA G
met pro thr asp leu pro pro ala ser glu asp ala gly pro
5′ gtcccaaccatgcccaccgatcttccgcctgcttctgaag **ATG** CGG GCC CAG
met arg ala gln
The amino acid sequences in the region of overlap are not the same.

(e) One might argue for the conservation of DNA by having the same region code for a multiple of products, and that might be the case in viruses and prokaryotes where genomic efficiency is more of an issue. However, eukaryotes appear to be much less likely to evolve strategies that conserve DNA sequences per se. However, if functionally and/or structurally related products can be conveniently regulated by such an arrangement, then perhaps an evolutionary advantage exists. The most obvious disadvantage is that if a mutation occurs in the common region, then two gene products are altered instead of one.

30. (a)

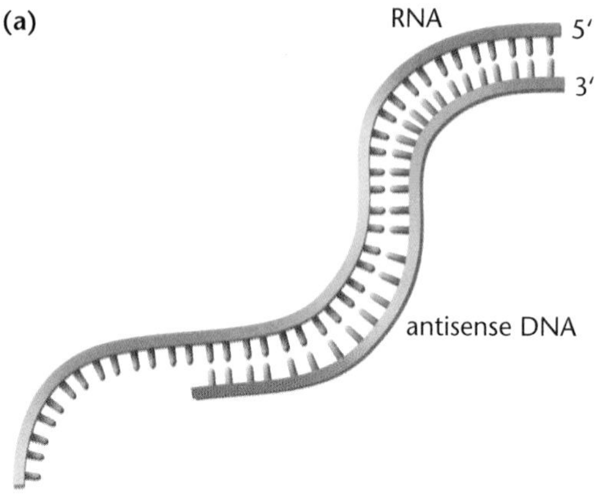

(b)

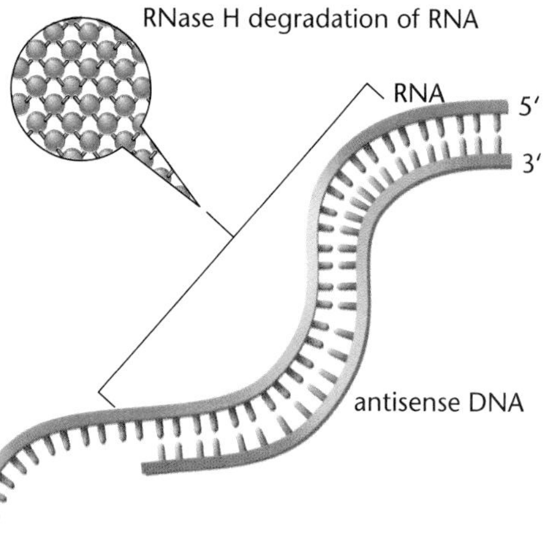

(c) Since the antisense strand is relatively short, under physiological conditions, nonspecific binding of the antisense strand to other RNAs in the cell may cause the degradation of nontargeted RNAs.

(d) Since the exact behavior of antisense DNA in a cell is not completely known, the answer to this question is elusive. However, given the complexity of intracellular events, it is likely that reduction of expression of a repressor gene group may induce other genes by eliminating such repressors.

(e) One of the major drawbacks to antisense therapy is the lack of specificity described here and the relative unknown behavior of DNA oligodeoxynucleotides in living cells.

Chapter 14

2. Transfer RNAs are "adaptor" molecules in that they provide a way for amino acids to interact with sequences of bases in nucleic acids. Amino acids are specifically and individually attached to the 3′ end of tRNAs, which possess a three-base sequence (the anticodon) to base-pair with three bases of mRNA. Messenger RNA, on the other hand, contains a copy of the triplet codes that are stored in DNA. The sequences of bases in mRNA interact, three at a time, with the anticodons of tRNAs.

Enzymes involved in transcription include the following: RNA polymerase (*E. coli*), and RNA polymerase I, II, III (eukaryotes). Those involved in translation include the following: aminoacyl tRNA synthetases, peptidyl transferase, and GTP-dependent release factors.

4. The sequence of base triplets in mRNA constitutes the sequence of codons. A three-base portion of the tRNA constitutes the anticodon.

6. The steps involved in tRNA charging are outlined in the text. An amino acid in the presence of ATP, Mg^{++}, and a specific aminoacyl synthetase produces an amino acid-AMP enzyme complex (+ PP_i).This complex interacts with a specific tRNA to produce the aminoacyl tRNA.

8. Phenylalanine is an amino acid that, like other amino acids, is required for protein synthesis. While too much phenylalanine and its derivatives cause PKU in phenylketonurics, too little will restrict protein synthesis.

10. Tyrosine is a precursor to melanin, skin pigment. Individuals with PKU fail to convert phenyalanine to tyrosine and even though tyrosine is obtained from the diet, at the population level, individuals with PKU have a tendency for less skin pigmentation.

12.

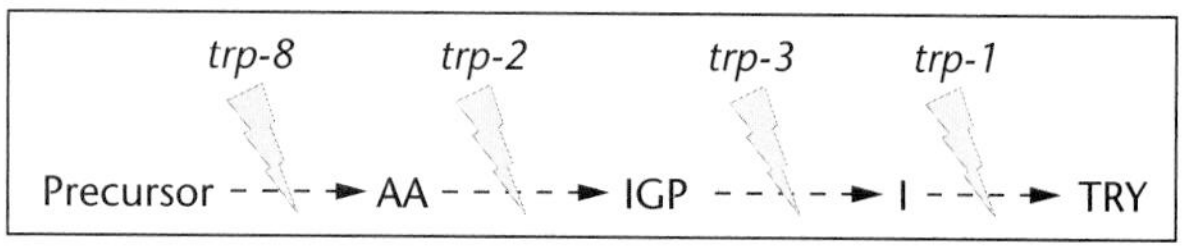

14. The fact that enzymes are a subclass of the general term *protein*, a *one-gene:one-protein* statement might seem to be more appropriate. However, some proteins are made up of subunits, each different type of subunit (polypeptide chain) being under the control of a different gene. Under this circumstance, one-gene:one-polypeptide might be more reasonable; however, many additional complexities exist and a simple statement regarding the relationship of a stretch of DNA to its physical product is difficult to justify.

16. The following types of normal hemoglobin:

Hemoglobin	***Polypeptide chains***
HbA	$2\alpha 2\beta$ (alpha, beta)
HbA_2	$2\alpha 2\delta$ (alpha, delta)
HbF	$2\alpha 2\gamma$ (alpha, gamma)
Gower 1	$2\zeta 2\varepsilon$ (zeta, epsilon)

The *alpha* and *beta* chains contain 141 and 146 amino acids, respectively. The *zeta* chain is similar to the *alpha* chain while the other chains are like the *beta* chain.

18. In the late 1940's, Pauling demonstrated a difference in the electrophoretic mobility of HbA and HbS (sickle-cell hemoglobin) and concluded that the difference had a chemical basis. Ingram determined that the chemical change occurs in the primary structure of the globin portion of the molecule using the fingerprinting technique. He found a change in the sixth amino acid in the β chain.

20. *Colinearity* refers to the sequential arrangement of subunits, amino acids, and nitrogenous bases, in proteins and DNA, respectively. Sequencing of genes and products in MS2 phage and studies on mutations in the A subunit of the *tryptophan synthetase* gene indicate a colinear relationship.

24. Enzymes function to regulate catabolic and anabolic activities of cells. They influence (lower) the *energy of activation,* thus allowing chemical reactions to occur under conditions that are compatible with living systems. Enzymes possess active sites and/or other domains that are sensitive to the environment. The active site is considered to be a crevice, or pit, that binds reactants, thus enhancing their interaction. The other domains mentioned above may influence the conformation and therefore function of the active site.

26. All of the substitutions involve one base change.

28. One can conclude that the amino acid is not involved in recognition of the codon.

30. Because cross **(a, b)** is essentially a monohybrid cross, there would be no difference in the results if crossing over occurred (or did not occur) between the *a* and *b* loci.

32. (a) outcomes from the last cross with its 9:4:3 ratio suggest two gene pairs.

(b) orange = *Y_R_*
yellow = *yyrr, yyR_*
red = *Y_rr*

(c) white– > yellow—***y***-> red—***r***-> orange (pathway V)

34.

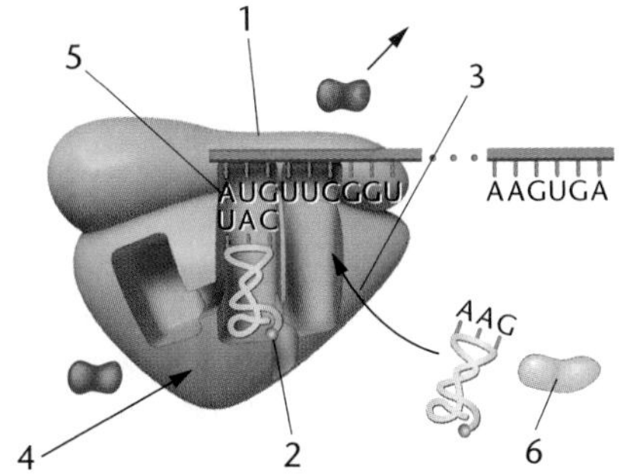

36. (a) Since protein synthesis is dependent on the passage of mRNA from DNA to ribosomes, any circumstance that compromises that flow will cause a reduction in protein synthesis. The more specific the binding of the antisense oligonucleotide to the target mRNA, the more specific the influence on protein synthesis. The ideal situation would be where a particular species of antisense oligonucleotide impacted on one and only one protein population.

(b) Clearly, a length of around 15-16 nucleotides is most effective in causing RNA degradation.

(c) A number of factors, including length of the oligonucleotide, are probably involved *in vivo*. It is likely that stability of the oligonucleotide is dependent on its base composition and length. The oligonucleotide must be small enough to diffuse effectively throughout the cell in order to "locate" the targeted mRNA and it must not assume a folded conformation that blocks opportunities for base pairing. Since the oligonucleotide is so much smaller than the target mRNA, it is also likely that the actual location of binding to the target is important in mRNA degradation. One of the main problems of antisense therapy is the introduction of the oligonucleotide into the interior of target cells.

Chapter 15

2. Mutagenized conidia can be cultured on complete, then minimal media to isolate strains that are nutritionally deficient (auxotrophs). Individual transfer of such isolates to minimal medium plus various supplements allows one to isolate those mutations that can be "repaired."

4. Each gene and its product function in an environment that has also evolved, or coevolved. A coordinated output of each gene product is required for life. Deviations from the norm, caused by mutation, are likely to be disruptive because of the complex and interactive environment in which each gene product must function. However, on occasion a beneficial variation occurs.

6. A *conditional* mutation is one that produces a wild type phenotype under one environmental condition and a mutant phenotype under a different condition.

8. All three of the agents are mutagenic because they cause base substitutions. Deaminating agents oxidatively deaminate bases such that cytosine is converted to uracil and adenine is converted to hypoxanthine. Uracil pairs with adenine and hypoxanthine pairs with cytosine. Alkylating agents donate an alkyl group to the amino or keto groups of nucleotides, thus altering base-pairing affinities. 6-ethyl guanine acts like adenine, thus pairing with thymine. Base analogues such as 5-bromouracil and 2-amino purine are incorporated as thymine and adenine, respectively, yet they base pair with guanine and cytosine, respectively.

10. X-rays are of higher energy and shorter wavelength than UV light. They have greater penetrating ability and can create more disruption of DNA.

12. *Photoreactivation* can lead to repair of UV-induced damage. An enzyme, photoreactivation enzyme, will absorb a photon of light to cleave thymine dimers. *Excision repair* involves the products of several genes, DNA polymerase I, and DNA ligase to clip out the UV-induced dimer, fill in, and join the phosphodiester backbone in the resulting gap. The excision repair process can be activated by damage that distorts the DNA helix. *Recombinational repair* is a system that responds to DNA that has escaped other repair mechanisms at the time of replication. If a gap is created on one of the newly synthesized strands, a rescue operation or "SOS response" allows the gap to be filled. Many different gene products are involved in this repair process: *rec*A, *lex*A. In SOS repair, the proofreading by DNA polymerase III is suppressed. This is therefore called an "error-prone system."

14. Each involves amplification of trinucleotide repeats. As the degree of amplification increases, so does the degree of expression of the abnormal phenotype. See the text for a detailed description of the role of trinucleotide repeats in a variety of human diseases. Genetic anticipation is the occurrence of an earlier age of onset of a genetic disease in successive generations.

16. *Xeroderma pigmentosum* is a form of human skin cancer caused by perhaps several rare autosomal genes that interfere with the repair of damaged DNA. Studies with heterokaryons provided evidence for complementation, indicating that there may be as many as seven different genes involved. The photoreactivation repair enzyme appears to be involved.

18. Each organism mentioned in the problem possesses a variety of transposable elements. Bacteria possess insertion sequences (about 800 to 1500 base pairs in length) as well as transposons, which are larger. Both are mobile in bacterial, viral, and plasmid DNAs and both have repeated base sequences at their ends. In maize, Barbara McClintock described the genetic behavior of mobile elements (*Ds* and *Ac*). *Ds* can move if *Ac* is present, thus *transposable controlling elements* exist. An *Ac* element is 4563 base pairs long and similar in structure to some bacterial transposons. Transposons often code for transposase enzymes, which are essential for transposition. *Copia* elements in *Drosophila* may be present in numerous copies in the genome and contain direct and inverted terminal repeats. *P* elements, also in *Drosophila,* are responsible for a phenomenon called hybrid dysgenesis.

Humans possess a variety of transposable elements including the *Alu* family of short interspersed elements (SINES) which are between 200 to 300 base pairs long and may exist in 300,000 copies per genome. Long interspersed elements (LINES) also occur in the human genome and seem to be capable of movement. Such elements share common structural features, are often mobile, and may influence gene activity.

20. It is likely that the reverse transcriptase, in making DNA, provides a DNA segment that is capable of integrating into the yeast chromosome as other types of DNA are known to do.

22. It is possible that through the reduction of certain environmental agents that cause mutations, mutation rates might be reduced. On the other hand, certain industrial and medical activities actually concentrate mutagens (radioactive agents and hazardous chemicals). Unless human populations are protected from such agents, mutation rates might actually increase. If one asks about the accumulation of mutations (not rates) in human populations as a result of improved living conditions and medical care, then it is likely that as the environment becomes less harsh (through improvements), more mutations will be tolerated as selection pressure decreases. However, as individuals live longer and have children at a later age, some studies indicate that older males accumulate more gametic mutations.

24. Both major forms of muscular dystrophy include muscular wasting of differing severity and age of onset. Both forms are caused by mutations in the *dystrophin* gene, which is very large composed of 97 exons and 2.6 Mb in length. Given the size of this gene and the number of exons/introns, many opportunities exist for mutational upset.

26. There are several ways in which an unexpected mutant gene may enter a pedigree. If a gene is incompletely penetrant, it may be present in a population and only express itself under certain conditions. It is unlikely that the gene for hemophilia behaved in this manner. If a gene's expression is suppressed by another mutation in an individual, it is possible that offspring may inherit a given gene and not inherit its suppressor. Such offspring would have hemophilia. Since all genetic variations must arise at some point, it is possible that the mutation in Queen Victoria's family was new, arising in the father. Lastly, is it possible that the mother was heterozygous and by chance, no other individuals in her family were unlucky enough to receive the mutant gene.

28. Your study should include examination of the following short-term aspects: immediate assessment of radiation amounts distributed in a matrix of the bomb sites as well as a control area not receiving bomb-induced radiation; radiation exposure as measured by radiation sickness and evidence of radiation poisoning from tissue samples; abortion rates; birthing rates; and chromosomal studies. Long-term assessment should include: sex-ratio distortion (males being more influenced by X-linked recessive lethals than females), chromosomal studies, birth and abortion rates, cancer frequency and type, and genetic disorders. In each case, data should be compared to the control site to see if changes are bomb related. In addition, to attempt to determine cause-effect, it is often helpful to show a dose response. Thus, by comparing the location of individuals at the time of exposure to the matrix of radiation amounts, one may be able to determine whether those most exposed to radiation suffer the most physiologically and genetically. If a positive correlation is observed, then statistically significant conclusions may be possible.

30. (a) For those organisms that generate energy by aerobic respiration, a process occurs that involves the reduction of molecular oxygen. Partially reduced species are produced as intermediates and by-products of such molecular action: O_2^-, H_2O_2, and OH^-. These species are potent electrophilic oxidants that escape mitochondria and attack numerous cellular components. Collectively, these are called reactive oxygen species (ROS).

(b)

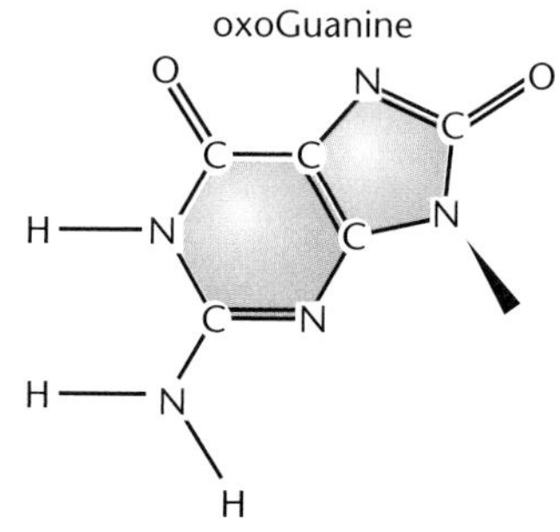

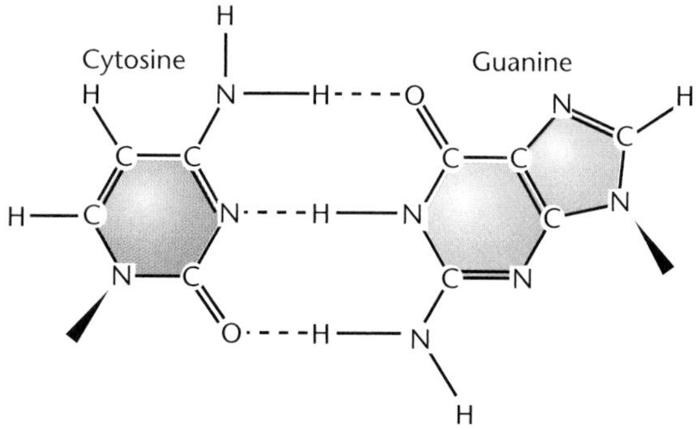

When casually examining the structures in the above diagrams, it is not immediately obvious that oxoG:A pairs should occur. However, hydrogen bonding can occur to any other base, including self pairs. Homopurine (A:A, G:G) and heteropurine (A:G) pairs represent anomalous base pairing possibilities even with nonaltered bases. While G:C is undoubtedly the most stable, several mispairs are actually stronger than the A—T pair. Base pairing is complicated by the fact that the purines possess two H-bonding faces, the Watson–Crick face, involving ring positions 1 and 6 for Adenine and, 1, 2, and 6 for Guanine, and the Hoogsteen face involving ring positions 6 and 7. The typical pairing mode is indicated as *wc*, where pairing occurs on the Watson–Crick face in the normal orientation, even for the mispair A:G. Alteration of pairing and favoring of the Hoogsteen face can occur with the alteration generated by oxoGuanine. Indeed, triple helix configurations commonly involve the Hoogsteen face.

(c) If not repaired (see below), the first round of replication involves the pairing of oxoG to Adenine (see above) while in the next round of replication, Adenine pairs with its normal Thymine. Therefore, if one starts with a G:C pair, one ends up with an A:T pair.

(d) It turns out that G:G>T:A transversions are quite commonly found in human cancers and are especially prevalent in the tumor suppressor gene *p53*. Thus, the cellular defense system has been extensively studied. One component is a triphosphatase that cleanses the nucleotide precursor pool by removing the two outermost phosphates from oxo-dGTP.

Another involves a DNA glycosylase that initiates repair of misreplicated oxoG:A by hydrolyzing the glycosidic bond linking the adenine base to the sugar. Another is a DNA glycosylase/lyase system that recognizes oxoG opposite cytosine. Of the three systems, the DNA glycosylases are probably the most effective.

32. Individuals with *xeroderma pigmentosum* (XP) are much more likely to contract skin cancer in youth than non-XP individuals. By age 20, approximately 80% of the XP population has skin cancer compared with approximately 4% in the non-XP group. XP individuals lack one or more genes involved in DNA repair.

Chapter 16

2. Under *negative* control, the regulatory molecule interferes with transcription, while in *positive* control, the regulatory molecule stimulates transcription.

4. (a) Due to the deletion of a base early in the *lac* Z gene there will be "frameshift" of all the reading frames downstream from the deletion. It is likely that either premature chain termination of translation will occur (from the introduction of a nonsense triplet in a reading frame) or the normal chain termination will be ignored. Regardless, a mutant condition for the *Z* gene will be likely. If such a cell is placed on a lactose medium, it will be incapable of growth because β-galactosidase is not available.
(b) If the deletion occurs early in the *A* gene, one might expect impaired function of the *A* gene product, but it will not influence the use of lactose as a carbon source.

6. $I^+O^+Z^+$ = Because of the function of the active repressor from the I^+ gene, and no lactose to influence its function, there will be **No Enzyme Made**.
$I^+O^cZ^+$ = There will be a **Functional Enzyme Made** because of the constitutive operator is in *cis* with a *Z* gene. The lactose in the medium will have no influence because of the constitutive operator. The repressor cannot bind to the mutant operator.
$I^-O^+Z^-$ = There will be a **Nonfunctional Enzyme Made**, because with I^- the system is constitutive but the *Z* gene is mutant. The absence of lactose in the medium will have no influence because of the nonfunctional repressor. The mutant repressor can not bind to the operator.
$I^-O^+Z^-$ = There will be a **Nonfunctional Enzyme Made** because with I^- the system is constitutive but the *Z* gene is mutant. The lactose in the medium will have no influence because of the nonfunctional repressor. The mutant repressor can not bind to the operator.
$I^-O^+Z^+/F'I^+$ = There will be **No Enzyme Made** because in the absence of lactose, the repressor product of the I^+ gene will bind to the operator and inhibit transcription.
$I^+O^cZ^+/F'O^+$ = Because there is a constitutive operator in *cis* with a normal *Z* gene, there will be **Functional Enzyme Made**. The lactose in the medium will have no influence because of the mutant operator.
$I^+O^+Z^-/F'I^+O^+Z^+$ = Because there is lactose in the medium, the repressor protein will not bind to the operator and transcription will occur. The presence of a normal *Z* gene allows a **Functional and Nonfunctional Enzyme to be Made**. The repressor protein is diffusable, working in *trans*.
$I^-O^+Z^-/F'I^+O^+Z^+$ = Because there is no lactose in the medium, the repressor protein (from I^+) will repress the operators and there will be **No Enzyme Made**.
$I^sO^+Z^+/F'O^+$ = With the product of I^S there is binding of the repressor to the operator and therefore **No Enzyme Made**. The lack of lactose in the medium is of no consequence because the mutant repressor is insensitive to lactose.
$I^+O^cZ^+/F'O^+Z^+$ = The arrangement of the constitutive operator (O^c) with the Z gene will cause a **Functional Enzyme to be Made**.

8. A single *E. coli* cell contains very few molecules of the *lac* repressor. However, the *lac* I^q mutation causes a 10× increase in repressor protein production, thus facilitating its isolation. With the use of dialysis against a radioactive gratuitous inducer (IPTG), Gilbert and Muller–Hill were able to identify the repressor protein in certain extracts of *lac* I^q cells. The material that bound the labeled IPTG was purified and shown to be heat labile and have other characteristics of protein. Extracts of *lac* I^- cells did not bind the labeled IPTG.

10. (a) With no lactose and no glucose, the operon is off because the *lac* repressor is bound to the operator and although CAP is bound to its binding site, it will not override the action of the repressor.
(b) With lactose added to the medium, the *lac* repressor is inactivated and the operon is transcribing the structural genes. With no glucose, the CAP is bound to its binding site, thus enhancing transcription.
(c) With no lactose present in the medium, the *lac* repressor is bound to the operator region, and since glucose inhibits adenyl cyclase, the CAP protein will not interact with its binding site. The operon is therefore "off."
(d) With lactose present, the *lac* repressor is inactivated, however since glucose is also present, CAP will not interact with its binding site. Under this condition transcription is severely diminished and the operon can be considered to be "off."

12. Attenuation functions to reduce the synthesis of tryptophan when it is in full supply. It does so by reducing transcription of the *tryptophan* operon. The same phenomenon is observed when tryptophan activates the repressor to shut off transcription of the *tryptophan* operon.

14. Neelaredoxin appears to be a protein that defends anaerobic and perhaps aerobic organisms from oxidative stress brought on by the metabolism of oxygen. The generation of oxygen-free radicals (creates the oxidative stress) is dependent on several molecular species including O_2 and H_2O_2. Apparently, relatively high levels of neelaredoxin are produced at all times (*constitutively expressed*) even when potential inducers of gene expression are not added to the system. Additional neelaredoxin gene expression is not responsive (*induced*) as a result of O_2 and H_2O_2 treatment.

16. Since a substance supplied in the medium (the antibiotic) causes the synthesis of the efflux pump components, two situations seem appropriate. Under a *negative control* system the antibiotic would interrupt the repressor to bring about induction (this would be an inducible system). Under *positive control* the antibiotic would activate an activator (this again would be an inducible system).

18. Because the deletion of the regulatory gene causes a loss of synthesis of the enzymes, the regulatory gene product can be viewed as one exerting *positive control*. When *tis* is present, no enzymes are made, therefore, *tis* must inactivate the positive regulatory protein. When *tis* is absent, the regulatory protein is free to exert its positive influence on transcription. Mutations in the operator negate the positive action of the regulator. The model below illustrates these points.

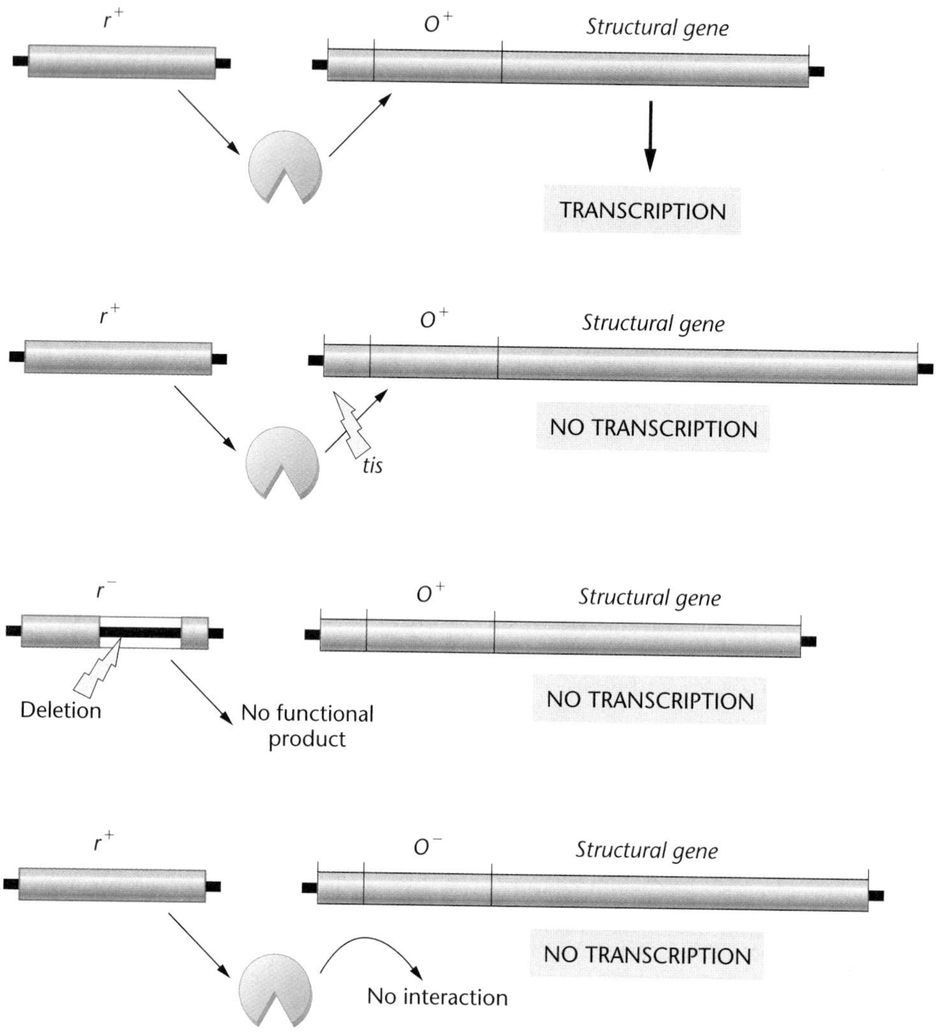

Model of regulatory system described in problem 18. This is an example of *positive* control.

20. (a) Call one constitutive mutation *lex*A$^-$ (mutation in the repressor gene product) and the other O^{uvrA-} (mutation in the operator).

(b) One can make partial diploid strains using F′. O^{uvrA^-} will be dominant to O^{uvrA^+} and *lex*A$^-$ will be recessive to *lex*A$^+$. O^{uvrA^-} will act in *cis*.

22. To get started, find the CACUUCC sequence. It pairs, with one mismatch, with a second region. Hint: The third region is composed of seven bases and starts with an AG.

Chapter 17

2. Your list should include the following:

chromatin remodeling
gene amplification
transcription
processing and transport
translation

See the text for a complete description of each topic.

4. Transcription factors are proteins that are *necessary* for the initiation of transcription. However, they are not *sufficient* for the initiation of transcription. To be activated, RNA polymerase II requires a number of transcription factors. Transcription factors contain at least two functional domains: one binds to the DNA sequences of promoters and/or enhancers, the other interacts with RNA polymerase or other transcription factors.

6. Both the *lac* and *gal* systems are influenced by catabolite repression, however, the *lac* system is under negative control whereas the *gal* system is under positive control. Both systems are inducible.

8. The model that depicts binding of factors to the nascent polypeptide chain is supported. There are a variety of experiments that could be used to substantiate such a model. One might stabilize the proposed MREI-protein complex with "crosslinkers," treat with RNAse to digest mRNA and to break up polysomes, then isolate individual ribosomes. One may use some specific antibody or other method to determine whether tubulin subunits contaminate the ribosome population.

10.

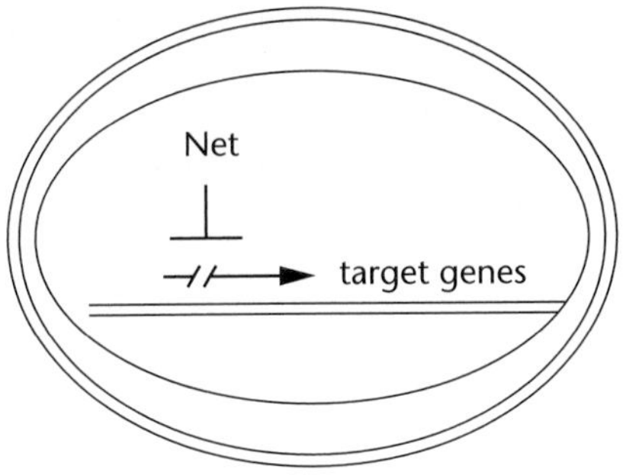

Neutral Conditions

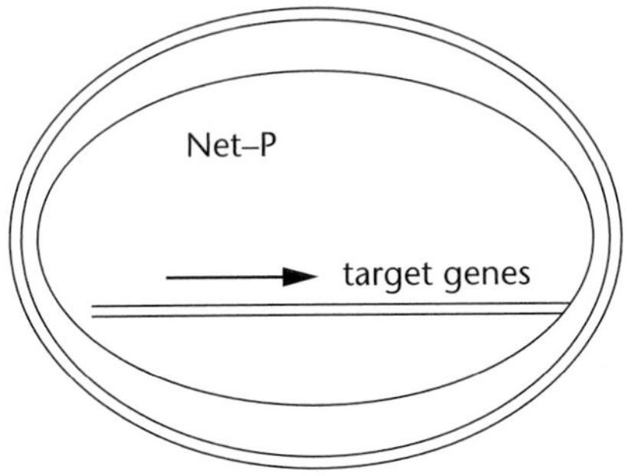

Phosphorylated Net

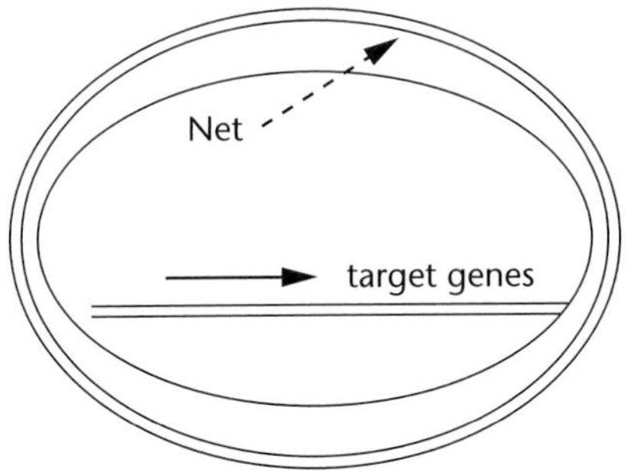

UV and Heat Shock

Sketches modified from Ducret et al. *Molecular and Cellular Biology* 1999 19:7076–7087.

12. The inherent flaw is that the genome in all nonlymphoid cells is the same, so how would DNA folding vary in such a way to trigger specific gene expression patterns in different cells? However, the same can be said for any form of eukaryotic genetic regulation. Highly differentiated cells have the same genetic material, so how are different genes expressed in different cells? In support of differential folding of chromatin contributing to genetic regulation, the same principles that apply to regulate gene activity as we presently view it (cell-specific factors, proteins) could also be applied to the 3-D organization within the nucleus.

14. Methylation of CpGs causes a reduction in luciferase expression that is somewhat proportional to the amount of methylation and patch size. Methylation within the transcription unit more drastically reduces luciferase expression compared to methylation outside the transcription unit. A high degree of methylation outside the transcription unit (593 CpGs) has as great of an impact on depressing transcription as the same degree of methylation within the transcription unit.

16. Criteria for determining the conservation of alternative splicing patterns would likely include the following:

(a) Similar mRNA length
(b) Conservation of splice junctions
(c) Positioning of homologous introns
(d) Size of homologous introns
(e) Exon nucleotide sequence homology
(f) Predicted amino acid physico-chemical similarity
(g) Same 5′–3′ orientation
(h) Conserved use of alternative stop codons in frame-shift splicing events
(i) Conserved use of alternative frames of translation

18. When splice specificity is lost one might observe several classes of altered RNAs: (1) a variety of nonspecific variants producing RNA pools with many lengths and combinations of exons and introns, (2) incomplete splicing where introns and exons are erroneously included or excluded in the mRNA product, and (3) a variety of nonsense products that result in premature RNA decay or truncated protein products. It is presently unknown as to whether cancer-specific splices initiate or result from tumorigenesis. Given the complexity of cancer induction and the maintenance of the transformed cellular state, gene products that are significant in regulating the cell cycle may certainly be influenced by alternative splicing and thus contribute to cancer.

Chapter 18

2. The G1 stage begins after mitosis and is involved in the synthesis of many cytoplasmic elements. In the S phase DNA synthesis occurs. G2 is a period of growth and preparation for mitosis. Most cell cycle time variation is caused by changes in the duration of G1. G0 is the nondividing state.

4. Kinases regulate other proteins by adding phosphate groups. Cyclins bind to the kinases, switching them on and off. CDK4 binds to cyclin D, moving cells from G1 to S. At the G2/mitosis border a CDK1 (cyclin dependent kinase) combines with another cyclin (cyclin B). Phosphorylation occurs, bringing about a series of changes in the nuclear membrane *via* caldesmon, cytoskeleton, and histone H1.

6. To say that a particular trait is inherited conveys the assumption that when a particular genetic circumstance is present, it will be revealed in the phenotype. When one discusses an inherited predisposition, one usually refers to situations where a particular phenotype is expressed in families in some consistent pattern. However, the phenotype may not always be expressed or may manifest itself in different ways. In retinoblastoma, the gene is inherited as an autosomal dominant and those that inherit the mutant *RB* allele are predisposed to develop eye tumors. However, approximately 10% of the people known to inherit the gene don't actually express it and in some cases expression involves only one eye rather than two.

8. Apoptosis or programmed cell death is a genetically controlled process that leads to death of a cell. It is a natural process involved in morphogenesis and a protective mechanism against cancer formation. During apoptosis, nuclear DNA becomes fragmented, cellular structures are disrupted, and the cells dissolve. Caspases are involved in the initiation and progress of apoptosis.

10. There are a number of ways in which protooncogenes are converted to oncogenes: point mutations in which a mutant gene acts as a positive "switch" in the cell cycle, translocations where a hybrid gene might be formed, and overexpression where a gene might acquire a new promoter and/or enhancer. In the case of RSV, an oncogene (*c-src*) was captured from the chicken genome.

12. Various kinases can be activated by breaks in DNA. One kinase, called ATM and/or a kinase called Chk2, phosphorylates BRCA1 and p53. The activated p53 arrests replication during the S phase to facilitate DNA repair. The activated BRCA1 protein, in conjunction with BRCA2, mRAD51, and other nuclear proteins is involved in repairing the DNA.
14. Protooncogenes are those that normally function to promote or maintain cell division. In the mutant state (oncogenes), they induce or maintain uncontrolled cell division, that is, there is a gain-of-function. Generally this gain-of-function takes the form of increased or abnormally continuous gene output. On the other hand, loss-of-function is generally attributed to tumor suppressor genes that function to halt passage through the cell cycle. When such genes are mutant, they have lost their capacity to halt the cell cycle.
16. Unfortunately, it is common to spend enormous amounts of money on dealing with diseases after they occur rather than concentrating on disease prevention. Too often pressure from special interest groups or lack of political stimulus retards advances in education and prevention. Obviously, it is less expensive, both in terms of human suffering and money, to seek preventive measures for as many diseases as possible. However, having gained some understanding of the mechanisms of disease, in this case cancer, it must also be stated that no matter what preventive measures are taken it will be impossible to completely eliminate disease from the human population. It is extremely important, however, that we increase efforts to educate and protect the human population from as many hazardous environmental agents as possible.
18. Normal cells are often capable of withstanding mutational assault because they have checkpoints and DNA repair mechanisms in place. When such mechanisms fail, cancer may be a result. Through mutation, such protective mechanisms are compromised in cancer cells and as a result they show higher than normal rates of mutation, chromosomal abnormalities, and genomic instability.
20. Certain environmental agents such as chemicals and X-rays cause mutations. Since genes control the cell cycle, mutations in cell cycle control genes, or those that impact on cell cycle control, can lead to cancer.
22. No, she will still have the general population risk of about 10%. In addition, it is possible that genetic tests won't detect all breast cancer mutations.
24. A benign tumor is a multicellular cell mass that is usually localized to a giving anatomical site. Malignant tumors are those generated by cells that have migrated to one or more secondary sites. Malignant tumors are more difficult to treat and can be life-threatening.
26. As with many forms of cancer, a single gene alteration is not the only requirement. The authors (Bose et al.) state "but only infrequently do the cells acquire the additional changes necessary to produce leukemia in humans." Some studies indicate variations (often deletions) in the region of the breakpoints may influence expression of CML.
28. Since cell cycle control is achieved by gene (DNA) products—proteins—any agent that causes damage to DNA is a potential carcinogen. Since cigarette smoke is known to contain an agent that changes DNA, in this case transversions, numerous modified gene products (including cell cycle controlling proteins) are likely to be produced. The fact that many cancer patients have such transversions in *p53* strongly suggests that cancer is caused by agents in cigarette smoke.
30. (a, b) Even though there are changes in the *BRCA1* gene, they don't always have physiological consequences. Such neutral polymorphisms make screening difficult in that one can't always be certain that a mutation will cause problems for the patient.
    (c) The polymorphism in *PM2* is probably a silent mutation because the third base of the codon is involved.
    (d) The polymorphism in *PM3* is probably a neutral missense mutation because the first base is involved.

Chapter 19

2. Eukaryotic mRNAs typically have a 3′ polyA, tail as indicated in the diagram below. The poly-dT segment provides a double-stranded section that serves to prime the production of the complementary strand.

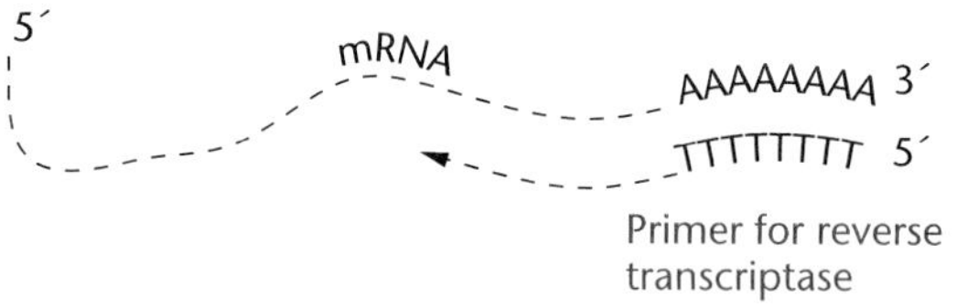

4. It is believed that the protein interacts with the major groove of the DNA helix. This information comes from the structure of the proteins that have been sufficiently well studied to suggest that the DNA major groove and "fingers" or extensions of the protein form the basis of interaction.
6. This segment contains the palindromic sequence of GGATCC, which is recognized by the restriction enzyme *Bam*HI. The double-stranded sequence is the following:

   CCTAGG
   GGATCC

8. Plasmids were the first to be used as cloning vectors and they are still routinely used to clone relatively small fragments of DNA. Because of their small size, they are relatively easy to separate from the host bacterial chromosome and they have relatively few restriction sites. They can be engineered fairly easily (i.e., polylinkers and reporter genes added). For cloning larger pieces of DNA such as entire eukaryotic genes, cosmids are often used. For instance, when modifications are made in the bacterial virus lambda (λ) relatively large inserts of about 20 kb can cloned. This is an important advantage when one needs to clone a large gene or generate a genomic library from a eukaryote. In addition, some cosmids will only accept inserts of a limited size, which means that small, less meaningful perhaps, fragments will not be cloned unnecessarily. Both plasmids and cosmids suffer from the limitation that they can only use bacteria as hosts.

   YACs (yeast artificial chromosomes) contain telomeres, an origin of replication, and a centromere and are extensively used to clone DNA in yeast. With selectable markers (TRP1 and URA3) and a cluster of restriction sites, DNA inserts ranging from 100 kb to 1000 kb can be cloned and inserted into yeast. Since yeast, being eukaryotes, undergo many of the typical RNA and protein processing steps of other, more complex eukaryotes, the advantages are numerous when working with eukaryotic genes.
10. This problem can be solved by the following expressions:

*Not*I	4^8
*Hin*fI	$4 \times 4 \times 1 \times 4 \times 4$
*Xho*II	$2 \times 4 \times 4 \times 4 \times 4 \times 2$

The reason for using a "1" in the *Hin*fI portion is that any of the four bases can be inserted for the "N" whereas only two bases can be used for "Pu" and "Py" in the *Xho*II portion.

12. (a) Because the *Drosophila* DNA has been cloned into the *Bam*H1 site in the ampicillin resistance gene of the plasmid, the gene will be mutated and any bacterium with the recombinant plasmid will be ampicillin sensitive. The tetracycline resistance gene remains active, however. Bacteria that have been transformed with the recombinant plasmid will be resistant to tetracycline and therefore tetracycline should be added to the medium.
(b) Colonies that grow on a tetracycline medium should be tested for growth on an ampicillin medium either by replica plating or some similar controlled transfer method. Those bacteria that do not grow on the ampicillin medium probably contain the *Drosophila* DNA insert.
(c) Resistance to both antibiotics by a transformed bacterium could be explained in several ways. First, if cleavage with the *Bam*H1 was incomplete, then no change in biological properties of the uncut plasmids would be expected. Also, it is possible that the cut ends of the plasmid were ligated together in the original form with no insert.

14. Given that there is only one site for the action of *Hin*dIII, then the following will occur. Cuts will be made such that a four-base single-stranded set of sticky ends will be produced. For the antibiotic resistance to be present, the ligation will reform the plasmid into its original form. However, two of the plasmids can join to form a dimer as indicated in the diagram below.

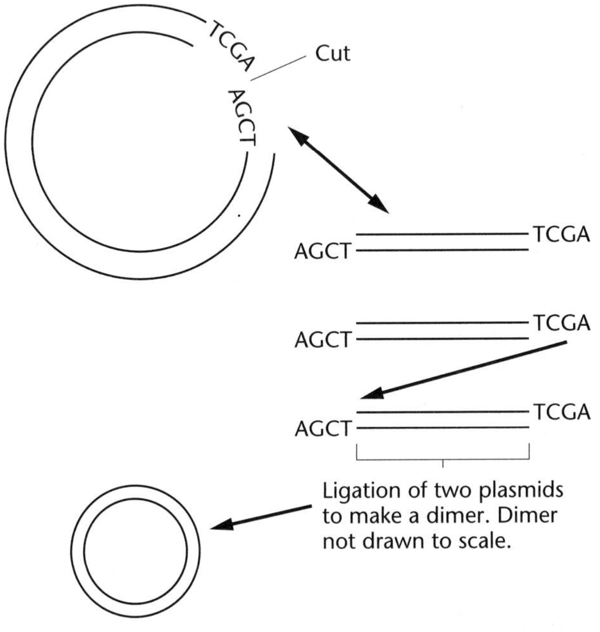

Ligation of two plasmids to make a dimer. Dimer not drawn to scale.

16. Because of complementary base pairing, the 3′ end of the DNA strand often loops back onto itself, thereby providing a primer for DNA polymerase I.

18. A filter is used to bind the DNA from the colonies containing recombinant plasmids. A labeled probe is constructed for the protein sequence of EF-1a. Since it is highly conserved, it should show considerable complementation to the human EF-1a cDNA. It is used to detect, through hybridization, the DNA of interest. Cells with the desired clone are then picked from the original plate and the plasmid is isolated from the cells.

20. The problem can be best solved by drawing out the strands, then placing the restriction sites in the appropriate positions as follows:

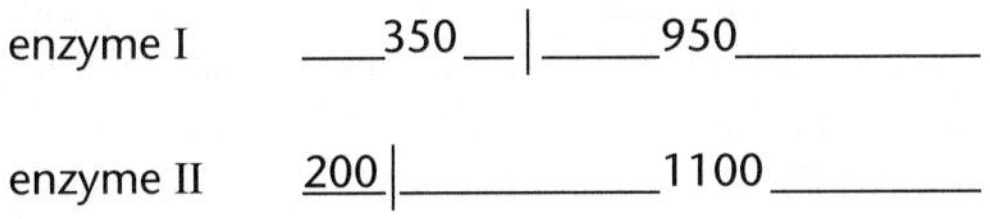

To determine the orientation of the restriction sites to each other, examine the results of the double digested DNA and note that there is a 150 bp fragment meaning that enzyme II cuts within the 350 bp fragment of enzyme I. Therefore the final map is as follows:

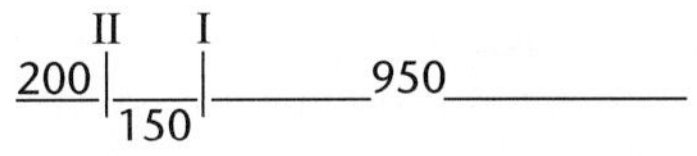

22. Option (b) fits the expectation because the thick band in the offspring probably represents the bands at approximately the same position in both parents. The likelihood of such a match is expected to be low in the general population.

24. (a) The overall size of the fragment is 12 kb. From the A + N digest, sites A and N must be 1 kb apart. N must be 2 kb from an E site. Pattern #5 is the likely choice. Notice that digest A + N breaks up the 6 kb E fragment.
(b) By drawing lines though sections that hybridize to the probe, one can see that the only place of consistent overlap to the probe is the 1 kb fragment between A and N.

26. Taking the number of bases recognized by *Bam*HI as 6, there would be approximately 4096 base pairs between sites. Given that lambda DNA contains approximately 48,500 base pairs, there would be about 11.8 sites (48,500/4096).

28. $T_m(°C) = 81.5 + 0.41\ (\%GC) - (675/N) = 81.5 + 0.41(33.3) - (675/21) =$ about 63°C. Subtracting 5°C gives us a good starting point of about 58°C for PCR with this primer. Notice that as the %GC and length increase, the T_m(°C) increases. GC pairs contain three hydrogen bonds rather than two as between AT pairs.

30. Since there is no 3′-hydroxyl group, chain elongation cannot take place and resulting fragments are formed that can be separated by electrophoresis. Where the ddNTP was incorporated, the length of each strand and therefore the position of the particular ddNTP is established and used to eventually provide the base sequence of the DNA.

Chapter 20

2. Knowing the sequence of DNA in an organism is only the beginning. Annotating the DNA is a significant challenge. Even at that, knowing how gene products interact in time and space (proteomics) will take additional rounds of technological advances as yet unconsidered. The work of Haas et al. identified variation in intron/exon splice sites, micro-exons, and alternative transcription start sites. Correlating various transcriptional and translational schemes with the phenotype will be an interesting adventure.

4. (a) Assuming an average gene size of 5,000 base pairs there would be about 6.7×10^7 base pairs comprising genes. Subtracting this value from 116.8 Mb gives 49.8 Mb between genes. Dividing 49.8 Mb by 13,379 genes gives about 3700 bases between genes.
(b) 54,934/13,379 = 4.11 exons
(c) 48,257/13,379 = 3.61 introns
(d) There is a marked increase in the number of genes involved in alternative transcripts.

6. First, calculate the number of base pairs in heterochromatic regions of the human genome:

$$0.15 \times 3000\ \text{Mbp} = 450 \times 10^6\ \text{bp}$$

Since there is one gene for 69,697 bp in the *Drosophila* there must be:

$$450{,}000{,}000/69{,}697 = 6{,}456.5$$

genes in the heterochromatic region of humans. In all likelihood this number is an overestimate because overall gene density in *Drosophila* is approximately 10× higher than in humans. Applying this information to the above calculation, perhaps as many as 650 genes are located in human heterochromatin. If these simple calculations are correct, then a considerable number of genes are yet to be identified. In addition, because these genes are located in heterochromatin, they may possess unique properties.

8. Open reading frames are identified by computer programs based on identification of start (ATG) and stop (TAA, TGA, TAG) codons. Notice that the percentage of GC pairs compared to AT pairs is quite low in such punctuation triplets. Therefore, when scanning DNA sequences for ORFs with high AT content, many short sequences are obtained that are clearly not likely to be involved in protein production. However, when DNA is GC rich, the likelihood of long ORFs similar to protein-coding size is increased. Therefore, the likelihood of falsely considering a sequence "protein-coding" increases with increasing GC content as indicated in the figure.

10. Functional genomics seeks to understand functional similarities of genomes across phylogenetic and evolutionary distances. Comparative genomics analyzes the arrangement and organization of families of genes within and among genomes.

12. Operons are genomic entities in which clusters of functionally related (often) genes are under coordinate control. In *A. aeolicus*, one operon contains six genes with widely varying functions. This finding and others like it causes one to reconsider the classical definition of an operon.

14. General similarities and differences:

Yeast	***Bacteria***
DNA	DNA
double-stranded	double-stranded
chromosomes	circular (*E. coli*)
12.1 Mb	naked nucleic acid
6200 genes	4.6 MB
	4397 genes

M. jannaschii (Archaea) has a circular, double-stranded DNA genome of about 1.7 Mb with about 1738 protein-coding genes. It contains three chromosomes and most genes (58%) in this organism don't match any other known genes.

16. Bacterial genes are densely packed in the chromosome. The protein-coding genes are mostly organized in polycistronic transcription units without introns. Eukaryotic genes are less densely packed in chromosomes and protein-coding genes are mostly organized as single transcription units with introns.

18. Gene duplication is a common phenomenon revealed by careful analysis of genome sequence data. The relatively small compact genome of *Arabidopsis* resembles those of *Drosophila* and *C. elegans* but not plants. Both coding and noncoding DNA loss can account for the compact genome of *Arabidopsis*. Compared with rice, maize, and barley, there are fewer "gene-empty" repetitive DNA regions in intergenic areas and many copies of transposable elements (intergenic also) have been lost.

20. The human genome is composed of over 3 billion nucleotides, about 5% of which code for genes. At least 50% of the genome is derived of transposable elements. Genes are unevenly distributed over chromosomes with clusters of gene-rich separated by gene-poor ones (deserts). Human genes tend to be larger and contain more and larger introns than in invertebrates such as *Drosophila*. Hundreds of genes have been transferred from bacteria into vertebrates. Duplicated regions are common, which may facilitate chromosomal rearrangement. The human genome appears to contain approximately 30,000 to 33,000 protein-coding genes.

22. Assuming that the APS strain is the ancestral strain, the remaining strains appear to have smaller genome sizes indicating genome reduction. The smallest *Buchnera* genome is approximately 448 kb compared to a genome size of *M. genitalium* of about 600 kb with about 480 protein-coding genes. The APS genome codes for about 564 genes in its 641 kb genome. A gene is coded every 1136 bp (641,000/564) for the APS strain and every 1250 bp (600,000/480) for *M. genitalium.* Given these data, the CCE species should code for approximately (448,000/1193) = 375.5 genes. (Note: 1193 was obtained as the average gene spacing of the two bacterial species mentioned above.) Using these calculations, the CCE strain would contain fewer genes than *M. genitalium*.

There are other possible approaches to determine minimum genome size to sustain life. Among them would include computational studies whereby one might estimate the number of essential chemical reactions that are needed for life. Another would be to take an organism with a small number of genes, then systematically mutate genes to see if elimination of genes caused reduced survival. By eliminating individual and groups of genes by mutation, the minimum number might be obtainable.

24. In general, one would expect certain factors (such as heat or salt) to favor evolution to increase protein stability: distribution of ionic interactions on the surface, density of hydrophobic residues and interactions, number of hydrogen and disulfide bonds. By examining the codon table, a high GC ratio would favor amino acids Ala, Gly, Pro, Arg, and Trp and minimize the use of Ile, Phe, Lys, Asn, and Tyr. How codon bias influences actual protein stability is not yet understood.

Most genomic sequences change by relatively gradual responses to mild selection over long periods of time. They strongly resemble their patterns of common descent. While the same can be said for organisms adapted to extreme environments, extraordinary physiological demands may dictate unexpected sequence bias.

26. While the β-globin gene family is a relatively large (60 kb) sequence and restriction analyses show that it is composed of six genes, one is a pseudogene and therefore does not produce a product. The five functional genes each contain two similarly sized introns which, when included with noncoding flanking regions (5′ and 3′) and spacer DNA between genes, account for the 95% mentioned in the question.

Chapter 21

2. (a) *Control of kidney development* could be studied most directly in humans and/or mice, since both are similar physiologically and anatomically. Since mice, rather than humans, can be genetically manipulated, the study of the genetics of kidney development would probably be more productive. While *Drosophila* has excretory organs (Malpighian tubules) they are not closely related to the mammalian kidney. Yeast do not have excretory organs; however, both yeast and *Drosophila* engage in membrane transport processes and the genetics of such processes will be fundamental and most likely apply to all organisms.

(b) Cancer can be studied in virtually all model organisms because it involves fundamentals of cell cycle control. However, much of what we know about cell cycle control derives from work in yeast and given that cell cycle systems are highly conserved processes, information obtained from any organism is often universally applicable. Yeast are well understood genetically and they can be more easily manipulated (genetically and environmentally) than the other organisms listed. Mammalian cell lines are often used to discover regulatory elements homologous to those seen in yeast.

(c) Cystic fibrosis is caused by mutations in the cystic fibrosis transmembrane conductance regulator gene that normally funnels chloride ions out of a cell, leading to a saltier cellular external environment that in turn draws water out of the cell by osmosis. The human *CFTR* gene has been cloned and expressed in yeast, so a viable way to look at the gene, its regulation, and product is available in a yeast model. In addition, *Drosophila* cell lines have been developed to study vertebrate expression systems for how an ion channels proteins. Using these transgenic systems one could expand knowledge of the transport systems in general. Seemingly, the most appropriate system to study human cystic fibrosis from a genetic standpoint would be the mouse; however, while strains have been generated with numerous mutations in the *CFTR* gene, mice don't develop the most serious symptom of human cystic fibrosis, namely chronic lung infections by *Pseudomonas aeruginosa*. Recently, this problem has been overcome by the development of a mouse strain that is hypersusceptible to this bacterium (Coleman *et al.* 2003) and holds promise as a model organism for CF. Depending on the experimental approach, either human cell lines, transgenic *Drosophila* or yeast, or new mouse models may be most appropriate.

(d) Molecular aspects of purine metabolism are highly conserved throughout the animal and plant kingdom. To study the genetics of purine metabolism it might be easiest to take advantage of the ease of mutant screens in yeast. In addition, certain strains of *Drosophila* respond differently to purine nutritional supplements, therefore one might use *Drosophila* to screen or select strains with altered purine metabolism.

(e) Genetic analyses of vertebrate immune function could be studied most directly in mouse models where genetic manipulation is possible. However, a relatively large number of forkhead-box (FOX) transcription factors have been found and characterized in *Drosophila melanogaster*. These transcription factors are members of a broader FOX family and have crucial roles in various aspects of immune regulation ranging from lymphocyte survival to thymic development. It would be possible therefore to study the immune system in a variety of organisms, taking advantage of the peculiarities of each.

(f) Like the study of cancer, the genetics of cell division might be most directly studied using yeast. The vast genetic understanding coupled with relative ease of genetic manipulation have already greatly enhanced our understanding of cell division processes. Because cell cycle mutations hamper the proliferation of the organism, conditional mutants are often successfully employed. Cultured mouse and human cells are also useful for studying specific aspects of cell division; however, their genetic manipulation is more difficult than yeast cells. In most cases, knockout/rescue strategies are useful.

4. The basic procedure in oligonucleotide site-directed mutagenesis is to synthesize an appropriate oligonucleotide that is homologous to the gene or portion of gene of interest. The oligonucleotide contains a desired mutation that may involve one or more nucleotides. The oligonucleotide is hybridized to M13 DNA that contains a cloned gene (or gene fragment) of interest and anneals to a homologous region. DNA polymerase is added along with other *in vitro* DNA polymerizing components to yield a double-stranded M13 molecule that is then transformed into *E. coli* where replication occurs. Semiconservative replication provides mutant molecules that are then screened or selected using a variety of standard techniques.

6. Balancer chromosomes contain multiple overlapping inversions that greatly reduce the recovery of crossover chromatids. By using such chromosomes, the mutagenized chromosome(s) remains intact. Such chromosomes are most useful when they contain dominant marker genes such as wing or eye shape or eye color. Balancer chromosomes usually contain a recessive lethal gene so that the balancer chromosome will not exist in the homozygous state.

8. Coat color markers are useful in generating knockout mice because when embryonic stem cells are injected into blastocysts the resulting chimeras will have patches of different fur colors, making them easy to recognize. Theoretically, any other surface marker (hair or hairless, etc.) would be usable as long as expression and penetrance of the dominant marker is high. Any test, molecular, biochemical, or visible that allows one to determine the presence of a gene in an organism would be usable, however, clearly visible tests are much more efficient. One could imagine using a gene that provided some needed substance for survival that circulates through body fluids. If the blastocyst is deficient for this vital substance, only the chimeras would survive.

10. (a) While autosomal and dominant mutations may be detected with the *ClB* technique it is specifically designed to detect recessive mutations, especially but not exclusively lethals, which are X-linked.

(b) The recessive, X-linked *l* gene is used to exclude hemizygous *ClB*/Y flies (males) in the F_2 generation. If no males appear in the F_2, then one (or more) X-linked recessive lethal was introduced on the mutagenized male.

(c) The multiply inverted X chromosome (*C*) suppresses the recovery of crossing over products on the X, thereby leaving the mutagenized X chromosome intact.

(d) *C* is a balancer chromosome that contains multiple overlapping inversions to greatly reduce the recovery of crossover products (chromatids).

12. Temperature-sensitive mutations allow one to collect mutations that impact on the vital functions of cells. To select such mutations, one conducts a mutagenesis experiment in which the yeast cells are grown at 23°C for example. In general, the mutagen selected is one that causes minor changes in DNA, such as base analogs, ultraviolet light, or nitrosoguanidine. The resulting colonies are then replica plated at the permissive temperature (23°C for example) and one sample is incubated at the restrictive temperature (36°C for example). Colonies that do not grow at the restrictive temperature are likely to contain temperature-sensitive lethal mutations. Returning to the original plate (permissive temperature) allows one to select cells for further analysis. Since yeast cells produce buds, the size of which characterizes the stage of the cell cycle, it is possible to isolate cell division cycle (*cdc*) defects by the morphology of the cells. In addition, if a gene is controlling a specific point in the cell cycle, a uniform morphology is likely.

14. Tissue-specific and/or temperal-specific patterns of gene expression are often studied by *in situ* hybridization that may reveal the pattern of mRNA in an organism. A labeled cDNA is used as a probe to hydrogen bond with the RNA. One may also use the northern blot technique that involves purification of mRNA from a tissue(s) of interest. Lastly, immunostaining may be used to generate a protein's expression profile.

16. P-elements containing a visible marker in the host fly are modified to contain a gene of interest. The recombinant P-element is then injected into *Drosophila* eggs along with a helper plasmid that contains the transposase gene that is transcribed and translated in the embryo thus enabling the P-element to insert into the embryo's DNA. Recombinant adult flies can be identified by expression of the visible marker. Offspring from such a fly may contain the gene of interest. To generate a transgenic mouse, DNA is injected into the haploid nucleus of the egg, which is then placed into the oviduct of a pseudopregnant female mouse. Resulting offspring are then screened for evidence of transgenesis. The difference between the two techniques resides in the type of DNA used and cell in which the DNA is injected.

18. If transgenesis occurs in a vital gene, the organism may not survive, and that could seriously complicate a study. In addition, genes are often influenced by position effects whereby their function is altered by their neighbors or broad chromosomal location (in heterochromatin, for example). If such position effects occur, accurate interpretation of gene action will be compromised.

20. Selecting for mutants occurs when one creates conditions that remove irrelevant organisms, leaving only the mutants of interest. Screening for mutants is often more labor intensive and involves visual examination of organisms.

22. Genetic complementation allows researchers to determine whether two or more mutations are in the same gene. Complementation groups are determined. Functional complementation allows one to identify the function of a cloned gene in a cDNA library by transforming mutant strains until one cDNA clone restores the mutant to wild type. This approach is often called a rescue experiment.

24. First, one should compare the two genes of interest against DNA sequences available in databases such as GenBank. This would allow the identification of genes or regions of homology to genes in the database. If homologous genes are identified in other organisms, their function may be known, or hints may be provided as to their function. Second, one should translate the DNA sequences into likely amino acid sequences. Since the locations of the introns and exons are known, this should be straightforward. Once the amino acid sequences are obtained, they can be compared to amino acid sequences in databases such as SwissProt or BLAST. From such searches, one may discover homologies to other proteins with known function. Third, one should analyze tissue-specific and temporal-specific patterns of gene expression using any number of techniques, such as *in situ* hybridization, northern blots, immunostaining, or microarray technology.

26. It is possible that yeast may process the candidate gene in a form that is similar to that seen in humans. Because abnormal accumulation of a protein is involved, yeast may accumulate the same protein if transformed with the candidate gene. It would be worth the effort to examine this possibility. For instance, Dr. Susan Lindquist at the Whitehead Institute has discovered a yeast protein that forms amyloid fibers similar to those found in Alzheimer's patients. Perhaps transforming yeast with the human gene would provide a gene product that, ideally, would be the same as that seen in humans, or at least, give some idea of the gene's behavior.

28. The first step in gaining insight into the functionality of a DNA sequence is to annotate that sequence. Given that an interesting motif is possibly produced, suggests, but does not prove, that the sequence is not "junk" DNA. Annotating the sequence involves use of a variety of annotation tools that scan the sequence in search of open reading frames (ORFs) that begin with an initiation sequence and end with a termination sequence. In addition, other landmarks are typically present in genes: intron/exon topology, intron/exon junction sequences, codon bias, upstream regulatory sequences, a 3′ polyadenylation signal, and in some cases CpG islands. Once a sequence is identified as a likely gene, comparisons with other known, well-studied genes may provide clues as to function. From that DNA sequence, the amino acid sequences can be predicted that can be compared to amino acid sequences in databases such as SwissProt or BLAST. From such searches, one may discover homologies to other proteins with known function. One could eventually analyze tissue-specific and temporal-specific patterns of gene expression using any number of techniques, such as *in situ* hybridization, northern blots, immunostaining, or microarray technology.

30. The fact that cDNA probes would detect an RNA pool in one place or another contrary to an immunofluorescence assay, which locates proteins, validates the complexity of genetic regulation in multicellular eukaryotes. It is likely that significant tissue-specific RNA processing or elimination is causing the results mentioned. Since the immunofluorescence assay does detect HOX protein in the testes, the researchers might wish to concentrate on locating the mRNA responsible for its synthesis. Assuming that it did not turn up in the cDNA study, perhaps something quite interesting is masking its detection. It might be equally interesting to tease out the reason for cDNA probes detecting an RNA but no protein product. Such exceptions to expectation are often the starting points for monumental discoveries.

Chapter 22

2. Kleter and Peijnenburg used the BLAST tool from the *http://www.ncbi.nlm.nih.gov/BLAST* Web site to conduct a series of alignment comparisons of transgenic sequences with sequences of known allergenic proteins. Of 33 transgenic proteins screened for identities of at least six contiguous amino acids found in allergenic proteins, 22 gave positive results.

4. (a) Both the saline and column extracts of Lkt50 appear to be capable of inducing at least 50% neutralization of toxicity when injected into rabbits.

(b) In order for a successful edible vaccine to be developed, numerous hurdles must be overcome. First, the immunogen must be stably incorporated into the host plant hereditary material and the host must express only that immunogen. During feeding, the immunogen must be transported across the intestinal wall unaltered, or altered in such a way as to stimulate the desired immune response. There must be guarantees that potentially harmful by products of transgenesis have not been produced. In other words, broad ecological and environmental issues must be addressed to prevent a transgenic plant from becoming an unintended vector for harm to the environment or any organisms feeding on the plant (directly or indirectly).

6. Enhancement therapy using gene products benefits from the application of modern biotechnology without being burdened by alteration of the genome. Enhancement gene therapy, however, opens the door to a variety of ethical issues. What limits can/should be imposed on individuals or institutions seeking to improve human qualities? What qualities should be open for enhancement? Gene therapy is not without medical and ethical risks.

8. Somatic gene therapy involves attempts to alter the genetic material in non–germ line cells. Clinical trials are currently underway. Germ line therapy, while certainly being more efficient (although perhaps more difficult technically) alters the germ line and is transmitted to offspring. There are considerable ethical problems associated with germ plasm therapy. It recalls previous attempts of the eugenics movements of past decades, which involved the use of selective breeding to purify the human stock. Some present-day biologists have said publicly that germ line gene therapy will *not* be conducted. Enhancement gene therapy raises a considerable ethical dilemma. Should genetic techniques be used to enhance human potential? It is generally felt that enhancement gene therapy, like germ line therapy is unacceptable.

10. p53 and pRB are tumor-suppressor proteins and are required by the cell to effectively monitor the cell cycle. Reduction in their activity would diminish normal cell cycle controls and most likely lead to cancer. It would be especially important if such viral-vectors are intended to treat cancer, where cell cycle control is likely already compromised.

12. (a) A microarray is a solid support containing an orderly arrangement of DNA samples. A typical array contains thousands of DNA spots, which may be small oligonucleotides, cDNAs, or short genomic sequences. Labeled sequences hybridize to the immobilized DNAs by standard base pairing. Such technology allows a method for monitoring RNA expression levels of thousands of genes in virtually any cell population.

(b) Using microarray technology, researchers can observe the overall behavior of the genome in cancer and normal cells and by comparison, determine which genes are active or inactive under various circumstances. It is possible to identify the set of genes whose expression or lack thereof defines the properties of each tumor type. This application can therefore lead to precise diagnosis and refine possible therapies. In addition, microarray profiling can be used to determine the efficacy of particular therapies. For instance, one can monitor responses to radiation and/or chemotherapy to determine the degree to which cells are responding.

14. *Drosophila* is a unique experimental organism in that there is a vast knowledge of its genetics, it is easily cultured and genetically manipulated, and it contains unique chromosomes, polytene chromosomes, which allow visual landmarks. Coupled with probe-labeling (sequence tagged sites), the visible landmarks (chromomeres) and ease of manipulation, one can actually see where important genes are located in chromosomes. *Drosophila* also contains P elements that allow sequence markers to be inserted into the genome. Microdissection of chromosomes is also useful in developing specific clones for sequencing. In addition, techniques have been developed (*in situ* hybridization) to allow scientists to actually determine the distributions of gene activities in all tissues of the organism.

16. Positional cloning relies on segregation (Mendelian) and linkage analysis. It is a departure from traditional methods of mapping and is used to isolate, map, and clone a gene with no knowledge of its gene product. Single gene disorders with simple, preferably dominant, expression would be most easily studied using positional cloning.

18. Positional cloning relies on segregation (Mendelian) and linkage analysis. Given the numerous limitations associated with such analyses in human populations (family and sample size, etc.) it is unlikely that this technique will be successfully applied to genetically complex traits in the near future.

20. Since both mutations occur in the CF gene, children who possess both alleles will suffer from CF. With both parents heterozygous, each child born will have a 25% chance of developing CF.

22. In general, bacteria do not process eukaryotic proteins in the same manner as eukaryotes. Transgenic eukaryotes are more likely to correctly process eukaryotic proteins, thus increasing the likelihood of their normal biological activity.

24. The answer provided here is based on the condition that individual I-2 is a carrier and the son, II-4, has the disorder. The 3 kb fragment occurs in the normal I-1 father and the normal son II-1. The affected son, II-4, has the 4 kb fragment. One daughter, II-2, is a carrier while the other daughter, II-3, is not a carrier.

26. The two major problems described here are common concerns related to genetic engineering. The first is the localization of the introduced DNA into the target tissue and target location in the genome. Inappropriate targeting may have serious consequences. In addition, it is often difficult to control the output of introduced DNA. Genetic regulation is complicated and subject to a number of factors including upstream and downstream signals as well as various posttranscriptional processing schemes. Artificial control of these factors will prove difficult.

28. Using restriction enzyme analysis to detect point mutations in humans is a tedious trial-and-error process. Given the size of the human genome in terms of base sequences and the relatively low number of unique restriction enzymes, the likelihood of matching a specific point mutation—separate from other normal sequence variations—to a desired gene is low.

30. (a) Y-linked excluded, X-linked recessive excluded, autosomal recessive possible but unlikely because the gene is stated as being rare, X-linked dominant possible if heterozygous, autosomal dominant possible.

(b) Chromosome 21 with the *B1* marker probably contains the mutation.

(c) The disease gene is segregating with some certainty with the B1 RFLP marker in the family. Since the mother also has the B3 marker, the offspring could be tested. If the child carries the B3 marker, then he/she does not carry the B1 marker that has been segregating with the defective gene. However, this prediction is not completely accurate because a crossover in the mother could put the undesirable gene with the B3 marker.

(d) The possibilities would include: a crossover between the restriction sites in the father giving a B1 chromosome, or a mutation eliminating either the B2 or B3 restriction site.

Chapter 23

2. The fact that nuclei from almost any source remain transcriptionally and translationally active substantiates the fact that the genetic code and the ancillary processes of transcription and translation are compatible throughout the animal and plant kingdoms. Because the egg represents an isolated, "closed" system that can be mechanically, environmentally, and to some extent biochemically manipulated, various conditions may be developed that allow one to study facets of gene regulation. For instance, the influence of transcriptional enhancers and suppressors may be studied along with factors that affect translational and post-translational processes. Combinations of injected nuclei may reveal nuclear–nuclear interactions that could not normally be studied by other methods.

4. Imaginal discs are small clusters of cells formed during larval stages. There are 12 bilaterally paired discs composed of eye-antennal, leg, wing, etc. and one unpaired genital disc.
6. (a-d) Genes that control early development are often dependent on the deposition of their products (mRNA, transcription factors, various structural proteins, etc.) in the egg by the mother. Such maternal effect genes control early events such as defining anterior–posterior polarity. Such products are placed in eggs during oogenesis and are activated immediately after fertilization. The phenotypes of maternal effect mutations vary from determination of general body plan to eye pigmentation and direction of shell coiling.
8. (a, b) Zygotic genes are activated or repressed depending on their response to maternal-effect gene products. Three subsets of zygotic genes divide the embryo into segments. These segmentation genes are normally transcribed in the developing embryo and their mutations have embryonic lethal phenotypes. The maternal genotype contains zygotic genes and these are passed to the embryo as with any other gene.
10. Because the polar cytoplasm contains information to form germ cells, one would expect such a transplantation procedure to generate germ cells in the anterior region. Work done by Illmensee and Mahowald in 1974 verified this expectation.
12. There are a variety of approaches to determine the level of control of a particular gene. First, one may determine whether levels of hnRNA are consistent among various cell types of interest. This is often accomplished by either direct isolation of the RNA and assessment by northern blotting or by use of *in situ* hybridization. If the hnRNA pools for a given gene are consistent in various cell types, then transcriptional control can be eliminated as a possibility. Support for translational control can be achieved directly by determining, in different cell types, the presence of a variety of mRNA species with common sequences. This can be accomplished only in cases where sufficient knowledge exists for specific mRNA trapping or labeling. Clues as to translational control via alternative splicing can sometimes be achieved by examining the amino acid sequence of proteins. Similarities in certain structural/functional motifs may indicate alternative RNA processing.
14. The "gain-of-function" *Antp* mutation causes the wild type *Antennapedia* gene to be expressed in the head and mutant flies have legs on the head in place of antenna. Such mutations are often dominant.
16. Because of the regulatory nature of *homeotic* genes in fundamental cellular activities of determination and differentiation, it would be difficult to ignore their possible impact on oncogenesis. Homeotic genes encode DNA binding domains that influence gene expression, and any factor that influences gene expression may, under some circumstances, influence cell cycle control.

    However attractive this model may be, there have been no homeotic transformations noted in mammary glands, so the typical expression of mutant homeotic genes in insects is not revealed in mammary tissue according to Lewis (2000). A substantial number of experiments will be needed to establish a functional link between homeotic gene mutation and cancer induction. Mutagenesis and transgenesis experiments are likely to be most productive in establishing a cause-effect relationship.
18. Two coupled approaches might be used. First, one could make transgenic flies that contain a series of deletions spanning all segments of the *bicoid* mRNA; the coding region, 5′ and 3′ untranslated regions. Comparison of stabilities of individual, deleted mRNAs with controls would indicate whether a particular segment of the mRNA contains a degradation signal sequence. If a degradation-sensitive region or signal sequence is located by deletion, that same intact region, when ligated to a noninvolved, nondegraded mRNA (like a ribosomal protein or tubulin mRNA) should foster degradation in a manner similar to the *bicoid* mRNA.
20.

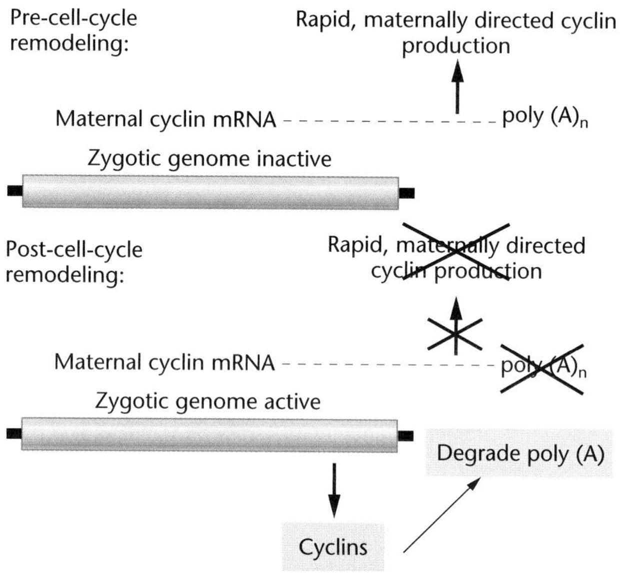

22. Given the information in the problem, it is likely that this gene normally controls the expression of *BX-C* genes in all body segments. The wild type product of *esc* stored in the egg may be required to interpret the information correctly stored in the egg cortex.
24. Three classes of flower *homeotic* genes are known that are activated in an overlapping pattern to specify various floral organs. Class *A* genes give rise to sepals. Expression of *A* and *B* class genes specify petals, *B* and *C* genes control stamen formation, and expression of *C* genes gives rise to carpels.
26. Because signal-receptor interactions depend on membrane-bound structures, the pathway can only work with adjacent cells. The advantage of such a system is that only cells in a certain location will be influenced—those in contact. A disadvantage would occur if large groups of cells are to be induced into a particular developmental pathway or if cells not in contact need to be induced.
28. If the *her-1*$^+$ product acts as a negative regulator, then when the gene is mutant, suppression over *tra-1*$^+$ is lost and hermaphroditism would be the result. This hypothesis fits the information provided. The double mutant should be male because even though there is no suppression from *her-1*$^-$, there is no *tra-1*$^+$ product to support hermaphrodite development.

Chapter 24

2. (a) *Polygenes* are genes that are involved in determining continuously varying or multiple factor traits.
   (b) *Additive alleles* are alleles that account for the hereditary influence on the phenotype in an additive way.
   (c) *Correlation* is a statistic that varies from -1 to $+1$ and describes the extent to which variation in one trait is associated with variation in another. It does not imply that a cause-and-effect relationship exists between two traits.

(d) *Monozygotic twins* are derived from a single fertilized egg and are thus genetically identical to each other. They provide a method for determining the influence of genetics and environment on certain traits. *Dizygotic twins* arise from two eggs fertilized by two sperm cells. They have the same genetic relationship as siblings. The role of genetics and the role of the environment can be studied by comparing the expression of traits in monozygotic and dizygotic twins. The higher concordance value for monozygotic twins as compared to the value for dizygotic twins indicates a significant genetic component for a given trait.

(e) *Heritability* is a measure of the degree to which the phenotypic variation of a given trait is due to genetic factors. A high heritability indicates that genetic factors are major contributers to phenotypic variation while environmental factors have little impact.

(f) QLT stands for Quantitative Trait Loci, which are situations where multiple genes contribute to a quantitative trait.

4. If you add the numbers given for the ratio, you obtain the value of 16, which is indicative of a dihybrid cross. The distribution is that of a dihybrid cross with additive effects.

(a) Because a dihybrid result has been identified, there are two loci involved in the production of color. There are two alleles at each locus for a total of four alleles.

(b, c) Because the description of red, medium-red, etc., gives us no indication of a *quantity* of color in any form of units, we would not be able to actually quantify a unit amount for each change in color. We can say that each gene (additive allele) provides an equal unit amount to the phenotype and the colors differ from each other in multiples of that unit amount. The number of additive alleles needed to produce each phenotype is given below:

1/16	= dark red	= *AABB*
4/16	= medium-dark red	= 2*AABb*
		2*AaBB*
6/16	= medium red	= *AAbb*
		4*AaBb*
		aaBB
4/16	= light red	= 2*aaBb*
		2*Aabb*
1/16	= white	= *aabb*

(d) F_1 = all light red
F_2 = 1/4 medium red
2/4 light red
1/4 white

6. As you read this question, notice that the strains are inbred, therefore homozygous, and that approximately 1/250 represent the shortest and tallest groups in the F_2 generation. See $1/4^n$ formula in the text.

(a, b) Referring to the text, see that where four gene pairs act additively, the proportion of one of the extreme phenotypes to the total number of offspring is 1/256 (add the numbers in each phenotypic class). The same may be said for the other extreme type. The extreme types in this problem are the 12 cm and 36 cm plants. From this observation, one would suggest that there are four gene pairs involved.

(c) If there are four gene pairs, there are nine $(2n + 1)$ phenotypic categories and eight increments between these categories. Since there is a difference of 24 cm between the extremes, 24 cm/8 = 3 cm for each increment (each of the additive alleles).

(d) A typical F_1 cross that produces a "typical" F_2 distribution would be where all gene pairs are heterozygous (*AaBbCcDd*), independently assorting, and additive. There are many possible sets of parents that would give an F_1 of this type.

The limitation is that each parent has genotypes that give a height of 24 cm as stated in the problem. Because the parents are inbred, it is expected that they are fully homozygous. An example:

AABBccdd × *aabbCCDD*

(e) Since the *aabbccdd* genotype gives a height of 12 cm and each uppercase allele adds 3 cm to the height, there are many possibilities for an 18 cm plant:

AAbbccdd,
AaBbccdd,
aaBbCcdd, etc.

Any plant with seven uppercase letters will be 33 cm tall:

AABBCCDd,
AABBCcDD,
AABbCCDD, for examples.
AABbcc × *aabbcc*

Gametes (18 cm tail)	***Gamete*** (6 cm tail)	***Offspring***
ABc	*abc*	*AaBbcc* (14 cm)
Abc		*Aabbcc* (10 cm)

8. For height, notice that average differences between MZ twins reared together (1.7 cm) and those MZ twins reared apart (1.8 cm) are similar (meaning little environmental influence) and considerably less than differences of DZ twins (4.4 cm) or sibs (4.5) reared together. These data indicate that genetics plays a major role in determining height. However, for weight, notice that MZ twins reared together have a much smaller (1.9 kg) difference than MZ twins reared apart, indicating that the environment has a considerable impact on weight. By comparing the weight differences of MZ twins reared apart with DZ twins and sibs reared together one can conclude that the environment has almost as much an influence on weight as genetics. For ridge count, the differences between MZ twins reared together and those reared apart are small. For the data in the table, it would appear that ridge count and height have the highest heritability values.

10. Many traits, especially those that we view as quantitative are likely to be determined by a polygenic mode with possible environmental influences. The following are some common examples: height; general body structure; skin color; and perhaps most common, behavioral traits, including intelligence.

12. (a) Using the following equations, H^2 and h^2 can be calculated as follows.

For back fat:
Broad sense heritability = H^2 = 12.2/30.6 = .398
Narrow sense heritability = h^2 = 8.44/30.6 = .276

For body length:
Broad sense heritability = H^2 = 26.4/52.4 = .504
Narrow sense heritability = h^2 = 11.7/52.4 = .223

(b) In animal and plant breeding, a measure of potential response to selection based on additive variance and dominance variance is termed narrow heritability (h^2). A relatively high narrow heritability is a prediction of the impact selection may have in altering an initial randomly breeding population. Therefore, of the two traits, selection for back fat would produce more response.

14. (a) For vitamin A

$$h_A{}^2 = V_A/V_P = V_A/(V_E + V_A + V_D) = 0.097$$

For cholesterol $h_A{}^2 = 0.223$

(b) Cholesterol content should be influenced to a greater extent by selection.

16. Given that both narrow sense heritability values are relatively high, it is likely that a farmer would be able to alter both milk protein content and butterfat by selection. The value of 0.91 for the correlation coefficient between protein content and butterfat suggests that if one selects for butterfat, protein content will increase. However, correlation coefficients describe the extent to which variation in one quantitative trait is associated with variation in another and does not reveal the underlying causes of such variation. Assuming that these dairy cows had been selected for high butterfat in the past and increased protein content followed that selection (for butterfat), it is likely that selection for butterfat would continue to correlate with increased protein content. However, there may well be a point where physiological circumstances change and selection for high butterfat may be at the expense of protein content.

18. Given the realized heritability value of 0.4 it is unlikely that selection experiments would cause a rapid and/or significant response to selection. A minor response might result from intense selection.

20. Since the rice plants are genetically identical, V_G is zero and $H^2 = V_G/V_P$ = zero. Broad-sense heritability is a measure to which the phenotypic variance is due to genetic factors. In this case, with genetically identical plants, H^2 is zero, and the variance observed in grain yield is due to the environment. Selection would not be effective in this strain of rice.

22. (a) The most direct explanation would involve two gene pairs, with each additive gene contributing about 1.2 mm to the phenotype.

(b) The fit to this backcross supports the original hypothesis.

(c) These data do not support the simple hypothesis provided in part (a).

(d, e) With these data, one can see no distinct phenotypic classes suggesting that the environment may play a role in eye development or that there are more genes involved.

24. For traits including blood type, eye color, and mental retardation, there is a fairly significant difference between MZ and DZ groups and therefore a high genetic component. However, for measles, the difference is not as significant, indicating a greater role of the environment. Hair color has a significant genetic component as do idiopathic epilepsy, schizophrenia, diabetes, allergies, cleft lip, and club foot. The genetic component to mammary cancer is present but minimal according to these data.

26. As with many traits that are caused by numerous loci acting additively, some genes have more influence on expression than others. In addition, environmental factors may play a role in the expression of some polygenic traits. In the case of brachydactyly, there are numerous modifier genes in the genome that can influence brachydactyly expression. Examination of OMIM (*Online Mendelian Inheritance of Man*) through *http://www.ncbi.nlm.nih.gov/* will illustrate this point.

28. It is likely that the flies maintained in the *Drosophila* repository are more highly inbred and less heterozygous than those recently obtained from the wild. Response to selection is dependent on genetic variation. The greater the genetic variation in a species, the more likely and dramatic the response to selection. Therefore, one would expect a greater response to selection in the wild population.

30. Breeders attempt to "select" out this disorder by first maintaining complete and detailed breeding records of afflicted strains. Second, they avoid breeding dogs whose close relatives are afflicted. The molecular-developmental mechanism that causes the "month of birth" effect in canine hip dysplasia is unknown. However, with many, perhaps all, quantitative traits, it is clear that there is a significant environmental influence on both the penetrance and/or expression of the phenotype. With many genes acting in various ways to influence a phenotype, there are opportunities for varied molecular and developmental intraorganismic microenvironments. Stated another way, the longer and more complex the molecular distance from the genome to the phenotype, the greater the likelihood for environmental factors to be involved in expression.

Chapter 25

2.

$$\begin{aligned} p &= \text{frequency of A} \\ &= 0.2 + .3 \\ &= 0.5 \\ q &= 1 - p = 0.5 \\ \text{Frequency of } AA &= p^2 \\ &= (.5)^2 \\ &= .25 \text{ or } 25\% \\ \text{Frequency of } Aa &= 2pq \\ &= 2(.5)(.5) \\ &= .5 \text{ or } 50\% \\ \text{Frequency of } aa &= q^2 \\ &= (.5)^2 \\ &= .25 \text{ or } 25\% \end{aligned}$$

4. In order for the Hardy–Weinberg equations to apply, the population must be in equilibrium.

6. (a) The equilibrium values will be as follows:

$$\begin{aligned} \text{Frequency of } l/l &= p^2 = (.7755)^2 \\ &= .6014 \text{ or } 60.14\% \\ \text{Frequency of } l/\Delta 32 &= 2pq \\ &= 2(.7755)(.2245) \\ &= .3482 \text{ or } 34.82\% \\ \text{Frequency of } \Delta 32/\Delta 32 &= q^2 = (.2245)^2 \\ &= .0504 \text{ or } 5.04\% \end{aligned}$$

Comparing these equilibrium values with the observed values strongly suggests that the observed values are drawn from a population in equilibrium.

(b) The equilibrium values will be as follows:

$$\begin{aligned} \text{Frequency of } AA &= p^2 = (.877)^2 \\ &= .7691 \text{ or } 76.91\% \\ \text{Frequency of } AS &= 2pq = 2(.877)(.123) \\ &= .2157 \text{ or } 21.57\% \\ \text{Frequency of } SS &= q^2 = (.123)^2 \\ &= .0151 \text{ or } 1.51\% \end{aligned}$$

Comparing these equilibrium values with the observed values suggests that the observed values may be drawn from a population that is not in equilibrium. Notice that there are more heterozygotes than predicted, and fewer *SS* types.

To test for a Hardy–Weinberg equilibrium, apply the chi-square test as follows.

$$\chi^2 = \frac{\Sigma(o - e)^2}{e}$$

$$(75.6 - 76.9)^2/76.9 + (24.2 - 21.6)^2/21.6 + (0.2 - 1.51)^2/1.5 = 1.47$$

In calculating degrees of freedom in a test of gene frequencies, the "free variables" are reduced by an additional degree of freedom because one estimated a parameter (p or q) used in determining the expected values. Therefore, there is one degree of freedom even though there are three classes. Checking the χ^2 table with one degree of freedom gives a value of 3.84 at the 0.05 probability level.

Since the χ^2 value calculated here is smaller, the null hypothesis (the observed values fluctuate from the equilibrium values by chance and chance alone) should not be rejected. Thus the frequencies of *AA*, *AS*, *SS* sampled a population that is in equilibrium.

8. The following formula calculates the frequency of an allele in the next generation for any selection scenario, given the frequencies of *a* and *A* in this generation and the fitness of all three genotypes.

$$q_{g+1} = [w_{Aa}p_g q_g + w_{aa}q_g{}^2]/[w_{AA}p_g{}^2 + w_{Aa}2p_g q_g + w_{aa}q_g{}^2]$$

where q_{g+1} is the frequency of the a allele in the next generation, q_g is the frequency of the a allele in this generation, p_g is the frequency of the A allele in this generation, and each "w" represents the fitness of their respective genotypes.

(a) $q_{g+1} = [.9(.7)(.3) + .8(.3)^2]/[1(.7)^2 + .9(2)(.7)(.3) + .8(.3)^2]$

$q_{g+1} = .278 \quad p_{g+1} = .722$

(b) $q_{g+1} = .289 \quad p_{g+1} = .711$

(c) $q_{g+1} = .298 \quad p_{g+1} = .702$

(d) $q_{g+1} = .319 \quad p_{g+1} = .681$

10. Since a dominant lethal gene is highly selected against, it is unlikely that it will exist at too high of frequency, if at all. However, if the gene shows incomplete penetrance or late age of onset (after reproductive age) it may remain in a population.

12. What one must do is predict the probability of one of the grandparents being heterozygous in this problem. Given the frequency of the disorder in the population as 1 in 10,000 individuals (0.0001), then $q^2 = 0.0001$, and $q = 0.01$. The frequency of heterozygosity is $2pq$ or approximately .02 as also stated in the problem. The probability for one of the grandparents to be heterozygous would therefore be $0.02 + 0.02$ or 0.04 or 1/25. If one of the grandparents is a carrier, then the probability of the offspring from a first-cousin mating being homozygous for the recessive gene is 1/16. Multiplying the two probabilities together gives $1/16 \times 1/25 = 1/400$.

Following the same analysis for the second-cousin mating gives $1/64 \times 1/25 = 1/1600$. Notice that the population at large has a frequency of homozygotes of 1/10,000; therefore, one can easily see how inbreeding increases the likelihood of homozygosity.

14. Because heterozygosity tends to mask expression of recessive genes, which may be desirable in a domesticated animal or plant, inbreeding schemes are often used to render strains homozygous so that such recessive genes can be expressed. In addition, assume that a particularly desirable trait occurs in a domesticated plant or animal. The best way to increase the frequency of individuals with that trait is by self-fertilization (not often possible) or by matings to blood relatives (inbreeding). In theory, one increases the likelihood of a gene "meeting itself" by various inbreeding schemes. There are disadvantages to increasing the degree of homozygosity by inbreeding. *Inbreeding depression* is a reduction in fitness often associated with an increase in homozygosity.

16. The quickest way to generate a homozygous line of an organism is to *self-fertilize* that organism. Because this is not always possible, brother-sister matings are often used.

18. If there were two males with hemophilia, $q = 2/2000$ and $p = 1998/2000$. The frequency of heterozygous females would be $2pq$ or $2(1998/2000 \times 2/2000) = .001998$. The number of heterozygous females would be 3.996, or approximately 4.

20. Because three of the affected infants had affected parents, only two "new" genes, from mutation, enter into the problem. The gene is dominant; therefore, each new case of achondroplasia arose from a single new mutation. There are 50,000 births, therefore 100,000 gametes (genes) are involved. The frequency of mutation is therefore given as follows: 2/100,000 or 2×10^{-5}.

22. Since $r1 = 0.81$ and $r2 = 0.19$ the expected frequency of heterozygotes would be $2pq \times 125$ or $2(0.81 \times 0.19) \times 125 = 38.475$. Given the following equation and substituting the values:

$$F = (H_e - H_o)/H_e$$
$$F = (38.475 - 20)/38.475 = 0.48$$

24. The following distribution of genotypes occurs among the 50 desert bighorn sheep in which the normal dominant *C* allele produces straight coats.

CC = 29 = straight coats
Cc = 17 = straight coats
cc = 4 = curled coats

Computing, $p = .75$ and $q = .25$ and $2pq = 0.375$ for the expected frequency of heterozygotes. Since 17/50 or (0.34) are observed as heterozygotes, the following equation applies:

$$F = (H_e - H_o)/H_e$$
$$F = (.375 - .34)/.375 = 0.093$$

This problem could also be solved using the actual numbers of sheep in each category where there would be 18.75 heterozygotes expected $(2pq)(50)$:

$$F = (18.75 - 17)/18.75 = 0.093$$

26. (a) The gene is most likely recessive because all affected individuals have unaffected parents and the condition clearly runs in families. For the population, since $q^2 = .002$, then $q = .045$, $p = .955$, and $2(pq) = 0.086$. For the community, since $q^2 = .005$, $q = .07$, $p = .93$, and $2(pq) = 0.13$.

(b) The "founder effect" is probably operating here. Relatively small, local populations that are relatively isolated in a reproductive sense tend to show differences in gene frequencies when compared to larger populations. In such small populations, homozygosity is increased because a gene has a higher probability of "meeting itself."

Chapter 26

2. A species is a group of interbreeding or potentially interbreeding populations that is reproductively isolated from all other such groups. Speciation is the process that leads to the formation of species. Evolution is the change in a population over time. Speciation is one of many results of evolution.

4. Organisms may appear to be similar but be reproductively isolated for a sufficient period to justify their species identity. If significant genetic differences occur, they can be considered separate species.

6. The subterranean niche is less broad and less dynamic than above ground. The underground microhabitat consists of a narrower range of climatic changes, thus genetic polymorphism is not selected for. This conclusion has been supported by additional studies that indicate that genetic diversity is positively correlated with niche-width.

8. Results from laboratory studies indicated that there was a selective advantage in having the two inversions present rather than either one. Thus, natural selection favored the maintenance of both inversions over the loss of either.

10. (a) Missense mutations cause amino acid changes.

(b) Horizontal transfer refers to the process of passing genetic information from one organism to another without producing offspring. In bacteria, plasmid transfer is an example of horizontal transfer.

(c) The fact that none of the isolates shared identical nucleotide changes indicates that there is little genetic exchange among different strains. Each alteration is unique, most likely originating in an ancestral strain and maintained in descendents of that strain only.

12. The approximate similarity of mutation rates among genes and lineages should provide more credible estimates of divergence times of species and allow for broader interpretations of sequence comparisons. It also provides for increased understanding of the mutational processes that govern evolution among mammalian genomes. For instance, if the rate of mutation is fairly constant among lineages or cells, which have a more rapid turnover, it indicates that replication-related errors do not make a significant contribution to mutation rates.

14. In general, speciation involves the gradual accumulation of genetic changes to a point where reproductive isolation occurs. Depending on environmental or geographic conditions, genetic changes may occur slowly or rapidly. They can involve point or chromosomal changes.

16. Reproductive isolating mechanisms are grouped into Prezygotic and Postzygotic, and include those listed below. See the text for specific examples and illustrations.

Geographic or ecological
Seasonal or temporal
Behavorial
Mechanical
Physiological
Hybrid inviability or weakness
Developmental hybrid sterility
Segregational hybrid sterility
F_2 breakdown

18. Polyploid plants that result from hybridization of two species would be expected to be more heterozygous than the diploid parental species because two distinct genomes are combined. Generally, genetic variation is an advantage unless a significant degree of that variation is outside acceptable physiological tolerance.

20. Somatic gene therapy, like any therapy, allows some individuals to live more normal lives than those not receiving therapy. As such, the ability of such individuals to contribute to the gene pool increases the likelihood that less fit genes will enter and be maintained in the gene pool. This is a normal consequence of therapy, genetic or not, and in the face of disease control and prevention, societies have generally accepted this consequence. Germ-line therapy could, if successful, lead to limited, isolated, and infrequent removal of a gene from a gene lineage. However, given the present state of the science, its impact on the course of human evolution will be diluted and negated by a host of other factors that afflict mankind.

22. Approach this problem by writing the possible codons for all the amino acids (except Arg and Asp, which show no change) in the human cytochrome c chain. Then determine the minimum number of nucleotide substitutions required for each changed amino acid in the various organisms. Once listed, count up the numbers for each organism: horse, 3; pig, 2; dog, 3; chicken, 3; bullfrog, 2; fungus, 6.

24. The classification of organisms into different species is based on evidence (morphological, genetic, ecological, etc.) that they are reproductively isolated. That is, there must be evidence that gene flow does not occur among the groups being called different species. Classifications above the species level (genus, family, etc.) are not based on such empirical data. Indeed, classification above the species level is somewhat arbitrary and based on traditions that extend far beyond DNA sequence information. In addition, recall that DNA sequence divergence is not always directly proportional to morphological, behavioral, or ecological divergence. While the genus classifications provided in this problem seem to be invalid, other factors, well beyond simple DNA sequence comparison, must be considered in classification practices. As more information is gained on the meaning of DNA sequence differences (ΔT_m) in comparison to morphological factors, many phylogenetic relationships will be reconsidered and it is possible that adjustments will be needed in some classification schemes.

26. (a) Since noncoding genomic regions are probably silent genetically, it is likely that they contribute little, if anything, to the phenotype. Selection acts on the phenotype, therefore, such noncoding regions are probably selectively neutral.

(b) These polymorphism data indicate that all the Lake Victoria area (lake and contributing rivers) cichlids are related by recent ancestry, whereas those from neighboring lakes are more distantly related. In addition, since Lake Victoria dried out about 14,000 years ago, it is likely that it was repopulated by a relatively small sample of cichlids.

28. The pattern of genetic distances through time indicates that from the present to about 25,000 years ago, modern humans and Cro-Magnons show an approximately constant number of differences. Conversely, there is an abrupt increase in genetic distance seen in comparing modern humans and Cro-Magnons with Neanderthals. The results indicate a clear discontinuity between modern humans, Cro-Magnons, and Neanderthals with respect to genetic variation in the mitochondrial DNAs sampled. Assuming that the sampling and analytical techniques used to generate the data are valid, it appears that Neanderthals made little, if any, genetic contributions to the Cro-Magnon or modern European gene pool. It could be argued that the absence of Neanderthal mtDNA lineages in living humans is a consequence of random drift or lineage extinction since the disappearance of Neanderthals. However, the examination of ancient Cro-Magnon mtDNA shows no evidence of a historical relationship and suggests that Neanderthals were not genetically related to the ancestors of modern humans.

Chapter 27

2. The frequency of the lethal gene in the captive population $(q^2 = 5/169$ and $q = .172)$ is approximately double that in the gene pool as a whole $(q = 0.09)$. Applying the formula

$$q_n = q_o/(1 + nq_o)$$

one can estimate that it would take 10 generations to reduce the lethal gene's frequency to .063 in the captive population with no intervention (random mating assumed). Since condors produce very few eggs per year, a more proactive approach seems justified.

(a, b) First, if detailed records are kept of the breeding partners of the captive birds, then knowledge of heterozygotes should be available. Breeding programs could be established to restrict matings between those carrying the lethal gene. Such "kinship management" is often used in captive populations. If kinship records are not available, it is often possible to establish kinship using genetic markers such as DNA microsatellite polymorphisms. Using such markers, one can often identify mating partners and link them to their offspring.

By coupling knowledge of mating partners with the likelihood of producing a lethal genetic combination, selective matings can often be used to minimize the influence of a deleterious gene. In addition, such markers can be used to establish matings that optimize genetic mixing, thus reducing inbreeding depression.

4. Both genetic drift and inbreeding tend to drive populations toward homozygosity. Genetic drift is more common when the effective breeding size of the population is low. When this condition prevails, inbreeding is also much more likely. They are different in that inbreeding can occur when certain population structures or behaviors favor matings between relatives, regardless of the effective size of the population. Inbreeding tends to increase the frequency of both homozygous classes at the expense of the heterozygotes. Genetic drift can lead to fixation of one allele or the other, thus producing a single homozygous class.

6. Inbreeding depression, over time, reduces the level of heterozygosity, usually a selectively advantageous quality of a species. When homozygosity increases (through loss of heterozygosity) deleterious alleles are likely to become more of a load on a population. Outbreeding depression occurs when there is a reduction in fitness of progeny from genetically diverse individuals. It is usually attributed to offspring being less well-adapted to the local environmental conditions of the parents. Even though forced outbreeding may be necessary to save a threatened species, where population numbers are low, it significantly and permanently changes the genetic make-up of the species.

8. Often, molecular assays of overall heterozygosity can indicate the degree of inbreeding and/or genetic drift. As inbreeding (and genetic drift for that matter) occurs, the degree of heterozygosity decreases. An allele whose frequency is dictated by inbreeding will not be uniquely influenced. That is, other alleles would be characterized by decreased heterozygosity as well. So, if the genome in general has a relatively high degree of heterozygosity, the gene is probably influenced by selection rather than inbreeding and/or genetic drift.

10. Generally, threatened species are captured and bred in an artificial environment until sufficient population numbers are achieved to ensure species survival. Next, genetic management strategies are applied to breed individuals in such a way as to increase genetic heterozygosity as much as possible. If plants are involved, seed banks are often used to maintain and facilitate long-term survival.

12. Allozymes are variants of a given allele often detected by electrophoresis. Such variation may or may not impact on the fitness of an individual. The greater the allozyme variation, the more genetically heterogeneous the individual. It is generally agreed that such genetic diversity is essential for long-term survival. All other factors being equal, allozyme variation is more likely to reflect physiological variation than RFLP variation because RFLP regions are not necessarily found in protein-coding regions of the genome. RFLP allows one to detect very small amounts of genetic diversity in a population and is unlikely to encounter an organism that is not in some way variable in terms RFLP with respect to other organisms (within and among species).

14. (a) The probability of being a heterozygote is $2pq = 2(.99)(.01) = 0.0198$. Multiplying this value by 20 gives the probability of being heterozygous:

$$0.0198 \times 20 = 0.396$$

(b) To determine N_e use the expression:

$$\begin{aligned} N_e/N &= .42 \\ N_e &= .42 \times 50 = 21 \\ H_t &= (1 - 1/2N_e)^t H_o \\ &= (1 - 1/42)^5 \times 0.0198 \\ &= 0.01755 \\ H_t/H_o &= .01755/.0198 = 0.886 \end{aligned}$$

Therefore there is a loss of approximately 11.4% heterozygosity after five generations.

16. (a) First, it will be necessary to determine whether the native habitat in the Asian steppes of the 1920s is suitable to any introduction. If the original range is supportive of reintroduction, care must be taken to introduce horses with maximum genetic diversity possible. To do so, you might monitor RFLP (fragment length polymorphisms) patterns or other indicators of genetic diversity.

(b) Since the founder breeding group included a domestic mare, it may be desirable to select those for reintroduction that are least like the domestic mare genetically. It might be desirable to release reasonably sized breeding groups in separate locations within the range to enhance eventual genetic diversity.

18. From a physiological standpoint, cryogenic preservation in liquid nitrogen can allow 100 years or more of seed storage for some species; however, such elaborate storage can only be offered to a small faction of the world's seeds. Thus, seeds of most species undergo storage loss, which decreases genetic diversity. Seeds of tropical plants are somewhat intolerant to cold storage and must be regenerated frequently, a practice that is prone to a loss of genetic diversity arising from genetic drift. Only a finite number of seeds can be used in each regeneration procedure and the restriction of sample size (often fewer than 100 plants) reduces genetic diversity. To somewhat counteract this problem, plants are grown under optimum conditions to reduce selection. Another problem with preserved seeds is the accumulation of deleterious mutations both as a result of seed storage and regeneration. Some studies indicate increased frequencies of chromosomal and mtDNA lesions, chlorophyll deficiency mutations, and decreased DNA polymerase activity associated with long-term seed storage.

20. The longest bottleneck-to-present interval occurred with cheetahs, and one would expect cheetahs to show the highest degree of microsatellite polymorphism. The shortest bottleneck-to-present interval occurred with the Gir Forest lions, so it would be expected to have the least polymorphism. Data from Driscoll et al. (2002 *Genome Research* 12:414–423) include the following estimates of microsatellite polymorphism in the three feline groups mentioned above: cheetahs (84.1%), pumas (42.9%), and Gir Forest lions (19.3%).

22. (a) The species with the greatest genetic variability, as estimated by these markers, is the domesticated cat. Domesticated cats share an immense and variable gene pool. Their staggering numbers and outbreeding behaviors allow them to maintain a high degree of genetic variability.

(b) The lion has the least genetic variability.

(c) Since allozymes code for proteins and proteins often provide a significant function in an organism, selection is stronger and mutations are less tolerated. Selection would be expected to be more harsh on DNA segments that are related to function. In addition, by their very nature, microsatellites and minisatellites are more mutable.

24. While flagship species (often large mammals) may make it possible to gather considerable public support and funding, they may reduce support for species that may have a greater impact on a community of species. Primary producers (plants) are a necessary component of a diverse and supportive habitat. If one focuses on a flagship species within an area, it is possible that other areas will suffer more dramatically because foundational species are lost. Using umbrella species to protect a large geographic area in hopes of protecting other species in that area is a reasonable approach. However, the size of an area is not necessarily a primary factor in determining species success. Diversity and productivity of a habitat are major contributors to species success. Since land is at a premium, it may be wiser in the long run to select umbrella species in diverse and productive habitats rather than on the basis of land size. By selecting sets of species that show considerable biodiversity, one increases the likelihood of protecting a sufficiently rich habitat to support many species. Such habitats are often of considerable economic value thereby making their availability limited.

Credits

Front Matter

p. ii, iii Richard Megna/Fundamental Photographs; **p. ix** Dr. Andrew S. Bajer, University of Oregon; **p. x top** Dr. Andrew S. Bajer, University of Oregon; **p. x bottom** Malcolm Gutter/Visuals Unlimited; **p. xi top** B. John Cabisco/Visuals Unlimited; **p. xi bottom** Dr. L. Caro/Science Photo Library/Photo Researchers, Inc.; **p. xii top** M. Wurtz/Biozentrum, University of Basel/Science Photo Library/Photo Researchers, Inc.; **p. xii bottom** Marc Henrie © Dorling Kindersley; **p. xiii top** Sovereign/Phototake NYC; **p. xiii bottom** Tania Midgley/Corbis/Bettmann; **p. xiv top** Jean Claude Revy/Phototake NYC; **p. xiv bottom** Ventana Medical Systems Inc.; **p. xv top** Dr. Gopal Murti/Science Photo Library/Photo Researchers, Inc.; **p. xv bottom** Image courtesy of Brian Harmon and John Sedat, University of California, San Francisco; **p. xvi top** MaizeGDB/Courtesy M.G. Neuffer; **p. xvi bottom** Reprinted from the front cover of Science, Vol. 292, May 4, 2001 Crystal structure of a Thermus thermophilus 70S ribosome containing three bound transfer RNAs(top) and exploded views showing its different molecular components (middle and bottom). Image provided by Dr. Albion Baucom (baucom@biology.ucsc.edu). Copyright American Association for the Advancement of Science; **p. xvii top** Dr. Andrew S. Bajer, University of Oregon; **p. xvii bottom** CNRI/Science Photo Library/Photo Researchers, Inc.; **p. xviii top** Biozentrum, University of Basel/Science Photo Library/Photo Researchers, Inc.; **p. xviii bottom** Dr. Gopal Murti/Photo Researchers, Inc.; **p. xix top** SPL/Photo Researchers, Inc.; **p. xix bottom** Dr. Andrew S. Bajer, University of Michael Gabridge/Visuals Unlimited; **p. xx** Alfred Pasieka/SPL/Photo Researchers, Inc.; **p. xxi top** Dr. Stanley Flegler/Visuals Unlimited; **p. xxi bottom** Photo courtesy of Roslin Institute; **p. xxii** Edward B. Lewis, California Institute of Technology; **p. xxiii** Edward S. Ross, California Academy of Sciences.

Chapter 1: Introduction to Genetics

p. 1, CO1 Omikron/Science Source/Photo Researchers, Inc.; **p. 3, 1.1** Malcolm Gutter/Visuals Unlimited; **p. 3, 1.2** Biophoto Associates/Science Source/Photo Researchers, Inc.; **p. 3, 1.3** Sovereign/Phototake NYC; **p. 3, 1.4** Micrograph by Conly L. Rieder, Division of Molecular Medicine, Wadsworth Center, Albany, New York 12201-0509; **p. 4, 1.6 (top and bottom)** Carolina Biological Supply Company/Phototake NYC; **p. 5, 1.7** Biozentrum, University of Basel/Science Photo Library/Photo Researchers, Inc.; **p. 6, 1.10** Manuel C. Peitsch/Corbis/Bettmann; **p. 7, 1.11** Photo Researchers, Inc.; **p. 7, 1.13** Oliver Meckes & Nicole Ottawa/Photo Researchers, Inc.; **p. 9, 1.15** Luis de la Maza, Ph.D., M.D./Phototake NYC; **p. 10, 1.16** Photo courtesy of Roslin Institute; **p. 10, 1.17** Science Museum/Science & Society Picture Library; **p. 11, 1.19** Alfred Pasieka/SPL/Photo Researchers, Inc.; **p. 12, 1.20(a)** John Paul Endress/Pearson Learning Photo Studio; **p. 12, 1.20(b)** Gilbert S. Grant/Photo Researchers, Inc.; **p. 12, 1.20(c)** Dr. David M. Phillips/Visuals Unlimited; **p. 12, 1.21(a)** Dr. Jeremy Burgess/Photo Researchers, Inc.; **p. 12, 1.21(b)** © Dr. David Phillips/Visuals Unlimited; **p. 12, 1.21(c)** From Genes & Development (1992) volume 6, pp. 1052–1057 as figure 1A. Matthew L. Springer/Cold Spring Harbor Press; **p. 13, 1.22(a)** Sinclair Stammers/Photo Researchers, Inc.; **p. 13, 1.22(b)** Wally Eberhart/Visuals Unlimited; **p. 13, 1.22(c)** © Mark Smith/Photo Researchers, Inc.

Chapter 2: Mitosis and Meiosis

p. 17, CO2 Dr. Andrew S. Bajer, University of Oregon; **p. 21, 2.2** CNRI/Science Photo Library/Photo Researchers, Inc.; **p. 21, 2.4(a)** Cytographics/Visuals Unlimited; **p. 21, 2.4(b)** L. Lisco/Don W. Fawcett/Visuals Unlimited; **p. 24, 2.7(a–d)** Dr. Andrew S. Bajer, University of Oregon; **p. 25, 2.7(e–g)** Dr. Andrew S. Bajer, University of Oregon; **p. 34, 2.13(a)** Biophoto Associates/Photo Researchers, Inc.; **p. 34, 2.13(b)** Andrew Syred/SPL/Photo Researchers, Inc.; **p. 34, 2.13(c)** Biophoto Associates/Science Source/Photo Researchers, Inc.

Chapter 3: Mendelian Genetics

p. 39, CO3 Archiv/Photo Researchers, Inc.

Chapter 4: Extensions of Mendelian Genetics

p. 66, CO4.1 a, c Dr. Ralph Somes; **p. 66, CO4.1 b, d** J. James Bitgood/University of Wisconsin, Animal Sciences Dept.; **p. 69, 4.1(a)** John D. Cunningham/Visuals Unlimited; **p. 73, 4.4(a)** Photo courtesy of Stanton K. Short (The Jackson Laboratory, Bar Harbor, ME); **p. 73, 4.4(b)** Fred Habegger/Grant Heilman Photography, Inc.; **p. 79, 4.9** Irene Vandermolen/Animals Animals/Earth Scenes; **p. 81, 4.11(R, L)** Carolina Biological Supply Company/Phototake NYC; **p. 83, 4.13** Mary Teresa Giancoli; **p. 84, 4.14** Hans Reinhard/Bruce Coleman Inc.; **p. 85, 4.15** Debra P. Hershkowitz/Bruce Coleman Inc.; **p. 86, 4.16(B)** Tanya Wolff, Washington University School of Medicine; **p. 86, 4.16(M)** Joel C. Eissenberg, Ph.D., Dept. of Biochemistry and Molecular Biology, St. Louis University Medical Center; **p. 86, 4.16(T)** Tanya Wolff, Washington University School of Medicine; **p. 86, 4.17(a)** Tanya Wolff, Washington University School of Medicine; **p. 86, 4.17(b)** Dr. Steven Henikoff, Howard Hughes Medical Institute, Fred Hutchinson Cancer Research Center, Seattle, Washington; **p. 87, 4.18(a)** Jane Burton/Bruce Coleman Inc.; **p. 87, 4.18(b)** Dr. William S. Klug; **p. 94, Problem 14, top left** Dusty L. Perrin; **p. 94, top right,** Corbis/Bettmann; **p. 94, bottom,** John Daniels/Ardea London Limited; **p. 96, Problem 39** Dale C. Spartas/DCS Photo, Inc.; **p. 97, Problem 41, top** Nigel J.H. Smith/Animals Animals/Earth Scenes; **p. 97, bottom** Francois Gohier/Photo Researchers, Inc.; **p. 98, Problem 44** Hans Reinhard/Bruce Coleman Inc.; **p. 92, 4.1 a, c (IS)** Dr. Ralph Somes; **p. 92, 4.1 b, d (IS)** J. James Bitgood/University of Wisconsin, Animal Sciences Dept.

Chapter 5: Chromosome Mapping in Eukaryotes

p. 100, CO5 B. John Cabisco/Visuals Unlimited; **p. 107, 5.7** ©2002 Clare A. Hasenkampf/Biological Photo Service; **p. 120, 5.17** Dr. Sheldon Wolff & Judy Bodycote/Laboratory of Radiobiology and Environmental Health, University of California, San Francisco; **p. 121, 5.18(b)** Biophoto Assoc./Photo Researchers, Inc.; **p. 122, 5.19(b)** John D. Cunningham/Visuals Unlimited; **p. 122, 5.19(c)** Science Source/Photo Researchers, Inc.; **p. 123, 5.20** James W. Richardson/Visuals Unlimited.

Chapter 6: Genetic Analysis and Mapping in Bacteria and Bacteriophages

p. 137, CO6 Dr. L. Caro/Science Photo Library/Photo Researchers, Inc.; **p. 139, 6.2** Michael G. Gabridge/Visuals Unlimited; **p. 141, 6.15** Dennis Kunkel/Phototake NYC; **p. 146, 6.12** K.G. Murti/Visuals Unlimited; **p. 149, 6.14** M. Wurtz/Biozentrum, University of Basel/Science Photo Library/Photo Researchers, Inc.; **p. 150, 6.16** Bruce Iverson/Bruce Iverson, Photomicrography.

Chapter 7: Sex Determination and Sex Chromosomes

p. 165, CO7 Wellcome Trust Medical Photographic Library; **p. 167, 7.1** Biophoto Assoc./Photo Researchers, Inc.; **p. 168, 7.3** Bill Beatty/Visuals Unlimited; **p. 169, 7.4** Dr. Maria Gallegos, University of California, San Francisco; **p. 171, 7.6(a, b)** Courtesy of the Greenwood Genetic Center, Greenwood, SC; **p. 172, 7.7(a, b)** Catherine G. Palmer, Indiana University; **p. 176, 7.9(T, B)** Stuart Kenter Associates; **p. 177, 7.11(a)** W. Layer/Okapia/Photo Researchers, Inc.; **p. 177, 7.11(b)** Marc Henrie © Dorling Kindersley; **p. 186, Problem 31** Associated Press/Courtesy of Texas A & M University.

Chapter 8: Chromosome Mutations: Variation in Chromosome Number and Arrangement

p. 187, CO8 National Institutes of Health/Evelin Schrock, Stan du Manoir and Tom Reid, NIH; **p. 190, 8.2** Ray Clarke; **p. 191, 8.3(b)** Richard Shiell/Animals Animals/Earth Scenes; **p. 192, 8.5(a)** Courtesy of the Greenwood Genetic Center, Greenwood, SC; **p. 192, 8.5(b)** William McCoy/Rainbow; **p. 193, 8.7** David D. Weaver, M.D., Indiana University; **p. 194, 8.8** David D. Weaver, M.D., Indiana University; **p. 196, 8.12** Ken Wagner/Phototake NYC; **p. 197, 8.13(b)** Pfizer, Inc./Phototake NYC; **p. 202, 8.18(a–c)** Mary Lilly/Carnegie Institution of Washington; **p. 208, 8.26** Science VU/Visuals Unlimited.

Chapter 9: Extranuclear Inheritance

p. 214, CO9 Tania Midgley/Corbis/Bettmann; **p. 215, 9.1** Grant Heilman Photography, Inc.; **p. 216, 9.2** RMF/Visuals Unlimited; **p. 217, 9.3** John D. Cunningham/Visuals Unlimited; **p. 217, 9.4(a, b)** Dr. Ronald A. Butow, Department of Molecular Biology and Oncology, University of Texas Southwestern Medical Center; **p. 219, 9.6** Dr. Richard D. Kolodnar/Dana-Farber Cancer Institute; **p. 219, 9.7** Don W. Fawcett/Kahri/Dawid/Science Source/Photo Researchers, Inc.; **p. 222, 9.9(a, b)** Dr. Alan Pestronk, Dept. of Neurology, Washington University School of Medicine, St. Louis; **p. 223, 9.10** John D. Cunningham/Visuals Unlimited; **p. 225, 9.13** Robert & Linda Mitchell Photography.

Chapter 10: DNA Structure and Analysis

p. 231, CO10 Jean Claude Revy/Phototake NYC; **p. 234, 10.3(a, b)** Bruce Iverson/Bruce Iverson, Photomicrography; **p. 237, 10.5(b)** Oliver Meckes/Max-Planck-Institut-Tubingen/Photo Researchers, Inc.; **p. 241, 10.8(b)** Runk/Schoenberger/Grant Heilman Photography, Inc.; **p. 241, 10.8(c)** Biology Media/Photo Researchers, Inc.; **p. 245, 10.13** Photo by M.H.F. Wilkins. Courtesy of Biophysics Department, King's College, London, England; **p. 249, 10.17** Ken Eward/Science Source/Photo Researchers, Inc.; **p. 255, 10.22** Ventana Medical Systems Inc.; **p. 257, 10.27** Dr. William S. Klug.

Chapter 11: DNA Replication and Recombination

p. 263, CO11 Dr. Gopal Murti/Science Photo Library/Photo Researchers, Inc.; **p. 267, 11.5(a)** Walter H. Hodge/Peter Arnold, Inc.; **p. 267, 11.5(b, c)** Figure from "Molecular Genetics," Pt. 1 pp. 74–75, J.H. Taylor (ed). Copyright ©1963 and renewed 1991, reproduced with permission from Elsevier Science Ltd.; **p. 268, 11.6** Reprinted from CELL, Vol. 25, 1981, pp 659, Sundin and Varshavsky, (1 figure), with permission from Elsevier Science. Courtesy of A. Varshavsky; **p. 275, 11.14** Reproduced by permission from H.J. Kreigstein and D.S. Hogness, Proceedings of the National Academy of Sciences 71:136 (1974), p. 137, Fig. 2; **p. 276, 11.15** Dr. Harold Weintraub, Howard Hughes Medical Institute, Fred Hutchinson Cancer Center/"Essential Molecular Biology" 2e, Freifelder & Malachinski, Jones & Bartlett, Fig. 7-24, pp. 141.

Chapter 12: DNA Organization in Chromosomes

p. 286, CO12 Science VU/BMRL/Visuals Unlimited; **p. 287, 12.1(a)** Dr. M. Wurtz/Biozentrum, University of Basel/Science Photo Library/Photo Researchers, Inc.; **p. 288, 12.2** Science Source/Photo Researchers, Inc.; **p. 288, 12.3** Dr. Gopal Murti/Science Photo Library/Photo Researchers, Inc.; **p. 290, 12.5** Image courtesy of Brian Harmon and John Sedat, University of California, San Francisco; **p. 291, 12.6** Science Source/Photo Researchers, Inc.; **p. 291, 12.7(b)** Omikron/Photo Researchers, Inc.; **p. 296, 12.11** Dr. David Adler/University of Washington Department of Pathology; **p. 296, 12.12** Douglas Chapman/University of Washington Department of Pathology.

Chapter 13: The Genetic Code and Transcription

p. 306, CO13 Prof. Oscar L. Miller/Science Photo Library/Photo Researchers, Inc.; **p. 323, 13.11** Bert W. O'Malley, M.D., Baylor College of Medicine; **p. 327, 13.15(b)** O.L. Miller, Jr. B.R. Beatty, Journal of Cellular Physiology, Vol. 74 (1969). Reprinted by permission of Wiley-Liss, Inc., a subsidiary of John Wiley & Sons, Inc.

Chapter 14: Translation and Proteins

p. 341, 14.9(a) "The Structure and Function of Polyribosomes." Alexander Rich, Jonathan R. Warner and Howard M. Goodman, 1963. Reproduced by permission of the Cold Spring Harbor Laboratory Press. Cold Spring Harbor Symp. Quant. Biol. 28 (1963) fig. 4C (top), p. 273 ©1964; **p. 341, 14.9(b)** E.V. Kiseleva; **p. 347, 14.13(a)** Dennis Kunkel/Phototake NYC; **p. 347, 14.13(b)** Francis Leroy/Biocosmos/Science Photo Library/Photo Researchers, Inc.; **p. 352, 14.19** Kenneth Eward/BioGrafx/Science Source/Photo Researchers, Inc.

Chapter 15: Gene Mutation, DNA Repair, and Transposition

p. 361, CO15 Francis Leroy/Biocosmos/Science Photo Library/Photo Researchers, Inc.; **p. 375, 15.11** Sue Ford/Science Photo Library/Photo Researchers, Inc.; **p. 376, 15.12(a)** Mary Evans Picture Library/Photo Researchers, Inc.; **p. 381, 15.18(a, b)** W. Clark Lambert, M.D., Ph. D./University of Medicine & Dentistry of New Jersey; **p. 383, 15.20** Stanley Cohen/Science Photo Library/Photo Researchers, Inc.; **p. 384, 15.21(a)** Courtesy of the Barbara McClintock Papers, American Philosophical Society; **p. 384, 15.21(b)** MaizeGDB/Courtesy M.G. Neuffer.

Chapter 16: Regulation of Gene Expression in Prokaryotes

p. 392, CO16 Nature vol 401 Sept 16, 1999 (Cover). Article: "Structure of the trp RNA-binding attenuation protein, TRAP, bound to

RNA" by Alfred A. Antson, Eleanor J. Dodson, Guy Dodson, Richard B. Greaves, Xiao-ping Chen & Paul Gollnick; **p. 400, 16.11(a–c)** Johnson Research Foundation/Science Lewis, et al/Johnson Research Foundation.

Chapter 17: Regulation of Gene Expression in Eukaryotes

p. 413, 17.2(b) From: Qumsiyeh, Mazin B. 1999 "Structure and function of the nucleus: anatomy and physiology of chromatin" Cellular Molecular Life Science Vol. 55, pp. 1129–1140, Fig. 1C, pg. 1132; **p. 418, 17.12** Dr. Paul B. Sigler; **p. 425, 17.23** From Black, D.L. 2000. Protein diversity from alternative splicing: A challenge for bioinformatics and post-genome biology. Cell **103**: 367-370, Figure 1, p.368; **p. 425, 17.24** From Hannon, G.J. 2002. RNA Interference. Nature 418: 244-251, Figure 2, p.246; **p. 426, 17.25** From Matzke, M. et all. 2004. Genetic analysis of RNA mediated transcriptional gene silencing. Biochim. Biophys. Acta 1677: 127-141, Figure 1, p.1678.

Chapter 18: Cell Cycle and Cancer

p. 434, CO18 SPL/Photo Researchers, Inc.; **p. 435, 18.1(a)** Courtesy of Hesed M. Padilla-Nash, Antonio Fargiano, and Thomas Ried. Affiliation is Section of Cancer Genomics, Genetics Branch, Center for Research, National Cancer Institute, National Institutes of Health, Bethesda, MD 20892; **p. 435, 18.1(b)** Courtesy of Hesed M. Padilla-Nash and Thomas Ried. Affiliation is Section of Cancer Genomics, Genetics Branch, Center for Research, National Cancer Institute, National Institutes of Health, Bethesda, MD 20892; **p. 437, 18.3(e–g)** SPL/Photo Researchers, Inc.; **p. 438, 18-4(a, b)** Courtesy of Professor Manfred Schwab, DKFZ, Heidelberg, Germany; **p. 441, 18.10(a)** Dr. Gopal Murti/Photo Researchers, Inc.; **p. 445, 18.13** Sung-Hou Kim, University of California, Department of Chemistry and Lawrence Berkeley National Laboratory, Berkeley, California; **p. 445, 18.14** Reproduced with permission from Y. Cho, Gorina, Jeffrey and Pavletich, Science 265: pp. 346–355, fig. 6b, pg. 352. Copyright 1994. American Association for the Advancement of Science; **p. 447, 18.17** Du Cane Medical Imaging Ltd./Photo Researchers, Inc.; **p. 448, 18.18** Dr. David M. Martin/Photo Researchers, Inc.; **p. 450, 18.21** VLA/Photo Researchers, Inc.

Chapter 19: Recombinant DNA Technology

p. 457, CO19 Michael Gabridge/Visuals Unlimited; **p. 459, 19.1** Riken Biomolecules Gallery/Kyushu Institute of Technology; **p. 461, 19.5** K.G. Murti/Visuals Unlimited; **p. 461, 19.7** Michael Gabridge/Visuals Unlimited; **p. 462, 19.8** Dr. M. Wurtz/Biozentrum, University of Basel/Science Photo Library/Photo Researchers, Inc.; **p. 464, 19.12** Reprinted with permission from Dr. Michael Blaber Florida State University; **p. 466, 19.16** Jon Gordon/Phototake NYC; **p. 469, 19.19** Bio-Rad Laboratories Diagnostics Group; **p. 472, 19.22** National Institutes of Health/Custom Medical Stock Photo, Inc.; **p. 476, 19.25** Reprinted with permission from Journal of Food Protection, Vol. 59, No. 6, 1996, Pages 573. Copyright held by the International Association for Food Protection, Des Moines, Iowa, U.S.A. Courtesy of Dr. Pina M. Fratamico; **p. 477, 19.27** Dr. Suzanne McCutcheon.

Chapter 20: Genomics and Proteomics

p. 484, CO20 Dr. Jeremy Burgess/Science Photo Library/Photo Researchers, Inc.; **p. 493, 20.7** B. Boonyartanakornkit/D. S. Clark, Vrdoljak/Visuals Unlimited; **p. 495, 20.11** Holt Studios International/Photo Researchers, Inc.; **p. 500, 20.18** The Institute for Genomic Research; **p. 501, 20.19** Reprinted from Genomics Vol 61, N.2, Breen, M. et.al., Reciprocal chromosome painting reveals detailed regions of conserved synteny between the karyotypes of the domestic dog (conis familiaris) and human, pp145-155, © 1999, with permission from Elsevier; **p. 501, 20.20** From Maglich J.M., Sluder A.E., et al., Beyond the human genome: examples of nuclear receptor analysis in model organisms and potential for drug discovery, American Journal of PharmacoGenomics, 2003; Volume 3, Issue 5: 345-353. Figure 1 on page 345; **p. 507, 20.27** Dr. Carl Merril/Laboratory of Biochemical Genetics of the National Institute of Mental Health, NIH; **p. 507, 20.28** Geoff Tompkinson/Science Photo Library/Photo Researchers, Inc.; **p. 508, 20.29(a)** ©BioPhoto/Photo Researchers, Inc.; **p. 508, 20.29(b)** Dr. David Pulak/Cellnucleus.com; **p. 509, 20.30** From: Aebersold, R. and Mann, M. 2003. Mass spectrometry-based proteomics. Reprinted with permission from Nature 422: 198–207, Fig. 4, pg 203. Copyright Macmillan Magazines Limited.

Chapter 21: Dissection of Gene Function: Mutational Analysis in Model Organisms

p. 516, CO21 Jean Claude Revy - ISM/Phototake NYC; **p. 518, 21.2(a)** Dr. Stanley Flegler/Visuals Unlimited; **p. 518, 21.2(b)** A.B. Dowsett/Science Photo Library/Photo Researchers, Inc.; **p. 519, 21.3(a)** The Provincial Museum of Alberta; **p. 519, 21.3(b)** Dr. Kimberly A. Hughes and Dr. Kevin A. Dixon/Department of Animal Biology/University of Illinois at Urbana-Champaign; **p. 522, 21.6(a)** Hank Morgan; **p. 522, 21.6(b)** Dr. R.L. Brinster/Peter Arnold, Inc.; **p. 532, 21.14(a–c)** Dr. Mario Capecchi - Figure 3a in: Hostikka SL, Capecchi MR, "The mouse Hoxc11 gene: genomic structure and expression pattern." Mech Dev. 1998 Jan;70(1-2):133–45; **p. 532, 21.15** Reproduced by permission of the American Society for Cell Biology from Molecular & Cellular Biology (20:8536-47). Copyright by the American Society for Cell Biology. Image courtesy of Dr. Gideon Dreyfuss; **p. 533, 21.16(a1, a2, b1, b2)** Reproduced with permission from Figure 2 from: Hjalt TA, Semina EV, Amendt BA, Murray JC (2000) The Pitx2 protein in mouse development. Developmental Dynamics 218:195–200). Copyright © John Wiley & Sons, Inc.; **p. 539, 21.22(a)** Dr. Lee Hartwell/Fred Hutchinson Cancer Research Center; **p. 539, 21.22(b)** Gabi Seethaler/Dr. Tim Hunt/Cancer Research UK; **p. 539, 21.22(c)** Dr. Paul Nurse; **p. 539, 21.23(a)** Figure 1 from Hartwell, L. et al. 1974. Genetic control of the cell division cycle in yeast. Science 183:46–51. Copyright American Association for the Advancement of Science; **p. 540, 21.24(a)** Adapted from Figure 3 from "Hartwell, L. et al. 1974. Genetic control of the cell division cycle in yeast. Science 183:46-51". Copyright American Association for the Advancement of Science; **p. 540, 21.24(b)** Figure 5 from "Hartwell, L. et al. 1974. Genetic control of the cell division cycle in yeast. Science 183:46–51." Copyright American Association for the Advancement of Science; **p. 541, 21.25** Christiane Nusslein-Volhard, Max Planck Institute of Developmental Biology, Tubingen, Germany; **p. 543, 21.27** Christiane Nusslein-Volhard, Max Planck Institute of Developmental Biology, Tubingen, Germany; **p. 543, 21.28(a)** Corbis/Bettmann; **p. 543, 21.28(b)** The Scotsman/Corbis/Sygma; **p. 544, 21.29** Courtesy of Tim Miller and Don Cleveland, University of California, San Diego.

Chapter 22: Applications and Ethics of Biotechnology

p. 549, CO22 Affymetrix, Inc.; **p. 552, 22.3** Reprinted with permission from "New Genes Boost Rice Nutrients" by I. Potrykus and P. Beyer, Science, August 13, 1999, Vol. 285, pp. 994; **p. 554, 22.5** Hank Morgan/Photo Researchers, Inc.; **p. 554, 22.6** Charles J. Arntzen, Florence Ely Nelson Distinguished Professor and Founding Director, Arizona Biomedical Institute at Arizona State University; **p. 551, 22.2** Reprinted with permission from International Service for the Acquisition of Agri-biotech Applications (ISAAA) Inc c/o IRRI los Banos, Laguna, Philippines; **p. 555, 22.7** Reprinted with permission from

Biodesign Institute Arizona State university Box 5001 Tempe AZ 85281-5001; **p. 558, 22.13(a)** Cancer Genetics Branch/National Human Genome Research Institute/NIH; **p. 559, 22.15** Reprinted with permission from Ash Alizadeh, Nature Magazine 2000: 403, pgs. 503–511, figure 4 left panel. Copyright 2000 Macmillian Magazines Limited; **p. 560, 22.17** GeneChip(R) Human Genome U133 Plus 2.0 Array. Courtesy of Affymetrix, Inc.; **p. 561, 22.19** Van De Silva; **p. 567, 22.26(a)** Dorothy Warburton/Phototake NYC; **p. 567, 22.26(b)** Courtesy of Werner Schempp; Glaser, Yer and Schempp et al (1998), with kind permission of Kluwer Academic Publishers. From Chromosome Research 6:481–486 (1998), Figure 4, published by Kluwer Academic Publishers: www.kluweronline.nl; **p. 568, 22.28** Courtesy of Cellmark.

Chapter 23: Developmental Genetics of Model Organisms

p. 575, CO23 Edward B. Lewis, California Institute of Technology; **p. 576, 23.1(a)** F.R. Turner/Visuals Unlimited; **p. 576, 23.1(b)** R. Calentine/Visuals Unlimited; **p. 578, 23.2(a)** Darwin Dale/Photo Researchers, Inc.; **p. 578, 23.2(b)** Tanya Wolff, Washington University School of Medicine; **p. 579, 23.3(a)** Urs Kloter/Georg Halder, University of Basel, Switzerland; **p. 580, 23.5** Dr. William S. Klug, The College of New Jersey; **p. 581, 23.7(a)** F. Rudolf Turner, Indiana University; **p. 583, 23.11** Jim Langeland, Stephen Paddock, and Sean Carroll, University of Wisconsin at Madison; **p. 584, 23.13(a)** Peter A. Lawrence and P. Johnston, "Development", 105, 761–767 (1989); **p. 584, 23.13(b)** Peter A. Lawrence "The Making of a Fly", Blackwells Scientific, 1992; **p. 584, 23.14** Jim Langeland, Stephen Paddock, and Sean Carroll, University of Wisconsin at Madison; **p. 585, 23.15(a, b)** Reproduced by permission from T. Kaufmann, et al., Advanced Genetics 27:309–362, 1990. Image courtesy of F. Rudolf Turner, Indiana University; **p. 586, 23.18** Courtesy of Dr. Alexey Veraksa and Dr. William McGinnis; **p. 587, 23.19** Reproduced with permission from [Figure 1a on page 332 from "Kmita, M. and Duboule, D. 2003. Organizing Axes in Time and Space; 25 Years of Colinear Tinkering. Science 2003 301:331–333".]. Copyright American Association for the Advancement of Science; **p. 587, 23.20** P. Barber/Custom Medical Stock Photo, Inc.; **p. 589, 23.23** Elliot M. Meyerowitz, California Institute of Technolgy, Division of Biology; **p. 589, 23.24(a, b)** Max-Planck-Institut fur Entwicklungsbiologie; **p. 590, 23.26(a–d)** Dr. Jose Luis Riechmann, Division of Biology, California Institute of Technolgy. From Science 2002: 295, pp. 1482–85; **p. 592, 23.28** James King-Holmes/Photo Researchers, Inc.

Chapter 24: Quantitative Genetics and Multifactorial Traits

p. 599, CO24 Ed Reschke/Peter Arnold, Inc.; **p. 610, 24.8** Courtesy of Dr. Steven D. Tanksley, Cornell University.

Chapter 25: Population Genetics

p. 617, CO25 Edward S. Ross, California Academy of Sciences; **p. 621, 25.4** From Leibert, F.et al., 1998. The CCR5 mutation conferring protection against HIV-1 in Caucasion populations has a single and recent origin in northeastern Europe. Human Molecular Genetics 7: 399-406, by permission of Oxford University Press; **p. 621, 25.3** Michel Samson, Frederick Libert, et al., "Resistance to HIV-1 infection in Caucasian individuals bearing mutant alleles of the CCR-5 chemokine receptor gene." Reprinted with permission from Nature [vol. 382, 22 August 1996, p. 725, Fig. 3]. Copyright 1996 Macmillan Magazines Limited; **p. 627, 25.9** Larsh K. Bristol/Visuals Unlimited; **p. 627, 25.10** Hans Pfletschinger/Peter Arnold, Inc.

Chapter 26: Evolutionary Genetics

p. 640, CO26 Breck P. Kent/Animals Animals/Earth Scenes; **p. 644, 26.3** Reprinted from Trends in Genetics 8: 392-98 Tsui L., The spectrum of cystic fibrosis mutations, copyright 1992, with permission from Elsevier; **p. 646, 26.8** From Powers, D.A. and Schulte, P.M. 1998. Evolutionary adaptations of gene structure and expression in natural populations in relation to a changing environment, J. Exp. Zool. Reprinted by permission of Wiley-Liss, Inc., a subsidiary of John Wiley & Sons, Inc; and reprinted with permission from the Annual Review of Genetics, Volume 25 © 2001 by Annual Reviews *www.annualreviews.org*; **p. 649, 26.9** Carl C. Hansen/Nancy Knowlton/Smithsonian Institution Photo Services; **p. 650, 26.11(a, b)** Kenneth Y. Kaneshiro, Ph.D., Director/University of Hawaii/CCRT; **p. 651, 26.13(a, b)** Courtesy of Toby Bradshaw and Doug W. Schemske, University of Washington. Photo by Jordan Rehm; **p. 652, 26.15(a, b)** Dr. Paul V. Loiselle; **p. 652, 26.16** Fig. 5 in Molecular Biology Evolution 15 (4) pages 391–407, Kazuhiko Takahasi/Norihiro Okada. The Society for Molecular Biology and Evolution; **p. 653, 26.17** C.B. & D.W. Frith/Bruce Coleman Inc.; **p. 655, 26.19** From Sibley, C.G. and Ahlquist, J.E. 1987. DNA hybridization evidence of hominoid phylogeny: results from an expanded data set. J.Mol.Evol. 26:99-121.Table 1, p101. with permission from Springer-Verlag GmbH; **p. 656, 26.20** From Fitch, W.M. et al., 1991. Positive Darwinian evolution in human influenza A viruses. Proc Natl. Acad.Sci. USA 88: 4270-73; **p. 657, 26.21** Reprinted from Current Biology Vol 7, No.3, Hillis, Phylogenic Analysis, pp R129-R131, Figure 1, copyright 1998 with permission from Elsevier.

Chapter 27: Conservation Genetics

p. 663, CO27 Scott Bauer/USDA/ARS/Agricultural Research Service; **p. 665, 27.2(a)** David Austen; **p. 665, 27.2(b)** Sarah Ward/Department of Soil and Crop Services/Colorado State University; **p. 666, 27.3** Sarah Ward/Department of Soil and Crop Services/Colorado State University; **p. 667, 27.4** Johnny Johnson/DRK Photo; **p. 669, 27.6** Art Wolfe/Getty Images Inc. - Stone Allstock; **p. 669, 27.7** Tim Thompson/Corbis/Bettmann; **p. 671, 27.10** Jim Brandenburg/Minden Pictures; **p. 673, 27.11** Lynn Stone/Animals Animals/Earth Scenes.

Index

Note: **Bold** page number indicates a figure; *italic* page number indicates a table.

B

C

D

E

F

G

H

I

J

K

L

N

O

P

Q

R

S

T

U

V